World Mortality in 2000: Life Tables for 191 Countries

A.D. LOPEZ

O.B. AHMAD

M. GUILLOT

B.D. FERGUSON

J.A. SALOMON

C.J.L. MURRAY

K.H. HILL

World Health Organization

Typeset and printed in Canada

WHO Library Cataloguing-in-Publication Data

World Mortality in 2000: Life Tables for 191 Countries / A. D. Lopez ... [et al.].

1.Life tables 2.Vital statistics 3.Data collection – methods

4.World Health Organization I. Lopez, Alan D.

ISBN 92 4 156204 8 (NLM classification: HB 1322)

Contents

Contributors

Alan D. LOPEZ, World Health Organization, Evidence and Information for Policy (EIP) – Epidemiology and Burden of Disease (EBD)

Omar B. AHMAD, School of Public Health, University of Ghana, P.O. Box LG13, Legon, Accra, Ghana

Michel GUILLOT, Center for Demography and Ecology, University of Wisconsin-Madison, 1180 Observatory Drive, Madison, WI 53706 USA

Brodie D. FERGUSON, World Health Organization, EIP/EBD

Joshua A. SALOMON, World Health Organization, EIP/EBD

Christopher J.L. MURRAY, World Health Organization, Executive Director – EIP

Kenneth H. HILL, Director, Hopkins Population Center. Acting Director, Hopkins Center on the Demography of Aging. Department of Population and Family Health Sciences, Johns Hopkins University, School of Hygiene and Public Health, 615 North Wolfe Street, RmW4027, Baltimore MD 21205. khill@jhsph.edu

The authors would especially like to express their gratitude to Mie Inoue for the countless hours she spent collecting mortality statistics from various sources and for her extraordinary efforts in formatting the tables and graphs presented here. She, more than anyone, deserves the credit for ensuring that this book has appeared.

Glossary of terms

DHS	Demographic and Health Surveys
DSP	Disease Surveillance Points
HALE	Healthy Life Expectancy
MICS	Multiple Indicator Child Survey
SRS	Sample Registration System

Introduction

The main purpose of this report is to compile a complete set of life tables for all WHO Member States for the year 2000 in order to compare current patterns of age/sex-specific mortality among countries based on comparable methods. Basic demographic information on the number of deaths by age and sex in a population is a critical input for the determination and evaluation of health policies and programmes. Beginning with the year 1999, WHO began making annual life tables for all Member States. These life tables have several uses and form the basis of all WHO's estimates about mortality patterns and levels worldwide. A key use of these life tables is in the construction of healthy life expectancy (HALE) which is the basic indicator of population health levels used by WHO and published each year in the World Health Report.

The construction of a life table requires reliable data on a population's mortality rates, by age and sex. The most reliable source of such data is a functioning vital registration system where all deaths are registered. Deaths at each age are related to the size of the population in that age group, usually estimated from population censuses, or continuous registration of all births, deaths and migrations. The resulting age/sex-specific death rates are then used to calculate a life table.

While the legal requirement for the registration of deaths is virtually universal, the cost of establishing and maintaining a system to record births and deaths implies that reliable data from routine registration are generally only available in the more economically advanced countries. Reasonably complete national data to calculate life tables in the late 1990s were only available for 74 countries, covering about one-quarter of the deaths estimated to have occurred in 2000 (see Table 1). In the absence of complete vital registration, sample registration or reliable information on mortality in childhood has been used, together with indirect demographic methods, to estimate life tables. This approach has been greatly facilitated by the availability of reliable estimates of child mortality in many countries of the developing world during the 1980s and 1990s from the Demographic and Health Surveys (DHS) Program, and more recently by the Multiple Indicator Child Survey (MICS) Programme led by UNICEF.

Several international agencies and other demographic centres routinely prepare national mortality estimates or life table compilations as part of their focus on sectoral monitoring. Thus, UNICEF has periodically reviewed available data on child mortality to assess progress with child survival targets and to evaluate interventions (Hill et al. 1999). A recent update of trends in child mortality during the 1990s has also just been completed (Ahmad et al. 2000). Three agencies or organizations, the United Nations Population Division, the World Bank and the United States Census Bureau have all produced international compilations of life tables, and in the case of the Population Division, at least, tables continue to be updated biennially (UN 2001; US Bureau of the Census; World Bank 2001). These various studies generally rely on the same data sources – censuses, surveys and vital registration – but can produce quite different results due to differences in the timing of data availability, differences in judgement about whether or how the basic data should be adjusted, and differences in estimation techniques and choice of models.

A comparative review of these various exercises highlights the variability in results from different procedures and judgement. For example, in India, adjusting the Sample Registration System (SRS) for underreporting of adult mortality, estimated at 13–14% in 1999–2000 (Mari Bhat 2000), yields an estimate of 9.8 million deaths in 2000, or 1 million more than the 2000 United Nations Population Assessment (UN 2001).

Differences such as these are not insignificant and have major implications for the monitoring, evaluation and reorientation of public health programmes in countries as well as at a global level. While it would obviously be desirable to develop a single set of life tables for all countries of the world, technical judgement, data availability and the timing of periodic assessments will continue to vary. Given WHO's needs for annual life table estimates as part of the continuous assessment of health system performance, and a preference for a model life table system based on a modification of the Brass logit system rather than other families of model life tables (Murray et al. 2001), WHO has constructed a new set of life tables, the results of which, for 2000, are reported in the Appendices.

This report provides details on the methods and data sources used to prepare these life tables. We begin with a brief review of the sources, types and quality of the data available. We examine the different sources of data and the problems and difficulties involved in using them in generating life tables. We also provide a brief review of the two main approaches used by WHO to estimate the parameters of the Brass logit system (α, β) for each country. For countries with a long series of vital registration data, lagged-time series analysis was used. For all other countries, α and β were estimated from either shorter time series of vital registration data or from survey or surveillance data on child and adult mortality. This is followed by a discussion of how the basic demographic input for the method, levels of $_5q_0$ and $_{45}q_{15}$, were estimated for countries. A brief summary of the major findings is provided and detailed country-specific and region-specific life tables for WHO's 191 Member States and 14 subregions[1] are given in an Appendix. Given the dominant role that child deaths play in overall mortality patterns and levels in many countries, plots of the data available to estimate child mortality for 2000 are shown individually.

Data sources and adjustments

Vital registration data

Ideally, life tables should be constructed from a long historical series of mortality data from vital registration where the deaths and population of the *de facto* (or occasionally *de jure*) population-at-risk are entirely covered by the system. In order to compute life tables for a given year (i.e. 2000) for which vital registration of deaths is not yet available for administrative reasons, short-term projections are required from the latest available year. This will require an adequate time series of data, with at least 15–20 years of mortality statistics. Table 9 shows the availability of vital registration data on mortality to the World Health Organization that could be used for life table estimation.

Firstly, vital registration data since 1980 were systematically evaluated for completeness using an array of demographic techniques including the Brass Growth-Balance method (Brass 1971), the Generalised Growth Balance method (Hill 1987) and the Bennett–Horiuchi technique (Bennett and Horiuchi 1984). These latter two methods require data on the age–sex distribution of the population from two adjacent censuses, as well as registered intercensal deaths. As a result, the simple Growth-Balance method was more commonly used to evaluate the completeness of death reporting since the basic data requirements (deaths and population by age and sex for a given year or period) were more easily met. On the other hand, the technique assumes that the population of interest is stable, which is unlikely to be the case in many developing countries, and involves a certain degree of arbitrariness in fitting the points used to estimate the extent of underreporting (Preston 1984).

Application of these methods to each country with vital registration data resulted in selecting 74 countries for which the vital registration data were judged to be sufficiently complete to compute life tables (see Table 1). These countries are listed under Category I in Table 9. For each of these countries, the last year (or last few years for small populations) were used to establish the "standard" pattern of l_x values by age, for each sex separately. Using this standard, a time series in α and β, the two parameters of the Brass Logit system, were generated. Time series techniques were then used to project the values of α and β to the year 2000, from which the 2000 life table was then obtained. These are more fully described in the methods section. For small populations (e.g. Cook Islands) three or four year moving averages were used for time series analysis rather than single-year data.

[1] To aid in demographic analysis, the 191 Member States have been divided into five mortality strata on the basis of their level of child (5q0) and adult male mortality (45q15) as follows: A = Very low child, very low adult; B = Low child, low adult; C = Low child, high adult; D = High child, high adult; E = High child, very high adult.
The matrix defined by the six WHO Regions (Afr = Africa; Amr = The Americas; Emr = Eastern Mediterranean; Eur = Europe; Sear = South-East Asia; Wpr = Western Pacific) and the five mortality strata leads to 14 subregions, since not every mortality stratum is represented in every Region, namely: AfrD, AfrE, AmrA, AmrB, AmrD, EmrB, EmrD, EurA, EurB, EurC, SearB, SearD, WprA, WprB.

Table 1. Mortality data sources (number of countries) for WHO subregions, 2000

Subregion	**Category I**: Complete vital statistics (coverage 95%+)	**Category II**: Incomplete vital statistics	**Category III**: Sample registration and surveillance systems	**Category IV**: Child mortality estimated from surveys and censuses	**Category V**: No recent data on child or adult mortality	Number of countries
AfrD	2	2	0	18	4	26
AfrE	0	2	1	14	3	20
AmrA	3	0	0	0	0	3
AmrB	17	9	0	0	0	26
AmrD	0	4	0	2	0	6
EmrB	4	3	0	6	0	13
EmrD	0	2	0	5	2	9
EurA	26	0	0	0	0	26
EurB	7	9	0	0	0	16
EurC	8	1	0	0	0	9
SearB	1	1	0	1	0	3
SearD	0	1	2	3	1	7
WprA	4	1	0	0	0	5
WprB	2	11	1	8	0	22
World	74	46	4	57	10	191

Table 2. Mortality data sources (% of deaths covered) for WHO subregions, 2000

Subregion	**Category I**: Complete vital statistics (coverage 95%+)	**Category II**: Incomplete vital statistics	**Category III**: Sample registration and surveillance systems	**Category IV**: Child mortality estimated from surveys and censuses	**Category V**: No recent data on child or adult mortality	Deaths as % of world total (WHO estimates)
AfrD	0%	4%	0%	89%	7%	8%
AfrE	0%	13%	9%	71%	7%	11%
AmrA	100%	0%	0%	0%	0%	5%
AmrB	39%	61%	0%	0%	0%	5%
AmrD	0%	65%	0%	35%	0%	1%
EmrB	2%	70%	0%	27%	0%	1%
EmrD	0%	18%	0%	68%	14%	6%
EurA	100%	0%	0%	0%	0%	7%
EurB	51%	49%	0%	0%	0%	4%
EurC	96%	4%	0%	0%	0%	7%
SearB	6%	19%	0%	76%	0%	4%
SearD	0%	0%	92%	7%	2%	21%
WprA	100%	0%	0%	0%	0%	2%
WprB	0%	9%	84%	7%	0%	18%
World	24%	12%	36%	25%	2%	100%

For Bosnia and Herzegovina, vital registration data were not available after 1991. The 1989/1991 data were judged to be complete and, in the absence of new information, were averaged and assumed to apply in 2000, since any background improvements in mortality were probably counteracted by the effects of the conflict in the 1990s.

For a second group of countries with vital registration (or sample vital registration in the case of Bangladesh, China, India and Tanzania), application of these indirect demographic methods to assess completeness suggested that some correction was required to the vital registration data. These 50 countries are listed under Categories II and III in Table 9. The extent of underreporting varied considerably but was typically of the order of 20–25% in the Central Asian Republics, and somewhat higher (30–40%) in several developing countries. For example, adult mortality was judged to be 84% complete for Egypt, 92% complete in Guatemala, 84–86% in India, 78–84% in Kazakhstan, 82–85% in Republic of Korea, and 73–75% in the Philippines. These completeness ratios were then applied to the vital registration data to correct for underreporting.

For child mortality, all available data points from child mortality surveys such as the DHS or the MICS programme of UNICEF were used to correct the vital registration data on child mortality. This analysis benefited greatly from the comprehensive country evaluation of data carried out by Hill et al. (1999) based on census and survey data available as of 1996. We have updated these country plots with more recent information and adjusted the 2000 predictions of Hill et al. where recent data suggested this was necessary (see Appendix). Where recent survey data were not available, child mortality rates were corrected, based on the correction factors suggested from the above techniques for adult mortality and on levels estimated for neighbouring countries. To the extent that child deaths are more likely to be underreported than adult deaths, this will underestimate levels of child mortality.

For those countries with a sufficiently long time series of vital registration, or sample registration data, the corrected time series were then analysed (as for Category I countries). For the remainder, levels of child ($_5q_0$) and adult ($_{45}q_{15}$) mortality were projected forward to 2000 using available evidence on the speed of mortality decline, or by assuming a pattern of mortality change in the 1990s consistent with economic growth. These projected values of child and adult mortality were then applied to the new Modified Logit Life Table system (see Murray et al. 2001, and methods section) to generate a full life table. In some cases (e.g. South Africa, Tanzania, Cambodia, Dominican Republic, Haiti and Myanmar), death rates were increased by adding on estimated mortality from HIV/AIDS. These methods and data sources for estimating HIV/AIDS mortality are described in a later section.

Multi-source approaches for specific populations

In two large developing countries, India and China, several data sources, including vital registration, surveillance systems and surveys are available to estimate mortality rates. None of these systems alone is sufficiently reliable to produce life tables for these countries without adjustments, but all are useful to estimate child and adult mortality. The data sources used and the adjustments made to them are as follows:

China

Three sources of mortality data were used to estimate the life table.

Disease Surveillance Points (DSP)

This is a nationally representative system of 145 epidemiological surveillance points operated by the Chinese Academy of Preventive Medicine and covering a population of 10 million people throughout China. Data on the age, sex and cause of 50,000–60,000 deaths are recorded each year. Periodic evaluations of the DSP data by re-surveying households at random suggest a level of underreporting of deaths of about 15% (Chinese Academy of Preventive Medicine 1997), although Growth-Balance of the data since 1991 suggests an average adjustment factor about twice this level. Annual data for the period 1991–1998 were used, with corrections, to estimate the trend in $_5q_0$ and $_{45}q_{15}$.

Vital registration

Data on the age, sex and cause of 725,000 deaths are collected annually from the vital registration system operated by the Ministry of Health, covering a population of 121 million (66 million in urban areas, 55 million in rural areas). While the data are not representative of mortality conditions throughout China, they are useful for suggesting trends in mortality, given the number of deaths covered. Trends in $_{45}q_{15}$ for the rural and urban coverage areas separately are shown in Figure 1. While underreporting yields implausibly low levels of ${}_{45}q_{15}$, these data suggest that there has been only a very modest decline in adult mortality during the 1990s (4–5% for males – both areas – and for females in rural areas, and 14% for females in urban areas).

Figure 1. Trends in 45q15 from vital registration data – China, 1990-1998

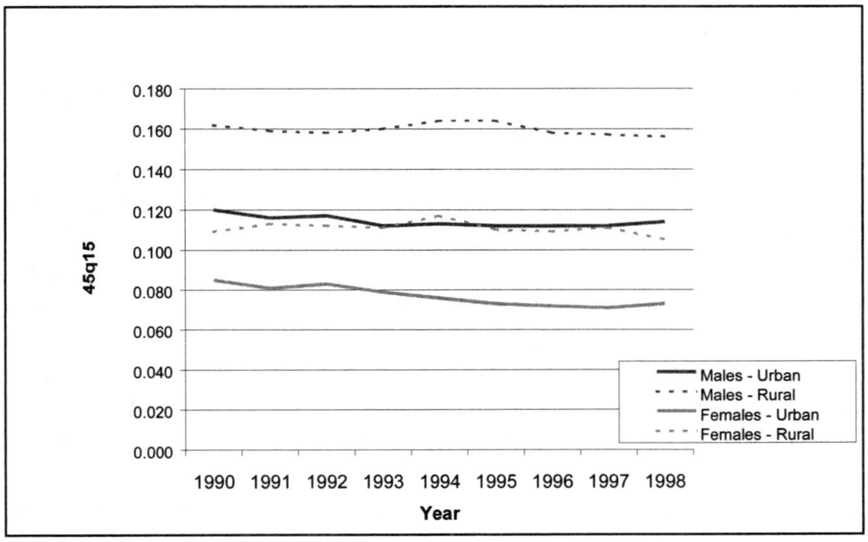

Survey data

Survey data is obtained from the annual 1 per 1000 household survey asking about deaths in the past 12 months. For example, the 1997 survey covered a population of 1,243,000 people spread over 864 counties (3164 townships, 4438 villages) in 31 provinces and recorded a total of 7,845 deaths. While this is a nationally representative sample, Growth-Balance methods suggest substantial underreporting of deaths (27% and 29% for males and females, respectively). Trends in the implied unadjusted ${}_{45}q_{15}$ from the surveys in the 1990s are shown in Figure 2 and suggest a somewhat more substantial decline although the much smaller number of deaths compared with vital registration make trend assessment difficult.

In all three systems, data were available for the period 1991–1998. Since Growth-Balance analyses suggested that underreporting had remained relatively constant during the 1990s, the average annual decline in ${}_{5}q_{0}$ and ${}_{45}q_{15}$ suggested by these three data sources was first calculated and applied to the 1990 Chinese life table based on the census to project death rates to 2000. Uncertainty around death rates in 2000 was generated from more optimistic and pessimistic assumptions about the rate of decline during the 1990s.

Figure 2. Trends in 45q15 (unadjusted) based on estimates from the Sample Survey of Population Change – China, 1991–1998

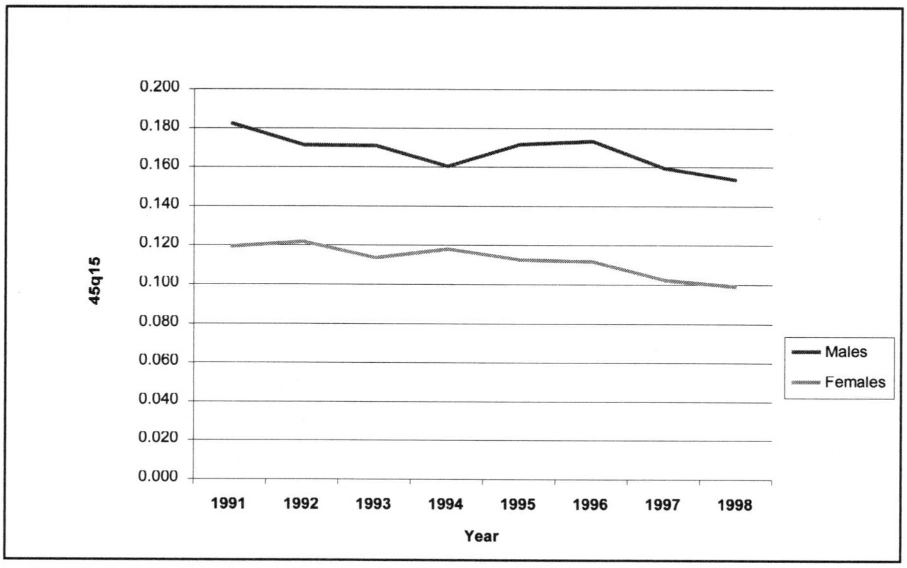

India

The most representative and reliable data on mortality rates by age and sex in India come from the Sample Registration System which has been in operation for several decades. We used data for the period 1990–1998 (latest year available) to compute annual life tables. Data are collected on vital events in 4436 rural and 2235 urban sampling units with a population of about 6 million people covering almost all States and Territories. Comparison of ${}_5q_0$ from the SRS with the rate reported from the DHS (National Family Health Survey) conducted in 1992–93 yield very similar results, suggesting that underreporting of child deaths is minimal. On the other hand, underreporting of adult deaths in the SRS during the 1990s has probably increased to around 15% based on the Bennett–Horiuchi variable -r methodology (Mari Bhat 2000). We therefore corrected the SRS death rates at age five and over by 14% for males and 16% for females, and projected these forward to 2000. Uncertainty intervals around age-specific death rates were generated from plausible projections to 2000 of the trend lines in these rates.

Census and survey data

For the remaining 67 countries (see Table 1), no reliable estimates of adult mortality were available. However, extensive direct and indirect estimates of child (${}_5q_0$) mortality were available for recent years from various census and survey programmes. These estimates were systematically evaluated and projected to 2000, along with uncertainty intervals (Ahmad et al. 2000). These estimates, with uncertainty ranges, were then applied to the Modified Logit Life Table System using the "global" standard to estimate a corresponding life table. More detail on the procedure is provided in the methods section.

In only a handful of countries in this group (Afghanistan, Angola, Burundi, Democratic People's Republic of Korea, Djibouti, Equatorial Guinea, Haiti, Liberia, Malawi, Sao Tome and Principe, and Swaziland) was there insufficient evidence in the 1990s to establish estimates of child mortality with reasonable confidence. The life tables for these countries are consequently based on estimated child mortality levels with wide uncertainty. Substantial caution in their use is essential until more recent evidence on child mortality levels becomes available.

For many countries in this category, mortality from HIV/AIDS is likely to be substantial. This would not be captured from the model life table application since the models were built from mortality data in the pre-AIDS era. For these countries, primarily in sub-Saharan Africa, AIDS-free life tables were first estimated by estimating child mortality levels excluding HIV. Age/sex-specific death rates from HIV were then estimated separately (Salomon and Murray 2001) and added to the model life table age-specific rates with appropriate adjustment for competing risks.

Methods

Life table construction

Standard life table methods were used to construct the time series of life tables for countries in Categories I and II, where time series data were available. The basic Brass Logit system was used to generate the trend in the two parameters (α, β) from these life tables. This system rests on the assumption that two distinct age-patterns of mortality can be related to each other by a linear transformation of the logit of their respective survivorship probabilities. Thus for any two observed series of survivorship values, l_x and l^s_x, where the latter is the standard, it is possible to find constants α and β such that

$$logit(l_x) = \alpha + \beta logit(l^s_x)$$

$$if\ logit(l_x) = 0.5\, ln\left(\frac{(1.0 - l_x)}{l_x}\right)$$

Then

$$0.5\, ln\left(\frac{(1.0 - l_x)}{l_x}\right) = \alpha + 0.5\beta ln\left(\frac{(1.0 - l^s_x)}{l^s_x}\right)$$

for all ages x between 1 and T and where l_0(radix) = 1. If the above equation holds for every pair of life tables, then any life table can be generated from a single standard life table by changing the pairs of (α,β) values used. In reality, the assumption of linearity is only approximately satisfied by pairs of actual life tables. However, the approximation is close enough to warrant the use of the model to study and fit observed mortality schedules. The parameter α varies the mortality *level* of the standard, while β varies the *slope* of the standard, i.e., it governs the relationship between the mortality in children and adults.

In circumstances where an historical sequence of life tables is available it is possible to generate a time series of α,β pairs using a country-specific standard. A plot of α and β, separately, against time should produce a trajectory of points (Lopez et al. 2000). If the plot of points for each parameter falls along a fairly straight line, that line could theoretically be projected forward to forecast estimates for any time in the future. These α,β estimates can then be substituted into the appropriate logit equations to obtain the corresponding life tables. Where, on the other hand, the trend in the points was erratic, the system could not provide an adequate forecast. In such situations, suitable techniques must then be applied to project mortality, given this pattern. As a result, three models were developed to accommodate different scenarios. In the first model the parameter at time t is assumed to be a simple linear function of time t:

$$\hat{\alpha}_t = \gamma_1 + \gamma_1 t$$
$$\hat{\beta}_t = \phi_1 + \phi_2 t$$

This model is suited to situations where the trend in α or β is clearly linear. In the second model, the α and β parameters at time t are assumed to be lagged linear functions of the parameters in the preceding periods. Thus the parameters for time $T+1$ are based on lag 1 model, those for time $T+2$ are based on lag 2 model, etc., where T corresponds to the time location of the standard life table. The following equations summarize these relationships:

$$\hat{\alpha}_{T+1} = \gamma_{11} + \gamma_{21}\alpha_T \qquad \hat{\beta}_{T+1} = \phi_{11} + \phi_{21}\beta_T \quad \text{1st forecast point}$$
$$\hat{\alpha}_{T+2} = \gamma_{12} + \gamma_{22}\alpha_T \qquad \hat{\beta}_{T+2} = \phi_{12} + \phi_{22}\beta_T \quad \text{2nd forecast point}$$
$$\hat{\alpha}_{T+3} = \gamma_{13} + \gamma_{23}\alpha_T \qquad \hat{\beta}_{T+3} = \phi_{13} + \phi_{23}\beta_T \quad \text{3rd forecast point}$$
$$\cdots\cdots \qquad \cdots\cdots$$
$$\hat{\alpha}_{T+n} = \gamma_{1n} + \gamma_{2n}\alpha_T \qquad \hat{\beta}_{T+n} = \phi_{1n} + \phi_{2n}\beta_T \quad \text{last forecast point}$$

This model is likely to be more suitable in situations where there are clear linear trends, but also regular oscillations in parameter values over time. The third approach combines the above two models:

$$\hat{\alpha}_{T+1} = \gamma_{11} + \gamma_{21}\alpha_T + \gamma_{31}(T+1) \qquad \hat{\beta}_{T+1} = \phi_{11} + \phi_{21}\beta_T + \phi_{31}(T+1) \quad \text{1st forecast point}$$
$$\hat{\alpha}_{T+2} = \gamma_{12} + \gamma_{22}\alpha_T + \gamma_{32}(T+2) \qquad \hat{\beta}_{T+2} = \phi_{12} + \phi_{22}\beta_T + \phi_{32}(T+2) \quad \text{2nd forecast point}$$
$$\hat{\alpha}_{T+3} = \gamma_{13} + \gamma_{23}\alpha_T + \gamma_{33}(T+3) \qquad \hat{\beta}_{T+3} = \phi_{13} + \phi_{23}\beta_T + \phi_{33}(T+3) \quad \text{3rd forecast point}$$
$$\bullet\bullet\bullet\bullet\bullet\bullet \qquad \bullet\bullet\bullet\bullet\bullet\bullet$$
$$\hat{\alpha}_{T+n} = \gamma_{1n} + \gamma_{2n}\alpha_T + \gamma_{3n}(T+n) \qquad \beta_{T+n} = \phi_{1n} + \phi_{2n}\beta_T + \phi_{3n}(T+4) \quad \text{last forecast point}$$

This model is suitable in situations where there are complex linear trends. In each country, all three models were used to forecast parameter estimates. The model that yielded time series of estimates which best fitted the historical trend was deemed adequate for that country.

To estimate the life table out to age 100+, the Coale-Guo procedure (Coale and Guo 1989) was used. It has long been observed that mortality rates at ages over 75 or 80 increase with age at a diminishing rate rather than at the constant Gompertz rate (Perks 1932). Using data from seven populations with relatively reliable mortality data at older ages (Austria, France, Germany, Japan, Netherlands, Norway and Sweden), Coale and Guo demonstrated that the relative increase from one age group to the next decreases above age 80 or 85. Using these findings, they developed a method of closing out life tables above age 80. Their technique incorporates an assumption of steady rather than Gompertzian constancy in the rate of increase in mortality with age above age 80. More specifically, the logarithm of the ratio of the mortality rate in the interval (x+5, x+10) to the ratio from (x, x+5) is assumed to decline by a constant increment as x rises above 80.

Estimating adult mortality

Measuring adult mortality is inherently more difficult than measuring child mortality. In part, this is due to the relative scarcity of adult deaths, particularly in the age range 15 to 60, relative to deaths in childhood or old age. Thus obtaining precise measures of adult mortality requires large numbers of observations typically covering long reference periods. It is also due, however, to the fact that, whereas mothers' reports of the survival of their children provide generally satisfactory estimates of child mortality, event histories that provide generally adequate and timely estimates of adult mortality have not been found. In contrast to child mortality estimation where information is easily collected from affected mothers, it is often difficult to identify the right informant to provide information on deceased adults. This often results in problems of under-counting and multiple reporting. Often the informant does not know the age of the deceased and birth certificates are often not available for older people in most developing countries. As a result, errors in the reporting of age can severely limit the ability to obtain good estimates of adult mortality.

The most widely used method of measuring adult mortality from survey data is that using information on the survival of mother and father to estimate adult female and male survivorship, respectively. Other methods include those using information on (a) survival of first husbands to estimate male adult survivorship, (b) survival of first wives to estimate female adult survivorship, (c) survival of siblings. These methods, although theoretically sound have proved difficult to apply in practice, often leading to underestimation of true levels of adult mortality by 15–60% in countries where it has been possible to validate them against registration data (Stanton et al. 2000).

An attempt to systematically review all available direct (primarily from Demographic Surveillance Systems) and indirect estimates on adult mortality levels in Africa concluded that only a relatively small fraction of them yielded plausible estimates, and many of these referred to years well before 2000 (Lopez et al. 2000). As a result, for all countries in Categories IV and V, a new modified logit life table system was used. Full details of the rationale and evaluation of the method are given elsewhere (Murray et al. 2001). Essentially the approach was based on matching the estimated level of child mortality to a plausible range of levels of l_{60} derived from an empirical database of over 1800 life tables with a wide range of life expectancies. In most cases, the mean value of l_{60} from among those within this plausible range was chosen as a basis for estimating adult mortality levels. In relatively few cases, values other than the mean value were chosen based on an assessment of available evidence about levels of adult mortality.

Estimating mortality from HIV/AIDS

For countries with the most extensive HIV/AIDS epidemics and no vital registration data, AIDS mortality was estimated separately and then incorporated into life tables that excluded AIDS. For sub-Saharan African countries, the total number of adult AIDS deaths was derived from backcalculation models using sentinel surveillance data on prevalence in pregnant women, with updates of previously published models (Salomon and Murray 2001) where more recent data has become available. In order to estimate age and sex-specific mortality, we have analysed registration and surveillance data on AIDS mortality from the following sources: the Adult Morbidity and Mortality Project in three districts of Tanzania; vital registration data from urban and rural South Africa; and Zimbabwe vital registration. These data provide the only reliable sources of population-based information on cause-specific mortality in continental sub-Saharan Africa available to WHO. In Figure 3 we have plotted the relative age and sex pattern of mortality rates from each of these sources, normalizing on the highest observed rate in each site. There is remarkable consistency in the pattern across these different sources, with the main differences appearing at the youngest and oldest ages. Based on these sources, we have developed a regional standard age pattern by taking the weighted average of these sources. The regional standard appears as a thick line in Figure 3. Using this standard, a given estimate of total adult deaths may be translated into age-specific death rates by applying the standard pattern of rates to the population age structure and then rescaling all of the rates such that the total number of deaths matches the specified figure.

Figure 3. Age pattern of HIV mortality

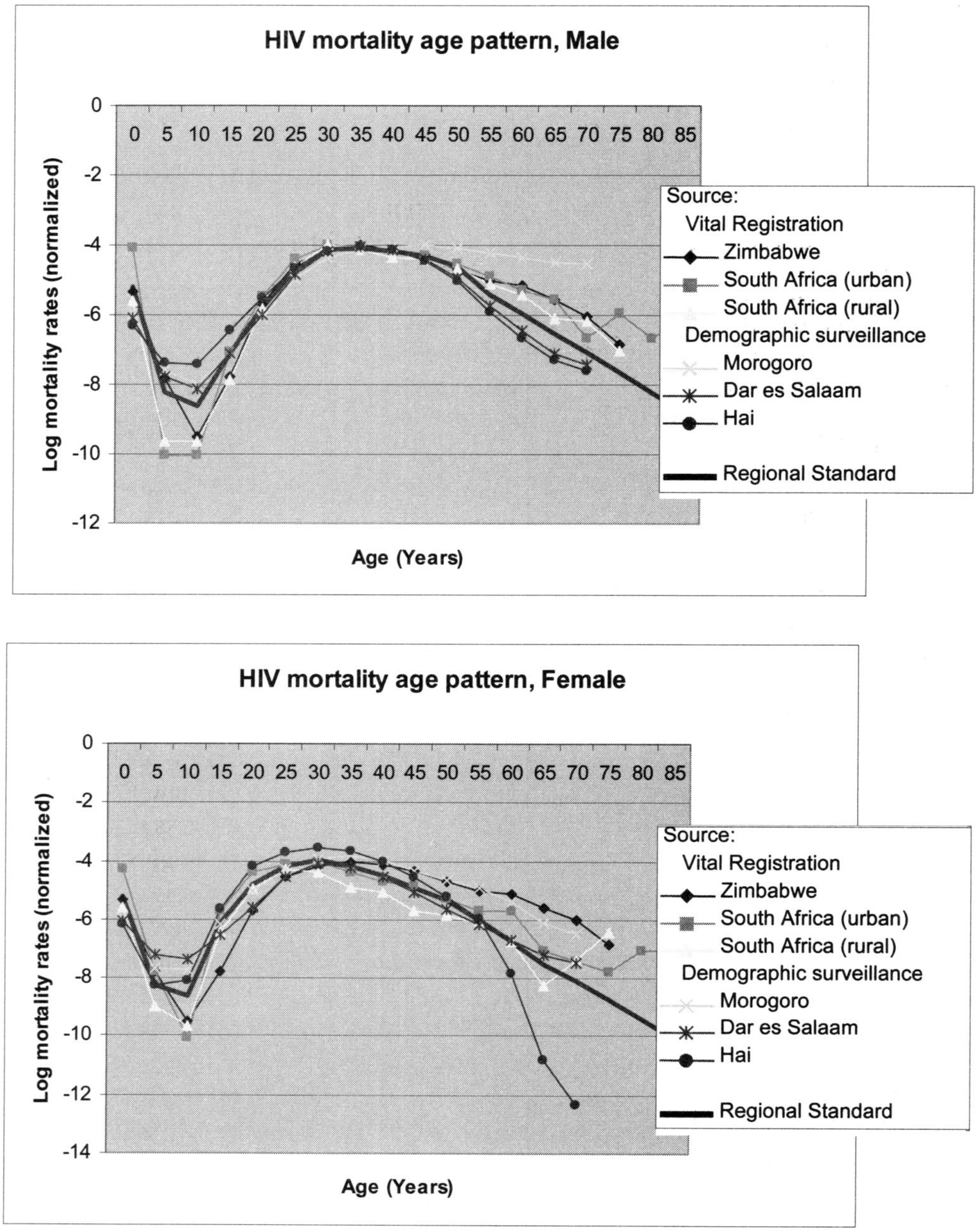

Given the dearth of data from which to estimate AIDS mortality directly and the uncertainties introduced by the backcalculation approach, it is important to try to quantify the level of uncertainty around the mortality estimates that result from these methods. Where enough data were available to undertake a maximum likelihood estimation approach in the backcalculation models (i.e. about 20 countries), the results included a measure of uncertainty around mortality estimates in each year. For the remaining countries, uncertainty intervals were derived based on an assessment of the coverage and representativeness of sentinel surveillance sites in each country. Probability distributions around the total number of deaths were then translated into distributions around age and sex-specific mortality rates using numerical simulation methods. Monte Carlo methods were used to sample from these distributions in order to incorporate uncertainty around AIDS mortality into the life table uncertainty estimates.

For four countries outside of sub-Saharan Africa (Cambodia, Dominican Republic, Haiti and Myanmar) where no vital registration data existed or where AIDS was inadequately reflected in registered deaths, we have also merged separate estimates of AIDS mortality into life tables excluding AIDS. The age pattern of AIDS mortality in these cases was based on regional reference patterns from countries with available vital statistics (Argentina, Brazil, Chile, Cuba, Mexico and Uruguay in the case of Haiti and Dominican Republic; and Thailand in the case of Myanmar and Cambodia).

Uncertainty bounds

There are several sources of uncertainty around the final values of α and β obtained from these models, including model uncertainty as to the correct specification as well as estimation uncertainty in identifying values for the regression coefficients. A detailed discussion of the sources of uncertainty and methods for uncertainty analysis for life tables may be found in Salomon et al. (2001). The level of uncertainty around estimates of α and β depends in part on the uncertainty around the regression coefficients γ_{ij} and ϕ_{ij}, and in turn implies some level of uncertainty around the life tables that are computed from these parameters. Because a complete life table is a complicated nonlinear function of the uncertain parameters, we have used Monte Carlo simulation techniques to develop numerical estimates of the ranges of uncertainty around the life tables. This uncertainty is captured by taking random draws of the regression coefficients γ_{ij} and ϕ_{ij} from normal distributions with means equal to the estimated coefficients and variances derived from the standard errors in the regression. In each of 1000 iterations, the draws of γ_{ij} and ϕ_{ij} are used to generate α and β estimates, which are then translated into complete life tables. Thus probability distributions may be defined around life table estimates by analysing the 1000 different simulated life tables. For example, a range may be defined around the estimate for life expectancy at birth by sorting the 1000 different estimates of e(0) in the simulated life tables and then identifying the 25th and 975th values as the bounds of an approximate 95% confidence interval.

In the absence of historical data (i.e., for countries in Categories IV and V), a modified logit system was used (Murray et al. 2001). Like the simple logit system, the new model is anchored on the relationship between two life tables. Unlike the simple logit system, however, the new formulation includes two additional parameters which correct for the nonlinearity in the original Brass method. The main input to the system is an estimated range around ${}_5q_0$. For any given ${}_5q_0$ value, the logit life table system provides a range of plausible values for ${}_{45}q_{15}$. These ranges have formed the basis for a set of Monte Carlo simulations of different possible life tables consistent with the knowledge and uncertainty regarding adult and child mortality. The range of simulated life tables allow calculation of uncertainty ranges around the various key indicators such as q_x, l_x and e_x.

Results

Overall life expectancy at birth (both sexes combined) in 2000 is estimated to range from 81.1 years in Japan (84.7 females, 77.5 for males) to 37.5 years in Malawi (see Table 3). For males, the next highest life expectancy was estimated for Sweden (77.3 years), followed by Andorra (77.2), Iceland (77.1), Monaco (76.8) and Switzerland (76.7). Male life expectancy exceeded 75.0 years in 19 countries in 2000 (see Table 4).

Table 3. Life expectancy at birth (years), top 10 and bottom 10 countries, 2000

Top 10 countries		Bottom 10 countries	
1 Japan	81.1	1 Malawi	37.5
2 Monaco	80.6	2 Sierra Leone	37.9
3 Andorra	80.5	3 Mozambique	38.7
4 San Marino	80.0	4 Zambia	39.4
5 Sweden	79.6	5 Rwanda	39.5
6 Switzerland	79.6	6 Burundi	41.0
7 Iceland	79.5	7 Central African Republic	42.0
8 Australia	79.3	8 Lesotho	42.1
9 Italy	79.2	9 Namibia	42.7
10 France	79.1	10 Dem. Rep. of the Congo	42.8

Table 4. Countries with male life expectancy greater than 75.0 years, 2000

Country	e0 (years)	Country	e0 (years)
Japan	77.5	Italy	76.0
Sweden	77.3	New Zealand	75.9
Andorra	77.2	Norway	75.7
Iceland	77.1	Netherlands	75.4
Monaco	76.8	Singapore	75.4
Switzerland	76.7	Greece	75.4
Israel	76.6	Malta	75.4
Australia	76.6	Spain	75.4
San Marino	76.1	France	75.2
Canada	76.0		

Table 5. Countries with female life expectancy greater than 80.0 years, 2000

Country	e0 (years)	Country	e0 (years)
Japan	84.7	Norway	81.4
Monaco	84.4	Austria	81.4
San Marino	83.8	Netherlands	81.0
Andorra	83.8	New Zealand	80.9
France	83.1	Belgium	80.9
Switzerland	82.5	Finland	80.9
Italy	82.4	Luxembourg	80.8
Spain	82.3	Greece	80.8
Australia	82.1	Malta	80.7
Sweden	82.0	Germany	80.6
Iceland	81.8	Israel	80.6
Canada	81.5	Singapore	80.2

Among females, the second highest life expectancy was estimated for Monaco (84.4 years), followed by San Marino and Andorra (83.8 years), and France (83.1). Twenty-four countries had an estimated life expectancy of 80 years or more for females in 2000 (see Table 5). Female life expectancy exceeded 75.0 years in 57 countries, or about one-third of WHO's Member States.

Given the extraordinary impact of the HIV/AIDS epidemic in sub-Saharan Africa, it is perhaps not surprising that the countries with the lowest life expectancy in 2000 are all from this region. Indeed, 37 of the 40 countries with the lowest life expectancy are in sub-Saharan Africa. While countries in this region suffer disproportionately from many of the factors which cause child death and the premature mortality of adults, including acute respiratory infection, diarrhoeal diseases and tuberculosis, the lack of progress in achieving further gains in life expectancy can largely be attributed to the effects of the HIV/AIDS epidemic over the last 15 years or so. If deaths from HIV/AIDS were to be excluded, life expectancy at birth in some countries of the region would be 15 to 20 years higher (see Table 6) This is particular true of countries of southern Africa (Botswana, Lesotho, Namibia, South Africa, Swaziland, Zambia, Zimbabwe), but reductions in life expectancy of the order of 5–10 years due to AIDS mortality in 2000 are common in many other African countries as well. On average, life expectancy at birth in sub-Saharan Africa is 6 years lower for males and slightly more than 7 years lower for females compared to what would have been the case in 2000 in the absence of HIV/AIDS mortality.

Table 6. Difference in life expectancy at birth for all possible causes of death and causes excluding AIDS, by sex, 2000 (years)

Males			
Namibia	18.5	Djibouti	5.9
Botswana	17.3	Haiti	5.4
Zambia	15.2	Eritrea	5.4
Zimbabwe	14.9	Nigeria	4.7
Lesotho	14.8	Somalia	4.7
Swaziland	14.2	Ghana	3.9
South Africa	12.1	Bahamas	3.9
Central African Republic	10.3	Gabon	3.8
Malawi	10.1	Dem. Rep. of the Congo	3.5
Kenya	9.8	Guyana	3.0
Burundi	9.8	Liberia	2.7
Côte d'Ivoire	9.0	Cambodia	2.7
Mozambique	8.3	Myanmar	2.5
Rwanda	7.8	Chad	2.3
Uganda	7.8	Honduras	2.2
United Republic of Tanzania	7.2	Sierra Leone	2.1
Ethiopia	7.2	Dominican Republic	2.1
Burkina Faso	6.6	Benin	2.1
Togo	6.3	Gambia	2.0
Cameroon	6.2	Angola	2.0
Congo	6.0		
Females			
Namibia	21.5	Congo	7.3
Botswana	20.2	Cameroon	7.3
Zambia	18.1	Djibouti	7.0
Zimbabwe	17.6	Eritrea	6.4
Lesotho	17.4	Somalia	5.7
Swaziland	16.8	Nigeria	5.6
South Africa	15.1	Ghana	4.7

Malawi	12.4	Dem. Rep. of the Congo	4.4
Central African Republic	12.2	Gabon	4.4
Burundi	11.7	Liberia	3.3
Kenya	11.6	Chad	2.8
Côte d'Ivoire	10.9	Sierra Leone	2.6
Mozambique	10.3	Angola	2.6
Rwanda	9.9	Gambia	2.4
Uganda	9.2	Benin	2.4
Ethiopia	8.8	Guinea-Bissau	2.1
United Republic of Tanzania	8.6	Haiti	2.1
Burkina Faso	7.8	Senegal	2.1
Togo	7.5		

Large sex differences in life expectancy persist in the more developed countries. At the beginning of the 20th century, female life expectancy exceeded that of males by 2 to 3 years, on average, at least in Europe, North America and Australia (Lopez 1983). In 2000, the female advantage had widened to 10 or more years in Kazakhstan (10.4 years), Lithuania (10.4), Ukraine (10.7), Estonia (11.0), Latvia (11.3) and Belarus (12.0), and was highest of all countries in the Russian Federation (12.6). Conversely, the differential was only half a year or less in countries such as Bangladesh, Lesotho and Zambia. In only a handful of countries including Botswana, Maldives, Namibia and Nepal was male life expectancy estimated to exceed that of females (by 0.2 to 0.5 years). Figure 4 shows the distribution of the female-male gap in life expectancy at birth across the 191 Member States of WHO in 2000. The extreme values observed in some Eastern European countries are clear from the tail of the distribution. In about one-quarter of countries, the male–female differential is between 5 and 6 years in favour of females. Since sex differentials in mortality are typically lower in developing countries, the overall global difference in male–female life expectancy at birth is slightly lower than this (4.7 years).

Figure 4. Distribution of female–male difference in life expectancy at birth, WHO Member States, 2000

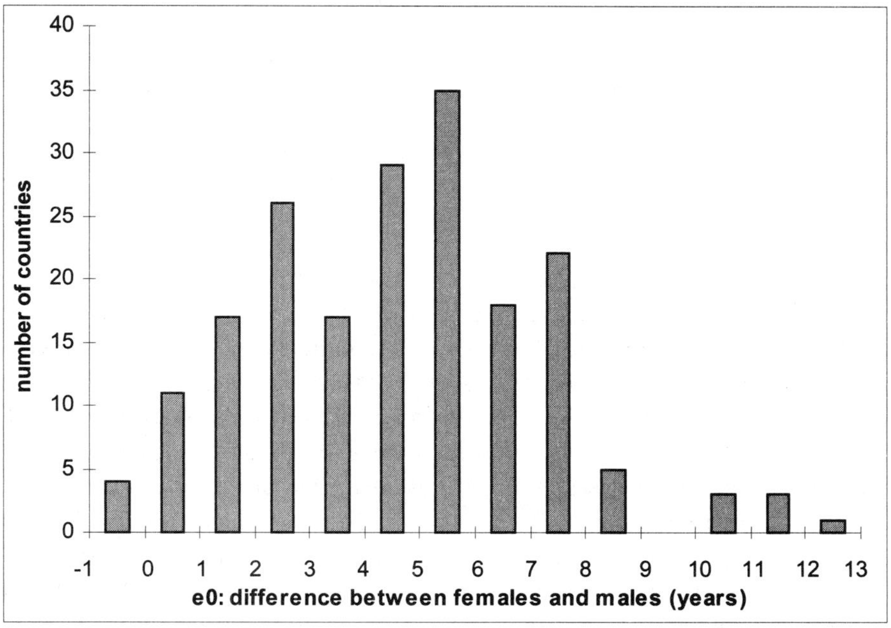

Useful summary indicators of prevailing mortality risks in a population are the probability of dying between birth and age 5, as an overall measure of health conditions among children, and the probability of dying between ages 15 and 60, as a measure of premature mortality among adults. These are shown for the various WHO Regions, and within them, the various mortality-based subregions, in Figure 5.

Figure 5. Chances of dying in childhood (0–4 years) and adulthood (15–59 years), by subregion, 2000

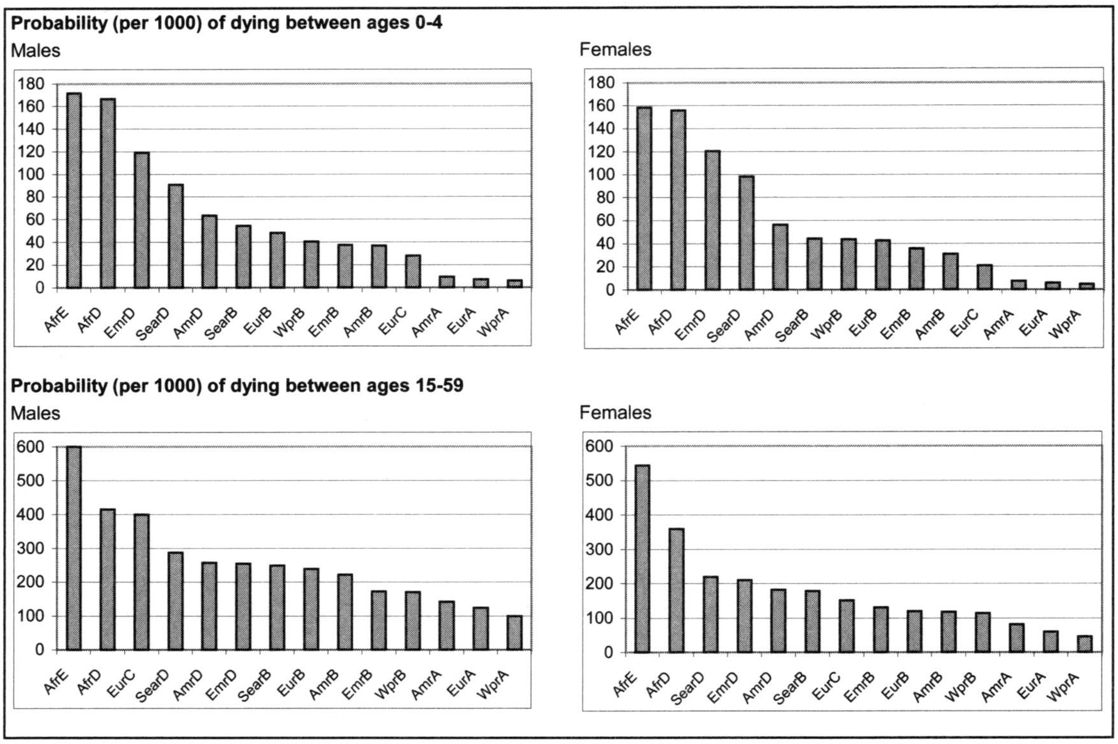

Differences in levels of child mortality remain vast. Of the 10.9 million deaths below age 5 estimated to have occurred in 2000 (Table 7), 99% of them were in developing regions. In industrialized countries (i.e. those classified into the A Regional Strata) the probability of child death ($_5q_0$) is typically less than 1% (0.4% in Sweden and Singapore), but it rises to almost 300 per 1000 in Sierra Leone. Levels of child mortality well in excess of 10% (100 per 1000) are still common throughout Africa and in parts of Asia (Afghanistan, Cambodia, Laos, Myanmar, Nepal and Pakistan).

Table 7. Total deaths by sex, age and WHO subregion, 2000

Sex	WHO subregions		all (000)	0-4 (000)	5-14 (000)	15-29 (000)	30-44 (000)	45-59 (000)	60-69 (000)	70-79 (000)	80+ (000)
Both sexes	World		55,619	10,893	1,438	3,619	5,062	6,993	8,006	10,281	9,328
	Afr	All	10,572	4,244	432	1,304	1,683	1,039	719	739	412
		D	4,244	1,928	182	396	484	383	319	348	204
		E	6,327	2,316	250	907	1,198	656	400	391	208
	Amr	All	5,872	469	64	297	438	758	838	1,280	1,729
		A	2,778	37	9	55	129	308	371	687	1,183
		B	2,584	313	38	194	254	386	404	517	477
		D	511	119	17	48	55	64	63	76	68

Sex	WHO subregions		all (000)	0-4 (000)	5-14 (000)	15-29 (000)	30-44 (000)	45-59 (000)	60-69 (000)	70-79 (000)	80+ (000)
	Emr	All	4,028	1,549	154	257	294	444	467	530	332
		B	682	127	20	49	51	101	111	132	91
		D	3,346	1,421	134	208	243	343	357	398	242
	Eur	All	9,668	254	48	241	541	1,178	1,704	2,602	3,098
		A	4,080	27	7	49	120	348	575	1,126	1,828
		B	1,952	170	24	67	126	264	370	510	420
		C	3,636	57	18	125	295	567	759	967	850
	Sear	All	14,091	3,309	536	1,030	1,325	2,029	2,211	2,301	1,350
		B	2,143	301	55	191	249	331	370	404	242
		D	11,948	3,008	481	838	1,076	1,699	1,841	1,896	1,107
	Wpr	All	11,389	1,068	203	491	780	1,544	2,066	2,828	2,408
		A	1,150	8	2	16	30	117	179	294	504
		B	10,238	1,060	201	475	750	1,427	1,888	2,534	1,903
Males	World		29,641	5,638	728	1,874	3,014	4,339	4,792	5,498	3,759
	Afr	All	5,417	2,257	211	522	890	609	387	364	177
	Afr	D	2,189	1,022	89	164	257	222	171	173	90
		E	3,228	1,235	122	357	633	387	216	191	87
	Amr	All	3,148	260	37	217	293	468	490	689	694
		A	1,382	21	5	40	84	192	220	371	448
		B	1,485	175	22	148	175	240	235	277	213
		D	282	65	9	29	34	36	35	41	33
	Emr	All	2,132	790	77	129	164	261	263	279	168
		B	382	67	11	31	31	62	64	71	45
		D	1,750	723	66	98	134	198	198	209	123
	Eur	All	4,947	142	30	180	397	817	1,072	1,288	1,020
		A	2,037	15	4	37	82	233	379	627	660
		B	1,053	93	15	46	86	177	224	253	159
		C	1,857	33	12	97	229	406	470	409	202
	Sear	All	7,661	1,655	258	540	789	1,208	1,272	1,256	683
		B	1,182	170	31	122	152	188	203	207	108
		D	6,479	1,484	226	418	637	1,020	1,069	1,049	575
	Wpr	All	6,337	534	115	287	479	977	1,307	1,621	1,017
		A	625	5	1	11	20	81	124	178	205

Sex	WHO subregions		all (000)	0-4 (000)	5-14 (000)	15-29 (000)	30-44 (000)	45-59 (000)	60-69 (000)	70-79 (000)	80+ (000)
		B	5,712	529	114	275	459	896	1,184	1,443	812
Females	World		25,978	5,255	709	1,745	2,048	2,654	3,215	4,784	5,569
	Afr	All	5,155	1,987	220	782	793	430	332	375	235
	Afr	D	2,056	906	92	232	227	161	148	175	114
		E	3,099	1,081	128	550	566	269	184	200	121
	Amr	All	2,725	209	27	80	145	290	348	591	1,035
		A	1,396	16	4	15	45	116	151	315	735
		B	1,099	138	15	46	79	147	169	241	264
		D	229	55	8	18	22	28	28	35	36
	Emr	All	1,896	758	77	128	129	184	205	251	164
		B	300	60	10	18	20	39	46	61	46
		D	1,596	698	68	110	109	145	159	190	119
	Eur	All	4,721	112	18	61	144	362	632	1,314	2,078
		A	2,043	12	3	12	38	115	197	499	1,168
		B	900	77	9	22	40	87	146	257	261
		C	1,779	23	6	28	66	160	289	558	648
	Sear	All	6,430	1,654	279	490	536	821	939	1,045	666
		B	961	131	23	69	97	142	167	197	134
		D	5,469	1,523	255	421	439	679	772	848	532
	Wpr	All	5,051	534	88	204	301	567	759	1,207	1,391
		A	525	3	1	4	10	37	55	116	299
		B	4,526	531	87	200	291	530	704	1,091	1,092

However, perhaps the widest disparities in mortality occur at the adult ages 15–59 years. In some southern African countries such as Botswana, Malawi, Namibia and Zambia, where HIV/AIDS is now a major public health problem, 65% or more of adults who survive to age 15 can be expected to die before age 60 on current mortality rates. In several others (e.g. Burundi, Lesotho, Mozambique, Rwanda, Swaziland and Zimbabwe) the risk exceeds 60%. The dramatic increase in $_{45}q_{15}$ in South Africa is also noteworthy, with estimated levels of 567 per 1000 and 502 per 1000 for males and females respectively in 2000, compared with levels around 250 to 350 per 1000 females and males respectively in 1990 (Timaeus 1999). At the other extreme, $_{45}q_{15}$ levels of 90–130 per 1000 are common in most developed countries for men, with risks as low as half this again for women.

The increasing disparity between levels of child and adult mortality in recent years is apparent from Figure 6 which contrasts average levels of $_{5}q_{0}$ with levels of $_{45}q_{15}$ for males and females separately based on the estimated country-specific life tables for 2000. Estimates of uncertainty are included for each point. The systematic departure of patterns of mortality in Eastern European countries, particularly males, from expected levels given the traditional relationship between child and adult mortality is clear, as is the vast uncertainty around estimated levels of adult mortality in several, primarily sub-Saharan African countries. By and large, uncertainty around child mortality levels is much less than around adult mortality, reflecting the knowledge gained from the extensive global survey programmes on child mortality levels and determinants over the past decades.

Figure 6. Adult mortality versus child mortality for 191 WHO Member States for the year 2000

Males

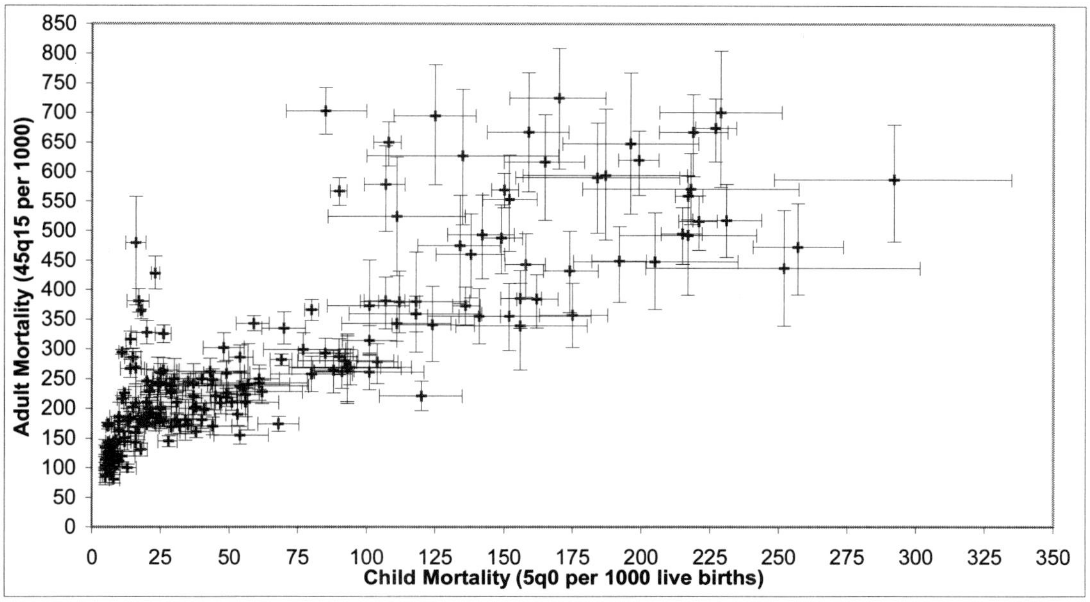

Females

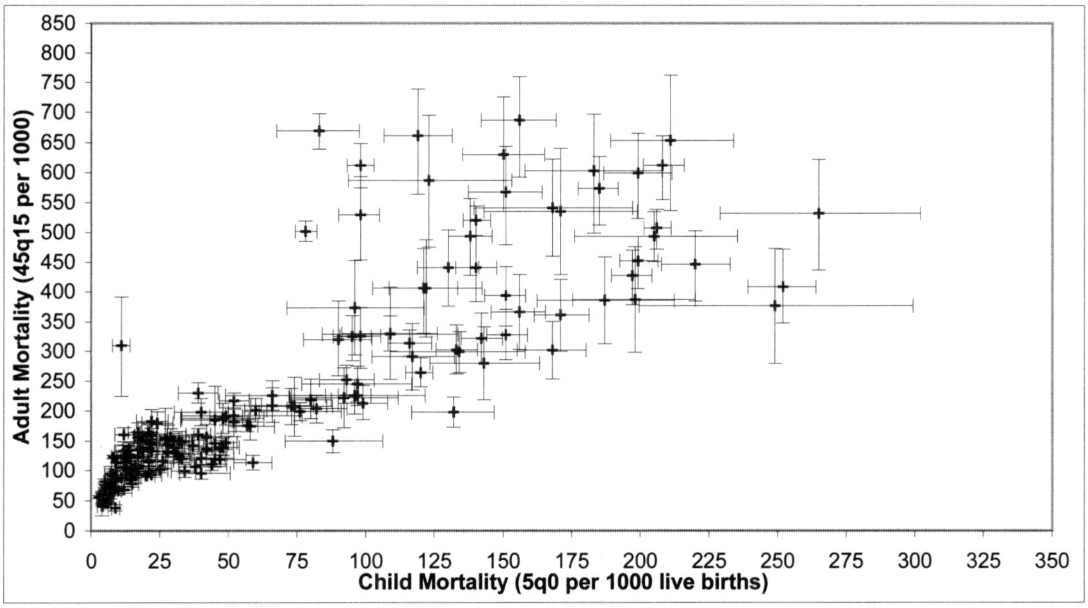

This is reflected in the age-pattern of uncertainty intervals shown for each country and region in the Appendix. Uncertainty bounds around q_x and l_x tend to widen around young adult ages, particularly in sub-Saharan Africa where reliable direct evidence on adult mortality levels is rare. Indeed, the width of these uncertainty intervals conveys as important a finding as the estimated central value, and argues for urgent investment in measuring adult mortality levels in those countries where the bounds remain unacceptably wide.

An alternative summary index of mortality conditions in a population is the average age of death, related to the average age of the population. In countries where child mortality is high, both the average age at death as well as the average age of the population will be comparatively low, and will increase more or less in tandem as health development improves. This is clear from Figure 7 which shows the relationship separately for males and females based on the life tables for each country in 2000. Interestingly, beyond an average age of the population of about 30, the average age at death rises somewhat more slowly than at lower average age levels, particularly for females. This attenuation is in part due to declining fertility in low mortality countries over the past 50 years but also reflects the fact that by the time average age at death reaches 60 years or so, much of the progress in reducing child and adult mortality has been achieved, and hence further reductions will only lead to progressively slower increases in average population age. As the figure also shows, there remain extraordinary large differences in average age at death among WHO Member States in 2000, ranging from as low as 15.5 in Niger (both sexes combined), 19.9 in Afghanistan, 20.3 and 20.5 in Somalia and Sierra Leone, respectively, to almost 80 years in Sweden (78.5), Norway (77.1) and Greece (76.8). This fourfold difference is significantly larger than the range in life expectancy reported earlier.

Figure 7. Average age of death and population, 2000, 191 countries

Males

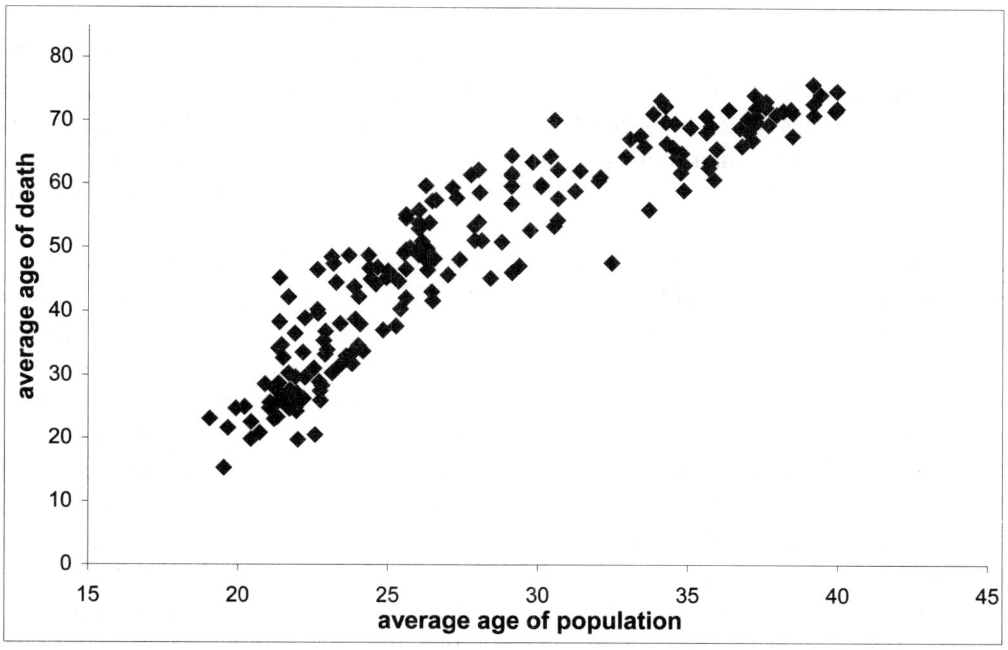

Females

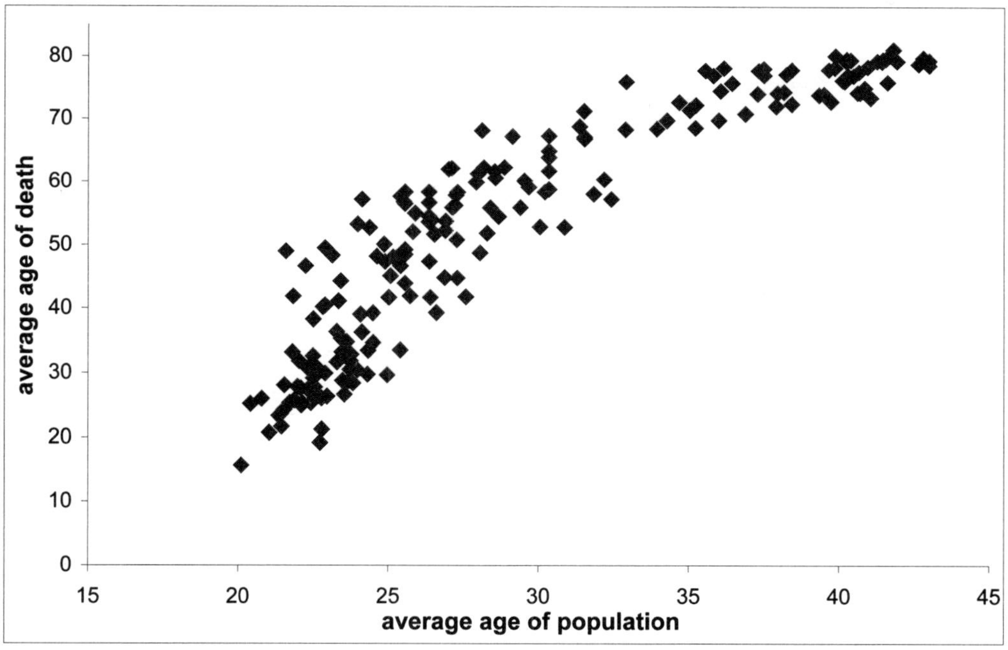

Just under 56 million people are estimated to have died in 2000, almost 30 million of whom were males (see Table 7). Worldwide, 10.9 million children below age 5 died in that year, about 3/4 of them in developing regions of Africa and south Asia. Another 10 million or so died in each of the oldest age groups 70–79 and 80+, the vast majority in the lower mortality A, B and C subregions. There were many more deaths in young adults ages 15–59 years (15.6 million), than in children ages 0-14 years (12.3 million), emphasizing the need for a more expansive view of policies to prevent premature death to address leading causes of young adult death as well.

Worldwide, life expectancy at birth in 2000 was estimated at 64.9 years (62.7 for males, 67.2 for females) (see Table 8). The very substantial differences in age patterns of mortality among subregions evident from Table 7 translate into wide variations in average regional life expectancy, with levels ranging from just over 44 years (both sexes) in AfrE where HIV/AIDS is most prevalent to an average level almost twice as high (80.9 years) in WprA. Average life expectancy attained 70 years or more in six subregions, interestingly not in EurB or C. In particular, average male life expectancy at birth in EurC was barely 60 years, among the lowest of any subregion in the world outside of Africa.

Table 8. Life expectancy at birth (years) by WHO subregion, 2000

Sub-Region	males	females	both sexes
AfrD	50.3	52.4	51.3
AfrE	43.5	45.2	44.3
AmrA	74.1	79.6	76.9
AmrB	67.5	74.5	70.9
AmrD	63.5	68.4	65.9
EmrB	68.6	71.6	70.0
EmrD	59.1	61.0	60.0
EurA	74.8	81.2	78.0
EurB	65.5	72.3	68.8
EurC	60.3	72.1	66.0
SearB	64.1	68.9	66.4
SearD	59.7	62.4	61.0
WprA	77.3	84.2	80.9
WprB	68.2	72.7	70.4
World	**62.7**	**67.2**	**64.9**

Discussion

The life table, and its associated parameters, is a key input into the assessment of how well, or poorly, heath systems are performing. Life tables have numerous other uses in epidemiology, demography and economics and the availability of annual, current life tables for all 191 WHO Member States should improve the quality and relevance of the analytical base for such uses. Careful evaluation of the data upon which life tables are to be constructed is essential if the results are to be used with any confidence. This report has tried to set out the countries for which vital registration appears to be working sufficiently well to prepare life tables with no (or very minimal) adjustments to data. For countries where this is not the case, we have outlined the procedures used to adjust the data, but the results obtained are very much dependent on a considered judgement of the evidence. This degree of subjectivity is unavoidable given current demographic practices for adjusting data; and further work is required to explore how formal statistical methods might be better applied to reduce this subjectivity in judging the extent of under-enumeration of deaths.

This analysis has also highlighted the vast degree of uncertainty and ignorance that exists with respect to levels and patterns of adult mortality in developing countries. For one-third of WHO's Member States, probably accounting for a similar proportion of deaths, little or nothing is reliably known about levels of adult mortality, particularly among younger adults below age 60. For these countries, there is little alternative but to follow the classical demographic approach of constructing adult mortality levels from child mortality rates on the assumption that the two are linked in some predictable fashion. However, evidence from the past 30 years or so suggests that this is not necessarily the case as new hazards such as HIV have emerged, or old ones, such as alcohol abuse, have become more extreme in some populations. We have attempted to avoid such distortions by creating a new model life table system to estimate adult death rates, but the results will remain uncertain until verification is eventually possible from systems which reliably capture deaths.

It is unlikely that such systems will be established in the near future in the countries where they are needed. Yet good estimates of adult mortality are required now for planning and monitoring in the health sector. Questions on deaths occurring in households, when asked with sufficient care and rigour, can yield useful, current data on adult mortality levels and patterns and should be routinely added to censuses and surveys. The forthcoming WHO World Health Survey will provide an ideal opportunity to expand our knowledge of adult mortality. At the same time, a focus on collecting data on child survival in the handful of countries (Category V) where no recent data are available will greatly assist programmes concerned with child health promotion (Hill and Yazbeck 1994).

References

Ahmad OB, Lopez AD, Inoue M (2000) Trends in child mortality: a reappraisal. *Bulletin of the World Health Organization,* **78**(10):1175-1191.

Bennett NG, Horiuchi S (1984) Mortality estimation from registered deaths in less developed countries. *Demography,* **21**(2):217-233.

Brass W (1971) On the scale of mortality. In: *Biological aspects of demography.* Brass W, ed. Taylor & Francis, London, UK.

Chinese Academy of Preventive Medicine (1997) *Annual report on the disease surveillance system, 1996.* Beijing.

Coale A, Guo G (1989) Revised regional model life tables at very low levels of mortality. *Population Index,* **55** (4):613-643.

Hill K (1987) Estimating census and death registration completeness. *Asian and Pacific Population Forum,* **1**(3):8-24.

Hill K, Pande R, Mahy M, Jones G (1999) *Trends in child mortality in the developing world: 1960-1996.* UNICEF, Division of Evaluation and Planning, New York, NY.

Hill K, Yazbeck A (1994) *Trends in under-five mortality, 1960-90. Estimates for 84 developing countries.* World Bank, Washington, DC.

Lopez AD (1983) The sex mortality differential in developed countries. In: *Sex differentials in mortality: trends, determinants and consequences.* Lopez AD, Ruzicka LT, eds. Australian National University Press, Canberra.

Lopez AD, Salomon JA, Ahmad OB, Murray CJL, Mafat D (2000) *Life tables for 191 countries: data, methods, results*. GPE discussion paper no 9. World Health Organization/ Global Programme on Evidence for Health Policy, Geneva.

Mari Bhat PN (2000) *Estimating life tables for India in the 1990s*. Unpublished paper prepared for GPE. World Health Organization/Global Programme on Evidence for Health Policy, Geneva.

Martin LG (1980) A modification for use in destabilized populations of Brass' technique for estimation completeness of registration. *Population Studies,* **34**(2):381-395.

Murray CJL, Ferguson BD, Lopez AD, Guillot M, Salomon JA, Ahmad OB (2001) *Modified logit life table system: principles, empirical validation and application*. GPE discussion paper no 39. World Health Organization/Global Programme on Evidence for Health Policy, Geneva.

Perks W (1932) On some experiments on the graduation of mortality statistics. *Journal of the Institute of Actuaries,* **63**:12-57.

Preston SH (1984) Use of direct and indirect techniques for estimating the completeness of death registration systems. In: *Data bases for mortality measurement.* UN, ed. United Nations, New York, NY.

Salomon JA, Mathers CD, Murray CJL, Ferguson B (2001) *Methods for life expectancy and healthy life expectancy uncertainty analysis*. GPE discussion paper no 10. World Health Organization/Global Programme on Evidence for Health Policy, Geneva.

Salomon JA, Murray CJL (2001) Modelling HIV/AIDS epidemics in sub-Saharan Africa using seroprevalence data from antenatal clinics. *Bulletin of the World Health Organization,* **79**(7):586-692.

Stanton E, Abderrahim A, Hill K (2000) An assessment of DHS maternal mortality indicators. *Studies in Family Planning,* **31**(2):111-124.

Timaeus IM (1999) Mortality in sub-Saharan Africa. In: *Health and mortality. Issues of global concern.* Proceedings of the symposium on health and mortality, November 1997, Brussels. Chamie J, Cliquet RL, eds. United Nations Population Division, New York, NY.

UN (2001) *World population prospects: the 2000 revision*. United Nations, New York, NY.

US Bureau of the Census, *International data base (IDB)*. United States Bureau of the Census, International Programs Center (IPC), Washington, DC. www.census.gov/ipc/www/idbnew.html

World Bank (2001) *World development indicators 2001*. The World Bank, Washington, DC.

Table 9. Availability of vital registration data on mortality in the WHO database, 1980-2000 (as of 15 September 2001)

Country	80	81	82	83	84	85	86	87	88	89	90	91	92	93	94	95	96	97	98	99	00
Category I: Complete Vital Statistics (coverage 95%+)																					
Andorra													x		x	x	x	x	x		x
Antigua and Barbuda				x			x	x			x	x	x	x	x	x					
Argentina	x	x	x	x	x	x	x	x	x	x	x	x	x	x	x	x	x			x	
Australia	x	x	x	x	x	x	x	x	x	x	x	x	x	x	x	x	x	x	x	x	
Austria	x	x	x	x	x	x	x	x	x	x	x	x	x	x	x	x	x	x	x	x	x
Bahamas	x	x	x	x	x	x	x	x	x	x	x	x	x	x	x	x	x	x			
Bahrain	x	x	x		x	x	x	x	x	x	x	x	x	x	x	x	x	x	x	x	x
Barbados	x	x	x	x	x	x	x	x	x	x	x	x	x	x	x	x					
Belarus		x	x	x	x	x	x	x	x	x	x	x	x	x	x	x	x	x	x	x	
Belgium	x	x	x	x	x	x	x	x	x	x	x	x	x	x	x	x	x	x	x		
Bosnia and Herzegovina						x	x	x	x	x	x	x									
Bulgaria	x	x	x	x	x	x	x	x	x	x	x	x	x	x	x	x	x	x	x	x	
Canada	x	x	x	x	x	x	x	x	x	x	x	x	x	x	x	x	x	x			
Chile	x	x	x	x	x	x	x	x	x	x	x	x	x	x	x	x	x	x	x		
Cook Islands	x	x	x	x	x	x	x	x	x	x	x	x	x	x	x	x	x	x	x	x	
Costa Rica	x	x	x	x	x	x	x	x	x	x	x	x	x	x	x	x		x		x	
Croatia			x	x	x	x	x	x	x	x	x	x	x	x	x	x	x	x	x	x	
Cuba	x	x	x	x	x	x	x	x	x	x	x	x	x	x	x	x	x	x	x	x	
Cyprus	x	x	x	x	x	x	x	x	x	x	x	x	x	x	x	x	x	x	x	x	
Czech Republic	x	x	x	x	x	x	x	x	x	x	x	x	x	x	x	x	x	x	x	x	x
Denmark	x	x	x	x	x	x	x	x	x	x	x	x	x	x	x	x	x	x	x		
Dominica	x	x	x	x	x	x	x	x	x	x	x	x	x	x	x						
Estonia		x	x	x	x	x	x	x	x	x	x	x	x	x	x	x	x	x	x	x	
Fiji	x	x	x	x	x	x	x	x					x	x	x	x	x	x	x	x	
Finland	x	x	x	x	x	x	x	x	x	x	x	x	x	x	x	x	x	x	x		
France	x	x	x	x	x	x	x	x	x	x	x	x	x	x	x	x	x	x	x		
Germany	x	x	x	x	x	x	x	x	x	x	x	x	x	x	x	x	x	x	x	x	
Greece	x	x	x	x	x	x	x	x	x	x	x	x	x	x	x	x	x	x	x		
Grenada					x				x						x	x	x				
Hungary	x	x	x	x	x	x	x	x	x	x	x	x	x	x	x	x	x	x	x	x	x

Country	80	81	82	83	84	85	86	87	88	89	90	91	92	93	94	95	96	97	98	99	00
Iceland	x	x	x	x	x	x	x	x	x	x	x	x	x	x	x	x	x	x	x		
Ireland	x	x	x	x	x	x	x	x	x	x	x	x	x	x	x	x	x	x	x		
Israel	x	x	x	x	x	x	x	x	x	x	x	x	x	x	x	x	x	x	x		
Italy	x	x	x	x	x	x	x	x	x	x	x	x	x	x	x	x	x	x	x		
Jamaica	x	x	x	x	x	x	x	x	x	x	x	x									
Japan	x	x	x	x	x	x	x	x	x	x	x	x	x	x	x	x	x	x	x		
Kuwait	x	x	x	x	x	x	x	x	x	x		x	x	x	x	x	x	x	x	x	
Latvia	x	x	x	x	x	x	x	x	x	x	x	x	x	x	x	x	x	x	x	x	
Lithuania		x	x	x	x	x	x	x	x	x	x	x	x	x	x	x	x	x	x	x	
Luxembourg	x	x	x	x	x	x	x	x	x	x	x	x	x	x	x	x	x	x	x	x	x
Malta	x	x	x	x	x	x	x	x	x	x	x	x	x	x	x	x	x	x	x	x	
Mauritius	x	x	x	x	x	x	x	x	x	x	x	x	x	x	x	x	x	x	x	x	
Mexico	x	x	x	x	x	x	x	x	x	x	x	x	x	x	x	x	x	x			x
Monaco		x	x	x			x	x													
Netherlands	x	x	x	x	x	x	x	x	x	x	x	x	x	x	x	x	x	x	x	x	
New Zealand	x	x	x	x	x	x	x	x	x	x	x	x	x	x	x	x	x	x	x		
Norway	x	x	x	x	x	x	x	x	x	x	x	x	x	x	x	x	x	x	x		
Poland	x	x	x	x	x	x	x	x	x	x	x	x	x	x	x	x	x	x	x	x	
Portugal	x	x	x	x	x	x	x	x	x	x	x	x	x	x	x	x	x	x	x	x	
Qatar		x	x	x		x	x	x	x	x	x		x	x	x	x	x	x			
Republic of Moldova		x	x	x	x	x	x	x	x	x	x	x	x	x	x	x	x	x	x	x	
Romania	x	x	x	x	x	x	x	x	x	x	x	x	x	x	x	x	x	x	x	x	
Russian Federation	x	x	x	x	x	x	x	x	x	x	x	x	x	x	x	x	x	x	x	x	
Saint Kitts and Nevis	x	x	x	x	x	x	x	x	x	x	x	x	x	x	x	x					
Saint Lucia	x	x	x	x	x	x	x	x	x	x		x	x	x	x	x					
Saint Vincent and the Grenadines	x		x	x	x	x	x	x	x				x			x	x				
San Marino	x	x	x	x	x	x	x	x	x	x	x	x	x	x	x	x	x	x	x		
Seychelles	x	x	x	x	x	x	x	x	x	x	x	x	x	x	x	x	x	x	x		
Singapore	x	x	x	x	x	x	x	x	x	x	x	x	x	x	x	x	x	x	x	x	x
Slovakia			x	x	x	x	x	x	x	x	x	x	x	x	x	x	x	x	x	x	
Slovenia			x	x	x	x	x	x	x	x	x	x	x	x	x	x	x	x	x	x	
Spain	x	x	x	x	x	x	x	x	x	x	x	x	x	x	x	x	x	x	x		
Sri Lanka	x	x	x	x	x	x	x	x	x	x		x	x	x	x	x	x				
Suriname	x	x	x	x	x	x	x	x	x	x	x	x	x	x	x	x	x	x			
Sweden	x	x	x	x	x	x	x	x	x	x	x	x	x	x	x	x	x	x	x	x	x

Country	80	81	82	83	84	85	86	87	88	89	90	91	92	93	94	95	96	97	98	99	00
Switzerland	x	x	x	x	x	x	x	x	x	x	x	x	x	x	x	x	x	x	x		
TFYR Macedonia			x	x	x	x	x	x	x	x	x	x	x	x	x	x	x	x			
Trinidad and Tobago	x	x	x	x	x	x	x	x	x	x	x	x	x	x	x	x					
Ukraine		x	x	x	x	x	x	x	x	x	x	x	x	x	x	x	x	x	x	x	x
United Kingdom	x	x	x	x	x	x	x	x	x	x	x	x	x	x	x	x	x	x	x	x	
United States of America	x	x	x	x	x	x	x	x	x	x	x	x	x	x	x	x	x	x	x	x	
Uruguay	x	x	x	x	x	x	x	x	x	x	x	x	x	x		x	x	x	x		
Venezuela	x	x	x	x	x	x	x	x	x	x	x	x	x	x	x						
Yugoslavia			x	x	x	x	x	x	x	x	x	x	x	x	x	x	x	x			
Category II: Incomplete Vital Statistics																					
Albania	x				x	x	x	x	x	x	x	x	x	x	x	x	x	x	x		
Algeria	x	x	x			x	x														
Armenia		x	x			x	x	x	x	x	x	x	x	x	x	x	x	x	x	x	
Azerbaijan		x	x	x	x	x	x	x	x	x	x	x	x	x	x	x	x	x	x	x	x
Belize	x	x	x	x	x	x	x	x	x	x	x	x	x	x	x	x	x				
Brazil	x	x	x	x	x	x	x	x	x	x	x	x	x	x	x	x					
Brunei Darussalam			x	x	x	x	x	x	x	x	x	x	x	x	x	x	x	x	x	x	x
Cape Verde	x	x	x	x	x	x					x	x							x		
Colombia			x	x	x	x	x	x	x	x	x	x	x	x	x	x					
Dominican Republic	x	x	x	x	x	x												x	x	x	
Ecuador	x	x	x	x	x	x	x	x	x	x	x	x	x	x	x	x	x	x	x		
Egypt	x	x	x	x	x	x	x	x	x	x	x	x	x	x	x	x				x	
El Salvador	x	x	x	x	x	x	x	x	x	x	x	x	x	x		x	x	x	x		
Georgia		x	x	x	x	x	x	x	x	x	x	x	x		x	x	x				
Guatemala	x	x	x	x	x	x	x	x	x			x	x	x		x	x	x			
Guyana					x									x	x	x	x				
Honduras	x	x	x	x	x																
Iran (Islamic Republic of)				x	x	x	x	x				x			x	x	x	x	x	x	
Kazakhstan		x	x	x	x	x	x	x	x	x	x	x	x	x	x	x	x	x	x	x	
Kyrgyzstan		x	x	x	x	x	x	x	x	x	x	x	x	x	x	x	x	x	x	x	
Malaysia											x	x	x	x	x	x	x	x	x		
Maldives					x	x	x	x	x	x	x	x	x	x		x	x	x	x		
Marshall Islands							x	x	x	x	x	x	x	x	x	x	x	x			
Mongolia								x	x	x		x	x	x	x	x	x	x	x	x	
Morocco								x	x	x	x	x	x	x	x	x	x	x	x		

Country	80	81	82	83	84	85	86	87	88	89	90	91	92	93	94	95	96	97	98	99	00
Nicaragua								x	x	x	x	x	x	x	x		x	x	x		
Niue	x	x	x	x	x	x	x	x	x	x	x	x	x	x	x	x	x	x	x		
Palau						x		x	x	x	x	x	x	x	x	x	x	x	x	x	
Panama	x	x	x	x	x	x	x	x	x	x	x	x	x	x	x	x	x	x			
Papua New Guinea	x	x	x	x	x			x	x	x	x	x	x	x	x	x	x	x	x		
Paraguay	x	x	x	x	x	x	x	x	x	x	x	x	x		x						
Peru	x	x	x	x	x	x	x	x	x	x											
Philippines	x	x	x	x	x	x	x	x	x	x	x	x	x	x	x	x	x				
Republic of Korea	x	x	x	x	x	x	x	x	x	x	x	x	x	x	x	x	x	x	x		
Samoa	x												x	x							
South Africa														x	x	x					
Syrian Arab Republic				x	x	x													x		x
Tajikistan		x	x			x	x	x	x	x	x	x	x	x	x	x					
Thailand	x	x	x	x	x	x	x	x	x	x	x	x	x	x	x	x	x	x	x	x	
Tonga																	x	x	x		
Tunisia	x							x	x	x				x	x	x		x	x		
Turkey													x	x	x	x	x	x			
Turkmenistan		x	x			x	x	x	x	x	x	x	x	x	x	x	x	x	x		
Tuvalu												x	x	x	x						
Uzbekistan		x	x	x	x	x	x	x	x	x	x	x	x	x	x	x	x	x	x		
Zimbabwe			x								x	x	x	x		x		x			
Category III: Sample Registration and Surveillance Systems																					
Bangladesh											x	x	x	x	x	x	x	x	x	x	
China								x	x	x	x	x	x	x	x	x	x	x	x	x	
India											x	x	x	x	x	x	x	x	x		
United Rep. of Tanzania													x	x	x	x	x	x	x	x	

Mortalité mondiale en 2000 : tables de mortalité pour 191 pays

A.D. LOPEZ

O.B. AHMAD

M. GUILLOT

B.D. FERGUSON

J.A. SALOMON

C.J.L. MURRAY

K.H. HILL

Organisation mondiale de la Santé

Les désignations employées et la présentation du matériel dans cette publication ne traduisent pas l'expression d'une quelconque opinion de la part du Secrétariat de l'Organisation mondiale de la Santé sur le statut juridique d'un pays, d'un territoire, d'une ville ou d'une région, sur leurs services officiels ou sur la délimitation de leurs frontières ou de leurs limites.

La référence à des entreprises spécifiques ou à des produits de certains fabricants ne signifie pas qu'elles ou ils soient approuvé(e)s ou recommandé(e)s par l'Organisation mondiale de la Santé de préférence à d'autres de nature semblable non mentionné(e)s. Sauf erreur ou omission, les produits brevetés se distinguent par une capitale à la première lettre de leur nom.

Composé et imprimé au Canada

Données sur le catalogage avant publication de la bibliothèque de l'OMS

Mortalité mondiale en 2000 : Tables de mortalité pour 191 pays / A. D. Lopez ... [et coll.].

1.Tables de mortalité 2.Statistiques de l'état civil 3.Collecte de données – méthodes

4.Organisation mondiale de la Santé I. Lopez, Alan D.

ISBN 92 4 156204 8 (classement NLM : HB 1322)

Table des matières

Collaborateurs

Alan D. LOPEZ, Organisation mondiale de la Santé, Bases factuelles et Information à l'appui des politiques (BFI) – Épidémiologie et charge de morbidité (EBD)

Omar B. AHMAD, École de santé publique, Université du Ghana, B.P. LG13, Legon, Accra, Ghana

Michel GUILLOT, Centre de démographie et d'écologie, Université du Wisconsin à Madison, 1180, promenade de l'Observatoire, Madison (WI) 53706 États-Unis d'Amérique

Brodie D. FERGUSON, Organisation mondiale de la Santé, BFI/EBD

Joshua A. SALOMON, Organisation mondiale de la Santé, BFI/EBD

Christopher J.L. MURRAY, Organisation mondiale de la Santé, Directeur exécutif– EIP

Kenneth H. HILL, directeur du Centre Hopkins sur la population, directeur intérimaire du Centre Hopkins sur la démographie du vieillissement, département des sciences de la population et de la santé familiale, Université Johns Hopkins, École d'hygiène et de santé publique, 615 Nord, rue Wolfe, pièce W4027, Baltimore (MD) 21205 É.-U. khill@jhsph.edu

Les auteurs tiennent à remercier tout particulièrement Mie Inoue pour les longues heures passées à rassembler des statistiques de mortalité de diverses sources et à présenter les tableaux et les graphiques qui figurent dans le présent ouvrage, qui n'aurait pu être publié sans sa généreuse contribution.

Glossaire

EDS	Enquêtes démographiques et sanitaire
PSÉ	Points de surveillance épidémiologique
EVAES	Espérance de vie en bonne santé
MICS	Enquête à indicateur multiple sur les enfants
SE	Système d'enregistrement

Introduction

L'objectif principal de ce rapport est de compiler un ensemble complet de tables de mortalité de tous les États Membres de l'OMS pour l'année 2000 afin de comparer les profils actuels de mortalité par âge et par sexe entre les pays en utilisant des méthodes comparables. Les informations démographiques de base sur le nombre de décès par âge et par sexe dans une population constituent un facteur essentiel pour la détermination et l'évaluation des politiques et des programmes de santé. Depuis 1999, l'OMS a entrepris de calculer les tables de mortalité pour tous les États Membres. Ces tables de mortalité ont plusieurs utilités et forment la base de toutes les estimations de l'OMS au sujet des profils et des niveaux de mortalité à l'échelle mondiale. Une des principales utilités de ces tables de mortalité réside dans l'élaboration de l'espérance de vie ajustée de l'état de santé (EVAES), qui constitue l'indicateur principal des niveaux de santé de la population utilisé par l'OMS qui sont publiés chaque année dans le Rapport sur la santé dans le monde.

L'élaboration d'une table de mortalité nécessite des données fiables sur les taux de mortalité d'une population par âge et par sexe. La source la plus fiable de données de ce genre est un registre d'état civil où tous les décès sont enregistrés. Les décès à chaque âge sont reliés à la taille de la population dans ce groupe d'âge, qui est habituellement estimé à partir des recensements, ou au fichier permanent de l'ensemble des naissances, des décès et des migrations. Les taux de mortalité par âge et par sexe qui en résultent sont ensuite utilisés pour calculer une table de mortalité.

Bien que des lois prescrivent, presque partout dans le monde, l'enregistrement des décès, le coût de l'élaboration et de l'entretien d'un système d'enregistrement des naissances et des décès implique que des données fiables sur l'enregistrement courant sont généralement disponibles seulement dans les pays plus développés au niveau économique. Des données nationales assez complètes pour calculer les tables de mortalité étaient seulement disponibles vers la fin des années 1990 pour 74 pays, ce qui couvre environ le quart des décès estimés en 2000 (voir Tableau 1). En l'absence de registre d'état civil complet, l'enregistrement par sondage ou des renseignements fiables sur la mortalité juvénile sont utilisés pour évaluer les tables de mortalité en s'aidant de méthodes démographiques indirectes. Dans de nombreux pays en développement, cette méthode a été grandement facilitée par la disponibilité d'estimations fiables sur la mortalité juvénile portant sur les années 1980 et 1990 à partir du programme des enquêtes démographiques et sanitaires, et plus récemment grâce au programme d'enquête à indicateur multiple sur les enfants (MICS), qui est dirigé par l'UNICEF.

Plusieurs organismes internationaux et autres centres d'études démographiques préparent en routine les estimations de mortalité nationales ou les compilations de tables de mortalité dans le cadre de leur activité de suivi du secteur. Ainsi, l'UNICEF passe périodiquement en revue les données disponibles sur la mortalité juvénile pour évaluer les interventions et le degré de réalisation des objectifs de survie des enfants (Hill et al. 1999). Une récente mise à jour des tendances au niveau de la mortalité juvénile durant les années 1990 vient aussi d'être tout juste terminée (Ahmad et al. 2000). Trois organismes ou organisations, la Division de la population de l'Organisation des Nations Unies, la Banque mondiale et le United States Census Bureau ont tous produit des compilations internationales de tables de mortalité et dans le cas de la Division de la population, au moins, les tables continuent d'être mises à jour tous les deux ans (UN 2001; US Bureau of the Census 0 AD; World Bank 2001). Ces diverses études s'appuient généralement sur les mêmes sources de données, soit les recensements, les enquêtes et les registres d'état civil. Mais elles peuvent produire des résultats très différents en raison de différences quant au moment de la disponibilité des données, au jugement sur la nécessité d'adapter les données de base et la méthodologie pour ce faire, et aux techniques d'évaluation et de choix des modèles.

Une analyse comparative de ces divers exercices met en relief la variabilité des résultats découlant de procédures et de jugements différents. Par exemple, en Inde, l'ajustement des données d'état civil pour la sous-déclaration de la mortalité adulte, estimée à 13–14 % en 1999–2000 (Mari Bhat 2000), produit une estimation de 9,8 millions de décès en 2000, soit 1 million de plus que l'estimation démographique l'Organisation des Nations Unies (UN 2001).

De telles différences ne sont pas sans importance et ont des conséquences importantes sur le suivi, l'évaluation et la réorientation des programmes de santé publique dans les pays de même qu'au niveau mondial. Bien qu'il soit assurément souhaitable de développer un ensemble unique de tables de mortalité pour tous les pays du monde, le jugement des experts, la disponibilité des données et le moment des évaluations périodiques continueront de varier. Étant donné que l'OMS a besoin d'estimations annuelles pour les tables de mortalité dans le cadre de l'évaluation continue de la performance des systèmes de santé, et qu'il y a une préférence pour un système de table type de mortalité inspiré d'une modification du système logit de Brass plutôt que des autres familles de tables type de mortalité (Murray et al. 2001), l'OMS a construit un nouvel ensemble de tables de mortalité dont les résultats, pour 2000, sont présentés dans les Annexes.

Le présent rapport fournit des détails sur les méthodes et les sources de données utilisées pour préparer ces tables de mortalité. Nous commençons par une brève analyse des sources, des types et de la qualité des données disponibles. Nous examinons les différentes sources de données et les problèmes et les difficultés que comporte leur utilisation pour produire des tables de mortalité. Nous effectuons aussi une brève analyse des approches principales utilisées par l'OMS pour évaluer les paramètres du système logit de Brass (α, β) pour chaque pays. Pour les pays qui possèdent une longue série de données d'état civil, une analyse chronologique différée a aussi été utilisée. Pour tous les autres pays, α et β ont été estimés soit à partir de séries chronologiques plus courtes de données d'état civil, soit à partir d'enquêtes ou de données de la surveillance sur la mortalité juvénile et adulte. Cela est suivi d'une discussion sur la façon dont les données démographiques de base pour la méthode, aux niveaux ${}_5q_0$ et ${}_{45}q_{15}$, ont été évalués pour les pays. Un court résumé des principaux résultats est fourni et des tables de mortalité détaillées par pays et par région pour les 191 États Membres de l'OMS et 14 sous-régions[1] sont présentées en annexe. Étant donné la place prépondérante qu'occupe la mortalité juvénile dans les modèles et niveaux de mortalité globaux de nombreux pays, les graphes des données disponibles pour évaluer la mortalité juvénile de 2000 sont illustrées séparément.

Sources de données et ajustements

Données d'état civil

Idéalement, les tables de mortalité devraient être construites à partir d'une longue série historique de données sur la mortalité provenant des regitres d'état civil où les décès et la population de la population à risque de fait (et à l'occasion de droit) sont entièrement couverts par le système. Pour calculer les tables de mortalité d'une année en particulier (c.-à-d. 2000) pour laquelle les données 'état civil sur les décès ne sont pas encore disponibles pour des raisons administratives, des extrapolations à court terme de la dernière année disponible sont nécessaires. Ce calcul nécessite une série chronologique adéquate de données, avec au moins 15 à 20 années de statistiques de mortalité. Le Tableau 9 présente les données d'état civil sur la mortalité dont l'Organisation mondiale de la Santé dispose pour l'évaluation des tables de mortalité.

Premièrement, depuis 1980, les données d'état civil ont été systématiquement évaluées quant à leur complétude, en utilisant une série de techniques démographiques dont la méthode croissance-équilibre de Brass (Brass 1971), la méthode croissance-équilibre généralisée (Hill 1987) et la technique Bennett et Horiuchi (Bennett and Horiuchi 1984). Ces deux dernières méthodes nécessitent des données sur la distribution par âge et par sexe de la population provenant de deux recensements contigus, ainsi que sur les décès enregistrés entre les recensements. Ainsi, la méthode croissance-équilibre simple a été plus couramment utilisée pour évaluer la complétude de l'enregistrement des décès puisque les données de base requises (décès et population par âge et par sexe pour une année ou une période donnée) étaient plus faciles à trouver. D'un autre côté, la technique suppose que la population en cause est stable, ce qui est improbable dans le cas de nombreux pays en développement, et elle est jusqu'à un certain point arbitraire pour fixer les points utilisés pour évaluer l'ampleur de la sous-déclaration (Preston 1984).

[1] Afin d'assister l'analyse démographique, les 191 États Membres ont été divisés en cinq strates de mortalité selon leur niveau de mortalité juvénile (5q0) et de mortalité des hommes adultes (45q15), comme suit : A = mortalité juvénile très faible, mortalité adulte très faible ; B = mortalité juvénile faible, mortalité adulte faible ; C = mortalité juvénile faible, mortalité adulte élevée ; D = mortalité juvénile élevée, mortalité adulte élevée ; E = mortalité juvénile élevée, mortalité adulte très élevée.
La matrice définie par les six régions de l'OMS (Afr = Afrique ; Amr = les Amériques ; Emr = Méditérannée orientale ; Eur = Europe ; Sear = Asie du Sud-Est ; Wpr = Pacifique occidental) et les cinq strates de mortalité conduisent à 14 sous-régions, puisque toutes les strates de mortalité ne sont pas représentées dans chaque région, c'est-à-dire : AfrD, AfrE, AmrA, AmrB, AmrD, EmrB, EmrD, EurA, EurB, EurC, SearB, SearD, WprA, WprB.

L'application de ces méthodes à chaque pays ayant des données des statistiques d'état civil a conduit au choix de 74 pays dont les données des statistiques d'état civil ont été jugées suffisamment complètes pour calculer les tables de mortalité (voir Tableau 1). Ces pays sont inscrits sous la Catégorie 1 dans le Tableau 9. Pour chacun de ces pays, la dernière année (ou les dernières années pour les petites populations) a été utilisée pour étabir la distribution normalisée des valeurs l_x par âge et par sexe, séparément. En utilisant cette norme, une série chronologique en α et β, les deux paramètres du système logit de Brass, a été produite. Les techniques de séries chronologiques ont ensuite été utilisées pour extrapoler les valeurs de α et β vers l'année 2000, à partir de laquelle la table de mortalité de 2000 a été obtenue. Elles sont décrites plus en détail dans la section des méthodes. Pour ce qui est des petites populations, par ex. les Îles Cook, des moyennes mobiles de trois ou quatre ans ont été utilisées pour l'analyse des séries chronologiques plutôt que les données d'une seule année.

Tableau 1. Sources des données sur la mortalité (nombre de pays) pour les sous-régions de l'OMS en 2000

Sous-région	**Catégorie I** : Statistiques de l'état civil complètes (couverture de 95%+)	**Catégorie II** : Statistiques de l'état civil incomplètes	**Catégorie III** : Systèmes d'enregistrement et de surveillance	**Catégorie IV** : Mortalité juvénile estimée à partir des enquêtes et des recensements	**Catégorie V** : Aucune donnée récente sur la mortalité juvénile ou adulte	Nombre de pays
AfrD	2	2	0	18	4	26
AfrE	0	2	1	14	3	20
AmrA	3	0	0	0	0	3
AmrB	17	9	0	0	0	26
AmrD	0	4	0	2	0	6
EmrB	4	3	0	6	0	13
EmrD	0	2	0	5	2	9
EurA	26	0	0	0	0	26
EurB	7	9	0	0	0	16
EurC	8	1	0	0	0	9
SearB	1	1	0	1	0	3
SearD	0	1	2	3	1	7
WprA	4	1	0	0	0	5
WprB	2	11	1	8	0	22
Monde	74	46	4	57	10	191

Tableau 2. Sources des données sur la mortalité (% des décès couverts) pour les sous-régions de l'OMS en 2000

Sous-région	**Catégorie I** : Statistiques de l'état civil complètes (couverture de 95%+)	**Catégorie II** : Statistiques de l'état civil incomplètes	**Catégorie III** : Systèmes d'enregistrement et de surveillance	**Catégorie IV** : Mortalité juvénile estimée à partir des enquêtes et des recensements	**Catégorie V** : Aucune donnée récente sur la mortalité juvénile ou adulte	Décès en tant En tant que % du total mondial (estimations de l'OMS)
AfrD	0 %	4 %	0 %	89 %	7 %	8 %
AfrE	0 %	13 %	9 %	71 %	7 %	11 %
AmrA	100 %	0 %	0 %	0 %	0 %	5 %
AmrB	39 %	61 %	0 %	0 %	0 %	5 %
AmrD	0 %	65 %	0 %	35 %	0 %	1 %
EmrB	2 %	70 %	0 %	27 %	0 %	1 %
EmrD	0 %	18 %	0 %	68 %	14 %	6 %
EurA	100 %	0 %	0 %	0 %	0 %	7 %
EurB	51 %	49 %	0 %	0 %	0 %	4 %
EurC	96 %	4 %	0 %	0 %	0 %	7 %
SearB	6 %	19 %	0 %	76 %	0 %	4 %
SearD	0 %	0 %	92 %	7 %	2 %	21 %
WprA	100 %	0 %	0 %	0 %	0 %	2 %
WprB	0 %	9 %	84 %	7 %	0 %	18 %
Monde	24 %	12 %	36 %	25 %	2 %	100 %

En ce qui concerne la Bosnie-Herzégovine, les données d'état civil n'étaient pas disponibles après 1991. Les données de 1989/1991 ont été jugées complètes et, en l'absence de nouveaux renseignements, elles ont été moyennées et appliquées à l'année 2000, étant donné que toute amélioration de fond au niveau de la mortalité a probablement été contrebalancée par les conséquences du conflit au cours des années 1990.

Pour ce qui est d'un second groupe de pays possédant des données d'état civil (ou un échantillon de données d'état civil dans le cas du Bangladesh, de la Chine, de l'Inde et de la Tanzanie), l'utilisation de ces méthodes démographiques indirectes pour évaluer la complétude a donné à penser que certaines corrections devaient être apportées aux données d'état civil. Ces 50 pays sont inscrits dans les Catégories II et III au Tableau 9. Le degré de la sous-déclaration a varié considérablement, mais était généralement de l'ordre de 20–25 % dans les républiques d'Asie centrale et légèrement plus élevé (30–40 %) dans plusieurs pays en développement. Par exemple, la mortalité adulte a été jugée complète à 84 % pour l'Égypte, à 92 % pour le Guatemala, à 84–86 % pour l'Inde, à 78–84 % pour le Kazakhstan, à 82–85 % pour la République de Corée et à 73–75 % pour les Philippines. Ces ratios de complétude ont par la suite été appliqués aux données d'état civil pour compenser la sous-déclaration.

En ce qui concerne la mortalité juvénile, tous les points de données disponibles provenant des enquêtes sur la mortalité juvénile, comme les enquêtes démographiques et sanitaires, ou le programme d'enquête à indicateurs multiples sur les enfants de l'UNICEF, ont été utilisés pour corriger les données des données d'état civil sur la mortalité juvénile. Cette analyse a grandement profité de l'évaluation complète des données des pays qui fut menée par Hill et al. en 1999 en se fondant sur les données des recensements et des enquêtes disponibles, en 1996. Nous avons mis à jour ces graphiques nationaux à l'aide de renseignements plus récents et nous avons ajusté les prédictions de Hill et coll. pour 2000 lorsque les données récentes ont donné à penser que cela était nécessaire (voir l'Annexe). Là où des données d'enquête récentes n'étaient pas disponibles, les taux de mortalité juvénile ont été corrigés selon des facteurs de correction suggérés par les techniques ci-dessus pour la mortalité adulte et à des niveaux estimés pour les pays voisins. En supposant que les décès d'enfants sont plus susceptibles d'être sous-déclarés que les décès d'adultes, cela sous-estimera les niveaux de mortalité juvénile.

Pour les pays possédant une série chronologique suffisamment longue de données d'état civil ou un échantillon de données d'état civil, les séries chronologiques corrigées ont ensuite été analysées (comme pour les pays de la Catégorie I). Pour le reste, les niveaux de mortalité juvénile ($_5q_0$) et adulte ($_{45}q_{15}$) ont été projetés pour l'an 2000 en utilisant les bases factuelles disponibles sur le rythme du déclin de la mortalité ou en supposant un modèle de changement de la mortalité dans les années 1990 conséquent avec la croissance économique. Ces valeurs projetées des mortalités juvénile et adulte ont par la suite été appliquées au nouveau système de table de mortalité par la méthode logit modifiée (voir Murray et al. 2001 et la section des méthodes) pour produire une table de mortalité complète. Dans certains cas, (par ex. l'Afrique du Sud, la Tanzanie, le Cambodge, la République dominicaine, Haïti et le Myanmar), les taux de mortalité ont été augmentés par l'ajout de la mortalité estimée due au VIH/SIDA. Ces méthodes et ces sources de données servant à estimer la mortalité due au VIH/SIDA sont décrites dans une section ultérieure.

Approches multi-sources pour des populations spécifiques

Dans deux grands pays en développement, l'Inde et la Chine, plusieurs sources de données, dont les registres d'état civil, des systèmes de surveillance et des enquêtes, sont disponibles pour estimer les taux de mortalité. Aucun de ces systèmes utilisé seul n'est suffisamment fiable pour produire des tables de mortalité pour ces pays sans ajustements, mais ils sont tous utiles pour estimer la mortalité juvénile et adulte. Voici les sources de données et les ajustements apportés.

Chine

Trois sources de données sur la mortalité ont été utilisées pour estimer la table de mortalité.

Points de surveillance des maladies (PSM)

Ce système composé de 145 points de surveillance épidémiologique représentatifs du pays est utilisé par l'Académie chinoise de médecine préventive; il couvre une population de 10 millions de personnes à travers la Chine. Les données sur l'âge, le sexe et la cause des 50000 à 60000 décès sont enregistrées chaque année. Les évaluations périodiques des données de points de surveillance des maladies faites en ré-interrogeant les foyers de façon aléatoire révèlent un niveau de sous-déclaration des décès d'environ 15 % (Chinese Academy of Preventive Medicine 1997), bien que la croissance et l'équilibre des données depuis 1991 révèlent un facteur d'ajustement de la moyenne qui est d'environ deux fois celui de ce niveau. Les données annuelles pour la période 1991-1998 ont été utilisées, avec des corrections, pour estimer la tendance en $_5q_0$ et $_{45}q_{15}$.

Registres d'état civil

Les données portant sur l'âge, le sexe et la cause de 725 000 décès sont collectées annuellement à partir des registres d'état civil du ministère de la Santé. Elles couvrent une population de 121 millions de personnes (66 millions dans les régions urbaines, 55 millions dans les régions rurales). Bien que les données ne soient pas représentatives des conditions de mortalité dans toute la Chine, elles sont utiles pour suggérer des tendances, étant donné le nombre de décès couverts. Les tendances en $_{45}q_{15}$ pour la couverture des régions rurales et urbaines sont illustrées à la Figure 1. Bien que la sous-déclaration produit des niveaux de $_{45}q_{15}$ invraisemblablement faibles, ces données révèlent seulement un déclin très modeste de la mortalité adulte au cours des années 1990 (4-5 % pour les hommes – dans les deux régions – et pour les femmes dans les régions rurales, et 14 % pour les femmes – dans les régions urbaines).

Figure 1. Tendances en 45q15 provenant des données des données d'état civil – Chine, 1990-1998

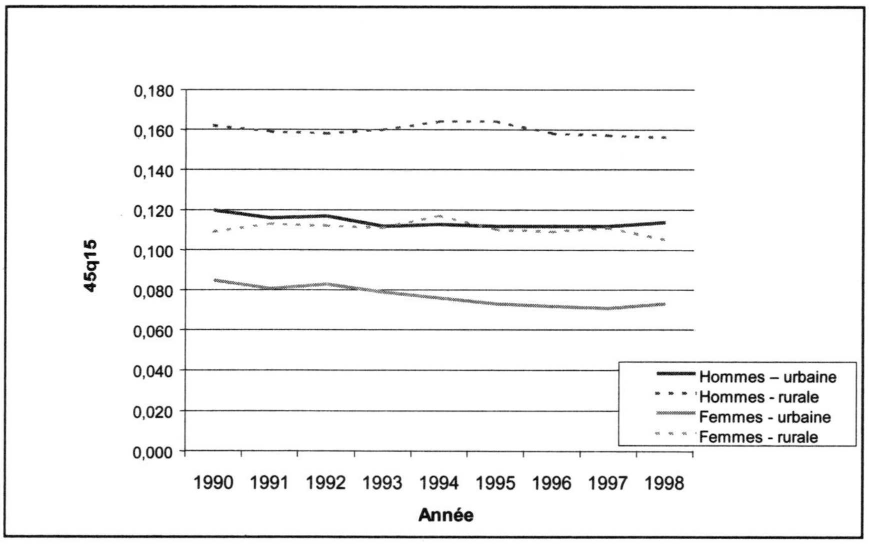

Données d'enquête

Les données d'enquête sont obtenues à partir de l'enquête annuelle réalisée auprès d'un foyer sur 1 000 concernant les décès au cours des 12 derniers mois. Par exemple, l'enquête de 1997 a couvert une population de 1 243 000 personnes réparties dans 864 comtés (3164 cantons, 4438 villages) dans 31 provinces et a enregistré un total de 7845 décès. Bien que cet échantillon soit représentatif, les méthodes croissance-équilibre révèlent une sous-déclaration importante des décès (27 % pour les hommes et et 29 % pour les femmes). Les tendances de la valeur ${}_{45}q_{15}$ implicite et non ajustée provenant des enquêtes dans les années 1990 sont illustrées à la Figure 2 et révèlent un déclin un peu plus important quoique le nombre beaucoup plus petit de décès par comparaison avec les registres d'état civil rendent difficile l'évaluation de la tendance.

Des données étaient disponibles pour la période 1991-1998 dans les trois systèmes. Puisque les analyses de croissance et d'équilibre révèlent que la sous-déclaration est demeurée relativement constante durant les années 1990, le déclin annuel moyen en ${}_{5}q_{0}$ et en ${}_{45}q_{15}$ révélé par ces trois sources de données a pour la première fois été calculé et appliqué à la table de mortalité chinoise de 1990 qui était appuyée sur le recensement pour faire des projections de taux de mortalité en 2000. Une incertitude autour des taux de mortalité en 2000 a été générée à partir des conceptions optimistes et pessimistes à propos du taux du déclin au cours des années 1990.

Figure 2. Tendances en 45q15 (non ajustées) fondées sur les estimations provenant des Enquêtes échantillons sur les mouvements démographiques– Chine, 1991–1998

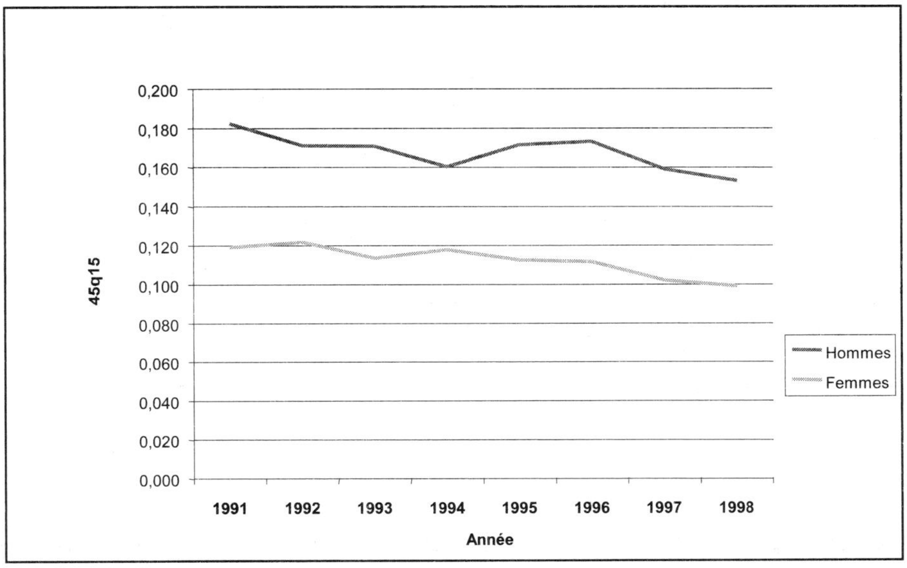

Inde

Les données les plus représentatives et les plus fiables sur les taux de mortalité par âge et par sexe proviennent des registres témoins en place depuis plusieurs décennies. Nous avons utilisé des données pour la période 1990–1998 (la dernière année disponible) pour calculer les tables de mortalité annuelles. Les données sont collectées à partir d'événements démographiques dans 4 436 unités d'échantillonnage rurales et dans 2 235 unités d'échantillonnage urbaines avec une population d'environ 6 millions de personnes couvrant presque tous les États et Territoires. La comparaison de ${}_5q_0$ provenant des registres témoins avec le taux révélé par les enquêtes démographiques et sanitaires (National Family Health Survey) effectuées en 1992–93 produit des résultats très similaires, évoquant que la sous-déclaration des décès d'enfants est minimale. D'un autre côté, la sous-déclaration des décès d'adultes dans les registres témoins au cours des années 1990 a probablement augmenté de 15 % d'après la méthodologie à variable -r de Bennett et Horiuchi (Mari Bhat 2000). Nous avons par conséquent corrigé les taux de décès des registres témoins à l'âge de cinq ans et nous l'avons augmenté de 14 % pour les hommes et de 16 % pour les femmes, et nous les avons extrapolés pour 2000. Les intervalles d'incertitude sur les taux de décès par âge ont été produits à partir d'extrapolations plausibles pour 2000 des tendances dans ces taux.

Données de recensement et d'enquête

Pour les 67 pays restants (voir Tableau 1), aucune estimation fiable de la mortalité adulte n'était disponible. Cependant, de nombreuses estimations directes et indirectes de la mortalité juvénile (${}_5q_0$) étaient disponibles pour les dernières années à partir de différents recensements et programmes d'enquêtes. Ces estimations ont été systématiquement évaluées et extrapolées pour 2000 avec des intervalles d'incertitude (Ahmad et al. 2000). Ces estimations, avec des intervalles d'incertitude, ont par la suite été appliquées au système de table de mortalité par la méthode logit modifiée en utilisant la norme « mondiale » pour estimer une table de mortalité correspondante. Des détails supplémentaires sur les procédures sont présentés à la section des méthodes.

Dans quelques pays de ce groupe (Afghanistan, Angola, Burundi, République populaire démocratique de Corée, Djibouti, Guinée équatoriale, Haïti, Libéria, Malawi, Sao Tome et Principe et Swaziland), les bases factuelles se sont révélées insuffisantes dans les années 1990 pour établir des estimations de la mortalité juvénile avec une confiance raisonnable. Les tables de mortalité pour ces pays sont en conséquence fondées sur les niveaux de mortalité juvénile avec une grande incertitude. Une grande précaution est essentielle lors de leur utilisation jusqu'à ce que des bases factuelles plus récentes sur les niveaux de mortalité juvénile soient disponibles.

Pour de nombreux pays dans cette catégorie, la mortalité causée par le VIH/SIDA est susceptible d'être élevée. Ce fait n'apparaît pas dans les tables types de mortalité puisque les distributions ont été conçues à partir de données sur la mortalité provenant de l'époque précédant l'arrivée du sida. Pour ces pays, principalement les pays d'Afrique subsaharienne, les tables de mortalité qui ne prennent pas le sida en considération ont pour la première fois été estimées à partir de l'estimation des niveaux de mortalité juvénile hors VIH. Les taux de mortalité par âge et par sexe causés par le VIH ont ensuite été estimés séparément (Salomon and Murray 2001) et ajoutés aux taux par âge de la table type de mortalité avec un ajustement approprié en prévision des risques possibles.

Méthodes

Élaboration d'une table de mortalité

Des méthodes normalisées de table de mortalité ont été utilisées pour élaborer la série chronologique des tables de mortalité pour les pays dans les Catégories I et II, lorsque les données sur les séries chronologiques étaient disponibles. Le système logit de Brass de base a été utilisé pour produire la tendance dans les deux paramètres (α, β) à partir de ces tables de mortalité. Ce système repose sur l'hypothèse que deux distributions distinctes de mortalité par âge peuvent être reliées l'une à l'autre par une transformation linéaire du logit de leurs probabilités de survie respectives. Il en est ainsi pour les deux séries de valeurs de survie observées, l_x et l^s_x.
Lorsque la dernière est la norme, il est possible de trouver des constantes α et β telles que

$$logit(l_x) = \alpha + \beta logit(l^s_x)$$

$$if\ logit(l_x) = 0.5\, ln\left(\frac{(1.0 - l_x)}{l_x}\right)$$

Then

$$0.5\, ln\left(\frac{(1.0 - l_x)}{l_x}\right) = \alpha + 0.5\beta ln\left(\frac{(1.0 - l^s_x}{l^s_x}\right)$$

pour tous les âges x entre 1 et T et où l_0(base) = 1. Si l'équation ci-dessus vaut pour chaque paire de tables de mortalité, alors toutes les tables de mortalité peuvent être produites à partir d'une seule table de mortalité normalisée en changeant les paires de valeurs (α, β) utilisées. En réalité, la supposition de linéarité est seulement approximativement respectée par les paires de tables de mortalité concrètes. Mais l'approximation est suffisamment exacte pour garantir l'utilisation du modèle permettant d'étudier et d'adapter les bulletins de mortalité observés. Le paramètre α fait varier le *niveau* de mortalité de la norme, tandis que β fait varier la *pente* de la norme, c'est-à-dire qu'il régit la relation entre la mortalité chez les enfants et les adultes.

Dans des circonstances où un historique de l'exécution des tables de mortalité est disponible, il est possible de produire une série chronologique de paires α,β en utilisant une norme propre à chaque pays. Un graphe de α et β, séparément, en fonction du temps devrait produire une trajectoire de points (Lopez et al. 2000). Si le graphe de points pour chaque paramètre se transforme en une ligne passablement directe, cette ligne pourrait théoriquement être extrapolée afin de prévoir les estimations pour chaque période ultérieure. Ces estimations α,β peuvent ensuite être transformées en équations logit adéquates afin d'obtenir les tables de mortalité correspondantes. Lorsque, par ailleurs, la tendance des points était erratique, le système ne pouvait pas fournir une prévision adéquate. Dans de telles situations, des techniques adaptées doivent ensuite être appliquées pour extrapoler la mortalité, en raison de cette distribution. Ainsi, trois modèles ont été conçus pour s'adapter à différents scénarios. Dans le premier modèle, le paramètre au temps t est supposé être une simple fonction linéaire de la période t :

$$\hat{\alpha}_t = \gamma_1 + \gamma_1 t$$
$$\hat{\beta}_t = \phi_1 + \phi_2 t$$

Ce modèle convient à des situations où la tendance en α ou β est clairement linéaire. Dans le second modèle, les paramètres α et β au temps t sont censés être des fonctions linéaires décalées des paramètres dans les périodes précédentes. Ainsi, les paramètres pour la période $T+1$ sont basés sur le modèle à pas 1, ceux pour la période $T+2$ sont basés sur la distribution à pas 2, etc., T correspondant à un temps donné de la table de mortalité normalisée. Les équations suivantes résument ces relations :

$$\hat{\alpha}_{T+1} = \gamma_{11} + \gamma_{21}\alpha_T \qquad \hat{\beta}_{T+1} = \phi_{11} + \phi_{21}\beta_T \quad \textit{1st forecast point}$$
$$\hat{\alpha}_{T+2} = \gamma_{12} + \gamma_{22}\alpha_T \qquad \hat{\beta}_{T+2} = \phi_{12} + \phi_{22}\beta_T \quad \textit{2nd forecast point}$$
$$\hat{\alpha}_{T+3} = \gamma_{13} + \gamma_{23}\alpha_T \qquad \hat{\beta}_{T+3} = \phi_{13} + \phi_{23}\beta_T \quad \textit{3rd forecast point}$$
$$\ldots\ldots\ldots\ldots \qquad \ldots\ldots\ldots\ldots$$
$$\hat{\alpha}_{T+n} = \gamma_{1n} + \gamma_{2n}\alpha_T \qquad \hat{\beta}_{T+n} = \phi_{1n} + \phi_{2n}\beta_T \quad \textit{last forecast point}$$

Ce modèle sera probablement plus approprié dans les situations où il existe des tendances linéaires claires et aussi des oscillations régulières dans les valeurs des paramètres au fil du temps. La troisième méthode combine les deux distributions ci-dessus :

$$\hat{\alpha}_{T+1} = \gamma_{11} + \gamma_{21}\alpha_T + \gamma_{31}(T+1) \qquad \hat{\beta}_{T+1} = \phi_{11} + \phi_{21}\beta_T + \phi_{31}(T+1) \quad \textit{1st forecast point}$$
$$\hat{\alpha}_{T+2} = \gamma_{12} + \gamma_{22}\alpha_T + \gamma_{32}(T+2) \qquad \hat{\beta}_{T+2} = \phi_{12} + \phi_{22}\beta_T + \phi_{32}(T+2) \quad \textit{2nd forecast point}$$
$$\hat{\alpha}_{T+3} = \gamma_{13} + \gamma_{23}\alpha_T + \gamma_{33}(T+3) \qquad \hat{\beta}_{T+3} = \phi_{13} + \phi_{23}\beta_T + \phi_{33}(T+3) \quad \textit{3rd forecast point}$$
$$\bullet\bullet\bullet\bullet\bullet\bullet\bullet\bullet \qquad \bullet\bullet\bullet\bullet\bullet\bullet\bullet\bullet$$
$$\hat{\alpha}_{T+n} = \gamma_{1n} + \gamma_{2n}\alpha_T + \gamma_{3n}(T+n) \qquad \beta_{T+n} = \phi_{1n} + \phi_{2n}\beta_T + \phi_{3n}(T+4) \quad \textit{last forecast point}$$

Ce modèle est approprié dans les situations où il existe des tendances linéaires complexes. Dans chaque pays, les trois modèles ont été utilisés pour prévoir les estimations des paramètres. Le modèle qui a produit les séries chronologiques d'estimations convenant le mieux à la tendance historique a été considéré adéquat pour ce pays.

Pour estimer la table de mortalité jusqu'à l'âge de 100+, la procédure de Coale et Guo (Coale and Guo 1989) a été utilisée. On observe depuis longtemps que les taux de mortalité à des âges dépassant 75 ou 80 ans augmentent avec l'âge à un taux décroissant plutôt que constant selon la loi de Gompertz (Perks 1932). En utilisant des données provenant de sept populations ayant des données sur la mortalité relativement fiables à des âges plus avancés (Autriche, France, Allemagne, Japon, Pays-Bas, Norvège et Suède), Coale et Guo ont démontré que l'augmentation relative d'un groupe d'âge à l'autre diminue au-delà de 80 ou 85 ans. À l'aide de ces découvertes, ils ont élaboré une méthode de rapprochement des tables de mortalité au-delà de 80 ans. Leur technique intègre une hypothèse de constance régulière plutôt que gompertzienne dans le cas du taux d'augmentation de la mortalité à un âge plus élevé que 80 ans. Plus précisément, le logarithme du rapport du taux de mortalité dans l'intervalle (x+5, x+10) avec le rapport de (x, x+5) est censé décliner d'une valeur constante lorsque x s'élève au-dessus de 80 ans.

Estimation de la mortalité adulte

La mesure de la mortalité adulte est par nature plus difficile que la mesure de la mortalité juvénile. Cela est en partie causé par la rareté relative des décès d'adultes, en particulier dans la plage d'âge de 15 à 60 ans, relativement aux décès dans l'enfance ou à un âge avancé. Ainsi, l'obtention de mesures précises de la mortalité adulte nécessite un grand nombre d'observations couvrant habituellement de longues périodes de référence. Cela est aussi causé, cependant, par le fait que, puisque les rapports faits par les mères au sujet de la survie de leurs enfants fournissent généralement des estimations satisfaisantes de la mortalité juvénile, les antécédents d'événements qui fournissent des estimations généralement adéquates et opportunes de la mortalité adulte n'ont pas été trouvés. Contrairement à l'estimation de la mortalité juvénile où l'information est facilement obtenue des mères concernées, il est souvent difficile de déterminer l'informateur approprié pour fournir des renseignements sur les adultes décédés. Cela aboutit souvent à des problèmes de sous-déclaration et de déclarations multiples. Souvent l'informateur ne connaît pas l'âge de la personne décédée et les certificats de naissance ne sont souvent pas disponibles pour les personnes plus âgées dans la plupart des pays en développement. En conséquence, les erreurs dans la déclaration de l'âge peuvent sérieusement restreindre la capacité d'obtenir de bonnes estimations de la mortalité adulte.

La méthode la plus couramment utilisée pour mesurer la mortalité adulte à partir des données d'enquêtes consiste à utiliser les renseignements sur la survie de la mère et du père pour estimer la survie des femmes et des hommes adultes, respectivement. Les autres méthodes comprennent celles qui utilisent les renseignements sur (a) la survie des premiers époux pour estimer la survie des hommes adultes, (b) la survie des premières épouses pour estimer la survie des femmes adultes, (c) la survie de la fratrie. Ces méthodes, bien que théoriquement fondées, se sont révélées difficiles en pratique, conduisant souvent à une sous-estimation des véritables niveaux de la mortalité adulte de 15 à 60 % dans les pays où il a été possible de les valider par rapport aux registres d'état civil (Stanton et al. 2000).

Une tentative visant à réviser systématiquement toutes les estimations directes (tirées principalement des systèmes de surveillance démographique) et indirectes sur les niveaux de mortalité adulte en Afrique a conclu que seule une fraction relativement petite de ces estimations a produit des estimations plausibles et qu'un grand nombre d'entre elles faisaient référence à des années bien avant 2000 (Lopez et al. 2000). Ainsi, pour tous les pays des Catégories IV et V, un nouveau système de table de mortalité à logit modifié a été utilisé. Les détails complets de la justification et de l'évaluation de la méthode sont donnés ailleurs (Murray et al. 2001). Essentiellement, la méthode a été basée sur la mise en rapport du niveau estimé de mortalité juvénile avec un degré plausible de niveaux de l_{60} provenant d'une base de données empirique comprenant plus de 1 800 tables de mortalité composées d'une vaste gamme d'espérances de vie. Dans la plupart des cas, la valeur de l_{60} a été choisie parmi cette gamme plausible comme base pour estimer les niveaux de mortalité adulte. Dans relativement peu de cas, la valeurs autres que la valeur moyenne ont été choisies d'après une évaluation des bases factuelles disponibles au sujet des niveaux de mortalité adulte.

Estimation de la mortalité par le VIH/SIDA

Pour les pays qui possèdent les épidémies de VIH/SIDA les plus étendues mais qui n'ont pas de données d'état civil, la mortalité causée par le sida a été estimée séparément, puis intégrée dans des tables de mortalité qui avaient exclu le sida. Pour les pays d'Afrique subsaharienne, le nombre total de décès d'adultes causés par le sida provient des modèles d'extrapolation utilisant des données de surveillance sentinelle sur la fréquence pour les femmes enceintes avec des mises à jour sur les distributions précédemment publiées (Salomon and Murray 2001) où un plus grand nombre de données est devenu disponible. Dans le but d'évaluer la mortalité par âge et par sexe, nous avons analysé les données d'état civil et de surveillance sur la mortalité due au sida à partir des sources suivantes : le projet sur la morbidité et la mortalité adulte dans trois districts de Tanzanie ; les données d'état civil provenant des régions urbaines et rurales d'Afrique du Sud et les données d'état civil du Zimbabwe. Ces données offrent les seules sources fiables de renseignements représentatifs sur la mortalité par cause en Afrique subsaharienne continentale dont dispose l'OMS. À la Figure 3, nous avons tracé les diagrammes par âge et par sexe à partir de chacune de ces sources en normalisant sur le plus haut taux observé sur chaque site. Il existe une uniformité remarquable dans le modèle entre ces différentes sources, les principales différences apparaissant aux âges les plus jeunes et les plus avancés. En s'appuyant sur ces sources, nous avons élaboré un modèle d'âge régional normalisé en prenant la moyenne pondérée de ces sources. La norme régionale apparaît en trait fort à la Figure 3. En utilisant cette norme, une estimation précise de la totalité des décès d'adultes peut être transformée en taux de mortalité par âge en utilisant le modèle normalisé des taux à la structure par âge de la population pour ensuite remettre tous les taux à l'échelle de façon à ce que le nombre total de décès corresponde à au nombre spécifié.

Figure 3. Distribution par âge de la mortalité causée par le VIH

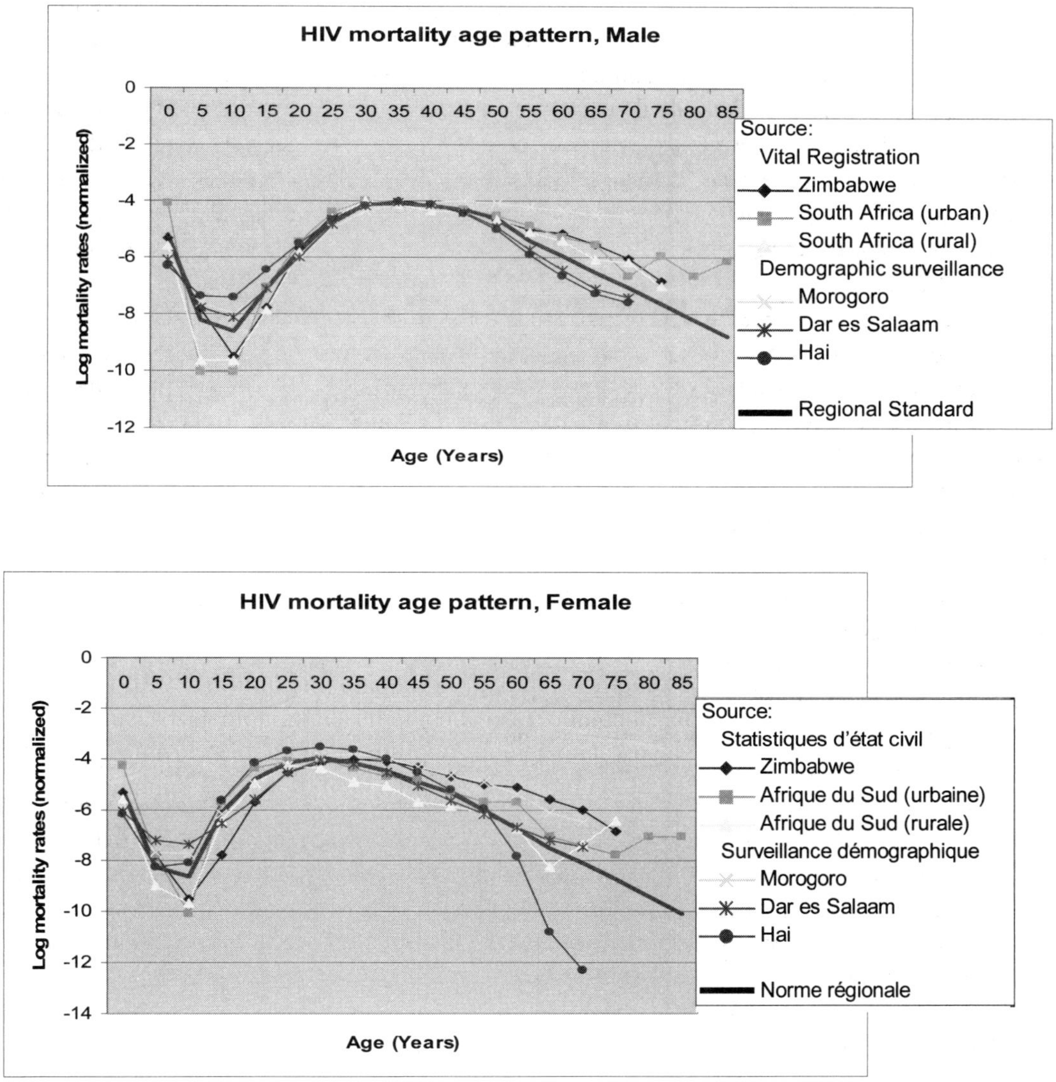

Étant donné le manque de données pour estimer la mortalité causée par le sida directement et les incertitudes introduites par la méthode d'extrapolation, il est important d'essayer de quantifier le niveau d'incertitude qui entoure les estimations de mortalité résultant de ces méthodes. Lorsque des données suffisantes étaient disponibles pour entreprendre une approche d'estimation de vraisemblance maximale dans les modèles d'extrapolation (c.-à-d. pour environ 20 pays), les résultats comprenaient une mesure d'incertitude autour des estimations de mortalité à chaque année. Pour les pays restants, les intervalles d'incertitude provenaient d'une évaluation de la couverture et de la représentativité des sites de surveillance sentinelle dans chaque pays. Les distributions de probabilités autour du nombre total de décès ont ensuite été transformées en distributions autour des taux de mortalité par âge et par sexe en utilisant des méthodes de simulation numérique. Les méthodes Monte Carlo ont été utilisées pour échantillonner à partir de ces distributions afin d'intégrer l'incertitude qui entoure la mortalité causée par le sida dans les estimations d'incertitude de la table de mortalité.

Pour quatre pays situés en dehors de l'Afrique subsaharienne (Cambodge, République dominicaine, Haïti et Myanmar) où il n'a jamais existé de données sur les données d'état civil ou dans lesquels le sida a été dénoté de façon inappropriée dans les décès inscrits, nous avons aussi combiné des estimations séparées de la mortalité causée par le sida dans des tables de mortalité hors sida. La distribution par âge de la mortalité causée par le sida dans ces cas a été fondée sur les distributions de références régionales provenant des pays ayant des données d'état civil disponibles (Argentine, Brésil, Chili, Cuba, Mexique et Uruguay dans le cas d'Haïti et de la République dominicaine ; et la Thaïlande dans le cas du Myanmar et du Cambodge).

Limites d'incertitude

Il y a plusieurs sources d'incertitude autour des valeurs finales de α et β obtenues de ces modèles, y compris l'incertitude du modèle quant à la spécification correcte ainsi que l'incertitude de l'estimation dans la détermination des valeurs pour les coefficients de régression. Une discussion détaillée des sources d'incertitude et des méthodes pour l'analyse d'incertitude pour les tables de mortalité est présentée dans Salomon et al. 2001. Le niveau d'incertitude autour des estimations de α et β dépend en partie de l'incertitude autour des coefficients de régression γ_{ij} et ϕ_{ij}, ce qui entraîne à son tour un certain niveau d'incertitude autour des tables de mortalité qui sont calculées à partir de ces paramètres. Étant donné qu'une table de mortalité complète est une fonction non linéaire compliquée des paramètres d'incertitude, nous avons utilisé les techniques de simulation Monte Carlo pour concevoir des estimations numériques des degrés d'incertitude autour des tables de mortalité. Cette incertitude est rendue en prenant des sélections aléatoires des coefficients de régression γ_{ij} et ϕ_{ij} à partir de distributions normales ayant des moyennes égales aux coefficients estimés et des variances provenant des erreurs-types dans la régression. Dans chacune des 1 000 itérations, les sélections de γ_{ij} et ϕ_{ij} sont utilisées pour produire des estimations α et β, qui sont ensuites transformées en tables de mortalité complexes. Ainsi, les distributions de probabilités peuvent être définies autour des estimations des tables de mortalité en analysant les 1 000 différentes tables de mortalité simulées. Par exemple, une plage peut être définie autour de l'estimation de l'espérance de vie à la naissance en classant les 1 000 estimations différentes de e(0) dans les tables de mortalité simulées et ensuite déterminer les 25e et 975e valeurs en tant que limites d'un intervalle de confiance de près de 95 %.

En l'absence de données historiques (c.-à-d. pour les pays des Catégories IV et V), un système à logit modifié a été utilisé (Murray et al. 2001). Comme le système à logit simple, le nouveau modèle est basé sur la relation entre deux tables de mortalité. Contrairement au système à logit simple, cependant, la nouvelle formulation comprend deux paramètres additionnels qui corrigent la non-linéarité de la méthode originale de Brass. Les principales données d'entrée du système sont une plage estimée autour de ${}_5q_0$. Pour toute valeur ${}_5q_0$, le système de table de mortalité à logit offre une plage de valeurs plausibles pour ${}_{45}q_{15}$. Ces plages constituent la fondation d'un ensemble de simulations Monte Carlo au sujet des différentes tables de mortalité possibles compatibles avec les connaissances et l'incertitude concernant la mortalité adulte et juvénile. La plage de tables de mortalité simulées permet le calcul de plages d'incertitude autour des divers indicateurs principaux tels que q_x, l_x et e_x.

Résultats

On estime que l'espérance de vie générale à la naissance (les deux sexes combinés) en 2000 varie de 81,1 ans au Japon (84,7 pour les femmes, 77,5 pour les hommes) à 37,5 ans au Malawi (voir Tableau 3). Pour les hommes, la deuxième espérance de vie la plus élevée a été estimée pour la Suède (77,3 ans), suivie par l'Andorre (77,2), l'Islande (77,1), Monaco (76,8) et la Suisse (76,7). L'espérance de vie des hommes a dépassé 75,0 ans dans 19 pays en 2000 (voir Tableau 4).

Tableau 3. Espérance de vie à la naissance (années), 10 premiers et 10 derniers pays, 2000

10 premiers pays		10 derniers pays	
1 Japon	81,1	1 Malawi	37,5
2 Monaco	80,6	2 Sierra Leone	37,9
3 Andorre	80,5	3 Mozambique	38,7
4 Saint-Marin	80,0	4 Zambie	39,4
5 Suède	79,6	5 Rwanda	39,5
6 Suisse	79,6	6 Burundi	41,0

7 Islande	79,5	7 République centrafricaine	42,0
8 Australie	79,3	8 Lesotho	42,1
9 Italie	79,2	9 Namibie	42,7
10 France	79,1	10 Rép. dém. du Congo	42,8

Tableau 4. Pays où l'espérance de vie des hommes est supérieure à 75,0 ans en 2000

Pays	e0 (années)	Pays	e0 (années)
Japon	77,5	Italie	76,0
Suède	77,3	Nouvelle-Zélande	75,9
Andorre	77,2	Norvège	75,7
Islande	77,1	Pays-Bas	75,4
Monaco	76,8	Singapour	75,4
Suisse	76,7	Grèce	75,4
Israël	76,6	Malte	75,4
Australie	76,6	Espagne	75,4
Saint-Marin	76,1	France	75,2
Canada	76,0		

Tableau 5. Pays où l'espérance de vie des femmes est supérieure à 80,0 ans en 2000

Pays	e0 (années)	Pays	e0 (années)
Japon	84,7	Norvège	81,4
Monaco	84,4	Autriche	81,4
Saint-Marin	83,8	Pays-Bas	81,0
Andorre	83,8	Nouvelle-Zélande	80,9
France	83,1	Belgique	80,9
Suisse	82,5	Finlande	80,9
Italie	82,4	Luxembourg	80,8
Espagne	82,3	Grèce	80,8
Australie	82,1	Malte	80,7
Suède	82,0	Allemagne	80,6
Islande	81,8	Israël	80,6
Canada	81,5	Singapour	80,2

Parmi les femmes, la deuxième espérance de vie la plus élevée a été estimée pour Monaco (84,4 ans), suivie par Saint-Marin et Andorre (83,8 ans) et la France (83,1). Vingt-quatre pays avaient une espérance de vie estimée à 80 ans ou plus pour les femmes en 2000 (voir Tableau 5). L'espérance de vie des femmes a dépassé 75,0 ans dans 57 pays, soit environ dans le tiers des États Membres de l'OMS.

Étant donné les répercussions phénoménales de l'épidémie de VIH/SIDA en Afrique subsaharienne, il n'est peut-être pas surprenant que les pays ayant la plus faible espérance de vie en 2000 proviennent tous de cette région. En effet, 37 des 40 pays ayant la plus faible espérance de vie proviennent de l'Afrique subsaharienne. Tandis que les pays de cette région souffrent de façon disproportionnée des facteurs causant la mortalité juvénile et la mortalité prématurée d'adultes, y compris les infections respiratoires aiguës, les maladies diarrhéiques et la tuberculose, le manque de progrès de l'espérance de vie peut en grande partie être attribué aux conséquences de l'épidémie de VIH/SIDA des quelque 15 dernières années. Si on excluait les décès attribués au VIH/SIDA, l'espérance de vie à la naissance dans certains pays de la région serait de 15 à 20 ans plus élevée (voir Tableau 6). Cela est particulièrement vrai dans les pays d'Afrique du Sud (Botswana, Lesotho, Namibie, Afrique du Sud, Swaziland, Zambie, Zimbabwe), mais des réductions de l'espérance de vie de l'ordre de 5 à 10 ans causées par la mortalité due au sida en 2000 sont aussi courantes dans de nombreux autres pays d'Afrique. En moyenne, l'espérance de vie à la naissance en Afrique subsaharienne est de 6 ans plus basse pour les hommes et un peu plus de 7 ans plus basse pour les femmes comparativement à ce que cela aurait été en 2000 en l'absence de la mortalité causée par le VIH/SIDA.

Tableau 6. Différence, en nombre d'années, au niveau de l'espérance de vie à la naissance lorsque le sida est exclu de l'estimation de toutes les maladies possibles pour l'année 2000.

Hommes			
Namibie	18,5	Djibouti	5,9
Botswana	17,3	Haïti	5,4
Zambie	15,2	Erythrée	5,4
Zimbabwe	14,9	Niger	4,7
Lesotho	14,8	Somalie	4,7
Swaziland	14,2	Ghana	3,9
Afrique du Sud	12,1	Bahamas	3,9
République centrafricaine	10,3	Gabon	3,8
Malawi	10,1	Rép. dém. du Congo	3,5
Kenya	9,8	Guyane	3,0
Burundi	9,8	Liberia	2,7
Côte d'Ivoire	9,0	Cambodge	2,7
Mozambique	8,3	Myanmar	2,5
Rwanda	7,8	Tchad	2,3
Ouganda	7,8	Honduras	2,2
République-Unie deTanzanie	7,2	Sierra Leone	2,1
Éthiopie	7,2	République dominicaine	2,1
Burkina Faso	6,6	Bénin	2,1
Togo	6,3	Gambie	2,0
Cameroun	6,2	Angola	2,0
Congo	6,0		
Femmes			
Namibie	21,5	Congo	7,3
Botswana	20,2	Cameroun	7,3
Zambie	18,1	Djibouti	7,0
Zimbabwe	17,6	Érythrée	6,4
Lesotho	17,4	Somalie	5,7
Swaziland	16,8	Nigeria	5,6
Afrique du Sud	15,1	Ghana	4,7
Malawi	12,4	Rép. dém. du Congo	4,4
République centrafricaine	12,2	Gabon	4,4
Burundi	11,7	Liberia	3,3
Kenya	11,6	Tchad	2,8
Côte d'Ivoire	10,9	Sierra Leone	2,6
Mozambique	10,3	Angola	2,6
Rwanda	9,9	Gambie	2,4
Ouganda	9,2	Bénin	2,4
Éthiopie	8,8	Guinée-Bissau	2,1
République-Unie deTanzanie	8,6	Haïti	2,1
Burkina Faso	7,8	Sénégal	2,1
Togo	7,5		

De grandes différences subsistent entre les sexes en matière d'espérance de vie dans les pays plus développés. Au début du 20e siècle, l'espérance de vie des femmes dépassait celle des hommes de 2 à 3 ans, en moyenne, en Europe, en Amérique du Nord et en Australie (Lopez 1983). En 2000, l'avantage des femmes a augmenté à 10 ans au Kazakhstan (10,4 ans), en Lithuanie (10,4), en Ukraine (10,7), en Estonie (11,0), en Lettonie (11,3) et au Bélarus (12,0), et il a atteint son plus haut niveau d'entre tous les pays dans la Fédération de Russie (12,6). Inversement, la différence était seulement d'une demi-année ou moins dans des pays tels que le Bangladesh, le

Lesotho et la Zambie. Seulement quelques pays, dont le Botswana, les Maldives, la Namibie et le Népal, ont vu l'espérance de vie des hommes dépasser celle des femmes (de 0,2 à 0,5 ans). La Figure 4 illustre la distribution de l'écart entre les hommes et les femmes en matière d'espérance de vie à la naissance pour les 191 États Membres de l'OMS en 2000. Les valeurs extrêmes observées dans certains pays d'Europe de l'Est se retrouvent complètement à la queue de la distribution. Dans environ le quart des pays, la différence entre les hommes et les femmes se situe entre 5 et 6 ans en faveur des femmes. Puisque les différences entre les sexes concernant la mortalité sont habituellement moins grandes dans les pays en développement, la différence générale dans le monde au niveau de l'espérance de vie à la naissance est légèrement inférieure à la dernière (4,7 ans).

Figure 4. Distribution de l'écart entre les hommes et les femmes en matière d'espérance de vie à la naissance, États Membres de l'OMS, 2000

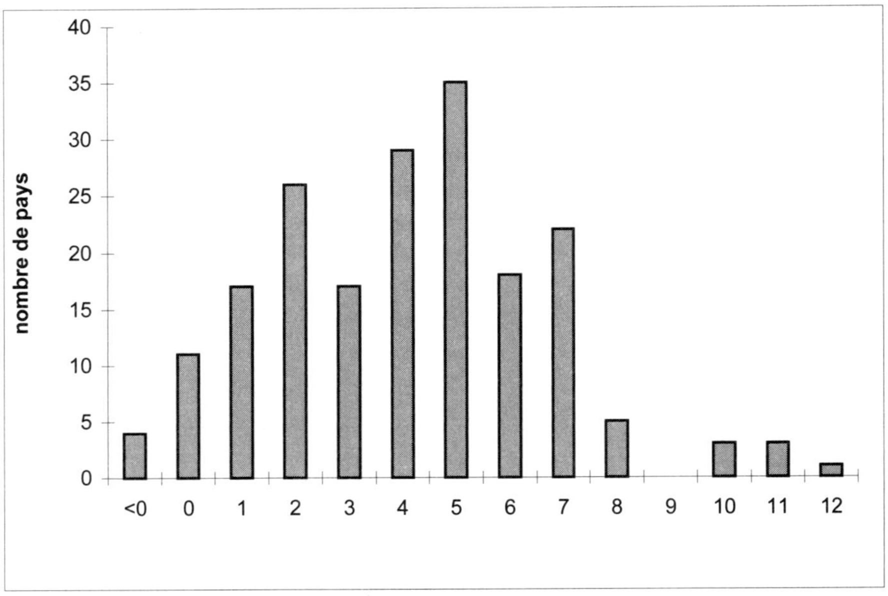

La probabilité de décès entre la naissance et l'âge de 5 ans, en tant que mesure générale des conditions de santé chez les enfants, et la probabilité de décès entre les âges de 15 et 60 ans, en tant que mesure de la mortalité prématurée chez les adultes sont des indicateurs sommaires utiles des risques de mortalité prédominants d'une population. Ils sont illustrés pour les différentes régions de l'OMS, et à l'intérieur d'elles, et pour les sous-régions en fonction de la mortalité, à la Figure 5.

Figure 5. Risques de décès pendant l'enfance (0 à 4 ans) et l'âge adulte (15 à 59 ans), par sous-région, 2000

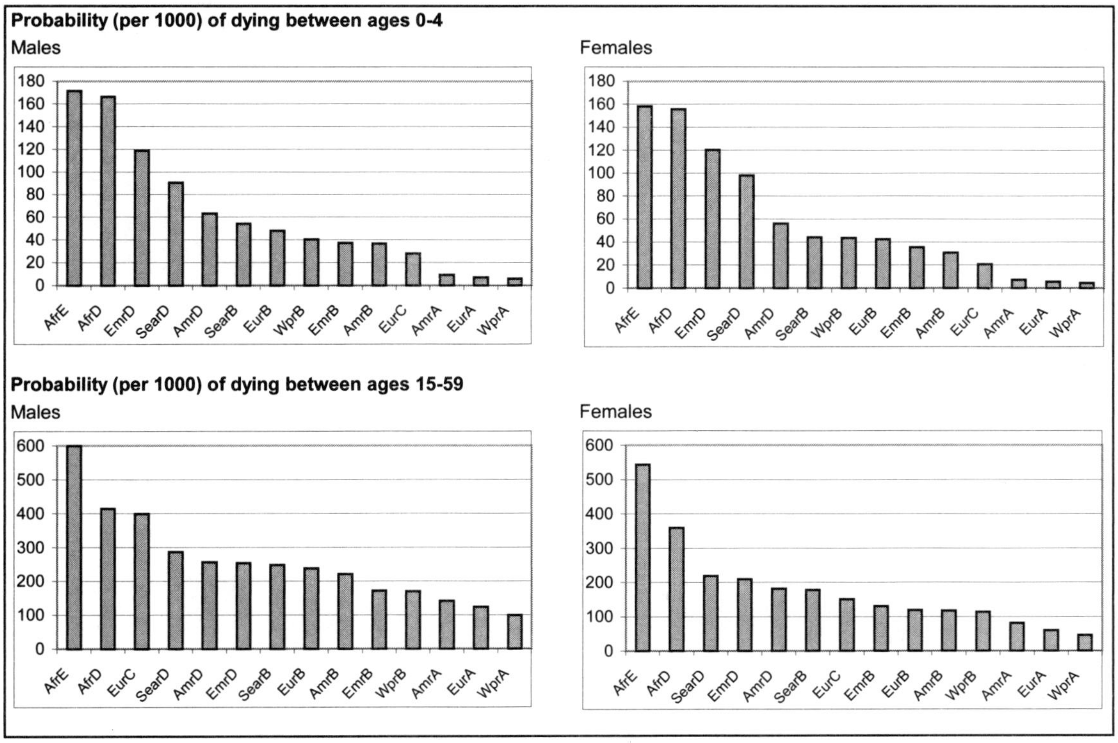

Les différences entre les niveaux de mortalité juvénile demeurent énormes. Des 10,9 millions de décès avant l'âge de 5 ans qui ont eu lieu en 2000 (Tableau 7), 99 % étaient dans des pays en développement. Dans les pays industrialisés (c.-à-d. ceux classés dans la strate régionale A), la probabilité de décès d'enfants ($_5q_0$) est habituellement inférieure à 1 % (0,4% en Suède et à Singapour), mais elle augmente à près de 300 pour 1000 en Sierra Leone. Les niveaux de mortalité juvénile qui dépassent 10 % (100 pour 1000) sont encore communs à travers l'Afrique et dans certaines parties d'Asie (Afghanistan, Cambodge, Laos, Myanmar, Népal et Pakistan).

Tableau 7. Nombre total de décès par âge, sexe et sous-région de l'OMS, 2000

Sexe	Sous-régions de l'OMS		tous (000)	0-4 (000)	5-14 (000)	15-29 (000)	30-44 (000)	45-59 (000)	60-69 (000)	70-79 (000)	80+ (000)
Deux sexes	Monde		55 619	10 893	1 438	3 619	5 062	6 993	8 006	10 281	9 328
	Afr	Tous	10 572	4 244	432	1 304	1 683	1 039	719	739	412
		D	4 244	1 928	182	396	484	383	319	348	204
		E	6 327	2 316	250	907	1 198	656	400	391	208
	Amr	Tous	5 872	469	64	297	438	758	838	1 280	1 729
		A	2 778	37	9	55	129	308	371	687	1 183
		B	2 584	313	38	194	254	386	404	517	477

Sexe	Sous-régions de l'OMS		tous (000)	0-4 (000)	5-14 (000)	15-29 (000)	30-44 (000)	45-59 (000)	60-69 (000)	70-79 (000)	80+ (000)
		D	511	119	17	48	55	64	63	76	68
	Emr	Tous	4 028	1 549	154	257	294	444	467	530	332
		B	682	127	20	49	51	101	111	132	91
		D	3 346	1 421	134	208	243	343	357	398	242
	Eur	Tous	9 668	254	48	241	541	1 178	1 704	2 602	3 098
		A	4 080	27	7	49	120	348	575	1 126	1 828
		B	1 952	170	24	67	126	264	370	510	420
		C	3 636	57	18	125	295	567	759	967	850
	Sear	Tous	14 091	3 309	536	1 030	1 325	2 029	2 211	2 301	1 350
		B	2 143	301	55	191	249	331	370	404	242
		D	11 948	3 008	481	838	1 076	1 699	1 841	1 896	1 107
	Wpr	Tous	11 389	1 068	203	491	780	1 544	2 066	2 828	2 408
		A	1 150	8	2	16	30	117	179	294	504
		B	10 238	1 060	201	475	750	1 427	1 888	2 534	1 903
Hommes	Monde		29 641	5 638	728	1 874	3 014	4 339	4 792	5 498	3 759
	Afr	Tous	5 417	2 257	211	522	890	609	387	364	177
	Afr	D	2 189	1 022	89	164	257	222	171	173	90
		E	3 228	1 235	122	357	633	387	216	191	87
	Amr	Tous	3 148	260	37	217	293	468	490	689	694
		A	1 382	21	5	40	84	192	220	371	448
		B	1 485	175	22	148	175	240	235	277	213
		D	282	65	9	29	34	36	35	41	33
	Emr	Tous	2 132	790	77	129	164	261	263	279	168
		B	382	67	11	31	31	62	64	71	45
		D	1 750	723	66	98	134	198	198	209	123
	Eur	Tous	4 947	142	30	180	397	817	1 072	1 288	1 020
		A	2 037	15	4	37	82	233	379	627	660
		B	1 053	93	15	46	86	177	224	253	159
		C	1 857	33	12	97	229	406	470	409	202
	Sear	Tous	7 661	1 655	258	540	789	1 208	1 272	1 256	683
		B	1 182	170	31	122	152	188	203	207	108
		D	6 479	1 484	226	418	637	1 020	1 069	1 049	575
	Wpr	Tous	6 337	534	115	287	479	977	1 307	1 621	1 017

Sexe	Sous-régions de l'OMS		tous (000)	0-4 (000)	5-14 (000)	15-29 (000)	30-44 (000)	45-59 (000)	60-69 (000)	70-79 (000)	80+ (000)
		A	625	5	1	11	20	81	124	178	205
		B	5 712	529	114	275	459	896	1 184	1 443	812
Femmes	Monde		25 978	5 255	709	1 745	2 048	2 654	3 215	4 784	5 569
	Afr	Tous	5 155	1 987	220	782	793	430	332	375	235
	Afr	D	2 056	906	92	232	227	161	148	175	114
		E	3 099	1 081	128	550	566	269	184	200	121
	Amr	Tous	2 725	209	27	80	145	290	348	591	1 035
		A	1 396	16	4	15	45	116	151	315	735
		B	1 099	138	15	46	79	147	169	241	264
		D	229	55	8	18	22	28	28	35	36
	Emr	Tous	1 896	758	77	128	129	184	205	251	164
		B	300	60	10	18	20	39	46	61	46
		D	1 596	698	68	110	109	145	159	190	119
	Eur	Tous	4 721	112	18	61	144	362	632	1 314	2 078
		A	2 043	12	3	12	38	115	197	499	1 168
		B	900	77	9	22	40	87	146	257	261
		C	1 779	23	6	28	66	160	289	558	648
	Sear	Tous	6 430	1 654	279	490	536	821	939	1 045	666
		B	961	131	23	69	97	142	167	197	134
		D	5 469	1 523	255	421	439	679	772	848	532
	Wpr	Tous	5 051	534	88	204	301	567	759	1 207	1 391
		A	525	3	1	4	10	37	55	116	299
		B	4 526	531	87	200	291	530	704	1 091	1 092

Cependant, les plus grandes disparités au niveau de la mortalité surviennent entre les âges de 15 et 59 ans. Dans certains pays d'Afrique australe, tels le Botswana, le Malawi, la Namibie et la Zambie, où le sida constitue maintenant un problème de santé publique important, 65 % ou plus des adultes qui survivent à l'âge de 15 ans mourront probablement avant l'âge de 60 ans selon les taux de mortalité actuels. Dans plusieurs autres pays (par ex. le Burundi, le Lesotho, le Mozambique, le Rwanda, le Swaziland et le Zimbabwe) le risque dépasse 60 %. L'augmentation remarquable de ${}_{45}q_{15}$ en Afrique du Sud est aussi digne de mention, avec des niveaux estimés à 567 pour 1 000 pour les hommes et et 502 pour1 000 pour les femmes en 1990 (Timaeus 1999). À l'autre extrême, les niveaux ${}_{45}q_{15}$ 90–130 pour 1 000 sont habituels dans la plupart des pays développés pour les hommes, avec des risques réduits de près de la moitié pour les femmes encore une fois.

La disparité croissante entre les niveaux de mortalité juvénile et adulte au cours des dernières années est évidente à la Figure 6, qui compare les niveaux moyens de ${}_5q_0$ avec les niveaux de ${}_{45}q_{15}$ pour les hommes et les femmes séparément, conformément aux tables de mortalité par pays de 2000. Les estimations d'incertitude sont comprises pour chaque point. La déviation systématique des distributions de la mortalité dans les pays d'Europe de l'Est est claire, particulièrement pour les hommes, par rapport aux niveaux prévus, étant donné la relation traditionnelle entre la mortalité juvénile et adulte. Une grande incertitude entoure aussi des niveaux estimés de mortalité adulte dans plusieurs pays, surtout de l'Afrique subsaharienne. Généralement, l'incertitude entourant les niveaux de mortalité juvénile est beaucoup moins grande que la mortalité adulte, dénotant les connaissances acquises à partir des vastes programmes d'enquête mondiaux sur les niveaux et les facteurs déterminants de la mortalité juvénile au cours des dernières décennies.

Figure 6. La mortalité adulte par opposition à la mortalité juvénile pour 191 États Membres de l'OMS pour l'année 2000

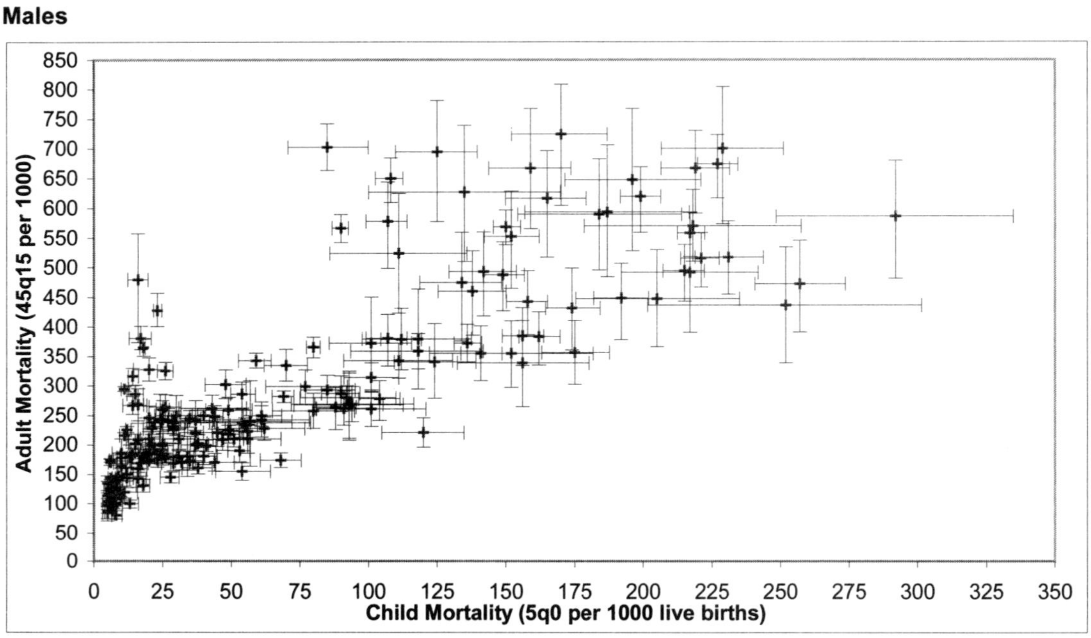

Females

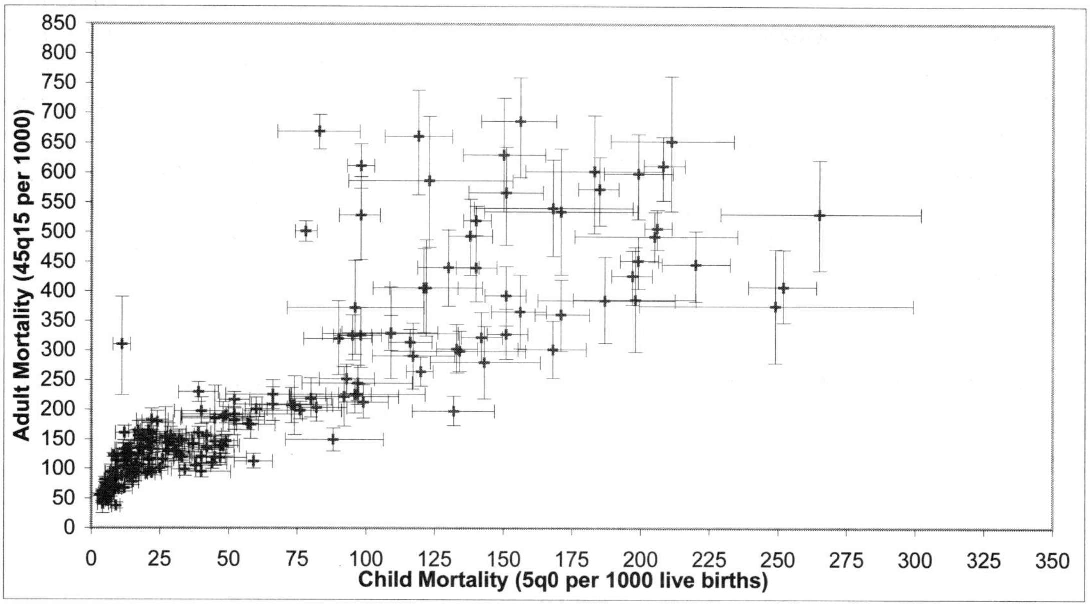

Cela est exprimé dans la distribution par âge des intervalles de mortalité illustrée pour chaque pays et région dans l'Annexe. Les limites d'incertitude autour de q_x et l_x ont tendance à s'élargir vers les âges adultes, en particulier en Afrique subsaharienne où les bases factuelles directes et fiables sur les niveaux de mortalité adulte sont rares. En effet, la dimension de ces intervalles d'incertitude montre un point important, qui est la valeur centrale; celle-ci exige un investissement urgent afin de mesurer les niveaux de mortalité adulte dans les pays où les limites sont encore inadmissiblement larges.

La moyenne d'âge des décès, par rapport à l'âge moyen d'une population, est un indice sommaire alternatif des conditions de mortalité dans une population. Dans les pays où la mortalité juvénile est élevée, l'âge moyen des décès et l'âge moyen de la population seront comparativement faibles et ils augmenteront plus ou moins ensemble lorsque le développement de la santé s'améliorera. La Figure 7 démontre clairement la relation, pour les hommes et les femmes séparément, en s'appuyant sur les tables de mortalité pour chaque pays en 2000. De façon intéressante, au-delà d'un âge moyen de la population d'environ 30 ans, l'âge moyen des décès augmente quelque peu plus lentement que pour les niveaux plus bas d'âge moyen, en particulier pour les femmes. Cette atténuation est en partie causée par le déclin de la fécondité dans les pays ayant un faible taux de mortalité au cours des 50 dernières années, mais elle révèle aussi le fait que pendant le temps où l'âge moyen des décès atteint environ 60 ans, une grande partie du progrès visant à réduire la mortalité juvénile et adulte a été réussie et, par conséquent, des réductions supplémentaires conduiront seulement à des augmentations progressivement plus lentes de l'âge moyen de la population. Comme le démontre aussi la figure, il demeure des différences extrêmement grandes au niveau de l'âge moyen de décès parmi les États Membres de l'OMS en 2000, allant d'un niveau aussi bas que 15,5 au Niger (les deux sexes combinés), 19,9 en Afghanistan, 20,3 et 20,5 en Somalie et en Sierra Leone, respectivement, jusqu'à près de 80 ans en Suède (78,5), en Norvège (77,1) et en Grèce (76,8). Cette différence d'un facteur 4 est beaucoup plus grande que l'intervalle de variation de l'espérance de vie cité précédemment.

Figure 7. Âge moyen des décès et de la population, 2000, 191 pays

Males

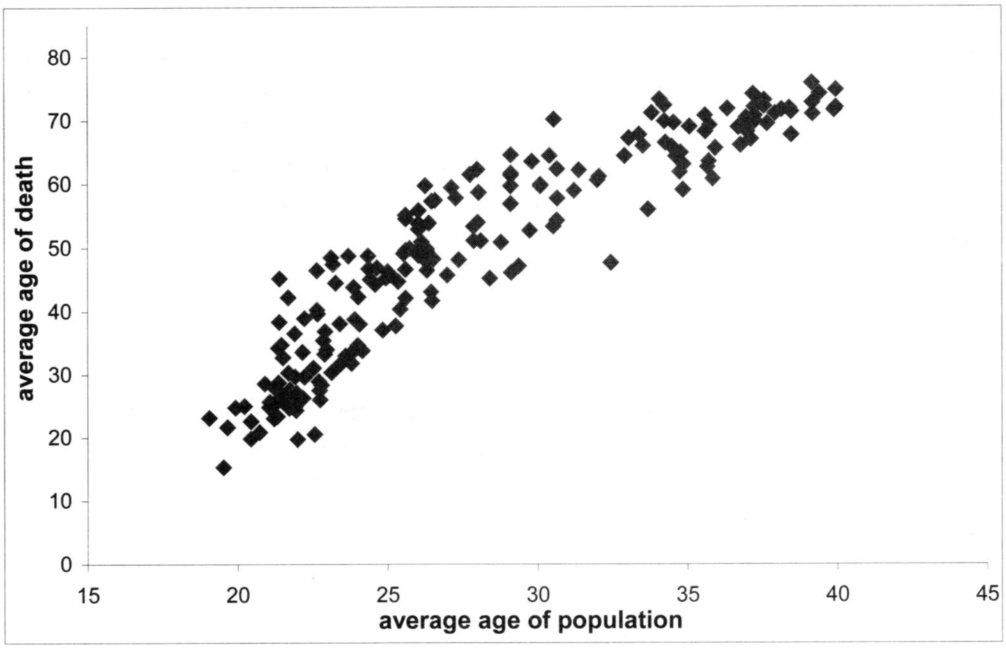

Females

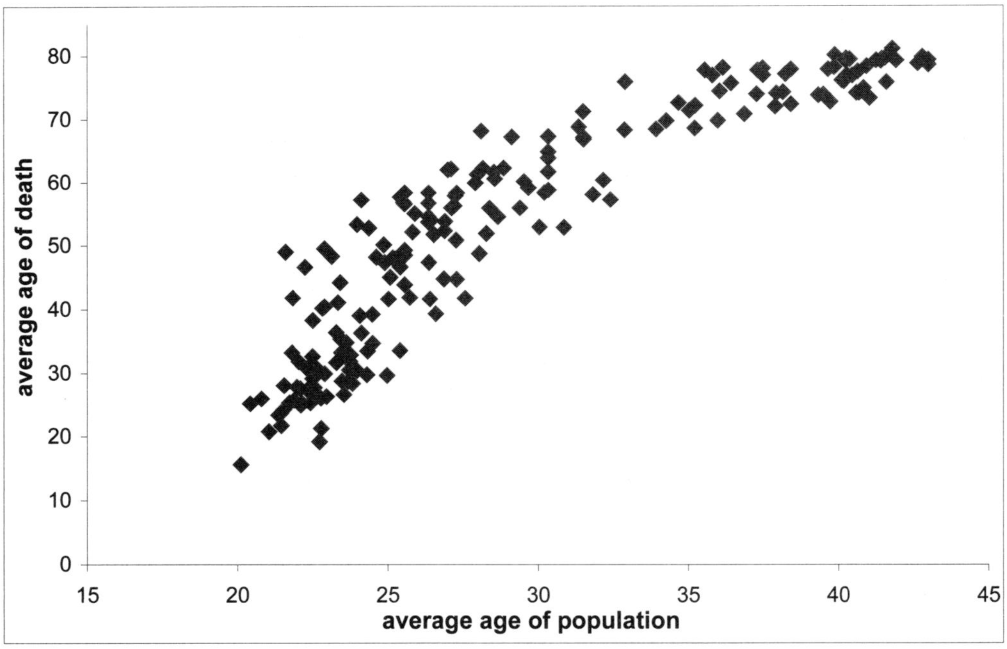

On estime à un peu moins de 56 millions le nombre de décès en 2000, desquels presque 30 millions étaient des hommes (voir Tableau 7). À travers le monde, 10,9 millions d'enfants de moins de 5 ans sont morts cette année là, le 3/4 d'entre eux provenant de régions en développement en Afrique et Asie du Sud. Quelque 10 autres millions meurent dans chacun des deux groupes d'âge les plus élevés, 70–79 et 80+, la vaste majorité dans les sous-régions à plus faible mortalité A, B et C. Il y a eu beaucoup plus de décès dans les groupes de jeunes adultes âgés entre 15 et 59 ans (15,6 millions) que dans les groupes d'enfants âgés de 0 à 14 ans (12,3 millions), ce qui montre le besoin d'une vue plus expansive des politiques visant à empêcher les décès prématurés dans le but de se pencher sur les principales causes de décès des jeunes adultes.

À l'échelle du monde, l'espérance de vie à la naissance en 2000 a été estimée à 64,9 ans (62.7 pour les hommes, 67,2 pour les femmes) (voir Tableau 8). Les différences les plus substantielles dans les distributions par âge de la mortalité au sein des sous-régions sont évidentes au Tableau 7 et elles expliquent les grandes variations au sujet de l'espérance de vie régionale, avec des niveaux se situant entre 44 ans (les deux sexes) en AfrE où le VIH/SIDA est le plus présent et un niveau moyen de près du double (80,9 ans) en WprA. L'espérance de vie moyenne a atteint 70 ans ou plus dans six sous-régions, qui, curieusement, ne sont pas en EurB ou C. En particulier, l'espérance de vie moyenne des hommes à la naissance en EurC était à peine de 60 ans, parmi les plus basses dans toutes les sous-régions du monde hors Afrique.

Tableau 8. Espérance de vie à la naissance (années) par sous-région de l'OMS, 2000

Sous-région	hommes	femmes	deux sexes
AfrD	50,3	52,4	51,3
AfrE	43,5	45,2	44,3
AmrA	74,1	79,6	76,9
AmrB	67,5	74,5	70,9
AmrD	63,5	68,4	65,9
EmrB	68,6	71,6	70,0
EmrD	59,1	61,0	60,0
EurA	74,8	81,2	78,0
EurB	65,5	72,3	68,8
EurC	60,3	72,1	66,0
SearB	64,1	68,9	66,4
SearD	59,7	62,4	61,0
WprA	77,3	84,2	80,9
WprB	68,2	72,7	70,4
Monde	**62,7**	**67,2**	**64,9**

Discussion

La table de mortalité et ses paramètres associés sont des renseignements importants pour évaluer la performance des systèmes de santé. Les tables de mortalité ont de nombreuses autres utilités en épidémiologie, en démographie et en économie, et la disponibilité de tables de mortalité courantes annuelles pour les 191 États Membres de l'OMS devrait améliorer la qualité et la pertinence de la base analytique pour de telles utilisations. Une évaluation minutieuse des données sur lesquelles les tables de mortalité sont construites est essentielle pour que les résultats soient utilisés en toute confiance. Ce rapport a essaié de répertorier les pays pour lesquels les données d'état civil semblent fonctionner suffisamment bien pour préparer les tables de mortalité sans apporter (ou très peu) de changements aux données. Dans les pays où la situation est différente, nous avons décrit les procédures utilisées pour ajuster les données, mais les résultats obtenus dépendent énormément d'un jugement motivé des bases factuelles. Ce degré de subjectivité est inévitable étant donné les pratiques démographiques actuelles pour l'ajustement des données ; un travail supplémentaire est nécessaire pour explorer de nouvelles façons de mieux utiliser les méthodes statistiques formelles pour réduire cette subjectivité dans l'appréciation de la portée de la sous-déclaration des décès.

Cette analyse met aussi en lumière le vaste degré d'incertitude et d'ignorance qui existent à l'égard des niveaux et des distributions de mortalité adulte dans les pays en développement. Pour le tiers des États Membres de l'OMS, comptant probablement pour une portion semblable de décès, peu de choses sont connues de façon fiable à propos des niveaux de mortalité adulte, particulièrement parmi les plus jeunes adultes de moins de 60 ans. Pour ces pays, on n'a guère d'autre choix que d'adopter la méthode démographique classique servant à élaborer des niveaux de mortalité adulte à partir des taux de mortalité juvénile, en supposant que les deux sont liés d'une manière prévisible. Cependant, les bases factuelles des quelque 30 dernières années révèlent que ce n'est pas nécessairement le cas puisque de nouveaux dangers, tels que le VIH, ont émergé et que de plus anciens, tels l'abus d'alcool, sont devenus extrêmes dans certaines populations. Nous avons tenté d'éviter de telles altérations en créant un nouveau système de table type de mortalité pour estimer les taux de mortalité adulte, mais les résultats demeureront incertains jusqu'à ce qu'une vérification soit possible à partir de registres d'état civil fiables. Il est peu probable que de tels systèmes soient établis dans un avenir rapproché dans les pays qui en ont besoin. Cependant, de bonnes estimations de la mortalité adulte sont nécessaires maintenant pour planifier et assurer le suivi dans le secteur de la santé. Les questions sur les décès survenant dans les foyers, lorsqu'elles sont posées avec suffisamment d'attention et de rigueur, peuvent produire des données courantes utiles sur les niveaux et les distributions de la mortalité adulte, et elles devraient être couramment ajoutées aux recensements et aux enquêtes. La prochaine enquête sur la santé dans le monde de l'OMS offrira une occasion idéale d'accroître nos connaissances sur la mortalité adulte. Au même moment, une priorité placée sur la collecte de données sur la survie des enfants dans une poignée de pays (Catégorie V) où aucune donnée récente n'est disponible, aidera grandement les programmes concernés par la promotion de la santé des enfants (Hill and Yazbeck 1994).

Références

Ahmad OB, Lopez AD, Inoue M (2000) Trends in child mortality: a reappraisal. *Bulletin of the World Health Organization,* **78**(10):1175-1191.

Bennett NG, Horiuchi S (1984) Mortality estimation from registered deaths in less developed countries. *Demography,* **21**(2):217-234.

Brass W (1971) On the scale of mortality. In: *Biological aspects of demography*, Brass W, ed., Taylor & Francis, London.

Chinese Academy of Preventive Medicine (1997) *Annual report on the disease surveillance system, 1996.* Beijing.

Coale A, Guo G (1989) Revised regional model life tables at very low levels of mortality. *Population Index,* **55**(4):613-643.

Hill K (1987) Estimating census and death registration completeness. *Asian and Pacific Population Forum,* **1**(3):8-24.

Hill K, Yazbeck A (1994) *Trends in under- five mortality, 1960-90. Estimates for 84 developing countries.* World Bank, Washington.

Hill K, Pande R, Mahy M, Jones G (1999) *Trends in child mortality in the developing world: 1960-1996.* UNICEF, Division of Evaluation and Planning, New York.

Lopez AD (1983) The sex mortality differential in developed countries. In: *Sex differentials in mortality: trends, determinants and consequences*, Lopez AD, Ruzicka LT, eds., Australian National University Press, Canberra.

Lopez AD, Salomon JA, Ahmad OB, Murray CJL, Mafat D (2000) *Life tables for 191 countries: data, methods, results*. World Health Organization/ Global Programme on Evidence for Health Policy, Geneva. (GPE discussion paper no 9)

Mari Bhat PN (2000) *Estimating life tables for India in the 1990s*. World Health Organization/Global Programme on Evidence for Health Policy, Geneva. (Unpublished paper prepared for GPE)

Murray CJL, Ferguson BD, Lopez AD, Guillot M, Salomon JA, Ahmad OB (2001) *Modified logit life table system: principles, empirical validation and application*. World Health Organization/Global Programme on Evidence for Health Policy, Geneva. (GPE discussion paper no 39)

Perks W (1932) On some experiments on the graduation of mortality statistics. *Journal of the Institute of Actuaries,* **63**:12-57.

Preston SH (1984) Use of direct and indirect techniques for estimating the completeness of death registration systems. In: *Data bases for mortality measurement*, UN, pp. 143-153, United Nations, New York.

Salomon JA, Mathers CD, Murray CJL, Ferguson B (2001) *Methods for life expectancy and healthy life expectancy uncertainty analysis*. World Health Organization/Global Programme on Evidence for Health Policy, Geneva. (GPE discussion paper no 10)

Salomon JA, Murray CJL (2001) Modelling HIV/AIDS epidemics in sub-Saharan Africa using seroprevalence data from antenatal clinics. *Bulletin of the World Health Organization,* **79**(7):586-692.

Stanton E, Abderrahim A, Hill K (2000) An assessment of DHS maternal mortality indicators. *Studies in Family Planning,* **31**(2):111-124.

Timaeus IM (1999) Mortality in sub-Saharan Africa. In: *Health and mortality. Issues of global concern*, pp.110-131, Chamie J, Cliquet RL, eds., United Nations Population Division; Population and Family Study Centre, New York; Brussels.

UN (2001) *World population prospects: the 2000 revision*. United Nations, New York.

US Bureau of the Census () *International data base (IDB)*. United States Bureau of the Census, International Programs Center (IPC), www.census.gov/ipc/www/idbnew.html

World Bank (2001) *World development indicators 2001*. World Bank, Washington.

Tableau 9. Disponibilité des données d'état civil dans la base de données de l'OMS, 1980-2000 (au 15 septembre 2001)

Pays	80	81	82	83	84	85	86	87	88	89	90	91	92	93	94	95	96	97	98	99	00
Catégorie I : Données d'état civil complètes (couverture de 95 % +)																					
Andorre													x		x	x	x	x	x		x
Antigua-et-Barbuda				x			x	x			x	x	x	x	x	x					
Argentine	x	x	x	x	x	x	x	x	x	x	x	x	x	x	x	x	x			x	
Australie	x	x	x	x	x	x	x	x	x	x	x	x	x	x	x	x	x	x	x	x	
Autriche	x	x	x	x	x	x	x	x	x	x	x	x	x	x	x	x	x	x	x	x	x
Bahamas	x	x	x	x	x	x	x	x	x	x	x	x	x	x	x	x	x	x			
Bahrein	x	x	x		x	x	x	x	x	x	x	x	x	x	x	x	x	x	x	x	x
Barbade	x	x	x	x	x	x	x	x	x	x	x	x	x	x	x	x					
Bélarus		x	x	x	x	x	x	x	x	x	x	x	x	x	x	x	x	x	x	x	
Belgique	x	x	x	x	x	x	x	x	x	x	x	x	x	x	x	x	x	x	x		
Bosnie-Herzégovine						x	x	x	x	x	x	x									
Bulgarie	x	x	x	x	x	x	x	x	x	x	x	x	x	x	x	x	x	x	x	x	
Canada	x	x	x	x	x	x	x	x	x	x	x	x	x	x	x	x	x	x			
Chili	x	x	x	x	x	x	x	x	x	x	x	x	x	x	x	x	x	x	x		
Îles Cook	x	x	x	x	x	x	x	x	x	x	x	x	x	x	x	x	x	x	x	x	
Costa-Rica	x	x	x	x	x	x	x	x	x	x	x	x	x	x	x	x		x		x	
Croatie			x	x	x	x	x	x	x	x	x	x	x	x	x	x	x	x	x	x	
Cuba	x	x	x	x	x	x	x	x	x	x	x	x	x	x	x	x	x	x	x	x	
Chypre	x	x	x	x	x	x	x	x	x	x	x	x	x	x	x	x	x	x	x	x	
République Tchèque	x	x	x	x	x	x	x	x	x	x	x	x	x	x	x	x	x	x	x	x	x
Danemark	x	x	x	x	x	x	x	x	x	x	x	x	x	x	x	x	x	x	x		
Dominique	x	x	x	x	x	x	x	x	x	x	x	x	x	x	x						
Estonie		x	x	x	x	x	x	x	x	x	x	x	x	x	x	x	x	x	x	x	
Fidji	x	x	x	x	x	x	x	x					x	x	x	x	x	x	x	x	
Finlande	x	x	x	x	x	x	x	x	x	x	x	x	x	x	x	x	x	x	x		
France	x	x	x	x	x	x	x	x	x	x	x	x	x	x	x	x	x	x	x		
Allemagne	x	x	x	x	x	x	x	x	x	x	x	x	x	x	x	x	x	x	x	x	
Grèce	x	x	x	x	x	x	x	x	x	x	x	x	x	x	x	x	x	x	x		
Granada					x				x						x	x	x				
Hongrie	x	x	x	x	x	x	x	x	x	x	x	x	x	x	x	x	x	x	x	x	x
Islande	x	x	x	x	x	x	x	x	x	x	x	x	x	x	x	x	x	x	x		

Pays	80	81	82	83	84	85	86	87	88	89	90	91	92	93	94	95	96	97	98	99	00
Irlande	x	x	x	x	x	x	x	x	x	x	x	x	x	x	x	x	x	x	x		
Israël	x	x	x	x	x	x	x	x	x	x	x	x	x	x	x	x	x	x	x		
Italie	x	x	x	x	x	x	x	x	x	x	x	x	x	x	x	x	x	x	x		
Jamaïque	x	x	x	x	x	x	x	x	x	x	x	x									
Japon	x	x	x	x	x	x	x	x	x	x	x	x	x	x	x	x	x	x	x		
Koweit	x	x	x	x	x	x	x	x	x	x		x	x	x	x	x	x	x	x	x	
Lettonie	x	x	x	x	x	x	x	x	x	x	x	x	x	x	x	x	x	x	x	x	
Lithuanie		x	x	x	x	x	x	x	x	x	x	x	x	x	x	x	x	x	x	x	
Luxembourg	x	x	x	x	x	x	x	x	x	x	x	x	x	x	x	x	x	x	x	x	x
Malte	x	x	x	x	x	x	x	x	x	x	x	x	x	x	x	x	x	x	x	x	
Île Maurice	x	x	x	x	x	x	x	x	x	x	x	x	x	x	x	x	x	x	x	x	
Mexique	x	x	x	x	x	x	x	x	x	x	x	x	x	x	x	x	x	x			x
Monaco		x	x	x			x	x													
Pays-Bas	x	x	x	x	x	x	x	x	x	x	x	x	x	x	x	x	x	x	x	x	
Nouvelle-Zélande	x	x	x	x	x	x	x	x	x	x	x	x	x	x	x	x	x	x	x		
Norvège	x	x	x	x	x	x	x	x	x	x	x	x	x	x	x	x	x	x	x		
Pologne	x	x	x	x	x	x	x	x	x	x	x	x	x	x	x	x	x	x	x	x	
Portugal	x	x	x	x	x	x	x	x	x	x	x	x	x	x	x	x	x	x	x	x	
Qatar		x	x	x		x	x	x	x	x	x		x	x	x	x	x	x			
République de Moldova		x	x	x	x	x	x	x	x	x	x	x	x	x	x	x	x	x	x	x	
Roumanie	x	x	x	x	x	x	x	x	x	x	x	x	x	x	x	x	x	x	x	x	
Fédération de Russie	x	x	x	x	x	x	x	x	x	x	x	x	x	x	x	x	x	x	x	x	
Saint-Kitts-et-Nevis	x	x	x	x	x	x	x	x	x	x	x	x	x	x	x	x					
Sainte-Lucie	x	x	x	x	x	x	x	x	x	x		x	x	x	x	x					
Saint-Vincent-et-les-Grenadines	x		x	x	x	x	x	x	x				x			x	x				
Saint-Marin	x	x	x	x	x	x	x	x	x	x	x	x	x	x	x	x	x	x	x		
Seychelles	x	x	x	x	x	x	x	x	x	x	x	x	x	x	x	x	x	x	x		
Singapour	x	x	x	x	x	x	x	x	x	x	x	x	x	x	x	x	x	x	x	x	x
Slovaquie			x	x	x	x	x	x	x	x	x	x	x	x	x	x	x	x	x	x	
Slovénie			x	x	x	x	x	x	x	x	x	x	x	x	x	x	x	x	x	x	
Espagne	x	x	x	x	x	x	x	x	x	x	x	x	x	x	x	x	x	x	x		
Sri Lanka	x	x	x	x	x	x	x	x	x	x		x	x	x	x	x	x				
Surinam	x	x	x	x	x	x	x	x	x	x	x	x	x	x	x	x	x	x			
Suède	x	x	x	x	x	x	x	x	x	x	x	x	x	x	x	x	x	x	x	x	x
Suisse	x	x	x	x	x	x	x	x	x	x	x	x	x	x	x	x	x	x	x		

Pays	80	81	82	83	84	85	86	87	88	89	90	91	92	93	94	95	96	97	98	99	00
TFYR Macédoine			x	x	x	x	x	x	x	x	x	x	x	x	x	x	x	x			
Trinidad et Tobago	x	x	x	x	x	x	x	x	x	x	x	x	x	x	x	x					
Ukraine		x	x	x	x	x	x	x	x	x	x	x	x	x	x	x	x	x	x	x	x
Royaume-Uni	x	x	x	x	x	x	x	x	x	x	x	x	x	x	x	x	x	x	x	x	
États-Unis d'Amérique	x	x	x	x	x	x	x	x	x	x	x	x	x	x	x	x	x	x	x	x	
Uruguay	x	x	x	x	x	x	x	x	x	x	x	x	x	x		x	x	x	x		
Venezuela	x	x	x	x	x	x	x	x	x	x	x	x	x	x	x						
Yugoslavie			x	x	x	x	x	x	x	x	x	x	x	x	x	x	x	x			
Catégorie II : Données d'état civil incomplètes																					
Albanie	x				x	x	x	x	x	x	x	x	x	x	x	x	x	x	x		
Algérie	x	x	x			x	x														
Arménie		x	x			x	x	x	x	x	x	x	x	x	x	x	x	x	x	x	
Azerbaïdjan		x	x	x	x	x	x	x	x	x	x	x	x	x	x	x	x	x	x	x	x
Belize	x	x	x	x	x	x	x	x	x	x	x	x	x	x	x	x	x				
Brésil	x	x	x	x	x	x	x	x	x	x	x	x	x	x	x	x					
Brunéi Darussalam			x	x	x	x	x	x	x	x	x	x	x	x	x	x	x	x	x	x	x
Cap-Vert	x	x	x	x	x	x					x	x							x		
Colombie			x	x	x	x	x	x	x	x	x	x	x	x	x	x					
République dominicaine	x	x	x	x	x	x												x	x	x	
Équateur	x	x	x	x	x	x	x	x	x	x	x	x	x	x	x	x	x	x	x		
Égypte	x	x	x	x	x	x	x	x	x	x	x	x	x	x	x	x				x	
El Salvador	x	x	x	x	x	x	x	x	x	x	x	x	x	x		x	x	x	x		
Géorgie		x	x	x	x	x	x	x	x	x	x	x	x		x	x	x				
Guatemala	x	x	x	x	x	x	x	x	x			x	x	x		x	x	x			
Guyana					x									x	x	x	x				
Honduras	x	x	x	x	x																
Iran (République islamique d')				x	x	x	x	x				x			x	x	x	x	x	x	
Kazakhstan		x	x	x	x	x	x	x	x	x	x	x	x	x	x	x	x	x	x	x	
Kyrgyzstan		x	x	x	x	x	x	x	x	x	x	x	x	x	x	x	x	x	x	x	
Malaysia											x	x	x	x	x	x	x	x	x		
Maldives					x	x	x	x	x	x	x	x	x	x		x	x	x	x		
Îles Marshall							x	x	x	x	x	x	x	x	x	x	x	x			
Mongolie								x	x	x		x	x	x	x	x	x	x	x	x	
Maroc								x	x	x	x	x	x	x	x	x	x	x	x		

Pays	80	81	82	83	84	85	86	87	88	89	90	91	92	93	94	95	96	97	98	99	00
Nicaragua								x	x	x	x	x	x	x	x		x	x	x		
Niue	x	x	x	x	x	x	x	x	x	x	x	x	x	x	x	x	x	x	x		
Palaos						x		x	x	x	x	x	x	x	x	x	x	x	x	x	
Panama	x	x	x	x	x	x	x	x	x	x	x	x	x	x	x	x	x	x			
Papouasie-Nouvelle-Guinée	x	x	x	x	x			x	x	x	x	x	x	x	x	x	x	x	x		
Paraguay	x	x	x	x	x	x	x	x	x	x	x	x	x		x						
Pérou	x	x	x	x	x	x	x	x	x	x											
Philippines	x	x	x	x	x	x	x	x	x	x	x	x	x	x	x	x	x				
République de Corée	x	x	x	x	x	x	x	x	x	x	x	x	x	x	x	x	x	x	x		
Samoa	x												x	x							
Afrique du Sud														x	x	x					
République arabe de Syrie				x	x	x													x		x
Tadjikistan		x	x			x	x	x	x	x	x	x	x	x	x	x					
Thaïlande	x	x	x	x	x	x	x	x	x	x	x	x	x	x	x	x	x	x	x	x	
Tonga																	x	x	x		
Tunisie	x							x	x	x				x	x	x		x	x		
Turquie													x	x	x	x	x	x			
Turkménistan		x	x			x	x	x	x	x	x	x	x	x	x	x	x	x	x		
Tuvalu												x	x	x	x						
Ouzbékistan		x	x	x	x	x	x	x	x	x	x	x	x	x	x	x	x	x	x		
Zimbabwe			x								x	x	x	x		x		x			
Catégorie III : Systèmes de surveillance et registres témoins																					
Bangladesh											x	x	x	x	x	x	x	x	x	x	
Chine								x	x	x	x	x	x	x	x	x	x	x	x	x	
Inde											x	x	x	x	x	x	x	x	x		
République-Unie de Tanzanie													x	x	x	x	x	x	x	x	

Mortalidad Mundial en 2000: Tablas de Mortalidad en 191 Países

A.D. LOPEZ

O.B. AHMAD

M. GUILLOT

B.D. FERGUSON

J.A. SALOMON

C.J.L. MURRAY

K.H. HILL

Organización Mundial de la Salud

Tipografiado e impreso en Canadá

Información de Publicaciones del Catálogo de la Biblioteca de la OMS

Mortalidad Mundial en 2000: Tablas de Mortalidad en 191 Países / A. D. López. [y otros.].

1.Tablas de Mortalidad 2.Estadísticas Demográficas 3.Recolección de datos – métodos

4.Organizaciones Mundial de la Salud I. Lopez, Alan D.

ISBN 92 4 156204 8 (NLM clasificación: HB 1322)

Indice

Aportes

Alan D. LOPEZ, Organización Mundial de la Salud, Pruebas Científicas para las Políticas de Salud (EIP) – Epidemiología y Carga de Morbilidad (EBD)

Omar B. AHMAD, School of Public Health, University of Ghana, P.O. Box LG13, Legon, Accra, Ghana

Michel GUILLOT, Center for Demography and Ecology, University of Wisconsin-Madison, 1180 Observatory Drive, Madison, WI 53706 EE.UU.

Brodie D. FERGUSON, Organización Mundial de la Salud, EIP/EBD

Joshua A. SALOMON, Organización Mundial de la Salud, EIP/EBD

Christopher J.L. MURRAY, Organización Mundial de la Salud, Director Ejecutivo – EIP

Kenneth H. HILL, Director, Hopkins Population Center. Subdirector, Hopkins Center on the Demography of Aging. Department of Population and Family Health Sciences, Johns Hopkins University, School of Hygiene and Public Health, 615 North Wolfe Street, RmW4027, Baltimore MD 21205. khill@jhsph.edu

Los autores quieren expresar especialmente su gratitud con Mie Inoue por las incontables horas en recolectar estadísticas de mortalidad de varias fuentes y por sus extraordinarios esfuerzos en formatear las tablas y gráficas presentadas aquí. Ella mas que nadie mas, merece todo el crédito por asegurar la publicación de este libro.

Glosario de términos

DHS	Demographic and Health Surveys / Encuestas Demográficas y de Salud
DSP	Disease Surveillance Points / Puntos de Vigilancia de Enfermedades
HALE	Healthy Life Expectancy / (EVAS) Esperanza de Vida Sana
MICS	Multiple Indicator Child Survey / Encuesta de Indicadores Infantiles Múltiples
SRS	Sample Registration System / Sistema de Registro por Muestreo

Introducción

La finalidad del presente informe es compilar una serie completa de tablas de mortalidad en todos los países miembros de la OMS, el año 2000, a fin de comparar, con métodos análogos, los patrones actuales de mortalidad relacionada con la edad y el sexo entre países. La información demográfica básica acerca del número de defunciones por edad y por sexo de una población es una base de datos fundamental para determinar y evaluar las políticas y los programas de salud. Comenzando por el año 1999, la OMS procedió a la confección de tablas de mortalidad de todos los Estados Miembros. Dichas listas de mortalidad sirven para varios usos y constituyen la base de todos los cálculos de la OMS relativos a patrones y niveles de mortalidad en todo el mundo. Uno de los usos estratégicos de dichas tablas de mortalidad es la elaboración del indicador de esperanza de vida sana (HALE), indicador básico de los niveles de salud de la población que utiliza la OMS y que cada año se publica en el Informe sobre la Salud en el Mundo.

La confección de una tabla de mortalidad requiere datos fiables de las tasas demográficas de mortalidad, por edad y por sexo. La fuente más fidedigna para la obtención de dichos datos es un sistema operativo de registro civil o de estadísticas demográficas en que se registren todas las defunciones. Las defunciones correspondientes a cada edad se relacionan en función de la población del grupo de personas de esa edad que, por lo general, se calcula a partir de los censos demográficos o del registro continuo de todos los nacimientos, defunciones y migraciones. Las tasas de mortalidad específicas por edad y por sexo obtenidas se usan para calcular la tabla de mortalidad.

Aunque las exigencias jurídicas para el registro de defunciones son prácticamente universales, el costo para establecer y mantener un sistema de registro de nacimientos y defunciones implica que los datos fidedignos de los registros corrientes sólo se pueden obtener, por lo general, en los países más avanzados económicamente. Sólo se pudieron recabar datos nacionales, lo suficientemente completos para el cálculo de las tablas de mortalidad, a finales del decenio de 1990, en 74 países, que abarcan aproximadamente 25% de las defunciones acaecidas, según cálculos en 2000 (véase la Tabla 1). Ante la falta de estadísticas demográficas completas, para calcular las tablas de mortalidad se utilizaron muestras de información de los registros o datos fidedignos relativos a la mortalidad infantil, además de métodos demográficos indirectos. Este método fue sumamente fácil, debido a la disponibilidad de cálculos fidedignos de mortalidad infantil en muchos países en desarrollo, durante los decenios de 1980 y 1990, gracias al Programa de Encuestas Demográficas y de la Salud (DHS) y, recientemente, al Programa de Encuestas Agrupadas de Indicadores Múltiples para la Infancia (MICS) que dirige el UNICEF.

Varias organizaciones internacionales y otros centros demográficos preparan, con regularidad, estimaciones de mortalidad nacional o compilaciones de tablas de mortalidad como parte de la labor de comprobación sectorial. Así pues, UNICEF revisa periódicamente los datos disponibles sobre mortalidad infantil, a fin de determinar las mejoras necesarias en relación con los objetivos de supervivencia infantil y evaluar las intervenciones (Hill y otros. 1999). Asimismo, se acaba de publicar una versión actualizada de las tendencias de mortalidad infantil en la década de los 90 (Ahmad y otros. 2000). Tres agencias u organizaciones, a saber, la División de las Naciones Unidas para la Población, el Banco Mundial y la Oficina del Censo de los Estados Unidos han preparado, por separado, compilaciones internacionales de tablas de mortalidad y, por lo que respecta a la División para la Población, las tablas se siguen actualizando al menos cada dos años (ONU2001; Oficina del Censo de los EE.UU.; Banco Mundial 2001). Estos diversos estudios se basan, por lo general, en fuentes análogas de datos – censos, encuestas y registros demográficos – pero pueden arrojar resultados muy diferentes, debido a discrepancias en la sincronización de la disponibilidad de datos, diferencias de juicio en cuanto a la idoneidad o a la manera en que se han de componer los datos básicos, así como a diferencias en las técnicas de estimación y en la elección de modelos.

Una revisión comparativa de estos ejercicios varios evidencia la variabilidad de resultados a partir de diferentes procedimientos y juicios. Por ejemplo, en India, al realizarse un ajuste del Sistema de Registro por Muestreo (SRS) ante la falta de declaraciones de defunciones de adultos, calculada en un 13–14%, los años 1999–2000 (Mari Bhat 2000), arroja una estimación de 9,8 millones de defunciones en el año 2000, es decir, 1 millón más que la Evaluación Demográfica de las Naciones Unidas para el año 2000 (ONU 2001).

Diferencias como ésta no son de desdeñar y tienen considerables repercusiones en la comprobación, evaluación y reorientación de programas nacionales de salud pública, así como a nivel mundial. Aunque es obvia la conveniencia de elaborar una serie única de tablas de mortalidad para todos los países del mundo, seguirá habiendo discrepancias en el juicio técnico, la disponibilidad de datos y la sincronización de evaluaciones periódicas. Habida cuenta de las necesidades de la OMS en cuanto a evaluaciones anuales de tablas de mortalidad como parte de la evaluación continua del desempeño del sistema de salud, y una preferencia por el modelo de sistema de tablas de mortalidad, basado en una modificación del sistema de logit de Brass, en lugar de otras familias de modelos de tablas de mortalidad (Murray y otros. 2001), la OMS ha confeccionado una nueva serie de tablas de mortalidad, cuyos resultados, para el año 2000, figuran en los Anexos.

Este informe proporciona detalles acerca de los métodos y fuentes de información utilizados en la preparación de estas tablas. Comenzamos con una somera revisión de las fuentes, tipos y calidad de los datos disponibles. Examinamos las diferentes fuentes de información, así como los problemas y las dificultades que plantea su uso en la confección de tablas de mortalidad. Asimismo, ofrecemos una breve reseña de los dos enfoques principales que utiliza la OMS en el cálculo de parámetros del sistema de logit de Brass (α, β) para cada país. Se utilizó el análisis de series de tiempo retardado para los países con largas series de registro de datos demográficos; en cuanto a los demás países, α y β se calcularon, bien a partir de series de tiempo más cortas en cuanto al registro de datos demográficos o a partir de datos de encuestas o de vigilancia de mortalidad infantil y de adultos. Sigue después un debate acerca del método empleado en el cálculo del insumo demográfico básico de los países, en los niveles de ${}_5q_0$ y ${}_{45}q_{15}$. En uno de los Anexos se ofrece un resumen sucinto de las principales recomendaciones, así como tablas de mortalidad detalladas por país y por región, para los 191 Estados Miembros de la OMS y las 14 subregiones[1]. Habida cuenta del papel preponderante que desempeña la mortalidad infantil en las pautas y niveles de mortalidad generales en muchos países, se publican, por separado, diagramas de datos disponibles para el cálculo de mortalidad infantil en el año 2000.

Fuentes de datos y ajustes

Registros Civiles

A la hora de confeccionar tablas de mortalidad, lo idóneo sería hacerlo a partir de una larga serie histórica de datos de mortalidad, obtenidos del registro civil de defunciones y del porcentaje de población en riesgo *de facto* (u ocasionalmente *de jure*), que el sistema abarque en su totalidad. Para realizar el cálculo de tablas de mortalidad en un determinado año (Ej. : 2000) en que todavía no se disponga de datos demográficos de defunciones, por razones de índole administrativa, se requieren proyecciones a corto plazo del año anterior o del último que se disponga. Esto exige una serie cronológica adecuada de datos que incluya, por lo menos, entre 15 y 20 años de estadísticas de mortalidad. La Tabla 9 evidencia la disponibilidad de datos relativos a las defunciones de los registros civiles para la OMS, que se podría utilizar en el cálculo de tablas de mortalidad.

En primer lugar, se evaluaron sistemáticamente los datos demográficos del registro, desde 1980, para verificar si estaban completos, mediante el uso de una amplia gama de técnicas demográficas, de las cuales

[1] Como ayuda para el análisis demográfico, los 191 Estados Miembros se han dividido en cinco estratos de mortalidad sobre la base del nivel de mortalidad infantil (5q0) y de varones adultos (45q15) de cada uno, a saber: A = infantil muy baja, adulta muy baja; B = infantil baja, adulta baja; C = infantil baja, adulta alta; D = infantil alta, adulta alta; E = infantil alta, adulta muy alta.
La matriz que definen las seis Regiones de la OMS (Afr = Africa; Amr = Las Américas; Emr = Mediterránea oriental; Eur = Europa; Sear = Asia Sudoriental; Wpr = Pácifico Occidental) y los cinco estratos de mortalidad configuran 14 subregiones, ya que no todos los estratos de mortalidad están representados en cada Región, a saber: AfrD, AfrE, AmrA, AmrB, AmrD, EmrB, EmrD, EurA, EurB, EurC, SearB, SearD, WprA, WprB.

el método de Equilibrio-Crecimiento de Brass (Brass 1971), el Método Generalizado de Crecimiento y Equilibrio (Hill 1987) y la técnica Bennett–Horiuchi (Bennett y Horiuchi,1984). Estos dos últimos métodos requieren datos referentes a la distribución de la población por edad y sexo de dos censos seguidos, así como de las defunciones registradas entre ambos censos. Por ello, se utilizó con más frecuencia el método más sencillo de Crecimiento-Equilibrio, a fin de evaluar el estado completo del registro de defunciones en que se cumplían con más facilidad los requisitos básicos de datos (defunciones y población por edad y sexo en un año o periodo determinados). Por el contrario, la técnica supone que la población considerada es estable, lo que probablemente no ocurre en muchos de los países en desarrollo y requiere cierto grado de arbitrariedad para ceñirse a los puntos que se utilizan en el cálculo del alcance de las defunciones sin declarar (Preston 1984).

La utilización de estos métodos en los registros civiles de cada país dio lugar a seleccionar 74 países en que se consideró suficientemente completo el registro de datos demográficos para el cálculo de las tablas de mortalidad (véase la Tabla 1). Estos países se enumeran en la Categoría I de la Tabla 9. En cada uno de estos países se tomó el ultimo año (o los últimos años en el caso de poblaciones reducidas) para establecer el patrón "estándar" de valores l_x por edad, y para cada sexo por separado. Utilizando este estándar, se generó una serie cronológica en α y β, los dos parámetros del sistema de Logit de Brass. Luego, se utilizaron técnicas de series cronológicas para proyectar los valores α y β al año 2000, de donde se obtuvo la tabla de mortalidad para el 2000. Hay una descripción más pormenorizada de las mismas en la sección de métodos. Por lo que atañe a poblaciones más reducidas (Ej. Islas Cook) se utilizaron movimientos demográficos medios de tres o cuatro años para el análisis de series cronológicas, en lugar de datos de un solo año.

Tabla 1. Fuentes de datos de mortalidad (número de países) por subregiones de la OMS, año 2000

Subregión	**Categoría I**: Estadísticas vitales completas (cobertura 95%+)	**Categoría II**: Estadísticas vitales incompletas	**Categoría III**: Registro por Muestreo y Sistemas de Vigilancia	**Categoría IV**: Mortalidad infantil calculada a partir de encuestas y censos	**Categoría V**: Sin datos recientes de mortalidad infantil o de adultos	Número de países
AfrD	2	2	0	18	4	26
AfrE	0	2	1	14	3	20
AmrA	3	0	0	0	0	3
AmrB	17	9	0	0	0	26
AmrD	0	4	0	2	0	6
EmrB	4	3	0	6	0	13
EmrD	0	2	0	5	2	9
EurA	26	0	0	0	0	26
EurB	7	9	0	0	0	16
EurC	8	1	0	0	0	9
SearB	1	1	0	1	0	3
SearD	0	1	2	3	1	7
WprA	4	1	0	0	0	5
WprB	2	11	1	8	0	22
Mundial	74	46	4	57	10	191

Tabla 2. Fuentes de datos de mortalidad (% de defunciones que abarca) por subregiones de la OMS, año 2000

Subregión	**Categoría I**: Estadísticas vitales completas (cobertura 95%+)	**Categoría II**: Estadísticas vitales incompletas	**Categoría III**: Registro por Muestreo y Sistemas de Vigilancia	**Categoría IV**: Mortalidad infantil calculada a partir de encuestas y censos	**Categoría V**: Sin datos recientes de mortalidad infantil o de adultos	Fallecimientos como % del total mundial (estimaciones de la OMS)
AfrD	0%	4%	0%	89%	7%	8%
AfrE	0%	13%	9%	71%	7%	11%
AmrA	100%	0%	0%	0%	0%	5%
AmrB	39%	61%	0%	0%	0%	5%
AmrD	0%	65%	0%	35%	0%	1%
EmrB	2%	70%	0%	27%	0%	1%
EmrD	0%	18%	0%	68%	14%	6%
EurA	100%	0%	0%	0%	0%	7%
EurB	51%	49%	0%	0%	0%	4%
EurC	96%	4%	0%	0%	0%	7%
SearB	6%	19%	0%	76%	0%	4%
SearD	0%	0%	92%	7%	2%	21%
WprA	100%	0%	0%	0%	0%	2%
WprB	0%	9%	84%	7%	0%	18%
Mundial	24%	12%	36%	25%	2%	100%

En Bosnia y Herzegovina, no pudieron recopilarse datos de los registros civiles desde 1991. Se consideraron completos los datos relativos a los años 1989/1991 y, ante la falta de nuevos datos, se calculó el promedio y se utilizó como supuesto para el año 2000, debido a que las secuelas del conflicto en el decenio de 1990 contrarrestaron cualquier incidencia positiva en las tasas de mortalidad.

Para un segundo grupo de países con registro civil (o con registro por muestreo de datos demográficos como en el caso de Bangla Desh, China, India y Tanzania), la aplicación de estos métodos demográficos indirectos para determinar si los datos son completos evidenció la necesidad de un cierto ajuste de los datos de los registros civiles. Estos 50 países figuran en las Categorías II y III de la Tabla 9. Hubo notables diferencias por lo que atañe al número de defunciones sin declarar pero, por lo general éste oscila entre 20 y 25%, en las Repúblicas del Asia Central, y algo más elevado (30–40%) en varios países en desarrollo. Por ejemplo, la mortalidad de adultos en Egipto se consideró completa en 84%, completa en 92% en Guatemala, 84–86% en India, 78–84% en Kazajstán, 82–85% en la República de Corea y 73–75% en Filipinas. Estas razones de completitud se aplicaron luego a los datos de registro civil para corregir la cifra de defunciones sin declarar.

Por lo que respecta a mortalidad infantil, se utilizaron todos los datos de puntos de las encuestas de mortalidad infantil como el DHS o el programa MICS de UNICEF para corregir los datos de los registros civiles relativos a mortalidad infantil. Este análisis se benefició sobremanera de la exhaustiva evaluación de datos por países que realizaron Hill y otros. (1999) sobre la base de información de censos y encuestas disponible a partir de 1996. Hemos actualizado los diagramas de estos países con información más reciente y realizado el ajuste de las predicciones de Hill y otros para el 2000, cuando fue oportuno, según datos recientes (véase el Anexo). Cuando no se dispuso de datos recientes, se efectuaron correcciones en las tasas de mortalidad infantil, sobre la base de los factores de corrección que sugieren las técnicas anteriormente mencionadas para la mortalidad de adultos y de los niveles estimados en los países limítrofes. Como es probable que el número de declaraciones de defunción de niños sea inferior al de adultos, se subestimarán los niveles de mortalidad infantil.

Para aquellos países con una serie cronológica de datos de registro civil lo suficientemente extensa, o con datos de registro por muestreo, se analizaron las series cronológicas (como en los países de la Categoría I). En cuanto al resto, se realizó una proyección para el año 2000 de los niveles de mortalidad infantil ($_5q_0$) y de adultos ($_{45}q_{15}$), utilizando las pruebas disponibles en relación con la velocidad del descenso en la mortalidad o mediante el patrón de cambio de mortalidad en el decenio de 1990, coherente con el desarrollo económico. Estos valores proyectados de mortalidad infantil y de adultos se aplicaron luego al nuevo Sistema Modificado de Tablas de Mortalidad de Logit (véase Murray y otros. 2001, y la sección de métodos) para confeccionar una tabla de mortalidad completa. En algunos casos (Ej. Sudáfrica, Tanzania, Camboya, República Dominicana, Haití y Myanmar), se aumentaron las tasas de mortalidad añadiendo la mortalidad estimada de VIH/SIDA. Estos métodos y Fuentes de datos para el cálculo de mortalidad por VIH/SIDA se analizan en una sección ulterior.

Procedimiento de fuentes múltiples para determinadas poblaciones

En dos de los países en desarrollo de gran extensión, India y China, se dispone de varias fuentes de datos, incluidas las de registro civil, las de los sistemas de vigilancia, así como las de encuestas, para calcular las tasas de mortalidad. Ninguno de estos sistemas es lo suficientemente fiable por sí solo para generar tablas de mortalidad de estos países, sin necesidad de ajustes, pero todos son útiles para el cálculo de mortalidad infantil y adulta. Se utilizaron las siguientes fuentes de datos y ajustes de los mismos:

China

Tres fuentes de datos de mortalidad sirvieron para el cálculo de la tabla de mortalidad.

Puntos de Vigilancia de Enfermedades (DSP)

Este es un sistema nacional representativo de 145 puntos de vigilancia epidemiológica, de los que se encarga la Academia China de Medicina Preventiva y que abarcan una población de 10 millones de personas a través de China. Cada año se registran datos relativos a edad, sexo y causa de 50,000–60,000 defunciones. De las evaluaciones periódicas de los datos DSP realizadas mediante nuevas encuestas, al azar, en los hogares se deduce que el porcentaje de defunciones sin declarar es aproximadamente 15% (Academia China de Medicina Preventiva, 1997), aunque el Crecimiento-Equilibrio de los datos, a partir de 1991 sugiere un factor de ajuste medio de aproximadamente el doble de este nivel. Se utilizaron los datos anuales para el periodo 1991–1998, con correcciones para calcular la tendencia en $_5q_0$ y $_{45}q_{15}$.

Registros civiles

Los datos relativos a edad, sexo y causa de 725.000 defunciones se recaban anualmente del sistema de registros civiles dependiente del Ministerio de Salud y que abarca una población de 121 millones (66 millones en zonas urbanas y 55 millones en zonas rurales). Aunque los datos no reflejen las condiciones de mortalidad en toda China, son útiles para observar las tendencias en mortalidad, habida cuenta del número de defunciones que abarcan. Las tendencias en $_{45}q_{15}$ para abarcar zonas rurales y urbanas por separado se observan en la Figura 1. Aunque el porcentaje de defunciones sin declarar dé lugar a niveles inverosímiles de $_{45}q_{15}$, estos datos sugieren que sólo ha habido un modesto descenso en la mortalidad adulta en el decenio de 1990 (4–5% para los varones – en ambas zonas – y para las mujeres en las zonas rurales, así como 14% para las mujeres en las zonas urbanas).

Figura 1. Tendencias en 45q15 a partir de datos de registro civil – China, 1990-1998

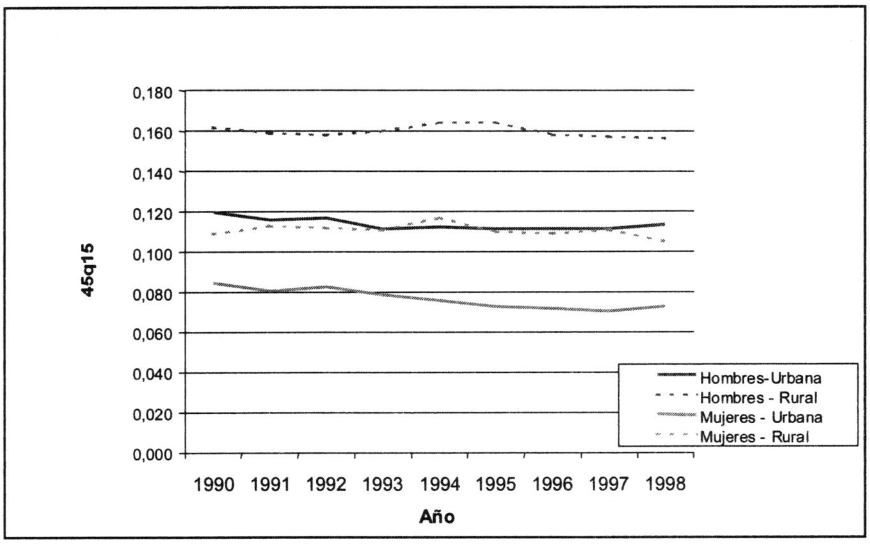

Datos de Encuestas

Los datos de encuestas se obtienen a partir de la encuesta anual de 1 de cada 1000 hogares, recabando información acerca de las defunciones acaecidas en los últimos 12 meses. Por ejemplo, la encuesta de 1997 abarcó una población de 1 243 000 personas, en más de 864 cabezas de partido (3164 municipios, 4438 poblados) en 31 provincias y se registró la cifra total de 7.845 defunciones. Aunque ésta sea una muestra representativa nacional, los métodos de Crecimiento-Equilibrio evidencian un porcentaje considerable de defunciones sin declarar (27% y 29% para varones y mujeres, respectivamente). Las tendencias en el implícito $_{45}q_{15}$ no ajustado de las encuestas del decenio de 1990 constan en la Figura 2 y reflejan un descenso algo más considerable, aunque el mucho más reducido número de defunciones, comparado con los datos de registro civil, dificulte la evaluación de tendencias.

En cada uno de estos tres sistemas, había datos disponibles para el periodo 1991–1998. Como los análisis de Crecimiento-Equilibrio evidenciaban que el porcentaje de defunciones sin declarar se mantuvo relativamente constante durante el decenio de 1990, el promedio de disminución anual en $_5q_0$ y $_{45}q_{15}$ que sugerían estas tres fuentes de datos se calculó primero y se aplicó a la tabla de mortalidad china correspondiente a 1990, sobre la base del censo, para proyectar las tasas de mortalidad al año 2000. La incertidumbre acerca de las tasas de mortalidad en 2000 obedeció a presunciones más optimistas y pesimistas acerca de la tasa de disminución en el decenio de 1990.

Figura 2. Tendencias en 45q15 (no ajustado) basadas en cálculos de la Encuesta Tipo de Cambios Demográficos – China, 1991–1998

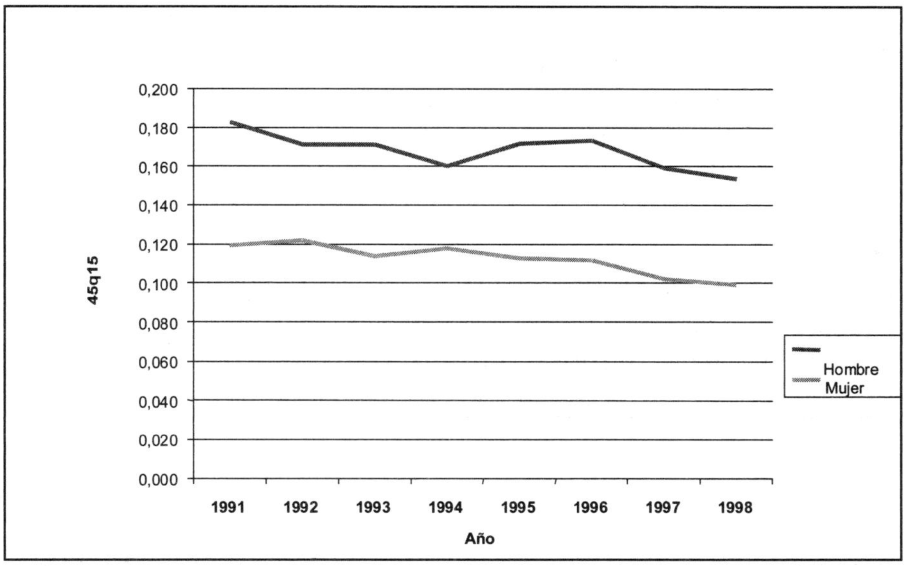

India

Los datos más representativos y fiables relativos a tasas de mortalidad por edad y sexo en la India proceden del Sistema de Registro de Muestras que lleva funcionando varios decenios. Utilizamos datos para el periodo 1990–1998 (ultimo año disponible) para el cómputo de tablas de mortalidad anual. Los datos se recaban de sucesos demográficos en unidades 4436unidades de muestreo rurales y 2235 urbanas con una población de aproximadamente 6 millones de personas que se extiende a casi todos los Estados y Territorios. Al comparar el $_5q_0$ del SRS con la tasa obtenida del DHS (Encuesta Nacional de Salud de la Familia) realizado en 1992–93 se observa que ambas arrojan resultados análogos, lo que sugiere que la insuficiencia de declaraciones de defunción infantil es mínima. No obstante, puede que haya aumentado en un 15% la insuficiencia de declaraciones de defunción de adultos en el SRS en los 90 sobre la base de la metodología de la variable -r de Bennett–Horiuchi (Mari Bhat 2000). Por consiguiente, hemos corregido las tasas de mortalidad SRS para edades de cinco años o mayores en 14% para los hombres y 16% para las mujeres y las proyectamos al año 2000. Se generaron intervalos de incertidumbre en relación con tasas de mortalidad específicas por edad a partir de proyecciones plausibles al 2000 de las líneas de tendencia en estas tasas.

Datos de Censos y de Encuestas

En los 67 países restantes (véase la Tabla 1), no se dispuso de estimaciones fiables de mortalidad adulta. Sin embargo, se pudieron utilizar amplias estimaciones, directas e indirectas, de mortalidad infantil ($_5q_0$) en años recientes, a partir de varios censos y programas de encuestas. Estas estimaciones se evaluaron sistemáticamente y se proyectaron al año 2000, con intervalos de incertidumbre (Ahmad y otros. 2000). Estas estimaciones, con campos de variabilidad de incertidumbre, se aplicaron luego al Sistema Modificado de Tablas de Mortalidad de Logit, utilizando el estándar "global" para calcular la correspondiente tabla de mortalidad. En la sección relativa a los métodos se ofrecen más pormenores al respecto.

Unicamente en un reducido número de países de este grupo (Afganistán, Angola, Burundi, República Popular Democrática de Corea, Djibouti, Guinea Ecuatorial, Haití, Liberia, Malawi, Santo Tomé y Príncipe y Swazilandia) hubo pruebas insuficientes en el decenio de 1990 para establecer estimaciones de mortalidad infantil con suficiente confianza. Las tablas de mortalidad para estos países se basan, por consiguiente, en los niveles de mortalidad infantil estimados con amplia incertidumbre. Se recomienda suma cautela en su uso, hasta que se disponga de pruebas más recientes relativas a niveles de mortalidad infantil.

En muchos de los países de esta categoría, es probable que la mortalidad por VIH/SIDA sea considerable. Esto no está captado en la aplicación de la tabla de mortalidad del modelo, ya que los modelos se confeccionaron a partir de datos anteriores a la época del SIDA. En estos países, principalmente en el Africa al sur del Sáhara, las tablas de mortalidad sin SIDA se calcularon primero mediante estimaciones de niveles de mortalidad infantil, excluyendo el VIH. Las tasas de mortalidad por edad y por sexo del VIH se estimaron luego por separado (Salomon y Murray 2001) y se añadieron a las tasas específicas por edad de la tabla de mortalidad del modelo, con el ajuste apropiado para los riesgos concurrentes.

Métodos

Confección de la tabla de mortalidad

Se utilizaron métodos normalizados de tablas de mortalidad para confeccionar las series cronológicas de tablas de mortalidad por lo que atañe a países de las Categorías I y II, en que se disponía de datos relativos a las series cronológicas. Se utilizó el sistema básico de Logit de Brass para generar la tendencia en los dos parámetros (α, β) de estas tablas de mortalidad. Este sistema se funda en el supuesto de que se puedan relacionar dos patrones de edad de mortalidad distintos entre sí por medio de una transformación lineal del logit de las probabilidades de supervivencia respectivas. Así pues, de cada dos series de valores de supervivencia observados, l_x y l^s_x, en que este último es la norma, es posible encontrar las constantes α y β de modo que

$$logit(l_x) = \alpha + \beta logit(l^s_x)$$

$$if\ logit(l_x) = 0.5 \ln\left(\frac{(1.0 - l_x)}{l_x}\right)$$

$$Then$$

$$0.5 \ln\left(\frac{(1.0 - l_x)}{l_x}\right) = \alpha + 0.5\beta \ln\left(\frac{(1.0 - l^s_x}{l^s_x}\right)$$

para todas las edades x entre 1 y T y cuando l_0(base numérica) = 1. Si se verifica la ecuación de arriba por cada par de tablas de mortalidad, se puede luego confeccionar cualquier tabla de mortalidad a partir de una sola tabla de mortalidad estándar, cambiando los pares de valores (α, β) utilizados. En realidad, el supuesto de proporcionalidad sólo se logra aproximadamente con pares de tablas de mortalidad reales. Sin embargo, la aproximación se acerca lo bastante para garantizar el uso del modelo de estudio y ceñirse a las proyecciones de mortalidad observadas. El parámetro α varía el *nivel* de mortalidad del estándar, mientras que β varía la *inclinación* del estándar, es decir, rige la relación entre mortalidad infantil y adulta.

En casos en que se disponga de una secuencia histórica de tablas de mortalidad, se puede generar una serie cronológica de pares α,β, utilizando un estándar específico para el país. Un diagrama de α y β, por separado, comparados con el tiempo ha de dar lugar a una trayectoria de puntos (López y otros. 2000). Si el diagrama de puntos para cada parámetro queda dentro de una línea bastante recta, en teoría, se podría proyectar dicha línea hacia delante para pronosticar estimaciones relativas a cualquier periodo en el futuro. Estas estimaciones α,β se pueden luego substituir en las apropiadas ecuaciones logit, para obtener las correspondientes tablas de mortalidad. Por otra parte, cuando sea errática la tendencia de los puntos, el sistema no puede proporcionar un pronóstico adecuado. En estas circunstancias, cabe aplicar técnicas adecuadas para proyectar la mortalidad, habida cuenta de la tendencia. Por ello se confeccionaron tres modelos para ajustarse a diferentes escenarios. En el primer modelo, se supone que el parámetro en el tiempo *t* es una función lineal simple del tiempo *t*:

$$\hat{\alpha}_t = \gamma_1 + \gamma_1 t$$

$$\hat{\beta}_t = \phi_1 + \phi_2 t$$

Este modelo es aplicable en circunstancias en que la tendencia en α o en β sea claramente lineal. En el segundo modelo, se supone que los parámetros α y β en el tiempo *t* son funciones lineales de desfase de los parámetros de los anteriores periodos. Así pues, los parámetros para el tiempo *T+1* se basan en el modelo de desfase 1, las correspondientes al tiempo *T+2* se basan en el modelo de desfase 2, etc., mientras que *T* corresponde a la localización del tiempo en la tabla de mortalidad estándar. En las siguientes ecuaciones se resumen estas relaciones:

$$\hat{\alpha}_{T+1} = \gamma_{11} + \gamma_{21}\alpha_T \qquad \hat{\beta}_{T+1} = \phi_{11} + \phi_{21}\beta_T \quad \textit{1st forecast point}$$

$$\hat{\alpha}_{T+2} = \gamma_{12} + \gamma_{22}\alpha_T \qquad \hat{\beta}_{T+2} = \phi_{12} + \phi_{22}\beta_T \quad \textit{2nd forecast point}$$

$$\hat{\alpha}_{T+3} = \gamma_{13} + \gamma_{23}\alpha_T \qquad \hat{\beta}_{T+3} = \phi_{13} + \phi_{23}\beta_T \quad \textit{3rd forecast point}$$

....

$$\hat{\alpha}_{T+n} = \gamma_{1n} + \gamma_{2n}\alpha_T \qquad \hat{\beta}_{T+n} = \phi_{1n} + \phi_{2n}\beta_T \quad \textit{last forecast point}$$

Este modelo es probablemente más adecuado en circunstancias con tendencias lineales claras, pero también con oscilaciones regulares en los valores de los parámetros con el tiempo. El tercer enfoque es una combinación de los modelos anteriores:

$$\hat{\alpha}_{T+1} = \gamma_{11} + \gamma_{21}\alpha_T + \gamma_{31}(T+1) \qquad \hat{\beta}_{T+1} = \phi_{11} + \phi_{21}\beta_T + \phi_{31}(T+1) \quad \textit{1st forecast point}$$

$$\hat{\alpha}_{T+2} = \gamma_{12} + \gamma_{22}\alpha_T + \gamma_{32}(T+2) \qquad \hat{\beta}_{T+2} = \phi_{12} + \phi_{22}\beta_T + \phi_{32}(T+2) \quad \textit{2nd forecast point}$$

$$\hat{\alpha}_{T+3} = \gamma_{13} + \gamma_{23}\alpha_T + \gamma_{33}(T+3) \qquad \hat{\beta}_{T+3} = \phi_{13} + \phi_{23}\beta_T + \phi_{33}(T+3) \quad \textit{3rd forecast point}$$

....

$$\hat{\alpha}_{T+n} = \gamma_{1n} + \gamma_{2n}\alpha_T + \gamma_{3n}(T+n) \qquad \beta_{T+n} = \phi_{1n} + \phi_{2n}\beta_T + \phi_{3n}(T+4) \quad \textit{last forecast point}$$

Este modelo es adecuado en circunstancias en que se den tendencias lineales complejas. Se utilizaron los tres modelos para pronosticar las estimaciones de parámetros en cada país. El modelo que generó las series cronológicas de estimaciones que más se ceñían a la tendencia histórica se consideró el más idóneo para el país en cuestión.

Para el cálculo de la tabla de mortalidad hasta llegar a la edad de 100+, se utilizó el procedimiento de Coale-Guo (Coale y Guo 1989). Hace tiempo que se viene observando que las tasas de mortalidad en edades de más de 75 u 80 años aumentan con la edad, con una tasa de disminución en lugar de la tasa constante de Gompertz (Perks 1932). Utilizando información relativa a siete grupos demográficos con datos de mortalidad relativamente fiables a edades más avanzadas (Austria, Francia, Alemania, Japón, Países Bajos, Noruega y Suecia), Coale y Guo demostraron que el aumento relativo de un grupo de la misma edad a otro disminuye al sobrepasar la edad de 80 u 85 años. Basándose en este descubrimiento, desarrollaron un método para completar las tablas de mortalidad en edades superiores a los 80 años. En su técnica se incorpora un supuesto de constancia uniforme en vez de la Gompertziana por lo que respecta a la tasa de aumento de mortalidad en edades superiores a los 80 años. Más concretamente, el logaritmo de la razón de la tasa de mortalidad en el intervalo (x+5, x+10) para la razón a partir de (x, x+5) se supone que disminuye con incremento uniforme al aumentar x por encima de 80.

Cálculo de la mortalidad adulta

El cálculo de la mortalidad en adultos es inherentemente más difícil que el de la infantil. Esto obedece en parte, a la relativa escasez de fallecimientos de adultos, sobre todo por lo que respecta al periodo comprendido en edades de 15 a 60 años, en relación con los fallecimientos ocurridos en la infancia o en la senectud. Así pues, con objeto de lograr cálculos precisos de mortalidad en adultos son necesarias muchas observaciones que típicamente abarcan periodos de larga referencia. Sin embargo, esto se debe también al hecho de que, aunque los informes de las madres relativos a la supervivencia de sus hijos proporcionan, por lo general, estimaciones satisfactorias acerca de la mortalidad infantil, no se han encontrado historiales con resultados que proporcionen estimaciones, por general, adecuadas ni oportunas, acerca de la mortalidad de adultos. Al contrario del cálculo de mortalidad infantil en que se recaba información fácilmente de las madres afectadas, es a menudo difícil dar con el declarante adecuado para obtener datos de adultos fallecidos. Esto conlleva, a menudo, problemas de omisiones o de declaraciones múltiples de un mismo fallecimiento. Con frecuencia, el declarante desconoce la edad del fallecido y, a menudo, no se pueden obtener certificaciones de nacimiento de las personas mayores, en casi todos los países en desarrollo. Por ello, los errores al declarar la edad pueden limitar, sobremanera, la capacidad para obtener buenas estimaciones de mortalidad en adultos.

El método más frecuentemente utilizado en el cálculo de la mortalidad en adultos a partir de datos de encuestas es el uso de los datos de supervivencia de los padres para calcular la supervivencia de varones y de mujeres adultas, respectivamente. Otro método incluye el uso de datos referentes a (a) la supervivencia de los primeros esposos para calcular la supervivencia de hombres adultos, (b) la supervivencia de las primeras esposas para calcular la supervivencia de mujeres adultas, (c) la supervivencia de hermanos. Se ha demostrado que estos métodos, aunque buenos en teoría, son difíciles de aplicar en la práctica, pues a menudo resultan en una subestimación de los auténticos niveles de mortalidad en adultos de 15 a 60%, en países en que ha sido imposible cotejarlos con datos de registro (Stanton y otros. 2000).

En uno de los intentos realizados para revisar sistemáticamente todas las estimaciones directas (principalmente de Sistemas de Vigilancia Demográfica) e indirectas de niveles de mortalidad adulta en Africa se llegó a la conclusión de que sólo una fracción relativamente pequeña de las mismas arrojaba cálculos admisibles y que muchas de ellas se referían a años muy anteriores a 2000 (López y otros. 2000). Por ello, se utilizó un nuevo sistema modificado de logit , para todos los países de las Categorías IV y V. Todos los pormenores de la base lógica y de la evaluación del método constan en otra parte del informe (Murray y otros. 2001). El enfoque se fundaba, principalmente, en armonizar el nivel estimado de mortalidad infantil con una variación verosímil de niveles de l_{60} ,derivados a partir de una base de datos empírica de más de 1800 tablas de mortalidad, con un amplio conjunto de valores relativos a la esperanza de vida. En la mayoría de los casos, se tomó el valor de la media de l_{60} de los valores verosímiles del conjunto como base para el cálculo de los niveles de mortalidad en adultos. Fueron relativamente pocos los casos en que se tomaron valores diferentes al valor de la media, sobre la base de una evaluación de pruebas disponibles relativas a los niveles de mortalidad en adultos.

Cálculo de mortalidad por VIH/SIDA

Por lo que respecta a los países con las mayores epidemias de VIH/SIDA y sin datos de registro civil, la mortalidad por SIDA se calculó por separado y luego se incluyó en las tablas de mortalidad que excluían el SIDA. En cuanto a los países de la región al sur del Sáhara, el número total de fallecimientos de adultos por SIDA se derivó de modelos de cálculo retrotraído, en que se utilizan datos de vigilancia centinela, principalmente de mujeres embarazadas, con actualizaciones de modelos publicados anteriormente (Salomon y Murray 2001) donde se dispone ahora de datos más recientes. A fin de calcular la mortalidad específica por edad y por sexo, hemos analizado datos de registro y de vigilancia relativos a la mortalidad por SIDA de las siguientes fuentes: el Proyecto de Morbilidad y Mortalidad de Adultos en tres distritos de Tanzania; datos de registros civiles de zonas rurales y urbanas en Sudáfrica y datos demográficos de Zimbabwe. Estos datos constituyen las únicas fuentes fiables de información demográfica relativas a mortalidad por causas específicas en el Africa continental subsahariana de que dispone la OMS. En la Figura 3 hemos preparado un diagrama de patrones relativos a las tasas de mortalidad por edad y por sexo de cada una de estas fuentes, usando como estándar la tasa más elevada que se observa en cada lugar. Hay una notable consistencia en los patrones observados en las tres diferentes fuentes, en que las diferencias más notables se dan en las edades más tempranas y en las más avanzadas. Sobre la base de estas fuentes, hemos desarrollado un patrón regional de edad estándar, tomando la media ponderada de estas fuentes. El patrón regional aparece en forma de línea gruesa en la Figura 3. Sirviéndonos de este estándar, se puede traducir una estimación dada del total de fallecimientos de adultos en tasas de mortalidad específicas por edad, aplicando el patrón estándar de tasas a la estructura por edades de la población y readaptando luego el escalamiento de las tasas, de manera que el número total de fallecimientos se ajuste a la cifra especificada.

Figura 3. Patrón por edad de mortalidad por VIH

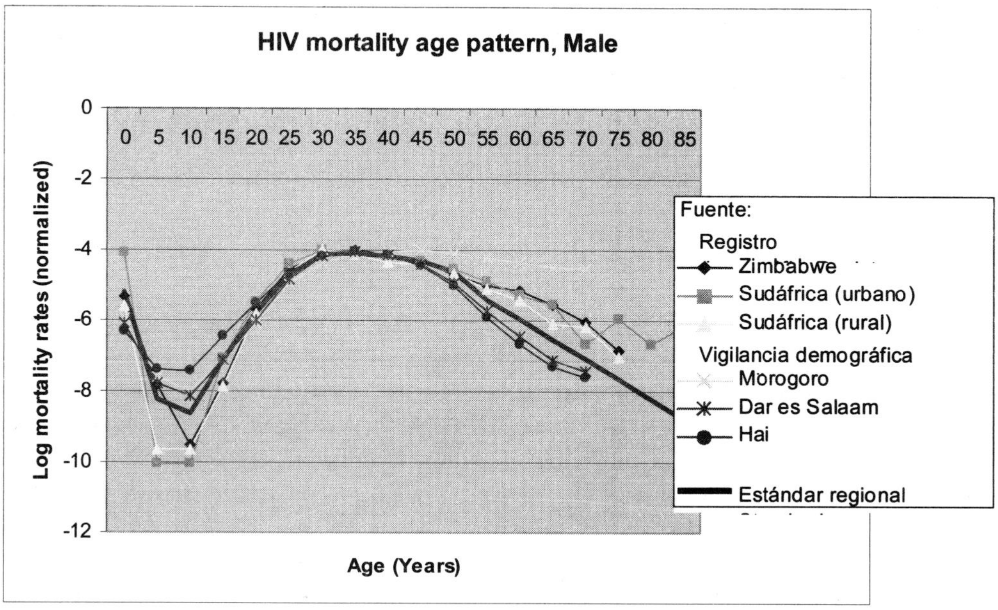

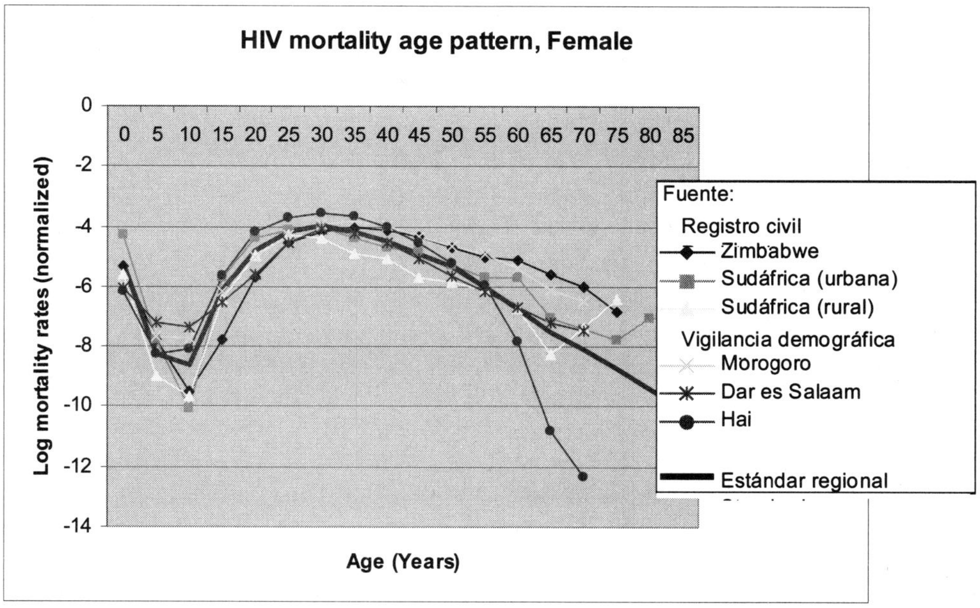

Habida cuenta de la falta de datos para poder calcular la mortalidad por SIDA directamente y de las incertidumbres que plantea el enfoque del cálculo retrotraído, es importante tratar de cuantificar el nivel de incertidumbre en las estimaciones de mortalidad resultantes de estos métodos. En los casos en que se disponía de datos suficientes para lograr un enfoque de estimación de máximas probabilidades en los modelos de cálculo retrotraído (Ej. en cerca de 20 países), los resultados incluían una cálculo de incertidumbre relativo a las estimaciones anuales de mortalidad. En los restantes países, los intervalos de incertidumbre se derivaron sobre la base de una evaluación de la cobertura y de la representatividad de los sitios de vigilancia centinela en cada país. Se tradujeron luego las probabilidades de distribución en número total de fallecimientos a distribuciones de tasas de mortalidad por edad y por sexo, utilizando métodos de simulación numérica. Se utilizaron los métodos de Monte Carlo para muestreo de estas distribuciones, a fin de incorporar la incertidumbre a la mortalidad por SIDA en las estimaciones de incertidumbre de la tabla de mortalidad.

Para los países fuera del Africa al sur del Sáhara (Camboya, República Dominicana, Haití y Myanmar), en que no había registros civiles o donde el SIDA no se reflejaba adecuadamente en las defunciones registradas, hemos combinado también las estimaciones separadas de mortalidad por SIDA en tablas de mortalidad que excluían el SIDA. El patrón de edad de la mortalidad por SIDA en estos casos se basó en patrones regionales de referencia de países que disponen de registros civiles (Argentina, Brasil, Chile, Cuba, México y Uruguay en el caso de Haití y de la República Dominicana; así como de Tailandia en el caso de Myanmar y de Camboya).

Límites de incertidumbre

Hay varias fuentes de incertidumbre en torno a los valores finales de α y β obtenidos a partir de estos modelos, incluida la incertidumbre del modelo en cuanto a la especificación correcta, así como el cálculo de la incertidumbre en valores de identificación para los coeficientes de regresión. Se puede encontrar una exhaustiva explicación de las fuentes de incertidumbre y de los métodos para el análisis de incertidumbre de las tablas de mortalidad en Salomon y otros. (2001). El nivel de incertidumbre en torno a las estimaciones de α y β depende, en parte, de la incertidumbre en torno a los coeficientes de regresión γ_{ij} y ϕ_{ij}, y ,a su vez,implica cierto nivel de incertidumbre en torno a las tablas de mortalidad que se computan a partir de estos parámetros. Habida cuenta de que una tabla de mortalidad es una complicada función no lineal de ciertos parámetros inciertos, utilizamos las técnicas de simulación de Monte Carlo para elaborar estimaciones numéricas de los campos de incertidumbre en torno a las tablas de mortalidad. Esta incertidumbre se capta tomando coeficientes de regresión al azar γ_{ij} y ϕ_{ij} a partir de distribuciones normales con medios idénticos a los de los coeficientes estimados y con variancias derivadas de los errores comunes en la regresión. En una de cada 1000 iteraciones, las deducciones de γ_{ij} y ϕ_{ij} se usan para generar estimaciones de α y β , que luego se traducen en tablas de mortalidad completas. Así pues, las probabilidades de distribución se pueden definir en torno a las estimaciones de las tablas de mortalidad, mediante el análisis de 1000 tablas de mortalidad simuladas y diferentes. Por ejemplo, se puede definir un campo en torno al cálculo de la esperanza de vida al nacer, seleccionando 1000 estimaciones diferentes de e(0) en las tablas de mortalidad simuladas e identificando luego los valores 25 y 975 como límites de un intervalo de confianza de aproximadamente 95%.

A falta de datos de historial (ej., para los países de las Categorías IV y V), se utilizó un sistema de logit modificado (Murray y otros. 2001). Al igual que el sistema de logit simple, el nuevo modelo estriba en la relación entre dos tablas de mortalidad. No obstante, a diferencia del sistema de logit simple, la nueva formulación incluye dos parámetros adicionales que corrigen la alinealidad en el método de Brass original. El insumo principal para el sistema es un campo estimado en torno a $_5q_0$. Para cualquier valor dado de $_5q_0$, el sistema de logit de tablas de mortalidad proporciona una escala de valores verosímiles en cuanto a $_{45}q_{15}$. Estas escalas constituyen la base de una serie de simulaciones de Monte Carlo de potenciales tablas de mortalidad diferentes, consecuentes con lo conocido y con lo incierto, acerca de mortalidad infantil y en adultos. La serie de tablas de mortalidad simuladas posibilita el cálculo de campos de incertidumbre en torno a varios indicadores clave como q_x, l_x y e_x.

Resultados

La esperanza total de vida al nacer (combinada para ambos sexos) en 2000 se calcula que oscilará entre 81,1 años en Japón (84,7 en mujeres, 77,5 en hombres) hasta 37,5 años en Malawi (véase la Tabla 3). En cuanto a hombres, la siguiente esperanza de vida al nacer más alta se estimó en Suecia (77,3 años), seguida de Andorra (77,2), Islandia (77,1), Mónaco (76,8) y Suiza (76,7). La esperanza de vida para hombres sobrepasó los 75,0 años en 19 países el año 2000 (véase la Tabla 4).

Tabla 3. Esperanza de vida al nacer (años), los 10 primeros y los 10 últimos países, en 2000

Los 10 primeros países		Los 10 últimos países	
1 Japón	81,1	1 Malawi	37,5
2 Mónaco	80,6	2 Sierra Leona	37,9
3 Andorra	80,5	3 Mozambique	38,7
4 San Marino	80,0	4 Zambia	39,4
5 Suecia	79,6	5 Rwanda	39,5
6 Suiza	79,6	6 Burundi	41,0
7 Islandia	79,5	7 República Centroafricana	42,0
8 Australia	79,3	8 Lesotho	42,1
9 Italia	79,2	9 Namibia	42,7
10 Francia	79,1	10 República Democrática del Congo	42,8

Tabla 4. Países con esperanza de vida para varones superior a 75,0 años, en 2000

País	e0 (años)	País	e0 (años)
Japón	77,5	Italia	76,0
Suecia	77,3	Nueva Zelandia	75,9
Andorra	77,2	Noruega	75,7
Islandia	77,1	Países Bajos	75,4
Mónaco	76,8	Singapur	75,4
Suiza	76,7	Grecia	75,4
Israel	76,6	Malta	75,4
Australia	76,6	España	75,4
San Marino	76,1	Francia	75,2
Canadá	76,0		

Tabla 5. Países con esperanza de vida para mujeres superior a 80,0 años, en 2000

País	E0 (años)	País	e0 (años)
Japón	84,7	Noruega	81,4
Mónaco	84,4	Austria	81,4
San Marino	83,8	Países Bajos	81,0
Andorra	83,8	Nueva Zelandia	80,9
Francia	83,1	Bélgica	80,9
Suiza	82,5	Finlandia	80,9
Italia	82,4	Luxemburgo	80,8
España	82,3	Grecia	80,8
Australia	82,1	Malta	80,7
Suecia	82,0	Alemania	80,6
Islandia	81,8	Israel	80,6
Canadá	81,5	Singapur	80,2

Entre las mujeres, la segunda tasa de esperanza de vida más alta se calculó para Mónaco (84,4 años), seguida de San Marino y de Andorra (83,8 años), así como Francia (83,1). En veinticuatro países hay una esperanza de vida estimada de 80 años o más para las mujeres en 2000 (véase la Tabla 5). La esperanza de vida para las mujeres sobrepasó los 75,0 años en 57 países, aproximadamente un tercio de los Estados Miembros de la OMS.

Habida cuenta de las extraordinarias repercusiones de la epidemia de VIH/SIDA en el Africa al sur de Sáhara, tal vez no sorprenda que los países con la esperanza de vida más baja en 2000 sean todos de esta Región. En efecto, 37 de los 40 países con la esperanza de vida más baja están en el Africa al sur del Sáhara. Aunque los países de esta Región adolecen, en forma desproporcionada, de muchos de los factores causantes de mortalidad infantil, así como de mortalidad prematura en adultos, de los cuales infecciones graves de las vías respiratorias, enfermedades diarreicas y tuberculosis, la falta de adelantos para lograr ganar más terreno en el campo de la esperanza de vida se puede atribuir, en gran parte, a los efectos de la epidemia del VIH/SIDA en los últimos 15 años, aproximadamente. Si se excluyesen los fallecimientos a consecuencia del VIH/SIDA, la esperanza de vida al nacer en algunos países de la región sería de 15 a 20 años más (véase la Tabla 6). Esto es particularmente cierto en países del Africa meridional (Botswana, Lesotho, Namibia, Sudáfrica, Swazilandia, Zambia, Zimbabwe), pero las disminuciones de la esperanza de vida de 5 a10 años, debido a la mortalidad por SIDA en 2000 son también comunes en muchos otros países africanos. En promedio, la esperanza de vida al nacer en el Africa al sur del Sáhara es de 6 años menos para los varones y poco más de 7 para las mujeres, comparada con lo que hubiese sido en 2000, de no haberse registrado la mortalidad por VIH/SIDA.

Tabla 6. Diferencia en esperanza de vida al nacer para todas las causas posibles de fallecimiento, excluido el SIDA, por sexo, en 2000 (años)

Varones			
Namibia	18,5	Djibouti	5,9
Botswana	17,3	Haití	5,4
Zambia	15,2	Eritrea	5,4
Zimbabwe	14,9	Nigeria	4,7
Lesotho	14,8	Somalia	4,7
Swazilandia	14,2	Ghana	3,9
Sudáfrica	12,1	Bahamas	3,9
República Centroafricana	10,3	Gabón	3,8
Malawi	10,1	República Dem. del Congo	3,5
Kenya	9,8	Guyana	3,0
Burundi	9,8	Liberia	2,7
Côte d'Ivoire	9,0	Camboya	2,7
Mozambique	8,3	Myanmar	2,5
Rwanda	7,8	Chad	2,3
Uganda	7,8	Honduras	2,2
República Unida de Tanzania	7,2	Sierra Leona	2,1
Etiopía	7,2	República Dominicana	2,1
Burkina Faso	6,6	Benin	2,1
Togo	6,3	Gambia	2,0
Camerún	6,2	Angola	2,0
Congo	6,0		
Mujeres			
Namibia	21,5	Congo	7,3
Botswana	20,2	Camerún	7,3
Zambia	18,1	Djibouti	7,0
Zimbabwe	17,6	Eritrea	6,4

Lesotho	17,4	Somalia	5,7
Swazilandia	16,8	Nigeria	5,6
Sudáfrica	15,1	Ghana	4,7
Malawi	12,4	República Dem. del Congo	4,4
República Centroafricana	12,2	Gabón	4,4
Burundi	11,7	Liberia	3,3
Kenya	11,6	Chad	2,8
Côte d'Ivoire	10,9	Sierra Leona	2,6
Mozambique	10,3	Angola	2,6
Rwanda	9,9	Gambia	2,4
Uganda	9,2	Benin	2,4
Etiopía	8,8	Guinea-Bissau	2,1
República Unida de Tanzanía	8,6	Haití	2,1
Burkina Faso	7,8	Senegal	2,1
Togo	7,5		

En los países más desarrollados persisten considerables diferencias de sexo en la esperanza de vida. A comienzos del siglo XX, la esperanza de vida de las mujeres superaba a la de los varones en 2 ó 3 años, en promedio, por lo menos en Europa, Norteamérica y Australia (López, 1983). En 2000, la ventaja de las mujeres registró un incremento hasta situarse en 10 o más años en Kazajstán (10,4 años), Lituania (10,4), Ucrania (10,7), Estonia (11,0), Latvia (11,3) y Belarús (12,0), y la más alta de todas se registró en la Federación de Rusia (12,6). A la inversa, la diferencia fue sólo de medio año o de menos en países como Bangladesh, Lesotho y Zambia. Solamente en un número muy reducido de países, de los cuales Botswana, Maldivas, Namibia y Nepal, el cálculo de la esperanza de vida de los hombres sobrepasó al de las mujeres. (de 0,2 a 0,5 años). En la Figura 4 se observa la distribución del desfase entre hombres y mujeres por lo que respecta a la esperanza de vida al nacer, entre los 191 Estados Miembros de la OMS en 2000. Los valores extremos que se observan en algunos países de Europa Oriental quedan claros en el límite de la distribución. En una cuarta parte de los países, aproximadamente, el diferencial varón-mujer es de entre 5 y 6 años, siendo más favorable para las mujeres. Como los diferenciales de mortalidad por sexo son, por lo general, más bajos en los países en desarrollo, el total de la diferencia en la esperanza de vida al nacer para varones y para mujeres es ligeramente inferior a ésta (4,7 años).

Figura 4. Distribución de la diferencia varón-mujer en la esperanza de vida al nacer, Estados Miembros de la OMS, 2000

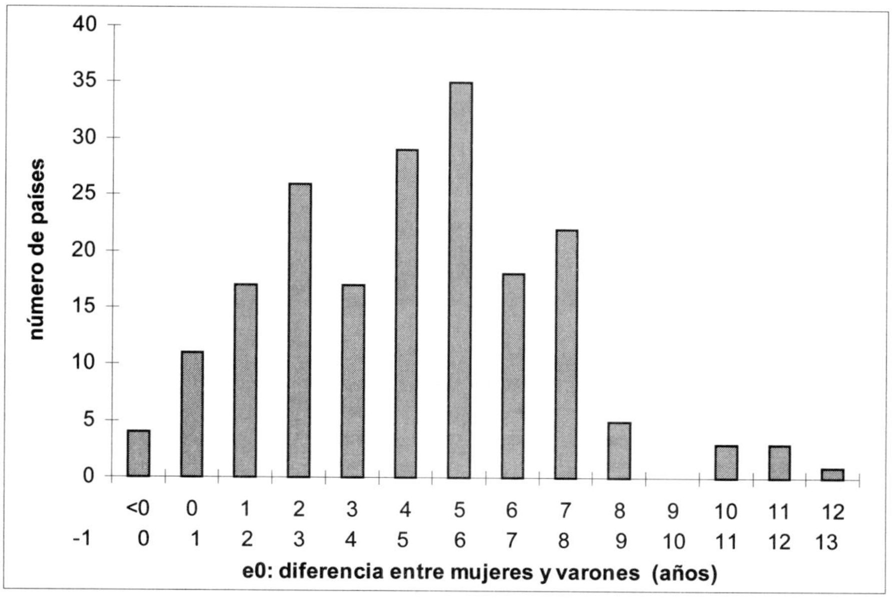

Los indicadores compendiados útiles de riesgos prevalecientes de mortalidad para la población son la probabilidad de morir entre el nacimiento y la edad de 5 años, como medida general de condiciones de salud infantil y la probabilidad de morir a edades comprendidas entre los 15 y los 60 años, como medida de mortalidad prematura en adultos. Estos figuran en las diferentes Regiones de la OMS, y dentro de las mismas, las diversas subregiones por tasas mortalidad.en la Figura 5.

Figura 5. Probabilidades de fallecimiento durante la infancia (0–4 años) y edad adulta (15–59 años), por subregión, 2000

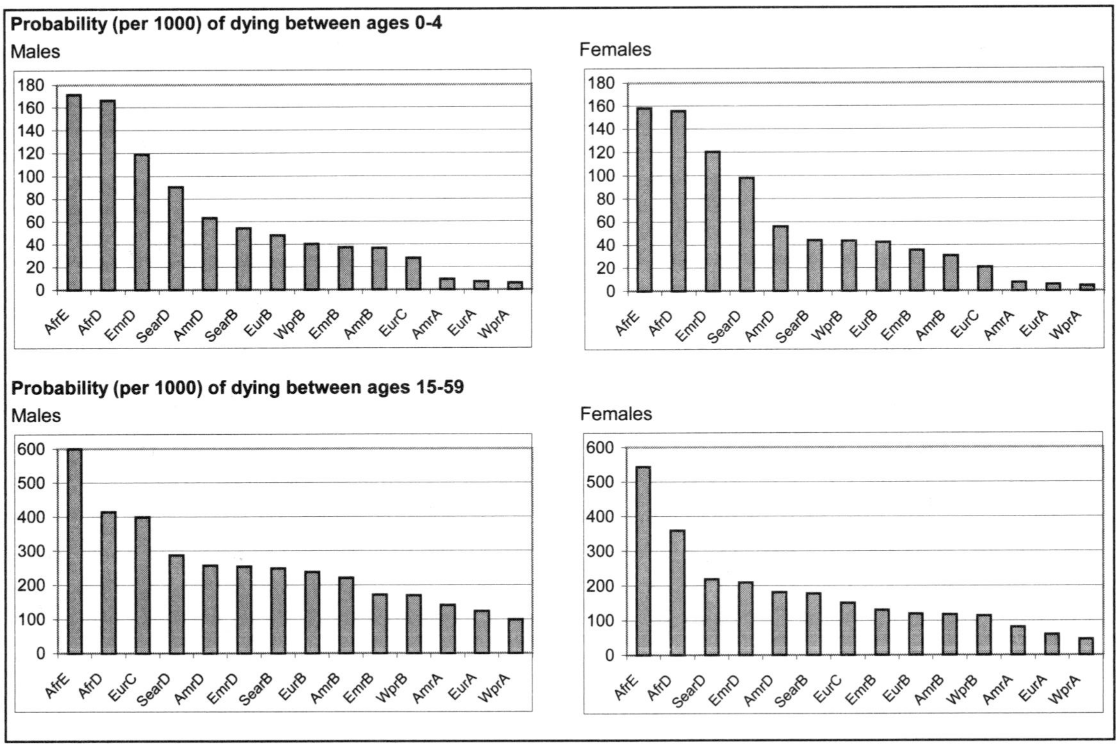

Sigue habiendo muchas diferencias en los niveles de mortalidad infantil. De los 10,9 millones de fallecimientos de menores de 5 años acaecidos en 2000 (véase laTabla 7), 99% se registró en regiones en desarrollo. En los países industrializados (Ej. : los clasificados en el Estrato Regional A) la probabilidad de mortalidad infantil ($_5q_0$) es típicamente inferior a 1% (0,4% en Suecia y Singapur), pero asciende a casi 300 por 1000 en Sierra Leona. Los niveles de mortalidad infantil muy por encima de 10% (100 por 1000) son todavía corrientes en Africa y en partes de Asia (Afganistán, Camboya, Laos, Myanmar, Nepal y Pakistán).

Tabla 7. Total de fallecimientos por sexo, edad y subregión de la OMS, 2000

Sexo	Subregiones OMS		Todas (000)	0-4 (000)	5-14 (000)	15-29 (000)	30-44 (000)	45-59 (000)	60-69 (000)	70-79 (000)	80+ (000)
Ambos sexos	Mundo		55 619	10 893	1 438	3 619	5 062	6 993	8 006	10 281	9 328
	Afr	Todas	10 572	4 244	432	1 304	1 683	1 039	719	739	412
		D	4 244	1 928	182	396	484	383	319	348	204
		E	6 327	2 316	250	907	1 198	656	400	391	208
	Amr	Todas	5 872	469	64	297	438	758	838	1 280	1 729
		A	2 778	37	9	55	129	308	371	687	1 183
		B	2 584	313	38	194	254	386	404	517	477

Sexo	Subregiones OMS		Todas (000)	0-4 (000)	5-14 (000)	15-29 (000)	30-44 (000)	45-59 (000)	60-69 (000)	70-79 (000)	80+ (000)
		D	511	119	17	48	55	64	63	76	68
	Emr	Todas	4 028	1 549	154	257	294	444	467	530	332
		B	682	127	20	49	51	101	111	132	91
		D	3 346	1 421	134	208	243	343	357	398	242
	Eur	Todas	9 668	254	48	241	541	1 178	1 704	2 602	3 098
		A	4 080	27	7	49	120	348	575	1 126	1 828
		B	1 952	170	24	67	126	264	370	510	420
		C	3 636	57	18	125	295	567	759	967	850
	Sear	Todas	14 091	3 309	536	1 030	1 325	2 029	2 211	2 301	1 350
		B	2 143	301	55	191	249	331	370	404	242
		D	11 948	3 008	481	838	1 076	1 699	1 841	1 896	1 107
	Wpr	Todas	11 389	1 068	203	491	780	1 544	2 066	2 828	2 408
		A	1 150	8	2	16	30	117	179	294	504
		B	10 238	1 060	201	475	750	1 427	1 888	2 534	1 903
Varones	Mundo		29 641	5 638	728	1 874	3 014	4 339	4 792	5 498	3 759
	Afr	Todas	5 417	2 257	211	522	890	609	387	364	177
	Afr	D	2 189	1 022	89	164	257	222	171	173	90
		E	3 228	1 235	122	357	633	387	216	191	87
	Amr	Todas	3 148	260	37	217	293	468	490	689	694
		A	1 382	21	5	40	84	192	220	371	448
		B	1 485	175	22	148	175	240	235	277	213
		D	282	65	9	29	34	36	35	41	33
	Emr	Todas	2 132	790	77	129	164	261	263	279	168
		B	382	67	11	31	31	62	64	71	45
		D	1 750	723	66	98	134	198	198	209	123
	Eur	Todas	4 947	142	30	180	397	817	1 072	1 288	1 020
		A	2 037	15	4	37	82	233	379	627	660
		B	1 053	93	15	46	86	177	224	253	159
		C	1 857	33	12	97	229	406	470	409	202
	Sear	Todas	7 661	1 655	258	540	789	1 208	1 272	1 256	683
		B	1 182	170	31	122	152	188	203	207	108
		D	6 479	1 484	226	418	637	1 020	1 069	1 049	575
	Wpr	Todas	6 337	534	115	287	479	977	1 307	1 621	1 017

Sexo	Subregiones OMS		Todas (000)	0-4 (000)	5-14 (000)	15-29 (000)	30-44 (000)	45-59 (000)	60-69 (000)	70-79 (000)	80+ (000)
		A	625	5	1	11	20	81	124	178	205
		B	5 712	529	114	275	459	896	1,184	1,443	812
Mujeres	Mundo		25 978	5 255	709	1 745	2 048	2 654	3 215	4 784	5 569
	Afr	Todas	5 155	1 987	220	782	793	430	332	375	235
	Afr	D	2 056	906	92	232	227	161	148	175	114
		E	3 099	1 081	128	550	566	269	184	200	121
	Amr	Todas	2 725	209	27	80	145	290	348	591	1 035
		A	1 396	16	4	15	45	116	151	315	735
		B	1 099	138	15	46	79	147	169	241	264
		D	229	55	8	18	22	28	28	35	36
	Emr	Todas	1 896	758	77	128	129	184	205	251	164
		B	300	60	10	18	20	39	46	61	46
		D	1 596	698	68	110	109	145	159	190	119
	Eur	Todas	4 721	112	18	61	144	362	632	1 314	2 078
		A	2 043	12	3	12	38	115	197	499	1 168
		B	900	77	9	22	40	87	146	257	261
		C	1 779	23	6	28	66	160	289	558	648
	Sear	Todas	6 430	1 654	279	490	536	821	939	1 045	666
		B	961	131	23	69	97	142	167	197	134
		D	5 469	1 523	255	421	439	679	772	848	532
	Wpr	Todas	5 051	534	88	204	301	567	759	1 207	1 391
		A	525	3	1	4	10	37	55	116	299
		B	4 526	531	87	200	291	530	704	1 091	1 092

Sin embargo, tal vez las mayores disparidades en la mortalidad ocurren en la edad adulta comprendida entre los 15 y los 59 años. En algunos países del Africa meridional como Botswana, Malawi, Namibia y Zambia, donde el VIH/SIDA constituye actualmente un grave problema de salud, se espera que 65% o más de adultos que sobreviven hasta la edad de 15 años morirán a la edad de 60 sobre la base de las tasas actuales de mortalidad. En muchos otros países (Ej. Burundi, Lesotho, Mozambique, Rwanda, Swazilandia y Zimbabwe) el riesgo sobrepasa 60%. Asimismo, cabe destacar el espectacular aumento del $_{45}q_{15}$ en Sudáfrica, con niveles estimados de 567 por 1000 y de 502 por 1000 para varones y mujeres respectivamente en 2000, comparado con niveles de aproximadamente entre 250 y 350 por cada 1000 mujeres y hombres respectivamente en 1990 (Timaeus, 1999). A la inversa, son muy corrientes en la mayoría de los países desarrollados niveles de $_{45}q_{15}$ de 90–130 por 1000 para varones, con la mitad de riesgos para las mujeres.

La creciente disparidad entre niveles de mortalidad infantil y adulta en los últimos años es evidente en la Figura 6, donde se contrastan niveles medios de $_{5}q_{o}$ con niveles de $_{45}q_{15}$ para varones y mujeres, por separado, sobre la base de las tablas de mortalidad específicas para cada país en 2000. Se incluyen estimaciones de incertidumbre para cada punto. Es patente el alejamiento sistemático de patrones de mortalidad en los países de Europa oriental, particularmente los relativos a varones, de los niveles esperados, habida cuenta de la relación tradicional entre la mortalidad infantil y la adulta, al igual que la considerable incertidumbre acerca de los niveles de mortalidad adulta en varios países del Africa, principalmente al sur del Sáhara. Por lo general, la incertidumbre acerca de los niveles de mortalidad infantil es muy inferior a la de los de mortalidad adulta, reflejando los conocimientos adquiridos a partir de los extensos programas de encuestas globales sobre niveles de mortalidad infantil y factores determinantes en los últimos decenios.

Figura 6. Comparación de mortalidad adulta y mortalidad infantil en 191 Estados Miembros de la OMS, el año 2000

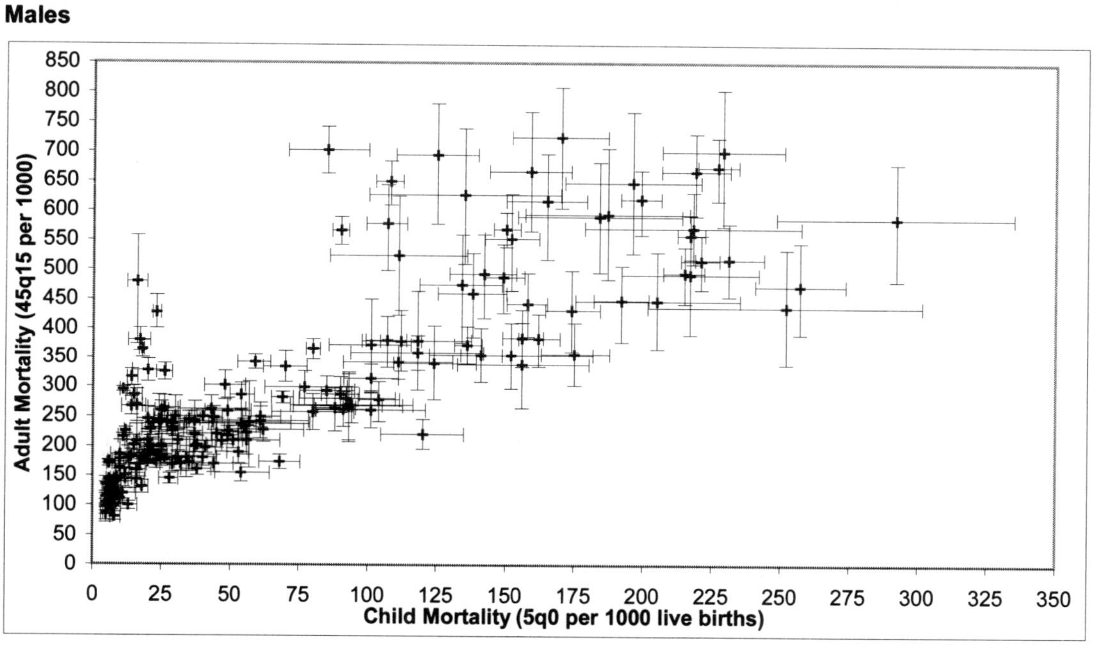

(Varones - Mortalidad en Adultos $_{45}q_{15}$ por 1000 – Mortalidad infantil $_{5}q_{0}$ por 1000 nacimientos vivos)

Females

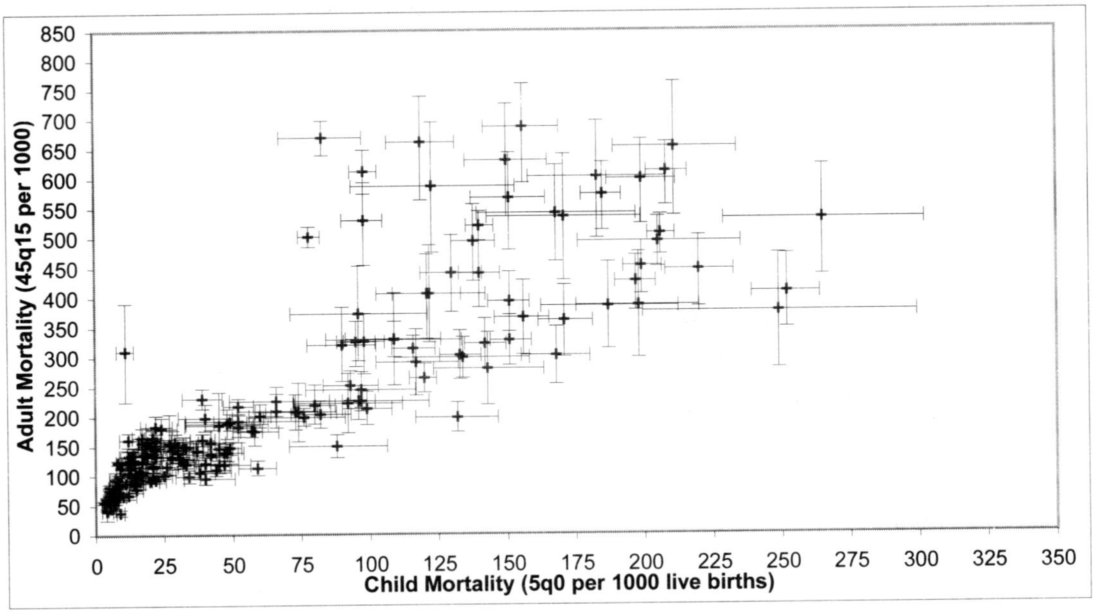

(Mujeres - Mortalidad en Adultos $_{45}q_{15}$ por 100 – Mortalidad infantil $_{5}q_{0}$ por 1000 nacimientos vivos)

Esto se refleja en el patrón-edad de los intervalos de incertidumbre para cada país en la Región que figuran en el Anexo. Los límites de incertidumbre en torno a q_x y a l_x tienden a extenderse en torno a edades adultas jóvenes, en particular por lo que respecta al Africa al sur del Sáhara, donde escasean las pruebas fiables relativas a los niveles de mortalidad adulta. En efecto, la extensión de estos intervalos de incertidumbre conlleva un importante hallazgo como principal valor estimado y plantea la urgencia de invertir en el cálculo de los niveles de mortalidad en esos países en que los límites continúan siendo inexcusablemente extensos.

Un índice compendiado alternativo de condiciones demográficas de mortalidad es la edad media de fallecimiento, en relación con la edad media de la población. En países con alta mortalidad infantil, tanto la edad media de fallecimiento como la edad media de la población son comparativamente bajas y aumentarán más o menos en tandem, en la medida en que se produzcan mejoras en el desarrollo de la salud. Esto es patente en la Figura 7, donde se demuestra la relación por separado para varones y mujeres sobre la base de las tablas de mortalidad para cada país en 2000. Cabe destacar que, más allá de una edad media de la población de aproximadamente 30 años, la edad media de fallecimiento aumenta algo más despacio que en los niveles de edad media inferiores, particularmente por lo que respecta a las mujeres. Esta disminución obedece en parte, a un descenso en la fertilidad en países con mortalidad baja, durante los últimos 50 años, pero evidencia también el hecho de que cuando la edad media de fallecimiento alcanza los 60 años, aproximadamente, ya se han conseguido casi todos los logros para disminuir la mortalidad infantil y adulta y de aquí que las reducciones ulteriores resulten sólo en aumentos más lentos en la edad media de la población. Como también se puede observar en la Figura, sigue habiendo extraordinarias diferencias considerables en la edad media de fallecimiento entre Estados Miembros de la OMS en 2000, que oscilan de una tan baja como 15,5 en Níger (ambos sexos combinados), 19,9 en Afganistán, 20,3 y 20,5 en Somalia y Sierra Leona, respectivamente, hasta casi 80 en Suecia (78,5), Noruega (77,1) y Grecia (76,8). Esta diferencia cuádruple es considerablemente más extensa que la escala en la esperanza de vida que anteriormente se presenta en este informe.

Figura 7. Edad media de fallecimiento y población para 2000 en 191 países

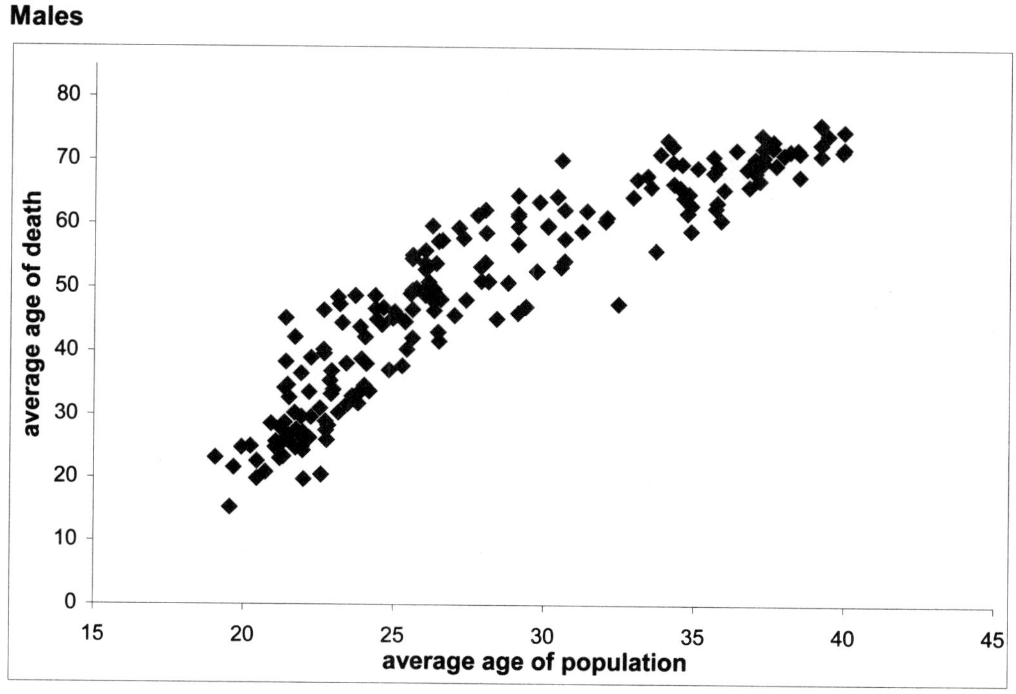

(Varones – edad media de fallecimiento- edad media de la población)

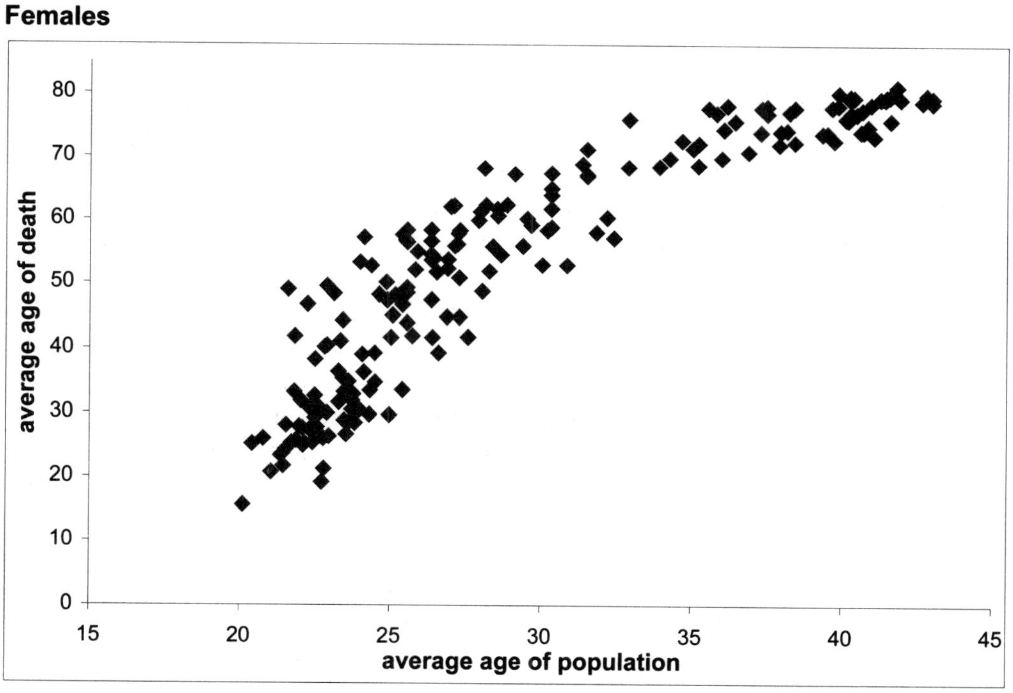

(Mujeres – edad media del fallecimiento – edad media de la población)

Se calcula que la cifra estimada de fallecimientos en 2000 fue poco menos de 56 millones de personas, de los cuales casi 30 millones eran varones (véase la Tabla 7). Ese mismo año murieron en el mundo 10,9 millones de niños menores de 5 años, de los cuales cerca de 3/4 en regiones en desarrollo de Africa y del Asia meridional. Otros 10 millones aproximadamente fallecieron en cada uno de los grupos de edad más avanzada, es decir, de 70–79 y 80+, la gran mayoría de los cuales en las subregiones A, B y C de mortalidad inferior. Hubo muchos más fallecimientos entre adultos jóvenes de edades comprendidas entre los 15 y los 59 años (15,6 millones), que entre los niños de edades entre 0 y 14 años (12,3 millones), poniendo de manifiesto la necesidad de un enfoque más amplio en las políticas de prevención de fallecimientos prematuros y de ocuparse, asimismo, de las causas principales de mortalidad en adultos jóvenes.

A escala mundial, la esperanza de vida al nacer en 2000 se calculó en 64,9 años (62,7 para los varones y 67,2 para las mujeres) (véase la Tabla 8). Las marcadísimas diferencias en patrones de edad de mortalidad entre las subregiones patentes en la Tabla 7 se traducen en extensas variaciones del promedio regional de esperanza de vida, con niveles que oscilan desde sólo poco más de 44 años (en ambos sexos) en AfrE donde más predomina el VIH/SIDA a un nivel medio casi el doble de alto (80,9 años) en WprA. El promedio de esperanza de vida alcanzó los 70 años o más en seis subregiones, sorprendentemente no en EurB o en C. En particular, el promedio de esperanza de vida al nacer para los varones en EurC fue de a penas 60 años, entre los más bajos de cualquier subregión en el mundo, salvo en Africa.

Tabla 8. Esperanza de vida al nacer (años) por subregión OMS, 2000

Subregión	varones	mujeres	ambos sexos
AfrD	50,3	52,4	51,3
AfrE	43,5	45,2	44,3
AmrA	74,1	79,6	76,9
AmrB	67,5	74,5	70,9
AmrD	63,5	68,4	65,9
EmrB	68,6	71,6	70,0
EmrD	59,1	61,0	60,0
EurA	74,8	81,2	78,0
EurB	65,5	72,3	68,8
EurC	60,3	72,1	66,0
SearB	64,1	68,9	66,4
SearD	59,7	62,4	61,0
WprA	77,3	84,2	80,9
WprB	68,2	72,7	70,4
Mundial	**62,7**	**67,2**	**64,9**

Debate

La tabla de mortalidad y los parámetros relacionados con la misma constituyen un insumo clave en la evaluación de lo bien o mal que funcionan los sistemas de salud. Las tablas de mortalidad tienen muchas otras aplicaciones en epidemiología, demografía y economía, y la disponibilidad de tablas anuales y actualizadas de mortalidad para los 191 Estados Miembros de la OMS han servir para mejorar la calidad e importancia de la base analítica para tales usos. Es fundamental llevar a cabo una detenida evaluación de los datos que han de servir para confeccionar las tablas de mortalidad, a fin de poder utilizar los resultados con cierta seguridad. La finalidad del presente informe es ayudar a que los países en que, al parecer, funcionan bastante bien los registros civiles inicien la preparación de tablas de mortalidad con pocos (o insignificantes) cambios o ajustes o ninguno en los datos. Para los países en que no sea este el caso, hemos perfilado los procedimientos que se utilizan para el ajuste de datos, pero los resultados que se obtengan dependen, en grado sumo, de un juicio considerado de las pruebas. Este grado de subjetividad es inevitable, habida cuenta de las actuales prácticas demográficas para el ajuste de datos, y requiere ampliar la labor para explorar cómo los métodos estadísticos formales se podrían aplicar mejor para reducir esta subjetividad al juzgar el alcance de la falta de registro de datos relativos al número de defunciones.

En el presente análisis se pone, asimismo, de manifiesto el amplio grado de incertidumbre y desconocimiento que existe con respecto a los niveles y patrones de mortalidad adulta en los países en desarrollo. En relación con un tercio de los Estados Miembros de la OMS, en que probablemente se haya registrado una proporción análoga de fallecimientos, poco o nada se sabe con certeza acerca de los niveles de mortalidad adulta, en particular por lo que se refiere a adultos más jóvenes menores de 60 años. En estos países, la única alternativa es atenerse al enfoque demográfico clásico de confeccionar los niveles de mortalidad adulta a partir de las tasas de mortalidad infantil, bajo el supuesto de que haya una relación entre ambas de alguna manera predecible. Sin embargo, los descubrimientos realizados en los últimos 30 años aproximadamente sugieren que no es éste realmente el caso al surgir nuevos peligros como el VIH, o viejos peligros como el alcoholismo, en grado sumo en ciertos segmentos demográficos. Hemos tratado de evitar tales deformaciones como éstas, confeccionando un modelo nuevo de sistemas de tablas de mortalidad para el cálculo de las tasas de mortalidad adulta, mas perdura la incertidumbre en los resultados hasta que sea posible la comprobación de los mismos mediante sistemas fidedignos en que se registren los fallecimientos.

Son escasas las probabilidades de que dichos sistemas se establezcan en un futuro próximo en países donde hagan falta. No obstante, ahora es cuando se requieren estimaciones fiables de mortalidad adulta para la planificación y control del sector de salud. Preguntas acerca de fallecimientos acaecidos en hogares, cuando se plantean con suficiente cautela y rigor, pueden dar lugar a datos actualizados útiles de niveles y patrones de mortalidad adulta y deberían incluirse periódicamente en censos y encuestas. La Encuesta de la OMS relativa a la Salud Mundial, que está a punto de publicarse, proporcionará una oportunidad idónea para ampliar nuestros conocimientos de mortalidad adulta. Al mismo tiempo, cabe destacar la conveniencia de recabar datos relativos a supervivencia infantil en los pocos países (Categoría V) en que no se disponga de información reciente, lo que servirá de gran ayuda para los programas relativos a la promoción de la salud infantil (Hill y Yazbeck 1994).

Referencias

Ahmad OB, López AD, Inoue M (2000) Trends in child mortality: a reappraisal. *Boletín de la Organización Mundial de la Salud,* **78**(10):1175-1191.

Bennett NG, Horiuchi S (1984) Mortality estimation from registered deaths in less developed countries. *Demography,* **21**(2):217-234.

Brass W (1971) On the scale of mortality. In: *Biological aspects of demography*, Brass W, ed., Taylor & Francis, Londres.

Chinese Academy of Preventive Medicine (1997) *Annual report on the disease surveillance system, 1996.* Beijing.

Coale A, Guo G (1989) Revised regional model life tables at very low levels of mortality. *Population Index,* **55**(4):613-643.

Hill K (1987) Estimating census and death registration completeness. *Asian and Pacific Population Forum,* **1**(3):8-24.

Hill K, Yazbeck A (1994) *Trends in under-five mortality, 1960-90. Estimates for 84 developing countries. Banco Mundial*, Washington.

Hill K, Pande R, Mahy M, Jones G (1999) *Trends in child mortality in the developing world: 1960-1996.* UNICEF, División de Evaluación y Planificación, Nueva York.

López AD (1983) The sex mortality differential in developed countries. In: *Sex differentials in mortality: trends, determinants and consequences*, Lopez AD, Ruzicka LT, eds., Australian National University Press, Canberra.

López AD, Salomon JA, Ahmad OB, Murray CJL, Mafat D (2000) *Life tables for 191 countries: data, methods, results.* Organización Mundial de la Salud/ Programa Global de Pruebas Científicas para las Políticas de Salud, Ginebra. (GPE discussion paper no 9)

Mari Bhat PN (2000) *Estimating life tables for India in the 1990s.* Organización Mundial de la Salud/Programa Global de Pruebas Científicas para las Políticas de Salud. (Informe inédito preparado para el GPE)

Murray CJL, Ferguson BD, López AD, Guillot M, Salomon JA, Ahmad OB (2001) *Modified logit life table system: principles, empirical validation and application.* Organización Mundial de la Salud/Programa Global de Pruebas Científicas para las Políticas de Salud, Ginebra. (GPE discussion paper no 39)

Perks W (1932) On some experiments on the graduation of mortality statistics. *Journal of the Institute of Actuaries,* **63**:12-57.

Preston SH (1984) Use of direct and indirect techniques for estimating the completeness of death registration systems. In: *Data bases for mortality measurement*, UN, pp. 143-153, Naciones Unidas, Nueva York.

Salomon JA, Mathers CD, Murray CJL, Ferguson B (2001) *Methods for life expectancy and healthy life expectancy uncertainty analysis*. Organización Mundial de la Salud/Programa Global de Pruebas Científicas para las Políticas de Salud, Ginebra. (GPE discussion paper no 10)

Salomon JA, Murray CJL (2001) Modelling HIV/AIDS epidemics in sub-Saharan Africa using seroprevalence data from antenatal clinics. *Boletín de la Organización Mundial de la Salud,* **79**(7):586-692.

Stanton E, Abderrahim A, Hill K (2000) An assessment of DHS maternal mortality indicators. *Studies in Family Planning,* **31**(2):111-124.

Timaeus IM (1999) Mortality in sub-Saharan Africa. In: *Health and mortality. Issues of global concern*, pp.110-131, Chamie J, Cliquet RL, eds., Naciones Unidas, División de Población; Population and Family Study Centre, New York; Brussels.

UN (2001) *World population prospects: the 2000 revision.* Naciones Unidas, Nueva York.

US Bureau of the Census () *International data base (IDB).* Oficina del Censo de los Estados Unidos, Centro de Programas Internacionales (IPC), www.census.gov/ipc/www/idbnew.html

World Bank (2001) *World development indicators 2001.* Banco Mundial, Washington.

Tabla 9. Disponibilidad de datos de registro demográfico relativos a mortalidad en la base de datos de la OMS, entre 1980 y 2000 (a 15 de septiembre de 2001)

País	80	81	82	83	84	85	86	87	88	89	90	91	92	93	94	95	96	97	98	99	00
Categoría I: Estadística demográfica completa (cobertura 95%+)																					
Andorra													x		x	x	x	x	x		x
Antigua y Barbuda				x			x	x			x	x	x	x	x	x					
Argentina	x	x	x	x	x	x	x	x	x	x	x	x	x	x	x	x	x			x	
Alemania	x	x	x	x	x	x	x	x	x	x	x	x	x	x	x	x	x	x	x	x	
Australia	x	x	x	x	x	x	x	x	x	x	x	x	x	x	x	x	x	x	x	x	
Austria	x	x	x	x	x	x	x	x	x	x	x	x	x	x	x	x	x	x	x	x	x
Bahamas	x	x	x	x	x	x	x	x	x	x	x	x	x	x	x	x	x	x			
Bahrain	x	x	x		x	x	x	x	x	x	x	x	x	x	x	x	x	x	x	x	x
Barbados	x	x	x	x	x	x	x	x	x	x	x	x	x	x	x	x					
Belarús		x	x	x	x	x	x	x	x	x	x	x	x	x	x	x	x	x	x	x	
Bélgica	x	x	x	x	x	x	x	x	x	x	x	x	x	x	x	x	x	x	x		
Bosnia y Herzegovina						x	x	x	x	x	x	x									
Bulgaria	x	x	x	x	x	x	x	x	x	x	x	x	x	x	x	x	x	x	x	x	
Canadá	x	x	x	x	x	x	x	x	x	x	x	x	x	x	x	x	x	x			
Chile	x	x	x	x	x	x	x	x	x	x	x	x	x	x	x	x	x	x	x		
Chipre	x	x	x	x	x	x	x	x	x	x	x	x	x	x	x	x	x	x	x	x	
Islas Cook	x	x	x	x	x	x	x	x	x	x	x	x	x	x	x	x	x	x	x	x	
Costa Rica	x	x	x	x	x	x	x	x	x	x	x	x	x	x	x	x		x		x	
Croacia			x	x	x	x	x	x	x	x	x	x	x	x	x	x	x	x	x	x	
Cuba	x	x	x	x	x	x	x	x	x	x	x	x	x	x	x	x	x	x	x	x	
Dinamarca	x	x	x	x	x	x	x	x	x	x	x	x	x	x	x	x	x	x	x		
Dominica	x	x	x	x	x	x	x	x	x	x	x	x	x	x	x						
Eslovaquia			x	x	x	x	x	x	x	x	x	x	x	x	x	x	x	x	x	x	
Eslovenia			x	x	x	x	x	x	x	x	x	x	x	x	x	x	x	x	x	x	
Estonia		x	x	x	x	x	x	x	x	x	x	x	x	x	x	x	x	x	x	x	
España	x	x	x	x	x	x	x	x	x	x	x	x	x	x	x	x	x	x	x		
Estados Unidos de América	x	x	x	x	x	x	x	x	x	x	x	x	x	x	x	x	x	x	x	x	
Ex Rep.Yugoslava de Macedonia			x	x	x	x	x	x	x	x	x	x	x	x	x	x	x	x			
Federación de Rusia	x	x	x	x	x	x	x	x	x	x	x	x	x	x	x	x	x	x	x	x	

País	80	81	82	83	84	85	86	87	88	89	90	91	92	93	94	95	96	97	98	99	00
Fiji	x	x	x	x	x	x	x	x					x	x	x	x	x	x	x	x	
Finlandia	x	x	x	x	x	x	x	x	x	x	x	x	x	x	x	x	x	x	x		
Francia	x	x	x	x	x	x	x	x	x	x	x	x	x	x	x	x	x	x	x		
Grecia	x	x	x	x	x	x	x	x	x	x	x	x	x	x	x	x	x	x	x		
Granada					x				x						x	x	x				
Hungría	x	x	x	x	x	x	x	x	x	x	x	x	x	x	x	x	x	x	x	x	x
Islandia	x	x	x	x	x	x	x	x	x	x	x	x	x	x	x	x	x	x	x		
Irlanda	x	x	x	x	x	x	x	x	x	x	x	x	x	x	x	x	x	x	x		
Israel	x	x	x	x	x	x	x	x	x	x	x	x	x	x	x	x	x	x	x		
Italia	x	x	x	x	x	x	x	x	x	x	x	x	x	x	x	x	x	x	x		
Jamaica	x	x	x	x	x	x	x	x	x	x	x	x									
Japón	x	x	x	x	x	x	x	x	x	x	x	x	x	x	x	x	x	x	x		
Kuwait	x	x	x	x	x	x	x	x	x	x		x	x	x	x	x	x	x	x	x	
Latvia	x	x	x	x	x	x	x	x	x	x	x	x	x	x	x	x	x	x	x	x	
Lituania		x	x	x	x	x	x	x	x	x	x	x	x	x	x	x	x	x	x	x	
Luxemburgo	x	x	x	x	x	x	x	x	x	x	x	x	x	x	x	x	x	x	x	x	x
Malta	x	x	x	x	x	x	x	x	x	x	x	x	x	x	x	x	x	x	x	x	
Mauricio	x	x	x	x	x	x	x	x	x	x	x	x	x	x	x	x	x	x	x	x	
México	x	x	x	x	x	x	x	x	x	x	x	x	x	x	x	x	x	x			x
Mónaco		x	x	x			x	x													
Nueva Zelandia	x	x	x	x	x	x	x	x	x	x	x	x	x	x	x	x	x	x	x		
Noruega	x	x	x	x	x	x	x	x	x	x	x	x	x	x	x	x	x	x	x		
Países Bajos	x	x	x	x	x	x	x	x	x	x	x	x	x	x	x	x	x	x	x	x	
Polonia	x	x	x	x	x	x	x	x	x	x	x	x	x	x	x	x	x	x	x	x	
Reino Unido	x	x	x	x	x	x	x	x	x	x	x	x	x	x	x	x	x	x	x	x	
Portugal	x	x	x	x	x	x	x	x	x	x	x	x	x	x	x	x	x	x	x	x	
Qatar		x	x	x		x	x	x	x	x	x		x	x	x	x	x	x			
República Checa	x	x	x	x	x	x	x	x	x	x	x	x	x	x	x	x	x	x	x	x	x
República de Moldova		x	x	x	x	x	x	x	x	x	x	x	x	x	x	x	x	x	x	x	
Rumania	x	x	x	x	x	x	x	x	x	x	x	x	x	x	x	x	x	x	x	x	
Saint Kitts y Nevis	x	x	x	x	x	x	x	x	x	x	x	x	x	x	x	x					
Santa Lucía	x	x	x	x	x	x	x	x	x	x		x	x	x	x	x					
San Vicente y las Granadinas	x		x	x	x	x	x	x	x				x			x	x				
San Marino	x	x	x	x	x	x	x	x	x	x	x	x	x	x	x	x	x	x	x		
Seychelles	x	x	x	x	x	x	x	x	x	x	x	x	x	x	x	x	x	x	x		

País	80	81	82	83	84	85	86	87	88	89	90	91	92	93	94	95	96	97	98	99	00
Singapur	x	x	x	x	x	x	x	x	x	x	x	x	x	x	x	x	x	x	x	x	x
Sri Lanka	x	x	x	x	x	x	x	x	x	x		x	x	x	x	x	x				
Suriname	x	x	x	x	x	x	x	x	x	x	x	x	x	x	x	x	x	x			
Suecia	x	x	x	x	x	x	x	x	x	x	x	x	x	x	x	x	x	x	x	x	x
Suiza	x	x	x	x	x	x	x	x	x	x	x	x	x	x	x	x	x	x	x		
Trinidad y Tobago	x	x	x	x	x	x	x	x	x	x	x	x	x	x	x	x					
Ucrania		x	x	x	x	x	x	x	x	x	x	x	x	x	x	x	x	x	x	x	x
Uruguay	x	x	x	x	x	x	x	x	x	x	x	x	x	x		x	x	x	x		
Venezuela	x	x	x	x	x	x	x	x	x	x	x	x	x	x	x						
Yugoslavia			x	x	x	x	x	x	x	x	x	x	x	x	x	x	x	x			
Categoría II: Datos demográficos incompletos																					
Albania	x				x	x	x	x	x	x	x	x	x	x	x	x	x	x	x		
Argelia	x	x	x			x	x														
Armenia		x	x			x	x	x	x	x	x	x	x	x	x	x	x	x	x	x	
Azerbaiyán		x	x	x	x	x	x	x	x	x	x	x	x	x	x	x	x	x	x	x	x
Belize	x	x	x	x	x	x	x	x	x	x	x	x	x	x	x	x	x				
Brasil	x	x	x	x	x	x	x	x	x	x	x	x	x	x	x	x					
Brunei Darussalam			x	x	x	x	x	x	x	x	x	x	x	x	x	x	x	x	x	x	x
Cabo Verde	x	x	x	x	x	x					x	x							x		
Colombia			x	x	x	x	x	x	x	x	x	x	x	x	x	x					
Ecuador	x	x	x	x	x	x	x	x	x	x	x	x	x	x	x	x	x	x	x		
Egipto	x	x	x	x	x	x	x	x	x	x	x	x	x	x	x	x				x	
El Salvador	x	x	x	x	x	x	x	x	x	x	x	x	x	x		x	x	x	x		
Filipinas	x	x	x	x	x	x	x	x	x	x	x	x	x	x	x	x	x				
Georgia		x	x	x	x	x	x	x	x	x	x	x	x		x	x	x				
Guatemala	x	x	x	x	x	x	x	x	x			x	x	x		x	x	x			
Guyana					x									x	x	x	x				
Honduras	x	x	x	x	x																
Irán (República Islámica del)				x	x	x	x	x				x			x	x	x	x	x	x	
Kazajstán		x	x	x	x	x	x	x	x	x	x	x	x	x	x	x	x	x	x	x	
Kirguistán		x	x	x	x	x	x	x	x	x	x	x	x	x	x	x	x	x	x	x	
Malasia											x	x	x	x	x	x	x	x	x		
Maldivas					x	x	x	x	x	x	x	x	x	x		x	x	x	x		
Islas Marshall							x	x	x	x	x	x	x	x	x	x	x	x			

País	80	81	82	83	84	85	86	87	88	89	90	91	92	93	94	95	96	97	98	99	00
Mongolia								x	x	x		x	x	x	x	x	x	x	x	x	
Marruecos								x	x	x	x	x	x	x	x	x	x	x	x		
Nicaragua								x	x	x	x	x	x	x	x		x	x	x		
Niue	x	x	x	x	x	x	x	x	x	x	x	x	x	x	x	x	x	x	x		
Palau						x		x	x	x	x	x	x	x	x	x	x	x	x	x	
Panamá	x	x	x	x	x	x	x	x	x	x	x	x	x	x	x	x	x	x			
Papua Nueva Guinea	x	x	x	x	x			x	x	x	x	x	x	x	x	x	x	x	x		
Paraguay	x	x	x	x	x	x	x	x	x	x	x	x	x		x						
Perú	x	x	x	x	x	x	x	x	x	x											
República Árabe Siria				x	x	x													x		x
República de Corea	x	x	x	x	x	x	x	x	x	x	x	x	x	x	x	x	x	x	x		
República Dominicana			x	x	x	x	x	x												x	x
Samoa	x												x	x							
Sudáfrica														x	x	x					
Tayikistán		x	x			x	x	x	x	x	x	x	x	x	x	x					
Tailandia	x	x	x	x	x	x	x	x	x	x	x	x	x	x	x	x	x	x	x	x	
Tonga																	x	x	x		
Túnez	x							x	x	x				x	x	x		x	x		
Turquía													x	x	x	x	x	x			
Turkmenistán		x	x			x	x	x	x	x	x	x	x	x	x	x	x	x	x		
Tuvalu												x	x	x	x						
Uzbekistán		x	x	x	x	x	x	x	x	x	x	x	x	x	x	x	x	x	x		
Zimbabwe			x								x	x	x	x		x		x			
Categoría III: Sistemas de Vigilancia y de Registo por Muestreo																					
Bangladesh											x	x	x	x	x	x	x	x	x	x	
China								x	x	x	x	x	x	x	x	x	x	x	x	x	
India											x	x	x	x	x	x	x	x	x		
República Unida de Tanzania													x	x	x	x	x	x	x	x	

APPENDIX A:
Life Tables by WHO sub-Region

APPENDICE A:
Tables de Mortalité détaillées par sous-région de L'OMS

Anexo A:
Tablas de Mortalidad detalladas por subregiones de la OMS

DEFINITIONS

The column abbreviations for each abridged life table in Appendix A and B are defined as follows:

Age represents the age interval x to $x+n$, where x= *0, 1, 5,.....,95,100,* and n is the width of the age interval (in years).

$_nM_x$ represents the age-specific death rate at ages x to $x+n$. At the very highest ages and for small populations, death rates will not necessarily match the death and population estimates due to rounding.

$_nq_x$ is defined as the probability of dying between exact ages x and $x+n$.

l_x represents the number of people alive at exact age x out of a hypothetical cohort of 100 000 at birth.

$_nd_x$ represents the number of deaths in the hypothetical cohort which occur in the age interval x to $x+n$.

The $_nL_x$ function is defined as the total number of person-years lived (contributed by those alive and those who died) between exact ages x and $x+n$. Each person who remains alive throughout an age interval contributes n years (n is the length of the interval).

e_x is defined as the average expected number of years of life left (life expectancy) for a person who reaches age x. e_0 is life expectancy at birth.

* * *

Actual deaths are the number of deaths by age and sex estimated to have occurred in the population in 2000. They are obtained by applying the $_nM_x$ to the ***Actual population*** for 2000 as estimated by the United Nations Population Division (Source: *World Population Prospects: the 2000 revision*. New York, United Nations, 2001). For small populations, the age-sex structure has been estimated using regional patterns.

* * *

COMPOSITION OF REGIONS

For analytical purposes, the 191 Member States of WHO have been divided into 5 mortality strata on the basis of their level of child (5q0) and adult (male) mortality (45q15) as follows: **A** = *Very low child, very low adult mortality;* **B** = *Low child, low adult mortality;* **C** = *Low child, high adult mortality;* **D** = *High child, high adult mortality;* **E** = *High child, very high adult mortality.*
The matrix defined by the six WHO Regions (**Afr** = *Africa;* **Amr** = *The Americas;* **Emr** = *Eastern Mediterranean;* **Eur** = *Europe;* **Sear** = *South-East Asia;* **Wpr** = *Western Pacific*) and the 5 mortality strata yields 14 sub-Regions, since not every mortality stratum is represented in every Region.
The following table gives the composition of each sub-Region.

WHO Region	Stratum	WHO Member States
Afr	D	Algeria, Angola, Benin, Burkina Faso, Cameroon, Cape Verde, Chad, Comoros, Equatorial Guinea, Gabon, Gambia, Ghana, Guinea, Guinea-Bissau, Liberia, Madagascar, Mali, Mauritania, Mauritius, Niger, Nigeria, Sao Tome and Principe, Senegal, Seychelles, Sierra Leone, Togo
	E	Botswana, Burundi, Central African Republic, Congo, Côte d'Ivoire, Democratic Republic of the Congo, Eritrea, Ethiopia, Kenya, Lesotho, Malawi, Mozambique, Namibia, Rwanda, South Africa, Swaziland, Uganda, United Republic of Tanzania, Zambia, Zimbabwe
Amr	A	Canada, Cuba, United States of America
	B	Antigua and Barbuda, Argentina, Bahamas, Barbados, Belize, Brazil, Chile, Colombia, Costa Rica, Dominica, Dominican Republic, El Salvador, Grenada, Guyana, Honduras, Jamaica, Mexico, Panama, Paraguay, Saint Kitts and Nevis, Saint Lucia, Saint Vincent and the Grenadines, Suriname, Trinidad and Tobago, Uruguay, Venezuela
	D	Bolivia, Ecuador, Guatemala, Haiti, Nicaragua, Peru
Emr	B	Bahrain, Cyprus, Iran (Islamic Republic of), Jordan, Kuwait, Lebanon, Libyan Arab Jamahiriya, Oman, Qatar, Saudi Arabia, Syrian Arab Republic, Tunisia, United Arab Emirates
	D	Afghanistan, Djibouti, Egypt, Iraq, Morocco, Pakistan, Somalia, Sudan, Yemen
Eur	A	Andorra, Austria, Belgium, Croatia, Czech Republic, Denmark, Finland, France, Germany, Greece, Iceland, Ireland, Israel, Italy, Luxembourg, Malta, Monaco, Netherlands, Norway, Portugal, San Marino, Slovenia, Spain, Sweden, Switzerland, United Kingdom
	B	Albania, Armenia, Azerbaijan, Bosnia and Herzegovina, Bulgaria, Georgia, Kyrgyzstan, Poland, Romania, Slovakia, Tajikistan, The Former Yugoslav Republic of Macedonia, Turkey, Turkmenistan, Uzbekistan, Yugoslavia
	C	Belarus, Estonia, Hungary, Kazakhstan, Latvia, Lithuania, Republic of Moldova, Russian Federation, Ukraine
Sear	B	Indonesia, Sri Lanka, Thailand
	D	Bangladesh, Bhutan, Democratic People's Republic of Korea, India, Maldives, Myanmar, Nepal
Wpr	A	Australia, Brunei Darussalam, Japan, New Zealand, Singapore
	B	Cambodia, China, Cook Islands, Fiji, Kiribati, Lao People's Democratic Republic, Malaysia, Marshall Islands, Micronesia (Federated States of), Mongolia, Nauru, Niue, Palau, Papua New Guinea, Philippines, Republic of Korea, Samoa, Solomon Islands, Tonga, Tuvalu, Vanuatu, Viet Nam

World / Monde / Mundo

Males, 2000 / *Hommes, 2000* / *Hombres, 2000*

Age	$_nM_x$	$_nq_x$			l_x			$_nd_x$	$_nL_x$	e_x			Actual deaths	Actual population
			percentiles			percentiles					percentiles			
		mean	2 1/2	97 1/2	mean	2 1/2	97 1/2			mean	2 1/2	97 1/2	(000)	(000)
<1	0.0655	0.0618	0.0584	0.0645	100,000	100,000	100,000	6,182	94,436	62.7	61.5	64.3	4,229	64,610
1-4	0.0056	0.0223	0.0212	0.0230	93,818	93,550	94,163	2,089	370,258	65.8	65.4	66.3	1,409	249,646
5-9	0.0014	0.0070	0.0067	0.0072	91,729	91,413	92,151	642	457,040	63.3	62.9	63.7	433	308,296
10-14	0.0010	0.0048	0.0046	0.0049	91,087	90,759	91,532	436	454,346	58.7	58.3	59.1	295	307,690
15-19	0.0015	0.0075	0.0072	0.0077	90,651	90,316	91,111	676	451,565	54.0	53.6	54.4	425	283,527
20-24	0.0025	0.0122	0.0118	0.0125	89,975	89,633	90,456	1,096	447,134	49.4	49.0	49.8	635	259,193
25-29	0.0032	0.0159	0.0153	0.0163	88,879	88,529	89,388	1,412	440,865	44.9	44.6	45.4	814	254,329
30-34	0.0039	0.0193	0.0186	0.0199	87,467	87,110	88,008	1,688	433,115	40.6	40.2	41.0	935	239,964
35-39	0.0045	0.0225	0.0217	0.0231	85,779	85,417	86,353	1,927	424,078	36.4	36.0	36.8	979	215,472
40-44	0.0059	0.0289	0.0279	0.0297	83,852	83,478	84,458	2,420	413,211	32.1	31.8	32.5	1,099	187,708
45-49	0.0076	0.0373	0.0360	0.0385	81,432	81,017	82,077	3,038	399,566	28.0	27.7	28.4	1,269	166,904
50-54	0.0108	0.0525	0.0505	0.0544	78,394	77,925	79,113	4,117	381,679	24.0	23.7	24.3	1,438	133,349
55-59	0.0157	0.0757	0.0726	0.0786	74,277	73,728	75,085	5,620	357,337	20.2	19.9	20.5	1,632	103,747
60-64	0.0245	0.1155	0.1111	0.1196	68,657	67,966	69,562	7,930	323,460	16.7	16.4	16.9	2,215	90,361
65-69	0.0359	0.1649	0.1588	0.1707	60,727	59,846	61,705	10,015	278,598	13.5	13.3	13.7	2,576	71,672
70-74	0.0547	0.2407	0.2332	0.2483	50,712	49,696	51,832	12,207	223,045	10.7	10.5	10.9	2,890	52,816
75-79	0.0811	0.3373	0.3280	0.3465	38,506	37,385	39,736	12,988	160,058	8.3	8.1	8.4	2,607	32,128
80-84	0.1271	0.4823	0.4728	0.4917	25,518	24,431	26,657	12,308	96,819	6.2	6.1	6.3	1,957	15,392
85-89	0.1882	0.6399	0.6331	0.6462	13,210	12,418	13,998	8,453	44,919	4.7	4.6	4.7	1,223	6,500
90-94	0.2622	0.7338	0.7292	0.7378	4,757	4,394	5,129	3,491	13,315	3.5	3.5	3.5	458	1,747
95-99	0.3629	0.8688	0.8658	0.8703	1,267	1,152	1,386	1,100	3,032	2.7	2.6	2.7	106	291
100+	0.5079	1	1	1	166	150	186	166	327	2.0	2.0	2.0	15	30
All ages													*29,641*	*3,045,372*

$\ln(_nq_x)$ with uncertainty interval

l_x with uncertainty interval

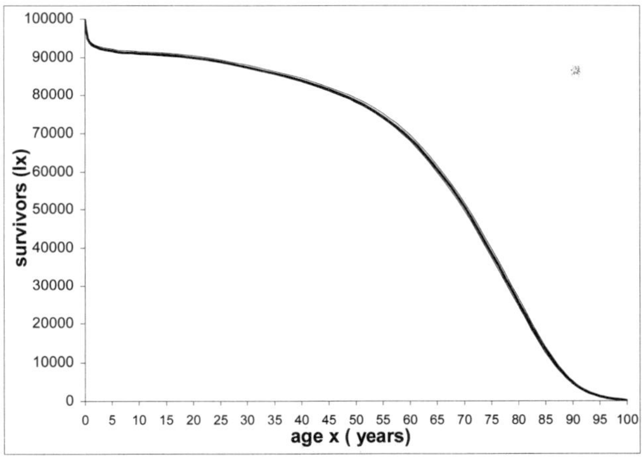

e_x with uncertainty interval

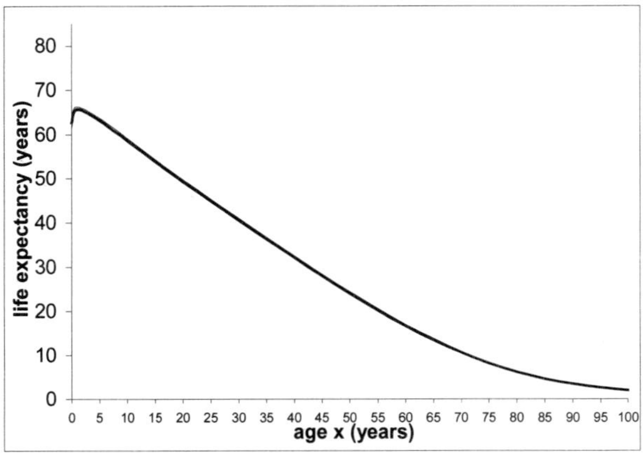

Age distribution of deaths with uncertainty interval

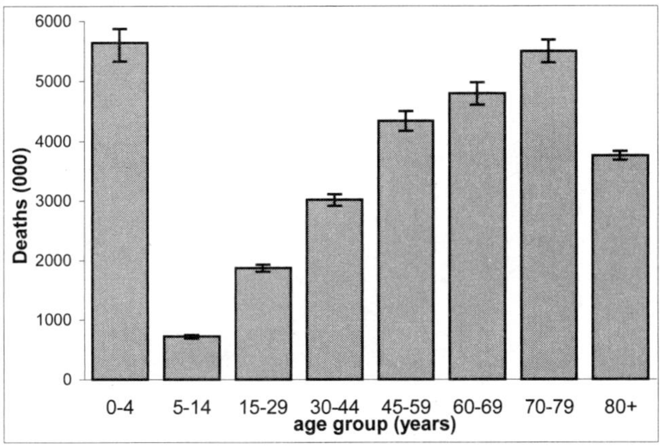

World — Monde — Mundo

Females, 2000 — *Femmes, 2000* — *Mujeres, 2000*

Age	$_nM_x$	$_nq_x$			l_x			$_nd_x$	$_nL_x$	e_x			Actual deaths	Actual population
			percentiles			percentiles					percentiles			
		mean	2½	97½	mean	2½	97½			mean	2½	97½	(000)	(000)
<1	0.0608	0.0577	0.0543	0.0602	100,000	100,000	100,000	5,768	94,809	67.2	65.9	69.2	3,735	61,399
1-4	0.0064	0.0253	0.0238	0.0262	94,232	93,978	94,570	2,385	371,204	70.3	69.7	70.9	1,520	236,463
5-9	0.0015	0.0075	0.0071	0.0078	91,847	91,539	92,285	689	457,512	68.1	67.5	68.6	439	291,531
10-14	0.0009	0.0046	0.0044	0.0048	91,158	90,836	91,621	422	454,733	63.6	63.1	64.1	270	291,099
15-19	0.0016	0.0078	0.0075	0.0081	90,735	90,406	91,211	711	451,899	58.9	58.4	59.4	424	269,573
20-24	0.0025	0.0122	0.0117	0.0127	90,024	89,686	90,518	1,102	447,366	54.3	53.8	54.8	611	247,988
25-29	0.0029	0.0144	0.0138	0.0150	88,922	88,576	89,441	1,282	441,405	50.0	49.5	50.4	709	244,147
30-34	0.0031	0.0153	0.0146	0.0159	87,640	87,284	88,188	1,338	434,852	45.6	45.2	46.1	713	231,716
35-39	0.0032	0.0157	0.0151	0.0164	86,301	85,907	86,883	1,358	428,112	41.3	40.9	41.8	662	208,575
40-44	0.0037	0.0184	0.0176	0.0192	84,943	84,513	85,566	1,562	420,813	36.9	36.5	37.4	673	181,392
45-49	0.0047	0.0232	0.0221	0.0242	83,382	82,912	84,061	1,932	412,080	32.6	32.2	33.0	766	163,376
50-54	0.0065	0.0321	0.0307	0.0336	81,450	80,930	82,202	2,614	400,715	28.3	27.9	28.7	865	132,644
55-59	0.0096	0.0470	0.0446	0.0495	78,836	78,211	79,679	3,708	384,910	24.2	23.8	24.5	1,023	106,206
60-64	0.0146	0.0705	0.0668	0.0741	75,128	74,337	76,090	5,295	362,403	20.2	19.9	20.6	1,412	96,674
65-69	0.0224	0.1062	0.1005	0.1116	69,833	68,826	70,933	7,413	330,635	16.6	16.2	16.9	1,802	80,382
70-74	0.0351	0.1612	0.1538	0.1685	62,421	61,147	63,686	10,060	286,952	13.2	13.0	13.5	2,295	65,459
75-79	0.0540	0.2380	0.2287	0.2476	52,360	50,886	53,728	12,461	230,650	10.3	10.1	10.5	2,489	46,064
80-84	0.0887	0.3629	0.3512	0.3746	39,900	38,396	41,381	14,481	163,296	7.7	7.6	7.9	2,255	25,430
85-89	0.1399	0.5184	0.5090	0.5272	25,419	24,114	26,790	13,176	94,154	5.7	5.6	5.8	1,906	13,619
90-94	0.2086	0.6415	0.6346	0.6476	12,243	11,400	13,141	7,853	37,654	4.2	4.1	4.2	1,006	4,825
95-99	0.3069	0.7989	0.7945	0.8023	4,390	4,020	4,801	3,507	11,427	3.0	3.0	3.1	334	1,089
100+	0.4548	1	1	1	883	797	986	883	1,941	2.2	2.2	2.3	68	149
All ages													*25,978*	*2,999,800*

$\ln(_nq_x)$ with uncertainty interval

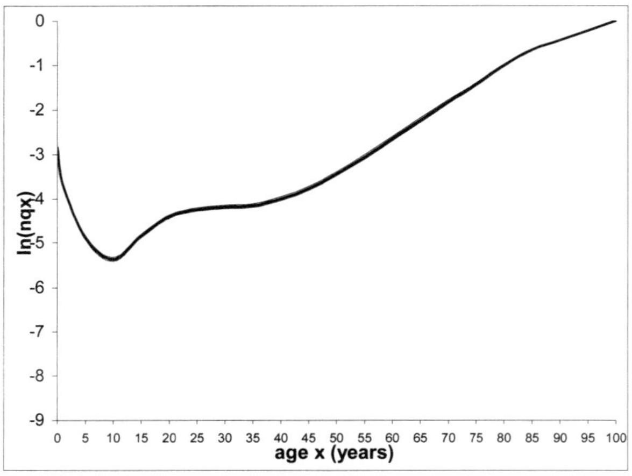

l_x with uncertainty interval

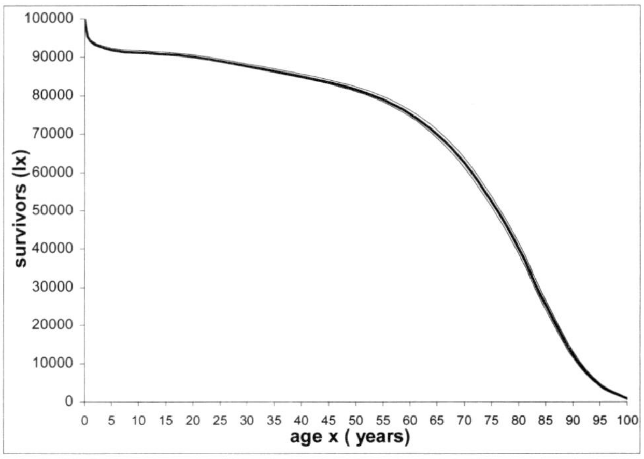

e_x with uncertainty interval

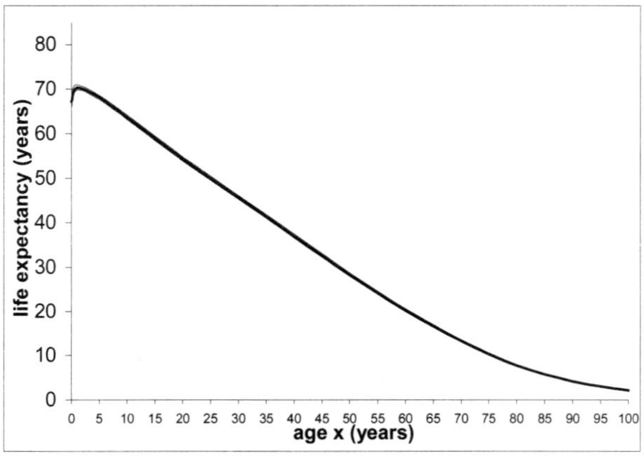

Age distribution of deaths with uncertainty interval

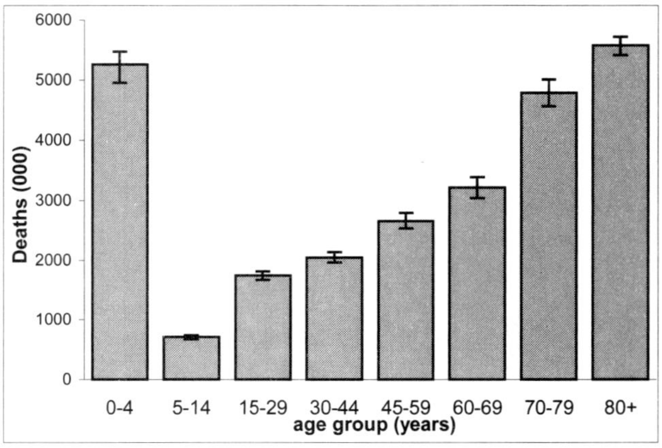

AfrD **AfrD** **AfrD**

Males, 2000 *Hommes, 2000* *Hombres, 2000*

Age	$_nM_x$	$_nq_x$			l_x			$_nd_x$	$_nL_x$	e_x			Actual deaths	Actual population
			percentiles			percentiles					percentiles			
		mean	2½	97½	mean	2½	97½			mean	2½	97½	(000)	(000)
<1	0.1425	0.1263	0.1203	0.1315	100,000	100,000	100,000	12,628	88,635	50.3	48.7	52.4	793	5,563
1-4	0.0117	0.0456	0.0432	0.0480	87,372	86,847	87,972	3,988	339,917	56.6	55.4	57.7	230	19,566
5-9	0.0027	0.0132	0.0125	0.0140	83,384	82,780	84,023	1,104	414,159	55.2	54.0	56.3	57	21,396
10-14	0.0017	0.0085	0.0080	0.0090	82,280	81,629	82,948	698	409,652	50.9	49.8	52.0	32	18,797
15-19	0.0019	0.0096	0.0091	0.0102	81,581	80,903	82,281	787	405,939	46.3	45.2	47.4	32	16,399
20-24	0.0041	0.0202	0.0190	0.0214	80,794	80,089	81,527	1,631	399,894	41.7	40.6	42.8	56	13,661
25-29	0.0070	0.0343	0.0320	0.0367	79,163	78,397	79,967	2,713	389,033	37.5	36.5	38.6	77	10,988
30-34	0.0100	0.0488	0.0454	0.0526	76,450	75,578	77,370	3,728	372,929	33.8	32.7	34.8	89	8,913
35-39	0.0117	0.0569	0.0528	0.0616	72,722	71,707	73,795	4,136	353,270	30.4	29.4	31.4	87	7,402
40-44	0.0135	0.0653	0.0606	0.0709	68,586	67,411	69,869	4,482	331,726	27.1	26.2	28.0	81	6,019
45-49	0.0157	0.0754	0.0697	0.0818	64,104	62,697	65,629	4,832	308,441	23.8	23.0	24.6	77	4,937
50-54	0.0190	0.0906	0.0837	0.0983	59,272	57,580	61,030	5,368	282,942	20.5	19.8	21.3	72	3,809
55-59	0.0239	0.1128	0.1041	0.1225	53,905	51,950	55,877	6,080	254,323	17.3	16.7	18.0	72	3,032
60-64	0.0343	0.1579	0.1460	0.1708	47,824	45,637	49,995	7,552	220,242	14.2	13.7	14.8	82	2,405
65-69	0.0488	0.2174	0.2022	0.2334	40,272	37,915	42,599	8,756	179,471	11.4	11.0	11.9	89	1,823
70-74	0.0737	0.3111	0.2916	0.3308	31,516	29,064	33,886	9,806	133,066	8.9	8.5	9.3	92	1,244
75-79	0.1116	0.4362	0.4135	0.4581	21,710	19,449	24,006	9,471	84,875	6.8	6.5	7.1	81	725
80-84	0.1670	0.5891	0.5666	0.6094	12,239	10,540	14,080	7,211	43,170	5.1	4.9	5.3	56	337
85-89	0.2431	0.7560	0.7381	0.7713	5,029	4,117	6,102	3,801	15,639	3.8	3.7	3.9	25	105
90-94	0.3455	0.8482	0.8341	0.8603	1,227	943	1,598	1,041	3,013	2.8	2.7	2.8	7	20
95-99	0.4761	0.9803	0.9770	0.9825	186	132	265	183	383	2.1	2.1	2.1	1	3
100+	0.6284	1	1	1	4	2	6	4	6	1.6	1.6	1.6	0	0
All ages													*2,189*	*147,143*

ln($_nq_x$) with uncertainty interval

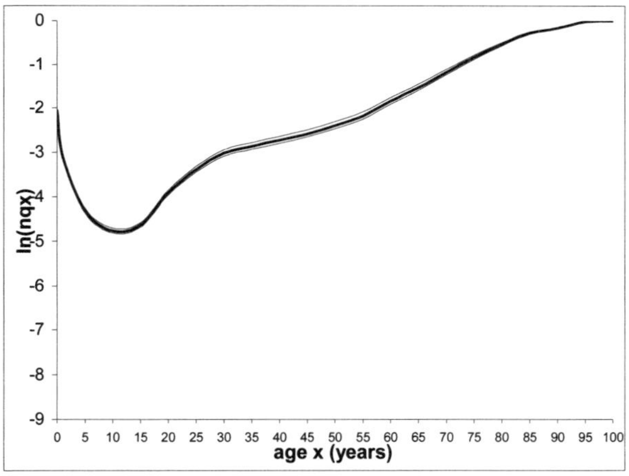

l_x with uncertainty interval

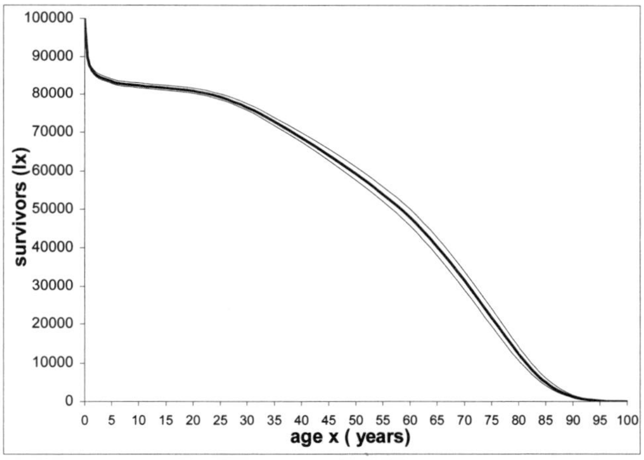

e_x with uncertainty interval

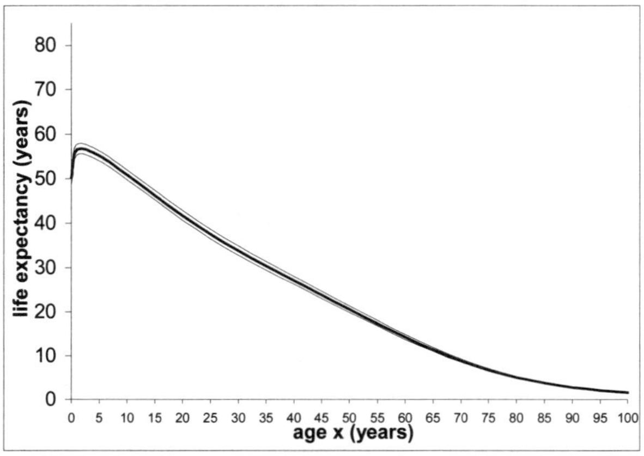

Age distribution of deaths with uncertainty interval

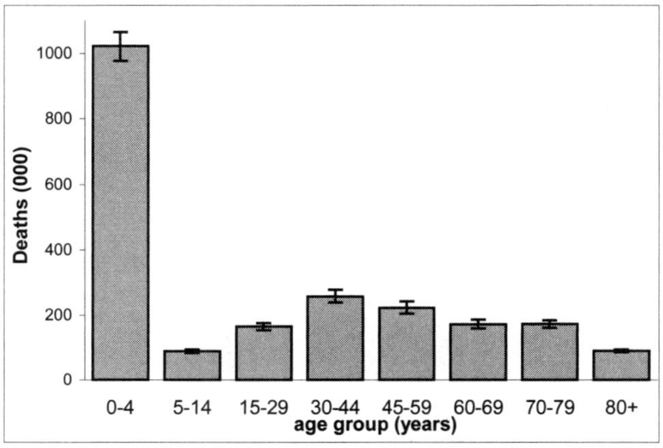

AfrD **AfrD** **AfrD**

Females, 2000 *Femmes, 2000* *Mujeres, 2000*

Age	$_nM_x$	$_nq_x$			l_x			$_nd_x$	$_nL_x$	e_x			Actual deaths	Actual population
			percentiles			percentiles					percentiles			
		mean	2½	97½	mean	2½	97½			mean	2½	97½	(000)	(000)
<1	0.1227	0.1105	0.1054	0.1161	100,000	100,000	100,000	11,049	90,056	52.4	50.7	54.5	655	5,337
1-4	0.0131	0.0507	0.0481	0.0530	88,951	88,394	89,459	4,511	344,977	57.9	56.7	59.2	251	19,216
5-9	0.0028	0.0139	0.0131	0.0147	84,440	83,943	85,059	1,176	419,261	56.9	55.8	58.1	59	20,919
10-14	0.0018	0.0091	0.0086	0.0096	83,264	82,751	83,907	761	414,418	52.7	51.6	53.8	34	18,388
15-19	0.0032	0.0160	0.0151	0.0170	82,503	81,968	83,160	1,321	409,211	48.2	47.1	49.3	52	16,073
20-24	0.0062	0.0303	0.0284	0.0323	81,182	80,610	81,862	2,461	399,754	43.9	42.8	45.0	83	13,475
25-29	0.0089	0.0437	0.0407	0.0468	78,720	78,078	79,435	3,442	384,995	40.2	39.2	41.3	97	10,876
30-34	0.0103	0.0501	0.0464	0.0539	75,278	74,497	76,146	3,774	366,954	36.9	36.0	37.9	91	8,895
35-39	0.0102	0.0496	0.0458	0.0534	71,504	70,481	72,553	3,545	348,656	33.7	32.8	34.7	76	7,453
40-44	0.0098	0.0480	0.0442	0.0517	67,959	66,720	69,189	3,262	331,639	30.4	29.6	31.2	60	6,114
45-49	0.0104	0.0509	0.0469	0.0549	64,697	63,269	66,102	3,291	315,256	26.8	26.0	27.6	53	5,082
50-54	0.0130	0.0628	0.0580	0.0677	61,406	59,804	63,003	3,858	297,383	23.1	22.4	23.8	52	3,983
55-59	0.0171	0.0821	0.0756	0.0885	57,547	55,767	59,337	4,724	275,928	19.4	18.8	20.1	56	3,288
60-64	0.0250	0.1175	0.1083	0.1265	52,824	50,853	54,824	6,206	248,603	16.0	15.4	16.5	68	2,707
65-69	0.0379	0.1732	0.1602	0.1856	46,618	44,458	48,832	8,072	212,908	12.7	12.3	13.3	80	2,108
70-74	0.0606	0.2630	0.2450	0.2798	38,545	36,258	40,925	10,137	167,385	9.9	9.5	10.3	90	1,486
75-79	0.0944	0.3819	0.3600	0.4020	28,409	26,178	30,869	10,849	114,919	7.5	7.3	7.8	85	904
80-84	0.1445	0.5309	0.5088	0.5518	17,559	15,686	19,756	9,322	64,491	5.6	5.4	5.8	64	443
85-89	0.2137	0.6965	0.6780	0.7133	8,237	7,042	9,704	5,737	26,844	4.2	4.1	4.3	34	161
90-94	0.3082	0.8007	0.7871	0.8139	2,500	2,031	3,104	2,002	6,495	3.1	3.0	3.1	12	39
95-99	0.4299	0.9388	0.9293	0.9463	498	382	654	468	1,088	2.3	2.3	2.3	3	7
100+	0.5810	1	1	1	31	21	44	31	53	1.7	1.7	1.8	1	1
All ages													*2,056*	*146,956*

ln($_nq_x$) with uncertainty interval

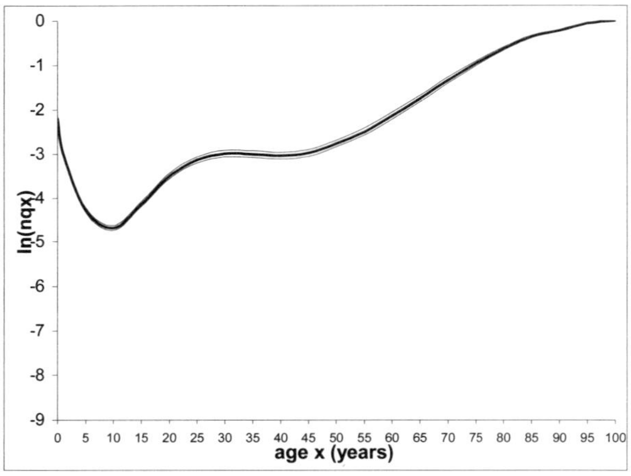

l_x with uncertainty interval

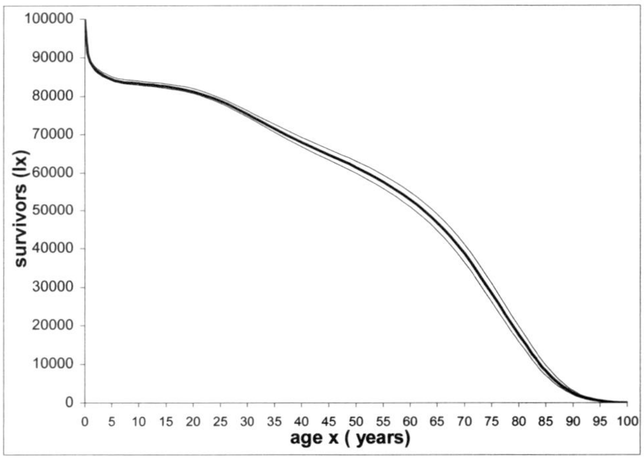

e_x with uncertainty interval

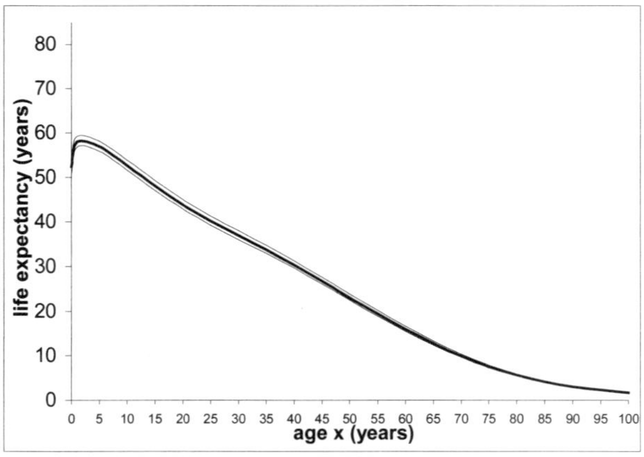

Age distribution of deaths with uncertainty interval

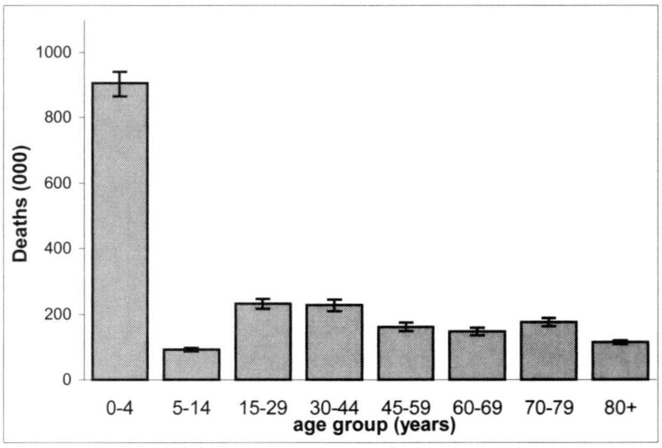

AfrE **AfrE** **AfrE**

Males, 2000 *Hommes, 2000* *Hombres, 2000*

Age	$_nM_x$	$_nq_x$			l_x			$_nd_x$	$_nL_x$	e_x			Actual deaths	Actual population
			percentiles			percentiles					percentiles			
		mean	2½	97½	mean	2½	97½			mean	2½	97½	(000)	(000)
<1	0.1417	0.1256	0.1147	0.1374	100,000	100,000	100,000	12,564	88,693	43.5	41.3	46.5	925	6,528
1-4	0.0136	0.0525	0.0489	0.0564	87,436	86,256	88,525	4,592	338,725	48.8	47.5	50.1	310	22,890
5-9	0.0030	0.0151	0.0140	0.0162	82,844	81,545	84,197	1,248	411,103	47.4	46.2	48.7	76	25,049
10-14	0.0021	0.0103	0.0096	0.0111	81,597	80,226	83,011	843	405,877	43.1	41.9	44.4	46	22,220
15-19	0.0030	0.0151	0.0142	0.0159	80,754	79,340	82,206	1,218	400,725	38.5	37.3	39.8	58	19,074
20-24	0.0072	0.0353	0.0331	0.0374	79,536	78,083	81,027	2,805	390,668	34.0	32.9	35.3	114	15,899
25-29	0.0143	0.0692	0.0652	0.0735	76,731	75,194	78,305	5,312	370,377	30.2	29.1	31.5	185	12,926
30-34	0.0222	0.1053	0.0989	0.1122	71,420	69,763	73,104	7,522	338,292	27.2	26.1	28.5	234	10,503
35-39	0.0251	0.1179	0.1098	0.1261	63,897	62,112	65,680	7,533	300,655	25.2	24.1	26.4	215	8,587
40-44	0.0263	0.1234	0.1143	0.1326	56,365	54,363	58,275	6,957	264,430	23.2	22.2	24.3	184	6,991
45-49	0.0273	0.1280	0.1180	0.1382	49,408	47,258	51,525	6,323	231,229	21.1	20.2	22.1	155	5,652
50-54	0.0281	0.1314	0.1205	0.1426	43,084	40,866	45,344	5,661	201,268	18.8	18.0	19.7	127	4,526
55-59	0.0295	0.1375	0.1257	0.1495	37,423	35,103	39,777	5,145	174,251	16.3	15.6	17.1	105	3,559
60-64	0.0398	0.1809	0.1653	0.1959	32,277	29,932	34,662	5,838	146,792	13.5	12.9	14.2	108	2,711
65-69	0.0541	0.2383	0.2190	0.2562	26,439	24,159	28,770	6,302	116,443	10.9	10.5	11.5	108	1,996
70-74	0.0794	0.3314	0.3070	0.3525	20,138	18,062	22,467	6,674	84,004	8.6	8.2	9.0	105	1,317
75-79	0.1176	0.4543	0.4259	0.4770	13,464	11,696	15,569	6,117	52,027	6.6	6.3	6.9	87	738
80-84	0.1724	0.6025	0.5740	0.6235	7,347	6,117	8,938	4,426	25,669	5.0	4.8	5.2	57	329
85-89	0.2482	0.7658	0.7424	0.7816	2,920	2,305	3,808	2,237	9,011	3.7	3.6	3.9	23	94
90-94	0.3450	0.8477	0.8312	0.8592	684	505	981	580	1,680	2.8	2.7	2.9	6	16
95-99	0.4653	0.9711	0.9662	0.9741	104	72	166	101	217	2.1	2.1	2.2	1	2
100+	0.6074	1	1	1	3	2	5	3	5	1.6	1.6	1.7	0	0
All ages													*3,228*	*171,608*

$\ln({}_nq_x)$ with uncertainty interval

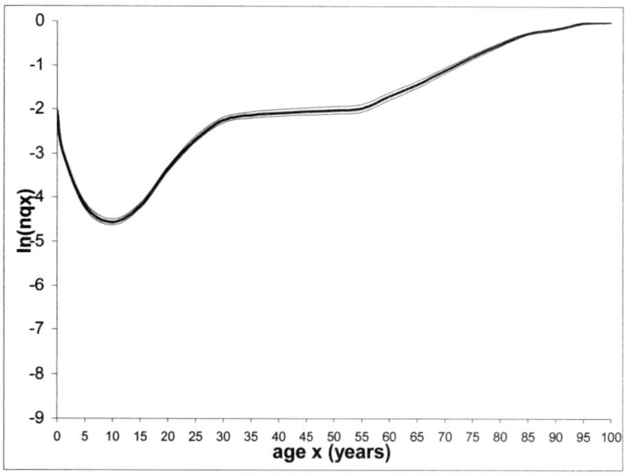

l_x with uncertainty interval

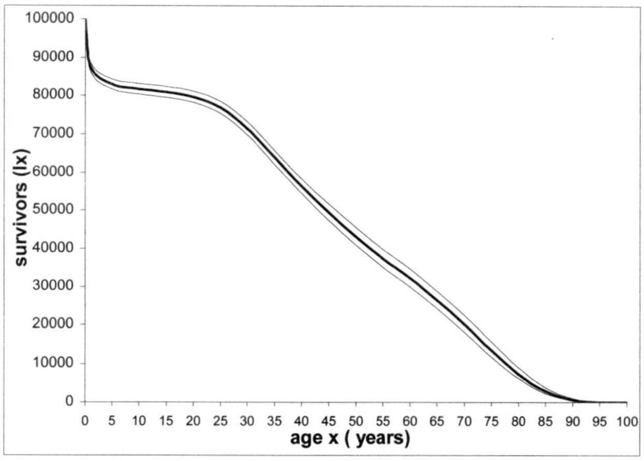

e_x with uncertainty interval

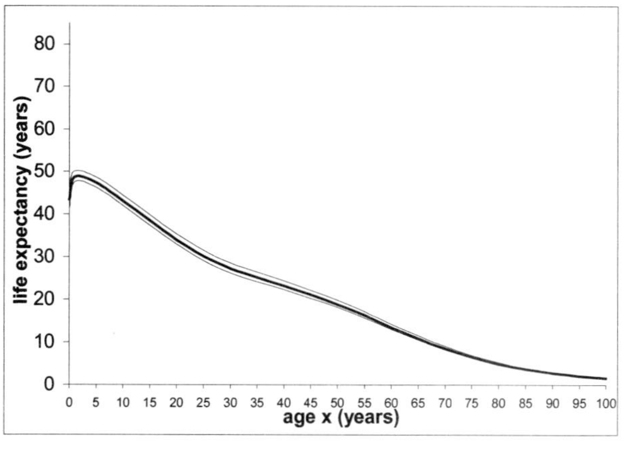

Age distribution of deaths with uncertainty interval

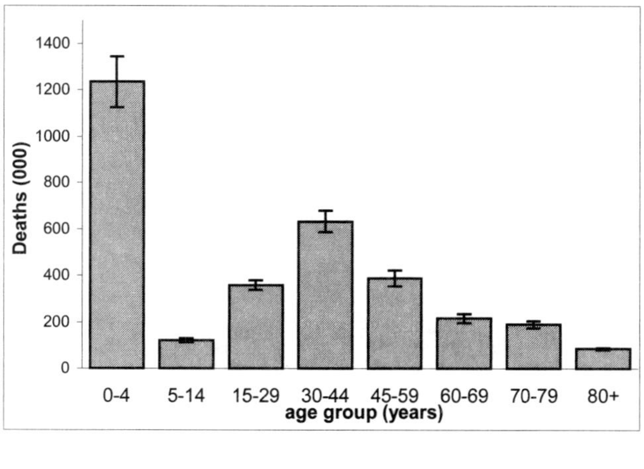

AfrE AfrE AfrE

Females, 2000 *Femmes, 2000* *Mujeres, 2000*

Age	$_nM_x$	$_nq_x$			l_x			$_nd_x$	$_nL_x$	e_x			Actual deaths	Actual population
			percentiles			percentiles					percentiles			
		mean	2½	97½	mean	2½	97½			mean	2½	97½	(000)	(000)
<1	0.1182	0.1068	0.0980	0.1158	100,000	100,000	100,000	10,684	90,384	45.2	43.0	48.6	743	6,282
1-4	0.0148	0.0573	0.0532	0.0613	89,316	88,416	90,202	5,120	344,975	49.6	48.1	51.3	338	22,774
5-9	0.0032	0.0158	0.0146	0.0169	84,196	83,142	85,309	1,329	417,658	48.5	47.1	50.2	79	24,875
10-14	0.0022	0.0110	0.0101	0.0117	82,867	81,787	84,059	909	412,063	44.2	42.8	45.9	49	22,149
15-19	0.0053	0.0261	0.0243	0.0279	81,958	80,853	83,199	2,143	404,433	39.7	38.3	41.3	101	19,088
20-24	0.0122	0.0590	0.0549	0.0631	79,815	78,614	81,139	4,713	387,293	35.7	34.3	37.3	194	15,923
25-29	0.0197	0.0940	0.0872	0.1008	75,102	73,710	76,553	7,057	357,869	32.8	31.5	34.3	255	12,936
30-34	0.0235	0.1111	0.1026	0.1195	68,045	66,400	69,642	7,563	321,318	30.9	29.7	32.3	247	10,493
35-39	0.0214	0.1014	0.0930	0.1094	60,482	58,616	62,408	6,133	287,078	29.5	28.4	30.7	184	8,622
40-44	0.0189	0.0902	0.0825	0.0974	54,349	52,345	56,506	4,903	259,489	27.5	26.6	28.6	135	7,124
45-49	0.0169	0.0813	0.0740	0.0879	49,446	47,252	51,837	4,020	237,182	25.0	24.2	26.0	100	5,886
50-54	0.0180	0.0862	0.0783	0.0931	45,426	43,096	47,999	3,916	217,341	22.0	21.3	22.9	88	4,865
55-59	0.0206	0.0980	0.0889	0.1056	41,510	39,086	44,241	4,069	197,379	18.8	18.2	19.6	82	3,975
60-64	0.0278	0.1298	0.1180	0.1395	37,442	34,960	40,306	4,859	175,061	15.6	15.1	16.3	88	3,154
65-69	0.0403	0.1831	0.1676	0.1957	32,583	30,087	35,549	5,965	148,003	12.5	12.1	13.1	97	2,405
70-74	0.0626	0.2705	0.2504	0.2870	26,618	24,202	29,591	7,201	115,088	9.8	9.5	10.3	104	1,663
75-79	0.0953	0.3849	0.3614	0.4038	19,417	17,270	22,183	7,473	78,402	7.5	7.2	7.8	96	1,007
80-84	0.1446	0.5311	0.5076	0.5499	11,944	10,317	14,166	6,344	43,860	5.6	5.5	5.9	72	498
85-89	0.2138	0.6965	0.6766	0.7122	5,600	4,659	6,976	3,901	18,248	4.2	4.1	4.3	35	165
90-94	0.3063	0.7981	0.7834	0.8096	1,699	1,347	2,256	1,356	4,428	3.1	3.0	3.2	11	36
95-99	0.4262	0.9352	0.9281	0.9400	343	258	489	321	753	2.3	2.3	2.3	2	5
100+	0.5742	1	1	1	22	16	35	22	39	1.7	1.7	1.8	0	0
All ages													*3,099*	*173,924*

$\ln({}_nq_x)$ with uncertainty interval

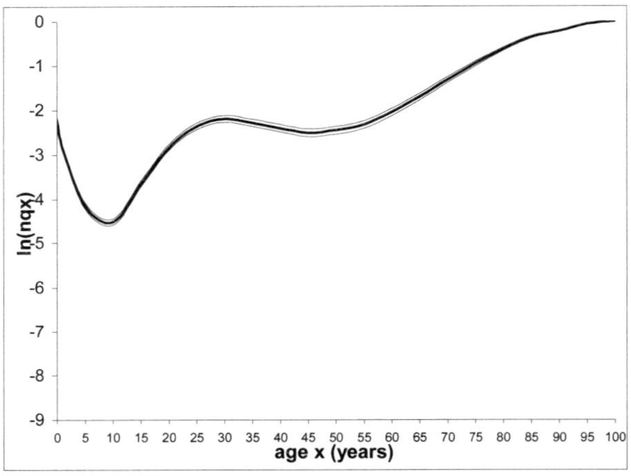

l_x with uncertainty interval

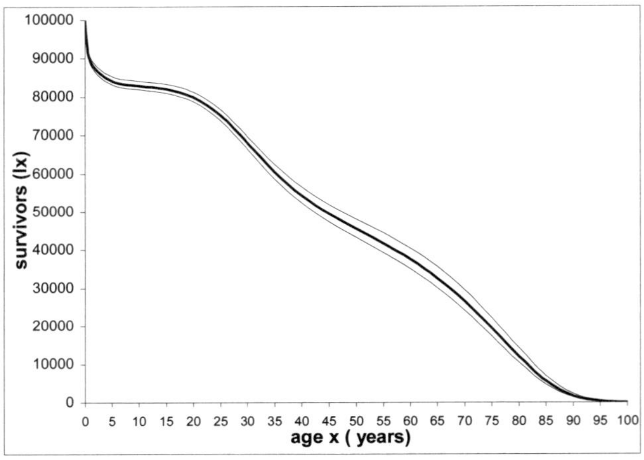

e_x with uncertainty interval

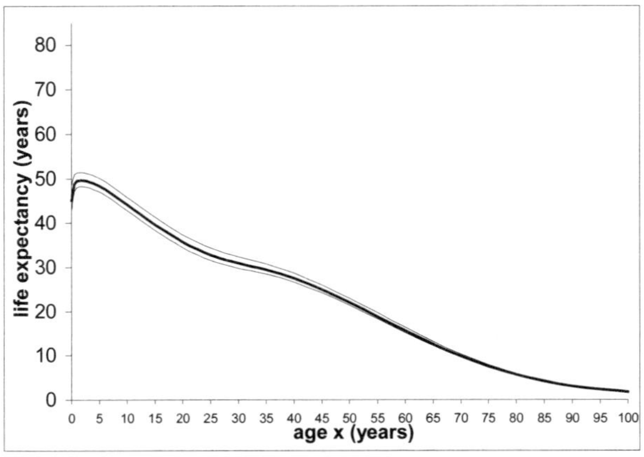

Age distribution of deaths with uncertainty interval

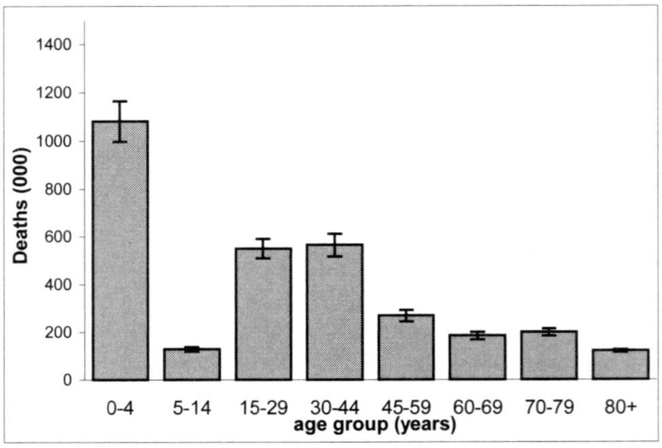

AmrA **AmrA** **AmrA**

Males, 2000 *Hommes, 2000* *Hombres, 2000*

Age	$_nM_x$	$_nq_x$			l_x			$_nd_x$	$_nL_x$	e_x			Actual deaths	Actual population
			percentiles			percentiles					percentiles			
		mean	2½	97½	mean	2½	97½			mean	2½	97½	(000)	(000)
<1	0.0077	0.0076	0.0072	0.0080	100,000	100,000	100,000	761	99,315	74.1	73.9	74.4	17	2,262
1-4	0.0004	0.0015	0.0014	0.0016	99,239	99,197	99,282	149	396,597	73.7	73.5	73.9	3	9,291
5-9	0.0002	0.0010	0.0009	0.0010	99,090	99,041	99,139	94	495,211	69.8	69.6	70.0	2	12,163
10-14	0.0002	0.0012	0.0012	0.0013	98,995	98,943	99,048	123	494,666	64.8	64.6	65.1	3	12,006
15-19	0.0009	0.0046	0.0044	0.0048	98,872	98,814	98,929	457	493,215	59.9	59.7	60.1	11	11,567
20-24	0.0013	0.0065	0.0063	0.0068	98,414	98,339	98,488	642	490,467	55.2	55.0	55.4	14	10,876
25-29	0.0013	0.0066	0.0064	0.0068	97,772	97,674	97,868	644	487,252	50.5	50.3	50.8	15	11,128
30-34	0.0015	0.0077	0.0075	0.0079	97,128	97,010	97,245	748	483,773	45.9	45.6	46.1	19	12,086
35-39	0.0020	0.0100	0.0098	0.0103	96,381	96,241	96,519	968	479,485	41.2	41.0	41.4	27	13,581
40-44	0.0029	0.0144	0.0140	0.0147	95,413	95,248	95,577	1,371	473,638	36.6	36.4	36.8	38	13,256
45-49	0.0044	0.0215	0.0210	0.0220	94,042	93,844	94,237	2,024	465,151	32.1	31.9	32.3	51	11,712
50-54	0.0063	0.0310	0.0303	0.0316	92,018	91,780	92,254	2,852	452,960	27.7	27.5	27.9	64	10,120
55-59	0.0100	0.0486	0.0476	0.0496	89,166	88,882	89,446	4,330	435,003	23.5	23.3	23.7	77	7,747
60-64	0.0156	0.0749	0.0733	0.0766	84,835	84,502	85,171	6,358	408,282	19.6	19.4	19.8	95	6,080
65-69	0.0244	0.1149	0.1123	0.1175	78,477	78,099	78,893	9,014	369,850	16.0	15.8	16.2	125	5,147
70-74	0.0378	0.1727	0.1687	0.1764	69,463	68,954	69,987	11,995	317,327	12.7	12.6	12.9	171	4,523
75-79	0.0574	0.2511	0.2452	0.2563	57,468	56,789	58,131	14,429	251,267	9.9	9.8	10.0	200	3,491
80-84	0.0944	0.3818	0.3738	0.3891	43,039	42,257	43,840	16,432	174,115	7.4	7.3	7.5	197	2,090
85-89	0.1503	0.5462	0.5377	0.5544	26,607	25,881	27,390	14,533	96,702	5.4	5.3	5.4	144	961
90-94	0.2325	0.6849	0.6776	0.6919	12,074	11,570	12,662	8,269	35,561	3.8	3.8	3.9	75	321
95-99	0.3479	0.8511	0.8450	0.8547	3,804	3,568	4,082	3,238	9,308	2.7	2.7	2.8	26	75
100+	0.5036	1	1	1	566	518	633	566	1,125	2.0	2.0	2.1	5	11
All ages													*1,382*	*160,495*

ln($_nq_x$) with uncertainty interval

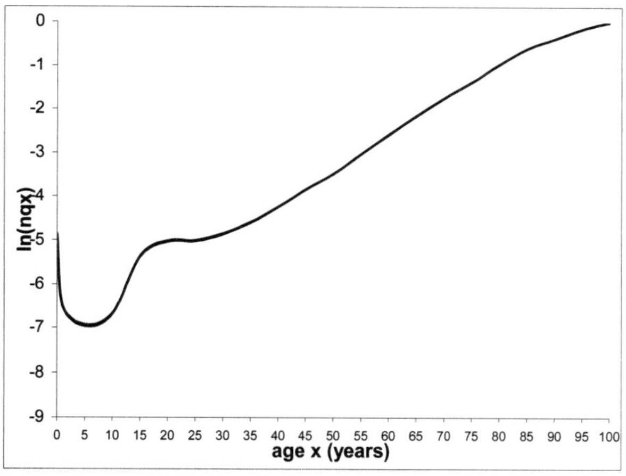

l_x with uncertainty interval

e_x with uncertainty interval

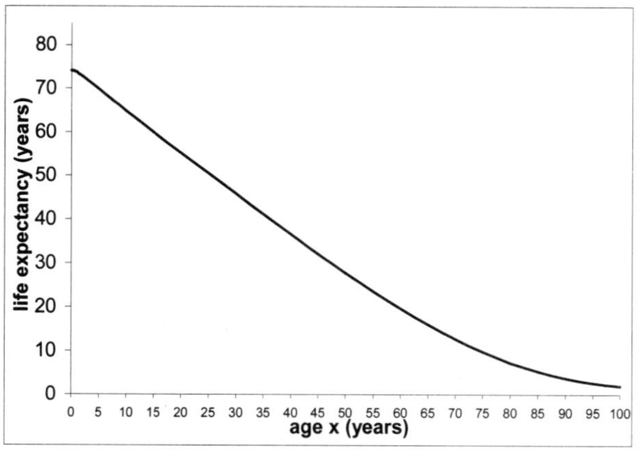

Age distribution of deaths with uncertainty interval

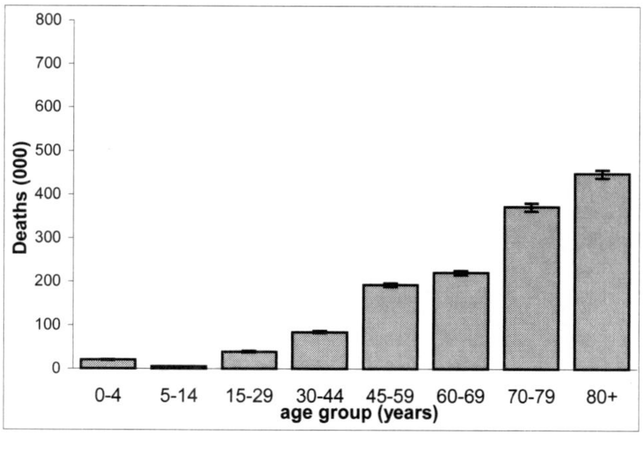

AmrA **AmrA** **AmrA**

Females, 2000 *Femmes, 2000* *Mujeres, 2000*

Age	$_nM_x$	$_nq_x$			l_x			$_nd_x$	$_nL_x$	e_x			Actual deaths	Actual population
			percentiles			percentiles					percentiles			
		mean	2½	97½	mean	2½	97½			mean	2½	97½	(000)	(000)
<1	0.0062	0.0061	0.0059	0.0064	100,000	100,000	100,000	612	99,450	79.6	79.4	79.7	13	2,150
1-4	0.0003	0.0012	0.0011	0.0012	99,388	99,361	99,412	118	397,271	79.1	79.0	79.2	3	8,851
5-9	0.0002	0.0008	0.0007	0.0008	99,271	99,239	99,298	75	496,166	75.2	75.1	75.3	2	11,593
10-14	0.0002	0.0008	0.0008	0.0008	99,196	99,163	99,225	80	495,781	70.2	70.1	70.4	2	11,450
15-19	0.0004	0.0020	0.0019	0.0021	99,116	99,080	99,148	198	495,085	65.3	65.2	65.4	4	11,034
20-24	0.0005	0.0023	0.0022	0.0023	98,918	98,876	98,955	226	494,025	60.4	60.3	60.5	5	10,477
25-29	0.0005	0.0027	0.0026	0.0028	98,692	98,644	98,735	265	492,798	55.6	55.4	55.7	6	10,895
30-34	0.0007	0.0036	0.0035	0.0037	98,427	98,373	98,476	357	491,243	50.7	50.6	50.8	9	11,882
35-39	0.0011	0.0055	0.0053	0.0056	98,070	98,008	98,128	536	489,010	45.9	45.8	46.0	15	13,312
40-44	0.0016	0.0081	0.0079	0.0083	97,534	97,460	97,603	789	485,697	41.1	41.0	41.2	21	13,109
45-49	0.0024	0.0121	0.0119	0.0123	96,745	96,656	96,830	1,170	480,801	36.4	36.3	36.5	29	11,791
50-54	0.0037	0.0184	0.0181	0.0187	95,575	95,466	95,681	1,760	473,477	31.8	31.7	32.0	38	10,351
55-59	0.0060	0.0296	0.0291	0.0300	93,815	93,677	93,947	2,773	462,145	27.4	27.3	27.5	49	8,092
60-64	0.0096	0.0469	0.0463	0.0476	91,042	90,867	91,208	4,269	444,540	23.1	23.1	23.3	63	6,546
65-69	0.0151	0.0728	0.0718	0.0737	86,773	86,552	86,975	6,315	418,080	19.2	19.1	19.3	88	5,829
70-74	0.0236	0.1116	0.1101	0.1129	80,459	80,174	80,699	8,980	379,842	15.5	15.4	15.6	132	5,583
75-79	0.0377	0.1722	0.1700	0.1739	71,478	71,136	71,783	12,305	326,628	12.1	12.0	12.2	183	4,867
80-84	0.0644	0.2773	0.2740	0.2798	59,173	58,791	59,577	16,406	254,849	9.1	9.0	9.2	222	3,454
85-89	0.1071	0.4223	0.4184	0.4260	42,767	42,377	43,229	18,060	168,683	6.6	6.6	6.7	221	2,065
90-94	0.1757	0.5753	0.5710	0.5793	24,707	24,381	25,121	14,213	80,892	4.6	4.6	4.7	170	965
95-99	0.2820	0.7638	0.7602	0.7671	10,493	10,278	10,761	8,015	28,421	3.2	3.2	3.3	92	325
100+	0.4460	1	1	1	2,478	2,393	2,571	2,478	5,557	2.2	2.2	2.3	30	68
All ages													*1,396*	*164,691*

ln($_nq_x$) with uncertainty interval

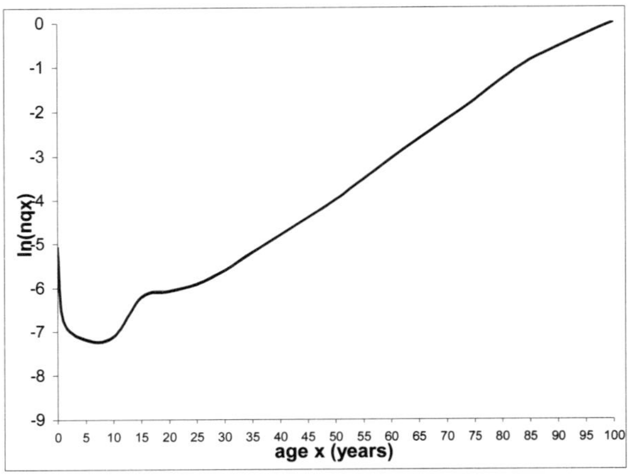

l_x with uncertainty interval

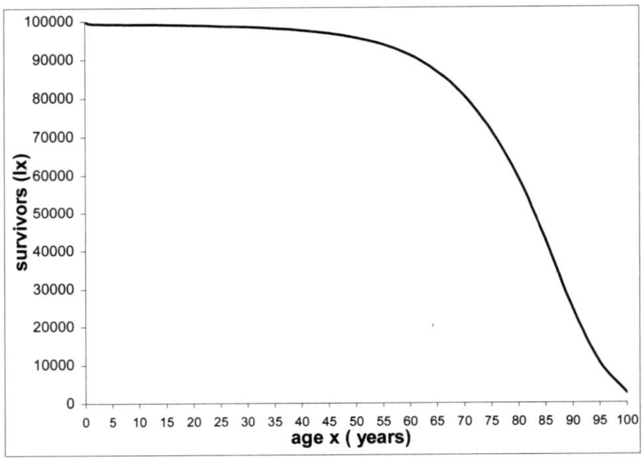

e_x with uncertainty interval

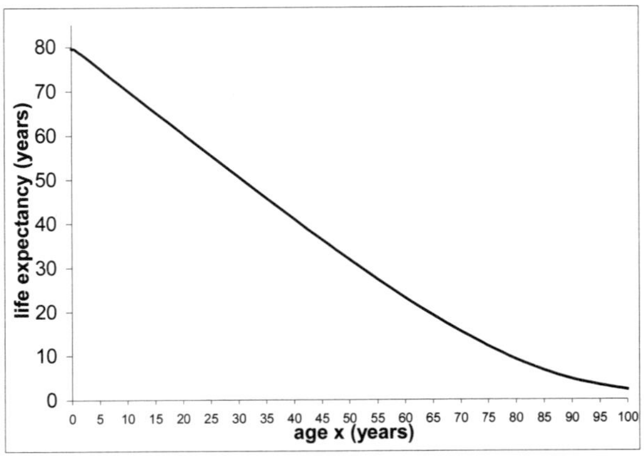

Age distribution of deaths with uncertainty interval

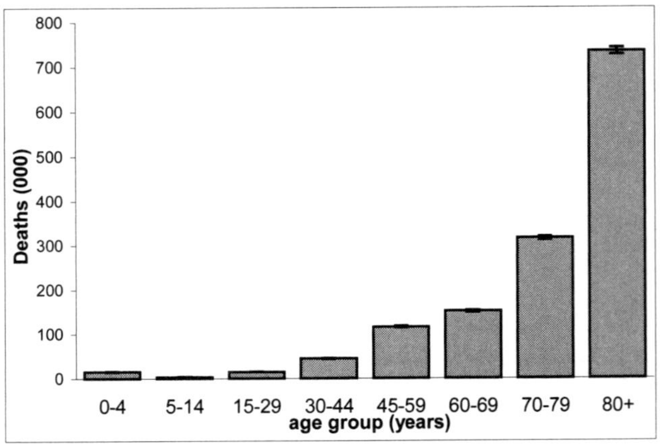

AmrB **AmrB** **AmrB**

Males, 2000 *Hommes, 2000* *Hombres, 2000*

Age	$_nM_x$	$_nq_x$			l_x			$_nd_x$	$_nL_x$	e_x			Actual deaths	Actual population
			percentiles			percentiles					percentiles			
		mean	2 1/2	97 1/2	mean	2 1/2	97 1/2			mean	2 1/2	97 1/2	(000)	(000)
<1	0.0306	0.0298	0.0260	0.0336	100,000	100,000	100,000	2,981	97,317	67.5	66.6	68.6	142	4,631
1-4	0.0018	0.0072	0.0066	0.0078	97,019	96,640	97,396	698	386,400	68.6	67.9	69.3	33	18,192
5-9	0.0005	0.0024	0.0022	0.0025	96,321	95,884	96,754	227	481,036	65.0	64.4	65.7	11	22,618
10-14	0.0005	0.0026	0.0024	0.0028	96,094	95,643	96,541	250	479,843	60.2	59.6	60.9	12	22,421
15-19	0.0017	0.0087	0.0080	0.0093	95,843	95,377	96,307	832	477,137	55.3	54.7	56.0	38	21,861
20-24	0.0027	0.0133	0.0124	0.0143	95,012	94,494	95,533	1,264	471,899	50.8	50.2	51.5	55	20,633
25-29	0.0030	0.0148	0.0139	0.0159	93,748	93,165	94,351	1,391	465,263	46.5	45.9	47.1	55	18,369
30-34	0.0033	0.0164	0.0154	0.0175	92,357	91,713	93,041	1,512	458,005	42.1	41.5	42.8	54	16,498
35-39	0.0039	0.0192	0.0180	0.0204	90,845	90,141	91,605	1,743	449,867	37.8	37.2	38.4	58	14,911
40-44	0.0050	0.0248	0.0233	0.0263	89,102	88,301	89,942	2,208	439,990	33.5	32.9	34.1	63	12,587
45-49	0.0067	0.0328	0.0309	0.0345	86,894	85,982	87,818	2,846	427,355	29.3	28.6	29.9	70	10,481
50-54	0.0096	0.0469	0.0441	0.0495	84,048	83,013	85,052	3,939	410,392	25.2	24.6	25.8	80	8,297
55-59	0.0140	0.0677	0.0636	0.0715	80,109	78,939	81,217	5,420	386,994	21.3	20.7	21.8	90	6,436
60-64	0.0210	0.0997	0.0935	0.1060	74,689	73,390	75,939	7,446	354,828	17.6	17.1	18.2	108	5,150
65-69	0.0321	0.1487	0.1395	0.1589	67,243	65,843	68,676	9,998	311,218	14.3	13.8	14.8	127	3,939
70-74	0.0475	0.2123	0.1994	0.2272	57,245	55,621	59,008	12,151	255,847	11.4	10.9	11.8	140	2,957
75-79	0.0724	0.3065	0.2903	0.3261	45,094	43,243	47,107	13,820	190,921	8.8	8.4	9.1	136	1,882
80-84	0.1172	0.4531	0.4345	0.4751	31,274	29,168	33,430	14,170	120,948	6.5	6.3	6.8	110	937
85-89	0.1793	0.6190	0.6030	0.6376	17,105	15,310	18,905	10,587	59,056	4.9	4.7	5.0	70	388
90-94	0.2399	0.6976	0.6838	0.7127	6,518	5,548	7,495	4,546	18,949	3.8	3.7	3.9	26	109
95-99	0.3278	0.8263	0.8137	0.8393	1,971	1,594	2,365	1,629	4,970	2.9	2.8	3.0	6	19
100+	0.4794	1	1	1	342	261	436	342	714	2.1	2.1	2.2	1	2
All ages													*1,485*	*213,318*

ln($_nq_x$) with uncertainty interval

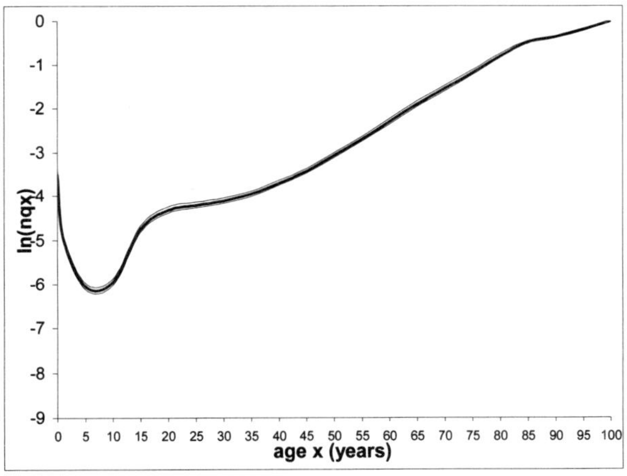

l_x with uncertainty interval

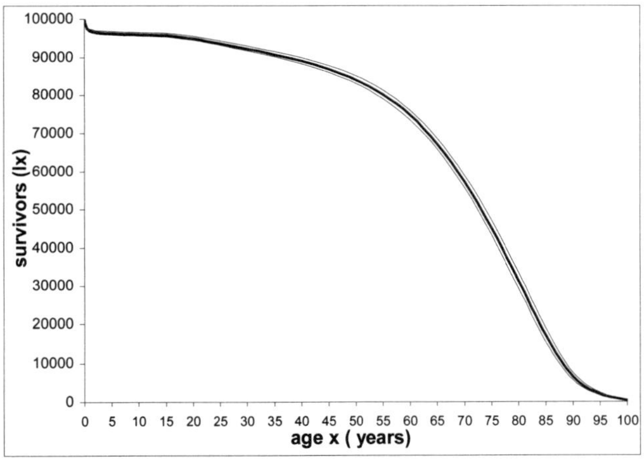

e_x with uncertainty interval

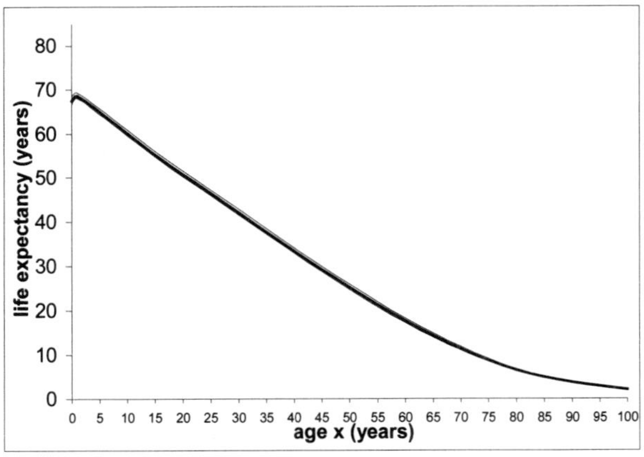

Age distribution of deaths with uncertainty interval

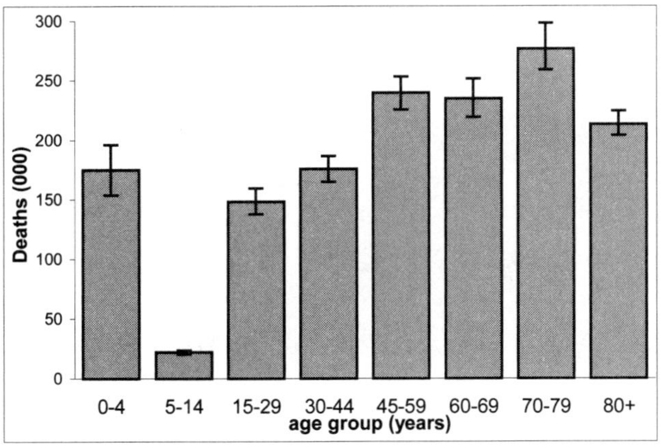

AmrB **AmrB** **AmrB**

Females, 2000 *Femmes, 2000* *Mujeres, 2000*

Age	$_nM_x$	$_nq_x$			l_x			$_nd_x$	$_nL_x$	e_x			Actual deaths (000)	Actual population (000)
		mean	percentiles 2½	97½	mean	percentiles 2½	97½			mean	percentiles 2½	97½		
<1	0.0252	0.0246	0.0220	0.0272	100,000	100,000	100,000	2,463	97,783	74.5	73.4	75.5	111	4,417
1-4	0.0015	0.0062	0.0056	0.0067	97,537	97,277	97,804	600	388,706	75.4	74.6	76.1	27	17,533
5-9	0.0004	0.0018	0.0017	0.0019	96,936	96,635	97,245	172	484,252	71.8	71.1	72.6	8	21,799
10-14	0.0004	0.0018	0.0016	0.0019	96,764	96,452	97,081	170	483,397	66.9	66.2	67.7	8	21,667
15-19	0.0006	0.0031	0.0028	0.0033	96,594	96,270	96,920	296	482,230	62.1	61.3	62.8	13	21,262
20-24	0.0008	0.0038	0.0035	0.0041	96,298	95,953	96,641	364	480,580	57.2	56.5	58.0	16	20,506
25-29	0.0009	0.0047	0.0044	0.0050	95,934	95,563	96,304	451	478,542	52.4	51.7	53.2	18	18,617
30-34	0.0012	0.0060	0.0056	0.0065	95,483	95,081	95,885	575	475,977	47.7	46.9	48.4	20	16,970
35-39	0.0017	0.0082	0.0076	0.0089	94,908	94,466	95,351	782	472,583	43.0	42.2	43.7	26	15,510
40-44	0.0024	0.0122	0.0112	0.0131	94,126	93,631	94,626	1,144	467,767	38.3	37.6	39.0	32	13,274
45-49	0.0036	0.0179	0.0165	0.0192	92,981	92,411	93,564	1,665	460,743	33.7	33.0	34.4	40	11,120
50-54	0.0055	0.0272	0.0252	0.0290	91,316	90,647	92,008	2,481	450,378	29.3	28.6	30.0	49	8,880
55-59	0.0082	0.0403	0.0374	0.0430	88,835	88,028	89,676	3,578	435,232	25.0	24.4	25.7	58	7,015
60-64	0.0128	0.0619	0.0574	0.0668	85,257	84,259	86,284	5,279	413,089	21.0	20.3	21.6	74	5,823
65-69	0.0203	0.0967	0.0896	0.1047	79,978	78,745	81,277	7,730	380,566	17.2	16.6	17.8	95	4,665
70-74	0.0307	0.1426	0.1314	0.1553	72,248	70,772	73,868	10,301	335,488	13.8	13.3	14.3	114	3,715
75-79	0.0496	0.2206	0.2038	0.2389	61,947	59,834	64,038	13,667	275,568	10.7	10.2	11.1	127	2,554
80-84	0.0824	0.3416	0.3198	0.3638	48,280	45,589	50,782	16,493	200,167	8.0	7.7	8.3	114	1,387
85-89	0.1337	0.5011	0.4793	0.5219	31,787	29,024	34,434	15,930	119,109	5.8	5.6	6.0	87	648
90-94	0.2078	0.6401	0.6235	0.6554	15,857	13,875	17,930	10,150	48,836	4.2	4.1	4.3	45	217
95-99	0.3124	0.8064	0.7968	0.8147	5,707	4,781	6,751	4,602	14,730	3.0	3.0	3.1	15	47
100+	0.4618	1	1	1	1,105	886	1,372	1,105	2,393	2.2	2.1	2.2	3	8
All ages													*1,099*	*217,633*

ln($_nq_x$) with uncertainty interval

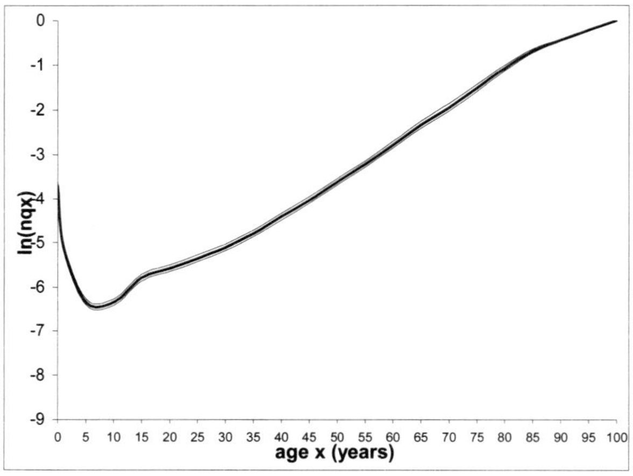

l_x with uncertainty interval

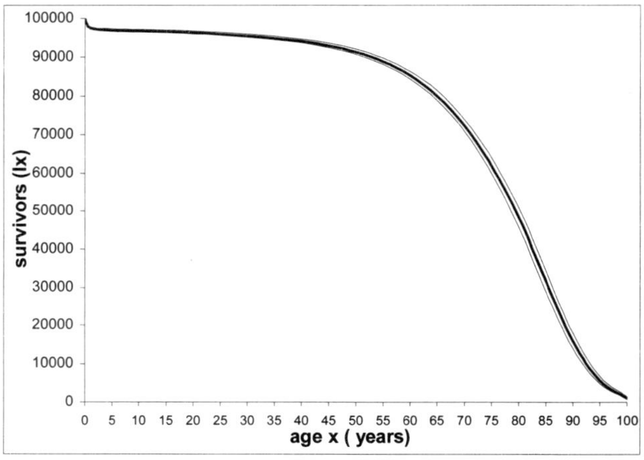

e_x with uncertainty interval

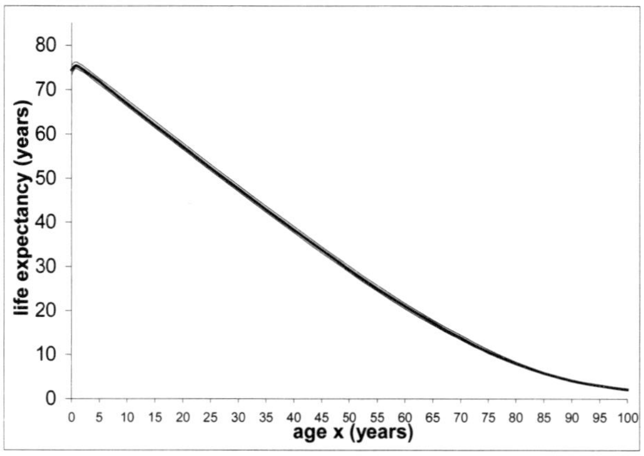

Age distribution of deaths with uncertainty interval

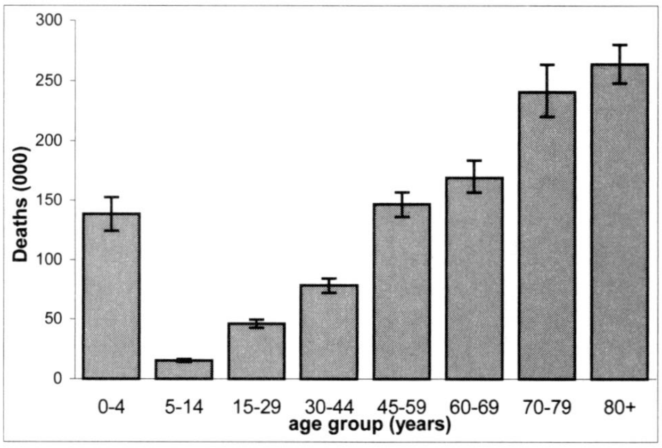

AmrD **AmrD** **AmrD**

Males, 2000 *Hommes, 2000* *Hombres, 2000*

Age	$_nM_x$	$_nq_x$			l_x			$_nd_x$	$_nL_x$	e_x			Actual deaths	Actual population
			percentiles			percentiles					percentiles			
		mean	2½	97½	mean	2½	97½			mean	2½	97½	(000)	(000)
<1	0.0473	0.0453	0.0408	0.0500	100,000	100,000	100,000	4,533	95,920	63.5	62.5	65.6	47	991
1-4	0.0047	0.0186	0.0174	0.0200	95,467	95,001	95,916	1,778	377,598	65.5	64.2	66.6	18	3,776
5-9	0.0011	0.0055	0.0051	0.0060	93,688	93,124	94,233	515	467,153	62.7	61.5	63.8	5	4,498
10-14	0.0010	0.0048	0.0045	0.0052	93,173	92,579	93,743	448	464,744	58.0	56.8	59.1	4	4,254
15-19	0.0018	0.0090	0.0084	0.0097	92,725	92,107	93,318	835	461,537	53.3	52.1	54.4	7	3,930
20-24	0.0031	0.0152	0.0141	0.0166	91,890	91,232	92,522	1,398	455,956	48.7	47.6	49.8	11	3,465
25-29	0.0039	0.0193	0.0176	0.0215	90,492	89,762	91,190	1,745	448,098	44.5	43.4	45.5	11	2,918
30-34	0.0046	0.0229	0.0207	0.0259	88,747	87,927	89,561	2,029	438,663	40.3	39.3	41.3	11	2,426
35-39	0.0056	0.0277	0.0253	0.0312	86,718	85,736	87,704	2,405	427,578	36.2	35.3	37.1	11	2,032
40-44	0.0065	0.0320	0.0292	0.0356	84,313	83,118	85,476	2,697	414,823	32.1	31.3	33.0	11	1,679
45-49	0.0082	0.0402	0.0368	0.0442	81,616	80,157	82,955	3,279	399,881	28.1	27.4	28.9	12	1,404
50-54	0.0108	0.0525	0.0480	0.0574	78,337	76,614	79,853	4,109	381,409	24.2	23.5	24.9	12	1,103
55-59	0.0147	0.0709	0.0648	0.0773	74,227	72,215	75,933	5,266	357,970	20.4	19.8	21.1	13	885
60-64	0.0223	0.1057	0.0967	0.1142	68,961	66,633	70,864	7,288	326,584	16.8	16.2	17.4	16	725
65-69	0.0337	0.1553	0.1427	0.1666	61,673	59,027	63,944	9,575	284,427	13.4	13.0	14.0	19	564
70-74	0.0515	0.2280	0.2112	0.2432	52,098	49,203	54,819	11,878	230,795	10.4	10.1	10.9	20	397
75-79	0.0849	0.3500	0.3287	0.3687	40,220	37,279	43,243	14,078	165,904	7.8	7.5	8.1	21	244
80-84	0.1407	0.5205	0.4976	0.5396	26,142	23,602	29,030	13,606	96,694	5.6	5.5	5.8	17	120
85-89	0.2219	0.7136	0.6957	0.7284	12,536	10,929	14,583	8,946	40,314	4.0	4.0	4.2	11	48
90-94	0.3272	0.8256	0.8128	0.8367	3,590	2,993	4,438	2,964	9,058	2.9	2.8	3.0	4	12
95-99	0.4539	0.9610	0.9558	0.9647	626	491	831	602	1,325	2.2	2.2	2.2	1	2
100+	0.6202	1	1	1	24	17	36	24	39	1.6	1.6	1.8	0	0
All ages													*282*	*35,474*

ln($_nq_x$) with uncertainty interval

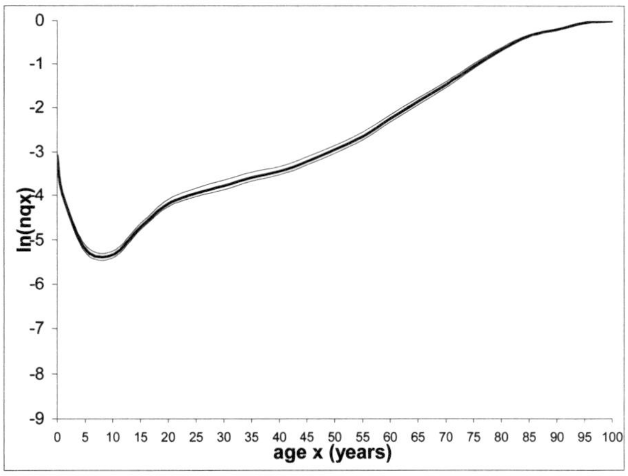

l_x with uncertainty interval

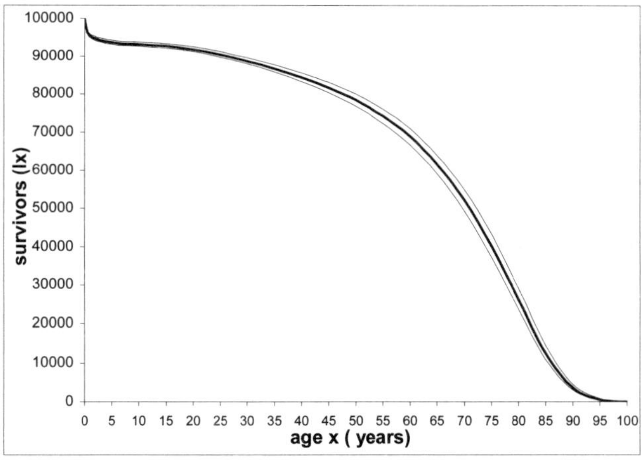

e_x with uncertainty interval

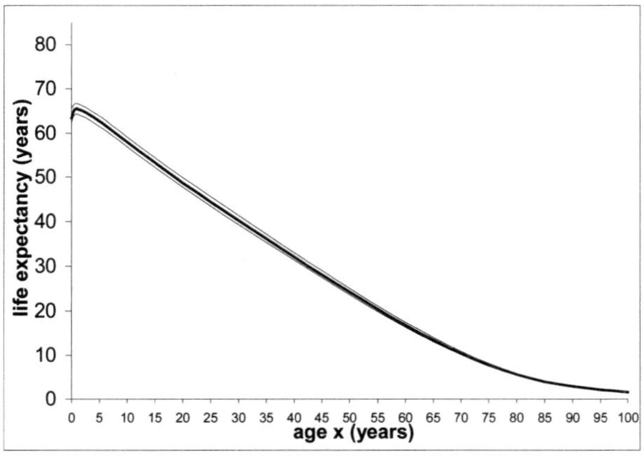

Age distribution of deaths with uncertainty interval

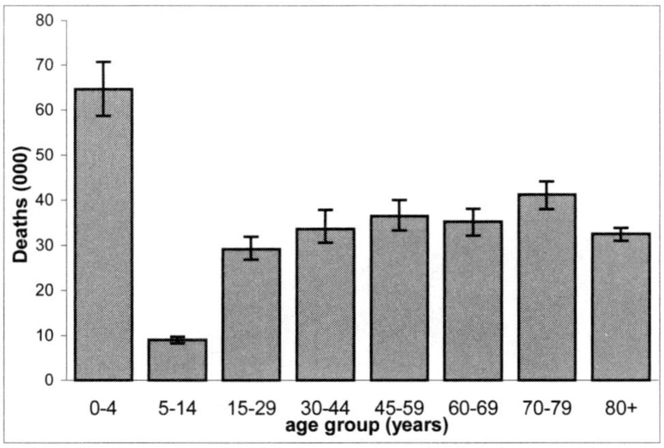

AmrD AmrD AmrD

Females, 2000 *Femmes, 2000* *Mujeres, 2000*

Age	$_nM_x$	$_nq_x$ mean	$_nq_x$ percentiles 2½	$_nq_x$ percentiles 97½	l_x mean	l_x percentiles 2½	l_x percentiles 97½	$_nd_x$	$_nL_x$	e_x mean	e_x percentiles 2½	e_x percentiles 97½	Actual deaths (000)	Actual population (000)
<1	0.0394	0.0381	0.0337	0.0425	100,000	100,000	100,000	3,808	96,573	68.4	67.2	70.6	37	945
1-4	0.0048	0.0188	0.0174	0.0203	96,192	95,753	96,634	1,811	380,421	70.1	69.0	71.2	17	3,642
5-9	0.0010	0.0052	0.0047	0.0057	94,381	93,806	94,943	489	470,681	67.4	66.3	68.5	5	4,346
10-14	0.0009	0.0044	0.0041	0.0049	93,892	93,281	94,486	418	468,414	62.8	61.7	63.8	4	4,126
15-19	0.0014	0.0071	0.0065	0.0078	93,474	92,835	94,092	667	465,703	58.0	56.9	59.1	5	3,831
20-24	0.0019	0.0095	0.0086	0.0104	92,807	92,126	93,460	881	461,833	53.4	52.4	54.4	7	3,427
25-29	0.0022	0.0108	0.0097	0.0120	91,926	91,187	92,622	995	457,142	48.9	47.9	49.9	6	2,951
30-34	0.0027	0.0133	0.0121	0.0147	90,931	90,103	91,701	1,213	451,623	44.4	43.4	45.4	7	2,517
35-39	0.0035	0.0173	0.0157	0.0191	89,718	88,783	90,594	1,551	444,714	40.0	39.0	40.9	7	2,145
40-44	0.0042	0.0206	0.0189	0.0228	88,167	87,109	89,165	1,819	436,288	35.7	34.7	36.5	7	1,787
45-49	0.0056	0.0274	0.0252	0.0302	86,348	85,164	87,471	2,369	425,818	31.4	30.5	32.2	8	1,496
50-54	0.0078	0.0384	0.0353	0.0424	83,979	82,656	85,241	3,225	411,832	27.2	26.4	27.9	9	1,175
55-59	0.0107	0.0523	0.0479	0.0578	80,754	79,273	82,175	4,220	393,218	23.2	22.4	23.9	10	946
60-64	0.0157	0.0754	0.0692	0.0831	76,533	74,812	78,126	5,768	368,246	19.3	18.6	19.9	12	787
65-69	0.0245	0.1152	0.1059	0.1262	70,765	68,684	72,599	8,155	333,437	15.7	15.1	16.2	15	629
70-74	0.0375	0.1716	0.1581	0.1866	62,610	60,014	64,898	10,746	286,184	12.4	11.9	12.9	17	461
75-79	0.0587	0.2560	0.2375	0.2756	51,864	48,813	54,597	13,277	226,127	9.4	9.0	9.8	17	298
80-84	0.1037	0.4116	0.3874	0.4369	38,587	35,412	41,583	15,883	153,229	6.8	6.5	7.1	16	157
85-89	0.1752	0.6091	0.5854	0.6324	22,704	20,092	25,467	13,829	78,951	4.8	4.6	5.0	12	69
90-94	0.2685	0.7435	0.7253	0.7607	8,876	7,401	10,558	6,599	24,581	3.4	3.3	3.5	6	21
95-99	0.4027	0.9119	0.9031	0.9194	2,276	1,771	2,892	2,076	5,155	2.4	2.4	2.5	2	4
100+	0.5839	1	1	1	201	143	280	201	344	1.7	1.7	2.0	0	0
All ages													*229*	*35,762*

ln($_nq_x$) with uncertainty interval

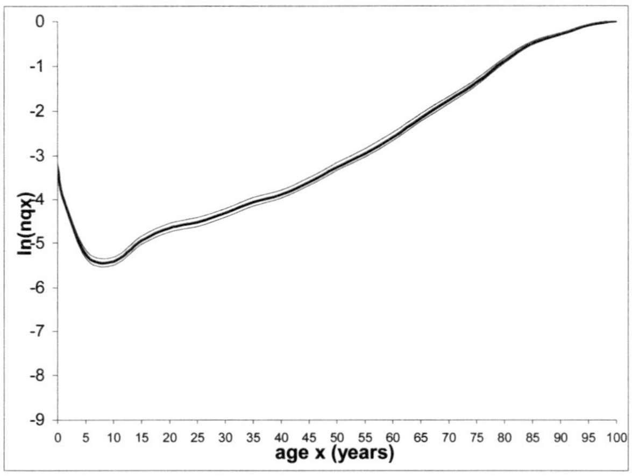

l_x with uncertainty interval

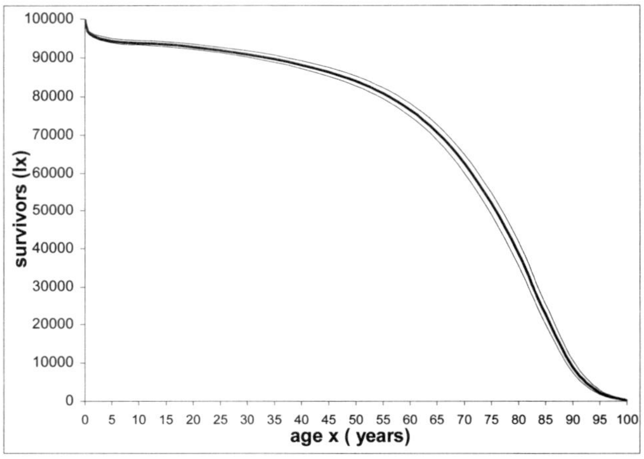

e_x with uncertainty interval

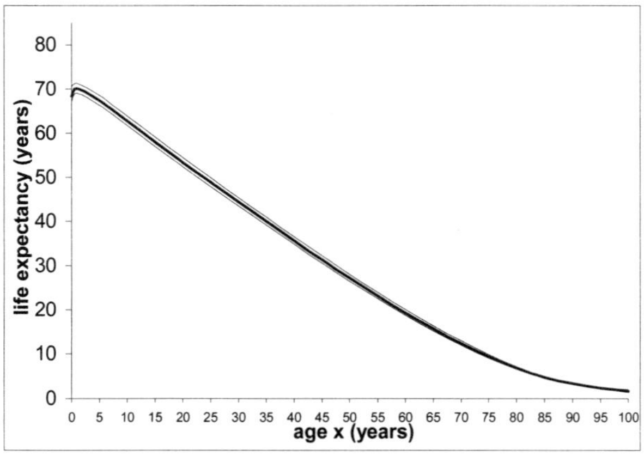

Age distribution of deaths with uncertainty interval

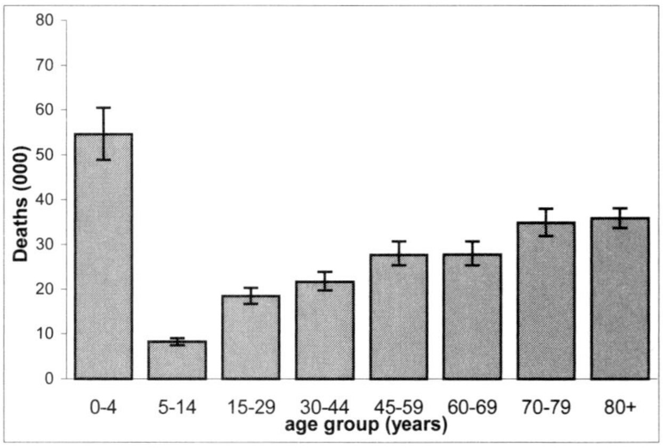

EmrB **EmrB** **EmrB**

Males, 2000 *Hommes, 2000* *Hombres, 2000*

Age	$_nM_x$	$_nq_x$			l_x			$_nd_x$	$_nL_x$	e_x			Actual deaths	Actual population
			percentiles			percentiles					percentiles			
		mean	2½	97½	mean	2½	97½			mean	2½	97½	(000)	(000)
<1	0.0326	0.0317	0.0281	0.0353	100,000	100,000	100,000	3,166	97,151	68.6	67.7	69.5	57	1,754
1-4	0.0015	0.0059	0.0054	0.0064	96,834	96,475	97,194	573	385,962	69.8	69.3	70.3	10	6,690
5-9	0.0006	0.0030	0.0028	0.0033	96,261	95,857	96,663	292	480,576	66.2	65.7	66.7	5	8,956
10-14	0.0006	0.0029	0.0026	0.0031	95,969	95,541	96,395	275	479,158	61.4	60.9	61.9	5	9,375
15-19	0.0012	0.0059	0.0054	0.0064	95,694	95,243	96,144	569	477,048	56.6	56.1	57.1	10	8,443
20-24	0.0017	0.0086	0.0079	0.0093	95,125	94,629	95,621	818	473,581	51.9	51.4	52.4	11	6,582
25-29	0.0017	0.0085	0.0078	0.0092	94,307	93,752	94,864	803	469,527	47.3	46.9	47.8	9	5,511
30-34	0.0018	0.0090	0.0084	0.0096	93,504	92,893	94,120	843	465,412	42.7	42.3	43.2	9	4,793
35-39	0.0022	0.0110	0.0102	0.0117	92,661	91,998	93,333	1,017	460,762	38.1	37.6	38.5	10	4,363
40-44	0.0031	0.0152	0.0143	0.0161	91,644	90,923	92,377	1,391	454,742	33.5	33.0	33.9	12	4,012
45-49	0.0050	0.0245	0.0230	0.0258	90,253	89,461	91,060	2,207	445,748	29.0	28.5	29.4	17	3,449
50-54	0.0081	0.0398	0.0377	0.0419	88,046	87,150	88,961	3,505	431,469	24.6	24.2	25.0	21	2,616
55-59	0.0129	0.0625	0.0595	0.0657	84,541	83,501	85,609	5,287	409,491	20.5	20.1	20.9	24	1,859
60-64	0.0214	0.1018	0.0969	0.1070	79,255	78,043	80,502	8,067	376,107	16.7	16.4	17.1	30	1,385
65-69	0.0339	0.1562	0.1491	0.1641	71,188	69,791	72,627	11,117	328,148	13.4	13.0	13.7	35	1,026
70-74	0.0540	0.2378	0.2270	0.2501	60,071	58,540	61,630	14,283	264,649	10.4	10.1	10.7	37	684
75-79	0.0853	0.3515	0.3361	0.3679	45,788	44,025	47,403	16,096	188,701	7.8	7.6	8.1	34	397
80-84	0.1379	0.5128	0.4949	0.5310	29,692	27,944	31,308	15,226	110,395	5.7	5.5	5.9	26	185
85-89	0.2183	0.7061	0.6899	0.7206	14,466	13,124	15,763	10,214	46,795	4.1	4.0	4.2	14	65
90-94	0.3262	0.8243	0.8149	0.8337	4,252	3,685	4,888	3,505	10,745	2.9	2.9	3.0	4	13
95-99	0.4699	0.9750	0.9547	0.9776	747	617	905	728	1,550	2.1	2.1	2.2	1	2
100+	0.5966	1	1	1	19	12	41	19	31	1.7	1.7	1.8	0	0
All ages													*382*	*72,161*

ln($_nq_x$) with uncertainty interval

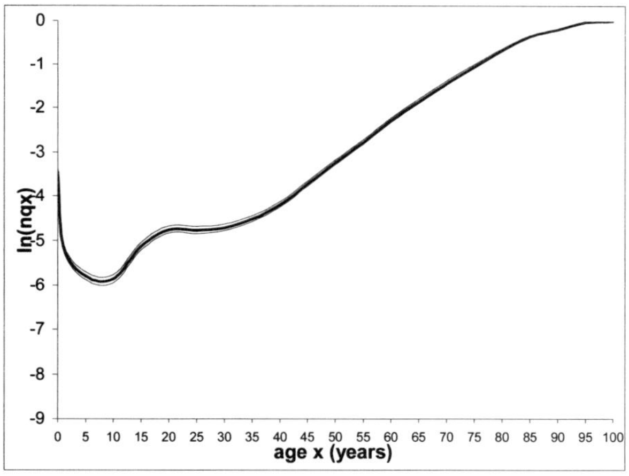

l_x with uncertainty interval

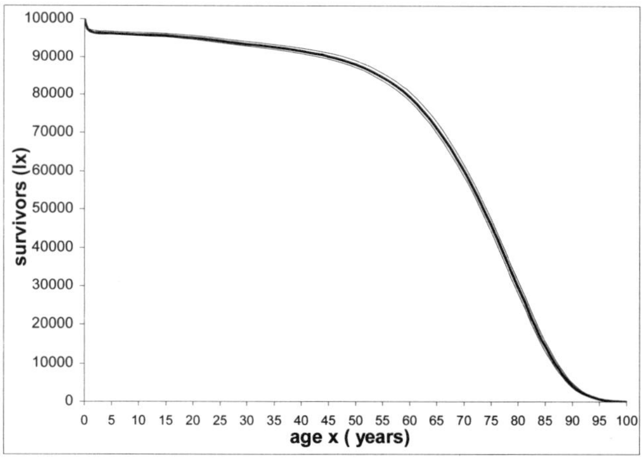

e_x with uncertainty interval

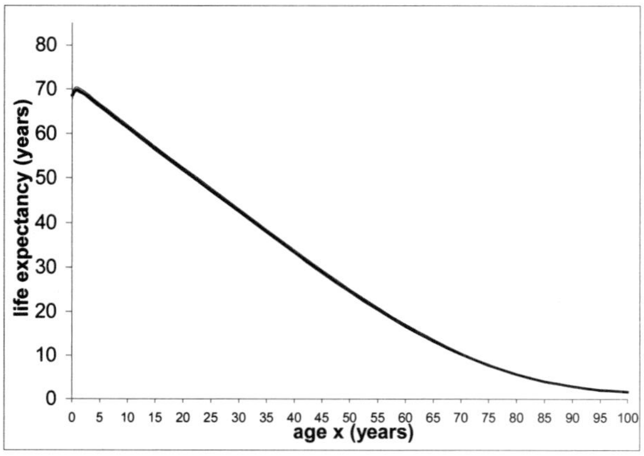

Age distribution of deaths with uncertainty interval

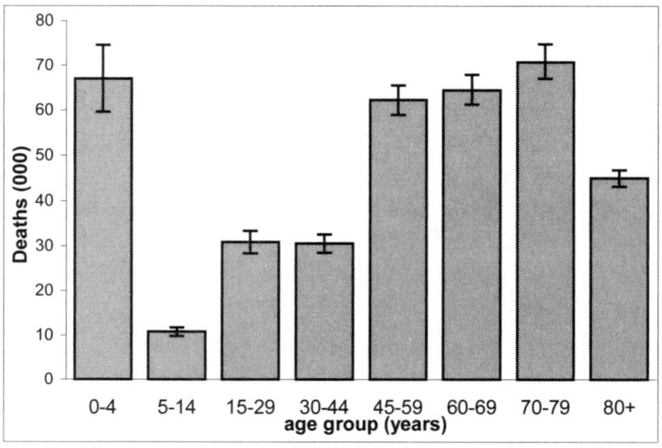

EmrB EmrB EmrB

Females, 2000 *Femmes, 2000* *Mujeres, 2000*

Age	$_nM_x$	$_nq_x$			l_x			$_nd_x$	$_nL_x$	e_x			Actual deaths	Actual population
			percentiles			percentiles					percentiles			
		mean	2½	97½	mean	2½	97½			mean	2½	97½	(000)	(000)
<1	0.0301	0.0293	0.0257	0.0329	100,000	100,000	100,000	2,931	97,363	71.6	70.7	72.5	50	1,677
1-4	0.0016	0.0062	0.0057	0.0067	97,069	96,708	97,428	602	386,832	72.7	72.2	73.3	10	6,342
5-9	0.0005	0.0027	0.0025	0.0030	96,467	96,059	96,869	262	481,682	69.2	68.6	69.8	5	8,502
10-14	0.0006	0.0028	0.0026	0.0031	96,206	95,775	96,628	270	480,353	64.4	63.8	64.9	5	8,922
15-19	0.0007	0.0037	0.0034	0.0040	95,936	95,481	96,380	357	478,786	59.5	59.0	60.1	6	8,118
20-24	0.0010	0.0050	0.0046	0.0054	95,579	95,097	96,050	474	476,709	54.7	54.2	55.3	6	6,311
25-29	0.0011	0.0056	0.0052	0.0060	95,105	94,587	95,609	529	474,202	50.0	49.5	50.6	6	5,189
30-34	0.0013	0.0063	0.0059	0.0068	94,576	94,021	95,114	600	471,380	45.3	44.8	45.8	6	4,436
35-39	0.0017	0.0082	0.0077	0.0088	93,976	93,384	94,548	772	467,948	40.5	40.1	41.1	6	3,864
40-44	0.0025	0.0122	0.0114	0.0130	93,203	92,566	93,814	1,139	463,170	35.9	35.4	36.4	8	3,304
45-49	0.0038	0.0187	0.0175	0.0199	92,064	91,362	92,727	1,718	456,027	31.3	30.8	31.8	10	2,726
50-54	0.0061	0.0300	0.0281	0.0320	90,347	89,551	91,080	2,711	444,957	26.8	26.4	27.3	13	2,150
55-59	0.0098	0.0478	0.0449	0.0508	87,636	86,707	88,464	4,190	427,706	22.6	22.2	23.0	16	1,609
60-64	0.0156	0.0749	0.0704	0.0793	83,446	82,352	84,450	6,252	401,601	18.6	18.2	19.0	20	1,281
65-69	0.0261	0.1223	0.1148	0.1288	77,194	75,893	78,465	9,442	362,364	14.9	14.5	15.3	26	1,011
70-74	0.0426	0.1923	0.1815	0.2022	67,752	66,204	69,324	13,030	306,182	11.6	11.3	11.9	30	711
75-79	0.0698	0.2973	0.2828	0.3106	54,721	52,916	56,685	16,267	232,939	8.8	8.5	9.0	31	440
80-84	0.1164	0.4507	0.4337	0.4667	38,454	36,695	40,628	17,330	148,947	6.4	6.2	6.6	25	214
85-89	0.1834	0.6287	0.6135	0.6436	21,124	19,639	22,914	13,281	72,419	4.6	4.5	4.7	15	80
90-94	0.2852	0.7685	0.7579	0.7789	7,843	7,083	8,734	6,027	21,133	3.2	3.2	3.3	5	17
95-99	0.4096	0.9189	0.9141	0.9234	1,816	1,580	2,105	1,668	4,073	2.4	2.4	2.4	1	2
100+	0.5442	1	1	1	147	123	180	147	271	1.8	1.8	2.0	0	0
All ages													*300*	*66,909*

$\ln(_nq_x)$ with uncertainty interval

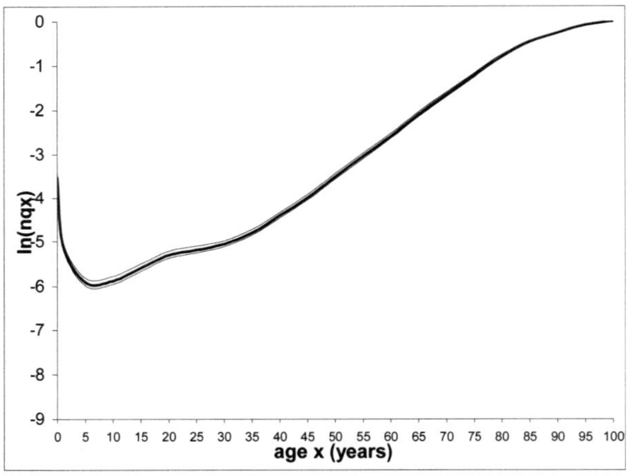

l_x with uncertainty interval

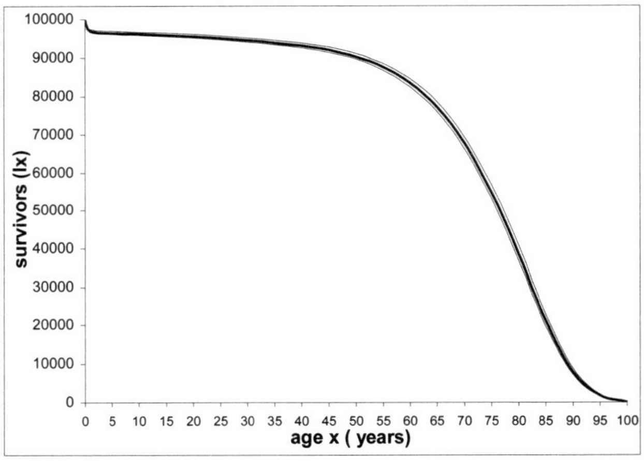

e_x with uncertainty interval

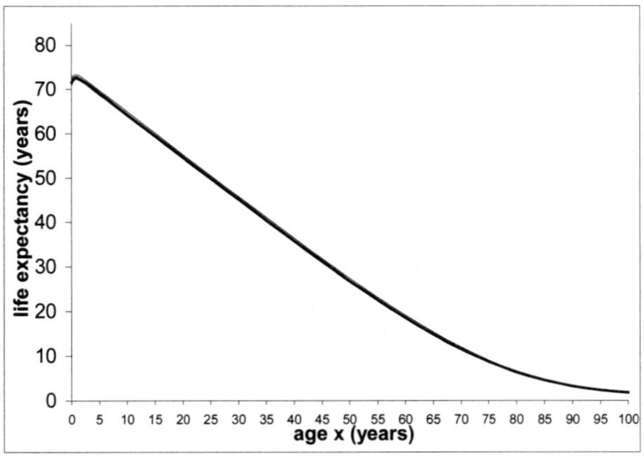

Age distribution of deaths with uncertainty interval

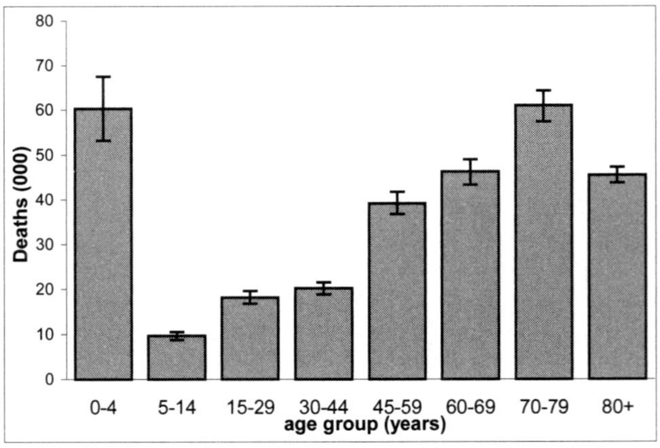

EmrD **EmrD** **EmrD**

Males, 2000 *Hommes, 2000* *Hombres, 2000*

Age	$_nM_x$	$_nq_x$			l_x			$_nd_x$	$_nL_x$	e_x			Actual deaths	Actual population
			percentiles			percentiles					percentiles			
		mean	2½	97½	mean	2½	97½			mean	2½	97½	(000)	(000)
<1	0.1037	0.0949	0.0836	0.1061	100,000	100,000	100,000	9,488	91,461	59.1	57.4	61.4	584	5,628
1-4	0.0067	0.0263	0.0241	0.0286	90,512	89,393	91,639	2,384	356,328	64.3	63.1	65.3	140	20,856
5-9	0.0017	0.0086	0.0079	0.0093	88,128	86,998	89,329	757	438,748	61.9	60.9	63.0	40	23,302
10-14	0.0012	0.0061	0.0057	0.0066	87,371	86,209	88,587	537	435,512	57.5	56.5	58.5	26	21,239
15-19	0.0015	0.0076	0.0070	0.0082	86,834	85,650	88,062	659	432,524	52.8	51.8	53.8	28	18,687
20-24	0.0021	0.0104	0.0097	0.0113	86,176	84,969	87,416	899	428,630	48.2	47.2	49.2	33	15,765
25-29	0.0028	0.0138	0.0127	0.0150	85,276	84,044	86,527	1,177	423,438	43.7	42.7	44.6	37	13,208
30-34	0.0035	0.0175	0.0160	0.0191	84,099	82,835	85,384	1,470	416,821	39.2	38.3	40.2	40	11,380
35-39	0.0045	0.0220	0.0202	0.0242	82,629	81,300	84,003	1,821	408,593	34.9	34.0	35.8	44	9,872
40-44	0.0058	0.0286	0.0263	0.0314	80,808	79,386	82,287	2,310	398,265	30.6	29.8	31.5	50	8,568
45-49	0.0082	0.0401	0.0371	0.0439	78,498	76,970	80,094	3,150	384,615	26.5	25.7	27.3	58	7,083
50-54	0.0120	0.0583	0.0539	0.0637	75,348	73,702	77,070	4,391	365,764	22.5	21.7	23.3	65	5,433
55-59	0.0182	0.0868	0.0799	0.0948	70,957	69,166	72,817	6,162	339,382	18.7	18.0	19.5	75	4,131
60-64	0.0272	0.1274	0.1169	0.1382	64,796	62,668	66,768	8,255	303,341	15.2	14.6	15.9	91	3,342
65-69	0.0429	0.1939	0.1772	0.2103	56,541	54,008	58,824	10,964	255,292	12.1	11.5	12.7	107	2,496
70-74	0.0656	0.2816	0.2585	0.3042	45,576	42,763	48,127	12,835	195,793	9.4	8.9	9.9	111	1,689
75-79	0.1033	0.4104	0.3804	0.4386	32,741	29,999	35,452	13,437	130,112	7.1	6.7	7.5	98	948
80-84	0.1554	0.5597	0.5262	0.5889	19,304	17,003	21,954	10,804	69,509	5.3	5.0	5.6	71	457
85-89	0.2357	0.7416	0.7089	0.7677	8,500	6,989	10,362	6,303	26,740	3.9	3.7	4.1	37	156
90-94	0.3475	0.8506	0.8220	0.8724	2,196	1,624	2,985	1,868	5,377	2.8	2.6	2.9	12	35
95-99	0.4752	0.9795	0.9706	0.9849	328	208	528	321	676	2.1	2.1	2.1	2	5
100+	0.7539	1	1	1	7	3	15	7	9	1.3	1.3	1.6	0	1
All ages													*1,750*	*174,279*

$\ln(_nq_x)$ with uncertainty interval

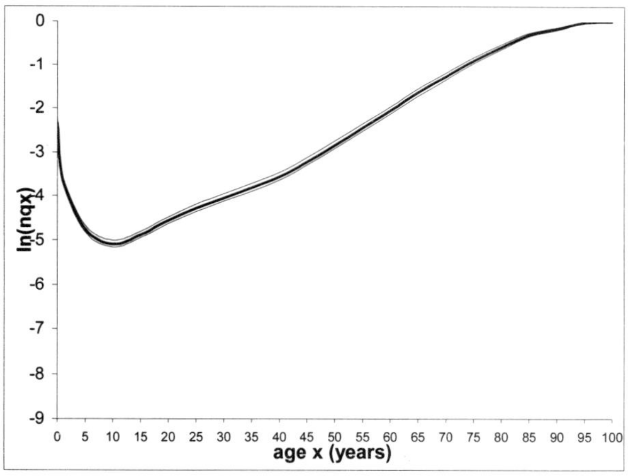

l_x with uncertainty interval

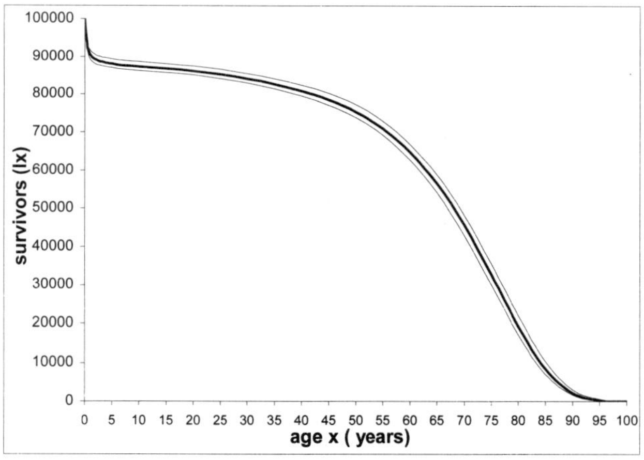

e_x with uncertainty interval

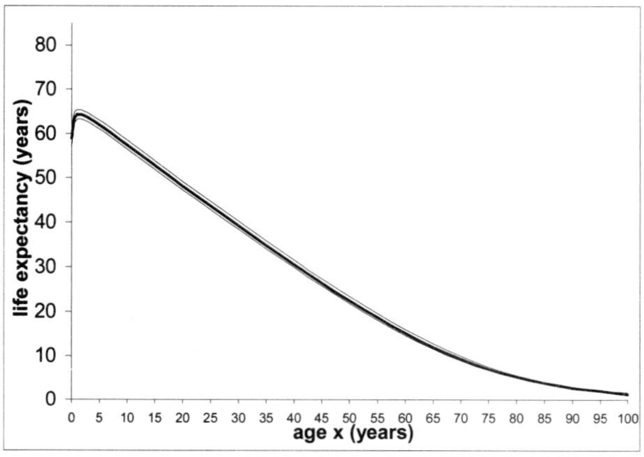

Age distribution of deaths with uncertainty interval

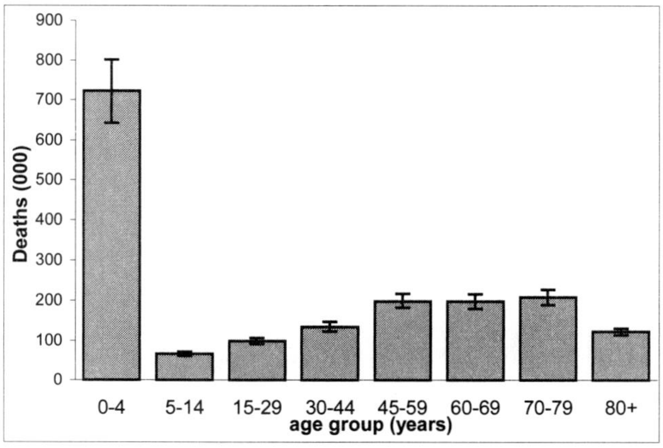

EmrD EmrD EmrD

Females, 2000 *Femmes, 2000* *Mujeres, 2000*

Age	$_nM_x$	$_nq_x$ mean	$_nq_x$ percentiles 2½	$_nq_x$ percentiles 97½	l_x mean	l_x percentiles 2½	l_x percentiles 97½	$_nd_x$	$_nL_x$	e_x mean	e_x percentiles 2½	e_x percentiles 97½	Actual deaths (000)	Actual population (000)
<1	0.1006	0.0922	0.0818	0.1036	100,000	100,000	100,000	9,224	91,698	61.0	59.2	63.5	543	5,399
1-4	0.0078	0.0307	0.0274	0.0337	90,776	89,636	91,823	2,787	356,414	66.2	65.1	67.5	155	19,825
5-9	0.0018	0.0092	0.0082	0.0101	87,989	86,805	89,143	810	437,920	64.3	63.2	65.4	41	22,129
10-14	0.0013	0.0066	0.0059	0.0072	87,179	85,938	88,357	574	434,459	59.8	58.8	60.9	27	20,168
15-19	0.0017	0.0085	0.0076	0.0093	86,605	85,324	87,795	736	431,183	55.2	54.2	56.2	30	17,719
20-24	0.0026	0.0127	0.0113	0.0139	85,869	84,544	87,076	1,091	426,617	50.7	49.7	51.7	38	15,041
25-29	0.0032	0.0159	0.0141	0.0174	84,778	83,395	86,008	1,348	420,522	46.3	45.3	47.2	41	12,775
30-34	0.0034	0.0168	0.0149	0.0184	83,430	81,987	84,768	1,402	413,647	42.0	41.0	42.9	37	11,033
35-39	0.0037	0.0186	0.0165	0.0204	82,028	80,538	83,505	1,523	406,334	37.7	36.7	38.5	36	9,494
40-44	0.0044	0.0217	0.0195	0.0237	80,505	78,974	82,129	1,745	398,164	33.3	32.4	34.2	36	8,248
45-49	0.0058	0.0288	0.0262	0.0314	78,760	77,180	80,528	2,270	388,128	29.0	28.1	29.8	40	6,819
50-54	0.0089	0.0436	0.0398	0.0475	76,491	74,850	78,421	3,337	374,112	24.8	23.9	25.6	48	5,356
55-59	0.0134	0.0646	0.0591	0.0707	73,154	71,432	75,304	4,729	353,948	20.8	20.0	21.5	57	4,269
60-64	0.0202	0.0960	0.0880	0.1058	68,425	66,580	70,852	6,567	325,709	17.1	16.3	17.8	72	3,574
65-69	0.0320	0.1483	0.1360	0.1644	61,858	59,715	64,587	9,175	286,356	13.6	12.9	14.3	86	2,699
70-74	0.0512	0.2269	0.2082	0.2512	52,684	50,242	55,672	11,952	233,539	10.6	10.0	11.1	100	1,958
75-79	0.0835	0.3452	0.3192	0.3757	40,732	37,846	43,744	14,062	168,502	8.0	7.5	8.4	89	1,072
80-84	0.1354	0.5057	0.4728	0.5388	26,669	23,629	29,310	13,486	99,632	5.8	5.5	6.2	70	515
85-89	0.2083	0.6848	0.6522	0.7144	13,184	10,898	15,238	9,028	43,348	4.2	4.0	4.5	36	172
90-94	0.3103	0.8035	0.7751	0.8293	4,156	3,112	5,289	3,339	10,761	3.0	2.9	3.2	11	35
95-99	0.4493	0.9568	0.9440	0.9664	817	531	1,189	781	1,739	2.2	2.2	2.3	2	4
100+	0.6381	1	1	1	35	18	66	35	55	1.6	1.6	1.7	0	0
All ages													*1,596*	*168,305*

$\ln(_nq_x)$ with uncertainty interval

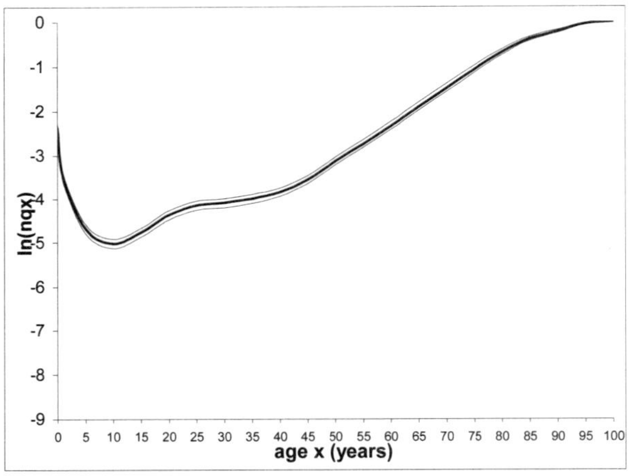

l_x with uncertainty interval

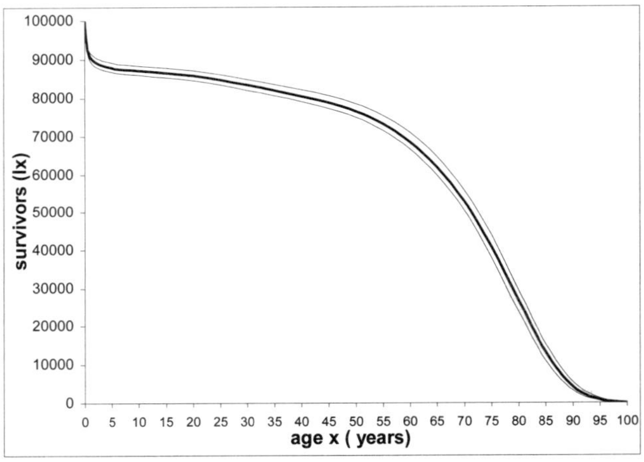

e_x with uncertainty interval

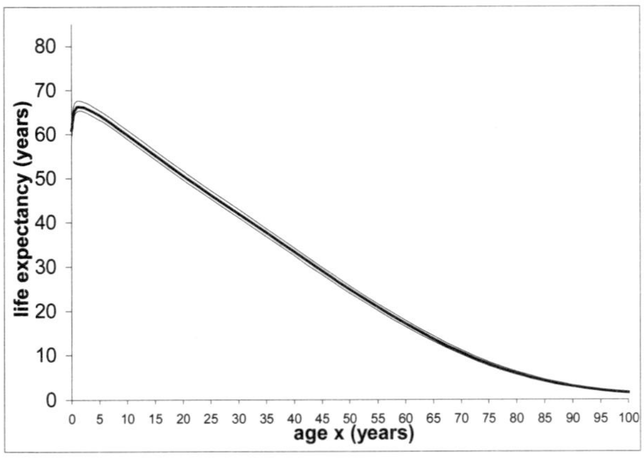

Age distribution of deaths with uncertainty interval

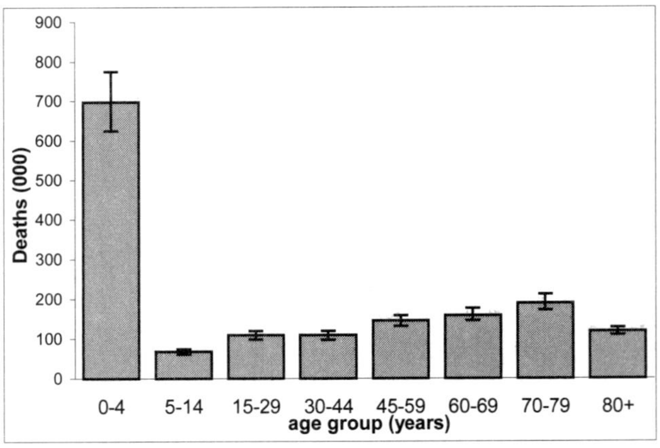

EurA **EurA** **EurA**

Males, 2000 *Hommes, 2000* *Hombres, 2000*

Age	$_nM_x$	$_nq_x$			l_x			$_nd_x$	$_nL_x$	e_x			Actual deaths	Actual population
			percentiles			percentiles					percentiles			
		mean	2½	97½	mean	2½	97½			mean	2½	97½	(000)	(000)
<1	0.0057	0.0057	0.0054	0.0059	100,000	100,000	100,000	567	99,490	74.8	74.5	75.2	12	2,189
1-4	0.0003	0.0013	0.0013	0.0013	99,433	99,409	99,458	129	397,421	74.3	74.1	74.4	3	9,035
5-9	0.0002	0.0008	0.0008	0.0008	99,304	99,276	99,333	79	496,320	70.3	70.2	70.5	2	12,015
10-14	0.0002	0.0009	0.0009	0.0010	99,225	99,194	99,257	92	495,892	65.4	65.3	65.5	2	12,515
15-19	0.0006	0.0032	0.0031	0.0033	99,132	99,099	99,167	315	494,875	60.5	60.3	60.6	8	12,834
20-24	0.0010	0.0049	0.0048	0.0051	98,818	98,774	98,863	487	492,870	55.6	55.5	55.8	13	13,444
25-29	0.0010	0.0051	0.0049	0.0052	98,330	98,272	98,389	499	490,404	50.9	50.8	51.0	15	14,912
30-34	0.0012	0.0060	0.0059	0.0062	97,831	97,761	97,902	588	487,687	46.2	46.0	46.3	20	16,449
35-39	0.0016	0.0079	0.0077	0.0080	97,243	97,159	97,328	764	484,307	41.4	41.3	41.5	26	16,579
40-44	0.0024	0.0121	0.0118	0.0123	96,480	96,381	96,581	1,163	479,491	36.7	36.6	36.8	36	15,043
45-49	0.0038	0.0187	0.0183	0.0191	95,317	95,201	95,439	1,782	472,128	32.1	32.0	32.3	53	13,956
50-54	0.0057	0.0283	0.0278	0.0288	93,534	93,382	93,680	2,643	461,064	27.7	27.6	27.8	77	13,504
55-59	0.0090	0.0442	0.0435	0.0449	90,891	90,695	91,074	4,017	444,413	23.4	23.3	23.6	103	11,408
60-64	0.0148	0.0712	0.0702	0.0722	86,874	86,624	87,110	6,188	418,901	19.4	19.3	19.5	158	10,711
65-69	0.0241	0.1135	0.1118	0.1151	80,686	80,375	80,985	9,155	380,544	15.7	15.6	15.8	220	9,157
70-74	0.0384	0.1751	0.1724	0.1779	71,531	71,165	71,891	12,524	326,345	12.4	12.3	12.5	291	7,595
75-79	0.0612	0.2655	0.2617	0.2698	59,007	58,593	59,432	15,667	255,868	9.5	9.4	9.6	335	5,476
80-84	0.1028	0.4088	0.4034	0.4144	43,340	42,806	43,842	17,717	172,410	7.0	7.0	7.1	268	2,606
85-89	0.1632	0.5796	0.5738	0.5852	25,624	25,067	26,114	14,852	90,988	5.2	5.1	5.2	249	1,526
90-94	0.2337	0.6869	0.6819	0.6915	10,772	10,403	11,113	7,399	31,661	3.8	3.8	3.9	113	482
95-99	0.3349	0.8353	0.8318	0.8386	3,373	3,213	3,535	2,817	8,412	2.8	2.8	2.9	27	81
100+	0.4743	1	1	1	556	520	595	556	1,171	2.1	2.1	2.2	3	7
All ages													*2,037*	*201,524*

$ln(_nq_x)$ with uncertainty interval

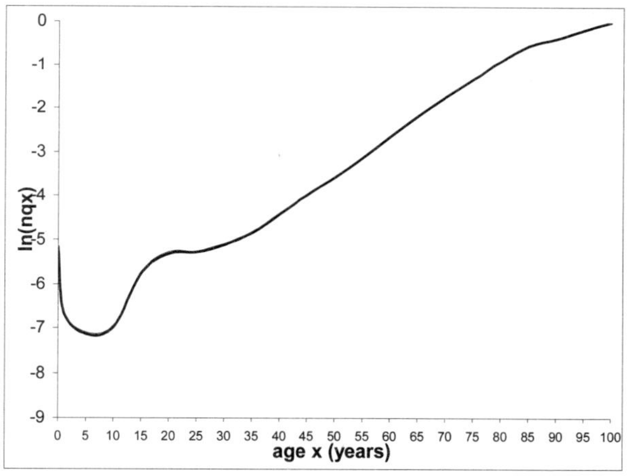

l_x with uncertainty interval

e_x with uncertainty interval

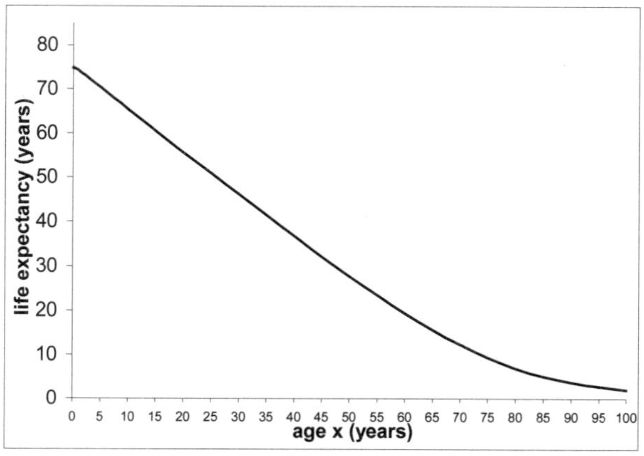

Age distribution of deaths with uncertainty interval

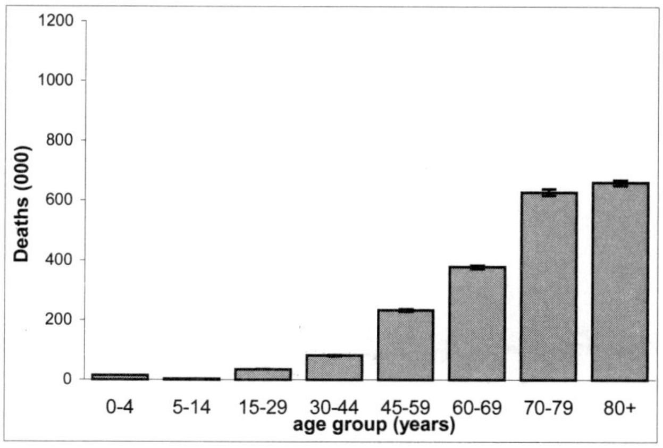

EurA | EurA | EurA

Females, 2000 | *Femmes, 2000* | *Mujeres, 2000*

Age	$_nM_x$	$_nq_x$			l_x			$_nd_x$	$_nL_x$	e_x			Actual deaths	Actual population
			percentiles			percentiles					percentiles			
		mean	2½	97½	mean	2½	97½			mean	2½	97½	(000)	(000)
<1	0.0045	0.0045	0.0043	0.0047	100,000	100,000	100,000	448	99,596	81.2	80.8	81.6	9	2,080
1-4	0.0003	0.0010	0.0010	0.0011	99,552	99,534	99,571	104	397,957	80.5	80.4	80.7	2	8,550
5-9	0.0001	0.0006	0.0005	0.0006	99,448	99,426	99,470	55	497,099	76.6	76.5	76.8	1	11,406
10-14	0.0001	0.0006	0.0006	0.0006	99,392	99,369	99,417	59	496,812	71.7	71.5	71.8	1	11,899
15-19	0.0003	0.0013	0.0013	0.0014	99,333	99,308	99,359	131	496,336	66.7	66.6	66.8	3	12,217
20-24	0.0003	0.0016	0.0015	0.0016	99,202	99,173	99,232	154	495,623	61.8	61.6	61.9	4	12,831
25-29	0.0004	0.0018	0.0017	0.0018	99,047	99,015	99,082	175	494,799	56.9	56.7	57.0	5	14,366
30-34	0.0005	0.0024	0.0024	0.0025	98,872	98,835	98,912	241	493,758	52.0	51.8	52.1	8	15,863
35-39	0.0008	0.0038	0.0037	0.0038	98,631	98,588	98,677	370	492,230	47.1	47.0	47.2	12	16,128
40-44	0.0012	0.0060	0.0058	0.0061	98,261	98,209	98,315	588	489,833	42.3	42.1	42.4	18	14,823
45-49	0.0019	0.0093	0.0091	0.0095	97,673	97,609	97,740	906	486,097	37.5	37.4	37.6	26	13,920
50-54	0.0029	0.0143	0.0140	0.0146	96,766	96,687	96,853	1,382	480,377	32.8	32.7	33.0	39	13,553
55-59	0.0042	0.0210	0.0206	0.0214	95,385	95,283	95,496	2,004	471,911	28.3	28.1	28.4	50	11,682
60-64	0.0068	0.0334	0.0328	0.0341	93,380	93,249	93,526	3,123	459,092	23.8	23.7	23.9	78	11,473
65-69	0.0113	0.0548	0.0538	0.0559	90,257	90,067	90,459	4,947	438,918	19.6	19.5	19.7	119	10,531
70-74	0.0195	0.0932	0.0916	0.0949	85,310	85,034	85,587	7,949	406,679	15.6	15.4	15.7	193	9,884
75-79	0.0351	0.1615	0.1591	0.1642	77,361	76,967	77,731	12,495	355,570	11.9	11.8	12.0	306	8,714
80-84	0.0673	0.2879	0.2838	0.2927	64,867	64,336	65,341	18,676	277,643	8.7	8.6	8.8	333	4,947
85-89	0.1202	0.4621	0.4560	0.4687	46,190	45,598	46,729	21,344	177,592	6.2	6.2	6.3	442	3,680
90-94	0.1909	0.6070	0.6013	0.6137	24,846	24,273	25,387	15,082	78,985	4.4	4.4	4.5	284	1,488
95-99	0.2953	0.7829	0.7785	0.7880	9,764	9,377	10,118	7,644	25,887	3.1	3.1	3.2	94	319
100+	0.4506	1	1	1	2,120	1,988	2,237	2,120	4,705	2.2	2.2	2.4	15	33
All ages													*2,043*	*210,386*

ln($_nq_x$) with uncertainty interval

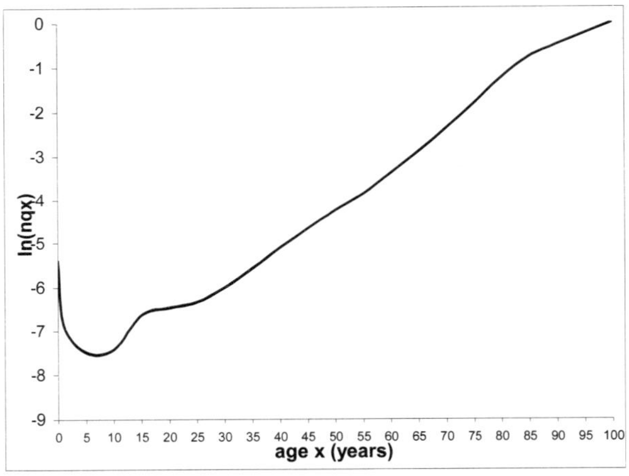

l_x with uncertainty interval

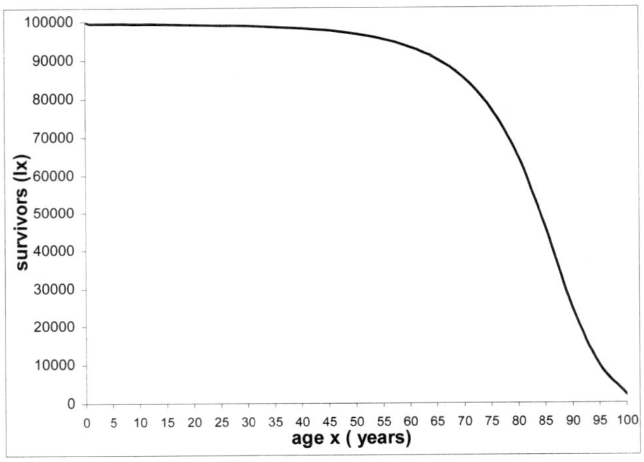

e_x with uncertainty interval

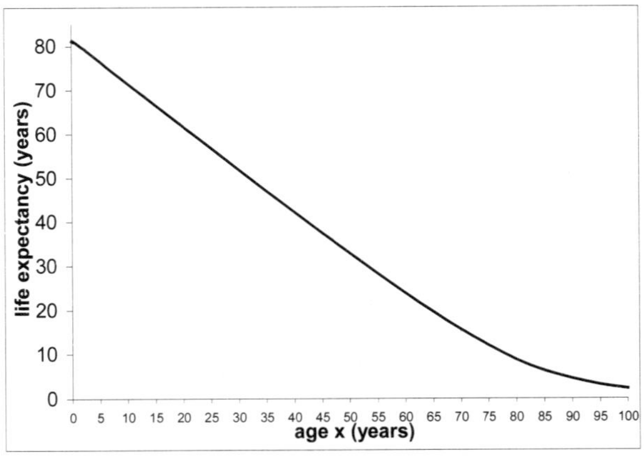

Age distribution of deaths with uncertainty interval

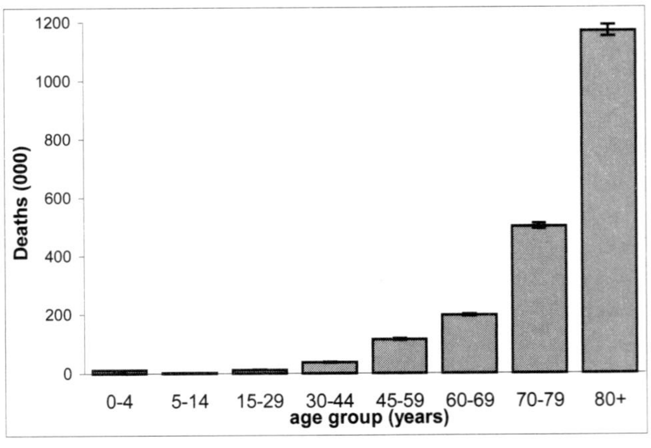

EurB **EurB** **EurB**

Males, 2000 *Hommes, 2000* *Hombres, 2000*

Age	$_nM_x$	$_nq_x$			l_x			$_nd_x$	$_nL_x$	e_x			Actual deaths	Actual population
			percentiles			percentiles					percentiles			
		mean	2½	97½	mean	2½	97½			mean	2½	97½	(000)	(000)
<1	0.0385	0.0372	0.0349	0.0397	100,000	100,000	100,000	3,719	96,653	65.5	65.0	66.5	72	1,873
1-4	0.0029	0.0114	0.0108	0.0119	96,281	96,029	96,512	1,094	382,497	67.1	66.8	67.4	21	7,273
5-9	0.0008	0.0038	0.0036	0.0039	95,186	94,895	95,464	359	475,036	63.8	63.6	64.1	8	9,982
10-14	0.0007	0.0035	0.0033	0.0036	94,828	94,524	95,121	328	473,320	59.1	58.8	59.4	7	10,101
15-19	0.0011	0.0057	0.0055	0.0060	94,500	94,186	94,808	542	471,147	54.3	54.0	54.6	12	10,372
20-24	0.0016	0.0081	0.0078	0.0084	93,959	93,629	94,287	764	467,883	49.6	49.3	49.9	16	10,003
25-29	0.0020	0.0097	0.0093	0.0101	93,195	92,847	93,550	906	463,709	44.9	44.7	45.2	18	9,001
30-34	0.0025	0.0123	0.0118	0.0127	92,289	91,919	92,676	1,135	458,605	40.4	40.1	40.7	20	7,922
35-39	0.0035	0.0175	0.0168	0.0180	91,153	90,759	91,578	1,591	451,789	35.8	35.6	36.1	27	7,567
40-44	0.0054	0.0267	0.0258	0.0274	89,563	89,142	90,034	2,392	441,833	31.4	31.2	31.7	40	7,376
45-49	0.0081	0.0395	0.0383	0.0404	87,171	86,719	87,703	3,444	427,243	27.2	27.0	27.5	52	6,486
50-54	0.0121	0.0586	0.0570	0.0600	83,726	83,240	84,333	4,906	406,368	23.2	23.0	23.5	60	5,000
55-59	0.0179	0.0856	0.0831	0.0877	78,821	78,272	79,518	6,746	377,238	19.5	19.3	19.8	65	3,609
60-64	0.0268	0.1255	0.1218	0.1283	72,074	71,444	72,879	9,045	337,760	16.1	15.9	16.4	100	3,731
65-69	0.0384	0.1752	0.1698	0.1794	63,030	62,345	63,899	11,041	287,546	13.1	12.9	13.3	124	3,240
70-74	0.0567	0.2484	0.2413	0.2544	51,989	51,191	52,898	12,913	227,661	10.3	10.2	10.5	139	2,442
75-79	0.0856	0.3527	0.3438	0.3606	39,076	38,247	40,101	13,782	160,923	7.9	7.8	8.0	114	1,334
80-84	0.1364	0.5085	0.4985	0.5171	25,294	24,558	26,316	12,863	94,311	5.8	5.7	5.9	66	487
85-89	0.2087	0.6858	0.6772	0.6935	12,431	11,859	13,197	8,525	40,841	4.3	4.2	4.3	60	286
90-94	0.2978	0.7864	0.7802	0.7921	3,906	3,635	4,259	3,071	10,314	3.2	3.1	3.2	26	87
95-99	0.4083	0.9176	0.9144	0.9208	834	756	936	766	1,875	2.4	2.4	2.4	6	15
100+	0.5332	1	1	1	69	61	80	69	129	1.9	1.9	1.9	1	2
All ages													*1,053*	*108,189*

ln($_nq_x$) with uncertainty interval

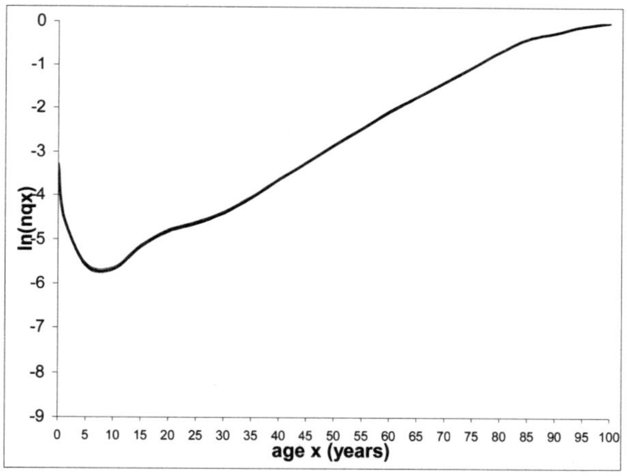

l_x with uncertainty interval

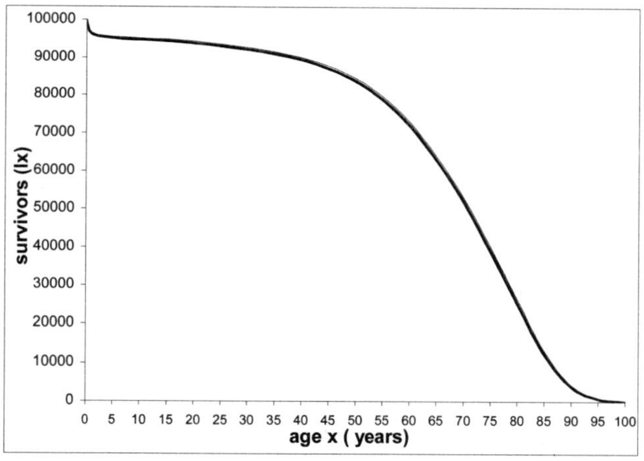

e_x with uncertainty interval

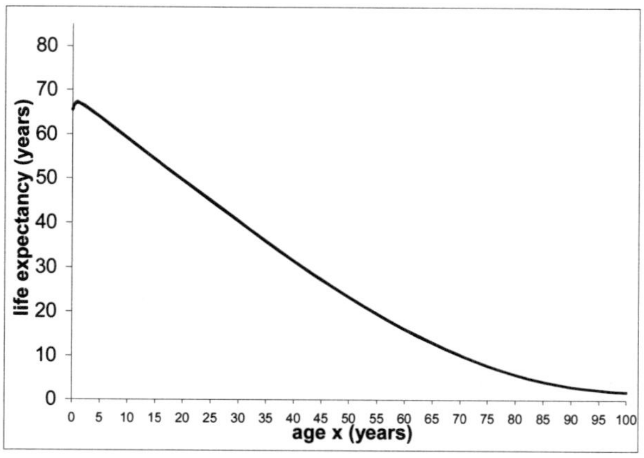

Age distribution of deaths with uncertainty interval

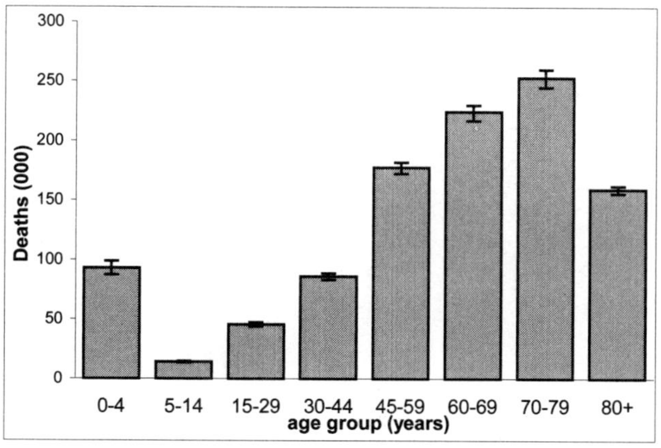

EurB **EurB** **EurB**

Females, 2000 *Femmes, 2000* *Mujeres, 2000*

Age	$_nM_x$	$_nq_x$			l_x			$_nd_x$	$_nL_x$	e_x			Actual deaths	Actual population
			percentiles			percentiles					percentiles			
		mean	2½	97½	mean	2½	97½			mean	2½	97½	(000)	(000)
<1	0.0331	0.0322	0.0301	0.0342	100,000	100,000	100,000	3,217	97,105	72.3	71.2	73.1	59	1,779
1-4	0.0026	0.0105	0.0100	0.0111	96,783	96,577	96,986	1,019	384,686	73.7	73.3	74.0	19	6,993
5-9	0.0005	0.0027	0.0026	0.0028	95,764	95,508	96,014	259	478,173	70.4	70.1	70.8	5	9,585
10-14	0.0004	0.0022	0.0021	0.0023	95,505	95,237	95,768	207	477,008	65.6	65.3	66.0	4	9,674
15-19	0.0006	0.0031	0.0029	0.0032	95,298	95,021	95,569	293	475,758	60.8	60.4	61.1	6	9,953
20-24	0.0008	0.0039	0.0037	0.0040	95,005	94,717	95,289	367	474,108	55.9	55.6	56.3	7	9,570
25-29	0.0009	0.0046	0.0044	0.0048	94,638	94,335	94,936	435	472,104	51.2	50.8	51.5	8	8,716
30-34	0.0012	0.0058	0.0056	0.0061	94,203	93,883	94,517	547	469,649	46.4	46.1	46.7	9	7,787
35-39	0.0017	0.0084	0.0081	0.0087	93,656	93,315	93,989	786	466,315	41.6	41.3	42.0	13	7,600
40-44	0.0025	0.0123	0.0118	0.0128	92,870	92,502	93,228	1,141	461,498	37.0	36.7	37.3	18	7,379
45-49	0.0037	0.0183	0.0177	0.0191	91,729	91,327	92,120	1,679	454,448	32.4	32.1	32.7	24	6,600
50-54	0.0055	0.0271	0.0261	0.0282	90,050	89,605	90,480	2,437	444,156	28.0	27.7	28.3	29	5,214
55-59	0.0086	0.0421	0.0405	0.0439	87,613	87,116	88,093	3,688	428,844	23.7	23.4	24.0	34	3,901
60-64	0.0138	0.0667	0.0642	0.0693	83,925	83,329	84,475	5,599	405,628	19.6	19.4	19.9	60	4,355
65-69	0.0221	0.1046	0.1008	0.1084	78,326	77,554	79,013	8,194	371,146	15.8	15.6	16.1	86	3,894
70-74	0.0364	0.1667	0.1610	0.1721	70,132	69,147	70,972	11,694	321,428	12.4	12.2	12.6	122	3,342
75-79	0.0617	0.2674	0.2588	0.2763	58,439	57,261	59,469	15,629	253,122	9.3	9.2	9.5	135	2,192
80-84	0.1052	0.4164	0.4043	0.4277	42,810	41,592	44,007	17,827	169,481	6.8	6.7	7.0	99	937
85-89	0.1716	0.6005	0.5874	0.6121	24,983	23,885	26,191	15,001	87,410	4.9	4.8	5.1	100	580
90-94	0.2540	0.7208	0.7096	0.7305	9,982	9,268	10,772	7,195	28,324	3.6	3.6	3.7	49	194
95-99	0.3445	0.8470	0.8412	0.8531	2,787	2,516	3,127	2,361	6,853	2.8	2.8	2.9	12	35
100+	0.4206	1	1	1	426	371	495	426	1,014	2.4	2.3	2.4	2	5
All ages													*900*	*110,284*

$ln(_nq_x)$ with uncertainty interval

l_x with uncertainty interval

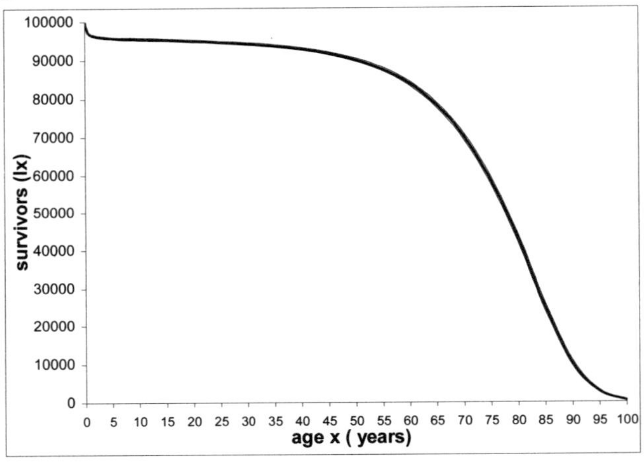

e_x with uncertainty interval

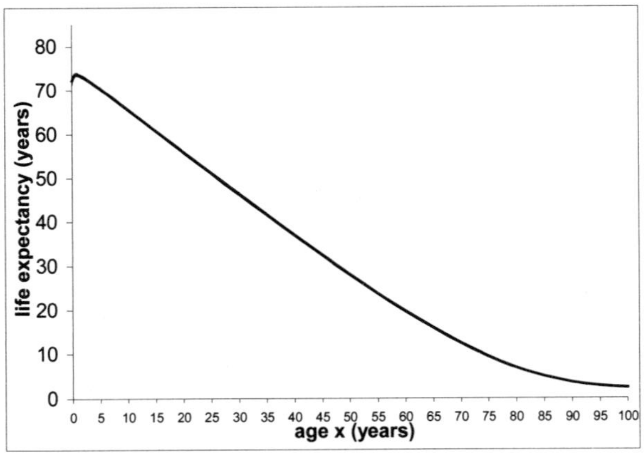

Age distribution of deaths with uncertainty interval

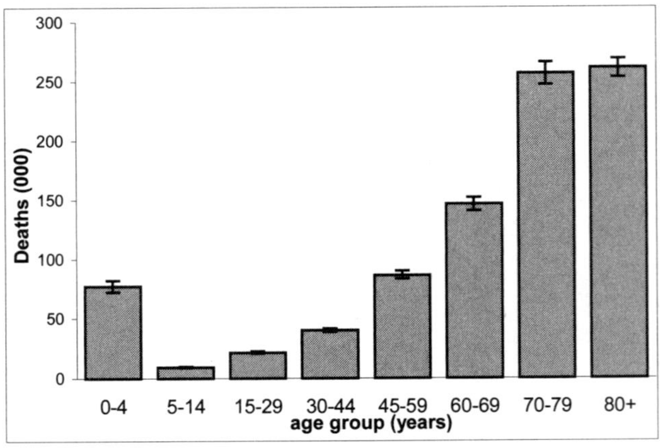

EurC

Males, 2000

EurC

Hommes, 2000

EurC

Hombres, 2000

Age	$_nM_x$	$_nq_x$ mean	$_nq_x$ percentiles 2½	$_nq_x$ percentiles 97½	l_x mean	l_x percentiles 2½	l_x percentiles 97½	$_nd_x$	$_nL_x$	e_x mean	e_x percentiles 2½	e_x percentiles 97½	Actual deaths (000)	Actual population (000)
<1	0.0230	0.0225	0.0215	0.0236	100,000	100,000	100,000	2,251	97,974	60.3	59.5	61.4	27	1,174
1-4	0.0014	0.0055	0.0052	0.0057	97,749	97,640	97,850	534	389,711	60.7	60.0	61.4	6	4,649
5-9	0.0007	0.0033	0.0031	0.0034	97,214	97,089	97,335	319	485,272	57.0	56.3	57.7	5	7,327
10-14	0.0007	0.0034	0.0032	0.0035	96,895	96,759	97,027	325	483,662	52.2	51.5	52.9	7	9,999
15-19	0.0019	0.0094	0.0089	0.0099	96,570	96,422	96,714	903	480,592	47.4	46.6	48.1	19	9,961
20-24	0.0040	0.0197	0.0186	0.0209	95,667	95,488	95,843	1,885	473,622	42.8	42.1	43.5	37	9,243
25-29	0.0047	0.0232	0.0220	0.0246	93,782	93,500	94,039	2,179	463,463	38.6	37.9	39.3	42	8,859
30-34	0.0059	0.0293	0.0276	0.0311	91,603	91,203	91,965	2,681	451,313	34.5	33.8	35.1	48	8,099
35-39	0.0079	0.0385	0.0364	0.0410	88,922	88,379	89,423	3,428	436,040	30.4	29.8	31.1	74	9,385
40-44	0.0109	0.0531	0.0501	0.0566	85,494	84,758	86,170	4,542	416,115	26.5	26.0	27.2	107	9,783
45-49	0.0154	0.0741	0.0699	0.0788	80,952	79,968	81,852	6,001	389,756	22.9	22.4	23.5	134	8,711
50-54	0.0213	0.1009	0.0950	0.1070	74,951	73,674	76,121	7,562	355,847	19.5	19.0	20.0	141	6,617
55-59	0.0299	0.1393	0.1323	0.1463	67,388	65,807	68,845	9,386	313,476	16.4	16.0	16.9	131	4,388
60-64	0.0412	0.1867	0.1769	0.1962	58,002	56,233	59,669	10,827	262,943	13.7	13.3	14.1	249	6,052
65-69	0.0570	0.2494	0.2377	0.2611	47,175	45,243	49,018	11,765	206,464	11.2	10.9	11.6	220	3,868
70-74	0.0767	0.3218	0.3089	0.3347	35,410	33,494	37,367	11,394	148,567	9.1	8.9	9.4	264	3,443
75-79	0.1007	0.4023	0.3912	0.4145	24,016	22,367	25,823	9,662	95,927	7.3	7.1	7.5	145	1,438
80-84	0.1495	0.5442	0.5329	0.5565	14,355	13,126	15,722	7,812	52,243	5.5	5.4	5.6	88	592
85-89	0.2173	0.7041	0.6947	0.7141	6,543	5,822	7,343	4,606	21,197	4.2	4.1	4.2	77	353
90-94	0.2999	0.7894	0.7818	0.7979	1,936	1,664	2,242	1,528	5,096	3.1	3.1	3.2	29	96
95-99	0.4179	0.9271	0.9226	0.9320	408	337	489	378	905	2.4	2.3	2.4	7	16
100+	0.5373	1	1	1	30	23	38	30	55	1.9	1.8	1.9	1	2
All ages													*1,857*	*114,055*

ln($_nq_x$) with uncertainty interval

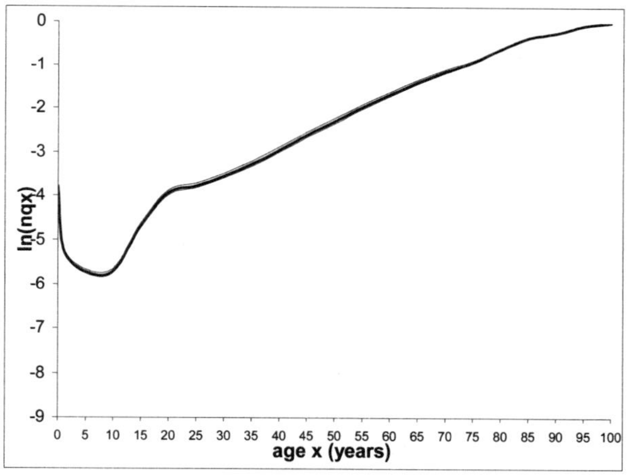

l_x with uncertainty interval

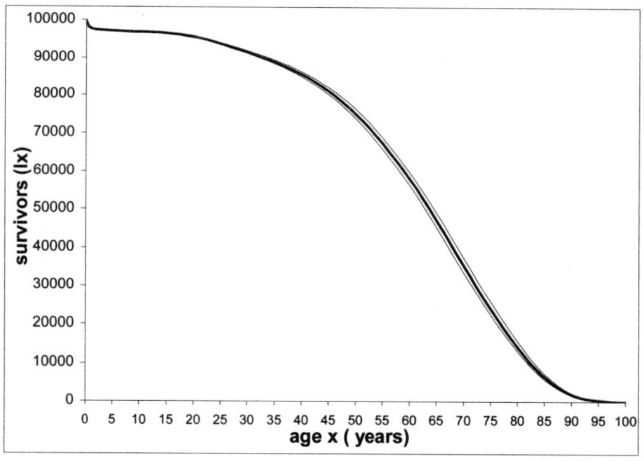

e_x with uncertainty interval

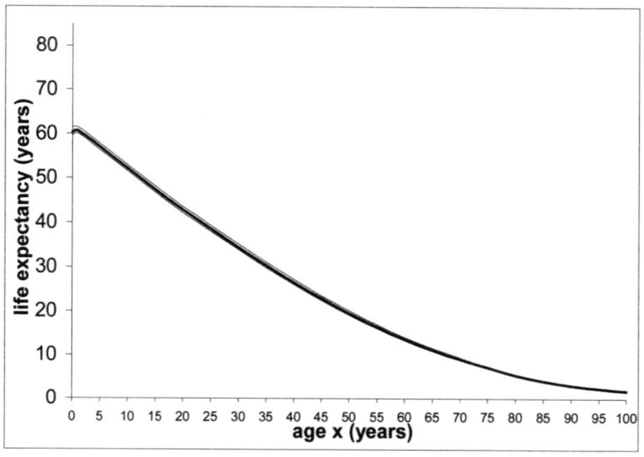

Age distribution of deaths with uncertainty interval

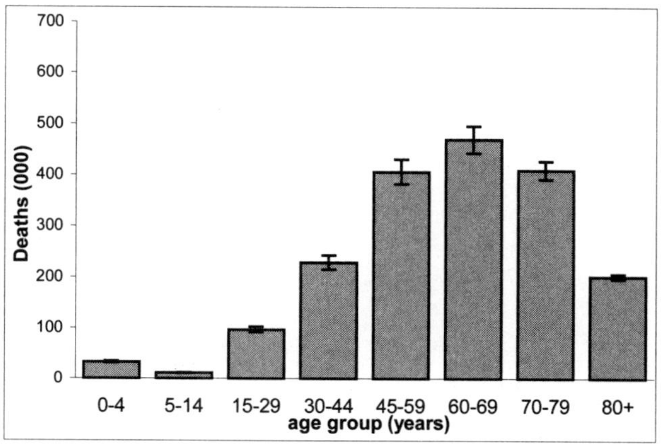

EurC **EurC** **EurC**

Females, 2000 *Femmes, 2000* *Mujeres, 2000*

Age	$_nM_x$	$_nq_x$			l_x			$_nd_x$	$_nL_x$	e_x			Actual deaths	Actual population
			percentiles			percentiles					percentiles			
		mean	2½	97½	mean	2½	97½			mean	2½	97½	(000)	(000)
<1	0.0166	0.0163	0.0151	0.0175	100,000	100,000	100,000	1,635	98,529	72.1	71.6	73.0	18	1,111
1-4	0.0011	0.0043	0.0040	0.0046	98,365	98,253	98,487	423	392,446	72.3	71.8	72.8	5	4,447
5-9	0.0004	0.0020	0.0019	0.0021	97,942	97,801	98,086	195	489,224	68.6	68.1	69.1	3	6,976
10-14	0.0003	0.0017	0.0016	0.0018	97,747	97,594	97,901	167	488,321	63.7	63.3	64.2	3	9,617
15-19	0.0008	0.0038	0.0036	0.0041	97,581	97,416	97,743	371	486,976	58.8	58.4	59.3	7	9,681
20-24	0.0010	0.0052	0.0049	0.0055	97,209	97,020	97,388	503	484,791	54.1	53.6	54.5	9	9,081
25-29	0.0013	0.0063	0.0059	0.0067	96,707	96,485	96,907	607	482,017	49.3	48.9	49.8	11	8,585
30-34	0.0016	0.0080	0.0076	0.0085	96,100	95,841	96,328	767	478,582	44.6	44.2	45.1	13	7,952
35-39	0.0022	0.0108	0.0103	0.0115	95,333	95,030	95,600	1,033	474,083	40.0	39.5	40.4	21	9,593
40-44	0.0032	0.0157	0.0150	0.0166	94,300	93,941	94,617	1,484	467,790	35.4	34.9	35.8	33	10,306
45-49	0.0048	0.0237	0.0225	0.0249	92,816	92,383	93,197	2,201	458,576	30.9	30.5	31.4	46	9,550
50-54	0.0071	0.0351	0.0334	0.0369	90,615	90,084	91,082	3,181	445,121	26.6	26.2	27.0	55	7,647
55-59	0.0107	0.0523	0.0499	0.0548	87,434	86,783	88,019	4,569	425,746	22.5	22.1	22.9	60	5,584
60-64	0.0159	0.0763	0.0726	0.0803	82,865	82,090	83,628	6,324	398,512	18.6	18.2	19.0	134	8,418
65-69	0.0258	0.1210	0.1147	0.1271	76,540	75,558	77,541	9,262	359,546	14.9	14.6	15.2	156	6,041
70-74	0.0417	0.1887	0.1790	0.1977	67,278	65,967	68,560	12,697	304,646	11.6	11.3	11.9	276	6,615
75-79	0.0689	0.2940	0.2809	0.3060	54,581	52,977	56,210	16,046	232,789	8.7	8.5	9.0	282	4,090
80-84	0.1195	0.4601	0.4440	0.4751	38,535	36,858	40,421	17,730	148,349	6.3	6.1	6.5	235	1,964
85-89	0.1930	0.6509	0.6369	0.6639	20,805	19,424	22,474	13,542	70,169	4.5	4.4	4.6	267	1,384
90-94	0.2786	0.7588	0.7481	0.7687	7,263	6,529	8,157	5,511	19,781	3.3	3.3	3.4	116	416
95-99	0.3851	0.8934	0.8861	0.8995	1,752	1,510	2,048	1,565	4,063	2.5	2.5	2.6	28	74
100+	0.5146	1	1	1	187	152	233	187	363	1.9	1.9	2.1	2	5
All ages													*1,779*	*129,137*

ln($_nq_x$) with uncertainty interval

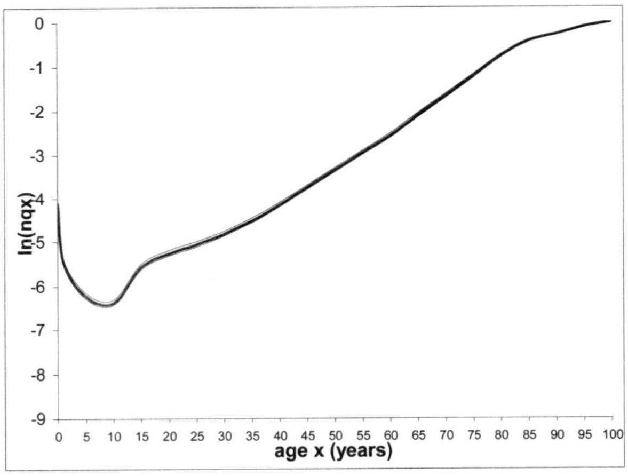

l_x with uncertainty interval

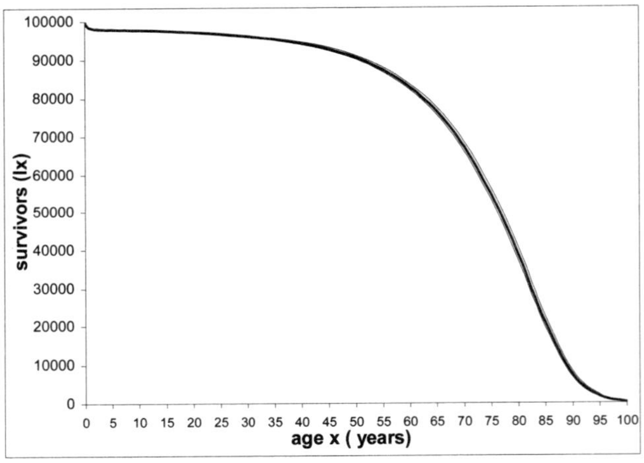

e_x with uncertainty interval

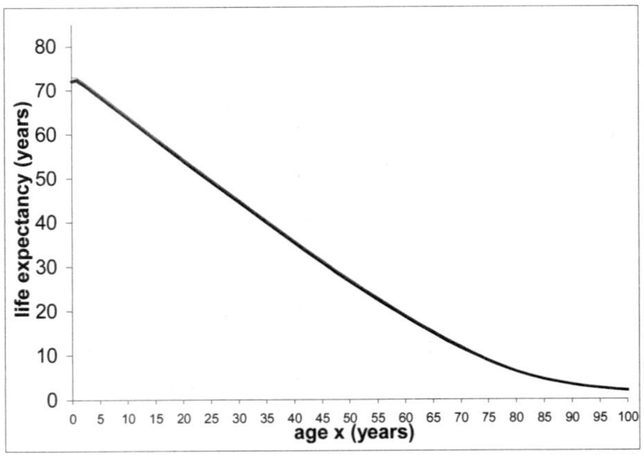

Age distribution of deaths with uncertainty interval

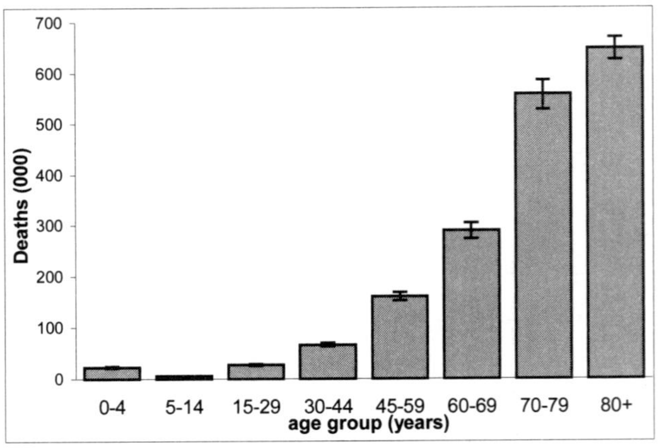

SearB **SearB** **SearB**

Males, 2000 *Hommes, 2000* *Hombres, 2000*

Age	$_nM_x$	$_nq_x$			l_x			$_nd_x$	$_nL_x$	e_x			Actual deaths	Actual population
			percentiles			percentiles					percentiles			
		mean	2 1/2	97 1/2	mean	2 1/2	97 1/2			mean	2 1/2	97 1/2	(000)	(000)
<1	0.0451	0.0433	0.0386	0.0484	100,000	100,000	100,000	4,333	96,101	64.1	63.0	65.3	136	3,024
1-4	0.0029	0.0115	0.0105	0.0125	95,667	95,156	96,141	1,096	380,039	66.0	65.3	66.7	34	11,792
5-9	0.0012	0.0058	0.0054	0.0063	94,571	93,998	95,130	551	471,480	62.8	62.1	63.4	17	14,607
10-14	0.0010	0.0049	0.0045	0.0052	94,021	93,420	94,617	457	468,961	58.1	57.4	58.8	14	14,745
15-19	0.0020	0.0098	0.0091	0.0105	93,564	92,944	94,189	917	465,527	53.4	52.7	54.0	29	14,809
20-24	0.0029	0.0145	0.0136	0.0154	92,647	91,988	93,325	1,345	459,872	48.9	48.2	49.5	42	14,335
25-29	0.0038	0.0190	0.0179	0.0202	91,302	90,574	92,049	1,736	452,168	44.6	43.9	45.2	51	13,298
30-34	0.0042	0.0207	0.0195	0.0220	89,565	88,754	90,383	1,853	443,195	40.4	39.8	41.0	52	12,411
35-39	0.0046	0.0226	0.0213	0.0240	87,713	86,819	88,595	1,982	433,607	36.2	35.6	36.8	49	10,654
40-44	0.0055	0.0274	0.0258	0.0290	85,730	84,756	86,680	2,346	422,786	32.0	31.4	32.5	51	9,273
45-49	0.0075	0.0369	0.0347	0.0391	83,384	82,317	84,442	3,081	409,219	27.8	27.2	28.3	55	7,353
50-54	0.0106	0.0517	0.0486	0.0548	80,303	79,125	81,494	4,155	391,129	23.8	23.2	24.3	59	5,593
55-59	0.0157	0.0754	0.0709	0.0799	76,148	74,849	77,493	5,743	366,385	19.9	19.4	20.4	73	4,677
60-64	0.0242	0.1143	0.1076	0.1210	70,406	68,969	71,928	8,048	331,909	16.3	15.9	16.8	96	3,945
65-69	0.0368	0.1683	0.1588	0.1784	62,358	60,757	64,145	10,498	285,546	13.1	12.7	13.5	107	2,921
70-74	0.0572	0.2502	0.2367	0.2649	51,860	50,096	53,835	12,975	226,865	10.3	9.9	10.6	113	1,974
75-79	0.0860	0.3538	0.3355	0.3727	38,886	36,924	40,910	13,756	160,038	7.9	7.6	8.2	94	1,095
80-84	0.1361	0.5078	0.4876	0.5276	25,129	23,301	27,036	12,760	93,747	5.8	5.6	6.0	66	482
85-89	0.2088	0.6860	0.6679	0.7027	12,369	11,035	13,711	8,485	40,634	4.2	4.1	4.4	33	156
90-94	0.3093	0.8022	0.7882	0.8147	3,884	3,281	4,547	3,116	10,074	3.0	3.0	3.1	9	28
95-99	0.4317	0.9405	0.9332	0.9466	768	608	963	723	1,674	2.3	2.3	2.3	1	3
100+	0.6032	1	1	1	46	32	64	46	76	1.7	1.6	1.8	0	0
All ages													*1,182*	*147,174*

ln($_nq_x$) with uncertainty interval

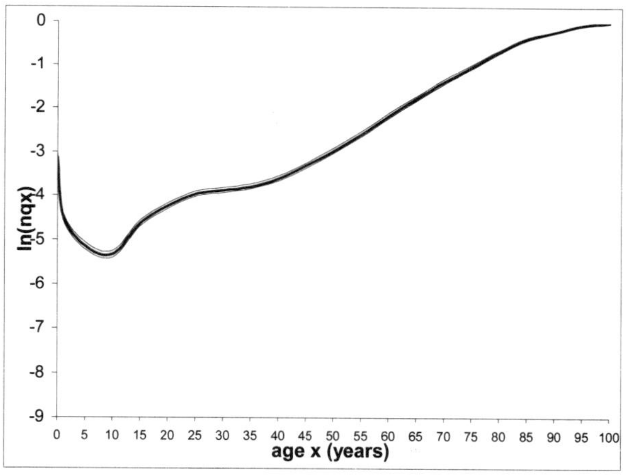

l_x with uncertainty interval

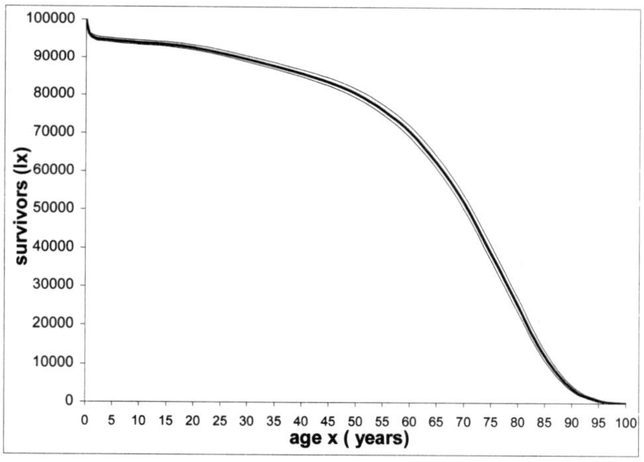

e_x with uncertainty interval

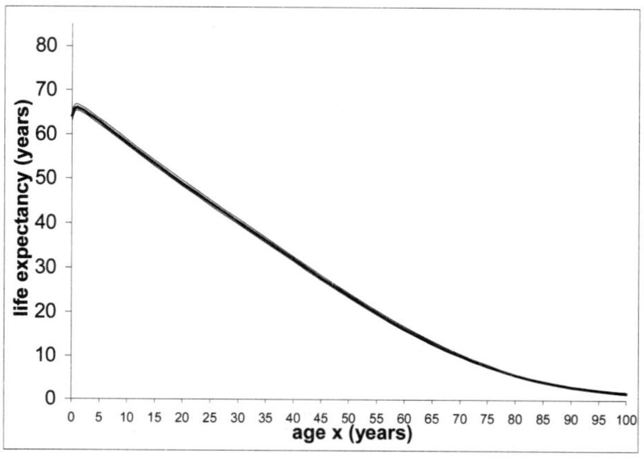

Age distribution of deaths with uncertainty interval

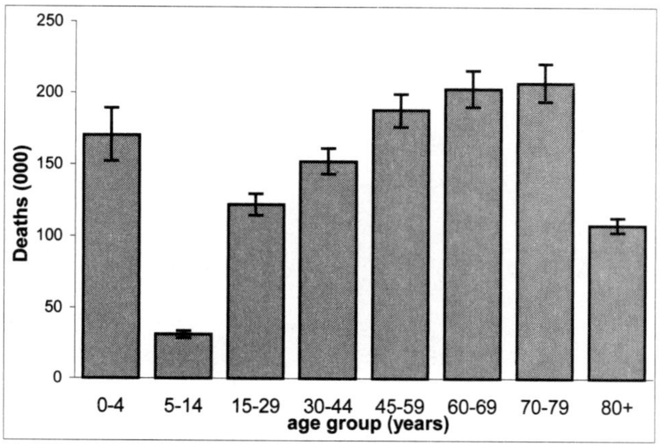

SearB

SearB

SearB

Females, 2000 *Femmes, 2000* *Mujeres, 2000*

Age	$_nM_x$	$_nq_x$			l_x			$_nd_x$	$_nL_x$	e_x			Actual deaths	Actual population
			percentiles			percentiles					percentiles			
		mean	2½	97½	mean	2½	97½			mean	2½	97½	(000)	(000)
<1	0.0348	0.0337	0.0296	0.0377	100,000	100,000	100,000	3,371	96,966	68.9	67.8	70.0	100	2,883
1-4	0.0027	0.0108	0.0097	0.0118	96,629	96,226	97,037	1,040	384,020	70.3	69.5	71.0	31	11,424
5-9	0.0009	0.0046	0.0041	0.0050	95,589	95,096	96,071	437	476,852	67.0	66.3	67.7	13	14,146
10-14	0.0007	0.0037	0.0033	0.0040	95,152	94,623	95,661	348	474,890	62.3	61.6	63.0	10	14,330
15-19	0.0012	0.0060	0.0055	0.0065	94,804	94,247	95,339	570	472,595	57.5	56.8	58.2	17	14,458
20-24	0.0017	0.0084	0.0077	0.0091	94,234	93,632	94,819	793	469,186	52.9	52.2	53.5	24	14,053
25-29	0.0021	0.0106	0.0097	0.0115	93,440	92,783	94,088	991	464,724	48.3	47.6	48.9	28	13,089
30-34	0.0024	0.0120	0.0110	0.0129	92,449	91,730	93,167	1,105	459,483	43.8	43.1	44.4	29	12,211
35-39	0.0030	0.0149	0.0137	0.0160	91,344	90,546	92,142	1,356	453,329	39.3	38.6	39.8	31	10,478
40-44	0.0039	0.0195	0.0180	0.0209	89,988	89,096	90,882	1,752	445,558	34.8	34.2	35.4	36	9,128
45-49	0.0055	0.0270	0.0250	0.0289	88,236	87,231	89,249	2,383	435,222	30.5	29.9	31.0	40	7,272
50-54	0.0079	0.0387	0.0360	0.0413	85,853	84,709	87,016	3,321	420,962	26.3	25.7	26.8	45	5,744
55-59	0.0114	0.0554	0.0518	0.0591	82,532	81,219	83,882	4,568	401,239	22.2	21.7	22.7	57	5,034
60-64	0.0173	0.0830	0.0781	0.0886	77,964	76,459	79,530	6,469	373,647	18.4	17.9	18.8	76	4,369
65-69	0.0269	0.1260	0.1190	0.1343	71,495	69,793	73,290	9,011	334,948	14.8	14.4	15.3	91	3,382
70-74	0.0439	0.1976	0.1861	0.2092	62,484	60,599	64,493	12,349	281,548	11.6	11.2	12.0	104	2,360
75-79	0.0703	0.2990	0.2818	0.3144	50,135	47,919	52,220	14,991	213,197	8.8	8.5	9.1	94	1,334
80-84	0.1142	0.4443	0.4222	0.4652	35,144	32,906	37,127	15,614	136,684	6.5	6.3	6.8	76	665
85-89	0.1803	0.6213	0.5993	0.6428	19,530	17,803	21,367	12,134	67,315	4.7	4.5	4.9	42	233
90-94	0.2784	0.7585	0.7410	0.7760	7,396	6,493	8,562	5,610	20,149	3.3	3.2	3.4	14	49
95-99	0.4038	0.9130	0.9040	0.9224	1,786	1,455	2,218	1,630	4,038	2.4	2.4	2.5	2	6
100+	0.5588	1	1	1	155	113	213	155	278	1.8	1.8	1.9	0	0
All ages													*961*	*146,647*

$\ln(_nq_x)$ with uncertainty interval

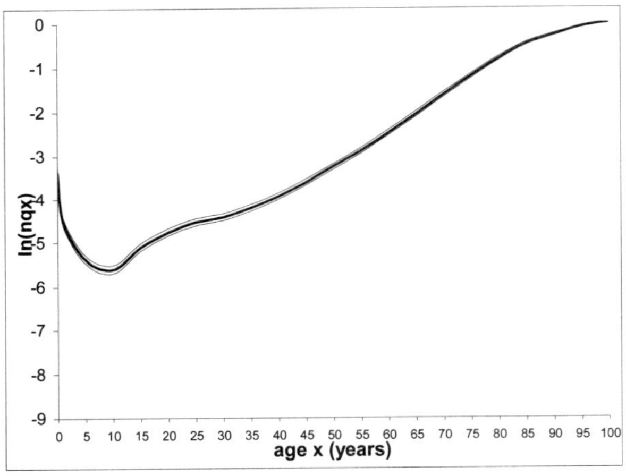

l_x with uncertainty interval

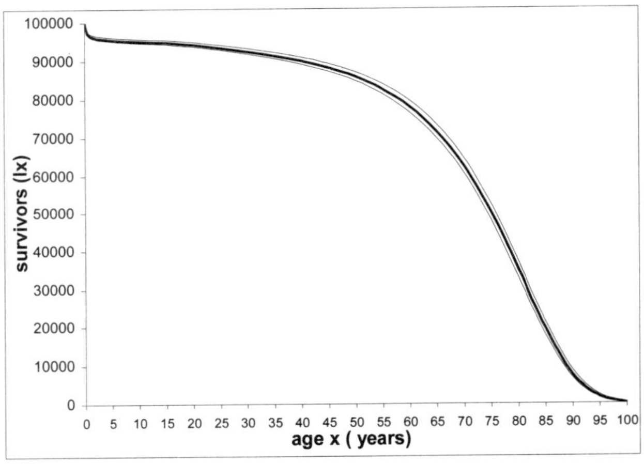

e_x with uncertainty interval

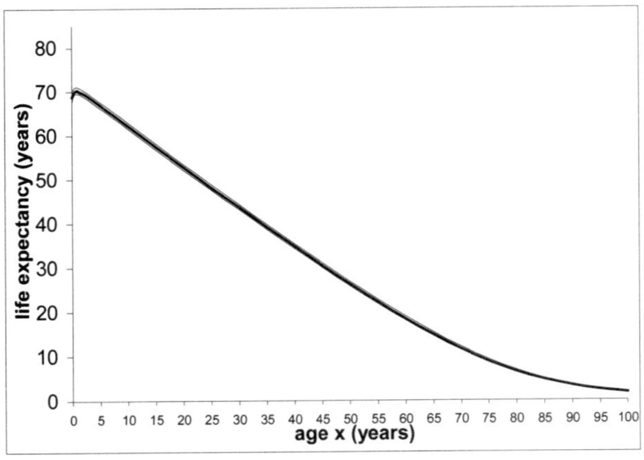

Age distribution of deaths with uncertainty interval

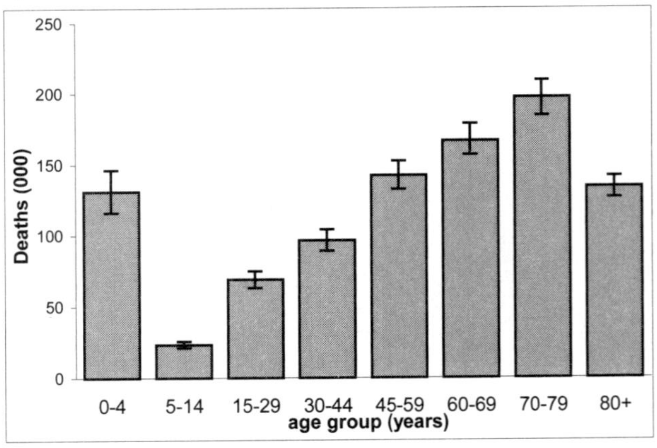

SearD   SearD   SearD

Males, 2000   *Hommes, 2000*   *Hombres, 2000*

Age	$_nM_x$	$_nq_x$			l_x			$_nd_x$	$_nL_x$	e_x			Actual deaths	Actual population
			percentiles			percentiles					percentiles			
		mean	2½	97½	mean	2½	97½			mean	2½	97½	(000)	(000)
<1	0.0659	0.0622	0.0565	0.0678	100,000	100,000	100,000	6,220	94,402	59.7	58.2	61.7	1,023	15,524
1-4	0.0077	0.0303	0.0278	0.0327	93,780	93,225	94,345	2,842	368,300	62.7	61.3	64.2	461	59,804
5-9	0.0019	0.0096	0.0088	0.0104	90,938	90,249	91,624	869	452,519	60.6	59.2	62.1	141	73,334
10-14	0.0012	0.0061	0.0056	0.0066	90,069	89,334	90,784	549	448,973	56.2	54.8	57.7	85	69,764
15-19	0.0015	0.0075	0.0069	0.0082	89,520	88,757	90,267	672	445,920	51.5	50.2	53.0	98	64,992
20-24	0.0024	0.0117	0.0107	0.0128	88,848	88,051	89,646	1,042	441,633	46.9	45.5	48.4	135	57,389
25-29	0.0034	0.0169	0.0154	0.0185	87,806	86,962	88,682	1,486	435,312	42.4	41.1	43.9	184	54,031
30-34	0.0041	0.0201	0.0183	0.0220	86,319	85,376	87,307	1,736	427,256	38.1	36.8	39.6	197	48,550
35-39	0.0046	0.0228	0.0207	0.0250	84,583	83,502	85,701	1,929	418,093	33.8	32.6	35.3	197	42,646
40-44	0.0066	0.0326	0.0295	0.0359	82,654	81,426	83,915	2,696	406,531	29.5	28.3	31.0	243	36,685
45-49	0.0089	0.0435	0.0392	0.0482	79,958	78,523	81,421	3,479	391,094	25.5	24.3	26.8	267	30,021
50-54	0.0143	0.0689	0.0612	0.0770	76,479	74,788	78,193	5,273	369,215	21.5	20.4	22.8	339	23,719
55-59	0.0216	0.1025	0.0899	0.1151	71,207	69,145	73,279	7,295	337,795	17.9	16.9	19.2	414	19,185
60-64	0.0334	0.1540	0.1340	0.1733	63,911	61,322	66,489	9,842	294,951	14.7	13.8	15.8	517	15,506
65-69	0.0465	0.2082	0.1808	0.2332	54,069	50,870	57,252	11,255	242,207	11.9	11.1	12.9	552	11,872
70-74	0.0691	0.2947	0.2582	0.3264	42,814	39,006	46,743	12,619	182,523	9.3	8.7	10.2	560	8,100
75-79	0.1030	0.4097	0.3660	0.4463	30,195	26,294	34,673	12,371	120,050	7.2	6.7	7.8	489	4,741
80-84	0.1526	0.5523	0.5070	0.5887	17,824	14,620	21,983	9,844	64,512	5.5	5.1	5.9	345	2,263
85-89	0.2189	0.7074	0.6685	0.7378	7,980	6,071	10,838	5,645	25,787	4.1	3.9	4.4	169	771
90-94	0.3044	0.7955	0.7644	0.8200	2,335	1,598	3,593	1,857	6,102	3.1	3.0	3.3	51	169
95-99	0.4097	0.9190	0.8989	0.9339	477	288	846	439	1,071	2.4	2.3	2.5	9	22
100+	0.5350	1	1	1	39	19	85	39	72	1.9	1.8	1.9	1	1
All ages													*6,479*	*639,091*

ln($_nq_x$) with uncertainty interval

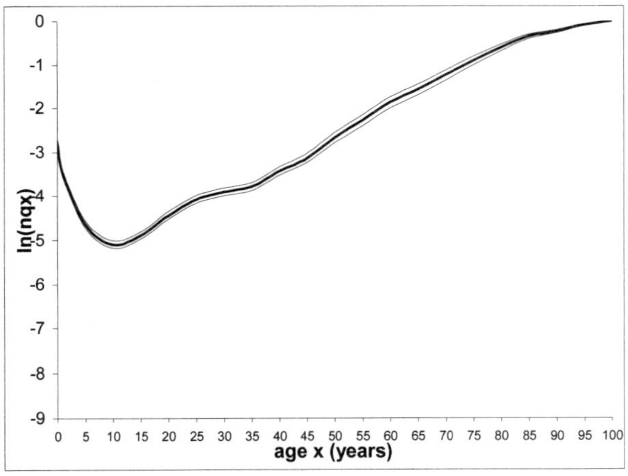

l_x with uncertainty interval

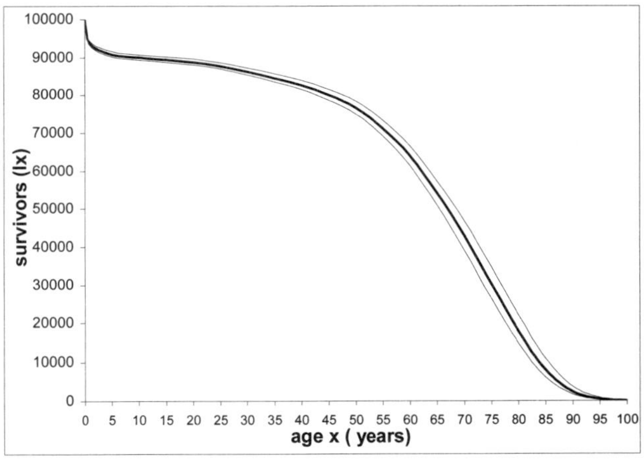

e_x with uncertainty interval

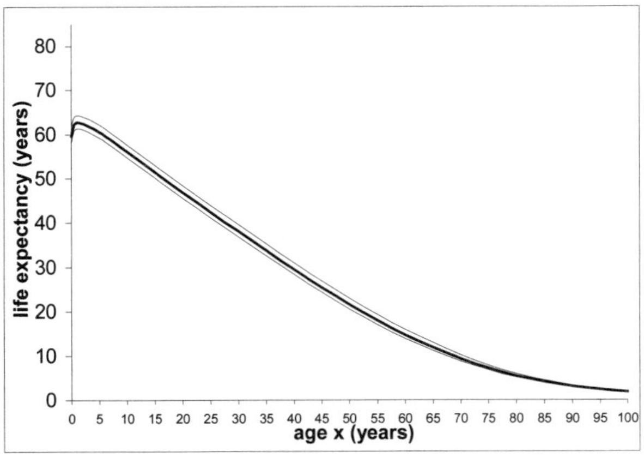

Age distribution of deaths with uncertainty interval

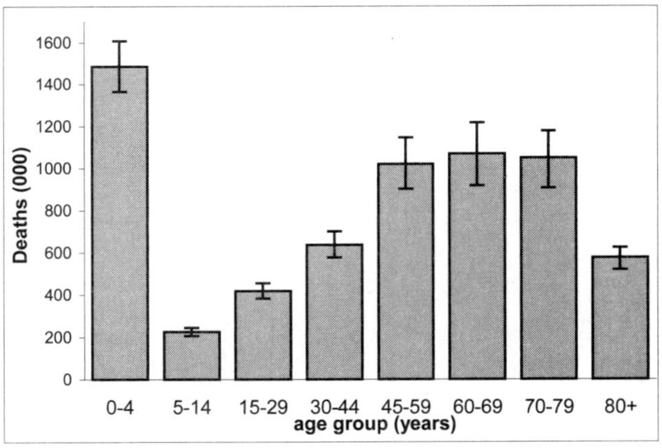

SearD **SearD** **SearD**

Females, 2000 *Femmes, 2000* *Mujeres, 2000*

Age	$_nM_x$	$_nq_x$			l_x			$_nd_x$	$_nL_x$	e_x			Actual deaths	Actual population
			percentiles			percentiles					percentiles			
		mean	2½	97½	mean	2½	97½			mean	2½	97½	(000)	(000)
<1	0.0670	0.0632	0.0561	0.0710	100,000	100,000	100,000	6,316	94,316	62.4	60.5	64.9	989	14,776
1-4	0.0095	0.0371	0.0332	0.0407	93,684	92,897	94,388	3,479	366,388	65.6	63.9	67.6	534	56,224
5-9	0.0025	0.0123	0.0109	0.0136	90,205	89,401	91,052	1,108	448,258	64.1	62.4	66.1	170	68,730
10-14	0.0013	0.0065	0.0058	0.0072	89,098	88,237	90,032	581	444,037	59.9	58.2	61.8	85	65,136
15-19	0.0020	0.0102	0.0090	0.0114	88,517	87,609	89,511	902	440,328	55.2	53.6	57.2	124	60,329
20-24	0.0028	0.0138	0.0122	0.0154	87,615	86,634	88,704	1,212	435,043	50.8	49.2	52.7	148	53,198
25-29	0.0030	0.0149	0.0132	0.0167	86,403	85,314	87,620	1,290	428,790	46.5	44.9	48.3	149	49,473
30-34	0.0032	0.0159	0.0141	0.0178	85,113	83,908	86,464	1,354	422,181	42.1	40.6	44.0	143	44,461
35-39	0.0035	0.0174	0.0154	0.0195	83,759	82,436	85,245	1,457	415,153	37.8	36.3	39.6	137	39,101
40-44	0.0047	0.0234	0.0206	0.0263	82,302	80,857	83,929	1,924	406,699	33.4	31.9	35.2	159	33,666
45-49	0.0067	0.0329	0.0289	0.0371	80,378	78,740	82,196	2,643	395,280	29.1	27.7	30.9	186	27,814
50-54	0.0093	0.0454	0.0396	0.0515	77,734	75,821	79,817	3,530	379,847	25.0	23.7	26.7	215	23,097
55-59	0.0142	0.0688	0.0593	0.0787	74,204	71,939	76,626	5,104	358,263	21.1	19.8	22.7	278	19,526
60-64	0.0215	0.1019	0.0867	0.1174	69,101	66,356	71,983	7,041	327,900	17.5	16.3	19.0	350	16,277
65-69	0.0333	0.1538	0.1297	0.1775	62,059	58,738	65,505	9,545	286,433	14.2	13.2	15.6	423	12,684
70-74	0.0500	0.2222	0.1872	0.2541	52,514	48,572	56,556	11,667	233,400	11.3	10.5	12.5	443	8,863
75-79	0.0748	0.3151	0.2681	0.3538	40,846	36,232	45,779	12,870	172,056	8.8	8.2	9.8	405	5,411
80-84	0.1120	0.4375	0.3812	0.4799	27,976	23,413	33,414	12,240	109,281	6.8	6.3	7.5	306	2,734
85-89	0.1658	0.5859	0.5288	0.6263	15,736	12,204	20,651	9,221	55,630	5.1	4.7	5.6	160	967
90-94	0.2419	0.7009	0.6521	0.7341	6,516	4,595	9,731	4,567	18,878	3.7	3.5	4.1	54	222
95-99	0.3478	0.8511	0.8158	0.8744	1,949	1,238	3,385	1,659	4,769	2.7	2.6	2.9	11	31
100+	0.4938	1	1	1	290	157	623	290	588	2.0	2.0	2.1	1	3
All ages													*5,469*	*602,722*

ln($_nq_x$) with uncertainty interval

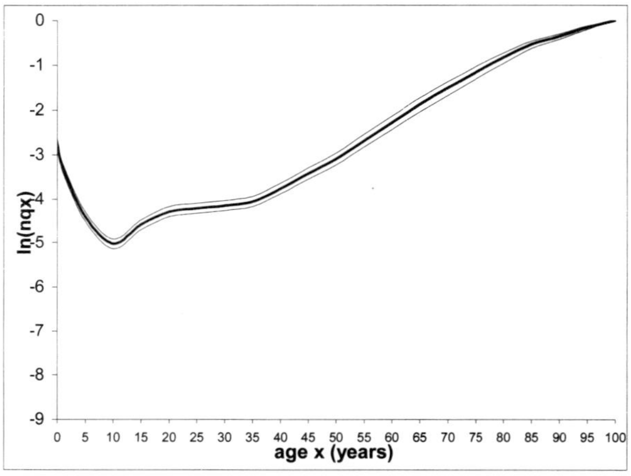

l_x with uncertainty interval

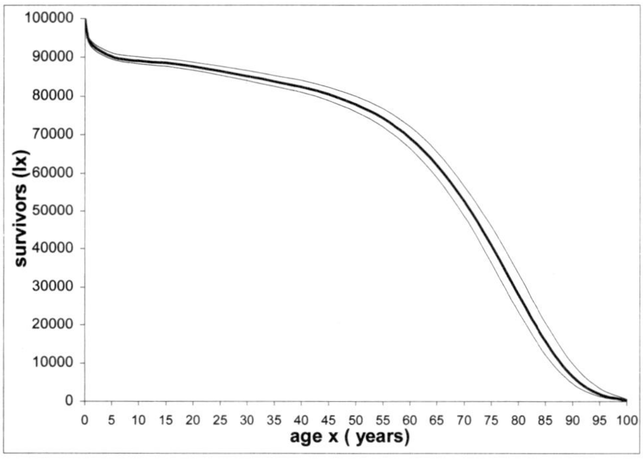

e_x with uncertainty interval

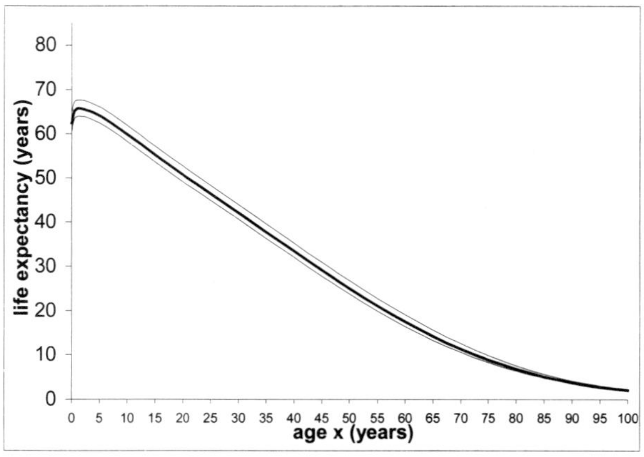

Age distribution of deaths with uncertainty interval

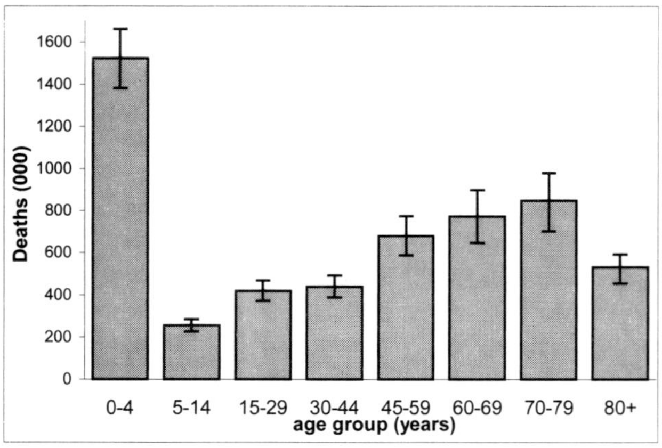

WprA **WprA** **WprA**

Males, 2000 *Hommes, 2000* *Hombres, 2000*

Age	$_nM_x$	$_nq_x$			l_x			$_nd_x$	$_nL_x$	e_x			Actual deaths	Actual population
			percentiles			percentiles					percentiles			
		mean	2½	97½	mean	2½	97½			mean	2½	97½	(000)	(000)
<1	0.0044	0.0044	0.0041	0.0046	100,000	100,000	100,000	436	99,608	77.3	77.1	77.6	4	823
1-4	0.0004	0.0015	0.0014	0.0015	99,564	99,539	99,590	144	397,911	76.7	76.5	76.8	1	3,288
5-9	0.0002	0.0009	0.0008	0.0009	99,420	99,389	99,451	85	496,888	72.8	72.6	73.0	1	4,100
10-14	0.0002	0.0008	0.0008	0.0009	99,335	99,301	99,370	82	496,470	67.8	67.7	68.0	1	4,340
15-19	0.0005	0.0027	0.0026	0.0028	99,253	99,215	99,290	271	495,588	62.9	62.8	63.1	3	4,818
20-24	0.0008	0.0038	0.0037	0.0039	98,982	98,934	99,029	374	493,976	58.1	57.9	58.2	4	5,358
25-29	0.0008	0.0039	0.0038	0.0040	98,608	98,547	98,667	385	492,078	53.3	53.1	53.4	5	6,090
30-34	0.0009	0.0045	0.0044	0.0046	98,223	98,149	98,293	444	490,006	48.5	48.3	48.6	5	5,490
35-39	0.0012	0.0059	0.0057	0.0060	97,779	97,693	97,861	573	487,465	43.7	43.5	43.9	6	5,224
40-44	0.0018	0.0089	0.0087	0.0091	97,207	97,106	97,301	861	483,880	38.9	38.8	39.1	9	5,006
45-49	0.0029	0.0145	0.0142	0.0148	96,345	96,225	96,455	1,398	478,233	34.3	34.1	34.4	16	5,458
50-54	0.0049	0.0241	0.0237	0.0245	94,948	94,803	95,079	2,290	469,014	29.7	29.6	29.9	30	6,118
55-59	0.0071	0.0346	0.0341	0.0352	92,658	92,484	92,817	3,210	455,263	25.4	25.3	25.6	35	4,953
60-64	0.0120	0.0582	0.0573	0.0592	89,447	89,245	89,642	5,209	434,214	21.2	21.1	21.4	51	4,261
65-69	0.0192	0.0914	0.0897	0.0930	84,238	84,001	84,482	7,699	401,945	17.4	17.2	17.5	73	3,791
70-74	0.0288	0.1345	0.1320	0.1368	76,540	76,211	76,864	10,297	356,956	13.9	13.7	14.0	88	3,040
75-79	0.0481	0.2147	0.2105	0.2187	66,243	65,802	66,699	14,224	295,653	10.6	10.5	10.8	90	1,870
80-84	0.0842	0.3479	0.3415	0.3540	52,019	51,460	52,631	18,096	214,853	7.9	7.8	8.0	88	1,047
85-89	0.1412	0.5217	0.5145	0.5290	33,923	33,309	34,605	17,698	125,368	5.7	5.7	5.8	75	532
90-94	0.2043	0.6333	0.6271	0.6397	16,224	15,699	16,798	10,275	50,296	4.2	4.2	4.3	33	161
95-99	0.2996	0.7889	0.7843	0.7937	5,949	5,656	6,264	4,693	15,665	3.1	3.1	3.2	8	27
100+	0.4487	1	1	1	1,256	1,167	1,351	1,256	2,798	2.2	2.2	2.4	1	2
All ages													*625*	*75,799*

ln($_nq_x$) with uncertainty interval

l_x with uncertainty interval

e_x with uncertainty interval

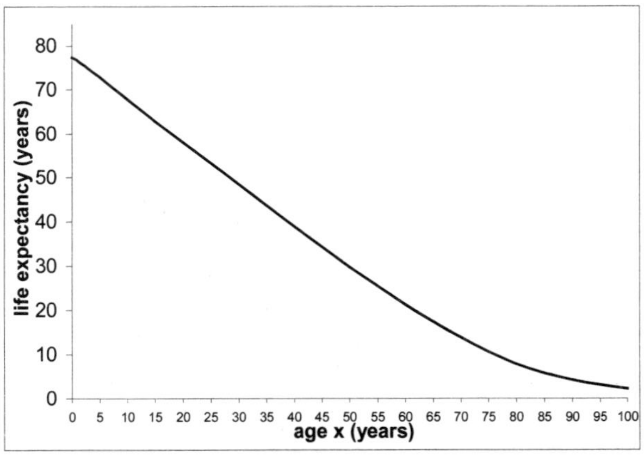

Age distribution of deaths with uncertainty interval

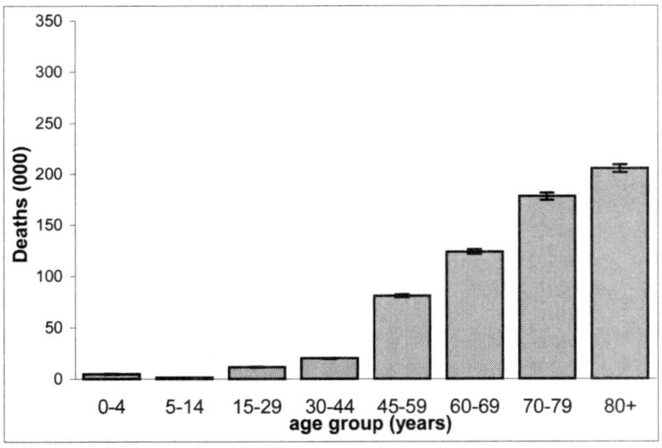

WprA **WprA** **WprA**

Females, 2000 *Femmes, 2000* *Mujeres, 2000*

Age	$_nM_x$	$_nq_x$			l_x			$_nd_x$	$_nL_x$	e_x			Actual deaths	Actual population
			percentiles			percentiles					percentiles			
		mean	2½	97½	mean	2½	97½			mean	2½	97½	(000)	(000)
<1	0.0032	0.0032	0.0029	0.0035	100,000	100,000	100,000	320	99,712	84.2	83.9	84.6	3	784
1-4	0.0003	0.0011	0.0010	0.0012	99,680	99,648	99,714	112	398,450	83.5	83.2	83.8	1	3,111
5-9	0.0001	0.0005	0.0005	0.0006	99,568	99,527	99,611	54	497,704	79.6	79.3	79.9	0	3,896
10-14	0.0001	0.0005	0.0005	0.0005	99,514	99,470	99,562	51	497,443	74.6	74.3	75.0	0	4,128
15-19	0.0002	0.0011	0.0010	0.0012	99,463	99,415	99,515	110	497,039	69.7	69.4	70.0	1	4,592
20-24	0.0003	0.0014	0.0013	0.0015	99,353	99,298	99,412	139	496,416	64.7	64.4	65.1	1	5,105
25-29	0.0003	0.0015	0.0015	0.0016	99,214	99,150	99,283	154	495,684	59.8	59.5	60.2	2	5,895
30-34	0.0004	0.0021	0.0020	0.0022	99,060	98,988	99,139	209	494,779	54.9	54.6	55.3	2	5,398
35-39	0.0006	0.0029	0.0028	0.0031	98,851	98,769	98,942	288	493,537	50.0	49.7	50.4	3	5,137
40-44	0.0009	0.0045	0.0043	0.0047	98,563	98,468	98,669	448	491,696	45.2	44.9	45.5	5	4,949
45-49	0.0015	0.0073	0.0070	0.0076	98,115	98,001	98,242	716	488,786	40.4	40.1	40.7	8	5,431
50-54	0.0021	0.0103	0.0099	0.0107	97,399	97,262	97,553	1,004	484,486	35.6	35.4	36.0	13	6,108
55-59	0.0032	0.0157	0.0152	0.0163	96,395	96,232	96,578	1,514	478,191	31.0	30.7	31.3	16	5,084
60-64	0.0052	0.0257	0.0250	0.0267	94,881	94,690	95,096	2,443	468,298	26.4	26.2	26.8	23	4,499
65-69	0.0075	0.0366	0.0354	0.0381	92,438	92,195	92,710	3,384	453,730	22.1	21.8	22.4	31	4,206
70-74	0.0128	0.0622	0.0598	0.0647	89,054	88,687	89,388	5,537	431,428	17.8	17.6	18.1	47	3,657
75-79	0.0242	0.1141	0.1093	0.1186	83,517	82,952	83,985	9,530	393,760	13.8	13.6	14.1	69	2,870
80-84	0.0471	0.2105	0.2014	0.2183	73,987	73,114	74,749	15,576	330,995	10.3	10.1	10.6	91	1,933
85-89	0.0891	0.3643	0.3505	0.3769	58,411	57,179	59,691	21,278	238,861	7.4	7.2	7.6	106	1,189
90-94	0.1498	0.5169	0.5013	0.5314	37,133	35,744	38,771	19,193	128,085	5.2	5.1	5.3	70	467
95-99	0.2477	0.7105	0.6973	0.7230	17,940	16,812	19,289	12,746	51,460	3.6	3.5	3.7	27	109
100+	0.4180	1	1	1	5,194	4,682	5,799	5,194	12,425	2.4	2.4	2.6	5	13
All ages													*525*	*78,560*

ln($_nq_x$) with uncertainty interval

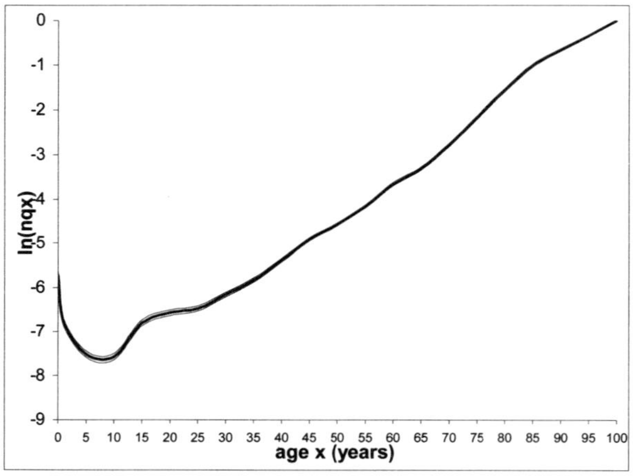

l_x with uncertainty interval

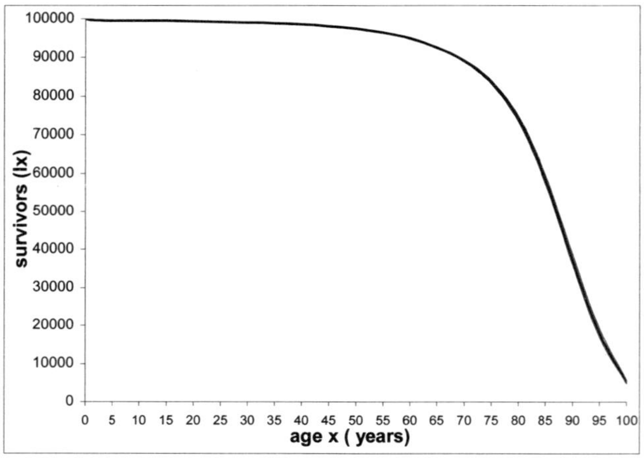

e_x with uncertainty interval

Age distribution of deaths with uncertainty interval

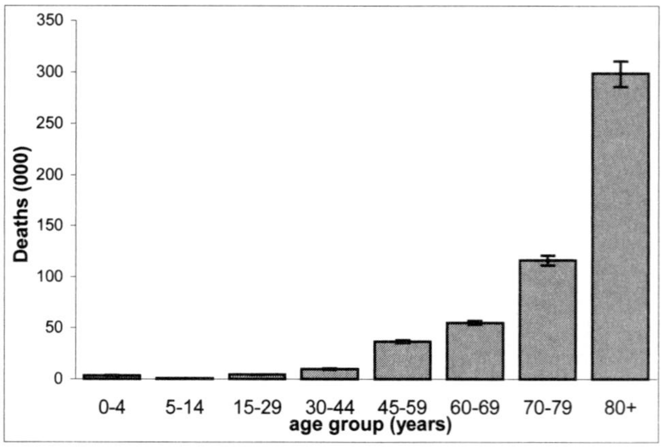

WprB | WprB | WprB

Males, 2000 | *Hommes, 2000* | *Hombres, 2000*

Age	$_nM_x$	$_nq_x$			l_x			$_nd_x$	$_nL_x$	e_x			Actual deaths	Actual population
			percentiles			percentiles					percentiles			
		mean	2½	97½	mean	2½	97½			mean	2½	97½	(000)	(000)
<1	0.0309	0.0301	0.0265	0.0338	100,000	100,000	100,000	3,008	97,293	68.2	67.4	69.1	391	12,647
1-4	0.0026	0.0105	0.0096	0.0113	96,992	96,615	97,349	1,015	385,533	69.3	68.7	70.0	138	52,544
5-9	0.0009	0.0046	0.0042	0.0049	95,977	95,532	96,417	440	478,786	66.1	65.4	66.7	63	68,949
10-14	0.0007	0.0033	0.0030	0.0035	95,537	95,068	96,010	315	476,897	61.3	60.7	62.0	50	75,913
15-19	0.0011	0.0054	0.0050	0.0059	95,222	94,737	95,718	519	474,811	56.5	55.9	57.2	72	65,778
20-24	0.0015	0.0074	0.0068	0.0080	94,703	94,192	95,240	702	471,758	51.8	51.2	52.5	93	62,540
25-29	0.0015	0.0075	0.0070	0.0081	94,001	93,453	94,589	708	468,233	47.2	46.6	47.8	111	73,089
30-34	0.0018	0.0092	0.0085	0.0098	93,292	92,703	93,930	859	464,315	42.5	41.9	43.2	138	74,445
35-39	0.0024	0.0118	0.0110	0.0126	92,434	91,794	93,130	1,091	459,440	37.9	37.3	38.5	149	62,669
40-44	0.0034	0.0167	0.0156	0.0177	91,342	90,640	92,107	1,522	452,908	33.3	32.8	33.9	173	51,431
45-49	0.0050	0.0248	0.0233	0.0263	89,821	89,037	90,671	2,229	443,531	28.9	28.3	29.5	252	50,201
50-54	0.0079	0.0386	0.0363	0.0411	87,592	86,703	88,552	3,383	429,502	24.5	24.0	25.1	291	36,892
55-59	0.0127	0.0614	0.0578	0.0655	84,209	83,190	85,296	5,172	408,116	20.4	19.9	21.0	353	27,877
60-64	0.0211	0.1002	0.0939	0.1071	79,037	77,881	80,239	7,922	375,382	16.6	16.1	17.1	514	24,358
65-69	0.0338	0.1557	0.1453	0.1661	71,116	69,696	72,491	11,072	327,898	13.2	12.7	13.6	670	19,831
70-74	0.0567	0.2482	0.2317	0.2636	60,044	58,119	61,760	14,903	262,961	10.1	9.7	10.5	760	13,411
75-79	0.0882	0.3612	0.3392	0.3835	45,141	42,837	47,405	16,305	184,941	7.6	7.3	8.0	683	7,751
80-84	0.1449	0.5318	0.5074	0.5575	28,836	26,646	31,209	15,335	105,840	5.6	5.3	5.8	501	3,459
85-89	0.2236	0.7170	0.6958	0.7386	13,500	11,968	15,326	9,680	43,302	4.0	3.9	4.2	237	1,058
90-94	0.3240	0.8215	0.8060	0.8371	3,820	3,128	4,643	3,138	9,686	2.9	2.9	3.0	64	197
95-99	0.4403	0.9486	0.9385	0.9535	682	510	894	647	1,469	2.2	2.2	2.3	9	21
100+	0.5680	1	1	1	35	24	55	35	62	1.8	1.7	1.9	1	1
All ages													*5,712*	*785,062*

ln($_nq_x$) with uncertainty interval

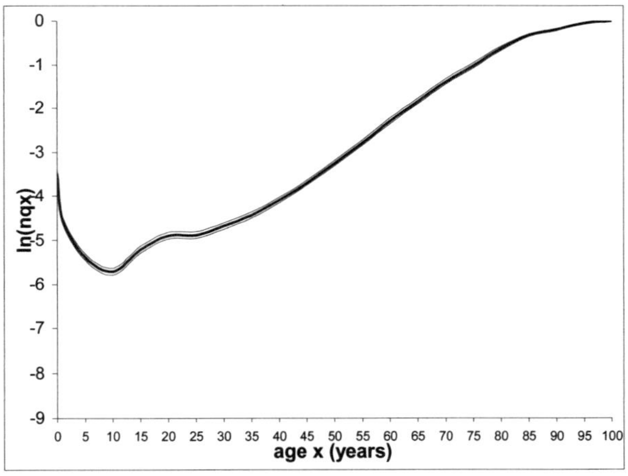

l_x with uncertainty interval

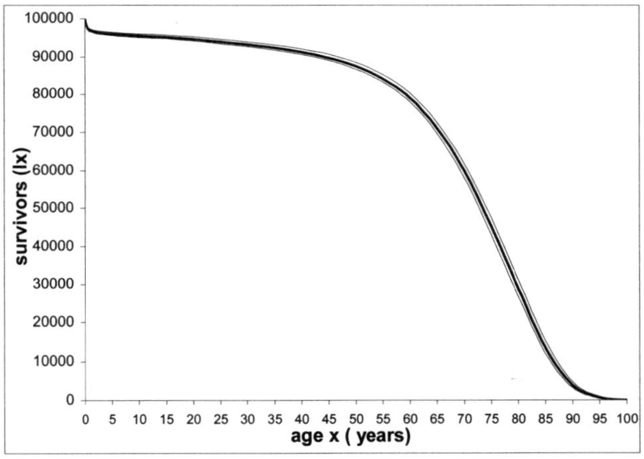

e_x with uncertainty interval

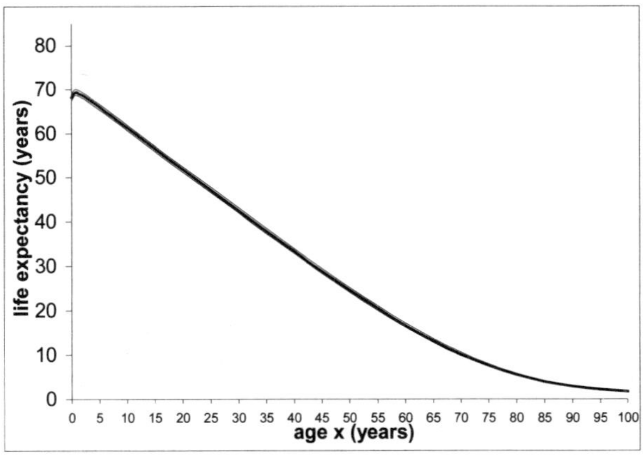

Age distribution of deaths with uncertainty interval

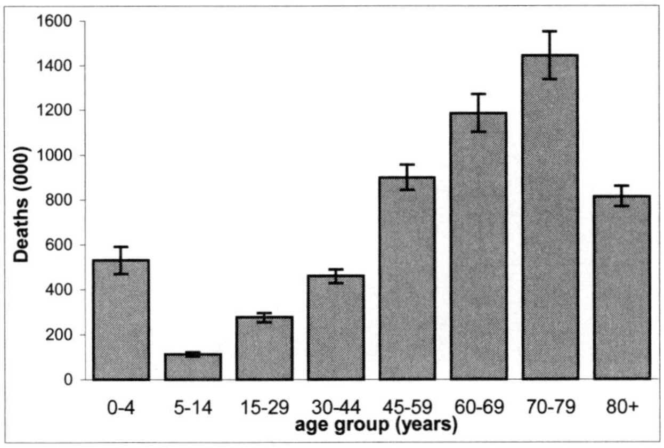

Females, 2000 *Femmes, 2000* *Mujeres, 2000*

Age	$_nM_x$	$_nq_x$			l_x			$_nd_x$	$_nL_x$	e_x			Actual deaths	Actual population
			percentiles			percentiles					percentiles			
		mean	2½	97½	mean	2½	97½			mean	2½	97½	(000)	(000)
<1	0.0343	0.0332	0.0295	0.0372	100,000	100,000	100,000	3,323	97,010	72.7	71.6	73.8	403	11,781
1-4	0.0027	0.0106	0.0098	0.0116	96,677	96,280	97,054	1,028	384,241	74.2	73.1	75.3	127	47,531
5-9	0.0008	0.0039	0.0036	0.0042	95,649	95,209	96,107	376	477,304	71.0	69.9	72.1	49	62,629
10-14	0.0005	0.0027	0.0025	0.0029	95,273	94,809	95,753	259	475,716	66.2	65.2	67.4	38	69,448
15-19	0.0009	0.0043	0.0040	0.0047	95,014	94,530	95,512	411	474,041	61.4	60.4	62.5	53	61,218
20-24	0.0012	0.0058	0.0053	0.0063	94,603	94,085	95,132	547	471,645	56.7	55.6	57.8	68	58,992
25-29	0.0011	0.0056	0.0051	0.0061	94,055	93,493	94,626	524	468,966	52.0	50.9	53.1	78	69,786
30-34	0.0013	0.0064	0.0059	0.0069	93,531	92,925	94,142	597	466,163	47.3	46.2	48.4	92	71,814
35-39	0.0016	0.0078	0.0072	0.0085	92,934	92,280	93,590	727	462,855	42.5	41.5	43.7	94	60,138
40-44	0.0022	0.0108	0.0099	0.0117	92,208	91,497	92,916	995	458,552	37.9	36.8	39.0	105	48,181
45-49	0.0033	0.0161	0.0148	0.0175	91,213	90,428	91,988	1,472	452,385	33.2	32.2	34.4	156	47,867
50-54	0.0050	0.0248	0.0227	0.0269	89,741	88,848	90,615	2,229	443,132	28.8	27.7	29.9	174	34,521
55-59	0.0077	0.0376	0.0343	0.0410	87,512	86,468	88,521	3,295	429,322	24.4	23.4	25.5	201	26,199
60-64	0.0126	0.0611	0.0551	0.0675	84,217	82,976	85,398	5,148	408,215	20.3	19.3	21.3	295	23,411
65-69	0.0201	0.0959	0.0851	0.1076	79,069	77,574	80,583	7,584	376,385	16.4	15.5	17.4	409	20,295
70-74	0.0345	0.1588	0.1394	0.1798	71,485	69,625	73,523	11,352	329,044	12.9	12.1	13.8	523	15,160
75-79	0.0551	0.2421	0.2122	0.2737	60,133	57,176	62,954	14,559	264,266	9.9	9.2	10.6	568	10,310
80-84	0.0954	0.3850	0.3430	0.4265	45,574	41,551	49,589	17,546	184,005	7.2	6.7	7.8	532	5,581
85-89	0.1567	0.5628	0.5172	0.6039	28,028	23,834	32,581	15,776	100,702	5.2	4.9	5.6	349	2,227
90-94	0.2436	0.7037	0.6650	0.7362	12,253	9,440	15,730	8,623	35,395	3.7	3.5	4.0	161	660
95-99	0.3595	0.8648	0.8383	0.8849	3,630	2,491	5,267	3,139	8,733	2.7	2.6	2.8	44	122
100+	0.5047	1	1	1	491	287	851	491	973	2.0	1.9	2.1	6	12
All ages													*4,526*	*747,885*

ln($_nq_x$) with uncertainty interval

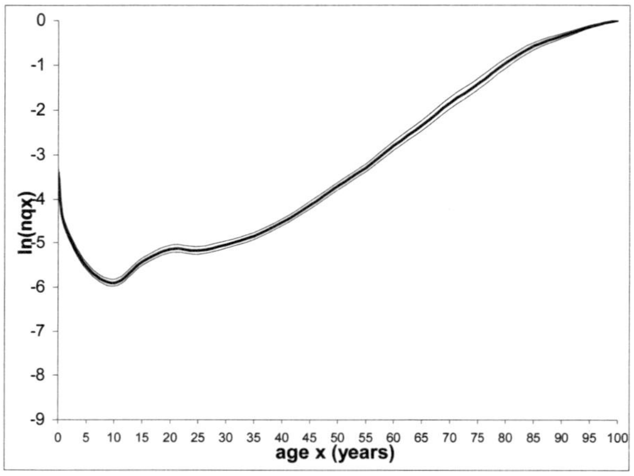

l_x with uncertainty interval

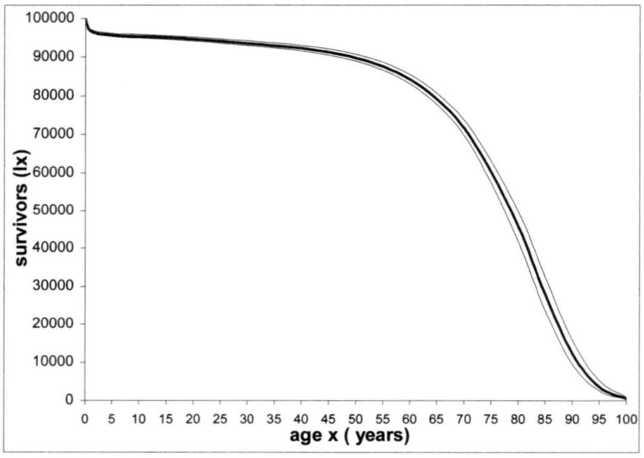

e_x with uncertainty interval

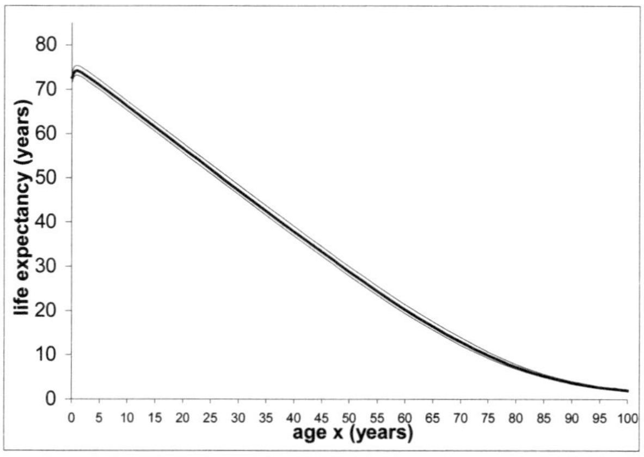

Age distribution of deaths with uncertainty interval

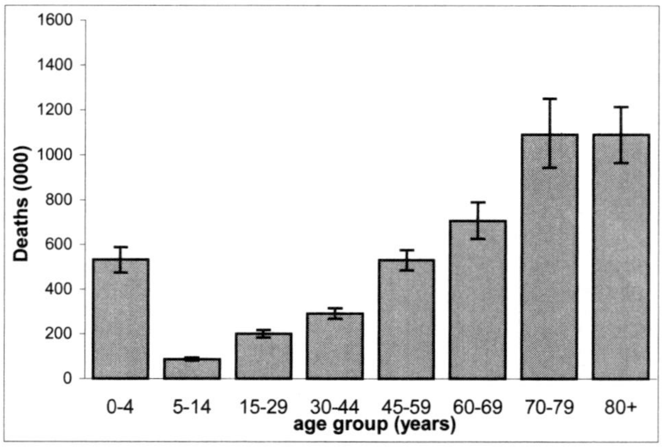

APPENDIX B:
Life Tables by Country

APPENDICE B:
Tables de Mortalité détaillées par pays

Anexo B:
Tablas de Mortalidad detalladas por país

Afghanistan | Afghanistan | Afganistán

Males, 2000 | *Hommes, 2000* | *Hombres, 2000*

Age	$_nM_x$	$_nq_x$			l_x			$_nd_x$	$_nL_x$	e_x			Actual deaths	Actual population
			percentiles			percentiles					percentiles			
		mean	2½	97½	mean	2½	97½			mean	2½	97½		
<1	0.2326	0.1923	0.1380	0.2509	100,000	100,000	100,000	19,231	82,692	44.2	38.5	50.3	106,106	456,248
1-4	0.0192	0.0734	0.0550	0.0952	80,769	74,911	86,201	5,925	308,855	53.7	48.5	59.1	28,742	1,498,200
5-9	0.0042	0.0210	0.0155	0.0278	74,844	69,852	79,843	1,573	370,285	53.8	49.5	58.9	6,649	1,564,615
10-14	0.0024	0.0117	0.0086	0.0156	73,270	67,982	78,480	859	364,203	49.9	45.8	54.9	3,204	1,358,220
15-19	0.0027	0.0134	0.0098	0.0178	72,411	66,934	77,790	969	359,633	45.5	41.5	50.5	3,186	1,182,458
20-24	0.0042	0.0206	0.0150	0.0274	71,442	65,755	77,010	1,470	353,535	41.0	37.1	46.0	4,151	998,362
25-29	0.0066	0.0324	0.0236	0.0431	69,972	63,977	75,822	2,266	344,193	36.8	32.9	41.7	5,514	837,356
30-34	0.0087	0.0425	0.0310	0.0565	67,705	61,257	73,976	2,879	331,331	33.0	29.1	37.8	6,215	715,326
35-39	0.0113	0.0549	0.0400	0.0729	64,827	57,846	71,606	3,562	315,229	29.3	25.5	34.0	6,774	599,516
40-44	0.0142	0.0686	0.0500	0.0907	61,265	53,699	68,630	4,203	295,816	25.9	22.2	30.4	7,096	499,379
45-49	0.0174	0.0833	0.0600	0.1110	57,061	48,920	65,047	4,754	273,423	22.6	19.1	26.9	7,071	406,693
50-54	0.0219	0.1037	0.0730	0.1419	52,308	43,609	60,966	5,426	247,975	19.5	16.2	23.4	7,293	333,321
55-59	0.0280	0.1307	0.0900	0.1820	46,882	37,722	56,320	6,126	219,096	16.4	13.5	20.1	7,497	268,129
60-64	0.0383	0.1749	0.1189	0.2450	40,756	31,397	50,856	7,129	185,957	13.5	10.9	16.8	7,878	205,493
65-69	0.0539	0.2374	0.1609	0.3292	33,627	24,496	44,133	7,984	148,175	10.9	8.7	13.7	7,858	145,841
70-74	0.0811	0.3371	0.2322	0.4519	25,643	17,390	36,003	8,646	106,602	8.5	6.7	10.9	7,405	91,301
75-79	0.1205	0.4629	0.3316	0.5889	16,998	9,925	26,223	7,868	65,318	6.5	5.2	8.4	5,575	46,277
80-84	0.1755	0.6099	0.4652	0.7294	9,129	4,080	17,069	5,568	31,728	4.9	4.0	6.3	2,920	16,640
85-89	0.2491	0.7675	0.6336	0.8616	3,562	1,104	9,122	2,734	10,974	3.7	3.2	4.7	884	3,547
90-94	0.3446	0.8471	0.7345	0.9194	828	153	3,342	701	2,036	2.8	2.4	3.4	137	398
95-99	0.4644	0.9702	0.9189	0.9909	127	12	887	123	264	2.1	2.0	2.4	10	21
100+	0.6099	1	1	1	4	0	72	4	6	1.6	1.6	1.8	1	1
All ages													*232,164*	*11,227,342*

ln($_nq_x$) with uncertainty interval

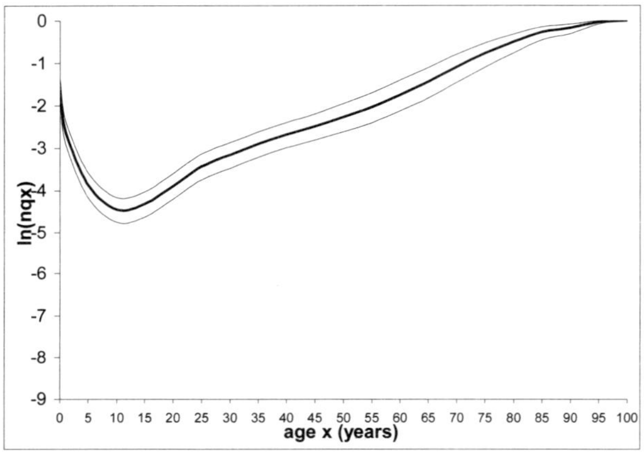

l_x with uncertainty interval

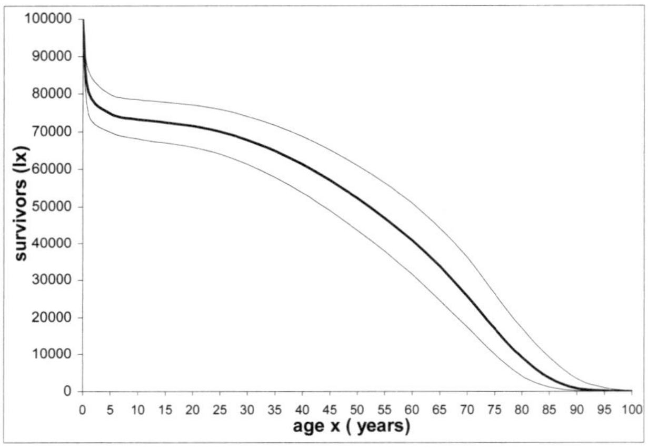

e_x with uncertainty interval

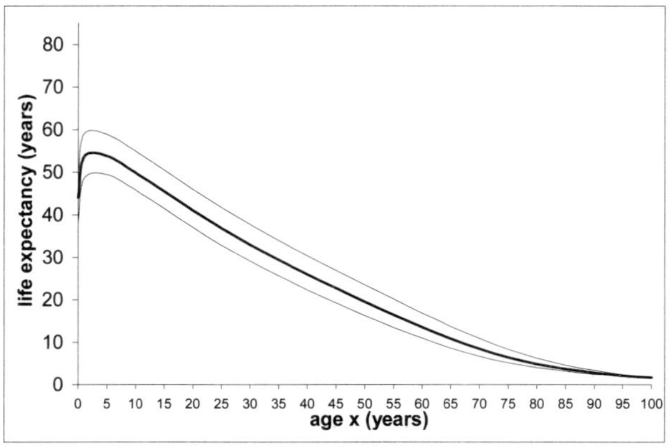

Distribution of deaths by age with uncertainty interval

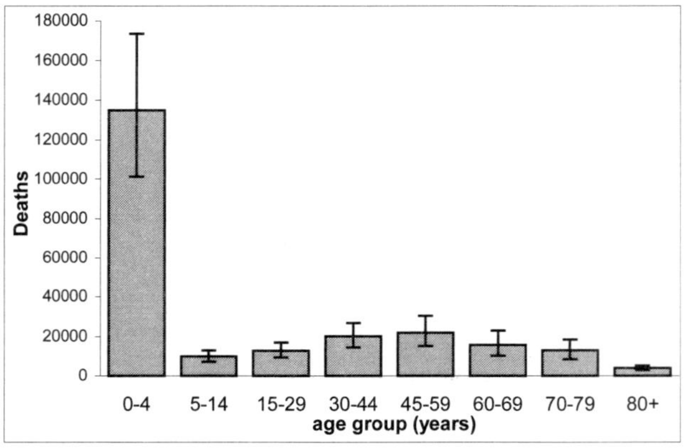

Afghanistan | Afghanistan | Afganistán

Females, 2000 | *Femmes, 2000* | *Mujeres, 2000*

Age	$_nM_x$	$_nq_x$			l_x			$_nd_x$	$_nL_x$	e_x			Actual deaths	Actual population
			percentiles			percentiles					percentiles			
		mean	2½	97½	mean	2½	97½			mean	2½	97½		
<1	0.2081	0.1753	0.1231	0.2299	100,000	100,000	100,000	17,529	84,224	45.1	39.0	51.7	90,867	436,601
1-4	0.0237	0.0898	0.0670	0.1168	82,471	77,014	87,694	7,404	312,115	53.7	47.8	60.2	33,585	1,415,753
5-9	0.0052	0.0258	0.0186	0.0343	75,067	70,068	80,056	1,933	370,503	54.8	49.8	60.6	7,667	1,469,442
10-14	0.0034	0.0166	0.0119	0.0223	73,134	67,932	78,406	1,216	362,630	51.2	46.5	56.7	4,249	1,266,926
15-19	0.0047	0.0233	0.0167	0.0313	71,918	66,518	77,359	1,677	355,397	47.0	42.5	52.4	5,174	1,096,436
20-24	0.0070	0.0344	0.0245	0.0462	70,241	64,472	75,957	2,419	345,157	43.1	38.8	48.3	6,477	924,253
25-29	0.0079	0.0386	0.0273	0.0520	67,822	61,494	73,920	2,621	332,558	39.5	35.3	44.5	6,114	775,792
30-34	0.0082	0.0400	0.0282	0.0540	65,201	58,303	71,686	2,611	319,480	36.0	32.0	40.8	5,408	661,831
35-39	0.0097	0.0474	0.0332	0.0640	62,591	55,167	69,623	2,969	305,531	32.4	28.6	37.0	5,387	554,374
40-44	0.0095	0.0464	0.0323	0.0626	59,622	51,657	67,253	2,765	291,196	28.9	25.3	33.2	4,391	462,489
45-49	0.0113	0.0547	0.0379	0.0740	56,857	48,448	65,016	3,111	276,507	25.2	21.8	29.3	4,273	379,843
50-54	0.0152	0.0731	0.0503	0.1003	53,746	44,912	62,461	3,930	258,905	21.5	18.3	25.4	4,790	315,593
55-59	0.0208	0.0990	0.0678	0.1374	49,816	40,563	59,166	4,934	236,746	18.0	15.1	21.6	5,404	259,330
60-64	0.0310	0.1436	0.0974	0.2002	44,882	35,304	54,999	6,447	208,294	14.7	12.1	18.0	6,311	203,873
65-69	0.0463	0.2076	0.1400	0.2871	38,435	28,788	49,368	7,980	172,224	11.7	9.5	14.7	6,870	148,260
70-74	0.0717	0.3039	0.2070	0.4083	30,455	21,262	41,902	9,255	129,135	9.1	7.3	11.7	6,801	94,890
75-79	0.1068	0.4216	0.2965	0.5405	21,200	13,358	32,365	8,938	83,653	7.0	5.6	9.1	5,261	49,240
80-84	0.1582	0.5667	0.4228	0.6850	12,262	6,179	21,924	6,949	43,936	5.3	4.4	6.9	2,871	18,150
85-89	0.2282	0.7266	0.5868	0.8250	5,313	1,954	12,190	3,860	16,912	4.0	3.4	5.1	916	4,013
90-94	0.3211	0.8177	0.6980	0.8949	1,452	345	5,037	1,188	3,699	3.0	2.5	3.7	158	492
95-99	0.4403	0.9485	0.8804	0.9799	265	37	1,521	251	570	2.2	2.1	2.6	14	31
100+	0.5885	1	1	1	14	1	182	14	23	1.7	1.7	1.9	1	1
All ages													*212,987*	*10,537,613*

ln($_nq_x$) with uncertainty interval

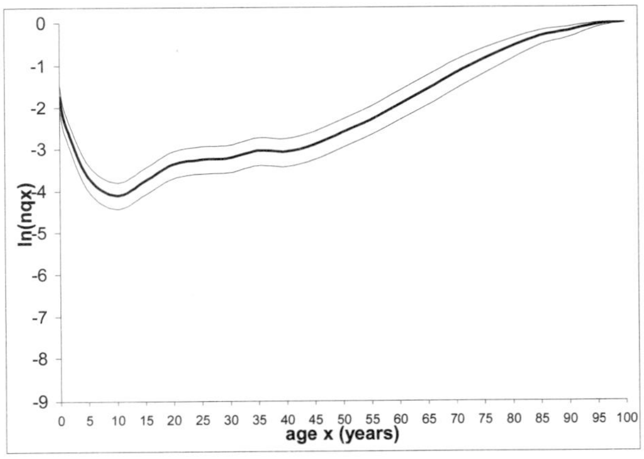

l_x with uncertainty interval

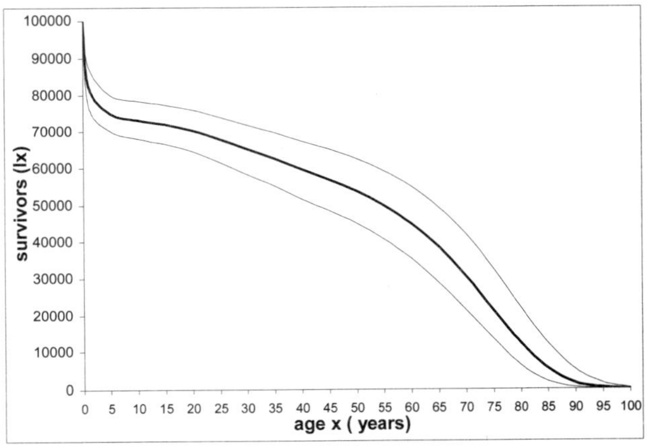

e_x with uncertainty interval

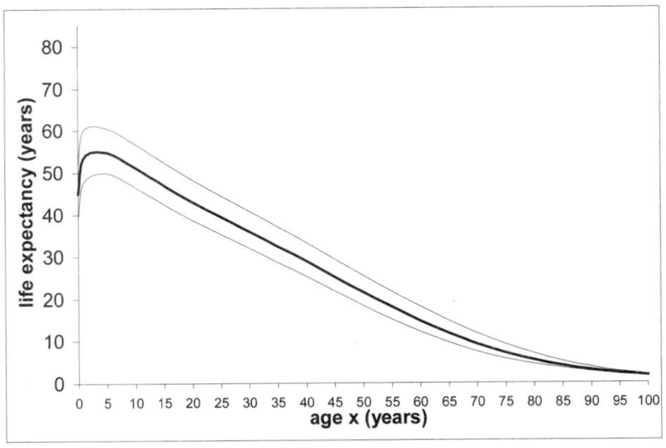

Distribution of deaths by age with uncertainty interval

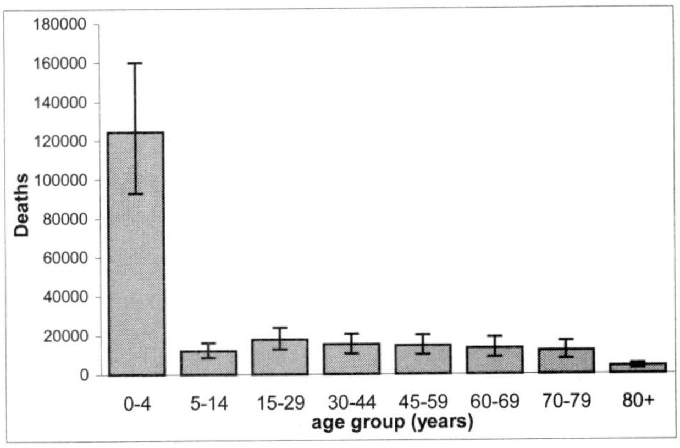

Albania Albanie Albania

Males, 2000 *Hommes, 2000* *Hombres, 2000*

Age	$_nM_x$	$_nq_x$ mean	$_nq_x$ percentiles 2½	$_nq_x$ percentiles 97½	l_x mean	l_x percentiles 2½	l_x percentiles 97½	$_nd_x$	$_nL_x$	e_x mean	e_x percentiles 2½	e_x percentiles 97½	Actual deaths	Actual population
<1	0.0353	0.0342	0.0260	0.0423	100,000	100,000	100,000	3,419	96,923	64.3	62.9	65.7	1,127	31,958
1-4	0.0033	0.0132	0.0109	0.0155	96,581	95,767	97,400	1,272	383,271	65.6	64.5	66.9	421	126,944
5-9	0.0014	0.0069	0.0058	0.0080	95,309	94,282	96,339	655	474,907	62.5	61.5	63.7	227	164,487
10-14	0.0013	0.0063	0.0054	0.0073	94,654	93,527	95,782	598	471,776	57.9	56.9	59.1	204	161,135
15-19	0.0027	0.0136	0.0117	0.0157	94,056	92,843	95,268	1,283	467,075	53.2	52.3	54.4	402	146,398
20-24	0.0046	0.0228	0.0200	0.0259	92,774	91,388	94,150	2,115	458,582	48.9	48.0	50.1	638	138,392
25-29	0.0052	0.0257	0.0229	0.0288	90,659	89,024	92,263	2,328	447,475	45.0	44.2	46.1	695	133,540
30-34	0.0032	0.0160	0.0144	0.0178	88,331	86,464	90,151	1,415	438,116	41.1	40.4	42.1	418	129,352
35-39	0.0034	0.0170	0.0153	0.0187	86,916	84,926	88,851	1,474	430,894	36.8	36.1	37.7	438	128,162
40-44	0.0045	0.0223	0.0201	0.0245	85,442	83,335	87,492	1,907	422,443	32.4	31.7	33.2	487	107,843
45-49	0.0056	0.0278	0.0250	0.0304	83,535	81,289	85,733	2,320	411,877	28.0	27.5	28.9	457	81,111
50-54	0.0069	0.0338	0.0303	0.0369	81,216	78,819	83,593	2,742	399,223	23.8	23.3	24.5	419	61,059
55-59	0.0106	0.0515	0.0462	0.0560	78,474	75,912	81,058	4,038	382,273	19.5	19.0	20.2	643	60,913
60-64	0.0227	0.1072	0.0965	0.1159	74,436	71,659	77,313	7,980	352,228	15.4	15.0	16.1	1,157	51,090
65-69	0.0404	0.1834	0.1663	0.1959	66,456	63,354	69,850	12,186	301,812	12.0	11.6	12.5	1,506	37,306
70-74	0.0612	0.2653	0.2438	0.2838	54,269	50,944	58,231	14,397	235,353	9.1	8.8	9.5	1,384	22,616
75-79	0.1091	0.4287	0.4016	0.4552	39,872	36,488	44,035	17,093	156,627	6.5	6.2	6.8	1,536	14,074
80-84	0.1919	0.6484	0.6186	0.6776	22,779	19,879	26,352	14,769	76,971	4.5	4.3	4.8	919	4,789
85-89	0.2904	0.8413	0.8198	0.8624	8,010	6,409	10,052	6,738	23,203	3.3	3.2	3.4	420	1,445
90-94	0.3960	0.9049	0.8896	0.9199	1,272	882	1,812	1,151	2,906	2.5	2.4	2.6	65	165
95-99	0.5280	0.9800	0.9747	0.9850	121	71	200	118	249	2.1	2.1	2.1	3	5
100+	0.7560	1	1	1	2	1	5	2	3	1.3	1.3	2.0	0	0
All ages													*13,567*	*1,602,784*

ln($_nq_x$) with uncertainty interval

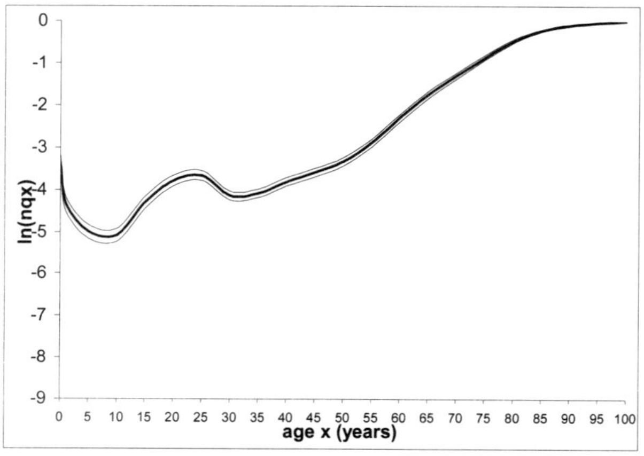

l_x with uncertainty interval

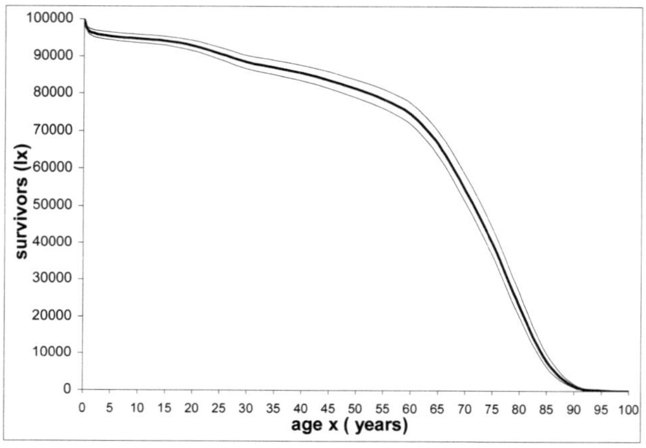

e_x with uncertainty interval

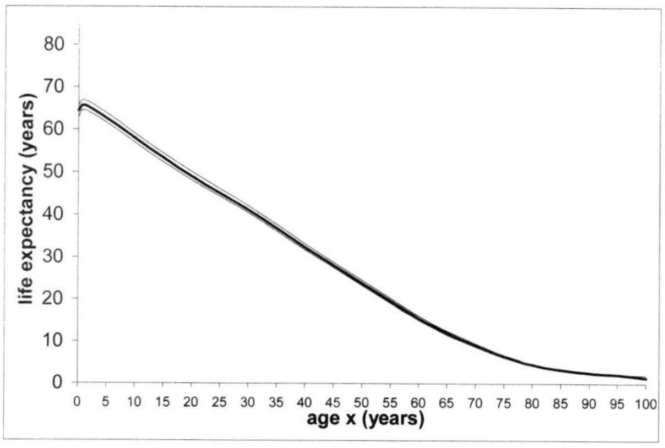

Distribution of deaths by age with uncertainty interval

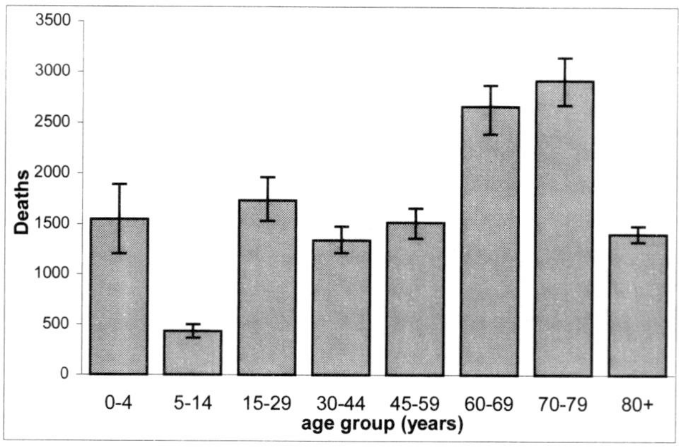

Albania | Albanie | Albania

Females, 2000 | *Femmes, 2000* | *Mujeres, 2000*

Age	$_nM_x$	$_nq_x$			l_x			$_nd_x$	$_nL_x$	e_x			Actual deaths	Actual population
			percentiles			percentiles					percentiles			
		mean	2½	97½	mean	2½	97½			mean	2½	97½		
<1	0.0310	0.0302	0.0237	0.0382	100,000	100,000	100,000	3,017	97,284	72.9	71.4	73.9	934	30,120
1-4	0.0027	0.0106	0.0089	0.0128	96,983	96,181	97,633	1,029	385,462	74.2	73.2	75.0	319	119,567
5-9	0.0007	0.0036	0.0030	0.0042	95,954	94,952	96,761	341	478,918	71.0	70.2	71.8	111	155,522
10-14	0.0005	0.0027	0.0023	0.0032	95,613	94,548	96,468	254	477,431	66.2	65.5	67.0	79	149,339
15-19	0.0011	0.0057	0.0049	0.0067	95,359	94,249	96,248	540	475,447	61.4	60.7	62.2	153	134,423
20-24	0.0012	0.0061	0.0053	0.0072	94,819	93,616	95,779	579	472,650	56.7	56.0	57.5	160	130,573
25-29	0.0011	0.0055	0.0048	0.0065	94,241	92,943	95,272	517	469,910	52.0	51.4	52.8	139	126,295
30-34	0.0010	0.0051	0.0045	0.0059	93,723	92,344	94,814	476	467,426	47.3	46.7	48.0	125	122,294
35-39	0.0015	0.0074	0.0065	0.0086	93,247	91,796	94,391	686	464,521	42.5	42.0	43.3	180	121,621
40-44	0.0020	0.0100	0.0089	0.0115	92,561	91,010	93,776	923	460,500	37.8	37.3	38.5	205	102,397
45-49	0.0026	0.0131	0.0118	0.0150	91,638	89,961	92,942	1,203	455,185	33.2	32.7	33.9	200	75,824
50-54	0.0036	0.0180	0.0163	0.0205	90,436	88,608	91,846	1,629	448,107	28.6	28.1	29.3	203	55,794
55-59	0.0058	0.0284	0.0259	0.0319	88,807	86,793	90,348	2,518	437,741	24.1	23.6	24.7	332	57,725
60-64	0.0106	0.0518	0.0474	0.0572	86,289	84,028	88,004	4,471	420,269	19.7	19.3	20.3	490	46,026
65-69	0.0194	0.0927	0.0847	0.1000	81,818	79,219	83,830	7,584	390,132	15.6	15.3	16.2	725	37,309
70-74	0.0364	0.1667	0.1530	0.1775	74,235	71,301	76,727	12,378	340,227	12.0	11.6	12.5	1,052	28,919
75-79	0.0639	0.2755	0.2556	0.2909	61,856	58,642	64,987	17,039	266,684	8.9	8.6	9.3	1,367	21,401
80-84	0.1214	0.4657	0.4415	0.4895	44,817	41,583	48,378	20,871	171,909	6.3	6.1	6.6	1,203	9,907
85-89	0.1840	0.6301	0.6045	0.6553	23,946	21,230	27,021	15,088	82,011	4.6	4.5	4.8	946	5,142
90-94	0.2778	0.7576	0.7360	0.7798	8,858	7,318	10,687	6,711	24,159	3.3	3.2	3.4	318	1,144
95-99	0.4179	0.9272	0.9161	0.9384	2,147	1,612	2,821	1,991	4,764	2.3	2.3	2.4	39	93
100+	0.6266	1	1	1	156	99	237	156	250	1.6	1.6	2.0	2	3
All ages													*9,281*	*1,531,438*

$\ln({}_nq_x)$ with uncertainty interval

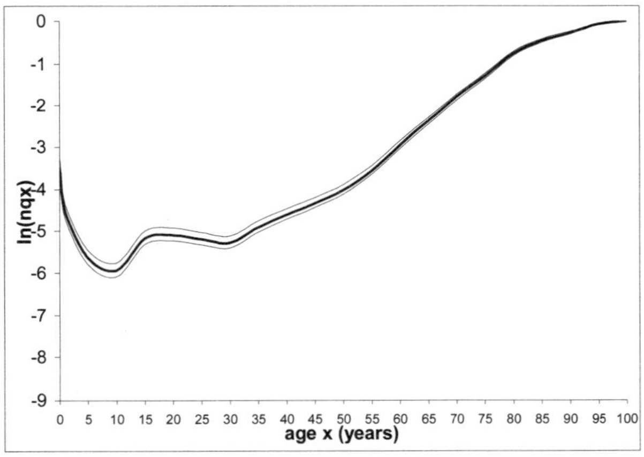

l_x with uncertainty interval

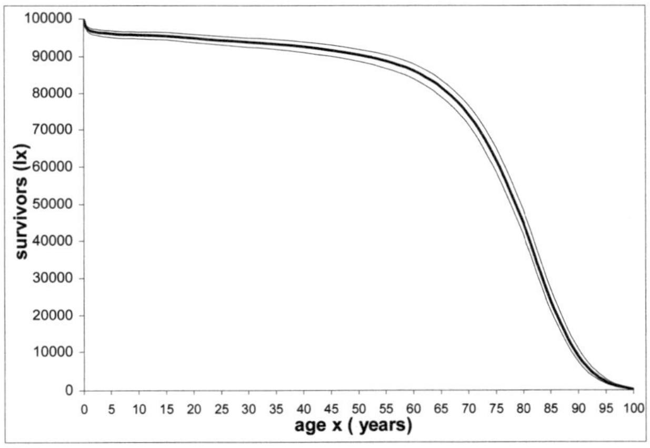

e_x with uncertainty interval

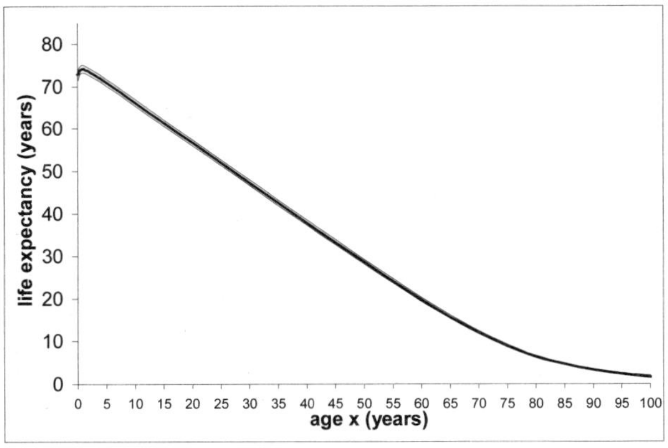

Distribution of deaths by age with uncertainty interval

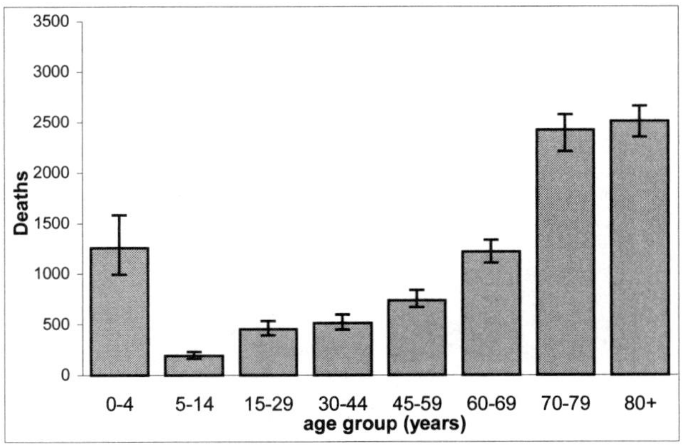

Algeria — Algérie — Argelia

Males, 2000 — *Hommes, 2000* — *Hombres, 2000*

Age	$_nM_x$	$_nq_x$ mean	$_nq_x$ percentiles 2½	$_nq_x$ percentiles 97½	l_x mean	l_x percentiles 2½	l_x percentiles 97½	$_nd_x$	$_nL_x$	e_x mean	e_x percentiles 2½	e_x percentiles 97½	Actual deaths	Actual population
<1	0.0464	0.0446	0.0360	0.0532	100,000	100,000	100,000	4,455	95,990	68.1	66.8	69.5	17,270	372,111
1-4	0.0026	0.0102	0.0087	0.0117	95,545	94,675	96,399	979	379,831	70.3	69.3	71.4	3,675	1,426,299
5-9	0.0009	0.0043	0.0037	0.0049	94,566	93,572	95,563	408	471,812	67.0	66.1	68.0	1,547	1,790,362
10-14	0.0007	0.0037	0.0032	0.0042	94,159	93,116	95,211	349	469,920	62.3	61.4	63.3	1,346	1,810,452
15-19	0.0012	0.0060	0.0052	0.0068	93,809	92,727	94,909	565	467,635	57.5	56.6	58.5	2,134	1,767,283
20-24	0.0014	0.0070	0.0061	0.0079	93,245	92,098	94,418	657	464,581	52.9	52.0	53.8	2,250	1,592,302
25-29	0.0014	0.0069	0.0060	0.0077	92,588	91,368	93,844	638	461,346	48.2	47.3	49.1	1,886	1,364,450
30-34	0.0014	0.0072	0.0062	0.0080	91,950	90,664	93,284	658	458,107	43.5	42.7	44.4	1,677	1,167,823
35-39	0.0018	0.0091	0.0080	0.0102	91,292	89,940	92,702	830	454,387	38.8	38.0	39.7	1,873	1,025,054
40-44	0.0027	0.0136	0.0120	0.0151	90,462	89,032	91,963	1,228	449,241	34.2	33.4	35.0	2,094	766,080
45-49	0.0044	0.0217	0.0193	0.0240	89,234	87,698	90,863	1,934	441,338	29.6	28.8	30.4	2,841	648,395
50-54	0.0073	0.0356	0.0321	0.0393	87,301	85,616	89,111	3,109	428,732	25.2	24.5	26.0	3,397	468,571
55-59	0.0120	0.0581	0.0524	0.0639	84,192	82,314	86,251	4,891	408,732	21.0	20.4	21.8	3,688	308,140
60-64	0.0194	0.0926	0.0839	0.1016	79,301	77,149	81,654	7,346	378,139	17.2	16.6	17.9	5,396	277,777
65-69	0.0313	0.1453	0.1320	0.1585	71,955	69,457	74,563	10,457	333,631	13.7	13.1	14.3	7,397	236,004
70-74	0.0513	0.2274	0.2074	0.2464	61,498	58,703	64,429	13,983	272,531	10.6	10.0	11.2	8,519	166,035
75-79	0.0840	0.3473	0.3184	0.3762	47,514	44,321	50,878	16,500	196,322	7.9	7.5	8.4	7,610	90,549
80-84	0.1349	0.5044	0.4677	0.5388	31,014	28,030	34,373	15,645	115,960	5.8	5.5	6.2	6,413	47,531
85-89	0.2085	0.6853	0.6499	0.7159	15,369	13,190	18,002	10,532	50,516	4.2	4.0	4.5	3,727	17,876
90-94	0.3074	0.7996	0.7718	0.8227	4,837	3,747	6,205	3,868	12,582	3.1	2.9	3.2	928	3,018
95-99	0.4324	0.9412	0.9274	0.9519	969	664	1,416	912	2,110	2.3	2.2	2.3	127	293
100+	0.5804	1	1	1	57	32	103	57	98	1.7	1.7	1.8	9	16
All ages													*85,804*	*15,346,421*

ln($_nq_x$) with uncertainty interval

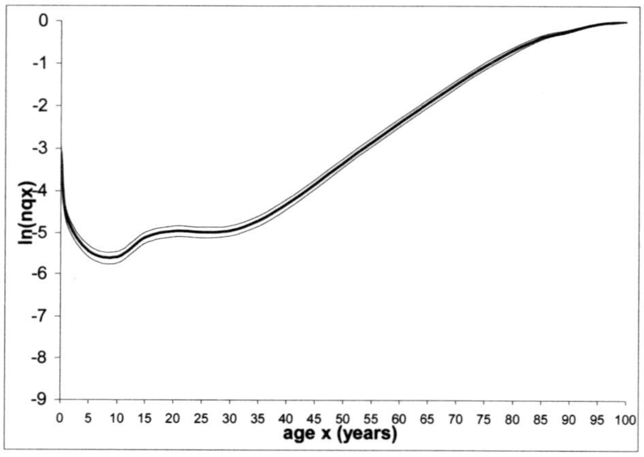

l_x with uncertainty interval

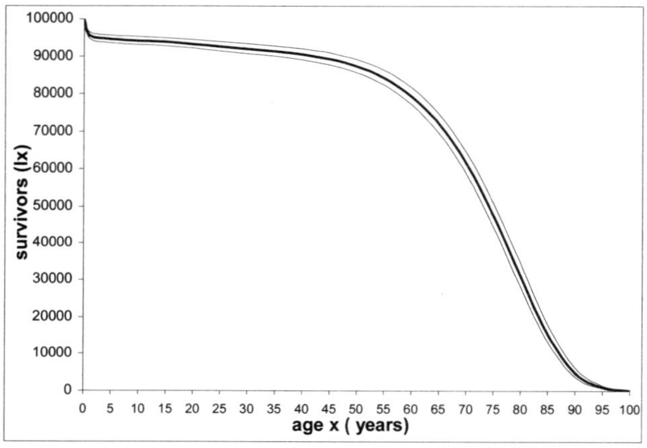

e_x with uncertainty interval

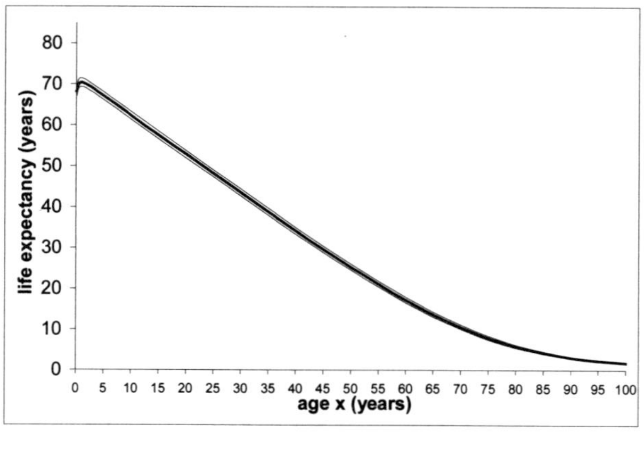

Distribution of deaths by age with uncertainty interval

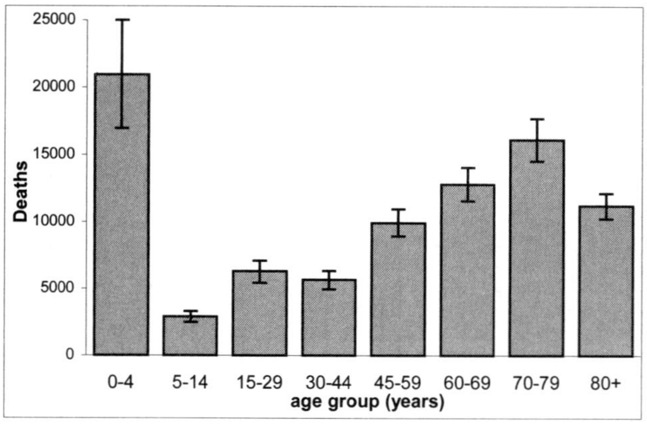

Algeria — Algérie — Argelia

Females, 2000 — *Femmes, 2000* — *Mujeres, 2000*

Age	$_nM_x$	$_nq_x$			l_x			$_nd_x$	$_nL_x$	e_x			Actual deaths	Actual population
			percentiles			percentiles					percentiles			
		mean	2½	97½	mean	2½	97½			mean	2½	97½		
<1	0.0396	0.0382	0.0301	0.0466	100,000	100,000	100,000	3,823	96,559	71.2	69.9	72.5	14,025	354,223
1-4	0.0023	0.0092	0.0077	0.0107	96,177	95,343	96,992	882	382,590	73.1	72.2	74.1	3,150	1,366,317
5-9	0.0006	0.0031	0.0026	0.0036	95,295	94,353	96,245	295	475,736	69.7	68.9	70.7	1,060	1,708,279
10-14	0.0004	0.0022	0.0019	0.0025	95,000	94,026	95,993	208	474,479	64.9	64.1	65.9	755	1,726,279
15-19	0.0006	0.0030	0.0025	0.0034	94,792	93,796	95,815	281	473,258	60.1	59.3	61.0	1,005	1,693,380
20-24	0.0008	0.0040	0.0034	0.0046	94,511	93,480	95,573	373	471,622	55.2	54.4	56.2	1,213	1,532,347
25-29	0.0010	0.0050	0.0043	0.0057	94,138	93,054	95,251	467	469,521	50.4	49.7	51.4	1,283	1,288,867
30-34	0.0013	0.0062	0.0054	0.0071	93,671	92,524	94,844	585	466,890	45.7	44.9	46.6	1,426	1,137,906
35-39	0.0017	0.0086	0.0075	0.0098	93,085	91,863	94,332	805	463,416	40.9	40.2	41.9	1,755	1,010,645
40-44	0.0024	0.0120	0.0105	0.0136	92,281	90,958	93,622	1,105	458,642	36.3	35.6	37.2	1,789	742,816
45-49	0.0036	0.0179	0.0158	0.0202	91,176	89,724	92,637	1,634	451,795	31.7	31.0	32.6	2,193	606,420
50-54	0.0055	0.0272	0.0243	0.0303	89,542	87,915	91,160	2,434	441,625	27.2	26.6	28.1	2,527	458,421
55-59	0.0084	0.0413	0.0371	0.0456	87,108	85,251	88,923	3,595	426,551	22.9	22.3	23.7	2,755	326,909
60-64	0.0138	0.0666	0.0598	0.0729	83,513	81,376	85,572	5,564	403,653	18.8	18.3	19.5	4,205	305,056
65-69	0.0233	0.1099	0.0987	0.1194	77,948	75,498	80,289	8,570	368,317	15.0	14.5	15.6	6,251	268,642
70-74	0.0411	0.1863	0.1683	0.1998	69,378	66,658	71,907	12,922	314,587	11.5	11.1	12.0	7,812	190,186
75-79	0.0721	0.3055	0.2805	0.3266	56,456	53,580	59,756	17,246	239,167	8.6	8.2	9.0	8,725	121,003
80-84	0.1214	0.4657	0.4363	0.4941	39,210	36,344	42,993	18,261	150,400	6.2	5.9	6.6	7,830	64,491
85-89	0.1938	0.6528	0.6223	0.6801	20,950	18,881	24,138	13,676	70,559	4.5	4.3	4.7	6,217	32,075
90-94	0.2931	0.7798	0.7547	0.8006	7,274	6,059	8,897	5,672	19,354	3.2	3.1	3.3	2,555	8,717
95-99	0.4198	0.9290	0.9156	0.9393	1,602	1,208	2,122	1,488	3,545	2.3	2.3	2.4	729	1,737
100+	0.5696	1	1	1	114	73	178	114	200	1.8	1.7	1.9	116	204
All ages													*79,377*	*14,944,920*

$ln(_nq_x)$ with uncertainty interval

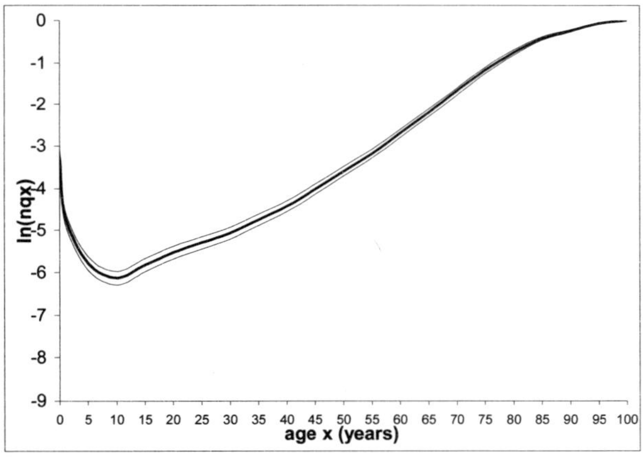

l_x with uncertainty interval

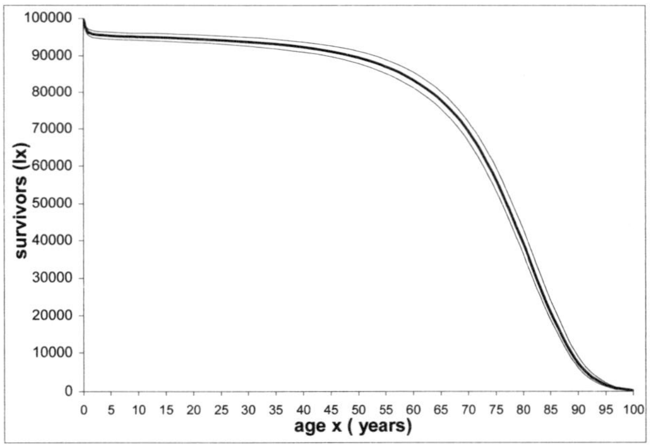

e_x with uncertainty interval

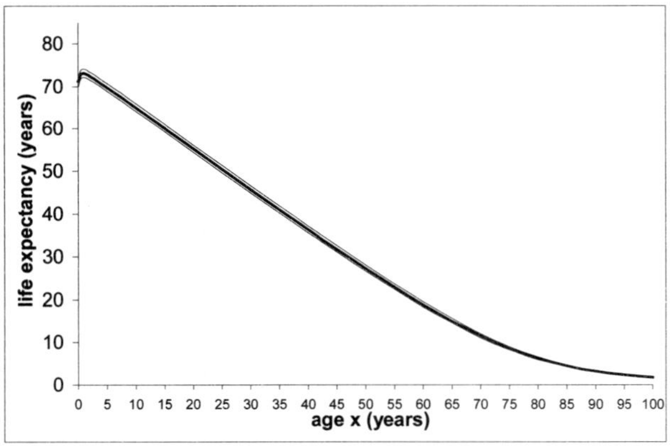

Distribution of deaths by age with uncertainty interval

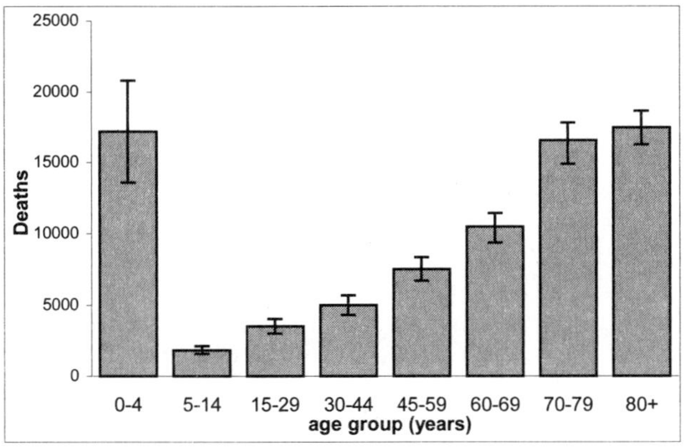

Andorra — Andorre — Andorra

Males, 2000 — *Hommes, 2000* — *Hombres, 2000*

Age	$_nM_x$	$_nq_x$ mean	$_nq_x$ percentiles 2½	$_nq_x$ percentiles 97½	l_x mean	l_x percentiles 2½	l_x percentiles 97½	$_nd_x$	$_nL_x$	e_x mean	e_x percentiles 2½	e_x percentiles 97½	Actual deaths	Actual population
<1	0.0042	0.0042	0.0025	0.0060	100,000	100,000	100,000	421	99,621	77.2	74.0	81.7	2	426
1-4	0.0002	0.0009	0.0006	0.0012	99,579	99,398	99,748	91	398,099	76.5	73.3	81.1	0	1,896
5-9	0.0002	0.0009	0.0006	0.0012	99,488	99,288	99,688	88	497,221	72.6	69.4	77.1	0	2,347
10-14	0.0002	0.0009	0.0006	0.0012	99,400	99,171	99,629	86	496,786	67.6	64.5	72.2	0	2,210
15-19	0.0005	0.0026	0.0017	0.0036	99,314	99,056	99,571	262	495,917	62.7	59.5	67.3	1	2,460
20-24	0.0008	0.0040	0.0026	0.0054	99,053	98,702	99,396	396	494,272	57.8	54.7	62.4	2	3,116
25-29	0.0009	0.0046	0.0030	0.0062	98,656	98,171	99,131	458	492,135	53.1	49.9	57.6	4	4,088
30-34	0.0013	0.0065	0.0043	0.0086	98,198	97,561	98,830	638	489,393	48.3	45.2	52.8	6	4,782
35-39	0.0016	0.0079	0.0052	0.0105	97,559	96,718	98,409	775	485,858	43.6	40.6	48.1	7	4,599
40-44	0.0021	0.0103	0.0068	0.0136	96,784	95,703	97,894	994	481,435	38.9	36.0	43.3	8	3,942
45-49	0.0031	0.0154	0.0102	0.0206	95,790	94,415	97,227	1,471	475,271	34.3	31.4	38.6	11	3,416
50-54	0.0048	0.0235	0.0156	0.0321	94,318	92,531	96,231	2,220	466,043	29.8	27.0	34.0	14	2,844
55-59	0.0070	0.0344	0.0220	0.0476	92,099	89,734	94,704	3,164	452,583	25.5	22.9	29.5	14	2,068
60-64	0.0118	0.0571	0.0354	0.0797	88,935	85,583	92,481	5,078	431,979	21.3	18.9	25.1	21	1,804
65-69	0.0183	0.0873	0.0529	0.1216	83,857	78,758	88,797	7,319	400,986	17.4	15.3	20.9	30	1,634
70-74	0.0295	0.1376	0.0830	0.1877	76,538	69,183	84,080	10,530	356,363	13.8	12.0	17.0	40	1,338
75-79	0.0495	0.2202	0.1364	0.2880	66,008	56,200	77,105	14,534	293,705	10.7	9.2	13.3	47	957
80-84	0.0835	0.3455	0.2299	0.4304	51,474	40,065	66,591	17,783	212,912	8.0	6.8	10.0	46	550
85-89	0.1343	0.5028	0.3753	0.5891	33,691	23,209	51,280	16,939	126,105	5.8	5.1	7.2	44	331
90-94	0.2058	0.6363	0.5315	0.7064	16,751	9,927	32,033	10,658	51,783	4.2	3.7	5.1	39	189
95-99	0.3005	0.7901	0.7193	0.8394	6,093	2,975	15,007	4,815	16,023	3.1	2.8	3.5	8	25
100+	0.4180	1	1	1	1,279	478	4,182	1,279	3,059	2.4	2.2	2.5	0	0
All ages													*346*	*45,024*

ln($_nq_x$) with uncertainty interval

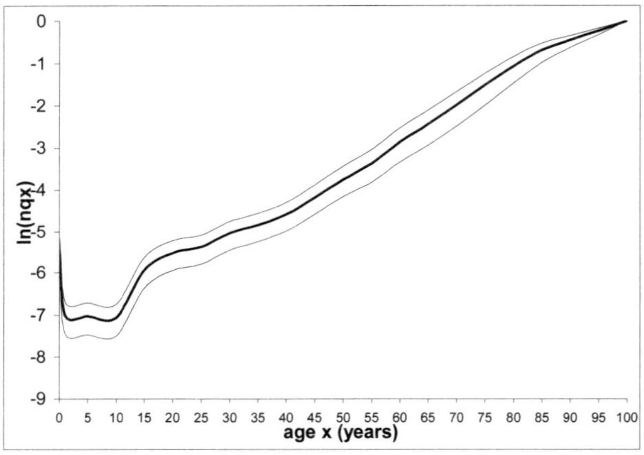

l_x with uncertainty interval

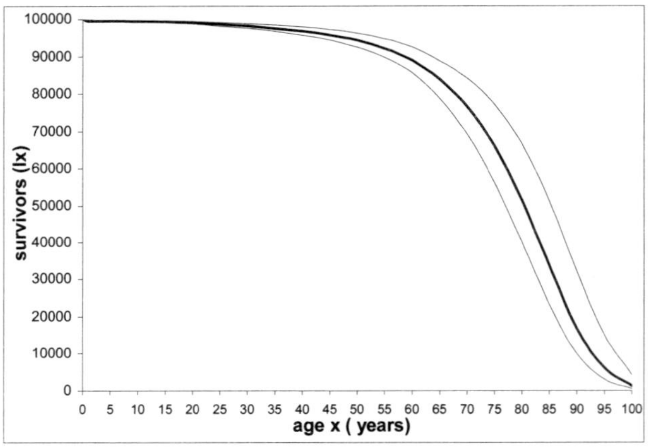

e_x with uncertainty interval

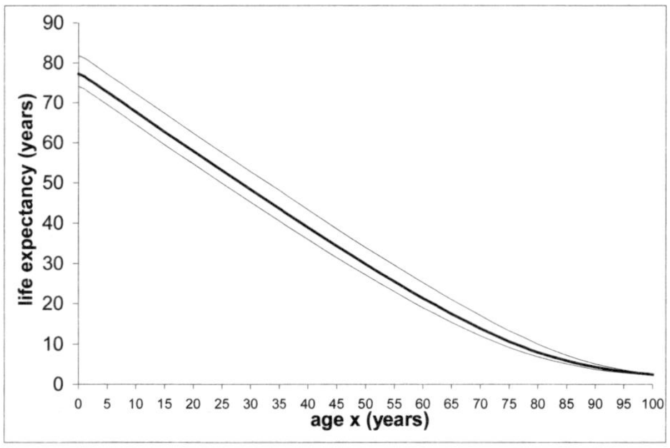

Distribution of deaths by age with uncertainty interval

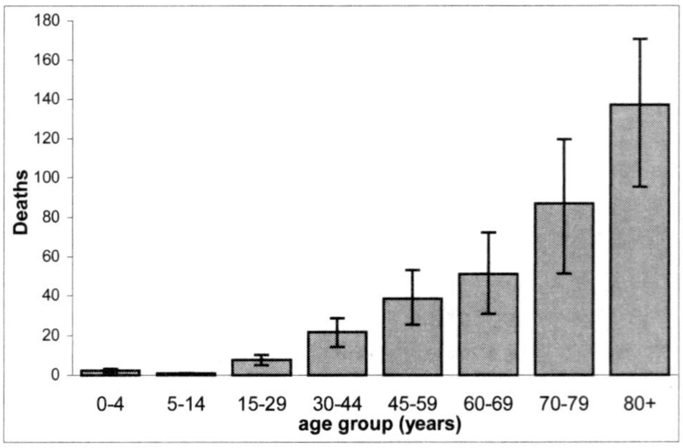

Andorra　　Andorre　　Andorra

Females, 2000　　*Femmes, 2000*　　*Mujeres, 2000*

Age	$_nM_x$	$_nq_x$			l_x			$_nd_x$	$_nL_x$	e_x			Actual deaths	Actual population
			percentiles			percentiles					percentiles			
		mean	2 1/2	97 1/2	mean	2 1/2	97 1/2			mean	2 1/2	97 1/2		
<1	0.0034	0.0034	0.0018	0.0051	100,000	100,000	100,000	342	99,692	83.8	80.6	87.3	1	396
1-4	0.0002	0.0008	0.0005	0.0012	99,658	99,487	99,822	84	398,430	83.1	80.0	86.6	0	1,691
5-9	0.0001	0.0005	0.0003	0.0007	99,574	99,374	99,773	50	497,743	79.2	76.1	82.7	0	2,179
10-14	0.0001	0.0006	0.0003	0.0009	99,524	99,306	99,744	59	497,471	74.2	71.2	77.7	0	2,063
15-19	0.0002	0.0012	0.0006	0.0017	99,465	99,226	99,711	115	497,037	69.2	66.2	72.8	1	2,302
20-24	0.0002	0.0011	0.0006	0.0017	99,350	99,070	99,647	113	496,467	64.3	61.3	67.8	1	3,041
25-29	0.0003	0.0015	0.0009	0.0022	99,237	98,912	99,582	152	495,802	59.4	56.4	62.9	1	4,110
30-34	0.0004	0.0021	0.0012	0.0030	99,084	98,695	99,495	206	494,907	54.5	51.5	58.0	2	4,427
35-39	0.0006	0.0030	0.0017	0.0044	98,878	98,402	99,375	292	493,662	49.6	46.7	53.1	2	3,961
40-44	0.0009	0.0045	0.0026	0.0066	98,586	97,980	99,205	440	491,832	44.7	41.9	48.2	3	3,264
45-49	0.0013	0.0064	0.0038	0.0096	98,146	97,331	98,943	628	489,162	39.9	37.1	43.4	4	2,742
50-54	0.0019	0.0094	0.0057	0.0142	97,518	96,399	98,565	917	485,299	35.2	32.5	38.6	4	2,331
55-59	0.0026	0.0131	0.0081	0.0198	96,601	95,035	98,001	1,263	479,850	30.5	27.9	33.8	5	1,857
60-64	0.0044	0.0218	0.0138	0.0330	95,339	93,157	97,208	2,075	471,505	25.8	23.4	29.1	7	1,575
65-69	0.0074	0.0363	0.0231	0.0548	93,263	90,086	95,869	3,387	457,849	21.4	19.1	24.5	11	1,498
70-74	0.0136	0.0659	0.0416	0.0977	89,876	85,152	93,597	5,923	434,573	17.1	15.1	20.1	18	1,295
75-79	0.0268	0.1256	0.0774	0.1787	83,953	76,831	89,452	10,545	393,402	13.1	11.5	15.8	26	956
80-84	0.0539	0.2373	0.1486	0.3124	73,408	63,101	81,962	17,422	323,483	9.6	8.2	11.9	33	609
85-89	0.1002	0.4005	0.2722	0.4993	55,986	43,388	69,702	22,425	223,867	6.8	5.8	8.6	38	375
90-94	0.1725	0.5683	0.4390	0.6604	33,561	22,748	50,730	19,072	110,589	4.7	4.0	5.9	38	219
95-99	0.2749	0.7532	0.6633	0.8169	14,489	8,216	28,460	10,913	39,703	3.3	3.0	3.9	10	35
100+	0.4056	1	1	1	3,575	1,505	9,583	3,575	8,816	2.5	2.4	2.6	0	1
All ages													*204*	*40,927*

ln($_nq_x$) with uncertainty interval

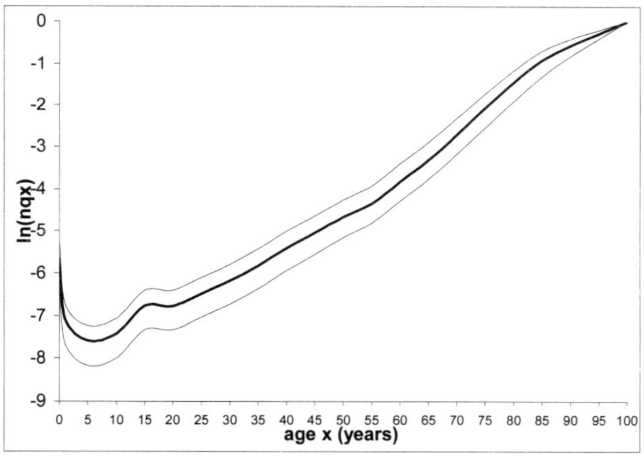

l_x with uncertainty interval

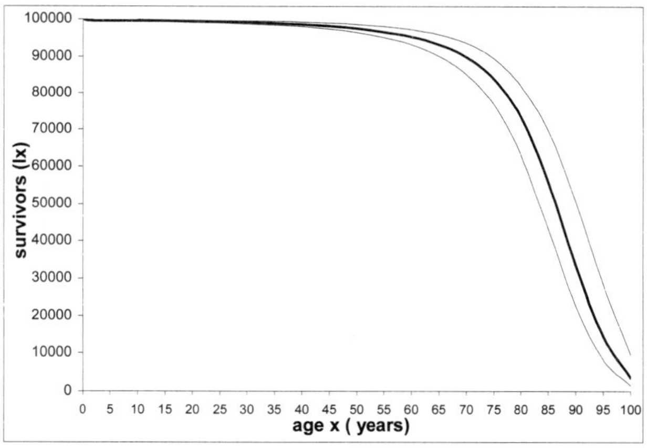

e_x with uncertainty interval

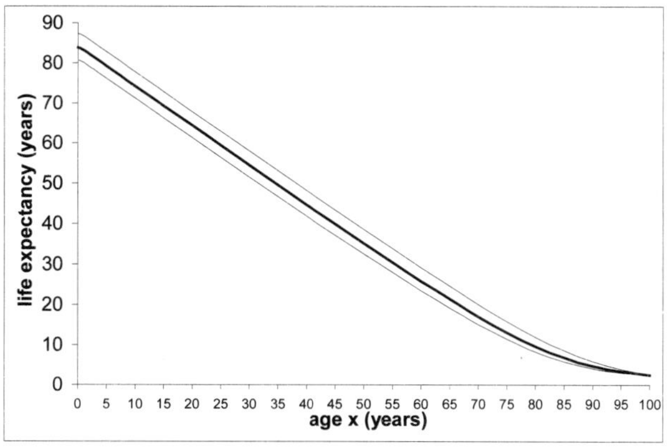

Distribution of deaths by age with uncertainty interval

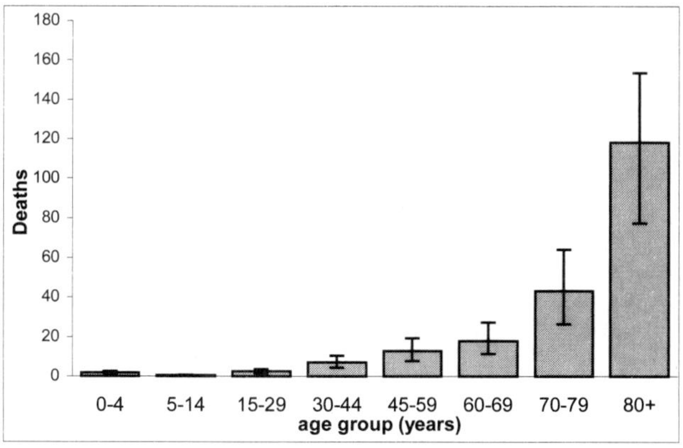

Angola — Angola — Angola

Males, 2000 — *Hommes, 2000* — *Hombres, 2000*

Age	$_nM_x$	$_nq_x$			l_x			$_nd_x$	$_nL_x$	e_x			Actual deaths	Actual population
			percentiles			percentiles					percentiles			
		mean	2½	97½	mean	2½	97½			mean	2½	97½		
<1	0.1935	0.1648	0.1375	0.1943	100,000	100,000	100,000	16,479	85,169	44.3	40.2	49.1	57,624	297,812
1-4	0.0162	0.0625	0.0501	0.0753	83,521	80,571	86,250	5,217	321,562	52.0	47.3	57.2	16,256	1,001,963
5-9	0.0040	0.0198	0.0155	0.0244	78,304	75,804	80,797	1,551	387,642	51.3	47.0	56.3	4,095	1,023,794
10-14	0.0029	0.0144	0.0112	0.0179	76,753	74,007	79,401	1,107	380,997	47.3	43.1	52.2	2,450	842,991
15-19	0.0026	0.0127	0.0098	0.0158	75,646	72,726	78,468	962	375,823	43.0	38.8	47.8	1,751	683,941
20-24	0.0058	0.0286	0.0219	0.0358	74,683	71,614	77,657	2,138	368,072	38.5	34.4	43.2	3,235	557,007
25-29	0.0097	0.0473	0.0359	0.0599	72,546	69,126	75,853	3,429	354,155	34.6	30.6	39.2	4,279	441,887
30-34	0.0129	0.0624	0.0469	0.0802	69,116	65,148	72,948	4,311	334,804	31.1	27.3	35.6	4,541	352,664
35-39	0.0154	0.0741	0.0551	0.0965	64,805	60,138	69,412	4,800	312,025	28.1	24.4	32.3	4,516	293,535
40-44	0.0177	0.0847	0.0626	0.1115	60,005	54,512	65,515	5,082	287,319	25.1	21.8	29.0	4,279	241,892
45-49	0.0199	0.0950	0.0692	0.1258	54,923	48,676	61,415	5,218	261,568	22.2	19.1	25.8	3,941	197,554
50-54	0.0232	0.1095	0.0790	0.1459	49,705	42,842	57,122	5,444	234,913	19.3	16.5	22.5	3,793	163,644
55-59	0.0282	0.1317	0.0944	0.1754	44,261	36,773	52,518	5,830	206,728	16.3	13.9	19.2	3,713	131,677
60-64	0.0389	0.1773	0.1275	0.2337	38,431	30,400	47,406	6,813	175,122	13.4	11.4	16.0	3,996	102,718
65-69	0.0548	0.2411	0.1754	0.3115	31,618	23,409	41,121	7,622	139,037	10.8	9.1	13.0	4,118	75,123
70-74	0.0825	0.3421	0.2561	0.4270	23,997	16,257	33,552	8,210	99,457	8.4	7.1	10.2	4,059	49,168
75-79	0.1245	0.4747	0.3720	0.5659	15,786	9,438	24,479	7,494	60,198	6.4	5.5	7.8	3,372	27,084
80-84	0.1715	0.6002	0.4955	0.6857	8,293	4,097	15,371	4,978	29,020	5.0	4.3	6.0	1,939	11,303
85-89	0.2448	0.7594	0.6663	0.8285	3,315	1,288	7,755	2,518	10,282	3.8	3.4	4.4	699	2,855
90-94	0.3402	0.8419	0.7655	0.8959	798	221	2,587	671	1,974	2.8	2.5	3.2	132	388
95-99	0.4603	0.9666	0.9325	0.9844	126	23	607	122	265	2.2	2.1	2.3	12	25
100+	0.6061	1	1	1	4	0	41	4	7	1.6	1.6	1.7	1	1
All ages													*132,799*	*6,499,026*

ln($_nq_x$) with uncertainty interval

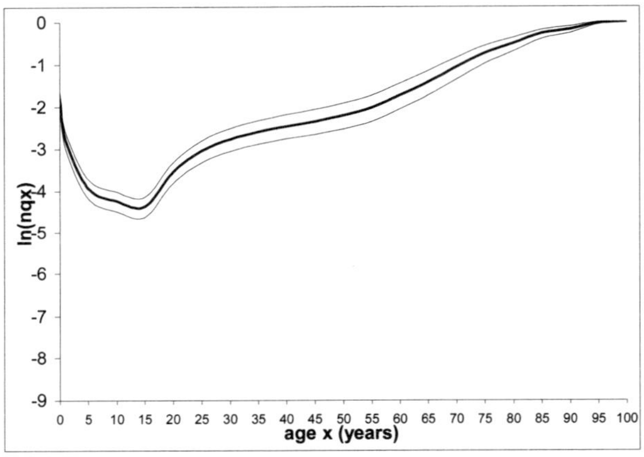

l_x with uncertainty interval

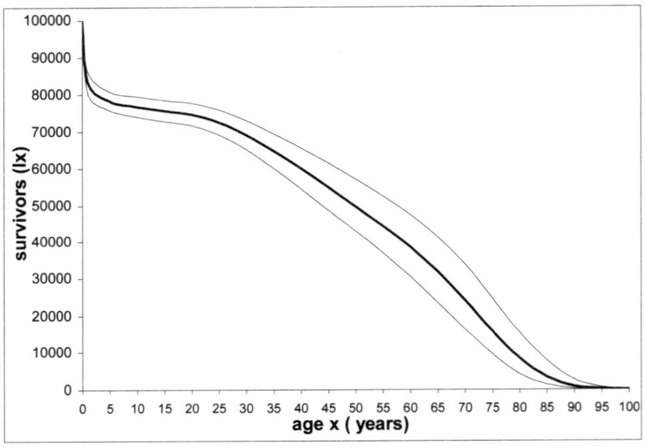

e_x with uncertainty interval

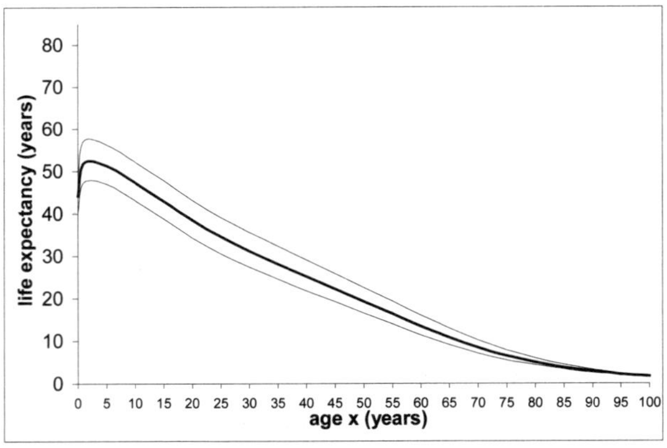

Distribution of deaths by age with uncertainty interval

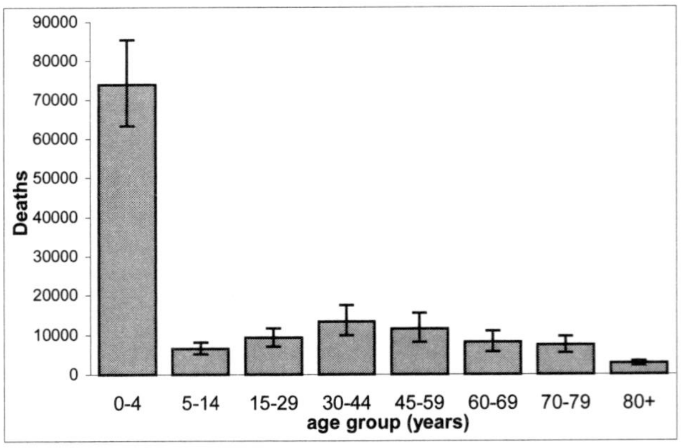

Angola

Angola

Angola

Females, 2000

Femmes, 2000

Mujeres, 2000

Age	$_nM_x$	$_nq_x$			l_x			$_nd_x$	$_nL_x$	e_x			Actual deaths	Actual population
			percentiles			percentiles					percentiles			
		mean	2½	97½	mean	2½	97½			mean	2½	97½		
<1	0.1609	0.1406	0.1163	0.1640	100,000	100,000	100,000	14,057	87,348	48.3	43.8	53.3	46,085	286,358
1-4	0.0174	0.0667	0.0536	0.0813	85,943	83,602	88,375	5,729	330,020	55.2	50.2	60.8	17,462	1,005,851
5-9	0.0038	0.0187	0.0145	0.0235	80,213	77,989	82,463	1,502	397,311	55.1	50.5	60.1	3,865	1,022,311
10-14	0.0025	0.0123	0.0095	0.0156	78,711	76,282	81,092	971	391,130	51.1	46.7	56.0	2,096	844,483
15-19	0.0039	0.0193	0.0147	0.0245	77,741	75,168	80,208	1,497	384,961	46.7	42.4	51.5	2,672	687,156
20-24	0.0069	0.0339	0.0257	0.0434	76,244	73,452	78,991	2,581	374,766	42.5	38.4	47.2	3,879	563,184
25-29	0.0094	0.0459	0.0343	0.0594	73,663	70,337	76,907	3,378	359,869	38.9	35.0	43.4	4,223	449,909
30-34	0.0106	0.0515	0.0381	0.0673	70,285	66,162	74,177	3,621	342,370	35.7	32.1	39.8	3,825	361,572
35-39	0.0111	0.0538	0.0393	0.0707	66,663	61,711	71,240	3,585	324,355	32.5	29.1	36.3	3,354	303,516
40-44	0.0105	0.0513	0.0372	0.0677	63,079	57,350	68,314	3,235	307,306	29.2	26.1	32.7	2,672	253,797
45-49	0.0114	0.0556	0.0401	0.0735	59,844	53,467	65,755	3,326	290,903	25.6	22.7	29.0	2,422	211,831
50-54	0.0145	0.0700	0.0503	0.0924	56,518	49,538	63,119	3,954	272,703	22.0	19.3	25.1	2,555	176,174
55-59	0.0193	0.0920	0.0662	0.1223	52,563	44,964	59,942	4,835	250,729	18.5	16.0	21.3	2,810	145,694
60-64	0.0286	0.1333	0.0964	0.1773	47,728	39,535	55,976	6,364	222,732	15.1	12.9	17.7	3,343	116,992
65-69	0.0433	0.1955	0.1422	0.2575	41,365	32,700	50,579	8,085	186,609	12.0	10.2	14.3	3,835	88,500
70-74	0.0681	0.2910	0.2153	0.3735	33,279	24,588	43,327	9,685	142,183	9.3	7.8	11.3	4,128	60,605
75-79	0.1037	0.4117	0.3155	0.5067	23,594	15,809	33,806	9,714	93,685	7.1	6.0	8.7	3,681	35,502
80-84	0.1541	0.5562	0.4498	0.6501	13,880	7,985	22,810	7,720	50,098	5.4	4.6	6.5	2,502	16,234
85-89	0.2242	0.7184	0.6199	0.7967	6,159	2,793	12,191	4,425	19,734	4.0	3.5	4.8	1,080	4,815
90-94	0.3174	0.8129	0.7318	0.8744	1,734	568	4,634	1,410	4,442	3.0	2.6	3.5	271	854
95-99	0.4370	0.9455	0.9014	0.9722	325	71	1,243	307	702	2.3	2.1	2.5	36	83
100+	0.5855	1	1	1	18	2	122	18	30	1.7	1.7	1.8	3	5
All ages													*116,796*	*6,635,426*

$\ln(_nq_x)$ with uncertainty interval

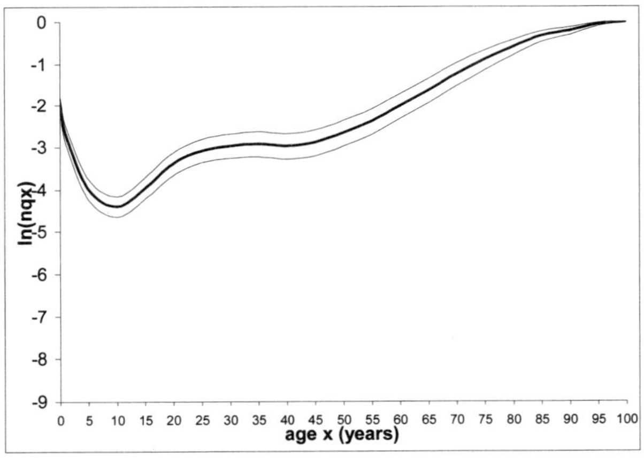

l_x with uncertainty interval

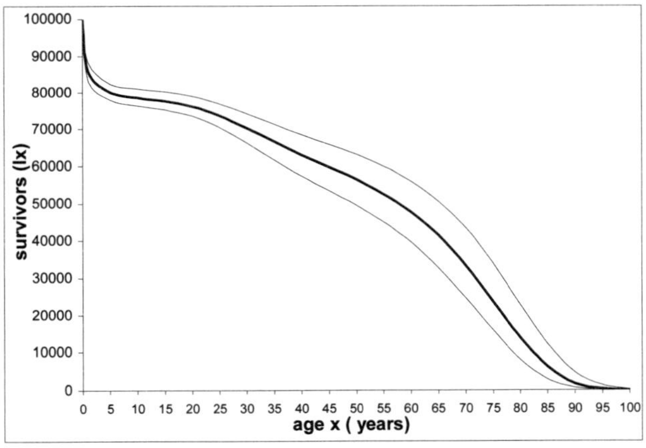

e_x with uncertainty interval

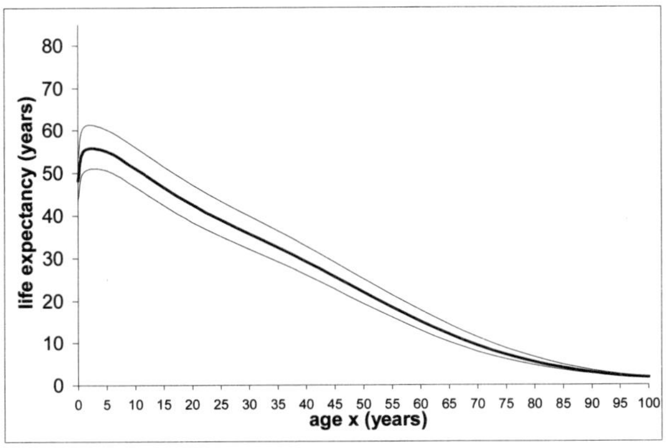

Distribution of deaths by age with uncertainty interval

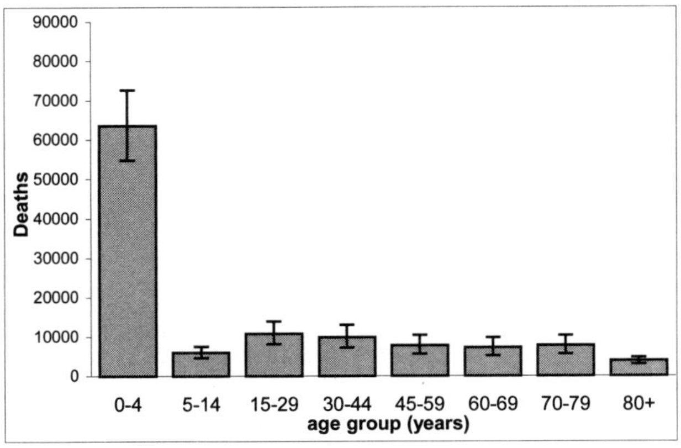

Antigua and Barbuda · Antigua-et-Barbuda · Antigua y Barbuda

Males, 2000 · *Hommes, 2000* · *Hombres, 2000*

Age	$_nM_x$	$_nq_x$ mean	$_nq_x$ percentiles 2½	$_nq_x$ percentiles 97½	l_x mean	l_x percentiles 2½	l_x percentiles 97½	$_nd_x$	$_nL_x$	e_x mean	e_x percentiles 2½	e_x percentiles 97½	Actual deaths	Actual population
<1	0.0240	0.0235	0.0183	0.0276	100,000	100,000	100,000	2,346	97,889	71.8	70.5	73.1	15	643
1-4	0.0005	0.0020	0.0016	0.0023	97,654	97,237	98,166	195	390,148	72.5	71.4	73.8	1	2,548
5-9	0.0004	0.0021	0.0018	0.0024	97,459	97,014	98,005	208	486,774	68.7	67.6	69.9	1	3,211
10-14	0.0006	0.0030	0.0025	0.0034	97,251	96,777	97,833	291	485,525	63.8	62.7	65.1	2	3,346
15-19	0.0019	0.0093	0.0078	0.0106	96,959	96,445	97,590	902	482,542	59.0	57.9	60.2	6	3,114
20-24	0.0028	0.0140	0.0119	0.0159	96,057	95,424	96,832	1,349	476,915	54.5	53.5	55.7	8	2,776
25-29	0.0036	0.0177	0.0152	0.0198	94,709	93,909	95,679	1,673	469,359	50.2	49.3	51.4	10	2,787
30-34	0.0017	0.0083	0.0072	0.0093	93,035	92,046	94,222	774	463,243	46.1	45.2	47.3	4	2,606
35-39	0.0021	0.0103	0.0090	0.0114	92,262	91,189	93,540	946	458,944	41.5	40.6	42.6	5	2,354
40-44	0.0036	0.0180	0.0159	0.0200	91,316	90,147	92,699	1,643	452,473	36.9	36.0	38.0	6	1,781
45-49	0.0057	0.0282	0.0253	0.0311	89,673	88,347	91,224	2,530	442,041	32.5	31.7	33.6	9	1,577
50-54	0.0083	0.0407	0.0366	0.0448	87,143	85,598	88,917	3,550	426,842	28.4	27.7	29.4	11	1,366
55-59	0.0107	0.0522	0.0468	0.0572	83,593	81,765	85,661	4,362	407,061	24.5	23.8	25.4	12	1,132
60-64	0.0141	0.0682	0.0611	0.0745	79,231	77,088	81,656	5,405	382,645	20.7	20.1	21.5	13	925
65-69	0.0201	0.0956	0.0857	0.1037	73,826	71,349	76,670	7,057	351,491	17.0	16.5	17.7	14	703
70-74	0.0332	0.1533	0.1382	0.1646	66,770	63,952	70,103	10,234	308,265	13.6	13.0	14.2	19	561
75-79	0.0499	0.2219	0.2022	0.2381	56,536	53,429	60,416	12,544	251,320	10.6	10.1	11.0	19	382
80-84	0.0844	0.3485	0.3237	0.3728	43,992	40,709	48,201	15,330	181,634	7.8	7.5	8.2	20	241
85-89	0.1373	0.5110	0.4838	0.5395	28,662	25,535	32,597	14,645	106,696	5.7	5.4	6.0	16	116
90-94	0.2147	0.6529	0.6282	0.6788	14,017	11,758	16,826	9,152	42,628	4.1	3.9	4.2	8	37
95-99	0.3229	0.8201	0.8019	0.8387	4,865	3,777	6,256	3,990	12,355	2.9	2.8	3.0	2	7
100+	0.4672	1	1	1	875	609	1,239	875	1,873	2.1	2.1	2.2	0	1
All ages													*204*	*32,213*

ln($_nq_x$) with uncertainty interval

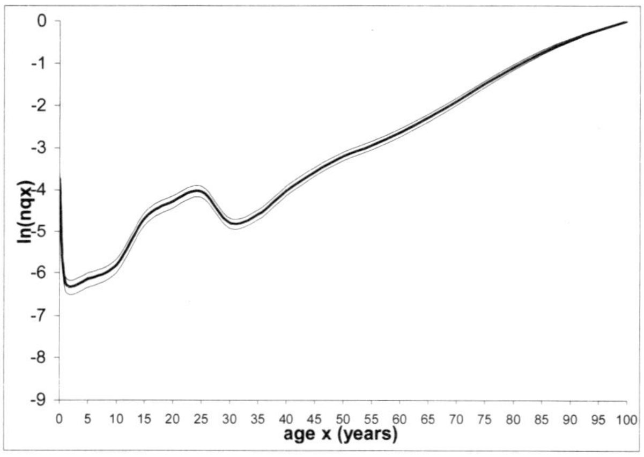

l_x with uncertainty interval

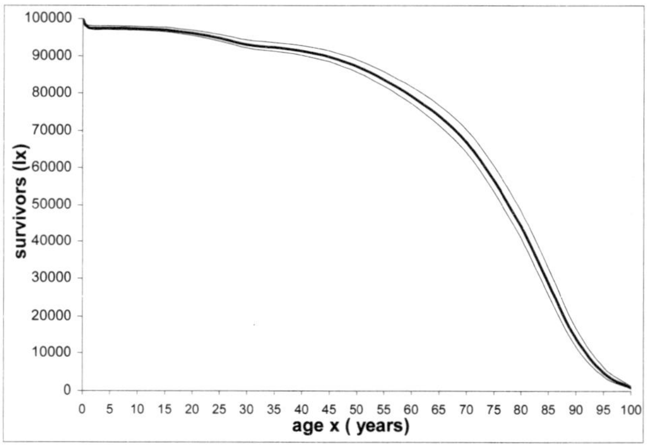

e_x with uncertainty interval

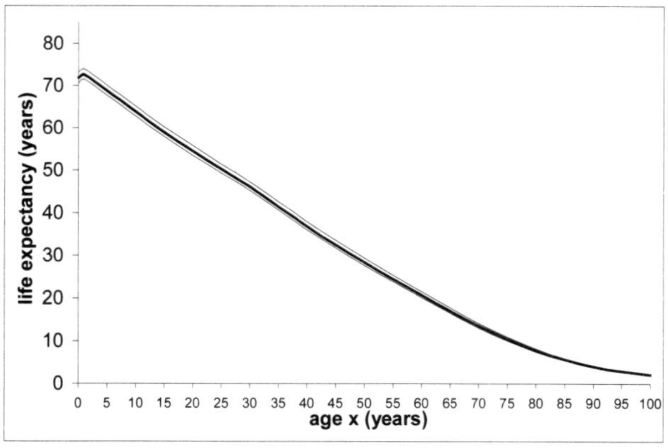

Distribution of deaths by age with uncertainty interval

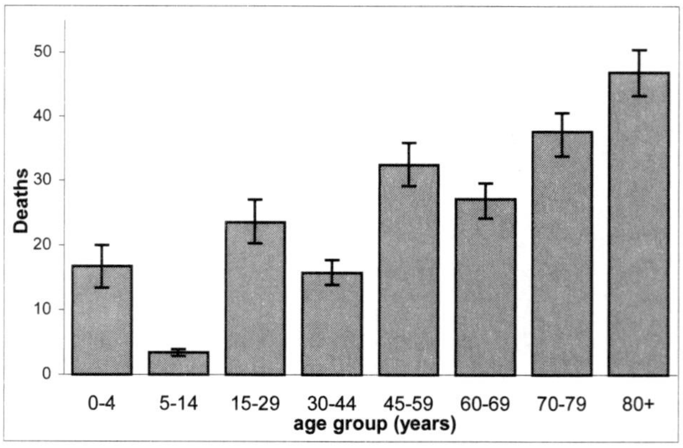

Antigua and Barbuda | Antigua-et-Barbuda | Antigua y Barbuda

Females, 2000 | *Femmes, 2000* | *Mujeres, 2000*

Age	$_nM_x$	$_nq_x$			l_x			$_nd_x$	$_nL_x$	e_x			Actual deaths	Actual population
			percentiles			percentiles					percentiles			
		mean	2½	97½	mean	2½	97½			mean	2½	97½		
<1	0.0186	0.0183	0.0144	0.0225	100,000	100,000	100,000	1,831	98,352	76.6	75.4	77.9	12	622
1-4	0.0010	0.0039	0.0032	0.0046	98,169	97,750	98,556	380	391,765	77.1	76.0	78.3	2	2,440
5-9	0.0004	0.0019	0.0016	0.0022	97,789	97,305	98,241	184	488,486	73.3	72.3	74.6	1	3,091
10-14	0.0005	0.0025	0.0021	0.0029	97,605	97,091	98,087	244	487,418	68.5	67.4	69.7	2	3,238
15-19	0.0009	0.0043	0.0036	0.0050	97,362	96,809	97,883	419	485,762	63.6	62.6	64.9	3	3,014
20-24	0.0013	0.0063	0.0053	0.0072	96,943	96,326	97,529	608	483,195	58.9	57.9	60.1	3	2,700
25-29	0.0014	0.0070	0.0060	0.0080	96,335	95,631	97,012	678	479,979	54.3	53.3	55.5	4	2,734
30-34	0.0018	0.0091	0.0078	0.0103	95,657	94,863	96,430	866	476,118	49.6	48.7	50.8	5	2,593
35-39	0.0024	0.0119	0.0103	0.0134	94,790	93,888	95,679	1,124	471,141	45.1	44.1	46.2	6	2,435
40-44	0.0033	0.0164	0.0144	0.0184	93,666	92,632	94,695	1,538	464,485	40.6	39.7	41.7	6	1,922
45-49	0.0046	0.0226	0.0200	0.0251	92,128	90,928	93,333	2,082	455,436	36.2	35.4	37.3	8	1,705
50-54	0.0058	0.0284	0.0255	0.0313	90,046	88,643	91,464	2,554	443,846	32.0	31.2	33.0	9	1,480
55-59	0.0072	0.0356	0.0320	0.0391	87,492	85,867	89,136	3,111	429,682	27.9	27.1	28.8	9	1,227
60-64	0.0109	0.0531	0.0476	0.0582	84,381	82,508	86,287	4,480	410,703	23.8	23.1	24.7	11	1,013
65-69	0.0162	0.0778	0.0696	0.0849	79,900	77,702	82,179	6,214	383,966	20.0	19.4	20.8	13	796
70-74	0.0244	0.1150	0.1032	0.1246	73,686	71,107	76,457	8,476	347,239	16.5	15.9	17.2	16	651
75-79	0.0348	0.1600	0.1443	0.1715	65,210	62,249	68,567	10,432	299,969	13.3	12.8	13.9	16	455
80-84	0.0536	0.2364	0.2158	0.2537	54,778	51,575	58,670	12,947	241,522	10.3	9.9	10.8	16	299
85-89	0.0845	0.3487	0.3243	0.3731	41,831	38,490	46,007	14,588	172,685	7.7	7.4	8.1	13	152
90-94	0.1361	0.4832	0.4536	0.5111	27,243	24,129	31,089	13,164	96,722	5.5	5.3	5.8	7	55
95-99	0.2242	0.6702	0.6411	0.6960	14,079	11,797	16,987	9,435	42,087	3.9	3.7	4.0	3	12
100+	0.3775	1	1	1	4,643	3,587	6,096	4,643	12,298	2.6	2.5	2.7	1	2
All ages													*164*	*32,635*

$\ln(_nq_x)$ with uncertainty interval

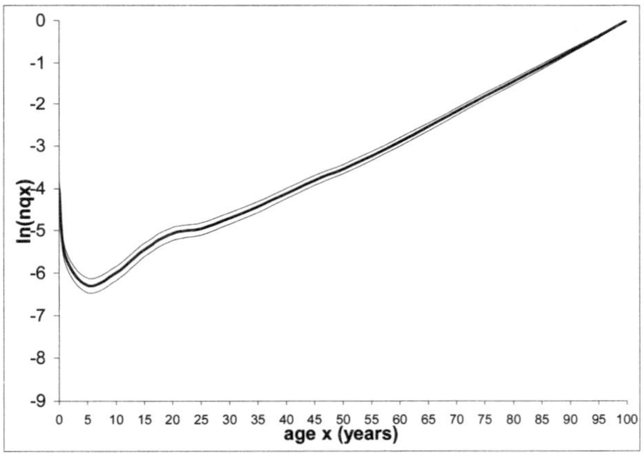

l_x with uncertainty interval

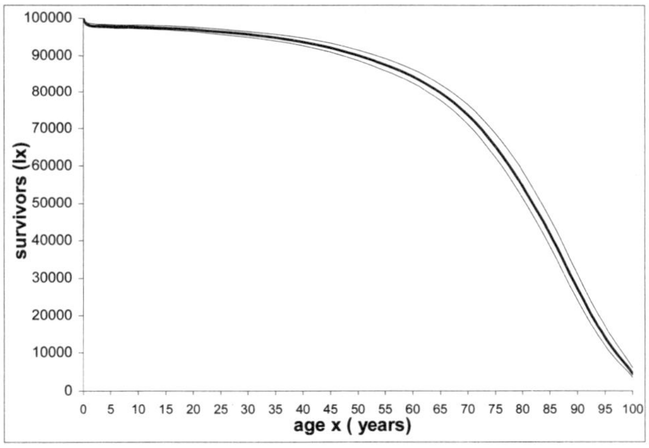

e_x with uncertainty interval

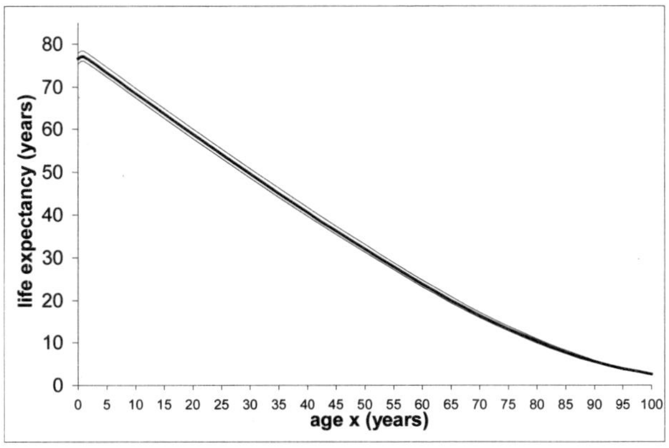

Distribution of deaths by age with uncertainty interval

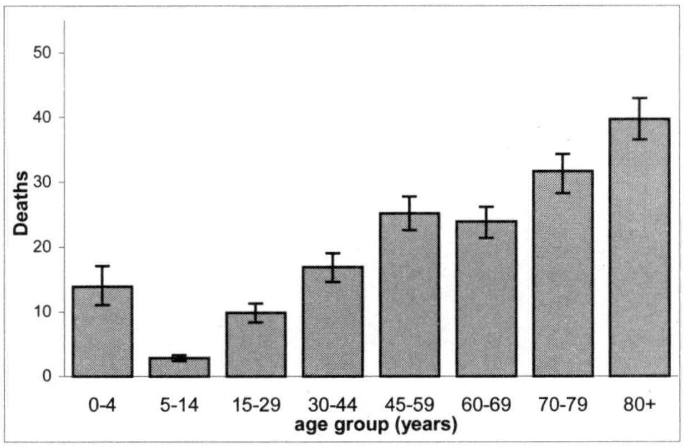

Argentina | Argentine | Argentina

Males, 2000 | *Hommes, 2000* | *Hombres, 2000*

Age	$_nM_x$	$_nq_x$			l_x			$_nd_x$	$_nL_x$	e_x			Actual deaths	Actual population
			percentiles			percentiles					percentiles			
		mean	2½	97½	mean	2½	97½			mean	2½	97½		
<1	0.0211	0.0207	0.0180	0.0234	100,000	100,000	100,000	2,071	98,136	70.2	69.8	70.7	7,620	361,085
1-4	0.0009	0.0035	0.0032	0.0038	97,929	97,660	98,198	343	390,891	70.7	70.2	71.3	1,245	1,417,460
5-9	0.0003	0.0016	0.0014	0.0017	97,585	97,287	97,885	154	487,542	66.9	66.4	67.6	550	1,738,363
10-14	0.0004	0.0018	0.0017	0.0020	97,431	97,120	97,744	179	486,710	62.0	61.5	62.7	624	1,697,758
15-19	0.0011	0.0053	0.0049	0.0057	97,253	96,927	97,580	515	484,974	57.1	56.6	57.8	1,776	1,671,350
20-24	0.0015	0.0072	0.0067	0.0078	96,737	96,373	97,104	700	481,937	52.4	51.9	53.1	2,462	1,695,904
25-29	0.0017	0.0087	0.0081	0.0092	96,038	95,627	96,452	833	478,106	47.8	47.3	48.5	2,469	1,417,648
30-34	0.0019	0.0095	0.0089	0.0101	95,205	94,745	95,670	905	473,762	43.2	42.7	43.9	2,364	1,237,264
35-39	0.0024	0.0118	0.0112	0.0125	94,300	93,790	94,816	1,115	468,712	38.6	38.1	39.3	2,745	1,154,192
40-44	0.0035	0.0174	0.0165	0.0182	93,185	92,622	93,758	1,618	461,880	34.0	33.5	34.7	3,723	1,062,588
45-49	0.0057	0.0283	0.0272	0.0294	91,567	90,940	92,209	2,594	451,349	29.6	29.0	30.3	5,729	996,928
50-54	0.0091	0.0443	0.0425	0.0460	88,973	88,270	89,698	3,940	435,014	25.4	24.8	26.1	7,912	873,610
55-59	0.0138	0.0666	0.0630	0.0697	85,033	84,262	85,833	5,665	411,002	21.4	20.8	22.1	10,222	741,624
60-64	0.0208	0.0988	0.0921	0.1041	79,368	78,566	80,193	7,843	377,232	17.8	17.2	18.5	12,994	625,004
65-69	0.0319	0.1476	0.1362	0.1574	71,525	70,683	72,386	10,556	331,235	14.4	13.9	15.1	17,060	535,325
70-74	0.0462	0.2071	0.1902	0.2223	60,969	59,702	62,441	12,625	273,282	11.5	11.0	12.1	20,032	433,617
75-79	0.0662	0.2840	0.2615	0.3046	48,344	46,641	50,567	13,731	207,393	8.9	8.5	9.3	19,261	290,934
80-84	0.1156	0.4485	0.4191	0.4752	34,613	32,432	37,343	15,525	134,254	6.4	6.1	6.7	16,968	146,728
85-89	0.1882	0.6399	0.6113	0.6652	19,088	17,020	21,692	12,214	64,907	4.6	4.4	4.8	10,263	54,540
90-94	0.2853	0.7686	0.7468	0.7880	6,874	5,698	8,431	5,284	18,521	3.3	3.1	3.4	3,070	10,760
95-99	0.4029	0.9121	0.8997	0.9229	1,591	1,208	2,135	1,451	3,601	2.4	2.4	2.5	323	802
100+	0.5302	1	1	1	140	93	214	140	264	1.9	1.9	2.0	8	15
All ages													*149,421*	*18,163,499*

$\ln(_nq_x)$ with uncertainty interval

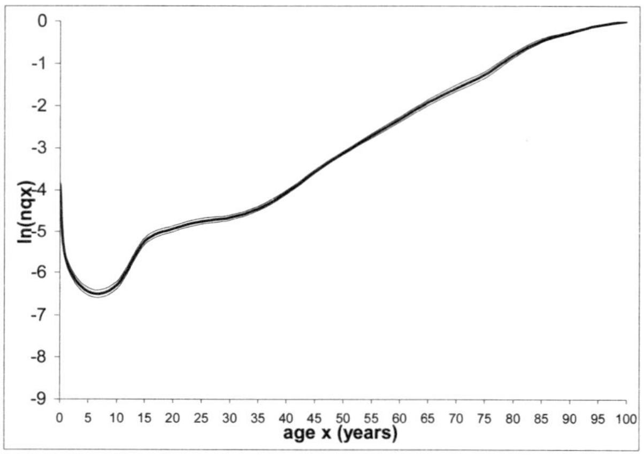

l_x with uncertainty interval

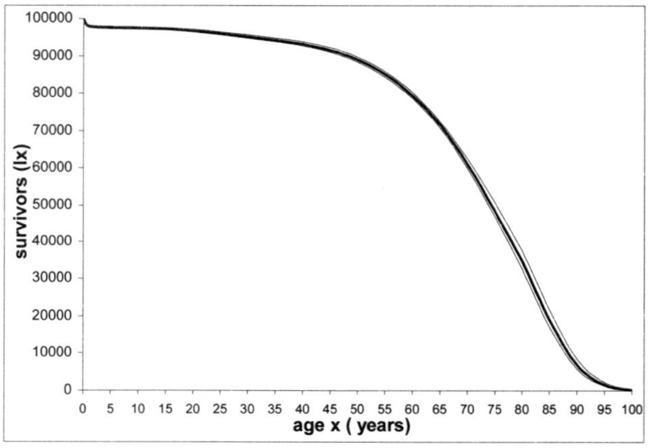

e_x with uncertainty interval

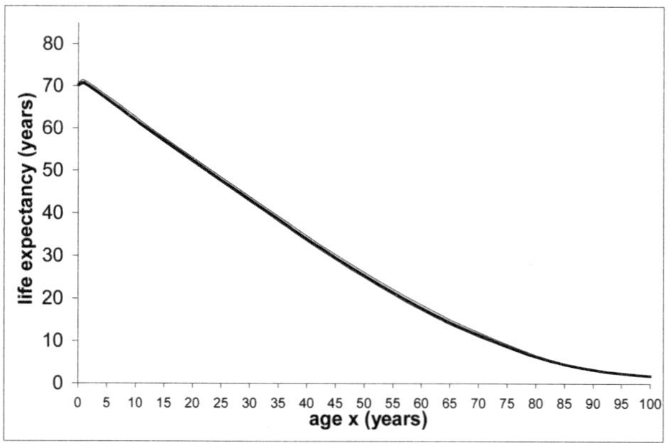

Distribution of deaths by age with uncertainty interval

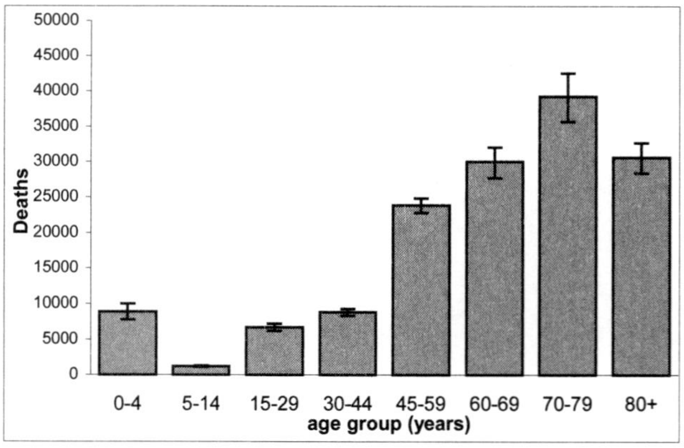

Argentina / Argentine / Argentina

Females, 2000 / *Femmes, 2000* / *Mujeres, 2000*

Age	$_nM_x$	$_nq_x$			l_x			$_nd_x$	$_nL_x$	e_x			Actual deaths	Actual population
			percentiles			percentiles					percentiles			
		mean	2½	97½	mean	2½	97½			mean	2½	97½		
<1	0.0178	0.0175	0.0149	0.0199	100,000	100,000	100,000	1,749	98,426	77.8	77.3	78.3	6,129	344,875
1-4	0.0008	0.0030	0.0027	0.0033	98,251	98,005	98,513	296	392,292	78.2	77.7	78.7	1,039	1,375,382
5-9	0.0002	0.0012	0.0011	0.0013	97,954	97,681	98,245	120	489,471	74.4	73.9	74.9	414	1,683,360
10-14	0.0003	0.0013	0.0012	0.0014	97,834	97,550	98,136	129	488,848	69.5	69.0	70.0	434	1,646,635
15-19	0.0004	0.0022	0.0020	0.0024	97,705	97,410	98,019	216	487,984	64.6	64.0	65.1	721	1,625,620
20-24	0.0006	0.0031	0.0029	0.0034	97,489	97,176	97,821	307	486,676	59.7	59.2	60.2	1,050	1,665,367
25-29	0.0008	0.0037	0.0035	0.0040	97,182	96,846	97,538	364	484,999	54.9	54.4	55.4	1,051	1,400,008
30-34	0.0010	0.0048	0.0044	0.0051	96,818	96,455	97,200	461	482,936	50.1	49.5	50.6	1,173	1,228,300
35-39	0.0013	0.0066	0.0062	0.0070	96,357	95,963	96,770	636	480,193	45.3	44.8	45.9	1,543	1,164,819
40-44	0.0020	0.0100	0.0095	0.0106	95,721	95,289	96,170	958	476,208	40.6	40.1	41.2	2,243	1,115,430
45-49	0.0029	0.0146	0.0139	0.0152	94,763	94,282	95,256	1,380	470,365	36.0	35.4	36.5	3,069	1,046,247
50-54	0.0043	0.0214	0.0204	0.0224	93,383	92,845	93,923	2,000	461,916	31.5	30.9	32.0	4,022	929,047
55-59	0.0059	0.0292	0.0278	0.0306	91,383	90,785	91,959	2,667	450,249	27.1	26.6	27.7	4,766	804,550
60-64	0.0088	0.0429	0.0406	0.0452	88,716	88,071	89,299	3,802	434,076	22.8	22.3	23.4	6,297	718,919
65-69	0.0140	0.0677	0.0637	0.0718	84,914	84,251	85,562	5,746	410,208	18.8	18.3	19.3	9,273	662,037
70-74	0.0220	0.1041	0.0970	0.1115	79,169	78,312	80,026	8,241	375,241	14.9	14.5	15.5	12,861	585,597
75-79	0.0355	0.1631	0.1504	0.1755	70,928	69,690	72,141	11,565	325,725	11.4	11.0	11.9	15,847	446,306
80-84	0.0728	0.3079	0.2841	0.3298	59,362	57,598	61,096	18,280	251,112	8.1	7.8	8.5	19,217	263,980
85-89	0.1359	0.5073	0.4754	0.5359	41,082	38,602	43,623	20,841	153,309	5.6	5.4	5.9	16,678	122,687
90-94	0.2312	0.6826	0.6543	0.7079	20,241	17,992	22,886	13,817	59,754	3.8	3.6	4.0	7,954	34,399
95-99	0.3583	0.8634	0.8461	0.8787	6,424	5,302	7,912	5,546	15,480	2.7	2.6	2.8	1,625	4,536
100+	0.5056	1	1	1	878	645	1,217	878	1,735	2.0	2.0	2.1	108	214
All ages													*117,515*	*18,868,315*

$\ln(_nq_x)$ with uncertainty interval

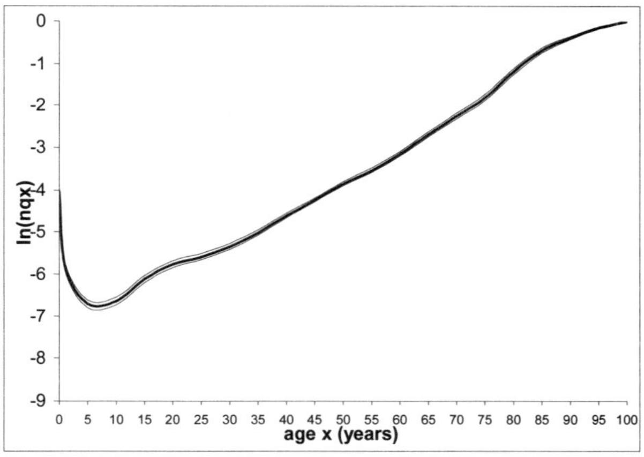

l_x with uncertainty interval

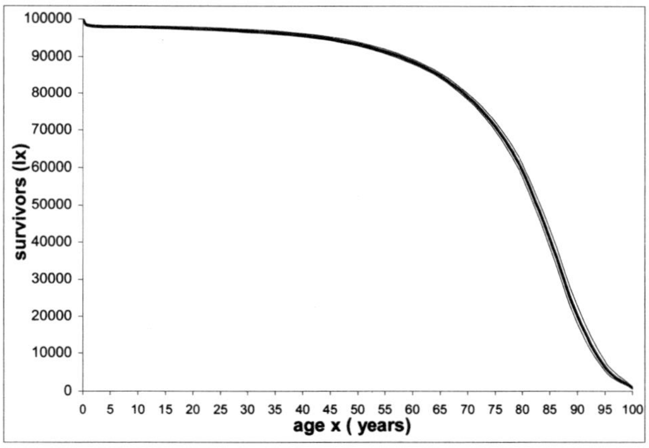

e_x with uncertainty interval

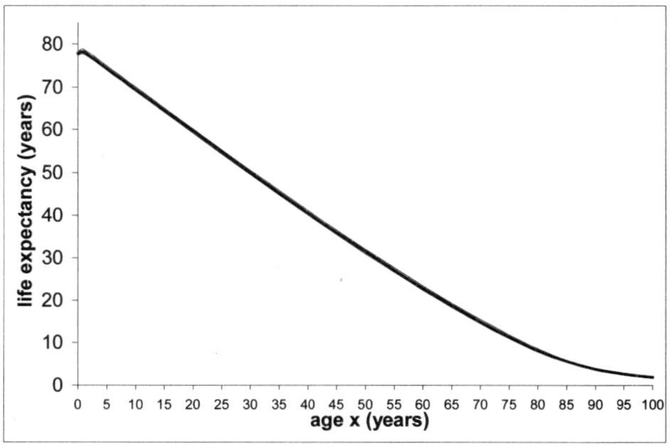

Distribution of deaths by age with uncertainty interval

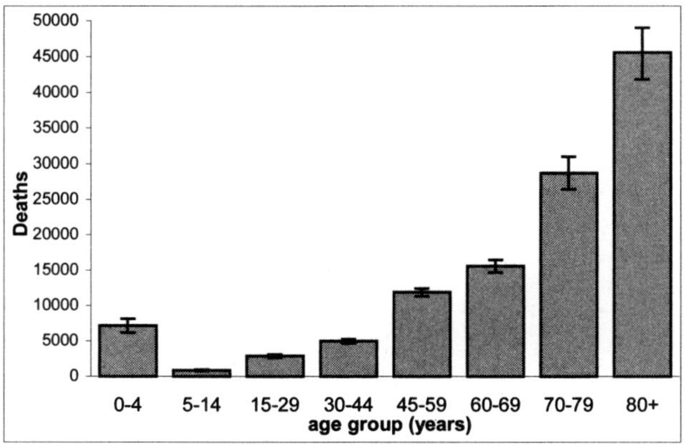

Armenia **Arménie** **Armenia**

Males, 2000 *Hommes, 2000* *Hombres, 2000*

Age	$_nM_x$	$_nq_x$			l_x			$_nd_x$	$_nL_x$	e_x			Actual deaths	Actual population
			percentiles			percentiles					percentiles			
		mean	2½	97½	mean	2½	97½			mean	2½	97½		
<1	0.0499	0.0477	0.0421	0.0528	100,000	100,000	100,000	4,774	95,703	64.4	63.8	65.0	1,060	21,254
1-4	0.0023	0.0090	0.0082	0.0097	95,226	94,720	95,794	855	378,851	66.6	66.2	67.2	191	84,453
5-9	0.0002	0.0012	0.0011	0.0013	94,371	93,800	95,005	113	471,571	63.2	62.8	63.8	39	161,258
10-14	0.0003	0.0014	0.0013	0.0015	94,258	93,679	94,900	128	470,969	58.3	57.9	58.8	53	194,551
15-19	0.0009	0.0043	0.0040	0.0047	94,130	93,542	94,782	409	469,628	53.4	53.0	53.9	164	188,271
20-24	0.0009	0.0044	0.0041	0.0048	93,721	93,104	94,402	415	467,567	48.6	48.2	49.1	145	163,117
25-29	0.0009	0.0046	0.0043	0.0050	93,306	92,661	94,014	431	465,452	43.8	43.4	44.3	134	144,826
30-34	0.0018	0.0089	0.0083	0.0095	92,875	92,201	93,611	827	462,308	39.0	38.6	39.5	242	135,434
35-39	0.0024	0.0120	0.0113	0.0128	92,048	91,322	92,834	1,106	457,476	34.3	34.0	34.8	359	148,367
40-44	0.0041	0.0204	0.0192	0.0217	90,942	90,151	91,790	1,858	450,066	29.7	29.4	30.2	617	149,325
45-49	0.0075	0.0369	0.0350	0.0389	89,084	88,193	90,024	3,289	437,199	25.3	25.0	25.7	816	108,505
50-54	0.0116	0.0565	0.0535	0.0592	85,795	84,759	86,871	4,845	416,866	21.2	20.9	21.6	863	74,296
55-59	0.0203	0.0968	0.0917	0.1010	80,951	79,742	82,220	7,834	385,170	17.3	17.0	17.7	973	47,821
60-64	0.0332	0.1532	0.1455	0.1592	73,117	71,684	74,680	11,204	337,576	13.9	13.6	14.2	2,534	76,365
65-69	0.0520	0.2303	0.2198	0.2382	61,913	60,273	63,815	14,258	273,923	10.9	10.7	11.2	2,910	55,911
70-74	0.0800	0.3333	0.3210	0.3450	47,656	45,916	49,788	15,885	198,566	8.4	8.2	8.7	4,031	50,384
75-79	0.1300	0.4906	0.4758	0.5049	31,771	30,076	33,805	15,586	119,889	6.4	6.2	6.6	2,326	17,891
80-84	0.1700	0.5965	0.5820	0.6118	16,185	14,890	17,722	9,654	56,789	5.2	5.0	5.3	1,025	6,029
85-89	0.2250	0.7200	0.7070	0.7340	6,531	5,781	7,408	4,702	20,898	4.1	4.0	4.2	923	4,100
90-94	0.3000	0.7895	0.7783	0.8016	1,829	1,538	2,170	1,444	4,812	3.1	3.1	3.2	360	1,201
95-99	0.4200	0.9292	0.9228	0.9357	385	305	481	358	852	2.3	2.3	2.4	73	173
100+	0.5400	1	1	1	27	20	37	27	50	1.9	1.7	1.9	5	9
All ages													*19,841*	*1,833,541*

ln($_nq_x$) with uncertainty interval

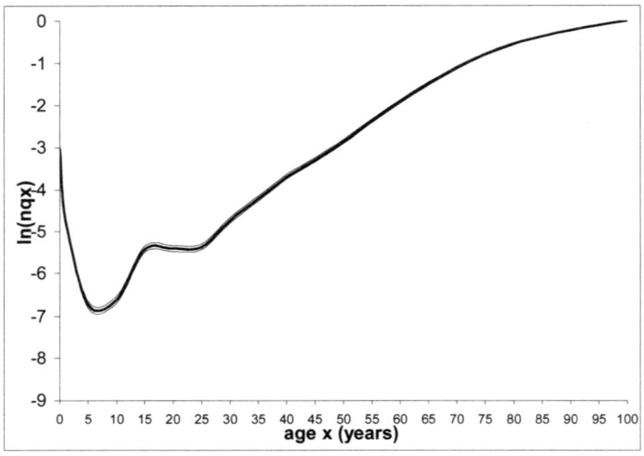

l_x with uncertainty interval

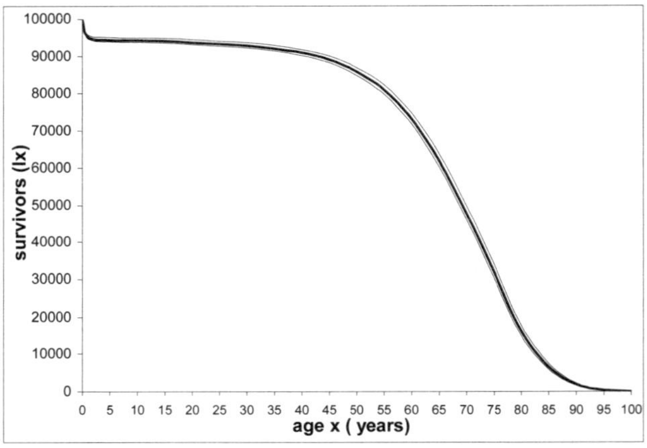

e_x with uncertainty interval

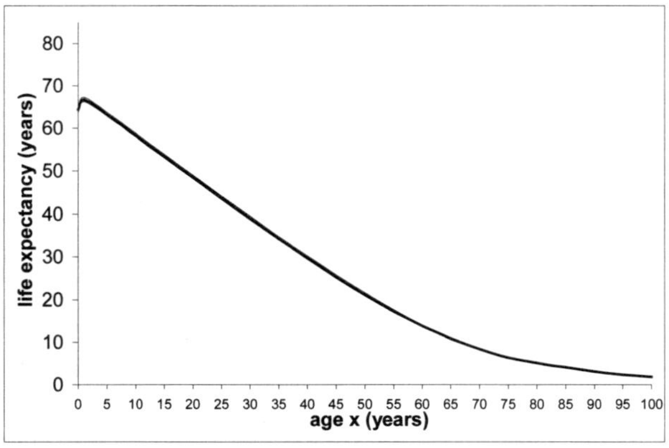

Distribution of deaths by age with uncertainty interval

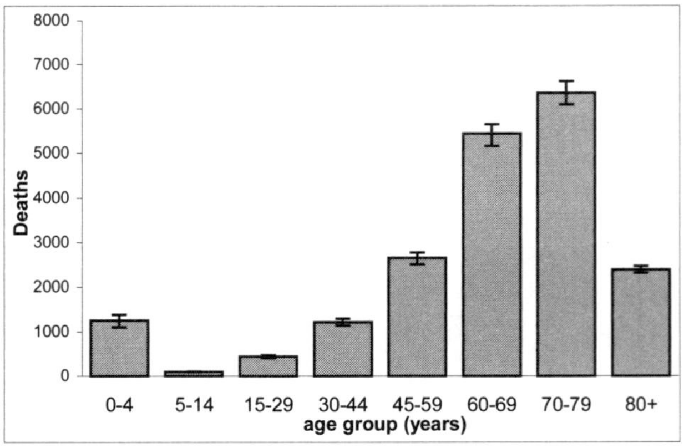

Armenia — Arménie — Armenia

Females, 2000 — *Femmes, 2000* — *Mujeres, 2000*

Age	$_nM_x$	$_nq_x$ mean	$_nq_x$ percentiles 2½	$_nq_x$ percentiles 97½	l_x mean	l_x percentiles 2½	l_x percentiles 97½	$_nd_x$	$_nL_x$	e_x mean	e_x percentiles 2½	e_x percentiles 97½	Actual deaths	Actual population
<1	0.0316	0.0307	0.0258	0.0359	100,000	100,000	100,000	3,068	97,239	71.2	70.2	72.3	644	20,417
1-4	0.0018	0.0073	0.0063	0.0083	96,932	96,413	97,423	704	386,036	72.5	71.7	73.4	147	80,458
5-9	0.0001	0.0007	0.0007	0.0009	96,227	95,615	96,810	72	480,956	69.0	68.3	69.9	23	151,210
10-14	0.0002	0.0011	0.0009	0.0012	96,155	95,533	96,747	104	480,516	64.1	63.3	65.0	40	184,632
15-19	0.0002	0.0011	0.0009	0.0012	96,051	95,416	96,656	104	479,997	59.1	58.4	60.0	39	180,512
20-24	0.0003	0.0013	0.0011	0.0015	95,948	95,300	96,565	123	479,431	54.2	53.5	55.1	40	157,244
25-29	0.0006	0.0029	0.0025	0.0033	95,825	95,161	96,458	275	478,435	49.2	48.5	50.2	81	141,575
30-34	0.0007	0.0036	0.0031	0.0041	95,549	94,851	96,215	344	476,889	44.4	43.7	45.3	101	140,735
35-39	0.0011	0.0054	0.0047	0.0061	95,206	94,465	95,913	515	474,743	39.5	38.8	40.4	182	167,627
40-44	0.0017	0.0086	0.0075	0.0096	94,691	93,889	95,458	811	471,429	34.7	34.1	35.6	291	169,088
45-49	0.0032	0.0160	0.0142	0.0179	93,880	92,984	94,738	1,506	465,638	30.0	29.4	30.9	411	127,161
50-54	0.0048	0.0236	0.0212	0.0263	92,375	91,315	93,391	2,184	456,415	25.5	24.8	26.3	420	87,845
55-59	0.0099	0.0483	0.0436	0.0532	90,191	88,914	91,413	4,360	440,054	21.0	20.4	21.8	584	58,987
60-64	0.0171	0.0819	0.0736	0.0897	85,831	84,184	87,424	7,031	411,575	17.0	16.5	17.7	1,615	94,555
65-69	0.0327	0.1510	0.1359	0.1636	78,799	76,634	80,988	11,903	364,241	13.3	12.8	13.9	2,292	70,137
70-74	0.0560	0.2454	0.2229	0.2619	66,897	64,093	69,985	16,419	293,436	10.2	9.7	10.7	3,711	66,329
75-79	0.0984	0.3950	0.3668	0.4210	50,478	47,307	54,382	19,937	202,546	7.7	7.3	8.1	3,031	30,790
80-84	0.1366	0.5091	0.4818	0.5378	30,541	27,389	34,435	15,549	113,830	6.0	5.7	6.3	1,837	13,451
85-89	0.1886	0.6408	0.6148	0.6670	14,991	12,658	17,846	9,606	50,940	4.7	4.5	4.9	1,559	8,268
90-94	0.2590	0.7287	0.7062	0.7506	5,385	4,215	6,875	3,924	15,152	3.6	3.4	3.7	563	2,173
95-99	0.3538	0.8582	0.8418	0.8739	1,461	1,051	2,020	1,254	3,543	2.7	2.6	2.8	90	255
100+	0.4808	1	1	1	207	133	320	207	431	2.1	2.0	2.1	3	7
All ages													*17,707*	*1,953,456*

ln($_nq_x$) with uncertainty interval

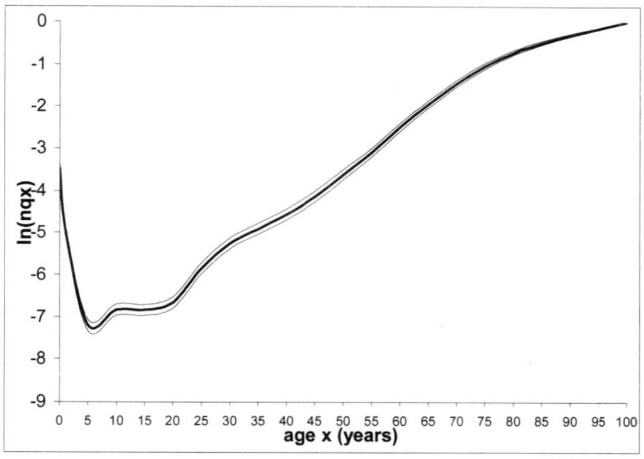

l_x with uncertainty interval

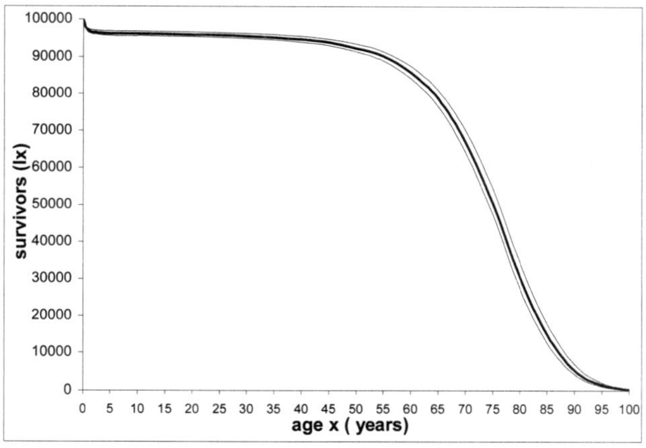

e_x with uncertainty interval

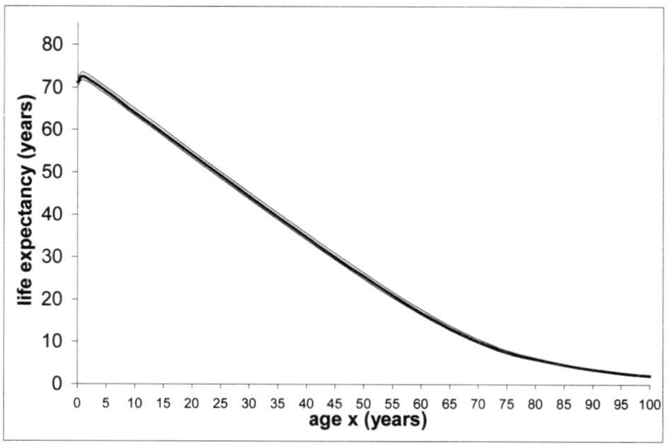

Distribution of deaths by age with uncertainty interval

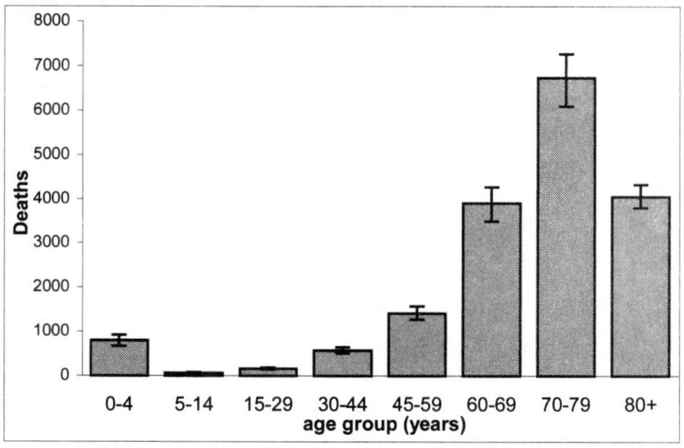

Australia **Australie** **Australia**

Males, 2000 *Hommes, 2000* *Hombres, 2000*

Age	$_nM_x$	$_nq_x$			l_x			$_nd_x$	$_nL_x$	e_x			Actual deaths	Actual population
			percentiles			percentiles					percentiles			
		mean	2½	97½	mean	2½	97½			mean	2½	97½		
<1	0.0060	0.0060	0.0052	0.0068	100,000	100,000	100,000	597	99,463	76.6	76.1	77.1	760	126,635
1-4	0.0003	0.0011	0.0010	0.0012	99,403	99,321	99,483	111	397,346	76.0	75.5	76.5	146	521,444
5-9	0.0001	0.0006	0.0006	0.0007	99,292	99,200	99,383	65	496,299	72.1	71.6	72.6	89	686,626
10-14	0.0002	0.0007	0.0007	0.0008	99,227	99,130	99,324	74	495,951	67.2	66.7	67.7	102	680,974
15-19	0.0007	0.0037	0.0034	0.0040	99,153	99,049	99,257	366	494,850	62.2	61.7	62.7	514	694,906
20-24	0.0011	0.0056	0.0052	0.0061	98,787	98,652	98,923	557	492,543	57.4	56.9	58.0	777	687,850
25-29	0.0013	0.0065	0.0060	0.0069	98,230	98,053	98,409	636	489,560	52.7	52.3	53.3	968	744,625
30-34	0.0013	0.0066	0.0062	0.0070	97,594	97,375	97,816	642	486,364	48.1	47.6	48.6	932	705,902
35-39	0.0014	0.0067	0.0064	0.0071	96,952	96,696	97,212	652	483,129	43.4	42.9	43.9	1,010	747,925
40-44	0.0018	0.0088	0.0084	0.0092	96,300	96,010	96,594	848	479,377	38.7	38.2	39.2	1,279	722,355
45-49	0.0024	0.0120	0.0115	0.0125	95,451	95,123	95,785	1,148	474,386	34.0	33.5	34.5	1,621	669,867
50-54	0.0040	0.0198	0.0190	0.0207	94,303	93,932	94,681	1,867	466,847	29.4	28.9	29.9	2,527	631,748
55-59	0.0070	0.0344	0.0329	0.0361	92,436	92,009	92,869	3,180	454,230	24.9	24.5	25.4	3,408	486,876
60-64	0.0110	0.0535	0.0509	0.0565	89,256	88,768	89,746	4,778	434,336	20.7	20.3	21.2	4,305	391,367
65-69	0.0190	0.0907	0.0856	0.0959	84,478	83,875	85,065	7,662	403,238	16.7	16.3	17.2	6,245	328,670
70-74	0.0340	0.1567	0.1473	0.1657	76,817	75,828	77,663	12,036	353,995	13.1	12.8	13.6	9,989	293,793
75-79	0.0540	0.2379	0.2241	0.2507	64,781	63,263	66,219	15,410	285,379	10.1	9.8	10.5	11,656	215,844
80-84	0.0920	0.3740	0.3541	0.3920	49,371	47,424	51,381	18,464	200,693	7.5	7.2	7.8	10,721	116,528
85-89	0.1510	0.5481	0.5259	0.5683	30,907	28,962	33,104	16,940	112,184	5.5	5.3	5.7	8,349	55,289
90-94	0.2120	0.6479	0.6289	0.6654	13,967	12,604	15,597	9,049	42,686	4.1	4.0	4.3	3,433	16,192
95-99	0.3050	0.7963	0.7821	0.8094	4,917	4,231	5,789	3,916	12,839	3.1	3.0	3.2	1,011	3,316
100+	0.4490	1	1	1	1,001	806	1,261	1,001	2,230	2.2	2.2	2.4	212	473
All ages													*70,053*	*9,529,205*

ln($_nq_x$) with uncertainty interval

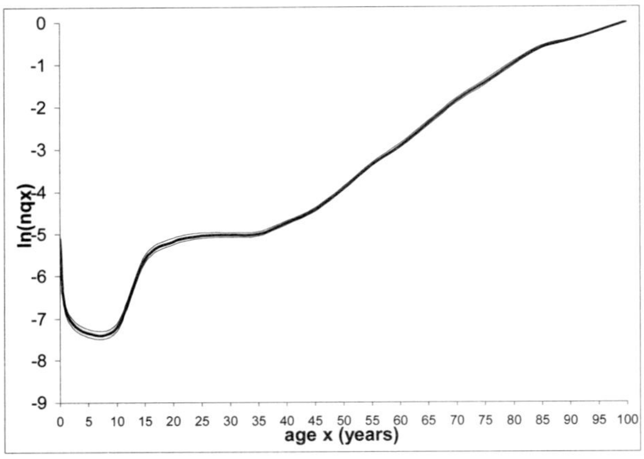

l_x with uncertainty interval

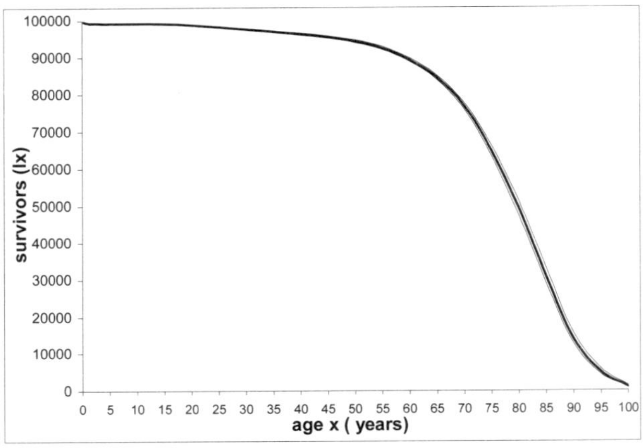

e_x with uncertainty interval

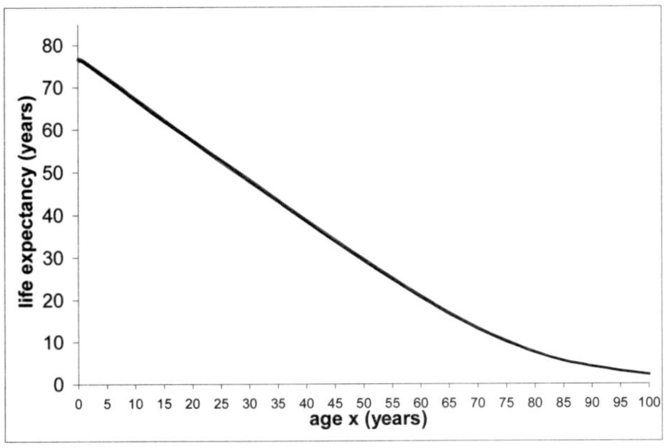

Distribution of deaths by age with uncertainty interval

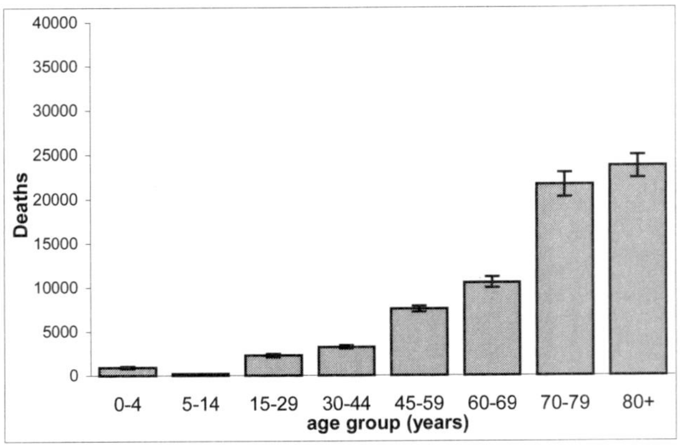

Australia **Australie** **Australia**

Females, 2000 *Femmes, 2000* *Mujeres, 2000*

Age	$_nM_x$	$_nq_x$ mean	$_nq_x$ percentiles 2½	$_nq_x$ percentiles 97½	l_x mean	l_x percentiles 2½	l_x percentiles 97½	$_nd_x$	$_nL_x$	e_x mean	e_x percentiles 2½	e_x percentiles 97½	Actual deaths	Actual population
<1	0.0040	0.0040	0.0033	0.0047	100,000	100,000	100,000	399	99,641	82.1	81.7	82.4	486	121,589
1-4	0.0002	0.0010	0.0008	0.0011	99,601	99,528	99,673	96	398,176	81.4	81.1	81.8	118	493,491
5-9	0.0001	0.0005	0.0004	0.0006	99,506	99,419	99,591	50	497,405	77.5	77.2	77.8	65	649,857
10-14	0.0001	0.0006	0.0006	0.0007	99,456	99,362	99,548	65	497,119	72.5	72.2	72.9	84	645,961
15-19	0.0003	0.0015	0.0014	0.0017	99,392	99,289	99,492	154	496,573	67.6	67.2	67.9	205	661,233
20-24	0.0004	0.0019	0.0017	0.0021	99,238	99,116	99,358	188	495,717	62.7	62.3	63.0	250	657,844
25-29	0.0004	0.0020	0.0018	0.0022	99,049	98,906	99,192	198	494,751	57.8	57.5	58.1	293	732,733
30-34	0.0005	0.0027	0.0024	0.0030	98,851	98,687	99,016	267	493,590	52.9	52.6	53.2	382	707,708
35-39	0.0007	0.0033	0.0030	0.0037	98,585	98,394	98,777	330	492,100	48.0	47.7	48.4	501	747,386
40-44	0.0011	0.0053	0.0048	0.0057	98,255	98,033	98,479	519	489,977	43.2	42.9	43.5	767	723,784
45-49	0.0016	0.0078	0.0072	0.0084	97,736	97,471	98,004	764	486,768	38.4	38.1	38.8	1,057	673,489
50-54	0.0023	0.0113	0.0105	0.0121	96,971	96,652	97,297	1,094	482,121	33.7	33.4	34.0	1,398	615,951
55-59	0.0040	0.0198	0.0186	0.0210	95,877	95,493	96,269	1,899	474,639	29.0	28.8	29.4	1,890	472,509
60-64	0.0060	0.0296	0.0279	0.0312	93,978	93,500	94,458	2,778	462,948	24.6	24.3	24.9	2,335	389,123
65-69	0.0100	0.0488	0.0465	0.0512	91,201	90,609	91,799	4,449	444,882	20.2	20.0	20.6	3,446	344,566
70-74	0.0180	0.0861	0.0823	0.0897	86,752	85,987	87,524	7,471	415,081	16.1	15.9	16.5	5,959	331,067
75-79	0.0310	0.1439	0.1370	0.1493	79,281	78,292	80,257	11,405	367,891	12.4	12.2	12.7	8,841	285,198
80-84	0.0610	0.2646	0.2525	0.2747	67,876	66,717	69,027	17,963	294,472	9.1	8.9	9.4	11,328	185,698
85-89	0.1130	0.4405	0.4237	0.4585	49,913	48,581	51,527	21,989	194,593	6.5	6.3	6.7	12,800	113,270
90-94	0.1800	0.5844	0.5664	0.6041	27,924	26,715	29,651	16,319	90,663	4.6	4.5	4.8	7,752	43,067
95-99	0.2800	0.7609	0.7460	0.7767	11,605	10,581	12,746	8,830	31,535	3.3	3.2	3.4	3,114	11,122
100+	0.4390	1	1	1	2,775	2,362	3,214	2,775	6,321	2.3	2.2	2.5	788	1,794
All ages													*63,859*	*9,608,440*

ln($_nq_x$) with uncertainty interval

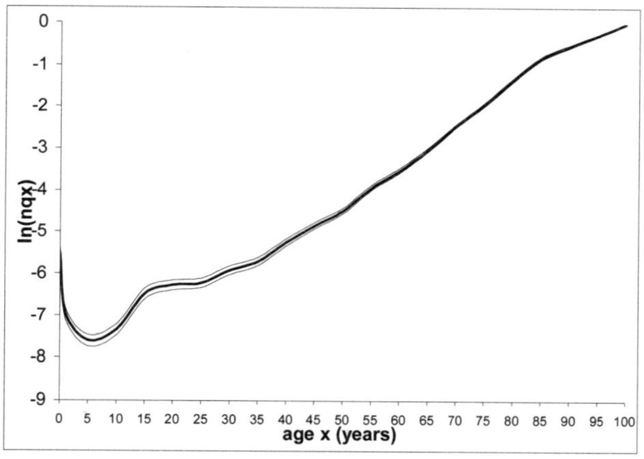

l_x with uncertainty interval

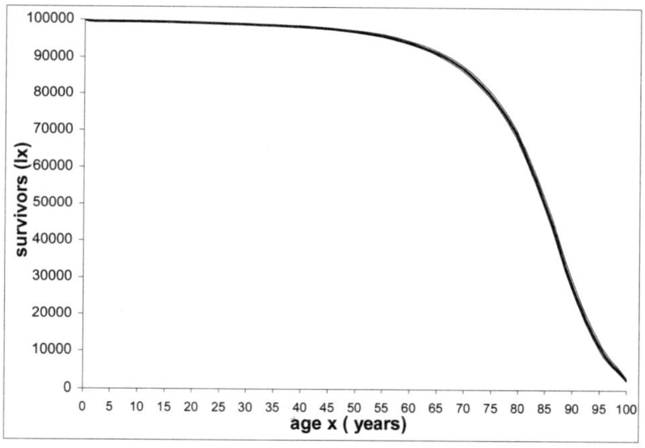

e_x with uncertainty interval

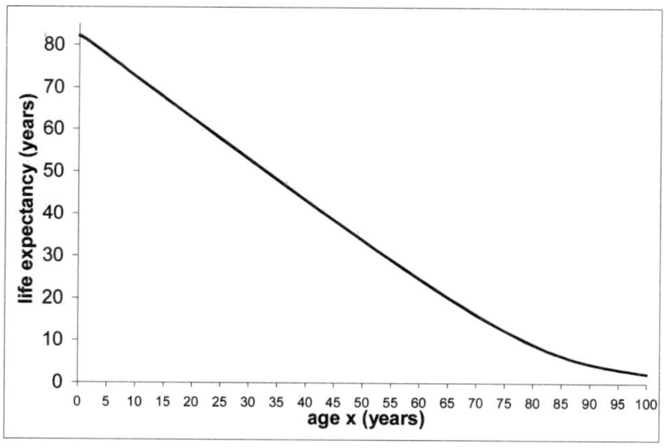

Distribution of deaths by age with uncertainty interval

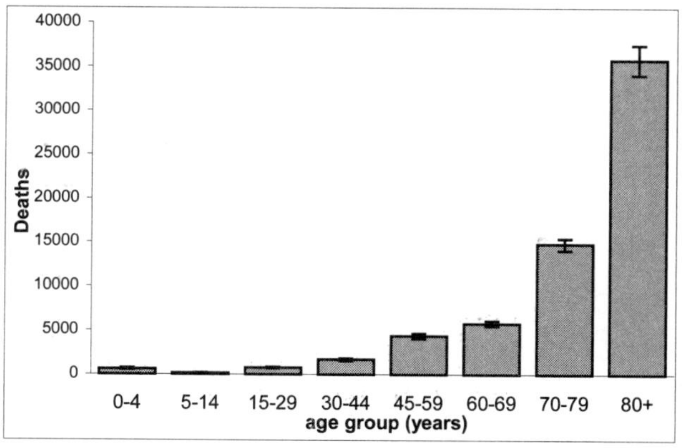

Austria | Autriche | Austria

Males, 2000 | *Hommes, 2000* | *Hombres, 2000*

Age	$_nM_x$	$_nq_x$ mean	$_nq_x$ percentiles 2½	$_nq_x$ percentiles 97½	l_x mean	l_x percentiles 2½	l_x percentiles 97½	$_nd_x$	$_nL_x$	e_x mean	e_x percentiles 2½	e_x percentiles 97½	Actual deaths	Actual population
<1	0.0042	0.0042	0.0034	0.0052	100,000	100,000	100,000	421	99,621	74.9	74.3	75.5	166	39,175
1-4	0.0004	0.0015	0.0013	0.0018	99,579	99,477	99,659	147	397,961	74.2	73.6	74.8	63	169,424
5-9	0.0001	0.0006	0.0005	0.0007	99,431	99,302	99,533	57	497,014	70.3	69.7	70.9	28	241,631
10-14	0.0001	0.0007	0.0006	0.0009	99,374	99,236	99,484	74	496,686	65.3	64.8	66.0	36	238,876
15-19	0.0008	0.0040	0.0034	0.0046	99,300	99,149	99,421	392	495,520	60.4	59.8	61.0	195	246,639
20-24	0.0011	0.0053	0.0047	0.0060	98,908	98,697	99,079	527	493,220	55.6	55.1	56.2	254	237,743
25-29	0.0008	0.0042	0.0037	0.0047	98,380	98,101	98,611	411	490,873	50.9	50.4	51.5	238	283,825
30-34	0.0010	0.0048	0.0044	0.0054	97,969	97,642	98,241	474	488,661	46.1	45.6	46.7	349	360,336
35-39	0.0015	0.0077	0.0070	0.0084	97,495	97,120	97,811	749	485,604	41.3	40.8	41.9	570	369,572
40-44	0.0024	0.0119	0.0109	0.0128	96,746	96,303	97,123	1,148	480,862	36.6	36.1	37.2	755	316,153
45-49	0.0036	0.0177	0.0165	0.0189	95,599	95,068	96,055	1,694	473,757	32.0	31.5	32.6	945	264,336
50-54	0.0061	0.0302	0.0283	0.0321	93,904	93,273	94,469	2,840	462,421	27.6	27.1	28.1	1,544	251,306
55-59	0.0094	0.0459	0.0429	0.0486	91,064	90,319	91,768	4,176	444,879	23.3	22.9	23.9	2,280	242,916
60-64	0.0150	0.0723	0.0677	0.0768	86,888	86,063	87,727	6,280	418,738	19.4	18.9	19.8	3,037	202,485
65-69	0.0249	0.1171	0.1099	0.1242	80,608	79,568	81,699	9,436	379,448	15.7	15.3	16.1	3,955	159,026
70-74	0.0384	0.1751	0.1647	0.1850	71,172	69,732	72,720	12,461	324,705	12.4	12.0	12.8	5,328	138,833
75-79	0.0616	0.2667	0.2515	0.2815	58,711	56,834	60,694	15,660	254,403	9.5	9.2	9.9	6,056	98,386
80-84	0.1014	0.4044	0.3839	0.4255	43,050	40,961	45,244	17,409	171,731	7.1	6.8	7.3	4,283	42,251
85-89	0.1600	0.5715	0.5498	0.5944	25,642	23,839	27,720	14,653	91,577	5.2	5.0	5.4	4,651	29,065
90-94	0.2421	0.7012	0.6818	0.7207	10,989	9,737	12,420	7,705	31,829	3.7	3.6	3.8	2,003	8,276
95-99	0.3510	0.8549	0.8413	0.8680	3,284	2,720	3,932	2,807	7,997	2.7	2.7	2.8	432	1,230
100+	0.4879	1	1	1	476	359	617	476	977	2.0	2.0	2.1	40	82
All ages													*37,207*	*3,941,566*

ln($_nq_x$) with uncertainty interval

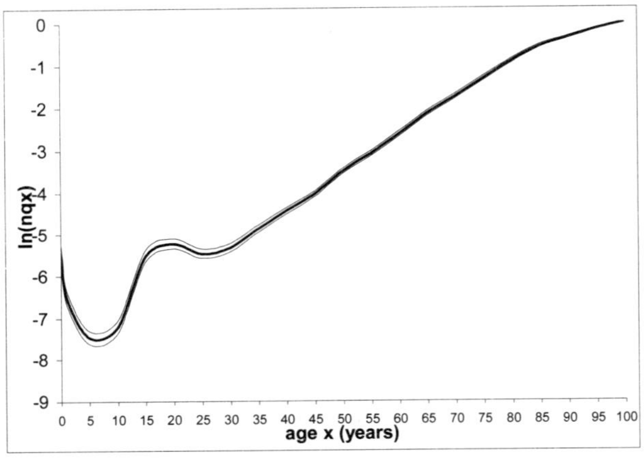

l_x with uncertainty interval

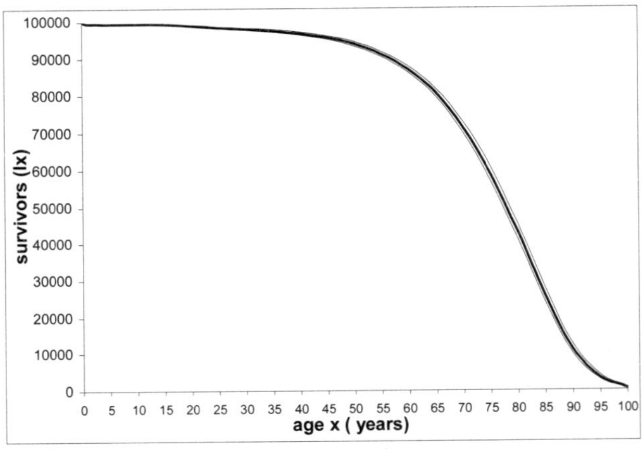

e_x with uncertainty interval

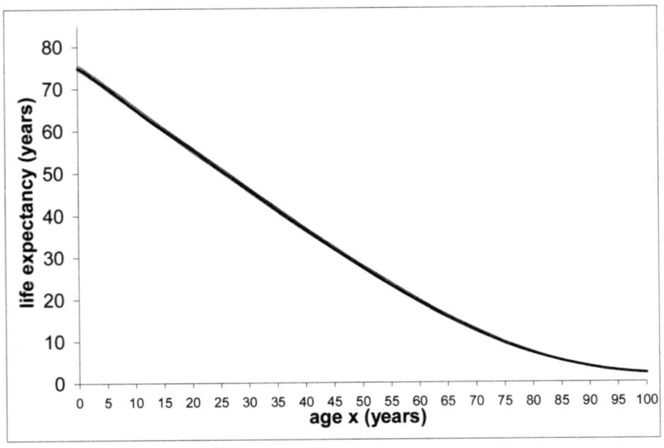

Distribution of deaths by age with uncertainty interval

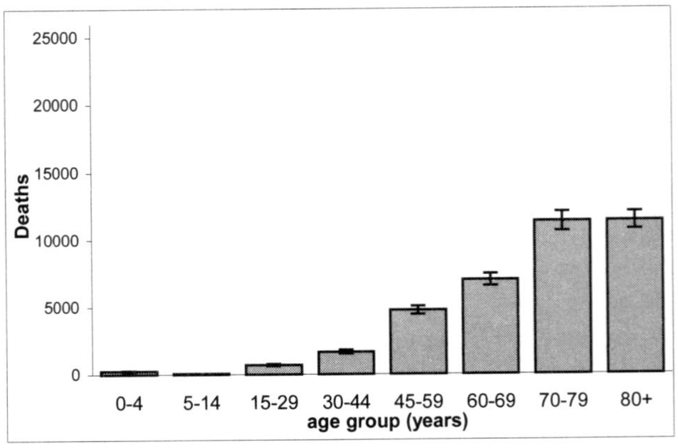

Austria **Autriche** **Austria**

Females, 2000 *Femmes, 2000* *Mujeres, 2000*

Age	$_nM_x$	$_nq_x$			l_x			$_nd_x$	$_nL_x$	e_x			Actual deaths	Actual population
			percentiles			percentiles					percentiles			
		mean	2½	97½	mean	2½	97½			mean	2½	97½		
<1	0.0039	0.0039	0.0032	0.0048	100,000	100,000	100,000	392	99,647	81.4	80.8	81.9	146	37,045
1-4	0.0002	0.0007	0.0006	0.0008	99,608	99,517	99,677	67	398,270	80.7	80.1	81.3	27	160,861
5-9	0.0001	0.0005	0.0004	0.0005	99,541	99,439	99,619	46	497,588	76.7	76.1	77.3	21	229,452
10-14	0.0001	0.0007	0.0006	0.0008	99,495	99,386	99,579	73	497,290	71.8	71.2	72.4	33	226,500
15-19	0.0003	0.0016	0.0014	0.0018	99,421	99,302	99,515	155	496,720	66.8	66.2	67.4	73	234,993
20-24	0.0003	0.0014	0.0013	0.0016	99,267	99,127	99,378	143	495,974	61.9	61.3	62.5	65	224,336
25-29	0.0003	0.0015	0.0014	0.0017	99,123	98,966	99,250	150	495,242	57.0	56.4	57.6	84	279,196
30-34	0.0004	0.0021	0.0019	0.0023	98,974	98,801	99,115	211	494,340	52.1	51.5	52.7	147	343,218
35-39	0.0007	0.0035	0.0032	0.0038	98,762	98,569	98,923	346	492,946	47.2	46.6	47.8	244	347,348
40-44	0.0012	0.0058	0.0054	0.0062	98,416	98,193	98,605	570	490,655	42.4	41.8	43.0	350	301,319
45-49	0.0019	0.0093	0.0088	0.0099	97,846	97,585	98,074	913	486,947	37.6	37.0	38.2	484	258,243
50-54	0.0031	0.0152	0.0144	0.0159	96,933	96,630	97,208	1,470	480,990	32.9	32.3	33.5	760	248,599
55-59	0.0042	0.0207	0.0195	0.0220	95,463	95,124	95,788	1,977	472,373	28.4	27.8	29.0	1,052	251,294
60-64	0.0062	0.0307	0.0289	0.0328	93,486	93,107	93,843	2,870	460,253	23.9	23.4	24.5	1,351	216,585
65-69	0.0103	0.0502	0.0468	0.0541	90,615	90,158	91,084	4,550	441,701	19.6	19.1	20.2	1,924	186,740
70-74	0.0187	0.0895	0.0825	0.0969	86,065	85,307	86,788	7,703	411,067	15.5	15.0	16.1	3,613	192,787
75-79	0.0339	0.1562	0.1431	0.1693	78,362	77,038	79,491	12,244	361,199	11.8	11.4	12.3	6,613	195,091
80-84	0.0675	0.2888	0.2656	0.3104	66,118	63,998	68,091	19,095	282,853	8.5	8.1	9.0	6,419	95,079
85-89	0.1243	0.4741	0.4413	0.5020	47,023	44,176	50,004	22,293	179,384	6.0	5.7	6.3	9,581	77,096
90-94	0.2115	0.6469	0.6158	0.6716	24,730	22,183	27,771	15,999	75,656	4.1	3.9	4.3	5,805	27,452
95-99	0.3326	0.8325	0.8119	0.8483	8,732	7,309	10,604	7,269	21,852	2.8	2.8	3.0	1,551	4,663
100+	0.4836	1	1	1	1,463	1,116	1,994	1,463	3,025	2.1	2.1	2.2	212	439
All ages													*40,555*	*4,138,336*

ln($_nq_x$) with uncertainty interval

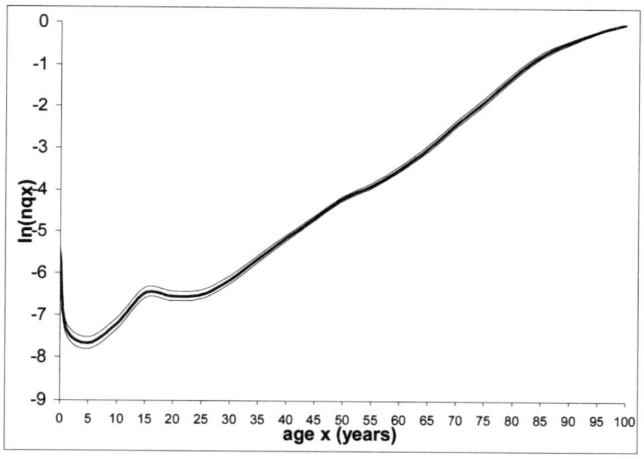

l_x with uncertainty interval

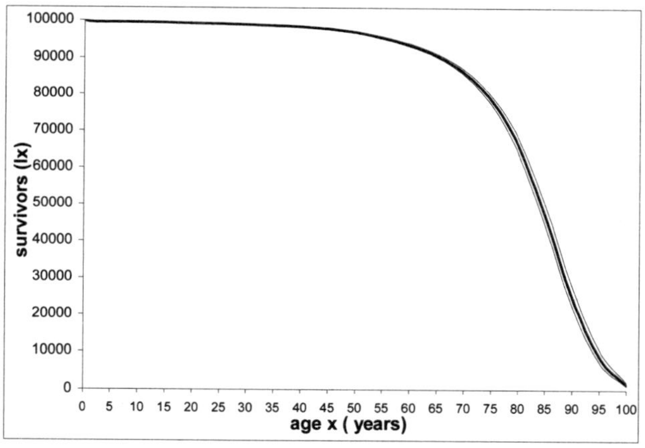

e_x with uncertainty interval

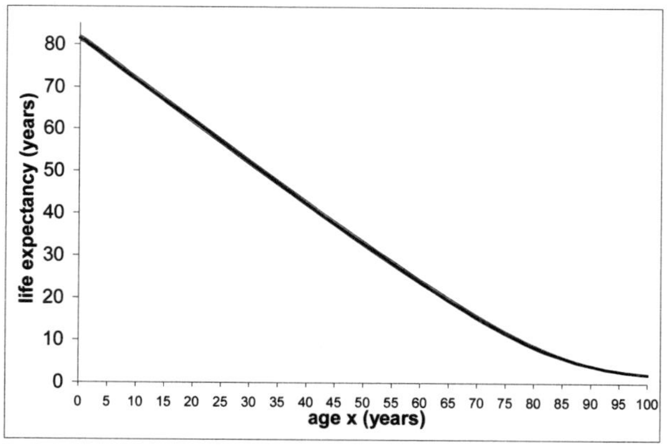

Distribution of deaths by age with uncertainty interval

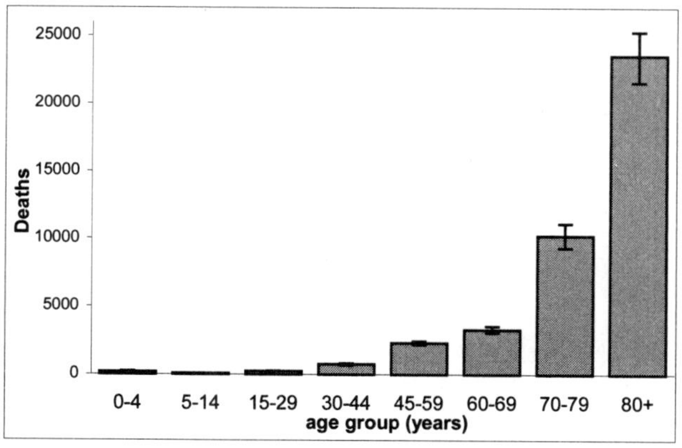

## Azerbaijan		Azerbaïdjan		Azerbaiyán

Males, 2000		*Hommes, 2000*		*Hombres, 2000*

Age	$_nM_x$	$_nq_x$ mean	$_nq_x$ percentiles 2½	$_nq_x$ percentiles 97½	l_x mean	l_x percentiles 2½	l_x percentiles 97½	$_nd_x$	$_nL_x$	e_x mean	e_x percentiles 2½	e_x percentiles 97½	Actual deaths	Actual population
<1	0.0800	0.0746	0.0592	0.0917	100,000	100,000	100,000	7,464	93,282	61.7	59.2	64.2	4,824	60,290
1-4	0.0072	0.0285	0.0241	0.0333	92,536	90,833	94,084	2,636	363,816	65.6	63.8	67.6	1,835	253,219
5-9	0.0015	0.0077	0.0066	0.0090	89,900	87,805	91,815	692	447,767	63.5	61.9	65.2	680	440,204
10-14	0.0010	0.0050	0.0043	0.0058	89,207	87,017	91,210	449	444,914	59.0	57.4	60.7	450	445,355
15-19	0.0020	0.0100	0.0086	0.0116	88,758	86,509	90,817	886	441,576	54.3	52.7	55.9	813	405,089
20-24	0.0027	0.0136	0.0117	0.0157	87,872	85,509	90,039	1,197	436,368	49.8	48.3	51.4	955	347,859
25-29	0.0025	0.0126	0.0109	0.0145	86,675	84,167	88,986	1,094	430,639	45.4	44.0	47.0	839	330,278
30-34	0.0033	0.0163	0.0140	0.0186	85,581	82,948	88,020	1,395	424,417	41.0	39.6	42.6	1,094	332,890
35-39	0.0046	0.0227	0.0196	0.0259	84,186	81,404	86,785	1,915	416,143	36.6	35.3	38.2	1,476	320,785
40-44	0.0071	0.0347	0.0300	0.0393	82,271	79,299	85,081	2,853	404,224	32.4	31.2	33.9	1,983	280,924
45-49	0.0095	0.0466	0.0405	0.0526	79,418	76,181	82,527	3,698	387,845	28.5	27.3	29.9	1,811	189,879
50-54	0.0127	0.0617	0.0540	0.0693	75,720	72,175	79,184	4,669	366,926	24.8	23.6	26.1	1,533	120,457
55-59	0.0160	0.0769	0.0679	0.0861	71,051	67,176	74,910	5,465	341,590	21.2	20.1	22.5	1,314	82,133
60-64	0.0263	0.1235	0.1092	0.1371	65,585	61,393	69,826	8,100	307,678	17.8	16.7	19.0	3,521	133,755
65-69	0.0348	0.1602	0.1420	0.1780	57,486	52,975	62,198	9,208	264,408	14.9	13.9	16.1	3,509	100,754
70-74	0.0520	0.2301	0.2046	0.2562	48,277	43,543	53,365	11,108	213,617	12.3	11.4	13.3	3,616	69,541
75-79	0.0686	0.2927	0.2607	0.3245	37,169	32,389	42,447	10,879	158,650	10.2	9.5	11.1	1,790	26,105
80-84	0.0871	0.3576	0.3209	0.3927	26,291	21,880	31,380	9,402	107,948	8.4	7.8	9.2	847	9,722
85-89	0.1144	0.4448	0.4044	0.4815	16,888	13,288	21,312	7,512	65,663	6.8	6.3	7.3	728	6,365
90-94	0.1560	0.5313	0.4905	0.5672	9,377	6,890	12,694	4,982	31,937	5.2	4.8	5.6	396	2,536
95-99	0.2340	0.6874	0.6483	0.7200	4,394	2,982	6,467	3,021	12,910	3.7	3.5	4.0	194	829
100+	0.3900	1	1	1	1,374	835	2,274	1,374	3,522	2.6	2.4	2.6	78	199
All ages													*34,284*	*3,959,168*

ln($_nq_x$) with uncertainty interval

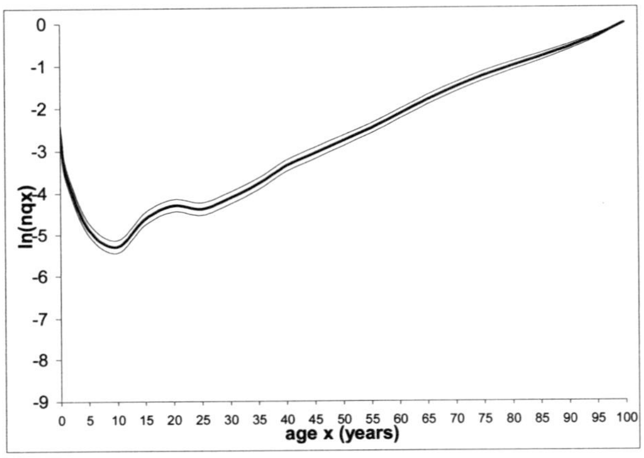

l_x with uncertainty interval

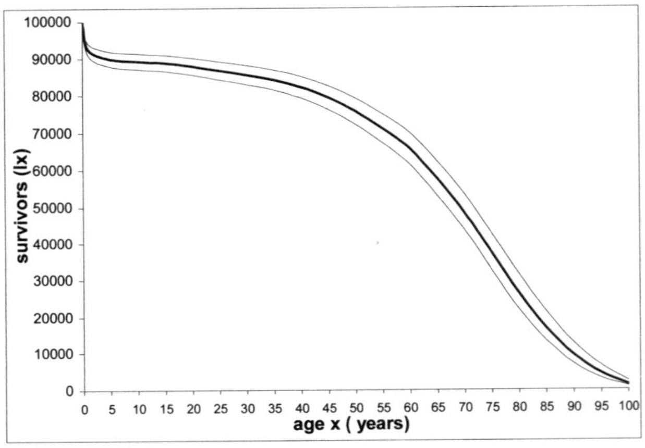

e_x with uncertainty interval

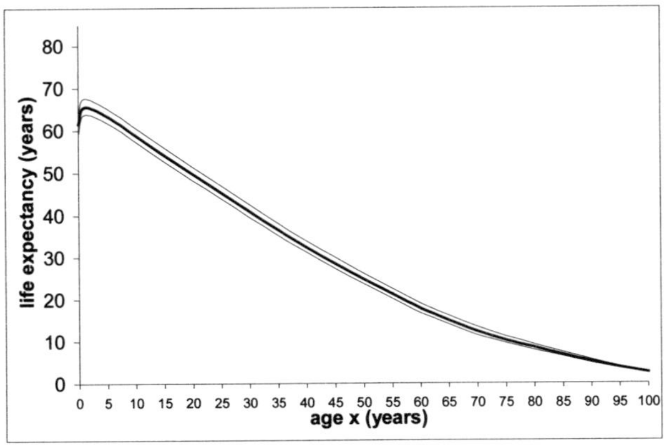

Distribution of deaths by age with uncertainty interval

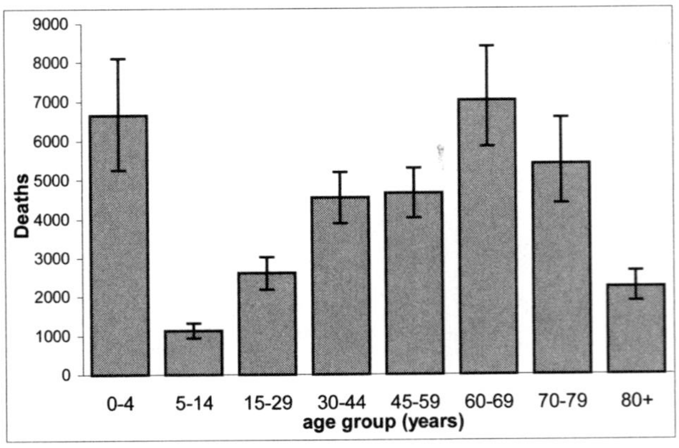

Azerbaijan | Azerbaïdjan | Azerbaiyán

Females, 2000 | *Femmes, 2000* | *Mujeres, 2000*

Age	$_nM_x$	$_nq_x$ mean	$_nq_x$ percentiles 2½	$_nq_x$ percentiles 97½	l_x mean	l_x percentiles 2½	l_x percentiles 97½	$_nd_x$	$_nL_x$	e_x mean	e_x percentiles 2½	e_x percentiles 97½	Actual deaths	Actual population
<1	0.0698	0.0657	0.0508	0.0812	100,000	100,000	100,000	6,571	94,086	68.8	66.4	71.3	4,020	57,556
1-4	0.0062	0.0243	0.0203	0.0286	93,429	91,876	94,919	2,271	368,264	72.7	70.7	74.8	1,468	238,095
5-9	0.0012	0.0059	0.0049	0.0069	91,158	89,370	92,931	534	454,453	70.5	68.6	72.6	485	412,443
10-14	0.0006	0.0031	0.0026	0.0037	90,624	88,767	92,466	285	452,404	65.9	64.0	68.0	267	423,208
15-19	0.0008	0.0041	0.0035	0.0048	90,338	88,444	92,217	370	450,765	61.1	59.2	63.2	317	386,254
20-24	0.0013	0.0066	0.0056	0.0077	89,968	88,025	91,893	595	448,352	56.3	54.4	58.4	447	337,032
25-29	0.0014	0.0070	0.0059	0.0081	89,373	87,351	91,369	624	445,305	51.7	49.8	53.8	421	300,874
30-34	0.0015	0.0077	0.0065	0.0089	88,749	86,647	90,817	684	442,037	47.0	45.2	49.1	510	329,793
35-39	0.0022	0.0111	0.0094	0.0128	88,066	85,877	90,209	974	437,892	42.3	40.5	44.4	803	360,816
40-44	0.0033	0.0164	0.0140	0.0189	87,091	84,786	89,341	1,430	431,881	37.8	36.0	39.9	1,012	305,690
45-49	0.0047	0.0233	0.0200	0.0267	85,661	83,196	88,054	1,998	423,311	33.4	31.6	35.4	980	207,527
50-54	0.0066	0.0325	0.0279	0.0370	83,663	80,995	86,233	2,716	411,527	29.1	27.4	31.1	894	135,520
55-59	0.0105	0.0512	0.0440	0.0590	80,948	78,021	83,779	4,141	394,385	25.0	23.4	27.0	991	94,334
60-64	0.0153	0.0737	0.0624	0.0859	76,807	73,536	80,027	5,659	369,884	21.2	19.7	23.1	2,497	163,177
65-69	0.0220	0.1043	0.0867	0.1227	71,147	67,528	74,805	7,418	337,191	17.7	16.2	19.5	2,733	124,230
70-74	0.0340	0.1567	0.1289	0.1846	63,729	59,847	67,798	9,985	293,682	14.5	13.1	16.1	3,208	94,346
75-79	0.0480	0.2143	0.1758	0.2535	53,744	49,211	58,374	11,517	239,928	11.7	10.5	13.1	2,427	50,562
80-84	0.0700	0.2979	0.2470	0.3505	42,227	36,872	47,778	12,578	179,691	9.2	8.2	10.4	1,904	27,204
85-89	0.1020	0.4064	0.3457	0.4680	29,649	24,250	35,975	12,049	118,123	7.1	6.3	7.9	1,874	18,372
90-94	0.1500	0.5172	0.4547	0.5792	17,600	13,217	23,539	9,104	60,691	5.2	4.6	5.8	1,418	9,456
95-99	0.2455	0.7068	0.6484	0.7602	8,497	5,562	12,834	6,006	24,466	3.6	3.3	4.0	933	3,801
100+	0.4000	1	1	1	2,491	1,334	4,486	2,491	6,227	2.5	2.3	2.6	728	1,820
All ages													*30,337*	*4,082,110*

ln($_nq_x$) with uncertainty interval

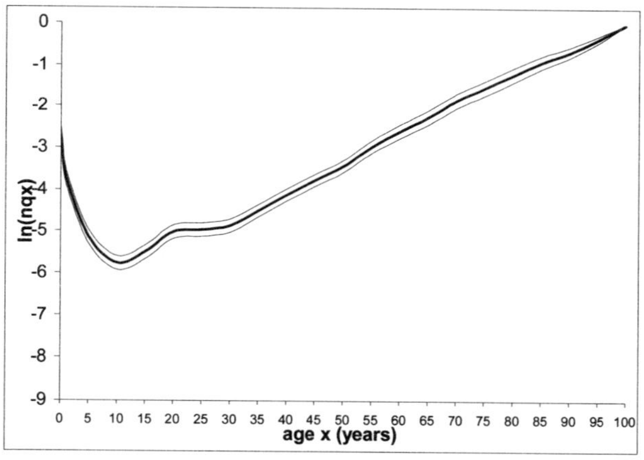

l_x with uncertainty interval

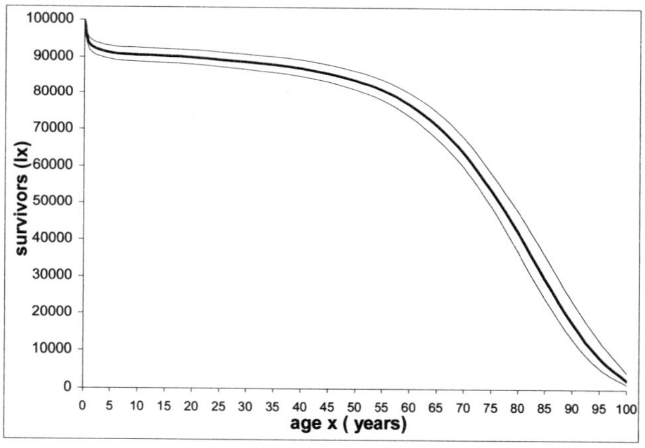

e_x with uncertainty interval

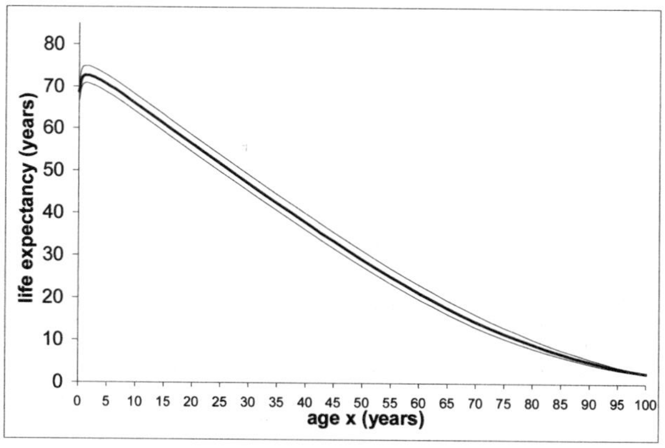

Distribution of deaths by age with uncertainty interval

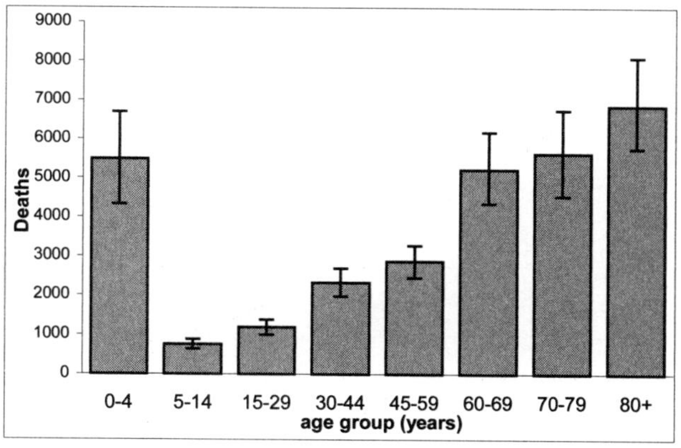

Bahamas

Males, 2000

Bahamas

Hommes, 2000

Bahamas

Hombres, 2000

Age	$_nM_x$	$_nq_x$			l_x			$_nd_x$	$_nL_x$	e_x			Actual deaths	Actual population
			percentiles			percentiles					percentiles			
		mean	2½	97½	mean	2½	97½			mean	2½	97½		
<1	0.0109	0.0108	0.0076	0.0153	100,000	100,000	100,000	1,077	99,031	68.0	67.1	68.9	33	2,993
1-4	0.0009	0.0036	0.0028	0.0047	98,923	98,466	99,236	359	394,829	67.7	66.8	68.8	11	12,471
5-9	0.0009	0.0045	0.0036	0.0057	98,563	98,002	98,955	446	491,702	64.0	63.0	65.1	14	15,832
10-14	0.0011	0.0054	0.0044	0.0067	98,117	97,440	98,599	534	489,251	59.2	58.3	60.4	16	14,458
15-19	0.0018	0.0089	0.0074	0.0108	97,583	96,783	98,162	873	485,734	54.5	53.6	55.7	27	14,847
20-24	0.0031	0.0155	0.0132	0.0182	96,711	95,737	97,434	1,495	479,814	50.0	49.0	51.2	42	13,332
25-29	0.0036	0.0179	0.0156	0.0205	95,215	93,996	96,152	1,705	471,812	45.8	44.8	47.0	50	13,903
30-34	0.0049	0.0243	0.0217	0.0272	93,510	92,066	94,648	2,275	461,861	41.6	40.5	42.8	63	12,881
35-39	0.0058	0.0288	0.0263	0.0315	91,235	89,561	92,590	2,629	449,602	37.5	36.5	38.8	73	12,423
40-44	0.0066	0.0325	0.0302	0.0348	88,606	86,742	90,151	2,878	435,835	33.6	32.6	34.8	62	9,388
45-49	0.0095	0.0463	0.0429	0.0495	85,728	83,728	87,432	3,970	418,715	29.6	28.6	30.9	66	6,952
50-54	0.0110	0.0538	0.0492	0.0581	81,758	79,583	83,678	4,395	397,802	25.9	24.9	27.1	59	5,382
55-59	0.0157	0.0755	0.0683	0.0822	77,363	74,963	79,562	5,842	372,210	22.3	21.3	23.4	74	4,684
60-64	0.0193	0.0920	0.0822	0.1008	71,521	68,801	74,128	6,579	341,158	18.9	17.9	20.0	71	3,663
65-69	0.0273	0.1279	0.1134	0.1415	64,942	61,867	68,036	8,308	303,941	15.5	14.6	16.5	72	2,646
70-74	0.0414	0.1874	0.1656	0.2092	56,634	53,114	60,323	10,615	256,634	12.4	11.6	13.3	75	1,822
75-79	0.0612	0.2653	0.2354	0.2963	46,019	42,004	50,332	12,209	199,574	9.7	9.1	10.5	69	1,124
80-84	0.0966	0.3890	0.3506	0.4292	33,810	29,559	38,483	13,152	136,171	7.3	6.8	7.9	76	782
85-89	0.1492	0.5433	0.5008	0.5862	20,658	16,872	24,992	11,224	75,230	5.4	5.1	5.9	56	372
90-94	0.2255	0.6725	0.6325	0.7102	9,434	6,983	12,477	6,344	28,137	3.9	3.7	4.2	26	114
95-99	0.3334	0.8334	0.8031	0.8601	3,090	2,024	4,585	2,575	7,724	2.8	2.7	3.0	9	27
100+	0.4822	1	1	1	515	283	903	515	1,067	2.1	2.0	2.2	3	6
All ages													*1,045*	*150,102*

$\ln({}_nq_x)$ with uncertainty interval

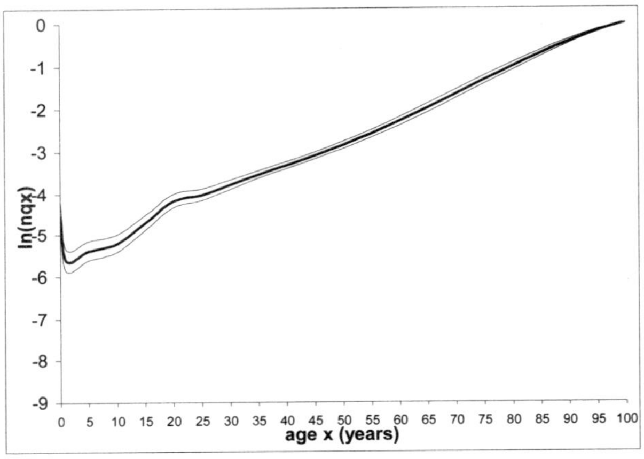

l_x with uncertainty interval

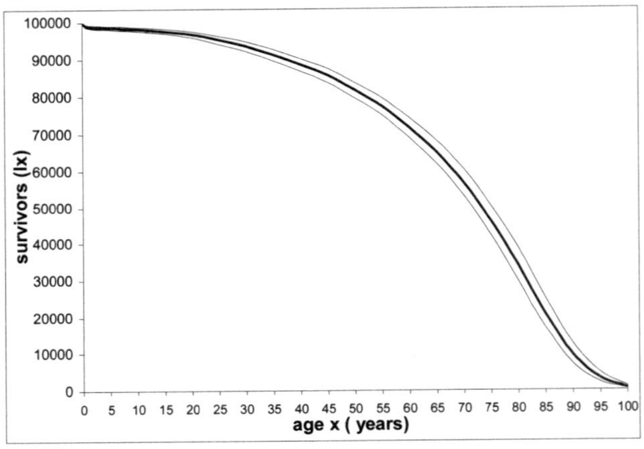

e_x with uncertainty interval

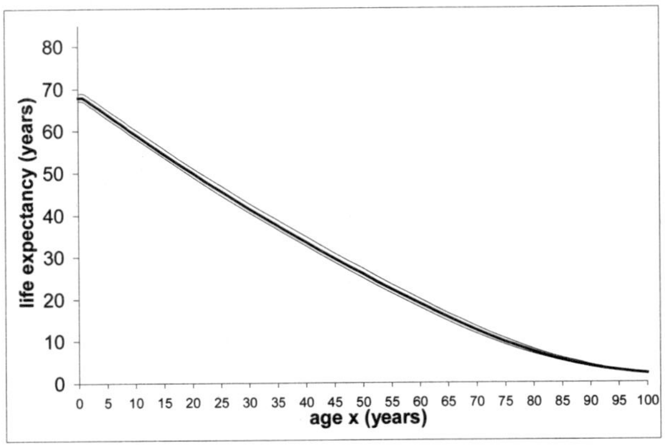

Distribution of deaths by age with uncertainty interval

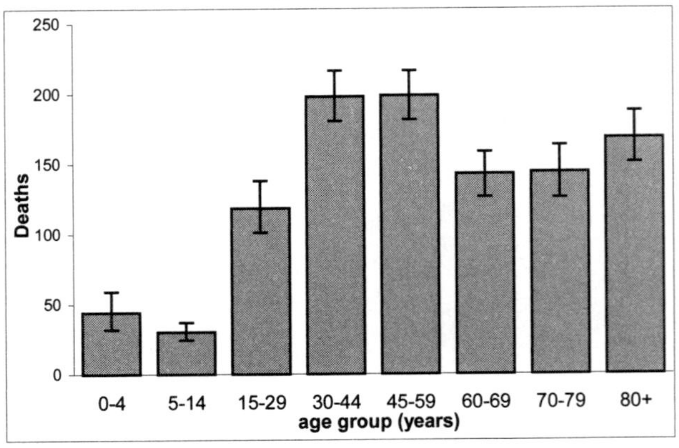

Bahamas **Bahamas** **Bahamas**

Females, 2000 *Femmes, 2000* *Mujeres, 2000*

Age	$_nM_x$	$_nq_x$ mean	$_nq_x$ percentiles 2½	$_nq_x$ percentiles 97½	l_x mean	l_x percentiles 2½	l_x percentiles 97½	$_nd_x$	$_nL_x$	e_x mean	e_x percentiles 2½	e_x percentiles 97½	Actual deaths	Actual population
<1	0.0105	0.0104	0.0073	0.0150	100,000	100,000	100,000	1,045	99,060	74.8	73.8	75.9	30	2,847
1-4	0.0005	0.0020	0.0015	0.0025	98,955	98,497	99,266	194	395,355	74.6	73.6	75.8	6	12,058
5-9	0.0005	0.0027	0.0021	0.0034	98,761	98,246	99,115	263	493,148	70.7	69.7	71.9	8	15,291
10-14	0.0005	0.0026	0.0021	0.0033	98,498	97,912	98,907	259	491,843	65.9	64.9	67.1	7	14,043
15-19	0.0006	0.0029	0.0024	0.0036	98,239	97,589	98,698	290	490,472	61.0	60.1	62.3	9	14,559
20-24	0.0012	0.0058	0.0048	0.0070	97,950	97,234	98,461	566	488,333	56.2	55.3	57.4	15	12,857
25-29	0.0017	0.0086	0.0073	0.0102	97,384	96,553	97,990	839	484,821	51.5	50.6	52.8	24	13,641
30-34	0.0022	0.0112	0.0097	0.0128	96,545	95,570	97,274	1,077	480,032	47.0	46.0	48.2	29	12,856
35-39	0.0031	0.0155	0.0137	0.0173	95,468	94,344	96,335	1,480	473,641	42.5	41.5	43.7	39	12,456
40-44	0.0044	0.0215	0.0195	0.0236	93,988	92,708	95,013	2,025	464,877	38.1	37.2	39.3	44	10,212
45-49	0.0057	0.0280	0.0257	0.0302	91,963	90,521	93,158	2,572	453,384	33.9	32.9	35.1	44	7,781
50-54	0.0070	0.0344	0.0314	0.0373	89,391	87,784	90,760	3,075	439,268	29.8	28.8	30.9	43	6,204
55-59	0.0092	0.0450	0.0406	0.0490	86,316	84,506	87,910	3,881	421,877	25.7	24.8	26.9	51	5,550
60-64	0.0122	0.0594	0.0531	0.0649	82,435	80,365	84,343	4,895	399,937	21.8	21.0	22.9	54	4,392
65-69	0.0183	0.0874	0.0775	0.0961	77,540	75,152	79,868	6,776	370,760	18.1	17.2	19.1	60	3,258
70-74	0.0288	0.1345	0.1189	0.1491	70,764	67,928	73,677	9,517	330,026	14.6	13.8	15.5	71	2,478
75-79	0.0414	0.1878	0.1660	0.2090	61,247	57,802	64,920	11,501	277,481	11.4	10.7	12.2	69	1,659
80-84	0.0750	0.3158	0.2807	0.3517	49,746	45,723	54,143	15,709	209,456	8.5	7.9	9.2	89	1,190
85-89	0.1200	0.4615	0.4180	0.5065	34,037	29,643	38,945	15,709	130,910	6.3	5.8	6.8	69	575
90-94	0.1900	0.6051	0.5606	0.6485	18,327	14,628	22,665	11,090	58,368	4.5	4.2	4.9	34	181
95-99	0.2800	0.7609	0.7230	0.7951	7,238	5,142	9,959	5,507	19,667	3.3	3.1	3.6	10	37
100+	0.4200	1	1	1	1,731	1,054	2,758	1,731	4,121	2.4	2.3	2.6	3	6
All ages													*809*	*154,131*

ln($_nq_x$) with uncertainty interval

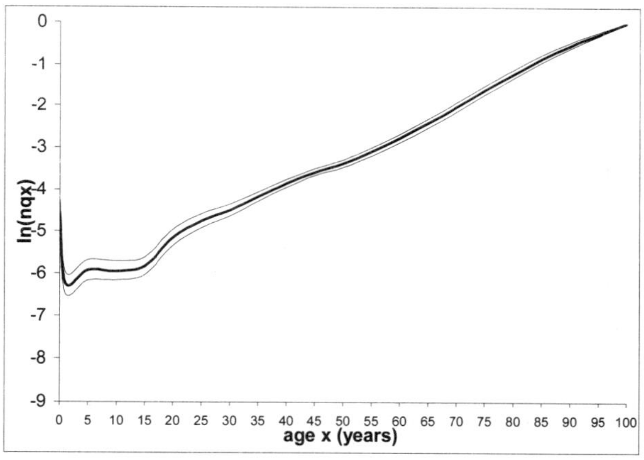

l_x with uncertainty interval

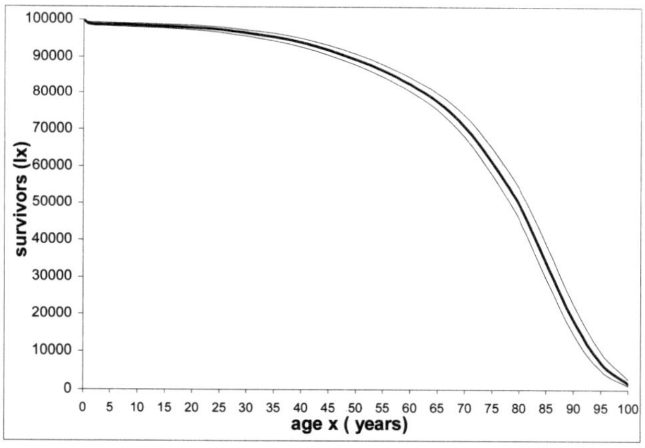

e_x with uncertainty interval

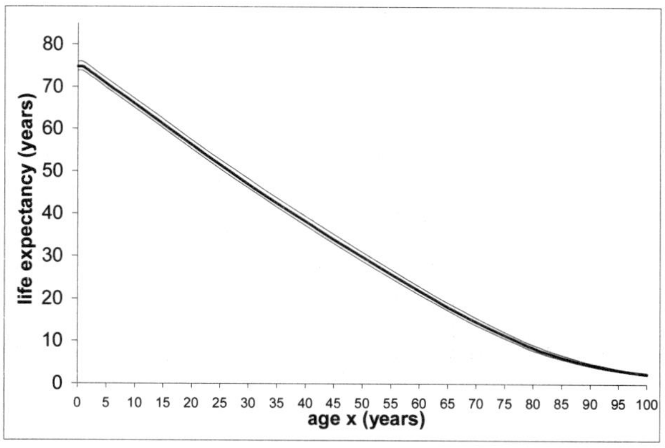

Distribution of deaths by age with uncertainty interval

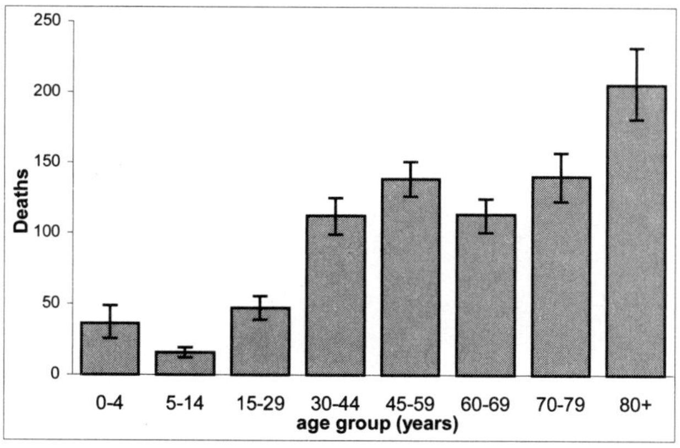

Bahrain / Bahreïn / Bahrein

Males, 2000 / *Hommes, 2000* / *Hombres, 2000*

Age	$_nM_x$	$_nq_x$ mean	$_nq_x$ percentiles 2½	$_nq_x$ percentiles 97½	l_x mean	l_x percentiles 2½	l_x percentiles 97½	$_nd_x$	$_nL_x$	e_x mean	e_x percentiles 2½	e_x percentiles 97½	Actual deaths	Actual population
<1	0.0094	0.0094	0.0059	0.0138	100,000	100,000	100,000	935	99,158	72.7	71.9	73.7	54	5,769
1-4	0.0004	0.0016	0.0011	0.0020	99,065	98,621	99,413	155	395,887	72.4	71.5	73.4	9	23,078
5-9	0.0002	0.0012	0.0008	0.0015	98,910	98,420	99,302	116	494,261	68.5	67.6	69.6	8	33,607
10-14	0.0002	0.0012	0.0009	0.0015	98,794	98,272	99,218	119	493,675	63.6	62.7	64.6	7	29,836
15-19	0.0007	0.0035	0.0026	0.0044	98,676	98,122	99,131	348	492,509	58.7	57.8	59.7	19	26,905
20-24	0.0010	0.0048	0.0037	0.0058	98,328	97,691	98,869	467	490,472	53.9	52.9	54.9	23	24,571
25-29	0.0009	0.0046	0.0036	0.0054	97,861	97,128	98,505	446	488,190	49.1	48.2	50.2	22	23,842
30-34	0.0010	0.0051	0.0042	0.0060	97,415	96,604	98,148	498	485,830	44.4	43.4	45.4	34	32,884
35-39	0.0013	0.0067	0.0056	0.0077	96,917	96,032	97,739	651	482,957	39.6	38.6	40.7	58	43,151
40-44	0.0021	0.0102	0.0088	0.0115	96,266	95,300	97,190	983	478,870	34.8	33.8	35.9	91	44,115
45-49	0.0034	0.0167	0.0149	0.0184	95,282	94,223	96,336	1,593	472,429	30.2	29.1	31.3	113	33,486
50-54	0.0057	0.0280	0.0255	0.0305	93,689	92,521	94,898	2,622	461,889	25.6	24.6	26.8	113	19,942
55-59	0.0096	0.0467	0.0422	0.0515	91,067	89,757	92,419	4,253	444,702	21.3	20.3	22.5	97	10,175
60-64	0.0155	0.0748	0.0667	0.0834	86,814	85,440	88,177	6,496	417,831	17.2	16.2	18.4	103	6,605
65-69	0.0324	0.1498	0.1301	0.1707	80,318	78,750	81,896	12,028	371,520	13.4	12.5	14.5	139	4,279
70-74	0.0527	0.2328	0.1977	0.2673	68,290	65,560	70,910	15,895	301,712	10.3	9.5	11.3	152	2,885
75-79	0.0856	0.3524	0.3005	0.4041	52,395	48,208	56,491	18,464	215,815	7.7	7.0	8.5	130	1,518
80-84	0.1400	0.5184	0.4562	0.5797	33,931	28,990	39,418	17,591	125,676	5.5	5.0	6.1	101	718
85-89	0.2425	0.7549	0.7015	0.8033	16,340	12,515	21,437	12,335	50,861	3.8	3.5	4.1	54	222
90-94	0.3678	0.8742	0.8392	0.9045	4,005	2,461	6,379	3,501	9,520	2.6	2.5	2.8	22	60
95-99	0.5369	0.9800	0.9693	0.9877	504	235	1,013	494	1,037	2.1	2.1	2.1	5	10
100+	0.7546	1	1	1	10	3	30	10	13	1.3	1.3	1.7	1	1
All ages													*1,354*	*367,659*

ln($_nq_x$) with uncertainty interval

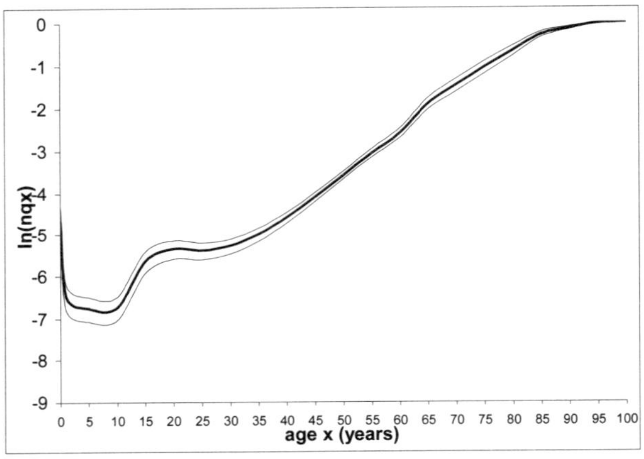

l_x with uncertainty interval

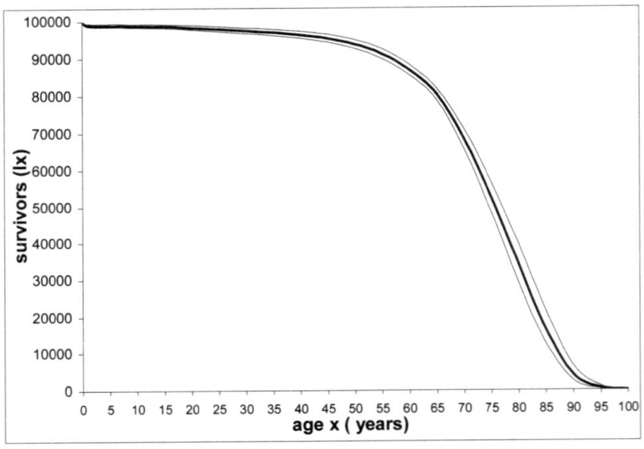

e_x with uncertainty interval

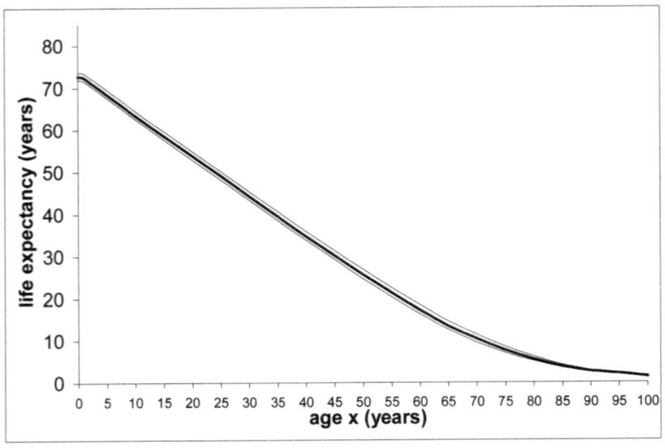

Distribution of deaths by age with uncertainty interval

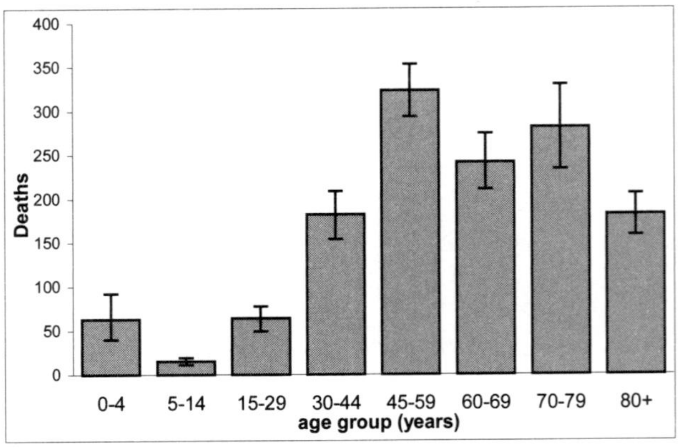

Bahrain **Bahreïn** **Bahrein**

Females, 2000 *Femmes, 2000* *Mujeres, 2000*

Age	$_nM_x$	$_nq_x$			l_x			$_nd_x$	$_nL_x$	e_x			Actual deaths	Actual population
			percentiles			percentiles					percentiles			
		mean	2½	97½	mean	2½	97½			mean	2½	97½		
<1	0.0068	0.0068	0.0039	0.0119	100,000	100,000	100,000	680	99,388	74.7	73.5	76.0	38	5,545
1-4	0.0004	0.0015	0.0010	0.0022	99,320	98,805	99,610	150	396,919	74.2	73.0	75.7	8	22,178
5-9	0.0002	0.0010	0.0007	0.0014	99,170	98,586	99,509	95	495,612	70.4	69.1	71.8	6	32,448
10-14	0.0002	0.0009	0.0006	0.0013	99,075	98,450	99,444	91	495,147	65.4	64.2	66.9	5	27,954
15-19	0.0004	0.0018	0.0013	0.0025	98,984	98,322	99,380	177	494,476	60.5	59.2	62.0	9	25,592
20-24	0.0004	0.0021	0.0016	0.0029	98,807	98,077	99,254	211	493,505	55.6	54.3	57.1	10	22,859
25-29	0.0005	0.0025	0.0018	0.0033	98,596	97,792	99,099	243	492,370	50.7	49.4	52.2	10	19,515
30-34	0.0007	0.0034	0.0026	0.0044	98,352	97,473	98,916	335	490,925	45.8	44.5	47.4	16	23,299
35-39	0.0010	0.0051	0.0041	0.0064	98,017	97,045	98,657	503	488,830	41.0	39.6	42.5	27	26,219
40-44	0.0017	0.0083	0.0069	0.0100	97,515	96,424	98,255	813	485,542	36.2	34.8	37.8	38	22,677
45-49	0.0025	0.0124	0.0107	0.0142	96,702	95,461	97,578	1,198	480,516	31.4	30.1	33.1	38	15,337
50-54	0.0043	0.0215	0.0191	0.0240	95,504	94,109	96,528	2,052	472,391	26.8	25.4	28.5	38	8,737
55-59	0.0080	0.0392	0.0342	0.0447	93,452	91,921	94,623	3,665	458,100	22.3	20.9	24.1	47	5,820
60-64	0.0140	0.0676	0.0577	0.0787	89,788	88,270	91,067	6,073	433,757	18.2	16.8	19.9	70	5,000
65-69	0.0273	0.1278	0.1050	0.1530	83,715	81,893	85,443	10,700	391,826	14.3	12.9	16.0	100	3,654
70-74	0.0466	0.2085	0.1641	0.2547	73,015	69,572	76,220	15,224	327,016	11.0	9.8	12.6	129	2,779
75-79	0.0725	0.3068	0.2386	0.3705	57,791	52,304	63,014	17,731	244,628	8.3	7.3	9.6	107	1,481
80-84	0.1302	0.4910	0.3970	0.5699	40,060	32,927	47,711	19,670	151,127	5.8	5.1	6.8	91	702
85-89	0.2209	0.7116	0.6227	0.7763	20,391	14,352	28,770	14,511	65,676	4.0	3.6	4.6	52	234
90-94	0.3546	0.8591	0.8014	0.8981	5,880	3,292	10,856	5,051	14,245	2.7	2.5	3.0	20	56
95-99	0.5380	0.9800	0.9635	0.9890	829	335	2,156	812	1,707	2.1	2.1	2.2	4	8
100+	0.7716	1	1	1	17	4	79	17	21	1.3	1.2	1.8	0	0
All ages													*864*	*272,094*

ln($_nq_x$) with uncertainty interval

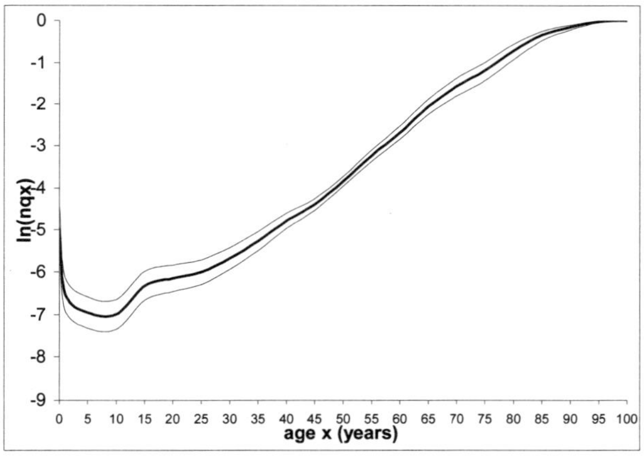

l_x with uncertainty interval

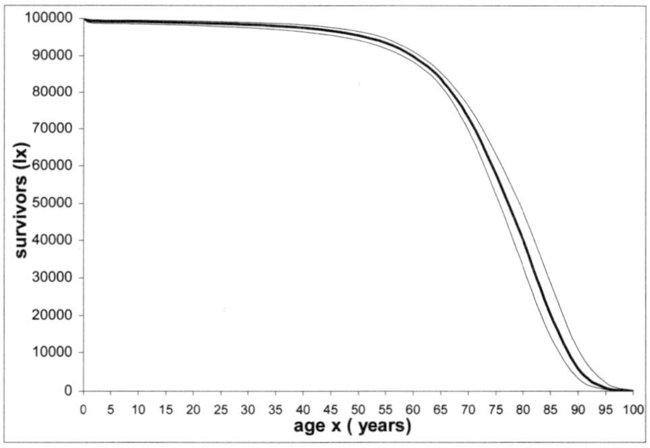

e_x with uncertainty interval

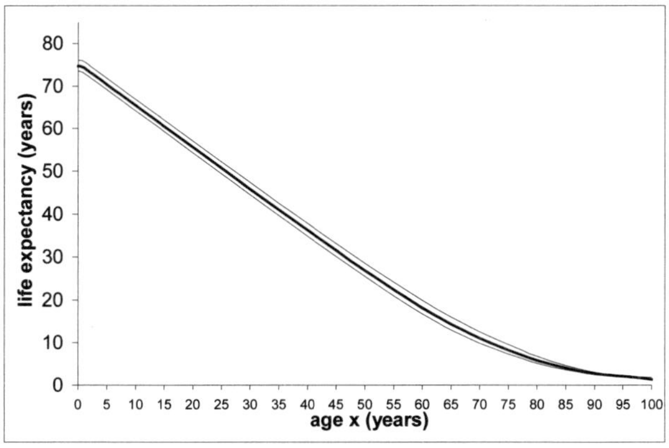

Distribution of deaths by age with uncertainty interval

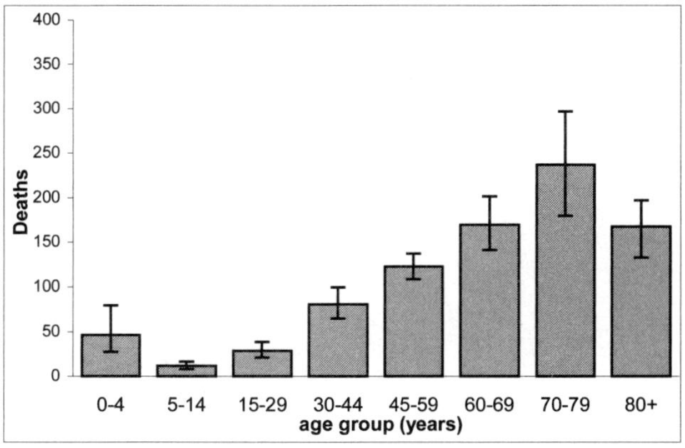

Bangladesh **Bangladesh** **Bangladesh**

Males, 2000 *Hommes, 2000* *Hombres, 2000*

Age	$_nM_x$	$_nq_x$			l_x			$_nd_x$	$_nL_x$	e_x			Actual deaths	Actual population
			percentiles			percentiles					percentiles			
		mean	2½	97½	mean	2½	97½			mean	2½	97½		
<1	0.0616	0.0583	0.0504	0.0666	100,000	100,000	100,000	5,834	94,749	60.4	58.5	62.4	126,894	2,060,755
1-4	0.0088	0.0346	0.0303	0.0387	94,166	93,343	94,958	3,253	368,854	63.1	61.5	64.9	66,164	7,501,107
5-9	0.0025	0.0124	0.0108	0.0140	90,912	89,914	91,912	1,129	451,738	61.3	59.9	62.8	22,891	9,156,390
10-14	0.0013	0.0067	0.0058	0.0076	89,783	88,700	90,916	600	447,413	57.1	55.7	58.5	11,565	8,616,951
15-19	0.0013	0.0065	0.0056	0.0073	89,182	88,033	90,387	578	444,467	52.4	51.1	53.9	10,127	7,790,355
20-24	0.0016	0.0080	0.0069	0.0090	88,605	87,391	89,877	706	441,258	47.8	46.4	49.2	10,667	6,667,014
25-29	0.0020	0.0100	0.0087	0.0113	87,899	86,608	89,254	875	437,306	43.1	41.8	44.5	11,947	5,973,688
30-34	0.0030	0.0149	0.0130	0.0169	87,024	85,638	88,481	1,296	431,881	38.5	37.3	39.9	15,493	5,164,224
35-39	0.0044	0.0218	0.0189	0.0247	85,728	84,204	87,334	1,866	423,978	34.1	32.9	35.4	19,643	4,464,237
40-44	0.0065	0.0320	0.0279	0.0363	83,863	82,143	85,680	2,682	412,609	29.8	28.7	31.1	23,555	3,623,833
45-49	0.0090	0.0440	0.0384	0.0499	81,181	79,191	83,293	3,573	396,972	25.7	24.6	26.9	26,139	2,904,285
50-54	0.0130	0.0630	0.0551	0.0712	77,608	75,241	80,094	4,886	375,826	21.7	20.8	22.9	25,231	1,940,828
55-59	0.0200	0.0952	0.0837	0.1072	72,722	69,885	75,679	6,926	346,297	18.0	17.1	19.1	31,771	1,588,529
60-64	0.0310	0.1439	0.1265	0.1604	65,796	62,395	69,329	9,465	305,320	14.7	13.9	15.7	38,609	1,245,455
65-69	0.0460	0.2063	0.1820	0.2291	56,331	52,388	60,447	11,620	252,608	11.7	11.0	12.6	43,381	943,072
70-74	0.0700	0.2979	0.2647	0.3284	44,712	40,506	49,169	13,318	190,262	9.1	8.5	9.8	44,272	632,453
75-79	0.1100	0.4314	0.3906	0.4680	31,393	27,492	35,843	13,542	123,111	6.9	6.5	7.5	38,948	354,073
80-84	0.1600	0.5714	0.5304	0.6074	17,851	14,876	21,623	10,201	63,754	5.3	5.0	5.6	25,631	160,193
85-89	0.2300	0.7302	0.6948	0.7599	7,650	5,903	10,126	5,586	24,287	3.9	3.8	4.2	13,271	57,699
90-94	0.3300	0.8291	0.8017	0.8518	2,064	1,421	3,076	1,712	5,187	2.9	2.7	3.0	3,822	11,582
95-99	0.4900	0.9919	0.9881	0.9944	353	212	605	350	714	2.0	2.0	2.1	447	912
100+	0.7500	1	1	1	3	1	7	3	4	1.3	1.3	1.6	37	49
All ages													*610,503*	*70,857,684*

ln($_nq_x$) with uncertainty interval

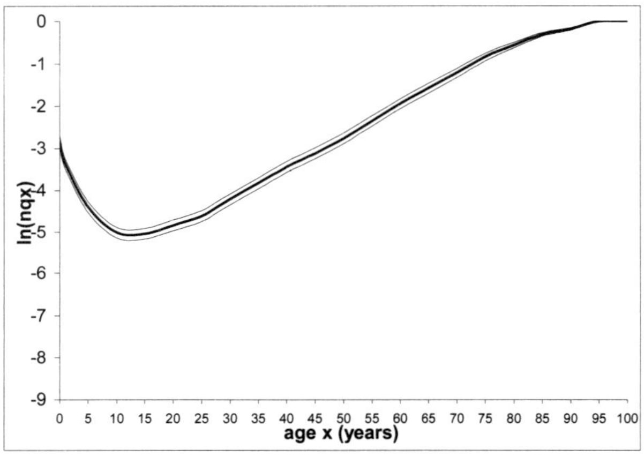

l_x with uncertainty interval

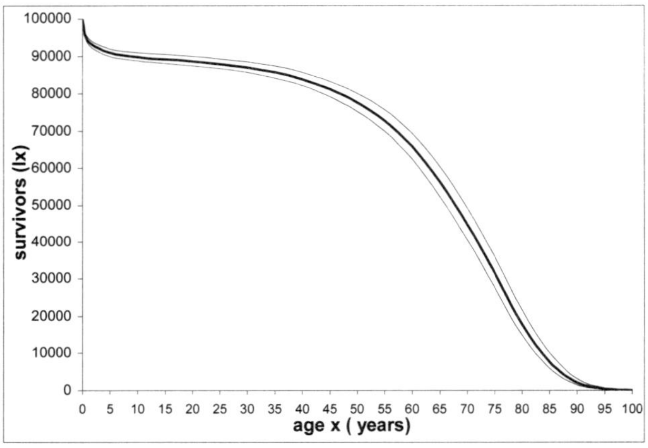

e_x with uncertainty interval

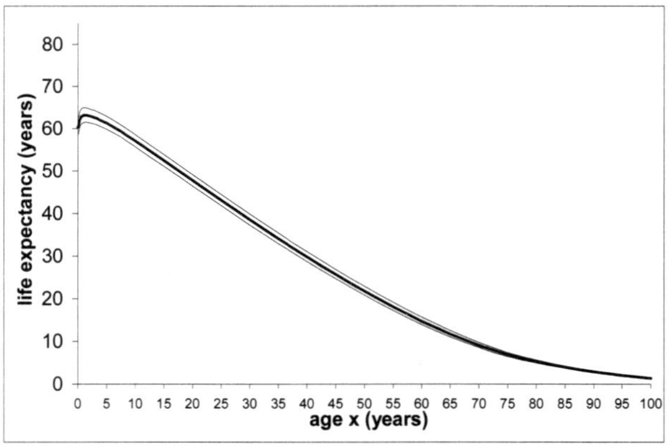

Distribution of deaths by age with uncertainty interval

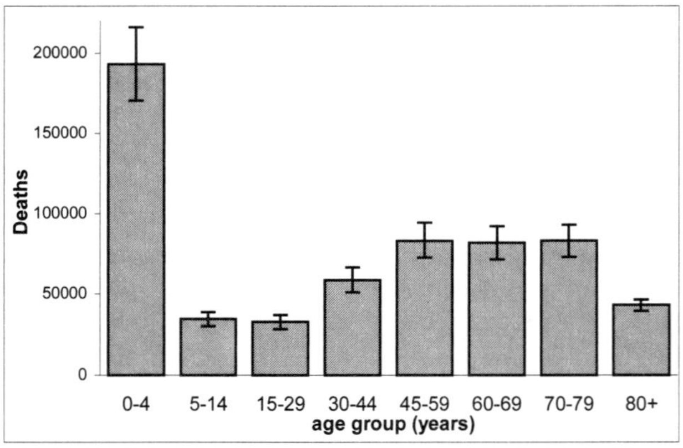

Females, 2000 *Femmes, 2000* *Mujeres, 2000*

Age	$_nM_x$	$_nq_x$			l_x			$_nd_x$	$_nL_x$	e_x			Actual deaths	Actual population
			percentiles			percentiles					percentiles			
		mean	2½	97½	mean	2½	97½			mean	2½	97½		
<1	0.0635	0.0601	0.0521	0.0689	100,000	100,000	100,000	6,009	94,592	60.8	59.2	62.7	126,122	1,985,313
1-4	0.0090	0.0351	0.0313	0.0387	93,991	93,114	94,789	3,299	368,045	63.7	62.2	65.3	63,691	7,104,728
5-9	0.0024	0.0121	0.0107	0.0134	90,691	89,692	91,691	1,096	450,718	62.0	60.6	63.4	21,039	8,655,308
10-14	0.0010	0.0050	0.0044	0.0055	89,596	88,510	90,672	447	446,862	57.7	56.4	59.1	8,107	8,109,576
15-19	0.0018	0.0092	0.0082	0.0102	89,149	88,028	90,259	821	443,694	53.0	51.7	54.4	13,500	7,298,742
20-24	0.0024	0.0120	0.0106	0.0133	88,328	87,145	89,506	1,058	438,996	48.5	47.2	49.8	15,050	6,243,606
25-29	0.0031	0.0155	0.0137	0.0172	87,270	86,009	88,532	1,351	432,972	44.0	42.8	45.3	17,380	5,568,616
30-34	0.0038	0.0187	0.0166	0.0209	85,919	84,565	87,283	1,609	425,571	39.7	38.6	40.9	18,075	4,780,239
35-39	0.0045	0.0224	0.0198	0.0250	84,310	82,851	85,817	1,887	416,830	35.4	34.3	36.6	18,487	4,082,855
40-44	0.0058	0.0285	0.0252	0.0318	82,422	80,829	84,114	2,346	406,246	31.1	30.1	32.3	18,879	3,269,085
45-49	0.0077	0.0379	0.0337	0.0424	80,076	78,300	81,990	3,039	392,785	27.0	26.0	28.1	20,018	2,587,712
50-54	0.0113	0.0550	0.0490	0.0614	77,038	74,998	79,228	4,241	374,587	22.9	22.0	24.0	21,550	1,903,548
55-59	0.0175	0.0838	0.0746	0.0931	72,797	70,395	75,347	6,098	348,739	19.1	18.3	20.2	28,006	1,601,525
60-64	0.0270	0.1265	0.1125	0.1397	66,699	63,840	69,705	8,435	312,405	15.7	14.9	16.6	34,003	1,259,372
65-69	0.0410	0.1859	0.1654	0.2053	58,264	54,922	61,766	10,834	264,234	12.6	11.8	13.4	37,904	924,481
70-74	0.0630	0.2721	0.2428	0.2994	47,430	43,757	51,299	12,908	204,881	9.9	9.3	10.6	39,143	621,322
75-79	0.0940	0.3806	0.3439	0.4142	34,522	30,938	38,503	13,138	139,767	7.6	7.1	8.2	32,990	350,962
80-84	0.1400	0.5185	0.4787	0.5542	21,384	18,400	25,054	11,088	79,201	5.7	5.4	6.2	22,821	163,010
85-89	0.2100	0.6885	0.6515	0.7201	10,296	8,281	13,024	7,089	33,758	4.2	4.0	4.5	11,873	56,539
90-94	0.3000	0.7895	0.7595	0.8146	3,207	2,324	4,521	2,532	8,439	3.1	3.0	3.3	4,079	13,595
95-99	0.4500	0.9574	0.9453	0.9666	675	433	1,079	646	1,437	2.2	2.2	2.3	608	1,350
100+	0.7000	1	1	1	29	14	58	29	41	1.4	1.4	1.7	65	93
All ages													*573,393*	*66,581,577*

ln($_nq_x$) with uncertainty interval

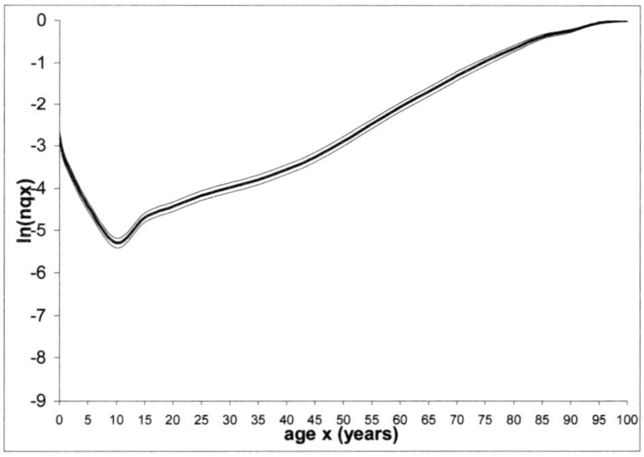

l_x with uncertainty interval

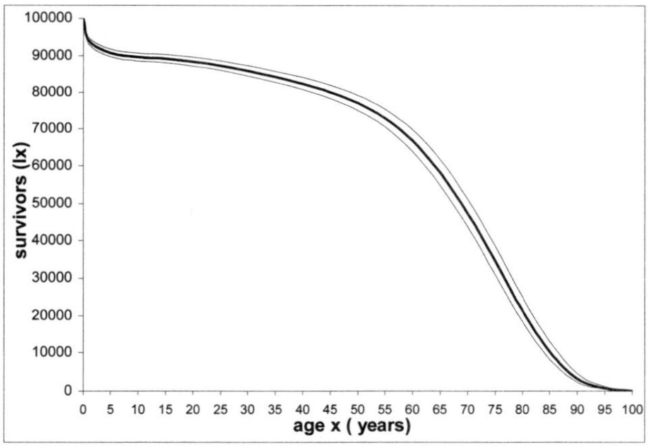

e_x with uncertainty interval

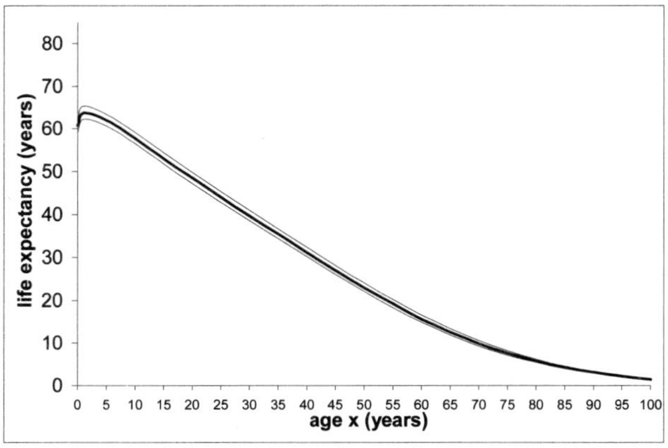

Distribution of deaths by age with uncertainty interval

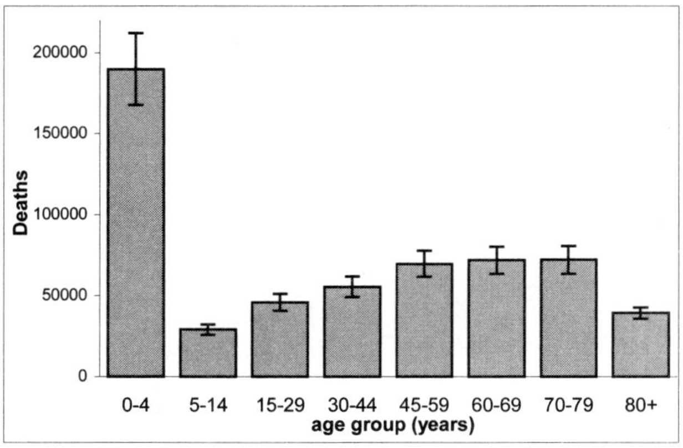

Barbados **Barbade** **Barbados**

Males, 2000 *Hommes, 2000* *Hombres, 2000*

Age	$_nM_x$	$_nq_x$			l_x			$_nd_x$	$_nL_x$	e_x			Actual deaths	Actual population
			percentiles			percentiles					percentiles			
		mean	2½	97½	mean	2½	97½			mean	2½	97½		
<1	0.0117	0.0115	0.0084	0.0146	100,000	100,000	100,000	1,155	98,961	71.6	70.4	72.8	21	1,816
1-4	0.0003	0.0013	0.0010	0.0015	98,845	98,545	99,158	125	395,081	71.4	70.4	72.7	2	6,760
5-9	0.0003	0.0015	0.0012	0.0018	98,720	98,395	99,060	152	493,220	67.5	66.5	68.7	3	9,214
10-14	0.0002	0.0012	0.0009	0.0014	98,568	98,213	98,941	117	492,546	62.6	61.6	63.8	2	10,232
15-19	0.0010	0.0048	0.0038	0.0056	98,451	98,075	98,849	468	491,084	57.7	56.7	58.9	10	10,756
20-24	0.0012	0.0061	0.0049	0.0071	97,983	97,524	98,477	596	488,425	52.9	51.9	54.2	13	10,494
25-29	0.0022	0.0111	0.0091	0.0129	97,387	96,832	97,995	1,082	484,232	48.2	47.3	49.5	27	12,242
30-34	0.0031	0.0152	0.0128	0.0175	96,306	95,585	97,099	1,467	477,861	43.8	42.8	44.9	36	11,862
35-39	0.0029	0.0146	0.0125	0.0167	94,839	93,913	95,855	1,386	470,730	39.4	38.5	40.6	36	12,307
40-44	0.0035	0.0173	0.0150	0.0196	93,453	92,348	94,654	1,614	463,230	34.9	34.1	36.1	38	10,956
45-49	0.0069	0.0339	0.0300	0.0381	91,839	90,542	93,234	3,117	451,403	30.5	29.7	31.6	60	8,677
50-54	0.0064	0.0315	0.0283	0.0350	88,722	87,096	90,435	2,796	436,621	26.5	25.8	27.5	42	6,574
55-59	0.0125	0.0608	0.0547	0.0668	85,926	84,045	87,872	5,227	416,564	22.3	21.6	23.3	50	4,019
60-64	0.0194	0.0927	0.0832	0.1011	80,699	78,433	83,069	7,483	384,789	18.6	17.9	19.5	67	3,426
65-69	0.0310	0.1440	0.1296	0.1554	73,216	70,507	76,159	10,544	339,721	15.2	14.6	16.0	91	2,925
70-74	0.0484	0.2157	0.1959	0.2306	62,672	59,548	66,290	13,521	279,560	12.3	11.8	13.0	137	2,842
75-79	0.0711	0.3018	0.2784	0.3234	49,152	45,817	53,302	14,835	208,671	10.0	9.5	10.6	151	2,117
80-84	0.0893	0.3649	0.3398	0.3895	34,317	31,001	38,463	12,523	140,275	8.3	7.9	8.8	132	1,480
85-89	0.1186	0.4574	0.4296	0.4836	21,793	18,925	25,394	9,969	84,044	6.6	6.3	7.0	85	719
90-94	0.1580	0.5360	0.5083	0.5611	11,824	9,773	14,486	6,337	40,110	5.1	4.9	5.4	30	188
95-99	0.2320	0.6840	0.6581	0.7067	5,487	4,290	7,123	3,753	16,176	3.8	3.6	3.9	5	23
100+	0.3820	1	1	1	1,734	1,258	2,436	1,734	4,540	2.6	2.4	2.7	0	1
All ages													*1,040*	*129,630*

ln($_nq_x$) with uncertainty interval

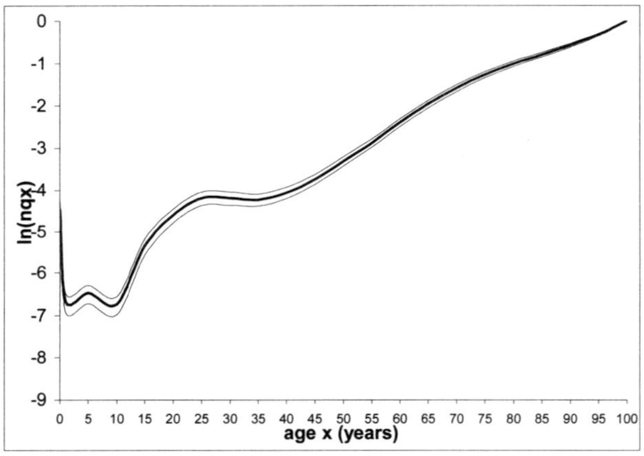

l_x with uncertainty interval

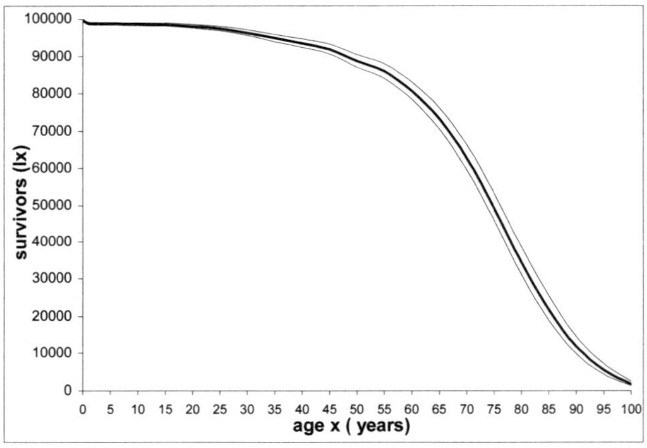

e_x with uncertainty interval

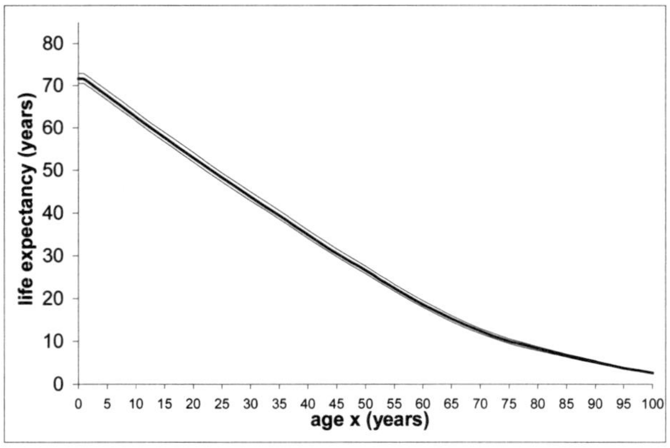

Distribution of deaths by age with uncertainty interval

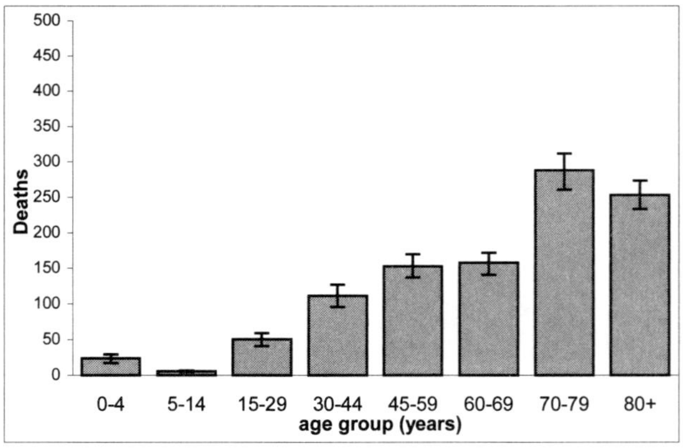

Barbados Barbade Barbados

Females, 2000 *Femmes, 2000* *Mujeres, 2000*

Age	$_nM_x$	$_nq_x$			l_x			$_nd_x$	$_nL_x$	e_x			Actual deaths	Actual population
			percentiles			percentiles					percentiles			
		mean	2½	97½	mean	2½	97½			mean	2½	97½		
<1	0.0114	0.0113	0.0087	0.0141	100,000	100,000	100,000	1,133	98,981	77.4	76.4	78.6	20	1,739
1-4	0.0004	0.0015	0.0013	0.0018	98,867	98,589	99,126	153	395,102	77.3	76.3	78.5	3	6,569
5-9	0.0003	0.0013	0.0010	0.0015	98,714	98,407	99,001	124	493,262	73.4	72.4	74.6	2	8,947
10-14	0.0003	0.0015	0.0012	0.0017	98,591	98,261	98,900	144	492,593	68.5	67.5	69.7	3	9,982
15-19	0.0006	0.0029	0.0024	0.0034	98,447	98,092	98,781	286	491,519	63.6	62.6	64.8	6	10,476
20-24	0.0006	0.0029	0.0024	0.0034	98,161	97,759	98,543	285	490,094	58.8	57.8	59.9	6	10,340
25-29	0.0008	0.0042	0.0035	0.0048	97,876	97,429	98,304	410	488,357	54.0	53.0	55.1	10	12,123
30-34	0.0011	0.0056	0.0048	0.0065	97,466	96,957	97,957	550	485,955	49.2	48.2	50.3	13	11,531
35-39	0.0016	0.0080	0.0069	0.0090	96,916	96,330	97,487	772	482,648	44.4	43.5	45.5	19	11,932
40-44	0.0024	0.0118	0.0103	0.0132	96,144	95,459	96,819	1,131	477,890	39.8	38.9	40.9	26	11,038
45-49	0.0037	0.0182	0.0161	0.0202	95,013	94,201	95,826	1,726	470,749	35.2	34.4	36.3	33	9,081
50-54	0.0057	0.0281	0.0252	0.0310	93,287	92,311	94,284	2,621	459,884	30.8	30.0	31.9	40	7,049
55-59	0.0080	0.0395	0.0357	0.0434	90,666	89,493	91,893	3,577	444,390	26.7	25.9	27.6	40	5,011
60-64	0.0132	0.0637	0.0573	0.0699	87,090	85,691	88,550	5,549	421,576	22.6	22.0	23.6	61	4,604
65-69	0.0186	0.0888	0.0797	0.0969	81,541	79,801	83,296	7,243	389,597	19.0	18.4	19.8	79	4,246
70-74	0.0281	0.1314	0.1182	0.1421	74,298	72,149	76,443	9,766	347,074	15.6	15.1	16.3	125	4,438
75-79	0.0412	0.1868	0.1691	0.1998	64,532	61,894	67,349	12,051	292,530	12.6	12.1	13.2	142	3,441
80-84	0.0593	0.2581	0.2366	0.2768	52,480	49,543	55,963	13,543	228,545	9.9	9.5	10.4	167	2,825
85-89	0.0911	0.3709	0.3466	0.3959	38,938	36,077	42,721	14,440	158,588	7.5	7.1	7.8	154	1,692
90-94	0.1374	0.4864	0.4608	0.5137	24,497	22,214	27,913	11,917	86,737	5.5	5.2	5.7	89	647
95-99	0.2333	0.6862	0.6612	0.7111	12,581	10,915	14,962	8,634	37,003	3.8	3.5	3.9	33	141
100+	0.3752	1	1	1	3,947	3,153	4,957	3,947	10,520	2.7	2.3	2.7	6	16
All ages													*1,077*	*137,868*

$\ln(_nq_x)$ with uncertainty interval

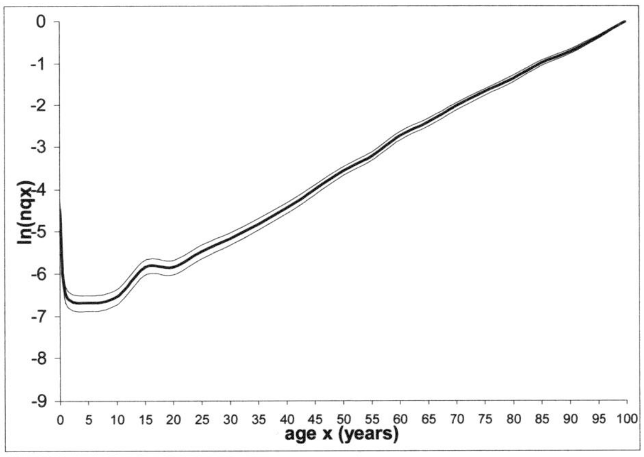

l_x with uncertainty interval

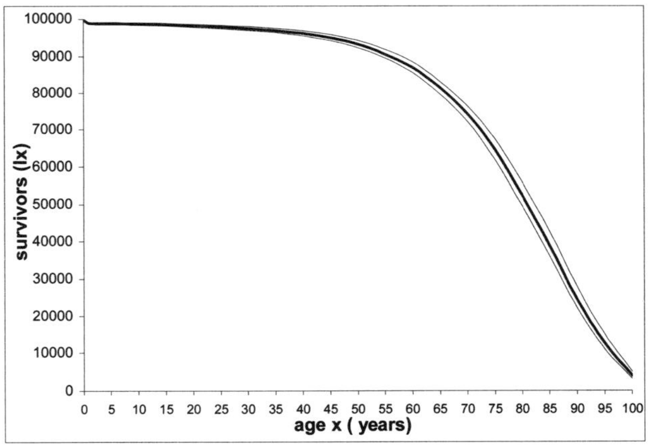

e_x with uncertainty interval

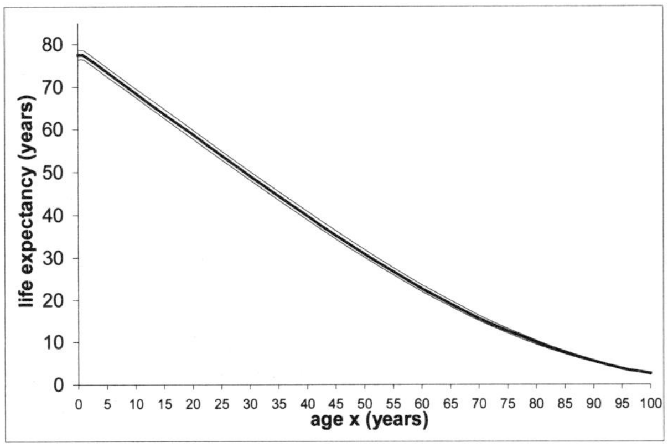

Distribution of deaths by age with uncertainty interval

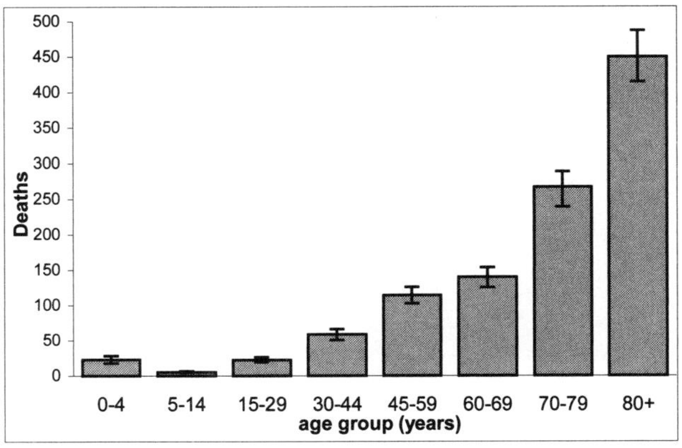

Belarus　　Bélarus　　Belarús

Males, 2000　　*Hommes, 2000*　　*Hombres, 2000*

Age	$_nM_x$	$_nq_x$ mean	$_nq_x$ percentiles 2½	$_nq_x$ percentiles 97½	l_x mean	l_x percentiles 2½	l_x percentiles 97½	$_nd_x$	$_nL_x$	e_x mean	e_x percentiles 2½	e_x percentiles 97½	Actual deaths	Actual population
<1	0.0130	0.0128	0.0096	0.0162	100,000	100,000	100,000	1,285	98,844	62.0	61.1	62.8	610	46,930
1-4	0.0010	0.0040	0.0032	0.0048	98,715	98,379	99,039	394	393,915	61.8	61.0	62.6	194	193,729
5-9	0.0004	0.0018	0.0015	0.0022	98,321	97,923	98,719	182	491,151	58.0	57.2	58.9	116	313,559
10-14	0.0004	0.0019	0.0016	0.0023	98,139	97,708	98,569	191	490,219	53.1	52.3	54.0	163	418,623
15-19	0.0010	0.0050	0.0041	0.0059	97,948	97,484	98,410	489	488,520	48.2	47.4	49.1	418	418,351
20-24	0.0030	0.0149	0.0125	0.0172	97,460	96,913	97,999	1,451	483,671	43.5	42.6	44.3	1,132	377,282
25-29	0.0040	0.0198	0.0170	0.0225	96,009	95,248	96,772	1,901	475,290	39.1	38.2	40.0	1,429	357,338
30-34	0.0050	0.0247	0.0217	0.0276	94,108	93,104	95,124	2,324	464,728	34.8	33.9	35.7	1,663	332,694
35-39	0.0070	0.0344	0.0309	0.0378	91,784	90,535	93,060	3,157	451,026	30.6	29.7	31.6	2,819	402,739
40-44	0.0100	0.0488	0.0447	0.0528	88,627	87,111	90,187	4,323	432,325	26.6	25.7	27.6	4,148	414,766
45-49	0.0140	0.0676	0.0632	0.0723	84,303	82,518	86,150	5,702	407,263	22.9	21.9	23.9	4,789	342,091
50-54	0.0210	0.0998	0.0923	0.1076	78,602	76,593	80,680	7,842	373,405	19.3	18.4	20.3	5,599	266,602
55-59	0.0310	0.1439	0.1316	0.1577	70,760	68,615	72,968	10,179	328,354	16.2	15.3	17.2	5,712	184,246
60-64	0.0420	0.1900	0.1719	0.2102	60,581	58,495	62,717	11,513	274,123	13.5	12.7	14.4	9,879	235,220
65-69	0.0560	0.2456	0.2215	0.2718	49,068	46,680	51,530	12,052	215,211	11.1	10.4	11.9	10,155	181,338
70-74	0.0770	0.3229	0.2926	0.3543	37,016	34,010	39,863	11,951	155,205	8.9	8.3	9.5	11,432	148,465
75-79	0.1070	0.4221	0.3874	0.4565	25,066	21,958	28,150	10,580	98,878	7.0	6.5	7.4	7,314	68,358
80-84	0.1640	0.5816	0.5452	0.6162	14,486	11,934	17,245	8,424	51,367	5.2	4.9	5.5	3,971	24,213
85-89	0.2390	0.7480	0.7174	0.7763	6,061	4,586	7,842	4,534	18,971	4.0	3.8	4.2	3,319	13,885
90-94	0.2890	0.7740	0.7470	0.7992	1,527	1,029	2,216	1,182	4,090	3.3	3.1	3.5	1,248	4,320
95-99	0.3710	0.8779	0.8576	0.8962	345	207	561	303	817	2.6	2.5	2.7	270	727
100+	0.4870	1	1	1	42	22	80	42	87	2.1	2.0	2.2	39	80
All ages													*76,419*	*4,745,556*

$ln(_nq_x)$ with uncertainty interval

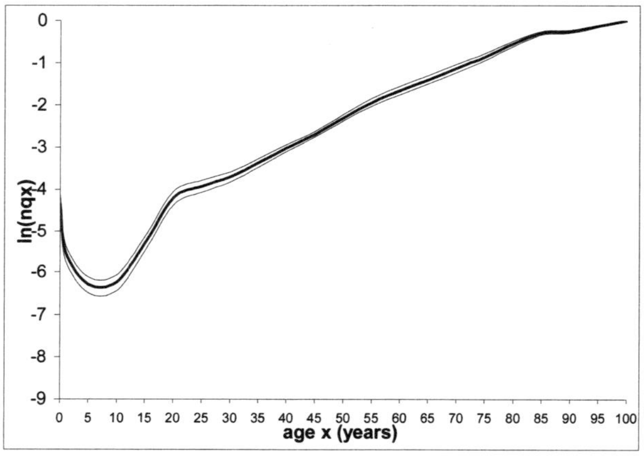

l_x with uncertainty interval

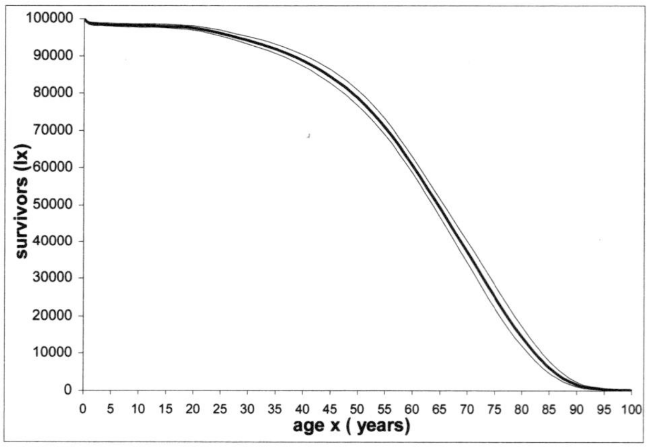

e_x with uncertainty interval

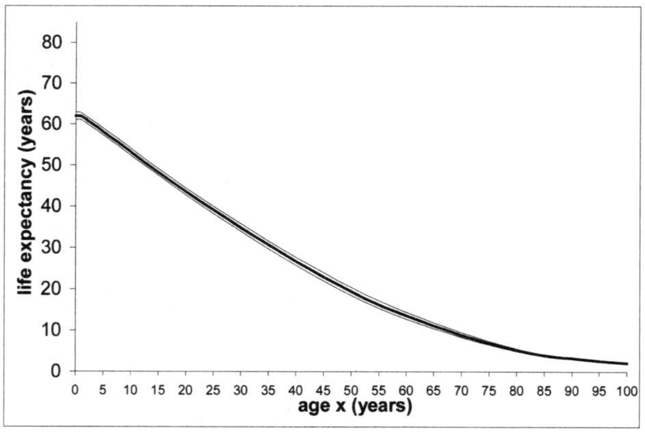

Distribution of deaths by age with uncertainty interval

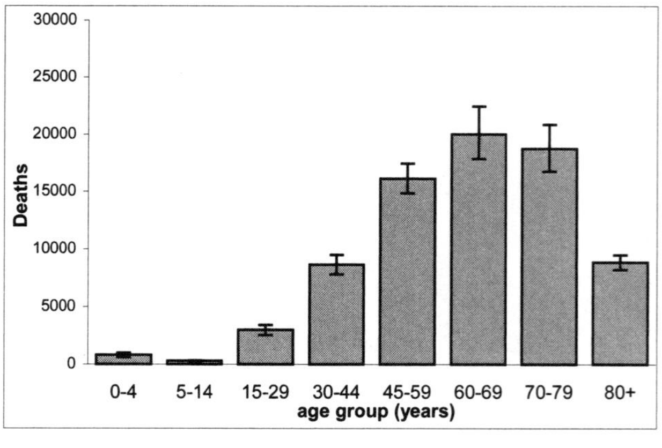

Belarus Bélarus Belarús

Females, 2000 *Femmes, 2000* *Mujeres, 2000*

Age	$_nM_x$	$_nq_x$			l_x			$_nd_x$	$_nL_x$	e_x			Actual deaths	Actual population
			percentiles			percentiles					percentiles			
		mean	2 1/2	97 1/2	mean	2 1/2	97 1/2			mean	2 1/2	97 1/2		
<1	0.0090	0.0089	0.0065	0.0114	100,000	100,000	100,000	893	99,197	74.0	73.1	74.8	403	44,802
1-4	0.0007	0.0027	0.0021	0.0032	99,107	98,860	99,346	265	395,792	73.6	72.7	74.5	122	182,369
5-9	0.0002	0.0009	0.0008	0.0011	98,842	98,544	99,135	94	493,976	69.8	68.9	70.7	57	299,447
10-14	0.0002	0.0011	0.0009	0.0013	98,748	98,434	99,060	109	493,470	64.9	64.0	65.8	89	404,379
15-19	0.0006	0.0028	0.0023	0.0033	98,640	98,306	98,972	276	492,509	60.0	59.1	60.9	230	410,580
20-24	0.0007	0.0036	0.0030	0.0041	98,364	97,985	98,747	353	490,935	55.1	54.2	56.0	270	375,559
25-29	0.0009	0.0044	0.0037	0.0051	98,010	97,579	98,453	435	488,964	50.3	49.4	51.2	321	361,019
30-34	0.0012	0.0058	0.0050	0.0066	97,575	97,085	98,087	569	486,453	45.5	44.6	46.4	409	349,579
35-39	0.0017	0.0084	0.0073	0.0093	97,006	96,447	97,599	811	483,002	40.8	39.9	41.7	702	417,707
40-44	0.0026	0.0127	0.0113	0.0140	96,195	95,551	96,891	1,223	477,914	36.1	35.2	37.0	1,120	437,388
45-49	0.0040	0.0198	0.0180	0.0215	94,971	94,224	95,800	1,881	470,154	31.5	30.7	32.4	1,497	374,263
50-54	0.0070	0.0344	0.0317	0.0374	93,091	92,225	94,080	3,202	457,447	27.1	26.3	28.0	2,143	306,139
55-59	0.0100	0.0488	0.0447	0.0534	89,888	88,871	91,058	4,385	438,480	23.0	22.2	23.9	2,307	230,696
60-64	0.0150	0.0723	0.0654	0.0797	85,504	84,351	86,780	6,181	412,065	19.0	18.3	19.9	4,981	332,074
65-69	0.0230	0.1087	0.0976	0.1201	79,323	77,795	80,776	8,626	375,048	15.3	14.7	16.1	6,657	289,453
70-74	0.0380	0.1735	0.1555	0.1908	70,696	68,452	72,697	12,267	322,815	11.9	11.3	12.6	10,856	285,684
75-79	0.0650	0.2796	0.2519	0.3066	58,430	55,391	61,395	16,335	251,310	8.9	8.3	9.4	11,985	184,377
80-84	0.1180	0.4556	0.4170	0.4947	42,094	38,587	45,885	19,178	162,527	6.3	5.9	6.8	9,112	77,220
85-89	0.1960	0.6577	0.6203	0.6956	22,916	19,966	26,544	15,072	76,900	4.5	4.3	4.9	10,777	54,984
90-94	0.2650	0.7382	0.7088	0.7687	7,844	6,122	10,015	5,790	21,849	3.5	3.3	3.7	5,086	19,191
95-99	0.3610	0.8665	0.8465	0.8872	2,054	1,416	2,917	1,780	4,930	2.7	2.6	2.8	1,492	4,132
100+	0.4900	1	1	1	274	160	448	274	559	2.0	2.0	2.3	195	398
All ages													*70,810*	*5,441,440*

ln($_nq_x$) with uncertainty interval

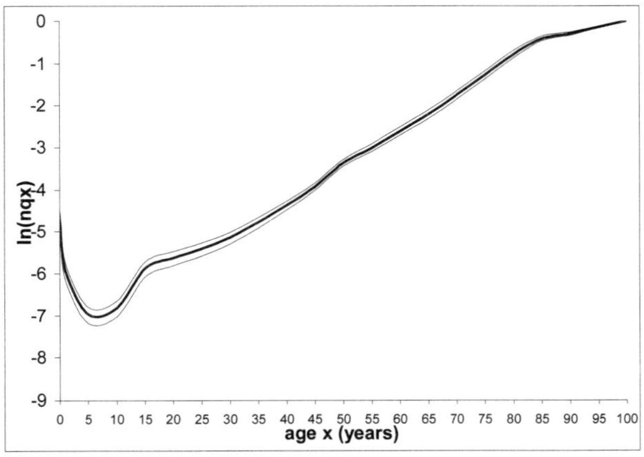

l_x with uncertainty interval

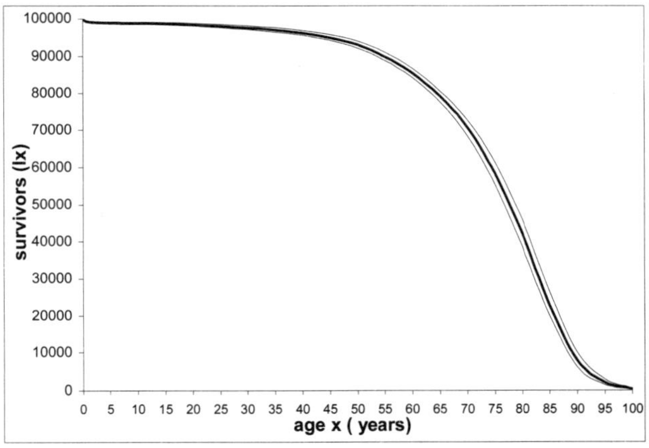

e_x with uncertainty interval

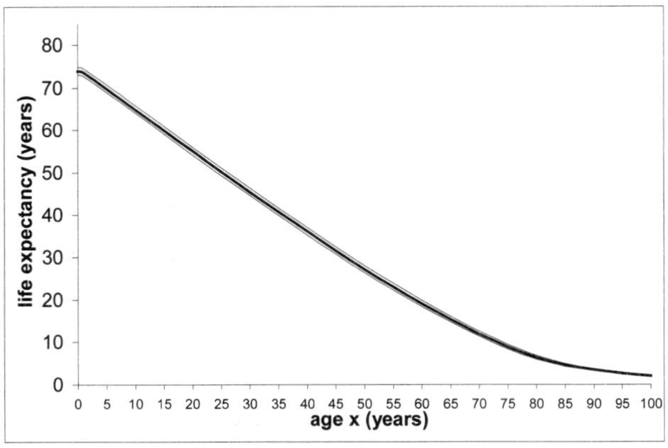

Distribution of deaths by age with uncertainty interval

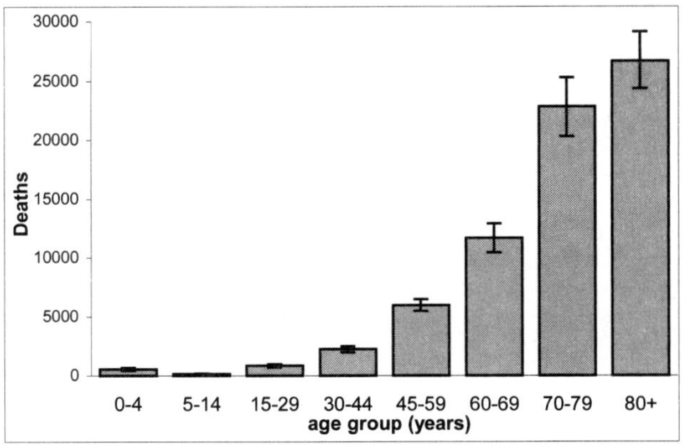

Belgium Belgique Bélgica

Males, 2000 *Hommes, 2000* *Hombres, 2000*

Age	$_nM_x$	$_nq_x$			l_x			$_nd_x$	$_nL_x$	e_x			Actual deaths	Actual population
			percentiles			percentiles					percentiles			
		mean	2½	97½	mean	2½	97½			mean	2½	97½		
<1	0.0070	0.0070	0.0059	0.0080	100,000	100,000	100,000	696	99,374	74.6	74.2	75.0	389	55,578
1-4	0.0004	0.0014	0.0012	0.0016	99,304	99,195	99,412	139	396,884	74.1	73.6	74.6	79	226,332
5-9	0.0002	0.0008	0.0008	0.0009	99,165	99,041	99,290	84	495,617	70.2	69.7	70.7	53	312,182
10-14	0.0002	0.0009	0.0008	0.0010	99,081	98,947	99,214	94	495,171	65.2	64.8	65.7	59	311,746
15-19	0.0008	0.0040	0.0036	0.0044	98,987	98,844	99,130	400	493,935	60.3	59.8	60.8	253	312,058
20-24	0.0013	0.0064	0.0059	0.0069	98,587	98,407	98,769	634	491,351	55.5	55.0	56.0	410	317,768
25-29	0.0012	0.0058	0.0054	0.0062	97,953	97,723	98,187	571	488,338	50.9	50.4	51.4	407	348,233
30-34	0.0012	0.0060	0.0056	0.0063	97,382	97,114	97,654	583	485,453	46.2	45.6	46.7	464	386,478
35-39	0.0015	0.0076	0.0072	0.0080	96,799	96,499	97,106	733	482,164	41.4	40.9	42.0	630	414,393
40-44	0.0025	0.0122	0.0116	0.0127	96,066	95,730	96,409	1,170	477,408	36.7	36.2	37.3	974	397,609
45-49	0.0040	0.0198	0.0191	0.0205	94,897	94,516	95,285	1,879	469,786	32.1	31.6	32.7	1,464	366,006
50-54	0.0060	0.0296	0.0284	0.0306	93,018	92,588	93,453	2,749	458,215	27.7	27.2	28.3	2,073	345,544
55-59	0.0090	0.0440	0.0420	0.0458	90,268	89,804	90,741	3,973	441,410	23.5	23.0	24.1	2,338	259,801
60-64	0.0140	0.0676	0.0642	0.0712	86,296	85,809	86,785	5,836	416,887	19.5	19.0	20.0	3,550	253,548
65-69	0.0230	0.1087	0.1022	0.1155	80,459	79,845	81,078	8,750	380,422	15.7	15.2	16.2	5,587	242,901
70-74	0.0380	0.1735	0.1619	0.1854	71,710	70,759	72,746	12,443	327,441	12.3	11.9	12.8	7,720	203,158
75-79	0.0600	0.2609	0.2434	0.2783	59,267	57,668	60,839	15,461	257,682	9.4	9.0	9.8	9,218	153,627
80-84	0.1060	0.4190	0.3949	0.4417	43,806	41,620	46,007	18,353	173,146	6.8	6.5	7.1	6,745	63,634
85-89	0.1770	0.6135	0.5889	0.6362	25,452	23,235	27,829	15,616	88,223	4.9	4.7	5.1	6,532	36,904
90-94	0.2430	0.7027	0.6829	0.7213	9,837	8,479	11,419	6,913	28,447	3.7	3.6	3.9	2,594	10,674
95-99	0.3420	0.8440	0.8302	0.8570	2,924	2,379	3,616	2,468	7,217	2.8	2.7	2.9	585	1,710
100+	0.4810	1	1	1	456	343	614	456	948	2.1	2.1	2.3	68	141
All ages													*52,192*	*5,020,025*

ln($_nq_x$) with uncertainty interval

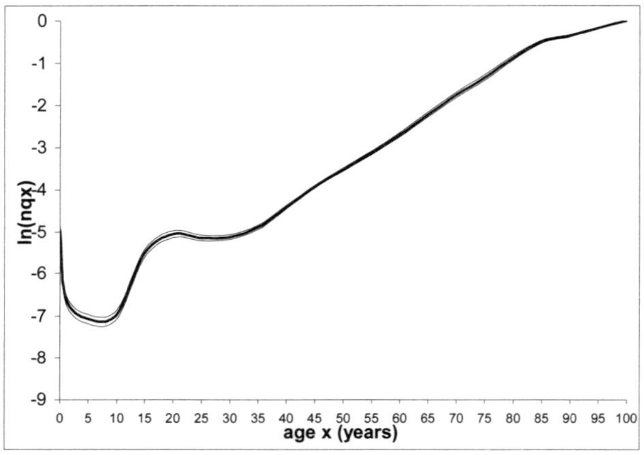

l_x with uncertainty interval

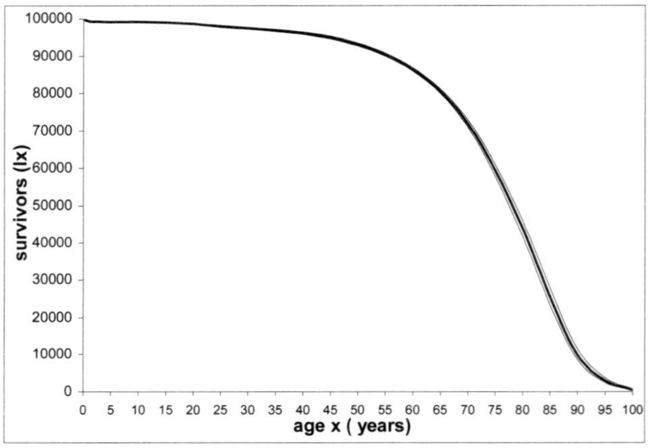

e_x with uncertainty interval

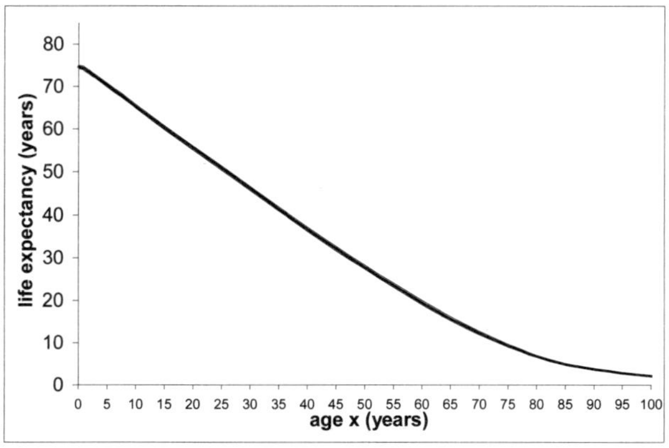

Distribution of deaths by age with uncertainty interval

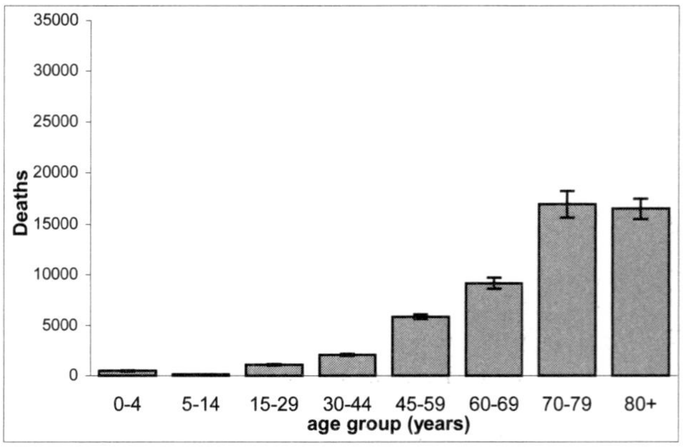

Belgium Belgique Bélgica

Females, 2000 *Femmes, 2000* *Mujeres, 2000*

Age	$_nM_x$	$_nq_x$ mean	$_nq_x$ percentiles 2½	$_nq_x$ percentiles 97½	l_x mean	l_x percentiles 2½	l_x percentiles 97½	$_nd_x$	$_nL_x$	e_x mean	e_x percentiles 2½	e_x percentiles 97½	Actual deaths	Actual population
<1	0.0050	0.0050	0.0044	0.0057	100,000	100,000	100,000	498	99,552	80.9	80.5	81.4	264	52,731
1-4	0.0003	0.0011	0.0010	0.0012	99,502	99,429	99,564	111	397,742	80.3	79.9	80.8	61	216,574
5-9	0.0002	0.0008	0.0008	0.0009	99,391	99,306	99,464	84	496,743	76.4	76.0	76.9	51	299,045
10-14	0.0002	0.0008	0.0007	0.0009	99,306	99,213	99,387	79	496,334	71.5	71.1	71.9	48	297,033
15-19	0.0003	0.0012	0.0011	0.0014	99,227	99,125	99,315	124	495,825	66.5	66.1	67.0	75	298,162
20-24	0.0003	0.0017	0.0015	0.0019	99,103	98,990	99,202	168	495,094	61.6	61.2	62.1	105	307,717
25-29	0.0004	0.0018	0.0016	0.0020	98,935	98,806	99,048	178	494,229	56.7	56.3	57.2	122	338,559
30-34	0.0005	0.0027	0.0025	0.0029	98,757	98,613	98,884	266	493,118	51.8	51.4	52.3	200	371,187
35-39	0.0009	0.0042	0.0039	0.0046	98,491	98,325	98,638	418	491,408	46.9	46.6	47.4	341	400,706
40-44	0.0015	0.0073	0.0068	0.0078	98,073	97,875	98,251	713	488,581	42.1	41.8	42.6	565	386,855
45-49	0.0020	0.0100	0.0093	0.0106	97,359	97,110	97,588	969	484,376	37.4	37.1	37.9	713	356,568
50-54	0.0030	0.0149	0.0140	0.0158	96,391	96,082	96,679	1,435	478,366	32.8	32.4	33.2	1,018	339,245
55-59	0.0050	0.0247	0.0234	0.0260	94,956	94,571	95,322	2,345	468,917	28.2	27.9	28.6	1,317	263,304
60-64	0.0070	0.0344	0.0326	0.0362	92,611	92,130	93,084	3,186	455,091	23.9	23.6	24.3	1,895	270,648
65-69	0.0110	0.0535	0.0507	0.0563	89,425	88,830	90,011	4,787	435,160	19.6	19.4	20.0	3,045	276,808
70-74	0.0180	0.0861	0.0814	0.0903	84,639	83,881	85,379	7,289	404,970	15.6	15.4	15.9	4,694	260,756
75-79	0.0340	0.1567	0.1483	0.1633	77,349	76,338	78,375	12,119	356,448	11.9	11.6	12.1	7,859	231,154
80-84	0.0670	0.2869	0.2737	0.2967	65,230	63,870	66,698	18,717	279,357	8.6	8.4	8.8	7,987	119,206
85-89	0.1260	0.4791	0.4637	0.4944	46,513	44,956	48,443	22,284	176,856	6.0	5.9	6.2	11,888	94,347
90-94	0.1980	0.6211	0.6069	0.6362	24,229	23,019	25,978	15,048	76,001	4.3	4.2	4.4	7,733	39,055
95-99	0.3020	0.7922	0.7812	0.8034	9,181	8,418	10,150	7,273	24,084	3.1	3.1	3.2	2,668	8,835
100+	0.4600	1	1	1	1,907	1,655	2,182	1,907	4,147	2.2	2.2	2.4	391	850
All ages													*53,036*	*5,229,345*

ln($_nq_x$) with uncertainty interval

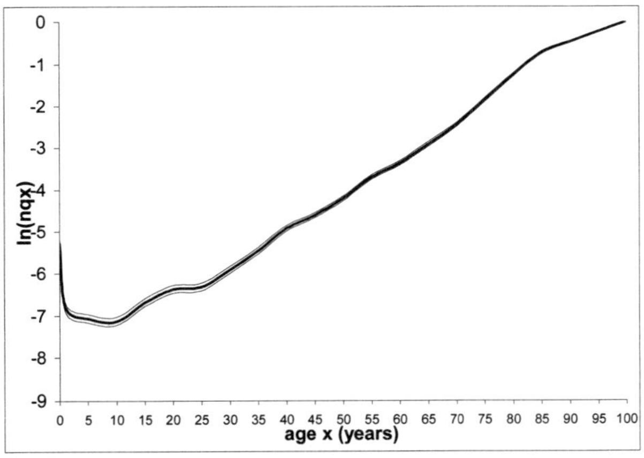

l_x with uncertainty interval

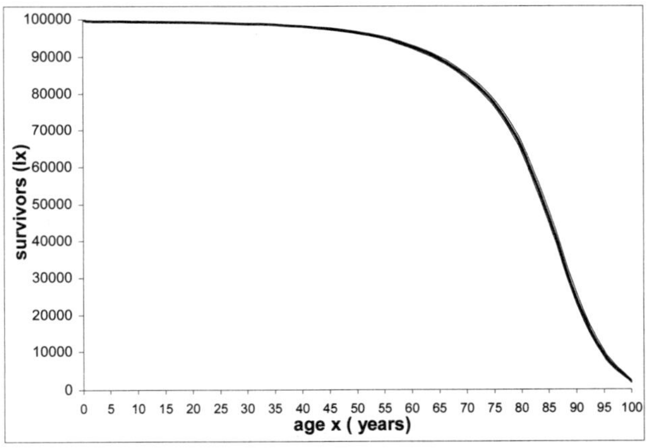

e_x with uncertainty interval

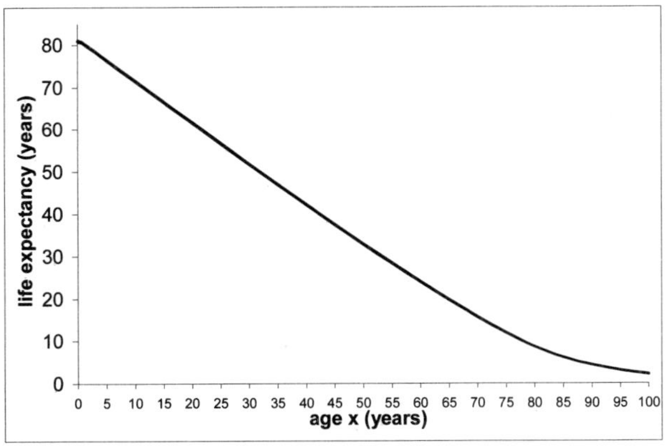

Distribution of deaths by age with uncertainty interval

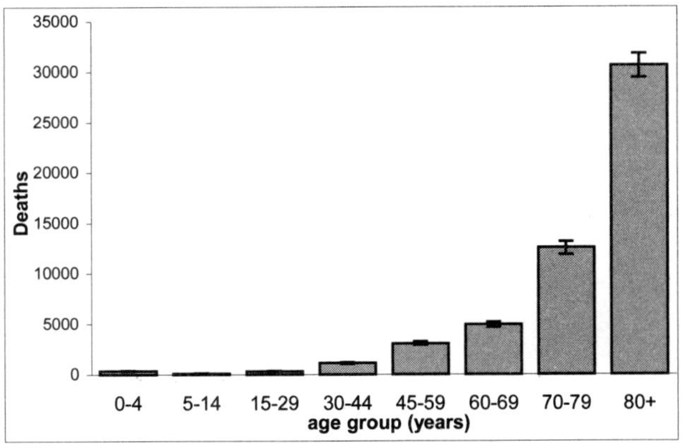

Belize Belize Belice

Males, 2000 *Hommes, 2000* *Hombres, 2000*

Age	$_nM_x$	$_nq_x$			l_x			$_nd_x$	$_nL_x$	e_x			Actual deaths	Actual population
			percentiles			percentiles					percentiles			
		mean	21/2	971/2	mean	21/2	971/2			mean	21/2	971/2		
<1	0.0307	0.0299	0.0243	0.0327	100,000	100,000	100,000	2,990	97,309	69.1	68.0	70.3	94	3,058
1-4	0.0018	0.0071	0.0061	0.0077	97,010	96,728	97,568	684	386,397	70.2	69.2	71.3	21	11,795
5-9	0.0007	0.0033	0.0029	0.0036	96,325	95,985	96,977	323	480,821	66.7	65.7	67.7	10	15,024
10-14	0.0005	0.0025	0.0022	0.0027	96,003	95,635	96,697	242	479,410	61.9	61.0	62.9	7	14,193
15-19	0.0017	0.0085	0.0075	0.0093	95,761	95,372	96,486	818	476,762	57.1	56.1	58.1	23	13,161
20-24	0.0026	0.0129	0.0114	0.0141	94,944	94,484	95,766	1,225	471,657	52.6	51.6	53.5	30	11,625
25-29	0.0032	0.0160	0.0143	0.0175	93,719	93,154	94,673	1,500	464,845	48.2	47.3	49.1	31	9,580
30-34	0.0023	0.0112	0.0101	0.0123	92,219	91,526	93,317	1,035	458,507	44.0	43.1	44.8	18	7,851
35-39	0.0031	0.0153	0.0138	0.0167	91,184	90,403	92,372	1,391	452,441	39.4	38.6	40.3	20	6,569
40-44	0.0058	0.0284	0.0259	0.0309	89,793	88,897	91,096	2,546	442,600	35.0	34.2	35.8	31	5,388
45-49	0.0070	0.0342	0.0314	0.0373	87,247	86,147	88,737	2,987	428,768	31.0	30.2	31.7	29	4,140
50-54	0.0077	0.0377	0.0345	0.0410	84,260	82,932	85,951	3,173	413,369	27.0	26.3	27.7	24	3,168
55-59	0.0115	0.0557	0.0512	0.0605	81,087	79,532	82,982	4,517	394,145	22.9	22.3	23.6	26	2,299
60-64	0.0174	0.0834	0.0768	0.0901	76,571	74,724	78,735	6,383	366,896	19.1	18.6	19.8	37	2,124
65-69	0.0266	0.1247	0.1151	0.1340	70,188	67,995	72,686	8,751	329,061	15.6	15.1	16.2	44	1,659
70-74	0.0415	0.1879	0.1738	0.2011	61,437	58,881	64,323	11,541	278,330	12.5	12.0	13.0	56	1,345
75-79	0.0546	0.2402	0.2225	0.2581	49,895	47,042	53,142	11,985	219,513	9.8	9.4	10.3	50	911
80-84	0.0924	0.3754	0.3512	0.4006	37,910	34,901	41,318	14,230	153,974	7.1	6.8	7.5	48	524
85-89	0.1640	0.5815	0.5547	0.6097	23,680	20,919	26,809	13,770	83,974	4.9	4.7	5.2	37	228
90-94	0.2725	0.7497	0.7286	0.7723	9,910	8,164	11,937	7,429	27,262	3.3	3.2	3.5	26	94
95-99	0.4242	0.9333	0.9237	0.9433	2,481	1,859	3,240	2,315	5,457	2.3	2.3	2.4	8	18
100+	0.6186	1	1	1	165	105	247	165	267	1.6	1.6	1.7	1	2
All ages													*671*	*114,756*

ln($_nq_x$) with uncertainty interval

l_x with uncertainty interval

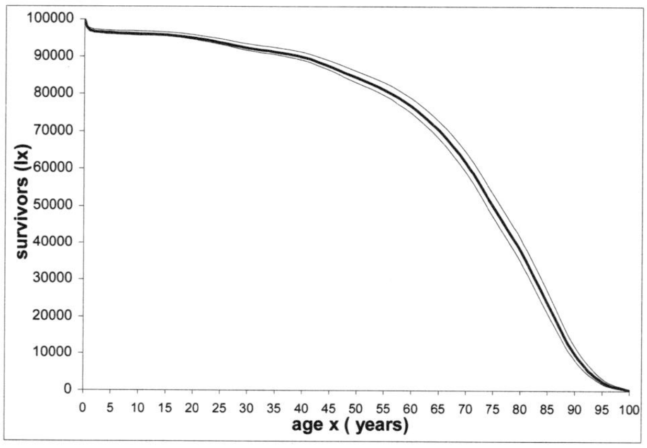

e_x with uncertainty interval

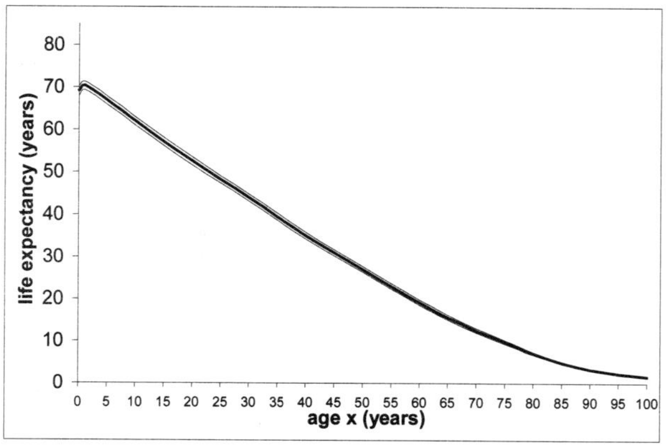

Distribution of deaths by age with uncertainty interval

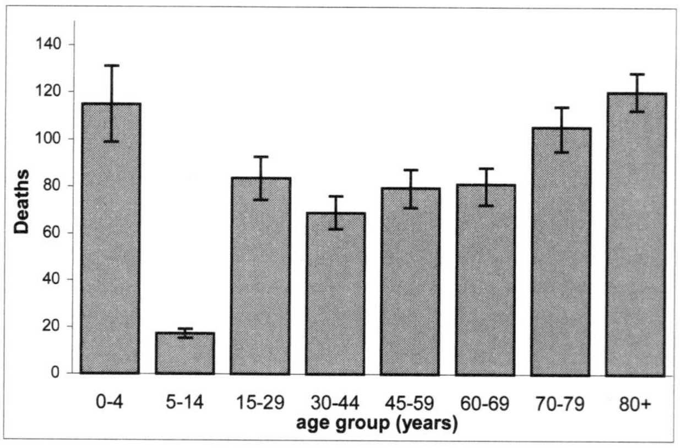

Belize　　Belize　　Belice

Females, 2000　　*Femmes, 2000*　　*Mujeres, 2000*

Age	$_nM_x$	$_nq_x$ mean	$_nq_x$ percentiles 2 1/2	$_nq_x$ percentiles 97 1/2	l_x mean	l_x percentiles 2 1/2	l_x percentiles 97 1/2	$_nd_x$	$_nL_x$	e_x mean	e_x percentiles 2 1/2	e_x percentiles 97 1/2	Actual deaths	Actual population
<1	0.0257	0.0251	0.0215	0.0278	100,000	100,000	100,000	2,513	97,738	74.7	74.0	75.2	76	2,969
1-4	0.0017	0.0069	0.0063	0.0075	97,487	97,221	97,852	673	388,331	75.6	74.9	76.2	20	11,451
5-9	0.0006	0.0028	0.0025	0.0030	96,813	96,494	97,240	266	483,401	72.1	71.4	72.7	8	14,573
10-14	0.0004	0.0021	0.0019	0.0022	96,547	96,207	96,996	200	482,236	67.3	66.6	67.9	6	13,745
15-19	0.0005	0.0027	0.0025	0.0029	96,347	95,992	96,812	259	481,088	62.5	61.8	63.1	7	12,849
20-24	0.0007	0.0036	0.0033	0.0039	96,088	95,714	96,574	347	479,572	57.6	56.9	58.2	8	11,135
25-29	0.0010	0.0051	0.0047	0.0054	95,741	95,342	96,252	484	477,493	52.8	52.1	53.4	10	9,619
30-34	0.0012	0.0059	0.0055	0.0064	95,257	94,824	95,802	566	474,868	48.1	47.4	48.7	9	7,838
35-39	0.0022	0.0110	0.0103	0.0117	94,691	94,219	95,272	1,038	470,859	43.4	42.7	44.0	14	6,573
40-44	0.0034	0.0168	0.0159	0.0179	93,653	93,113	94,292	1,574	464,330	38.8	38.2	39.4	17	5,110
45-49	0.0040	0.0198	0.0190	0.0211	92,079	91,442	92,789	1,823	455,837	34.4	33.8	35.0	15	3,838
50-54	0.0060	0.0297	0.0284	0.0316	90,256	89,517	91,027	2,677	444,587	30.1	29.5	30.7	18	2,983
55-59	0.0075	0.0367	0.0349	0.0392	87,579	86,684	88,444	3,210	429,870	25.9	25.4	26.5	16	2,147
60-64	0.0120	0.0583	0.0552	0.0625	84,369	83,282	85,359	4,915	409,557	21.8	21.3	22.4	23	1,918
65-69	0.0176	0.0841	0.0789	0.0904	79,454	78,074	80,650	6,679	380,573	18.0	17.5	18.6	27	1,563
70-74	0.0270	0.1265	0.1177	0.1358	72,775	71,018	74,283	9,203	340,866	14.4	14.0	14.9	34	1,268
75-79	0.0376	0.1720	0.1593	0.1842	63,572	61,374	65,537	10,937	290,517	11.2	10.8	11.6	36	957
80-84	0.0746	0.3143	0.2923	0.3363	52,635	50,068	55,095	16,543	221,819	8.0	7.6	8.3	43	575
85-89	0.1386	0.5148	0.4865	0.5441	36,092	33,231	38,991	18,580	134,011	5.5	5.2	5.7	39	282
90-94	0.2419	0.7009	0.6764	0.7266	17,512	15,151	20,022	12,274	50,739	3.6	3.5	3.8	34	139
95-99	0.3961	0.9050	0.8925	0.9179	5,238	4,142	6,478	4,741	11,970	2.4	2.4	2.5	13	32
100+	0.6086	1	1	1	498	340	697	498	818	1.6	1.6	1.9	3	5
All ages													*477*	*111,569*

$\ln(_nq_x)$ with uncertainty interval

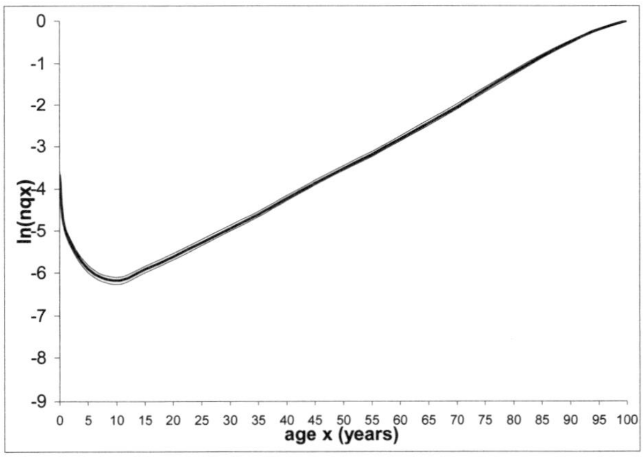

l_x with uncertainty interval

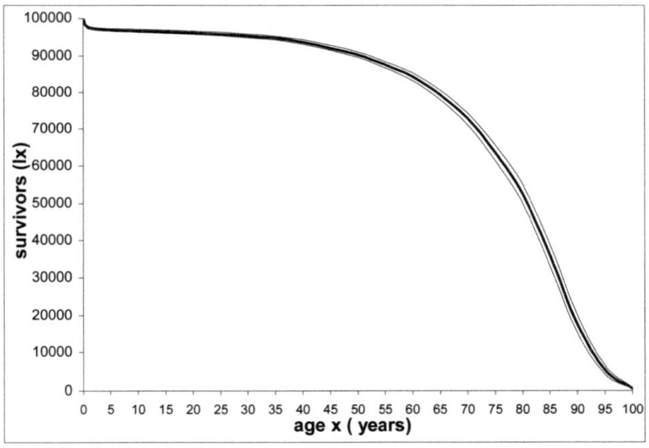

e_x with uncertainty interval

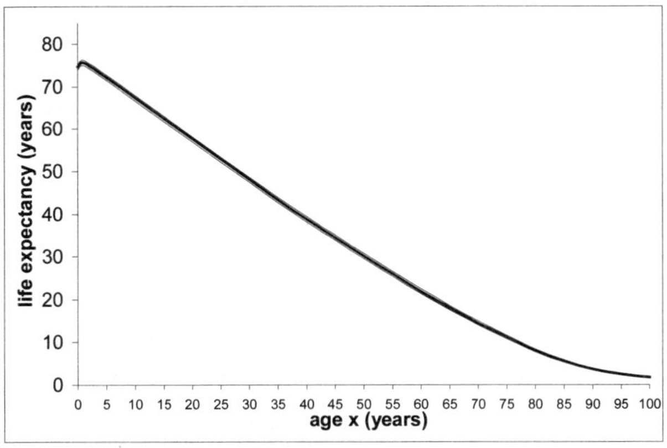

Distribution of deaths by age with uncertainty interval

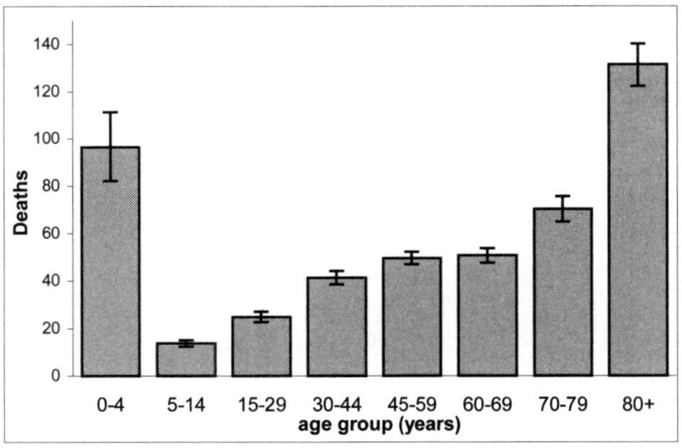

Benin   Bénin   Benin

Males, 2000   *Hommes, 2000*   *Hombres, 2000*

Age	$_nM_x$	$_nq_x$ mean	$_nq_x$ percentiles 2½	$_nq_x$ percentiles 97½	l_x mean	l_x percentiles 2½	l_x percentiles 97½	$_nd_x$	$_nL_x$	e_x mean	e_x percentiles 2½	e_x percentiles 97½	Actual deaths	Actual population
<1	0.1400	0.1243	0.1154	0.1335	100,000	100,000	100,000	12,433	88,810	51.7	49.6	54.0	17,313	123,670
1-4	0.0111	0.0433	0.0388	0.0478	87,567	86,650	88,463	3,795	341,161	58.0	55.9	60.6	4,821	433,473
5-9	0.0026	0.0129	0.0114	0.0144	83,772	83,024	84,521	1,080	416,161	56.6	54.6	59.0	1,243	478,859
10-14	0.0016	0.0081	0.0071	0.0090	82,692	81,858	83,510	668	411,791	52.3	50.4	54.6	683	421,430
15-19	0.0016	0.0079	0.0069	0.0088	82,024	81,119	82,893	648	408,502	47.7	45.8	50.0	571	360,136
20-24	0.0035	0.0174	0.0152	0.0195	81,376	80,402	82,293	1,415	403,343	43.0	41.2	45.3	956	272,542
25-29	0.0057	0.0280	0.0244	0.0315	79,961	78,832	81,032	2,238	394,210	38.7	37.0	41.0	1,172	206,434
30-34	0.0080	0.0393	0.0339	0.0444	77,723	76,346	79,058	3,051	380,989	34.8	33.1	37.0	1,414	176,614
35-39	0.0097	0.0475	0.0407	0.0540	74,672	72,953	76,375	3,548	364,492	31.1	29.4	33.2	1,434	147,293
40-44	0.0117	0.0566	0.0481	0.0647	71,125	69,012	73,259	4,029	345,550	27.5	26.0	29.5	1,366	117,159
45-49	0.0143	0.0692	0.0584	0.0792	67,095	64,549	69,716	4,646	323,863	24.0	22.6	25.8	1,346	93,806
50-54	0.0183	0.0874	0.0734	0.1003	62,450	59,437	65,614	5,460	298,600	20.6	19.3	22.3	1,464	80,059
55-59	0.0239	0.1129	0.0946	0.1300	56,990	53,491	60,751	6,434	268,865	17.4	16.2	18.9	1,270	53,060
60-64	0.0338	0.1557	0.1307	0.1792	50,556	46,595	54,938	7,871	233,102	14.3	13.2	15.6	1,427	42,265
65-69	0.0487	0.2170	0.1836	0.2480	42,685	38,362	47,659	9,261	190,272	11.4	10.6	12.5	1,652	33,937
70-74	0.0739	0.3118	0.2685	0.3510	33,424	29,018	38,908	10,420	141,069	8.9	8.2	9.8	1,759	23,809
75-79	0.1114	0.4356	0.3854	0.4796	23,004	18,958	28,463	10,021	89,965	6.8	6.3	7.5	1,745	15,667
80-84	0.1653	0.5849	0.5342	0.6278	12,983	9,866	17,495	7,593	45,929	5.1	4.8	5.6	1,368	8,277
85-89	0.2384	0.7469	0.7036	0.7819	5,389	3,678	8,139	4,025	16,883	3.9	3.6	4.1	616	2,584
90-94	0.3340	0.8341	0.7998	0.8614	1,364	805	2,397	1,138	3,407	2.9	2.7	3.1	185	554
95-99	0.4545	0.9615	0.9465	0.9720	226	112	475	218	479	2.2	2.1	2.3	28	61
100+	0.6007	1	1	1	9	3	25	9	15	1.7	1.6	1.7	3	5
All ages													*43,837*	*3,091,694*

ln($_nq_x$) with uncertainty interval

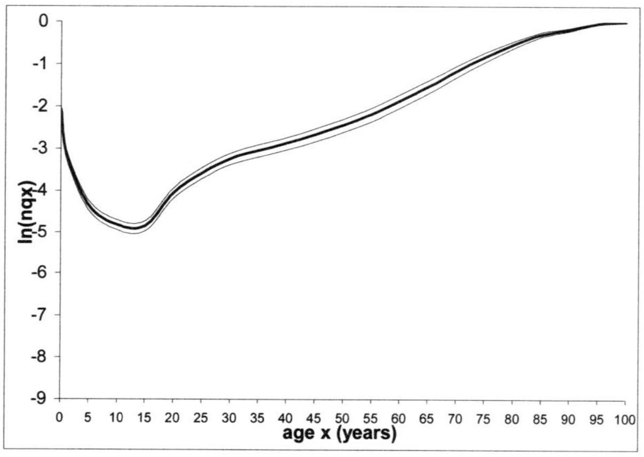

l_x with uncertainty interval

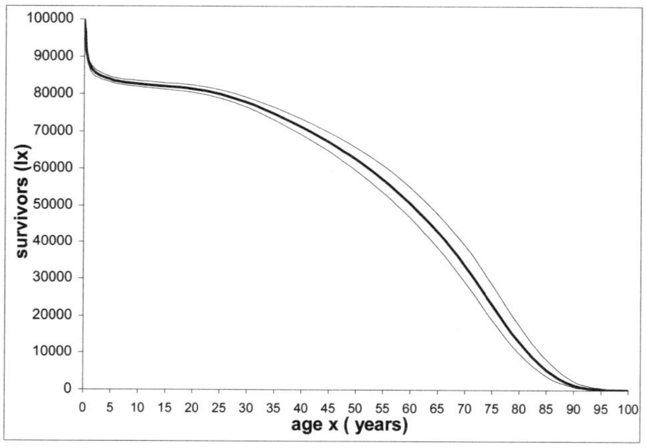

e_x with uncertainty interval

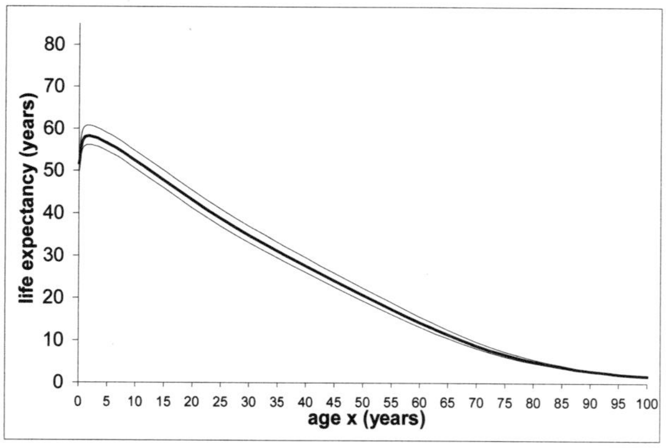

Distribution of deaths by age with uncertainty interval

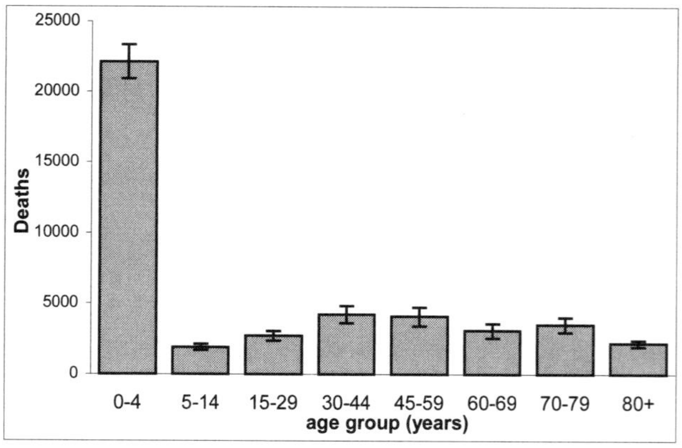

Benin | Bénin | Benin

Females, 2000 | *Femmes, 2000* | *Mujeres, 2000*

Age	$_nM_x$	$_nq_x$ mean	$_nq_x$ percentiles 2½	$_nq_x$ percentiles 97½	l_x mean	l_x percentiles 2½	l_x percentiles 97½	$_nd_x$	$_nL_x$	e_x mean	e_x percentiles 2½	e_x percentiles 97½	Actual deaths	Actual population
<1	0.1211	0.1092	0.1003	0.1173	100,000	100,000	100,000	10,918	90,174	53.8	51.7	56.0	14,335	118,401
1-4	0.0122	0.0474	0.0430	0.0520	89,082	88,275	89,969	4,223	346,194	59.4	57.0	62.0	5,276	432,517
5-9	0.0027	0.0134	0.0119	0.0149	84,859	84,109	85,596	1,135	421,459	58.3	56.0	60.8	1,285	477,207
10-14	0.0018	0.0087	0.0078	0.0098	83,724	82,938	84,523	731	416,794	54.0	51.8	56.5	740	421,619
15-19	0.0028	0.0140	0.0124	0.0157	82,993	82,165	83,831	1,160	412,065	49.5	47.3	51.9	1,022	363,090
20-24	0.0050	0.0249	0.0219	0.0281	81,833	80,939	82,749	2,035	404,078	45.1	43.0	47.6	1,391	276,275
25-29	0.0070	0.0344	0.0300	0.0392	79,798	78,756	80,927	2,745	392,130	41.2	39.2	43.6	1,464	209,234
30-34	0.0081	0.0396	0.0342	0.0455	77,054	75,709	78,477	3,053	377,635	37.6	35.7	39.8	1,518	187,796
35-39	0.0085	0.0418	0.0357	0.0484	74,000	72,285	75,755	3,095	362,265	34.1	32.3	36.2	1,492	174,614
40-44	0.0085	0.0417	0.0353	0.0484	70,906	68,790	72,995	2,955	347,142	30.4	28.8	32.4	1,200	140,989
45-49	0.0097	0.0472	0.0397	0.0551	67,951	65,457	70,357	3,210	331,731	26.6	25.1	28.5	1,008	104,209
50-54	0.0126	0.0612	0.0511	0.0716	64,741	61,850	67,481	3,963	313,800	22.8	21.4	24.6	1,021	80,865
55-59	0.0171	0.0822	0.0683	0.0962	60,779	57,422	63,947	4,997	291,401	19.2	17.9	20.8	970	56,580
60-64	0.0257	0.1208	0.1002	0.1409	55,782	51,898	59,525	6,736	262,068	15.7	14.5	17.1	1,266	49,269
65-69	0.0396	0.1802	0.1499	0.2089	49,045	44,584	53,478	8,839	223,129	12.5	11.5	13.7	1,383	34,915
70-74	0.0633	0.2734	0.2302	0.3135	40,206	35,283	45,386	10,994	173,547	9.7	8.8	10.7	1,663	26,251
75-79	0.0984	0.3947	0.3404	0.4431	29,212	24,301	34,939	11,531	117,234	7.4	6.7	8.2	1,503	15,279
80-84	0.1492	0.5434	0.4853	0.5926	17,681	13,651	23,045	9,609	64,386	5.5	5.1	6.1	988	6,620
85-89	0.2196	0.7089	0.6580	0.7503	8,073	5,634	11,861	5,723	26,058	4.1	3.8	4.5	774	3,522
90-94	0.3135	0.8077	0.7678	0.8400	2,350	1,407	4,056	1,898	6,056	3.0	2.8	3.2	215	686
95-99	0.4339	0.9426	0.9219	0.9576	452	225	942	426	982	2.3	2.2	2.4	40	93
100+	0.5825	1	1	1	26	10	74	26	45	1.7	1.7	1.8	4	7
All ages													*40,560*	*3,180,038*

ln($_nq_x$) with uncertainty interval

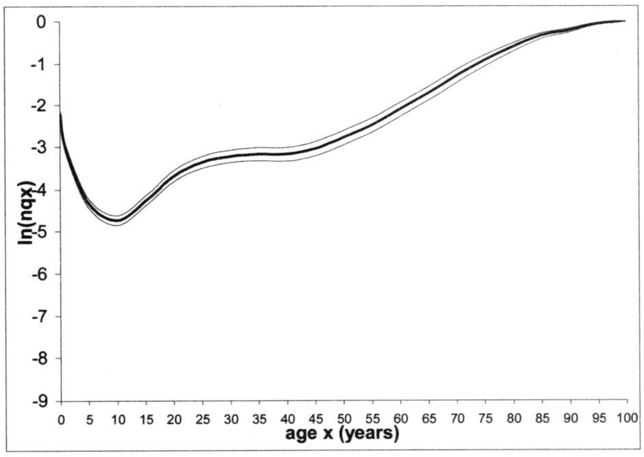

l_x with uncertainty interval

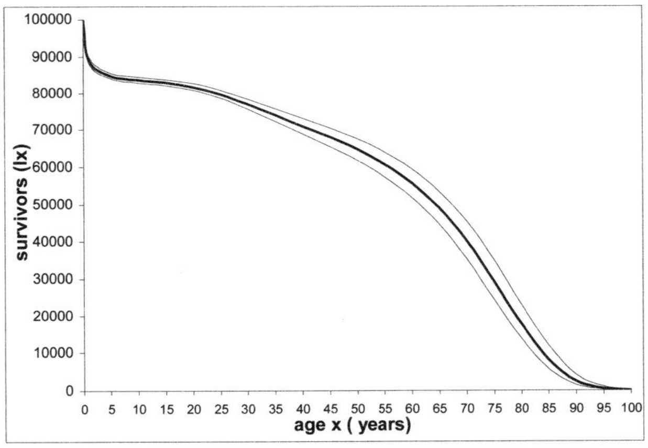

e_x with uncertainty interval

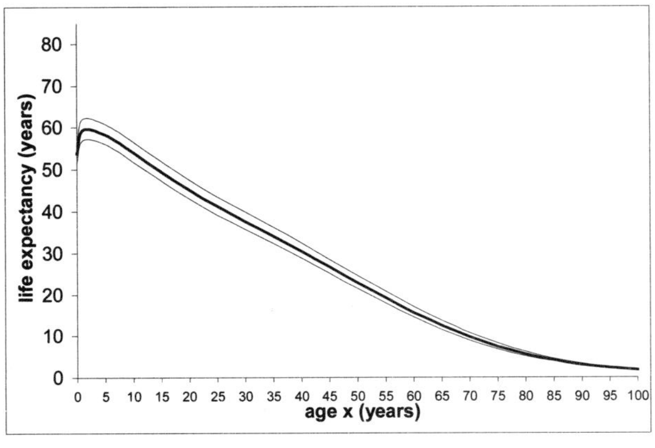

Distribution of deaths by age with uncertainty interval

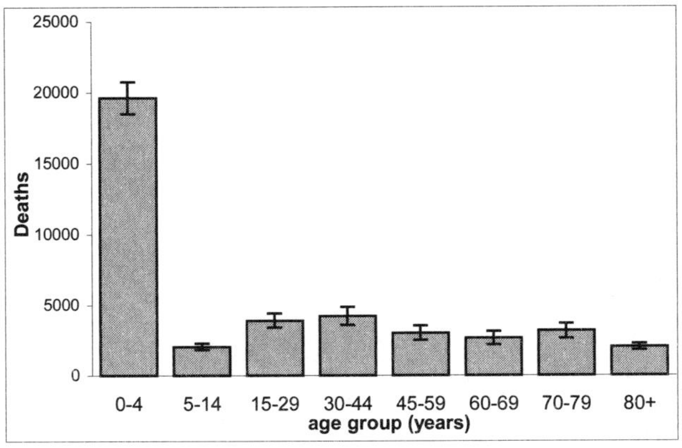

Bhutan   Bhoutan   Bhután

Males, 2000   *Hommes, 2000*   *Hombres, 2000*

Age	$_nM_x$	$_nq_x$			l_x			$_nd_x$	$_nL_x$	e_x			Actual deaths	Actual population
			percentiles			percentiles					percentiles			
		mean	2½	97½	mean	2½	97½			mean	2½	97½		
<1	0.0779	0.0728	0.0548	0.0914	100,000	100,000	100,000	7,280	93,448	60.4	57.1	64.5	2,845	36,522
1-4	0.0055	0.0215	0.0165	0.0265	92,720	90,859	94,519	1,996	366,090	64.1	61.2	67.4	712	130,597
5-9	0.0015	0.0077	0.0059	0.0095	90,724	88,730	92,718	699	451,874	61.5	58.8	64.6	234	151,192
10-14	0.0011	0.0056	0.0042	0.0069	90,025	87,922	92,112	500	448,877	57.0	54.3	60.0	151	135,958
15-19	0.0017	0.0083	0.0063	0.0103	89,525	87,344	91,715	745	445,765	52.3	49.6	55.3	190	113,747
20-24	0.0025	0.0125	0.0095	0.0154	88,781	86,486	91,136	1,108	441,132	47.7	45.1	50.7	233	92,857
25-29	0.0028	0.0141	0.0107	0.0174	87,672	85,184	90,273	1,235	435,273	43.3	40.8	46.2	210	74,019
30-34	0.0034	0.0168	0.0127	0.0208	86,437	83,705	89,309	1,449	428,562	38.9	36.5	41.7	205	60,520
35-39	0.0044	0.0219	0.0166	0.0272	84,988	81,977	88,174	1,858	420,293	34.5	32.2	37.2	229	51,843
40-44	0.0061	0.0303	0.0230	0.0378	83,129	79,773	86,714	2,516	409,358	30.2	28.0	32.8	276	44,876
45-49	0.0089	0.0434	0.0330	0.0544	80,614	76,814	84,723	3,501	394,314	26.1	24.0	28.6	341	38,357
50-54	0.0132	0.0637	0.0488	0.0799	77,112	72,743	81,923	4,914	373,276	22.1	20.2	24.5	417	31,707
55-59	0.0195	0.0928	0.0713	0.1158	72,198	67,133	77,926	6,698	344,247	18.5	16.7	20.7	540	27,768
60-64	0.0290	0.1353	0.1040	0.1670	65,500	59,488	72,330	8,865	305,341	15.1	13.4	17.2	659	22,696
65-69	0.0433	0.1954	0.1507	0.2411	56,636	49,554	64,611	11,068	255,509	12.1	10.6	13.9	747	17,234
70-74	0.0668	0.2862	0.2230	0.3516	45,568	37,853	54,370	13,043	195,233	9.4	8.1	10.9	798	11,942
75-79	0.1025	0.4078	0.3281	0.4873	32,525	25,181	41,578	13,265	129,463	7.2	6.2	8.3	733	7,155
80-84	0.1554	0.5595	0.4740	0.6393	19,260	13,547	27,479	10,777	69,358	5.4	4.7	6.2	523	3,364
85-89	0.2280	0.7262	0.6497	0.7911	8,483	4,912	14,351	6,160	27,016	4.0	3.6	4.5	235	1,030
90-94	0.3239	0.8213	0.7608	0.8708	2,323	1,026	4,990	1,908	5,891	2.9	2.7	3.3	62	191
95-99	0.4452	0.9531	0.9235	0.9731	415	133	1,175	396	889	2.2	2.1	2.4	8	18
100+	0.5922	1	1	1	19	4	87	19	33	1.7	1.6	1.8	1	1
All ages													*10,348*	*1,053,594*

ln($_nq_x$) with uncertainty interval

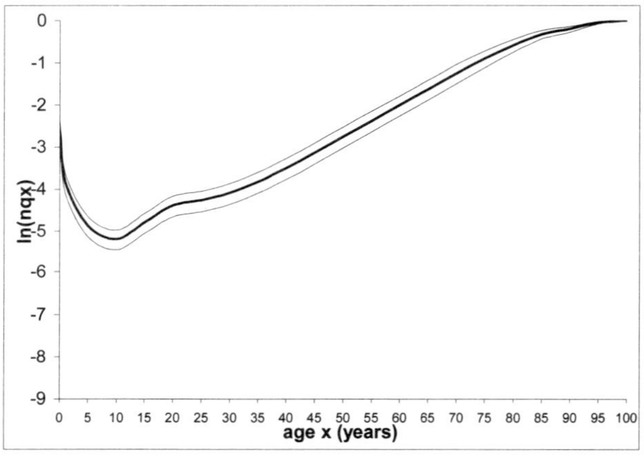

l_x with uncertainty interval

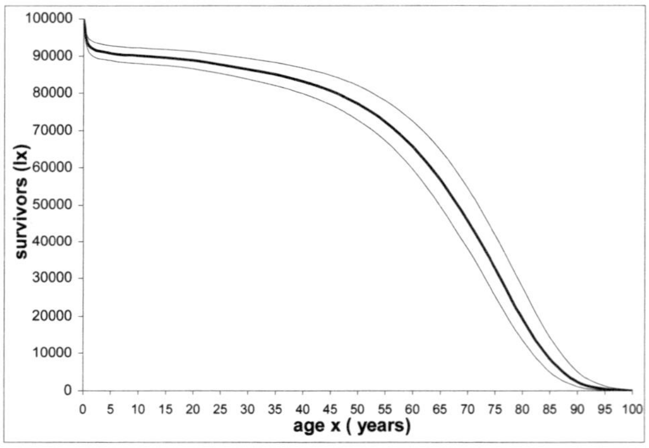

e_x with uncertainty interval

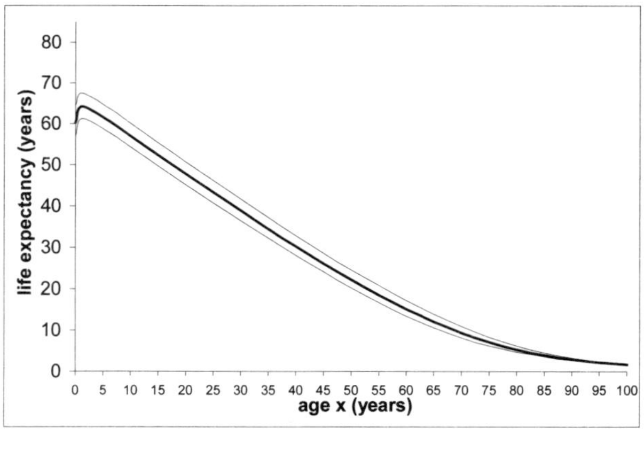

Distribution of deaths by age with uncertainty interval

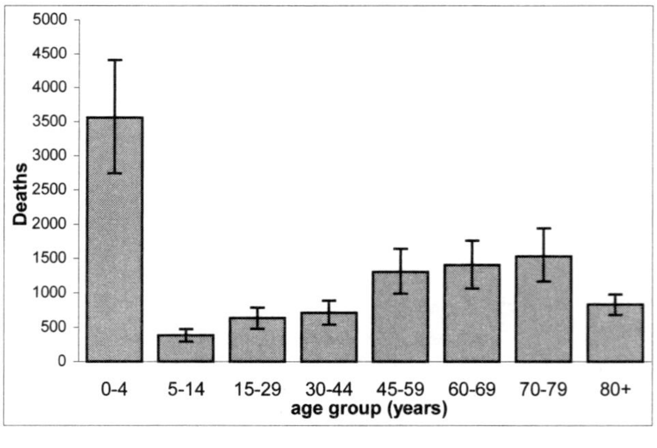

Bhutan　　　Bhoutan　　　Bhután

Females, 2000　　　*Femmes, 2000*　　　*Mujeres, 2000*

Age	$_nM_x$	$_nq_x$			l_x			$_nd_x$	$_nL_x$	e_x			Actual deaths	Actual population
			percentiles			percentiles					percentiles			
		mean	2½	97½	mean	2½	97½			mean	2½	97½		
<1	0.0727	0.0683	0.0504	0.0859	100,000	100,000	100,000	6,826	93,857	62.5	58.7	66.3	2,519	34,635
1-4	0.0065	0.0254	0.0196	0.0319	93,174	91,407	94,964	2,368	367,013	66.0	62.9	70.2	810	125,567
5-9	0.0016	0.0081	0.0062	0.0102	90,806	88,812	92,795	737	452,189	63.7	60.7	67.9	237	145,230
10-14	0.0011	0.0057	0.0044	0.0072	90,069	87,922	92,199	511	449,069	59.2	56.2	63.4	149	131,106
15-19	0.0017	0.0084	0.0064	0.0106	89,558	87,292	91,797	750	445,917	54.5	51.6	58.7	185	110,290
20-24	0.0023	0.0115	0.0089	0.0145	88,809	86,370	91,205	1,021	441,490	50.0	47.1	54.1	208	89,841
25-29	0.0027	0.0134	0.0103	0.0170	87,787	85,115	90,397	1,179	435,989	45.5	42.7	49.6	192	71,137
30-34	0.0032	0.0156	0.0120	0.0198	86,608	83,670	89,460	1,355	429,653	41.1	38.3	45.1	184	58,278
35-39	0.0041	0.0205	0.0157	0.0259	85,253	82,017	88,380	1,745	421,905	36.7	34.0	40.6	209	50,513
40-44	0.0051	0.0251	0.0192	0.0317	83,509	79,898	86,980	2,095	412,306	32.4	29.8	36.3	225	44,242
45-49	0.0070	0.0342	0.0257	0.0431	81,414	77,371	85,286	2,785	400,105	28.2	25.7	31.9	266	38,221
50-54	0.0100	0.0486	0.0356	0.0622	78,628	74,044	83,011	3,824	383,583	24.1	21.7	27.7	323	32,371
55-59	0.0143	0.0691	0.0490	0.0896	74,805	69,543	79,843	5,170	361,100	20.2	17.9	23.6	410	28,646
60-64	0.0220	0.1042	0.0719	0.1362	69,635	63,588	75,483	7,254	330,040	16.5	14.4	19.7	530	24,108
65-69	0.0345	0.1587	0.1073	0.2096	62,381	55,474	69,331	9,902	287,148	13.2	11.2	16.1	657	19,046
70-74	0.0563	0.2467	0.1665	0.3240	52,478	44,658	60,712	12,946	230,027	10.2	8.5	12.7	768	13,646
75-79	0.0898	0.3668	0.2552	0.4650	39,532	30,610	50,575	14,502	161,407	7.7	6.4	9.7	768	8,543
80-84	0.1401	0.5189	0.3865	0.6198	25,030	16,736	37,670	12,988	92,682	5.7	4.9	7.2	600	4,282
85-89	0.2108	0.6902	0.5649	0.7733	12,042	6,363	23,110	8,311	39,433	4.2	3.7	5.2	309	1,468
90-94	0.3057	0.7973	0.6986	0.8600	3,731	1,443	10,056	2,975	9,731	3.1	2.7	3.7	100	326
95-99	0.4275	0.9365	0.8822	0.9653	756	202	3,031	708	1,657	2.3	2.2	2.6	18	43
100+	0.5766	1	1	1	48	7	357	48	83	1.7	1.7	1.9	2	3
All ages													*9,668*	*1,031,542*

ln($_nq_x$) with uncertainty interval

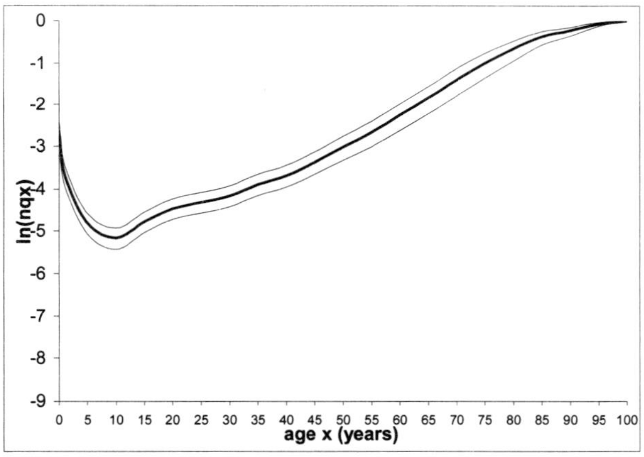

l_x with uncertainty interval

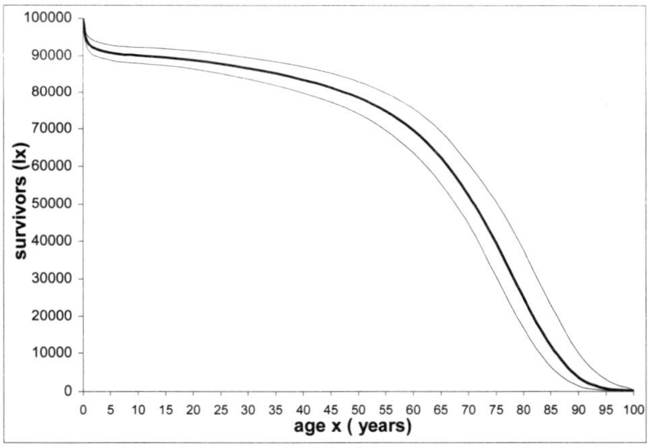

e_x with uncertainty interval

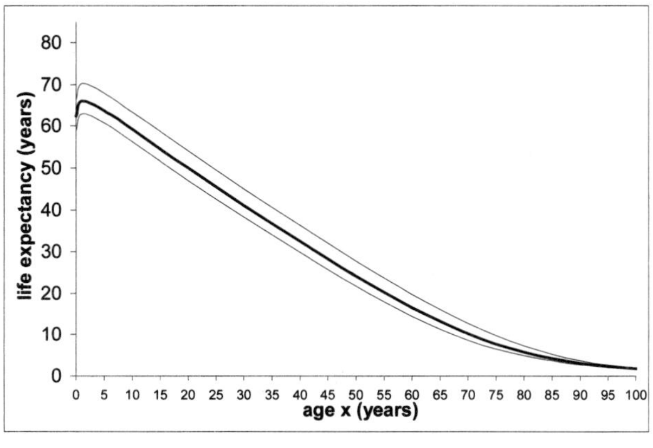

Distribution of deaths by age with uncertainty interval

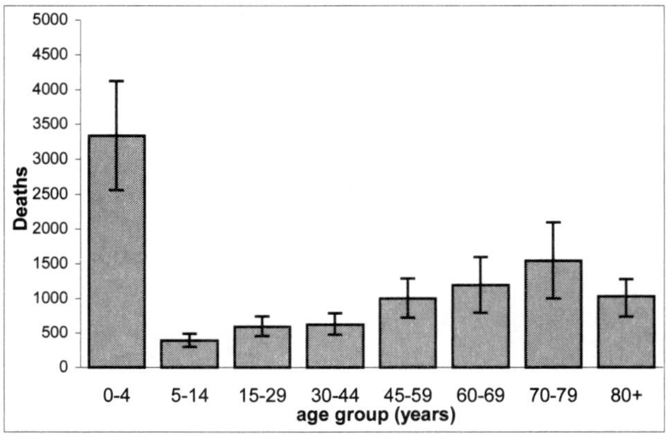

Bolivia **Bolivie** **Bolivia**

Males, 2000 *Hommes, 2000* *Hombres, 2000*

Age	$_nM_x$	$_nq_x$			l_x			$_nd_x$	$_nL_x$	e_x			Actual deaths	Actual population
			percentiles			percentiles					percentiles			
		mean	2½	97½	mean	2½	97½			mean	2½	97½		
<1	0.0737	0.0691	0.0622	0.0767	100,000	100,000	100,000	6,914	93,777	60.9	59.2	63.3	9,513	129,030
1-4	0.0051	0.0202	0.0179	0.0224	93,086	92,333	93,784	1,882	367,826	64.4	62.6	66.8	2,498	488,281
5-9	0.0015	0.0074	0.0064	0.0082	91,204	90,454	91,954	671	454,341	61.7	60.0	64.0	839	567,909
10-14	0.0011	0.0054	0.0047	0.0060	90,532	89,734	91,346	487	451,445	57.2	55.4	59.4	534	495,343
15-19	0.0017	0.0084	0.0072	0.0093	90,046	89,200	90,906	752	448,348	52.5	50.7	54.6	730	435,360
20-24	0.0025	0.0124	0.0107	0.0138	89,294	88,375	90,226	1,104	443,709	47.9	46.2	50.0	952	382,731
25-29	0.0028	0.0138	0.0118	0.0154	88,190	87,164	89,229	1,216	437,911	43.4	41.8	45.5	935	336,751
30-34	0.0033	0.0163	0.0139	0.0183	86,974	85,829	88,129	1,422	431,318	39.0	37.4	41.0	894	271,216
35-39	0.0043	0.0213	0.0180	0.0240	85,553	84,259	86,870	1,823	423,208	34.6	33.0	36.6	954	221,560
40-44	0.0060	0.0296	0.0249	0.0335	83,730	82,249	85,258	2,476	412,462	30.3	28.8	32.2	1,139	189,733
45-49	0.0087	0.0426	0.0356	0.0487	81,254	79,520	83,102	3,461	397,619	26.2	24.7	28.0	1,387	159,351
50-54	0.0129	0.0627	0.0521	0.0724	77,793	75,715	80,144	4,878	376,770	22.2	20.8	23.9	1,699	131,239
55-59	0.0192	0.0916	0.0757	0.1066	72,915	70,386	75,971	6,678	347,879	18.5	17.3	20.1	1,963	102,235
60-64	0.0287	0.1339	0.1109	0.1563	66,237	63,123	70,217	8,872	309,003	15.2	14.0	16.6	2,405	83,779
65-69	0.0429	0.1938	0.1615	0.2250	57,364	53,374	62,431	11,117	259,030	12.1	11.2	13.4	2,747	64,007
70-74	0.0662	0.2842	0.2390	0.3246	46,248	41,366	52,330	13,141	198,385	9.4	8.7	10.5	3,028	45,707
75-79	0.1017	0.4055	0.3493	0.4510	33,106	27,939	39,661	13,423	131,973	7.2	6.6	8.0	2,461	24,195
80-84	0.1545	0.5573	0.4984	0.6018	19,683	15,338	25,514	10,969	70,993	5.4	5.0	5.9	1,621	10,489
85-89	0.2271	0.7243	0.6745	0.7602	8,714	6,128	12,696	6,311	27,792	4.0	3.8	4.3	866	3,815
90-94	0.3229	0.8201	0.7821	0.8473	2,403	1,483	4,133	1,971	6,102	2.9	2.8	3.2	288	892
95-99	0.4443	0.9522	0.9344	0.9638	432	229	900	412	926	2.2	2.2	2.3	63	141
100+	0.5914	1	1	1	21	8	59	21	35	1.7	1.7	1.8	11	18
All ages													*37,527*	*4,143,782*

ln($_nq_x$) with uncertainty interval

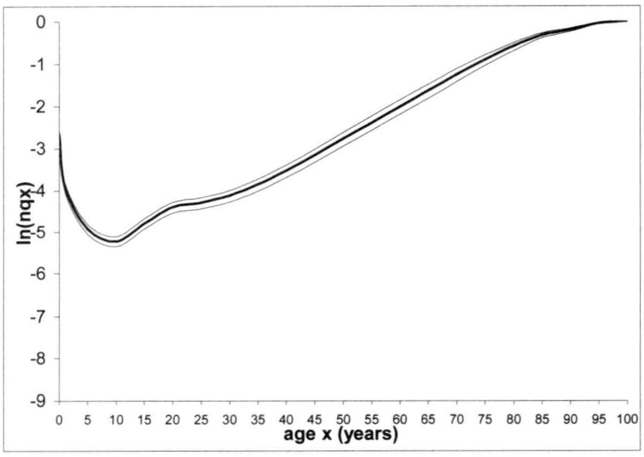

l_x with uncertainty interval

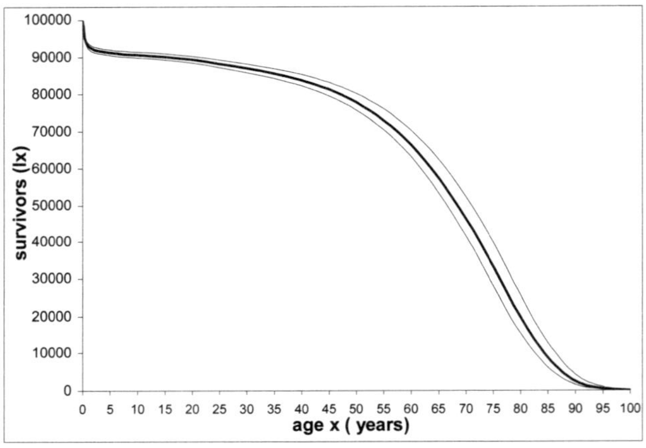

e_x with uncertainty interval

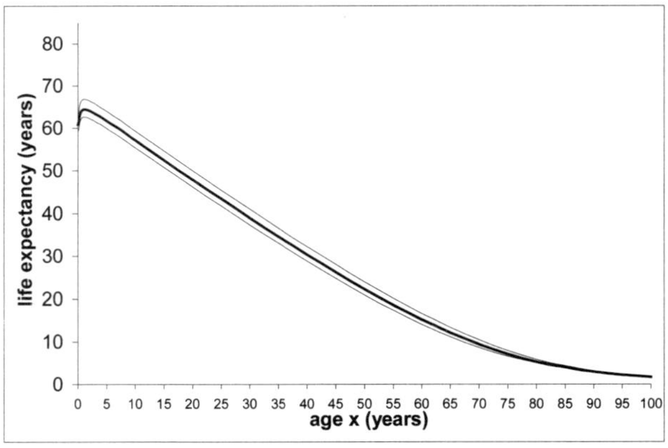

Distribution of deaths by age with uncertainty interval

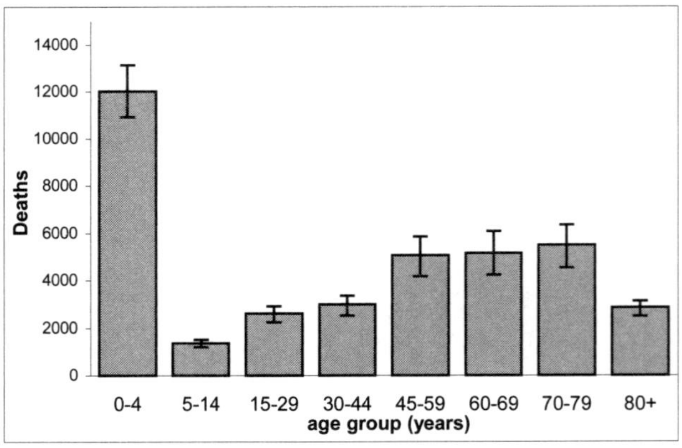

Bolivia **Bolivie** **Bolivia**

Females, 2000 *Femmes, 2000* *Mujeres, 2000*

Age	$_nM_x$	$_nq_x$			l_x			$_nd_x$	$_nL_x$	e_x			Actual deaths	Actual population
			percentiles			percentiles					percentiles			
		mean	2½	97½	mean	2½	97½			mean	2½	97½		
<1	0.0627	0.0593	0.0527	0.0667	100,000	100,000	100,000	5,933	94,661	63.6	61.3	65.8	7,702	122,886
1-4	0.0056	0.0220	0.0197	0.0249	94,067	93,325	94,729	2,067	371,308	66.6	64.4	68.9	2,619	470,398
5-9	0.0015	0.0073	0.0065	0.0084	92,000	91,250	92,746	675	458,312	64.1	62.0	66.3	805	546,362
10-14	0.0011	0.0053	0.0047	0.0061	91,325	90,500	92,128	486	455,409	59.5	57.5	61.7	512	479,870
15-19	0.0016	0.0081	0.0071	0.0094	90,839	89,960	91,683	740	452,345	54.8	52.9	57.0	697	426,284
20-24	0.0022	0.0109	0.0095	0.0127	90,099	89,137	91,004	986	448,030	50.2	48.3	52.4	839	381,104
25-29	0.0026	0.0127	0.0110	0.0148	89,113	88,040	90,112	1,132	442,734	45.8	43.9	47.9	873	341,335
30-34	0.0030	0.0150	0.0129	0.0175	87,981	86,738	89,089	1,319	436,605	41.3	39.5	43.4	844	279,162
35-39	0.0040	0.0198	0.0169	0.0232	86,661	85,219	87,895	1,717	429,013	36.9	35.2	38.9	925	231,097
40-44	0.0050	0.0249	0.0210	0.0293	84,944	83,237	86,339	2,113	419,438	32.6	31.0	34.5	1,011	200,724
45-49	0.0069	0.0341	0.0286	0.0403	82,831	80,799	84,433	2,823	407,100	28.4	26.8	30.2	1,190	171,675
50-54	0.0099	0.0483	0.0401	0.0572	80,009	77,544	81,900	3,863	390,387	24.3	22.9	26.0	1,404	141,910
55-59	0.0141	0.0683	0.0564	0.0810	76,146	73,106	78,613	5,201	367,727	20.4	19.1	22.0	1,588	112,242
60-64	0.0216	0.1024	0.0841	0.1210	70,945	67,183	74,180	7,264	336,564	16.7	15.5	18.2	2,051	95,020
65-69	0.0337	0.1554	0.1279	0.1824	63,681	59,052	67,938	9,896	293,664	13.3	12.3	14.6	2,546	75,537
70-74	0.0549	0.2414	0.2014	0.2806	53,785	48,313	59,246	12,983	236,465	10.3	9.4	11.4	3,073	55,966
75-79	0.0878	0.3598	0.3075	0.4087	40,801	34,939	47,317	14,681	167,303	7.8	7.1	8.7	2,741	31,241
80-84	0.1377	0.5121	0.4524	0.5620	26,120	20,949	32,715	13,376	97,161	5.8	5.3	6.4	1,999	14,518
85-89	0.2081	0.6843	0.6305	0.7253	12,744	9,213	17,744	8,721	41,918	4.3	4.0	4.7	1,192	5,730
90-94	0.3029	0.7935	0.7521	0.8239	4,023	2,531	6,428	3,192	10,537	3.1	2.9	3.3	459	1,516
95-99	0.4250	0.9340	0.9121	0.9488	831	446	1,594	776	1,825	2.3	2.2	2.4	120	282
100+	0.5743	1	1	1	55	23	140	55	95	1.7	1.7	1.8	25	44
All ages													*35,214*	*4,184,903*

ln($_nq_x$) with uncertainty interval

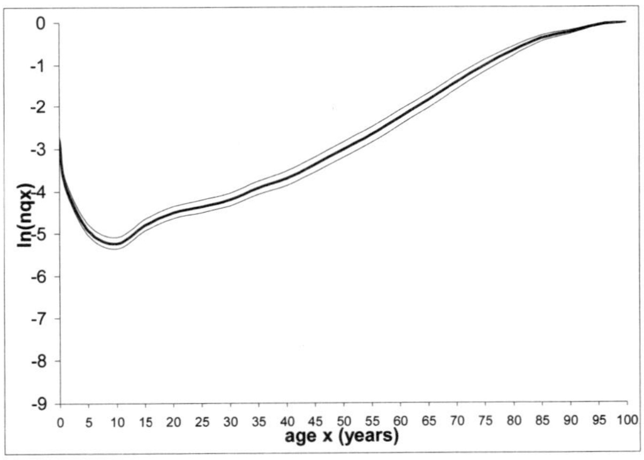

l_x with uncertainty interval

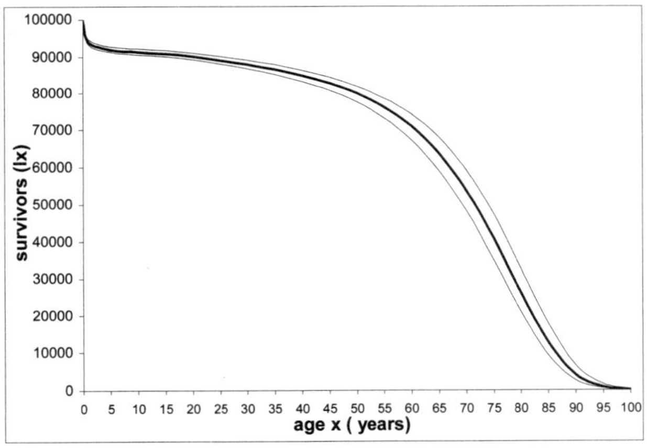

e_x with uncertainty interval

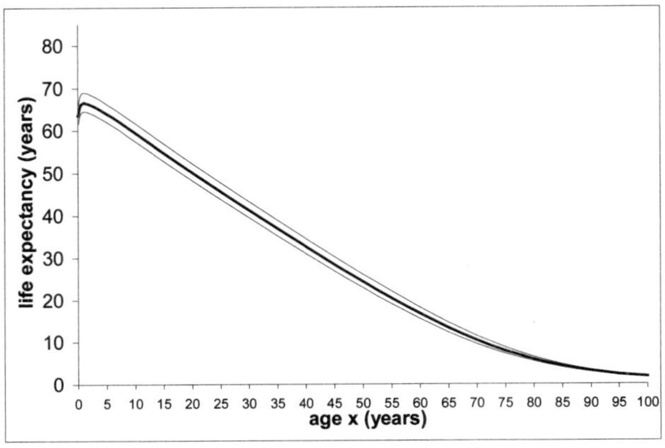

Distribution of deaths by age with uncertainty interval

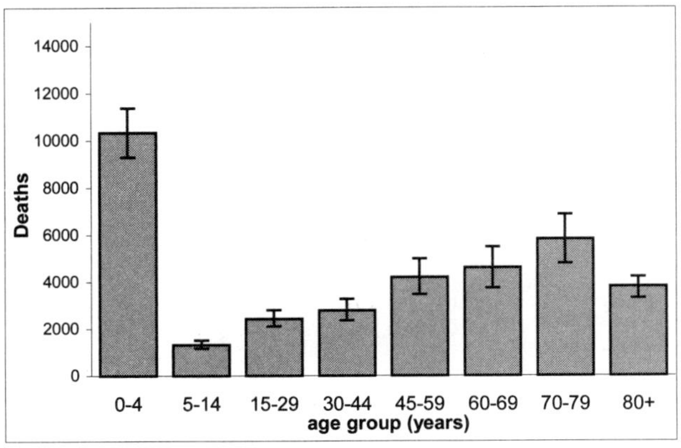

Bosnia and Herzegovina — Bosnie-Herzégovine — Bosnia y Herzegovina

Males, 2000 — *Hommes, 2000* — *Hombres, 2000*

Age	$_nM_x$	$_nq_x$			l_x			$_nd_x$	$_nL_x$	e_x			Actual deaths	Actual population
			percentiles			percentiles					percentiles			
		mean	2½	97½	mean	2½	97½			mean	2½	97½		
<1	0.0193	0.0190	0.0149	0.0232	100,000	100,000	100,000	1,899	98,291	68.7	67.0	70.4	381	19,693
1-4	0.0006	0.0024	0.0020	0.0029	98,101	97,681	98,511	239	391,828	69.0	67.5	70.6	53	85,934
5-9	0.0004	0.0018	0.0015	0.0021	97,861	97,413	98,311	175	488,869	65.2	63.7	66.8	44	123,703
10-14	0.0003	0.0017	0.0013	0.0020	97,686	97,206	98,164	161	488,029	60.3	58.8	61.9	53	159,256
15-19	0.0006	0.0028	0.0023	0.0034	97,525	97,015	98,027	275	486,938	55.4	53.9	57.0	90	159,774
20-24	0.0011	0.0054	0.0044	0.0064	97,250	96,690	97,793	522	484,944	50.5	49.1	52.1	159	147,587
25-29	0.0012	0.0062	0.0051	0.0074	96,728	96,076	97,359	599	482,139	45.8	44.3	47.4	192	154,647
30-34	0.0015	0.0077	0.0063	0.0092	96,128	95,374	96,859	741	478,789	41.1	39.6	42.7	259	167,116
35-39	0.0023	0.0112	0.0092	0.0133	95,387	94,500	96,236	1,072	474,255	36.4	34.9	37.9	407	180,049
40-44	0.0036	0.0180	0.0148	0.0212	94,315	93,240	95,337	1,695	467,339	31.8	30.4	33.3	630	173,803
45-49	0.0065	0.0318	0.0264	0.0373	92,620	91,262	93,923	2,945	455,740	27.3	25.9	28.8	1,007	155,848
50-54	0.0108	0.0527	0.0443	0.0616	89,676	87,858	91,440	4,727	436,561	23.1	21.8	24.6	1,179	108,896
55-59	0.0171	0.0820	0.0699	0.0955	84,949	82,482	87,392	6,964	407,334	19.2	18.1	20.7	1,323	77,383
60-64	0.0269	0.1260	0.1062	0.1466	77,985	74,740	81,286	9,825	365,362	15.7	14.7	17.0	2,465	91,661
65-69	0.0400	0.1818	0.1529	0.2098	68,160	63,964	72,381	12,392	309,821	12.6	11.7	13.7	3,068	76,718
70-74	0.0570	0.2495	0.2119	0.2859	55,768	50,797	60,642	13,913	244,058	9.9	9.2	10.8	2,877	50,470
75-79	0.0861	0.3541	0.3076	0.3980	41,855	36,458	47,774	14,821	172,222	7.4	6.8	8.0	1,855	21,556
80-84	0.1674	0.5900	0.5344	0.6366	27,034	22,203	33,055	15,950	95,296	5.0	4.7	5.5	1,172	7,004
85-89	0.2550	0.7786	0.7363	0.8121	11,084	8,133	15,226	8,630	33,846	3.7	3.5	4.0	1,261	4,946
90-94	0.3438	0.8462	0.8151	0.8711	2,454	1,544	3,946	2,077	6,040	2.8	2.7	3.0	528	1,536
95-99	0.4103	0.9196	0.8985	0.9359	377	202	729	347	846	2.4	2.3	2.5	95	232
100+	0.4897	1	1	1	30	13	74	30	62	2.0	2.0	2.4	11	22
All ages													*19,109*	*1,967,834*

$\ln(_nq_x)$ with uncertainty interval

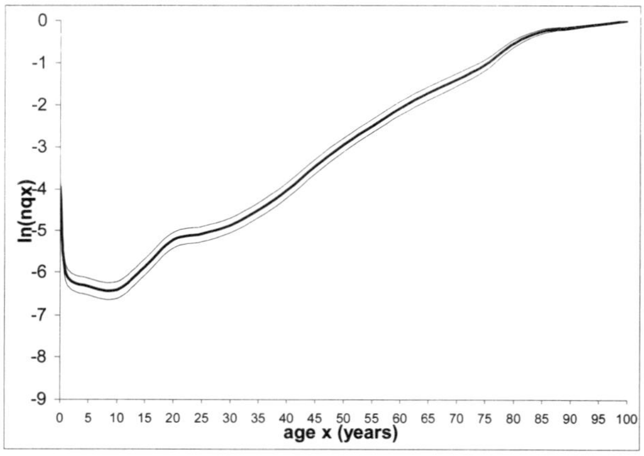

l_x with uncertainty interval

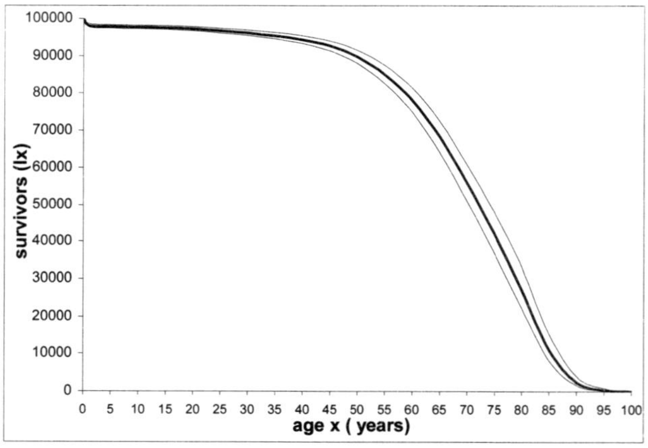

e_x with uncertainty interval

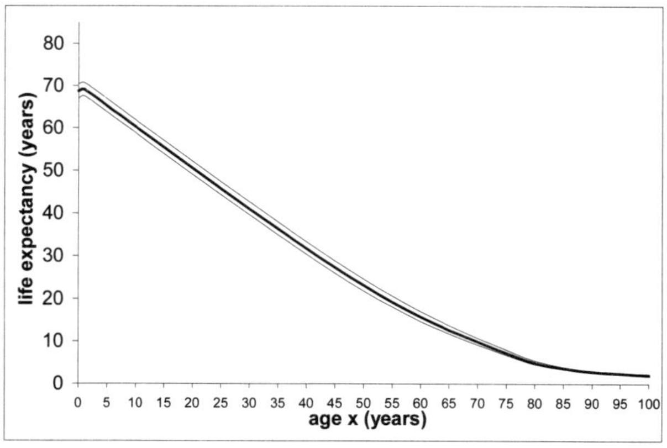

Distribution of deaths by age with uncertainty interval

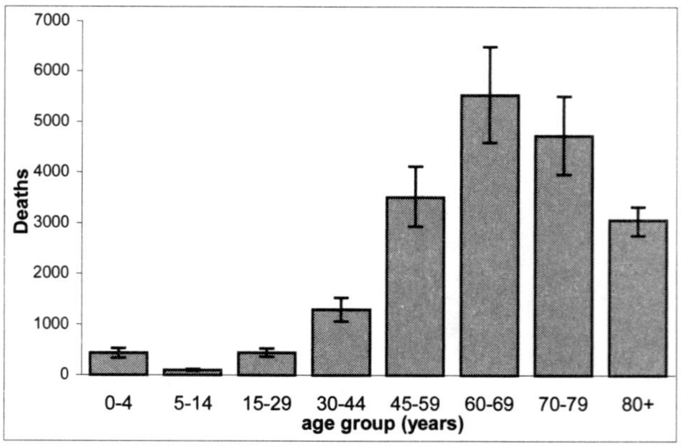

Bosnia and Herzegovina　　Bosnie-Herzégovine　　Bosnia y Herzegovina

Females, 2000　　*Femmes, 2000*　　*Mujeres, 2000*

Age	$_nM_x$	$_nq_x$			l_x			$_nd_x$	$_nL_x$	e_x			Actual deaths	Actual population
			percentiles			percentiles					percentiles			
		mean	2½	97½	mean	2½	97½			mean	2½	97½		
<1	0.0148	0.0146	0.0107	0.0185	100,000	100,000	100,000	1,462	98,684	74.7	73.4	76.2	275	18,552
1-4	0.0004	0.0018	0.0014	0.0022	98,538	98,148	98,927	176	393,729	74.8	73.7	76.3	36	80,386
5-9	0.0002	0.0010	0.0008	0.0012	98,362	97,937	98,784	99	491,561	70.9	69.8	72.4	23	116,088
10-14	0.0002	0.0012	0.0009	0.0014	98,263	97,819	98,703	114	491,027	66.0	64.9	67.5	35	149,717
15-19	0.0003	0.0015	0.0012	0.0019	98,148	97,680	98,611	151	490,365	61.1	60.0	62.5	46	150,459
20-24	0.0003	0.0017	0.0014	0.0021	97,998	97,499	98,490	169	489,564	56.2	55.1	57.6	48	139,677
25-29	0.0005	0.0025	0.0020	0.0031	97,828	97,295	98,353	248	488,520	51.3	50.2	52.7	75	147,062
30-34	0.0006	0.0029	0.0024	0.0035	97,580	96,998	98,153	287	487,181	46.4	45.3	47.8	95	161,884
35-39	0.0009	0.0043	0.0035	0.0051	97,292	96,655	97,919	414	485,426	41.5	40.5	42.9	150	176,209
40-44	0.0016	0.0079	0.0065	0.0094	96,878	96,164	97,578	768	482,469	36.7	35.6	38.1	275	172,490
45-49	0.0029	0.0146	0.0122	0.0171	96,109	95,261	96,939	1,403	477,039	32.0	30.9	33.3	456	155,025
50-54	0.0045	0.0224	0.0189	0.0259	94,706	93,633	95,748	2,123	468,222	27.4	26.4	28.7	514	113,374
55-59	0.0079	0.0388	0.0329	0.0447	92,583	91,210	93,903	3,592	453,932	23.0	22.1	24.3	705	89,088
60-64	0.0138	0.0668	0.0564	0.0767	88,990	87,199	90,716	5,947	430,084	18.8	18.0	20.0	1,498	108,364
65-69	0.0260	0.1220	0.1024	0.1384	83,043	80,741	85,450	10,130	389,892	15.0	14.2	16.1	2,371	91,268
70-74	0.0395	0.1797	0.1515	0.2011	72,913	69,572	76,444	13,100	331,816	11.7	11.0	12.6	2,726	69,046
75-79	0.0674	0.2884	0.2487	0.3206	59,813	55,630	64,842	17,251	255,938	8.7	8.1	9.4	2,830	41,984
80-84	0.1245	0.4748	0.4279	0.5212	42,562	38,202	48,713	20,209	162,288	6.2	5.7	6.7	1,910	15,335
85-89	0.1933	0.6517	0.6077	0.6956	22,353	18,972	27,867	14,568	75,346	4.6	4.2	4.9	1,946	10,063
90-94	0.2752	0.7537	0.7195	0.7882	7,785	5,819	10,734	5,868	21,324	3.4	3.2	3.6	740	2,689
95-99	0.3591	0.8643	0.8404	0.8883	1,918	1,232	2,907	1,657	4,616	2.7	2.6	2.9	167	464
100+	0.4685	1	1	1	260	138	464	260	555	2.1	2.1	2.4	22	48
All ages													*16,944*	*2,009,272*

ln($_nq_x$) with uncertainty interval

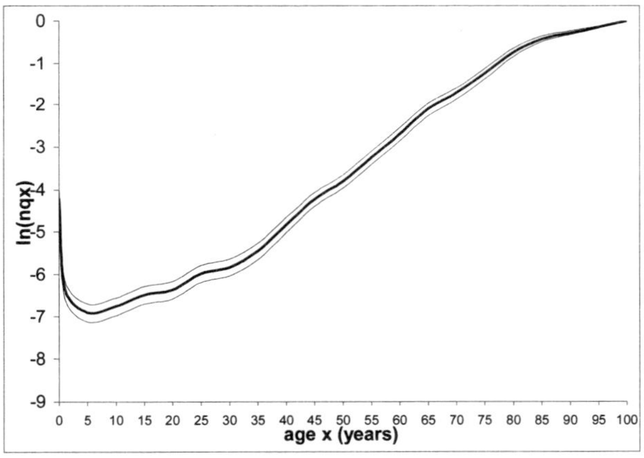

l_x with uncertainty interval

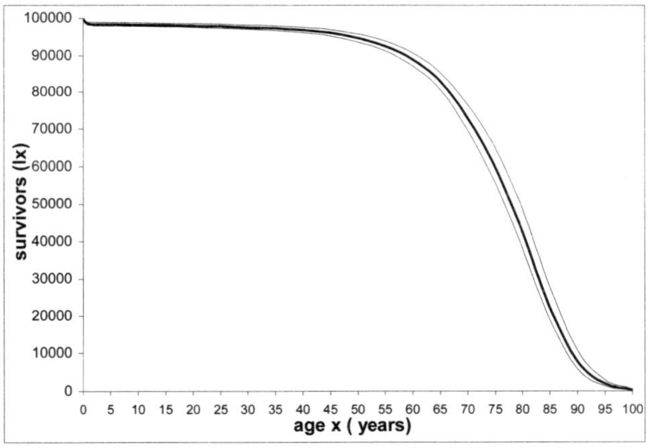

e_x with uncertainty interval

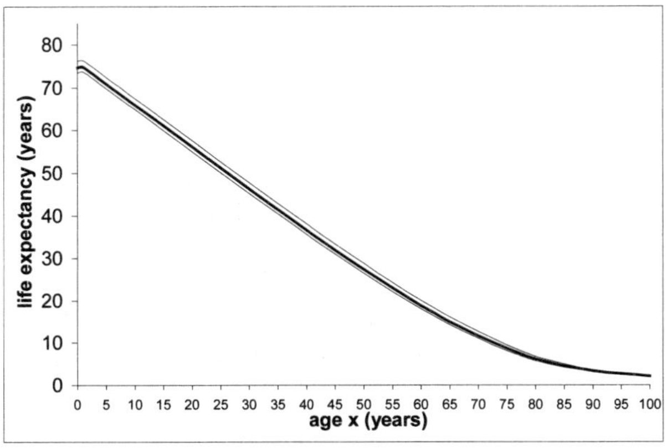

Distribution of deaths by age with uncertainty interval

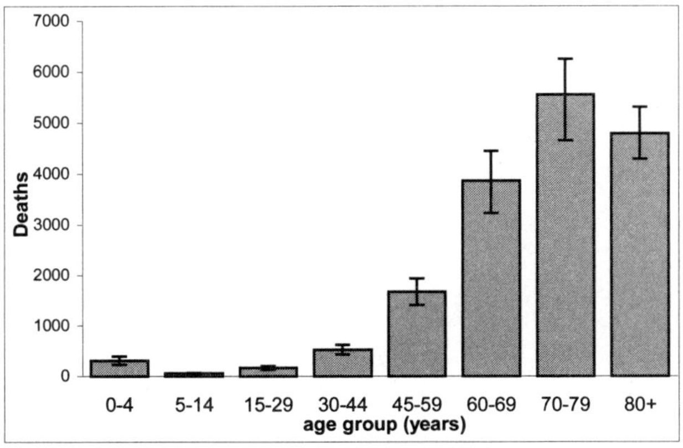

Botswana | Botswana | Botswana

Males, 2000 | *Hommes, 2000* | *Hombres, 2000*

Age	$_nM_x$	$_nq_x$			l_x			$_nd_x$	$_nL_x$	e_x			Actual deaths	Actual population
			percentiles			percentiles					percentiles			
		mean	2½	97½	mean	2½	97½			mean	2½	97½		
<1	0.0573	0.0545	0.0435	0.0656	100,000	100,000	100,000	5,445	95,099	44.6	42.1	46.9	1,372	23,954
1-4	0.0083	0.0327	0.0274	0.0381	94,555	93,437	95,653	3,089	370,806	46.1	43.9	48.2	742	89,103
5-9	0.0015	0.0075	0.0063	0.0088	91,466	90,009	92,939	685	455,619	43.6	41.6	45.6	165	109,692
10-14	0.0012	0.0060	0.0051	0.0070	90,781	89,232	92,353	545	452,543	39.0	37.0	40.9	125	103,986
15-19	0.0035	0.0174	0.0148	0.0204	90,236	88,614	91,884	1,573	447,246	34.2	32.2	36.1	332	94,471
20-24	0.0088	0.0429	0.0365	0.0499	88,663	86,840	90,525	3,801	433,811	29.7	27.8	31.6	693	79,076
25-29	0.0203	0.0965	0.0835	0.1111	84,862	82,605	87,199	8,185	403,846	26.0	24.1	27.8	1,233	60,824
30-34	0.0337	0.1553	0.1376	0.1760	76,676	73,427	79,857	11,906	353,618	23.5	21.7	25.3	1,621	48,153
35-39	0.0369	0.1690	0.1515	0.1885	64,771	60,503	68,871	10,947	296,487	22.3	20.6	24.1	1,455	39,411
40-44	0.0370	0.1693	0.1513	0.1891	53,824	49,096	58,339	9,110	246,346	21.3	19.7	23.0	1,115	30,160
45-49	0.0365	0.1673	0.1489	0.1877	44,714	39,959	49,247	7,481	204,869	20.2	18.8	21.6	812	22,249
50-54	0.0342	0.1575	0.1399	0.1768	37,233	32,699	41,550	5,863	171,508	18.7	17.5	20.0	523	15,296
55-59	0.0313	0.1453	0.1291	0.1631	31,370	27,162	35,472	4,558	145,455	16.8	15.8	17.8	382	12,201
60-64	0.0378	0.1726	0.1539	0.1928	26,812	22,951	30,687	4,627	122,492	14.2	13.3	15.1	394	10,433
65-69	0.0497	0.2211	0.1984	0.2452	22,185	18,761	25,732	4,905	98,663	11.6	10.9	12.4	326	6,566
70-74	0.0711	0.3019	0.2737	0.3312	17,280	14,318	20,423	5,216	73,360	9.2	8.7	9.9	324	4,559
75-79	0.1038	0.4122	0.3790	0.4458	12,064	9,637	14,782	4,973	47,888	7.1	6.7	7.6	279	2,687
80-84	0.1543	0.5568	0.5209	0.5922	7,091	5,341	9,179	3,948	25,585	5.4	5.1	5.8	215	1,396
85-89	0.2240	0.7180	0.6844	0.7500	3,143	2,181	4,395	2,256	10,073	4.0	3.8	4.3	92	409
90-94	0.3165	0.8118	0.7834	0.8383	886	547	1,383	720	2,273	3.0	2.8	3.2	31	98
95-99	0.4360	0.9445	0.9297	0.9572	167	89	298	158	361	2.3	2.2	2.3	5	11
100+	0.5846	1	1	1	9	4	21	9	16	1.7	1.7	1.8	1	1
All ages													*12,238*	*754,736*

$\ln(_nq_x)$ with uncertainty interval

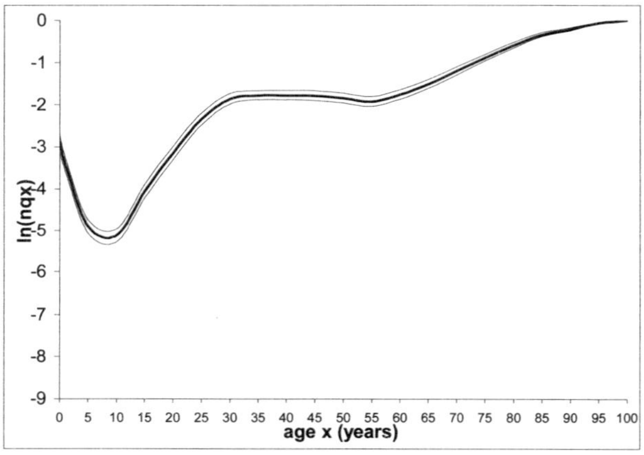

l_x with uncertainty interval

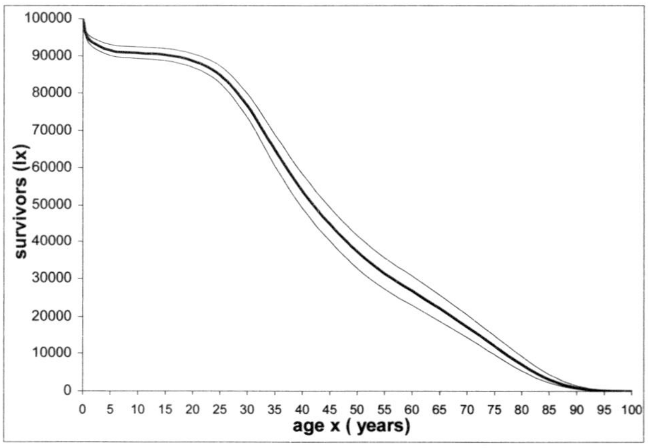

e_x with uncertainty interval

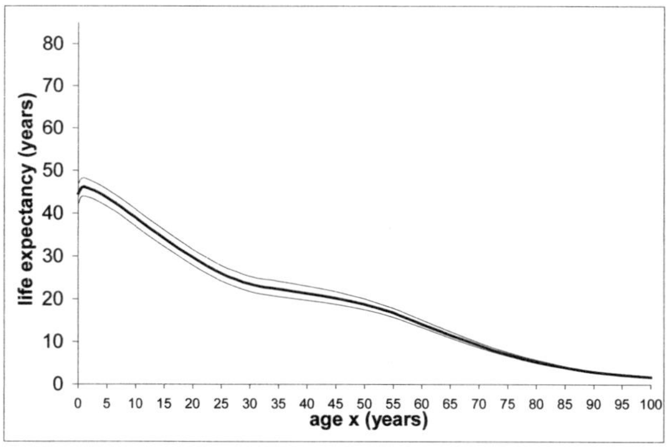

Distribution of deaths by age with uncertainty interval

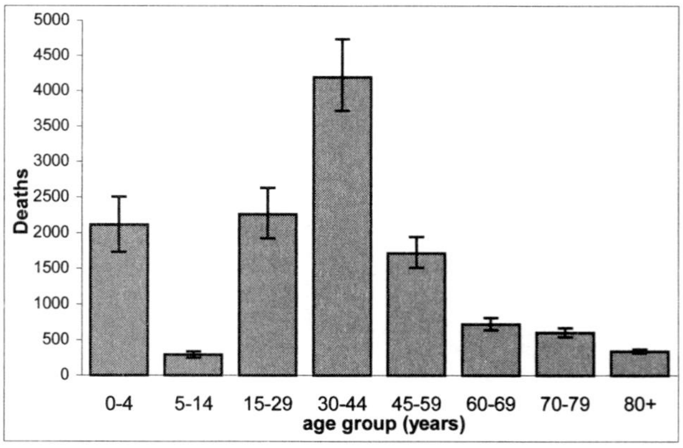

Botswana **Botswana** **Botswana**

Females, 2000 *Femmes, 2000* *Mujeres, 2000*

Age	$_nM_x$	$_nq_x$			l_x			$_nd_x$	$_nL_x$	e_x			Actual deaths	Actual population
			percentiles			percentiles					percentiles			
		mean	2½	97½	mean	2½	97½			mean	2½	97½		
<1	0.0511	0.0489	0.0384	0.0596	100,000	100,000	100,000	4,885	95,603	44.4	42.3	46.6	1,177	23,033
1-4	0.0090	0.0354	0.0300	0.0414	95,115	94,040	96,157	3,368	372,375	45.7	43.9	47.5	796	88,002
5-9	0.0017	0.0083	0.0070	0.0096	91,746	90,247	93,243	759	456,835	43.3	41.7	45.0	180	108,144
10-14	0.0014	0.0067	0.0058	0.0078	90,988	89,392	92,569	614	453,403	38.7	37.1	40.3	139	103,014
15-19	0.0061	0.0300	0.0258	0.0346	90,374	88,702	92,021	2,711	445,091	33.9	32.4	35.5	575	94,330
20-24	0.0173	0.0830	0.0726	0.0940	87,663	85,639	89,563	7,275	420,127	29.9	28.4	31.4	1,358	78,444
25-29	0.0309	0.1432	0.1286	0.1583	80,388	77,590	83,030	11,514	373,156	27.4	25.8	28.9	1,822	59,037
30-34	0.0379	0.1732	0.1594	0.1872	68,874	65,312	72,351	11,926	314,557	26.5	24.9	28.2	1,804	47,575
35-39	0.0326	0.1508	0.1385	0.1643	56,948	53,104	60,817	8,589	263,268	26.5	25.0	28.1	1,331	40,788
40-44	0.0281	0.1312	0.1194	0.1442	48,359	44,596	52,223	6,346	225,929	25.8	24.4	27.2	919	32,718
45-49	0.0232	0.1097	0.0993	0.1212	42,013	38,448	45,723	4,608	198,544	24.3	23.1	25.6	595	25,633
50-54	0.0222	0.1053	0.0951	0.1166	37,405	34,053	40,926	3,939	177,177	22.0	20.9	23.1	516	23,205
55-59	0.0225	0.1066	0.0961	0.1181	33,466	30,339	36,774	3,566	158,414	19.3	18.4	20.3	426	18,939
60-64	0.0273	0.1277	0.1153	0.1413	29,900	27,010	32,975	3,819	139,951	16.3	15.5	17.2	429	15,722
65-69	0.0368	0.1687	0.1526	0.1860	26,081	23,457	28,879	4,399	119,408	13.4	12.6	14.1	411	11,146
70-74	0.0550	0.2418	0.2200	0.2649	21,682	19,341	24,214	5,242	95,305	10.6	10.0	11.2	422	7,669
75-79	0.0826	0.3423	0.3144	0.3712	16,440	14,241	18,758	5,627	68,131	8.1	7.7	8.6	464	5,623
80-84	0.1296	0.4894	0.4562	0.5227	10,812	8,955	12,823	5,291	40,834	6.0	5.7	6.4	325	2,507
85-89	0.1967	0.6594	0.6262	0.6917	5,521	4,276	6,974	3,641	18,505	4.4	4.2	4.7	158	802
90-94	0.2886	0.7734	0.7447	0.8008	1,881	1,320	2,607	1,455	5,040	3.2	3.1	3.4	46	158
95-99	0.4093	0.9186	0.9012	0.9340	426	263	666	391	956	2.4	2.3	2.5	12	29
100+	0.5608	1	1	1	35	17	66	35	62	1.8	1.7	1.9	1	2
All ages													*13,904*	*786,520*

$\ln(_nq_x)$ with uncertainty interval

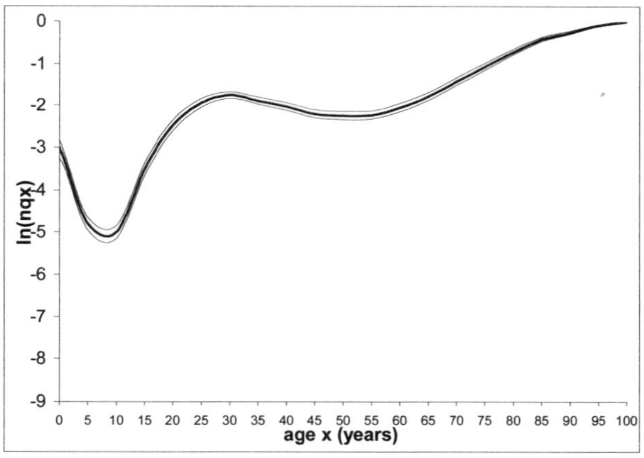

l_x with uncertainty interval

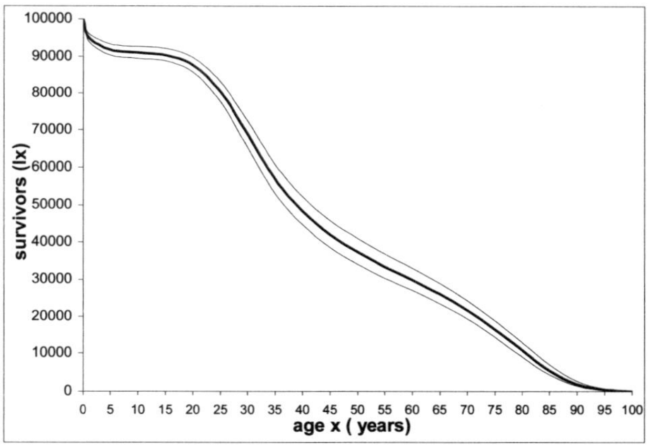

e_x with uncertainty interval

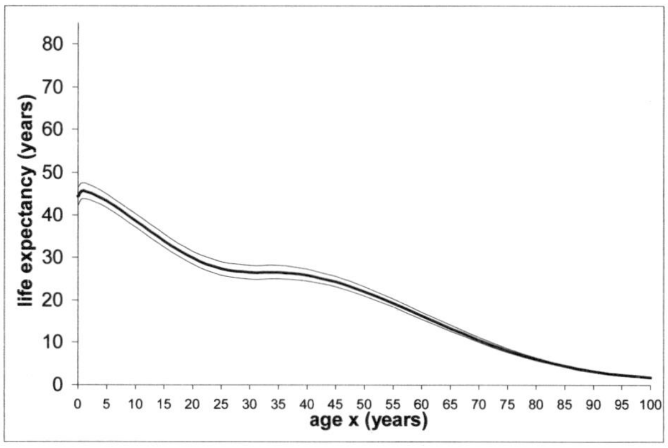

Distribution of deaths by age with uncertainty interval

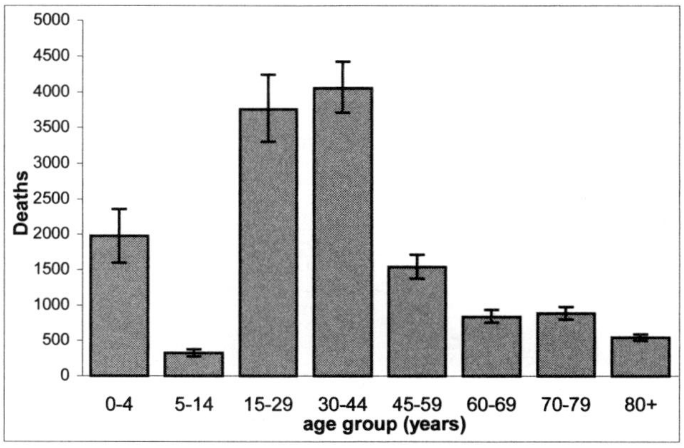

Brazil — Brésil — Brasil

Males, 2000 — *Hommes, 2000* — *Hombres, 2000*

Age	$_nM_x$	$_nq_x$ mean	$_nq_x$ percentiles 2½	$_nq_x$ percentiles 97½	l_x mean	l_x percentiles 2½	l_x percentiles 97½	$_nd_x$	$_nL_x$	e_x mean	e_x percentiles 2½	e_x percentiles 97½	Actual deaths	Actual population
<1	0.0406	0.0391	0.0316	0.0463	100,000	100,000	100,000	3,914	96,477	64.5	63.1	65.8	67,182	1,655,800
1-4	0.0026	0.0104	0.0090	0.0120	96,086	95,366	96,837	1,004	381,933	66.1	64.9	67.5	17,054	6,489,491
5-9	0.0004	0.0022	0.0019	0.0025	95,082	94,234	95,927	211	474,882	62.8	61.6	64.2	3,646	8,201,031
10-14	0.0005	0.0027	0.0024	0.0031	94,871	93,995	95,737	260	473,703	57.9	56.8	59.3	4,724	8,602,694
15-19	0.0020	0.0098	0.0085	0.0111	94,611	93,702	95,504	931	470,726	53.0	51.9	54.5	17,404	8,802,865
20-24	0.0030	0.0150	0.0131	0.0169	93,680	92,658	94,669	1,408	464,880	48.5	47.5	50.0	25,114	8,292,701
25-29	0.0035	0.0172	0.0152	0.0192	92,272	91,092	93,426	1,587	457,392	44.2	43.1	45.7	25,154	7,248,023
30-34	0.0038	0.0190	0.0169	0.0210	90,685	89,343	92,009	1,720	449,124	40.0	38.8	41.4	25,775	6,730,732
35-39	0.0045	0.0225	0.0201	0.0248	88,965	87,465	90,459	2,001	439,822	35.7	34.5	37.1	29,116	6,400,388
40-44	0.0061	0.0301	0.0272	0.0331	86,964	85,302	88,636	2,621	428,268	31.5	30.3	32.9	32,827	5,364,112
45-49	0.0082	0.0402	0.0364	0.0440	84,343	82,498	86,221	3,391	413,238	27.4	26.2	28.8	36,338	4,428,419
50-54	0.0118	0.0574	0.0512	0.0630	80,952	78,912	83,054	4,644	393,152	23.4	22.3	24.7	40,600	3,437,352
55-59	0.0170	0.0815	0.0717	0.0904	76,309	74,068	78,647	6,221	365,991	19.7	18.6	20.9	43,748	2,573,798
60-64	0.0247	0.1162	0.1010	0.1308	70,088	67,638	72,630	8,142	330,083	16.2	15.2	17.4	52,237	2,117,736
65-69	0.0379	0.1729	0.1493	0.1966	61,946	59,278	64,709	10,711	282,951	13.0	12.1	14.0	57,773	1,526,165
70-74	0.0555	0.2437	0.2108	0.2772	51,235	48,163	54,753	12,487	224,956	10.2	9.4	11.1	63,349	1,141,272
75-79	0.0862	0.3545	0.3113	0.3978	38,748	34,975	43,211	13,737	159,396	7.7	7.1	8.4	58,523	679,059
80-84	0.1460	0.5348	0.4848	0.5825	25,011	21,063	29,761	13,376	91,614	5.5	5.1	6.0	46,933	321,462
85-89	0.2330	0.7362	0.6942	0.7751	11,635	8,795	15,333	8,565	36,761	4.0	3.7	4.3	28,261	121,290
90-94	0.2940	0.7811	0.7473	0.8134	3,070	1,985	4,689	2,398	8,155	3.2	3.0	3.4	8,656	29,443
95-99	0.3860	0.8943	0.8714	0.9156	672	372	1,185	601	1,557	2.5	2.4	2.7	1,690	4,378
100+	0.5754	1	1	1	71	32	152	71	123	1.7	1.7	2.2	219	381
All ages													*686,323*	*84,168,592*

ln($_nq_x$) with uncertainty interval

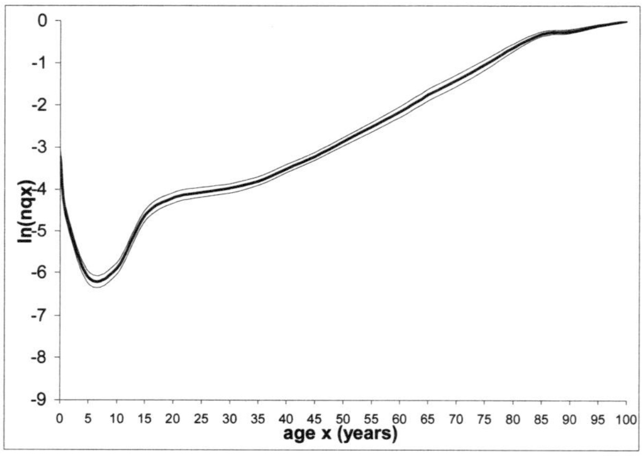

l_x with uncertainty interval

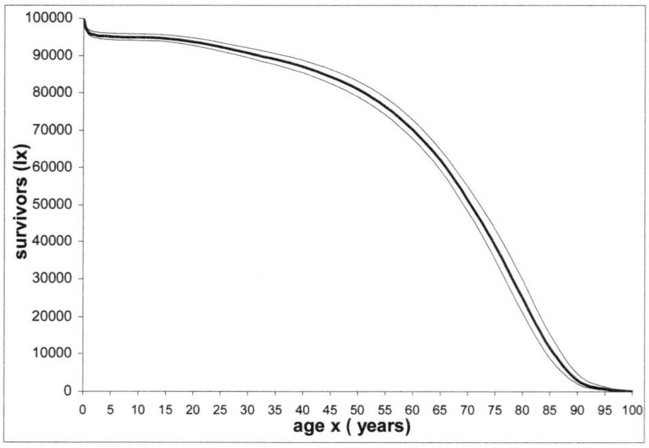

e_x with uncertainty interval

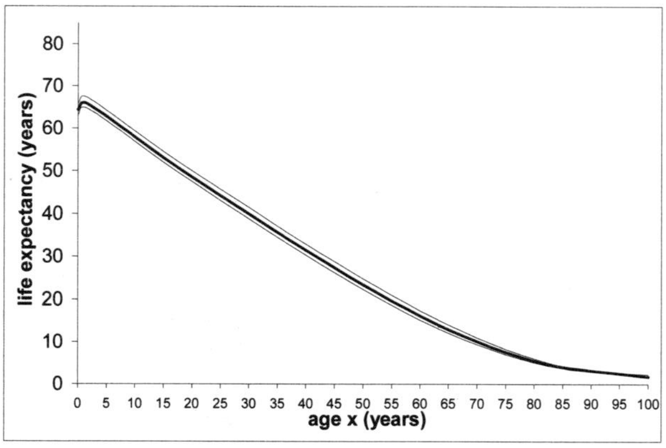

Distribution of deaths by age with uncertainty interval

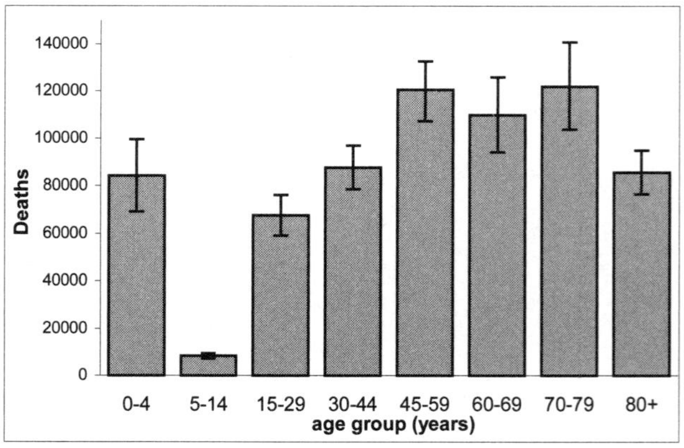

Brazil **Brésil** **Brasil**

Females, 2000 *Femmes, 2000* *Mujeres, 2000*

Age	$_nM_x$	$_nq_x$			l_x			$_nd_x$	$_nL_x$	e_x			Actual deaths	Actual population
			percentiles			percentiles					percentiles			
		mean	2½	97½	mean	2½	97½			mean	2½	97½		
<1	0.0348	0.0337	0.0276	0.0400	100,000	100,000	100,000	3,374	96,964	71.9	70.3	73.5	55,394	1,592,115
1-4	0.0023	0.0090	0.0076	0.0103	96,626	96,004	97,238	865	384,429	73.4	72.0	74.9	14,104	6,267,716
5-9	0.0003	0.0017	0.0015	0.0020	95,761	95,064	96,459	165	478,394	70.0	68.6	71.6	2,734	7,933,012
10-14	0.0004	0.0019	0.0016	0.0022	95,596	94,877	96,313	180	477,531	65.1	63.7	66.7	3,149	8,335,168
15-19	0.0007	0.0033	0.0028	0.0038	95,416	94,672	96,157	314	476,294	60.3	58.8	61.9	5,651	8,558,514
20-24	0.0008	0.0041	0.0035	0.0046	95,101	94,317	95,886	387	474,539	55.4	54.0	57.1	6,758	8,275,841
25-29	0.0011	0.0053	0.0045	0.0060	94,714	93,879	95,552	500	472,321	50.7	49.2	52.3	7,817	7,388,334
30-34	0.0014	0.0070	0.0060	0.0080	94,214	93,317	95,119	659	469,425	45.9	44.5	47.5	9,706	6,917,337
35-39	0.0019	0.0095	0.0081	0.0108	93,556	92,578	94,547	885	465,566	41.2	39.8	42.8	12,632	6,647,504
40-44	0.0030	0.0146	0.0126	0.0167	92,671	91,591	93,775	1,357	459,961	36.6	35.2	38.2	16,647	5,641,464
45-49	0.0043	0.0213	0.0183	0.0242	91,314	90,085	92,583	1,941	451,717	32.1	30.7	33.7	20,253	4,714,474
50-54	0.0065	0.0322	0.0279	0.0366	89,373	87,948	90,864	2,879	439,667	27.7	26.3	29.3	24,383	3,723,097
55-59	0.0097	0.0474	0.0410	0.0537	86,494	84,791	88,290	4,100	422,219	23.6	22.2	25.1	27,800	2,863,177
60-64	0.0148	0.0712	0.0612	0.0819	82,394	80,235	84,613	5,864	397,311	19.6	18.3	21.1	36,062	2,443,377
65-69	0.0238	0.1125	0.0951	0.1314	76,530	73,799	79,327	8,607	361,132	15.9	14.7	17.3	43,890	1,841,436
70-74	0.0361	0.1656	0.1382	0.1949	67,923	64,538	71,399	11,248	311,493	12.7	11.6	13.9	52,496	1,453,756
75-79	0.0605	0.2628	0.2198	0.3064	56,675	52,386	61,136	14,895	246,134	9.7	8.8	10.7	55,664	919,796
80-84	0.0972	0.3910	0.3338	0.4439	41,779	36,332	47,420	16,336	168,057	7.2	6.6	8.0	44,947	462,401
85-89	0.1535	0.5546	0.4919	0.6068	25,444	20,204	31,571	14,110	91,943	5.3	4.8	5.8	29,519	192,347
90-94	0.2401	0.6979	0.6443	0.7406	11,334	7,964	15,987	7,910	32,938	3.7	3.4	4.0	13,236	55,116
95-99	0.3722	0.8793	0.8461	0.9041	3,424	2,085	5,662	3,010	8,087	2.6	2.5	2.8	3,865	10,383
100+	0.5631	1	1	1	413	204	871	413	734	1.8	1.7	1.9	745	1,323
All ages													*487,451*	*86,237,688*

ln($_nq_x$) with uncertainty interval

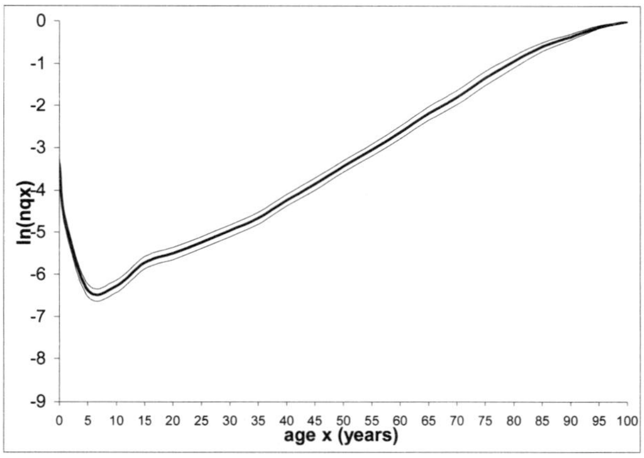

l_x with uncertainty interval

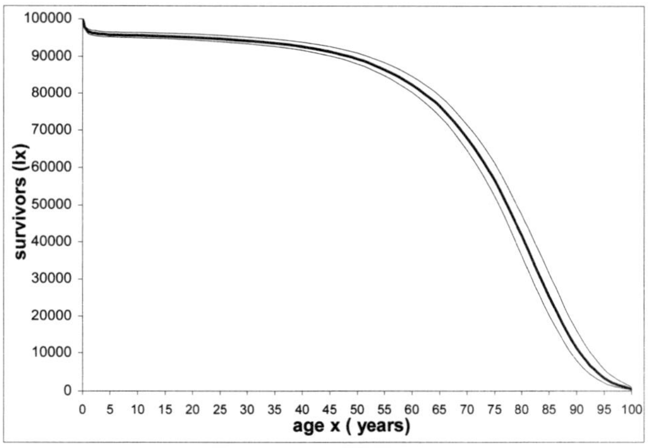

e_x with uncertainty interval

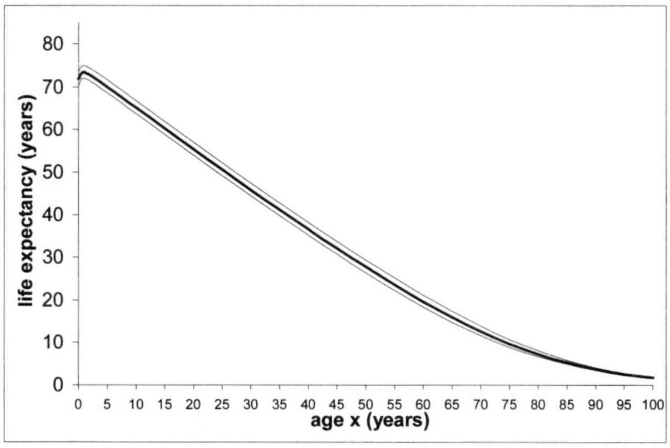

Distribution of deaths by age with uncertainty interval

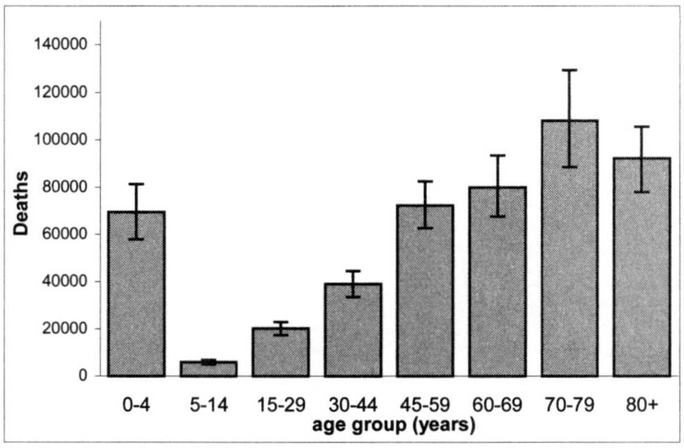

Brunei Darussalam   Brunéi Darussalam   Brunei Darussalam

Males, 2000   *Hommes, 2000*   *Hombres, 2000*

Age	$_nM_x$	$_nq_x$			l_x			$_nd_x$	$_nL_x$	e_x			Actual deaths	Actual population
			percentiles			percentiles					percentiles			
		mean	2½	97½	mean	2½	97½			mean	2½	97½		
<1	0.0091	0.0090	0.0075	0.0115	100,000	100,000	100,000	898	99,192	73.4	72.1	74.8	29	3,211
1-4	0.0007	0.0029	0.0025	0.0035	99,102	98,847	99,246	283	395,728	73.0	71.8	74.5	10	14,430
5-9	0.0004	0.0018	0.0016	0.0022	98,819	98,501	99,003	180	493,643	69.3	68.0	70.7	7	18,745
10-14	0.0005	0.0024	0.0021	0.0029	98,638	98,283	98,848	239	492,594	64.4	63.2	65.8	9	17,546
15-19	0.0010	0.0051	0.0044	0.0061	98,399	97,996	98,641	499	490,749	59.5	58.3	61.0	15	15,091
20-24	0.0015	0.0075	0.0065	0.0088	97,900	97,402	98,209	731	487,674	54.8	53.7	56.3	21	13,724
25-29	0.0014	0.0071	0.0062	0.0083	97,169	96,543	97,573	691	484,120	50.2	49.1	51.6	18	12,492
30-34	0.0014	0.0069	0.0061	0.0081	96,479	95,741	96,969	671	480,717	45.6	44.5	46.9	20	14,120
35-39	0.0020	0.0098	0.0086	0.0113	95,808	94,971	96,380	940	476,692	40.9	39.8	42.2	30	15,109
40-44	0.0032	0.0159	0.0140	0.0181	94,869	93,900	95,551	1,506	470,578	36.2	35.2	37.6	47	14,833
45-49	0.0045	0.0222	0.0196	0.0251	93,363	92,201	94,215	2,077	461,620	31.8	30.8	33.1	54	12,078
50-54	0.0065	0.0320	0.0282	0.0360	91,285	89,884	92,368	2,919	449,129	27.4	26.5	28.7	53	8,164
55-59	0.0097	0.0472	0.0415	0.0529	88,366	86,650	89,760	4,168	431,410	23.3	22.4	24.5	46	4,719
60-64	0.0168	0.0805	0.0705	0.0897	84,198	82,064	86,036	6,774	404,054	19.3	18.5	20.4	56	3,320
65-69	0.0311	0.1444	0.1268	0.1591	77,424	74,701	79,972	11,181	359,166	15.8	15.0	16.8	67	2,155
70-74	0.0424	0.1918	0.1696	0.2082	66,242	62,815	69,834	12,705	299,448	13.0	12.4	13.9	62	1,468
75-79	0.0650	0.2796	0.2515	0.3019	53,537	49,739	57,991	14,967	230,266	10.5	10.0	11.2	57	880
80-84	0.0853	0.3515	0.3228	0.3759	38,570	34,723	43,407	13,556	158,959	8.6	8.2	9.1	39	457
85-89	0.1099	0.4310	0.4011	0.4555	25,014	21,672	29,397	10,781	98,118	6.9	6.6	7.3	20	182
90-94	0.1528	0.5239	0.4943	0.5474	14,233	11,801	17,606	7,456	48,798	5.2	5.0	5.5	11	70
95-99	0.2294	0.6794	0.6520	0.7003	6,777	5,342	8,903	4,604	20,073	3.8	3.6	4.0	2	10
100+	0.3717	1	1	1	2,173	1,601	3,099	2,173	5,847	2.7	2.5	2.8	0	1
All ages													*673*	*172,805*

$\ln(_nq_x)$ with uncertainty interval

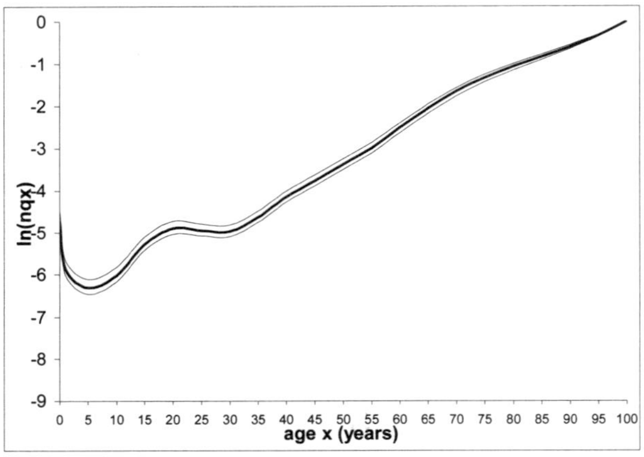

l_x with uncertainty interval

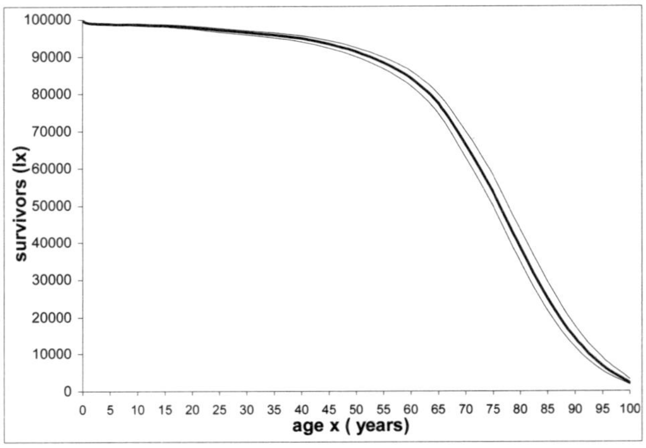

e_x with uncertainty interval

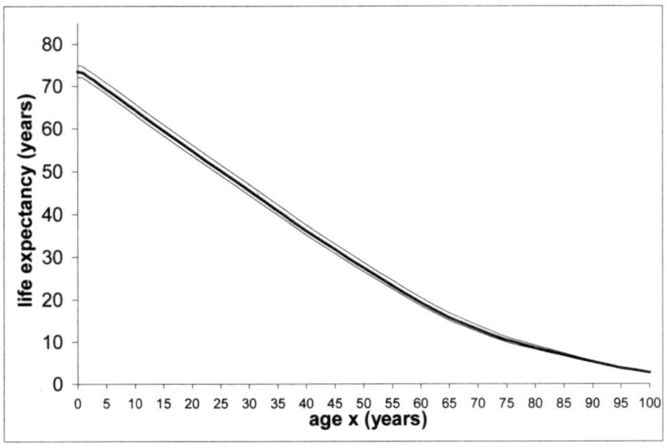

Distribution of deaths by age with uncertainty interval

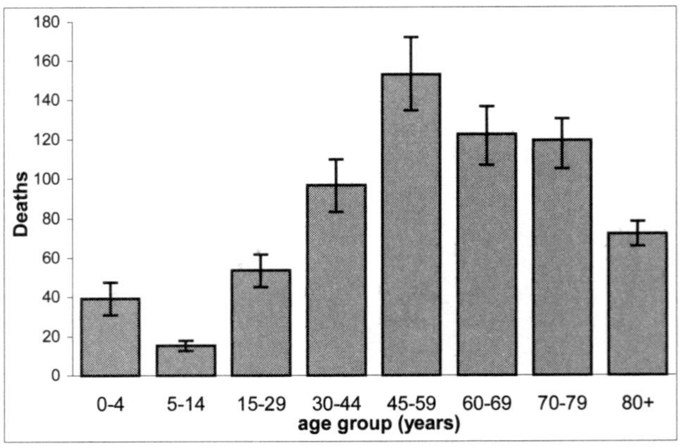

Brunei Darussalam

Brunéi Darussalam

Brunei Darussalam

Females, 2000

Femmes, 2000

Mujeres, 2000

Age	$_nM_x$	$_nq_x$			l_x			$_nd_x$	$_nL_x$	e_x			Actual deaths	Actual population
			percentiles			percentiles					percentiles			
		mean	21/2	971/2	mean	21/2	971/2			mean	21/2	971/2		
<1	0.0054	0.0054	0.0038	0.0070	100,000	100,000	100,000	535	99,518	78.6	77.5	80.2	16	3,037
1-4	0.0004	0.0017	0.0013	0.0021	99,465	99,305	99,623	167	397,459	78.0	76.9	79.6	6	13,684
5-9	0.0001	0.0003	0.0002	0.0004	99,298	99,099	99,496	30	496,415	74.2	73.0	75.8	1	17,801
10-14	0.0001	0.0006	0.0005	0.0008	99,268	99,063	99,474	61	496,188	69.2	68.1	70.8	2	16,235
15-19	0.0004	0.0019	0.0014	0.0023	99,207	98,988	99,427	185	495,573	64.2	63.1	65.8	5	14,435
20-24	0.0008	0.0041	0.0032	0.0049	99,022	98,764	99,284	408	494,091	59.3	58.2	60.9	10	12,673
25-29	0.0006	0.0029	0.0024	0.0035	98,614	98,275	98,961	289	492,346	54.6	53.5	56.2	7	11,361
30-34	0.0008	0.0040	0.0033	0.0048	98,325	97,934	98,728	398	490,629	49.7	48.6	51.3	11	13,142
35-39	0.0011	0.0056	0.0046	0.0066	97,927	97,468	98,404	552	488,255	44.9	43.8	46.5	15	13,131
40-44	0.0018	0.0087	0.0073	0.0101	97,375	96,830	97,947	849	484,754	40.2	39.1	41.7	22	12,369
45-49	0.0028	0.0139	0.0119	0.0159	96,526	95,862	97,230	1,339	479,285	35.5	34.4	37.0	27	9,516
50-54	0.0043	0.0213	0.0186	0.0241	95,188	94,347	96,073	2,031	470,862	31.0	29.9	32.5	27	6,163
55-59	0.0073	0.0357	0.0308	0.0400	93,157	92,086	94,271	3,324	457,476	26.6	25.6	28.0	26	3,607
60-64	0.0111	0.0538	0.0460	0.0605	89,833	88,450	91,249	4,835	437,080	22.5	21.6	23.8	33	2,978
65-69	0.0184	0.0881	0.0748	0.0990	84,999	83,212	86,771	7,490	406,268	18.6	17.8	19.9	43	2,321
70-74	0.0277	0.1293	0.1097	0.1447	77,509	75,200	80,213	10,024	362,484	15.2	14.4	16.3	38	1,368
75-79	0.0412	0.1867	0.1597	0.2066	67,485	64,340	71,415	12,597	305,932	12.1	11.4	13.0	37	890
80-84	0.0668	0.2861	0.2503	0.3145	54,888	51,045	60,008	15,705	235,179	9.2	8.7	10.0	30	450
85-89	0.1034	0.4109	0.3713	0.4465	39,184	35,183	44,989	16,100	155,667	6.9	6.5	7.5	23	223
90-94	0.1530	0.5243	0.4840	0.5600	23,083	19,853	28,283	12,102	79,111	5.0	4.8	5.4	14	94
95-99	0.2624	0.7341	0.6999	0.7638	10,981	8,831	14,441	8,061	30,722	3.4	3.3	3.6	5	20
100+	0.4355	1	1	1	2,920	2,097	4,203	2,920	6,705	2.3	2.2	2.5	1	2
All ages													*398*	*155,500*

$\ln(_nq_x)$ with uncertainty interval

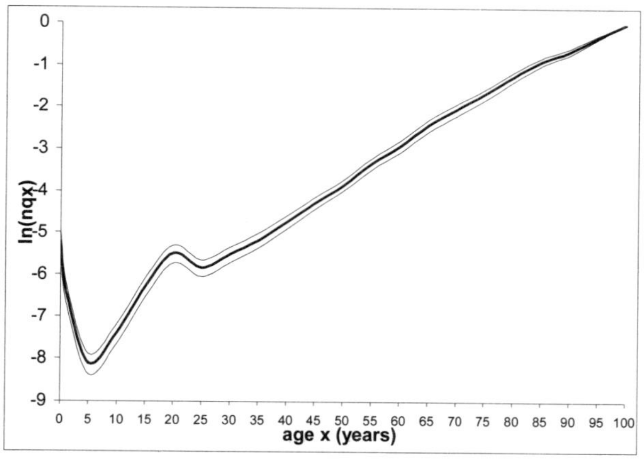

l_x with uncertainty interval

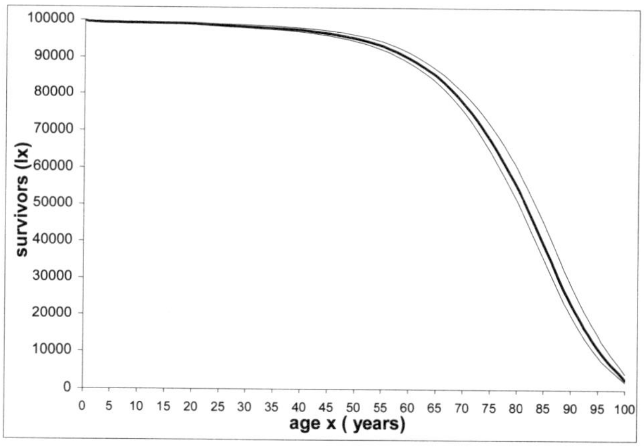

e_x with uncertainty interval

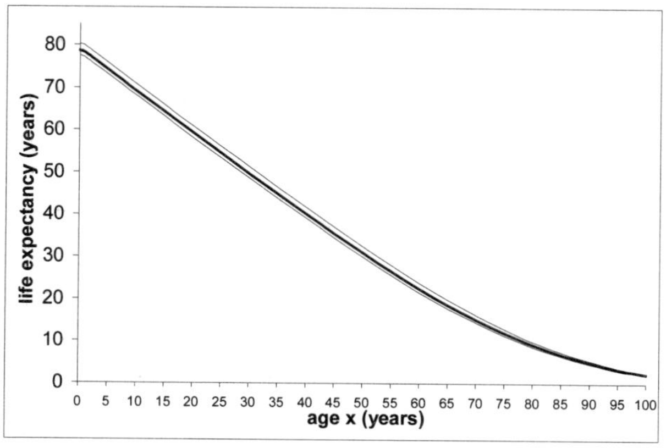

Distribution of deaths by age with uncertainty interval

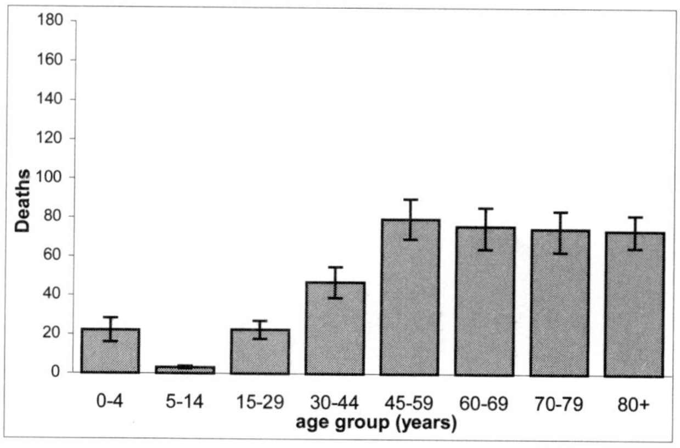

Bulgaria Bulgarie Bulgaria

Males, 2000 *Hommes, 2000* *Hombres, 2000*

Age	$_nM_x$	$_nq_x$			l_x			$_nd_x$	$_nL_x$	e_x			Actual deaths	Actual population
			percentiles			percentiles					percentiles			
		mean	2½	97½	mean	2½	97½			mean	2½	97½		
<1	0.0180	0.0177	0.0156	0.0203	100,000	100,000	100,000	1,771	98,406	67.4	67.0	67.7	581	32,286
1-4	0.0010	0.0040	0.0036	0.0044	98,229	97,972	98,441	392	391,974	67.6	67.2	68.0	129	129,499
5-9	0.0004	0.0021	0.0020	0.0024	97,837	97,542	98,084	210	488,658	63.8	63.5	64.2	93	216,604
10-14	0.0004	0.0019	0.0018	0.0021	97,627	97,312	97,891	190	487,658	59.0	58.6	59.4	102	262,058
15-19	0.0009	0.0045	0.0041	0.0049	97,436	97,105	97,717	437	486,088	54.1	53.7	54.5	254	281,936
20-24	0.0013	0.0062	0.0058	0.0067	96,999	96,631	97,313	604	483,484	49.3	49.0	49.7	391	312,870
25-29	0.0014	0.0067	0.0063	0.0072	96,395	95,981	96,752	648	480,352	44.6	44.3	45.0	399	295,658
30-34	0.0020	0.0100	0.0093	0.0106	95,746	95,288	96,147	953	476,349	39.9	39.5	40.3	540	269,994
35-39	0.0030	0.0149	0.0140	0.0158	94,793	94,279	95,252	1,411	470,439	35.3	34.9	35.7	808	269,459
40-44	0.0050	0.0247	0.0234	0.0259	93,382	92,797	93,917	2,306	461,146	30.8	30.4	31.2	1,370	273,969
45-49	0.0080	0.0392	0.0376	0.0408	91,076	90,405	91,712	3,572	446,453	26.5	26.1	26.9	2,248	280,948
50-54	0.0130	0.0630	0.0607	0.0652	87,505	86,753	88,253	5,509	423,752	22.4	22.1	22.8	3,557	273,591
55-59	0.0200	0.0952	0.0915	0.0985	81,996	81,145	82,862	7,809	390,457	18.8	18.5	19.1	4,363	218,160
60-64	0.0290	0.1352	0.1296	0.1401	74,187	73,301	75,116	10,030	345,859	15.5	15.2	15.8	5,888	203,025
65-69	0.0400	0.1818	0.1744	0.1886	64,157	63,166	65,155	11,665	291,622	12.5	12.2	12.8	8,215	205,379
70-74	0.0620	0.2684	0.2579	0.2788	52,492	51,367	53,785	14,089	227,238	9.8	9.5	10.0	10,220	164,842
75-79	0.1000	0.4000	0.3857	0.4142	38,403	37,170	39,882	15,361	153,613	7.4	7.2	7.6	11,103	111,030
80-84	0.1460	0.5348	0.5189	0.5499	23,042	21,870	24,401	12,323	84,403	5.7	5.6	5.9	5,388	36,904
85-89	0.2100	0.6885	0.6737	0.7019	10,719	9,869	11,660	7,380	35,145	4.4	4.3	4.5	4,368	20,799
90-94	0.2590	0.7288	0.7154	0.7404	3,339	2,942	3,783	2,433	9,394	3.6	3.5	3.7	1,129	4,361
95-99	0.3420	0.8440	0.8335	0.8529	906	764	1,074	764	2,235	2.8	2.8	2.9	133	389
100+	0.4690	1	1	1	141	112	178	141	301	2.1	2.1	2.2	9	20
All ages													*61,289*	*3,863,781*

$ln(_nq_x)$ with uncertainty interval

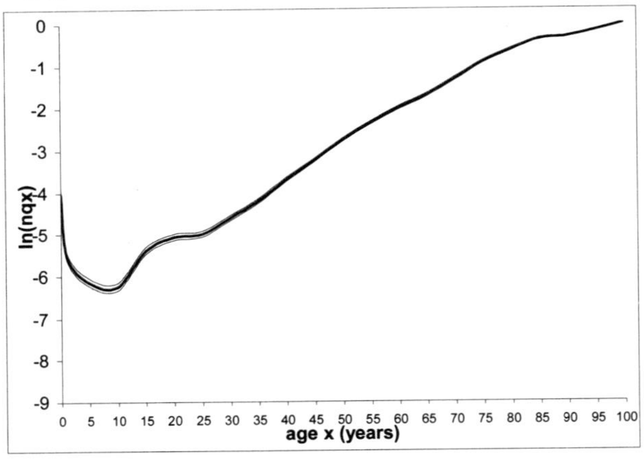

l_x with uncertainty interval

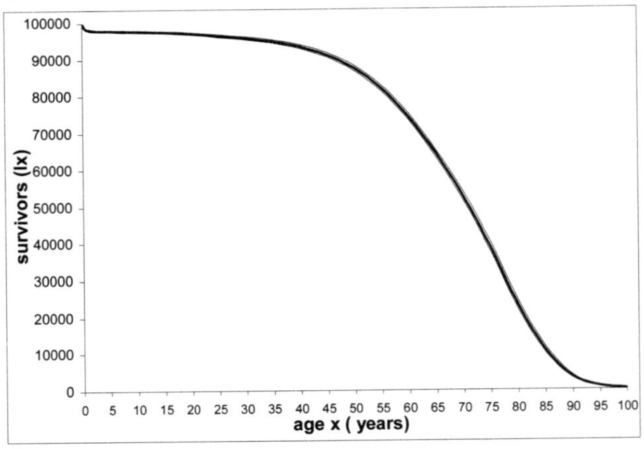

e_x with uncertainty interval

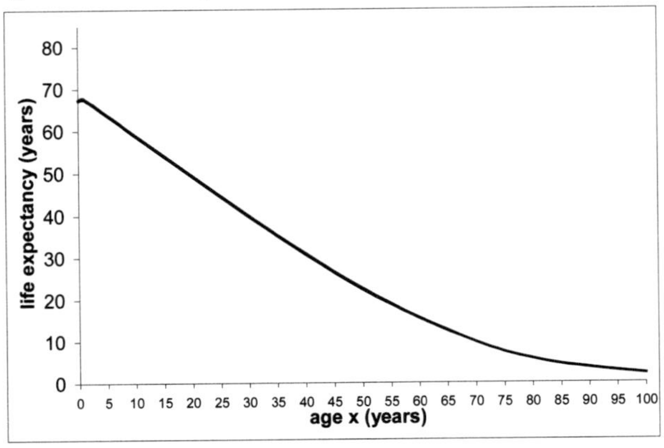

Distribution of deaths by age with uncertainty interval

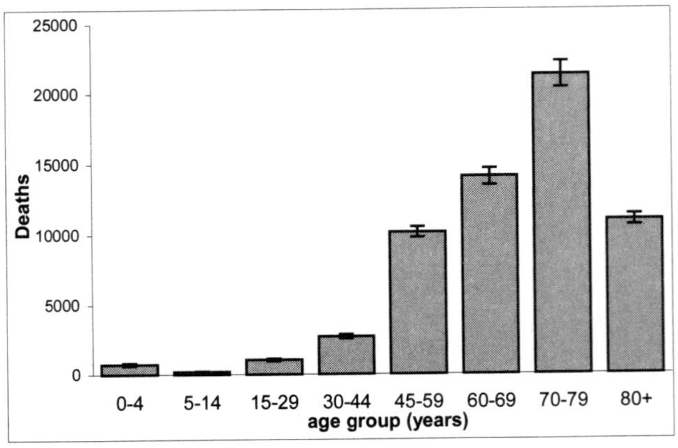

Bulgaria **Bulgarie** **Bulgaria**

Females, 2000 *Femmes, 2000* *Mujeres, 2000*

Age	$_nM_x$	$_nq_x$ mean	$_nq_x$ percentiles 2½	$_nq_x$ percentiles 97½	l_x mean	l_x percentiles 2½	l_x percentiles 97½	$_nd_x$	$_nL_x$	e_x mean	e_x percentiles 2½	e_x percentiles 97½	Actual deaths	Actual population
<1	0.0130	0.0128	0.0113	0.0147	100,000	100,000	100,000	1,285	98,844	74.9	74.5	75.4	398	30,603
1-4	0.0010	0.0040	0.0036	0.0044	98,715	98,527	98,870	394	393,915	74.9	74.5	75.4	124	124,302
5-9	0.0003	0.0014	0.0013	0.0016	98,321	98,092	98,512	142	491,249	71.2	70.8	71.6	60	205,945
10-14	0.0003	0.0015	0.0014	0.0016	98,179	97,936	98,381	147	490,525	66.3	65.9	66.7	75	250,604
15-19	0.0004	0.0021	0.0020	0.0023	98,031	97,775	98,247	211	489,631	61.4	61.0	61.8	115	266,819
20-24	0.0005	0.0023	0.0021	0.0025	97,821	97,546	98,053	225	488,543	56.5	56.1	57.0	138	299,114
25-29	0.0006	0.0030	0.0028	0.0032	97,596	97,302	97,846	292	487,250	51.6	51.3	52.1	172	286,340
30-34	0.0010	0.0048	0.0044	0.0052	97,304	96,986	97,576	466	485,355	46.8	46.4	47.2	254	264,384
35-39	0.0013	0.0066	0.0061	0.0071	96,838	96,485	97,143	637	482,597	42.0	41.6	42.4	355	268,759
40-44	0.0020	0.0100	0.0093	0.0106	96,201	95,805	96,548	957	478,611	37.3	36.9	37.7	563	281,375
45-49	0.0030	0.0149	0.0141	0.0157	95,244	94,790	95,647	1,418	472,673	32.6	32.3	33.0	884	294,535
50-54	0.0050	0.0247	0.0235	0.0259	93,826	93,301	94,302	2,317	463,337	28.1	27.7	28.5	1,472	294,384
55-59	0.0080	0.0392	0.0373	0.0411	91,509	90,898	92,080	3,589	448,573	23.7	23.4	24.1	1,939	242,340
60-64	0.0120	0.0583	0.0553	0.0610	87,920	87,228	88,595	5,122	426,798	19.6	19.3	19.9	2,848	237,368
65-69	0.0200	0.0952	0.0903	0.0995	82,799	82,021	83,579	7,886	394,280	15.6	15.4	16.0	5,015	250,741
70-74	0.0360	0.1651	0.1568	0.1716	74,913	73,870	75,971	12,371	343,639	12.0	11.8	12.3	7,786	216,289
75-79	0.0690	0.2942	0.2816	0.3040	62,542	61,194	64,032	18,403	266,705	8.9	8.7	9.1	11,495	166,598
80-84	0.1150	0.4466	0.4324	0.4607	44,140	42,660	46,002	19,713	171,416	6.5	6.4	6.7	6,718	58,417
85-89	0.1830	0.6278	0.6125	0.6422	24,427	23,266	26,098	15,335	83,797	4.8	4.7	4.9	6,615	36,146
90-94	0.2450	0.7061	0.6926	0.7181	9,092	8,367	10,007	6,419	26,202	3.7	3.6	3.8	2,266	9,248
95-99	0.3390	0.8404	0.8302	0.8490	2,673	2,359	3,025	2,246	6,625	2.8	2.8	2.9	392	1,156
100+	0.4760	1	1	1	427	356	512	427	896	2.1	2.1	2.3	40	83
All ages													*49,721*	*4,085,550*

ln($_nq_x$) with uncertainty interval

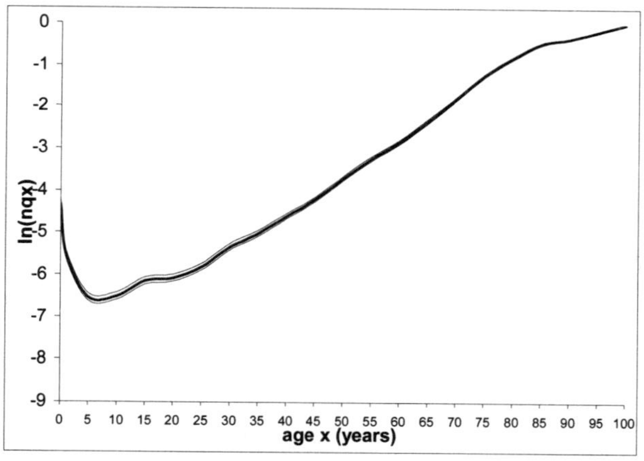

l_x with uncertainty interval

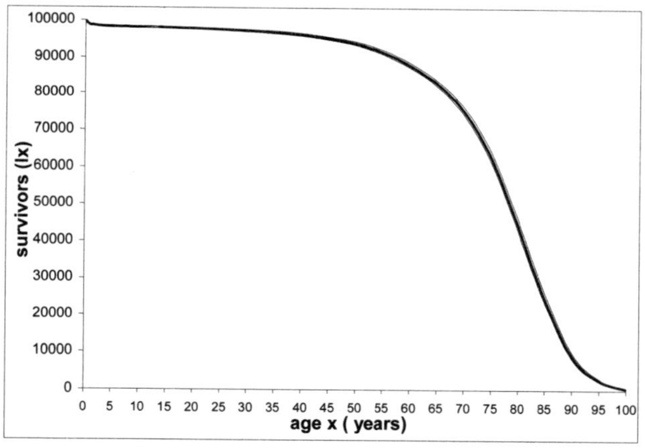

e_x with uncertainty interval

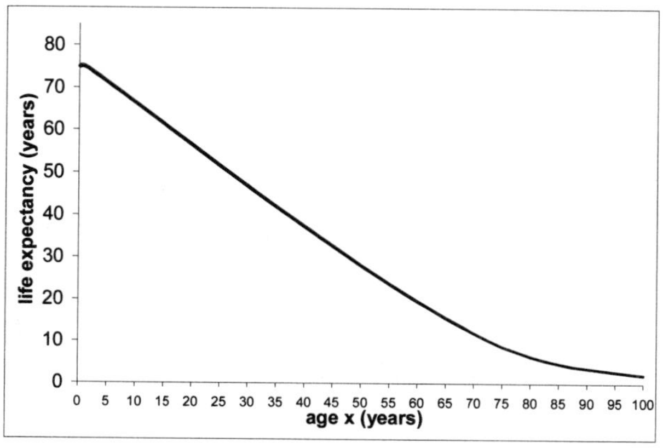

Distribution of deaths by age with uncertainty interval

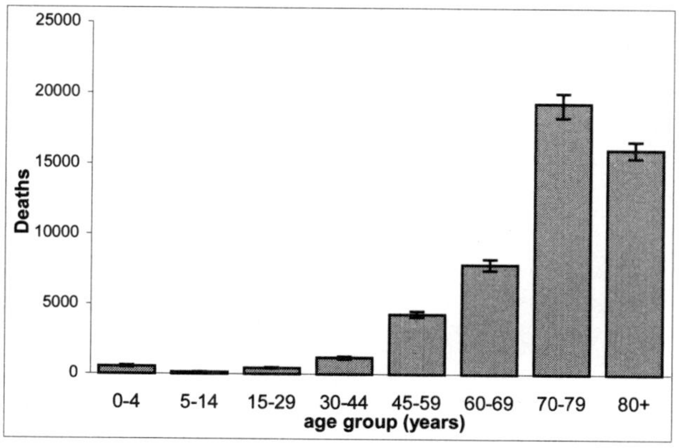

Burkina Faso

Burkina Faso

Burkina Faso

Males, 2000 — *Hommes, 2000* — *Hombres, 2000*

Age	$_nM_x$	$_nq_x$			l_x			$_nd_x$	$_nL_x$	e_x			Actual deaths	Actual population
			percentiles			percentiles					percentiles			
		mean	2½	97½	mean	2½	97½			mean	2½	97½		
<1	0.1910	0.1630	0.1561	0.1696	100,000	100,000	100,000	16,296	85,334	42.6	41.2	44.1	48,388	253,381
1-4	0.0169	0.0650	0.0608	0.0694	83,704	83,036	84,393	5,442	321,754	49.9	48.2	51.8	14,556	860,608
5-9	0.0035	0.0175	0.0162	0.0188	78,262	77,764	78,761	1,366	387,893	49.3	47.7	51.1	3,243	920,743
10-14	0.0021	0.0103	0.0095	0.0111	76,895	76,314	77,433	793	382,494	45.1	43.6	46.9	1,643	792,326
15-19	0.0018	0.0092	0.0085	0.0100	76,102	75,467	76,668	700	378,761	40.5	39.0	42.3	1,243	672,327
20-24	0.0058	0.0284	0.0261	0.0308	75,402	74,719	75,997	2,142	371,655	35.9	34.4	37.6	2,923	507,109
25-29	0.0121	0.0587	0.0536	0.0638	73,260	72,429	73,990	4,298	355,555	31.9	30.4	33.6	4,256	352,057
30-34	0.0189	0.0901	0.0817	0.0984	68,962	67,817	70,014	6,212	329,279	28.7	27.3	30.3	5,018	266,006
35-39	0.0216	0.1025	0.0923	0.1124	62,750	61,144	64,270	6,435	297,663	26.3	25.0	27.8	4,870	225,278
40-44	0.0232	0.1094	0.0980	0.1202	56,315	54,270	58,297	6,162	266,170	24.0	22.8	25.4	3,871	167,187
45-49	0.0247	0.1165	0.1041	0.1282	50,153	47,746	52,532	5,841	236,161	21.6	20.6	22.9	2,992	120,960
50-54	0.0264	0.1240	0.1107	0.1366	44,312	41,636	47,012	5,494	207,823	19.2	18.2	20.3	3,068	116,045
55-59	0.0290	0.1352	0.1208	0.1488	38,818	35,972	41,808	5,247	180,971	16.5	15.7	17.5	2,761	95,224
60-64	0.0379	0.1731	0.1553	0.1900	33,571	30,650	36,757	5,811	153,325	13.7	13.0	14.5	2,798	73,830
65-69	0.0525	0.2319	0.2096	0.2529	27,759	24,871	31,047	6,436	122,706	11.1	10.5	11.7	3,169	60,412
70-74	0.0783	0.3273	0.2993	0.3532	21,323	18,638	24,541	6,978	89,170	8.6	8.2	9.2	3,532	45,132
75-79	0.1165	0.4511	0.4192	0.4802	14,345	12,116	17,196	6,471	55,548	6.6	6.3	7.0	3,406	29,243
80-84	0.1708	0.5985	0.5665	0.6272	7,874	6,309	9,988	4,713	27,590	5.0	4.8	5.3	2,325	13,610
85-89	0.2440	0.7578	0.7306	0.7818	3,162	2,352	4,330	2,396	9,818	3.8	3.6	4.0	969	3,969
90-94	0.3393	0.8408	0.8189	0.8599	766	513	1,166	644	1,897	2.8	2.7	2.9	197	582
95-99	0.4594	0.9659	0.9570	0.9729	122	72	211	118	256	2.2	2.1	2.2	23	49
100+	0.6053	1	1	1	4	2	9	4	7	1.7	1.6	1.7	1	2
All ages													*115,251*	*5,576,080*

ln($_nq_x$) with uncertainty interval

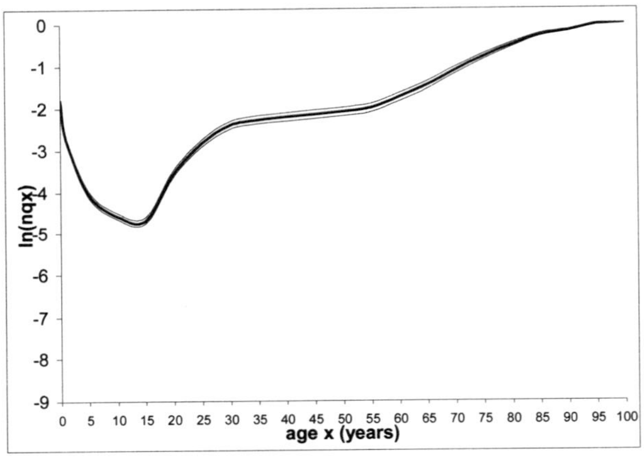

l_x with uncertainty interval

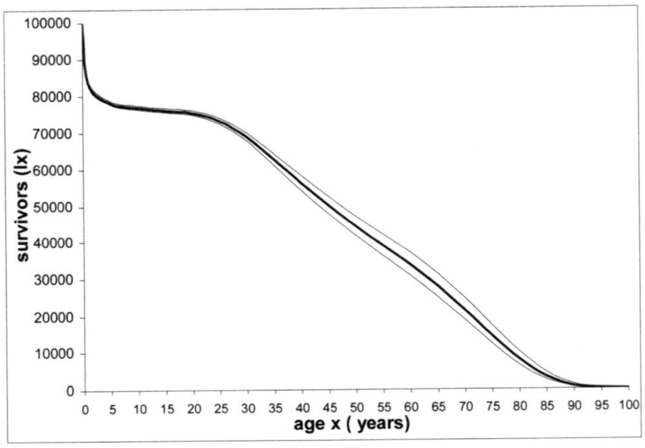

e_x with uncertainty interval

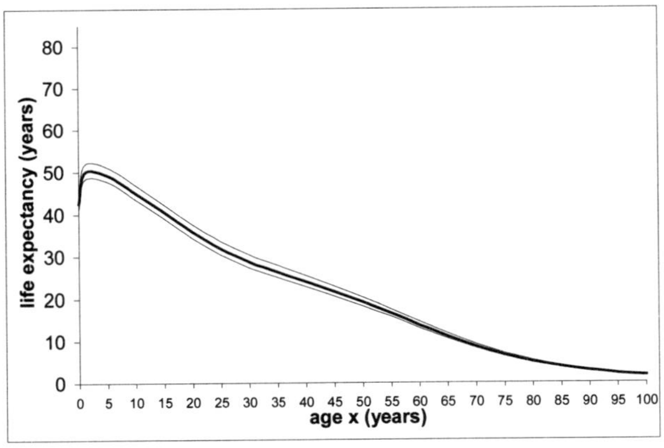

Distribution of deaths by age with uncertainty interval

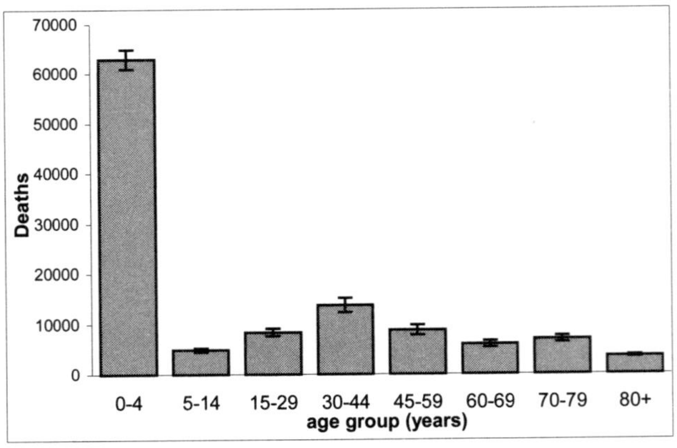

Burkina Faso **Burkina Faso** **Burkina Faso**

Females, 2000 *Femmes, 2000* *Mujeres, 2000*

Age	$_nM_x$	$_nq_x$			l_x			$_nd_x$	$_nL_x$	e_x			Actual deaths	Actual population
			percentiles			percentiles					percentiles			
		mean	2½	97½	mean	2½	97½			mean	2½	97½		
<1	0.1661	0.1445	0.1384	0.1506	100,000	100,000	100,000	14,447	86,998	43.6	42.3	45.2	40,322	242,818
1-4	0.0189	0.0722	0.0679	0.0759	85,553	84,942	86,165	6,181	327,378	49.9	48.3	51.9	16,105	852,998
5-9	0.0038	0.0190	0.0176	0.0201	79,372	78,872	79,866	1,511	393,084	49.7	48.1	51.5	3,491	908,341
10-14	0.0025	0.0123	0.0113	0.0130	77,861	77,298	78,421	954	386,921	45.6	44.0	47.4	1,937	785,409
15-19	0.0050	0.0245	0.0225	0.0260	76,907	76,294	77,518	1,882	379,830	41.1	39.6	42.9	3,343	674,540
20-24	0.0110	0.0536	0.0492	0.0573	75,025	74,311	75,736	4,022	365,071	37.1	35.6	38.8	6,000	544,624
25-29	0.0173	0.0829	0.0756	0.0892	71,003	70,075	71,932	5,886	340,302	34.0	32.6	35.7	7,307	422,449
30-34	0.0203	0.0968	0.0877	0.1049	65,117	63,884	66,468	6,302	309,831	31.9	30.5	33.4	6,578	323,394
35-39	0.0188	0.0899	0.0811	0.0979	58,815	57,282	60,638	5,288	280,855	30.0	28.8	31.4	4,877	259,030
40-44	0.0164	0.0789	0.0710	0.0863	53,527	51,775	55,720	4,225	257,073	27.7	26.7	29.0	3,231	196,588
45-49	0.0152	0.0731	0.0657	0.0801	49,302	47,404	51,764	3,606	237,495	24.9	24.0	26.0	2,352	154,916
50-54	0.0170	0.0813	0.0731	0.0890	45,696	43,629	48,361	3,715	219,191	21.7	20.8	22.6	2,325	137,165
55-59	0.0205	0.0974	0.0876	0.1066	41,981	39,745	44,828	4,089	199,680	18.4	17.6	19.2	2,567	125,338
60-64	0.0290	0.1350	0.1219	0.1474	37,892	35,509	40,899	5,116	176,668	15.1	14.4	15.8	3,175	109,624
65-69	0.0433	0.1953	0.1774	0.2121	32,776	30,276	35,913	6,402	147,874	12.0	11.5	12.7	3,811	88,037
70-74	0.0678	0.2899	0.2661	0.3120	26,374	23,855	29,542	7,647	112,754	9.4	8.9	9.9	4,556	67,183
75-79	0.1030	0.4097	0.3815	0.4356	18,728	16,412	21,681	7,673	74,456	7.2	6.8	7.6	4,136	40,133
80-84	0.1541	0.5561	0.5262	0.5828	11,055	9,265	13,409	6,148	39,906	5.4	5.2	5.7	2,971	19,284
85-89	0.2242	0.7183	0.6916	0.7416	4,907	3,870	6,343	3,525	15,725	4.0	3.9	4.2	1,325	5,910
90-94	0.3173	0.8128	0.7909	0.8316	1,383	1,002	1,948	1,124	3,542	3.0	2.9	3.1	344	1,083
95-99	0.4370	0.9455	0.9344	0.9544	259	169	405	245	560	2.3	2.2	2.3	53	121
100+	0.5854	1	1	1	14	8	26	14	24	1.7	1.7	1.8	4	7
All ages													*120,808*	*5,958,992*

ln($_nq_x$) with uncertainty interval

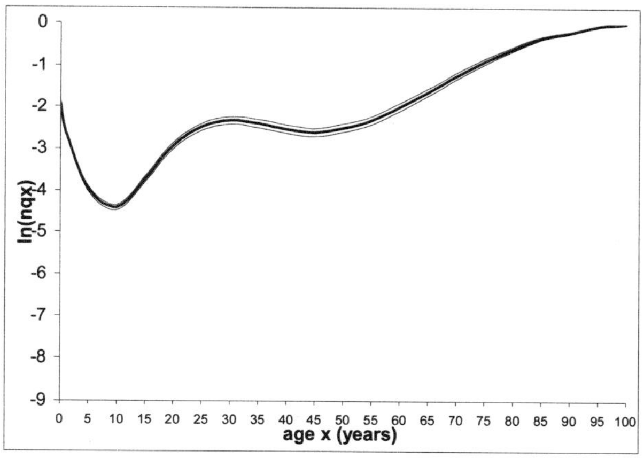

l_x with uncertainty interval

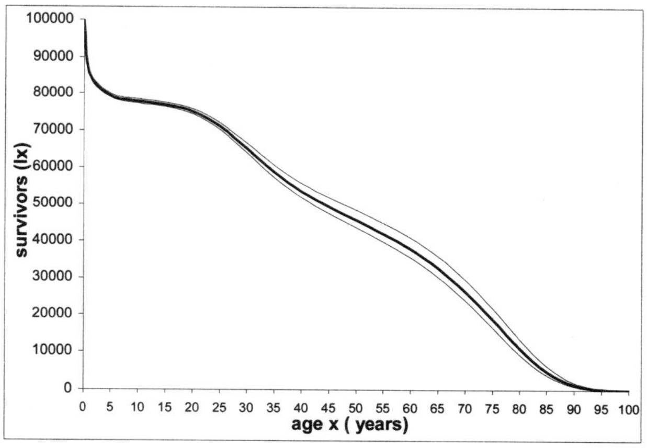

e_x with uncertainty interval

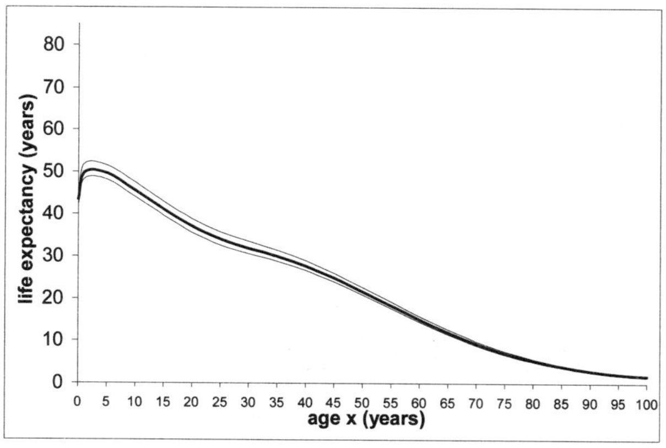

Distribution of deaths by age with uncertainty interval

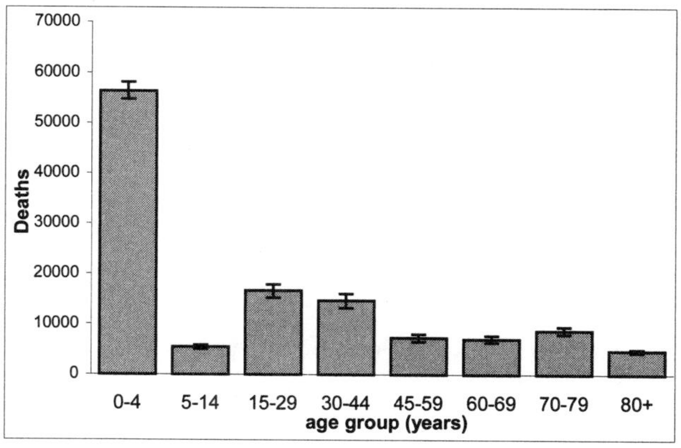

Burundi **Burundi** **Burundi**

Males, 2000 *Hommes, 2000* *Hombres, 2000*

Age	$_nM_x$	$_nq_x$			l_x			$_nd_x$	$_nL_x$	e_x			Actual deaths	Actual population
			percentiles			percentiles					percentiles			
		mean	2½	97½	mean	2½	97½			mean	2½	97½		
<1	0.1648	0.1435	0.1147	0.1715	100,000	100,000	100,000	14,350	87,085	40.6	35.4	46.0	20,693	125,583
1-4	0.0161	0.0618	0.0487	0.0785	85,650	82,854	88,528	5,295	329,894	46.4	40.6	52.2	6,951	433,100
5-9	0.0033	0.0166	0.0128	0.0218	80,356	77,909	82,854	1,333	398,446	45.4	40.0	50.8	1,659	495,955
10-14	0.0021	0.0105	0.0081	0.0140	79,023	76,435	81,652	833	393,032	41.1	35.8	46.5	972	458,624
15-19	0.0029	0.0142	0.0109	0.0190	78,190	75,381	80,946	1,114	388,166	36.5	31.3	41.8	1,076	375,095
20-24	0.0077	0.0379	0.0287	0.0509	77,076	73,969	80,067	2,920	378,081	32.0	26.8	37.3	2,086	270,095
25-29	0.0165	0.0792	0.0596	0.1075	74,156	70,248	77,766	5,876	356,091	28.1	23.1	33.3	3,164	191,758
30-34	0.0263	0.1234	0.0913	0.1687	68,280	62,698	73,128	8,427	320,335	25.3	20.4	30.3	4,114	156,377
35-39	0.0296	0.1376	0.1005	0.1886	59,853	52,122	66,384	8,236	278,677	23.6	18.8	28.0	4,170	141,095
40-44	0.0307	0.1425	0.1035	0.1983	51,618	42,294	59,587	7,354	239,703	21.9	17.6	25.9	3,850	125,502
45-49	0.0315	0.1458	0.1061	0.2043	44,264	34,079	53,259	6,455	205,180	20.2	16.4	23.6	3,200	101,705
50-54	0.0315	0.1459	0.1067	0.2043	37,808	27,374	47,433	5,515	175,255	18.2	14.9	21.2	1,965	62,437
55-59	0.0317	0.1468	0.1081	0.2046	32,293	22,052	42,292	4,740	149,618	15.8	13.1	18.4	1,482	46,770
60-64	0.0423	0.1913	0.1428	0.2615	27,554	17,795	37,719	5,270	124,594	13.1	10.8	15.4	1,522	35,994
65-69	0.0569	0.2490	0.1893	0.3315	22,284	13,420	32,313	5,548	97,549	10.6	8.8	12.5	1,561	27,453
70-74	0.0828	0.3431	0.2676	0.4396	16,736	9,181	26,142	5,743	69,322	8.3	6.9	9.9	1,658	20,010
75-79	0.1214	0.4657	0.3768	0.5693	10,993	5,145	19,019	5,120	42,166	6.4	5.3	7.6	1,550	12,762
80-84	0.1805	0.6220	0.5283	0.7209	5,873	2,216	11,673	3,653	20,234	4.8	4.1	5.6	1,076	5,958
85-89	0.2596	0.7872	0.7057	0.8640	2,220	618	5,339	1,748	6,732	3.6	3.2	4.1	444	1,709
90-94	0.3604	0.8658	0.7999	0.9236	472	84	1,529	409	1,135	2.7	2.4	3.0	86	239
95-99	0.4832	0.9863	0.9683	0.9958	63	6	303	63	129	2.1	2.0	2.1	9	19
100+	0.6252	1	1	1	1	0	10	1	1	1.6	1.6	1.7	1	1
All ages													*63,288*	*3,088,241*

ln($_nq_x$) with uncertainty interval

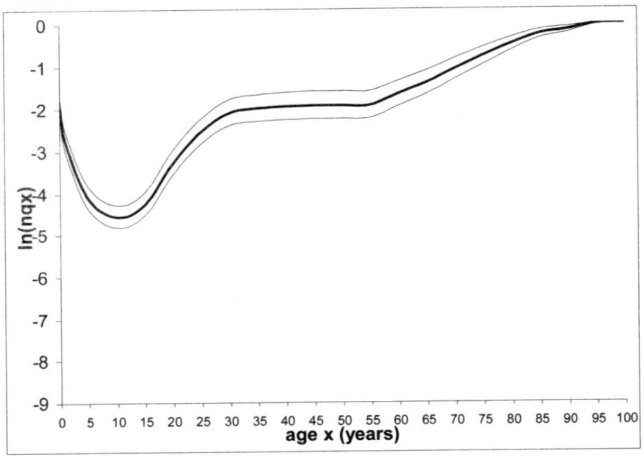

l_x with uncertainty interval

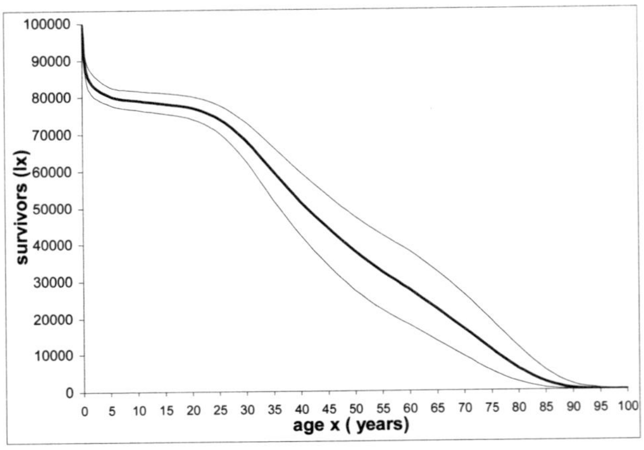

e_x with uncertainty interval

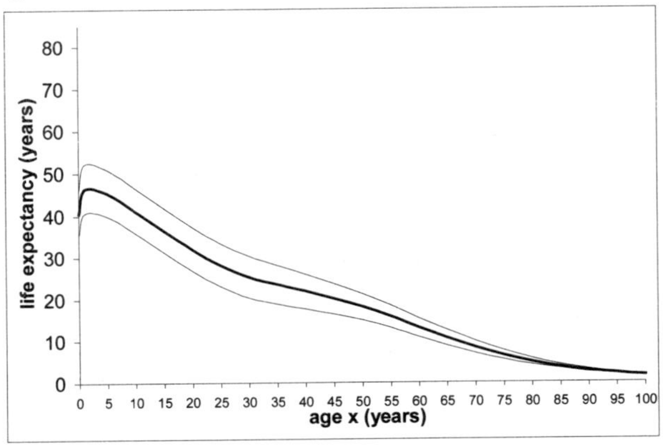

Distribution of deaths by age with uncertainty interval

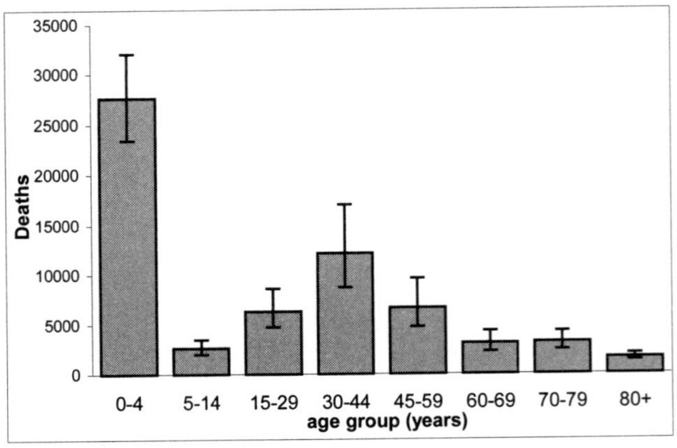

Burundi | **Burundi** | **Burundi**

Females, 2000 | *Femmes, 2000* | *Mujeres, 2000*

Age	$_nM_x$	$_nq_x$			l_x			$_nd_x$	$_nL_x$	e_x			Actual deaths	Actual population
			percentiles			percentiles					percentiles			
		mean	2½	97½	mean	2½	97½			mean	2½	97½		
<1	0.1391	0.1236	0.0988	0.1500	100,000	100,000	100,000	12,364	88,873	41.3	37.1	46.3	16,783	120,637
1-4	0.0176	0.0677	0.0554	0.0801	87,636	85,003	90,125	5,936	336,299	46.2	41.3	51.8	7,665	434,266
5-9	0.0037	0.0182	0.0146	0.0220	81,701	79,205	84,197	1,484	404,793	45.4	40.8	50.9	1,819	496,211
10-14	0.0025	0.0126	0.0101	0.0153	80,217	77,543	82,917	1,007	398,566	41.2	36.6	46.6	1,159	458,894
15-19	0.0062	0.0307	0.0245	0.0377	79,210	76,384	82,076	2,432	389,969	36.7	32.1	42.0	2,357	378,033
20-24	0.0147	0.0711	0.0558	0.0884	76,778	73,590	80,055	5,456	370,249	32.8	28.3	38.0	4,085	277,184
25-29	0.0242	0.1139	0.0877	0.1434	71,322	67,380	75,491	8,123	336,300	30.1	25.6	35.1	4,866	201,448
30-34	0.0289	0.1348	0.1021	0.1710	63,199	58,215	68,583	8,520	294,693	28.6	24.3	33.4	4,818	166,665
35-39	0.0260	0.1219	0.0914	0.1569	54,679	48,359	61,123	6,665	256,730	27.7	23.8	32.0	3,969	152,860
40-44	0.0225	0.1063	0.0793	0.1386	48,013	40,853	55,194	5,104	227,306	26.2	22.8	30.0	3,138	139,732
45-49	0.0196	0.0933	0.0692	0.1224	42,909	35,325	50,816	4,002	204,539	24.0	21.0	27.3	2,332	119,151
50-54	0.0202	0.0962	0.0711	0.1264	38,907	31,142	47,283	3,743	185,174	21.2	18.6	24.2	1,699	84,057
55-59	0.0225	0.1066	0.0788	0.1397	35,163	27,359	43,883	3,749	166,443	18.2	15.9	20.8	1,552	68,886
60-64	0.0300	0.1396	0.1038	0.1813	31,414	23,704	40,368	4,386	146,105	15.1	13.1	17.4	1,683	56,072
65-69	0.0434	0.1956	0.1471	0.2499	27,028	19,610	36,091	5,286	121,926	12.1	10.5	14.1	1,922	44,343
70-74	0.0668	0.2862	0.2201	0.3561	21,742	14,951	30,642	6,223	93,152	9.5	8.1	11.1	2,192	32,815
75-79	0.1007	0.4021	0.3199	0.4828	15,519	9,727	23,684	6,241	61,992	7.3	6.3	8.6	2,157	21,423
80-84	0.1511	0.5485	0.4570	0.6311	9,278	5,031	15,842	5,089	33,669	5.5	4.8	6.4	1,646	10,893
85-89	0.2208	0.7114	0.6251	0.7833	4,189	1,856	8,522	2,980	13,496	4.1	3.6	4.7	802	3,633
90-94	0.3137	0.8081	0.7353	0.8660	1,209	402	3,195	977	3,114	3.0	2.7	3.5	226	721
95-99	0.4334	0.9421	0.9013	0.9685	232	54	846	219	504	2.3	2.1	2.5	36	82
100+	0.5823	1	1	1	13	2	83	13	23	1.7	1.7	1.9	3	5
All ages													*66,909*	*3,268,011*

ln($_nq_x$) with uncertainty interval

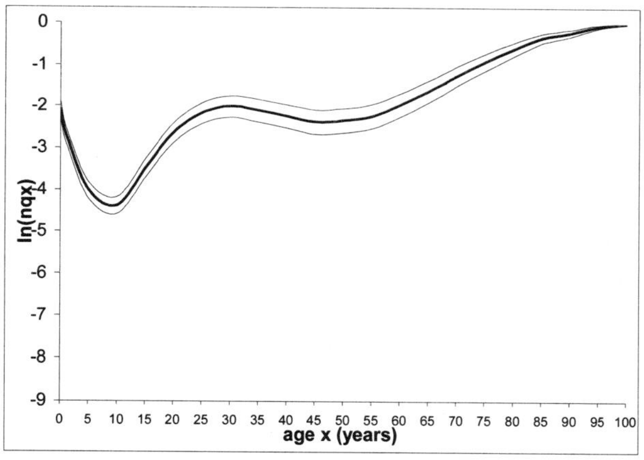

l_x with uncertainty interval

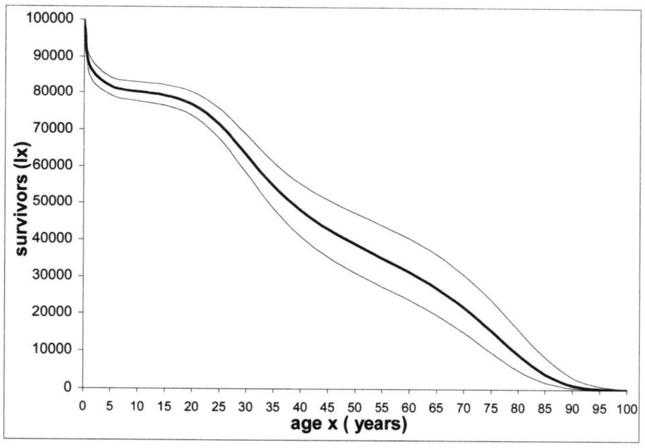

e_x with uncertainty interval

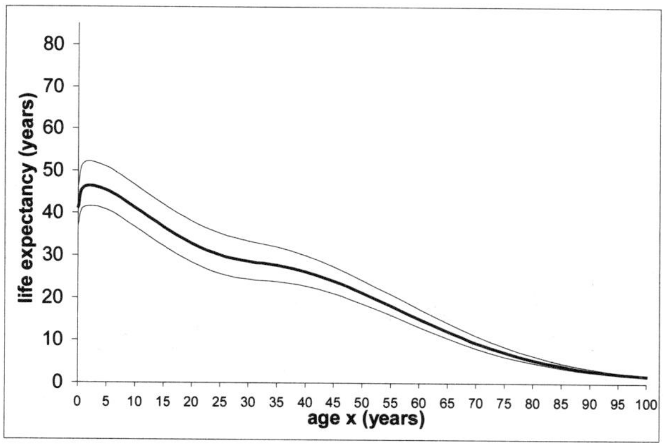

Distribution of deaths by age with uncertainty interval

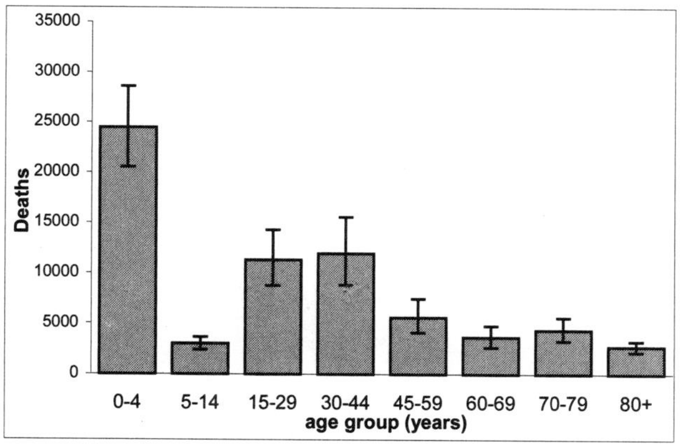

Cambodia

Cambodge

Camboya

Males, 2000

Hommes, 2000

Hombres, 2000

Age	$_nM_x$	$_nq_x$			l_x			$_nd_x$	$_nL_x$	e_x			Actual deaths	Actual population
			percentiles			percentiles					percentiles			
		mean	2½	97½	mean	2½	97½			mean	2½	97½		
<1	0.1162	0.1052	0.0999	0.1108	100,000	100,000	100,000	10,523	90,529	53.4	51.6	55.0	26,086	224,412
1-4	0.0089	0.0347	0.0323	0.0377	89,477	88,922	90,010	3,109	350,446	58.7	56.8	60.4	7,501	845,630
5-9	0.0022	0.0109	0.0100	0.0119	86,368	85,872	86,863	940	429,491	56.7	55.0	58.3	2,102	960,750
10-14	0.0014	0.0071	0.0066	0.0078	85,428	84,876	85,966	609	425,619	52.3	50.6	53.9	1,266	884,808
15-19	0.0016	0.0081	0.0074	0.0089	84,819	84,226	85,385	688	422,378	47.7	46.0	49.2	1,374	844,096
20-24	0.0044	0.0217	0.0199	0.0240	84,132	83,481	84,739	1,827	416,092	43.0	41.4	44.6	1,701	387,363
25-29	0.0082	0.0402	0.0366	0.0447	82,305	81,479	83,023	3,310	403,249	38.9	37.3	40.4	3,849	468,834
30-34	0.0090	0.0441	0.0399	0.0494	78,995	77,839	79,946	3,486	386,257	35.5	34.0	36.9	3,757	416,286
35-39	0.0098	0.0476	0.0429	0.0536	75,508	73,997	76,755	3,598	368,547	32.0	30.6	33.3	3,433	351,647
40-44	0.0098	0.0479	0.0429	0.0540	71,910	70,034	73,465	3,444	350,943	28.5	27.2	29.7	2,765	281,833
45-49	0.0118	0.0572	0.0511	0.0647	68,467	66,250	70,312	3,916	332,544	24.8	23.6	25.9	2,619	222,391
50-54	0.0161	0.0773	0.0690	0.0875	64,551	61,966	66,717	4,992	310,273	21.1	20.0	22.1	2,741	170,388
55-59	0.0225	0.1066	0.0952	0.1204	59,559	56,544	62,111	6,351	281,916	17.7	16.7	18.6	2,832	125,698
60-64	0.0324	0.1499	0.1343	0.1686	53,208	49,733	56,196	7,978	246,095	14.5	13.6	15.3	2,508	77,362
65-69	0.0472	0.2109	0.1903	0.2357	45,230	41,351	48,648	9,540	202,299	11.6	10.9	12.3	2,598	55,093
70-74	0.0718	0.3042	0.2768	0.3358	35,690	31,637	39,392	10,857	151,306	9.1	8.5	9.6	2,648	36,909
75-79	0.1084	0.4265	0.3937	0.4622	24,832	21,083	28,444	10,591	97,686	6.9	6.5	7.4	2,122	19,572
80-84	0.1621	0.5767	0.5431	0.6113	14,242	11,421	17,158	8,214	50,675	5.2	4.9	5.5	2,035	12,556
85-89	0.2351	0.7403	0.7119	0.7683	6,028	4,465	7,755	4,463	18,984	3.9	3.7	4.1	777	3,304
90-94	0.3308	0.8302	0.8078	0.8518	1,565	1,034	2,226	1,299	3,928	2.9	2.8	3.0	152	459
95-99	0.4516	0.9589	0.9491	0.9675	266	153	428	255	565	2.2	2.2	2.2	15	34
100+	0.5981	1	1	1	11	5	22	11	18	1.7	1.7	1.7	1	1
All ages													*74,883*	*6,389,426*

$\ln({}_nq_x)$ with uncertainty interval

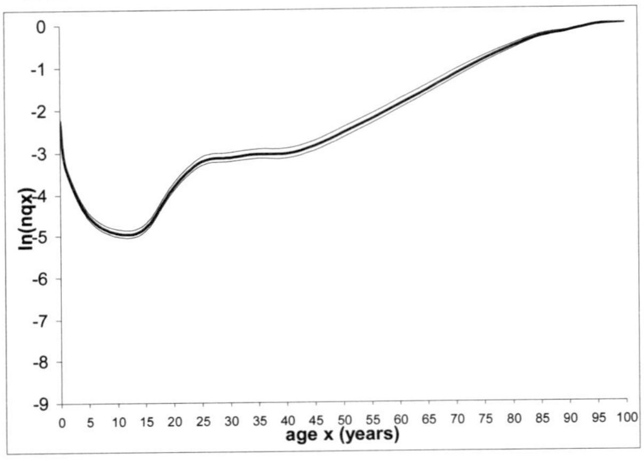

l_x with uncertainty interval

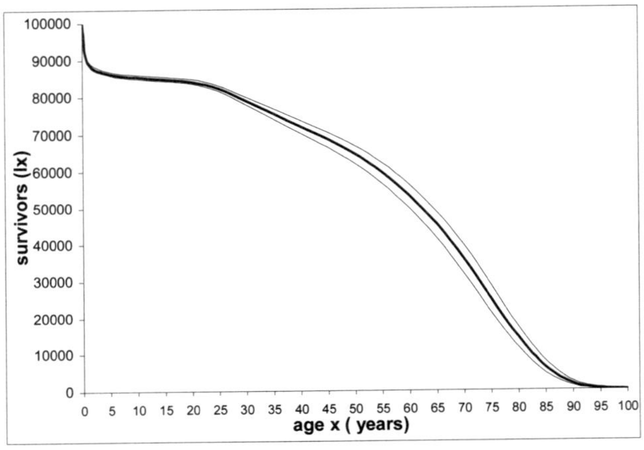

e_x with uncertainty interval

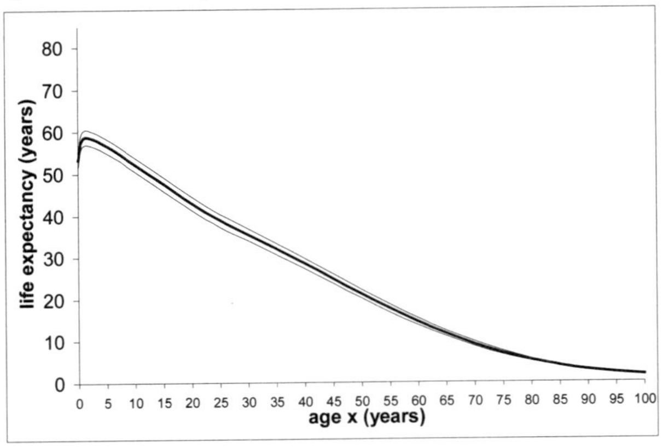

Distribution of deaths by age with uncertainty interval

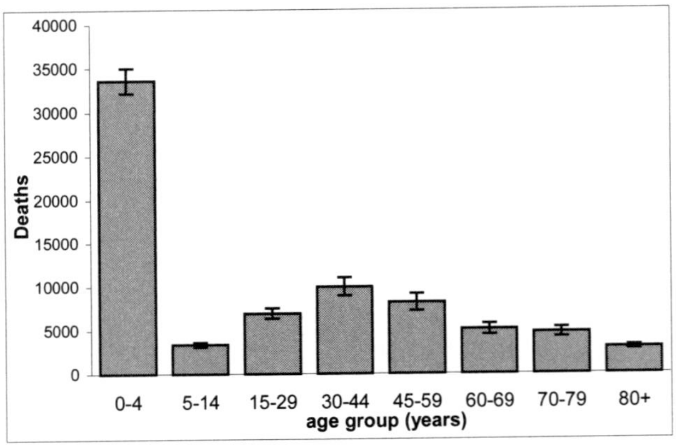

Cambodia Cambodge Camboya

Females, 2000 *Femmes, 2000* *Mujeres, 2000*

Age	$_nM_x$	$_nq_x$ mean	$_nq_x$ percentiles 2½	$_nq_x$ percentiles 97½	l_x mean	l_x percentiles 2½	l_x percentiles 97½	$_nd_x$	$_nL_x$	e_x mean	e_x percentiles 2½	e_x percentiles 97½	Actual deaths	Actual population
<1	0.0948	0.0873	0.0824	0.0926	100,000	100,000	100,000	8,734	92,139	58.5	57.2	60.0	20,260	213,726
1-4	0.0090	0.0353	0.0328	0.0375	91,266	90,735	91,761	3,225	357,323	63.1	61.6	64.8	7,429	823,025
5-9	0.0021	0.0105	0.0097	0.0113	88,041	87,551	88,539	926	437,888	61.4	59.9	63.0	1,976	934,833
10-14	0.0014	0.0071	0.0066	0.0077	87,115	86,605	87,664	620	434,024	57.0	55.6	58.6	1,232	862,168
15-19	0.0023	0.0113	0.0104	0.0122	86,495	85,960	87,086	978	430,029	52.4	51.0	53.9	1,876	824,746
20-24	0.0038	0.0190	0.0174	0.0206	85,517	84,924	86,174	1,621	423,532	48.0	46.6	49.5	1,489	389,115
25-29	0.0044	0.0219	0.0200	0.0239	83,896	83,206	84,676	1,835	414,893	43.9	42.5	45.3	2,065	467,028
30-34	0.0046	0.0227	0.0207	0.0250	82,061	81,260	82,984	1,865	405,643	39.8	38.5	41.2	2,031	441,716
35-39	0.0057	0.0279	0.0252	0.0307	80,196	79,281	81,265	2,235	395,392	35.7	34.4	37.0	2,239	396,170
40-44	0.0061	0.0303	0.0272	0.0336	77,961	76,912	79,206	2,360	383,905	31.6	30.5	32.9	2,040	331,805
45-49	0.0079	0.0389	0.0348	0.0433	75,601	74,392	77,031	2,941	370,653	27.5	26.4	28.8	2,124	267,685
50-54	0.0110	0.0535	0.0475	0.0598	72,660	71,171	74,318	3,890	353,575	23.5	22.5	24.7	2,426	220,486
55-59	0.0155	0.0747	0.0659	0.0837	68,770	66,915	70,721	5,135	331,012	19.7	18.8	20.8	2,737	176,437
60-64	0.0237	0.1117	0.0981	0.1252	63,635	61,317	65,949	7,108	300,403	16.1	15.3	17.1	3,004	126,942
65-69	0.0370	0.1692	0.1484	0.1891	56,526	53,639	59,316	9,565	258,719	12.8	12.1	13.7	3,834	103,697
70-74	0.0598	0.2603	0.2298	0.2878	46,961	43,498	50,427	12,223	204,248	9.9	9.3	10.7	4,277	71,461
75-79	0.0943	0.3816	0.3424	0.4143	34,738	30,980	38,807	13,256	140,550	7.5	7.1	8.1	3,743	39,688
80-84	0.1450	0.5321	0.4899	0.5653	21,482	18,143	25,519	11,430	78,834	5.6	5.3	6.0	2,573	17,744
85-89	0.2156	0.7004	0.6638	0.7283	10,052	7,901	13,018	7,040	32,658	4.2	4.0	4.4	1,136	5,272
90-94	0.3100	0.8031	0.7749	0.8245	3,012	2,159	4,377	2,419	7,802	3.0	2.9	3.2	249	803
95-99	0.4311	0.9399	0.9255	0.9502	593	382	985	557	1,293	2.3	2.2	2.4	23	54
100+	0.5799	1	1	1	36	19	73	36	61	1.7	1.7	1.8	2	3
All ages													*68,765*	*6,714,604*

$\ln(_nq_x)$ with uncertainty interval

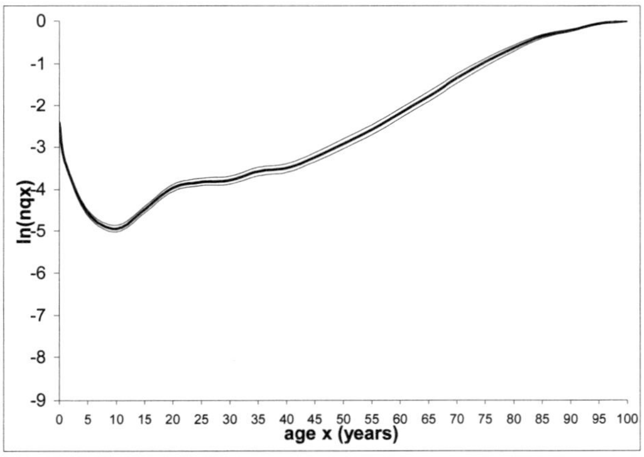

l_x with uncertainty interval

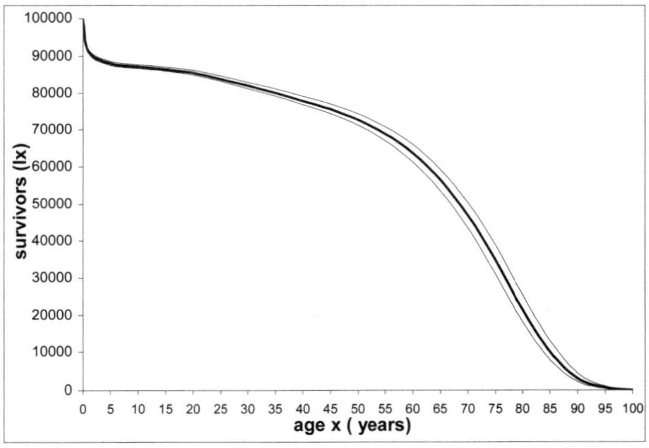

e_x with uncertainty interval

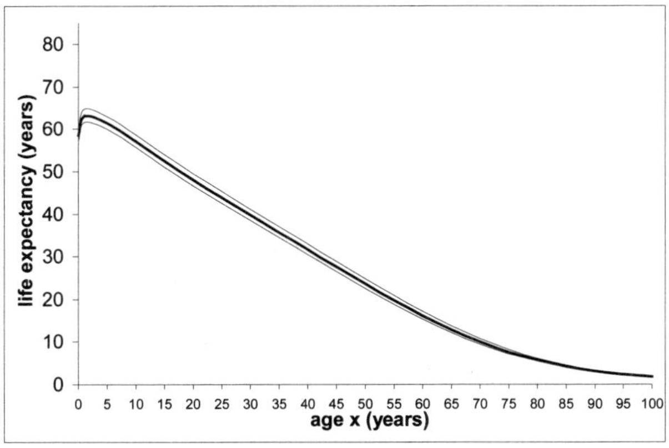

Distribution of deaths by age with uncertainty interval

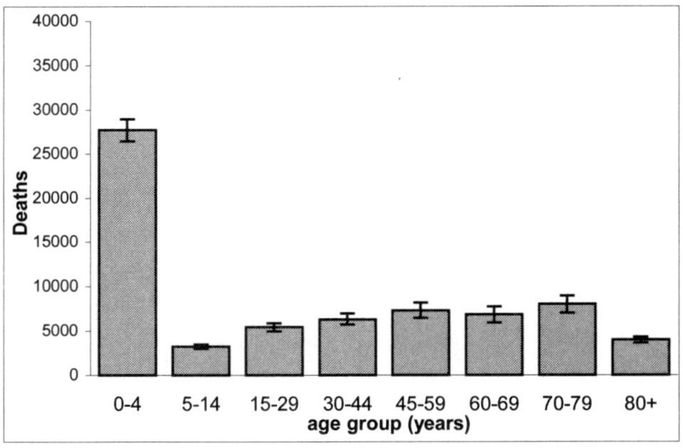

Cameroon Cameroun Camerún

Males, 2000 *Hommes, 2000* *Hombres, 2000*

Age	$_nM_x$	$_nq_x$			l_x			$_nd_x$	$_nL_x$	e_x			Actual deaths	Actual population
			percentiles			percentiles					percentiles			
		mean	2½	97½	mean	2½	97½			mean	2½	97½		
<1	0.1245	0.1119	0.1034	0.1197	100,000	100,000	100,000	11,194	89,926	49.0	46.7	51.5	32,356	259,935
1-4	0.0108	0.0421	0.0380	0.0463	88,806	88,031	89,662	3,740	346,250	54.2	51.6	57.1	9,954	921,628
5-9	0.0024	0.0118	0.0105	0.0132	85,067	84,321	85,814	1,002	422,828	52.5	50.0	55.3	2,554	1,077,804
10-14	0.0015	0.0077	0.0068	0.0086	84,065	83,272	84,867	648	418,703	48.1	45.6	50.9	1,497	967,727
15-19	0.0021	0.0104	0.0092	0.0117	83,417	82,587	84,255	865	414,922	43.4	41.0	46.2	1,758	843,336
20-24	0.0048	0.0239	0.0210	0.0271	82,552	81,672	83,443	1,975	407,822	38.8	36.5	41.6	3,372	696,243
25-29	0.0097	0.0473	0.0411	0.0539	80,577	79,554	81,677	3,809	393,362	34.7	32.4	37.5	5,416	559,395
30-34	0.0151	0.0729	0.0626	0.0840	76,768	75,322	78,281	5,599	369,842	31.3	29.1	34.0	6,727	444,326
35-39	0.0171	0.0822	0.0694	0.0956	71,169	69,006	73,298	5,849	341,221	28.6	26.5	31.1	6,126	357,363
40-44	0.0183	0.0875	0.0731	0.1023	65,319	62,411	68,074	5,715	312,310	25.9	24.1	28.2	5,260	287,436
45-49	0.0199	0.0946	0.0785	0.1108	59,604	56,027	62,997	5,640	283,922	23.2	21.5	25.3	4,770	240,126
50-54	0.0218	0.1034	0.0854	0.1210	53,964	49,817	57,992	5,579	255,873	20.4	18.9	22.2	4,378	200,787
55-59	0.0249	0.1172	0.0967	0.1368	48,385	43,788	52,967	5,671	227,748	17.4	16.1	19.0	4,181	167,913
60-64	0.0336	0.1551	0.1285	0.1804	42,714	37,801	47,771	6,625	197,008	14.4	13.3	15.8	4,534	134,826
65-69	0.0478	0.2133	0.1785	0.2460	36,089	31,013	41,631	7,698	161,199	11.6	10.6	12.8	4,918	102,975
70-74	0.0721	0.3054	0.2602	0.3465	28,391	23,449	34,202	8,670	120,277	9.0	8.3	10.0	5,180	71,857
75-79	0.1089	0.4279	0.3747	0.4745	19,720	15,416	25,302	8,438	77,506	6.9	6.3	7.6	4,712	43,284
80-84	0.1624	0.5776	0.5231	0.6234	11,282	8,193	15,821	6,516	40,120	5.2	4.8	5.7	3,316	20,414
85-89	0.2354	0.7409	0.6945	0.7788	4,766	3,101	7,544	3,531	15,002	3.9	3.7	4.2	1,444	6,136
90-94	0.3310	0.8303	0.7936	0.8601	1,235	686	2,305	1,025	3,098	2.9	2.7	3.1	353	1,066
95-99	0.4517	0.9590	0.9424	0.9708	210	96	476	201	445	2.2	2.1	2.3	44	97
100+	0.5982	1	1	1	9	3	27	9	14	1.7	1.6	1.8	3	5
All ages													*112,853*	*7,404,679*

ln($_nq_x$) with uncertainty interval

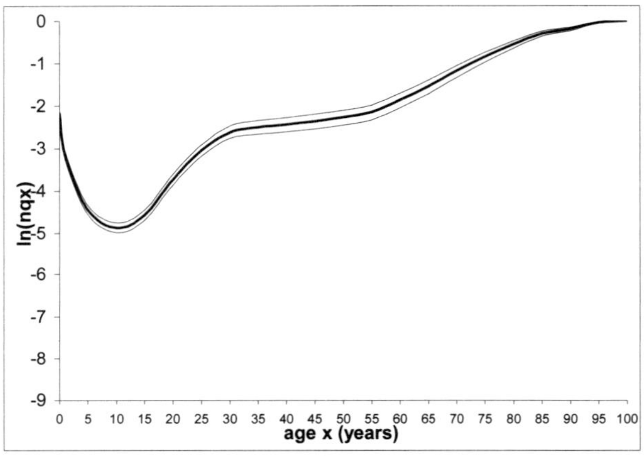

l_x with uncertainty interval

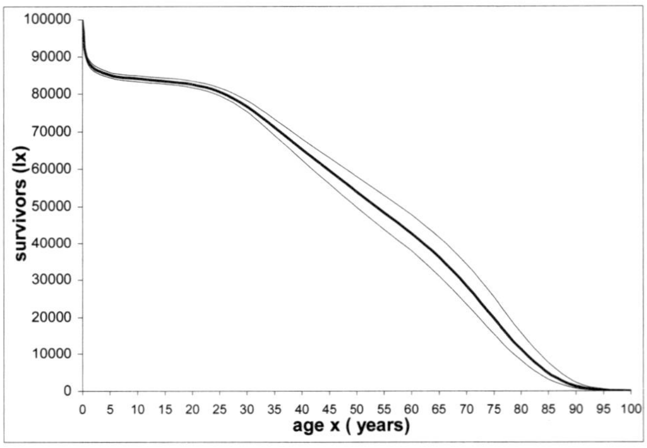

e_x with uncertainty interval

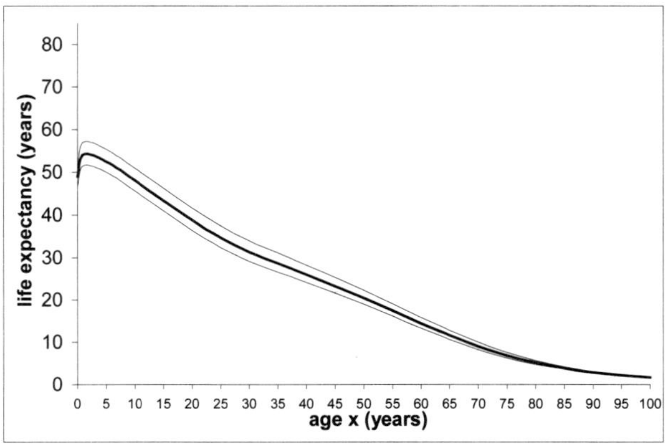

Distribution of deaths by age with uncertainty interval

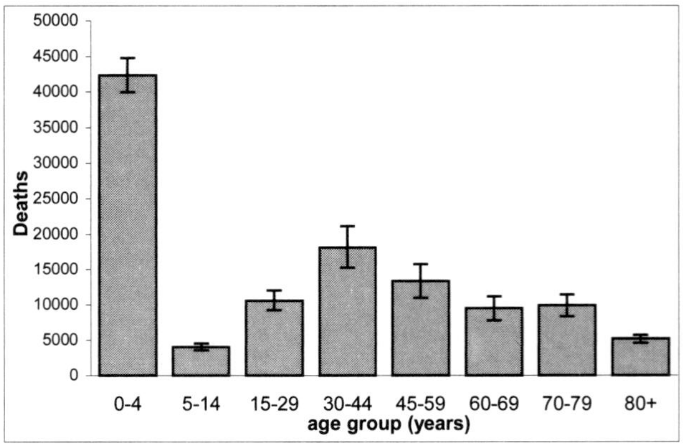

Cameroon Cameroun Camerún

Females, 2000 *Femmes, 2000* *Mujeres, 2000*

Age	$_nM_x$	$_nq_x$ mean	$_nq_x$ percentiles 2 1/2	$_nq_x$ percentiles 97 1/2	l_x mean	l_x percentiles 2 1/2	l_x percentiles 97 1/2	$_nd_x$	$_nL_x$	e_x mean	e_x percentiles 2 1/2	e_x percentiles 97 1/2	Actual deaths	Actual population
<1	0.1091	0.0993	0.0910	0.1070	100,000	100,000	100,000	9,933	91,060	50.4	47.8	53.1	27,198	249,338
1-4	0.0117	0.0454	0.0409	0.0499	90,067	89,303	90,903	4,085	350,465	54.9	52.1	58.1	10,644	913,225
5-9	0.0024	0.0120	0.0107	0.0134	85,982	85,244	86,724	1,032	427,333	53.5	50.7	56.5	2,567	1,063,443
10-14	0.0016	0.0080	0.0071	0.0090	84,951	84,158	85,751	679	423,057	49.1	46.4	52.1	1,536	957,455
15-19	0.0036	0.0179	0.0158	0.0203	84,272	83,438	85,111	1,511	417,582	44.4	41.8	47.5	3,033	838,066
20-24	0.0085	0.0414	0.0361	0.0471	82,761	81,829	83,731	3,425	405,242	40.2	37.6	43.2	5,861	693,422
25-29	0.0138	0.0668	0.0575	0.0769	79,336	78,070	80,685	5,297	383,437	36.8	34.3	39.7	7,690	556,666
30-34	0.0166	0.0796	0.0674	0.0927	74,039	72,092	75,988	5,893	355,461	34.3	32.0	37.0	7,338	442,595
35-39	0.0152	0.0732	0.0611	0.0859	68,145	65,411	70,754	4,990	328,252	32.0	30.0	34.5	5,465	359,476
40-44	0.0136	0.0658	0.0545	0.0775	63,156	59,793	66,297	4,155	305,390	29.4	27.6	31.6	4,006	294,470
45-49	0.0128	0.0619	0.0509	0.0730	59,000	55,160	62,649	3,649	285,879	26.3	24.6	28.3	3,212	251,636
50-54	0.0143	0.0692	0.0568	0.0816	55,351	51,135	59,419	3,832	267,175	22.8	21.4	24.7	3,092	215,597
55-59	0.0175	0.0838	0.0687	0.0986	51,519	46,961	55,995	4,317	246,801	19.4	18.0	21.0	3,228	184,553
60-64	0.0250	0.1177	0.0967	0.1377	47,202	42,332	52,086	5,554	222,124	15.9	14.7	17.4	3,802	152,054
65-69	0.0381	0.1739	0.1440	0.2023	41,648	36,507	46,988	7,244	190,130	12.7	11.7	13.9	4,549	119,400
70-74	0.0611	0.2650	0.2229	0.3039	34,404	29,160	40,221	9,118	149,223	9.8	9.0	10.9	5,263	86,134
75-79	0.0957	0.3861	0.3332	0.4328	25,285	20,379	31,255	9,761	102,023	7.5	6.8	8.3	5,197	54,320
80-84	0.1464	0.5358	0.4789	0.5839	15,524	11,659	20,841	8,318	56,825	5.6	5.1	6.1	4,017	27,445
85-89	0.2169	0.7032	0.6527	0.7443	7,206	4,917	10,860	5,067	23,362	4.1	3.9	4.5	2,019	9,311
90-94	0.3111	0.8046	0.7646	0.8369	2,139	1,257	3,771	1,721	5,532	3.0	2.8	3.3	612	1,968
95-99	0.4320	0.9408	0.9199	0.9561	418	205	888	393	910	2.3	2.2	2.4	105	244
100+	0.5807	1	1	1	25	9	71	25	43	1.7	1.7	1.8	9	16
All ages													*110,446*	*7,470,834*

$\ln(_nq_x)$ with uncertainty interval

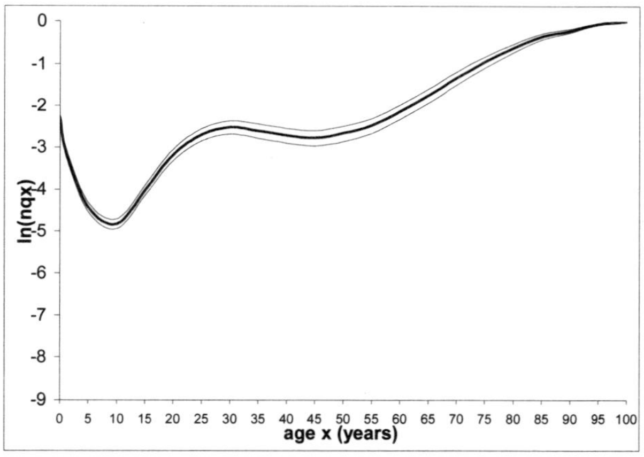

l_x with uncertainty interval

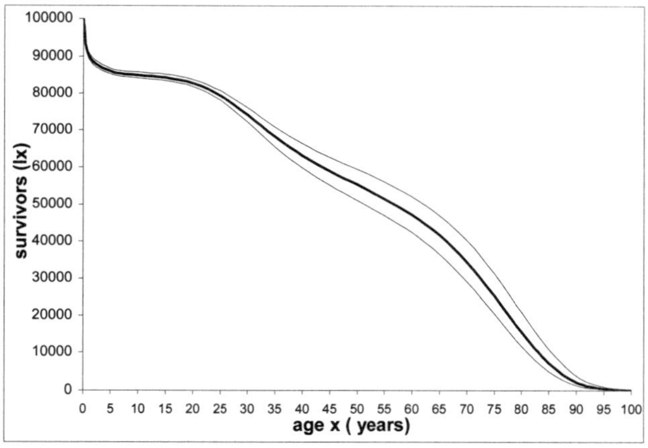

e_x with uncertainty interval

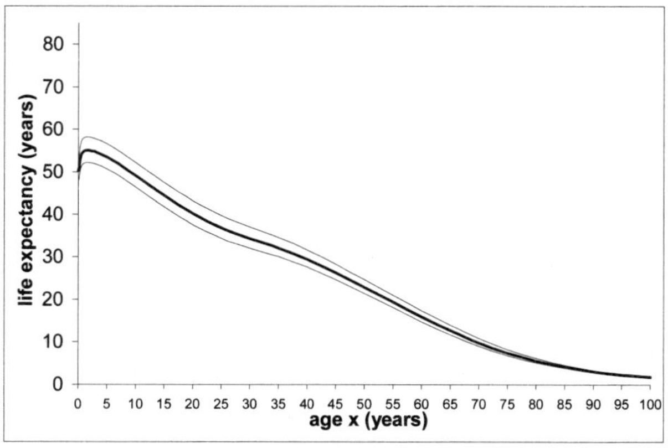

Distribution of deaths by age with uncertainty interval

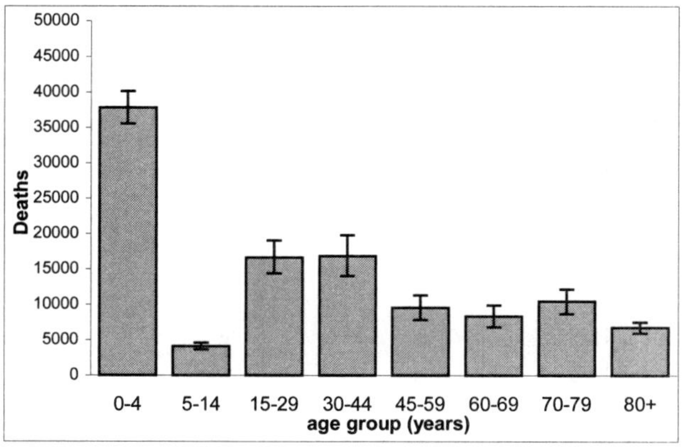

Canada **Canada** **Canadá**

Males, 2000 *Hommes, 2000* *Hombres, 2000*

Age	$_nM_x$	$_nq_x$			l_x			$_nd_x$	$_nL_x$	e_x			Actual deaths	Actual population
			percentiles			percentiles					percentiles			
		mean	2½	97½	mean	2½	97½			mean	2½	97½		
<1	0.0046	0.0046	0.0040	0.0053	100,000	100,000	100,000	458	99,588	76.0	75.6	76.5	818	177,976
1-4	0.0003	0.0010	0.0009	0.0012	99,542	99,466	99,602	102	397,924	75.4	74.9	75.9	191	742,201
5-9	0.0001	0.0007	0.0006	0.0008	99,440	99,351	99,510	69	497,027	71.4	71.0	71.9	146	1,043,260
10-14	0.0002	0.0010	0.0009	0.0011	99,371	99,274	99,448	99	496,607	66.5	66.1	67.0	209	1,051,576
15-19	0.0007	0.0033	0.0030	0.0037	99,272	99,164	99,359	331	495,531	61.5	61.1	62.0	705	1,054,523
20-24	0.0008	0.0041	0.0038	0.0045	98,941	98,799	99,057	404	493,692	56.7	56.3	57.2	858	1,047,955
25-29	0.0008	0.0039	0.0036	0.0042	98,536	98,358	98,685	381	491,727	52.0	51.6	52.5	825	1,063,965
30-34	0.0010	0.0049	0.0046	0.0053	98,155	97,947	98,331	485	489,561	47.2	46.8	47.6	1,139	1,150,008
35-39	0.0013	0.0063	0.0059	0.0067	97,670	97,428	97,877	614	486,813	42.4	42.0	42.9	1,716	1,359,737
40-44	0.0018	0.0091	0.0085	0.0096	97,056	96,774	97,301	881	483,074	37.6	37.3	38.1	2,386	1,307,662
45-49	0.0027	0.0133	0.0126	0.0140	96,174	95,840	96,471	1,280	477,670	32.9	32.6	33.4	3,098	1,155,735
50-54	0.0045	0.0220	0.0210	0.0231	94,894	94,496	95,256	2,091	469,242	28.4	28.0	28.8	4,553	1,021,716
55-59	0.0077	0.0378	0.0359	0.0396	92,803	92,324	93,256	3,508	455,245	23.9	23.6	24.4	5,976	775,607
60-64	0.0130	0.0632	0.0597	0.0662	89,295	88,728	89,863	5,641	432,374	19.8	19.5	20.2	8,068	618,431
65-69	0.0221	0.1049	0.0990	0.1096	83,654	82,924	84,367	8,778	396,327	15.9	15.7	16.3	12,216	551,548
70-74	0.0366	0.1677	0.1585	0.1743	74,876	73,850	75,991	12,553	342,999	12.5	12.3	12.8	16,790	458,775
75-79	0.0605	0.2626	0.2500	0.2723	62,323	60,986	63,944	16,368	270,697	9.5	9.3	9.8	20,317	336,009
80-84	0.1010	0.4032	0.3885	0.4176	45,955	44,477	47,958	18,529	183,455	7.1	6.9	7.3	18,913	187,254
85-89	0.1610	0.5739	0.5572	0.5898	27,426	26,131	29,326	15,741	97,781	5.1	5.0	5.3	14,775	91,784
90-94	0.2448	0.7058	0.6911	0.7192	11,686	10,789	12,874	8,247	33,687	3.7	3.6	3.8	6,750	27,571
95-99	0.3553	0.8599	0.8499	0.8688	3,438	3,030	3,903	2,957	8,322	2.7	2.7	2.8	1,944	5,471
100+	0.4920	1	1	1	482	398	585	482	979	2.0	2.0	2.1	289	588
All ages													*122,682*	*15,229,352*

$\ln(_nq_x)$ with uncertainty interval

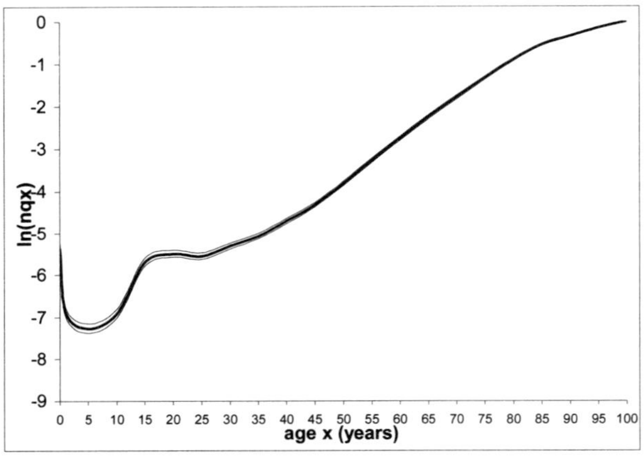

l_x with uncertainty interval

e_x with uncertainty interval

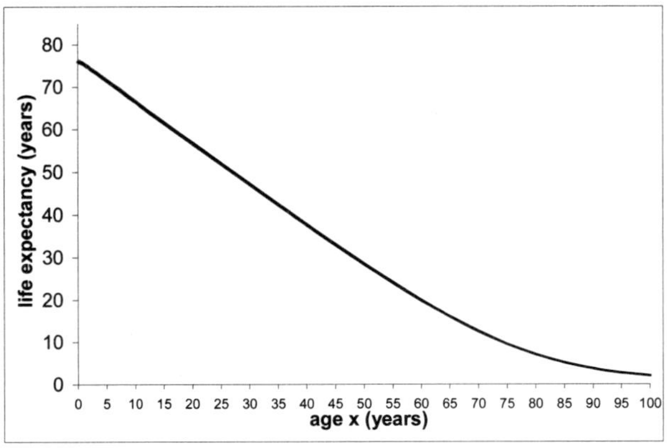

Distribution of deaths by age with uncertainty interval

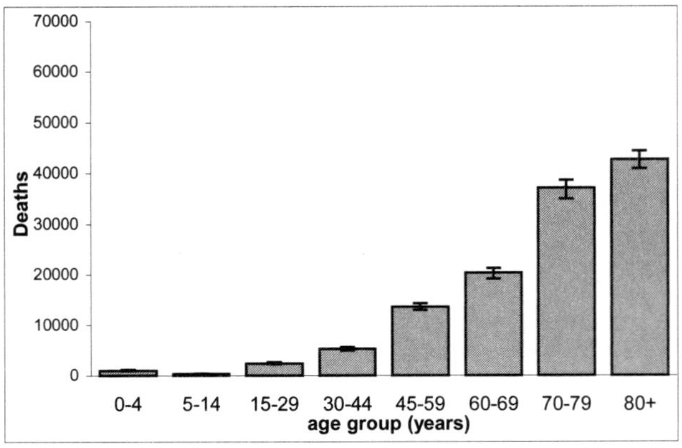

Females, 2000 *Femmes, 2000* *Mujeres, 2000*

Age	$_nM_x$	$_nq_x$			l_x			$_nd_x$	$_nL_x$	e_x			Actual deaths	Actual population
			percentiles			percentiles					percentiles			
		mean	2½	97½	mean	2½	97½			mean	2½	97½		
<1	0.0038	0.0038	0.0033	0.0045	100,000	100,000	100,000	382	99,656	81.5	81.0	81.9	655	170,720
1-4	0.0002	0.0009	0.0008	0.0010	99,618	99,550	99,671	88	398,261	80.8	80.4	81.3	155	703,539
5-9	0.0001	0.0006	0.0005	0.0007	99,530	99,451	99,592	58	497,506	76.9	76.4	77.3	115	993,020
10-14	0.0001	0.0007	0.0006	0.0007	99,472	99,386	99,541	65	497,199	71.9	71.5	72.4	131	999,889
15-19	0.0003	0.0014	0.0013	0.0016	99,407	99,313	99,482	140	496,688	66.9	66.5	67.4	282	1,002,223
20-24	0.0003	0.0013	0.0012	0.0015	99,268	99,157	99,356	133	496,006	62.0	61.6	62.5	268	1,000,597
25-29	0.0003	0.0016	0.0014	0.0018	99,135	99,009	99,236	158	495,280	57.1	56.7	57.6	330	1,036,986
30-34	0.0004	0.0020	0.0019	0.0022	98,977	98,836	99,092	202	494,380	52.2	51.8	52.7	462	1,129,902
35-39	0.0006	0.0032	0.0029	0.0035	98,775	98,614	98,907	313	493,092	47.3	46.9	47.8	849	1,337,018
40-44	0.0010	0.0052	0.0048	0.0056	98,462	98,274	98,619	511	491,032	42.5	42.1	42.9	1,359	1,305,799
45-49	0.0017	0.0084	0.0078	0.0089	97,951	97,724	98,144	821	487,704	37.7	37.3	38.1	1,962	1,165,938
50-54	0.0026	0.0129	0.0121	0.0137	97,130	96,851	97,374	1,255	482,514	33.0	32.6	33.4	2,681	1,030,549
55-59	0.0045	0.0223	0.0210	0.0235	95,875	95,533	96,183	2,134	474,041	28.4	28.0	28.8	3,559	790,682
60-64	0.0074	0.0363	0.0343	0.0382	93,741	93,326	94,145	3,398	460,210	23.9	23.6	24.4	4,759	644,464
65-69	0.0120	0.0581	0.0550	0.0613	90,343	89,794	90,889	5,252	438,585	19.8	19.5	20.2	7,112	593,942
70-74	0.0200	0.0951	0.0896	0.0999	85,091	84,340	85,851	8,089	405,233	15.8	15.6	16.2	10,941	548,128
75-79	0.0347	0.1597	0.1508	0.1668	77,002	75,922	78,095	12,301	354,259	12.2	12.0	12.5	16,413	472,714
80-84	0.0630	0.2721	0.2587	0.2820	64,702	63,255	66,269	17,604	279,498	9.1	8.8	9.3	19,657	312,089
85-89	0.1095	0.4297	0.4141	0.4449	47,098	45,456	49,126	20,237	184,896	6.5	6.3	6.7	20,765	189,722
90-94	0.1822	0.5891	0.5720	0.6052	26,861	25,506	28,784	15,823	86,834	4.5	4.4	4.7	13,890	76,225
95-99	0.2906	0.7763	0.7626	0.7887	11,038	10,150	12,205	8,568	29,483	3.2	3.1	3.3	5,889	20,263
100+	0.4441	1	1	1	2,469	2,145	2,840	2,469	5,561	2.3	2.2	2.3	1,304	2,937
All ages													*113,538*	*15,527,346*

ln($_nq_x$) with uncertainty interval

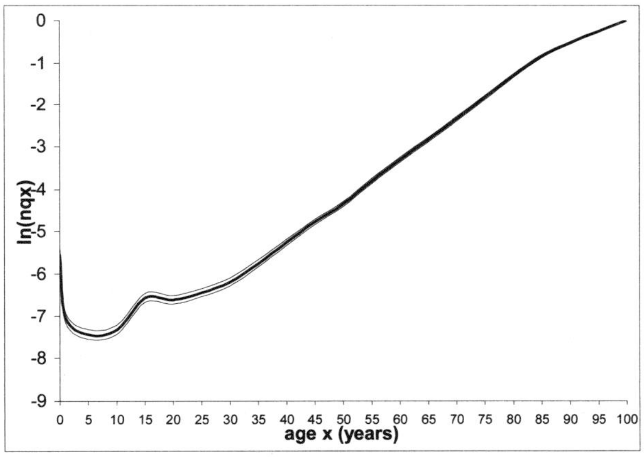

l_x with uncertainty interval

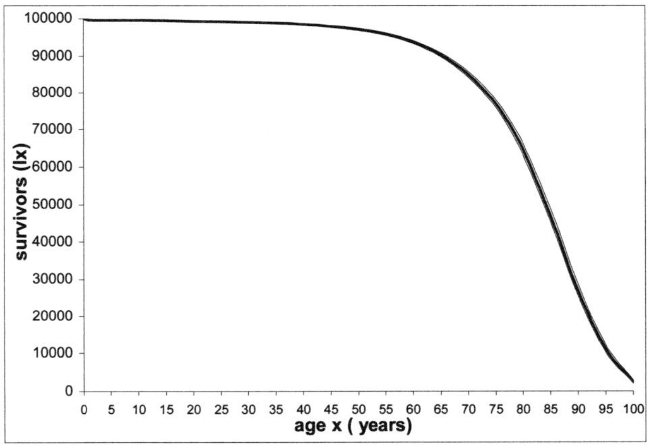

e_x with uncertainty interval

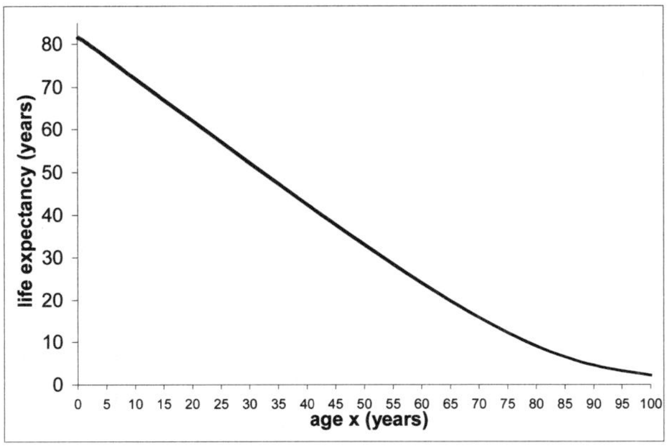

Distribution of deaths by age with uncertainty interval

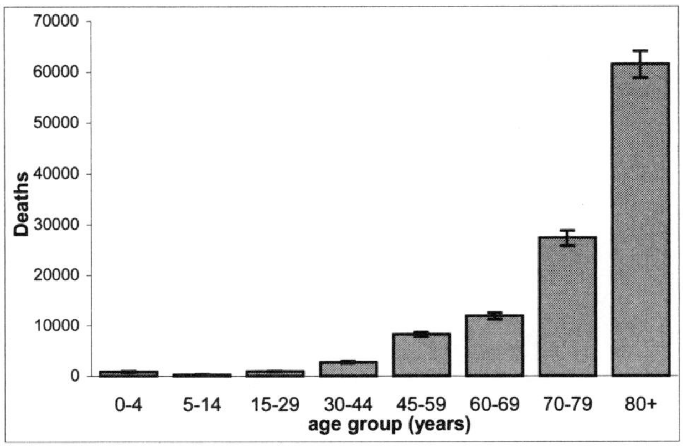

Cape Verde Cap-Vert Cabo Verde

Males, 2000 *Hommes, 2000* *Hombres, 2000*

Age	$_nM_x$	$_nq_x$			l_x			$_nd_x$	$_nL_x$	e_x			Actual deaths	Actual population
			percentiles			percentiles					percentiles			
		mean	2½	97½	mean	2½	97½			mean	2½	97½		
<1	0.0488	0.0468	0.0379	0.0586	100,000	100,000	100,000	4,676	95,791	66.5	64.4	68.0	317	6,484
1-4	0.0024	0.0094	0.0080	0.0114	95,324	94,137	96,215	899	379,138	68.7	67.3	69.9	57	24,036
5-9	0.0005	0.0024	0.0021	0.0029	94,425	93,067	95,448	229	471,552	65.3	64.0	66.5	14	28,191
10-14	0.0003	0.0015	0.0013	0.0018	94,196	92,795	95,251	142	470,625	60.5	59.2	61.7	8	26,008
15-19	0.0008	0.0041	0.0035	0.0049	94,054	92,628	95,129	383	469,315	55.6	54.3	56.8	20	24,468
20-24	0.0015	0.0073	0.0062	0.0087	93,672	92,176	94,800	681	466,657	50.8	49.5	52.0	31	21,056
25-29	0.0027	0.0137	0.0118	0.0162	92,991	91,376	94,210	1,270	461,781	46.2	44.9	47.3	47	16,955
30-34	0.0019	0.0095	0.0082	0.0112	91,721	89,896	93,102	872	456,427	41.8	40.6	42.9	33	17,231
35-39	0.0030	0.0147	0.0129	0.0173	90,850	88,887	92,335	1,340	450,899	37.1	36.0	38.2	42	14,123
40-44	0.0048	0.0238	0.0210	0.0277	89,510	87,350	91,147	2,132	442,220	32.7	31.6	33.7	30	6,208
45-49	0.0070	0.0344	0.0307	0.0396	87,378	84,929	89,234	3,006	429,375	28.4	27.4	29.4	16	2,324
50-54	0.0095	0.0464	0.0418	0.0527	84,372	81,569	86,498	3,915	412,075	24.3	23.4	25.3	11	1,182
55-59	0.0160	0.0769	0.0691	0.0868	80,458	77,270	82,885	6,189	386,815	20.4	19.6	21.3	23	1,409
60-64	0.0240	0.1132	0.1013	0.1264	74,269	70,565	77,157	8,408	350,323	16.9	16.1	17.7	60	2,483
65-69	0.0330	0.1524	0.1363	0.1678	65,861	61,649	69,342	10,039	304,207	13.7	13.0	14.5	81	2,457
70-74	0.0460	0.2063	0.1849	0.2274	55,822	51,303	59,888	11,515	250,323	10.7	10.1	11.4	95	2,058
75-79	0.0800	0.3333	0.3009	0.3658	44,307	39,637	48,813	14,769	184,613	7.8	7.4	8.4	84	1,052
80-84	0.1452	0.5328	0.4896	0.5724	29,538	25,138	34,125	15,736	108,349	5.5	5.2	5.9	113	780
85-89	0.2300	0.7302	0.6909	0.7630	13,802	10,750	17,417	10,077	43,815	4.0	3.8	4.2	101	441
90-94	0.3300	0.8291	0.7993	0.8532	3,724	2,547	5,383	3,088	9,357	2.9	2.8	3.1	54	164
95-99	0.4500	0.9574	0.9443	0.9670	636	374	1,080	609	1,354	2.2	2.2	2.3	11	24
100+	0.5600	1	1	1	27	12	60	27	48	1.8	1.8	1.9	1	2
All ages													*1,247*	*199,136*

ln($_nq_x$) with uncertainty interval

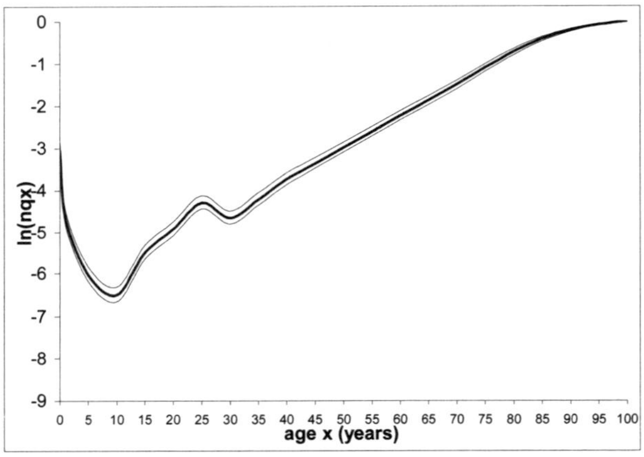

l_x with uncertainty interval

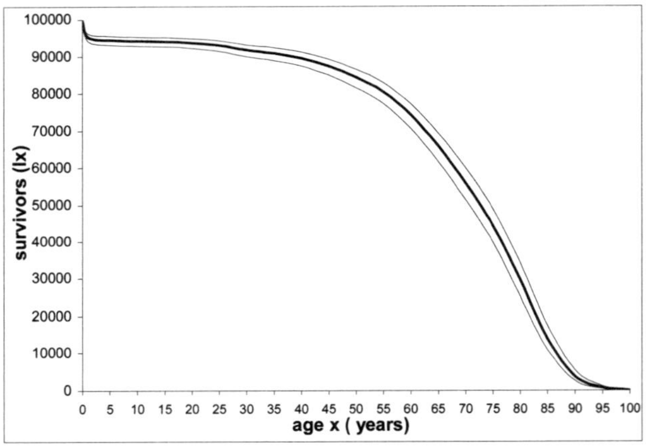

e_x with uncertainty interval

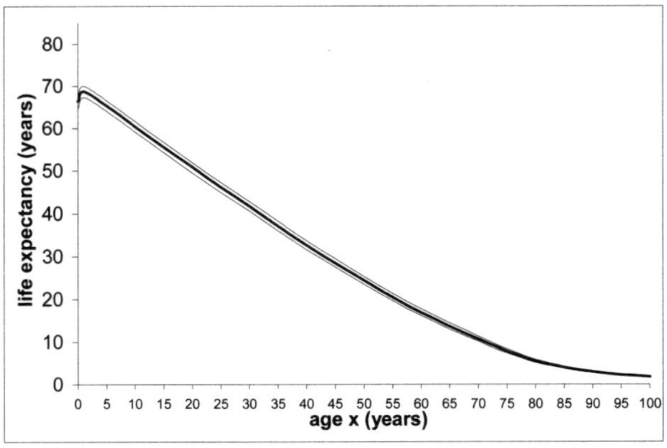

Distribution of deaths by age with uncertainty interval

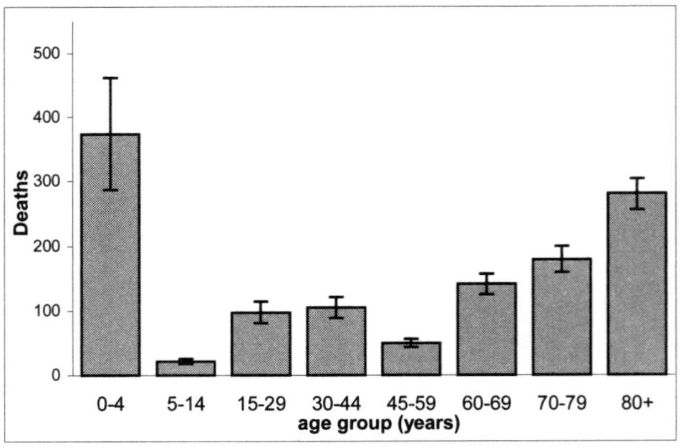

Cape Verde | Cap-Vert | Cabo Verde

Females, 2000 | *Femmes, 2000* | *Mujeres, 2000*

Age	$_nM_x$	$_nq_x$			l_x			$_nd_x$	$_nL_x$	e_x			Actual deaths	Actual population
			percentiles			percentiles					percentiles			
		mean	2½	97½	mean	2½	97½			mean	2½	97½		
<1	0.0346	0.0336	0.0247	0.0463	100,000	100,000	100,000	3,355	96,980	72.3	71.1	73.3	214	6,176
1-4	0.0017	0.0068	0.0056	0.0083	96,645	95,374	97,526	654	385,008	73.8	72.7	74.9	40	23,682
5-9	0.0006	0.0030	0.0025	0.0036	95,990	94,578	96,983	288	479,231	70.3	69.2	71.4	17	27,628
10-14	0.0004	0.0021	0.0017	0.0025	95,702	94,233	96,741	198	478,016	65.5	64.4	66.6	11	25,504
15-19	0.0005	0.0023	0.0020	0.0028	95,504	93,997	96,574	224	476,962	60.6	59.6	61.8	11	24,097
20-24	0.0007	0.0036	0.0031	0.0043	95,280	93,733	96,383	345	475,540	55.8	54.7	56.9	16	22,606
25-29	0.0008	0.0039	0.0033	0.0046	94,936	93,329	96,088	368	473,758	51.0	49.9	52.1	14	18,195
30-34	0.0011	0.0056	0.0049	0.0066	94,568	92,901	95,770	534	471,502	46.2	45.1	47.3	22	19,378
35-39	0.0016	0.0080	0.0070	0.0092	94,033	92,284	95,303	749	468,293	41.4	40.4	42.6	27	16,859
40-44	0.0024	0.0119	0.0106	0.0136	93,284	91,433	94,640	1,113	463,638	36.7	35.7	37.9	28	11,586
45-49	0.0040	0.0197	0.0178	0.0220	92,171	90,190	93,638	1,815	456,320	32.1	31.1	33.3	27	6,736
50-54	0.0056	0.0276	0.0252	0.0308	90,357	88,204	91,968	2,495	445,546	27.7	26.7	28.9	18	3,129
55-59	0.0090	0.0440	0.0398	0.0493	87,862	85,486	89,653	3,867	429,641	23.4	22.4	24.6	35	3,884
60-64	0.0150	0.0723	0.0645	0.0812	83,995	81,270	86,082	6,072	404,794	19.4	18.4	20.5	85	5,672
65-69	0.0220	0.1043	0.0913	0.1181	77,923	74,670	80,526	8,125	369,303	15.7	14.8	16.8	104	4,744
70-74	0.0330	0.1524	0.1311	0.1740	69,798	65,854	73,177	10,639	322,394	12.3	11.5	13.2	119	3,597
75-79	0.0600	0.2609	0.2237	0.2973	59,159	54,397	63,580	15,433	257,214	9.0	8.3	9.9	105	1,752
80-84	0.1150	0.4465	0.3888	0.5017	43,726	38,222	49,357	19,524	169,823	6.3	5.8	7.0	157	1,363
85-89	0.2000	0.6667	0.6024	0.7198	24,203	19,047	30,169	16,135	80,675	4.4	4.0	4.9	161	807
90-94	0.3000	0.7895	0.7382	0.8286	8,068	5,337	11,996	6,369	21,230	3.1	2.9	3.5	99	330
95-99	0.4000	0.9091	0.8774	0.9317	1,698	915	3,141	1,544	3,860	2.5	2.4	2.6	36	89
100+	0.5000	1	1	1	154	62	385	154	309	2.0	2.0	2.1	4	8
All ages													*1,349*	*227,822*

$\ln(_nq_x)$ with uncertainty interval

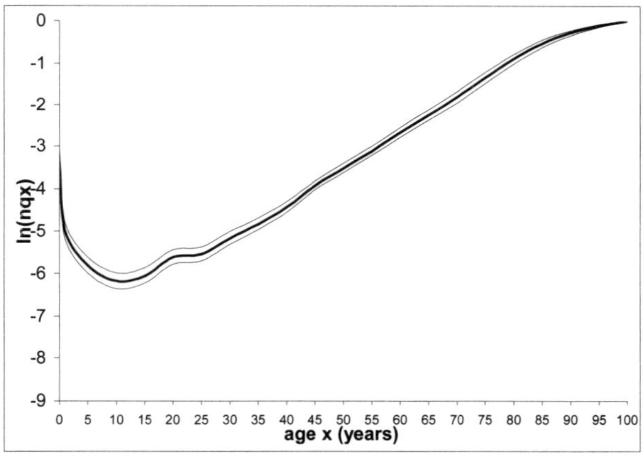

l_x with uncertainty interval

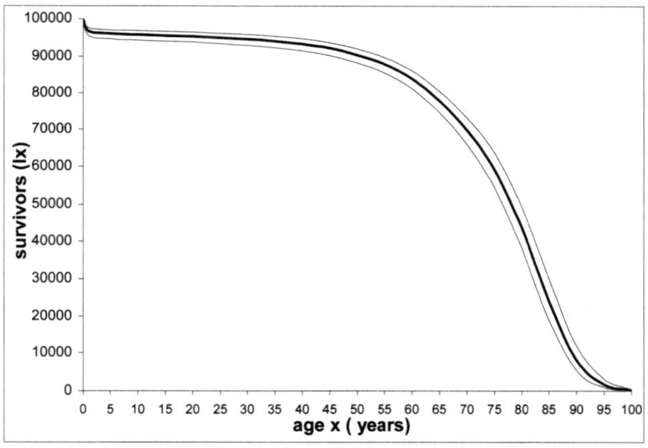

e_x with uncertainty interval

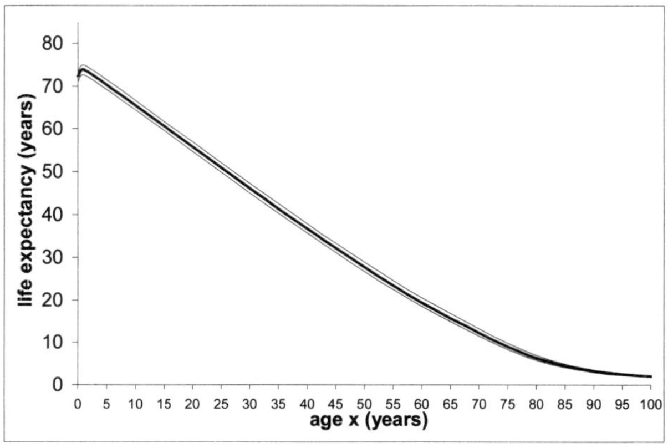

Distribution of deaths by age with uncertainty interval

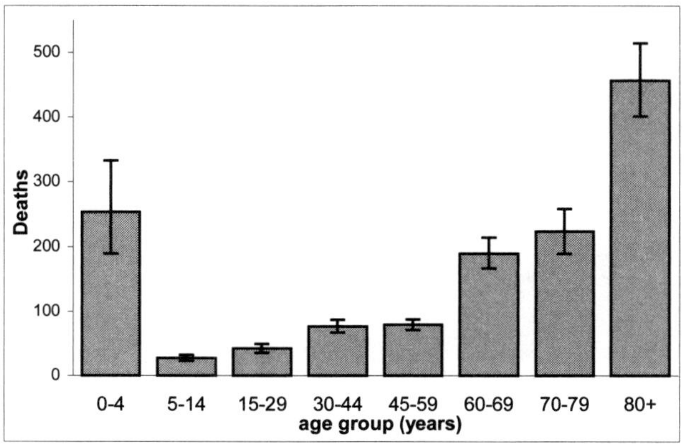

Central African Republic République centrafricaine República Centroafricana

Males, 2000 *Hommes, 2000* *Hombres, 2000*

Age	$_nM_x$	$_nq_x$			l_x			$_nd_x$	$_nL_x$	e_x			Actual deaths	Actual population
			percentiles			percentiles					percentiles			
		mean	2½	97½	mean	2½	97½			mean	2½	97½		
<1	0.1693	0.1469	0.1373	0.1567	100,000	100,000	100,000	14,690	86,779	41.6	39.6	44.0	11,487	67,855
1-4	0.0159	0.0611	0.0552	0.0667	85,310	84,326	86,272	5,212	328,729	47.8	45.5	50.6	3,744	236,125
5-9	0.0032	0.0157	0.0140	0.0174	80,097	79,363	80,830	1,259	397,338	46.8	44.6	49.5	839	264,755
10-14	0.0019	0.0097	0.0086	0.0107	78,838	77,986	79,629	762	392,284	42.5	40.3	45.2	445	228,924
15-19	0.0023	0.0116	0.0102	0.0128	78,076	77,151	78,945	903	388,123	37.9	35.7	40.6	454	195,428
20-24	0.0068	0.0335	0.0295	0.0372	77,173	76,162	78,137	2,582	379,413	33.3	31.1	35.9	1,131	166,174
25-29	0.0154	0.0740	0.0647	0.0827	74,592	73,327	75,831	5,521	359,156	29.3	27.2	32.0	1,957	127,280
30-34	0.0248	0.1169	0.1010	0.1311	69,071	67,264	70,911	8,073	325,170	26.5	24.4	29.0	2,557	103,005
35-39	0.0277	0.1296	0.1109	0.1469	60,998	58,443	63,702	7,905	285,226	24.7	22.7	27.0	2,450	88,411
40-44	0.0285	0.1330	0.1132	0.1517	53,093	49,921	56,565	7,062	247,809	23.0	21.2	25.0	2,184	76,631
45-49	0.0291	0.1357	0.1154	0.1550	46,031	42,449	50,077	6,246	214,540	21.1	19.6	22.9	1,894	65,063
50-54	0.0291	0.1357	0.1155	0.1550	39,785	35,982	44,269	5,399	185,429	19.0	17.7	20.6	1,464	50,270
55-59	0.0295	0.1375	0.1173	0.1567	34,386	30,518	39,158	4,729	160,110	16.6	15.5	17.9	1,253	42,424
60-64	0.0375	0.1715	0.1473	0.1943	29,658	25,840	34,564	5,086	135,575	13.8	12.9	15.0	1,295	34,518
65-69	0.0514	0.2278	0.1978	0.2557	24,572	20,828	29,472	5,598	108,865	11.2	10.4	12.1	1,369	26,620
70-74	0.0764	0.3207	0.2832	0.3549	18,974	15,503	23,642	6,084	79,659	8.8	8.2	9.5	1,435	18,783
75-79	0.1139	0.4434	0.3996	0.4821	12,890	10,003	16,947	5,715	50,161	6.7	6.3	7.3	1,266	11,116
80-84	0.1679	0.5912	0.5458	0.6299	7,175	5,187	10,163	4,242	25,269	5.1	4.8	5.5	868	5,173
85-89	0.2409	0.7519	0.7120	0.7846	2,933	1,924	4,598	2,205	9,152	3.8	3.6	4.1	398	1,651
90-94	0.3362	0.8370	0.8043	0.8632	728	416	1,314	609	1,812	2.8	2.7	3.0	105	312
95-99	0.4566	0.9633	0.9493	0.9732	119	57	254	114	250	2.2	2.1	2.2	15	32
100+	0.6027	1	1	1	4	2	13	4	7	1.7	1.6	1.7	1	2
All ages													*38,610*	*1,810,552*

ln($_nq_x$) with uncertainty interval

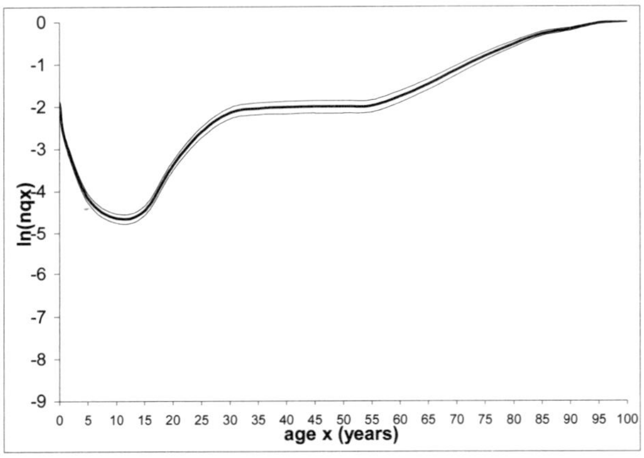

l_x with uncertainty interval

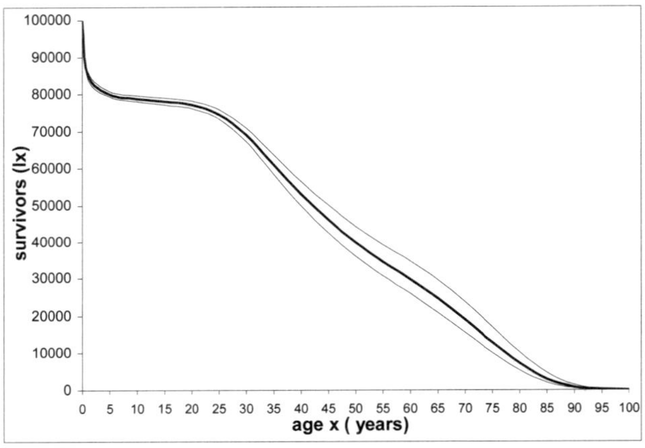

e_x with uncertainty interval

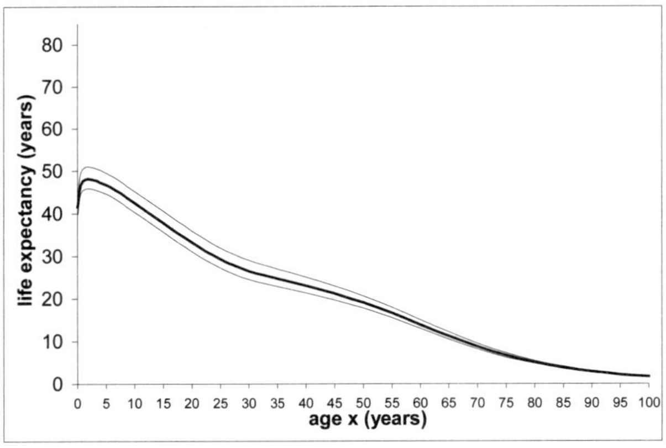

Distribution of deaths by age with uncertainty interval

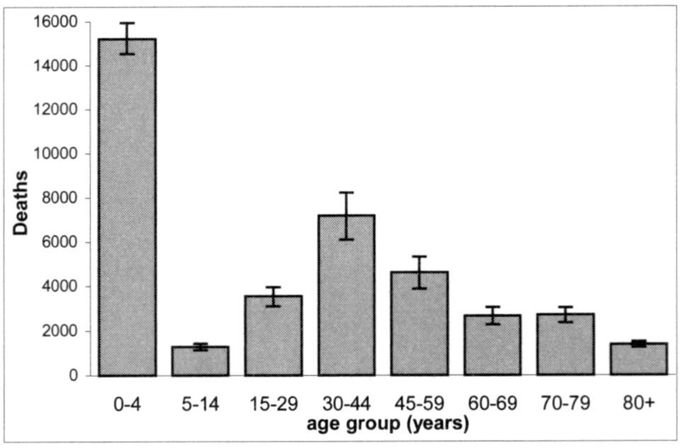

Central African Republic　　République centrafricaine　　República Centroafricana

Females, 2000　　*Femmes, 2000*　　*Mujeres, 2000*

Age	$_nM_x$	$_nq_x$			l_x			$_nd_x$	$_nL_x$	e_x			Actual deaths	Actual population
			percentiles			percentiles					percentiles			
		mean	2½	97½	mean	2½	97½			mean	2½	97½		
<1	0.1448	0.1281	0.1189	0.1382	100,000	100,000	100,000	12,807	88,473	42.5	39.9	45.2	9,445	65,246
1-4	0.0168	0.0647	0.0587	0.0712	87,193	86,184	88,113	5,645	335,222	47.7	44.9	50.9	4,021	238,779
5-9	0.0033	0.0162	0.0144	0.0181	81,547	80,814	82,272	1,318	404,441	46.9	44.2	50.0	861	264,071
10-14	0.0021	0.0105	0.0093	0.0118	80,229	79,360	81,018	844	399,035	42.6	40.0	45.7	493	232,901
15-19	0.0054	0.0264	0.0233	0.0298	79,385	78,427	80,221	2,097	391,682	38.0	35.4	41.1	1,082	202,069
20-24	0.0136	0.0658	0.0575	0.0746	77,288	76,106	78,294	5,083	373,733	34.0	31.4	37.1	2,354	173,100
25-29	0.0229	0.1083	0.0934	0.1236	72,205	70,453	73,785	7,817	341,484	31.2	28.7	34.2	3,106	135,682
30-34	0.0275	0.1285	0.1095	0.1475	64,388	61,742	66,858	8,273	301,259	29.7	27.4	32.4	2,974	108,295
35-39	0.0243	0.1144	0.0966	0.1317	56,116	52,635	59,471	6,417	264,535	28.7	26.7	31.1	2,278	93,897
40-44	0.0206	0.0979	0.0823	0.1128	49,698	45,703	53,655	4,864	236,332	27.1	25.3	29.2	1,718	83,462
45-49	0.0176	0.0842	0.0706	0.0972	44,834	40,549	49,175	3,774	214,736	24.8	23.3	26.6	1,282	72,928
50-54	0.0181	0.0864	0.0726	0.0998	41,060	36,618	45,684	3,549	196,426	21.8	20.5	23.4	1,047	57,937
55-59	0.0203	0.0967	0.0813	0.1114	37,511	32,982	42,369	3,626	178,489	18.6	17.5	20.0	1,022	50,298
60-64	0.0278	0.1298	0.1098	0.1489	33,885	29,333	38,923	4,397	158,432	15.4	14.4	16.6	1,170	42,154
65-69	0.0413	0.1872	0.1598	0.2129	29,488	25,005	34,651	5,519	133,641	12.3	11.4	13.3	1,386	33,565
70-74	0.0651	0.2801	0.2430	0.3141	23,969	19,739	29,114	6,713	103,060	9.5	8.8	10.4	1,601	24,575
75-79	0.1002	0.4005	0.3555	0.4405	17,255	13,613	22,038	6,911	69,000	7.3	6.8	7.9	1,589	15,865
80-84	0.1510	0.5481	0.5004	0.5893	10,345	7,690	14,204	5,670	37,548	5.5	5.1	5.9	1,247	8,259
85-89	0.2213	0.7123	0.6696	0.7482	4,674	3,164	7,096	3,330	15,048	4.1	3.8	4.4	644	2,912
90-94	0.3148	0.8095	0.7747	0.8384	1,345	797	2,345	1,089	3,458	3.0	2.8	3.2	206	653
95-99	0.4350	0.9436	0.9254	0.9572	256	129	528	242	556	2.3	2.2	2.4	38	87
100+	0.5835	1	1	1	14	6	39	14	25	1.7	1.7	1.8	4	6
All ages													*39,566*	*1,906,741*

ln($_nq_x$) with uncertainty interval

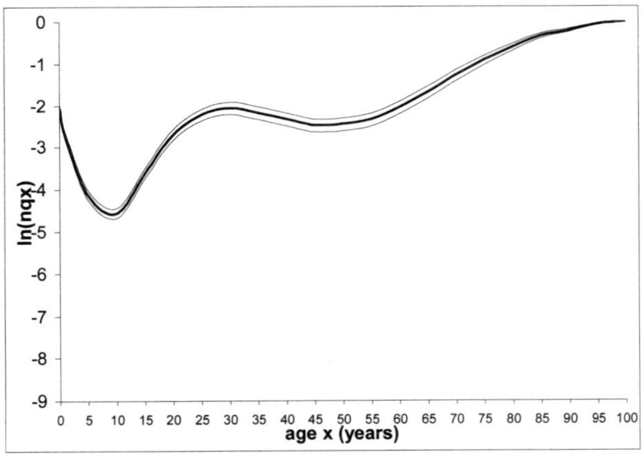

l_x with uncertainty interval

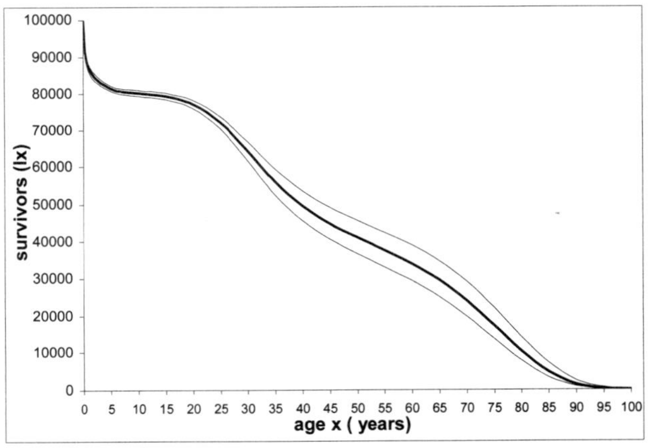

e_x with uncertainty interval

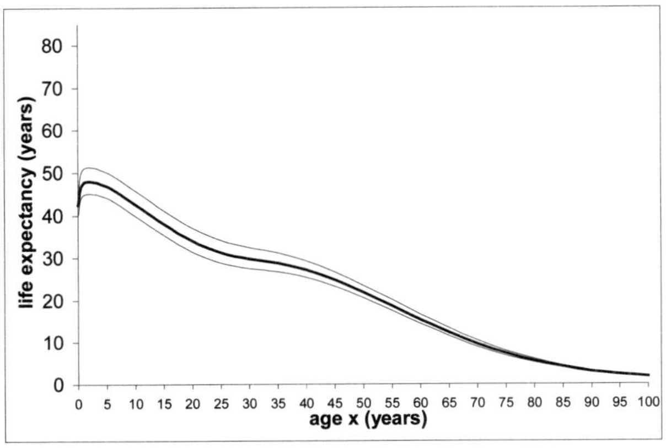

Distribution of deaths by age with uncertainty interval

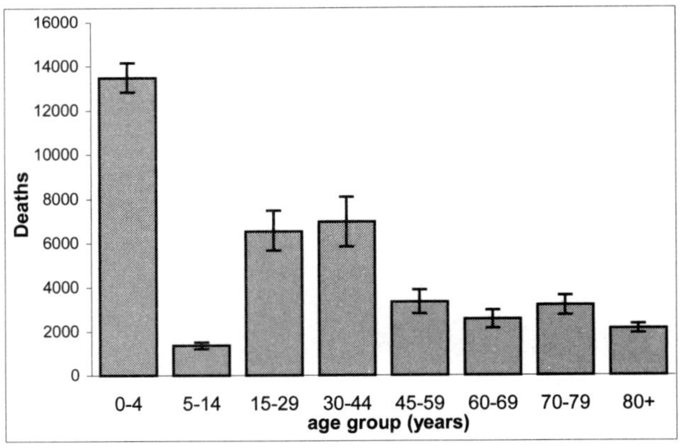

Chad — Tchad — Chad

Males, 2000 — *Hommes, 2000* — *Hombres, 2000*

Age	$_nM_x$	$_nq_x$			l_x			$_nd_x$	$_nL_x$	e_x			Actual deaths	Actual population
			percentiles			percentiles					percentiles			
		mean	2½	97½	mean	2½	97½			mean	2½	97½		
<1	0.1683	0.1462	0.1335	0.1594	100,000	100,000	100,000	14,618	86,844	47.4	44.8	50.5	28,246	167,807
1-4	0.0139	0.0536	0.0465	0.0605	85,382	84,064	86,651	4,580	330,535	54.5	51.7	58.1	8,049	580,850
5-9	0.0034	0.0166	0.0142	0.0191	80,802	79,802	81,801	1,344	400,648	53.5	50.9	56.8	1,995	594,568
10-14	0.0023	0.0116	0.0098	0.0134	79,458	78,296	80,577	923	394,980	49.3	46.8	52.6	1,150	492,012
15-19	0.0022	0.0107	0.0090	0.0123	78,535	77,250	79,737	840	390,573	44.9	42.4	48.1	877	407,988
20-24	0.0048	0.0236	0.0199	0.0273	77,695	76,297	79,009	1,836	383,883	40.3	37.8	43.5	1,637	342,257
25-29	0.0079	0.0390	0.0324	0.0453	75,859	74,211	77,439	2,955	371,905	36.2	33.8	39.4	2,235	281,237
30-34	0.0109	0.0529	0.0436	0.0619	72,903	70,850	74,922	3,860	354,868	32.6	30.3	35.6	2,443	224,594
35-39	0.0131	0.0632	0.0514	0.0743	69,044	66,465	71,641	4,366	334,305	29.3	27.1	32.1	2,373	181,744
40-44	0.0151	0.0730	0.0588	0.0862	64,678	61,526	67,925	4,720	311,591	26.1	24.0	28.7	2,278	150,383
45-49	0.0176	0.0841	0.0673	0.1002	59,958	56,233	63,885	5,045	287,179	23.0	21.1	25.4	2,177	123,948
50-54	0.0211	0.1003	0.0799	0.1200	54,913	50,648	59,523	5,509	260,794	19.8	18.1	22.0	2,045	96,793
55-59	0.0263	0.1235	0.0983	0.1480	49,404	44,658	54,685	6,103	231,763	16.8	15.3	18.7	2,146	81,510
60-64	0.0366	0.1677	0.1341	0.1999	43,301	38,182	49,208	7,261	198,352	13.8	12.5	15.5	2,367	64,661
65-69	0.0520	0.2302	0.1862	0.2713	36,040	30,734	42,611	8,295	159,461	11.1	10.0	12.5	2,522	48,482
70-74	0.0785	0.3282	0.2717	0.3788	27,745	22,588	34,676	9,104	115,962	8.6	7.8	9.8	2,543	32,384
75-79	0.1182	0.4562	0.3910	0.5123	18,640	14,031	25,254	8,504	71,940	6.6	6.0	7.5	2,168	18,340
80-84	0.1688	0.5935	0.5271	0.6473	10,136	6,850	15,379	6,015	35,640	5.1	4.6	5.7	1,351	8,003
85-89	0.2420	0.7539	0.6962	0.7980	4,120	2,423	7,246	3,106	12,836	3.8	3.5	4.2	511	2,111
90-94	0.3374	0.8384	0.7919	0.8730	1,014	492	2,180	850	2,520	2.8	2.6	3.1	104	307
95-99	0.4577	0.9643	0.9441	0.9768	164	63	447	158	345	2.2	2.1	2.3	10	22
100+	0.6037	1	1	1	6	1	25	6	10	1.7	1.6	1.7	1	1
All ages													*69,227*	*3,900,002*

$\ln({}_nq_x)$ with uncertainty interval

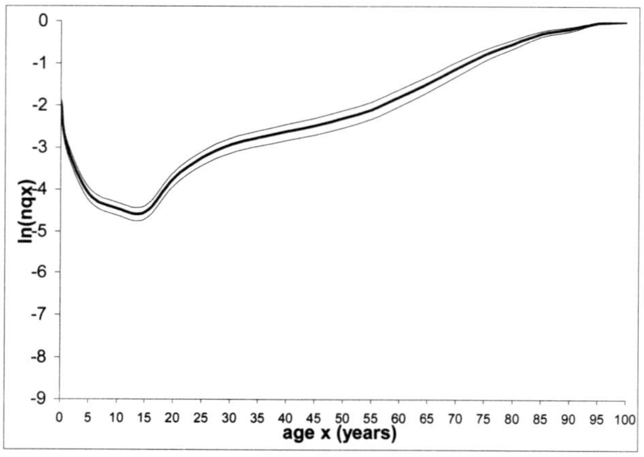

l_x with uncertainty interval

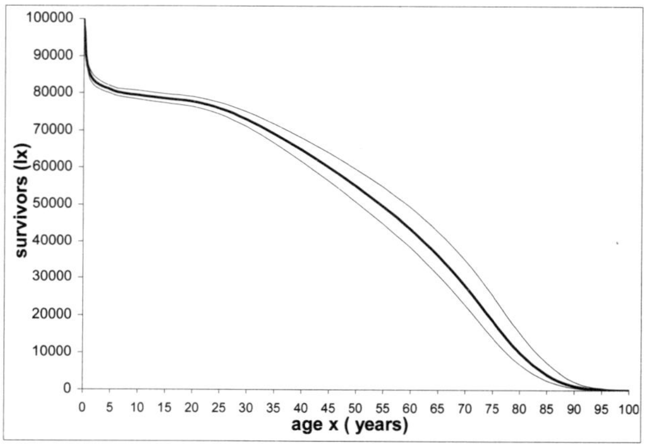

e_x with uncertainty interval

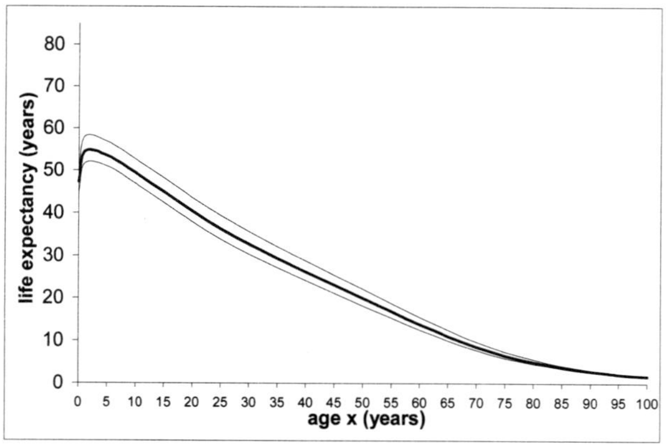

Distribution of deaths by age with uncertainty interval

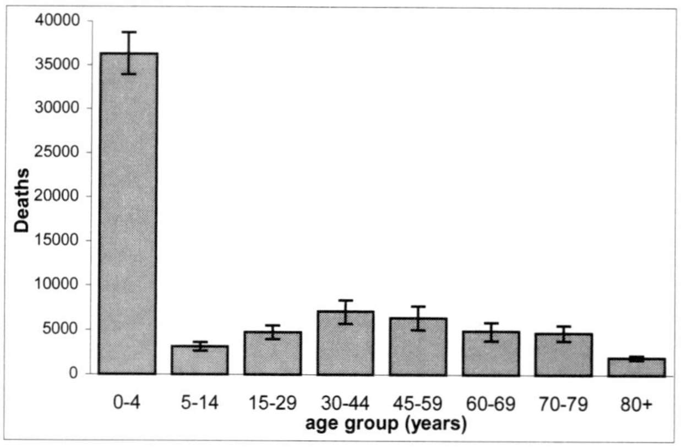

Chad Tchad Chad

Females, 2000 *Femmes, 2000* *Mujeres, 2000*

Age	$_nM_x$	$_nq_x$ mean	$_nq_x$ percentiles 2½	$_nq_x$ percentiles 97½	l_x mean	l_x percentiles 2½	l_x percentiles 97½	$_nd_x$	$_nL_x$	e_x mean	e_x percentiles 2½	e_x percentiles 97½	Actual deaths	Actual population
<1	0.1379	0.1227	0.1104	0.1351	100,000	100,000	100,000	12,269	88,958	51.1	48.1	54.3	22,469	162,920
1-4	0.0144	0.0555	0.0483	0.0633	87,731	86,486	88,955	4,873	339,231	57.3	53.9	61.0	8,328	579,744
5-9	0.0031	0.0156	0.0132	0.0181	82,858	81,869	83,856	1,291	411,066	56.5	53.5	60.1	1,862	593,073
10-14	0.0021	0.0102	0.0087	0.0120	81,568	80,435	82,634	836	405,750	52.4	49.4	55.9	1,014	492,300
15-19	0.0033	0.0164	0.0138	0.0193	80,732	79,476	81,855	1,326	400,346	47.9	45.0	51.3	1,360	410,615
20-24	0.0060	0.0296	0.0247	0.0351	79,406	77,950	80,632	2,354	391,147	43.7	40.8	47.0	2,082	346,087
25-29	0.0084	0.0412	0.0340	0.0492	77,053	75,236	78,600	3,176	377,324	39.9	37.2	43.1	2,409	286,211
30-34	0.0096	0.0471	0.0383	0.0567	73,877	71,555	75,923	3,480	360,685	36.5	34.0	39.6	2,218	229,850
35-39	0.0100	0.0488	0.0393	0.0592	70,397	67,497	72,999	3,438	343,391	33.2	30.9	36.0	1,875	187,286
40-44	0.0097	0.0473	0.0377	0.0576	66,959	63,502	70,109	3,166	326,883	29.8	27.7	32.4	1,521	157,015
45-49	0.0106	0.0518	0.0409	0.0632	63,794	59,846	67,441	3,302	310,715	26.2	24.2	28.6	1,402	131,959
50-54	0.0136	0.0656	0.0516	0.0801	60,492	56,063	64,645	3,968	292,541	22.4	20.7	24.7	1,427	105,252
55-59	0.0181	0.0867	0.0680	0.1058	56,524	51,569	61,259	4,902	270,366	18.8	17.2	20.9	1,649	90,938
60-64	0.0270	0.1264	0.0991	0.1532	51,622	46,114	57,021	6,526	241,796	15.4	14.0	17.3	2,007	74,349
65-69	0.0413	0.1871	0.1476	0.2248	45,096	39,049	51,257	8,437	204,389	12.3	11.0	13.9	2,377	57,574
70-74	0.0655	0.2815	0.2261	0.3324	36,659	30,311	43,661	10,319	157,500	9.5	8.5	10.8	2,641	40,315
75-79	0.1008	0.4027	0.3340	0.4624	26,341	20,340	33,789	10,607	105,187	7.3	6.5	8.3	2,445	24,248
80-84	0.1515	0.5494	0.4763	0.6089	15,734	11,066	22,504	8,644	57,061	5.5	4.9	6.2	1,707	11,271
85-89	0.2217	0.7133	0.6490	0.7631	7,090	4,394	11,785	5,057	22,808	4.1	3.8	4.5	784	3,536
90-94	0.3153	0.8101	0.7592	0.8491	2,033	1,041	4,137	1,647	5,224	3.0	2.8	3.3	214	678
95-99	0.4353	0.9439	0.9174	0.9618	386	157	996	364	837	2.3	2.2	2.4	31	72
100+	0.5838	1	1	1	22	6	82	22	37	1.7	1.7	1.8	2	4
All ages													*61,824*	*3,985,297*

ln($_nq_x$) with uncertainty interval

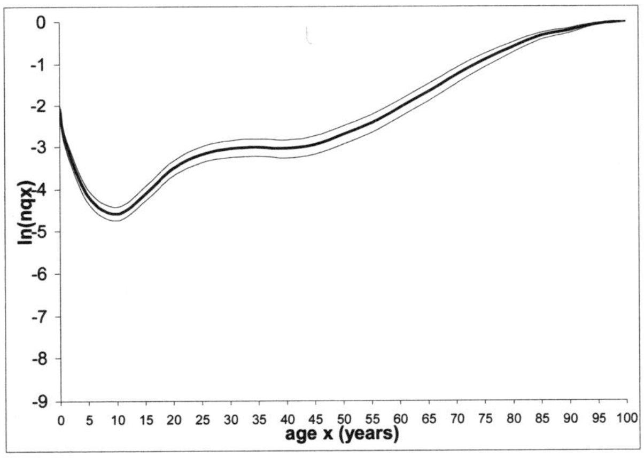

l_x with uncertainty interval

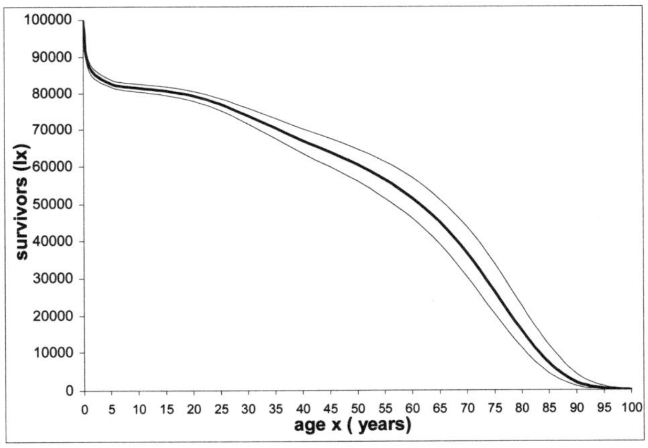

e_x with uncertainty interval

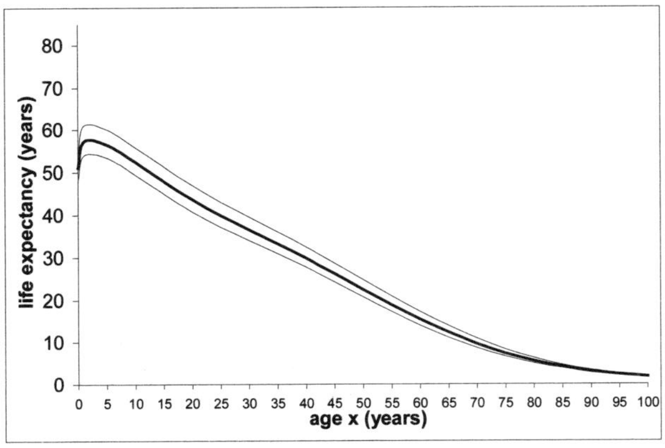

Distribution of deaths by age with uncertainty interval

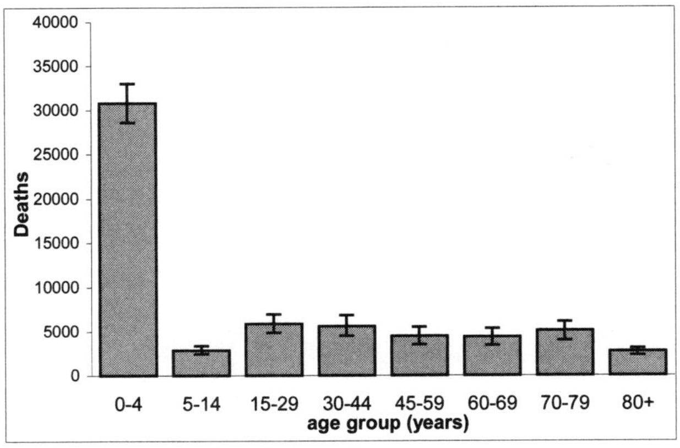

Chile | Chili | Chile

Males, 2000 | *Hommes, 2000* | *Hombres, 2000*

Age	$_nM_x$	$_nq_x$			l_x			$_nd_x$	$_nL_x$	e_x			Actual deaths	Actual population
			percentiles			percentiles					percentiles			
		mean	2½	97½	mean	2½	97½			mean	2½	97½		
<1	0.0100	0.0099	0.0078	0.0113	100,000	100,000	100,000	991	99,108	72.5	72.0	73.8	1,464	146,440
1-4	0.0006	0.0022	0.0018	0.0025	99,009	98,873	99,217	221	395,504	72.2	71.6	73.5	329	587,151
5-9	0.0003	0.0014	0.0012	0.0015	98,787	98,630	99,035	138	493,592	68.4	67.8	69.7	208	743,576
10-14	0.0003	0.0013	0.0011	0.0014	98,649	98,479	98,920	128	492,926	63.5	62.9	64.8	189	725,468
15-19	0.0008	0.0039	0.0032	0.0042	98,521	98,339	98,814	383	491,647	58.6	57.9	59.8	507	649,887
20-24	0.0013	0.0065	0.0055	0.0071	98,138	97,922	98,494	641	489,086	53.8	53.1	55.1	794	606,134
25-29	0.0016	0.0080	0.0068	0.0086	97,497	97,231	97,954	777	485,542	49.1	48.5	50.4	975	609,474
30-34	0.0019	0.0093	0.0079	0.0099	96,720	96,400	97,291	895	481,362	44.5	43.8	45.7	1,138	611,702
35-39	0.0021	0.0105	0.0091	0.0112	95,825	95,448	96,522	1,010	476,597	39.9	39.2	41.1	1,279	603,456
40-44	0.0030	0.0149	0.0129	0.0157	94,814	94,381	95,649	1,412	470,542	35.3	34.6	36.5	1,566	521,975
45-49	0.0040	0.0198	0.0172	0.0208	93,403	92,901	94,419	1,850	462,389	30.8	30.1	31.9	1,697	424,341
50-54	0.0070	0.0344	0.0302	0.0361	91,553	90,972	92,793	3,149	449,892	26.3	25.7	27.5	2,430	347,132
55-59	0.0110	0.0535	0.0469	0.0567	88,404	87,687	89,991	4,732	430,189	22.2	21.5	23.2	3,144	285,847
60-64	0.0160	0.0769	0.0671	0.0825	83,672	82,720	85,769	6,436	402,268	18.3	17.7	19.3	3,445	215,332
65-69	0.0280	0.1308	0.1143	0.1417	77,235	75,896	80,013	10,106	360,914	14.6	14.0	15.5	4,852	173,297
70-74	0.0440	0.1982	0.1747	0.2154	67,130	65,144	70,866	13,305	302,387	11.4	10.9	12.1	5,712	129,811
75-79	0.0670	0.2869	0.2575	0.3104	53,825	51,111	58,486	15,444	230,513	8.7	8.2	9.2	5,445	81,273
80-84	0.1230	0.4704	0.4362	0.5016	38,380	35,249	43,426	18,053	146,771	6.1	5.8	6.5	5,362	43,593
85-89	0.2080	0.6842	0.6562	0.7127	20,328	17,567	24,483	13,908	66,867	4.3	4.1	4.6	3,811	18,322
90-94	0.2780	0.7579	0.7366	0.7806	6,419	5,048	8,417	4,865	17,501	3.4	3.2	3.5	1,602	5,764
95-99	0.3760	0.8835	0.8696	0.8982	1,554	1,108	2,217	1,373	3,651	2.6	2.5	2.7	398	1,059
100+	0.5010	1	1	1	181	113	289	181	362	2.0	2.0	2.2	69	137
All ages													*46,417*	*7,531,171*

ln($_nq_x$) with uncertainty interval

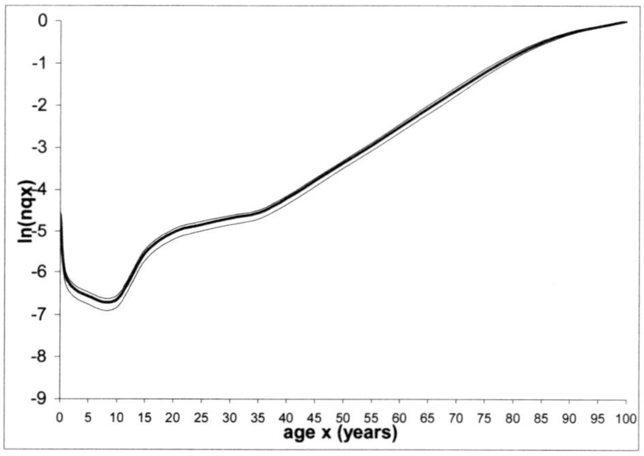

l_x with uncertainty interval

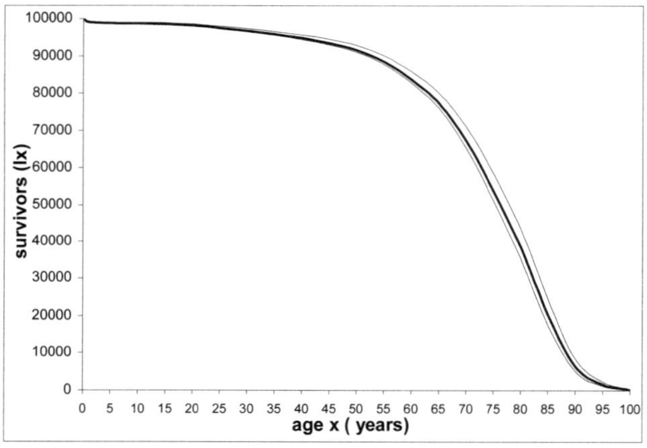

e_x with uncertainty interval

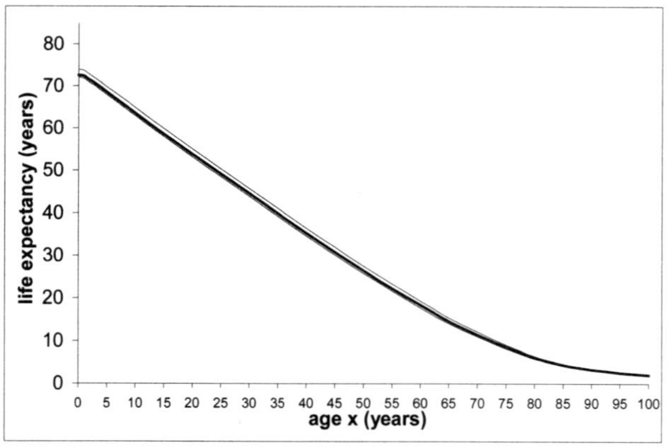

Distribution of deaths by age with uncertainty interval

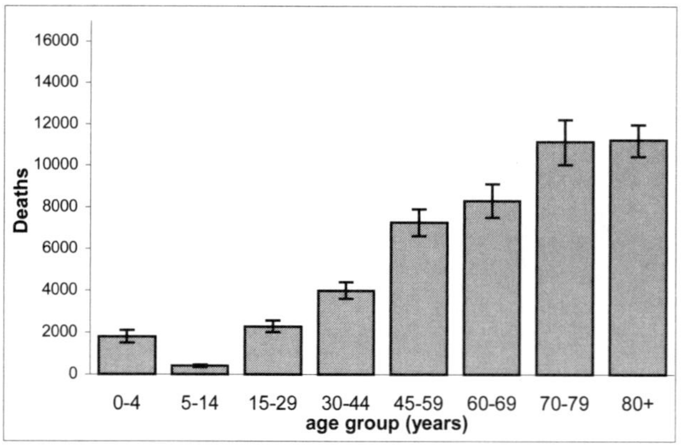

Chile | Chili | Chile

Females, 2000 | *Femmes, 2000* | *Mujeres, 2000*

Age	$_nM_x$	$_nq_x$ mean	$_nq_x$ percentiles 2½	$_nq_x$ percentiles 97½	l_x mean	l_x percentiles 2½	l_x percentiles 97½	$_nd_x$	$_nL_x$	e_x mean	e_x percentiles 2½	e_x percentiles 97½	Actual deaths	Actual population
<1	0.0080	0.0079	0.0067	0.0092	100,000	100,000	100,000	794	99,285	79.5	78.9	80.5	1,116	139,533
1-4	0.0004	0.0016	0.0014	0.0018	99,206	99,077	99,333	163	396,433	79.1	78.5	80.1	233	567,741
5-9	0.0002	0.0009	0.0008	0.0011	99,043	98,894	99,191	94	494,981	75.3	74.6	76.2	136	717,606
10-14	0.0002	0.0008	0.0007	0.0009	98,949	98,790	99,109	84	494,535	70.3	69.7	71.3	119	700,685
15-19	0.0003	0.0014	0.0013	0.0016	98,865	98,697	99,036	143	493,967	65.4	64.8	66.4	182	629,167
20-24	0.0003	0.0016	0.0014	0.0018	98,722	98,539	98,910	158	493,214	60.5	59.8	61.5	189	591,039
25-29	0.0004	0.0019	0.0017	0.0021	98,564	98,365	98,770	187	492,352	55.6	54.9	56.6	228	600,479
30-34	0.0006	0.0028	0.0025	0.0031	98,377	98,160	98,605	275	491,197	50.7	50.0	51.6	340	607,334
35-39	0.0009	0.0043	0.0039	0.0047	98,102	97,860	98,361	426	489,445	45.8	45.2	46.8	527	605,890
40-44	0.0012	0.0062	0.0055	0.0067	97,676	97,399	97,981	604	486,871	41.0	40.4	42.0	656	528,691
45-49	0.0022	0.0107	0.0096	0.0115	97,072	96,749	97,440	1,043	482,755	36.2	35.6	37.2	940	435,308
50-54	0.0030	0.0149	0.0134	0.0158	96,030	95,639	96,501	1,430	476,573	31.6	31.0	32.5	1,093	364,345
55-59	0.0050	0.0247	0.0223	0.0263	94,600	94,124	95,206	2,336	467,160	27.0	26.4	27.9	1,553	310,622
60-64	0.0080	0.0392	0.0351	0.0421	92,264	91,648	93,083	3,618	452,275	22.7	22.1	23.5	1,957	244,563
65-69	0.0140	0.0676	0.0600	0.0731	88,646	87,792	89,819	5,995	428,241	18.5	18.0	19.3	2,974	212,460
70-74	0.0230	0.1087	0.0962	0.1179	82,650	81,378	84,428	8,988	390,782	14.6	14.2	15.4	3,970	172,603
75-79	0.0390	0.1777	0.1578	0.1922	73,662	71,784	76,309	13,088	335,592	11.1	10.7	11.7	4,764	122,165
80-84	0.0770	0.3229	0.2924	0.3462	60,574	57,988	64,268	19,557	253,981	8.0	7.6	8.5	5,820	75,581
85-89	0.1430	0.5267	0.4937	0.5544	41,018	37,911	45,476	21,604	151,079	5.6	5.4	5.9	5,313	37,156
90-94	0.2160	0.6553	0.6289	0.6787	19,414	16,893	23,023	12,723	58,901	4.0	3.9	4.3	2,988	13,834
95-99	0.3200	0.8163	0.7979	0.8324	6,691	5,428	8,545	5,462	17,069	2.9	2.9	3.1	949	2,965
100+	0.4700	1	1	1	1,229	910	1,727	1,229	2,615	2.1	2.1	2.4	174	370
All ages													*36,223*	*7,680,137*

ln($_nq_x$) with uncertainty interval

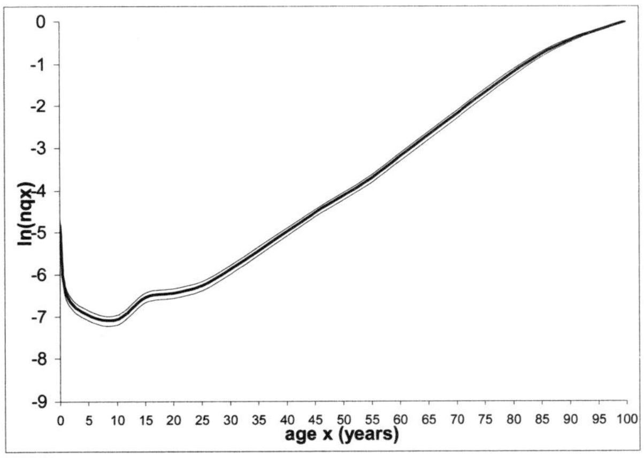

l_x with uncertainty interval

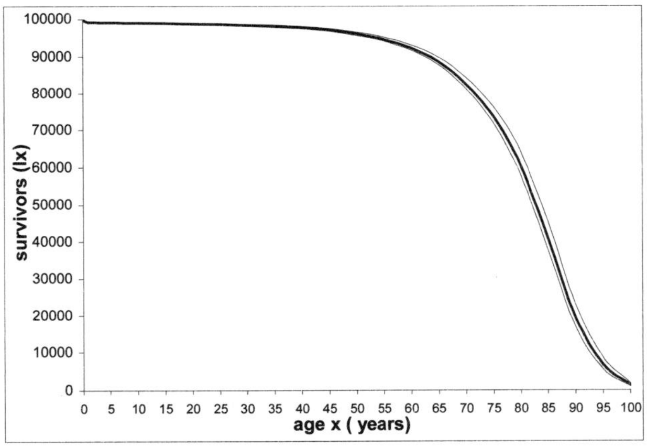

e_x with uncertainty interval

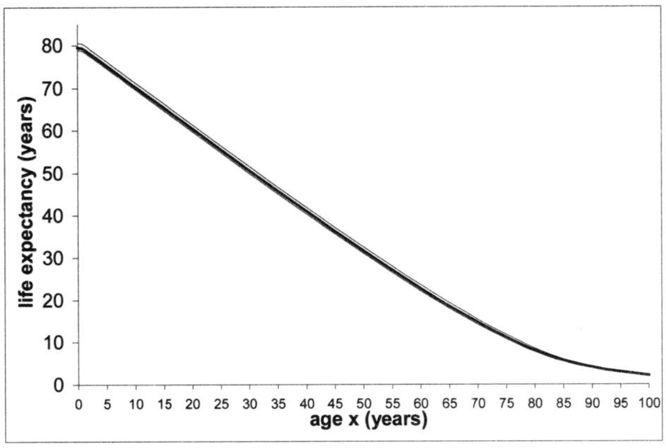

Distribution of deaths by age with uncertainty interval

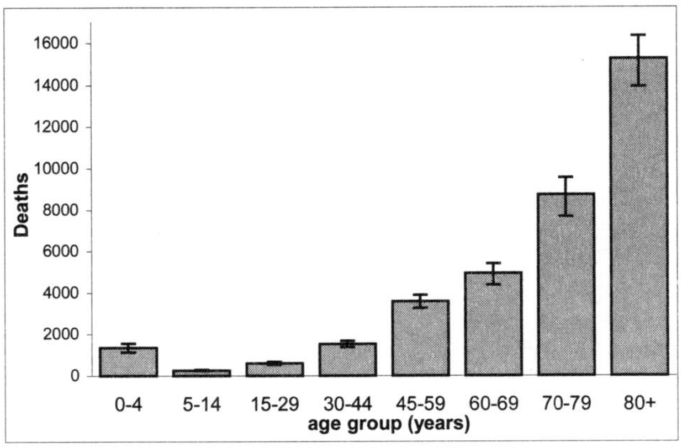

China | Chine | China

Males, 2000 | *Hommes, 2000* | *Hombres, 2000*

Age	$_nM_x$	$_nq_x$			l_x			$_nd_x$	$_nL_x$	e_x			Actual deaths	Actual population
			percentiles			percentiles					percentiles			
		mean	2 1/2	97 1/2	mean	2 1/2	97 1/2			mean	2 1/2	97 1/2		
<1	0.0295	0.0288	0.0243	0.0333	100,000	100,000	100,000	2,878	97,410	68.9	68.1	69.6	289,240	9,789,358
1-4	0.0023	0.0091	0.0080	0.0101	97,122	96,668	97,571	881	386,373	69.9	69.2	70.6	94,181	41,302,457
5-9	0.0008	0.0038	0.0033	0.0042	96,241	95,693	96,791	363	480,297	66.5	65.8	67.3	40,990	54,254,253
10-14	0.0006	0.0028	0.0025	0.0031	95,878	95,292	96,467	268	478,721	61.8	61.1	62.5	34,665	62,010,609
15-19	0.0010	0.0049	0.0044	0.0054	95,610	94,997	96,227	466	476,888	56.9	56.2	57.7	51,417	52,665,364
20-24	0.0013	0.0066	0.0059	0.0073	95,145	94,486	95,808	627	474,156	52.2	51.5	52.9	67,408	50,962,035
25-29	0.0013	0.0066	0.0059	0.0072	94,518	93,801	95,240	622	471,033	47.5	46.8	48.2	82,307	62,313,709
30-34	0.0017	0.0084	0.0076	0.0091	93,895	93,125	94,673	785	467,516	42.8	42.2	43.5	108,743	64,799,597
35-39	0.0021	0.0105	0.0096	0.0114	93,111	92,276	93,954	982	463,100	38.2	37.5	38.9	114,417	53,963,309
40-44	0.0030	0.0151	0.0139	0.0163	92,129	91,219	93,049	1,390	457,169	33.5	32.9	34.2	133,615	43,936,472
45-49	0.0047	0.0232	0.0216	0.0249	90,739	89,734	91,755	2,109	448,422	29.0	28.4	29.7	208,544	44,348,433
50-54	0.0076	0.0371	0.0346	0.0397	88,630	87,500	89,773	3,289	434,928	24.6	24.1	25.3	246,424	32,586,002
55-59	0.0123	0.0598	0.0557	0.0642	85,341	84,056	86,637	5,105	413,943	20.5	19.9	21.1	301,137	24,418,990
60-64	0.0208	0.0989	0.0915	0.1063	80,236	78,790	81,684	7,933	381,349	16.6	16.1	17.2	446,780	21,478,264
65-69	0.0335	0.1547	0.1429	0.1660	72,304	70,755	73,862	11,185	333,555	13.2	12.7	13.7	591,037	17,625,451
70-74	0.0566	0.2480	0.2300	0.2652	61,118	59,035	63,130	15,156	267,701	10.1	9.7	10.6	678,920	11,991,424
75-79	0.0881	0.3610	0.3366	0.3858	45,962	43,452	48,603	16,590	188,334	7.7	7.3	8.0	604,590	6,863,420
80-84	0.1442	0.5299	0.5023	0.5589	29,372	26,971	32,135	15,565	107,946	5.6	5.3	5.9	434,883	3,015,983
85-89	0.2214	0.7126	0.6887	0.7374	13,807	12,055	15,876	9,839	44,436	4.1	3.9	4.2	199,985	903,178
90-94	0.3190	0.8150	0.7969	0.8335	3,968	3,166	4,933	3,234	10,137	3.0	2.9	3.1	52,049	163,173
95-99	0.4311	0.9399	0.9306	0.9490	734	527	997	690	1,600	2.3	2.2	2.3	7,343	17,033
100+	0.5465	1	1	1	44	27	69	44	81	1.8	1.8	1.9	599	1,096
All ages													*4,789,275*	*659,409,610*

ln($_nq_x$) with uncertainty interval

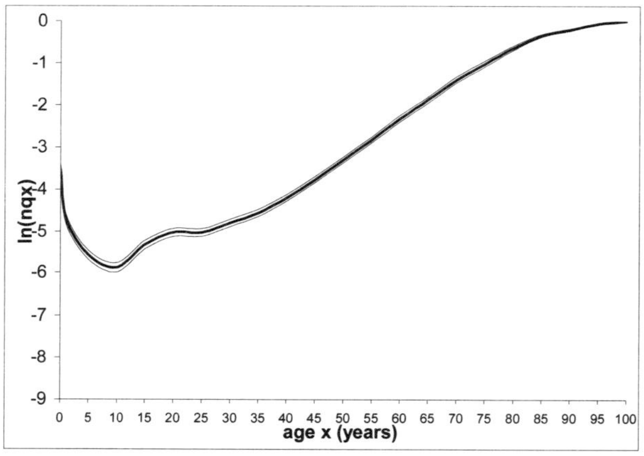

l_x with uncertainty interval

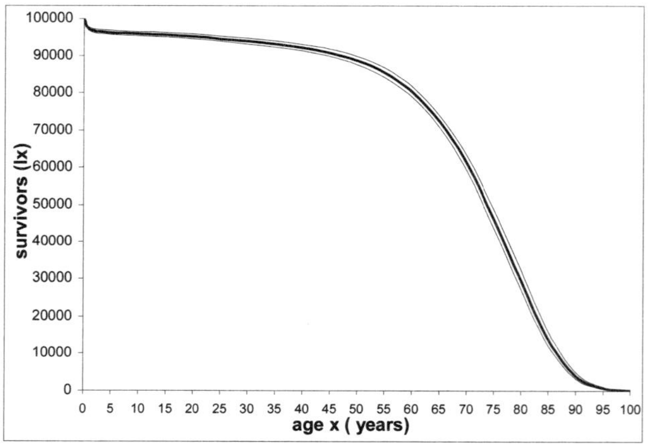

e_x with uncertainty interval

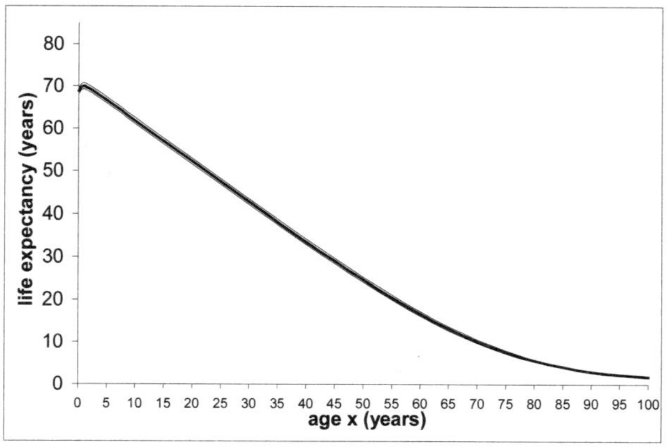

Distribution of deaths by age with uncertainty interval

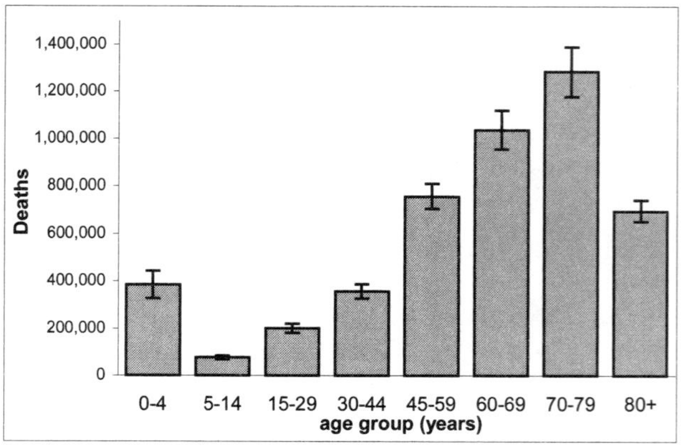

China | Chine | China

Females, 2000 | *Femmes, 2000* | *Mujeres, 2000*

Age	$_nM_x$	$_nq_x$			l_x			$_nd_x$	$_nL_x$	e_x			Actual deaths	Actual population
			percentiles			percentiles					percentiles			
		mean	2½	97½	mean	2½	97½			mean	2½	97½		
<1	0.0360	0.0349	0.0302	0.0399	100,000	100,000	100,000	3,489	96,860	73.0	71.9	74.2	327,161	9,081,978
1-4	0.0023	0.0093	0.0083	0.0103	96,511	96,010	96,981	899	383,885	74.6	73.5	75.9	86,238	36,810,399
5-9	0.0006	0.0028	0.0025	0.0031	95,611	95,062	96,159	268	477,387	71.3	70.1	72.6	27,297	48,638,859
10-14	0.0004	0.0021	0.0019	0.0024	95,344	94,774	95,914	203	476,210	66.5	65.3	67.8	23,946	56,168,606
15-19	0.0008	0.0040	0.0036	0.0045	95,141	94,553	95,729	385	474,741	61.7	60.5	63.0	39,365	48,594,588
20-24	0.0011	0.0056	0.0050	0.0061	94,756	94,135	95,380	527	472,462	56.9	55.7	58.2	53,297	47,770,057
25-29	0.0011	0.0053	0.0048	0.0058	94,229	93,564	94,901	499	469,896	52.2	51.0	53.5	62,886	59,211,993
30-34	0.0012	0.0061	0.0055	0.0067	93,730	93,018	94,446	567	467,231	47.5	46.3	48.8	75,517	62,217,774
35-39	0.0015	0.0073	0.0066	0.0080	93,163	92,399	93,928	681	464,111	42.7	41.5	44.0	75,365	51,378,028
40-44	0.0020	0.0101	0.0091	0.0111	92,482	91,658	93,304	933	460,076	38.0	36.8	39.3	82,394	40,608,874
45-49	0.0032	0.0156	0.0141	0.0171	91,548	90,644	92,444	1,431	454,164	33.4	32.2	34.7	132,174	41,938,969
50-54	0.0049	0.0243	0.0220	0.0266	90,117	89,097	91,119	2,189	445,114	28.9	27.7	30.2	148,237	30,147,872
55-59	0.0075	0.0370	0.0333	0.0407	87,928	86,744	89,093	3,255	431,505	24.5	23.4	25.8	170,412	22,592,422
60-64	0.0125	0.0604	0.0536	0.0677	84,674	83,273	86,057	5,113	410,585	20.4	19.3	21.6	252,304	20,259,508
65-69	0.0199	0.0948	0.0826	0.1080	79,560	77,847	81,302	7,543	378,945	16.5	15.5	17.7	352,129	17,690,857
70-74	0.0340	0.1568	0.1346	0.1804	72,018	69,866	74,271	11,292	331,858	13.0	12.1	14.1	452,380	13,294,807
75-79	0.0545	0.2398	0.2053	0.2753	60,726	57,413	63,962	14,564	267,217	10.0	9.2	10.9	493,336	9,051,362
80-84	0.0939	0.3802	0.3310	0.4269	46,161	41,609	50,747	17,548	186,935	7.3	6.7	8.0	459,879	4,898,837
85-89	0.1538	0.5555	0.5014	0.6020	28,613	23,848	33,949	15,893	103,330	5.3	4.9	5.8	300,783	1,955,549
90-94	0.2397	0.6972	0.6523	0.7335	12,719	9,492	16,926	8,868	36,992	3.7	3.5	4.0	141,852	591,702
95-99	0.3555	0.8601	0.8312	0.8824	3,851	2,530	5,885	3,312	9,318	2.7	2.6	2.9	40,111	112,841
100+	0.5014	1	1	1	539	297	993	539	1,075	2.0	2.0	2.1	5,770	11,508
All ages													*3,802,835*	*623,027,390*

$ln(_nq_x)$ with uncertainty interval

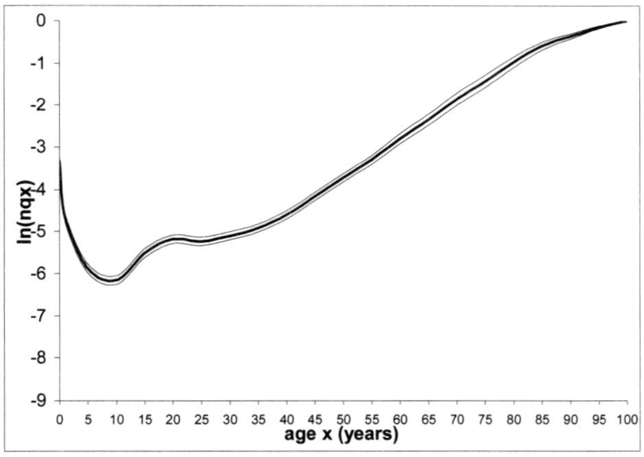

l_x with uncertainty interval

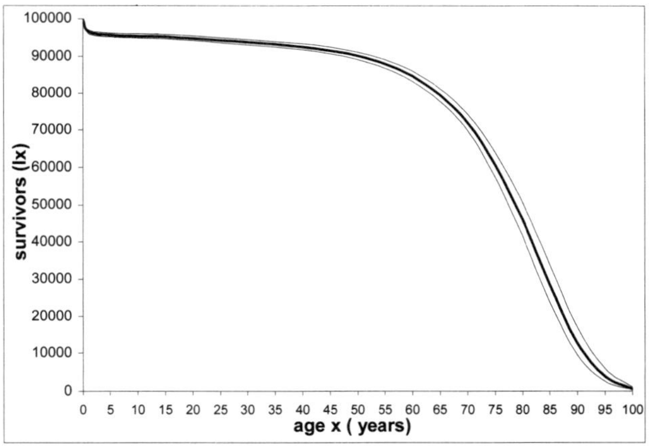

e_x with uncertainty interval

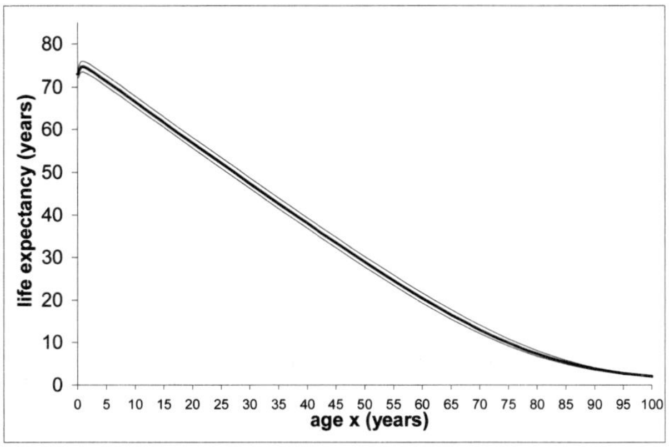

Distribution of deaths by age with uncertainty interval

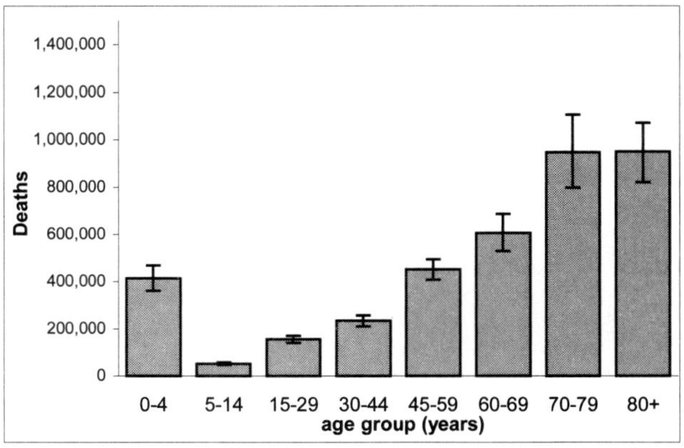

Colombia Colombie Colombia

Males, 2000 *Hommes, 2000* *Hombres, 2000*

Age	$_nM_x$	$_nq_x$			l_x			$_nd_x$	$_nL_x$	e_x			Actual deaths	Actual population
			percentiles			percentiles					percentiles			
		mean	2½	97½	mean	2½	97½			mean	2½	97½		
<1	0.0242	0.0237	0.0197	0.0277	100,000	100,000	100,000	2,366	97,871	67.2	66.3	68.1	11,775	487,129
1-4	0.0013	0.0053	0.0046	0.0060	97,634	97,234	98,029	519	389,291	67.8	67.0	68.9	2,590	1,941,744
5-9	0.0005	0.0027	0.0024	0.0030	97,115	96,665	97,565	262	484,922	64.2	63.3	65.3	1,295	2,401,172
10-14	0.0007	0.0033	0.0029	0.0037	96,854	96,382	97,334	322	483,463	59.4	58.5	60.5	1,471	2,207,066
15-19	0.0033	0.0162	0.0145	0.0179	96,531	96,026	97,047	1,565	478,746	54.5	53.6	55.7	6,834	2,091,100
20-24	0.0052	0.0258	0.0235	0.0281	94,967	94,309	95,641	2,450	468,708	50.4	49.5	51.6	10,206	1,952,251
25-29	0.0051	0.0253	0.0233	0.0272	92,516	91,661	93,396	2,336	456,741	46.7	45.7	47.9	8,771	1,714,620
30-34	0.0047	0.0234	0.0218	0.0250	90,180	89,169	91,216	2,107	445,634	42.8	41.8	44.0	7,927	1,676,982
35-39	0.0048	0.0239	0.0225	0.0254	88,073	86,946	89,224	2,108	435,097	38.8	37.8	40.0	7,179	1,481,772
40-44	0.0052	0.0256	0.0240	0.0272	85,965	84,740	87,208	2,203	424,319	34.7	33.7	35.8	6,315	1,216,249
45-49	0.0059	0.0292	0.0271	0.0312	83,762	82,453	85,082	2,447	412,695	30.5	29.5	31.7	5,912	997,276
50-54	0.0079	0.0386	0.0354	0.0414	81,316	79,932	82,699	3,137	398,736	26.4	25.4	27.5	5,993	761,774
55-59	0.0121	0.0586	0.0533	0.0632	78,179	76,726	79,633	4,583	379,436	22.3	21.4	23.4	6,757	559,401
60-64	0.0193	0.0919	0.0827	0.1003	73,596	72,078	75,108	6,762	351,074	18.6	17.7	19.6	8,152	423,260
65-69	0.0302	0.1404	0.1255	0.1549	66,834	65,245	68,420	9,384	310,710	15.2	14.3	16.1	9,811	324,874
70-74	0.0432	0.1950	0.1738	0.2163	57,450	55,424	59,736	11,205	259,240	12.2	11.5	13.1	10,384	240,266
75-79	0.0634	0.2736	0.2451	0.3024	46,246	43,660	49,351	12,653	199,596	9.6	9.0	10.3	10,484	165,387
80-84	0.0987	0.3960	0.3603	0.4316	33,593	30,456	37,254	13,302	134,709	7.3	6.8	7.8	8,568	86,765
85-89	0.1505	0.5467	0.5092	0.5832	20,291	17,310	23,831	11,092	73,722	5.4	5.1	5.8	5,970	39,675
90-94	0.2242	0.6703	0.6377	0.7018	9,198	7,215	11,696	6,165	27,495	3.9	3.7	4.2	2,996	13,362
95-99	0.3269	0.8252	0.8013	0.8479	3,033	2,152	4,238	2,503	7,656	2.9	2.8	3.0	1,068	3,268
100+	0.4661	1	1	1	530	327	842	530	1,138	2.1	2.1	2.2	288	619
All ages													*140,747*	*20,786,012*

$\ln(_nq_x)$ with uncertainty interval

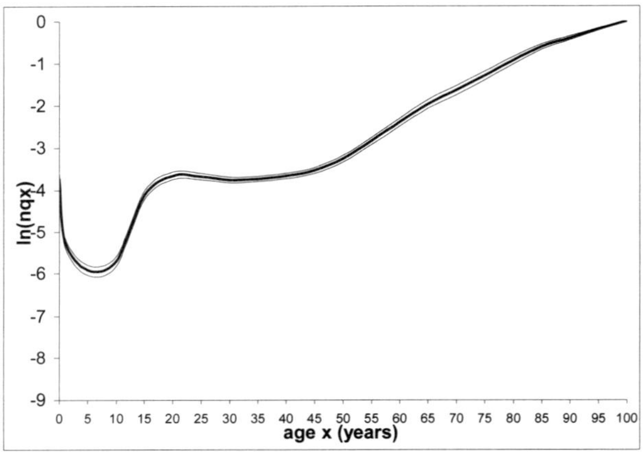

l_x with uncertainty interval

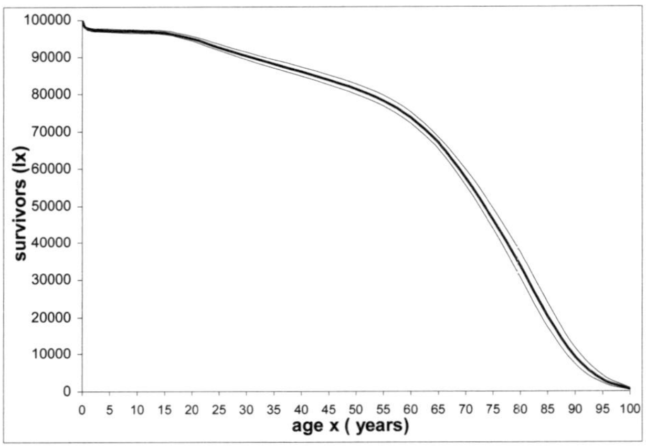

e_x with uncertainty interval

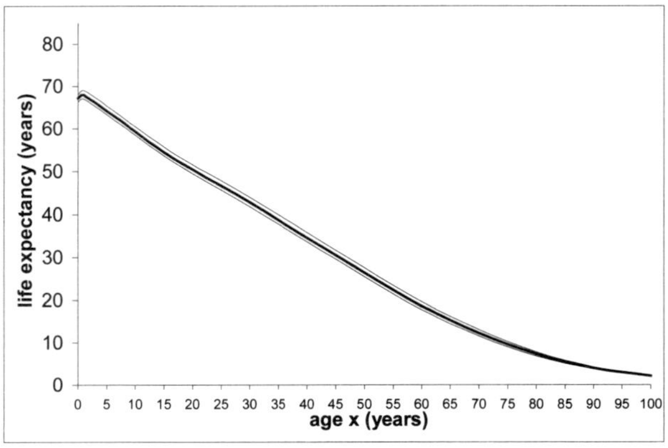

Distribution of deaths by age with uncertainty interval

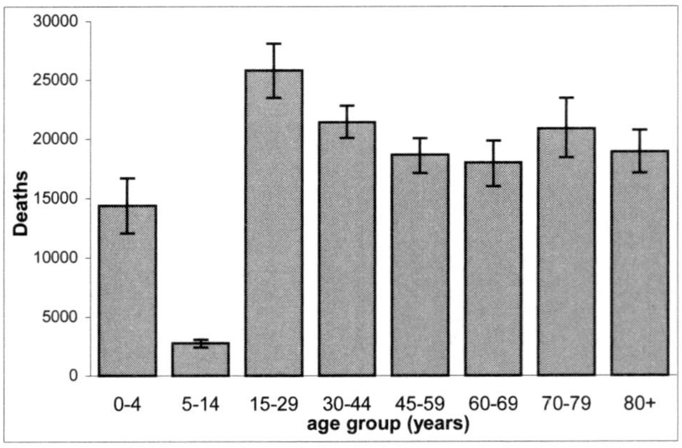

Colombia | Colombie | Colombia

Females, 2000 | *Femmes, 2000* | *Mujeres, 2000*

Age	$_nM_x$	$_nq_x$			l_x			$_nd_x$	$_nL_x$	e_x			Actual deaths	Actual population
			percentiles			percentiles					percentiles			
		mean	2½	97½	mean	2½	97½			mean	2½	97½		
<1	0.0166	0.0163	0.0136	0.0192	100,000	100,000	100,000	1,633	98,530	75.1	74.4	76.0	7,684	463,491
1-4	0.0011	0.0045	0.0039	0.0051	98,367	98,081	98,644	445	392,397	75.3	74.7	76.2	2,120	1,867,472
5-9	0.0004	0.0019	0.0016	0.0021	97,921	97,582	98,255	184	489,145	71.7	71.1	72.5	870	2,308,562
10-14	0.0004	0.0021	0.0018	0.0023	97,737	97,377	98,093	205	488,173	66.8	66.2	67.6	894	2,129,507
15-19	0.0009	0.0045	0.0040	0.0050	97,532	97,151	97,912	444	486,551	61.9	61.3	62.8	1,851	2,029,789
20-24	0.0009	0.0044	0.0039	0.0048	97,088	96,662	97,518	425	484,378	57.2	56.6	58.0	1,700	1,935,921
25-29	0.0010	0.0052	0.0046	0.0056	96,663	96,197	97,137	499	482,068	52.4	51.9	53.3	1,822	1,760,845
30-34	0.0011	0.0057	0.0052	0.0062	96,164	95,656	96,687	549	479,450	47.7	47.1	48.5	2,020	1,765,318
35-39	0.0016	0.0078	0.0071	0.0084	95,616	95,064	96,189	748	476,210	43.0	42.4	43.8	2,483	1,581,668
40-44	0.0021	0.0107	0.0098	0.0115	94,868	94,263	95,505	1,011	471,814	38.3	37.7	39.1	2,833	1,322,415
45-49	0.0032	0.0159	0.0147	0.0170	93,857	93,189	94,571	1,489	465,565	33.7	33.1	34.4	3,523	1,102,007
50-54	0.0052	0.0257	0.0239	0.0274	92,369	91,624	93,181	2,375	455,907	29.2	28.6	29.9	4,379	840,590
55-59	0.0083	0.0409	0.0378	0.0433	89,994	89,158	90,931	3,678	440,775	24.9	24.4	25.6	5,106	611,912
60-64	0.0143	0.0689	0.0635	0.0731	86,316	85,342	87,386	5,944	416,719	20.8	20.4	21.5	6,856	480,618
65-69	0.0229	0.1083	0.0997	0.1148	80,372	79,197	81,571	8,703	380,101	17.2	16.8	17.8	8,850	386,512
70-74	0.0320	0.1484	0.1369	0.1568	71,669	70,263	73,438	10,633	331,761	14.0	13.6	14.5	9,546	297,837
75-79	0.0499	0.2219	0.2065	0.2338	61,036	59,255	63,382	13,546	271,314	11.0	10.6	11.4	10,666	213,636
80-84	0.0782	0.3270	0.3083	0.3446	47,490	45,457	50,296	15,528	198,629	8.4	8.1	8.7	9,395	120,176
85-89	0.1214	0.4657	0.4433	0.4871	31,962	30,025	34,792	14,883	122,601	6.2	6.0	6.5	7,609	62,680
90-94	0.1870	0.5989	0.5763	0.6197	17,079	15,607	19,249	10,228	54,708	4.5	4.3	4.7	4,894	26,174
95-99	0.2856	0.7690	0.7504	0.7854	6,850	5,935	8,015	5,268	18,447	3.2	3.1	3.3	2,464	8,627
100+	0.4326	1	1	1	1,582	1,273	1,965	1,582	3,657	2.3	2.2	2.4	1,268	2,932
All ages													*98,831*	*21,318,689*

$\ln(_nq_x)$ with uncertainty interval

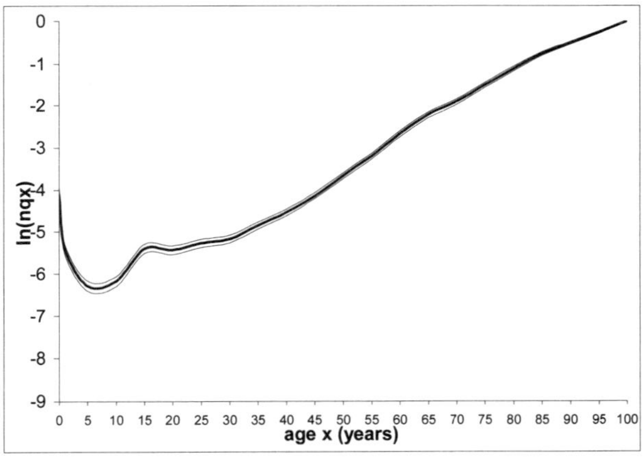

l_x with uncertainty interval

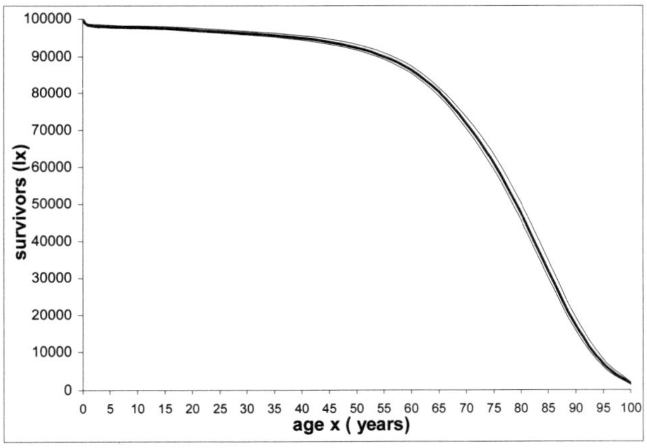

e_x with uncertainty interval

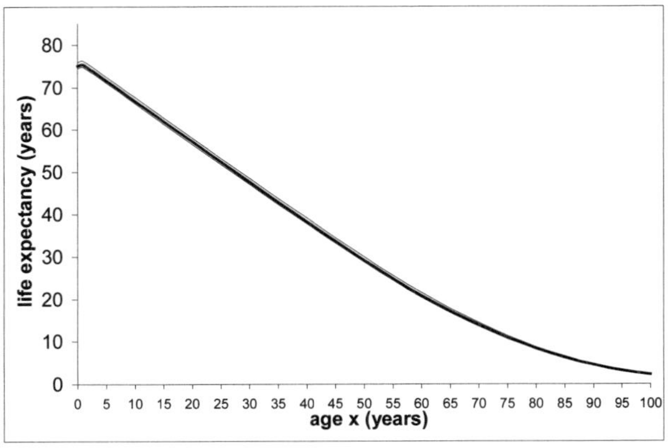

Distribution of deaths by age with uncertainty interval

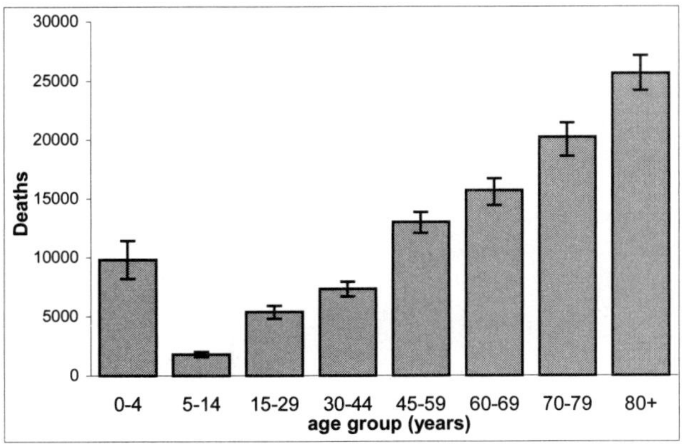

Comoros   Comores   Comoras

Males, 2000   *Hommes, 2000*   *Hombres, 2000*

Age	$_nM_x$	$_nq_x$			l_x			$_nd_x$	$_nL_x$	e_x			Actual deaths	Actual population
			percentiles			percentiles					percentiles			
		mean	2½	97½	mean	2½	97½			mean	2½	97½		
<1	0.0877	0.0813	0.0744	0.0885	100,000	100,000	100,000	8,129	92,684	55.3	53.6	57.7	1,154	13,153
1-4	0.0070	0.0275	0.0246	0.0300	91,871	91,148	92,564	2,524	361,426	59.2	57.3	61.6	324	46,346
5-9	0.0021	0.0104	0.0092	0.0114	89,347	88,602	90,097	926	444,421	56.8	55.0	59.1	104	49,864
10-14	0.0015	0.0076	0.0066	0.0083	88,421	87,603	89,238	668	440,436	52.4	50.6	54.6	68	44,837
15-19	0.0025	0.0123	0.0108	0.0136	87,753	86,876	88,630	1,082	436,060	47.8	46.0	50.0	105	42,163
20-24	0.0041	0.0204	0.0178	0.0225	86,671	85,698	87,654	1,770	428,928	43.3	41.6	45.5	141	34,180
25-29	0.0049	0.0243	0.0210	0.0269	84,901	83,771	86,060	2,060	419,352	39.2	37.5	41.2	138	28,064
30-34	0.0061	0.0301	0.0259	0.0336	82,840	81,530	84,205	2,494	407,966	35.1	33.5	37.1	142	23,206
35-39	0.0080	0.0393	0.0337	0.0442	80,347	78,822	81,981	3,160	393,833	31.1	29.5	33.0	141	17,525
40-44	0.0108	0.0523	0.0446	0.0594	77,187	75,395	79,221	4,040	375,833	27.3	25.8	29.1	150	13,951
45-49	0.0143	0.0691	0.0586	0.0788	73,146	71,032	75,690	5,051	353,104	23.7	22.3	25.4	159	11,097
50-54	0.0193	0.0919	0.0778	0.1054	68,095	65,607	71,258	6,258	324,833	20.2	18.9	21.8	171	8,889
55-59	0.0258	0.1214	0.1029	0.1394	61,838	58,883	65,715	7,507	290,422	17.0	15.9	18.5	179	6,929
60-64	0.0361	0.1654	0.1404	0.1892	54,331	50,675	58,952	8,984	249,195	14.0	13.0	15.3	189	5,232
65-69	0.0507	0.2250	0.1920	0.2550	45,347	41,088	50,635	10,205	201,222	11.3	10.5	12.4	188	3,714
70-74	0.0753	0.3168	0.2742	0.3528	35,142	30,610	40,789	11,133	147,876	8.9	8.2	9.7	178	2,370
75-79	0.1114	0.4358	0.3865	0.4749	24,009	19,810	29,390	10,463	93,888	6.8	6.4	7.5	145	1,297
80-84	0.1643	0.5822	0.5328	0.6199	13,546	10,414	17,851	7,887	48,014	5.2	4.9	5.6	91	557
85-89	0.2359	0.7419	0.7005	0.7725	5,659	3,977	8,340	4,198	17,800	3.9	3.7	4.2	38	161
90-94	0.3299	0.8290	0.7966	0.8529	1,461	913	2,498	1,211	3,671	2.9	2.8	3.1	10	29
95-99	0.4495	0.9570	0.9422	0.9669	250	136	508	239	532	2.2	2.2	2.3	1	3
100+	0.5967	1	1	1	11	5	29	11	18	1.7	1.6	1.8	0	0
All ages													*3,815*	*353,567*

ln($_nq_x$) with uncertainty interval

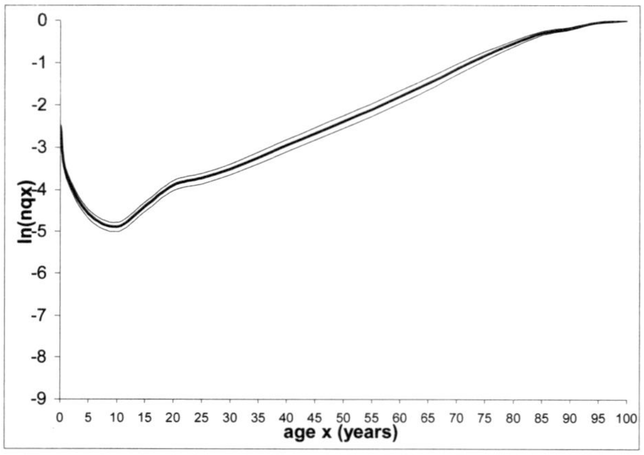

l_x with uncertainty interval

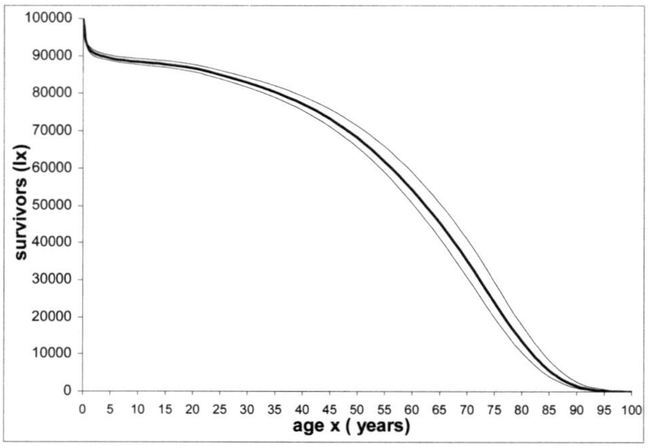

e_x with uncertainty interval

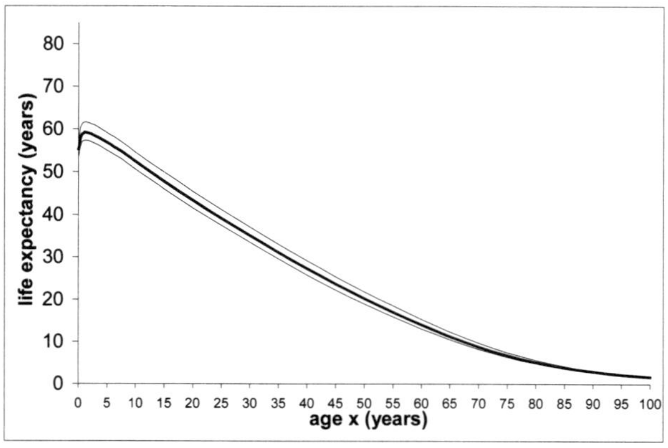

Distribution of deaths by age with uncertainty interval

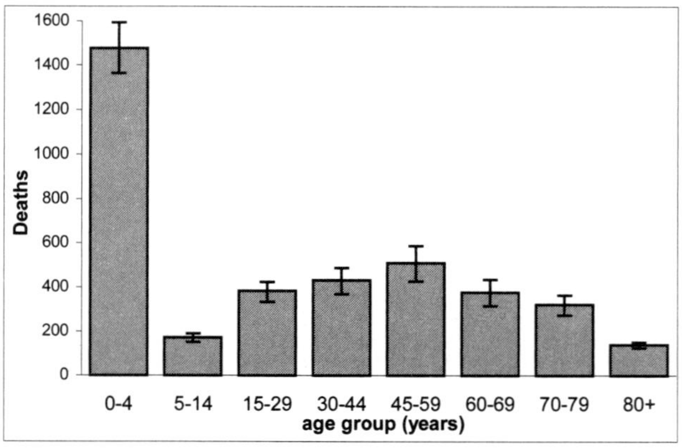

Comoros **Comores** **Comoras**

Females, 2000 *Femmes, 2000* *Mujeres, 2000*

Age	$_nM_x$	$_nq_x$			l_x			$_nd_x$	$_nL_x$	e_x			Actual deaths	Actual population
			percentiles			percentiles					percentiles			
		mean	2½	97½	mean	2½	97½			mean	2½	97½		
<1	0.0699	0.0658	0.0598	0.0725	100,000	100,000	100,000	6,576	94,082	58.1	56.4	60.4	873	12,485
1-4	0.0080	0.0314	0.0284	0.0342	93,424	92,753	94,023	2,935	366,652	61.1	59.3	63.5	359	44,815
5-9	0.0023	0.0115	0.0102	0.0126	90,489	89,790	91,185	1,042	449,842	59.1	57.3	61.3	112	48,436
10-14	0.0018	0.0090	0.0079	0.0098	89,447	88,694	90,219	802	445,232	54.7	52.9	56.9	79	43,846
15-19	0.0031	0.0152	0.0133	0.0167	88,645	87,833	89,482	1,343	439,870	50.2	48.4	52.4	126	41,407
20-24	0.0041	0.0201	0.0176	0.0222	87,302	86,380	88,249	1,758	432,116	45.9	44.2	48.0	137	33,738
25-29	0.0046	0.0227	0.0197	0.0251	85,544	84,476	86,635	1,944	422,861	41.8	40.1	43.9	128	27,834
30-34	0.0054	0.0265	0.0229	0.0295	83,600	82,362	84,884	2,218	412,458	37.7	36.1	39.7	123	22,920
35-39	0.0070	0.0346	0.0297	0.0388	81,383	79,953	82,887	2,816	399,873	33.7	32.1	35.6	125	17,782
40-44	0.0086	0.0421	0.0360	0.0476	78,567	76,896	80,418	3,311	384,557	29.8	28.3	31.6	122	14,181
45-49	0.0111	0.0542	0.0460	0.0617	75,256	73,313	77,526	4,078	366,085	26.0	24.6	27.7	128	11,467
50-54	0.0149	0.0718	0.0608	0.0821	71,178	68,924	73,956	5,110	343,116	22.4	21.1	23.9	141	9,451
55-59	0.0198	0.0945	0.0800	0.1085	66,068	63,432	69,459	6,241	314,739	18.9	17.7	20.4	151	7,627
60-64	0.0281	0.1312	0.1114	0.1506	59,827	56,680	63,906	7,849	279,513	15.6	14.6	16.9	165	5,890
65-69	0.0407	0.1845	0.1569	0.2106	51,978	48,146	56,788	9,593	235,909	12.6	11.7	13.7	182	4,471
70-74	0.0622	0.2691	0.2309	0.3029	42,386	38,007	47,814	11,407	183,410	9.9	9.2	10.8	189	3,042
75-79	0.0936	0.3791	0.3318	0.4178	30,978	26,494	36,608	11,745	125,530	7.6	7.1	8.3	168	1,793
80-84	0.1429	0.5264	0.4749	0.5662	19,234	15,424	24,212	10,124	70,859	5.7	5.3	6.2	120	843
85-89	0.2117	0.6921	0.6463	0.7261	9,110	6,711	12,654	6,305	29,786	4.2	4.0	4.6	57	270
90-94	0.3043	0.7953	0.7593	0.8219	2,805	1,852	4,475	2,231	7,331	3.1	2.9	3.3	17	57
95-99	0.4243	0.9334	0.9138	0.9468	574	333	1,077	536	1,263	2.3	2.3	2.4	3	7
100+	0.5741	1	1	1	38	18	93	38	67	1.7	1.7	1.8	0	0
All ages													*3,507*	*352,362*

ln($_nq_x$) with uncertainty interval

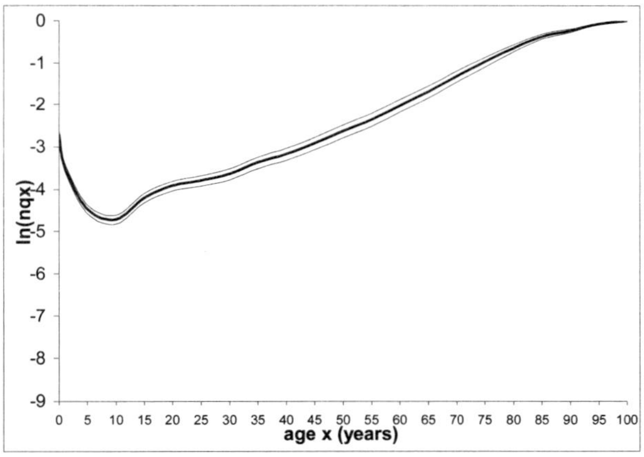

l_x with uncertainty interval

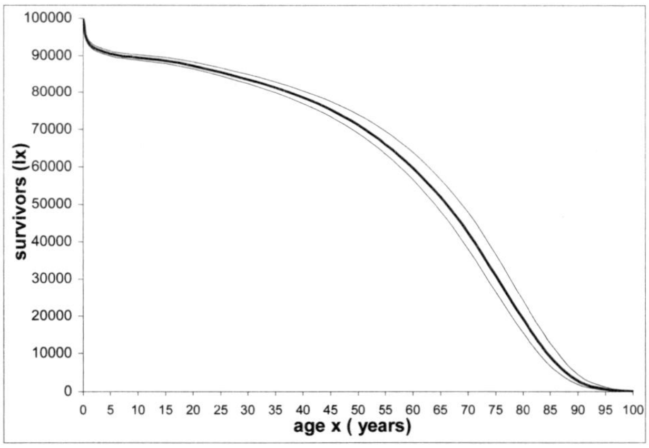

e_x with uncertainty interval

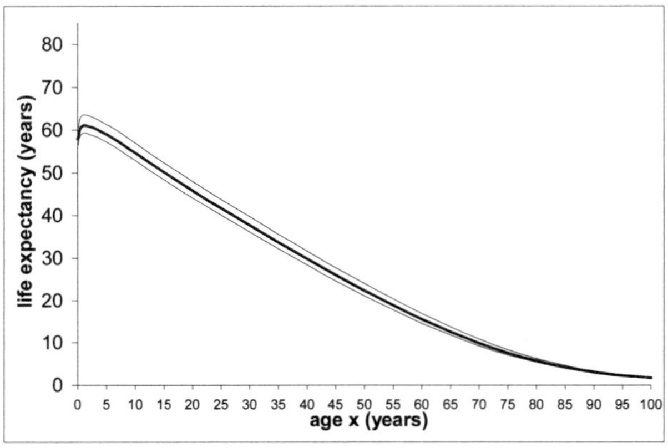

Distribution of deaths by age with uncertainty interval

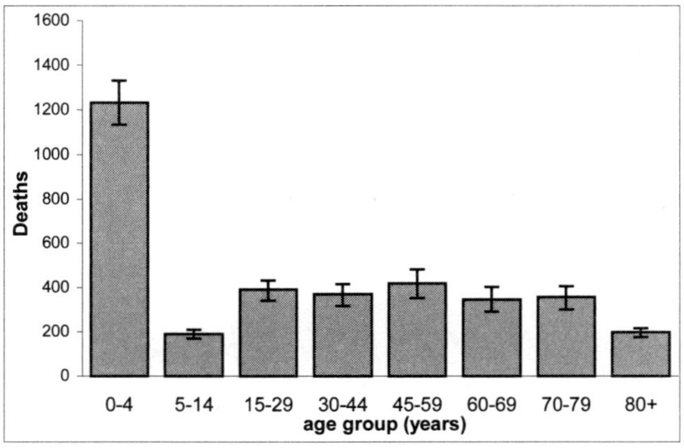

Congo

Congo

Congo

Males, 2000

Hommes, 2000

Hombres, 2000

Age	$_nM_x$	$_nq_x$			l_x			$_nd_x$	$_nL_x$	e_x			Actual deaths	Actual population
			percentiles			percentiles					percentiles			
		mean	21/2	971/2	mean	21/2	971/2			mean	21/2	971/2		
<1	0.1103	0.1003	0.0848	0.1157	100,000	100,000	100,000	10,033	90,970	50.1	46.7	55.2	6,860	62,202
1-4	0.0095	0.0370	0.0296	0.0435	89,967	88,427	91,517	3,331	351,875	54.7	50.8	59.8	2,080	219,779
5-9	0.0024	0.0117	0.0092	0.0139	86,637	85,151	88,134	1,016	430,644	52.8	49.0	57.5	536	227,108
10-14	0.0019	0.0093	0.0072	0.0110	85,621	83,974	87,316	792	426,123	48.4	44.6	53.0	346	186,339
15-19	0.0026	0.0131	0.0101	0.0157	84,828	83,047	86,688	1,114	421,358	43.8	40.1	48.4	406	153,660
20-24	0.0052	0.0256	0.0196	0.0308	83,715	81,744	85,808	2,144	413,213	39.3	35.7	43.9	678	130,579
25-29	0.0095	0.0466	0.0353	0.0569	81,570	79,233	84,123	3,799	398,354	35.3	31.7	39.7	1,016	106,545
30-34	0.0142	0.0684	0.0512	0.0849	77,771	74,792	81,151	5,317	375,564	31.9	28.5	36.1	1,220	86,145
35-39	0.0161	0.0772	0.0572	0.0973	72,454	68,621	76,996	5,596	348,281	29.1	25.9	33.0	1,116	69,462
40-44	0.0173	0.0829	0.0611	0.1054	66,858	62,215	72,590	5,543	320,432	26.3	23.5	30.0	967	55,899
45-49	0.0188	0.0897	0.0660	0.1145	61,315	55,937	68,158	5,502	292,819	23.4	20.9	26.8	854	45,433
50-54	0.0208	0.0988	0.0725	0.1260	55,813	49,532	63,661	5,517	265,272	20.5	18.2	23.5	766	36,821
55-59	0.0242	0.1142	0.0835	0.1448	50,296	43,290	59,025	5,742	237,125	17.5	15.5	20.2	737	30,414
60-64	0.0333	0.1539	0.1130	0.1940	44,554	37,020	54,029	6,857	205,627	14.4	12.7	16.8	811	24,308
65-69	0.0477	0.2131	0.1585	0.2650	37,697	29,881	47,780	8,035	168,399	11.6	10.1	13.6	878	18,393
70-74	0.0723	0.3060	0.2338	0.3707	29,662	22,060	39,952	9,077	125,620	9.0	7.9	10.7	919	12,719
75-79	0.1099	0.4310	0.3440	0.5034	20,586	14,018	30,217	8,873	80,744	6.9	6.1	8.1	828	7,535
80-84	0.1603	0.5721	0.4819	0.6419	11,712	7,089	19,614	6,700	41,810	5.2	4.7	6.1	560	3,496
85-89	0.2331	0.7364	0.6576	0.7938	5,012	2,561	10,162	3,691	15,832	3.9	3.6	4.5	248	1,063
90-94	0.3288	0.8276	0.7649	0.8724	1,321	528	3,479	1,093	3,325	2.9	2.7	3.3	64	194
95-99	0.4498	0.9572	0.9273	0.9747	228	67	818	218	485	2.2	2.1	2.3	9	20
100+	0.5964	1	1	1	10	2	59	10	16	1.7	1.7	1.8	1	1
All ages													*21,898*	*1,478,115*

ln($_nq_x$) with uncertainty interval

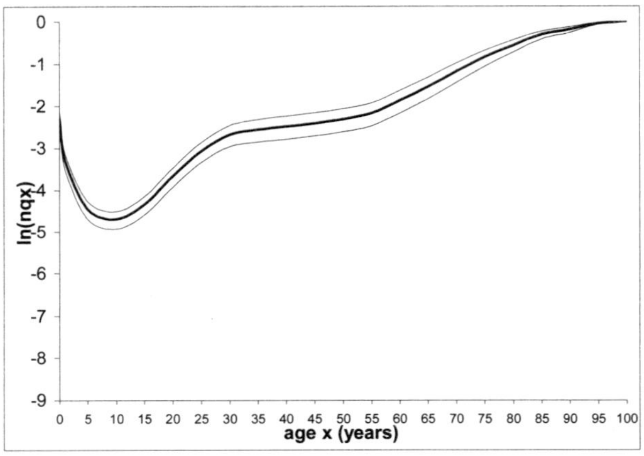

l_x with uncertainty interval

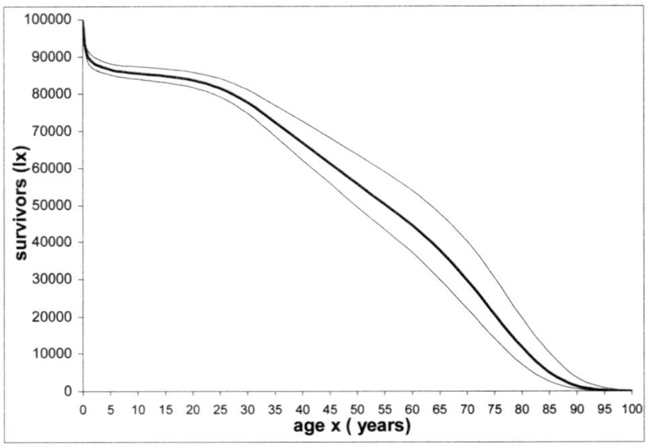

e_x with uncertainty interval

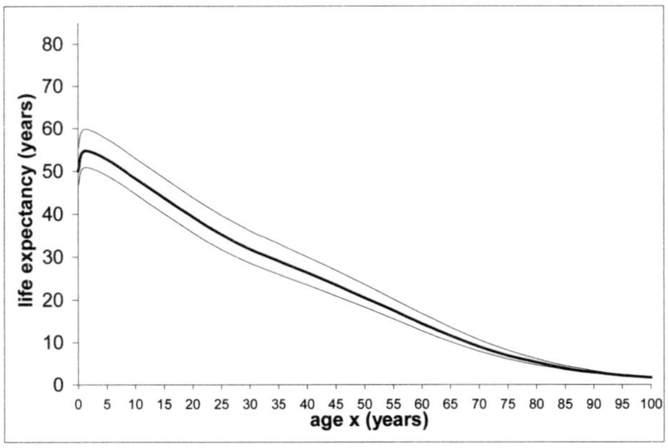

Distribution of deaths by age with uncertainty interval

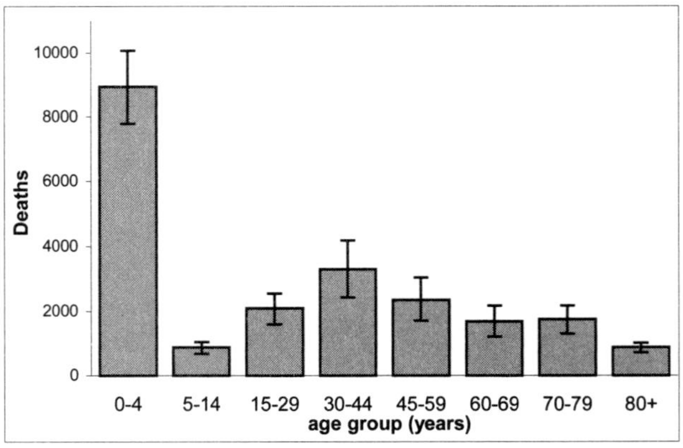

Congo

Females, 2000

Congo

Femmes, 2000

Congo

Mujeres, 2000

Age	$_nM_x$	$_nq_x$ mean	$_nq_x$ percentiles 2 1/2	$_nq_x$ percentiles 97 1/2	l_x mean	l_x percentiles 2 1/2	l_x percentiles 97 1/2	$_nd_x$	$_nL_x$	e_x mean	e_x percentiles 2 1/2	e_x percentiles 97 1/2	Actual deaths	Actual population
<1	0.0946	0.0872	0.0680	0.1072	100,000	100,000	100,000	8,721	92,151	52.9	49.0	57.5	5,715	60,390
1-4	0.0099	0.0385	0.0311	0.0466	91,279	89,279	93,202	3,519	356,672	56.9	52.7	61.8	2,175	220,484
5-9	0.0021	0.0106	0.0084	0.0129	87,761	85,762	89,748	926	436,487	55.2	51.1	59.9	488	229,681
10-14	0.0015	0.0073	0.0058	0.0089	86,834	84,672	88,966	635	432,584	50.7	46.7	55.4	279	190,323
15-19	0.0032	0.0161	0.0127	0.0196	86,199	83,917	88,431	1,388	427,527	46.1	42.1	50.8	514	158,342
20-24	0.0074	0.0361	0.0284	0.0441	84,812	82,268	87,259	3,062	416,403	41.8	37.8	46.4	997	135,610
25-29	0.0119	0.0580	0.0453	0.0718	81,750	78,643	84,656	4,739	396,900	38.3	34.5	42.8	1,321	110,627
30-34	0.0143	0.0693	0.0536	0.0874	77,010	73,085	80,763	5,334	371,717	35.5	31.9	39.7	1,287	89,668
35-39	0.0133	0.0644	0.0486	0.0824	71,676	66,930	76,415	4,618	346,836	32.9	29.7	36.8	974	73,185
40-44	0.0122	0.0591	0.0438	0.0762	67,058	61,701	72,601	3,961	325,388	30.0	27.1	33.6	734	60,305
45-49	0.0117	0.0569	0.0418	0.0737	63,097	57,298	69,281	3,591	306,507	26.7	24.1	30.0	591	50,428
50-54	0.0134	0.0649	0.0472	0.0841	59,506	53,277	66,225	3,860	287,879	23.2	20.8	26.2	552	41,174
55-59	0.0166	0.0797	0.0578	0.1039	55,646	48,867	62,884	4,437	267,139	19.6	17.4	22.4	574	34,589
60-64	0.0240	0.1130	0.0819	0.1471	51,209	43,834	58,962	5,787	241,579	16.1	14.2	18.6	681	28,407
65-69	0.0367	0.1681	0.1226	0.2170	45,422	37,489	53,694	7,635	208,024	12.9	11.2	15.0	818	22,290
70-74	0.0592	0.2579	0.1919	0.3254	37,787	29,549	46,804	9,744	164,576	10.0	8.6	11.8	961	16,237
75-79	0.0935	0.3788	0.2929	0.4601	28,043	20,220	37,823	10,623	113,660	7.6	6.5	9.0	979	10,478
80-84	0.1438	0.5288	0.4308	0.6126	17,421	11,091	26,746	9,212	64,073	5.6	4.9	6.7	795	5,530
85-89	0.2144	0.6978	0.6052	0.7692	8,209	4,309	15,170	5,728	26,723	4.2	3.7	4.9	432	2,014
90-94	0.3089	0.8016	0.7261	0.8572	2,480	1,002	5,913	1,988	6,437	3.0	2.7	3.5	147	475
95-99	0.4302	0.9391	0.8973	0.9647	492	145	1,584	462	1,074	2.3	2.2	2.5	30	69
100+	0.5790	1	1	1	30	5	161	30	52	1.7	1.7	1.9	3	5
All ages													*21,047*	*1,540,311*

ln($_nq_x$) with uncertainty interval

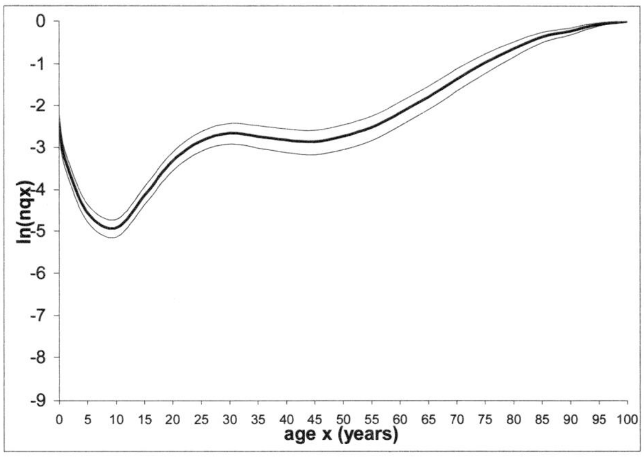

l_x with uncertainty interval

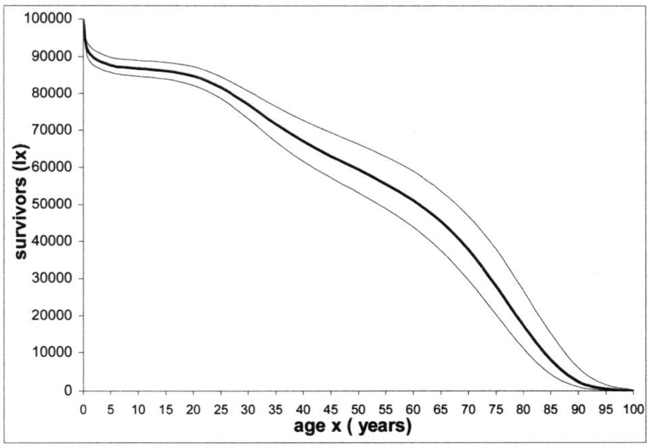

e_x with uncertainty interval

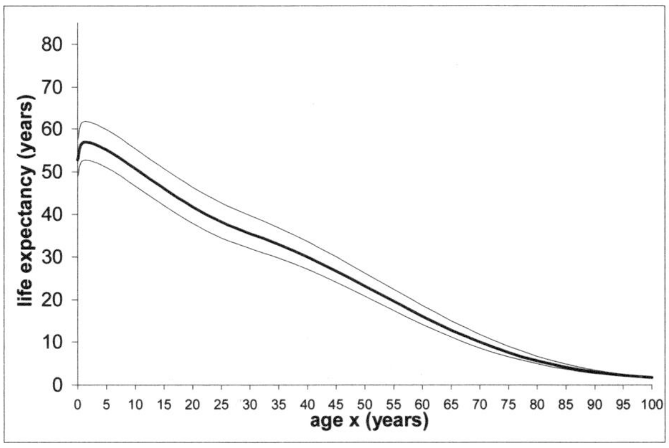

Distribution of deaths by age with uncertainty interval

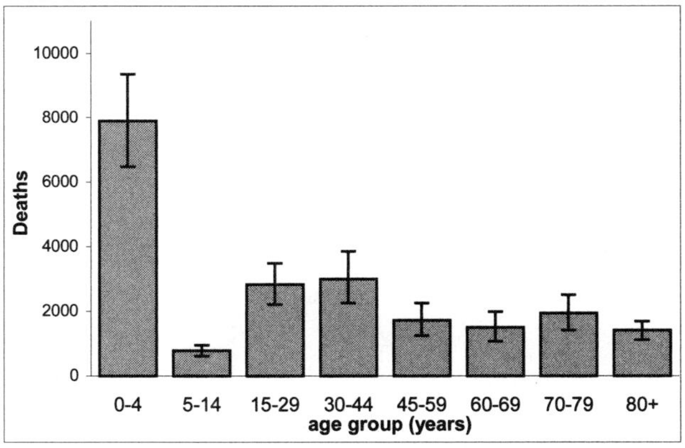

Cook Islands　　　Iles Cook　　　Islas Cook

Males, 2000　　　*Hommes, 2000*　　　*Hombres, 2000*

Age	$_nM_x$	$_nq_x$			l_x			$_nd_x$	$_nL_x$	e_x			Actual deaths	Actual population
			percentiles			percentiles					percentiles			
		mean	2 1/2	97 1/2	mean	2 1/2	97 1/2			mean	2 1/2	97 1/2		
<1	0.0169	0.0167	0.0125	0.0207	100,000	100,000	100,000	1,669	98,498	68.7	67.8	69.4	4	226
1-4	0.0016	0.0063	0.0051	0.0074	98,331	97,928	98,747	620	391,837	68.8	68.0	69.5	1	941
5-9	0.0006	0.0029	0.0024	0.0034	97,711	97,200	98,243	283	487,849	65.2	64.4	65.9	1	1,157
10-14	0.0009	0.0046	0.0039	0.0054	97,429	96,872	98,009	452	486,012	60.4	59.6	61.1	1	1,216
15-19	0.0011	0.0053	0.0045	0.0061	96,976	96,350	97,631	516	483,591	55.7	54.9	56.4	1	1,116
20-24	0.0015	0.0077	0.0065	0.0088	96,460	95,760	97,194	741	480,449	51.0	50.2	51.7	1	973
25-29	0.0015	0.0077	0.0066	0.0087	95,719	94,920	96,559	735	476,760	46.4	45.6	47.1	1	877
30-34	0.0022	0.0108	0.0094	0.0121	94,984	94,094	95,919	1,022	472,368	41.7	41.0	42.4	2	768
35-39	0.0029	0.0143	0.0127	0.0159	93,963	92,957	95,016	1,341	466,461	37.1	36.4	37.8	2	655
40-44	0.0042	0.0206	0.0186	0.0227	92,622	91,479	93,812	1,904	458,348	32.6	32.0	33.3	2	543
45-49	0.0054	0.0268	0.0247	0.0292	90,717	89,406	92,069	2,432	447,507	28.3	27.6	28.9	2	457
50-54	0.0078	0.0381	0.0353	0.0415	88,286	86,792	89,795	3,361	433,024	24.0	23.4	24.6	3	349
55-59	0.0120	0.0584	0.0543	0.0635	84,924	83,192	86,622	4,956	412,231	19.8	19.3	20.4	4	296
60-64	0.0201	0.0957	0.0885	0.1036	79,968	77,913	81,921	7,652	380,711	15.9	15.4	16.5	5	236
65-69	0.0387	0.1766	0.1628	0.1898	72,316	69,845	74,671	12,774	329,647	12.3	11.8	12.9	7	178
70-74	0.0600	0.2609	0.2400	0.2816	59,543	56,585	62,517	15,533	258,881	9.4	9.0	9.9	7	114
75-79	0.1020	0.4064	0.3767	0.4374	44,010	40,653	47,513	17,884	175,337	6.8	6.5	7.2	7	70
80-84	0.1740	0.6063	0.5740	0.6400	26,125	22,872	29,616	15,839	91,029	4.8	4.6	5.1	5	28
85-89	0.2736	0.8124	0.7884	0.8361	10,286	8,233	12,615	8,356	30,541	3.4	3.3	3.5	3	10
90-94	0.4406	0.9489	0.9384	0.9586	1,930	1,349	2,670	1,832	4,156	2.3	2.2	2.3	1	2
95-99	0.7154	0.9800	0.9743	0.9845	99	56	165	97	203	2.1	2.1	2.1	0	0
100+	0.9895	1	1	1	2	1	4	2	2	1.0	1.0	1.9	0	0
All ages													*60*	*10,212*

ln($_nq_x$) with uncertainty interval

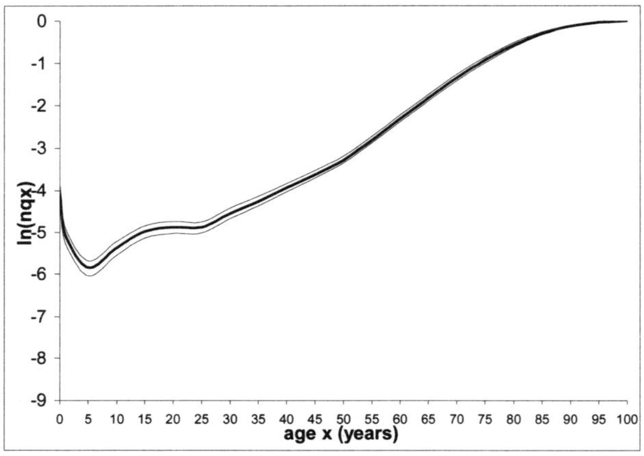

l_x with uncertainty interval

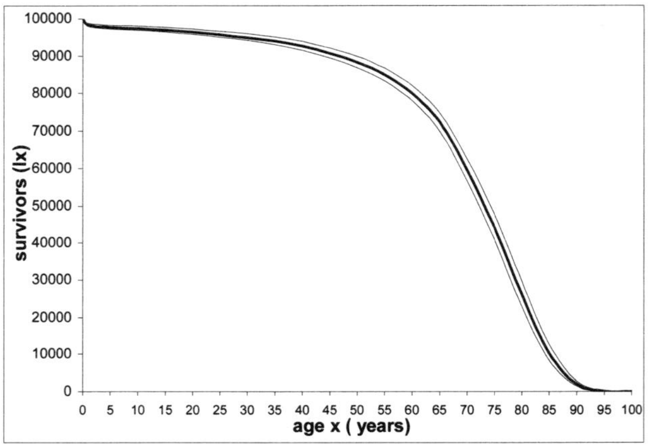

e_x with uncertainty interval

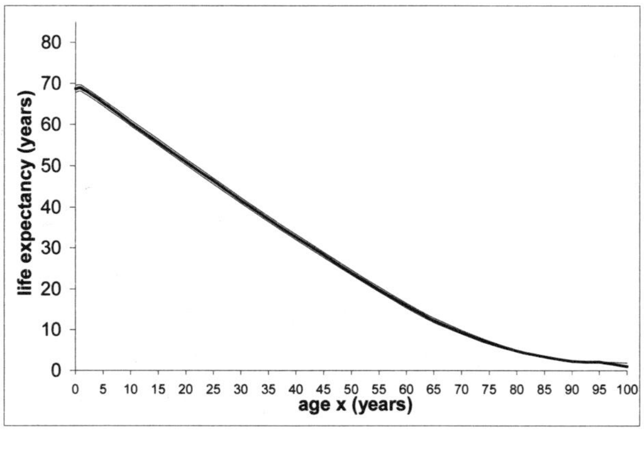

Distribution of deaths by age with uncertainty interval

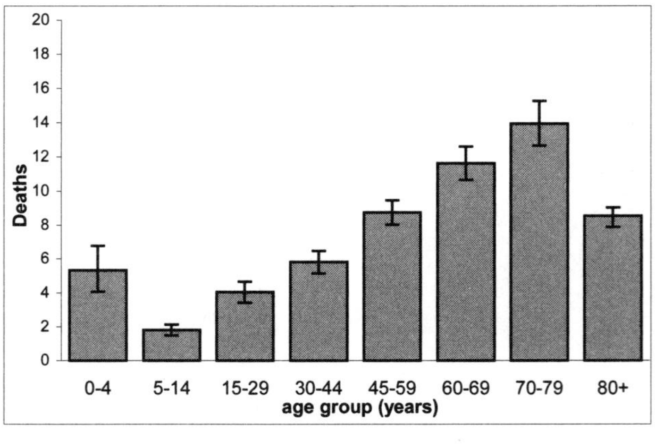

Cook Islands | Iles Cook | Islas Cook

Females, 2000 | *Femmes, 2000* | *Mujeres, 2000*

Age	$_nM_x$	$_nq_x$			l_x			$_nd_x$	$_nL_x$	e_x			Actual deaths	Actual population
			percentiles			percentiles					percentiles			
		mean	2½	97½	mean	2½	97½			mean	2½	97½		
<1	0.0194	0.0191	0.0148	0.0228	100,000	100,000	100,000	1,905	98,285	72.1	71.3	73.1	4	216
1-4	0.0003	0.0010	0.0009	0.0012	98,095	97,719	98,524	103	392,133	72.5	71.7	73.4	0	893
5-9	0.0005	0.0023	0.0019	0.0026	97,992	97,601	98,438	224	489,402	68.6	67.8	69.5	0	1,092
10-14	0.0008	0.0040	0.0033	0.0045	97,768	97,345	98,252	387	487,875	63.8	62.9	64.7	1	1,130
15-19	0.0006	0.0031	0.0026	0.0035	97,381	96,904	97,927	303	486,150	59.0	58.2	59.9	1	1,034
20-24	0.0019	0.0092	0.0079	0.0104	97,078	96,560	97,671	896	483,152	54.2	53.4	55.1	2	881
25-29	0.0019	0.0095	0.0082	0.0106	96,182	95,553	96,901	909	478,637	49.7	48.9	50.5	1	742
30-34	0.0020	0.0100	0.0088	0.0111	95,273	94,540	96,106	948	473,994	45.1	44.4	45.9	1	630
35-39	0.0026	0.0130	0.0116	0.0144	94,325	93,492	95,265	1,230	468,549	40.5	39.8	41.3	2	578
40-44	0.0030	0.0149	0.0135	0.0163	93,095	92,145	94,157	1,386	462,008	36.1	35.3	36.8	1	487
45-49	0.0047	0.0230	0.0211	0.0250	91,709	90,639	92,890	2,110	453,269	31.6	30.9	32.3	2	422
50-54	0.0063	0.0311	0.0285	0.0336	89,599	88,370	90,928	2,783	441,037	27.2	26.6	28.0	2	321
55-59	0.0099	0.0483	0.0443	0.0523	86,816	85,404	88,337	4,196	423,589	23.0	22.4	23.7	3	277
60-64	0.0175	0.0836	0.0769	0.0904	82,620	80,941	84,420	6,908	395,827	19.1	18.5	19.7	4	233
65-69	0.0250	0.1176	0.1085	0.1269	75,711	73,622	77,931	8,907	356,288	15.6	15.0	16.2	5	180
70-74	0.0417	0.1890	0.1743	0.2027	66,804	64,278	69,474	12,626	302,456	12.3	11.8	12.9	5	128
75-79	0.0585	0.2551	0.2354	0.2749	54,178	51,251	57,361	13,823	236,336	9.6	9.2	10.1	5	89
80-84	0.1004	0.4012	0.3738	0.4297	40,356	37,160	43,859	16,189	161,307	7.1	6.7	7.5	4	42
85-89	0.1629	0.5789	0.5497	0.6097	24,167	21,192	27,465	13,989	85,860	5.1	4.9	5.4	3	19
90-94	0.2420	0.7010	0.6757	0.7271	10,177	8,271	12,368	7,135	29,483	3.8	3.6	4.0	1	6
95-99	0.3300	0.8291	0.8102	0.8483	3,043	2,257	4,011	2,523	7,645	2.9	2.8	3.0	0	1
100+	0.4620	1	1	1	520	342	761	520	1,125	2.2	2.1	2.3	0	0
All ages													*49*	*9,400*

ln($_nq_x$) with uncertainty interval

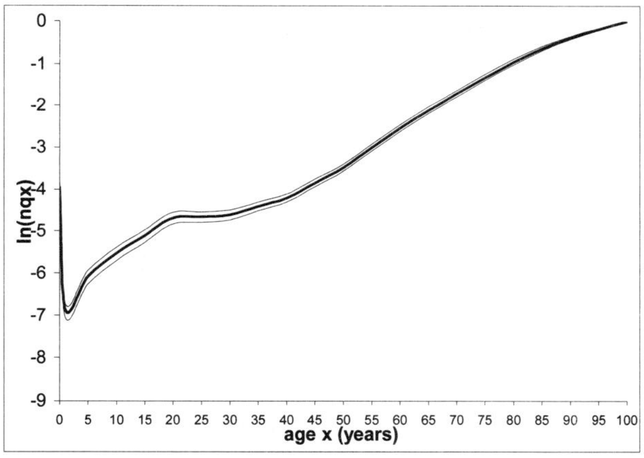

l_x with uncertainty interval

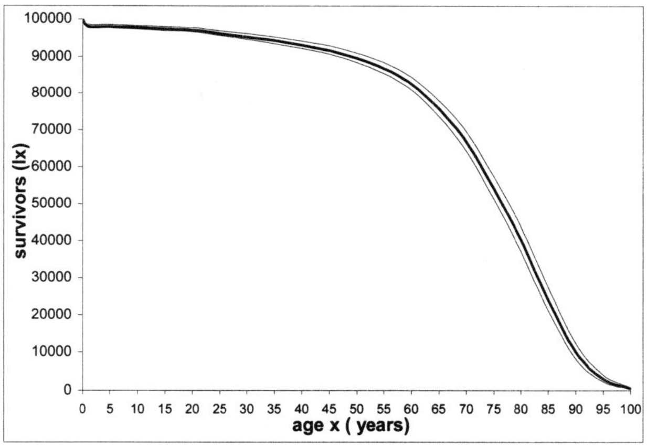

e_x with uncertainty interval

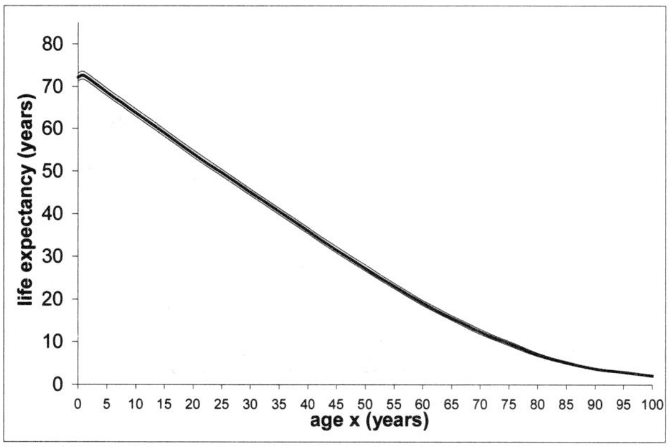

Distribution of deaths by age with uncertainty interval

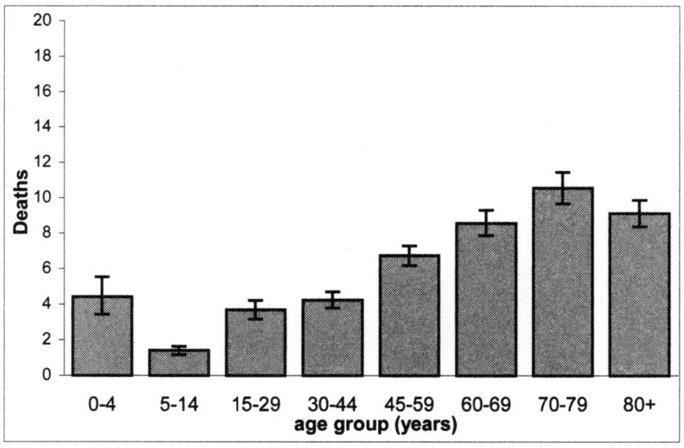

Costa Rica | Costa Rica | Costa Rica

Males, 2000 | *Hommes, 2000* | *Hombres, 2000*

Age	$_nM_x$	$_nq_x$			l_x			$_nd_x$	$_nL_x$	e_x			Actual deaths	Actual population
			percentiles			percentiles					percentiles			
		mean	21/2	971/2	mean	21/2	971/2			mean	21/2	971/2		
<1	0.0140	0.0138	0.0118	0.0158	100,000	100,000	100,000	1,383	98,756	73.4	72.7	74.5	642	45,857
1-4	0.0010	0.0040	0.0035	0.0045	98,617	98,416	98,824	394	393,525	73.5	72.8	74.5	180	179,946
5-9	0.0003	0.0013	0.0011	0.0014	98,224	97,978	98,481	128	490,800	69.8	69.1	70.8	57	218,426
10-14	0.0004	0.0019	0.0017	0.0022	98,096	97,836	98,370	191	490,004	64.8	64.2	65.9	87	222,240
15-19	0.0010	0.0050	0.0044	0.0055	97,905	97,624	98,203	488	488,305	60.0	59.3	61.0	207	206,898
20-24	0.0010	0.0050	0.0044	0.0055	97,417	97,086	97,775	486	485,870	55.2	54.6	56.3	189	189,267
25-29	0.0010	0.0050	0.0044	0.0055	96,931	96,553	97,348	483	483,446	50.5	49.8	51.5	161	161,020
30-34	0.0020	0.0100	0.0088	0.0109	96,448	96,025	96,921	960	479,839	45.8	45.1	46.8	320	160,123
35-39	0.0020	0.0100	0.0088	0.0108	95,488	94,983	96,070	950	475,064	41.2	40.5	42.2	301	150,605
40-44	0.0030	0.0149	0.0133	0.0161	94,538	93,958	95,224	1,408	469,170	36.6	35.9	37.5	382	127,421
45-49	0.0040	0.0198	0.0177	0.0212	93,130	92,450	93,962	1,844	461,041	32.1	31.5	33.0	402	100,499
50-54	0.0060	0.0296	0.0266	0.0316	91,286	90,490	92,297	2,698	449,685	27.7	27.1	28.6	464	77,275
55-59	0.0080	0.0392	0.0353	0.0420	88,588	87,630	89,839	3,474	434,255	23.5	22.9	24.3	469	58,631
60-64	0.0150	0.0723	0.0649	0.0776	85,114	83,952	86,668	6,153	410,188	19.3	18.8	20.1	698	46,530
65-69	0.0230	0.1087	0.0975	0.1169	78,961	77,437	81,045	8,587	373,339	15.6	15.1	16.3	862	37,487
70-74	0.0400	0.1818	0.1640	0.1947	70,374	68,385	73,142	12,795	319,883	12.2	11.8	12.8	1,097	27,417
75-79	0.0600	0.2609	0.2383	0.2784	57,579	55,067	61,145	15,021	250,343	9.4	9.0	9.8	1,044	17,398
80-84	0.1060	0.4190	0.3919	0.4431	42,558	39,737	46,574	17,831	168,215	6.8	6.5	7.2	939	8,862
85-89	0.1760	0.6111	0.5838	0.6356	24,728	22,129	28,324	15,111	85,860	4.9	4.7	5.2	605	3,436
90-94	0.2420	0.7010	0.6788	0.7215	9,616	8,063	11,788	6,741	27,857	3.7	3.6	3.9	221	913
95-99	0.3370	0.8379	0.8221	0.8525	2,875	2,246	3,786	2,409	7,148	2.8	2.8	3.0	59	175
100+	0.4730	1	1	1	466	331	674	466	985	2.1	2.1	2.3	10	22
All ages													*9,396*	*2,040,448*

$\ln(_nq_x)$ with uncertainty interval

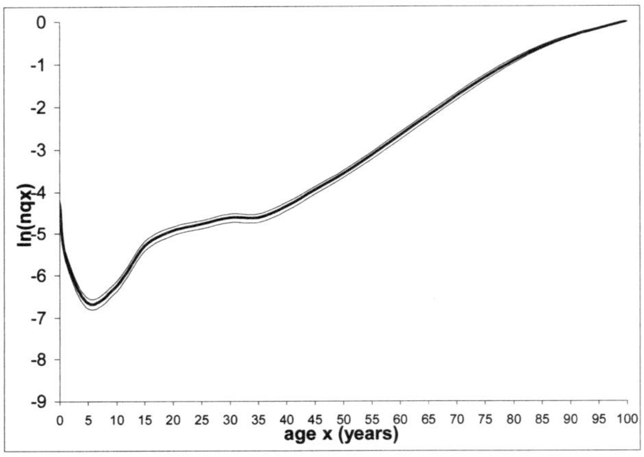

l_x with uncertainty interval

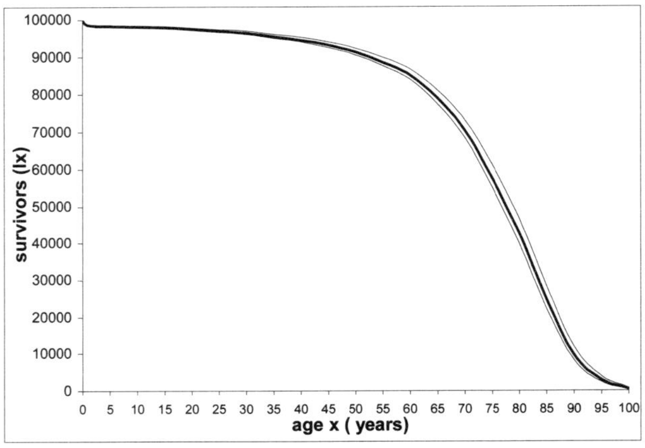

e_x with uncertainty interval

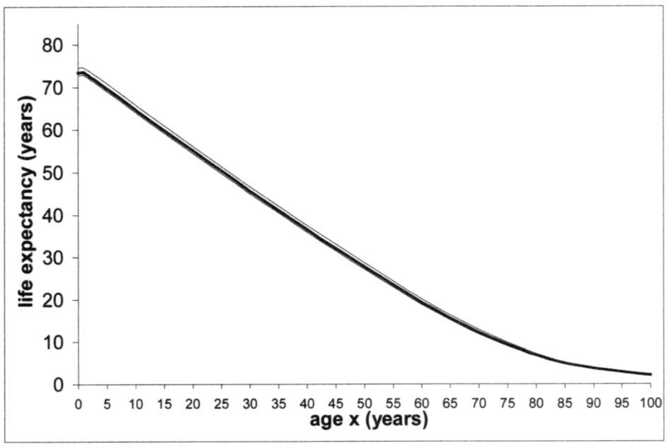

Distribution of deaths by age with uncertainty interval

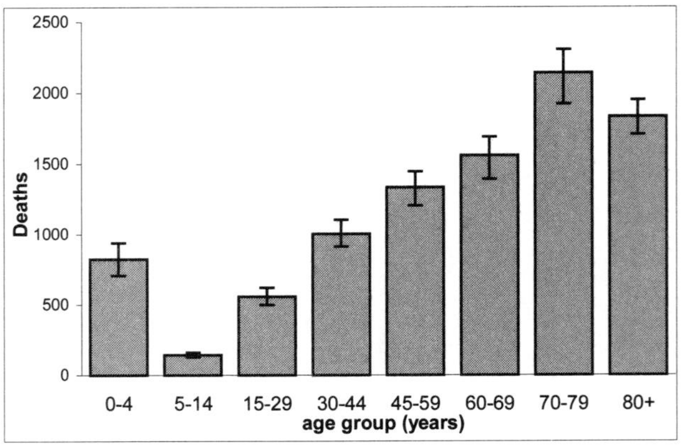

Costa Rica

Costa Rica

Costa Rica

Females, 2000

Femmes, 2000

Mujeres, 2000

Age	$_nM_x$	$_nq_x$			l_x			$_nd_x$	$_nL_x$	e_x			Actual deaths	Actual population
			percentiles			percentiles					percentiles			
		mean	21/2	971/2	mean	21/2	971/2			mean	21/2	971/2		
<1	0.0110	0.0109	0.0088	0.0128	100,000	100,000	100,000	1,089	99,020	78.8	78.1	79.8	481	43,694
1-4	0.0010	0.0040	0.0034	0.0045	98,911	98,725	99,123	395	394,696	78.6	78.0	79.6	172	172,085
5-9	0.0002	0.0010	0.0009	0.0012	98,516	98,286	98,784	103	492,322	74.9	74.3	75.9	44	208,591
10-14	0.0002	0.0011	0.0010	0.0012	98,413	98,172	98,695	108	491,793	70.0	69.4	71.0	47	211,391
15-19	0.0003	0.0017	0.0015	0.0019	98,305	98,053	98,601	167	491,105	65.1	64.5	66.1	67	195,765
20-24	0.0005	0.0023	0.0020	0.0025	98,138	97,870	98,456	225	490,124	60.2	59.6	61.2	82	179,201
25-29	0.0005	0.0022	0.0020	0.0024	97,912	97,623	98,258	220	489,010	55.3	54.7	56.3	69	153,197
30-34	0.0010	0.0050	0.0044	0.0054	97,692	97,384	98,064	487	487,242	50.5	49.8	51.4	154	154,298
35-39	0.0010	0.0050	0.0044	0.0054	97,205	96,858	97,632	485	484,812	45.7	45.1	46.7	147	146,845
40-44	0.0020	0.0100	0.0089	0.0106	96,720	96,338	97,199	962	481,194	40.9	40.3	41.9	250	125,105
45-49	0.0020	0.0100	0.0090	0.0105	95,758	95,316	96,329	953	476,406	36.3	35.7	37.2	199	99,328
50-54	0.0040	0.0198	0.0181	0.0210	94,805	94,311	95,458	1,877	469,331	31.6	31.0	32.5	308	76,992
55-59	0.0050	0.0247	0.0224	0.0262	92,927	92,335	93,729	2,295	458,901	27.2	26.6	28.1	294	58,867
60-64	0.0080	0.0392	0.0354	0.0419	90,633	89,915	91,628	3,554	444,279	22.8	22.3	23.7	384	48,051
65-69	0.0150	0.0723	0.0649	0.0775	87,079	86,151	88,385	6,295	419,656	18.7	18.2	19.5	594	39,604
70-74	0.0250	0.1176	0.1055	0.1264	80,784	79,472	82,646	9,504	380,159	14.9	14.5	15.6	759	30,351
75-79	0.0420	0.1900	0.1714	0.2036	71,280	69,425	73,927	13,546	322,533	11.6	11.1	12.2	886	21,106
80-84	0.0710	0.3015	0.2762	0.3225	57,733	55,289	61,256	17,406	245,153	8.7	8.3	9.2	832	11,714
85-89	0.1180	0.4556	0.4278	0.4842	40,328	37,458	44,340	18,373	155,705	6.4	6.1	6.8	600	5,082
90-94	0.1760	0.5759	0.5492	0.6042	21,954	19,320	25,370	12,644	71,840	4.7	4.5	5.0	263	1,497
95-99	0.2670	0.7413	0.7198	0.7643	9,311	7,647	11,438	6,901	25,848	3.4	3.3	3.6	72	269
100+	0.4230	1	1	1	2,409	1,803	3,205	2,409	5,695	2.4	2.3	2.5	9	21
All ages													*6,713*	*1,983,054*

$\ln(_nq_x)$ with uncertainty interval

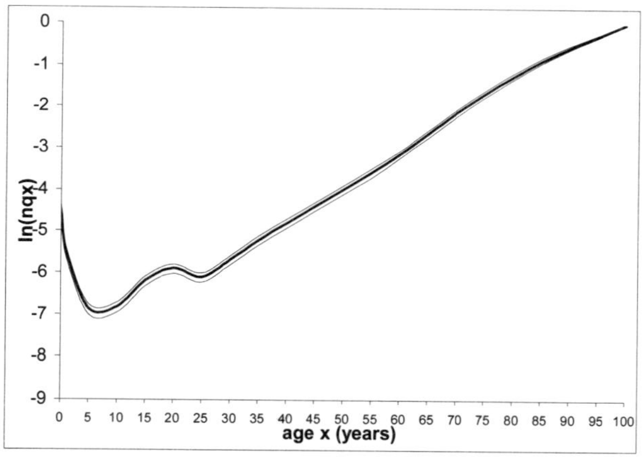

l_x with uncertainty interval

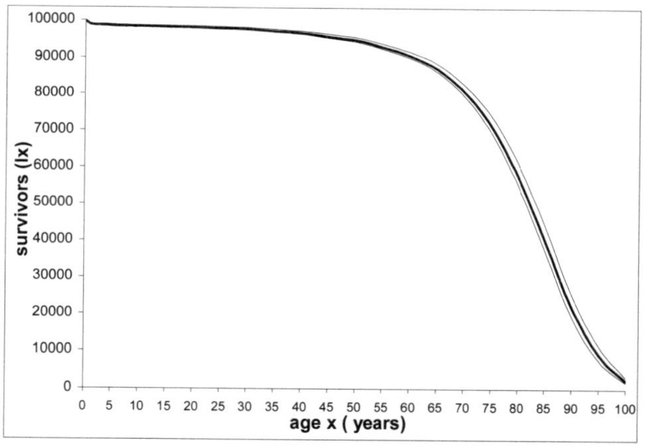

e_x with uncertainty interval

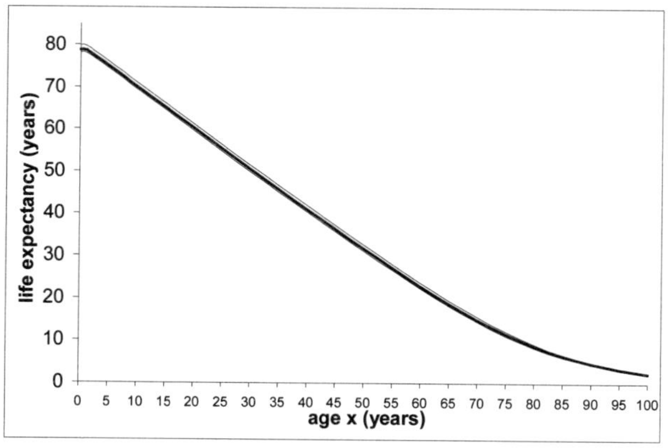

Distribution of deaths by age with uncertainty interval

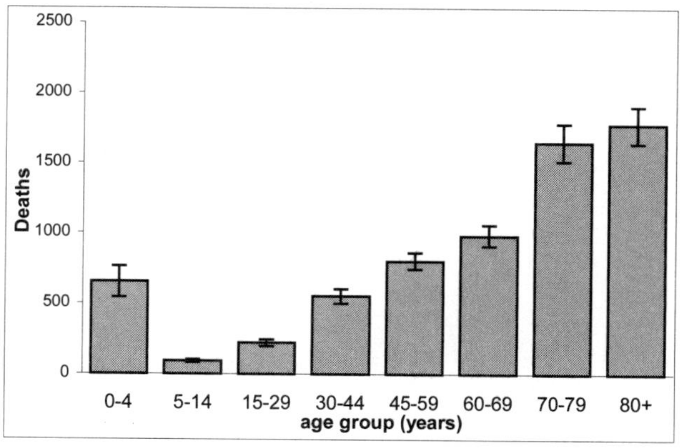

## Côte d'Ivoire		Côte d'Ivoire		Côte d'Ivoire

Males, 2000		*Hommes, 2000*		*Hombres, 2000*

Age	$_nM_x$	$_nq_x$			l_x			$_nd_x$	$_nL_x$	e_x			Actual deaths	Actual population
			percentiles			percentiles					percentiles			
		mean	2 1/2	97 1/2	mean	2 1/2	97 1/2			mean	2 1/2	97 1/2		
<1	0.1252	0.1125	0.1012	0.1228	100,000	100,000	100,000	11,249	89,876	46.4	43.2	49.9	33,736	269,551
1-4	0.0115	0.0447	0.0390	0.0507	88,751	87,719	89,883	3,968	345,481	51.2	47.7	55.5	10,906	949,456
5-9	0.0026	0.0129	0.0110	0.0149	84,783	83,784	85,780	1,093	421,183	49.6	46.2	53.7	2,817	1,085,778
10-14	0.0019	0.0093	0.0079	0.0109	83,690	82,607	84,742	780	416,500	45.2	41.8	49.3	2,025	1,080,651
15-19	0.0027	0.0133	0.0112	0.0156	82,910	81,766	84,031	1,101	411,796	40.6	37.3	44.6	2,576	963,225
20-24	0.0061	0.0302	0.0252	0.0356	81,809	80,573	83,081	2,467	402,877	36.1	32.8	40.1	4,661	761,305
25-29	0.0127	0.0615	0.0508	0.0735	79,342	77,831	80,960	4,882	384,505	32.1	28.9	36.1	7,521	592,311
30-34	0.0198	0.0945	0.0762	0.1143	74,460	72,156	76,779	7,033	354,717	29.1	26.0	32.9	9,717	490,042
35-39	0.0221	0.1046	0.0826	0.1278	67,427	63,909	70,751	7,052	319,502	26.9	24.1	30.4	9,499	430,344
40-44	0.0229	0.1081	0.0842	0.1326	60,374	55,741	64,714	6,525	285,559	24.7	22.2	27.9	8,699	380,683
45-49	0.0237	0.1120	0.0866	0.1373	53,849	48,350	59,171	6,030	254,171	22.4	20.2	25.3	7,656	322,730
50-54	0.0245	0.1156	0.0892	0.1414	47,819	41,712	53,940	5,526	225,282	19.9	18.0	22.4	6,238	254,319
55-59	0.0262	0.1231	0.0951	0.1505	42,293	35,823	49,044	5,207	198,449	17.2	15.5	19.4	5,467	208,345
60-64	0.0347	0.1595	0.1241	0.1935	37,086	30,466	44,382	5,916	170,641	14.2	12.8	16.1	5,615	161,962
65-69	0.0487	0.2171	0.1713	0.2598	31,170	24,632	38,875	6,767	138,934	11.5	10.3	13.1	5,639	115,767
70-74	0.0732	0.3095	0.2504	0.3623	24,403	18,321	32,214	7,553	103,134	9.0	8.0	10.2	5,444	74,343
75-79	0.1108	0.4340	0.3641	0.4931	16,850	11,790	24,147	7,312	65,971	6.9	6.2	7.8	4,549	41,041
80-84	0.1621	0.5768	0.5050	0.6346	9,538	6,072	15,356	5,501	33,937	5.2	4.7	5.9	2,888	17,819
85-89	0.2350	0.7402	0.6777	0.7883	4,037	2,223	7,601	2,988	12,714	3.9	3.6	4.3	1,187	5,049
90-94	0.3306	0.8299	0.7796	0.8679	1,049	470	2,450	870	2,633	2.9	2.7	3.2	285	862
95-99	0.4514	0.9587	0.9354	0.9735	178	62	540	171	379	2.2	2.1	2.3	36	80
100+	0.5979	1	1	1	7	2	35	7	12	1.7	1.7	1.8	2	4
All ages													*137,163*	*8,205,667*

$\ln(_nq_x)$ with uncertainty interval

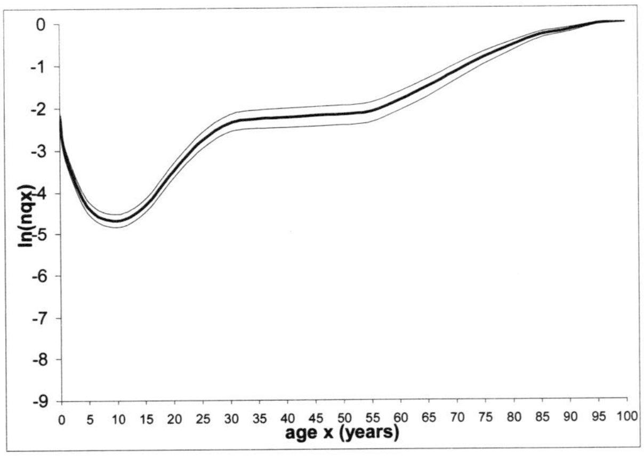

l_x with uncertainty interval

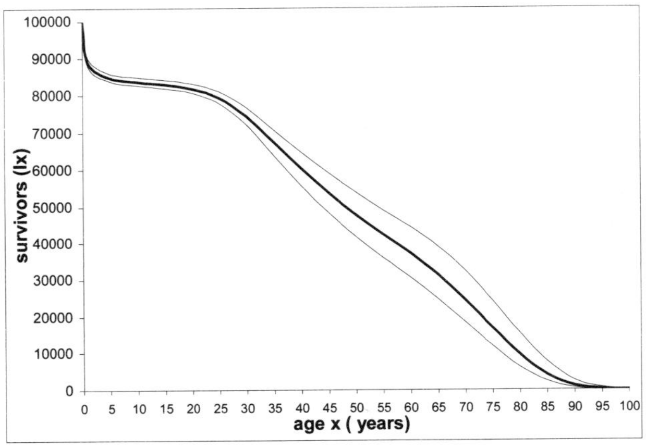

e_x with uncertainty interval

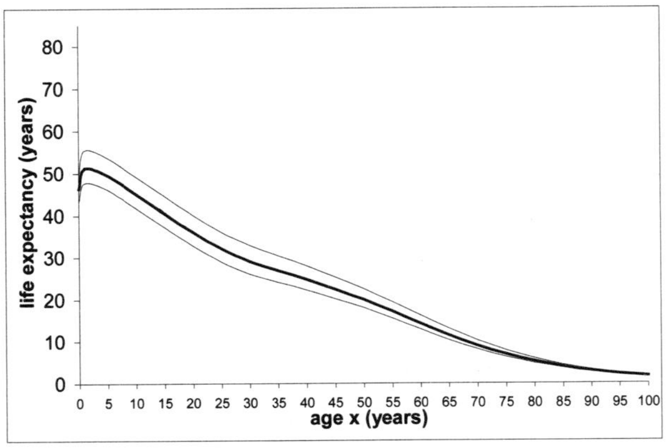

Distribution of deaths by age with uncertainty interval

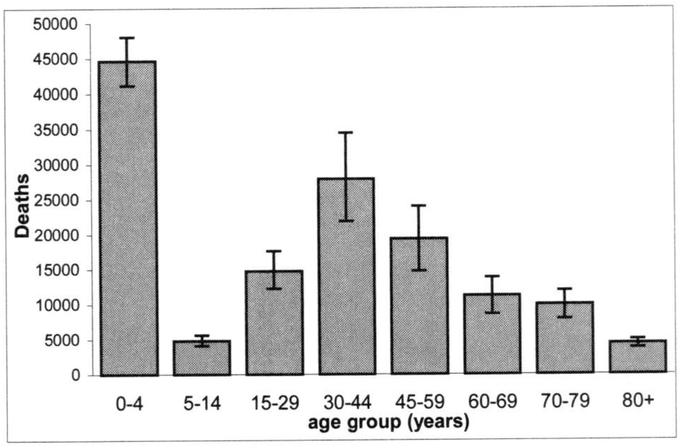

Côte d'Ivoire

Côte d'Ivoire

Côte d'Ivoire

Females, 2000 / *Femmes, 2000* / *Mujeres, 2000*

Age	$_nM_x$	$_nq_x$ mean	$_nq_x$ percentiles 2½	$_nq_x$ percentiles 97½	l_x mean	l_x percentiles 2½	l_x percentiles 97½	$_nd_x$	$_nL_x$	e_x mean	e_x percentiles 2½	e_x percentiles 97½	Actual deaths	Actual population
<1	0.1058	0.0966	0.0870	0.1056	100,000	100,000	100,000	9,661	91,306	48.4	45.5	51.4	27,383	258,811
1-4	0.0118	0.0457	0.0408	0.0503	90,339	89,435	91,303	4,131	351,443	52.5	49.4	56.2	11,091	943,503
5-9	0.0024	0.0118	0.0103	0.0132	86,208	85,412	87,004	1,014	428,507	51.0	47.9	54.6	2,556	1,080,080
10-14	0.0016	0.0080	0.0070	0.0090	85,194	84,315	86,093	681	424,269	46.6	43.5	50.1	1,729	1,077,228
15-19	0.0041	0.0205	0.0179	0.0232	84,513	83,574	85,489	1,730	418,242	41.9	38.9	45.5	3,975	960,846
20-24	0.0104	0.0506	0.0437	0.0581	82,783	81,691	83,960	4,188	403,448	37.7	34.7	41.2	7,803	751,779
25-29	0.0175	0.0841	0.0713	0.0979	78,596	77,133	80,272	6,607	376,462	34.6	31.7	38.0	9,943	566,597
30-34	0.0212	0.1006	0.0835	0.1186	71,989	69,700	74,461	7,242	341,841	32.6	29.9	35.7	9,447	445,958
35-39	0.0189	0.0901	0.0736	0.1071	64,748	61,432	68,054	5,834	309,153	30.9	28.6	33.8	7,060	374,147
40-44	0.0164	0.0789	0.0639	0.0942	58,914	54,851	62,831	4,650	282,943	28.7	26.7	31.3	5,270	320,639
45-49	0.0145	0.0701	0.0565	0.0838	54,263	49,685	58,698	3,805	261,805	26.0	24.2	28.2	3,882	267,102
50-54	0.0154	0.0742	0.0596	0.0885	50,458	45,523	55,329	3,742	242,937	22.8	21.1	24.8	3,190	207,099
55-59	0.0179	0.0855	0.0687	0.1018	46,716	41,495	51,971	3,994	223,597	19.4	17.9	21.2	3,106	173,910
60-64	0.0249	0.1172	0.0945	0.1392	42,723	37,277	48,345	5,009	201,092	16.0	14.7	17.6	3,476	139,564
65-69	0.0377	0.1721	0.1400	0.2027	37,714	32,115	43,774	6,492	172,340	12.7	11.7	14.1	3,897	103,443
70-74	0.0604	0.2624	0.2172	0.3041	31,222	25,662	37,644	8,193	135,628	9.9	9.0	11.0	4,197	69,480
75-79	0.0949	0.3834	0.3263	0.4337	23,029	17,950	29,468	8,830	93,071	7.5	6.8	8.4	3,860	40,687
80-84	0.1452	0.5327	0.4707	0.5847	14,199	10,270	19,852	7,565	52,085	5.6	5.1	6.2	2,784	19,169
85-89	0.2158	0.7008	0.6449	0.7457	6,635	4,311	10,508	4,650	21,549	4.2	3.9	4.6	1,310	6,071
90-94	0.3101	0.8033	0.7585	0.8388	1,985	1,096	3,731	1,594	5,141	3.0	2.8	3.3	375	1,208
95-99	0.4312	0.9400	0.9162	0.9570	390	177	901	367	851	2.3	2.2	2.4	61	142
100+	0.5800	1	1	1	23	8	76	23	40	1.7	1.7	1.8	5	9
All ages													*116,401*	*7,807,472*

ln($_nq_x$) with uncertainty interval

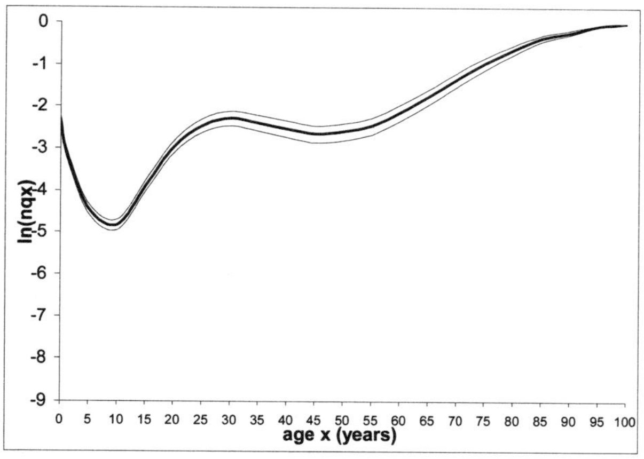

l_x with uncertainty interval

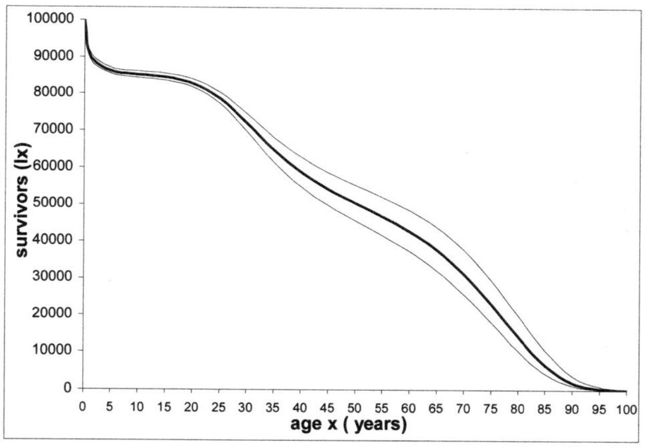

e_x with uncertainty interval

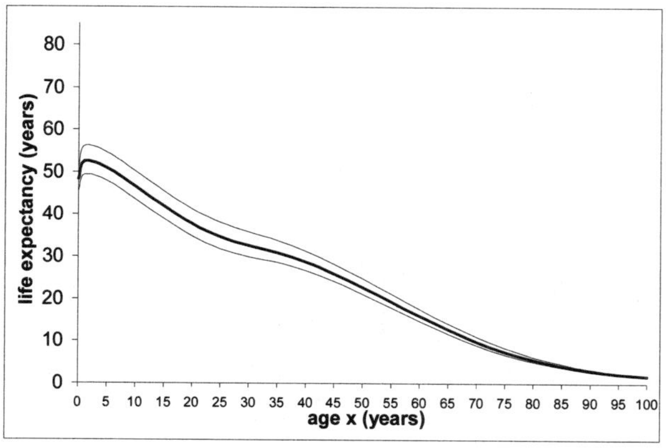

Distribution of deaths by age with uncertainty interval

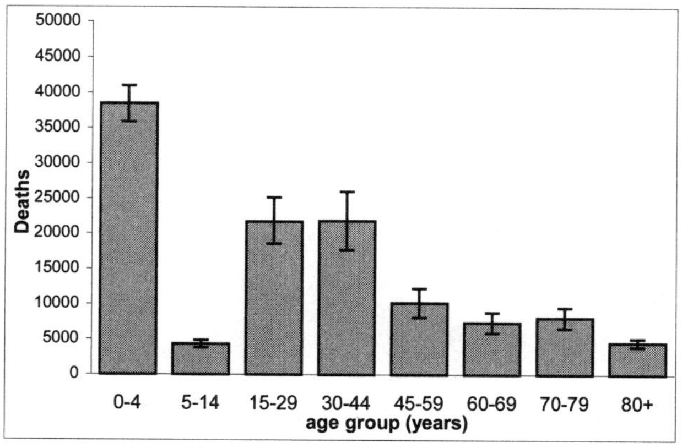

Croatia — Croatie — Croacia

Males, 2000 — *Hommes, 2000* — *Hombres, 2000*

Age	$_nM_x$	$_nq_x$ mean	$_nq_x$ percentiles 2½	$_nq_x$ percentiles 97½	l_x mean	l_x percentiles 2½	l_x percentiles 97½	$_nd_x$	$_nL_x$	e_x mean	e_x percentiles 2½	e_x percentiles 97½	Actual deaths	Actual population
<1	0.0092	0.0091	0.0075	0.0109	100,000	100,000	100,000	909	99,181	69.8	69.4	70.2	255	27,768
1-4	0.0003	0.0012	0.0010	0.0014	99,091	98,914	99,254	118	396,079	69.4	69.0	69.9	33	110,203
5-9	0.0002	0.0010	0.0009	0.0011	98,973	98,780	99,151	98	494,617	65.5	65.1	66.0	27	137,990
10-14	0.0002	0.0012	0.0011	0.0014	98,874	98,668	99,065	120	494,070	60.6	60.2	61.0	37	153,835
15-19	0.0009	0.0043	0.0038	0.0048	98,754	98,532	98,960	422	492,716	55.7	55.2	56.1	141	164,257
20-24	0.0012	0.0059	0.0053	0.0065	98,332	98,061	98,586	578	490,216	50.9	50.4	51.3	192	162,813
25-29	0.0011	0.0056	0.0051	0.0062	97,754	97,423	98,064	551	487,392	46.2	45.7	46.6	180	158,934
30-34	0.0014	0.0069	0.0064	0.0074	97,203	96,823	97,559	669	484,342	41.4	41.0	41.9	229	165,966
35-39	0.0021	0.0103	0.0097	0.0111	96,534	96,102	96,938	997	480,177	36.7	36.2	37.2	366	176,258
40-44	0.0035	0.0173	0.0164	0.0183	95,537	95,040	96,001	1,657	473,542	32.0	31.6	32.5	645	184,333
45-49	0.0060	0.0296	0.0285	0.0307	93,880	93,299	94,415	2,775	462,462	27.6	27.1	28.1	1,020	169,925
50-54	0.0092	0.0450	0.0434	0.0465	91,105	90,436	91,703	4,097	445,283	23.3	22.9	23.8	1,206	131,112
55-59	0.0138	0.0669	0.0640	0.0700	87,008	86,291	87,657	5,822	420,486	19.3	18.8	19.8	1,840	132,898
60-64	0.0230	0.1087	0.1029	0.1148	81,186	80,512	81,801	8,829	383,859	15.5	15.1	16.0	3,007	130,718
65-69	0.0400	0.1818	0.1705	0.1930	72,357	71,558	73,175	13,156	328,897	12.1	11.7	12.6	4,329	108,235
70-74	0.0700	0.2979	0.2788	0.3163	59,202	57,750	60,602	17,634	251,921	9.2	8.9	9.6	4,849	69,273
75-79	0.0990	0.3968	0.3737	0.4206	41,567	39,540	43,631	16,494	166,601	7.1	6.8	7.4	3,365	33,990
80-84	0.1600	0.5714	0.5470	0.5973	25,073	23,067	27,325	14,328	89,548	5.1	4.9	5.3	3,335	20,845
85-89	0.2600	0.7879	0.7690	0.8073	10,746	9,344	12,379	8,466	32,563	3.6	3.5	3.7	2,663	10,243
90-94	0.3600	0.8654	0.8518	0.8793	2,279	1,800	2,851	1,973	5,479	2.7	2.6	2.8	1,032	2,868
95-99	0.4800	0.9836	0.9802	0.9868	307	217	420	302	629	2.1	2.1	2.1	177	369
100+	0.6400	1	1	1	5	3	8	5	8	1.6	1.6	1.7	13	21
All ages													*28,942*	*2,252,854*

ln($_nq_x$) with uncertainty interval

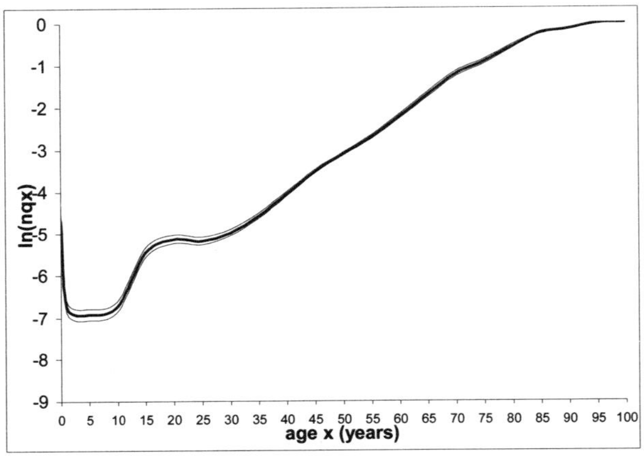

l_x with uncertainty interval

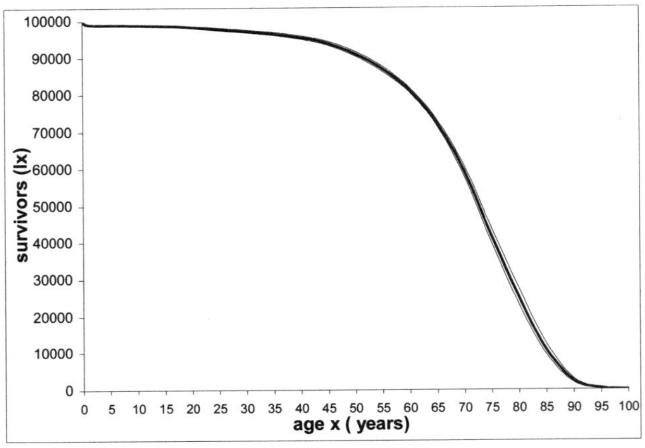

e_x with uncertainty interval

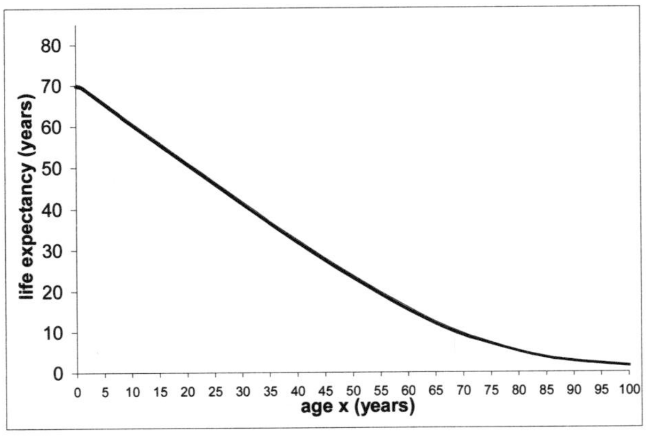

Distribution of deaths by age with uncertainty interval

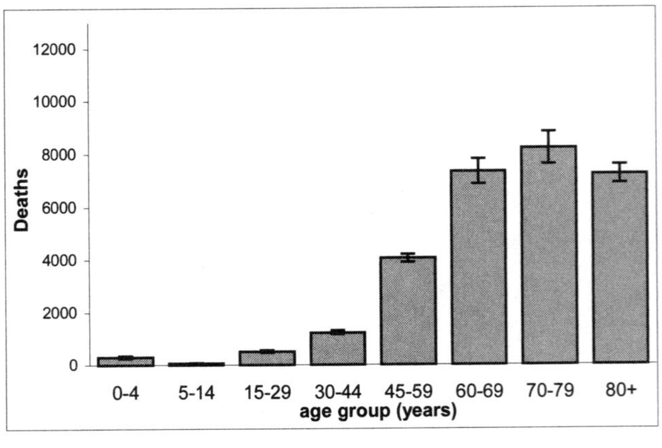

Croatia

Croatie

Croacia

Females, 2000

Femmes, 2000

Mujeres, 2000

Age	$_nM_x$	$_nq_x$			l_x			$_nd_x$	$_nL_x$	e_x			Actual deaths	Actual population
			percentiles			percentiles					percentiles			
		mean	2½	97½	mean	2½	97½			mean	2½	97½		
<1	0.0066	0.0065	0.0056	0.0077	100,000	100,000	100,000	654	99,411	77.7	77.3	78.2	173	26,283
1-4	0.0002	0.0009	0.0008	0.0011	99,346	99,227	99,440	94	397,157	77.2	76.8	77.7	25	104,875
5-9	0.0001	0.0006	0.0006	0.0007	99,252	99,120	99,356	64	496,099	73.3	72.9	73.7	17	131,806
10-14	0.0001	0.0007	0.0006	0.0007	99,188	99,048	99,300	65	495,778	68.4	68.0	68.8	19	146,888
15-19	0.0003	0.0016	0.0015	0.0018	99,123	98,974	99,242	163	495,209	63.4	63.0	63.8	52	157,175
20-24	0.0003	0.0015	0.0014	0.0017	98,961	98,791	99,097	152	494,422	58.5	58.1	58.9	48	156,956
25-29	0.0003	0.0017	0.0016	0.0019	98,808	98,621	98,961	171	493,613	53.6	53.2	54.0	54	154,943
30-34	0.0005	0.0026	0.0024	0.0029	98,637	98,430	98,806	257	492,542	48.7	48.3	49.1	86	164,111
35-39	0.0009	0.0044	0.0040	0.0048	98,380	98,145	98,574	431	490,820	43.8	43.4	44.2	153	174,247
40-44	0.0014	0.0071	0.0065	0.0078	97,948	97,671	98,180	698	487,996	39.0	38.6	39.4	256	179,193
45-49	0.0022	0.0109	0.0101	0.0118	97,250	96,912	97,539	1,064	483,592	34.2	33.9	34.7	365	165,909
50-54	0.0037	0.0184	0.0172	0.0196	96,186	95,769	96,550	1,770	476,508	29.6	29.2	30.0	499	134,229
55-59	0.0057	0.0279	0.0263	0.0296	94,417	93,894	94,886	2,638	465,487	25.1	24.8	25.5	805	141,996
60-64	0.0103	0.0502	0.0476	0.0529	91,778	91,145	92,372	4,607	447,373	20.7	20.5	21.1	1,545	149,987
65-69	0.0177	0.0850	0.0804	0.0891	87,171	86,396	87,959	7,405	417,341	16.7	16.5	17.1	2,466	138,980
70-74	0.0300	0.1395	0.1320	0.1457	79,765	78,727	80,792	11,130	371,002	13.0	12.8	13.3	3,565	118,845
75-79	0.0580	0.2533	0.2410	0.2621	68,635	67,356	69,976	17,384	299,718	9.7	9.5	10.0	4,038	69,616
80-84	0.0900	0.3673	0.3532	0.3804	51,252	49,703	53,110	18,827	209,191	7.2	7.0	7.4	4,269	47,430
85-89	0.1700	0.5965	0.5800	0.6122	32,425	31,010	34,353	19,341	113,770	4.9	4.8	5.0	4,574	26,908
90-94	0.2600	0.7303	0.7160	0.7432	13,084	12,167	14,323	9,555	36,752	3.5	3.4	3.6	2,319	8,921
95-99	0.4000	0.9091	0.9011	0.9158	3,528	3,124	3,987	3,207	8,019	2.4	2.4	2.5	596	1,489
100+	0.5800	1	1	1	321	263	392	321	553	1.7	1.7	1.7	63	108
All ages													*25,986*	*2,400,895*

ln($_nq_x$) with uncertainty interval

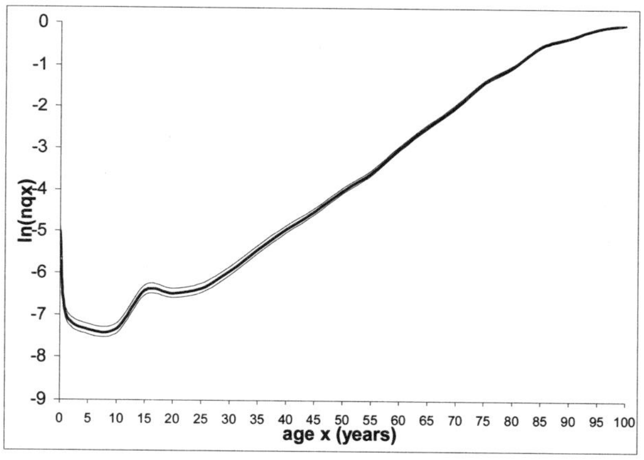

l_x with uncertainty interval

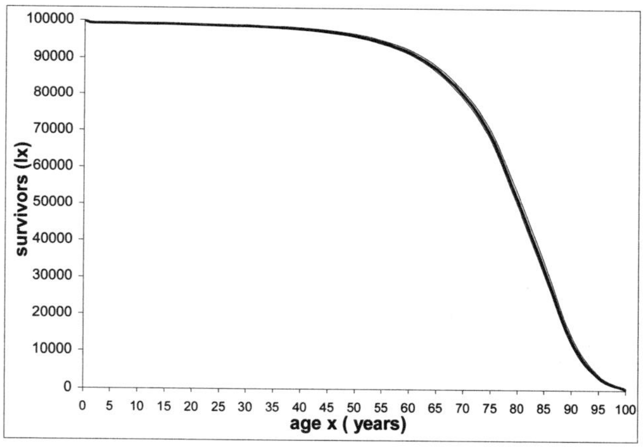

e_x with uncertainty interval

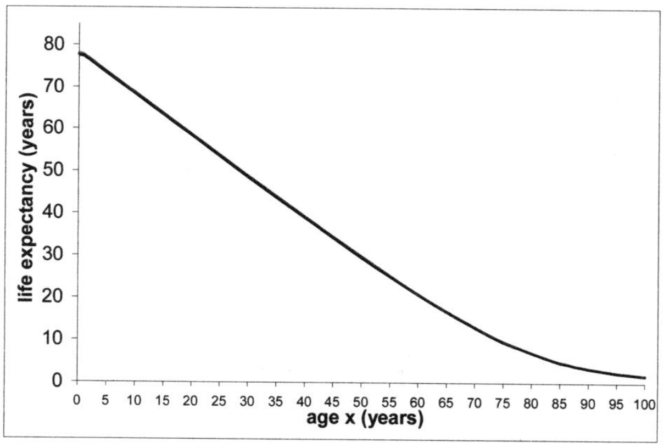

Distribution of deaths by age with uncertainty interval

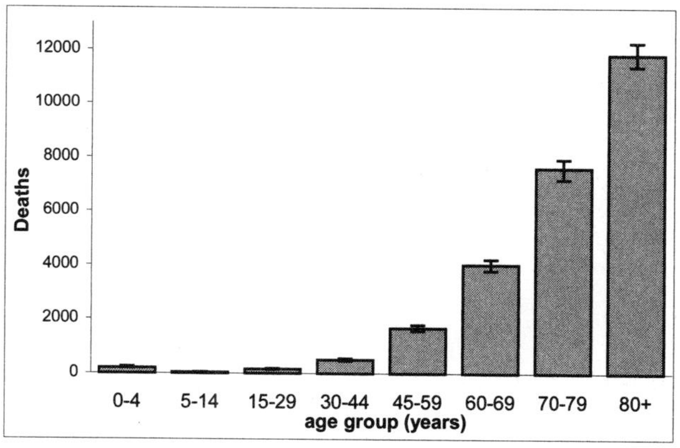

Cuba

Males, 2000

Cuba

Hommes, 2000

Cuba

Hombres, 2000

Age	$_nM_x$	$_nq_x$			l_x			$_nd_x$	$_nL_x$	e_x			Actual deaths	Actual population
			percentiles			percentiles					percentiles			
		mean	2½	97½	mean	2½	97½			mean	2½	97½		
<1	0.0075	0.0074	0.0066	0.0083	100,000	100,000	100,000	742	99,332	73.7	73.3	74.0	526	70,451
1-4	0.0005	0.0021	0.0019	0.0023	99,258	99,173	99,340	208	396,534	73.2	72.9	73.6	156	297,893
5-9	0.0003	0.0017	0.0016	0.0018	99,050	98,950	99,148	168	494,831	69.4	69.0	69.7	137	403,676
10-14	0.0003	0.0017	0.0016	0.0018	98,882	98,771	98,992	169	493,989	64.5	64.1	64.8	153	446,223
15-19	0.0008	0.0042	0.0040	0.0045	98,713	98,589	98,834	418	492,520	59.6	59.2	59.9	329	387,522
20-24	0.0012	0.0059	0.0056	0.0063	98,295	98,143	98,441	580	490,024	54.8	54.5	55.2	452	381,986
25-29	0.0014	0.0072	0.0068	0.0075	97,715	97,529	97,890	700	486,824	50.1	49.8	50.5	777	540,360
30-34	0.0016	0.0081	0.0078	0.0085	97,015	96,793	97,226	790	483,098	45.5	45.1	45.9	927	566,709
35-39	0.0022	0.0108	0.0103	0.0112	96,224	95,967	96,470	1,038	478,528	40.8	40.5	41.2	1,144	527,447
40-44	0.0031	0.0152	0.0147	0.0158	95,187	94,890	95,472	1,451	472,308	36.3	35.9	36.6	1,036	337,445
45-49	0.0044	0.0220	0.0213	0.0226	93,736	93,394	94,067	2,059	463,534	31.8	31.4	32.2	1,507	339,330
50-54	0.0062	0.0305	0.0295	0.0314	91,677	91,287	92,061	2,792	451,405	27.4	27.1	27.8	1,919	310,167
55-59	0.0099	0.0482	0.0466	0.0499	88,885	88,450	89,321	4,288	433,705	23.2	22.9	23.6	2,652	268,230
60-64	0.0149	0.0720	0.0693	0.0749	84,597	84,111	85,080	6,093	407,753	19.3	18.9	19.7	3,358	224,718
65-69	0.0229	0.1082	0.1033	0.1132	78,504	77,934	79,080	8,492	371,290	15.6	15.2	15.9	3,757	164,251
70-74	0.0352	0.1617	0.1534	0.1697	70,012	69,186	70,830	11,319	321,764	12.2	11.9	12.5	4,795	136,308
75-79	0.0600	0.2609	0.2473	0.2733	58,693	57,528	59,840	15,311	255,189	9.0	8.8	9.3	5,719	95,322
80-84	0.1150	0.4466	0.4270	0.4634	43,382	41,803	44,968	19,375	168,474	6.3	6.2	6.5	7,734	67,256
85-89	0.1905	0.6453	0.6263	0.6614	24,008	22,430	25,767	15,493	81,306	4.4	4.3	4.5	6,321	33,172
90-94	0.3236	0.8210	0.8081	0.8319	8,515	7,615	9,630	6,991	21,603	2.9	2.9	3.0	3,382	10,450
95-99	0.5129	0.9800	0.9771	0.9824	1,524	1,289	1,848	1,494	3,140	2.1	2.1	2.1	887	1,729
100+	0.7328	1	1	1	30	23	42	30	42	1.4	1.4	1.8	52	71
All ages													*47,720*	*5,610,716*

$\ln(_nq_x)$ with uncertainty interval

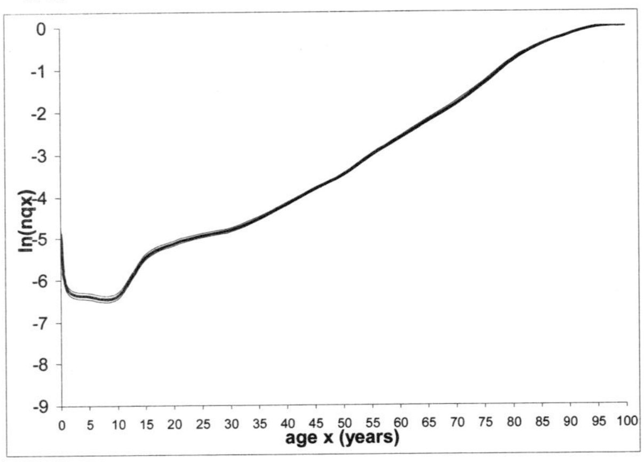

l_x with uncertainty interval

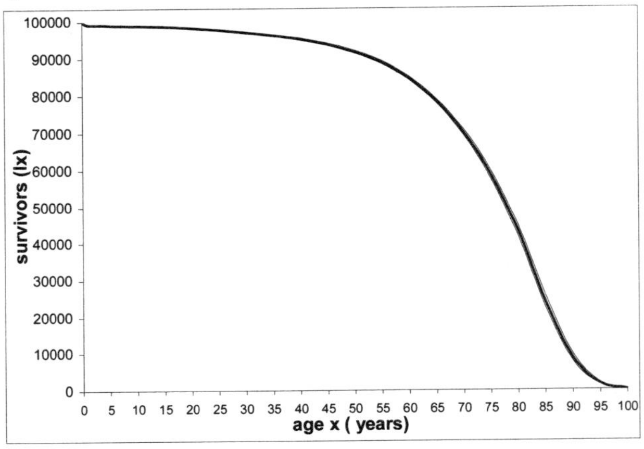

e_x with uncertainty interval

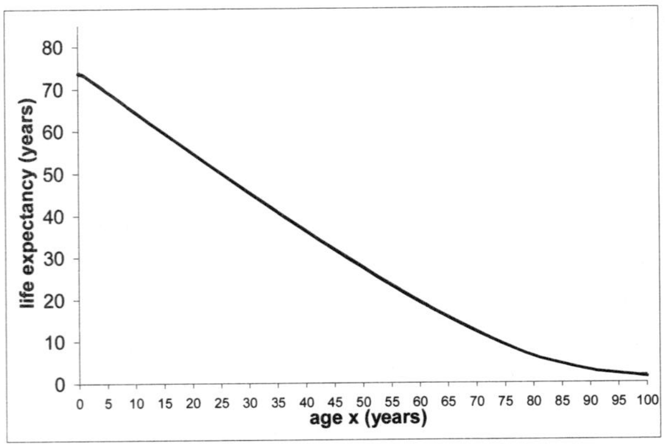

Distribution of deaths by age with uncertainty interval

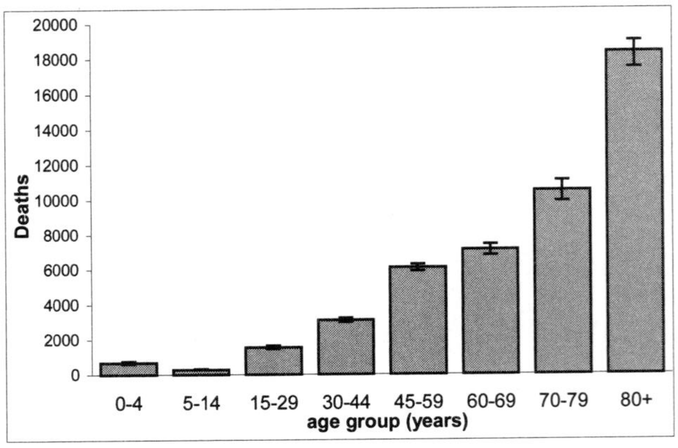

Cuba

Cuba

Cuba

Females, 2000 *Femmes, 2000* *Mujeres, 2000*

Age	$_nM_x$	$_nq_x$ mean	$_nq_x$ percentiles 2½	$_nq_x$ percentiles 97½	l_x mean	l_x percentiles 2½	l_x percentiles 97½	$_nd_x$	$_nL_x$	e_x mean	e_x percentiles 2½	e_x percentiles 97½	Actual deaths	Actual population
<1	0.0060	0.0060	0.0054	0.0065	100,000	100,000	100,000	596	99,463	77.5	77.1	78.0	406	67,774
1-4	0.0005	0.0020	0.0019	0.0022	99,404	99,345	99,463	202	397,130	77.0	76.6	77.5	144	282,053
5-9	0.0002	0.0011	0.0010	0.0012	99,202	99,127	99,276	109	495,735	73.1	72.8	73.6	84	383,903
10-14	0.0002	0.0010	0.0010	0.0011	99,093	99,010	99,175	102	495,208	68.2	67.8	68.7	87	425,214
15-19	0.0005	0.0024	0.0022	0.0026	98,991	98,901	99,080	237	494,361	63.3	62.9	63.7	178	370,393
20-24	0.0006	0.0031	0.0029	0.0033	98,754	98,648	98,859	302	493,014	58.4	58.1	58.9	224	365,185
25-29	0.0007	0.0035	0.0033	0.0037	98,452	98,326	98,576	347	491,393	53.6	53.2	54.1	368	521,955
30-34	0.0009	0.0043	0.0041	0.0046	98,105	97,959	98,251	427	489,460	48.8	48.4	49.2	477	547,417
35-39	0.0013	0.0062	0.0059	0.0066	97,679	97,507	97,848	610	486,868	44.0	43.6	44.4	661	527,377
40-44	0.0020	0.0101	0.0096	0.0106	97,068	96,865	97,270	983	482,885	39.2	38.9	39.7	709	348,086
45-49	0.0031	0.0152	0.0145	0.0158	96,085	95,835	96,333	1,457	476,784	34.6	34.3	35.1	1,067	349,232
50-54	0.0044	0.0218	0.0209	0.0228	94,628	94,317	94,935	2,063	467,984	30.1	29.8	30.5	1,432	324,924
55-59	0.0063	0.0311	0.0296	0.0325	92,565	92,184	92,945	2,880	455,628	25.7	25.4	26.1	1,763	278,938
60-64	0.0103	0.0504	0.0479	0.0526	89,686	89,227	90,154	4,519	437,131	21.5	21.2	21.9	2,408	232,921
65-69	0.0176	0.0845	0.0801	0.0880	85,167	84,564	85,759	7,193	407,851	17.5	17.2	17.8	3,098	175,645
70-74	0.0278	0.1298	0.1231	0.1348	77,974	77,120	78,871	10,123	364,560	13.9	13.7	14.2	4,187	150,787
75-79	0.0440	0.1982	0.1886	0.2050	67,850	66,722	69,161	13,448	305,632	10.6	10.4	10.8	4,615	104,891
80-84	0.0894	0.3654	0.3515	0.3765	54,402	53,067	56,115	19,878	222,318	7.6	7.4	7.7	6,937	77,590
85-89	0.1349	0.5044	0.4906	0.5171	34,525	33,222	36,388	17,414	129,089	5.5	5.4	5.6	5,327	39,489
90-94	0.2570	0.7256	0.7140	0.7366	17,111	16,143	18,507	12,416	48,306	3.5	3.4	3.6	3,263	12,695
95-99	0.4074	0.9166	0.9111	0.9219	4,695	4,267	5,246	4,303	10,563	2.4	2.4	2.4	783	1,921
100+	0.6310	1	1	1	391	334	463	391	620	1.6	1.6	2.0	44	70
All ages													*38,263*	*5,588,460*

ln($_nq_x$) with uncertainty interval

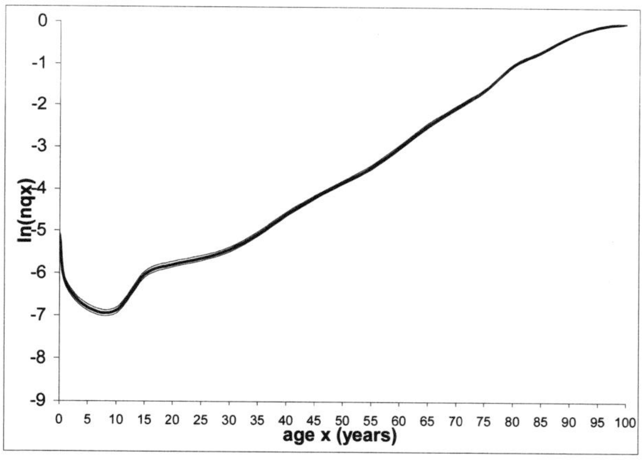

l_x with uncertainty interval

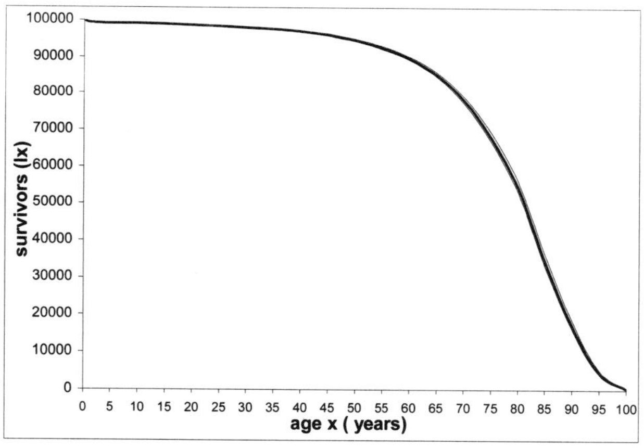

e_x with uncertainty interval

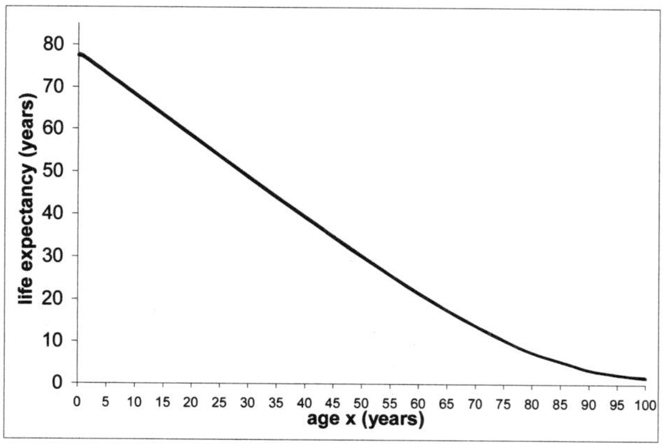

Distribution of deaths by age with uncertainty interval

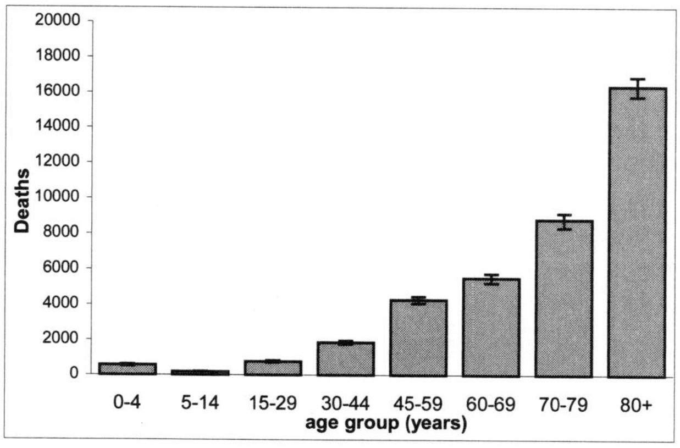

Cyprus

Chypre

Chipre

Males, 2000 *Hommes, 2000* *Hombres, 2000*

Age	$_nM_x$	$_nq_x$ mean	$_nq_x$ percentiles 2½	$_nq_x$ percentiles 97½	l_x mean	l_x percentiles 2½	l_x percentiles 97½	$_nd_x$	$_nL_x$	e_x mean	e_x percentiles 2½	e_x percentiles 97½	Actual deaths	Actual population
<1	0.0077	0.0077	0.0067	0.0080	100,000	100,000	100,000	769	99,308	74.8	74.3	75.6	41	5,308
1-4	0.0004	0.0014	0.0013	0.0015	99,231	99,199	99,329	141	396,585	74.4	73.8	75.2	8	22,238
5-9	0.0001	0.0004	0.0004	0.0004	99,090	99,052	99,202	41	495,347	70.5	69.9	71.3	3	32,358
10-14	0.0002	0.0008	0.0007	0.0009	99,049	99,009	99,166	82	495,039	65.5	65.0	66.3	6	33,757
15-19	0.0015	0.0077	0.0070	0.0080	98,967	98,923	99,091	761	492,932	60.6	60.0	61.4	50	32,726
20-24	0.0018	0.0089	0.0081	0.0093	98,206	98,130	98,401	874	488,846	56.0	55.5	56.8	53	29,479
25-29	0.0017	0.0083	0.0077	0.0087	97,332	97,217	97,601	812	484,633	51.5	51.0	52.3	45	26,797
30-34	0.0011	0.0056	0.0052	0.0059	96,521	96,368	96,853	544	481,245	46.9	46.4	47.7	32	27,996
35-39	0.0014	0.0068	0.0063	0.0071	95,977	95,799	96,351	654	478,252	42.1	41.6	42.9	38	28,159
40-44	0.0019	0.0093	0.0086	0.0098	95,324	95,115	95,746	890	474,394	37.4	36.9	38.2	53	28,524
45-49	0.0028	0.0140	0.0129	0.0147	94,434	94,183	94,921	1,320	468,870	32.7	32.3	33.5	72	25,632
50-54	0.0044	0.0218	0.0200	0.0229	93,114	92,800	93,695	2,029	460,498	28.2	27.7	28.9	101	23,007
55-59	0.0081	0.0397	0.0364	0.0420	91,085	90,671	91,817	3,619	446,380	23.7	23.3	24.4	164	20,189
60-64	0.0114	0.0554	0.0504	0.0589	87,467	86,859	88,479	4,844	425,223	19.6	19.2	20.3	179	15,689
65-69	0.0200	0.0953	0.0866	0.1015	82,623	81,747	84,017	7,873	393,431	15.6	15.3	16.2	259	12,946
70-74	0.0358	0.1644	0.1499	0.1749	74,750	73,450	76,741	12,289	343,027	12.0	11.7	12.5	364	10,154
75-79	0.0601	0.2613	0.2408	0.2767	62,461	60,605	65,238	16,318	271,510	8.9	8.6	9.3	472	7,858
80-84	0.1183	0.4564	0.4308	0.4776	46,143	43,837	49,531	21,059	178,067	6.1	5.9	6.4	576	4,869
85-89	0.2080	0.6843	0.6645	0.7031	25,084	22,900	28,194	17,165	82,506	4.2	4.1	4.3	506	2,434
90-94	0.3272	0.8256	0.8147	0.8381	7,919	6,798	9,458	6,538	19,980	2.9	2.8	3.0	305	933
95-99	0.4601	0.9665	0.9632	0.9705	1,381	1,101	1,753	1,335	2,901	2.2	2.1	2.2	92	200
100+	0.5784	1	1	1	46	32	65	46	80	1.7	1.7	1.9	17	29
All ages													*3,436*	*391,282*

$\ln(_nq_x)$ with uncertainty interval

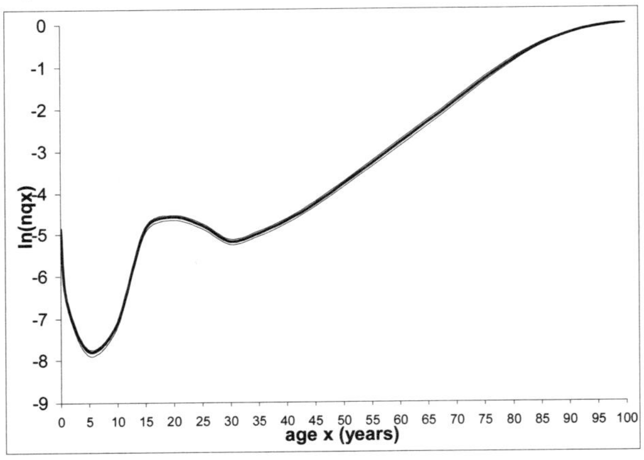

l_x with uncertainty interval

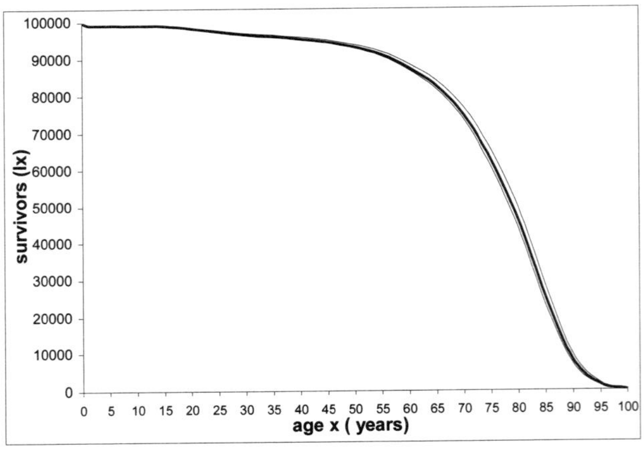

e_x with uncertainty interval

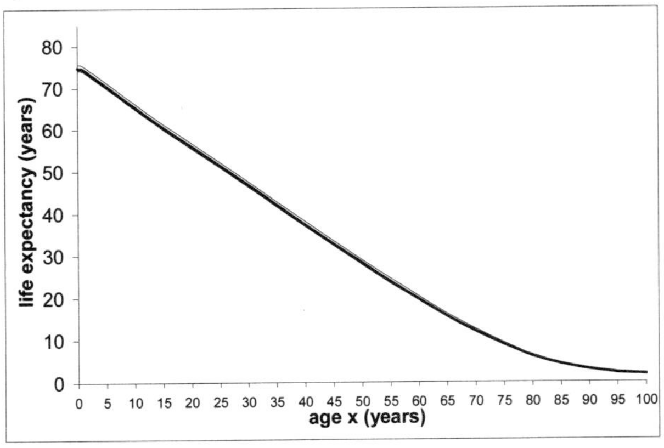

Distribution of deaths by age with uncertainty interval

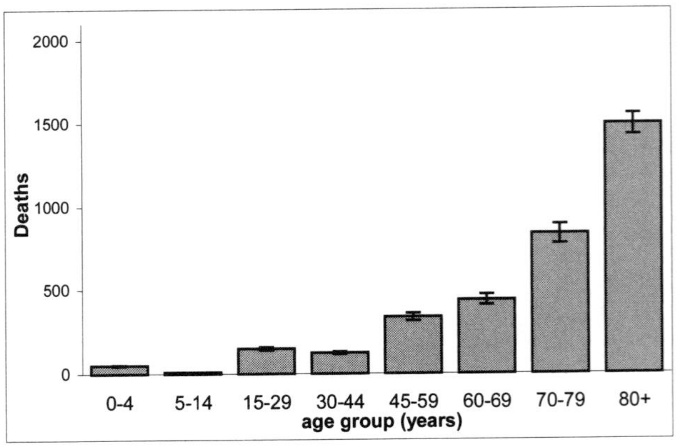

Cyprus   Chypre   Chipre

Females, 2000   *Femmes, 2000*   *Mujeres, 2000*

Age	$_nM_x$	$_nq_x$			l_x			$_nd_x$	$_nL_x$	e_x			Actual deaths	Actual population
			percentiles			percentiles					percentiles			
		mean	2½	97½	mean	2½	97½			mean	2½	97½		
<1	0.0066	0.0065	0.0054	0.0073	100,000	100,000	100,000	653	99,412	79.0	78.3	79.8	33	5,007
1-4	0.0002	0.0007	0.0006	0.0007	99,347	99,274	99,458	67	397,228	78.5	77.8	79.3	3	20,740
5-9	0.0001	0.0004	0.0003	0.0004	99,280	99,200	99,401	39	496,304	74.5	73.9	75.4	2	30,265
10-14	0.0001	0.0004	0.0003	0.0004	99,241	99,157	99,367	39	496,109	69.5	68.9	70.4	2	31,542
15-19	0.0003	0.0015	0.0012	0.0016	99,202	99,115	99,334	144	495,650	64.6	63.9	65.4	9	30,540
20-24	0.0003	0.0017	0.0014	0.0018	99,058	98,957	99,210	164	494,881	59.7	59.0	60.5	9	28,156
25-29	0.0005	0.0027	0.0024	0.0030	98,894	98,777	99,068	269	493,798	54.8	54.1	55.6	14	25,825
30-34	0.0007	0.0034	0.0030	0.0037	98,625	98,483	98,834	336	492,284	49.9	49.3	50.8	19	27,210
35-39	0.0008	0.0038	0.0033	0.0041	98,288	98,115	98,539	375	490,506	45.1	44.5	45.9	21	27,574
40-44	0.0011	0.0053	0.0047	0.0058	97,914	97,708	98,210	522	488,266	40.2	39.6	41.1	30	28,360
45-49	0.0017	0.0085	0.0075	0.0092	97,392	97,142	97,747	828	484,892	35.4	34.8	36.3	44	25,503
50-54	0.0024	0.0119	0.0105	0.0129	96,564	96,245	97,013	1,145	479,958	30.7	30.1	31.5	56	23,478
55-59	0.0044	0.0216	0.0191	0.0235	95,419	95,004	95,994	2,060	471,945	26.1	25.5	26.9	90	20,668
60-64	0.0070	0.0345	0.0306	0.0376	93,359	92,775	94,157	3,224	458,736	21.6	21.0	22.3	119	16,972
65-69	0.0146	0.0703	0.0624	0.0768	90,135	89,286	91,272	6,339	434,830	17.3	16.8	18.0	218	14,959
70-74	0.0228	0.1078	0.0951	0.1179	83,797	82,425	85,579	9,036	396,394	13.4	13.0	14.0	300	13,178
75-79	0.0460	0.2063	0.1831	0.2236	74,761	72,709	77,437	15,422	335,250	9.7	9.3	10.2	481	10,458
80-84	0.1060	0.4190	0.3828	0.4470	59,339	56,455	63,256	24,862	234,543	6.5	6.3	6.9	724	6,829
85-89	0.1900	0.6441	0.6130	0.6709	34,478	31,220	39,041	22,206	116,874	4.5	4.3	4.7	671	3,530
90-94	0.3100	0.8031	0.7855	0.8213	12,272	10,276	15,108	9,856	31,792	3.0	2.9	3.1	455	1,469
95-99	0.4400	0.9483	0.9415	0.9554	2,416	1,836	3,240	2,291	5,207	2.2	2.2	2.3	110	251
100+	0.5600	1	1	1	125	82	189	125	223	1.8	1.8	2.1	17	31
All ages													*3,430*	*392,545*

$\ln(_nq_x)$ with uncertainty interval

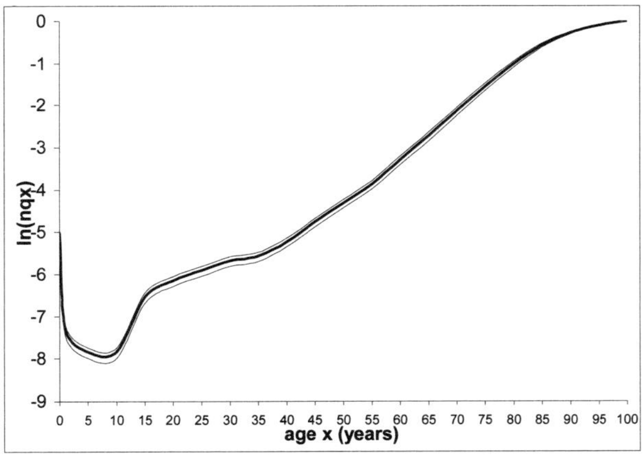

l_x with uncertainty interval

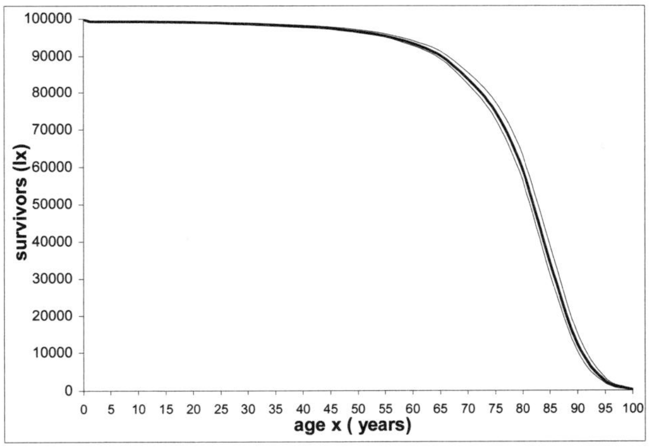

e_x with uncertainty interval

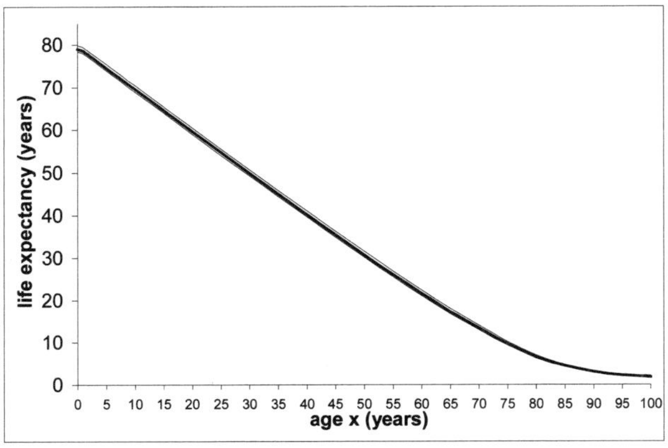

Distribution of deaths by age with uncertainty interval

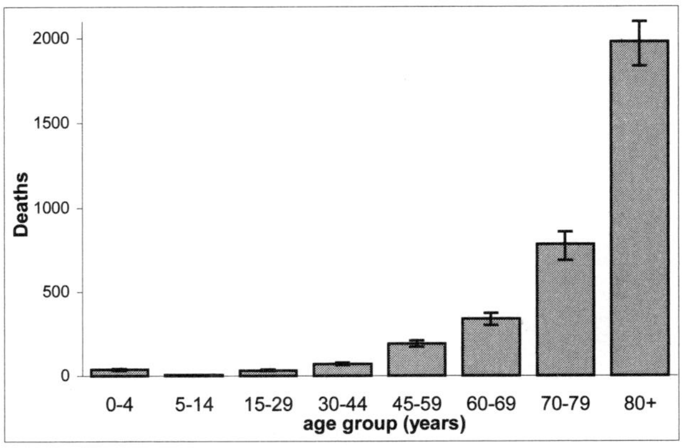

Czech Republic | République tchèque | República Checa

Males, 2000 | *Hommes, 2000* | *Hombres, 2000*

Age	$_nM_x$	$_nq_x$ mean	$_nq_x$ percentiles 2 1/2	$_nq_x$ percentiles 97 1/2	l_x mean	l_x percentiles 2 1/2	l_x percentiles 97 1/2	$_nd_x$	$_nL_x$	e_x mean	e_x percentiles 2 1/2	e_x percentiles 97 1/2	Actual deaths	Actual population
<1	0.0050	0.0050	0.0044	0.0057	100,000	100,000	100,000	498	99,552	71.5	71.3	71.8	225	45,072
1-4	0.0003	0.0012	0.0011	0.0014	99,502	99,429	99,561	123	397,713	70.9	70.6	71.1	58	186,012
5-9	0.0002	0.0010	0.0009	0.0011	99,379	99,292	99,449	99	496,646	67.0	66.7	67.2	60	302,130
10-14	0.0002	0.0012	0.0011	0.0013	99,280	99,183	99,358	119	496,100	62.0	61.8	62.3	79	331,201
15-19	0.0007	0.0036	0.0033	0.0039	99,161	99,052	99,248	356	494,912	57.1	56.8	57.4	255	354,300
20-24	0.0011	0.0052	0.0049	0.0056	98,804	98,664	98,918	517	492,728	52.3	52.0	52.6	467	444,710
25-29	0.0011	0.0052	0.0049	0.0056	98,287	98,108	98,433	515	490,148	47.6	47.3	47.8	452	430,185
30-34	0.0013	0.0066	0.0063	0.0070	97,772	97,559	97,946	648	487,241	42.8	42.5	43.1	465	349,399
35-39	0.0020	0.0100	0.0096	0.0104	97,124	96,875	97,327	966	483,205	38.1	37.8	38.3	691	345,276
40-44	0.0030	0.0149	0.0144	0.0155	96,158	95,864	96,397	1,432	477,210	33.4	33.1	33.7	1,052	350,643
45-49	0.0060	0.0296	0.0289	0.0303	94,726	94,382	95,006	2,800	466,631	28.9	28.6	29.2	2,390	398,358
50-54	0.0090	0.0440	0.0431	0.0449	91,926	91,522	92,251	4,046	449,518	24.7	24.4	25.0	3,554	394,867
55-59	0.0140	0.0676	0.0660	0.0693	87,881	87,450	88,235	5,944	424,544	20.7	20.4	21.0	4,234	302,398
60-64	0.0230	0.1087	0.1056	0.1120	81,937	81,551	82,279	8,910	387,409	17.0	16.8	17.3	4,940	214,770
65-69	0.0330	0.1524	0.1472	0.1575	73,027	72,584	73,511	11,131	337,306	13.8	13.5	14.1	6,456	195,633
70-74	0.0520	0.2301	0.2218	0.2379	61,896	61,175	62,622	14,241	273,874	10.8	10.6	11.1	8,506	163,578
75-79	0.0820	0.3402	0.3287	0.3519	47,654	46,627	48,675	16,214	197,735	8.3	8.1	8.5	9,475	115,548
80-84	0.1250	0.4762	0.4618	0.4902	31,440	30,288	32,662	14,971	119,771	6.3	6.2	6.5	4,916	39,328
85-89	0.1860	0.6348	0.6205	0.6482	16,468	15,520	17,536	10,454	56,206	4.8	4.7	4.9	4,673	25,125
90-94	0.2390	0.6960	0.6835	0.7075	6,014	5,460	6,615	4,186	17,513	3.8	3.7	3.9	1,480	6,191
95-99	0.3250	0.8228	0.8130	0.8317	1,828	1,597	2,093	1,504	4,629	2.9	2.9	3.0	214	657
100+	0.4590	1	1	1	324	269	391	324	706	2.2	2.2	2.3	14	30
All ages													*54,654*	*4,995,411*

ln($_nq_x$) with uncertainty interval

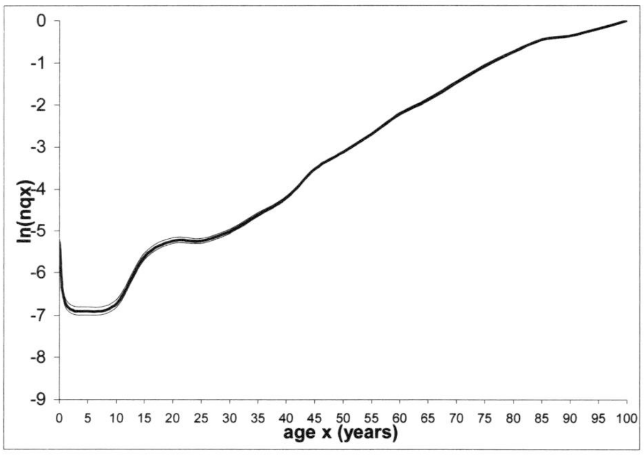

l_x with uncertainty interval

e_x with uncertainty interval

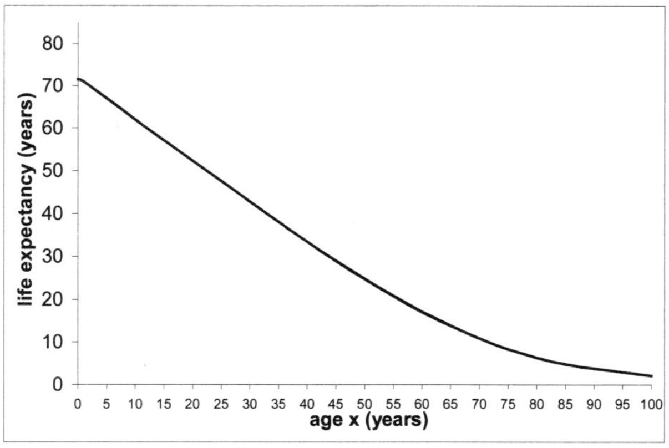

Distribution of deaths by age with uncertainty interval

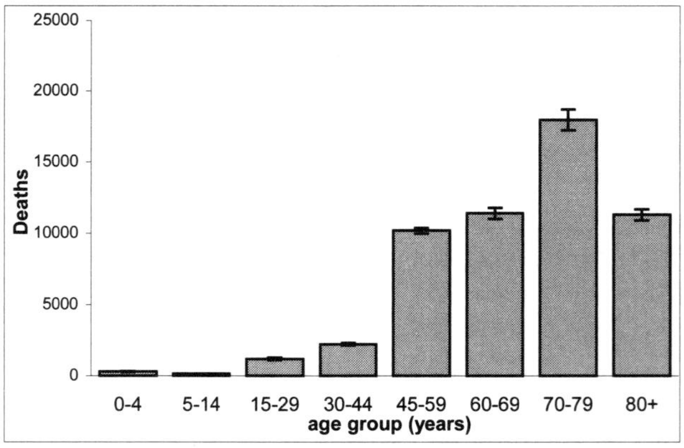

Czech Republic　　République tchèque　　República Checa

Females, 2000　　*Femmes, 2000*　　*Mujeres, 2000*

Age	$_nM_x$	$_nq_x$			l_x			$_nd_x$	$_nL_x$	e_x			Actual deaths	Actual population
			percentiles			percentiles					percentiles			
		mean	2½	97½	mean	2½	97½			mean	2½	97½		
<1	0.0050	0.0050	0.0045	0.0055	100,000	100,000	100,000	498	99,552	78.2	77.9	78.5	215	43,069
1-4	0.0002	0.0009	0.0008	0.0010	99,502	99,447	99,551	88	397,799	77.6	77.3	77.9	39	176,529
5-9	0.0002	0.0010	0.0009	0.0011	99,415	99,352	99,470	99	496,825	73.7	73.4	74.0	57	286,339
10-14	0.0001	0.0007	0.0007	0.0008	99,315	99,245	99,378	69	496,403	68.7	68.5	69.0	44	316,110
15-19	0.0003	0.0014	0.0013	0.0015	99,246	99,171	99,313	139	495,882	63.8	63.5	64.1	95	338,318
20-24	0.0003	0.0016	0.0015	0.0018	99,107	99,022	99,183	163	495,127	58.9	58.6	59.2	141	426,758
25-29	0.0003	0.0017	0.0016	0.0018	98,944	98,848	99,030	168	494,298	54.0	53.7	54.3	140	412,960
30-34	0.0005	0.0023	0.0022	0.0024	98,776	98,670	98,871	227	493,311	49.1	48.8	49.3	155	335,979
35-39	0.0008	0.0039	0.0037	0.0041	98,549	98,430	98,656	388	491,772	44.2	43.9	44.5	264	333,751
40-44	0.0014	0.0068	0.0065	0.0071	98,160	98,023	98,287	665	489,138	39.3	39.1	39.6	470	345,492
45-49	0.0020	0.0100	0.0096	0.0103	97,495	97,329	97,649	970	485,049	34.6	34.3	34.9	803	401,368
50-54	0.0040	0.0198	0.0193	0.0204	96,525	96,327	96,710	1,911	477,846	29.9	29.6	30.2	1,631	407,807
55-59	0.0060	0.0296	0.0288	0.0305	94,613	94,374	94,842	2,796	466,076	25.5	25.2	25.7	1,962	326,964
60-64	0.0100	0.0488	0.0474	0.0504	91,817	91,551	92,077	4,479	447,888	21.2	20.9	21.4	2,500	249,964
65-69	0.0170	0.0815	0.0788	0.0844	87,338	86,998	87,658	7,121	418,888	17.1	16.9	17.4	4,280	251,760
70-74	0.0290	0.1352	0.1302	0.1400	80,217	79,665	80,692	10,845	373,972	13.4	13.2	13.7	7,080	244,140
75-79	0.0520	0.2301	0.2217	0.2377	69,372	68,541	70,132	15,962	306,955	10.1	9.9	10.3	10,895	209,517
80-84	0.0920	0.3740	0.3616	0.3852	53,410	52,254	54,583	19,975	217,114	7.4	7.2	7.6	7,665	83,319
85-89	0.1570	0.5637	0.5498	0.5766	33,436	32,199	34,800	18,849	120,056	5.3	5.2	5.5	10,142	64,596
90-94	0.2240	0.6699	0.6579	0.6809	14,587	13,707	15,600	9,771	43,621	4.0	3.9	4.1	4,316	19,266
95-99	0.3220	0.8189	0.8102	0.8269	4,816	4,385	5,332	3,944	12,248	2.9	2.9	3.0	757	2,350
100+	0.4670	1	1	1	872	760	1,010	872	1,867	2.1	2.1	2.3	29	63
All ages													*53,679*	*5,276,419*

ln($_nq_x$) with uncertainty interval

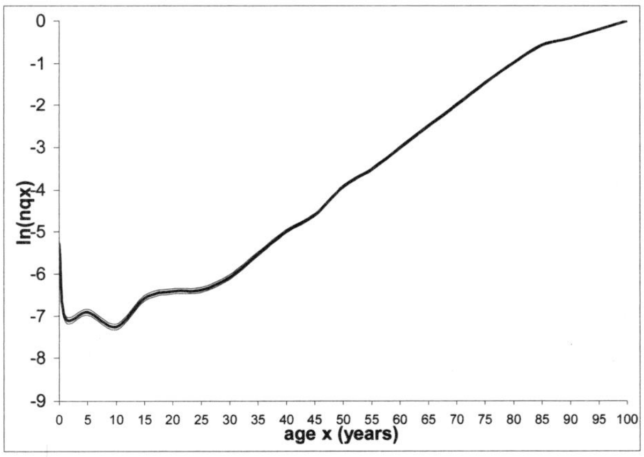

l_x with uncertainty interval

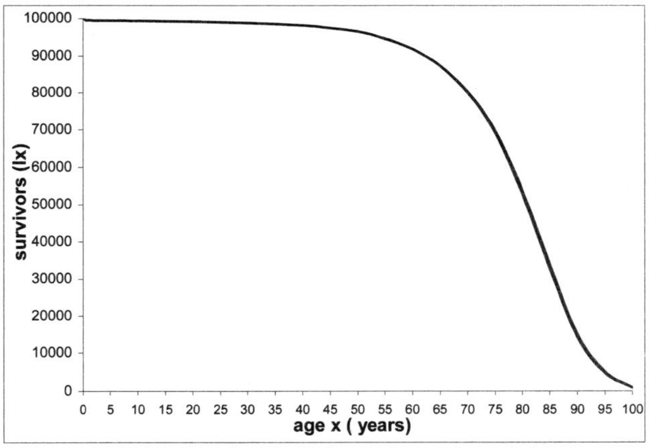

e_x with uncertainty interval

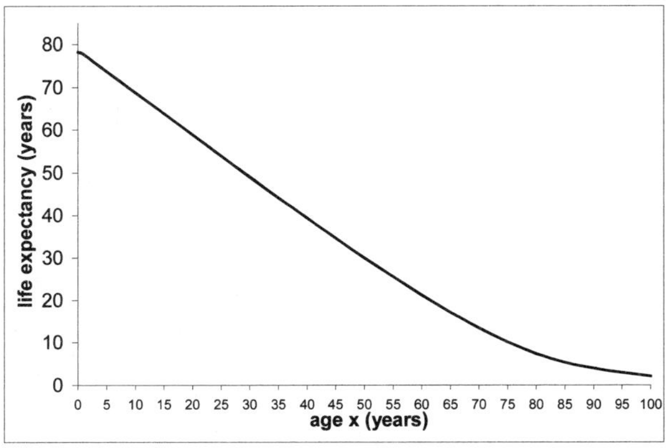

Distribution of deaths by age with uncertainty interval

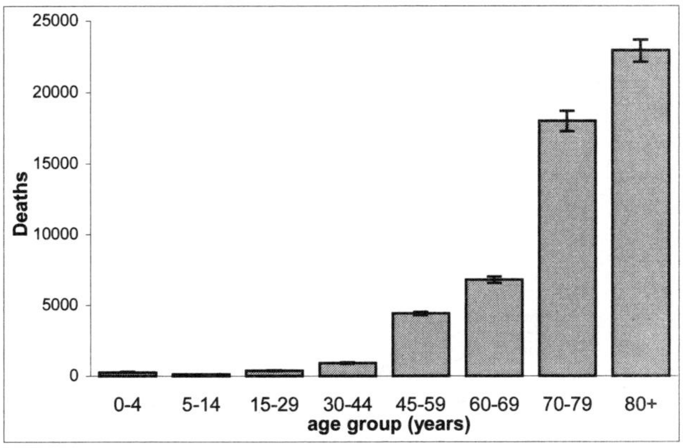

Democratic People's Republic of Korea
République populaire démocratique de Corée
República Popular Democrática de Corea

Males, 2000 — *Hommes, 2000* — *Hombres, 2000*

Age	$_nM_x$	$_nq_x$ mean	$_nq_x$ percentile 2½	$_nq_x$ percentile 97½	l_x mean	l_x percentile 2½	l_x percentile 97½	$_nd_x$	$_nL_x$	e_x mean	e_x percentile 2½	e_x percentile 97½	Actual deaths	Actual population
<1	0.0447	0.0430	0.0222	0.0657	100,000	100,000	100,000	4,301	96,129	64.5	62.2	66.4	8,681	194,043
1-4	0.0029	0.0115	0.0076	0.0151	95,699	93,425	97,781	1,098	380,162	66.4	64.6	68.7	2,290	792,973
5-9	0.0010	0.0049	0.0034	0.0064	94,601	92,109	97,039	467	471,838	63.1	61.3	65.6	1,042	1,051,769
10-14	0.0008	0.0040	0.0028	0.0052	94,134	91,521	96,709	379	469,723	58.4	56.5	60.9	789	978,432
15-19	0.0016	0.0081	0.0058	0.0102	93,755	91,049	96,436	759	466,879	53.7	51.7	56.2	1,484	913,072
20-24	0.0023	0.0113	0.0083	0.0140	92,997	90,116	95,874	1,048	462,363	49.1	47.1	51.7	1,862	821,661
25-29	0.0024	0.0118	0.0090	0.0144	91,949	88,854	95,067	1,082	457,039	44.6	42.5	47.3	2,496	1,054,231
30-34	0.0028	0.0137	0.0107	0.0164	90,867	87,578	94,200	1,245	451,223	40.1	38.0	42.8	3,044	1,103,710
35-39	0.0036	0.0178	0.0144	0.0210	89,622	86,140	93,164	1,600	444,134	35.6	33.4	38.4	3,116	865,235
40-44	0.0051	0.0253	0.0212	0.0290	88,023	84,332	91,774	2,223	434,556	31.2	28.9	34.0	3,881	758,642
45-49	0.0076	0.0373	0.0326	0.0420	85,800	81,887	89,763	3,197	421,005	27.0	24.6	29.8	3,511	462,284
50-54	0.0115	0.0560	0.0476	0.0647	82,602	78,481	86,826	4,628	401,441	22.9	20.5	25.7	7,230	627,138
55-59	0.0174	0.0835	0.0674	0.1012	77,974	73,739	82,318	6,512	373,592	19.2	16.7	21.9	9,672	554,874
60-64	0.0265	0.1242	0.0956	0.1594	71,462	67,348	75,547	8,878	335,117	15.7	13.3	18.3	11,746	443,364
65-69	0.0400	0.1820	0.1349	0.2434	62,584	58,946	66,072	11,389	284,449	12.5	10.4	14.9	10,507	262,428
70-74	0.0620	0.2686	0.1959	0.3616	51,195	45,396	56,167	13,750	221,602	9.8	7.9	11.9	9,330	150,369
75-79	0.0958	0.3865	0.2863	0.5028	37,446	29,135	45,155	14,472	151,047	7.4	6.0	9.2	8,180	85,373
80-84	0.1474	0.5387	0.4200	0.6566	22,973	14,486	32,229	12,375	83,928	5.5	4.6	6.8	5,927	40,198
85-89	0.2192	0.7079	0.5954	0.8036	10,598	4,974	18,693	7,503	34,234	4.1	3.5	4.9	3,266	14,902
90-94	0.3146	0.8093	0.7187	0.8815	3,095	977	7,564	2,505	7,962	3.0	2.6	3.5	1,090	3,465
95-99	0.4363	0.9448	0.8964	0.9751	590	116	2,128	558	1,279	2.3	2.1	2.5	215	492
100+	0.5843	1	1	1	33	3	220	33	56	1.7	1.7	1.9	23	39
All ages													*99,381*	*11,178,694*

$\ln(_nq_x)$ with uncertainty interval

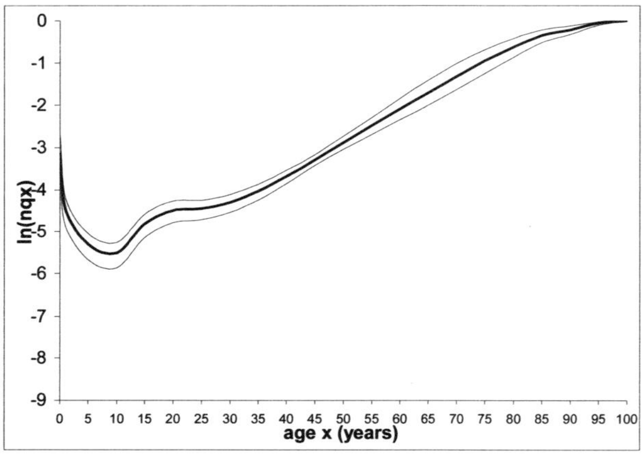

l_x with uncertainty interval

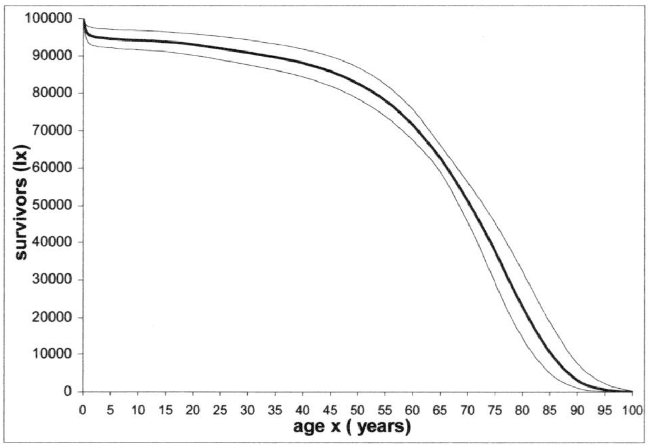

e_x with uncertainty interval

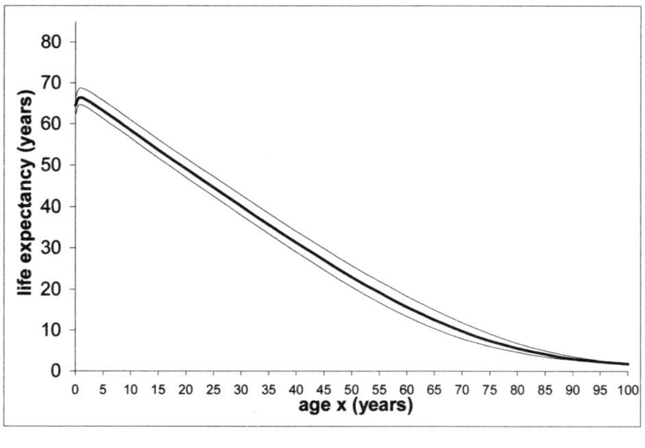

Distribution of deaths by age with uncertainty interval

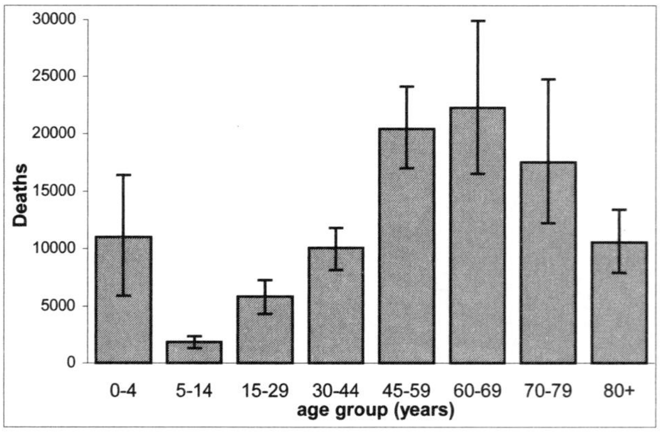

Democratic People's Republic of Korea

Females, 2000

République populaire démocratique de Corée

Femmes, 2000

República Popular Democrática de Corea

Mujeres, 2000

Age	$_nM_x$	$_nq_x$ mean	$_nq_x$ percentiles 2½	$_nq_x$ percentiles 97½	l_x mean	l_x percentiles 2½	l_x percentiles 97½	$_nd_x$	$_nL_x$	e_x mean	e_x percentiles 2½	e_x percentiles 97½	Actual deaths	Actual population
<1	0.0405	0.0391	0.0236	0.0600	100,000	100,000	100,000	3,908	96,482	67.2	64.8	69.2	7,451	183,927
1-4	0.0034	0.0134	0.0095	0.0180	96,092	93,998	97,644	1,290	381,271	68.9	67.3	70.8	2,574	760,905
5-9	0.0010	0.0051	0.0037	0.0066	94,802	92,311	96,716	479	472,811	65.8	64.2	67.7	1,018	1,004,881
10-14	0.0008	0.0039	0.0029	0.0051	94,323	91,700	96,362	371	470,684	61.2	59.5	63.1	738	934,867
15-19	0.0013	0.0065	0.0048	0.0083	93,951	91,233	96,084	607	468,237	56.4	54.7	58.3	1,133	873,740
20-24	0.0017	0.0083	0.0063	0.0105	93,344	90,478	95,622	777	464,777	51.8	50.1	53.7	1,319	788,945
25-29	0.0019	0.0096	0.0074	0.0119	92,567	89,527	95,020	889	460,612	47.2	45.5	49.1	1,954	1,012,344
30-34	0.0024	0.0118	0.0093	0.0144	91,678	88,459	94,316	1,080	455,689	42.6	40.9	44.5	2,510	1,058,735
35-39	0.0032	0.0160	0.0129	0.0192	90,598	87,185	93,442	1,453	449,356	38.1	36.4	40.0	2,695	833,348
40-44	0.0043	0.0215	0.0178	0.0251	89,145	85,509	92,235	1,913	440,940	33.7	32.0	35.6	3,197	736,840
45-49	0.0061	0.0302	0.0259	0.0347	87,231	83,359	90,596	2,637	429,564	29.3	27.7	31.2	2,797	455,597
50-54	0.0088	0.0432	0.0369	0.0496	84,594	80,484	88,253	3,658	413,825	25.2	23.6	27.0	5,476	619,491
55-59	0.0127	0.0615	0.0522	0.0718	80,936	76,635	84,862	4,980	392,230	21.2	19.7	23.0	7,491	589,995
60-64	0.0194	0.0923	0.0766	0.1092	75,956	71,600	80,017	7,010	362,255	17.4	16.1	19.1	9,259	478,460
65-69	0.0303	0.1406	0.1146	0.1676	68,946	64,811	72,829	9,695	320,490	13.9	12.7	15.5	9,243	305,545
70-74	0.0496	0.2208	0.1786	0.2627	59,250	54,978	63,789	13,081	263,549	10.8	9.7	12.2	10,036	202,196
75-79	0.0805	0.3350	0.2732	0.3968	46,169	40,595	51,803	15,469	192,175	8.2	7.2	9.4	10,610	131,814
80-84	0.1293	0.4885	0.4071	0.5606	30,701	24,838	37,554	14,996	116,013	6.0	5.3	7.0	9,398	72,708
85-89	0.1991	0.6647	0.5813	0.7290	15,705	11,225	21,975	10,439	52,424	4.4	4.0	5.1	6,595	33,117
90-94	0.2943	0.7815	0.7125	0.8307	5,265	3,042	8,901	4,115	13,981	3.2	2.9	3.6	2,947	10,015
95-99	0.4173	0.9265	0.8869	0.9510	1,150	515	2,559	1,066	2,554	2.3	2.2	2.6	820	1,965
100+	0.5675	1	1	1	85	25	289	85	149	1.8	1.7	1.9	141	248
All ages													*99,402*	*11,089,683*

ln($_nq_x$) with uncertainty interval

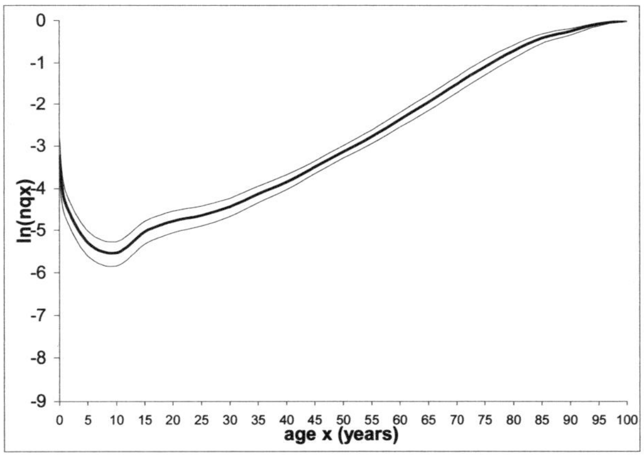

l_x with uncertainty interval

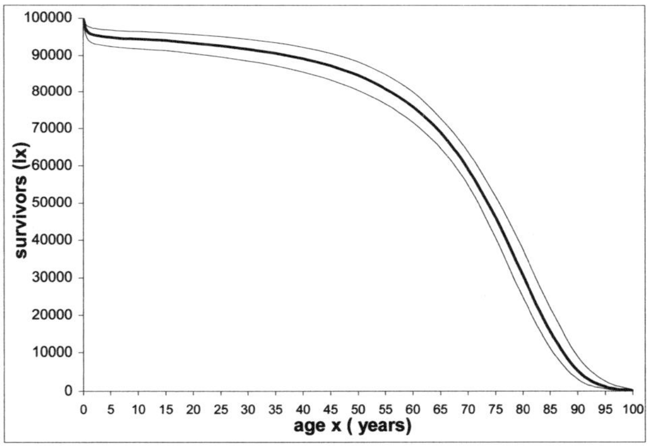

e_x with uncertainty interval

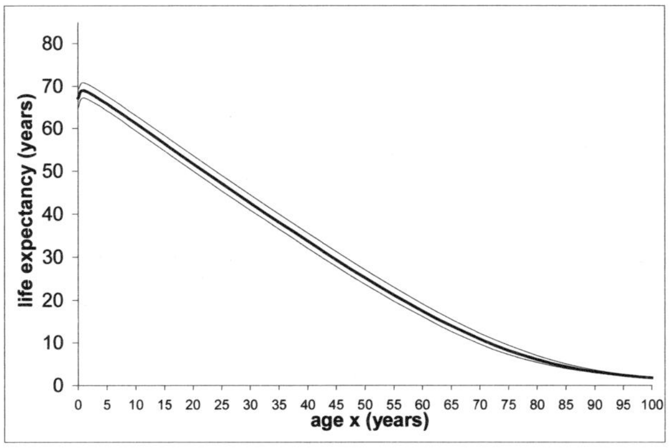

Distribution of deaths by age with uncertainty interval

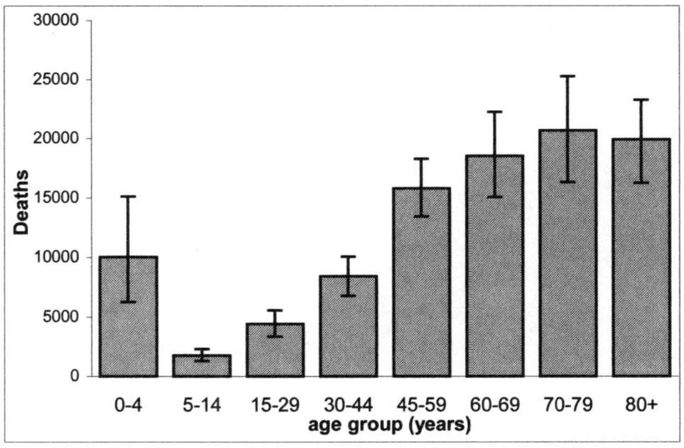

Democratic Republic of the Congo

Males, 2000

République démocratique du Congo

Hommes, 2000

República Democrática del Congo

Hombres, 2000

Age	$_nM_x$	$_nq_x$ mean	$_nq_x$ percentiles 2½	$_nq_x$ percentiles 97½	l_x mean	l_x percentiles 2½	l_x percentiles 97½	$_nd_x$	$_nL_x$	e_x mean	e_x percentiles 2½	e_x percentiles 97½	Actual deaths	Actual population
<1	0.1903	0.1625	0.1261	0.2029	100,000	100,000	100,000	16,248	85,377	41.6	37.8	45.5	211,044	1,108,938
1-4	0.0172	0.0661	0.0546	0.0793	83,752	79,714	87,392	5,534	321,725	48.7	45.7	51.8	67,668	3,933,861
5-9	0.0042	0.0206	0.0170	0.0248	78,218	74,253	82,143	1,611	387,061	48.0	45.4	50.8	17,101	4,109,159
10-14	0.0029	0.0144	0.0118	0.0173	76,607	72,521	80,675	1,100	380,284	44.0	41.4	46.6	9,601	3,318,238
15-19	0.0031	0.0153	0.0126	0.0184	75,507	71,316	79,711	1,155	374,645	39.6	37.1	42.2	8,310	2,695,091
20-24	0.0072	0.0356	0.0295	0.0425	74,351	70,049	78,695	2,645	365,145	35.2	32.7	37.7	15,395	2,125,426
25-29	0.0124	0.0603	0.0502	0.0715	71,707	67,072	76,346	4,322	347,728	31.4	29.0	34.0	21,147	1,701,464
30-34	0.0175	0.0840	0.0704	0.0987	67,385	62,278	72,444	5,659	322,776	28.2	26.0	30.8	24,290	1,385,386
35-39	0.0208	0.0987	0.0834	0.1157	61,725	56,134	67,214	6,090	293,403	25.6	23.5	28.1	23,225	1,118,969
40-44	0.0233	0.1100	0.0934	0.1290	55,636	49,699	61,429	6,122	262,872	23.1	21.2	25.5	21,197	910,095
45-49	0.0254	0.1196	0.1000	0.1404	49,513	43,417	55,610	5,920	232,767	20.7	18.9	22.9	18,712	735,799
50-54	0.0279	0.1303	0.1077	0.1531	43,594	37,524	49,871	5,682	203,763	18.1	16.6	20.2	16,870	604,951
55-59	0.0314	0.1457	0.1193	0.1707	37,911	32,037	44,197	5,523	175,749	15.5	14.2	17.3	15,495	493,035
60-64	0.0447	0.2009	0.1642	0.2332	32,388	26,862	38,509	6,507	145,672	12.7	11.6	14.3	17,077	382,285
65-69	0.0607	0.2633	0.2160	0.3026	25,881	20,949	31,562	6,815	112,367	10.2	9.3	11.6	16,800	276,991
70-74	0.0888	0.3632	0.3021	0.4107	19,066	14,780	23,990	6,924	78,018	8.0	7.3	9.2	16,209	182,636
75-79	0.1304	0.4918	0.4194	0.5458	12,141	8,834	16,106	5,971	45,781	6.2	5.6	7.0	13,413	102,842
80-84	0.1860	0.6348	0.5597	0.6879	6,171	4,068	9,304	3,917	21,061	4.7	4.3	5.3	8,424	45,289
85-89	0.2661	0.7990	0.7343	0.8412	2,253	1,303	4,096	1,801	6,766	3.5	3.3	3.9	3,341	12,555
90-94	0.3676	0.8741	0.8221	0.9063	453	215	1,088	396	1,077	2.6	2.5	2.9	718	1,952
95-99	0.4904	0.9922	0.9827	0.9961	57	21	194	57	115	2.0	2.0	2.1	86	175
100+	0.6316	1	1	1	0	0	3	0	1	1.6	1.6	1.6	6	9
All ages													*546,128*	*25,245,146*

$\ln(_nq_x)$ with uncertainty interval

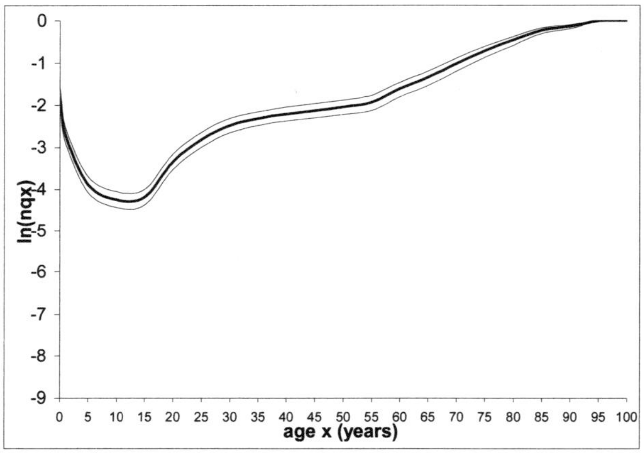

l_x with uncertainty interval

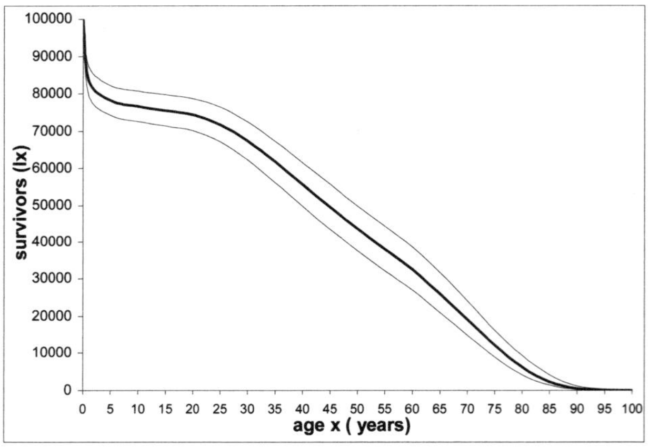

e_x with uncertainty interval

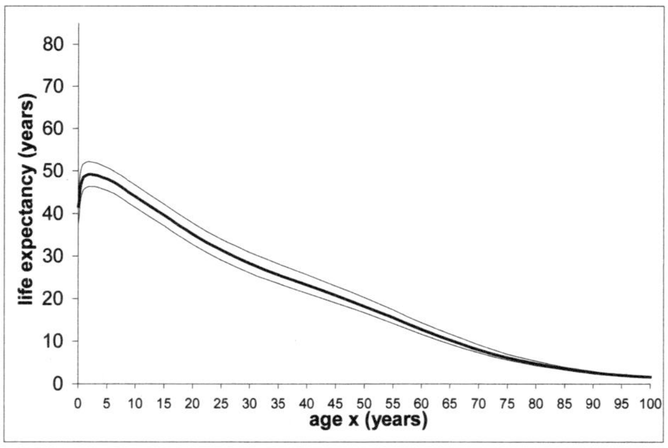

Distribution of deaths by age with uncertainty interval

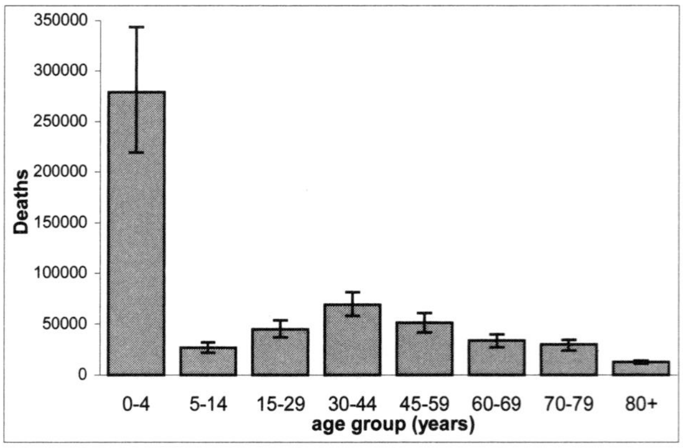

Democratic Republic of the Congo

Females, 2000

République démocratique du Congo

Femmes, 2000

República Democrática del Congo

Mujeres, 2000

Age	$_nM_x$	$_nq_x$			l_x			$_nd_x$	$_nL_x$	e_x			Actual deaths	Actual population
			percentiles			percentiles					percentiles			
		mean	2½	97½	mean	2½	97½			mean	2½	97½		
<1	0.1605	0.1402	0.1136	0.1707	100,000	100,000	100,000	14,024	87,378	44.0	40.8	47.1	171,142	1,066,287
1-4	0.0198	0.0758	0.0652	0.0873	85,976	82,929	88,641	6,514	328,269	50.2	47.6	52.7	77,739	3,917,568
5-9	0.0045	0.0224	0.0194	0.0259	79,462	76,476	82,402	1,783	392,850	50.1	47.9	52.3	18,538	4,084,443
10-14	0.0032	0.0157	0.0136	0.0181	77,679	74,610	80,778	1,217	385,351	46.2	44.2	48.4	10,445	3,307,687
15-19	0.0057	0.0280	0.0244	0.0323	76,462	73,317	79,682	2,144	376,950	41.9	39.9	44.1	15,320	2,694,001
20-24	0.0103	0.0502	0.0439	0.0576	74,318	71,011	77,740	3,729	362,267	38.1	36.2	40.2	21,897	2,127,045
25-29	0.0144	0.0696	0.0611	0.0794	70,589	66,924	74,325	4,914	340,659	35.0	33.2	37.0	24,676	1,710,685
30-34	0.0166	0.0795	0.0701	0.0901	65,675	61,612	69,754	5,221	315,322	32.4	30.7	34.4	23,168	1,399,174
35-39	0.0166	0.0798	0.0707	0.0898	60,454	56,063	64,805	4,824	290,208	30.0	28.4	31.9	18,954	1,140,156
40-44	0.0155	0.0746	0.0655	0.0836	55,629	51,027	60,147	4,152	267,768	27.3	25.9	29.1	14,574	940,037
45-49	0.0156	0.0752	0.0656	0.0846	51,478	46,765	56,072	3,872	247,708	24.3	23.1	26.0	12,114	774,881
50-54	0.0183	0.0875	0.0759	0.0986	47,605	42,850	52,214	4,168	227,608	21.1	19.9	22.7	12,232	668,020
55-59	0.0226	0.1069	0.0923	0.1205	43,438	38,703	48,002	4,645	205,577	17.9	16.8	19.3	12,855	568,933
60-64	0.0315	0.1462	0.1257	0.1647	38,793	34,163	43,335	5,671	179,787	14.8	13.8	16.0	14,552	461,352
65-69	0.0458	0.2054	0.1764	0.2307	33,122	28,739	37,547	6,804	148,600	11.9	11.0	13.0	16,036	350,224
70-74	0.0700	0.2977	0.2571	0.3313	26,318	22,407	30,412	7,836	112,000	9.3	8.6	10.2	17,082	244,157
75-79	0.1040	0.4128	0.3614	0.4528	18,482	15,360	21,950	7,629	73,336	7.1	6.6	7.9	15,375	147,793
80-84	0.1544	0.5571	0.4992	0.6006	10,853	8,569	13,889	6,046	39,149	5.4	5.0	6.0	11,101	71,885
85-89	0.2241	0.7182	0.6633	0.7575	4,807	3,466	6,868	3,452	15,404	4.0	3.8	4.4	5,249	23,419
90-94	0.3168	0.8121	0.7656	0.8444	1,355	851	2,313	1,100	3,473	3.0	2.8	3.3	1,495	4,720
95-99	0.4361	0.9447	0.9198	0.9598	255	135	542	240	551	2.3	2.2	2.4	255	584
100+	0.5848	1	1	1	14	6	43	14	24	1.7	1.7	1.8	23	39
All ages													*514,819*	*25,703,090*

$\ln(_nq_x)$ with uncertainty interval

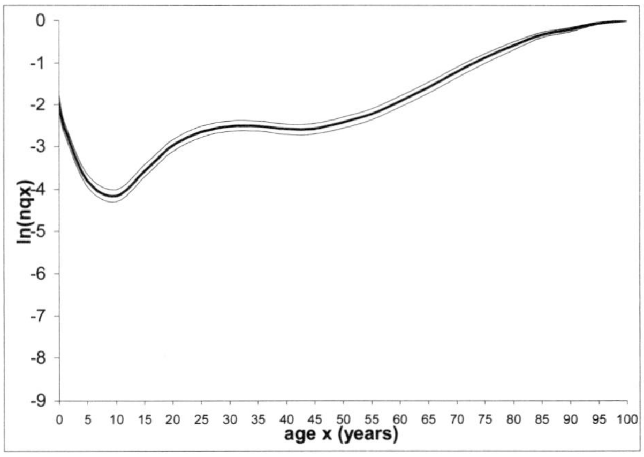

l_x with uncertainty interval

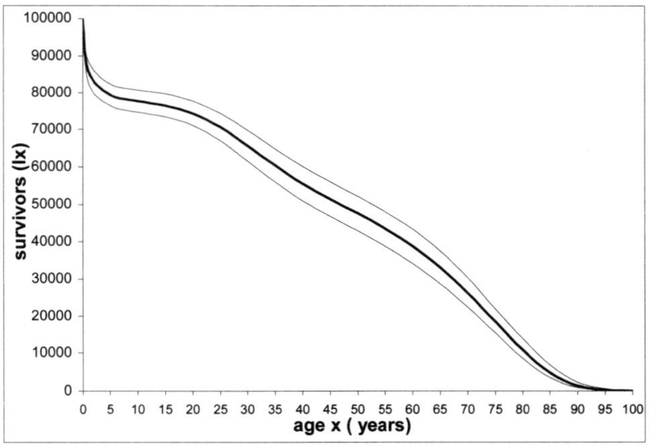

e_x with uncertainty interval

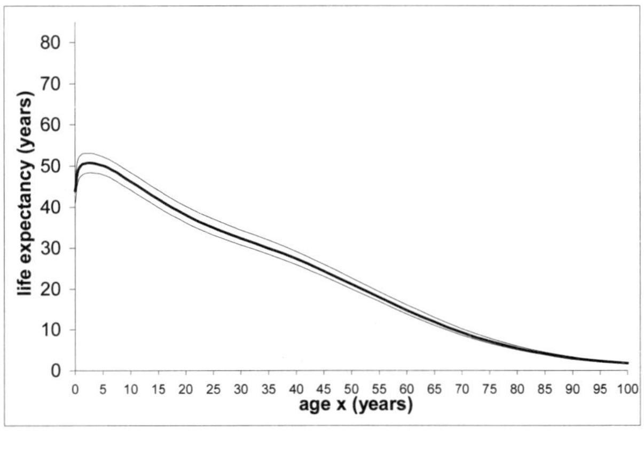

Distribution of deaths by age with uncertainty interval

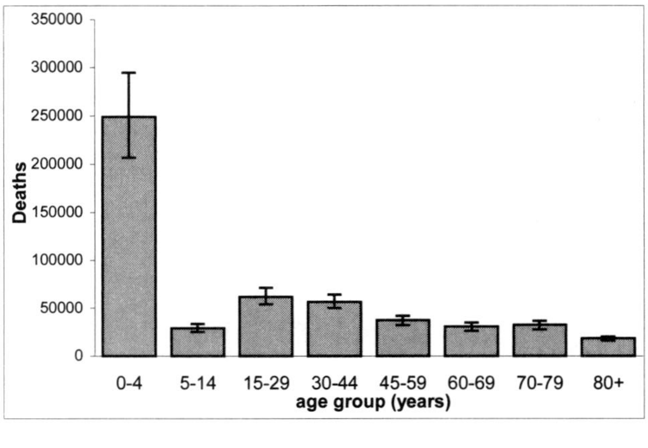

Denmark Danemark Dinamarca

Males, 2000 *Hommes, 2000* *Hombres, 2000*

Age	$_nM_x$	$_nq_x$			l_x			$_nd_x$	$_nL_x$	e_x			Actual deaths	Actual population
			percentiles			percentiles					percentiles			
		mean	2½	97½	mean	2½	97½			mean	2½	97½		
<1	0.0054	0.0054	0.0048	0.0060	100,000	100,000	100,000	538	99,515	74.2	73.9	74.6	179	33,020
1-4	0.0004	0.0017	0.0016	0.0019	99,462	99,400	99,522	170	397,438	73.6	73.3	74.1	58	135,584
5-9	0.0002	0.0009	0.0008	0.0009	99,291	99,217	99,366	86	496,242	69.7	69.4	70.2	30	175,642
10-14	0.0001	0.0007	0.0006	0.0008	99,205	99,124	99,287	70	495,852	64.8	64.4	65.3	22	154,202
15-19	0.0007	0.0034	0.0031	0.0036	99,136	99,049	99,222	332	494,847	59.8	59.5	60.3	94	140,437
20-24	0.0007	0.0033	0.0030	0.0035	98,803	98,693	98,914	322	493,213	55.0	54.7	55.5	108	166,212
25-29	0.0010	0.0048	0.0045	0.0051	98,482	98,351	98,614	475	491,223	50.2	49.9	50.7	186	192,785
30-34	0.0012	0.0059	0.0055	0.0062	98,007	97,848	98,168	575	488,599	45.4	45.1	45.9	245	208,208
35-39	0.0017	0.0086	0.0082	0.0090	97,432	97,243	97,625	838	485,066	40.7	40.3	41.2	360	208,331
40-44	0.0028	0.0141	0.0135	0.0147	96,594	96,367	96,826	1,362	479,566	36.0	35.7	36.5	542	190,849
45-49	0.0041	0.0202	0.0195	0.0209	95,232	94,954	95,517	1,923	471,354	31.5	31.1	32.0	759	186,129
50-54	0.0061	0.0300	0.0290	0.0310	93,309	92,976	93,651	2,797	459,554	27.1	26.7	27.6	1,231	202,196
55-59	0.0093	0.0455	0.0437	0.0472	90,512	90,121	90,908	4,115	442,272	22.9	22.5	23.3	1,631	175,310
60-64	0.0168	0.0804	0.0768	0.0836	86,397	85,947	86,840	6,950	414,607	18.8	18.5	19.2	2,187	130,452
65-69	0.0269	0.1261	0.1198	0.1314	79,446	78,824	80,056	10,016	372,191	15.3	14.9	15.6	2,821	104,827
70-74	0.0408	0.1850	0.1757	0.1934	69,430	68,496	70,446	12,844	315,040	12.1	11.8	12.4	3,603	88,363
75-79	0.0668	0.2861	0.2730	0.2984	56,586	55,358	58,067	16,188	242,458	9.3	9.0	9.6	4,630	69,347
80-84	0.1047	0.4148	0.3992	0.4296	40,397	38,895	42,214	16,758	160,093	7.0	6.8	7.2	4,439	42,405
85-89	0.1595	0.5703	0.5544	0.5853	23,640	22,185	25,344	13,481	84,496	5.2	5.1	5.3	3,344	20,960
90-94	0.2364	0.6916	0.6785	0.7043	10,159	9,218	11,250	7,026	29,715	3.8	3.7	3.9	1,504	6,362
95-99	0.3407	0.8424	0.8333	0.8512	3,133	2,738	3,599	2,639	7,746	2.8	2.8	2.8	349	1,025
100+	0.4772	1	1	1	494	409	600	494	1,034	2.1	2.1	2.1	43	90
All ages													*28,366*	*2,632,736*

ln($_nq_x$) with uncertainty interval

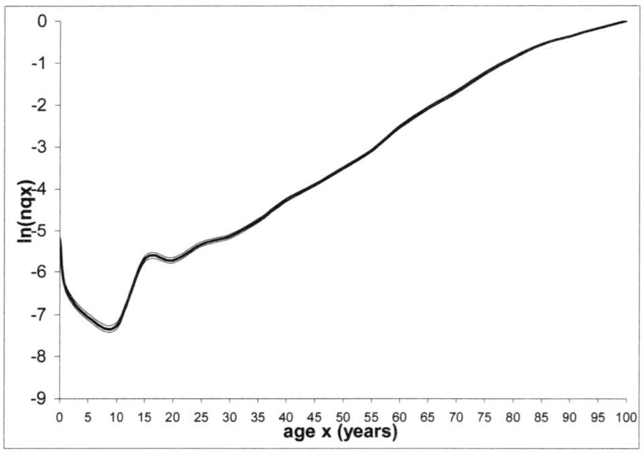

l_x with uncertainty interval

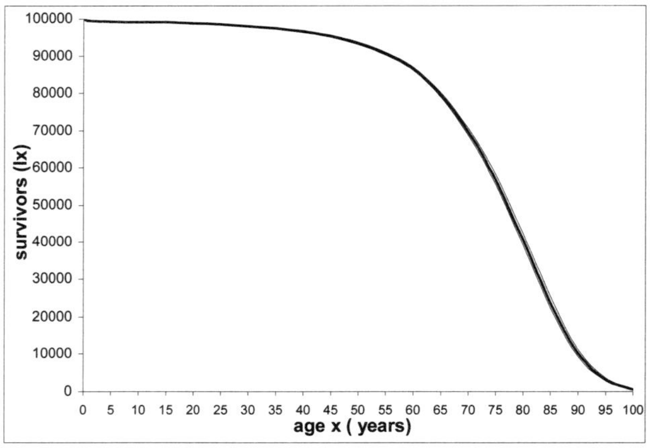

e_x with uncertainty interval

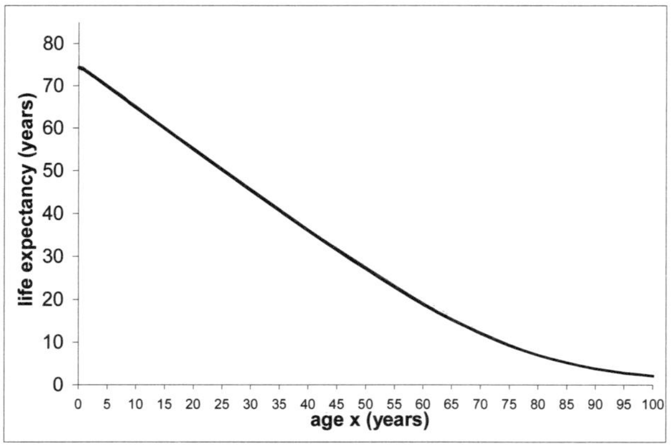

Distribution of deaths by age with uncertainty interval

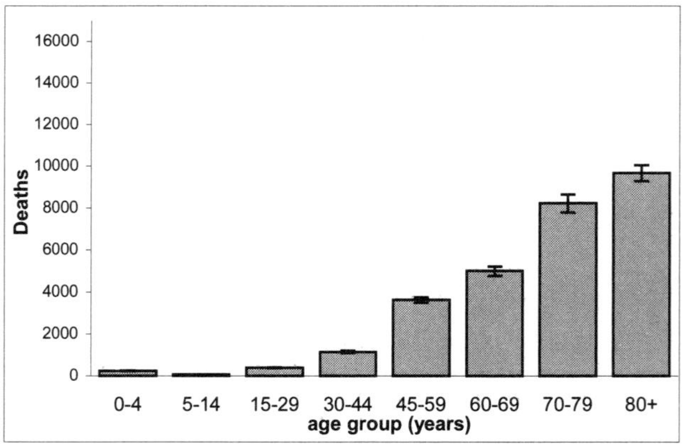

Denmark   Danemark   Dinamarca

Females, 2000   *Femmes, 2000*   *Mujeres, 2000*

Age	$_nM_x$	$_nq_x$ mean	$_nq_x$ percentiles 2½	$_nq_x$ percentiles 97½	l_x mean	l_x percentiles 2½	l_x percentiles 97½	$_nd_x$	$_nL_x$	e_x mean	e_x percentiles 2½	e_x percentiles 97½	Actual deaths	Actual population
<1	0.0043	0.0043	0.0037	0.0049	100,000	100,000	100,000	430	99,613	78.5	78.1	79.0	136	31,598
1-4	0.0002	0.0010	0.0009	0.0011	99,570	99,510	99,625	99	398,042	77.8	77.4	78.4	32	127,714
5-9	0.0001	0.0005	0.0004	0.0005	99,471	99,400	99,536	45	497,241	73.9	73.5	74.4	15	167,134
10-14	0.0001	0.0007	0.0007	0.0008	99,426	99,350	99,495	73	496,946	69.0	68.5	69.5	22	146,366
15-19	0.0002	0.0009	0.0008	0.0010	99,353	99,269	99,429	92	496,534	64.0	63.6	64.5	25	134,781
20-24	0.0003	0.0013	0.0012	0.0014	99,261	99,168	99,346	128	495,984	59.1	58.6	59.6	41	159,392
25-29	0.0003	0.0016	0.0015	0.0017	99,133	99,027	99,229	158	495,268	54.1	53.7	54.6	60	186,419
30-34	0.0005	0.0027	0.0024	0.0029	98,974	98,854	99,085	262	494,217	49.2	48.8	49.7	105	198,523
35-39	0.0009	0.0047	0.0043	0.0051	98,712	98,569	98,843	463	492,404	44.3	43.9	44.8	187	198,568
40-44	0.0017	0.0085	0.0079	0.0091	98,249	98,069	98,414	832	489,167	39.5	39.1	40.0	313	183,930
45-49	0.0024	0.0121	0.0114	0.0129	97,418	97,178	97,634	1,180	484,138	34.9	34.5	35.3	443	181,855
50-54	0.0039	0.0195	0.0185	0.0206	96,238	95,927	96,522	1,880	476,488	30.3	29.9	30.7	780	197,778
55-59	0.0068	0.0333	0.0316	0.0350	94,358	93,955	94,734	3,142	463,933	25.8	25.4	26.3	1,177	173,771
60-64	0.0118	0.0573	0.0544	0.0603	91,216	90,700	91,717	5,231	443,001	21.6	21.3	22.1	1,594	134,957
65-69	0.0198	0.0942	0.0892	0.0988	85,985	85,288	86,675	8,104	409,664	17.8	17.5	18.2	2,301	116,349
70-74	0.0301	0.1399	0.1323	0.1459	77,881	76,887	78,889	10,893	362,174	14.4	14.1	14.7	3,219	107,013
75-79	0.0457	0.2052	0.1949	0.2127	66,988	65,666	68,394	13,748	300,573	11.3	11.0	11.6	4,498	98,344
80-84	0.0741	0.3126	0.2996	0.3241	53,241	51,702	55,061	16,646	224,590	8.6	8.3	8.8	5,376	72,539
85-89	0.1182	0.4561	0.4421	0.4710	36,595	35,109	38,564	16,692	141,246	6.3	6.1	6.5	5,473	46,309
90-94	0.1854	0.5957	0.5822	0.6104	19,903	18,818	21,497	11,857	63,947	4.5	4.4	4.6	3,573	19,268
95-99	0.2862	0.7700	0.7593	0.7815	8,047	7,340	8,927	6,196	21,646	3.2	3.1	3.3	1,231	4,300
100+	0.4348	1	1	1	1,851	1,603	2,119	1,851	4,256	2.3	2.2	2.3	183	421
All ages													*30,783*	*2,687,329*

ln($_nq_x$) with uncertainty interval

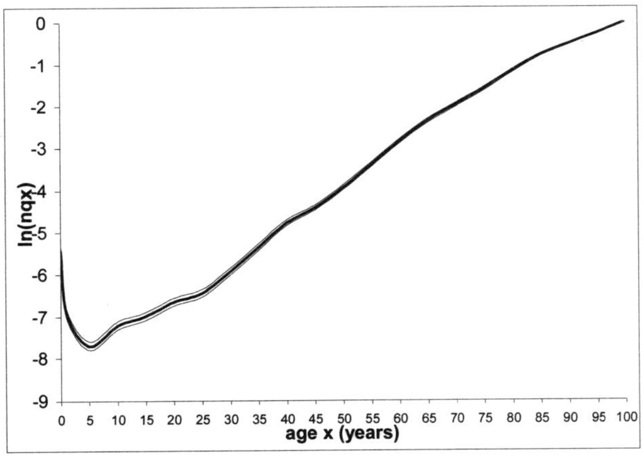

l_x with uncertainty interval

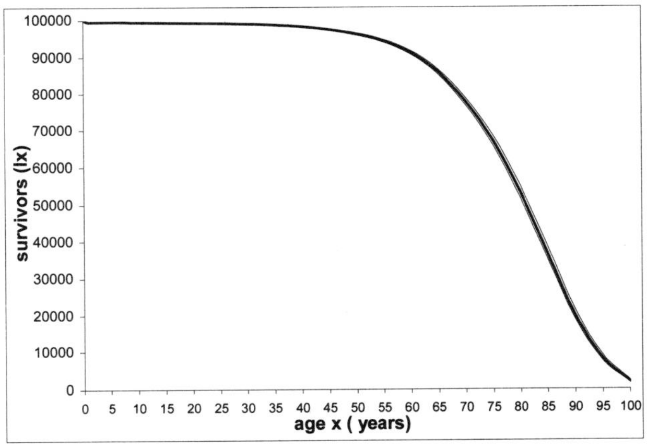

e_x with uncertainty interval

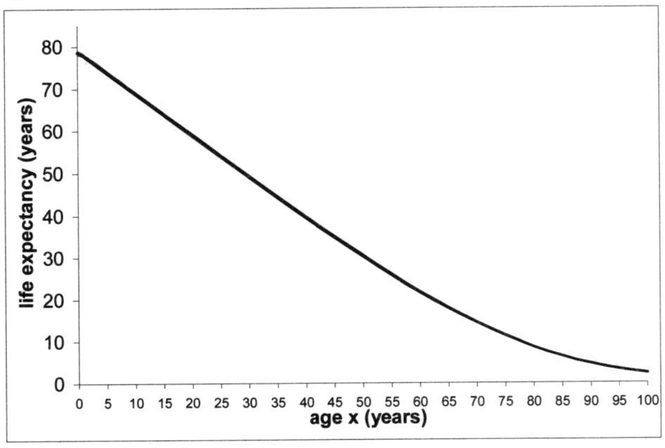

Distribution of deaths by age with uncertainty interval

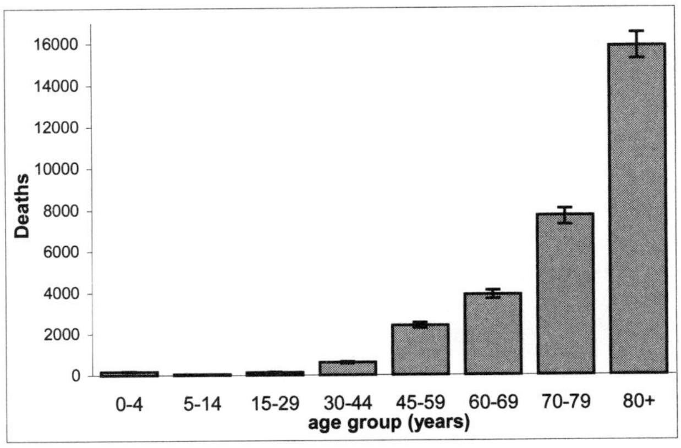

Djibouti   Djibouti   Djibouti

Males, 2000   *Hommes, 2000*   *Hombres, 2000*

Age	$_nM_x$	$_nq_x$			l_x			$_nd_x$	$_nL_x$	e_x			Actual deaths	Actual population
			percentiles			percentiles					percentiles			
		mean	2½	97½	mean	2½	97½			mean	2½	97½		
<1	0.1554	0.1363	0.1079	0.1662	100,000	100,000	100,000	13,633	87,730	43.5	39.4	48.3	1,807	11,631
1-4	0.0144	0.0555	0.0442	0.0671	86,367	83,375	89,214	4,795	333,961	49.3	44.9	54.0	574	40,010
5-9	0.0029	0.0145	0.0114	0.0178	81,572	78,605	84,571	1,183	404,904	48.1	44.0	52.5	135	46,340
10-14	0.0018	0.0091	0.0071	0.0112	80,389	77,329	83,541	730	400,123	43.8	39.7	48.1	72	39,562
15-19	0.0023	0.0114	0.0089	0.0141	79,660	76,524	82,918	905	396,036	39.1	35.1	43.4	76	33,234
20-24	0.0063	0.0310	0.0243	0.0386	78,755	75,483	82,143	2,442	387,668	34.6	30.6	38.8	174	27,664
25-29	0.0139	0.0673	0.0528	0.0844	76,312	72,645	80,060	5,138	368,718	30.6	26.7	34.8	263	18,883
30-34	0.0224	0.1061	0.0828	0.1341	71,175	66,741	75,833	7,550	336,999	27.6	23.8	31.7	337	15,021
35-39	0.0250	0.1178	0.0913	0.1499	63,625	58,130	69,508	7,498	299,380	25.6	22.0	29.4	305	12,165
40-44	0.0259	0.1216	0.0940	0.1560	56,127	49,673	63,044	6,824	263,576	23.7	20.5	27.2	277	10,708
45-49	0.0267	0.1253	0.0958	0.1615	49,303	41,995	56,954	6,176	231,077	21.6	18.8	24.8	272	10,171
50-54	0.0272	0.1273	0.0968	0.1641	43,128	35,340	51,234	5,491	201,910	19.4	16.9	22.1	267	9,804
55-59	0.0283	0.1320	0.1001	0.1695	37,637	29,688	45,952	4,968	175,763	16.8	14.7	19.2	229	8,088
60-64	0.0364	0.1669	0.1272	0.2117	32,669	24,809	41,159	5,452	149,712	14.0	12.2	16.1	219	6,021
65-69	0.0504	0.2237	0.1726	0.2784	27,216	19,730	35,700	6,088	120,861	11.3	9.9	13.1	202	4,006
70-74	0.0751	0.3162	0.2497	0.3828	21,128	14,428	29,231	6,681	88,939	8.8	7.7	10.3	175	2,336
75-79	0.1124	0.4388	0.3589	0.5131	14,447	8,912	21,528	6,340	56,386	6.8	5.9	7.9	129	1,148
80-84	0.1663	0.5872	0.5023	0.6607	8,107	4,349	13,700	4,761	28,635	5.1	4.5	5.9	73	441
85-89	0.2393	0.7486	0.6726	0.8101	3,347	1,485	6,818	2,505	10,470	3.8	3.5	4.4	26	110
90-94	0.3347	0.8350	0.7725	0.8837	841	286	2,232	702	2,099	2.9	2.6	3.2	6	17
95-99	0.4551	0.9621	0.9332	0.9797	139	34	508	134	293	2.2	2.1	2.3	0	1
100+	0.6013	1	1	1	5	1	34	5	9	1.7	1.6	1.8	0	0
All ages													*5,620*	*297,361*

ln($_nq_x$) with uncertainty interval

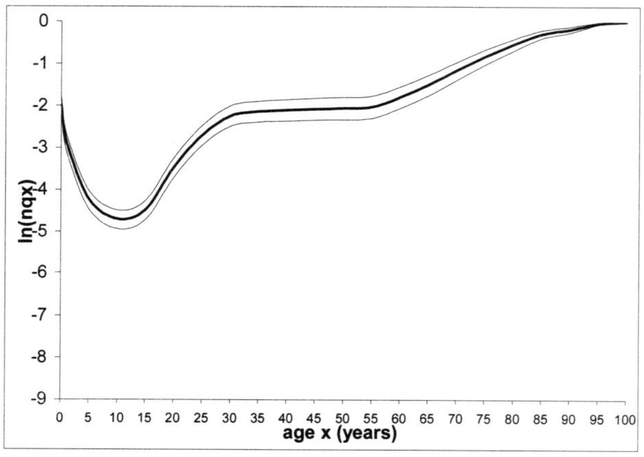

l_x with uncertainty interval

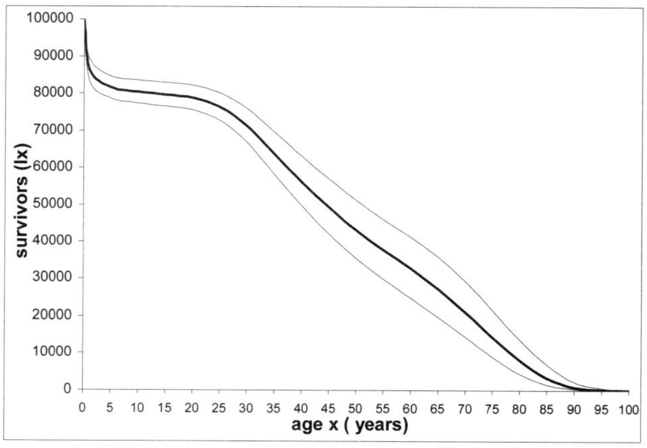

e_x with uncertainty interval

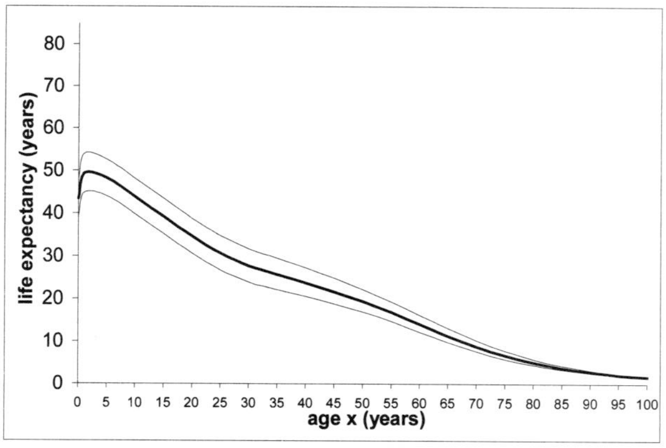

Distribution of deaths by age with uncertainty interval

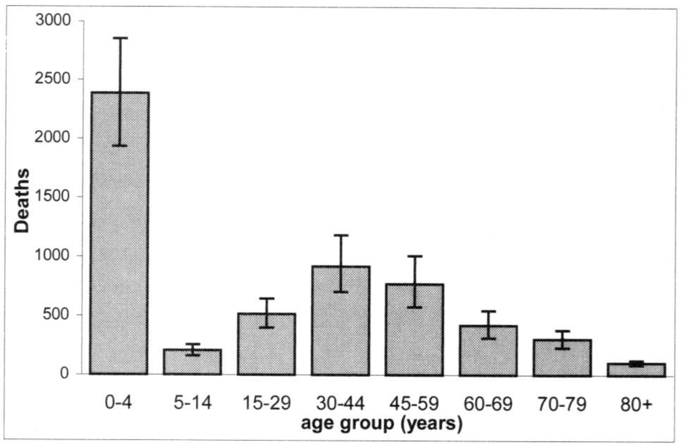

Djibouti **Djibouti** **Djibouti**

Females, 2000 *Femmes, 2000* *Mujeres, 2000*

Age	$_nM_x$	$_nq_x$ mean	$_nq_x$ percentiles 2½	$_nq_x$ percentiles 97½	l_x mean	l_x percentiles 2½	l_x percentiles 97½	$_nd_x$	$_nL_x$	e_x mean	e_x percentiles 2½	e_x percentiles 97½	Actual deaths	Actual population
<1	0.1310	0.1172	0.0899	0.1465	100,000	100,000	100,000	11,719	89,453	44.7	40.5	49.2	1,457	11,120
1-4	0.0149	0.0577	0.0467	0.0696	88,281	85,345	91,013	5,094	340,900	49.6	45.3	54.1	592	39,619
5-9	0.0029	0.0145	0.0117	0.0176	83,188	80,295	86,069	1,207	412,921	48.5	44.4	52.8	134	45,802
10-14	0.0019	0.0095	0.0076	0.0116	81,981	78,889	85,024	780	407,953	44.2	40.2	48.4	75	39,192
15-19	0.0048	0.0239	0.0192	0.0291	81,201	77,977	84,345	1,938	401,157	39.6	35.6	43.8	161	33,285
20-24	0.0122	0.0592	0.0478	0.0724	79,262	75,719	82,680	4,694	384,576	35.5	31.6	39.6	347	28,411
25-29	0.0205	0.0977	0.0785	0.1201	74,568	70,312	78,725	7,287	354,625	32.6	28.8	36.6	433	21,066
30-34	0.0247	0.1163	0.0927	0.1440	67,282	62,104	72,462	7,824	316,847	30.9	27.2	34.5	469	18,985
35-39	0.0219	0.1038	0.0823	0.1296	59,457	53,537	65,535	6,174	281,852	29.6	26.3	32.8	379	17,313
40-44	0.0188	0.0896	0.0706	0.1130	53,284	46,795	59,894	4,775	254,481	27.7	24.9	30.6	308	16,432
45-49	0.0163	0.0782	0.0614	0.0992	48,509	41,732	55,412	3,793	233,061	25.2	22.7	27.7	257	15,809
50-54	0.0170	0.0815	0.0639	0.1034	44,716	37,774	51,827	3,643	214,471	22.1	19.9	24.4	250	14,742
55-59	0.0194	0.0925	0.0727	0.1171	41,073	33,969	48,382	3,797	195,871	18.9	17.0	20.9	234	12,081
60-64	0.0267	0.1254	0.0990	0.1577	37,275	30,108	44,803	4,672	174,696	15.6	13.9	17.4	239	8,941
65-69	0.0401	0.1821	0.1451	0.2260	32,603	25,519	40,277	5,938	148,171	12.4	11.0	14.0	236	5,897
70-74	0.0636	0.2743	0.2226	0.3337	26,665	19,960	34,276	7,315	115,039	9.6	8.4	11.0	218	3,426
75-79	0.0985	0.3951	0.3291	0.4670	19,350	13,484	26,397	7,644	77,640	7.3	6.4	8.4	168	1,706
80-84	0.1493	0.5435	0.4693	0.6188	11,706	7,238	17,521	6,362	42,623	5.5	4.9	6.3	102	683
85-89	0.2196	0.7089	0.6390	0.7747	5,344	2,802	9,160	3,788	17,248	4.1	3.7	4.6	41	187
90-94	0.3134	0.8077	0.7488	0.8604	1,555	649	3,260	1,256	4,008	3.0	2.7	3.4	11	34
95-99	0.4339	0.9425	0.9103	0.9665	299	92	804	282	650	2.3	2.2	2.4	2	4
100+	0.5824	1	1	1	17	3	72	17	30	1.7	1.7	1.8	0	0
All ages													*6,113*	*334,735*

ln($_nq_x$) with uncertainty interval

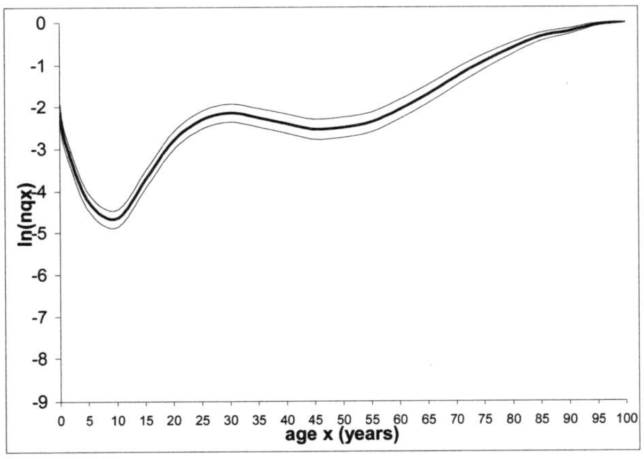

l_x with uncertainty interval

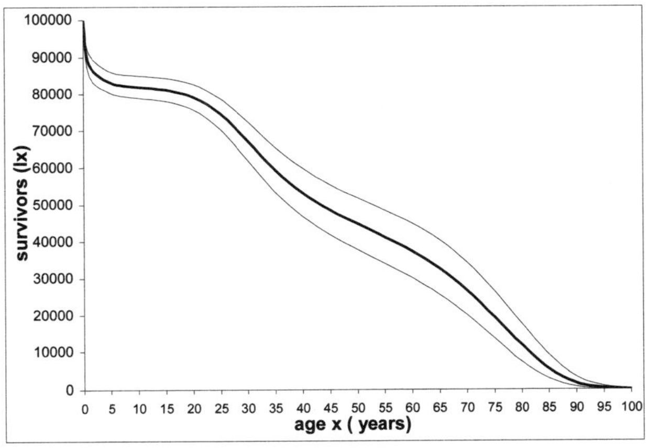

e_x with uncertainty interval

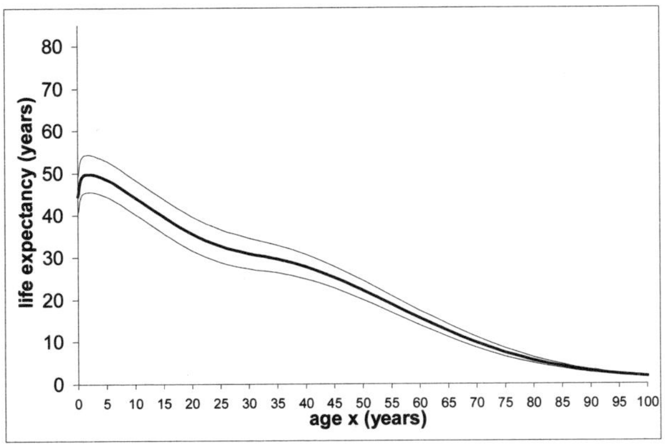

Distribution of deaths by age with uncertainty interval

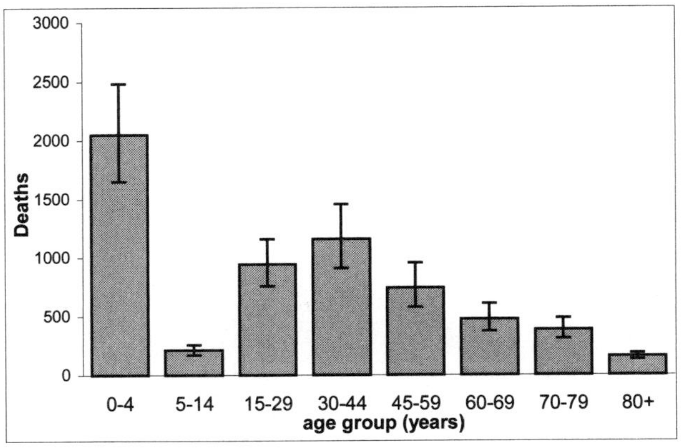

Dominica | Dominique | Dominica

Males, 2000 | *Hommes, 2000* | *Hombres, 2000*

Age	$_nM_x$	$_nq_x$ mean	$_nq_x$ percentiles 2½	$_nq_x$ percentiles 97½	l_x mean	l_x percentiles 2½	l_x percentiles 97½	$_nd_x$	$_nL_x$	e_x mean	e_x percentiles 2½	e_x percentiles 97½	Actual deaths	Actual population
<1	0.0124	0.0123	0.0084	0.0155	100,000	100,000	100,000	1,226	98,897	72.6	71.6	73.7	9	700
1-4	0.0005	0.0021	0.0016	0.0025	98,774	98,451	99,163	204	394,608	72.5	71.4	73.8	1	2,773
5-9	0.0005	0.0025	0.0019	0.0030	98,570	98,209	99,007	246	492,237	68.7	67.5	69.9	2	3,494
10-14	0.0005	0.0024	0.0019	0.0028	98,324	97,918	98,816	237	491,029	63.8	62.7	65.1	2	3,641
15-19	0.0007	0.0036	0.0029	0.0042	98,087	97,640	98,628	351	489,557	59.0	57.8	60.3	2	3,389
20-24	0.0010	0.0047	0.0039	0.0055	97,736	97,231	98,346	464	487,518	54.2	53.0	55.5	3	3,020
25-29	0.0013	0.0066	0.0055	0.0076	97,272	96,696	97,966	642	484,752	49.4	48.2	50.8	4	3,032
30-34	0.0017	0.0085	0.0072	0.0096	96,629	95,962	97,429	817	481,103	44.7	43.5	46.1	5	2,836
35-39	0.0030	0.0146	0.0128	0.0165	95,812	95,039	96,730	1,403	475,552	40.1	38.9	41.5	8	2,562
40-44	0.0041	0.0202	0.0181	0.0224	94,409	93,472	95,494	1,904	467,284	35.7	34.4	37.1	8	1,938
45-49	0.0061	0.0298	0.0277	0.0327	92,505	91,378	93,761	2,757	455,632	31.4	30.1	32.7	10	1,716
50-54	0.0095	0.0465	0.0430	0.0508	89,748	88,394	91,160	4,175	438,300	27.2	26.0	28.6	14	1,486
55-59	0.0130	0.0630	0.0569	0.0694	85,572	83,899	87,240	5,387	414,394	23.4	22.2	24.8	16	1,232
60-64	0.0173	0.0830	0.0737	0.0930	80,185	78,074	82,276	6,657	384,284	19.9	18.7	21.1	17	1,006
65-69	0.0250	0.1176	0.1030	0.1332	73,528	70,810	76,216	8,650	346,016	16.4	15.3	17.6	19	764
70-74	0.0401	0.1822	0.1586	0.2067	64,878	61,375	68,365	11,818	294,846	13.3	12.2	14.4	24	611
75-79	0.0575	0.2515	0.2195	0.2865	53,060	48,686	57,525	13,342	231,946	10.7	9.8	11.6	24	416
80-84	0.0814	0.3380	0.2986	0.3828	39,718	34,738	44,901	13,426	165,027	8.4	7.7	9.2	21	262
85-89	0.1177	0.4548	0.4106	0.5050	26,292	21,441	31,491	11,958	101,567	6.5	5.9	7.0	15	126
90-94	0.1744	0.5724	0.5298	0.6206	14,334	10,614	18,559	8,205	47,058	4.8	4.3	5.1	7	40
95-99	0.2642	0.7369	0.7013	0.7768	6,130	4,027	8,727	4,517	17,097	3.4	3.2	3.6	2	8
100+	0.4097	1	1	1	1,613	899	2,607	1,613	3,936	2.4	2.2	2.6	0	1
All ages													*214*	*35,053*

$\ln(_nq_x)$ with uncertainty interval

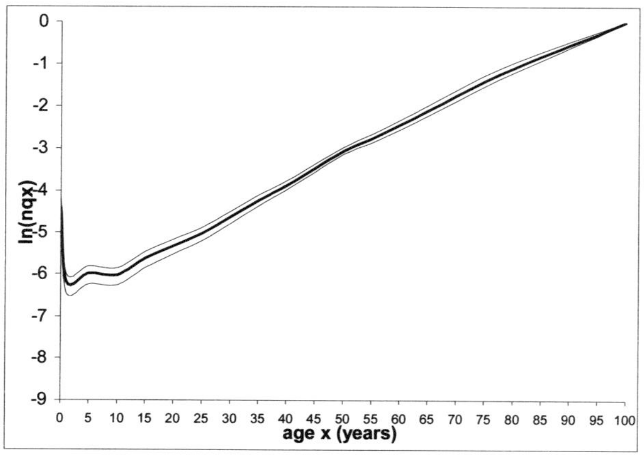

l_x with uncertainty interval

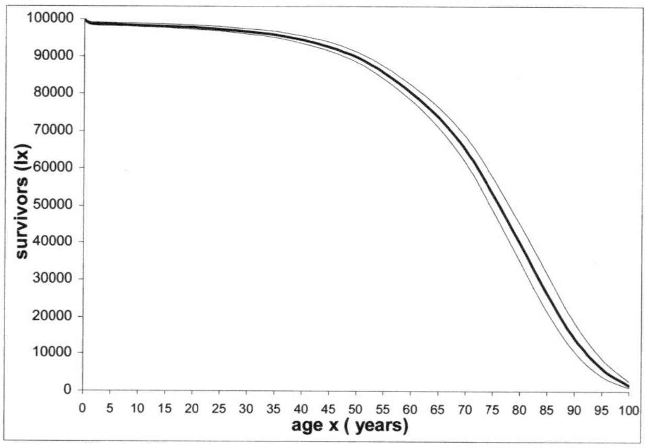

e_x with uncertainty interval

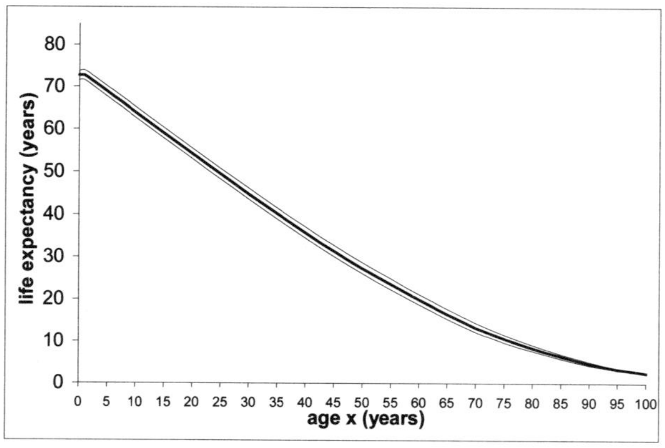

Distribution of deaths by age with uncertainty interval

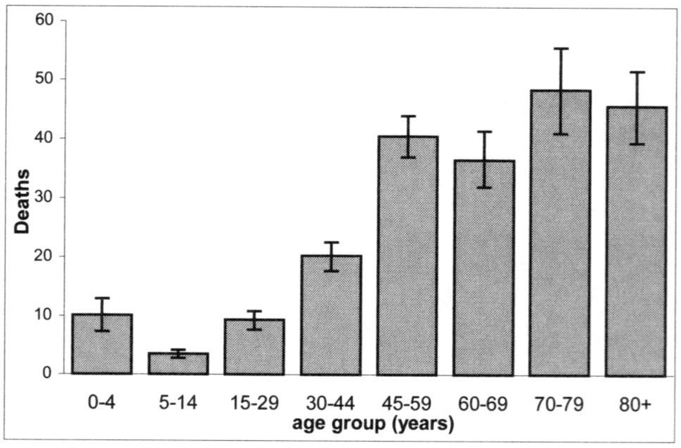

Dominica — Dominique — Dominica

Females, 2000 — *Femmes, 2000* — *Mujeres, 2000*

Age	$_nM_x$	$_nq_x$			l_x			$_nd_x$	$_nL_x$	e_x			Actual deaths	Actual population
			percentiles			percentiles					percentiles			
		mean	2½	97½	mean	2½	97½			mean	2½	97½		
<1	0.0106	0.0105	0.0075	0.0135	100,000	100,000	100,000	1,046	99,058	78.2	77.1	79.7	7	677
1-4	0.0006	0.0024	0.0018	0.0028	98,954	98,648	99,250	234	395,254	78.1	77.0	79.6	2	2,655
5-9	0.0002	0.0011	0.0009	0.0013	98,720	98,374	99,066	108	493,331	74.3	73.2	75.8	1	3,363
10-14	0.0003	0.0016	0.0013	0.0019	98,612	98,249	98,981	159	492,664	69.3	68.2	70.9	1	3,524
15-19	0.0005	0.0025	0.0020	0.0029	98,453	98,061	98,853	244	491,655	64.4	63.3	66.0	2	3,279
20-24	0.0003	0.0013	0.0011	0.0016	98,209	97,773	98,656	132	490,715	59.6	58.5	61.2	1	2,937
25-29	0.0006	0.0029	0.0024	0.0034	98,077	97,619	98,548	282	489,682	54.7	53.5	56.2	2	2,975
30-34	0.0011	0.0056	0.0047	0.0065	97,795	97,292	98,316	547	487,609	49.8	48.7	51.4	3	2,822
35-39	0.0019	0.0093	0.0079	0.0106	97,248	96,663	97,857	901	483,989	45.1	43.9	46.7	5	2,650
40-44	0.0027	0.0134	0.0117	0.0150	96,347	95,644	97,084	1,288	478,516	40.5	39.3	42.1	6	2,092
45-49	0.0035	0.0173	0.0155	0.0190	95,059	94,215	95,948	1,640	471,195	36.0	34.8	37.6	6	1,856
50-54	0.0048	0.0238	0.0213	0.0262	93,419	92,431	94,460	2,221	461,543	31.6	30.4	33.1	8	1,611
55-59	0.0069	0.0340	0.0300	0.0377	91,198	90,058	92,391	3,101	448,239	27.3	26.1	28.8	9	1,335
60-64	0.0100	0.0488	0.0422	0.0542	88,097	86,792	89,420	4,297	429,742	23.2	22.0	24.6	11	1,102
65-69	0.0151	0.0729	0.0621	0.0820	83,800	82,352	85,246	6,111	403,722	19.2	18.1	20.6	13	866
70-74	0.0243	0.1145	0.0965	0.1308	77,689	75,689	79,799	8,893	366,214	15.5	14.5	16.8	17	708
75-79	0.0398	0.1810	0.1524	0.2084	68,797	66,079	72,098	12,455	312,846	12.2	11.3	13.3	20	495
80-84	0.0650	0.2796	0.2385	0.3191	56,342	52,593	61,108	15,751	242,331	9.4	8.6	10.3	21	325
85-89	0.1000	0.4000	0.3499	0.4494	40,590	35,810	46,535	16,236	162,362	7.1	6.4	7.7	17	166
90-94	0.1514	0.5205	0.4681	0.5703	24,354	19,830	30,202	12,675	83,745	5.1	4.6	5.5	9	59
95-99	0.2692	0.7446	0.7002	0.7843	11,679	8,636	15,921	8,696	32,307	3.4	3.1	3.6	3	13
100+	0.3802	1	1	1	2,983	1,873	4,687	2,983	7,847	2.6	2.2	2.7	1	2
All ages													*164*	*35,511*

ln($_nq_x$) with uncertainty interval

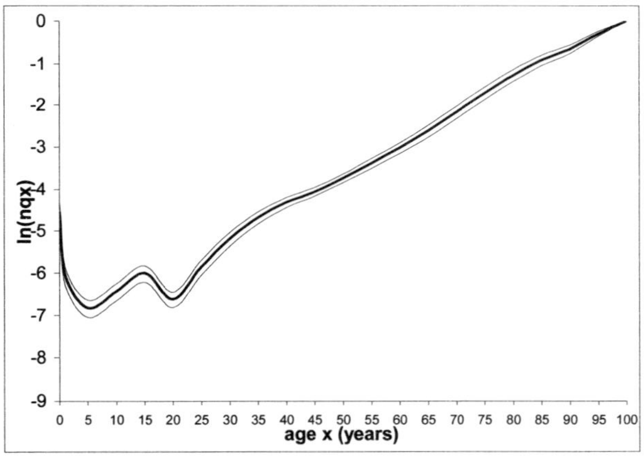

l_x with uncertainty interval

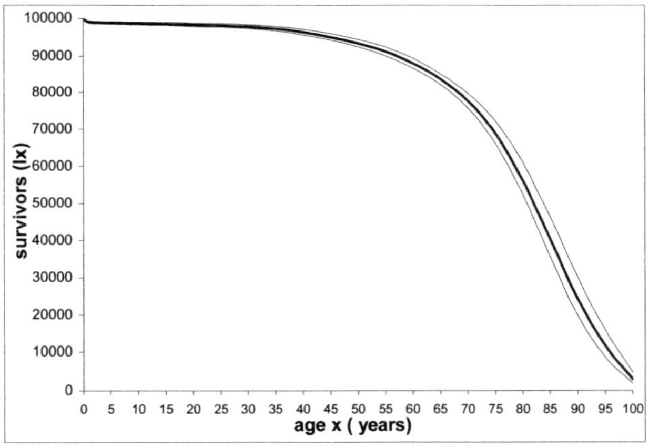

e_x with uncertainty interval

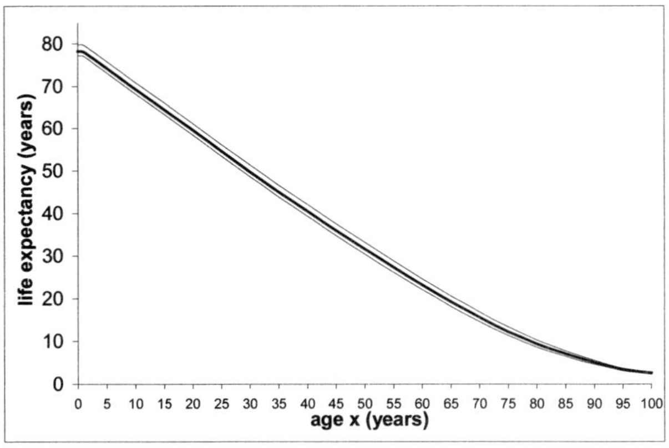

Distribution of deaths by age with uncertainty interval

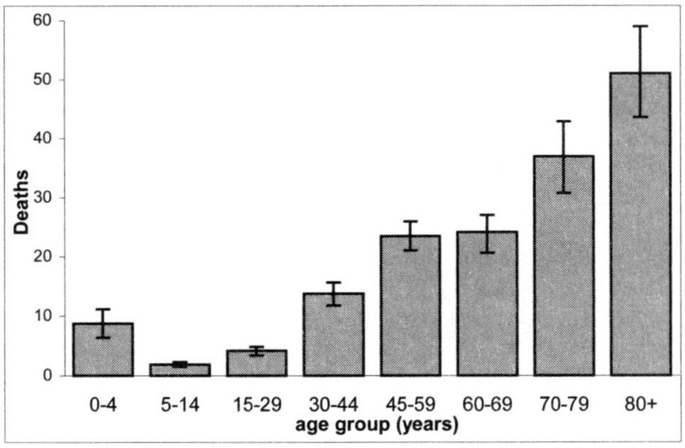

Dominican Republic République dominicaine República Dominicana

Males, 2000 *Hommes, 2000* *Hombres, 2000*

Age	$_nM_x$	$_nq_x$			l_x			$_nd_x$	$_nL_x$	e_x			Actual deaths	Actual population
			percentiles			percentiles					percentiles			
		mean	2½	97½	mean	2½	97½			mean	2½	97½		
<1	0.0489	0.0469	0.0423	0.0514	100,000	100,000	100,000	4,687	95,782	65.5	64.5	66.5	4,752	97,109
1-4	0.0020	0.0081	0.0074	0.0087	95,313	94,859	95,765	769	379,407	67.7	66.9	68.8	774	381,476
5-9	0.0008	0.0038	0.0035	0.0042	94,544	94,051	95,044	364	471,810	64.2	63.4	65.3	364	471,871
10-14	0.0007	0.0036	0.0033	0.0039	94,180	93,661	94,702	342	470,046	59.5	58.6	60.6	348	478,423
15-19	0.0019	0.0094	0.0087	0.0101	93,838	93,294	94,383	884	466,981	54.7	53.8	55.8	873	460,872
20-24	0.0038	0.0186	0.0172	0.0200	92,954	92,347	93,561	1,730	460,444	50.2	49.3	51.3	1,475	392,386
25-29	0.0049	0.0243	0.0226	0.0260	91,224	90,502	91,948	2,219	450,571	46.1	45.2	47.2	1,782	361,866
30-34	0.0060	0.0297	0.0278	0.0317	89,005	88,149	89,867	2,646	438,407	42.2	41.3	43.3	2,067	342,466
35-39	0.0059	0.0290	0.0272	0.0308	86,358	85,359	87,369	2,502	425,537	38.4	37.5	39.4	1,733	294,768
40-44	0.0060	0.0295	0.0276	0.0313	83,856	82,737	84,993	2,476	413,092	34.5	33.6	35.5	1,429	238,407
45-49	0.0062	0.0306	0.0285	0.0326	81,380	80,156	82,629	2,494	400,667	30.4	29.6	31.4	1,195	191,916
50-54	0.0078	0.0382	0.0352	0.0407	78,886	77,568	80,235	3,012	386,901	26.3	25.5	27.3	1,209	155,264
55-59	0.0107	0.0521	0.0478	0.0559	75,874	74,459	77,329	3,954	369,486	22.3	21.5	23.2	1,275	119,113
60-64	0.0187	0.0892	0.0813	0.0963	71,920	70,391	73,490	6,416	343,561	18.3	17.6	19.2	1,765	94,518
65-69	0.0293	0.1364	0.1236	0.1481	65,504	63,819	67,218	8,936	305,181	14.9	14.2	15.7	2,067	70,584
70-74	0.0472	0.2110	0.1910	0.2294	56,568	54,757	58,631	11,934	253,007	11.8	11.3	12.6	2,534	53,713
75-79	0.0706	0.2999	0.2732	0.3245	44,634	42,338	47,431	13,386	189,706	9.3	8.9	9.9	1,975	27,988
80-84	0.0988	0.3962	0.3655	0.4241	31,248	28,599	34,473	12,382	125,286	7.3	6.9	7.7	1,412	14,283
85-89	0.1464	0.5358	0.5039	0.5641	18,866	16,470	21,872	10,108	69,061	5.4	5.2	5.7	785	5,365
90-94	0.2274	0.6759	0.6479	0.7004	8,758	7,180	10,851	5,920	26,031	3.8	3.6	4.0	271	1,193
95-99	0.3728	0.8800	0.8627	0.8945	2,838	2,151	3,821	2,498	6,699	2.5	2.5	2.7	61	164
100+	0.6436	1	1	1	341	227	525	341	529	1.6	1.5	1.8	7	11
All ages													*30,150*	*4,253,756*

$\ln({}_nq_x)$ with uncertainty interval

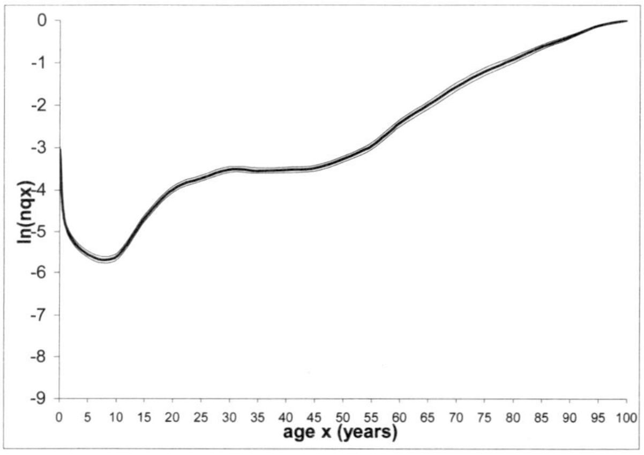

l_x with uncertainty interval

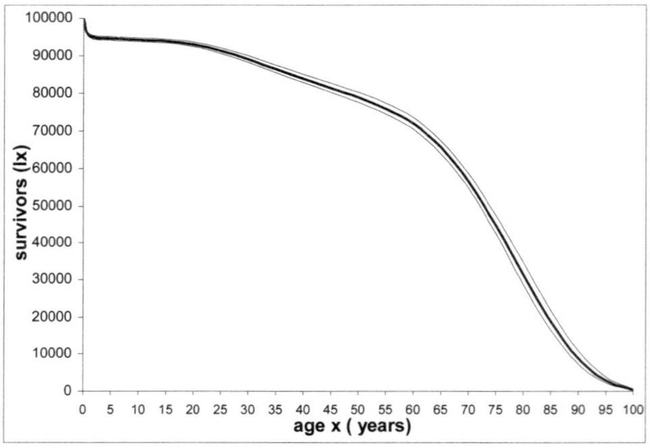

e_x with uncertainty interval

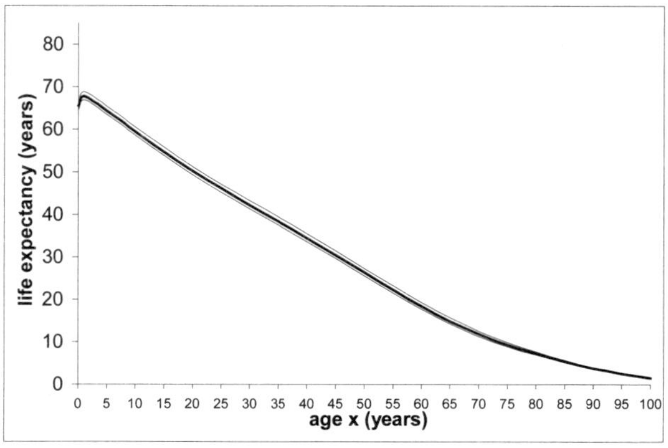

Distribution of deaths by age with uncertainty interval

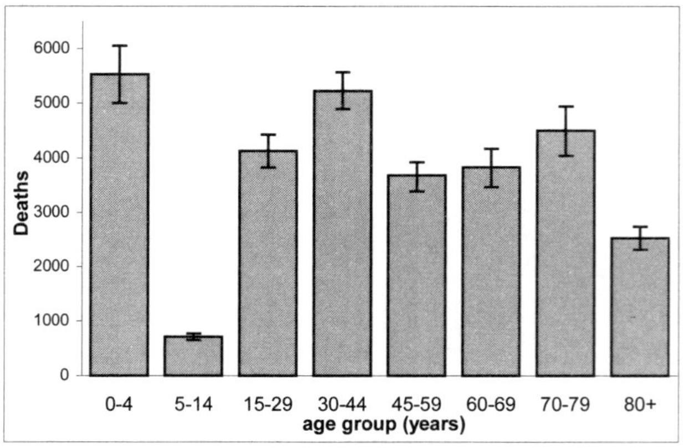

Dominican Republic / République dominicaine / República Dominicana

Females, 2000 / *Femmes, 2000* / *Mujeres, 2000*

Age	$_nM_x$	$_nq_x$			l_x			$_nd_x$	$_nL_x$	e_x			Actual deaths	Actual population
			percentiles			percentiles					percentiles			
		mean	2½	97½	mean	2½	97½			mean	2½	97½		
<1	0.0411	0.0396	0.0355	0.0438	100,000	100,000	100,000	3,963	96,433	71.6	70.6	72.7	3,765	91,612
1-4	0.0015	0.0061	0.0056	0.0066	96,037	95,621	96,450	585	382,743	73.6	72.7	74.5	564	369,411
5-9	0.0008	0.0039	0.0035	0.0042	95,452	95,004	95,901	369	476,338	70.0	69.1	71.0	352	455,126
10-14	0.0007	0.0035	0.0032	0.0038	95,083	94,606	95,555	333	474,584	65.3	64.4	66.2	323	460,328
15-19	0.0010	0.0050	0.0046	0.0054	94,750	94,248	95,249	470	472,576	60.5	59.6	61.4	438	440,260
20-24	0.0019	0.0094	0.0086	0.0102	94,280	93,742	94,815	882	469,197	55.8	54.9	56.7	696	370,531
25-29	0.0028	0.0139	0.0128	0.0151	93,398	92,796	94,000	1,299	463,745	51.3	50.5	52.2	951	339,482
30-34	0.0028	0.0138	0.0127	0.0150	92,100	91,401	92,794	1,275	457,312	47.0	46.2	47.9	877	314,632
35-39	0.0028	0.0140	0.0129	0.0152	90,825	90,028	91,605	1,274	450,940	42.6	41.8	43.5	801	283,722
40-44	0.0032	0.0160	0.0147	0.0173	89,551	88,660	90,413	1,433	444,173	38.2	37.4	39.0	782	242,296
45-49	0.0042	0.0207	0.0190	0.0223	88,118	87,125	89,066	1,821	436,038	33.8	33.0	34.6	808	193,441
50-54	0.0055	0.0271	0.0250	0.0293	86,297	85,181	87,347	2,342	425,631	29.4	28.7	30.2	848	154,142
55-59	0.0073	0.0357	0.0329	0.0386	83,955	82,691	85,136	2,998	412,283	25.2	24.5	25.9	875	120,313
60-64	0.0124	0.0603	0.0555	0.0652	80,958	79,520	82,328	4,880	392,588	21.0	20.4	21.7	1,215	97,736
65-69	0.0185	0.0885	0.0812	0.0958	76,077	74,390	77,728	6,734	363,552	17.2	16.6	17.9	1,404	75,825
70-74	0.0319	0.1479	0.1356	0.1595	69,344	67,383	71,318	10,255	321,079	13.6	13.1	14.2	1,812	56,721
75-79	0.0534	0.2356	0.2169	0.2531	59,088	56,747	61,436	13,921	260,638	10.6	10.1	11.1	1,612	30,177
80-84	0.0812	0.3374	0.3123	0.3603	45,167	42,445	48,045	15,238	187,741	8.1	7.7	8.5	1,254	15,452
85-89	0.1276	0.4836	0.4546	0.5099	29,929	27,324	32,931	14,474	113,461	5.9	5.6	6.2	751	5,885
90-94	0.2074	0.6392	0.6124	0.6632	15,455	13,519	17,816	9,879	47,638	4.1	3.9	4.2	320	1,544
95-99	0.3488	0.8523	0.8349	0.8674	5,576	4,577	6,906	4,752	13,623	2.7	2.6	2.8	96	276
100+	0.6069	1	1	1	824	607	1,140	824	1,357	1.6	1.6	1.9	16	27
All ages													*20,561*	*4,118,939*

ln($_nq_x$) with uncertainty interval

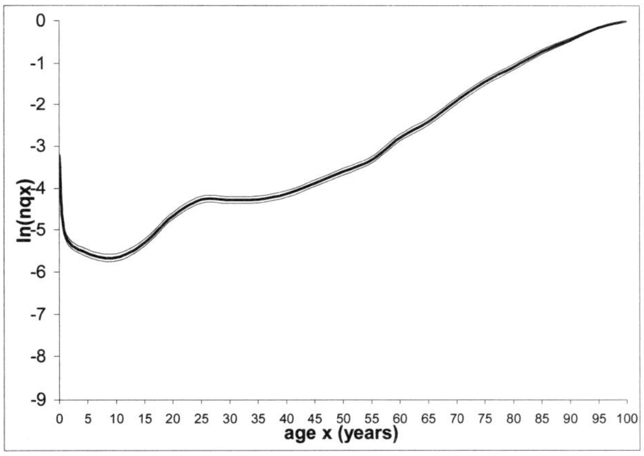

l_x with uncertainty interval

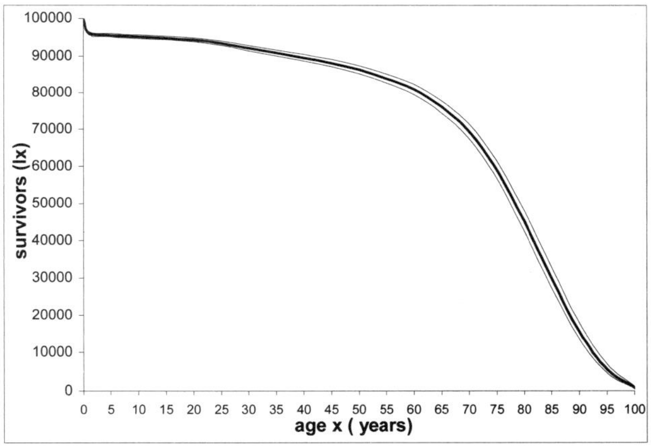

e_x with uncertainty interval

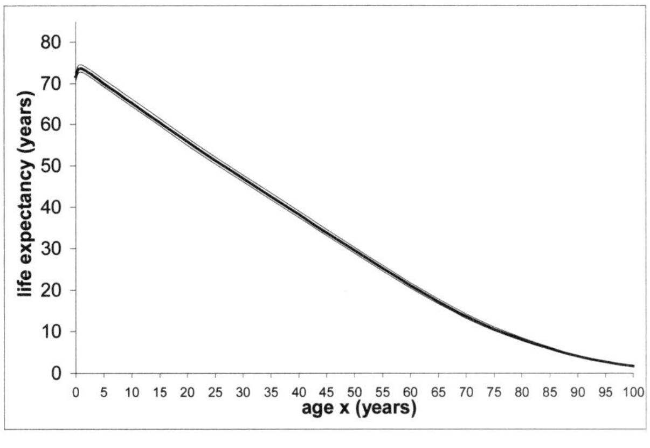

Distribution of deaths by age with uncertainty interval

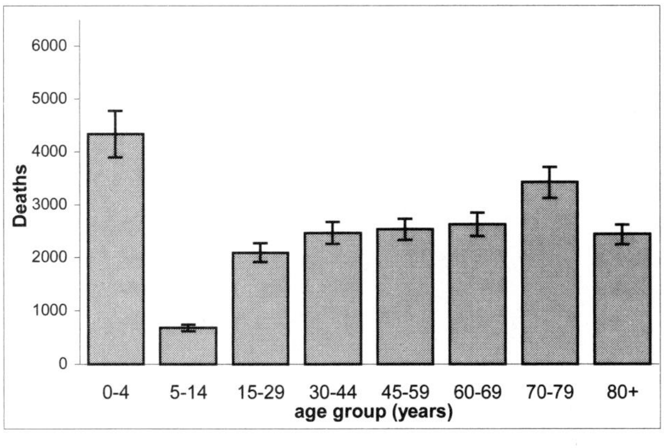

Ecuador | Equateur | Ecuador

Males, 2000 | *Hommes, 2000* | *Hombres, 2000*

Age	$_nM_x$	$_nq_x$			l_x			$_nd_x$	$_nL_x$	e_x			Actual deaths	Actual population
			percentiles			percentiles					percentiles			
		mean	2½	97½	mean	2½	97½			mean	2½	97½		
<1	0.0242	0.0237	0.0211	0.0264	100,000	100,000	100,000	2,367	97,869	68.2	67.5	69.0	3,690	152,549
1-4	0.0044	0.0173	0.0158	0.0187	97,633	97,360	97,892	1,690	386,476	68.9	68.2	69.5	2,598	594,258
5-9	0.0007	0.0037	0.0034	0.0040	95,943	95,549	96,341	353	478,834	66.1	65.5	66.7	537	728,656
10-14	0.0008	0.0038	0.0035	0.0041	95,590	95,174	96,015	362	477,048	61.3	60.7	61.9	531	700,358
15-19	0.0017	0.0086	0.0079	0.0092	95,229	94,789	95,680	816	474,104	56.5	55.9	57.1	1,165	676,584
20-24	0.0028	0.0139	0.0129	0.0148	94,413	93,925	94,921	1,310	468,788	52.0	51.4	52.6	1,749	625,896
25-29	0.0031	0.0153	0.0143	0.0164	93,102	92,535	93,696	1,429	461,940	47.7	47.2	48.2	1,715	554,561
30-34	0.0033	0.0164	0.0154	0.0174	91,674	91,022	92,351	1,502	454,614	43.4	42.9	43.9	1,600	484,155
35-39	0.0040	0.0197	0.0185	0.0208	90,172	89,439	90,930	1,774	446,423	39.1	38.6	39.5	1,633	410,780
40-44	0.0046	0.0227	0.0214	0.0240	88,397	87,576	89,241	2,007	436,969	34.8	34.4	35.2	1,579	343,854
45-49	0.0059	0.0290	0.0274	0.0306	86,390	85,480	87,318	2,507	425,684	30.6	30.2	31.0	1,697	288,123
50-54	0.0079	0.0389	0.0368	0.0411	83,883	82,875	84,902	3,265	411,255	26.4	26.0	26.8	1,678	211,359
55-59	0.0109	0.0532	0.0505	0.0561	80,619	79,503	81,733	4,293	392,361	22.4	22.0	22.7	1,846	168,757
60-64	0.0174	0.0835	0.0795	0.0877	76,326	75,088	77,535	6,375	365,692	18.5	18.2	18.8	2,387	136,944
65-69	0.0263	0.1234	0.1179	0.1289	69,951	68,580	71,328	8,632	328,175	14.9	14.6	15.2	2,785	105,878
70-74	0.0387	0.1766	0.1689	0.1837	61,319	59,745	62,910	10,828	279,525	11.7	11.4	11.9	2,968	76,625
75-79	0.0663	0.2845	0.2727	0.2962	50,491	48,792	52,214	14,363	216,547	8.6	8.4	8.9	3,291	49,618
80-84	0.1222	0.4679	0.4511	0.4843	36,128	34,511	37,779	16,904	138,379	6.1	5.9	6.3	3,210	26,277
85-89	0.2059	0.6795	0.6627	0.6951	19,224	18,057	20,574	13,063	63,460	4.3	4.2	4.4	2,359	11,461
90-94	0.3174	0.8129	0.8003	0.8242	6,160	5,510	6,876	5,008	15,778	3.0	2.9	3.0	968	3,049
95-99	0.4477	0.9554	0.9502	0.9598	1,153	969	1,372	1,101	2,459	2.2	2.2	2.2	218	486
100+	0.5779	1	1	1	51	39	68	51	89	1.7	1.7	1.9	24	42
All ages													*40,228*	*6,350,270*

ln($_nq_x$) with uncertainty interval

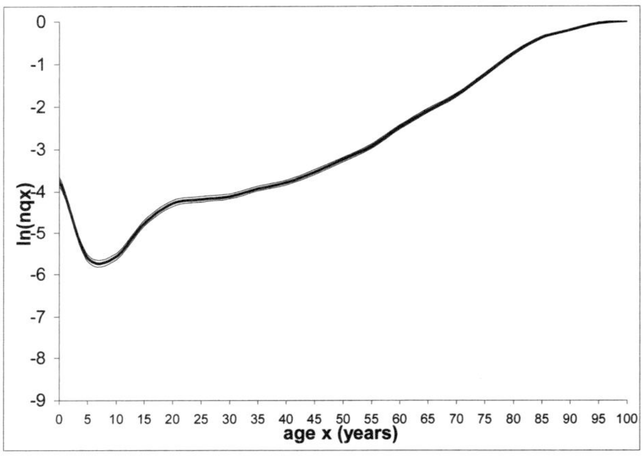

l_x with uncertainty interval

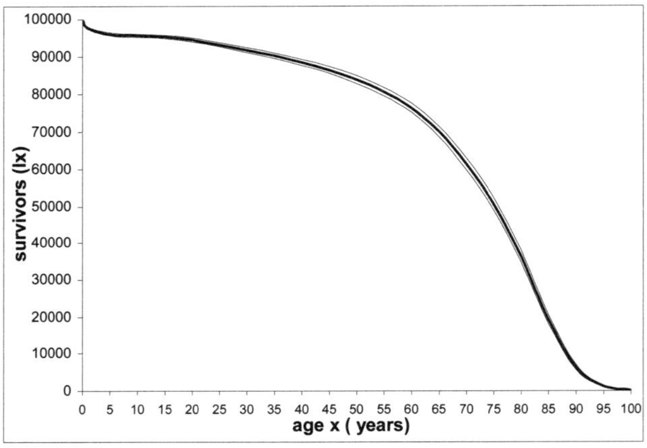

e_x with uncertainty interval

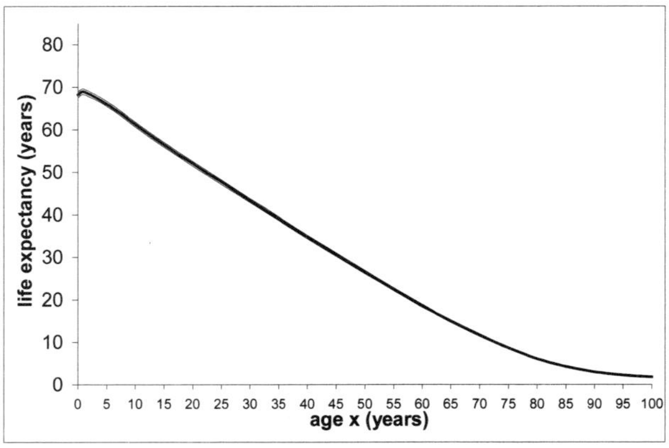

Distribution of deaths by age with uncertainty interval

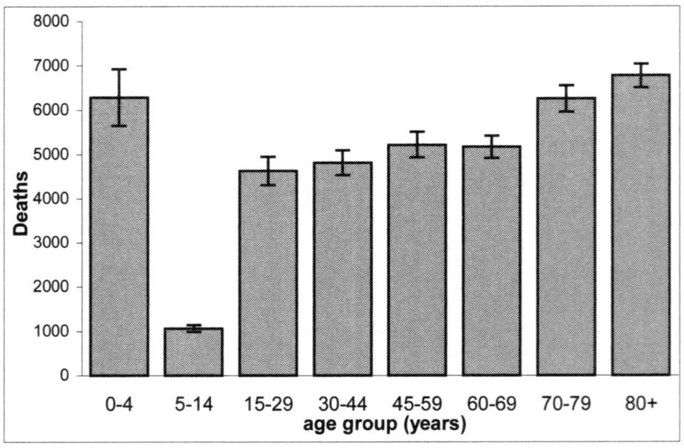

Ecuador **Equateur** **Ecuador**

Females, 2000 *Femmes, 2000* *Mujeres, 2000*

Age	$_nM_x$	$_nq_x$ mean	$_nq_x$ percentiles 2½	$_nq_x$ percentiles 97½	l_x mean	l_x percentiles 2½	l_x percentiles 97½	$_nd_x$	$_nL_x$	e_x mean	e_x percentiles 2½	e_x percentiles 97½	Actual deaths	Actual population
<1	0.0194	0.0190	0.0172	0.0211	100,000	100,000	100,000	1,904	98,286	74.2	73.5	74.8	2,815	145,285
1-4	0.0036	0.0142	0.0131	0.0154	98,096	97,891	98,275	1,391	389,045	74.6	74.1	75.2	2,052	574,094
5-9	0.0006	0.0029	0.0027	0.0031	96,705	96,405	96,988	277	482,833	71.7	71.2	72.3	404	704,198
10-14	0.0006	0.0031	0.0028	0.0033	96,428	96,107	96,731	294	481,405	66.9	66.4	67.5	415	678,537
15-19	0.0010	0.0051	0.0048	0.0055	96,134	95,791	96,457	494	479,435	62.1	61.6	62.7	677	657,398
20-24	0.0011	0.0055	0.0051	0.0059	95,640	95,260	95,996	523	476,891	57.4	57.0	57.9	670	610,651
25-29	0.0012	0.0059	0.0055	0.0063	95,117	94,700	95,506	559	474,186	52.7	52.3	53.2	641	543,983
30-34	0.0014	0.0072	0.0067	0.0077	94,558	94,102	94,982	680	471,090	48.0	47.6	48.5	690	477,976
35-39	0.0020	0.0097	0.0091	0.0104	93,878	93,377	94,342	914	467,106	43.3	42.9	43.8	799	408,486
40-44	0.0027	0.0132	0.0124	0.0141	92,964	92,405	93,480	1,229	461,750	38.7	38.3	39.2	918	344,969
45-49	0.0037	0.0181	0.0171	0.0193	91,736	91,101	92,317	1,661	454,526	34.2	33.8	34.7	1,068	292,347
50-54	0.0053	0.0262	0.0248	0.0278	90,075	89,346	90,738	2,359	444,478	29.8	29.5	30.3	1,148	216,415
55-59	0.0072	0.0356	0.0336	0.0375	87,716	86,866	88,483	3,118	430,785	25.5	25.2	26.0	1,270	175,454
60-64	0.0116	0.0563	0.0532	0.0591	84,598	83,606	85,481	4,762	411,084	21.4	21.1	21.8	1,679	144,955
65-69	0.0178	0.0853	0.0806	0.0893	79,836	78,668	80,860	6,809	382,156	17.5	17.3	17.9	2,052	115,158
70-74	0.0270	0.1264	0.1196	0.1317	73,027	71,679	74,205	9,230	342,059	13.9	13.7	14.2	2,345	86,900
75-79	0.0446	0.2007	0.1908	0.2080	63,797	62,323	65,316	12,807	286,967	10.6	10.3	10.8	2,683	60,123
80-84	0.0848	0.3498	0.3361	0.3624	50,990	49,440	52,853	17,836	210,359	7.6	7.4	7.8	2,953	34,828
85-89	0.1490	0.5427	0.5265	0.5584	33,154	31,725	35,091	17,994	120,783	5.3	5.2	5.5	2,555	17,153
90-94	0.2421	0.7012	0.6865	0.7147	15,160	14,184	16,521	10,630	43,909	3.7	3.6	3.8	1,316	5,437
95-99	0.3638	0.8698	0.8603	0.8782	4,530	4,046	5,092	3,940	10,829	2.6	2.6	2.7	393	1,081
100+	0.5057	1	1	1	590	493	707	590	1,167	2.0	2.0	2.1	50	98
All ages													*29,594*	*6,295,526*

$\ln(_nq_x)$ with uncertainty interval

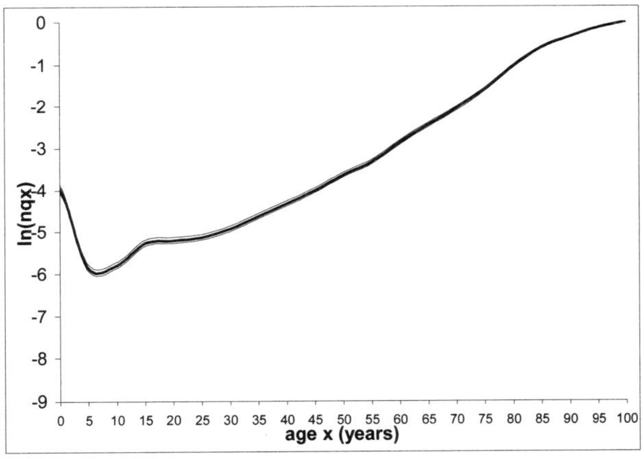

l_x with uncertainty interval

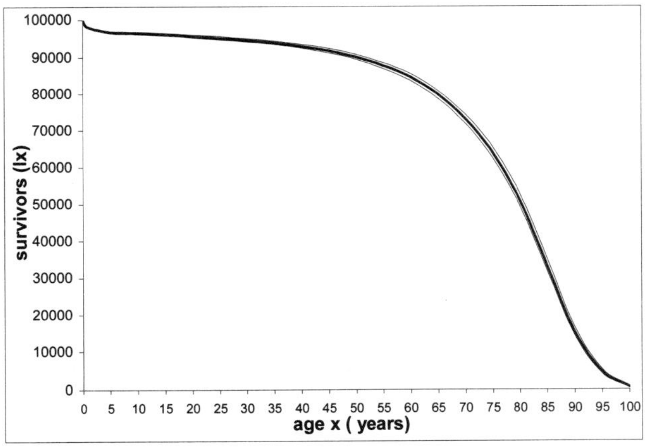

e_x with uncertainty interval

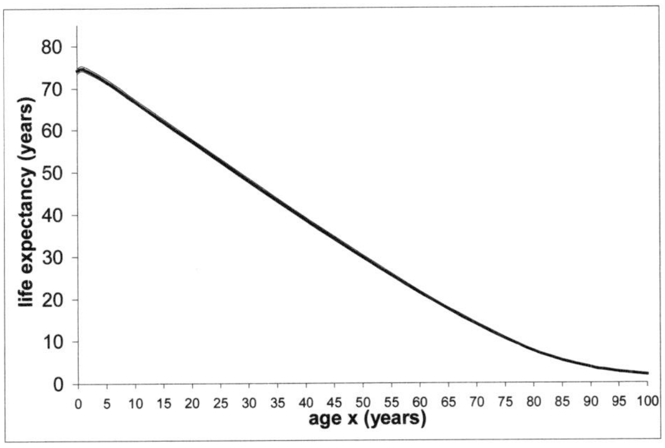

Distribution of deaths by age with uncertainty interval

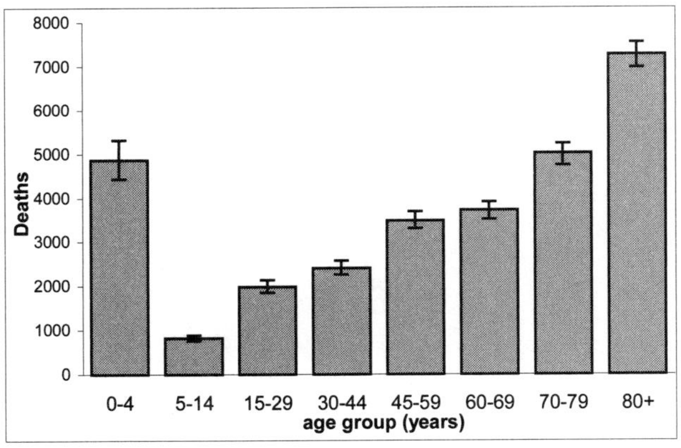

Egypt | Egypte | Egipto

Males, 2000 | *Hommes, 2000* | *Hombres, 2000*

Age	$_nM_x$	$_nq_x$			l_x			$_nd_x$	$_nL_x$	e_x			Actual deaths	Actual population
			percentiles			percentiles					percentiles			
		mean	2 1/2	97 1/2	mean	2 1/2	97 1/2			mean	2 1/2	97 1/2		
<1	0.0394	0.0381	0.0347	0.0414	100,000	100,000	100,000	3,806	96,574	65.4	64.6	66.3	33,066	838,920
1-4	0.0033	0.0129	0.0120	0.0139	96,194	95,856	96,531	1,245	381,787	67.0	66.3	67.7	10,617	3,257,091
5-9	0.0010	0.0049	0.0046	0.0053	94,949	94,551	95,349	469	473,572	63.8	63.1	64.6	3,951	3,989,177
10-14	0.0009	0.0044	0.0041	0.0047	94,480	94,049	94,909	418	471,355	59.2	58.5	59.8	3,718	4,193,258
15-19	0.0012	0.0059	0.0054	0.0063	94,062	93,603	94,517	553	468,928	54.4	53.7	55.1	4,572	3,878,862
20-24	0.0012	0.0062	0.0057	0.0066	93,509	93,013	94,001	580	466,097	49.7	49.1	50.4	4,014	3,227,983
25-29	0.0014	0.0070	0.0065	0.0075	92,930	92,395	93,462	650	463,022	45.0	44.4	45.7	3,766	2,681,008
30-34	0.0019	0.0094	0.0087	0.0101	92,279	91,703	92,855	868	459,227	40.3	39.7	41.0	4,670	2,472,089
35-39	0.0024	0.0121	0.0112	0.0129	91,412	90,781	92,046	1,104	454,298	35.7	35.1	36.3	5,275	2,170,157
40-44	0.0039	0.0193	0.0179	0.0206	90,307	89,610	91,014	1,740	447,187	31.1	30.5	31.7	7,820	2,009,883
45-49	0.0071	0.0349	0.0324	0.0373	88,568	87,770	89,384	3,088	435,117	26.6	26.1	27.2	11,906	1,677,626
50-54	0.0112	0.0546	0.0508	0.0584	85,479	84,519	86,478	4,666	415,733	22.5	22.0	23.1	13,478	1,200,975
55-59	0.0169	0.0810	0.0755	0.0866	80,814	79,641	82,059	6,542	387,713	18.7	18.2	19.2	14,208	841,980
60-64	0.0267	0.1251	0.1168	0.1334	74,271	72,812	75,809	9,291	348,130	15.1	14.6	15.6	18,453	691,421
65-69	0.0424	0.1915	0.1790	0.2031	64,980	63,097	66,927	12,447	293,786	11.9	11.5	12.3	22,597	533,361
70-74	0.0679	0.2904	0.2723	0.3068	52,534	50,293	54,823	15,254	224,535	9.1	8.8	9.5	24,314	357,895
75-79	0.1100	0.4314	0.4085	0.4523	37,280	34,990	39,696	16,082	146,196	6.8	6.6	7.1	23,247	211,335
80-84	0.1637	0.5809	0.5582	0.6017	21,198	19,377	23,361	12,314	75,206	5.1	4.9	5.2	16,006	97,750
85-89	0.2579	0.7840	0.7669	0.7998	8,884	7,771	10,288	6,965	27,007	3.6	3.5	3.7	7,432	28,816
90-94	0.3715	0.8785	0.8669	0.8892	1,919	1,559	2,398	1,686	4,537	2.6	2.6	2.7	1,388	3,736
95-99	0.4895	0.9915	0.9896	0.9930	233	173	319	231	472	2.0	2.0	2.0	108	220
100+	0.5898	1	1	1	2	1	3	2	3	1.7	1.6	1.7	3	5
All ages													*234,608*	*34,363,548*

ln($_nq_x$) with uncertainty interval

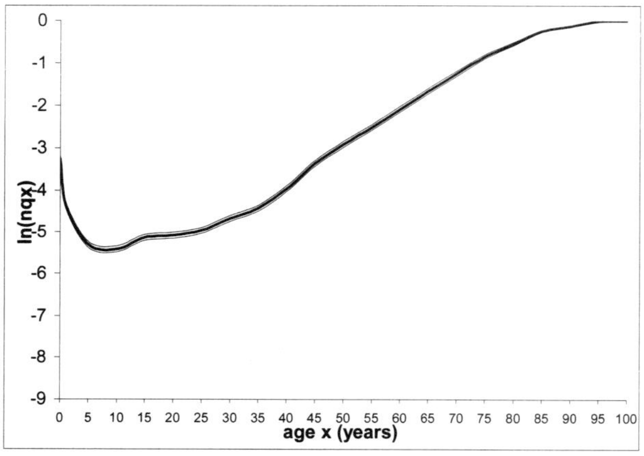

l_x with uncertainty interval

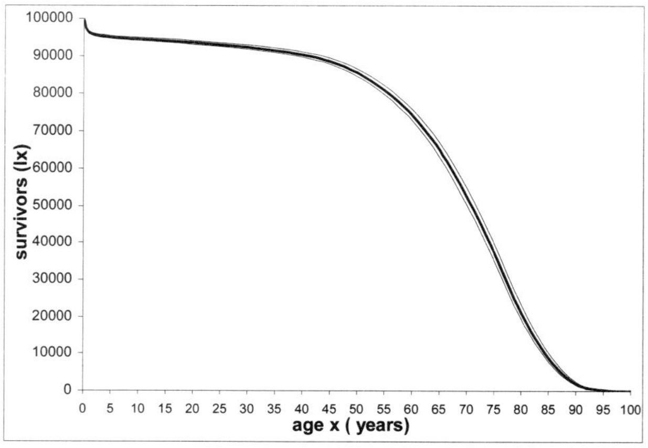

e_x with uncertainty interval

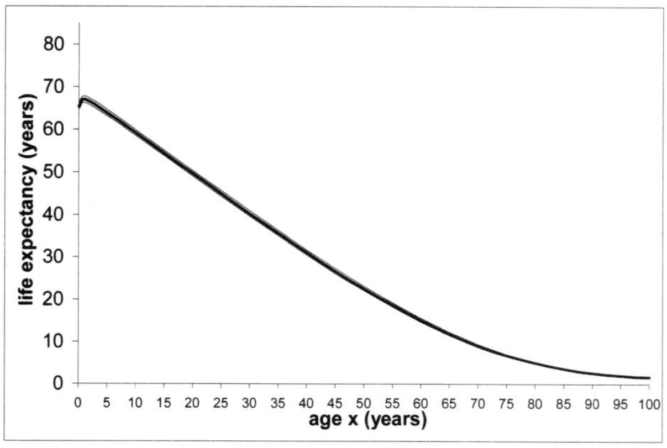

Distribution of deaths by age with uncertainty interval

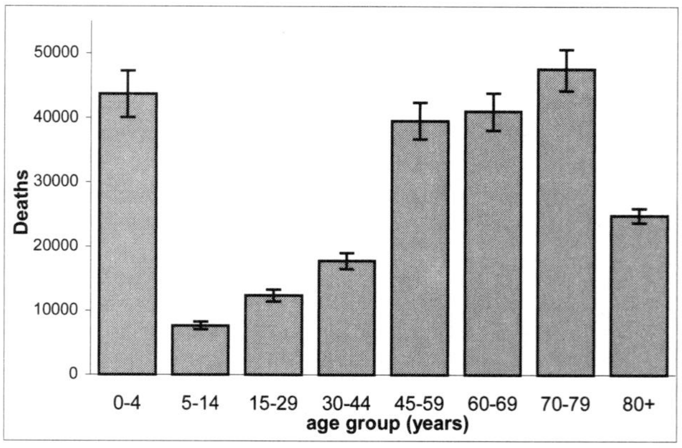

Egypt — Egypte — Egipto

Females, 2000 — *Femmes, 2000* — *Mujeres, 2000*

Age	$_nM_x$	$_nq_x$			l_x			$_nd_x$	$_nL_x$	e_x			Actual deaths	Actual population
			percentiles			percentiles					percentiles			
		mean	2½	97½	mean	2½	97½			mean	2½	97½		
<1	0.0370	0.0358	0.0325	0.0393	100,000	100,000	100,000	3,583	96,775	69.1	68.1	70.0	29,867	806,654
1-4	0.0035	0.0138	0.0128	0.0150	96,417	96,071	96,746	1,331	382,473	70.7	69.9	71.4	10,818	3,108,335
5-9	0.0008	0.0039	0.0036	0.0042	95,086	94,688	95,484	372	474,498	67.6	66.9	68.3	2,990	3,811,612
10-14	0.0007	0.0034	0.0031	0.0037	94,714	94,299	95,132	319	472,770	62.9	62.2	63.6	2,701	3,999,342
15-19	0.0008	0.0040	0.0037	0.0044	94,394	93,959	94,829	382	471,017	58.1	57.4	58.8	2,968	3,661,891
20-24	0.0008	0.0042	0.0039	0.0046	94,012	93,550	94,470	399	469,066	53.3	52.6	54.0	2,574	3,029,867
25-29	0.0011	0.0055	0.0051	0.0059	93,614	93,121	94,096	512	466,791	48.5	47.9	49.2	2,747	2,505,554
30-34	0.0012	0.0061	0.0057	0.0067	93,102	92,569	93,616	571	464,085	43.8	43.1	44.4	2,864	2,329,254
35-39	0.0017	0.0084	0.0078	0.0091	92,532	91,953	93,079	780	460,709	39.0	38.4	39.7	3,500	2,067,910
40-44	0.0025	0.0124	0.0115	0.0135	91,752	91,111	92,344	1,140	455,910	34.3	33.7	35.0	4,926	1,970,233
45-49	0.0041	0.0202	0.0187	0.0219	90,612	89,883	91,271	1,832	448,481	29.7	29.2	30.4	6,940	1,699,120
50-54	0.0074	0.0364	0.0336	0.0395	88,780	87,911	89,543	3,236	435,812	25.3	24.8	25.9	9,332	1,256,816
55-59	0.0120	0.0583	0.0538	0.0630	85,544	84,437	86,517	4,983	415,265	21.2	20.7	21.7	10,978	914,869
60-64	0.0180	0.0861	0.0797	0.0927	80,561	79,119	81,862	6,938	385,461	17.3	16.9	17.8	14,128	784,874
65-69	0.0290	0.1352	0.1256	0.1447	73,623	71,782	75,334	9,954	343,231	13.7	13.3	14.2	18,634	642,547
70-74	0.0479	0.2140	0.1993	0.2277	63,669	61,409	65,851	13,626	284,283	10.5	10.1	10.9	22,109	461,274
75-79	0.0855	0.3521	0.3299	0.3723	50,044	47,522	52,600	17,622	206,164	7.6	7.4	8.0	24,121	282,199
80-84	0.1479	0.5398	0.5136	0.5634	32,422	30,074	35,002	17,500	118,360	5.4	5.2	5.7	20,150	136,282
85-89	0.2352	0.7405	0.7193	0.7594	14,922	13,286	16,892	11,049	46,986	3.9	3.8	4.0	10,553	44,873
90-94	0.3439	0.8464	0.8319	0.8594	3,872	3,211	4,741	3,277	9,530	2.8	2.7	2.9	2,373	6,900
95-99	0.4625	0.9685	0.9634	0.9730	595	452	797	576	1,246	2.1	2.1	2.2	235	508
100+	0.5718	1	1	1	19	12	29	19	33	1.7	1.7	1.8	8	14
All ages													*205,513*	*33,520,928*

ln($_nq_x$) with uncertainty interval

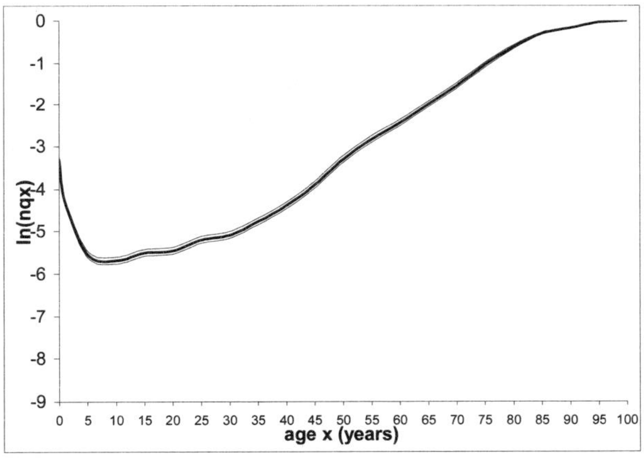

l_x with uncertainty interval

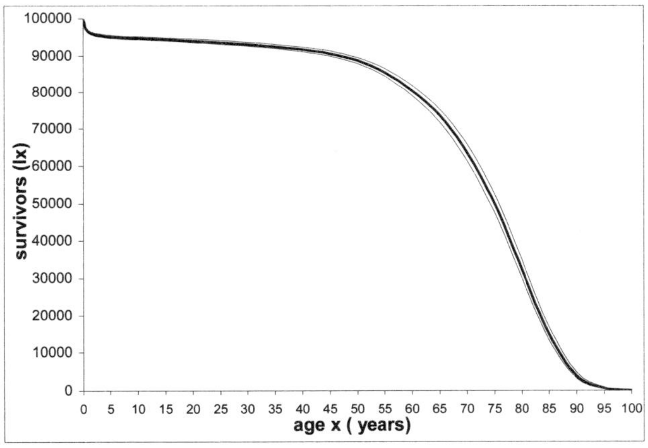

e_x with uncertainty interval

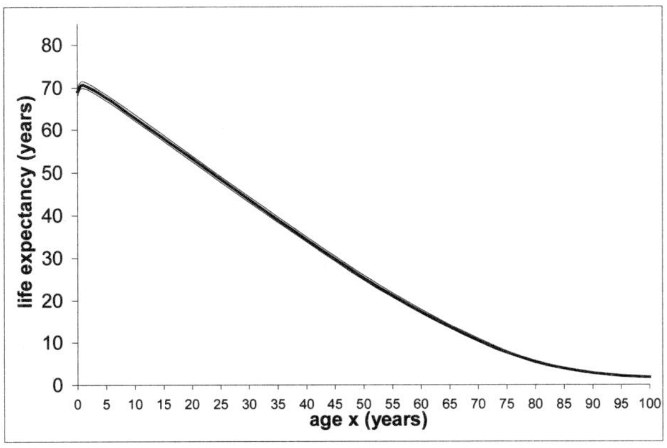

Distribution of deaths by age with uncertainty interval

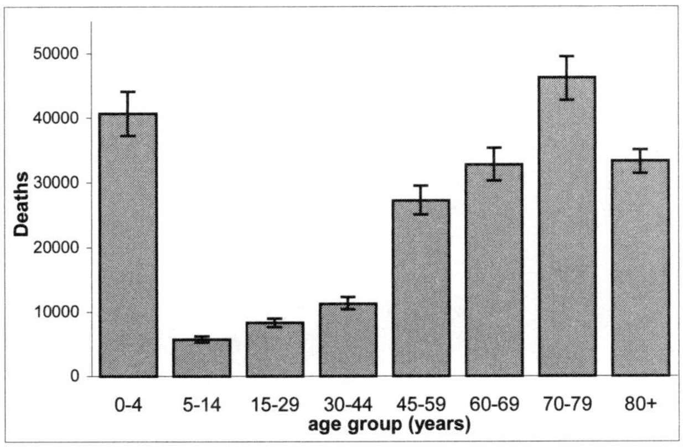

El Salvador **El Salvador** **El Salvador**

Males, 2000 *Hommes, 2000* *Hombres, 2000*

Age	$_nM_x$	$_nq_x$ mean	$_nq_x$ percentiles 2½	$_nq_x$ percentiles 97½	l_x mean	l_x percentiles 2½	l_x percentiles 97½	$_nd_x$	$_nL_x$	e_x mean	e_x percentiles 2½	e_x percentiles 97½	Actual deaths	Actual population
<1	0.0342	0.0332	0.0296	0.0368	100,000	100,000	100,000	3,320	97,012	66.3	65.5	67.1	2,848	83,222
1-4	0.0017	0.0070	0.0064	0.0076	96,680	96,318	97,037	673	385,104	67.5	66.8	68.3	566	323,796
5-9	0.0007	0.0036	0.0033	0.0038	96,007	95,591	96,417	341	479,181	64.0	63.2	64.8	272	382,350
10-14	0.0007	0.0034	0.0031	0.0037	95,666	95,223	96,102	324	477,520	59.2	58.5	60.0	236	348,269
15-19	0.0025	0.0124	0.0115	0.0134	95,342	94,875	95,802	1,187	473,743	54.4	53.7	55.2	835	333,311
20-24	0.0039	0.0193	0.0179	0.0207	94,155	93,603	94,701	1,819	466,229	50.1	49.3	50.9	1,327	340,194
25-29	0.0041	0.0205	0.0192	0.0219	92,336	91,664	93,002	1,896	456,941	46.0	45.2	46.8	1,164	280,606
30-34	0.0051	0.0251	0.0235	0.0266	90,440	89,656	91,219	2,267	446,533	41.9	41.2	42.7	1,051	206,997
35-39	0.0059	0.0291	0.0274	0.0308	88,173	87,270	89,074	2,566	434,451	37.9	37.2	38.7	906	153,397
40-44	0.0064	0.0316	0.0299	0.0333	85,607	84,584	86,631	2,706	421,270	34.0	33.3	34.8	865	134,699
45-49	0.0080	0.0393	0.0372	0.0416	82,901	81,766	84,042	3,259	406,358	30.0	29.3	30.8	922	115,004
50-54	0.0099	0.0483	0.0454	0.0513	79,642	78,368	80,920	3,847	388,593	26.1	25.5	26.9	989	99,859
55-59	0.0117	0.0568	0.0531	0.0607	75,795	74,344	77,246	4,308	368,206	22.3	21.8	23.0	918	78,434
60-64	0.0186	0.0890	0.0827	0.0953	71,487	69,834	73,142	6,363	341,529	18.5	18.0	19.2	1,218	65,373
65-69	0.0265	0.1240	0.1148	0.1330	65,124	63,181	67,090	8,079	305,426	15.1	14.6	15.7	1,429	54,043
70-74	0.0403	0.1829	0.1690	0.1958	57,046	54,776	59,387	10,431	259,152	11.9	11.5	12.4	1,593	39,587
75-79	0.0649	0.2791	0.2590	0.2974	46,615	44,051	49,350	13,008	200,555	9.0	8.7	9.4	1,630	25,126
80-84	0.1085	0.4266	0.4009	0.4500	33,607	30,949	36,566	14,337	132,194	6.5	6.3	6.8	1,335	12,310
85-89	0.1888	0.6413	0.6156	0.6645	19,270	17,021	21,906	12,358	65,458	4.5	4.3	4.7	769	4,073
90-94	0.3077	0.8000	0.7814	0.8170	6,913	5,710	8,421	5,530	17,974	3.0	2.9	3.1	276	896
95-99	0.4695	0.9746	0.9697	0.9788	1,383	1,045	1,841	1,348	2,871	2.1	2.1	2.1	37	78
100+	0.6707	1	1	1	35	22	56	35	52	1.5	1.5	1.6	1	2
All ages													*21,188*	*3,081,626*

ln($_nq_x$) with uncertainty interval

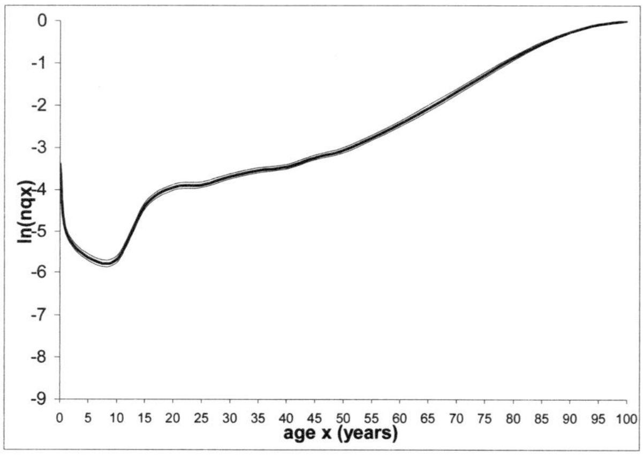

l_x with uncertainty interval

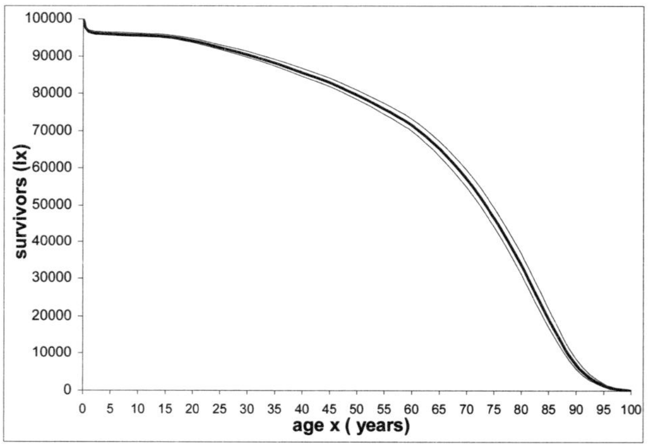

e_x with uncertainty interval

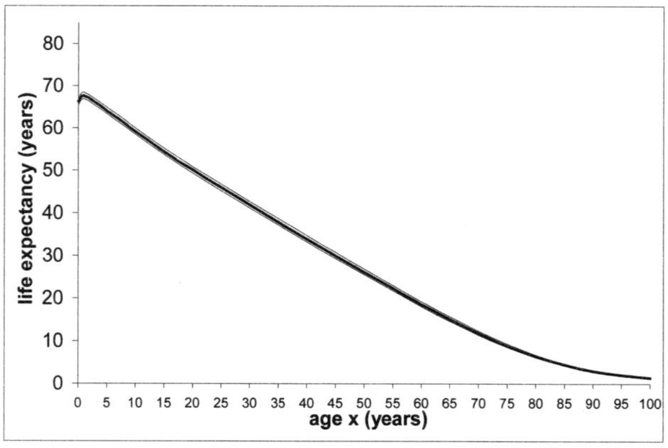

Distribution of deaths by age with uncertainty interval

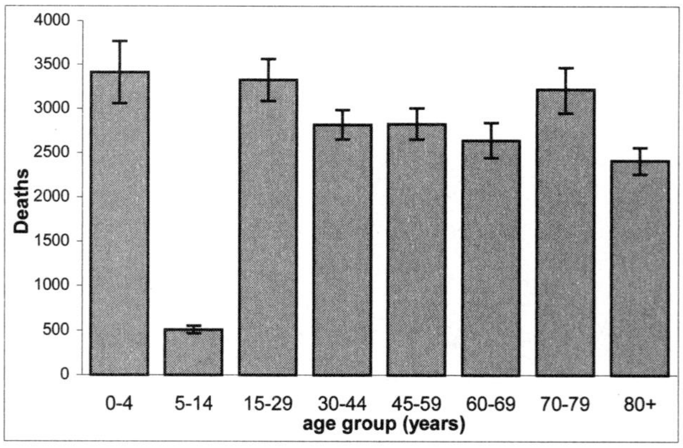

El Salvador

El Salvador

El Salvador

Females, 2000 *Femmes, 2000* *Mujeres, 2000*

Age	$_nM_x$	$_nq_x$			l_x			$_nd_x$	$_nL_x$	e_x			Actual deaths	Actual population
			percentiles			percentiles					percentiles			
		mean	2½	97½	mean	2½	97½			mean	2½	97½		
<1	0.0263	0.0257	0.0218	0.0284	100,000	100,000	100,000	2,571	97,686	73.3	72.4	74.4	2,086	79,259
1-4	0.0018	0.0074	0.0064	0.0080	97,429	97,164	97,819	718	387,994	74.2	73.4	75.3	575	311,172
5-9	0.0007	0.0034	0.0030	0.0037	96,711	96,389	97,190	332	482,727	70.7	69.9	71.8	254	368,871
10-14	0.0006	0.0029	0.0026	0.0032	96,379	96,031	96,898	282	481,192	66.0	65.2	67.1	198	337,685
15-19	0.0011	0.0057	0.0050	0.0061	96,098	95,728	96,650	545	479,125	61.2	60.4	62.2	370	324,800
20-24	0.0013	0.0064	0.0056	0.0069	95,552	95,141	96,165	607	476,244	56.5	55.7	57.6	428	335,942
25-29	0.0014	0.0068	0.0060	0.0073	94,945	94,489	95,625	642	473,120	51.8	51.1	52.9	391	288,280
30-34	0.0016	0.0081	0.0072	0.0087	94,303	93,800	95,050	763	469,606	47.2	46.4	48.2	378	232,479
35-39	0.0029	0.0145	0.0130	0.0156	93,540	92,983	94,364	1,357	464,304	42.5	41.8	43.5	534	182,805
40-44	0.0032	0.0156	0.0141	0.0168	92,182	91,533	93,134	1,442	457,306	38.1	37.5	39.1	498	158,004
45-49	0.0046	0.0228	0.0207	0.0245	90,740	89,998	91,818	2,069	448,529	33.7	33.0	34.6	598	129,682
50-54	0.0063	0.0310	0.0282	0.0333	88,671	87,797	89,916	2,750	436,483	29.4	28.8	30.3	695	110,408
55-59	0.0096	0.0471	0.0427	0.0506	85,922	84,874	87,382	4,047	419,493	25.3	24.7	26.1	836	86,666
60-64	0.0130	0.0628	0.0568	0.0675	81,875	80,577	83,652	5,143	396,518	21.4	20.9	22.2	958	73,884
65-69	0.0202	0.0960	0.0869	0.1029	76,732	75,138	78,900	7,365	365,247	17.7	17.2	18.4	1,278	63,371
70-74	0.0284	0.1325	0.1203	0.1415	69,367	67,403	72,047	9,193	323,851	14.3	13.8	14.9	1,396	49,189
75-79	0.0427	0.1928	0.1765	0.2055	60,174	57,862	63,379	11,604	271,858	11.1	10.7	11.6	1,455	34,089
80-84	0.0764	0.3209	0.2985	0.3416	48,569	45,974	52,193	15,585	203,886	8.1	7.8	8.5	1,459	19,082
85-89	0.1308	0.4929	0.4679	0.5192	32,985	30,268	36,611	16,260	124,275	5.8	5.6	6.1	1,019	7,786
90-94	0.2141	0.6518	0.6290	0.6761	16,725	14,552	19,483	10,901	50,924	4.0	3.9	4.2	523	2,443
95-99	0.3348	0.8351	0.8198	0.8517	5,824	4,713	7,228	4,864	14,530	2.8	2.8	2.9	120	357
100+	0.5004	1	1	1	960	699	1,302	960	1,919	2.0	2.0	2.1	9	17
All ages													*16,058*	*3,196,271*

$\ln({}_nq_x)$ with uncertainty interval

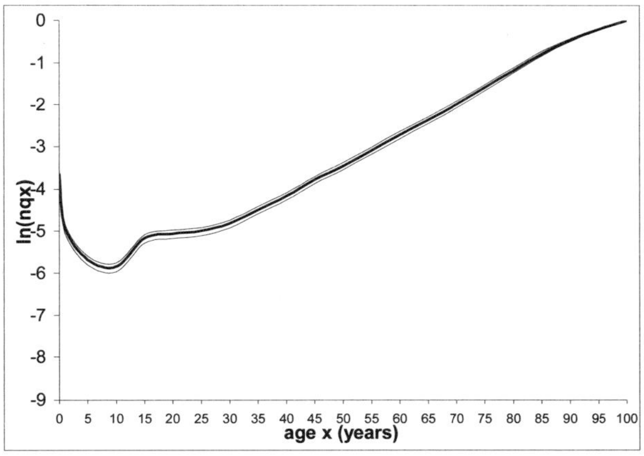

l_x with uncertainty interval

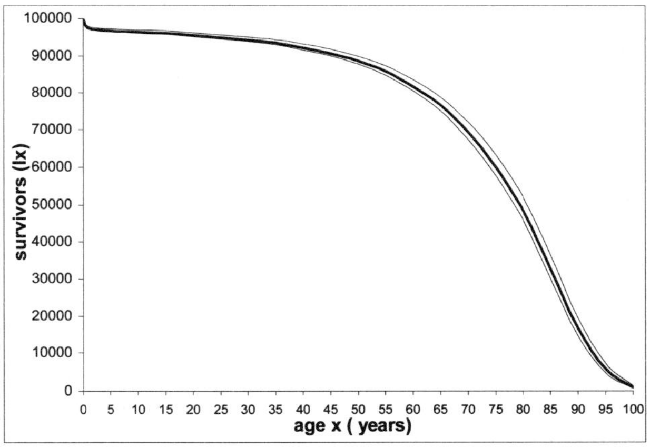

e_x with uncertainty interval

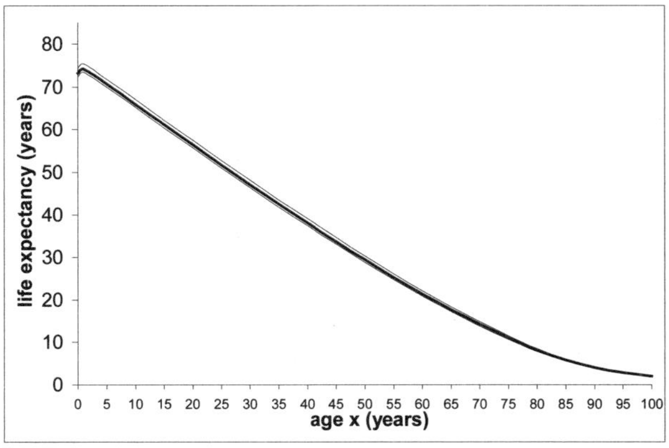

Distribution of deaths by age with uncertainty interval

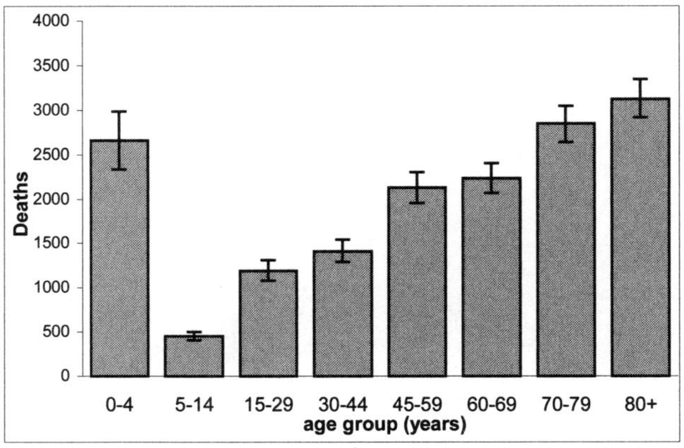

Equatorial Guinea   Guinée équatoriale   Guinea Ecuatorial

Males, 2000   *Hommes, 2000*   *Hombres, 2000*

Age	$_nM_x$	$_nq_x$ mean	$_nq_x$ percentiles 2½	$_nq_x$ percentiles 97½	l_x mean	l_x percentiles 2½	l_x percentiles 97½	$_nd_x$	$_nL_x$	e_x mean	e_x percentiles 2½	e_x percentiles 97½	Actual deaths	Actual population
<1	0.1354	0.1207	0.0953	0.1450	100,000	100,000	100,000	12,066	89,141	53.5	49.7	57.6	1,246	9,202
1-4	0.0104	0.0406	0.0330	0.0494	87,934	85,499	90,474	3,571	343,167	59.9	56.3	64.2	323	31,012
5-9	0.0025	0.0125	0.0100	0.0153	84,363	81,966	86,760	1,051	419,189	58.3	55.0	62.4	81	32,489
10-14	0.0016	0.0078	0.0062	0.0097	83,312	80,808	85,834	652	414,931	54.0	50.8	58.1	43	27,438
15-19	0.0015	0.0074	0.0058	0.0091	82,660	80,083	85,285	610	411,775	49.4	46.3	53.4	34	22,773
20-24	0.0030	0.0151	0.0119	0.0186	82,050	79,406	84,770	1,235	407,162	44.8	41.7	48.7	57	18,924
25-29	0.0043	0.0211	0.0166	0.0262	80,815	77,939	83,727	1,708	399,803	40.4	37.4	44.3	69	16,142
30-34	0.0056	0.0275	0.0214	0.0342	79,106	75,893	82,278	2,175	390,094	36.3	33.3	40.0	76	13,603
35-39	0.0071	0.0350	0.0270	0.0439	76,931	73,297	80,426	2,695	377,920	32.2	29.3	35.8	81	11,335
40-44	0.0092	0.0451	0.0341	0.0570	74,236	70,095	78,116	3,345	362,820	28.3	25.4	31.8	89	9,703
45-49	0.0122	0.0592	0.0441	0.0754	70,892	66,153	75,237	4,196	343,968	24.5	21.8	27.8	105	8,638
50-54	0.0167	0.0802	0.0589	0.1026	66,696	61,270	71,730	5,346	320,113	20.9	18.3	24.0	107	6,398
55-59	0.0231	0.1092	0.0794	0.1423	61,350	55,171	67,296	6,698	290,003	17.5	15.1	20.4	124	5,351
60-64	0.0332	0.1533	0.1109	0.2024	54,652	47,751	61,681	8,376	252,318	14.3	12.2	17.0	146	4,403
65-69	0.0482	0.2152	0.1562	0.2837	46,276	38,653	54,402	9,959	206,481	11.5	9.6	13.8	166	3,449
70-74	0.0735	0.3103	0.2283	0.3994	36,316	28,290	45,762	11,268	153,412	8.9	7.4	10.9	172	2,336
75-79	0.1110	0.4343	0.3314	0.5347	25,048	17,386	35,163	10,878	98,045	6.8	5.7	8.3	150	1,350
80-84	0.1649	0.5838	0.4723	0.6798	14,170	8,090	23,234	8,272	50,169	5.1	4.4	6.2	102	621
85-89	0.2380	0.7460	0.6467	0.8225	5,898	2,590	12,000	4,400	18,489	3.9	3.4	4.6	43	181
90-94	0.3336	0.8336	0.7539	0.8923	1,498	460	4,240	1,249	3,743	2.9	2.5	3.3	11	32
95-99	0.4541	0.9612	0.9236	0.9820	249	50	1,044	240	527	2.2	2.1	2.4	1	3
100+	0.6004	1	1	1	10	1	80	10	16	1.7	1.6	1.8	0	0
All ages													*3,226*	*225,383*

ln($_nq_x$) with uncertainty interval

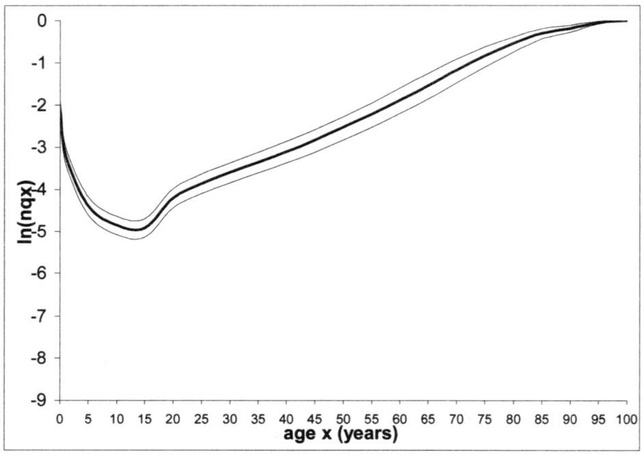

l_x with uncertainty interval

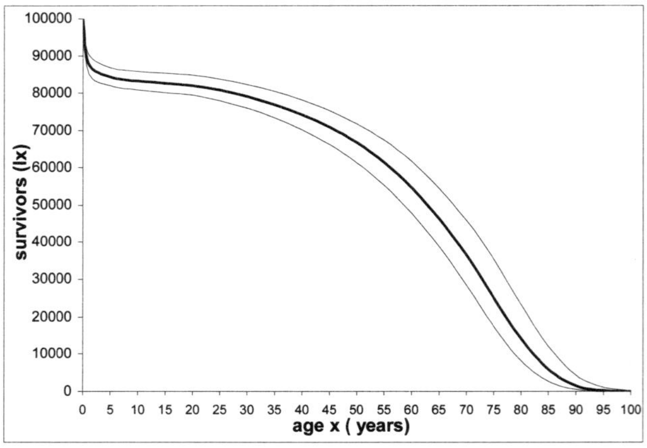

e_x with uncertainty interval

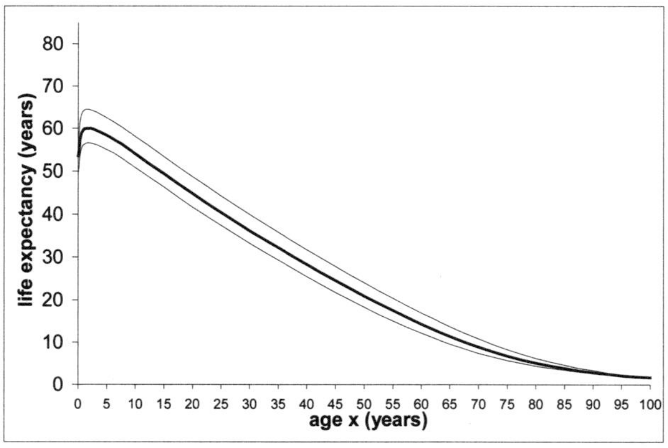

Distribution of deaths by age with uncertainty interval

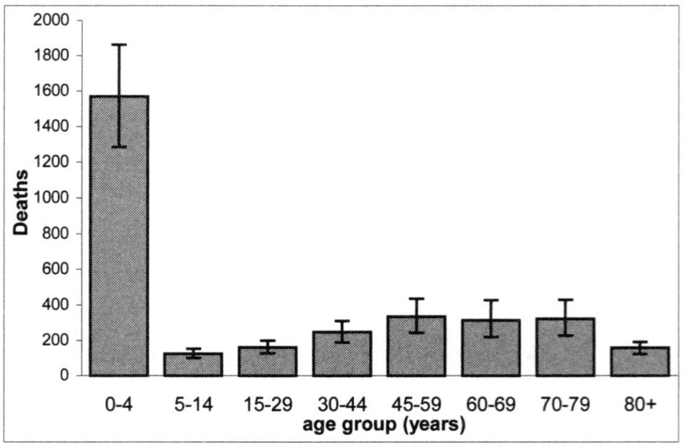

Equatorial Guinea | Guinée équatoriale | Guinea Ecuatorial

Females, 2000 | *Femmes, 2000* | *Mujeres, 2000*

Age	$_nM_x$	$_nq_x$			l_x			$_nd_x$	$_nL_x$	e_x			Actual deaths	Actual population
			percentiles			percentiles					percentiles			
		mean	2½	97½	mean	2½	97½			mean	2½	97½		
<1	0.1150	0.1042	0.0848	0.1271	100,000	100,000	100,000	10,421	90,621	56.2	52.6	60.0	1,027	8,934
1-4	0.0112	0.0437	0.0361	0.0524	89,579	87,286	91,523	3,914	348,921	61.7	58.1	66.0	346	30,866
5-9	0.0025	0.0127	0.0103	0.0154	85,665	83,665	87,655	1,085	425,612	60.5	57.1	64.5	82	32,348
10-14	0.0017	0.0083	0.0067	0.0101	84,580	82,444	86,726	702	421,146	56.2	53.0	60.1	46	27,410
15-19	0.0024	0.0120	0.0096	0.0146	83,878	81,641	86,127	1,003	416,886	51.7	48.5	55.5	55	22,854
20-24	0.0037	0.0185	0.0147	0.0226	82,876	80,495	85,271	1,530	410,555	47.2	44.1	51.0	71	19,150
25-29	0.0046	0.0229	0.0180	0.0282	81,346	78,700	83,961	1,859	402,082	43.1	40.1	46.7	76	16,479
30-34	0.0052	0.0257	0.0200	0.0320	79,487	76,489	82,365	2,044	392,325	39.0	36.2	42.5	73	14,061
35-39	0.0062	0.0304	0.0233	0.0382	77,443	74,065	80,609	2,358	381,320	35.0	32.3	38.4	73	11,807
40-44	0.0067	0.0329	0.0250	0.0417	75,085	71,278	78,589	2,473	369,242	31.0	28.5	34.2	69	10,236
45-49	0.0085	0.0414	0.0311	0.0528	72,612	68,373	76,500	3,008	355,540	27.0	24.6	30.0	78	9,262
50-54	0.0117	0.0570	0.0423	0.0730	69,604	64,866	73,980	3,971	338,093	23.1	20.8	25.9	82	6,966
55-59	0.0166	0.0795	0.0585	0.1020	65,633	60,292	70,624	5,218	315,122	19.3	17.2	22.0	101	6,070
60-64	0.0252	0.1186	0.0868	0.1520	60,416	54,393	66,271	7,166	284,164	15.7	13.8	18.2	126	4,992
65-69	0.0391	0.1781	0.1307	0.2283	53,250	46,247	60,365	9,484	242,540	12.5	10.8	14.8	157	4,019
70-74	0.0627	0.2711	0.2013	0.3414	43,766	35,849	52,476	11,867	189,163	9.7	8.3	11.6	180	2,871
75-79	0.0977	0.3926	0.2988	0.4764	31,899	23,917	41,834	12,523	128,190	7.4	6.3	8.9	173	1,771
80-84	0.1485	0.5416	0.4339	0.6259	19,377	12,803	29,048	10,494	70,647	5.5	4.8	6.7	129	866
85-89	0.2190	0.7075	0.6066	0.7777	8,882	4,801	16,044	6,285	28,700	4.1	3.7	4.9	63	286
90-94	0.3129	0.8070	0.7251	0.8612	2,598	1,075	6,228	2,096	6,699	3.0	2.7	3.5	20	63
95-99	0.4335	0.9421	0.8974	0.9665	501	151	1,712	472	1,090	2.3	2.2	2.5	3	8
100+	0.5821	1	1	1	29	5	176	29	50	1.7	1.7	1.9	1	1
All ages													*3,031*	*231,320*

$\ln(_nq_x)$ with uncertainty interval

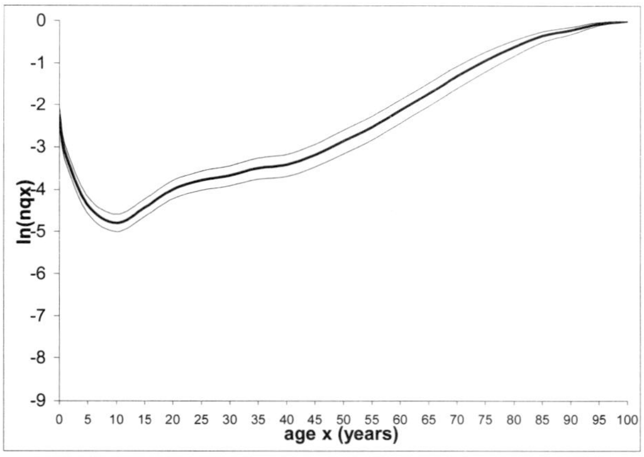

l_x with uncertainty interval

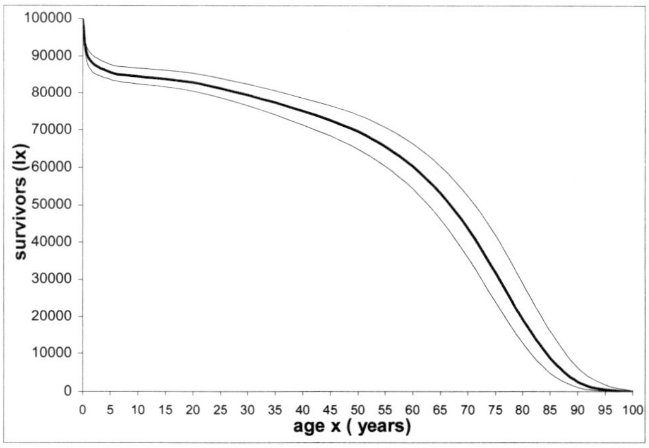

e_x with uncertainty interval

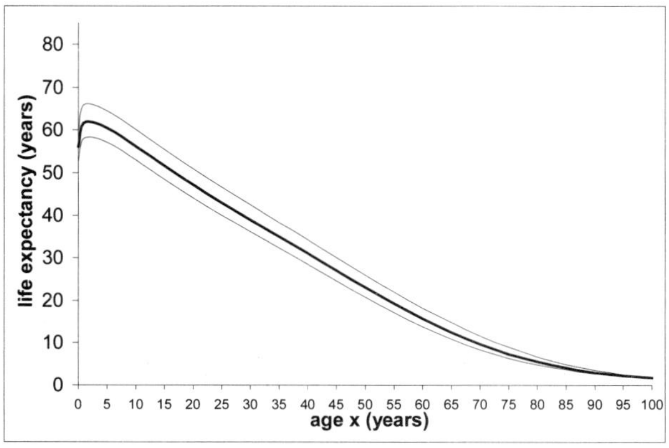

Distribution of deaths by age with uncertainty interval

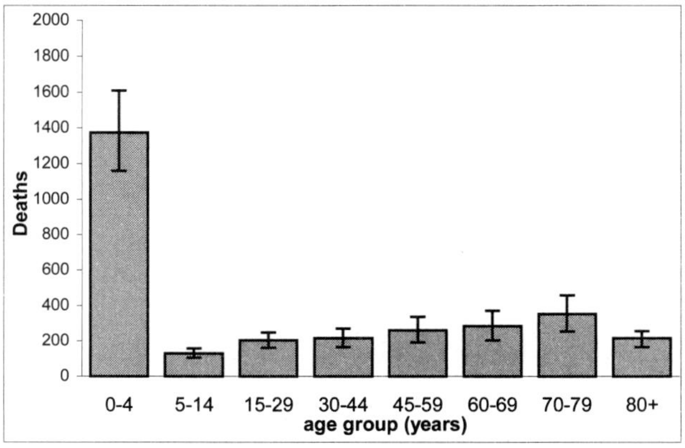

Eritrea　　Erythrée　　Eritrea

Males, 2000　　*Hommes, 2000*　　*Hombres, 2000*

Age	$_nM_x$	$_nq_x$			l_x			$_nd_x$	$_nL_x$	e_x			Actual deaths	Actual population
			percentiles			percentiles					percentiles			
		mean	2 1/2	97 1/2	mean	2 1/2	97 1/2			mean	2 1/2	97 1/2		
<1	0.1167	0.1056	0.0934	0.1186	100,000	100,000	100,000	10,560	90,496	49.1	46.3	52.0	7,831	67,110
1-4	0.0103	0.0403	0.0353	0.0451	89,440	88,141	90,662	3,605	349,106	53.8	50.8	57.4	2,517	243,732
5-9	0.0025	0.0122	0.0106	0.0139	85,835	84,614	87,057	1,049	426,551	52.0	49.1	55.5	656	266,613
10-14	0.0017	0.0083	0.0072	0.0095	84,786	83,476	86,083	704	422,169	47.6	44.8	51.0	383	229,435
15-19	0.0025	0.0123	0.0106	0.0142	84,082	82,708	85,461	1,038	417,814	43.0	40.2	46.4	471	189,506
20-24	0.0053	0.0260	0.0222	0.0301	83,044	81,577	84,549	2,155	409,830	38.5	35.7	41.9	847	161,129
25-29	0.0094	0.0458	0.0387	0.0537	80,888	79,217	82,656	3,705	395,178	34.5	31.7	37.8	1,289	137,420
30-34	0.0141	0.0683	0.0567	0.0810	77,183	75,101	79,400	5,271	372,737	31.0	28.4	34.2	1,628	115,092
35-39	0.0164	0.0786	0.0639	0.0943	71,912	69,277	74,752	5,650	345,434	28.1	25.7	31.1	1,553	94,934
40-44	0.0181	0.0865	0.0694	0.1045	66,262	62,836	69,727	5,733	316,976	25.3	23.1	28.1	1,401	77,448
45-49	0.0202	0.0963	0.0765	0.1166	60,529	56,271	64,651	5,829	288,072	22.5	20.5	25.0	1,230	60,793
50-54	0.0229	0.1084	0.0856	0.1310	54,700	49,712	59,513	5,927	258,682	19.6	17.8	21.8	1,228	53,591
55-59	0.0267	0.1253	0.0989	0.1509	48,773	43,198	54,325	6,112	228,583	16.7	15.1	18.6	1,119	41,861
60-64	0.0379	0.1730	0.1373	0.2072	42,660	36,682	48,856	7,382	194,847	13.7	12.3	15.4	1,182	31,206
65-69	0.0529	0.2335	0.1877	0.2763	35,279	29,128	42,146	8,239	155,796	11.0	9.9	12.5	1,156	21,857
70-74	0.0783	0.3274	0.2695	0.3793	27,040	21,168	34,234	8,853	113,067	8.6	7.8	9.8	1,075	13,733
75-79	0.1160	0.4497	0.3832	0.5063	18,187	13,252	25,008	8,179	70,487	6.6	6.0	7.4	855	7,369
80-84	0.1739	0.6060	0.5390	0.6600	10,008	6,642	15,426	6,064	34,879	4.9	4.5	5.5	536	3,082
85-89	0.2517	0.7724	0.7173	0.8150	3,944	2,265	7,111	3,046	12,103	3.7	3.4	4.0	214	852
90-94	0.3516	0.8556	0.8129	0.8880	898	419	2,011	768	2,184	2.7	2.6	3.0	50	143
95-99	0.4743	0.9788	0.9651	0.9871	130	47	376	127	268	2.1	2.1	2.2	7	14
100+	0.6174	1	1	1	3	1	13	3	4	1.6	1.6	1.7	1	1
All ages													*27,228*	*1,816,921*

ln($_nq_x$) with uncertainty interval

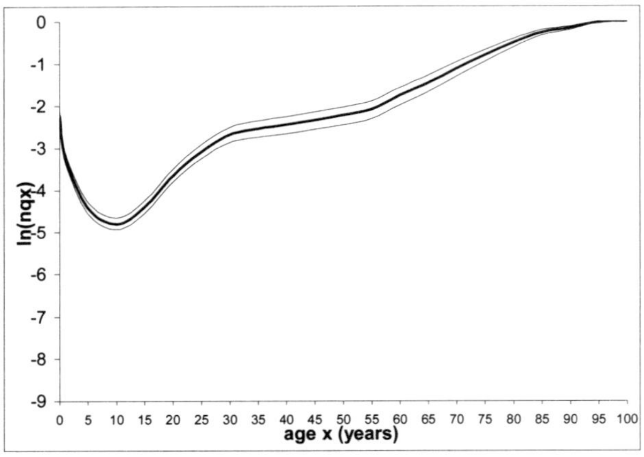

l_x with uncertainty interval

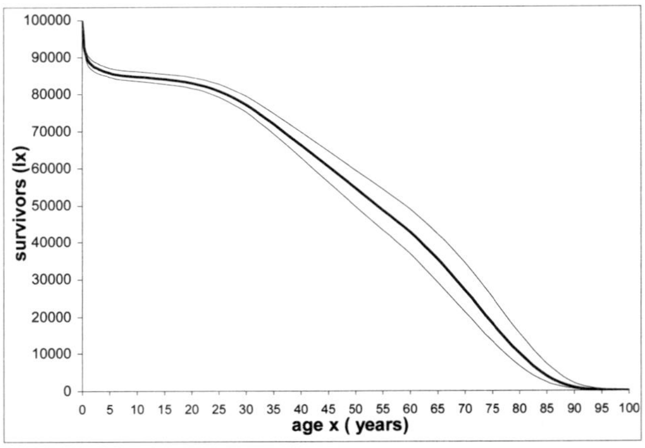

e_x with uncertainty interval

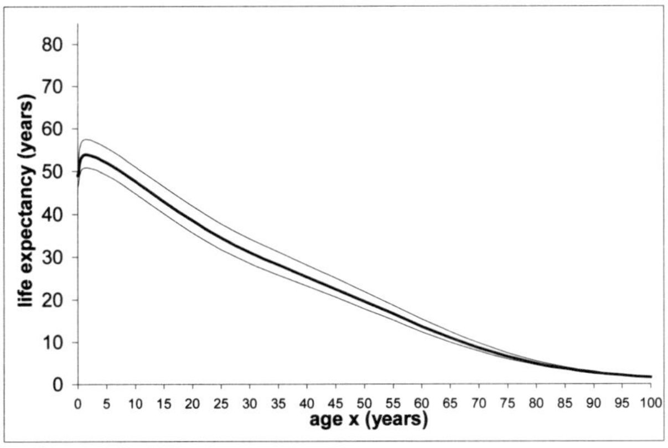

Distribution of deaths by age with uncertainty interval

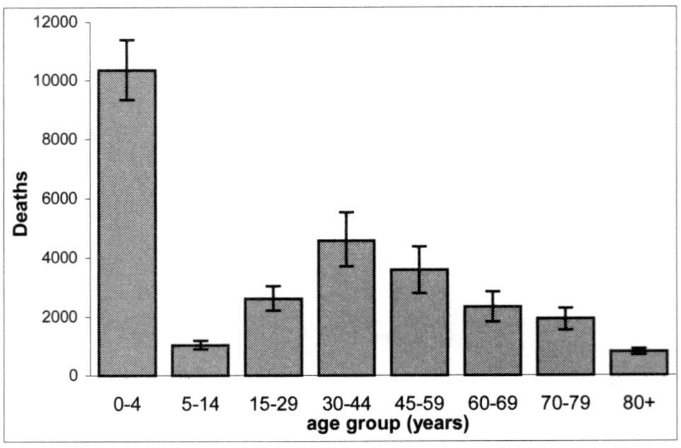

Eritrea | Erythrée | Eritrea

Females, 2000 | *Femmes, 2000* | *Mujeres, 2000*

Age	$_nM_x$	$_nq_x$			l_x			$_nd_x$	$_nL_x$	e_x			Actual deaths	Actual population
			percentiles			percentiles					percentiles			
		mean	2 1/2	97 1/2	mean	2 1/2	97 1/2			mean	2 1/2	97 1/2		
<1	0.0976	0.0897	0.0791	0.1013	100,000	100,000	100,000	8,971	91,926	51.0	48.1	54.0	6,285	64,405
1-4	0.0113	0.0441	0.0390	0.0497	91,029	89,872	92,093	4,016	354,479	55.0	51.8	58.6	2,743	242,084
5-9	0.0026	0.0130	0.0114	0.0148	87,013	85,896	88,130	1,133	432,235	53.4	50.3	57.0	695	265,053
10-14	0.0019	0.0093	0.0081	0.0107	85,881	84,661	87,094	802	427,398	49.1	46.0	52.6	430	229,185
15-19	0.0040	0.0197	0.0171	0.0226	85,079	83,770	86,374	1,679	421,195	44.5	41.5	48.0	758	190,247
20-24	0.0081	0.0399	0.0342	0.0461	83,399	81,876	84,867	3,325	408,684	40.4	37.4	43.9	1,318	161,929
25-29	0.0125	0.0607	0.0512	0.0712	80,074	78,124	81,908	4,864	388,210	37.0	34.1	40.3	1,724	137,562
30-34	0.0150	0.0722	0.0596	0.0858	75,210	72,650	77,620	5,428	362,480	34.2	31.5	37.4	1,736	115,942
35-39	0.0144	0.0695	0.0564	0.0834	69,782	66,588	72,810	4,849	336,788	31.7	29.2	34.6	1,390	96,580
40-44	0.0136	0.0657	0.0528	0.0793	64,933	61,233	68,482	4,268	313,997	28.8	26.7	31.5	1,089	80,112
45-49	0.0136	0.0656	0.0523	0.0794	60,665	56,505	64,699	3,982	293,372	25.7	23.8	28.1	873	64,350
50-54	0.0158	0.0758	0.0601	0.0917	56,684	52,016	61,146	4,295	272,680	22.3	20.6	24.5	879	55,786
55-59	0.0194	0.0925	0.0732	0.1116	52,388	47,246	57,410	4,844	249,830	18.9	17.4	20.9	872	44,950
60-64	0.0272	0.1273	0.1009	0.1528	47,544	41,972	53,131	6,051	222,591	15.6	14.2	17.4	951	34,979
65-69	0.0402	0.1825	0.1458	0.2175	41,492	35,569	47,698	7,573	188,529	12.5	11.3	14.1	1,026	25,551
70-74	0.0627	0.2712	0.2204	0.3179	33,919	27,884	40,745	9,197	146,603	9.8	8.8	11.0	1,072	17,094
75-79	0.0960	0.3869	0.3239	0.4421	24,722	19,121	31,763	9,566	99,696	7.5	6.7	8.4	944	9,842
80-84	0.1460	0.5349	0.4665	0.5914	15,156	10,785	21,474	8,107	55,513	5.6	5.1	6.3	656	4,493
85-89	0.2157	0.7006	0.6392	0.7490	7,049	4,465	11,456	4,939	22,898	4.2	3.8	4.6	303	1,403
90-94	0.3089	0.8017	0.7525	0.8399	2,110	1,121	4,134	1,692	5,476	3.0	2.8	3.3	84	273
95-99	0.4292	0.9381	0.9115	0.9566	419	179	1,023	393	915	2.3	2.2	2.4	15	34
100+	0.5784	1	1	1	26	8	91	26	45	1.7	1.7	1.8	1	2
All ages													*25,844*	*1,841,856*

ln($_nq_x$) with uncertainty interval

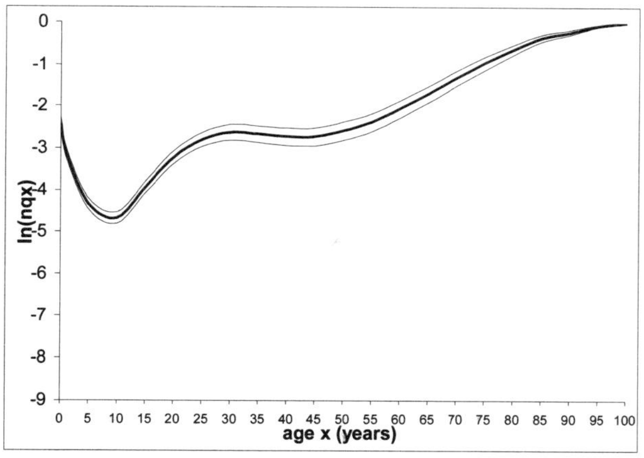

l_x with uncertainty interval

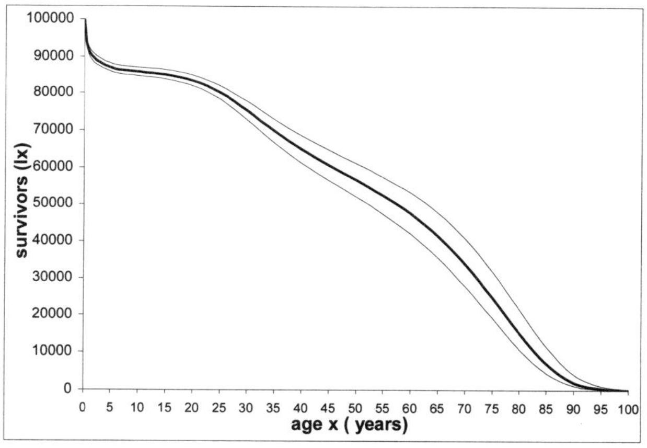

e_x with uncertainty interval

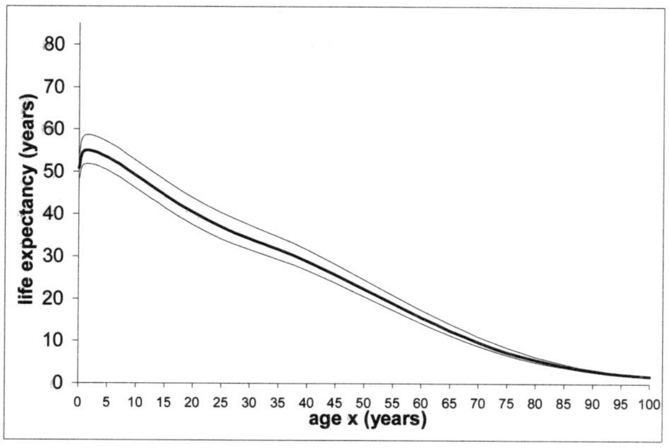

Distribution of deaths by age with uncertainty interval

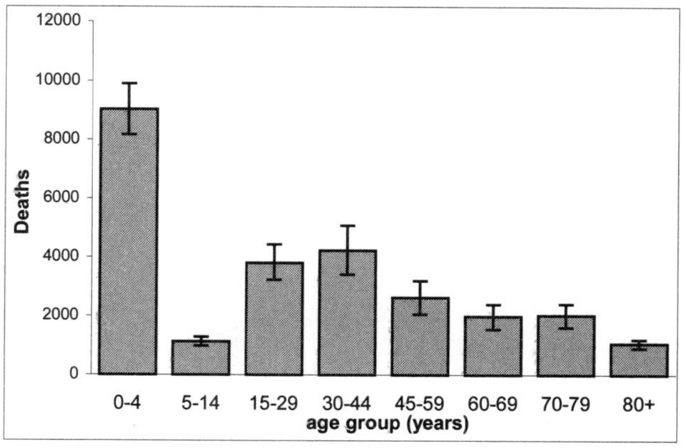

Estonia   Estonie   Estonia

Males, 2000   *Hommes, 2000*   *Hombres, 2000*

Age	$_nM_x$	$_nq_x$ mean	$_nq_x$ percentiles 2½	$_nq_x$ percentiles 97½	l_x mean	l_x percentiles 2½	l_x percentiles 97½	$_nd_x$	$_nL_x$	e_x mean	e_x percentiles 2½	e_x percentiles 97½	Actual deaths	Actual population
<1	0.0118	0.0117	0.0101	0.0133	100,000	100,000	100,000	1,172	98,945	65.4	64.8	66.1	76	6,413
1-4	0.0007	0.0026	0.0023	0.0029	98,828	98,665	98,988	260	394,685	65.2	64.6	65.9	16	24,622
5-9	0.0003	0.0013	0.0011	0.0014	98,567	98,377	98,755	125	492,524	61.4	60.8	62.1	10	39,945
10-14	0.0001	0.0005	0.0005	0.0006	98,442	98,240	98,643	53	492,079	56.4	55.9	57.2	6	55,253
15-19	0.0013	0.0064	0.0058	0.0071	98,389	98,181	98,595	631	490,368	51.5	50.9	52.2	69	53,552
20-24	0.0020	0.0101	0.0091	0.0111	97,758	97,484	98,023	984	486,330	46.8	46.2	47.5	103	50,952
25-29	0.0021	0.0106	0.0097	0.0116	96,774	96,405	97,124	1,028	481,300	42.2	41.7	42.9	109	50,972
30-34	0.0028	0.0140	0.0129	0.0152	95,746	95,286	96,181	1,343	475,373	37.7	37.1	38.4	143	50,562
35-39	0.0054	0.0265	0.0246	0.0286	94,403	93,835	94,940	2,504	465,755	33.2	32.7	33.8	267	49,621
40-44	0.0083	0.0407	0.0382	0.0435	91,899	91,153	92,601	3,742	450,141	29.0	28.5	29.6	414	49,781
45-49	0.0114	0.0554	0.0522	0.0586	88,158	87,192	89,062	4,883	428,581	25.1	24.7	25.7	526	46,180
50-54	0.0166	0.0797	0.0750	0.0838	83,275	82,089	84,377	6,639	399,778	21.5	21.1	22.0	661	39,800
55-59	0.0258	0.1213	0.1142	0.1270	76,636	75,247	77,914	9,296	359,942	18.1	17.7	18.6	825	31,934
60-64	0.0338	0.1560	0.1473	0.1624	67,341	65,796	68,829	10,504	310,442	15.3	14.9	15.7	1,145	33,828
65-69	0.0457	0.2050	0.1946	0.2134	56,836	55,227	58,691	11,652	255,049	12.6	12.3	13.0	1,201	26,278
70-74	0.0634	0.2737	0.2618	0.2850	45,184	43,518	47,268	12,365	195,005	10.2	9.9	10.5	1,316	20,748
75-79	0.0857	0.3530	0.3392	0.3665	32,819	31,296	34,892	11,585	135,131	8.1	7.9	8.4	895	10,439
80-84	0.1274	0.4831	0.4672	0.4982	21,234	20,044	23,000	10,258	80,525	6.2	6.0	6.4	563	4,421
85-89	0.1866	0.6363	0.6207	0.6506	10,976	10,077	12,158	6,984	37,420	4.6	4.5	4.8	496	2,660
90-94	0.2695	0.7452	0.7319	0.7572	3,992	3,521	4,538	2,975	11,037	3.4	3.3	3.5	214	793
95-99	0.3838	0.8920	0.8833	0.8997	1,017	855	1,212	907	2,364	2.5	2.5	2.6	51	133
100+	0.5388	1	1	1	110	86	141	110	204	1.9	1.8	1.9	8	14
All ages													*9,112*	*648,901*

ln($_nq_x$) with uncertainty interval

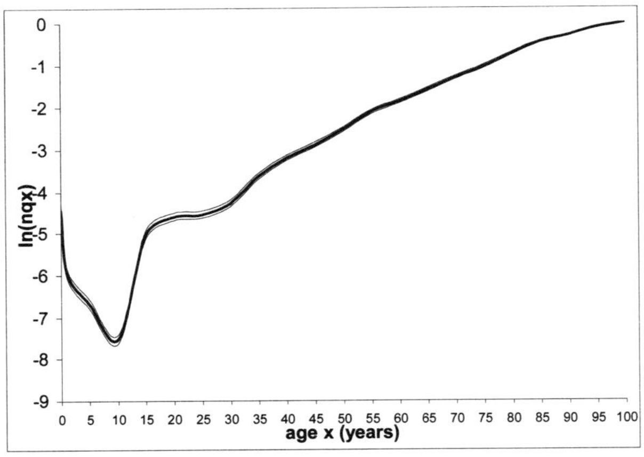

l_x with uncertainty interval

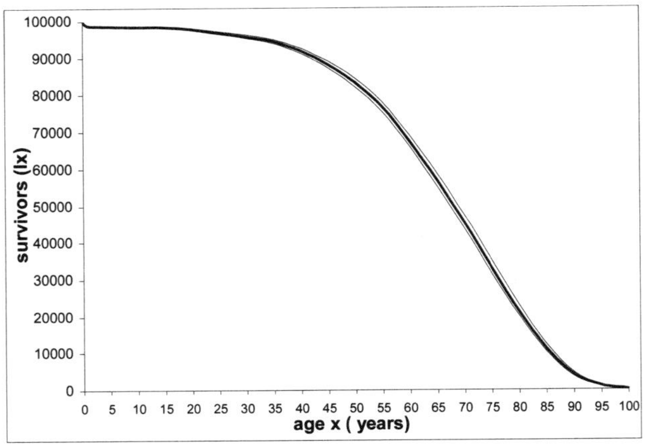

e_x with uncertainty interval

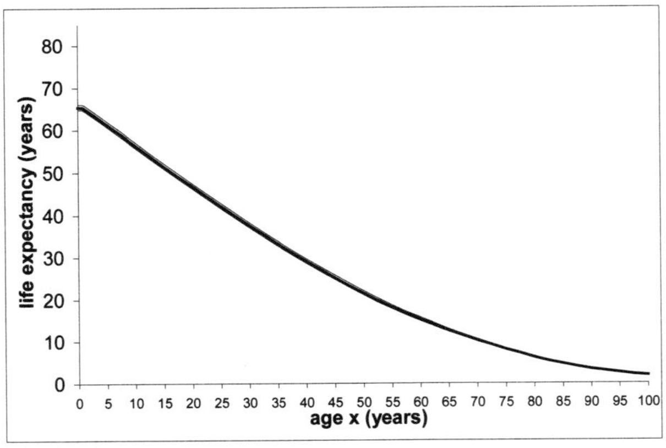

Distribution of deaths by age with uncertainty interval

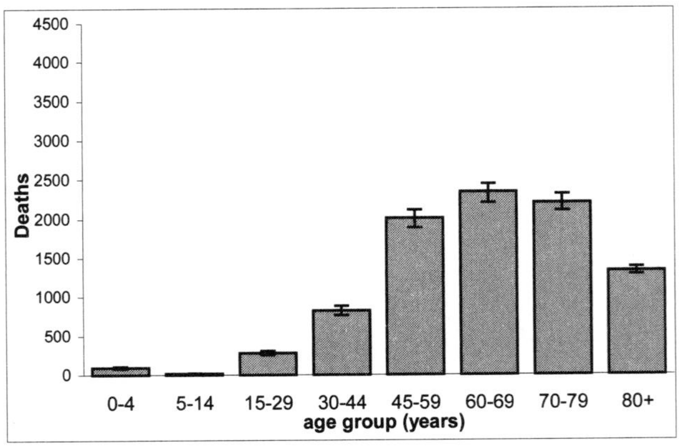

Estonia — Estonie — Estonia

Females, 2000 — *Femmes, 2000* — *Mujeres, 2000*

Age	$_nM_x$	$_nq_x$ mean	$_nq_x$ percentiles 2½	$_nq_x$ percentiles 97½	l_x mean	l_x percentiles 2½	l_x percentiles 97½	$_nd_x$	$_nL_x$	e_x mean	e_x percentiles 2½	e_x percentiles 97½	Actual deaths	Actual population
<1	0.0071	0.0071	0.0055	0.0087	100,000	100,000	100,000	708	99,363	76.5	75.5	77.6	43	6,090
1-4	0.0006	0.0024	0.0020	0.0028	99,292	99,131	99,452	239	396,594	76.0	75.0	77.2	14	23,466
5-9	0.0003	0.0013	0.0010	0.0015	99,053	98,854	99,253	124	494,954	72.2	71.2	73.4	9	37,910
10-14	0.0001	0.0007	0.0006	0.0008	98,929	98,710	99,148	68	494,475	67.3	66.3	68.5	7	52,834
15-19	0.0004	0.0019	0.0016	0.0022	98,861	98,631	99,091	186	493,841	62.3	61.3	63.6	20	52,201
20-24	0.0004	0.0022	0.0019	0.0025	98,675	98,417	98,933	216	492,837	57.4	56.4	58.7	22	49,470
25-29	0.0005	0.0023	0.0020	0.0027	98,460	98,170	98,747	229	491,726	52.5	51.6	53.8	23	49,538
30-34	0.0009	0.0046	0.0040	0.0053	98,231	97,909	98,551	456	490,015	47.7	46.7	48.9	44	46,878
35-39	0.0016	0.0080	0.0070	0.0090	97,775	97,394	98,157	782	486,921	42.9	41.9	44.1	82	50,785
40-44	0.0024	0.0119	0.0105	0.0134	96,993	96,519	97,470	1,156	482,075	38.2	37.2	39.5	128	53,331
45-49	0.0039	0.0191	0.0170	0.0211	95,837	95,244	96,439	1,828	474,615	33.6	32.7	34.9	200	51,928
50-54	0.0058	0.0287	0.0258	0.0315	94,009	93,231	94,778	2,694	463,311	29.2	28.3	30.5	277	47,647
55-59	0.0082	0.0404	0.0361	0.0447	91,315	90,304	92,310	3,687	447,360	25.0	24.1	26.2	334	40,568
60-64	0.0117	0.0570	0.0501	0.0635	87,629	86,359	88,879	4,991	425,665	21.0	20.0	22.1	553	47,127
65-69	0.0182	0.0871	0.0755	0.0972	82,637	81,104	84,215	7,195	395,199	17.1	16.2	18.1	748	41,069
70-74	0.0294	0.1370	0.1179	0.1546	75,442	73,253	77,578	10,338	351,366	13.5	12.7	14.4	1,142	38,825
75-79	0.0502	0.2231	0.1926	0.2522	65,104	62,080	68,434	14,524	289,210	10.2	9.5	11.0	1,423	28,335
80-84	0.0897	0.3662	0.3230	0.4072	50,580	46,840	55,256	18,522	206,595	7.4	6.9	8.0	1,191	13,283
85-89	0.1515	0.5493	0.5034	0.5917	32,058	27,769	37,408	17,609	116,267	5.3	4.9	5.7	1,419	9,369
90-94	0.2421	0.7012	0.6657	0.7340	14,449	11,339	18,577	10,132	41,849	3.7	3.5	3.9	791	3,268
95-99	0.3663	0.8725	0.8522	0.8915	4,317	3,016	6,210	3,767	10,285	2.6	2.5	2.7	216	589
100+	0.5243	1	1	1	550	327	918	550	1,049	1.9	1.9	2.0	30	58
All ages													*8,716*	*744,569*

ln($_nq_x$) with uncertainty interval

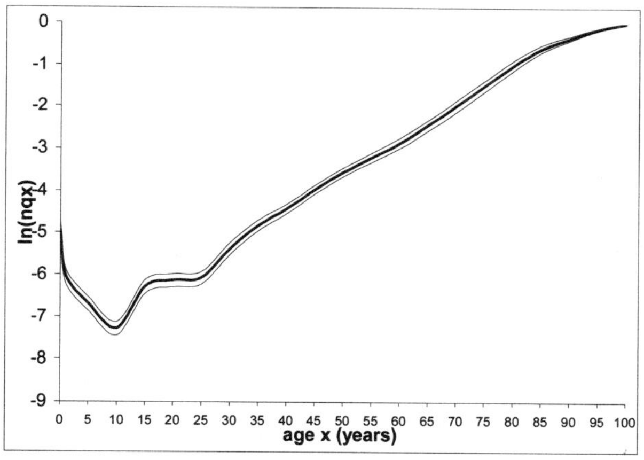

l_x with uncertainty interval

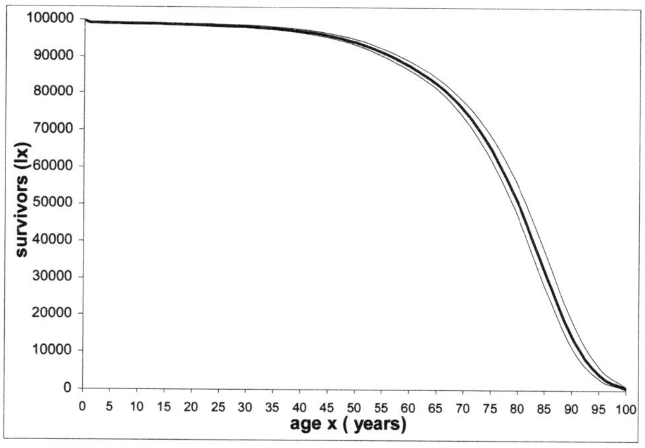

e_x with uncertainty interval

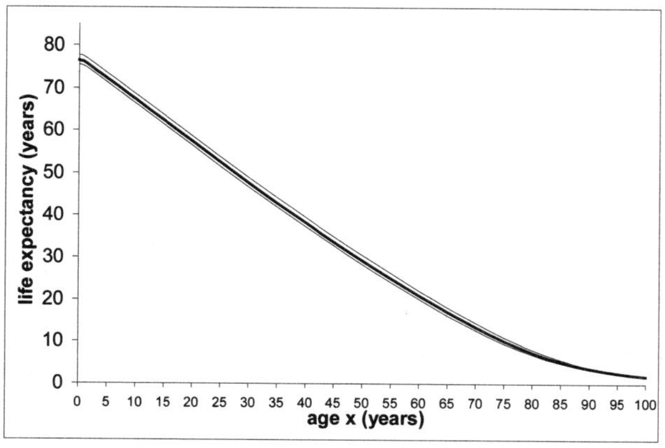

Distribution of deaths by age with uncertainty interval

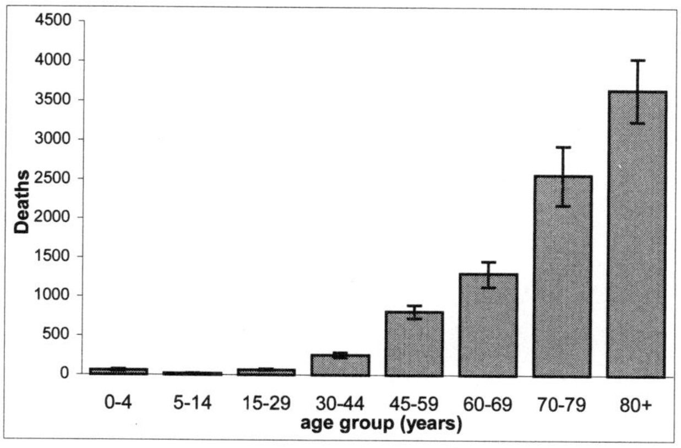

Ethiopia — Ethiopie — Etiopía

Males, 2000 — *Hommes, 2000* — *Hombres, 2000*

Age	$_nM_x$	$_nq_x$ mean	$_nq_x$ percentiles 2½	$_nq_x$ percentiles 97½	l_x mean	l_x percentiles 2½	l_x percentiles 97½	$_nd_x$	$_nL_x$	e_x mean	e_x percentiles 2½	e_x percentiles 97½	Actual deaths	Actual population
<1	0.1573	0.1378	0.1072	0.1689	100,000	100,000	100,000	13,782	87,597	42.8	37.9	48.6	200,875	1,276,764
1-4	0.0147	0.0569	0.0435	0.0713	86,218	83,113	89,277	4,909	333,093	48.7	43.4	54.1	64,118	4,350,901
5-9	0.0034	0.0166	0.0125	0.0214	81,310	78,320	84,305	1,353	403,167	47.5	42.6	52.5	15,559	4,637,394
10-14	0.0023	0.0113	0.0084	0.0146	79,957	76,785	83,244	902	397,530	43.3	38.5	48.2	9,059	3,991,981
15-19	0.0030	0.0147	0.0109	0.0191	79,055	75,740	82,544	1,160	392,374	38.7	34.0	43.6	9,605	3,247,796
20-24	0.0071	0.0347	0.0258	0.0456	77,895	74,399	81,642	2,706	382,709	34.3	29.6	39.0	19,499	2,758,131
25-29	0.0137	0.0661	0.0489	0.0875	75,189	71,291	79,538	4,969	363,521	30.4	25.8	35.1	31,137	2,277,761
30-34	0.0209	0.0993	0.0732	0.1323	70,220	65,231	75,652	6,972	333,669	27.4	22.9	31.9	39,283	1,880,169
35-39	0.0239	0.1126	0.0830	0.1506	63,248	56,600	70,112	7,121	298,437	25.1	21.0	29.4	37,551	1,573,699
40-44	0.0255	0.1201	0.0890	0.1623	56,127	48,077	64,290	6,740	263,785	23.0	19.2	27.0	32,606	1,276,185
45-49	0.0271	0.1268	0.0932	0.1724	49,387	40,378	58,571	6,261	231,283	20.8	17.5	24.4	28,435	1,050,320
50-54	0.0285	0.1329	0.0969	0.1806	43,126	33,604	53,025	5,730	201,305	18.4	15.6	21.6	24,823	872,111
55-59	0.0306	0.1419	0.1031	0.1918	37,396	27,758	47,707	5,307	173,712	15.9	13.4	18.7	21,206	694,088
60-64	0.0421	0.1904	0.1391	0.2526	32,089	22,666	42,531	6,110	145,169	13.1	11.0	15.6	22,158	526,474
65-69	0.0572	0.2503	0.1851	0.3236	25,979	17,212	36,206	6,502	113,641	10.6	8.9	12.7	21,805	381,132
70-74	0.0837	0.3461	0.2628	0.4313	19,477	11,815	28,933	6,741	80,535	8.3	7.0	10.0	20,875	249,390
75-79	0.1231	0.4705	0.3721	0.5610	12,736	6,719	20,648	5,993	48,700	6.4	5.4	7.7	17,019	138,291
80-84	0.1806	0.6221	0.5188	0.7072	6,743	2,949	12,777	4,195	23,230	4.8	4.2	5.7	10,578	58,575
85-89	0.2596	0.7871	0.6973	0.8531	2,549	864	6,100	2,006	7,728	3.6	3.2	4.2	4,024	15,502
90-94	0.3603	0.8657	0.7932	0.9155	543	127	1,847	470	1,304	2.7	2.4	3.1	803	2,228
95-99	0.4830	0.9861	0.9659	0.9948	73	11	382	72	149	2.1	2.0	2.2	75	156
100+	0.6251	1	1	1	1	0	13	1	2	1.6	1.6	1.7	3	5
All ages													*631,094*	*31,259,053*

ln($_nq_x$) with uncertainty interval

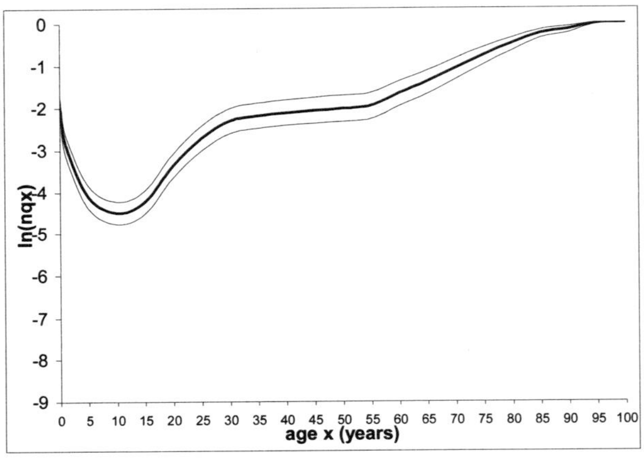

l_x with uncertainty interval

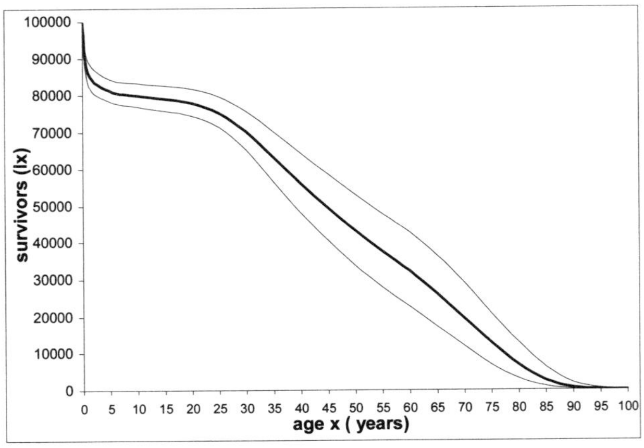

e_x with uncertainty interval

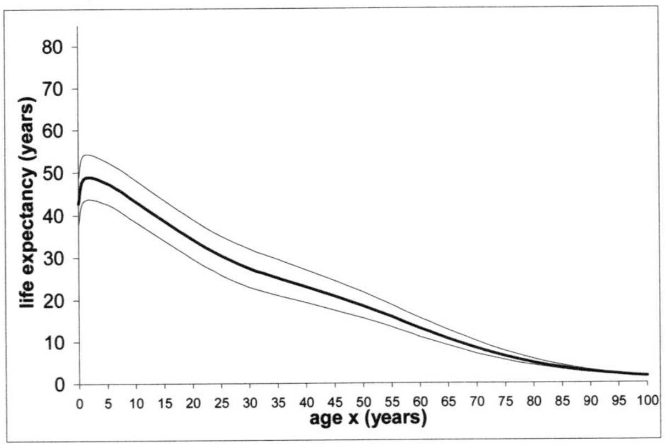

Distribution of deaths by age with uncertainty interval

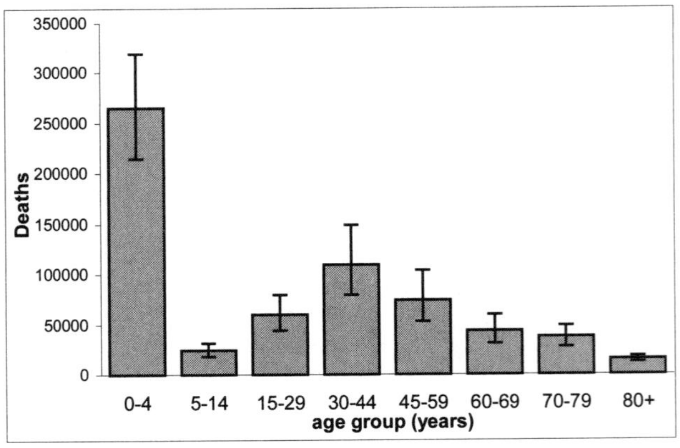

Ethiopia Ethiopie Etiopía

Females, 2000 *Femmes, 2000* *Mujeres, 2000*

Age	$_nM_x$	$_nq_x$ mean	$_nq_x$ percentiles 21/2	$_nq_x$ percentiles 971/2	l_x mean	l_x percentiles 21/2	l_x percentiles 971/2	$_nd_x$	$_nL_x$	e_x mean	e_x percentiles 21/2	e_x percentiles 971/2	Actual deaths	Actual population
<1	0.1301	0.1165	0.0912	0.1456	100,000	100,000	100,000	11,650	89,515	44.7	39.7	50.6	159,394	1,224,713
1-4	0.0160	0.0618	0.0483	0.0755	88,350	85,436	90,885	5,456	340,304	49.5	44.2	55.6	69,636	4,343,000
5-9	0.0035	0.0174	0.0133	0.0218	82,893	80,113	85,693	1,445	410,854	48.7	43.6	54.3	16,211	4,609,022
10-14	0.0025	0.0122	0.0093	0.0155	81,448	78,450	84,453	996	404,751	44.5	39.5	50.0	9,798	3,980,242
15-19	0.0054	0.0267	0.0201	0.0340	80,452	77,307	83,595	2,148	396,890	40.0	35.0	45.4	17,585	3,249,133
20-24	0.0117	0.0568	0.0424	0.0730	78,304	74,784	81,857	4,447	380,401	36.0	31.1	41.3	32,273	2,760,584
25-29	0.0184	0.0879	0.0651	0.1142	73,857	69,490	78,323	6,490	353,058	33.1	28.3	38.1	41,862	2,277,161
30-34	0.0219	0.1036	0.0763	0.1357	67,366	61,893	73,107	6,979	319,383	31.0	26.5	35.7	41,286	1,889,286
35-39	0.0203	0.0967	0.0710	0.1285	60,387	53,687	67,493	5,839	287,337	29.3	25.2	33.5	32,542	1,601,354
40-44	0.0182	0.0871	0.0639	0.1170	54,548	46,861	62,704	4,751	260,861	27.2	23.5	30.9	24,128	1,324,647
45-49	0.0169	0.0811	0.0594	0.1100	49,796	41,494	58,695	4,040	238,883	24.5	21.2	27.9	18,872	1,116,032
50-54	0.0185	0.0884	0.0647	0.1202	45,757	37,081	55,193	4,047	218,668	21.5	18.6	24.5	17,012	919,278
55-59	0.0217	0.1029	0.0754	0.1396	41,710	32,808	51,581	4,291	197,822	18.3	15.8	21.0	16,316	752,093
60-64	0.0296	0.1380	0.1017	0.1857	37,419	28,450	47,622	5,164	174,183	15.1	12.9	17.5	17,460	588,931
65-69	0.0431	0.1947	0.1451	0.2573	32,255	23,457	42,648	6,279	145,575	12.2	10.3	14.2	18,823	436,364
70-74	0.0666	0.2853	0.2175	0.3656	25,975	17,705	36,228	7,410	111,350	9.5	8.0	11.2	19,674	295,624
75-79	0.1003	0.4011	0.3159	0.4929	18,565	11,369	27,969	7,446	74,209	7.3	6.1	8.7	17,347	172,892
80-84	0.1506	0.5471	0.4510	0.6396	11,119	5,791	18,692	6,083	40,388	5.5	4.7	6.5	11,945	79,309
85-89	0.2202	0.7102	0.6187	0.7893	5,036	2,087	10,070	3,577	16,240	4.1	3.6	4.8	5,308	24,103
90-94	0.3131	0.8073	0.7301	0.8703	1,460	440	3,775	1,178	3,763	3.0	2.7	3.5	1,396	4,459
95-99	0.4329	0.9416	0.8980	0.9700	281	57	998	265	612	2.3	2.1	2.5	209	483
100+	0.5817	1	1	1	16	2	102	16	28	1.7	1.7	1.9	15	25
All ages													*589,091*	*31,648,735*

ln($_nq_x$) with uncertainty interval

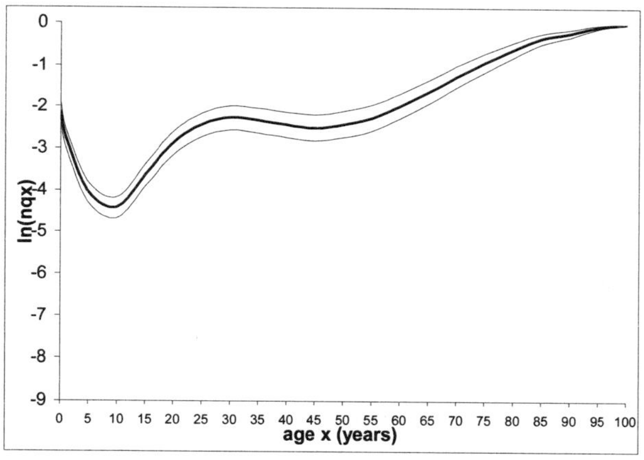

l_x with uncertainty interval

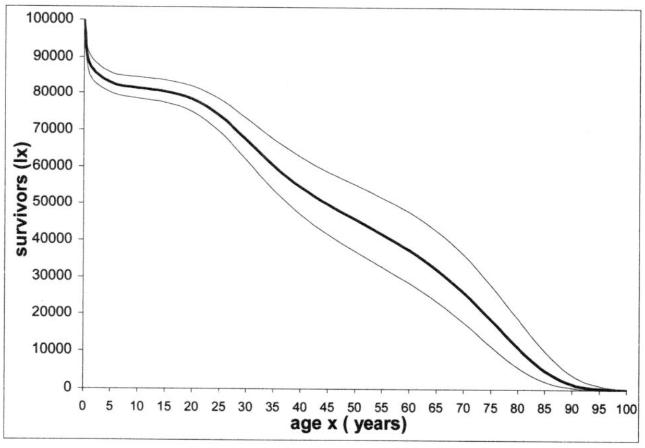

e_x with uncertainty interval

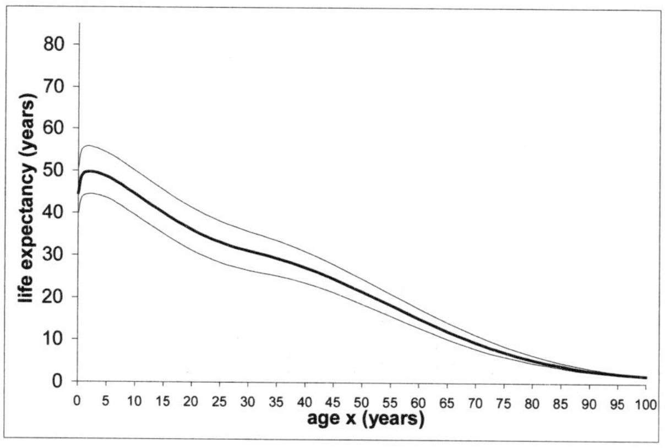

Distribution of deaths by age with uncertainty interval

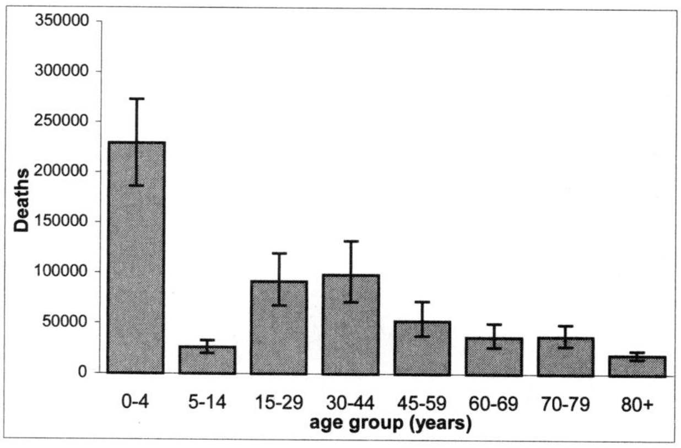

Fiji | Fidji | Fiji

Males, 2000 | *Hommes, 2000* | *Hombres, 2000*

Age	$_nM_x$	$_nq_x$			l_x			$_nd_x$	$_nL_x$	e_x			Actual deaths	Actual population
			percentiles			percentiles					percentiles			
		mean	2½	97½	mean	2½	97½			mean	2½	97½		
<1	0.0204	0.0201	0.0146	0.0257	100,000	100,000	100,000	2,007	98,194	66.9	65.7	68.1	203	9,921
1-4	0.0018	0.0071	0.0056	0.0085	97,993	97,428	98,540	692	390,310	67.3	66.3	68.4	71	39,884
5-9	0.0010	0.0051	0.0041	0.0060	97,301	96,595	97,990	491	485,274	63.8	62.8	64.8	47	46,326
10-14	0.0006	0.0029	0.0024	0.0035	96,809	96,012	97,592	285	483,333	59.1	58.1	60.1	26	43,701
15-19	0.0012	0.0062	0.0051	0.0073	96,524	95,675	97,359	599	481,125	54.3	53.3	55.2	56	44,903
20-24	0.0013	0.0067	0.0056	0.0079	95,926	94,973	96,864	643	478,020	49.6	48.7	50.5	54	40,059
25-29	0.0017	0.0082	0.0069	0.0096	95,282	94,225	96,324	784	474,451	44.9	44.0	45.8	54	32,662
30-34	0.0025	0.0125	0.0107	0.0144	94,498	93,321	95,657	1,181	469,537	40.3	39.4	41.1	75	29,746
35-39	0.0034	0.0167	0.0145	0.0191	93,317	91,974	94,635	1,556	462,693	35.7	34.9	36.6	97	28,910
40-44	0.0050	0.0247	0.0219	0.0279	91,760	90,221	93,263	2,271	453,124	31.3	30.5	32.1	133	26,472
45-49	0.0076	0.0374	0.0337	0.0416	89,489	87,702	91,221	3,351	439,070	27.0	26.3	27.8	157	20,575
50-54	0.0141	0.0682	0.0621	0.0747	86,139	84,054	88,149	5,876	416,003	23.0	22.3	23.7	233	16,528
55-59	0.0180	0.0863	0.0790	0.0939	80,263	77,777	82,673	6,924	384,005	19.5	18.9	20.2	222	12,321
60-64	0.0266	0.1245	0.1150	0.1344	73,339	70,477	76,140	9,131	343,867	16.1	15.5	16.7	236	8,895
65-69	0.0424	0.1917	0.1776	0.2047	64,208	61,007	67,382	12,306	290,275	13.0	12.5	13.6	246	5,809
70-74	0.0542	0.2387	0.2215	0.2559	51,902	48,518	55,412	12,388	228,542	10.5	10.1	11.0	198	3,655
75-79	0.0875	0.3588	0.3345	0.3829	39,515	36,102	43,141	14,179	162,126	8.0	7.6	8.5	170	1,940
80-84	0.1306	0.4922	0.4635	0.5201	25,336	22,280	28,709	12,469	95,506	6.1	5.8	6.4	154	1,178
85-89	0.1898	0.6436	0.6153	0.6726	12,867	10,692	15,404	8,281	43,630	4.6	4.4	4.9	40	210
90-94	0.2687	0.7439	0.7190	0.7694	4,585	3,501	5,926	3,411	12,694	3.4	3.3	3.6	14	51
95-99	0.3704	0.8772	0.8596	0.8947	1,174	807	1,665	1,030	2,781	2.6	2.5	2.7	1	4
100+	0.4972	1	1	1	144	85	234	144	290	2.0	1.9	2.1	0	0
All ages													*2,486*	*413,750*

ln($_nq_x$) with uncertainty interval

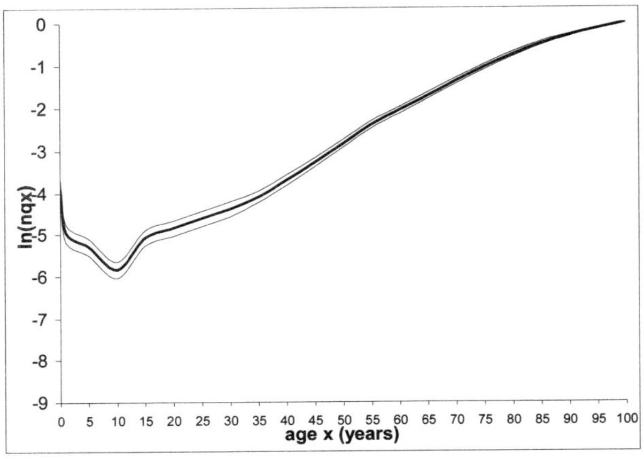

l_x with uncertainty interval

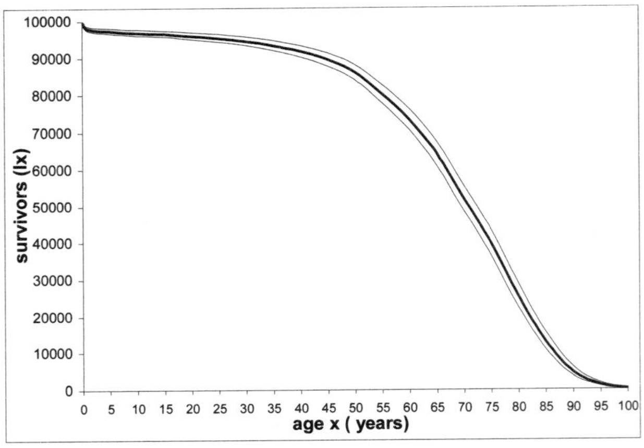

e_x with uncertainty interval

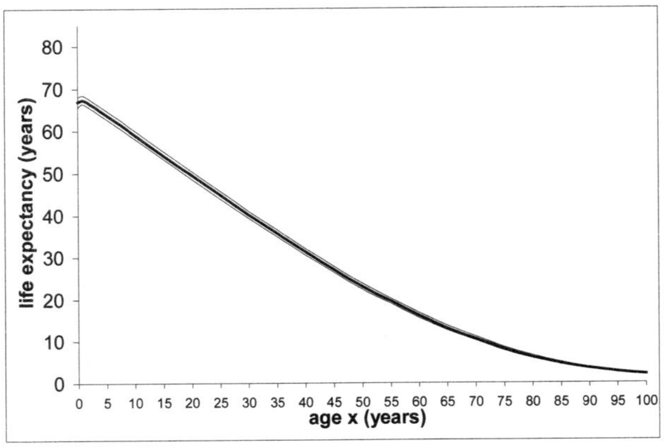

Distribution of deaths by age with uncertainty interval

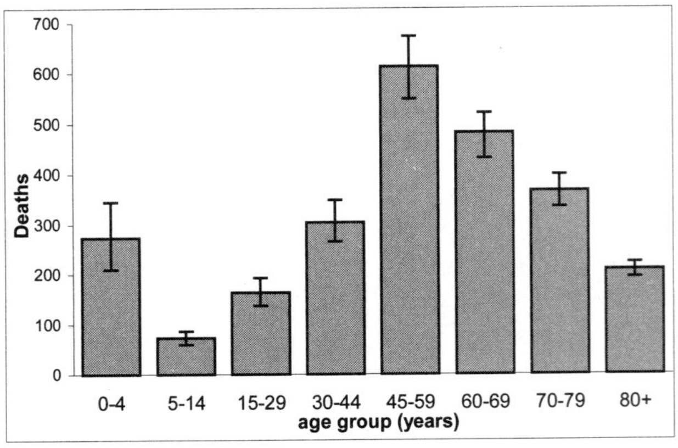

Fiji **Fidji** **Fiji**

Females, 2000 *Femmes, 2000* *Mujeres, 2000*

Age	$_nM_x$	$_nq_x$			l_x			$_nd_x$	$_nL_x$	e_x			Actual deaths	Actual population
			percentiles			percentiles					percentiles			
		mean	2½	97½	mean	2½	97½			mean	2½	97½		
<1	0.0191	0.0188	0.0142	0.0238	100,000	100,000	100,000	1,877	98,310	71.2	69.9	72.4	180	9,409
1-4	0.0013	0.0052	0.0042	0.0063	98,123	97,616	98,576	512	391,262	71.5	70.5	72.7	49	37,437
5-9	0.0005	0.0027	0.0022	0.0032	97,611	96,999	98,160	260	487,404	67.9	66.9	69.1	23	43,547
10-14	0.0004	0.0019	0.0016	0.0023	97,351	96,688	97,947	188	486,283	63.1	62.1	64.2	16	41,102
15-19	0.0008	0.0041	0.0034	0.0049	97,163	96,465	97,792	399	484,815	58.2	57.3	59.3	35	42,556
20-24	0.0010	0.0050	0.0041	0.0059	96,764	95,992	97,460	479	482,620	53.4	52.5	54.6	38	38,240
25-29	0.0015	0.0077	0.0065	0.0090	96,284	95,428	97,058	739	479,574	48.7	47.8	49.8	47	30,802
30-34	0.0017	0.0085	0.0072	0.0099	95,545	94,566	96,431	808	475,704	44.0	43.2	45.1	49	28,640
35-39	0.0023	0.0113	0.0098	0.0131	94,737	93,634	95,737	1,073	471,001	39.4	38.5	40.5	64	28,203
40-44	0.0035	0.0173	0.0151	0.0198	93,664	92,408	94,803	1,625	464,257	34.8	34.0	35.9	89	25,383
45-49	0.0058	0.0285	0.0253	0.0322	92,039	90,574	93,369	2,628	453,625	30.4	29.6	31.4	118	20,301
50-54	0.0096	0.0470	0.0424	0.0520	89,411	87,660	91,008	4,198	436,561	26.2	25.5	27.2	162	16,872
55-59	0.0133	0.0645	0.0581	0.0704	85,213	83,106	87,148	5,493	412,334	22.4	21.7	23.3	169	12,653
60-64	0.0192	0.0914	0.0823	0.0993	79,720	77,255	82,085	7,290	380,378	18.7	18.2	19.6	184	9,609
65-69	0.0321	0.1486	0.1343	0.1599	72,431	69,583	75,329	10,766	335,239	15.4	14.8	16.1	203	6,314
70-74	0.0433	0.1954	0.1779	0.2086	61,665	58,459	65,212	12,051	278,197	12.6	12.1	13.2	187	4,316
75-79	0.0590	0.2571	0.2369	0.2754	49,614	46,262	53,612	12,755	216,182	10.1	9.6	10.6	139	2,360
80-84	0.0900	0.3673	0.3429	0.3914	36,859	33,522	40,912	13,540	150,445	7.7	7.3	8.1	153	1,695
85-89	0.1400	0.5185	0.4898	0.5452	23,319	20,403	26,883	12,091	86,367	5.7	5.5	6.0	45	321
90-94	0.2100	0.6442	0.6165	0.6684	11,228	9,279	13,716	7,233	34,441	4.2	4.0	4.4	19	90
95-99	0.3000	0.7895	0.7665	0.8084	3,995	3,076	5,259	3,154	10,513	3.1	3.0	3.2	2	7
100+	0.4335	1	1	1	841	589	1,228	841	1,940	2.3	2.2	2.4	0	0
All ages													*1,970*	*399,857*

ln($_nq_x$) with uncertainty interval

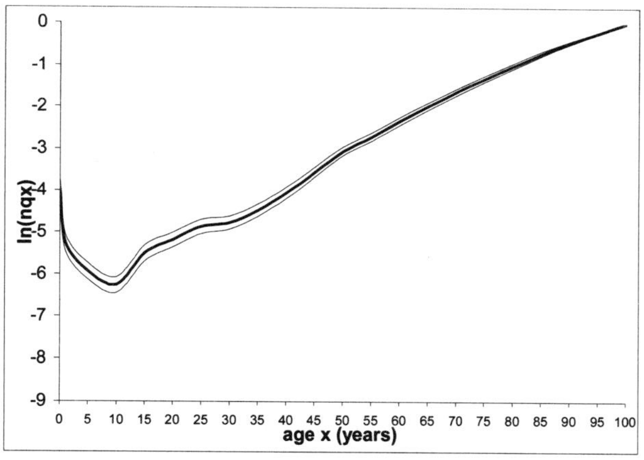

l_x with uncertainty interval

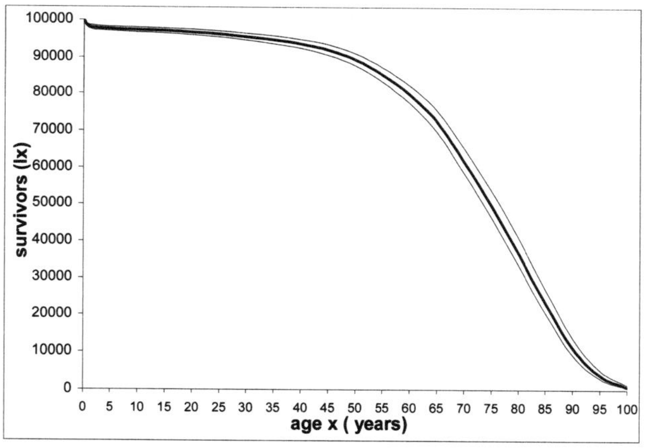

e_x with uncertainty interval

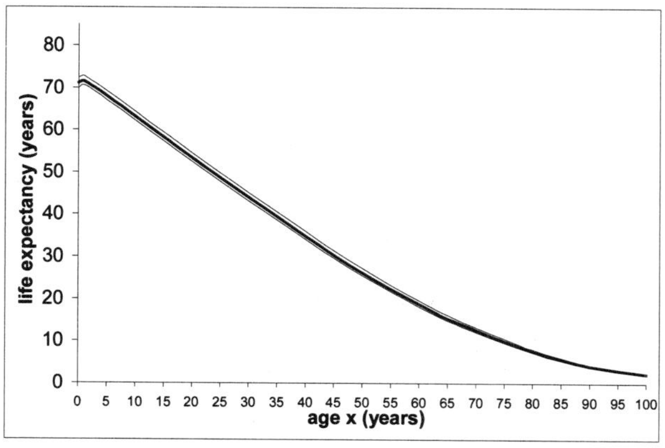

Distribution of deaths by age with uncertainty interval

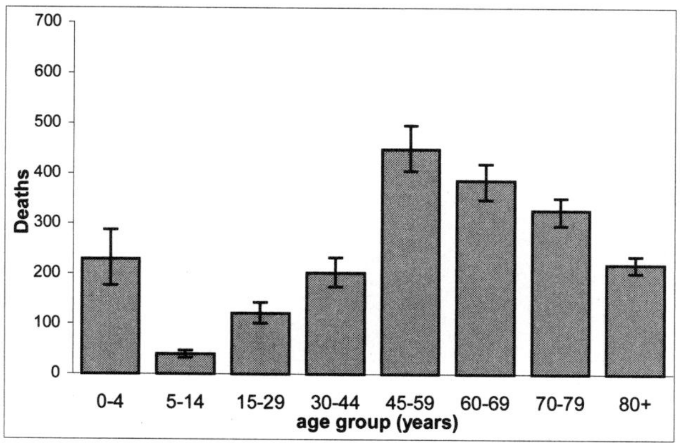

Finland | Finlande | Finlandia

Males, 2000 | *Hommes, 2000* | *Hombres, 2000*

Age	$_nM_x$	$_nq_x$			l_x			$_nd_x$	$_nL_x$	e_x			Actual deaths	Actual population
			percentiles			percentiles					percentiles			
		mean	2 1/2	97 1/2	mean	2 1/2	97 1/2			mean	2 1/2	97 1/2		
<1	0.0050	0.0050	0.0043	0.0056	100,000	100,000	100,000	498	99,552	73.7	73.4	74.0	144	28,797
1-4	0.0002	0.0008	0.0008	0.0009	99,502	99,436	99,569	84	397,808	73.1	72.7	73.4	25	119,273
5-9	0.0001	0.0005	0.0005	0.0006	99,419	99,344	99,493	55	496,957	69.1	68.8	69.5	18	166,758
10-14	0.0002	0.0008	0.0008	0.0009	99,364	99,284	99,444	84	496,609	64.2	63.8	64.5	28	161,776
15-19	0.0007	0.0033	0.0031	0.0036	99,280	99,192	99,368	332	495,568	59.2	58.9	59.6	113	169,392
20-24	0.0009	0.0046	0.0042	0.0049	98,948	98,832	99,064	454	493,603	54.4	54.1	54.8	154	167,240
25-29	0.0011	0.0055	0.0052	0.0059	98,493	98,345	98,644	545	491,105	49.6	49.3	50.0	172	155,301
30-34	0.0014	0.0072	0.0068	0.0076	97,948	97,767	98,133	703	487,985	44.9	44.6	45.3	258	179,361
35-39	0.0020	0.0100	0.0095	0.0104	97,246	97,028	97,467	968	483,809	40.2	39.9	40.6	388	193,839
40-44	0.0030	0.0149	0.0144	0.0154	96,278	96,020	96,541	1,433	477,807	35.6	35.2	36.0	586	195,377
45-49	0.0050	0.0247	0.0241	0.0253	94,845	94,541	95,152	2,342	468,368	31.1	30.7	31.5	1,018	203,588
50-54	0.0070	0.0344	0.0335	0.0352	92,503	92,152	92,852	3,182	454,559	26.8	26.4	27.2	1,507	215,215
55-59	0.0100	0.0488	0.0472	0.0502	89,321	88,944	89,684	4,357	435,711	22.7	22.3	23.1	1,476	147,571
60-64	0.0160	0.0769	0.0741	0.0798	84,964	84,600	85,300	6,536	408,479	18.7	18.4	19.1	1,984	124,004
65-69	0.0260	0.1221	0.1168	0.1274	78,428	78,002	78,918	9,573	368,207	15.1	14.7	15.5	2,698	103,776
70-74	0.0420	0.1900	0.1808	0.1989	68,855	68,118	69,652	13,086	311,560	11.8	11.5	12.2	3,690	87,856
75-79	0.0680	0.2906	0.2765	0.3034	55,769	54,569	56,940	16,206	238,330	9.0	8.8	9.3	3,877	57,022
80-84	0.1120	0.4375	0.4193	0.4541	39,563	38,011	41,154	17,309	154,542	6.7	6.5	6.9	3,181	28,406
85-89	0.1780	0.6159	0.5980	0.6324	22,254	20,792	23,896	13,707	77,004	4.9	4.8	5.1	2,481	13,941
90-94	0.2390	0.6960	0.6814	0.7099	8,547	7,686	9,605	5,949	24,891	3.8	3.7	3.9	983	4,112
95-99	0.3300	0.8291	0.8184	0.8394	2,599	2,232	3,060	2,155	6,529	2.9	2.8	3.0	177	537
100+	0.4670	1	1	1	444	359	556	444	951	2.1	2.1	2.3	15	32
All ages													*24,974*	*2,523,174*

$\ln(_nq_x)$ with uncertainty interval

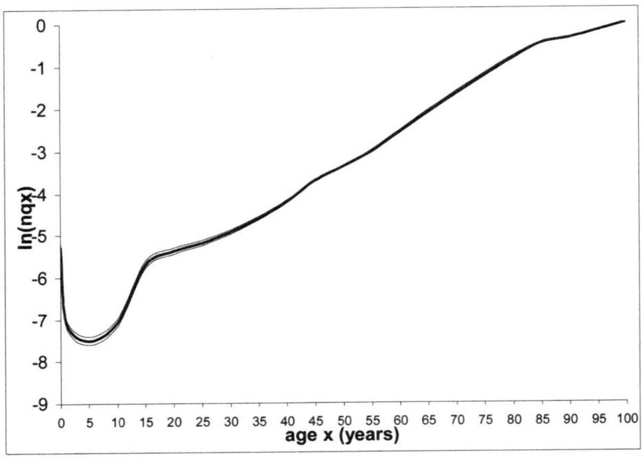

l_x with uncertainty interval

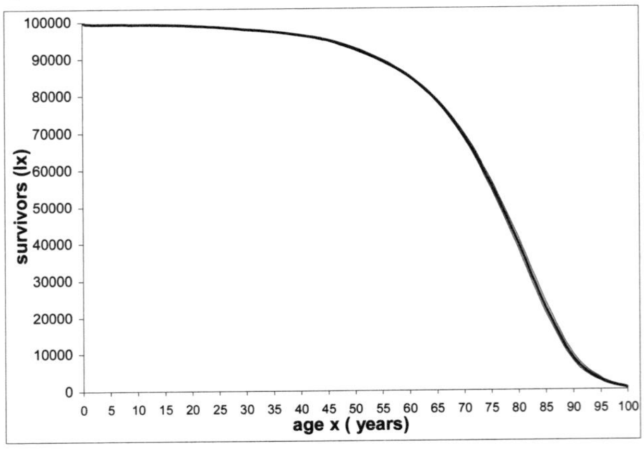

e_x with uncertainty interval

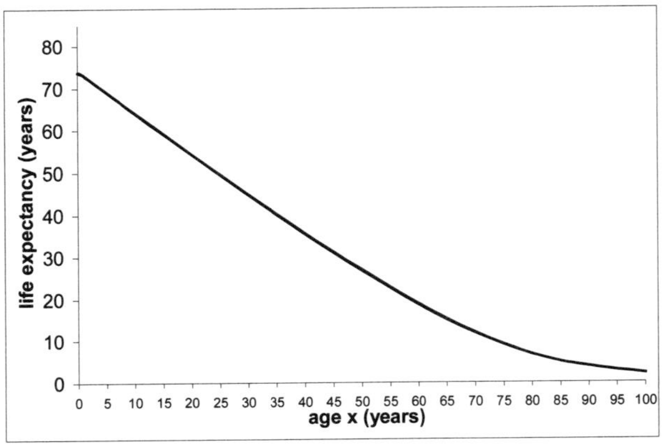

Distribution of deaths by age with uncertainty interval

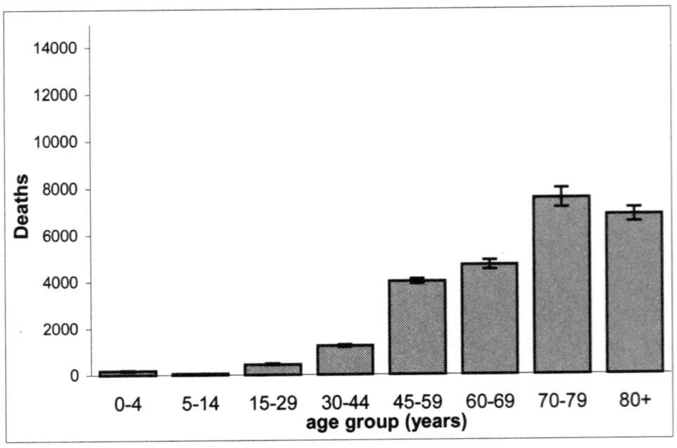

Finland | **Finlande** | **Finlandia**

Females, 2000 | *Femmes, 2000* | *Mujeres, 2000*

Age	$_nM_x$	$_nq_x$			l_x			$_nd_x$	$_nL_x$	e_x			Actual deaths	Actual population
			percentiles			percentiles					percentiles			
		mean	2½	97½	mean	2½	97½			mean	2½	97½		
<1	0.0040	0.0040	0.0035	0.0046	100,000	100,000	100,000	399	99,641	80.9	80.6	81.4	111	27,783
1-4	0.0001	0.0006	0.0005	0.0006	99,601	99,543	99,653	56	398,272	80.2	79.9	80.8	16	113,646
5-9	0.0001	0.0004	0.0004	0.0005	99,546	99,481	99,603	45	497,616	76.3	75.9	76.8	14	160,583
10-14	0.0001	0.0007	0.0006	0.0008	99,501	99,432	99,562	70	497,330	71.3	71.0	71.8	22	154,627
15-19	0.0003	0.0013	0.0012	0.0015	99,431	99,355	99,499	134	496,821	66.4	66.0	66.9	44	161,890
20-24	0.0004	0.0020	0.0019	0.0022	99,297	99,208	99,378	203	495,977	61.5	61.1	62.0	66	160,605
25-29	0.0003	0.0017	0.0016	0.0018	99,094	98,987	99,192	168	495,048	56.6	56.2	57.1	51	149,059
30-34	0.0006	0.0027	0.0025	0.0030	98,925	98,806	99,037	272	493,948	51.7	51.3	52.2	95	172,983
35-39	0.0008	0.0037	0.0035	0.0040	98,654	98,514	98,786	369	492,346	46.8	46.4	47.3	140	186,374
40-44	0.0012	0.0062	0.0058	0.0066	98,285	98,122	98,442	607	489,904	42.0	41.6	42.5	235	189,364
45-49	0.0020	0.0100	0.0094	0.0105	97,677	97,480	97,872	972	485,955	37.2	36.9	37.7	398	198,978
50-54	0.0030	0.0149	0.0142	0.0156	96,705	96,458	96,950	1,440	479,926	32.6	32.2	33.1	633	210,983
55-59	0.0040	0.0198	0.0188	0.0208	95,265	94,961	95,568	1,886	471,611	28.0	27.7	28.5	601	150,214
60-64	0.0060	0.0296	0.0279	0.0309	93,379	93,020	93,738	2,760	459,995	23.5	23.2	24.0	808	134,601
65-69	0.0110	0.0535	0.0503	0.0561	90,619	90,173	91,054	4,851	440,968	19.2	18.9	19.6	1,351	122,783
70-74	0.0210	0.0998	0.0936	0.1045	85,768	85,135	86,422	8,556	407,450	15.1	14.9	15.5	2,562	121,986
75-79	0.0380	0.1735	0.1630	0.1813	77,212	76,309	78,336	13,397	352,565	11.5	11.3	11.9	3,973	104,550
80-84	0.0720	0.3051	0.2892	0.3170	63,814	62,475	65,570	19,469	270,400	8.4	8.2	8.7	4,891	67,930
85-89	0.1290	0.4877	0.4702	0.5035	44,346	42,722	46,608	21,628	167,658	6.0	5.9	6.2	5,426	42,062
90-94	0.1960	0.6171	0.6012	0.6319	22,718	21,399	24,693	14,020	71,529	4.4	4.3	4.5	2,944	15,018
95-99	0.2950	0.7825	0.7702	0.7936	8,698	7,918	9,784	6,806	23,071	3.1	3.1	3.3	717	2,431
100+	0.4480	1	1	1	1,892	1,636	2,211	1,892	4,223	2.2	2.2	2.5	82	182
All ages													*25,177*	*2,648,632*

ln($_nq_x$) with uncertainty interval

l_x with uncertainty interval

e_x with uncertainty interval

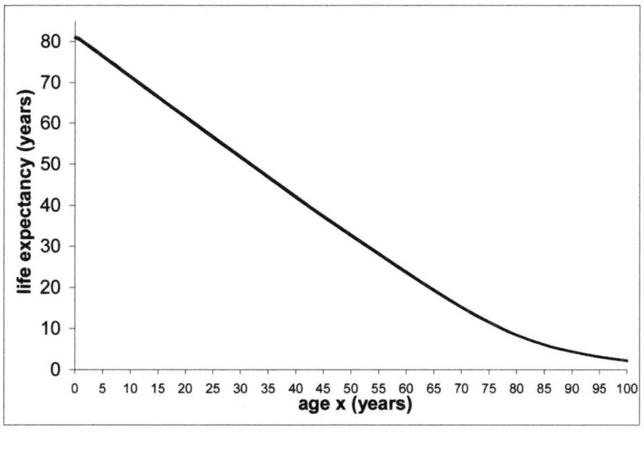

Distribution of deaths by age with uncertainty interval

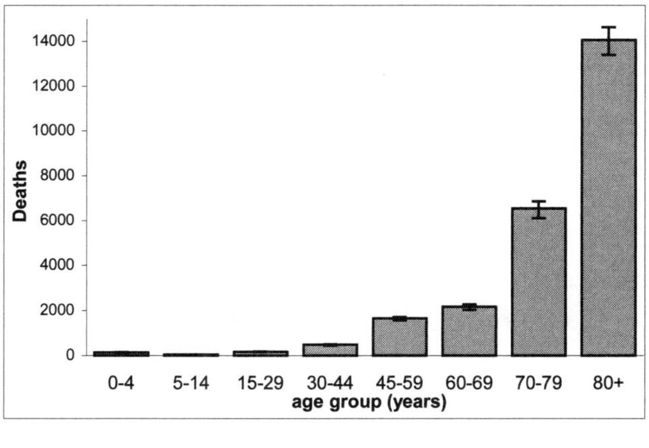

France **France** **Francia**

Males, 2000 *Hommes, 2000* *Hombres, 2000*

Age	$_nM_x$	$_nq_x$			l_x			$_nd_x$	$_nL_x$	e_x			Actual deaths	Actual population
			percentiles			percentiles					percentiles			
		mean	2½	97½	mean	2½	97½			mean	2½	97½		
<1	0.0067	0.0067	0.0063	0.0071	100,000	100,000	100,000	666	99,401	75.2	74.9	75.5	2,446	365,073
1-4	0.0004	0.0016	0.0015	0.0017	99,334	99,292	99,374	159	396,955	74.7	74.4	75.0	599	1,497,046
5-9	0.0002	0.0010	0.0010	0.0010	99,175	99,125	99,223	99	495,628	70.8	70.5	71.1	370	1,850,270
10-14	0.0002	0.0010	0.0010	0.0010	99,076	99,021	99,128	99	495,133	65.9	65.6	66.2	393	1,966,451
15-19	0.0007	0.0035	0.0033	0.0037	98,977	98,917	99,033	346	494,021	60.9	60.7	61.2	1,400	1,999,493
20-24	0.0013	0.0065	0.0062	0.0068	98,631	98,556	98,701	639	491,559	56.1	55.9	56.4	2,526	1,942,942
25-29	0.0013	0.0065	0.0062	0.0067	97,992	97,889	98,087	635	488,374	51.5	51.3	51.8	2,780	2,138,438
30-34	0.0016	0.0080	0.0077	0.0083	97,357	97,229	97,477	776	484,847	46.8	46.6	47.1	3,424	2,140,120
35-39	0.0021	0.0104	0.0101	0.0108	96,582	96,426	96,729	1,009	480,386	42.2	41.9	42.4	4,564	2,173,539
40-44	0.0032	0.0159	0.0154	0.0164	95,573	95,384	95,755	1,517	474,071	37.6	37.4	37.8	6,776	2,117,632
45-49	0.0047	0.0232	0.0225	0.0240	94,056	93,818	94,285	2,185	464,817	33.1	32.9	33.4	9,844	2,094,541
50-54	0.0066	0.0325	0.0316	0.0335	91,871	91,566	92,160	2,983	451,899	28.9	28.7	29.1	13,528	2,049,768
55-59	0.0095	0.0464	0.0452	0.0477	88,889	88,502	89,249	4,124	434,132	24.8	24.6	25.0	13,130	1,382,059
60-64	0.0142	0.0686	0.0667	0.0702	84,764	84,283	85,204	5,812	409,292	20.8	20.7	21.1	18,440	1,298,596
65-69	0.0210	0.0998	0.0970	0.1020	78,952	78,366	79,475	7,876	375,071	17.2	17.1	17.4	26,301	1,252,422
70-74	0.0315	0.1460	0.1421	0.1489	71,076	70,395	71,695	10,377	329,436	13.8	13.7	14.0	33,783	1,072,473
75-79	0.0468	0.2095	0.2045	0.2133	60,699	59,964	61,504	12,716	271,704	10.8	10.6	10.9	39,094	835,347
80-84	0.0822	0.3409	0.3344	0.3471	47,983	47,211	48,929	16,359	199,017	7.9	7.8	8.0	27,946	339,980
85-89	0.1394	0.5169	0.5098	0.5245	31,624	30,927	32,566	16,345	117,255	5.8	5.7	5.8	33,673	241,554
90-94	0.2029	0.6306	0.6243	0.6377	15,278	14,787	15,944	9,635	47,487	4.3	4.2	4.3	16,674	82,177
95-99	0.3006	0.7903	0.7854	0.7957	5,643	5,358	5,966	4,460	14,837	3.1	3.1	3.2	4,415	14,686
100+	0.4524	1	1	1	1,183	1,095	1,273	1,183	2,616	2.2	2.2	2.4	519	1,148
All ages													*262,626*	*28,855,755*

$\ln(_nq_x)$ with uncertainty interval

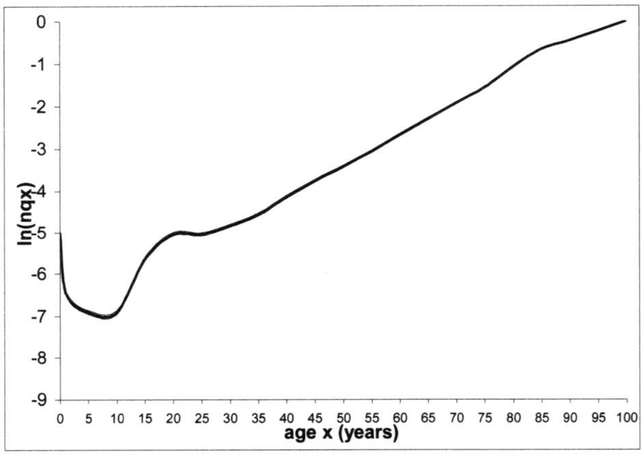

l_x with uncertainty interval

e_x with uncertainty interval

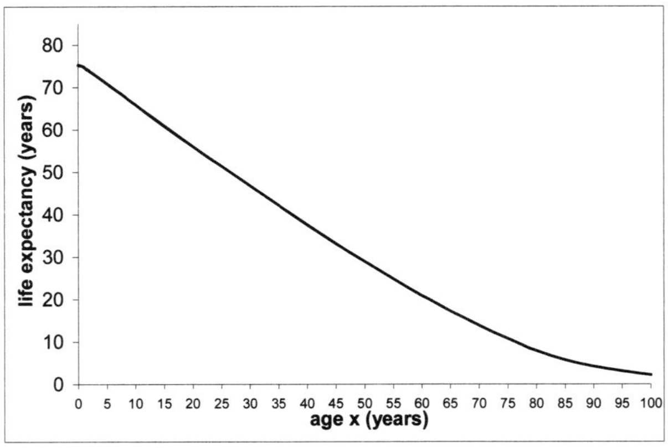

Distribution of deaths by age with uncertainty interval

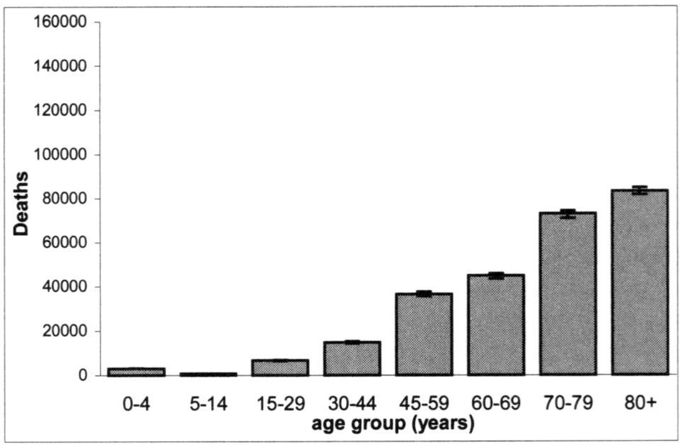

France | France | Francia

Females, 2000 | *Femmes, 2000* | *Mujeres, 2000*

Age	$_nM_x$	$_nq_x$ mean	$_nq_x$ percentiles 2½	$_nq_x$ percentiles 97½	l_x mean	l_x percentiles 2½	l_x percentiles 97½	$_nd_x$	$_nL_x$	e_x mean	e_x percentiles 2½	e_x percentiles 97½	Actual deaths	Actual population
<1	0.0046	0.0046	0.0042	0.0050	100,000	100,000	100,000	458	99,588	83.1	82.6	83.7	1,594	346,534
1-4	0.0003	0.0012	0.0011	0.0013	99,542	99,500	99,583	119	397,881	82.5	82.0	83.1	427	1,424,402
5-9	0.0001	0.0005	0.0005	0.0005	99,423	99,373	99,472	50	496,988	78.6	78.1	79.2	177	1,768,492
10-14	0.0001	0.0005	0.0005	0.0005	99,373	99,320	99,426	50	496,740	73.6	73.1	74.2	188	1,879,251
15-19	0.0003	0.0015	0.0014	0.0016	99,323	99,266	99,381	149	496,244	68.6	68.1	69.3	573	1,910,350
20-24	0.0004	0.0020	0.0018	0.0021	99,174	99,107	99,243	198	495,376	63.7	63.2	64.4	747	1,868,264
25-29	0.0004	0.0020	0.0019	0.0021	98,976	98,895	99,059	198	494,386	58.9	58.4	59.5	839	2,096,805
30-34	0.0006	0.0030	0.0028	0.0032	98,778	98,684	98,876	296	493,152	54.0	53.5	54.6	1,277	2,127,590
35-39	0.0009	0.0045	0.0042	0.0048	98,482	98,367	98,601	442	491,307	49.1	48.7	49.8	1,976	2,195,370
40-44	0.0013	0.0065	0.0060	0.0069	98,040	97,896	98,188	635	488,614	44.3	43.9	45.0	2,808	2,159,731
45-49	0.0020	0.0100	0.0093	0.0106	97,405	97,219	97,594	969	484,603	39.6	39.2	40.2	4,256	2,128,244
50-54	0.0028	0.0139	0.0130	0.0148	96,436	96,190	96,684	1,341	478,828	35.0	34.6	35.6	5,753	2,054,536
55-59	0.0038	0.0188	0.0176	0.0201	95,095	94,768	95,420	1,790	471,001	30.4	30.0	31.0	5,325	1,401,247
60-64	0.0054	0.0266	0.0249	0.0284	93,305	92,875	93,737	2,486	460,313	26.0	25.6	26.5	7,555	1,399,030
65-69	0.0085	0.0416	0.0386	0.0444	90,820	90,256	91,394	3,780	444,650	21.6	21.3	22.2	12,464	1,466,372
70-74	0.0137	0.0662	0.0613	0.0705	87,040	86,250	87,815	5,765	420,789	17.5	17.1	17.9	19,134	1,396,647
75-79	0.0238	0.1123	0.1038	0.1190	81,275	80,170	82,372	9,129	383,555	13.5	13.2	14.0	29,930	1,257,577
80-84	0.0494	0.2198	0.2043	0.2311	72,147	70,629	73,824	15,861	321,080	9.9	9.7	10.3	29,923	605,728
85-89	0.0967	0.3894	0.3679	0.4063	56,285	54,339	58,740	21,916	226,637	7.0	6.8	7.3	54,530	563,914
90-94	0.1633	0.5480	0.5285	0.5654	34,370	32,459	37,132	18,835	115,342	4.9	4.8	5.1	42,455	259,982
95-99	0.2676	0.7422	0.7293	0.7551	15,534	14,191	17,509	11,529	43,084	3.4	3.3	3.5	17,272	64,544
100+	0.4362	1	1	1	4,005	3,475	4,739	4,005	9,182	2.3	2.3	2.6	3,186	7,303
All ages													*242,389*	*30,381,913*

$\ln(_nq_x)$ with uncertainty interval

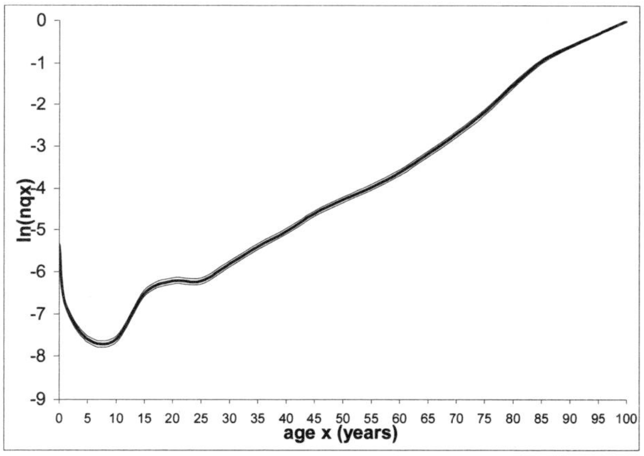

l_x with uncertainty interval

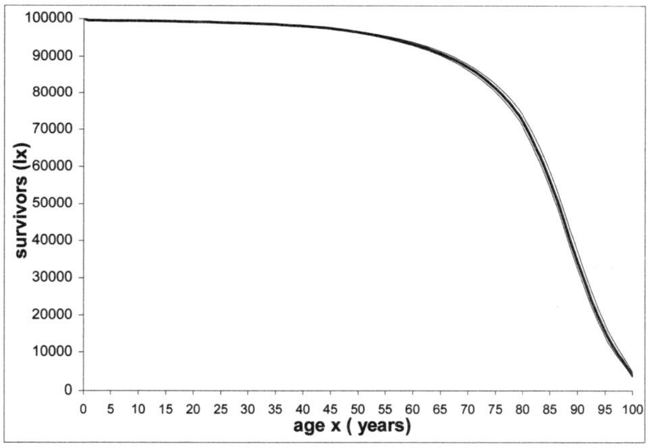

e_x with uncertainty interval

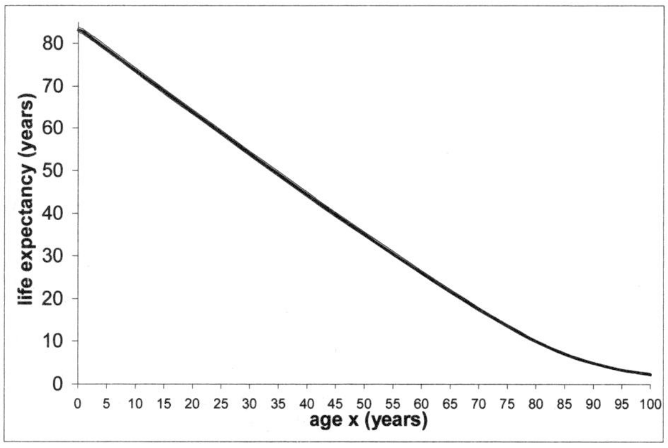

Distribution of deaths by age with uncertainty interval

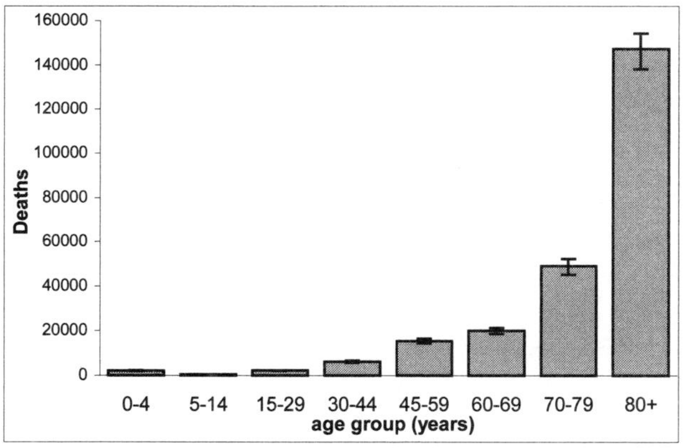

Gabon Gabon Gabón

Males, 2000 *Hommes, 2000* *Hombres, 2000*

Age	$_nM_x$	$_nq_x$			l_x			$_nd_x$	$_nL_x$	e_x			Actual deaths	Actual population
			percentiles			percentiles					percentiles			
		mean	2½	97½	mean	2½	97½			mean	2½	97½		
<1	0.0977	0.0898	0.0712	0.1097	100,000	100,000	100,000	8,979	91,919	54.6	50.3	59.2	2,076	21,258
1-4	0.0078	0.0306	0.0243	0.0379	91,021	89,033	92,883	2,784	357,403	59.0	54.9	63.6	601	77,106
5-9	0.0019	0.0094	0.0074	0.0119	88,237	86,247	90,228	832	439,104	56.8	53.0	61.3	154	81,133
10-14	0.0013	0.0065	0.0051	0.0082	87,405	85,291	89,540	567	435,606	52.3	48.6	56.7	90	68,865
15-19	0.0019	0.0094	0.0073	0.0119	86,838	84,628	89,078	815	432,152	47.6	44.0	52.0	106	56,210
20-24	0.0036	0.0178	0.0137	0.0227	86,023	83,676	88,414	1,534	426,279	43.0	39.4	47.4	177	49,057
25-29	0.0060	0.0298	0.0227	0.0382	84,489	81,878	87,162	2,517	416,152	38.8	35.1	43.1	255	42,224
30-34	0.0089	0.0436	0.0328	0.0563	81,972	78,752	85,100	3,575	400,924	34.9	31.3	39.1	316	35,414
35-39	0.0104	0.0506	0.0375	0.0657	78,397	74,320	82,152	3,968	382,068	31.4	27.9	35.4	324	31,193
40-44	0.0118	0.0574	0.0421	0.0746	74,430	69,440	78,892	4,269	361,478	27.9	24.6	31.7	335	28,339
45-49	0.0140	0.0675	0.0491	0.0888	70,161	64,259	75,486	4,735	338,969	24.5	21.3	28.0	366	26,167
50-54	0.0171	0.0821	0.0588	0.1103	65,426	58,624	71,675	5,370	313,706	21.1	18.1	24.3	395	23,075
55-59	0.0219	0.1037	0.0736	0.1413	60,056	52,401	67,395	6,229	284,706	17.7	15.1	20.7	433	19,792
60-64	0.0324	0.1497	0.1060	0.2043	53,827	45,438	62,301	8,060	248,982	14.5	12.2	17.1	528	16,322
65-69	0.0472	0.2109	0.1504	0.2842	45,766	36,879	55,433	9,653	204,698	11.6	9.6	13.8	625	13,250
70-74	0.0717	0.3041	0.2226	0.3965	36,113	27,052	46,650	10,984	153,105	9.0	7.5	10.9	673	9,381
75-79	0.1089	0.4279	0.3285	0.5299	25,129	16,326	35,968	10,752	98,766	6.8	5.7	8.2	654	6,009
80-84	0.1659	0.5862	0.4818	0.6826	14,377	7,675	24,152	8,428	50,815	5.1	4.4	6.1	474	2,859
85-89	0.2433	0.7564	0.6686	0.8309	5,949	2,436	12,515	4,500	18,495	3.8	3.3	4.4	224	922
90-94	0.3435	0.8459	0.7775	0.9013	1,449	412	4,147	1,226	3,568	2.8	2.5	3.2	67	196
95-99	0.4669	0.9724	0.9460	0.9880	223	41	916	217	465	2.1	2.1	2.3	10	21
100+	0.6107	1	1	1	6	0	49	6	10	1.6	1.6	1.7	1	2
All ages													*8,884*	*608,795*

$\ln(_nq_x)$ with uncertainty interval

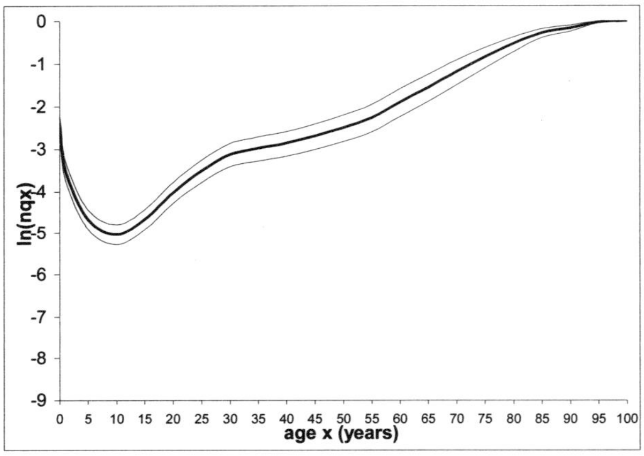

l_x with uncertainty interval

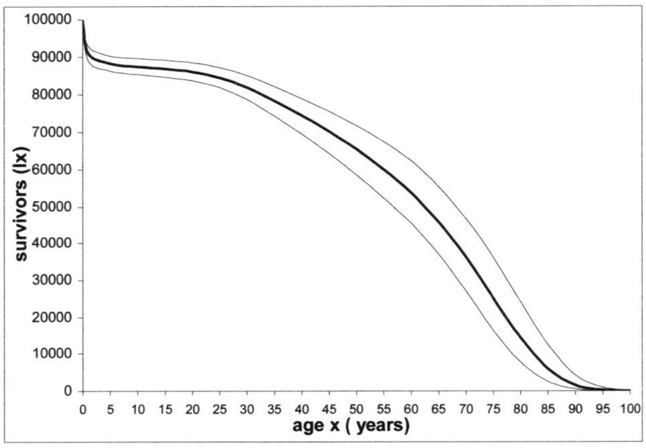

e_x with uncertainty interval

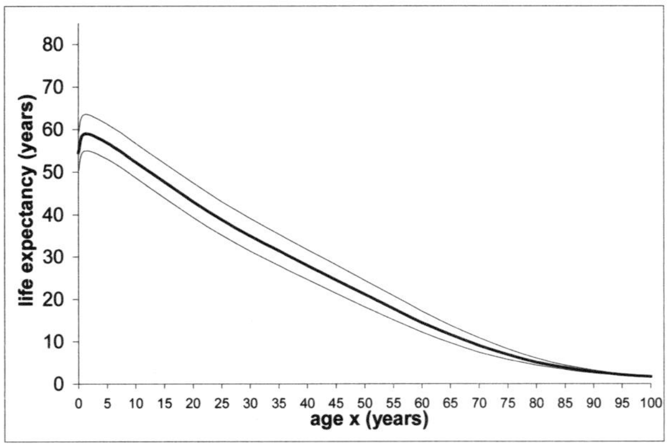

Distribution of deaths by age with uncertainty interval

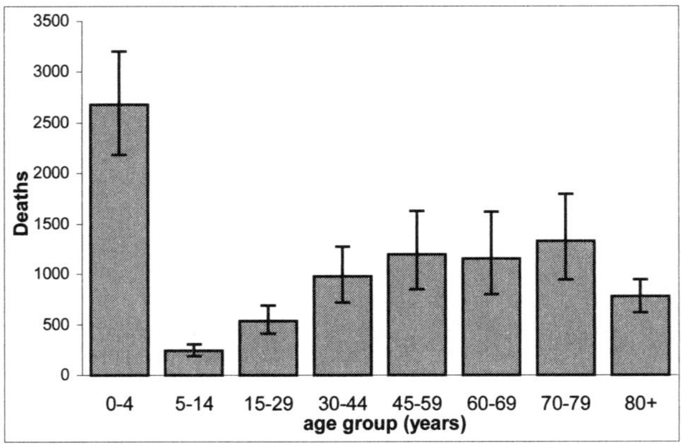

Gabon Gabon Gabón

Females, 2000 *Femmes, 2000* *Mujeres, 2000*

Age	$_nM_x$	$_nq_x$ mean	$_nq_x$ percentiles 2½	$_nq_x$ percentiles 97½	l_x mean	l_x percentiles 2½	l_x percentiles 97½	$_nd_x$	$_nL_x$	e_x mean	e_x percentiles 2½	e_x percentiles 97½	Actual deaths	Actual population
<1	0.0846	0.0787	0.0620	0.0959	100,000	100,000	100,000	7,865	92,921	56.9	52.9	61.6	1,726	20,391
1-4	0.0083	0.0327	0.0258	0.0398	92,135	90,411	93,800	3,009	361,318	60.8	56.6	65.8	638	76,596
5-9	0.0019	0.0093	0.0073	0.0115	89,126	87,393	90,872	833	443,548	58.8	54.7	63.6	151	80,408
10-14	0.0013	0.0064	0.0050	0.0079	88,293	86,387	90,165	566	440,051	54.3	50.3	59.1	88	68,467
15-19	0.0025	0.0125	0.0097	0.0155	87,727	85,702	89,699	1,097	435,892	49.6	45.7	54.4	141	56,088
20-24	0.0051	0.0252	0.0193	0.0314	86,630	84,376	88,796	2,183	427,691	45.2	41.3	49.9	252	49,281
25-29	0.0079	0.0385	0.0294	0.0483	84,447	81,737	87,040	3,255	414,095	41.3	37.6	45.9	335	42,603
30-34	0.0094	0.0460	0.0347	0.0586	81,191	77,816	84,486	3,737	396,614	37.9	34.3	42.3	337	35,789
35-39	0.0092	0.0452	0.0332	0.0585	77,454	73,348	81,548	3,500	378,520	34.6	31.3	38.7	292	31,579
40-44	0.0090	0.0440	0.0318	0.0578	73,954	69,209	78,780	3,258	361,626	31.1	28.1	34.9	261	28,951
45-49	0.0096	0.0467	0.0333	0.0618	70,696	65,404	76,184	3,304	345,222	27.4	24.6	31.0	259	27,064
50-54	0.0118	0.0575	0.0404	0.0764	67,392	61,554	73,522	3,874	327,277	23.6	21.0	27.0	287	24,229
55-59	0.0155	0.0747	0.0521	0.0994	63,519	57,055	70,360	4,748	305,723	19.9	17.5	23.0	341	21,989
60-64	0.0230	0.1087	0.0754	0.1445	58,771	51,660	66,409	6,387	277,886	16.3	14.1	19.1	428	18,601
65-69	0.0356	0.1635	0.1138	0.2170	52,384	44,258	60,937	8,562	240,514	13.0	11.1	15.5	529	14,857
70-74	0.0578	0.2524	0.1791	0.3286	43,822	34,822	53,773	11,060	191,459	10.1	8.5	12.2	658	11,389
75-79	0.0917	0.3730	0.2750	0.4661	32,762	23,702	44,137	12,219	133,262	7.6	6.4	9.3	664	7,240
80-84	0.1421	0.5244	0.4123	0.6191	20,543	13,001	31,939	10,772	75,783	5.7	4.9	6.9	555	3,905
85-89	0.2128	0.6945	0.5909	0.7733	9,770	4,952	18,665	6,785	31,889	4.2	3.7	5.0	314	1,478
90-94	0.3075	0.7998	0.7181	0.8601	2,985	1,123	7,588	2,387	7,763	3.1	2.7	3.6	104	337
95-99	0.4291	0.9380	0.8935	0.9657	598	157	2,139	561	1,307	2.3	2.2	2.5	20	47
100+	0.5780	1	1	1	37	5	228	37	64	1.7	1.7	1.9	2	4
All ages													*8,382*	*621,293*

ln($_nq_x$) with uncertainty interval

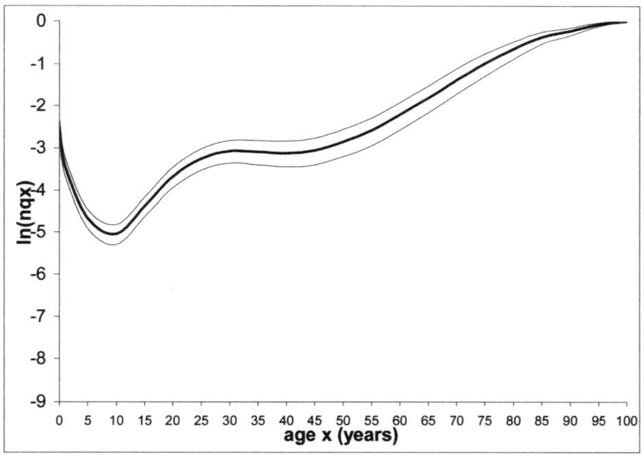

l_x with uncertainty interval

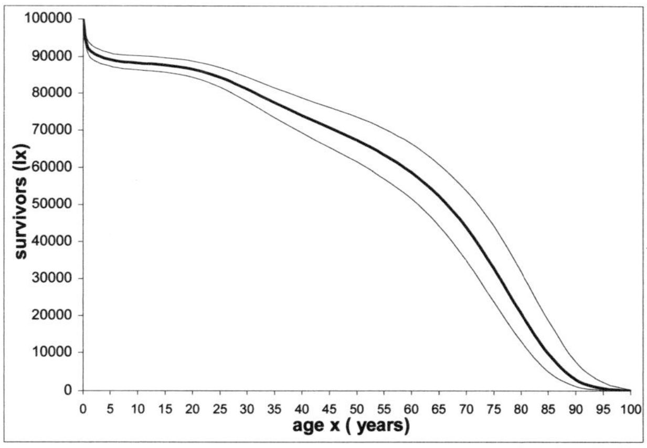

e_x with uncertainty interval

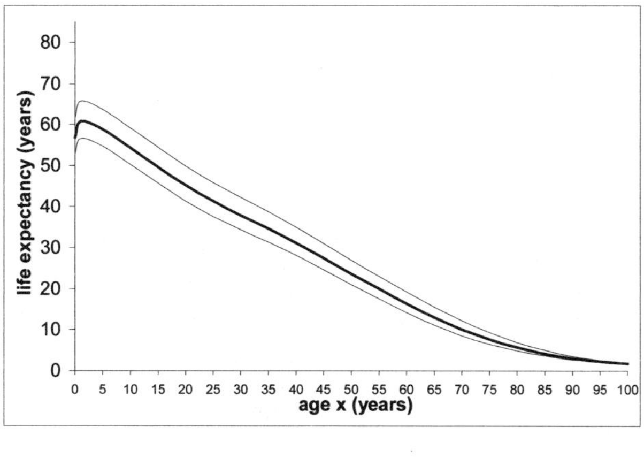

Distribution of deaths by age with uncertainty interval

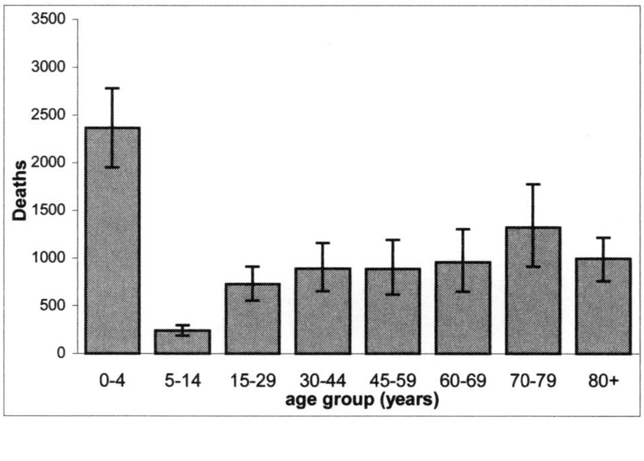

Gambia **Gambie** **Gambia**

Males, 2000 *Hommes, 2000* *Hombres, 2000*

Age	$_nM_x$	$_nq_x$			l_x			$_nd_x$	$_nL_x$	e_x			Actual deaths	Actual population
			percentiles			percentiles					percentiles			
		mean	2½	97½	mean	2½	97½			mean	2½	97½		
<1	0.0825	0.0768	0.0633	0.0916	100,000	100,000	100,000	7,679	93,089	55.9	52.5	59.9	1,916	23,225
1-4	0.0066	0.0259	0.0209	0.0310	92,321	90,839	93,672	2,387	363,553	59.5	56.0	63.8	526	80,096
5-9	0.0018	0.0090	0.0072	0.0109	89,933	88,433	91,429	811	447,640	57.1	53.7	61.2	156	85,963
10-14	0.0013	0.0066	0.0052	0.0080	89,123	87,516	90,721	585	444,150	52.6	49.2	56.7	96	73,174
15-19	0.0022	0.0109	0.0087	0.0133	88,537	86,823	90,224	969	440,265	47.9	44.5	52.0	135	61,253
20-24	0.0038	0.0188	0.0148	0.0229	87,569	85,672	89,413	1,645	433,730	43.4	40.1	47.5	207	54,453
25-29	0.0054	0.0265	0.0207	0.0325	85,923	83,717	88,035	2,276	423,927	39.2	36.0	43.2	265	49,300
30-34	0.0074	0.0363	0.0282	0.0448	83,647	81,017	86,198	3,035	410,649	35.2	32.1	39.1	332	44,930
35-39	0.0090	0.0439	0.0338	0.0549	80,612	77,428	83,766	3,538	394,217	31.4	28.4	35.1	360	40,066
40-44	0.0109	0.0532	0.0401	0.0675	77,075	73,263	80,924	4,097	375,131	27.7	24.9	31.3	365	33,463
45-49	0.0137	0.0661	0.0489	0.0850	72,978	68,491	77,613	4,827	352,821	24.1	21.6	27.5	370	27,008
50-54	0.0177	0.0845	0.0616	0.1094	68,151	62,936	73,666	5,762	326,348	20.7	18.3	23.8	389	22,045
55-59	0.0233	0.1101	0.0794	0.1425	62,389	56,323	68,869	6,866	294,778	17.4	15.3	20.2	405	17,378
60-64	0.0344	0.1583	0.1140	0.2029	55,522	48,512	62,996	8,787	255,646	14.2	12.4	16.7	447	12,991
65-69	0.0493	0.2196	0.1597	0.2772	46,736	38,668	55,388	10,263	208,022	11.4	9.9	13.5	440	8,927
70-74	0.0740	0.3122	0.2328	0.3838	36,473	27,997	46,541	11,386	153,900	8.9	7.7	10.6	407	5,503
75-79	0.1107	0.4335	0.3382	0.5120	25,087	17,380	35,670	10,876	98,244	6.8	5.9	8.1	322	2,909
80-84	0.1672	0.5895	0.4902	0.6634	14,211	8,619	23,527	8,377	50,112	5.1	4.5	6.0	199	1,188
85-89	0.2437	0.7571	0.6738	0.8145	5,834	2,948	11,902	4,417	18,128	3.8	3.4	4.3	78	322
90-94	0.3427	0.8449	0.7811	0.8881	1,417	547	3,878	1,197	3,494	2.8	2.6	3.2	18	53
95-99	0.4653	0.9710	0.9461	0.9842	220	61	849	213	459	2.1	2.1	2.3	2	4
100+	0.6096	1	1	1	6	1	46	6	10	1.6	1.6	1.7	0	0
All ages													*7,434*	*644,251*

$ln(_nq_x)$ with uncertainty interval

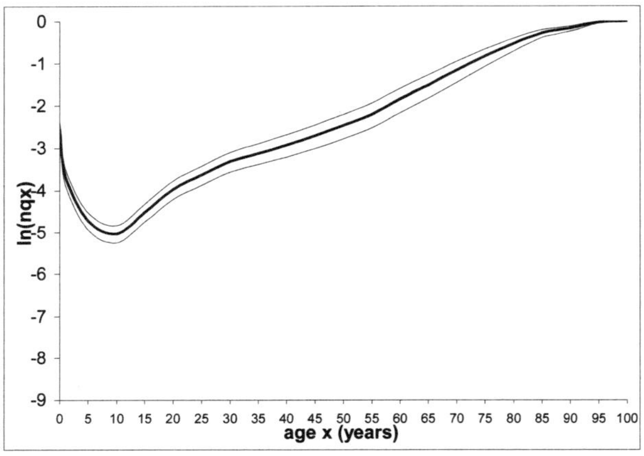

l_x with uncertainty interval

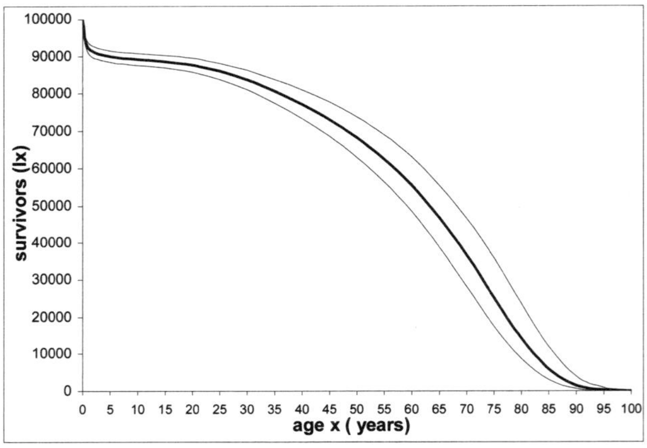

e_x with uncertainty interval

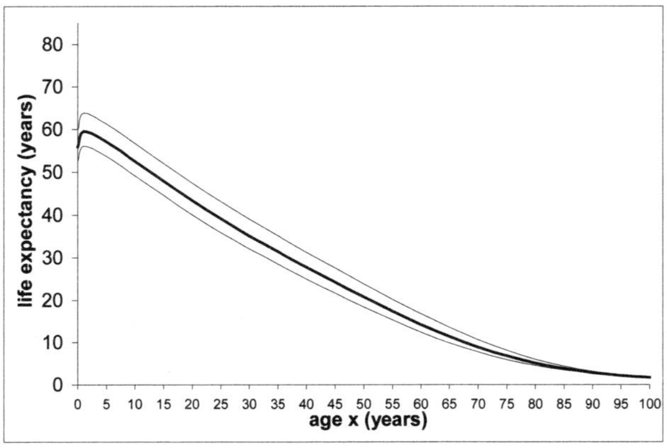

Distribution of deaths by age with uncertainty interval

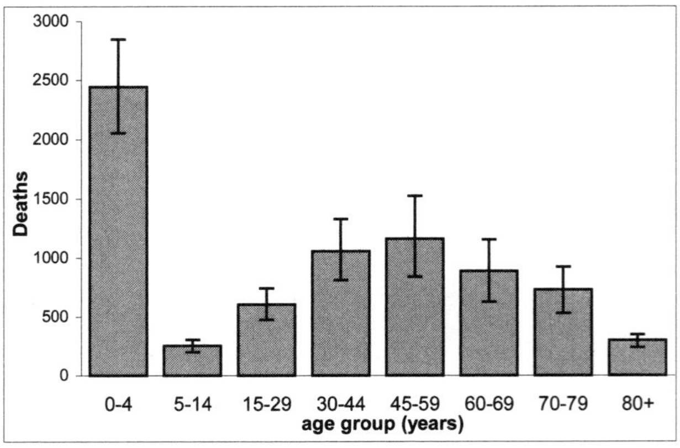

Gambia Gambie Gambia

Females, 2000 *Femmes, 2000* *Mujeres, 2000*

Age	$_nM_x$	$_nq_x$			l_x			$_nd_x$	$_nL_x$	e_x			Actual deaths	Actual population
			percentiles			percentiles					percentiles			
		mean	2½	97½	mean	2½	97½			mean	2½	97½		
<1	0.0672	0.0634	0.0528	0.0759	100,000	100,000	100,000	6,341	94,293	58.7	55.6	62.3	1,502	22,331
1-4	0.0071	0.0281	0.0234	0.0329	93,659	92,415	94,721	2,632	368,320	61.6	58.3	65.5	574	80,385
5-9	0.0019	0.0094	0.0077	0.0111	91,027	89,777	92,264	851	453,007	59.4	56.1	63.3	162	85,967
10-14	0.0014	0.0070	0.0058	0.0083	90,176	88,858	91,534	635	449,291	54.9	51.6	58.8	104	73,388
15-19	0.0026	0.0131	0.0107	0.0156	89,541	88,123	90,999	1,175	444,765	50.3	47.0	54.1	163	61,817
20-24	0.0044	0.0218	0.0176	0.0260	88,365	86,763	90,010	1,926	437,010	45.9	42.7	49.8	245	55,670
25-29	0.0061	0.0300	0.0241	0.0360	86,439	84,536	88,388	2,593	425,712	41.9	38.7	45.7	311	51,016
30-34	0.0073	0.0358	0.0286	0.0435	83,846	81,514	86,200	3,003	411,721	38.1	35.1	41.8	341	46,685
35-39	0.0079	0.0389	0.0308	0.0480	80,843	78,018	83,672	3,141	396,361	34.4	31.6	38.0	330	41,613
40-44	0.0086	0.0420	0.0327	0.0527	77,702	74,372	81,093	3,266	380,344	30.7	28.0	34.1	299	34,770
45-49	0.0101	0.0491	0.0374	0.0623	74,436	70,611	78,399	3,654	363,045	27.0	24.5	30.2	284	28,198
50-54	0.0130	0.0628	0.0471	0.0804	70,782	66,446	75,363	4,447	342,793	23.2	20.9	26.2	296	22,798
55-59	0.0172	0.0824	0.0608	0.1059	66,335	61,383	71,627	5,464	318,016	19.6	17.5	22.4	314	18,305
60-64	0.0248	0.1170	0.0854	0.1502	60,871	55,141	66,958	7,120	286,555	16.1	14.3	18.7	347	13,958
65-69	0.0372	0.1701	0.1238	0.2161	53,751	46,859	60,711	9,144	245,894	12.9	11.4	15.2	365	9,809
70-74	0.0586	0.2556	0.1882	0.3166	44,607	36,732	52,720	11,401	194,532	10.1	8.8	12.0	366	6,239
75-79	0.0908	0.3699	0.2814	0.4405	33,206	25,102	42,772	12,283	135,322	7.7	6.8	9.2	314	3,454
80-84	0.1403	0.5193	0.4183	0.5913	20,923	14,044	30,734	10,866	77,451	5.7	5.1	6.8	213	1,521
85-89	0.2097	0.6880	0.5944	0.7486	10,057	5,772	17,878	6,919	32,988	4.2	3.8	5.0	95	454
90-94	0.3033	0.7940	0.7199	0.8408	3,138	1,472	7,251	2,492	8,215	3.1	2.8	3.5	26	85
95-99	0.4242	0.9333	0.8919	0.9561	646	239	2,031	603	1,422	2.3	2.2	2.5	4	9
100+	0.5738	1	1	1	43	11	219	43	75	1.7	1.7	1.9	0	0
All ages													*6,653*	*658,472*

$\ln(_nq_x)$ with uncertainty interval

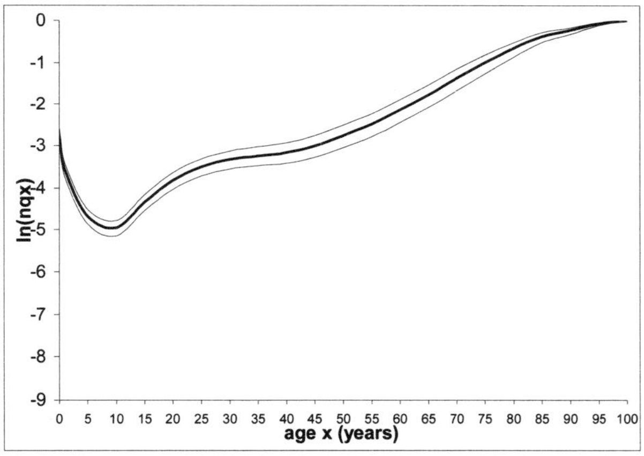

l_x with uncertainty interval

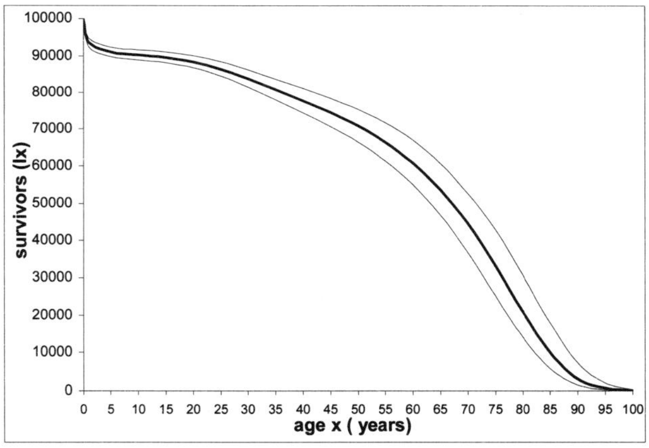

e_x with uncertainty interval

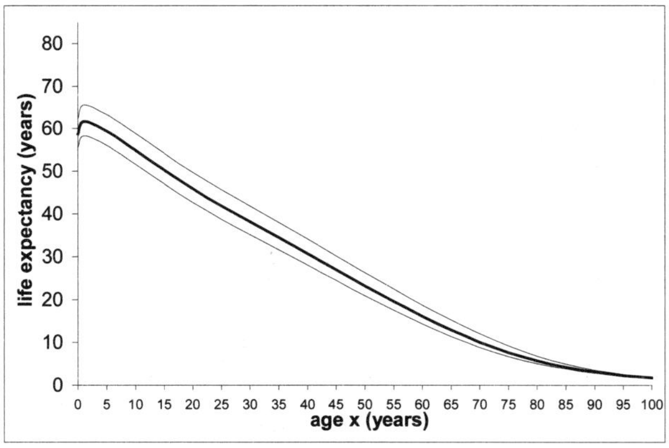

Distribution of deaths by age with uncertainty interval

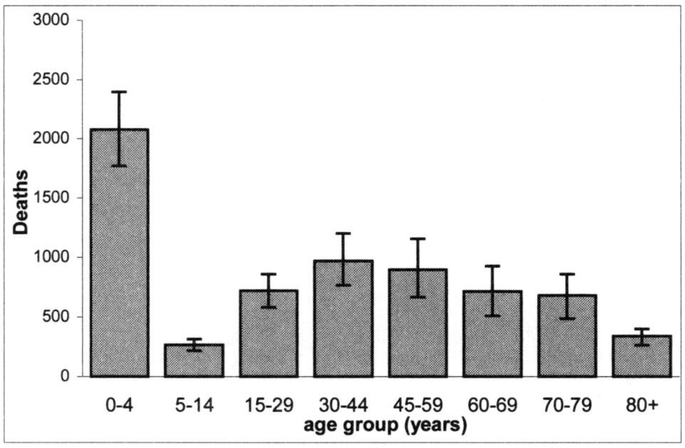

Georgia | Géorgie | Georgia

Males, 2000 | *Hommes, 2000* | *Hombres, 2000*

Age	$_nM_x$	$_nq_x$			l_x			$_nd_x$	$_nL_x$	e_x			Actual deaths	Actual population
			percentiles			percentiles					percentiles			
		mean	2½	97½	mean	2½	97½			mean	2½	97½		
<1	0.0262	0.0256	0.0187	0.0327	100,000	100,000	100,000	2,559	97,697	65.7	64.0	67.7	797	30,434
1-4	0.0011	0.0045	0.0035	0.0055	97,441	96,730	98,129	441	388,706	66.4	65.0	68.4	139	122,852
5-9	0.0003	0.0013	0.0010	0.0016	97,000	96,196	97,788	126	484,685	62.7	61.3	64.7	47	181,497
10-14	0.0003	0.0013	0.0010	0.0016	96,874	96,043	97,689	130	484,043	57.8	56.4	59.7	58	217,343
15-19	0.0006	0.0032	0.0025	0.0039	96,743	95,886	97,587	312	482,938	52.9	51.5	54.8	138	213,677
20-24	0.0013	0.0067	0.0052	0.0081	96,432	95,511	97,343	644	480,548	48.0	46.7	50.0	271	202,103
25-29	0.0021	0.0103	0.0081	0.0123	95,788	94,740	96,833	983	476,479	43.3	42.0	45.2	403	195,305
30-34	0.0028	0.0141	0.0113	0.0168	94,804	93,571	96,046	1,339	470,674	38.8	37.4	40.6	527	185,384
35-39	0.0040	0.0197	0.0160	0.0232	93,466	91,996	94,956	1,837	462,736	34.3	33.0	36.1	800	201,615
40-44	0.0061	0.0300	0.0249	0.0350	91,629	89,860	93,434	2,744	451,283	29.9	28.7	31.7	1,141	187,674
45-49	0.0079	0.0388	0.0323	0.0448	88,884	86,715	91,109	3,447	435,804	25.8	24.6	27.4	1,268	160,353
50-54	0.0140	0.0676	0.0560	0.0774	85,437	82,833	88,164	5,778	412,741	21.7	20.7	23.2	1,751	125,057
55-59	0.0186	0.0890	0.0734	0.1016	79,659	76,423	83,225	7,088	380,574	18.1	17.2	19.5	1,741	93,479
60-64	0.0276	0.1291	0.1069	0.1464	72,571	68,656	77,114	9,369	339,429	14.6	13.7	15.8	3,770	136,581
65-69	0.0507	0.2248	0.1895	0.2518	63,201	58,606	68,869	14,208	280,485	11.4	10.6	12.4	5,067	100,032
70-74	0.0667	0.2860	0.2468	0.3221	48,993	43,850	55,818	14,012	209,933	9.0	8.3	9.7	6,076	91,026
75-79	0.1125	0.4389	0.3928	0.4861	34,980	29,728	42,043	15,354	136,516	6.6	6.1	7.1	4,372	38,873
80-84	0.1797	0.6201	0.5737	0.6664	19,626	15,278	25,527	12,170	67,706	4.8	4.4	5.2	2,804	15,600
85-89	0.2700	0.8060	0.7708	0.8389	7,456	5,097	10,883	6,010	22,257	3.5	3.3	3.7	2,555	9,462
90-94	0.3600	0.8654	0.8384	0.8905	1,447	821	2,494	1,252	3,478	2.7	2.6	2.8	1,072	2,977
95-99	0.4800	0.9836	0.9765	0.9891	195	90	403	192	399	2.1	2.1	2.1	289	602
100+	0.6100	1	1	1	3	1	9	3	5	1.6	1.6	1.8	45	73
All ages													*35,132*	*2,511,999*

ln($_nq_x$) with uncertainty interval

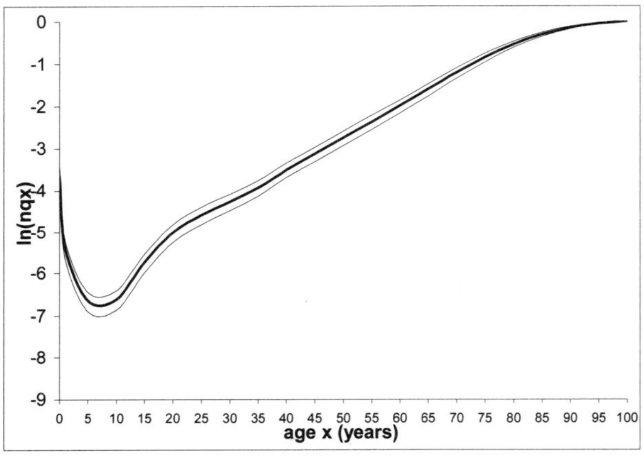

l_x with uncertainty interval

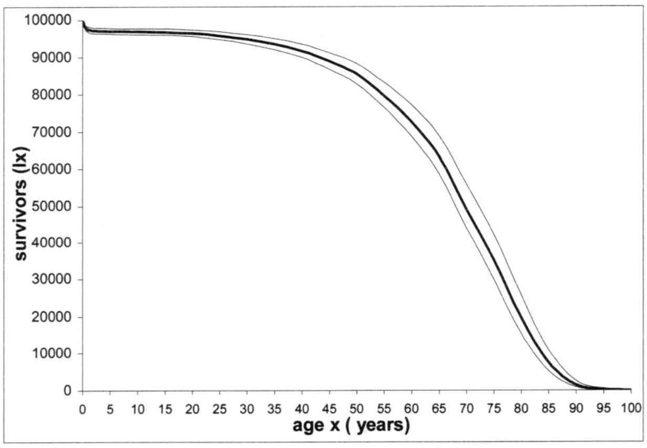

e_x with uncertainty interval

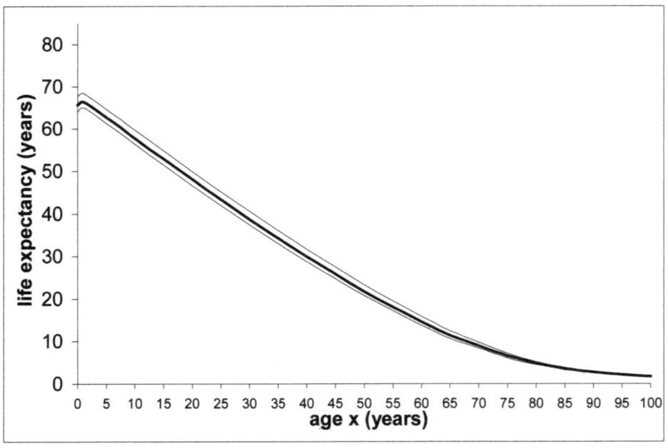

Distribution of deaths by age with uncertainty interval

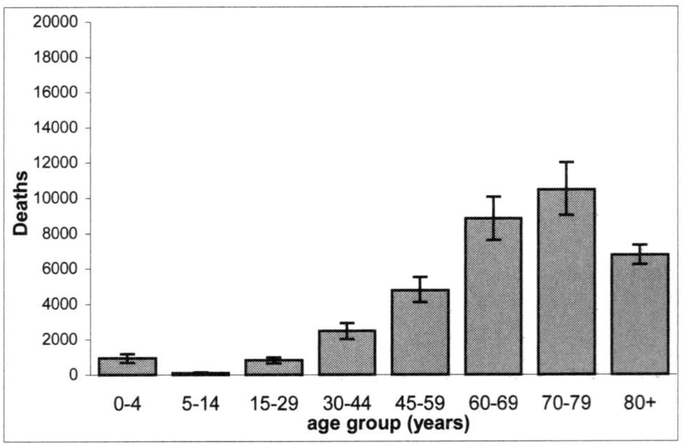

Georgia Géorgie Georgia

Females, 2000 *Femmes, 2000* *Mujeres, 2000*

Age	$_nM_x$	$_nq_x$			l_x			$_nd_x$	$_nL_x$	e_x			Actual deaths	Actual population
			percentiles			percentiles					percentiles			
		mean	2½	97½	mean	2½	97½			mean	2½	97½		
<1	0.0153	0.0151	0.0117	0.0202	100,000	100,000	100,000	1,507	98,644	71.8	70.3	74.2	441	28,875
1-4	0.0010	0.0040	0.0032	0.0050	98,493	97,978	98,828	390	393,035	71.9	70.4	74.3	116	116,856
5-9	0.0002	0.0012	0.0010	0.0015	98,102	97,488	98,514	118	490,216	68.2	66.7	70.5	42	172,766
10-14	0.0002	0.0008	0.0006	0.0010	97,984	97,340	98,419	77	489,728	63.3	61.8	65.6	32	206,302
15-19	0.0003	0.0016	0.0012	0.0019	97,907	97,244	98,357	153	489,154	58.3	56.8	60.7	64	206,762
20-24	0.0006	0.0028	0.0023	0.0035	97,755	97,056	98,234	272	488,092	53.4	51.9	55.7	109	195,891
25-29	0.0008	0.0039	0.0032	0.0049	97,482	96,718	98,013	384	486,451	48.5	47.1	50.9	145	183,923
30-34	0.0013	0.0065	0.0053	0.0080	97,098	96,246	97,701	629	483,918	43.7	42.3	46.0	239	184,161
35-39	0.0020	0.0100	0.0081	0.0121	96,469	95,480	97,188	960	479,945	39.0	37.5	41.3	442	221,137
40-44	0.0030	0.0149	0.0121	0.0179	95,509	94,323	96,400	1,422	473,990	34.4	32.9	36.6	618	205,848
45-49	0.0043	0.0213	0.0169	0.0252	94,087	92,634	95,238	2,001	465,432	29.8	28.5	32.0	767	178,424
50-54	0.0065	0.0320	0.0250	0.0377	92,086	90,297	93,627	2,943	453,071	25.4	24.1	27.5	934	143,845
55-59	0.0098	0.0480	0.0369	0.0567	89,143	86,895	91,289	4,282	435,009	21.2	19.9	23.1	1,128	114,625
60-64	0.0163	0.0781	0.0594	0.0929	84,861	81,967	87,917	6,631	407,729	17.1	16.0	18.9	2,755	169,437
65-69	0.0305	0.1418	0.1079	0.1685	78,230	74,350	82,690	11,095	363,414	13.4	12.3	15.0	3,958	129,625
70-74	0.0487	0.2170	0.1679	0.2545	67,135	61,822	73,764	14,568	299,257	10.2	9.3	11.5	6,384	131,144
75-79	0.0973	0.3914	0.3198	0.4477	52,568	46,088	61,377	20,573	211,405	7.3	6.6	8.3	7,718	79,307
80-84	0.1533	0.5540	0.4807	0.6090	31,994	25,455	41,746	17,725	115,660	5.4	4.9	6.1	6,637	43,311
85-89	0.2284	0.7269	0.6651	0.7679	14,270	9,954	21,677	10,373	45,415	4.0	3.7	4.4	6,016	26,340
90-94	0.3222	0.8192	0.7733	0.8482	3,896	2,310	7,260	3,192	9,907	3.0	2.8	3.2	2,900	9,001
95-99	0.4302	0.9390	0.9147	0.9533	705	351	1,646	662	1,538	2.3	2.2	2.4	947	2,202
100+	0.5435	1	1	1	43	16	140	43	79	1.8	1.8	1.9	146	269
All ages													*42,541*	*2,750,051*

ln($_nq_x$) with uncertainty interval

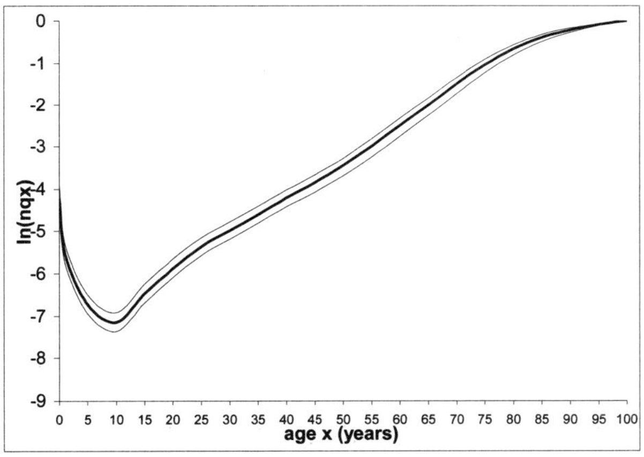

l_x with uncertainty interval

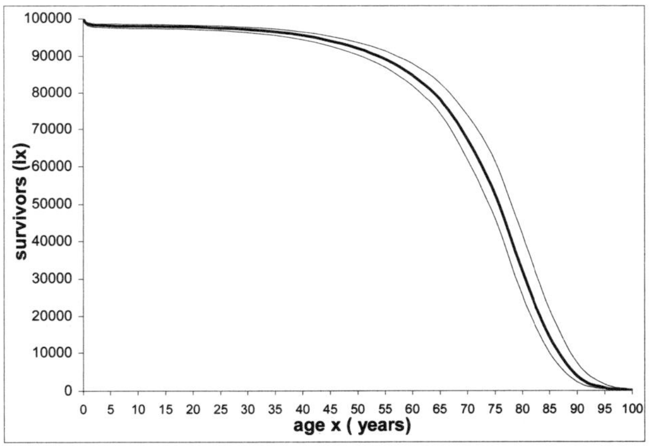

e_x with uncertainty interval

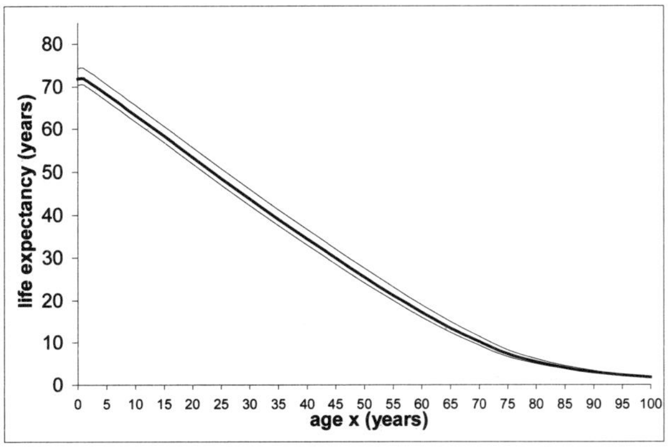

Distribution of deaths by age with uncertainty interval

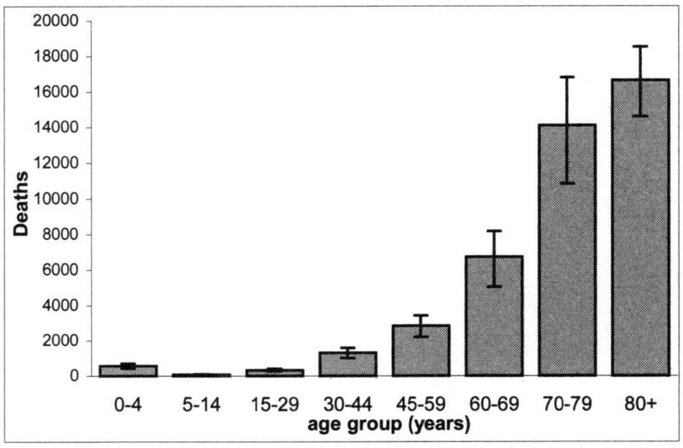

Germany   Allemagne   Alemania

Males, 2000   *Hommes, 2000*   *Hombres, 2000*

Age	$_nM_x$	$_nq_x$ mean	$_nq_x$ percentiles 2½	$_nq_x$ percentiles 97½	l_x mean	l_x percentiles 2½	l_x percentiles 97½	$_nd_x$	$_nL_x$	e_x mean	e_x percentiles 2½	e_x percentiles 97½	Actual deaths	Actual population
<1	0.0050	0.0050	0.0043	0.0056	100,000	100,000	100,000	498	99,552	74.3	73.9	74.7	1,879	375,747
1-4	0.0003	0.0011	0.0010	0.0012	99,502	99,438	99,566	107	397,751	73.7	73.3	74.1	429	1,589,276
5-9	0.0001	0.0006	0.0005	0.0007	99,395	99,320	99,469	60	496,825	69.7	69.3	70.2	258	2,148,706
10-14	0.0002	0.0008	0.0007	0.0009	99,335	99,255	99,414	79	496,478	64.8	64.4	65.2	389	2,433,799
15-19	0.0006	0.0030	0.0028	0.0033	99,256	99,168	99,342	302	495,523	59.8	59.4	60.3	1,442	2,363,275
20-24	0.0009	0.0043	0.0040	0.0046	98,954	98,840	99,064	425	493,706	55.0	54.6	55.4	1,979	2,300,662
25-29	0.0008	0.0039	0.0037	0.0042	98,529	98,384	98,670	388	491,674	50.2	49.8	50.7	2,047	2,590,542
30-34	0.0010	0.0048	0.0045	0.0051	98,141	97,969	98,306	475	489,515	45.4	45.0	45.9	3,457	3,563,783
35-39	0.0015	0.0074	0.0070	0.0078	97,666	97,465	97,857	720	486,528	40.6	40.2	41.1	5,539	3,742,607
40-44	0.0025	0.0126	0.0120	0.0132	96,946	96,707	97,173	1,223	481,670	35.9	35.5	36.4	8,274	3,257,519
45-49	0.0040	0.0198	0.0190	0.0206	95,722	95,429	96,006	1,895	473,872	31.3	30.9	31.8	11,541	2,885,149
50-54	0.0060	0.0296	0.0285	0.0307	93,827	93,467	94,181	2,773	462,200	26.9	26.5	27.4	14,570	2,428,273
55-59	0.0100	0.0488	0.0469	0.0509	91,053	90,615	91,486	4,442	444,163	22.7	22.3	23.1	25,840	2,584,035
60-64	0.0160	0.0769	0.0735	0.0805	86,612	86,096	87,121	6,662	416,403	18.7	18.4	19.1	43,996	2,749,728
65-69	0.0270	0.1265	0.1204	0.1324	79,949	79,273	80,603	10,111	374,470	15.1	14.7	15.4	51,866	1,920,957
70-74	0.0410	0.1859	0.1768	0.1944	69,839	68,773	70,817	12,986	316,729	11.9	11.6	12.2	61,950	1,510,980
75-79	0.0660	0.2833	0.2701	0.2953	56,853	55,405	58,294	16,104	244,004	9.0	8.8	9.3	61,694	934,752
80-84	0.1130	0.4405	0.4231	0.4565	40,749	39,098	42,522	17,952	158,864	6.6	6.4	6.8	44,656	395,186
85-89	0.1820	0.6254	0.6084	0.6412	22,797	21,388	24,441	14,258	78,340	4.8	4.7	5.0	48,923	268,810
90-94	0.2450	0.7061	0.6923	0.7192	8,539	7,717	9,544	6,029	24,608	3.7	3.6	3.8	21,347	87,129
95-99	0.3370	0.8379	0.8280	0.8474	2,510	2,169	2,937	2,103	6,241	2.8	2.8	2.9	5,299	15,725
100+	0.4710	1	1	1	407	332	505	407	864	2.1	2.1	2.3	654	1,388
All ages													*418,027*	*40,148,028*

ln($_nq_x$) with uncertainty interval

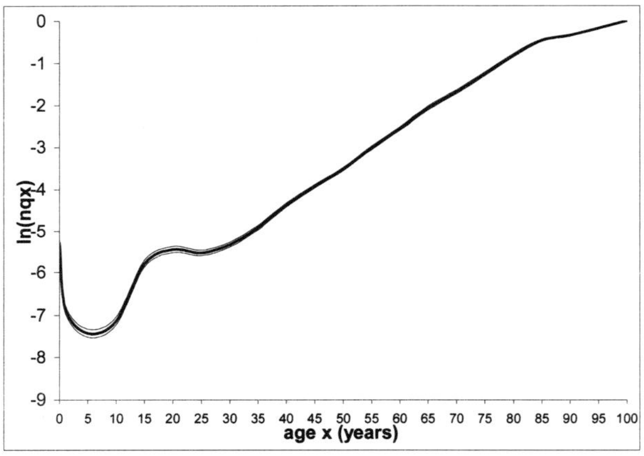

l_x with uncertainty interval

e_x with uncertainty interval

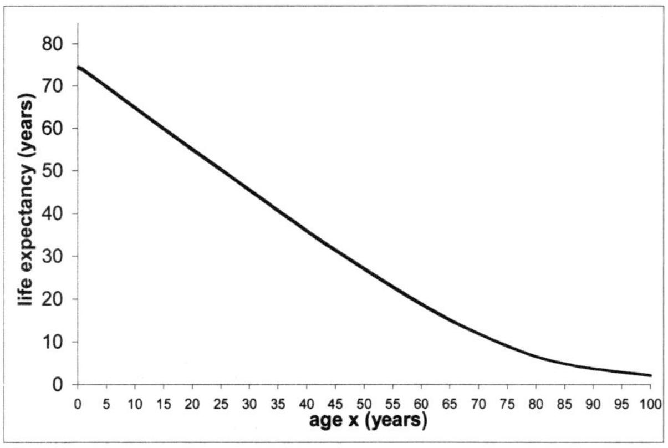

Distribution of deaths by age with uncertainty interval

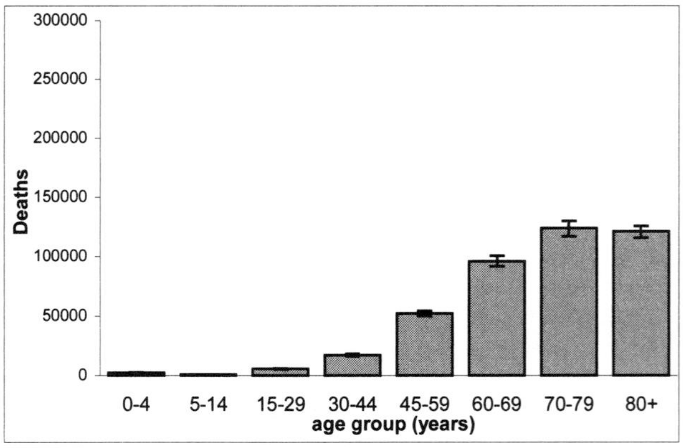

Germany | Allemagne | Alemania

Females, 2000 | *Femmes, 2000* | *Mujeres, 2000*

Age	$_nM_x$	$_nq_x$			l_x			$_nd_x$	$_nL_x$	e_x			Actual deaths	Actual population
			percentiles			percentiles					percentiles			
		mean	2½	97½	mean	2½	97½			mean	2½	97½		
<1	0.0040	0.0040	0.0035	0.0045	100,000	100,000	100,000	399	99,641	80.6	80.3	81.1	1,421	355,317
1-4	0.0002	0.0009	0.0008	0.0010	99,601	99,553	99,652	92	398,186	80.0	79.7	80.4	346	1,505,994
5-9	0.0001	0.0004	0.0004	0.0005	99,510	99,453	99,569	45	497,437	76.0	75.7	76.5	183	2,031,299
10-14	0.0001	0.0005	0.0005	0.0006	99,465	99,405	99,529	55	497,189	71.1	70.8	71.5	253	2,298,781
15-19	0.0003	0.0013	0.0012	0.0015	99,410	99,346	99,479	134	496,717	66.1	65.8	66.5	603	2,233,578
20-24	0.0003	0.0014	0.0013	0.0015	99,276	99,201	99,357	139	496,034	61.2	60.9	61.6	606	2,163,017
25-29	0.0003	0.0016	0.0015	0.0017	99,137	99,052	99,229	158	495,291	56.3	56.0	56.7	781	2,441,895
30-34	0.0004	0.0021	0.0019	0.0022	98,979	98,883	99,083	208	494,375	51.4	51.1	51.8	1,390	3,309,159
35-39	0.0007	0.0036	0.0034	0.0039	98,771	98,662	98,890	360	492,957	46.5	46.2	46.9	2,547	3,488,757
40-44	0.0013	0.0062	0.0059	0.0066	98,411	98,282	98,554	613	490,524	41.6	41.3	42.0	3,848	3,078,413
45-49	0.0020	0.0100	0.0095	0.0104	97,798	97,639	97,975	973	486,558	36.9	36.6	37.3	5,628	2,814,072
50-54	0.0030	0.0149	0.0143	0.0155	96,825	96,624	97,047	1,442	480,522	32.2	31.9	32.6	7,136	2,378,605
55-59	0.0040	0.0198	0.0190	0.0206	95,384	95,133	95,655	1,889	472,196	27.7	27.4	28.1	10,343	2,585,770
60-64	0.0070	0.0344	0.0328	0.0357	93,495	93,197	93,808	3,216	459,434	23.2	22.9	23.5	20,059	2,865,520
65-69	0.0120	0.0583	0.0553	0.0606	90,279	89,894	90,656	5,259	438,246	18.9	18.7	19.3	25,936	2,161,299
70-74	0.0210	0.0998	0.0945	0.1038	85,020	84,468	85,605	8,482	403,894	14.9	14.7	15.2	43,118	2,053,217
75-79	0.0380	0.1735	0.1647	0.1800	76,538	75,761	77,514	13,281	349,489	11.3	11.1	11.6	72,063	1,896,397
80-84	0.0750	0.3158	0.3022	0.3263	63,257	62,123	64,751	19,976	266,347	8.1	8.0	8.4	73,120	974,938
85-89	0.1370	0.5102	0.4956	0.5255	43,281	41,917	45,185	22,084	161,197	5.8	5.6	5.9	114,480	835,619
90-94	0.2090	0.6423	0.6291	0.6566	21,197	20,118	22,792	13,615	65,143	4.1	4.1	4.3	68,487	327,689
95-99	0.3110	0.8044	0.7948	0.8149	7,583	6,921	8,398	6,100	19,614	3.0	3.0	3.1	19,781	63,606
100+	0.4620	1	1	1	1,483	1,281	1,701	1,483	3,209	2.2	2.2	2.4	2,678	5,797
All ages													*474,807*	*41,868,739*

ln($_nq_x$) with uncertainty interval

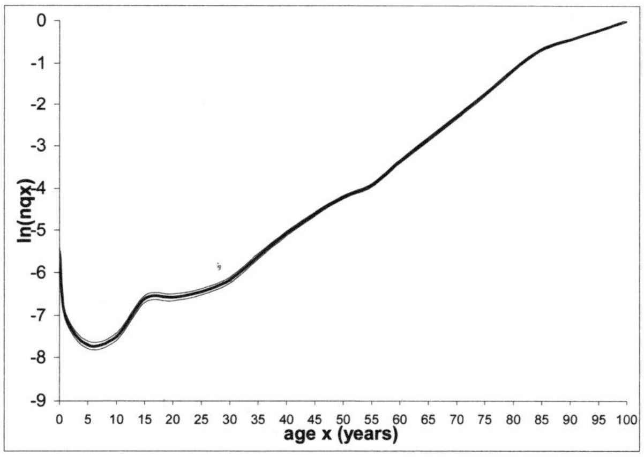

l_x with uncertainty interval

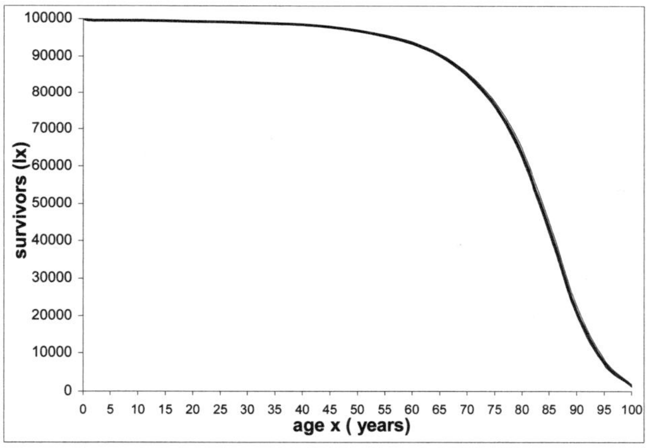

e_x with uncertainty interval

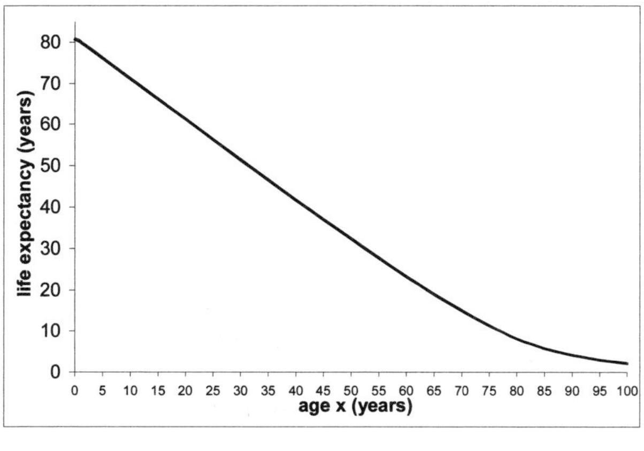

Distribution of deaths by age with uncertainty interval

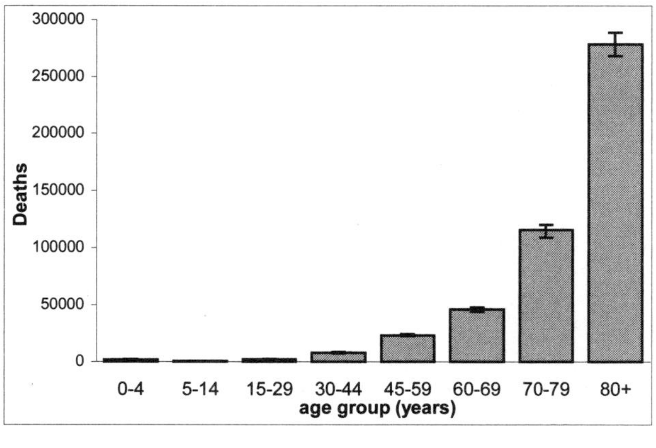

Ghana **Ghana** **Ghana**

Males, 2000 *Hommes, 2000* *Hombres, 2000*

Age	$_nM_x$	$_nq_x$			l_x			$_nd_x$	$_nL_x$	e_x			Actual deaths	Actual population
			percentiles			percentiles					percentiles			
		mean	2 1/2	97 1/2	mean	2 1/2	97 1/2			mean	2 1/2	97 1/2		
<1	0.0925	0.0854	0.0778	0.0930	100,000	100,000	100,000	8,535	92,318	55.0	52.9	57.3	28,523	308,508
1-4	0.0074	0.0290	0.0260	0.0320	91,465	90,697	92,221	2,656	359,485	59.1	56.7	61.9	8,221	1,112,927
5-9	0.0018	0.0090	0.0080	0.0100	88,809	88,071	89,551	802	442,041	56.9	54.5	59.6	2,404	1,325,060
10-14	0.0013	0.0063	0.0056	0.0070	88,007	87,199	88,800	555	438,649	52.3	50.0	55.1	1,556	1,230,224
15-19	0.0019	0.0096	0.0084	0.0107	87,453	86,588	88,304	836	435,173	47.7	45.3	50.4	2,189	1,139,446
20-24	0.0036	0.0179	0.0157	0.0201	86,617	85,666	87,557	1,552	429,204	43.1	40.8	45.8	3,368	931,326
25-29	0.0061	0.0299	0.0261	0.0340	85,065	83,954	86,173	2,545	418,961	38.8	36.6	41.5	4,647	764,968
30-34	0.0090	0.0439	0.0378	0.0506	82,520	81,143	83,908	3,624	403,537	35.0	32.8	37.5	5,441	605,820
35-39	0.0104	0.0508	0.0431	0.0593	78,895	77,143	80,686	4,009	384,454	31.4	29.4	33.9	5,265	504,938
40-44	0.0118	0.0573	0.0479	0.0676	74,887	72,732	77,118	4,295	363,696	28.0	26.1	30.3	4,901	415,002
45-49	0.0139	0.0673	0.0554	0.0799	70,592	68,032	73,313	4,747	341,091	24.6	22.8	26.7	4,676	335,949
50-54	0.0170	0.0815	0.0664	0.0972	65,844	62,614	69,102	5,366	315,807	21.1	19.5	23.2	4,630	272,475
55-59	0.0217	0.1027	0.0831	0.1227	60,478	56,528	64,312	6,214	286,857	17.8	16.4	19.6	4,657	215,009
60-64	0.0320	0.1481	0.1196	0.1759	54,265	49,595	58,786	8,038	251,228	14.5	13.3	16.2	5,308	165,897
65-69	0.0467	0.2090	0.1696	0.2452	46,227	40,870	51,635	9,659	206,985	11.6	10.6	13.0	5,721	122,589
70-74	0.0710	0.3016	0.2489	0.3465	36,568	30,848	42,836	11,030	155,262	9.1	8.3	10.2	5,903	83,087
75-79	0.1079	0.4250	0.3616	0.4747	25,537	20,158	32,172	10,853	100,552	6.9	6.3	7.7	5,307	49,168
80-84	0.1647	0.5832	0.5178	0.6313	14,684	10,602	20,538	8,564	52,009	5.1	4.7	5.7	3,766	22,873
85-89	0.2419	0.7536	0.6994	0.7916	6,120	3,932	9,904	4,612	19,069	3.8	3.6	4.2	1,667	6,890
90-94	0.3419	0.8439	0.8026	0.8725	1,508	829	2,978	1,272	3,721	2.8	2.6	3.0	417	1,219
95-99	0.4653	0.9710	0.9556	0.9801	235	108	588	228	491	2.1	2.1	2.2	53	114
100+	0.6093	1	1	1	7	2	26	7	11	1.6	1.6	1.7	4	6
All ages													*108,622*	*9,613,495*

ln($_nq_x$) with uncertainty interval

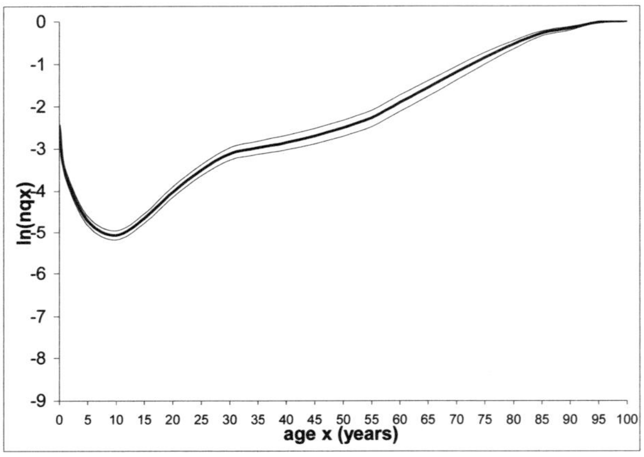

l_x with uncertainty interval

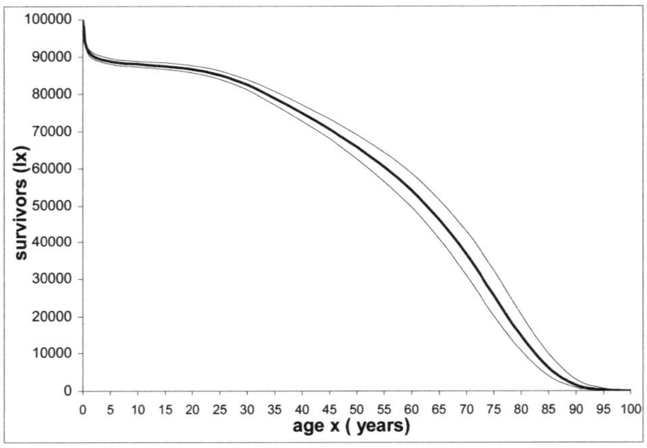

e_x with uncertainty interval

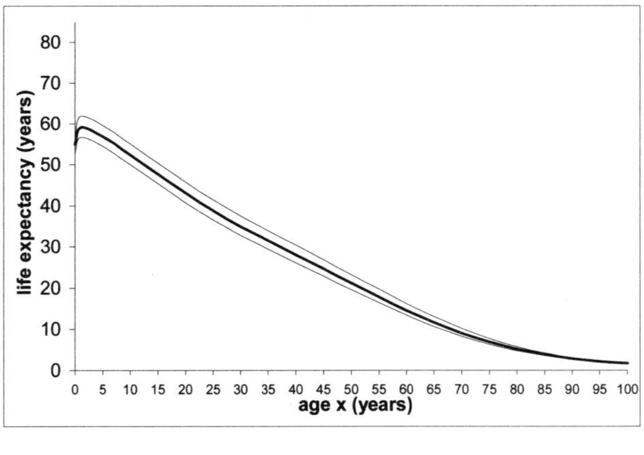

Distribution of deaths by age with uncertainty interval

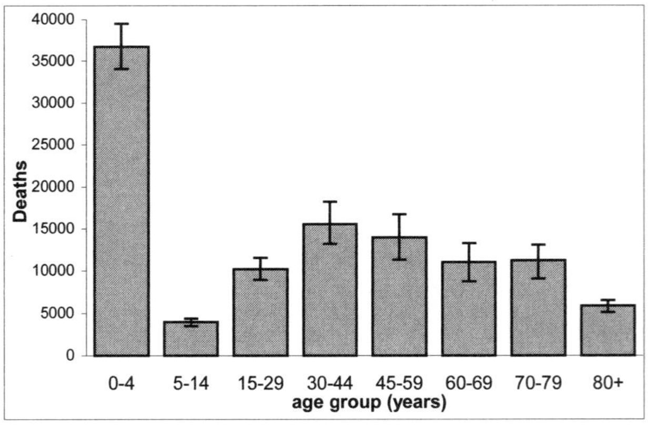

Ghana **Ghana** **Ghana**

Females, 2000 *Femmes, 2000* *Mujeres, 2000*

Age	$_nM_x$	$_nq_x$			l_x			$_nd_x$	$_nL_x$	e_x			Actual deaths	Actual population
			percentiles			percentiles					percentiles			
		mean	2½	97½	mean	2½	97½			mean	2½	97½		
<1	0.0757	0.0709	0.0640	0.0790	100,000	100,000	100,000	7,091	93,618	57.9	55.5	61.2	22,686	299,523
1-4	0.0074	0.0292	0.0253	0.0326	92,909	92,100	93,603	2,715	365,121	61.3	58.7	64.7	8,163	1,097,699
5-9	0.0017	0.0085	0.0072	0.0096	90,194	89,446	90,943	765	449,058	59.1	56.5	62.3	2,230	1,308,644
10-14	0.0012	0.0059	0.0050	0.0067	89,429	88,644	90,195	532	445,816	54.6	52.0	57.7	1,453	1,218,296
15-19	0.0024	0.0121	0.0102	0.0137	88,897	88,069	89,705	1,076	441,796	49.9	47.4	53.0	2,759	1,132,315
20-24	0.0050	0.0247	0.0206	0.0281	87,821	86,862	88,780	2,170	433,680	45.5	43.0	48.5	4,648	929,047
25-29	0.0078	0.0382	0.0315	0.0437	85,651	84,420	86,947	3,274	420,070	41.5	39.1	44.5	5,967	765,589
30-34	0.0094	0.0459	0.0374	0.0533	82,377	80,728	84,205	3,781	402,433	38.1	35.8	40.9	5,718	608,577
35-39	0.0092	0.0447	0.0361	0.0527	78,596	76,466	81,056	3,517	384,188	34.8	32.7	37.4	4,681	511,364
40-44	0.0089	0.0437	0.0350	0.0520	75,079	72,513	78,132	3,281	367,194	31.3	29.3	33.7	3,802	425,480
45-49	0.0094	0.0461	0.0367	0.0552	71,798	68,844	75,399	3,311	350,714	27.6	25.8	29.9	3,299	349,502
50-54	0.0116	0.0564	0.0448	0.0678	68,487	65,165	72,628	3,859	332,789	23.9	22.2	26.0	3,346	288,545
55-59	0.0151	0.0729	0.0579	0.0880	64,628	60,912	69,376	4,714	311,355	20.1	18.6	22.1	3,520	232,509
60-64	0.0223	0.1058	0.0843	0.1274	59,914	55,669	65,357	6,337	283,726	16.5	15.2	18.3	4,107	183,890
65-69	0.0346	0.1592	0.1270	0.1904	53,577	48,576	59,850	8,529	246,561	13.2	12.0	14.7	4,838	139,856
70-74	0.0563	0.2466	0.1987	0.2899	45,048	39,326	52,189	11,110	197,464	10.2	9.3	11.5	5,517	98,060
75-79	0.0897	0.3664	0.3027	0.4186	33,938	27,924	41,627	12,436	138,600	7.7	7.0	8.8	5,442	60,647
80-84	0.1400	0.5184	0.4466	0.5708	21,502	16,236	28,679	11,147	79,641	5.7	5.3	6.5	4,210	30,077
85-89	0.2106	0.6898	0.6246	0.7328	10,355	6,968	15,720	7,142	33,918	4.2	3.9	4.7	2,128	10,105
90-94	0.3055	0.7970	0.7461	0.8295	3,212	1,862	5,901	2,560	8,381	3.1	2.9	3.4	650	2,129
95-99	0.4273	0.9363	0.9092	0.9520	652	317	1,498	611	1,429	2.3	2.2	2.4	114	266
100+	0.5764	1	1	1	42	15	136	42	72	1.7	1.7	1.9	10	18
All ages													*99,288*	*9,692,138*

ln($_nq_x$) with uncertainty interval

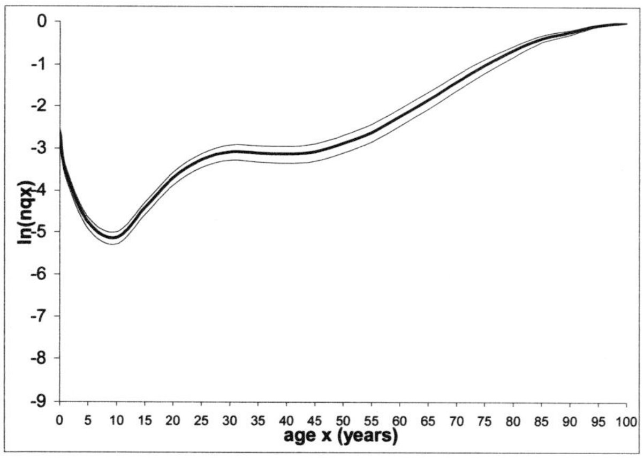

l_x with uncertainty interval

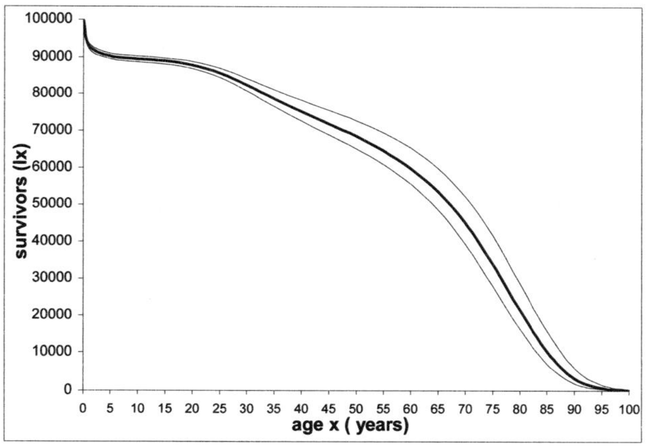

e_x with uncertainty interval

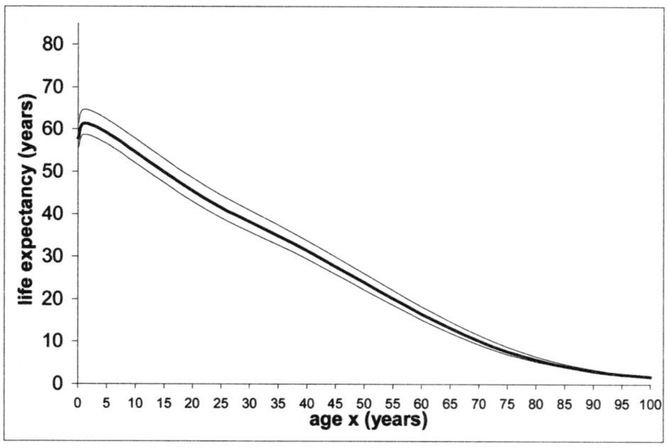

Distribution of deaths by age with uncertainty interval

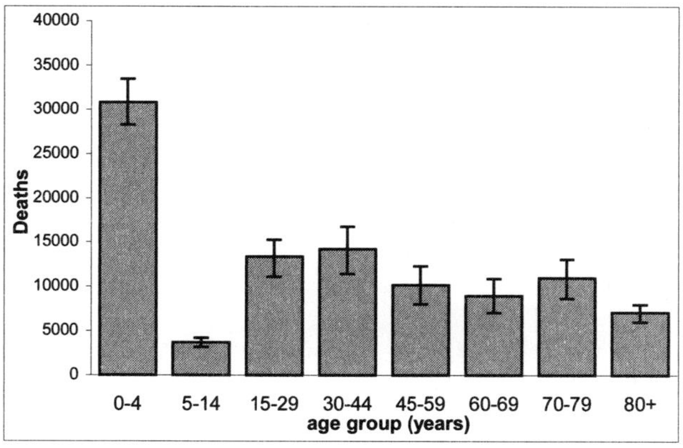

Greece　　　Grèce　　　Grecia

Males, 2000　　　*Hommes, 2000*　　　*Hombres, 2000*

Age	$_nM_x$	$_nq_x$			l_x			$_nd_x$	$_nL_x$	e_x			Actual deaths	Actual population
			percentiles			percentiles					percentiles			
		mean	2½	97½	mean	2½	97½			mean	2½	97½		
<1	0.0061	0.0061	0.0053	0.0070	100,000	100,000	100,000	609	99,452	75.4	75.1	75.7	320	52,194
1-4	0.0003	0.0010	0.0009	0.0011	99,391	99,303	99,466	100	397,325	74.8	74.5	75.2	52	206,466
5-9	0.0001	0.0007	0.0007	0.0008	99,291	99,193	99,375	72	496,277	70.9	70.6	71.3	39	270,520
10-14	0.0002	0.0008	0.0007	0.0009	99,220	99,114	99,310	77	495,905	66.0	65.6	66.3	46	294,356
15-19	0.0007	0.0033	0.0031	0.0036	99,142	99,029	99,239	331	494,886	61.0	60.7	61.4	237	355,187
20-24	0.0010	0.0052	0.0049	0.0057	98,812	98,668	98,935	516	492,768	56.2	55.9	56.6	417	397,693
25-29	0.0011	0.0054	0.0050	0.0058	98,295	98,110	98,454	529	490,155	51.5	51.2	51.8	440	407,765
30-34	0.0011	0.0057	0.0053	0.0060	97,767	97,545	97,956	553	487,451	46.8	46.4	47.1	464	408,963
35-39	0.0014	0.0068	0.0064	0.0071	97,214	96,959	97,431	657	484,427	42.0	41.7	42.4	517	381,422
40-44	0.0022	0.0108	0.0103	0.0113	96,557	96,267	96,805	1,043	480,179	37.3	37.0	37.6	799	367,788
45-49	0.0033	0.0166	0.0160	0.0172	95,514	95,179	95,805	1,585	473,611	32.7	32.3	33.0	1,164	347,942
50-54	0.0054	0.0265	0.0255	0.0275	93,930	93,541	94,274	2,491	463,422	28.2	27.9	28.5	1,761	327,616
55-59	0.0079	0.0389	0.0374	0.0403	91,439	90,995	91,849	3,559	448,297	23.9	23.6	24.2	2,300	289,662
60-64	0.0131	0.0636	0.0612	0.0660	87,880	87,403	88,348	5,591	425,423	19.7	19.5	20.0	3,895	296,374
65-69	0.0217	0.1029	0.0988	0.1069	82,289	81,719	82,877	8,467	390,279	15.9	15.7	16.2	6,224	286,915
70-74	0.0357	0.1638	0.1571	0.1702	73,823	73,022	74,668	12,095	338,875	12.4	12.2	12.7	8,466	237,186
75-79	0.0572	0.2503	0.2405	0.2595	61,727	60,626	62,856	15,447	270,018	9.4	9.2	9.6	8,471	148,078
80-84	0.1056	0.4179	0.4035	0.4326	46,280	44,929	47,694	19,339	183,052	6.7	6.5	6.9	8,742	82,745
85-89	0.1795	0.6195	0.6047	0.6353	26,941	25,674	28,381	16,690	92,981	4.7	4.6	4.8	8,476	47,220
90-94	0.2806	0.7617	0.7496	0.7741	10,251	9,416	11,176	7,809	27,830	3.3	3.2	3.4	5,295	18,871
95-99	0.4035	0.9127	0.9059	0.9194	2,443	2,127	2,786	2,229	5,525	2.4	2.4	2.5	1,805	4,474
100+	0.5339	1	1	1	213	172	260	213	399	1.9	1.9	2.0	314	588
All ages													*60,242*	*5,230,025*

ln($_nq_x$) with uncertainty interval

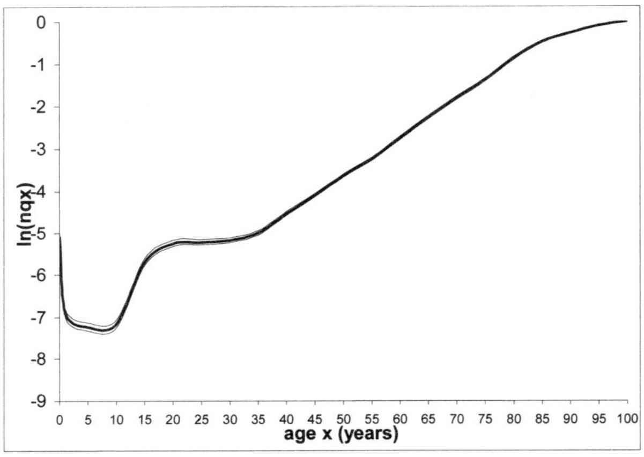

l_x with uncertainty interval

e_x with uncertainty interval

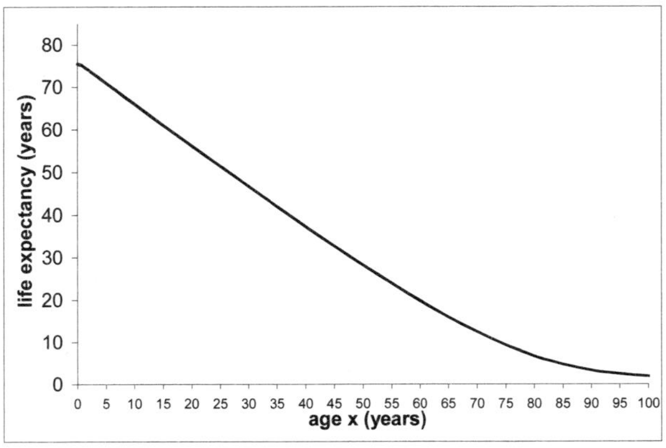

Distribution of deaths by age with uncertainty interval

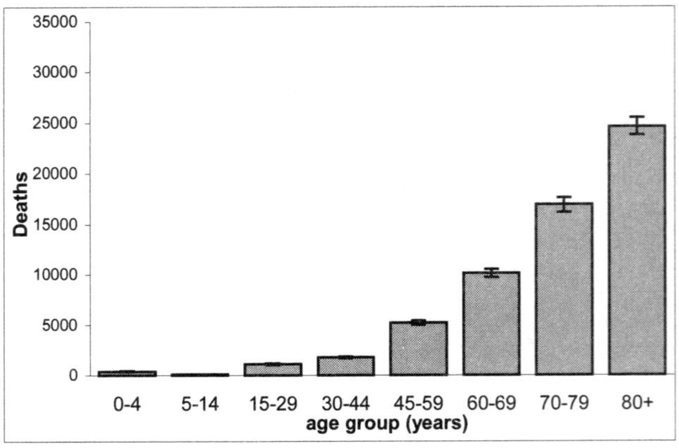

Greece | Grèce | Grecia

Females, 2000 | *Femmes, 2000* | *Mujeres, 2000*

Age	$_nM_x$	$_nq_x$			l_x			$_nd_x$	$_nL_x$	e_x			Actual deaths	Actual population
			percentiles			percentiles					percentiles			
		mean	2½	97½	mean	2½	97½			mean	2½	97½		
<1	0.0052	0.0051	0.0043	0.0060	100,000	100,000	100,000	514	99,537	80.8	80.1	81.6	253	49,078
1-4	0.0003	0.0011	0.0009	0.0012	99,486	99,399	99,571	105	397,693	80.2	79.6	81.0	51	192,887
5-9	0.0001	0.0005	0.0005	0.0006	99,381	99,282	99,480	54	496,772	76.2	75.6	77.1	27	253,691
10-14	0.0001	0.0006	0.0005	0.0006	99,328	99,223	99,434	56	496,499	71.3	70.7	72.1	31	278,781
15-19	0.0003	0.0013	0.0011	0.0015	99,272	99,160	99,385	129	496,036	66.3	65.7	67.1	88	335,962
20-24	0.0003	0.0015	0.0013	0.0016	99,143	99,015	99,271	145	495,349	61.4	60.8	62.2	111	377,134
25-29	0.0004	0.0018	0.0016	0.0020	98,997	98,853	99,143	180	494,536	56.5	55.9	57.3	142	391,243
30-34	0.0004	0.0021	0.0019	0.0024	98,817	98,654	98,982	211	493,558	51.6	51.0	52.4	172	401,733
35-39	0.0006	0.0029	0.0026	0.0032	98,606	98,421	98,793	287	492,312	46.7	46.1	47.5	224	384,080
40-44	0.0009	0.0047	0.0043	0.0052	98,319	98,106	98,535	464	490,434	41.8	41.3	42.6	352	372,063
45-49	0.0015	0.0074	0.0067	0.0080	97,855	97,600	98,112	722	487,470	37.0	36.4	37.8	515	348,142
50-54	0.0022	0.0111	0.0101	0.0120	97,133	96,820	97,449	1,075	482,979	32.3	31.7	33.1	738	331,763
55-59	0.0031	0.0152	0.0139	0.0165	96,058	95,662	96,452	1,462	476,637	27.6	27.1	28.4	928	302,425
60-64	0.0054	0.0269	0.0243	0.0291	94,596	94,100	95,087	2,542	466,627	23.0	22.5	23.7	1,768	324,642
65-69	0.0099	0.0484	0.0433	0.0526	92,054	91,411	92,687	4,453	449,140	18.6	18.1	19.3	3,204	323,178
70-74	0.0202	0.0963	0.0855	0.1049	87,602	86,647	88,610	8,435	416,921	14.4	13.9	15.0	5,816	287,460
75-79	0.0385	0.1756	0.1561	0.1905	79,167	77,613	81,033	13,905	361,073	10.6	10.3	11.2	7,681	199,464
80-84	0.0866	0.3558	0.3229	0.3804	65,262	62,827	68,382	23,219	268,263	7.4	7.1	7.8	10,455	120,796
85-89	0.1600	0.5714	0.5382	0.5992	42,043	39,038	46,302	24,024	150,153	5.1	4.9	5.3	11,423	71,396
90-94	0.2600	0.7303	0.7069	0.7516	18,018	15,845	21,381	13,159	50,613	3.5	3.4	3.7	6,910	26,578
95-99	0.3800	0.8879	0.8748	0.8997	4,859	3,949	6,224	4,314	11,353	2.6	2.5	2.7	2,478	6,522
100+	0.5200	1	1	1	545	400	762	545	1,048	1.9	1.9	2.3	478	919
All ages													*53,847*	*5,379,937*

ln($_nq_x$) with uncertainty interval

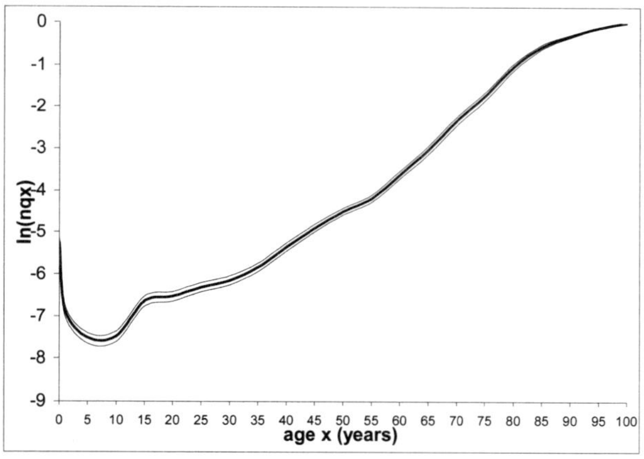

l_x with uncertainty interval

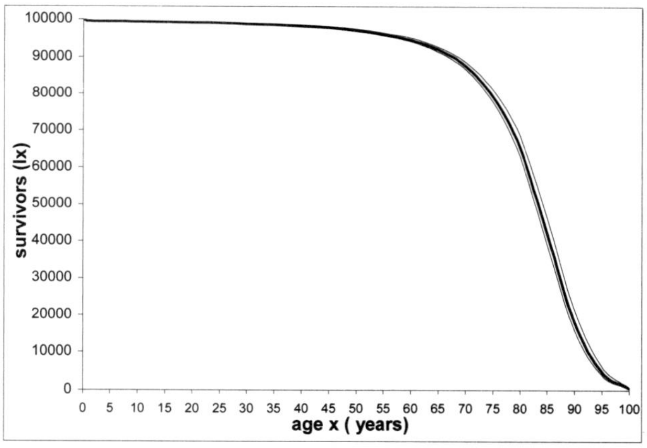

e_x with uncertainty interval

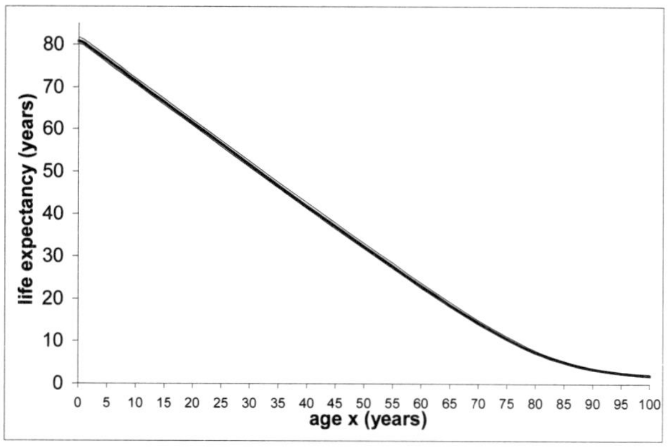

Distribution of deaths by age with uncertainty interval

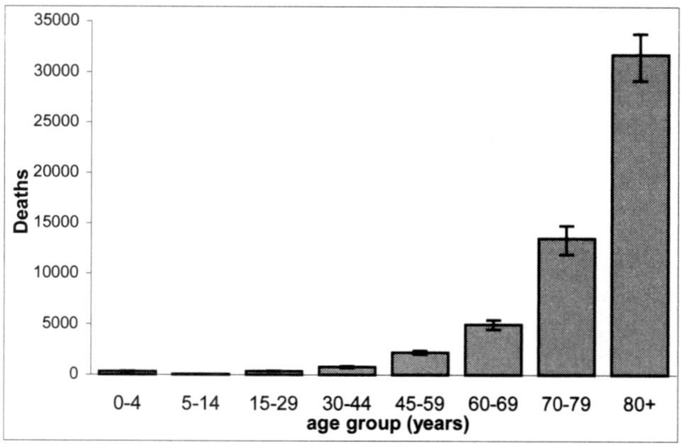

Grenada | Grenade | Granada

Males, 2000 | *Hommes, 2000* | *Hombres, 2000*

Age	$_nM_x$	$_nq_x$			l_x			$_nd_x$	$_nL_x$	e_x			Actual deaths	Actual population
			percentiles			percentiles					percentiles			
		mean	2½	97½	mean	2½	97½			mean	2½	97½		
<1	0.0187	0.0184	0.0131	0.0227	100,000	100,000	100,000	1,843	98,341	70.9	69.6	72.1	17	927
1-4	0.0016	0.0065	0.0051	0.0076	98,157	97,729	98,691	636	391,100	71.2	69.8	72.7	6	3,674
5-9	0.0003	0.0013	0.0010	0.0015	97,520	96,981	98,191	122	487,296	67.7	66.3	69.2	1	4,630
10-14	0.0003	0.0015	0.0012	0.0017	97,398	96,839	98,094	144	486,629	62.7	61.3	64.3	1	4,824
15-19	0.0006	0.0030	0.0024	0.0035	97,254	96,671	97,978	287	485,550	57.8	56.4	59.4	3	4,490
20-24	0.0010	0.0050	0.0040	0.0058	96,966	96,337	97,746	484	483,623	53.0	51.6	54.6	4	4,002
25-29	0.0014	0.0069	0.0057	0.0080	96,483	95,777	97,351	669	480,741	48.3	46.8	49.8	6	4,018
30-34	0.0024	0.0118	0.0099	0.0136	95,814	95,009	96,796	1,134	476,233	43.6	42.1	45.2	9	3,758
35-39	0.0035	0.0173	0.0149	0.0197	94,680	93,717	95,834	1,643	469,292	39.1	37.6	40.7	12	3,394
40-44	0.0050	0.0247	0.0218	0.0278	93,037	91,869	94,403	2,297	459,442	34.7	33.3	36.3	13	2,568
45-49	0.0070	0.0343	0.0313	0.0381	90,740	89,318	92,341	3,112	445,918	30.5	29.1	32.1	16	2,273
50-54	0.0100	0.0488	0.0441	0.0543	87,627	85,912	89,455	4,275	427,451	26.5	25.1	28.1	20	1,970
55-59	0.0143	0.0689	0.0609	0.0775	83,353	81,247	85,513	5,741	402,411	22.8	21.4	24.3	23	1,632
60-64	0.0190	0.0907	0.0787	0.1036	77,611	74,951	80,306	7,039	370,461	19.3	18.0	20.7	25	1,333
65-69	0.0259	0.1217	0.1044	0.1403	70,573	67,186	73,983	8,590	331,388	15.9	14.8	17.3	26	1,013
70-74	0.0415	0.1880	0.1606	0.2166	61,982	57,760	66,261	11,655	280,775	12.8	11.7	14.0	34	809
75-79	0.0661	0.2836	0.2442	0.3232	50,328	45,250	55,616	14,271	215,960	10.2	9.3	11.2	36	551
80-84	0.0877	0.3595	0.3140	0.4049	36,056	30,624	42,036	12,963	147,874	8.2	7.5	9.0	30	348
85-89	0.1210	0.4644	0.4155	0.5118	23,093	18,226	28,835	10,724	88,655	6.4	5.9	7.0	20	167
90-94	0.1736	0.5708	0.5241	0.6158	12,369	8,898	16,855	7,060	40,666	4.8	4.4	5.2	9	53
95-99	0.2592	0.7290	0.6891	0.7670	5,309	3,419	8,022	3,871	14,934	3.5	3.2	3.7	3	10
100+	0.4024	1	1	1	1,439	797	2,494	1,439	3,575	2.5	2.2	2.6	0	1
All ages													*315*	*46,447*

$\ln({}_nq_x)$ with uncertainty interval

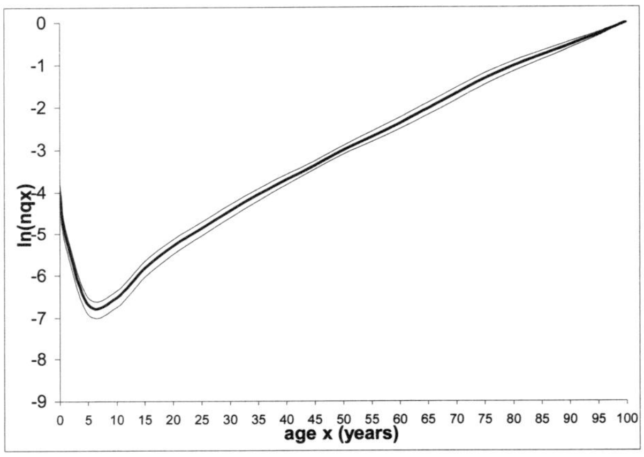

l_x with uncertainty interval

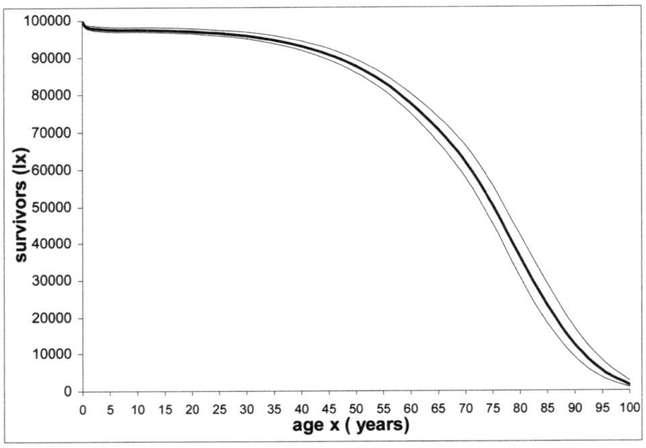

e_x with uncertainty interval

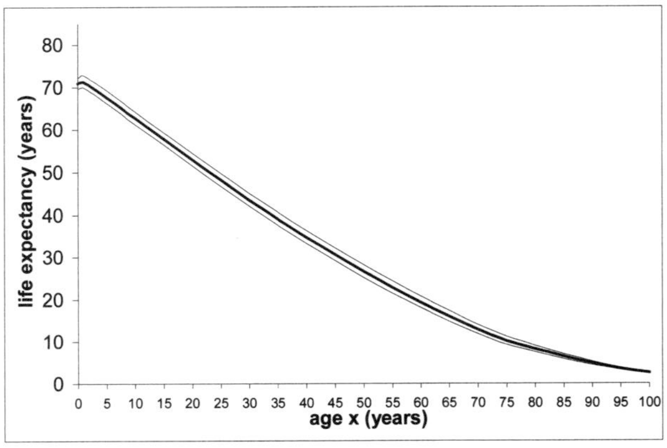

Distribution of deaths by age with uncertainty interval

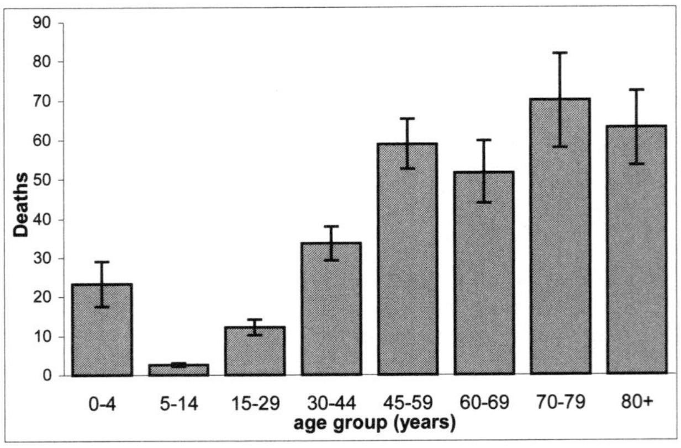

Grenada | Grenade | Granada

Females, 2000 | *Femmes, 2000* | *Mujeres, 2000*

Age	$_nM_x$	$_nq_x$			l_x			$_nd_x$	$_nL_x$	e_x			Actual deaths	Actual population
			percentiles			percentiles					percentiles			
		mean	2½	97½	mean	2½	97½			mean	2½	97½		
<1	0.0186	0.0183	0.0123	0.0230	100,000	100,000	100,000	1,825	98,357	73.2	72.1	74.6	17	897
1-4	0.0009	0.0035	0.0027	0.0042	98,175	97,705	98,767	347	391,865	73.6	72.2	75.1	3	3,518
5-9	0.0006	0.0031	0.0025	0.0037	97,827	97,295	98,499	308	488,366	69.9	68.5	71.4	3	4,456
10-14	0.0006	0.0030	0.0024	0.0036	97,519	96,936	98,258	297	486,853	65.1	63.7	66.6	3	4,669
15-19	0.0009	0.0045	0.0036	0.0052	97,222	96,592	98,021	433	485,028	60.3	58.8	61.9	4	4,345
20-24	0.0010	0.0050	0.0041	0.0057	96,789	96,094	97,671	483	482,738	55.5	54.1	57.1	4	3,892
25-29	0.0010	0.0050	0.0041	0.0057	96,306	95,543	97,274	480	480,331	50.8	49.3	52.4	4	3,942
30-34	0.0014	0.0069	0.0058	0.0078	95,826	95,000	96,873	664	477,469	46.0	44.6	47.7	5	3,739
35-39	0.0025	0.0124	0.0107	0.0139	95,162	94,256	96,310	1,182	472,853	41.3	39.9	43.0	9	3,511
40-44	0.0035	0.0173	0.0153	0.0192	93,980	92,946	95,284	1,630	465,822	36.8	35.3	38.5	10	2,771
45-49	0.0055	0.0271	0.0246	0.0295	92,349	91,165	93,829	2,505	455,483	32.4	30.9	34.1	14	2,459
50-54	0.0079	0.0387	0.0350	0.0422	89,844	88,479	91,525	3,479	440,522	28.3	26.8	29.9	17	2,134
55-59	0.0110	0.0535	0.0477	0.0597	86,365	84,741	88,326	4,623	420,266	24.3	22.9	25.9	19	1,769
60-64	0.0156	0.0750	0.0655	0.0852	81,742	79,682	84,117	6,130	393,385	20.5	19.2	22.1	23	1,460
65-69	0.0240	0.1132	0.0973	0.1304	75,612	72,892	78,604	8,560	356,661	17.0	15.7	18.5	28	1,147
70-74	0.0366	0.1677	0.1426	0.1943	67,052	63,386	70,959	11,245	307,149	13.8	12.6	15.2	34	938
75-79	0.0590	0.2571	0.2193	0.2981	55,807	51,073	60,838	14,350	243,160	11.1	10.0	12.3	39	656
80-84	0.0757	0.3181	0.2740	0.3681	41,457	35,846	47,499	13,188	174,315	9.1	8.2	10.0	33	431
85-89	0.1023	0.4074	0.3581	0.4633	28,269	22,649	34,484	11,517	112,553	7.2	6.5	7.9	22	220
90-94	0.1460	0.5077	0.4579	0.5635	16,752	12,157	22,134	8,505	58,247	5.4	4.9	5.9	11	79
95-99	0.2198	0.6623	0.6165	0.7129	8,248	5,306	11,999	5,462	24,851	4.0	3.5	4.2	4	17
100+	0.3491	1	1	1	2,785	1,523	4,602	2,785	7,978	2.9	2.4	3.0	1	3
All ages													*305*	*47,055*

ln($_nq_x$) with uncertainty interval

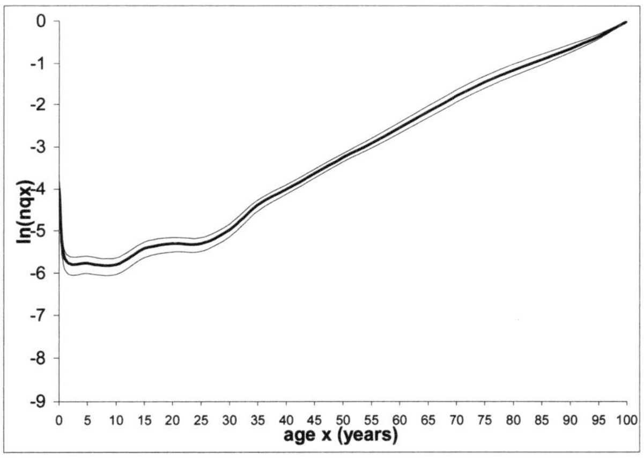

l_x with uncertainty interval

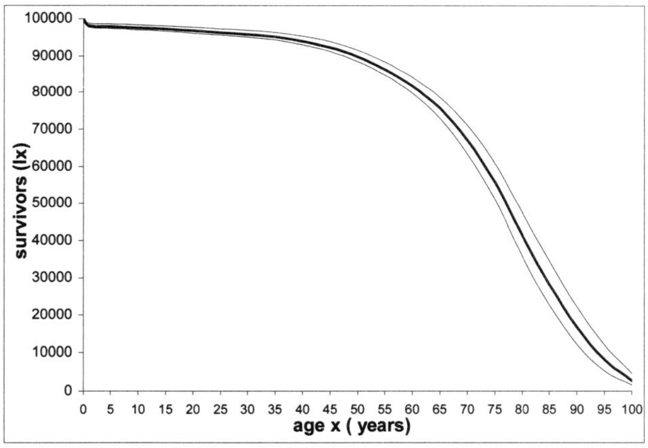

e_x with uncertainty interval

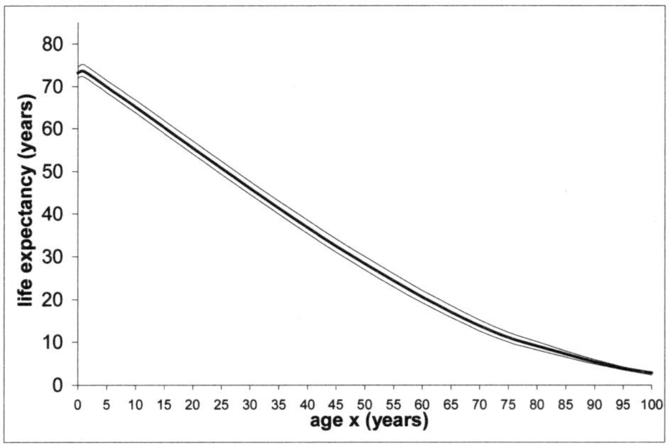

Distribution of deaths by age with uncertainty interval

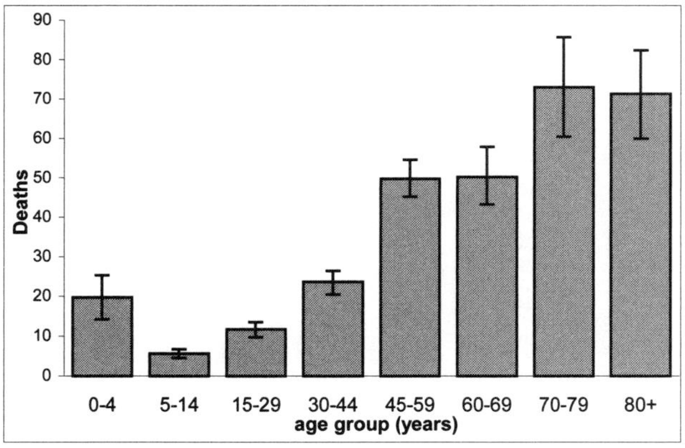

Guatemala

Males, 2000

Guatemala

Hommes, 2000

Guatemala

Hombres, 2000

Age	$_nM_x$	$_nq_x$			l_x			$_nd_x$	$_nL_x$	e_x			Actual deaths	Actual population
			percentiles			percentiles					percentiles			
		mean	2½	97½	mean	2½	97½			mean	2½	97½		
<1	0.0386	0.0373	0.0339	0.0411	100,000	100,000	100,000	3,727	96,646	63.5	62.2	65.1	7,696	199,586
1-4	0.0045	0.0177	0.0163	0.0193	96,273	95,887	96,613	1,701	381,010	65.0	63.8	66.5	3,314	742,253
5-9	0.0008	0.0041	0.0038	0.0045	94,572	94,038	95,039	391	471,882	62.1	61.0	63.6	702	847,094
10-14	0.0009	0.0042	0.0039	0.0046	94,181	93,613	94,679	399	469,906	57.4	56.3	58.8	634	745,612
15-19	0.0019	0.0096	0.0088	0.0105	93,782	93,179	94,311	904	466,647	52.6	51.6	54.1	1,278	659,210
20-24	0.0035	0.0176	0.0160	0.0192	92,877	92,196	93,482	1,632	460,305	48.1	47.1	49.5	1,935	545,578
25-29	0.0045	0.0225	0.0203	0.0246	91,245	90,424	91,989	2,050	451,099	43.9	43.0	45.3	1,967	432,649
30-34	0.0053	0.0263	0.0237	0.0288	89,195	88,203	90,119	2,349	440,101	39.9	39.0	41.2	1,802	337,558
35-39	0.0075	0.0368	0.0331	0.0402	86,846	85,665	87,982	3,200	426,229	35.9	35.0	37.1	2,036	271,266
40-44	0.0081	0.0397	0.0355	0.0432	83,646	82,221	85,072	3,321	409,928	32.2	31.4	33.3	1,776	219,190
45-49	0.0100	0.0489	0.0437	0.0531	80,325	78,665	82,049	3,926	391,812	28.4	27.7	29.4	1,860	185,599
50-54	0.0119	0.0576	0.0515	0.0624	76,400	74,487	78,466	4,404	370,989	24.7	24.1	25.7	1,721	144,980
55-59	0.0145	0.0698	0.0624	0.0754	71,996	69,836	74,426	5,027	347,414	21.1	20.5	21.9	1,706	117,905
60-64	0.0198	0.0942	0.0845	0.1015	66,969	64,568	69,780	6,312	319,068	17.5	16.9	18.2	1,947	98,444
65-69	0.0307	0.1428	0.1288	0.1532	60,658	58,016	63,883	8,659	281,642	14.0	13.5	14.7	2,472	80,402
70-74	0.0483	0.2153	0.1965	0.2303	51,999	49,125	55,652	11,196	232,005	11.0	10.5	11.5	2,756	57,107
75-79	0.0813	0.3378	0.3141	0.3590	40,803	37,813	44,715	13,782	169,561	8.3	7.9	8.7	2,759	33,945
80-84	0.1259	0.4788	0.4524	0.5028	27,021	24,238	30,671	12,937	102,762	6.2	6.0	6.5	1,918	15,233
85-89	0.1885	0.6406	0.6161	0.6627	14,084	12,051	16,796	9,022	47,864	4.6	4.5	4.8	1,061	5,626
90-94	0.2692	0.7446	0.7248	0.7625	5,062	4,065	6,448	3,769	14,002	3.4	3.3	3.5	340	1,263
95-99	0.3890	0.8976	0.8854	0.9084	1,293	965	1,775	1,161	2,983	2.5	2.5	2.6	75	193
100+	0.5888	1	1	1	132	88	203	132	225	1.7	1.7	1.9	10	17
All ages													*41,762*	*5,740,710*

ln($_nq_x$) with uncertainty interval

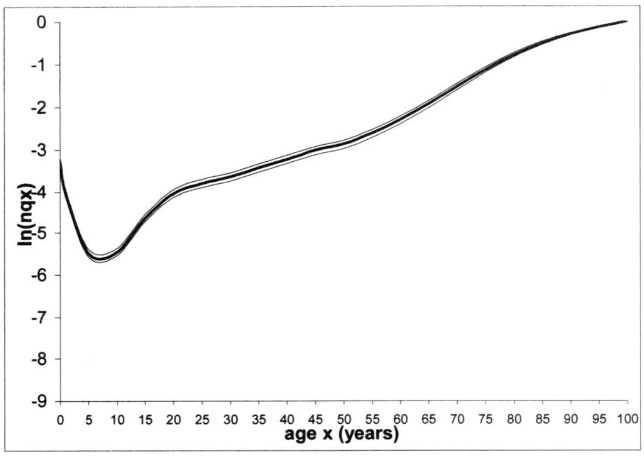

l_x with uncertainty interval

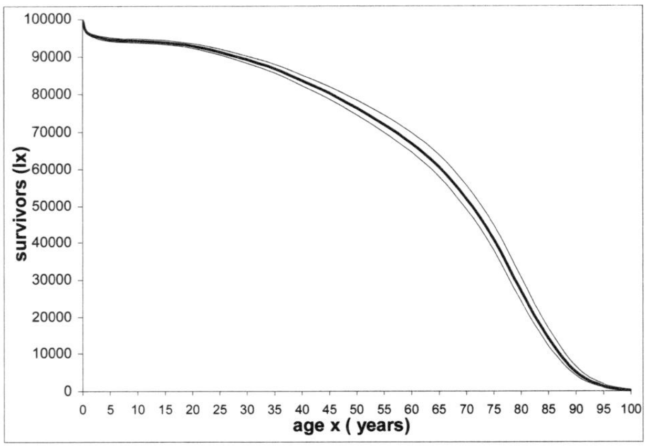

e_x with uncertainty interval

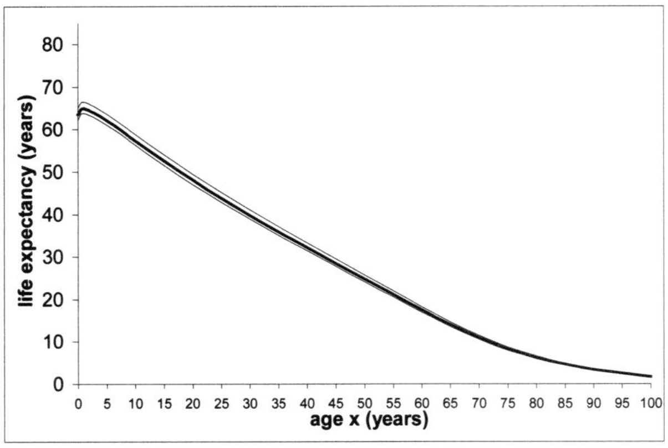

Distribution of deaths by age with uncertainty interval

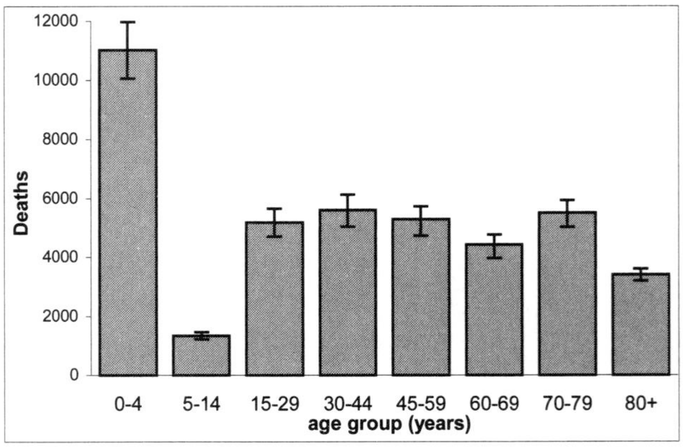

Guatemala

Guatemala

Guatemala

Females, 2000 *Femmes, 2000* *Mujeres, 2000*

Age	$_nM_x$	$_nq_x$			l_x			$_nd_x$	$_nL_x$	e_x			Actual deaths	Actual population
			percentiles			percentiles					percentiles			
		mean	21/2	971/2	mean	21/2	971/2			mean	21/2	971/2		
<1	0.0346	0.0336	0.0291	0.0366	100,000	100,000	100,000	3,360	96,976	68.6	67.2	70.6	6,582	189,991
1-4	0.0049	0.0196	0.0174	0.0210	96,640	96,345	97,090	1,890	382,025	69.9	68.6	71.9	3,530	713,483
5-9	0.0008	0.0041	0.0036	0.0044	94,750	94,326	95,404	391	472,775	67.3	66.0	69.2	671	812,189
10-14	0.0008	0.0041	0.0036	0.0044	94,360	93,907	95,059	383	470,841	62.6	61.3	64.4	581	715,008
15-19	0.0012	0.0060	0.0052	0.0065	93,977	93,494	94,721	561	468,481	57.8	56.6	59.6	758	633,078
20-24	0.0017	0.0084	0.0073	0.0092	93,416	92,889	94,226	786	465,112	53.1	51.9	54.9	898	531,073
25-29	0.0019	0.0094	0.0081	0.0102	92,629	92,038	93,535	868	460,976	48.6	47.4	50.3	807	428,393
30-34	0.0025	0.0125	0.0108	0.0137	91,761	91,097	92,774	1,152	455,927	44.0	42.9	45.7	867	343,206
35-39	0.0036	0.0177	0.0153	0.0195	90,610	89,845	91,769	1,607	449,032	39.5	38.4	41.2	1,005	280,742
40-44	0.0044	0.0216	0.0185	0.0239	89,003	88,093	90,369	1,926	440,199	35.2	34.1	36.8	996	227,638
45-49	0.0055	0.0271	0.0231	0.0301	87,077	85,988	88,696	2,362	429,480	30.9	29.9	32.4	1,053	191,520
50-54	0.0084	0.0410	0.0348	0.0456	84,715	83,403	86,646	3,471	414,899	26.7	25.8	28.1	1,249	149,244
55-59	0.0111	0.0540	0.0458	0.0602	81,244	79,604	83,631	4,390	395,246	22.8	21.9	24.1	1,322	119,051
60-64	0.0158	0.0762	0.0645	0.0849	76,854	74,809	79,805	5,853	369,638	18.9	18.1	20.1	1,584	100,034
65-69	0.0265	0.1244	0.1058	0.1383	71,001	68,455	74,660	8,833	332,923	15.3	14.5	16.3	2,233	84,174
70-74	0.0420	0.1901	0.1637	0.2102	62,168	58,987	66,757	11,818	281,296	12.1	11.4	12.9	2,581	61,439
75-79	0.0661	0.2835	0.2496	0.3114	50,350	46,587	55,828	14,275	216,064	9.3	8.8	10.0	2,485	37,611
80-84	0.1000	0.4000	0.3630	0.4341	36,075	32,082	41,894	14,430	144,301	7.0	6.6	7.5	1,748	17,478
85-89	0.1698	0.5961	0.5611	0.6311	21,645	18,155	26,688	12,902	75,971	5.0	4.7	5.3	1,182	6,959
90-94	0.2344	0.6882	0.6595	0.7173	8,743	6,697	11,713	6,017	25,666	3.8	3.6	4.0	438	1,867
95-99	0.3631	0.8689	0.8508	0.8874	2,727	1,894	3,988	2,369	6,525	2.7	2.5	2.7	139	383
100+	0.4898	1	1	1	357	213	595	357	730	2.0	1.8	2.1	24	49
All ages													*32,733*	*5,644,610*

$\ln(_nq_x)$ with uncertainty interval

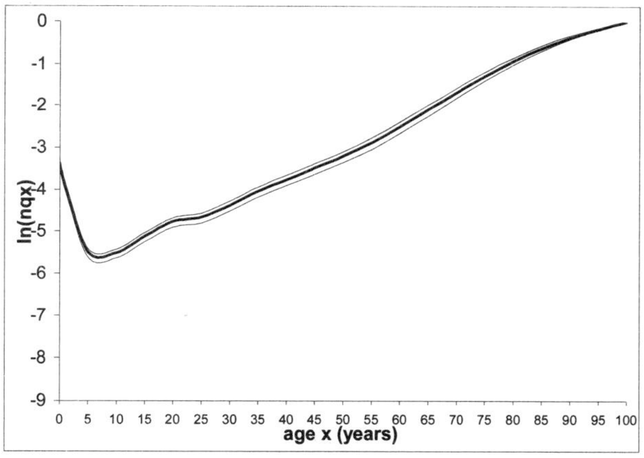

l_x with uncertainty interval

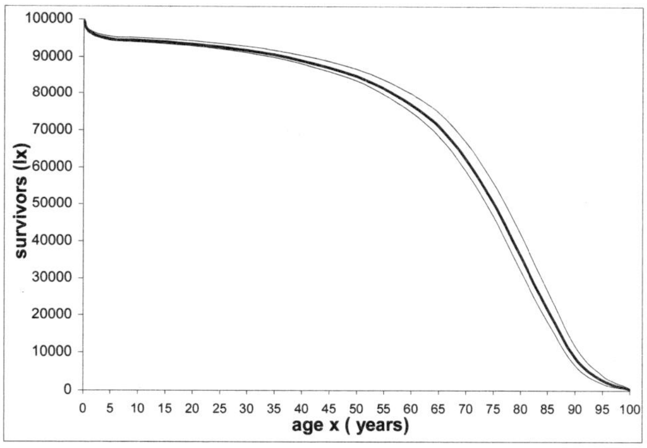

e_x with uncertainty interval

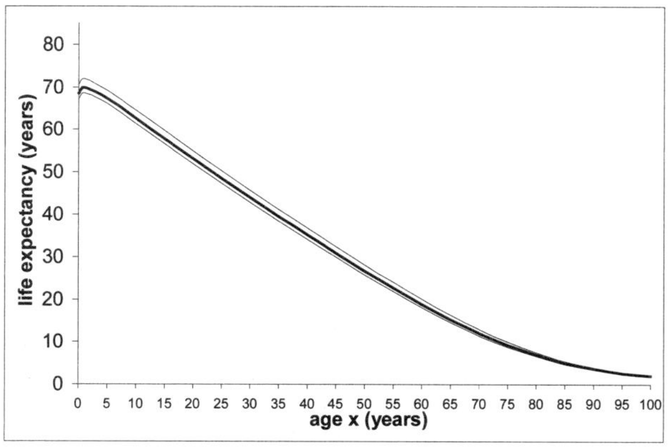

Distribution of deaths by age with uncertainty interval

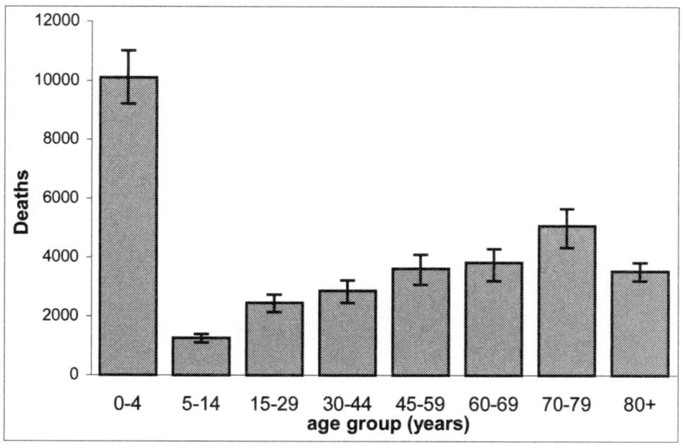

## Guinea	Guinée	Guinea

Males, 2000	*Hommes, 2000*	*Hombres, 2000*

Age	$_nM_x$	$_nq_x$ mean	$_nq_x$ percentiles 2 1/2	$_nq_x$ percentiles 97 1/2	l_x mean	l_x percentiles 2 1/2	l_x percentiles 97 1/2	$_nd_x$	$_nL_x$	e_x mean	e_x percentiles 2 1/2	e_x percentiles 97 1/2	Actual deaths	Actual population
<1	0.1500	0.1322	0.1200	0.1443	100,000	100,000	100,000	13,216	88,106	49.0	46.1	52.0	25,692	171,275
1-4	0.0125	0.0487	0.0422	0.0557	86,784	85,571	87,996	4,228	336,989	55.4	52.3	59.0	7,041	561,243
5-9	0.0031	0.0152	0.0129	0.0177	82,556	81,562	83,555	1,256	409,639	54.2	51.3	57.6	1,786	582,367
10-14	0.0019	0.0097	0.0082	0.0113	81,300	80,153	82,374	787	404,531	50.0	47.2	53.3	982	504,869
15-19	0.0021	0.0104	0.0088	0.0123	80,513	79,250	81,640	840	400,463	45.5	42.7	48.7	954	454,799
20-24	0.0045	0.0224	0.0187	0.0264	79,673	78,284	80,856	1,784	393,904	40.9	38.2	44.1	1,709	377,367
25-29	0.0067	0.0329	0.0273	0.0392	77,889	76,230	79,280	2,564	383,036	36.8	34.2	39.9	1,987	296,961
30-34	0.0091	0.0444	0.0364	0.0532	75,325	73,271	77,114	3,342	368,271	33.0	30.5	36.0	2,323	255,983
35-39	0.0113	0.0551	0.0445	0.0665	71,983	69,378	74,301	3,965	350,004	29.4	27.0	32.3	2,414	213,082
40-44	0.0139	0.0672	0.0537	0.0816	68,019	64,765	70,969	4,570	328,669	25.9	23.8	28.6	2,459	176,853
45-49	0.0170	0.0816	0.0646	0.0994	63,449	59,483	67,119	5,180	304,295	22.6	20.7	25.1	2,477	145,514
50-54	0.0213	0.1012	0.0796	0.1232	58,269	53,570	62,721	5,897	276,605	19.4	17.6	21.7	2,315	108,578
55-59	0.0271	0.1270	0.0996	0.1539	52,373	46,972	57,645	6,649	245,241	16.3	14.7	18.3	2,281	84,133
60-64	0.0399	0.1813	0.1429	0.2180	45,724	39,744	51,790	8,289	207,897	13.3	12.0	15.1	2,596	65,106
65-69	0.0556	0.2442	0.1949	0.2903	37,435	31,115	44,350	9,142	164,320	10.7	9.6	12.2	2,633	47,330
70-74	0.0823	0.3413	0.2792	0.3966	28,293	22,175	35,708	9,655	117,326	8.4	7.5	9.5	2,529	30,729
75-79	0.1212	0.4651	0.3948	0.5244	18,638	13,502	25,738	8,669	71,516	6.4	5.8	7.3	2,031	16,754
80-84	0.1805	0.6220	0.5526	0.6773	9,969	6,527	15,576	6,200	34,343	4.8	4.4	5.4	1,255	6,953
85-89	0.2596	0.7872	0.7315	0.8296	3,768	2,112	6,969	2,967	11,426	3.6	3.4	4.0	479	1,845
90-94	0.3605	0.8659	0.8233	0.8978	802	360	1,871	694	1,926	2.7	2.5	2.9	103	286
95-99	0.4832	0.9863	0.9757	0.9923	107	37	331	106	219	2.1	2.0	2.1	11	23
100+	0.6253	1	1	1	1	0	8	1	2	1.6	1.6	1.7	1	1
All ages													*66,058*	*4,102,051*

$\ln(_nq_x)$ with uncertainty interval

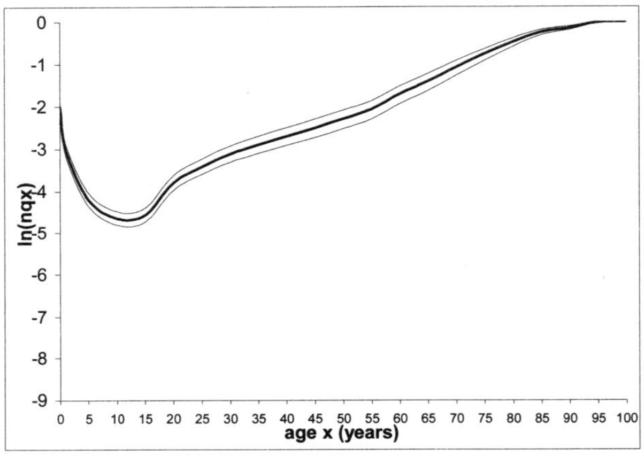

l_x with uncertainty interval

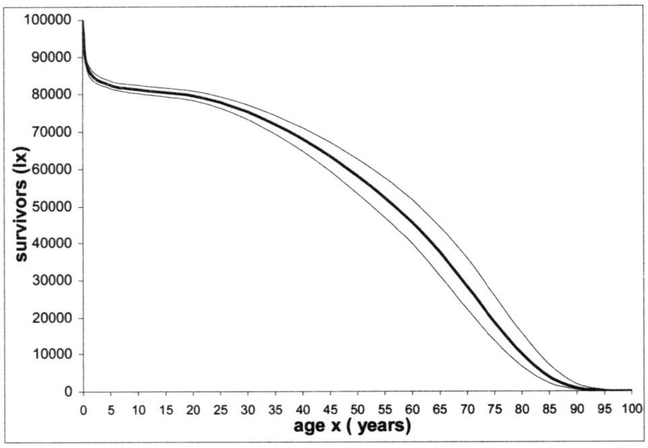

e_x with uncertainty interval

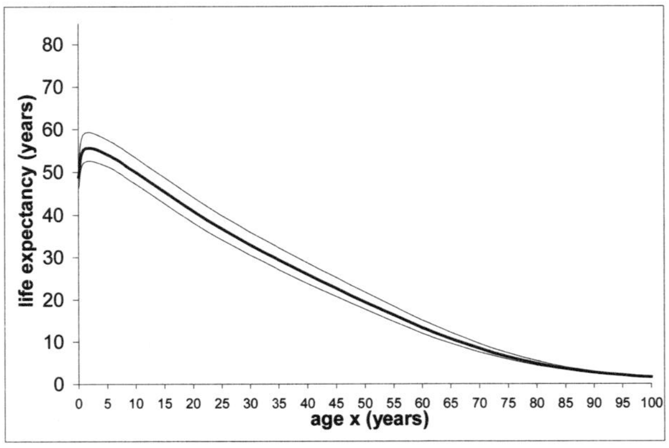

Distribution of deaths by age with uncertainty interval

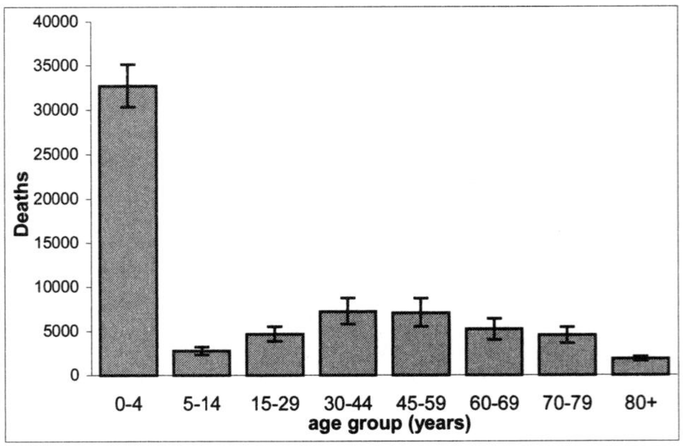

Guinea — Guinée — Guinea

Females, 2000 — *Femmes, 2000* — *Mujeres, 2000*

Age	$_nM_x$	$_nq_x$ mean	$_nq_x$ percentiles 2½	$_nq_x$ percentiles 97½	l_x mean	l_x percentiles 2½	l_x percentiles 97½	$_nd_x$	$_nL_x$	e_x mean	e_x percentiles 2½	e_x percentiles 97½	Actual deaths	Actual population
<1	0.1203	0.1086	0.0973	0.1189	100,000	100,000	100,000	10,856	90,230	52.0	49.0	55.2	19,719	163,900
1-4	0.0136	0.0527	0.0461	0.0596	89,144	88,106	90,272	4,696	345,306	57.3	54.0	61.1	7,500	551,461
5-9	0.0033	0.0162	0.0139	0.0188	84,448	83,452	85,444	1,371	418,815	56.4	53.3	60.0	1,853	566,275
10-14	0.0023	0.0114	0.0096	0.0133	83,078	81,981	84,167	943	413,030	52.3	49.3	55.8	1,120	490,569
15-19	0.0037	0.0185	0.0156	0.0217	82,134	80,943	83,352	1,519	406,876	47.9	44.9	51.4	1,649	441,757
20-24	0.0059	0.0289	0.0241	0.0342	80,616	79,270	82,042	2,328	397,259	43.7	40.8	47.1	2,157	368,026
25-29	0.0074	0.0365	0.0301	0.0437	78,288	76,623	80,041	2,857	384,297	39.9	37.2	43.2	2,166	291,368
30-34	0.0085	0.0414	0.0337	0.0500	75,431	73,309	77,592	3,125	369,342	36.4	33.8	39.5	2,132	252,006
35-39	0.0095	0.0465	0.0373	0.0565	72,306	69,648	74,914	3,361	353,127	32.8	30.4	35.8	2,000	210,111
40-44	0.0099	0.0484	0.0384	0.0592	68,945	65,712	72,030	3,335	336,389	29.3	27.1	32.1	1,733	174,779
45-49	0.0115	0.0561	0.0441	0.0689	65,610	61,824	69,158	3,681	318,848	25.7	23.6	28.3	1,675	145,051
50-54	0.0149	0.0720	0.0562	0.0885	61,929	57,566	65,987	4,459	298,498	22.0	20.2	24.5	1,683	112,680
55-59	0.0198	0.0944	0.0733	0.1158	57,470	52,471	62,220	5,423	273,792	18.6	16.9	20.8	1,791	90,413
60-64	0.0287	0.1338	0.1039	0.1631	52,047	46,393	57,574	6,963	242,828	15.2	13.7	17.2	2,057	71,753
65-69	0.0426	0.1924	0.1503	0.2326	45,084	38,824	51,468	8,673	203,738	12.2	10.9	13.9	2,280	53,550
70-74	0.0661	0.2836	0.2256	0.3369	36,411	29,839	43,702	10,325	156,241	9.5	8.5	10.9	2,369	35,842
75-79	0.0999	0.3996	0.3284	0.4614	26,086	19,899	33,842	10,424	104,368	7.3	6.5	8.4	2,041	20,430
80-84	0.1503	0.5463	0.4701	0.6082	15,661	10,851	22,728	8,557	56,915	5.5	5.0	6.3	1,356	9,018
85-89	0.2200	0.7096	0.6420	0.7616	7,105	4,318	12,044	5,042	22,919	4.1	3.8	4.6	587	2,667
90-94	0.3129	0.8070	0.7531	0.8478	2,063	1,029	4,312	1,665	5,320	3.0	2.8	3.3	157	502
95-99	0.4327	0.9414	0.9127	0.9605	398	157	1,065	375	866	2.3	2.2	2.4	24	55
100+	0.5816	1	1	1	23	6	93	23	40	1.7	1.7	1.8	2	3
All ages													*58,048*	*4,052,216*

ln($_nq_x$) with uncertainty interval

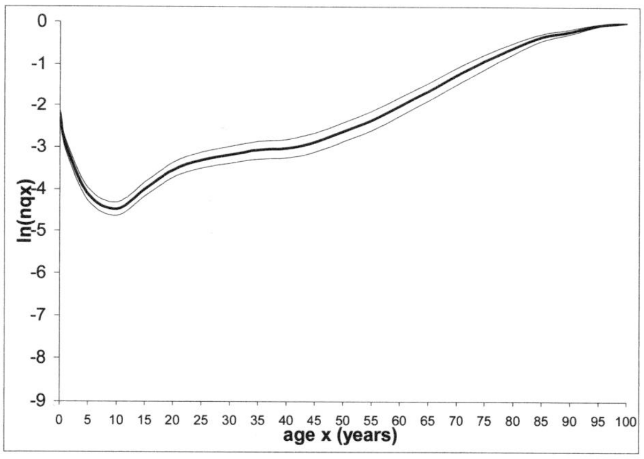

l_x with uncertainty interval

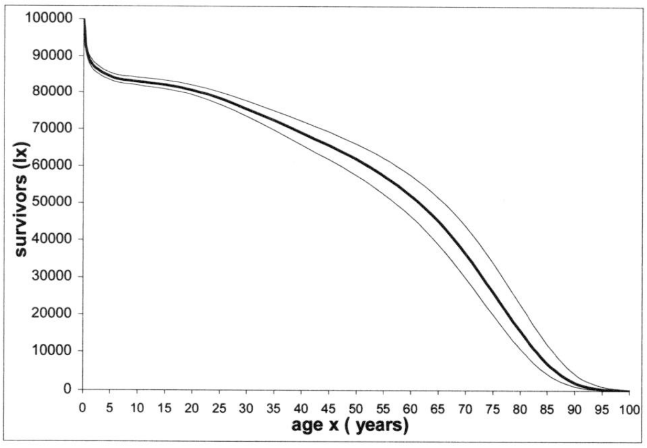

e_x with uncertainty interval

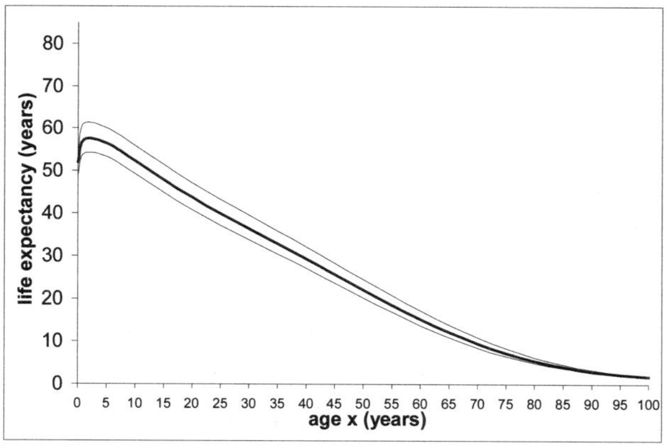

Distribution of deaths by age with uncertainty interval

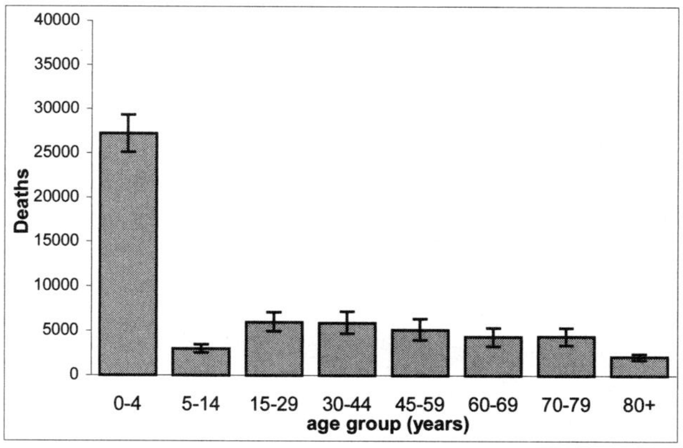

Guinea-Bissau | Guinée-Bissau | Guinea-Bissau

Males, 2000 | *Hommes, 2000* | *Hombres, 2000*

Age	$_nM_x$	$_nq_x$ mean	$_nq_x$ percentiles 2½	$_nq_x$ percentiles 97½	l_x mean	l_x percentiles 2½	l_x percentiles 97½	$_nd_x$	$_nL_x$	e_x mean	e_x percentiles 2½	e_x percentiles 97½	Actual deaths	Actual population
<1	0.1890	0.1615	0.1530	0.1711	100,000	100,000	100,000	16,153	85,462	44.5	43.0	46.6	4,453	23,558
1-4	0.0165	0.0634	0.0574	0.0677	83,847	82,887	84,704	5,313	322,637	52.0	50.0	54.5	1,342	81,504
5-9	0.0039	0.0191	0.0170	0.0207	78,534	77,785	79,282	1,498	388,927	51.4	49.6	53.8	325	84,291
10-14	0.0023	0.0116	0.0103	0.0126	77,036	76,197	77,884	894	382,946	47.4	45.6	49.7	166	71,263
15-19	0.0022	0.0107	0.0095	0.0117	76,142	75,243	77,051	817	378,669	42.9	41.1	45.2	130	60,246
20-24	0.0055	0.0271	0.0240	0.0298	75,325	74,373	76,291	2,044	371,516	38.3	36.6	40.6	281	51,032
25-29	0.0088	0.0429	0.0376	0.0473	73,281	72,193	74,397	3,142	358,550	34.3	32.6	36.5	372	42,472
30-34	0.0122	0.0592	0.0516	0.0659	70,139	68,845	71,536	4,152	340,316	30.8	29.1	32.8	441	36,153
35-39	0.0151	0.0725	0.0628	0.0813	65,987	64,430	67,846	4,786	317,972	27.6	26.0	29.5	472	31,369
40-44	0.0179	0.0856	0.0738	0.0966	61,202	59,361	63,586	5,241	292,905	24.5	23.1	26.3	456	25,456
45-49	0.0209	0.0991	0.0851	0.1122	55,960	53,780	58,897	5,545	265,939	21.6	20.3	23.2	458	21,957
50-54	0.0246	0.1161	0.0995	0.1317	50,415	47,786	53,887	5,851	237,448	18.7	17.5	20.1	424	17,214
55-59	0.0296	0.1379	0.1184	0.1564	44,564	41,493	48,524	6,146	207,453	15.8	14.8	17.0	422	14,239
60-64	0.0429	0.1937	0.1675	0.2184	38,417	35,004	42,778	7,443	173,481	12.9	12.0	14.0	488	11,371
65-69	0.0587	0.2558	0.2236	0.2857	30,975	27,360	35,613	7,924	135,064	10.4	9.7	11.3	505	8,603
70-74	0.0860	0.3541	0.3149	0.3895	23,051	19,542	27,652	8,162	94,849	8.2	7.6	8.8	494	5,737
75-79	0.1257	0.4783	0.4350	0.5167	14,889	11,931	18,943	7,122	56,640	6.3	5.8	6.8	406	3,231
80-84	0.1863	0.6354	0.5926	0.6718	7,767	5,772	10,694	4,935	26,497	4.7	4.4	5.0	256	1,374
85-89	0.2665	0.7996	0.7651	0.8277	2,832	1,899	4,341	2,264	8,498	3.5	3.4	3.7	93	348
90-94	0.3680	0.8746	0.8478	0.8959	567	329	1,011	496	1,348	2.6	2.5	2.8	15	40
95-99	0.4908	0.9926	0.9884	0.9952	71	34	152	71	144	2.0	2.0	2.1	1	3
100+	0.6320	1	1	1	1	0	2	1	1	1.6	1.6	1.6	0	0
All ages													*11,999*	*591,461*

ln($_nq_x$) with uncertainty interval

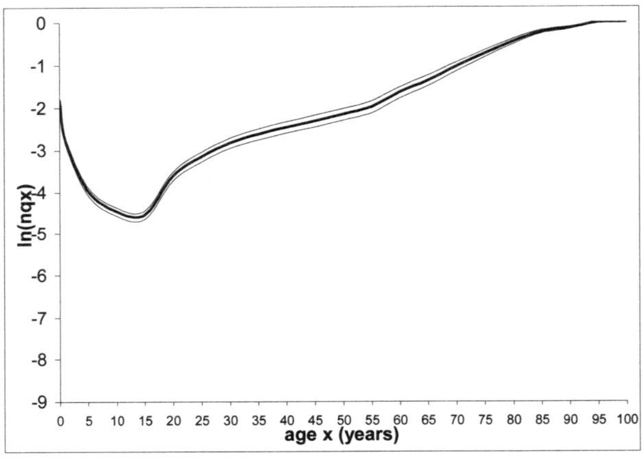

l_x with uncertainty interval

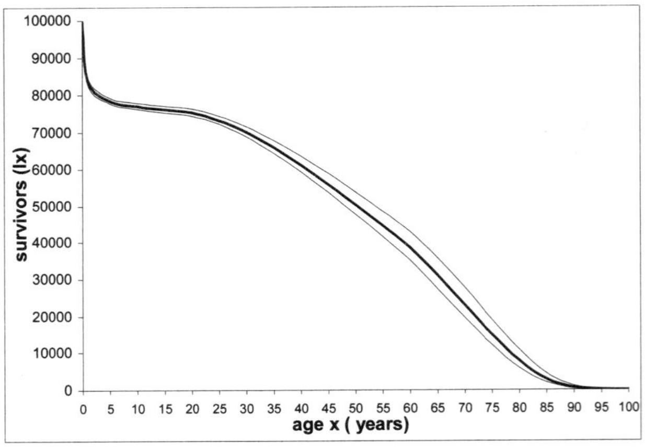

e_x with uncertainty interval

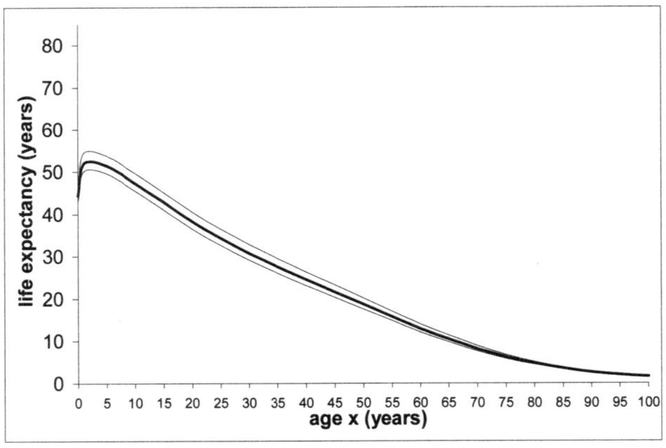

Distribution of deaths by age with uncertainty interval

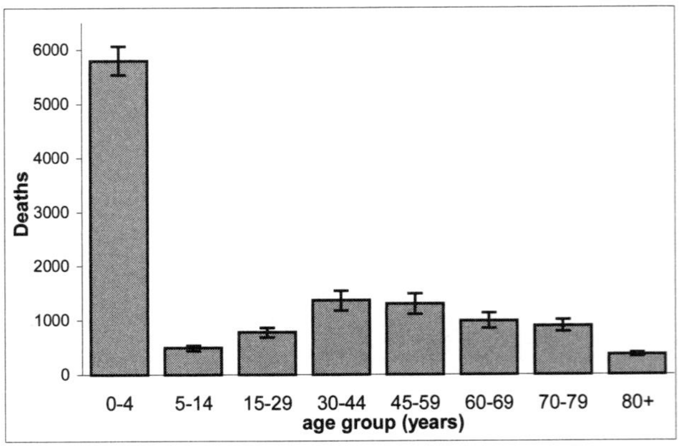

Guinea-Bissau | Guinée-Bissau | Guinea-Bissau

Females, 2000 | *Femmes, 2000* | *Mujeres, 2000*

Age	$_nM_x$	$_nq_x$			l_x			$_nd_x$	$_nL_x$	e_x			Actual deaths	Actual population
			percentiles			percentiles					percentiles			
		mean	2½	97½	mean	2½	97½			mean	2½	97½		
<1	0.1542	0.1354	0.1258	0.1453	100,000	100,000	100,000	13,540	87,814	46.9	45.1	49.3	3,511	22,772
1-4	0.0185	0.0710	0.0642	0.0765	86,460	85,466	87,423	6,137	331,109	53.2	51.1	56.1	1,515	81,755
5-9	0.0043	0.0214	0.0190	0.0233	80,322	79,575	81,066	1,719	397,313	53.2	51.1	55.8	366	84,522
10-14	0.0030	0.0148	0.0131	0.0162	78,603	77,746	79,491	1,161	390,112	49.3	47.2	51.9	214	71,774
15-19	0.0049	0.0243	0.0214	0.0266	77,442	76,504	78,453	1,881	382,508	45.0	43.0	47.5	299	60,847
20-24	0.0080	0.0391	0.0342	0.0431	75,561	74,490	76,776	2,954	370,422	41.1	39.1	43.5	413	51,848
25-29	0.0102	0.0499	0.0433	0.0556	72,607	71,290	74,147	3,622	353,981	37.6	35.7	40.0	445	43,466
30-34	0.0115	0.0559	0.0481	0.0628	68,985	67,370	70,925	3,857	335,283	34.5	32.7	36.7	428	37,227
35-39	0.0125	0.0604	0.0517	0.0683	65,128	63,205	67,491	3,934	315,805	31.4	29.7	33.4	405	32,525
40-44	0.0123	0.0595	0.0506	0.0677	61,194	58,975	63,976	3,638	296,874	28.2	26.7	30.1	327	26,702
45-49	0.0135	0.0652	0.0553	0.0744	57,555	55,084	60,708	3,750	278,401	24.8	23.4	26.5	316	23,457
50-54	0.0168	0.0807	0.0683	0.0924	53,805	51,101	57,315	4,342	258,170	21.4	20.1	22.9	308	18,337
55-59	0.0217	0.1030	0.0872	0.1180	49,463	46,528	53,349	5,093	234,582	18.1	16.9	19.4	340	15,644
60-64	0.0309	0.1435	0.1219	0.1640	44,370	41,077	48,632	6,367	205,931	14.8	13.8	16.1	399	12,896
65-69	0.0453	0.2033	0.1741	0.2307	38,003	34,342	42,690	7,725	170,699	11.9	11.1	13.0	456	10,074
70-74	0.0694	0.2956	0.2571	0.3310	30,277	26,419	35,258	8,951	129,009	9.3	8.6	10.2	487	7,023
75-79	0.1033	0.4106	0.3652	0.4512	21,327	17,675	26,194	8,757	84,741	7.2	6.6	7.8	435	4,211
80-84	0.1541	0.5562	0.5081	0.5975	12,570	9,703	16,629	6,991	45,371	5.4	5.0	5.9	303	1,968
85-89	0.2238	0.7175	0.6743	0.7530	5,579	3,912	8,164	4,003	17,886	4.1	3.8	4.4	132	588
90-94	0.3165	0.8117	0.7765	0.8401	1,576	970	2,643	1,279	4,042	3.0	2.8	3.2	29	93
95-99	0.4358	0.9444	0.9261	0.9577	297	156	585	280	643	2.3	2.2	2.4	3	8
100+	0.5845	1	1	1	17	7	43	17	28	1.7	1.7	1.8	0	0
All ages													*11,133*	*607,737*

ln($_nq_x$) with uncertainty interval

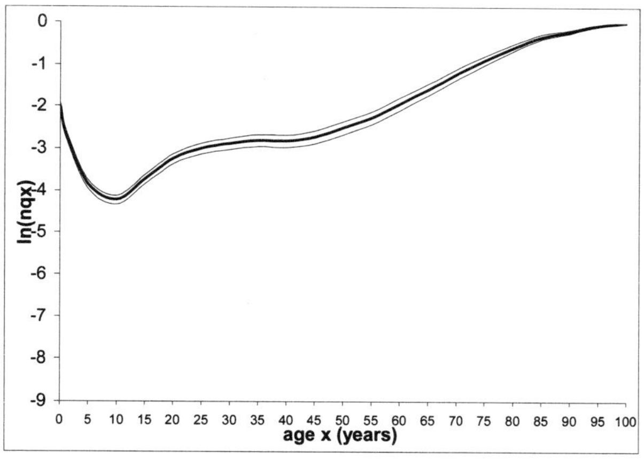

l_x with uncertainty interval

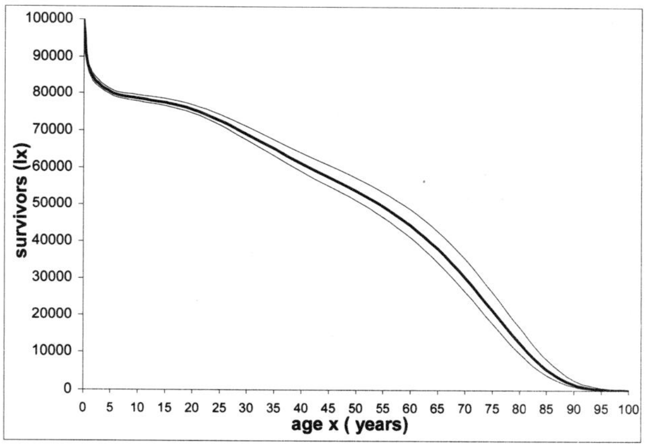

e_x with uncertainty interval

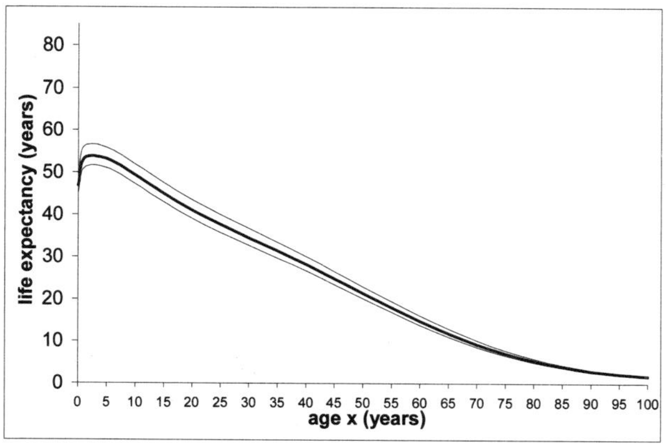

Distribution of deaths by age with uncertainty interval

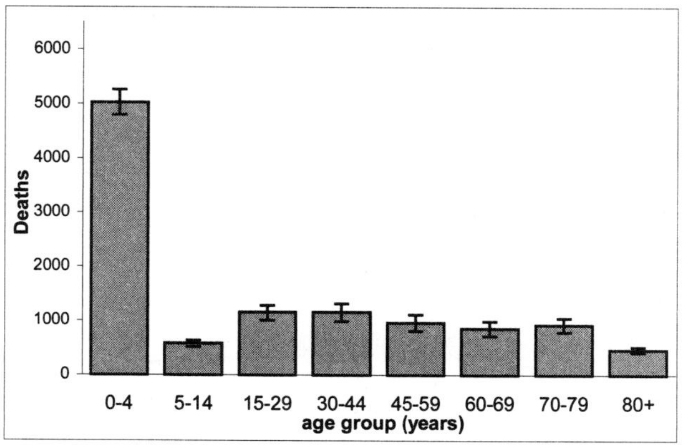

Guyana **Guyana** **Guyana**

Males, 2000 *Hommes, 2000* *Hombres, 2000*

Age	$_nM_x$	$_nq_x$ mean	$_nq_x$ percentiles 2 1/2	$_nq_x$ percentiles 97 1/2	l_x mean	l_x percentiles 2 1/2	l_x percentiles 97 1/2	$_nd_x$	$_nL_x$	e_x mean	e_x percentiles 2 1/2	e_x percentiles 97 1/2	Actual deaths	Actual population
<1	0.0720	0.0676	0.0535	0.0812	100,000	100,000	100,000	6,760	93,916	61.5	59.2	63.3	599	8,316
1-4	0.0026	0.0105	0.0089	0.0121	93,240	91,876	94,651	975	370,619	64.9	63.2	66.2	87	33,036
5-9	0.0008	0.0039	0.0033	0.0045	92,265	90,764	93,812	356	460,435	61.6	59.9	62.9	30	38,876
10-14	0.0005	0.0024	0.0020	0.0028	91,909	90,358	93,504	218	458,999	56.8	55.2	58.1	18	37,495
15-19	0.0012	0.0058	0.0049	0.0067	91,691	90,109	93,315	528	457,135	51.9	50.3	53.3	48	41,867
20-24	0.0026	0.0127	0.0109	0.0147	91,163	89,508	92,856	1,155	452,928	47.2	45.7	48.5	101	39,553
25-29	0.0037	0.0185	0.0160	0.0214	90,008	88,195	91,844	1,662	445,886	42.8	41.3	44.1	135	36,160
30-34	0.0047	0.0231	0.0202	0.0267	88,346	86,311	90,373	2,041	436,629	38.5	37.1	39.9	138	29,619
35-39	0.0059	0.0289	0.0256	0.0333	86,306	84,008	88,545	2,493	425,295	34.4	33.1	35.7	144	24,583
40-44	0.0080	0.0392	0.0351	0.0450	83,812	81,210	86,280	3,287	410,845	30.3	29.2	31.6	160	19,955
45-49	0.0100	0.0488	0.0443	0.0558	80,526	77,552	83,248	3,928	392,808	26.5	25.4	27.7	159	15,881
50-54	0.0140	0.0676	0.0611	0.0768	76,598	73,226	79,563	5,181	370,036	22.7	21.7	23.9	170	12,154
55-59	0.0210	0.0998	0.0893	0.1122	71,417	67,601	74,701	7,125	339,273	19.2	18.2	20.3	177	8,410
60-64	0.0300	0.1395	0.1240	0.1548	64,292	60,018	68,032	8,971	299,034	16.0	15.1	17.0	197	6,577
65-69	0.0400	0.1818	0.1611	0.2024	55,321	50,727	59,597	10,058	251,460	13.2	12.4	14.1	230	5,758
70-74	0.0530	0.2340	0.2079	0.2608	45,263	40,461	49,995	10,591	199,836	10.6	9.9	11.4	252	4,752
75-79	0.0800	0.3333	0.2977	0.3682	34,672	29,908	39,601	11,557	144,465	8.1	7.6	8.7	260	3,254
80-84	0.1300	0.4906	0.4462	0.5310	23,114	18,895	27,812	11,339	87,224	5.9	5.5	6.3	213	1,636
85-89	0.2127	0.6943	0.6511	0.7304	11,775	8,862	15,402	8,175	38,438	4.1	3.9	4.4	158	741
90-94	0.3519	0.8559	0.8257	0.8796	3,600	2,389	5,373	3,081	8,756	2.7	2.6	2.9	56	158
95-99	0.5014	0.9800	0.9713	0.9857	519	288	937	508	1,068	2.1	2.1	2.1	15	29
100+	0.6152	1	1	1	10	4	27	10	17	1.6	1.6	1.7	2	3
All ages													*3,347*	*368,813*

$\ln(_nq_x)$ with uncertainty interval

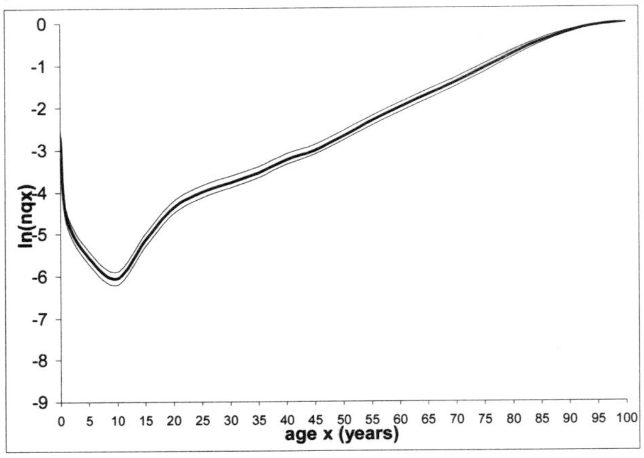

l_x with uncertainty interval

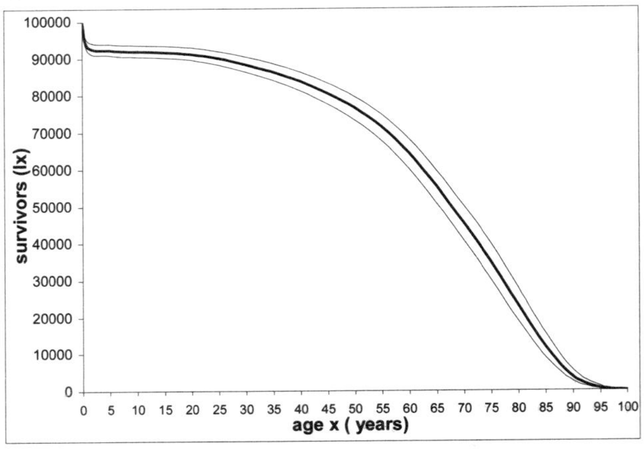

e_x with uncertainty interval

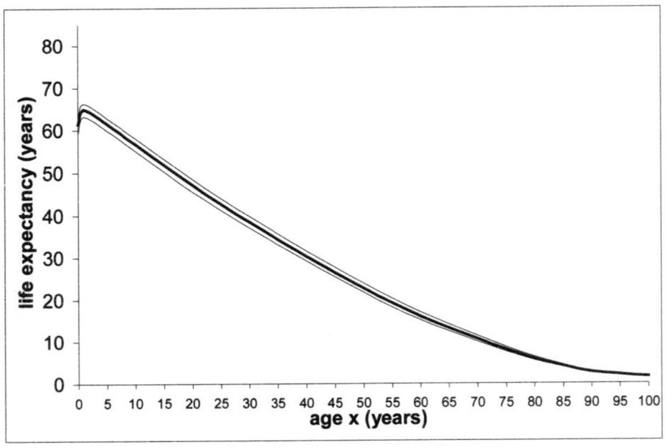

Distribution of deaths by age with uncertainty interval

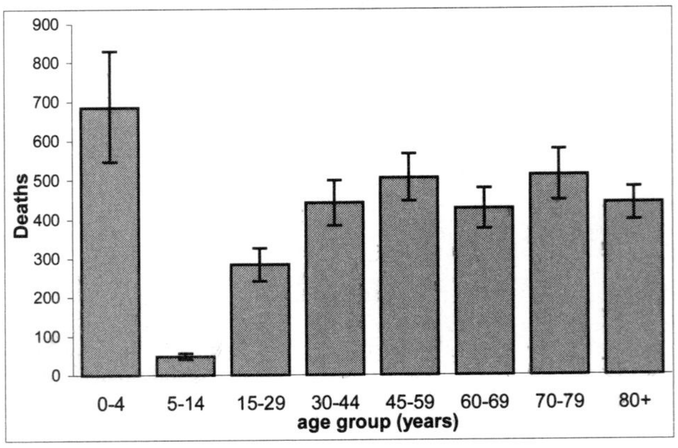

Guyana

Females, 2000

Guyana

Femmes, 2000

Guyana

Mujeres, 2000

Age	$_nM_x$	$_nq_x$ mean	$_nq_x$ percentiles 2½	$_nq_x$ percentiles 97½	l_x mean	l_x percentiles 2½	l_x percentiles 97½	$_nd_x$	$_nL_x$	e_x mean	e_x percentiles 2½	e_x percentiles 97½	Actual deaths	Actual population
<1	0.0616	0.0583	0.0438	0.0714	100,000	100,000	100,000	5,834	94,749	67.0	64.8	69.7	481	7,808
1-4	0.0020	0.0081	0.0065	0.0096	94,166	92,863	95,616	762	374,832	70.2	68.6	72.6	66	32,411
5-9	0.0007	0.0035	0.0028	0.0041	93,403	91,968	94,995	325	466,202	66.7	65.2	69.1	27	37,983
10-14	0.0005	0.0025	0.0020	0.0030	93,078	91,589	94,728	233	464,806	61.9	60.4	64.3	18	36,629
15-19	0.0011	0.0054	0.0044	0.0064	92,845	91,317	94,536	502	462,969	57.1	55.6	59.4	44	41,002
20-24	0.0022	0.0110	0.0090	0.0130	92,343	90,734	94,121	1,019	459,168	52.4	50.9	54.7	89	39,940
25-29	0.0032	0.0157	0.0130	0.0184	91,324	89,556	93,270	1,437	453,027	47.9	46.5	50.2	119	37,592
30-34	0.0037	0.0182	0.0153	0.0212	89,887	87,907	92,053	1,639	445,337	43.7	42.3	45.8	119	32,326
35-39	0.0028	0.0141	0.0117	0.0163	88,248	86,044	90,649	1,241	438,139	39.4	38.1	41.5	83	29,342
40-44	0.0037	0.0184	0.0153	0.0212	87,007	84,644	89,586	1,600	431,038	35.0	33.7	36.9	90	24,369
45-49	0.0060	0.0296	0.0245	0.0338	85,408	82,853	88,217	2,524	420,728	30.6	29.3	32.5	108	17,958
50-54	0.0094	0.0458	0.0379	0.0520	82,883	80,050	86,056	3,798	404,922	26.4	25.2	28.2	134	14,321
55-59	0.0148	0.0715	0.0590	0.0810	79,085	75,891	82,798	5,652	381,297	22.6	21.5	24.2	155	10,479
60-64	0.0200	0.0952	0.0787	0.1078	73,433	69,746	77,914	6,994	349,682	19.1	18.1	20.6	163	8,125
65-69	0.0284	0.1327	0.1104	0.1495	66,440	62,226	71,782	8,817	310,156	15.9	14.9	17.1	206	7,247
70-74	0.0387	0.1763	0.1486	0.1987	57,623	52,922	63,855	10,157	262,721	12.9	12.0	14.0	232	5,998
75-79	0.0600	0.2609	0.2251	0.2942	47,466	42,404	54,367	12,382	206,372	10.2	9.4	11.0	255	4,249
80-84	0.0817	0.3392	0.3005	0.3796	35,083	29,927	42,130	11,901	145,663	7.9	7.3	8.5	195	2,388
85-89	0.1394	0.5169	0.4710	0.5635	23,182	18,568	29,469	11,982	85,955	5.6	5.2	6.1	165	1,183
90-94	0.2240	0.6699	0.6256	0.7113	11,200	8,104	15,590	7,503	33,490	3.9	3.6	4.2	63	282
95-99	0.3392	0.8406	0.8080	0.8689	3,697	2,339	5,836	3,107	9,161	2.8	2.6	2.9	20	60
100+	0.4838	1	1	1	589	307	1,120	589	1,218	2.1	1.7	2.1	4	8
All ages													*2,837*	*391,700*

ln($_nq_x$) with uncertainty interval

l_x with uncertainty interval

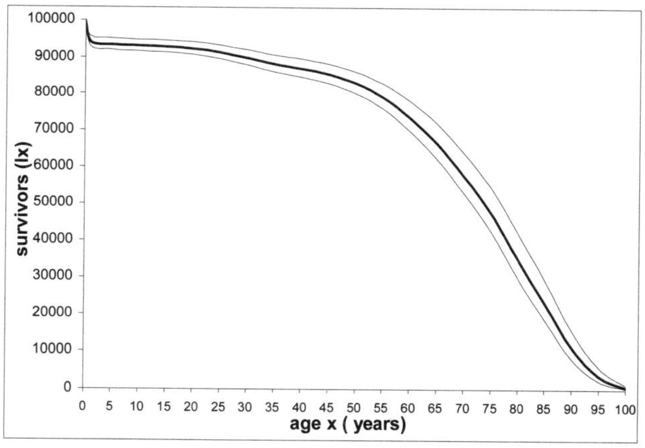

e_x with uncertainty interval

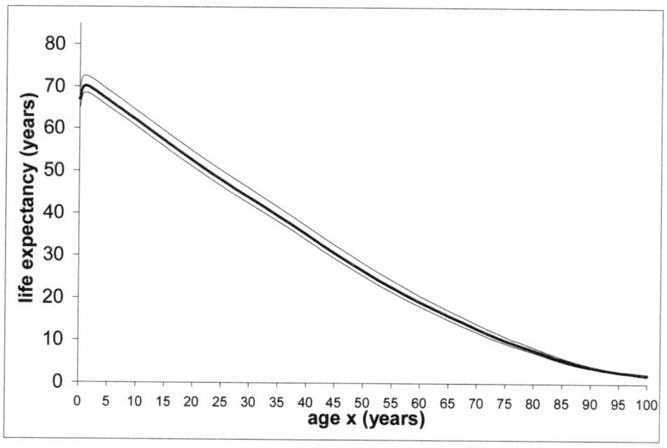

Distribution of deaths by age with uncertainty interval

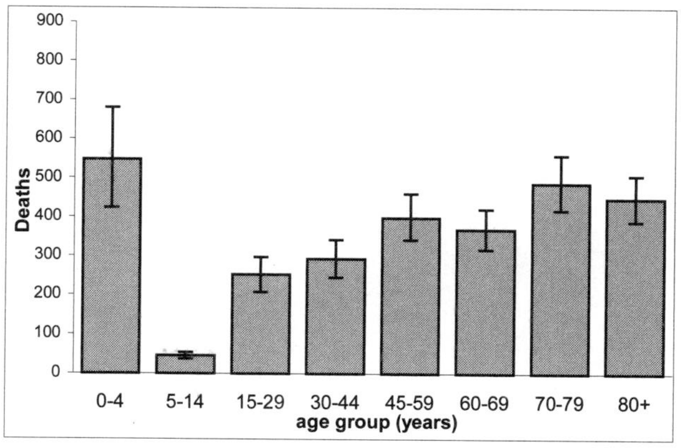

Haiti — Haïti — Haití

Males, 2000 — *Hommes, 2000* — *Hombres, 2000*

Age	$_nM_x$	$_nq_x$ mean	$_nq_x$ percentiles 2½	$_nq_x$ percentiles 97½	l_x mean	l_x percentiles 2½	l_x percentiles 97½	$_nd_x$	$_nL_x$	e_x mean	e_x percentiles 2½	e_x percentiles 97½	Actual deaths	Actual population
<1	0.0911	0.0842	0.0609	0.1081	100,000	100,000	100,000	8,423	92,419	49.7	44.6	55.0	11,309	124,084
1-4	0.0074	0.0291	0.0218	0.0372	91,577	89,194	93,910	2,668	359,906	53.2	48.4	58.0	3,361	453,493
5-9	0.0022	0.0111	0.0082	0.0144	88,910	86,415	91,402	985	442,085	50.8	46.2	55.5	1,189	533,826
10-14	0.0016	0.0079	0.0059	0.0103	87,925	85,279	90,612	697	437,882	46.3	41.9	51.0	899	565,314
15-19	0.0028	0.0139	0.0103	0.0182	87,228	84,431	90,058	1,213	433,107	41.7	37.3	46.3	1,383	493,612
20-24	0.0066	0.0324	0.0240	0.0426	86,015	82,959	89,131	2,787	423,108	37.2	32.9	41.9	2,575	391,026
25-29	0.0118	0.0575	0.0427	0.0760	83,228	79,600	86,993	4,784	404,182	33.4	29.3	38.0	3,485	294,455
30-34	0.0164	0.0790	0.0589	0.1046	78,445	73,623	83,282	6,194	376,738	30.2	26.4	34.7	3,948	240,129
35-39	0.0179	0.0857	0.0644	0.1134	72,251	65,922	78,375	6,193	345,771	27.6	24.1	31.8	3,596	200,774
40-44	0.0191	0.0914	0.0677	0.1201	66,058	58,446	73,326	6,035	315,201	25.0	21.6	28.8	3,062	159,908
45-49	0.0205	0.0977	0.0714	0.1281	60,023	51,424	68,239	5,861	285,460	22.2	19.2	25.8	2,701	131,549
50-54	0.0238	0.1125	0.0817	0.1486	54,161	44,884	63,116	6,093	255,573	19.4	16.6	22.6	2,538	106,457
55-59	0.0292	0.1362	0.0989	0.1809	48,068	38,398	57,573	6,549	223,968	16.5	14.0	19.5	2,580	88,218
60-64	0.0389	0.1774	0.1288	0.2345	41,519	31,797	51,333	7,366	189,181	13.7	11.6	16.3	2,766	71,044
65-69	0.0528	0.2333	0.1708	0.3036	34,153	24,845	43,910	7,968	150,846	11.2	9.4	13.4	2,891	54,738
70-74	0.0770	0.3228	0.2419	0.4075	26,185	17,904	36,002	8,454	109,791	8.8	7.4	10.6	2,940	38,187
75-79	0.1127	0.4396	0.3427	0.5320	17,731	11,052	27,097	7,795	69,169	6.8	5.7	8.2	2,697	23,930
80-84	0.1648	0.5837	0.4804	0.6731	9,936	5,172	17,557	5,800	35,183	5.2	4.4	6.2	1,987	12,053
85-89	0.2376	0.7453	0.6524	0.8186	4,137	1,690	9,015	3,083	12,975	3.9	3.4	4.5	1,118	4,705
90-94	0.3303	0.8295	0.7529	0.8874	1,053	307	3,133	874	2,646	2.9	2.6	3.3	431	1,306
95-99	0.4499	0.9574	0.9194	0.9794	180	35	774	172	382	2.2	2.1	2.4	110	244
100+	0.5970	1	1	1	8	1	62	8	13	1.7	1.7	1.8	18	30
All ages													*57,585*	*3,989,082*

ln($_nq_x$) with uncertainty interval

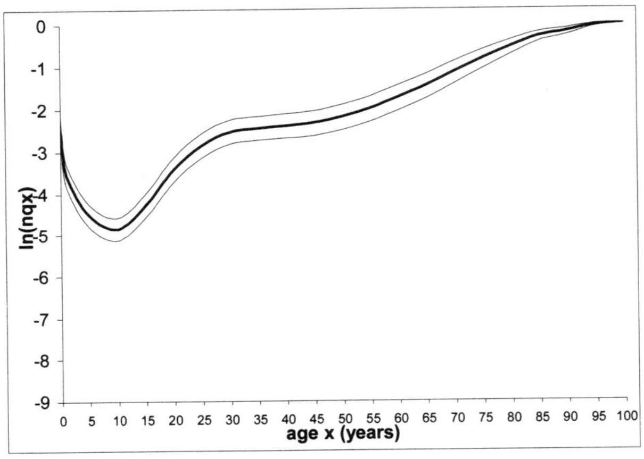

l_x with uncertainty interval

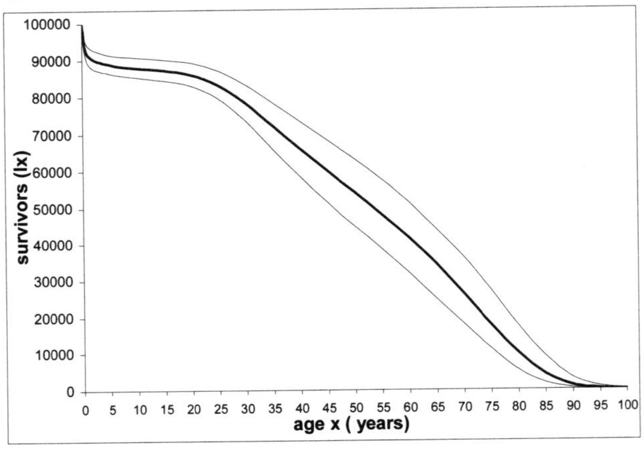

e_x with uncertainty interval

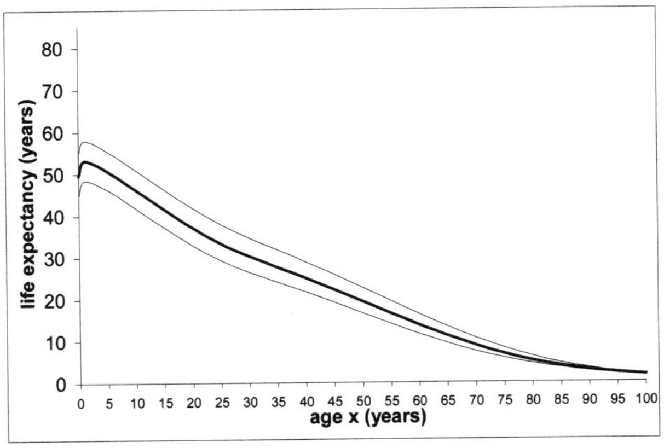

Distribution of deaths by age with uncertainty interval

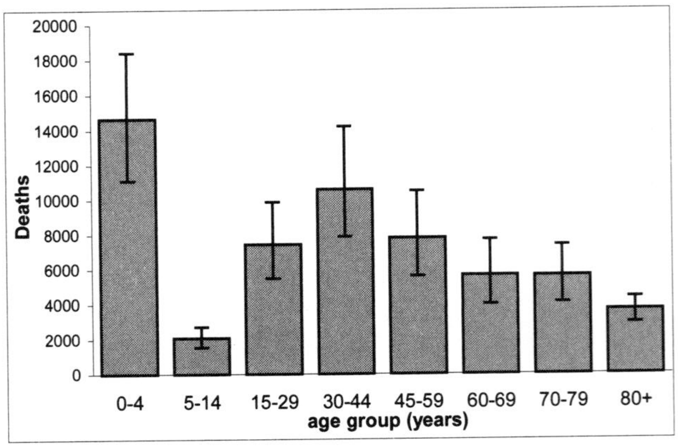

Haiti **Haïti** **Haití**

Females, 2000 *Femmes, 2000* *Mujeres, 2000*

Age	$_nM_x$	$_nq_x$ mean	$_nq_x$ percentiles 2½	$_nq_x$ percentiles 97½	l_x mean	l_x percentiles 2½	l_x percentiles 97½	$_nd_x$	$_nL_x$	e_x mean	e_x percentiles 2½	e_x percentiles 97½	Actual deaths	Actual population
<1	0.0705	0.0663	0.0450	0.0896	100,000	100,000	100,000	6,628	94,035	56.1	52.0	61.2	8,330	118,176
1-4	0.0081	0.0319	0.0234	0.0405	93,372	91,035	95,496	2,983	366,328	59.1	55.2	63.7	3,572	438,733
5-9	0.0024	0.0119	0.0087	0.0151	90,389	87,890	92,883	1,074	449,260	57.0	53.2	61.6	1,241	518,995
10-14	0.0018	0.0091	0.0067	0.0115	89,315	86,597	92,034	814	444,540	52.6	48.9	57.2	1,013	552,869
15-19	0.0032	0.0159	0.0117	0.0201	88,501	85,598	91,419	1,409	438,982	48.1	44.4	52.6	1,555	484,318
20-24	0.0051	0.0253	0.0187	0.0318	87,092	83,879	90,350	2,203	429,950	43.8	40.2	48.3	2,020	394,165
25-29	0.0072	0.0351	0.0260	0.0443	84,888	81,215	88,664	2,982	416,988	39.9	36.3	44.3	2,203	308,074
30-34	0.0082	0.0403	0.0298	0.0510	81,907	77,657	86,354	3,301	401,282	36.3	32.8	40.6	2,154	261,935
35-39	0.0096	0.0469	0.0348	0.0596	78,606	73,785	83,756	3,687	383,812	32.7	29.4	36.8	2,200	229,012
40-44	0.0106	0.0517	0.0385	0.0666	74,919	69,540	80,798	3,875	364,907	29.2	26.2	33.1	2,036	191,706
45-49	0.0125	0.0606	0.0444	0.0789	71,044	65,171	77,623	4,305	344,456	25.6	22.8	29.3	2,030	162,397
50-54	0.0160	0.0769	0.0554	0.1007	66,738	60,425	74,009	5,130	320,866	22.1	19.5	25.6	2,109	131,877
55-59	0.0209	0.0993	0.0705	0.1301	61,608	54,662	69,592	6,115	292,753	18.8	16.3	21.9	2,252	107,823
60-64	0.0287	0.1338	0.0944	0.1758	55,493	47,944	64,297	7,427	258,898	15.6	13.3	18.4	2,487	86,682
65-69	0.0409	0.1856	0.1312	0.2442	48,066	39,929	57,572	8,919	218,034	12.6	10.6	15.0	2,724	66,601
70-74	0.0624	0.2699	0.1944	0.3492	39,147	30,431	48,993	10,565	169,324	9.9	8.3	11.9	2,906	46,568
75-79	0.0936	0.3792	0.2838	0.4734	28,582	20,215	39,269	10,838	115,816	7.6	6.4	9.2	2,773	29,632
80-84	0.1429	0.5263	0.4189	0.6233	17,744	11,063	28,125	9,339	65,371	5.7	4.8	6.9	2,190	15,327
85-89	0.2116	0.6920	0.5893	0.7748	8,405	4,167	16,342	5,816	27,482	4.2	3.7	5.0	1,309	6,186
90-94	0.3040	0.7950	0.7100	0.8600	2,588	938	6,660	2,058	6,768	3.1	2.7	3.6	568	1,868
95-99	0.4241	0.9332	0.8838	0.9643	531	131	1,898	495	1,167	2.3	2.2	2.6	165	390
100+	0.5739	1	1	1	35	5	213	35	62	1.7	1.7	1.9	32	55
All ages													*47,867*	*4,153,389*

ln($_nq_x$) with uncertainty interval

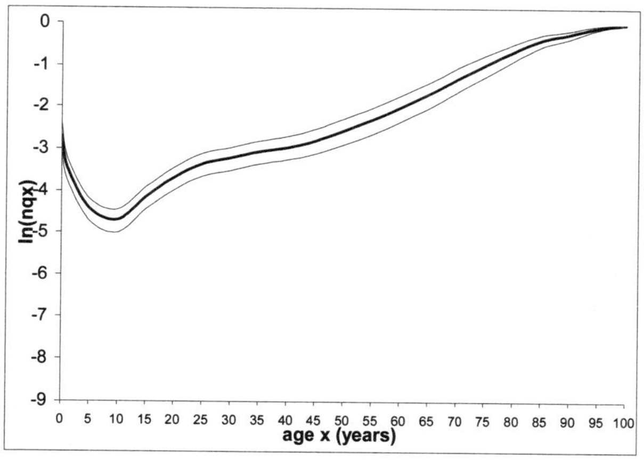

l_x with uncertainty interval

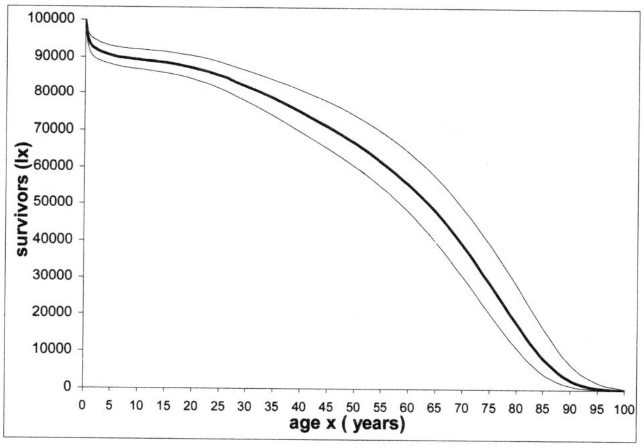

e_x with uncertainty interval

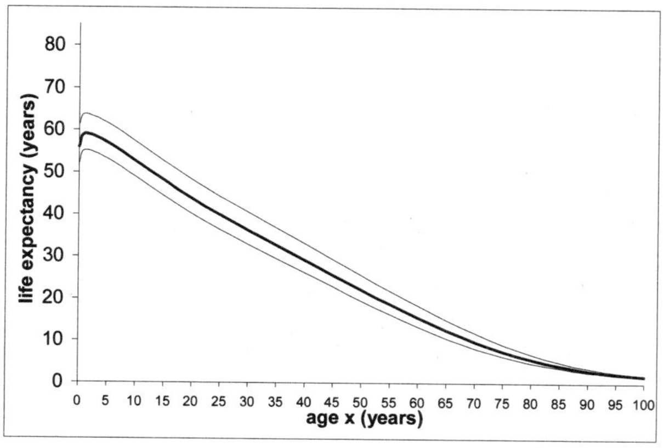

Distribution of deaths by age with uncertainty interval

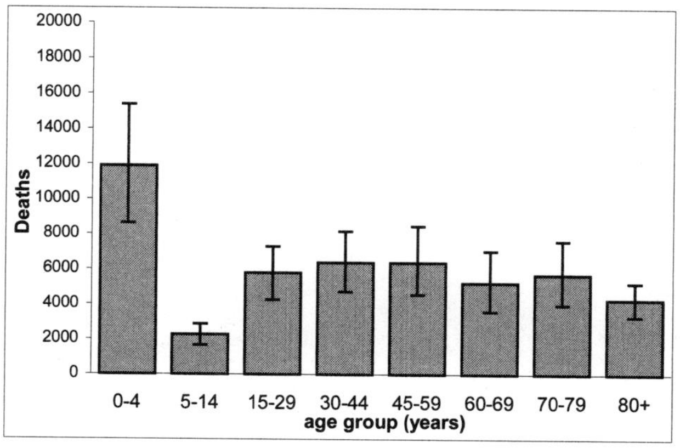

Honduras

Honduras

Honduras

Males, 2000 *Hommes, 2000* *Hombres, 2000*

Age	$_nM_x$	$_nq_x$			l_x			$_nd_x$	$_nL_x$	e_x			Actual deaths	Actual population
			percentiles			percentiles					percentiles			
		mean	2½	97½	mean	2½	97½			mean	2½	97½		
<1	0.0285	0.0278	0.0239	0.0317	100,000	100,000	100,000	2,778	97,500	66.3	64.5	67.9	2,862	100,464
1-4	0.0045	0.0177	0.0158	0.0199	97,222	96,833	97,608	1,719	384,765	67.2	65.4	68.7	1,745	390,627
5-9	0.0015	0.0073	0.0065	0.0082	95,504	94,911	96,069	692	475,788	64.4	62.7	65.8	672	461,448
10-14	0.0009	0.0045	0.0040	0.0050	94,811	94,134	95,449	422	473,000	59.9	58.2	61.2	369	413,044
15-19	0.0015	0.0076	0.0068	0.0086	94,389	93,660	95,070	719	470,147	55.1	53.5	56.4	560	366,036
20-24	0.0034	0.0166	0.0148	0.0188	93,670	92,853	94,424	1,556	464,459	50.5	48.9	51.8	1,045	311,922
25-29	0.0041	0.0203	0.0182	0.0231	92,114	91,106	93,025	1,874	455,884	46.3	44.8	47.6	1,056	256,861
30-34	0.0044	0.0217	0.0193	0.0247	90,240	88,998	91,336	1,956	446,309	42.3	40.8	43.5	943	215,078
35-39	0.0046	0.0227	0.0202	0.0260	88,284	86,797	89,571	2,004	436,408	38.1	36.8	39.4	792	172,540
40-44	0.0059	0.0292	0.0261	0.0335	86,279	84,542	87,758	2,523	425,089	34.0	32.7	35.2	812	136,749
45-49	0.0060	0.0293	0.0262	0.0336	83,756	81,710	85,469	2,456	412,641	29.9	28.8	31.1	644	108,198
50-54	0.0082	0.0400	0.0358	0.0459	81,300	78,962	83,233	3,256	398,363	25.7	24.7	26.9	649	79,429
55-59	0.0120	0.0583	0.0521	0.0668	78,045	75,337	80,253	4,553	378,839	21.7	20.7	22.8	772	64,219
60-64	0.0186	0.0887	0.0786	0.1009	73,491	70,308	76,069	6,515	351,168	17.9	17.0	18.9	984	53,062
65-69	0.0300	0.1395	0.1229	0.1573	66,976	63,212	70,094	9,346	311,517	14.4	13.6	15.3	1,246	41,532
70-74	0.0444	0.1999	0.1760	0.2243	57,631	53,268	61,476	11,520	259,354	11.3	10.7	12.1	1,263	28,433
75-79	0.0700	0.2979	0.2648	0.3301	46,111	41,322	50,654	13,735	196,216	8.5	8.0	9.1	1,265	18,067
80-84	0.1200	0.4615	0.4195	0.4990	32,376	27,682	37,241	14,943	124,522	6.1	5.7	6.5	1,014	8,449
85-89	0.2100	0.6885	0.6490	0.7217	17,433	13,868	21,618	12,003	57,158	4.2	4.0	4.4	573	2,729
90-94	0.3400	0.8416	0.8152	0.8631	5,430	3,860	7,588	4,570	13,441	2.8	2.7	2.9	194	572
95-99	0.4800	0.9836	0.9778	0.9878	860	528	1,402	846	1,763	2.1	2.1	2.1	31	65
100+	0.6374	1	1	1	14	6	31	14	22	1.6	1.5	1.7	3	4
All ages													*19,493*	*3,229,528*

ln($_nq_x$) with uncertainty interval

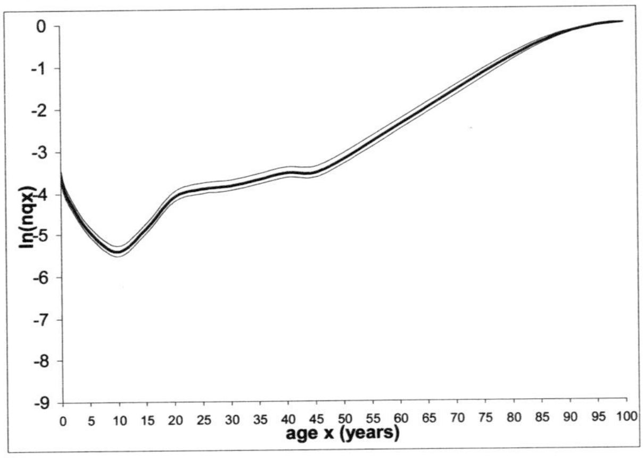

l_x with uncertainty interval

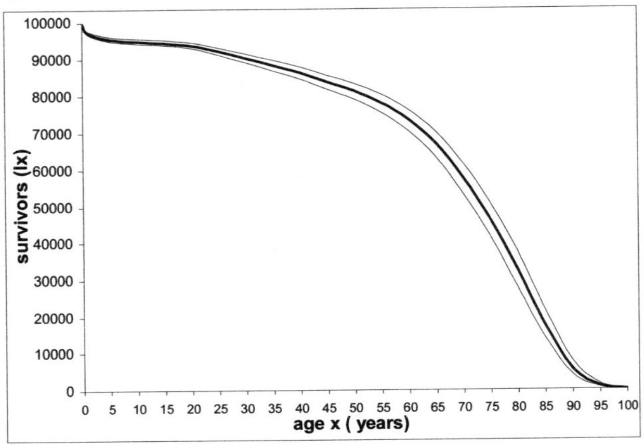

e_x with uncertainty interval

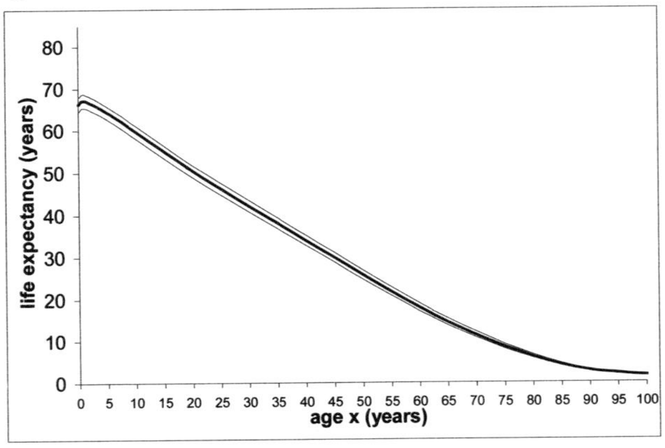

Distribution of deaths by age with uncertainty interval

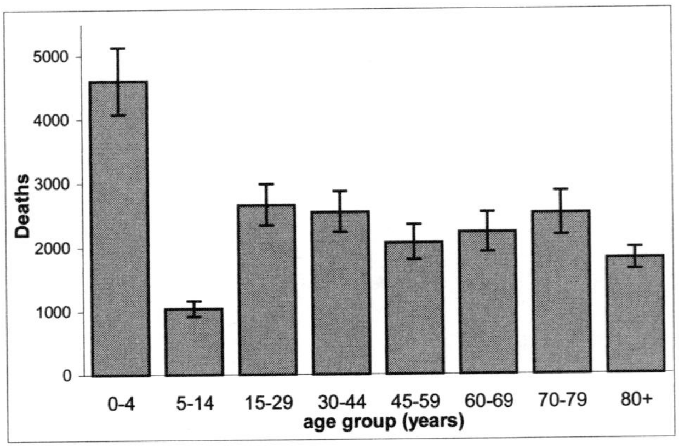

Honduras **Honduras** **Honduras**

Females, 2000 *Femmes, 2000* *Mujeres, 2000*

Age	$_nM_x$	$_nq_x$			l_x			$_nd_x$	$_nL_x$	e_x			Actual deaths	Actual population
			percentiles			percentiles					percentiles			
		mean	2½	97½	mean	2½	97½			mean	2½	97½		
<1	0.0250	0.0244	0.0213	0.0288	100,000	100,000	100,000	2,441	97,803	71.0	69.2	72.5	2,411	96,600
1-4	0.0047	0.0185	0.0166	0.0210	97,559	97,117	97,874	1,807	385,901	71.7	70.1	73.3	1,759	375,840
5-9	0.0011	0.0056	0.0050	0.0064	95,753	95,077	96,249	539	477,416	69.1	67.5	70.6	502	445,154
10-14	0.0007	0.0035	0.0031	0.0040	95,214	94,469	95,765	335	475,231	64.4	62.9	65.9	281	399,009
15-19	0.0009	0.0047	0.0042	0.0053	94,879	94,091	95,464	444	473,282	59.7	58.2	61.1	333	354,518
20-24	0.0019	0.0093	0.0082	0.0105	94,434	93,589	95,065	874	469,986	54.9	53.5	56.4	564	303,463
25-29	0.0022	0.0110	0.0098	0.0125	93,560	92,605	94,281	1,029	465,229	50.4	49.0	51.9	558	252,073
30-34	0.0028	0.0139	0.0123	0.0158	92,531	91,446	93,358	1,282	459,450	46.0	44.6	47.4	594	212,981
35-39	0.0033	0.0165	0.0147	0.0189	91,249	90,002	92,209	1,510	452,469	41.6	40.3	42.9	580	173,659
40-44	0.0041	0.0204	0.0181	0.0234	89,739	88,298	90,856	1,833	444,112	37.2	36.0	38.5	574	139,099
45-49	0.0040	0.0200	0.0177	0.0230	87,906	86,230	89,213	1,759	435,135	33.0	31.8	34.2	449	111,222
50-54	0.0061	0.0300	0.0265	0.0345	86,148	84,247	87,635	2,584	424,279	28.6	27.5	29.8	501	82,268
55-59	0.0087	0.0426	0.0373	0.0491	83,564	81,340	85,314	3,562	408,914	24.4	23.4	25.5	592	67,958
60-64	0.0143	0.0690	0.0600	0.0792	80,002	77,349	82,128	5,517	386,216	20.4	19.4	21.4	818	57,248
65-69	0.0195	0.0932	0.0806	0.1065	74,485	71,224	77,201	6,939	355,077	16.7	15.9	17.7	898	45,935
70-74	0.0321	0.1484	0.1284	0.1678	67,546	63,639	70,979	10,023	312,671	13.1	12.4	14.0	1,042	32,504
75-79	0.0476	0.2128	0.1848	0.2393	57,523	52,959	61,865	12,242	257,007	10.0	9.4	10.7	1,032	21,670
80-84	0.0900	0.3673	0.3236	0.4052	45,280	40,284	50,430	16,633	184,817	7.0	6.6	7.6	978	10,864
85-89	0.1800	0.6207	0.5694	0.6593	28,647	23,960	34,112	17,781	98,781	4.6	4.4	5.0	708	3,934
90-94	0.3000	0.7895	0.7518	0.8152	10,866	8,164	14,687	8,578	28,595	3.1	3.0	3.3	290	965
95-99	0.4400	0.9483	0.9330	0.9581	2,288	1,508	3,646	2,169	4,930	2.2	2.2	2.3	59	134
100+	0.5923	1	1	1	118	63	244	118	200	1.7	1.7	1.8	5	9
All ages													*15,529*	*3,187,107*

ln($_nq_x$) with uncertainty interval

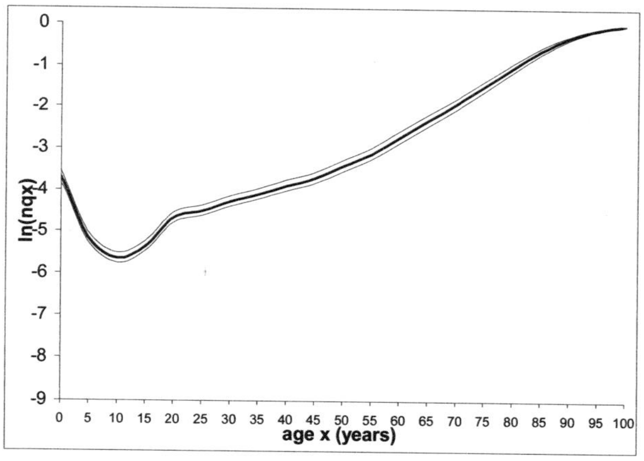

l_x with uncertainty interval

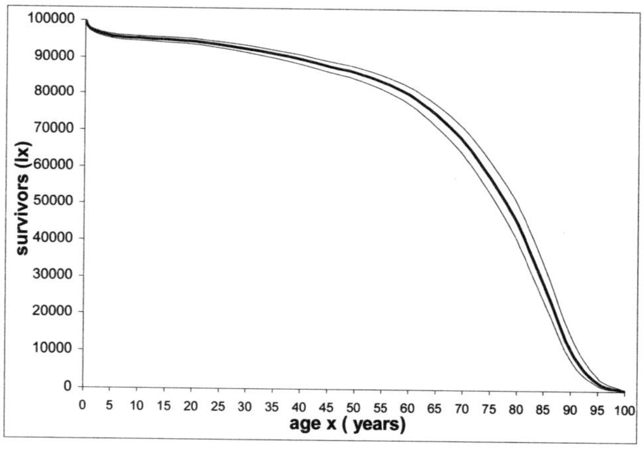

e_x with uncertainty interval

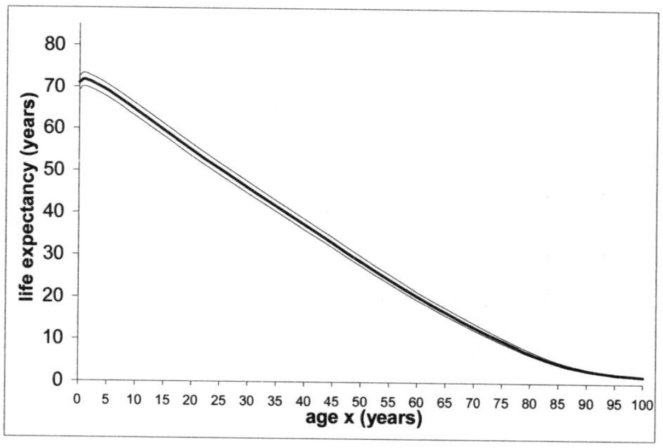

Distribution of deaths by age with uncertainty interval

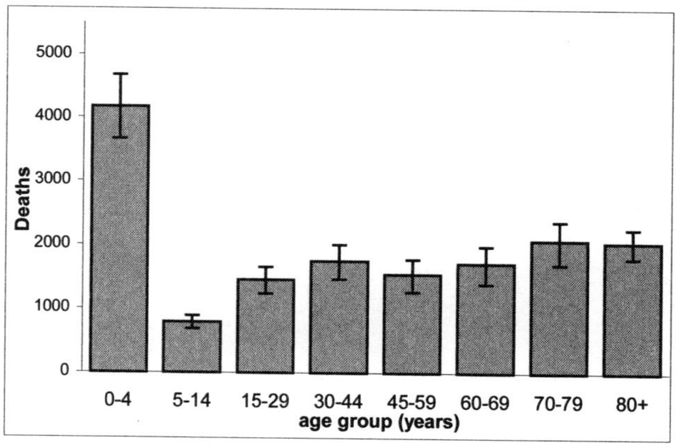

Hungary | Hongrie | Hungría

Males, 2000 | *Hommes, 2000* | *Hombres, 2000*

Age	$_nM_x$	$_nq_x$ mean	$_nq_x$ percentiles 2½	$_nq_x$ percentiles 97½	l_x mean	l_x percentiles 2½	l_x percentiles 97½	$_nd_x$	$_nL_x$	e_x mean	e_x percentiles 2½	e_x percentiles 97½	Actual deaths	Actual population
<1	0.0093	0.0092	0.0081	0.0105	100,000	100,000	100,000	920	99,172	66.3	66.1	66.5	457	49,266
1-4	0.0005	0.0019	0.0017	0.0021	99,080	98,949	99,186	190	395,864	65.9	65.7	66.2	97	201,619
5-9	0.0002	0.0010	0.0009	0.0011	98,890	98,741	99,013	102	494,197	62.0	61.8	62.3	62	301,136
10-14	0.0003	0.0016	0.0014	0.0017	98,789	98,630	98,920	154	493,558	57.1	56.9	57.3	98	312,326
15-19	0.0006	0.0029	0.0027	0.0031	98,635	98,462	98,778	285	492,460	52.2	51.9	52.4	190	327,963
20-24	0.0010	0.0049	0.0045	0.0052	98,349	98,153	98,514	477	490,553	47.3	47.1	47.6	411	422,438
25-29	0.0012	0.0062	0.0058	0.0066	97,872	97,640	98,070	607	487,842	42.5	42.3	42.8	496	398,659
30-34	0.0021	0.0103	0.0097	0.0109	97,265	96,993	97,500	998	483,828	37.8	37.5	38.1	711	344,677
35-39	0.0043	0.0213	0.0203	0.0223	96,266	95,940	96,555	2,047	476,214	33.2	32.9	33.4	1,293	300,735
40-44	0.0081	0.0399	0.0386	0.0411	94,219	93,802	94,593	3,756	461,708	28.8	28.5	29.1	2,892	355,548
45-49	0.0121	0.0586	0.0574	0.0597	90,464	89,944	90,933	5,301	439,067	24.9	24.6	25.2	4,722	391,095
50-54	0.0170	0.0814	0.0798	0.0831	85,163	84,592	85,688	6,936	408,475	21.3	21.0	21.7	5,429	319,752
55-59	0.0235	0.1111	0.1082	0.1142	78,227	77,673	78,770	8,690	369,411	18.0	17.6	18.3	6,405	272,268
60-64	0.0337	0.1553	0.1503	0.1606	69,537	69,071	70,011	10,799	320,689	14.9	14.6	15.3	7,296	216,674
65-69	0.0457	0.2051	0.1974	0.2132	58,738	58,154	59,333	12,048	263,573	12.2	11.9	12.5	8,879	194,245
70-74	0.0671	0.2871	0.2758	0.2987	46,691	45,803	47,527	13,406	199,937	9.7	9.4	10.0	10,672	159,150
75-79	0.0966	0.3891	0.3747	0.4033	33,284	32,177	34,339	12,951	134,044	7.6	7.4	7.8	10,911	112,933
80-84	0.1393	0.5166	0.5007	0.5318	20,333	19,212	21,437	10,505	75,404	5.9	5.7	6.0	6,042	43,368
85-89	0.1970	0.6600	0.6450	0.6738	9,828	8,996	10,704	6,487	32,924	4.5	4.4	4.6	5,075	25,759
90-94	0.2760	0.7549	0.7422	0.7665	3,341	2,934	3,799	2,522	9,139	3.3	3.3	3.4	1,498	5,428
95-99	0.3900	0.8986	0.8904	0.9058	819	685	980	736	1,887	2.5	2.5	2.5	181	464
100+	0.5200	1	1	1	83	65	107	83	160	1.9	1.8	1.9	2	3
All ages													*73,818*	*4,755,506*

$ln(_nq_x)$ with uncertainty interval

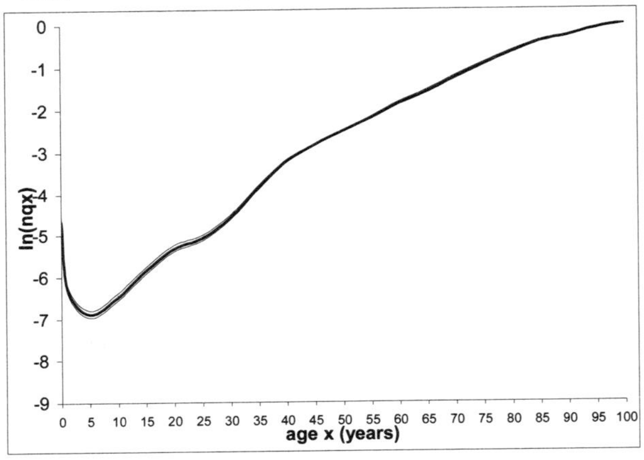

l_x with uncertainty interval

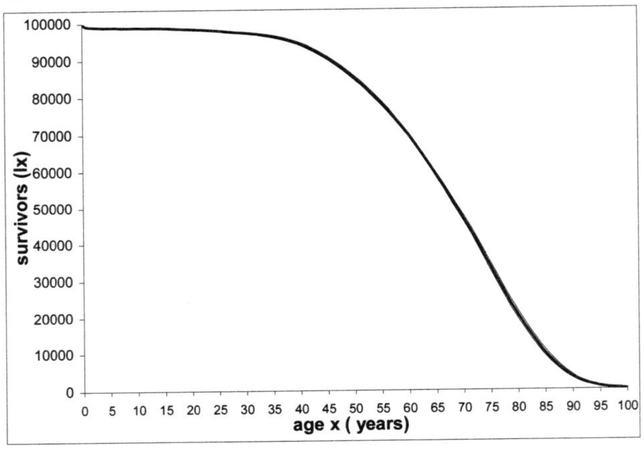

e_x with uncertainty interval

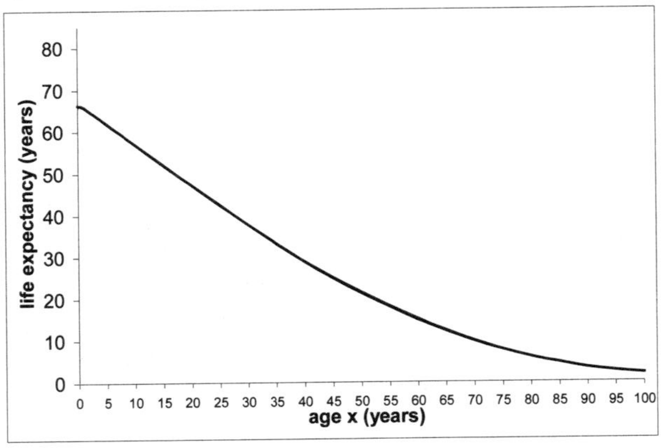

Distribution of deaths by age with uncertainty interval

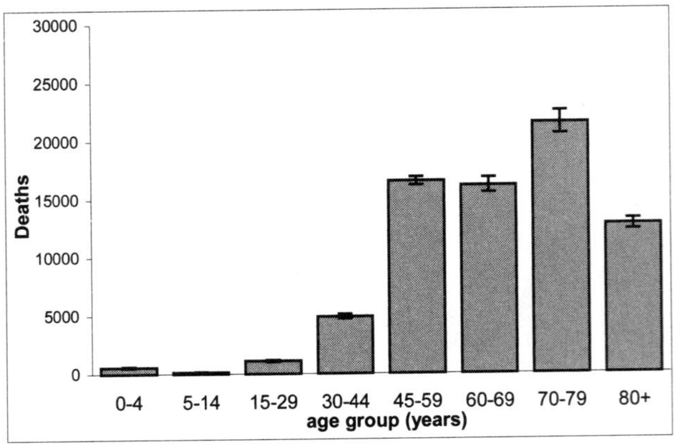

Hungary   Hongrie   Hungría

Females, 2000   *Femmes, 2000*   *Mujeres, 2000*

Age	$_nM_x$	$_nq_x$			l_x			$_nd_x$	$_nL_x$	e_x			Actual deaths	Actual population
			percentiles			percentiles					percentiles			
		mean	2½	97½	mean	2½	97½			mean	2½	97½		
<1	0.0072	0.0072	0.0062	0.0081	100,000	100,000	100,000	716	99,356	75.2	74.9	75.5	341	47,303
1-4	0.0003	0.0012	0.0011	0.0013	99,284	99,195	99,378	122	396,844	74.7	74.4	75.1	59	191,550
5-9	0.0002	0.0008	0.0007	0.0008	99,162	99,062	99,268	78	495,617	70.8	70.5	71.2	45	287,012
10-14	0.0002	0.0008	0.0007	0.0009	99,085	98,978	99,197	79	495,225	65.9	65.6	66.2	48	298,558
15-19	0.0003	0.0014	0.0013	0.0016	99,006	98,893	99,125	143	494,669	60.9	60.6	61.3	91	314,067
20-24	0.0003	0.0016	0.0014	0.0017	98,862	98,738	98,995	154	493,927	56.0	55.7	56.4	125	403,708
25-29	0.0004	0.0021	0.0019	0.0022	98,709	98,574	98,854	207	493,026	51.1	50.8	51.5	160	381,045
30-34	0.0007	0.0036	0.0033	0.0038	98,502	98,353	98,663	355	491,621	46.2	45.9	46.6	240	333,010
35-39	0.0017	0.0083	0.0078	0.0087	98,147	97,977	98,333	814	488,700	41.3	41.0	41.7	497	298,649
40-44	0.0031	0.0152	0.0145	0.0158	97,333	97,122	97,569	1,478	482,970	36.7	36.3	37.0	1,114	363,799
45-49	0.0045	0.0224	0.0216	0.0230	95,855	95,591	96,158	2,143	473,916	32.2	31.9	32.6	1,868	413,034
50-54	0.0062	0.0305	0.0296	0.0313	93,712	93,402	94,077	2,854	461,424	27.9	27.5	28.2	2,168	350,582
55-59	0.0091	0.0444	0.0428	0.0456	90,858	90,503	91,263	4,030	444,214	23.7	23.3	24.0	2,910	320,711
60-64	0.0132	0.0641	0.0615	0.0661	86,828	86,452	87,228	5,566	420,223	19.6	19.3	20.0	3,852	290,787
65-69	0.0217	0.1030	0.0985	0.1068	81,261	80,824	81,730	8,373	385,374	15.8	15.5	16.1	6,062	278,997
70-74	0.0358	0.1642	0.1567	0.1709	72,888	72,254	73,683	11,969	334,518	12.3	12.1	12.6	9,208	257,340
75-79	0.0621	0.2689	0.2573	0.2800	60,919	60,006	62,138	16,382	263,640	9.3	9.0	9.5	13,068	210,311
80-84	0.1067	0.4213	0.4053	0.4378	44,537	43,225	46,142	18,763	175,777	6.8	6.6	7.0	9,684	90,719
85-89	0.1727	0.6031	0.5868	0.6206	25,774	24,414	27,365	15,545	90,008	4.9	4.7	5.0	10,942	63,359
90-94	0.2632	0.7353	0.7220	0.7496	10,229	9,301	11,265	7,522	28,581	3.5	3.4	3.6	4,176	15,867
95-99	0.3777	0.8853	0.8771	0.8941	2,708	2,329	3,126	2,397	6,347	2.6	2.5	2.6	617	1,633
100+	0.5106	1	1	1	310	247	383	310	608	2.0	1.9	2.0	4	8
All ages													*67,278*	*5,212,049*

$\ln(_nq_x)$ with uncertainty interval

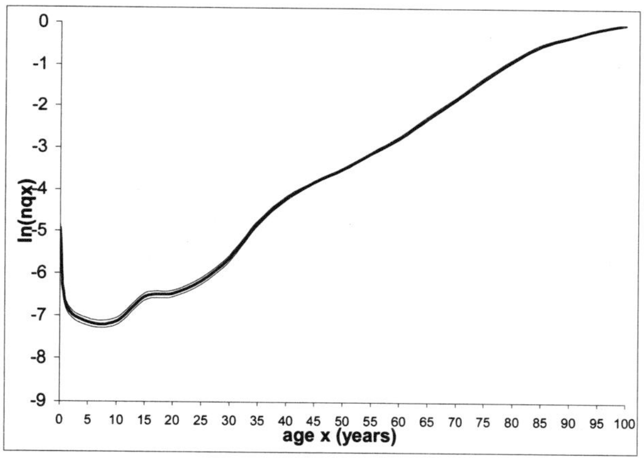

l_x with uncertainty interval

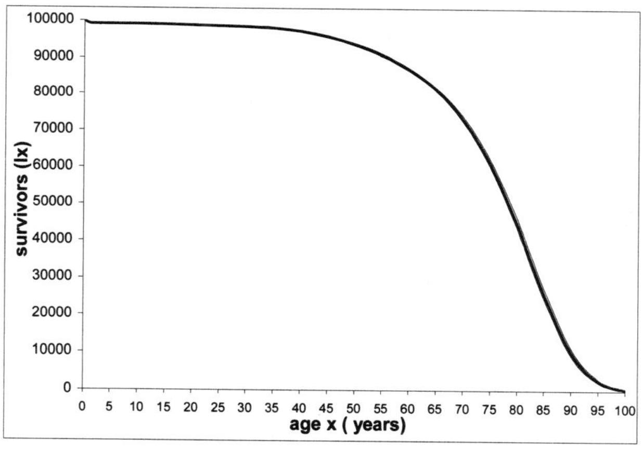

e_x with uncertainty interval

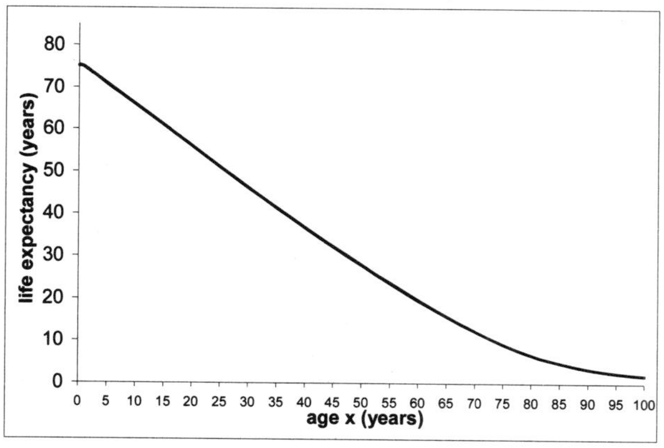

Distribution of deaths by age with uncertainty interval

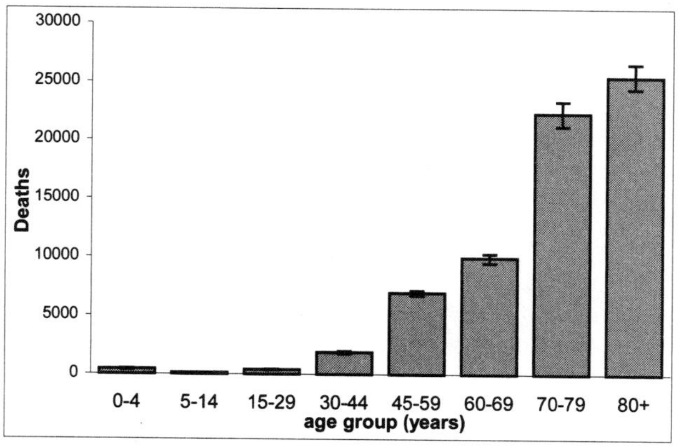

Iceland / Islande / Islandia

Males, 2000 — *Hommes, 2000* — *Hombres, 2000*

Age	$_nM_x$	$_nq_x$ mean	$_nq_x$ percentiles 2½	$_nq_x$ percentiles 97½	l_x mean	l_x percentiles 2½	l_x percentiles 97½	$_nd_x$	$_nL_x$	e_x mean	e_x percentiles 2½	e_x percentiles 97½	Actual deaths	Actual population
<1	0.0040	0.0040	0.0024	0.0066	100,000	100,000	100,000	398	99,642	77.1	75.7	78.6	9	2,335
1-4	0.0002	0.0010	0.0007	0.0014	99,602	99,341	99,758	96	398,178	76.4	75.0	78.0	2	8,423
5-9	0.0002	0.0008	0.0006	0.0012	99,506	99,203	99,692	83	497,323	72.5	71.1	74.1	2	11,658
10-14	0.0001	0.0006	0.0004	0.0009	99,423	99,085	99,634	61	496,964	67.6	66.1	69.1	1	10,993
15-19	0.0003	0.0016	0.0012	0.0023	99,362	99,001	99,591	163	496,405	62.6	61.2	64.2	4	10,844
20-24	0.0005	0.0027	0.0020	0.0036	99,200	98,778	99,472	264	495,337	57.7	56.3	59.3	6	10,655
25-29	0.0008	0.0041	0.0032	0.0053	98,935	98,427	99,273	407	493,659	52.8	51.4	54.4	9	10,557
30-34	0.0009	0.0045	0.0036	0.0057	98,528	97,906	98,957	448	491,520	48.1	46.6	49.6	9	10,129
35-39	0.0012	0.0062	0.0051	0.0074	98,080	97,352	98,596	605	488,886	43.3	41.8	44.9	13	10,737
40-44	0.0015	0.0075	0.0064	0.0088	97,475	96,629	98,094	735	485,536	38.5	37.1	40.1	16	10,503
45-49	0.0018	0.0092	0.0080	0.0104	96,740	95,780	97,466	885	481,486	33.8	32.4	35.4	17	9,469
50-54	0.0035	0.0173	0.0152	0.0196	95,854	94,787	96,687	1,663	475,115	29.1	27.7	30.7	29	8,195
55-59	0.0070	0.0344	0.0302	0.0396	94,192	92,933	95,214	3,240	462,858	24.5	23.2	26.1	43	6,189
60-64	0.0111	0.0541	0.0464	0.0634	90,952	89,256	92,342	4,918	442,463	20.3	19.0	21.9	53	4,761
65-69	0.0194	0.0926	0.0773	0.1098	86,034	83,598	88,060	7,963	410,261	16.4	15.2	17.8	90	4,656
70-74	0.0343	0.1582	0.1300	0.1876	78,071	74,418	81,255	12,347	359,486	12.8	11.7	14.1	139	4,044
75-79	0.0634	0.2737	0.2263	0.3191	65,724	60,461	70,693	17,992	283,639	9.7	8.8	10.9	185	2,919
80-84	0.0970	0.3904	0.3263	0.4478	47,732	41,165	54,695	18,636	192,068	7.4	6.6	8.4	168	1,732
85-89	0.1462	0.5354	0.4642	0.5951	29,095	22,731	36,846	15,578	106,532	5.5	5.0	6.3	124	849
90-94	0.2171	0.6574	0.5939	0.7082	13,517	9,203	19,742	8,886	40,927	4.0	3.7	4.5	68	314
95-99	0.3176	0.8132	0.7662	0.8494	4,631	2,685	8,016	3,766	11,857	3.0	2.8	3.2	28	87
100+	0.4578	1	1	1	865	404	1,874	865	1,890	2.2	2.1	2.2	5	10
All ages													*1,020*	*140,059*

ln($_nq_x$) with uncertainty interval

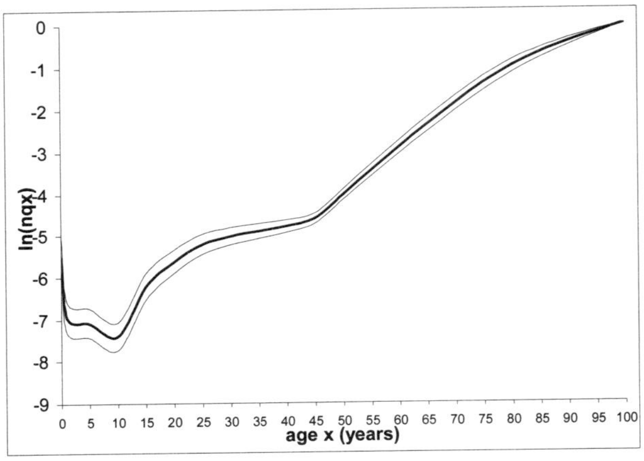

l_x with uncertainty interval

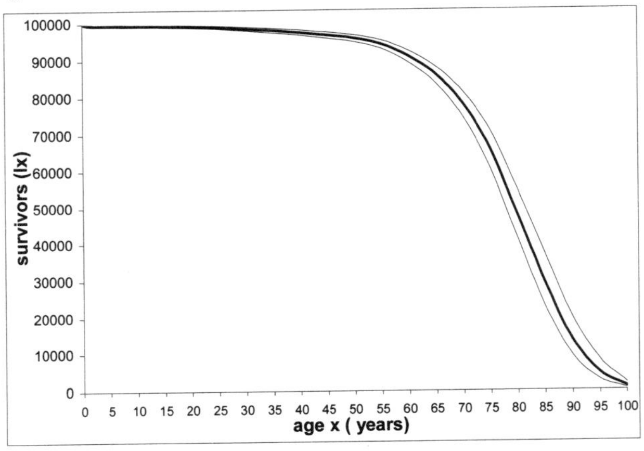

e_x with uncertainty interval

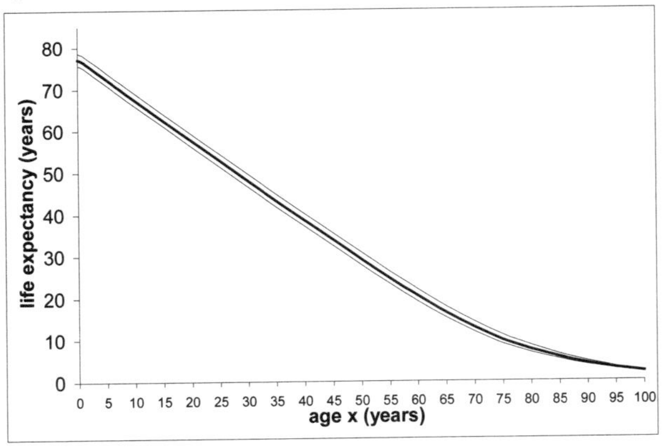

Distribution of deaths by age with uncertainty interval

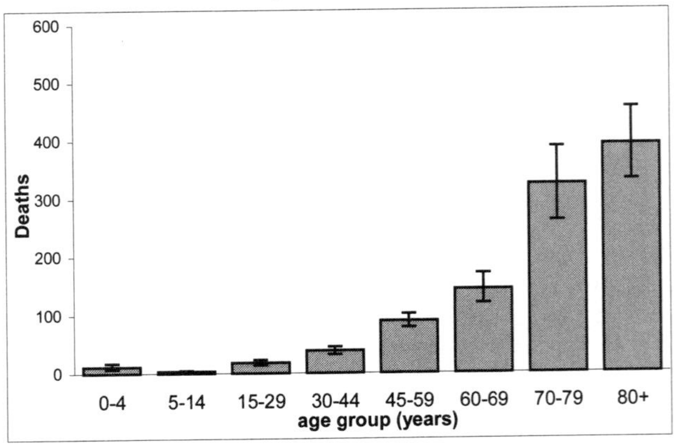

Iceland   Islande   Islandia

Females, 2000   *Femmes, 2000*   *Mujeres, 2000*

Age	$_nM_x$	$_nq_x$			l_x			$_nd_x$	$_nL_x$	e_x			Actual deaths	Actual population
			percentiles			percentiles					percentiles			
		mean	2½	97½	mean	2½	97½			mean	2½	97½		
<1	0.0029	0.0029	0.0018	0.0037	100,000	100,000	100,000	291	99,739	81.8	80.5	83.9	6	2,214
1-4	0.0005	0.0022	0.0015	0.0026	99,709	99,628	99,816	215	398,322	81.1	79.7	83.2	4	7,954
5-9	0.0002	0.0009	0.0006	0.0010	99,494	99,370	99,663	87	497,254	77.2	75.9	79.4	2	11,059
10-14	0.0002	0.0011	0.0008	0.0013	99,407	99,266	99,600	107	496,767	72.3	71.0	74.4	2	10,391
15-19	0.0004	0.0021	0.0016	0.0025	99,300	99,139	99,521	206	495,983	67.4	66.1	69.5	4	10,344
20-24	0.0006	0.0029	0.0022	0.0034	99,093	98,894	99,366	286	494,753	62.5	61.2	64.6	6	10,324
25-29	0.0003	0.0017	0.0013	0.0020	98,808	98,560	99,147	168	493,617	57.7	56.4	59.8	3	10,245
30-34	0.0004	0.0019	0.0015	0.0022	98,639	98,363	99,015	190	492,720	52.8	51.5	54.9	4	10,094
35-39	0.0005	0.0026	0.0021	0.0030	98,449	98,142	98,865	256	491,605	47.9	46.6	50.0	6	10,602
40-44	0.0008	0.0039	0.0032	0.0045	98,193	97,845	98,661	386	490,002	43.0	41.7	45.1	8	10,211
45-49	0.0013	0.0065	0.0053	0.0075	97,808	97,402	98,348	634	487,454	38.2	36.9	40.2	12	9,067
50-54	0.0022	0.0109	0.0090	0.0125	97,174	96,676	97,824	1,063	483,212	33.4	32.1	35.4	17	7,705
55-59	0.0040	0.0198	0.0161	0.0229	96,111	95,467	96,942	1,903	475,796	28.7	27.5	30.7	25	6,159
60-64	0.0065	0.0320	0.0254	0.0374	94,208	93,285	95,379	3,019	463,492	24.3	23.0	26.2	32	4,923
65-69	0.0112	0.0543	0.0421	0.0639	91,189	89,796	92,953	4,949	443,572	20.0	18.8	21.8	55	4,967
70-74	0.0185	0.0884	0.0675	0.1049	86,240	84,061	89,041	7,625	412,138	16.0	14.9	17.7	84	4,536
75-79	0.0326	0.1509	0.1150	0.1790	78,615	75,244	83,032	11,860	363,426	12.3	11.3	13.8	119	3,642
80-84	0.0619	0.2680	0.2095	0.3125	66,755	61,772	73,484	17,888	289,055	9.0	8.2	10.2	153	2,474
85-89	0.1109	0.4342	0.3603	0.4923	48,867	42,466	58,087	21,216	191,295	6.4	5.8	7.3	165	1,491
90-94	0.1878	0.6007	0.5358	0.6541	27,651	21,559	37,161	16,609	88,427	4.4	4.0	4.9	119	636
95-99	0.3006	0.7904	0.7500	0.8275	11,042	7,457	17,251	8,727	29,028	3.1	2.9	3.3	52	172
100+	0.4547	1	1	1	2,315	1,286	4,313	2,315	5,091	2.2	2.2	2.3	10	22
All ages													*889*	*139,232*

ln($_nq_x$) with uncertainty interval

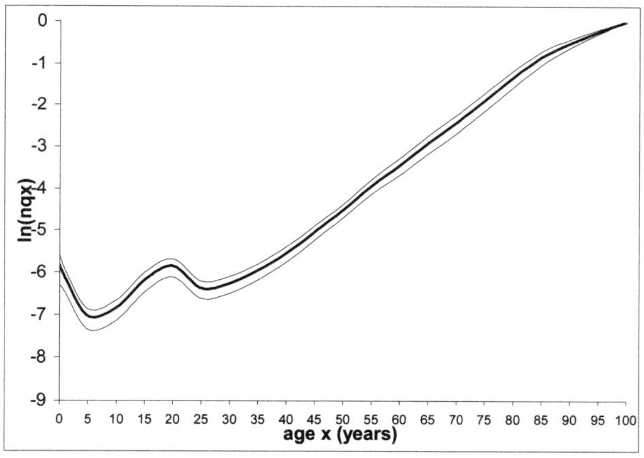

l_x with uncertainty interval

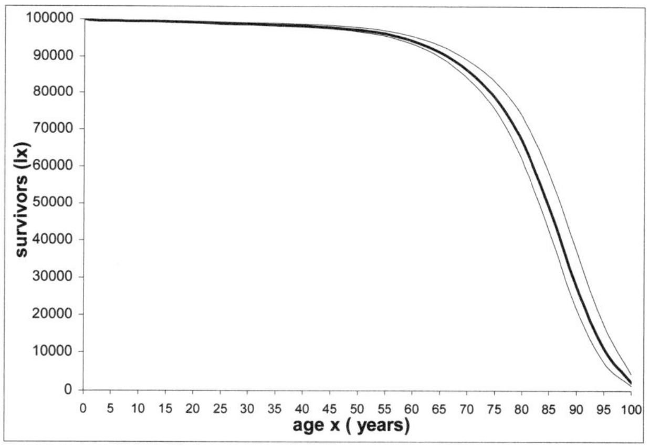

e_x with uncertainty interval

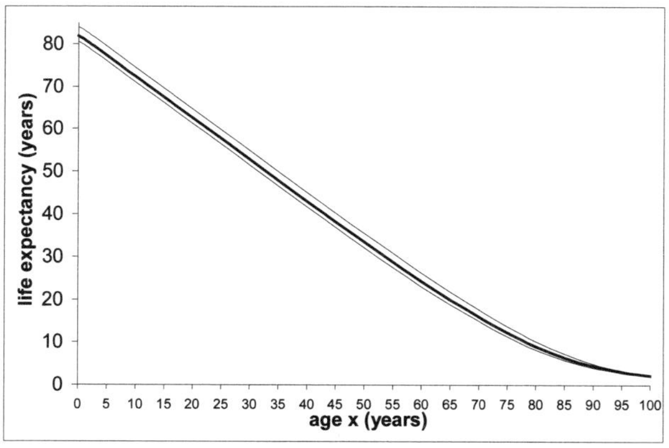

Distribution of deaths by age with uncertainty interval

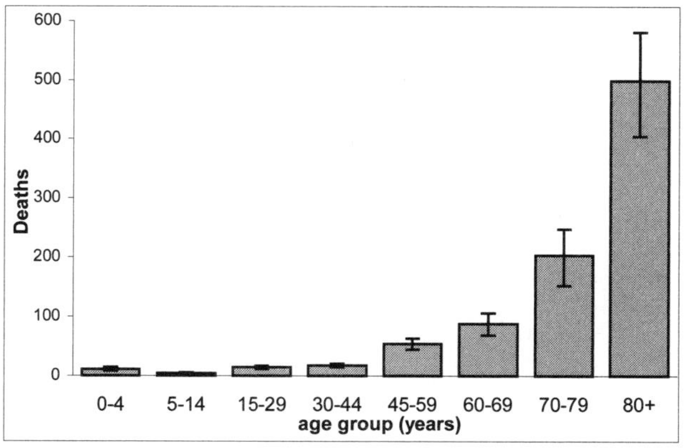

India

Inde

India

Males, 2000 — *Hommes, 2000* — *Hombres, 2000*

Age	$_nM_x$	$_nq_x$ mean	$_nq_x$ percentiles 2½	$_nq_x$ percentiles 97½	l_x mean	l_x percentiles 2½	l_x percentiles 97½	$_nd_x$	$_nL_x$	e_x mean	e_x percentiles 2½	e_x percentiles 97½	Actual deaths	Actual population
<1	0.0660	0.0623	0.0562	0.0690	100,000	100,000	100,000	6,231	94,392	59.8	58.2	61.6	809,778	12,267,518
1-4	0.0075	0.0295	0.0267	0.0323	93,769	93,103	94,382	2,762	368,448	62.8	61.2	64.6	357,897	47,746,577
5-9	0.0018	0.0090	0.0081	0.0099	91,007	90,308	91,704	820	452,986	60.6	59.1	62.4	106,163	58,611,961
10-14	0.0012	0.0058	0.0052	0.0064	90,187	89,425	90,955	527	449,618	56.1	54.6	57.9	65,672	56,056,390
15-19	0.0015	0.0074	0.0066	0.0082	89,660	88,853	90,476	663	446,643	51.5	50.0	53.3	77,899	52,487,646
20-24	0.0024	0.0117	0.0104	0.0129	88,997	88,128	89,873	1,040	442,386	46.8	45.3	48.6	109,091	46,397,872
25-29	0.0034	0.0170	0.0151	0.0188	87,957	86,989	88,926	1,494	436,051	42.3	40.9	44.1	150,441	43,917,606
30-34	0.0040	0.0200	0.0177	0.0221	86,463	85,354	87,566	1,728	427,997	38.0	36.6	39.8	160,028	39,639,034
35-39	0.0045	0.0220	0.0195	0.0245	84,736	83,463	85,991	1,867	419,009	33.8	32.4	35.5	156,352	35,083,215
40-44	0.0066	0.0323	0.0285	0.0360	82,868	81,423	84,287	2,677	407,647	29.5	28.1	31.1	199,794	30,419,921
45-49	0.0088	0.0431	0.0380	0.0484	80,191	78,504	81,857	3,456	392,314	25.4	24.1	27.0	220,749	25,059,645
50-54	0.0145	0.0699	0.0609	0.0790	76,735	74,750	78,748	5,366	370,261	21.4	20.2	23.0	287,310	19,826,261
55-59	0.0220	0.1043	0.0895	0.1185	71,369	68,963	73,899	7,441	338,243	17.8	16.7	19.3	352,140	16,006,381
60-64	0.0341	0.1571	0.1336	0.1787	63,928	61,036	67,104	10,043	294,532	14.6	13.6	15.9	444,645	13,039,797
65-69	0.0469	0.2100	0.1781	0.2377	53,885	50,280	57,752	11,316	241,133	11.8	11.0	13.0	472,897	10,076,816
70-74	0.0696	0.2963	0.2538	0.3320	42,569	38,329	47,188	12,613	181,310	9.3	8.6	10.3	479,147	6,887,505
75-79	0.1031	0.4099	0.3589	0.4509	29,955	25,660	35,212	12,278	119,080	7.2	6.7	7.9	416,361	4,037,984
80-84	0.1526	0.5524	0.4994	0.5930	17,677	14,179	22,574	9,765	63,971	5.5	5.1	6.0	294,803	1,931,288
85-89	0.2184	0.7063	0.6604	0.7402	7,912	5,835	11,300	5,588	25,589	4.2	3.9	4.5	142,830	654,032
90-94	0.3019	0.7921	0.7552	0.8194	2,324	1,516	3,838	1,841	6,096	3.1	3.0	3.4	43,478	143,999
95-99	0.4034	0.9126	0.8891	0.9290	483	274	940	441	1,093	2.4	2.3	2.5	7,618	18,883
100+	0.5209	1	1	1	42	19	104	42	81	1.9	1.8	2.0	670	1,287
All ages													*5,355,764*	*520,311,618*

ln($_nq_x$) with uncertainty interval

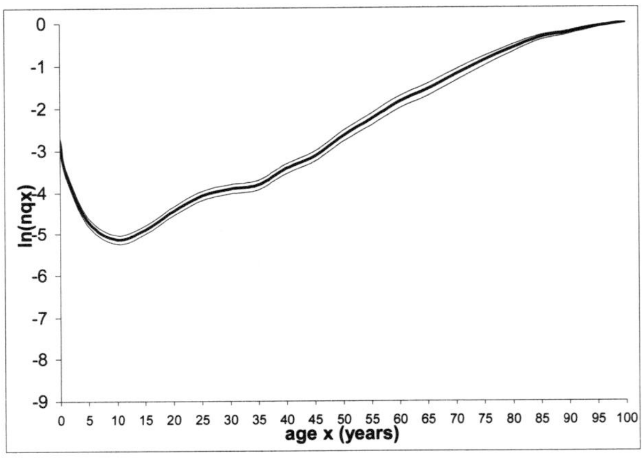

l_x with uncertainty interval

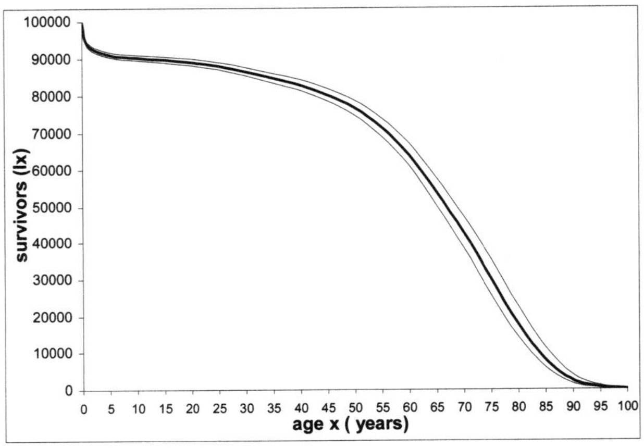

e_x with uncertainty interval

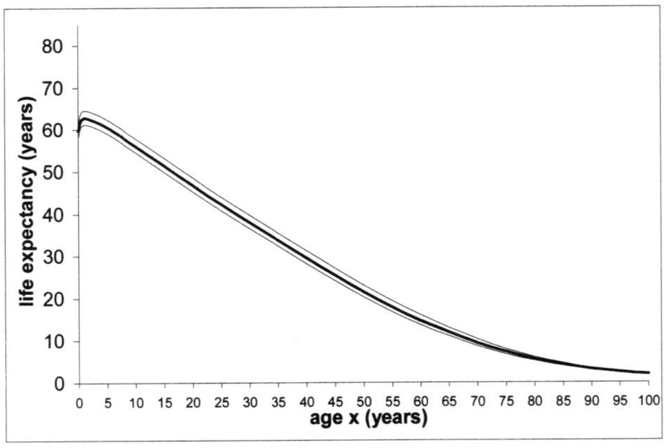

Distribution of deaths by age with uncertainty interval

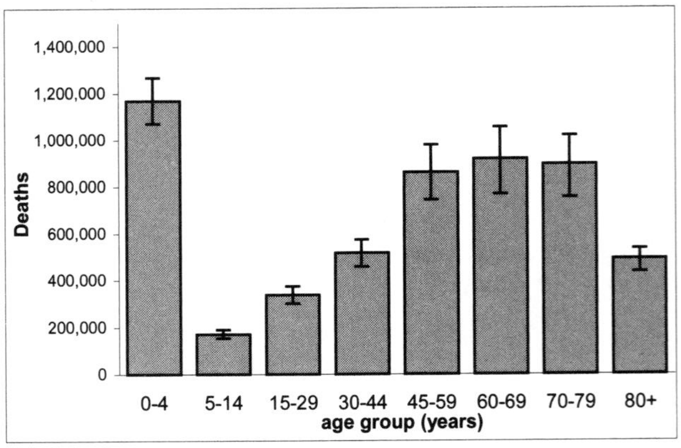

India   Inde   India

Females, 2000   *Femmes, 2000*   *Mujeres, 2000*

Age	$_nM_x$	$_nq_x$			l_x			$_nd_x$	$_nL_x$	e_x			Actual deaths	Actual population
			percentiles			percentiles					percentiles			
		mean	2½	97½	mean	2½	97½			mean	2½	97½		
<1	0.0676	0.0638	0.0558	0.0728	100,000	100,000	100,000	6,377	94,261	62.7	60.7	65.0	788,471	11,655,599
1-4	0.0096	0.0376	0.0333	0.0419	93,623	92,723	94,421	3,525	366,035	65.9	63.8	68.3	430,704	44,728,827
5-9	0.0025	0.0125	0.0109	0.0140	90,099	89,199	90,999	1,128	447,675	64.5	62.4	66.8	137,870	54,725,561
10-14	0.0013	0.0066	0.0058	0.0074	88,971	87,977	89,980	588	443,384	60.2	58.2	62.6	69,159	52,128,201
15-19	0.0021	0.0103	0.0089	0.0116	88,383	87,328	89,451	907	439,645	55.6	53.6	57.9	100,055	48,475,620
20-24	0.0028	0.0139	0.0120	0.0156	87,475	86,326	88,636	1,213	434,345	51.2	49.2	53.5	119,360	42,745,630
25-29	0.0029	0.0145	0.0125	0.0164	86,263	84,988	87,546	1,247	428,194	46.9	44.9	49.1	116,085	39,848,328
30-34	0.0030	0.0151	0.0130	0.0171	85,015	83,605	86,438	1,281	421,874	42.5	40.6	44.7	109,226	35,983,667
35-39	0.0032	0.0159	0.0137	0.0181	83,735	82,183	85,316	1,334	415,339	38.1	36.3	40.3	102,540	31,934,500
40-44	0.0045	0.0222	0.0191	0.0255	82,401	80,703	84,148	1,833	407,421	33.7	31.9	35.8	124,848	27,744,053
45-49	0.0065	0.0320	0.0274	0.0370	80,568	78,670	82,541	2,577	396,396	29.4	27.7	31.5	150,316	23,125,487
50-54	0.0090	0.0440	0.0373	0.0513	77,991	75,818	80,281	3,432	381,374	25.3	23.7	27.3	172,910	19,212,249
55-59	0.0139	0.0673	0.0563	0.0792	74,559	72,035	77,260	5,020	360,243	21.4	19.8	23.3	226,553	16,258,203
60-64	0.0210	0.0998	0.0823	0.1180	69,539	66,546	72,812	6,937	330,350	17.7	16.3	19.5	288,036	13,715,985
65-69	0.0328	0.1514	0.1237	0.1791	62,601	58,968	66,588	9,475	289,318	14.4	13.2	16.0	354,148	10,813,430
70-74	0.0488	0.2173	0.1774	0.2550	53,126	48,425	57,889	11,545	236,767	11.5	10.5	13.0	368,365	7,554,310
75-79	0.0727	0.3075	0.2543	0.3537	41,581	36,077	47,324	12,786	175,938	9.0	8.2	10.2	335,430	4,615,527
80-84	0.1083	0.4262	0.3626	0.4764	28,795	23,315	35,291	12,273	113,290	6.9	6.3	7.8	252,665	2,332,329
85-89	0.1593	0.5696	0.5044	0.6180	16,522	12,213	22,495	9,411	59,080	5.2	4.8	5.9	130,278	817,817
90-94	0.2311	0.6824	0.6254	0.7234	7,110	4,690	11,148	4,852	20,996	3.9	3.6	4.3	42,440	183,658
95-99	0.3307	0.8300	0.7872	0.8597	2,258	1,310	4,176	1,875	5,669	2.9	2.7	3.1	8,199	24,796
100+	0.4668	1	1	1	384	187	889	384	823	2.1	2.0	2.2	915	1,961
All ages													*4,428,572*	*488,625,738*

ln($_nq_x$) with uncertainty interval

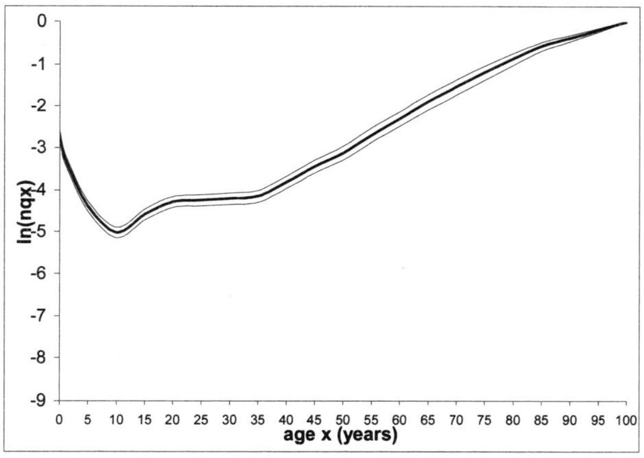

l_x with uncertainty interval

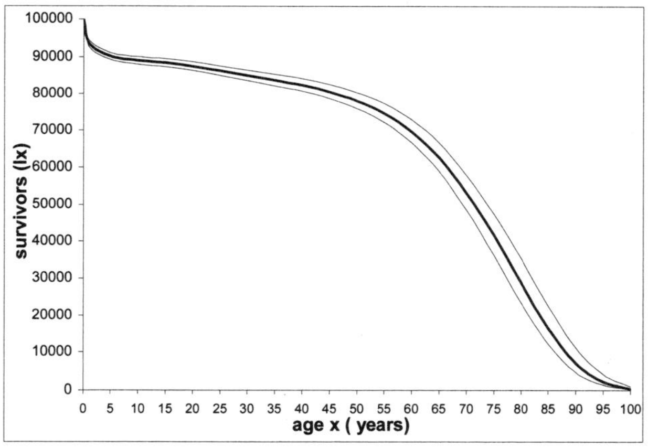

e_x with uncertainty interval

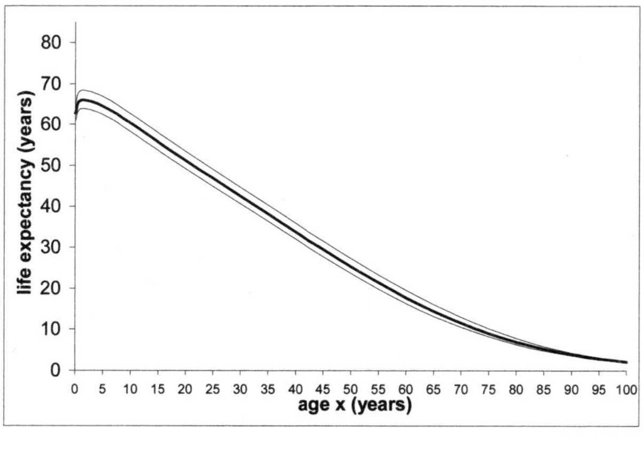

Distribution of deaths by age with uncertainty interval

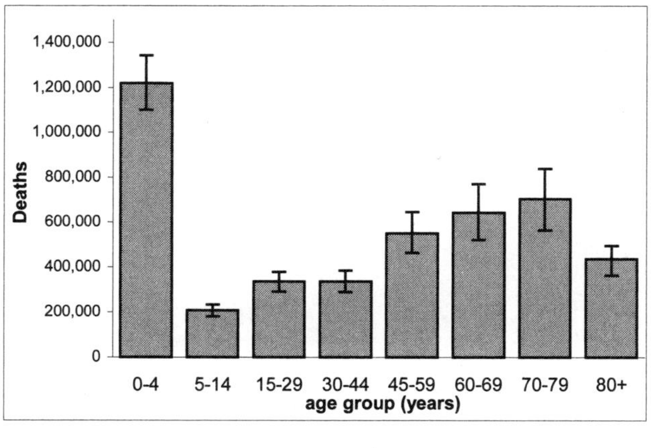

Indonesia | Indonésie | Indonesia

Males, 2000 | *Hommes, 2000* | *Hombres, 2000*

Age	$_nM_x$	$_nq_x$			l_x			$_nd_x$	$_nL_x$	e_x			Actual deaths	Actual population
			percentiles			percentiles					percentiles			
		mean	2½	97½	mean	2½	97½			mean	2½	97½		
<1	0.0508	0.0486	0.0426	0.0547	100,000	100,000	100,000	4,858	95,628	63.4	62.3	64.6	115,106	2,265,987
1-4	0.0034	0.0134	0.0121	0.0147	95,142	94,529	95,735	1,280	377,499	65.6	64.7	66.6	29,921	8,827,779
5-9	0.0012	0.0060	0.0054	0.0065	93,863	93,169	94,562	559	467,918	62.5	61.6	63.4	13,148	11,015,331
10-14	0.0010	0.0051	0.0046	0.0056	93,304	92,568	94,050	474	465,336	57.8	57.0	58.7	11,279	11,066,826
15-19	0.0018	0.0091	0.0082	0.0099	92,830	92,059	93,613	844	462,041	53.1	52.3	54.0	19,948	10,924,150
20-24	0.0025	0.0125	0.0114	0.0136	91,986	91,156	92,834	1,151	457,054	48.6	47.8	49.4	26,427	10,491,127
25-29	0.0027	0.0134	0.0122	0.0145	90,835	89,931	91,769	1,216	451,136	44.2	43.4	44.9	25,932	9,622,946
30-34	0.0031	0.0153	0.0139	0.0165	89,619	88,631	90,651	1,367	444,679	39.7	39.0	40.5	27,132	8,825,898
35-39	0.0040	0.0196	0.0179	0.0212	88,252	87,169	89,388	1,732	436,932	35.3	34.6	36.0	29,162	7,357,771
40-44	0.0055	0.0271	0.0248	0.0293	86,521	85,326	87,781	2,347	426,736	30.9	30.3	31.7	35,545	6,463,446
45-49	0.0080	0.0392	0.0359	0.0423	84,174	82,840	85,590	3,297	412,626	26.7	26.1	27.4	40,548	5,074,543
50-54	0.0120	0.0580	0.0534	0.0624	80,877	79,373	82,488	4,693	392,650	22.7	22.2	23.4	45,043	3,768,203
55-59	0.0179	0.0856	0.0793	0.0917	76,183	74,484	78,026	6,525	364,604	19.0	18.4	19.6	58,710	3,280,630
60-64	0.0271	0.1269	0.1182	0.1354	69,658	67,710	71,794	8,842	326,187	15.5	15.0	16.1	75,430	2,782,660
65-69	0.0409	0.1854	0.1730	0.1971	60,816	58,564	63,291	11,273	275,898	12.4	11.9	12.9	84,271	2,062,393
70-74	0.0633	0.2732	0.2556	0.2910	49,543	47,118	52,226	13,535	213,877	9.7	9.3	10.1	86,500	1,366,834
75-79	0.0978	0.3928	0.3692	0.4159	36,008	33,681	38,582	14,144	144,679	7.4	7.0	7.7	71,930	735,764
80-84	0.1491	0.5430	0.5163	0.5678	21,864	19,977	24,108	11,871	79,640	5.5	5.3	5.8	48,093	322,630
85-89	0.2210	0.7118	0.6878	0.7330	9,992	8,666	11,534	7,113	32,179	4.1	3.9	4.2	22,856	103,404
90-94	0.3166	0.8119	0.7929	0.8283	2,880	2,313	3,576	2,338	7,384	3.0	2.9	3.1	5,944	18,773
95-99	0.4383	0.9467	0.9374	0.9542	542	397	740	513	1,170	2.3	2.2	2.3	742	1,692
100+	0.5860	1	1	1	29	18	46	29	49	1.7	1.7	1.8	45	76
All ages													*873,710*	*106,378,863*

$\ln(_nq_x)$ with uncertainty interval

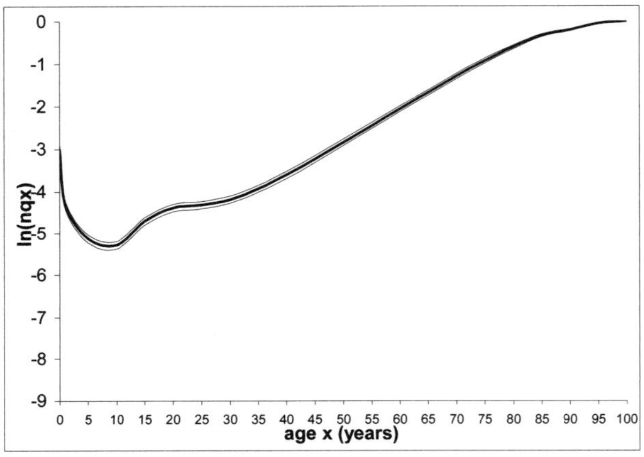

l_x with uncertainty interval

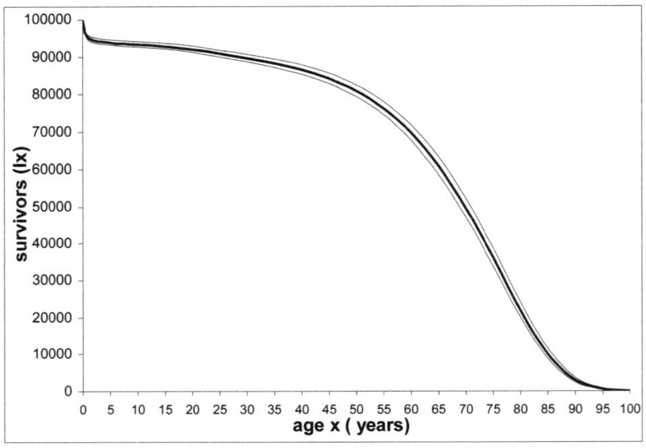

e_x with uncertainty interval

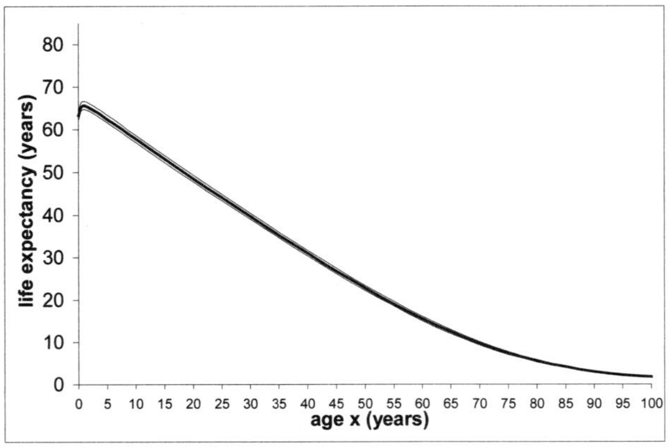

Distribution of deaths by age with uncertainty interval

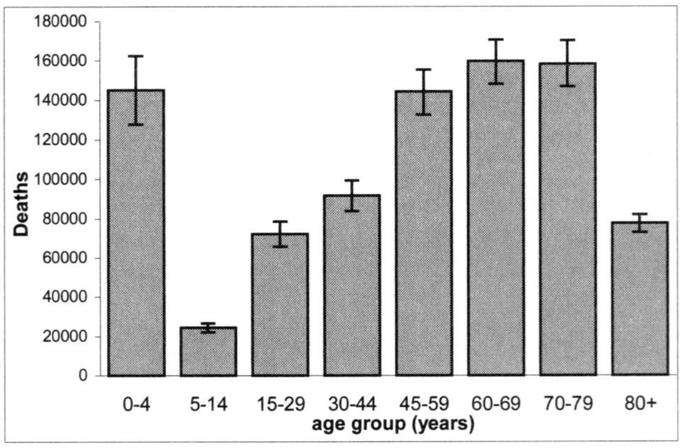

Females, 2000 *Femmes, 2000* *Mujeres, 2000*

Age	$_nM_x$	$_nq_x$			l_x			$_nd_x$	$_nL_x$	e_x			Actual deaths	Actual population
			percentiles			percentiles					percentiles			
		mean	2½	97½	mean	2½	97½			mean	2½	97½		
<1	0.0382	0.0369	0.0321	0.0420	100,000	100,000	100,000	3,692	96,677	67.4	66.4	68.5	82,414	2,158,083
1-4	0.0032	0.0127	0.0113	0.0139	96,308	95,799	96,793	1,220	382,304	69.0	68.2	69.9	27,225	8,530,054
5-9	0.0010	0.0050	0.0045	0.0055	95,088	94,489	95,685	475	474,251	65.9	65.1	66.7	10,669	10,643,491
10-14	0.0008	0.0041	0.0036	0.0044	94,612	93,972	95,250	384	472,102	61.2	60.4	62.0	8,723	10,723,998
15-19	0.0013	0.0066	0.0059	0.0072	94,228	93,556	94,901	621	469,590	56.5	55.7	57.2	14,069	10,639,975
20-24	0.0017	0.0084	0.0076	0.0092	93,607	92,884	94,339	787	466,071	51.8	51.1	52.6	17,239	10,212,852
25-29	0.0020	0.0097	0.0088	0.0106	92,821	92,037	93,624	902	461,849	47.2	46.5	48.0	18,334	9,386,349
30-34	0.0024	0.0118	0.0107	0.0129	91,919	91,069	92,802	1,087	456,875	42.7	42.0	43.4	20,457	8,595,340
35-39	0.0032	0.0161	0.0146	0.0174	90,831	89,909	91,807	1,459	450,509	38.1	37.5	38.8	23,195	7,163,030
40-44	0.0043	0.0215	0.0197	0.0233	89,372	88,361	90,465	1,922	442,058	33.7	33.1	34.4	27,369	6,295,924
45-49	0.0061	0.0302	0.0277	0.0326	87,451	86,316	88,687	2,639	430,656	29.4	28.8	30.1	30,373	4,956,098
50-54	0.0088	0.0430	0.0398	0.0464	84,812	83,515	86,226	3,647	414,941	25.3	24.7	25.9	35,288	4,015,040
55-59	0.0126	0.0610	0.0566	0.0657	81,165	79,676	82,793	4,955	393,436	21.3	20.8	21.9	45,884	3,643,377
60-64	0.0192	0.0915	0.0851	0.0980	76,210	74,513	78,073	6,974	363,615	17.5	17.0	18.0	60,227	3,140,354
65-69	0.0299	0.1393	0.1297	0.1484	69,236	67,271	71,330	9,644	322,072	14.0	13.6	14.5	72,191	2,410,931
70-74	0.0491	0.2187	0.2043	0.2325	59,592	57,312	61,865	13,033	265,379	10.9	10.5	11.3	80,377	1,636,589
75-79	0.0797	0.3324	0.3121	0.3532	46,559	44,132	49,049	15,478	194,101	8.2	7.9	8.6	73,241	918,491
80-84	0.1281	0.4852	0.4613	0.5105	31,081	28,859	33,544	15,082	117,701	6.1	5.8	6.3	57,315	447,289
85-89	0.1979	0.6619	0.6388	0.6853	15,999	14,338	17,943	10,590	53,521	4.4	4.3	4.6	31,318	158,274
90-94	0.2930	0.7797	0.7609	0.7979	5,409	4,512	6,458	4,217	14,393	3.2	3.1	3.3	9,828	33,544
95-99	0.4160	0.9253	0.9150	0.9348	1,192	912	1,530	1,103	2,651	2.4	2.3	2.4	1,598	3,841
100+	0.5665	1	1	1	89	59	128	89	157	1.8	1.8	1.9	134	237
All ages													*747,469*	*105,713,161*

ln($_nq_x$) with uncertainty interval

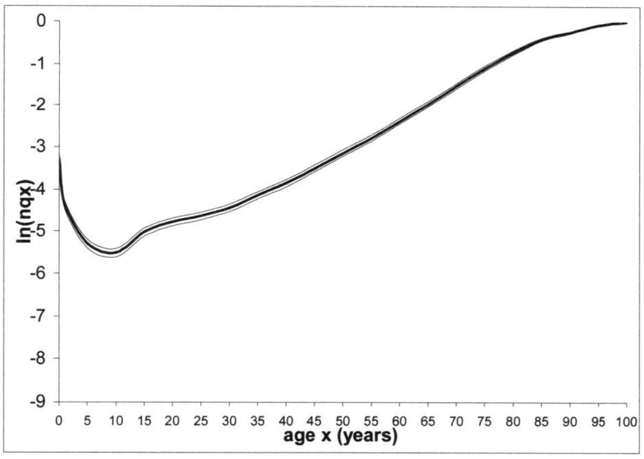

l_x with uncertainty interval

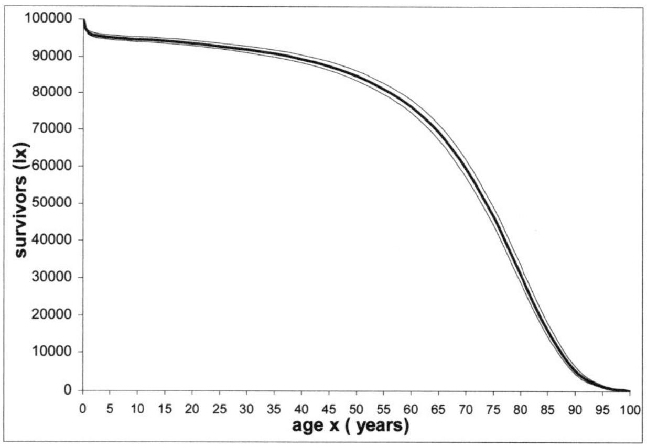

e_x with uncertainty interval

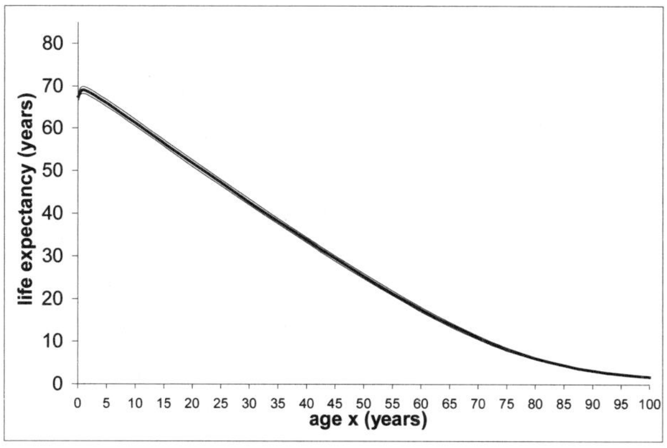

Distribution of deaths by age with uncertainty interval

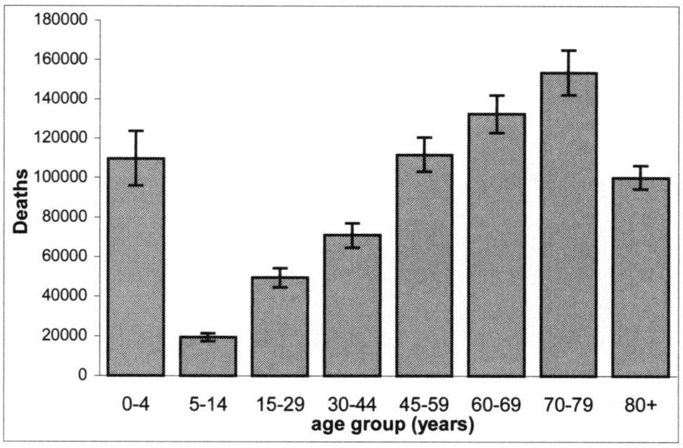

Iran (Islamic Republic of) | Iran (République islamique d') | Irán (República Islámica del)

Males, 2000 | *Hommes, 2000* | *Hombres, 2000*

Age	$_nM_x$	$_nq_x$			l_x			$_nd_x$	$_nL_x$	e_x			Actual deaths	Actual population
			percentiles			percentiles					percentiles			
		mean	2½	97½	mean	2½	97½			mean	2½	97½		
<1	0.0395	0.0381	0.0315	0.0449	100,000	100,000	100,000	3,812	96,569	68.1	67.0	69.3	31,581	800,039
1-4	0.0014	0.0058	0.0050	0.0065	96,188	95,515	96,850	555	383,420	69.8	68.8	70.7	4,462	3,082,343
5-9	0.0006	0.0032	0.0027	0.0036	95,633	94,889	96,370	303	477,408	66.2	65.2	67.1	2,929	4,620,759
10-14	0.0006	0.0032	0.0028	0.0036	95,330	94,549	96,108	307	475,884	61.4	60.4	62.3	3,240	5,019,441
15-19	0.0012	0.0062	0.0054	0.0070	95,023	94,206	95,840	591	473,639	56.6	55.7	57.5	5,698	4,567,310
20-24	0.0019	0.0093	0.0081	0.0104	94,432	93,548	95,323	874	469,977	51.9	51.0	52.8	6,076	3,267,632
25-29	0.0019	0.0095	0.0083	0.0106	93,558	92,580	94,553	886	465,578	47.4	46.5	48.3	5,218	2,743,001
30-34	0.0019	0.0093	0.0082	0.0103	92,673	91,607	93,767	859	461,216	42.8	42.0	43.7	4,460	2,393,908
35-39	0.0022	0.0107	0.0095	0.0118	91,813	90,668	92,999	983	456,611	38.2	37.4	39.1	4,619	2,146,359
40-44	0.0028	0.0139	0.0124	0.0153	90,831	89,601	92,116	1,261	451,001	33.6	32.8	34.4	5,257	1,879,523
45-49	0.0047	0.0232	0.0209	0.0254	89,569	88,240	90,973	2,077	442,654	29.0	28.2	29.9	7,497	1,597,504
50-54	0.0080	0.0393	0.0357	0.0431	87,492	86,020	89,072	3,436	428,870	24.7	23.9	25.5	9,843	1,228,381
55-59	0.0128	0.0620	0.0568	0.0681	84,056	82,362	85,881	5,213	407,247	20.6	19.9	21.3	11,217	876,358
60-64	0.0217	0.1029	0.0938	0.1127	78,843	76,893	80,945	8,115	373,928	16.8	16.1	17.5	14,107	650,040
65-69	0.0343	0.1579	0.1436	0.1717	70,728	68,470	73,069	11,167	325,725	13.4	12.8	14.0	16,572	483,397
70-74	0.0540	0.2379	0.2172	0.2579	59,562	56,711	62,210	14,169	262,386	10.4	9.9	11.0	17,724	328,223
75-79	0.0850	0.3505	0.3222	0.3795	45,393	42,213	48,553	15,911	187,187	7.9	7.5	8.4	16,496	194,069
80-84	0.1350	0.5047	0.4722	0.5385	29,482	26,585	32,784	14,879	110,213	5.8	5.5	6.1	11,989	88,806
85-89	0.2100	0.6885	0.6591	0.7186	14,603	12,490	17,154	10,055	47,879	4.2	4.0	4.4	5,505	26,213
90-94	0.3200	0.8163	0.7944	0.8383	4,549	3,515	5,837	3,713	11,603	2.9	2.8	3.0	1,509	4,717
95-99	0.4800	0.9836	0.9789	0.9877	835	568	1,193	822	1,712	2.1	2.1	2.1	205	428
100+	0.6900	1	1	1	14	7	25	14	20	1.4	1.4	1.6	12	17
All ages													*186,215*	*35,998,468*

ln($_nq_x$) with uncertainty interval

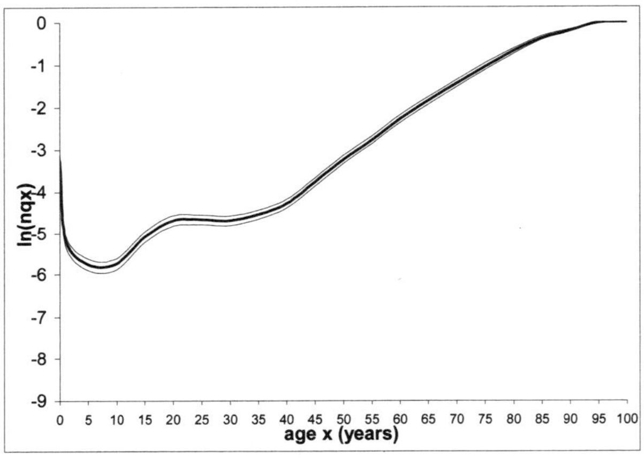

l_x with uncertainty interval

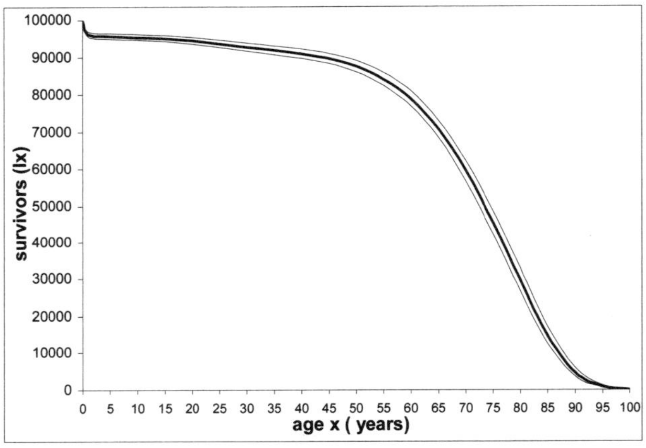

e_x with uncertainty interval

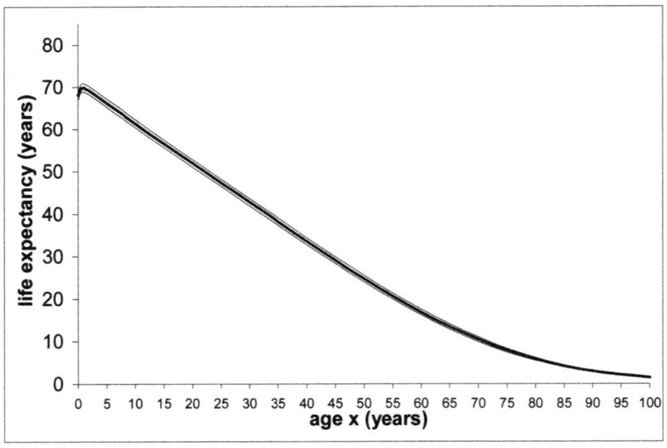

Distribution of deaths by age with uncertainty interval

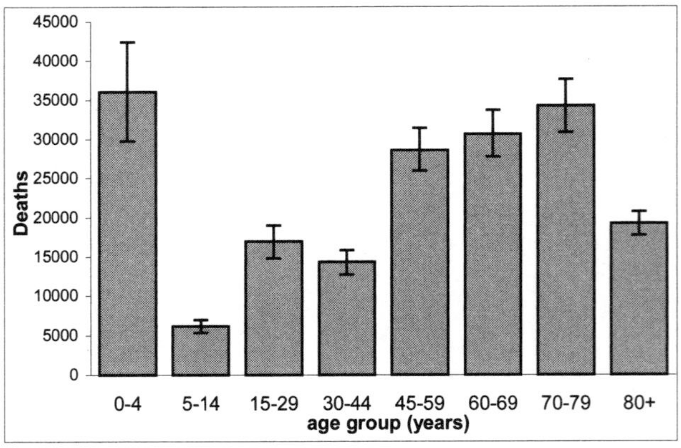

Iran (Islamic Republic of) — Iran (République islamique d') — Irán (República Islámica del)

Females, 2000 — *Femmes, 2000* — *Mujeres, 2000*

Age	$_nM_x$	$_nq_x$ mean	$_nq_x$ percentiles 2½	$_nq_x$ percentiles 97½	l_x mean	l_x percentiles 2½	l_x percentiles 97½	$_nd_x$	$_nL_x$	e_x mean	e_x percentiles 2½	e_x percentiles 97½	Actual deaths	Actual population
<1	0.0418	0.0403	0.0336	0.0472	100,000	100,000	100,000	4,031	96,372	69.9	68.8	71.0	32,175	769,268
1-4	0.0016	0.0065	0.0057	0.0073	95,969	95,284	96,642	622	382,384	71.8	71.0	72.8	4,724	2,903,119
5-9	0.0006	0.0031	0.0027	0.0035	95,347	94,600	96,096	295	475,998	68.3	67.4	69.2	2,701	4,361,481
10-14	0.0007	0.0035	0.0031	0.0040	95,052	94,278	95,836	337	474,418	63.5	62.6	64.4	3,374	4,745,465
15-19	0.0008	0.0041	0.0036	0.0046	94,715	93,905	95,537	389	472,602	58.7	57.9	59.6	3,627	4,407,764
20-24	0.0012	0.0059	0.0052	0.0066	94,326	93,471	95,191	557	470,237	54.0	53.1	54.9	3,718	3,137,392
25-29	0.0013	0.0065	0.0057	0.0072	93,769	92,852	94,693	605	467,331	49.3	48.4	50.1	3,377	2,608,580
30-34	0.0013	0.0065	0.0058	0.0072	93,164	92,182	94,149	605	464,306	44.6	43.8	45.4	2,953	2,265,298
35-39	0.0016	0.0080	0.0072	0.0089	92,558	91,514	93,602	737	460,949	39.8	39.0	40.7	3,208	2,005,305
40-44	0.0024	0.0121	0.0109	0.0134	91,821	90,704	92,931	1,109	456,333	35.1	34.4	36.0	4,222	1,736,895
45-49	0.0038	0.0188	0.0170	0.0207	90,712	89,492	91,916	1,704	449,299	30.5	29.8	31.4	5,589	1,473,524
50-54	0.0065	0.0320	0.0290	0.0350	89,008	87,642	90,340	2,846	437,923	26.1	25.3	26.9	7,602	1,169,530
55-59	0.0109	0.0533	0.0486	0.0582	86,161	84,584	87,672	4,591	419,330	21.8	21.2	22.6	9,337	852,945
60-64	0.0175	0.0838	0.0765	0.0914	81,571	79,725	83,302	6,838	390,758	17.9	17.3	18.7	11,493	656,725
65-69	0.0295	0.1374	0.1249	0.1496	74,732	72,621	76,817	10,266	347,997	14.4	13.8	15.0	15,058	510,451
70-74	0.0470	0.2103	0.1916	0.2278	64,467	61,983	67,004	13,557	288,441	11.2	10.7	11.8	16,998	361,665
75-79	0.0750	0.3158	0.2891	0.3426	50,910	48,005	53,884	16,077	214,357	8.6	8.1	9.1	16,698	222,644
80-84	0.1200	0.4615	0.4290	0.4953	34,833	31,829	38,211	16,077	133,973	6.4	6.0	6.7	12,614	105,117
85-89	0.1800	0.6207	0.5893	0.6533	18,756	16,395	21,644	11,642	64,677	4.7	4.4	4.9	5,747	31,930
90-94	0.2845	0.7675	0.7423	0.7928	7,114	5,692	8,862	5,460	19,192	3.2	3.1	3.4	1,677	5,896
95-99	0.4200	0.9292	0.9164	0.9414	1,654	1,179	2,274	1,537	3,660	2.3	2.3	2.4	236	563
100+	0.5700	1	1	1	117	69	188	117	205	1.8	1.7	1.8	16	28
All ages													*167,146*	*34,331,585*

ln($_nq_x$) with uncertainty interval

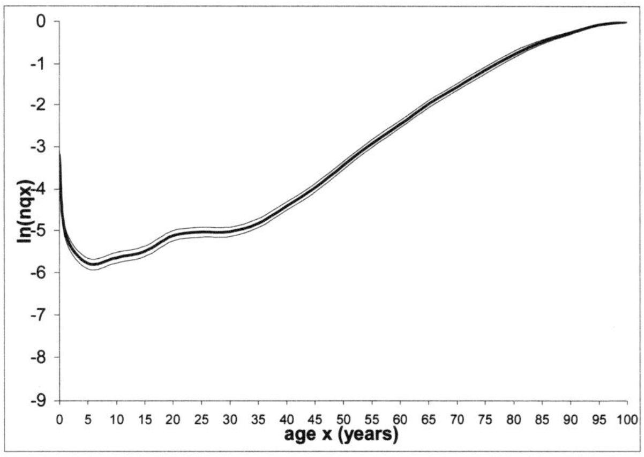

l_x with uncertainty interval

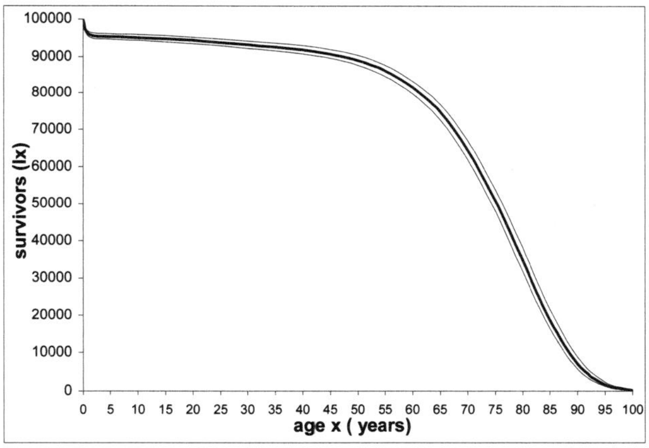

e_x with uncertainty interval

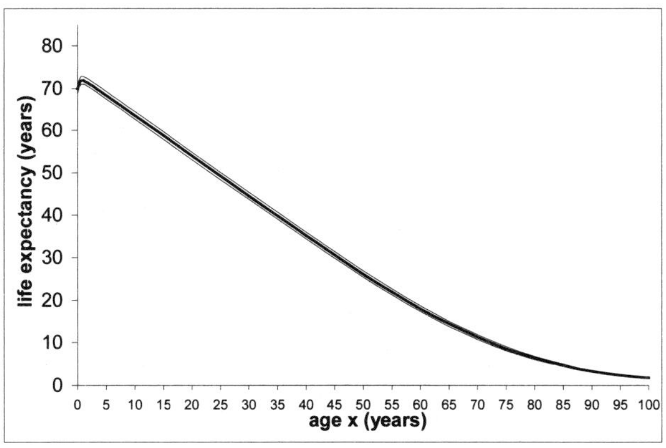

Distribution of deaths by age with uncertainty interval

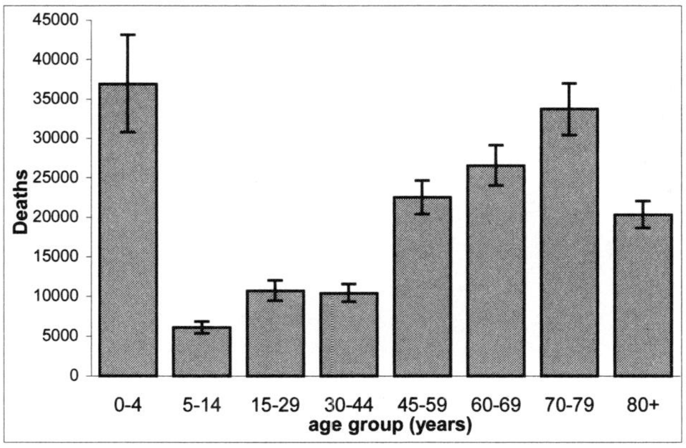

Iraq | **Iraq** | **Iraq**

Males, 2000 | *Hommes, 2000* | *Hombres, 2000*

Age	$_nM_x$	$_nq_x$			l_x			$_nd_x$	$_nL_x$	e_x			Actual deaths	Actual population
			percentiles			percentiles					percentiles			
		mean	2½	97½	mean	2½	97½			mean	2½	97½		
<1	0.0668	0.0630	0.0497	0.0768	100,000	100,000	100,000	6,303	94,327	61.7	59.7	64.0	26,469	396,117
1-4	0.0046	0.0181	0.0150	0.0210	93,697	92,321	95,028	1,694	370,723	64.9	63.2	66.8	6,494	1,421,190
5-9	0.0014	0.0068	0.0057	0.0079	92,003	90,504	93,499	625	458,453	62.1	60.4	64.0	2,128	1,561,102
10-14	0.0010	0.0051	0.0042	0.0059	91,378	89,799	92,944	464	455,731	57.5	55.8	59.4	1,529	1,502,118
15-19	0.0017	0.0084	0.0070	0.0097	90,914	89,277	92,539	761	452,667	52.7	51.1	54.6	2,151	1,278,742
20-24	0.0024	0.0121	0.0102	0.0140	90,153	88,425	91,884	1,095	448,027	48.2	46.6	50.1	2,651	1,085,291
25-29	0.0027	0.0133	0.0112	0.0152	89,058	87,207	90,943	1,184	442,331	43.7	42.1	45.6	2,582	964,541
30-34	0.0032	0.0157	0.0133	0.0179	87,874	85,899	89,922	1,377	435,928	39.3	37.7	41.1	2,464	780,104
35-39	0.0041	0.0204	0.0174	0.0232	86,497	84,389	88,727	1,765	428,072	34.9	33.3	36.7	2,630	637,736
40-44	0.0058	0.0285	0.0244	0.0323	84,732	82,440	87,183	2,411	417,630	30.5	29.1	32.3	3,006	520,726
45-49	0.0084	0.0412	0.0357	0.0468	82,320	79,802	85,053	3,395	403,114	26.4	24.9	28.1	3,549	421,322
50-54	0.0126	0.0611	0.0526	0.0699	78,925	76,127	82,014	4,819	382,579	22.4	21.0	24.1	4,233	336,094
55-59	0.0188	0.0896	0.0760	0.1036	74,106	70,963	77,623	6,643	353,926	18.7	17.3	20.3	4,889	260,479
60-64	0.0282	0.1316	0.1101	0.1530	67,464	63,986	71,425	8,880	315,118	15.3	14.0	16.7	5,432	192,759
65-69	0.0422	0.1910	0.1590	0.2238	58,584	54,687	62,883	11,192	264,938	12.2	11.1	13.5	5,727	135,566
70-74	0.0653	0.2806	0.2348	0.3268	47,392	42,775	52,193	13,300	203,709	9.5	8.6	10.6	5,689	87,142
75-79	0.1004	0.4013	0.3411	0.4571	34,092	29,081	39,879	13,682	136,254	7.2	6.5	8.1	5,018	49,976
80-84	0.1530	0.5534	0.4863	0.6097	20,410	15,888	26,147	11,294	73,814	5.4	4.9	6.1	3,706	24,221
85-89	0.2254	0.7209	0.6611	0.7675	9,116	6,208	13,259	6,572	29,150	4.0	3.7	4.4	1,890	8,385
90-94	0.3213	0.8180	0.7709	0.8536	2,544	1,450	4,467	2,081	6,478	3.0	2.8	3.2	680	2,118
95-99	0.4427	0.9508	0.9279	0.9659	463	214	1,023	440	995	2.2	2.2	2.3	155	349
100+	0.5900	1	1	1	23	7	74	23	39	1.7	1.7	1.8	28	48
All ages													*93,102*	*11,666,126*

$\ln(_nq_x)$ with uncertainty interval

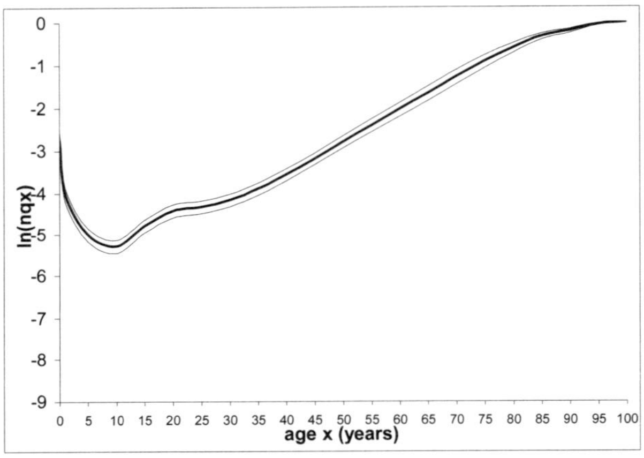

l_x with uncertainty interval

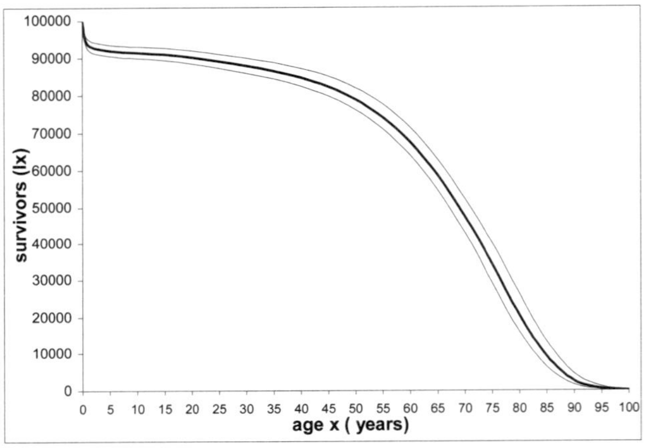

e_x with uncertainty interval

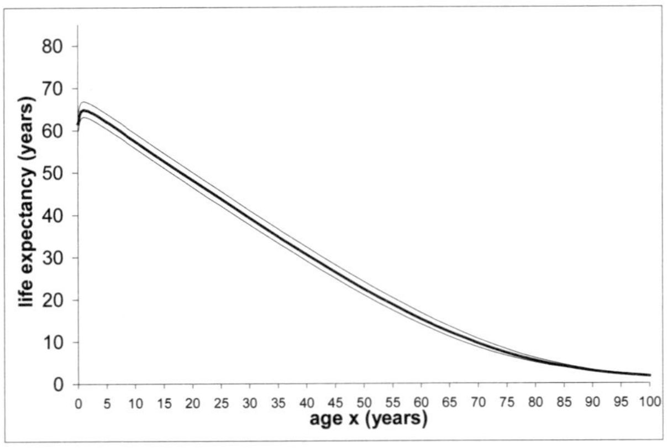

Distribution of deaths by age with uncertainty interval

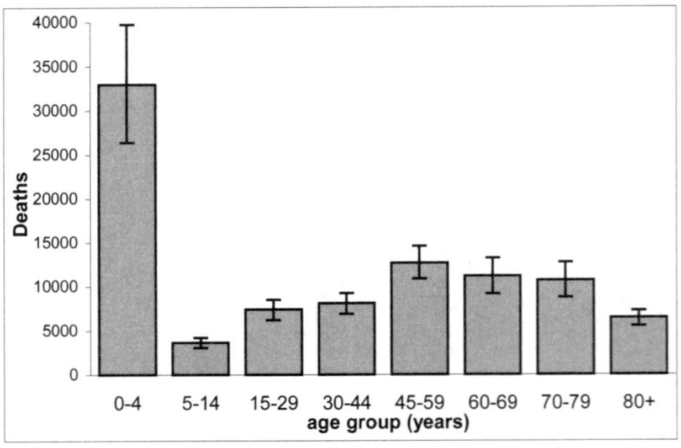

Females, 2000 *Femmes, 2000* *Mujeres, 2000*

Age	$_nM_x$	$_nq_x$ mean	$_nq_x$ percentiles 2½	$_nq_x$ percentiles 97½	l_x mean	l_x percentiles 2½	l_x percentiles 97½	$_nd_x$	$_nL_x$	e_x mean	e_x percentiles 2½	e_x percentiles 97½	Actual deaths	Actual population
<1	0.0573	0.0545	0.0422	0.0676	100,000	100,000	100,000	5,453	95,092	64.7	62.3	67.2	21,562	376,001
1-4	0.0049	0.0195	0.0160	0.0233	94,547	93,242	95,777	1,843	373,764	67.4	65.5	69.5	6,720	1,362,659
5-9	0.0013	0.0066	0.0054	0.0079	92,704	91,204	94,179	614	461,982	64.7	63.0	66.8	1,989	1,495,896
10-14	0.0010	0.0048	0.0040	0.0058	92,089	90,492	93,650	446	459,331	60.1	58.5	62.1	1,397	1,439,307
15-19	0.0015	0.0075	0.0061	0.0088	91,643	89,972	93,265	683	456,508	55.4	53.8	57.4	1,818	1,214,368
20-24	0.0020	0.0099	0.0082	0.0117	90,960	89,176	92,670	904	452,540	50.8	49.2	52.8	2,062	1,031,995
25-29	0.0023	0.0116	0.0096	0.0136	90,056	88,129	91,877	1,041	447,677	46.3	44.7	48.2	2,167	931,620
30-34	0.0028	0.0138	0.0114	0.0162	89,015	86,929	90,955	1,226	442,009	41.8	40.3	43.7	2,096	755,920
35-39	0.0037	0.0183	0.0152	0.0215	87,789	85,523	89,907	1,609	434,924	37.3	35.9	39.2	2,277	615,675
40-44	0.0047	0.0233	0.0195	0.0273	86,180	83,692	88,536	2,012	425,872	33.0	31.5	34.8	2,377	503,072
45-49	0.0066	0.0323	0.0272	0.0376	84,168	81,421	86,808	2,719	414,044	28.7	27.3	30.4	2,697	410,678
50-54	0.0094	0.0460	0.0387	0.0533	81,449	78,386	84,450	3,749	397,875	24.6	23.3	26.2	3,116	330,693
55-59	0.0135	0.0655	0.0547	0.0753	77,701	74,261	81,156	5,087	375,787	20.7	19.4	22.2	3,515	259,690
60-64	0.0207	0.0986	0.0820	0.1143	72,614	68,739	76,603	7,159	345,173	16.9	15.7	18.4	4,178	201,432
65-69	0.0326	0.1505	0.1253	0.1750	65,455	61,105	70,033	9,852	302,645	13.5	12.4	14.8	4,812	147,826
70-74	0.0534	0.2354	0.1970	0.2739	55,603	50,828	60,668	13,087	245,298	10.5	9.5	11.6	5,345	100,180
75-79	0.0859	0.3536	0.3003	0.4079	42,516	37,598	48,634	15,033	174,999	7.9	7.1	8.8	5,183	60,341
80-84	0.1357	0.5066	0.4449	0.5678	27,483	22,586	33,771	13,922	102,611	5.9	5.3	6.5	4,071	30,005
85-89	0.2062	0.6802	0.6229	0.7339	13,561	9,950	18,483	9,224	44,745	4.3	3.9	4.7	2,115	10,257
90-94	0.3013	0.7913	0.7458	0.8325	4,337	2,648	6,948	3,432	11,389	3.1	2.9	3.4	649	2,155
95-99	0.4237	0.9328	0.9082	0.9529	905	444	1,749	844	1,992	2.3	2.2	2.4	134	316
100+	0.5732	1	1	1	61	21	157	61	106	1.7	1.7	1.9	19	33
All ages													*80,299*	*11,280,119*

ln($_nq_x$) with uncertainty interval

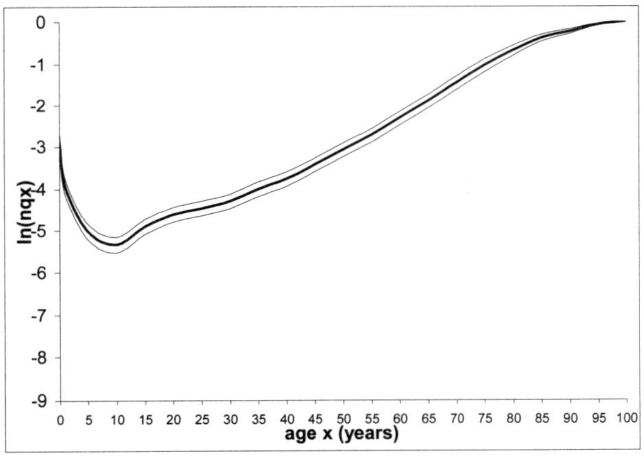

l_x with uncertainty interval

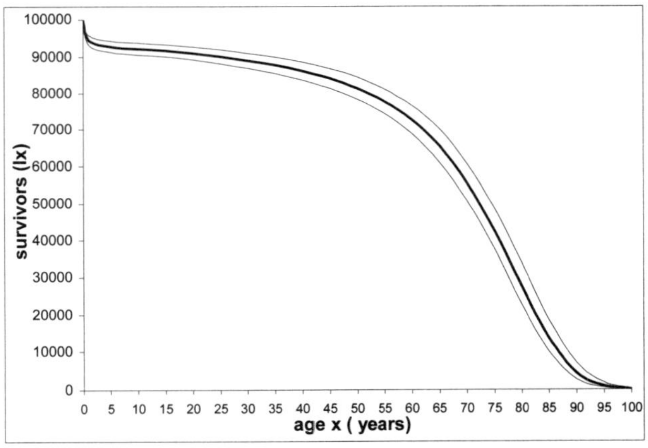

e_x with uncertainty interval

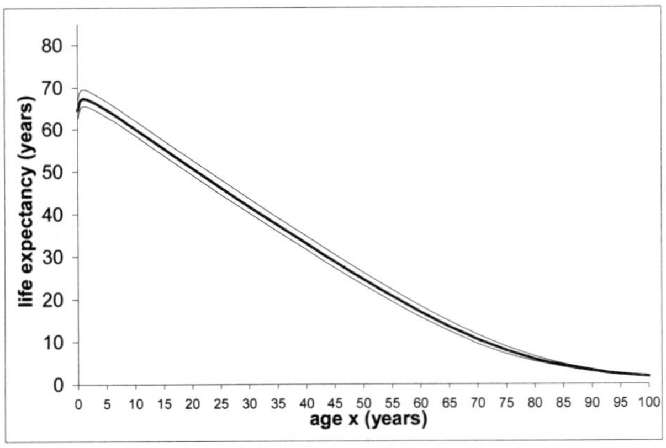

Distribution of deaths by age with uncertainty interval

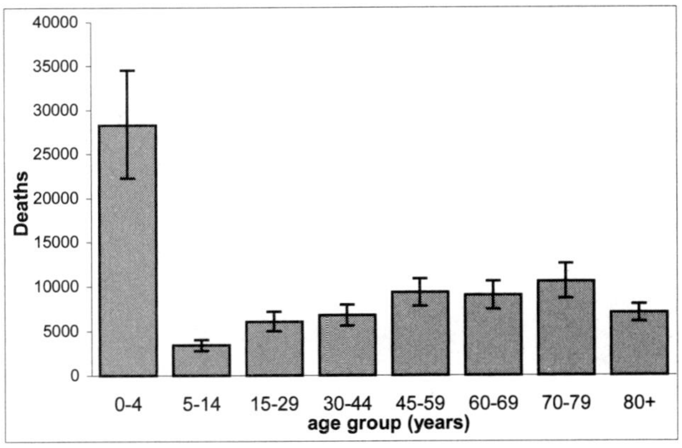

# Ireland 	Irlande 	Irlanda

Males, 2000 	*Hommes, 2000* 	*Hombres, 2000*

Age	$_nM_x$	$_nq_x$ mean	$_nq_x$ percentiles 2½	$_nq_x$ percentiles 97½	l_x mean	l_x percentiles 2½	l_x percentiles 97½	$_nd_x$	$_nL_x$	e_x mean	e_x percentiles 2½	e_x percentiles 97½	Actual deaths	Actual population
<1	0.0056	0.0056	0.0047	0.0065	100,000	100,000	100,000	559	99,497	74.1	73.6	74.6	156	27,771
1-4	0.0002	0.0008	0.0007	0.0009	99,441	99,351	99,528	79	397,574	73.5	73.0	74.0	22	108,646
5-9	0.0001	0.0005	0.0004	0.0006	99,362	99,262	99,459	51	496,682	69.5	69.1	70.0	14	134,746
10-14	0.0002	0.0010	0.0009	0.0011	99,311	99,205	99,415	101	496,302	64.6	64.1	65.1	31	150,844
15-19	0.0007	0.0033	0.0029	0.0036	99,210	99,092	99,326	324	495,239	59.6	59.2	60.1	114	174,858
20-24	0.0013	0.0063	0.0056	0.0069	98,886	98,731	99,039	621	492,877	54.8	54.4	55.3	222	176,409
25-29	0.0012	0.0062	0.0056	0.0068	98,265	98,047	98,481	609	489,804	50.1	49.7	50.6	187	150,811
30-34	0.0012	0.0061	0.0056	0.0066	97,656	97,383	97,928	597	486,790	45.4	45.0	45.9	159	129,854
35-39	0.0013	0.0063	0.0058	0.0068	97,060	96,737	97,381	609	483,776	40.7	40.3	41.2	161	127,836
40-44	0.0017	0.0085	0.0079	0.0091	96,451	96,082	96,818	819	480,206	35.9	35.6	36.4	215	126,304
45-49	0.0027	0.0135	0.0126	0.0144	95,632	95,206	96,055	1,293	474,927	31.2	30.9	31.7	327	120,238
50-54	0.0043	0.0213	0.0200	0.0226	94,339	93,835	94,840	2,009	466,673	26.6	26.3	27.1	487	113,021
55-59	0.0084	0.0412	0.0390	0.0435	92,330	91,724	92,930	3,806	452,136	22.2	21.8	22.6	761	90,458
60-64	0.0155	0.0745	0.0703	0.0785	88,524	87,775	89,258	6,597	426,129	18.0	17.7	18.4	1,138	73,534
65-69	0.0284	0.1328	0.1252	0.1392	81,927	80,982	82,859	10,877	382,443	14.2	14.0	14.6	1,764	62,018
70-74	0.0476	0.2126	0.2012	0.2211	71,050	69,759	72,370	15,107	317,482	11.0	10.8	11.4	2,393	50,288
75-79	0.0785	0.3280	0.3134	0.3405	55,943	54,335	57,808	18,348	233,843	8.4	8.1	8.6	2,967	37,819
80-84	0.1250	0.4762	0.4609	0.4922	37,594	36,003	39,691	17,901	143,220	6.2	6.0	6.4	2,706	21,651
85-89	0.1903	0.6447	0.6300	0.6602	19,693	18,548	21,379	12,697	66,724	4.6	4.5	4.7	1,856	9,751
90-94	0.2769	0.7562	0.7441	0.7686	6,996	6,310	7,840	5,291	19,109	3.3	3.3	3.4	668	2,414
95-99	0.3850	0.8933	0.8855	0.9009	1,705	1,460	1,974	1,523	3,957	2.5	2.5	2.6	135	351
100+	0.5117	1	1	1	182	145	224	182	356	2.0	1.9	2.0	17	34
All ages													*16,502*	*1,889,656*

ln($_nq_x$) with uncertainty interval

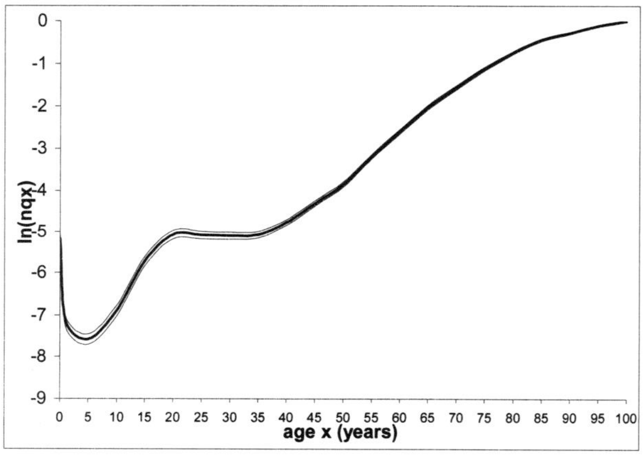

l_x with uncertainty interval

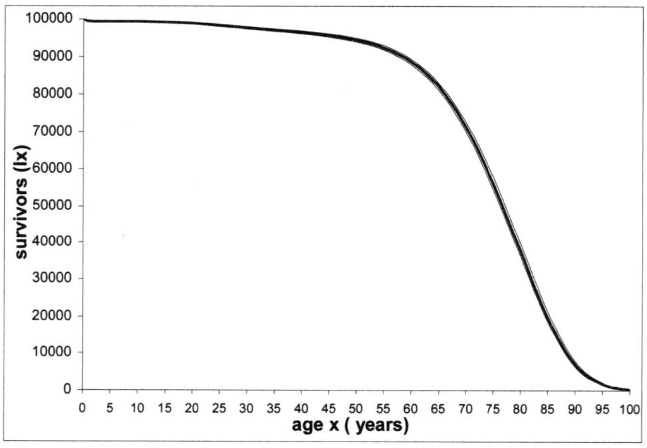

e_x with uncertainty interval

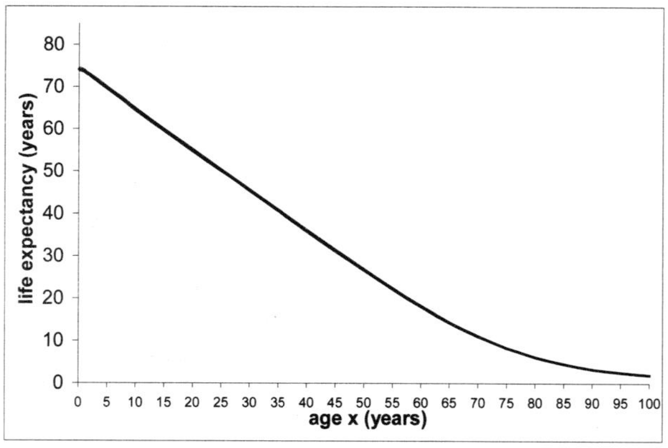

Distribution of deaths by age with uncertainty interval

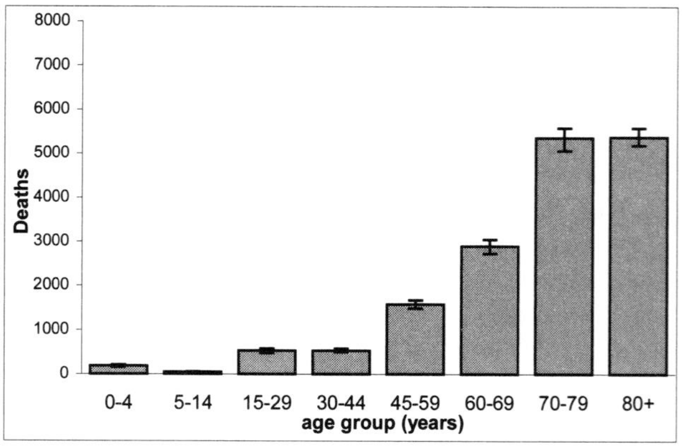

Ireland   Irlande   Irlanda

Females, 2000   *Femmes, 2000*   *Mujeres, 2000*

Age	$_nM_x$	$_nq_x$			l_x			$_nd_x$	$_nL_x$	e_x			Actual deaths	Actual population
			percentiles			percentiles					percentiles			
		mean	2½	97½	mean	2½	97½			mean	2½	97½		
<1	0.0043	0.0043	0.0039	0.0048	100,000	100,000	100,000	432	99,611	79.7	79.3	80.1	114	26,187
1-4	0.0002	0.0008	0.0007	0.0008	99,568	99,524	99,610	77	398,087	79.0	78.7	79.4	20	101,445
5-9	0.0001	0.0003	0.0002	0.0003	99,491	99,441	99,539	27	497,388	75.1	74.7	75.5	7	126,875
10-14	0.0001	0.0006	0.0006	0.0007	99,464	99,412	99,515	65	497,160	70.1	69.8	70.5	19	143,188
15-19	0.0003	0.0013	0.0012	0.0014	99,400	99,343	99,455	132	496,669	65.1	64.8	65.6	44	165,722
20-24	0.0003	0.0015	0.0014	0.0016	99,268	99,202	99,332	148	495,971	60.2	59.9	60.6	50	168,809
25-29	0.0004	0.0020	0.0018	0.0021	99,120	99,045	99,194	194	495,117	55.3	55.0	55.7	57	146,266
30-34	0.0005	0.0023	0.0022	0.0025	98,926	98,839	99,012	232	494,052	50.4	50.1	50.8	60	129,019
35-39	0.0007	0.0037	0.0035	0.0039	98,695	98,593	98,794	362	492,569	45.5	45.2	45.9	99	134,590
40-44	0.0011	0.0054	0.0052	0.0057	98,333	98,212	98,451	534	490,328	40.7	40.4	41.1	141	129,327
45-49	0.0019	0.0094	0.0090	0.0098	97,798	97,651	97,942	917	486,699	35.9	35.6	36.3	228	120,822
50-54	0.0030	0.0147	0.0141	0.0153	96,881	96,694	97,064	1,427	480,839	31.2	30.9	31.6	331	111,587
55-59	0.0048	0.0236	0.0226	0.0245	95,454	95,218	95,688	2,253	471,641	26.6	26.3	27.0	423	88,536
60-64	0.0083	0.0407	0.0388	0.0423	93,202	92,905	93,498	3,795	456,523	22.2	21.9	22.6	620	74,576
65-69	0.0151	0.0727	0.0690	0.0756	89,407	89,021	89,801	6,502	430,782	18.0	17.8	18.4	1,000	66,253
70-74	0.0258	0.1213	0.1149	0.1261	82,905	82,334	83,594	10,059	389,380	14.3	14.0	14.6	1,580	61,151
75-79	0.0458	0.2056	0.1954	0.2130	72,847	71,953	73,986	14,979	326,787	10.9	10.7	11.2	2,469	53,862
80-84	0.0796	0.3318	0.3181	0.3424	57,868	56,625	59,531	19,202	241,337	8.1	7.9	8.3	2,896	36,396
85-89	0.1322	0.4968	0.4824	0.5100	38,667	37,305	40,592	19,209	145,311	5.8	5.7	6.0	2,687	20,327
90-94	0.2102	0.6446	0.6317	0.6566	19,458	18,410	20,992	12,542	59,663	4.1	4.0	4.2	1,444	6,870
95-99	0.3200	0.8163	0.8073	0.8247	6,916	6,344	7,680	5,645	17,643	2.9	2.9	3.0	465	1,453
100+	0.4661	1	1	1	1,271	1,112	1,464	1,271	2,726	2.1	2.1	2.2	78	168
All ages													*14,831*	*1,913,429*

$ln(_nq_x)$ with uncertainty interval

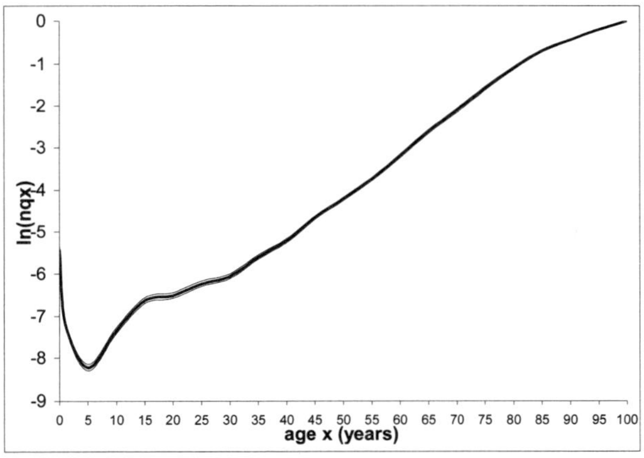

l_x with uncertainty interval

e_x with uncertainty interval

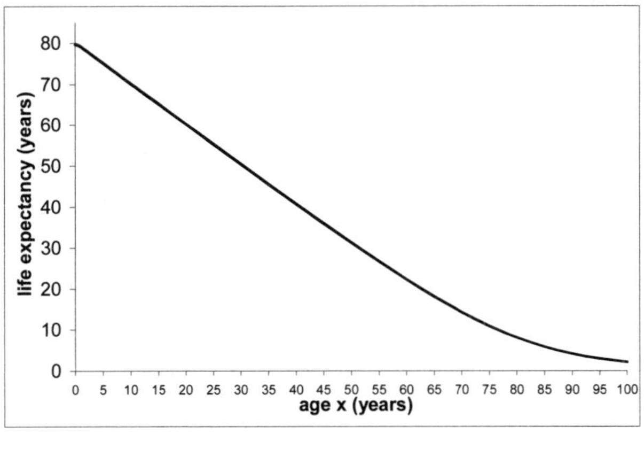

Distribution of deaths by age with uncertainty interval

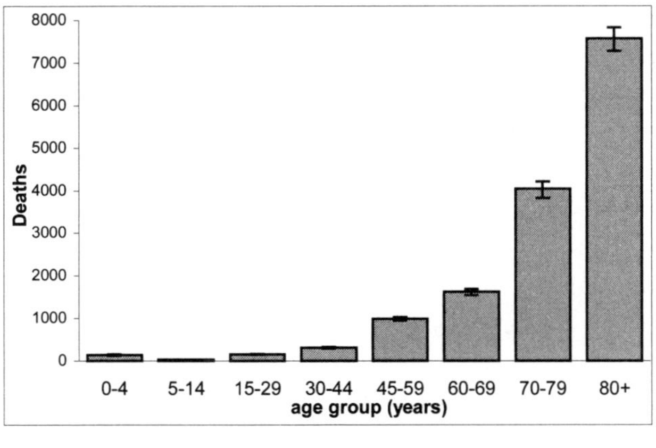

Israel **Israël** **Israel**

Males, 2000 *Hommes, 2000* *Hombres, 2000*

Age	$_nM_x$	$_nq_x$			l_x			$_nd_x$	$_nL_x$	e_x			Actual deaths	Actual population
			percentiles			percentiles					percentiles			
		mean	2½	97½	mean	2½	97½			mean	2½	97½		
<1	0.0060	0.0060	0.0056	0.0066	100,000	100,000	100,000	597	99,463	76.6	76.3	76.9	378	63,007
1-4	0.0004	0.0016	0.0015	0.0017	99,403	99,335	99,439	159	397,232	76.1	75.8	76.4	101	252,653
5-9	0.0002	0.0008	0.0008	0.0009	99,244	99,163	99,288	79	496,023	72.2	71.9	72.5	46	284,561
10-14	0.0002	0.0008	0.0008	0.0009	99,165	99,077	99,212	79	495,627	67.3	67.0	67.6	44	275,384
15-19	0.0005	0.0025	0.0024	0.0027	99,086	98,992	99,136	247	494,810	62.3	62.0	62.6	135	269,089
20-24	0.0008	0.0039	0.0038	0.0042	98,838	98,726	98,900	390	493,217	57.5	57.2	57.8	208	262,861
25-29	0.0007	0.0036	0.0035	0.0039	98,449	98,311	98,526	359	491,346	52.7	52.4	53.0	176	241,531
30-34	0.0009	0.0043	0.0041	0.0045	98,090	97,931	98,181	421	489,397	47.9	47.6	48.2	171	198,925
35-39	0.0012	0.0061	0.0059	0.0064	97,669	97,488	97,775	599	486,848	43.1	42.8	43.4	226	183,439
40-44	0.0018	0.0087	0.0084	0.0091	97,070	96,861	97,197	846	483,237	38.3	38.0	38.6	312	178,522
45-49	0.0026	0.0127	0.0123	0.0131	96,225	95,981	96,378	1,219	478,075	33.6	33.3	34.0	444	174,060
50-54	0.0044	0.0217	0.0210	0.0224	95,005	94,721	95,195	2,063	469,871	29.0	28.7	29.4	685	156,004
55-59	0.0080	0.0392	0.0380	0.0405	92,943	92,597	93,193	3,645	455,602	24.6	24.3	24.9	795	99,371
60-64	0.0120	0.0583	0.0563	0.0602	89,298	88,844	89,657	5,202	433,485	20.5	20.3	20.8	1,108	92,314
65-69	0.0210	0.0998	0.0962	0.1034	84,096	83,492	84,606	8,390	399,506	16.6	16.4	16.9	1,653	78,725
70-74	0.0330	0.1524	0.1467	0.1579	75,706	74,863	76,467	11,540	349,683	13.2	13.0	13.5	2,226	67,461
75-79	0.0530	0.2340	0.2254	0.2419	64,167	63,042	65,252	15,015	283,298	10.1	9.9	10.4	2,832	53,432
80-84	0.0920	0.3740	0.3616	0.3850	49,152	47,793	50,542	18,382	199,805	7.5	7.3	7.7	2,581	28,054
85-89	0.1530	0.5533	0.5395	0.5648	30,770	29,395	32,268	17,026	111,284	5.4	5.3	5.6	2,349	15,351
90-94	0.2170	0.6572	0.6452	0.6665	13,744	12,792	14,859	9,032	41,622	4.1	4.0	4.2	1,052	4,846
95-99	0.3120	0.8058	0.7967	0.8124	4,712	4,266	5,272	3,797	12,168	3.0	3.0	3.1	266	851
100+	0.4570	1	1	1	915	800	1,072	915	2,002	2.2	2.2	2.4	23	50
All ages													*17,809*	*2,980,491*

ln($_nq_x$) with uncertainty interval

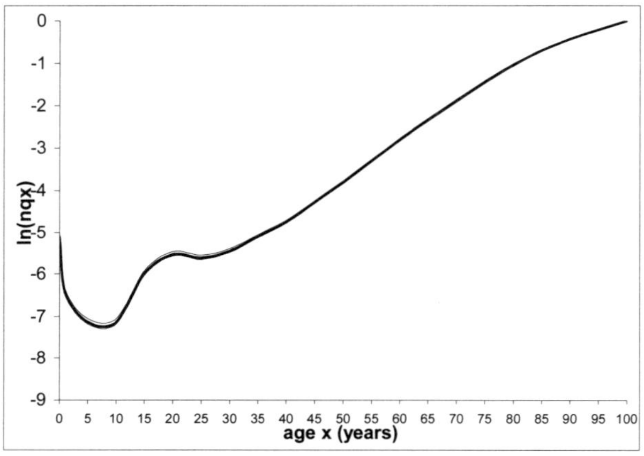

l_x with uncertainty interval

e_x with uncertainty interval

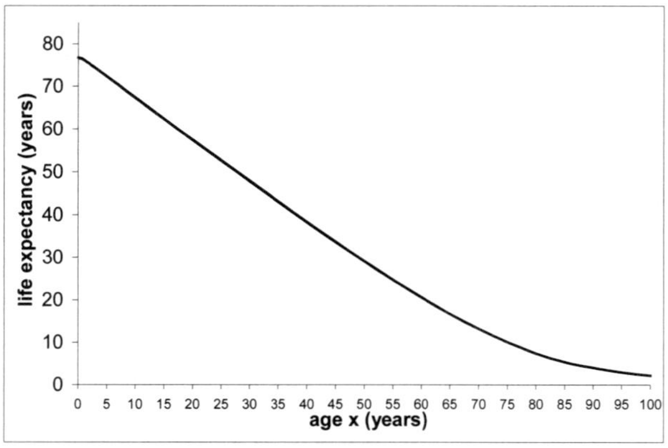

Distribution of deaths by age with uncertainty interval

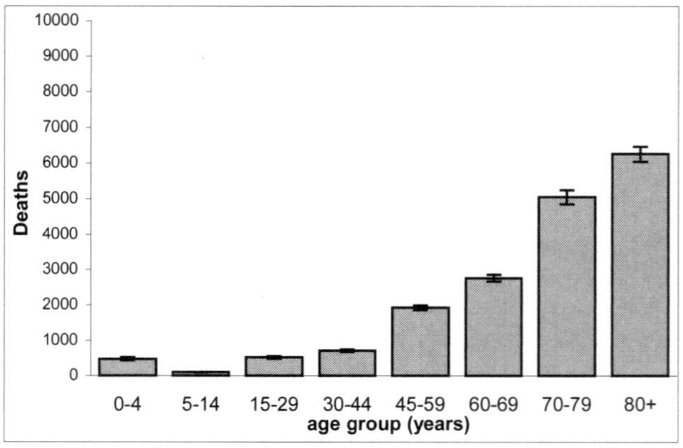

Israel

Israël

Israel

Females, 2000

Femmes, 2000

Mujeres, 2000

Age	$_nM_x$	$_nq_x$			l_x			$_nd_x$	$_nL_x$	e_x			Actual deaths	Actual population
			percentiles			percentiles					percentiles			
		mean	2½	97½	mean	2½	97½			mean	2½	97½		
<1	0.0050	0.0050	0.0044	0.0055	100,000	100,000	100,000	498	99,552	80.6	80.3	81.0	301	60,121
1-4	0.0003	0.0011	0.0010	0.0012	99,502	99,446	99,561	107	397,751	80.0	79.7	80.4	64	237,770
5-9	0.0001	0.0006	0.0006	0.0007	99,395	99,330	99,464	65	496,813	76.1	75.8	76.5	35	270,145
10-14	0.0001	0.0005	0.0005	0.0005	99,330	99,260	99,404	50	496,527	71.1	70.8	71.5	26	262,782
15-19	0.0002	0.0011	0.0010	0.0012	99,281	99,208	99,360	109	496,130	66.1	65.8	66.6	56	254,776
20-24	0.0002	0.0010	0.0009	0.0011	99,171	99,090	99,259	99	495,609	61.2	60.9	61.6	50	247,975
25-29	0.0003	0.0015	0.0014	0.0016	99,072	98,984	99,167	153	494,978	56.3	56.0	56.7	73	234,368
30-34	0.0004	0.0022	0.0020	0.0023	98,919	98,821	99,026	217	494,051	51.3	51.1	51.8	89	201,660
35-39	0.0007	0.0032	0.0030	0.0034	98,702	98,590	98,824	320	492,707	46.5	46.2	46.9	123	189,967
40-44	0.0010	0.0047	0.0045	0.0050	98,381	98,251	98,523	466	490,741	41.6	41.3	42.0	180	189,312
45-49	0.0017	0.0085	0.0080	0.0089	97,915	97,761	98,084	829	487,503	36.8	36.5	37.2	314	184,703
50-54	0.0030	0.0149	0.0142	0.0155	97,086	96,895	97,297	1,445	481,818	32.1	31.8	32.5	499	166,492
55-59	0.0040	0.0198	0.0189	0.0206	95,641	95,394	95,917	1,894	473,470	27.5	27.3	27.9	427	106,631
60-64	0.0080	0.0392	0.0374	0.0407	93,747	93,429	94,101	3,676	459,544	23.0	22.8	23.4	848	106,061
65-69	0.0130	0.0630	0.0598	0.0654	90,071	89,626	90,579	5,670	436,177	18.9	18.6	19.2	1,269	97,652
70-74	0.0220	0.1043	0.0989	0.1083	84,400	83,763	85,159	8,800	400,001	15.0	14.8	15.3	2,022	91,929
75-79	0.0380	0.1735	0.1648	0.1799	75,600	74,687	76,738	13,118	345,207	11.4	11.2	11.7	2,924	76,950
80-84	0.0730	0.3087	0.2956	0.3189	62,482	61,254	64,089	19,286	264,196	8.3	8.1	8.5	3,060	41,914
85-89	0.1330	0.4991	0.4848	0.5139	43,196	41,722	45,146	21,558	162,087	5.9	5.7	6.0	3,621	27,226
90-94	0.2020	0.6289	0.6160	0.6430	21,639	20,283	23,261	13,608	67,368	4.2	4.2	4.4	1,889	9,350
95-99	0.3050	0.7963	0.7867	0.8068	8,030	7,240	8,933	6,395	20,967	3.1	3.0	3.2	596	1,954
100+	0.4600	1	1	1	1,635	1,398	1,905	1,635	3,555	2.2	2.2	2.4	91	198
All ages													*18,558*	*3,059,936*

ln($_nq_x$) with uncertainty interval

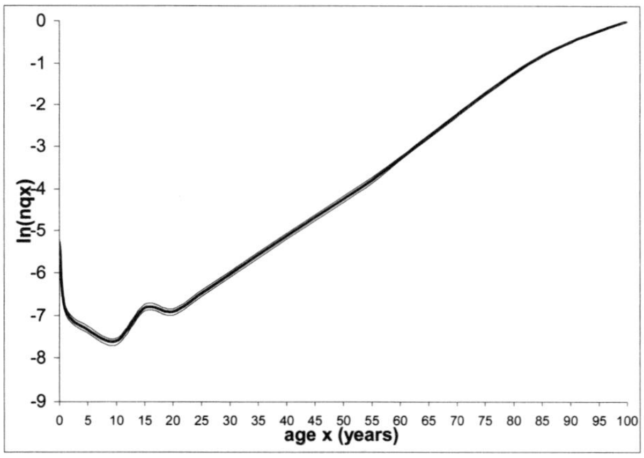

l_x with uncertainty interval

e_x with uncertainty interval

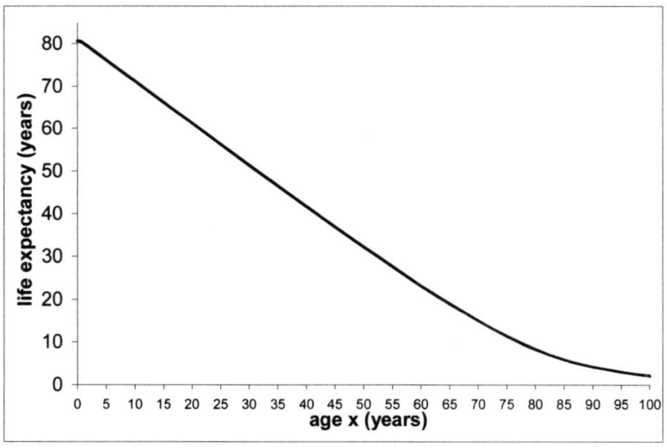

Distribution of deaths by age with uncertainty interval

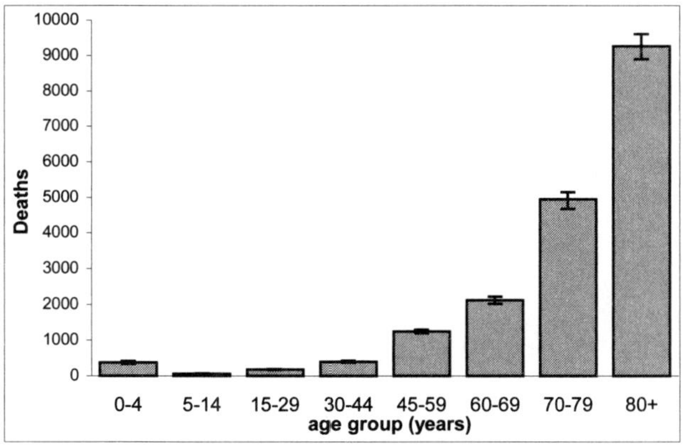

Italy   Italie   Italia

Males, 2000   *Hommes, 2000*   *Hombres, 2000*

Age	$_nM_x$	$_nq_x$ mean	$_nq_x$ percentiles 2½	$_nq_x$ percentiles 97½	l_x mean	l_x percentiles 2½	l_x percentiles 97½	$_nd_x$	$_nL_x$	e_x mean	e_x percentiles 2½	e_x percentiles 97½	Actual deaths	Actual population
<1	0.0060	0.0060	0.0054	0.0066	100,000	100,000	100,000	597	99,463	76.0	75.6	76.3	1,568	261,400
1-4	0.0003	0.0011	0.0010	0.0012	99,403	99,339	99,462	107	397,355	75.4	75.1	75.7	294	1,088,198
5-9	0.0002	0.0008	0.0008	0.0009	99,296	99,224	99,363	84	496,269	71.5	71.2	71.8	243	1,427,856
10-14	0.0002	0.0010	0.0010	0.0011	99,212	99,133	99,285	104	495,798	66.5	66.2	66.9	303	1,441,835
15-19	0.0007	0.0032	0.0030	0.0035	99,107	99,022	99,188	322	494,733	61.6	61.3	62.0	1,014	1,559,576
20-24	0.0010	0.0047	0.0045	0.0050	98,786	98,680	98,887	468	492,759	56.8	56.5	57.1	1,759	1,851,183
25-29	0.0010	0.0049	0.0047	0.0052	98,318	98,186	98,447	485	490,375	52.1	51.7	52.4	2,232	2,254,590
30-34	0.0012	0.0062	0.0059	0.0065	97,832	97,678	97,986	605	487,650	47.3	47.0	47.7	2,939	2,370,010
35-39	0.0015	0.0073	0.0070	0.0076	97,228	97,047	97,409	712	484,358	42.6	42.3	42.9	3,373	2,294,490
40-44	0.0020	0.0100	0.0096	0.0103	96,516	96,307	96,725	960	480,177	37.9	37.6	38.2	4,023	2,011,726
45-49	0.0030	0.0149	0.0144	0.0153	95,555	95,315	95,797	1,423	474,220	33.2	32.9	33.6	5,608	1,869,181
50-54	0.0050	0.0247	0.0240	0.0254	94,133	93,855	94,413	2,324	464,852	28.7	28.4	29.0	9,673	1,934,569
55-59	0.0080	0.0392	0.0381	0.0405	91,808	91,487	92,132	3,600	450,041	24.4	24.1	24.7	13,042	1,630,217
60-64	0.0130	0.0630	0.0609	0.0651	88,208	87,847	88,567	5,553	427,157	20.3	20.0	20.6	21,337	1,641,329
65-69	0.0220	0.1043	0.1005	0.1080	82,655	82,218	83,082	8,618	391,730	16.5	16.2	16.8	31,495	1,431,606
70-74	0.0360	0.1651	0.1589	0.1710	74,037	73,338	74,659	12,226	339,619	13.1	12.8	13.4	43,463	1,207,300
75-79	0.0590	0.2571	0.2478	0.2656	61,811	60,796	62,796	15,890	269,327	10.2	10.0	10.4	52,708	893,351
80-84	0.0900	0.3673	0.3551	0.3785	45,920	44,670	47,230	16,869	187,430	7.8	7.7	8.0	34,697	385,517
85-89	0.1370	0.5102	0.4969	0.5226	29,052	27,849	30,403	14,823	108,200	5.9	5.8	6.1	35,607	259,908
90-94	0.1900	0.6051	0.5929	0.6164	14,228	13,352	15,240	8,609	45,313	4.5	4.4	4.6	14,411	75,848
95-99	0.2760	0.7549	0.7449	0.7642	5,619	5,134	6,200	4,242	15,369	3.3	3.3	3.4	3,251	11,780
100+	0.4260	1	1	1	1,377	1,212	1,578	1,377	3,232	2.3	2.3	2.4	424	995
All ages													*283,463*	*27,902,465*

$\ln(_nq_x)$ with uncertainty interval

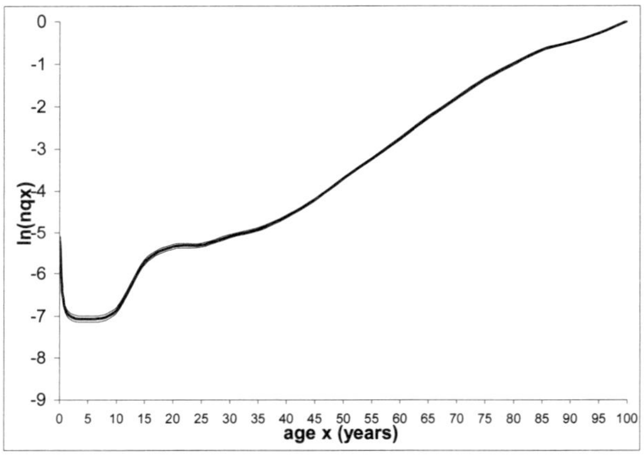

l_x with uncertainty interval

e_x with uncertainty interval

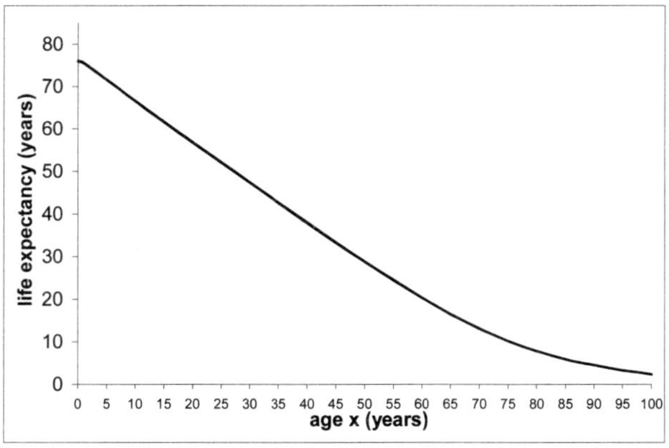

Distribution of deaths by age with uncertainty interval

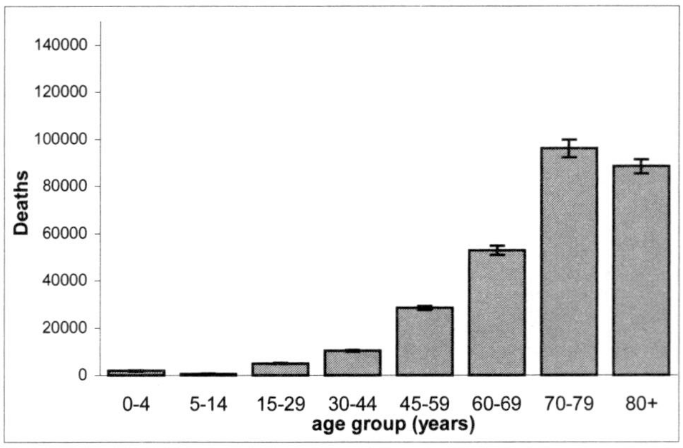

Italy | Italie | Italia

Females, 2000 | *Femmes, 2000* | *Mujeres, 2000*

Age	$_nM_x$	$_nq_x$			l_x			$_nd_x$	$_nL_x$	e_x			Actual deaths	Actual population
			percentiles			percentiles					percentiles			
		mean	2½	97½	mean	2½	97½			mean	2½	97½		
<1	0.0050	0.0050	0.0044	0.0056	100,000	100,000	100,000	498	99,552	82.4	82.0	82.7	1,237	247,420
1-4	0.0003	0.0012	0.0011	0.0013	99,502	99,438	99,565	123	397,713	81.8	81.4	82.2	317	1,023,252
5-9	0.0001	0.0006	0.0006	0.0007	99,379	99,304	99,453	65	496,733	77.9	77.5	78.3	176	1,352,095
10-14	0.0001	0.0006	0.0005	0.0006	99,314	99,234	99,394	60	496,423	72.9	72.5	73.3	165	1,373,827
15-19	0.0002	0.0010	0.0010	0.0011	99,255	99,170	99,340	104	496,014	68.0	67.6	68.4	314	1,493,492
20-24	0.0003	0.0013	0.0012	0.0014	99,151	99,058	99,244	134	495,419	63.0	62.6	63.4	481	1,782,314
25-29	0.0003	0.0016	0.0015	0.0018	99,017	98,916	99,120	163	494,676	58.1	57.7	58.5	721	2,183,905
30-34	0.0005	0.0024	0.0023	0.0026	98,854	98,742	98,969	242	493,663	53.2	52.8	53.6	1,140	2,326,328
35-39	0.0007	0.0033	0.0031	0.0035	98,612	98,487	98,743	330	492,234	48.3	47.9	48.7	1,531	2,284,346
40-44	0.0010	0.0048	0.0045	0.0050	98,282	98,139	98,433	471	490,233	43.5	43.1	43.9	1,939	2,019,969
45-49	0.0015	0.0076	0.0073	0.0079	97,811	97,645	97,987	745	487,193	38.7	38.3	39.1	2,903	1,897,291
50-54	0.0025	0.0123	0.0119	0.0128	97,066	96,870	97,273	1,196	482,339	34.0	33.6	34.4	4,923	1,985,133
55-59	0.0040	0.0198	0.0190	0.0206	95,870	95,636	96,114	1,898	474,603	29.3	29.0	29.8	6,884	1,721,050
60-64	0.0060	0.0296	0.0284	0.0309	93,971	93,700	94,249	2,777	462,913	24.9	24.5	25.3	10,842	1,807,010
65-69	0.0100	0.0488	0.0468	0.0511	91,194	90,872	91,513	4,448	444,848	20.6	20.2	21.0	16,870	1,686,960
70-74	0.0170	0.0815	0.0776	0.0857	86,745	86,227	87,220	7,073	416,045	16.5	16.2	16.9	26,647	1,567,463
75-79	0.0330	0.1524	0.1444	0.1602	79,673	78,852	80,369	12,144	368,003	12.7	12.4	13.1	44,929	1,361,493
80-84	0.0580	0.2533	0.2403	0.2652	67,528	66,270	68,700	17,103	294,884	9.6	9.3	9.9	40,422	696,927
85-89	0.1010	0.4032	0.3845	0.4201	50,425	48,703	52,195	20,331	201,298	7.0	6.8	7.2	57,013	564,487
90-94	0.1590	0.5383	0.5186	0.5562	30,094	28,368	32,045	16,198	101,876	5.0	4.9	5.2	33,281	209,316
95-99	0.2520	0.7175	0.7013	0.7324	13,896	12,677	15,341	9,971	39,567	3.5	3.5	3.7	10,066	39,946
100+	0.4150	1	1	1	3,925	3,402	4,583	3,925	9,458	2.4	2.4	2.6	1,456	3,509
All ages													*264,257*	*29,627,533*

ln($_nq_x$) with uncertainty interval

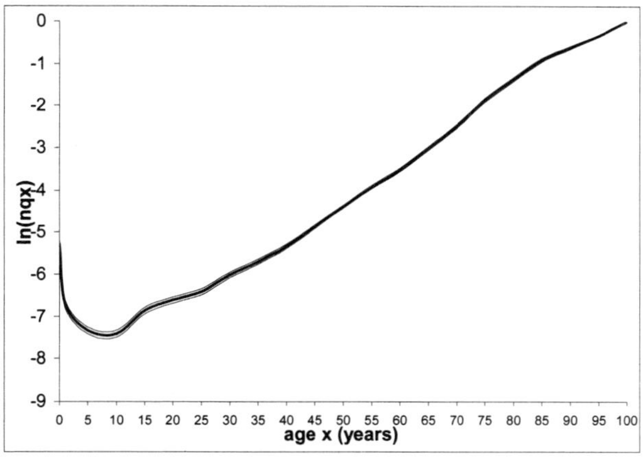

l_x with uncertainty interval

e_x with uncertainty interval

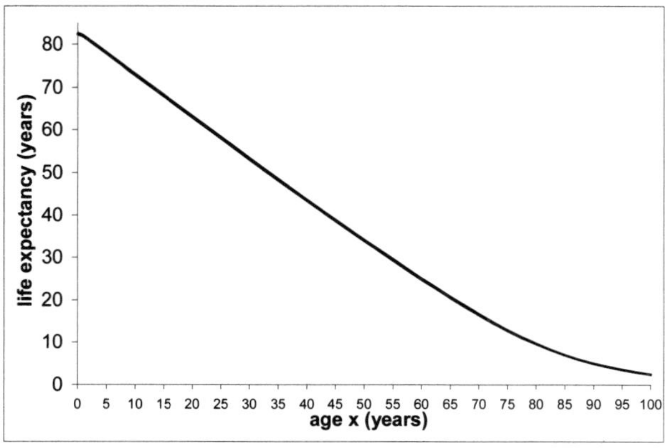

Distribution of deaths by age with uncertainty interval

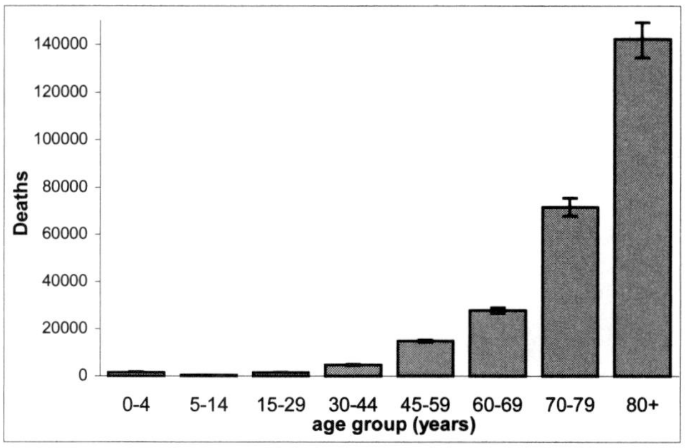

Jamaica **Jamaïque** **Jamaica**

Males, 2000 *Hommes, 2000* *Hombres, 2000*

Age	$_nM_x$	$_nq_x$			l_x			$_nd_x$	$_nL_x$	e_x			Actual deaths	Actual population
			percentiles			percentiles					percentiles			
		mean	2½	97½	mean	2½	97½			mean	2½	97½		
<1	0.0138	0.0137	0.0114	0.0165	100,000	100,000	100,000	1,367	98,770	72.8	72.0	74.7	366	26,484
1-4	0.0008	0.0031	0.0026	0.0035	98,633	98,348	98,861	301	393,810	72.9	72.0	74.7	82	107,665
5-9	0.0003	0.0017	0.0015	0.0020	98,332	98,002	98,604	168	491,240	69.1	68.2	70.9	48	140,747
10-14	0.0004	0.0019	0.0016	0.0021	98,164	97,811	98,461	183	490,364	64.2	63.3	66.0	51	137,059
15-19	0.0011	0.0057	0.0049	0.0065	97,982	97,604	98,304	560	488,508	59.3	58.4	61.1	158	137,591
20-24	0.0020	0.0099	0.0084	0.0110	97,422	96,973	97,825	960	484,710	54.6	53.7	56.4	245	123,650
25-29	0.0024	0.0121	0.0104	0.0134	96,462	95,905	97,000	1,168	479,390	50.1	49.2	51.9	265	108,589
30-34	0.0025	0.0124	0.0106	0.0135	95,294	94,621	95,993	1,180	473,518	45.7	44.8	47.5	230	92,216
35-39	0.0027	0.0132	0.0113	0.0144	94,113	93,339	94,974	1,246	467,453	41.3	40.3	43.0	219	82,302
40-44	0.0038	0.0190	0.0163	0.0205	92,868	91,998	93,896	1,765	459,927	36.8	35.8	38.4	260	67,779
45-49	0.0048	0.0237	0.0204	0.0255	91,103	90,113	92,365	2,163	450,107	32.5	31.5	34.0	252	52,454
50-54	0.0072	0.0355	0.0304	0.0382	88,940	87,816	90,481	3,159	436,801	28.2	27.2	29.7	313	43,288
55-59	0.0104	0.0507	0.0432	0.0546	85,781	84,462	87,727	4,348	418,033	24.1	23.2	25.6	381	36,597
60-64	0.0161	0.0773	0.0657	0.0843	81,432	79,847	83,938	6,295	391,426	20.3	19.4	21.6	481	29,909
65-69	0.0221	0.1047	0.0890	0.1153	75,138	73,115	78,425	7,869	356,017	16.8	16.0	17.9	546	24,721
70-74	0.0360	0.1651	0.1416	0.1824	67,269	64,685	71,442	11,109	308,574	13.4	12.7	14.4	808	22,447
75-79	0.0550	0.2418	0.2109	0.2661	56,160	52,885	61,324	13,577	246,859	10.6	10.0	11.4	931	16,924
80-84	0.0855	0.3522	0.3158	0.3824	42,583	38,815	48,391	14,997	175,424	8.2	7.7	8.8	925	10,820
85-89	0.1226	0.4693	0.4305	0.5005	27,587	23,971	33,110	12,945	105,569	6.3	6.0	6.8	707	5,769
90-94	0.1799	0.5842	0.5475	0.6130	14,641	11,974	18,855	8,553	47,546	4.7	4.4	5.0	361	2,006
95-99	0.2699	0.7457	0.7155	0.7690	6,088	4,634	8,532	4,540	16,820	3.4	3.2	3.6	120	444
100+	0.4142	1	1	1	1,548	1,071	2,427	1,548	3,738	2.4	2.4	2.5	28	67
All ages													*7,777*	*1,269,528*

ln($_nq_x$) with uncertainty interval

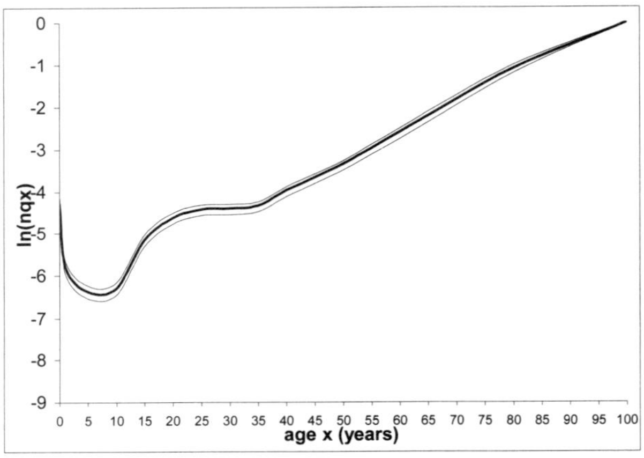

l_x with uncertainty interval

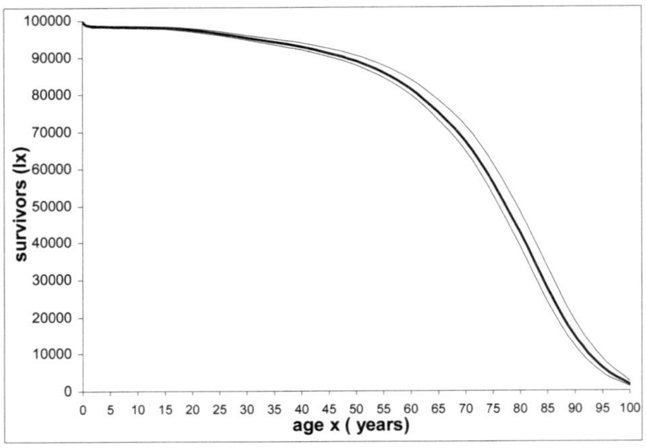

e_x with uncertainty interval

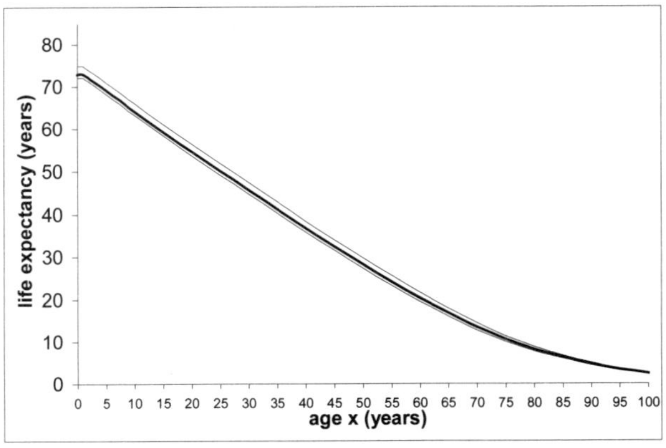

Distribution of deaths by age with uncertainty interval

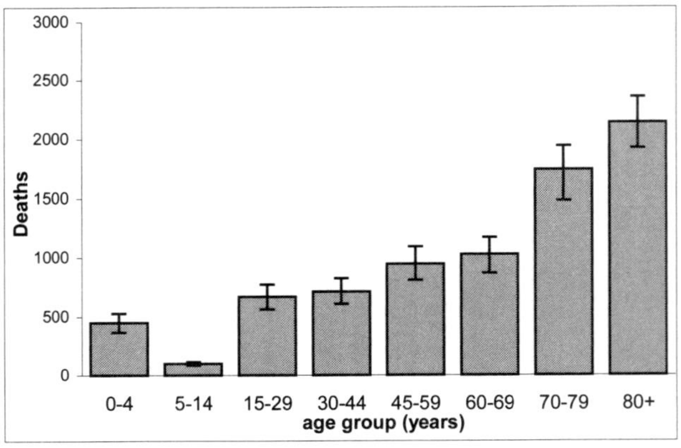

Jamaica | Jamaïque | Jamaica

Females, 2000 | *Femmes, 2000* | *Mujeres, 2000*

Age	$_nM_x$	$_nq_x$			l_x			$_nd_x$	$_nL_x$	e_x			Actual deaths	Actual population
			percentiles			percentiles					percentiles			
		mean	2½	97½	mean	2½	97½			mean	2½	97½		
<1	0.0117	0.0116	0.0088	0.0138	100,000	100,000	100,000	1,160	98,956	76.6	75.3	78.3	299	25,465
1-4	0.0007	0.0028	0.0022	0.0032	98,840	98,615	99,121	277	394,694	76.5	75.2	78.2	72	103,074
5-9	0.0003	0.0013	0.0010	0.0015	98,563	98,295	98,899	127	492,495	72.7	71.4	74.4	35	134,986
10-14	0.0002	0.0011	0.0009	0.0013	98,435	98,148	98,796	112	491,896	67.8	66.5	69.5	31	134,941
15-19	0.0003	0.0017	0.0014	0.0020	98,323	98,019	98,705	170	491,190	62.8	61.6	64.5	47	135,191
20-24	0.0007	0.0034	0.0028	0.0039	98,153	97,824	98,567	335	489,929	57.9	56.7	59.6	84	122,380
25-29	0.0013	0.0063	0.0052	0.0072	97,818	97,441	98,293	613	487,559	53.1	51.9	54.8	137	109,287
30-34	0.0016	0.0079	0.0066	0.0090	97,205	96,744	97,786	769	484,104	48.5	47.3	50.1	157	98,646
35-39	0.0021	0.0106	0.0090	0.0120	96,436	95,874	97,142	1,026	479,616	43.8	42.7	45.5	197	92,128
40-44	0.0025	0.0125	0.0107	0.0140	95,410	94,722	96,269	1,191	474,072	39.3	38.1	40.9	193	76,872
45-49	0.0038	0.0189	0.0163	0.0211	94,219	93,394	95,242	1,781	466,641	34.7	33.6	36.3	218	57,211
50-54	0.0058	0.0286	0.0247	0.0320	92,438	91,424	93,691	2,644	455,578	30.3	29.3	31.9	259	44,679
55-59	0.0089	0.0435	0.0373	0.0488	89,794	88,502	91,373	3,908	439,196	26.2	25.2	27.6	331	37,217
60-64	0.0130	0.0630	0.0536	0.0707	85,885	84,181	87,965	5,407	415,908	22.2	21.3	23.6	415	31,941
65-69	0.0180	0.0861	0.0730	0.0966	80,478	78,227	83,252	6,931	385,063	18.6	17.8	19.8	504	27,998
70-74	0.0275	0.1287	0.1094	0.1435	73,547	70,668	77,171	9,462	344,080	15.1	14.3	16.1	724	26,313
75-79	0.0415	0.1878	0.1615	0.2076	64,085	60,528	68,725	12,036	290,335	11.9	11.3	12.8	879	21,202
80-84	0.0674	0.2883	0.2536	0.3165	52,049	47,960	57,627	15,007	222,726	9.1	8.6	9.8	954	14,156
85-89	0.1083	0.4263	0.3876	0.4617	37,042	32,783	43,015	15,789	145,735	6.8	6.4	7.3	900	8,306
90-94	0.1600	0.5405	0.5052	0.5754	21,252	17,648	26,343	11,488	71,799	5.0	4.7	5.4	564	3,527
95-99	0.2500	0.7143	0.6843	0.7438	9,765	7,494	13,036	6,975	27,899	3.6	3.4	3.8	200	800
100+	0.4050	1	1	1	2,790	1,920	4,116	2,790	6,889	2.5	2.5	2.6	96	237
All ages													*7,296*	*1,306,557*

$\ln(_nq_x)$ with uncertainty interval

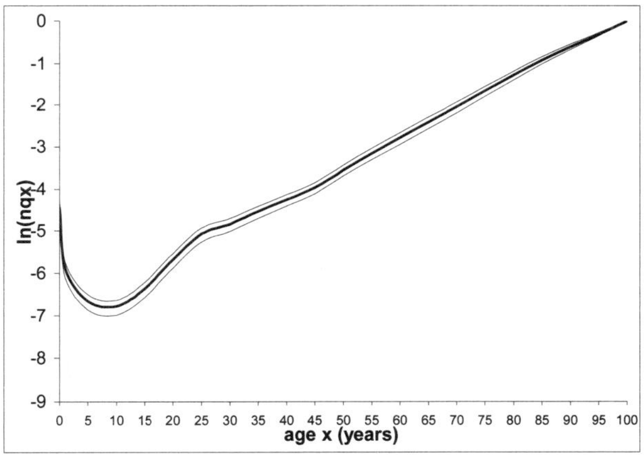

l_x with uncertainty interval

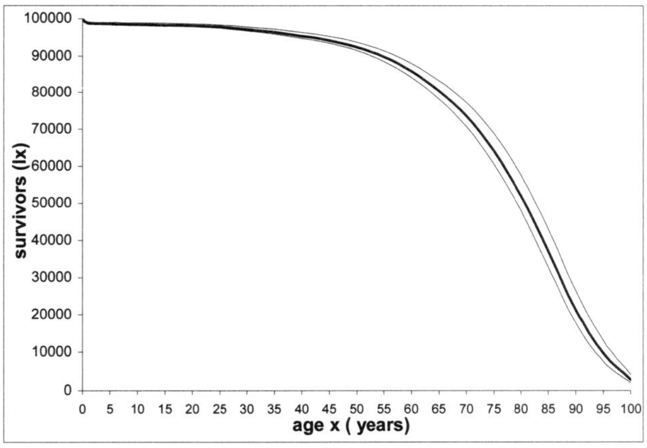

e_x with uncertainty interval

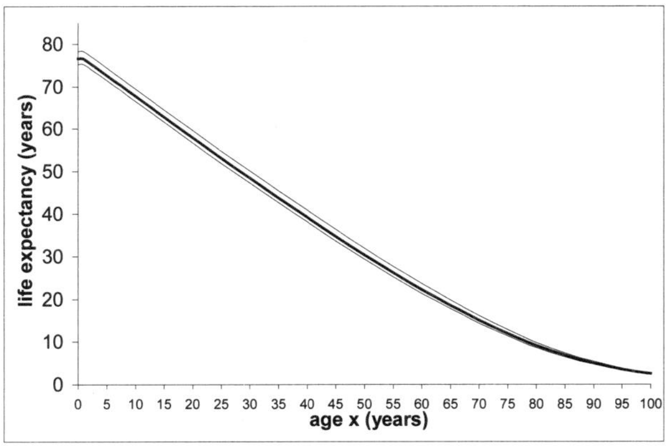

Distribution of deaths by age with uncertainty interval

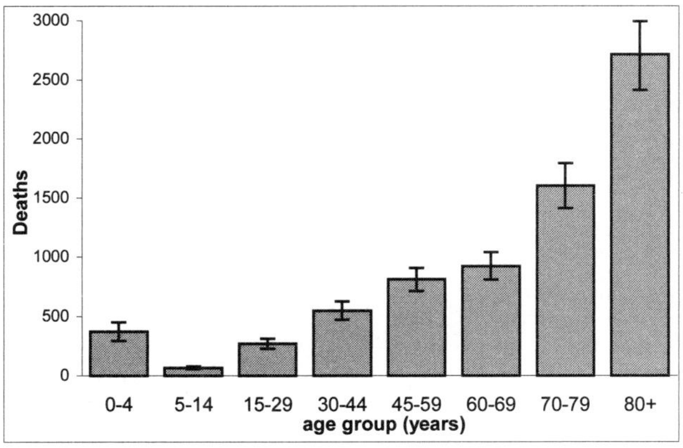

Japan Japon Japón

Males, 2000 *Hommes, 2000* *Hombres, 2000*

Age	$_nM_x$	$_nq_x$			l_x			$_nd_x$	$_nL_x$	e_x			Actual deaths	Actual population
			percentiles			percentiles					percentiles			
		mean	2½	97½	mean	2½	97½			mean	2½	97½		
<1	0.0040	0.0040	0.0037	0.0042	100,000	100,000	100,000	399	99,641	77.5	77.4	77.7	2,553	638,200
1-4	0.0004	0.0015	0.0015	0.0016	99,601	99,575	99,625	151	398,043	76.9	76.7	77.0	958	2,520,472
5-9	0.0002	0.0009	0.0009	0.0009	99,450	99,417	99,481	89	497,027	73.0	72.8	73.1	553	3,072,056
10-14	0.0002	0.0008	0.0008	0.0008	99,361	99,324	99,395	79	496,605	68.0	67.9	68.2	536	3,352,322
15-19	0.0005	0.0025	0.0024	0.0026	99,281	99,241	99,319	248	495,787	63.1	62.9	63.3	1,919	3,838,596
20-24	0.0007	0.0033	0.0032	0.0035	99,033	98,983	99,080	331	494,339	58.2	58.1	58.4	2,949	4,401,886
25-29	0.0007	0.0034	0.0033	0.0035	98,702	98,641	98,760	335	492,673	53.4	53.2	53.6	3,435	5,052,032
30-34	0.0008	0.0041	0.0040	0.0042	98,367	98,296	98,434	402	490,829	48.6	48.4	48.8	3,654	4,456,501
35-39	0.0011	0.0056	0.0054	0.0057	97,965	97,883	98,042	547	488,456	43.8	43.6	44.0	4,602	4,108,773
40-44	0.0018	0.0088	0.0086	0.0089	97,418	97,323	97,508	854	484,954	39.0	38.8	39.2	6,913	3,928,118
45-49	0.0030	0.0149	0.0146	0.0151	96,564	96,454	96,671	1,438	479,226	34.3	34.2	34.5	13,447	4,482,353
50-54	0.0050	0.0247	0.0243	0.0251	95,126	94,996	95,256	2,349	469,760	29.8	29.7	30.0	26,153	5,230,675
55-59	0.0070	0.0344	0.0339	0.0350	92,778	92,621	92,934	3,191	455,910	25.5	25.4	25.7	30,038	4,291,110
60-64	0.0120	0.0583	0.0572	0.0594	89,586	89,409	89,762	5,219	434,885	21.3	21.2	21.5	44,737	3,728,083
65-69	0.0190	0.0907	0.0889	0.0925	84,368	84,146	84,587	7,651	402,709	17.5	17.4	17.7	63,602	3,347,491
70-74	0.0280	0.1308	0.1280	0.1335	76,716	76,370	77,025	10,038	358,487	14.0	13.9	14.1	74,268	2,652,425
75-79	0.0470	0.2103	0.2059	0.2143	66,679	66,174	67,156	14,022	298,338	10.7	10.6	10.9	74,760	1,590,648
80-84	0.0830	0.3437	0.3371	0.3497	52,657	51,995	53,330	18,097	218,040	7.9	7.8	8.0	74,377	896,112
85-89	0.1400	0.5185	0.5107	0.5258	34,559	33,851	35,331	17,920	127,997	5.8	5.7	5.8	64,413	460,096
90-94	0.2030	0.6308	0.6239	0.6374	16,640	16,100	17,245	10,497	51,708	4.3	4.2	4.3	28,315	139,481
95-99	0.2980	0.7867	0.7814	0.7917	6,143	5,849	6,483	4,833	16,217	3.1	3.1	3.2	6,812	22,859
100+	0.4480	1	1	1	1,310	1,219	1,416	1,310	2,925	2.2	2.2	2.4	763	1,704
All ages													*529,760*	*62,211,993*

$\ln(_nq_x)$ with uncertainty interval

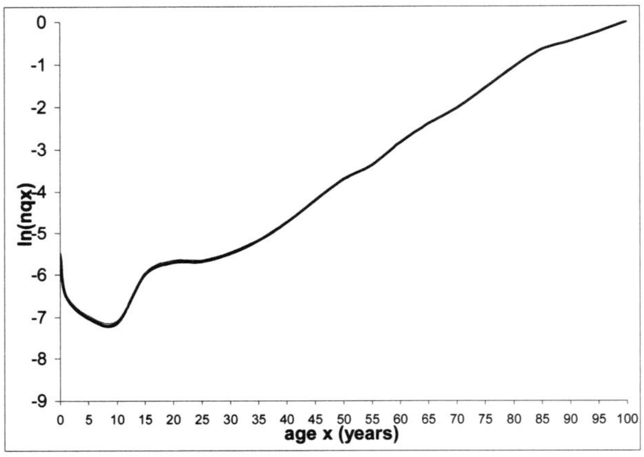

l_x with uncertainty interval

e_x with uncertainty interval

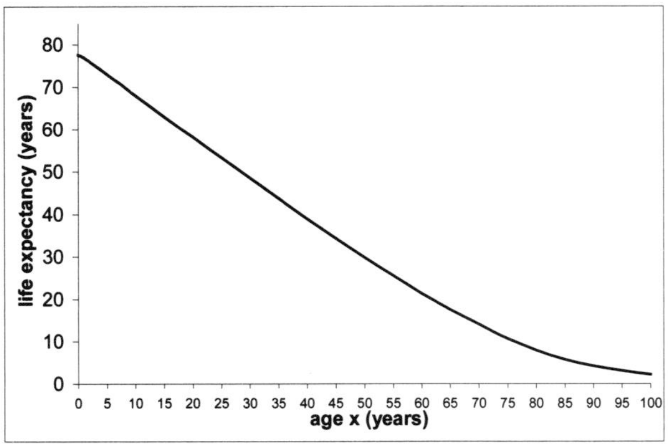

Distribution of deaths by age with uncertainty interval

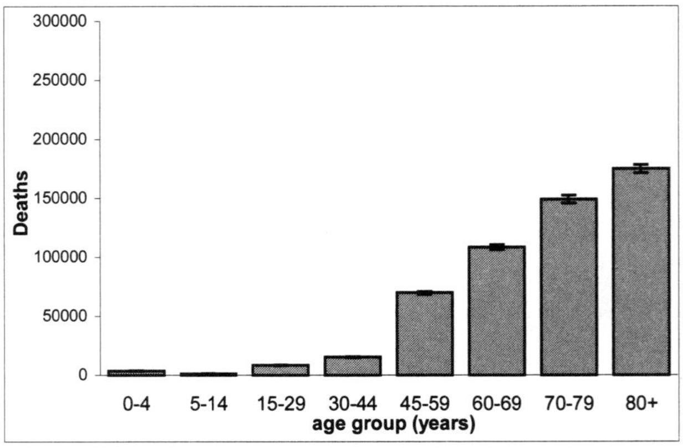

Japan

Japon

Japón

Females, 2000 — *Femmes, 2000* — *Mujeres, 2000*

Age	$_nM_x$	$_nq_x$ mean	$_nq_x$ percentiles 2½	$_nq_x$ percentiles 97½	l_x mean	l_x percentiles 2½	l_x percentiles 97½	$_nd_x$	$_nL_x$	e_x mean	e_x percentiles 2½	e_x percentiles 97½	Actual deaths	Actual population
<1	0.0030	0.0030	0.0026	0.0034	100,000	100,000	100,000	299	99,731	84.7	84.3	85.0	1,820	606,654
1-4	0.0003	0.0012	0.0011	0.0013	99,701	99,664	99,735	116	398,526	83.9	83.6	84.3	692	2,386,615
5-9	0.0001	0.0005	0.0005	0.0006	99,585	99,538	99,630	55	497,789	80.0	79.7	80.4	322	2,925,921
10-14	0.0001	0.0004	0.0004	0.0005	99,530	99,479	99,579	45	497,540	75.1	74.7	75.4	287	3,191,778
15-19	0.0002	0.0009	0.0009	0.0010	99,486	99,431	99,538	94	497,192	70.1	69.7	70.5	696	3,663,899
20-24	0.0003	0.0013	0.0012	0.0014	99,391	99,330	99,451	129	496,633	65.1	64.8	65.5	1,091	4,194,583
25-29	0.0003	0.0014	0.0014	0.0015	99,262	99,192	99,330	144	495,951	60.2	59.9	60.6	1,413	4,871,638
30-34	0.0004	0.0019	0.0018	0.0021	99,118	99,040	99,196	193	495,109	55.3	55.0	55.7	1,695	4,346,011
35-39	0.0006	0.0028	0.0026	0.0029	98,925	98,836	99,014	277	493,934	50.4	50.1	50.8	2,248	4,014,248
40-44	0.0009	0.0043	0.0041	0.0045	98,649	98,546	98,752	423	492,185	45.6	45.2	45.9	3,328	3,869,188
45-49	0.0014	0.0071	0.0069	0.0074	98,225	98,106	98,348	700	489,377	40.7	40.4	41.1	6,372	4,455,709
50-54	0.0020	0.0100	0.0096	0.0103	97,525	97,386	97,674	970	485,201	36.0	35.7	36.4	10,483	5,241,494
55-59	0.0030	0.0149	0.0144	0.0155	96,555	96,393	96,730	1,438	479,182	31.3	31.0	31.7	13,313	4,437,747
60-64	0.0050	0.0247	0.0238	0.0257	95,118	94,931	95,317	2,349	469,716	26.8	26.5	27.2	19,820	3,964,074
65-69	0.0070	0.0344	0.0332	0.0359	92,769	92,524	93,005	3,191	455,867	22.4	22.1	22.8	26,177	3,739,588
70-74	0.0120	0.0583	0.0558	0.0610	89,578	89,204	89,913	5,218	434,844	18.1	17.8	18.5	38,649	3,220,733
75-79	0.0230	0.1087	0.1035	0.1141	84,360	83,766	84,857	9,174	398,864	14.1	13.8	14.4	57,578	2,503,373
80-84	0.0450	0.2022	0.1921	0.2118	75,186	74,236	75,979	15,206	337,914	10.5	10.2	10.8	76,119	1,691,541
85-89	0.0860	0.3539	0.3382	0.3687	59,980	58,545	61,383	21,227	246,830	7.5	7.3	7.8	89,728	1,043,351
90-94	0.1460	0.5076	0.4902	0.5239	38,752	37,026	40,624	19,673	134,744	5.3	5.1	5.4	59,969	410,746
95-99	0.2430	0.7027	0.6878	0.7163	19,080	17,711	20,678	13,408	55,176	3.6	3.6	3.8	22,993	94,621
100+	0.4140	1	1	1	5,672	5,034	6,416	5,672	13,701	2.4	2.4	2.6	4,475	10,809
All ages													*439,267*	*64,884,321*

ln($_nq_x$) with uncertainty interval

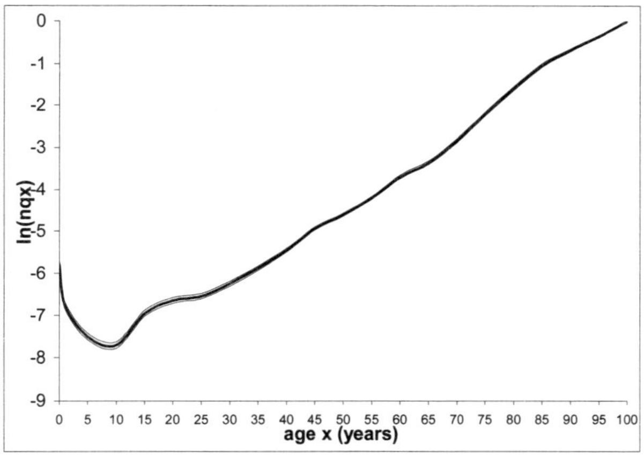

l_x with uncertainty interval

e_x with uncertainty interval

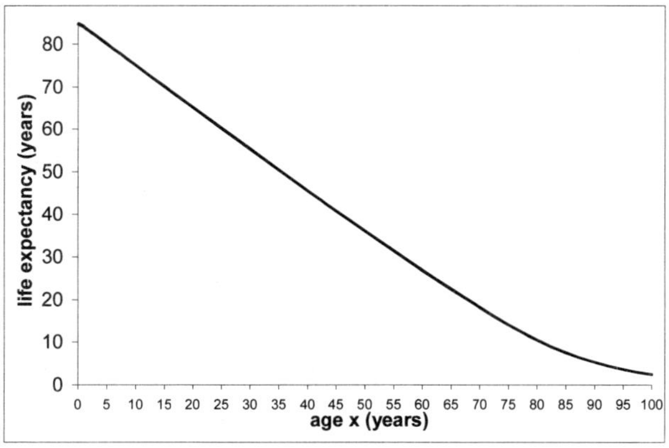

Distribution of deaths by age with uncertainty interval

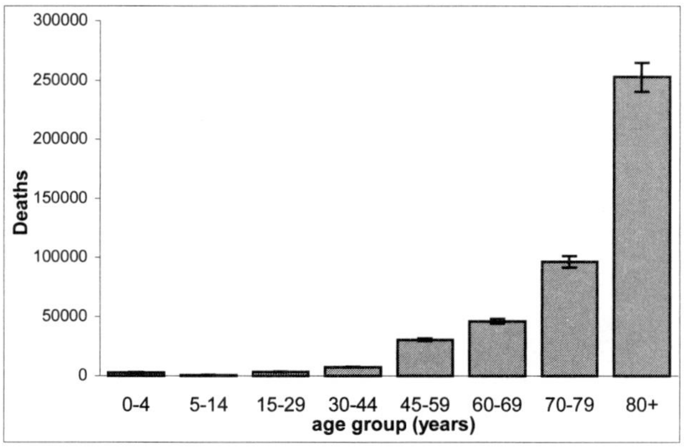

Jordan / Jordanie / Jordania

Males, 2000 / *Hommes, 2000* / *Hombres, 2000*

Age	$_nM_x$	$_nq_x$			l_x			$_nd_x$	$_nL_x$	e_x			Actual deaths	Actual population
			percentiles			percentiles					percentiles			
		mean	2½	97½	mean	2½	97½			mean	2½	97½		
<1	0.0208	0.0204	0.0169	0.0240	100,000	100,000	100,000	2,040	98,164	68.5	67.3	69.8	1,703	81,954
1-4	0.0012	0.0047	0.0040	0.0053	97,960	97,604	98,308	460	390,736	68.9	67.6	70.4	362	307,657
5-9	0.0005	0.0026	0.0023	0.0030	97,500	97,101	97,899	257	486,859	65.3	63.9	66.7	170	322,183
10-14	0.0005	0.0024	0.0021	0.0028	97,244	96,816	97,675	237	485,626	60.4	59.0	61.9	144	295,774
15-19	0.0012	0.0062	0.0054	0.0070	97,007	96,554	97,469	603	483,528	55.6	54.2	57.0	345	276,597
20-24	0.0017	0.0087	0.0076	0.0098	96,404	95,888	96,942	835	479,933	50.9	49.5	52.3	445	255,594
25-29	0.0018	0.0088	0.0077	0.0100	95,569	94,957	96,206	844	475,736	46.3	44.9	47.8	439	247,519
30-34	0.0021	0.0103	0.0091	0.0116	94,725	94,022	95,458	976	471,186	41.7	40.3	43.1	440	212,442
35-39	0.0027	0.0136	0.0120	0.0153	93,749	92,931	94,585	1,277	465,554	37.1	35.7	38.5	404	147,317
40-44	0.0040	0.0199	0.0177	0.0223	92,472	91,509	93,441	1,839	457,766	32.6	31.2	34.0	398	99,048
45-49	0.0062	0.0304	0.0270	0.0339	90,634	89,470	91,792	2,752	446,289	28.2	26.8	29.6	450	73,051
50-54	0.0096	0.0471	0.0417	0.0526	87,882	86,441	89,299	4,140	429,059	24.0	22.6	25.4	601	62,262
55-59	0.0150	0.0724	0.0632	0.0827	83,742	81,929	85,501	6,063	403,551	20.1	18.7	21.4	887	59,020
60-64	0.0234	0.1107	0.0956	0.1287	77,679	75,417	79,852	8,598	366,899	16.4	15.2	17.6	1,056	45,065
65-69	0.0360	0.1653	0.1420	0.1939	69,081	66,347	71,894	11,418	316,861	13.2	12.1	14.2	1,125	31,215
70-74	0.0562	0.2463	0.2126	0.2871	57,663	53,560	61,603	14,201	252,813	10.3	9.4	11.2	1,049	18,679
75-79	0.0873	0.3583	0.3150	0.4084	43,462	38,183	48,438	15,571	178,381	7.8	7.1	8.5	871	9,980
80-84	0.1368	0.5098	0.4615	0.5609	27,891	22,590	33,136	14,217	103,910	5.8	5.4	6.3	646	4,724
85-89	0.2067	0.6815	0.6391	0.7232	13,673	9,919	17,787	9,318	45,071	4.3	4.0	4.6	572	2,766
90-94	0.3012	0.7911	0.7592	0.8219	4,355	2,746	6,374	3,445	11,439	3.1	2.9	3.3	248	825
95-99	0.4229	0.9321	0.9154	0.9472	910	489	1,535	848	2,005	2.3	2.3	2.4	67	159
100+	0.5726	1	1	1	62	26	130	62	108	1.7	1.7	1.8	15	27
All ages													*12,438*	*2,553,858*

$\ln(_nq_x)$ with uncertainty interval

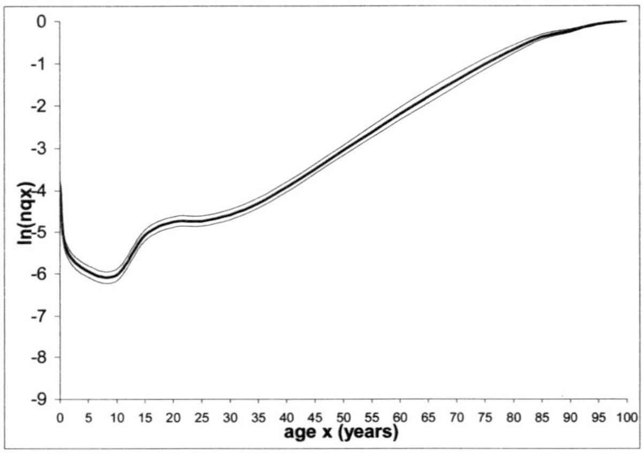

l_x with uncertainty interval

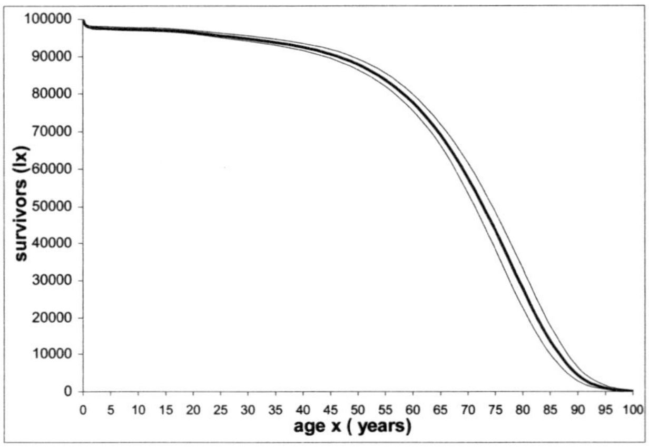

e_x with uncertainty interval

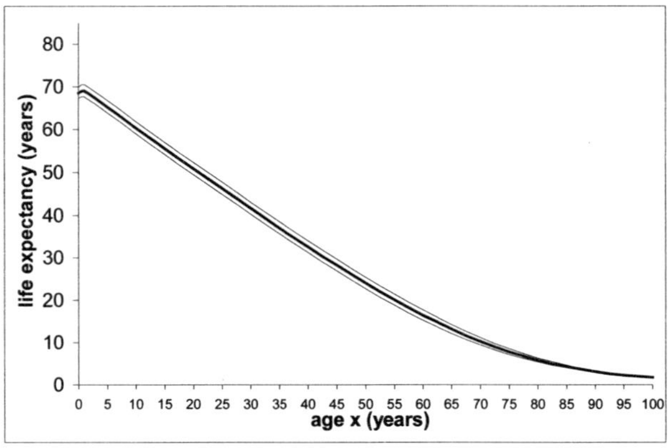

Distribution of deaths by age with uncertainty interval

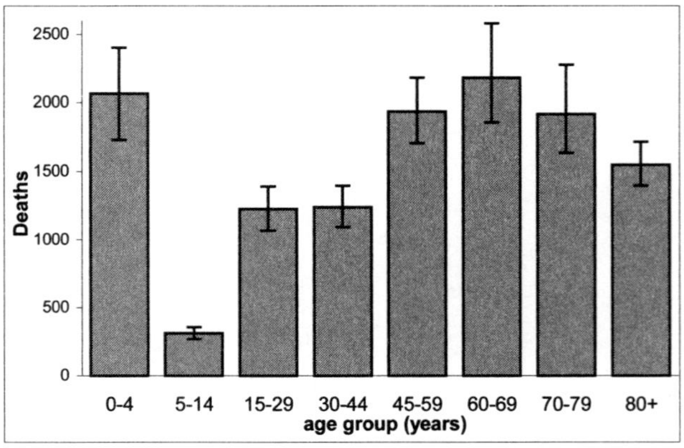

Jordan — Jordanie — Jordania

Females, 2000 — *Femmes, 2000* — *Mujeres, 2000*

Age	$_nM_x$	$_nq_x$ mean	$_nq_x$ percentiles 2½	$_nq_x$ percentiles 97½	l_x mean	l_x percentiles 2½	l_x percentiles 97½	$_nd_x$	$_nL_x$	e_x mean	e_x percentiles 2½	e_x percentiles 97½	Actual deaths	Actual population
<1	0.0173	0.0170	0.0137	0.0203	100,000	100,000	100,000	1,699	98,471	72.5	71.7	73.3	1,346	78,052
1-4	0.0013	0.0051	0.0044	0.0058	98,301	97,969	98,625	501	392,003	72.7	71.9	73.5	375	293,620
5-9	0.0005	0.0024	0.0021	0.0028	97,800	97,400	98,192	239	488,404	69.1	68.3	69.9	150	305,957
10-14	0.0004	0.0021	0.0019	0.0024	97,562	97,131	97,983	209	487,285	64.2	63.5	65.0	121	282,410
15-19	0.0008	0.0039	0.0034	0.0044	97,353	96,897	97,799	381	485,809	59.4	58.6	60.2	207	263,167
20-24	0.0010	0.0048	0.0043	0.0054	96,971	96,471	97,461	468	483,685	54.6	53.8	55.4	229	236,778
25-29	0.0011	0.0055	0.0049	0.0062	96,503	95,951	97,042	536	481,177	49.8	49.1	50.6	238	213,456
30-34	0.0015	0.0072	0.0065	0.0080	95,968	95,362	96,559	695	478,100	45.1	44.4	45.9	250	172,239
35-39	0.0021	0.0104	0.0094	0.0114	95,273	94,602	95,931	989	473,889	40.4	39.7	41.2	268	128,458
40-44	0.0031	0.0154	0.0141	0.0167	94,283	93,530	95,029	1,450	467,791	35.8	35.1	36.6	282	90,987
45-49	0.0046	0.0229	0.0212	0.0247	92,833	91,977	93,692	2,127	458,850	31.3	30.6	32.1	316	68,250
50-54	0.0069	0.0339	0.0313	0.0364	90,707	89,714	91,700	3,071	445,856	27.0	26.3	27.8	440	63,891
55-59	0.0101	0.0493	0.0457	0.0532	87,636	86,494	88,774	4,318	427,385	22.9	22.2	23.6	526	52,088
60-64	0.0155	0.0745	0.0686	0.0807	83,318	82,033	84,580	6,211	401,063	18.9	18.3	19.6	642	41,423
65-69	0.0243	0.1144	0.1048	0.1239	77,107	75,647	78,585	8,818	363,491	15.2	14.7	15.9	730	30,073
70-74	0.0401	0.1824	0.1671	0.1967	68,289	66,342	70,182	12,454	310,312	11.9	11.4	12.4	670	16,689
75-79	0.0665	0.2851	0.2624	0.3075	55,835	53,301	58,358	15,919	239,379	9.0	8.5	9.5	827	12,438
80-84	0.1118	0.4369	0.4070	0.4677	39,916	37,091	42,993	17,439	155,984	6.6	6.2	6.9	524	4,685
85-89	0.1791	0.6185	0.5887	0.6496	22,477	20,110	25,339	13,902	77,631	4.7	4.5	5.0	604	3,374
90-94	0.2733	0.7508	0.7267	0.7755	8,575	7,083	10,372	6,439	23,560	3.4	3.2	3.5	268	982
95-99	0.3974	0.9063	0.8924	0.9200	2,137	1,590	2,825	1,936	4,874	2.5	2.4	2.5	84	212
100+	0.5504	1	1	1	200	127	302	200	364	1.8	1.8	1.9	15	28
All ages													*9,113*	*2,359,257*

ln($_nq_x$) with uncertainty interval

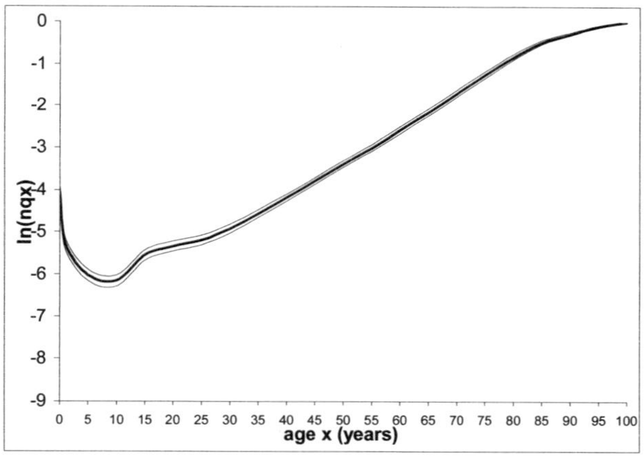

l_x with uncertainty interval

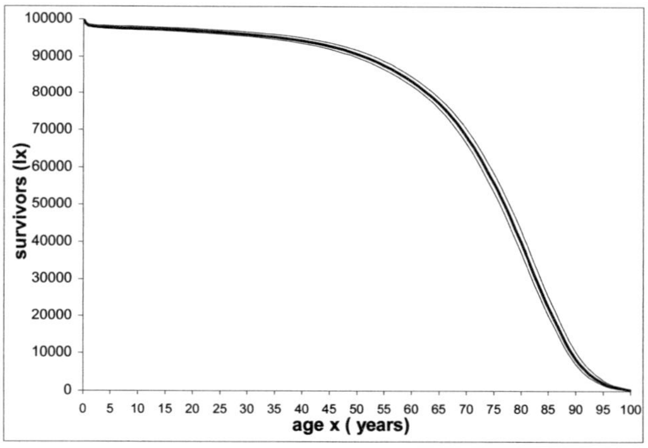

e_x with uncertainty interval

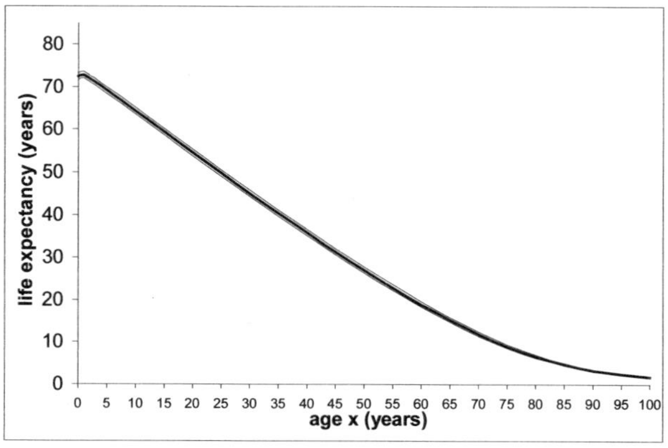

Distribution of deaths by age with uncertainty interval

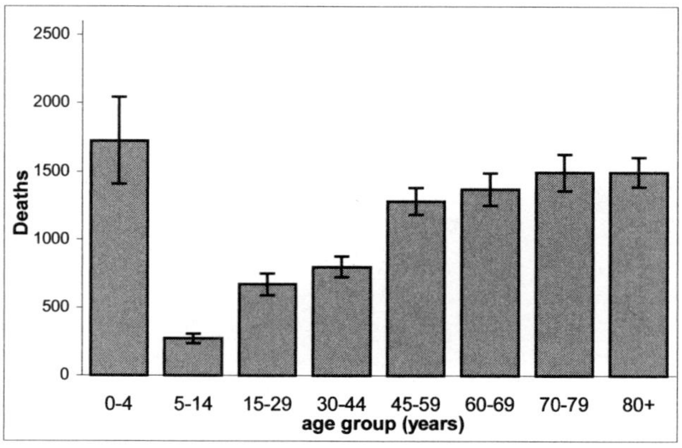

Kazakhstan / Kazakhstan / Kazajstán

Males, 2000 / *Hommes, 2000* / *Hombres, 2000*

Age	$_nM_x$	$_nq_x$ mean	$_nq_x$ percentiles 2½	$_nq_x$ percentiles 97½	l_x mean	l_x percentiles 2½	l_x percentiles 97½	$_nd_x$	$_nL_x$	e_x mean	e_x percentiles 2½	e_x percentiles 97½	Actual deaths	Actual population
<1	0.0716	0.0672	0.0647	0.0696	100,000	100,000	100,000	6,724	93,948	58.0	57.1	59.0	9,406	131,414
1-4	0.0034	0.0136	0.0131	0.0143	93,276	93,039	93,525	1,272	370,050	61.2	60.3	62.2	1,778	517,052
5-9	0.0012	0.0057	0.0055	0.0060	92,004	91,754	92,253	529	458,696	58.0	57.2	59.0	831	720,797
10-14	0.0012	0.0059	0.0057	0.0062	91,475	91,209	91,737	543	456,018	53.4	52.5	54.3	1,010	849,390
15-19	0.0027	0.0136	0.0130	0.0143	90,932	90,648	91,214	1,236	451,573	48.7	47.8	49.6	2,124	776,371
20-24	0.0052	0.0258	0.0245	0.0272	89,697	89,353	90,023	2,312	442,703	44.3	43.5	45.2	3,711	710,379
25-29	0.0063	0.0310	0.0294	0.0327	87,384	86,924	87,793	2,707	430,155	40.4	39.6	41.3	4,054	644,307
30-34	0.0064	0.0314	0.0296	0.0332	84,678	84,079	85,183	2,655	416,750	36.6	35.9	37.4	3,677	577,197
35-39	0.0076	0.0373	0.0351	0.0395	82,023	81,288	82,620	3,059	402,465	32.7	32.0	33.5	4,801	631,585
40-44	0.0090	0.0440	0.0413	0.0467	78,963	78,073	79,704	3,475	386,130	28.9	28.2	29.6	5,382	598,050
45-49	0.0120	0.0583	0.0546	0.0618	75,488	74,426	76,410	4,397	366,448	25.1	24.5	25.8	5,710	475,800
50-54	0.0165	0.0792	0.0742	0.0841	71,091	69,824	72,237	5,633	341,373	21.5	20.9	22.2	5,553	336,553
55-59	0.0253	0.1189	0.1114	0.1258	65,458	63,954	66,877	7,781	307,840	18.1	17.6	18.8	5,026	198,866
60-64	0.0348	0.1599	0.1501	0.1691	57,678	55,907	59,429	9,224	265,330	15.3	14.8	15.8	10,539	303,176
65-69	0.0486	0.2169	0.2038	0.2287	48,454	46,478	50,505	10,508	216,002	12.7	12.3	13.2	7,344	150,968
70-74	0.0623	0.2696	0.2543	0.2830	37,947	35,909	40,172	10,231	164,155	10.5	10.2	10.9	8,516	136,638
75-79	0.0841	0.3474	0.3300	0.3621	27,716	25,826	29,887	9,629	114,506	8.5	8.2	8.8	4,185	49,766
80-84	0.1200	0.4615	0.4427	0.4772	18,087	16,514	19,943	8,348	69,564	6.6	6.4	6.9	2,412	20,100
85-89	0.1649	0.5839	0.5658	0.5988	9,739	8,641	11,100	5,687	34,478	5.2	5.0	5.3	1,869	11,334
90-94	0.2321	0.6841	0.6681	0.6973	4,052	3,474	4,819	2,772	11,945	3.9	3.8	4.0	740	3,187
95-99	0.3149	0.8096	0.7967	0.8202	1,280	1,055	1,599	1,036	3,292	3.0	2.9	3.1	191	606
100+	0.4120	1	1	1	244	191	325	244	592	2.4	2.2	2.5	30	73
All ages													*88,889*	*7,843,609*

ln($_nq_x$) with uncertainty interval

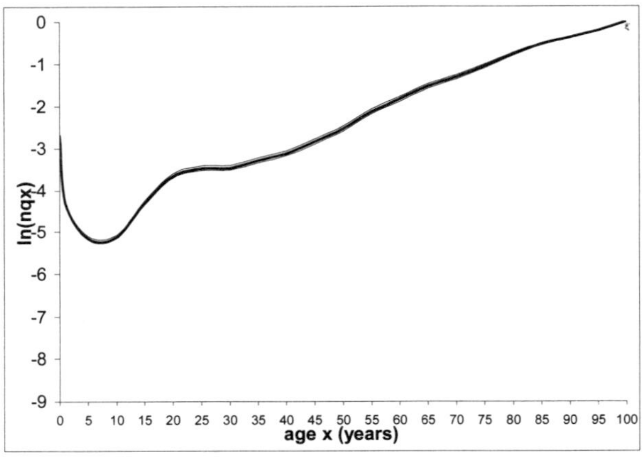

l_x with uncertainty interval

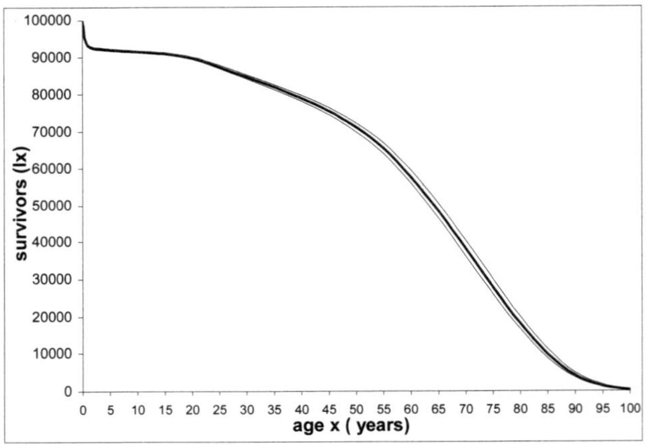

e_x with uncertainty interval

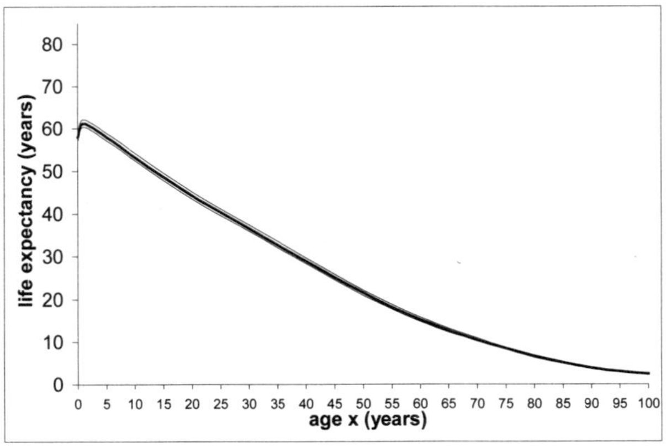

Distribution of deaths by age with uncertainty interval

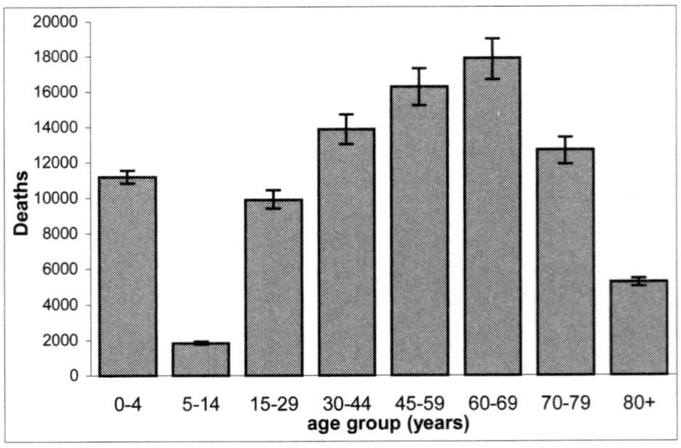

Kazakhstan **Kazakhstan** **Kazajstán**

Females, 2000 *Femmes, 2000* *Mujeres, 2000*

Age	$_nM_x$	$_nq_x$			l_x			$_nd_x$	$_nL_x$	e_x			Actual deaths	Actual population
			percentiles			percentiles					percentiles			
		mean	2 1/2	97 1/2	mean	2 1/2	97 1/2			mean	2 1/2	97 1/2		
<1	0.0515	0.0492	0.0455	0.0532	100,000	100,000	100,000	4,920	95,572	68.4	66.8	69.9	6,355	123,451
1-4	0.0029	0.0114	0.0105	0.0123	95,080	94,682	95,454	1,083	377,721	70.9	69.4	72.5	1,439	501,526
5-9	0.0008	0.0038	0.0035	0.0041	93,997	93,597	94,396	355	469,096	67.7	66.3	69.3	525	693,702
10-14	0.0007	0.0033	0.0031	0.0036	93,642	93,222	94,064	310	467,433	63.0	61.5	64.5	548	826,229
15-19	0.0013	0.0065	0.0060	0.0071	93,332	92,892	93,774	609	465,136	58.2	56.7	59.7	999	763,114
20-24	0.0018	0.0090	0.0083	0.0099	92,723	92,240	93,209	839	461,515	53.6	52.1	55.1	1,283	705,690
25-29	0.0026	0.0131	0.0120	0.0144	91,883	91,339	92,431	1,203	456,410	49.0	47.6	50.5	1,648	625,103
30-34	0.0029	0.0146	0.0134	0.0161	90,680	90,046	91,315	1,327	450,085	44.6	43.3	46.1	1,606	544,750
35-39	0.0036	0.0177	0.0161	0.0196	89,354	88,621	90,084	1,584	442,808	40.3	39.0	41.7	2,329	650,843
40-44	0.0046	0.0226	0.0204	0.0250	87,769	86,909	88,613	1,979	433,899	35.9	34.7	37.3	2,841	622,770
45-49	0.0063	0.0311	0.0279	0.0347	85,790	84,733	86,777	2,672	422,270	31.7	30.5	33.1	3,260	515,199
50-54	0.0085	0.0414	0.0369	0.0462	83,118	81,792	84,299	3,439	406,992	27.7	26.5	29.0	3,224	381,484
55-59	0.0131	0.0635	0.0563	0.0711	79,679	78,010	81,101	5,060	385,743	23.7	22.7	25.0	3,200	243,954
60-64	0.0177	0.0848	0.0748	0.0949	74,619	72,467	76,457	6,325	357,281	20.2	19.2	21.3	6,994	395,076
65-69	0.0248	0.1166	0.1027	0.1301	68,294	65,592	70,739	7,962	321,562	16.8	15.9	17.8	5,331	215,298
70-74	0.0371	0.1697	0.1501	0.1886	60,331	57,061	63,474	10,240	276,057	13.7	12.9	14.6	9,327	251,442
75-79	0.0559	0.2452	0.2191	0.2706	50,091	46,342	53,949	12,284	219,747	11.0	10.3	11.7	7,762	138,858
80-84	0.0780	0.3264	0.2954	0.3556	37,807	33,930	42,131	12,339	158,190	8.8	8.2	9.3	5,259	67,418
85-89	0.1089	0.4279	0.3936	0.4584	25,469	22,034	29,652	10,899	100,096	6.8	6.4	7.2	4,881	44,833
90-94	0.1654	0.5527	0.5186	0.5813	14,570	11,966	17,899	8,052	48,693	5.0	4.7	5.3	2,380	14,390
95-99	0.2467	0.7088	0.6793	0.7326	6,518	5,010	8,534	4,620	18,729	3.7	3.4	3.8	828	3,356
100+	0.3614	1	1	1	1,898	1,340	2,720	1,898	5,252	2.8	2.3	2.9	136	375
All ages													*72,153*	*8,328,861*

$\ln(_nq_x)$ with uncertainty interval

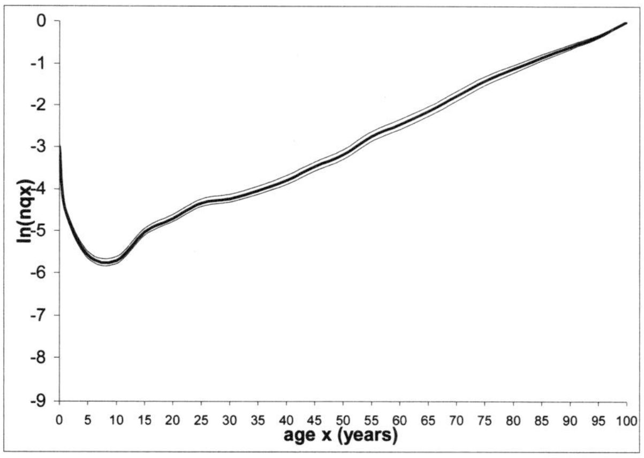

l_x with uncertainty interval

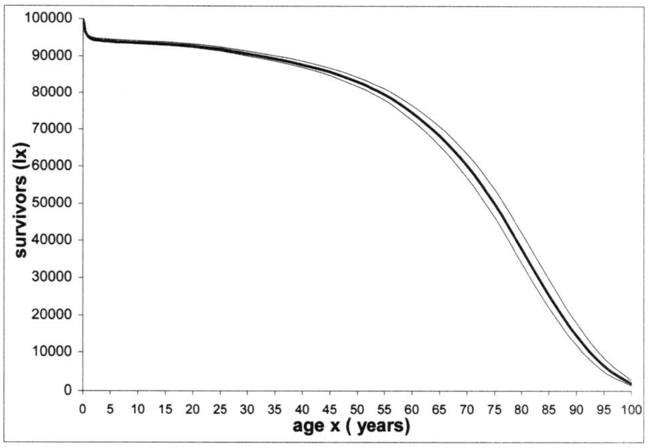

e_x with uncertainty interval

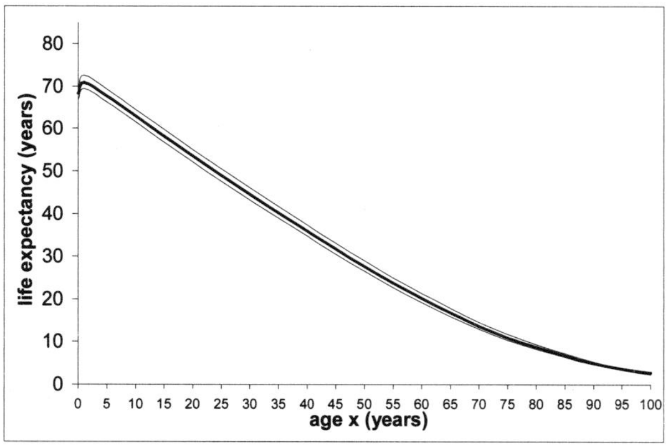

Distribution of deaths by age with uncertainty interval

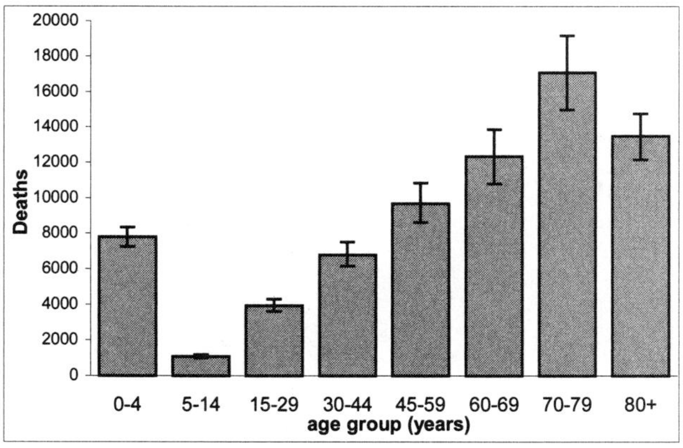

Kenya **Kenya** **Kenya**

Males, 2000 *Hommes, 2000* *Hombres, 2000*

Age	$_nM_x$	$_nq_x$			l_x			$_nd_x$	$_nL_x$	e_x			Actual deaths	Actual population
			percentiles			percentiles					percentiles			
		mean	2 1/2	97 1/2	mean	2 1/2	97 1/2			mean	2 1/2	97 1/2		
<1	0.0815	0.0759	0.0680	0.0837	100,000	100,000	100,000	7,591	93,168	48.2	45.6	52.0	42,562	522,374
1-4	0.0085	0.0332	0.0287	0.0366	92,409	91,632	93,196	3,067	362,274	51.2	48.1	55.0	15,615	1,844,226
5-9	0.0018	0.0091	0.0077	0.0101	89,342	88,602	90,088	810	444,684	48.9	45.9	52.6	3,928	2,157,866
10-14	0.0014	0.0071	0.0060	0.0079	88,532	87,731	89,365	631	441,084	44.3	41.3	48.0	3,111	2,175,583
15-19	0.0029	0.0144	0.0121	0.0161	87,901	87,051	88,827	1,267	436,339	39.6	36.6	43.3	5,557	1,913,482
20-24	0.0065	0.0320	0.0266	0.0362	86,634	85,656	87,750	2,771	426,243	35.1	32.2	38.8	10,193	1,567,950
25-29	0.0142	0.0687	0.0565	0.0792	83,863	82,588	85,412	5,764	404,906	31.2	28.3	34.8	17,299	1,215,196
30-34	0.0230	0.1089	0.0883	0.1280	78,099	76,209	80,587	8,504	369,235	28.3	25.5	31.7	21,284	924,092
35-39	0.0252	0.1187	0.0955	0.1414	69,595	66,821	73,468	8,262	327,320	26.5	23.9	29.7	18,631	738,159
40-44	0.0254	0.1194	0.0958	0.1430	61,333	57,374	66,455	7,322	288,360	24.7	22.4	27.6	14,948	588,688
45-49	0.0254	0.1195	0.0955	0.1431	54,011	49,172	60,090	6,454	253,920	22.7	20.7	25.3	11,751	462,361
50-54	0.0248	0.1166	0.0929	0.1393	47,557	42,135	54,318	5,545	223,924	20.5	18.8	22.7	8,257	333,446
55-59	0.0247	0.1163	0.0927	0.1384	42,012	36,264	49,204	4,887	197,844	17.8	16.4	19.8	5,614	227,287
60-64	0.0320	0.1480	0.1187	0.1745	37,125	31,245	44,556	5,495	171,890	14.8	13.7	16.5	6,214	194,411
65-69	0.0448	0.2013	0.1634	0.2341	31,631	25,793	39,141	6,368	142,233	12.0	11.0	13.4	7,464	166,705
70-74	0.0673	0.2882	0.2387	0.3286	25,262	19,755	32,564	7,280	108,111	9.4	8.6	10.5	7,956	118,150
75-79	0.1027	0.4085	0.3487	0.4549	17,982	13,270	24,541	7,345	71,548	7.2	6.6	8.0	7,427	72,345
80-84	0.1535	0.5545	0.4915	0.6013	10,637	7,253	15,922	5,899	38,438	5.4	5.0	6.0	5,508	35,892
85-89	0.2258	0.7217	0.6656	0.7615	4,738	2,910	8,097	3,420	15,143	4.0	3.8	4.4	2,752	12,186
90-94	0.3214	0.8182	0.7731	0.8497	1,319	702	2,707	1,079	3,357	3.0	2.8	3.2	788	2,451
95-99	0.4429	0.9509	0.9286	0.9646	240	107	614	228	515	2.2	2.2	2.3	111	251
100+	0.5901	1	1	1	12	4	44	12	20	1.7	1.7	1.8	9	15
All ages													*216,981*	*15,273,116*

ln($_nq_x$) with uncertainty interval

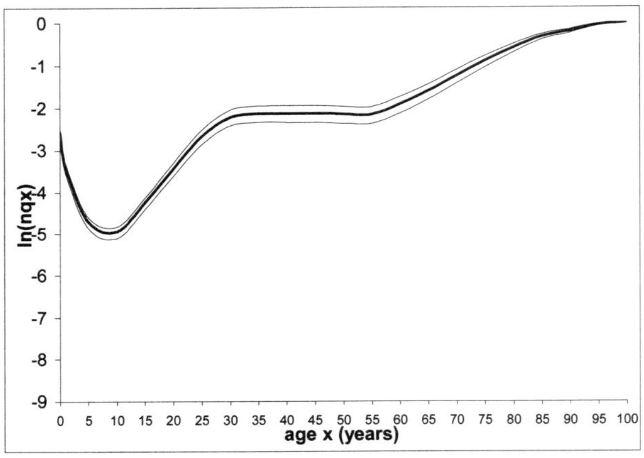

l_x with uncertainty interval

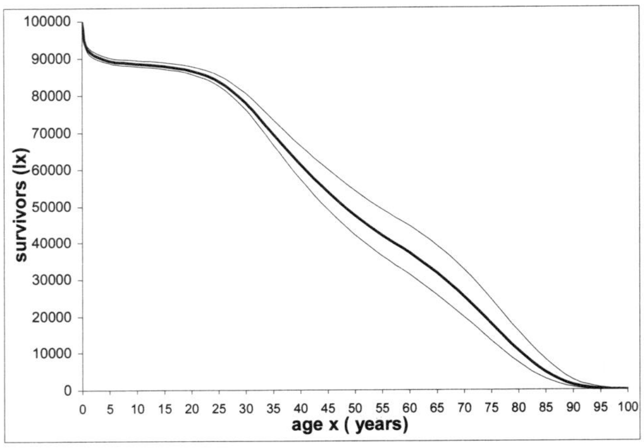

e_x with uncertainty interval

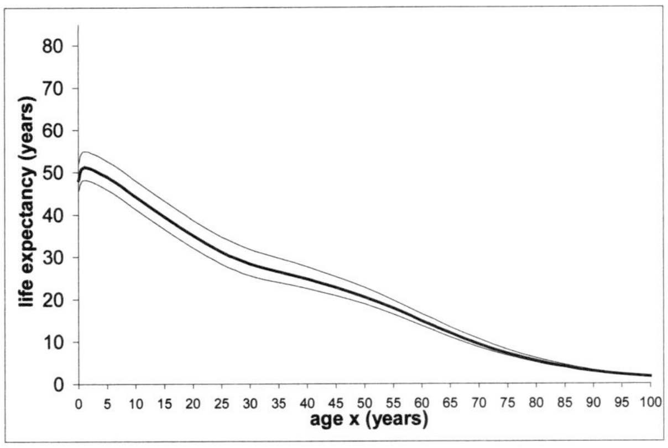

Distribution of deaths by age with uncertainty interval

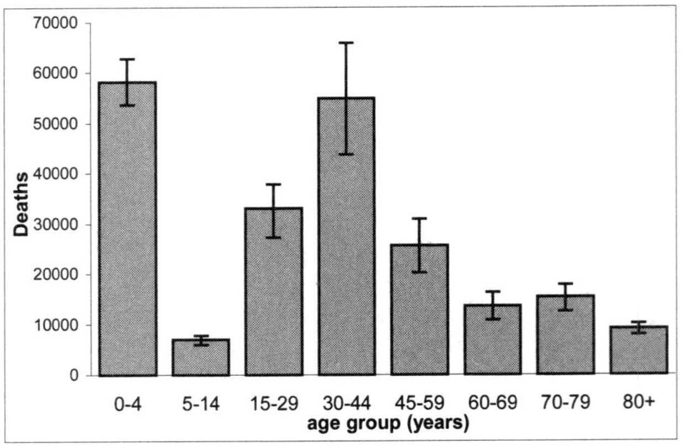

Kenya

Females, 2000

Kenya

Femmes, 2000

Kenya

Mujeres, 2000

Age	$_nM_x$	$_nq_x$ mean	$_nq_x$ percentiles 2½	$_nq_x$ percentiles 97½	l_x mean	l_x percentiles 2½	l_x percentiles 97½	$_nd_x$	$_nL_x$	e_x mean	e_x percentiles 2½	e_x percentiles 97½	Actual deaths	Actual population
<1	0.0702	0.0660	0.0593	0.0740	100,000	100,000	100,000	6,603	94,058	49.6	46.8	53.5	35,006	498,687
1-4	0.0086	0.0338	0.0298	0.0372	93,397	92,597	94,066	3,154	366,020	52.1	48.9	56.2	15,779	1,831,204
5-9	0.0016	0.0082	0.0071	0.0091	90,244	89,494	90,991	740	449,368	49.9	46.6	53.9	3,517	2,136,703
10-14	0.0012	0.0060	0.0051	0.0067	89,504	88,690	90,331	536	446,180	45.2	42.0	49.2	2,598	2,164,121
15-19	0.0043	0.0211	0.0180	0.0237	88,968	88,108	89,859	1,876	440,150	40.5	37.3	44.5	8,158	1,914,084
20-24	0.0118	0.0572	0.0480	0.0648	87,092	86,070	88,212	4,984	423,000	36.3	33.1	40.3	18,507	1,570,598
25-29	0.0208	0.0989	0.0812	0.1144	82,108	80,597	83,853	8,120	390,239	33.4	30.2	37.2	25,216	1,211,884
30-34	0.0254	0.1195	0.0967	0.1413	73,988	71,504	76,958	8,843	347,832	31.8	28.8	35.2	23,475	923,375
35-39	0.0219	0.1040	0.0834	0.1244	65,145	61,715	69,513	6,772	308,794	30.7	28.1	33.7	16,434	749,329
40-44	0.0187	0.0895	0.0716	0.1076	58,373	54,206	63,715	5,222	278,809	29.0	26.8	31.6	11,441	610,835
45-49	0.0156	0.0750	0.0600	0.0904	53,151	48,371	59,155	3,986	255,788	26.6	24.7	28.8	7,677	492,588
50-54	0.0154	0.0740	0.0593	0.0892	49,164	43,997	55,609	3,639	236,725	23.6	21.9	25.5	5,529	359,718
55-59	0.0166	0.0796	0.0640	0.0957	45,526	40,073	52,312	3,626	218,564	20.3	18.8	22.0	4,164	251,023
60-64	0.0222	0.1050	0.0850	0.1254	41,900	36,237	48,963	4,398	198,504	16.8	15.5	18.4	4,891	220,761
65-69	0.0331	0.1530	0.1249	0.1809	37,502	31,692	44,804	5,739	173,162	13.5	12.4	14.8	6,214	187,505
70-74	0.0536	0.2364	0.1957	0.2745	31,763	25,957	39,187	7,507	140,046	10.4	9.6	11.6	7,041	131,354
75-79	0.0860	0.3537	0.3003	0.4003	24,256	18,833	31,444	8,580	99,827	7.9	7.2	8.8	6,870	79,927
80-84	0.1354	0.5059	0.4450	0.5546	15,675	11,294	21,860	7,930	58,552	5.9	5.4	6.5	5,682	41,953
85-89	0.2059	0.6795	0.6228	0.7218	7,746	5,030	11,958	5,263	25,569	4.3	4.0	4.7	3,186	15,476
90-94	0.3010	0.7908	0.7451	0.8239	2,482	1,399	4,467	1,963	6,522	3.1	2.9	3.4	1,153	3,832
95-99	0.4234	0.9325	0.9073	0.9491	519	246	1,139	484	1,143	2.3	2.2	2.5	245	579
100+	0.5729	1	1	1	35	13	106	35	61	1.7	1.7	1.9	26	45
All ages													*212,810*	*15,395,581*

ln($_nq_x$) with uncertainty interval

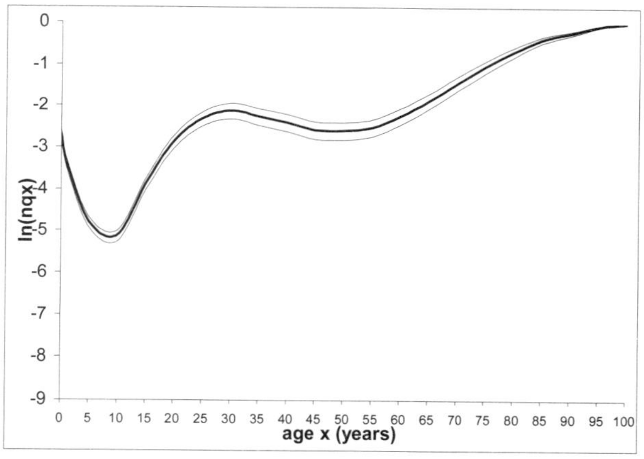

l_x with uncertainty interval

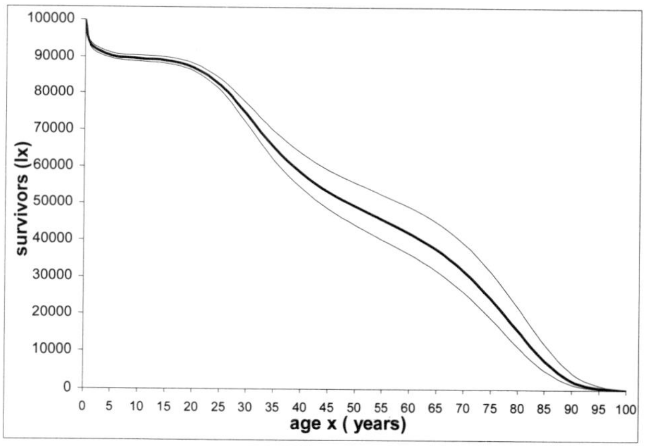

e_x with uncertainty interval

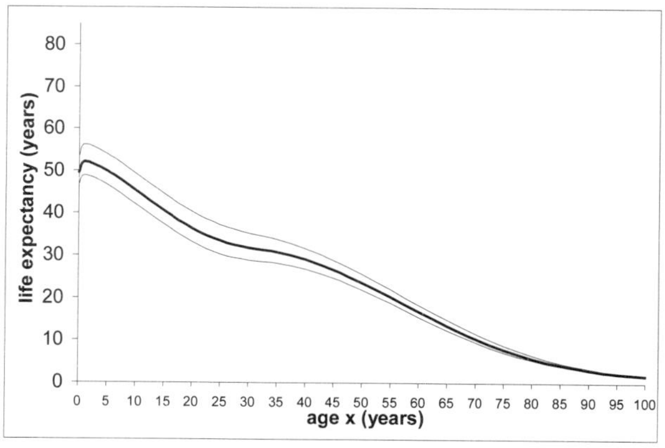

Distribution of deaths by age with uncertainty interval

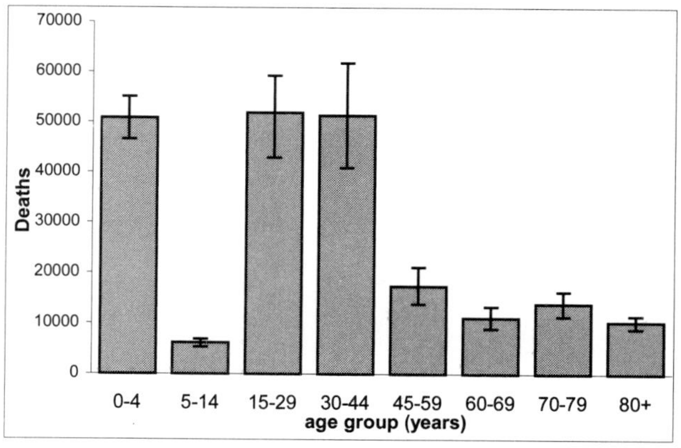

Kiribati **Kiribati** **Kiribati**

Males, 2000 *Hommes, 2000* *Hombres, 2000*

Age	$_nM_x$	$_nq_x$			l_x			$_nd_x$	$_nL_x$	e_x			Actual deaths	Actual population
			percentiles			percentiles					percentiles			
		mean	2½	97½	mean	2½	97½			mean	2½	97½		
<1	0.0781	0.0730	0.0611	0.0854	100,000	100,000	100,000	7,295	93,434	60.4	57.7	64.2	95	1,213
1-4	0.0055	0.0216	0.0178	0.0253	92,705	91,462	93,889	2,000	366,018	64.1	61.3	68.0	27	4,868
5-9	0.0015	0.0077	0.0062	0.0091	90,704	89,477	91,953	700	451,772	61.5	58.8	65.2	9	5,670
10-14	0.0011	0.0056	0.0045	0.0066	90,004	88,691	91,361	501	448,770	57.0	54.3	60.5	5	4,752
15-19	0.0017	0.0083	0.0066	0.0098	89,504	88,113	90,939	745	445,658	52.3	49.6	55.8	7	4,133
20-24	0.0025	0.0125	0.0099	0.0148	88,759	87,254	90,310	1,109	441,025	47.7	45.0	51.2	9	3,386
25-29	0.0028	0.0141	0.0110	0.0168	87,651	85,975	89,374	1,236	435,163	43.3	40.7	46.6	9	3,075
30-34	0.0034	0.0168	0.0130	0.0201	86,415	84,550	88,329	1,451	428,447	38.9	36.3	42.1	11	3,142
35-39	0.0044	0.0219	0.0168	0.0263	84,964	82,867	87,102	1,860	420,171	34.5	32.0	37.7	12	2,675
40-44	0.0062	0.0303	0.0230	0.0368	83,104	80,704	85,525	2,518	409,228	30.2	27.8	33.3	14	2,350
45-49	0.0089	0.0435	0.0326	0.0536	80,587	77,783	83,425	3,503	394,176	26.1	23.7	29.0	18	2,006
50-54	0.0132	0.0638	0.0473	0.0798	77,084	73,732	80,707	4,915	373,130	22.1	19.9	25.0	22	1,662
55-59	0.0195	0.0928	0.0682	0.1176	72,168	68,107	76,894	6,699	344,095	18.5	16.5	21.1	23	1,191
60-64	0.0290	0.1354	0.0994	0.1723	65,470	60,504	71,647	8,864	305,188	15.1	13.3	17.5	28	969
65-69	0.0433	0.1955	0.1439	0.2466	56,605	50,270	64,525	11,066	255,363	12.1	10.6	14.2	32	745
70-74	0.0668	0.2863	0.2131	0.3517	45,540	37,874	55,163	13,038	195,102	9.4	8.2	11.2	39	591
75-79	0.1025	0.4079	0.3149	0.4805	32,501	24,553	43,093	13,259	129,360	7.2	6.3	8.6	35	337
80-84	0.1554	0.5596	0.4588	0.6300	19,243	12,756	28,975	10,769	69,291	5.4	4.8	6.3	21	136
85-89	0.2281	0.7262	0.6383	0.7824	8,474	4,753	15,556	6,154	26,984	4.0	3.6	4.6	11	49
90-94	0.3239	0.8214	0.7534	0.8639	2,320	1,051	5,627	1,905	5,882	2.9	2.7	3.3	3	10
95-99	0.4452	0.9531	0.9201	0.9705	414	146	1,387	395	887	2.2	2.1	2.4	1	2
100+	0.5923	1	1	1	19	4	111	19	33	1.7	1.6	1.8	0	0
All ages													*430*	*42,962*

$\ln(_nq_x)$ with uncertainty interval

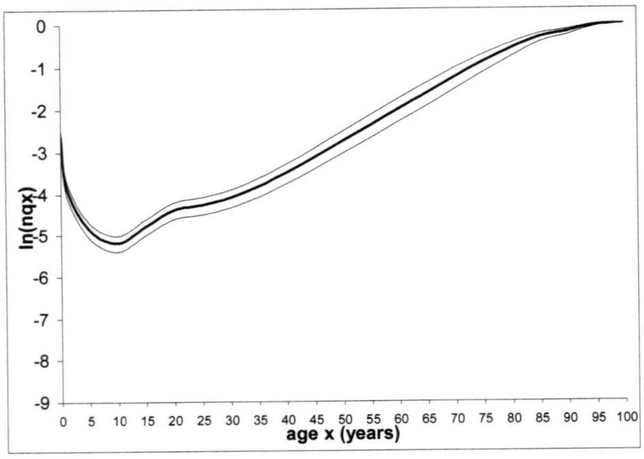

l_x with uncertainty interval

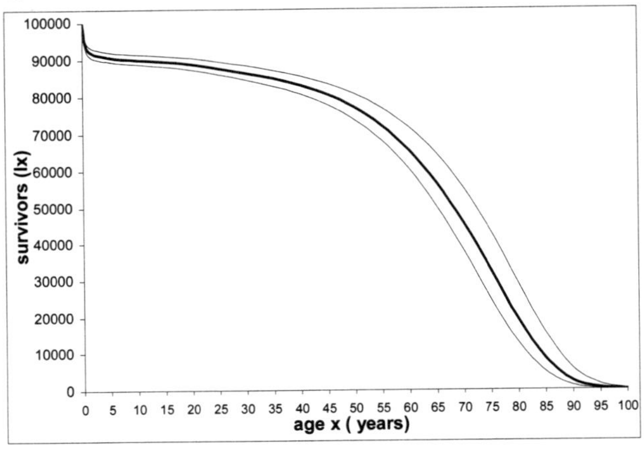

e_x with uncertainty interval

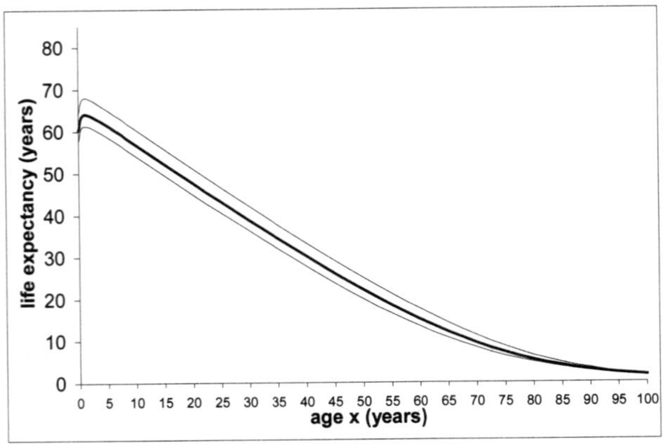

Distribution of deaths by age with uncertainty interval

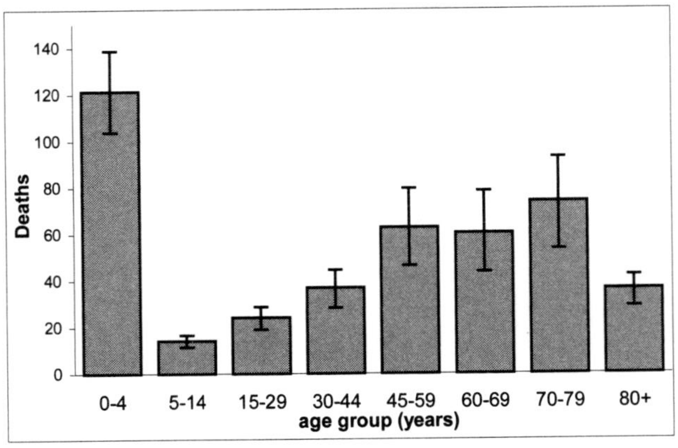

Kiribati **Kiribati** **Kiribati**

Females, 2000 *Femmes, 2000* *Mujeres, 2000*

Age	$_nM_x$	$_nq_x$			l_x			$_nd_x$	$_nL_x$	e_x			Actual deaths	Actual population
			percentiles			percentiles					percentiles			
		mean	2½	97½	mean	2½	97½			mean	2½	97½		
<1	0.0582	0.0553	0.0440	0.0664	100,000	100,000	100,000	5,526	95,027	64.5	61.2	68.4	67	1,148
1-4	0.0050	0.0198	0.0160	0.0240	94,474	93,363	95,599	1,870	373,407	67.3	64.3	71.3	23	4,579
5-9	0.0013	0.0067	0.0054	0.0082	92,604	91,413	93,801	621	461,467	64.6	61.9	68.5	7	5,338
10-14	0.0010	0.0049	0.0039	0.0060	91,983	90,713	93,289	449	458,791	60.1	57.3	63.8	4	4,477
15-19	0.0015	0.0075	0.0060	0.0093	91,534	90,190	92,924	687	455,951	55.3	52.6	59.1	6	3,869
20-24	0.0020	0.0100	0.0079	0.0124	90,847	89,367	92,365	910	451,959	50.7	48.0	54.4	7	3,254
25-29	0.0023	0.0117	0.0091	0.0145	89,937	88,255	91,626	1,048	447,064	46.2	43.5	49.8	7	2,898
30-34	0.0028	0.0139	0.0107	0.0174	88,889	86,971	90,774	1,232	441,362	41.7	39.1	45.2	7	2,578
35-39	0.0037	0.0184	0.0140	0.0232	87,656	85,460	89,773	1,616	434,241	37.3	34.7	40.7	9	2,440
40-44	0.0047	0.0234	0.0175	0.0296	86,040	83,478	88,458	2,017	425,161	33.0	30.4	36.2	10	2,163
45-49	0.0066	0.0324	0.0239	0.0411	84,024	81,006	86,812	2,722	413,313	28.7	26.2	31.9	12	1,827
50-54	0.0094	0.0462	0.0335	0.0586	81,301	77,680	84,581	3,753	397,125	24.6	22.2	27.6	14	1,479
55-59	0.0136	0.0657	0.0470	0.0841	77,549	73,131	81,481	5,091	375,015	20.6	18.4	23.4	14	1,062
60-64	0.0208	0.0989	0.0701	0.1284	72,457	67,049	77,612	7,165	344,374	16.9	14.9	19.5	19	936
65-69	0.0327	0.1510	0.1070	0.1965	65,292	58,714	72,170	9,857	301,820	13.5	11.7	15.8	24	742
70-74	0.0535	0.2360	0.1703	0.3020	55,436	47,760	64,449	13,082	244,472	10.4	9.0	12.4	31	573
75-79	0.0861	0.3543	0.2668	0.4355	42,353	33,626	53,473	15,008	174,246	7.9	6.8	9.5	30	350
80-84	0.1359	0.5073	0.4034	0.5900	27,345	18,982	39,179	13,872	102,046	5.8	5.1	7.0	23	166
85-89	0.2064	0.6808	0.5834	0.7498	13,473	7,808	23,082	9,173	44,434	4.3	3.8	5.0	12	58
90-94	0.3016	0.7917	0.7160	0.8435	4,301	1,981	9,285	3,405	11,289	3.1	2.8	3.6	4	12
95-99	0.4240	0.9331	0.8919	0.9577	896	317	2,634	836	1,972	2.3	2.2	2.5	1	2
100+	0.5734	1	1	1	60	14	285	60	105	1.7	1.7	1.9	0	0
All ages													*331*	*39,949*

ln($_nq_x$) with uncertainty interval

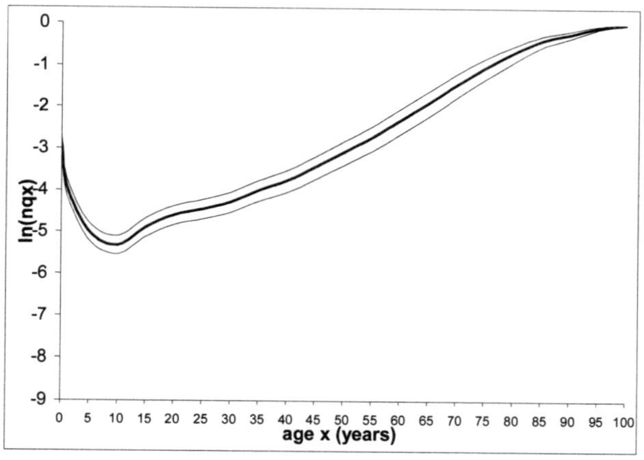

l_x with uncertainty interval

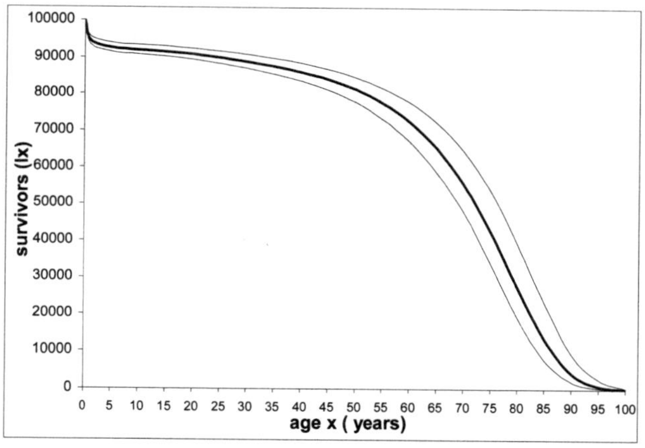

e_x with uncertainty interval

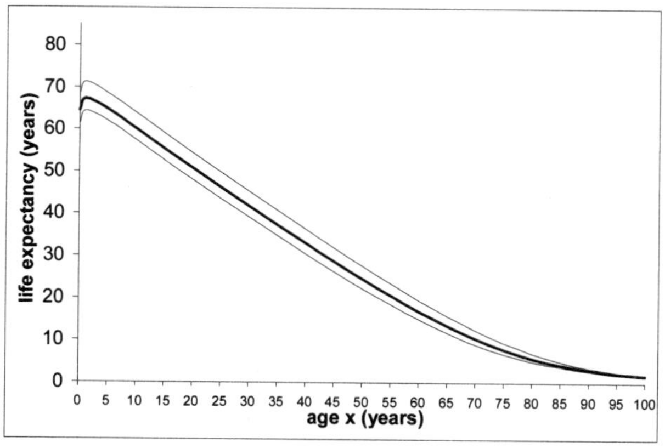

Distribution of deaths by age with uncertainty interval

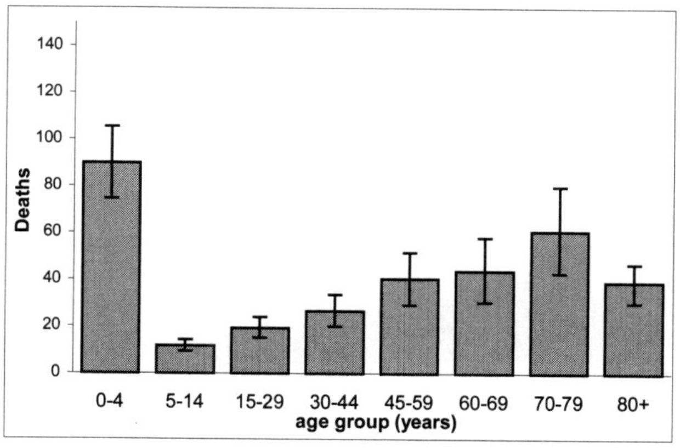

Kuwait   Koweït   Kuwait

Males, 2000   *Hommes, 2000*   *Hombres, 2000*

Age	$_nM_x$	$_nq_x$ mean	$_nq_x$ percentiles 2½	$_nq_x$ percentiles 97½	l_x mean	l_x percentiles 2½	l_x percentiles 97½	$_nd_x$	$_nL_x$	e_x mean	e_x percentiles 2½	e_x percentiles 97½	Actual deaths	Actual population
<1	0.0108	0.0107	0.0086	0.0135	100,000	100,000	100,000	1,065	99,041	74.2	73.4	75.0	151	14,084
1-4	0.0007	0.0028	0.0024	0.0033	98,935	98,647	99,135	273	395,084	74.0	73.2	74.8	41	58,855
5-9	0.0004	0.0022	0.0019	0.0026	98,662	98,323	98,900	215	492,771	70.2	69.4	71.0	39	90,610
10-14	0.0004	0.0018	0.0016	0.0021	98,447	98,072	98,714	175	491,797	65.4	64.6	66.2	50	140,773
15-19	0.0009	0.0046	0.0040	0.0053	98,272	97,869	98,561	450	490,232	60.5	59.7	61.3	120	130,250
20-24	0.0015	0.0074	0.0066	0.0084	97,821	97,353	98,164	728	487,286	55.7	54.9	56.6	149	99,887
25-29	0.0007	0.0037	0.0034	0.0042	97,093	96,533	97,513	363	484,559	51.1	50.4	52.0	43	57,275
30-34	0.0008	0.0041	0.0037	0.0046	96,730	96,129	97,185	400	482,650	46.3	45.5	47.2	41	49,322
35-39	0.0010	0.0050	0.0046	0.0055	96,330	95,687	96,822	484	480,440	41.5	40.7	42.3	104	102,993
40-44	0.0016	0.0079	0.0072	0.0086	95,846	95,157	96,379	756	477,342	36.7	35.9	37.5	172	108,730
45-49	0.0025	0.0126	0.0117	0.0136	95,091	94,336	95,682	1,198	472,458	32.0	31.2	32.8	227	89,645
50-54	0.0041	0.0205	0.0191	0.0221	93,893	93,053	94,565	1,928	464,644	27.3	26.6	28.2	274	66,073
55-59	0.0077	0.0380	0.0353	0.0411	91,965	91,000	92,757	3,491	451,097	22.9	22.2	23.7	385	49,723
60-64	0.0149	0.0720	0.0661	0.0785	88,474	87,262	89,480	6,371	426,443	18.7	18.0	19.4	450	30,121
65-69	0.0253	0.1189	0.1080	0.1300	82,103	80,414	83,563	9,761	386,113	14.9	14.3	15.6	398	15,758
70-74	0.0465	0.2082	0.1888	0.2267	72,342	69,958	74,537	15,065	324,050	11.6	11.0	12.3	304	6,533
75-79	0.0708	0.3006	0.2737	0.3253	57,278	54,097	60,465	17,219	243,339	9.0	8.5	9.6	186	2,624
80-84	0.1105	0.4329	0.3989	0.4637	40,058	36,498	43,917	17,342	156,936	6.8	6.4	7.2	122	1,101
85-89	0.1674	0.5900	0.5558	0.6203	22,716	19,576	26,399	13,403	80,072	5.0	4.8	5.3	46	273
90-94	0.2460	0.7076	0.6784	0.7324	9,313	7,433	11,725	6,590	26,793	3.7	3.5	3.9	19	79
95-99	0.3506	0.8543	0.8338	0.8711	2,723	1,989	3,771	2,326	6,635	2.7	2.7	2.8	4	12
100+	0.4847	1	1	1	397	256	626	397	818	2.1	2.0	2.1	0	1
All ages													*3,326*	*1,114,722*

ln($_nq_x$) with uncertainty interval

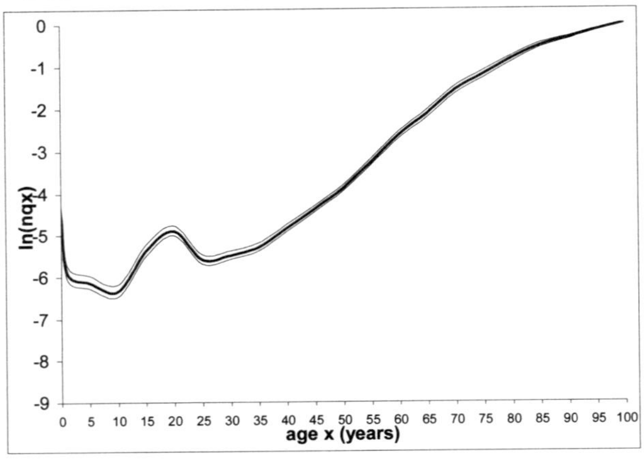

l_x with uncertainty interval

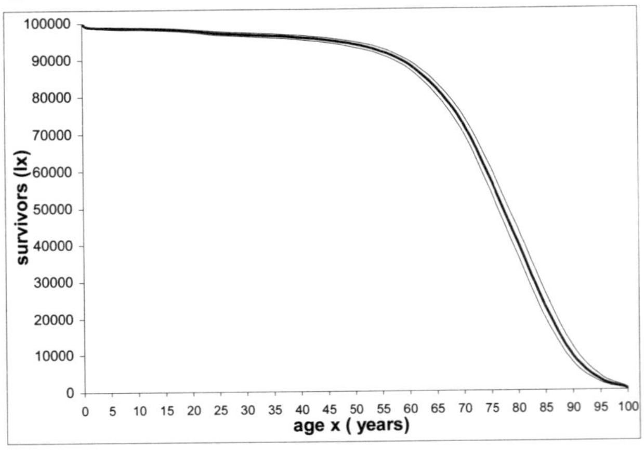

e_x with uncertainty interval

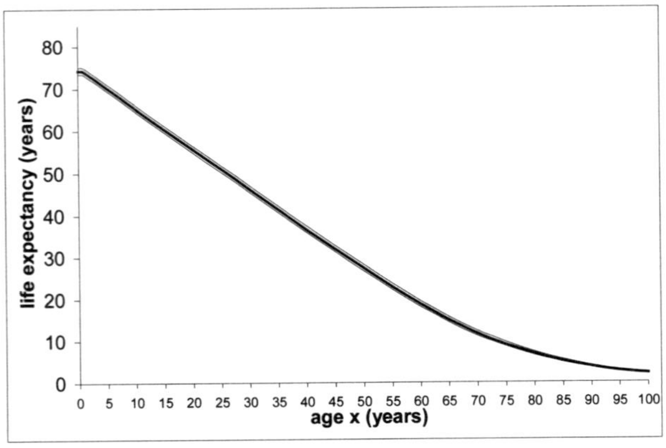

Distribution of deaths by age with uncertainty interval

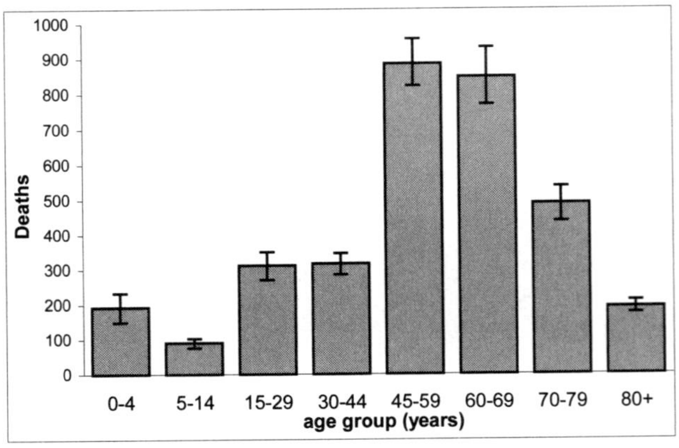

Kuwait

Koweït

Kuwait

Females, 2000 — *Femmes, 2000* — *Mujeres, 2000*

Age	$_nM_x$	$_nq_x$ mean	$_nq_x$ percentiles 2½	$_nq_x$ percentiles 97½	l_x mean	l_x percentiles 2½	l_x percentiles 97½	$_nd_x$	$_nL_x$	e_x mean	e_x percentiles 2½	e_x percentiles 97½	Actual deaths	Actual population
<1	0.0099	0.0098	0.0079	0.0123	100,000	100,000	100,000	979	99,119	76.8	76.1	77.7	135	13,673
1-4	0.0006	0.0022	0.0019	0.0026	99,021	98,770	99,207	221	395,553	76.6	75.9	77.4	32	57,286
5-9	0.0002	0.0012	0.0010	0.0014	98,800	98,511	99,018	116	493,709	72.7	72.1	73.6	21	88,054
10-14	0.0002	0.0012	0.0011	0.0014	98,684	98,376	98,918	120	493,120	67.8	67.2	68.7	33	136,018
15-19	0.0002	0.0012	0.0010	0.0013	98,564	98,237	98,814	113	492,538	62.9	62.2	63.8	29	124,471
20-24	0.0004	0.0018	0.0015	0.0020	98,451	98,107	98,716	173	491,821	58.0	57.3	58.8	32	90,156
25-29	0.0004	0.0022	0.0019	0.0025	98,277	97,907	98,564	212	490,856	53.1	52.4	53.9	21	47,820
30-34	0.0005	0.0023	0.0020	0.0026	98,065	97,666	98,377	225	489,763	48.2	47.5	49.0	16	34,775
35-39	0.0008	0.0041	0.0036	0.0046	97,840	97,411	98,179	399	488,201	43.3	42.7	44.1	37	44,862
40-44	0.0011	0.0054	0.0048	0.0061	97,441	96,963	97,824	527	485,885	38.4	37.8	39.3	51	46,631
45-49	0.0016	0.0082	0.0074	0.0090	96,913	96,376	97,351	794	482,581	33.6	33.0	34.5	66	40,350
50-54	0.0033	0.0164	0.0149	0.0179	96,119	95,505	96,632	1,573	476,664	28.9	28.3	29.7	91	27,695
55-59	0.0057	0.0281	0.0258	0.0306	94,546	93,800	95,192	2,660	466,081	24.3	23.8	25.1	114	20,006
60-64	0.0130	0.0630	0.0572	0.0684	91,886	90,929	92,736	5,785	444,970	20.0	19.4	20.7	157	12,053
65-69	0.0210	0.0998	0.0901	0.1079	86,102	84,711	87,432	8,590	409,034	16.1	15.7	16.9	142	6,755
70-74	0.0390	0.1777	0.1608	0.1904	77,512	75,569	79,556	13,772	353,129	12.7	12.2	13.3	164	4,207
75-79	0.0588	0.2563	0.2343	0.2729	63,740	61,182	66,767	16,340	277,850	9.8	9.4	10.3	149	2,537
80-84	0.0952	0.3844	0.3584	0.4091	47,400	44,485	51,125	18,218	191,455	7.4	7.0	7.8	134	1,407
85-89	0.1491	0.5430	0.5132	0.5709	29,182	26,285	32,803	15,846	106,294	5.4	5.2	5.7	82	547
90-94	0.2261	0.6736	0.6461	0.6981	13,336	11,279	15,969	8,984	39,728	3.9	3.7	4.1	53	235
95-99	0.3321	0.8318	0.8116	0.8489	4,352	3,405	5,651	3,620	10,901	2.9	2.8	3.0	38	115
100+	0.4722	1	1	1	732	514	1,065	732	1,551	2.1	2.1	2.2	14	29
All ages													*1,609*	*799,682*

ln($_nq_x$) with uncertainty interval

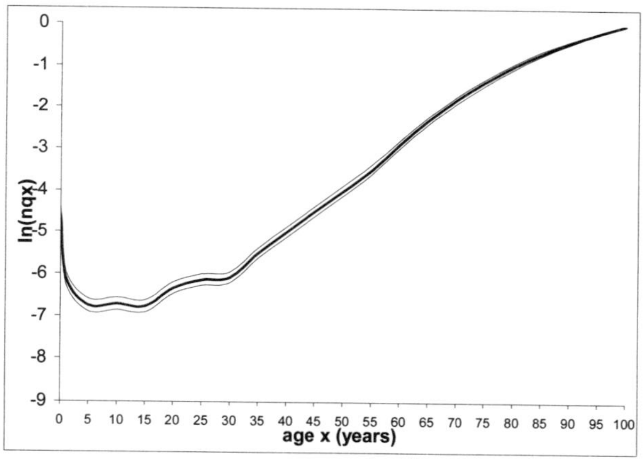

l_x with uncertainty interval

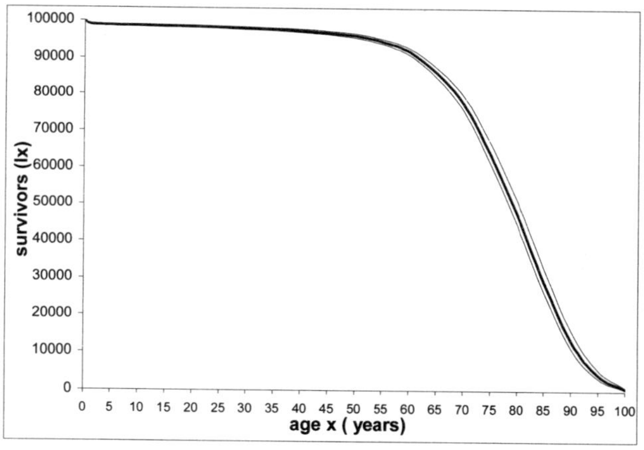

e_x with uncertainty interval

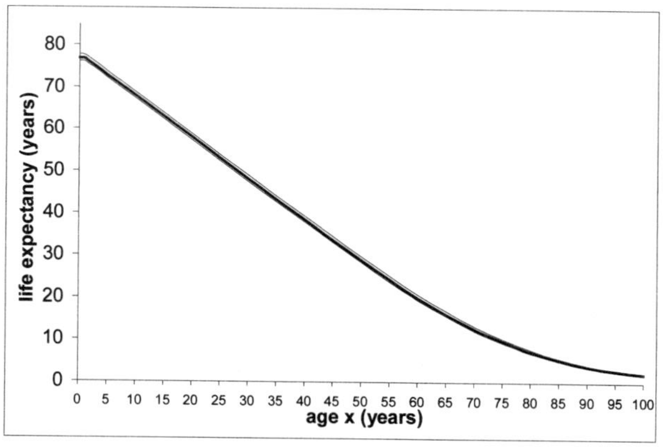

Distribution of deaths by age with uncertainty interval

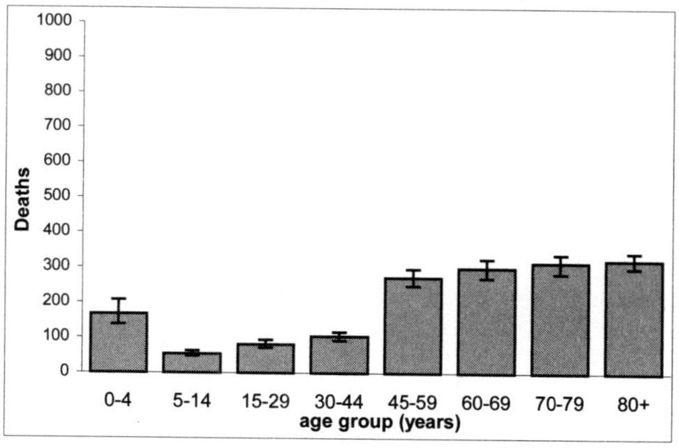

Kyrgyzstan Kirghizistan Kirguistán

Males, 2000 *Hommes, 2000* *Hombres, 2000*

Age	$_nM_x$	$_nq_x$			l_x			$_nd_x$	$_nL_x$	e_x			Actual deaths	Actual population
			percentiles			percentiles					percentiles			
		mean	2½	97½	mean	2½	97½			mean	2½	97½		
<1	0.0512	0.0490	0.0428	0.0555	100,000	100,000	100,000	4,898	95,592	60.0	58.4	61.6	2,788	54,423
1-4	0.0056	0.0219	0.0196	0.0244	95,102	94,454	95,721	2,087	375,399	62.1	60.8	63.5	1,169	210,247
5-9	0.0009	0.0044	0.0039	0.0049	93,015	92,145	93,848	410	464,048	59.5	58.3	60.7	251	283,568
10-14	0.0009	0.0044	0.0039	0.0049	92,605	91,692	93,480	408	462,002	54.7	53.5	55.9	261	295,563
15-19	0.0016	0.0079	0.0070	0.0088	92,196	91,241	93,112	728	459,161	50.0	48.8	51.1	408	257,376
20-24	0.0027	0.0134	0.0120	0.0150	91,468	90,438	92,457	1,228	454,271	45.3	44.2	46.5	595	220,109
25-29	0.0039	0.0195	0.0174	0.0218	90,240	89,085	91,351	1,763	446,794	40.9	39.8	42.0	783	198,376
30-34	0.0047	0.0233	0.0208	0.0259	88,477	87,146	89,758	2,058	437,243	36.7	35.7	37.7	798	169,617
35-39	0.0068	0.0335	0.0300	0.0372	86,420	84,891	87,892	2,894	424,865	32.5	31.5	33.5	1,142	167,678
40-44	0.0085	0.0418	0.0375	0.0463	83,526	81,734	85,256	3,489	408,909	28.5	27.7	29.5	1,269	148,749
45-49	0.0127	0.0614	0.0555	0.0677	80,037	77,953	82,056	4,916	387,897	24.7	23.9	25.6	1,419	111,980
50-54	0.0181	0.0864	0.0785	0.0947	75,121	72,672	77,506	6,491	359,380	21.1	20.4	22.0	1,316	72,840
55-59	0.0225	0.1067	0.0974	0.1160	68,631	65,791	71,419	7,320	324,853	17.9	17.2	18.7	1,013	44,946
60-64	0.0348	0.1601	0.1466	0.1729	61,310	58,158	64,462	9,816	282,013	14.7	14.1	15.4	2,296	65,970
65-69	0.0477	0.2130	0.1953	0.2309	51,495	48,104	55,014	10,970	230,048	12.0	11.5	12.7	2,207	46,278
70-74	0.0646	0.2780	0.2561	0.3007	40,525	36,998	44,272	11,264	174,463	9.6	9.2	10.1	2,553	39,544
75-79	0.0921	0.3744	0.3487	0.4017	29,261	25,871	32,935	10,956	118,913	7.4	7.0	7.7	1,396	15,156
80-84	0.1548	0.5580	0.5297	0.5880	18,305	15,477	21,451	10,215	65,986	5.3	5.0	5.5	872	5,633
85-89	0.2407	0.7514	0.7275	0.7758	8,090	6,377	10,089	6,079	25,252	3.8	3.7	4.0	777	3,228
90-94	0.3465	0.8495	0.8318	0.8674	2,011	1,429	2,749	1,708	4,929	2.8	2.7	2.9	351	1,014
95-99	0.4617	0.9679	0.9611	0.9743	303	190	462	293	634	2.2	2.1	2.2	124	269
100+	0.5694	1	1	1	10	5	18	10	17	1.8	1.7	1.8	30	52
All ages													*23,819*	*2,412,616*

ln($_nq_x$) with uncertainty interval

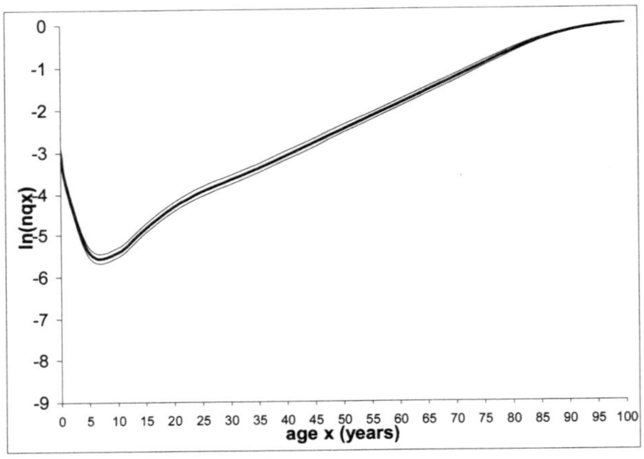

l_x with uncertainty interval

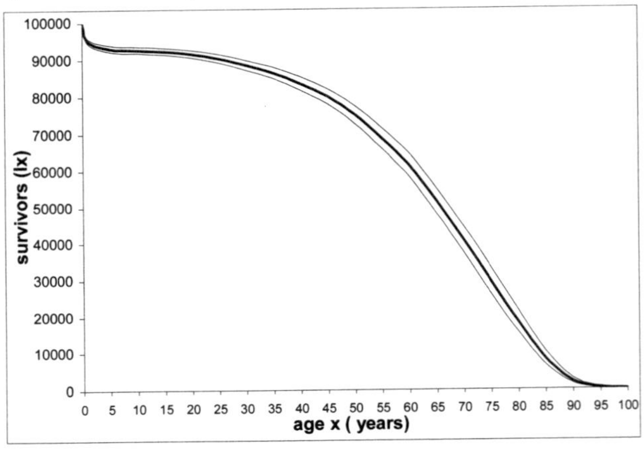

e_x with uncertainty interval

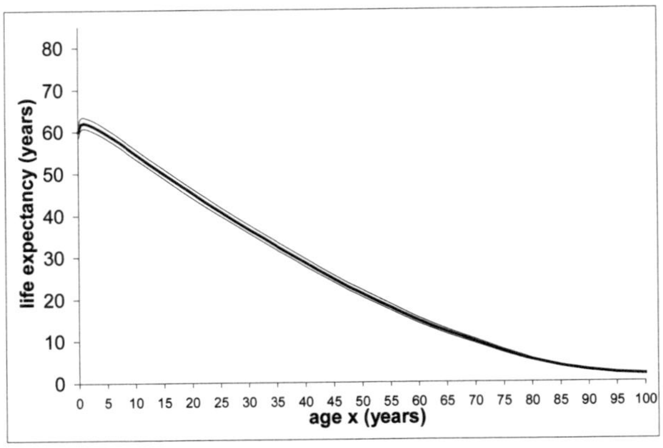

Distribution of deaths by age with uncertainty interval

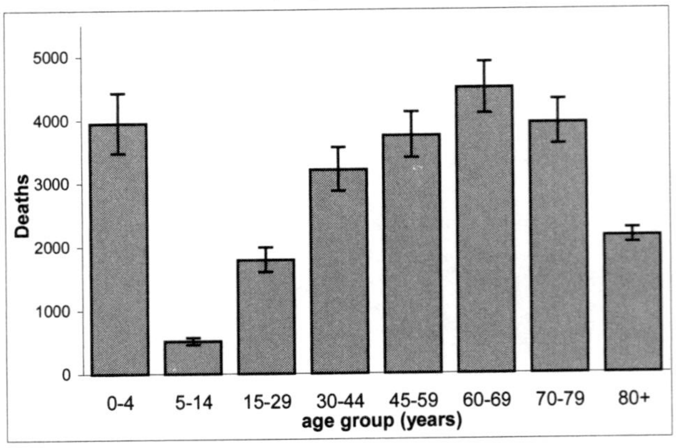

Kyrgyzstan

Kirghizistan

Kirguistán

Females, 2000

Femmes, 2000

Mujeres, 2000

Age	$_nM_x$	$_nq_x$			l_x			$_nd_x$	$_nL_x$	e_x			Actual deaths	Actual population
			percentiles			percentiles					percentiles			
		mean	2½	97½	mean	2½	97½			mean	2½	97½		
<1	0.0413	0.0398	0.0329	0.0465	100,000	100,000	100,000	3,981	96,418	68.8	66.9	70.9	2,170	52,557
1-4	0.0049	0.0195	0.0167	0.0224	96,019	95,352	96,707	1,868	379,595	70.7	69.0	72.6	1,017	206,776
5-9	0.0006	0.0031	0.0027	0.0036	94,152	93,214	95,094	292	470,028	68.0	66.6	69.8	173	278,682
10-14	0.0005	0.0026	0.0022	0.0030	93,860	92,881	94,841	245	468,686	63.2	61.8	65.0	151	288,138
15-19	0.0007	0.0037	0.0031	0.0042	93,615	92,600	94,629	344	467,214	58.4	57.0	60.2	187	253,666
20-24	0.0011	0.0056	0.0048	0.0065	93,271	92,207	94,331	526	465,040	53.6	52.2	55.4	247	217,974
25-29	0.0017	0.0086	0.0074	0.0100	92,745	91,606	93,875	802	461,721	48.9	47.6	50.6	344	198,430
30-34	0.0021	0.0106	0.0091	0.0122	91,943	90,691	93,179	975	457,280	44.3	43.0	46.0	352	165,136
35-39	0.0029	0.0142	0.0122	0.0164	90,968	89,580	92,331	1,291	451,614	39.7	38.5	41.4	503	175,816
40-44	0.0035	0.0172	0.0148	0.0199	89,677	88,113	91,206	1,546	444,521	35.3	34.1	36.9	543	156,186
45-49	0.0055	0.0269	0.0230	0.0310	88,131	86,361	89,855	2,372	434,726	30.9	29.8	32.4	654	119,929
50-54	0.0082	0.0404	0.0344	0.0463	85,759	83,685	87,785	3,462	420,142	26.6	25.6	28.1	665	80,708
55-59	0.0127	0.0616	0.0523	0.0702	82,298	79,810	84,765	5,065	398,824	22.7	21.7	24.0	614	48,367
60-64	0.0177	0.0848	0.0721	0.0960	77,232	74,205	80,331	6,553	369,778	19.0	18.2	20.2	1,410	79,580
65-69	0.0264	0.1239	0.1057	0.1385	70,679	67,082	74,543	8,760	331,495	15.5	14.7	16.6	1,598	60,487
70-74	0.0398	0.1810	0.1561	0.2000	61,919	57,790	66,663	11,205	281,583	12.3	11.6	13.2	2,409	60,534
75-79	0.0628	0.2712	0.2388	0.3004	50,714	46,234	56,260	13,755	219,184	9.5	8.9	10.2	2,201	35,070
80-84	0.1008	0.4026	0.3651	0.4412	36,959	32,345	42,825	14,878	147,601	7.1	6.6	7.7	1,694	16,809
85-89	0.1565	0.5626	0.5232	0.6031	22,081	18,075	27,189	12,422	79,351	5.3	4.9	5.6	1,516	9,683
90-94	0.2350	0.6892	0.6545	0.7240	9,659	7,174	12,964	6,658	28,324	3.8	3.6	4.1	708	3,013
95-99	0.3412	0.8431	0.8181	0.8672	3,002	1,980	4,479	2,531	7,417	2.8	2.7	2.9	220	644
100+	0.4790	1	1	1	471	263	815	471	983	2.1	2.0	2.1	22	46
All ages													*19,399*	*2,508,231*

$\ln(_nq_x)$ with uncertainty interval

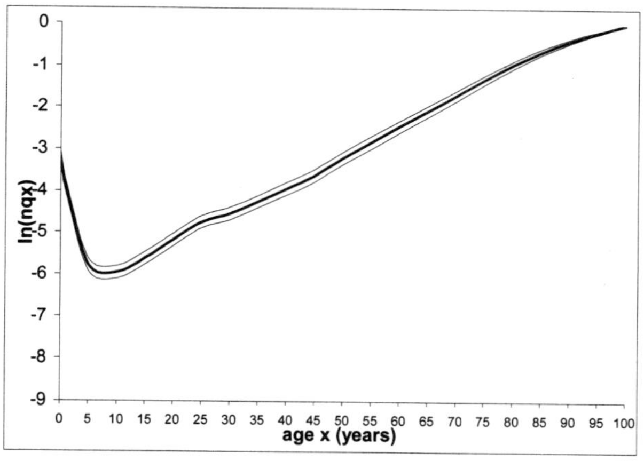

l_x with uncertainty interval

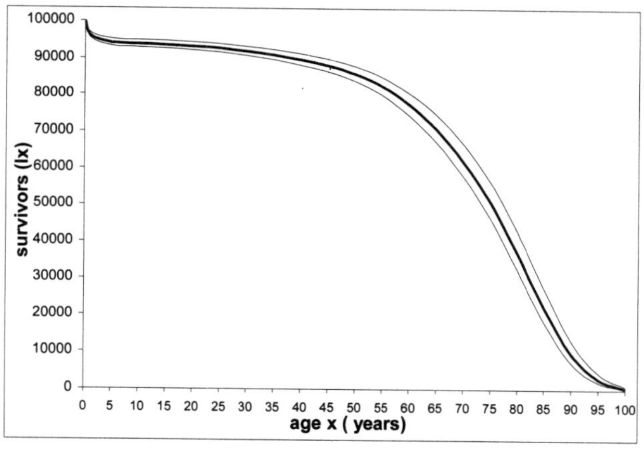

e_x with uncertainty interval

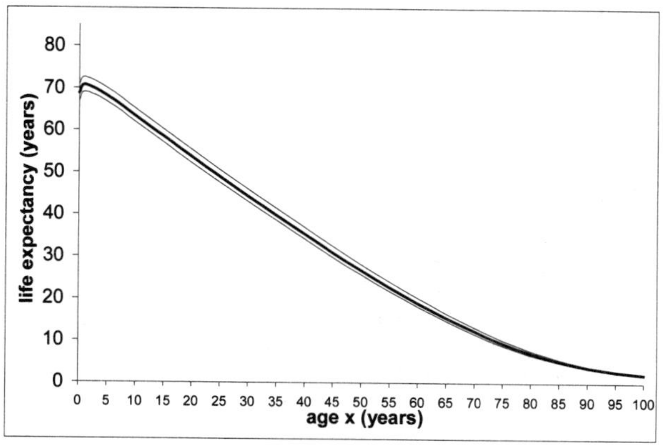

Distribution of deaths by age with uncertainty interval

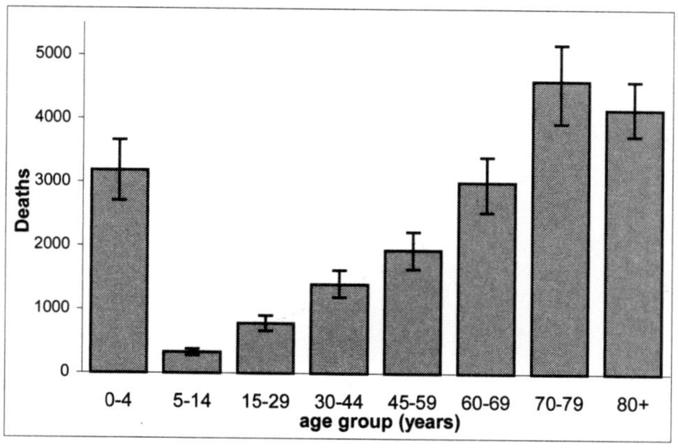

Lao People's Democratic Republic | République démocratique populaire lao | República Democrática Popular Lao

Males, 2000 | *Hommes, 2000* | *Hombres, 2000*

Age	$_nM_x$	$_nq_x$ mean	$_nq_x$ percentiles 2½	$_nq_x$ percentiles 97½	l_x mean	l_x percentiles 2½	l_x percentiles 97½	$_nd_x$	$_nL_x$	e_x mean	e_x percentiles 2½	e_x percentiles 97½	Actual deaths	Actual population
<1	0.0655	0.0619	0.0473	0.0862	100,000	100,000	100,000	6,187	94,432	52.2	47.5	55.0	6,136	93,654
1-4	0.0255	0.0960	0.0827	0.1205	93,813	91,382	95,270	9,004	353,643	54.7	50.4	57.0	8,457	332,166
5-9	0.0073	0.0357	0.0310	0.0449	84,809	80,368	87,388	3,031	416,468	56.3	53.1	58.2	2,760	379,294
10-14	0.0037	0.0184	0.0160	0.0231	81,778	76,757	84,679	1,506	405,126	53.3	50.5	55.1	1,274	342,600
15-19	0.0040	0.0199	0.0174	0.0249	80,272	74,982	83,322	1,599	397,363	49.3	46.7	51.1	1,130	280,863
20-24	0.0039	0.0195	0.0170	0.0243	78,673	73,113	81,874	1,532	389,535	45.2	42.8	47.0	933	237,278
25-29	0.0050	0.0247	0.0217	0.0307	77,141	71,336	80,479	1,905	380,943	41.1	38.7	42.8	1,006	201,284
30-34	0.0060	0.0296	0.0260	0.0365	75,236	69,148	78,734	2,224	370,622	37.0	34.8	38.8	968	161,325
35-39	0.0070	0.0344	0.0304	0.0423	73,012	66,621	76,684	2,511	358,784	33.1	31.0	34.8	967	138,078
40-44	0.0090	0.0440	0.0391	0.0537	70,501	63,804	74,352	3,103	344,748	29.2	27.1	30.9	1,086	120,707
45-49	0.0120	0.0583	0.0519	0.0704	67,398	60,378	71,446	3,926	327,176	25.4	23.4	27.1	1,041	86,720
50-54	0.0166	0.0797	0.0705	0.0951	63,472	56,127	67,736	5,062	304,706	21.8	19.9	23.5	1,233	74,205
55-59	0.0240	0.1132	0.0994	0.1346	58,410	50,787	62,960	6,612	275,521	18.5	16.7	20.1	1,176	49,019
60-64	0.0320	0.1481	0.1283	0.1772	51,798	43,953	56,703	7,671	239,812	15.5	13.9	17.0	1,681	52,555
65-69	0.0460	0.2063	0.1772	0.2464	44,127	36,164	49,427	9,102	197,878	12.8	11.4	14.2	1,434	31,181
70-74	0.0640	0.2759	0.2365	0.3260	35,024	27,254	40,670	9,662	150,967	10.5	9.3	11.7	1,931	30,175
75-79	0.0800	0.3333	0.2869	0.3880	25,363	18,368	31,054	8,454	105,677	8.5	7.6	9.5	1,071	13,389
80-84	0.1200	0.4615	0.4049	0.5221	16,908	11,240	22,144	7,804	65,032	6.5	5.9	7.3	997	8,309
85-89	0.1674	0.5900	0.5305	0.6481	9,104	5,372	13,179	5,371	32,094	5.0	4.5	5.6	388	2,320
90-94	0.2493	0.7132	0.6580	0.7637	3,733	1,890	6,187	2,662	10,678	3.6	3.3	4.0	126	504
95-99	0.3570	0.8619	0.8204	0.8967	1,071	447	2,116	923	2,585	2.7	2.5	2.9	38	106
100+	0.4914	1	1	1	148	46	380	148	301	2.0	1.9	2.1	2	4
All ages													*35,836*	*2,635,736*

ln($_nq_x$) with uncertainty interval

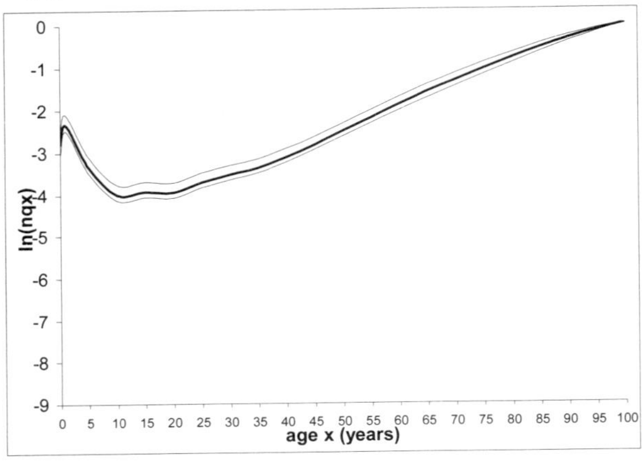

l_x with uncertainty interval

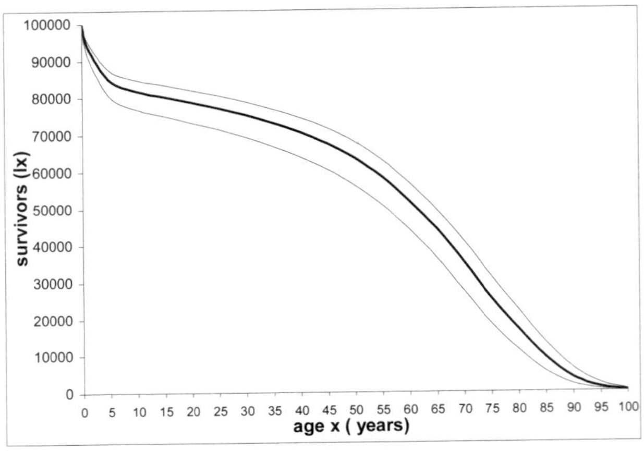

e_x with uncertainty interval

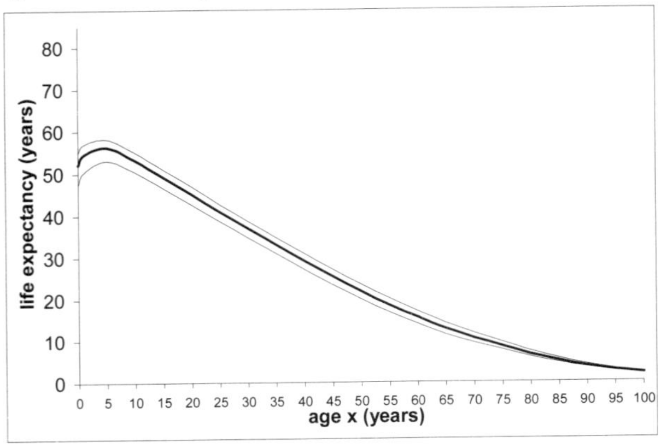

Distribution of deaths by age with uncertainty interval

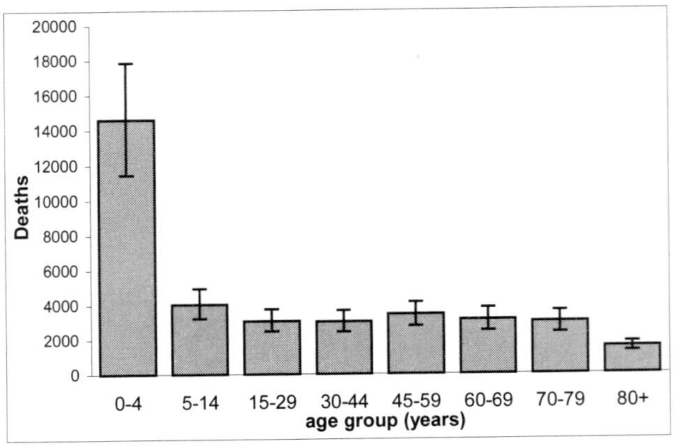

Lao People's Democratic Republic | République démocratique populaire lao | República Democrática Popular Lao

Females, 2000 | *Femmes, 2000* | *Mujeres, 2000*

Age	$_nM_x$	$_nq_x$			l_x			$_nd_x$	$_nL_x$	e_x			Actual deaths	Actual population
			percentiles			percentiles					percentiles			
		mean	2½	97½	mean	2½	97½			mean	2½	97½		
<1	0.0461	0.0443	0.0348	0.0646	100,000	100,000	100,000	4,431	96,012	56.1	52.1	58.4	4,116	89,195
1-4	0.0248	0.0937	0.0835	0.1163	95,569	93,544	96,524	8,950	360,796	57.6	54.2	59.9	7,969	321,214
5-9	0.0069	0.0337	0.0300	0.0410	86,619	82,666	88,463	2,921	425,793	59.4	56.9	61.5	2,513	366,432
10-14	0.0033	0.0162	0.0143	0.0195	83,698	79,281	85,812	1,353	415,109	56.4	54.0	58.4	1,081	331,750
15-19	0.0040	0.0199	0.0177	0.0239	82,345	77,734	84,581	1,643	407,620	52.3	50.0	54.3	1,102	273,401
20-24	0.0035	0.0173	0.0153	0.0206	80,703	75,875	83,085	1,395	400,026	48.3	46.1	50.4	812	232,847
25-29	0.0042	0.0205	0.0182	0.0244	79,308	74,309	81,810	1,629	392,466	44.1	42.0	46.2	824	198,454
30-34	0.0048	0.0239	0.0212	0.0282	77,679	72,498	80,319	1,855	383,756	40.0	37.9	42.0	830	171,663
35-39	0.0061	0.0302	0.0268	0.0353	75,824	70,455	78,616	2,287	373,401	35.9	33.9	37.9	885	144,474
40-44	0.0076	0.0371	0.0330	0.0432	73,537	67,967	76,509	2,728	360,862	32.0	30.0	33.9	947	125,212
45-49	0.0103	0.0501	0.0444	0.0586	70,808	65,027	73,982	3,546	345,178	28.1	26.2	30.0	951	92,570
50-54	0.0127	0.0617	0.0542	0.0724	67,263	61,219	70,700	4,148	325,945	24.4	22.7	26.3	1,005	78,978
55-59	0.0177	0.0849	0.0736	0.0997	63,115	56,789	66,866	5,358	302,180	20.9	19.2	22.7	1,051	59,268
60-64	0.0255	0.1197	0.1023	0.1409	57,757	51,128	61,944	6,913	271,502	17.6	16.1	19.3	1,504	59,072
65-69	0.0360	0.1653	0.1397	0.1946	50,844	43,927	55,606	8,403	233,210	14.6	13.3	16.2	1,254	34,807
70-74	0.0500	0.2222	0.1868	0.2599	42,440	35,380	47,839	9,431	188,624	12.0	10.9	13.4	1,734	34,680
75-79	0.0650	0.2796	0.2353	0.3231	33,009	26,184	38,905	9,228	141,975	9.8	8.9	10.9	946	14,556
80-84	0.0950	0.3838	0.3279	0.4338	23,781	17,725	29,750	9,128	96,084	7.6	6.9	8.5	915	9,634
85-89	0.1400	0.5185	0.4549	0.5706	14,653	10,035	19,995	7,598	54,270	5.8	5.3	6.5	491	3,505
90-94	0.2000	0.6250	0.5627	0.6727	7,055	4,309	10,900	4,409	22,047	4.3	3.9	4.8	167	837
95-99	0.3000	0.7895	0.7366	0.8271	2,646	1,410	4,766	2,089	6,962	3.1	2.9	3.4	79	264
100+	0.4500	1	1	1	557	244	1,255	557	1,238	2.2	2.1	2.4	6	14
All ages													*31,182*	*2,642,827*

$ln(_nq_x)$ with uncertainty interval

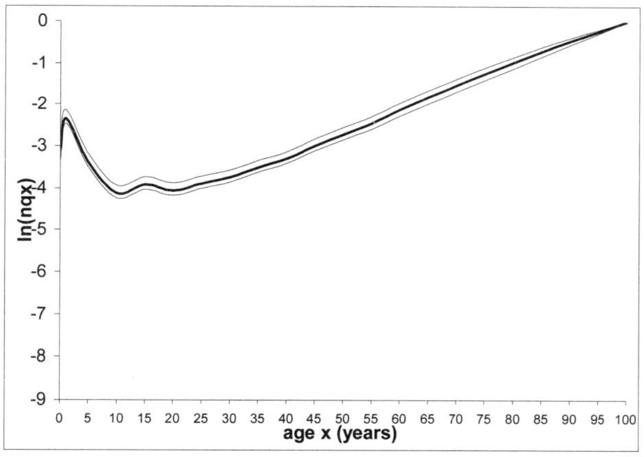

l_x with uncertainty interval

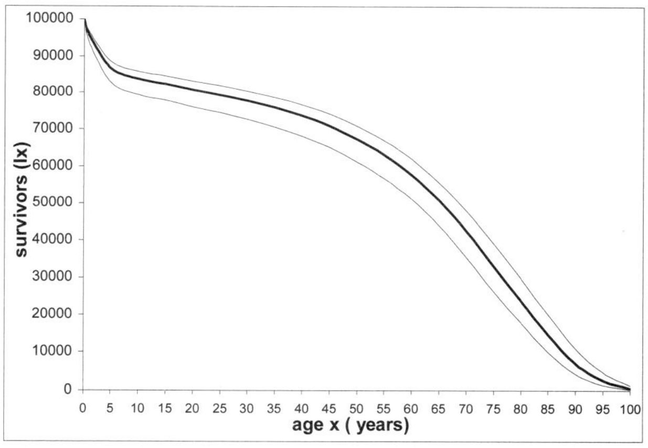

e_x with uncertainty interval

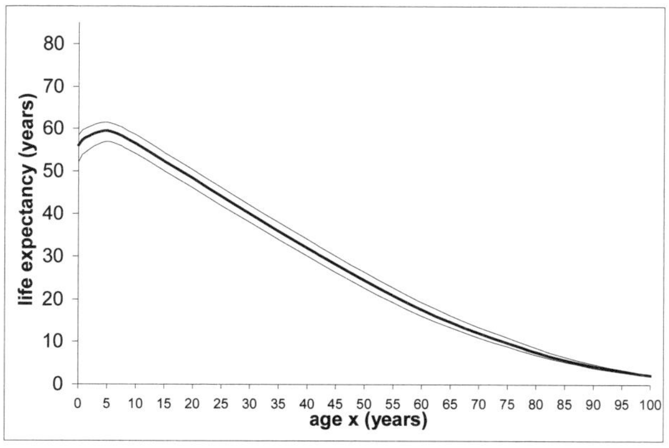

Distribution of deaths by age with uncertainty interval

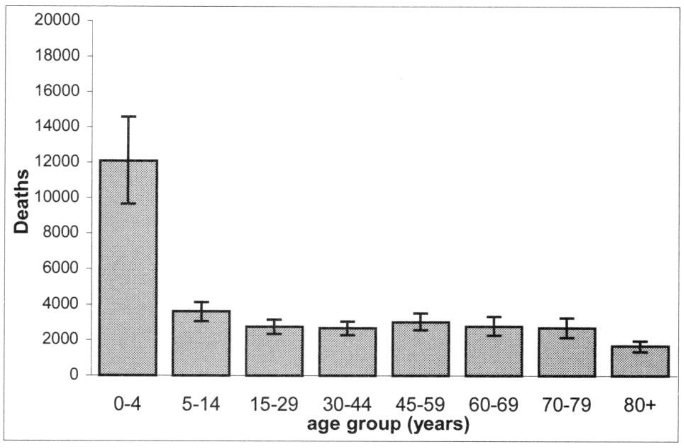

Latvia — Lettonie — Letonia

Males, 2000 — *Hommes, 2000* — *Hombres, 2000*

Age	$_nM_x$	$_nq_x$ mean	$_nq_x$ percentile 2½	$_nq_x$ percentile 97½	l_x mean	l_x percentile 2½	l_x percentile 97½	$_nd_x$	$_nL_x$	e_x mean	e_x percentile 2½	e_x percentile 97½	Actual deaths	Actual population
<1	0.0167	0.0165	0.0115	0.0207	100,000	100,000	100,000	1,647	98,518	64.2	62.9	64.9	161	9,660
1-4	0.0009	0.0036	0.0028	0.0044	98,353	97,926	98,848	356	392,559	64.3	63.0	65.0	35	38,164
5-9	0.0008	0.0038	0.0029	0.0045	97,998	97,500	98,575	369	489,064	60.5	59.3	61.2	54	71,148
10-14	0.0006	0.0029	0.0022	0.0034	97,628	97,061	98,287	278	487,446	55.7	54.5	56.5	55	96,359
15-19	0.0016	0.0080	0.0064	0.0094	97,350	96,733	98,066	781	484,796	50.9	49.7	51.6	150	93,055
20-24	0.0037	0.0184	0.0152	0.0214	96,569	95,819	97,437	1,774	478,407	46.2	45.1	47.1	305	82,302
25-29	0.0040	0.0197	0.0168	0.0226	94,794	93,773	95,956	1,867	469,304	42.1	40.9	42.9	343	86,099
30-34	0.0043	0.0212	0.0186	0.0241	92,927	91,654	94,340	1,973	459,704	37.9	36.8	38.7	375	87,485
35-39	0.0060	0.0297	0.0267	0.0333	90,955	89,445	92,581	2,698	448,027	33.6	32.6	34.5	539	89,465
40-44	0.0080	0.0394	0.0365	0.0437	88,256	86,466	90,105	3,481	432,579	29.6	28.6	30.5	705	87,605
45-49	0.0112	0.0545	0.0518	0.0600	84,775	82,684	86,812	4,623	412,318	25.7	24.7	26.6	849	75,720
50-54	0.0176	0.0842	0.0790	0.0926	80,152	77,723	82,314	6,748	383,889	22.0	21.1	22.9	1,161	66,040
55-59	0.0230	0.1088	0.1008	0.1194	73,404	70,523	75,813	7,984	347,057	18.8	17.9	19.7	1,412	61,386
60-64	0.0307	0.1426	0.1311	0.1562	65,419	62,100	68,169	9,330	303,772	15.8	14.9	16.6	1,917	62,417
65-69	0.0425	0.1922	0.1753	0.2124	56,089	52,398	59,231	10,780	253,496	13.0	12.2	13.8	1,987	46,732
70-74	0.0555	0.2438	0.2219	0.2702	45,309	41,271	48,846	11,045	198,932	10.5	9.9	11.2	1,938	34,905
75-79	0.0828	0.3431	0.3142	0.3774	34,264	30,120	38,010	11,754	141,934	8.1	7.6	8.6	1,353	16,343
80-84	0.1296	0.4895	0.4553	0.5288	22,510	18,753	26,067	11,018	85,003	6.1	5.7	6.5	842	6,497
85-89	0.1946	0.6545	0.6215	0.6910	11,492	8,836	14,199	7,521	38,656	4.5	4.2	4.8	680	3,496
90-94	0.2801	0.7610	0.7335	0.7909	3,971	2,731	5,374	3,022	10,789	3.3	3.1	3.5	259	926
95-99	0.3867	0.8950	0.8770	0.9139	949	571	1,432	849	2,197	2.5	2.4	2.6	50	130
100+	0.5119	1	1	1	100	49	176	100	195	2.0	1.9	2.0	5	9
All ages													*15,176*	*1,115,943*

ln($_nq_x$) with uncertainty interval

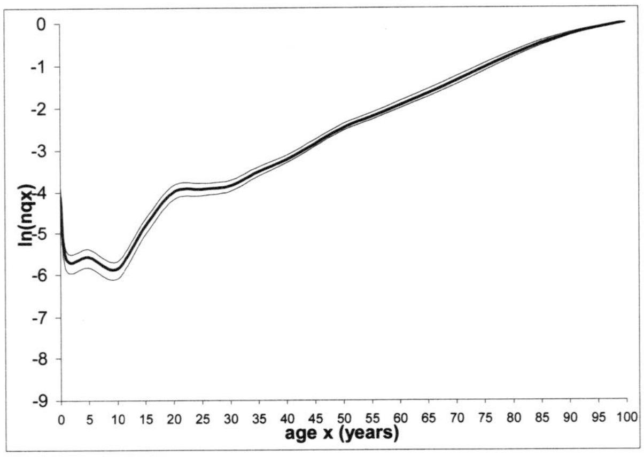

l_x with uncertainty interval

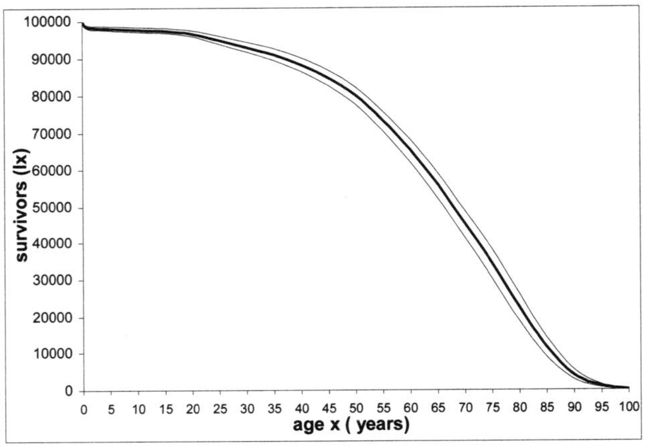

e_x with uncertainty interval

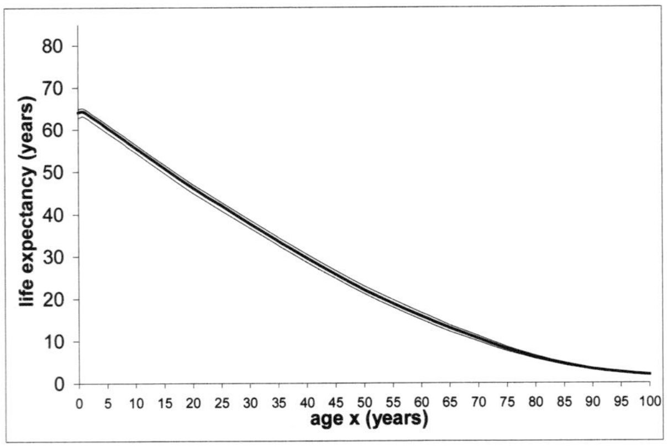

Distribution of deaths by age with uncertainty interval

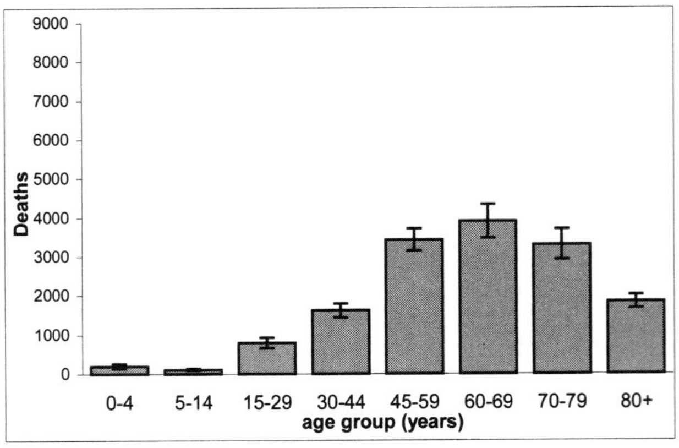

Latvia — Lettonie — Letonia

Females, 2000 — *Femmes, 2000* — *Mujeres, 2000*

Age	$_nM_x$	$_nq_x$			l_x			$_nd_x$	$_nL_x$	e_x			Actual deaths	Actual population
			percentiles			percentiles					percentiles			
		mean	2½	97½	mean	2½	97½			mean	2½	97½		
<1	0.0123	0.0121	0.0093	0.0137	100,000	100,000	100,000	1,212	98,909	75.5	74.5	76.5	112	9,169
1-4	0.0005	0.0021	0.0017	0.0023	98,788	98,625	99,068	207	394,654	75.5	74.4	76.4	19	36,335
5-9	0.0003	0.0014	0.0012	0.0016	98,581	98,394	98,897	142	492,548	71.6	70.6	72.6	20	68,035
10-14	0.0002	0.0010	0.0008	0.0011	98,439	98,235	98,779	96	491,953	66.7	65.7	67.7	18	92,069
15-19	0.0006	0.0029	0.0025	0.0033	98,343	98,128	98,698	290	490,988	61.8	60.8	62.7	53	89,912
20-24	0.0008	0.0041	0.0035	0.0046	98,053	97,804	98,453	399	489,267	57.0	56.0	57.9	66	80,582
25-29	0.0008	0.0040	0.0034	0.0044	97,654	97,358	98,113	387	487,303	52.2	51.2	53.1	66	83,478
30-34	0.0011	0.0057	0.0050	0.0064	97,267	96,926	97,779	556	484,945	47.4	46.4	48.3	95	83,094
35-39	0.0016	0.0079	0.0069	0.0088	96,711	96,306	97,294	760	481,653	42.6	41.7	43.6	145	92,065
40-44	0.0027	0.0132	0.0118	0.0148	95,951	95,461	96,623	1,270	476,579	38.0	37.0	38.9	251	94,032
45-49	0.0041	0.0201	0.0182	0.0223	94,681	94,051	95,482	1,899	468,656	33.4	32.5	34.4	346	85,414
50-54	0.0060	0.0295	0.0269	0.0327	92,782	91,950	93,742	2,736	457,068	29.1	28.2	30.0	471	78,625
55-59	0.0084	0.0411	0.0374	0.0457	90,046	88,941	91,222	3,701	440,976	24.9	24.1	25.7	654	77,942
60-64	0.0120	0.0583	0.0528	0.0649	86,345	84,873	87,811	5,030	419,150	20.8	20.1	21.7	1,025	85,395
65-69	0.0176	0.0843	0.0758	0.0939	81,315	79,361	83,173	6,858	389,430	17.0	16.3	17.7	1,278	72,585
70-74	0.0292	0.1360	0.1217	0.1506	74,457	71,906	76,868	10,128	346,965	13.3	12.7	14.0	2,049	70,191
75-79	0.0524	0.2317	0.2082	0.2538	64,329	61,077	67,511	14,905	284,382	10.0	9.5	10.6	2,712	51,740
80-84	0.0940	0.3805	0.3480	0.4124	49,424	45,575	53,452	18,804	200,109	7.2	6.8	7.7	2,337	24,869
85-89	0.1578	0.5657	0.5326	0.6007	30,620	26,781	34,851	17,322	109,794	5.2	4.9	5.4	3,051	19,336
90-94	0.2480	0.7111	0.6862	0.7400	13,298	10,694	16,290	9,456	38,121	3.6	3.5	3.8	1,927	7,769
95-99	0.3652	0.8713	0.8569	0.8893	3,842	2,781	5,113	3,348	9,167	2.6	2.6	2.7	662	1,814
100+	0.5035	1	1	1	494	308	732	494	982	2.0	2.0	2.1	77	152
All ages													*17,433*	*1,304,603*

ln($_nq_x$) with uncertainty interval

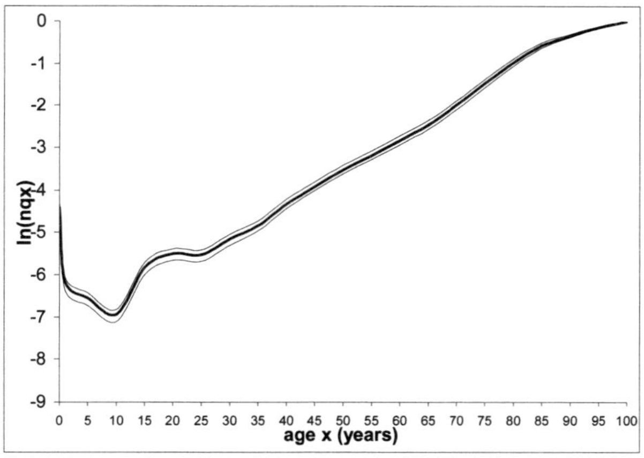

l_x with uncertainty interval

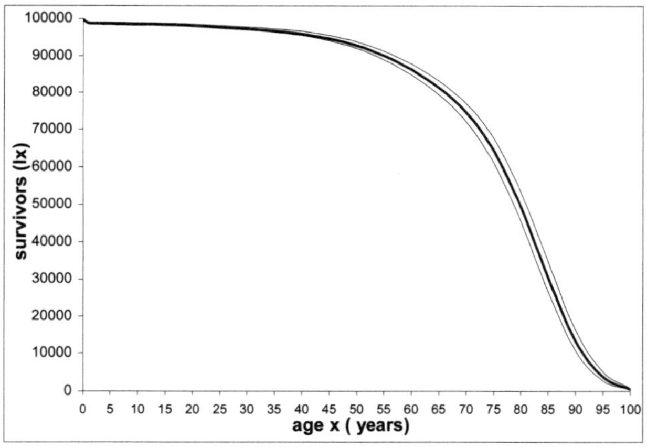

e_x with uncertainty interval

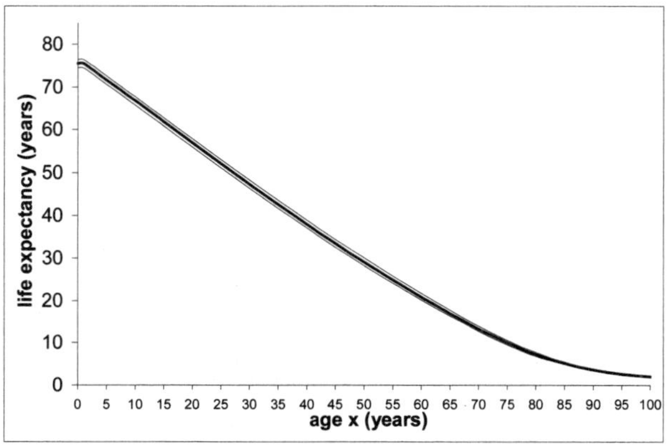

Distribution of deaths by age with uncertainty interval

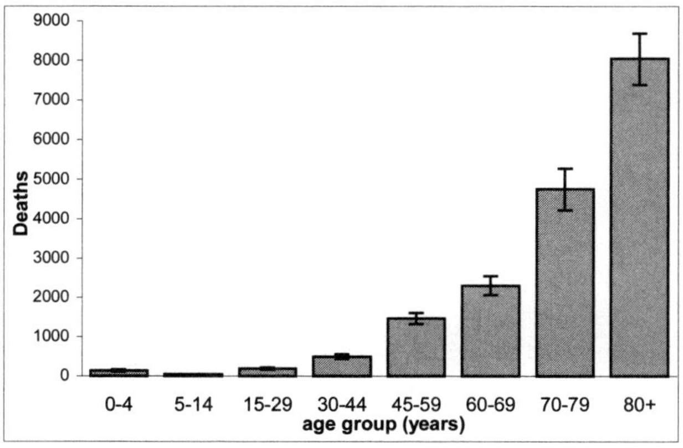

Lebanon | Liban | Líbano

Males, 2000 | *Hommes, 2000* | *Hombres, 2000*

Age	$_nM_x$	$_nq_x$ mean	$_nq_x$ percentiles 2½	$_nq_x$ percentiles 97½	l_x mean	l_x percentiles 2½	l_x percentiles 97½	$_nd_x$	$_nL_x$	e_x mean	e_x percentiles 2½	e_x percentiles 97½	Actual deaths	Actual population
<1	0.0183	0.0180	0.0137	0.0223	100,000	100,000	100,000	1,804	98,377	69.1	67.8	70.6	622	33,931
1-4	0.0010	0.0040	0.0032	0.0048	98,196	97,769	98,626	396	391,834	69.4	68.1	70.7	139	137,459
5-9	0.0005	0.0024	0.0019	0.0028	97,800	97,308	98,293	231	488,423	65.6	64.4	67.0	92	195,011
10-14	0.0004	0.0022	0.0018	0.0026	97,569	97,035	98,099	217	487,303	60.8	59.6	62.1	84	188,846
15-19	0.0012	0.0058	0.0048	0.0069	97,352	96,779	97,922	568	485,339	55.9	54.7	57.2	200	170,657
20-24	0.0016	0.0082	0.0068	0.0096	96,784	96,112	97,454	791	481,941	51.2	50.0	52.5	261	158,853
25-29	0.0017	0.0083	0.0069	0.0097	95,993	95,191	96,796	799	477,966	46.6	45.5	47.9	281	168,307
30-34	0.0020	0.0097	0.0082	0.0113	95,194	94,268	96,124	927	473,652	42.0	40.9	43.2	316	161,679
35-39	0.0026	0.0129	0.0110	0.0148	94,267	93,205	95,337	1,216	468,296	37.4	36.3	38.6	303	116,866
40-44	0.0038	0.0189	0.0163	0.0216	93,051	91,823	94,292	1,762	460,849	32.9	31.8	34.0	302	78,930
45-49	0.0059	0.0291	0.0252	0.0329	91,289	89,840	92,758	2,658	449,798	28.4	27.4	29.6	402	67,945
50-54	0.0093	0.0455	0.0396	0.0512	88,631	86,890	90,405	4,030	433,078	24.2	23.3	25.3	504	54,136
55-59	0.0146	0.0703	0.0617	0.0792	84,601	82,495	86,760	5,949	408,131	20.3	19.4	21.3	598	41,015
60-64	0.0229	0.1081	0.0954	0.1214	78,652	76,147	81,274	8,503	372,003	16.6	15.8	17.5	922	40,319
65-69	0.0353	0.1621	0.1434	0.1804	70,149	67,184	73,428	11,373	322,313	13.3	12.5	14.2	1,363	38,622
70-74	0.0551	0.2422	0.2152	0.2685	58,776	55,066	62,644	14,233	258,298	10.4	9.7	11.2	1,674	30,370
75-79	0.0858	0.3531	0.3166	0.3905	44,543	40,466	48,862	15,729	183,391	7.9	7.4	8.5	1,473	17,180
80-84	0.1349	0.5044	0.4626	0.5474	28,814	25,164	33,075	14,535	107,732	5.9	5.5	6.3	900	6,672
85-89	0.2045	0.6766	0.6390	0.7149	14,279	11,641	17,586	9,661	47,242	4.3	4.1	4.6	660	3,226
90-94	0.2987	0.7877	0.7590	0.8173	4,618	3,319	6,348	3,637	12,176	3.1	3.0	3.3	204	683
95-99	0.4205	0.9297	0.9144	0.9447	980	607	1,530	911	2,167	2.3	2.3	2.4	31	73
100+	0.5705	1	1	1	69	34	131	69	121	1.8	1.7	1.8	2	4
All ages													*11,332*	*1,710,784*

ln($_nq_x$) with uncertainty interval

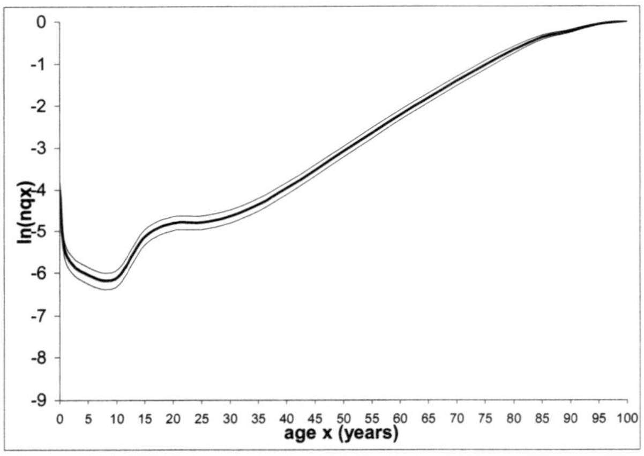

l_x with uncertainty interval

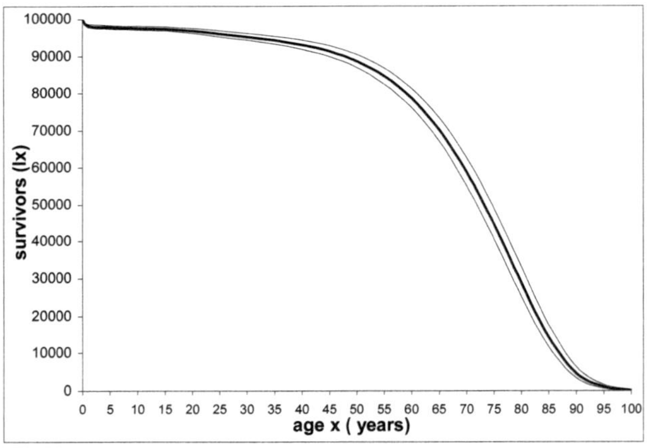

e_x with uncertainty interval

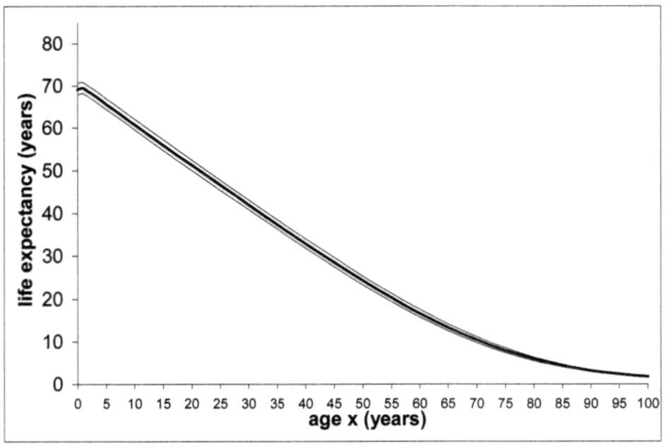

Distribution of deaths by age with uncertainty interval

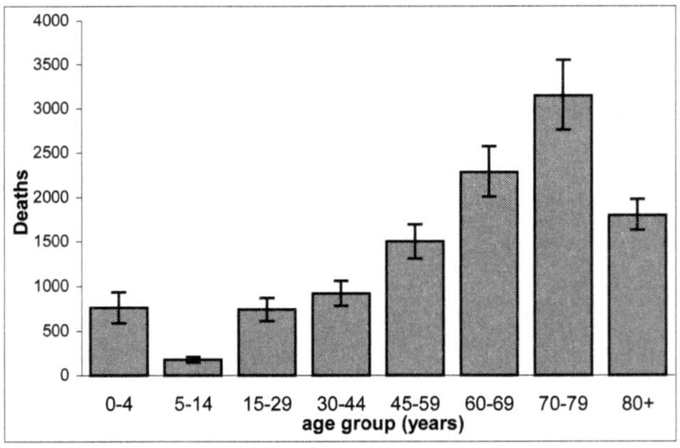

Lebanon   Liban   Líbano

Females, 2000   *Femmes, 2000*   *Mujeres, 2000*

Age	$_nM_x$	$_nq_x$			l_x			$_nd_x$	$_nL_x$	e_x			Actual deaths	Actual population
			percentiles			percentiles					percentiles			
		mean	2½	97½	mean	2½	97½			mean	2½	97½		
<1	0.0141	0.0140	0.0103	0.0176	100,000	100,000	100,000	1,395	98,744	73.3	72.3	74.6	458	32,408
1-4	0.0010	0.0041	0.0033	0.0049	98,605	98,235	98,967	405	393,448	73.4	72.4	74.6	135	131,528
5-9	0.0004	0.0021	0.0017	0.0025	98,200	97,755	98,644	204	490,490	69.7	68.7	70.9	78	187,348
10-14	0.0004	0.0019	0.0015	0.0022	97,996	97,514	98,478	183	489,521	64.8	63.9	66.1	68	182,282
15-19	0.0007	0.0035	0.0029	0.0041	97,812	97,299	98,329	343	488,206	59.9	59.0	61.2	117	166,920
20-24	0.0009	0.0043	0.0035	0.0050	97,470	96,897	98,047	419	486,301	55.1	54.2	56.4	135	157,138
25-29	0.0010	0.0049	0.0041	0.0058	97,051	96,410	97,698	480	484,054	50.4	49.5	51.6	164	165,302
30-34	0.0013	0.0065	0.0055	0.0076	96,571	95,856	97,297	631	481,275	45.6	44.7	46.8	208	158,453
35-39	0.0019	0.0095	0.0081	0.0109	95,939	95,134	96,764	910	477,422	40.9	40.0	42.1	252	132,304
40-44	0.0029	0.0143	0.0124	0.0163	95,029	94,104	95,984	1,362	471,743	36.3	35.4	37.4	326	113,102
45-49	0.0044	0.0216	0.0190	0.0243	93,668	92,583	94,798	2,026	463,275	31.7	31.0	32.9	374	85,515
50-54	0.0066	0.0322	0.0285	0.0359	91,642	90,341	92,996	2,954	450,826	27.4	26.6	28.5	404	61,699
55-59	0.0097	0.0472	0.0417	0.0523	88,688	87,127	90,302	4,184	432,979	23.2	22.5	24.2	484	50,101
60-64	0.0148	0.0715	0.0628	0.0789	84,504	82,665	86,381	6,040	407,417	19.2	18.6	20.2	692	46,672
65-69	0.0232	0.1096	0.0963	0.1200	78,463	76,290	80,657	8,598	370,822	15.5	15.0	16.4	1,006	43,406
70-74	0.0383	0.1748	0.1546	0.1898	69,866	67,255	72,756	12,215	318,791	12.1	11.6	12.8	1,347	35,161
75-79	0.0636	0.2743	0.2465	0.2972	57,651	54,534	61,509	15,812	248,724	9.2	8.7	9.7	1,357	21,343
80-84	0.1078	0.4247	0.3925	0.4572	41,839	38,639	46,346	17,768	164,773	6.7	6.3	7.1	970	8,995
85-89	0.1741	0.6066	0.5742	0.6402	24,070	21,449	28,157	14,601	83,850	4.8	4.6	5.1	826	4,745
90-94	0.2677	0.7423	0.7158	0.7691	9,470	7,786	11,857	7,029	26,261	3.4	3.3	3.6	300	1,122
95-99	0.3917	0.9004	0.8848	0.9156	2,440	1,798	3,274	2,197	5,610	2.5	2.4	2.6	59	150
100+	0.5456	1	1	1	243	152	373	243	445	1.8	1.8	1.9	6	11
All ages													*9,769*	*1,785,705*

ln($_nq_x$) with uncertainty interval

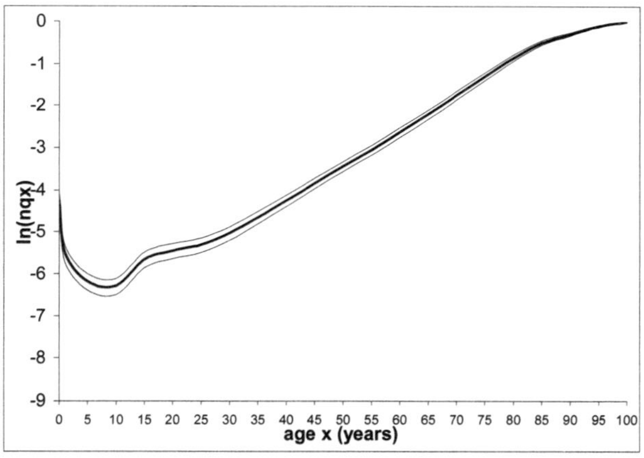

l_x with uncertainty interval

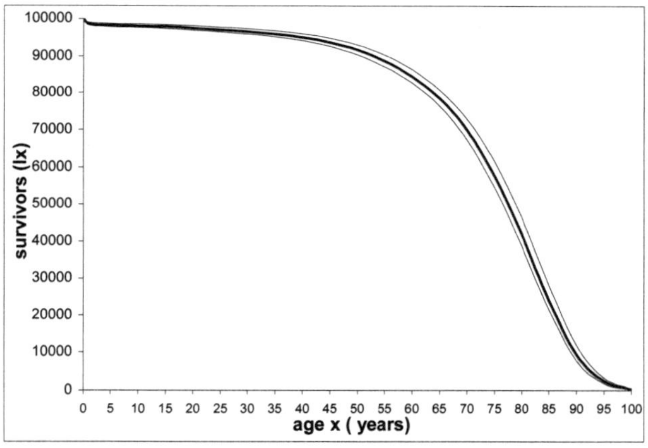

e_x with uncertainty interval

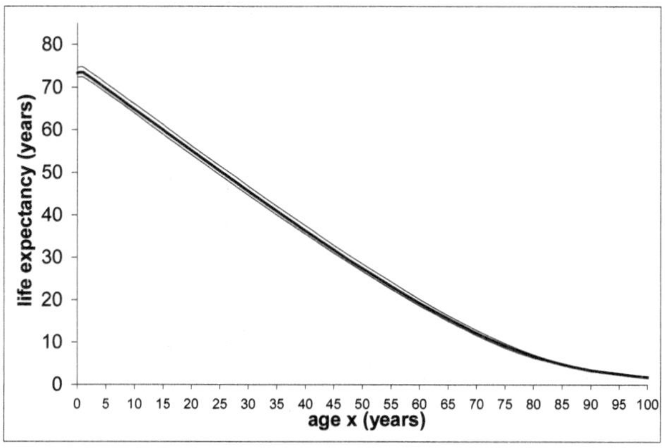

Distribution of deaths by age with uncertainty interval

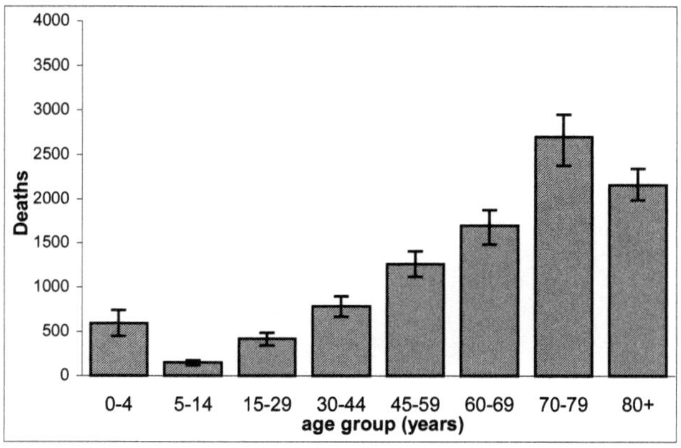

Males, 2000 *Hommes, 2000* *Hombres, 2000*

Age	$_nM_x$	$_nq_x$			l_x			$_nd_x$	$_nL_x$	e_x			Actual deaths	Actual population
			percentiles			percentiles					percentiles			
		mean	2½	97½	mean	2½	97½			mean	2½	97½		
<1	0.1265	0.1136	0.0977	0.1316	100,000	100,000	100,000	11,360	89,776	42.0	37.6	46.4	4,034	31,880
1-4	0.0131	0.0509	0.0423	0.0608	88,640	86,840	90,226	4,508	343,742	46.4	41.5	51.2	1,523	116,144
5-9	0.0025	0.0123	0.0100	0.0152	84,132	82,635	85,617	1,036	418,071	44.8	40.0	49.5	332	134,005
10-14	0.0017	0.0082	0.0067	0.0102	83,096	81,485	84,677	685	413,769	40.3	35.6	45.0	201	121,773
15-19	0.0029	0.0145	0.0117	0.0182	82,412	80,726	84,054	1,199	409,061	35.6	31.0	40.3	318	108,640
20-24	0.0078	0.0384	0.0307	0.0486	81,213	79,364	83,066	3,119	398,267	31.1	26.5	35.7	737	94,063
25-29	0.0186	0.0887	0.0701	0.1142	78,094	75,653	80,514	6,926	373,155	27.2	22.7	31.8	1,449	78,058
30-34	0.0307	0.1426	0.1109	0.1861	71,168	67,061	74,868	10,147	330,473	24.6	20.3	29.0	1,994	64,952
35-39	0.0336	0.1549	0.1188	0.2036	61,021	54,578	66,513	9,454	281,469	23.3	19.3	27.3	1,875	55,819
40-44	0.0334	0.1543	0.1176	0.2020	51,567	43,468	58,494	7,955	237,946	22.1	18.6	25.6	1,550	46,365
45-49	0.0329	0.1519	0.1159	0.1988	43,612	34,686	51,465	6,623	201,503	20.7	17.6	23.8	1,230	37,437
50-54	0.0309	0.1433	0.1098	0.1871	36,989	27,845	45,343	5,300	171,696	19.0	16.4	21.6	1,027	33,266
55-59	0.0288	0.1344	0.1035	0.1748	31,689	22,710	40,315	4,260	147,795	16.7	14.6	18.9	804	27,894
60-64	0.0371	0.1698	0.1321	0.2176	27,429	18,816	36,141	4,659	125,497	13.9	12.1	15.8	822	22,150
65-69	0.0507	0.2248	0.1774	0.2823	22,770	14,809	31,353	5,119	101,053	11.3	9.8	12.9	819	16,163
70-74	0.0749	0.3153	0.2547	0.3844	17,651	10,728	25,751	5,566	74,340	8.8	7.7	10.1	817	10,914
75-79	0.1120	0.4376	0.3655	0.5142	12,085	6,703	19,116	5,288	47,205	6.7	5.9	7.7	721	6,436
80-84	0.1694	0.5949	0.5185	0.6703	6,797	3,296	12,028	4,043	23,876	5.0	4.4	5.7	407	2,405
85-89	0.2472	0.7639	0.6969	0.8252	2,753	1,087	5,697	2,103	8,509	3.7	3.4	4.2	143	578
90-94	0.3475	0.8507	0.7968	0.8978	650	190	1,706	553	1,591	2.8	2.5	3.1	23	66
95-99	0.4708	0.9758	0.9565	0.9884	97	19	347	95	201	2.1	2.1	2.2	2	4
100+	0.6142	1	1	1	2	0	15	2	4	1.6	1.6	1.7	0	0
All ages													*20,829*	*1,009,012*

ln($_nq_x$) with uncertainty interval

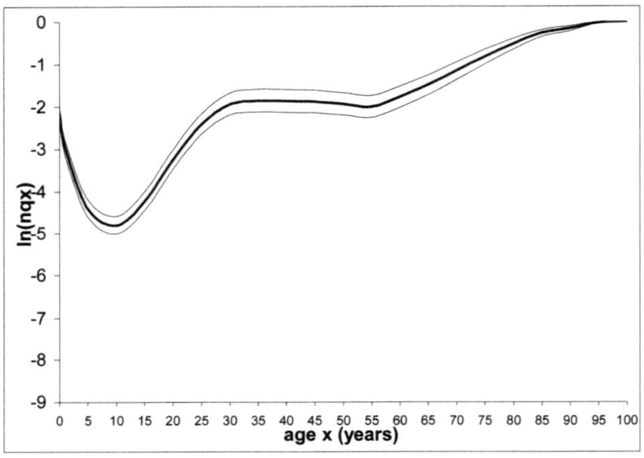

l_x with uncertainty interval

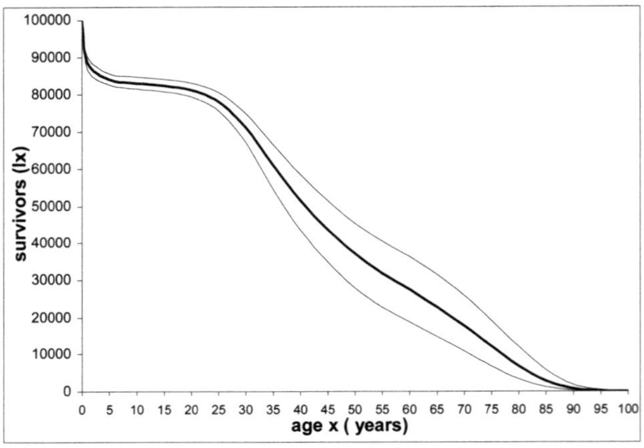

e_x with uncertainty interval

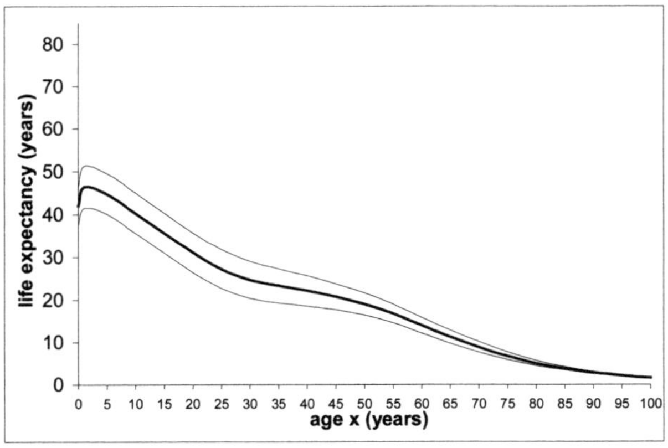

Distribution of deaths by age with uncertainty interval

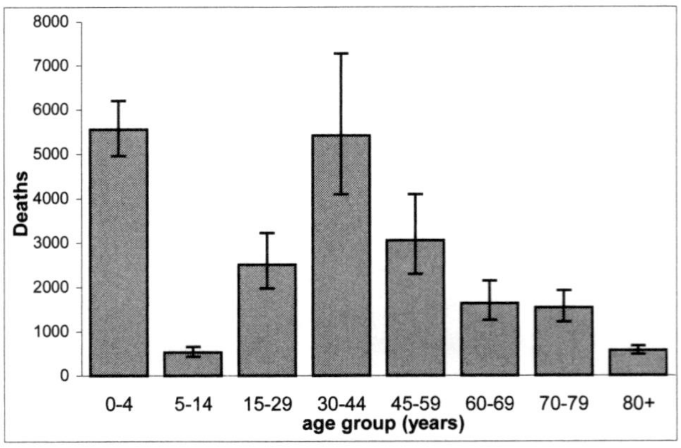

Lesotho **Lesotho** **Lesotho**

Females, 2000 *Femmes, 2000* *Mujeres, 2000*

Age	$_nM_x$	$_nq_x$			l_x			$_nd_x$	$_nL_x$	e_x			Actual deaths	Actual population
			percentiles			percentiles					percentiles			
		mean	21/2	971/2	mean	21/2	971/2			mean	21/2	971/2		
<1	0.1123	0.1020	0.0857	0.1180	100,000	100,000	100,000	10,199	90,821	42.2	37.7	46.9	3,442	30,654
1-4	0.0139	0.0536	0.0459	0.0630	89,801	88,205	91,432	4,817	347,644	46.0	40.9	51.1	1,579	113,982
5-9	0.0025	0.0124	0.0103	0.0150	84,984	83,486	86,484	1,055	422,283	44.5	39.6	49.5	327	130,948
10-14	0.0017	0.0084	0.0069	0.0102	83,929	82,354	85,530	703	417,889	40.1	35.1	45.0	201	119,273
15-19	0.0057	0.0283	0.0232	0.0347	83,226	81,570	84,912	2,355	410,243	35.4	30.5	40.3	615	107,035
20-24	0.0162	0.0779	0.0629	0.0974	80,871	78,902	82,841	6,303	388,596	31.3	26.5	36.3	1,502	92,627
25-29	0.0286	0.1336	0.1058	0.1701	74,568	71,370	77,401	9,964	347,928	28.8	24.0	33.6	2,195	76,629
30-34	0.0348	0.1601	0.1251	0.2055	64,604	59,345	69,001	10,343	297,159	27.8	23.3	32.4	2,252	64,702
35-39	0.0298	0.1386	0.1066	0.1796	54,260	47,147	60,368	7,522	252,496	27.7	23.7	31.7	1,701	57,100
40-44	0.0248	0.1167	0.0890	0.1521	46,738	38,725	53,876	5,456	220,051	26.7	23.3	30.2	1,207	48,693
45-49	0.0199	0.0947	0.0718	0.1237	41,282	32,910	49,005	3,909	196,638	24.9	22.0	27.9	805	40,512
50-54	0.0189	0.0902	0.0683	0.1177	37,373	28,912	45,409	3,371	178,437	22.2	19.8	24.9	702	37,172
55-59	0.0196	0.0934	0.0707	0.1214	34,002	25,585	42,225	3,174	162,074	19.2	17.1	21.5	623	31,804
60-64	0.0256	0.1202	0.0916	0.1548	30,828	22,559	39,146	3,704	144,878	15.9	14.1	18.0	661	25,841
65-69	0.0378	0.1725	0.1329	0.2188	27,123	19,164	35,438	4,680	123,918	12.8	11.2	14.5	763	20,205
70-74	0.0602	0.2616	0.2058	0.3232	22,444	15,097	30,547	5,872	97,539	9.9	8.6	11.4	865	14,375
75-79	0.0944	0.3818	0.3101	0.4554	16,572	10,350	24,015	6,328	67,041	7.5	6.5	8.7	847	8,977
80-84	0.1449	0.5319	0.4510	0.6084	10,244	5,666	16,517	5,449	37,599	5.6	4.9	6.5	566	3,905
85-89	0.2154	0.7001	0.6236	0.7672	4,795	2,233	9,030	3,357	15,584	4.2	3.7	4.7	226	1,051
90-94	0.3098	0.8028	0.7390	0.8566	1,438	520	3,397	1,154	3,726	3.0	2.7	3.4	49	157
95-99	0.4310	0.9398	0.9043	0.9648	284	75	886	266	618	2.3	2.2	2.5	6	13
100+	0.5797	1	1	1	17	3	85	17	29	1.7	1.7	1.9	0	0
All ages													*21,135*	*1,025,655*

$\ln(_nq_x)$ with uncertainty interval

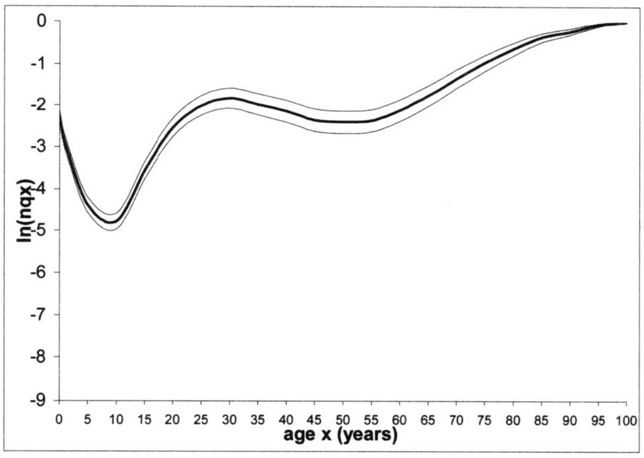

l_x with uncertainty interval

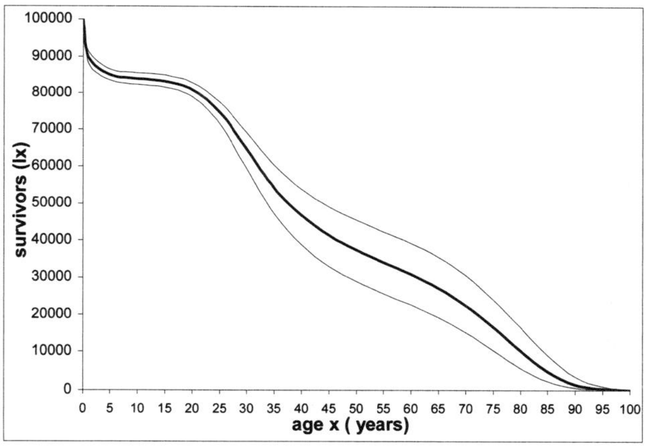

e_x with uncertainty interval

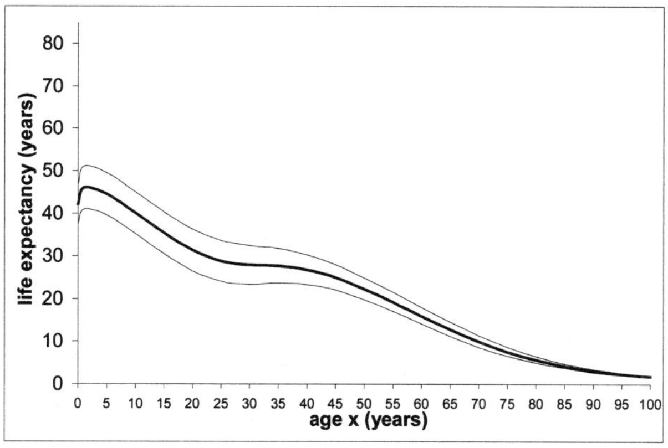

Distribution of deaths by age with uncertainty interval

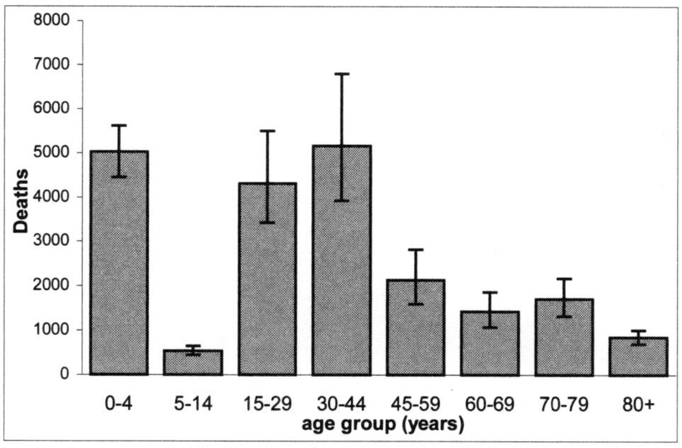

Liberia — Libéria — Liberia

Males, 2000 — *Hommes, 2000* — *Hombres, 2000*

Age	$_nM_x$	$_nq_x$			l_x			$_nd_x$	$_nL_x$	e_x			Actual deaths	Actual population
			percentiles			percentiles					percentiles			
		mean	2½	97½	mean	2½	97½			mean	2½	97½		
<1	0.1814	0.1560	0.1268	0.1900	100,000	100,000	100,000	15,596	85,963	46.6	42.8	50.9	10,038	55,328
1-4	0.0151	0.0584	0.0477	0.0702	84,404	80,997	87,322	4,928	325,788	54.3	50.3	58.4	3,346	221,224
5-9	0.0033	0.0165	0.0132	0.0202	79,476	76,484	82,466	1,310	394,106	53.5	49.9	57.5	553	166,447
10-14	0.0020	0.0097	0.0078	0.0120	78,166	74,978	81,317	760	388,932	49.4	45.9	53.3	353	180,868
15-19	0.0015	0.0073	0.0059	0.0091	77,406	74,101	80,665	568	385,612	44.8	41.4	48.7	285	193,561
20-24	0.0042	0.0209	0.0167	0.0259	76,838	73,447	80,177	1,607	380,174	40.1	36.8	44.0	727	171,971
25-29	0.0076	0.0371	0.0295	0.0462	75,231	71,600	78,793	2,793	369,174	35.9	32.7	39.8	1,009	133,393
30-34	0.0110	0.0535	0.0423	0.0670	72,438	68,411	76,372	3,876	352,501	32.2	29.1	35.9	944	85,851
35-39	0.0132	0.0641	0.0503	0.0807	68,562	64,022	72,979	4,392	331,830	28.9	26.0	32.4	731	55,205
40-44	0.0153	0.0739	0.0573	0.0934	64,170	59,031	69,081	4,742	308,994	25.7	23.0	29.0	638	41,541
45-49	0.0179	0.0858	0.0656	0.1087	59,428	53,738	65,024	5,101	284,385	22.6	20.1	25.6	628	35,036
50-54	0.0215	0.1022	0.0771	0.1301	54,326	47,961	60,634	5,551	257,755	19.5	17.2	22.3	743	34,496
55-59	0.0266	0.1247	0.0934	0.1589	48,776	41,786	55,747	6,082	228,673	16.4	14.4	18.9	785	29,522
60-64	0.0392	0.1785	0.1337	0.2252	42,694	35,268	50,228	7,619	194,421	13.4	11.7	15.6	924	23,586
65-69	0.0551	0.2420	0.1829	0.3001	35,075	27,520	43,040	8,487	154,156	10.7	9.4	12.6	936	17,002
70-74	0.0820	0.3403	0.2633	0.4097	26,588	19,503	34,589	9,047	110,322	8.4	7.3	9.9	886	10,800
75-79	0.1215	0.4660	0.3729	0.5409	17,541	11,649	25,345	8,174	67,268	6.4	5.6	7.6	705	5,804
80-84	0.1815	0.6242	0.5254	0.6948	9,366	5,358	15,839	5,846	32,216	4.8	4.3	5.6	448	2,471
85-89	0.2614	0.7905	0.7055	0.8450	3,520	1,645	7,439	2,782	10,644	3.6	3.3	4.1	215	824
90-94	0.3630	0.8689	0.8015	0.9097	738	259	2,142	641	1,765	2.7	2.5	3.0	70	192
95-99	0.4862	0.9888	0.9725	0.9950	97	24	420	96	197	2.1	2.0	2.1	12	24
100+	0.6277	1	1	1	1	0	12	1	2	1.6	1.6	1.7	2	3
All ages													*24,979*	*1,465,149*

ln($_nq_x$) with uncertainty interval

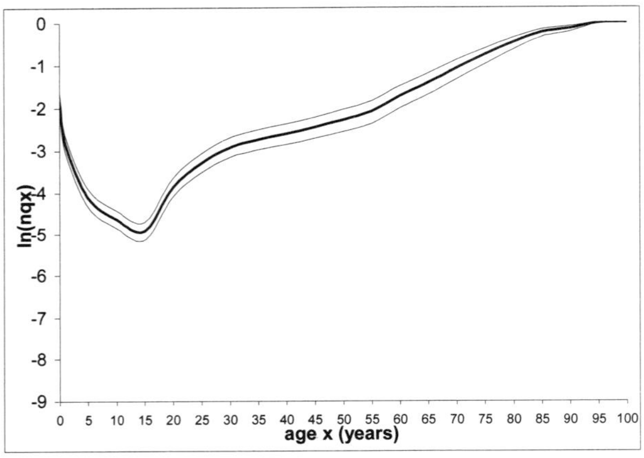

l_x with uncertainty interval

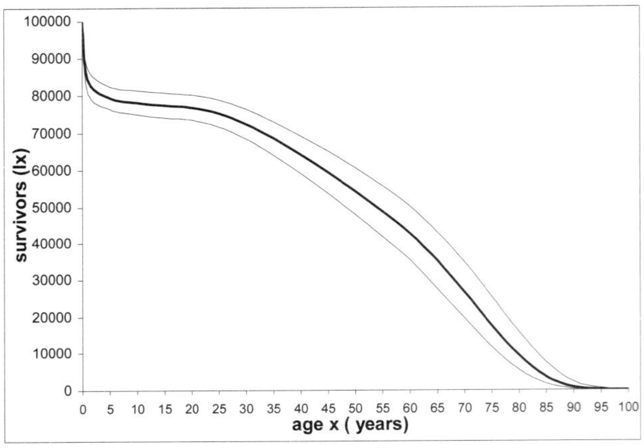

e_x with uncertainty interval

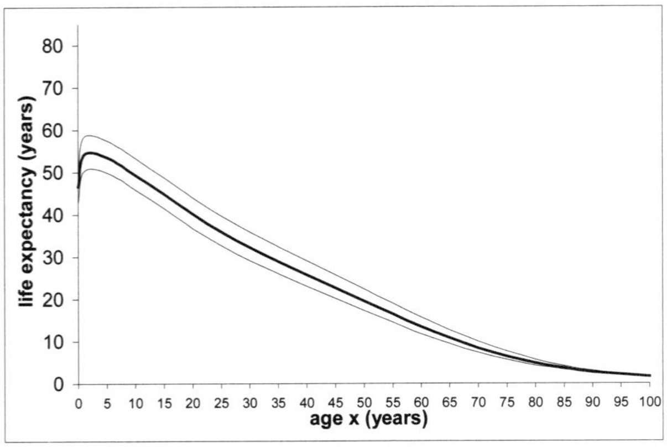

Distribution of deaths by age with uncertainty interval

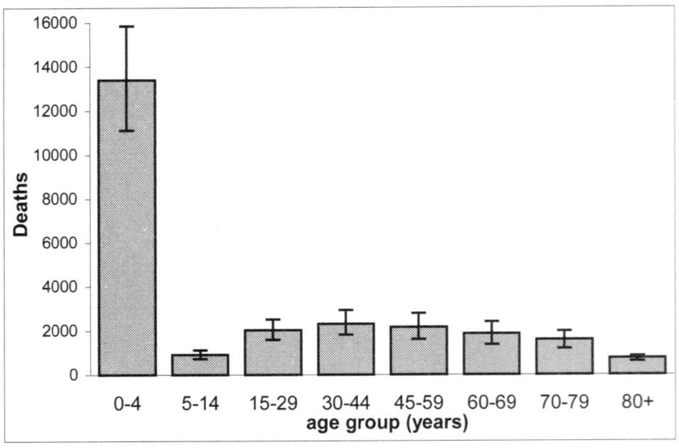

Liberia　　Libéria　　Liberia

Females, 2000　　*Femmes, 2000*　　*Mujeres, 2000*

Age	$_nM_x$	$_nq_x$			l_x			$_nd_x$	$_nL_x$	e_x			Actual deaths	Actual population
			percentiles			percentiles					percentiles			
		mean	2½	97½	mean	2½	97½			mean	2½	97½		
<1	0.1512	0.1331	0.1084	0.1585	100,000	100,000	100,000	13,312	88,019	49.1	45.4	53.3	8,046	53,200
1-4	0.0163	0.0626	0.0516	0.0737	86,688	84,148	89,160	5,429	333,724	55.7	51.6	60.1	3,588	220,574
5-9	0.0035	0.0172	0.0140	0.0207	81,260	78,764	83,751	1,397	402,805	55.3	51.6	59.4	578	166,766
10-14	0.0022	0.0111	0.0090	0.0134	79,863	77,213	82,546	885	397,102	51.2	47.6	55.2	399	179,274
15-19	0.0037	0.0181	0.0146	0.0220	78,978	76,219	81,803	1,430	391,316	46.8	43.2	50.7	705	192,992
20-24	0.0068	0.0334	0.0268	0.0410	77,548	74,589	80,601	2,594	381,257	42.6	39.1	46.4	1,155	169,724
25-29	0.0096	0.0468	0.0372	0.0579	74,954	71,644	78,413	3,511	365,994	39.0	35.7	42.6	1,235	128,751
30-34	0.0110	0.0535	0.0420	0.0666	71,443	67,688	75,436	3,822	347,660	35.8	32.7	39.1	879	79,971
35-39	0.0112	0.0545	0.0423	0.0684	67,621	63,401	72,167	3,686	328,889	32.6	29.8	35.8	561	50,098
40-44	0.0106	0.0516	0.0398	0.0650	63,935	59,266	68,980	3,297	311,432	29.4	26.7	32.2	401	37,906
45-49	0.0113	0.0551	0.0423	0.0695	60,638	55,523	66,096	3,341	294,836	25.8	23.4	28.5	371	32,701
50-54	0.0143	0.0688	0.0527	0.0868	57,297	51,661	63,137	3,944	276,623	22.2	19.9	24.7	483	33,896
55-59	0.0188	0.0900	0.0689	0.1141	53,353	47,177	59,592	4,802	254,759	18.6	16.6	20.9	576	30,539
60-64	0.0278	0.1301	0.1000	0.1652	48,551	41,843	55,249	6,316	226,963	15.2	13.4	17.3	709	25,485
65-69	0.0423	0.1913	0.1480	0.2415	42,235	35,066	49,724	8,080	190,973	12.1	10.6	14.0	828	19,565
70-74	0.0668	0.2861	0.2243	0.3538	34,154	26,853	42,326	9,771	146,345	9.4	8.1	11.0	897	13,428
75-79	0.1020	0.4065	0.3281	0.4859	24,384	17,704	32,682	9,911	97,141	7.2	6.2	8.4	794	7,779
80-84	0.1530	0.5534	0.4671	0.6331	14,473	9,298	21,703	8,009	52,341	5.4	4.7	6.3	545	3,563
85-89	0.2232	0.7163	0.6372	0.7833	6,463	3,411	11,289	4,630	20,743	4.1	3.6	4.6	291	1,302
90-94	0.3165	0.8117	0.7471	0.8644	1,834	739	4,091	1,489	4,704	3.0	2.7	3.4	106	336
95-99	0.4363	0.9448	0.9103	0.9683	345	100	1,035	326	748	2.3	2.1	2.4	25	57
100+	0.5848	1	1	1	19	3	93	19	33	1.7	1.7	1.8	5	8
All ages													*23,177*	*1,447,915*

ln($_nq_x$) with uncertainty interval

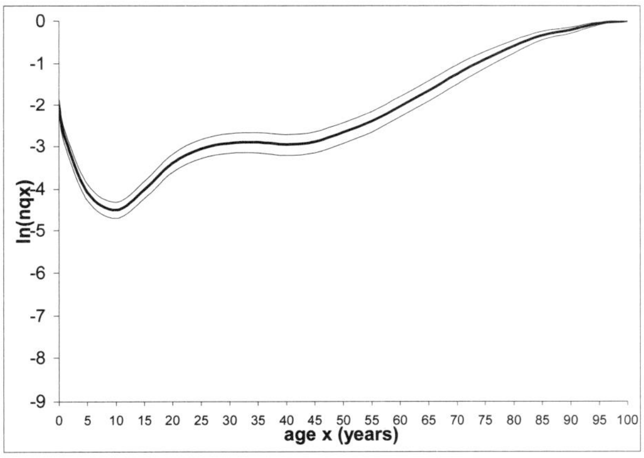

l_x with uncertainty interval

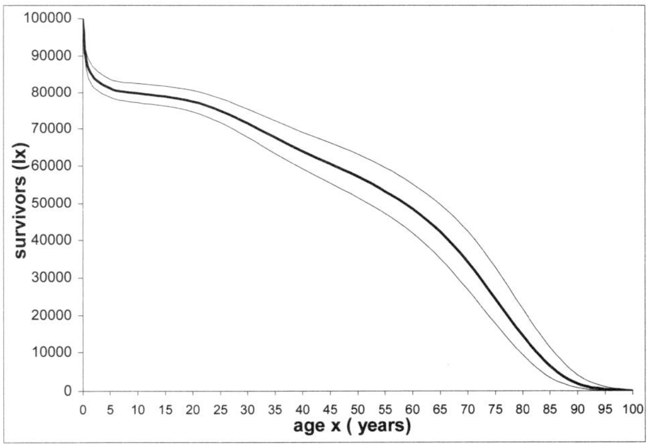

e_x with uncertainty interval

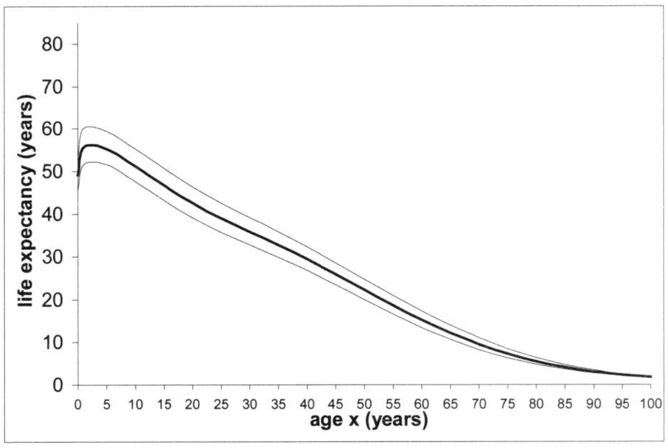

Distribution of deaths by age with uncertainty interval

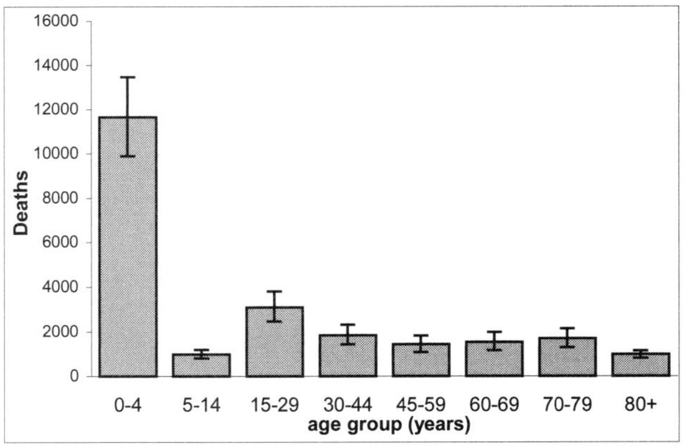

Libyan Arab Jamahiriya | Jamahiriya arabe libyenne | Jamahiriya Arabe Libia

Males, 2000 | *Hommes, 2000* | *Hombres, 2000*

Age	$_nM_x$	$_nq_x$			l_x			$_nd_x$	$_nL_x$	e_x			Actual deaths	Actual population
			percentiles			percentiles					percentiles			
		mean	2 1/2	97 1/2	mean	2 1/2	97 1/2			mean	2 1/2	97 1/2		
<1	0.0257	0.0251	0.0201	0.0303	100,000	100,000	100,000	2,512	97,739	67.5	66.4	68.6	1,820	70,800
1-4	0.0015	0.0060	0.0051	0.0070	97,488	96,975	97,986	588	388,542	68.3	67.3	69.3	391	258,256
5-9	0.0006	0.0031	0.0027	0.0036	96,900	96,303	97,487	304	483,741	64.7	63.7	65.7	174	276,055
10-14	0.0006	0.0028	0.0024	0.0032	96,596	95,955	97,226	272	482,299	59.9	58.9	60.8	177	313,163
15-19	0.0014	0.0068	0.0059	0.0078	96,324	95,645	96,992	656	479,980	55.0	54.1	56.0	480	351,542
20-24	0.0019	0.0094	0.0082	0.0107	95,668	94,902	96,422	903	476,084	50.4	49.5	51.3	541	285,126
25-29	0.0019	0.0096	0.0085	0.0108	94,765	93,888	95,630	912	471,547	45.8	45.0	46.8	465	240,359
30-34	0.0023	0.0112	0.0099	0.0125	93,853	92,872	94,821	1,051	466,638	41.3	40.4	42.1	433	192,078
35-39	0.0030	0.0147	0.0131	0.0164	92,802	91,710	93,878	1,366	460,594	36.7	35.9	37.6	439	147,843
40-44	0.0043	0.0213	0.0191	0.0236	91,436	90,214	92,638	1,948	452,309	32.2	31.4	33.1	571	132,687
45-49	0.0065	0.0322	0.0292	0.0354	89,488	88,104	90,847	2,882	440,234	27.9	27.1	28.7	801	122,376
50-54	0.0102	0.0495	0.0450	0.0540	86,605	85,022	88,155	4,289	422,305	23.7	23.0	24.5	1,091	107,453
55-59	0.0157	0.0754	0.0687	0.0822	82,317	80,512	84,150	6,209	396,060	19.8	19.2	20.5	1,345	85,803
60-64	0.0243	0.1144	0.1047	0.1243	76,107	73,981	78,237	8,707	358,771	16.2	15.6	16.9	1,550	63,852
65-69	0.0371	0.1699	0.1556	0.1833	67,401	64,978	69,819	11,450	308,379	13.0	12.4	13.6	1,659	44,681
70-74	0.0578	0.2524	0.2320	0.2725	55,951	53,188	58,853	14,123	244,447	10.1	9.6	10.7	1,513	26,183
75-79	0.0896	0.3661	0.3393	0.3938	41,828	38,878	44,943	15,314	170,854	7.7	7.3	8.1	1,240	13,831
80-84	0.1398	0.5179	0.4882	0.5491	26,514	23,948	29,501	13,732	98,237	5.7	5.5	6.0	846	6,053
85-89	0.2103	0.6891	0.6611	0.7165	12,781	10,931	14,993	8,807	41,888	4.2	4.0	4.4	435	2,069
90-94	0.3050	0.7964	0.7739	0.8176	3,974	3,098	5,051	3,165	10,375	3.1	3.0	3.2	151	496
95-99	0.4268	0.9358	0.9240	0.9462	809	565	1,128	757	1,774	2.3	2.3	2.4	32	76
100+	0.5759	1	1	1	52	30	84	52	90	1.7	1.7	1.8	5	8
All ages													*16,157*	*2,740,790*

ln($_nq_x$) with uncertainty interval

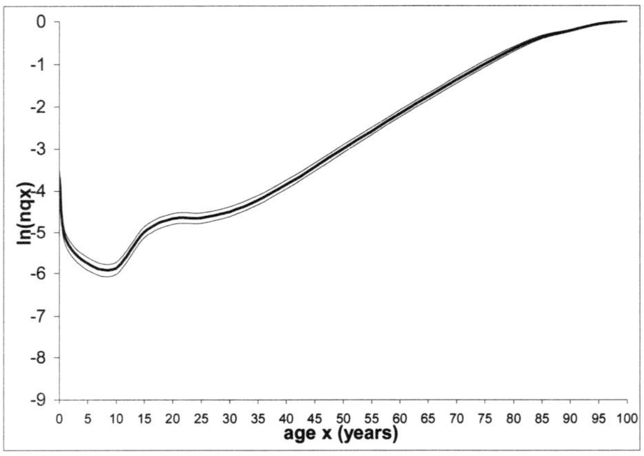

l_x with uncertainty interval

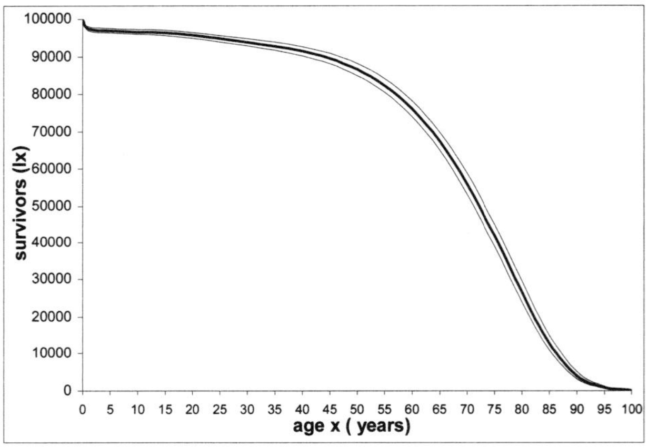

e_x with uncertainty interval

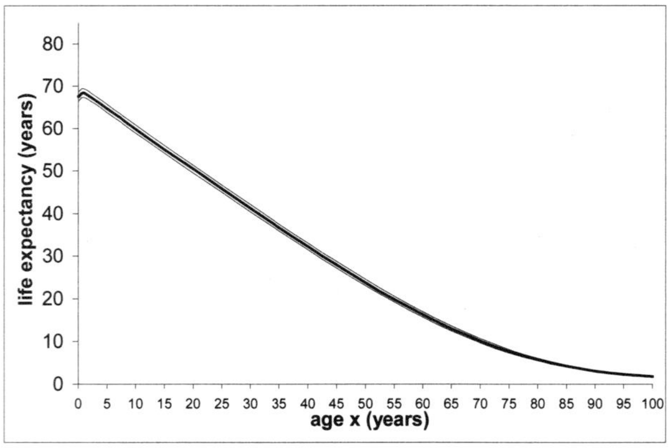

Distribution of deaths by age with uncertainty interval

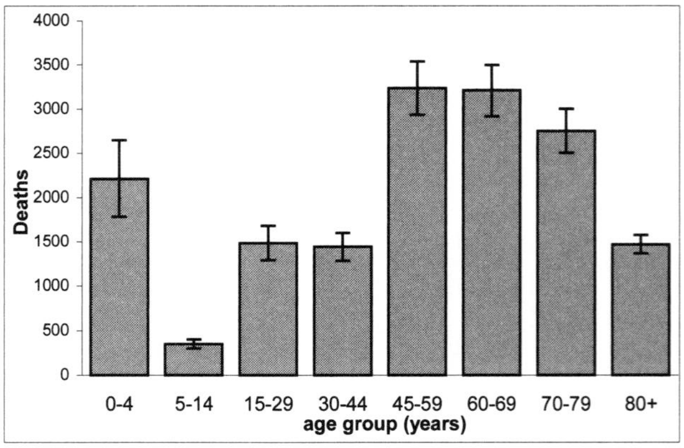

Libyan Arab Jamahiriya   Jamahiriya arabe libyenne   Jamahiriya Arabe Libia

Females, 2000   *Femmes, 2000*   *Mujeres, 2000*

Age	$_nM_x$	$_nq_x$ mean	$_nq_x$ percentiles 2 1/2	$_nq_x$ percentiles 97 1/2	l_x mean	l_x percentiles 2 1/2	l_x percentiles 97 1/2	$_nd_x$	$_nL_x$	e_x mean	e_x percentiles 2 1/2	e_x percentiles 97 1/2	Actual deaths	Actual population
<1	0.0227	0.0222	0.0176	0.0270	100,000	100,000	100,000	2,223	97,999	71.0	69.9	72.3	1,530	67,428
1-4	0.0017	0.0069	0.0059	0.0079	97,777	97,300	98,237	676	389,484	71.6	70.3	73.1	428	246,246
5-9	0.0006	0.0031	0.0026	0.0035	97,100	96,555	97,645	297	484,759	68.1	66.8	69.7	161	263,069
10-14	0.0005	0.0026	0.0022	0.0029	96,803	96,225	97,383	251	483,388	63.3	61.9	64.9	156	299,517
15-19	0.0009	0.0046	0.0040	0.0052	96,552	95,948	97,162	443	481,653	58.5	57.1	60.1	312	339,163
20-24	0.0011	0.0057	0.0050	0.0064	96,109	95,462	96,774	548	479,175	53.8	52.3	55.3	316	276,092
25-29	0.0013	0.0066	0.0057	0.0074	95,561	94,866	96,293	627	476,238	49.0	47.6	50.6	307	233,064
30-34	0.0017	0.0084	0.0074	0.0094	94,934	94,191	95,739	798	472,675	44.4	42.9	45.9	310	183,788
35-39	0.0024	0.0119	0.0105	0.0132	94,136	93,340	95,027	1,117	467,888	39.7	38.3	41.3	333	139,350
40-44	0.0034	0.0171	0.0152	0.0189	93,019	92,110	94,020	1,586	461,132	35.2	33.7	36.7	415	120,661
45-49	0.0051	0.0250	0.0225	0.0274	91,433	90,373	92,572	2,282	451,462	30.7	29.3	32.3	521	103,132
50-54	0.0074	0.0365	0.0326	0.0404	89,151	87,897	90,474	3,251	437,629	26.4	25.1	28.0	611	82,226
55-59	0.0108	0.0527	0.0461	0.0597	85,900	84,407	87,458	4,524	418,193	22.4	21.0	23.9	677	62,561
60-64	0.0166	0.0795	0.0680	0.0920	81,377	79,616	83,186	6,468	390,712	18.5	17.2	19.9	780	47,137
65-69	0.0260	0.1219	0.1022	0.1432	74,908	72,872	76,981	9,128	351,722	14.8	13.7	16.2	894	34,439
70-74	0.0429	0.1938	0.1610	0.2284	65,780	62,566	68,817	12,745	297,040	11.5	10.5	12.7	1,018	23,731
75-79	0.0708	0.3007	0.2518	0.3493	53,035	48,276	57,671	15,948	225,306	8.7	7.9	9.7	1,070	15,113
80-84	0.1174	0.4538	0.3917	0.5094	37,087	31,412	43,152	16,828	143,365	6.4	5.8	7.1	957	8,152
85-89	0.1858	0.6343	0.5742	0.6828	20,259	15,410	26,248	12,850	69,169	4.6	4.3	5.1	583	3,138
90-94	0.2806	0.7618	0.7157	0.7972	7,409	4,889	11,177	5,644	20,113	3.3	3.1	3.6	228	812
95-99	0.4045	0.9137	0.8882	0.9323	1,765	991	3,177	1,613	3,987	2.4	2.3	2.5	46	113
100+	0.5565	1	1	1	152	67	355	152	274	1.8	1.8	1.9	4	8
All ages													*11,655*	*2,548,940*

ln($_nq_x$) with uncertainty interval

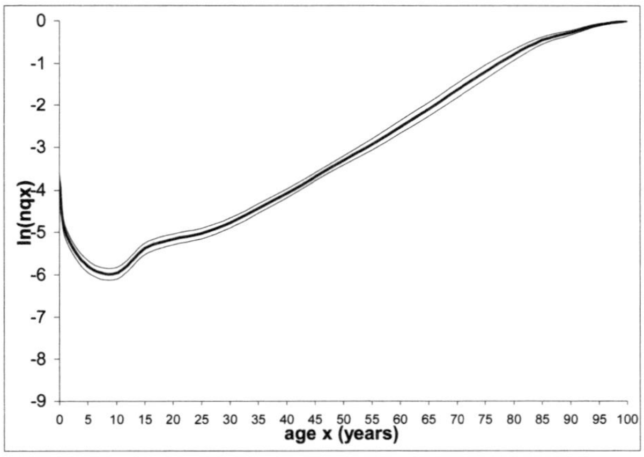

l_x with uncertainty interval

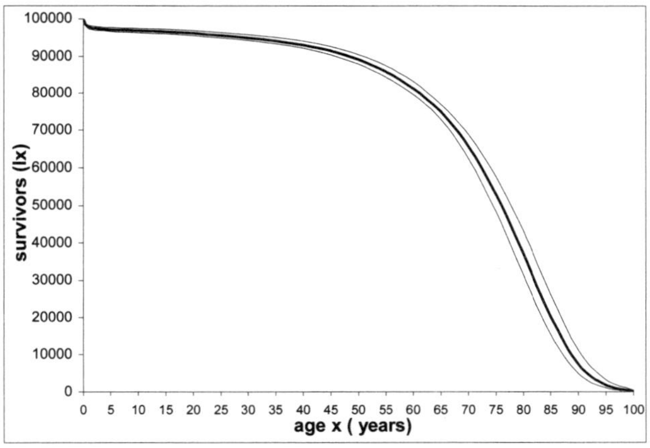

e_x with uncertainty interval

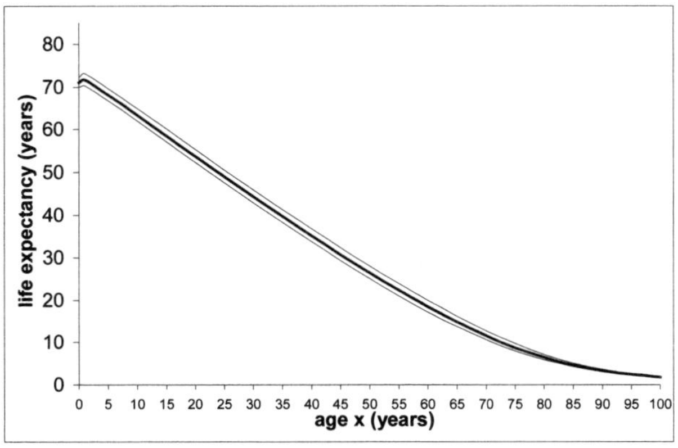

Distribution of deaths by age with uncertainty interval

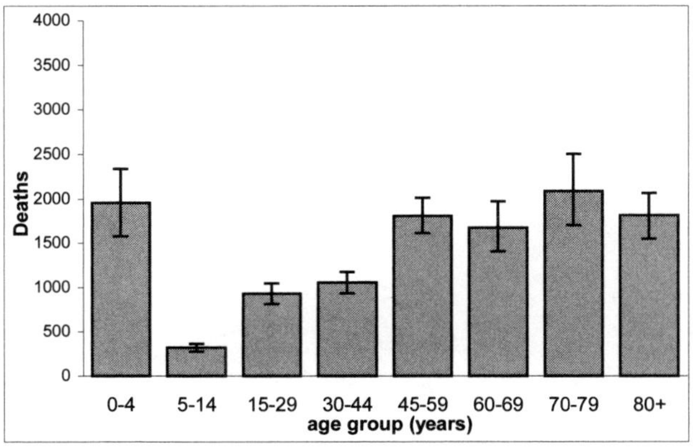

Lithuania — Lituanie — Lituania

Males, 2000 — *Hommes, 2000* — *Hombres, 2000*

Age	$_nM_x$	$_nq_x$			l_x			$_nd_x$	$_nL_x$	e_x			Actual deaths	Actual population
			percentiles			percentiles					percentiles			
		mean	2½	97½	mean	2½	97½			mean	2½	97½		
<1	0.0117	0.0115	0.0091	0.0140	100,000	100,000	100,000	1,154	98,962	66.9	66.1	67.6	216	18,520
1-4	0.0009	0.0035	0.0029	0.0040	98,846	98,602	99,089	343	394,563	66.6	65.9	67.4	67	76,940
5-9	0.0004	0.0021	0.0018	0.0024	98,504	98,206	98,804	206	492,003	62.9	62.1	63.6	52	124,575
10-14	0.0005	0.0025	0.0021	0.0029	98,297	97,970	98,630	248	490,868	58.0	57.2	58.7	75	147,786
15-19	0.0015	0.0074	0.0063	0.0084	98,050	97,687	98,421	724	488,439	53.1	52.4	53.9	206	139,267
20-24	0.0027	0.0133	0.0115	0.0149	97,326	96,872	97,802	1,295	483,393	48.5	47.8	49.3	350	130,580
25-29	0.0031	0.0155	0.0137	0.0172	96,031	95,437	96,674	1,492	476,426	44.1	43.4	44.9	444	141,850
30-34	0.0035	0.0175	0.0156	0.0191	94,539	93,811	95,348	1,650	468,570	39.8	39.0	40.6	498	141,403
35-39	0.0050	0.0249	0.0227	0.0270	92,889	92,025	93,858	2,316	458,656	35.4	34.7	36.2	759	150,274
40-44	0.0072	0.0353	0.0327	0.0379	90,573	89,538	91,730	3,201	444,864	31.3	30.5	32.1	945	131,310
45-49	0.0107	0.0522	0.0490	0.0554	87,372	86,146	88,732	4,557	425,469	27.3	26.6	28.1	1,167	108,986
50-54	0.0146	0.0706	0.0665	0.0749	82,815	81,390	84,374	5,844	399,467	23.7	23.0	24.5	1,326	90,624
55-59	0.0190	0.0905	0.0847	0.0964	76,971	75,390	78,661	6,964	367,447	20.3	19.7	21.1	1,692	89,288
60-64	0.0266	0.1248	0.1159	0.1336	70,007	68,356	71,709	8,738	328,191	17.1	16.4	17.8	2,155	80,938
65-69	0.0358	0.1643	0.1520	0.1761	61,269	59,554	63,052	10,070	281,172	14.2	13.6	14.8	2,461	68,708
70-74	0.0500	0.2222	0.2057	0.2389	51,200	49,234	53,159	11,378	227,554	11.5	10.9	12.0	2,454	49,084
75-79	0.0720	0.3051	0.2840	0.3270	39,822	37,560	42,153	12,149	168,737	9.0	8.6	9.4	2,008	27,887
80-84	0.1100	0.4314	0.4065	0.4577	27,673	25,412	30,181	11,937	108,521	6.9	6.6	7.2	1,297	11,795
85-89	0.1600	0.5714	0.5465	0.5979	15,736	13,896	17,911	8,992	56,198	5.2	5.0	5.4	1,298	8,115
90-94	0.2300	0.6805	0.6583	0.7037	6,744	5,588	8,116	4,589	19,952	3.8	3.7	4.0	1,002	4,356
95-99	0.3600	0.8654	0.8503	0.8807	2,155	1,656	2,766	1,865	5,180	2.6	2.6	2.7	249	692
100+	0.5500	1	1	1	290	198	411	290	527	1.8	1.8	2.0	50	90
All ages													*20,771*	*1,743,068*

ln($_nq_x$) with uncertainty interval

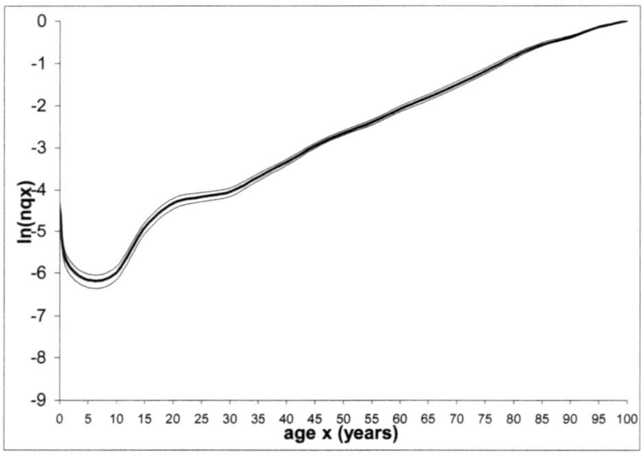

l_x with uncertainty interval

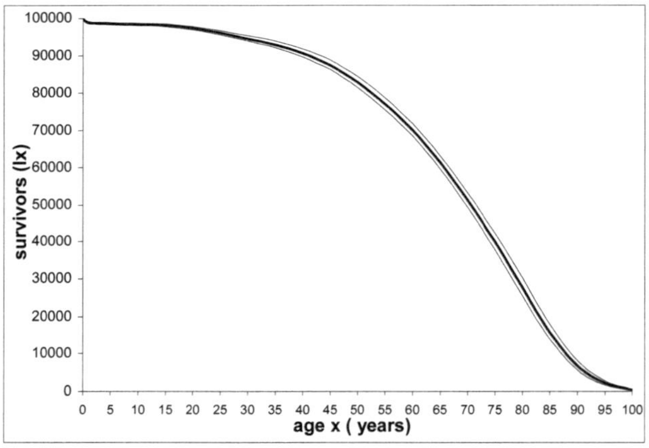

e_x with uncertainty interval

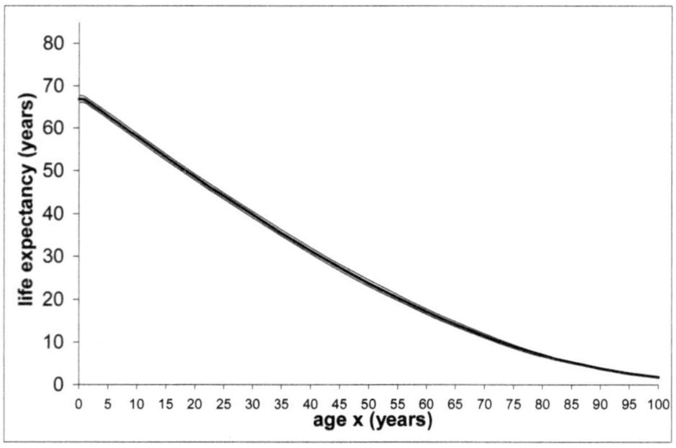

Distribution of deaths by age with uncertainty interval

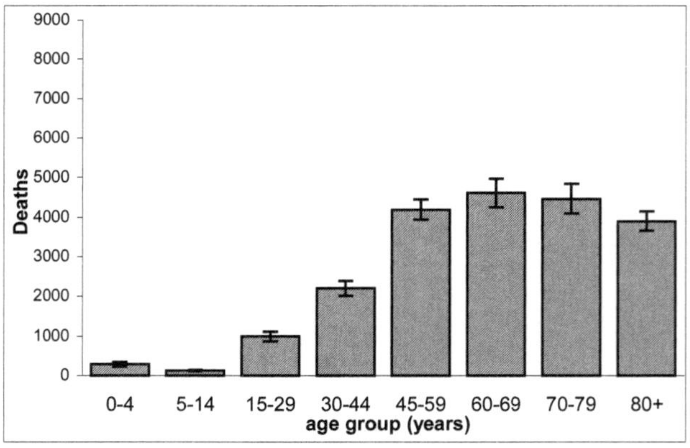

Lithuania | **Lituanie** | **Lituania**

Females, 2000 | *Femmes, 2000* | *Mujeres, 2000*

Age	$_nM_x$	$_nq_x$ mean	$_nq_x$ percentiles 2½	$_nq_x$ percentiles 97½	l_x mean	l_x percentiles 2½	l_x percentiles 97½	$_nd_x$	$_nL_x$	e_x mean	e_x percentiles 2½	e_x percentiles 97½	Actual deaths	Actual population
<1	0.0099	0.0098	0.0077	0.0120	100,000	100,000	100,000	983	99,115	77.2	76.5	78.0	174	17,571
1-4	0.0005	0.0022	0.0018	0.0025	99,017	98,799	99,231	216	395,548	77.0	76.2	77.8	40	73,121
5-9	0.0003	0.0014	0.0012	0.0016	98,801	98,551	99,049	136	493,664	73.2	72.4	74.0	33	118,546
10-14	0.0003	0.0013	0.0011	0.0015	98,665	98,396	98,933	130	492,999	68.3	67.5	69.1	37	142,128
15-19	0.0006	0.0029	0.0025	0.0033	98,535	98,248	98,821	285	491,962	63.3	62.6	64.2	78	134,212
20-24	0.0006	0.0029	0.0025	0.0032	98,250	97,927	98,574	284	490,539	58.5	57.7	59.4	74	128,455
25-29	0.0008	0.0039	0.0035	0.0044	97,966	97,610	98,324	385	488,866	53.7	52.9	54.5	108	137,097
30-34	0.0010	0.0051	0.0046	0.0056	97,581	97,184	97,982	496	486,663	48.9	48.1	49.8	139	136,446
35-39	0.0015	0.0073	0.0066	0.0079	97,085	96,641	97,535	707	483,655	44.1	43.3	45.0	221	151,422
40-44	0.0023	0.0117	0.0108	0.0125	96,377	95,876	96,888	1,125	479,075	39.4	38.6	40.3	326	138,726
45-49	0.0035	0.0172	0.0161	0.0182	95,252	94,676	95,839	1,636	472,172	34.9	34.0	35.8	420	121,306
50-54	0.0053	0.0261	0.0244	0.0277	93,616	92,960	94,279	2,440	461,981	30.4	29.6	31.3	570	107,954
55-59	0.0068	0.0333	0.0310	0.0357	91,176	90,448	91,901	3,037	448,289	26.2	25.3	27.1	756	111,670
60-64	0.0103	0.0503	0.0465	0.0544	88,139	87,350	88,911	4,437	429,604	22.0	21.2	22.9	1,143	110,686
65-69	0.0155	0.0746	0.0681	0.0815	83,702	82,836	84,585	6,246	402,897	18.0	17.2	18.9	1,628	104,990
70-74	0.0260	0.1219	0.1097	0.1348	77,456	76,123	78,746	9,441	363,678	14.3	13.5	15.1	2,415	93,030
75-79	0.0495	0.2204	0.1970	0.2444	68,015	66,030	69,945	14,988	302,604	10.9	10.3	11.6	3,049	61,551
80-84	0.0790	0.3297	0.2964	0.3625	53,027	50,021	55,967	17,486	221,420	8.3	7.8	8.9	2,389	30,250
85-89	0.1240	0.4734	0.4336	0.5096	35,541	31,891	39,379	16,825	135,644	6.1	5.8	6.6	2,828	22,800
90-94	0.1919	0.6090	0.5715	0.6427	18,716	15,650	22,280	11,398	59,387	4.4	4.1	4.7	1,723	8,976
95-99	0.2926	0.7791	0.7505	0.8044	7,318	5,630	9,518	5,701	19,487	3.2	3.0	3.3	571	1,951
100+	0.4394	1	1	1	1,617	1,117	2,372	1,617	3,680	2.3	2.2	2.3	60	137
All ages													*18,783*	*1,953,025*

$\ln({}_nq_x)$ with uncertainty interval

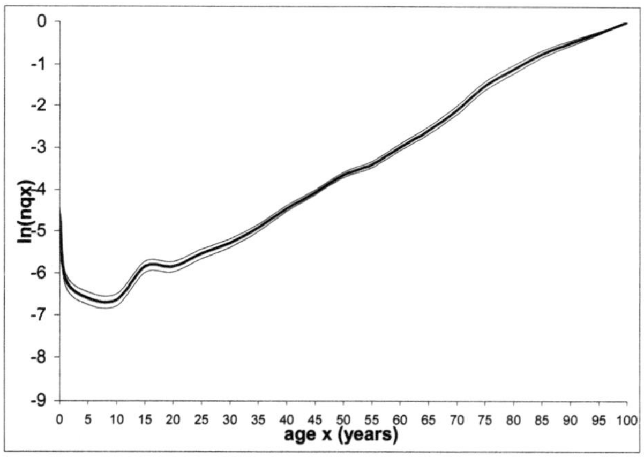

l_x with uncertainty interval

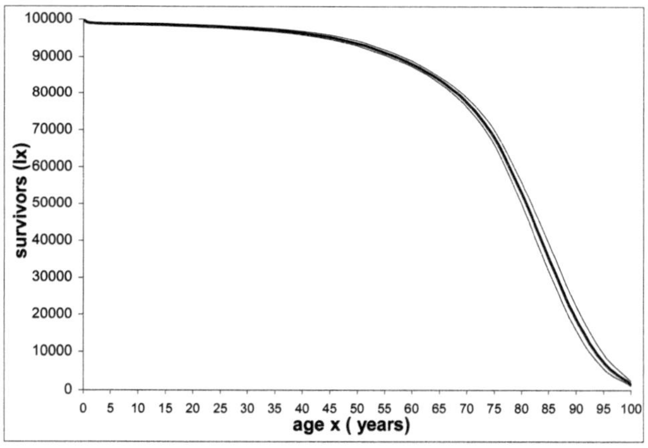

e_x with uncertainty interval

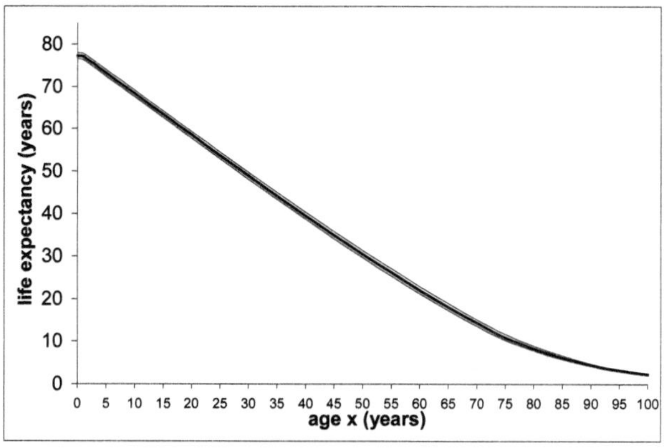

Distribution of deaths by age with uncertainty interval

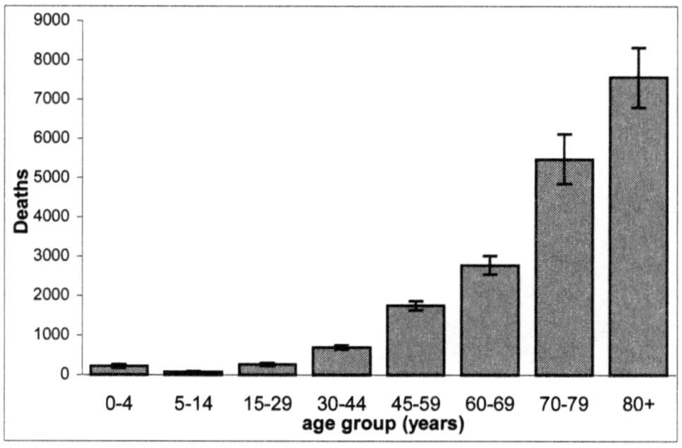

Luxembourg | Luxembourg | Luxemburgo

Males, 2000 | *Hommes, 2000* | *Hombres, 2000*

Age	$_nM_x$	$_nq_x$			l_x			$_nd_x$	$_nL_x$	e_x			Actual deaths	Actual population
			percentiles			percentiles					percentiles			
		mean	2½	97½	mean	2½	97½			mean	2½	97½		
<1	0.0045	0.0045	0.0034	0.0060	100,000	100,000	100,000	447	99,598	73.9	73.0	74.9	12	2,625
1-4	0.0002	0.0008	0.0007	0.0010	99,553	99,402	99,664	83	398,015	73.2	72.4	74.2	2	11,560
5-9	0.0001	0.0006	0.0005	0.0007	99,471	99,298	99,598	59	497,204	69.3	68.4	70.3	2	14,547
10-14	0.0002	0.0010	0.0008	0.0013	99,411	99,224	99,550	101	496,804	64.3	63.5	65.3	3	13,044
15-19	0.0004	0.0021	0.0017	0.0026	99,311	99,099	99,468	210	496,028	59.4	58.5	60.4	6	13,170
20-24	0.0008	0.0042	0.0035	0.0050	99,101	98,842	99,296	412	494,472	54.5	53.7	55.5	12	13,968
25-29	0.0012	0.0062	0.0052	0.0073	98,688	98,347	98,954	613	491,909	49.7	48.9	50.7	19	15,064
30-34	0.0015	0.0073	0.0062	0.0084	98,075	97,627	98,435	712	488,597	45.0	44.2	46.0	26	17,536
35-39	0.0019	0.0093	0.0081	0.0106	97,364	96,806	97,823	907	484,549	40.3	39.6	41.3	35	18,510
40-44	0.0029	0.0143	0.0126	0.0160	96,456	95,778	97,031	1,381	478,829	35.7	34.9	36.7	50	17,246
45-49	0.0035	0.0175	0.0157	0.0193	95,075	94,242	95,806	1,663	471,220	31.2	30.5	32.1	56	15,813
50-54	0.0068	0.0333	0.0300	0.0366	93,413	92,420	94,306	3,110	459,288	26.7	26.0	27.6	98	14,468
55-59	0.0100	0.0487	0.0436	0.0535	90,302	89,034	91,474	4,399	440,515	22.5	21.9	23.4	118	11,788
60-64	0.0164	0.0788	0.0701	0.0863	85,904	84,267	87,489	6,773	412,587	18.5	18.0	19.3	174	10,620
65-69	0.0244	0.1152	0.1024	0.1253	79,131	76,993	81,353	9,116	372,864	14.9	14.4	15.6	227	9,292
70-74	0.0387	0.1766	0.1578	0.1905	70,015	67,349	73,026	12,365	319,160	11.5	11.1	12.1	303	7,829
75-79	0.0679	0.2902	0.2637	0.3123	57,649	54,519	61,502	16,732	246,418	8.4	8.1	8.9	285	4,195
80-84	0.1288	0.4871	0.4561	0.5189	40,918	37,494	45,287	19,932	154,757	5.9	5.6	6.2	276	2,146
85-89	0.2199	0.7094	0.6805	0.7380	20,985	18,039	24,632	14,886	67,710	4.0	3.9	4.2	257	1,171
90-94	0.3377	0.8388	0.8180	0.8581	6,099	4,727	7,869	5,116	15,147	2.8	2.7	2.9	94	279
95-99	0.4668	0.9724	0.9656	0.9780	983	671	1,432	956	2,048	2.1	2.1	2.2	16	35
100+	0.5807	1	1	1	27	15	49	27	47	1.7	1.7	1.9	1	1
All ages													*2,071*	*214,907*

$\ln(_nq_x)$ with uncertainty interval

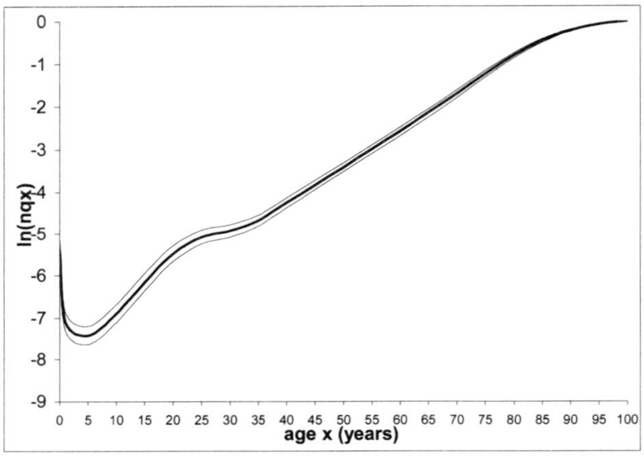

l_x with uncertainty interval

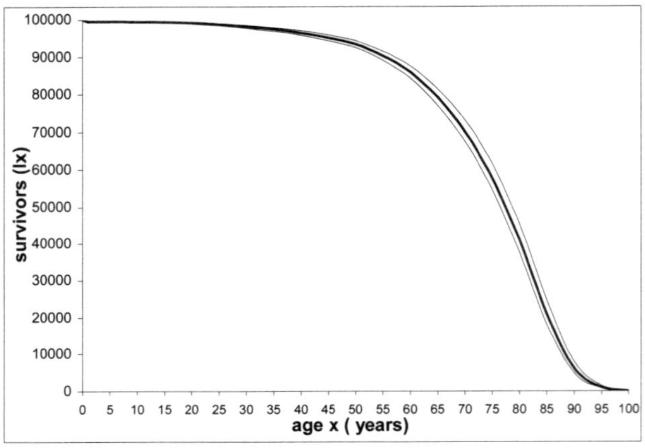

e_x with uncertainty interval

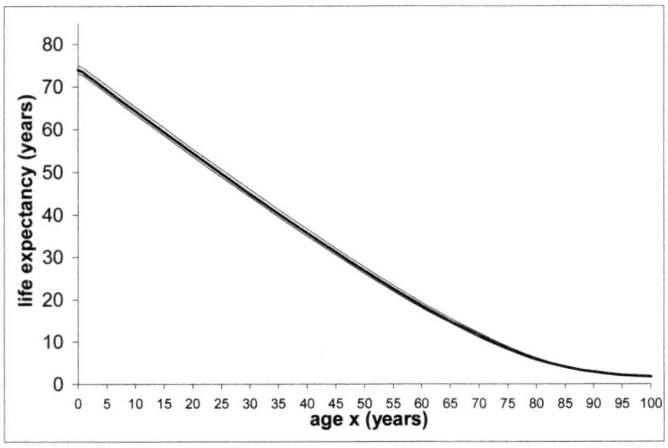

Distribution of deaths by age with uncertainty interval

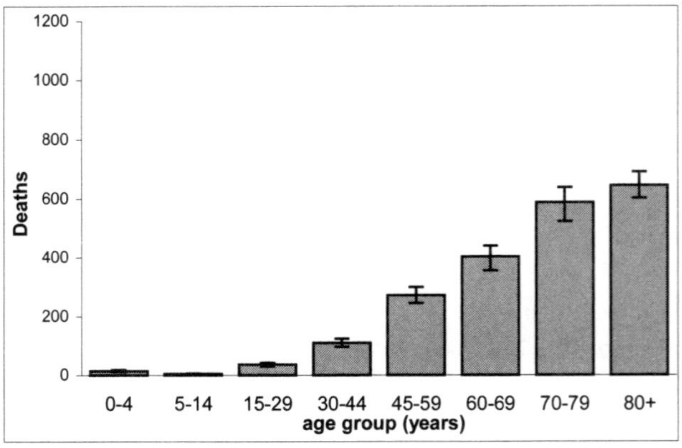

Luxembourg | Luxembourg | Luxemburgo

Females, 2000 | *Femmes, 2000* | *Mujeres, 2000*

Age	$_nM_x$	$_nq_x$			l_x			$_nd_x$	$_nL_x$	e_x			Actual deaths	Actual population
			percentiles			percentiles					percentiles			
		mean	2½	97½	mean	2½	97½			mean	2½	97½		
<1	0.0036	0.0036	0.0024	0.0041	100,000	100,000	100,000	359	99,677	80.8	79.7	82.0	9	2,481
1-4	0.0004	0.0017	0.0013	0.0019	99,641	99,589	99,756	171	398,155	80.1	79.1	81.3	5	10,909
5-9	0.0001	0.0004	0.0003	0.0004	99,470	99,395	99,628	36	497,262	76.2	75.2	77.4	1	13,780
10-14	0.0001	0.0004	0.0003	0.0005	99,435	99,356	99,602	41	497,071	71.2	70.2	72.5	1	12,534
15-19	0.0002	0.0011	0.0008	0.0012	99,394	99,309	99,570	109	496,696	66.2	65.3	67.5	3	12,583
20-24	0.0003	0.0015	0.0012	0.0017	99,285	99,186	99,486	149	496,051	61.3	60.3	62.6	4	13,375
25-29	0.0004	0.0021	0.0017	0.0024	99,136	99,018	99,369	211	495,152	56.4	55.4	57.7	6	14,553
30-34	0.0005	0.0025	0.0020	0.0028	98,925	98,780	99,200	247	494,008	51.5	50.6	52.8	9	17,066
35-39	0.0007	0.0033	0.0027	0.0037	98,678	98,502	98,999	322	492,585	46.6	45.7	47.9	12	18,275
40-44	0.0013	0.0066	0.0056	0.0074	98,356	98,139	98,732	647	490,163	41.8	40.9	43.0	22	17,052
45-49	0.0019	0.0095	0.0083	0.0108	97,709	97,412	98,181	933	486,214	37.0	36.1	38.3	29	15,241
50-54	0.0034	0.0170	0.0151	0.0191	96,776	96,365	97,363	1,640	479,780	32.4	31.5	33.6	46	13,476
55-59	0.0045	0.0222	0.0197	0.0252	95,136	94,520	95,893	2,117	470,387	27.9	27.1	29.1	52	11,544
60-64	0.0078	0.0383	0.0334	0.0433	93,019	92,140	94,007	3,559	456,198	23.5	22.7	24.6	86	11,023
65-69	0.0119	0.0577	0.0498	0.0651	89,460	88,151	90,869	5,158	434,406	19.3	18.6	20.4	123	10,389
70-74	0.0218	0.1034	0.0888	0.1158	84,302	82,414	86,347	8,716	399,721	15.3	14.7	16.3	214	9,824
75-79	0.0385	0.1755	0.1513	0.1941	75,586	72,870	78,683	13,269	344,759	11.8	11.2	12.6	305	7,918
80-84	0.0676	0.2892	0.2540	0.3167	62,317	58,728	66,780	18,020	266,536	8.8	8.3	9.4	311	4,606
85-89	0.1145	0.4450	0.4047	0.4845	44,297	40,131	49,819	19,711	172,209	6.4	5.9	6.8	401	3,501
90-94	0.1868	0.5985	0.5649	0.6382	24,586	20,689	29,657	14,714	78,789	4.5	4.2	4.7	278	1,486
95-99	0.2937	0.7806	0.7589	0.8103	9,872	7,484	12,904	7,706	26,242	3.2	3.0	3.3	79	269
100+	0.4450	1	1	1	2,166	1,420	3,111	2,166	4,867	2.2	2.2	2.3	12	26
All ages													*2,007*	*221,911*

ln($_nq_x$) with uncertainty interval

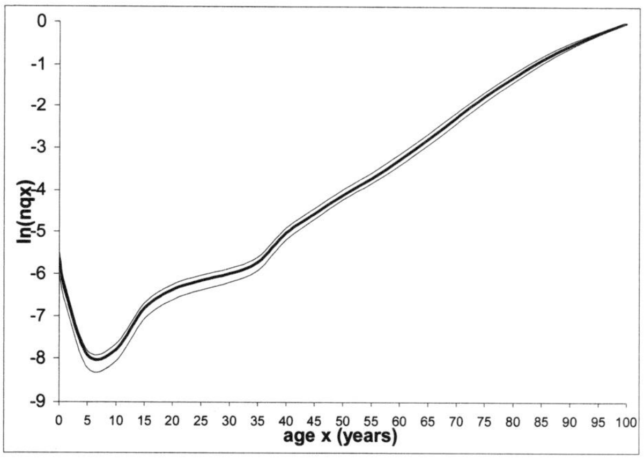

l_x with uncertainty interval

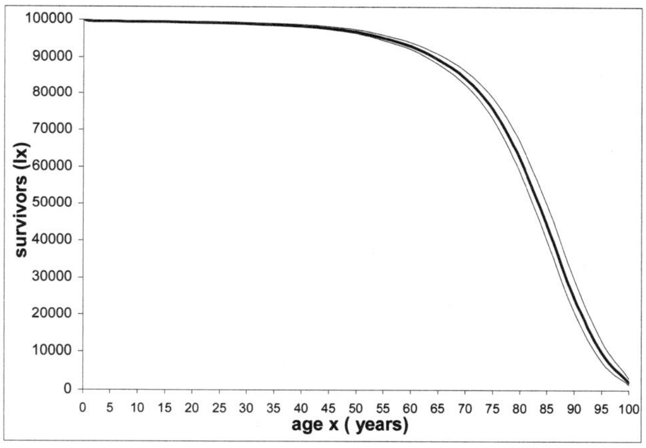

e_x with uncertainty interval

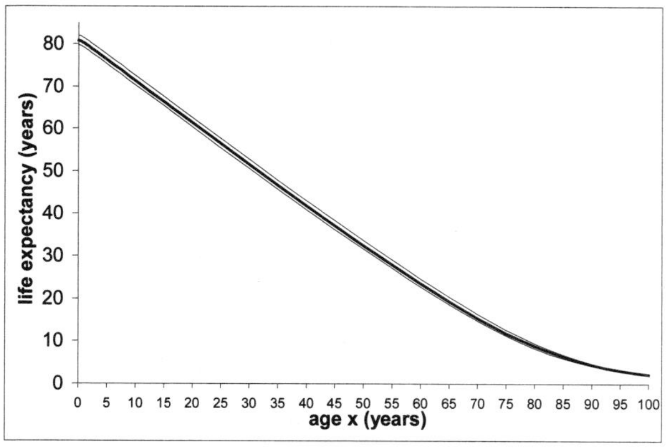

Distribution of deaths by age with uncertainty interval

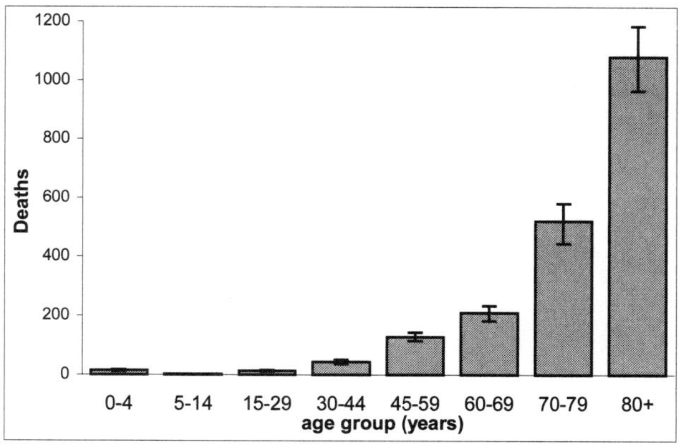

Madagascar

Males, 2000

Madagascar

Hommes, 2000

Madagascar

Hombres, 2000

Age	$_nM_x$	$_nq_x$			l_x			$_nd_x$	$_nL_x$	e_x			Actual deaths	Actual population
			percentiles			percentiles					percentiles			
		mean	2 1/2	97 1/2	mean	2 1/2	97 1/2			mean	2 1/2	97 1/2		
<1	0.1335	0.1192	0.1107	0.1268	100,000	100,000	100,000	11,920	89,272	51.7	49.6	53.9	42,011	314,648
1-4	0.0108	0.0420	0.0379	0.0464	88,080	87,320	88,930	3,702	343,438	57.7	55.4	60.3	12,091	1,121,809
5-9	0.0028	0.0137	0.0122	0.0153	84,379	83,630	85,129	1,154	419,011	56.2	54.0	58.6	3,197	1,161,151
10-14	0.0018	0.0089	0.0079	0.0100	83,225	82,411	84,040	741	414,275	51.9	49.8	54.3	1,750	978,591
15-19	0.0020	0.0101	0.0089	0.0114	82,484	81,615	83,376	835	410,334	47.3	45.3	49.7	1,698	833,923
20-24	0.0040	0.0196	0.0172	0.0222	81,649	80,716	82,627	1,600	404,245	42.8	40.8	45.1	2,785	703,605
25-29	0.0052	0.0256	0.0223	0.0292	80,049	78,994	81,197	2,050	395,120	38.6	36.7	40.9	3,033	584,683
30-34	0.0066	0.0324	0.0280	0.0372	77,999	76,713	79,367	2,529	383,674	34.6	32.7	36.8	3,219	488,458
35-39	0.0086	0.0419	0.0359	0.0484	75,470	73,879	77,114	3,162	369,448	30.6	28.9	32.7	3,531	412,594
40-44	0.0112	0.0543	0.0461	0.0630	72,309	70,311	74,298	3,925	351,730	26.9	25.2	28.9	3,870	346,801
45-49	0.0145	0.0700	0.0588	0.0816	68,383	65,880	70,796	4,785	329,955	23.3	21.7	25.1	4,099	282,634
50-54	0.0193	0.0920	0.0768	0.1075	63,598	60,505	66,508	5,850	303,368	19.8	18.4	21.6	3,841	199,175
55-59	0.0258	0.1210	0.1005	0.1413	57,749	54,002	61,329	6,990	271,268	16.6	15.3	18.1	4,236	164,380
60-64	0.0385	0.1754	0.1459	0.2033	50,759	46,372	55,082	8,904	231,533	13.5	12.4	14.9	5,005	130,144
65-69	0.0543	0.2389	0.2002	0.2750	41,855	36,944	46,945	9,999	184,276	10.8	9.9	12.0	5,229	96,369
70-74	0.0806	0.3356	0.2865	0.3800	31,856	26,845	37,545	10,690	132,554	8.5	7.7	9.4	5,196	64,434
75-79	0.1192	0.4593	0.4036	0.5076	21,166	16,748	26,787	9,721	81,528	6.5	6.0	7.2	4,406	36,951
80-84	0.1781	0.6161	0.5615	0.6615	11,445	8,347	15,975	7,051	39,598	4.9	4.5	5.3	2,943	16,526
85-89	0.2567	0.7818	0.7383	0.8168	4,394	2,840	7,005	3,435	13,382	3.6	3.4	3.9	1,260	4,908
90-94	0.3573	0.8622	0.8291	0.8885	959	520	1,833	827	2,314	2.7	2.6	2.9	314	879
95-99	0.4800	0.9836	0.9747	0.9893	132	58	313	130	271	2.1	2.0	2.1	40	83
100+	0.6224	1	1	1	2	1	8	2	3	1.6	1.6	1.7	2	4
All ages													*113,756*	*7,942,750*

ln($_nq_x$) with uncertainty interval

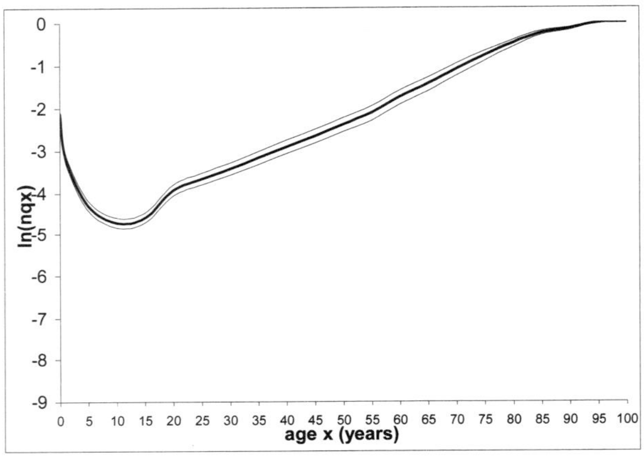

l_x with uncertainty interval

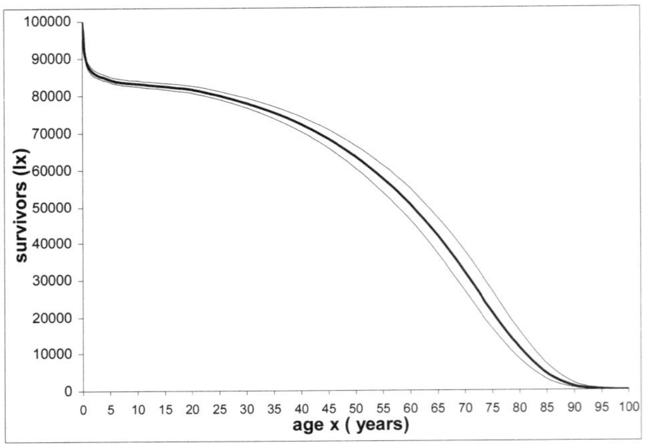

e_x with uncertainty interval

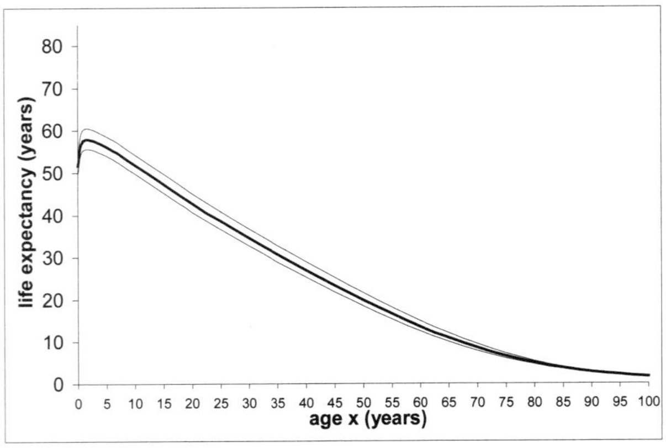

Distribution of deaths by age with uncertainty interval

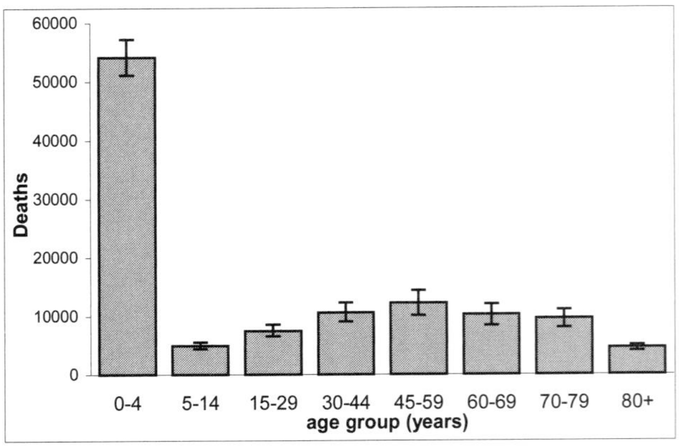

Madagascar

Females, 2000

Madagascar

Femmes, 2000

Madagascar

Mujeres, 2000

Age	$_nM_x$	$_nq_x$ mean	$_nq_x$ percentiles 2½	$_nq_x$ percentiles 97½	l_x mean	l_x percentiles 2½	l_x percentiles 97½	$_nd_x$	$_nL_x$	e_x mean	e_x percentiles 2½	e_x percentiles 97½	Actual deaths	Actual population
<1	0.1100	0.1001	0.0913	0.1082	100,000	100,000	100,000	10,011	90,990	54.6	52.5	56.8	33,611	305,484
1-4	0.0120	0.0468	0.0424	0.0512	89,989	89,176	90,874	4,210	349,850	59.7	57.3	62.3	13,533	1,124,508
5-9	0.0030	0.0148	0.0132	0.0165	85,778	85,035	86,527	1,270	425,718	58.5	56.3	61.0	3,462	1,160,968
10-14	0.0021	0.0104	0.0093	0.0117	84,509	83,706	85,325	881	420,342	54.4	52.2	56.8	2,046	976,116
15-19	0.0032	0.0160	0.0141	0.0180	83,628	82,784	84,501	1,338	414,795	49.9	47.8	52.3	2,676	829,765
20-24	0.0045	0.0225	0.0197	0.0255	82,290	81,386	83,249	1,849	406,827	45.7	43.6	48.1	3,197	703,202
25-29	0.0052	0.0258	0.0225	0.0295	80,441	79,396	81,566	2,077	397,010	41.7	39.7	44.0	3,080	588,672
30-34	0.0059	0.0289	0.0249	0.0332	78,363	77,098	79,715	2,262	386,162	37.7	35.8	39.9	2,882	492,084
35-39	0.0073	0.0358	0.0307	0.0414	76,102	74,559	77,704	2,727	373,692	33.8	32.0	35.9	3,033	415,690
40-44	0.0082	0.0401	0.0340	0.0466	73,375	71,482	75,280	2,943	359,519	29.9	28.2	32.0	2,862	349,731
45-49	0.0103	0.0503	0.0423	0.0587	70,433	68,151	72,663	3,542	343,308	26.1	24.5	28.0	2,949	285,829
50-54	0.0140	0.0674	0.0563	0.0789	66,891	64,150	69,506	4,509	323,180	22.3	20.9	24.1	2,938	210,566
55-59	0.0190	0.0908	0.0754	0.1065	62,382	59,087	65,470	5,665	297,745	18.8	17.4	20.4	3,405	178,943
60-64	0.0279	0.1304	0.1079	0.1525	56,716	52,796	60,479	7,395	265,095	15.4	14.2	16.9	4,046	145,043
65-69	0.0417	0.1888	0.1566	0.2191	49,322	44,747	53,862	9,310	223,334	12.3	11.3	13.6	4,592	110,168
70-74	0.0650	0.2795	0.2349	0.3208	40,012	34,955	45,344	11,184	172,098	9.6	8.8	10.7	4,921	75,727
75-79	0.0987	0.3958	0.3408	0.4446	28,827	23,822	34,694	11,410	115,613	7.3	6.7	8.2	4,439	44,984
80-84	0.1491	0.5430	0.4844	0.5925	17,418	13,345	22,871	9,457	63,446	5.5	5.1	6.1	3,180	21,331
85-89	0.2187	0.7070	0.6556	0.7487	7,960	5,514	11,793	5,628	25,732	4.1	3.8	4.5	1,543	7,057
90-94	0.3118	0.8055	0.7650	0.8381	2,332	1,386	4,062	1,879	6,026	3.0	2.8	3.3	479	1,535
95-99	0.4317	0.9405	0.9192	0.9560	454	224	955	427	989	2.3	2.2	2.4	85	197
100+	0.5806	1	1	1	27	10	77	27	47	1.7	1.7	1.8	8	14
All ages													*102,968*	*8,027,614*

$\ln(_nq_x)$ with uncertainty interval

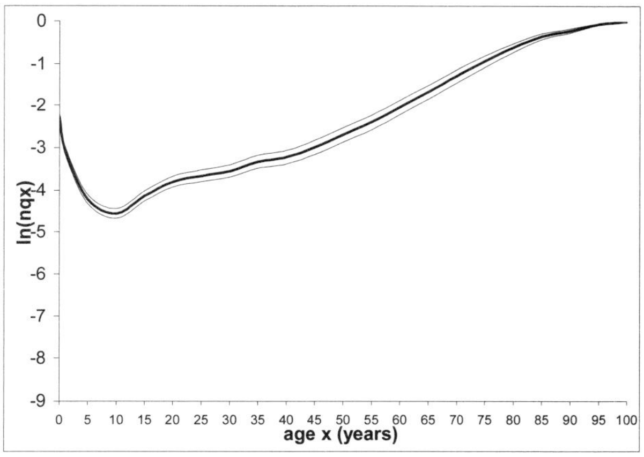

l_x with uncertainty interval

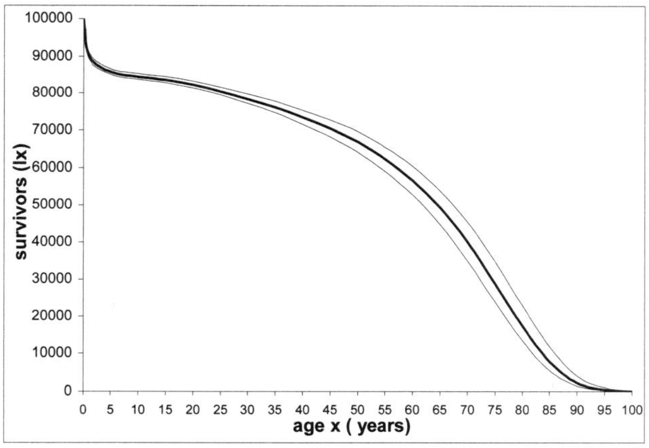

e_x with uncertainty interval

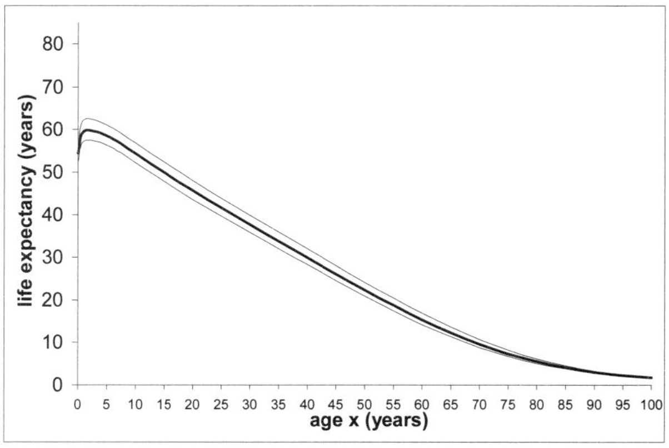

Distribution of deaths by age with uncertainty interval

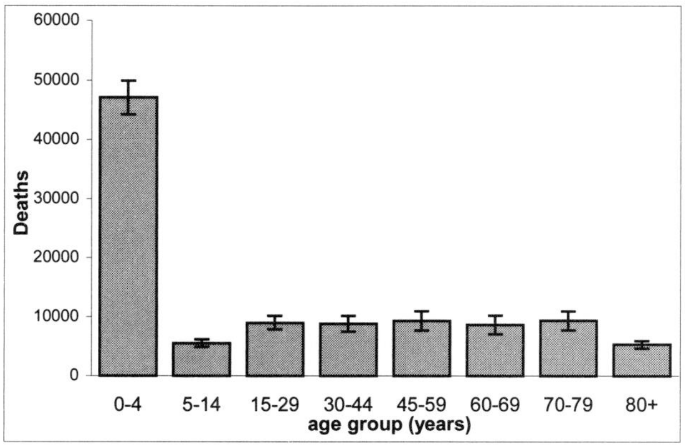

Malawi **Malawi** **Malawi**

Males, 2000 *Hommes, 2000* *Hombres, 2000*

Age	$_nM_x$	$_nq_x$			l_x			$_nd_x$	$_nL_x$	e_x			Actual deaths	Actual population
			percentiles			percentiles					percentiles			
		mean	2½	97½	mean	2½	97½			mean	2½	97½		
<1	0.1958	0.1665	0.1416	0.1930	100,000	100,000	100,000	16,648	85,017	37.1	33.5	42.8	46,802	239,004
1-4	0.0195	0.0746	0.0580	0.0885	83,352	80,703	85,840	6,216	318,490	43.5	38.9	49.8	15,603	799,482
5-9	0.0040	0.0199	0.0150	0.0241	77,136	74,888	79,331	1,536	381,842	42.9	38.3	48.8	3,500	870,204
10-14	0.0025	0.0123	0.0092	0.0150	75,600	73,150	78,140	931	375,675	38.7	34.1	44.5	1,831	738,888
15-19	0.0031	0.0155	0.0115	0.0190	74,670	72,067	77,421	1,155	370,461	34.2	29.6	39.9	1,937	621,454
20-24	0.0090	0.0442	0.0326	0.0548	73,515	70,724	76,531	3,248	359,454	29.7	25.1	35.3	4,695	519,585
25-29	0.0195	0.0928	0.0677	0.1173	70,267	66,957	74,034	6,518	335,039	25.9	21.4	31.4	7,649	393,126
30-34	0.0310	0.1438	0.1039	0.1862	63,749	59,353	69,018	9,170	295,818	23.3	18.9	28.6	9,943	320,748
35-39	0.0349	0.1605	0.1151	0.2127	54,579	48,874	61,847	8,761	250,992	21.8	17.6	26.8	8,955	256,550
40-44	0.0362	0.1658	0.1187	0.2218	45,818	39,069	54,729	7,597	210,098	20.5	16.7	25.0	7,433	205,581
45-49	0.0366	0.1677	0.1194	0.2243	38,221	30,618	48,222	6,408	175,086	19.1	15.8	23.0	5,974	163,235
50-54	0.0357	0.1640	0.1168	0.2186	31,813	23,950	42,391	5,219	146,020	17.4	14.6	20.8	5,114	143,100
55-59	0.0346	0.1594	0.1138	0.2112	26,595	18,760	37,334	4,240	122,373	15.3	13.0	18.2	3,923	113,232
60-64	0.0455	0.2042	0.1478	0.2657	22,355	14,797	32,965	4,565	100,360	12.7	10.8	15.2	3,938	86,563
65-69	0.0599	0.2605	0.1924	0.3314	17,789	10,865	27,925	4,635	77,360	10.4	8.8	12.4	3,814	63,655
70-74	0.0864	0.3552	0.2710	0.4374	13,155	7,265	22,340	4,673	54,092	8.2	6.9	9.8	3,651	42,266
75-79	0.1255	0.4776	0.3805	0.5661	8,482	4,087	16,029	4,051	32,282	6.3	5.4	7.5	3,038	24,205
80-84	0.1856	0.6338	0.5329	0.7187	4,431	1,773	9,808	2,808	15,133	4.7	4.1	5.6	2,262	12,189
85-89	0.2654	0.7978	0.7110	0.8638	1,623	499	4,581	1,295	4,877	3.5	3.2	4.1	833	3,138
90-94	0.3667	0.8730	0.8029	0.9228	328	68	1,324	286	781	2.6	2.4	3.0	182	495
95-99	0.4895	0.9915	0.9768	0.9973	42	5	261	41	84	2.0	2.0	2.1	27	55
100+	0.6308	1	1	1	0	0	6	0	1	1.6	1.6	1.7	2	3
All ages													*141,104*	*5,616,758*

$\ln(_nq_x)$ with uncertainty interval

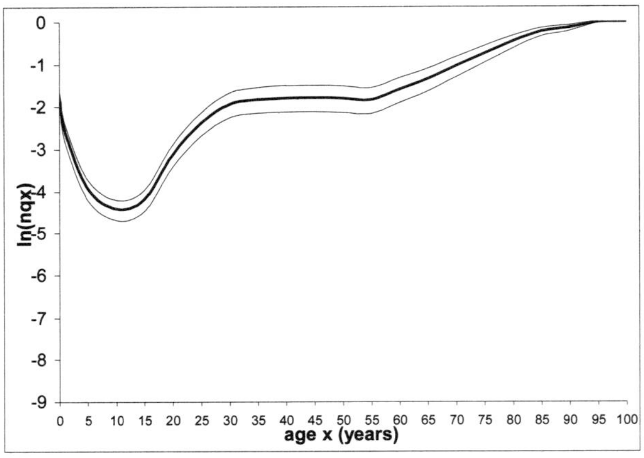

l_x with uncertainty interval

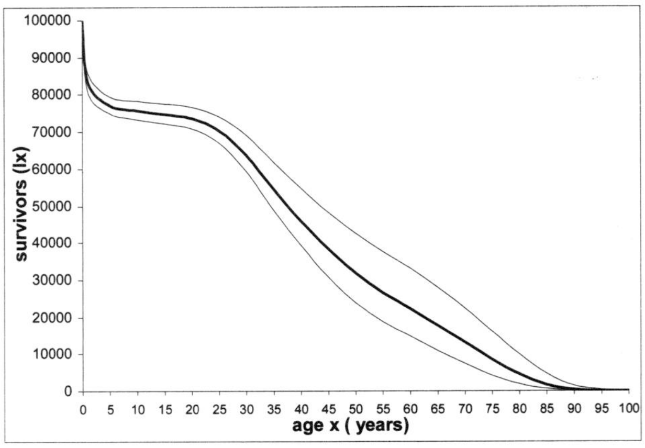

e_x with uncertainty interval

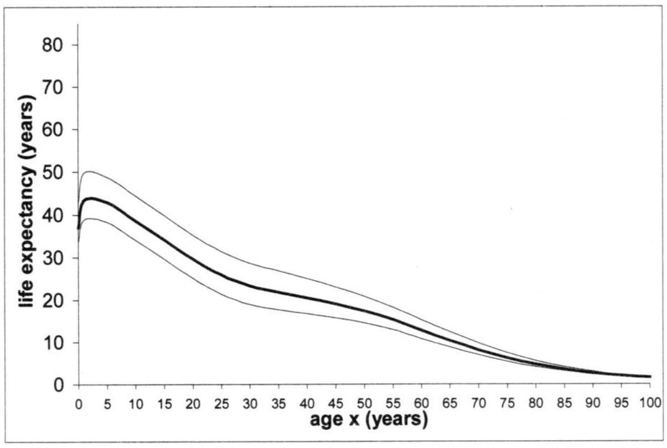

Distribution of deaths by age with uncertainty interval

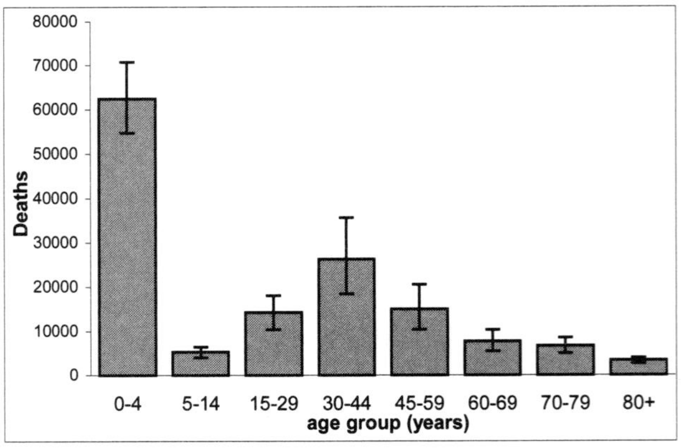

Malawi **Malawi** **Malawi**

Females, 2000 *Femmes, 2000* *Mujeres, 2000*

Age	$_nM_x$	$_nq_x$			l_x			$_nd_x$	$_nL_x$	e_x			Actual deaths	Actual population
			percentiles			percentiles					percentiles			
		mean	2½	97½	mean	2½	97½			mean	2½	97½		
<1	0.1622	0.1415	0.1158	0.1661	100,000	100,000	100,000	14,152	87,263	37.8	32.6	43.0	37,271	229,811
1-4	0.0214	0.0814	0.0672	0.1009	85,848	83,394	88,421	6,991	326,612	43.0	37.2	49.2	16,818	785,704
5-9	0.0044	0.0220	0.0175	0.0283	78,856	76,614	81,092	1,734	389,947	42.7	37.2	48.8	3,786	851,341
10-14	0.0031	0.0152	0.0120	0.0197	77,122	74,447	79,533	1,171	382,683	38.6	33.2	44.7	2,216	724,165
15-19	0.0074	0.0365	0.0285	0.0476	75,951	72,981	78,534	2,772	372,824	34.2	28.8	40.2	4,531	609,355
20-24	0.0174	0.0832	0.0641	0.1093	73,179	69,505	76,165	6,088	350,673	30.4	25.0	36.3	8,830	508,584
25-29	0.0282	0.1317	0.0999	0.1740	67,090	61,907	70,977	8,839	313,356	27.9	22.6	33.7	11,139	394,912
30-34	0.0336	0.1550	0.1150	0.2051	58,252	51,137	63,698	9,031	268,682	26.8	21.7	32.2	10,887	323,901
35-39	0.0301	0.1402	0.1021	0.1890	49,221	40,677	56,184	6,899	228,858	26.2	21.7	31.1	8,002	265,457
40-44	0.0258	0.1212	0.0874	0.1649	42,322	33,171	50,328	5,130	198,785	25.1	21.1	29.3	5,726	221,905
45-49	0.0221	0.1048	0.0752	0.1430	37,192	27,902	45,812	3,897	176,219	23.2	19.8	26.9	4,125	186,531
50-54	0.0224	0.1061	0.0762	0.1444	33,296	24,092	42,259	3,533	157,644	20.6	17.7	23.8	3,696	164,902
55-59	0.0244	0.1151	0.0829	0.1557	29,762	20,788	38,925	3,426	140,246	17.8	15.3	20.6	3,272	133,957
60-64	0.0320	0.1480	0.1075	0.1976	26,336	17,726	35,574	3,898	121,937	14.8	12.6	17.2	3,364	105,259
65-69	0.0455	0.2041	0.1507	0.2670	22,439	14,421	31,590	4,580	100,743	11.9	10.1	14.0	3,533	77,703
70-74	0.0692	0.2950	0.2237	0.3741	17,859	10,791	26,635	5,269	76,120	9.3	7.9	11.1	3,704	53,513
75-79	0.1030	0.4096	0.3225	0.4994	12,589	6,755	20,676	5,156	50,057	7.2	6.1	8.5	3,302	32,058
80-84	0.1536	0.5550	0.4590	0.6466	7,433	3,382	14,008	4,125	26,853	5.4	4.6	6.4	2,561	16,672
85-89	0.2233	0.7164	0.6260	0.7961	3,308	1,195	7,578	2,370	10,616	4.1	3.5	4.7	1,087	4,870
90-94	0.3159	0.8110	0.7343	0.8753	938	244	2,834	761	2,408	3.0	2.6	3.5	274	866
95-99	0.4353	0.9439	0.9010	0.9724	177	30	753	167	385	2.3	2.1	2.5	52	120
100+	0.5840	1	1	1	10	1	75	10	17	1.7	1.7	1.9	5	8
All ages													*138,183*	*5,691,594*

ln($_nq_x$) with uncertainty interval

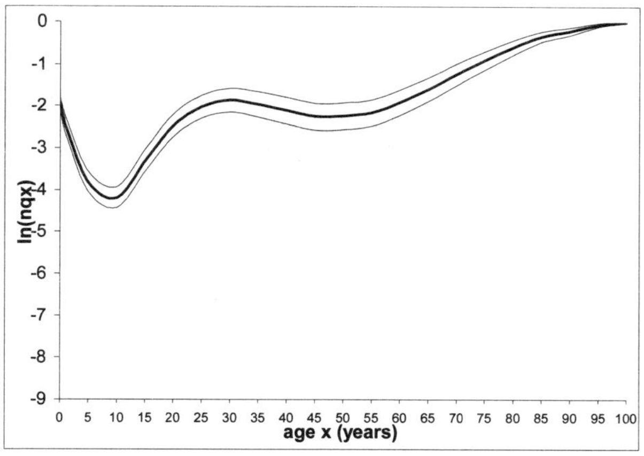

l_x with uncertainty interval

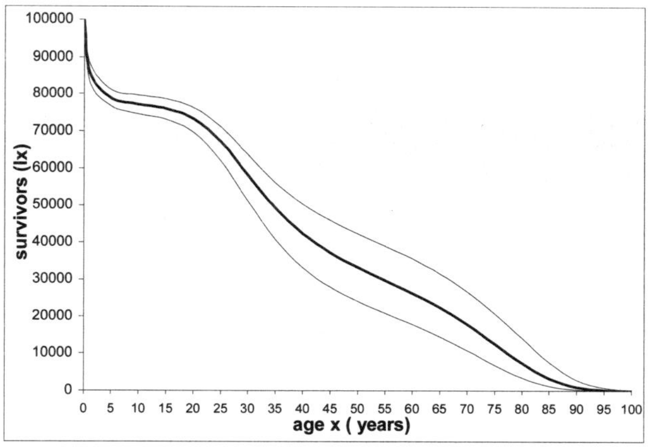

e_x with uncertainty interval

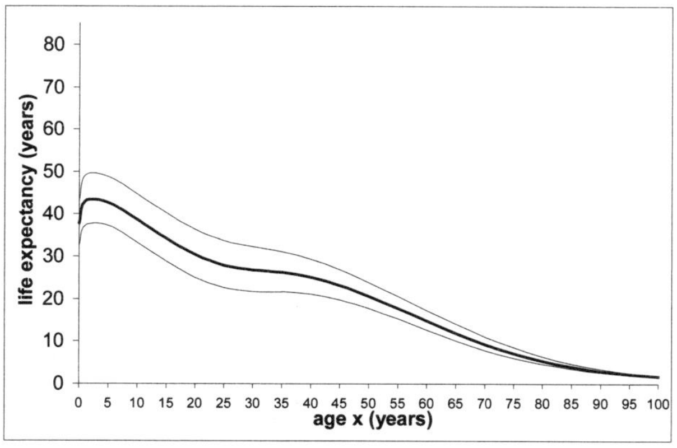

Distribution of deaths by age with uncertainty interval

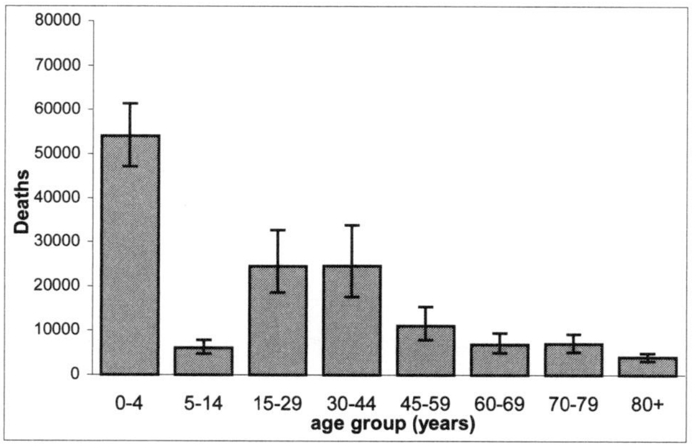

## Malaysia		Malaisie		Malasia

Males, 2000		*Hommes, 2000*		*Hombres, 2000*

Age	${}_nM_x$	${}_nq_x$			l_x			${}_nd_x$	${}_nL_x$	e_x			Actual deaths	Actual population
			percentiles			percentiles					percentiles			
		mean	2½	97½	mean	2½	97½			mean	2½	97½		
<1	0.0115	0.0114	0.0094	0.0135	100,000	100,000	100,000	1,142	98,972	68.3	67.4	69.3	3,121	270,343
1-4	0.0009	0.0036	0.0031	0.0041	98,858	98,651	99,061	357	394,575	68.1	67.2	69.1	976	1,080,047
5-9	0.0005	0.0025	0.0021	0.0028	98,501	98,252	98,751	243	491,899	64.4	63.4	65.3	681	1,379,609
10-14	0.0006	0.0029	0.0025	0.0033	98,258	97,976	98,540	287	490,574	59.5	58.6	60.5	680	1,160,980
15-19	0.0016	0.0080	0.0070	0.0090	97,971	97,651	98,289	781	487,903	54.7	53.8	55.6	1,847	1,153,112
20-24	0.0020	0.0098	0.0086	0.0110	97,190	96,772	97,600	952	483,569	50.1	49.2	51.0	1,956	993,494
25-29	0.0021	0.0104	0.0092	0.0116	96,238	95,711	96,758	1,003	478,682	45.6	44.7	46.5	1,849	882,589
30-34	0.0026	0.0128	0.0114	0.0141	95,235	94,602	95,862	1,214	473,140	41.0	40.1	42.0	2,066	804,838
35-39	0.0032	0.0159	0.0143	0.0175	94,021	93,270	94,770	1,496	466,364	36.5	35.6	37.5	2,494	777,404
40-44	0.0042	0.0208	0.0188	0.0228	92,525	91,642	93,414	1,923	457,817	32.1	31.2	33.0	2,956	703,665
45-49	0.0056	0.0277	0.0252	0.0302	90,602	89,566	91,654	2,508	446,740	27.7	26.8	28.6	3,255	579,824
50-54	0.0093	0.0453	0.0414	0.0495	88,094	86,886	89,334	3,987	430,502	23.4	22.6	24.4	4,217	455,312
55-59	0.0146	0.0705	0.0642	0.0773	84,107	82,652	85,609	5,926	405,719	19.4	18.6	20.3	4,694	321,395
60-64	0.0244	0.1148	0.1035	0.1266	78,181	76,413	79,987	8,973	368,474	15.7	14.9	16.5	6,569	269,762
65-69	0.0384	0.1750	0.1565	0.1943	69,209	66,951	71,391	12,114	315,758	12.4	11.7	13.2	6,767	176,399
70-74	0.0622	0.2691	0.2410	0.2972	57,095	54,188	59,930	15,363	247,065	9.5	8.9	10.1	7,670	123,338
75-79	0.1000	0.4000	0.3635	0.4344	41,731	38,232	45,307	16,693	166,925	7.1	6.7	7.6	6,470	64,698
80-84	0.1650	0.5841	0.5460	0.6181	25,039	21,624	28,835	14,624	88,633	5.1	4.9	5.5	6,262	37,949
85-89	0.2400	0.7500	0.7206	0.7761	10,414	8,273	13,091	7,811	32,545	3.8	3.7	4.0	3,490	14,541
90-94	0.3400	0.8416	0.8205	0.8605	2,604	1,865	3,658	2,191	6,445	2.8	2.7	2.9	1,593	4,686
95-99	0.4700	0.9751	0.9686	0.9805	412	263	657	402	856	2.1	2.1	2.2	344	731
100+	0.6500	1	1	1	10	5	21	10	16	1.5	1.5	1.8	64	99
All ages													*70,017*	*11,254,815*

ln(${}_nq_x$) with uncertainty interval

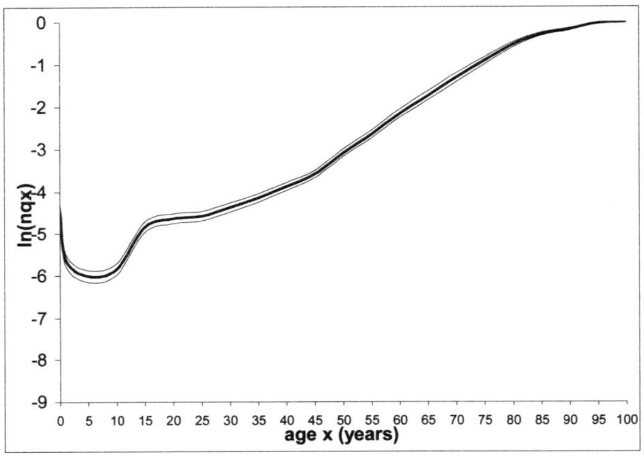

l_x with uncertainty interval

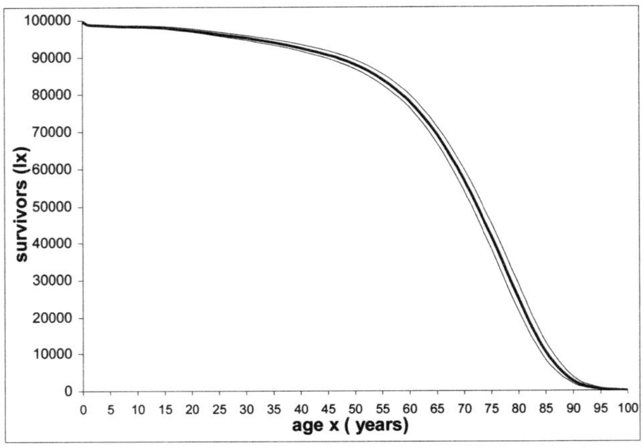

e_x with uncertainty interval

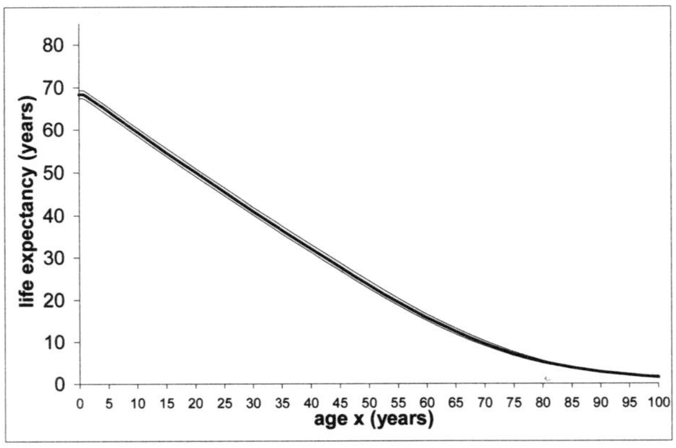

Distribution of deaths by age with uncertainty interval

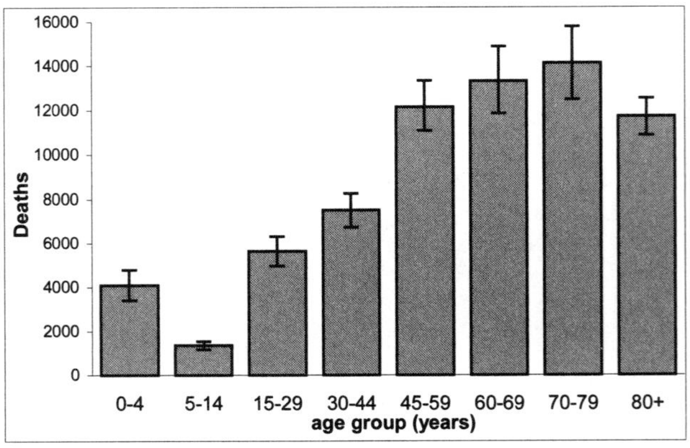

Malaysia　　Malaisie　　Malasia

Females, 2000　　*Femmes, 2000*　　*Mujeres, 2000*

Age	$_nM_x$	$_nq_x$ mean	$_nq_x$ percentiles 2½	$_nq_x$ percentiles 97½	l_x mean	l_x percentiles 2½	l_x percentiles 97½	$_nd_x$	$_nL_x$	e_x mean	e_x percentiles 2½	e_x percentiles 97½	Actual deaths	Actual population
<1	0.0090	0.0089	0.0069	0.0108	100,000	100,000	100,000	888	99,201	74.1	73.2	75.4	2,272	253,844
1-4	0.0008	0.0031	0.0026	0.0037	99,112	98,917	99,307	310	395,703	73.8	72.9	75.1	805	1,025,738
5-9	0.0004	0.0018	0.0015	0.0021	98,802	98,554	99,049	174	493,573	70.0	69.1	71.3	461	1,308,283
10-14	0.0004	0.0020	0.0017	0.0024	98,628	98,352	98,903	200	492,638	65.1	64.2	66.4	446	1,096,334
15-19	0.0005	0.0025	0.0021	0.0029	98,427	98,121	98,733	245	491,525	60.3	59.3	61.6	546	1,095,078
20-24	0.0006	0.0030	0.0025	0.0034	98,183	97,841	98,524	292	490,183	55.4	54.5	56.7	569	957,005
25-29	0.0007	0.0034	0.0029	0.0039	97,891	97,508	98,273	335	488,617	50.6	49.7	51.9	598	872,465
30-34	0.0010	0.0050	0.0042	0.0056	97,556	97,126	97,988	483	486,572	45.7	44.8	47.0	798	804,044
35-39	0.0014	0.0069	0.0059	0.0078	97,073	96,578	97,574	666	483,701	41.0	40.1	42.2	1,066	774,897
40-44	0.0020	0.0101	0.0088	0.0114	96,407	95,829	96,999	973	479,605	36.2	35.4	37.5	1,400	690,442
45-49	0.0032	0.0160	0.0140	0.0179	95,435	94,743	96,150	1,525	473,361	31.6	30.7	32.8	1,820	564,924
50-54	0.0054	0.0268	0.0237	0.0299	93,910	93,061	94,803	2,518	463,253	27.0	26.2	28.2	2,399	441,192
55-59	0.0091	0.0447	0.0391	0.0495	91,391	90,309	92,541	4,086	446,741	22.7	21.9	23.8	2,818	308,075
60-64	0.0160	0.0772	0.0669	0.0856	87,305	85,856	88,796	6,736	419,686	18.7	18.0	19.7	4,424	275,637
65-69	0.0250	0.1176	0.1017	0.1303	80,569	78,638	82,620	9,479	379,150	15.0	14.4	15.9	4,831	193,237
70-74	0.0410	0.1859	0.1618	0.2040	71,091	68,559	74,221	13,219	322,406	11.7	11.1	12.5	6,004	146,451
75-79	0.0700	0.2979	0.2645	0.3239	57,872	54,574	62,214	17,238	246,264	8.8	8.3	9.4	5,604	80,061
80-84	0.1200	0.4615	0.4241	0.4951	40,633	37,057	45,760	18,754	156,283	6.4	6.1	6.9	5,985	49,875
85-89	0.1800	0.6207	0.5845	0.6526	21,880	19,070	26,355	13,580	75,447	4.8	4.6	5.1	3,366	18,700
90-94	0.2500	0.7143	0.6849	0.7406	8,299	6,700	10,795	5,928	23,712	3.6	3.5	3.8	1,584	6,336
95-99	0.3600	0.8654	0.8458	0.8827	2,371	1,742	3,311	2,052	5,700	2.7	2.6	2.8	330	916
100+	0.5000	1	1	1	319	205	509	319	638	2.0	2.0	2.2	68	136
All ages													*48,194*	*10,963,670*

ln($_nq_x$) with uncertainty interval

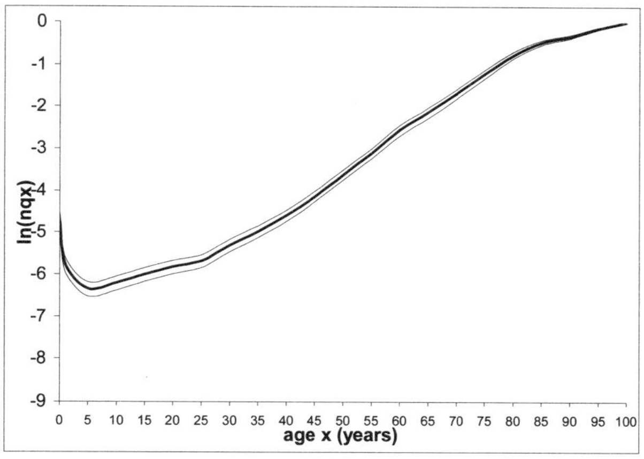

l_x with uncertainty interval

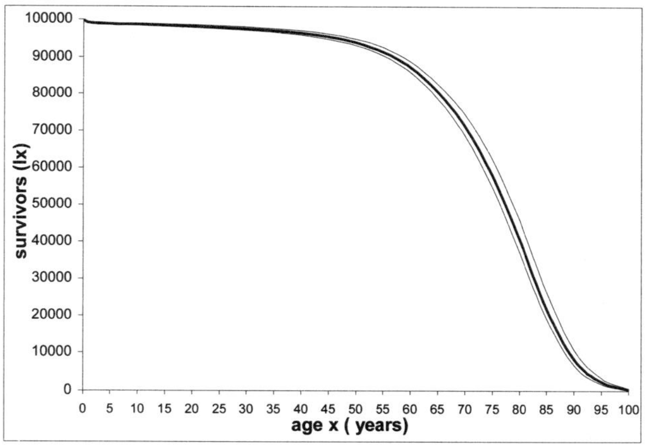

e_x with uncertainty interval

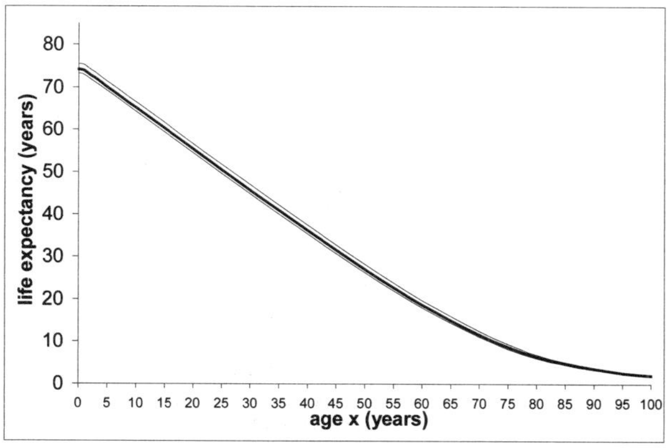

Distribution of deaths by age with uncertainty interval

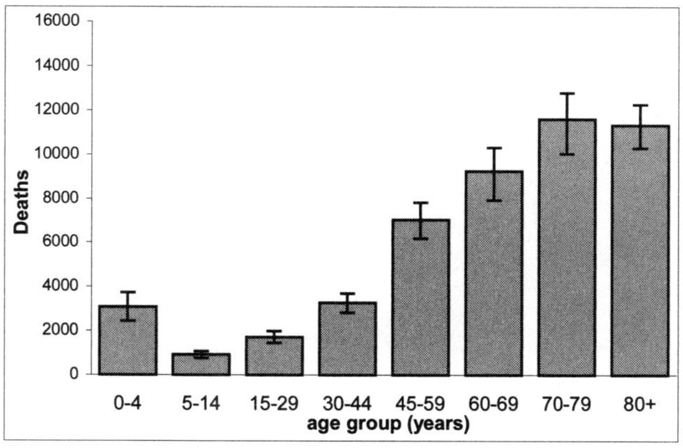

Maldives | **Maldives** | **Maldivas**

Males, 2000 | *Hommes, 2000* | *Hombres, 2000*

Age	$_nM_x$	$_nq_x$ mean	$_nq_x$ percentiles 2½	$_nq_x$ percentiles 97½	l_x mean	l_x percentiles 2½	l_x percentiles 97½	$_nd_x$	$_nL_x$	e_x mean	e_x percentiles 2½	e_x percentiles 97½	Actual deaths	Actual population
<1	0.0479	0.0459	0.0341	0.0580	100,000	100,000	100,000	4,589	95,870	64.6	62.9	66.2	253	5,292
1-4	0.0044	0.0173	0.0142	0.0206	95,411	94,201	96,593	1,648	377,690	66.7	65.6	67.8	83	19,043
5-9	0.0016	0.0082	0.0068	0.0096	93,763	92,259	95,224	767	466,899	63.9	62.8	64.9	34	20,896
10-14	0.0010	0.0049	0.0041	0.0057	92,996	91,370	94,573	451	463,854	59.4	58.4	60.4	19	19,983
15-19	0.0009	0.0043	0.0037	0.0051	92,545	90,849	94,186	402	461,721	54.7	53.7	55.7	15	16,913
20-24	0.0010	0.0047	0.0040	0.0055	92,143	90,388	93,839	437	459,625	49.9	48.9	50.9	13	13,221
25-29	0.0016	0.0079	0.0068	0.0091	91,707	89,889	93,459	722	456,729	45.1	44.2	46.1	17	10,939
30-34	0.0022	0.0111	0.0096	0.0128	90,985	89,068	92,828	1,007	452,408	40.4	39.5	41.4	20	9,095
35-39	0.0034	0.0169	0.0147	0.0193	89,978	87,932	91,939	1,517	446,099	35.9	35.0	36.8	27	7,907
40-44	0.0050	0.0247	0.0218	0.0279	88,461	86,239	90,586	2,184	436,847	31.4	30.6	32.4	31	6,237
45-49	0.0076	0.0373	0.0335	0.0415	86,277	83,834	88,610	3,217	423,343	27.2	26.4	28.1	39	5,174
50-54	0.0120	0.0583	0.0529	0.0641	83,060	80,354	85,644	4,838	403,203	23.1	22.4	24.0	39	3,288
55-59	0.0180	0.0861	0.0786	0.0942	78,221	75,204	81,110	6,737	374,265	19.4	18.7	20.2	58	3,218
60-64	0.0260	0.1221	0.1112	0.1328	71,485	68,118	74,738	8,726	335,609	16.0	15.4	16.7	72	2,781
65-69	0.0404	0.1837	0.1678	0.1986	62,759	59,074	66,423	11,527	284,976	12.9	12.3	13.5	90	2,226
70-74	0.0560	0.2456	0.2248	0.2665	51,232	47,341	55,277	12,583	224,700	10.2	9.7	10.8	95	1,688
75-79	0.0900	0.3673	0.3381	0.3961	38,648	34,722	42,850	14,197	157,749	7.7	7.3	8.2	82	914
80-84	0.1378	0.5124	0.4779	0.5459	24,451	20,969	28,361	12,529	90,934	5.8	5.4	6.1	49	354
85-89	0.2100	0.6885	0.6555	0.7216	11,923	9,523	14,807	8,209	39,090	4.2	4.0	4.4	27	127
90-94	0.3200	0.8163	0.7900	0.8422	3,714	2,651	5,102	3,031	9,473	2.9	2.8	3.1	8	26
95-99	0.4800	0.9836	0.9777	0.9884	682	418	1,071	671	1,398	2.1	2.1	2.1	1	3
100+	0.7879	1	1	1	11	5	24	11	14	1.3	1.3	1.6	0	0
All ages													*1,074*	*149,325*

ln($_nq_x$) with uncertainty interval

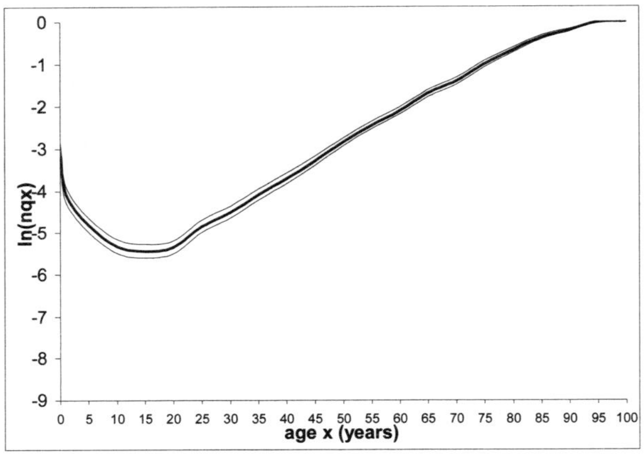

l_x with uncertainty interval

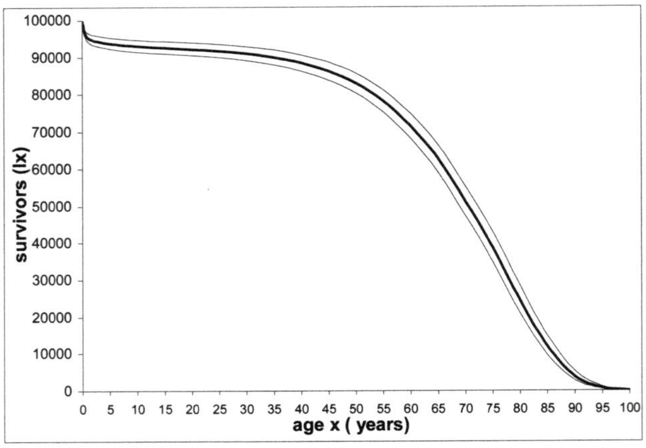

e_x with uncertainty interval

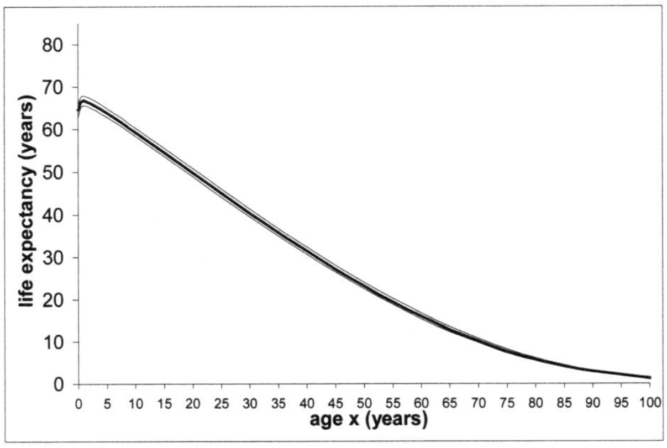

Distribution of deaths by age with uncertainty interval

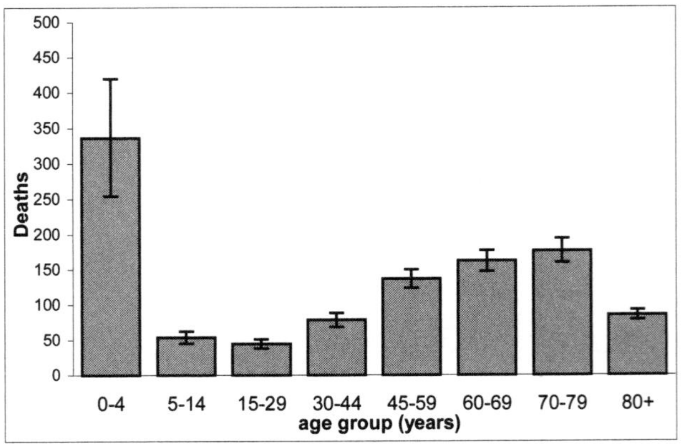

Maldives **Maldives** **Maldivas**

Females, 2000 *Femmes, 2000* *Mujeres, 2000*

Age	$_nM_x$	$_nq_x$ mean	$_nq_x$ percentiles 2½	$_nq_x$ percentiles 97½	l_x mean	l_x percentiles 2½	l_x percentiles 97½	$_nd_x$	$_nL_x$	e_x mean	e_x percentiles 2½	e_x percentiles 97½	Actual deaths	Actual population
<1	0.0501	0.0479	0.0321	0.0638	100,000	100,000	100,000	4,793	95,686	64.4	62.4	66.7	253	5,057
1-4	0.0047	0.0185	0.0140	0.0226	95,207	93,625	96,787	1,765	376,593	66.7	65.3	68.4	85	18,041
5-9	0.0015	0.0075	0.0058	0.0090	93,442	91,505	95,431	698	465,467	63.9	62.6	65.6	29	19,654
10-14	0.0008	0.0040	0.0031	0.0048	92,744	90,682	94,879	370	462,796	59.4	58.1	61.0	15	19,181
15-19	0.0009	0.0045	0.0035	0.0054	92,374	90,249	94,584	415	460,833	54.6	53.3	56.2	15	16,647
20-24	0.0015	0.0075	0.0059	0.0089	91,959	89,766	94,250	687	458,079	49.8	48.5	51.4	19	12,662
25-29	0.0024	0.0119	0.0096	0.0140	91,272	88,971	93,692	1,089	453,639	45.2	43.9	46.8	24	10,128
30-34	0.0034	0.0169	0.0137	0.0196	90,183	87,724	92,797	1,520	447,117	40.7	39.4	42.3	28	8,343
35-39	0.0039	0.0193	0.0160	0.0221	88,663	86,008	91,526	1,712	439,036	36.3	35.1	37.9	28	7,097
40-44	0.0046	0.0227	0.0191	0.0257	86,951	84,104	90,066	1,977	429,812	32.0	30.8	33.5	28	6,149
45-49	0.0075	0.0368	0.0316	0.0411	84,974	81,941	88,346	3,128	417,050	27.7	26.5	29.1	40	5,287
50-54	0.0110	0.0535	0.0468	0.0595	81,846	78,577	85,556	4,381	398,277	23.7	22.5	25.0	38	3,467
55-59	0.0160	0.0769	0.0668	0.0860	77,465	73,904	81,555	5,959	372,428	19.9	18.8	21.2	43	2,662
60-64	0.0295	0.1373	0.1190	0.1544	71,506	67,548	76,109	9,821	332,978	16.3	15.3	17.5	70	2,378
65-69	0.0377	0.1723	0.1496	0.1943	61,685	57,116	67,050	10,630	281,849	13.5	12.5	14.6	73	1,940
70-74	0.0550	0.2418	0.2104	0.2736	51,055	46,021	57,022	12,343	224,416	10.8	10.0	11.8	77	1,400
75-79	0.0790	0.3299	0.2883	0.3709	38,712	33,428	45,026	12,769	161,636	8.4	7.8	9.2	74	937
80-84	0.1200	0.4615	0.4102	0.5086	25,942	21,031	32,043	11,973	99,779	6.3	5.9	7.0	61	510
85-89	0.1800	0.6207	0.5676	0.6654	13,969	10,335	18,901	8,670	48,169	4.6	4.3	5.1	13	74
90-94	0.2900	0.7754	0.7313	0.8103	5,299	3,458	8,174	4,109	14,167	3.2	3.0	3.4	5	18
95-99	0.4500	0.9574	0.9395	0.9691	1,190	656	2,197	1,139	2,532	2.2	2.1	2.3	1	2
100+	0.7000	1	1	1	51	20	133	51	72	1.4	1.4	1.6	0	0
All ages													*1,021*	*141,634*

$\ln(_nq_x)$ with uncertainty interval

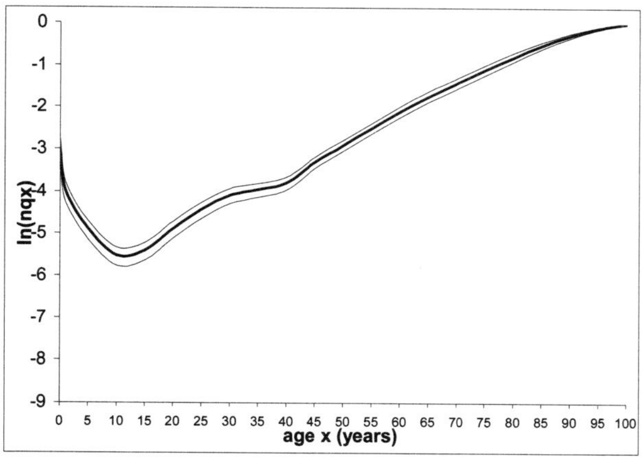

l_x with uncertainty interval

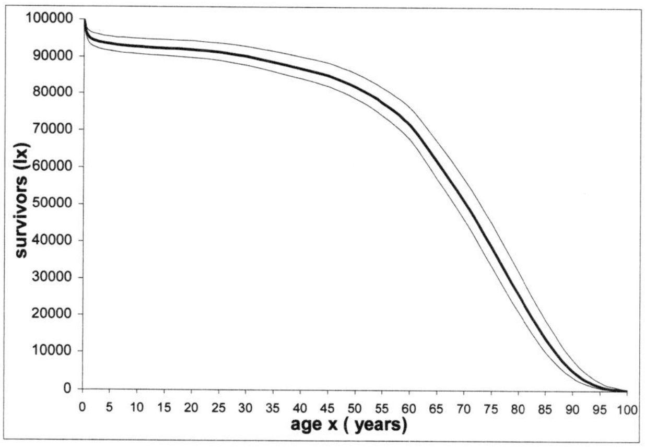

e_x with uncertainty interval

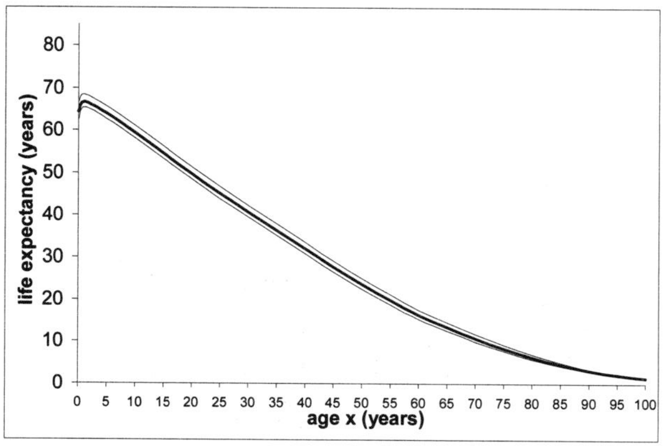

Distribution of deaths by age with uncertainty interval

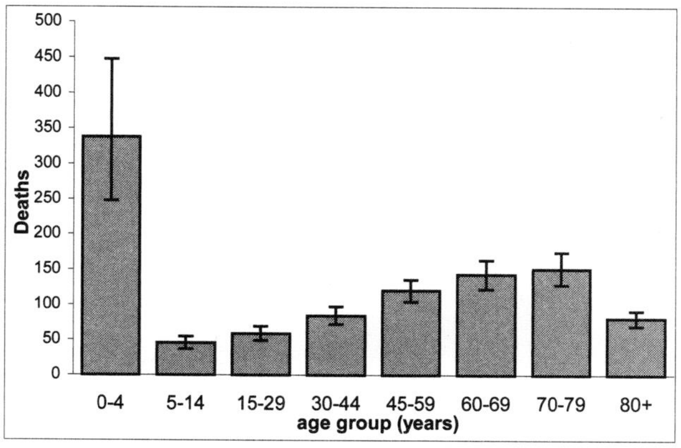

Mali | Mali | Malí

Males, 2000 | *Hommes, 2000* | *Hombres, 2000*

Age	$_nM_x$	$_nq_x$ mean	$_nq_x$ percentiles 2½	$_nq_x$ percentiles 97½	l_x mean	l_x percentiles 2½	l_x percentiles 97½	$_nd_x$	$_nL_x$	e_x mean	e_x percentiles 2½	e_x percentiles 97½	Actual deaths	Actual population
<1	0.2062	0.1740	0.1598	0.1881	100,000	100,000	100,000	17,395	84,344	42.7	40.6	45.3	49,388	239,472
1-4	0.0181	0.0695	0.0614	0.0767	82,605	81,188	84,019	5,740	316,643	50.6	48.0	53.7	15,258	841,659
5-9	0.0043	0.0214	0.0185	0.0238	76,865	75,616	78,113	1,642	380,217	50.3	47.7	53.1	3,631	840,652
10-14	0.0027	0.0135	0.0117	0.0152	75,222	73,841	76,573	1,018	373,566	46.3	43.8	49.1	1,940	711,685
15-19	0.0024	0.0120	0.0103	0.0136	74,204	72,731	75,652	892	368,790	41.9	39.5	44.6	1,462	604,277
20-24	0.0061	0.0302	0.0259	0.0342	73,312	71,759	74,849	2,217	361,016	37.4	35.0	40.1	3,151	513,003
25-29	0.0097	0.0472	0.0402	0.0539	71,095	69,345	72,852	3,357	347,081	33.5	31.1	36.1	4,004	414,014
30-34	0.0132	0.0637	0.0539	0.0733	67,738	65,684	69,825	4,315	327,902	30.0	27.7	32.5	4,257	323,530
35-39	0.0163	0.0783	0.0659	0.0912	63,423	60,966	66,000	4,967	304,697	26.9	24.7	29.2	4,153	254,795
40-44	0.0194	0.0924	0.0773	0.1087	58,456	55,575	61,654	5,400	278,782	24.0	21.9	26.1	3,859	199,259
45-49	0.0222	0.1053	0.0879	0.1248	53,057	49,777	56,887	5,587	251,315	21.2	19.3	23.1	3,515	158,115
50-54	0.0258	0.1213	0.1013	0.1444	47,469	43,823	51,884	5,760	222,946	18.4	16.7	20.1	3,316	128,337
55-59	0.0306	0.1423	0.1188	0.1694	41,709	37,584	46,627	5,936	193,706	15.6	14.1	17.1	3,297	107,613
60-64	0.0444	0.1997	0.1677	0.2361	35,773	31,324	41,069	7,145	161,004	12.7	11.5	14.1	3,998	90,090
65-69	0.0603	0.2620	0.2221	0.3060	28,628	24,030	34,136	7,501	124,387	10.3	9.3	11.4	4,346	72,072
70-74	0.0882	0.3614	0.3120	0.4135	21,127	16,676	26,474	7,635	86,546	8.0	7.2	8.9	4,714	53,431
75-79	0.1287	0.4868	0.4312	0.5425	13,492	9,780	18,103	6,568	51,039	6.2	5.6	6.8	4,564	35,465
80-84	0.1884	0.6405	0.5855	0.6929	6,924	4,474	10,182	4,435	23,533	4.7	4.3	5.1	3,914	20,770
85-89	0.2691	0.8044	0.7599	0.8445	2,489	1,374	4,216	2,003	7,440	3.5	3.3	3.8	2,669	9,918
90-94	0.3711	0.8780	0.8433	0.9081	487	214	1,012	427	1,152	2.6	2.5	2.8	1,450	3,909
95-99	0.4939	0.9951	0.9909	0.9976	59	20	159	59	120	2.0	2.0	2.0	604	1,222
100+	0.6347	1	1	1	0	0	1	0	0	1.6	1.6	1.6	190	299
All ages													*127,682*	*5,623,587*

ln($_nq_x$) with uncertainty interval

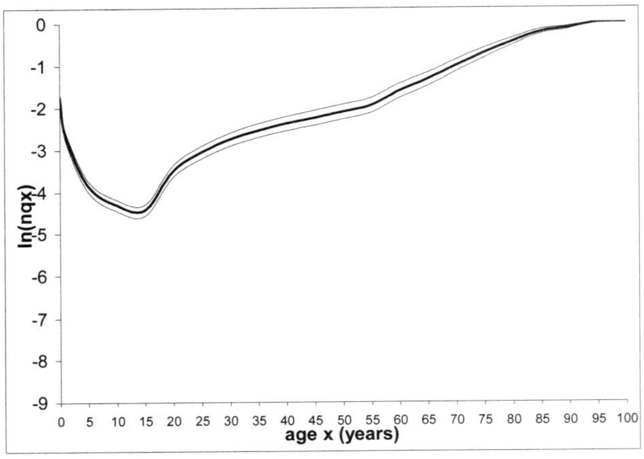

l_x with uncertainty interval

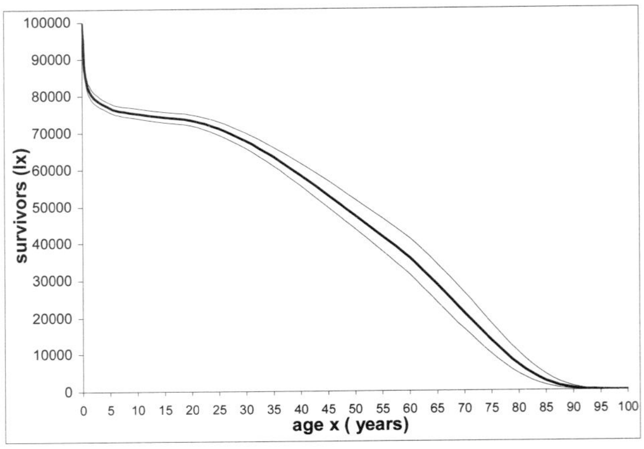

e_x with uncertainty interval

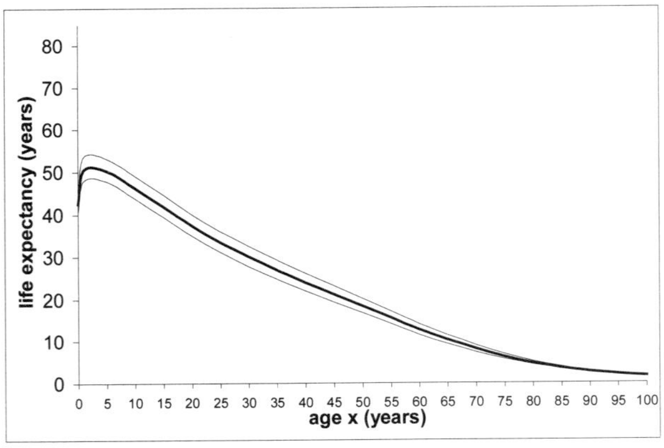

Distribution of deaths by age with uncertainty interval

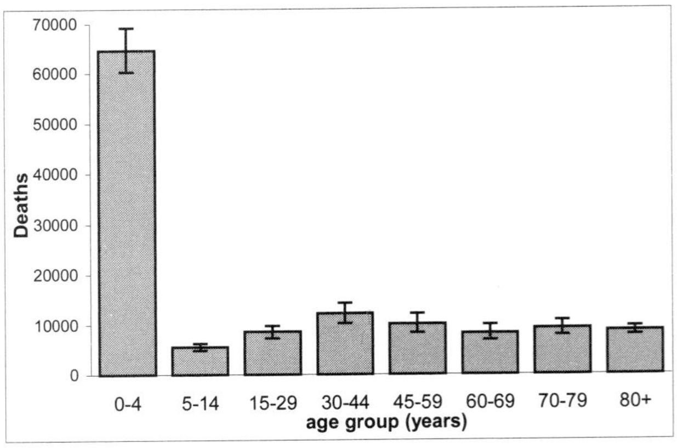

Mali

Mali

Malí

Females, 2000

Femmes, 2000

Mujeres, 2000

Age	$_nM_x$	$_nq_x$			l_x			$_nd_x$	$_nL_x$	e_x			Actual deaths	Actual population
			percentiles			percentiles					percentiles			
		mean	21/2	971/2	mean	21/2	971/2			mean	21/2	971/2		
<1	0.1742	0.1506	0.1361	0.1654	100,000	100,000	100,000	15,056	86,450	44.6	42.0	47.8	40,101	230,262
1-4	0.0215	0.0819	0.0719	0.0916	84,944	83,455	86,394	6,961	323,072	51.5	48.7	55.2	17,905	831,030
5-9	0.0050	0.0248	0.0213	0.0283	77,984	76,737	79,233	1,938	385,074	52.0	49.3	55.4	4,184	831,408
10-14	0.0035	0.0172	0.0147	0.0197	76,046	74,599	77,441	1,308	376,960	48.3	45.6	51.5	2,459	708,856
15-19	0.0056	0.0277	0.0235	0.0317	74,738	73,133	76,304	2,068	368,522	44.1	41.4	47.2	3,368	600,302
20-24	0.0088	0.0432	0.0364	0.0496	72,670	70,815	74,509	3,137	355,509	40.2	37.6	43.3	4,532	513,477
25-29	0.0110	0.0533	0.0445	0.0615	69,533	67,300	71,788	3,707	338,399	36.9	34.4	39.9	4,529	413,496
30-34	0.0121	0.0587	0.0486	0.0680	65,827	63,160	68,571	3,865	319,470	33.9	31.5	36.6	3,907	322,947
35-39	0.0133	0.0643	0.0529	0.0752	61,962	58,868	65,205	3,983	299,849	30.8	28.6	33.3	3,418	257,258
40-44	0.0130	0.0628	0.0514	0.0741	57,978	54,478	61,715	3,640	280,791	27.8	25.7	30.1	2,698	208,105
45-49	0.0142	0.0685	0.0559	0.0814	54,338	50,507	58,499	3,722	262,385	24.5	22.6	26.6	2,453	172,925
50-54	0.0176	0.0845	0.0689	0.1007	50,616	46,491	55,177	4,275	242,391	21.1	19.4	23.0	2,576	146,063
55-59	0.0227	0.1072	0.0876	0.1281	46,341	41,943	51,312	4,968	219,284	17.8	16.3	19.5	2,854	125,974
60-64	0.0321	0.1486	0.1220	0.1770	41,373	36,757	46,773	6,147	191,497	14.6	13.3	16.1	3,470	108,091
65-69	0.0466	0.2088	0.1732	0.2468	35,226	30,506	41,068	7,356	157,739	11.8	10.6	13.0	4,142	88,825
70-74	0.0710	0.3016	0.2550	0.3505	27,870	23,225	33,954	8,404	118,338	9.2	8.3	10.2	4,824	67,921
75-79	0.1050	0.4158	0.3612	0.4717	19,466	15,084	25,297	8,093	77,094	7.1	6.4	7.9	4,931	46,972
80-84	0.1558	0.5607	0.5028	0.6172	11,372	7,969	16,159	6,376	40,920	5.4	4.9	5.9	4,521	29,011
85-89	0.2256	0.7213	0.6690	0.7696	4,996	3,050	8,015	3,603	15,971	4.0	3.7	4.4	3,354	14,864
90-94	0.3183	0.8141	0.7712	0.8529	1,392	703	2,633	1,134	3,561	3.0	2.8	3.2	2,055	6,455
95-99	0.4376	0.9460	0.9237	0.9636	259	103	595	245	560	2.3	2.2	2.4	1,004	2,294
100+	0.5861	1	1	1	14	4	45	14	24	1.7	1.7	1.8	396	675
All ages													*123,678*	*5,727,211*

ln($_nq_x$) with uncertainty interval

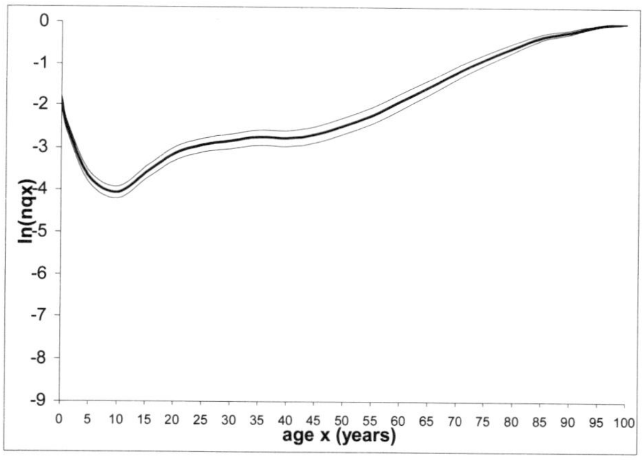

l_x with uncertainty interval

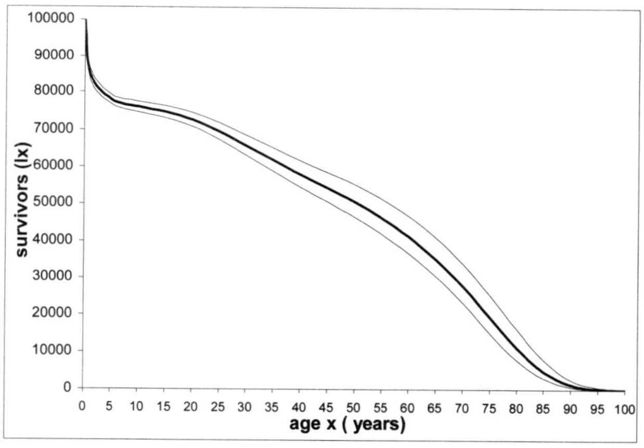

e_x with uncertainty interval

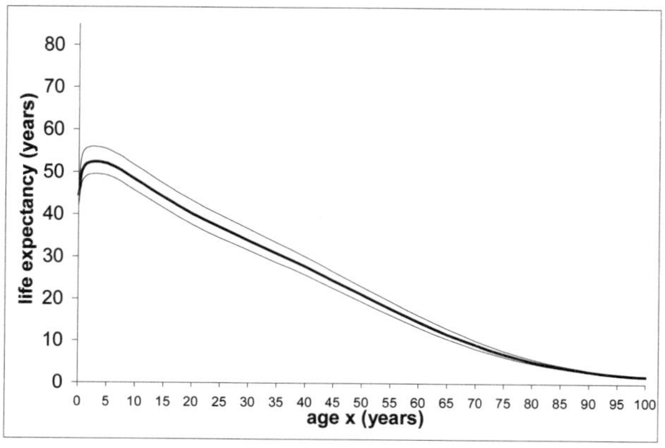

Distribution of deaths by age with uncertainty interval

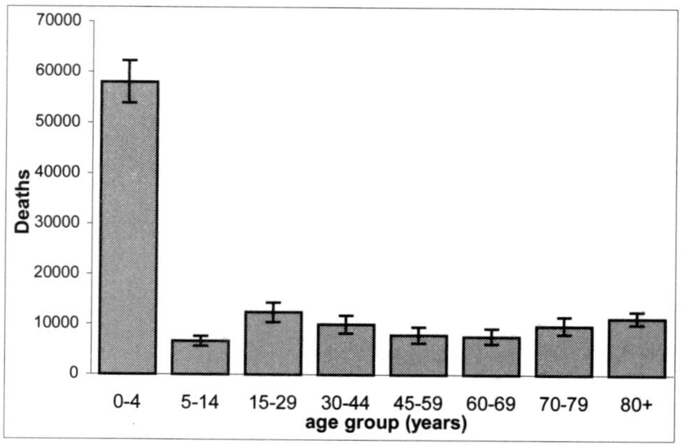

Malta Malte Malta

Males, 2000 *Hommes, 2000* *Hombres, 2000*

Age	$_nM_x$	$_nq_x$			l_x			$_nd_x$	$_nL_x$	e_x			Actual deaths	Actual population
			percentiles			percentiles					percentiles			
		mean	2½	97½	mean	2½	97½			mean	2½	97½		
<1	0.0088	0.0088	0.0074	0.0102	100,000	100,000	100,000	876	99,211	75.4	74.6	76.2	21	2,378
1-4	0.0002	0.0008	0.0007	0.0009	99,124	98,984	99,263	83	396,297	75.0	74.2	75.9	2	9,991
5-9	0.0001	0.0004	0.0003	0.0004	99,041	98,891	99,191	37	495,113	71.1	70.3	72.0	1	13,508
10-14	0.0003	0.0014	0.0013	0.0016	99,004	98,849	99,158	142	494,664	66.1	65.3	67.0	4	14,580
15-19	0.0004	0.0020	0.0018	0.0023	98,862	98,689	99,034	201	493,806	61.2	60.4	62.1	6	15,005
20-24	0.0005	0.0024	0.0021	0.0027	98,661	98,464	98,857	236	492,713	56.3	55.5	57.2	7	15,208
25-29	0.0013	0.0063	0.0056	0.0070	98,425	98,202	98,648	618	490,579	51.5	50.6	52.3	17	13,754
30-34	0.0007	0.0034	0.0031	0.0038	97,807	97,518	98,097	333	488,202	46.8	46.0	47.7	8	11,809
35-39	0.0009	0.0046	0.0041	0.0050	97,474	97,153	97,797	447	486,251	41.9	41.1	42.8	12	13,566
40-44	0.0019	0.0096	0.0088	0.0105	97,026	96,663	97,392	935	482,796	37.1	36.3	38.0	29	15,045
45-49	0.0033	0.0162	0.0149	0.0176	96,092	95,649	96,538	1,560	476,559	32.4	31.6	33.3	47	14,243
50-54	0.0055	0.0271	0.0251	0.0291	94,532	93,971	95,096	2,562	466,253	27.9	27.2	28.8	84	15,221
55-59	0.0091	0.0446	0.0413	0.0483	91,969	91,247	92,692	4,100	449,596	23.7	22.9	24.5	95	10,459
60-64	0.0145	0.0702	0.0643	0.0767	87,869	86,949	88,793	6,168	423,926	19.6	18.9	20.5	123	8,446
65-69	0.0232	0.1095	0.0993	0.1201	81,701	80,477	82,932	8,948	386,136	15.9	15.3	16.7	163	7,046
70-74	0.0367	0.1680	0.1517	0.1837	72,753	70,814	74,454	12,223	333,207	12.6	12.0	13.2	203	5,538
75-79	0.0639	0.2755	0.2505	0.2992	60,530	57,803	63,159	16,677	260,956	9.6	9.1	10.2	267	4,171
80-84	0.0984	0.3947	0.3648	0.4233	43,852	40,592	47,335	17,309	175,989	7.3	7.0	7.8	204	2,078
85-89	0.1487	0.5420	0.5109	0.5700	26,543	23,587	30,062	14,386	96,750	5.5	5.2	5.8	123	828
90-94	0.2208	0.6641	0.6378	0.6871	12,157	10,173	14,644	8,074	36,563	4.0	3.8	4.2	52	234
95-99	0.3222	0.8191	0.8003	0.8351	4,083	3,184	5,259	3,344	10,381	2.9	2.8	3.0	7	23
100+	0.4617	1	1	1	738	525	1,049	738	1,599	2.2	2.1	2.2	0	1
All ages													*1,477*	*193,132*

ln($_nq_x$) with uncertainty interval

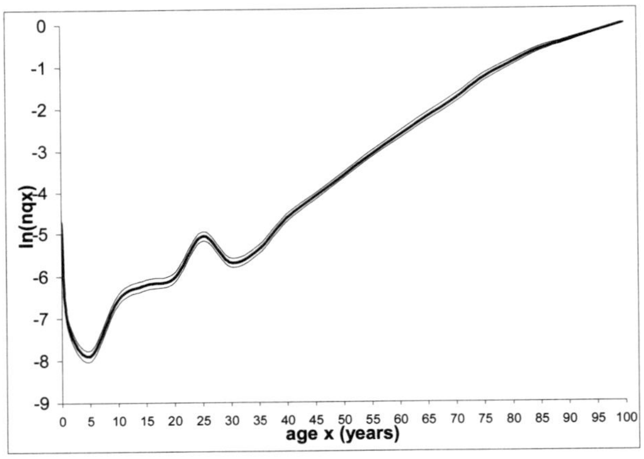

l_x with uncertainty interval

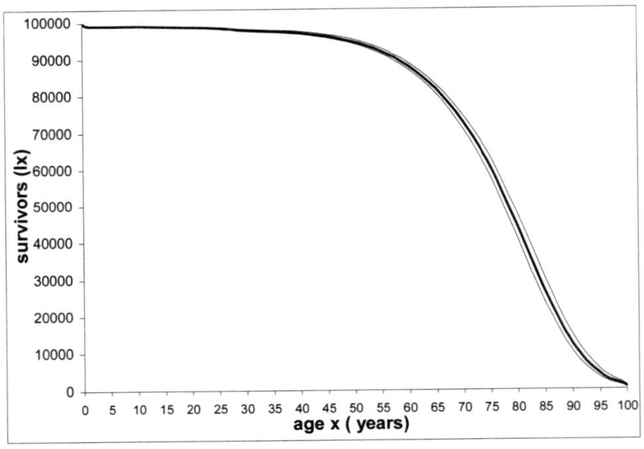

e_x with uncertainty interval

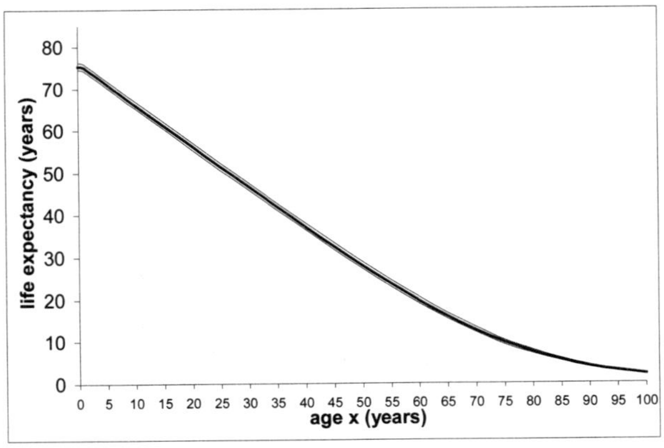

Distribution of deaths by age with uncertainty interval

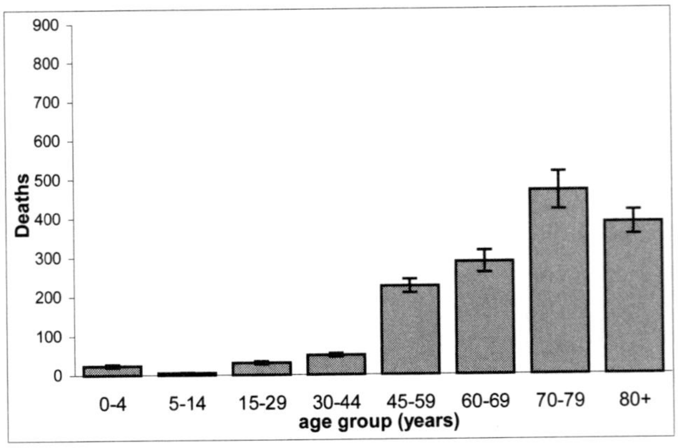

Malta | Malte | Malta

Females, 2000 | *Femmes, 2000* | *Mujeres, 2000*

Age	$_nM_x$	$_nq_x$ mean	$_nq_x$ percentiles 2½	$_nq_x$ percentiles 97½	l_x mean	l_x percentiles 2½	l_x percentiles 97½	$_nd_x$	$_nL_x$	e_x mean	e_x percentiles 2½	e_x percentiles 97½	Actual deaths	Actual population
<1	0.0050	0.0049	0.0036	0.0063	100,000	100,000	100,000	494	99,555	80.7	79.5	82.2	11	2,246
1-4	0.0002	0.0007	0.0005	0.0009	99,506	99,371	99,640	70	397,856	80.1	79.0	81.7	2	9,431
5-9	0.0001	0.0003	0.0002	0.0004	99,436	99,288	99,586	30	497,104	76.2	75.0	77.7	1	12,680
10-14	0.0002	0.0012	0.0009	0.0014	99,406	99,252	99,562	119	496,732	71.2	70.1	72.7	3	13,773
15-19	0.0003	0.0014	0.0011	0.0017	99,287	99,109	99,469	141	496,083	66.3	65.1	67.8	4	14,256
20-24	0.0003	0.0014	0.0012	0.0017	99,146	98,941	99,358	143	495,373	61.4	60.2	62.9	4	14,243
25-29	0.0003	0.0013	0.0011	0.0016	99,003	98,770	99,244	132	494,685	56.5	55.3	58.0	3	12,978
30-34	0.0003	0.0016	0.0013	0.0019	98,871	98,613	99,138	160	493,956	51.6	50.4	53.0	4	11,194
35-39	0.0005	0.0023	0.0019	0.0028	98,711	98,423	99,008	232	492,976	46.6	45.5	48.1	6	13,028
40-44	0.0008	0.0039	0.0032	0.0046	98,479	98,148	98,819	385	491,435	41.7	40.6	43.2	12	14,794
45-49	0.0014	0.0072	0.0059	0.0084	98,095	97,696	98,501	703	488,716	36.9	35.8	38.4	20	14,000
50-54	0.0021	0.0106	0.0089	0.0124	97,392	96,878	97,909	1,035	484,371	32.1	31.1	33.6	33	15,229
55-59	0.0035	0.0171	0.0145	0.0200	96,357	95,687	97,021	1,649	477,662	27.5	26.4	28.9	38	11,089
60-64	0.0062	0.0304	0.0258	0.0357	94,708	93,811	95,599	2,881	466,337	22.9	21.9	24.3	59	9,494
65-69	0.0115	0.0561	0.0467	0.0656	91,827	90,534	93,098	5,151	446,259	18.5	17.6	19.9	102	8,853
70-74	0.0250	0.1176	0.0969	0.1361	86,676	84,593	88,565	10,197	407,888	14.5	13.7	15.7	184	7,349
75-79	0.0440	0.1982	0.1637	0.2249	76,479	73,079	79,859	15,158	344,500	11.1	10.4	12.1	261	5,927
80-84	0.0800	0.3333	0.2835	0.3717	61,321	56,686	66,783	20,440	255,504	8.2	7.6	9.0	272	3,398
85-89	0.1250	0.4762	0.4241	0.5188	40,881	36,038	47,849	19,467	155,736	6.1	5.6	6.6	238	1,904
90-94	0.2000	0.6250	0.5819	0.6627	21,414	17,709	27,556	13,384	66,918	4.3	4.0	4.6	153	763
95-99	0.3000	0.7895	0.7582	0.8165	8,030	5,993	11,480	6,340	21,132	3.1	3.0	3.3	50	165
100+	0.4400	1	1	1	1,691	1,110	2,715	1,691	3,842	2.3	2.2	2.5	7	15
All ages													*1,465*	*196,809*

ln($_nq_x$) with uncertainty interval

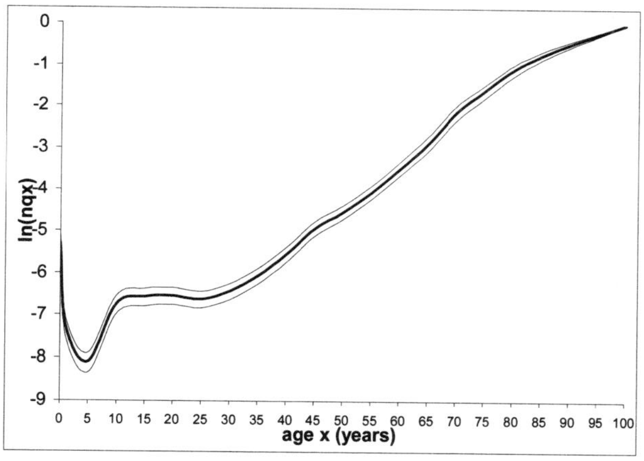

l_x with uncertainty interval

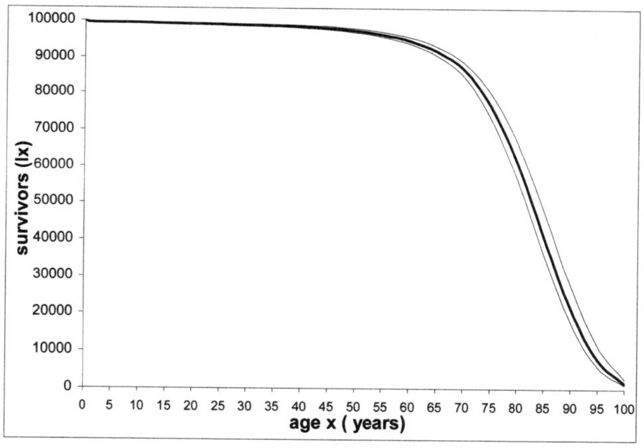

e_x with uncertainty interval

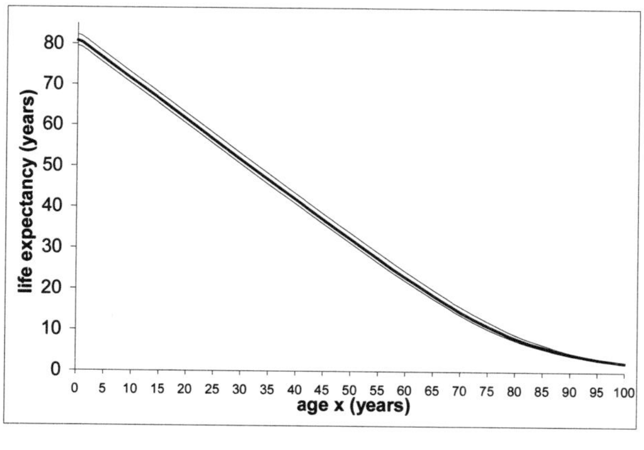

Distribution of deaths by age with uncertainty interval

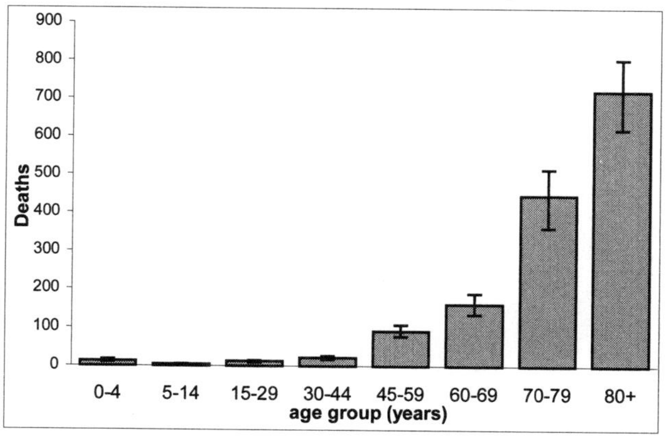

Marshall Islands **Iles Marshall** **Islas Marshall**

Males, 2000 *Hommes, 2000* *Hombres, 2000*

Age	$_nM_x$	$_nq_x$			l_x			$_nd_x$	$_nL_x$	e_x			Actual deaths	Actual population
			percentiles			percentiles					percentiles			
		mean	2½	97½	mean	2½	97½			mean	2½	97½		
<1	0.0407	0.0392	0.0335	0.0468	100,000	100,000	100,000	3,923	96,469	62.8	61.4	64.3	30	748
1-4	0.0023	0.0091	0.0080	0.0105	96,077	95,322	96,649	876	382,207	64.3	63.3	65.7	7	3,003
5-9	0.0006	0.0031	0.0027	0.0035	95,202	94,321	95,877	291	475,281	60.9	59.8	62.3	2	3,497
10-14	0.0004	0.0022	0.0019	0.0025	94,911	93,990	95,621	207	474,036	56.1	55.0	57.4	1	2,931
15-19	0.0009	0.0043	0.0038	0.0050	94,704	93,755	95,438	410	472,493	51.2	50.1	52.5	2	2,549
20-24	0.0024	0.0119	0.0105	0.0136	94,293	93,289	95,075	1,125	468,654	46.4	45.3	47.8	5	2,088
25-29	0.0030	0.0149	0.0131	0.0169	93,168	92,021	94,078	1,387	462,373	41.9	40.9	43.3	6	1,897
30-34	0.0050	0.0247	0.0218	0.0277	91,781	90,470	92,845	2,266	453,240	37.5	36.5	38.8	10	1,938
35-39	0.0065	0.0320	0.0284	0.0356	89,515	87,961	90,819	2,863	440,417	33.4	32.4	34.7	11	1,650
40-44	0.0074	0.0363	0.0324	0.0400	86,652	84,832	88,242	3,148	425,391	29.4	28.4	30.7	11	1,450
45-49	0.0120	0.0583	0.0523	0.0635	83,504	81,436	85,384	4,864	405,360	25.4	24.5	26.6	15	1,237
50-54	0.0150	0.0723	0.0645	0.0791	78,640	76,267	80,918	5,685	378,987	21.9	20.9	22.9	15	1,025
55-59	0.0197	0.0941	0.0836	0.1037	72,955	70,234	75,702	6,864	347,615	18.4	17.5	19.4	15	735
60-64	0.0300	0.1395	0.1240	0.1542	66,091	62,952	69,374	9,222	307,399	15.0	14.3	15.9	18	598
65-69	0.0450	0.2022	0.1803	0.2229	56,869	53,247	60,771	11,502	255,590	12.1	11.4	12.9	21	459
70-74	0.0650	0.2796	0.2505	0.3063	45,367	41,379	49,813	12,683	195,128	9.5	8.9	10.1	24	364
75-79	0.1000	0.4000	0.3640	0.4326	32,684	28,703	37,335	13,074	130,736	7.2	6.8	7.7	21	208
80-84	0.1600	0.5714	0.5333	0.6051	19,610	16,285	23,745	11,206	70,037	5.3	5.0	5.7	13	84
85-89	0.2200	0.7097	0.6765	0.7381	8,404	6,430	11,081	5,964	27,111	4.0	3.9	4.3	7	30
90-94	0.3400	0.8416	0.8172	0.8619	2,440	1,684	3,585	2,053	6,040	2.8	2.7	3.0	2	6
95-99	0.4500	0.9574	0.9464	0.9661	387	233	655	370	822	2.2	2.2	2.3	0	1
100+	0.8000	1	1	1	16	8	35	16	21	1.3	1.2	1.8	0	0
All ages													*235*	*26,498*

ln($_nq_x$) with uncertainty interval

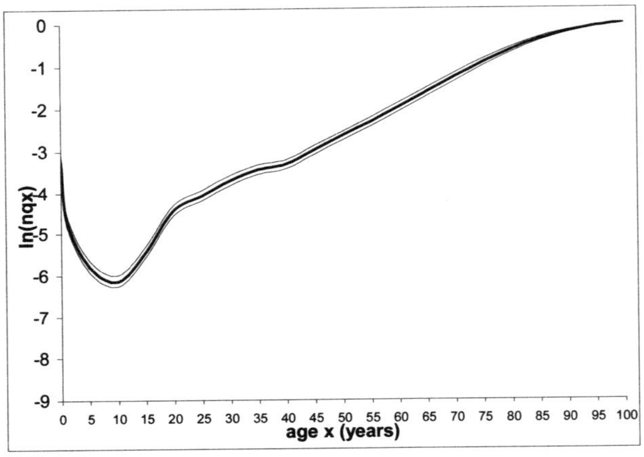

l_x with uncertainty interval

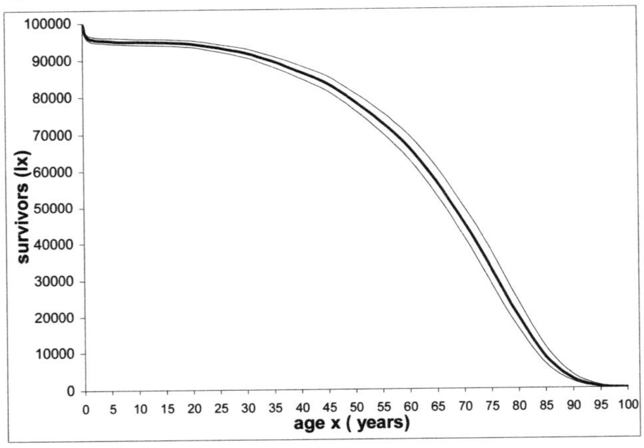

e_x with uncertainty interval

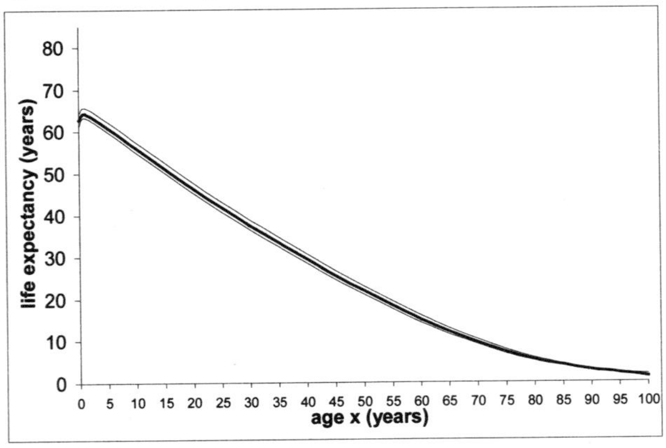

Distribution of deaths by age with uncertainty interval

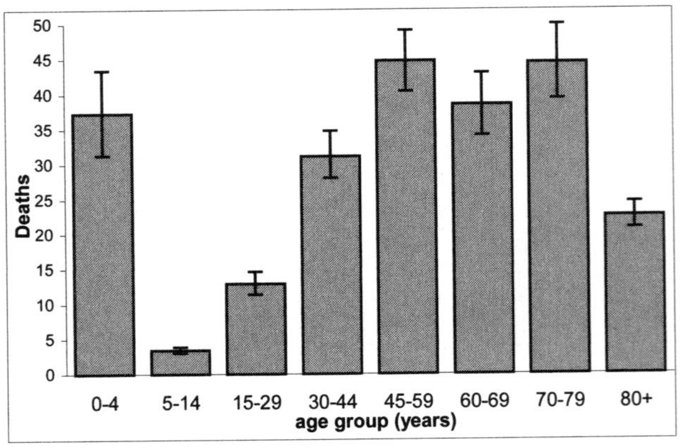

Marshall Islands | **Iles Marshall** | **Islas Marshall**

Females, 2000 | *Femmes, 2000* | *Mujeres, 2000*

Age	$_nM_x$	$_nq_x$			l_x			$_nd_x$	$_nL_x$	e_x			Actual deaths	Actual population
			percentiles			percentiles					percentiles			
		mean	21/2	971/2	mean	21/2	971/2			mean	21/2	971/2		
<1	0.0343	0.0332	0.0273	0.0403	100,000	100,000	100,000	3,324	97,008	67.8	66.6	69.0	24	708
1-4	0.0015	0.0058	0.0050	0.0068	96,676	95,975	97,270	564	385,350	69.2	68.2	70.1	4	2,824
5-9	0.0006	0.0029	0.0025	0.0034	96,112	95,324	96,784	283	479,854	65.5	64.6	66.5	2	3,292
10-14	0.0004	0.0022	0.0019	0.0025	95,829	94,999	96,539	211	478,621	60.7	59.8	61.7	1	2,761
15-19	0.0005	0.0027	0.0023	0.0031	95,619	94,759	96,357	259	477,448	55.9	55.0	56.8	1	2,386
20-24	0.0014	0.0070	0.0060	0.0080	95,360	94,463	96,131	663	475,144	51.0	50.1	51.9	3	2,007
25-29	0.0024	0.0120	0.0105	0.0137	94,697	93,710	95,550	1,139	470,639	46.3	45.5	47.2	4	1,787
30-34	0.0033	0.0164	0.0145	0.0184	93,558	92,428	94,543	1,531	463,964	41.9	41.0	42.7	5	1,590
35-39	0.0044	0.0218	0.0195	0.0243	92,027	90,724	93,173	2,003	455,130	37.5	36.7	38.3	7	1,505
40-44	0.0055	0.0271	0.0246	0.0299	90,025	88,523	91,358	2,442	444,018	33.3	32.6	34.1	7	1,334
45-49	0.0083	0.0404	0.0371	0.0441	87,583	85,876	89,112	3,540	429,063	29.2	28.5	29.9	9	1,127
50-54	0.0110	0.0535	0.0493	0.0579	84,043	82,089	85,804	4,499	408,967	25.3	24.7	26.0	10	912
55-59	0.0154	0.0741	0.0686	0.0799	79,544	77,334	81,572	5,898	382,976	21.6	21.0	22.2	10	655
60-64	0.0220	0.1043	0.0971	0.1117	73,646	71,154	75,975	7,679	349,035	18.1	17.6	18.7	13	578
65-69	0.0330	0.1524	0.1424	0.1620	65,968	63,206	68,599	10,055	304,700	14.9	14.4	15.5	15	458
70-74	0.0484	0.2159	0.2020	0.2295	55,912	52,970	58,834	12,070	249,386	12.2	11.7	12.7	17	353
75-79	0.0621	0.2686	0.2517	0.2857	43,842	40,816	46,949	11,777	189,769	9.8	9.4	10.3	13	216
80-84	0.0942	0.3812	0.3589	0.4028	32,066	29,153	35,132	12,222	129,772	7.5	7.2	7.9	10	102
85-89	0.1424	0.5250	0.4999	0.5484	19,843	17,409	22,522	10,418	73,172	5.6	5.4	5.9	5	36
90-94	0.2144	0.6524	0.6282	0.6736	9,425	7,862	11,264	6,149	28,680	4.1	3.9	4.2	2	7
95-99	0.3216	0.8184	0.7993	0.8341	3,276	2,566	4,188	2,681	8,338	2.9	2.8	3.0	0	1
100+	0.4804	1	1	1	595	426	840	595	1,239	2.1	2.0	2.1	0	0
All ages													*164*	*24,639*

ln($_nq_x$) with uncertainty interval

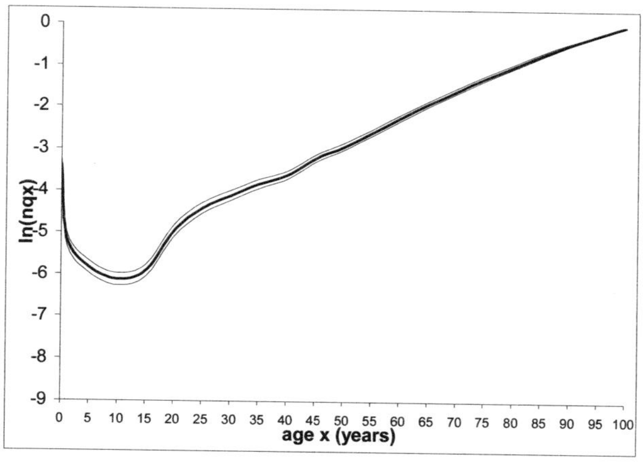

l_x with uncertainty interval

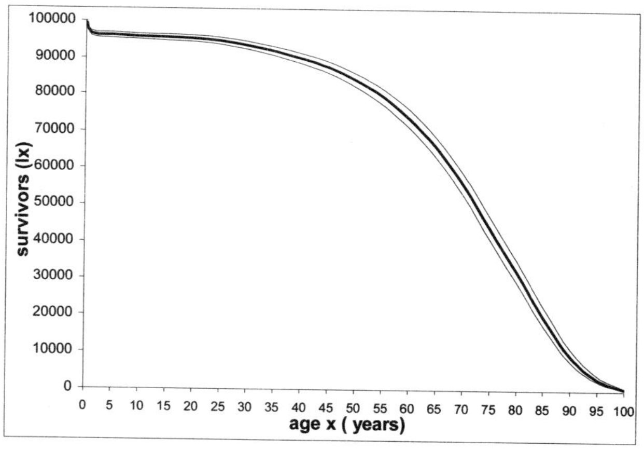

e_x with uncertainty interval

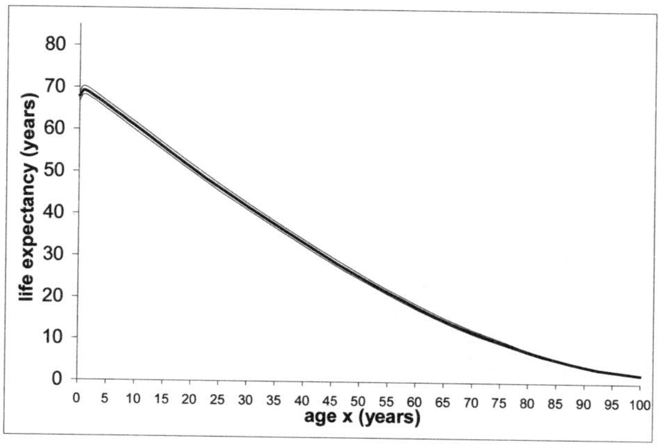

Distribution of deaths by age with uncertainty interval

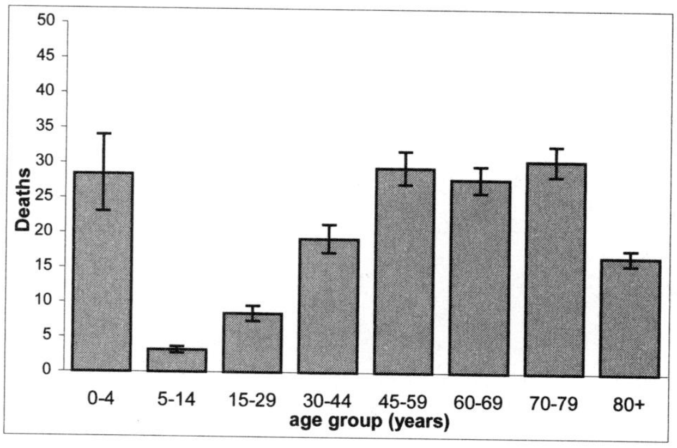

Mauritania **Mauritanie** **Mauritania**

Males, 2000 *Hommes, 2000* *Hombres, 2000*

Age	$_nM_x$	$_nq_x$			l_x			$_nd_x$	$_nL_x$	e_x			Actual deaths	Actual population
			percentiles			percentiles					percentiles			
		mean	2½	97½	mean	2½	97½			mean	2½	97½		
<1	0.1536	0.1349	0.1223	0.1482	100,000	100,000	100,000	13,495	87,855	51.7	49.1	54.3	8,026	52,249
1-4	0.0120	0.0467	0.0410	0.0529	86,505	85,178	87,766	4,042	336,319	58.7	56.0	61.8	2,214	184,206
5-9	0.0028	0.0140	0.0121	0.0161	82,463	81,218	83,695	1,154	409,428	57.5	55.0	60.4	540	191,539
10-14	0.0017	0.0085	0.0073	0.0099	81,308	79,988	82,578	693	404,809	53.3	50.9	56.1	278	162,271
15-19	0.0014	0.0069	0.0059	0.0080	80,615	79,245	81,930	558	401,681	48.7	46.4	51.5	200	143,770
20-24	0.0032	0.0158	0.0135	0.0184	80,057	78,646	81,412	1,268	397,116	44.0	41.7	46.8	380	118,931
25-29	0.0047	0.0231	0.0196	0.0270	78,789	77,258	80,264	1,820	389,397	39.7	37.5	42.4	460	98,411
30-34	0.0061	0.0303	0.0255	0.0356	76,969	75,267	78,671	2,330	379,022	35.6	33.4	38.2	498	80,970
35-39	0.0079	0.0386	0.0322	0.0457	74,639	72,678	76,633	2,884	365,986	31.6	29.6	34.1	509	64,632
40-44	0.0101	0.0493	0.0407	0.0586	71,755	69,360	74,111	3,538	349,931	27.8	25.9	30.1	579	57,267
45-49	0.0132	0.0637	0.0521	0.0760	68,217	65,296	71,011	4,346	330,221	24.1	22.3	26.3	622	47,242
50-54	0.0177	0.0848	0.0688	0.1015	63,871	60,332	67,186	5,418	305,811	20.6	19.0	22.6	639	36,037
55-59	0.0241	0.1138	0.0917	0.1359	58,453	54,210	62,482	6,649	275,641	17.2	15.8	19.1	654	27,106
60-64	0.0343	0.1581	0.1274	0.1877	51,803	46,841	56,666	8,191	238,540	14.1	12.9	15.8	681	19,839
65-69	0.0495	0.2204	0.1788	0.2596	43,613	38,049	49,318	9,612	194,034	11.3	10.2	12.7	885	17,861
70-74	0.0752	0.3165	0.2617	0.3662	34,001	28,228	40,498	10,760	143,104	8.8	8.0	10.0	633	8,419
75-79	0.1131	0.4409	0.3766	0.4965	23,241	18,002	29,898	10,247	90,587	6.7	6.1	7.6	746	6,593
80-84	0.1673	0.5897	0.5237	0.6437	12,994	9,184	18,639	7,663	45,813	5.1	4.7	5.7	453	2,711
85-89	0.2404	0.7509	0.6940	0.7945	5,331	3,310	8,856	4,003	16,648	3.8	3.6	4.2	194	806
90-94	0.3360	0.8366	0.7913	0.8705	1,328	684	2,686	1,111	3,308	2.8	2.7	3.1	45	134
95-99	0.4563	0.9631	0.9434	0.9756	217	89	553	209	458	2.2	2.1	2.3	5	11
100+	0.6024	1	1	1	8	2	31	8	13	1.7	1.6	1.7	0	0
All ages													*19,239*	*1,321,005*

ln($_nq_x$) with uncertainty interval

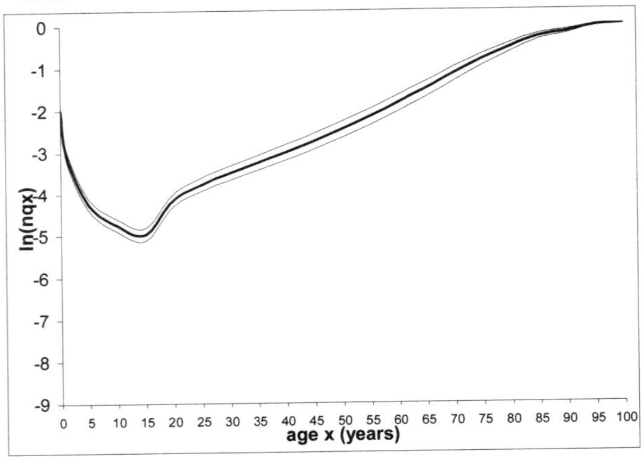

l_x with uncertainty interval

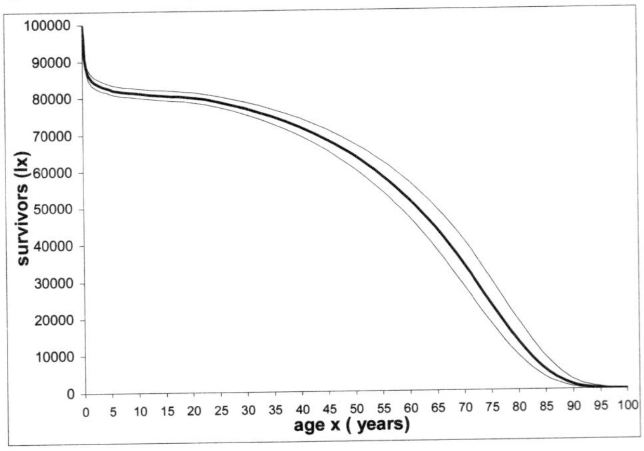

e_x with uncertainty interval

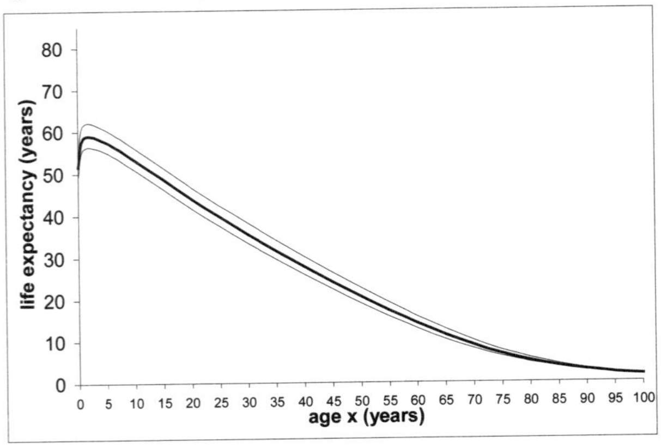

Distribution of deaths by age with uncertainty interval

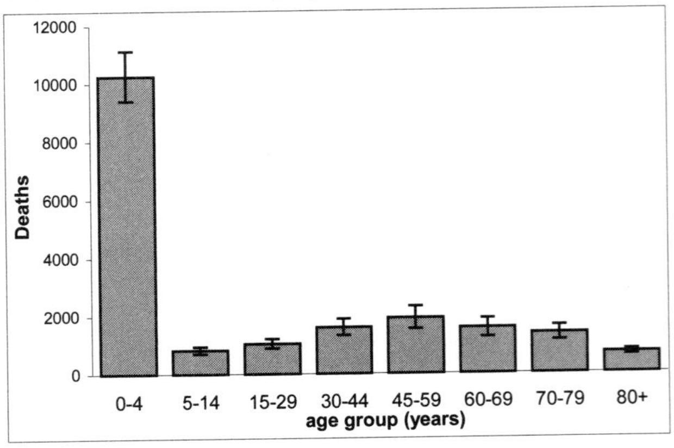

Mauritania

Mauritanie

Mauritania

Females, 2000 *Femmes, 2000* *Mujeres, 2000*

Age	$_nM_x$	$_nq_x$ mean	$_nq_x$ percentiles 21/2	$_nq_x$ percentiles 971/2	l_x mean	l_x percentiles 21/2	l_x percentiles 971/2	$_nd_x$	$_nL_x$	e_x mean	e_x percentiles 21/2	e_x percentiles 971/2	Actual deaths	Actual population
<1	0.1356	0.1209	0.1080	0.1347	100,000	100,000	100,000	12,089	89,120	53.5	50.9	56.4	6,884	50,752
1-4	0.0137	0.0532	0.0464	0.0602	87,911	86,529	89,199	4,678	340,417	59.9	57.0	63.1	2,517	183,162
5-9	0.0031	0.0152	0.0131	0.0175	83,233	81,986	84,479	1,263	413,008	59.2	56.6	62.1	582	190,388
10-14	0.0020	0.0098	0.0084	0.0114	81,970	80,628	83,375	803	407,844	55.0	52.5	57.9	318	161,630
15-19	0.0028	0.0139	0.0118	0.0162	81,167	79,765	82,674	1,128	403,016	50.6	48.1	53.4	398	142,285
20-24	0.0044	0.0215	0.0182	0.0252	80,039	78,536	81,691	1,724	395,886	46.2	43.8	49.0	517	118,724
25-29	0.0053	0.0263	0.0221	0.0310	78,315	76,642	80,189	2,058	386,430	42.2	39.9	44.9	529	99,278
30-34	0.0059	0.0290	0.0242	0.0344	76,257	74,383	78,396	2,214	375,749	38.3	36.1	40.9	480	81,441
35-39	0.0069	0.0342	0.0282	0.0407	74,043	71,843	76,467	2,529	363,893	34.3	32.3	36.8	486	69,883
40-44	0.0073	0.0359	0.0294	0.0429	71,514	68,921	74,261	2,564	351,161	30.5	28.6	32.8	428	58,675
45-49	0.0091	0.0444	0.0360	0.0533	68,950	65,962	72,018	3,058	337,106	26.5	24.7	28.7	445	49,070
50-54	0.0125	0.0607	0.0490	0.0731	65,892	62,448	69,331	4,001	319,458	22.6	21.0	24.7	470	37,512
55-59	0.0176	0.0842	0.0675	0.1014	61,891	57,883	65,793	5,210	296,432	18.9	17.4	20.8	539	30,680
60-64	0.0267	0.1250	0.0999	0.1499	56,682	52,014	61,145	7,082	265,702	15.4	14.1	17.2	589	22,096
65-69	0.0410	0.1861	0.1493	0.2211	49,599	44,217	54,934	9,231	224,918	12.3	11.1	13.8	831	20,238
70-74	0.0653	0.2806	0.2285	0.3286	40,368	34,449	46,628	11,326	173,525	9.5	8.6	10.8	911	13,954
75-79	0.1005	0.4016	0.3368	0.4583	29,042	23,217	35,973	11,663	116,054	7.3	6.5	8.2	833	8,285
80-84	0.1515	0.5493	0.4804	0.6066	17,379	12,705	23,857	9,547	63,029	5.5	5.0	6.1	594	3,919
85-89	0.2217	0.7133	0.6510	0.7615	7,832	5,093	12,393	5,587	25,196	4.1	3.8	4.5	279	1,258
90-94	0.3153	0.8101	0.7600	0.8478	2,246	1,221	4,297	1,819	5,771	3.0	2.8	3.3	83	262
95-99	0.4353	0.9439	0.9177	0.9612	426	187	1,018	403	925	2.3	2.2	2.4	13	29
100+	0.5838	1	1	1	24	7	82	24	41	1.7	1.7	1.8	1	2
All ages													*18,726*	*1,343,523*

ln($_nq_x$) with uncertainty interval

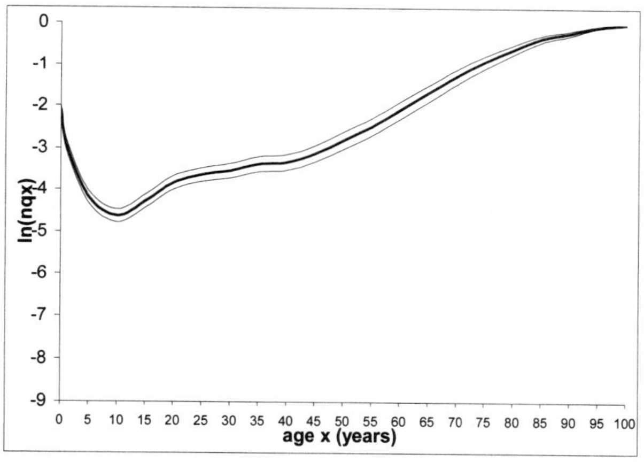

l_x with uncertainty interval

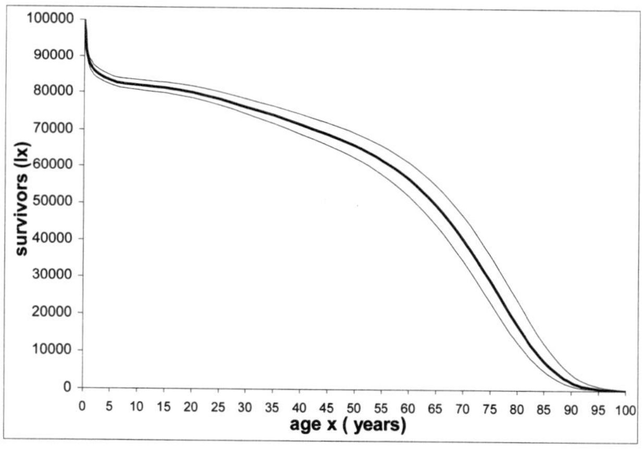

e_x with uncertainty interval

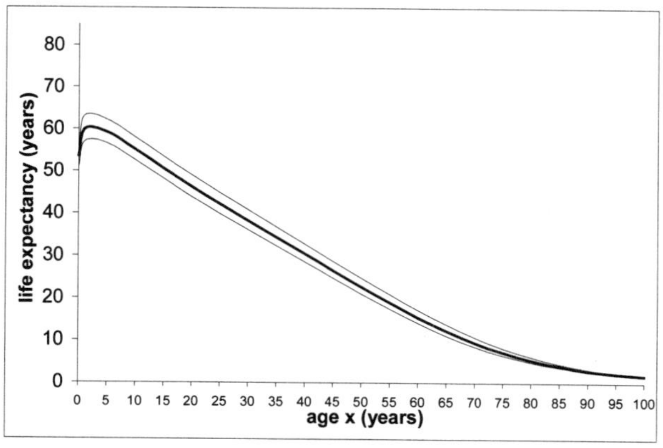

Distribution of deaths by age with uncertainty interval

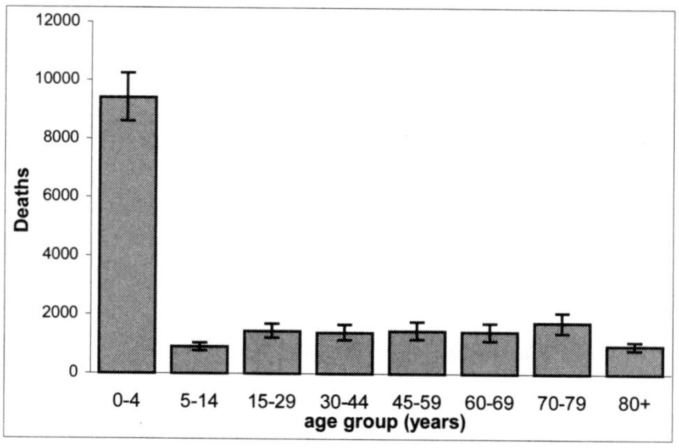

Mauritius   Maurice   Mauricio

Males, 2000   *Hommes, 2000*   *Hombres, 2000*

Age	$_nM_x$	$_nq_x$ mean	$_nq_x$ percentiles 2 1/2	$_nq_x$ percentiles 97 1/2	l_x mean	l_x percentiles 2 1/2	l_x percentiles 97 1/2	$_nd_x$	$_nL_x$	e_x mean	e_x percentiles 2 1/2	e_x percentiles 97 1/2	Actual deaths	Actual population
<1	0.0195	0.0192	0.0183	0.0234	100,000	100,000	100,000	1,917	98,275	67.6	67.0	68.1	191	9,788
1-4	0.0004	0.0018	0.0017	0.0021	98,083	97,656	98,173	176	391,911	68.0	67.3	68.6	17	38,439
5-9	0.0002	0.0009	0.0009	0.0011	97,907	97,453	98,005	89	489,313	64.1	63.5	64.7	10	54,581
10-14	0.0003	0.0015	0.0015	0.0017	97,818	97,351	97,919	148	488,719	59.1	58.5	59.8	15	48,053
15-19	0.0005	0.0027	0.0026	0.0031	97,670	97,181	97,777	263	487,689	54.2	53.6	54.8	28	51,750
20-24	0.0010	0.0051	0.0048	0.0057	97,406	96,881	97,525	493	485,799	49.4	48.8	50.0	56	55,510
25-29	0.0014	0.0069	0.0066	0.0078	96,913	96,325	97,052	670	482,891	44.6	44.0	45.2	65	46,647
30-34	0.0019	0.0096	0.0092	0.0106	96,243	95,577	96,410	921	478,914	39.9	39.3	40.5	95	49,488
35-39	0.0034	0.0171	0.0164	0.0187	95,322	94,562	95,526	1,627	472,543	35.3	34.7	35.9	175	50,726
40-44	0.0055	0.0272	0.0261	0.0293	93,695	92,795	93,963	2,547	462,108	30.8	30.3	31.5	247	44,757
45-49	0.0086	0.0419	0.0403	0.0446	91,148	90,080	91,511	3,823	446,183	26.6	26.1	27.3	328	38,228
50-54	0.0117	0.0568	0.0542	0.0602	87,325	86,064	87,822	4,964	424,215	22.7	22.2	23.3	317	27,069
55-59	0.0176	0.0844	0.0796	0.0895	82,361	80,882	83,064	6,949	394,432	18.9	18.4	19.5	330	18,705
60-64	0.0311	0.1444	0.1349	0.1529	75,412	73,644	76,456	10,893	349,826	15.4	14.9	16.0	459	14,732
65-69	0.0408	0.1853	0.1718	0.1961	64,519	62,384	66,145	11,957	292,702	12.6	12.2	13.1	467	11,434
70-74	0.0618	0.2677	0.2483	0.2822	52,562	50,148	54,783	14,071	227,634	9.9	9.6	10.4	575	9,306
75-79	0.0932	0.3778	0.3520	0.3948	38,492	35,997	41,179	14,543	156,100	7.5	7.3	8.0	545	5,846
80-84	0.1453	0.5329	0.5027	0.5499	23,948	21,784	26,682	12,761	87,839	5.6	5.5	5.9	361	2,485
85-89	0.2159	0.7011	0.6734	0.7148	11,187	9,804	13,268	7,843	36,329	4.2	4.1	4.4	181	840
90-94	0.3057	0.7973	0.7751	0.8076	3,344	2,797	4,333	2,666	8,722	3.1	3.0	3.2	53	174
95-99	0.4125	0.9218	0.9086	0.9276	678	538	975	625	1,515	2.4	2.4	2.5	9	21
100+	0.5304	1	1	1	53	39	89	53	100	1.9	1.9	1.9	1	1
All ages													*4,523*	*578,580*

ln($_nq_x$) with uncertainty interval

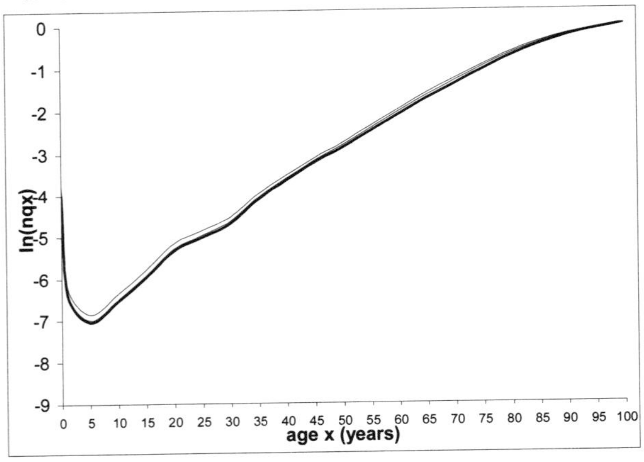

l_x with uncertainty interval

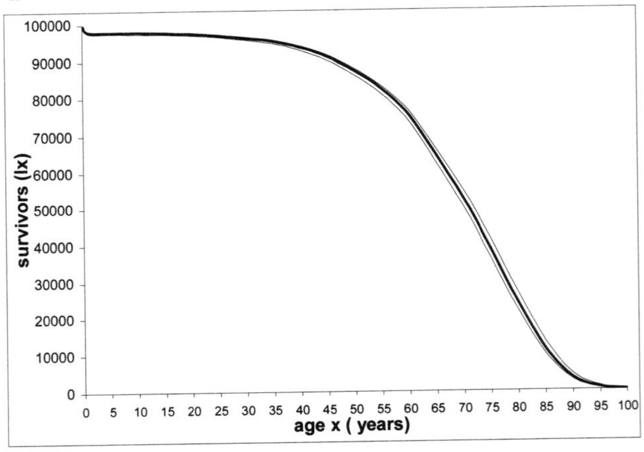

e_x with uncertainty interval

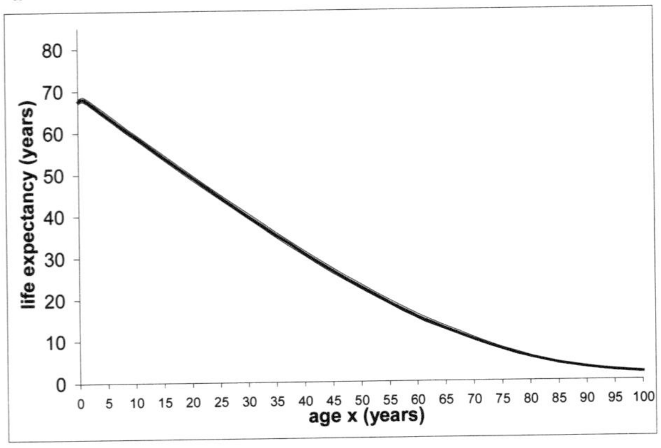

Distribution of deaths by age with uncertainty interval

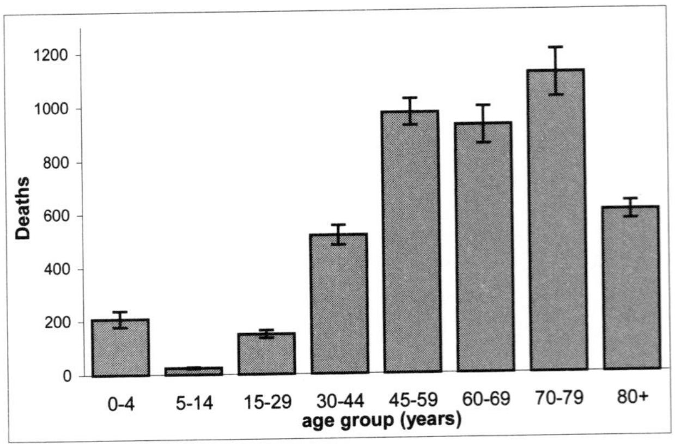

Mauritius | **Maurice** | **Mauricio**

Females, 2000 | *Femmes, 2000* | *Mujeres, 2000*

Age	$_nM_x$	$_nq_x$			l_x			$_nd_x$	$_nL_x$	e_x			Actual deaths	Actual population
			percentiles			percentiles					percentiles			
		mean	2½	97½	mean	2½	97½			mean	2½	97½		
<1	0.0144	0.0142	0.0140	0.0161	100,000	100,000	100,000	1,420	98,722	74.6	74.2	75.1	135	9,376
1-4	0.0005	0.0020	0.0020	0.0022	98,580	98,394	98,602	200	393,838	74.7	74.3	75.2	19	37,345
5-9	0.0003	0.0013	0.0013	0.0014	98,379	98,173	98,405	128	491,577	70.9	70.5	71.3	14	52,864
10-14	0.0003	0.0013	0.0013	0.0014	98,252	98,033	98,281	128	490,939	66.0	65.6	66.4	12	47,159
15-19	0.0003	0.0016	0.0016	0.0018	98,124	97,893	98,156	162	490,215	61.0	60.7	61.5	17	50,149
20-24	0.0004	0.0021	0.0020	0.0023	97,962	97,716	97,998	206	489,297	56.1	55.8	56.6	23	54,184
25-29	0.0006	0.0028	0.0028	0.0031	97,757	97,493	97,798	278	488,087	51.2	50.9	51.7	26	44,928
30-34	0.0008	0.0040	0.0039	0.0043	97,478	97,189	97,527	389	486,419	46.4	46.0	46.8	38	46,981
35-39	0.0012	0.0058	0.0057	0.0063	97,089	96,767	97,149	567	484,029	41.6	41.2	42.0	58	49,128
40-44	0.0020	0.0099	0.0095	0.0106	96,522	96,154	96,599	951	480,233	36.8	36.5	37.2	87	43,695
45-49	0.0032	0.0157	0.0152	0.0169	95,571	95,132	95,679	1,503	474,097	32.1	31.8	32.5	120	37,985
50-54	0.0054	0.0266	0.0255	0.0283	94,068	93,527	94,229	2,500	464,088	27.6	27.3	28.0	154	28,527
55-59	0.0093	0.0453	0.0433	0.0477	91,567	90,881	91,824	4,146	447,472	23.3	23.0	23.7	195	21,062
60-64	0.0150	0.0724	0.0690	0.0755	87,421	86,544	87,846	6,325	421,293	19.3	19.0	19.6	262	17,450
65-69	0.0252	0.1186	0.1131	0.1232	81,096	80,007	81,782	9,615	381,441	15.6	15.4	15.9	348	13,823
70-74	0.0368	0.1687	0.1612	0.1744	71,481	70,154	72,536	12,057	327,260	12.3	12.2	12.6	455	12,359
75-79	0.0603	0.2619	0.2517	0.2692	59,423	57,921	60,847	15,562	258,211	9.3	9.2	9.5	526	8,732
80-84	0.1057	0.4181	0.4059	0.4268	43,861	42,329	45,534	18,340	173,456	6.8	6.7	6.9	464	4,386
85-89	0.1736	0.6053	0.5915	0.6127	25,521	24,262	27,052	15,448	88,987	4.8	4.8	5.0	342	1,971
90-94	0.2667	0.7408	0.7291	0.7459	10,074	9,396	11,050	7,463	27,979	3.4	3.4	3.5	154	577
95-99	0.3835	0.8917	0.8842	0.8944	2,611	2,388	2,994	2,328	6,070	2.5	2.5	2.6	39	102
100+	0.5161	1	1	1	283	252	347	283	548	1.9	1.9	2.0	4	8
All ages													*3,491*	*582,791*

ln($_nq_x$) with uncertainty interval

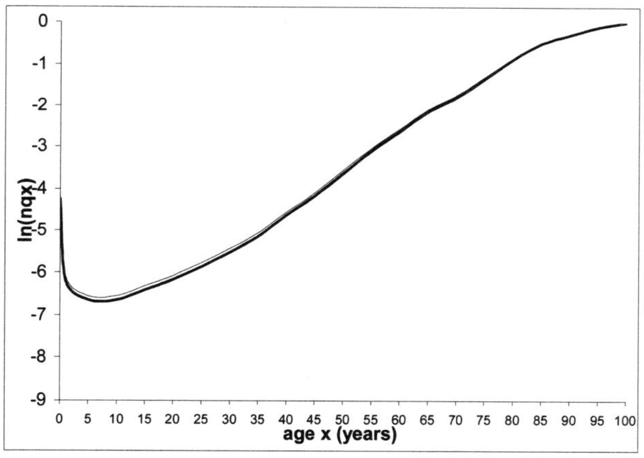

l_x with uncertainty interval

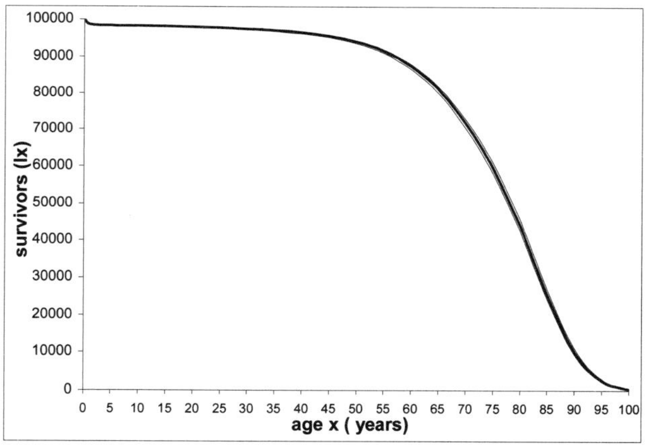

e_x with uncertainty interval

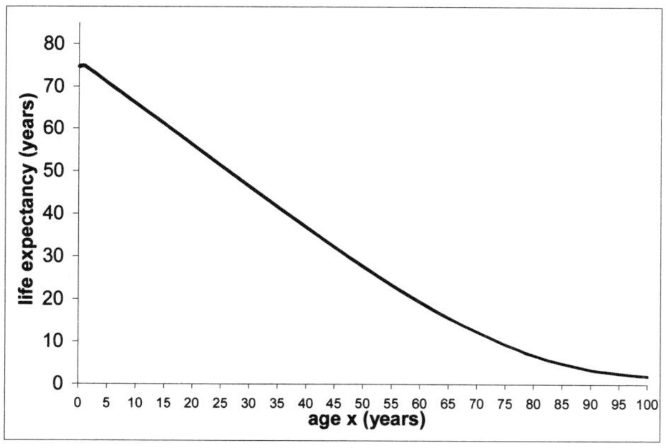

Distribution of deaths by age with uncertainty interval

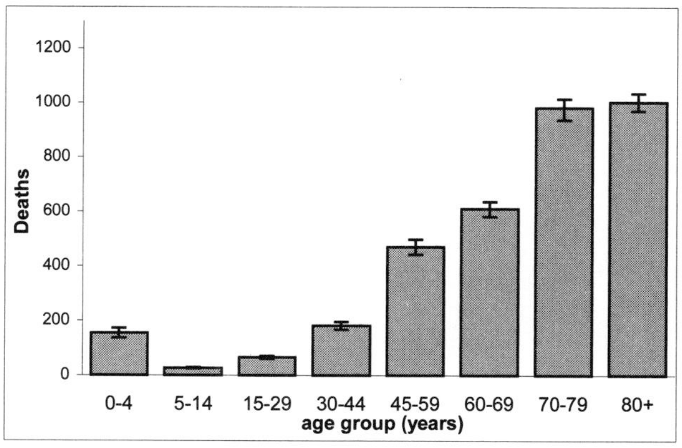

Mexico — Mexique — México

Males, 2000 — *Hommes, 2000* — *Hombres, 2000*

Age	$_nM_x$	$_nq_x$ mean	$_nq_x$ percentiles 2½	$_nq_x$ percentiles 97½	l_x mean	l_x percentiles 2½	l_x percentiles 97½	$_nd_x$	$_nL_x$	e_x mean	e_x percentiles 2½	e_x percentiles 97½	Actual deaths	Actual population
<1	0.0271	0.0265	0.0221	0.0310	100,000	100,000	100,000	2,647	97,618	71.0	70.2	71.8	31,466	1,160,612
1-4	0.0013	0.0052	0.0045	0.0058	97,353	96,898	97,791	503	388,206	71.9	71.0	72.9	5,892	4,544,128
5-9	0.0005	0.0023	0.0020	0.0025	96,850	96,351	97,349	218	483,705	68.3	67.3	69.3	2,528	5,606,424
10-14	0.0005	0.0024	0.0021	0.0026	96,632	96,115	97,155	230	482,584	63.4	62.5	64.4	2,579	5,405,598
15-19	0.0011	0.0055	0.0049	0.0061	96,402	95,868	96,951	532	480,677	58.6	57.6	59.6	5,661	5,110,474
20-24	0.0017	0.0082	0.0074	0.0090	95,869	95,293	96,475	788	477,375	53.9	52.9	54.9	8,083	4,895,677
25-29	0.0020	0.0100	0.0090	0.0109	95,081	94,431	95,765	948	473,036	49.3	48.3	50.3	9,212	4,598,406
30-34	0.0023	0.0116	0.0106	0.0126	94,133	93,402	94,903	1,094	467,932	44.8	43.8	45.8	8,736	3,737,123
35-39	0.0031	0.0155	0.0143	0.0167	93,040	92,224	93,898	1,444	461,587	40.3	39.3	41.3	9,560	3,055,549
40-44	0.0040	0.0200	0.0186	0.0214	91,595	90,681	92,556	1,831	453,400	35.9	34.9	36.9	10,425	2,581,690
45-49	0.0056	0.0276	0.0259	0.0293	89,764	88,743	90,834	2,478	442,626	31.5	30.6	32.6	11,746	2,097,761
50-54	0.0081	0.0396	0.0371	0.0423	87,286	86,146	88,472	3,460	427,781	27.4	26.4	28.4	13,436	1,661,315
55-59	0.0118	0.0571	0.0528	0.0616	83,826	82,564	85,125	4,786	407,166	23.4	22.5	24.4	15,693	1,334,963
60-64	0.0173	0.0830	0.0759	0.0905	79,040	77,675	80,415	6,560	378,800	19.7	18.8	20.7	17,895	1,033,346
65-69	0.0263	0.1234	0.1115	0.1354	72,480	71,074	73,893	8,943	340,044	16.2	15.4	17.1	21,058	800,741
70-74	0.0386	0.1759	0.1579	0.1934	63,538	61,605	65,379	11,178	289,743	13.1	12.4	14.0	22,108	573,044
75-79	0.0617	0.2672	0.2402	0.2919	52,360	49,691	54,978	13,990	226,823	10.4	9.8	11.1	24,099	390,726
80-84	0.0841	0.3476	0.3153	0.3772	38,370	35,192	41,704	13,337	158,505	8.3	7.8	8.9	16,545	196,628
85-89	0.1194	0.4599	0.4244	0.4918	25,032	21,987	28,555	11,512	96,381	6.4	6.0	6.8	11,683	97,812
90-94	0.1764	0.5768	0.5429	0.6074	13,520	11,252	16,435	7,799	44,203	4.7	4.4	5.0	6,310	35,763
95-99	0.2712	0.7476	0.7195	0.7728	5,721	4,446	7,512	4,277	15,774	3.3	3.2	3.5	2,074	7,648
100+	0.4337	1	1	1	1,444	1,013	2,107	1,444	3,329	2.3	2.1	2.4	337	778
All ages													*257,127*	*48,926,206*

$\ln({}_nq_x)$ with uncertainty interval

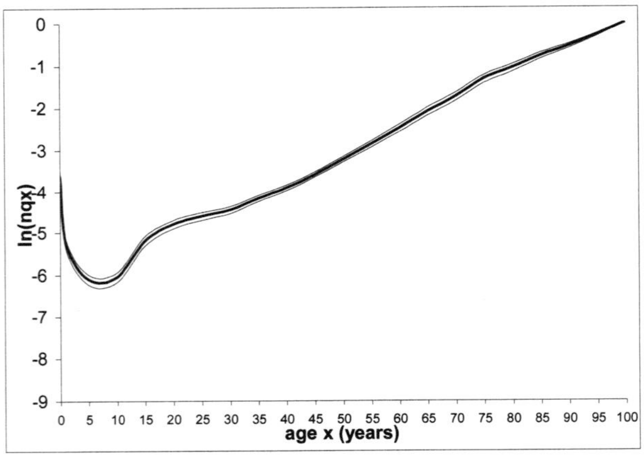

l_x with uncertainty interval

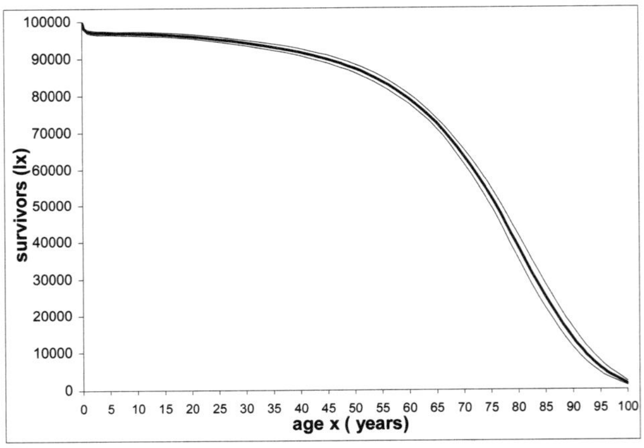

e_x with uncertainty interval

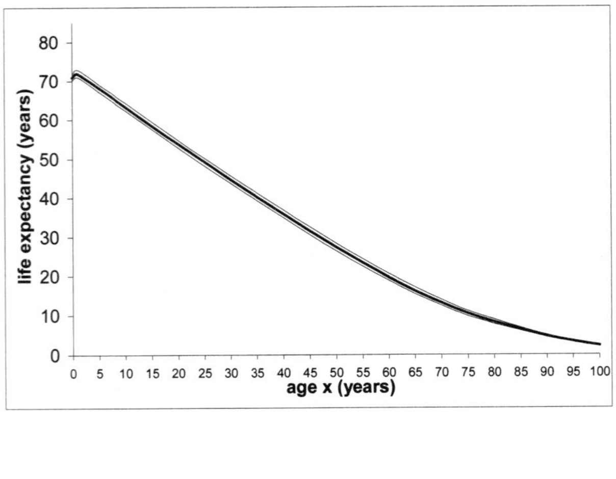

Distribution of deaths by age with uncertainty interval

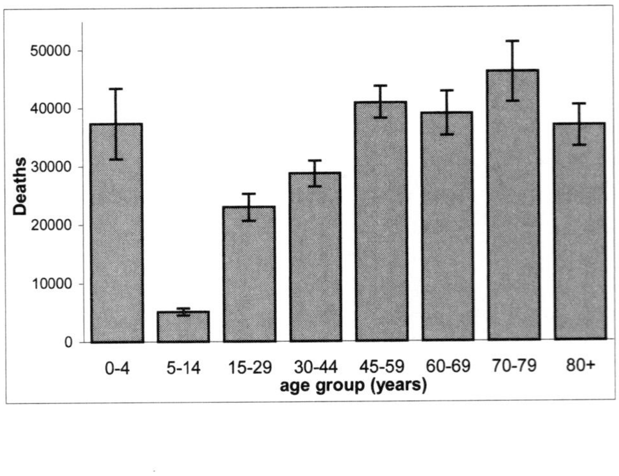

Mexico | Mexique | México

Females, 2000 | *Femmes, 2000* | *Mujeres, 2000*

Age	$_nM_x$	$_nq_x$ mean	$_nq_x$ percentiles 2½	$_nq_x$ percentiles 97½	l_x mean	l_x percentiles 2½	l_x percentiles 97½	$_nd_x$	$_nL_x$	e_x mean	e_x percentiles 2½	e_x percentiles 97½	Actual deaths	Actual population
<1	0.0215	0.0211	0.0188	0.0238	100,000	100,000	100,000	2,107	98,104	76.2	75.7	76.8	23,536	1,095,951
1-4	0.0010	0.0040	0.0037	0.0044	97,893	97,625	98,119	393	390,629	76.9	76.4	77.4	4,396	4,367,144
5-9	0.0003	0.0015	0.0014	0.0016	97,500	97,198	97,757	148	487,130	73.2	72.7	73.7	1,636	5,380,773
10-14	0.0003	0.0014	0.0013	0.0015	97,352	97,038	97,620	137	486,417	68.3	67.8	68.8	1,468	5,209,260
15-19	0.0004	0.0022	0.0020	0.0023	97,215	96,891	97,493	209	485,551	63.4	62.9	63.9	2,156	5,004,543
20-24	0.0005	0.0026	0.0025	0.0028	97,006	96,665	97,298	257	484,386	58.5	58.1	59.0	2,621	4,944,015
25-29	0.0006	0.0032	0.0030	0.0034	96,749	96,390	97,059	309	482,971	53.6	53.2	54.2	3,038	4,745,722
30-34	0.0008	0.0042	0.0039	0.0045	96,440	96,059	96,771	403	481,191	48.8	48.4	49.3	3,320	3,965,697
35-39	0.0012	0.0061	0.0057	0.0065	96,037	95,630	96,393	589	478,710	44.0	43.6	44.5	4,022	3,266,035
40-44	0.0019	0.0093	0.0088	0.0099	95,447	95,006	95,839	889	475,014	39.3	38.9	39.8	5,196	2,776,743
45-49	0.0031	0.0152	0.0144	0.0160	94,558	94,069	94,999	1,438	469,196	34.6	34.2	35.1	6,917	2,256,528
50-54	0.0049	0.0244	0.0232	0.0257	93,120	92,567	93,631	2,276	459,911	30.1	29.7	30.6	8,812	1,780,973
55-59	0.0077	0.0379	0.0361	0.0398	90,844	90,215	91,445	3,444	445,611	25.8	25.4	26.3	11,171	1,445,157
60-64	0.0121	0.0590	0.0559	0.0618	87,400	86,701	88,100	5,152	424,119	21.7	21.4	22.1	13,846	1,139,754
65-69	0.0198	0.0943	0.0893	0.0985	82,248	81,450	83,077	7,753	391,855	17.9	17.6	18.3	17,953	907,401
70-74	0.0298	0.1385	0.1314	0.1441	74,494	73,440	75,564	10,320	346,673	14.5	14.3	14.9	20,318	682,542
75-79	0.0476	0.2128	0.2029	0.2201	64,175	62,859	65,609	13,657	286,729	11.5	11.2	11.7	23,655	496,632
80-84	0.0716	0.3036	0.2917	0.3143	50,517	49,040	52,299	15,335	214,248	8.9	8.7	9.1	18,863	263,527
85-89	0.1094	0.4295	0.4156	0.4432	35,182	33,788	37,044	15,112	138,129	6.7	6.5	6.8	15,494	141,619
90-94	0.1701	0.5631	0.5483	0.5770	20,070	19,033	21,590	11,301	66,445	4.8	4.7	4.9	9,987	58,717
95-99	0.2690	0.7443	0.7313	0.7558	8,769	8,060	9,659	6,526	24,265	3.4	3.3	3.4	4,098	15,237
100+	0.4326	1	1	1	2,243	1,968	2,553	2,243	5,184	2.3	2.2	2.3	888	2,054
All ages													*203,392*	*49,946,024*

ln($_nq_x$) with uncertainty interval

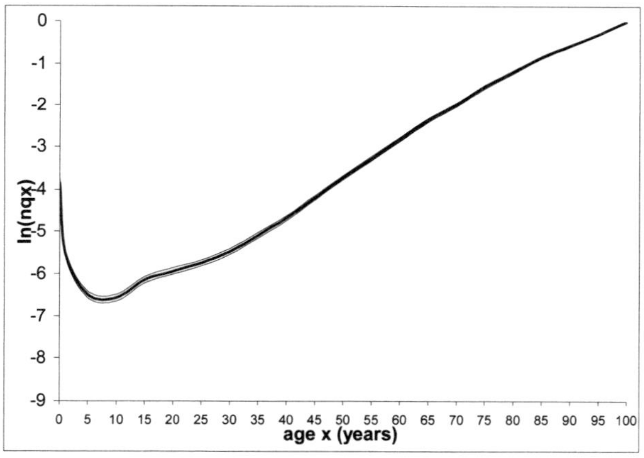

l_x with uncertainty interval

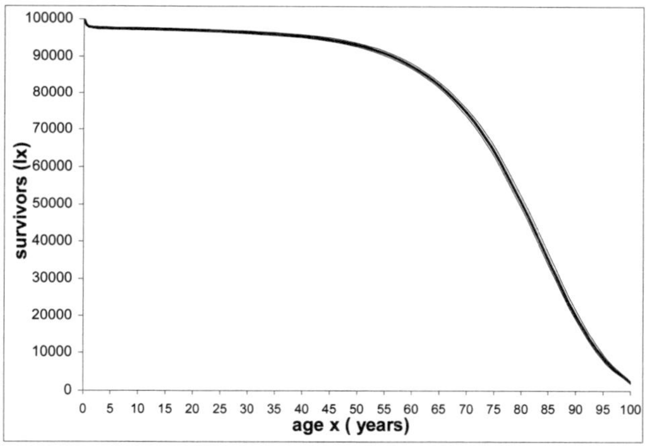

e_x with uncertainty interval

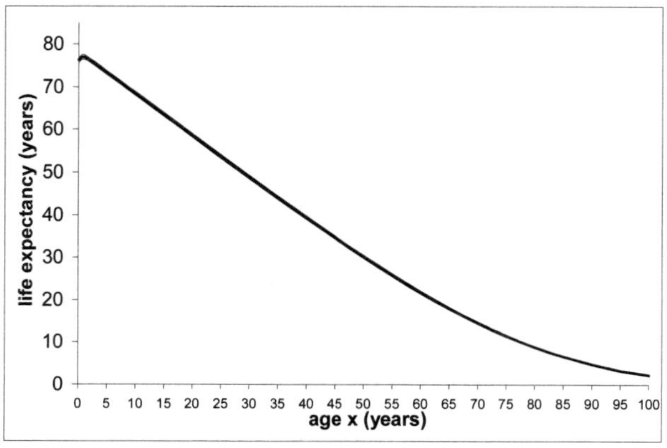

Distribution of deaths by age with uncertainty interval

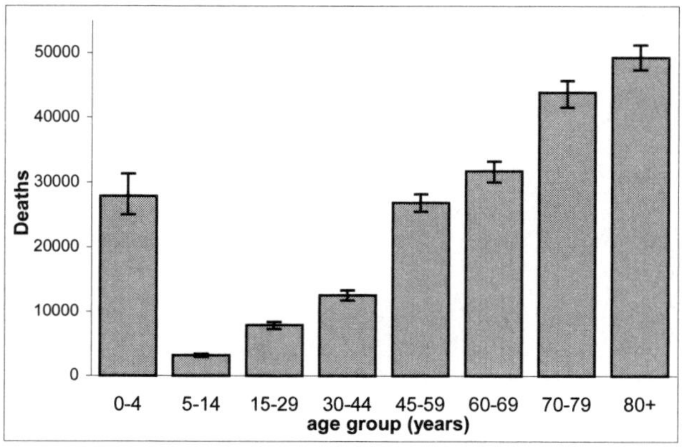

Micronesia (Federated States of) Micronésie (Etats fédérés de) Micronesia (Estados Federados de)

Males, 2000 *Hommes, 2000* *Hombres, 2000*

Age	$_nM_x$	$_nq_x$			l_x			$_nd_x$	$_nL_x$	e_x			Actual deaths	Actual population
			percentiles			percentiles					percentiles			
		mean	2½	97½	mean	2½	97½			mean	2½	97½		
<1	0.0506	0.0484	0.0338	0.0644	100,000	100,000	100,000	4,842	95,642	63.7	61.5	65.9	91	1,796
1-4	0.0033	0.0132	0.0102	0.0162	95,158	93,556	96,623	1,256	377,616	66.0	64.4	67.7	24	7,206
5-9	0.0011	0.0054	0.0043	0.0066	93,902	92,152	95,636	511	468,231	62.8	61.3	64.5	9	8,393
10-14	0.0009	0.0043	0.0034	0.0053	93,391	91,593	95,227	404	465,943	58.2	56.7	59.8	6	7,034
15-19	0.0017	0.0082	0.0065	0.0100	92,987	91,116	94,901	766	463,019	53.4	51.9	55.1	10	6,117
20-24	0.0023	0.0115	0.0093	0.0138	92,221	90,209	94,275	1,065	458,442	48.8	47.4	50.5	12	5,012
25-29	0.0025	0.0122	0.0099	0.0145	91,156	88,959	93,392	1,110	453,004	44.4	42.9	46.0	11	4,552
30-34	0.0029	0.0142	0.0117	0.0168	90,046	87,670	92,455	1,279	447,030	39.9	38.4	41.6	13	4,650
35-39	0.0037	0.0185	0.0154	0.0216	88,766	86,200	91,357	1,642	439,727	35.4	34.0	37.1	15	3,959
40-44	0.0053	0.0261	0.0220	0.0302	87,124	84,335	89,921	2,271	429,945	31.1	29.7	32.7	18	3,479
45-49	0.0078	0.0383	0.0329	0.0439	84,854	81,794	87,913	3,248	416,147	26.8	25.5	28.4	23	2,969
50-54	0.0118	0.0573	0.0495	0.0651	81,605	78,226	85,022	4,678	396,330	22.8	21.5	24.3	29	2,461
55-59	0.0178	0.0851	0.0735	0.0975	76,927	73,208	80,734	6,548	368,264	19.0	17.8	20.5	31	1,763
60-64	0.0269	0.1262	0.1073	0.1456	70,379	66,385	74,513	8,881	329,689	15.6	14.4	16.9	39	1,434
65-69	0.0406	0.1844	0.1554	0.2128	61,497	57,332	65,719	11,342	279,131	12.4	11.4	13.7	45	1,103
70-74	0.0629	0.2719	0.2300	0.3146	50,155	45,878	54,891	13,638	216,680	9.7	8.8	10.8	55	875
75-79	0.0971	0.3907	0.3335	0.4455	36,517	31,653	41,793	14,269	146,912	7.4	6.7	8.3	48	498
80-84	0.1491	0.5430	0.4762	0.6006	22,248	17,879	27,705	12,080	81,040	5.5	5.0	6.2	30	201
85-89	0.2210	0.7118	0.6499	0.7601	10,168	7,210	14,270	7,238	32,745	4.1	3.8	4.5	16	73
90-94	0.3166	0.8119	0.7623	0.8489	2,930	1,729	4,911	2,379	7,513	3.0	2.8	3.3	5	15
95-99	0.4383	0.9467	0.9215	0.9630	551	261	1,167	522	1,190	2.3	2.2	2.4	1	2
100+	0.5860	1	1	1	29	10	92	29	50	1.7	1.7	1.8	0	0
All ages													*532*	*63,592*

ln($_nq_x$) with uncertainty interval

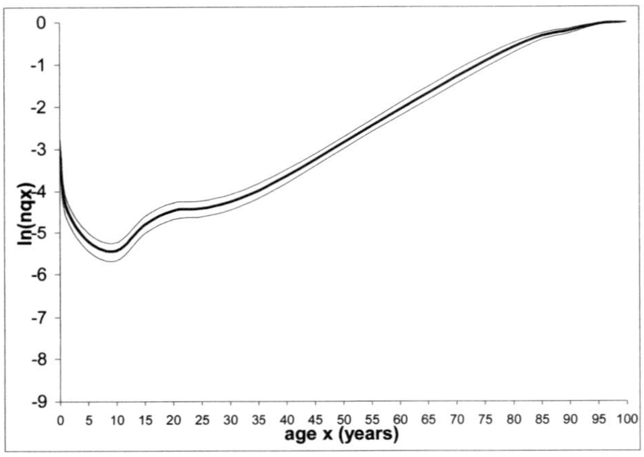

l_x with uncertainty interval

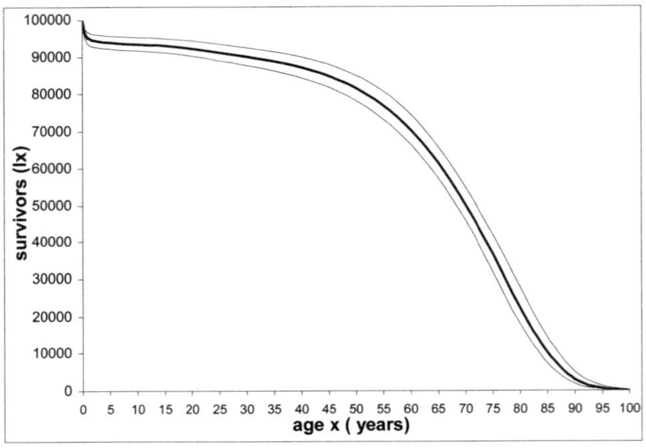

e_x with uncertainty interval

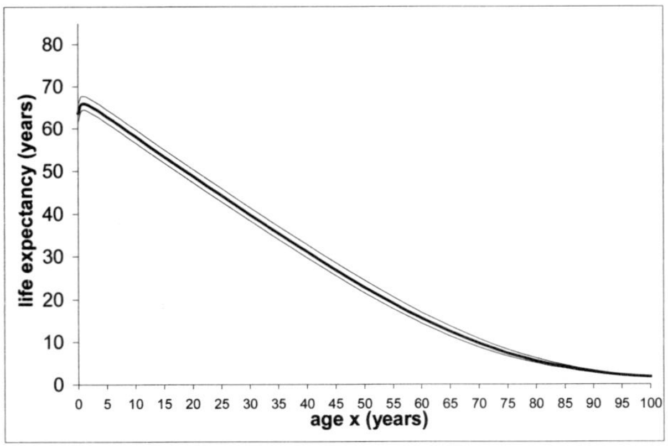

Distribution of deaths by age with uncertainty interval

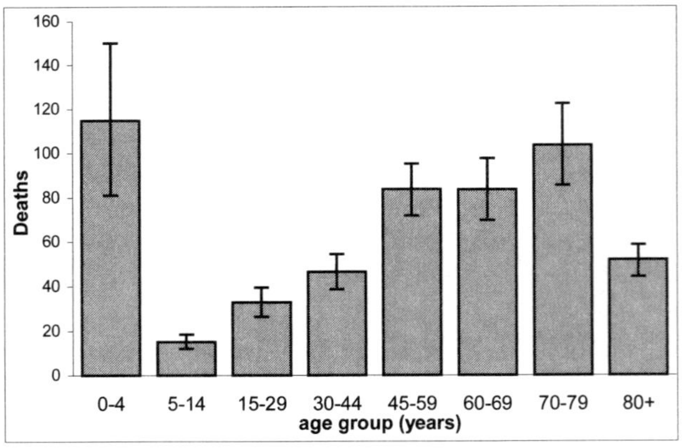

Micronesia (Federated States of) — Micronésie (Etats fédérés de) — Micronesia (Estados Federados de)

Females, 2000 — *Femmes, 2000* — *Mujeres, 2000*

Age	$_nM_x$	$_nq_x$ mean	$_nq_x$ percentiles 2½	$_nq_x$ percentiles 97½	l_x mean	l_x percentiles 2½	l_x percentiles 97½	$_nd_x$	$_nL_x$	e_x mean	e_x percentiles 2½	e_x percentiles 97½	Actual deaths	Actual population
<1	0.0373	0.0361	0.0240	0.0490	100,000	100,000	100,000	3,612	96,749	67.7	65.7	69.6	63	1,700
1-4	0.0031	0.0123	0.0093	0.0154	96,388	95,096	97,603	1,186	382,704	69.3	67.7	71.1	21	6,777
5-9	0.0010	0.0047	0.0037	0.0059	95,201	93,708	96,695	452	474,877	66.1	64.6	67.9	8	7,901
10-14	0.0008	0.0038	0.0029	0.0046	94,749	93,158	96,342	355	472,859	61.4	60.0	63.2	5	6,626
15-19	0.0013	0.0062	0.0049	0.0076	94,394	92,728	96,062	589	470,497	56.6	55.2	58.4	7	5,727
20-24	0.0016	0.0080	0.0063	0.0097	93,805	92,021	95,591	749	467,151	52.0	50.6	53.7	8	4,817
25-29	0.0019	0.0092	0.0074	0.0111	93,056	91,129	94,983	857	463,136	47.4	46.0	49.1	8	4,290
30-34	0.0023	0.0114	0.0093	0.0135	92,199	90,120	94,276	1,048	458,374	42.8	41.5	44.5	9	3,816
35-39	0.0031	0.0156	0.0129	0.0183	91,151	88,900	93,395	1,418	452,210	38.2	36.9	39.9	11	3,612
40-44	0.0043	0.0210	0.0177	0.0245	89,733	87,270	92,178	1,887	443,948	33.8	32.5	35.4	14	3,201
45-49	0.0060	0.0298	0.0254	0.0342	87,846	85,136	90,517	2,614	432,696	29.5	28.2	31.0	16	2,704
50-54	0.0087	0.0426	0.0365	0.0489	85,232	82,234	88,164	3,633	417,079	25.3	24.1	26.8	19	2,189
55-59	0.0125	0.0607	0.0517	0.0693	81,599	78,295	84,822	4,951	395,619	21.3	20.2	22.7	20	1,571
60-64	0.0191	0.0909	0.0768	0.1038	76,648	73,070	80,150	6,970	365,816	17.5	16.5	18.9	26	1,386
65-69	0.0298	0.1385	0.1167	0.1592	69,678	65,939	73,443	9,648	324,270	14.0	13.0	15.3	33	1,098
70-74	0.0488	0.2175	0.1844	0.2499	60,030	55,878	64,448	13,057	267,508	10.9	10.0	12.0	41	847
75-79	0.0793	0.3307	0.2830	0.3793	46,973	42,523	52,422	15,536	196,027	8.2	7.5	9.1	41	518
80-84	0.1278	0.4841	0.4264	0.5420	31,437	26,717	37,382	15,220	119,137	6.1	5.5	6.7	31	246
85-89	0.1974	0.6609	0.6039	0.7134	16,217	12,518	21,110	10,719	54,291	4.4	4.1	4.9	17	86
90-94	0.2925	0.7790	0.7319	0.8194	5,499	3,588	8,298	4,284	14,643	3.2	3.0	3.5	5	18
95-99	0.4156	0.9249	0.8982	0.9454	1,215	648	2,180	1,124	2,704	2.4	2.3	2.5	1	2
100+	0.5661	1	1	1	91	35	214	91	161	1.8	1.7	1.9	0	0
All ages													*405*	*59,133*

$\ln(_nq_x)$ with uncertainty interval

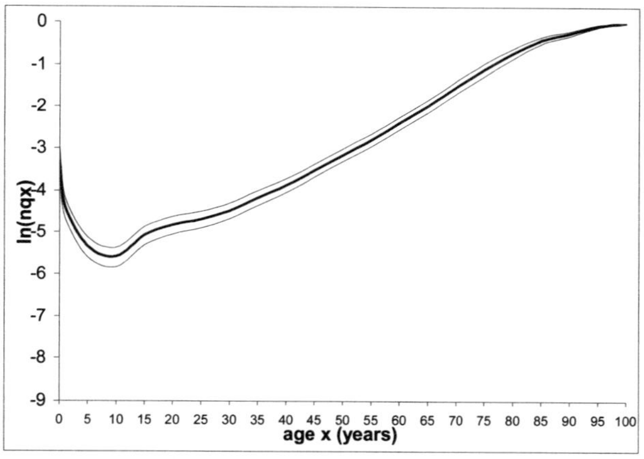

l_x with uncertainty interval

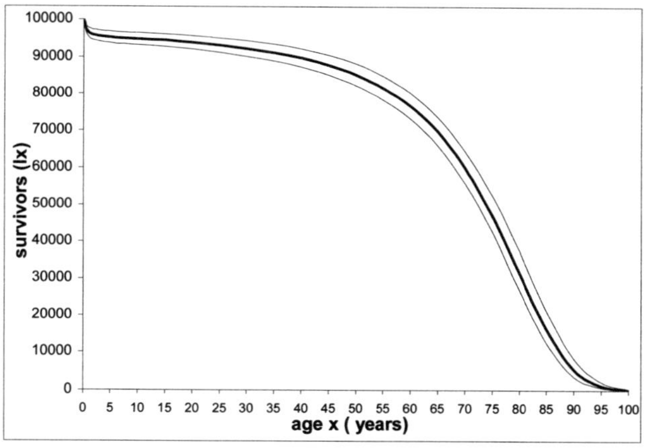

e_x with uncertainty interval

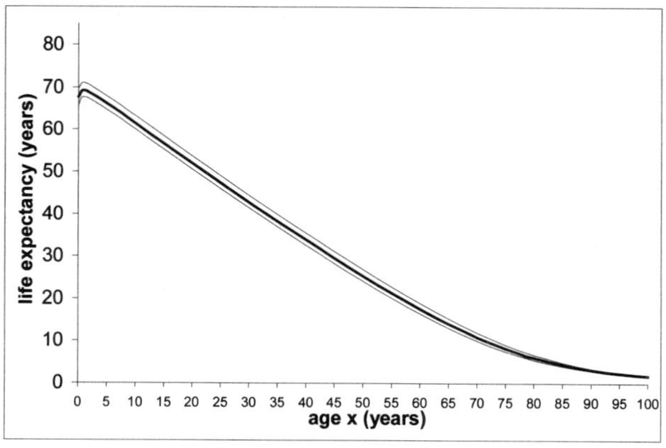

Distribution of deaths by age with uncertainty interval

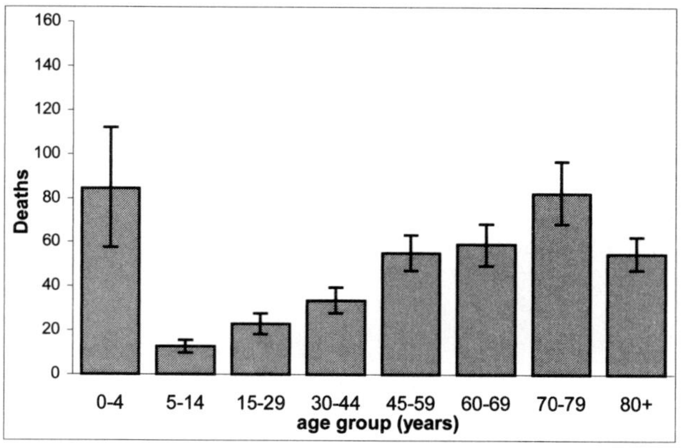

Monaco **Monaco** **Mónaco**

Males, 2000 *Hommes, 2000* *Hombres, 2000*

Age	$_nM_x$	$_nq_x$			l_x			$_nd_x$	$_nL_x$	e_x			Actual deaths	Actual population
			percentiles			percentiles					percentiles			
		mean	2½	97½	mean	2½	97½			mean	2½	97½		
<1	0.0057	0.0057	0.0028	0.0084	100,000	100,000	100,000	571	99,486	76.8	75.2	79.8	1	206
1-4	0.0003	0.0012	0.0007	0.0016	99,429	99,163	99,718	120	397,430	76.2	74.6	79.3	0	845
5-9	0.0002	0.0008	0.0005	0.0011	99,309	99,001	99,648	82	496,341	72.3	70.7	75.3	0	1,044
10-14	0.0002	0.0009	0.0005	0.0012	99,227	98,892	99,600	89	495,913	67.4	65.8	70.4	0	1,109
15-19	0.0006	0.0031	0.0019	0.0041	99,138	98,774	99,547	309	494,917	62.4	60.8	65.5	1	1,128
20-24	0.0011	0.0054	0.0034	0.0069	98,829	98,372	99,359	535	492,805	57.6	56.0	60.6	1	1,096
25-29	0.0011	0.0055	0.0036	0.0069	98,293	97,694	99,019	545	490,104	52.9	51.3	55.9	1	1,206
30-34	0.0013	0.0066	0.0044	0.0080	97,748	97,021	98,660	641	487,140	48.2	46.6	51.2	2	1,207
35-39	0.0018	0.0089	0.0062	0.0107	97,108	96,245	98,223	869	483,367	43.5	42.0	46.4	2	1,226
40-44	0.0027	0.0134	0.0096	0.0157	96,239	95,216	97,611	1,286	477,981	38.9	37.4	41.8	3	1,195
45-49	0.0040	0.0197	0.0147	0.0227	94,953	93,724	96,670	1,874	470,081	34.4	32.9	37.2	5	1,182
50-54	0.0056	0.0276	0.0208	0.0319	93,079	91,600	95,251	2,566	458,981	30.0	28.5	32.7	6	1,156
55-59	0.0080	0.0394	0.0294	0.0461	90,513	88,679	93,266	3,569	443,643	25.8	24.3	28.4	6	780
60-64	0.0121	0.0587	0.0431	0.0691	86,944	84,594	90,522	5,106	421,955	21.8	20.3	24.1	9	733
65-69	0.0181	0.0867	0.0631	0.1022	81,838	78,749	86,617	7,093	391,457	18.0	16.6	20.1	13	707
70-74	0.0276	0.1291	0.0941	0.1546	74,745	70,701	81,151	9,647	349,607	14.4	13.2	16.3	17	605
75-79	0.0419	0.1895	0.1406	0.2280	65,098	59,768	73,514	12,336	294,650	11.2	10.0	12.7	20	471
80-84	0.0756	0.3179	0.2471	0.3809	52,762	46,143	63,180	16,774	221,874	8.2	7.2	9.4	15	192
85-89	0.1294	0.4888	0.4093	0.5677	35,988	28,566	47,571	17,592	135,958	5.9	5.1	6.7	18	136
90-94	0.2099	0.6440	0.5770	0.7151	18,395	12,350	28,102	11,847	56,436	4.1	3.6	4.6	10	46
95-99	0.3228	0.8200	0.7731	0.8682	6,548	3,519	11,886	5,369	16,634	2.9	2.7	3.2	3	8
100+	0.4705	1	1	1	1,179	464	2,697	1,179	2,506	2.1	2.1	2.3	0	1
All ages													*132*	*16,279*

ln($_nq_x$) with uncertainty interval

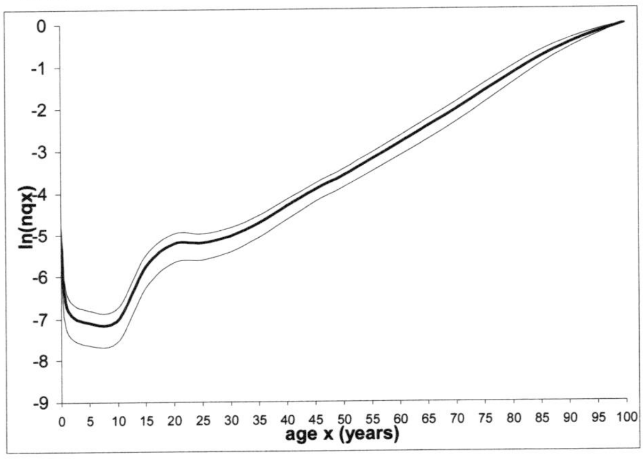

l_x with uncertainty interval

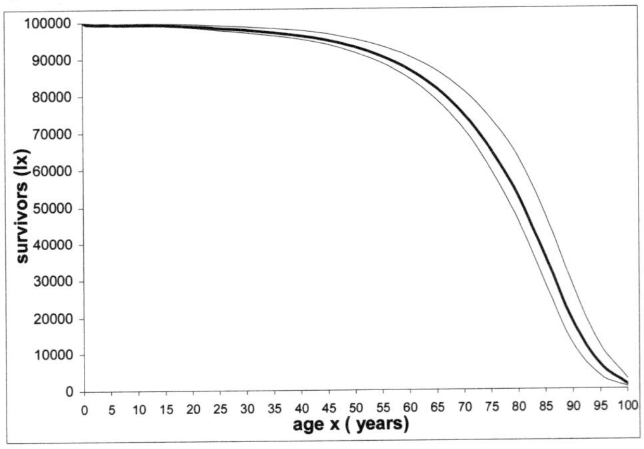

e_x with uncertainty interval

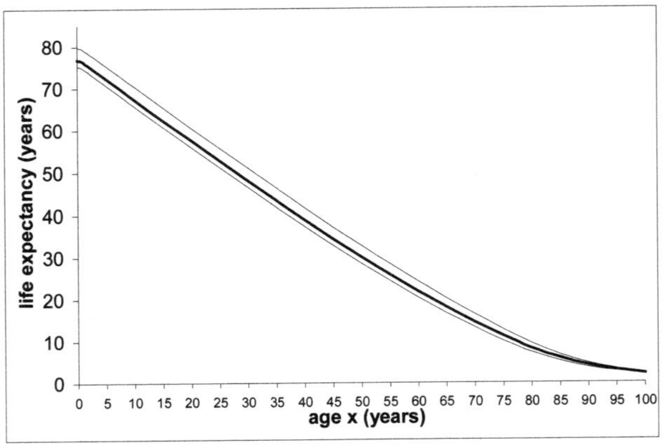

Distribution of deaths by age with uncertainty interval

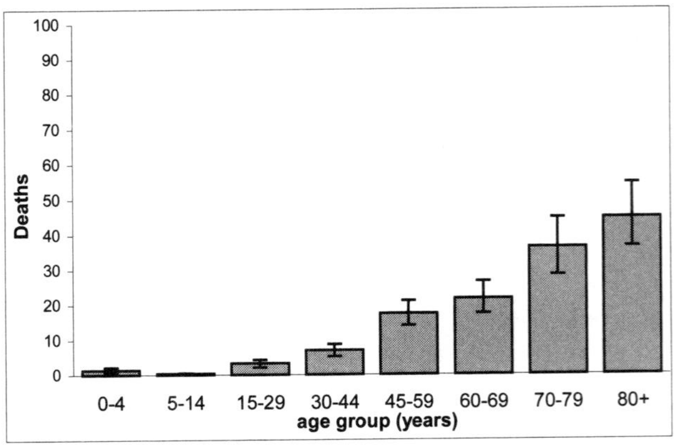

Females, 2000 *Femmes, 2000* *Mujeres, 2000*

Age	$_nM_x$	$_nq_x$			l_x			$_nd_x$	$_nL_x$	e_x			Actual deaths	Actual population
			percentiles			percentiles					percentiles			
		mean	2½	97½	mean	2½	97½			mean	2½	97½		
<1	0.0040	0.0040	0.0024	0.0081	100,000	100,000	100,000	403	99,637	84.4	81.6	86.4	1	196
1-4	0.0002	0.0010	0.0006	0.0017	99,597	99,186	99,762	95	398,157	83.8	81.0	85.9	0	804
5-9	0.0001	0.0005	0.0004	0.0009	99,501	99,022	99,698	52	497,375	79.9	77.1	82.0	0	998
10-14	0.0001	0.0006	0.0004	0.0010	99,449	98,935	99,663	61	497,093	74.9	72.2	77.0	0	1,060
15-19	0.0003	0.0013	0.0009	0.0020	99,388	98,834	99,621	125	496,630	69.9	67.2	72.1	0	1,078
20-24	0.0003	0.0017	0.0012	0.0027	99,264	98,632	99,534	168	495,897	65.0	62.4	67.1	0	1,054
25-29	0.0004	0.0018	0.0013	0.0028	99,095	98,364	99,412	180	495,026	60.1	57.5	62.3	0	1,183
30-34	0.0005	0.0027	0.0020	0.0041	98,915	98,085	99,279	266	493,912	55.2	52.6	57.4	1	1,200
35-39	0.0008	0.0039	0.0030	0.0058	98,650	97,684	99,079	384	492,287	50.4	47.8	52.5	1	1,239
40-44	0.0012	0.0059	0.0047	0.0085	98,265	97,119	98,780	579	489,879	45.6	43.1	47.7	1	1,218
45-49	0.0018	0.0087	0.0072	0.0122	97,686	96,295	98,313	853	486,300	40.8	38.4	43.0	2	1,201
50-54	0.0025	0.0125	0.0104	0.0174	96,834	95,118	97,603	1,207	481,150	36.2	33.8	38.3	3	1,159
55-59	0.0034	0.0168	0.0140	0.0235	95,626	93,463	96,586	1,605	474,120	31.6	29.4	33.7	3	791
60-64	0.0048	0.0236	0.0196	0.0330	94,021	91,269	95,235	2,218	464,562	27.1	25.0	29.2	4	789
65-69	0.0076	0.0371	0.0303	0.0517	91,803	88,258	93,368	3,407	450,499	22.7	20.8	24.7	6	827
70-74	0.0123	0.0596	0.0469	0.0819	88,396	83,699	90,542	5,265	428,818	18.5	16.8	20.4	10	788
75-79	0.0215	0.1022	0.0782	0.1370	83,131	76,844	86,293	8,499	394,407	14.5	13.1	16.4	15	709
80-84	0.0455	0.2044	0.1545	0.2596	74,632	66,314	79,544	15,253	335,027	10.8	9.6	12.6	16	342
85-89	0.0746	0.3143	0.2393	0.3846	59,379	49,100	67,251	18,663	250,236	8.0	7.0	9.5	24	318
90-94	0.1364	0.4840	0.3822	0.5631	40,715	30,218	51,155	19,707	144,457	5.5	4.8	6.7	20	147
95-99	0.2322	0.6843	0.5845	0.7472	21,009	13,203	31,604	14,375	61,918	3.7	3.5	4.6	8	36
100+	0.3981	1	1	1	6,633	3,338	13,132	6,633	16,662	2.5	2.4	3.3	2	4
All ages													*117*	*17,141*

ln($_nq_x$) with uncertainty interval

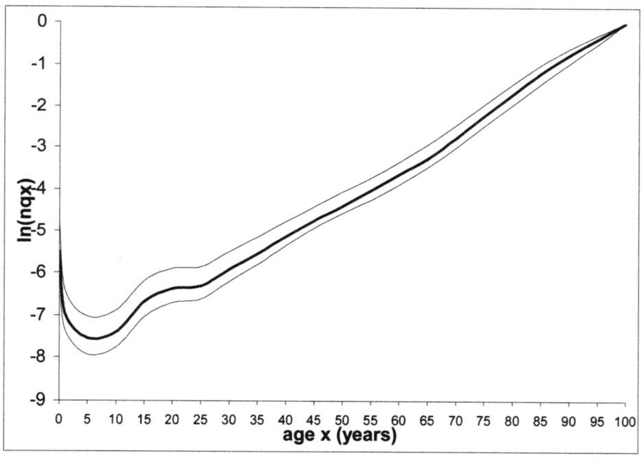

l_x with uncertainty interval

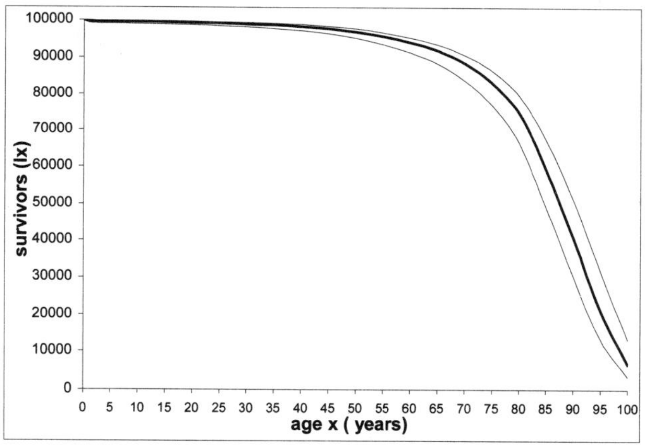

e_x with uncertainty interval

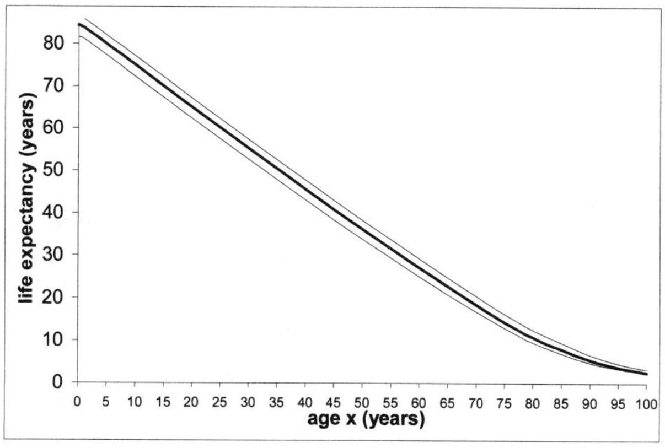

Distribution of deaths by age with uncertainty interval

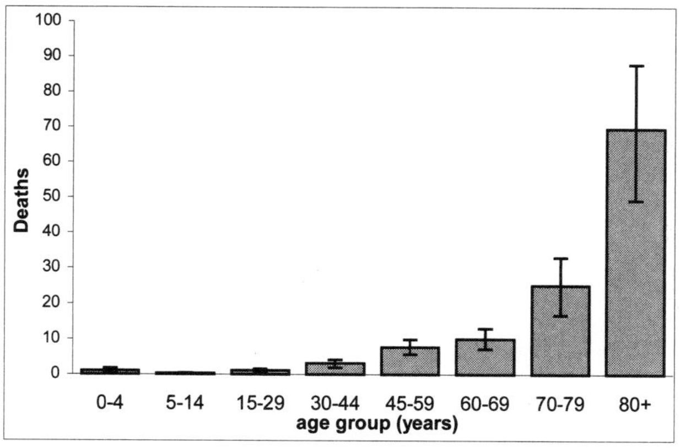

Mongolia

Mongolie

Mongolia

Males, 2000

Hommes, 2000

Hombres, 2000

Age	$_nM_x$	$_nq_x$			l_x			$_nd_x$	$_nL_x$	e_x			Actual deaths	Actual population
			percentiles			percentiles					percentiles			
		mean	2½	97½	mean	2½	97½			mean	2½	97½		
<1	0.0809	0.0754	0.0598	0.0920	100,000	100,000	100,000	7,543	93,211	61.2	59.3	62.9	2,358	29,143
1-4	0.0047	0.0185	0.0158	0.0212	92,457	90,801	94,015	1,707	365,731	65.1	63.8	66.6	511	109,486
5-9	0.0012	0.0059	0.0051	0.0067	90,750	89,003	92,493	536	452,410	62.3	60.9	63.9	177	149,276
10-14	0.0010	0.0050	0.0044	0.0057	90,214	88,406	92,008	454	449,935	57.7	56.3	59.3	167	165,926
15-19	0.0015	0.0073	0.0063	0.0082	89,760	87,901	91,599	651	447,171	53.0	51.6	54.6	207	142,355
20-24	0.0028	0.0137	0.0120	0.0154	89,109	87,181	91,018	1,219	442,495	48.3	46.9	50.0	354	128,514
25-29	0.0044	0.0218	0.0194	0.0244	87,889	85,841	89,921	1,918	434,652	44.0	42.5	45.6	512	116,060
30-34	0.0044	0.0219	0.0196	0.0243	85,972	83,753	88,173	1,879	425,159	39.9	38.5	41.6	442	99,958
35-39	0.0053	0.0259	0.0234	0.0286	84,092	81,729	86,438	2,182	415,006	35.7	34.3	37.5	475	90,396
40-44	0.0070	0.0342	0.0312	0.0373	81,910	79,402	84,397	2,799	402,555	31.6	30.2	33.3	475	68,322
45-49	0.0100	0.0488	0.0447	0.0530	79,112	76,443	81,738	3,858	385,913	27.7	26.3	29.4	446	44,620
50-54	0.0125	0.0607	0.0550	0.0669	75,254	72,423	78,007	4,568	364,849	23.9	22.6	25.6	386	30,850
55-59	0.0180	0.0861	0.0764	0.0962	70,686	67,740	73,540	6,088	338,209	20.3	19.0	22.0	534	29,647
60-64	0.0278	0.1299	0.1131	0.1465	64,598	61,613	67,520	8,388	302,019	17.0	15.7	18.6	615	22,145
65-69	0.0400	0.1818	0.1559	0.2068	56,210	53,347	59,034	10,220	255,498	14.2	13.0	15.7	715	17,875
70-74	0.0538	0.2370	0.2018	0.2715	45,990	42,937	48,939	10,901	202,697	11.8	10.7	13.1	602	11,197
75-79	0.0725	0.3068	0.2621	0.3500	35,089	31,407	38,761	10,764	148,535	9.7	8.7	10.8	528	7,292
80-84	0.0940	0.3807	0.3293	0.4291	24,325	20,596	28,600	9,261	98,473	7.8	7.1	8.7	344	3,660
85-89	0.1288	0.4873	0.4308	0.5382	15,064	11,889	19,182	7,340	56,970	6.1	5.6	6.8	146	1,133
90-94	0.1863	0.5976	0.5415	0.6464	7,724	5,490	10,919	4,616	24,772	4.5	4.1	5.0	28	151
95-99	0.2845	0.7674	0.7186	0.8077	3,108	1,941	5,006	2,385	8,385	3.2	3.0	3.5	3	9
100+	0.4585	1	1	1	723	373	1,409	723	1,577	2.2	2.0	2.3	0	0
All ages													*10,027*	*1,268,015*

ln($_nq_x$) with uncertainty interval

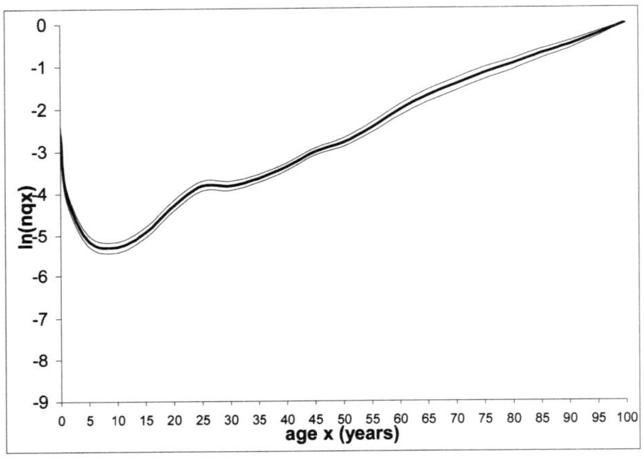

l_x with uncertainty interval

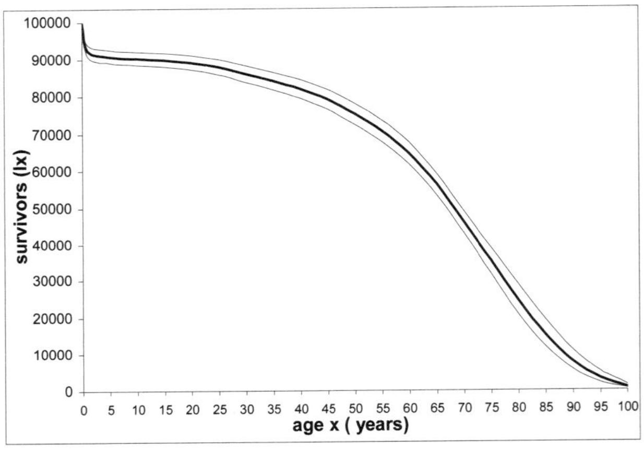

e_x with uncertainty interval

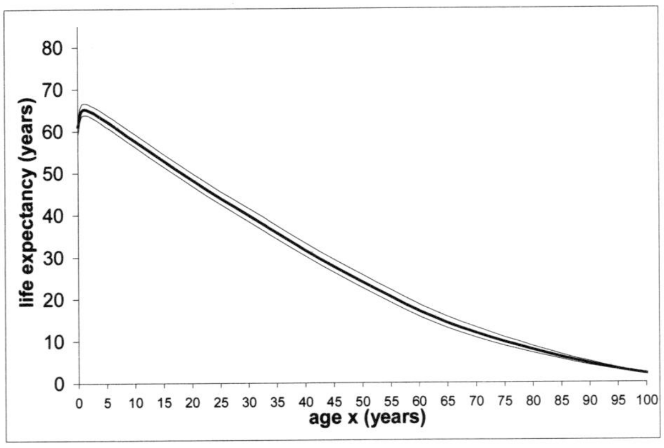

Distribution of deaths by age with uncertainty interval

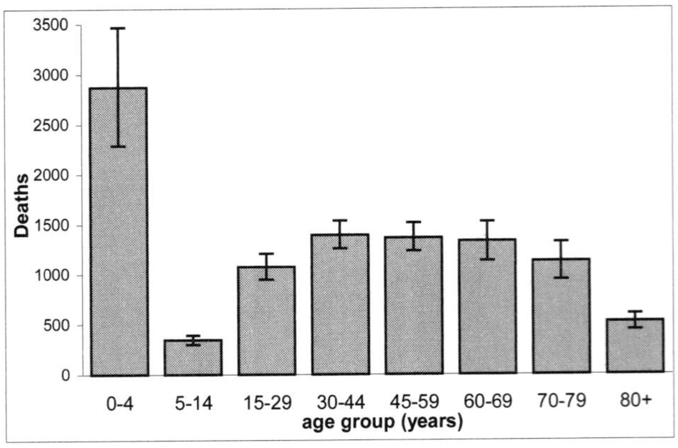

Mongolia   Mongolie   Mongolia

Females, 2000   *Femmes, 2000*   *Mujeres, 2000*

Age	$_nM_x$	$_nq_x$			l_x			$_nd_x$	$_nL_x$	e_x			Actual deaths	Actual population
			percentiles			percentiles					percentiles			
		mean	2½	97½	mean	2½	97½			mean	2½	97½		
<1	0.0620	0.0587	0.0462	0.0717	100,000	100,000	100,000	5,868	94,719	66.9	65.4	68.4	1,717	27,716
1-4	0.0047	0.0185	0.0157	0.0212	94,132	92,831	95,380	1,741	372,349	70.1	68.8	71.4	492	105,134
5-9	0.0009	0.0046	0.0040	0.0052	92,391	90,894	93,881	427	460,886	67.4	66.1	68.8	132	142,892
10-14	0.0006	0.0029	0.0025	0.0033	91,964	90,426	93,508	268	459,150	62.7	61.4	64.1	95	162,316
15-19	0.0008	0.0040	0.0035	0.0045	91,696	90,133	93,274	368	457,560	57.9	56.6	59.2	112	139,374
20-24	0.0013	0.0063	0.0055	0.0071	91,328	89,732	92,950	578	455,195	53.1	51.8	54.5	161	126,939
25-29	0.0024	0.0120	0.0105	0.0134	90,750	89,106	92,438	1,088	451,030	48.4	47.2	49.8	274	113,621
30-34	0.0022	0.0111	0.0098	0.0123	89,662	87,936	91,467	993	445,827	44.0	42.7	45.3	223	100,023
35-39	0.0036	0.0180	0.0161	0.0199	88,669	86,871	90,572	1,598	439,351	39.4	38.2	40.8	335	92,171
40-44	0.0049	0.0242	0.0219	0.0265	87,071	85,141	89,114	2,107	430,087	35.1	33.9	36.4	342	69,824
45-49	0.0069	0.0339	0.0311	0.0368	84,964	82,884	87,163	2,882	417,615	30.9	29.7	32.2	316	45,828
50-54	0.0092	0.0450	0.0411	0.0490	82,082	79,838	84,446	3,691	401,183	26.9	25.8	28.2	294	31,944
55-59	0.0130	0.0630	0.0571	0.0692	78,391	75,996	80,894	4,935	379,618	23.1	22.0	24.3	387	29,775
60-64	0.0200	0.0952	0.0854	0.1053	73,456	70,948	76,035	6,996	349,791	19.4	18.4	20.6	467	23,331
65-69	0.0300	0.1395	0.1241	0.1548	66,460	63,936	68,982	9,274	309,117	16.2	15.2	17.3	635	21,181
70-74	0.0433	0.1955	0.1733	0.2173	57,187	54,598	59,749	11,179	257,986	13.4	12.5	14.5	598	13,812
75-79	0.0563	0.2466	0.2191	0.2748	46,008	42,780	49,247	11,347	201,670	11.1	10.3	12.0	572	10,174
80-84	0.0756	0.3180	0.2834	0.3521	34,660	31,175	38,428	11,023	145,744	8.9	8.3	9.7	482	6,377
85-89	0.1063	0.4200	0.3788	0.4583	23,637	20,400	27,455	9,928	93,367	6.9	6.4	7.5	254	2,393
90-94	0.1563	0.5321	0.4886	0.5705	13,710	11,145	16,924	7,295	46,664	5.1	4.8	5.5	66	424
95-99	0.2404	0.6983	0.6575	0.7323	6,415	4,787	8,535	4,479	18,636	3.7	3.4	3.9	8	34
100+	0.3865	1	1	1	1,935	1,282	2,913	1,935	5,007	2.6	2.3	2.7	0	1
All ages													*7,965*	*1,265,284*

ln($_nq_x$) with uncertainty interval

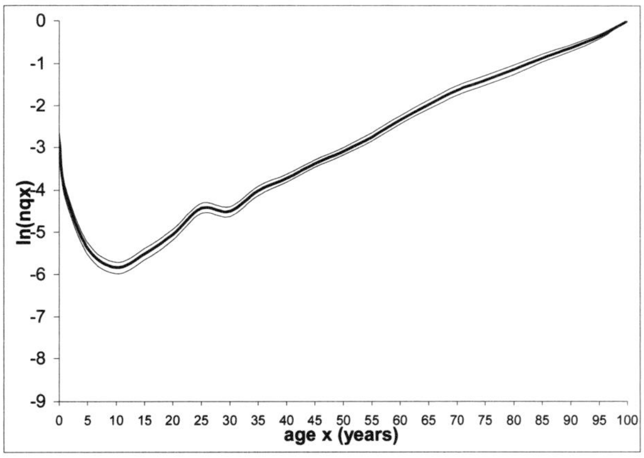

l_x with uncertainty interval

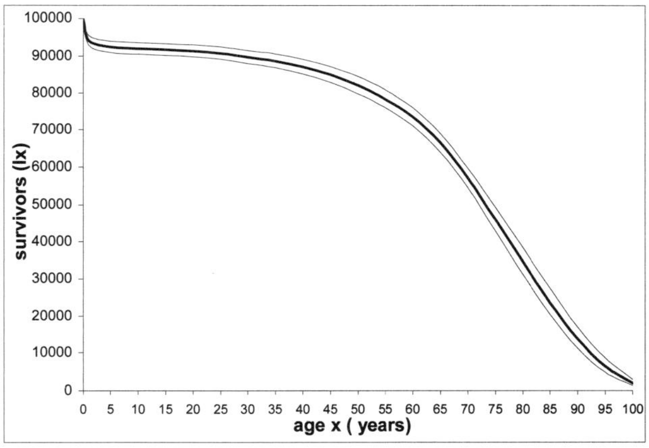

e_x with uncertainty interval

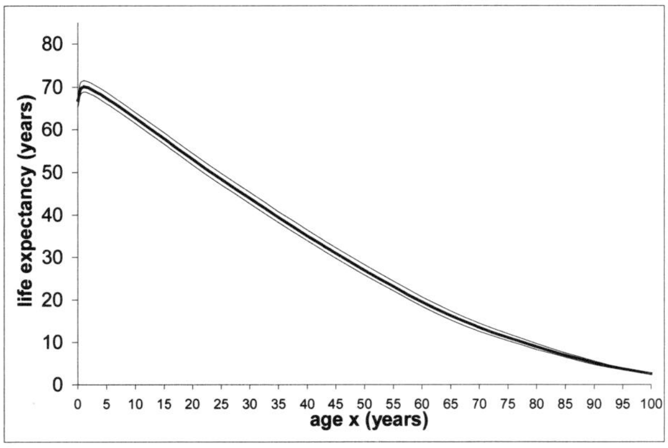

Distribution of deaths by age with uncertainty interval

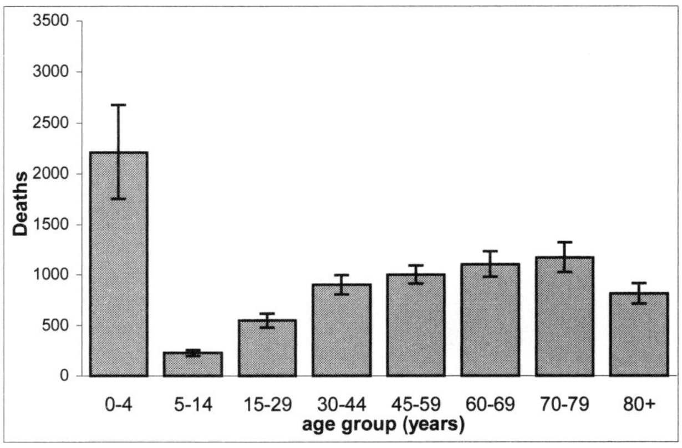

Morocco **Maroc** **Marruecos**

Males, 2000 *Hommes, 2000* *Hombres, 2000*

Age	$_nM_x$	$_nq_x$			l_x			$_nd_x$	$_nL_x$	e_x			Actual deaths	Actual population
			percentiles			percentiles					percentiles			
		mean	2½	97½	mean	2½	97½			mean	2½	97½		
<1	0.0580	0.0551	0.0486	0.0618	100,000	100,000	100,000	5,512	95,039	66.1	65.0	67.1	21,195	365,464
1-4	0.0034	0.0136	0.0124	0.0150	94,488	93,820	95,140	1,289	374,860	68.9	68.1	69.7	5,044	1,466,885
5-9	0.0010	0.0049	0.0045	0.0054	93,199	92,452	93,947	460	464,847	65.8	65.1	66.6	1,721	1,739,240
10-14	0.0007	0.0037	0.0034	0.0041	92,739	91,955	93,519	343	462,839	61.2	60.4	61.9	1,262	1,701,351
15-19	0.0012	0.0058	0.0052	0.0063	92,396	91,586	93,199	533	460,650	56.4	55.6	57.1	1,883	1,628,560
20-24	0.0014	0.0072	0.0066	0.0079	91,864	91,011	92,700	661	457,666	51.7	51.0	52.4	2,170	1,502,959
25-29	0.0014	0.0072	0.0066	0.0078	91,203	90,296	92,078	655	454,377	47.0	46.3	47.8	1,911	1,325,947
30-34	0.0016	0.0080	0.0073	0.0087	90,548	89,589	91,460	720	450,941	42.4	41.7	43.1	1,835	1,149,765
35-39	0.0021	0.0105	0.0096	0.0114	89,828	88,814	90,780	942	446,785	37.7	37.0	38.4	1,998	947,306
40-44	0.0032	0.0157	0.0144	0.0171	88,886	87,803	89,909	1,398	440,935	33.1	32.4	33.7	2,699	851,186
45-49	0.0051	0.0254	0.0234	0.0275	87,488	86,306	88,611	2,223	431,882	28.5	27.9	29.2	3,378	656,201
50-54	0.0085	0.0414	0.0383	0.0446	85,265	83,940	86,537	3,532	417,495	24.2	23.7	24.9	3,925	464,018
55-59	0.0137	0.0662	0.0613	0.0713	81,733	80,209	83,219	5,414	395,131	20.2	19.6	20.8	4,233	308,986
60-64	0.0222	0.1050	0.0974	0.1128	76,319	74,546	78,079	8,013	361,563	16.4	15.9	17.0	6,382	287,934
65-69	0.0351	0.1615	0.1499	0.1725	68,306	66,260	70,357	11,033	313,947	13.0	12.6	13.6	7,965	226,634
70-74	0.0566	0.2481	0.2309	0.2648	57,273	54,945	59,657	14,207	250,848	10.1	9.7	10.5	10,167	179,521
75-79	0.0909	0.3703	0.3471	0.3942	43,066	40,534	45,649	15,949	175,458	7.6	7.2	7.9	8,294	91,242
80-84	0.1451	0.5323	0.5067	0.5594	27,117	24,887	29,677	14,435	99,497	5.6	5.3	5.8	7,519	51,826
85-89	0.2211	0.7119	0.6885	0.7351	12,682	11,089	14,550	9,029	40,837	4.1	3.9	4.2	3,515	15,897
90-94	0.3217	0.8185	0.8004	0.8359	3,653	2,937	4,506	2,990	9,295	2.9	2.9	3.0	878	2,730
95-99	0.4469	0.9547	0.9467	0.9618	663	482	889	633	1,416	2.2	2.2	2.3	131	293
100+	0.5929	1	1	1	30	18	47	30	51	1.7	1.7	1.8	9	15
All ages													*98,112*	*14,963,960*

$\ln(_nq_x)$ with uncertainty interval

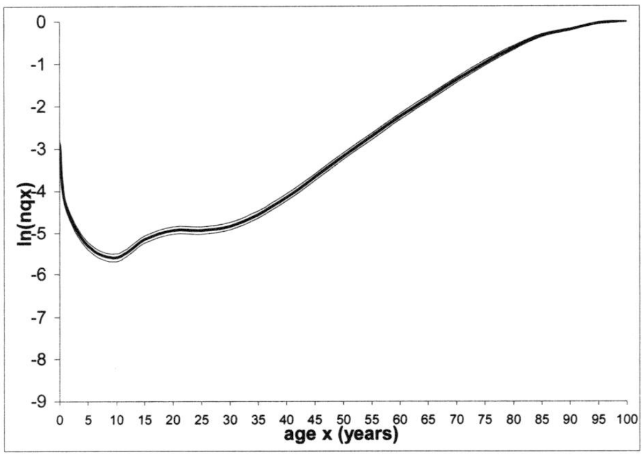

l_x with uncertainty interval

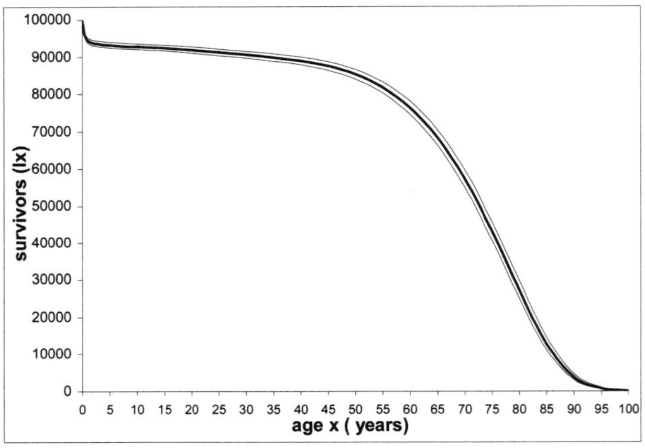

e_x with uncertainty interval

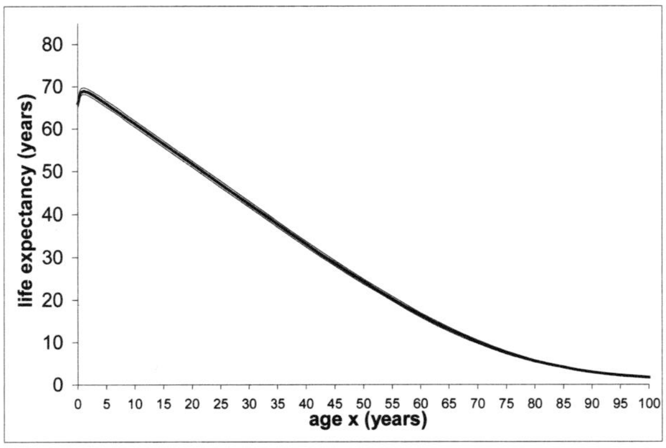

Distribution of deaths by age with uncertainty interval

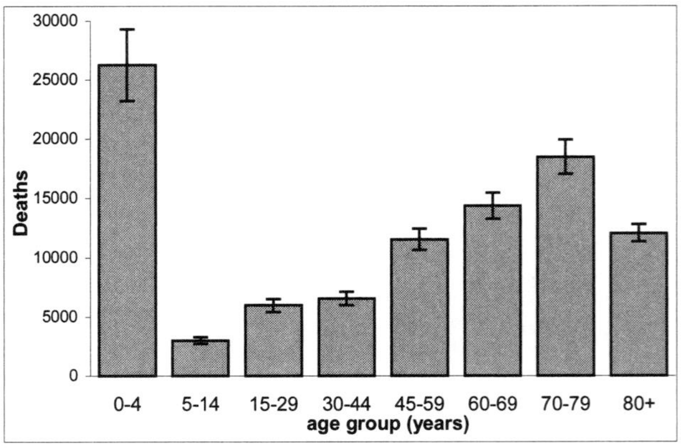

Morocco · Maroc · Marruecos

Females, 2000 · *Femmes, 2000* · *Mujeres, 2000*

Age	$_nM_x$	$_nq_x$			l_x			$_nd_x$	$_nL_x$	e_x			Actual deaths	Actual population
			percentiles			percentiles					percentiles			
		mean	2½	97½	mean	2½	97½			mean	2½	97½		
<1	0.0505	0.0483	0.0420	0.0548	100,000	100,000	100,000	4,832	95,651	70.4	69.1	71.8	17,751	351,408
1-4	0.0028	0.0112	0.0099	0.0125	95,168	94,523	95,799	1,067	378,112	73.0	71.9	74.4	3,987	1,412,590
5-9	0.0006	0.0032	0.0029	0.0036	94,101	93,405	94,799	305	469,744	69.8	68.7	71.2	1,087	1,676,083
10-14	0.0004	0.0020	0.0018	0.0022	93,797	93,077	94,511	186	468,518	65.0	64.0	66.4	652	1,642,171
15-19	0.0004	0.0022	0.0020	0.0025	93,611	92,878	94,343	208	467,532	60.2	59.1	61.5	701	1,576,019
20-24	0.0007	0.0034	0.0030	0.0038	93,402	92,653	94,156	314	466,226	55.3	54.2	56.6	978	1,450,179
25-29	0.0009	0.0046	0.0040	0.0051	93,088	92,310	93,872	425	464,378	50.5	49.4	51.8	1,172	1,280,797
30-34	0.0012	0.0058	0.0051	0.0065	92,663	91,846	93,487	536	461,976	45.7	44.6	47.0	1,291	1,113,087
35-39	0.0016	0.0081	0.0071	0.0090	92,127	91,263	93,002	742	458,781	40.9	39.9	42.3	1,599	988,041
40-44	0.0022	0.0110	0.0097	0.0123	91,385	90,457	92,327	1,003	454,416	36.2	35.2	37.5	1,909	864,682
45-49	0.0034	0.0168	0.0149	0.0188	90,382	89,371	91,413	1,522	448,103	31.6	30.6	32.9	2,243	660,346
50-54	0.0053	0.0262	0.0231	0.0292	88,859	87,721	90,038	2,325	438,486	27.1	26.1	28.4	2,579	486,513
55-59	0.0083	0.0405	0.0353	0.0453	86,535	85,158	87,939	3,502	423,919	22.8	21.8	24.0	3,007	363,981
60-64	0.0138	0.0665	0.0575	0.0746	83,033	81,313	84,786	5,524	401,354	18.6	17.7	19.8	5,217	379,089
65-69	0.0236	0.1115	0.0956	0.1260	77,509	75,298	79,773	8,643	365,937	14.8	13.9	15.8	6,150	260,358
70-74	0.0422	0.1908	0.1635	0.2167	68,866	66,027	71,859	13,139	311,481	11.3	10.6	12.3	9,645	228,654
75-79	0.0746	0.3145	0.2718	0.3534	55,727	52,066	60,111	17,525	234,823	8.4	7.8	9.2	7,308	97,925
80-84	0.1251	0.4766	0.4218	0.5211	38,202	33,804	43,657	18,208	145,492	6.1	5.7	6.7	7,309	58,404
85-89	0.1988	0.6639	0.6119	0.7026	19,995	16,190	24,931	13,275	66,787	4.4	4.1	4.8	3,886	19,549
90-94	0.2990	0.7881	0.7492	0.8163	6,720	4,837	9,623	5,296	17,713	3.1	3.0	3.4	1,191	3,982
95-99	0.4260	0.9350	0.9157	0.9481	1,424	898	2,413	1,332	3,126	2.3	2.3	2.4	233	548
100+	0.5748	1	1	1	93	47	203	93	161	1.7	1.7	1.9	21	37
All ages													*79,916*	*14,914,443*

ln($_nq_x$) with uncertainty interval

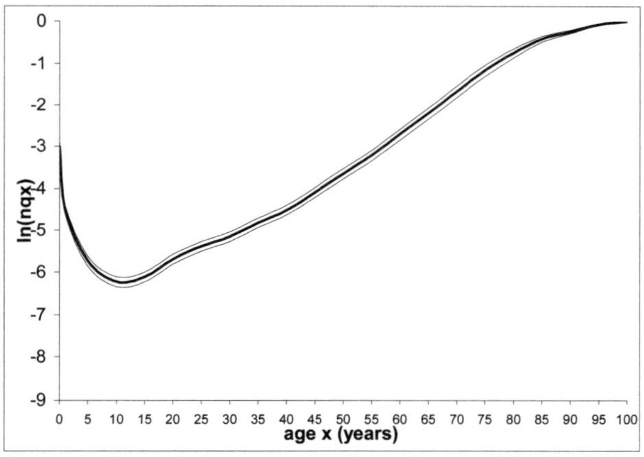

l_x with uncertainty interval

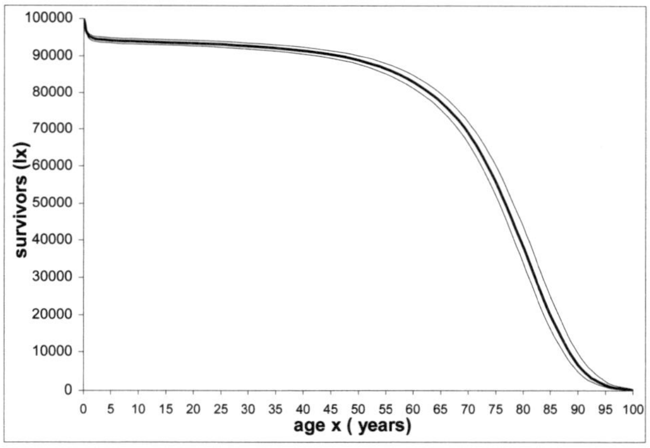

e_x with uncertainty interval

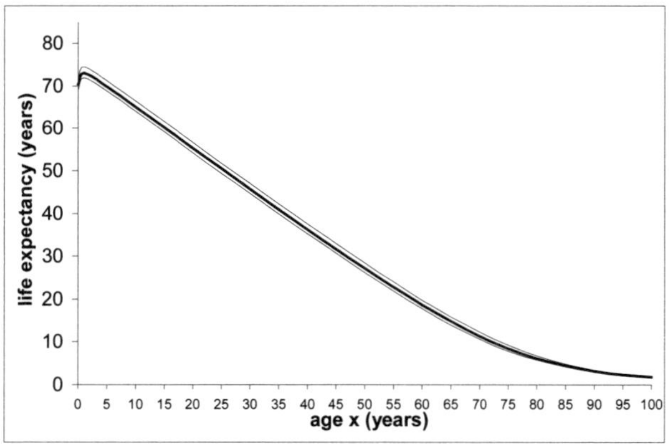

Distribution of deaths by age with uncertainty interval

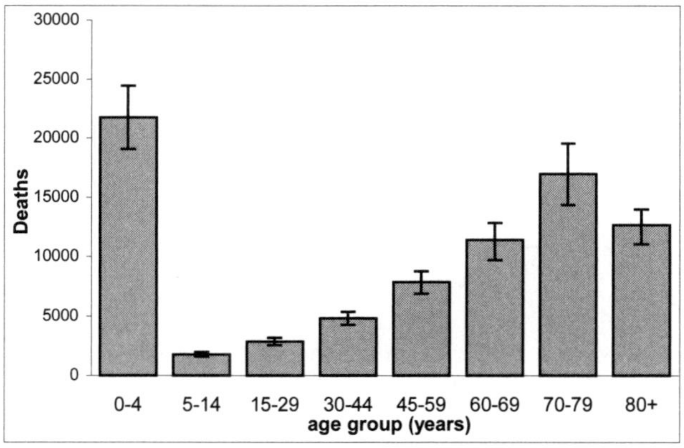

Mozambique | **Mozambique** | **Mozambique**

Males, 2000 | *Hommes, 2000* | *Hombres, 2000*

Age	$_nM_x$	$_nq_x$			l_x			$_nd_x$	$_nL_x$	e_x			Actual deaths	Actual population
			percentiles			percentiles					percentiles			
		mean	2 1/2	97 1/2	mean	2 1/2	97 1/2			mean	2 1/2	97 1/2		
<1	0.1959	0.1665	0.1558	0.1760	100,000	100,000	100,000	16,650	85,015	37.9	36.3	40.2	70,281	358,845
1-4	0.0191	0.0729	0.0660	0.0780	83,350	82,396	84,424	6,076	318,815	44.5	42.3	47.2	23,449	1,230,342
5-9	0.0042	0.0209	0.0186	0.0227	77,273	76,544	78,022	1,613	382,333	43.8	41.7	46.4	5,510	1,305,790
10-14	0.0028	0.0140	0.0124	0.0153	75,660	74,890	76,529	1,060	375,649	39.7	37.6	42.3	3,161	1,119,842
15-19	0.0034	0.0167	0.0148	0.0183	74,600	73,774	75,557	1,246	369,885	35.3	33.1	37.8	3,251	965,394
20-24	0.0089	0.0433	0.0381	0.0477	73,354	72,463	74,415	3,176	358,831	30.8	28.7	33.3	7,382	834,055
25-29	0.0179	0.0856	0.0748	0.0951	70,178	69,106	71,508	6,006	335,877	27.1	25.0	29.5	12,084	675,846
30-34	0.0276	0.1289	0.1118	0.1449	64,173	62,684	66,161	8,271	300,186	24.4	22.3	26.7	15,283	554,710
35-39	0.0313	0.1452	0.1253	0.1651	55,902	53,780	58,763	8,118	259,215	22.6	20.7	24.7	14,242	454,739
40-44	0.0330	0.1523	0.1312	0.1743	47,784	45,010	51,399	7,280	220,720	21.1	19.3	22.9	12,135	367,916
45-49	0.0339	0.1561	0.1346	0.1789	40,504	37,257	44,654	6,321	186,718	19.4	17.8	21.0	9,939	293,593
50-54	0.0338	0.1559	0.1348	0.1787	34,183	30,685	38,644	5,330	157,590	17.5	16.2	18.9	8,618	254,816
55-59	0.0340	0.1568	0.1361	0.1793	28,853	25,259	33,433	4,524	132,956	15.3	14.2	16.4	6,970	204,851
60-64	0.0457	0.2051	0.1797	0.2328	24,329	20,731	28,882	4,991	109,169	12.7	11.7	13.6	7,278	159,200
65-69	0.0607	0.2635	0.2335	0.2960	19,338	15,906	23,692	5,096	83,952	10.3	9.5	11.1	7,122	117,324
70-74	0.0879	0.3603	0.3243	0.3986	14,242	11,198	18,160	5,132	58,382	8.1	7.5	8.7	6,823	77,616
75-79	0.1283	0.4858	0.4454	0.5275	9,110	6,735	12,265	4,426	34,487	6.2	5.8	6.7	5,524	43,038
80-84	0.1859	0.6347	0.5943	0.6750	4,684	3,182	6,788	2,973	15,989	4.7	4.4	5.0	3,368	18,115
85-89	0.2659	0.7987	0.7652	0.8308	1,711	1,034	2,741	1,367	5,140	3.5	3.3	3.7	1,264	4,755
90-94	0.3673	0.8737	0.8472	0.8984	345	175	638	301	820	2.6	2.5	2.8	252	687
95-99	0.4901	0.9919	0.9875	0.9951	44	18	96	43	88	2.0	2.0	2.1	24	48
100+	0.6313	1	1	1	0	0	1	0	1	1.6	1.6	1.6	1	2
All ages													*223,961*	*9,041,524*

ln($_nq_x$) with uncertainty interval

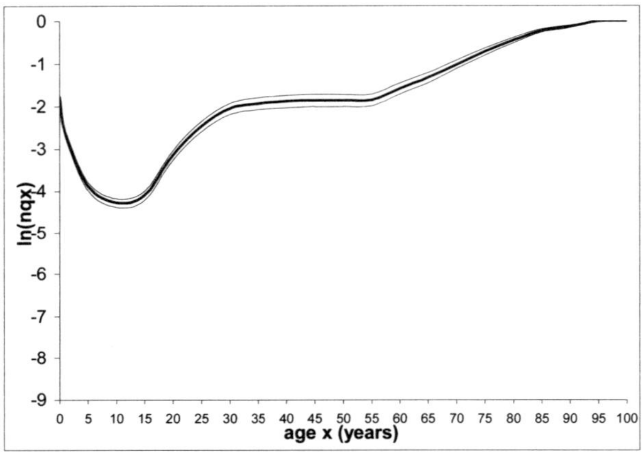

l_x with uncertainty interval

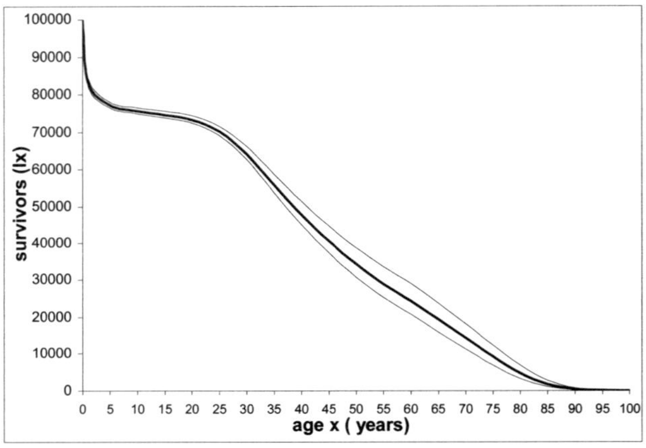

e_x with uncertainty interval

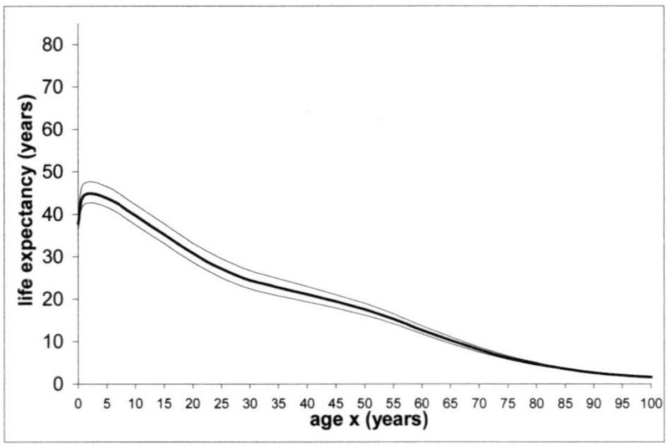

Distribution of deaths by age with uncertainty interval

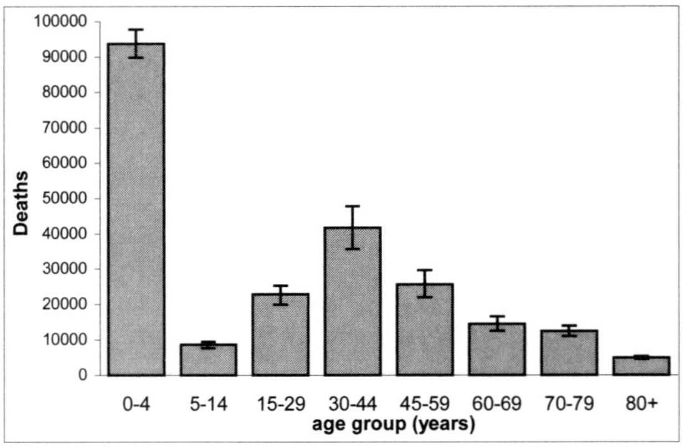

Mozambique **Mozambique** **Mozambique**

Females, 2000 *Femmes, 2000* *Mujeres, 2000*

Age	$_nM_x$	$_nq_x$ mean	$_nq_x$ percentiles 2½	$_nq_x$ percentiles 97½	l_x mean	l_x percentiles 2½	l_x percentiles 97½	$_nd_x$	$_nL_x$	e_x mean	e_x percentiles 2½	e_x percentiles 97½	Actual deaths	Actual population
<1	0.1606	0.1403	0.1309	0.1500	100,000	100,000	100,000	14,034	87,370	39.5	37.7	42.1	55,798	347,381
1-4	0.0208	0.0793	0.0723	0.0852	85,966	85,001	86,906	6,816	327,506	44.9	42.5	48.0	25,837	1,241,448
5-9	0.0045	0.0220	0.0197	0.0240	79,150	78,401	79,898	1,745	391,388	44.7	42.3	47.7	5,842	1,310,231
10-14	0.0031	0.0154	0.0136	0.0167	77,405	76,558	78,286	1,188	384,055	40.6	38.2	43.6	3,476	1,123,354
15-19	0.0069	0.0340	0.0300	0.0373	76,217	75,276	77,191	2,595	374,597	36.2	33.9	39.1	6,715	969,272
20-24	0.0153	0.0736	0.0644	0.0815	73,622	72,475	74,803	5,420	354,558	32.4	30.1	35.3	12,785	836,299
25-29	0.0242	0.1140	0.0988	0.1276	68,201	66,628	69,813	7,775	321,570	29.8	27.5	32.5	16,349	676,206
30-34	0.0286	0.1336	0.1148	0.1509	60,427	58,281	62,910	8,071	281,955	28.3	26.2	30.8	15,986	558,462
35-39	0.0262	0.1229	0.1051	0.1397	52,355	49,715	55,687	6,436	245,688	27.3	25.4	29.5	12,180	464,984
40-44	0.0228	0.1079	0.0920	0.1230	45,920	42,895	49,835	4,955	217,212	25.7	24.1	27.7	8,743	383,275
45-49	0.0202	0.0963	0.0821	0.1099	40,965	37,617	45,249	3,945	194,962	23.5	22.1	25.2	6,348	313,739
50-54	0.0212	0.1007	0.0860	0.1148	37,020	33,482	41,536	3,729	175,776	20.8	19.5	22.2	5,939	279,904
55-59	0.0239	0.1127	0.0965	0.1282	33,290	29,637	37,963	3,752	157,072	17.8	16.8	19.1	5,531	231,553
60-64	0.0318	0.1474	0.1270	0.1668	29,538	25,839	34,299	4,355	136,804	14.8	13.8	15.9	5,897	185,241
65-69	0.0456	0.2045	0.1779	0.2293	25,183	21,530	29,942	5,149	113,042	11.9	11.1	12.8	6,401	140,520
70-74	0.0694	0.2958	0.2614	0.3274	20,034	16,594	24,615	5,926	85,353	9.3	8.7	10.1	6,698	96,464
75-79	0.1033	0.4106	0.3698	0.4472	14,107	11,162	18,181	5,793	56,055	7.2	6.7	7.8	5,854	56,652
80-84	0.1538	0.5553	0.5115	0.5932	8,315	6,176	11,450	4,617	30,030	5.4	5.1	5.8	4,020	26,143
85-89	0.2234	0.7167	0.6765	0.7502	3,697	2,518	5,577	2,650	11,862	4.1	3.8	4.3	1,798	8,046
90-94	0.3161	0.8112	0.7776	0.8385	1,047	631	1,792	850	2,688	3.0	2.8	3.2	479	1,516
95-99	0.4355	0.9440	0.9264	0.9571	198	102	395	187	429	2.3	2.2	2.4	68	157
100+	0.5842	1	1	1	11	4	29	11	19	1.7	1.7	1.8	6	11
All ages													*212,751*	*9,250,858*

ln($_nq_x$) with uncertainty interval

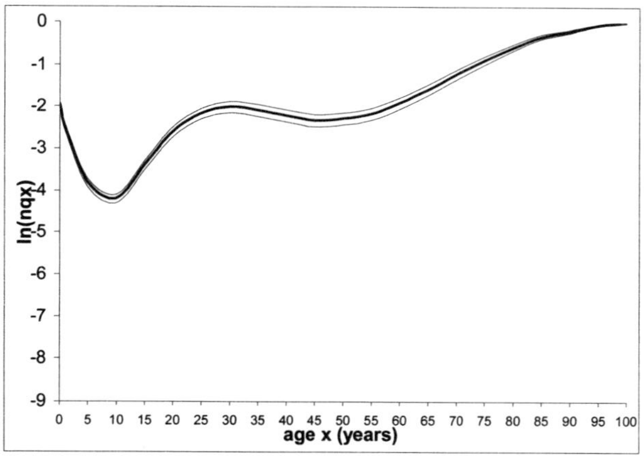

l_x with uncertainty interval

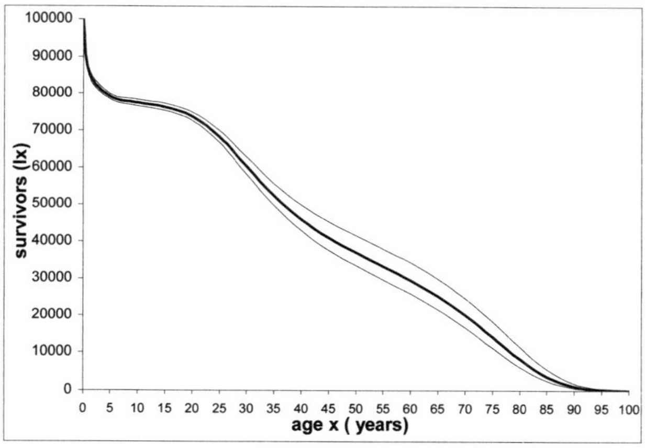

e_x with uncertainty interval

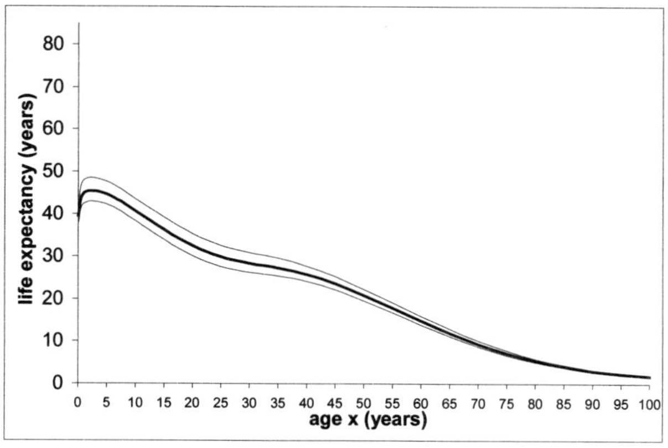

Distribution of deaths by age with uncertainty interval

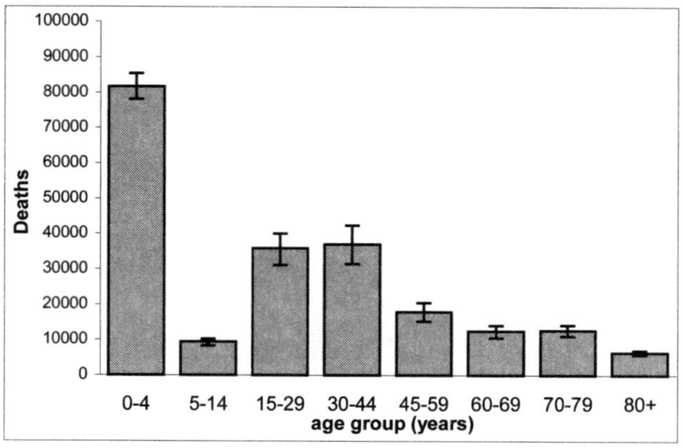

Myanmar

Males, 2000

Myanmar

Hommes, 2000

Myanmar

Hombres, 2000

Age	$_nM_x$	$_nq_x$			l_x			$_nd_x$	$_nL_x$	e_x			Actual deaths	Actual population
			percentiles			percentiles					percentiles			
		mean	2½	97½	mean	2½	97½			mean	2½	97½		
<1	0.0931	0.0859	0.0691	0.1036	100,000	100,000	100,000	8,589	92,270	56.2	54.0	58.5	53,302	572,598
1-4	0.0069	0.0272	0.0232	0.0312	91,411	89,636	93,088	2,483	359,685	60.5	58.9	62.1	14,961	2,167,480
5-9	0.0018	0.0090	0.0077	0.0103	88,928	86,932	90,912	798	442,647	58.1	56.7	59.7	4,889	2,713,544
10-14	0.0012	0.0062	0.0053	0.0071	88,131	86,045	90,202	548	439,283	53.6	52.2	55.2	3,158	2,532,101
15-19	0.0017	0.0085	0.0073	0.0097	87,583	85,439	89,712	746	436,050	48.9	47.5	50.5	4,167	2,437,034
20-24	0.0040	0.0199	0.0172	0.0226	86,837	84,616	89,042	1,726	429,870	44.3	42.9	45.9	9,407	2,342,589
25-29	0.0071	0.0351	0.0306	0.0396	85,111	82,726	87,492	2,989	418,083	40.2	38.8	41.8	15,049	2,105,148
30-34	0.0078	0.0382	0.0336	0.0428	82,122	79,496	84,776	3,137	402,768	36.6	35.2	38.1	14,111	1,811,749
35-39	0.0084	0.0412	0.0365	0.0459	78,985	76,149	81,875	3,256	386,786	32.9	31.6	34.4	12,850	1,526,442
40-44	0.0086	0.0419	0.0375	0.0466	75,729	72,683	78,839	3,176	370,705	29.2	28.0	30.6	11,284	1,316,965
45-49	0.0105	0.0513	0.0457	0.0569	72,553	69,339	75,885	3,722	353,461	25.4	24.2	26.7	11,718	1,112,976
50-54	0.0147	0.0709	0.0628	0.0790	68,831	65,451	72,378	4,883	331,948	21.6	20.5	22.9	13,233	899,497
55-59	0.0210	0.0999	0.0879	0.1120	63,948	60,404	67,703	6,385	303,776	18.1	17.0	19.3	14,375	683,849
60-64	0.0307	0.1424	0.1251	0.1611	57,562	53,921	61,468	8,200	267,313	14.8	13.8	16.0	15,330	499,775
65-69	0.0451	0.2027	0.1767	0.2298	49,363	45,728	53,262	10,004	221,804	11.9	11.0	12.9	17,378	385,295
70-74	0.0690	0.2942	0.2570	0.3315	39,359	35,810	43,160	11,581	167,841	9.2	8.5	10.1	20,277	293,863
75-79	0.1050	0.4157	0.3680	0.4603	27,778	24,270	31,862	11,547	110,022	7.1	6.5	7.7	19,598	186,737
80-84	0.1582	0.5668	0.5145	0.6121	16,231	13,098	19,855	9,200	58,153	5.3	4.9	5.8	15,420	97,466
85-89	0.2310	0.7323	0.6856	0.7698	7,030	5,081	9,595	5,148	22,282	3.9	3.7	4.3	7,815	33,825
90-94	0.3269	0.8252	0.7878	0.8543	1,882	1,170	3,016	1,553	4,752	2.9	2.8	3.1	2,698	8,253
95-99	0.4480	0.9556	0.9382	0.9676	329	170	640	314	702	2.2	2.2	2.3	579	1,292
100+	0.5947	1	1	1	15	6	40	15	25	1.7	1.7	1.8	67	113
All ages													*281,665*	*23,728,591*

ln($_nq_x$) with uncertainty interval

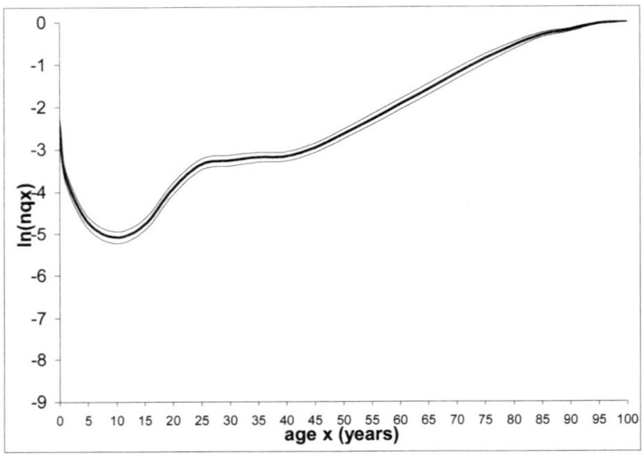

l_x with uncertainty interval

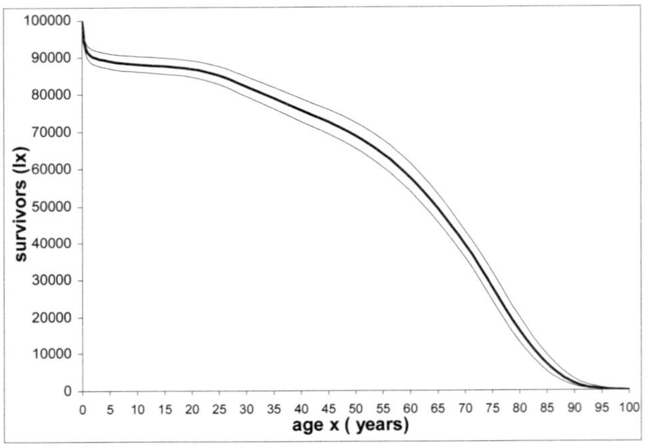

e_x with uncertainty interval

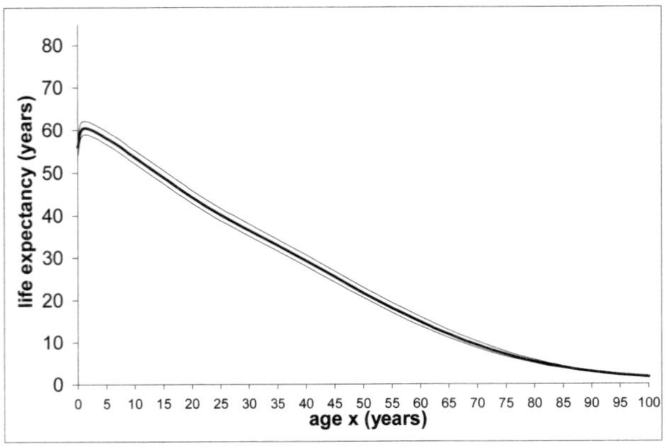

Distribution of deaths by age with uncertainty interval

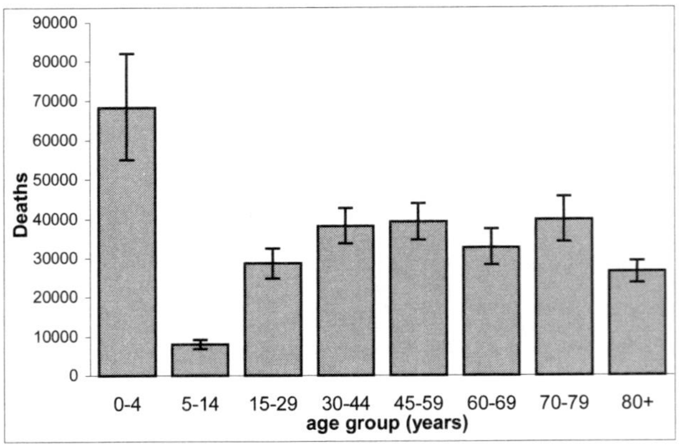

Myanmar **Myanmar** **Myanmar**

Females, 2000 *Femmes, 2000* *Mujeres, 2000*

Age	$_nM_x$	$_nq_x$			l_x			$_nd_x$	$_nL_x$	e_x			Actual deaths	Actual population
			percentiles			percentiles					percentiles			
		mean	2½	97½	mean	2½	97½			mean	2½	97½		
<1	0.0761	0.0712	0.0545	0.0889	100,000	100,000	100,000	7,120	93,592	61.1	58.6	63.6	41,292	542,747
1-4	0.0070	0.0276	0.0228	0.0326	92,880	91,109	94,552	2,559	365,377	64.8	63.0	66.8	14,875	2,123,993
5-9	0.0017	0.0086	0.0071	0.0101	90,321	88,322	92,321	774	449,669	62.6	60.8	64.6	4,576	2,659,691
10-14	0.0012	0.0060	0.0050	0.0071	89,547	87,435	91,629	538	446,390	58.1	56.4	60.1	3,005	2,494,066
15-19	0.0020	0.0100	0.0084	0.0117	89,009	86,821	91,145	887	442,829	53.4	51.7	55.4	4,836	2,415,097
20-24	0.0033	0.0165	0.0140	0.0193	88,122	85,814	90,339	1,458	436,966	49.0	47.3	50.9	7,795	2,335,572
25-29	0.0038	0.0191	0.0162	0.0221	86,664	84,169	89,058	1,652	429,190	44.7	43.1	46.7	8,142	2,115,184
30-34	0.0040	0.0200	0.0171	0.0231	85,012	82,317	87,614	1,702	420,804	40.6	39.0	42.4	7,431	1,836,894
35-39	0.0050	0.0248	0.0213	0.0284	83,310	80,417	86,112	2,069	411,377	36.3	34.8	38.2	7,839	1,558,800
40-44	0.0057	0.0279	0.0240	0.0319	81,241	78,130	84,267	2,265	400,544	32.2	30.7	34.0	7,669	1,356,505
45-49	0.0074	0.0364	0.0314	0.0415	78,976	75,652	82,223	2,872	387,702	28.0	26.6	29.8	8,670	1,170,283
50-54	0.0103	0.0502	0.0431	0.0576	76,104	72,546	79,597	3,822	370,967	24.0	22.6	25.6	9,736	945,111
55-59	0.0146	0.0702	0.0595	0.0811	72,283	68,473	76,046	5,075	348,727	20.1	18.8	21.7	10,655	732,229
60-64	0.0222	0.1052	0.0884	0.1219	67,208	63,163	71,235	7,070	318,365	16.5	15.2	17.9	12,160	547,532
65-69	0.0348	0.1602	0.1339	0.1885	60,138	55,937	64,460	9,634	276,604	13.1	11.9	14.4	15,056	432,283
70-74	0.0568	0.2486	0.2088	0.2934	50,504	46,128	55,205	12,557	221,127	10.2	9.1	11.3	19,364	340,992
75-79	0.0905	0.3690	0.3135	0.4288	37,947	33,096	43,350	14,003	154,726	7.7	6.9	8.6	20,544	226,997
80-84	0.1409	0.5209	0.4559	0.5857	23,944	19,197	29,656	12,472	88,539	5.7	5.1	6.4	17,636	125,201
85-89	0.2115	0.6917	0.6311	0.7474	11,472	7,954	15,966	7,936	37,521	4.2	3.8	4.7	9,825	46,454
90-94	0.3064	0.7982	0.7499	0.8411	3,536	2,009	5,831	2,823	9,214	3.1	2.8	3.4	3,786	12,359
95-99	0.4281	0.9370	0.9112	0.9574	714	319	1,458	669	1,562	2.3	2.2	2.4	917	2,142
100+	0.5771	1	1	1	45	14	129	45	78	1.7	1.7	1.8	125	216
All ages													*235,934*	*24,020,348*

ln($_nq_x$) with uncertainty interval

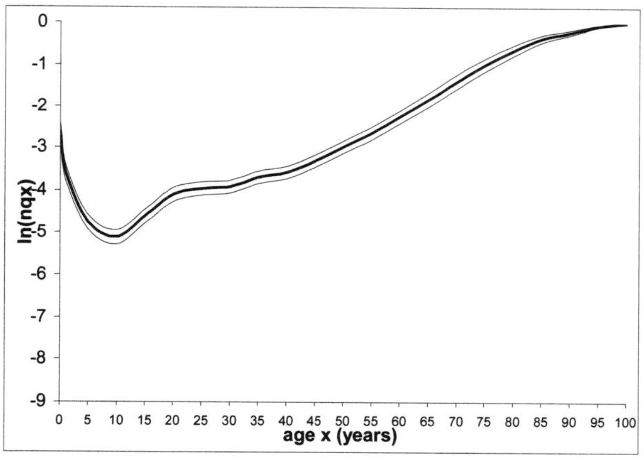

l_x with uncertainty interval

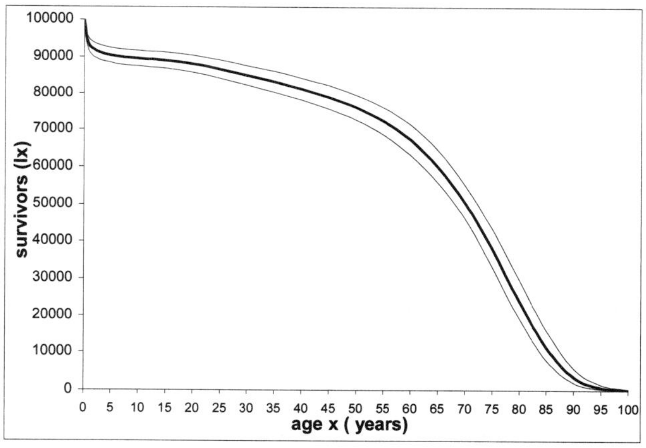

e_x with uncertainty interval

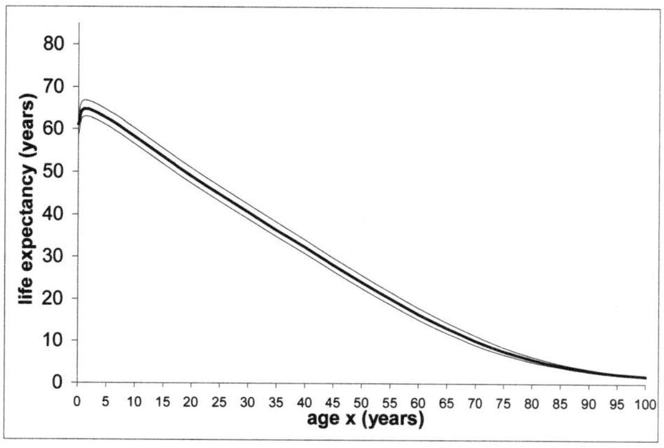

Distribution of deaths by age with uncertainty interval

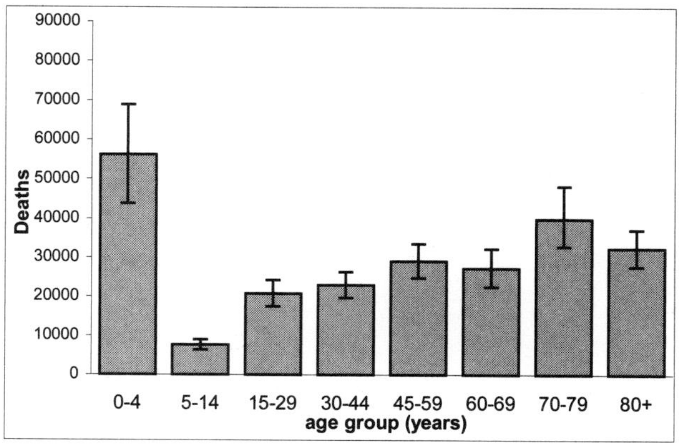

Namibia — Namibie — Namibia

Males, 2000 — *Hommes, 2000* — *Hombres, 2000*

Age	$_nM_x$	$_nq_x$ mean	$_nq_x$ percentiles 2½	$_nq_x$ percentiles 97½	l_x mean	l_x percentiles 2½	l_x percentiles 97½	$_nd_x$	$_nL_x$	e_x mean	e_x percentiles 2½	e_x percentiles 97½	Actual deaths	Actual population
<1	0.0928	0.0857	0.0716	0.1001	100,000	100,000	100,000	8,569	92,288	42.8	39.0	48.4	2,856	30,760
1-4	0.0110	0.0428	0.0343	0.0503	91,431	89,988	92,840	3,912	356,335	45.8	41.6	51.6	1,218	110,919
5-9	0.0019	0.0097	0.0075	0.0115	87,519	86,031	89,017	847	435,478	43.8	39.6	49.4	255	131,003
10-14	0.0014	0.0070	0.0054	0.0084	86,672	85,078	88,224	607	431,842	39.2	35.0	44.7	160	113,896
15-19	0.0033	0.0163	0.0125	0.0195	86,065	84,395	87,733	1,402	426,819	34.4	30.3	40.0	306	93,204
20-24	0.0085	0.0418	0.0318	0.0505	84,663	82,773	86,632	3,539	414,467	30.0	25.9	35.5	684	80,117
25-29	0.0207	0.0986	0.0737	0.1220	81,124	78,603	83,877	7,998	385,624	26.2	22.1	31.7	1,422	68,562
30-34	0.0347	0.1597	0.1178	0.2013	73,126	69,233	77,696	11,676	336,439	23.8	19.8	29.1	1,915	55,177
35-39	0.0376	0.1719	0.1263	0.2185	61,449	55,865	68,540	10,562	280,844	22.8	19.1	27.8	1,645	43,755
40-44	0.0368	0.1685	0.1224	0.2136	50,888	43,660	59,886	8,575	233,004	22.0	18.7	26.4	1,225	33,293
45-49	0.0353	0.1624	0.1177	0.2060	42,313	34,333	52,437	6,870	194,392	21.0	18.1	24.8	895	25,320
50-54	0.0318	0.1474	0.1070	0.1867	35,443	27,319	46,091	5,226	164,152	19.5	17.1	22.8	680	21,370
55-59	0.0279	0.1305	0.0949	0.1648	30,217	22,290	40,975	3,943	141,230	17.5	15.5	20.2	472	16,891
60-64	0.0336	0.1549	0.1140	0.1935	26,275	18,683	36,914	4,070	121,198	14.7	13.0	17.0	492	14,661
65-69	0.0455	0.2043	0.1530	0.2511	22,205	15,140	32,502	4,536	99,683	12.0	10.6	13.9	525	11,531
70-74	0.0674	0.2885	0.2223	0.3462	17,669	11,420	27,268	5,097	75,601	9.4	8.3	11.0	574	8,513
75-79	0.1020	0.4064	0.3260	0.4728	12,571	7,553	21,111	5,109	50,085	7.2	6.3	8.4	541	5,308
80-84	0.1543	0.5567	0.4695	0.6248	7,463	4,030	14,229	4,155	26,927	5.4	4.8	6.3	391	2,537
85-89	0.2266	0.7234	0.6435	0.7823	3,308	1,512	7,549	2,393	10,558	4.0	3.6	4.6	185	817
90-94	0.3221	0.8191	0.7532	0.8663	915	329	2,691	750	2,327	2.9	2.7	3.3	52	161
95-99	0.4435	0.9515	0.9177	0.9714	166	44	664	158	355	2.2	2.1	2.4	7	15
100+	0.5907	1	1	1	8	1	55	8	14	1.7	1.7	1.8	1	1
All ages													*16,501*	*867,811*

ln($_nq_x$) with uncertainty interval

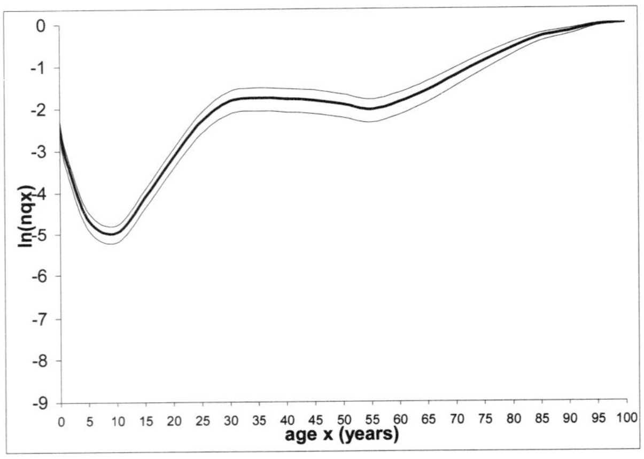

l_x with uncertainty interval

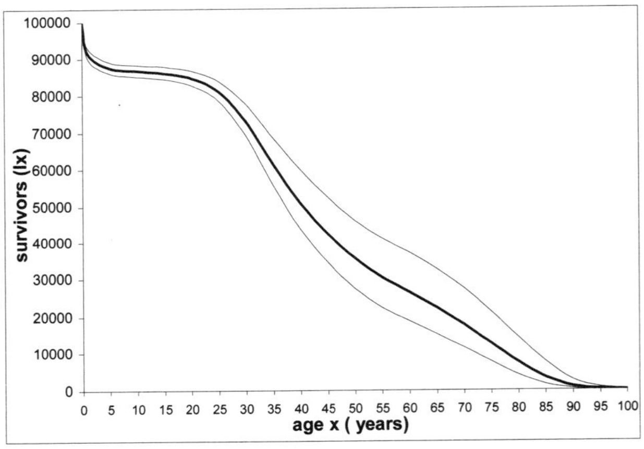

e_x with uncertainty interval

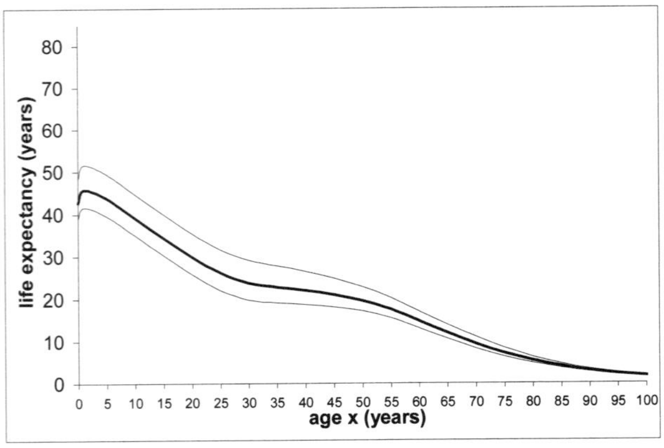

Distribution of deaths by age with uncertainty interval

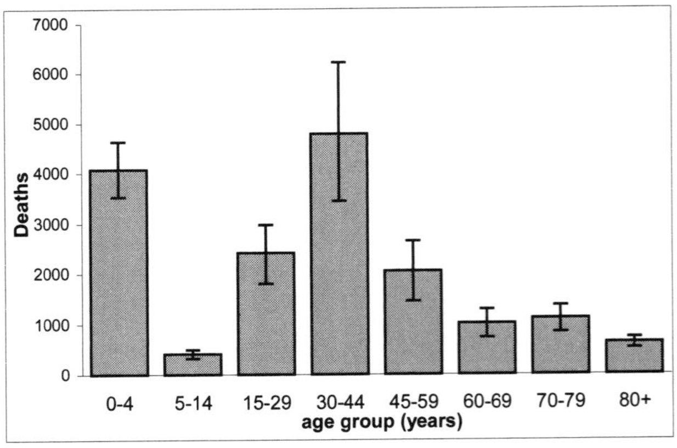

Namibia Namibie Namibia

Females, 2000 *Femmes, 2000* *Mujeres, 2000*

Age	$_nM_x$	$_nq_x$			l_x			$_nd_x$	$_nL_x$	e_x			Actual deaths	Actual population
			percentiles			percentiles					percentiles			
		mean	2½	97½	mean	2½	97½			mean	2½	97½		
<1	0.0838	0.0779	0.0659	0.0898	100,000	100,000	100,000	7,790	92,989	42.6	39.0	47.8	2,484	29,648
1-4	0.0114	0.0445	0.0372	0.0509	92,210	91,017	93,412	4,107	358,981	45.2	41.2	50.6	1,257	109,885
5-9	0.0019	0.0096	0.0078	0.0111	88,102	86,856	89,338	845	438,400	43.3	39.2	48.5	250	129,473
10-14	0.0014	0.0068	0.0055	0.0079	87,258	85,929	88,551	596	434,798	38.7	34.6	43.9	155	112,812
15-19	0.0060	0.0296	0.0238	0.0346	86,662	85,270	87,994	2,567	426,892	33.9	29.9	39.1	556	92,543
20-24	0.0180	0.0859	0.0678	0.1018	84,095	82,412	85,893	7,223	402,417	29.9	25.8	35.0	1,421	79,180
25-29	0.0324	0.1499	0.1164	0.1819	76,872	74,109	80,068	11,521	355,556	27.4	23.5	32.5	2,160	66,653
30-34	0.0396	0.1802	0.1393	0.2211	65,351	61,067	70,744	11,779	297,306	26.8	23.1	31.6	2,150	54,270
35-39	0.0334	0.1540	0.1178	0.1897	53,572	47,715	60,893	8,252	247,228	27.2	23.9	31.4	1,503	45,014
40-44	0.0275	0.1287	0.0978	0.1589	45,319	38,662	53,653	5,832	212,018	26.7	23.9	30.3	1,000	36,364
45-49	0.0213	0.1010	0.0765	0.1251	39,488	32,543	48,306	3,989	187,467	25.3	22.9	28.3	624	29,336
50-54	0.0192	0.0917	0.0695	0.1134	35,499	28,507	44,514	3,254	169,360	22.8	20.8	25.4	498	25,925
55-59	0.0187	0.0894	0.0680	0.1103	32,245	25,312	41,325	2,882	154,020	19.9	18.1	22.1	415	22,177
60-64	0.0234	0.1105	0.0846	0.1355	29,363	22,560	38,420	3,246	138,700	16.6	15.1	18.5	428	18,293
65-69	0.0341	0.1572	0.1218	0.1905	26,117	19,551	35,048	4,106	120,322	13.3	12.1	15.0	490	14,349
70-74	0.0547	0.2407	0.1906	0.2862	22,011	15,891	30,607	5,298	96,811	10.3	9.3	11.8	601	10,987
75-79	0.0873	0.3582	0.2929	0.4146	16,713	11,426	24,525	5,987	68,597	7.8	7.0	9.0	590	6,762
80-84	0.1371	0.5104	0.4358	0.5713	10,726	6,776	17,311	5,475	39,943	5.8	5.2	6.6	487	3,556
85-89	0.2075	0.6832	0.6123	0.7381	5,251	2,921	9,768	3,588	17,287	4.3	3.9	4.8	259	1,250
90-94	0.3026	0.7930	0.7344	0.8374	1,664	765	3,787	1,319	4,360	3.1	2.8	3.5	82	271
95-99	0.4248	0.9339	0.9007	0.9558	344	124	1,006	322	757	2.3	2.2	2.5	15	36
100+	0.5742	1	1	1	23	5	100	23	40	1.7	1.7	1.9	1	2
All ages													*17,427*	*888,786*

ln($_nq_x$) with uncertainty interval

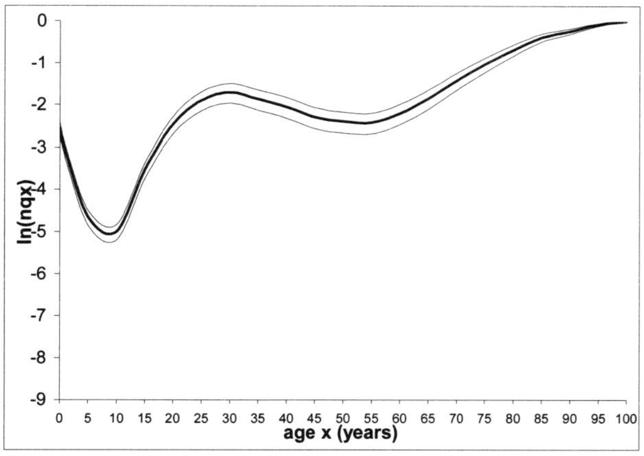

l_x with uncertainty interval

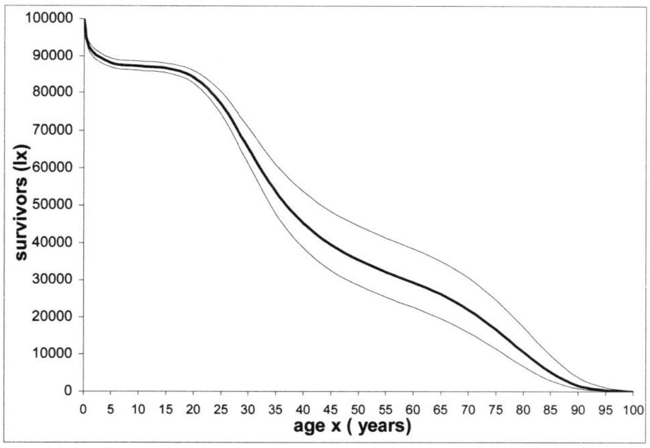

e_x with uncertainty interval

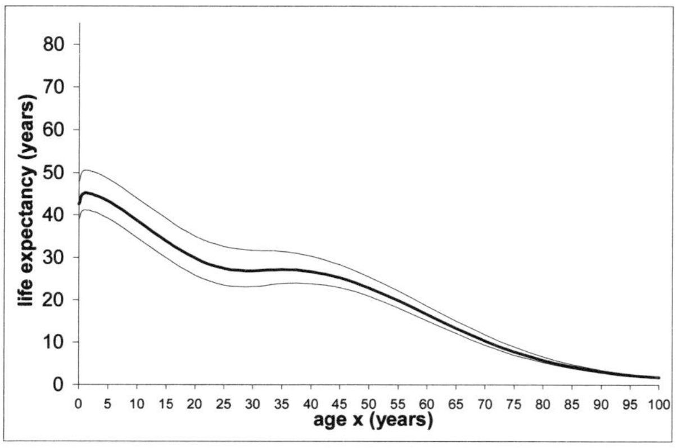

Distribution of deaths by age with uncertainty interval

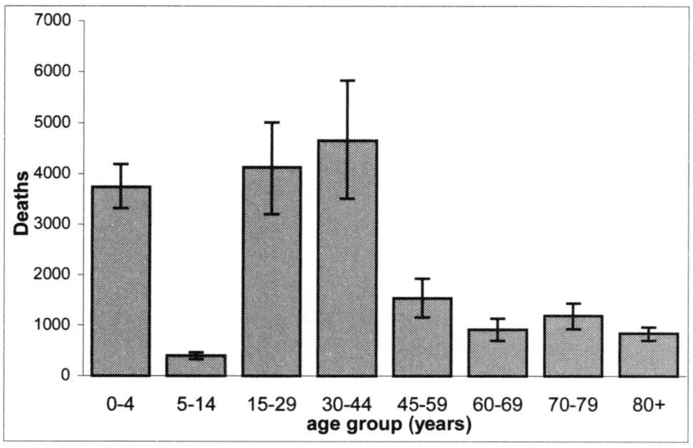

Nauru **Nauru** **Nauru**

Males, 2000 *Hommes, 2000* *Hombres, 2000*

Age	$_nM_x$	$_nq_x$			l_x			$_nd_x$	$_nL_x$	e_x			Actual deaths	Actual population
			percentiles			percentiles					percentiles			
		mean	2½	97½	mean	2½	97½			mean	2½	97½		
<1	0.0122	0.0120	0.0100	0.0161	100,000	100,000	100,000	1,204	98,917	58.8	55.3	63.0	2	179
1-4	0.0010	0.0040	0.0034	0.0053	98,796	98,393	98,997	396	394,235	58.5	55.2	62.7	1	718
5-9	0.0006	0.0028	0.0024	0.0037	98,400	97,877	98,661	276	491,312	54.7	51.4	58.9	0	836
10-14	0.0006	0.0028	0.0023	0.0037	98,124	97,516	98,427	274	489,937	49.8	46.6	54.0	0	700
15-19	0.0017	0.0087	0.0071	0.0115	97,850	97,156	98,196	849	487,129	45.0	41.8	49.2	1	609
20-24	0.0032	0.0160	0.0128	0.0213	97,001	96,040	97,496	1,553	481,124	40.4	37.2	44.5	2	499
25-29	0.0043	0.0211	0.0163	0.0281	95,448	93,996	96,252	2,010	472,215	36.0	33.0	40.0	2	453
30-34	0.0062	0.0306	0.0230	0.0408	93,438	91,356	94,686	2,864	460,030	31.7	28.9	35.7	3	463
35-39	0.0096	0.0470	0.0345	0.0623	90,574	87,626	92,505	4,261	442,218	27.6	25.0	31.4	4	394
40-44	0.0151	0.0730	0.0526	0.0953	86,313	82,168	89,311	6,299	415,819	23.9	21.5	27.5	5	346
45-49	0.0222	0.1052	0.0753	0.1345	80,014	74,335	84,609	8,420	379,021	20.5	18.5	23.9	7	296
50-54	0.0302	0.1405	0.1010	0.1745	71,594	64,337	78,239	10,062	332,816	17.7	15.8	20.6	7	245
55-59	0.0379	0.1730	0.1262	0.2093	61,532	53,111	70,338	10,642	281,055	15.1	13.6	17.6	7	176
60-64	0.0493	0.2195	0.1647	0.2600	50,890	41,993	61,461	11,169	226,528	12.8	11.6	14.8	7	143
65-69	0.0618	0.2677	0.2083	0.3092	39,721	31,076	51,338	10,632	172,026	10.7	9.7	12.3	7	110
70-74	0.0827	0.3425	0.2791	0.3859	29,089	21,468	40,643	9,964	120,535	8.6	7.9	9.8	7	87
75-79	0.1126	0.4395	0.3761	0.4829	19,125	13,183	29,299	8,405	74,612	6.8	6.3	7.7	6	50
80-84	0.1623	0.5772	0.5142	0.6178	10,720	6,817	18,279	6,188	38,131	5.2	4.9	5.8	3	20
85-89	0.2297	0.7295	0.6755	0.7628	4,533	2,606	8,881	3,306	14,397	4.0	3.8	4.4	2	7
90-94	0.3194	0.8155	0.7720	0.8421	1,226	618	2,881	1,000	3,131	3.0	2.8	3.2	0	1
95-99	0.4364	0.9449	0.9218	0.9573	226	98	657	214	490	2.3	2.2	2.4	0	0
100+	0.5858	1	1	1	12	4	51	12	21	1.7	1.7	1.8	0	0
All ages													*73*	*6,332*

$\ln(_nq_x)$ with uncertainty interval

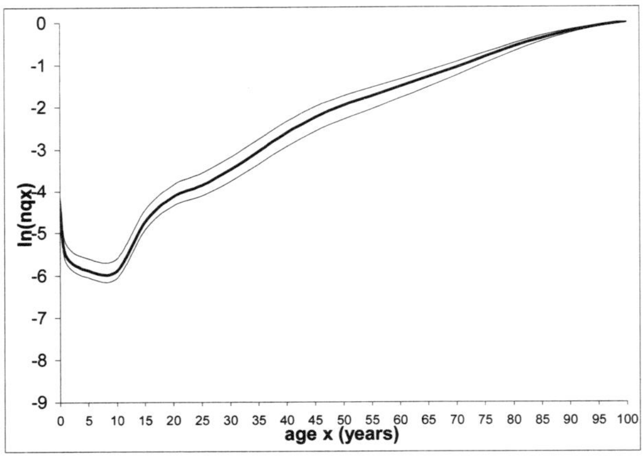

l_x with uncertainty interval

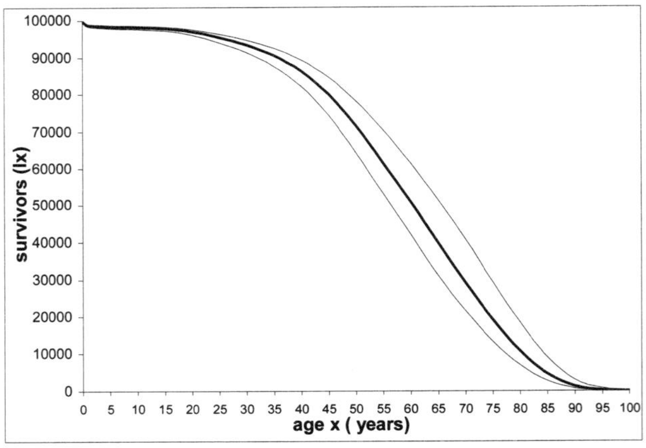

e_x with uncertainty interval

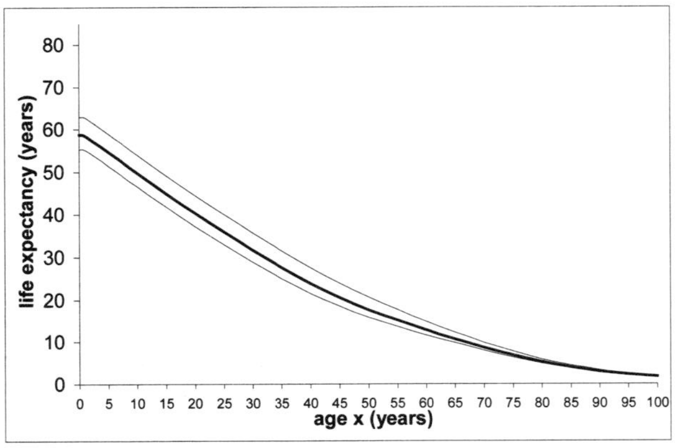

Distribution of deaths by age with uncertainty interval

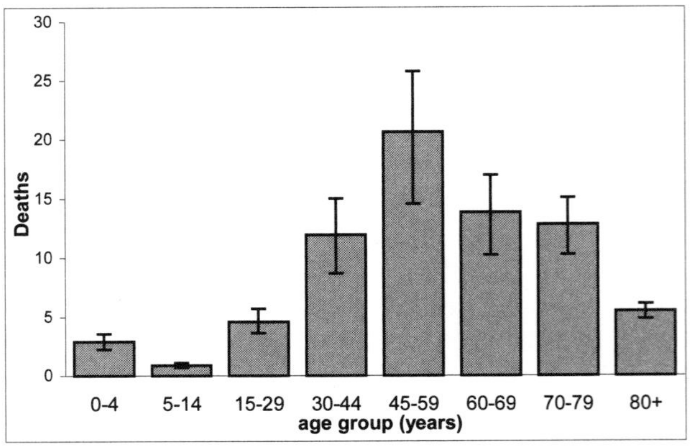

Nauru

Females, 2000 — *Femmes, 2000* — *Mujeres, 2000*

Age	$_nM_x$	$_nq_x$ mean	$_nq_x$ percentiles 2½	$_nq_x$ percentiles 97½	l_x mean	l_x percentiles 2½	l_x percentiles 97½	$_nd_x$	$_nL_x$	e_x mean	e_x percentiles 2½	e_x percentiles 97½	Actual deaths	Actual population
<1	0.0070	0.0070	0.0052	0.0099	100,000	100,000	100,000	696	99,374	66.6	62.5	71.3	1	169
1-4	0.0010	0.0041	0.0031	0.0055	99,304	99,015	99,478	403	396,248	66.0	62.1	70.8	1	675
5-9	0.0005	0.0025	0.0019	0.0034	98,901	98,471	99,165	249	493,882	62.3	58.4	67.0	0	787
10-14	0.0005	0.0026	0.0020	0.0036	98,652	98,132	98,972	258	492,615	57.4	53.6	62.2	0	660
15-19	0.0012	0.0060	0.0046	0.0082	98,394	97,780	98,773	588	490,499	52.6	48.8	57.3	1	570
20-24	0.0017	0.0084	0.0064	0.0117	97,806	96,974	98,320	824	486,969	47.9	44.2	52.6	1	480
25-29	0.0021	0.0104	0.0080	0.0146	96,982	95,841	97,687	1,013	482,378	43.3	39.7	47.9	1	427
30-34	0.0030	0.0149	0.0112	0.0209	95,969	94,446	96,910	1,433	476,265	38.7	35.2	43.3	1	380
35-39	0.0048	0.0236	0.0172	0.0330	94,537	92,476	95,827	2,234	467,098	34.2	30.9	38.8	2	360
40-44	0.0078	0.0384	0.0271	0.0534	92,303	89,420	94,180	3,542	452,659	30.0	26.9	34.4	2	319
45-49	0.0121	0.0589	0.0405	0.0810	88,761	84,643	91,627	5,224	430,744	26.1	23.3	30.3	3	269
50-54	0.0177	0.0848	0.0574	0.1144	83,537	77,786	87,914	7,080	399,983	22.6	20.1	26.4	4	218
55-59	0.0237	0.1120	0.0755	0.1470	76,456	68,891	82,870	8,565	360,869	19.5	17.3	22.9	4	156
60-64	0.0306	0.1423	0.0966	0.1806	67,891	58,763	76,617	9,662	315,302	16.6	14.8	19.6	4	138
65-69	0.0382	0.1742	0.1207	0.2161	58,229	48,149	69,214	10,145	265,786	13.9	12.5	16.4	4	109
70-74	0.0512	0.2268	0.1627	0.2739	48,085	37,743	60,863	10,908	213,155	11.3	10.2	13.3	4	84
75-79	0.0698	0.2970	0.2236	0.3470	37,177	27,404	50,962	11,043	158,278	8.9	8.1	10.4	4	52
80-84	0.1118	0.4369	0.3535	0.4918	26,134	17,895	39,565	11,418	102,127	6.6	6.1	7.7	3	24
85-89	0.1735	0.6050	0.5298	0.6548	14,717	9,095	25,580	8,904	51,323	4.9	4.5	5.5	1	9
90-94	0.2607	0.7315	0.6748	0.7706	5,813	3,139	12,028	4,252	16,308	3.5	3.3	3.9	0	2
95-99	0.3794	0.8872	0.8522	0.9108	1,561	720	3,912	1,385	3,650	2.5	2.4	2.7	0	0
100+	0.5347	1	1	1	176	64	578	176	329	1.9	1.9	2.0	0	0
All ages													*42*	*5,888*

ln($_nq_x$) with uncertainty interval

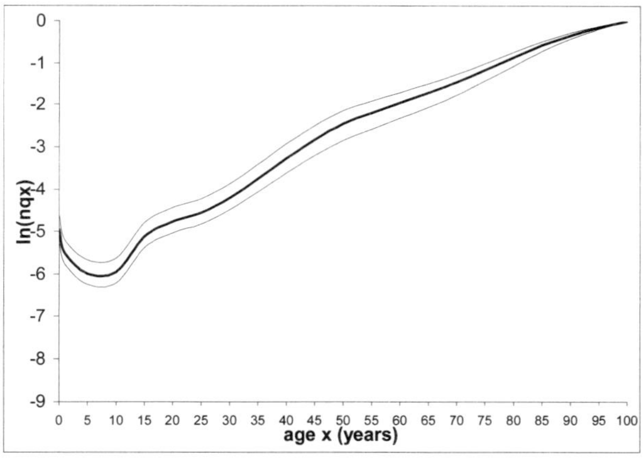

l_x with uncertainty interval

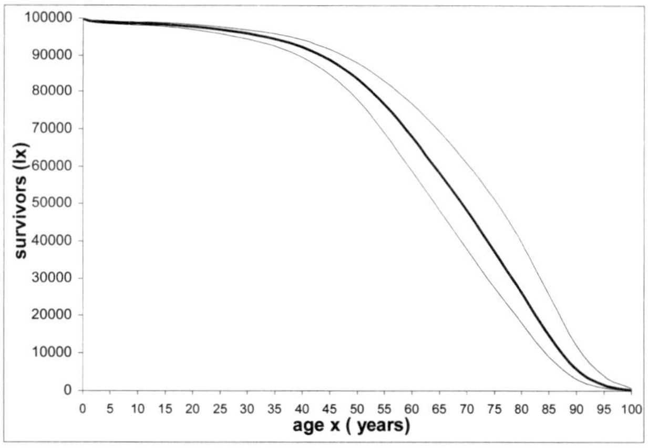

e_x with uncertainty interval

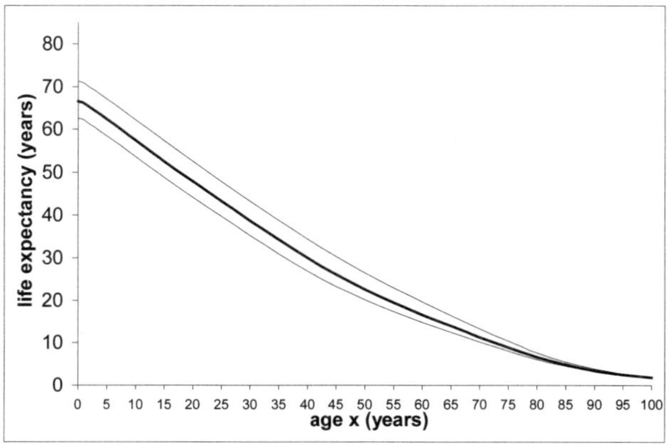

Distribution of deaths by age with uncertainty interval

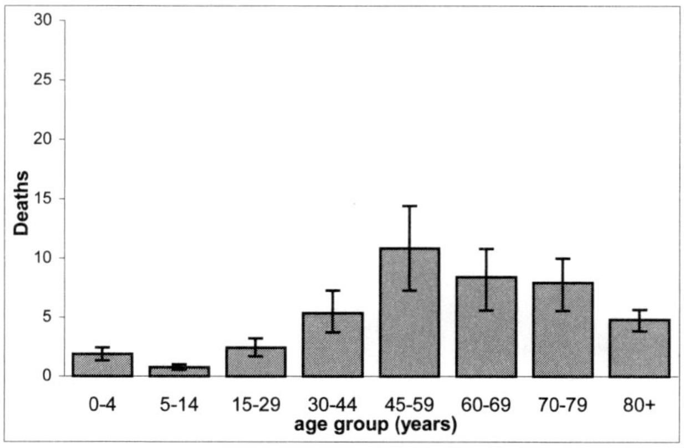

Nepal **Népal** **Nepal**

Males, 2000 *Hommes, 2000* *Hombres, 2000*

Age	$_nM_x$	$_nq_x$			l_x			$_nd_x$	$_nL_x$	e_x			Actual deaths	Actual population
			percentiles			percentiles					percentiles			
		mean	2½	97½	mean	2½	97½			mean	2½	97½		
<1	0.0543	0.0518	0.0451	0.0589	100,000	100,000	100,000	5,179	95,339	58.5	56.7	60.4	21,015	386,898
1-4	0.0134	0.0519	0.0469	0.0568	94,821	94,110	95,487	4,925	367,465	60.7	59.0	62.4	19,389	1,446,547
5-9	0.0034	0.0169	0.0153	0.0186	89,896	88,898	90,892	1,523	445,673	59.9	58.5	61.4	5,563	1,627,949
10-14	0.0028	0.0139	0.0125	0.0152	88,373	87,262	89,486	1,229	438,794	55.9	54.5	57.4	3,989	1,424,483
15-19	0.0033	0.0165	0.0149	0.0181	87,145	85,943	88,358	1,438	432,127	51.7	50.3	53.1	4,107	1,233,729
20-24	0.0039	0.0195	0.0176	0.0215	85,706	84,403	87,033	1,675	424,344	47.5	46.2	48.9	4,160	1,054,185
25-29	0.0048	0.0237	0.0214	0.0261	84,031	82,617	85,484	1,993	415,175	43.4	42.2	44.7	4,300	895,837
30-34	0.0058	0.0286	0.0258	0.0314	82,039	80,501	83,634	2,345	404,330	39.4	38.2	40.6	4,415	761,241
35-39	0.0070	0.0344	0.0311	0.0378	79,694	78,023	81,446	2,741	391,614	35.5	34.4	36.7	4,531	647,351
40-44	0.0087	0.0423	0.0383	0.0465	76,952	75,143	78,880	3,259	376,614	31.6	30.6	32.8	4,452	514,486
45-49	0.0104	0.0506	0.0459	0.0555	73,694	71,656	75,857	3,728	359,149	27.9	26.9	29.0	4,547	438,136
50-54	0.0135	0.0653	0.0592	0.0714	69,966	67,682	72,378	4,566	338,416	24.3	23.3	25.3	5,269	390,542
55-59	0.0181	0.0864	0.0783	0.0941	65,400	62,851	68,081	5,648	312,883	20.8	19.9	21.8	5,791	320,835
60-64	0.0253	0.1188	0.1079	0.1292	59,753	56,937	62,707	7,098	281,019	17.5	16.7	18.4	6,365	252,017
65-69	0.0361	0.1656	0.1505	0.1806	52,655	49,609	55,853	8,717	241,481	14.5	13.8	15.3	6,670	184,760
70-74	0.0496	0.2208	0.2014	0.2409	43,937	40,783	47,267	9,701	195,435	11.9	11.3	12.6	6,061	122,108
75-79	0.0677	0.2895	0.2660	0.3143	34,237	31,205	37,550	9,910	146,408	9.6	9.1	10.1	4,664	68,912
80-84	0.0993	0.3977	0.3702	0.4270	24,327	21,709	27,451	9,674	97,448	7.5	7.1	7.9	3,021	30,434
85-89	0.1400	0.5185	0.4899	0.5488	14,652	12,555	17,181	7,598	54,268	5.8	5.5	6.1	1,257	8,976
90-94	0.2000	0.6250	0.5981	0.6531	7,055	5,664	8,757	4,409	22,047	4.3	4.1	4.5	319	1,594
95-99	0.2900	0.7754	0.7529	0.7984	2,646	1,965	3,510	2,051	7,074	3.3	3.1	3.3	44	151
100+	0.3800	1	1	1	594	396	863	594	1,564	2.6	2.2	2.7	3	7
All ages													*119,932*	*11,811,178*

$\ln(_nq_x)$ with uncertainty interval

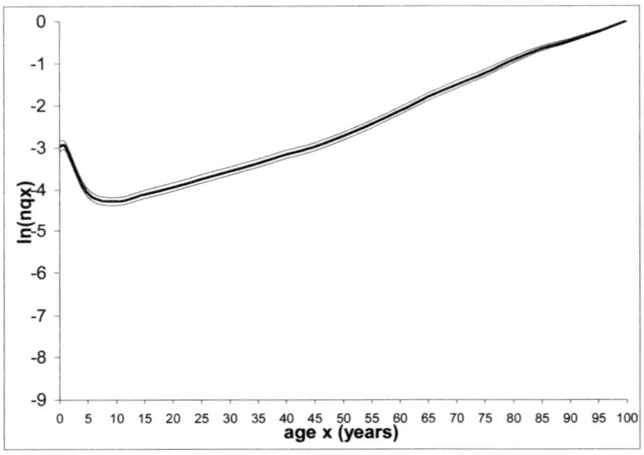

l_x with uncertainty interval

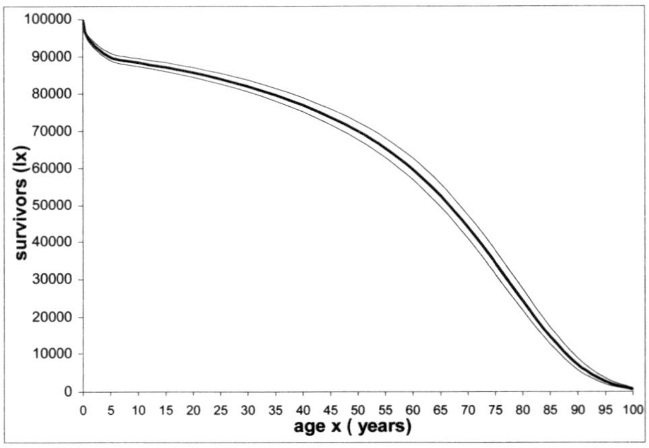

e_x with uncertainty interval

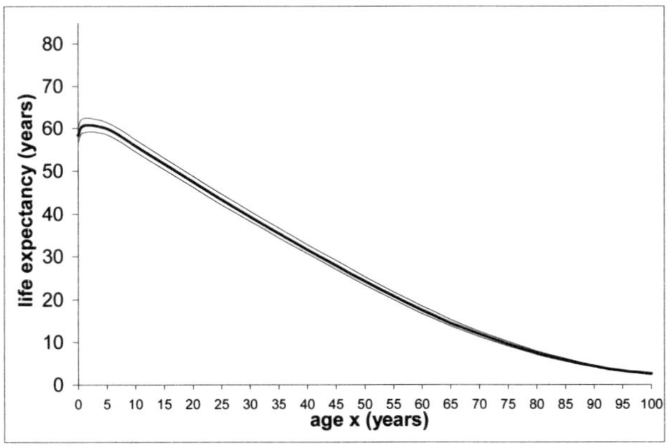

Distribution of deaths by age with uncertainty interval

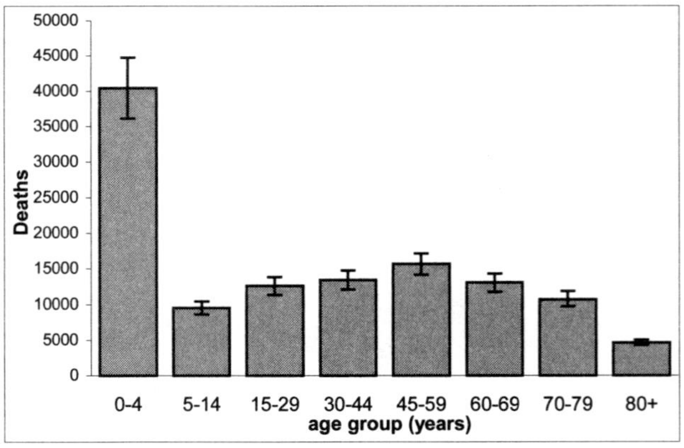

Nepal

Népal

Nepal

Females, 2000 — *Femmes, 2000* — *Mujeres, 2000*

Age	$_nM_x$	$_nq_x$			l_x			$_nd_x$	$_nL_x$	e_x			Actual deaths	Actual population
			percentiles			percentiles					percentiles			
		mean	2½	97½	mean	2½	97½			mean	2½	97½		
<1	0.0634	0.0600	0.0538	0.0661	100,000	100,000	100,000	5,997	94,603	58.0	56.3	59.6	23,358	368,474
1-4	0.0155	0.0597	0.0554	0.0645	94,003	93,388	94,624	5,613	362,542	60.7	59.1	62.3	21,082	1,361,695
5-9	0.0033	0.0164	0.0152	0.0179	88,390	87,594	89,184	1,453	438,318	60.5	59.1	61.9	5,040	1,520,174
10-14	0.0031	0.0153	0.0142	0.0167	86,937	86,046	87,809	1,334	431,351	56.5	55.1	57.9	4,077	1,318,679
15-19	0.0034	0.0169	0.0156	0.0184	85,603	84,610	86,552	1,447	424,398	52.3	51.0	53.7	3,886	1,139,293
20-24	0.0045	0.0223	0.0206	0.0243	84,156	83,053	85,186	1,875	416,092	48.2	46.9	49.5	4,422	981,313
25-29	0.0059	0.0291	0.0269	0.0317	82,281	81,038	83,426	2,396	405,415	44.2	43.1	45.5	5,005	847,034
30-34	0.0070	0.0344	0.0317	0.0375	79,885	78,468	81,182	2,748	392,556	40.4	39.4	41.7	5,145	735,015
35-39	0.0086	0.0421	0.0387	0.0459	77,137	75,530	78,600	3,251	377,559	36.8	35.8	37.9	5,460	634,130
40-44	0.0087	0.0427	0.0391	0.0464	73,886	72,066	75,535	3,153	361,550	33.3	32.3	34.4	4,441	509,248
45-49	0.0090	0.0440	0.0403	0.0478	70,734	68,723	72,549	3,113	345,885	29.7	28.8	30.7	3,883	431,432
50-54	0.0121	0.0588	0.0538	0.0638	67,621	65,438	69,588	3,978	328,157	25.9	25.0	26.9	4,611	380,345
55-59	0.0160	0.0769	0.0704	0.0831	63,642	61,266	65,782	4,896	305,972	22.4	21.6	23.3	5,010	313,115
60-64	0.0220	0.1043	0.0955	0.1126	58,747	56,175	61,063	6,125	278,420	19.1	18.3	19.9	5,485	249,338
65-69	0.0300	0.1395	0.1279	0.1511	52,621	49,876	55,181	7,343	244,751	16.0	15.3	16.7	5,630	187,667
70-74	0.0410	0.1859	0.1708	0.2015	45,279	42,437	48,102	8,419	205,346	13.2	12.5	13.8	5,277	128,706
75-79	0.0570	0.2495	0.2304	0.2695	36,860	34,068	39,833	9,195	161,311	10.6	10.1	11.1	4,338	76,100
80-84	0.0840	0.3471	0.3239	0.3719	27,665	25,141	30,568	9,603	114,318	8.3	7.9	8.7	2,995	35,656
85-89	0.1200	0.4615	0.4363	0.4887	18,062	15,999	20,556	8,336	69,470	6.4	6.1	6.7	1,376	11,463
90-94	0.1800	0.5844	0.5594	0.6108	9,726	8,183	11,566	5,684	31,577	4.7	4.5	4.9	423	2,351
95-99	0.2600	0.7303	0.7084	0.7531	4,042	3,184	5,087	2,952	11,354	3.6	3.3	3.6	73	280
100+	0.3500	1	1	1	1,090	786	1,477	1,090	3,114	2.9	2.4	2.9	6	18
All ages													*121,023*	*11,231,526*

ln($_nq_x$) with uncertainty interval

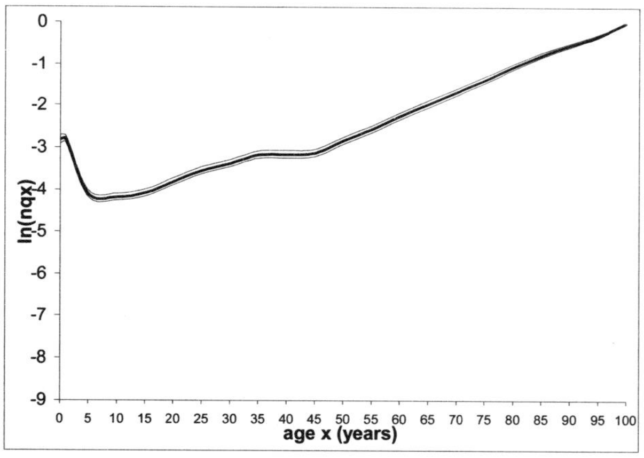

l_x with uncertainty interval

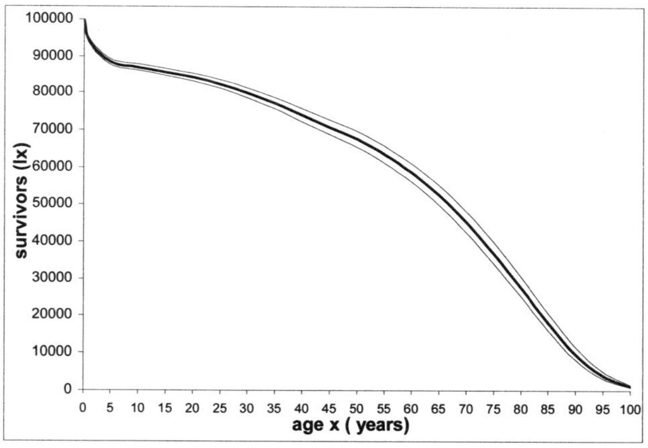

e_x with uncertainty interval

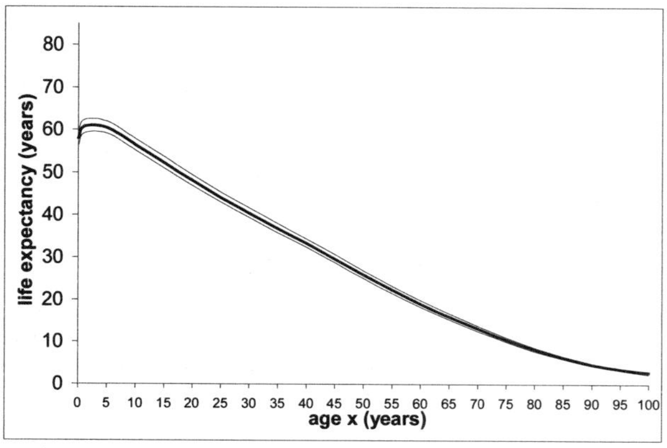

Distribution of deaths by age with uncertainty interval

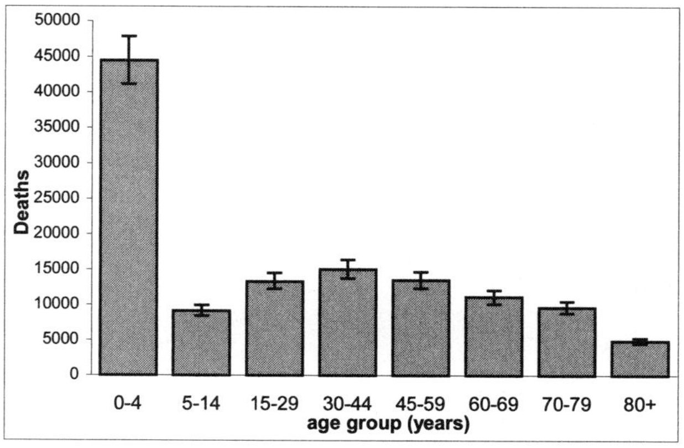

Netherlands   Pays-Bas   Países Bajos

Males, 2000   *Hommes, 2000*   *Hombres, 2000*

Age	$_nM_x$	$_nq_x$			l_x			$_nd_x$	$_nL_x$	e_x			Actual deaths	Actual population
			percentiles			percentiles					percentiles			
		mean	2½	97½	mean	2½	97½			mean	2½	97½		
<1	0.0050	0.0050	0.0044	0.0056	100,000	100,000	100,000	498	99,552	75.4	75.0	76.0	463	92,507
1-4	0.0004	0.0017	0.0015	0.0018	99,502	99,441	99,562	167	397,608	74.8	74.4	75.4	163	387,489
5-9	0.0002	0.0007	0.0007	0.0008	99,335	99,261	99,410	74	496,490	70.9	70.5	71.5	77	510,351
10-14	0.0001	0.0007	0.0006	0.0008	99,261	99,180	99,342	69	496,130	66.0	65.6	66.6	69	494,254
15-19	0.0004	0.0022	0.0020	0.0024	99,191	99,106	99,278	218	495,412	61.0	60.6	61.6	206	467,732
20-24	0.0007	0.0033	0.0030	0.0035	98,973	98,871	99,076	326	494,051	56.2	55.7	56.8	315	477,198
25-29	0.0007	0.0032	0.0030	0.0035	98,647	98,522	98,774	320	492,436	51.3	50.9	51.9	374	575,542
30-34	0.0008	0.0037	0.0035	0.0040	98,327	98,182	98,475	368	490,716	46.5	46.0	47.1	511	681,674
35-39	0.0009	0.0044	0.0042	0.0047	97,959	97,791	98,131	435	488,708	41.7	41.2	42.3	608	683,403
40-44	0.0017	0.0083	0.0078	0.0087	97,524	97,330	97,722	806	485,606	36.8	36.4	37.5	1,040	626,553
45-49	0.0030	0.0149	0.0142	0.0156	96,718	96,481	96,958	1,440	479,990	32.1	31.6	32.7	1,743	581,073
50-54	0.0050	0.0247	0.0236	0.0258	95,278	94,976	95,583	2,353	470,509	27.6	27.1	28.2	2,972	594,365
55-59	0.0070	0.0344	0.0326	0.0360	92,926	92,539	93,310	3,196	456,637	23.2	22.7	23.8	3,085	440,675
60-64	0.0140	0.0676	0.0634	0.0710	89,729	89,259	90,192	6,069	433,474	18.9	18.4	19.5	5,078	362,704
65-69	0.0250	0.1176	0.1093	0.1245	83,660	83,003	84,330	9,842	393,696	15.1	14.7	15.7	7,642	305,674
70-74	0.0410	0.1859	0.1721	0.1984	73,818	72,750	75,066	13,726	334,776	11.8	11.4	12.3	10,030	244,644
75-79	0.0680	0.2906	0.2705	0.3095	60,092	58,537	62,144	17,463	256,805	8.9	8.6	9.3	12,261	180,304
80-84	0.1140	0.4436	0.4195	0.4668	42,630	40,420	45,332	18,910	165,874	6.6	6.3	6.9	10,934	95,909
85-89	0.1820	0.6254	0.6039	0.6465	23,720	21,554	26,317	14,835	81,512	4.8	4.7	5.0	8,026	44,097
90-94	0.2440	0.7044	0.6883	0.7208	8,885	7,620	10,425	6,258	25,649	3.7	3.6	3.8	3,172	13,001
95-99	0.3360	0.8367	0.8256	0.8482	2,626	2,128	3,250	2,197	6,540	2.8	2.8	2.9	770	2,292
100+	0.4690	1	1	1	429	323	567	429	915	2.1	2.1	2.3	104	222
All ages													*69,642*	*7,861,663*

ln($_nq_x$) with uncertainty interval

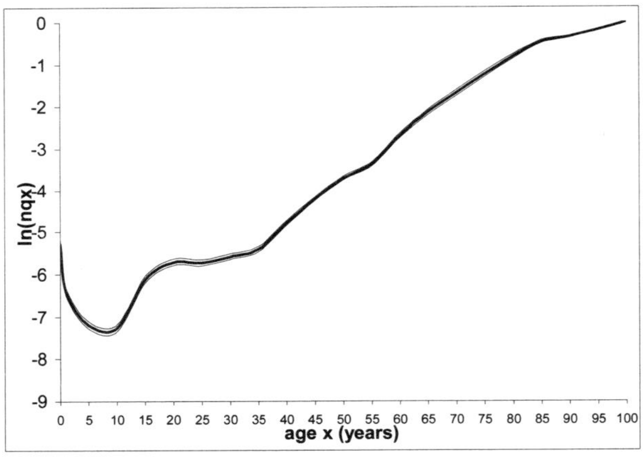

l_x with uncertainty interval

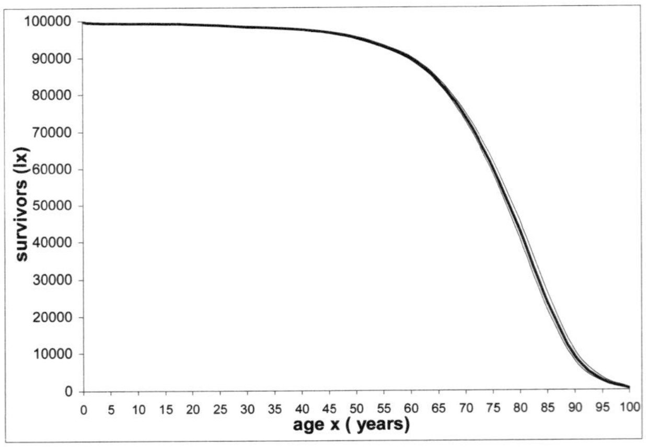

e_x with uncertainty interval

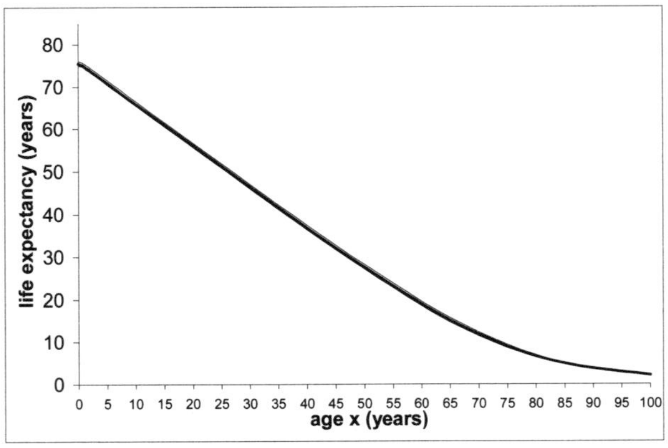

Distribution of deaths by age with uncertainty interval

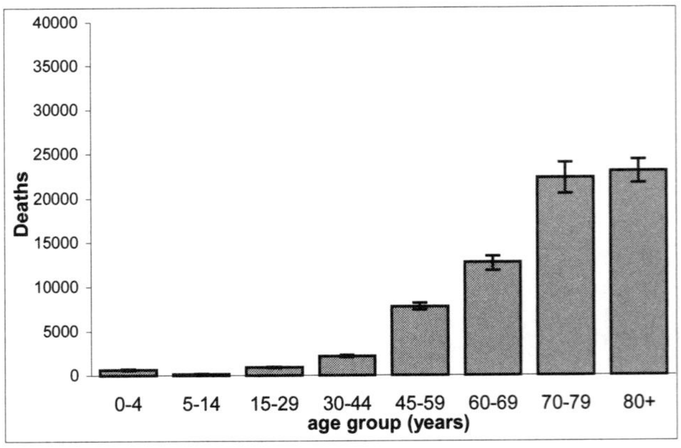

Netherlands | Pays-Bas | Países Bajos

Females, 2000 | *Femmes, 2000* | *Mujeres, 2000*

Age	$_nM_x$	$_nq_x$ mean	$_nq_x$ percentiles 2½	$_nq_x$ percentiles 97½	l_x mean	l_x percentiles 2½	l_x percentiles 97½	$_nd_x$	$_nL_x$	e_x mean	e_x percentiles 2½	e_x percentiles 97½	Actual deaths	Actual population
<1	0.0040	0.0040	0.0034	0.0046	100,000	100,000	100,000	399	99,641	81.0	80.5	81.4	352	88,060
1-4	0.0003	0.0013	0.0011	0.0014	99,601	99,536	99,663	127	398,100	80.3	79.8	80.7	118	369,075
5-9	0.0001	0.0004	0.0004	0.0005	99,474	99,393	99,550	45	497,258	76.4	75.9	76.8	44	487,045
10-14	0.0001	0.0005	0.0004	0.0006	99,429	99,344	99,510	50	497,022	71.4	70.9	71.9	47	473,023
15-19	0.0002	0.0011	0.0010	0.0013	99,380	99,289	99,466	114	496,612	66.5	66.0	66.9	103	448,104
20-24	0.0002	0.0012	0.0011	0.0013	99,265	99,162	99,364	119	496,029	61.5	61.0	62.0	110	457,561
25-29	0.0003	0.0016	0.0015	0.0018	99,146	99,032	99,256	163	495,323	56.6	56.1	57.0	184	558,523
30-34	0.0005	0.0023	0.0021	0.0025	98,983	98,853	99,108	227	494,346	51.7	51.2	52.1	298	646,773
35-39	0.0007	0.0036	0.0033	0.0039	98,755	98,607	98,900	355	492,890	46.8	46.3	47.2	469	650,783
40-44	0.0014	0.0067	0.0063	0.0072	98,401	98,226	98,573	662	490,348	42.0	41.5	42.4	819	606,609
45-49	0.0020	0.0100	0.0094	0.0105	97,739	97,522	97,955	973	486,262	37.2	36.8	37.7	1,129	564,387
50-54	0.0030	0.0149	0.0142	0.0156	96,766	96,502	97,035	1,441	480,229	32.6	32.1	33.0	1,722	573,975
55-59	0.0050	0.0247	0.0235	0.0260	95,325	95,011	95,650	2,354	470,743	28.0	27.6	28.5	2,145	428,991
60-64	0.0080	0.0392	0.0372	0.0415	92,972	92,612	93,351	3,646	455,744	23.7	23.3	24.1	2,963	370,313
65-69	0.0120	0.0583	0.0552	0.0617	89,326	88,831	89,806	5,203	433,620	19.5	19.2	20.0	4,063	338,568
70-74	0.0200	0.0952	0.0901	0.1009	84,122	83,347	84,853	8,012	400,582	15.6	15.3	16.0	6,190	309,520
75-79	0.0350	0.1609	0.1518	0.1700	76,111	74,934	77,194	12,248	349,934	12.0	11.7	12.3	9,643	275,513
80-84	0.0670	0.2869	0.2718	0.3010	63,863	62,197	65,375	18,325	273,503	8.8	8.5	9.1	12,430	185,520
85-89	0.1190	0.4586	0.4376	0.4777	45,538	43,527	47,526	20,883	175,485	6.3	6.1	6.6	13,523	113,637
90-94	0.1840	0.5928	0.5725	0.6113	24,656	22,953	26,625	14,615	79,432	4.5	4.4	4.7	8,357	45,418
95-99	0.2810	0.7623	0.7468	0.7768	10,040	8,986	11,335	7,654	27,239	3.3	3.2	3.4	2,722	9,688
100+	0.4370	1	1	1	2,386	2,010	2,870	2,386	5,460	2.3	2.3	2.5	436	998
All ages													*67,866*	*8,002,084*

ln($_nq_x$) with uncertainty interval

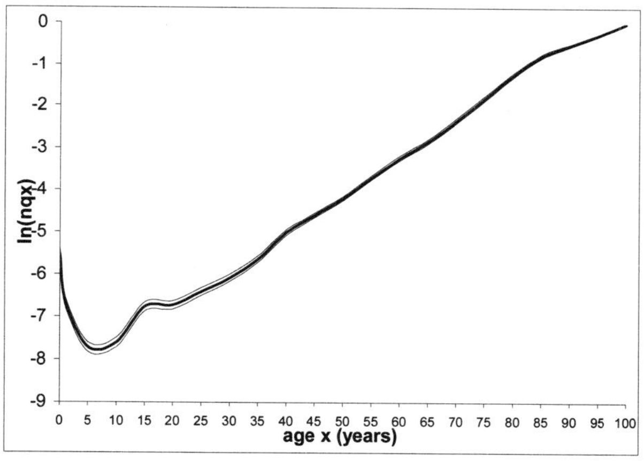

l_x with uncertainty interval

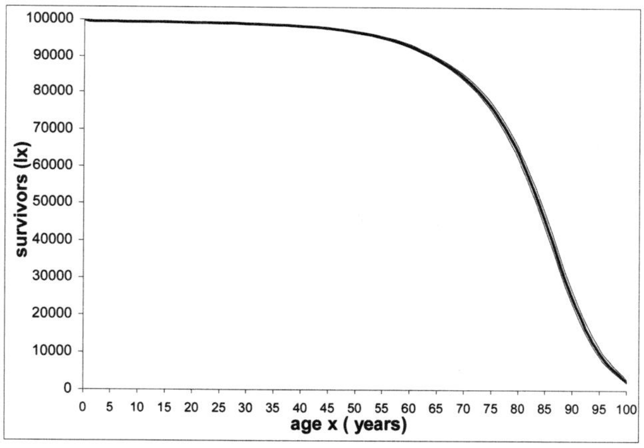

e_x with uncertainty interval

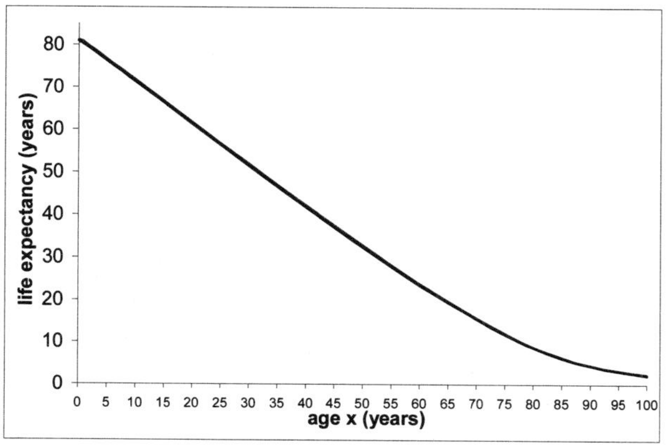

Distribution of deaths by age with uncertainty interval

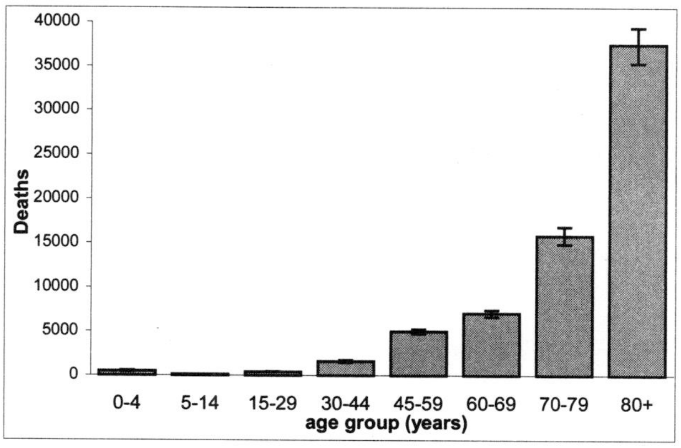

New Zealand — Nouvelle-Zélande — Nueva Zelandia

Males, 2000 — *Hommes, 2000* — *Hombres, 2000*

Age	$_nM_x$	$_nq_x$ mean	$_nq_x$ percentiles 2 1/2	$_nq_x$ percentiles 97 1/2	l_x mean	l_x percentiles 2 1/2	l_x percentiles 97 1/2	$_nd_x$	$_nL_x$	e_x mean	e_x percentiles 2 1/2	e_x percentiles 97 1/2	Actual deaths	Actual population
<1	0.0056	0.0055	0.0051	0.0067	100,000	100,000	100,000	553	99,502	75.9	75.2	76.7	159	28,567
1-4	0.0004	0.0015	0.0014	0.0017	99,447	99,325	99,490	147	397,435	75.3	74.6	76.1	42	113,881
5-9	0.0002	0.0010	0.0009	0.0011	99,300	99,155	99,354	95	496,262	71.4	70.7	72.2	30	154,539
10-14	0.0003	0.0014	0.0013	0.0016	99,205	99,045	99,265	141	495,673	66.5	65.8	67.3	42	147,950
15-19	0.0009	0.0044	0.0041	0.0050	99,064	98,884	99,134	436	494,229	61.6	60.9	62.4	122	138,579
20-24	0.0012	0.0062	0.0058	0.0069	98,628	98,390	98,729	614	491,603	56.9	56.1	57.6	158	126,466
25-29	0.0014	0.0067	0.0063	0.0074	98,013	97,707	98,159	661	488,413	52.2	51.5	53.0	173	127,556
30-34	0.0012	0.0058	0.0054	0.0063	97,352	96,982	97,544	568	485,339	47.5	46.8	48.3	156	133,229
35-39	0.0015	0.0074	0.0068	0.0080	96,784	96,367	97,017	716	482,129	42.8	42.1	43.6	222	149,460
40-44	0.0020	0.0099	0.0092	0.0106	96,068	95,598	96,353	951	477,961	38.1	37.4	38.9	282	141,580
45-49	0.0026	0.0131	0.0121	0.0139	95,117	94,587	95,471	1,247	472,467	33.5	32.8	34.2	334	126,501
50-54	0.0042	0.0208	0.0192	0.0222	93,870	93,273	94,314	1,956	464,461	28.9	28.2	29.6	495	117,628
55-59	0.0079	0.0387	0.0356	0.0415	91,914	91,201	92,500	3,554	450,685	24.4	23.8	25.2	714	90,588
60-64	0.0135	0.0655	0.0603	0.0708	88,360	87,416	89,205	5,788	427,331	20.3	19.7	21.0	980	72,368
65-69	0.0216	0.1025	0.0941	0.1111	82,572	81,230	83,826	8,465	391,699	16.6	16.0	17.2	1,343	62,149
70-74	0.0337	0.1553	0.1420	0.1682	74,107	72,202	75,936	11,509	341,763	13.2	12.7	13.8	1,870	55,532
75-79	0.0550	0.2417	0.2219	0.2597	62,598	60,059	65,153	15,131	275,162	10.1	9.8	10.7	2,174	39,529
80-84	0.0902	0.3682	0.3401	0.3894	47,467	44,464	50,698	17,475	193,647	7.6	7.3	8.0	1,938	21,473
85-89	0.1441	0.5298	0.4982	0.5497	29,992	27,148	33,457	15,889	110,237	5.5	5.3	5.8	1,485	10,306
90-94	0.2240	0.6699	0.6424	0.6847	14,103	12,225	16,789	9,448	42,172	3.9	3.8	4.1	644	2,875
95-99	0.3389	0.8403	0.8216	0.8492	4,655	3,855	6,003	3,912	11,541	2.8	2.8	2.9	150	442
100+	0.4990	1	1	1	744	581	1,071	744	1,490	2.0	2.0	2.1	11	23
All ages													*13,524*	*1,861,221*

ln($_nq_x$) with uncertainty interval

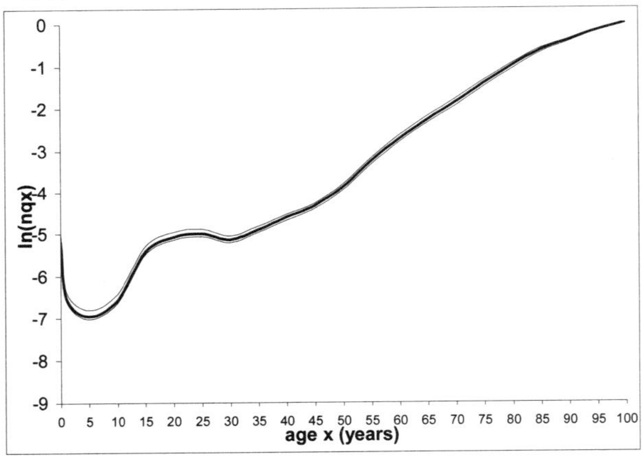

l_x with uncertainty interval

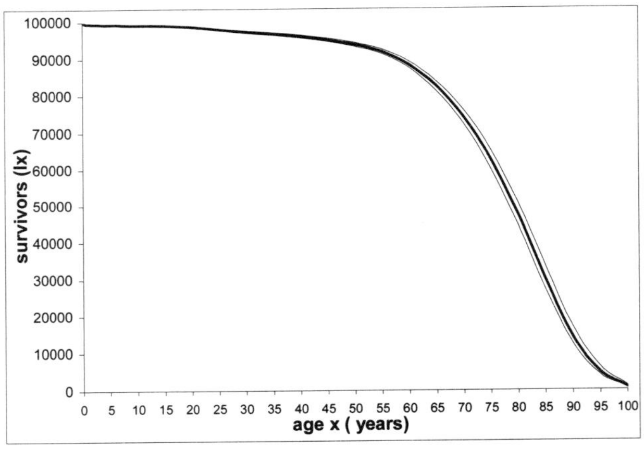

e_x with uncertainty interval

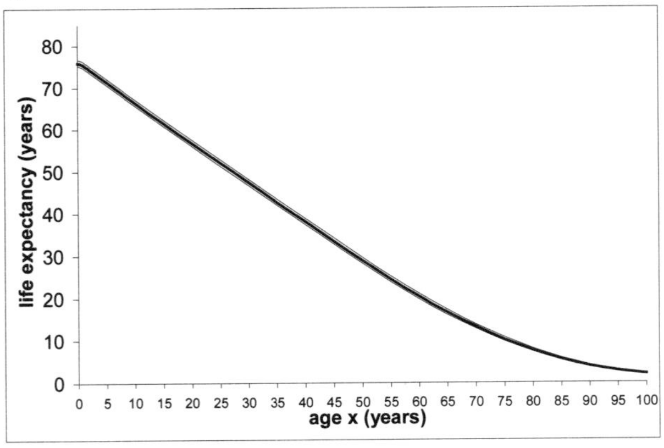

Distribution of deaths by age with uncertainty interval

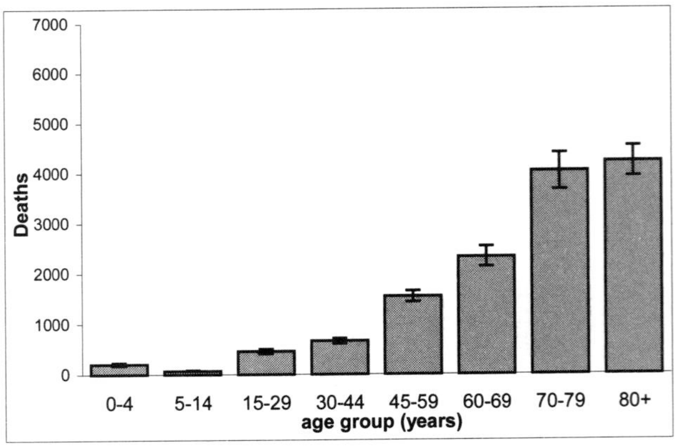

New Zealand — Nouvelle-Zélande — Nueva Zelandia

Females, 2000 — *Femmes, 2000* — *Mujeres, 2000*

Age	$_nM_x$	$_nq_x$ mean	$_nq_x$ percentiles 2½	$_nq_x$ percentiles 97½	l_x mean	l_x percentiles 2½	l_x percentiles 97½	$_nd_x$	$_nL_x$	e_x mean	e_x percentiles 2½	e_x percentiles 97½	Actual deaths	Actual population
<1	0.0043	0.0043	0.0033	0.0049	100,000	100,000	100,000	427	99,616	80.9	79.9	82.1	116	27,065
1-4	0.0003	0.0012	0.0010	0.0014	99,573	99,506	99,666	123	397,997	80.3	79.3	81.4	33	107,669
5-9	0.0001	0.0006	0.0005	0.0007	99,450	99,367	99,564	63	497,092	76.4	75.4	77.5	19	145,798
10-14	0.0002	0.0011	0.0009	0.0012	99,387	99,296	99,511	106	496,667	71.4	70.5	72.6	30	141,726
15-19	0.0006	0.0029	0.0025	0.0033	99,280	99,176	99,421	291	495,674	66.5	65.5	67.6	77	130,635
20-24	0.0004	0.0018	0.0015	0.0020	98,989	98,849	99,173	173	494,513	61.7	60.7	62.8	42	119,830
25-29	0.0004	0.0019	0.0017	0.0022	98,816	98,656	99,023	192	493,600	56.8	55.9	57.9	51	132,144
30-34	0.0007	0.0034	0.0030	0.0038	98,624	98,441	98,856	339	492,271	51.9	51.0	53.0	101	146,032
35-39	0.0007	0.0037	0.0033	0.0041	98,285	98,063	98,558	361	490,519	47.1	46.2	48.2	117	158,988
40-44	0.0014	0.0069	0.0062	0.0077	97,923	97,660	98,236	674	487,930	42.2	41.3	43.3	203	146,585
45-49	0.0018	0.0090	0.0082	0.0101	97,249	96,907	97,629	875	484,057	37.5	36.6	38.6	232	128,501
50-54	0.0031	0.0154	0.0140	0.0173	96,374	95,928	96,831	1,482	478,163	32.8	32.0	33.9	364	117,314
55-59	0.0053	0.0260	0.0234	0.0293	94,891	94,270	95,477	2,471	468,279	28.3	27.5	29.4	480	91,012
60-64	0.0084	0.0411	0.0365	0.0462	92,420	91,510	93,243	3,800	452,601	24.0	23.3	25.0	627	74,663
65-69	0.0129	0.0626	0.0552	0.0701	88,620	87,285	89,836	5,551	429,224	19.9	19.3	20.8	838	64,773
70-74	0.0197	0.0938	0.0823	0.1043	83,069	81,165	84,879	7,792	395,867	16.1	15.5	16.9	1,216	61,755
75-79	0.0341	0.1570	0.1381	0.1722	75,277	72,703	77,894	11,817	346,844	12.5	12.0	13.2	1,817	53,318
80-84	0.0598	0.2601	0.2322	0.2811	63,460	60,181	67,134	16,507	276,034	9.4	8.9	9.9	2,203	36,837
85-89	0.1025	0.4079	0.3742	0.4366	46,953	43,262	51,545	19,150	186,891	6.8	6.4	7.2	2,240	21,862
90-94	0.1714	0.5660	0.5361	0.5958	27,803	24,373	32,256	15,735	91,810	4.7	4.5	5.0	1,422	8,297
95-99	0.2798	0.7606	0.7409	0.7835	12,068	9,851	14,964	9,179	32,801	3.3	3.1	3.4	508	1,816
100+	0.4461	1	1	1	2,889	2,133	3,877	2,889	6,476	2.2	2.2	2.3	73	163
All ages													*12,808*	*1,916,783*

ln($_nq_x$) with uncertainty interval

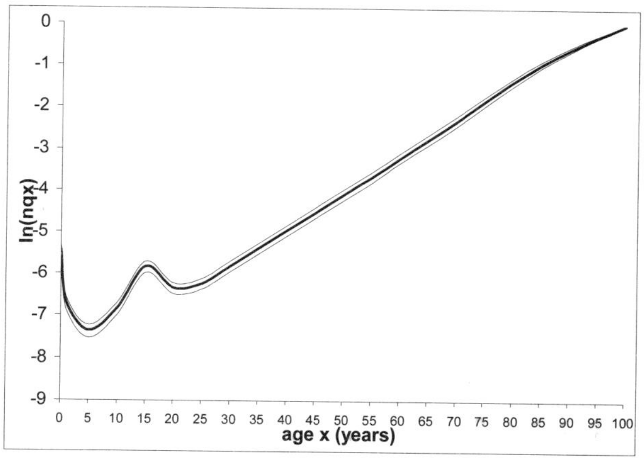

l_x with uncertainty interval

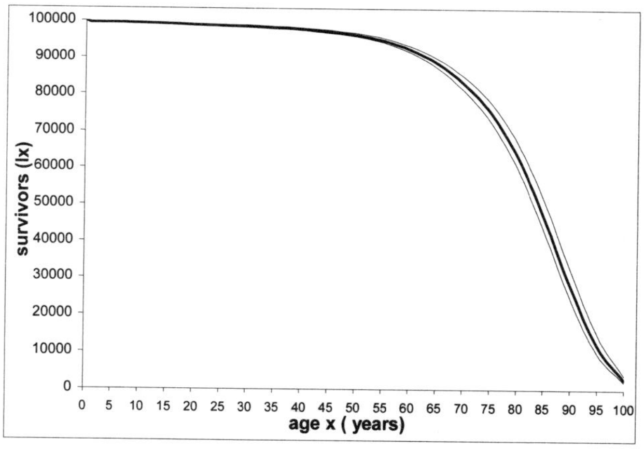

e_x with uncertainty interval

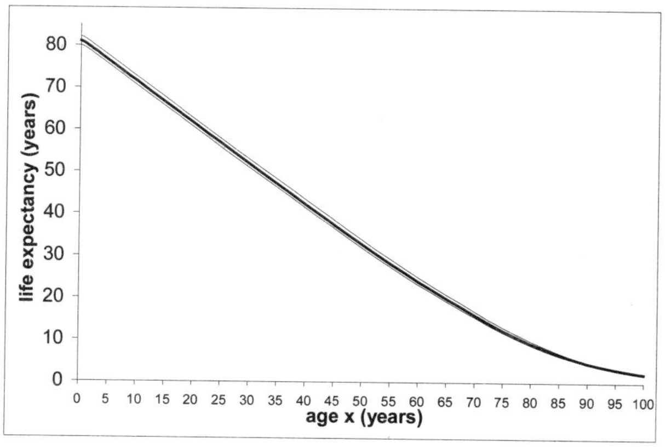

Distribution of deaths by age with uncertainty interval

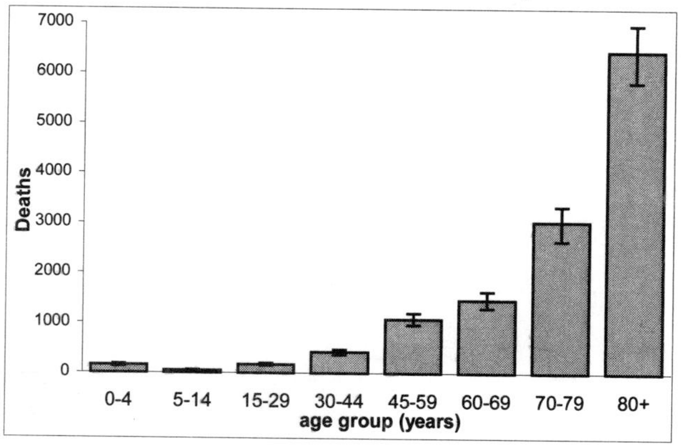

Nicaragua **Nicaragua** **Nicaragua**

Males, 2000 *Hommes, 2000* *Hombres, 2000*

Age	$_nM_x$	$_nq_x$			l_x			$_nd_x$	$_nL_x$	e_x			Actual deaths	Actual population
			percentiles			percentiles					percentiles			
		mean	2½	97½	mean	2½	97½			mean	2½	97½		
<1	0.0415	0.0400	0.0363	0.0436	100,000	100,000	100,000	3,999	96,401	66.4	65.4	67.5	3,549	85,547
1-4	0.0023	0.0092	0.0085	0.0099	96,001	95,641	96,367	882	381,887	68.2	67.1	69.3	745	322,530
5-9	0.0010	0.0048	0.0044	0.0052	95,119	94,725	95,519	455	474,458	64.8	63.7	65.9	356	371,690
10-14	0.0012	0.0059	0.0054	0.0064	94,664	94,244	95,090	559	471,925	60.1	59.0	61.2	378	319,705
15-19	0.0020	0.0098	0.0091	0.0106	94,106	93,648	94,570	923	468,221	55.5	54.4	56.5	599	304,058
20-24	0.0026	0.0131	0.0121	0.0142	93,183	92,655	93,710	1,220	462,864	51.0	49.9	52.0	635	240,929
25-29	0.0029	0.0146	0.0135	0.0158	91,963	91,344	92,569	1,342	456,459	46.6	45.6	47.7	577	196,363
30-34	0.0035	0.0175	0.0162	0.0189	90,621	89,904	91,316	1,587	449,137	42.3	41.2	43.3	555	157,015
35-39	0.0048	0.0236	0.0219	0.0254	89,034	88,205	89,833	2,098	439,925	38.0	37.0	39.0	609	127,714
40-44	0.0057	0.0281	0.0261	0.0303	86,936	85,963	87,867	2,446	428,565	33.8	32.8	34.9	593	103,959
45-49	0.0067	0.0327	0.0302	0.0353	84,490	83,359	85,566	2,763	415,541	29.8	28.8	30.7	555	83,522
50-54	0.0095	0.0464	0.0426	0.0505	81,727	80,429	82,955	3,792	399,153	25.7	24.7	26.6	569	59,916
55-59	0.0133	0.0644	0.0588	0.0705	77,935	76,429	79,354	5,016	377,133	21.8	20.9	22.7	603	45,306
60-64	0.0194	0.0923	0.0838	0.1018	72,919	71,176	74,554	6,734	347,760	18.1	17.3	19.0	705	36,389
65-69	0.0290	0.1350	0.1218	0.1492	66,185	64,193	68,125	8,936	308,585	14.7	14.0	15.5	790	27,293
70-74	0.0450	0.2022	0.1823	0.2229	57,249	54,749	59,681	11,578	257,298	11.6	11.0	12.3	896	19,903
75-79	0.0690	0.2942	0.2669	0.3210	45,670	42,547	48,667	13,438	194,756	8.9	8.5	9.5	898	13,018
80-84	0.1093	0.4293	0.3959	0.4599	32,232	28,889	35,605	13,839	126,564	6.6	6.3	7.0	649	5,939
85-89	0.1783	0.6167	0.5833	0.6450	18,394	15,603	21,510	11,343	63,609	4.7	4.5	5.0	364	2,040
90-94	0.2800	0.7609	0.7349	0.7819	7,050	5,540	8,964	5,364	19,158	3.3	3.2	3.5	117	419
95-99	0.4000	0.9091	0.8943	0.9206	1,686	1,208	2,376	1,533	3,832	2.4	2.4	2.5	16	41
100+	0.6000	1	1	1	153	96	251	153	255	1.7	1.6	1.8	1	2
All ages													*14,761*	*2,523,298*

ln($_nq_x$) with uncertainty interval

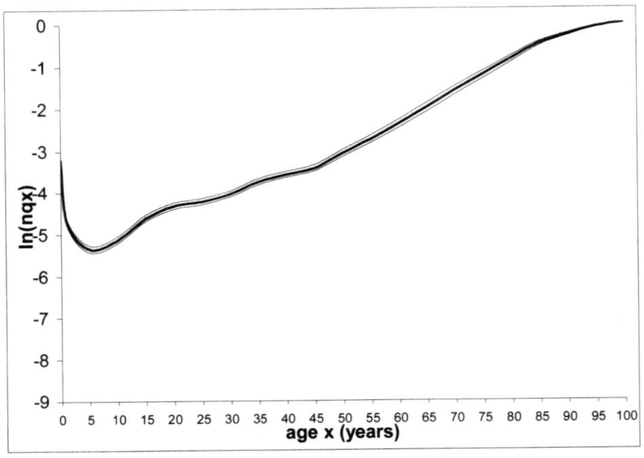

l_x with uncertainty interval

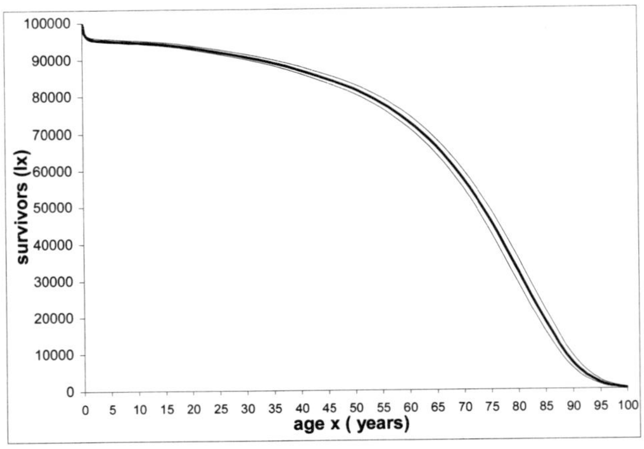

e_x with uncertainty interval

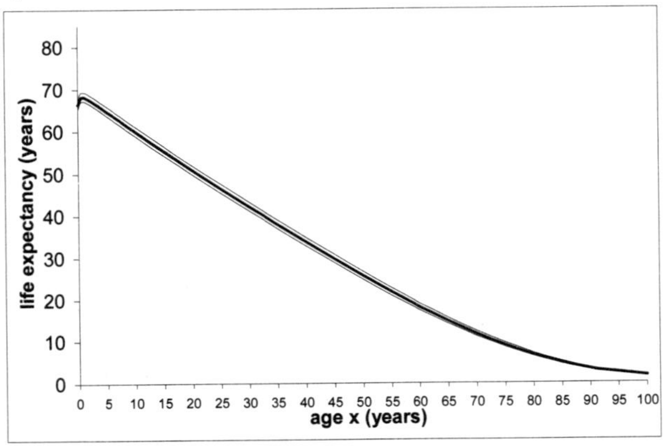

Distribution of deaths by age with uncertainty interval

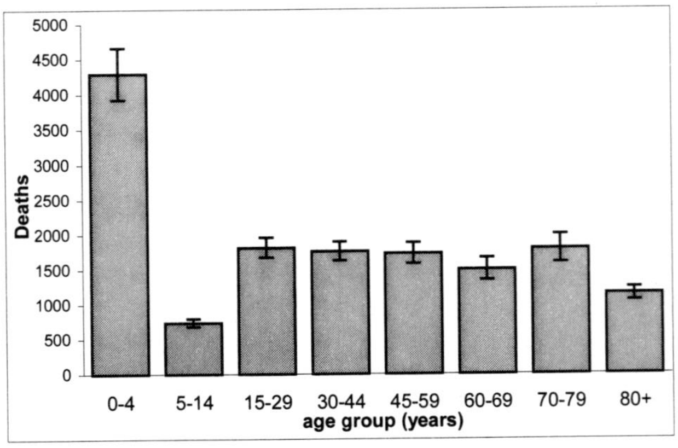

Nicaragua

Females, 2000

Nicaragua

Femmes, 2000

Nicaragua

Mujeres, 2000

Age	$_nM_x$	$_nq_x$			l_x			$_nd_x$	$_nL_x$	e_x			Actual deaths	Actual population
			percentiles			percentiles					percentiles			
		mean	21/2	971/2	mean	21/2	971/2			mean	21/2	971/2		
<1	0.0304	0.0296	0.0260	0.0335	100,000	100,000	100,000	2,963	97,333	71.1	70.2	71.9	2,455	80,628
1-4	0.0024	0.0096	0.0087	0.0105	97,037	96,653	97,396	936	385,901	72.2	71.5	73.0	757	312,372
5-9	0.0010	0.0050	0.0046	0.0055	96,101	95,651	96,549	485	479,292	68.9	68.2	69.7	364	359,214
10-14	0.0010	0.0052	0.0047	0.0056	95,616	95,133	96,107	496	476,839	64.3	63.6	65.0	323	310,056
15-19	0.0016	0.0079	0.0072	0.0085	95,120	94,596	95,653	751	473,720	59.6	58.9	60.3	473	298,244
20-24	0.0017	0.0085	0.0078	0.0092	94,368	93,788	94,964	802	469,838	55.0	54.4	55.7	412	241,642
25-29	0.0016	0.0079	0.0073	0.0085	93,567	92,929	94,224	741	465,981	50.5	49.9	51.1	322	202,394
30-34	0.0022	0.0107	0.0099	0.0115	92,826	92,138	93,538	997	461,636	45.9	45.3	46.5	358	165,711
35-39	0.0028	0.0141	0.0130	0.0151	91,829	91,079	92,611	1,291	455,916	41.3	40.8	42.0	387	136,553
40-44	0.0035	0.0171	0.0159	0.0183	90,537	89,713	91,403	1,549	448,816	36.9	36.3	37.5	385	111,451
45-49	0.0050	0.0248	0.0231	0.0265	88,989	88,083	89,949	2,207	439,427	32.5	32.0	33.1	450	89,589
50-54	0.0072	0.0355	0.0332	0.0378	86,782	85,776	87,866	3,081	426,207	28.3	27.8	28.8	465	64,291
55-59	0.0096	0.0469	0.0439	0.0498	83,701	82,558	84,933	3,924	408,694	24.2	23.8	24.7	475	49,456
60-64	0.0140	0.0676	0.0634	0.0718	79,777	78,469	81,161	5,396	385,395	20.3	19.9	20.8	569	40,614
65-69	0.0225	0.1065	0.1001	0.1128	74,381	72,885	75,916	7,918	352,111	16.6	16.2	17.0	722	32,096
70-74	0.0340	0.1567	0.1474	0.1653	66,463	64,772	68,179	10,414	306,281	13.2	12.9	13.6	846	24,881
75-79	0.0500	0.2222	0.2095	0.2347	56,049	54,135	58,068	12,455	249,109	10.2	9.9	10.6	831	16,624
80-84	0.0878	0.3599	0.3414	0.3790	43,594	41,537	45,754	15,691	178,743	7.5	7.2	7.7	714	8,134
85-89	0.1528	0.5528	0.5318	0.5752	27,903	26,061	29,984	15,426	100,952	5.2	5.1	5.4	499	3,264
90-94	0.2463	0.7082	0.6893	0.7272	12,478	11,181	13,964	8,836	35,879	3.6	3.5	3.8	199	806
95-99	0.3676	0.8740	0.8618	0.8858	3,641	3,050	4,319	3,183	8,659	2.6	2.6	2.7	37	100
100+	0.5080	1	1	1	459	348	590	459	903	2.0	2.0	2.0	3	5
All ages													*12,042*	*2,548,125*

$ln(_nq_x)$ with uncertainty interval

l_x with uncertainty interval

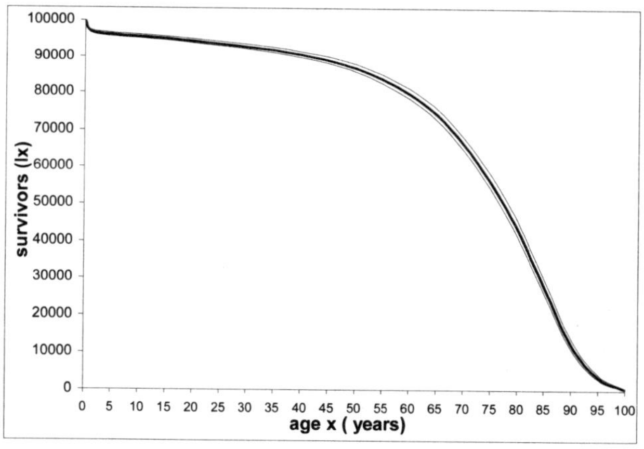

e_x with uncertainty interval

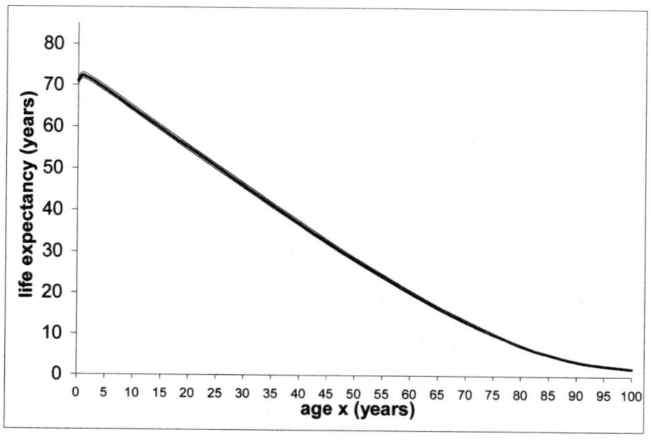

Distribution of deaths by age with uncertainty interval

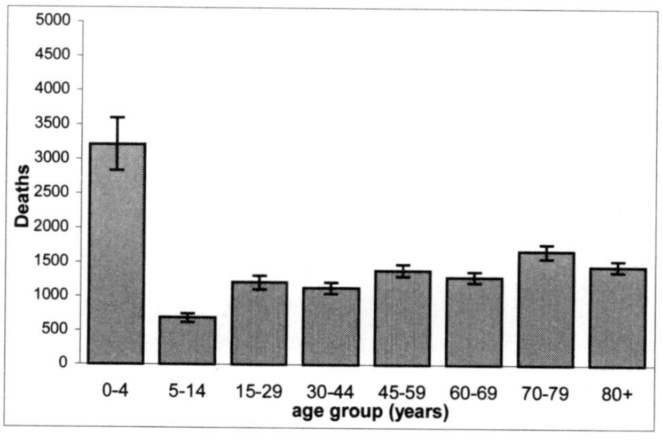

Niger

Niger

Níger

Males, 2000

Hommes, 2000

Hombres, 2000

Age	$_nM_x$	$_nq_x$			l_x			$_nd_x$	$_nL_x$	e_x			Actual deaths	Actual population
			percentiles			percentiles					percentiles			
		mean	2 1/2	97 1/2	mean	2 1/2	97 1/2			mean	2 1/2	97 1/2		
<1	0.2380	0.1960	0.1751	0.2180	100,000	100,000	100,000	19,599	82,361	42.7	40.2	46.5	62,771	263,784
1-4	0.0199	0.0761	0.0634	0.0862	80,401	78,204	82,487	6,121	306,915	52.1	48.8	56.5	17,808	892,964
5-9	0.0044	0.0216	0.0175	0.0250	74,281	72,643	75,925	1,601	367,399	52.3	49.1	56.2	3,810	873,969
10-14	0.0024	0.0120	0.0097	0.0140	72,679	70,877	74,515	873	361,212	48.4	45.3	52.2	1,717	710,351
15-19	0.0034	0.0169	0.0136	0.0198	71,806	69,912	73,759	1,216	355,990	43.9	40.9	47.7	2,015	590,047
20-24	0.0046	0.0228	0.0183	0.0268	70,590	68,570	72,709	1,609	348,929	39.7	36.7	43.3	2,225	482,624
25-29	0.0078	0.0385	0.0307	0.0456	68,981	66,796	71,381	2,653	338,275	35.5	32.6	39.0	3,017	384,696
30-34	0.0108	0.0527	0.0417	0.0630	66,329	63,877	69,192	3,498	322,898	31.8	29.0	35.2	3,309	305,477
35-39	0.0136	0.0658	0.0517	0.0795	62,831	60,045	66,303	4,137	303,812	28.5	25.8	31.6	3,326	244,258
40-44	0.0164	0.0788	0.0616	0.0962	58,694	55,546	62,873	4,625	281,909	25.3	22.8	28.3	3,220	196,303
45-49	0.0193	0.0921	0.0717	0.1135	54,069	50,359	59,003	4,981	257,895	22.3	20.0	25.0	3,017	156,188
50-54	0.0233	0.1101	0.0855	0.1362	49,088	44,755	54,774	5,403	231,934	19.3	17.2	21.8	2,567	110,168
55-59	0.0287	0.1339	0.1042	0.1659	43,685	38,757	50,088	5,849	203,803	16.3	14.5	18.6	2,452	85,426
60-64	0.0388	0.1770	0.1376	0.2179	37,836	32,453	44,865	6,696	172,442	13.5	11.9	15.5	2,479	63,852
65-69	0.0542	0.2388	0.1869	0.2902	31,141	25,534	38,629	7,436	137,112	10.8	9.5	12.5	2,459	45,349
70-74	0.0814	0.3383	0.2700	0.4013	23,704	18,141	31,272	8,020	98,472	8.4	7.4	9.8	2,368	29,074
75-79	0.1208	0.4640	0.3824	0.5327	15,684	10,861	22,604	7,277	60,230	6.5	5.7	7.6	1,894	15,677
80-84	0.1759	0.6108	0.5256	0.6765	8,407	5,075	13,689	5,135	29,199	4.9	4.4	5.7	1,123	6,388
85-89	0.2495	0.7683	0.6935	0.8212	3,272	1,642	6,390	2,514	10,076	3.7	3.4	4.2	415	1,663
90-94	0.3450	0.8476	0.7863	0.8891	758	294	1,959	643	1,863	2.8	2.6	3.1	87	251
95-99	0.4648	0.9706	0.9454	0.9836	116	33	419	112	241	2.1	2.1	2.3	9	19
100+	0.6102	1	1	1	3	1	23	3	6	1.6	1.6	1.7	1	1
All ages													*122,087*	*5,458,529*

ln($_nq_x$) with uncertainty interval

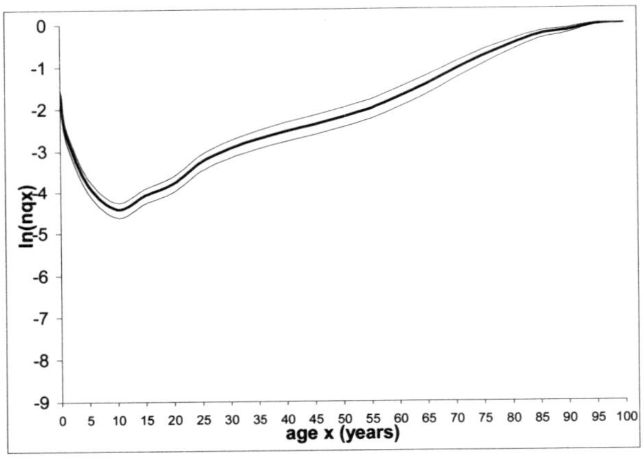

l_x with uncertainty interval

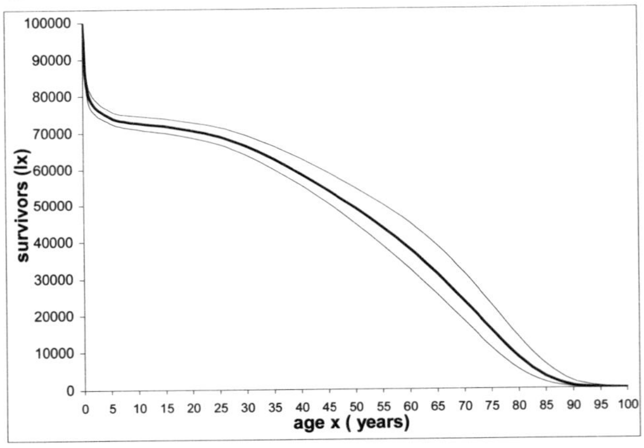

e_x with uncertainty interval

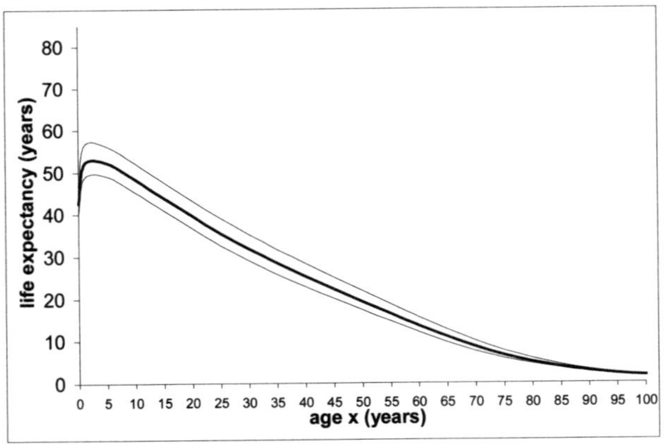

Distribution of deaths by age with uncertainty interval

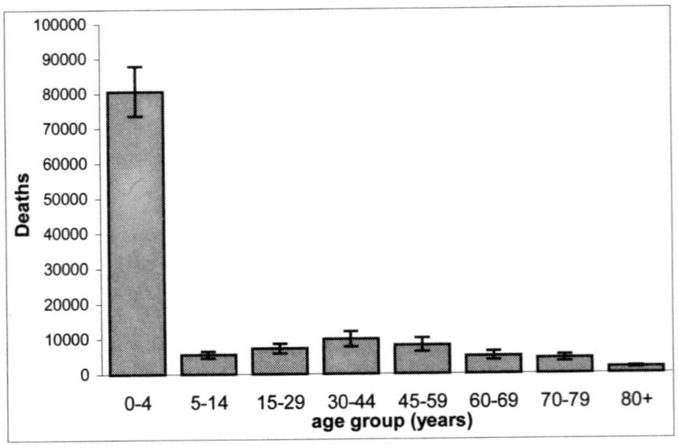

Niger **Niger** **Níger**

Females, 2000 *Femmes, 2000* *Mujeres, 2000*

Age	$_nM_x$	$_nq_x$ mean	$_nq_x$ percentiles 2½	$_nq_x$ percentiles 97½	l_x mean	l_x percentiles 2½	l_x percentiles 97½	$_nd_x$	$_nL_x$	e_x mean	e_x percentiles 2½	e_x percentiles 97½	Actual deaths	Actual population
<1	0.2098	0.1765	0.1601	0.1932	100,000	100,000	100,000	17,649	84,116	43.9	41.4	47.0	52,536	250,388
1-4	0.0241	0.0913	0.0797	0.1020	82,351	80,678	83,987	7,518	311,360	52.3	48.9	55.9	21,185	877,361
5-9	0.0053	0.0260	0.0222	0.0298	74,833	73,604	76,081	1,943	369,307	53.4	50.2	56.8	4,452	846,414
10-14	0.0034	0.0168	0.0142	0.0194	72,890	71,492	74,395	1,222	361,395	49.7	46.6	53.0	2,321	686,085
15-19	0.0050	0.0247	0.0209	0.0288	71,668	70,136	73,335	1,772	353,910	45.5	42.5	48.8	2,847	568,723
20-24	0.0080	0.0392	0.0331	0.0460	69,896	68,173	71,800	2,742	342,625	41.6	38.7	44.8	3,736	466,841
25-29	0.0097	0.0476	0.0399	0.0562	67,154	65,136	69,426	3,195	327,781	38.2	35.4	41.2	3,654	374,786
30-34	0.0105	0.0510	0.0422	0.0608	63,959	61,605	66,657	3,259	311,646	35.0	32.4	37.8	3,133	299,597
35-39	0.0116	0.0562	0.0462	0.0677	60,700	58,020	63,825	3,410	294,973	31.8	29.3	34.4	2,793	241,578
40-44	0.0109	0.0531	0.0433	0.0645	57,289	54,169	60,847	3,042	278,842	28.5	26.2	30.9	2,141	196,209
45-49	0.0122	0.0590	0.0479	0.0722	54,247	50,744	58,173	3,202	263,231	25.0	22.9	27.2	1,927	158,448
50-54	0.0158	0.0759	0.0613	0.0932	51,045	47,166	55,337	3,876	245,535	21.4	19.4	23.4	1,861	117,879
55-59	0.0212	0.1005	0.0810	0.1235	47,169	42,833	51,865	4,740	223,994	17.9	16.2	19.8	2,020	95,467
60-64	0.0311	0.1443	0.1164	0.1769	42,429	37,632	47,554	6,124	196,834	14.7	13.1	16.3	2,295	73,773
65-69	0.0464	0.2080	0.1688	0.2524	36,305	31,104	41,845	7,550	162,648	11.7	10.4	13.2	2,495	53,750
70-74	0.0717	0.3041	0.2508	0.3618	28,754	23,380	34,531	8,744	121,911	9.1	8.1	10.3	2,518	35,100
75-79	0.1069	0.4217	0.3571	0.4874	20,010	14,921	25,807	8,439	78,954	7.0	6.2	8.0	2,077	19,433
80-84	0.1582	0.5668	0.4978	0.6323	11,572	7,649	16,592	6,559	41,460	5.3	4.8	6.0	1,314	8,306
85-89	0.2283	0.7267	0.6646	0.7820	5,013	2,813	8,332	3,643	15,957	4.0	3.6	4.4	546	2,391
90-94	0.3211	0.8178	0.7669	0.8618	1,370	613	2,795	1,120	3,489	3.0	2.7	3.3	141	438
95-99	0.4403	0.9486	0.9223	0.9679	250	85	652	237	538	2.2	2.2	2.4	20	46
100+	0.5885	1	1	1	13	3	51	13	22	1.7	1.7	1.8	2	3
All ages													*116,013*	*5,373,016*

ln($_nq_x$) with uncertainty interval

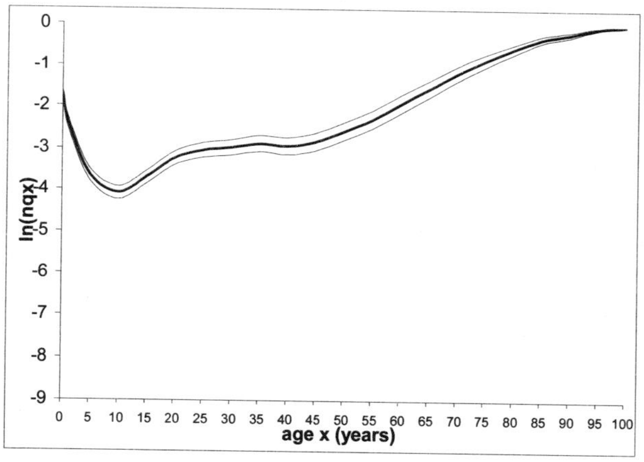

l_x with uncertainty interval

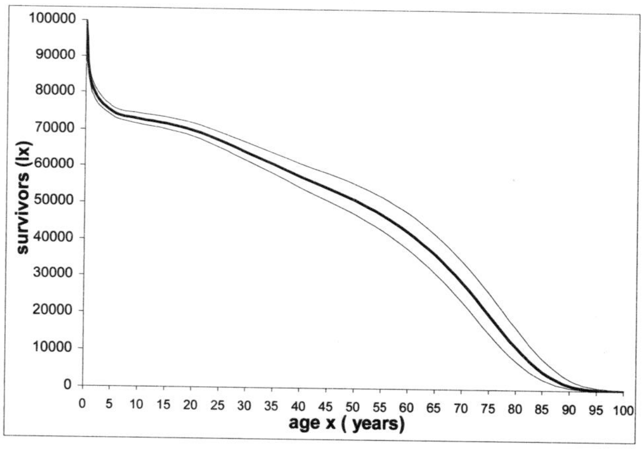

e_x with uncertainty interval

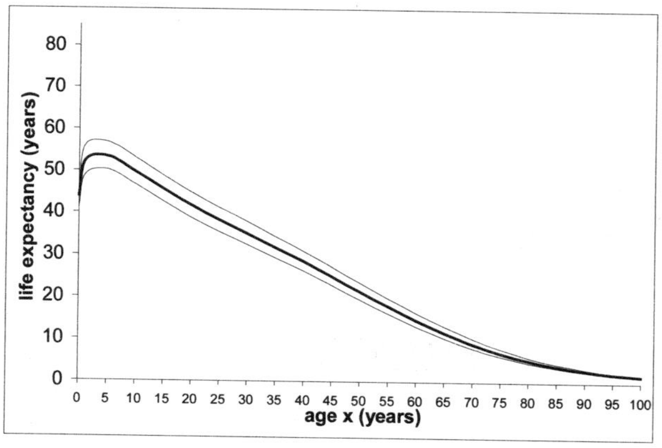

Distribution of deaths by age with uncertainty interval

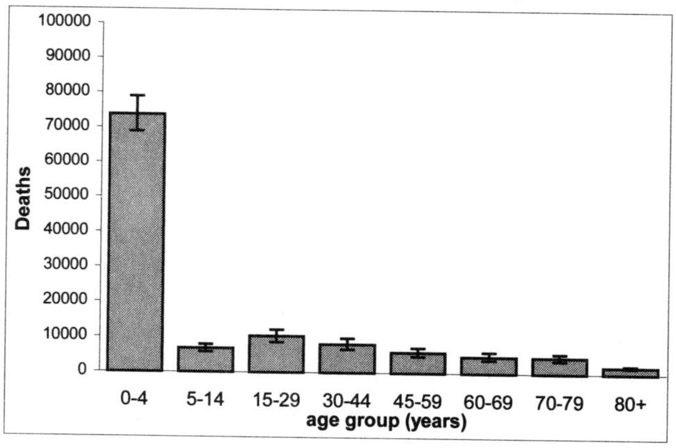

Nigeria

Nigéria

Nigeria

Males, 2000 — *Hommes, 2000* — *Hombres, 2000*

Age	$_nM_x$	$_nq_x$ mean	$_nq_x$ percentiles 2½	$_nq_x$ percentiles 97½	l_x mean	l_x percentiles 2½	l_x percentiles 97½	$_nd_x$	$_nL_x$	e_x mean	e_x percentiles 2½	e_x percentiles 97½	Actual deaths	Actual population
<1	0.1338	0.1194	0.1110	0.1269	100,000	100,000	100,000	11,941	89,253	49.8	47.6	52.2	295,969	2,212,210
1-4	0.0112	0.0435	0.0392	0.0479	88,059	87,310	88,896	3,827	343,051	55.6	53.1	58.3	86,834	7,783,505
5-9	0.0025	0.0125	0.0111	0.0139	84,232	83,484	84,975	1,049	418,536	54.0	51.7	56.7	21,573	8,604,565
10-14	0.0016	0.0079	0.0070	0.0089	83,182	82,378	83,990	661	414,260	49.7	47.4	52.3	11,909	7,463,260
15-19	0.0018	0.0091	0.0081	0.0103	82,521	81,667	83,396	754	410,722	45.1	42.8	47.6	11,738	6,391,461
20-24	0.0042	0.0209	0.0184	0.0236	81,767	80,857	82,719	1,707	404,567	40.5	38.3	43.0	22,206	5,261,634
25-29	0.0078	0.0384	0.0335	0.0437	80,060	79,021	81,190	3,073	392,616	36.3	34.1	38.8	32,658	4,172,090
30-34	0.0118	0.0575	0.0496	0.0660	76,986	75,600	78,444	4,423	373,874	32.6	30.6	35.0	39,607	3,347,581
35-39	0.0137	0.0663	0.0564	0.0768	72,563	70,628	74,496	4,812	350,785	29.5	27.5	31.7	37,808	2,756,071
40-44	0.0152	0.0734	0.0618	0.0855	67,751	65,206	70,195	4,975	326,317	26.4	24.6	28.5	35,168	2,306,651
45-49	0.0173	0.0831	0.0695	0.0970	62,776	59,633	65,725	5,219	300,833	23.3	21.7	25.2	32,981	1,901,259
50-54	0.0202	0.0963	0.0801	0.1125	57,557	53,847	61,108	5,544	273,927	20.1	18.7	21.9	29,201	1,442,836
55-59	0.0245	0.1156	0.0959	0.1347	52,013	47,791	56,151	6,014	245,033	17.0	15.8	18.6	29,501	1,202,041
60-64	0.0357	0.1640	0.1366	0.1901	46,000	41,355	50,683	7,545	211,137	13.9	12.8	15.3	33,327	932,639
65-69	0.0510	0.2260	0.1899	0.2598	38,455	33,518	43,725	8,691	170,546	11.2	10.3	12.3	35,109	688,919
70-74	0.0766	0.3215	0.2752	0.3635	29,763	24,875	35,420	9,568	124,896	8.7	8.0	9.6	36,083	470,993
75-79	0.1149	0.4463	0.3928	0.4929	20,195	15,929	25,672	9,013	78,443	6.6	6.1	7.3	31,842	277,137
80-84	0.1733	0.6045	0.5511	0.6493	11,182	8,171	15,588	6,760	39,011	4.9	4.6	5.4	21,780	125,685
85-89	0.2519	0.7729	0.7296	0.8080	4,422	2,879	6,997	3,418	13,566	3.7	3.5	4.0	8,986	35,667
90-94	0.3528	0.8570	0.8239	0.8835	1,004	553	1,892	861	2,439	2.7	2.6	2.9	2,115	5,993
95-99	0.4761	0.9803	0.9705	0.9869	144	64	333	141	296	2.1	2.1	2.1	253	532
100+	0.6188	1	1	1	3	1	10	3	5	1.6	1.6	1.7	17	28
All ages													*856,664*	*57,382,757*

$\ln(_nq_x)$ with uncertainty interval

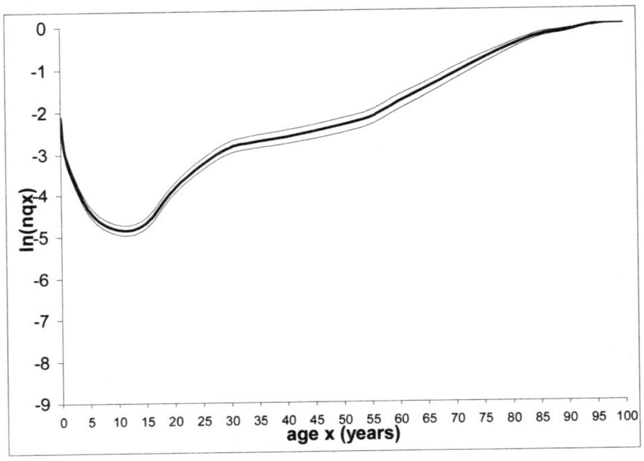

l_x with uncertainty interval

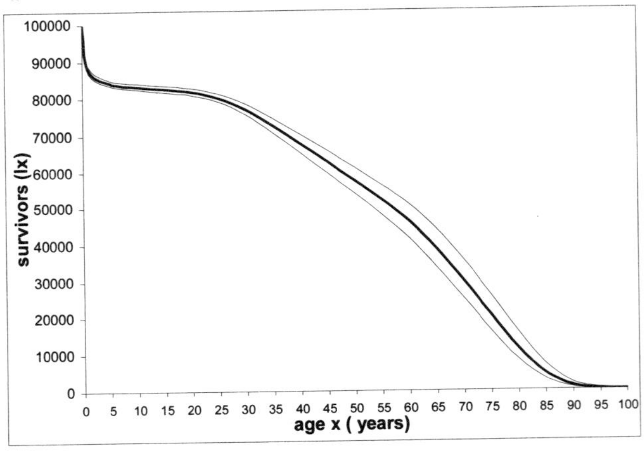

e_x with uncertainty interval

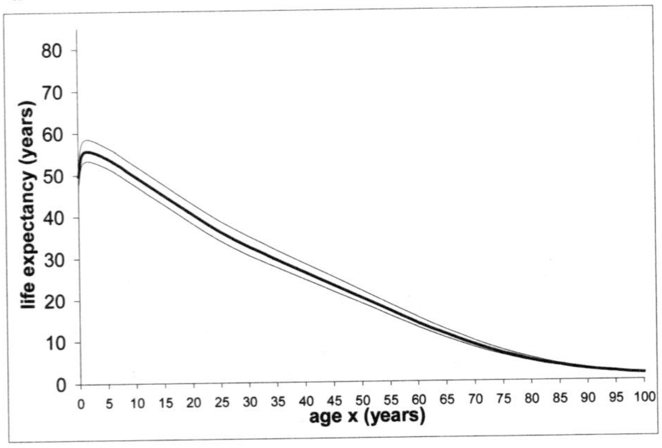

Distribution of deaths by age with uncertainty interval

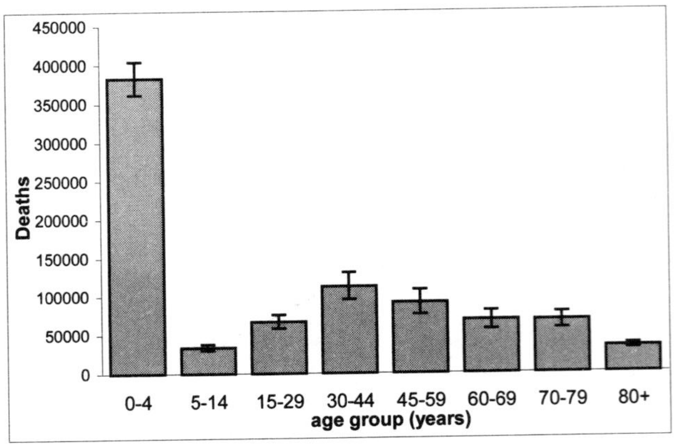

Nigeria

Nigéria

Nigeria

Females, 2000 *Femmes, 2000* *Mujeres, 2000*

Age	$_nM_x$	$_nq_x$ mean	$_nq_x$ 2 1/2 percentile	$_nq_x$ 97 1/2 percentile	l_x mean	l_x 2 1/2 percentile	l_x 97 1/2 percentile	$_nd_x$	$_nL_x$	e_x mean	e_x 2 1/2 percentile	e_x 97 1/2 percentile	Actual deaths	Actual population
<1	0.1192	0.1076	0.0990	0.1154	100,000	100,000	100,000	10,763	90,313	51.4	49.0	53.9	252,292	2,116,947
1-4	0.0125	0.0484	0.0437	0.0532	89,237	88,456	90,096	4,316	346,589	56.6	53.9	59.6	94,270	7,570,655
5-9	0.0026	0.0131	0.0117	0.0147	84,921	84,175	85,670	1,113	421,823	55.4	52.9	58.3	21,969	8,327,716
10-14	0.0017	0.0086	0.0076	0.0097	83,808	82,997	84,605	720	417,240	51.1	48.6	53.9	12,468	7,220,993
15-19	0.0033	0.0164	0.0144	0.0185	83,088	82,235	83,931	1,360	412,039	46.5	44.1	49.3	20,431	6,189,574
20-24	0.0070	0.0342	0.0299	0.0388	81,728	80,787	82,714	2,793	401,657	42.3	39.9	45.0	35,529	5,109,875
25-29	0.0107	0.0523	0.0453	0.0600	78,935	77,740	80,221	4,128	384,355	38.7	36.4	41.3	43,581	4,057,744
30-34	0.0127	0.0616	0.0526	0.0713	74,807	73,109	76,546	4,609	362,513	35.7	33.5	38.2	41,673	3,277,811
35-39	0.0122	0.0592	0.0499	0.0690	70,198	67,897	72,443	4,158	340,595	32.9	30.9	35.1	33,219	2,720,971
40-44	0.0113	0.0550	0.0460	0.0644	66,040	63,209	68,729	3,633	321,118	29.8	28.0	31.9	26,079	2,305,190
45-49	0.0114	0.0555	0.0461	0.0652	62,407	59,136	65,474	3,463	303,378	26.3	24.8	28.3	22,271	1,950,932
50-54	0.0137	0.0661	0.0547	0.0777	58,944	55,283	62,425	3,897	284,977	22.7	21.3	24.5	20,332	1,486,781
55-59	0.0175	0.0839	0.0693	0.0985	55,047	50,990	58,968	4,620	263,683	19.2	17.9	20.8	22,123	1,262,606
60-64	0.0256	0.1204	0.0994	0.1406	50,427	45,970	54,826	6,072	236,953	15.7	14.6	17.2	26,014	1,015,214
65-69	0.0392	0.1787	0.1484	0.2073	44,355	39,509	49,286	7,924	201,963	12.5	11.5	13.8	30,257	771,128
70-74	0.0628	0.2712	0.2286	0.3107	36,430	31,350	41,959	9,881	157,449	9.7	8.9	10.7	33,969	541,282
75-79	0.0977	0.3925	0.3392	0.4398	26,549	21,694	32,369	10,420	106,697	7.4	6.8	8.2	32,298	330,719
80-84	0.1485	0.5415	0.4845	0.5897	16,129	12,262	21,390	8,733	58,814	5.5	5.1	6.1	23,734	159,833
85-89	0.2189	0.7074	0.6572	0.7483	7,396	5,093	11,028	5,232	23,900	4.1	3.8	4.5	11,220	51,250
90-94	0.3129	0.8069	0.7673	0.8390	2,164	1,282	3,780	1,746	5,581	3.0	2.8	3.3	3,265	10,435
95-99	0.4334	0.9421	0.9215	0.9572	418	206	880	394	908	2.3	2.2	2.4	544	1,255
100+	0.5820	1	1	1	24	9	69	24	42	1.7	1.7	1.8	49	85
All ages													*807,586*	*56,478,996*

ln($_nq_x$) with uncertainty interval

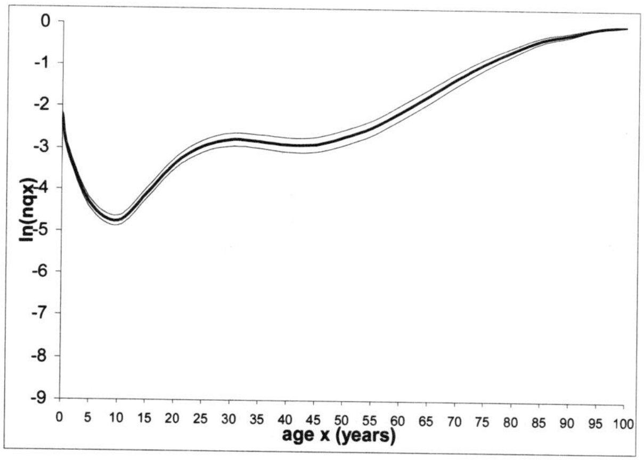

l_x with uncertainty interval

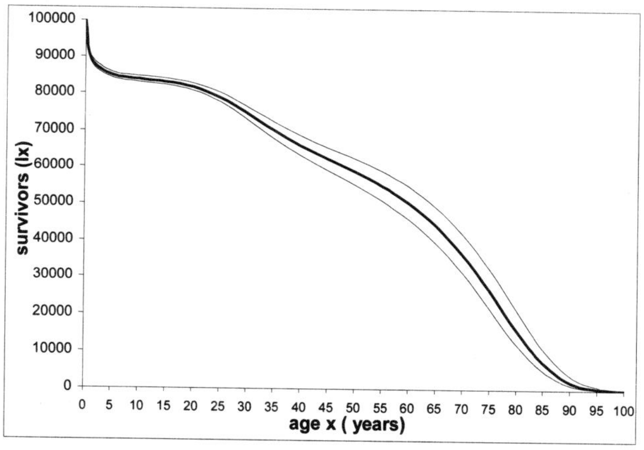

e_x with uncertainty interval

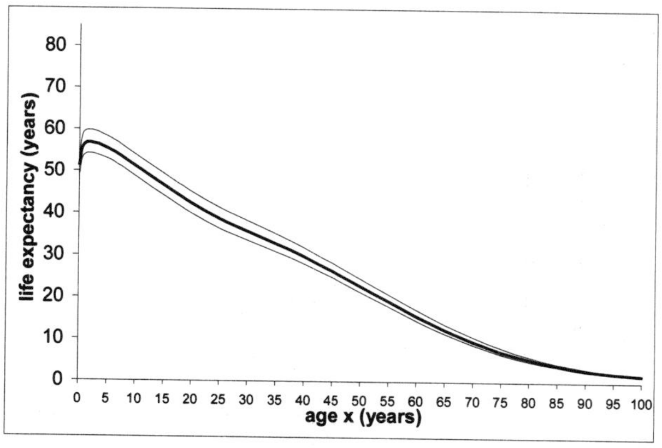

Distribution of deaths by age with uncertainty interval

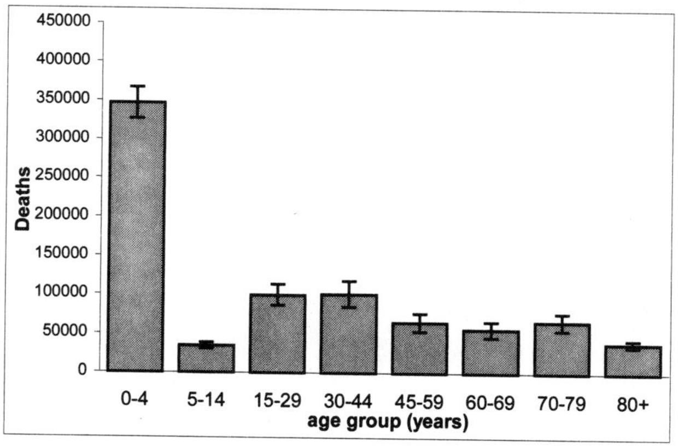

Niue

Nioué

Niue

Males, 2000 — *Hommes, 2000* — *Hombres, 2000*

Age	$_nM_x$	$_nq_x$			l_x			$_nd_x$	$_nL_x$	e_x			Actual deaths	Actual population
			percentiles			percentiles					percentiles			
		mean	2½	97½	mean	2½	97½			mean	2½	97½		
<1	0.0306	0.0297	0.0207	0.0408	100,000	100,000	100,000	2,974	97,324	69.5	66.7	71.9	1	23
1-4	0.0011	0.0044	0.0033	0.0057	97,026	95,918	97,927	428	387,079	70.6	68.2	72.8	0	96
5-9	0.0008	0.0039	0.0029	0.0050	96,599	95,370	97,606	376	482,053	66.9	64.6	69.0	0	118
10-14	0.0006	0.0032	0.0024	0.0041	96,223	94,891	97,322	307	480,345	62.2	59.9	64.3	0	124
15-19	0.0012	0.0058	0.0044	0.0074	95,915	94,501	97,088	555	478,189	57.4	55.1	59.4	0	114
20-24	0.0021	0.0105	0.0081	0.0134	95,361	93,801	96,661	1,002	474,297	52.7	50.5	54.7	0	99
25-29	0.0023	0.0116	0.0090	0.0146	94,358	92,549	95,881	1,092	469,062	48.2	46.1	50.2	0	90
30-34	0.0016	0.0082	0.0064	0.0102	93,266	91,200	95,016	761	464,429	43.8	41.8	45.7	0	78
35-39	0.0028	0.0138	0.0110	0.0172	92,505	90,267	94,406	1,273	459,345	39.1	37.2	41.0	0	67
40-44	0.0037	0.0185	0.0149	0.0231	91,233	88,713	93,372	1,685	451,950	34.6	32.8	36.5	0	55
45-49	0.0050	0.0247	0.0203	0.0308	89,547	86,664	91,980	2,211	442,209	30.2	28.5	32.1	0	47
50-54	0.0081	0.0396	0.0325	0.0491	87,336	83,995	90,116	3,458	428,037	25.9	24.3	27.7	0	36
55-59	0.0130	0.0630	0.0519	0.0774	83,878	79,868	87,184	5,280	406,190	21.9	20.4	23.6	0	30
60-64	0.0220	0.1041	0.0865	0.1258	78,598	73,688	82,661	8,182	372,533	18.2	16.9	19.8	1	24
65-69	0.0286	0.1335	0.1104	0.1580	70,415	64,416	75,508	9,399	328,580	15.0	13.8	16.5	1	18
70-74	0.0466	0.2086	0.1737	0.2428	61,017	54,237	67,170	12,729	273,261	11.9	10.8	13.2	1	12
75-79	0.0634	0.2738	0.2285	0.3202	48,288	41,070	55,504	13,220	208,390	9.4	8.5	10.5	0	7
80-84	0.1059	0.4185	0.3585	0.4792	35,068	27,919	42,821	14,676	138,649	7.0	6.3	7.9	0	3
85-89	0.1472	0.5380	0.4765	0.5988	20,392	14,540	27,471	10,971	74,530	5.3	4.8	5.9	0	1
90-94	0.2457	0.7072	0.6516	0.7586	9,420	5,834	14,380	6,662	27,114	3.6	3.3	3.9	0	0
95-99	0.4102	0.9195	0.8901	0.9433	2,758	1,408	5,011	2,536	6,182	2.4	2.3	2.5	0	0
100+	0.6850	1	1	1	222	80	551	222	324	1.5	1.4	1.8	0	0
All ages													*6*	*1,042*

ln($_nq_x$) with uncertainty interval

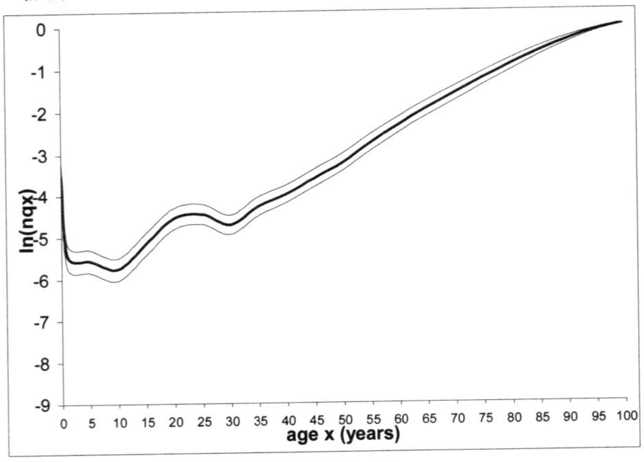

l_x with uncertainty interval

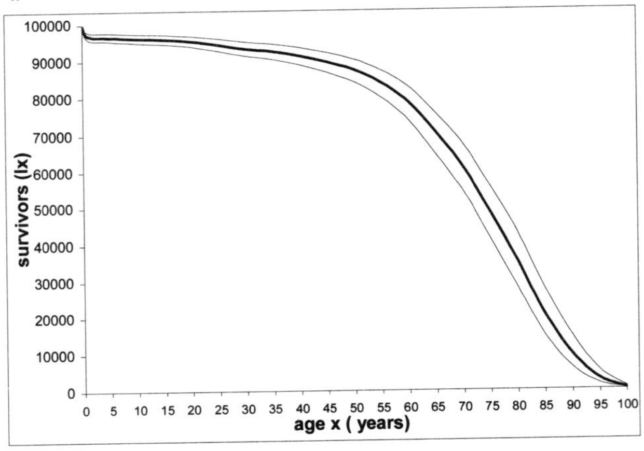

e_x with uncertainty interval

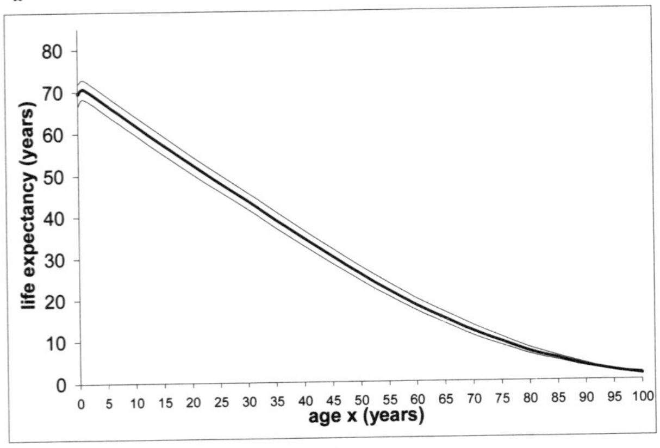

Distribution of deaths by age with uncertainty interval

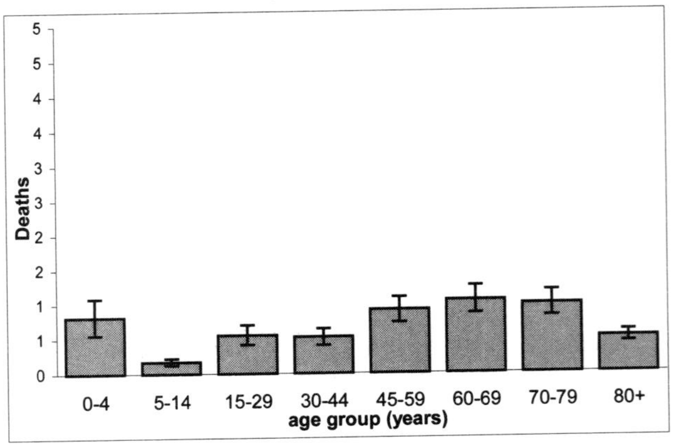

Niue | Nioué | Niue

Females, 2000 | *Femmes, 2000* | *Mujeres, 2000*

Age	$_nM_x$	$_nq_x$			l_x			$_nd_x$	$_nL_x$	e_x			Actual deaths	Actual population
			percentiles			percentiles					percentiles			
		mean	2 1/2	97 1/2	mean	2 1/2	97 1/2			mean	2 1/2	97 1/2		
<1	0.0266	0.0260	0.0165	0.0367	100,000	100,000	100,000	2,598	97,662	72.8	71.5	75.6	1	22
1-4	0.0006	0.0024	0.0017	0.0031	97,402	96,334	98,354	234	389,047	73.7	72.4	76.5	0	91
5-9	0.0006	0.0028	0.0019	0.0035	97,168	96,040	98,192	272	485,161	69.9	68.6	72.7	0	111
10-14	0.0006	0.0030	0.0021	0.0037	96,896	95,701	98,001	291	483,752	65.1	63.7	67.8	0	115
15-19	0.0008	0.0041	0.0029	0.0051	96,605	95,343	97,796	400	482,022	60.3	58.9	63.0	0	106
20-24	0.0013	0.0064	0.0046	0.0078	96,204	94,854	97,510	616	479,482	55.5	54.2	58.2	0	90
25-29	0.0012	0.0058	0.0042	0.0070	95,589	94,116	97,065	558	476,550	50.9	49.5	53.5	0	76
30-34	0.0017	0.0084	0.0062	0.0099	95,031	93,457	96,655	800	473,157	46.2	44.8	48.8	0	64
35-39	0.0022	0.0110	0.0082	0.0128	94,232	92,529	96,059	1,036	468,569	41.5	40.1	44.1	0	59
40-44	0.0032	0.0161	0.0122	0.0183	93,196	91,348	95,273	1,500	462,229	37.0	35.6	39.5	0	50
45-49	0.0045	0.0222	0.0173	0.0248	91,696	89,676	94,108	2,038	453,382	32.5	31.1	34.9	0	43
50-54	0.0065	0.0320	0.0252	0.0356	89,657	87,452	92,482	2,867	441,119	28.2	26.8	30.5	0	33
55-59	0.0098	0.0476	0.0371	0.0533	86,790	84,340	90,151	4,130	423,625	24.1	22.7	26.2	0	28
60-64	0.0159	0.0763	0.0592	0.0869	82,660	79,845	86,803	6,311	397,522	20.1	18.8	22.2	0	24
65-69	0.0220	0.1043	0.0808	0.1206	76,349	72,906	81,664	7,961	361,844	16.6	15.4	18.4	0	18
70-74	0.0385	0.1756	0.1378	0.2041	68,389	64,116	75,068	12,009	311,921	13.2	12.1	14.8	1	13
75-79	0.0550	0.2418	0.1943	0.2819	56,380	51,029	64,724	13,630	247,823	10.5	9.5	11.8	1	9
80-84	0.0848	0.3498	0.2931	0.4050	42,749	36,642	52,145	14,955	176,360	8.1	7.3	9.0	0	4
85-89	0.1210	0.4645	0.4037	0.5255	27,795	21,801	36,862	12,910	106,697	6.1	5.5	6.8	0	2
90-94	0.1980	0.6211	0.5611	0.6784	14,884	10,344	21,979	9,244	46,689	4.2	3.8	4.7	0	1
95-99	0.3300	0.8291	0.7847	0.8678	5,640	3,326	9,646	4,676	14,171	2.8	2.7	3.1	0	0
100+	0.5500	1	1	1	964	440	2,077	964	1,752	1.8	1.8	2.1	0	0
All ages													*5*	*960*

ln($_nq_x$) with uncertainty interval

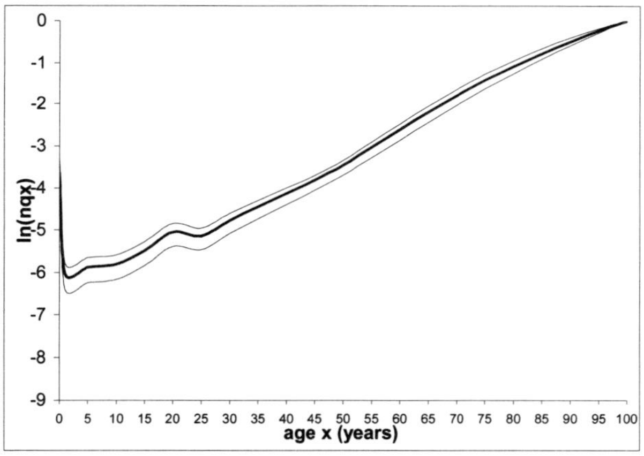

l_x with uncertainty interval

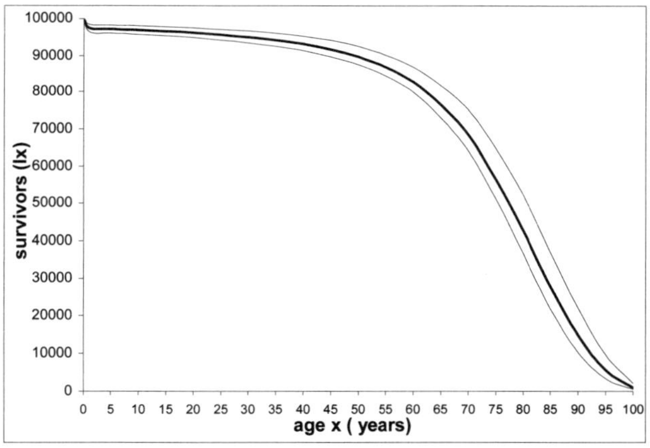

e_x with uncertainty interval

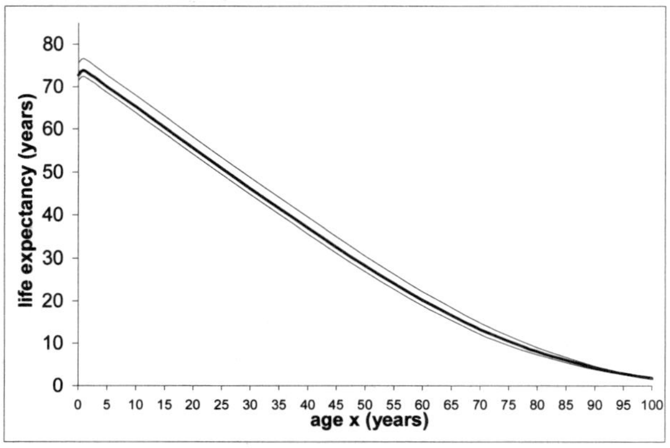

Distribution of deaths by age with uncertainty interval

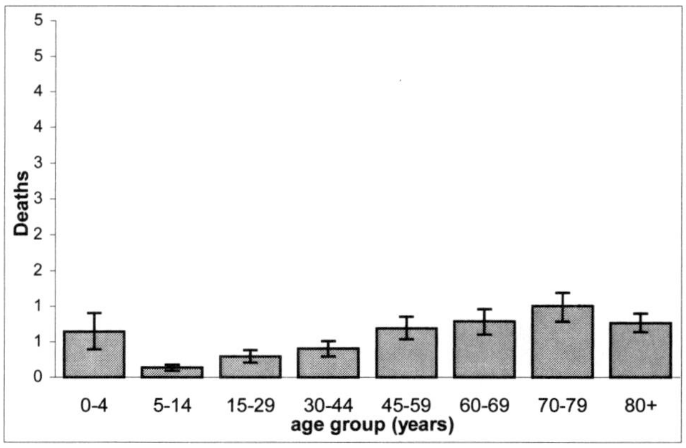

Norway　　　　Norvège　　　　Noruega

Males, 2000　　　　*Hommes, 2000*　　　　*Hombres, 2000*

Age	$_nM_x$	$_nq_x$			l_x			$_nd_x$	$_nL_x$	e_x			Actual deaths	Actual population
			percentiles			percentiles					percentiles			
		mean	2½	97½	mean	2½	97½			mean	2½	97½		
<1	0.0043	0.0043	0.0040	0.0048	100,000	100,000	100,000	433	99,611	75.7	75.5	76.0	128	29,548
1-4	0.0002	0.0009	0.0008	0.0010	99,567	99,525	99,603	89	398,055	75.0	74.8	75.3	27	119,040
5-9	0.0001	0.0007	0.0006	0.0007	99,478	99,429	99,520	67	497,222	71.1	70.9	71.4	21	157,434
10-14	0.0002	0.0008	0.0007	0.0008	99,411	99,356	99,457	77	496,862	66.2	65.9	66.4	23	147,252
15-19	0.0005	0.0027	0.0026	0.0029	99,334	99,274	99,385	272	495,990	61.2	61.0	61.4	74	134,921
20-24	0.0011	0.0055	0.0052	0.0059	99,062	98,983	99,129	546	493,945	56.4	56.1	56.6	153	138,600
25-29	0.0010	0.0051	0.0048	0.0054	98,516	98,402	98,612	501	491,329	51.7	51.4	51.9	167	163,686
30-34	0.0011	0.0056	0.0053	0.0059	98,015	97,873	98,136	548	488,706	46.9	46.7	47.2	199	177,166
35-39	0.0014	0.0070	0.0067	0.0073	97,467	97,297	97,611	682	485,629	42.2	42.0	42.4	237	168,504
40-44	0.0020	0.0099	0.0094	0.0103	96,785	96,582	96,957	954	481,537	37.4	37.2	37.7	323	163,079
45-49	0.0028	0.0139	0.0134	0.0145	95,830	95,588	96,042	1,335	475,813	32.8	32.6	33.0	438	156,225
50-54	0.0047	0.0231	0.0223	0.0240	94,495	94,204	94,757	2,187	467,007	28.2	28.0	28.4	744	158,872
55-59	0.0075	0.0369	0.0358	0.0381	92,308	91,952	92,641	3,405	453,027	23.8	23.7	24.0	919	122,294
60-64	0.0130	0.0629	0.0611	0.0646	88,903	88,465	89,329	5,589	430,542	19.6	19.5	19.8	1,185	91,279
65-69	0.0215	0.1020	0.0991	0.1044	83,314	82,751	83,852	8,497	395,326	15.8	15.7	16.0	1,707	79,408
70-74	0.0367	0.1681	0.1635	0.1717	74,817	74,135	75,453	12,579	342,637	12.3	12.2	12.5	2,729	74,345
75-79	0.0627	0.2711	0.2646	0.2760	62,238	61,475	63,053	16,870	269,015	9.3	9.2	9.4	4,032	64,291
80-84	0.1063	0.4199	0.4125	0.4273	45,368	44,564	46,370	19,052	179,211	6.8	6.7	6.9	4,259	40,061
85-89	0.1705	0.5978	0.5895	0.6058	26,316	25,655	27,241	15,732	92,252	4.9	4.9	5.0	3,278	19,224
90-94	0.2589	0.7285	0.7214	0.7351	10,584	10,151	11,127	7,711	29,789	3.5	3.5	3.6	1,570	6,067
95-99	0.3718	0.8788	0.8741	0.8830	2,873	2,689	3,069	2,525	6,791	2.6	2.6	2.6	444	1,194
100+	0.5054	1	1	1	348	315	386	348	689	2.0	2.0	2.0	74	147
All ages													*22,732*	*2,212,637*

$\ln(_nq_x)$ with uncertainty interval

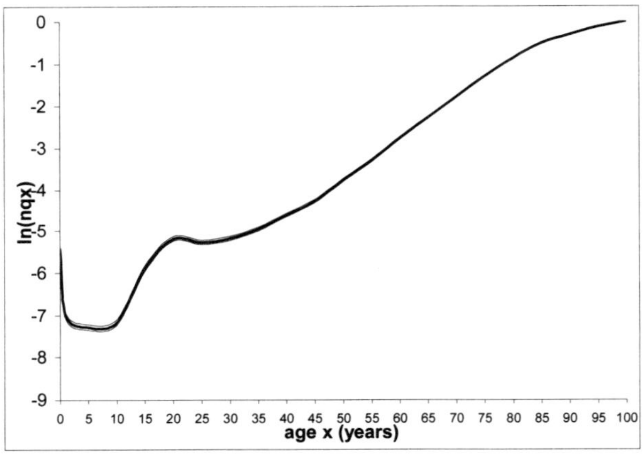

l_x with uncertainty interval

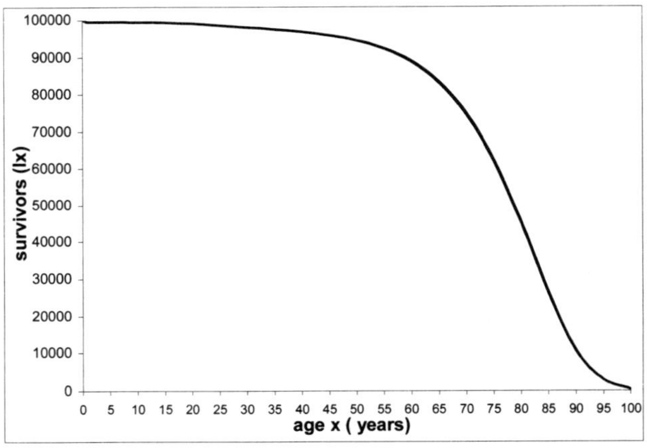

e_x with uncertainty interval

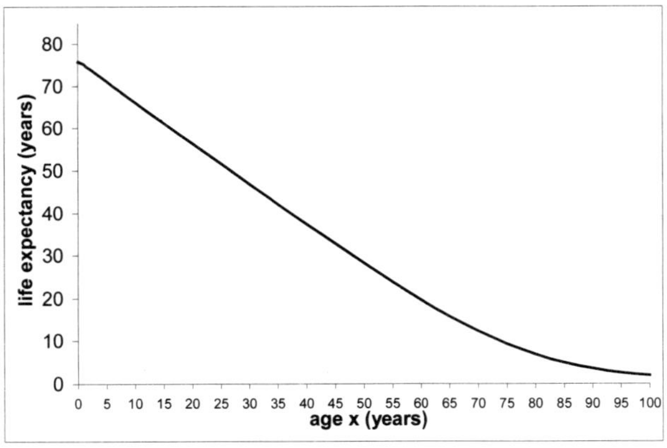

Distribution of deaths by age with uncertainty interval

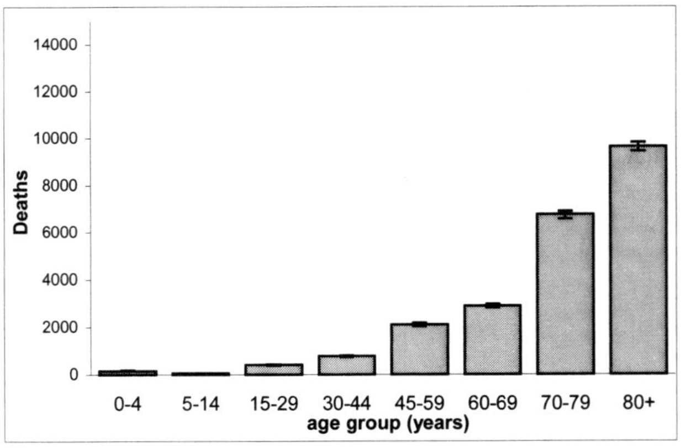

Norway | Norvège | Noruega

Females, 2000 | *Femmes, 2000* | *Mujeres, 2000*

Age	$_nM_x$	$_nq_x$			l_x			$_nd_x$	$_nL_x$	e_x			Actual deaths	Actual population
			percentiles			percentiles					percentiles			
		mean	2½	97½	mean	2½	97½			mean	2½	97½		
<1	0.0033	0.0033	0.0029	0.0037	100,000	100,000	100,000	330	99,703	81.4	80.9	82.0	92	27,915
1-4	0.0003	0.0010	0.0009	0.0011	99,670	99,629	99,710	101	398,438	80.6	80.2	81.3	28	112,797
5-9	0.0001	0.0004	0.0003	0.0004	99,569	99,519	99,619	37	497,754	76.7	76.2	77.4	11	148,814
10-14	0.0001	0.0006	0.0006	0.0007	99,532	99,479	99,586	64	497,502	71.7	71.3	72.4	18	140,019
15-19	0.0003	0.0015	0.0014	0.0016	99,468	99,410	99,527	149	496,969	66.8	66.3	67.4	39	128,432
20-24	0.0003	0.0017	0.0015	0.0018	99,319	99,250	99,391	167	496,178	61.9	61.4	62.5	45	132,721
25-29	0.0003	0.0016	0.0015	0.0017	99,152	99,070	99,237	157	495,369	57.0	56.5	57.6	50	158,695
30-34	0.0005	0.0027	0.0025	0.0029	98,996	98,901	99,092	271	494,299	52.1	51.6	52.7	93	169,458
35-39	0.0007	0.0037	0.0034	0.0039	98,724	98,611	98,839	365	492,709	47.2	46.7	47.8	119	160,616
40-44	0.0012	0.0061	0.0057	0.0065	98,359	98,222	98,497	599	490,298	42.4	41.9	43.0	192	156,824
45-49	0.0017	0.0085	0.0080	0.0089	97,760	97,587	97,933	827	486,733	37.6	37.2	38.2	255	150,042
50-54	0.0030	0.0149	0.0141	0.0157	96,933	96,717	97,149	1,443	481,058	32.9	32.5	33.5	456	152,066
55-59	0.0042	0.0209	0.0196	0.0221	95,490	95,211	95,770	2,001	472,449	28.4	28.0	29.0	509	120,144
60-64	0.0069	0.0337	0.0314	0.0356	93,490	93,140	93,841	3,153	459,566	23.9	23.5	24.5	653	95,116
65-69	0.0110	0.0536	0.0496	0.0568	90,337	89,861	90,833	4,844	439,574	19.7	19.3	20.2	969	87,980
70-74	0.0189	0.0901	0.0831	0.0955	85,493	84,770	86,323	7,704	408,205	15.7	15.3	16.1	1,705	90,356
75-79	0.0356	0.1635	0.1510	0.1727	77,789	76,689	79,146	12,715	357,158	12.0	11.7	12.4	3,252	91,349
80-84	0.0659	0.2827	0.2639	0.2967	65,074	63,444	67,191	18,400	279,372	8.8	8.5	9.1	4,574	69,446
85-89	0.1156	0.4485	0.4267	0.4665	46,675	44,664	49,458	20,932	181,043	6.3	6.1	6.5	4,830	41,777
90-94	0.1926	0.6103	0.5924	0.6268	25,743	23,998	28,354	15,712	81,577	4.4	4.2	4.5	3,339	17,335
95-99	0.3045	0.7956	0.7831	0.8067	10,031	8,974	11,549	7,980	26,212	3.1	3.0	3.1	1,235	4,057
100+	0.4567	1	1	1	2,050	1,734	2,482	2,050	4,490	2.2	2.2	2.3	198	434
All ages													*22,662*	*2,256,393*

ln($_nq_x$) with uncertainty interval

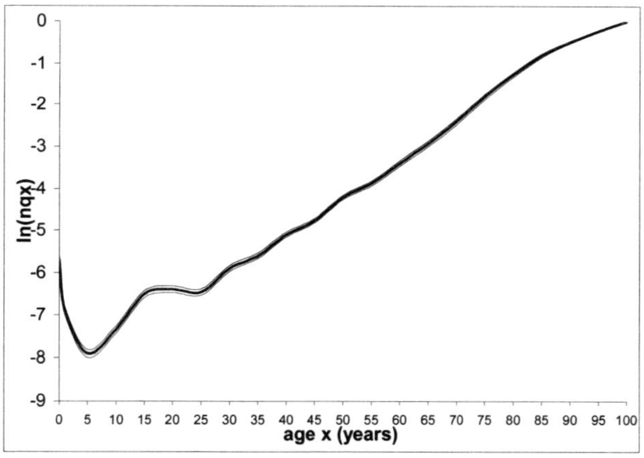

l_x with uncertainty interval

e_x with uncertainty interval

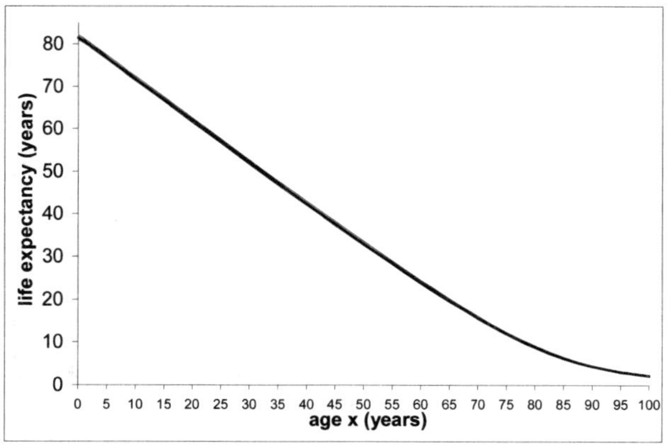

Distribution of deaths by age with uncertainty interval

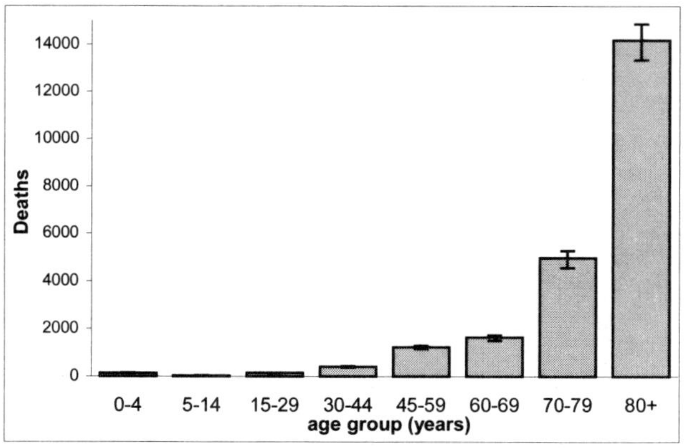

Oman — Oman — Omán

Males, 2000 — *Hommes, 2000* — *Hombres, 2000*

Age	$_nM_x$	$_nq_x$ mean	$_nq_x$ percentiles 2½	$_nq_x$ percentiles 97½	l_x mean	l_x percentiles 2½	l_x percentiles 97½	$_nd_x$	$_nL_x$	e_x mean	e_x percentiles 2½	e_x percentiles 97½	Actual deaths	Actual population
<1	0.0167	0.0165	0.0122	0.0208	100,000	100,000	100,000	1,645	98,519	69.5	68.5	70.7	737	44,152
1-4	0.0009	0.0036	0.0029	0.0043	98,355	97,924	98,782	355	392,568	69.7	68.7	70.9	146	161,968
5-9	0.0004	0.0022	0.0018	0.0026	98,000	97,504	98,496	214	489,466	65.9	64.9	67.2	83	189,131
10-14	0.0004	0.0021	0.0017	0.0025	97,786	97,253	98,323	203	488,422	61.0	60.1	62.3	72	173,785
15-19	0.0011	0.0056	0.0046	0.0065	97,583	97,015	98,157	544	486,554	56.2	55.2	57.4	155	138,592
20-24	0.0016	0.0078	0.0065	0.0091	97,039	96,383	97,709	760	483,294	51.5	50.5	52.7	171	108,963
25-29	0.0016	0.0080	0.0068	0.0092	96,279	95,510	97,071	768	479,473	46.9	45.9	48.1	138	86,275
30-34	0.0019	0.0094	0.0080	0.0107	95,510	94,638	96,415	894	475,318	42.2	41.3	43.5	154	81,962
35-39	0.0025	0.0124	0.0108	0.0141	94,617	93,636	95,641	1,176	470,143	37.6	36.6	38.8	180	71,820
40-44	0.0037	0.0183	0.0162	0.0205	93,441	92,320	94,604	1,713	462,921	33.0	32.1	34.2	255	68,911
45-49	0.0057	0.0283	0.0254	0.0312	91,728	90,427	93,066	2,597	452,147	28.6	27.6	29.8	442	76,903
50-54	0.0091	0.0444	0.0399	0.0487	89,131	87,602	90,679	3,957	435,762	24.4	23.4	25.5	494	54,402
55-59	0.0143	0.0690	0.0612	0.0759	85,174	83,380	86,937	5,873	411,186	20.4	19.5	21.5	496	34,705
60-64	0.0225	0.1064	0.0936	0.1177	79,301	77,258	81,305	8,439	375,407	16.7	15.8	17.7	525	23,337
65-69	0.0348	0.1600	0.1403	0.1784	70,862	68,584	73,199	11,340	325,959	13.4	12.6	14.3	471	13,537
70-74	0.0544	0.2394	0.2109	0.2671	59,522	56,598	62,925	14,248	261,989	10.5	9.7	11.3	517	9,498
75-79	0.0847	0.3495	0.3107	0.3892	45,274	41,755	49,629	15,825	186,805	8.0	7.4	8.6	462	5,449
80-84	0.1336	0.5007	0.4557	0.5465	29,448	25,745	34,036	14,744	110,383	5.9	5.5	6.4	335	2,507
85-89	0.2029	0.6731	0.6323	0.7141	14,705	11,860	18,309	9,897	48,780	4.3	4.1	4.6	154	757
90-94	0.2970	0.7852	0.7540	0.8169	4,807	3,391	6,731	3,775	12,712	3.1	3.0	3.3	42	140
95-99	0.4188	0.9280	0.9112	0.9441	1,032	621	1,656	958	2,288	2.3	2.3	2.4	5	13
100+	0.5689	1	1	1	74	35	147	74	131	1.8	1.7	1.9	1	1
All ages													*6,033*	*1,346,808*

ln($_nq_x$) with uncertainty interval

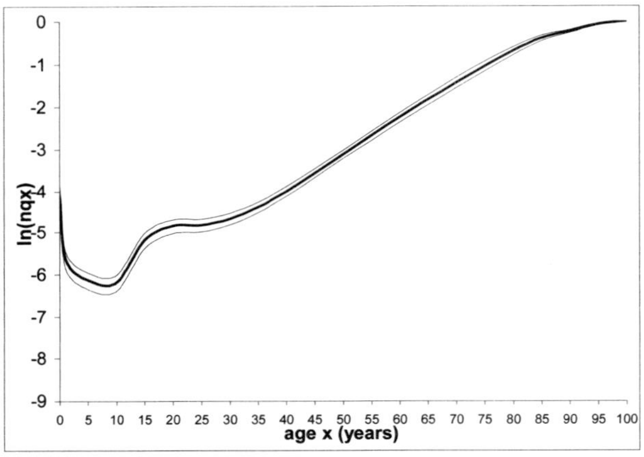

l_x with uncertainty interval

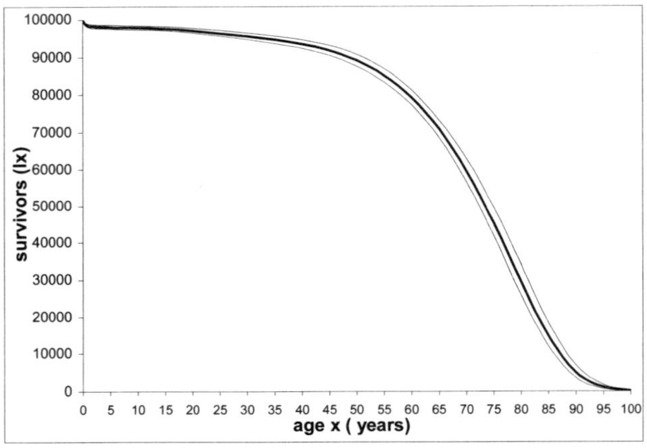

e_x with uncertainty interval

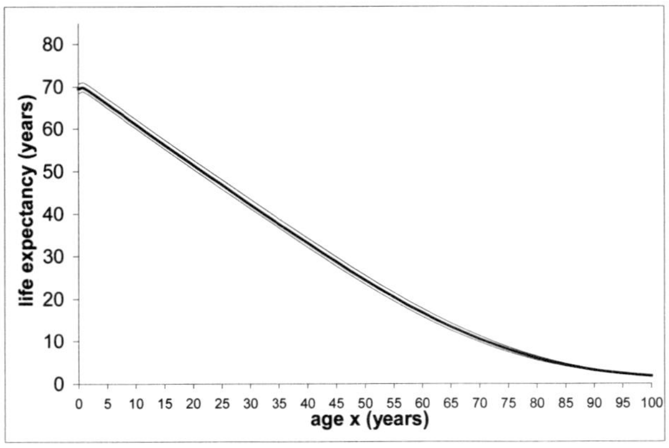

Distribution of deaths by age with uncertainty interval

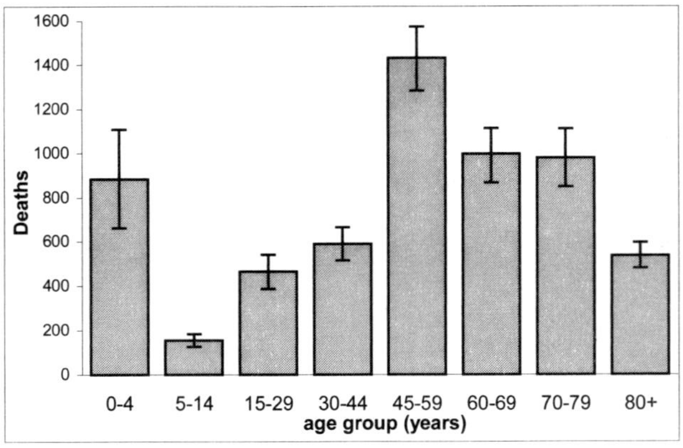

Oman Oman Omán

Females, 2000 *Femmes, 2000* *Mujeres, 2000*

Age	$_nM_x$	$_nq_x$ mean	$_nq_x$ percentiles 2½	$_nq_x$ percentiles 97½	l_x mean	l_x percentiles 2½	l_x percentiles 97½	$_nd_x$	$_nL_x$	e_x mean	e_x percentiles 2½	e_x percentiles 97½	Actual deaths	Actual population
<1	0.0137	0.0136	0.0104	0.0169	100,000	100,000	100,000	1,357	98,779	73.5	72.6	74.6	578	42,089
1-4	0.0010	0.0040	0.0033	0.0047	98,643	98,309	98,961	393	393,629	73.5	72.7	74.6	155	155,520
5-9	0.0004	0.0020	0.0017	0.0024	98,250	97,851	98,640	200	490,751	69.8	69.0	70.9	74	182,843
10-14	0.0004	0.0018	0.0015	0.0021	98,050	97,621	98,474	180	489,801	64.9	64.1	66.0	62	169,349
15-19	0.0007	0.0035	0.0029	0.0040	97,870	97,415	98,324	338	488,506	60.0	59.2	61.1	94	136,044
20-24	0.0008	0.0042	0.0036	0.0049	97,532	97,031	98,040	413	486,630	55.2	54.4	56.3	91	107,274
25-29	0.0010	0.0049	0.0042	0.0056	97,119	96,563	97,689	473	484,415	50.4	49.7	51.5	80	81,877
30-34	0.0013	0.0064	0.0056	0.0073	96,646	96,029	97,283	623	481,674	45.7	44.9	46.7	92	70,880
35-39	0.0019	0.0094	0.0082	0.0105	96,023	95,328	96,742	900	477,866	41.0	40.2	42.0	101	53,792
40-44	0.0029	0.0142	0.0126	0.0158	95,123	94,323	95,950	1,351	472,240	36.3	35.6	37.3	130	45,605
45-49	0.0043	0.0215	0.0193	0.0236	93,773	92,832	94,746	2,013	463,831	31.8	31.1	32.8	171	39,454
50-54	0.0065	0.0320	0.0289	0.0349	91,760	90,641	92,916	2,940	451,449	27.4	26.8	28.4	191	29,330
55-59	0.0096	0.0469	0.0420	0.0509	88,820	87,495	90,182	4,168	433,680	23.3	22.7	24.2	247	25,648
60-64	0.0147	0.0711	0.0635	0.0772	84,652	83,119	86,207	6,019	408,212	19.3	18.7	20.1	297	20,141
65-69	0.0230	0.1090	0.0972	0.1180	78,633	76,911	80,430	8,569	371,743	15.6	15.1	16.3	281	12,202
70-74	0.0381	0.1739	0.1560	0.1870	70,064	68,039	72,613	12,181	319,869	12.2	11.7	12.8	373	9,807
75-79	0.0632	0.2728	0.2483	0.2930	57,883	55,313	61,288	15,791	249,940	9.2	8.8	9.7	328	5,186
80-84	0.1073	0.4230	0.3946	0.4517	42,092	39,281	46,070	17,805	165,949	6.7	6.4	7.1	309	2,884
85-89	0.1734	0.6049	0.5760	0.6348	24,287	21,948	27,892	14,691	84,708	4.8	4.6	5.1	203	1,172
90-94	0.2669	0.7411	0.7163	0.7650	9,596	8,081	11,701	7,111	26,646	3.4	3.3	3.6	61	230
95-99	0.3909	0.8995	0.8846	0.9132	2,485	1,899	3,220	2,235	5,719	2.5	2.4	2.6	10	25
100+	0.5449	1	1	1	250	165	364	250	458	1.8	1.8	1.9	1	1
All ages													*3,931*	*1,191,353*

ln($_nq_x$) with uncertainty interval

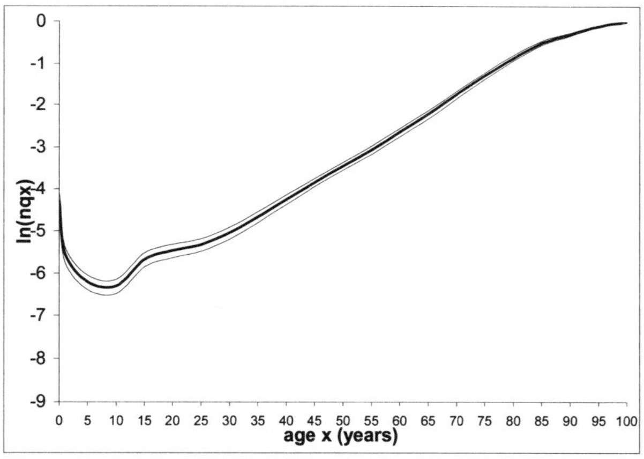

l_x with uncertainty interval

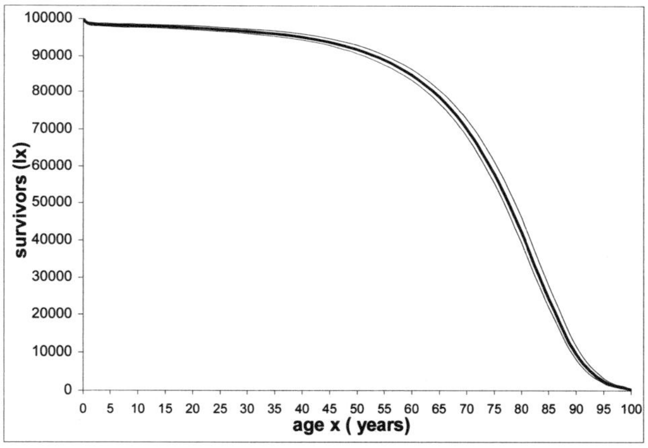

e_x with uncertainty interval

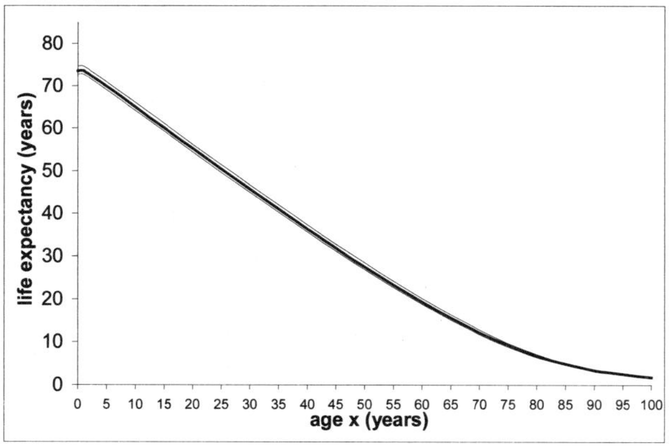

Distribution of deaths by age with uncertainty interval

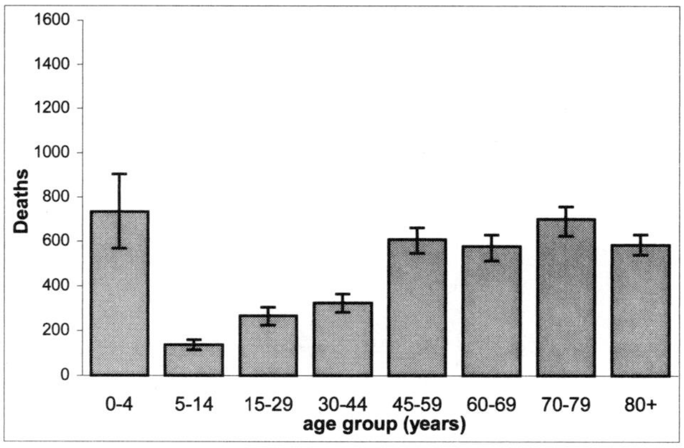

Pakistan Pakistan Pakistán

Males, 2000 *Hommes, 2000* *Hombres, 2000*

Age	$_nM_x$	$_nq_x$ mean	$_nq_x$ percentiles 2½	$_nq_x$ percentiles 97½	l_x mean	l_x percentiles 2½	l_x percentiles 97½	$_nd_x$	$_nL_x$	e_x mean	e_x percentiles 2½	e_x percentiles 97½	Actual deaths	Actual population
<1	0.1094	0.0996	0.0853	0.1139	100,000	100,000	100,000	9,960	91,036	60.1	58.1	62.3	262,647	2,400,649
1-4	0.0057	0.0225	0.0196	0.0254	90,040	88,609	91,468	2,023	355,305	65.8	64.3	67.5	51,390	9,026,576
5-9	0.0016	0.0081	0.0071	0.0091	88,017	86,519	89,516	713	438,303	63.3	61.7	65.0	16,394	10,075,998
10-14	0.0013	0.0064	0.0056	0.0072	87,304	85,773	88,857	557	435,129	58.8	57.2	60.4	11,395	8,908,883
15-19	0.0015	0.0075	0.0065	0.0084	86,748	85,178	88,341	648	432,117	54.1	52.6	55.8	11,467	7,644,458
20-24	0.0018	0.0090	0.0078	0.0101	86,099	84,479	87,742	771	428,568	49.5	48.0	51.2	11,331	6,295,057
25-29	0.0023	0.0112	0.0098	0.0126	85,328	83,635	87,056	955	424,253	44.9	43.4	46.6	11,478	5,101,228
30-34	0.0027	0.0134	0.0117	0.0151	84,373	82,581	86,205	1,131	419,038	40.4	38.9	42.0	11,740	4,348,016
35-39	0.0036	0.0178	0.0156	0.0201	83,242	81,333	85,195	1,485	412,497	35.9	34.5	37.5	14,210	3,947,156
40-44	0.0048	0.0237	0.0208	0.0268	81,757	79,701	83,865	1,939	403,938	31.5	30.1	33.1	17,092	3,560,814
45-49	0.0067	0.0329	0.0289	0.0372	79,818	77,578	82,121	2,630	392,516	27.2	25.8	28.8	20,362	3,039,179
50-54	0.0101	0.0491	0.0428	0.0553	77,188	74,715	79,744	3,788	376,472	23.1	21.7	24.6	23,613	2,347,022
55-59	0.0166	0.0798	0.0691	0.0912	73,401	70,624	76,291	5,859	352,355	19.1	17.8	20.6	30,728	1,847,921
60-64	0.0250	0.1176	0.1006	0.1364	67,541	64,374	70,880	7,946	317,842	15.6	14.4	17.0	37,236	1,489,433
65-69	0.0420	0.1900	0.1602	0.2218	59,595	56,005	63,396	11,326	269,663	12.3	11.2	13.6	46,507	1,107,312
70-74	0.0625	0.2704	0.2270	0.3144	48,270	44,247	52,602	13,054	208,714	9.6	8.7	10.7	46,514	743,720
75-79	0.0990	0.3968	0.3381	0.4514	35,216	30,520	40,185	13,973	141,146	7.3	6.6	8.1	41,493	419,117
80-84	0.1500	0.5455	0.4798	0.6010	21,243	16,744	26,387	11,587	77,246	5.4	4.9	6.0	31,497	209,977
85-89	0.2300	0.7302	0.6717	0.7756	9,656	6,687	13,727	7,050	30,653	3.9	3.6	4.3	18,958	82,426
90-94	0.3500	0.8537	0.8115	0.8849	2,606	1,507	4,507	2,224	6,355	2.7	2.6	3.0	8,028	22,937
95-99	0.4800	0.9836	0.9727	0.9900	381	175	850	375	781	2.1	2.0	2.1	1,967	4,098
100+	0.7800	1	1	1	6	2	23	6	8	1.3	1.3	1.6	384	492
All ages													*726,430*	*72,622,469*

$ln(_nq_x)$ with uncertainty interval

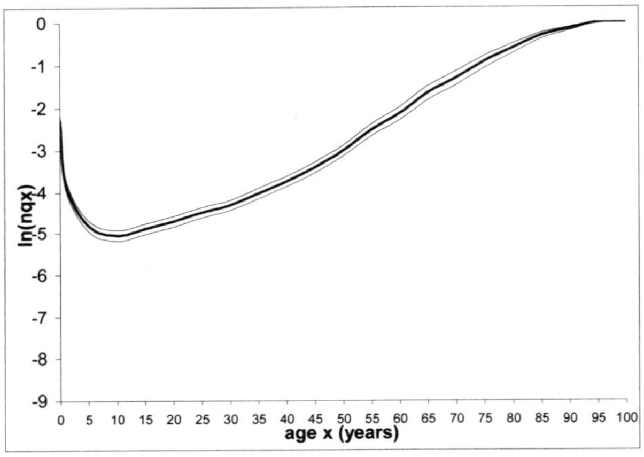

l_x with uncertainty interval

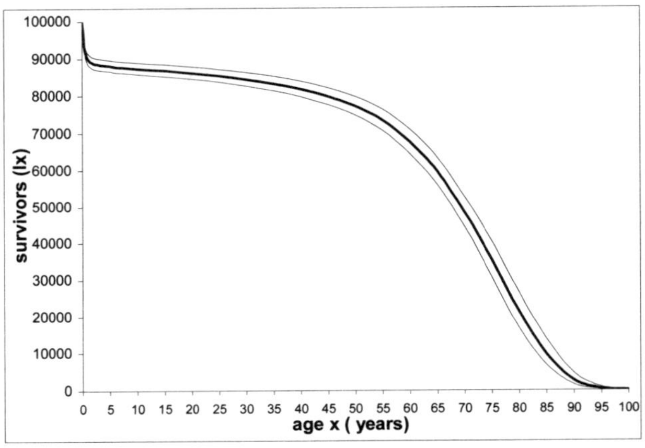

e_x with uncertainty interval

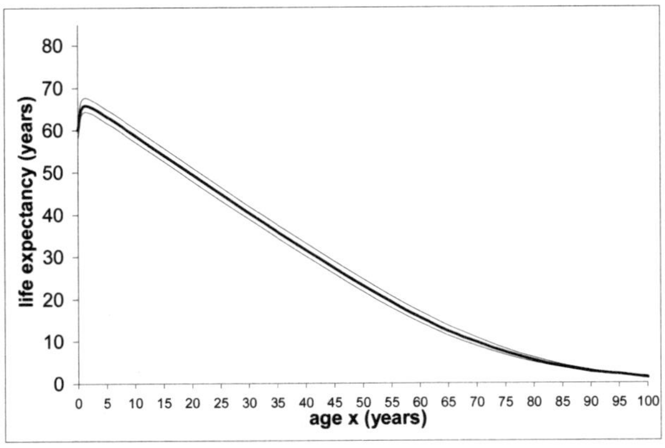

Distribution of deaths by age with uncertainty interval

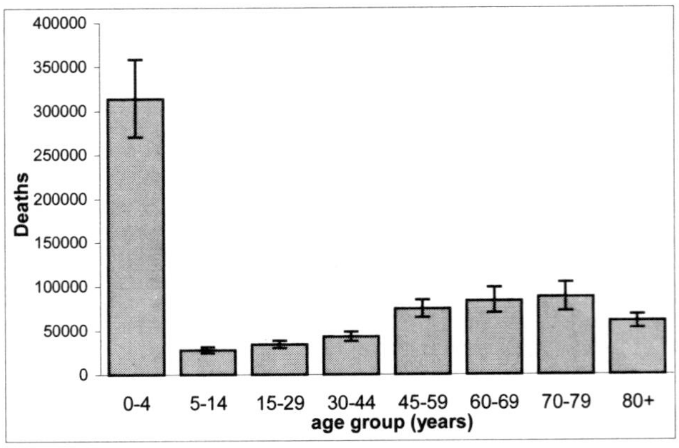

Pakistan **Pakistan** **Pakistán**

Females, 2000 *Femmes, 2000* *Mujeres, 2000*

Age	$_nM_x$	$_nq_x$			l_x			$_nd_x$	$_nL_x$	e_x			Actual deaths	Actual population
			percentiles			percentiles					percentiles			
		mean	2½	97½	mean	2½	97½			mean	2½	97½		
<1	0.1180	0.1067	0.0924	0.1221	100,000	100,000	100,000	10,667	90,400	60.7	58.4	63.2	272,381	2,308,317
1-4	0.0072	0.0281	0.0243	0.0320	89,333	87,789	90,763	2,512	351,303	67.0	65.1	69.3	60,595	8,474,834
5-9	0.0019	0.0095	0.0082	0.0108	86,821	85,329	88,317	821	432,053	64.9	63.0	67.2	17,972	9,459,185
10-14	0.0015	0.0075	0.0064	0.0085	86,000	84,422	87,596	643	428,394	60.5	58.6	62.8	12,550	8,366,687
15-19	0.0017	0.0085	0.0073	0.0097	85,358	83,702	87,031	722	424,982	55.9	54.0	58.2	12,234	7,196,582
20-24	0.0025	0.0124	0.0107	0.0142	84,635	82,893	86,396	1,051	420,547	51.4	49.5	53.6	15,050	6,019,888
25-29	0.0030	0.0149	0.0128	0.0170	83,584	81,717	85,470	1,244	414,807	47.0	45.1	49.2	15,071	5,023,657
30-34	0.0029	0.0144	0.0124	0.0164	82,339	80,328	84,373	1,185	408,733	42.7	40.8	44.8	12,420	4,282,693
35-39	0.0030	0.0149	0.0128	0.0170	81,154	79,007	83,326	1,208	402,749	38.2	36.4	40.4	11,133	3,711,043
40-44	0.0040	0.0198	0.0171	0.0226	79,946	77,664	82,256	1,583	395,771	33.8	32.0	35.9	12,735	3,183,837
45-49	0.0055	0.0271	0.0234	0.0310	78,363	75,909	80,851	2,126	386,499	29.4	27.7	31.5	14,596	2,653,783
50-54	0.0085	0.0416	0.0355	0.0478	76,237	73,561	78,958	3,173	373,253	25.2	23.5	27.2	17,912	2,107,306
55-59	0.0130	0.0630	0.0527	0.0735	73,064	70,078	76,115	4,600	353,822	21.1	19.5	23.1	23,148	1,780,619
60-64	0.0200	0.0952	0.0782	0.1131	68,465	65,078	71,956	6,520	326,022	17.4	15.9	19.2	29,204	1,460,215
65-69	0.0320	0.1481	0.1195	0.1781	61,944	58,099	65,968	9,177	286,778	14.0	12.6	15.7	35,175	1,099,232
70-74	0.0500	0.2222	0.1778	0.2676	52,767	48,514	57,335	11,726	234,521	11.0	9.8	12.4	40,030	800,603
75-79	0.0770	0.3229	0.2603	0.3828	41,041	35,902	46,376	13,250	172,080	8.4	7.4	9.6	32,411	420,918
80-84	0.1240	0.4733	0.3941	0.5416	27,791	22,158	34,305	13,153	106,072	6.2	5.5	7.1	24,396	196,738
85-89	0.1900	0.6441	0.5642	0.7056	14,638	10,157	20,784	9,428	49,620	4.5	4.1	5.1	13,001	68,427
90-94	0.3000	0.7895	0.7255	0.8352	5,210	2,990	9,057	4,113	13,711	3.1	2.8	3.4	4,953	16,510
95-99	0.4600	0.9664	0.9435	0.9794	1,097	493	2,486	1,060	2,304	2.2	2.1	2.3	1,103	2,397
100+	0.6700	1	1	1	37	10	140	37	55	1.5	1.5	1.7	165	246
All ages													*678,236*	*68,633,717*

ln($_nq_x$) with uncertainty interval

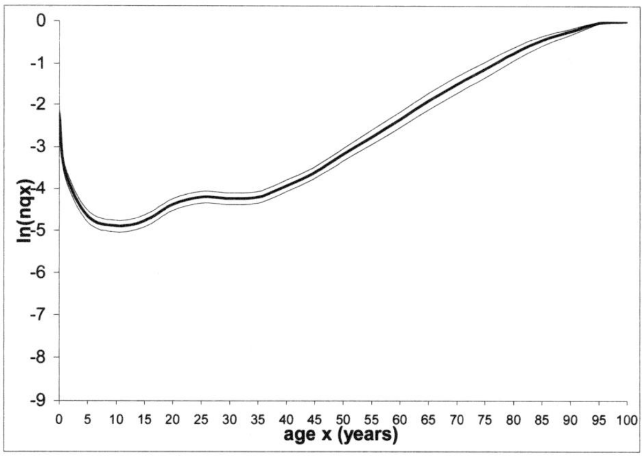

l_x with uncertainty interval

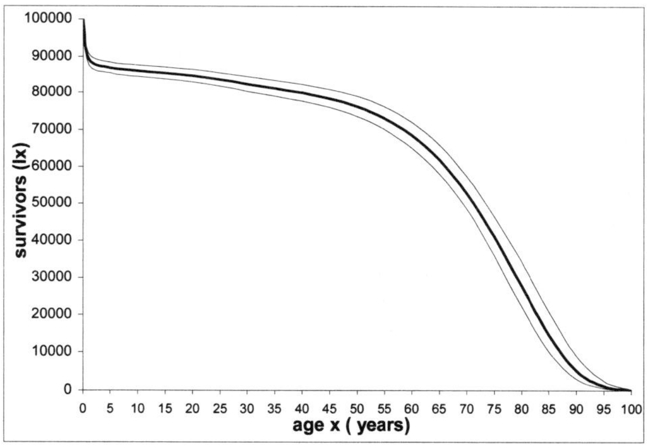

e_x with uncertainty interval

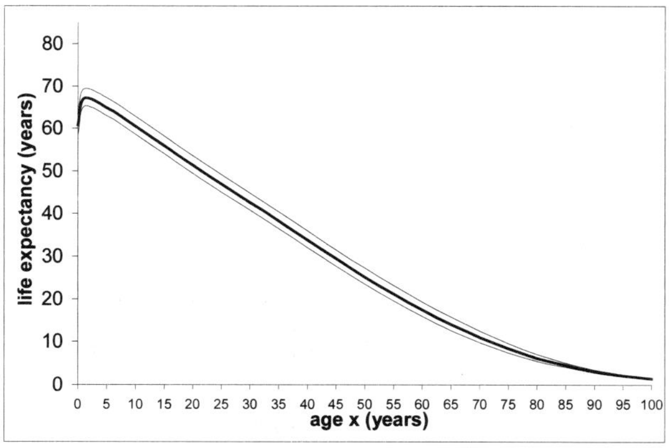

Distribution of deaths by age with uncertainty interval

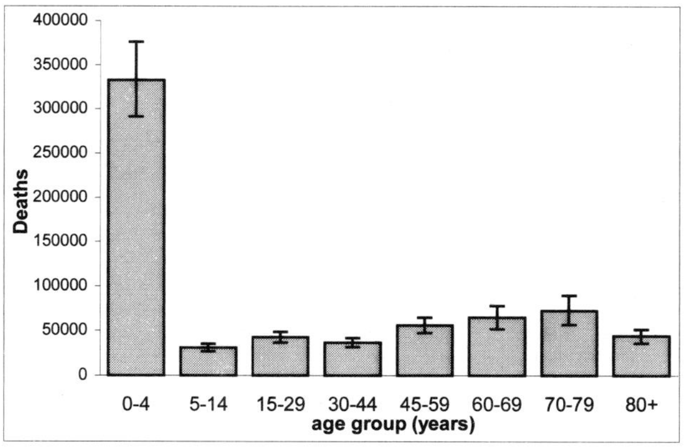

Palau   Palaos   Palau

Males, 2000   *Hommes, 2000*   *Hombres, 2000*

Age	$_nM_x$	$_nq_x$ mean	$_nq_x$ percentiles 2½	$_nq_x$ percentiles 97½	l_x mean	l_x percentiles 2½	l_x percentiles 97½	$_nd_x$	$_nL_x$	e_x mean	e_x percentiles 2½	e_x percentiles 97½	Actual deaths	Actual population
<1	0.0213	0.0209	0.0172	0.0259	100,000	100,000	100,000	2,091	98,119	64.7	63.6	65.2	6	281
1-4	0.0013	0.0052	0.0045	0.0061	97,909	97,408	98,279	505	390,427	65.1	64.2	65.6	1	1,126
5-9	0.0004	0.0020	0.0018	0.0024	97,405	96,814	97,839	196	486,535	61.4	60.5	62.0	1	1,312
10-14	0.0003	0.0014	0.0012	0.0016	97,209	96,584	97,668	134	485,712	56.5	55.7	57.1	0	1,099
15-19	0.0013	0.0067	0.0059	0.0078	97,076	96,429	97,550	647	483,761	51.6	50.8	52.2	1	956
20-24	0.0025	0.0123	0.0110	0.0142	96,429	95,678	96,977	1,183	479,187	46.9	46.1	47.5	2	783
25-29	0.0037	0.0184	0.0167	0.0210	95,246	94,321	95,914	1,753	471,849	42.5	41.7	43.1	3	711
30-34	0.0031	0.0154	0.0142	0.0174	93,494	92,341	94,311	1,441	463,864	38.2	37.5	38.9	2	727
35-39	0.0039	0.0195	0.0181	0.0217	92,052	90,734	92,972	1,791	455,783	33.8	33.1	34.4	2	619
40-44	0.0055	0.0271	0.0256	0.0300	90,261	88,761	91,286	2,449	445,184	29.4	28.8	30.0	3	544
45-49	0.0080	0.0392	0.0375	0.0429	87,812	86,100	88,949	3,444	430,453	25.2	24.6	25.8	4	464
50-54	0.0115	0.0561	0.0536	0.0609	84,369	82,405	85,610	4,734	410,009	21.1	20.5	21.7	4	385
55-59	0.0218	0.1032	0.0977	0.1112	79,635	77,391	81,020	8,218	377,629	17.2	16.7	17.8	6	276
60-64	0.0368	0.1686	0.1583	0.1798	71,417	68,784	73,109	12,039	326,986	13.9	13.4	14.4	8	224
65-69	0.0579	0.2528	0.2369	0.2680	59,377	56,413	61,534	15,010	259,362	11.2	10.7	11.7	10	172
70-74	0.0735	0.3105	0.2916	0.3295	44,367	41,295	46,954	13,774	187,400	9.1	8.7	9.6	10	137
75-79	0.1192	0.4591	0.4345	0.4827	30,593	27,688	33,263	14,045	117,851	7.1	6.8	7.5	9	78
80-84	0.1428	0.5261	0.5018	0.5486	16,548	14,323	18,809	8,706	60,974	6.0	5.8	6.3	4	31
85-89	0.1782	0.6163	0.5928	0.6370	7,842	6,465	9,372	4,833	27,126	5.0	4.8	5.2	2	11
90-94	0.2316	0.6833	0.6614	0.7019	3,009	2,347	3,816	2,056	8,876	3.9	3.8	4.1	1	2
95-99	0.3136	0.8080	0.7895	0.8232	953	700	1,292	770	2,455	3.0	2.9	3.1	0	0
100+	0.4424	1	1	1	183	124	272	183	414	2.3	2.0	2.3	0	0
All ages													*81*	*9,937*

$\ln(_nq_x)$ with uncertainty interval

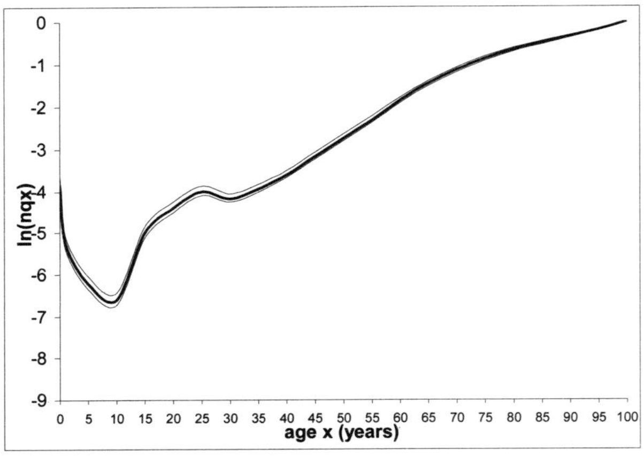

l_x with uncertainty interval

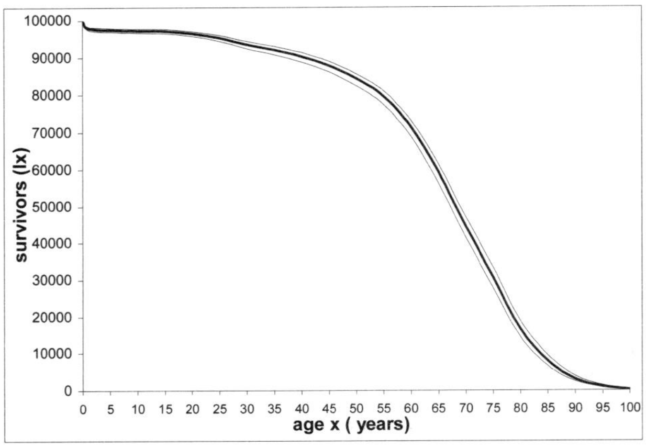

e_x with uncertainty interval

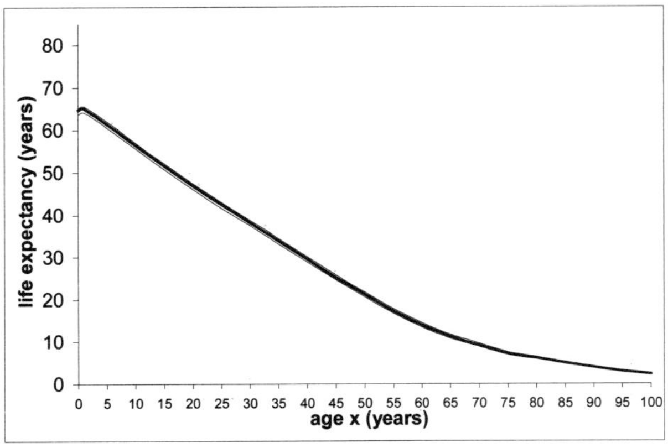

Distribution of deaths by age with uncertainty interval

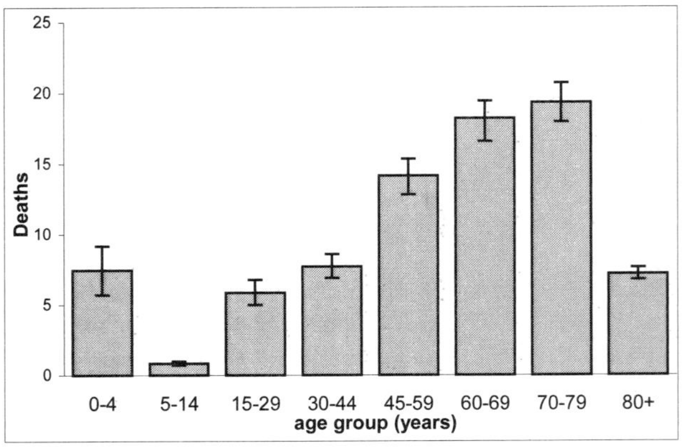

Palau | Palaos | Palau

Females, 2000 | *Femmes, 2000* | *Mujeres, 2000*

Age	$_nM_x$	$_nq_x$ mean	$_nq_x$ percentiles 2½	$_nq_x$ percentiles 97½	l_x mean	l_x percentiles 2½	l_x percentiles 97½	$_nd_x$	$_nL_x$	e_x mean	e_x percentiles 2½	e_x percentiles 97½	Actual deaths	Actual population
<1	0.0180	0.0177	0.0129	0.0227	100,000	100,000	100,000	1,773	98,404	69.3	68.3	70.3	5	266
1-4	0.0011	0.0046	0.0036	0.0055	98,227	97,727	98,711	447	391,834	69.5	68.6	70.5	1	1,059
5-9	0.0003	0.0015	0.0012	0.0018	97,780	97,191	98,358	150	488,525	65.8	65.0	66.8	0	1,235
10-14	0.0003	0.0013	0.0010	0.0015	97,630	97,013	98,238	124	487,840	60.9	60.1	61.9	0	1,035
15-19	0.0007	0.0033	0.0026	0.0039	97,506	96,866	98,139	320	486,729	56.0	55.2	57.0	1	895
20-24	0.0008	0.0038	0.0031	0.0045	97,186	96,489	97,880	368	485,009	51.2	50.4	52.2	1	753
25-29	0.0013	0.0065	0.0053	0.0076	96,818	96,059	97,580	627	482,522	46.3	45.6	47.3	1	670
30-34	0.0016	0.0080	0.0067	0.0093	96,191	95,331	97,062	770	479,028	41.6	40.9	42.6	1	596
35-39	0.0023	0.0116	0.0098	0.0133	95,420	94,445	96,414	1,106	474,337	37.0	36.2	37.9	1	564
40-44	0.0041	0.0203	0.0176	0.0230	94,314	93,189	95,469	1,918	466,777	32.4	31.6	33.3	2	500
45-49	0.0047	0.0230	0.0203	0.0256	92,396	91,045	93,790	2,125	456,670	28.0	27.3	28.9	2	423
50-54	0.0092	0.0450	0.0405	0.0493	90,271	88,711	91,884	4,059	441,209	23.6	22.9	24.5	3	342
55-59	0.0158	0.0760	0.0687	0.0828	86,212	84,341	88,163	6,550	414,686	19.6	19.0	20.4	4	246
60-64	0.0211	0.1003	0.0905	0.1087	79,662	77,355	82,109	7,987	378,343	16.0	15.5	16.7	5	217
65-69	0.0459	0.2060	0.1873	0.2203	71,675	68,950	74,681	14,768	321,454	12.5	12.0	13.1	8	172
70-74	0.0643	0.2770	0.2549	0.2955	56,907	53,762	60,695	15,764	245,122	10.1	9.6	10.6	9	132
75-79	0.0946	0.3824	0.3578	0.4068	41,142	37,876	45,226	15,734	166,375	8.0	7.6	8.4	8	81
80-84	0.1295	0.4892	0.4614	0.5163	25,408	22,468	29,045	12,429	95,967	6.3	6.0	6.7	5	38
85-89	0.1758	0.6106	0.5831	0.6389	12,979	10,869	15,644	7,925	45,084	5.0	4.8	5.3	2	13
90-94	0.2300	0.6805	0.6551	0.7068	5,054	3,924	6,522	3,439	14,953	3.9	3.7	4.1	1	3
95-99	0.3100	0.8031	0.7819	0.8248	1,615	1,150	2,249	1,297	4,184	3.1	2.9	3.2	0	0
100+	0.4168	1	1	1	318	202	491	318	763	2.4	2.2	2.5	0	0
All ages													*59*	*9,241*

ln($_nq_x$) with uncertainty interval

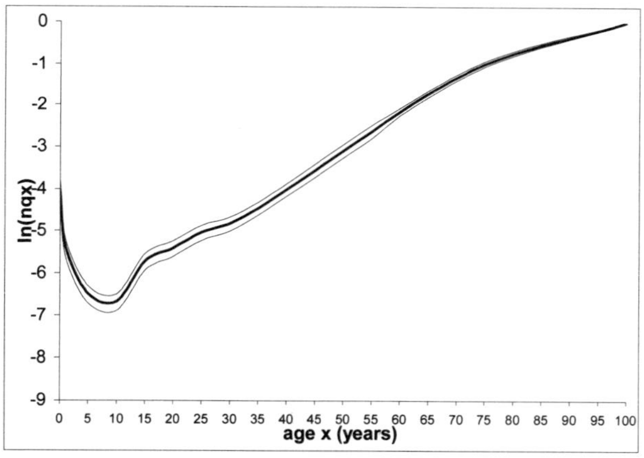

l_x with uncertainty interval

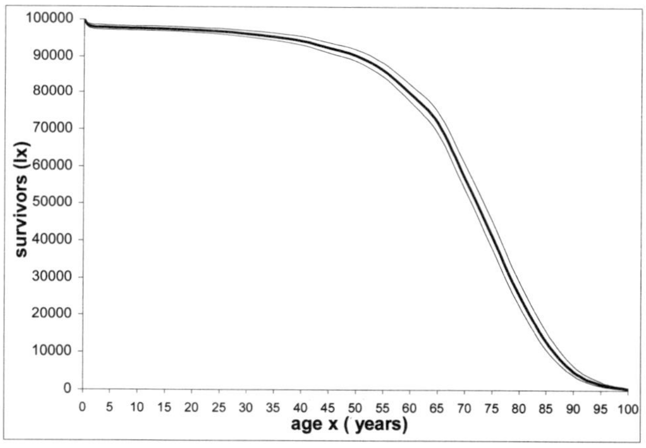

e_x with uncertainty interval

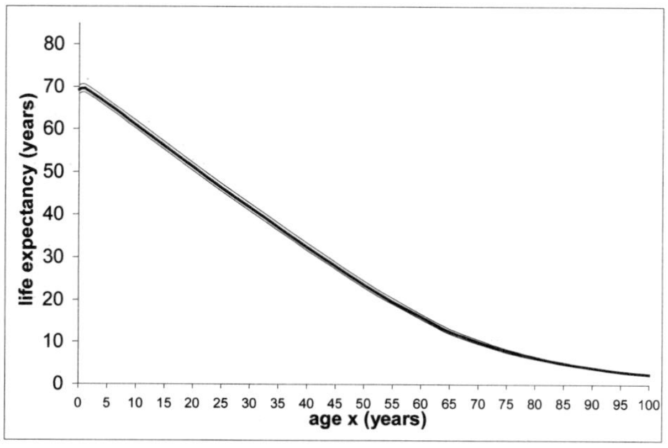

Distribution of deaths by age with uncertainty interval

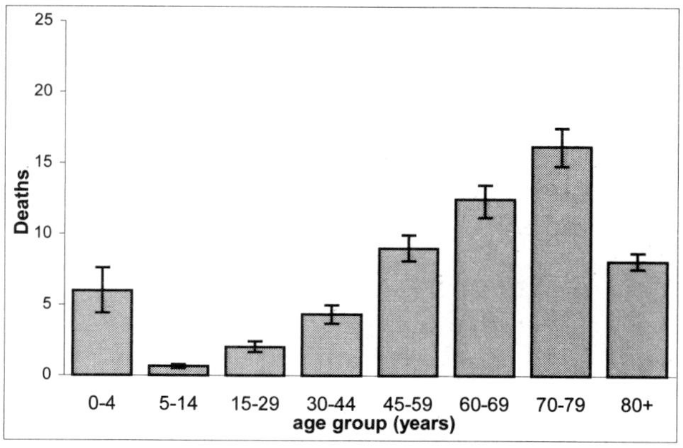

Panama | Panama | Panamá

Males, 2000 | *Hommes, 2000* | *Hombres, 2000*

Age	$_nM_x$	$_nq_x$			l_x			$_nd_x$	$_nL_x$	e_x			Actual deaths	Actual population
			percentiles			percentiles					percentiles			
		mean	2 1/2	97 1/2	mean	2 1/2	97 1/2			mean	2 1/2	97 1/2		
<1	0.0212	0.0208	0.0179	0.0237	100,000	100,000	100,000	2,082	98,126	71.5	70.7	72.2	653	30,770
1-4	0.0017	0.0069	0.0062	0.0076	97,918	97,634	98,211	674	390,054	72.0	71.2	72.8	213	123,376
5-9	0.0005	0.0026	0.0023	0.0029	97,244	96,891	97,604	253	485,588	68.5	67.7	69.3	80	153,985
10-14	0.0004	0.0019	0.0018	0.0021	96,991	96,613	97,375	189	484,483	63.7	62.9	64.5	58	148,578
15-19	0.0010	0.0052	0.0047	0.0057	96,802	96,406	97,203	503	482,753	58.8	58.0	59.6	144	137,804
20-24	0.0016	0.0081	0.0074	0.0089	96,299	95,855	96,745	781	479,542	54.1	53.3	54.9	211	129,239
25-29	0.0021	0.0105	0.0096	0.0115	95,518	95,003	96,029	1,001	475,086	49.5	48.8	50.3	265	125,921
30-34	0.0027	0.0135	0.0125	0.0147	94,517	93,915	95,104	1,273	469,402	45.0	44.3	45.8	314	115,884
35-39	0.0029	0.0144	0.0134	0.0156	93,244	92,539	93,920	1,343	462,862	40.6	39.9	41.3	294	101,421
40-44	0.0029	0.0144	0.0135	0.0156	91,901	91,094	92,660	1,325	456,192	36.2	35.5	36.9	246	84,701
45-49	0.0042	0.0206	0.0193	0.0222	90,576	89,675	91,410	1,868	448,210	31.6	31.0	32.4	293	70,371
50-54	0.0058	0.0285	0.0265	0.0308	88,708	87,682	89,646	2,529	437,218	27.3	26.7	27.9	337	58,279
55-59	0.0080	0.0390	0.0361	0.0422	86,179	84,984	87,268	3,365	422,483	23.0	22.4	23.6	377	47,379
60-64	0.0157	0.0753	0.0692	0.0816	82,814	81,395	84,117	6,237	398,479	18.8	18.3	19.4	574	36,664
65-69	0.0216	0.1025	0.0938	0.1112	76,577	74,751	78,293	7,848	363,266	15.2	14.7	15.7	599	27,712
70-74	0.0341	0.1570	0.1436	0.1697	68,729	66,442	70,952	10,793	316,663	11.6	11.2	12.1	733	21,495
75-79	0.0650	0.2797	0.2572	0.3004	57,936	55,164	60,763	16,204	249,170	8.3	8.0	8.7	921	14,159
80-84	0.1407	0.5204	0.4875	0.5497	41,732	38,594	45,134	21,715	154,372	5.5	5.3	5.8	1,170	8,315
85-89	0.2350	0.7402	0.7129	0.7641	20,017	17,378	23,132	14,817	63,040	3.8	3.7	4.0	855	3,639
90-94	0.3640	0.8700	0.8533	0.8846	5,199	4,100	6,642	4,524	12,426	2.7	2.6	2.7	345	947
95-99	0.4900	0.9919	0.9895	0.9937	676	473	975	670	1,368	2.0	2.0	2.1	74	150
100+	0.6400	1	1	1	5	3	10	5	9	1.6	1.5	2.0	8	12
All ages													*8,763*	*1,440,801*

$\ln(_nq_x)$ with uncertainty interval

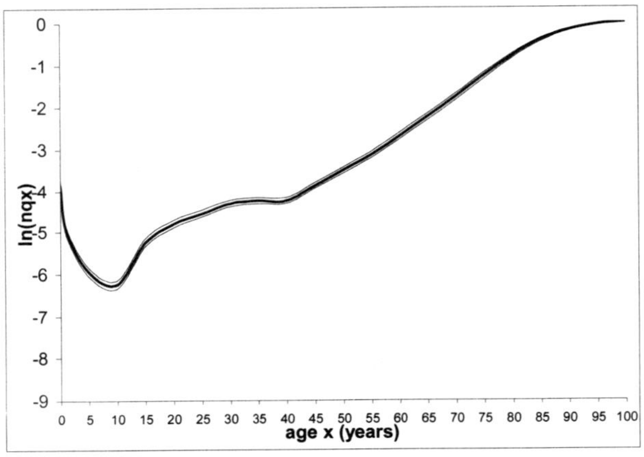

l_x with uncertainty interval

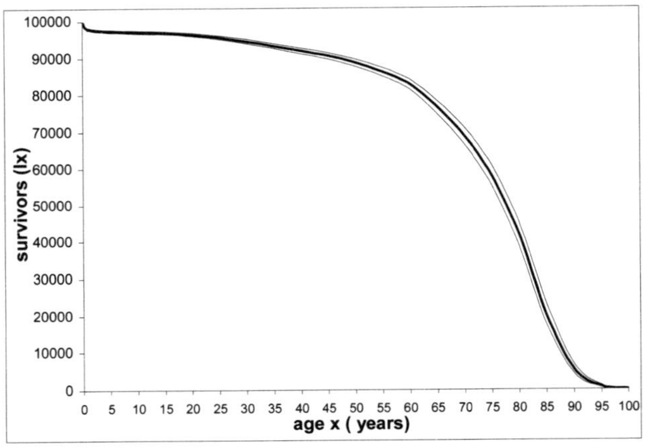

e_x with uncertainty interval

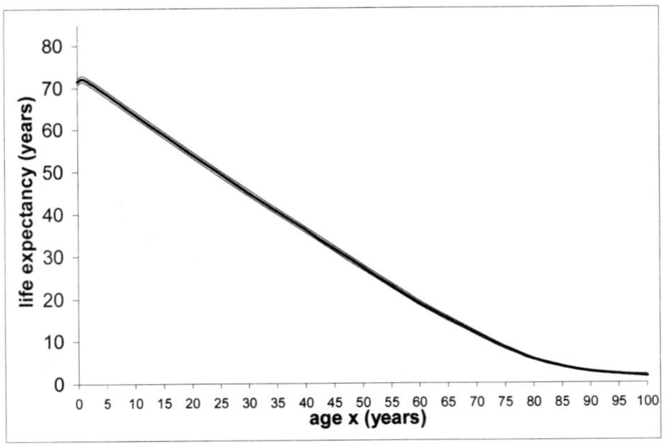

Distribution of deaths by age with uncertainty interval

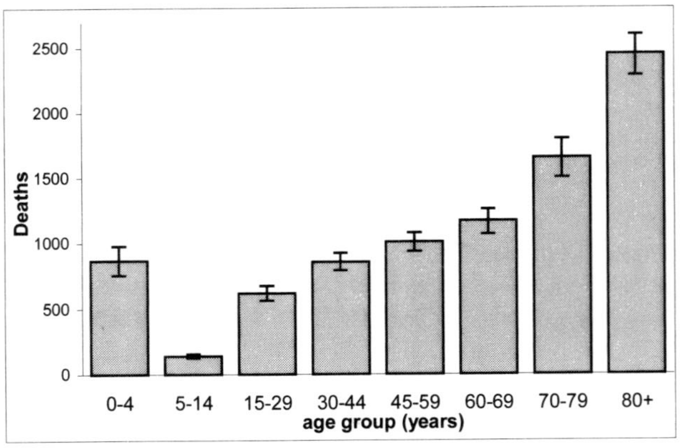

Panama

Panama

Panamá

Females, 2000

Femmes, 2000

Mujeres, 2000

Age	$_nM_x$	$_nq_x$ mean	$_nq_x$ percentiles 2½	$_nq_x$ percentiles 97½	l_x mean	l_x percentiles 2½	l_x percentiles 97½	$_nd_x$	$_nL_x$	e_x mean	e_x percentiles 2½	e_x percentiles 97½	Actual deaths	Actual population
<1	0.0165	0.0162	0.0152	0.0189	100,000	100,000	100,000	1,624	98,538	76.3	75.8	76.7	483	29,305
1-4	0.0016	0.0062	0.0059	0.0069	98,376	98,115	98,483	612	392,034	76.6	76.1	77.0	184	117,928
5-9	0.0003	0.0016	0.0016	0.0018	97,764	97,439	97,901	159	488,420	73.1	72.6	73.5	48	146,864
10-14	0.0003	0.0017	0.0016	0.0018	97,604	97,263	97,748	164	487,613	68.2	67.7	68.6	48	142,911
15-19	0.0006	0.0028	0.0027	0.0031	97,441	97,085	97,592	273	486,521	63.3	62.8	63.7	75	133,879
20-24	0.0008	0.0041	0.0039	0.0044	97,168	96,787	97,331	397	484,846	58.5	58.0	58.9	103	125,537
25-29	0.0009	0.0047	0.0045	0.0051	96,771	96,357	96,951	454	482,718	53.7	53.2	54.1	117	124,228
30-34	0.0010	0.0051	0.0049	0.0055	96,317	95,867	96,515	495	480,346	48.9	48.5	49.4	118	114,949
35-39	0.0014	0.0069	0.0066	0.0074	95,822	95,336	96,038	660	477,459	44.2	43.7	44.6	141	102,026
40-44	0.0022	0.0107	0.0104	0.0114	95,162	94,632	95,402	1,022	473,255	39.5	39.0	39.9	185	85,478
45-49	0.0027	0.0136	0.0131	0.0144	94,140	93,551	94,414	1,281	467,498	34.9	34.4	35.3	193	70,313
50-54	0.0043	0.0212	0.0205	0.0224	92,859	92,205	93,173	1,973	459,363	30.3	29.9	30.7	245	57,064
55-59	0.0056	0.0276	0.0265	0.0292	90,886	90,139	91,265	2,510	448,157	25.9	25.5	26.3	259	46,305
60-64	0.0098	0.0477	0.0455	0.0505	88,377	87,509	88,851	4,217	431,340	21.6	21.2	22.0	356	36,423
65-69	0.0166	0.0798	0.0757	0.0845	84,159	83,089	84,808	6,719	404,000	17.5	17.2	17.9	458	27,523
70-74	0.0248	0.1167	0.1103	0.1233	77,441	76,070	78,387	9,039	364,607	13.8	13.5	14.1	555	22,404
75-79	0.0402	0.1827	0.1727	0.1920	68,402	66,690	69,741	12,497	310,767	10.3	10.1	10.6	647	16,091
80-84	0.0896	0.3661	0.3482	0.3822	55,905	53,883	57,700	20,468	228,354	7.1	6.9	7.3	865	9,656
85-89	0.1745	0.6076	0.5844	0.6248	35,437	33,289	37,609	21,531	123,357	4.7	4.6	4.9	779	4,465
90-94	0.2969	0.7852	0.7667	0.7968	13,906	12,489	15,632	10,919	36,774	3.1	3.1	3.2	392	1,320
95-99	0.4413	0.9495	0.9420	0.9536	2,987	2,538	3,646	2,836	6,427	2.2	2.2	2.3	92	209
100+	0.5730	1	1	1	151	118	211	151	263	1.7	1.7	2.0	9	15
All ages													*6,353*	*1,414,893*

ln($_nq_x$) with uncertainty interval

l_x with uncertainty interval

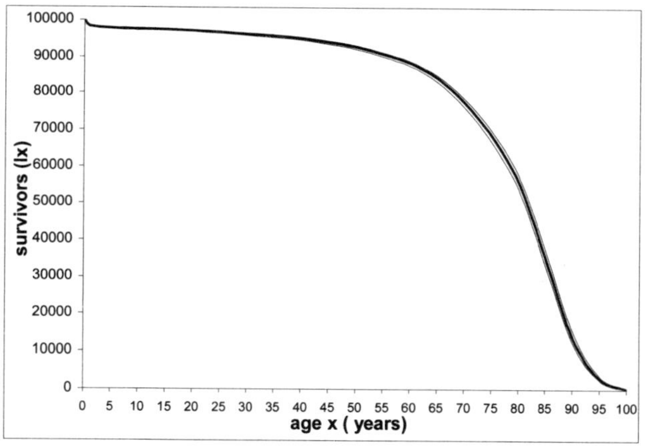

e_x with uncertainty interval

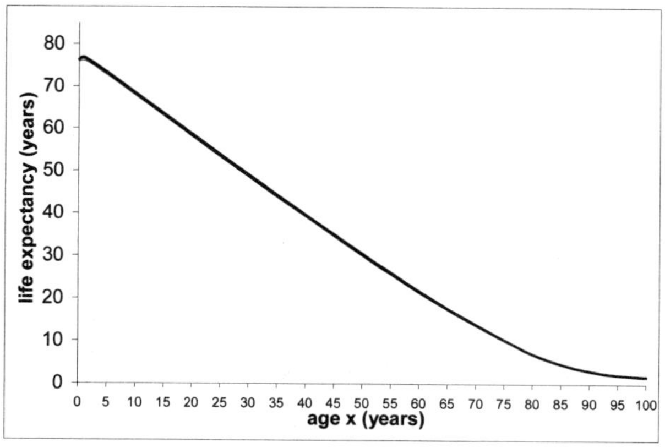

Distribution of deaths by age with uncertainty interval

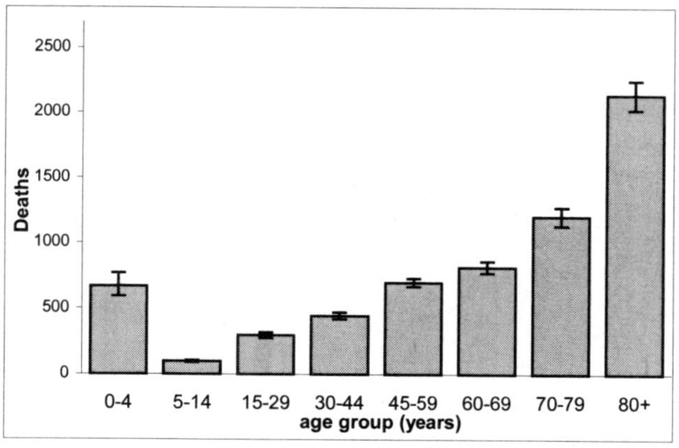

Papua New Guinea **Papouasie-Nouvelle-Guinée** **Papua Nueva Guinea**

Males, 2000 *Hommes, 2000* *Hombres, 2000*

Age	$_nM_x$	$_nq_x$			l_x			$_nd_x$	$_nL_x$	e_x			Actual deaths	Actual population
			percentiles			percentiles					percentiles			
		mean	2 1/2	97 1/2	mean	2 1/2	97 1/2			mean	2 1/2	97 1/2		
<1	0.0882	0.0818	0.0586	0.0967	100,000	100,000	100,000	8,176	92,642	55.1	52.6	57.5	6,777	76,797
1-4	0.0100	0.0390	0.0321	0.0449	91,824	90,330	94,136	3,586	358,692	59.0	57.0	60.4	2,893	289,435
5-9	0.0031	0.0154	0.0129	0.0177	88,239	86,277	91,114	1,356	437,803	57.3	55.6	58.7	1,078	347,866
10-14	0.0023	0.0115	0.0097	0.0132	86,883	84,752	89,939	999	431,916	53.2	51.5	54.6	664	287,093
15-19	0.0030	0.0148	0.0126	0.0170	85,884	83,631	89,064	1,270	426,245	48.8	47.2	50.2	866	290,739
20-24	0.0043	0.0212	0.0183	0.0244	84,614	82,211	87,939	1,797	418,579	44.5	42.9	45.9	1,036	241,428
25-29	0.0045	0.0223	0.0195	0.0255	82,817	80,209	86,326	1,845	409,475	40.4	38.9	41.8	962	213,553
30-34	0.0054	0.0267	0.0237	0.0305	80,973	78,163	84,645	2,164	399,455	36.2	34.8	37.7	904	166,833
35-39	0.0075	0.0367	0.0330	0.0418	78,809	75,776	82,643	2,895	386,809	32.2	30.8	33.6	1,120	149,616
40-44	0.0109	0.0531	0.0483	0.0601	75,914	72,606	79,917	4,028	369,502	28.3	27.0	29.7	1,301	119,356
45-49	0.0143	0.0688	0.0631	0.0775	71,886	68,239	76,060	4,946	347,068	24.8	23.5	26.1	1,374	96,405
50-54	0.0171	0.0820	0.0741	0.0920	66,941	62,952	71,260	5,489	320,982	21.4	20.2	22.7	1,191	69,665
55-59	0.0220	0.1042	0.0931	0.1167	61,452	57,159	65,977	6,405	291,247	18.1	17.0	19.3	1,251	56,902
60-64	0.0315	0.1459	0.1291	0.1637	55,047	50,490	59,833	8,029	255,163	14.9	13.9	16.1	1,317	41,860
65-69	0.0512	0.2271	0.2004	0.2547	47,018	42,225	52,107	10,678	208,395	12.0	11.1	13.1	1,351	26,363
70-74	0.0687	0.2931	0.2590	0.3271	36,340	31,468	41,666	10,653	155,067	9.8	9.1	10.7	1,265	18,410
75-79	0.0968	0.3896	0.3480	0.4294	25,687	21,176	30,875	10,007	103,416	7.8	7.2	8.6	865	8,936
80-84	0.1285	0.4863	0.4414	0.5312	15,680	12,083	20,132	7,624	59,337	6.3	5.8	6.8	457	3,553
85-89	0.1777	0.6151	0.5705	0.6605	8,055	5,665	11,245	4,955	27,889	4.8	4.5	5.2	243	1,366
90-94	0.2559	0.7238	0.6838	0.7645	3,100	1,923	4,829	2,244	8,769	3.5	3.3	3.8	73	287
95-99	0.3838	0.8919	0.8654	0.9172	856	453	1,527	764	1,990	2.5	2.4	2.7	14	36
100+	0.5994	1	1	1	93	38	206	93	154	1.7	1.6	1.8	1	2
All ages													*27,002*	*2,506,501*

ln($_nq_x$) with uncertainty interval

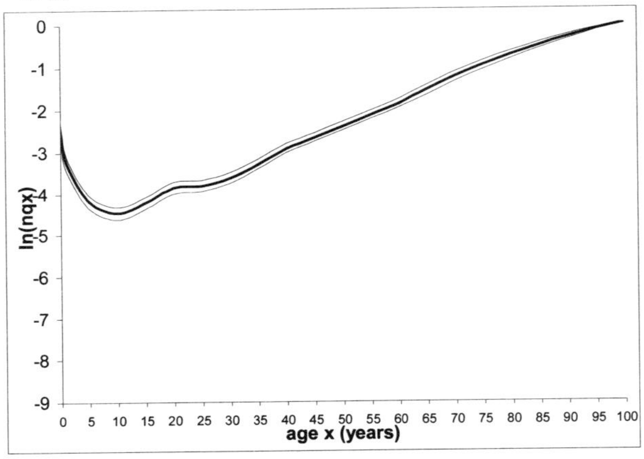

l_x with uncertainty interval

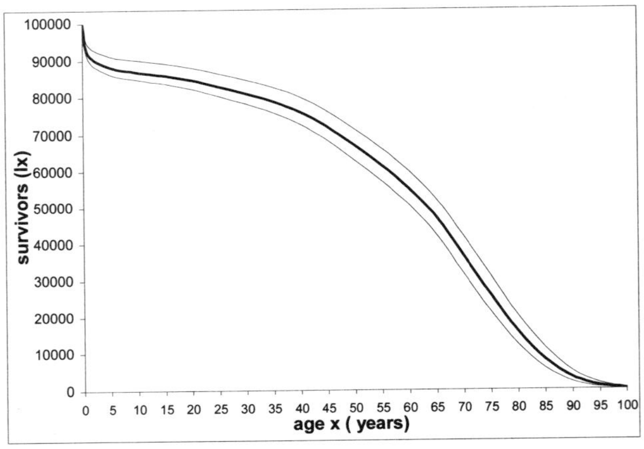

e_x with uncertainty interval

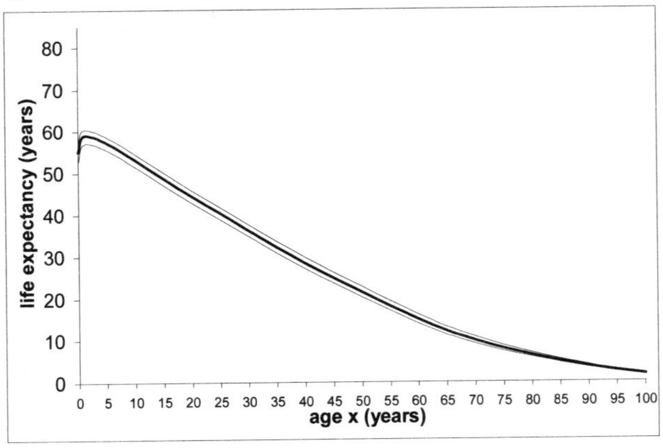

Distribution of deaths by age with uncertainty interval

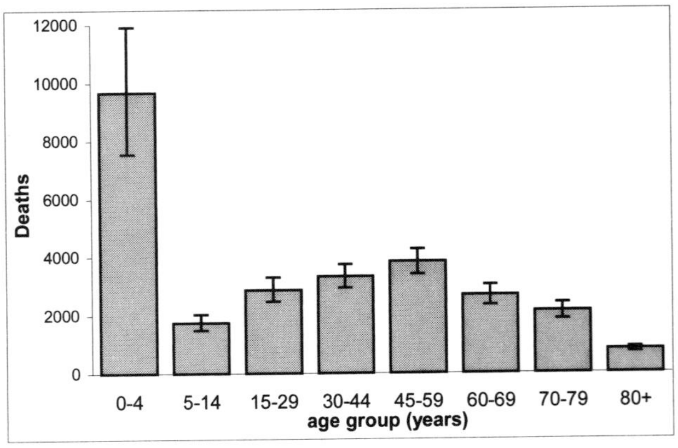

Papua New Guinea **Papouasie-Nouvelle-Guinée** **Papua Nueva Guinea**

Females, 2000 *Femmes, 2000* *Mujeres, 2000*

Age	$_nM_x$	$_nq_x$			l_x			$_nd_x$	$_nL_x$	e_x			Actual deaths	Actual population
			percentiles			percentiles					percentiles			
		mean	2½	97½	mean	2½	97½			mean	2½	97½		
<1	0.0790	0.0737	0.0484	0.0877	100,000	100,000	100,000	7,371	93,366	57.5	54.8	60.0	5,747	72,794
1-4	0.0097	0.0379	0.0296	0.0436	92,629	91,225	95,164	3,513	362,084	61.0	58.6	62.3	2,634	271,530
5-9	0.0033	0.0164	0.0134	0.0190	89,116	87,246	92,345	1,464	441,920	59.4	57.1	60.6	1,083	327,094
10-14	0.0027	0.0136	0.0113	0.0158	87,652	85,592	91,109	1,196	435,269	55.3	53.2	56.5	705	256,405
15-19	0.0026	0.0131	0.0110	0.0151	86,456	84,244	90,080	1,128	429,458	51.0	49.0	52.3	665	253,081
20-24	0.0040	0.0199	0.0170	0.0230	85,327	82,974	89,094	1,700	422,387	46.7	44.7	47.9	807	200,523
25-29	0.0051	0.0251	0.0218	0.0289	83,628	81,068	87,583	2,099	412,889	42.6	40.7	43.8	934	183,789
30-34	0.0062	0.0303	0.0268	0.0349	81,528	78,721	85,674	2,470	401,467	38.6	36.8	39.8	973	158,157
35-39	0.0078	0.0383	0.0347	0.0442	79,059	75,978	83,375	3,031	387,714	34.7	33.0	36.0	1,151	147,180
40-44	0.0097	0.0474	0.0439	0.0547	76,027	72,619	80,479	3,603	371,127	31.0	29.4	32.2	1,130	116,412
45-49	0.0111	0.0538	0.0498	0.0620	72,424	68,650	76,947	3,894	352,385	27.4	25.9	28.6	1,040	94,125
50-54	0.0145	0.0697	0.0642	0.0803	68,530	64,393	73,114	4,779	330,704	23.9	22.4	25.0	985	68,158
55-59	0.0190	0.0907	0.0830	0.1039	63,751	59,225	68,420	5,782	304,302	20.5	19.1	21.6	1,048	55,151
60-64	0.0250	0.1176	0.1065	0.1348	57,970	53,070	62,744	6,820	272,798	17.3	16.0	18.3	999	39,976
65-69	0.0367	0.1679	0.1508	0.1932	51,150	45,914	56,064	8,589	234,276	14.2	13.1	15.2	945	25,790
70-74	0.0496	0.2208	0.1974	0.2538	42,561	37,041	47,610	9,398	189,310	11.6	10.6	12.5	929	18,712
75-79	0.0747	0.3149	0.2826	0.3576	33,163	27,640	38,210	10,442	139,709	9.2	8.3	9.9	663	8,873
80-84	0.1095	0.4299	0.3904	0.4811	22,721	17,756	27,412	9,767	89,184	7.2	6.5	7.8	378	3,449
85-89	0.1400	0.5185	0.4774	0.5731	12,953	9,214	16,711	6,716	47,975	5.8	5.2	6.2	172	1,232
90-94	0.2000	0.6250	0.5850	0.6783	6,237	3,933	8,733	3,898	19,490	4.3	3.9	4.6	49	243
95-99	0.3000	0.7895	0.7559	0.8330	2,339	1,266	3,624	1,846	6,155	3.1	2.9	3.3	11	37
100+	0.4500	1	1	1	492	211	885	492	1,094	2.2	2.2	2.4	1	3
All ages													*23,051*	*2,302,714*

ln($_nq_x$) with uncertainty interval

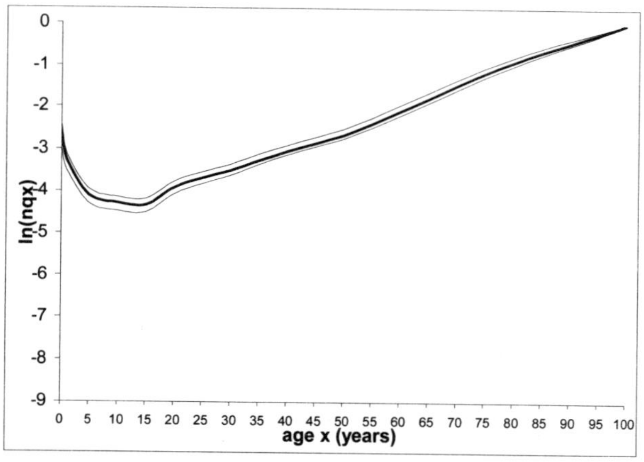

l_x with uncertainty interval

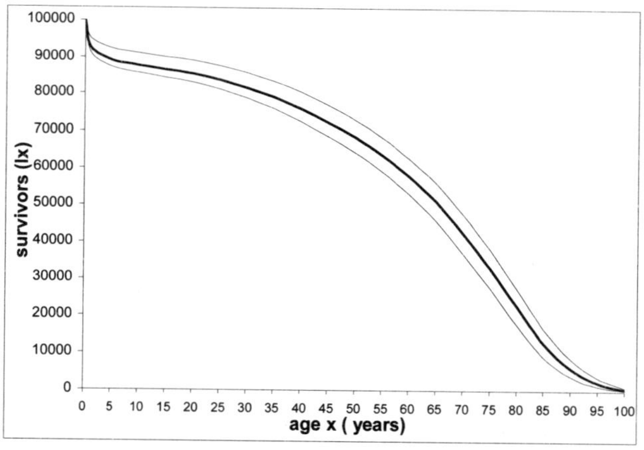

e_x with uncertainty interval

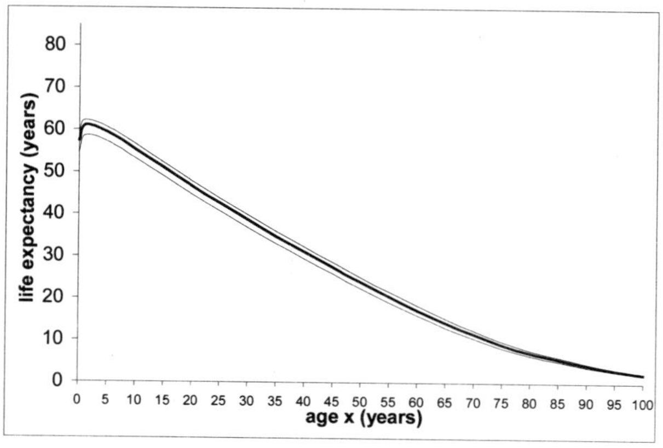

Distribution of deaths by age with uncertainty interval

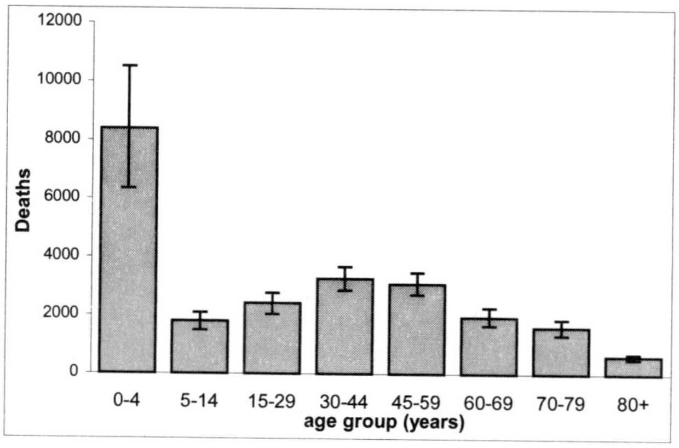

Paraguay | Paraguay | Paraguay

Males, 2000 | *Hommes, 2000* | *Hombres, 2000*

Age	$_nM_x$	$_nq_x$			l_x			$_nd_x$	$_nL_x$	e_x			Actual deaths	Actual population
			percentiles			percentiles					percentiles			
		mean	2 1/2	97 1/2	mean	2 1/2	97 1/2			mean	2 1/2	97 1/2		
<1	0.0276	0.0269	0.0212	0.0298	100,000	100,000	100,000	2,690	97,579	70.2	69.0	71.6	2,267	82,241
1-4	0.0020	0.0081	0.0068	0.0088	97,310	97,021	97,877	787	387,350	71.2	70.0	72.6	634	312,110
5-9	0.0005	0.0024	0.0020	0.0026	96,523	96,165	97,215	232	482,034	67.7	66.6	69.2	178	370,720
10-14	0.0005	0.0026	0.0022	0.0028	96,291	95,913	97,018	246	480,838	62.9	61.8	64.3	175	340,607
15-19	0.0014	0.0069	0.0059	0.0076	96,044	95,644	96,809	664	478,561	58.0	57.0	59.5	423	304,657
20-24	0.0019	0.0092	0.0079	0.0101	95,380	94,921	96,239	881	474,699	53.4	52.4	54.8	452	243,889
25-29	0.0019	0.0097	0.0084	0.0106	94,499	93,961	95,476	916	470,208	48.9	47.8	50.3	415	212,846
30-34	0.0019	0.0096	0.0084	0.0105	93,584	92,962	94,675	897	465,676	44.4	43.3	45.7	371	192,687
35-39	0.0021	0.0106	0.0094	0.0117	92,687	91,984	93,883	987	460,967	39.8	38.7	41.1	363	169,711
40-44	0.0035	0.0172	0.0152	0.0189	91,700	90,908	93,004	1,575	454,563	35.2	34.1	36.5	508	146,485
45-49	0.0058	0.0286	0.0257	0.0314	90,125	89,192	91,586	2,573	444,191	30.7	29.7	32.0	738	127,312
50-54	0.0080	0.0391	0.0349	0.0430	87,552	86,392	89,234	3,420	429,209	26.6	25.5	27.8	610	76,513
55-59	0.0115	0.0559	0.0495	0.0617	84,132	82,680	86,119	4,703	408,902	22.5	21.5	23.7	728	63,254
60-64	0.0205	0.0975	0.0856	0.1080	79,429	77,583	81,859	7,746	377,778	18.7	17.8	19.8	1,006	49,079
65-69	0.0285	0.1328	0.1161	0.1484	71,683	69,206	74,849	9,522	334,609	15.5	14.6	16.5	893	31,375
70-74	0.0406	0.1842	0.1614	0.2060	62,161	58,939	66,158	11,451	282,178	12.5	11.7	13.3	920	22,660
75-79	0.0582	0.2542	0.2251	0.2826	50,710	46,797	55,478	12,890	221,326	9.7	9.1	10.4	835	14,338
80-84	0.0979	0.3933	0.3558	0.4311	37,820	33,571	42,988	14,874	151,916	7.2	6.7	7.7	750	7,660
85-89	0.1570	0.5639	0.5245	0.6032	22,946	19,100	27,692	12,938	82,385	5.2	4.9	5.6	463	2,951
90-94	0.2403	0.6983	0.6657	0.7315	10,008	7,579	13,166	6,988	29,075	3.7	3.5	4.0	157	653
95-99	0.3510	0.8548	0.8331	0.8769	3,020	2,035	4,401	2,581	7,355	2.7	2.6	2.9	26	75
100+	0.4890	1	1	1	438	250	735	438	897	2.0	2.0	2.1	2	4
All ages													*12,913*	*2,771,827*

ln($_nq_x$) with uncertainty interval

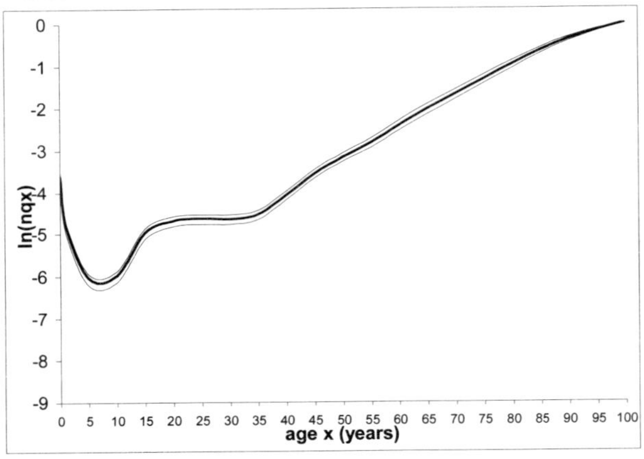

l_x with uncertainty interval

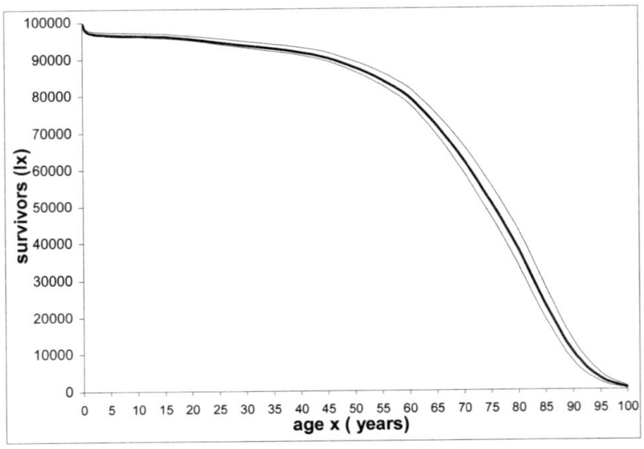

e_x with uncertainty interval

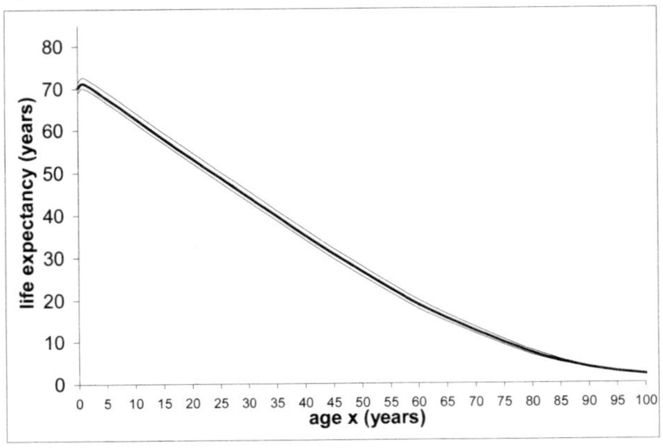

Distribution of deaths by age with uncertainty interval

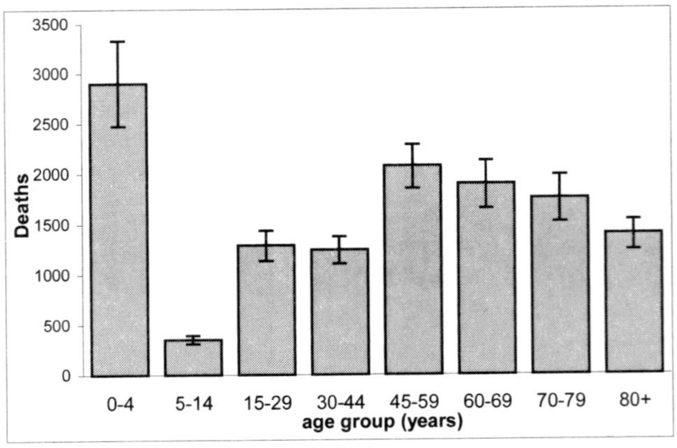

Paraguay

Females, 2000

Paraguay

Femmes, 2000

Paraguay

Mujeres, 2000

Age	$_nM_x$	$_nq_x$			l_x			$_nd_x$	$_nL_x$	e_x			Actual deaths	Actual population
			percentiles			percentiles					percentiles			
		mean	2 1/2	97 1/2	mean	2 1/2	97 1/2			mean	2 1/2	97 1/2		
<1	0.0249	0.0244	0.0192	0.0280	100,000	100,000	100,000	2,440	97,804	74.2	72.5	75.8	1,963	78,700
1-4	0.0017	0.0068	0.0057	0.0077	97,560	97,200	98,078	660	388,655	75.0	73.4	76.8	511	300,665
5-9	0.0004	0.0022	0.0018	0.0025	96,900	96,453	97,523	210	483,973	71.5	69.9	73.3	155	357,902
10-14	0.0004	0.0021	0.0018	0.0024	96,690	96,214	97,345	206	482,933	66.7	65.1	68.4	141	329,783
15-19	0.0009	0.0044	0.0037	0.0050	96,483	95,980	97,170	420	481,367	61.8	60.2	63.6	258	295,573
20-24	0.0010	0.0048	0.0041	0.0055	96,064	95,504	96,813	457	479,175	57.1	55.5	58.8	226	236,981
25-29	0.0011	0.0056	0.0048	0.0064	95,606	94,983	96,420	539	476,685	52.3	50.8	54.1	234	207,249
30-34	0.0013	0.0063	0.0054	0.0073	95,068	94,371	95,957	601	473,835	47.6	46.1	49.3	238	187,287
35-39	0.0019	0.0095	0.0082	0.0109	94,466	93,687	95,436	896	470,093	42.9	41.4	44.6	315	165,494
40-44	0.0027	0.0135	0.0117	0.0155	93,571	92,668	94,656	1,263	464,695	38.3	36.8	40.0	387	142,276
45-49	0.0048	0.0237	0.0208	0.0272	92,307	91,230	93,544	2,187	456,068	33.8	32.3	35.4	582	121,428
50-54	0.0061	0.0298	0.0262	0.0343	90,120	88,744	91,597	2,689	443,877	29.5	28.2	31.1	454	74,895
55-59	0.0080	0.0392	0.0340	0.0453	87,431	85,699	89,196	3,429	428,582	25.4	24.1	26.9	502	62,689
60-64	0.0134	0.0646	0.0553	0.0750	84,002	81,819	86,166	5,430	406,435	21.3	20.1	22.8	701	52,486
65-69	0.0210	0.0997	0.0844	0.1159	78,572	75,683	81,401	7,835	373,274	17.6	16.5	19.0	832	39,620
70-74	0.0295	0.1375	0.1158	0.1592	70,737	66,914	74,530	9,729	329,365	14.3	13.3	15.5	938	31,759
75-79	0.0464	0.2080	0.1762	0.2388	61,009	56,264	65,897	12,687	273,324	11.2	10.3	12.2	966	20,812
80-84	0.0805	0.3351	0.2903	0.3793	48,321	42,830	54,288	16,194	201,120	8.4	7.7	9.3	969	12,040
85-89	0.1127	0.4397	0.3905	0.4885	32,127	26,583	38,527	14,125	125,322	6.4	5.9	7.0	589	5,227
90-94	0.1833	0.5913	0.5432	0.6380	18,002	13,596	23,483	10,645	58,074	4.5	4.2	5.0	274	1,495
95-99	0.2887	0.7735	0.7365	0.8088	7,357	4,921	10,726	5,690	19,712	3.2	3.0	3.5	74	255
100+	0.4402	1	1	1	1,666	941	2,826	1,666	3,785	2.3	2.2	2.5	11	24
All ages													*11,320*	*2,724,640*

ln($_nq_x$) with uncertainty interval

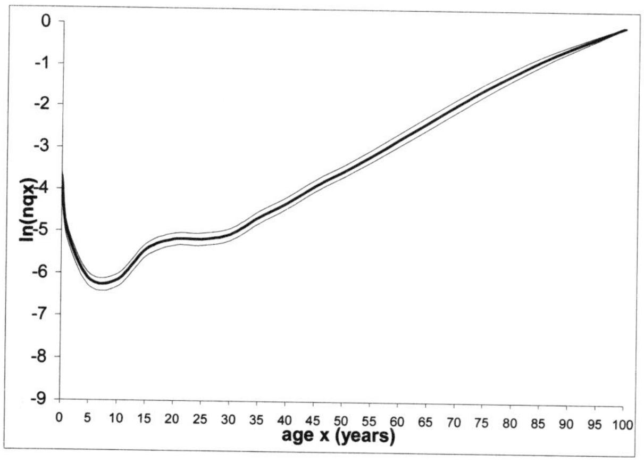

l_x with uncertainty interval

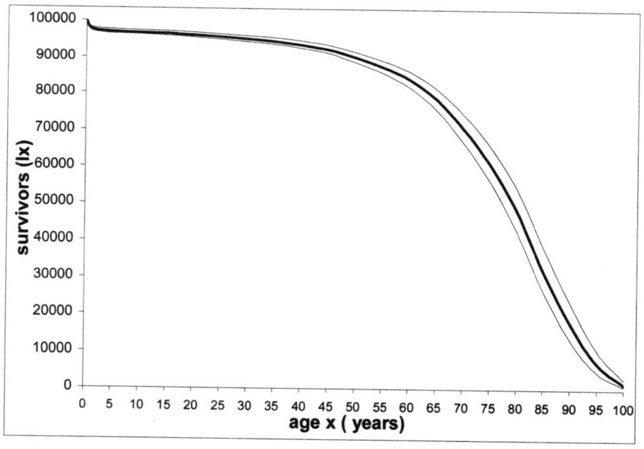

e_x with uncertainty interval

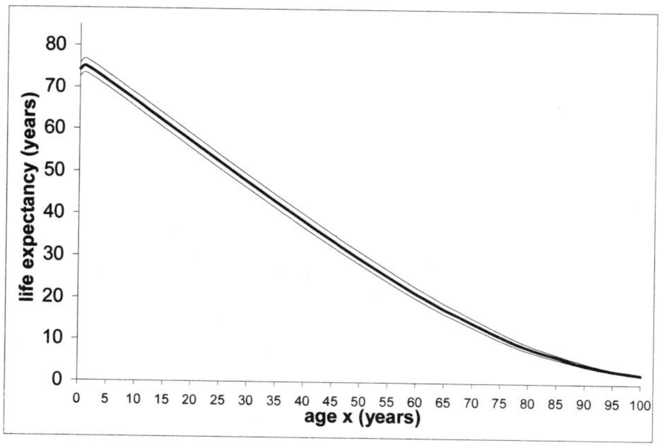

Distribution of deaths by age with uncertainty interval

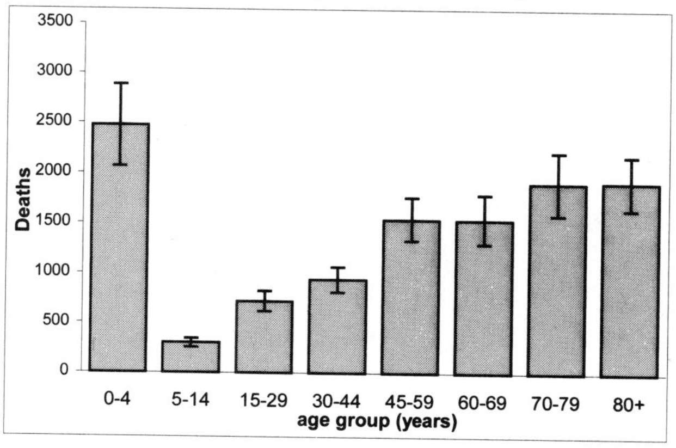

Peru   Pérou   Perú

Males, 2000   *Hommes, 2000*   *Hombres, 2000*

Age	$_nM_x$	$_nq_x$			l_x			$_nd_x$	$_nL_x$	e_x			Actual deaths	Actual population
			percentiles			percentiles					percentiles			
		mean	2½	97½	mean	2½	97½			mean	2½	97½		
<1	0.0369	0.0357	0.0324	0.0391	100,000	100,000	100,000	3,571	96,786	66.7	65.3	68.0	11,060	299,791
1-4	0.0045	0.0177	0.0163	0.0192	96,429	96,092	96,763	1,711	381,612	68.2	66.8	69.6	5,267	1,175,055
5-9	0.0009	0.0046	0.0042	0.0050	94,719	94,320	95,119	436	472,504	65.4	64.0	66.8	1,338	1,448,747
10-14	0.0008	0.0039	0.0036	0.0043	94,283	93,857	94,712	372	470,484	60.7	59.4	62.0	1,128	1,428,096
15-19	0.0014	0.0071	0.0065	0.0078	93,911	93,463	94,365	671	467,878	55.9	54.6	57.2	1,953	1,361,581
20-24	0.0022	0.0108	0.0099	0.0119	93,240	92,746	93,738	1,006	463,684	51.3	50.0	52.6	2,775	1,278,541
25-29	0.0024	0.0121	0.0111	0.0134	92,234	91,669	92,807	1,116	458,378	46.8	45.6	48.1	2,686	1,102,921
30-34	0.0026	0.0129	0.0117	0.0143	91,118	90,465	91,777	1,172	452,657	42.3	41.2	43.6	2,425	936,369
35-39	0.0033	0.0161	0.0146	0.0180	89,945	89,198	90,694	1,451	446,100	37.9	36.7	39.1	2,601	800,052
40-44	0.0042	0.0207	0.0186	0.0231	88,495	87,632	89,352	1,831	437,897	33.4	32.4	34.7	2,769	662,442
45-49	0.0060	0.0294	0.0262	0.0330	86,664	85,612	87,660	2,546	426,957	29.1	28.1	30.3	3,316	556,157
50-54	0.0082	0.0401	0.0355	0.0452	84,119	82,789	85,316	3,377	412,150	24.9	23.9	26.0	3,681	449,282
55-59	0.0119	0.0579	0.0509	0.0653	80,741	79,046	82,198	4,673	392,023	20.8	19.9	21.9	4,321	362,450
60-64	0.0200	0.0952	0.0833	0.1074	76,068	73,880	77,899	7,244	362,229	17.0	16.1	17.9	5,965	298,268
65-69	0.0315	0.1460	0.1275	0.1639	68,824	65,942	71,412	10,051	318,992	13.5	12.8	14.3	7,287	231,280
70-74	0.0492	0.2190	0.1925	0.2445	58,773	55,132	62,304	12,873	261,682	10.4	9.8	11.0	7,855	159,683
75-79	0.0866	0.3559	0.3193	0.3910	45,900	41,735	50,307	16,338	188,654	7.6	7.1	8.1	8,590	99,189
80-84	0.1500	0.5455	0.5039	0.5823	29,562	25,624	34,242	16,125	107,498	5.4	5.1	5.7	7,497	49,978
85-89	0.2400	0.7500	0.7171	0.7771	13,437	10,771	16,911	10,078	41,991	3.8	3.6	4.0	4,865	20,272
90-94	0.3500	0.8537	0.8316	0.8716	3,359	2,401	4,725	2,868	8,193	2.7	2.6	2.9	1,890	5,400
95-99	0.4800	0.9836	0.9784	0.9874	492	308	796	484	1,007	2.1	2.1	2.1	369	769
100+	0.6849	1	1	1	8	4	17	8	12	1.5	1.5	1.8	38	55
All ages													*89,676*	*12,726,378*

ln($_nq_x$) with uncertainty interval

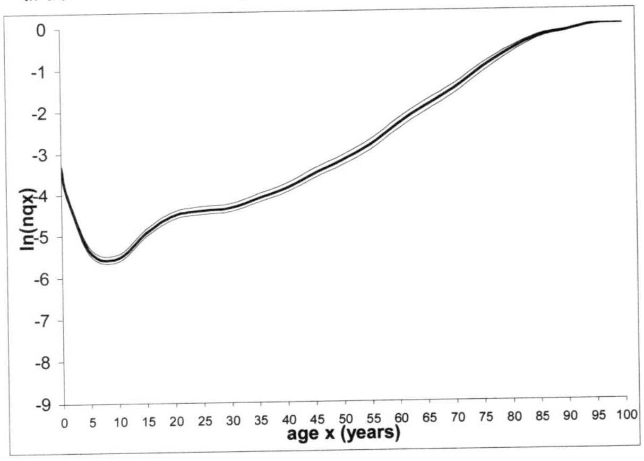

l_x with uncertainty interval

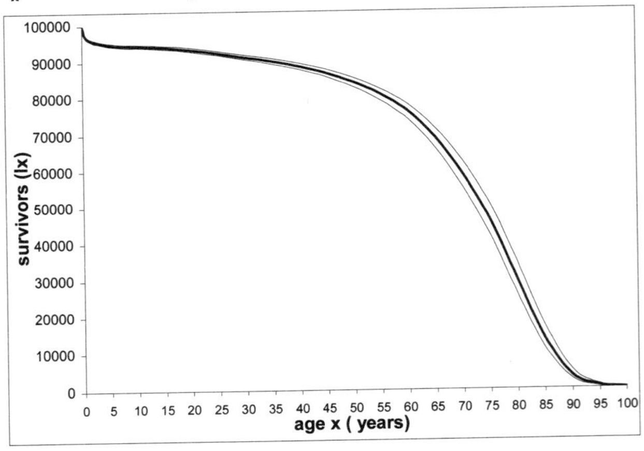

e_x with uncertainty interval

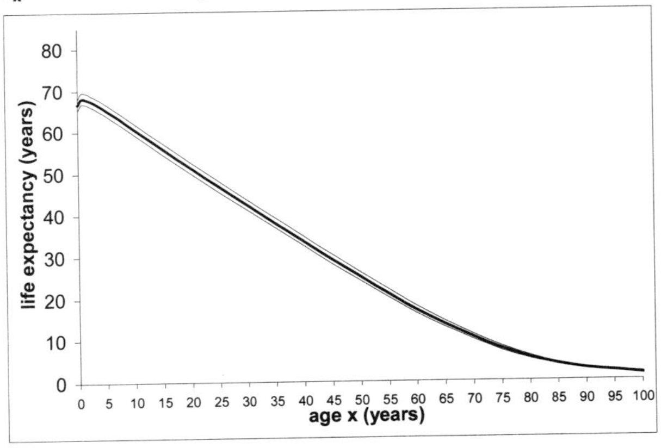

Distribution of deaths by age with uncertainty interval

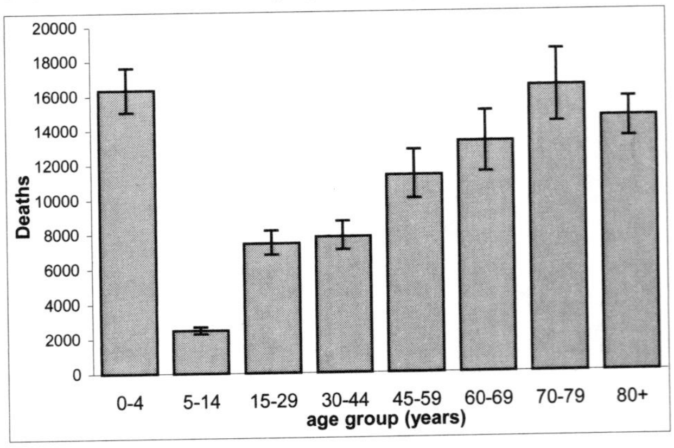

Peru / Pérou / Perú

Females, 2000 / *Femmes, 2000* / *Mujeres, 2000*

Age	$_nM_x$	$_nq_x$ mean	$_nq_x$ percentiles 2 1/2	$_nq_x$ percentiles 97 1/2	l_x mean	l_x percentiles 2 1/2	l_x percentiles 97 1/2	$_nd_x$	$_nL_x$	e_x mean	e_x percentiles 2 1/2	e_x percentiles 97 1/2	Actual deaths	Actual population
<1	0.0326	0.0316	0.0284	0.0352	100,000	100,000	100,000	3,165	97,152	71.6	70.1	73.0	9,390	288,261
1-4	0.0042	0.0168	0.0153	0.0186	96,835	96,482	97,156	1,627	383,437	72.9	71.5	74.4	4,808	1,133,151
5-9	0.0007	0.0037	0.0033	0.0041	95,208	94,811	95,608	349	475,169	70.2	68.8	71.6	1,033	1,404,734
10-14	0.0006	0.0030	0.0027	0.0033	94,859	94,433	95,279	285	473,583	65.4	64.1	66.8	836	1,389,414
15-19	0.0010	0.0050	0.0045	0.0055	94,574	94,118	95,018	469	471,699	60.6	59.3	62.0	1,324	1,331,837
20-24	0.0013	0.0067	0.0061	0.0075	94,105	93,598	94,590	628	468,956	55.9	54.6	57.3	1,700	1,268,424
25-29	0.0014	0.0070	0.0063	0.0078	93,477	92,899	94,016	653	465,753	51.2	50.0	52.6	1,579	1,126,801
30-34	0.0019	0.0093	0.0084	0.0104	92,824	92,173	93,420	862	461,965	46.6	45.4	47.9	1,847	989,497
35-39	0.0025	0.0125	0.0114	0.0141	91,962	91,213	92,631	1,152	456,930	42.0	40.8	43.3	2,166	859,296
40-44	0.0030	0.0147	0.0133	0.0166	90,810	89,929	91,578	1,337	450,709	37.5	36.3	38.8	2,108	710,788
45-49	0.0043	0.0213	0.0190	0.0240	89,474	88,438	90,354	1,903	442,609	33.0	31.9	34.3	2,530	588,278
50-54	0.0060	0.0296	0.0262	0.0334	87,570	86,315	88,608	2,588	431,381	28.7	27.6	29.9	2,826	471,008
55-59	0.0085	0.0416	0.0365	0.0473	84,982	83,431	86,228	3,537	416,069	24.5	23.5	25.6	3,249	382,234
60-64	0.0124	0.0601	0.0523	0.0684	81,445	79,503	82,961	4,893	394,994	20.4	19.5	21.5	3,966	320,180
65-69	0.0200	0.0952	0.0824	0.1085	76,552	74,103	78,469	7,291	364,535	16.6	15.8	17.6	5,114	255,715
70-74	0.0300	0.1395	0.1205	0.1582	69,262	66,156	72,002	9,664	322,147	13.1	12.4	13.9	5,549	184,976
75-79	0.0487	0.2172	0.1888	0.2445	59,597	55,719	63,328	12,943	265,628	9.8	9.2	10.5	5,978	122,688
80-84	0.1000	0.4000	0.3573	0.4404	46,654	42,177	51,371	18,662	186,616	6.8	6.4	7.3	6,705	67,046
85-89	0.1800	0.6207	0.5755	0.6595	27,992	23,845	33,018	17,375	96,525	4.7	4.4	5.0	5,380	29,887
90-94	0.2800	0.7609	0.7271	0.7884	10,618	8,165	13,921	8,079	28,853	3.3	3.1	3.5	2,579	9,212
95-99	0.4300	0.9389	0.9243	0.9499	2,539	1,728	3,757	2,384	5,544	2.3	2.2	2.4	725	1,687
100+	0.6569	1	1	1	155	87	284	155	236	1.5	1.5	2.0	123	187
All ages													*71,515*	*12,935,301*

$\ln(_nq_x)$ with uncertainty interval

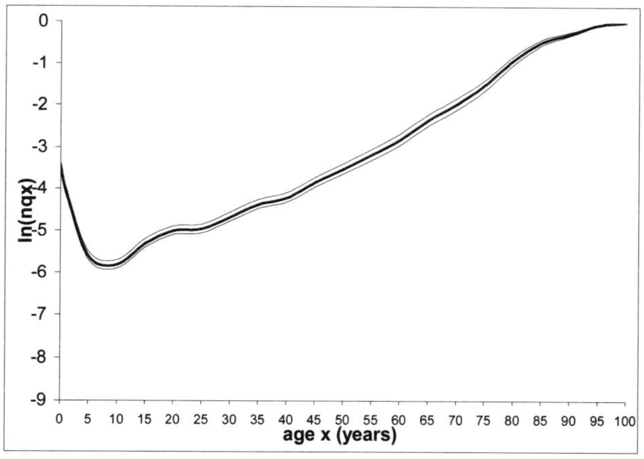

l_x with uncertainty interval

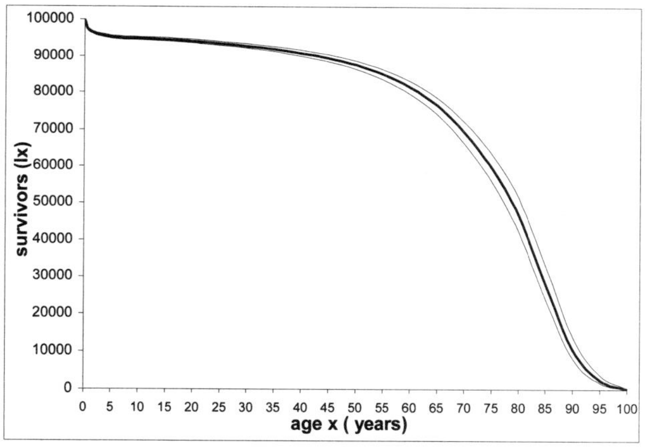

e_x with uncertainty interval

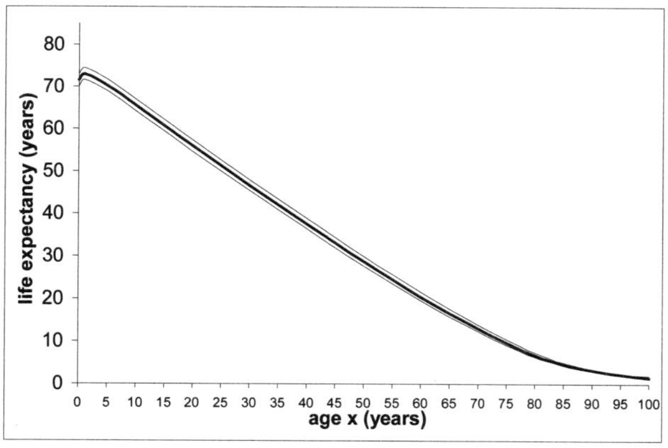

Distribution of deaths by age with uncertainty interval

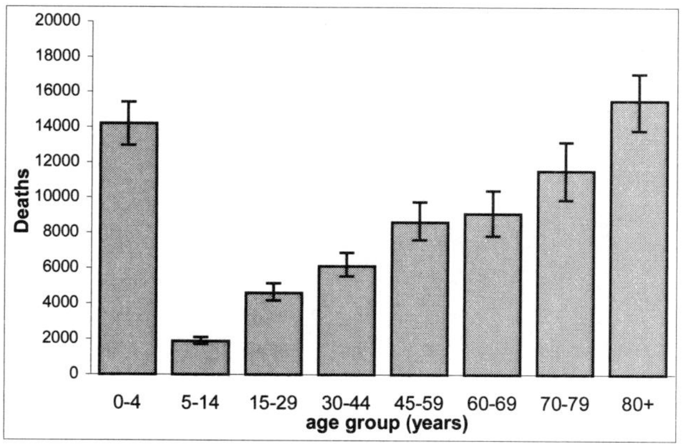

Philippines **Philippines** **Filipinas**

Males, 2000 *Hommes, 2000* *Hombres, 2000*

Age	$_nM_x$	$_nq_x$			l_x			$_nd_x$	$_nL_x$	e_x			Actual deaths	Actual population
			percentiles			percentiles					percentiles			
		mean	21/2	971/2	mean	21/2	971/2			mean	21/2	971/2		
<1	0.0323	0.0314	0.0284	0.0343	100,000	100,000	100,000	3,137	97,176	64.6	63.6	65.5	33,158	1,027,051
1-4	0.0032	0.0125	0.0117	0.0135	96,863	96,574	97,159	1,214	384,537	65.7	64.8	66.5	12,641	4,004,065
5-9	0.0013	0.0062	0.0058	0.0067	95,649	95,274	96,027	597	476,751	62.5	61.6	63.3	6,153	4,913,722
10-14	0.0010	0.0050	0.0047	0.0054	95,052	94,636	95,470	476	474,068	57.9	57.0	58.6	4,581	4,560,381
15-19	0.0015	0.0077	0.0072	0.0083	94,576	94,126	95,025	729	471,055	53.2	52.3	53.9	6,419	4,145,284
20-24	0.0024	0.0121	0.0113	0.0130	93,846	93,346	94,344	1,139	466,385	48.6	47.7	49.3	8,964	3,671,984
25-29	0.0033	0.0162	0.0150	0.0174	92,708	92,129	93,280	1,498	459,794	44.2	43.3	44.9	10,282	3,156,207
30-34	0.0037	0.0184	0.0171	0.0198	91,210	90,527	91,877	1,679	451,852	39.8	39.1	40.6	10,060	2,707,908
35-39	0.0048	0.0239	0.0223	0.0259	89,531	88,731	90,302	2,144	442,297	35.5	34.8	36.3	11,115	2,293,396
40-44	0.0061	0.0299	0.0280	0.0324	87,388	86,436	88,286	2,617	430,396	31.4	30.7	32.1	11,694	1,923,354
45-49	0.0083	0.0404	0.0378	0.0438	84,771	83,637	85,818	3,428	415,284	27.2	26.6	27.9	13,219	1,601,380
50-54	0.0111	0.0539	0.0506	0.0583	81,343	79,977	82,570	4,386	395,751	23.3	22.7	24.0	13,468	1,215,365
55-59	0.0161	0.0772	0.0723	0.0834	76,957	75,311	78,395	5,941	369,935	19.5	18.9	20.1	15,552	968,450
60-64	0.0237	0.1119	0.1040	0.1204	71,017	69,031	72,731	7,943	335,226	15.9	15.4	16.5	17,282	729,328
65-69	0.0400	0.1818	0.1684	0.1946	63,073	60,721	65,169	11,468	286,697	12.6	12.1	13.1	21,672	541,792
70-74	0.0600	0.2609	0.2419	0.2782	51,606	48,904	54,195	13,462	224,372	9.8	9.5	10.2	18,847	314,121
75-79	0.0860	0.3539	0.3306	0.3753	38,143	35,299	41,085	13,499	156,968	7.4	7.1	7.7	16,470	191,515
80-84	0.1550	0.5585	0.5317	0.5827	24,644	22,051	27,502	13,763	88,812	5.1	4.9	5.3	13,876	89,537
85-89	0.2784	0.8208	0.8020	0.8373	10,881	9,202	12,880	8,931	32,077	3.4	3.3	3.5	8,487	30,485
90-94	0.4307	0.9395	0.9306	0.9472	1,950	1,497	2,551	1,832	4,254	2.3	2.3	2.4	2,574	5,977
95-99	0.5737	0.9800	0.9756	0.9830	118	79	177	116	243	2.1	2.1	2.1	310	541
100+	0.7000	1	1	1	2	1	4	2	3	1.4	1.4	1.9	17	24
All ages													*256,839*	*38,091,867*

ln($_nq_x$) with uncertainty interval

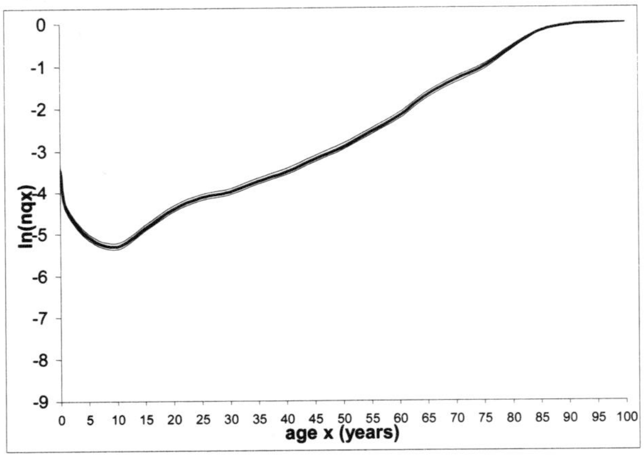

l_x with uncertainty interval

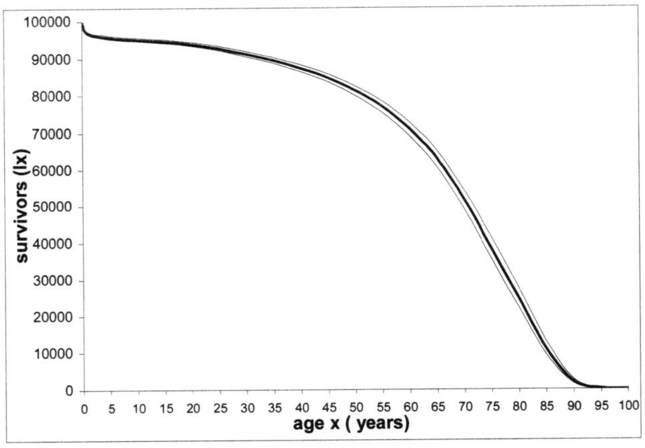

e_x with uncertainty interval

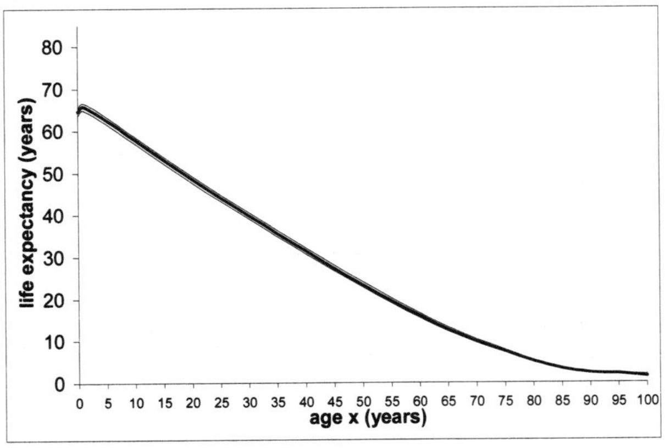

Distribution of deaths by age with uncertainty interval

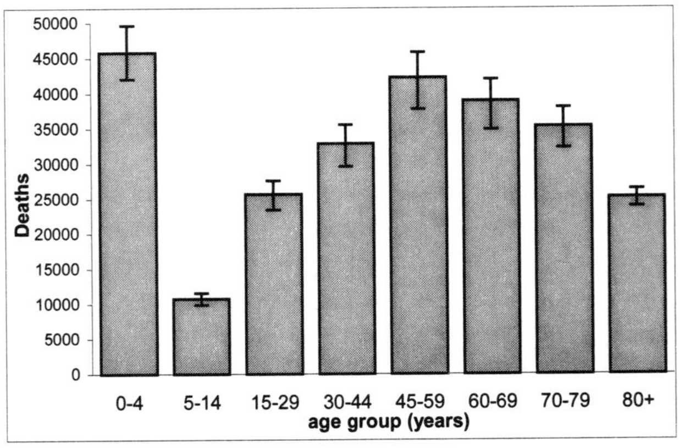

Philippines | Philippines | Filipinas

Females, 2000 | *Femmes, 2000* | *Mujeres, 2000*

Age	$_nM_x$	$_nq_x$			l_x			$_nd_x$	$_nL_x$	e_x			Actual deaths	Actual population
			percentiles			percentiles					percentiles			
		mean	2½	97½	mean	2½	97½			mean	2½	97½		
<1	0.0264	0.0258	0.0223	0.0280	100,000	100,000	100,000	2,578	97,680	71.1	70.1	72.7	25,533	967,547
1-4	0.0030	0.0120	0.0106	0.0128	97,422	97,197	97,774	1,167	386,887	72.0	71.0	73.5	11,566	3,832,653
5-9	0.0011	0.0055	0.0049	0.0059	96,255	95,950	96,738	531	479,946	68.8	67.9	70.3	5,208	4,707,497
10-14	0.0008	0.0039	0.0034	0.0042	95,724	95,380	96,265	369	477,697	64.2	63.3	65.6	3,382	4,382,094
15-19	0.0009	0.0046	0.0041	0.0050	95,355	94,983	95,937	439	475,678	59.5	58.5	60.9	3,691	3,999,428
20-24	0.0012	0.0060	0.0053	0.0064	94,916	94,511	95,545	565	473,167	54.7	53.8	56.1	4,254	3,559,862
25-29	0.0015	0.0076	0.0068	0.0083	94,351	93,902	95,040	720	469,955	50.0	49.1	51.4	4,719	3,081,656
30-34	0.0019	0.0092	0.0082	0.0100	93,631	93,125	94,396	864	465,995	45.4	44.5	46.7	4,955	2,672,020
35-39	0.0026	0.0127	0.0112	0.0139	92,767	92,192	93,624	1,182	460,880	40.8	39.9	42.1	5,867	2,287,922
40-44	0.0033	0.0164	0.0144	0.0179	91,585	90,914	92,572	1,502	454,169	36.3	35.4	37.5	6,420	1,940,895
45-49	0.0044	0.0219	0.0191	0.0239	90,083	89,288	91,239	1,972	445,482	31.8	31.0	33.1	7,248	1,637,036
50-54	0.0060	0.0297	0.0258	0.0325	88,110	87,152	89,493	2,617	434,008	27.5	26.7	28.7	7,255	1,203,054
55-59	0.0089	0.0435	0.0377	0.0477	85,493	84,321	87,181	3,716	418,174	23.3	22.6	24.3	8,855	996,418
60-64	0.0142	0.0684	0.0591	0.0752	81,777	80,298	83,898	5,590	394,908	19.2	18.6	20.2	11,280	796,915
65-69	0.0240	0.1132	0.0980	0.1245	76,187	74,260	78,941	8,625	359,372	15.4	14.8	16.3	14,502	604,235
70-74	0.0400	0.1818	0.1588	0.1988	67,562	65,017	71,203	12,284	307,099	12.1	11.6	12.8	16,207	405,186
75-79	0.0549	0.2415	0.2140	0.2632	55,278	52,093	59,897	13,350	243,015	9.2	8.8	9.8	15,452	281,279
80-84	0.1091	0.4286	0.3926	0.4606	41,928	38,383	47,077	17,972	164,712	6.4	6.1	6.7	15,460	141,693
85-89	0.2000	0.6667	0.6356	0.6976	23,957	20,704	28,596	15,971	79,856	4.3	4.1	4.5	10,334	51,672
90-94	0.3400	0.8416	0.8221	0.8611	7,986	6,261	10,421	6,721	19,766	2.8	2.7	2.9	3,744	11,012
95-99	0.5059	0.9800	0.9756	0.9846	1,265	869	1,854	1,240	2,606	2.1	2.1	2.1	622	1,229
100+	0.6500	1	1	1	25	13	45	25	39	1.5	1.5	1.8	57	87
All ages													*186,611*	*37,561,390*

ln($_nq_x$) with uncertainty interval

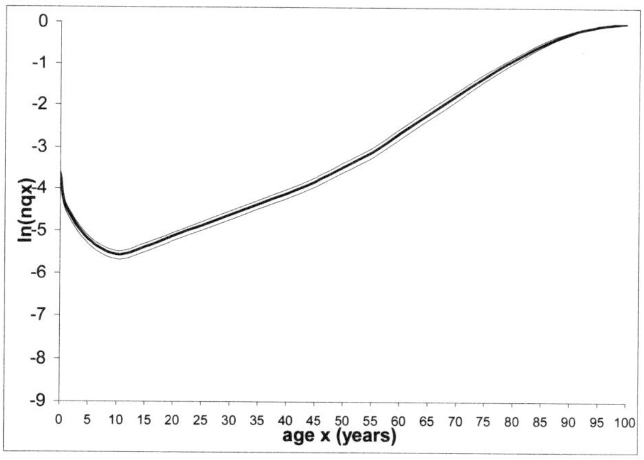

l_x with uncertainty interval

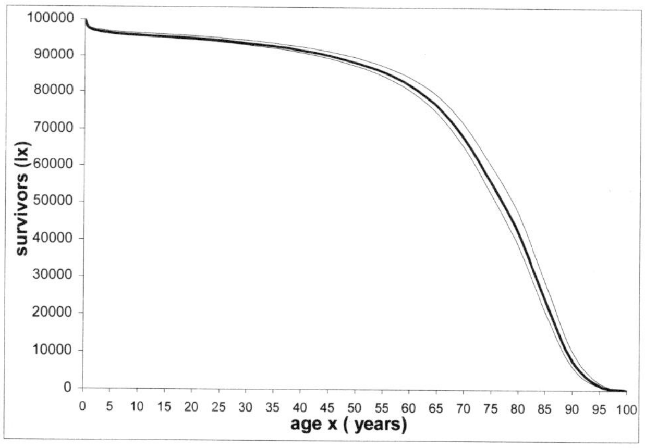

e_x with uncertainty interval

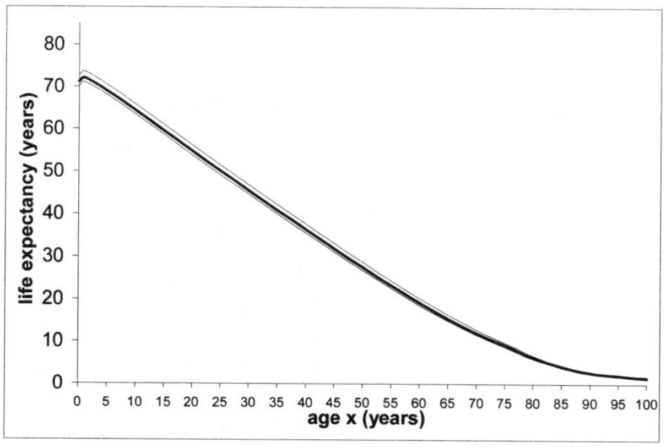

Distribution of deaths by age with uncertainty interval

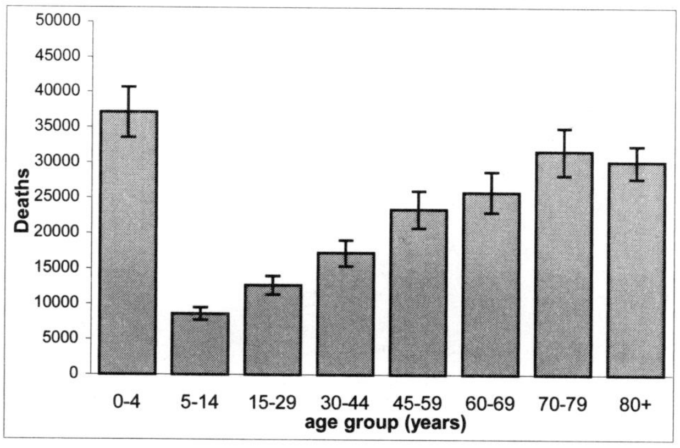

Poland　　　Pologne　　　Polonia

Males, 2000　　　*Hommes, 2000*　　　*Hombres, 2000*

Age	$_nM_x$	$_nq_x$			l_x			$_nd_x$	$_nL_x$	e_x			Actual deaths	Actual population
			percentiles			percentiles					percentiles			
		mean	2½	97½	mean	2½	97½			mean	2½	97½		
<1	0.0110	0.0109	0.0100	0.0118	100,000	100,000	100,000	1,094	99,015	69.2	68.9	69.5	2,294	207,679
1-4	0.0004	0.0016	0.0015	0.0017	98,906	98,816	98,997	155	395,253	69.0	68.6	69.4	320	816,525
5-9	0.0002	0.0012	0.0011	0.0013	98,751	98,652	98,851	119	493,459	65.1	64.7	65.5	306	1,268,131
10-14	0.0003	0.0014	0.0013	0.0015	98,632	98,526	98,739	136	492,821	60.2	59.8	60.5	414	1,499,101
15-19	0.0009	0.0045	0.0042	0.0047	98,496	98,381	98,611	442	491,376	55.3	54.9	55.6	1,543	1,716,197
20-24	0.0014	0.0068	0.0064	0.0071	98,054	97,916	98,194	666	488,608	50.5	50.1	50.9	2,219	1,628,860
25-29	0.0015	0.0073	0.0070	0.0077	97,389	97,217	97,561	716	485,154	45.8	45.4	46.2	2,104	1,426,666
30-34	0.0020	0.0100	0.0096	0.0104	96,673	96,470	96,878	967	480,949	41.1	40.8	41.5	2,483	1,235,170
35-39	0.0033	0.0164	0.0157	0.0170	95,706	95,466	95,952	1,566	474,617	36.5	36.1	36.9	4,361	1,321,642
40-44	0.0051	0.0253	0.0244	0.0261	94,140	93,845	94,444	2,379	464,753	32.1	31.7	32.5	8,120	1,586,016
45-49	0.0079	0.0387	0.0376	0.0397	91,761	91,397	92,136	3,550	449,931	27.9	27.5	28.3	12,078	1,531,000
50-54	0.0116	0.0566	0.0550	0.0582	88,211	87,773	88,666	4,990	428,581	23.9	23.5	24.3	14,133	1,213,873
55-59	0.0175	0.0839	0.0811	0.0869	83,221	82,717	83,747	6,985	398,643	20.2	19.8	20.6	13,204	753,516
60-64	0.0255	0.1199	0.1152	0.1247	76,236	75,675	76,803	9,141	358,328	16.8	16.4	17.2	19,659	770,649
65-69	0.0368	0.1686	0.1612	0.1756	67,095	66,388	67,792	11,312	307,197	13.7	13.4	14.1	25,551	693,895
70-74	0.0530	0.2341	0.2237	0.2436	55,783	54,728	56,814	13,056	246,276	11.0	10.8	11.3	28,377	535,275
75-79	0.0760	0.3194	0.3062	0.3310	42,727	41,397	44,066	13,648	179,517	8.6	8.4	8.9	24,935	327,985
80-84	0.1184	0.4567	0.4413	0.4699	29,080	27,696	30,573	13,281	112,196	6.5	6.3	6.7	15,325	129,460
85-89	0.1781	0.6163	0.6016	0.6289	15,799	14,691	17,080	9,736	54,653	4.8	4.7	4.9	13,151	73,820
90-94	0.2591	0.7289	0.7170	0.7394	6,063	5,466	6,804	4,419	17,055	3.5	3.5	3.6	5,699	21,994
95-99	0.3643	0.8703	0.8622	0.8773	1,643	1,431	1,925	1,430	3,926	2.7	2.6	2.7	1,217	3,342
100+	0.4950	1	1	1	213	176	265	213	431	2.0	2.0	2.0	118	239
All ages													*197,611*	*18,761,035*

$\ln(_nq_x)$ with uncertainty interval

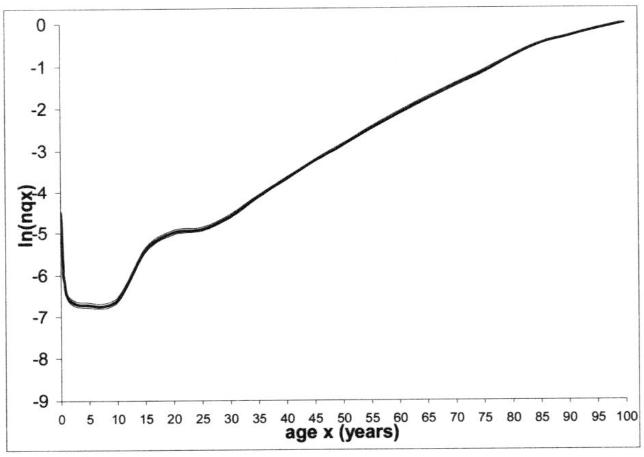

l_x with uncertainty interval

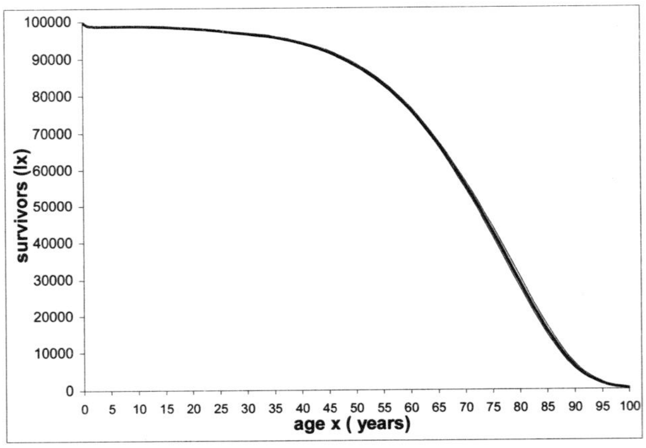

e_x with uncertainty interval

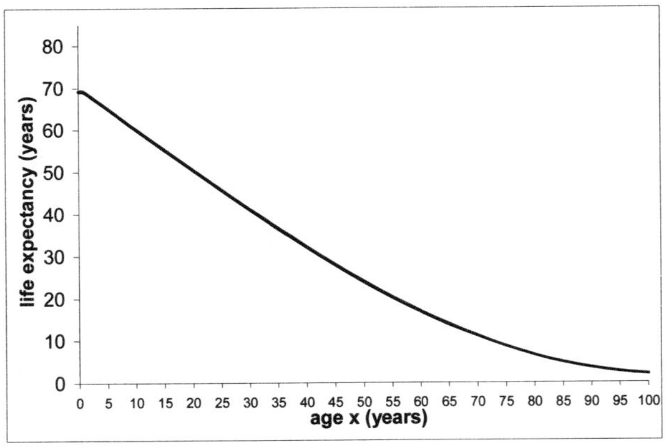

Distribution of deaths by age with uncertainty interval

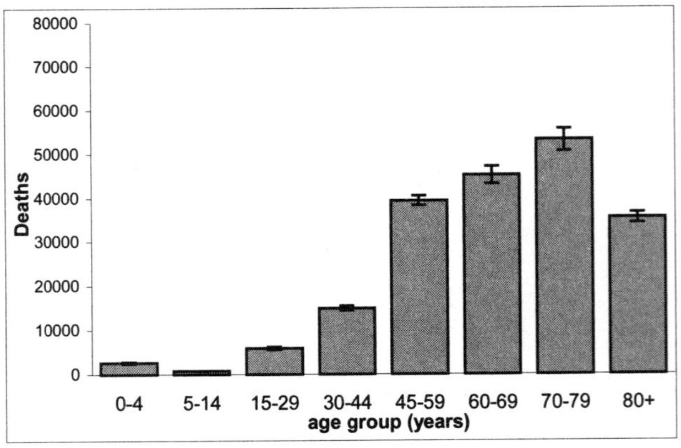

Poland   Pologne   Polonia

Females, 2000   *Femmes, 2000*   *Mujeres, 2000*

Age	$_nM_x$	$_nq_x$ mean	$_nq_x$ percentiles 2½	$_nq_x$ percentiles 97½	l_x mean	l_x percentiles 2½	l_x percentiles 97½	$_nd_x$	$_nL_x$	e_x mean	e_x percentiles 2½	e_x percentiles 97½	Actual deaths	Actual population
<1	0.0095	0.0094	0.0080	0.0107	100,000	100,000	100,000	939	99,155	77.7	77.2	78.2	1,873	197,790
1-4	0.0003	0.0013	0.0012	0.0014	99,061	98,925	99,196	129	395,935	77.4	76.9	78.0	251	771,854
5-9	0.0002	0.0009	0.0008	0.0010	98,932	98,784	99,081	89	494,438	73.5	73.0	74.1	218	1,206,737
10-14	0.0002	0.0008	0.0008	0.0009	98,843	98,686	99,000	82	494,009	68.6	68.1	69.2	238	1,427,589
15-19	0.0003	0.0016	0.0015	0.0017	98,761	98,596	98,926	158	493,407	63.6	63.1	64.2	528	1,647,157
20-24	0.0004	0.0018	0.0016	0.0019	98,602	98,424	98,782	176	492,573	58.7	58.2	59.3	559	1,566,995
25-29	0.0004	0.0020	0.0018	0.0021	98,427	98,233	98,622	192	491,653	53.8	53.3	54.4	540	1,379,404
30-34	0.0006	0.0029	0.0027	0.0032	98,234	98,026	98,445	289	490,450	48.9	48.4	49.5	701	1,191,070
35-39	0.0011	0.0054	0.0050	0.0058	97,946	97,716	98,179	526	488,414	44.1	43.6	44.7	1,402	1,301,093
40-44	0.0019	0.0093	0.0087	0.0099	97,420	97,155	97,689	904	484,839	39.3	38.8	39.9	2,959	1,587,876
45-49	0.0030	0.0148	0.0140	0.0155	96,516	96,202	96,837	1,426	479,015	34.6	34.2	35.2	4,691	1,575,820
50-54	0.0045	0.0223	0.0212	0.0233	95,090	94,712	95,471	2,116	470,160	30.1	29.6	30.7	5,818	1,292,879
55-59	0.0063	0.0312	0.0297	0.0329	92,974	92,522	93,414	2,904	457,610	25.7	25.3	26.3	5,429	855,316
60-64	0.0100	0.0490	0.0461	0.0518	90,070	89,553	90,576	4,414	439,313	21.5	21.0	22.1	9,481	943,582
65-69	0.0163	0.0784	0.0730	0.0830	85,656	85,038	86,274	6,712	411,497	17.5	17.0	18.0	15,067	923,661
70-74	0.0275	0.1286	0.1191	0.1368	78,943	77,977	79,861	10,154	369,332	13.7	13.3	14.2	22,875	832,063
75-79	0.0482	0.2150	0.1992	0.2296	68,790	67,354	70,349	14,792	306,970	10.4	10.0	10.8	29,506	612,348
80-84	0.0871	0.3577	0.3350	0.3793	53,998	52,089	56,334	19,316	221,700	7.6	7.3	7.9	23,853	273,764
85-89	0.1479	0.5398	0.5140	0.5634	34,682	32,375	37,464	18,720	126,609	5.4	5.2	5.6	27,342	184,927
90-94	0.2355	0.6900	0.6683	0.7101	15,962	14,146	18,147	11,013	46,770	3.8	3.6	3.9	14,665	62,277
95-99	0.3520	0.8560	0.8422	0.8688	4,949	4,125	5,964	4,236	12,035	2.7	2.7	2.8	3,370	9,576
100+	0.4937	1	1	1	713	547	940	713	1,443	2.0	2.0	2.1	313	634
All ages													*171,679*	*19,844,412*

ln($_nq_x$) with uncertainty interval

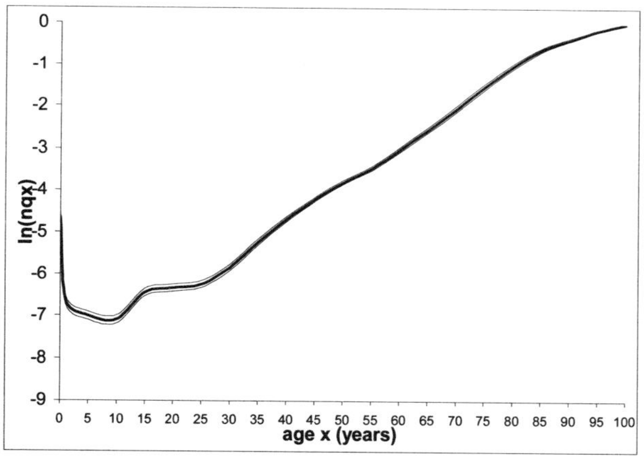

l_x with uncertainty interval

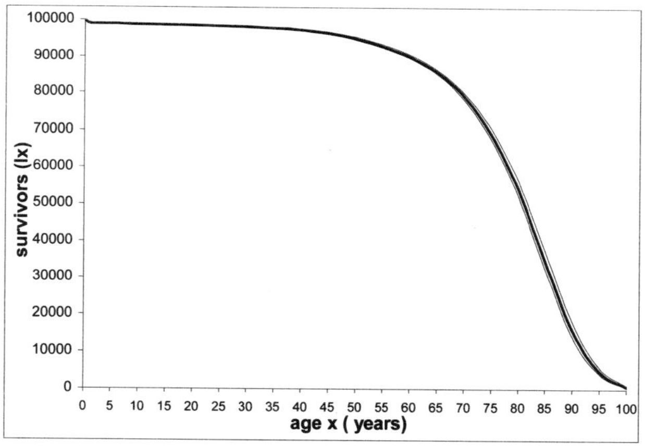

e_x with uncertainty interval

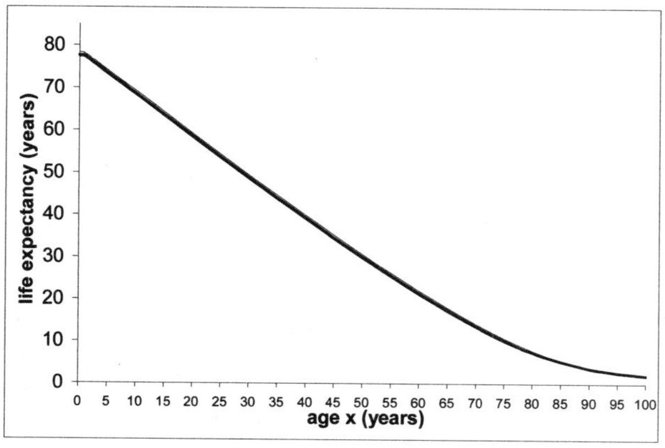

Distribution of deaths by age with uncertainty interval

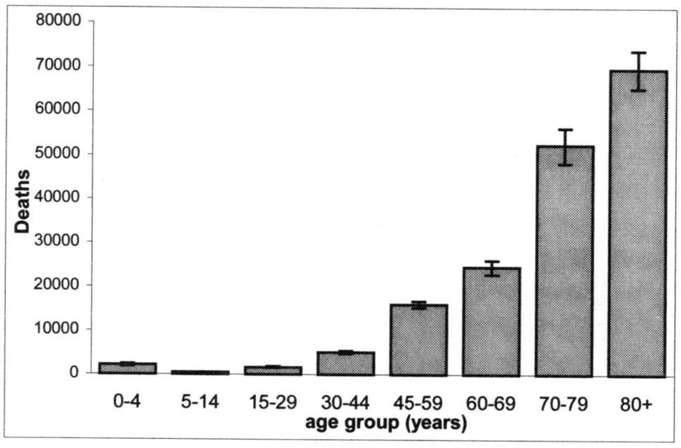

Portugal

Portugal

Portugal

Males, 2000 | *Hommes, 2000* | *Hombres, 2000*

Age	$_nM_x$	$_nq_x$			l_x			$_nd_x$	$_nL_x$	e_x			Actual deaths	Actual population
			percentiles			percentiles					percentiles			
		mean	2½	97½	mean	2½	97½			mean	2½	97½		
<1	0.0060	0.0060	0.0050	0.0070	100,000	100,000	100,000	597	99,463	71.7	71.4	72.1	343	57,169
1-4	0.0010	0.0040	0.0035	0.0044	99,403	99,303	99,504	397	396,661	71.2	70.8	71.5	231	231,261
5-9	0.0003	0.0014	0.0012	0.0015	99,007	98,862	99,155	139	494,687	67.4	67.1	67.8	80	286,781
10-14	0.0004	0.0020	0.0018	0.0022	98,868	98,710	99,032	198	493,846	62.5	62.2	62.9	113	281,768
15-19	0.0010	0.0050	0.0046	0.0054	98,671	98,495	98,855	497	492,110	57.7	57.3	58.0	341	337,947
20-24	0.0015	0.0075	0.0069	0.0080	98,173	97,958	98,403	738	489,021	52.9	52.6	53.3	607	402,135
25-29	0.0019	0.0097	0.0090	0.0102	97,435	97,172	97,722	941	484,824	48.3	47.9	48.7	813	418,815
30-34	0.0023	0.0116	0.0110	0.0122	96,494	96,184	96,840	1,122	479,666	43.8	43.4	44.2	886	378,469
35-39	0.0030	0.0149	0.0143	0.0155	95,372	95,019	95,773	1,420	473,311	39.3	38.8	39.7	1,062	353,974
40-44	0.0040	0.0198	0.0192	0.0204	93,952	93,549	94,404	1,860	465,110	34.8	34.4	35.2	1,330	332,407
45-49	0.0050	0.0247	0.0240	0.0253	92,092	91,642	92,585	2,274	454,774	30.5	30.1	30.9	1,569	313,737
50-54	0.0070	0.0344	0.0333	0.0355	89,818	89,333	90,332	3,090	441,365	26.2	25.8	26.6	2,059	294,190
55-59	0.0100	0.0488	0.0470	0.0506	86,728	86,225	87,231	4,231	423,065	22.0	21.6	22.4	2,517	251,736
60-64	0.0160	0.0769	0.0736	0.0804	82,498	81,997	82,994	6,346	396,623	18.0	17.6	18.4	3,856	240,992
65-69	0.0270	0.1265	0.1201	0.1329	76,152	75,616	76,697	9,630	356,682	14.3	14.0	14.7	5,978	221,398
70-74	0.0450	0.2022	0.1911	0.2128	66,521	65,596	67,426	13,454	298,972	11.0	10.7	11.4	8,260	183,551
75-79	0.0720	0.3051	0.2886	0.3215	53,068	51,635	54,474	16,190	224,862	8.2	7.9	8.5	9,499	131,933
80-84	0.1370	0.5102	0.4887	0.5327	36,877	35,117	38,712	18,816	137,346	5.7	5.5	5.9	9,030	65,912
85-89	0.2340	0.7382	0.7200	0.7573	18,061	16,567	19,793	13,332	56,975	4.0	3.8	4.1	6,494	27,754
90-94	0.3070	0.7991	0.7849	0.8140	4,729	4,021	5,537	3,779	12,308	3.1	3.0	3.2	1,927	6,277
95-99	0.3860	0.8943	0.8846	0.9045	950	748	1,187	850	2,202	2.5	2.5	2.6	247	641
100+	0.4930	1	1	1	100	71	136	100	204	2.0	2.0	2.3	28	56
All ages													*57,270*	*4,818,903*

ln($_nq_x$) with uncertainty interval

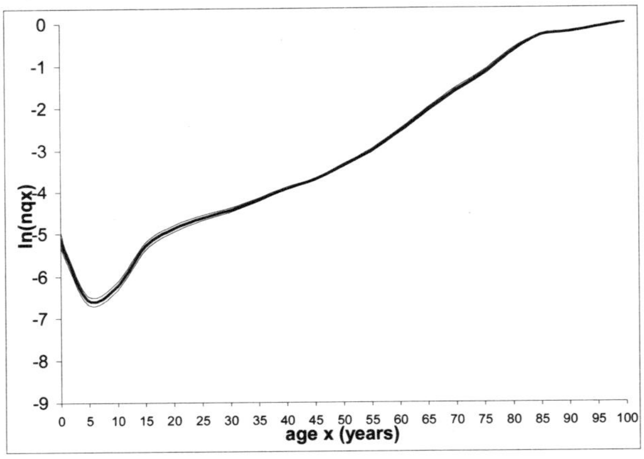

l_x with uncertainty interval

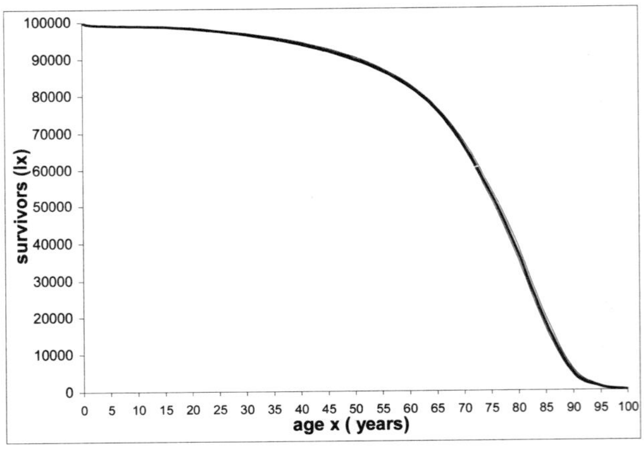

e_x with uncertainty interval

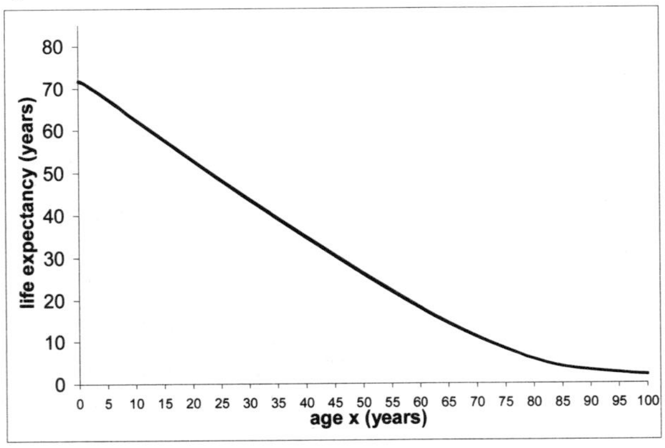

Distribution of deaths by age with uncertainty interval

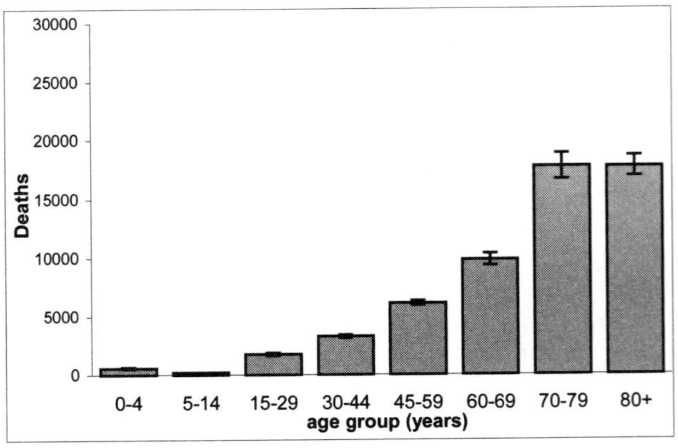

Portugal

Portugal

Portugal

Females, 2000

Femmes, 2000

Mujeres, 2000

Age	$_nM_x$	$_nq_x$			l_x			$_nd_x$	$_nL_x$	e_x			Actual deaths	Actual population
			percentiles			percentiles					percentiles			
		mean	2½	97½	mean	2½	97½			mean	2½	97½		
<1	0.0050	0.0050	0.0042	0.0058	100,000	100,000	100,000	498	99,552	79.3	78.8	79.8	271	54,189
1-4	0.0005	0.0018	0.0016	0.0020	99,502	99,423	99,581	179	397,580	78.7	78.1	79.2	98	218,296
5-9	0.0003	0.0013	0.0012	0.0014	99,323	99,225	99,423	129	496,294	74.8	74.3	75.4	71	272,336
10-14	0.0002	0.0010	0.0009	0.0011	99,194	99,084	99,307	104	495,711	69.9	69.4	70.5	57	269,892
15-19	0.0003	0.0016	0.0015	0.0018	99,090	98,970	99,214	163	495,043	65.0	64.4	65.5	107	323,618
20-24	0.0005	0.0024	0.0022	0.0026	98,927	98,793	99,066	242	494,029	60.1	59.5	60.6	192	391,933
25-29	0.0006	0.0027	0.0025	0.0030	98,685	98,532	98,846	271	492,746	55.2	54.7	55.8	229	416,034
30-34	0.0007	0.0036	0.0034	0.0039	98,414	98,243	98,597	359	491,172	50.4	49.8	50.9	276	378,488
35-39	0.0011	0.0053	0.0049	0.0056	98,055	97,863	98,266	518	488,980	45.5	45.0	46.1	390	367,579
40-44	0.0014	0.0070	0.0066	0.0074	97,537	97,314	97,781	680	485,983	40.8	40.2	41.3	493	351,796
45-49	0.0020	0.0100	0.0095	0.0104	96,856	96,599	97,138	964	481,873	36.0	35.5	36.6	672	335,801
50-54	0.0030	0.0149	0.0142	0.0157	95,893	95,596	96,217	1,428	475,895	31.4	30.9	31.9	963	320,890
55-59	0.0040	0.0198	0.0189	0.0210	94,465	94,117	94,835	1,871	467,649	26.8	26.3	27.4	1,144	286,105
60-64	0.0070	0.0344	0.0325	0.0367	92,594	92,198	93,000	3,185	455,010	22.3	21.8	22.8	1,989	284,180
65-69	0.0120	0.0583	0.0544	0.0625	89,409	88,937	89,876	5,208	434,026	18.0	17.6	18.5	3,328	277,374
70-74	0.0230	0.1087	0.1007	0.1171	84,201	83,379	84,884	9,157	398,114	14.0	13.6	14.5	5,779	251,264
75-79	0.0420	0.1900	0.1752	0.2042	75,044	73,616	76,279	14,262	339,568	10.4	10.0	10.8	8,351	198,836
80-84	0.0900	0.3673	0.3423	0.3909	60,783	58,581	62,911	22,328	248,092	7.2	6.9	7.6	10,469	116,324
85-89	0.1700	0.5965	0.5687	0.6219	38,454	35,786	41,379	22,938	134,927	5.0	4.8	5.2	10,394	61,139
90-94	0.2520	0.7175	0.6961	0.7373	15,517	13,630	17,802	11,134	44,182	3.6	3.5	3.8	4,485	17,799
95-99	0.3580	0.8631	0.8497	0.8754	4,383	3,593	5,369	3,783	10,566	2.7	2.7	2.8	902	2,519
100+	0.4940	1	1	1	600	451	806	600	1,215	2.0	2.0	2.3	104	210
All ages													*50,763*	*5,196,602*

ln($_nq_x$) with uncertainty interval

l_x with uncertainty interval

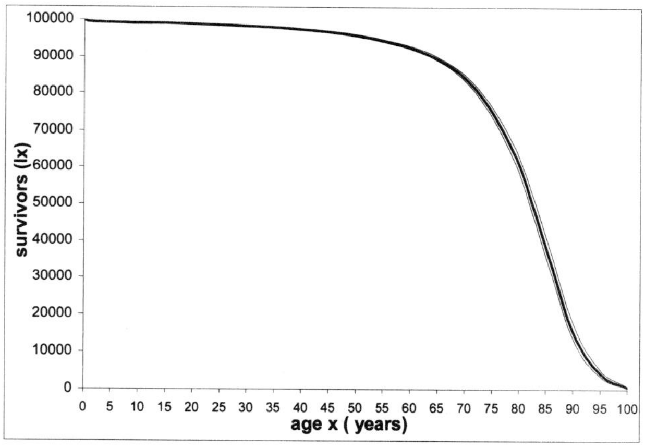

e_x with uncertainty interval

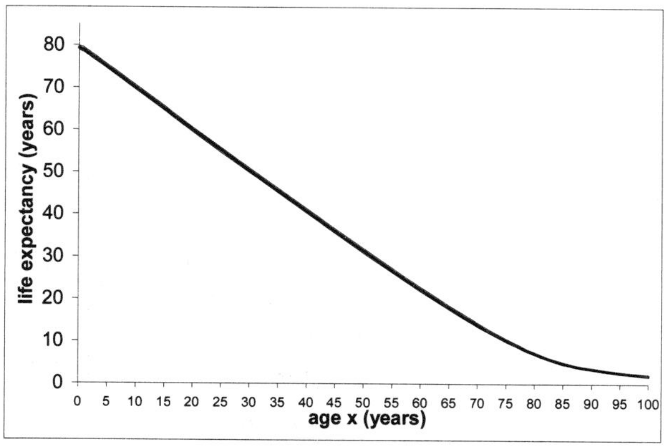

Distribution of deaths by age with uncertainty interval

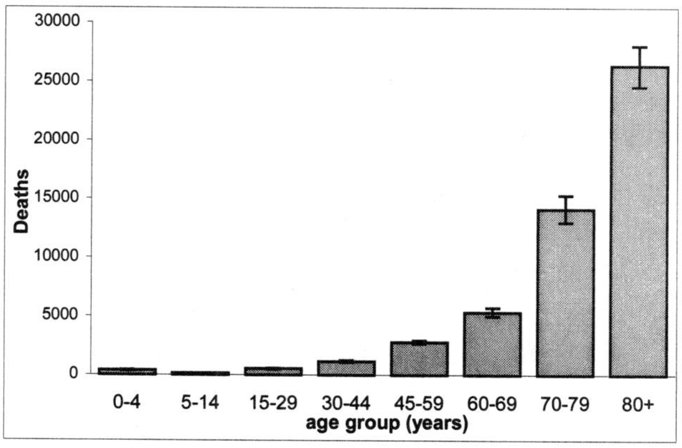

Qatar **Qatar** **Qatar**

Males, 2000 *Hommes, 2000* *Hombres, 2000*

Age	$_nM_x$	$_nq_x$			l_x			$_nd_x$	$_nL_x$	e_x			Actual deaths	Actual population
			percentiles			percentiles					percentiles			
		mean	2½	97½	mean	2½	97½			mean	2½	97½		
<1	0.0176	0.0174	0.0149	0.0205	100,000	100,000	100,000	1,735	98,438	70.4	70.1	70.7	103	5,856
1-4	0.0006	0.0026	0.0023	0.0029	98,265	97,949	98,506	253	392,453	70.6	70.3	71.1	14	21,355
5-9	0.0006	0.0031	0.0028	0.0034	98,012	97,667	98,276	303	489,302	66.8	66.5	67.3	16	25,729
10-14	0.0004	0.0018	0.0016	0.0020	97,709	97,331	97,999	176	488,105	62.0	61.7	62.5	9	24,305
15-19	0.0013	0.0064	0.0059	0.0070	97,533	97,138	97,838	624	486,106	57.1	56.8	57.6	26	20,289
20-24	0.0017	0.0087	0.0081	0.0094	96,909	96,458	97,259	844	482,437	52.5	52.1	53.0	29	16,520
25-29	0.0012	0.0060	0.0057	0.0064	96,065	95,551	96,467	578	478,882	47.9	47.5	48.4	30	24,887
30-34	0.0015	0.0075	0.0071	0.0079	95,487	94,936	95,918	713	475,655	43.2	42.8	43.7	62	41,622
35-39	0.0025	0.0124	0.0119	0.0130	94,774	94,184	95,237	1,177	470,929	38.5	38.1	39.0	126	50,248
40-44	0.0035	0.0175	0.0169	0.0182	93,597	92,955	94,100	1,641	463,882	34.0	33.5	34.5	172	48,740
45-49	0.0060	0.0296	0.0288	0.0303	91,956	91,263	92,496	2,718	452,984	29.5	29.1	30.1	227	37,793
50-54	0.0084	0.0410	0.0398	0.0421	89,238	88,506	89,819	3,657	437,045	25.3	24.9	25.9	193	23,097
55-59	0.0118	0.0572	0.0549	0.0589	85,580	84,860	86,180	4,892	415,673	21.3	20.9	21.9	155	13,202
60-64	0.0200	0.0952	0.0905	0.0989	80,689	80,065	81,248	7,685	384,232	17.5	17.0	18.0	136	6,803
65-69	0.0350	0.1609	0.1517	0.1688	73,004	72,474	73,529	11,748	335,651	14.0	13.6	14.5	116	3,312
70-74	0.0500	0.2222	0.2086	0.2351	61,256	60,399	62,352	13,613	272,250	11.2	10.8	11.7	75	1,490
75-79	0.0750	0.3158	0.2973	0.3345	47,644	46,337	49,343	15,045	200,606	8.7	8.4	9.1	48	640
80-84	0.1200	0.4615	0.4373	0.4848	32,598	30,925	34,668	15,045	125,379	6.6	6.4	6.9	27	228
85-89	0.1600	0.5714	0.5469	0.5940	17,553	15,943	19,433	10,030	62,689	5.2	5.0	5.4	13	79
90-94	0.2400	0.6977	0.6757	0.7172	7,523	6,472	8,734	5,248	21,868	3.7	3.6	3.9	2	10
95-99	0.3500	0.8537	0.8380	0.8671	2,274	1,830	2,822	1,941	5,547	2.8	2.7	2.8	1	2
100+	0.4700	1	1	1	333	243	457	333	708	2.1	2.1	2.2	0	0
All ages													*1,580*	*366,207*

ln($_nq_x$) with uncertainty interval

l_x with uncertainty interval

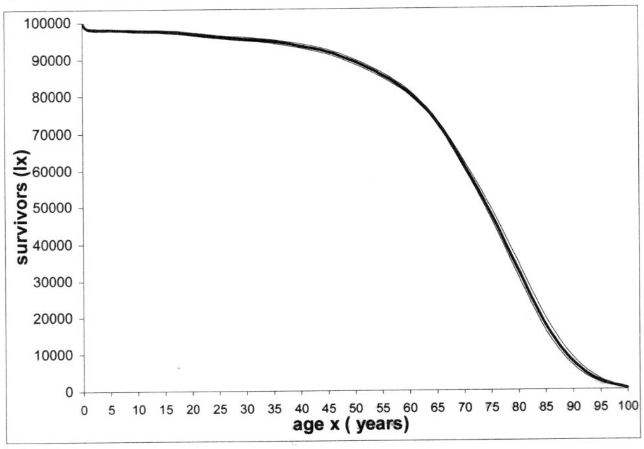

e_x with uncertainty interval

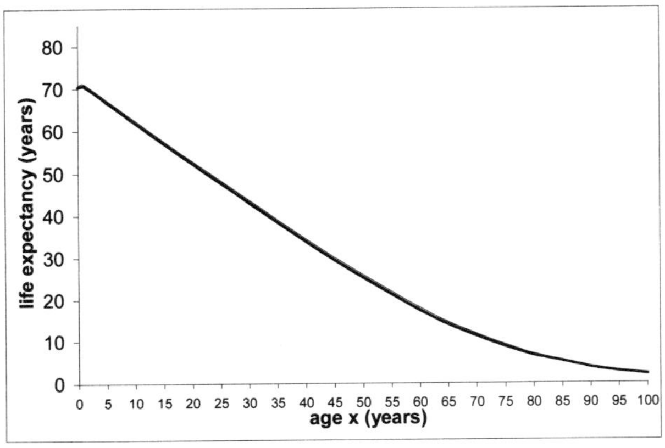

Distribution of deaths by age with uncertainty interval

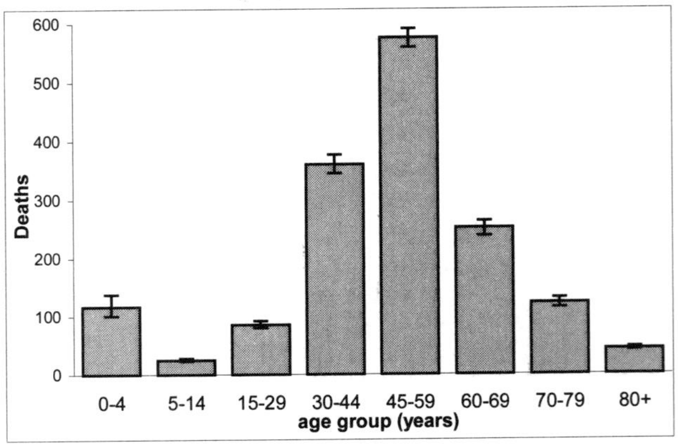

Qatar

Qatar

Qatar

Females, 2000

Femmes, 2000

Mujeres, 2000

Age	$_nM_x$	$_nq_x$			l_x			$_nd_x$	$_nL_x$	e_x			Actual deaths	Actual population
			percentiles			percentiles					percentiles			
		mean	2½	97½	mean	2½	97½			mean	2½	97½		
<1	0.0106	0.0105	0.0089	0.0122	100,000	100,000	100,000	1,054	99,051	75.0	74.6	75.4	60	5,630
1-4	0.0006	0.0023	0.0020	0.0025	98,946	98,781	99,113	225	395,243	74.8	74.3	75.2	12	20,412
5-9	0.0003	0.0014	0.0013	0.0016	98,721	98,534	98,914	141	493,251	71.0	70.5	71.4	7	24,711
10-14	0.0004	0.0019	0.0017	0.0020	98,580	98,379	98,788	182	492,442	66.0	65.6	66.5	9	23,020
15-19	0.0003	0.0016	0.0014	0.0017	98,397	98,181	98,624	155	491,598	61.2	60.7	61.6	6	19,289
20-24	0.0004	0.0020	0.0018	0.0022	98,242	98,013	98,484	195	490,722	56.3	55.8	56.7	6	14,406
25-29	0.0006	0.0029	0.0026	0.0031	98,047	97,802	98,306	282	489,528	51.4	50.9	51.8	7	12,402
30-34	0.0009	0.0047	0.0043	0.0050	97,765	97,499	98,049	455	487,686	46.5	46.0	47.0	13	14,042
35-39	0.0015	0.0075	0.0069	0.0079	97,310	97,015	97,629	726	484,735	41.7	41.2	42.2	28	18,697
40-44	0.0027	0.0136	0.0128	0.0142	96,584	96,250	96,953	1,312	479,640	37.0	36.5	37.5	50	18,445
45-49	0.0041	0.0202	0.0194	0.0209	95,272	94,885	95,710	1,924	471,549	32.5	32.0	33.0	50	12,283
50-54	0.0058	0.0287	0.0278	0.0296	93,348	92,915	93,854	2,681	460,037	28.1	27.6	28.6	40	6,846
55-59	0.0094	0.0457	0.0441	0.0473	90,667	90,210	91,220	4,146	442,970	23.9	23.4	24.4	37	3,930
60-64	0.0164	0.0787	0.0754	0.0821	86,521	86,028	87,064	6,807	415,586	19.9	19.4	20.4	38	2,338
65-69	0.0252	0.1185	0.1125	0.1247	79,714	79,178	80,279	9,449	374,946	16.4	15.9	16.9	36	1,420
70-74	0.0360	0.1651	0.1555	0.1746	70,265	69,435	71,180	11,603	322,316	13.2	12.8	13.7	29	809
75-79	0.0576	0.2517	0.2366	0.2664	58,662	57,314	59,984	14,768	256,388	10.4	10.0	10.8	20	342
80-84	0.0850	0.3505	0.3310	0.3706	43,894	42,083	45,712	15,385	181,004	8.0	7.7	8.3	12	146
85-89	0.1300	0.4906	0.4686	0.5139	28,508	26,614	30,581	13,985	107,578	6.0	5.7	6.2	8	58
90-94	0.2000	0.6250	0.6046	0.6468	14,523	13,033	16,250	9,077	45,385	4.3	4.1	4.4	1	5
95-99	0.3000	0.7895	0.7735	0.8062	5,446	4,603	6,419	4,300	14,332	3.1	3.0	3.2	0	1
100+	0.4200	1	1	1	1,147	892	1,449	1,147	2,730	2.4	2.1	2.4	0	0
All ages													*469*	*199,232*

ln($_nq_x$) with uncertainty interval

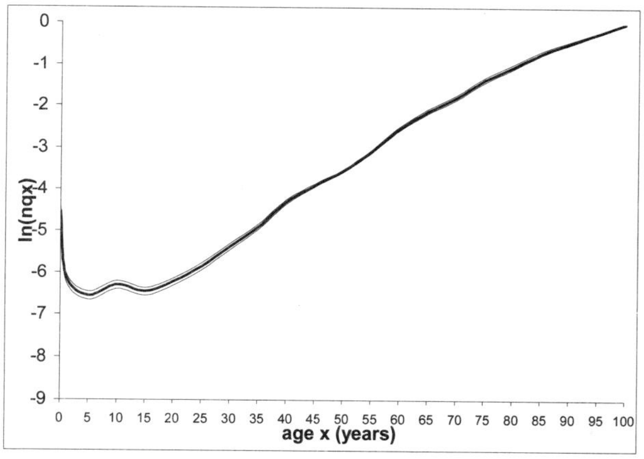

l_x with uncertainty interval

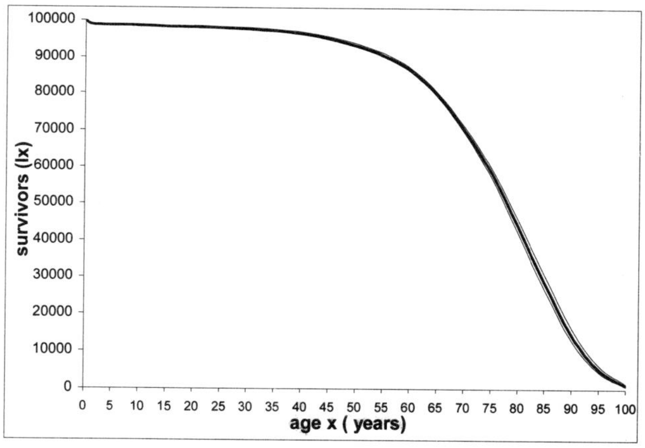

e_x with uncertainty interval

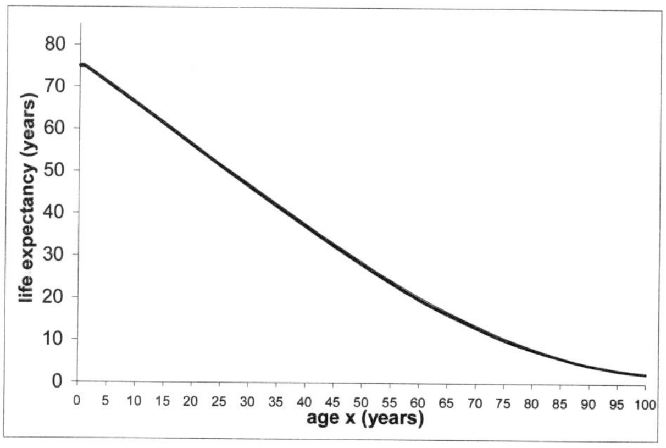

Distribution of deaths by age with uncertainty interval

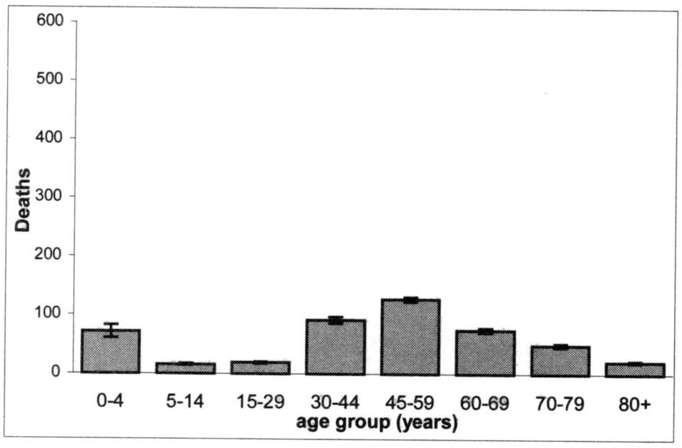

Republic of Korea　　République de Corée　　República de Corea

Males, 2000　　*Hommes, 2000*　　*Hombres, 2000*

Age	$_nM_x$	$_nq_x$ mean	$_nq_x$ percentiles 2½	$_nq_x$ percentiles 97½	l_x mean	l_x percentiles 2½	l_x percentiles 97½	$_nd_x$	$_nL_x$	e_x mean	e_x percentiles 2½	e_x percentiles 97½	Actual deaths	Actual population
<1	0.0078	0.0078	0.0054	0.0111	100,000	100,000	100,000	777	99,301	70.5	69.2	72.3	2,501	319,743
1-4	0.0006	0.0022	0.0016	0.0030	99,223	98,893	99,462	221	396,363	70.0	68.7	71.8	731	1,311,418
5-9	0.0004	0.0019	0.0014	0.0025	99,002	98,600	99,298	191	494,533	66.2	64.9	68.0	698	1,803,961
10-14	0.0003	0.0015	0.0011	0.0020	98,811	98,349	99,155	150	493,680	61.3	60.0	63.1	520	1,712,685
15-19	0.0009	0.0044	0.0034	0.0057	98,661	98,154	99,041	434	492,221	56.4	55.1	58.2	1,731	1,963,724
20-24	0.0011	0.0054	0.0042	0.0069	98,227	97,598	98,710	534	489,801	51.6	50.4	53.4	2,199	2,016,446
25-29	0.0013	0.0066	0.0052	0.0082	97,693	96,927	98,295	644	486,855	46.9	45.7	48.7	2,989	2,258,436
30-34	0.0017	0.0083	0.0066	0.0101	97,049	96,130	97,786	801	483,242	42.2	41.0	43.9	3,583	2,161,303
35-39	0.0026	0.0132	0.0106	0.0158	96,248	95,157	97,144	1,266	478,073	37.5	36.3	39.2	5,636	2,127,794
40-44	0.0041	0.0201	0.0166	0.0236	94,982	93,651	96,111	1,906	470,142	33.0	31.8	34.7	8,190	2,019,730
45-49	0.0064	0.0317	0.0267	0.0365	93,075	91,438	94,520	2,949	458,002	28.6	27.5	30.2	9,521	1,478,485
50-54	0.0089	0.0437	0.0367	0.0497	90,126	88,104	91,993	3,939	440,780	24.5	23.5	26.0	10,668	1,193,759
55-59	0.0141	0.0682	0.0568	0.0777	86,186	83,724	88,613	5,876	416,242	20.5	19.5	21.9	14,220	1,007,239
60-64	0.0214	0.1015	0.0844	0.1154	80,310	77,218	83,577	8,155	381,164	16.8	16.0	18.1	18,625	870,552
65-69	0.0327	0.1513	0.1263	0.1705	72,155	68,304	76,526	10,916	333,486	13.4	12.7	14.5	19,282	589,046
70-74	0.0521	0.2307	0.1959	0.2575	61,239	56,657	66,860	14,126	270,880	10.4	9.7	11.2	17,854	342,358
75-79	0.0875	0.3588	0.3149	0.3960	47,113	42,069	53,762	16,906	193,299	7.7	7.2	8.4	18,684	213,628
80-84	0.1421	0.5242	0.4769	0.5657	30,207	25,412	36,832	15,836	111,446	5.7	5.3	6.1	13,744	96,722
85-89	0.2173	0.7041	0.6621	0.7392	14,371	11,037	19,267	10,119	46,560	4.1	3.9	4.4	6,257	28,791
90-94	0.3129	0.8070	0.7747	0.8338	4,253	2,878	6,511	3,432	10,968	3.0	2.9	3.2	1,757	5,614
95-99	0.4241	0.9332	0.9156	0.9470	821	478	1,467	766	1,806	2.3	2.3	2.4	187	441
100+	0.5411	1	1	1	55	25	124	55	101	1.8	1.8	1.9	10	18
All ages													*159,585*	*23,521,893*

ln($_nq_x$) with uncertainty interval

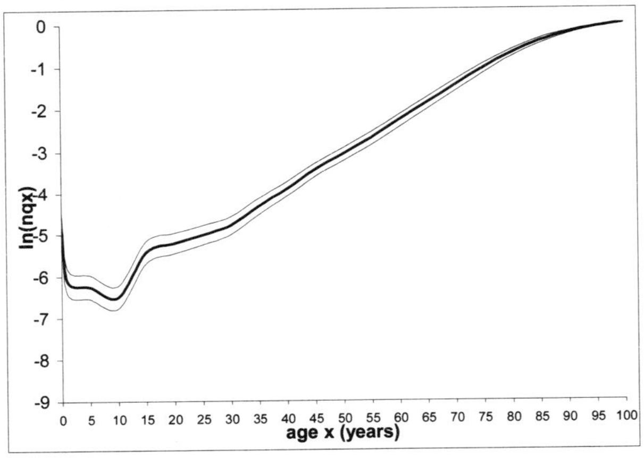

l_x with uncertainty interval

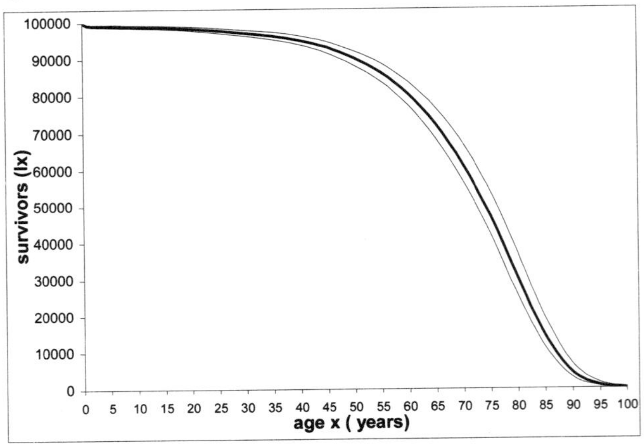

e_x with uncertainty interval

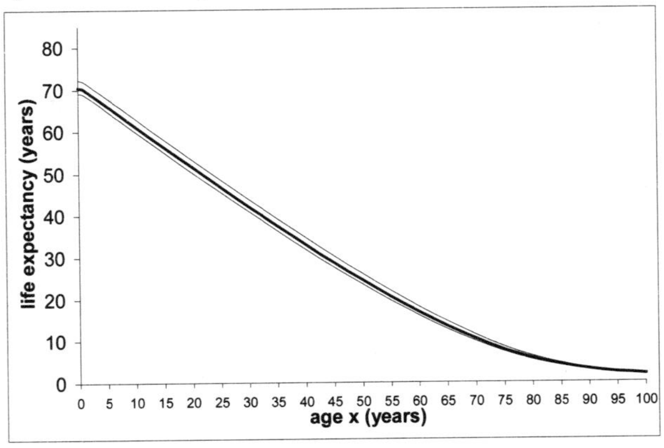

Distribution of deaths by age with uncertainty interval

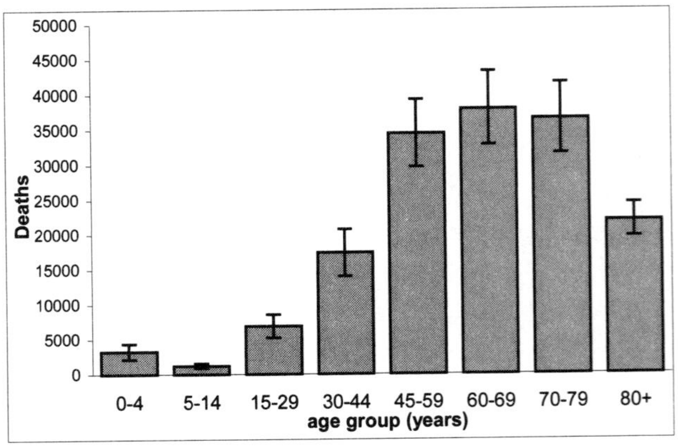

Republic of Korea / République de Corée / República de Corea

Females, 2000 / *Femmes, 2000* / *Mujeres, 2000*

Age	$_nM_x$	$_nq_x$			l_x			$_nd_x$	$_nL_x$	e_x			Actual deaths	Actual population
			percentiles			percentiles					percentiles			
		mean	2½	97½	mean	2½	97½			mean	2½	97½		
<1	0.0071	0.0071	0.0054	0.0104	100,000	100,000	100,000	710	99,361	78.3	76.9	79.9	2,117	296,333
1-4	0.0005	0.0019	0.0015	0.0026	99,290	98,958	99,457	190	396,705	77.9	76.5	79.5	564	1,177,562
5-9	0.0002	0.0011	0.0009	0.0015	99,100	98,702	99,303	113	495,219	74.0	72.7	75.6	362	1,586,336
10-14	0.0002	0.0010	0.0008	0.0014	98,987	98,552	99,212	101	494,685	69.1	67.8	70.7	312	1,532,019
15-19	0.0004	0.0022	0.0018	0.0028	98,886	98,419	99,130	213	493,900	64.2	62.8	65.8	793	1,837,498
20-24	0.0005	0.0024	0.0019	0.0031	98,673	98,141	98,956	235	492,780	59.3	58.0	60.9	906	1,900,426
25-29	0.0006	0.0029	0.0024	0.0037	98,439	97,839	98,764	282	491,488	54.5	53.1	56.0	1,231	2,145,299
30-34	0.0007	0.0036	0.0030	0.0045	98,157	97,480	98,531	352	489,903	49.6	48.3	51.2	1,481	2,060,342
35-39	0.0010	0.0052	0.0043	0.0064	97,805	97,037	98,239	504	487,762	44.8	43.5	46.3	2,139	2,069,308
40-44	0.0014	0.0069	0.0058	0.0085	97,300	96,412	97,818	673	484,820	40.0	38.8	41.5	2,743	1,977,014
45-49	0.0021	0.0104	0.0088	0.0126	96,628	95,592	97,254	1,006	480,622	35.3	34.1	36.8	3,001	1,433,306
50-54	0.0031	0.0155	0.0131	0.0186	95,621	94,385	96,400	1,483	474,399	30.6	29.5	32.1	3,702	1,183,979
55-59	0.0048	0.0239	0.0202	0.0286	94,138	92,627	95,135	2,253	465,059	26.0	25.0	27.5	5,015	1,035,221
60-64	0.0080	0.0391	0.0325	0.0466	91,885	89,977	93,212	3,592	450,447	21.6	20.6	23.0	7,618	955,277
65-69	0.0142	0.0685	0.0563	0.0812	88,293	85,780	90,179	6,047	426,349	17.4	16.5	18.7	10,956	772,443
70-74	0.0264	0.1239	0.1014	0.1445	82,246	78,818	85,098	10,190	385,757	13.5	12.8	14.6	14,590	552,336
75-79	0.0511	0.2267	0.1879	0.2568	72,057	67,425	76,467	16,333	319,451	10.1	9.4	11.0	19,585	383,076
80-84	0.0930	0.3773	0.3245	0.4187	55,724	50,113	62,100	21,025	226,057	7.3	6.8	8.0	19,554	210,237
85-89	0.1578	0.5658	0.5113	0.6082	34,699	29,129	41,949	19,631	124,416	5.2	4.8	5.6	13,563	85,959
90-94	0.2496	0.7137	0.6674	0.7471	15,068	11,414	20,501	10,753	43,079	3.6	3.4	3.9	5,487	21,981
95-99	0.3683	0.8748	0.8454	0.8946	4,315	2,886	6,818	3,774	10,249	2.6	2.5	2.8	821	2,229
100+	0.5066	1	1	1	540	304	1,054	540	1,066	2.0	2.0	2.1	34	67
All ages													*116,572*	*23,218,248*

ln($_nq_x$) with uncertainty interval

l_x with uncertainty interval

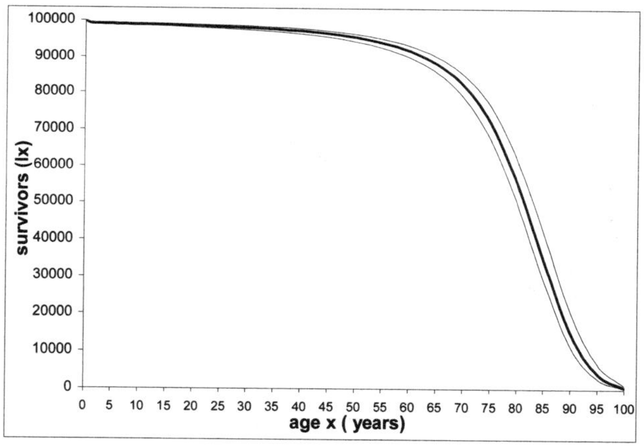

e_x with uncertainty interval

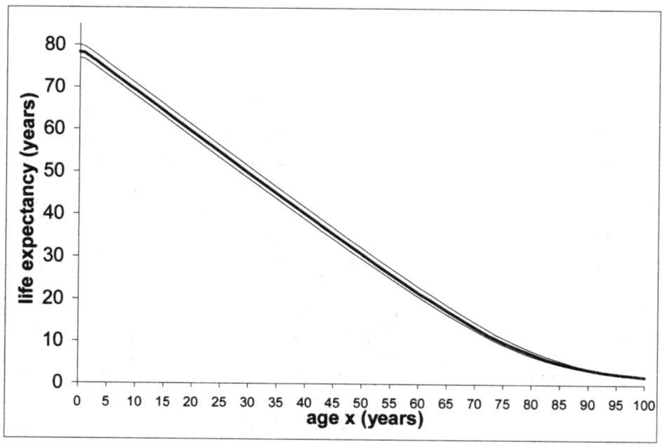

Distribution of deaths by age with uncertainty interval

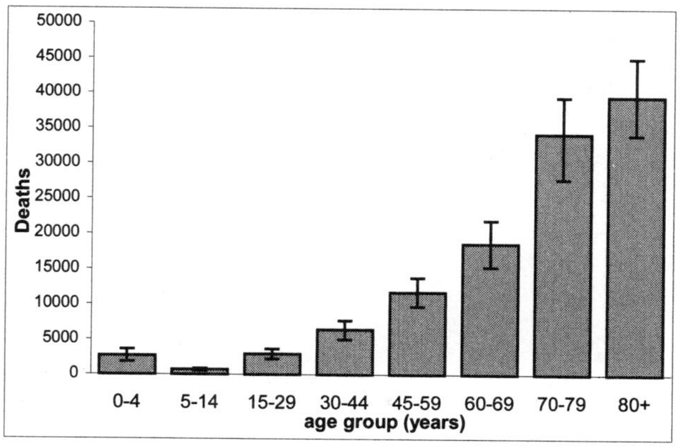

Republic of Moldova — République de Moldova — República de Moldova

Males, 2000 — *Hommes, 2000* — *Hombres, 2000*

Age	$_nM_x$	$_nq_x$			l_x			$_nd_x$	$_nL_x$	e_x			Actual deaths	Actual population
			percentiles			percentiles					percentiles			
		mean	2½	97½	mean	2½	97½			mean	2½	97½		
<1	0.0216	0.0212	0.0186	0.0238	100,000	100,000	100,000	2,116	98,095	63.1	62.4	63.8	580	26,864
1-4	0.0012	0.0049	0.0044	0.0054	97,884	97,621	98,136	480	390,383	63.5	62.8	64.2	130	105,722
5-9	0.0006	0.0029	0.0026	0.0032	97,404	97,095	97,700	284	486,309	59.8	59.1	60.5	96	164,824
10-14	0.0005	0.0026	0.0024	0.0029	97,120	96,785	97,442	255	484,962	54.9	54.3	55.7	110	209,937
15-19	0.0012	0.0058	0.0053	0.0063	96,865	96,507	97,209	558	482,931	50.1	49.4	50.8	235	203,381
20-24	0.0019	0.0097	0.0089	0.0105	96,307	95,900	96,698	930	479,210	45.3	44.7	46.1	352	181,135
25-29	0.0030	0.0150	0.0138	0.0163	95,377	94,891	95,842	1,432	473,304	40.8	40.1	41.5	469	154,966
30-34	0.0039	0.0194	0.0179	0.0209	93,945	93,347	94,518	1,822	465,169	36.3	35.6	37.1	530	135,343
35-39	0.0052	0.0258	0.0240	0.0277	92,123	91,394	92,823	2,378	454,669	32.0	31.3	32.8	794	151,886
40-44	0.0085	0.0414	0.0388	0.0442	89,745	88,862	90,595	3,717	439,430	27.8	27.1	28.6	1,401	165,596
45-49	0.0120	0.0582	0.0549	0.0615	86,027	84,938	87,082	5,004	417,628	23.9	23.2	24.7	1,763	147,115
50-54	0.0169	0.0812	0.0762	0.0864	81,024	79,711	82,304	6,580	388,668	20.2	19.6	21.0	1,736	102,522
55-59	0.0259	0.1217	0.1131	0.1303	74,444	72,821	76,032	9,060	349,569	16.8	16.2	17.5	1,953	75,356
60-64	0.0372	0.1703	0.1573	0.1829	65,384	63,330	67,430	11,136	299,081	13.8	13.2	14.4	2,973	79,856
65-69	0.0551	0.2420	0.2234	0.2591	54,248	51,749	56,825	13,128	238,422	11.1	10.6	11.6	3,346	60,769
70-74	0.0723	0.3061	0.2839	0.3265	41,120	38,342	44,130	12,587	174,135	8.8	8.4	9.2	3,430	47,450
75-79	0.1136	0.4422	0.4157	0.4664	28,534	25,823	31,602	12,619	111,122	6.6	6.3	6.9	2,995	26,378
80-84	0.1780	0.6159	0.5899	0.6393	15,915	13,778	18,464	9,802	55,070	4.8	4.7	5.0	1,765	9,914
85-89	0.2619	0.7913	0.7717	0.8087	6,113	4,970	7,571	4,837	18,471	3.6	3.5	3.7	1,083	4,136
90-94	0.3615	0.8671	0.8527	0.8800	1,276	951	1,728	1,106	3,060	2.7	2.6	2.8	154	427
95-99	0.4684	0.9737	0.9683	0.9778	170	114	255	165	352	2.1	2.1	2.2	3	7
100+	0.5696	1	1	1	4	3	8	4	8	1.8	1.8	1.8	0	0
All ages													*25,898*	*2,053,584*

$\ln(_nq_x)$ with uncertainty interval

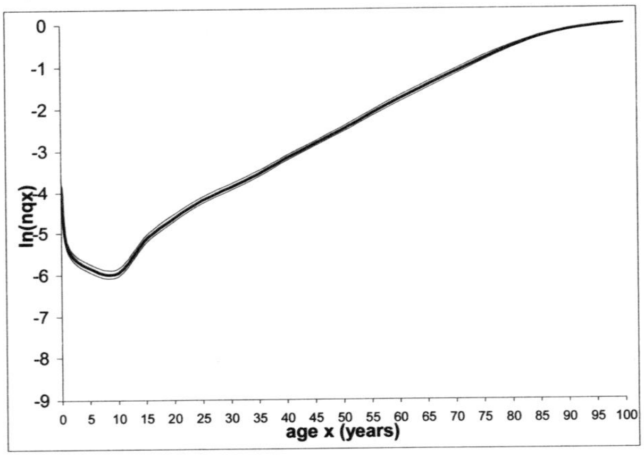

l_x with uncertainty interval

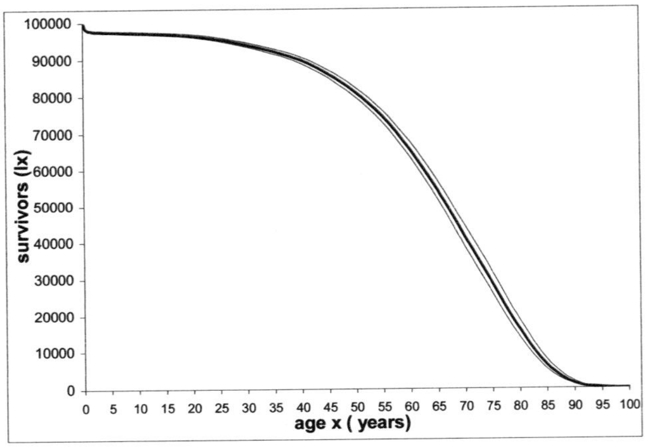

e_x with uncertainty interval

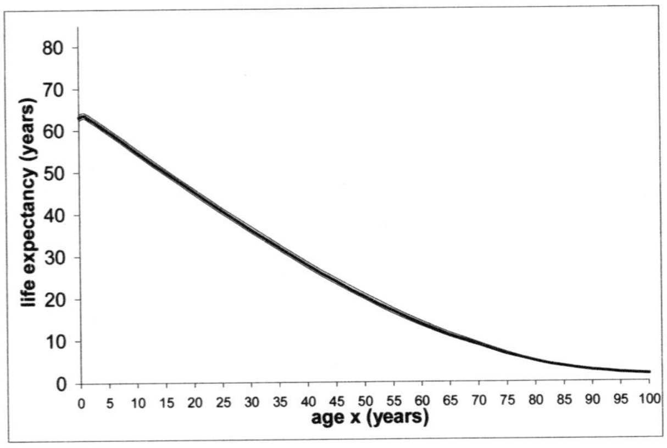

Distribution of deaths by age with uncertainty interval

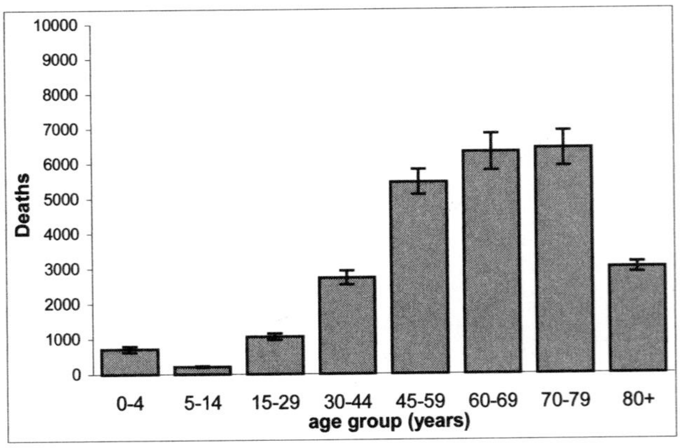

Republic of Moldova — République de Moldova — República de Moldova

Females, 2000 — *Femmes, 2000* — *Mujeres, 2000*

Age	$_nM_x$	$_nq_x$ mean	$_nq_x$ percentiles 2½	$_nq_x$ percentiles 97½	l_x mean	l_x percentiles 2½	l_x percentiles 97½	$_nd_x$	$_nL_x$	e_x mean	e_x percentiles 2½	e_x percentiles 97½	Actual deaths	Actual population
<1	0.0170	0.0168	0.0143	0.0194	100,000	100,000	100,000	1,679	98,489	70.5	69.6	71.4	432	25,343
1-4	0.0011	0.0043	0.0037	0.0048	98,321	98,058	98,568	419	392,280	70.7	69.8	71.7	107	100,451
5-9	0.0005	0.0025	0.0022	0.0028	97,902	97,588	98,200	245	488,900	66.9	66.1	68.0	79	157,716
10-14	0.0004	0.0018	0.0016	0.0020	97,658	97,316	97,984	177	487,844	62.1	61.3	63.1	74	202,240
15-19	0.0005	0.0025	0.0022	0.0028	97,480	97,117	97,826	246	486,785	57.2	56.4	58.3	100	198,236
20-24	0.0006	0.0029	0.0026	0.0033	97,234	96,844	97,608	287	485,453	52.4	51.6	53.4	105	177,816
25-29	0.0009	0.0044	0.0039	0.0049	96,947	96,525	97,353	429	483,663	47.5	46.7	48.5	140	157,436
30-34	0.0013	0.0066	0.0059	0.0073	96,518	96,051	96,969	640	480,990	42.7	41.9	43.7	181	136,237
35-39	0.0020	0.0101	0.0090	0.0111	95,878	95,346	96,394	965	476,978	38.0	37.2	39.0	338	167,093
40-44	0.0030	0.0146	0.0132	0.0161	94,913	94,289	95,522	1,390	471,091	33.3	32.5	34.4	536	181,780
45-49	0.0056	0.0276	0.0251	0.0301	93,523	92,775	94,259	2,580	461,166	28.8	28.0	29.8	920	164,525
50-54	0.0082	0.0402	0.0368	0.0436	90,943	89,986	91,892	3,653	445,584	24.5	23.7	25.5	978	119,325
55-59	0.0140	0.0675	0.0610	0.0735	87,290	86,058	88,512	5,894	421,716	20.5	19.7	21.4	1,333	95,351
60-64	0.0211	0.1004	0.0897	0.1095	81,396	79,736	83,112	8,168	386,559	16.8	16.0	17.6	2,269	107,405
65-69	0.0342	0.1576	0.1403	0.1734	73,228	71,008	75,657	11,539	337,292	13.3	12.7	14.1	2,980	87,099
70-74	0.0519	0.2296	0.2052	0.2528	61,689	58,693	65,046	14,166	273,030	10.4	9.8	11.0	4,053	78,123
75-79	0.0855	0.3522	0.3200	0.3830	47,523	43,853	51,697	16,737	195,774	7.7	7.3	8.2	4,472	52,306
80-84	0.1439	0.5291	0.4918	0.5620	30,786	27,057	35,155	16,288	113,212	5.6	5.3	5.9	2,990	20,784
85-89	0.2251	0.7202	0.6888	0.7469	14,498	11,852	17,865	10,441	46,389	4.0	3.8	4.2	2,385	10,597
90-94	0.3274	0.8258	0.8033	0.8448	4,057	3,000	5,560	3,350	10,234	2.9	2.8	3.0	592	1,809
95-99	0.4426	0.9507	0.9401	0.9589	707	466	1,094	672	1,518	2.2	2.2	2.3	83	187
100+	0.5563	1	1	1	35	19	65	35	63	1.8	1.8	1.9	6	10
All ages													*25,154*	*2,241,869*

ln($_nq_x$) with uncertainty interval

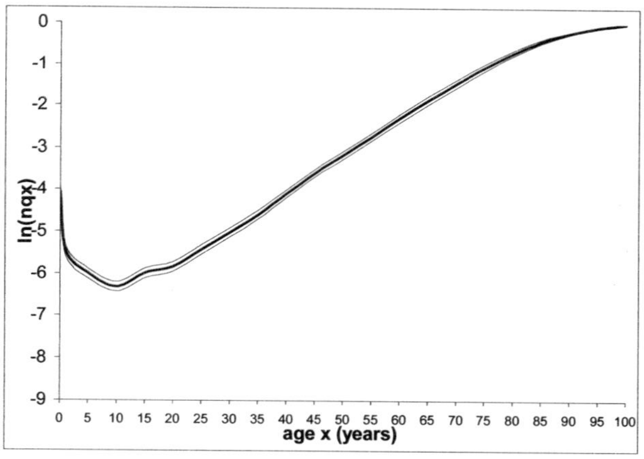

l_x with uncertainty interval

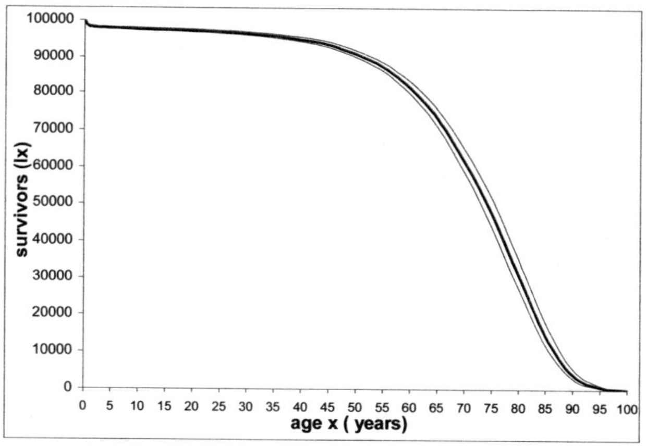

e_x with uncertainty interval

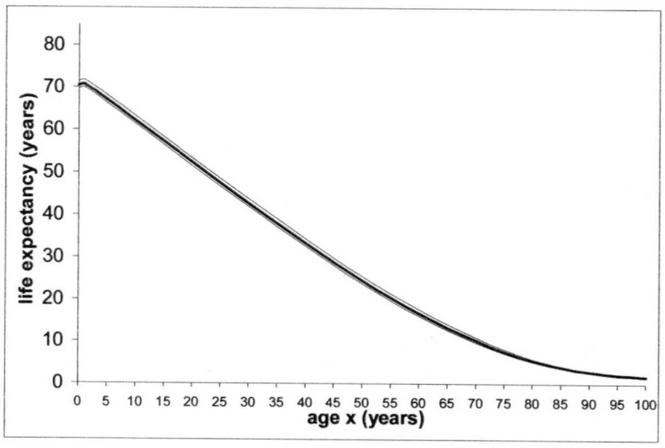

Distribution of deaths by age with uncertainty interval

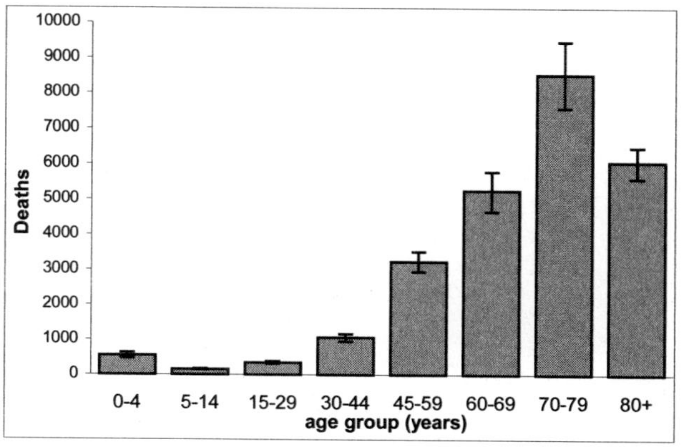

Romania   Roumanie   Rumania

Males, 2000   *Hommes, 2000*   *Hombres, 2000*

Age	$_nM_x$	$_nq_x$ mean	$_nq_x$ percentiles 2½	$_nq_x$ percentiles 97½	l_x mean	l_x percentiles 2½	l_x percentiles 97½	$_nd_x$	$_nL_x$	e_x mean	e_x percentiles 2½	e_x percentiles 97½	Actual deaths	Actual population
<1	0.0208	0.0204	0.0182	0.0227	100,000	100,000	100,000	2,045	98,160	66.2	65.5	67.0	2,452	117,703
1-4	0.0012	0.0047	0.0042	0.0051	97,955	97,733	98,177	456	390,727	66.6	65.9	67.4	544	466,313
5-9	0.0007	0.0037	0.0034	0.0040	97,499	97,250	97,749	358	486,602	62.9	62.2	63.7	460	625,505
10-14	0.0008	0.0038	0.0034	0.0041	97,141	96,867	97,418	365	484,794	58.2	57.4	58.9	667	885,166
15-19	0.0009	0.0045	0.0042	0.0049	96,776	96,475	97,079	438	482,785	53.4	52.6	54.1	769	846,717
20-24	0.0011	0.0057	0.0052	0.0061	96,338	96,006	96,675	548	480,318	48.6	47.9	49.4	1,137	995,763
25-29	0.0015	0.0077	0.0071	0.0083	95,790	95,418	96,169	734	477,112	43.9	43.1	44.6	1,412	917,603
30-34	0.0021	0.0103	0.0096	0.0111	95,055	94,629	95,489	981	472,825	39.2	38.4	40.0	1,938	934,257
35-39	0.0038	0.0187	0.0174	0.0201	94,075	93,579	94,576	1,759	465,975	34.6	33.8	35.4	2,444	647,243
40-44	0.0064	0.0316	0.0294	0.0338	92,315	91,702	92,932	2,915	454,289	30.2	29.4	31.0	5,097	794,280
45-49	0.0098	0.0479	0.0447	0.0512	89,400	88,612	90,186	4,283	436,293	26.1	25.3	26.9	7,710	785,351
50-54	0.0140	0.0677	0.0632	0.0721	85,117	84,108	86,115	5,761	411,184	22.3	21.5	23.0	9,006	642,869
55-59	0.0205	0.0976	0.0907	0.1047	79,357	78,069	80,645	7,742	377,427	18.7	18.0	19.4	10,301	502,170
60-64	0.0296	0.1378	0.1271	0.1488	71,614	69,998	73,211	9,865	333,408	15.4	14.8	16.1	16,774	566,884
65-69	0.0418	0.1894	0.1742	0.2047	61,749	59,852	63,649	11,693	279,513	12.5	12.0	13.1	20,505	490,176
70-74	0.0614	0.2661	0.2457	0.2862	50,056	47,753	52,348	13,320	216,982	9.9	9.4	10.3	23,505	382,895
75-79	0.0915	0.3724	0.3473	0.3960	36,736	34,087	39,474	13,682	149,477	7.5	7.2	7.9	21,430	234,125
80-84	0.1480	0.5401	0.5127	0.5647	23,054	20,589	25,752	12,452	84,142	5.5	5.3	5.8	11,417	77,145
85-89	0.2250	0.7200	0.6970	0.7402	10,602	8,973	12,529	7,634	33,928	4.0	3.9	4.2	11,221	49,870
90-94	0.3217	0.8185	0.8013	0.8337	2,969	2,341	3,789	2,430	7,554	3.0	2.9	3.1	4,242	13,187
95-99	0.4325	0.9412	0.9324	0.9487	539	392	753	507	1,173	2.3	2.2	2.3	736	1,703
100+	0.5500	1	1	1	32	20	51	32	58	1.8	1.8	1.9	54	99
All ages													*153,820*	*10,977,024*

ln($_nq_x$) with uncertainty interval

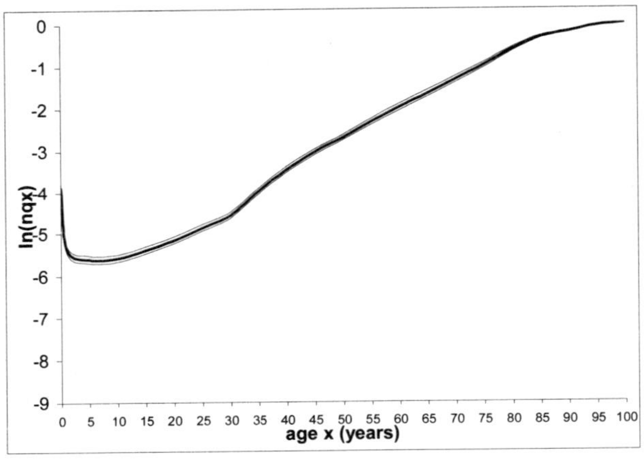

l_x with uncertainty interval

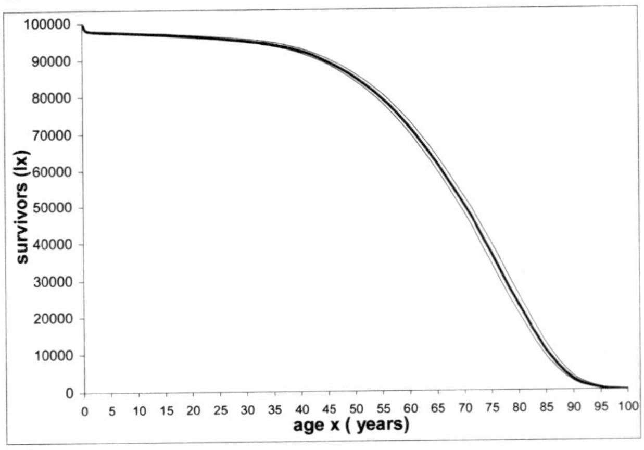

e_x with uncertainty interval

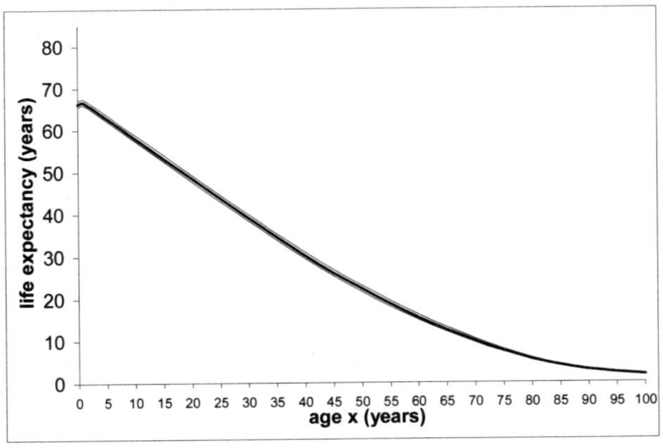

Distribution of deaths by age with uncertainty interval

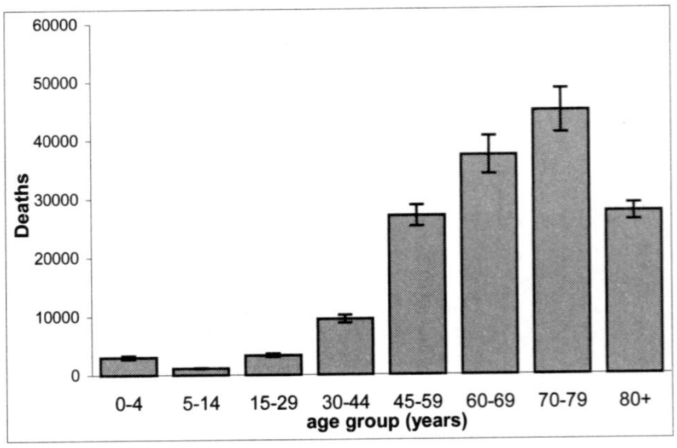

Romania / Roumanie / Rumania

Females, 2000 / *Femmes, 2000* / *Mujeres, 2000*

Age	$_nM_x$	$_nq_x$ mean	$_nq_x$ percentiles 2½	$_nq_x$ percentiles 97½	l_x mean	l_x percentiles 2½	l_x percentiles 97½	$_nd_x$	$_nL_x$	e_x mean	e_x percentiles 2½	e_x percentiles 97½	Actual deaths	Actual population
<1	0.0174	0.0172	0.0150	0.0194	100,000	100,000	100,000	1,717	98,455	73.5	72.6	74.6	1,946	111,620
1-4	0.0010	0.0039	0.0035	0.0043	98,283	98,060	98,500	381	392,220	73.7	72.9	74.9	428	441,464
5-9	0.0005	0.0024	0.0021	0.0026	97,903	97,654	98,151	232	488,934	70.0	69.2	71.2	282	595,428
10-14	0.0005	0.0024	0.0021	0.0026	97,671	97,400	97,942	230	487,778	65.2	64.3	66.4	402	851,803
15-19	0.0004	0.0022	0.0020	0.0024	97,441	97,148	97,735	216	486,662	60.3	59.5	61.5	361	813,449
20-24	0.0005	0.0023	0.0021	0.0026	97,224	96,912	97,541	228	485,554	55.5	54.6	56.7	448	955,645
25-29	0.0006	0.0031	0.0027	0.0034	96,997	96,664	97,337	297	484,243	50.6	49.7	51.8	545	890,291
30-34	0.0009	0.0044	0.0040	0.0048	96,700	96,340	97,070	426	482,437	45.7	44.8	46.9	803	909,380
35-39	0.0015	0.0076	0.0068	0.0083	96,274	95,877	96,685	728	479,552	40.9	40.0	42.1	970	639,048
40-44	0.0025	0.0126	0.0114	0.0137	95,547	95,086	96,026	1,201	474,729	36.2	35.3	37.4	2,030	802,268
45-49	0.0040	0.0197	0.0179	0.0215	94,345	93,788	94,934	1,858	467,080	31.6	30.7	32.8	3,215	808,021
50-54	0.0058	0.0285	0.0258	0.0311	92,487	91,785	93,233	2,637	455,841	27.2	26.3	28.4	3,924	678,308
55-59	0.0087	0.0424	0.0377	0.0464	89,849	88,962	90,801	3,810	439,724	23.0	22.1	24.1	4,833	557,907
60-64	0.0135	0.0651	0.0571	0.0715	86,040	84,876	87,253	5,600	416,199	18.9	18.0	19.9	9,007	669,394
65-69	0.0228	0.1078	0.0935	0.1205	80,440	78,894	82,007	8,674	380,515	15.0	14.2	16.0	13,815	606,062
70-74	0.0395	0.1798	0.1553	0.2027	71,766	69,625	74,284	12,903	326,574	11.5	10.8	12.3	20,065	507,842
75-79	0.0709	0.3010	0.2632	0.3371	58,863	55,856	62,748	17,720	250,015	8.5	7.9	9.2	25,641	361,770
80-84	0.1252	0.4769	0.4307	0.5200	41,143	37,029	46,230	19,621	156,661	6.1	5.6	6.5	17,601	140,536
85-89	0.2043	0.6762	0.6369	0.7120	21,522	17,772	26,320	14,552	71,229	4.3	4.1	4.6	18,896	92,492
90-94	0.3076	0.7999	0.7738	0.8245	6,970	5,118	9,557	5,575	18,123	3.1	2.9	3.2	7,541	24,513
95-99	0.4276	0.9366	0.9243	0.9480	1,394	898	2,161	1,306	3,054	2.3	2.3	2.4	1,383	3,234
100+	0.5487	1	1	1	88	47	164	88	161	1.8	1.8	1.9	116	211
All ages													*134,255*	*11,460,686*

ln($_nq_x$) with uncertainty interval

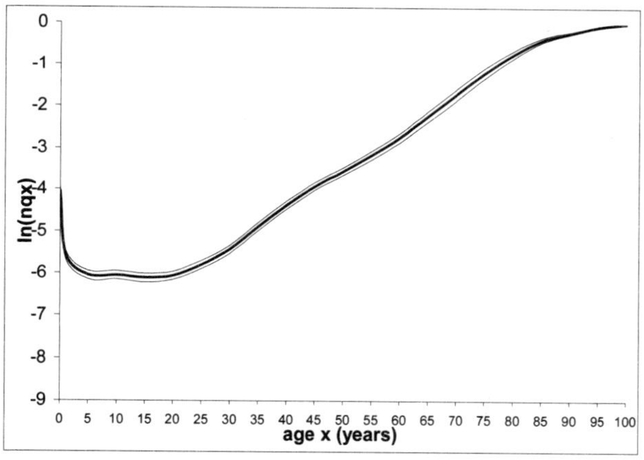

l_x with uncertainty interval

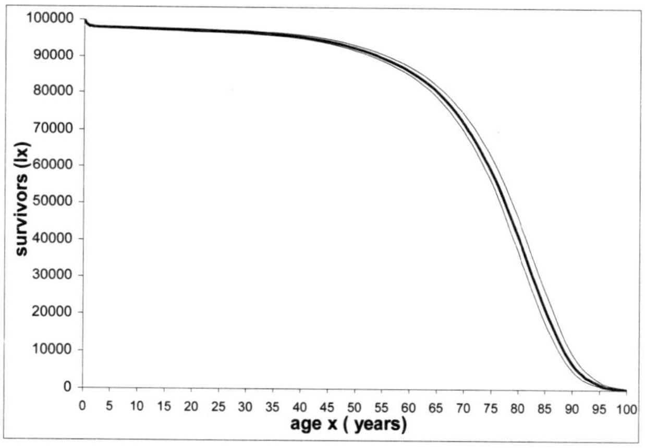

e_x with uncertainty interval

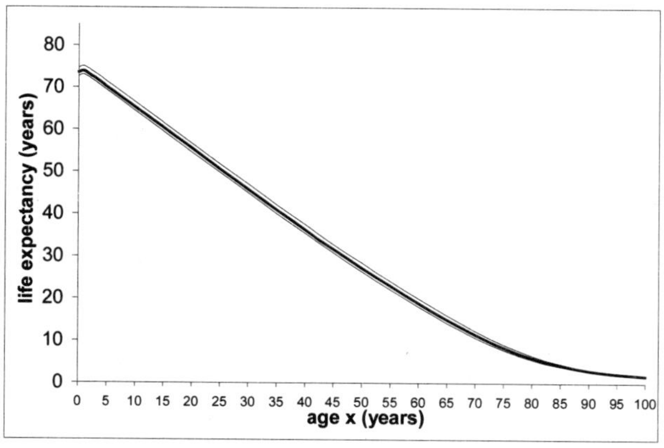

Distribution of deaths by age with uncertainty interval

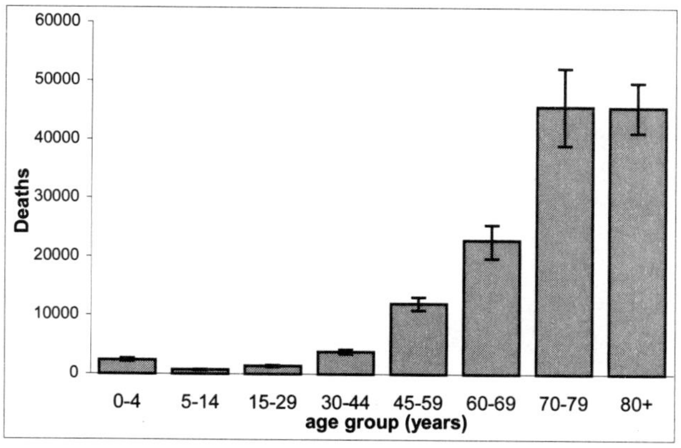

Russian Federation | Fédération de Russie | Federación de Rusia

Males, 2000 | *Hommes, 2000* | *Hombres, 2000*

Age	$_nM_x$	$_nq_x$			l_x			$_nd_x$	$_nL_x$	e_x			Actual deaths	Actual population
			percentiles			percentiles					percentiles			
		mean	2½	97½	mean	2½	97½			mean	2½	97½		
<1	0.0187	0.0184	0.0171	0.0197	100,000	100,000	100,000	1,836	98,348	59.4	58.2	60.5	12,385	663,372
1-4	0.0012	0.0049	0.0046	0.0052	98,164	98,033	98,295	478	391,508	59.5	58.3	60.6	3,164	2,590,910
5-9	0.0007	0.0036	0.0033	0.0038	97,686	97,536	97,836	348	487,559	55.8	54.6	56.9	2,927	4,097,544
10-14	0.0007	0.0036	0.0033	0.0039	97,338	97,166	97,506	350	485,813	51.0	49.8	52.1	4,329	6,005,263
15-19	0.0021	0.0107	0.0099	0.0115	96,987	96,789	97,174	1,036	482,347	46.1	45.0	47.3	12,970	6,038,019
20-24	0.0047	0.0231	0.0214	0.0251	95,951	95,671	96,197	2,220	474,206	41.6	40.5	42.8	25,624	5,473,320
25-29	0.0054	0.0267	0.0246	0.0291	93,731	93,272	94,133	2,505	462,395	37.5	36.5	38.7	28,178	5,202,029
30-34	0.0068	0.0332	0.0305	0.0362	91,227	90,562	91,815	3,029	448,560	33.5	32.5	34.6	32,515	4,814,711
35-39	0.0086	0.0423	0.0389	0.0463	88,197	87,289	89,011	3,731	431,659	29.6	28.6	30.7	49,815	5,763,178
40-44	0.0118	0.0575	0.0529	0.0630	84,466	83,269	85,552	4,859	410,182	25.8	24.9	26.8	72,972	6,159,502
45-49	0.0167	0.0801	0.0732	0.0876	79,607	78,045	81,029	6,373	382,100	22.2	21.4	23.2	91,164	5,465,569
50-54	0.0229	0.1084	0.0988	0.1182	73,233	71,210	75,058	7,941	346,316	18.9	18.2	19.8	94,838	4,136,186
55-59	0.0325	0.1502	0.1369	0.1626	65,293	62,795	67,537	9,805	301,952	15.9	15.2	16.7	77,317	2,381,040
60-64	0.0438	0.1973	0.1805	0.2121	55,488	52,585	58,122	10,946	250,076	13.3	12.7	13.9	158,980	3,632,257
65-69	0.0608	0.2641	0.2438	0.2821	44,542	41,458	47,629	11,762	193,308	10.9	10.4	11.5	139,858	2,298,646
70-74	0.0808	0.3360	0.3140	0.3557	32,781	29,856	36,015	11,014	136,369	8.9	8.6	9.3	165,752	2,052,215
75-79	0.1040	0.4128	0.3909	0.4325	21,767	19,355	24,706	8,985	86,371	7.2	6.9	7.5	79,786	766,968
80-84	0.1529	0.5530	0.5320	0.5722	12,782	11,048	15,048	7,069	46,236	5.5	5.3	5.7	50,247	328,666
85-89	0.2180	0.7055	0.6880	0.7216	5,713	4,736	7,043	4,030	18,489	4.2	4.0	4.3	42,578	195,330
90-94	0.3008	0.7906	0.7767	0.8037	1,683	1,319	2,197	1,330	4,423	3.1	3.0	3.2	16,280	54,121
95-99	0.4200	0.9292	0.9215	0.9363	352	259	491	327	780	2.3	2.3	2.4	4,202	10,004
100+	0.5400	1	1	1	25	17	39	25	46	1.9	1.8	1.9	667	1,235
All ages													*1,166,546*	*68,130,085*

ln($_nq_x$) with uncertainty interval

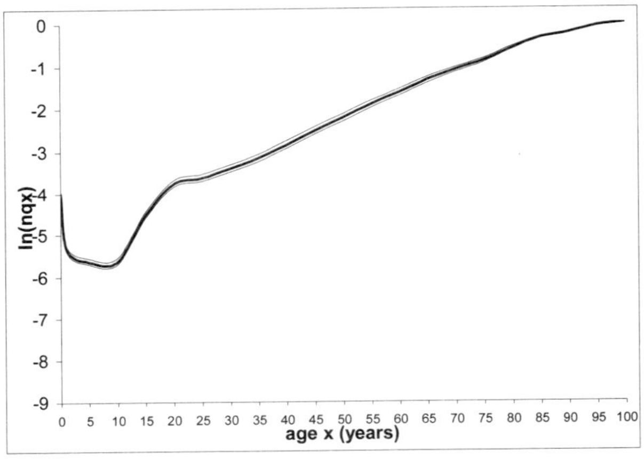

l_x with uncertainty interval

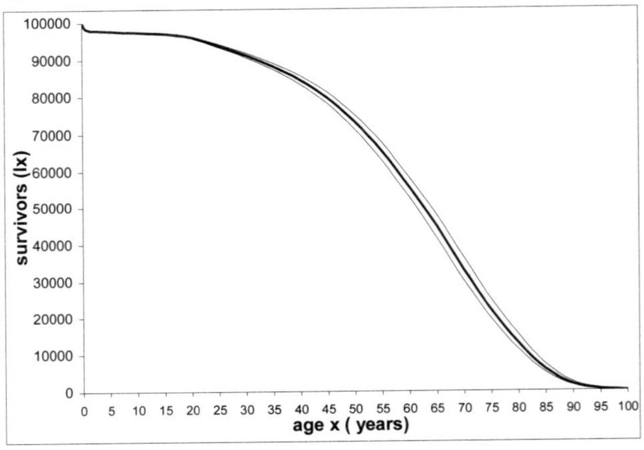

e_x with uncertainty interval

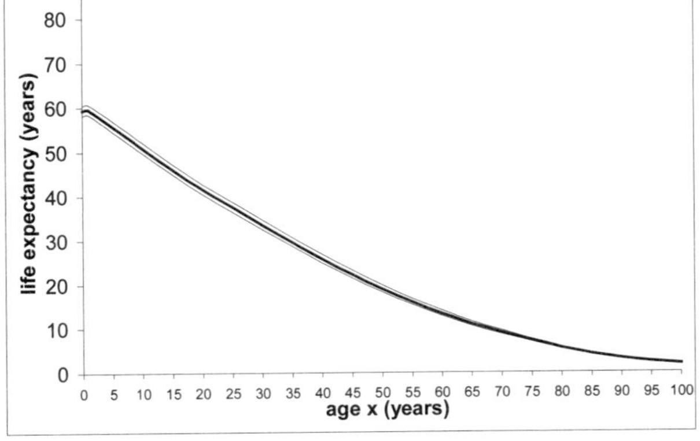

Distribution of deaths by age with uncertainty interval

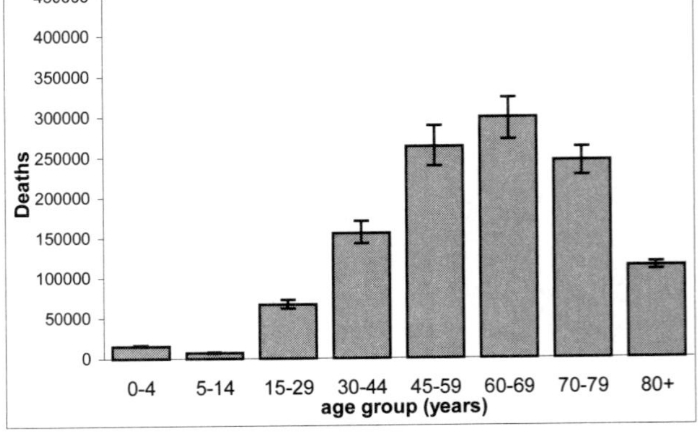

Russian Federation — Fédération de Russie — Federación de Rusia

Females, 2000 — *Femmes, 2000* — *Mujeres, 2000*

Age	$_nM_x$	$_nq_x$			l_x			$_nd_x$	$_nL_x$	e_x			Actual deaths	Actual population
			percentiles			percentiles					percentiles			
		mean	2½	97½	mean	2½	97½			mean	2½	97½		
<1	0.0136	0.0134	0.0122	0.0148	100,000	100,000	100,000	1,344	98,791	72.0	71.3	72.7	8,537	627,599
1-4	0.0009	0.0037	0.0034	0.0040	98,656	98,523	98,784	364	393,752	72.0	71.2	72.7	2,292	2,480,562
5-9	0.0004	0.0021	0.0019	0.0022	98,292	98,143	98,442	203	490,954	68.2	67.5	69.0	1,611	3,892,156
10-14	0.0003	0.0017	0.0016	0.0018	98,089	97,925	98,252	168	490,025	63.4	62.6	64.1	1,980	5,765,338
15-19	0.0008	0.0042	0.0039	0.0045	97,921	97,744	98,095	409	488,583	58.5	57.7	59.2	4,905	5,865,794
20-24	0.0012	0.0058	0.0054	0.0062	97,512	97,305	97,715	565	486,149	53.7	53.0	54.5	6,246	5,375,839
25-29	0.0013	0.0066	0.0061	0.0071	96,947	96,700	97,188	641	483,136	49.0	48.3	49.7	6,667	5,027,663
30-34	0.0017	0.0084	0.0078	0.0090	96,307	96,014	96,592	813	479,502	44.3	43.6	45.1	7,951	4,690,640
35-39	0.0023	0.0113	0.0105	0.0120	95,494	95,147	95,832	1,075	474,782	39.7	39.0	40.4	13,217	5,835,230
40-44	0.0032	0.0161	0.0150	0.0172	94,419	94,003	94,825	1,520	468,294	35.1	34.4	35.8	20,924	6,448,190
45-49	0.0049	0.0243	0.0226	0.0260	92,899	92,391	93,396	2,258	458,850	30.6	30.0	31.3	29,348	5,962,997
50-54	0.0074	0.0362	0.0337	0.0388	90,641	90,007	91,276	3,280	445,003	26.3	25.7	27.0	35,062	4,756,380
55-59	0.0111	0.0541	0.0503	0.0581	87,360	86,563	88,194	4,724	424,991	22.2	21.6	22.9	33,887	3,048,370
60-64	0.0161	0.0776	0.0717	0.0836	82,636	81,600	83,731	6,409	397,158	18.3	17.8	19.0	82,647	5,121,518
65-69	0.0268	0.1254	0.1155	0.1352	76,227	74,805	77,626	9,558	357,240	14.7	14.2	15.2	98,357	3,676,205
70-74	0.0432	0.1949	0.1796	0.2091	66,669	64,760	68,511	12,996	300,854	11.4	11.0	11.9	174,335	4,035,697
75-79	0.0710	0.3015	0.2802	0.3217	53,673	51,219	56,148	16,180	227,914	8.6	8.2	9.0	170,419	2,400,588
80-84	0.1224	0.4685	0.4428	0.4926	37,493	34,842	40,414	17,567	143,548	6.2	6.0	6.5	146,386	1,196,213
85-89	0.1967	0.6593	0.6358	0.6801	19,926	17,799	22,489	13,137	66,788	4.5	4.3	4.7	165,808	842,950
90-94	0.2800	0.7609	0.7432	0.7763	6,789	5,693	8,142	5,166	18,449	3.3	3.2	3.4	72,586	259,235
95-99	0.3900	0.8986	0.8880	0.9077	1,623	1,273	2,083	1,459	3,741	2.5	2.5	2.6	18,911	48,489
100+	0.5357	1	1	1	165	117	233	165	307	1.9	1.8	2.1	1,836	3,428
All ages													*1,103,912*	*77,361,081*

ln($_nq_x$) with uncertainty interval

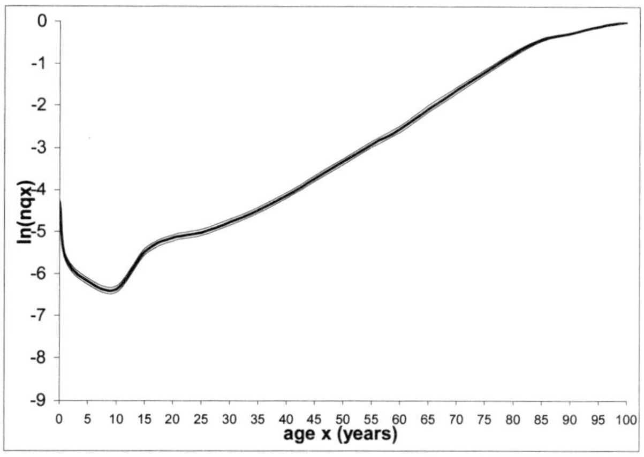

l_x with uncertainty interval

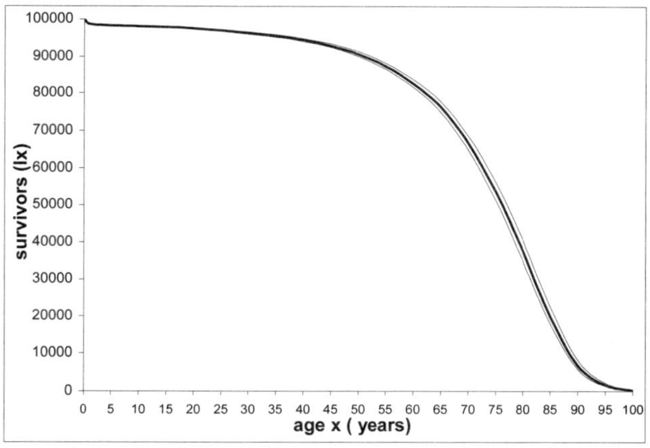

e_x with uncertainty interval

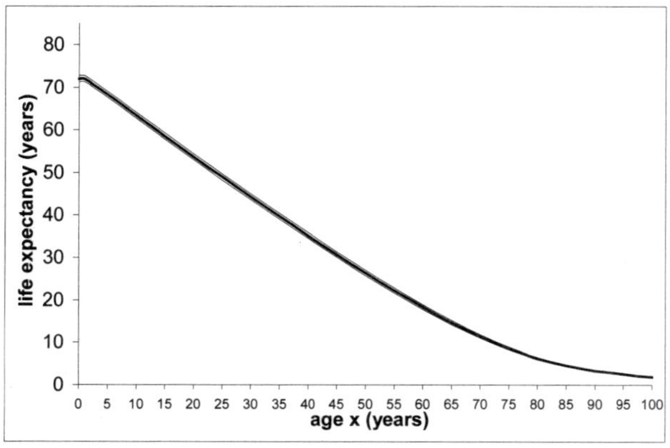

Distribution of deaths by age with uncertainty interval

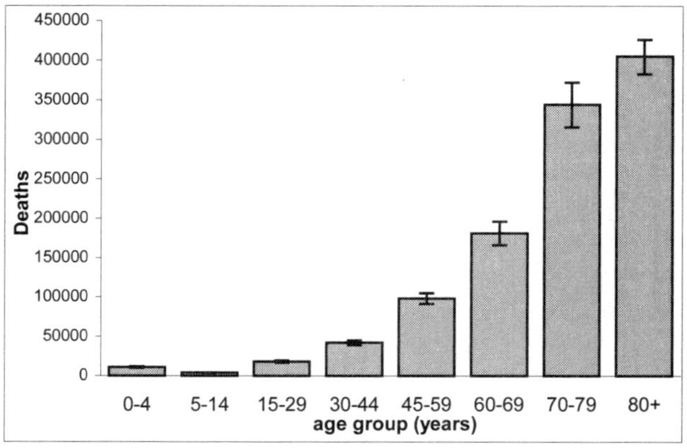

Rwanda — Rwanda — Rwanda

Males, 2000 — *Hommes, 2000* — *Hombres, 2000*

Age	$_nM_x$	$_nq_x$			l_x			$_nd_x$	$_nL_x$	e_x			Actual deaths	Actual population
			percentiles			percentiles					percentiles			
		mean	2 1/2	97 1/2	mean	2 1/2	97 1/2			mean	2 1/2	97 1/2		
<1	0.1876	0.1605	0.1460	0.1752	100,000	100,000	100,000	16,048	85,556	38.5	36.0	41.6	25,584	136,390
1-4	0.0182	0.0697	0.0610	0.0782	83,952	82,476	85,396	5,855	321,753	44.9	42.1	48.5	9,194	505,241
5-9	0.0041	0.0205	0.0176	0.0233	78,096	76,848	79,345	1,598	386,485	44.1	41.4	47.6	2,249	543,868
10-14	0.0029	0.0142	0.0121	0.0162	76,498	75,099	77,878	1,087	379,773	40.0	37.3	43.4	1,420	496,330
15-19	0.0034	0.0171	0.0145	0.0196	75,411	73,880	76,896	1,288	373,836	35.5	32.8	38.9	1,585	459,790
20-24	0.0087	0.0427	0.0361	0.0491	74,123	72,435	75,777	3,169	362,693	31.1	28.4	34.4	3,312	379,100
25-29	0.0175	0.0838	0.0701	0.0964	70,954	68,882	73,030	5,943	339,915	27.4	24.7	30.6	5,206	297,804
30-34	0.0268	0.1256	0.1041	0.1459	65,012	62,240	67,876	8,168	304,639	24.7	22.0	27.7	6,146	229,230
35-39	0.0305	0.1419	0.1167	0.1673	56,844	53,219	60,734	8,065	264,056	22.8	20.4	25.6	5,369	175,782
40-44	0.0323	0.1493	0.1224	0.1775	48,778	44,507	53,546	7,285	225,681	21.2	19.0	23.7	4,341	134,498
45-49	0.0332	0.1533	0.1258	0.1827	41,494	36,861	46,910	6,359	191,571	19.5	17.5	21.7	3,433	103,424
50-54	0.0332	0.1533	0.1262	0.1826	35,134	30,393	41,010	5,386	162,206	17.6	15.9	19.4	3,029	91,209
55-59	0.0337	0.1553	0.1285	0.1845	29,748	25,087	35,835	4,621	137,189	15.3	13.9	16.8	2,428	72,093
60-64	0.0455	0.2044	0.1711	0.2404	25,127	20,476	31,231	5,136	112,796	12.7	11.5	14.0	2,511	55,140
65-69	0.0608	0.2638	0.2242	0.3061	19,991	15,554	25,889	5,274	86,769	10.3	9.3	11.3	2,411	39,662
70-74	0.0883	0.3618	0.3141	0.4114	14,717	10,793	20,085	5,324	60,273	8.0	7.3	8.9	2,243	25,395
75-79	0.1299	0.4903	0.4365	0.5440	9,393	6,352	13,773	4,605	35,451	6.2	5.6	6.8	1,722	13,256
80-84	0.1848	0.6319	0.5783	0.6834	4,788	2,897	7,743	3,026	16,374	4.7	4.3	5.2	965	5,220
85-89	0.2646	0.7962	0.7512	0.8371	1,762	917	3,246	1,403	5,303	3.5	3.3	3.8	398	1,506
90-94	0.3657	0.8720	0.8362	0.9033	359	149	799	313	856	2.6	2.5	2.8	90	247
95-99	0.4885	0.9907	0.9838	0.9951	46	14	129	46	93	2.0	2.0	2.1	11	22
100+	0.6300	1	1	1	0	0	2	0	1	1.6	1.6	1.6	1	1
All ages													*83,649*	*3,765,208*

ln($_nq_x$) with uncertainty interval

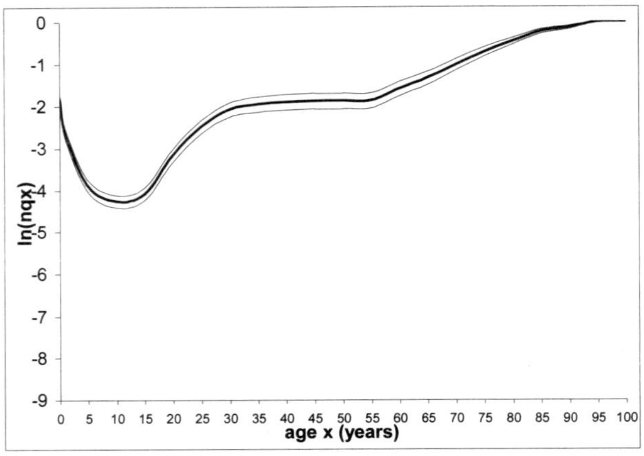

l_x with uncertainty interval

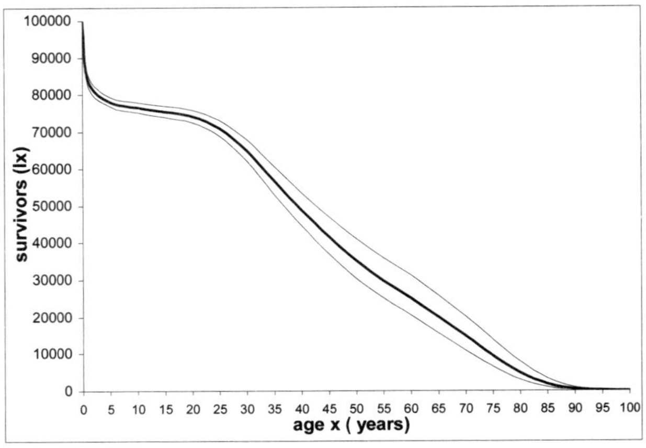

e_x with uncertainty interval

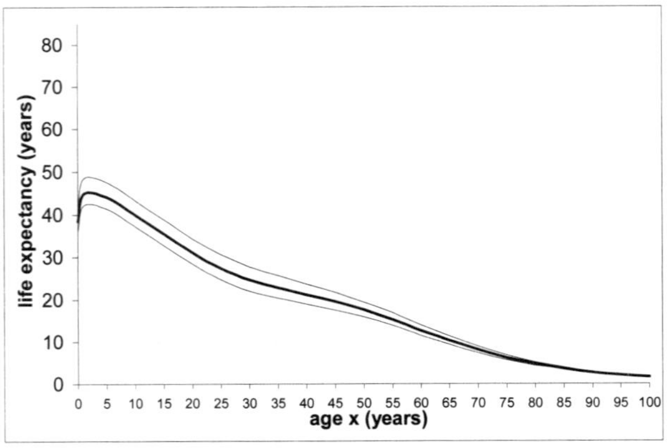

Distribution of deaths by age with uncertainty interval

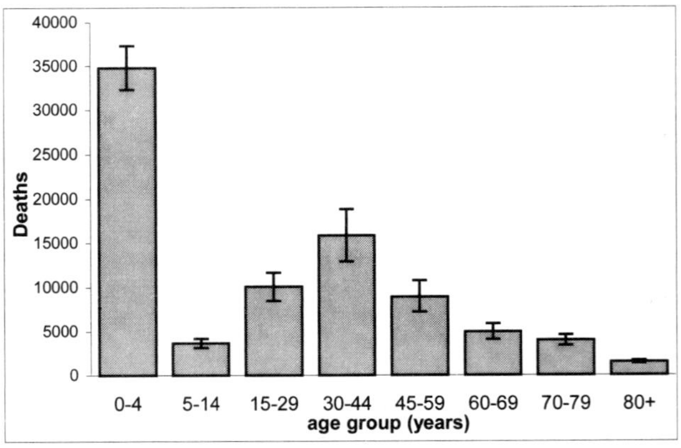

Rwanda

Females, 2000

Rwanda

Femmes, 2000

Rwanda

Mujeres, 2000

Age	$_nM_x$	$_nq_x$ mean	$_nq_x$ percentiles 2½	$_nq_x$ percentiles 97½	l_x mean	l_x percentiles 2½	l_x percentiles 97½	$_nd_x$	$_nL_x$	e_x mean	e_x percentiles 2½	e_x percentiles 97½	Actual deaths	Actual population
<1	0.1528	0.1343	0.1188	0.1481	100,000	100,000	100,000	13,430	87,913	40.5	37.4	44.0	20,111	131,650
1-4	0.0196	0.0748	0.0658	0.0845	86,570	85,185	88,122	6,474	330,742	45.8	42.4	49.9	9,982	509,973
5-9	0.0042	0.0208	0.0179	0.0240	80,096	78,846	81,341	1,667	396,313	45.4	42.1	49.3	2,294	545,420
10-14	0.0029	0.0145	0.0124	0.0169	78,429	77,071	79,830	1,141	389,295	41.3	38.1	45.2	1,469	501,426
15-19	0.0066	0.0324	0.0275	0.0377	77,289	75,781	78,814	2,502	380,188	36.9	33.7	40.7	3,074	467,059
20-24	0.0146	0.0704	0.0591	0.0824	74,786	72,944	76,587	5,262	360,775	33.0	29.9	36.8	5,620	385,313
25-29	0.0232	0.1094	0.0907	0.1290	69,524	66,932	71,899	7,609	328,597	30.3	27.2	34.0	6,893	297,700
30-34	0.0275	0.1285	0.1053	0.1520	61,915	58,300	65,302	7,958	289,681	28.8	25.8	32.1	6,205	225,886
35-39	0.0252	0.1184	0.0961	0.1412	53,957	49,440	58,340	6,386	253,821	27.6	25.0	30.6	4,383	174,212
40-44	0.0220	0.1043	0.0844	0.1254	47,571	42,511	52,639	4,962	225,452	26.0	23.6	28.6	3,011	136,837
45-49	0.0196	0.0935	0.0756	0.1129	42,610	37,265	48,129	3,985	203,087	23.7	21.7	26.0	2,143	109,225
50-54	0.0207	0.0982	0.0795	0.1186	38,625	33,150	44,492	3,794	183,641	20.9	19.1	22.9	2,042	98,863
55-59	0.0234	0.1105	0.0897	0.1331	34,831	29,322	40,955	3,847	164,539	17.9	16.4	19.7	1,901	81,307
60-64	0.0313	0.1452	0.1189	0.1739	30,984	25,535	37,280	4,500	143,672	14.9	13.5	16.3	2,016	64,382
65-69	0.0450	0.2022	0.1676	0.2395	26,485	21,238	32,849	5,356	119,034	12.0	10.8	13.2	2,162	48,054
70-74	0.0688	0.2935	0.2482	0.3413	21,129	16,326	27,343	6,202	90,140	9.3	8.4	10.4	2,272	33,025
75-79	0.1028	0.4088	0.3549	0.4640	14,927	10,846	20,556	6,103	59,378	7.2	6.5	8.0	2,043	19,873
80-84	0.1530	0.5533	0.4950	0.6101	8,824	5,813	13,255	4,882	31,915	5.4	4.9	6.0	1,458	9,531
85-89	0.2226	0.7151	0.6613	0.7649	3,942	2,267	6,672	2,819	12,662	4.1	3.7	4.5	721	3,240
90-94	0.3153	0.8102	0.7652	0.8506	1,123	533	2,241	910	2,885	3.0	2.8	3.3	208	660
95-99	0.4348	0.9434	0.9193	0.9623	213	80	520	201	462	2.3	2.2	2.4	34	79
100+	0.5835	1	1	1	12	3	41	12	21	1.7	1.7	1.8	3	5
All ages													*80,049*	*3,843,720*

$\ln(_nq_x)$ with uncertainty interval

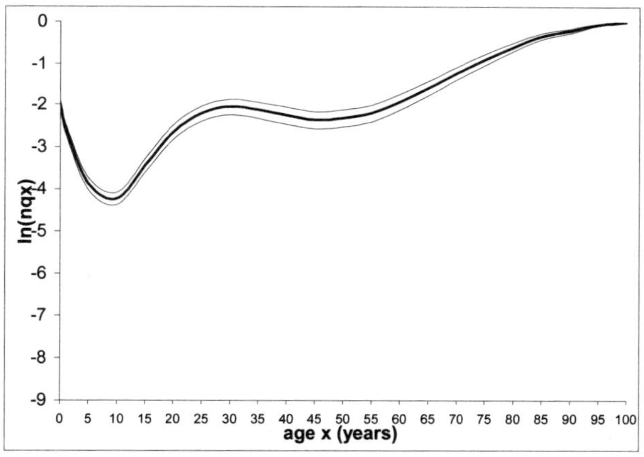

l_x with uncertainty interval

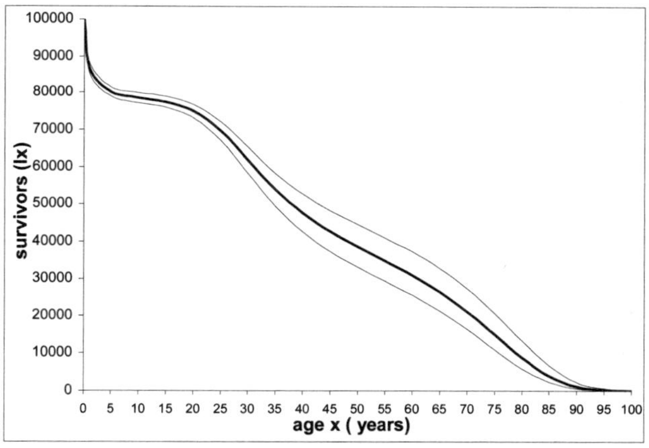

e_x with uncertainty interval

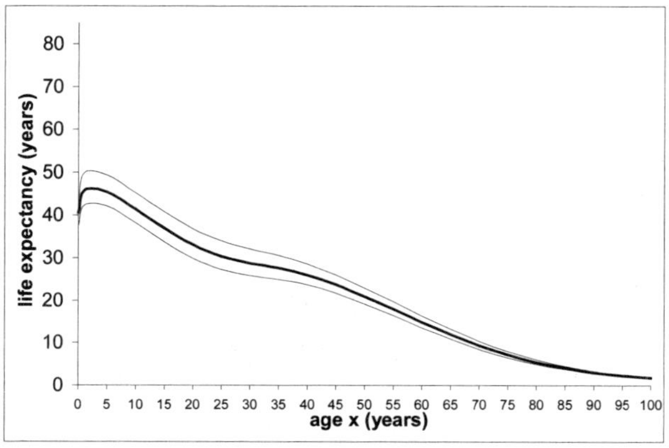

Distribution of deaths by age with uncertainty interval

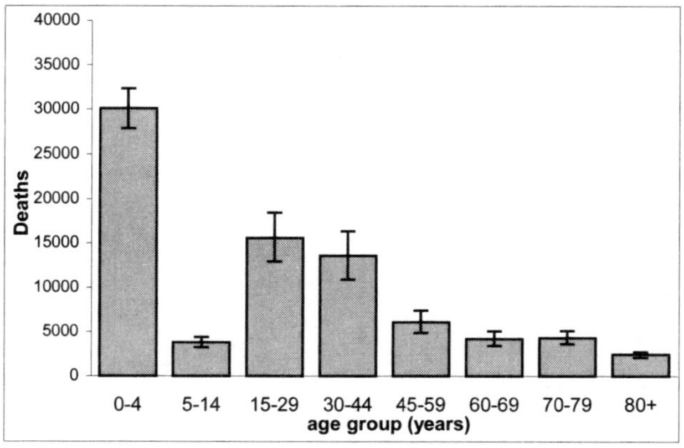

Saint Kitts and Nevis   Saint-Kitts-et-Nevis   Saint Kitts y Nevis

Males, 2000   *Hommes, 2000*   *Hombres, 2000*

Age	$_nM_x$	$_nq_x$ mean	$_nq_x$ percentiles 2½	$_nq_x$ percentiles 97½	l_x mean	l_x percentiles 2½	l_x percentiles 97½	$_nd_x$	$_nL_x$	e_x mean	e_x percentiles 2½	e_x percentiles 97½	Actual deaths	Actual population
<1	0.0235	0.0230	0.0199	0.0255	100,000	100,000	100,000	2,298	97,932	66.1	65.3	67.4	9	381
1-4	0.0006	0.0024	0.0021	0.0026	97,702	97,452	98,006	234	390,248	66.6	65.9	67.9	1	1,512
5-9	0.0003	0.0015	0.0014	0.0017	97,468	97,197	97,798	150	486,966	62.8	62.1	64.0	1	1,905
10-14	0.0005	0.0023	0.0021	0.0026	97,318	97,033	97,664	228	486,021	57.9	57.2	59.1	1	1,985
15-19	0.0008	0.0041	0.0037	0.0045	97,090	96,785	97,462	402	484,448	53.0	52.3	54.2	2	1,848
20-24	0.0011	0.0056	0.0050	0.0061	96,689	96,348	97,104	544	482,083	48.2	47.5	49.4	2	1,647
25-29	0.0019	0.0092	0.0082	0.0100	96,144	95,756	96,618	886	478,506	43.4	42.8	44.7	3	1,653
30-34	0.0024	0.0117	0.0104	0.0127	95,258	94,796	95,826	1,119	473,493	38.8	38.1	40.0	4	1,546
35-39	0.0045	0.0222	0.0198	0.0240	94,139	93,589	94,826	2,095	465,460	34.3	33.6	35.4	6	1,397
40-44	0.0061	0.0299	0.0266	0.0321	92,045	91,340	92,949	2,748	453,354	30.0	29.3	31.1	6	1,057
45-49	0.0074	0.0361	0.0321	0.0386	89,297	88,410	90,478	3,227	438,417	25.8	25.2	26.9	7	935
50-54	0.0136	0.0656	0.0581	0.0700	86,070	84,994	87,570	5,643	416,243	21.7	21.0	22.7	11	810
55-59	0.0180	0.0861	0.0761	0.0922	80,427	79,043	82,484	6,927	384,820	18.1	17.4	19.0	12	672
60-64	0.0294	0.1371	0.1215	0.1477	73,501	71,755	76,208	10,078	342,308	14.5	13.9	15.3	16	549
65-69	0.0526	0.2323	0.2084	0.2502	63,423	61,155	66,946	14,733	280,280	11.4	10.9	12.1	22	417
70-74	0.0737	0.3112	0.2841	0.3337	48,689	45,852	52,994	15,151	205,569	9.1	8.7	9.6	25	333
75-79	0.0950	0.3838	0.3575	0.4088	33,538	30,549	37,936	12,873	135,509	7.1	6.8	7.4	22	227
80-84	0.1584	0.5674	0.5408	0.5939	20,665	18,059	24,373	11,725	74,013	5.0	4.8	5.2	23	143
85-89	0.2852	0.8325	0.8144	0.8501	8,940	7,334	11,192	7,443	26,094	3.3	3.2	3.4	20	69
90-94	0.4408	0.9490	0.9409	0.9566	1,497	1,099	2,078	1,421	3,224	2.3	2.2	2.3	10	22
95-99	0.5846	0.9800	0.9763	0.9841	76	48	123	75	157	2.1	2.1	2.1	3	4
100+	0.6654	1	1	1	2	1	3	2	2	1.5	1.5	1.8	0	1
All ages													*203*	*19,112*

ln($_nq_x$) with uncertainty interval

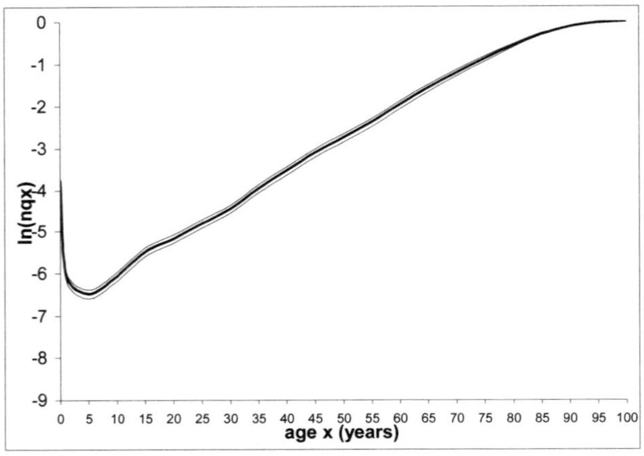

l_x with uncertainty interval

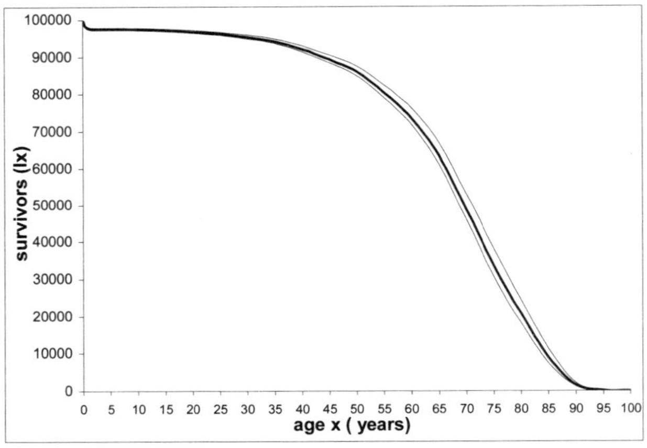

e_x with uncertainty interval

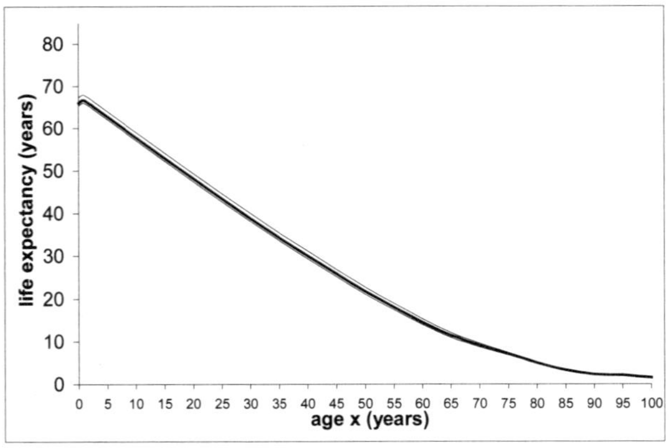

Distribution of deaths by age with uncertainty interval

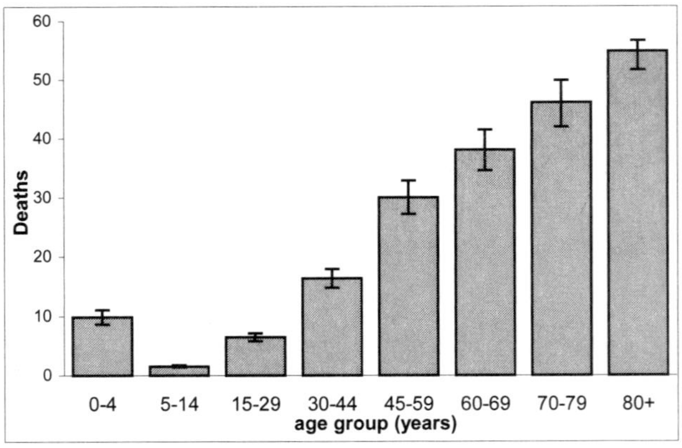

Saint Kitts and Nevis — Saint-Kitts-et-Nevis — Saint Kitts y Nevis

Females, 2000 — *Femmes, 2000* — *Mujeres, 2000*

Age	$_nM_x$	$_nq_x$			l_x			$_nd_x$	$_nL_x$	e_x			Actual deaths	Actual population
			percentiles			percentiles					percentiles			
		mean	21/2	971/2	mean	21/2	971/2			mean	21/2	971/2		
<1	0.0174	0.0172	0.0138	0.0197	100,000	100,000	100,000	1,715	98,456	72.0	70.9	73.3	6	369
1-4	0.0012	0.0049	0.0042	0.0055	98,285	98,032	98,618	483	391,981	72.3	71.1	73.7	2	1,448
5-9	0.0001	0.0006	0.0005	0.0007	97,802	97,493	98,204	59	488,864	68.6	67.5	70.1	0	1,834
10-14	0.0001	0.0007	0.0006	0.0008	97,744	97,428	98,154	71	488,540	63.7	62.5	65.1	0	1,921
15-19	0.0004	0.0022	0.0019	0.0024	97,672	97,348	98,092	211	487,835	58.7	57.6	60.1	1	1,788
20-24	0.0007	0.0035	0.0030	0.0039	97,461	97,113	97,909	341	486,456	53.8	52.7	55.3	1	1,602
25-29	0.0012	0.0060	0.0052	0.0067	97,121	96,735	97,613	581	484,152	49.0	47.9	50.4	2	1,622
30-34	0.0014	0.0072	0.0063	0.0080	96,540	96,090	97,104	696	480,960	44.3	43.2	45.7	2	1,539
35-39	0.0022	0.0109	0.0096	0.0121	95,844	95,319	96,491	1,049	476,599	39.6	38.5	41.0	3	1,445
40-44	0.0030	0.0149	0.0132	0.0165	94,796	94,162	95,562	1,411	470,450	35.0	33.9	36.4	3	1,140
45-49	0.0046	0.0228	0.0203	0.0252	93,384	92,608	94,302	2,127	461,604	30.5	29.4	31.9	5	1,012
50-54	0.0066	0.0327	0.0291	0.0364	91,258	90,275	92,390	2,982	448,833	26.1	25.0	27.5	6	878
55-59	0.0117	0.0568	0.0496	0.0636	88,276	86,988	89,705	5,013	428,847	21.9	20.9	23.3	9	728
60-64	0.0194	0.0923	0.0794	0.1037	83,263	81,454	85,252	7,685	397,102	18.1	17.1	19.3	12	601
65-69	0.0291	0.1357	0.1157	0.1542	75,578	73,011	78,482	10,255	352,252	14.7	13.8	15.8	14	472
70-74	0.0456	0.2046	0.1747	0.2333	65,323	61,756	69,406	13,366	293,201	11.6	10.8	12.5	18	386
75-79	0.0604	0.2623	0.2263	0.2970	51,957	47,351	57,281	13,626	225,721	8.9	8.3	9.6	16	270
80-84	0.1164	0.4509	0.4030	0.4957	38,331	33,285	44,316	17,282	148,451	6.2	5.8	6.7	21	177
85-89	0.2023	0.6717	0.6291	0.7116	21,049	16,786	26,458	14,139	69,898	4.3	4.0	4.6	18	90
90-94	0.3167	0.8120	0.7843	0.8395	6,910	4,841	9,813	5,611	17,718	3.0	2.8	3.1	10	32
95-99	0.4467	0.9544	0.9438	0.9650	1,299	777	2,117	1,240	2,776	2.2	2.2	2.3	3	7
100+	0.5677	1	1	1	59	27	119	59	104	1.8	1.8	1.9	1	1
All ages													*153*	*19,361*

ln($_nq_x$) with uncertainty interval

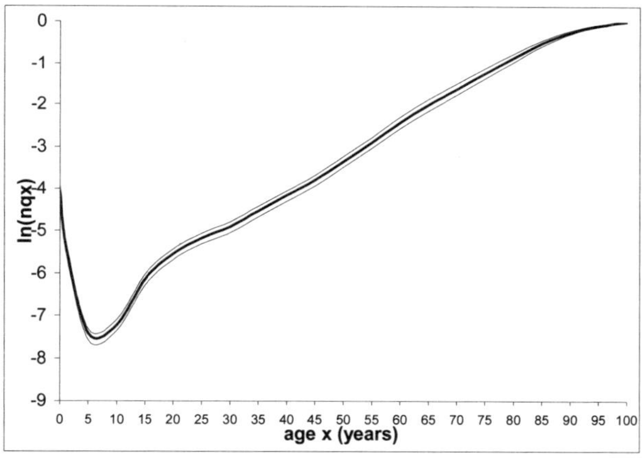

l_x with uncertainty interval

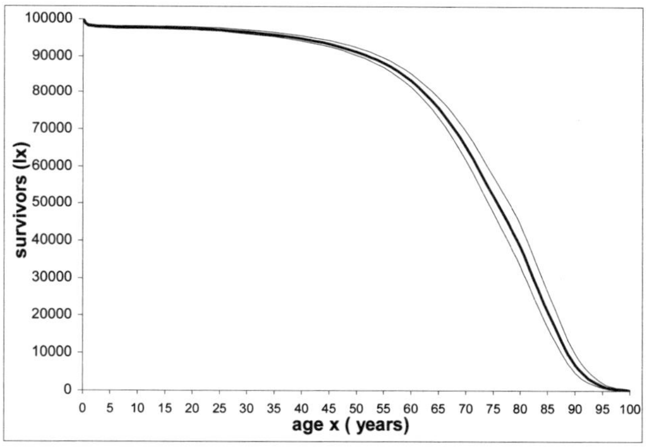

e_x with uncertainty interval

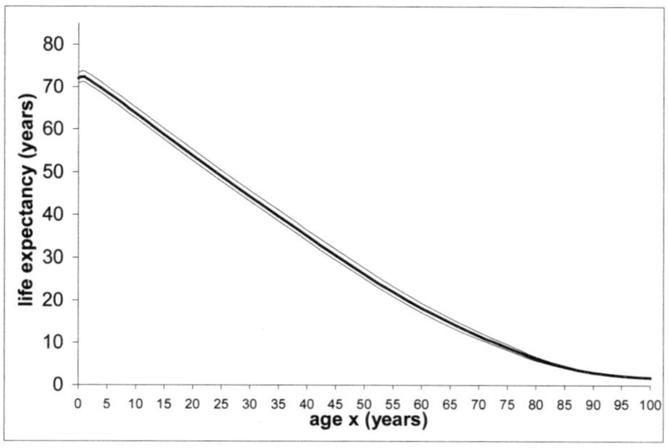

Distribution of deaths by age with uncertainty interval

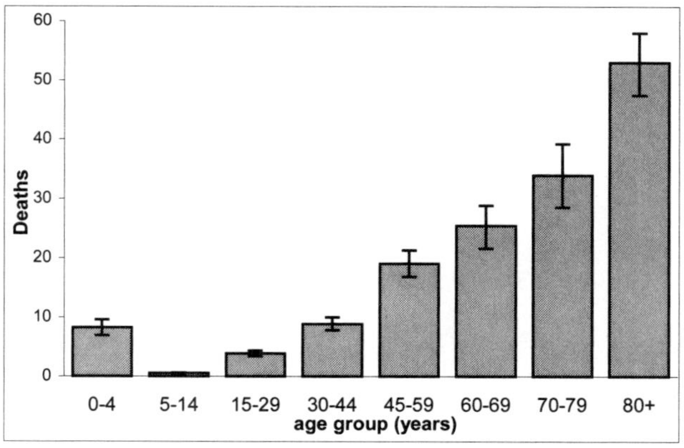

Saint Lucia   Sainte-Lucie   Santa Lucía

Males, 2000   *Hommes, 2000*   *Hombres, 2000*

Age	$_nM_x$	$_nq_x$			l_x			$_nd_x$	$_nL_x$	e_x			Actual deaths	Actual population
			percentiles			percentiles					percentiles			
		mean	2½	97½	mean	2½	97½			mean	2½	97½		
<1	0.0174	0.0171	0.0142	0.0204	100,000	100,000	100,000	1,710	98,461	69.2	68.3	69.9	30	1,732
1-4	0.0008	0.0033	0.0029	0.0038	98,290	97,961	98,581	321	392,390	69.4	68.4	70.4	6	6,866
5-9	0.0004	0.0021	0.0019	0.0024	97,969	97,594	98,299	207	489,328	65.6	64.6	66.6	4	8,349
10-14	0.0006	0.0028	0.0025	0.0032	97,762	97,358	98,116	274	488,125	60.7	59.7	61.8	4	7,174
15-19	0.0009	0.0046	0.0041	0.0053	97,488	97,047	97,873	452	486,310	55.9	54.8	57.0	7	7,964
20-24	0.0014	0.0069	0.0062	0.0078	97,036	96,536	97,469	668	483,509	51.2	50.1	52.2	10	7,473
25-29	0.0023	0.0114	0.0103	0.0128	96,368	95,788	96,867	1,101	479,087	46.5	45.4	47.6	15	6,677
30-34	0.0026	0.0129	0.0118	0.0143	95,267	94,565	95,865	1,231	473,258	42.0	40.9	43.1	15	5,694
35-39	0.0032	0.0161	0.0149	0.0177	94,036	93,212	94,732	1,513	466,399	37.5	36.4	38.7	16	4,884
40-44	0.0047	0.0231	0.0216	0.0252	92,523	91,563	93,323	2,141	457,265	33.1	32.0	34.2	17	3,650
45-49	0.0068	0.0333	0.0315	0.0360	90,383	89,252	91,306	3,012	444,384	28.8	27.7	30.0	20	2,913
50-54	0.0116	0.0562	0.0522	0.0612	87,371	86,035	88,429	4,912	424,574	24.7	23.7	25.9	27	2,293
55-59	0.0137	0.0660	0.0600	0.0731	82,459	80,769	83,813	5,443	398,685	21.1	20.0	22.2	26	1,909
60-64	0.0214	0.1015	0.0908	0.1137	77,016	74,867	78,783	7,818	365,533	17.4	16.4	18.4	28	1,331
65-69	0.0308	0.1431	0.1264	0.1613	69,197	66,357	71,629	9,900	321,237	14.0	13.2	15.0	33	1,061
70-74	0.0488	0.2174	0.1914	0.2444	59,297	55,654	62,574	12,890	264,262	11.0	10.3	11.8	45	924
75-79	0.0772	0.3235	0.2870	0.3588	46,407	42,051	50,600	15,012	194,507	8.3	7.8	9.0	57	742
80-84	0.1344	0.5030	0.4580	0.5424	31,395	26,964	36,075	15,792	117,498	6.1	5.7	6.6	60	447
85-89	0.1801	0.6209	0.5800	0.6542	15,604	12,339	19,552	9,688	53,799	4.8	4.5	5.1	39	217
90-94	0.2673	0.7417	0.7086	0.7676	5,916	4,267	8,212	4,388	16,416	3.5	3.3	3.7	17	64
95-99	0.3664	0.8727	0.8501	0.8898	1,528	992	2,393	1,333	3,640	2.6	2.6	2.8	5	14
100+	0.5129	1	1	1	195	109	359	195	379	1.9	1.9	2.3	2	3
All ages													*483*	*72,381*

ln($_nq_x$) with uncertainty interval

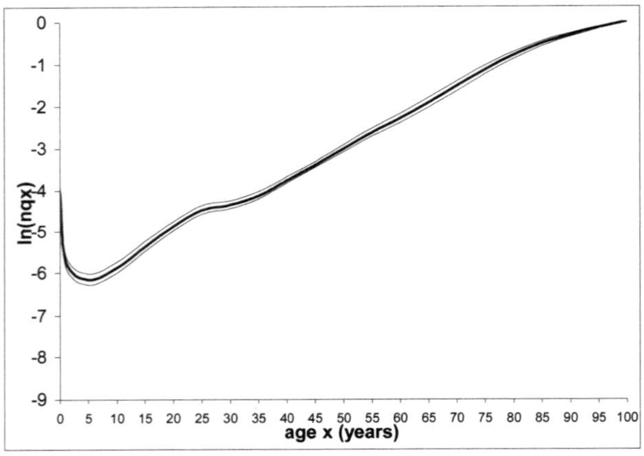

l_x with uncertainty interval

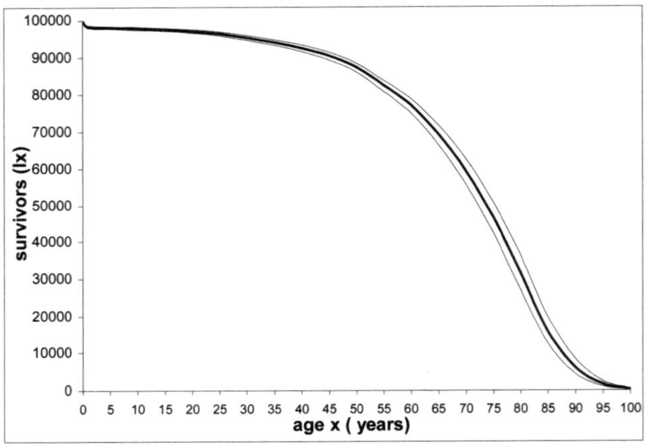

e_x with uncertainty interval

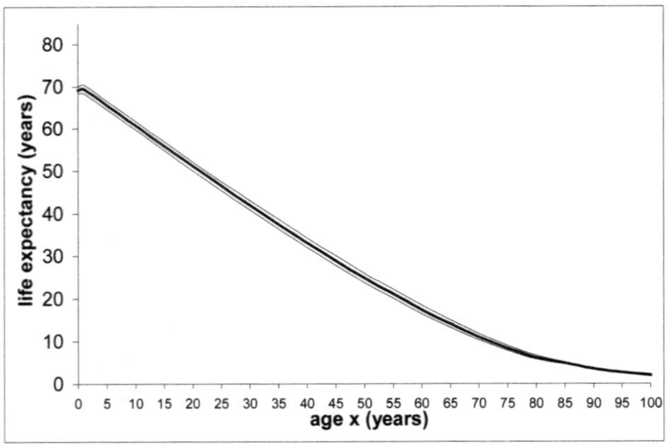

Distribution of deaths by age with uncertainty interval

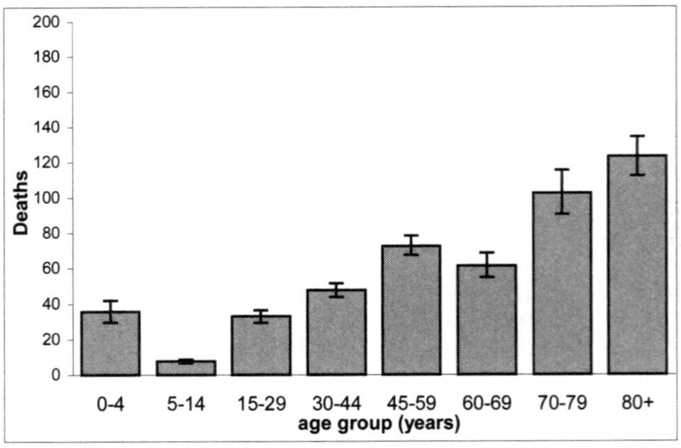

Saint Lucia

Sainte-Lucie

Santa Lucía

Females, 2000 *Femmes, 2000* *Mujeres, 2000*

Age	$_nM_x$	$_nq_x$ mean	$_nq_x$ percentiles 2 1/2	$_nq_x$ percentiles 97 1/2	l_x mean	l_x percentiles 2 1/2	l_x percentiles 97 1/2	$_nd_x$	$_nL_x$	e_x mean	e_x percentiles 2 1/2	e_x percentiles 97 1/2	Actual deaths	Actual population
<1	0.0119	0.0118	0.0105	0.0156	100,000	100,000	100,000	1,177	98,940	74.2	73.1	75.2	20	1,644
1-4	0.0007	0.0027	0.0025	0.0034	98,823	98,437	98,952	272	394,638	74.1	73.0	75.2	4	6,447
5-9	0.0003	0.0015	0.0014	0.0019	98,551	98,100	98,704	149	492,382	70.3	69.2	71.4	2	7,858
10-14	0.0002	0.0011	0.0010	0.0014	98,402	97,918	98,569	113	491,727	65.4	64.3	66.5	2	7,309
15-19	0.0004	0.0019	0.0017	0.0023	98,289	97,780	98,465	187	490,976	60.5	59.4	61.5	3	8,039
20-24	0.0005	0.0026	0.0024	0.0032	98,101	97,552	98,294	258	489,862	55.6	54.5	56.7	4	7,381
25-29	0.0009	0.0044	0.0040	0.0053	97,843	97,241	98,058	432	488,137	50.7	49.6	51.8	6	6,875
30-34	0.0012	0.0058	0.0053	0.0069	97,412	96,728	97,663	566	485,644	45.9	44.9	47.0	7	5,955
35-39	0.0021	0.0107	0.0098	0.0125	96,846	96,063	97,143	1,035	481,641	41.2	40.1	42.3	11	5,241
40-44	0.0028	0.0140	0.0129	0.0161	95,811	94,865	96,189	1,341	475,699	36.6	35.6	37.7	12	4,220
45-49	0.0038	0.0190	0.0175	0.0214	94,469	93,341	94,949	1,794	467,862	32.1	31.1	33.2	12	3,187
50-54	0.0071	0.0348	0.0321	0.0388	92,676	91,344	93,287	3,222	455,323	27.7	26.7	28.7	18	2,516
55-59	0.0102	0.0496	0.0451	0.0555	89,454	87,801	90,292	4,434	436,184	23.6	22.7	24.6	21	2,043
60-64	0.0155	0.0745	0.0669	0.0835	85,020	82,925	86,222	6,334	409,265	19.7	18.9	20.6	26	1,709
65-69	0.0223	0.1057	0.0941	0.1182	78,686	75,997	80,451	8,317	372,638	16.1	15.4	16.9	32	1,415
70-74	0.0356	0.1633	0.1452	0.1810	70,369	67,013	72,877	11,492	323,115	12.7	12.1	13.5	44	1,232
75-79	0.0603	0.2618	0.2340	0.2861	58,877	54,882	62,298	15,416	255,844	9.7	9.2	10.3	67	1,116
80-84	0.1001	0.4003	0.3621	0.4303	43,461	39,183	47,719	17,398	173,808	7.2	6.8	7.7	68	676
85-89	0.1531	0.5536	0.5129	0.5825	26,063	22,321	30,440	14,428	94,244	5.3	5.1	5.7	52	339
90-94	0.2333	0.6863	0.6506	0.7093	11,635	9,318	14,828	7,985	34,220	3.8	3.7	4.1	32	139
95-99	0.3333	0.8333	0.8074	0.8490	3,650	2,708	5,181	3,042	9,125	2.9	2.8	3.0	17	50
100+	0.4677	1	1	1	608	409	998	608	1,301	2.1	2.1	2.2	5	11
All ages													*465*	*75,402*

ln($_nq_x$) with uncertainty interval

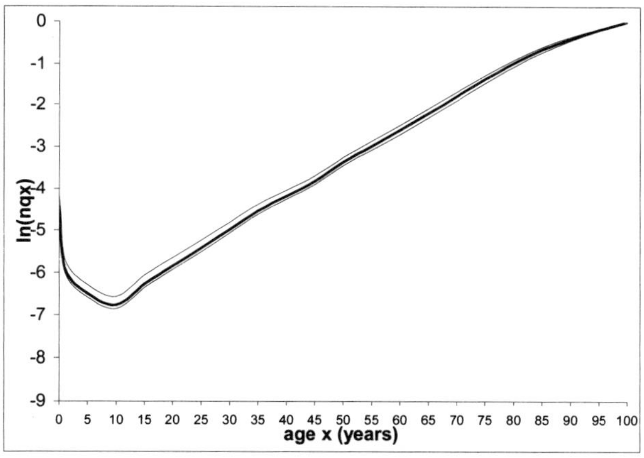

l_x with uncertainty interval

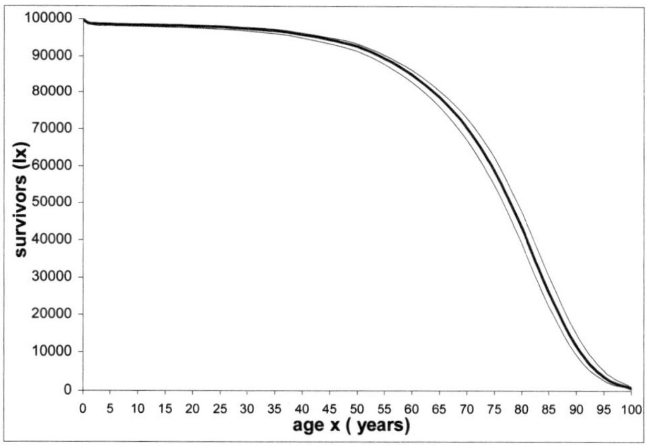

e_x with uncertainty interval

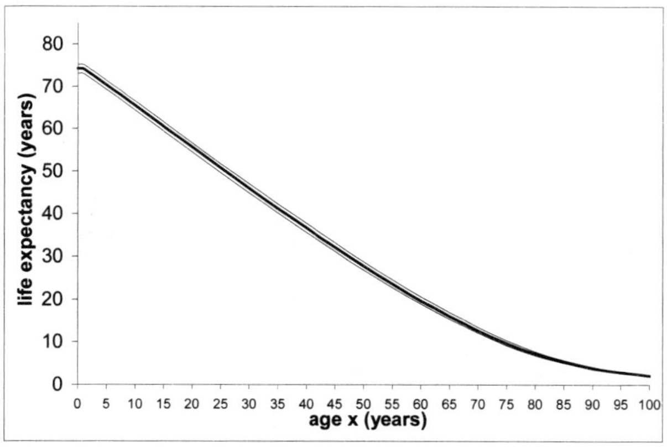

Distribution of deaths by age with uncertainty interval

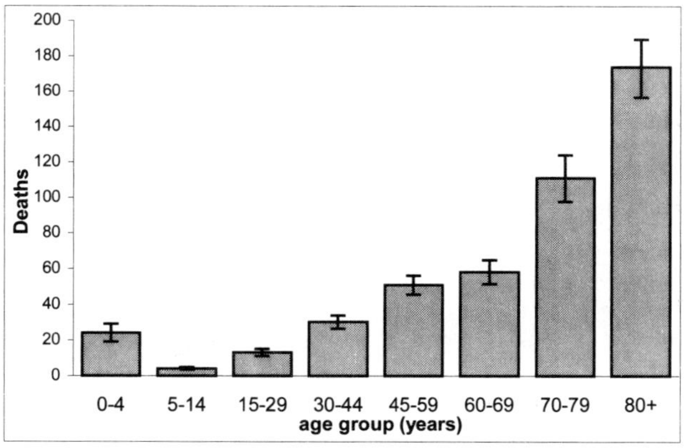

Saint Vincent and the Grenadines   Saint-Vincent-et-Grenadines   San Vicente y las Granadinas

Males, 2000   *Hommes, 2000*   *Hombres, 2000*

Age	$_nM_x$	$_nq_x$ mean	$_nq_x$ percentiles 2½	$_nq_x$ percentiles 97½	l_x mean	l_x percentiles 2½	l_x percentiles 97½	$_nd_x$	$_nL_x$	e_x mean	e_x percentiles 2½	e_x percentiles 97½	Actual deaths	Actual population
<1	0.0167	0.0164	0.0139	0.0208	100,000	100,000	100,000	1,644	98,520	67.7	66.5	68.7	19	1,123
1-4	0.0009	0.0036	0.0032	0.0043	98,356	97,918	98,611	352	392,578	67.9	66.7	68.9	4	4,451
5-9	0.0004	0.0020	0.0018	0.0024	98,003	97,498	98,300	196	489,527	64.1	62.9	65.2	2	5,609
10-14	0.0003	0.0013	0.0011	0.0015	97,807	97,265	98,126	126	488,721	59.2	58.1	60.3	2	5,844
15-19	0.0007	0.0033	0.0029	0.0039	97,681	97,117	98,013	318	487,610	54.3	53.2	55.4	4	5,440
20-24	0.0016	0.0079	0.0071	0.0093	97,363	96,742	97,729	769	484,891	49.5	48.3	50.5	8	4,849
25-29	0.0018	0.0091	0.0082	0.0106	96,593	95,847	97,038	882	480,762	44.8	43.7	45.9	9	4,868
30-34	0.0032	0.0160	0.0145	0.0183	95,712	94,833	96,240	1,531	474,730	40.2	39.2	41.3	15	4,553
35-39	0.0047	0.0234	0.0214	0.0266	94,180	93,095	94,845	2,205	465,389	35.8	34.8	36.9	19	4,112
40-44	0.0065	0.0320	0.0294	0.0359	91,975	90,623	92,819	2,941	452,523	31.6	30.7	32.7	20	3,111
45-49	0.0085	0.0416	0.0384	0.0463	89,034	87,374	90,093	3,705	435,907	27.6	26.7	28.6	23	2,754
50-54	0.0115	0.0559	0.0512	0.0620	85,329	83,331	86,630	4,769	414,721	23.7	22.9	24.6	27	2,386
55-59	0.0180	0.0861	0.0784	0.0949	80,559	78,168	82,194	6,938	385,452	19.9	19.2	20.8	36	1,978
60-64	0.0270	0.1265	0.1147	0.1384	73,621	70,748	75,751	9,310	344,831	16.6	15.9	17.4	44	1,616
65-69	0.0378	0.1727	0.1567	0.1877	64,311	60,959	67,064	11,108	293,785	13.6	13.0	14.4	46	1,227
70-74	0.0530	0.2340	0.2128	0.2535	53,203	49,515	56,552	12,449	234,893	11.0	10.4	11.6	52	981
75-79	0.0793	0.3310	0.3035	0.3564	40,754	36,965	44,518	13,492	170,041	8.5	8.1	9.1	53	668
80-84	0.1188	0.4580	0.4265	0.4865	27,262	23,792	31,008	12,486	105,096	6.5	6.2	6.9	50	421
85-89	0.1742	0.6067	0.5746	0.6336	14,776	12,217	17,782	8,964	51,469	4.9	4.7	5.2	35	202
90-94	0.2499	0.7142	0.6858	0.7367	5,812	4,476	7,564	4,151	16,607	3.6	3.5	3.8	16	65
95-99	0.3511	0.8550	0.8341	0.8708	1,661	1,178	2,376	1,420	4,045	2.7	2.6	2.8	4	13
100+	0.4828	1	1	1	241	152	394	241	499	2.1	2.0	2.1	1	1
All ages													*489*	*56,272*

ln($_nq_x$) with uncertainty interval

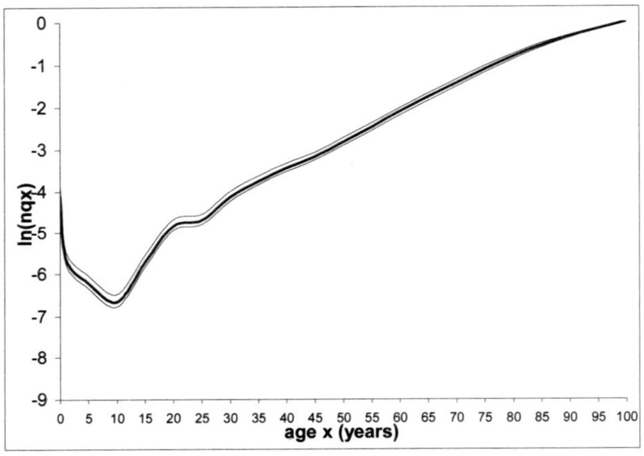

l_x with uncertainty interval

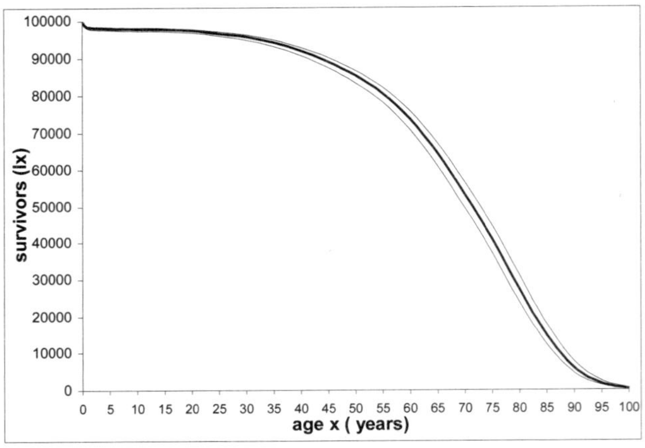

e_x with uncertainty interval

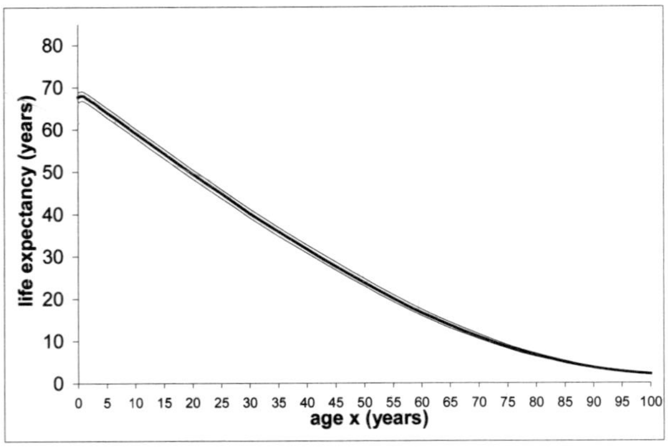

Distribution of deaths by age with uncertainty interval

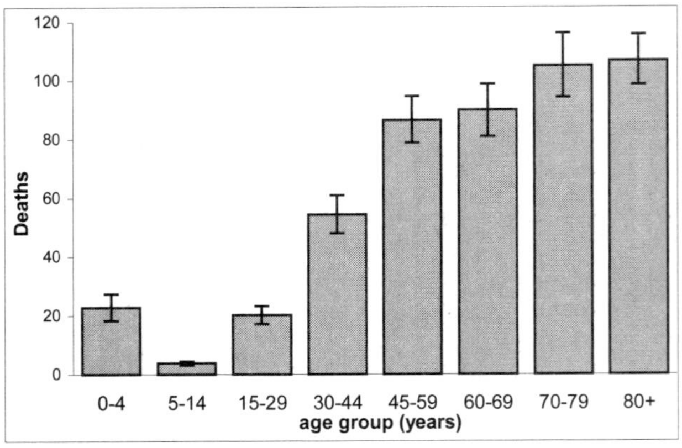

Saint Vincent and the Grenadines   Saint-Vincent-et-Grenadines   San Vicente y las Granadinas

Females, 2000   *Femmes, 2000*   *Mujeres, 2000*

Age	$_nM_x$	$_nq_x$			l_x			$_nd_x$	$_nL_x$	e_x			Actual deaths	Actual population
			percentiles			percentiles					percentiles			
		mean	21/2	971/2	mean	21/2	971/2			mean	21/2	971/2		
<1	0.0137	0.0136	0.0119	0.0179	100,000	100,000	100,000	1,357	98,778	73.3	72.1	74.8	15	1,087
1-4	0.0009	0.0034	0.0031	0.0042	98,643	98,210	98,811	339	393,756	73.3	72.1	74.9	4	4,262
5-9	0.0003	0.0016	0.0014	0.0019	98,303	97,794	98,507	155	491,129	69.6	68.3	71.1	2	5,399
10-14	0.0004	0.0021	0.0019	0.0026	98,148	97,606	98,369	207	490,224	64.7	63.5	66.2	2	5,657
15-19	0.0005	0.0023	0.0021	0.0028	97,941	97,357	98,185	229	489,133	59.8	58.6	61.4	2	5,264
20-24	0.0011	0.0053	0.0047	0.0063	97,712	97,083	97,980	517	487,267	54.9	53.7	56.5	5	4,716
25-29	0.0013	0.0066	0.0059	0.0078	97,195	96,470	97,519	639	484,377	50.2	49.0	51.8	6	4,776
30-34	0.0018	0.0091	0.0081	0.0106	96,556	95,722	96,948	877	480,588	45.5	44.4	47.1	8	4,530
35-39	0.0030	0.0146	0.0131	0.0168	95,679	94,710	96,163	1,402	474,892	40.9	39.8	42.5	13	4,254
40-44	0.0042	0.0206	0.0184	0.0232	94,278	93,120	94,906	1,945	466,527	36.5	35.4	38.0	14	3,358
45-49	0.0058	0.0284	0.0254	0.0314	92,333	90,956	93,156	2,619	455,117	32.2	31.1	33.7	17	2,979
50-54	0.0075	0.0368	0.0326	0.0409	89,714	88,102	90,789	3,302	440,314	28.1	27.0	29.5	19	2,586
55-59	0.0110	0.0535	0.0468	0.0598	86,412	84,496	87,825	4,625	420,494	24.1	23.0	25.4	24	2,143
60-64	0.0150	0.0722	0.0624	0.0808	81,786	79,446	83,713	5,906	394,167	20.3	19.3	21.6	27	1,769
65-69	0.0220	0.1043	0.0895	0.1170	75,880	73,028	78,487	7,912	359,623	16.7	15.7	17.9	31	1,390
70-74	0.0365	0.1674	0.1436	0.1884	67,969	64,486	71,467	11,376	311,404	13.3	12.5	14.4	42	1,137
75-79	0.0540	0.2379	0.2049	0.2665	56,593	52,337	61,205	13,465	249,303	10.5	9.8	11.4	43	795
80-84	0.0850	0.3505	0.3055	0.3862	43,128	38,387	48,663	15,117	177,848	8.0	7.4	8.7	44	522
85-89	0.1317	0.4954	0.4437	0.5324	28,011	23,562	33,797	13,878	105,361	5.9	5.6	6.5	35	266
90-94	0.2010	0.6269	0.5789	0.6583	14,133	11,018	18,801	8,860	44,088	4.3	4.0	4.7	19	95
95-99	0.3018	0.7920	0.7553	0.8147	5,274	3,765	7,916	4,177	13,838	3.1	3.0	3.3	6	21
100+	0.4464	1	1	1	1,097	698	1,937	1,097	2,457	2.2	2.2	2.3	2	3
All ages													*379*	*57,007*

ln($_nq_x$) with uncertainty interval

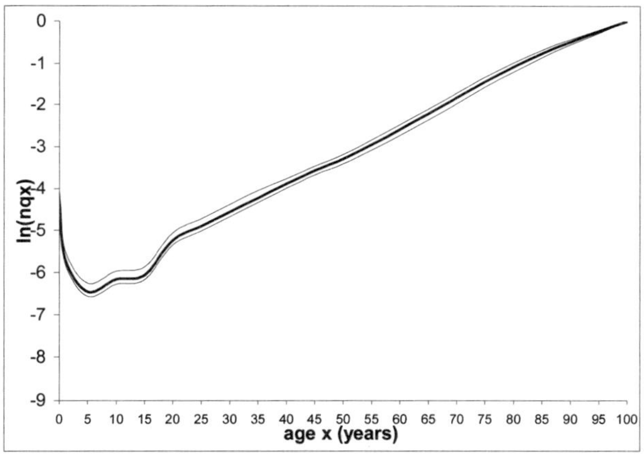

l_x with uncertainty interval

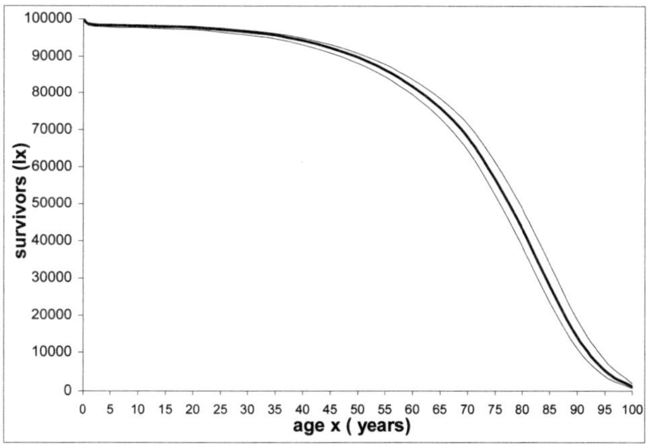

e_x with uncertainty interval

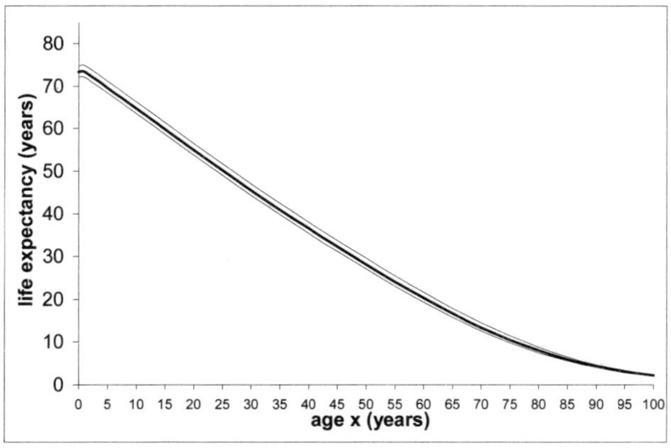

Distribution of deaths by age with uncertainty interval

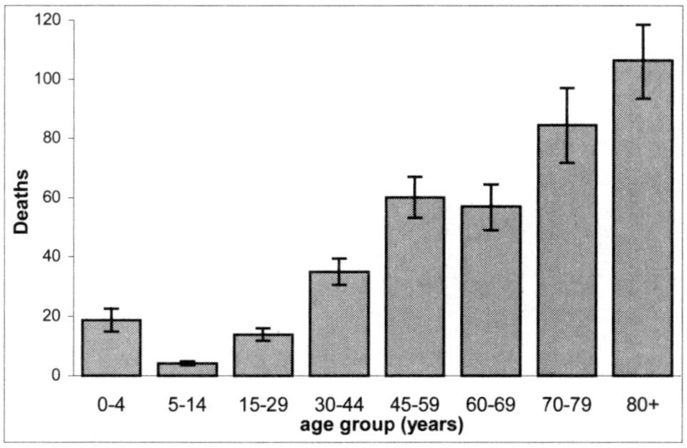

Samoa

Samoa

Samoa

Males, 2000 — *Hommes, 2000* — *Hombres, 2000*

Age	$_nM_x$	$_nq_x$ mean	$_nq_x$ percentiles 21/2	$_nq_x$ percentiles 971/2	l_x mean	l_x percentiles 21/2	l_x percentiles 971/2	$_nd_x$	$_nL_x$	e_x mean	e_x percentiles 21/2	e_x percentiles 971/2	Actual deaths	Actual population
<1	0.0296	0.0288	0.0243	0.0341	100,000	100,000	100,000	2,883	97,405	66.7	65.5	67.7	63	2,121
1-4	0.0022	0.0089	0.0077	0.0101	97,117	96,588	97,567	861	386,401	67.7	66.7	68.6	20	9,021
5-9	0.0012	0.0062	0.0054	0.0070	96,256	95,612	96,812	594	479,797	64.3	63.3	65.2	14	11,395
10-14	0.0007	0.0035	0.0031	0.0040	95,663	94,943	96,287	339	477,466	59.7	58.7	60.6	8	11,165
15-19	0.0015	0.0073	0.0065	0.0083	95,324	94,563	95,986	700	474,869	54.9	53.9	55.8	15	10,202
20-24	0.0016	0.0078	0.0069	0.0088	94,624	93,778	95,361	738	471,273	50.3	49.4	51.2	14	8,962
25-29	0.0019	0.0094	0.0084	0.0105	93,886	92,955	94,699	882	467,222	45.6	44.8	46.6	14	7,648
30-34	0.0028	0.0140	0.0126	0.0156	93,003	91,975	93,902	1,299	461,771	41.1	40.2	42.0	15	5,223
35-39	0.0037	0.0182	0.0165	0.0202	91,705	90,541	92,721	1,665	454,361	36.6	35.8	37.5	14	3,771
40-44	0.0053	0.0262	0.0240	0.0289	90,039	88,714	91,193	2,357	444,304	32.2	31.4	33.1	16	3,097
45-49	0.0078	0.0383	0.0354	0.0420	87,682	86,149	89,008	3,359	430,013	28.0	27.3	28.9	19	2,409
50-54	0.0139	0.0671	0.0623	0.0729	84,323	82,533	85,853	5,654	407,478	24.0	23.4	24.8	27	1,941
55-59	0.0170	0.0815	0.0754	0.0882	78,669	76,518	80,502	6,413	377,310	20.6	20.0	21.3	29	1,723
60-64	0.0241	0.1138	0.1050	0.1226	72,256	69,767	74,434	8,224	340,717	17.2	16.6	17.9	39	1,625
65-69	0.0373	0.1706	0.1576	0.1828	64,031	61,215	66,619	10,926	292,842	14.1	13.5	14.7	48	1,282
70-74	0.0466	0.2089	0.1931	0.2246	53,105	50,027	56,116	11,093	237,795	11.5	11.0	12.0	42	906
75-79	0.0746	0.3145	0.2920	0.3371	42,013	38,792	45,279	13,215	177,027	8.8	8.4	9.3	49	651
80-84	0.1123	0.4385	0.4104	0.4654	28,798	25,715	32,057	12,629	112,418	6.8	6.4	7.1	27	236
85-89	0.1658	0.5862	0.5566	0.6129	16,169	13,747	18,902	9,478	57,151	5.1	4.9	5.4	13	76
90-94	0.2402	0.6980	0.6716	0.7210	6,691	5,321	8,382	4,670	19,446	3.8	3.6	3.9	5	20
95-99	0.3412	0.8430	0.8234	0.8596	2,021	1,485	2,752	1,704	4,994	2.8	2.7	2.9	1	3
100+	0.4755	1	1	1	317	208	486	317	667	2.1	2.0	2.1	0	0
All ages													*491*	*83,477*

ln($_nq_x$) with uncertainty interval

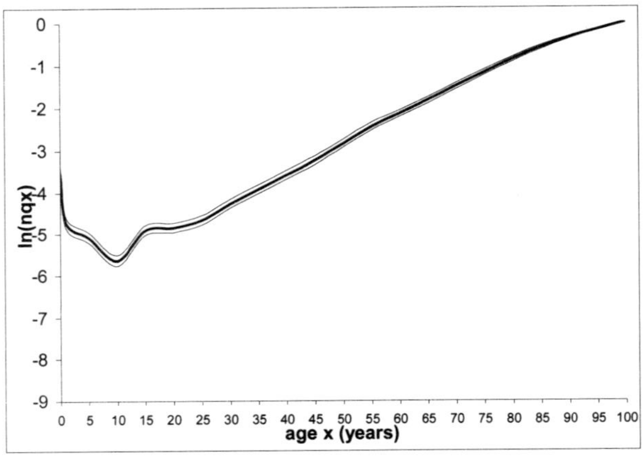

l_x with uncertainty interval

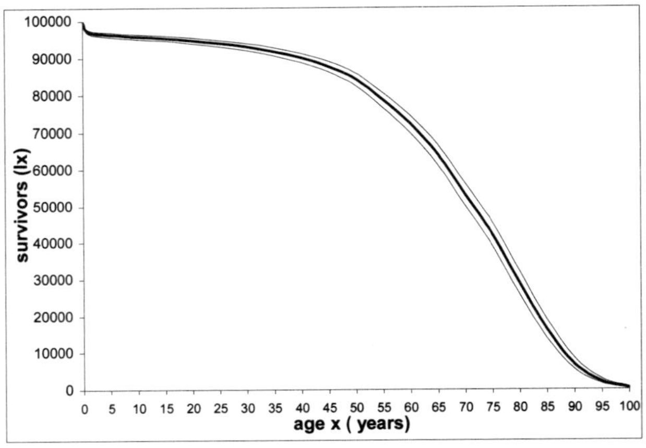

e_x with uncertainty interval

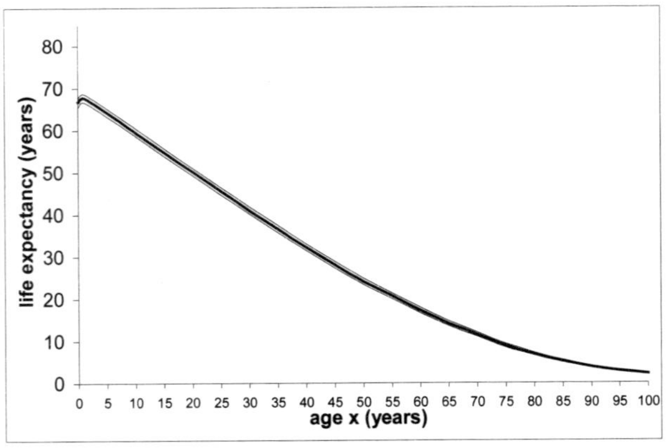

Distribution of deaths by age with uncertainty interval

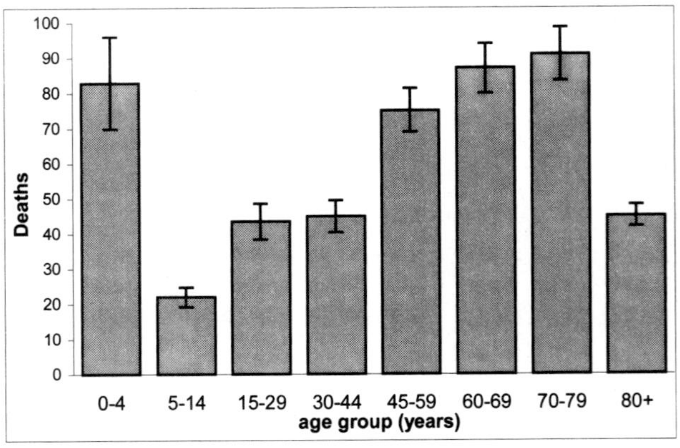

Samoa

Females, 2000

Samoa

Femmes, 2000

Samoa

Mujeres, 2000

Age	$_nM_x$	$_nq_x$ mean	$_nq_x$ percentiles 2½	$_nq_x$ percentiles 97½	l_x mean	l_x percentiles 2½	l_x percentiles 97½	$_nd_x$	$_nL_x$	e_x mean	e_x percentiles 2½	e_x percentiles 97½	Actual deaths	Actual population
<1	0.0249	0.0244	0.0192	0.0299	100,000	100,000	100,000	2,440	97,804	72.9	71.8	74.0	50	2,010
1-4	0.0014	0.0054	0.0046	0.0063	97,560	97,008	98,085	527	388,976	73.8	72.6	74.9	12	8,536
5-9	0.0005	0.0027	0.0023	0.0031	97,033	96,397	97,638	259	484,518	70.1	69.0	71.3	6	10,803
10-14	0.0004	0.0019	0.0016	0.0022	96,774	96,099	97,417	185	483,408	65.3	64.2	66.5	4	10,309
15-19	0.0008	0.0040	0.0034	0.0046	96,589	95,887	97,258	384	481,986	60.4	59.3	61.6	7	9,003
20-24	0.0009	0.0047	0.0040	0.0054	96,205	95,449	96,926	450	479,902	55.7	54.6	56.9	7	7,633
25-29	0.0014	0.0070	0.0061	0.0080	95,755	94,938	96,534	674	477,093	50.9	49.8	52.1	8	5,647
30-34	0.0019	0.0095	0.0083	0.0107	95,082	94,180	95,942	899	473,161	46.3	45.2	47.4	6	3,253
35-39	0.0025	0.0124	0.0111	0.0139	94,183	93,175	95,143	1,170	467,989	41.7	40.6	42.8	7	2,899
40-44	0.0035	0.0173	0.0157	0.0192	93,013	91,881	94,088	1,614	461,030	37.2	36.2	38.3	10	2,748
45-49	0.0046	0.0227	0.0209	0.0250	91,399	90,116	92,611	2,078	451,802	32.8	31.8	33.9	11	2,397
50-54	0.0072	0.0354	0.0323	0.0389	89,321	87,863	90,677	3,159	438,709	28.5	27.5	29.6	15	2,050
55-59	0.0098	0.0478	0.0433	0.0528	86,162	84,444	87,747	4,121	420,508	24.5	23.6	25.5	19	1,898
60-64	0.0150	0.0723	0.0649	0.0801	82,041	79,985	83,949	5,931	395,379	20.6	19.7	21.5	28	1,851
65-69	0.0240	0.1132	0.1011	0.1253	76,110	73,581	78,500	8,616	359,011	17.0	16.2	17.9	35	1,461
70-74	0.0320	0.1481	0.1321	0.1634	67,494	64,358	70,564	9,999	312,473	13.8	13.1	14.6	38	1,173
75-79	0.0500	0.2222	0.1989	0.2451	57,495	53,839	61,243	12,777	255,533	10.8	10.1	11.5	43	867
80-84	0.0800	0.3333	0.3008	0.3666	44,718	40,642	49,061	14,906	186,326	8.1	7.6	8.7	32	401
85-89	0.1300	0.4906	0.4520	0.5304	29,812	25,741	34,303	14,625	112,499	6.0	5.6	6.4	21	159
90-94	0.2000	0.6250	0.5888	0.6619	15,187	12,089	18,797	9,492	47,460	4.3	4.0	4.6	10	48
95-99	0.3000	0.7895	0.7610	0.8174	5,695	4,087	7,729	4,496	14,988	3.1	2.9	3.3	3	9
100+	0.4300	1	1	1	1,199	747	1,847	1,199	2,788	2.3	2.1	2.4	0	1
All ages													*370*	*75,156*

ln($_nq_x$) with uncertainty interval

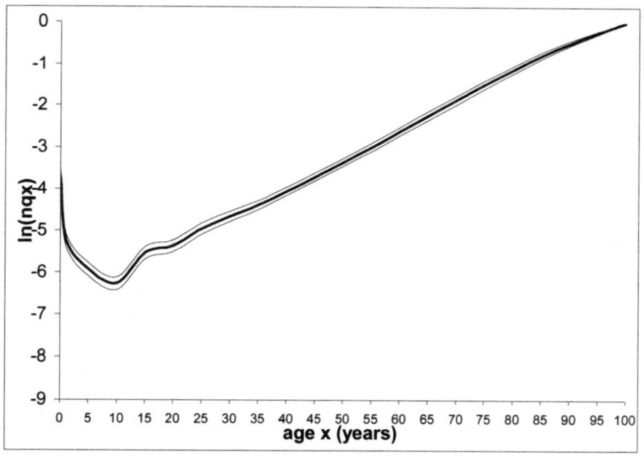

l_x with uncertainty interval

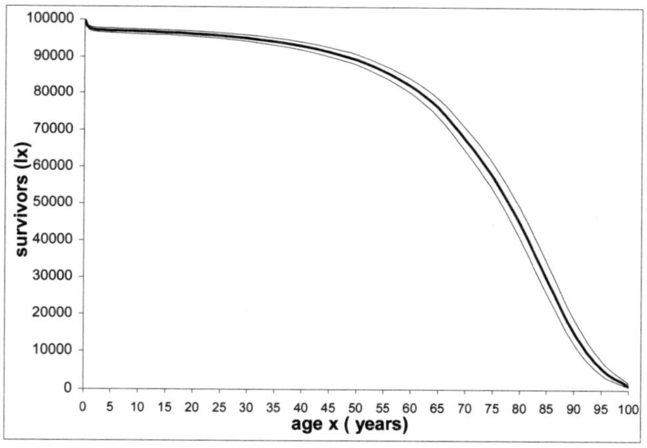

e_x with uncertainty interval

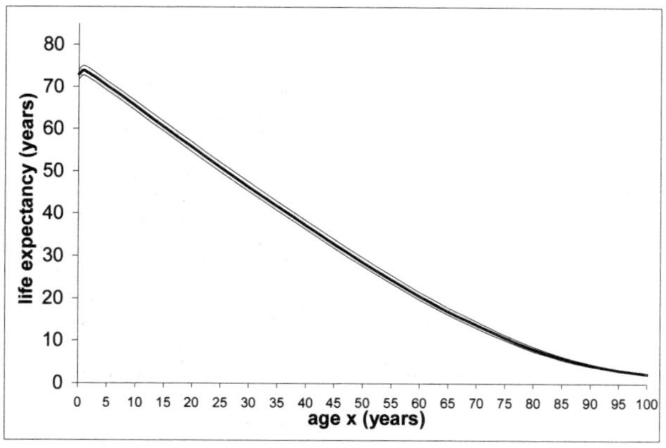

Distribution of deaths by age with uncertainty interval

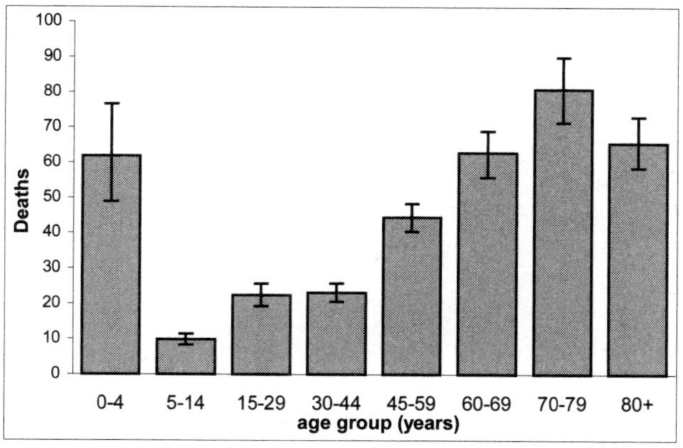

San Marino   Saint-Marin   San Marino

Males, 2000   *Hommes, 2000*   *Hombres, 2000*

Age	$_nM_x$	$_nq_x$			l_x			$_nd_x$	$_nL_x$	e_x			Actual deaths	Actual population
			percentiles			percentiles					percentiles			
		mean	2½	97½	mean	2½	97½			mean	2½	97½		
<1	0.0077	0.0076	0.0062	0.0097	100,000	100,000	100,000	763	99,314	76.1	75.4	76.9	1	120
1-4	0.0002	0.0008	0.0007	0.0010	99,237	99,026	99,384	79	396,759	75.7	75.0	76.4	0	502
5-9	0.0002	0.0008	0.0007	0.0010	99,158	98,931	99,316	79	495,592	71.8	71.0	72.5	0	658
10-14	0.0002	0.0011	0.0009	0.0013	99,079	98,836	99,248	105	495,132	66.8	66.1	67.5	0	665
15-19	0.0008	0.0040	0.0035	0.0047	98,974	98,712	99,158	395	493,882	61.9	61.2	62.6	1	719
20-24	0.0007	0.0035	0.0031	0.0040	98,579	98,249	98,815	344	492,034	57.2	56.4	57.8	1	853
25-29	0.0006	0.0030	0.0027	0.0034	98,235	97,853	98,513	294	490,437	52.3	51.6	53.0	1	1,039
30-34	0.0006	0.0032	0.0028	0.0036	97,940	97,518	98,251	313	488,919	47.5	46.8	48.2	1	1,092
35-39	0.0008	0.0040	0.0036	0.0045	97,627	97,165	97,972	390	487,162	42.6	41.9	43.3	1	1,057
40-44	0.0013	0.0065	0.0059	0.0072	97,238	96,729	97,621	630	484,613	37.8	37.1	38.5	1	927
45-49	0.0019	0.0095	0.0086	0.0104	96,608	96,031	97,050	913	480,754	33.0	32.3	33.7	2	861
50-54	0.0039	0.0191	0.0175	0.0209	95,694	95,028	96,213	1,826	473,907	28.3	27.6	29.0	3	892
55-59	0.0063	0.0312	0.0286	0.0340	93,869	93,046	94,525	2,924	462,032	23.8	23.2	24.5	5	751
60-64	0.0126	0.0611	0.0560	0.0669	90,944	89,886	91,818	5,555	440,834	19.5	18.9	20.2	10	756
65-69	0.0193	0.0921	0.0845	0.1010	85,390	83,875	86,673	7,868	407,279	15.6	15.1	16.2	13	660
70-74	0.0397	0.1807	0.1649	0.1968	77,522	75,402	79,347	14,008	352,590	11.9	11.5	12.5	22	556
75-79	0.0650	0.2796	0.2561	0.3016	63,514	60,564	66,265	17,757	273,179	9.0	8.6	9.5	27	412
80-84	0.1069	0.4218	0.3897	0.4507	45,758	42,298	49,297	19,300	180,539	6.6	6.3	6.9	19	178
85-89	0.1883	0.6402	0.6071	0.6692	26,458	23,233	30,085	16,938	89,944	4.5	4.3	4.8	23	120
90-94	0.2994	0.7886	0.7637	0.8095	9,520	7,684	11,820	7,508	25,075	3.1	3.0	3.3	10	35
95-99	0.4296	0.9385	0.9267	0.9477	2,012	1,464	2,793	1,888	4,396	2.3	2.3	2.3	2	5
100+	0.5562	1	1	1	124	77	205	124	223	1.8	1.7	1.8	0	0
All ages													*141*	*12,859*

ln($_nq_x$) with uncertainty interval

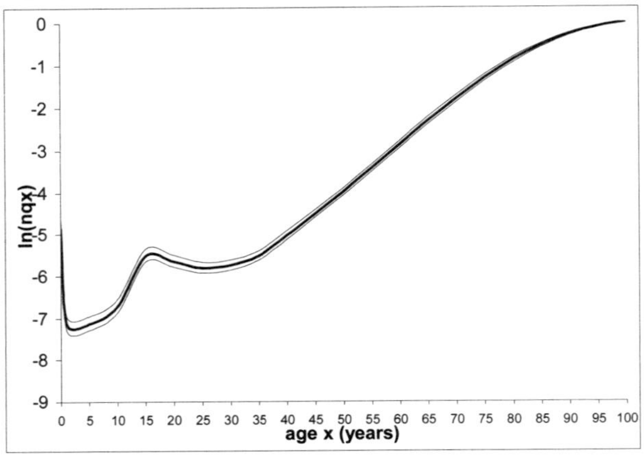

l_x with uncertainty interval

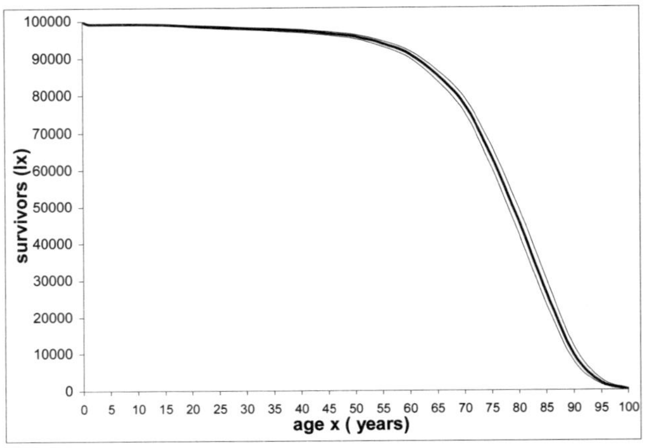

e_x with uncertainty interval

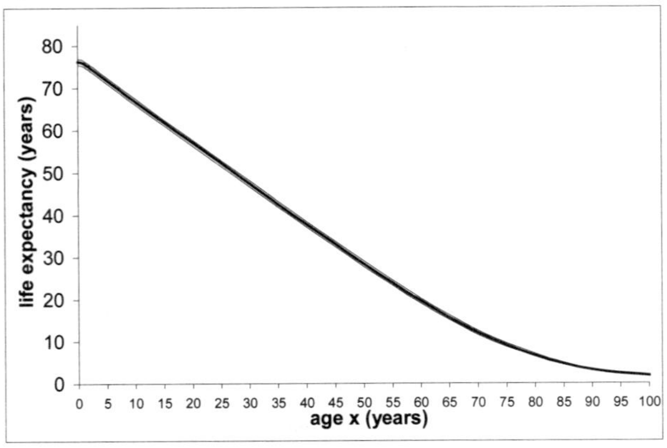

Distribution of deaths by age with uncertainty interval

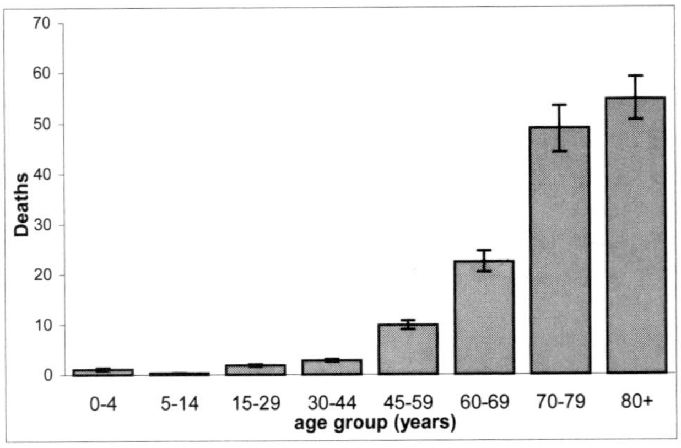

San Marino Saint-Marin San Marino

Females, 2000 *Femmes, 2000* *Mujeres, 2000*

Age	$_nM_x$	$_nq_x$ mean	$_nq_x$ percentiles 2½	$_nq_x$ percentiles 97½	l_x mean	l_x percentiles 2½	l_x percentiles 97½	$_nd_x$	$_nL_x$	e_x mean	e_x percentiles 2½	e_x percentiles 97½	Actual deaths	Actual population
<1	0.0079	0.0079	0.0063	0.0089	100,000	100,000	100,000	788	99,291	83.8	82.9	84.8	1	114
1-4	0.0003	0.0011	0.0009	0.0012	99,212	99,110	99,371	106	396,595	83.5	82.6	84.5	0	472
5-9	0.0001	0.0007	0.0006	0.0007	99,106	98,992	99,281	66	495,367	79.6	78.7	80.5	0	623
10-14	0.0001	0.0006	0.0005	0.0006	99,040	98,919	99,225	57	495,059	74.6	73.8	75.6	0	633
15-19	0.0002	0.0011	0.0009	0.0012	98,983	98,856	99,176	104	494,657	69.7	68.8	70.6	0	688
20-24	0.0003	0.0013	0.0011	0.0014	98,879	98,740	99,086	127	494,078	64.7	63.9	65.7	0	821
25-29	0.0004	0.0018	0.0016	0.0021	98,752	98,599	98,976	183	493,304	59.8	59.0	60.8	0	1,007
30-34	0.0004	0.0019	0.0017	0.0022	98,570	98,396	98,817	190	492,373	54.9	54.1	55.9	0	1,072
35-39	0.0005	0.0025	0.0022	0.0028	98,380	98,185	98,649	246	491,284	50.0	49.2	51.0	1	1,053
40-44	0.0007	0.0035	0.0031	0.0039	98,134	97,910	98,431	343	489,812	45.1	44.4	46.1	1	931
45-49	0.0011	0.0056	0.0050	0.0062	97,791	97,528	98,123	545	487,591	40.3	39.5	41.2	1	874
50-54	0.0015	0.0075	0.0068	0.0084	97,246	96,919	97,629	727	484,411	35.5	34.8	36.4	1	915
55-59	0.0028	0.0137	0.0124	0.0154	96,519	96,106	96,963	1,320	479,295	30.7	30.0	31.7	2	793
60-64	0.0050	0.0247	0.0222	0.0278	95,199	94,630	95,758	2,351	470,118	26.1	25.5	27.0	4	833
65-69	0.0075	0.0368	0.0328	0.0413	92,848	92,004	93,629	3,418	455,698	21.7	21.1	22.6	6	777
70-74	0.0150	0.0723	0.0638	0.0809	89,431	88,201	90,562	6,465	430,991	17.5	16.9	18.3	11	722
75-79	0.0239	0.1129	0.0992	0.1252	82,966	81,070	84,784	9,367	391,410	13.6	13.2	14.4	15	627
80-84	0.0481	0.2147	0.1901	0.2339	73,598	70,918	76,375	15,805	328,479	10.1	9.6	10.7	15	321
85-89	0.0912	0.3712	0.3366	0.3982	57,793	54,332	61,859	21,455	235,328	7.1	6.8	7.6	24	260
90-94	0.1628	0.5469	0.5137	0.5781	36,338	32,696	41,035	19,875	122,065	4.9	4.6	5.1	16	96
95-99	0.2741	0.7520	0.7303	0.7766	16,463	13,795	19,957	12,380	45,174	3.3	3.2	3.5	5	18
100+	0.4347	1	1	1	4,083	3,082	5,382	4,083	9,391	2.3	2.3	2.4	1	2
All ages													*105*	*13,655*

ln($_nq_x$) with uncertainty interval

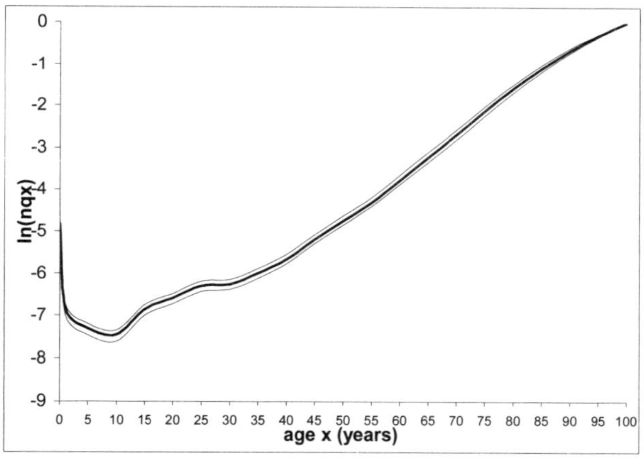

l_x with uncertainty interval

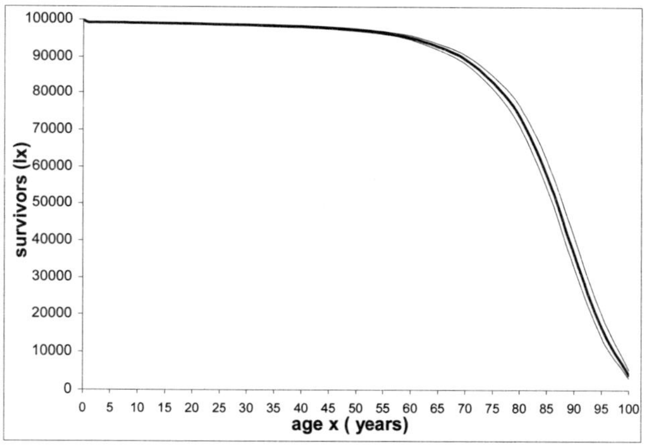

e_x with uncertainty interval

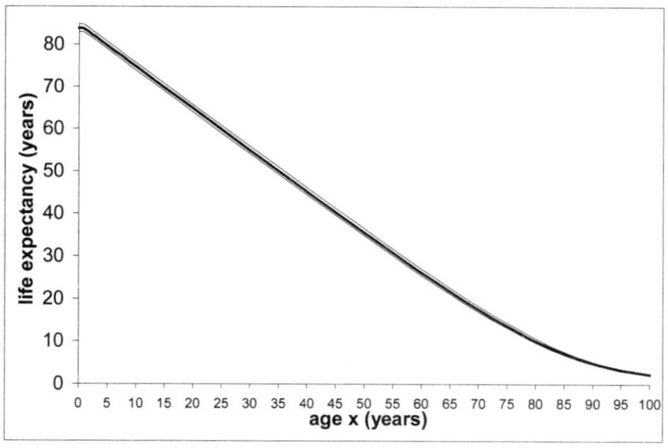

Distribution of deaths by age with uncertainty interval

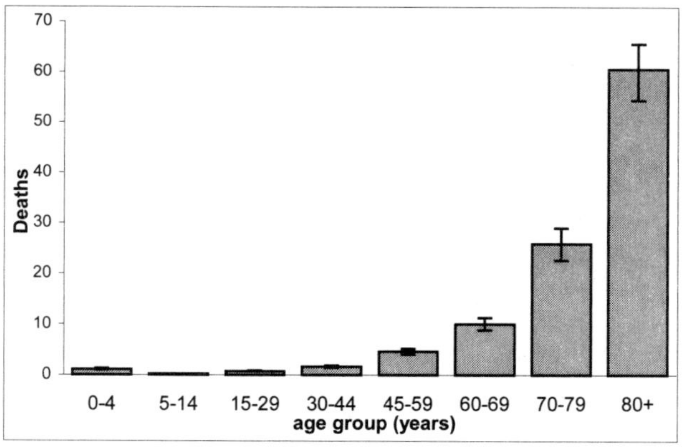

Sao Tome and Principe | Sao Tomé-et-Principe | Santo Tomé y Príncipe

Males, 2000 | *Hommes, 2000* | *Hombres, 2000*

Age	$_nM_x$	$_nq_x$			l_x			$_nd_x$	$_nL_x$	e_x			Actual deaths	Actual population
			percentiles			percentiles					percentiles			
		mean	2 1/2	97 1/2	mean	2 1/2	97 1/2			mean	2 1/2	97 1/2		
<1	0.0790	0.0737	0.0552	0.0935	100,000	100,000	100,000	7,371	93,366	60.3	58.0	62.7	165	2,091
1-4	0.0055	0.0219	0.0176	0.0258	92,629	90,646	94,481	2,024	365,656	64.1	62.5	65.9	43	7,752
5-9	0.0016	0.0078	0.0064	0.0092	90,604	88,356	92,816	706	451,258	61.5	60.0	63.2	14	9,092
10-14	0.0011	0.0056	0.0046	0.0066	89,899	87,571	92,226	503	448,236	56.9	55.5	58.6	9	8,388
15-19	0.0017	0.0083	0.0069	0.0097	89,396	87,016	91,802	743	445,120	52.2	50.9	53.9	13	7,892
20-24	0.0025	0.0125	0.0104	0.0145	88,653	86,201	91,172	1,110	440,489	47.7	46.3	49.3	17	6,791
25-29	0.0029	0.0142	0.0119	0.0164	87,543	84,996	90,223	1,240	434,615	43.2	41.9	44.8	16	5,469
30-34	0.0034	0.0169	0.0143	0.0194	86,303	83,664	89,150	1,456	427,873	38.8	37.6	40.3	19	5,558
35-39	0.0045	0.0220	0.0188	0.0252	84,846	82,083	87,874	1,867	419,564	34.4	33.2	35.9	20	4,555
40-44	0.0062	0.0304	0.0263	0.0346	82,979	80,023	86,216	2,526	408,580	30.2	29.0	31.6	12	2,002
45-49	0.0089	0.0436	0.0381	0.0494	80,453	77,269	83,934	3,512	393,486	26.0	24.9	27.4	7	750
50-54	0.0132	0.0640	0.0557	0.0721	76,941	73,499	80,692	4,923	372,401	22.1	21.0	23.4	5	381
55-59	0.0195	0.0931	0.0807	0.1045	72,019	68,316	76,025	6,703	343,338	18.4	17.4	19.6	9	454
60-64	0.0291	0.1357	0.1178	0.1528	65,316	61,421	69,471	8,862	304,426	15.1	14.1	16.2	23	801
65-69	0.0434	0.1958	0.1706	0.2209	56,454	52,406	60,769	11,055	254,631	12.1	11.2	13.0	34	792
70-74	0.0669	0.2867	0.2511	0.3229	45,399	41,276	49,977	13,017	194,450	9.4	8.6	10.2	44	664
75-79	0.1026	0.4084	0.3612	0.4533	32,381	28,442	37,261	13,226	128,843	7.1	6.6	7.8	35	339
80-84	0.1556	0.5601	0.5073	0.6061	19,156	15,814	23,428	10,729	68,956	5.4	5.0	5.9	39	252
85-89	0.2283	0.7266	0.6794	0.7648	8,427	6,245	11,319	6,123	26,826	4.0	3.7	4.3	32	142
90-94	0.3241	0.8216	0.7840	0.8510	2,304	1,469	3,593	1,893	5,840	2.9	2.8	3.1	17	53
95-99	0.4454	0.9533	0.9354	0.9656	411	219	776	392	879	2.2	2.2	2.3	3	8
100+	0.5924	1	1	1	19	8	50	19	32	1.7	1.7	1.8	0	1
All ages													*579*	*64,227*

ln($_nq_x$) with uncertainty interval

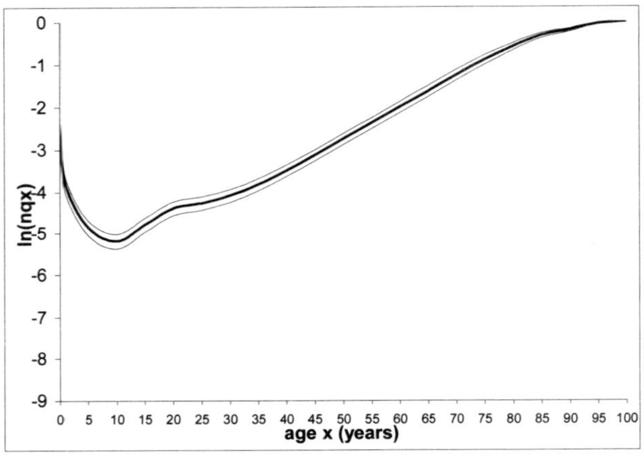

l_x with uncertainty interval

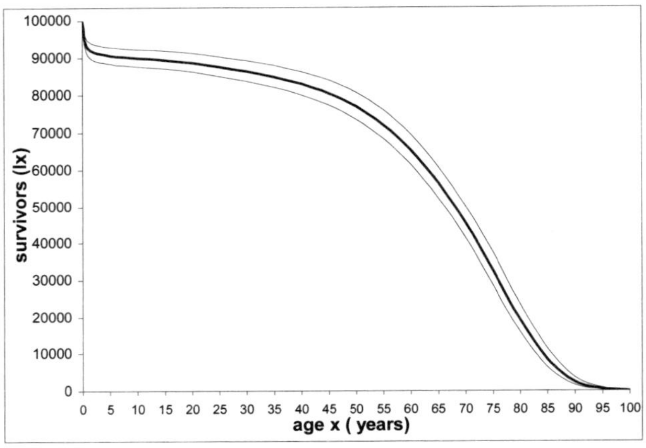

e_x with uncertainty interval

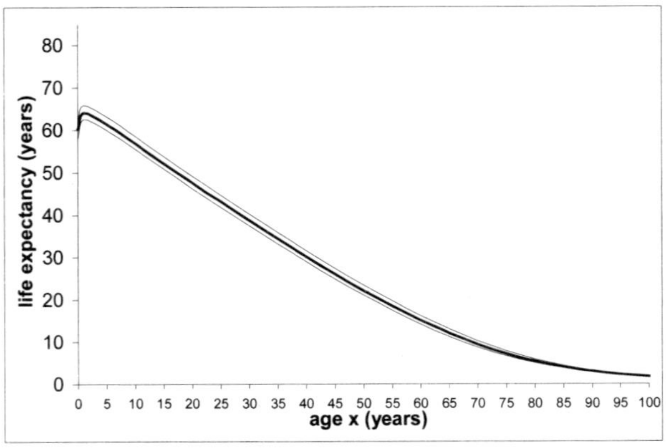

Distribution of deaths by age with uncertainty interval

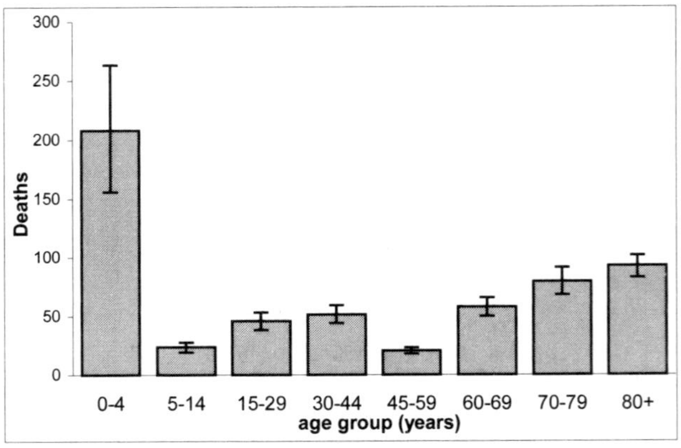

Sao Tome and Principe | Sao Tomé-et-Principe | Santo Tomé y Príncipe

Females, 2000 | *Femmes, 2000* | *Mujeres, 2000*

Age	$_nM_x$	$_nq_x$			l_x			$_nd_x$	$_nL_x$	e_x			Actual deaths	Actual population
			percentiles			percentiles					percentiles			
		mean	2½	97½	mean	2½	97½			mean	2½	97½		
<1	0.0768	0.0718	0.0512	0.0939	100,000	100,000	100,000	7,183	93,535	61.9	59.7	63.8	153	1,992
1-4	0.0069	0.0270	0.0221	0.0319	92,817	90,610	94,879	2,510	365,243	65.7	64.5	67.0	52	7,638
5-9	0.0017	0.0085	0.0072	0.0099	90,307	87,839	92,778	769	449,611	63.5	62.1	64.9	15	8,911
10-14	0.0012	0.0059	0.0050	0.0068	89,538	86,988	92,115	529	446,366	59.0	57.6	60.4	10	8,226
15-19	0.0017	0.0086	0.0074	0.0099	89,009	86,400	91,654	768	443,125	54.3	53.0	55.7	13	7,772
20-24	0.0024	0.0119	0.0103	0.0136	88,241	85,552	90,978	1,053	438,572	49.8	48.4	51.2	18	7,291
25-29	0.0028	0.0140	0.0122	0.0158	87,188	84,397	90,038	1,217	432,897	45.3	44.0	46.8	16	5,868
30-34	0.0033	0.0162	0.0143	0.0182	85,971	83,076	88,935	1,389	426,382	40.9	39.6	42.4	20	6,250
35-39	0.0043	0.0211	0.0188	0.0234	84,582	81,586	87,655	1,781	418,457	36.6	35.2	38.0	23	5,438
40-44	0.0052	0.0256	0.0231	0.0281	82,801	79,701	85,985	2,117	408,714	32.3	30.9	33.7	19	3,737
45-49	0.0071	0.0347	0.0317	0.0378	80,684	77,458	83,958	2,802	396,417	28.1	26.7	29.5	15	2,173
50-54	0.0101	0.0493	0.0446	0.0544	77,882	74,533	81,212	3,842	379,806	24.0	22.7	25.4	10	1,009
55-59	0.0145	0.0701	0.0623	0.0789	74,040	70,595	77,345	5,189	357,229	20.1	18.8	21.5	18	1,253
60-64	0.0223	0.1056	0.0923	0.1211	68,851	65,392	72,095	7,273	326,074	16.5	15.2	17.8	41	1,829
65-69	0.0350	0.1608	0.1386	0.1871	61,578	58,279	64,681	9,902	283,136	13.1	12.0	14.4	54	1,530
70-74	0.0570	0.2494	0.2120	0.2909	51,676	48,842	54,540	12,890	226,154	10.1	9.1	11.3	66	1,160
75-79	0.0908	0.3699	0.3146	0.4276	38,785	34,638	42,596	14,348	158,058	7.7	6.9	8.6	51	565
80-84	0.1412	0.5217	0.4537	0.5868	24,438	19,912	28,964	12,749	90,316	5.7	5.1	6.4	62	440
85-89	0.2118	0.6924	0.6262	0.7502	11,689	8,338	15,606	8,093	38,211	4.2	3.8	4.7	55	260
90-94	0.3066	0.7986	0.7443	0.8438	3,596	2,113	5,834	2,871	9,364	3.1	2.8	3.4	33	106
95-99	0.4283	0.9373	0.9077	0.9587	724	331	1,492	679	1,585	2.3	2.2	2.4	12	29
100+	0.5773	1	1	1	45	14	138	45	79	1.7	1.7	1.8	1	3
All ages													*760*	*73,480*

ln($_nq_x$) with uncertainty interval

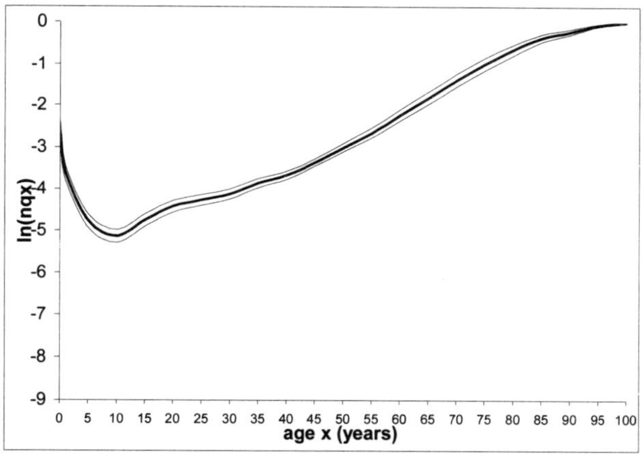

l_x with uncertainty interval

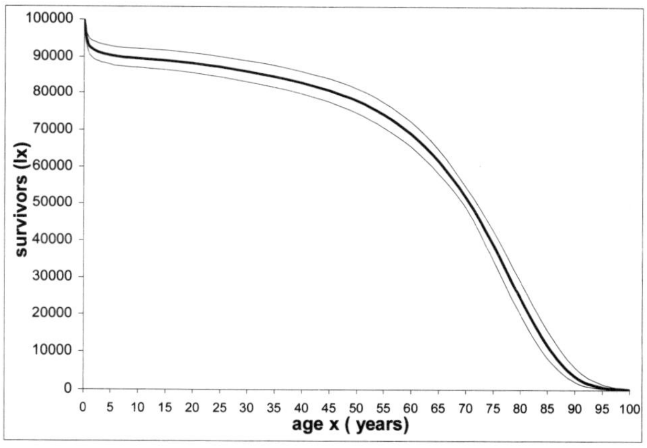

e_x with uncertainty interval

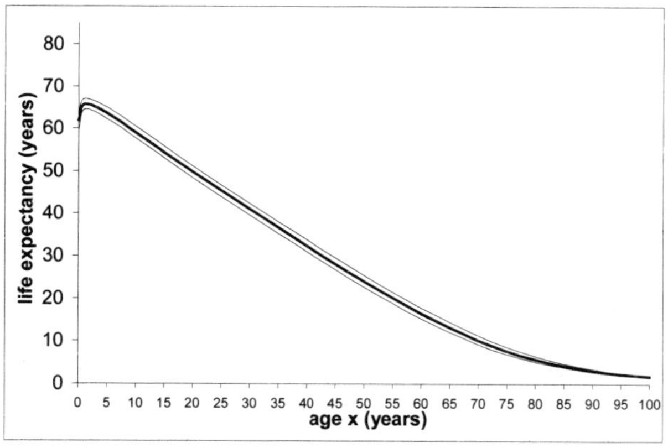

Distribution of deaths by age with uncertainty interval

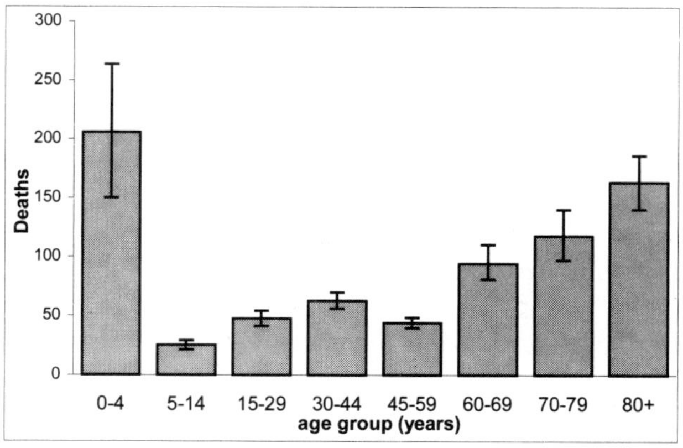

Saudi Arabia **Arabie saoudite** **Arabia Saudita**

Males, 2000 *Hommes, 2000* *Hombres, 2000*

Age	$_nM_x$	$_nq_x$			l_x			$_nd_x$	$_nL_x$	e_x			Actual deaths	Actual population
			percentiles			percentiles					percentiles			
		mean	2½	97½	mean	2½	97½			mean	2½	97½		
<1	0.0336	0.0326	0.0283	0.0369	100,000	100,000	100,000	3,264	97,063	68.1	67.0	69.1	11,581	344,430
1-4	0.0019	0.0076	0.0068	0.0085	96,736	96,310	97,172	736	385,179	69.4	68.5	70.3	2,458	1,286,012
5-9	0.0007	0.0035	0.0031	0.0039	96,000	95,500	96,494	336	479,162	65.9	65.0	66.7	1,047	1,494,617
10-14	0.0006	0.0030	0.0027	0.0033	95,665	95,133	96,189	285	477,610	61.1	60.3	61.9	800	1,340,188
15-19	0.0013	0.0064	0.0058	0.0071	95,379	94,821	95,931	612	475,368	56.3	55.4	57.1	1,379	1,072,011
20-24	0.0017	0.0083	0.0075	0.0092	94,768	94,154	95,375	787	471,872	51.6	50.8	52.4	1,542	924,896
25-29	0.0016	0.0081	0.0073	0.0089	93,981	93,298	94,656	760	468,005	47.0	46.3	47.8	1,113	685,240
30-34	0.0018	0.0090	0.0081	0.0099	93,221	92,472	93,958	841	464,003	42.4	41.6	43.2	952	525,394
35-39	0.0024	0.0117	0.0106	0.0128	92,380	91,556	93,183	1,079	459,203	37.8	37.0	38.5	1,260	535,871
40-44	0.0034	0.0171	0.0155	0.0187	91,301	90,381	92,191	1,557	452,612	33.2	32.5	33.9	2,304	669,863
45-49	0.0054	0.0266	0.0242	0.0290	89,744	88,693	90,762	2,385	442,757	28.7	28.0	29.4	3,535	656,330
50-54	0.0086	0.0422	0.0386	0.0460	87,359	86,123	88,562	3,686	427,580	24.4	23.8	25.1	4,217	489,189
55-59	0.0137	0.0664	0.0610	0.0722	83,673	82,181	85,132	5,558	404,471	20.4	19.8	21.0	4,469	325,279
60-64	0.0219	0.1037	0.0955	0.1120	78,115	76,308	79,898	8,098	370,333	16.7	16.1	17.3	4,659	213,094
65-69	0.0344	0.1582	0.1461	0.1699	70,018	67,821	72,208	11,080	322,390	13.3	12.8	13.8	4,897	142,497
70-74	0.0548	0.2410	0.2232	0.2582	58,938	56,331	61,494	14,202	259,187	10.3	9.9	10.8	4,794	87,493
75-79	0.0871	0.3577	0.3334	0.3825	44,737	41,992	47,574	16,001	183,681	7.8	7.4	8.2	4,308	49,458
80-84	0.1386	0.5146	0.4872	0.5434	28,736	26,301	31,503	14,787	106,712	5.8	5.5	6.0	3,049	22,001
85-89	0.2113	0.6913	0.6668	0.7167	13,949	12,180	16,040	9,642	45,639	4.2	4.0	4.4	1,441	6,822
90-94	0.3087	0.8014	0.7826	0.8208	4,307	3,451	5,336	3,451	11,179	3.0	2.9	3.2	392	1,271
95-99	0.4325	0.9412	0.9319	0.9503	855	618	1,155	805	1,861	2.3	2.2	2.3	61	140
100+	0.5806	1	1	1	50	31	78	50	87	1.7	1.7	1.8	3	6
All ages													*60,263*	*10,872,102*

$\ln(_nq_x)$ with uncertainty interval

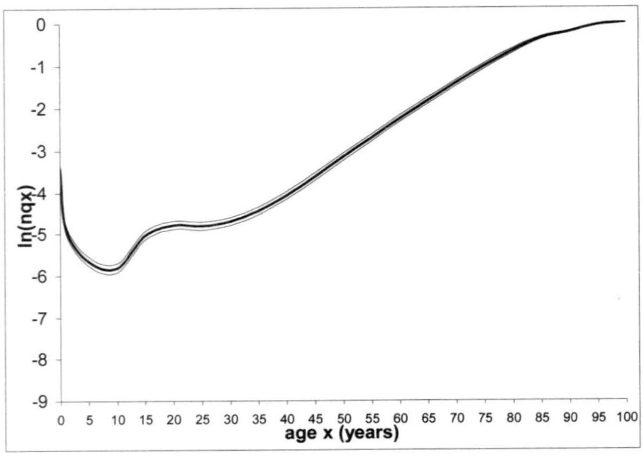

l_x with uncertainty interval

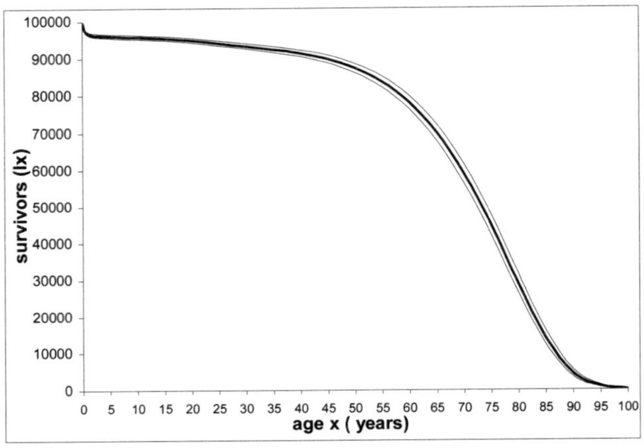

e_x with uncertainty interval

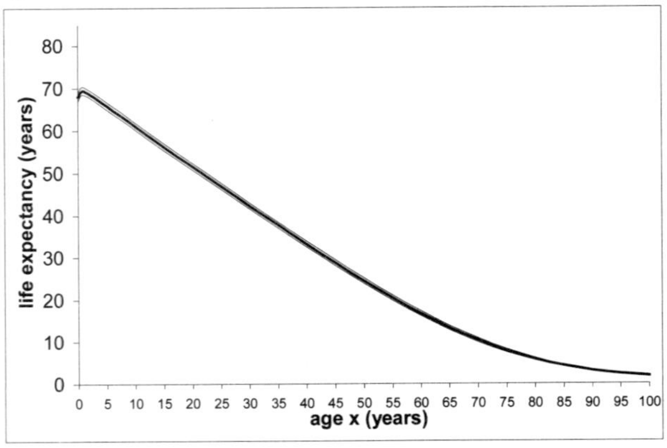

Distribution of deaths by age with uncertainty interval

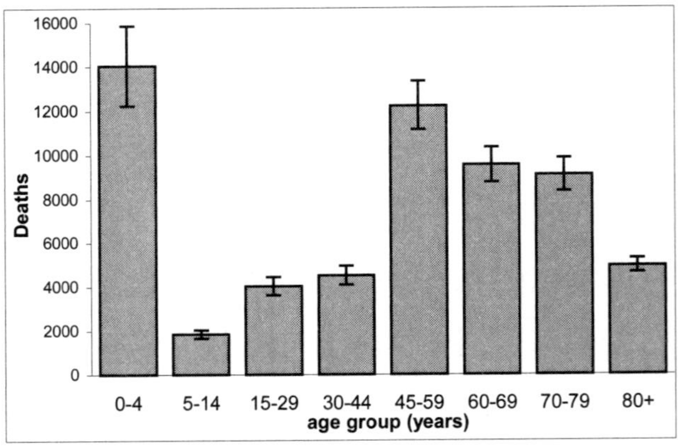

Saudi Arabia — Arabie saoudite — Arabia Saudita

Females, 2000 — *Femmes, 2000* — *Mujeres, 2000*

Age	$_nM_x$	$_nq_x$ mean	$_nq_x$ percentiles 2 1/2	$_nq_x$ percentiles 97 1/2	l_x mean	l_x percentiles 2 1/2	l_x percentiles 97 1/2	$_nd_x$	$_nL_x$	e_x mean	e_x percentiles 2 1/2	e_x percentiles 97 1/2	Actual deaths	Actual population
<1	0.0214	0.0210	0.0176	0.0245	100,000	100,000	100,000	2,098	98,112	73.5	72.5	74.8	7,015	328,029
1-4	0.0013	0.0051	0.0045	0.0058	97,902	97,554	98,240	502	390,401	74.1	73.2	75.3	1,581	1,228,359
5-9	0.0004	0.0022	0.0019	0.0025	97,399	96,999	97,798	218	486,452	70.5	69.6	71.7	640	1,430,637
10-14	0.0004	0.0018	0.0016	0.0021	97,182	96,757	97,607	178	485,463	65.6	64.8	66.8	470	1,282,494
15-19	0.0006	0.0031	0.0027	0.0035	97,004	96,557	97,451	297	484,275	60.7	59.9	61.9	628	1,022,395
20-24	0.0007	0.0037	0.0033	0.0042	96,706	96,224	97,189	362	482,627	55.9	55.1	57.1	679	905,806
25-29	0.0009	0.0044	0.0039	0.0049	96,345	95,819	96,869	422	480,669	51.1	50.3	52.3	581	661,861
30-34	0.0011	0.0057	0.0050	0.0064	95,923	95,346	96,495	546	478,249	46.3	45.5	47.5	583	509,915
35-39	0.0016	0.0081	0.0072	0.0091	95,377	94,736	96,008	772	474,953	41.6	40.8	42.7	706	434,281
40-44	0.0024	0.0119	0.0106	0.0133	94,605	93,876	95,316	1,123	470,215	36.9	36.1	38.0	922	385,937
45-49	0.0036	0.0179	0.0160	0.0199	93,481	92,631	94,301	1,671	463,228	32.3	31.6	33.4	1,213	336,217
50-54	0.0054	0.0269	0.0239	0.0297	91,810	90,792	92,777	2,468	452,880	27.9	27.1	28.9	1,529	280,622
55-59	0.0082	0.0402	0.0354	0.0442	89,342	88,098	90,501	3,592	437,730	23.6	22.9	24.6	1,760	214,468
60-64	0.0131	0.0633	0.0556	0.0696	85,750	84,220	87,139	5,430	415,174	19.5	18.8	20.4	2,105	160,952
65-69	0.0215	0.1020	0.0893	0.1119	80,320	78,443	82,201	8,191	381,120	15.6	15.0	16.4	2,592	120,601
70-74	0.0372	0.1702	0.1499	0.1854	72,129	69,895	74,857	12,277	329,951	12.1	11.6	12.8	3,054	82,084
75-79	0.0646	0.2781	0.2491	0.3004	59,852	56,963	63,632	16,644	257,649	9.1	8.6	9.6	3,388	52,443
80-84	0.1109	0.4341	0.4004	0.4637	43,208	39,969	47,780	18,755	169,152	6.6	6.2	7.0	2,842	25,636
85-89	0.1801	0.6210	0.5878	0.6503	24,453	21,760	28,649	15,185	84,302	4.7	4.5	5.0	1,649	9,156
90-94	0.2770	0.7564	0.7309	0.7793	9,268	7,704	11,670	7,010	25,308	3.3	3.2	3.5	547	1,975
95-99	0.4032	0.9124	0.8985	0.9246	2,257	1,705	3,056	2,060	5,108	2.4	2.4	2.5	100	247
100+	0.5556	1	1	1	198	129	310	198	356	1.8	1.8	1.9	8	14
All ages													*34,592*	*9,474,129*

$\ln(_nq_x)$ with uncertainty interval

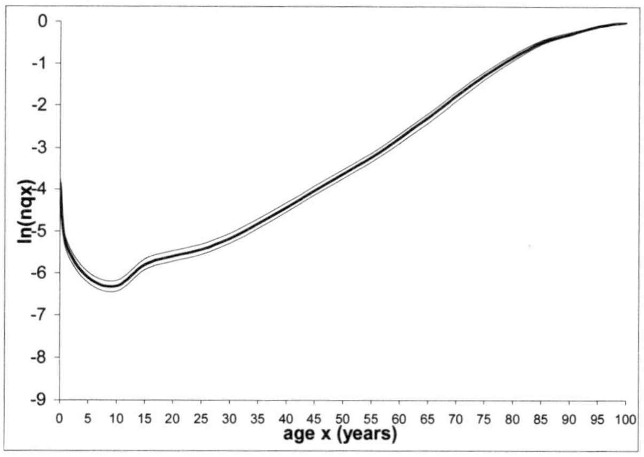

l_x with uncertainty interval

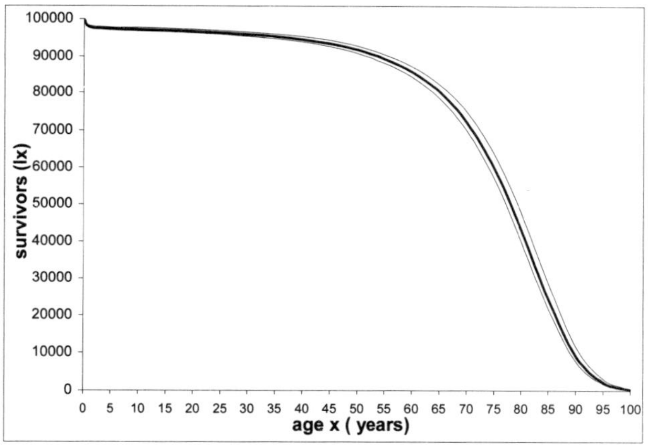

e_x with uncertainty interval

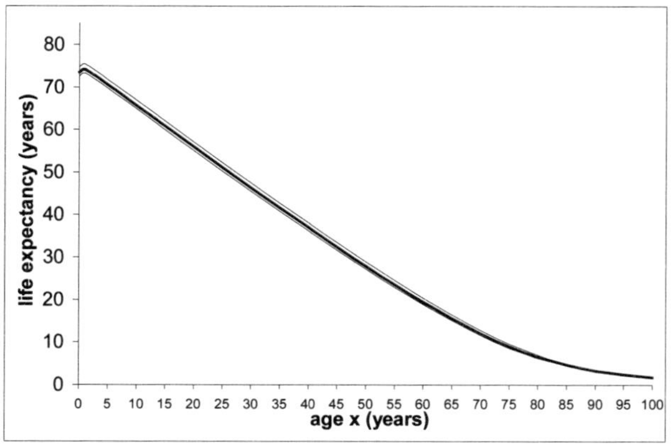

Distribution of deaths by age with uncertainty interval

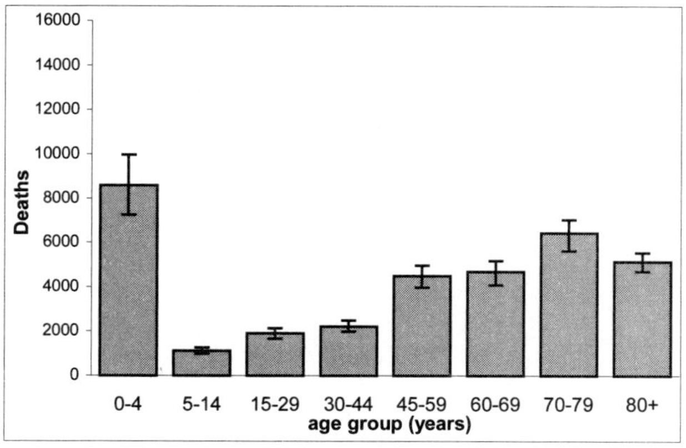

Senegal　　　Sénégal　　　Senegal

Males, 2000　　　*Hommes, 2000*　　　*Hombres, 2000*

Age	$_nM_x$	$_nq_x$			l_x			$_nd_x$	$_nL_x$	e_x			Actual deaths	Actual population
			percentiles			percentiles					percentiles			
		mean	2½	97½	mean	2½	97½			mean	2½	97½		
<1	0.1204	0.1086	0.1000	0.1167	100,000	100,000	100,000	10,863	90,224	54.0	52.0	56.2	22,032	182,993
1-4	0.0093	0.0365	0.0330	0.0402	89,137	88,333	90,002	3,258	348,731	59.6	57.3	62.1	5,814	622,375
5-9	0.0023	0.0113	0.0101	0.0125	85,880	85,131	86,627	966	426,983	57.8	55.6	60.2	1,579	697,670
10-14	0.0015	0.0073	0.0065	0.0082	84,913	84,118	85,714	621	423,015	53.4	51.3	55.8	888	605,154
15-19	0.0017	0.0082	0.0073	0.0092	84,293	83,468	85,134	693	419,730	48.8	46.7	51.2	842	509,823
20-24	0.0033	0.0161	0.0142	0.0182	83,599	82,742	84,486	1,349	414,625	44.2	42.1	46.6	1,409	433,189
25-29	0.0050	0.0245	0.0214	0.0278	82,251	81,331	83,226	2,016	406,213	39.9	37.8	42.2	1,750	352,637
30-34	0.0069	0.0338	0.0293	0.0387	80,235	79,159	81,398	2,713	394,391	35.8	33.9	38.1	1,990	289,213
35-39	0.0084	0.0411	0.0353	0.0474	77,522	76,134	78,987	3,187	379,640	32.0	30.1	34.1	2,027	241,405
40-44	0.0102	0.0498	0.0423	0.0579	74,334	72,543	76,158	3,704	362,413	28.2	26.5	30.3	2,046	200,189
45-49	0.0129	0.0624	0.0524	0.0730	70,631	68,342	72,870	4,409	342,130	24.6	23.0	26.5	2,082	161,583
50-54	0.0169	0.0810	0.0674	0.0950	66,221	63,355	68,945	5,361	317,704	21.0	19.6	22.9	2,113	125,213
55-59	0.0227	0.1072	0.0887	0.1258	60,860	57,338	64,181	6,524	287,991	17.7	16.3	19.3	2,215	97,798
60-64	0.0324	0.1499	0.1237	0.1752	54,337	50,123	58,421	8,144	251,324	14.5	13.3	16.0	2,411	74,408
65-69	0.0471	0.2108	0.1749	0.2447	46,193	41,342	51,084	9,740	206,616	11.6	10.6	12.9	2,426	51,466
70-74	0.0719	0.3046	0.2566	0.3485	36,453	31,261	42,113	11,102	154,512	9.0	8.2	10.1	2,260	31,456
75-79	0.1089	0.4279	0.3707	0.4781	25,351	20,463	31,307	10,849	99,635	6.9	6.3	7.7	1,618	14,864
80-84	0.1626	0.5779	0.5195	0.6267	14,503	10,793	19,702	8,382	51,559	5.2	4.8	5.7	749	4,605
85-89	0.2356	0.7413	0.6922	0.7808	6,121	4,071	9,466	4,537	19,261	3.9	3.6	4.2	214	907
90-94	0.3312	0.8307	0.7925	0.8612	1,584	892	2,914	1,315	3,971	2.9	2.7	3.1	30	92
95-99	0.4520	0.9592	0.9422	0.9712	268	124	605	257	569	2.2	2.1	2.3	2	5
100+	0.5984	1	1	1	11	4	35	11	18	1.7	1.6	1.8	0	0
All ages													*56,497*	*4,697,045*

ln($_nq_x$) with uncertainty interval

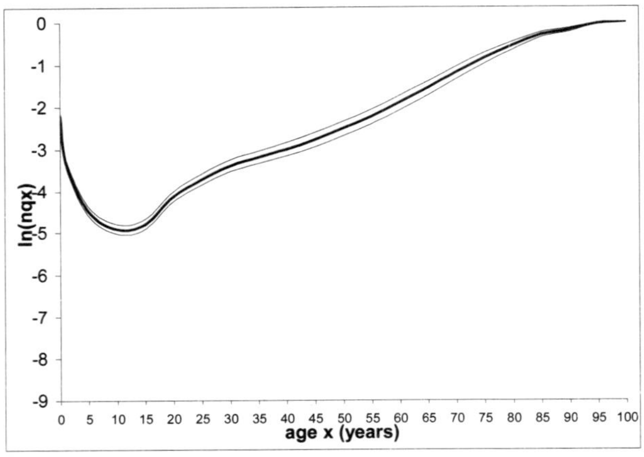

l_x with uncertainty interval

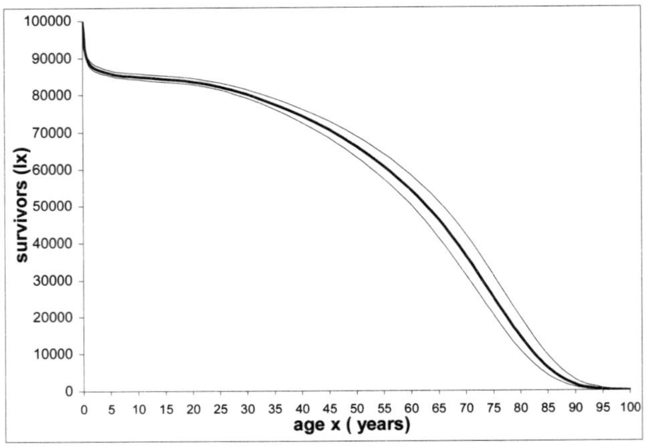

e_x with uncertainty interval

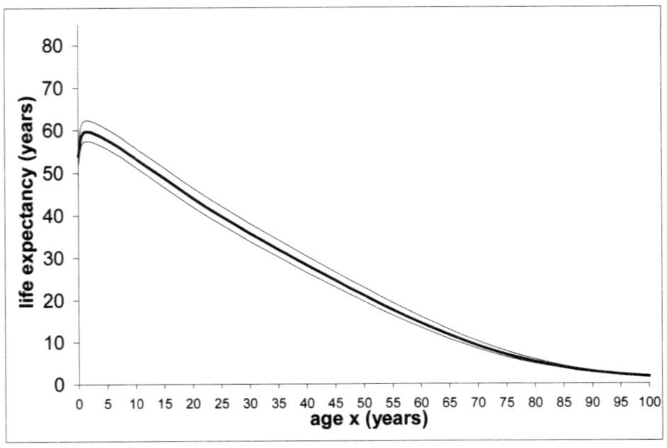

Distribution of deaths by age with uncertainty interval

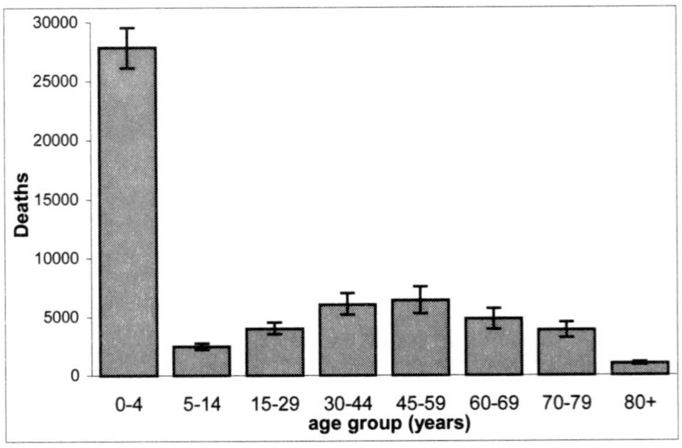

Senegal Sénégal Senegal

Females, 2000 *Femmes, 2000* *Mujeres, 2000*

Age	$_nM_x$	$_nq_x$ mean	$_nq_x$ percentiles 2½	$_nq_x$ percentiles 97½	l_x mean	l_x percentiles 2½	l_x percentiles 97½	$_nd_x$	$_nL_x$	e_x mean	e_x percentiles 2½	e_x percentiles 97½	Actual deaths	Actual population
<1	0.1058	0.0966	0.0892	0.1050	100,000	100,000	100,000	9,663	91,303	56.1	54.0	58.4	18,623	175,955
1-4	0.0103	0.0404	0.0362	0.0442	90,337	89,497	91,079	3,648	352,592	61.1	58.7	63.7	6,321	611,010
5-9	0.0023	0.0116	0.0103	0.0130	86,689	85,940	87,439	1,007	430,927	59.6	57.3	62.1	1,599	684,293
10-14	0.0015	0.0077	0.0068	0.0086	85,682	84,858	86,505	660	426,761	55.3	53.1	57.8	922	596,365
15-19	0.0025	0.0123	0.0109	0.0139	85,022	84,141	85,917	1,048	422,491	50.7	48.5	53.2	1,251	504,291
20-24	0.0043	0.0214	0.0188	0.0243	83,974	83,001	84,984	1,799	415,372	46.3	44.1	48.7	1,877	433,368
25-29	0.0060	0.0295	0.0256	0.0337	82,175	81,043	83,386	2,421	404,819	42.2	40.2	44.6	2,118	354,110
30-34	0.0070	0.0342	0.0294	0.0394	79,753	78,408	81,241	2,727	391,948	38.4	36.5	40.7	2,026	291,227
35-39	0.0075	0.0367	0.0313	0.0427	77,026	75,444	78,828	2,830	378,055	34.7	32.9	36.9	1,833	244,904
40-44	0.0077	0.0377	0.0318	0.0441	74,196	72,294	76,324	2,798	363,985	31.0	29.2	33.0	1,576	205,039
45-49	0.0090	0.0439	0.0368	0.0516	71,398	69,108	73,845	3,137	349,148	27.1	25.5	29.0	1,510	168,101
50-54	0.0119	0.0578	0.0480	0.0681	68,261	65,542	71,058	3,943	331,449	23.2	21.7	25.0	1,550	130,267
55-59	0.0163	0.0784	0.0646	0.0926	64,318	61,080	67,538	5,041	308,989	19.5	18.1	21.2	1,737	106,470
60-64	0.0246	0.1158	0.0951	0.1365	59,277	55,427	62,998	6,865	279,224	15.9	14.7	17.4	2,113	85,946
65-69	0.0381	0.1739	0.1429	0.2036	52,412	47,859	56,864	9,117	239,270	12.7	11.6	14.0	2,480	65,084
70-74	0.0613	0.2659	0.2209	0.3059	43,295	38,117	48,604	11,512	187,696	9.8	9.0	10.9	2,555	41,659
75-79	0.0961	0.3873	0.3298	0.4339	31,783	26,456	37,867	12,309	128,143	7.4	6.8	8.3	1,810	18,843
80-84	0.1468	0.5370	0.4747	0.5841	19,474	14,978	25,380	10,458	71,227	5.6	5.2	6.2	799	5,445
85-89	0.2173	0.7041	0.6493	0.7434	9,017	6,252	13,331	6,349	29,211	4.1	3.9	4.5	211	973
90-94	0.3115	0.8051	0.7624	0.8354	2,668	1,619	4,675	2,148	6,896	3.0	2.9	3.3	35	111
95-99	0.4323	0.9411	0.9190	0.9554	520	270	1,111	489	1,132	2.3	2.2	2.4	5	11
100+	0.5810	1	1	1	31	12	90	31	53	1.7	1.7	1.8	1	1
All ages													*52,953*	*4,723,473*

ln($_nq_x$) with uncertainty interval

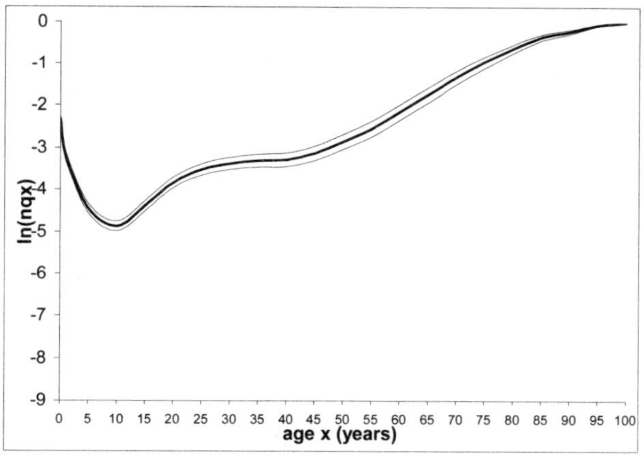

l_x with uncertainty interval

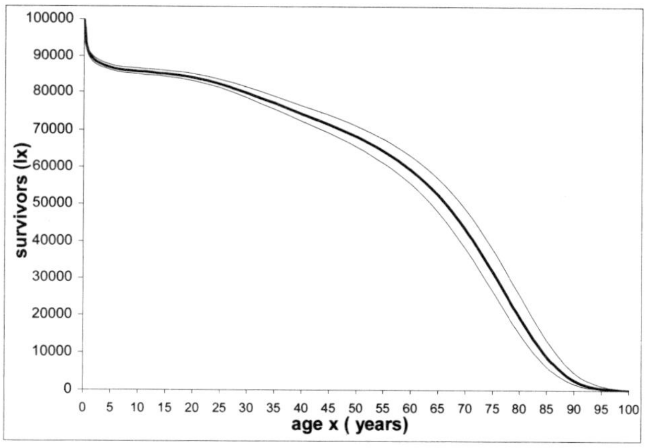

e_x with uncertainty interval

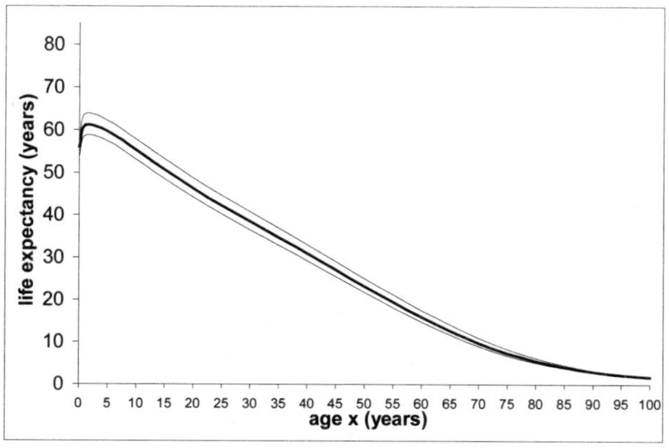

Distribution of deaths by age with uncertainty interval

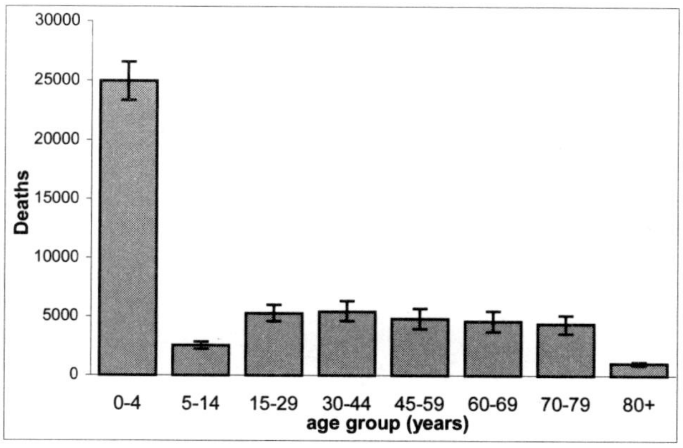

Seychelles **Seychelles** **Seychelles**

Males, 2000 *Hommes, 2000* *Hombres, 2000*

Age	$_nM_x$	$_nq_x$ mean	$_nq_x$ percentiles 2½	$_nq_x$ percentiles 97½	l_x mean	l_x percentiles 2½	l_x percentiles 97½	$_nd_x$	$_nL_x$	e_x mean	e_x percentiles 2½	e_x percentiles 97½	Actual deaths	Actual population
<1	0.0130	0.0129	0.0110	0.0144	100,000	100,000	100,000	1,285	98,843	66.5	65.3	67.9	9	678
1-4	0.0008	0.0033	0.0029	0.0036	98,715	98,558	98,896	323	394,084	66.4	65.2	67.8	2	2,661
5-9	0.0003	0.0015	0.0014	0.0017	98,392	98,199	98,613	152	491,581	62.6	61.4	64.0	1	3,779
10-14	0.0005	0.0023	0.0021	0.0026	98,240	98,029	98,480	230	490,627	57.7	56.5	59.1	2	3,327
15-19	0.0005	0.0024	0.0021	0.0026	98,010	97,773	98,277	231	489,476	52.8	51.7	54.2	2	3,583
20-24	0.0015	0.0076	0.0066	0.0085	97,780	97,515	98,074	740	487,050	47.9	46.8	49.3	6	3,843
25-29	0.0020	0.0098	0.0086	0.0110	97,040	96,688	97,422	946	482,836	43.3	42.2	44.7	6	3,230
30-34	0.0034	0.0168	0.0148	0.0188	96,094	95,629	96,586	1,610	476,446	38.7	37.6	40.1	12	3,426
35-39	0.0041	0.0203	0.0179	0.0229	94,484	93,828	95,160	1,920	467,620	34.3	33.3	35.6	14	3,512
40-44	0.0061	0.0299	0.0264	0.0337	92,564	91,682	93,455	2,771	455,891	30.0	29.0	31.2	19	3,099
45-49	0.0107	0.0522	0.0461	0.0586	89,792	88,595	90,985	4,691	437,235	25.8	24.9	27.0	28	2,647
50-54	0.0135	0.0653	0.0572	0.0729	85,101	83,403	86,791	5,555	411,620	22.1	21.3	23.2	25	1,874
55-59	0.0207	0.0986	0.0863	0.1093	79,547	77,323	81,823	7,842	378,128	18.5	17.8	19.5	27	1,295
60-64	0.0295	0.1373	0.1206	0.1505	71,704	68,874	74,759	9,843	333,915	15.2	14.6	16.1	30	1,020
65-69	0.0421	0.1904	0.1688	0.2067	61,862	58,511	65,743	11,776	279,868	12.2	11.7	12.9	33	792
70-74	0.0601	0.2614	0.2355	0.2823	50,086	46,419	54,644	13,093	217,696	9.5	9.1	10.1	39	644
75-79	0.0982	0.3941	0.3641	0.4197	36,993	33,314	41,773	14,579	148,517	7.0	6.7	7.4	40	405
80-84	0.1700	0.5965	0.5681	0.6224	22,414	19,331	26,563	13,370	78,646	5.0	4.8	5.2	29	172
85-89	0.2600	0.7879	0.7673	0.8066	9,044	7,299	11,472	7,126	27,407	3.6	3.5	3.7	15	58
90-94	0.3500	0.8537	0.8385	0.8677	1,918	1,411	2,670	1,638	4,679	2.8	2.7	2.8	4	12
95-99	0.4600	0.9664	0.9605	0.9718	281	187	431	271	590	2.2	2.1	2.2	1	1
100+	0.6795	1	1	1	9	5	17	9	14	1.5	1.5	1.9	0	0
All ages													*344*	*40,060*

ln($_nq_x$) with uncertainty interval

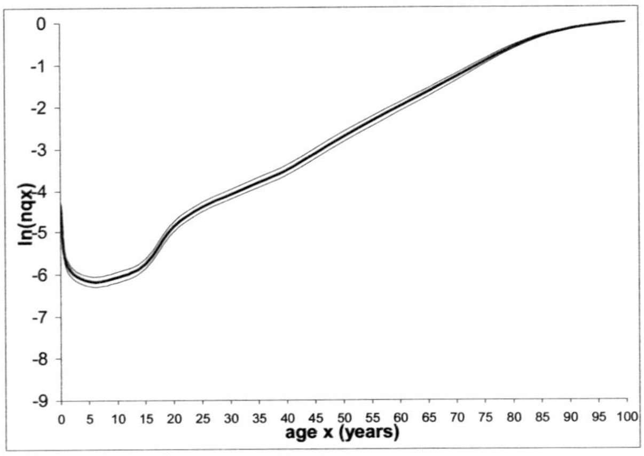

l_x with uncertainty interval

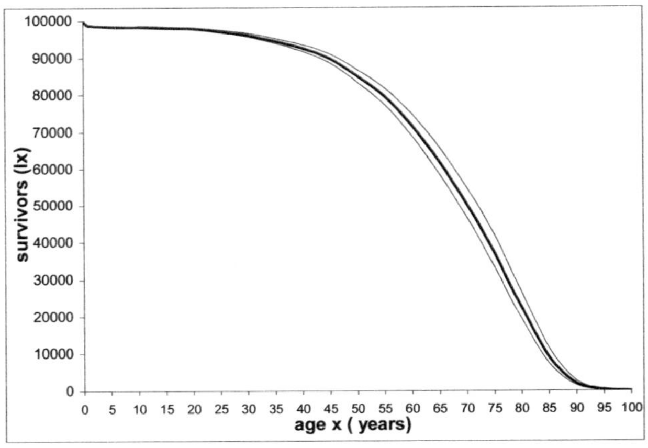

e_x with uncertainty interval

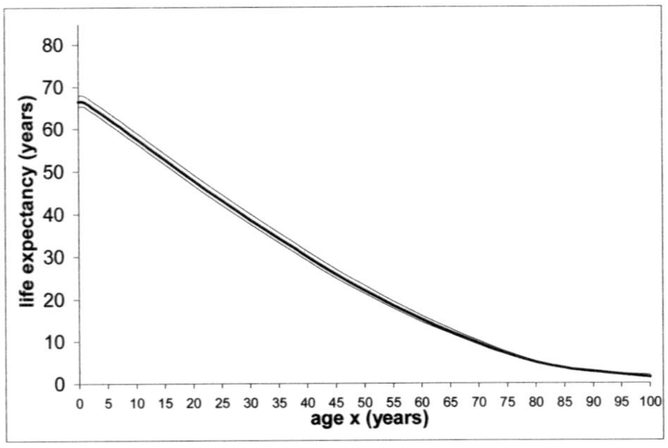

Distribution of deaths by age with uncertainty interval

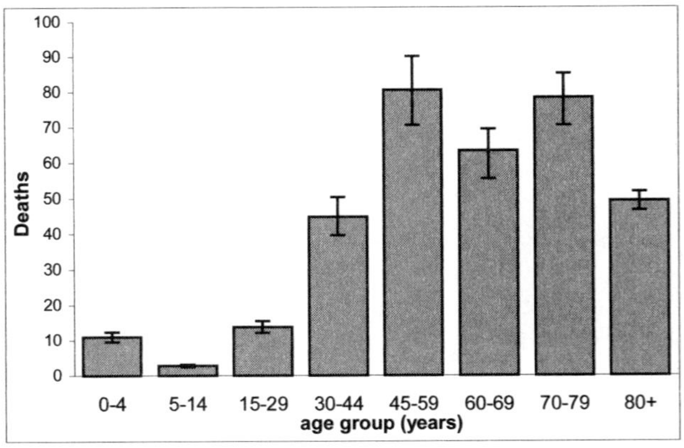

Seychelles

Females, 2000

Seychelles

Femmes, 2000

Seychelles

Mujeres, 2000

Age	${}_nM_x$	${}_nq_x$			l_x			${}_nd_x$	${}_nL_x$	e_x			Actual deaths	Actual population
			percentiles			percentiles					percentiles			
		mean	2½	97½	mean	2½	97½			mean	2½	97½		
<1	0.0105	0.0104	0.0087	0.0123	100,000	100,000	100,000	1,040	99,064	74.2	72.4	76.2	7	649
1-4	0.0004	0.0015	0.0013	0.0018	98,960	98,769	99,131	149	395,484	74.0	72.3	76.0	1	2,586
5-9	0.0004	0.0018	0.0015	0.0021	98,811	98,592	99,003	174	493,622	70.1	68.4	72.1	1	3,660
10-14	0.0003	0.0014	0.0012	0.0017	98,637	98,385	98,854	143	492,830	65.2	63.6	67.2	1	3,265
15-19	0.0005	0.0024	0.0021	0.0029	98,495	98,216	98,731	236	491,884	60.3	58.7	62.3	2	3,472
20-24	0.0004	0.0019	0.0017	0.0023	98,259	97,935	98,527	190	490,819	55.4	53.8	57.4	1	3,752
25-29	0.0007	0.0035	0.0030	0.0042	98,069	97,709	98,363	345	489,481	50.5	49.0	52.5	2	3,111
30-34	0.0012	0.0060	0.0052	0.0072	97,724	97,297	98,065	585	487,158	45.7	44.2	47.7	4	3,253
35-39	0.0017	0.0085	0.0073	0.0102	97,139	96,597	97,559	822	483,641	41.0	39.5	42.9	6	3,402
40-44	0.0025	0.0124	0.0106	0.0149	96,317	95,614	96,849	1,193	478,602	36.3	34.8	38.2	8	3,025
45-49	0.0035	0.0174	0.0147	0.0210	95,124	94,186	95,822	1,652	471,488	31.7	30.3	33.6	9	2,630
50-54	0.0060	0.0297	0.0247	0.0358	93,471	92,210	94,415	2,777	460,415	27.2	25.9	29.0	12	1,975
55-59	0.0094	0.0461	0.0378	0.0554	90,695	88,906	92,082	4,177	443,031	23.0	21.8	24.7	14	1,458
60-64	0.0160	0.0768	0.0623	0.0915	86,518	83,983	88,605	6,644	415,979	19.0	17.9	20.6	19	1,208
65-69	0.0239	0.1128	0.0911	0.1325	79,874	76,297	83,086	9,012	376,838	15.4	14.4	16.8	23	957
70-74	0.0427	0.1927	0.1574	0.2206	70,862	66,186	75,513	13,658	320,162	12.0	11.2	13.2	37	856
75-79	0.0647	0.2784	0.2330	0.3141	57,203	51,589	63,628	15,923	246,208	9.3	8.6	10.2	39	605
80-84	0.1103	0.4324	0.3789	0.4761	41,280	35,384	48,803	17,849	161,776	6.9	6.4	7.6	34	304
85-89	0.1589	0.5687	0.5225	0.6088	23,431	18,536	30,310	13,325	83,841	5.2	4.9	5.6	22	136
90-94	0.2319	0.6839	0.6498	0.7159	10,106	7,252	14,474	6,911	29,796	3.8	3.6	4.1	9	40
95-99	0.3430	0.8452	0.8227	0.8663	3,195	2,060	5,069	2,700	7,873	2.8	2.7	2.9	2	7
100+	0.5138	1	1	1	495	275	899	495	962	1.9	1.9	2.3	0	1
All ages													*252*	*40,351*

$\ln({}_nq_x)$ with uncertainty interval

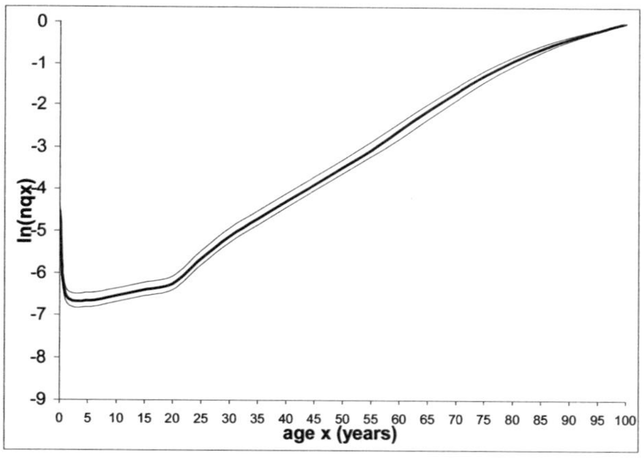

l_x with uncertainty interval

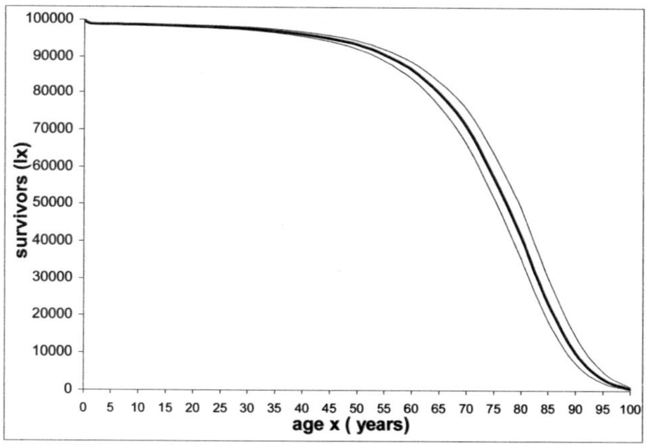

e_x with uncertainty interval

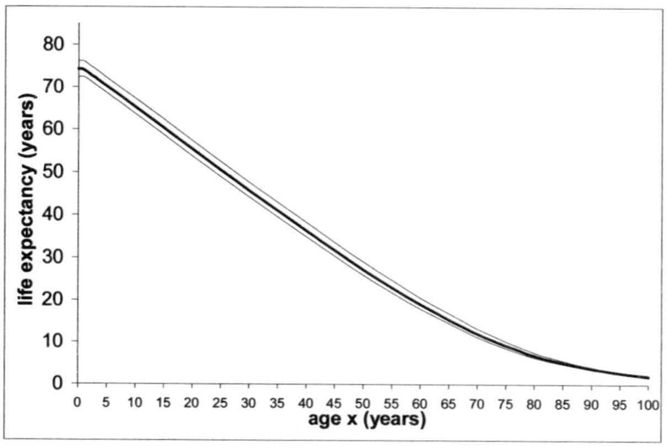

Distribution of deaths by age with uncertainty interval

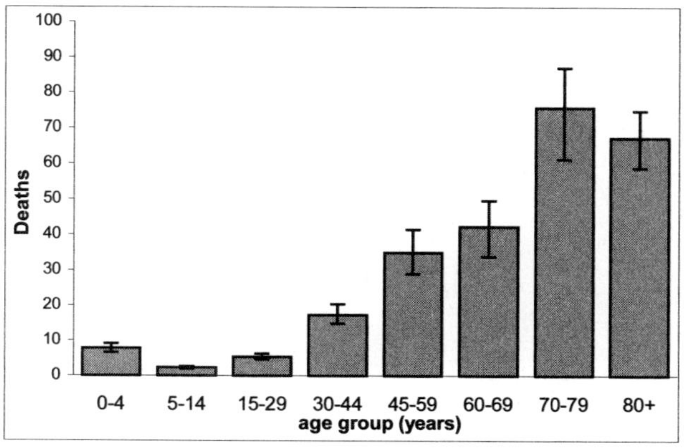

Sierra Leone | **Sierra Leone** | **Sierra Leona**

Males, 2000 | *Hommes, 2000* | *Hombres, 2000*

Age	$_nM_x$	$_nq_x$ mean	$_nq_x$ percentiles 2½	$_nq_x$ percentiles 97½	l_x mean	l_x percentiles 2½	l_x percentiles 97½	$_nd_x$	$_nL_x$	e_x mean	e_x percentiles 2½	e_x percentiles 97½	Actual deaths	Actual population
<1	0.2750	0.2205	0.1783	0.2707	100,000	100,000	100,000	22,046	80,159	37.0	32.6	41.9	25,704	93,460
1-4	0.0241	0.0912	0.0728	0.1118	77,954	72,926	82,169	7,107	294,760	46.4	41.8	51.7	7,454	309,167
5-9	0.0056	0.0277	0.0216	0.0347	70,847	66,502	75,161	1,965	349,325	46.9	42.8	51.9	1,717	305,260
10-14	0.0037	0.0184	0.0142	0.0231	68,883	64,358	73,304	1,265	341,252	43.2	39.2	48.0	973	262,391
15-19	0.0045	0.0221	0.0171	0.0278	67,618	62,987	72,221	1,493	334,357	39.0	35.0	43.7	1,005	224,938
20-24	0.0075	0.0370	0.0285	0.0467	66,125	61,377	70,937	2,447	324,507	34.8	30.9	39.5	1,459	193,541
25-29	0.0132	0.0637	0.0490	0.0804	63,678	58,760	68,821	4,059	308,244	31.0	27.3	35.6	2,130	161,785
30-34	0.0180	0.0861	0.0659	0.1086	59,619	54,270	65,269	5,133	285,266	28.0	24.4	32.3	2,410	133,970
35-39	0.0216	0.1026	0.0781	0.1301	54,487	48,379	60,696	5,593	258,453	25.4	22.0	29.5	2,392	110,536
40-44	0.0244	0.1152	0.0865	0.1470	48,894	42,133	55,631	5,633	230,389	23.0	20.0	26.7	2,246	91,867
45-49	0.0264	0.1237	0.0919	0.1582	43,261	36,058	50,732	5,350	202,933	20.6	18.0	24.0	1,988	75,411
50-54	0.0287	0.1339	0.0990	0.1711	37,912	30,514	45,925	5,076	176,868	18.2	15.9	21.2	1,745	60,794
55-59	0.0323	0.1493	0.1103	0.1898	32,835	25,478	41,187	4,901	151,923	15.6	13.6	18.3	1,573	48,750
60-64	0.0425	0.1921	0.1429	0.2413	27,934	20,839	36,410	5,367	126,252	12.9	11.3	15.2	1,597	37,555
65-69	0.0584	0.2547	0.1921	0.3138	22,567	16,033	30,896	5,747	98,465	10.4	9.0	12.3	1,564	26,802
70-74	0.0875	0.3590	0.2784	0.4294	16,819	11,185	24,562	6,038	69,003	8.1	7.0	9.7	1,469	16,785
75-79	0.1312	0.4940	0.3989	0.5697	10,782	6,382	17,399	5,326	40,592	6.2	5.4	7.5	1,119	8,526
80-84	0.1787	0.6176	0.5184	0.6892	5,455	2,746	10,453	3,369	18,853	4.9	4.3	5.8	555	3,103
85-89	0.2526	0.7742	0.6838	0.8325	2,086	853	5,005	1,615	6,392	3.7	3.3	4.3	170	672
90-94	0.3482	0.8515	0.7752	0.8975	471	143	1,560	401	1,152	2.8	2.5	3.2	25	71
95-99	0.4680	0.9734	0.9422	0.9867	70	15	343	68	145	2.1	2.1	2.3	2	4
100+	0.6131	1	1	1	2	0	19	2	3	1.6	1.6	1.6	0	0
All ages													*59,295*	*2,165,388*

ln($_nq_x$) with uncertainty interval

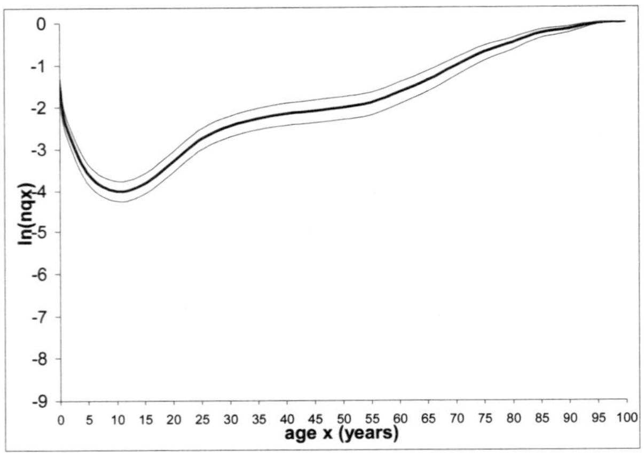

l_x with uncertainty interval

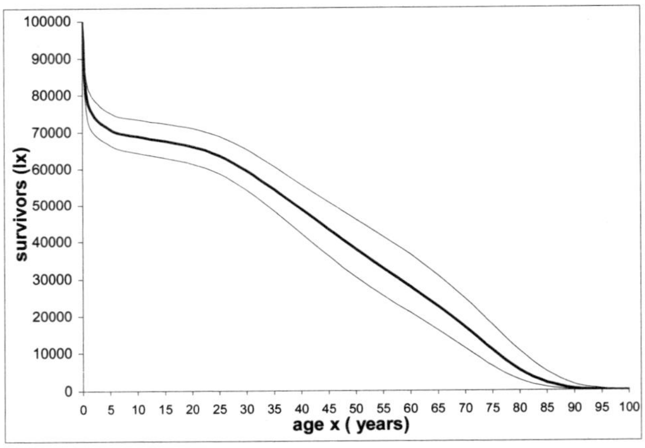

e_x with uncertainty interval

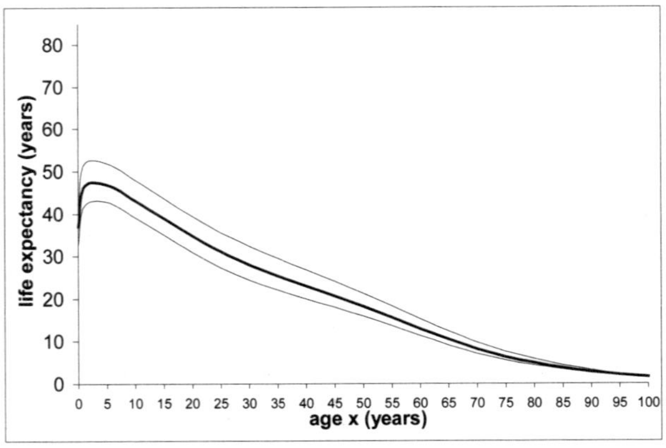

Distribution of deaths by age with uncertainty interval

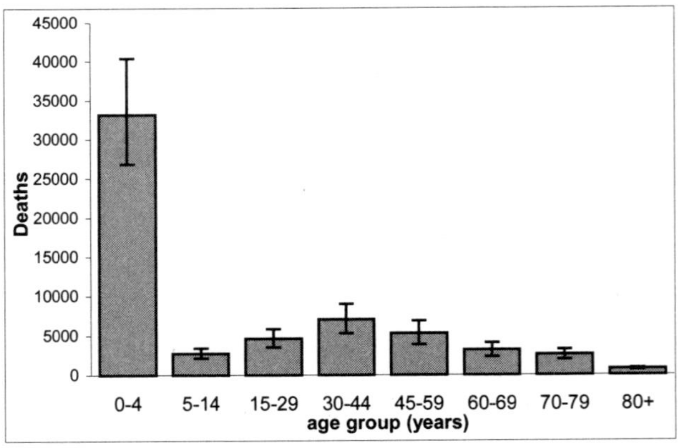

Sierra Leone **Sierra Leone** **Sierra Leona**

Females, 2000 *Femmes, 2000* *Mujeres, 2000*

Age	$_nM_x$	$_nq_x$			l_x			$_nd_x$	$_nL_x$	e_x			Actual deaths	Actual population
			percentiles			percentiles					percentiles			
		mean	2½	97½	mean	2½	97½			mean	2½	97½		
<1	0.2128	0.1786	0.1440	0.2202	100,000	100,000	100,000	17,861	83,925	38.8	34.3	43.9	19,126	89,866
1-4	0.0282	0.1057	0.0858	0.1279	82,139	77,975	85,600	8,683	307,715	46.3	41.3	51.9	8,848	313,576
5-9	0.0066	0.0324	0.0254	0.0403	73,456	69,801	77,097	2,377	361,335	47.5	43.1	52.6	2,036	309,405
10-14	0.0046	0.0228	0.0178	0.0285	71,078	67,216	74,941	1,618	351,347	44.1	39.7	48.9	1,223	265,583
15-19	0.0077	0.0379	0.0294	0.0476	69,460	65,406	73,488	2,630	340,728	40.0	35.8	44.8	1,760	228,058
20-24	0.0125	0.0605	0.0467	0.0762	66,831	62,439	71,108	4,044	324,044	36.5	32.4	41.0	2,460	197,148
25-29	0.0158	0.0760	0.0584	0.0959	62,787	57,910	67,478	4,775	301,996	33.7	29.7	38.0	2,624	165,942
30-34	0.0174	0.0834	0.0637	0.1053	58,012	52,394	63,361	4,838	277,964	31.3	27.6	35.2	2,407	138,312
35-39	0.0183	0.0875	0.0666	0.1114	53,174	46,877	59,281	4,654	254,235	28.9	25.5	32.5	2,105	114,974
40-44	0.0170	0.0815	0.0620	0.1046	48,520	41,703	55,330	3,956	232,709	26.4	23.4	29.7	1,647	96,904
45-49	0.0173	0.0829	0.0631	0.1068	44,564	37,421	51,900	3,696	213,580	23.5	20.9	26.5	1,409	81,443
50-54	0.0203	0.0967	0.0735	0.1244	40,868	33,528	48,626	3,952	194,460	20.4	18.0	23.1	1,351	66,485
55-59	0.0249	0.1173	0.0893	0.1505	36,916	29,488	45,035	4,332	173,751	17.3	15.3	19.7	1,378	55,287
60-64	0.0344	0.1582	0.1211	0.2009	32,584	25,214	40,973	5,154	150,036	14.3	12.5	16.4	1,518	44,189
65-69	0.0489	0.2178	0.1688	0.2722	27,430	20,357	35,933	5,975	122,214	11.5	10.0	13.4	1,601	32,742
70-74	0.0735	0.3105	0.2456	0.3780	21,455	15,059	29,737	6,663	90,620	9.1	7.8	10.6	1,579	21,479
75-79	0.1074	0.4235	0.3444	0.4987	14,793	9,489	22,220	6,264	58,303	7.0	6.1	8.2	1,256	11,691
80-84	0.1582	0.5668	0.4798	0.6422	8,529	4,757	14,284	4,834	30,558	5.3	4.7	6.2	757	4,785
85-89	0.2284	0.7269	0.6451	0.7914	3,695	1,702	7,292	2,686	11,759	4.0	3.6	4.6	291	1,275
90-94	0.3213	0.8181	0.7487	0.8699	1,009	355	2,545	825	2,569	3.0	2.7	3.4	62	192
95-99	0.4406	0.9489	0.9117	0.9712	184	46	630	174	395	2.2	2.1	2.4	7	15
100+	0.5888	1	1	1	9	1	56	9	16	1.7	1.7	1.8	1	1
All ages													*55,446*	*2,239,352*

ln($_nq_x$) with uncertainty interval

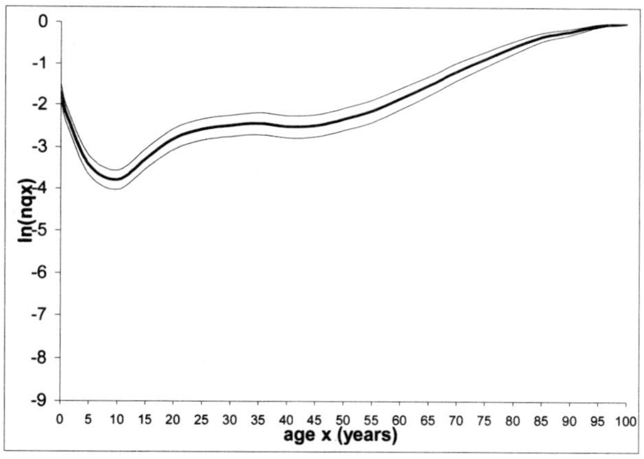

l_x with uncertainty interval

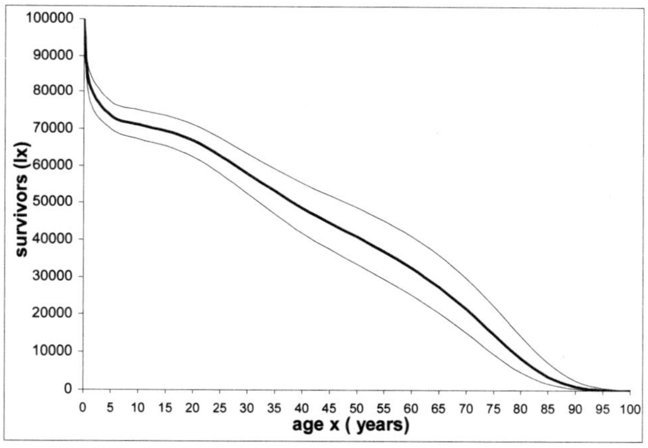

e_x with uncertainty interval

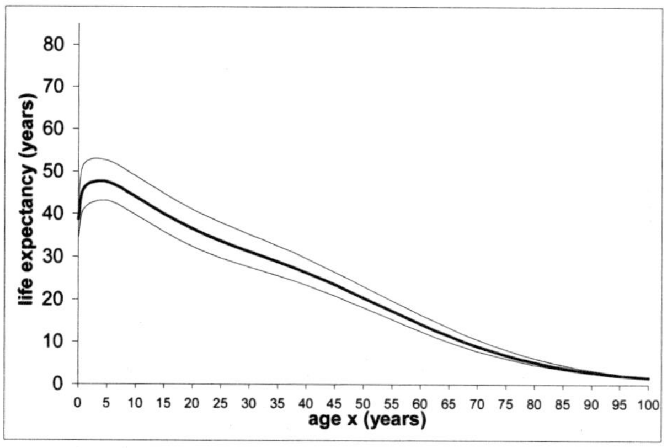

Distribution of deaths by age with uncertainty interval

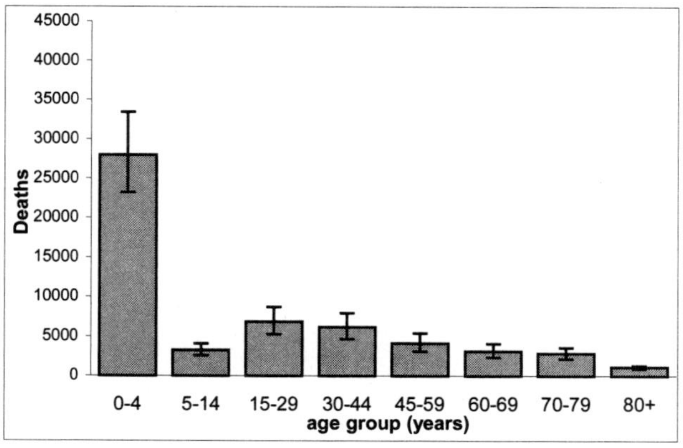

Singapore — Singapour — Singapur

Males, 2000 — *Hommes, 2000* — *Hombres, 2000*

Age	$_nM_x$	$_nq_x$ mean	$_nq_x$ percentiles 2½	$_nq_x$ percentiles 97½	l_x mean	l_x percentiles 2½	l_x percentiles 97½	$_nd_x$	$_nL_x$	e_x mean	e_x percentiles 2½	e_x percentiles 97½	Actual deaths	Actual population
<1	0.0037	0.0037	0.0031	0.0043	100,000	100,000	100,000	373	99,664	75.4	74.8	76.0	101	26,871
1-4	0.0003	0.0013	0.0011	0.0014	99,627	99,567	99,686	126	398,206	74.7	74.1	75.3	37	117,705
5-9	0.0001	0.0006	0.0005	0.0007	99,501	99,426	99,576	62	497,351	70.8	70.2	71.4	21	168,514
10-14	0.0002	0.0011	0.0009	0.0012	99,439	99,358	99,521	106	496,931	65.8	65.2	66.5	30	141,532
15-19	0.0004	0.0022	0.0020	0.0025	99,333	99,241	99,426	223	496,109	60.9	60.3	61.6	59	130,900
20-24	0.0012	0.0060	0.0054	0.0065	99,111	98,997	99,226	594	494,068	56.0	55.4	56.7	155	128,545
25-29	0.0011	0.0056	0.0052	0.0060	98,517	98,353	98,684	553	491,200	51.3	50.8	52.0	172	153,005
30-34	0.0012	0.0058	0.0054	0.0062	97,963	97,761	98,174	569	488,394	46.6	46.1	47.3	211	180,676
35-39	0.0013	0.0067	0.0063	0.0071	97,394	97,155	97,645	653	485,339	41.9	41.3	42.6	273	202,968
40-44	0.0020	0.0098	0.0092	0.0103	96,741	96,462	97,033	943	481,349	37.1	36.6	37.8	390	198,998
45-49	0.0030	0.0147	0.0139	0.0154	95,798	95,469	96,143	1,405	475,478	32.5	31.9	33.2	493	166,865
50-54	0.0050	0.0245	0.0233	0.0257	94,393	94,000	94,802	2,312	466,185	27.9	27.3	28.6	644	129,769
55-59	0.0090	0.0442	0.0418	0.0466	92,081	91,607	92,571	4,072	450,225	23.5	23.0	24.2	721	79,750
60-64	0.0158	0.0758	0.0709	0.0803	88,009	87,444	88,585	6,675	423,358	19.5	19.0	20.2	1,039	65,919
65-69	0.0268	0.1255	0.1164	0.1332	81,334	80,489	82,114	10,207	381,153	15.9	15.4	16.5	1,350	50,410
70-74	0.0409	0.1855	0.1716	0.1981	71,127	69,789	72,544	13,194	322,651	12.8	12.4	13.4	1,510	36,919
75-79	0.0573	0.2506	0.2329	0.2674	57,933	56,119	60,092	14,519	253,369	10.2	9.8	10.6	1,304	22,764
80-84	0.0890	0.3640	0.3422	0.3850	43,414	41,222	46,100	15,804	177,562	7.8	7.5	8.1	1,134	12,738
85-89	0.1363	0.5083	0.4848	0.5301	27,610	25,351	30,320	14,034	102,967	5.8	5.6	6.0	811	5,949
90-94	0.2058	0.6361	0.6155	0.6555	13,577	11,916	15,580	8,636	41,974	4.2	4.1	4.4	432	2,100
95-99	0.3062	0.7980	0.7828	0.8124	4,940	4,116	5,956	3,942	12,874	3.1	3.0	3.1	115	375
100+	0.4493	1	1	1	998	777	1,291	998	2,221	2.2	2.2	2.3	29	64
All ages													*11,030*	*2,023,336*

ln($_nq_x$) with uncertainty interval

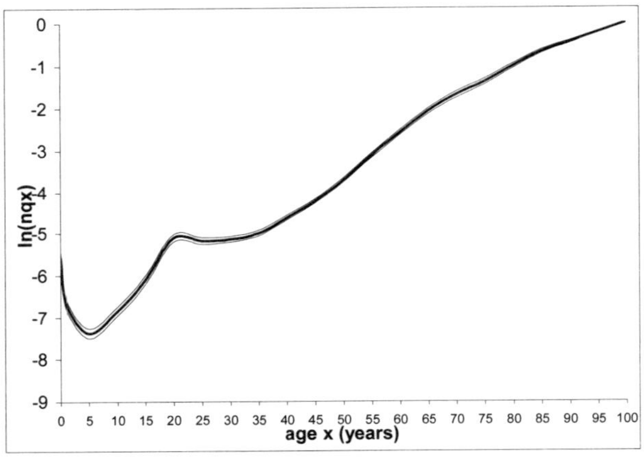

l_x with uncertainty interval

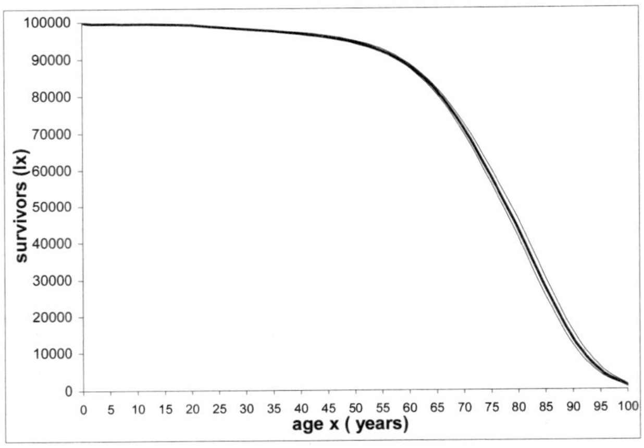

e_x with uncertainty interval

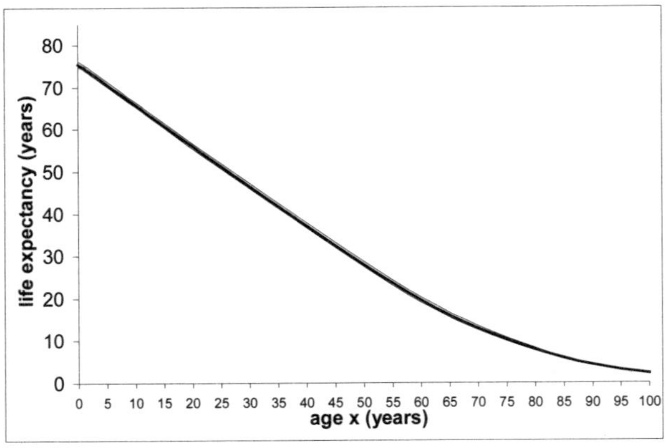

Distribution of deaths by age with uncertainty interval

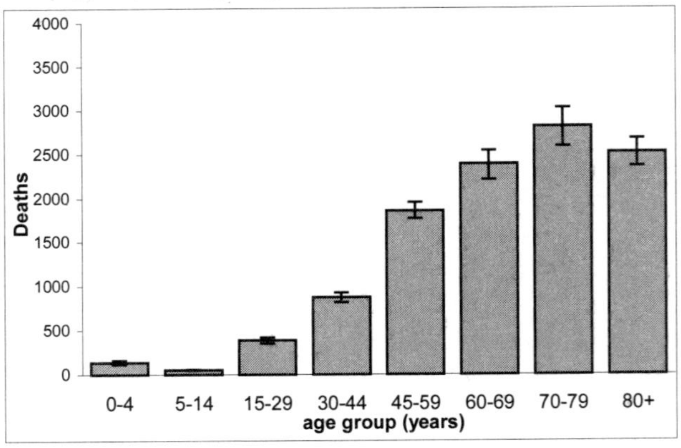

Singapore

Singapour

Singapur

Females, 2000 *Femmes, 2000* *Mujeres, 2000*

Age	$_nM_x$	$_nq_x$			l_x			$_nd_x$	$_nL_x$	e_x			Actual deaths	Actual population
			percentiles			percentiles					percentiles			
		mean	2½	97½	mean	2½	97½			mean	2½	97½		
<1	0.0030	0.0030	0.0024	0.0036	100,000	100,000	100,000	304	99,726	80.2	79.6	81.1	77	25,315
1-4	0.0002	0.0010	0.0008	0.0011	99,696	99,637	99,755	96	398,554	79.5	78.8	80.4	26	109,457
5-9	0.0001	0.0004	0.0004	0.0005	99,600	99,527	99,674	43	497,895	75.5	74.9	76.4	13	157,108
10-14	0.0001	0.0007	0.0006	0.0008	99,558	99,478	99,638	70	497,613	70.6	69.9	71.5	19	131,802
15-19	0.0003	0.0015	0.0013	0.0017	99,487	99,398	99,578	150	497,061	65.6	65.0	66.5	37	121,442
20-24	0.0003	0.0015	0.0013	0.0017	99,337	99,228	99,448	150	496,311	60.7	60.1	61.6	36	119,693
25-29	0.0004	0.0021	0.0019	0.0024	99,187	99,060	99,318	210	495,412	55.8	55.1	56.7	62	147,329
30-34	0.0005	0.0023	0.0020	0.0025	98,978	98,827	99,133	223	494,330	50.9	50.3	51.8	84	185,539
35-39	0.0006	0.0029	0.0026	0.0032	98,755	98,580	98,935	286	493,058	46.0	45.4	46.9	118	202,882
40-44	0.0010	0.0048	0.0043	0.0053	98,469	98,266	98,679	473	491,161	41.1	40.5	42.0	190	196,912
45-49	0.0016	0.0081	0.0074	0.0087	97,996	97,751	98,251	790	488,004	36.3	35.7	37.2	265	163,762
50-54	0.0030	0.0150	0.0139	0.0161	97,206	96,900	97,526	1,459	482,382	31.6	31.0	32.5	385	127,433
55-59	0.0049	0.0243	0.0224	0.0261	95,747	95,342	96,162	2,328	472,913	27.1	26.4	27.9	390	79,149
60-64	0.0096	0.0468	0.0425	0.0503	93,419	92,892	93,940	4,370	456,168	22.7	22.1	23.5	649	67,754
65-69	0.0157	0.0758	0.0683	0.0818	89,049	88,297	89,832	6,746	428,376	18.7	18.1	19.4	868	55,089
70-74	0.0253	0.1188	0.1067	0.1281	82,302	81,136	83,679	9,776	387,071	15.0	14.5	15.7	1,071	42,424
75-79	0.0436	0.1965	0.1775	0.2105	72,526	70,747	74,747	14,252	327,000	11.7	11.2	12.2	1,203	27,598
80-84	0.0697	0.2966	0.2719	0.3158	58,274	55,854	61,477	17,284	248,159	8.9	8.5	9.3	1,279	18,367
85-89	0.1105	0.4329	0.4060	0.4571	40,990	38,332	44,763	17,743	160,590	6.6	6.3	6.9	1,104	9,994
90-94	0.1740	0.5716	0.5476	0.5956	23,246	21,080	26,589	13,287	76,371	4.7	4.5	4.9	757	4,352
95-99	0.2720	0.7488	0.7299	0.7677	9,959	8,593	11,995	7,458	27,423	3.3	3.2	3.4	300	1,104
100+	0.4220	1	1	1	2,501	1,998	3,196	2,501	5,928	2.4	2.3	2.4	115	273
All ages													*9,049*	*1,994,778*

$\ln({}_nq_x)$ with uncertainty interval

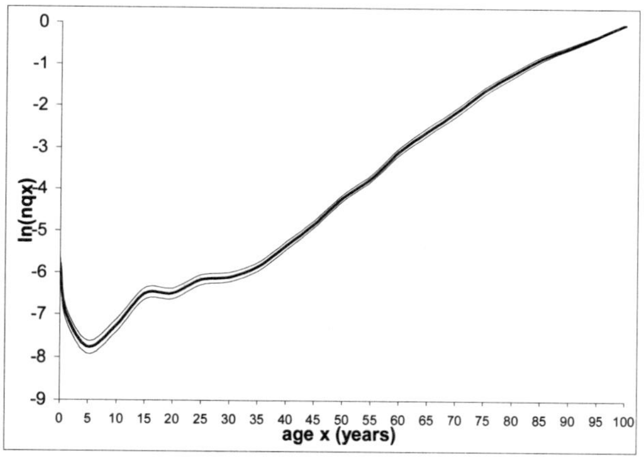

l_x with uncertainty interval

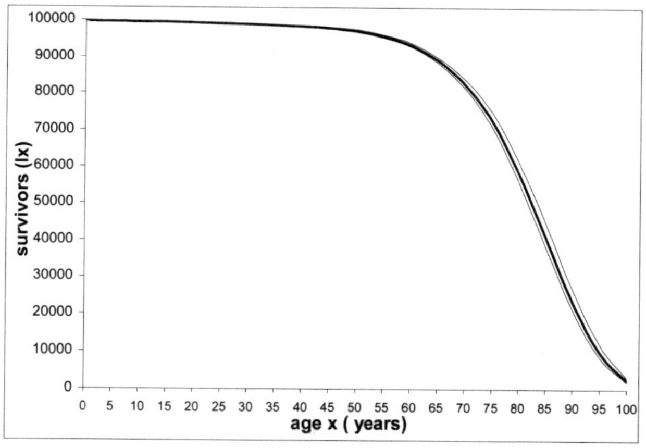

e_x with uncertainty interval

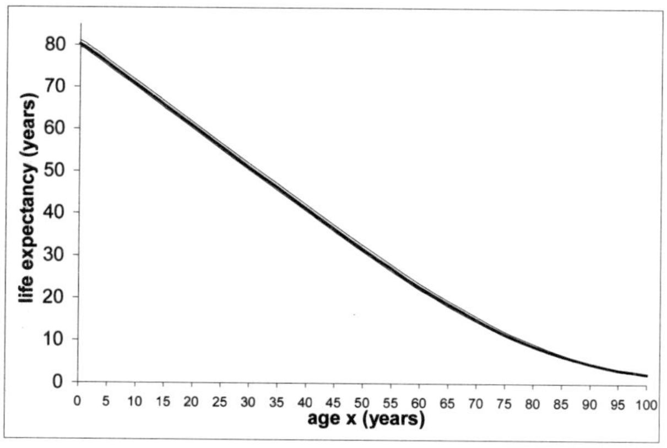

Distribution of deaths by age with uncertainty interval

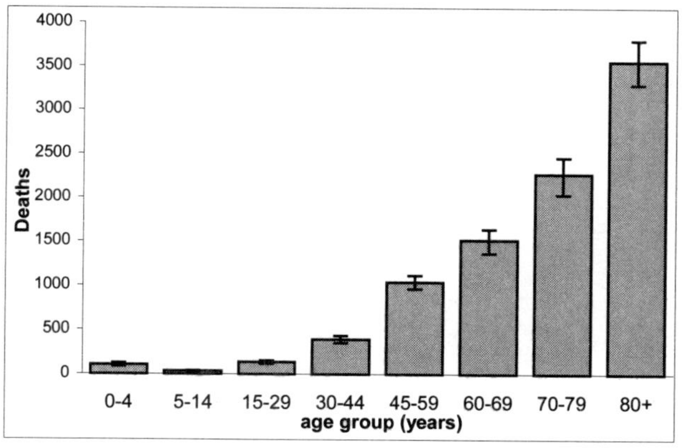

Slovakia **Slovaquie** **Eslovaquia**

Males, 2000 *Hommes, 2000* *Hombres, 2000*

Age	$_nM_x$	$_nq_x$			l_x			$_nd_x$	$_nL_x$	e_x			Actual deaths	Actual population
			percentiles			percentiles					percentiles			
		mean	2½	97½	mean	2½	97½			mean	2½	97½		
<1	0.0090	0.0089	0.0076	0.0102	100,000	100,000	100,000	893	99,197	69.2	68.8	69.7	265	29,440
1-4	0.0005	0.0019	0.0017	0.0021	99,107	98,977	99,237	190	395,973	68.9	68.4	69.4	57	118,295
5-9	0.0003	0.0013	0.0012	0.0014	98,917	98,768	99,067	129	494,265	65.0	64.5	65.5	48	183,472
10-14	0.0003	0.0015	0.0014	0.0016	98,789	98,627	98,951	148	493,573	60.1	59.6	60.6	62	207,585
15-19	0.0006	0.0028	0.0026	0.0031	98,641	98,466	98,816	281	492,501	55.1	54.7	55.7	130	227,796
20-24	0.0011	0.0054	0.0050	0.0059	98,360	98,162	98,559	535	490,463	50.3	49.9	50.9	263	241,701
25-29	0.0012	0.0059	0.0054	0.0063	97,825	97,588	98,064	575	487,688	45.6	45.1	46.1	256	216,951
30-34	0.0016	0.0081	0.0075	0.0086	97,250	96,974	97,530	785	484,288	40.8	40.4	41.4	301	185,852
35-39	0.0025	0.0122	0.0115	0.0129	96,465	96,140	96,797	1,175	479,390	36.1	35.7	36.7	482	196,881
40-44	0.0046	0.0228	0.0218	0.0239	95,291	94,903	95,688	2,176	471,013	31.5	31.0	32.2	957	207,193
45-49	0.0070	0.0344	0.0332	0.0356	93,115	92,638	93,605	3,203	457,566	27.2	26.7	27.9	1,429	204,174
50-54	0.0120	0.0583	0.0563	0.0602	89,912	89,349	90,492	5,238	436,465	23.1	22.6	23.7	1,893	157,771
55-59	0.0180	0.0861	0.0819	0.0898	84,674	84,043	85,315	7,293	405,139	19.4	18.8	20.0	2,103	116,817
60-64	0.0280	0.1308	0.1229	0.1377	77,382	76,751	78,024	10,125	361,596	16.0	15.4	16.6	2,675	95,528
65-69	0.0400	0.1818	0.1695	0.1929	67,257	66,340	68,185	12,229	305,713	13.0	12.5	13.5	3,421	85,518
70-74	0.0600	0.2609	0.2432	0.2771	55,028	53,647	56,626	14,355	239,254	10.3	9.9	10.8	4,088	68,137
75-79	0.0900	0.3673	0.3449	0.3882	40,673	38,864	42,856	14,941	166,013	8.1	7.7	8.5	4,387	48,741
80-84	0.1280	0.4848	0.4609	0.5072	25,732	23,788	28,076	12,476	97,470	6.3	6.1	6.6	2,239	17,489
85-89	0.1810	0.6231	0.6010	0.6438	13,256	11,738	15,136	8,259	45,631	4.9	4.8	5.1	2,123	11,727
90-94	0.2280	0.6770	0.6581	0.6950	4,997	4,190	6,039	3,383	14,836	4.0	3.8	4.1	790	3,467
95-99	0.3090	0.8018	0.7866	0.8162	1,614	1,278	2,065	1,294	4,188	3.0	3.0	3.1	197	637
100+	0.4450	1	1	1	320	235	441	320	719	2.2	2.2	2.3	38	86
All ages													*28,204*	*2,625,258*

$\ln(_nq_x)$ with uncertainty interval

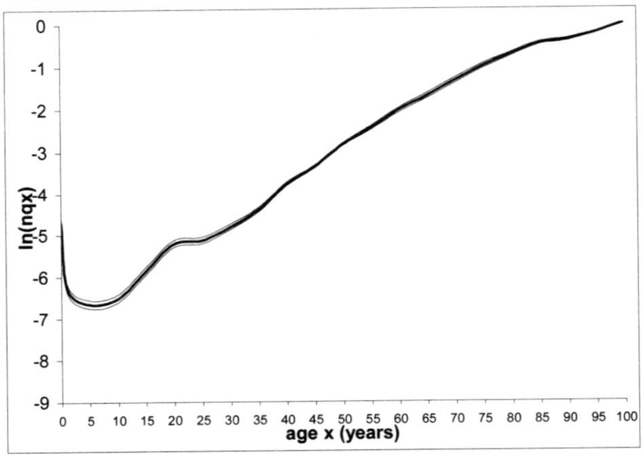

l_x with uncertainty interval

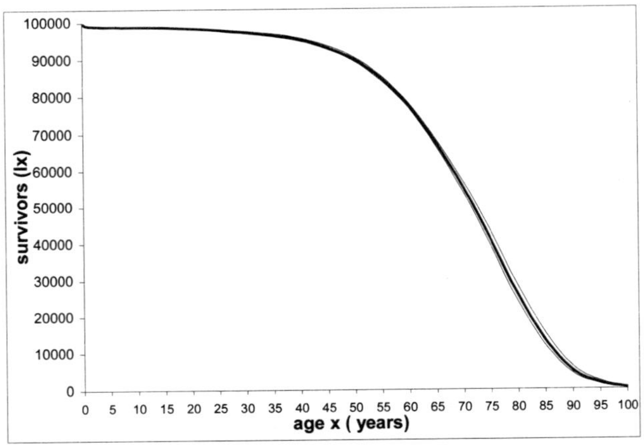

e_x with uncertainty interval

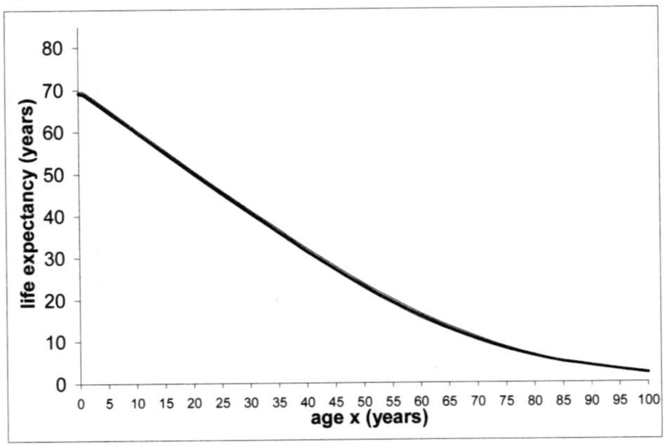

Distribution of deaths by age with uncertainty interval

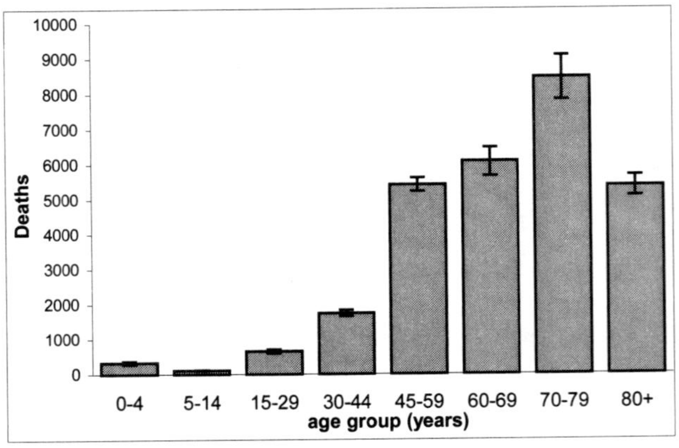

Slovakia **Slovaquie** **Eslovaquia**

Females, 2000 *Femmes, 2000* *Mujeres, 2000*

Age	$_nM_x$	$_nq_x$ mean	$_nq_x$ percentiles 2½	$_nq_x$ percentiles 97½	l_x mean	l_x percentiles 2½	l_x percentiles 97½	$_nd_x$	$_nL_x$	e_x mean	e_x percentiles 2½	e_x percentiles 97½	Actual deaths	Actual population
<1	0.0070	0.0070	0.0061	0.0078	100,000	100,000	100,000	696	99,374	77.5	77.2	78.0	198	28,335
1-4	0.0003	0.0013	0.0012	0.0014	99,304	99,220	99,388	127	396,913	77.1	76.7	77.6	36	112,809
5-9	0.0002	0.0011	0.0010	0.0012	99,177	99,081	99,273	109	495,614	73.2	72.8	73.7	39	175,007
10-14	0.0002	0.0009	0.0009	0.0010	99,068	98,963	99,173	94	495,107	68.3	67.9	68.7	38	199,075
15-19	0.0002	0.0011	0.0010	0.0012	98,974	98,861	99,086	109	494,599	63.3	63.0	63.8	48	218,204
20-24	0.0003	0.0015	0.0014	0.0016	98,865	98,744	98,986	148	493,957	58.4	58.0	58.9	70	232,994
25-29	0.0004	0.0018	0.0017	0.0020	98,717	98,585	98,849	182	493,130	53.5	53.1	54.0	78	210,655
30-34	0.0005	0.0025	0.0023	0.0027	98,535	98,389	98,680	246	492,059	48.6	48.2	49.1	91	181,613
35-39	0.0009	0.0047	0.0044	0.0050	98,289	98,126	98,450	461	490,292	43.7	43.3	44.2	182	193,415
40-44	0.0014	0.0071	0.0067	0.0075	97,828	97,637	98,017	697	487,397	38.9	38.5	39.4	294	205,313
45-49	0.0027	0.0133	0.0126	0.0139	97,131	96,903	97,356	1,288	482,434	34.1	33.8	34.6	555	207,998
50-54	0.0040	0.0198	0.0189	0.0206	95,843	95,559	96,125	1,898	474,469	29.6	29.2	30.0	677	169,237
55-59	0.0070	0.0344	0.0327	0.0358	93,945	93,601	94,294	3,232	461,646	25.1	24.8	25.6	962	137,494
60-64	0.0100	0.0488	0.0462	0.0509	90,713	90,287	91,145	4,425	442,505	20.9	20.6	21.3	1,221	122,081
65-69	0.0180	0.0861	0.0813	0.0899	86,288	85,764	86,855	7,432	412,863	16.9	16.6	17.2	2,132	118,434
70-74	0.0300	0.1395	0.1317	0.1454	78,857	78,100	79,796	11,003	366,776	13.2	13.0	13.5	3,224	107,469
75-79	0.0570	0.2495	0.2368	0.2585	67,854	66,746	69,289	16,926	296,952	9.9	9.7	10.2	5,011	87,921
80-84	0.0950	0.3838	0.3687	0.3979	50,927	49,497	52,880	19,548	205,767	7.4	7.2	7.6	3,188	33,555
85-89	0.1520	0.5507	0.5355	0.5670	31,379	29,994	33,385	17,281	113,694	5.5	5.3	5.6	3,604	23,713
90-94	0.2120	0.6479	0.6346	0.6624	14,098	13,107	15,462	9,134	43,087	4.1	4.0	4.2	1,517	7,157
95-99	0.3040	0.7950	0.7847	0.8060	4,964	4,425	5,599	3,946	12,980	3.1	3.0	3.2	281	923
100+	0.4490	1	1	1	1,018	858	1,192	1,018	2,266	2.2	2.2	2.4	15	33
All ages													*23,461*	*2,773,435*

ln($_nq_x$) with uncertainty interval

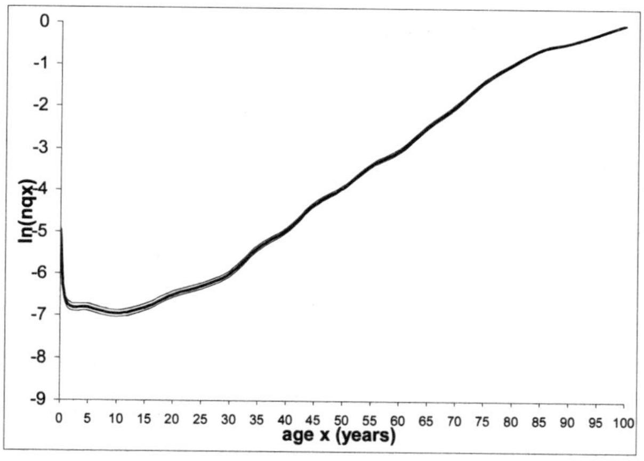

l_x with uncertainty interval

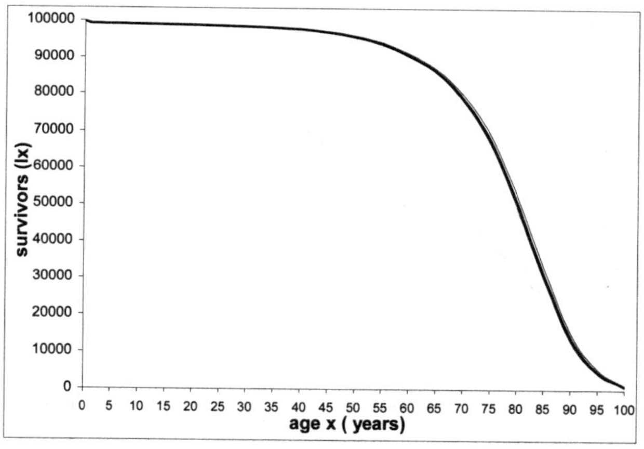

e_x with uncertainty interval

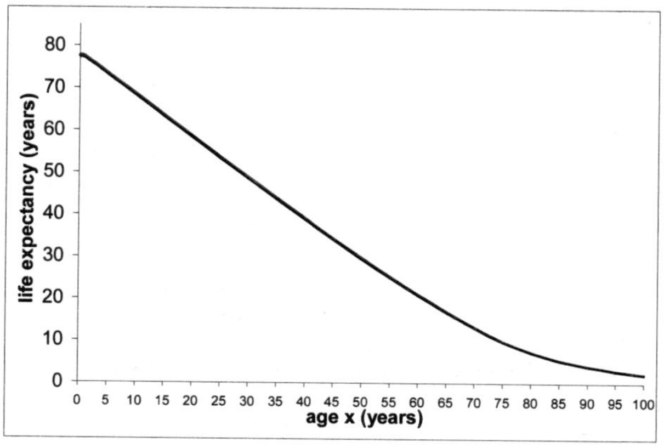

Distribution of deaths by age with uncertainty interval

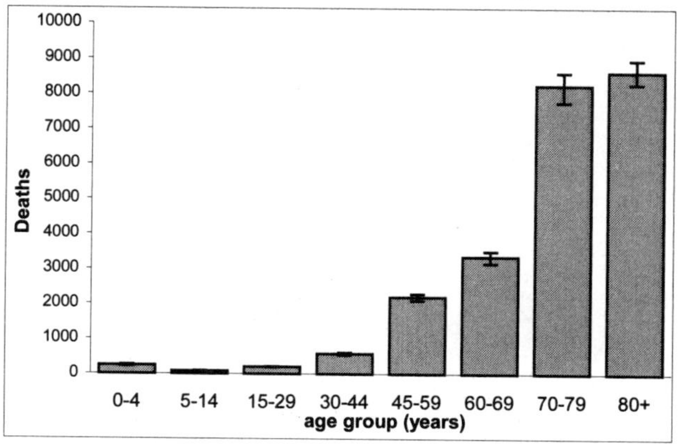

Slovenia   Slovénie   Eslovenia

Males, 2000   *Hommes, 2000*   *Hombres, 2000*

Age	$_nM_x$	$_nq_x$ mean	$_nq_x$ percentiles 2½	$_nq_x$ percentiles 97½	l_x mean	l_x percentiles 2½	l_x percentiles 97½	$_nd_x$	$_nL_x$	e_x mean	e_x percentiles 2½	e_x percentiles 97½	Actual deaths	Actual population
<1	0.0047	0.0047	0.0039	0.0055	100,000	100,000	100,000	471	99,576	71.9	71.5	72.3	43	9,150
1-4	0.0004	0.0015	0.0013	0.0017	99,529	99,449	99,609	149	397,760	71.2	70.8	71.6	14	37,127
5-9	0.0002	0.0010	0.0009	0.0012	99,380	99,281	99,480	103	496,644	67.3	66.9	67.7	11	52,067
10-14	0.0002	0.0012	0.0011	0.0014	99,277	99,166	99,390	124	496,077	62.4	61.9	62.8	16	63,445
15-19	0.0008	0.0042	0.0037	0.0046	99,153	99,028	99,281	414	494,732	57.4	57.0	57.9	60	71,409
20-24	0.0012	0.0060	0.0054	0.0065	98,739	98,572	98,912	593	492,215	52.7	52.2	53.1	93	77,637
25-29	0.0012	0.0062	0.0056	0.0066	98,147	97,928	98,375	604	489,223	48.0	47.5	48.5	90	73,079
30-34	0.0015	0.0074	0.0069	0.0079	97,543	97,280	97,820	722	485,907	43.3	42.8	43.7	110	73,749
35-39	0.0022	0.0107	0.0100	0.0113	96,820	96,513	97,150	1,038	481,505	38.6	38.1	39.1	166	77,023
40-44	0.0033	0.0166	0.0157	0.0173	95,782	95,421	96,173	1,587	474,942	34.0	33.5	34.5	264	79,134
45-49	0.0060	0.0294	0.0283	0.0304	94,195	93,771	94,661	2,770	464,051	29.5	29.0	30.0	487	81,609
50-54	0.0081	0.0397	0.0382	0.0410	91,425	90,925	91,978	3,626	448,062	25.3	24.8	25.8	529	65,369
55-59	0.0130	0.0631	0.0604	0.0657	87,799	87,251	88,404	5,537	425,154	21.2	20.8	21.8	702	53,927
60-64	0.0209	0.0993	0.0945	0.1044	82,262	81,705	82,852	8,172	390,882	17.5	17.0	18.0	1,033	49,393
65-69	0.0335	0.1547	0.1458	0.1636	74,090	73,359	74,880	11,463	341,793	14.2	13.7	14.6	1,423	42,430
70-74	0.0460	0.2061	0.1933	0.2182	62,627	61,417	63,847	12,909	280,862	11.3	10.9	11.7	1,402	30,513
75-79	0.0757	0.3183	0.2995	0.3355	49,718	48,013	51,405	15,827	209,022	8.6	8.3	8.9	1,292	17,060
80-84	0.1194	0.4598	0.4372	0.4805	33,891	31,921	35,984	15,582	130,499	6.4	6.2	6.7	757	6,339
85-89	0.1811	0.6233	0.6018	0.6431	18,309	16,641	20,253	11,413	63,012	4.8	4.6	4.9	641	3,538
90-94	0.2643	0.7370	0.7198	0.7531	6,896	5,974	8,064	5,082	19,233	3.5	3.4	3.6	423	1,600
95-99	0.3708	0.8777	0.8665	0.8883	1,814	1,477	2,260	1,592	4,293	2.6	2.6	2.7	65	175
100+	0.5006	1	1	1	222	166	302	222	443	2.0	2.0	2.0	4	7
All ages													*9,625*	*965,780*

ln($_nq_x$) with uncertainty interval

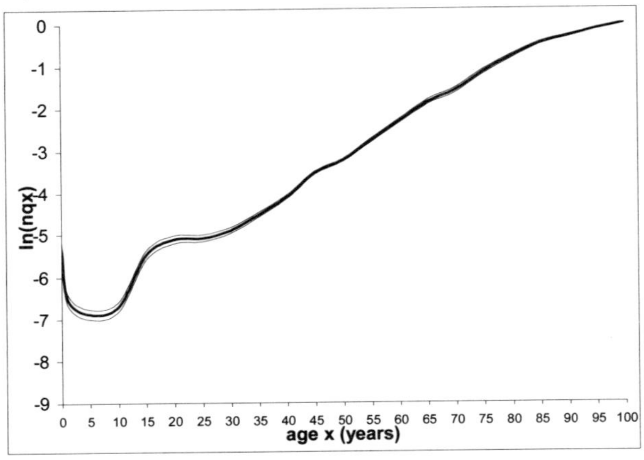

l_x with uncertainty interval

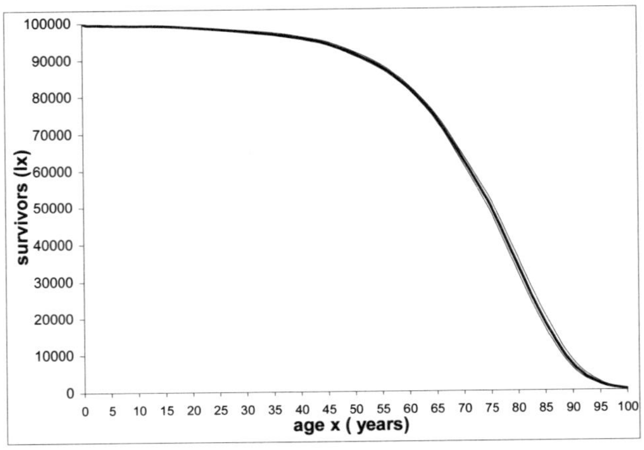

e_x with uncertainty interval

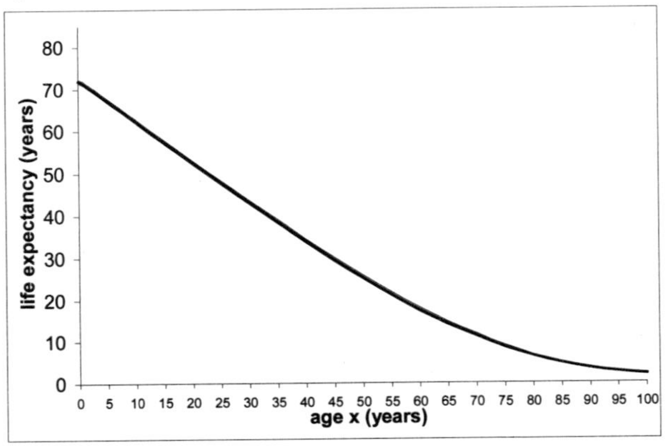

Distribution of deaths by age with uncertainty interval

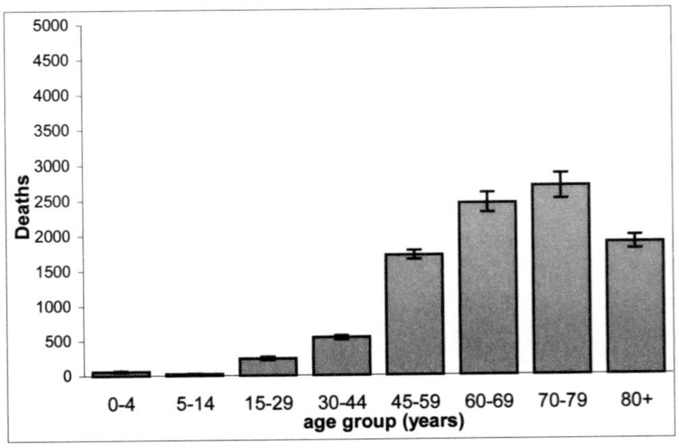

Slovenia　　　　Slovénie　　　　Eslovenia

Females, 2000　　　　*Femmes, 2000*　　　　*Mujeres, 2000*

Age	$_nM_x$	$_nq_x$ mean	$_nq_x$ percentiles 2½	$_nq_x$ percentiles 97½	l_x mean	l_x percentiles 2½	l_x percentiles 97½	$_nd_x$	$_nL_x$	e_x mean	e_x percentiles 2½	e_x percentiles 97½	Actual deaths	Actual population
<1	0.0040	0.0040	0.0033	0.0046	100,000	100,000	100,000	396	99,644	79.4	78.8	80.1	35	8,697
1-4	0.0002	0.0006	0.0006	0.0007	99,604	99,538	99,669	64	398,263	78.7	78.1	79.4	6	35,252
5-9	0.0000	0.0002	0.0002	0.0002	99,540	99,466	99,613	19	497,653	74.8	74.1	75.5	2	49,474
10-14	0.0001	0.0006	0.0005	0.0007	99,521	99,444	99,596	62	497,449	69.8	69.1	70.5	8	60,309
15-19	0.0003	0.0013	0.0011	0.0014	99,459	99,374	99,542	125	496,979	64.8	64.2	65.5	17	68,261
20-24	0.0004	0.0018	0.0016	0.0020	99,333	99,234	99,431	178	496,221	59.9	59.2	60.6	26	73,494
25-29	0.0004	0.0022	0.0020	0.0024	99,155	99,038	99,273	218	495,232	55.0	54.3	55.7	31	70,622
30-34	0.0006	0.0030	0.0027	0.0033	98,937	98,799	99,078	296	493,946	50.1	49.5	50.9	44	73,941
35-39	0.0009	0.0045	0.0041	0.0049	98,641	98,476	98,811	443	492,098	45.3	44.6	46.0	69	76,995
40-44	0.0016	0.0080	0.0073	0.0086	98,198	97,995	98,407	782	489,035	40.5	39.8	41.2	124	77,226
45-49	0.0024	0.0117	0.0109	0.0125	97,416	97,154	97,684	1,141	484,226	35.8	35.1	36.5	183	77,735
50-54	0.0038	0.0189	0.0177	0.0200	96,275	95,941	96,615	1,816	476,834	31.2	30.5	31.9	243	63,809
55-59	0.0055	0.0272	0.0255	0.0292	94,459	94,033	94,888	2,573	465,861	26.7	26.1	27.4	310	56,135
60-64	0.0076	0.0374	0.0349	0.0405	91,886	91,364	92,399	3,440	450,828	22.4	21.8	23.1	418	54,830
65-69	0.0138	0.0665	0.0612	0.0724	88,446	87,791	89,093	5,885	427,516	18.2	17.6	18.8	759	55,164
70-74	0.0242	0.1139	0.1037	0.1244	82,561	81,453	83,551	9,404	389,294	14.3	13.8	14.9	1,199	49,619
75-79	0.0405	0.1838	0.1667	0.2001	73,157	71,323	74,762	13,443	332,178	10.8	10.4	11.3	1,542	38,092
80-84	0.0823	0.3412	0.3131	0.3677	59,714	57,053	62,296	20,373	247,638	7.7	7.3	8.1	1,273	15,472
85-89	0.1515	0.5495	0.5177	0.5795	39,341	36,190	42,794	21,618	142,660	5.4	5.1	5.7	1,498	9,888
90-94	0.2300	0.6805	0.6561	0.7041	17,723	15,353	20,638	12,060	52,435	3.9	3.7	4.1	1,346	5,853
95-99	0.3300	0.8291	0.8137	0.8446	5,663	4,544	7,098	4,695	14,229	2.9	2.8	3.0	319	968
100+	0.4700	1	1	1	968	706	1,323	968	2,059	2.1	2.1	2.4	31	66
All ages													*9,484*	*1,021,902*

$\ln(_nq_x)$ with uncertainty interval

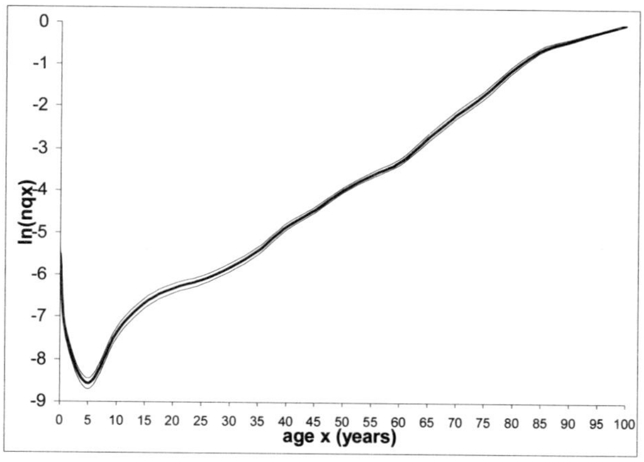

l_x with uncertainty interval

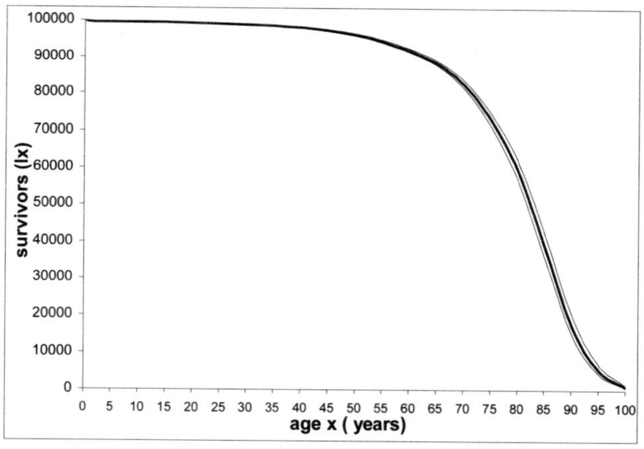

e_x with uncertainty interval

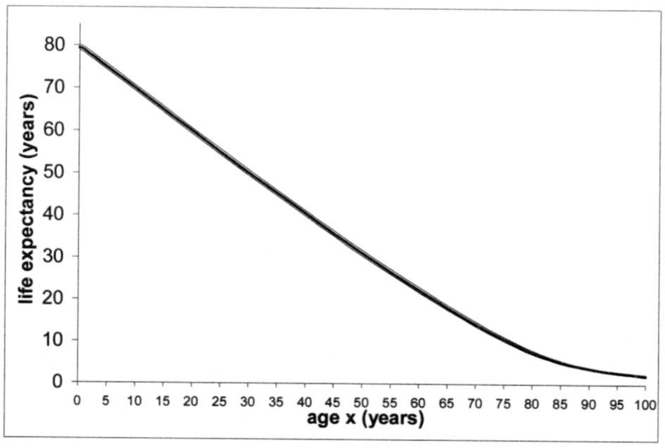

Distribution of deaths by age with uncertainty interval

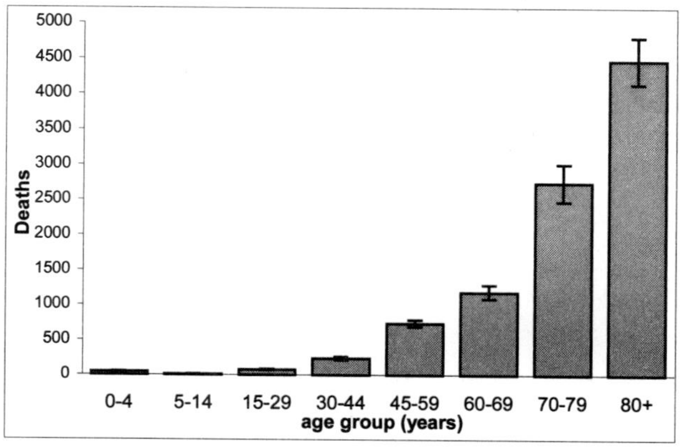

Solomon Islands **Iles Salomon** **Islas Salomón**

Males, 2000 *Hommes, 2000* *Hombres, 2000*

Age	$_nM_x$	$_nq_x$			l_x			$_nd_x$	$_nL_x$	e_x			Actual deaths	Actual population
			percentiles			percentiles					percentiles			
		mean	2½	97½	mean	2½	97½			mean	2½	97½		
<1	0.0306	0.0298	0.0246	0.0353	100,000	100,000	100,000	2,979	97,319	66.6	63.5	70.8	260	8,483
1-4	0.0019	0.0074	0.0058	0.0091	97,021	96,468	97,544	720	386,355	67.6	64.7	71.7	60	32,415
5-9	0.0007	0.0036	0.0028	0.0045	96,301	95,701	96,900	351	480,625	64.1	61.3	68.1	25	34,178
10-14	0.0006	0.0032	0.0024	0.0039	95,949	95,272	96,613	305	478,983	59.3	56.5	63.3	18	28,353
15-19	0.0015	0.0073	0.0056	0.0091	95,644	94,897	96,363	700	476,470	54.5	51.8	58.4	36	24,805
20-24	0.0020	0.0102	0.0076	0.0127	94,944	94,034	95,791	964	472,310	49.9	47.2	53.7	44	21,670
25-29	0.0021	0.0104	0.0077	0.0131	93,980	92,839	95,032	979	467,454	45.4	42.8	49.1	40	19,148
30-34	0.0024	0.0121	0.0089	0.0154	93,001	91,619	94,299	1,128	462,188	40.8	38.3	44.5	34	13,817
35-39	0.0032	0.0159	0.0115	0.0203	91,874	90,209	93,464	1,460	455,720	36.3	33.9	39.9	34	10,729
40-44	0.0046	0.0228	0.0163	0.0292	90,414	88,381	92,391	2,062	446,915	31.8	29.5	35.3	40	8,751
45-49	0.0070	0.0342	0.0240	0.0440	88,352	85,796	90,890	3,019	434,214	27.5	25.3	30.8	50	7,164
50-54	0.0107	0.0520	0.0362	0.0671	85,334	82,022	88,706	4,442	415,564	23.4	21.4	26.5	61	5,733
55-59	0.0164	0.0786	0.0542	0.1008	80,892	76,519	85,499	6,356	388,570	19.6	17.8	22.4	82	5,037
60-64	0.0251	0.1182	0.0816	0.1497	74,536	68,804	80,869	8,812	350,650	16.0	14.4	18.6	92	3,641
65-69	0.0382	0.1745	0.1224	0.2161	65,724	58,502	74,270	11,469	299,947	12.8	11.4	15.0	103	2,685
70-74	0.0593	0.2584	0.1884	0.3155	54,255	45,887	65,176	14,019	236,227	10.0	8.8	11.7	114	1,915
75-79	0.0918	0.3734	0.2897	0.4408	40,236	31,778	52,900	15,025	163,617	7.6	6.7	8.9	90	976
80-84	0.1425	0.5253	0.4418	0.5914	25,211	18,304	37,573	13,242	92,948	5.7	5.1	6.5	63	445
85-89	0.2133	0.6956	0.6318	0.7483	11,968	7,540	20,972	8,326	39,028	4.2	3.8	4.6	28	131
90-94	0.3083	0.8008	0.7555	0.8398	3,643	1,898	7,721	2,917	9,462	3.1	2.8	3.3	8	25
95-99	0.4300	0.9388	0.9162	0.9570	726	304	1,866	681	1,584	2.3	2.2	2.4	1	3
100+	0.5787	1	1	1	44	13	152	44	77	1.7	1.7	1.9	0	0
All ages													*1,284*	*230,104*

ln($_nq_x$) with uncertainty interval

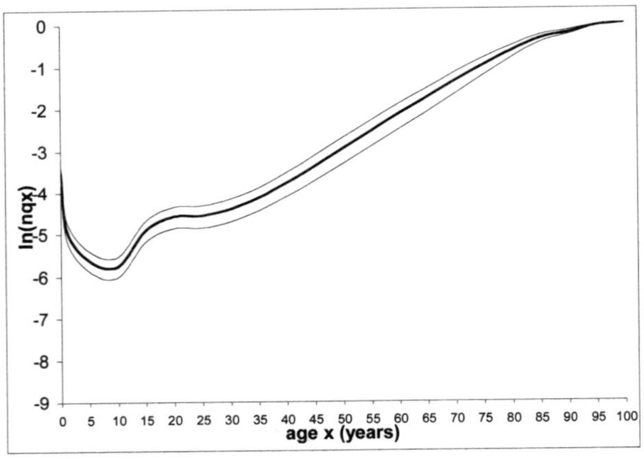

l_x with uncertainty interval

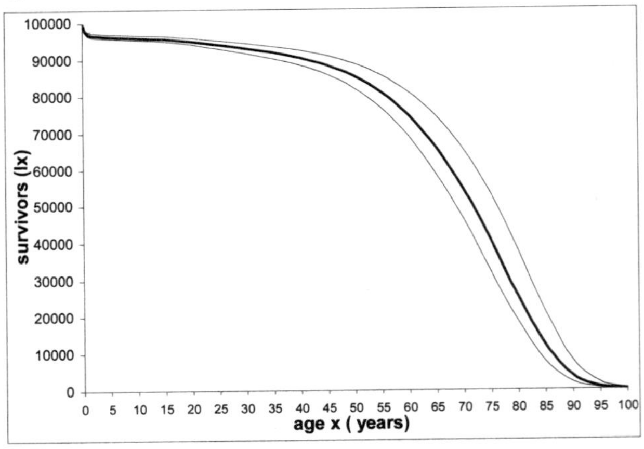

e_x with uncertainty interval

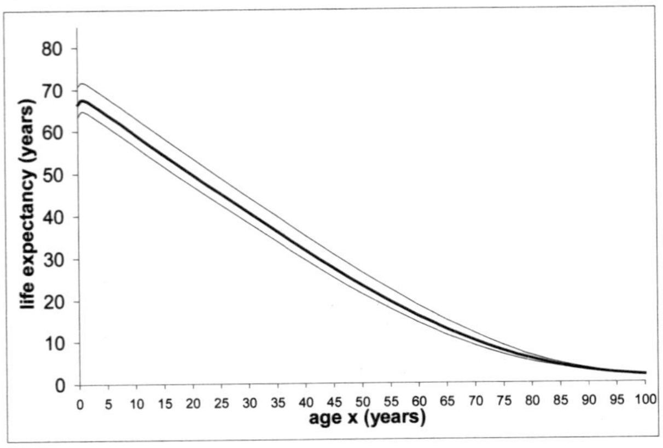

Distribution of deaths by age with uncertainty interval

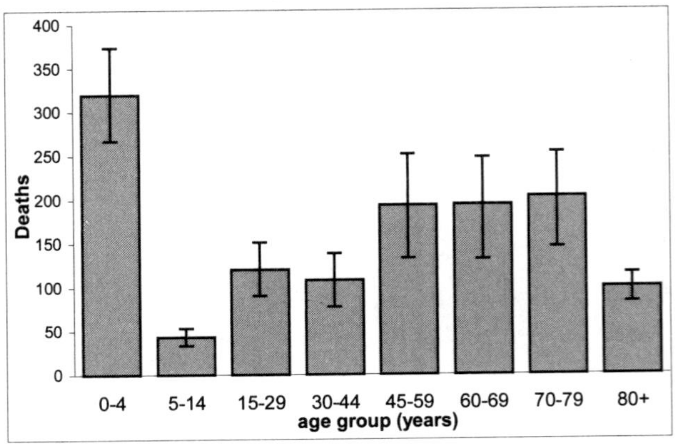

Solomon Islands — Iles Salomon — Islas Salomón

Females, 2000 — *Femmes, 2000* — *Mujeres, 2000*

Age	$_nM_x$	$_nq_x$ mean	$_nq_x$ percentiles 2 1/2	$_nq_x$ percentiles 97 1/2	l_x mean	l_x percentiles 2 1/2	l_x percentiles 97 1/2	$_nd_x$	$_nL_x$	e_x mean	e_x percentiles 2 1/2	e_x percentiles 97 1/2	Actual deaths	Actual population
<1	0.0211	0.0207	0.0158	0.0258	100,000	100,000	100,000	2,074	98,134	71.4	67.8	76.3	169	8,003
1-4	0.0016	0.0064	0.0049	0.0081	97,926	97,417	98,425	626	390,203	71.9	68.5	76.7	48	30,230
5-9	0.0006	0.0029	0.0022	0.0037	97,300	96,701	97,896	281	485,800	68.4	65.0	73.1	19	32,025
10-14	0.0005	0.0025	0.0018	0.0032	97,019	96,366	97,663	239	484,499	63.6	60.3	68.2	13	26,624
15-19	0.0009	0.0044	0.0032	0.0057	96,780	96,071	97,467	427	482,833	58.7	55.5	63.4	20	23,105
20-24	0.0011	0.0055	0.0039	0.0071	96,353	95,544	97,119	526	480,451	54.0	50.8	58.5	22	19,928
25-29	0.0013	0.0063	0.0044	0.0083	95,827	94,893	96,688	602	477,630	49.3	46.1	53.8	22	17,641
30-34	0.0016	0.0081	0.0056	0.0108	95,225	94,114	96,224	770	474,199	44.5	41.5	49.0	22	13,607
35-39	0.0023	0.0115	0.0078	0.0154	94,455	93,103	95,632	1,083	469,566	39.9	36.9	44.3	26	11,184
40-44	0.0033	0.0166	0.0111	0.0225	93,372	91,671	94,881	1,551	462,983	35.3	32.4	39.6	30	8,954
45-49	0.0049	0.0244	0.0160	0.0334	91,821	89,609	93,827	2,242	453,501	30.9	28.1	35.0	34	6,939
50-54	0.0073	0.0358	0.0229	0.0493	89,579	86,618	92,330	3,205	439,883	26.6	24.0	30.5	41	5,586
55-59	0.0106	0.0518	0.0324	0.0715	86,374	82,351	90,220	4,471	420,692	22.5	20.1	26.2	48	4,557
60-64	0.0163	0.0782	0.0482	0.1074	81,903	76,462	87,297	6,403	393,506	18.6	16.5	22.0	55	3,397
65-69	0.0255	0.1199	0.0738	0.1618	75,500	68,247	83,085	9,050	354,873	14.9	13.1	18.0	62	2,426
70-74	0.0422	0.1908	0.1201	0.2490	66,450	57,202	76,950	12,676	300,558	11.6	10.0	14.2	77	1,818
75-79	0.0697	0.2967	0.1975	0.3774	53,773	43,002	67,711	15,952	228,987	8.8	7.5	10.8	57	812
80-84	0.1159	0.4494	0.3304	0.5400	37,821	27,318	54,339	16,998	146,611	6.4	5.5	7.8	41	356
85-89	0.1841	0.6303	0.5236	0.7077	20,823	13,149	36,387	13,125	71,304	4.6	4.1	5.5	20	107
90-94	0.2788	0.7590	0.6863	0.8154	7,698	3,843	17,336	5,843	20,962	3.3	3.0	3.8	6	22
95-99	0.4027	0.9119	0.8725	0.9410	1,855	710	5,422	1,692	4,200	2.4	2.3	2.6	1	3
100+	0.5550	1	1	1	163	42	672	163	294	1.8	1.7	1.9	0	0
All ages													*834*	*217,324*

ln($_nq_x$) with uncertainty interval

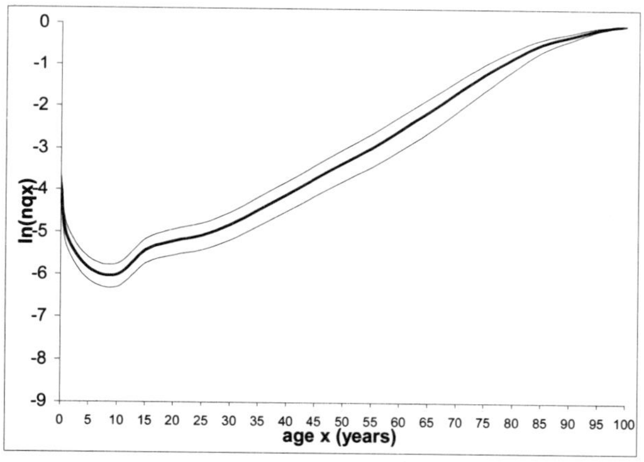

l_x with uncertainty interval

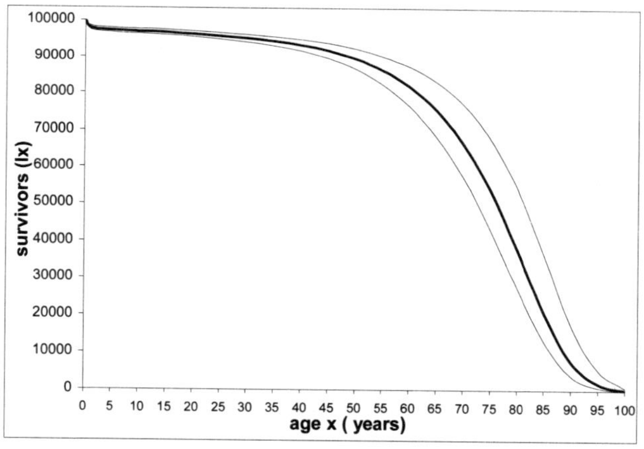

e_x with uncertainty interval

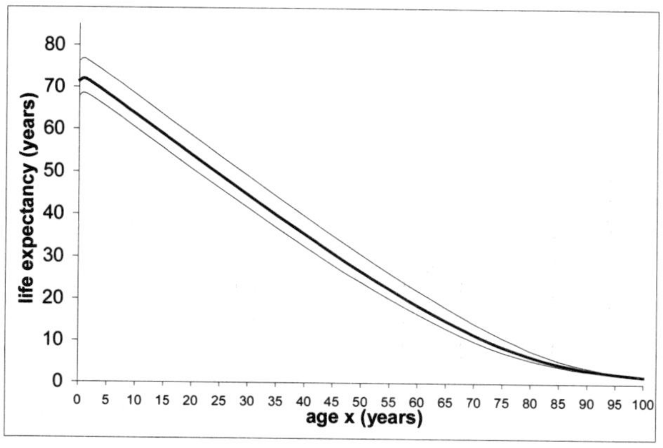

Distribution of deaths by age with uncertainty interval

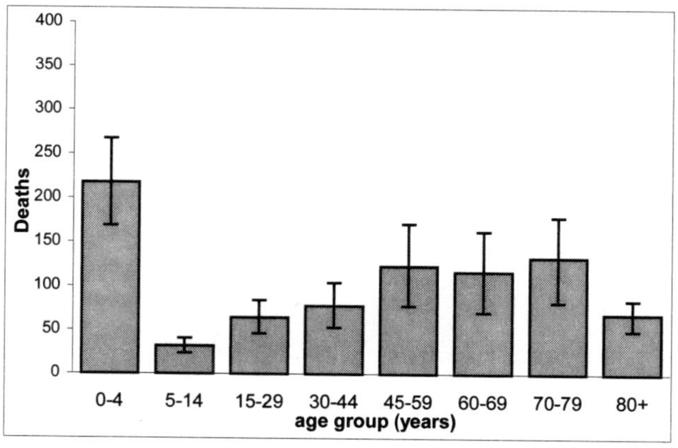

Somalia

Somalie

Somalia

Males, 2000 *Hommes, 2000* *Hombres, 2000*

Age	$_nM_x$	$_nq_x$			l_x			$_nd_x$	$_nL_x$	e_x			Actual deaths	Actual population
			percentiles			percentiles					percentiles			
		mean	2½	97½	mean	2½	97½			mean	2½	97½		
<1	0.1959	0.1665	0.1568	0.1752	100,000	100,000	100,000	16,655	85,011	43.8	42.4	45.8	39,772	203,006
1-4	0.0169	0.0649	0.0592	0.0691	83,345	82,480	84,318	5,407	320,404	51.5	49.6	53.9	11,710	693,888
5-9	0.0036	0.0178	0.0160	0.0192	77,938	77,239	78,638	1,389	386,217	51.0	49.1	53.2	2,370	658,833
10-14	0.0021	0.0104	0.0093	0.0113	76,549	75,799	77,321	797	380,752	46.9	45.0	49.1	1,165	556,797
15-19	0.0016	0.0081	0.0072	0.0088	75,752	74,971	76,586	613	377,229	42.3	40.5	44.5	755	464,877
20-24	0.0051	0.0253	0.0226	0.0276	75,140	74,335	76,020	1,903	370,939	37.7	35.8	39.8	1,953	380,546
25-29	0.0101	0.0494	0.0438	0.0542	73,236	72,359	74,265	3,620	357,131	33.6	31.8	35.7	3,131	308,940
30-34	0.0154	0.0740	0.0652	0.0818	69,616	68,555	70,953	5,155	335,195	30.2	28.4	32.2	3,825	248,686
35-39	0.0180	0.0861	0.0753	0.0960	64,461	63,067	66,325	5,548	308,438	27.4	25.7	29.3	3,649	202,845
40-44	0.0199	0.0947	0.0825	0.1066	58,914	57,149	61,329	5,579	280,620	24.8	23.2	26.4	3,322	167,076
45-49	0.0220	0.1041	0.0905	0.1179	53,335	51,136	56,271	5,553	252,790	22.1	20.7	23.6	2,996	136,404
50-54	0.0246	0.1158	0.1006	0.1316	47,781	45,174	51,180	5,533	225,075	19.4	18.1	20.7	2,486	101,121
55-59	0.0283	0.1322	0.1150	0.1504	42,249	39,294	46,033	5,586	197,280	16.6	15.5	17.7	2,168	76,572
60-64	0.0377	0.1722	0.1506	0.1953	36,663	33,460	40,739	6,314	167,530	13.7	12.8	14.7	2,309	61,267
65-69	0.0525	0.2322	0.2050	0.2613	30,349	27,016	34,603	7,047	134,126	11.0	10.2	11.9	2,284	43,468
70-74	0.0786	0.3286	0.2945	0.3649	23,302	19,957	27,511	7,657	97,364	8.6	8.0	9.3	2,264	28,788
75-79	0.1171	0.4529	0.4141	0.4936	15,644	12,675	19,408	7,086	60,507	6.6	6.1	7.1	1,859	15,877
80-84	0.1716	0.6003	0.5609	0.6402	8,559	6,419	11,366	5,138	29,948	5.0	4.7	5.4	1,145	6,675
85-89	0.2449	0.7594	0.7256	0.7924	3,421	2,310	4,977	2,598	10,610	3.8	3.6	4.0	451	1,840
90-94	0.3402	0.8418	0.8145	0.8680	823	479	1,356	693	2,037	2.8	2.7	3.0	97	285
95-99	0.4602	0.9666	0.9555	0.9759	130	63	249	126	273	2.2	2.1	2.2	11	23
100+	0.6061	1	1	1	4	2	11	4	7	1.7	1.6	1.7	1	1
All ages													*89,721*	*4,357,815*

ln($_nq_x$) with uncertainty interval

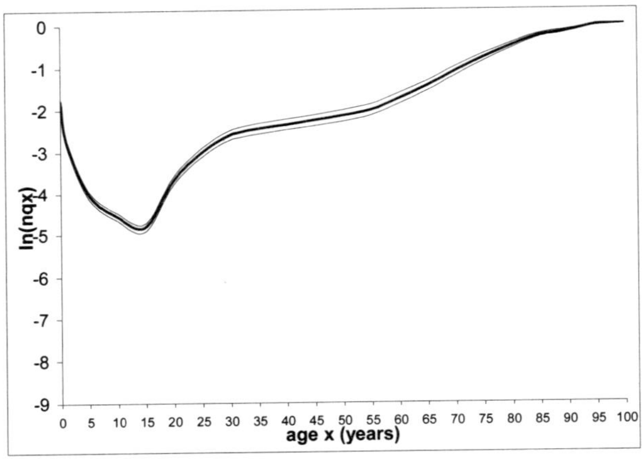

l_x with uncertainty interval

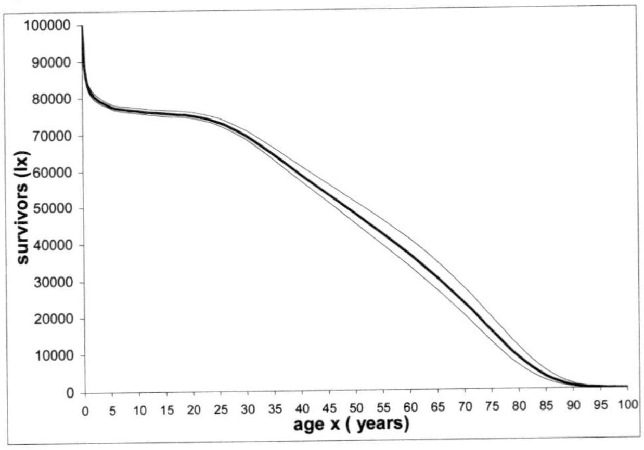

e_x with uncertainty interval

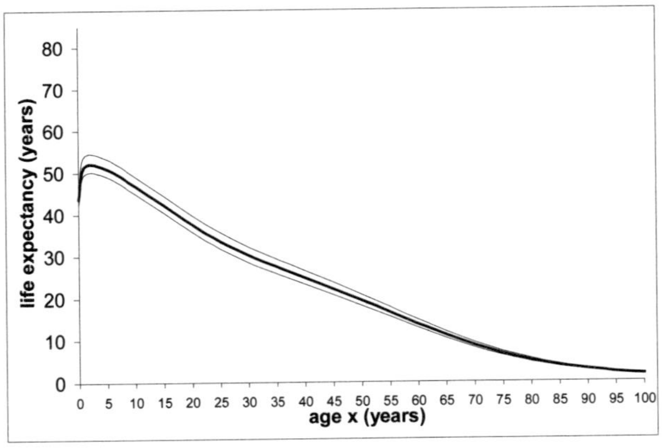

Distribution of deaths by age with uncertainty interval

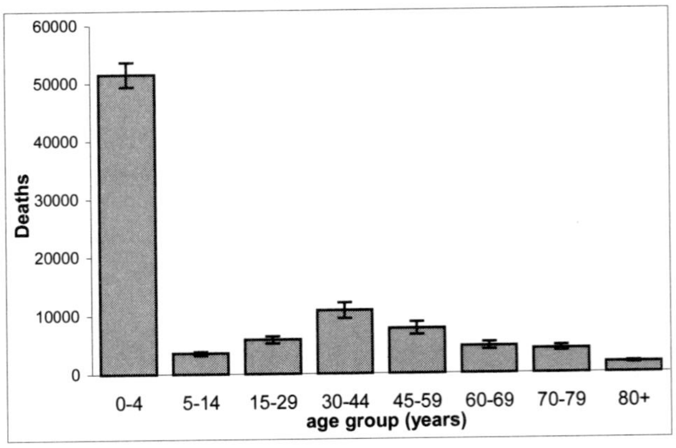

Somalia | Somalie | Somalia

Females, 2000 | *Femmes, 2000* | *Mujeres, 2000*

Age	$_nM_x$	$_nq_x$			l_x			$_nd_x$	$_nL_x$	e_x			Actual deaths	Actual population
			percentiles			percentiles					percentiles			
		mean	2½	97½	mean	2½	97½			mean	2½	97½		
<1	0.1610	0.1406	0.1319	0.1496	100,000	100,000	100,000	14,065	87,342	45.9	44.3	48.2	31,738	197,093
1-4	0.0178	0.0684	0.0622	0.0734	85,935	85,042	86,812	5,879	329,633	52.4	50.3	55.1	12,362	693,175
5-9	0.0037	0.0184	0.0165	0.0199	80,057	79,360	80,750	1,474	396,600	52.2	50.1	54.7	2,428	653,354
10-14	0.0024	0.0118	0.0106	0.0129	78,583	77,792	79,406	931	390,589	48.1	46.1	50.6	1,318	553,022
15-19	0.0044	0.0216	0.0192	0.0235	77,652	76,801	78,567	1,675	384,075	43.7	41.7	46.1	2,029	465,308
20-24	0.0090	0.0441	0.0390	0.0483	75,978	75,018	77,058	3,350	371,512	39.6	37.6	42.0	3,461	383,816
25-29	0.0136	0.0657	0.0577	0.0726	72,627	71,414	74,048	4,773	351,202	36.3	34.4	38.6	4,278	314,781
30-34	0.0158	0.0761	0.0662	0.0849	67,854	66,279	69,763	5,165	326,357	33.7	31.9	35.8	4,044	255,537
35-39	0.0152	0.0731	0.0631	0.0822	62,689	60,744	65,115	4,580	301,995	31.2	29.6	33.1	3,180	209,641
40-44	0.0136	0.0659	0.0566	0.0745	58,109	55,863	60,971	3,828	280,974	28.5	27.0	30.2	2,373	174,182
45-49	0.0134	0.0646	0.0554	0.0733	54,281	51,811	57,483	3,506	262,640	25.3	24.0	26.9	1,930	144,549
50-54	0.0157	0.0755	0.0647	0.0858	50,775	48,125	54,263	3,835	244,287	21.9	20.7	23.3	1,681	107,110
55-59	0.0198	0.0942	0.0808	0.1070	46,940	44,076	50,708	4,421	223,647	18.5	17.4	19.7	1,667	84,311
60-64	0.0285	0.1332	0.1146	0.1508	42,519	39,362	46,570	5,662	198,440	15.1	14.2	16.2	1,997	69,981
65-69	0.0430	0.1940	0.1682	0.2183	36,857	33,427	41,235	7,152	166,407	12.1	11.3	13.0	2,091	48,651
70-74	0.0675	0.2888	0.2540	0.3209	29,706	26,131	34,298	8,580	127,078	9.4	8.7	10.2	2,254	33,387
75-79	0.1028	0.4088	0.3672	0.4462	21,126	17,746	25,585	8,637	84,036	7.2	6.7	7.8	2,019	19,640
80-84	0.1538	0.5554	0.5113	0.5937	12,489	9,830	16,190	6,936	45,102	5.4	5.1	5.8	1,383	8,995
85-89	0.2239	0.7178	0.6782	0.7507	5,552	4,000	7,898	3,985	17,798	4.0	3.8	4.3	646	2,884
90-94	0.3171	0.8125	0.7803	0.8389	1,567	1,001	2,527	1,273	4,015	3.0	2.8	3.2	182	575
95-99	0.4368	0.9453	0.9288	0.9576	294	162	550	278	636	2.3	2.2	2.3	30	68
100+	0.5853	1	1	1	16	7	39	16	27	1.7	1.7	1.8	2	4
All ages													*83,092*	*4,420,064*

ln($_nq_x$) with uncertainty interval

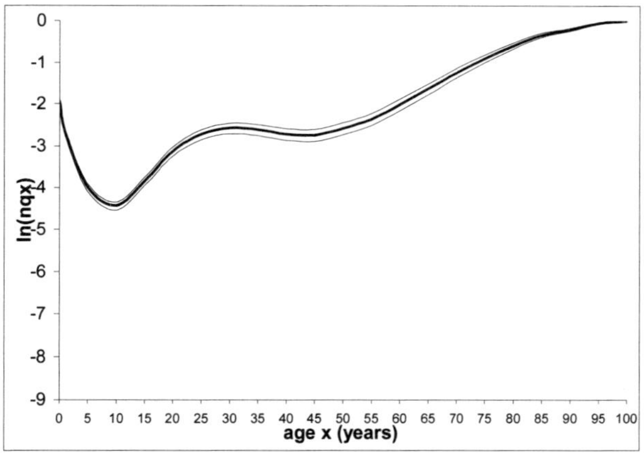

l_x with uncertainty interval

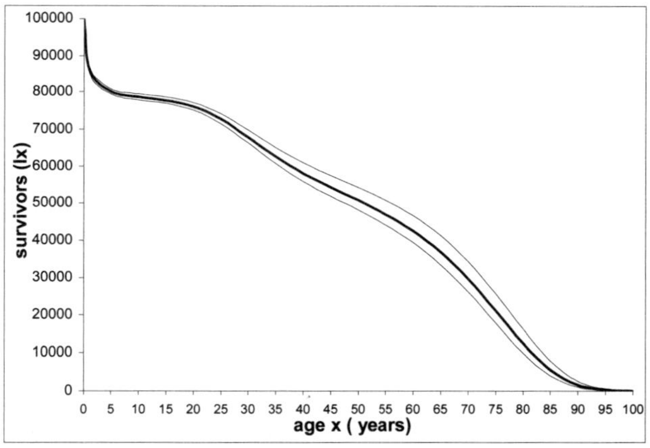

e_x with uncertainty interval

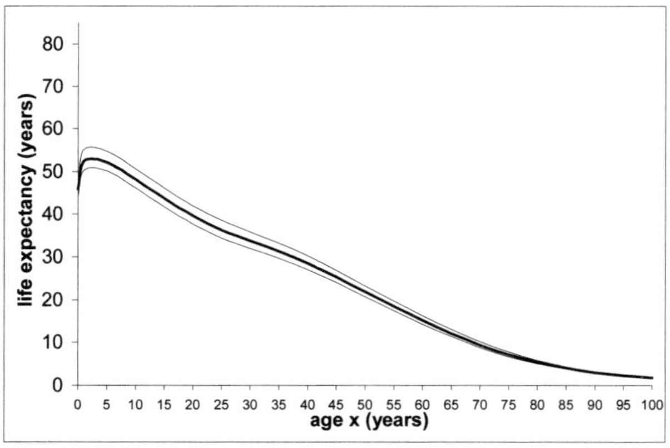

Distribution of deaths by age with uncertainty interval

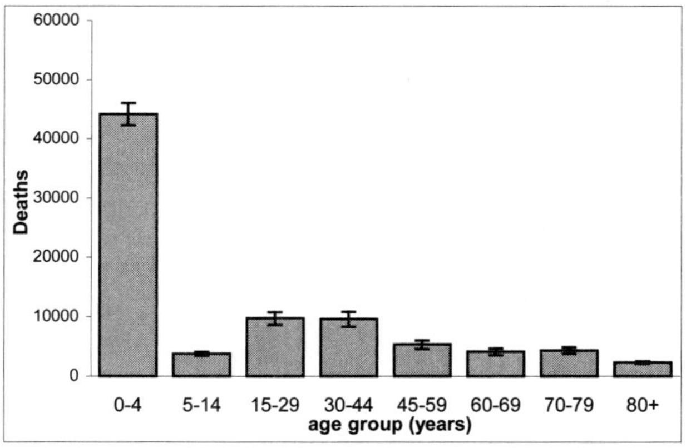

South Africa　　　Afrique du Sud　　　Sudáfrica

Males, 2000　　　*Hommes, 2000*　　　*Hombres, 2000*

Age	$_nM_x$	$_nq_x$ mean	$_nq_x$ percentiles 2½	$_nq_x$ percentiles 97½	l_x mean	l_x percentiles 2½	l_x percentiles 97½	$_nd_x$	$_nL_x$	e_x mean	e_x percentiles 2½	e_x percentiles 97½	Actual deaths	Actual population
<1	0.0668	0.0630	0.0604	0.0658	100,000	100,000	100,000	6,302	94,328	49.6	48.5	50.7	36,770	550,374
1-4	0.0073	0.0285	0.0274	0.0300	93,698	93,422	93,960	2,675	368,373	51.9	50.8	53.1	14,938	2,057,253
5-9	0.0016	0.0078	0.0074	0.0082	91,023	90,727	91,322	707	453,349	49.4	48.4	50.5	3,730	2,391,027
10-14	0.0012	0.0060	0.0057	0.0063	90,316	89,984	90,636	542	450,226	44.8	43.7	45.9	2,880	2,393,459
15-19	0.0028	0.0141	0.0134	0.0149	89,774	89,415	90,115	1,267	445,704	40.0	39.0	41.1	6,646	2,338,078
20-24	0.0063	0.0309	0.0293	0.0327	88,507	88,081	88,896	2,737	435,694	35.5	34.6	36.7	13,252	2,109,337
25-29	0.0133	0.0641	0.0604	0.0681	85,770	85,197	86,263	5,502	415,097	31.6	30.6	32.7	25,009	1,886,872
30-34	0.0214	0.1017	0.0953	0.1081	80,268	79,397	80,978	8,165	380,929	28.6	27.7	29.6	36,553	1,705,293
35-39	0.0237	0.1118	0.1045	0.1189	72,103	70,813	73,259	8,063	340,359	26.6	25.7	27.5	34,152	1,441,640
40-44	0.0242	0.1143	0.1067	0.1214	64,040	62,394	65,606	7,320	301,901	24.6	23.8	25.5	28,934	1,193,320
45-49	0.0249	0.1171	0.1093	0.1244	56,720	54,821	58,607	6,640	266,999	22.4	21.8	23.2	24,569	987,890
50-54	0.0250	0.1176	0.1097	0.1250	50,080	48,021	52,198	5,891	235,671	20.1	19.5	20.8	18,205	728,274
55-59	0.0257	0.1207	0.1125	0.1280	44,189	42,051	46,455	5,332	207,612	17.4	16.9	18.0	14,234	554,208
60-64	0.0342	0.1573	0.1471	0.1664	38,856	36,703	41,200	6,113	178,998	14.5	14.1	15.0	13,681	400,583
65-69	0.0475	0.2122	0.1992	0.2235	32,743	30,644	35,099	6,949	146,344	11.7	11.4	12.1	13,328	280,700
70-74	0.0703	0.2991	0.2827	0.3128	25,794	23,858	28,048	7,714	109,686	9.2	8.9	9.5	12,030	171,047
75-79	0.1052	0.4165	0.3977	0.4323	18,080	16,423	20,044	7,531	71,573	7.0	6.8	7.3	8,875	84,343
80-84	0.1596	0.5705	0.5512	0.5864	10,549	9,331	12,057	6,018	37,699	5.3	5.1	5.4	5,749	36,010
85-89	0.2343	0.7387	0.7223	0.7521	4,531	3,867	5,411	3,347	14,286	3.9	3.8	4.0	2,503	10,682
90-94	0.3319	0.8315	0.8186	0.8420	1,184	962	1,502	984	2,966	2.9	2.8	2.9	725	2,185
95-99	0.4541	0.9612	0.9558	0.9654	199	153	273	192	422	2.2	2.2	2.2	139	305
100+	0.5998	1	1	1	8	5	12	8	13	1.7	1.7	1.7	20	33
All ages													*316,920*	*21,322,913*

$\ln(_nq_x)$ with uncertainty interval

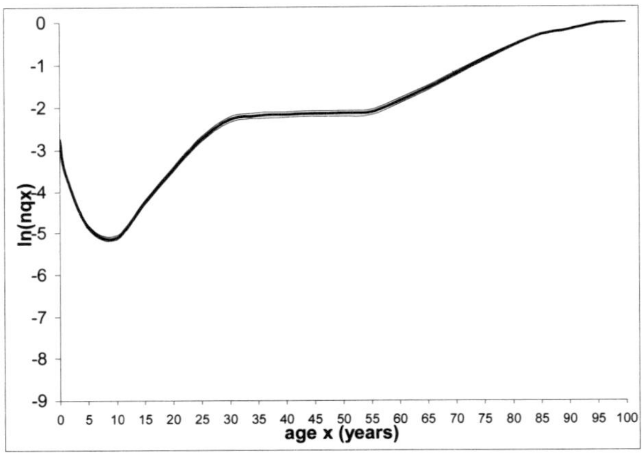

l_x with uncertainty interval

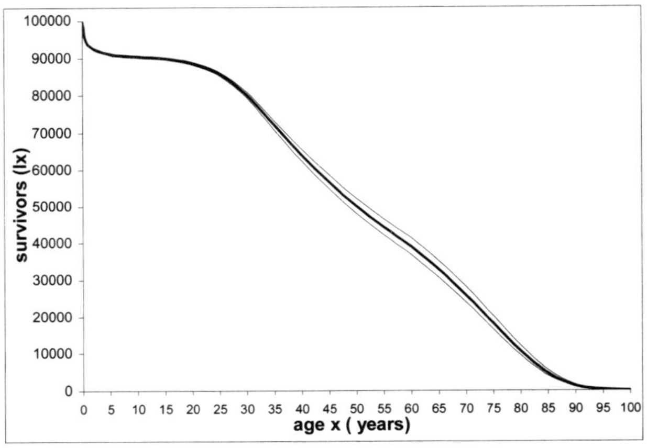

e_x with uncertainty interval

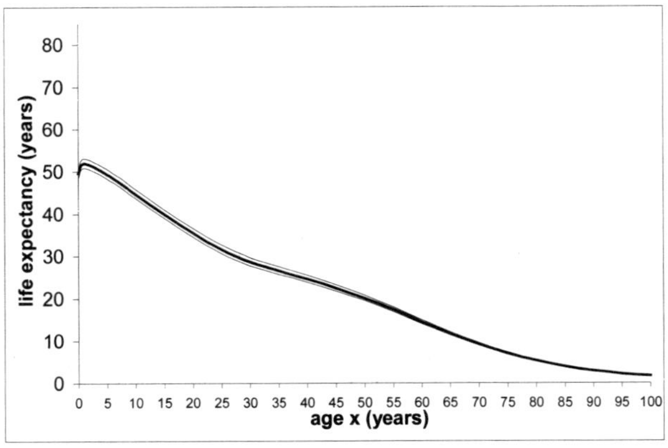

Distribution of deaths by age with uncertainty interval

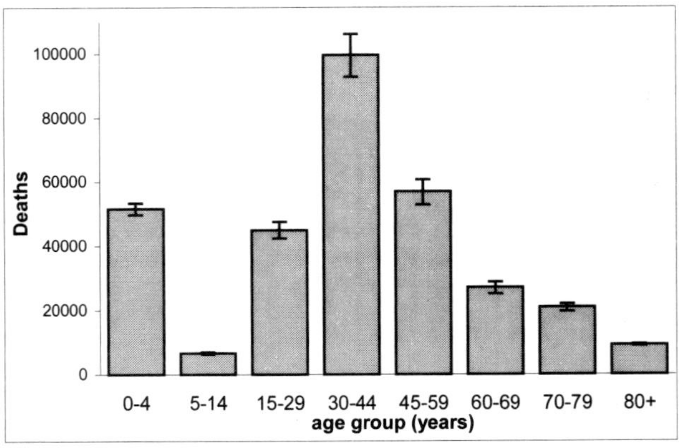

South Africa — Afrique du Sud — Sudáfrica

Females, 2000 — *Femmes, 2000* — *Mujeres, 2000*

Age	$_nM_x$	$_nq_x$ mean	$_nq_x$ percentiles 2½	$_nq_x$ percentiles 97½	l_x mean	l_x percentiles 2½	l_x percentiles 97½	$_nd_x$	$_nL_x$	e_x mean	e_x percentiles 2½	e_x percentiles 97½	Actual deaths	Actual population
<1	0.0545	0.0519	0.0487	0.0553	100,000	100,000	100,000	5,193	95,327	52.1	51.1	53.0	29,107	534,344
1-4	0.0070	0.0277	0.0263	0.0291	94,807	94,475	95,127	2,625	372,930	53.9	53.0	54.8	14,318	2,034,362
5-9	0.0014	0.0068	0.0065	0.0072	92,182	91,785	92,580	629	459,341	51.4	50.5	52.3	3,252	2,375,521
10-14	0.0010	0.0052	0.0049	0.0054	91,554	91,145	91,971	475	456,582	46.7	45.8	47.6	2,492	2,397,545
15-19	0.0039	0.0193	0.0183	0.0203	91,079	90,660	91,513	1,755	451,009	42.0	41.1	42.9	9,166	2,355,476
20-24	0.0107	0.0521	0.0495	0.0547	89,324	88,830	89,838	4,650	434,995	37.7	36.9	38.6	22,681	2,121,561
25-29	0.0189	0.0901	0.0858	0.0946	84,674	83,984	85,391	7,631	404,293	34.7	33.8	35.6	35,709	1,891,947
30-34	0.0231	0.1093	0.1043	0.1148	77,043	76,046	78,052	8,423	364,157	32.9	32.1	33.7	39,942	1,726,769
35-39	0.0200	0.0953	0.0906	0.1001	68,620	67,341	69,882	6,540	326,749	31.6	30.9	32.4	28,876	1,442,690
40-44	0.0173	0.0829	0.0786	0.0871	62,080	60,636	63,487	5,149	297,528	29.7	29.1	30.3	20,541	1,187,037
45-49	0.0146	0.0706	0.0668	0.0742	56,931	55,393	58,446	4,019	274,609	27.1	26.6	27.7	14,460	987,989
50-54	0.0146	0.0706	0.0667	0.0743	52,912	51,312	54,489	3,737	255,219	24.0	23.5	24.5	11,704	799,315
55-59	0.0160	0.0768	0.0726	0.0807	49,175	47,514	50,793	3,777	236,433	20.6	20.2	21.1	10,481	656,072
60-64	0.0213	0.1011	0.0956	0.1062	45,398	43,694	47,038	4,589	215,518	17.1	16.8	17.6	10,993	516,319
65-69	0.0316	0.1466	0.1388	0.1538	40,809	39,082	42,452	5,982	189,091	13.8	13.5	14.1	12,284	388,306
70-74	0.0510	0.2260	0.2149	0.2363	34,827	33,120	36,488	7,870	154,461	10.7	10.4	11.0	13,506	265,072
75-79	0.0817	0.3393	0.3249	0.3527	26,957	25,371	28,646	9,146	111,921	8.1	7.9	8.4	13,760	168,383
80-84	0.1304	0.4918	0.4754	0.5071	17,811	16,484	19,341	8,759	67,158	6.0	5.8	6.2	12,048	92,372
85-89	0.2001	0.6670	0.6515	0.6812	9,052	8,161	10,144	6,038	30,166	4.4	4.3	4.5	6,885	34,400
90-94	0.2950	0.7825	0.7699	0.7939	3,014	2,604	3,529	2,359	7,996	3.2	3.1	3.2	2,696	9,141
95-99	0.4176	0.9269	0.9200	0.9330	656	538	810	608	1,455	2.3	2.3	2.4	625	1,496
100+	0.5679	1	1	1	48	36	64	48	84	1.8	1.7	1.8	95	167
All ages													*315,621*	*21,986,284*

$\ln(_nq_x)$ with uncertainty interval

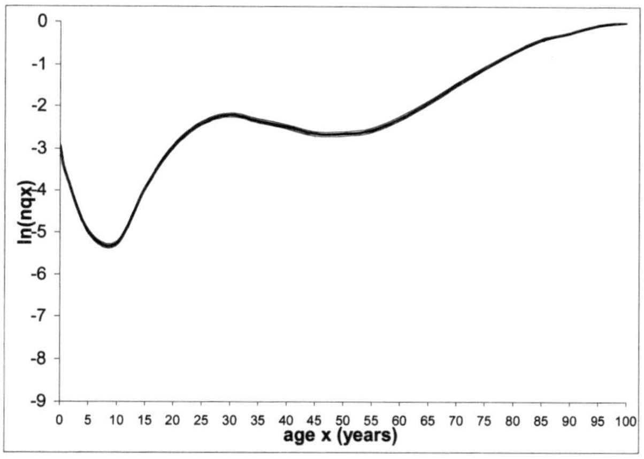

l_x with uncertainty interval

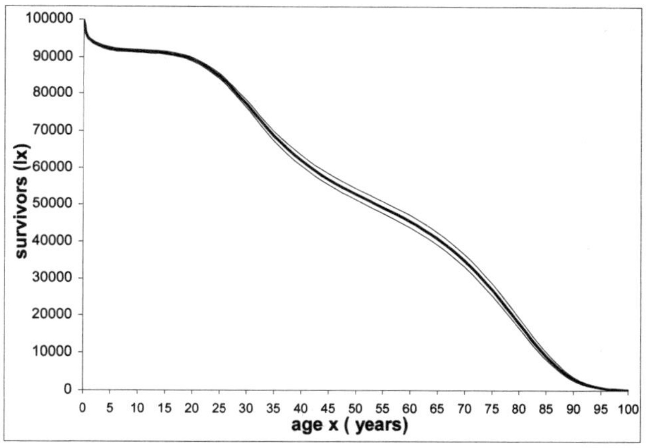

e_x with uncertainty interval

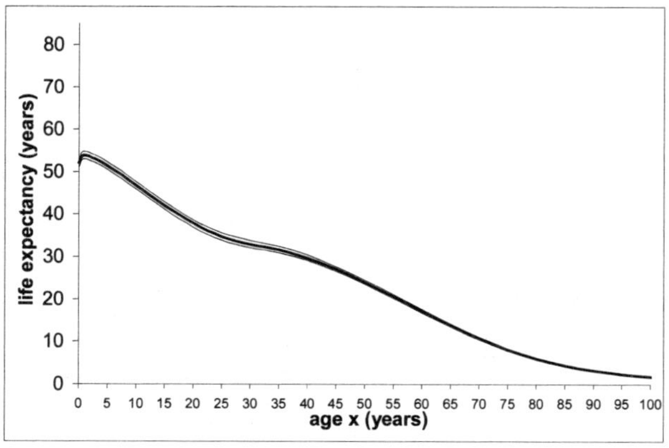

Distribution of deaths by age with uncertainty interval

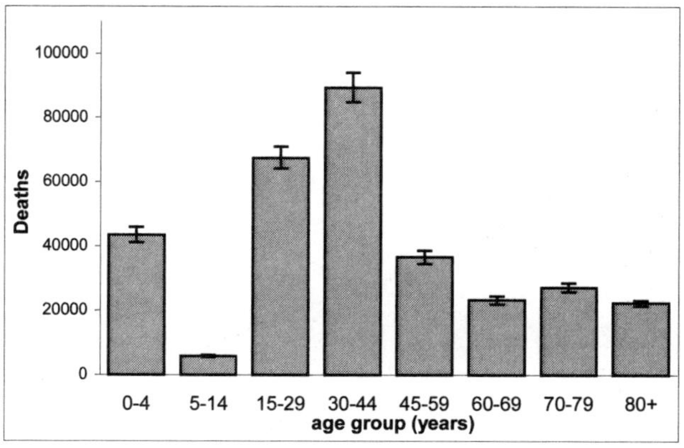

Spain Espagne España

Males, 2000 *Hommes, 2000* *Hombres, 2000*

Age	$_nM_x$	$_nq_x$			l_x			$_nd_x$	$_nL_x$	e_x			Actual deaths	Actual population
			percentiles			percentiles					percentiles			
		mean	2½	97½	mean	2½	97½			mean	2½	97½		
<1	0.0050	0.0049	0.0043	0.0056	100,000	100,000	100,000	493	99,556	75.4	74.9	76.0	918	185,141
1-4	0.0003	0.0011	0.0010	0.0012	99,507	99,442	99,569	107	397,770	74.7	74.3	75.4	202	753,967
5-9	0.0002	0.0010	0.0009	0.0011	99,400	99,325	99,473	103	496,741	70.8	70.3	71.5	211	1,015,070
10-14	0.0002	0.0010	0.0009	0.0011	99,297	99,212	99,380	101	496,232	65.9	65.4	66.5	218	1,073,234
15-19	0.0006	0.0031	0.0028	0.0034	99,196	99,103	99,288	307	495,213	60.9	60.5	61.6	816	1,317,536
20-24	0.0009	0.0047	0.0043	0.0051	98,889	98,770	99,008	464	493,286	56.1	55.6	56.8	1,531	1,627,552
25-29	0.0011	0.0055	0.0051	0.0058	98,425	98,271	98,578	537	490,784	51.4	50.9	52.0	1,872	1,712,114
30-34	0.0015	0.0076	0.0071	0.0081	97,889	97,698	98,078	746	487,577	46.6	46.1	47.3	2,537	1,657,239
35-39	0.0019	0.0093	0.0088	0.0099	97,142	96,904	97,374	905	483,448	42.0	41.5	42.7	2,927	1,563,013
40-44	0.0024	0.0120	0.0114	0.0127	96,237	95,949	96,515	1,159	478,287	37.4	36.8	38.0	3,426	1,413,619
45-49	0.0036	0.0180	0.0172	0.0188	95,078	94,734	95,411	1,711	471,110	32.8	32.3	33.4	4,529	1,246,548
50-54	0.0056	0.0276	0.0262	0.0289	93,366	92,954	93,768	2,573	460,398	28.3	27.8	29.0	6,606	1,181,845
55-59	0.0082	0.0402	0.0379	0.0422	90,793	90,299	91,275	3,649	444,841	24.1	23.5	24.7	8,416	1,025,959
60-64	0.0138	0.0667	0.0622	0.0701	87,144	86,570	87,704	5,810	421,194	20.0	19.5	20.6	12,584	912,296
65-69	0.0214	0.1017	0.0942	0.1079	81,334	80,573	82,102	8,268	385,998	16.2	15.7	16.8	20,498	956,971
70-74	0.0347	0.1595	0.1475	0.1704	73,066	71,969	74,361	11,657	336,186	12.8	12.3	13.3	27,565	794,973
75-79	0.0581	0.2535	0.2355	0.2705	61,409	59,920	63,391	15,569	268,120	9.7	9.4	10.1	33,101	570,044
80-84	0.0980	0.3936	0.3705	0.4158	45,839	43,710	48,465	18,043	184,088	7.2	6.9	7.5	29,730	303,319
85-89	0.1576	0.5654	0.5427	0.5874	27,796	25,533	30,509	15,715	99,693	5.2	5.0	5.4	22,986	145,820
90-94	0.2415	0.7003	0.6830	0.7176	12,081	10,534	13,953	8,460	35,027	3.7	3.6	3.8	11,586	47,967
95-99	0.3526	0.8568	0.8461	0.8677	3,621	2,975	4,423	3,103	8,799	2.7	2.7	2.8	2,257	6,402
100+	0.4905	1	1	1	519	394	681	519	1,058	2.0	2.0	2.1	52	106
All ages													*194,568*	*19,510,735*

$\ln({}_nq_x)$ with uncertainty interval

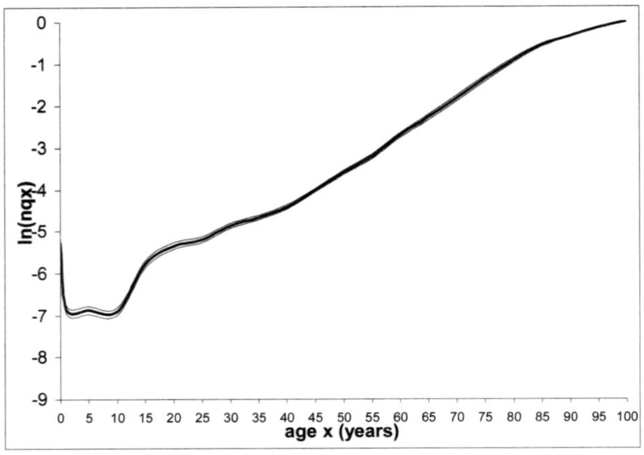

l_x with uncertainty interval

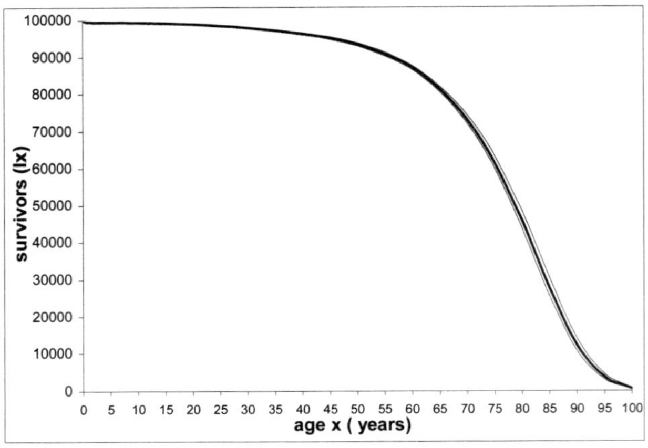

e_x with uncertainty interval

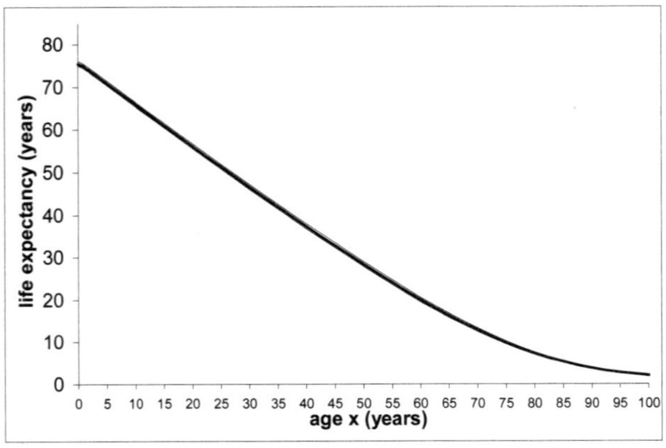

Distribution of deaths by age with uncertainty interval

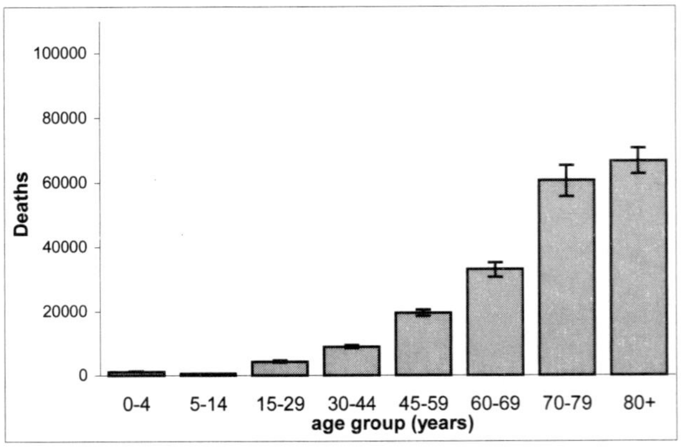

Spain | Espagne | España

Females, 2000 | *Femmes, 2000* | *Mujeres, 2000*

Age	$_nM_x$	$_nq_x$ mean	$_nq_x$ percentiles 2½	$_nq_x$ percentiles 97½	l_x mean	l_x percentiles 2½	l_x percentiles 97½	$_nd_x$	$_nL_x$	e_x mean	e_x percentiles 2½	e_x percentiles 97½	Actual deaths	Actual population
<1	0.0040	0.0040	0.0036	0.0045	100,000	100,000	100,000	401	99,639	82.3	82.1	82.7	703	174,497
1-4	0.0002	0.0010	0.0009	0.0011	99,599	99,551	99,641	99	398,158	81.7	81.4	82.1	174	702,614
5-9	0.0001	0.0006	0.0005	0.0006	99,500	99,444	99,550	59	497,354	77.8	77.5	78.1	112	951,915
10-14	0.0001	0.0007	0.0006	0.0008	99,441	99,380	99,495	69	497,034	72.8	72.5	73.2	142	1,017,721
15-19	0.0003	0.0014	0.0013	0.0015	99,372	99,306	99,431	134	496,525	67.8	67.6	68.2	339	1,254,442
20-24	0.0003	0.0013	0.0013	0.0014	99,238	99,161	99,306	133	495,857	62.9	62.7	63.3	418	1,558,174
25-29	0.0004	0.0018	0.0017	0.0019	99,105	99,019	99,182	179	495,078	58.0	57.7	58.4	595	1,649,510
30-34	0.0005	0.0024	0.0023	0.0026	98,926	98,829	99,014	241	494,029	53.1	52.8	53.5	788	1,614,790
35-39	0.0007	0.0035	0.0033	0.0037	98,685	98,573	98,787	342	492,571	48.2	48.0	48.6	1,075	1,548,918
40-44	0.0011	0.0052	0.0050	0.0055	98,343	98,213	98,462	515	490,429	43.4	43.1	43.8	1,491	1,420,245
45-49	0.0015	0.0075	0.0072	0.0078	97,828	97,675	97,970	734	487,306	38.6	38.3	39.0	1,899	1,260,665
50-54	0.0022	0.0110	0.0106	0.0115	97,094	96,912	97,264	1,071	482,795	33.9	33.6	34.3	2,697	1,216,124
55-59	0.0031	0.0153	0.0147	0.0159	96,024	95,811	96,228	1,471	476,440	29.2	29.0	29.6	3,330	1,078,394
60-64	0.0052	0.0255	0.0244	0.0265	94,553	94,304	94,798	2,411	466,736	24.7	24.4	25.0	5,220	1,010,606
65-69	0.0087	0.0425	0.0405	0.0441	92,142	91,829	92,453	3,915	450,923	20.2	20.0	20.6	9,669	1,113,713
70-74	0.0160	0.0769	0.0730	0.0799	88,227	87,783	88,660	6,784	424,175	16.0	15.8	16.3	16,094	1,006,256
75-79	0.0315	0.1458	0.1384	0.1514	81,443	80,804	82,182	11,875	377,525	12.1	12.0	12.4	25,785	819,720
80-84	0.0632	0.2729	0.2605	0.2817	69,567	68,605	70,809	18,984	300,377	8.8	8.6	9.0	34,400	544,304
85-89	0.1175	0.4542	0.4394	0.4669	50,583	49,280	52,364	22,977	195,474	6.2	6.0	6.3	37,336	317,625
90-94	0.2024	0.6296	0.6160	0.6422	27,606	26,407	29,357	17,382	85,886	4.2	4.1	4.3	24,231	119,728
95-99	0.3226	0.8196	0.8102	0.8283	10,224	9,506	11,219	8,380	25,981	2.9	2.9	3.0	6,173	19,137
100+	0.4759	1	1	1	1,844	1,635	2,098	1,844	3,875	2.1	2.1	2.2	210	441
All ages													*172,881*	*20,399,539*

ln($_nq_x$) with uncertainty interval

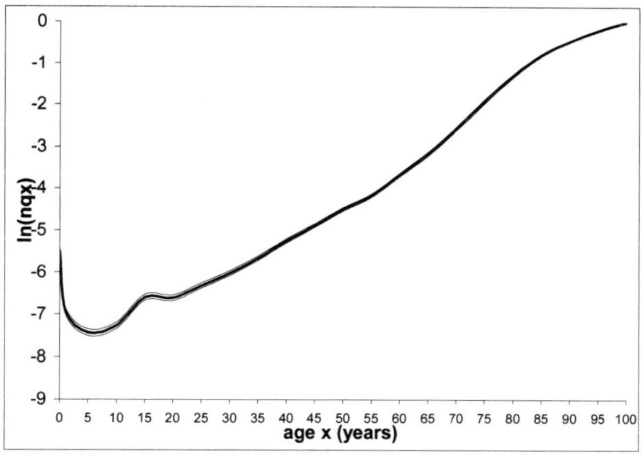

l_x with uncertainty interval

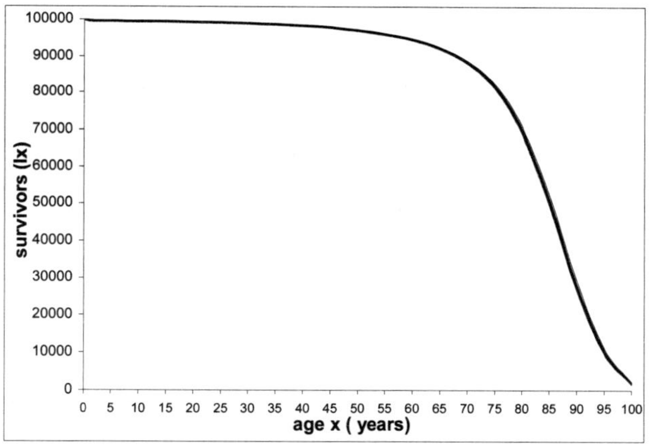

e_x with uncertainty interval

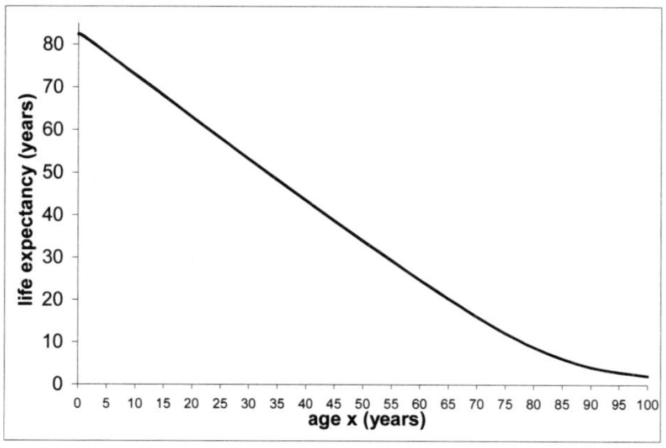

Distribution of deaths by age with uncertainty interval

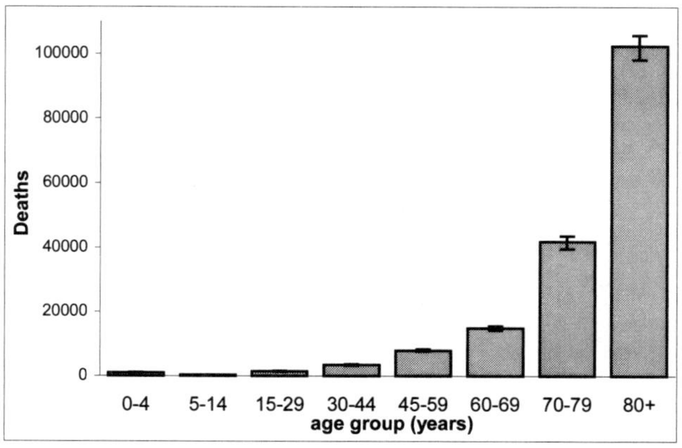

Sri Lanka | Sri Lanka | Sri Lanka

Males, 2000 | *Hommes, 2000* | *Hombres, 2000*

Age	$_nM_x$	$_nq_x$			l_x			$_nd_x$	$_nL_x$	e_x			Actual deaths	Actual population
			percentiles			percentiles					percentiles			
		mean	2 1/2	97 1/2	mean	2 1/2	97 1/2			mean	2 1/2	97 1/2		
<1	0.0200	0.0196	0.0166	0.0228	100,000	100,000	100,000	1,962	98,234	67.6	65.0	71.1	3,245	162,471
1-4	0.0011	0.0043	0.0036	0.0050	98,038	97,721	98,345	418	391,150	68.0	65.5	71.5	674	631,149
5-9	0.0009	0.0043	0.0036	0.0051	97,620	97,272	97,969	422	487,047	64.3	61.8	67.7	735	848,030
10-14	0.0010	0.0050	0.0042	0.0059	97,199	96,801	97,603	489	484,771	59.5	57.1	63.0	894	886,375
15-19	0.0020	0.0101	0.0083	0.0119	96,710	96,230	97,189	976	481,109	54.8	52.4	58.2	1,953	962,639
20-24	0.0039	0.0191	0.0155	0.0226	95,734	95,086	96,378	1,827	474,101	50.4	48.0	53.7	3,398	881,757
25-29	0.0039	0.0191	0.0153	0.0228	93,907	92,936	94,869	1,795	465,045	46.3	44.0	49.5	3,027	784,124
30-34	0.0037	0.0182	0.0143	0.0218	92,111	90,820	93,392	1,673	456,372	42.1	40.0	45.2	2,844	775,646
35-39	0.0049	0.0241	0.0188	0.0290	90,438	88,850	92,015	2,183	446,730	37.9	35.8	40.9	3,699	756,873
40-44	0.0061	0.0301	0.0232	0.0363	88,254	86,286	90,217	2,653	434,639	33.7	31.8	36.6	4,159	681,311
45-49	0.0081	0.0395	0.0302	0.0477	85,601	83,185	88,108	3,379	419,559	29.7	27.9	32.4	4,449	552,408
50-54	0.0096	0.0471	0.0358	0.0569	82,222	79,266	85,444	3,872	401,433	25.8	24.2	28.3	5,026	521,165
55-59	0.0137	0.0663	0.0504	0.0799	78,351	74,766	82,381	5,195	378,765	22.0	20.5	24.3	5,156	375,913
60-64	0.0200	0.0951	0.0724	0.1137	73,155	68,790	78,232	6,955	348,388	18.4	16.9	20.4	6,017	301,380
65-69	0.0315	0.1458	0.1123	0.1719	66,200	60,965	72,570	9,653	306,869	15.0	13.8	16.8	7,110	226,034
70-74	0.0460	0.2061	0.1622	0.2414	56,547	50,487	64,422	11,654	253,602	12.2	11.1	13.7	8,011	174,324
75-79	0.0638	0.2753	0.2237	0.3178	44,893	38,486	53,975	12,357	193,574	9.7	8.8	10.8	7,739	121,225
80-84	0.0967	0.3894	0.3315	0.4378	32,536	26,586	41,899	12,670	131,006	7.4	6.8	8.2	5,001	51,709
85-89	0.1460	0.5349	0.4801	0.5820	19,866	15,222	28,008	10,626	72,767	5.5	5.1	6.1	2,787	19,088
90-94	0.2171	0.6574	0.6143	0.6969	9,240	6,364	14,561	6,074	27,979	4.0	3.8	4.4	786	3,619
95-99	0.3178	0.8134	0.7825	0.8420	3,166	1,929	5,610	2,575	8,104	3.0	2.8	3.1	121	382
100+	0.4580	1	1	1	591	305	1,207	591	1,290	2.2	2.0	2.2	10	22
All ages													*76,841*	*9,717,644*

ln($_nq_x$) with uncertainty interval

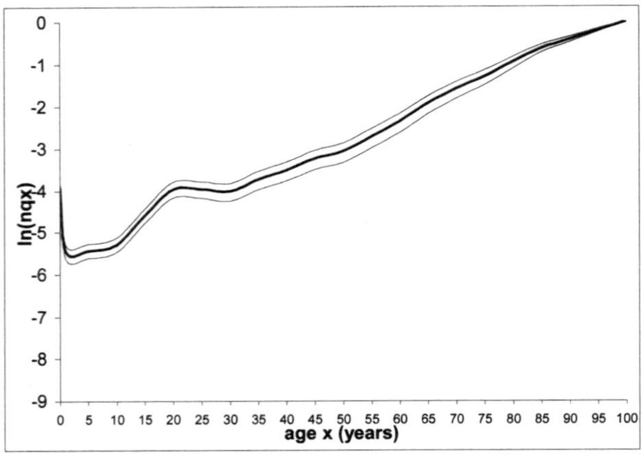

l_x with uncertainty interval

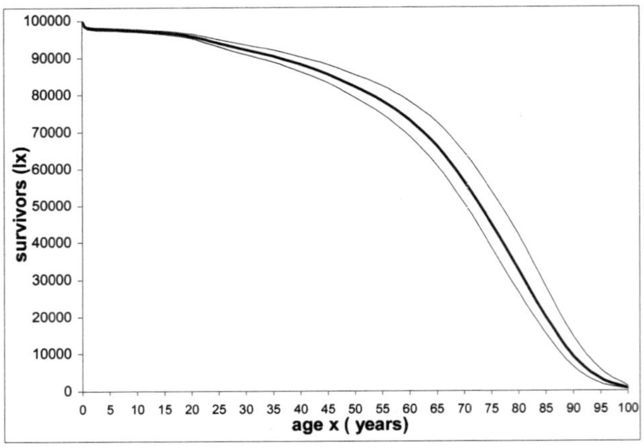

e_x with uncertainty interval

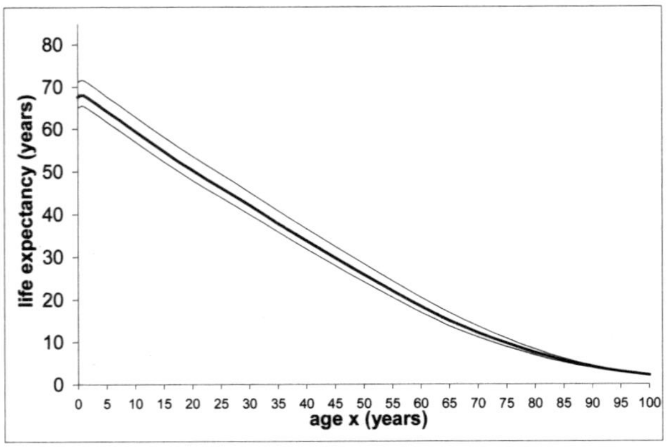

Distribution of deaths by age with uncertainty interval

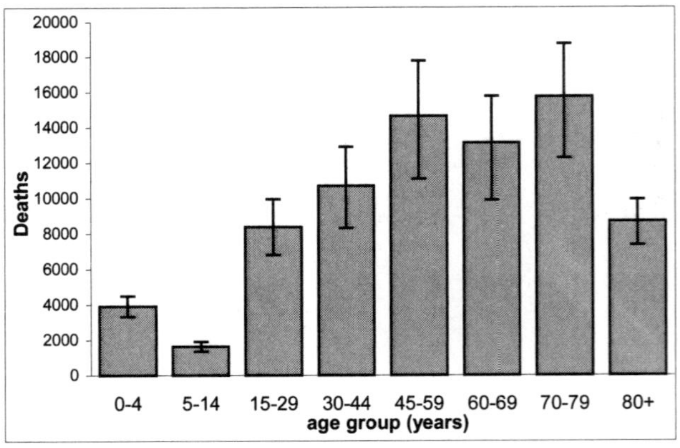

Sri Lanka — Sri Lanka — Sri Lanka

Females, 2000 — *Femmes, 2000* — *Mujeres, 2000*

Age	$_nM_x$	$_nq_x$ mean	$_nq_x$ percentiles 2½	$_nq_x$ percentiles 97½	l_x mean	l_x percentiles 2½	l_x percentiles 97½	$_nd_x$	$_nL_x$	e_x mean	e_x percentiles 2½	e_x percentiles 97½	Actual deaths	Actual population
<1	0.0140	0.0138	0.0113	0.0165	100,000	100,000	100,000	1,378	98,760	75.3	73.9	76.8	2,180	156,222
1-4	0.0009	0.0037	0.0031	0.0042	98,622	98,355	98,870	362	393,620	75.4	73.9	76.9	561	610,584
5-9	0.0006	0.0030	0.0026	0.0035	98,260	97,962	98,559	299	490,553	71.7	70.2	73.2	501	821,853
10-14	0.0006	0.0029	0.0024	0.0033	97,961	97,619	98,302	281	489,101	66.9	65.4	68.4	494	859,620
15-19	0.0011	0.0054	0.0046	0.0062	97,680	97,296	98,061	527	487,080	62.0	60.6	63.6	1,011	934,167
20-24	0.0013	0.0063	0.0054	0.0072	97,152	96,693	97,608	609	484,239	57.4	55.9	58.9	1,074	853,721
25-29	0.0013	0.0067	0.0058	0.0077	96,543	96,000	97,081	649	481,092	52.7	51.3	54.3	1,023	757,878
30-34	0.0014	0.0070	0.0061	0.0080	95,894	95,264	96,516	674	477,785	48.1	46.6	49.6	1,054	747,555
35-39	0.0019	0.0093	0.0081	0.0106	95,220	94,504	95,927	890	473,876	43.4	42.0	44.9	1,357	722,644
40-44	0.0025	0.0123	0.0107	0.0139	94,330	93,504	95,143	1,163	468,744	38.8	37.4	40.3	1,594	642,747
45-49	0.0034	0.0168	0.0147	0.0190	93,168	92,205	94,113	1,568	461,918	34.2	32.9	35.7	1,758	518,031
50-54	0.0055	0.0271	0.0237	0.0307	91,600	90,464	92,712	2,484	451,790	29.8	28.5	31.2	2,115	384,634
55-59	0.0081	0.0396	0.0345	0.0452	89,116	87,730	90,470	3,527	436,764	25.5	24.3	26.9	2,710	335,629
60-64	0.0123	0.0597	0.0518	0.0686	85,589	83,877	87,322	5,113	415,165	21.5	20.3	22.8	3,345	271,585
65-69	0.0191	0.0913	0.0782	0.1053	80,477	78,308	82,714	7,351	384,005	17.7	16.6	18.9	4,165	217,567
70-74	0.0312	0.1446	0.1231	0.1660	73,125	70,152	76,013	10,571	339,199	14.2	13.2	15.3	5,905	189,476
75-79	0.0483	0.2155	0.1844	0.2455	62,554	58,504	66,350	13,483	279,065	11.2	10.4	12.1	4,819	99,736
80-84	0.0740	0.3122	0.2718	0.3511	49,072	44,203	54,117	15,322	207,054	8.6	7.9	9.3	4,334	58,567
85-89	0.1220	0.4674	0.4189	0.5111	33,750	28,894	39,410	15,776	129,309	6.3	5.8	6.9	2,357	19,323
90-94	0.1800	0.5844	0.5391	0.6228	17,974	14,283	22,830	10,504	58,357	4.7	4.3	5.0	714	3,965
95-99	0.2700	0.7459	0.7096	0.7752	7,470	5,388	10,429	5,571	20,635	3.4	3.2	3.6	154	569
100+	0.3900	1	1	1	1,898	1,211	3,003	1,898	4,868	2.6	2.2	2.7	12	32
All ages													*43,238*	*9,206,105*

ln($_nq_x$) with uncertainty interval

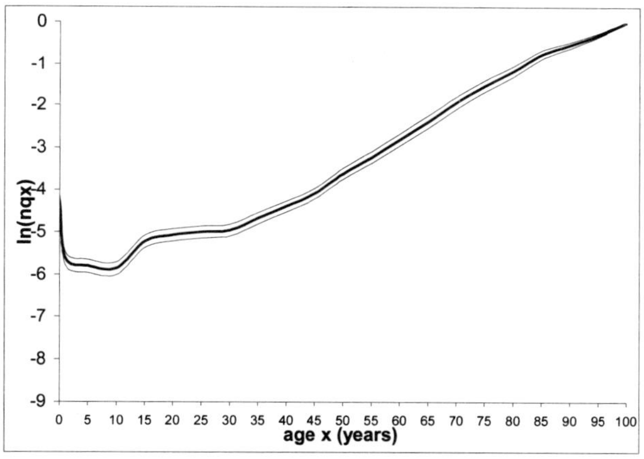

l_x with uncertainty interval

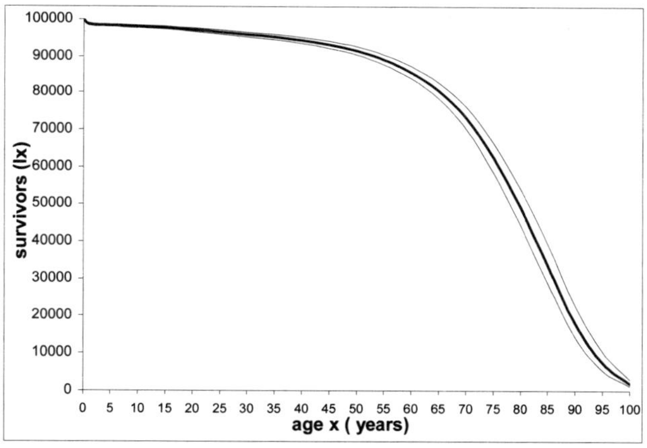

e_x with uncertainty interval

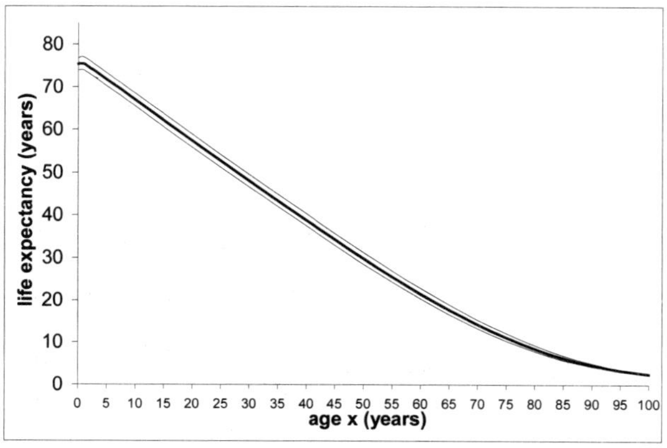

Distribution of deaths by age with uncertainty interval

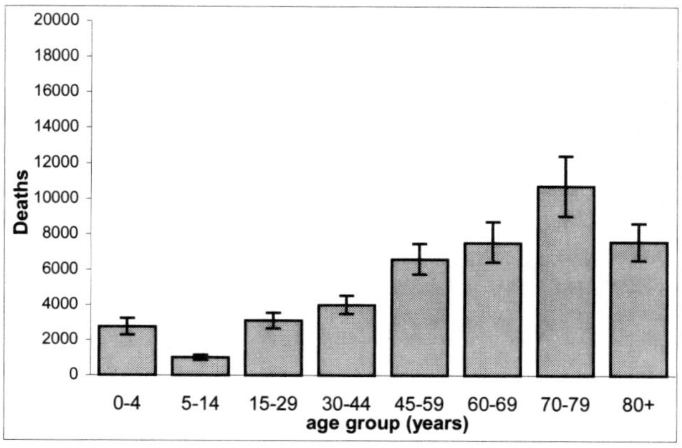

Sudan   Soudan   Sudán

Males, 2000   *Hommes, 2000*   *Hombres, 2000*

Age	$_nM_x$	$_nq_x$			l_x			$_nd_x$	$_nL_x$	e_x			Actual deaths	Actual population
			percentiles			percentiles					percentiles			
		mean	2½	97½	mean	2½	97½			mean	2½	97½		
<1	0.1048	0.0957	0.0816	0.1114	100,000	100,000	100,000	9,575	91,383	55.4	52.4	58.7	54,667	521,741
1-4	0.0080	0.0315	0.0267	0.0370	90,425	88,856	91,840	2,852	354,856	60.3	57.2	63.9	15,197	1,890,747
5-9	0.0020	0.0100	0.0083	0.0118	87,573	86,084	89,072	872	435,686	58.2	55.3	61.7	4,215	2,106,779
10-14	0.0013	0.0067	0.0056	0.0080	86,701	85,106	88,325	581	432,056	53.8	50.9	57.2	2,467	1,835,864
15-19	0.0017	0.0086	0.0071	0.0102	86,121	84,456	87,827	737	428,761	49.1	46.3	52.5	2,853	1,658,936
20-24	0.0032	0.0156	0.0129	0.0187	85,383	83,630	87,194	1,335	423,579	44.5	41.7	47.9	4,583	1,453,752
25-29	0.0047	0.0233	0.0191	0.0279	84,048	82,105	86,048	1,955	415,353	40.2	37.5	43.5	6,014	1,277,801
30-34	0.0065	0.0321	0.0261	0.0387	82,093	79,812	84,368	2,635	403,877	36.1	33.5	39.3	6,909	1,058,902
35-39	0.0079	0.0388	0.0312	0.0473	79,458	76,722	82,096	3,081	389,587	32.2	29.7	35.3	6,886	870,825
40-44	0.0096	0.0469	0.0372	0.0578	76,377	73,121	79,429	3,582	372,930	28.4	26.0	31.4	6,868	715,057
45-49	0.0122	0.0590	0.0461	0.0734	72,795	68,959	76,340	4,298	353,231	24.7	22.4	27.5	6,930	569,575
50-54	0.0160	0.0771	0.0593	0.0963	68,497	64,016	72,677	5,279	329,289	21.1	18.9	23.7	7,914	493,656
55-59	0.0217	0.1029	0.0783	0.1288	63,219	58,038	68,166	6,506	299,829	17.6	15.7	20.0	8,519	392,625
60-64	0.0328	0.1514	0.1146	0.1905	56,713	50,843	62,614	8,587	262,096	14.4	12.6	16.5	9,846	300,497
65-69	0.0479	0.2139	0.1627	0.2677	48,125	41,276	55,064	10,295	214,889	11.5	10.0	13.3	10,412	217,316
70-74	0.0729	0.3084	0.2395	0.3775	37,830	30,480	45,921	11,667	159,983	8.9	7.7	10.4	10,465	143,505
75-79	0.1105	0.4328	0.3497	0.5101	26,163	19,314	34,925	11,324	102,506	6.8	5.9	7.9	9,020	81,655
80-84	0.1679	0.5913	0.5054	0.6647	14,839	9,515	22,713	8,775	52,260	5.1	4.5	5.8	6,109	36,380
85-89	0.2457	0.7611	0.6902	0.8178	6,064	3,190	11,233	4,616	18,783	3.8	3.4	4.2	2,667	10,854
90-94	0.3462	0.8491	0.7955	0.8913	1,449	581	3,480	1,230	3,553	2.8	2.6	3.1	677	1,955
95-99	0.4696	0.9747	0.9557	0.9864	219	63	712	213	454	2.1	2.1	2.2	87	185
100+	0.6130	1	1	1	6	1	32	6	9	1.6	1.6	1.7	5	8
All ages													*183,310*	*15,638,615*

$\ln(_nq_x)$ with uncertainty interval

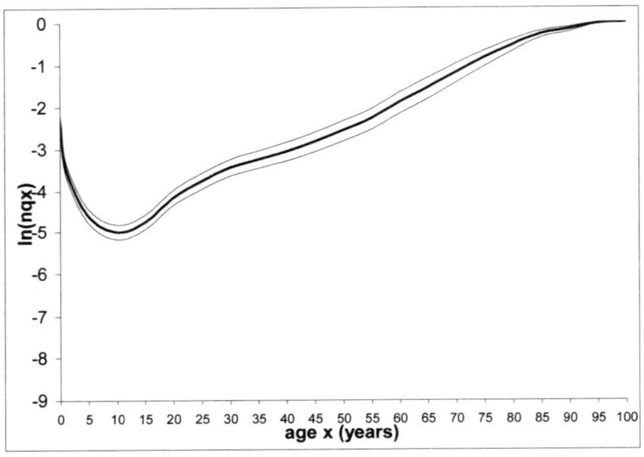

l_x with uncertainty interval

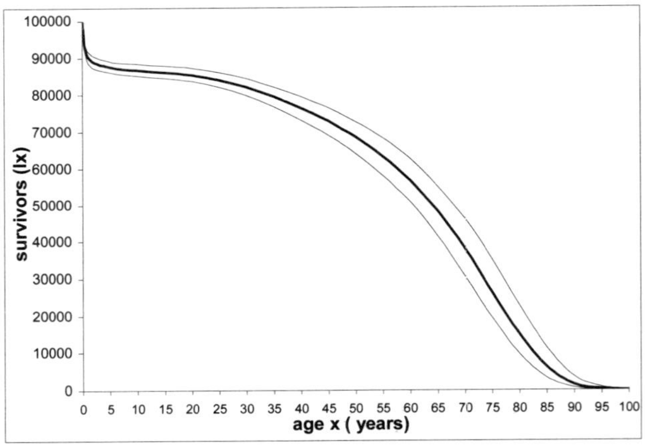

e_x with uncertainty interval

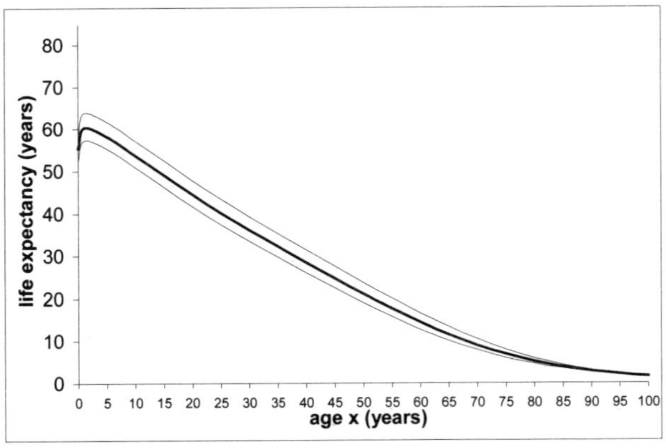

Distribution of deaths by age with uncertainty interval

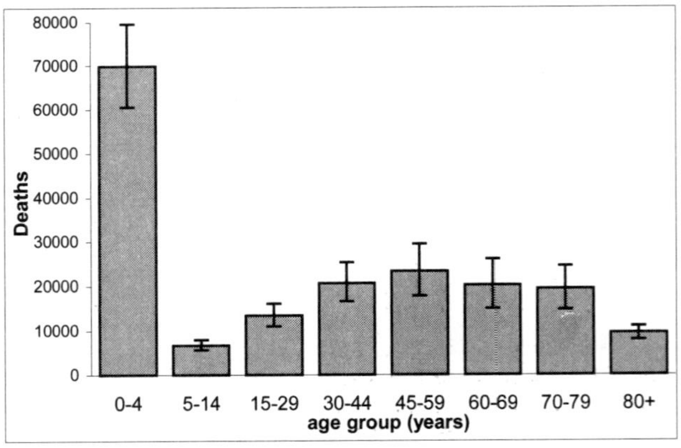

Sudan

Soudan

Sudán

Females, 2000

Femmes, 2000

Mujeres, 2000

Age	$_nM_x$	$_nq_x$			l_x			$_nd_x$	$_nL_x$	e_x			Actual deaths	Actual population
			percentiles			percentiles					percentiles			
		mean	2½	97½	mean	2½	97½			mean	2½	97½		
<1	0.0926	0.0855	0.0718	0.0994	100,000	100,000	100,000	8,547	92,307	57.8	54.6	61.4	45,772	494,307
1-4	0.0089	0.0348	0.0288	0.0406	91,453	90,057	92,821	3,182	358,173	62.2	59.0	65.9	16,184	1,821,694
5-9	0.0020	0.0102	0.0083	0.0120	88,271	86,771	89,769	896	439,113	60.3	57.3	64.0	4,142	2,029,929
10-14	0.0014	0.0068	0.0056	0.0081	87,375	85,790	88,971	598	435,376	55.9	53.0	59.5	2,438	1,773,412
15-19	0.0023	0.0113	0.0092	0.0135	86,776	85,131	88,473	985	431,419	51.3	48.4	54.8	3,667	1,606,718
20-24	0.0040	0.0199	0.0161	0.0238	85,792	84,014	87,658	1,707	424,690	46.9	44.0	50.3	5,701	1,418,234
25-29	0.0056	0.0278	0.0224	0.0334	84,084	82,080	86,247	2,341	414,570	42.8	40.0	46.1	7,110	1,259,337
30-34	0.0066	0.0327	0.0262	0.0395	81,744	79,410	84,315	2,671	402,041	38.9	36.3	42.2	6,984	1,051,243
35-39	0.0071	0.0349	0.0276	0.0427	79,073	76,281	82,107	2,759	388,467	35.2	32.6	38.2	6,181	870,406
40-44	0.0074	0.0362	0.0283	0.0448	76,314	73,059	79,821	2,763	374,664	31.3	29.0	34.2	5,304	719,269
45-49	0.0086	0.0422	0.0326	0.0526	73,552	69,848	77,522	3,106	359,993	27.4	25.2	30.2	4,992	578,628
50-54	0.0114	0.0554	0.0424	0.0694	70,446	66,263	74,920	3,904	342,467	23.5	21.4	26.1	5,840	512,246
55-59	0.0156	0.0752	0.0570	0.0945	66,541	61,804	71,618	5,003	320,199	19.8	17.8	22.1	6,498	415,861
60-64	0.0236	0.1112	0.0839	0.1395	61,538	56,179	67,325	6,845	290,579	16.2	14.4	18.3	7,692	326,533
65-69	0.0366	0.1678	0.1267	0.2109	54,693	48,488	61,311	9,179	250,519	12.9	11.3	14.8	8,939	243,974
70-74	0.0593	0.2583	0.1981	0.3203	45,514	38,382	53,220	11,756	198,182	10.0	8.7	11.6	9,909	167,054
75-79	0.0937	0.3794	0.3003	0.4558	33,759	26,357	42,676	12,809	136,769	7.5	6.6	8.8	9,392	100,285
80-84	0.1443	0.5302	0.4419	0.6080	20,949	14,668	29,814	11,107	76,979	5.6	4.9	6.5	6,879	47,680
85-89	0.2149	0.6989	0.6195	0.7637	9,842	5,750	16,560	6,879	32,014	4.2	3.7	4.8	3,401	15,825
90-94	0.3094	0.8023	0.7404	0.8519	2,963	1,359	6,268	2,377	7,684	3.0	2.8	3.4	1,066	3,444
95-99	0.4306	0.9395	0.9066	0.9624	586	201	1,627	550	1,278	2.3	2.2	2.5	188	437
100+	0.5794	1	1	1	35	8	152	35	61	1.7	1.7	1.9	18	31
All ages													*168,296*	*15,456,547*

ln($_nq_x$) with uncertainty interval

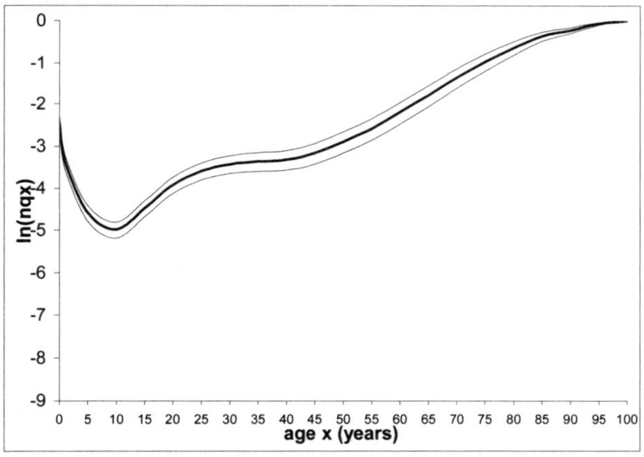

l_x with uncertainty interval

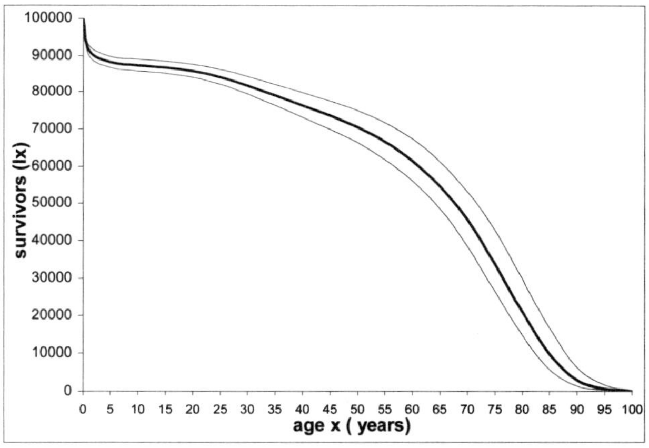

e_x with uncertainty interval

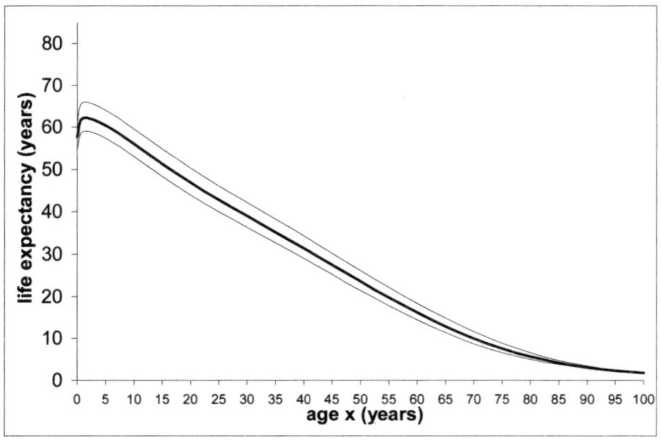

Distribution of deaths by age with uncertainty interval

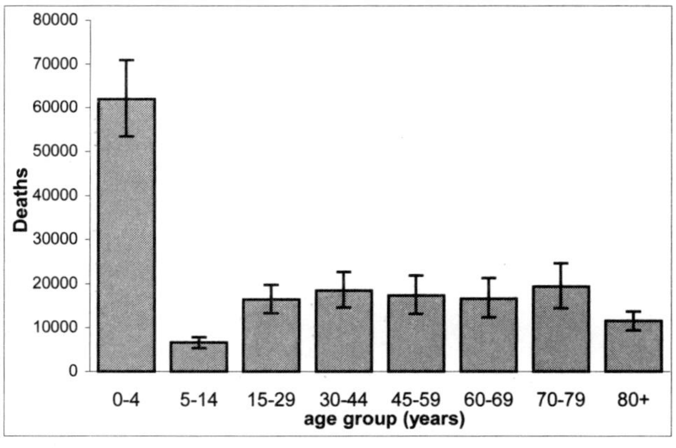

Suriname | Suriname | Suriname

Males, 2000 | *Hommes, 2000* | *Hombres, 2000*

Age	$_nM_x$	$_nq_x$			l_x			$_nd_x$	$_nL_x$	e_x			Actual deaths	Actual population
			percentiles			percentiles					percentiles			
		mean	2½	97½	mean	2½	97½			mean	2½	97½		
<1	0.0275	0.0268	0.0221	0.0315	100,000	100,000	100,000	2,679	97,589	68.0	66.4	69.7	109	3,982
1-4	0.0006	0.0025	0.0021	0.0029	97,321	96,850	97,790	240	388,708	68.9	67.5	70.5	10	16,323
5-9	0.0003	0.0017	0.0015	0.0020	97,081	96,584	97,579	170	484,980	65.1	63.7	66.7	7	21,323
10-14	0.0003	0.0016	0.0014	0.0018	96,911	96,391	97,430	155	484,168	60.2	58.8	61.8	7	23,146
15-19	0.0008	0.0042	0.0036	0.0048	96,756	96,214	97,292	405	482,768	55.3	53.9	56.9	21	25,481
20-24	0.0016	0.0082	0.0071	0.0094	96,351	95,752	96,932	788	479,784	50.5	49.2	52.1	33	20,116
25-29	0.0020	0.0097	0.0084	0.0112	95,563	94,857	96,238	932	475,484	45.9	44.6	47.5	39	19,789
30-34	0.0024	0.0120	0.0104	0.0137	94,631	93,798	95,413	1,133	470,323	41.3	40.1	42.9	44	18,406
35-39	0.0035	0.0175	0.0152	0.0199	93,498	92,517	94,413	1,635	463,404	36.8	35.6	38.3	59	16,688
40-44	0.0055	0.0271	0.0236	0.0308	91,863	90,680	92,971	2,486	453,103	32.4	31.3	33.8	55	10,025
45-49	0.0080	0.0392	0.0341	0.0446	89,378	87,911	90,759	3,505	438,126	28.2	27.2	29.6	53	6,629
50-54	0.0118	0.0573	0.0495	0.0649	85,873	84,051	87,603	4,921	417,060	24.3	23.3	25.6	61	5,181
55-59	0.0167	0.0803	0.0692	0.0903	80,951	78,626	83,185	6,497	388,515	20.6	19.8	21.8	79	4,731
60-64	0.0224	0.1061	0.0915	0.1182	74,455	71,529	77,286	7,898	352,527	17.2	16.4	18.2	105	4,670
65-69	0.0335	0.1545	0.1342	0.1698	66,556	63,074	70,155	10,282	307,077	13.9	13.3	14.8	140	4,178
70-74	0.0484	0.2158	0.1900	0.2360	56,274	52,365	60,739	12,143	251,015	11.0	10.4	11.7	141	2,911
75-79	0.0784	0.3278	0.2956	0.3557	44,131	40,157	49,198	14,467	184,489	8.3	7.9	8.9	137	1,751
80-84	0.1261	0.4794	0.4433	0.5119	29,664	26,242	34,654	14,222	112,765	6.2	5.8	6.6	100	793
85-89	0.1931	0.6512	0.6175	0.6811	15,442	13,065	19,192	10,056	52,070	4.5	4.3	4.8	80	415
90-94	0.2816	0.7632	0.7367	0.7869	5,386	4,200	7,201	4,110	14,598	3.3	3.2	3.5	45	160
95-99	0.3909	0.8995	0.8831	0.9139	1,275	897	1,855	1,147	2,935	2.5	2.4	2.6	20	50
100+	0.5166	1	1	1	128	78	217	128	248	1.9	1.9	2.0	11	21
All ages													*1,357*	*206,769*

$\ln(_nq_x)$ with uncertainty interval

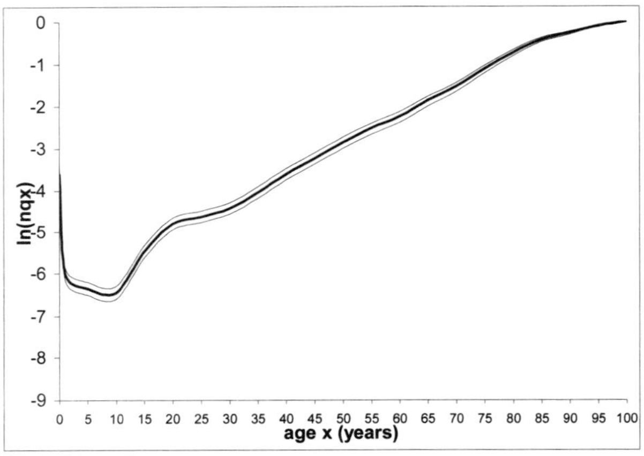

l_x with uncertainty interval

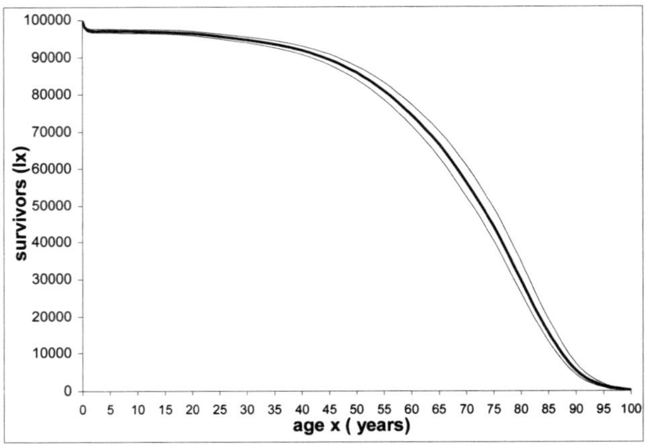

e_x with uncertainty interval

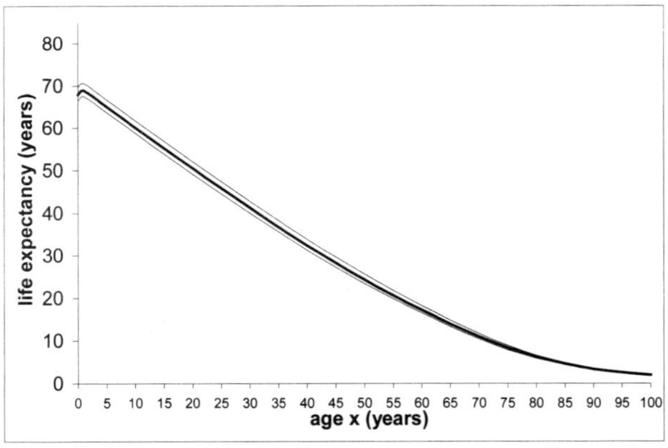

Distribution of deaths by age with uncertainty interval

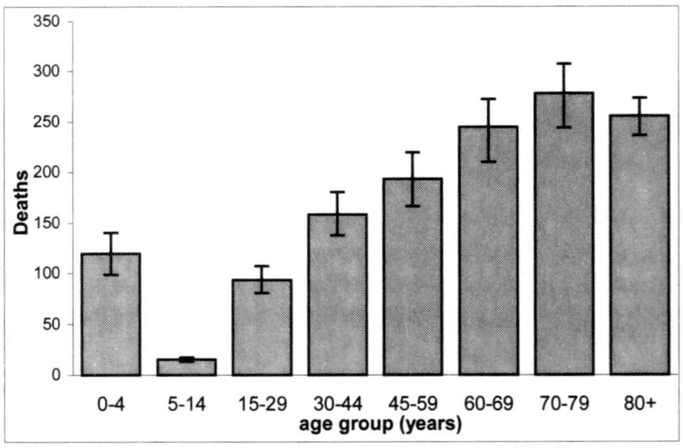

Suriname

Females, 2000

Suriname

Femmes, 2000

Suriname

Mujeres, 2000

Age	$_nM_x$	$_nq_x$			l_x			$_nd_x$	$_nL_x$	e_x			Actual deaths	Actual population
			percentiles			percentiles					percentiles			
		mean	2½	97½	mean	2½	97½			mean	2½	97½		
<1	0.0192	0.0189	0.0152	0.0226	100,000	100,000	100,000	1,887	98,302	73.5	71.9	75.4	73	3,814
1-4	0.0004	0.0017	0.0014	0.0020	98,113	97,736	98,484	172	392,041	73.9	72.5	75.8	7	15,724
5-9	0.0003	0.0014	0.0012	0.0017	97,942	97,542	98,337	139	489,359	70.1	68.6	71.9	6	20,523
10-14	0.0003	0.0013	0.0011	0.0016	97,802	97,385	98,218	131	488,682	65.2	63.7	67.0	6	22,289
15-19	0.0007	0.0035	0.0029	0.0041	97,671	97,237	98,105	345	487,491	60.3	58.8	62.1	17	24,499
20-24	0.0008	0.0041	0.0034	0.0047	97,326	96,849	97,811	395	485,642	55.5	54.1	57.3	16	19,527
25-29	0.0011	0.0057	0.0047	0.0066	96,931	96,396	97,473	552	483,276	50.7	49.3	52.5	22	19,310
30-34	0.0015	0.0075	0.0063	0.0087	96,379	95,764	96,998	725	480,084	45.9	44.6	47.7	27	18,148
35-39	0.0019	0.0096	0.0081	0.0112	95,654	94,929	96,388	923	475,965	41.3	40.0	43.0	32	16,724
40-44	0.0030	0.0147	0.0124	0.0172	94,732	93,872	95,611	1,396	470,169	36.7	35.4	38.4	32	10,946
45-49	0.0047	0.0234	0.0197	0.0272	93,336	92,281	94,429	2,182	461,223	32.2	31.0	33.8	36	7,705
50-54	0.0064	0.0317	0.0266	0.0369	91,153	89,814	92,569	2,886	448,554	27.9	26.8	29.5	40	6,261
55-59	0.0095	0.0463	0.0385	0.0537	88,268	86,510	90,085	4,086	431,126	23.7	22.7	25.2	58	6,093
60-64	0.0150	0.0722	0.0596	0.0833	84,182	81,860	86,520	6,075	405,723	19.7	18.8	21.1	89	5,924
65-69	0.0220	0.1043	0.0860	0.1190	78,107	75,044	81,120	8,143	370,178	16.1	15.2	17.3	112	5,098
70-74	0.0347	0.1595	0.1326	0.1800	69,964	66,111	74,123	11,157	321,929	12.7	11.9	13.7	126	3,625
75-79	0.0575	0.2513	0.2133	0.2809	58,807	54,273	64,297	14,775	257,099	9.6	9.0	10.4	131	2,278
80-84	0.1012	0.4039	0.3565	0.4448	44,032	39,330	50,586	17,783	175,703	7.0	6.5	7.5	110	1,088
85-89	0.1671	0.5893	0.5451	0.6313	26,249	22,304	32,553	15,468	92,577	5.0	4.6	5.3	85	509
90-94	0.2585	0.7280	0.6925	0.7613	10,782	8,378	14,774	7,849	30,360	3.5	3.3	3.8	57	219
95-99	0.3750	0.8824	0.8605	0.9023	2,932	2,006	4,450	2,587	6,899	2.6	2.5	2.7	24	65
100+	0.5100	1	1	1	345	196	594	345	676	2.0	1.9	2.1	11	21
All ages													*1,118*	*210,390*

$\ln({}_nq_x)$ with uncertainty interval

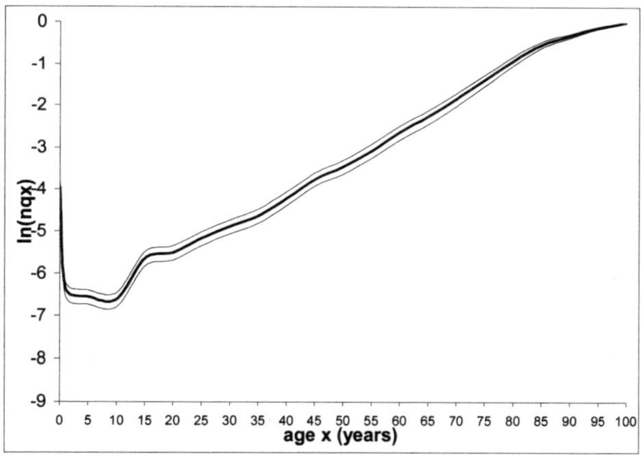

l_x with uncertainty interval

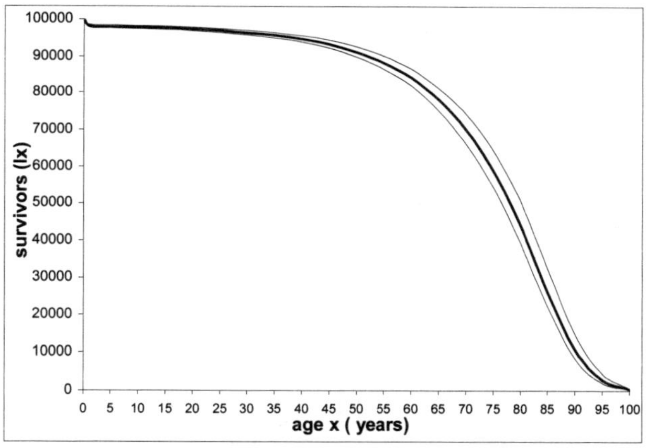

e_x with uncertainty interval

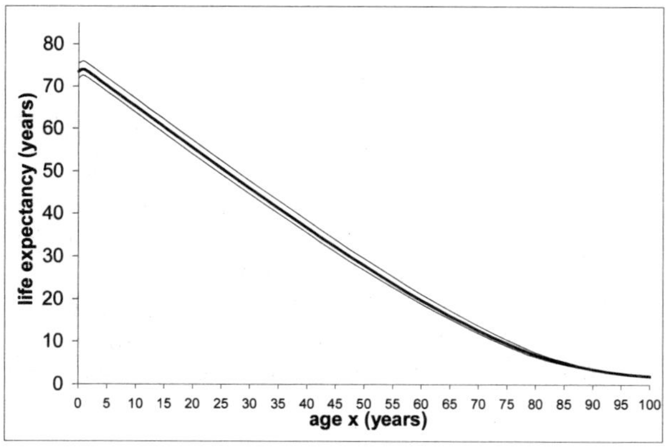

Distribution of deaths by age with uncertainty interval

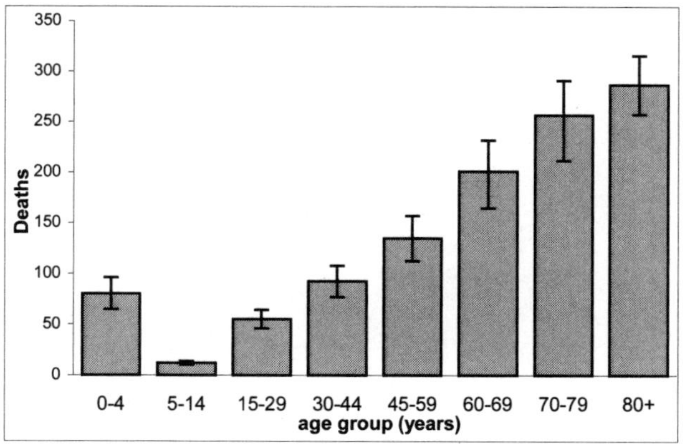

Swaziland **Swaziland** **Swazilandia**

Males, 2000 *Hommes, 2000* *Hombres, 2000*

Age	$_nM_x$	$_nq_x$			l_x			$_nd_x$	$_nL_x$	e_x			Actual deaths	Actual population
			percentiles			percentiles					percentiles			
		mean	2 1/2	97 1/2	mean	2 1/2	97 1/2			mean	2 1/2	97 1/2		
<1	0.1055	0.0963	0.0656	0.1286	100,000	100,000	100,000	9,634	91,329	44.7	38.8	51.2	1,581	14,987
1-4	0.0109	0.0426	0.0303	0.0571	90,366	87,140	93,435	3,850	352,224	48.5	42.8	54.2	599	54,833
5-9	0.0021	0.0105	0.0074	0.0144	86,516	83,016	89,997	911	430,304	46.6	41.3	52.1	136	64,356
10-14	0.0015	0.0073	0.0052	0.0101	85,605	81,932	89,292	627	426,459	42.1	36.9	47.5	86	58,387
15-19	0.0028	0.0141	0.0099	0.0194	84,978	81,153	88,817	1,201	421,888	37.4	32.2	42.8	141	49,452
20-24	0.0071	0.0350	0.0247	0.0482	83,777	79,667	87,910	2,931	411,558	32.9	27.8	38.3	319	44,816
25-29	0.0165	0.0791	0.0562	0.1090	80,846	75,952	85,741	6,396	388,241	29.0	24.0	34.4	597	36,229
30-34	0.0272	0.1271	0.0918	0.1736	74,450	67,675	80,920	9,465	348,588	26.2	21.5	31.6	788	29,017
35-39	0.0297	0.1382	0.1016	0.1865	64,985	55,928	73,493	8,984	302,464	24.7	20.1	29.8	709	23,878
40-44	0.0297	0.1380	0.0990	0.1879	56,001	45,505	66,026	7,730	260,679	23.2	19.0	27.9	579	19,524
45-49	0.0294	0.1368	0.0967	0.1888	48,271	37,174	59,195	6,605	224,840	21.6	17.8	25.7	465	15,814
50-54	0.0280	0.1308	0.0917	0.1815	41,665	30,547	53,000	5,451	194,700	19.6	16.4	23.2	356	12,729
55-59	0.0268	0.1257	0.0878	0.1743	36,214	25,424	47,599	4,551	169,695	17.2	14.4	20.3	270	10,070
60-64	0.0350	0.1607	0.1131	0.2201	31,663	21,389	42,860	5,090	145,592	14.3	11.9	17.0	280	7,999
65-69	0.0484	0.2157	0.1541	0.2888	26,574	17,139	37,277	5,732	118,539	11.5	9.6	13.8	290	5,997
70-74	0.0720	0.3051	0.2243	0.3946	20,842	12,620	30,730	6,359	88,314	9.0	7.5	10.9	300	4,164
75-79	0.1085	0.4267	0.3276	0.5270	14,483	7,760	23,707	6,180	56,968	6.9	5.7	8.3	264	2,438
80-84	0.1650	0.5840	0.4759	0.6830	8,304	3,671	15,940	4,849	29,396	5.1	4.3	6.2	193	1,167
85-89	0.2421	0.7540	0.6567	0.8344	3,455	1,163	8,354	2,605	10,761	3.8	3.3	4.5	80	332
90-94	0.3419	0.8439	0.7651	0.9053	850	193	2,868	717	2,097	2.8	2.5	3.2	20	58
95-99	0.4653	0.9710	0.9385	0.9887	133	18	674	129	277	2.1	2.1	2.3	2	5
100+	0.6093	1	1	1	4	0	41	4	6	1.6	1.6	1.8	0	0
All ages													*8,055*	*456,252*

$\ln(_nq_x)$ with uncertainty interval

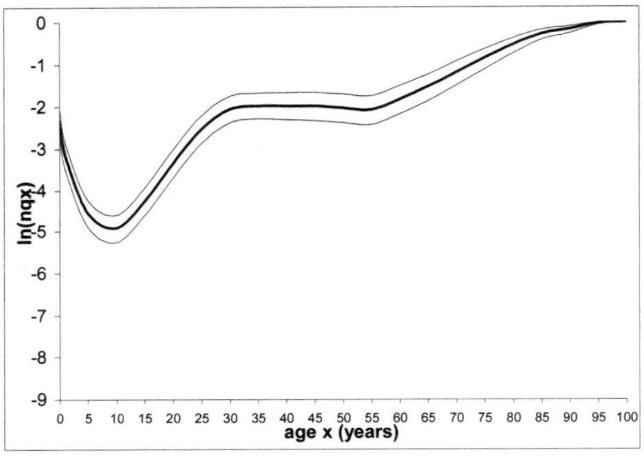

l_x with uncertainty interval

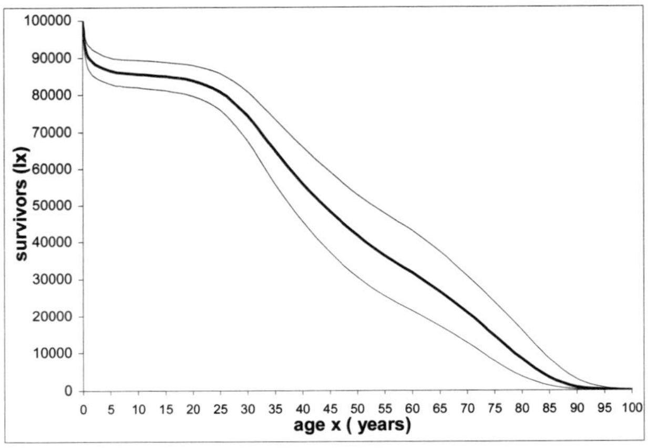

e_x with uncertainty interval

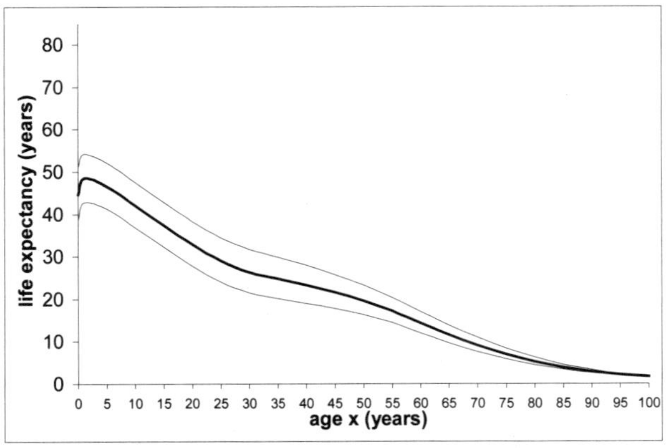

Distribution of deaths by age with uncertainty interval

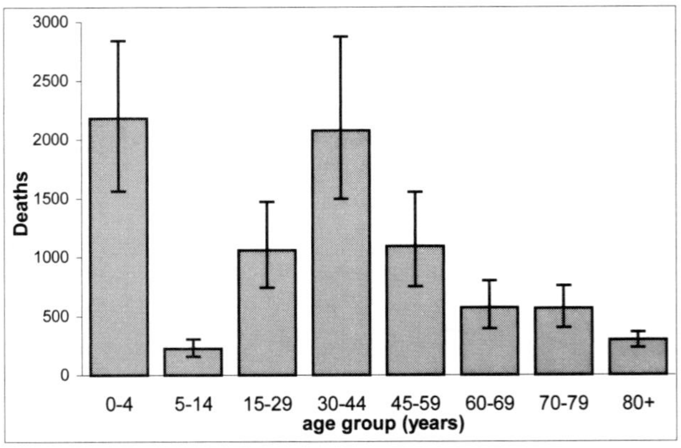

Swaziland — Swaziland — Swazilandia

Females, 2000 — *Femmes, 2000* — *Mujeres, 2000*

Age	$_nM_x$	$_nq_x$			l_x			$_nd_x$	$_nL_x$	e_x			Actual deaths	Actual population
			percentiles			percentiles					percentiles			
		mean	2 1/2	97 1/2	mean	2 1/2	97 1/2			mean	2 1/2	97 1/2		
<1	0.0904	0.0836	0.0584	0.1117	100,000	100,000	100,000	8,361	92,475	45.6	39.4	52.1	1,303	14,410
1-4	0.0111	0.0434	0.0315	0.0568	91,639	88,827	94,157	3,976	357,011	48.7	42.6	54.7	610	54,737
5-9	0.0020	0.0101	0.0073	0.0135	87,662	84,669	90,647	887	436,094	46.9	41.1	52.8	131	64,235
10-14	0.0014	0.0070	0.0051	0.0095	86,775	83,573	89,964	611	432,347	42.3	36.6	48.3	83	58,604
15-19	0.0050	0.0248	0.0178	0.0335	86,164	82,818	89,510	2,140	425,469	37.6	32.0	43.6	251	49,881
20-24	0.0142	0.0685	0.0495	0.0927	84,024	80,189	87,913	5,759	405,723	33.5	28.0	39.5	644	45,348
25-29	0.0251	0.1182	0.0863	0.1595	78,265	72,889	83,566	9,255	368,190	30.8	25.5	36.8	925	36,805
30-34	0.0306	0.1423	0.1057	0.1900	69,011	61,262	76,353	9,821	320,502	29.5	24.5	35.4	916	29,882
35-39	0.0263	0.1232	0.0887	0.1648	59,190	49,621	68,286	7,292	277,720	29.0	24.4	34.3	651	24,805
40-44	0.0221	0.1045	0.0736	0.1422	51,898	41,534	61,948	5,424	245,928	27.8	23.7	32.4	456	20,664
45-49	0.0179	0.0857	0.0596	0.1179	46,474	35,852	57,005	3,981	222,417	25.7	22.2	29.7	304	16,982
50-54	0.0172	0.0825	0.0570	0.1139	42,493	31,867	53,237	3,505	203,704	22.9	19.8	26.5	240	13,940
55-59	0.0181	0.0864	0.0596	0.1194	38,988	28,490	49,810	3,370	186,516	19.7	17.0	22.9	203	11,222
60-64	0.0238	0.1122	0.0775	0.1539	35,618	25,360	46,410	3,998	168,096	16.4	14.0	19.2	219	9,223
65-69	0.0354	0.1625	0.1133	0.2194	31,620	21,798	42,231	5,138	145,257	13.1	11.1	15.6	254	7,178
70-74	0.0569	0.2489	0.1774	0.3265	26,483	17,461	36,603	6,592	115,932	10.2	8.5	12.3	291	5,113
75-79	0.0903	0.3682	0.2724	0.4628	19,890	12,200	29,621	7,324	81,142	7.7	6.4	9.4	290	3,209
80-84	0.1405	0.5198	0.4070	0.6192	12,567	6,622	21,551	6,532	46,503	5.7	4.8	7.0	229	1,629
85-89	0.2111	0.6908	0.5803	0.7777	6,035	2,522	12,779	4,169	19,752	4.2	3.7	5.1	115	544
90-94	0.3059	0.7976	0.7045	0.8665	1,866	560	5,363	1,488	4,865	3.1	2.7	3.6	33	109
95-99	0.4277	0.9367	0.8826	0.9685	378	75	1,585	354	827	2.3	2.1	2.6	6	13
100+	0.5767	1	1	1	24	2	186	24	41	1.7	1.7	1.9	1	1
All ages													*8,152*	*468,534*

ln($_nq_x$) with uncertainty interval

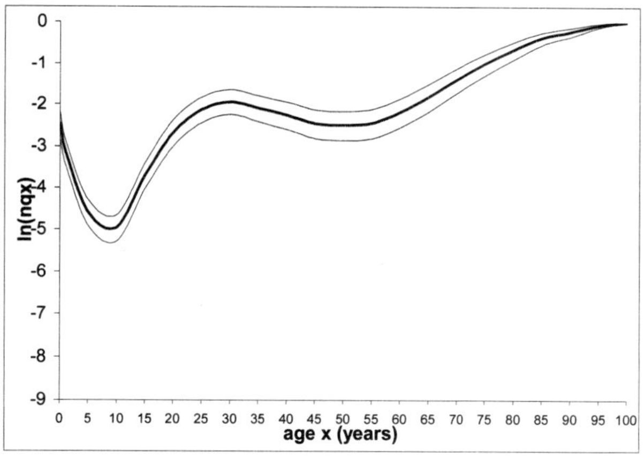

l_x with uncertainty interval

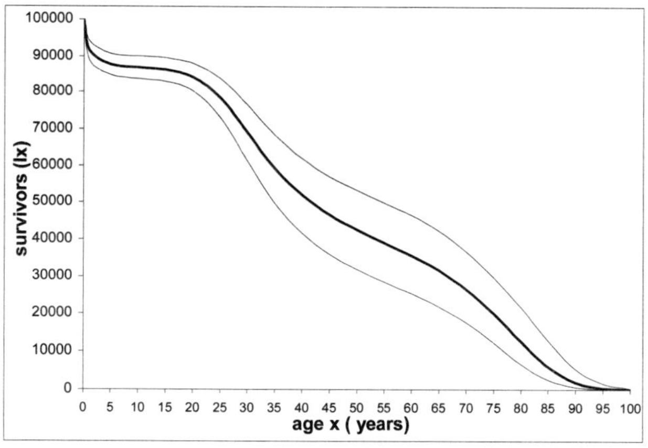

e_x with uncertainty interval

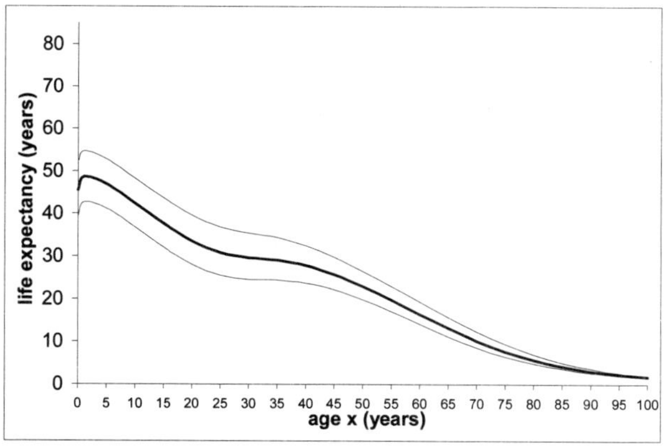

Distribution of deaths by age with uncertainty interval

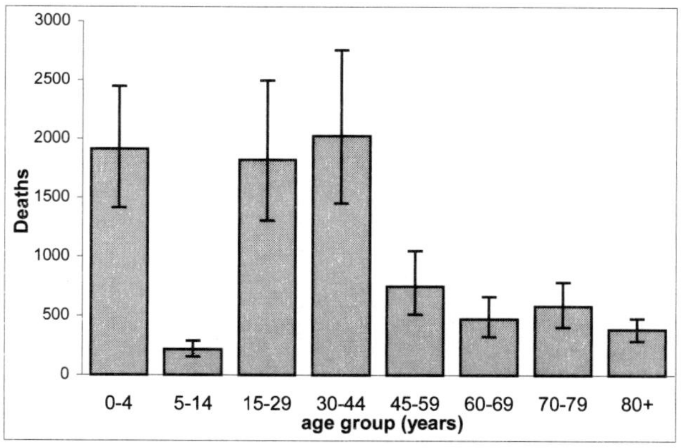

Sweden   Suède   Suecia

Males, 2000   *Hommes, 2000*   *Hombres, 2000*

Age	$_nM_x$	$_nq_x$ mean	$_nq_x$ percentiles 2½	$_nq_x$ percentiles 97½	l_x mean	l_x percentiles 2½	l_x percentiles 97½	$_nd_x$	$_nL_x$	e_x mean	e_x percentiles 2½	e_x percentiles 97½	Actual deaths	Actual population
<1	0.0040	0.0040	0.0035	0.0047	100,000	100,000	100,000	403	99,638	77.3	77.0	77.7	178	44,074
1-4	0.0001	0.0005	0.0005	0.0006	99,597	99,531	99,653	52	398,265	76.7	76.3	77.0	24	182,240
5-9	0.0001	0.0005	0.0004	0.0005	99,545	99,473	99,607	49	497,605	72.7	72.3	73.1	30	305,763
10-14	0.0001	0.0007	0.0006	0.0008	99,497	99,419	99,563	69	497,311	67.7	67.3	68.1	41	293,527
15-19	0.0005	0.0024	0.0021	0.0026	99,428	99,342	99,501	235	496,552	62.8	62.4	63.2	122	258,101
20-24	0.0008	0.0038	0.0035	0.0041	99,193	99,083	99,288	377	495,023	57.9	57.5	58.3	199	261,830
25-29	0.0007	0.0037	0.0034	0.0040	98,816	98,673	98,942	363	493,175	53.1	52.7	53.5	219	297,313
30-34	0.0008	0.0038	0.0035	0.0040	98,454	98,283	98,606	372	491,338	48.3	47.9	48.7	244	322,273
35-39	0.0010	0.0049	0.0046	0.0052	98,081	97,886	98,257	482	489,201	43.5	43.1	43.9	315	319,013
40-44	0.0015	0.0076	0.0072	0.0080	97,599	97,376	97,801	742	486,140	38.7	38.3	39.1	458	299,993
45-49	0.0026	0.0129	0.0124	0.0134	96,857	96,598	97,095	1,250	481,161	34.0	33.6	34.4	776	298,936
50-54	0.0040	0.0200	0.0193	0.0207	95,607	95,303	95,891	1,913	473,253	29.4	29.0	29.8	1,332	329,460
55-59	0.0063	0.0309	0.0298	0.0321	93,694	93,348	94,023	2,896	461,230	24.9	24.6	25.3	1,835	292,208
60-64	0.0109	0.0529	0.0508	0.0553	90,798	90,430	91,157	4,805	441,979	20.6	20.3	21.0	2,360	217,109
65-69	0.0185	0.0882	0.0841	0.0924	85,993	85,530	86,449	7,583	411,008	16.7	16.3	17.0	3,351	181,596
70-74	0.0318	0.1474	0.1400	0.1546	78,410	77,640	79,078	11,558	363,155	13.0	12.7	13.3	5,284	166,032
75-79	0.0542	0.2386	0.2268	0.2494	66,852	65,636	67,991	15,948	294,389	9.8	9.6	10.1	7,988	147,451
80-84	0.0935	0.3790	0.3620	0.3945	50,904	49,278	52,571	19,294	206,284	7.1	6.9	7.4	8,746	93,505
85-89	0.1639	0.5814	0.5624	0.5988	31,610	29,948	33,472	18,379	112,102	5.0	4.8	5.1	7,699	46,958
90-94	0.2635	0.7358	0.7204	0.7498	13,231	12,105	14,567	9,736	36,947	3.4	3.3	3.5	3,876	14,708
95-99	0.4198	0.9290	0.9218	0.9352	3,495	3,039	4,066	3,247	7,735	2.4	2.3	2.4	983	2,343
100+	0.5128	1	1	1	248	197	316	248	484	2.0	1.7	2.0	96	187
All ages													*46,154*	*4,374,620*

ln($_nq_x$) with uncertainty interval

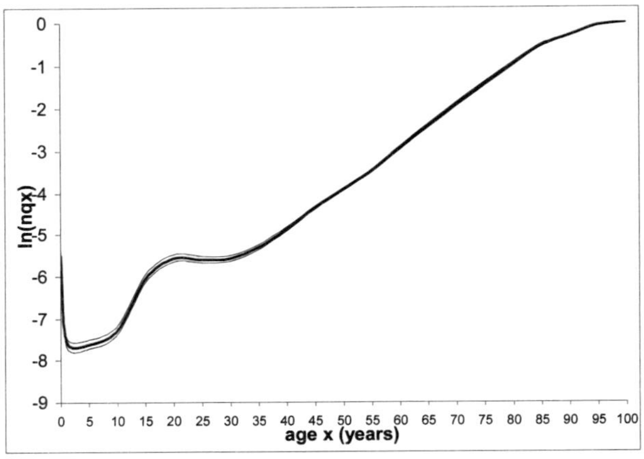

l_x with uncertainty interval

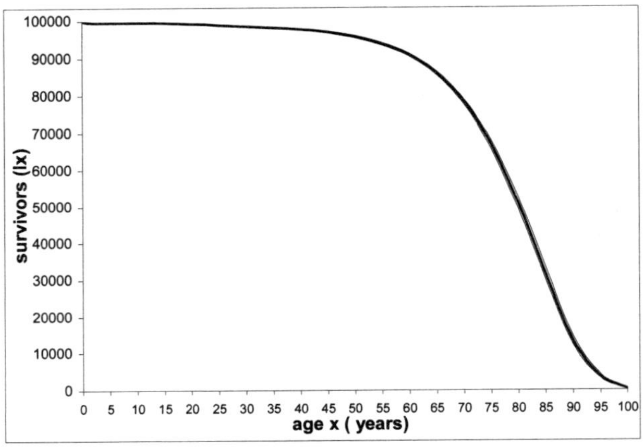

e_x with uncertainty interval

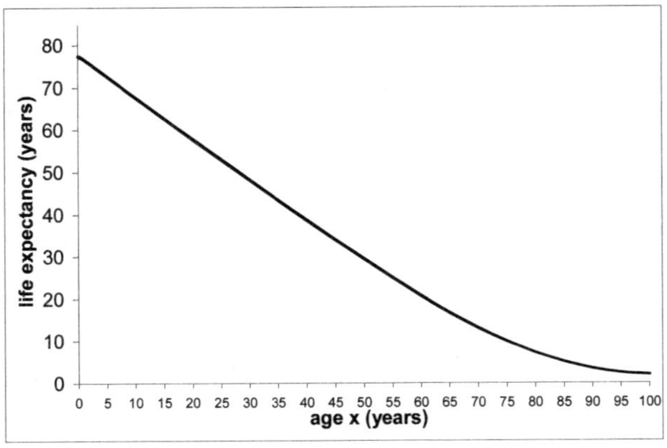

Distribution of deaths by age with uncertainty interval

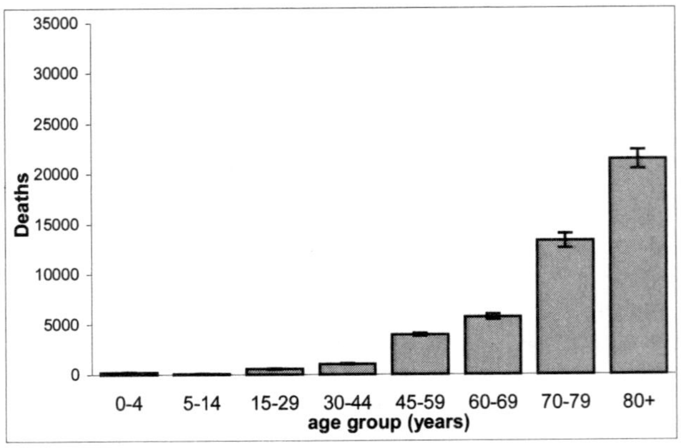

Sweden / Suède / Suecia

Females, 2000 / *Femmes, 2000* / *Mujeres, 2000*

Age	$_nM_x$	$_nq_x$			l_x			$_nd_x$	$_nL_x$	e_x			Actual deaths	Actual population
			percentiles			percentiles					percentiles			
		mean	2½	97½	mean	2½	97½			mean	2½	97½		
<1	0.0028	0.0028	0.0024	0.0032	100,000	100,000	100,000	282	99,746	82.0	81.6	82.4	118	41,796
1-4	0.0001	0.0005	0.0004	0.0005	99,718	99,682	99,756	46	398,762	81.2	80.9	81.6	20	172,079
5-9	0.0001	0.0005	0.0004	0.0005	99,672	99,631	99,715	46	498,244	77.2	76.9	77.6	27	290,566
10-14	0.0001	0.0006	0.0006	0.0007	99,626	99,580	99,673	64	497,969	72.3	71.9	72.7	36	278,522
15-19	0.0002	0.0012	0.0010	0.0013	99,562	99,511	99,615	115	497,521	67.3	67.0	67.7	57	244,460
20-24	0.0002	0.0012	0.0011	0.0013	99,447	99,386	99,511	123	496,925	62.4	62.1	62.8	62	249,364
25-29	0.0003	0.0013	0.0012	0.0014	99,323	99,253	99,399	128	496,299	57.5	57.1	57.9	74	286,864
30-34	0.0003	0.0014	0.0013	0.0015	99,196	99,116	99,282	139	495,632	52.5	52.2	52.9	86	307,848
35-39	0.0007	0.0033	0.0030	0.0035	99,057	98,968	99,153	324	494,473	47.6	47.3	48.0	199	302,923
40-44	0.0009	0.0047	0.0044	0.0049	98,732	98,623	98,852	461	492,510	42.7	42.4	43.1	269	287,663
45-49	0.0017	0.0085	0.0080	0.0089	98,271	98,138	98,420	831	489,280	37.9	37.6	38.3	492	289,840
50-54	0.0027	0.0135	0.0129	0.0141	97,441	97,268	97,632	1,319	483,905	33.2	32.9	33.6	878	322,101
55-59	0.0045	0.0222	0.0213	0.0232	96,121	95,899	96,365	2,137	475,265	28.7	28.4	29.0	1,277	284,042
60-64	0.0068	0.0336	0.0321	0.0351	93,985	93,703	94,284	3,159	462,026	24.3	24.0	24.6	1,513	221,241
65-69	0.0105	0.0510	0.0485	0.0533	90,826	90,447	91,218	4,635	442,543	20.0	19.8	20.3	2,093	199,857
70-74	0.0175	0.0839	0.0796	0.0875	86,191	85,628	86,749	7,235	412,869	16.0	15.7	16.3	3,492	199,248
75-79	0.0318	0.1473	0.1398	0.1529	78,956	78,135	79,808	11,631	365,703	12.2	12.0	12.5	6,277	197,370
80-84	0.0615	0.2667	0.2546	0.2755	67,325	66,187	68,653	17,956	291,736	8.9	8.7	9.1	9,113	148,058
85-89	0.1154	0.4479	0.4333	0.4620	49,369	48,008	51,173	22,114	191,561	6.2	6.0	6.3	10,858	94,055
90-94	0.2039	0.6325	0.6190	0.6472	27,255	26,062	29,002	17,240	84,557	4.1	4.0	4.2	8,073	39,597
95-99	0.3350	0.8354	0.8263	0.8451	10,015	9,250	10,997	8,367	24,977	2.8	2.8	2.9	3,030	9,046
100+	0.4776	1	1	1	1,649	1,433	1,881	1,649	3,452	2.1	2.1	2.1	446	934
All ages													*48,489*	*4,467,474*

ln($_nq_x$) with uncertainty interval

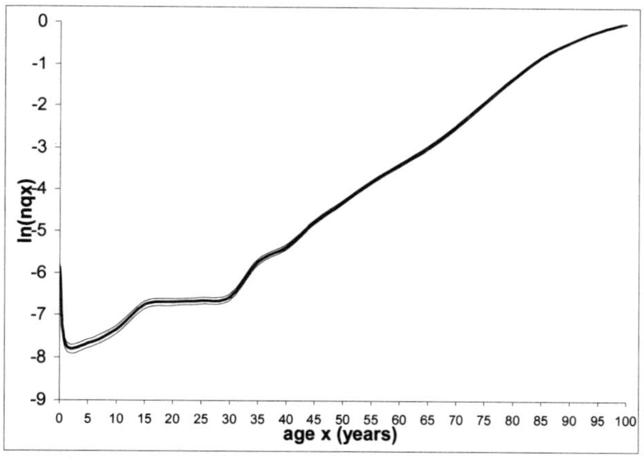

l_x with uncertainty interval

e_x with uncertainty interval

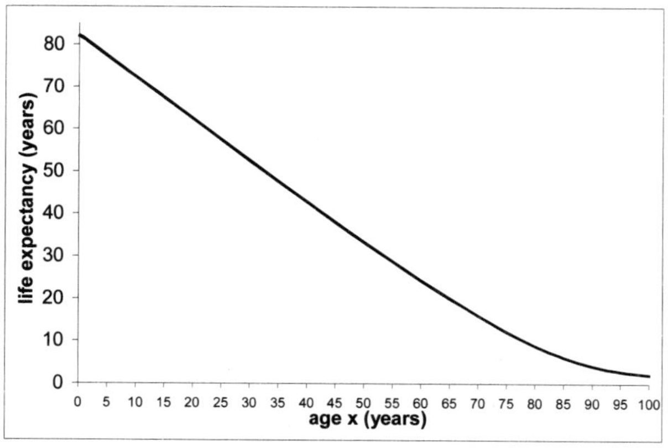

Distribution of deaths by age with uncertainty interval

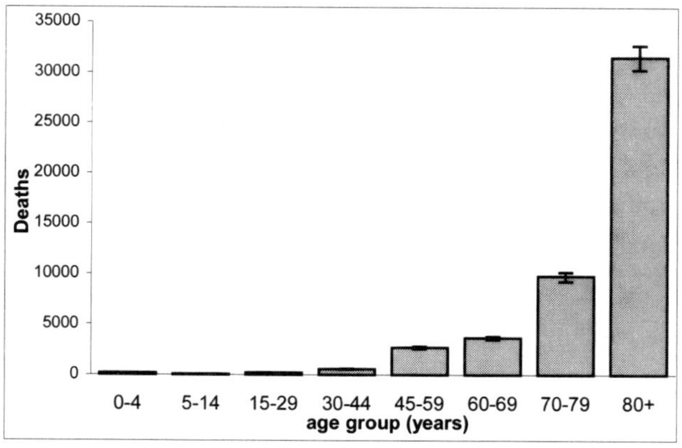

Switzerland | Suisse | Suiza

Males, 2000 | *Hommes, 2000* | *Hombres, 2000*

Age	$_nM_x$	$_nq_x$ mean	$_nq_x$ percentiles 2½	$_nq_x$ percentiles 97½	l_x mean	l_x percentiles 2½	l_x percentiles 97½	$_nd_x$	$_nL_x$	e_x mean	e_x percentiles 2½	e_x percentiles 97½	Actual deaths	Actual population
<1	0.0050	0.0050	0.0046	0.0054	100,000	100,000	100,000	498	99,552	76.7	76.3	77.1	181	36,263
1-4	0.0003	0.0011	0.0010	0.0012	99,502	99,458	99,545	107	397,751	76.1	75.7	76.5	41	151,121
5-9	0.0001	0.0006	0.0006	0.0007	99,395	99,345	99,445	65	496,813	72.2	71.8	72.6	28	213,020
10-14	0.0002	0.0010	0.0009	0.0011	99,330	99,276	99,385	99	496,403	67.2	66.8	67.6	42	210,816
15-19	0.0006	0.0030	0.0029	0.0032	99,231	99,171	99,292	302	495,399	62.3	61.9	62.7	123	201,246
20-24	0.0011	0.0054	0.0051	0.0057	98,929	98,850	99,008	538	493,300	57.4	57.0	57.8	214	196,199
25-29	0.0011	0.0053	0.0050	0.0056	98,391	98,283	98,500	520	490,655	52.7	52.3	53.2	235	221,907
30-34	0.0011	0.0056	0.0053	0.0058	97,871	97,737	98,005	547	487,989	48.0	47.6	48.4	315	281,545
35-39	0.0011	0.0055	0.0053	0.0058	97,324	97,166	97,483	539	485,276	43.3	42.9	43.7	353	317,990
40-44	0.0019	0.0093	0.0089	0.0096	96,786	96,605	96,967	896	481,689	38.5	38.1	38.9	552	296,694
45-49	0.0027	0.0136	0.0131	0.0141	95,890	95,675	96,105	1,300	476,199	33.8	33.4	34.2	727	266,306
50-54	0.0043	0.0215	0.0207	0.0223	94,590	94,329	94,848	2,031	467,873	29.3	28.9	29.7	1,137	261,993
55-59	0.0070	0.0344	0.0332	0.0358	92,559	92,238	92,873	3,184	454,837	24.8	24.5	25.2	1,653	236,173
60-64	0.0120	0.0583	0.0559	0.0608	89,375	88,979	89,763	5,206	433,861	20.6	20.3	21.0	2,224	185,304
65-69	0.0180	0.0861	0.0821	0.0902	84,169	83,646	84,680	7,249	402,723	16.8	16.5	17.1	2,867	159,271
70-74	0.0320	0.1481	0.1409	0.1551	76,920	76,121	77,626	11,396	356,111	13.1	12.8	13.4	3,984	124,497
75-79	0.0540	0.2379	0.2266	0.2483	65,524	64,314	66,670	15,587	288,654	10.0	9.7	10.2	5,007	92,720
80-84	0.0950	0.3838	0.3687	0.3984	49,937	48,360	51,561	19,168	201,766	7.3	7.1	7.5	5,200	54,740
85-89	0.1600	0.5714	0.5554	0.5863	30,769	29,183	32,551	17,582	109,891	5.3	5.1	5.4	4,408	27,551
90-94	0.2250	0.6716	0.6586	0.6835	13,187	12,105	14,441	8,857	39,364	4.0	3.9	4.1	2,080	9,246
95-99	0.3210	0.8176	0.8084	0.8259	4,330	3,831	4,909	3,540	11,029	2.9	2.9	3.0	548	1,707
100+	0.4640	1	1	1	790	667	940	790	1,702	2.2	2.1	2.3	83	179
All ages													*32,002*	*3,546,488*

$ln(_nq_x)$ with uncertainty interval

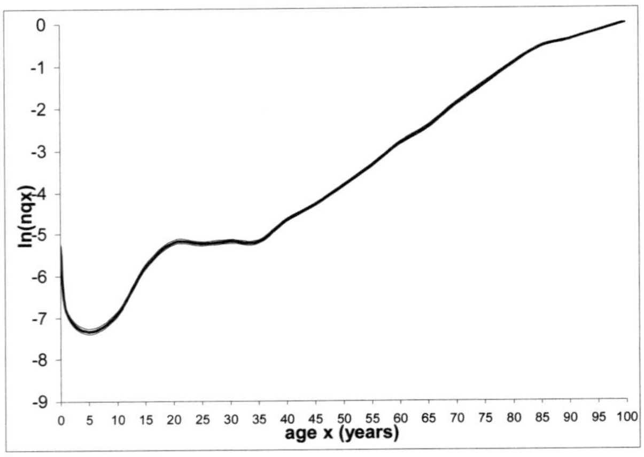

l_x with uncertainty interval

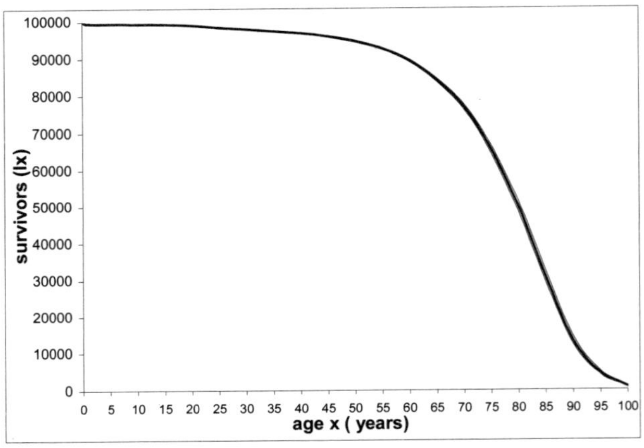

e_x with uncertainty interval

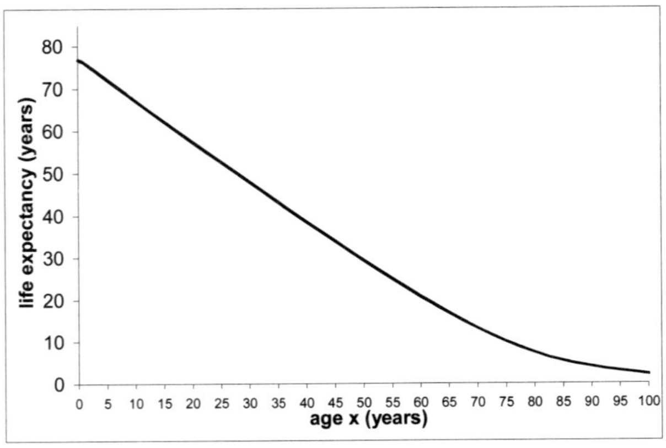

Distribution of deaths by age with uncertainty interval

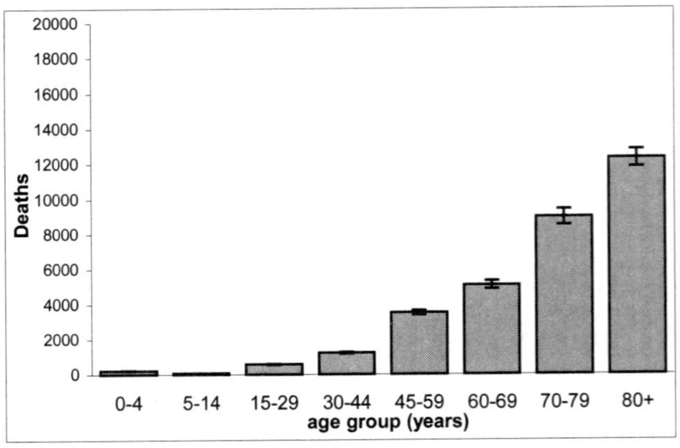

Switzerland **Suisse** **Suiza**

Females, 2000 *Femmes, 2000* *Mujeres, 2000*

Age	$_nM_x$	$_nq_x$ mean	$_nq_x$ 2½	$_nq_x$ 97½	l_x mean	l_x 2½	l_x 97½	$_nd_x$	$_nL_x$	e_x mean	e_x 2½	e_x 97½	Actual deaths	Actual population
<1	0.0050	0.0050	0.0044	0.0056	100,000	100,000	100,000	498	99,552	82.5	82.1	83.0	173	34,503
1-4	0.0002	0.0009	0.0008	0.0010	99,502	99,437	99,559	91	397,789	81.9	81.6	82.4	33	143,894
5-9	0.0001	0.0005	0.0005	0.0005	99,411	99,337	99,475	50	496,930	78.0	77.6	78.5	20	203,123
10-14	0.0001	0.0006	0.0006	0.0007	99,361	99,283	99,430	65	496,644	73.0	72.7	73.5	26	201,468
15-19	0.0002	0.0011	0.0011	0.0012	99,296	99,212	99,370	114	496,197	68.1	67.7	68.6	44	191,620
20-24	0.0004	0.0020	0.0019	0.0022	99,182	99,089	99,265	203	495,404	63.1	62.8	63.6	77	187,422
25-29	0.0004	0.0019	0.0018	0.0021	98,979	98,870	99,077	193	494,414	58.3	57.9	58.8	82	210,319
30-34	0.0005	0.0027	0.0025	0.0029	98,786	98,663	98,897	266	493,266	53.4	53.0	53.9	143	265,372
35-39	0.0007	0.0035	0.0033	0.0037	98,520	98,380	98,647	344	491,740	48.5	48.2	49.0	208	296,463
40-44	0.0012	0.0057	0.0054	0.0060	98,176	98,015	98,322	563	489,472	43.7	43.3	44.2	317	275,650
45-49	0.0017	0.0087	0.0083	0.0091	97,613	97,424	97,786	846	485,951	38.9	38.6	39.4	442	253,953
50-54	0.0028	0.0141	0.0135	0.0147	96,767	96,544	96,975	1,364	480,426	34.2	33.9	34.7	724	254,845
55-59	0.0040	0.0198	0.0189	0.0206	95,403	95,139	95,658	1,889	472,292	29.7	29.4	30.1	935	233,841
60-64	0.0060	0.0296	0.0280	0.0308	93,514	93,202	93,818	2,764	460,659	25.2	24.9	25.7	1,166	194,277
65-69	0.0090	0.0440	0.0415	0.0460	90,750	90,377	91,133	3,994	443,765	20.9	20.6	21.3	1,614	179,374
70-74	0.0150	0.0723	0.0680	0.0756	86,756	86,245	87,325	6,272	418,101	16.8	16.5	17.2	2,425	161,682
75-79	0.0290	0.1352	0.1272	0.1412	80,484	79,751	81,386	10,881	375,219	12.9	12.7	13.2	4,083	140,788
80-84	0.0560	0.2456	0.2324	0.2550	69,603	68,491	71,033	17,096	305,277	9.5	9.3	9.8	5,399	96,404
85-89	0.1040	0.4127	0.3959	0.4266	52,508	51,024	54,523	21,670	208,364	6.8	6.6	7.0	6,566	63,137
90-94	0.1680	0.5585	0.5425	0.5733	30,838	29,393	32,936	17,223	102,519	4.8	4.7	5.0	4,771	28,399
95-99	0.2690	0.7443	0.7313	0.7562	13,615	12,632	15,018	10,134	37,672	3.4	3.3	3.5	1,777	6,605
100+	0.4320	1	1	1	3,481	3,089	3,983	3,481	8,058	2.3	2.3	2.5	337	780
All ages													*31,362*	*3,623,919*

ln($_nq_x$) with uncertainty interval

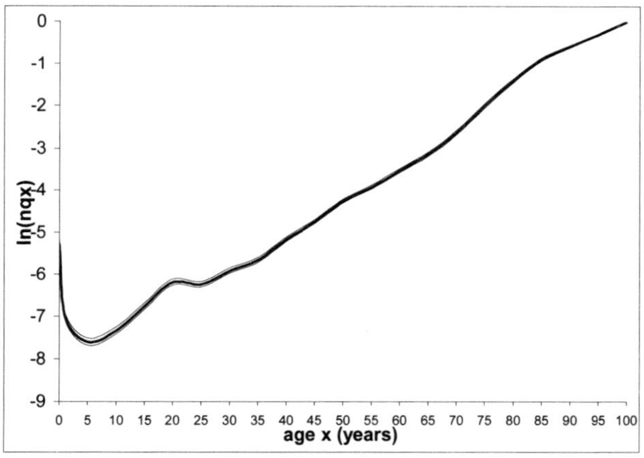

l_x with uncertainty interval

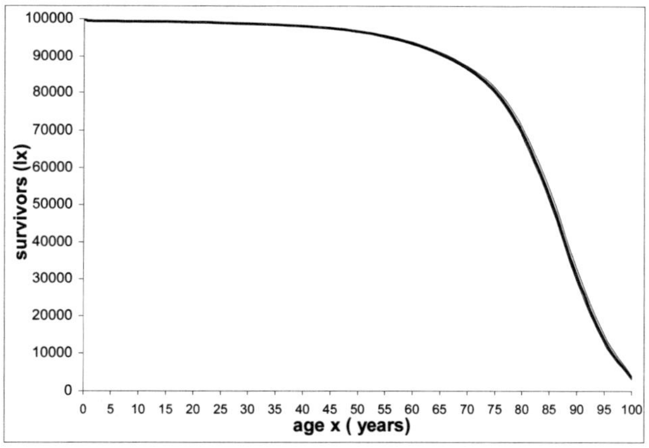

e_x with uncertainty interval

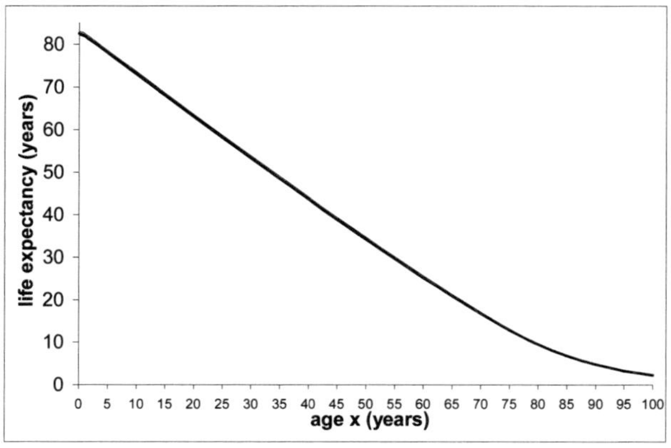

Distribution of deaths by age with uncertainty interval

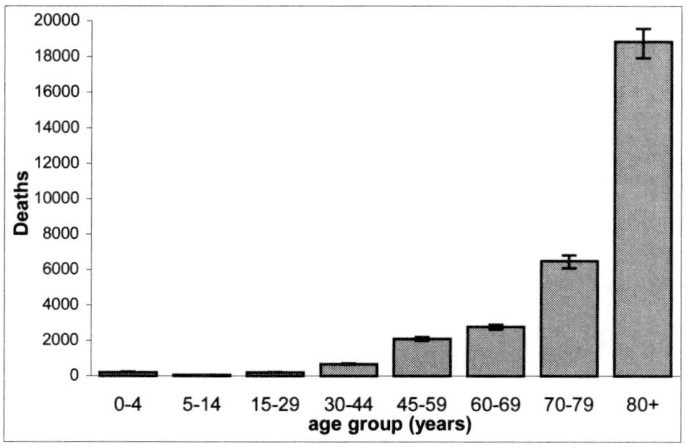

Syrian Arab Republic   République arabe syrienne   República Arabe Siria

Males, 2000   *Hommes, 2000*   *Hombres, 2000*

Age	$_nM_x$	$_nq_x$ mean	$_nq_x$ percentiles 2½	$_nq_x$ percentiles 97½	l_x mean	l_x percentiles 2½	l_x percentiles 97½	$_nd_x$	$_nL_x$	e_x mean	e_x percentiles 2½	e_x percentiles 97½	Actual deaths	Actual population
<1	0.0270	0.0263	0.0229	0.0297	100,000	100,000	100,000	2,633	97,631	69.3	68.4	70.2	6,466	239,811
1-4	0.0015	0.0058	0.0052	0.0064	97,367	97,027	97,710	568	388,107	70.2	69.4	71.1	1,325	905,836
5-9	0.0006	0.0029	0.0026	0.0032	96,800	96,403	97,198	281	483,296	66.6	65.8	67.5	625	1,074,564
10-14	0.0005	0.0026	0.0023	0.0028	96,519	96,095	96,943	249	481,971	61.8	61.0	62.6	595	1,153,863
15-19	0.0012	0.0060	0.0054	0.0066	96,270	95,824	96,717	574	479,915	56.9	56.2	57.8	1,200	1,004,071
20-24	0.0015	0.0077	0.0070	0.0085	95,696	95,199	96,195	738	476,634	52.2	51.5	53.1	1,266	816,956
25-29	0.0015	0.0074	0.0067	0.0081	94,958	94,398	95,521	706	473,023	47.6	47.0	48.5	996	667,439
30-34	0.0017	0.0083	0.0075	0.0090	94,252	93,632	94,876	779	469,310	43.0	42.3	43.8	963	580,079
35-39	0.0022	0.0107	0.0098	0.0116	93,472	92,788	94,161	1,001	464,860	38.3	37.7	39.1	1,027	476,881
40-44	0.0032	0.0157	0.0144	0.0170	92,471	91,707	93,239	1,454	458,721	33.7	33.1	34.5	1,087	342,809
45-49	0.0050	0.0247	0.0227	0.0267	91,017	90,145	91,893	2,248	449,465	29.2	28.6	29.9	1,234	246,659
50-54	0.0081	0.0395	0.0364	0.0427	88,769	87,747	89,808	3,510	435,070	24.9	24.3	25.6	1,542	191,161
55-59	0.0130	0.0628	0.0576	0.0677	85,259	84,034	86,523	5,356	412,906	20.8	20.3	21.4	1,786	137,662
60-64	0.0207	0.0984	0.0901	0.1054	79,903	78,437	81,447	7,860	379,866	17.0	16.6	17.6	2,523	121,906
65-69	0.0327	0.1513	0.1390	0.1608	72,043	70,203	73,884	10,897	332,972	13.6	13.2	14.1	3,281	100,268
70-74	0.0524	0.2315	0.2146	0.2445	61,146	58,917	63,614	14,157	270,337	10.6	10.2	11.0	3,896	74,405
75-79	0.0836	0.3458	0.3254	0.3652	46,989	44,588	49,962	16,247	194,327	8.0	7.7	8.3	3,268	39,087
80-84	0.1337	0.5011	0.4791	0.5242	30,742	28,627	33,705	15,403	115,201	5.9	5.7	6.1	2,592	19,388
85-89	0.2050	0.6777	0.6579	0.6987	15,338	13,848	17,495	10,395	50,704	4.3	4.2	4.5	1,247	6,082
90-94	0.3014	0.7914	0.7760	0.8075	4,943	4,172	5,902	3,912	12,980	3.1	3.0	3.2	391	1,299
95-99	0.4248	0.9338	0.9259	0.9418	1,031	803	1,302	963	2,267	2.3	2.3	2.4	76	178
100+	0.5740	1	1	1	68	47	96	68	119	1.7	1.7	1.8	10	18
All ages													*37,398*	*8,200,422*

ln($_nq_x$) with uncertainty interval

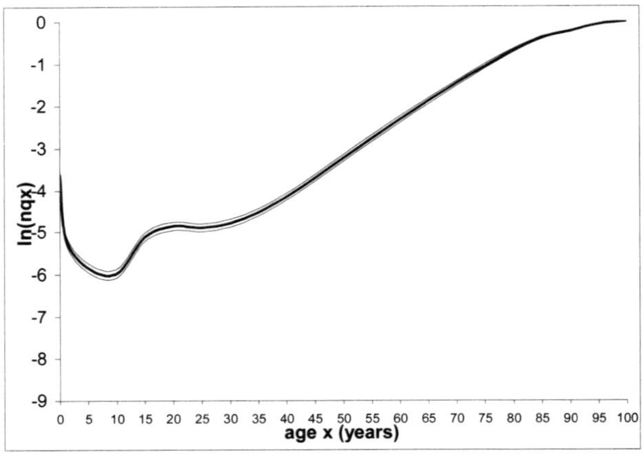

l_x with uncertainty interval

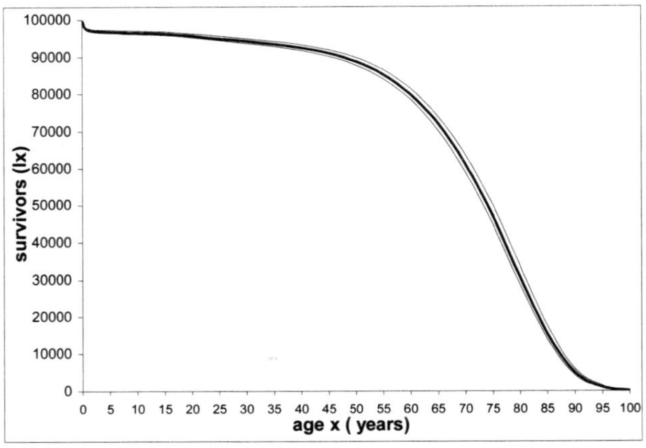

e_x with uncertainty interval

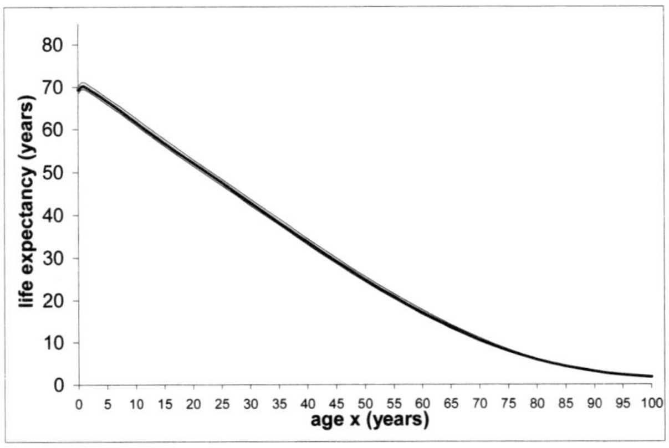

Distribution of deaths by age with uncertainty interval

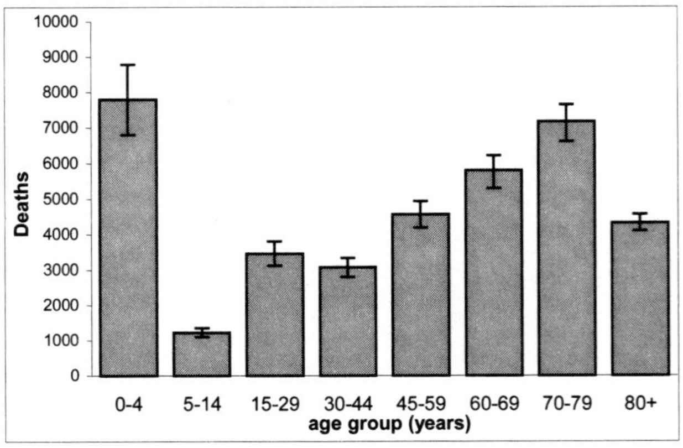

Syrian Arab Republic | République arabe syrienne | República Arabe Siria

Females, 2000 | *Femmes, 2000* | *Mujeres, 2000*

Age	$_nM_x$	$_nq_x$			l_x			$_nd_x$	$_nL_x$	e_x			Actual deaths	Actual population
			percentiles			percentiles					percentiles			
		mean	2½	97½	mean	2½	97½			mean	2½	97½		
<1	0.0226	0.0221	0.0188	0.0253	100,000	100,000	100,000	2,213	98,009	72.4	71.7	73.2	5,136	227,525
1-4	0.0015	0.0060	0.0053	0.0067	97,787	97,473	98,117	587	389,741	73.1	72.5	73.8	1,304	865,811
5-9	0.0005	0.0026	0.0023	0.0029	97,201	96,819	97,599	253	485,372	69.5	68.9	70.2	536	1,029,405
10-14	0.0004	0.0021	0.0019	0.0024	96,948	96,539	97,374	208	484,219	64.7	64.1	65.3	480	1,114,973
15-19	0.0007	0.0037	0.0033	0.0041	96,740	96,309	97,189	355	482,812	59.8	59.3	60.5	717	976,256
20-24	0.0009	0.0045	0.0040	0.0050	96,385	95,917	96,872	435	480,839	55.0	54.5	55.7	718	793,828
25-29	0.0010	0.0052	0.0047	0.0058	95,950	95,438	96,483	502	478,496	50.3	49.7	50.9	679	646,601
30-34	0.0014	0.0067	0.0061	0.0074	95,448	94,887	96,030	644	475,630	45.5	45.0	46.1	740	546,678
35-39	0.0019	0.0095	0.0086	0.0105	94,804	94,183	95,447	904	471,759	40.8	40.3	41.4	870	453,919
40-44	0.0028	0.0138	0.0126	0.0151	93,900	93,198	94,624	1,298	466,253	36.2	35.7	36.8	945	339,233
45-49	0.0042	0.0205	0.0189	0.0223	92,601	91,793	93,432	1,903	458,250	31.6	31.2	32.2	1,027	247,366
50-54	0.0062	0.0305	0.0283	0.0328	90,699	89,748	91,670	2,769	446,571	27.3	26.8	27.8	1,220	196,811
55-59	0.0092	0.0450	0.0421	0.0481	87,930	86,801	89,074	3,960	429,750	23.0	22.7	23.6	1,373	148,965
60-64	0.0145	0.0698	0.0651	0.0740	83,970	82,633	85,309	5,861	405,198	19.0	18.7	19.5	1,944	134,371
65-69	0.0234	0.1103	0.1029	0.1164	78,109	76,549	79,641	8,617	369,003	15.2	14.9	15.7	2,582	110,588
70-74	0.0397	0.1807	0.1692	0.1892	69,492	67,749	71,142	12,555	316,074	11.8	11.5	12.2	3,244	81,681
75-79	0.0676	0.2890	0.2734	0.3020	56,937	55,088	59,064	16,455	243,549	8.9	8.6	9.2	2,803	41,493
80-84	0.1142	0.4441	0.4257	0.4622	40,482	38,603	42,915	17,978	157,467	6.5	6.3	6.7	2,667	23,360
85-89	0.1833	0.6284	0.6087	0.6467	22,505	21,088	24,601	14,141	77,169	4.7	4.5	4.8	1,358	7,408
90-94	0.2794	0.7600	0.7439	0.7744	8,363	7,482	9,490	6,356	22,748	3.3	3.2	3.4	500	1,790
95-99	0.4047	0.9139	0.9049	0.9217	2,007	1,688	2,385	1,834	4,533	2.4	2.4	2.5	98	241
100+	0.5567	1	1	1	173	132	227	173	310	1.8	1.8	1.9	19	35
All ages													*30,959*	*7,988,338*

ln($_nq_x$) with uncertainty interval

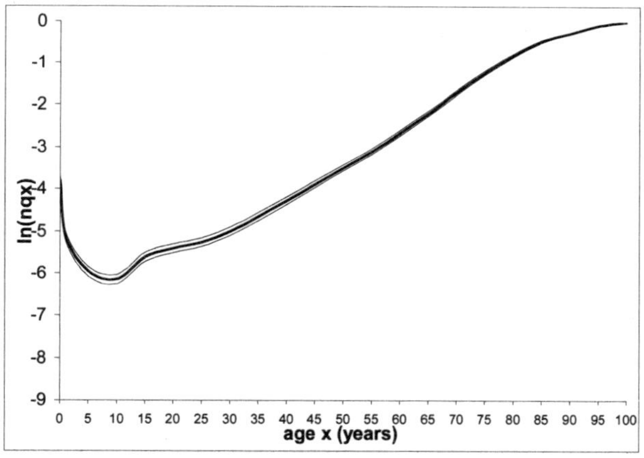

l_x with uncertainty interval

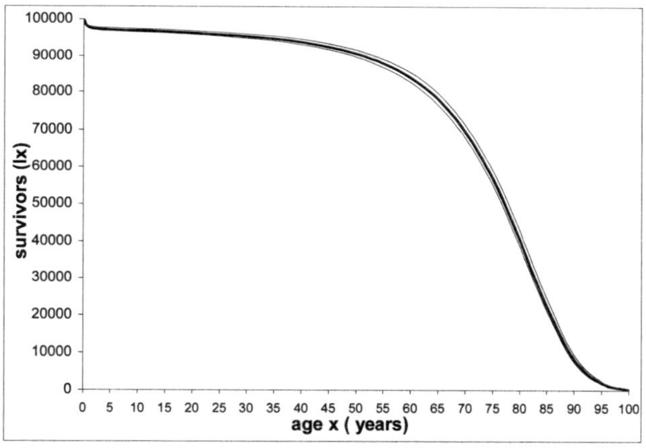

e_x with uncertainty interval

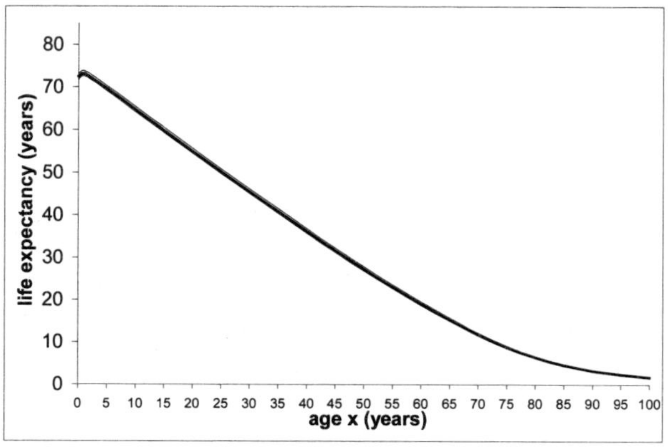

Distribution of deaths by age with uncertainty interval

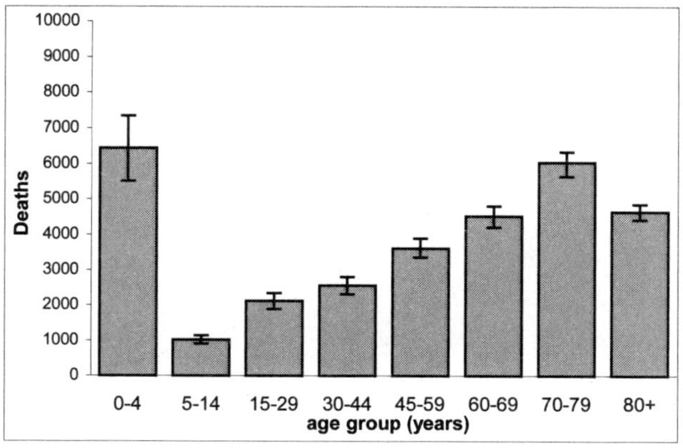

Tajikistan | Tadjikistan | Tayikistán

Males, 2000 | *Hommes, 2000* | *Hombres, 2000*

Age	$_nM_x$	$_nq_x$			l_x			$_nd_x$	$_nL_x$	e_x			Actual deaths	Actual population
			percentiles			percentiles					percentiles			
		mean	2½	97½	mean	2½	97½			mean	2½	97½		
<1	0.0549	0.0523	0.0457	0.0609	100,000	100,000	100,000	5,229	95,294	60.4	58.9	62.3	4,529	82,538
1-4	0.0089	0.0348	0.0309	0.0390	94,771	93,913	95,435	3,296	371,175	62.7	61.6	64.4	2,753	310,091
5-9	0.0009	0.0043	0.0038	0.0048	91,476	90,253	92,482	396	456,389	60.9	59.9	62.4	350	403,455
10-14	0.0007	0.0035	0.0031	0.0039	91,080	89,817	92,127	316	454,611	56.2	55.1	57.6	291	418,836
15-19	0.0013	0.0065	0.0058	0.0073	90,764	89,470	91,845	591	452,344	51.4	50.3	52.8	439	336,013
20-24	0.0028	0.0139	0.0123	0.0155	90,173	88,821	91,315	1,254	447,730	46.7	45.7	48.1	763	272,499
25-29	0.0037	0.0181	0.0160	0.0201	88,919	87,449	90,190	1,611	440,566	42.3	41.3	43.7	891	243,588
30-34	0.0049	0.0240	0.0213	0.0265	87,307	85,695	88,744	2,100	431,288	38.1	37.1	39.4	988	203,013
35-39	0.0058	0.0286	0.0253	0.0313	85,208	83,425	86,856	2,435	419,951	33.9	33.0	35.2	1,117	192,685
40-44	0.0075	0.0370	0.0327	0.0403	82,773	80,811	84,660	3,060	406,212	29.9	28.9	31.1	1,169	155,172
45-49	0.0100	0.0488	0.0433	0.0528	79,712	77,554	81,889	3,888	388,841	25.9	25.0	27.1	1,078	107,849
50-54	0.0136	0.0658	0.0585	0.0716	75,824	73,458	78,345	4,986	366,653	22.1	21.2	23.2	952	69,982
55-59	0.0198	0.0944	0.0841	0.1031	70,837	68,201	73,759	6,686	337,472	18.5	17.7	19.5	1,067	53,859
60-64	0.0309	0.1436	0.1281	0.1570	64,151	61,171	67,555	9,210	297,732	15.2	14.4	16.1	2,035	65,770
65-69	0.0442	0.1991	0.1781	0.2183	54,941	51,567	58,898	10,936	247,366	12.3	11.6	13.1	2,097	47,422
70-74	0.0677	0.2897	0.2609	0.3162	44,005	40,308	48,410	12,746	188,159	9.7	9.1	10.4	2,669	39,397
75-79	0.0983	0.3946	0.3604	0.4257	31,259	27,564	35,782	12,334	125,457	7.7	7.2	8.2	1,568	15,946
80-84	0.1368	0.5097	0.4745	0.5415	18,924	15,831	22,885	9,646	70,505	6.0	5.7	6.4	936	6,844
85-89	0.1892	0.6422	0.6090	0.6712	9,278	7,259	12,027	5,959	31,494	4.7	4.4	4.9	805	4,255
90-94	0.2600	0.7304	0.7015	0.7552	3,319	2,387	4,703	2,424	9,324	3.5	3.4	3.7	425	1,634
95-99	0.3551	0.8597	0.8382	0.8776	895	584	1,404	769	2,167	2.7	2.6	2.8	207	584
100+	0.4819	1	1	1	126	71	227	126	261	2.1	1.9	2.1	54	112
All ages													*27,183*	*3,031,544*

$\ln(_nq_x)$ with uncertainty interval

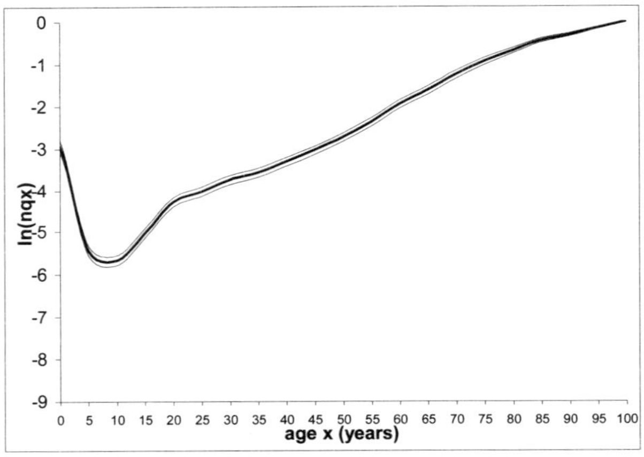

l_x with uncertainty interval

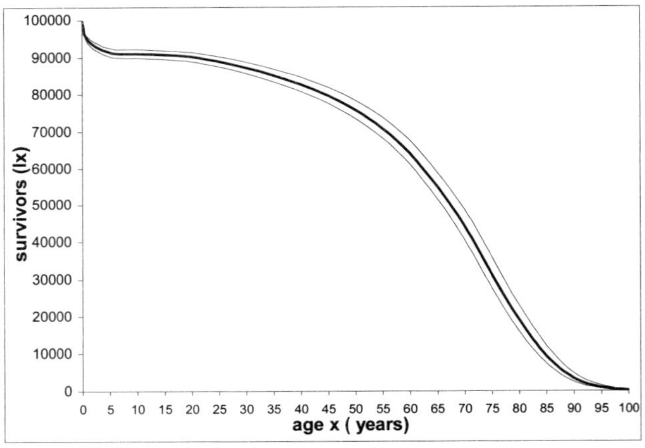

e_x with uncertainty interval

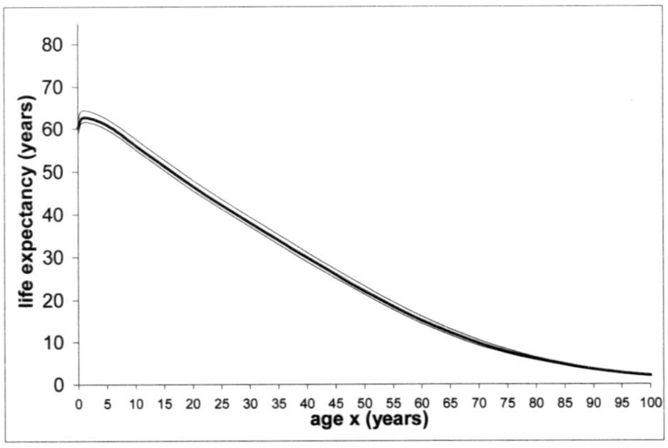

Distribution of deaths by age with uncertainty interval

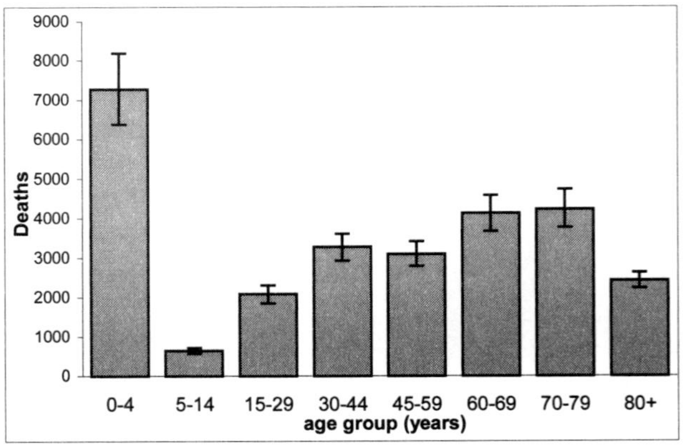

Tajikistan Tadjikistan Tayikistán

Females, 2000 *Femmes, 2000* *Mujeres, 2000*

Age	$_nM_x$	$_nq_x$			l_x			$_nd_x$	$_nL_x$	e_x			Actual deaths	Actual population
			percentiles			percentiles					percentiles			
		mean	21/2	971/2	mean	21/2	971/2			mean	21/2	971/2		
<1	0.0475	0.0455	0.0398	0.0563	100,000	100,000	100,000	4,551	95,904	64.7	62.9	66.9	3,737	78,757
1-4	0.0098	0.0384	0.0340	0.0444	95,449	94,374	96,025	3,669	372,991	66.8	65.3	68.8	2,968	301,748
5-9	0.0008	0.0039	0.0034	0.0045	91,780	90,181	92,760	359	458,003	65.4	64.1	67.2	308	392,342
10-14	0.0006	0.0030	0.0026	0.0034	91,421	89,776	92,443	273	456,424	60.6	59.4	62.4	244	409,112
15-19	0.0010	0.0049	0.0043	0.0056	91,148	89,469	92,202	448	454,622	55.8	54.6	57.6	324	328,592
20-24	0.0015	0.0076	0.0066	0.0086	90,700	88,967	91,808	686	451,787	51.1	49.9	52.8	407	268,064
25-29	0.0017	0.0085	0.0074	0.0097	90,014	88,200	91,203	767	448,155	46.5	45.3	48.1	409	238,995
30-34	0.0025	0.0122	0.0106	0.0138	89,248	87,347	90,530	1,089	443,515	41.8	40.7	43.5	522	212,400
35-39	0.0033	0.0165	0.0142	0.0185	88,159	86,141	89,573	1,451	437,166	37.3	36.2	38.9	668	201,424
40-44	0.0044	0.0218	0.0188	0.0244	86,708	84,544	88,300	1,891	428,811	32.9	31.9	34.4	702	159,130
45-49	0.0065	0.0320	0.0275	0.0357	84,816	82,478	86,641	2,712	417,302	28.6	27.6	30.0	728	112,091
50-54	0.0088	0.0433	0.0371	0.0483	82,104	79,533	84,262	3,553	401,640	24.4	23.5	25.8	636	71,885
55-59	0.0158	0.0760	0.0651	0.0847	78,552	75,691	81,139	5,970	377,833	20.4	19.5	21.7	756	47,818
60-64	0.0245	0.1152	0.0990	0.1277	72,581	69,283	75,854	8,364	341,998	16.9	16.1	18.1	1,722	70,412
65-69	0.0375	0.1714	0.1487	0.1884	64,218	60,438	68,341	11,005	293,576	13.8	13.0	14.8	2,125	56,697
70-74	0.0540	0.2379	0.2097	0.2604	53,213	49,054	58,181	12,659	234,418	11.1	10.5	11.9	2,567	47,540
75-79	0.0800	0.3333	0.3005	0.3622	40,554	36,282	45,979	13,518	168,976	8.8	8.3	9.5	2,340	29,247
80-84	0.1151	0.4469	0.4073	0.4788	27,036	23,140	32,163	12,082	104,975	7.0	6.5	7.5	1,735	15,076
85-89	0.1511	0.5482	0.5076	0.5790	14,954	12,060	19,064	8,198	54,274	5.6	5.3	6.0	1,321	8,747
90-94	0.2058	0.6361	0.5982	0.6640	6,756	5,077	9,386	4,297	20,886	4.3	4.0	4.6	730	3,546
95-99	0.2909	0.7767	0.7444	0.7994	2,458	1,706	3,771	1,909	6,563	3.2	3.1	3.4	415	1,428
100+	0.4268	1	1	1	549	342	964	549	1,286	2.3	2.3	2.5	166	388
All ages													*25,530*	*3,055,439*

$\ln(_nq_x)$ with uncertainty interval

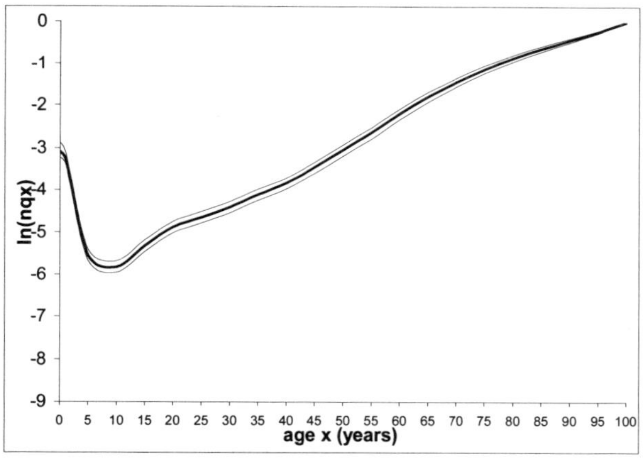

l_x with uncertainty interval

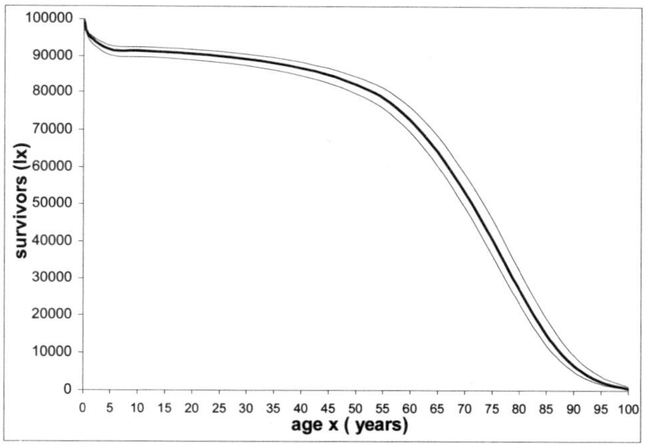

e_x with uncertainty interval

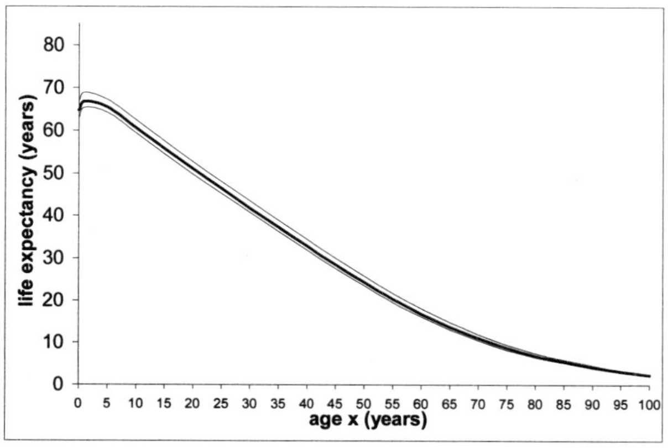

Distribution of deaths by age with uncertainty interval

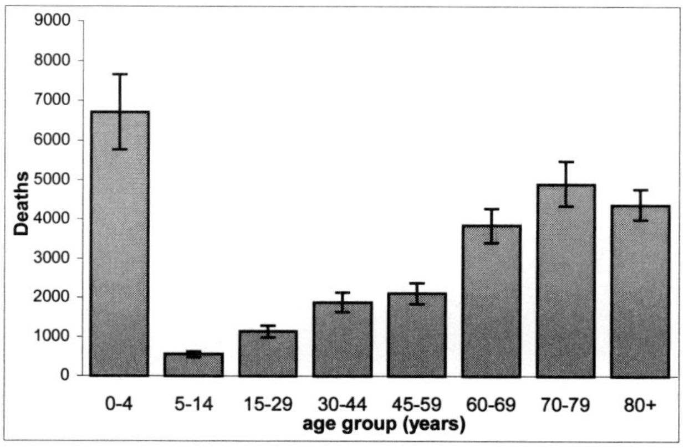

Thailand Thaïlande Tailandia

Males, 2000 *Hommes, 2000* *Hombres, 2000*

Age	$_nM_x$	$_nq_x$ mean	$_nq_x$ percentiles 2½	$_nq_x$ percentiles 97½	l_x mean	l_x percentiles 2½	l_x percentiles 97½	$_nd_x$	$_nL_x$	e_x mean	e_x percentiles 2½	e_x percentiles 97½	Actual deaths	Actual population
<1	0.0302	0.0294	0.0252	0.0324	100,000	100,000	100,000	2,940	97,354	66.0	65.0	67.0	17,988	595,638
1-4	0.0015	0.0058	0.0052	0.0063	97,060	96,762	97,477	566	386,881	67.0	66.1	68.1	3,413	2,332,818
5-9	0.0012	0.0058	0.0052	0.0062	96,494	96,150	96,971	557	481,077	63.4	62.5	64.4	3,175	2,743,607
10-14	0.0008	0.0039	0.0035	0.0042	95,937	95,551	96,470	376	478,745	58.8	57.9	59.8	2,194	2,792,247
15-19	0.0025	0.0124	0.0112	0.0133	95,561	95,146	96,131	1,181	474,852	54.0	53.1	55.0	7,270	2,922,675
20-24	0.0041	0.0202	0.0184	0.0218	94,380	93,878	95,059	1,908	467,129	49.6	48.8	50.6	12,101	2,962,345
25-29	0.0076	0.0375	0.0346	0.0403	92,472	91,836	93,309	3,470	453,683	45.6	44.8	46.6	22,108	2,890,557
30-34	0.0078	0.0382	0.0356	0.0410	89,002	88,133	90,081	3,404	436,500	42.3	41.5	43.2	21,907	2,809,470
35-39	0.0062	0.0307	0.0286	0.0329	85,598	84,518	86,875	2,630	421,415	38.9	38.1	39.7	15,848	2,539,183
40-44	0.0055	0.0272	0.0252	0.0292	82,968	81,735	84,393	2,259	409,191	35.0	34.3	35.8	11,748	2,127,749
45-49	0.0060	0.0296	0.0273	0.0317	80,709	79,352	82,264	2,387	397,575	30.9	30.3	31.7	10,364	1,726,060
50-54	0.0072	0.0352	0.0324	0.0376	78,321	76,840	80,017	2,757	384,713	26.8	26.2	27.5	9,341	1,303,308
55-59	0.0092	0.0452	0.0416	0.0484	75,564	73,948	77,422	3,415	369,281	22.7	22.1	23.3	9,435	1,020,126
60-64	0.0165	0.0792	0.0728	0.0850	72,148	70,371	74,204	5,716	346,451	18.6	18.1	19.3	14,205	860,904
65-69	0.0253	0.1190	0.1093	0.1279	66,432	64,389	68,802	7,904	312,401	15.0	14.5	15.6	16,004	632,569
70-74	0.0425	0.1920	0.1771	0.2065	58,528	56,154	61,279	11,239	264,545	11.7	11.2	12.2	18,400	433,112
75-79	0.0607	0.2635	0.2436	0.2837	47,290	44,560	50,428	12,460	205,298	8.9	8.5	9.3	14,426	237,693
80-84	0.1162	0.4503	0.4227	0.4791	34,830	31,918	38,141	15,683	134,941	6.2	5.9	6.5	12,504	107,592
85-89	0.2069	0.6818	0.6555	0.7093	19,147	16,627	22,020	13,054	63,099	4.2	4.0	4.4	6,936	33,524
90-94	0.3423	0.8444	0.8274	0.8625	6,093	4,834	7,585	5,145	15,029	2.8	2.7	2.9	2,030	5,930
95-99	0.4900	0.9919	0.9897	0.9939	948	665	1,310	940	1,919	2.0	2.0	2.1	272	556
100+	0.7530	1	1	1	8	4	13	8	10	1.3	1.3	1.7	23	30
All ages													*231,693*	*31,077,693*

ln($_nq_x$) with uncertainty interval

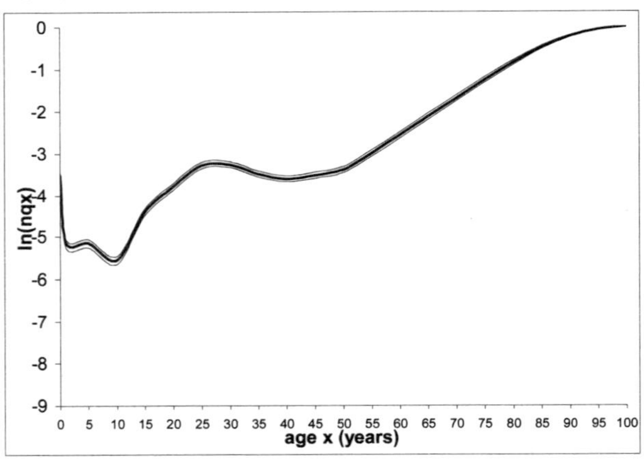

l_x with uncertainty interval

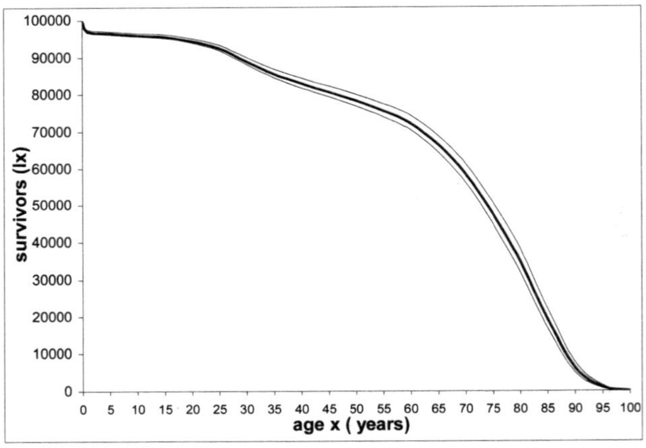

e_x with uncertainty interval

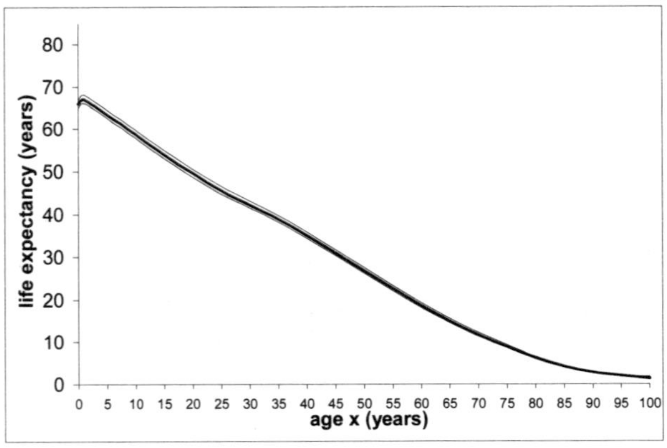

Distribution of deaths by age with uncertainty interval

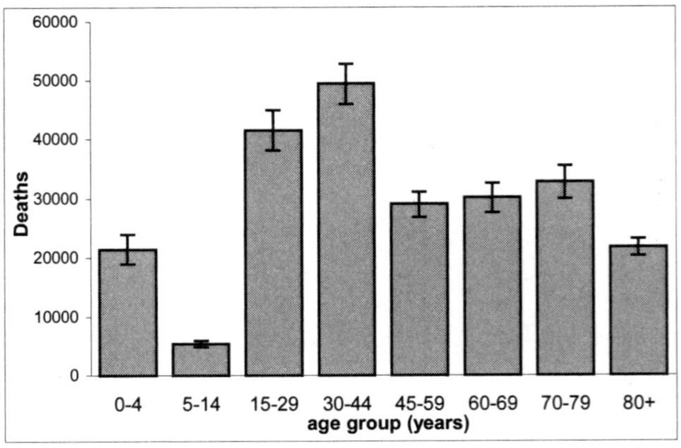

Thailand **Thaïlande** **Tailandia**

Females, 2000 *Femmes, 2000* *Mujeres, 2000*

Age	$_nM_x$	$_nq_x$			l_x			$_nd_x$	$_nL_x$	e_x			Actual deaths	Actual population
			percentiles			percentiles					percentiles			
		mean	2 1/2	97 1/2	mean	2 1/2	97 1/2			mean	2 1/2	97 1/2		
<1	0.0275	0.0268	0.0235	0.0297	100,000	100,000	100,000	2,684	97,585	72.4	71.1	74.2	15,630	568,357
1-4	0.0014	0.0055	0.0049	0.0060	97,316	97,029	97,654	533	387,986	73.4	72.2	75.1	3,139	2,283,105
5-9	0.0007	0.0034	0.0030	0.0037	96,783	96,447	97,178	325	483,104	69.7	68.6	71.5	1,801	2,680,495
10-14	0.0005	0.0023	0.0021	0.0026	96,458	96,091	96,887	224	481,731	65.0	63.8	66.7	1,280	2,745,952
15-19	0.0008	0.0041	0.0036	0.0045	96,234	95,846	96,687	393	480,187	60.1	59.0	61.8	2,361	2,883,900
20-24	0.0018	0.0091	0.0081	0.0100	95,841	95,416	96,336	871	477,027	55.4	54.2	57.1	5,452	2,986,241
25-29	0.0029	0.0144	0.0127	0.0159	94,970	94,461	95,561	1,370	471,423	50.8	49.7	52.5	8,562	2,945,183
30-34	0.0027	0.0136	0.0119	0.0150	93,599	92,958	94,349	1,274	464,812	46.6	45.5	48.1	7,860	2,868,307
35-39	0.0026	0.0130	0.0113	0.0144	92,326	91,563	93,231	1,203	458,620	42.2	41.1	43.7	6,802	2,592,664
40-44	0.0032	0.0157	0.0135	0.0173	91,122	90,243	92,178	1,431	452,036	37.7	36.7	39.2	6,928	2,189,162
45-49	0.0043	0.0211	0.0181	0.0234	89,692	88,678	90,932	1,895	443,722	33.2	32.3	34.7	7,678	1,797,723
50-54	0.0059	0.0290	0.0248	0.0321	87,797	86,607	89,284	2,547	432,618	28.9	28.0	30.3	7,913	1,344,327
55-59	0.0083	0.0405	0.0344	0.0450	85,250	83,826	87,074	3,451	417,623	24.7	23.8	26.0	8,718	1,054,826
60-64	0.0126	0.0611	0.0518	0.0681	81,799	80,056	84,079	4,999	396,498	20.6	19.9	21.8	12,066	957,109
65-69	0.0194	0.0926	0.0785	0.1032	76,800	74,606	79,724	7,110	366,226	16.8	16.1	17.9	14,633	753,715
70-74	0.0323	0.1493	0.1274	0.1656	69,690	66,910	73,467	10,405	322,437	13.3	12.6	14.2	17,228	533,854
75-79	0.0498	0.2216	0.1920	0.2445	59,285	55,828	64,105	13,137	263,582	10.2	9.6	10.9	15,724	315,500
80-84	0.0900	0.3673	0.3283	0.3993	46,148	42,176	51,797	16,952	188,360	7.3	7.0	7.8	14,355	159,495
85-89	0.1500	0.5455	0.5075	0.5787	29,196	25,335	34,794	15,925	106,166	5.2	4.9	5.5	8,236	54,905
90-94	0.2700	0.7459	0.7166	0.7717	13,271	10,673	17,138	9,898	36,660	3.3	3.2	3.5	3,153	11,679
95-99	0.4264	0.9354	0.9230	0.9460	3,373	2,436	4,857	3,155	7,399	2.3	2.3	2.4	552	1,295
100+	0.6000	1	1	1	218	132	374	218	363	1.7	1.7	1.8	52	87
All ages													*170,122*	*31,727,881*

ln($_nq_x$) with uncertainty interval

l_x with uncertainty interval

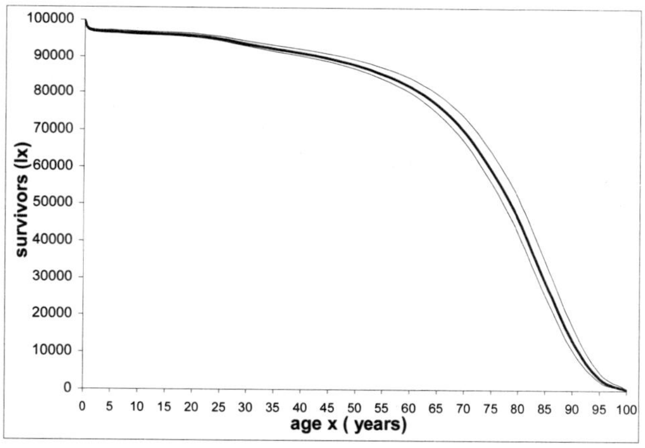

e_x with uncertainty interval

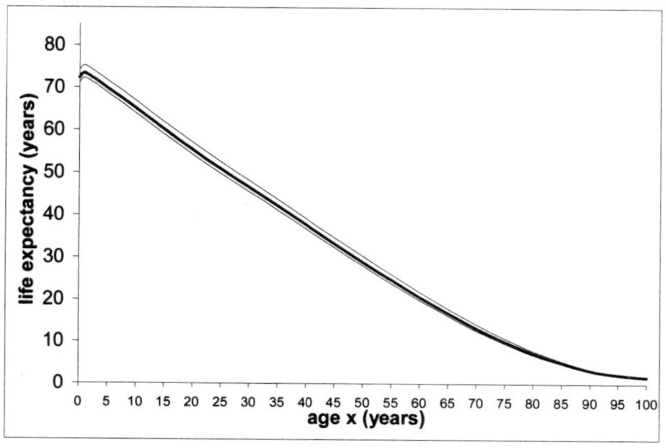

Distribution of deaths by age with uncertainty interval

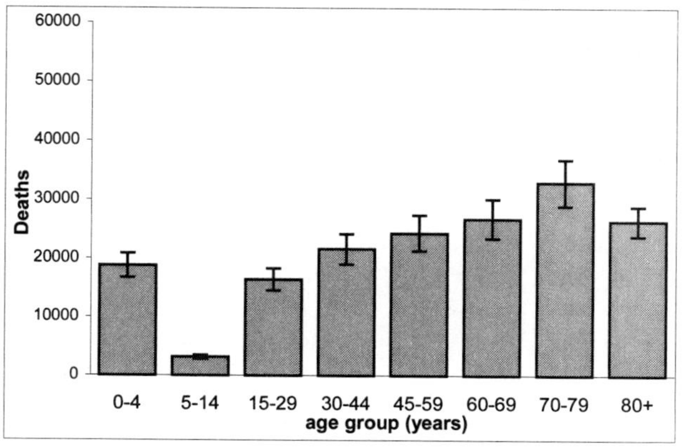

The former Yugoslav Republic of Macedonia | Ex-République yougoslave de Macédoine | Ex República Yugoslava de Macedonia

Males, 2000 | *Hommes, 2000* | *Hombres, 2000*

Age	$_nM_x$	$_nq_x$			l_x			$_nd_x$	$_nL_x$	e_x			Actual deaths	Actual population
			percentiles			percentiles					percentiles			
		mean	2½	97½	mean	2½	97½			mean	2½	97½		
<1	0.0141	0.0139	0.0112	0.0166	100,000	100,000	100,000	1,388	98,751	70.2	69.8	70.8	213	15,145
1-4	0.0006	0.0024	0.0021	0.0027	98,612	98,342	98,876	238	393,879	70.2	69.7	70.8	36	59,880
5-9	0.0003	0.0016	0.0014	0.0018	98,375	98,076	98,671	156	491,482	66.4	65.8	67.0	25	79,492
10-14	0.0002	0.0011	0.0009	0.0012	98,218	97,902	98,535	106	490,827	61.5	60.9	62.1	18	82,766
15-19	0.0006	0.0030	0.0026	0.0033	98,112	97,785	98,443	294	489,827	56.6	56.0	57.2	52	86,300
20-24	0.0007	0.0036	0.0032	0.0040	97,819	97,459	98,185	352	488,214	51.7	51.1	52.3	60	84,013
25-29	0.0006	0.0030	0.0027	0.0033	97,467	97,071	97,873	292	486,605	46.9	46.3	47.5	47	79,096
30-34	0.0010	0.0050	0.0045	0.0055	97,175	96,751	97,612	485	484,664	42.0	41.4	42.7	77	76,815
35-39	0.0016	0.0079	0.0072	0.0086	96,690	96,223	97,175	764	481,541	37.2	36.6	37.9	119	75,108
40-44	0.0028	0.0137	0.0126	0.0146	95,926	95,399	96,478	1,310	476,355	32.5	31.9	33.2	204	74,189
45-49	0.0048	0.0239	0.0225	0.0253	94,616	94,005	95,263	2,263	467,423	27.9	27.3	28.6	338	69,789
50-54	0.0088	0.0431	0.0410	0.0452	92,353	91,639	93,120	3,979	451,819	23.5	22.9	24.2	483	54,879
55-59	0.0141	0.0679	0.0638	0.0719	88,374	87,564	89,247	5,998	426,877	19.5	18.9	20.2	658	46,848
60-64	0.0221	0.1046	0.0971	0.1125	82,377	81,484	83,251	8,620	390,333	15.7	15.1	16.4	946	42,840
65-69	0.0387	0.1764	0.1615	0.1912	73,756	72,661	74,904	13,007	336,264	12.3	11.7	12.9	1,454	37,585
70-74	0.0634	0.2737	0.2494	0.2963	60,749	58,844	62,634	16,628	262,176	9.4	8.9	9.9	1,647	25,973
75-79	0.0966	0.3889	0.3573	0.4177	44,121	41,407	46,858	17,160	177,707	6.9	6.6	7.3	1,599	16,556
80-84	0.1752	0.6091	0.5743	0.6397	26,962	24,173	30,113	16,421	93,755	4.8	4.6	5.0	1,065	6,078
85-89	0.2843	0.8309	0.8078	0.8511	10,540	8,795	12,819	8,758	30,806	3.3	3.2	3.5	829	2,916
90-94	0.4130	0.9223	0.9091	0.9338	1,782	1,316	2,464	1,643	3,980	2.4	2.3	2.5	282	682
95-99	0.5368	0.9800	0.9748	0.9842	139	88	224	136	285	2.1	2.1	2.1	43	81
100+	0.6244	1	1	1	3	1	6	3	4	1.6	1.6	1.9	2	3
All ages													*10,198*	*1,017,034*

$ln(_nq_x)$ with uncertainty interval

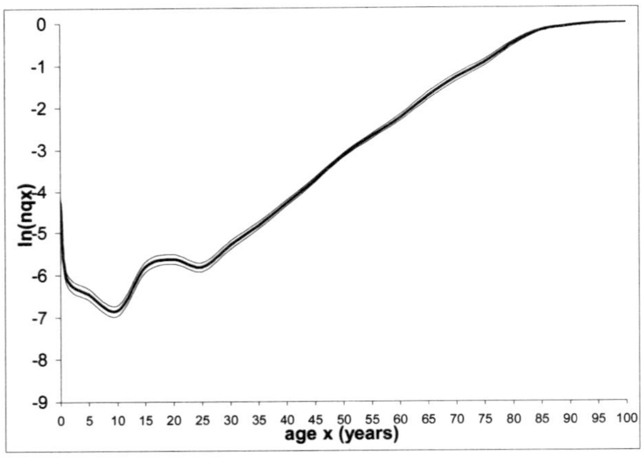

l_x with uncertainty interval

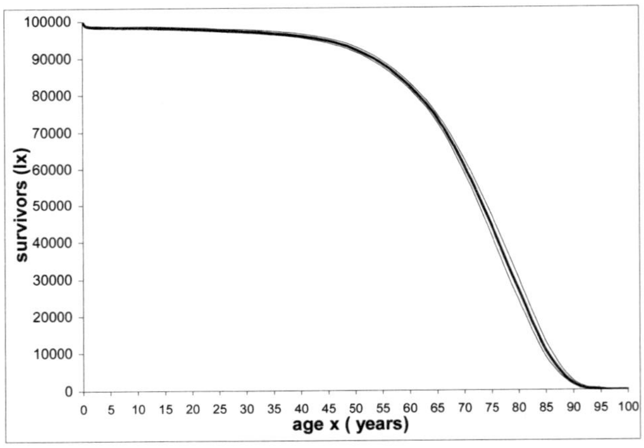

e_x with uncertainty interval

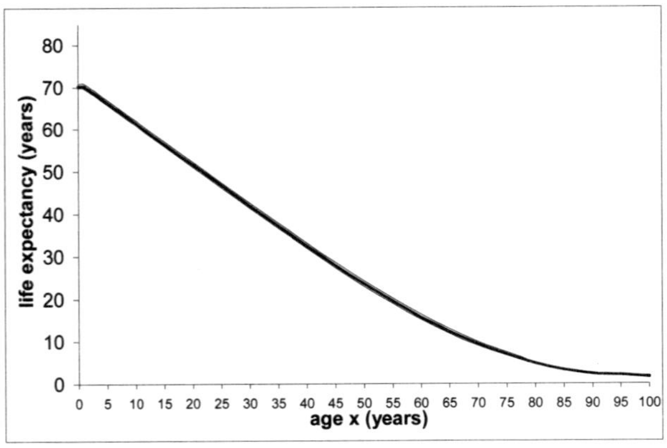

Distribution of deaths by age with uncertainty interval

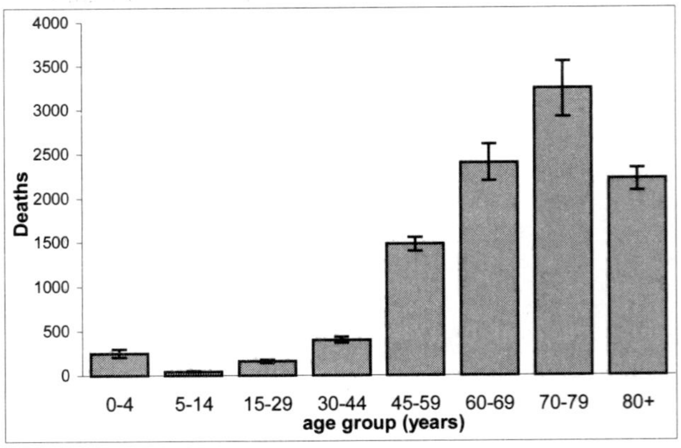

The former Yugoslav Republic of Macedonia

Females, 2000

Ex-République yougoslave de Macédoine

Femmes, 2000

Ex República Yugoslava de Macedonia

Mujeres, 2000

Age	$_nM_x$	$_nq_x$ mean	$_nq_x$ percentiles 21/2	$_nq_x$ percentiles 971/2	l_x mean	l_x percentiles 21/2	l_x percentiles 971/2	$_nd_x$	$_nL_x$	e_x mean	e_x percentiles 21/2	e_x percentiles 971/2	Actual deaths	Actual population
<1	0.0130	0.0128	0.0109	0.0154	100,000	100,000	100,000	1,281	98,847	74.8	74.4	75.2	183	14,154
1-4	0.0006	0.0024	0.0022	0.0028	98,719	98,458	98,910	242	394,297	74.8	74.3	75.2	34	55,429
5-9	0.0002	0.0012	0.0011	0.0014	98,478	98,186	98,692	119	492,091	71.0	70.5	71.4	18	74,671
10-14	0.0002	0.0008	0.0007	0.0009	98,359	98,053	98,585	79	491,597	66.0	65.6	66.5	13	78,303
15-19	0.0002	0.0010	0.0009	0.0011	98,280	97,965	98,513	98	491,154	61.1	60.6	61.6	16	81,906
20-24	0.0003	0.0012	0.0011	0.0014	98,182	97,856	98,423	123	490,602	56.1	55.7	56.6	20	80,622
25-29	0.0004	0.0018	0.0016	0.0020	98,059	97,720	98,311	176	489,855	51.2	50.7	51.7	27	75,940
30-34	0.0006	0.0030	0.0028	0.0033	97,883	97,526	98,149	293	488,681	46.3	45.8	46.8	45	74,782
35-39	0.0009	0.0047	0.0043	0.0051	97,590	97,205	97,879	455	486,811	41.4	41.0	41.9	69	73,568
40-44	0.0017	0.0083	0.0078	0.0089	97,135	96,713	97,457	808	483,654	36.6	36.1	37.1	120	71,528
45-49	0.0029	0.0143	0.0136	0.0150	96,327	95,849	96,697	1,374	478,199	31.9	31.4	32.4	198	69,098
50-54	0.0047	0.0231	0.0221	0.0240	94,953	94,407	95,386	2,193	469,282	27.3	26.8	27.9	268	57,237
55-59	0.0071	0.0349	0.0333	0.0366	92,760	92,154	93,253	3,238	455,702	22.9	22.4	23.5	355	49,890
60-64	0.0127	0.0617	0.0585	0.0651	89,521	88,905	90,039	5,527	433,788	18.7	18.2	19.2	599	47,008
65-69	0.0214	0.1017	0.0953	0.1083	83,994	83,402	84,594	8,540	398,618	14.7	14.2	15.2	927	43,267
70-74	0.0415	0.1880	0.1740	0.2020	75,454	74,417	76,469	14,186	341,802	11.1	10.7	11.6	1,349	32,500
75-79	0.0755	0.3175	0.2928	0.3403	61,267	59,580	62,945	19,454	257,703	8.1	7.7	8.5	1,713	22,693
80-84	0.1367	0.5093	0.4767	0.5385	41,814	39,306	44,371	21,296	155,828	5.7	5.4	6.0	1,184	8,661
85-89	0.2252	0.7205	0.6921	0.7450	20,518	18,217	23,221	14,783	65,631	4.0	3.8	4.2	988	4,387
90-94	0.3380	0.8391	0.8193	0.8555	5,735	4,683	7,144	4,812	14,238	2.8	2.7	2.9	375	1,111
95-99	0.4617	0.9678	0.9608	0.9733	923	677	1,284	893	1,935	2.2	2.1	2.2	82	177
100+	0.5741	1	1	1	30	18	50	30	52	1.7	1.7	1.8	5	9
All ages													*8,588*	*1,016,941*

$\ln(_nq_x)$ with uncertainty interval

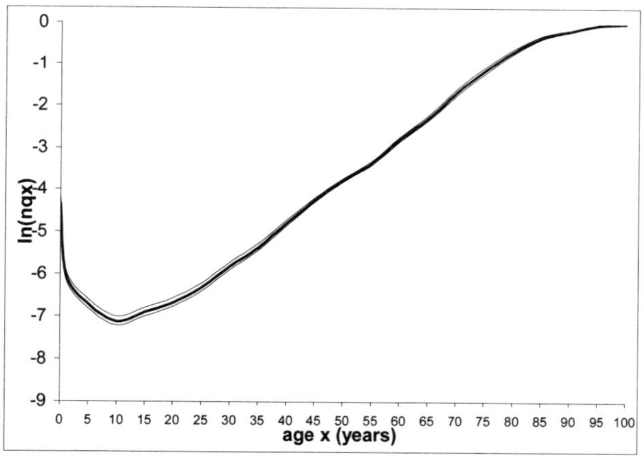

l_x with uncertainty interval

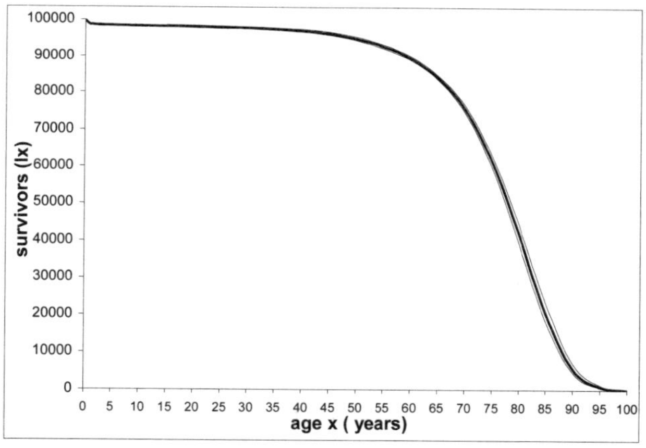

e_x with uncertainty interval

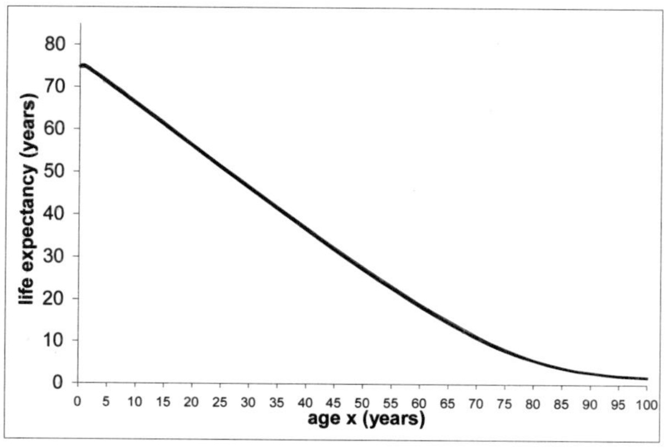

Distribution of deaths by age with uncertainty interval

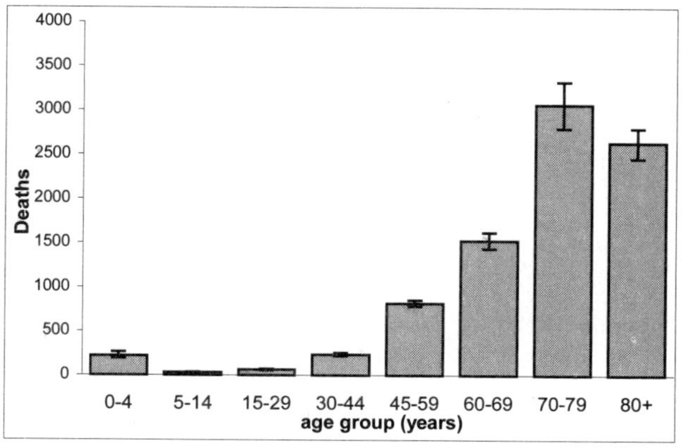

Togo Togo Togo

Males, 2000 *Hommes, 2000* *Hombres, 2000*

Age	$_nM_x$	$_nq_x$			l_x			$_nd_x$	$_nL_x$	e_x			Actual deaths	Actual population
			percentiles			percentiles					percentiles			
		mean	21/2	971/2	mean	21/2	971/2			mean	21/2	971/2		
<1	0.1142	0.1035	0.0907	0.1173	100,000	100,000	100,000	10,352	90,683	50.5	47.7	53.6	9,708	85,040
1-4	0.0098	0.0381	0.0332	0.0429	89,648	88,268	90,932	3,417	350,390	55.4	52.2	59.0	2,932	300,716
5-9	0.0022	0.0109	0.0094	0.0124	86,231	84,982	87,470	939	428,805	53.5	50.5	57.1	726	331,413
10-14	0.0015	0.0073	0.0062	0.0083	85,291	83,956	86,590	620	424,908	49.1	46.1	52.6	421	288,798
15-19	0.0021	0.0102	0.0087	0.0118	84,672	83,277	86,049	864	421,198	44.4	41.4	47.9	508	247,788
20-24	0.0045	0.0223	0.0190	0.0259	83,807	82,330	85,294	1,865	414,374	39.8	36.9	43.3	953	211,738
25-29	0.0087	0.0425	0.0357	0.0500	81,942	80,290	83,667	3,484	401,001	35.7	32.8	39.1	1,466	168,733
30-34	0.0135	0.0651	0.0537	0.0776	78,458	76,407	80,630	5,105	379,529	32.2	29.4	35.5	1,814	134,819
35-39	0.0153	0.0736	0.0595	0.0890	73,353	70,725	76,166	5,401	353,263	29.2	26.7	32.3	1,681	109,938
40-44	0.0165	0.0792	0.0630	0.0964	67,952	64,611	71,410	5,380	326,311	26.3	24.0	29.2	1,492	90,478
45-49	0.0182	0.0870	0.0684	0.1064	62,572	58,380	66,665	5,443	299,253	23.4	21.3	26.0	1,352	74,356
50-54	0.0204	0.0971	0.0758	0.1188	57,129	52,171	61,861	5,550	271,772	20.4	18.5	22.7	1,167	57,153
55-59	0.0239	0.1129	0.0878	0.1376	51,580	45,972	57,076	5,823	243,340	17.3	15.7	19.4	1,123	46,926
60-64	0.0343	0.1580	0.1234	0.1914	45,757	39,645	51,959	7,229	210,711	14.2	12.8	16.0	1,247	36,343
65-69	0.0491	0.2185	0.1725	0.2618	38,528	32,088	45,504	8,418	171,593	11.4	10.2	12.9	1,331	27,123
70-74	0.0740	0.3121	0.2522	0.3662	30,110	23,767	37,654	9,397	127,055	8.9	7.9	10.1	1,372	18,548
75-79	0.1115	0.4360	0.3652	0.4963	20,712	15,183	28,159	9,031	80,985	6.7	6.1	7.7	1,222	10,961
80-84	0.1691	0.5942	0.5216	0.6523	11,681	7,762	17,876	6,941	41,055	5.0	4.6	5.7	866	5,125
85-89	0.2470	0.7636	0.7032	0.8095	4,740	2,718	8,553	3,620	14,653	3.7	3.5	4.1	374	1,513
90-94	0.3475	0.8507	0.8043	0.8854	1,121	518	2,538	953	2,744	2.8	2.6	3.0	88	254
95-99	0.4708	0.9758	0.9598	0.9854	167	59	497	163	347	2.1	2.1	2.2	11	23
100+	0.6142	1	1	1	4	1	20	4	7	1.6	1.6	1.7	1	1
All ages													*31,855*	*2,247,787*

ln($_nq_x$) with uncertainty interval

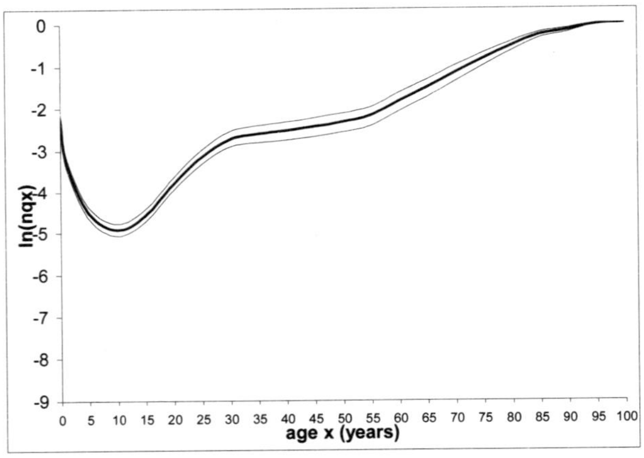

l_x with uncertainty interval

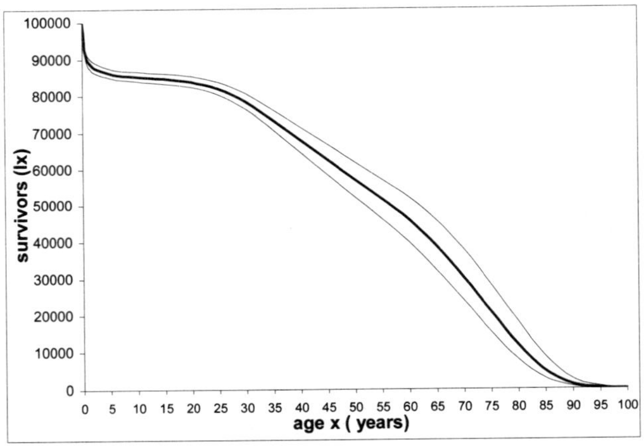

e_x with uncertainty interval

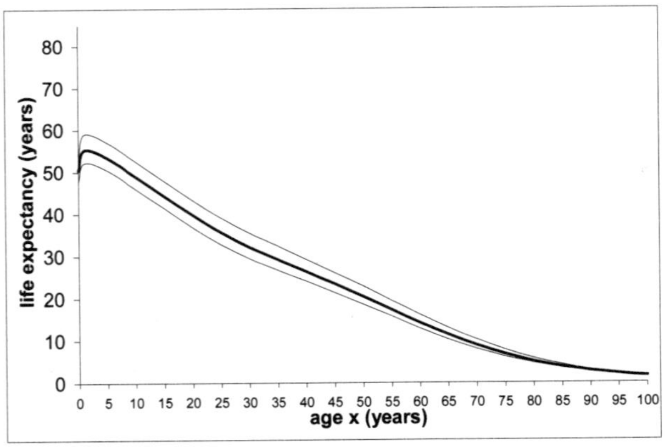

Distribution of deaths by age with uncertainty interval

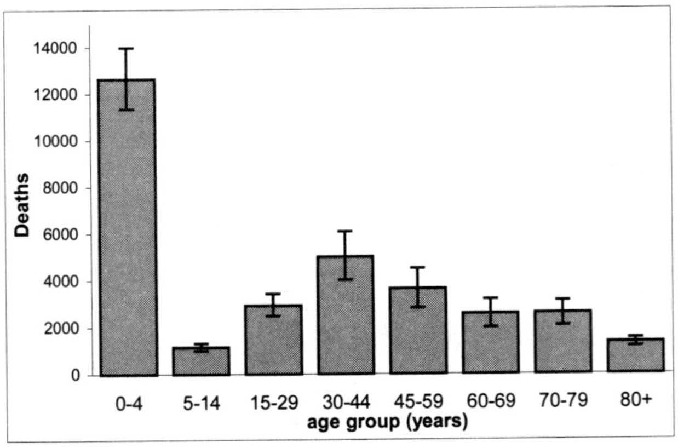

Togo

Females, 2000

Togo

Femmes, 2000

Togo

Mujeres, 2000

Age	$_nM_x$	$_nq_x$			l_x			$_nd_x$	$_nL_x$	e_x			Actual deaths	Actual population
			percentiles			percentiles					percentiles			
		mean	2½	97½	mean	2½	97½			mean	2½	97½		
<1	0.0934	0.0861	0.0743	0.0979	100,000	100,000	100,000	8,612	92,250	53.0	50.0	57.4	7,633	81,769
1-4	0.0098	0.0382	0.0318	0.0436	91,388	90,213	92,568	3,487	357,184	57.0	53.6	61.4	2,920	299,047
5-9	0.0021	0.0103	0.0084	0.0119	87,901	86,657	89,148	904	437,245	55.2	51.9	59.4	681	329,190
10-14	0.0014	0.0070	0.0057	0.0081	86,997	85,665	88,348	608	433,463	50.8	47.4	54.9	404	287,537
15-19	0.0032	0.0158	0.0128	0.0184	86,388	84,998	87,813	1,368	428,522	46.1	42.8	50.2	791	247,637
20-24	0.0074	0.0361	0.0289	0.0421	85,020	83,500	86,611	3,070	417,426	41.8	38.5	45.8	1,561	212,205
25-29	0.0120	0.0583	0.0461	0.0692	81,950	80,010	84,090	4,778	397,805	38.3	35.1	42.1	2,032	169,137
30-34	0.0145	0.0698	0.0545	0.0844	77,172	74,543	80,215	5,390	372,384	35.5	32.6	39.0	1,959	135,366
35-39	0.0134	0.0647	0.0500	0.0791	71,782	68,417	75,843	4,642	347,305	33.0	30.3	36.2	1,491	111,548
40-44	0.0122	0.0591	0.0454	0.0729	67,140	63,200	72,052	3,967	325,781	30.1	27.7	33.0	1,138	93,455
45-49	0.0117	0.0567	0.0435	0.0703	63,173	58,794	68,779	3,585	306,901	26.8	24.7	29.5	916	78,406
50-54	0.0133	0.0646	0.0495	0.0800	59,588	54,719	65,786	3,847	288,322	23.3	21.4	25.7	828	62,080
55-59	0.0165	0.0792	0.0608	0.0981	55,741	50,339	62,527	4,417	267,663	19.7	18.0	21.9	850	51,544
60-64	0.0238	0.1121	0.0860	0.1380	51,324	45,401	58,717	5,755	242,235	16.2	14.6	18.2	990	41,672
65-69	0.0364	0.1669	0.1287	0.2038	45,569	39,136	53,615	7,606	208,833	12.9	11.6	14.7	1,165	31,998
70-74	0.0588	0.2563	0.2010	0.3078	37,964	31,196	46,582	9,732	165,489	10.0	8.9	11.5	1,329	22,604
75-79	0.0929	0.3770	0.3053	0.4399	28,232	21,695	36,953	10,645	114,548	7.6	6.7	8.7	1,304	14,027
80-84	0.1435	0.5279	0.4481	0.5928	17,587	12,290	25,273	9,285	64,723	5.7	5.1	6.5	1,007	7,023
85-89	0.2141	0.6972	0.6249	0.7523	8,302	5,101	13,897	5,788	27,040	4.2	3.8	4.7	513	2,395
90-94	0.3086	0.8013	0.7440	0.8442	2,514	1,263	5,213	2,014	6,526	3.1	2.8	3.4	149	483
95-99	0.4300	0.9389	0.9081	0.9591	500	197	1,335	469	1,091	2.3	2.2	2.4	25	58
100+	0.5788	1	1	1	31	8	123	31	53	1.7	1.7	1.9	2	4
All ages													*29,688*	*2,279,185*

$\ln(_nq_x)$ with uncertainty interval

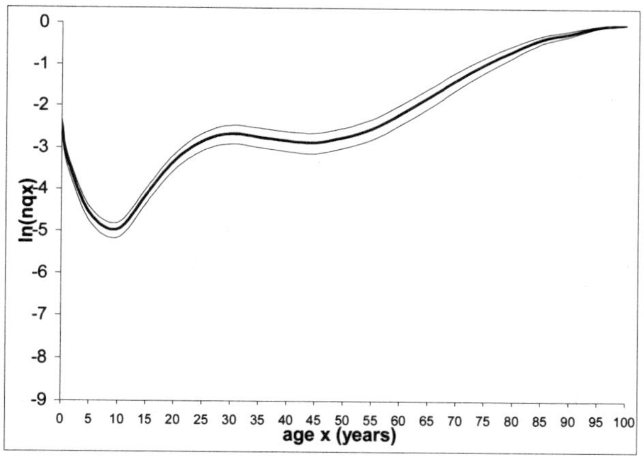

l_x with uncertainty interval

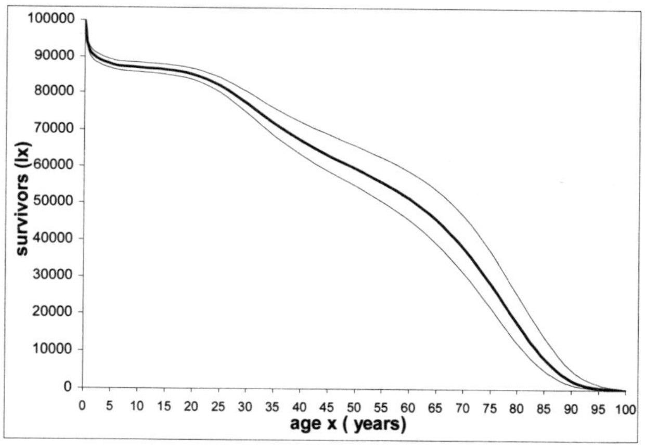

e_x with uncertainty interval

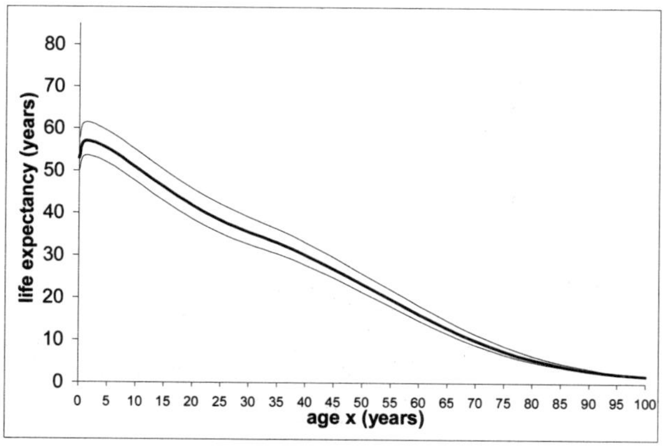

Distribution of deaths by age with uncertainty interval

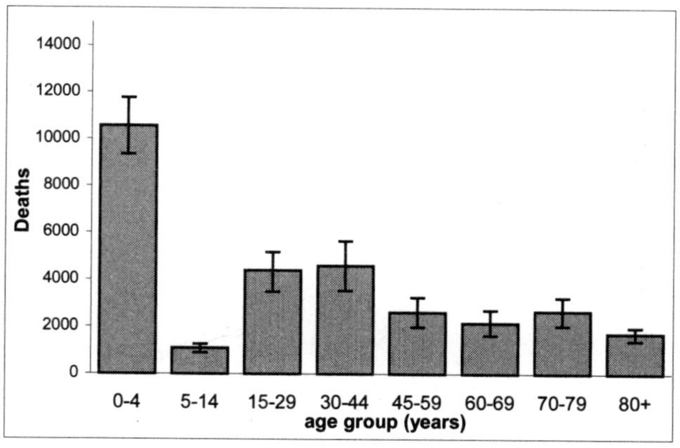

Tonga **Tonga** **Tonga**

Males, 2000 *Hommes, 2000* *Hombres, 2000*

Age	$_nM_x$	$_nq_x$			l_x			$_nd_x$	$_nL_x$	e_x			Actual deaths	Actual population
			percentiles			percentiles					percentiles			
		mean	2½	97½	mean	2½	97½			mean	2½	97½		
<1	0.0231	0.0226	0.0188	0.0264	100,000	100,000	100,000	2,259	97,967	67.4	66.8	68.3	26	1,144
1-4	0.0015	0.0061	0.0053	0.0068	97,741	97,364	98,119	592	389,543	68.0	67.3	68.8	7	4,750
5-9	0.0005	0.0023	0.0020	0.0026	97,149	96,704	97,599	223	485,185	64.4	63.7	65.2	3	5,842
10-14	0.0003	0.0014	0.0012	0.0016	96,925	96,456	97,402	136	484,287	59.5	58.9	60.3	2	6,143
15-19	0.0007	0.0034	0.0030	0.0038	96,789	96,306	97,282	330	483,122	54.6	54.0	55.4	4	5,635
20-24	0.0018	0.0088	0.0078	0.0097	96,459	95,942	96,990	847	480,179	49.8	49.1	50.6	9	4,912
25-29	0.0015	0.0074	0.0066	0.0081	95,612	95,015	96,234	708	476,291	45.2	44.5	46.0	7	4,429
30-34	0.0021	0.0106	0.0096	0.0116	94,904	94,245	95,597	1,010	471,993	40.5	39.9	41.3	8	3,879
35-39	0.0043	0.0213	0.0193	0.0229	93,893	93,154	94,682	2,000	464,466	35.9	35.3	36.7	14	3,309
40-44	0.0065	0.0320	0.0294	0.0340	91,893	91,020	92,852	2,939	452,118	31.6	31.0	32.4	18	2,744
45-49	0.0087	0.0426	0.0394	0.0448	88,954	87,927	90,123	3,787	435,304	27.6	27.0	28.4	20	2,309
50-54	0.0106	0.0515	0.0477	0.0544	85,167	83,992	86,569	4,387	414,868	23.7	23.1	24.4	19	1,760
55-59	0.0150	0.0723	0.0667	0.0770	80,780	79,422	82,441	5,840	389,302	19.9	19.3	20.5	22	1,497
60-64	0.0240	0.1132	0.1041	0.1213	74,941	73,308	76,941	8,484	353,494	16.2	15.7	16.8	29	1,190
65-69	0.0340	0.1567	0.1441	0.1683	66,457	64,419	68,929	10,413	306,253	13.0	12.5	13.5	31	899
70-74	0.0565	0.2475	0.2285	0.2662	56,044	53,575	58,995	13,871	245,545	9.9	9.5	10.4	32	574
75-79	0.0906	0.3693	0.3437	0.3957	42,174	39,314	45,515	15,575	171,932	7.4	7.0	7.7	32	353
80-84	0.1600	0.5714	0.5426	0.6016	26,599	23,759	29,871	15,199	94,997	5.2	5.0	5.5	22	139
85-89	0.2400	0.7500	0.7265	0.7741	11,400	9,466	13,663	8,550	35,624	3.8	3.7	4.0	12	49
90-94	0.3400	0.8416	0.8242	0.8592	2,850	2,139	3,737	2,398	7,054	2.8	2.7	2.9	4	11
95-99	0.4900	0.9919	0.9894	0.9939	451	301	657	448	914	2.0	2.0	2.1	1	1
100+	0.7412	1	1	1	4	2	7	4	5	1.3	1.3	1.8	0	0
All ages													*320*	*51,570*

ln($_nq_x$) with uncertainty interval

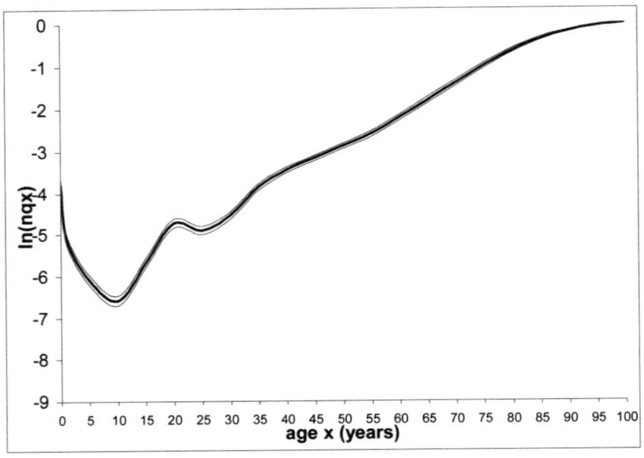

l_x with uncertainty interval

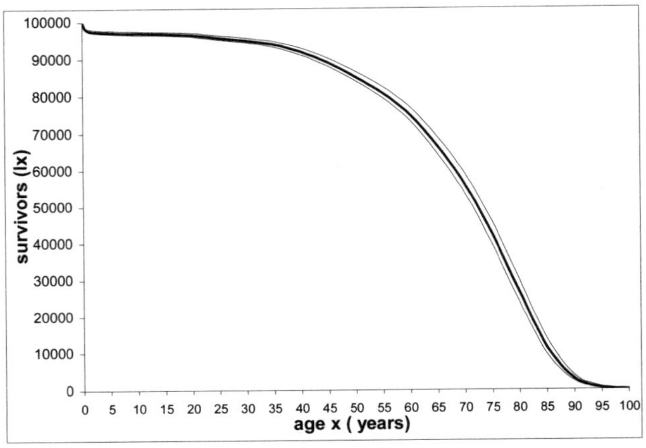

e_x with uncertainty interval

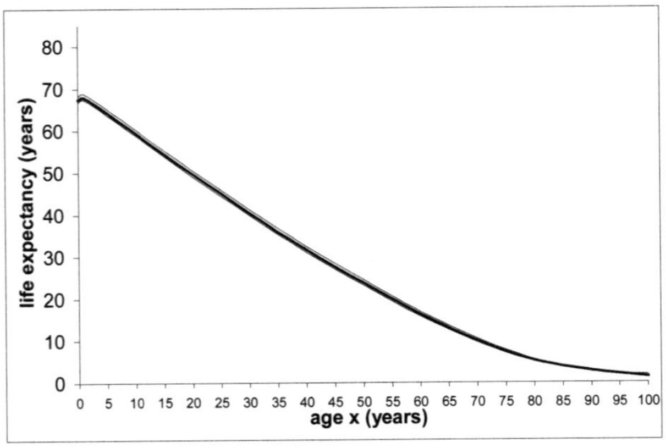

Distribution of deaths by age with uncertainty interval

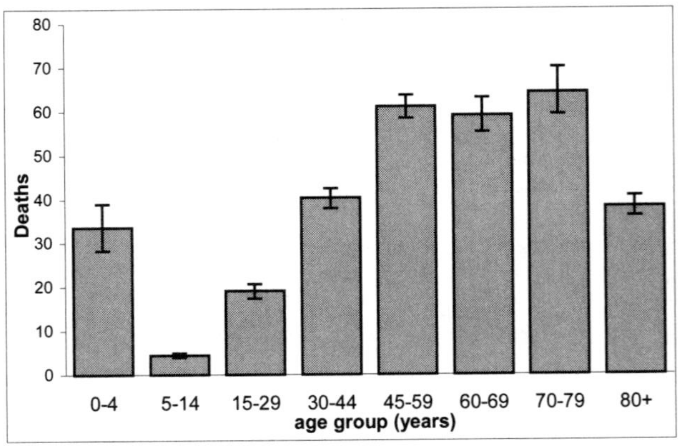

Tonga **Tonga** **Tonga**

Females, 2000 *Femmes, 2000* *Mujeres, 2000*

Age	$_nM_x$	$_nq_x$			l_x			$_nd_x$	$_nL_x$	e_x			Actual deaths	Actual population
			percentiles			percentiles					percentiles			
		mean	2½	97½	mean	2½	97½			mean	2½	97½		
<1	0.0130	0.0129	0.0103	0.0151	100,000	100,000	100,000	1,286	98,842	72.9	72.7	73.9	14	1,092
1-4	0.0015	0.0059	0.0050	0.0065	98,714	98,491	98,973	583	393,456	72.8	72.5	73.9	7	4,509
5-9	0.0004	0.0018	0.0015	0.0019	98,131	97,848	98,481	173	490,222	69.2	68.9	70.3	2	5,513
10-14	0.0004	0.0019	0.0017	0.0021	97,958	97,660	98,334	190	489,314	64.4	64.0	65.4	2	5,709
15-19	0.0006	0.0030	0.0026	0.0032	97,768	97,454	98,172	293	488,106	59.5	59.1	60.5	3	5,224
20-24	0.0009	0.0043	0.0037	0.0046	97,475	97,138	97,920	416	486,334	54.6	54.3	55.7	4	4,447
25-29	0.0010	0.0048	0.0041	0.0051	97,059	96,693	97,559	462	484,139	49.9	49.5	50.9	4	3,746
30-34	0.0017	0.0083	0.0073	0.0088	96,597	96,202	97,156	806	480,971	45.1	44.7	46.1	5	3,181
35-39	0.0030	0.0151	0.0133	0.0157	95,791	95,355	96,446	1,442	475,353	40.5	40.0	41.5	9	2,916
40-44	0.0043	0.0212	0.0190	0.0217	94,350	93,862	95,159	1,996	466,757	36.0	35.6	37.0	11	2,458
45-49	0.0057	0.0283	0.0257	0.0286	92,353	91,827	93,352	2,613	455,232	31.8	31.3	32.7	12	2,130
50-54	0.0073	0.0359	0.0325	0.0365	89,740	89,205	90,956	3,221	440,646	27.6	27.1	28.5	12	1,621
55-59	0.0101	0.0493	0.0445	0.0508	86,519	85,946	88,002	4,267	421,927	23.5	23.1	24.4	14	1,398
60-64	0.0150	0.0723	0.0650	0.0752	82,252	81,582	84,090	5,946	396,395	19.6	19.2	20.4	18	1,175
65-69	0.0230	0.1087	0.0979	0.1144	76,306	75,444	78,623	8,298	360,785	16.0	15.5	16.6	21	911
70-74	0.0340	0.1567	0.1417	0.1665	68,008	66,810	70,928	10,656	313,401	12.6	12.2	13.2	22	645
75-79	0.0570	0.2495	0.2284	0.2667	57,352	55,687	60,877	14,307	250,995	9.5	9.1	9.9	26	451
80-84	0.1000	0.4000	0.3743	0.4257	43,046	40,835	46,973	17,218	172,183	6.8	6.5	7.1	21	212
85-89	0.1800	0.6207	0.5941	0.6485	25,827	23,452	29,390	16,031	89,060	4.7	4.5	4.9	17	94
90-94	0.2800	0.7609	0.7399	0.7825	9,797	8,244	11,930	7,454	26,621	3.3	3.2	3.4	9	31
95-99	0.4200	0.9292	0.9189	0.9393	2,343	1,793	3,103	2,177	5,183	2.3	2.3	2.4	2	5
100+	0.6831	1	1	1	166	109	252	166	243	1.5	1.5	1.8	1	1
All ages													*234*	*47,469*

ln($_nq_x$) with uncertainty interval

l_x with uncertainty interval

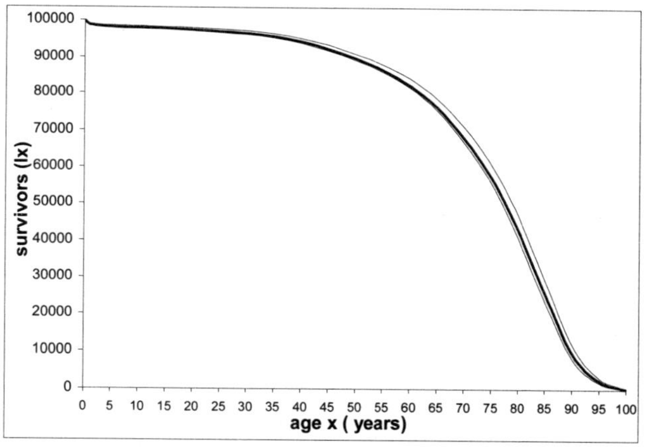

e_x with uncertainty interval

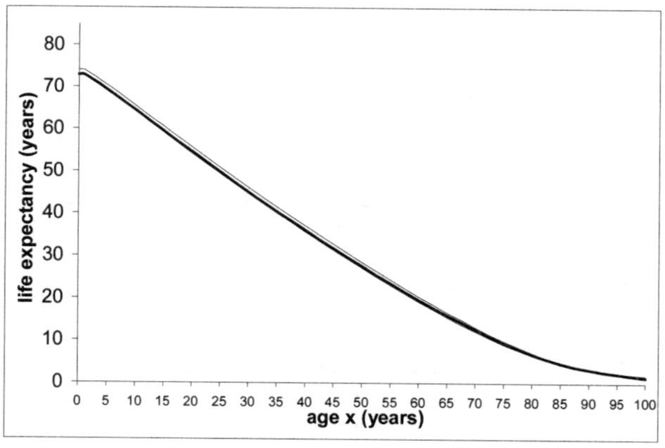

Distribution of deaths by age with uncertainty interval

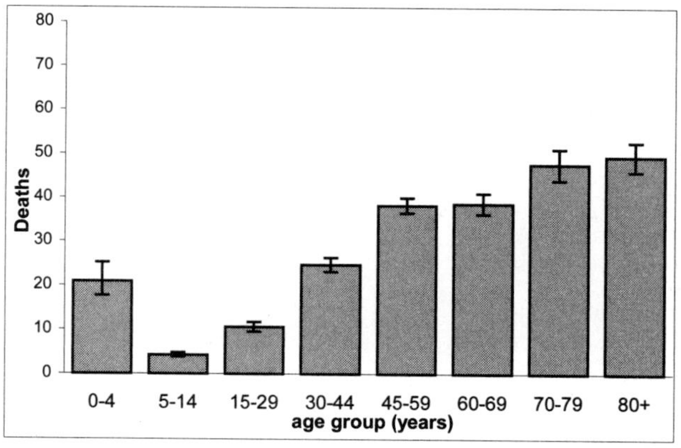

Trinidad and Tobago   Trinité-et-Tobago   Trinidad y Tabago

Males, 2000   *Hommes, 2000*   *Hombres, 2000*

Age	$_nM_x$	$_nq_x$			l_x			$_nd_x$	$_nL_x$	e_x			Actual deaths	Actual population
			percentiles			percentiles					percentiles			
		mean	2½	97½	mean	2½	97½			mean	2½	97½		
<1	0.0140	0.0139	0.0099	0.0179	100,000	100,000	100,000	1,386	98,752	68.5	67.1	70.2	130	9,257
1-4	0.0006	0.0024	0.0019	0.0030	98,614	98,206	99,011	242	393,874	68.5	67.1	70.1	22	35,294
5-9	0.0003	0.0015	0.0011	0.0018	98,372	97,923	98,821	147	491,493	64.7	63.2	66.2	16	53,503
10-14	0.0003	0.0014	0.0011	0.0017	98,225	97,747	98,704	137	490,782	59.8	58.3	61.3	18	65,746
15-19	0.0006	0.0031	0.0024	0.0038	98,088	97,581	98,594	308	489,669	54.8	53.4	56.4	46	73,057
20-24	0.0014	0.0071	0.0056	0.0086	97,780	97,211	98,346	694	487,165	50.0	48.6	51.5	88	61,704
25-29	0.0026	0.0129	0.0103	0.0155	97,086	96,376	97,798	1,254	482,295	45.3	44.0	46.8	139	53,598
30-34	0.0026	0.0131	0.0107	0.0156	95,832	94,881	96,789	1,260	476,010	40.9	39.6	42.3	125	47,139
35-39	0.0032	0.0157	0.0129	0.0185	94,572	93,400	95,756	1,485	469,148	36.4	35.1	37.8	163	51,438
40-44	0.0043	0.0211	0.0177	0.0246	93,087	91,676	94,518	1,968	460,517	32.0	30.7	33.3	191	44,682
45-49	0.0060	0.0295	0.0251	0.0340	91,120	89,422	92,846	2,687	448,881	27.6	26.4	28.9	222	37,140
50-54	0.0098	0.0477	0.0408	0.0548	88,433	86,392	90,516	4,217	431,622	23.4	22.2	24.6	301	30,772
55-59	0.0164	0.0787	0.0675	0.0904	84,216	81,726	86,765	6,624	404,519	19.4	18.4	20.6	381	23,250
60-64	0.0249	0.1173	0.1014	0.1342	77,592	74,553	80,670	9,105	365,195	15.8	14.9	16.9	454	18,225
65-69	0.0386	0.1761	0.1525	0.1989	68,486	64,832	72,253	12,061	312,278	12.6	11.8	13.6	511	13,242
70-74	0.0630	0.2721	0.2370	0.3047	56,425	51,934	61,097	15,355	243,736	9.8	9.1	10.6	661	10,492
75-79	0.0935	0.3788	0.3341	0.4193	41,070	36,313	46,166	15,559	166,450	7.5	6.9	8.2	750	8,020
80-84	0.1478	0.5396	0.4901	0.5826	25,510	21,503	30,430	13,766	93,136	5.5	5.1	6.0	676	4,572
85-89	0.2214	0.7127	0.6703	0.7487	11,744	9,126	15,358	8,370	37,797	4.1	3.8	4.4	431	1,946
90-94	0.3143	0.8088	0.7767	0.8368	3,375	2,311	5,063	2,730	8,685	3.0	2.9	3.2	177	563
95-99	0.4227	0.9318	0.9140	0.9465	645	380	1,130	601	1,422	2.3	2.3	2.4	43	101
100+	0.5385	1	1	1	44	21	97	44	82	1.9	1.8	1.9	6	11
All ages													*5,550*	*643,752*

$\ln(_nq_x)$ with uncertainty interval

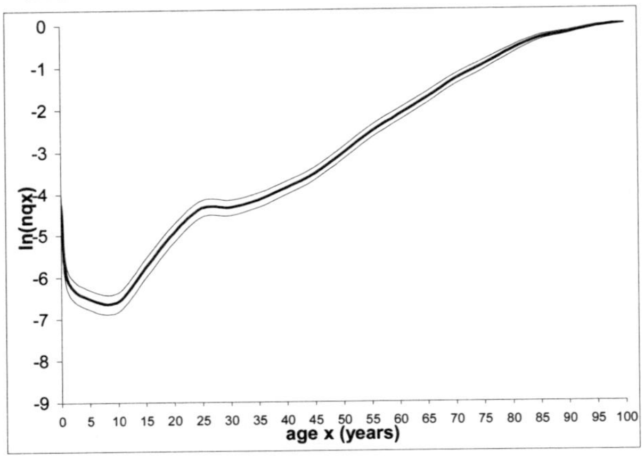

l_x with uncertainty interval

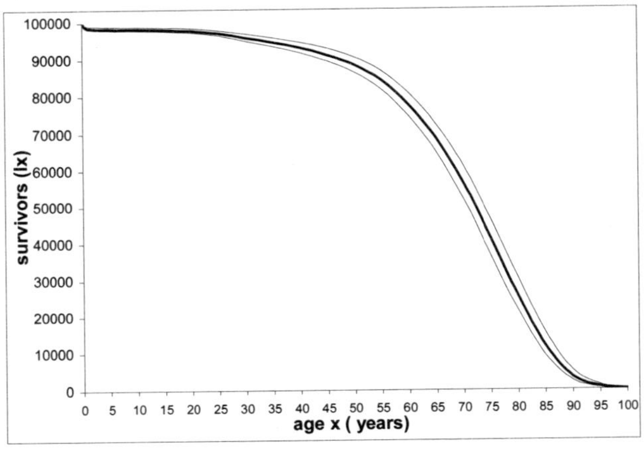

e_x with uncertainty interval

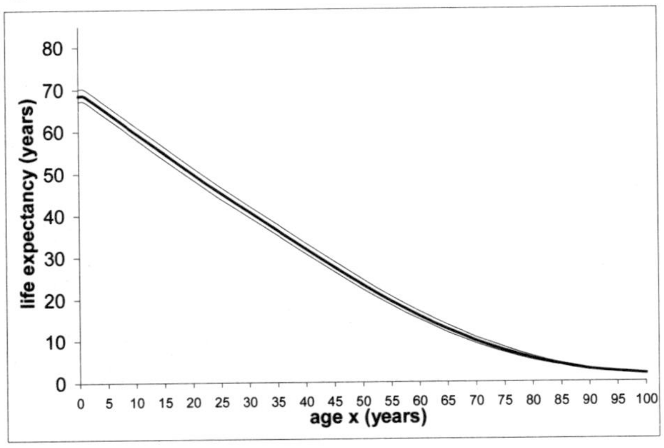

Distribution of deaths by age with uncertainty interval

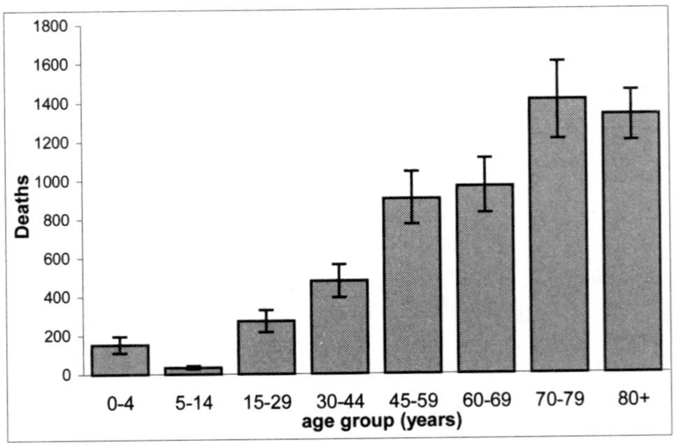

Trinidad and Tobago

Trinité-et-Tobago

Trinidad y Tabago

Females, 2000

Femmes, 2000

Mujeres, 2000

Age	$_nM_x$	$_nq_x$			l_x			$_nd_x$	$_nL_x$	e_x			Actual deaths	Actual population
			percentiles			percentiles					percentiles			
		mean	21/2	971/2	mean	21/2	971/2			mean	21/2	971/2		
<1	0.0110	0.0109	0.0076	0.0144	100,000	100,000	100,000	1,093	99,016	73.8	73.0	74.9	97	8,783
1-4	0.0005	0.0020	0.0015	0.0024	98,907	98,564	99,238	194	395,162	73.6	72.7	74.8	17	34,405
5-9	0.0002	0.0009	0.0007	0.0011	98,713	98,328	99,087	88	493,345	69.7	68.9	70.9	9	51,992
10-14	0.0003	0.0013	0.0010	0.0016	98,625	98,223	99,019	127	492,808	64.8	63.9	66.0	17	64,490
15-19	0.0005	0.0024	0.0019	0.0029	98,498	98,070	98,918	238	491,895	59.9	59.0	61.1	35	71,769
20-24	0.0007	0.0036	0.0029	0.0043	98,260	97,787	98,727	356	490,410	55.0	54.1	56.2	44	60,618
25-29	0.0010	0.0048	0.0040	0.0056	97,904	97,367	98,436	469	488,347	50.2	49.3	51.4	50	52,517
30-34	0.0015	0.0077	0.0065	0.0088	97,435	96,824	98,044	748	485,305	45.4	44.5	46.7	74	48,263
35-39	0.0020	0.0100	0.0087	0.0112	96,687	95,971	97,404	964	481,025	40.8	39.8	42.0	105	52,341
40-44	0.0026	0.0131	0.0117	0.0145	95,723	94,893	96,554	1,256	475,473	36.1	35.2	37.4	121	45,931
45-49	0.0041	0.0202	0.0184	0.0219	94,466	93,516	95,413	1,906	467,566	31.6	30.6	32.8	154	37,689
50-54	0.0062	0.0306	0.0278	0.0330	92,560	91,472	93,624	2,831	455,723	27.2	26.2	28.4	191	30,699
55-59	0.0100	0.0488	0.0437	0.0529	89,729	88,513	90,896	4,377	437,704	23.0	22.0	24.1	244	24,352
60-64	0.0160	0.0769	0.0678	0.0843	85,352	84,064	86,616	6,566	410,347	19.0	18.1	20.1	313	19,582
65-69	0.0260	0.1221	0.1065	0.1359	78,787	77,348	80,386	9,617	369,891	15.4	14.5	16.4	391	15,030
70-74	0.0413	0.1873	0.1632	0.2100	69,170	67,080	71,821	12,957	313,456	12.2	11.4	13.1	501	12,117
75-79	0.0615	0.2666	0.2333	0.2996	56,213	53,170	60,098	14,984	243,604	9.4	8.7	10.2	604	9,821
80-84	0.1000	0.4000	0.3553	0.4456	41,229	37,356	46,010	16,491	164,915	6.9	6.4	7.5	578	5,784
85-89	0.1700	0.5965	0.5481	0.6448	24,737	21,001	29,435	14,756	86,797	4.9	4.5	5.3	500	2,944
90-94	0.2700	0.7459	0.7074	0.7840	9,982	7,545	13,178	7,445	27,574	3.4	3.2	3.7	306	1,135
95-99	0.3800	0.8879	0.8641	0.9108	2,537	1,630	3,856	2,252	5,927	2.6	2.4	2.7	114	300
100+	0.5000	1	1	1	284	145	524	284	569	2.0	1.9	2.0	27	54
All ages													*4,493*	*650,616*

$ln(_nq_x)$ with uncertainty interval

l_x with uncertainty interval

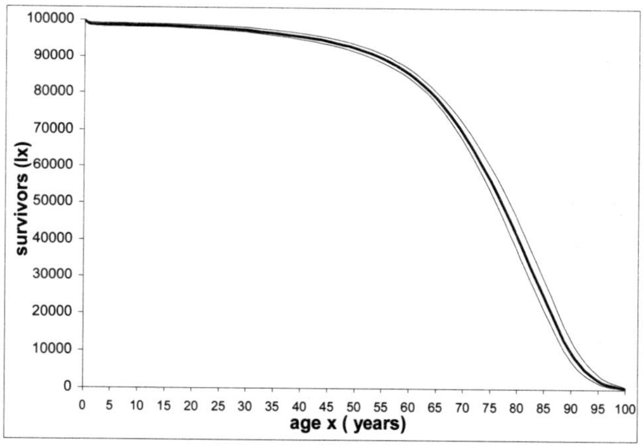

e_x with uncertainty interval

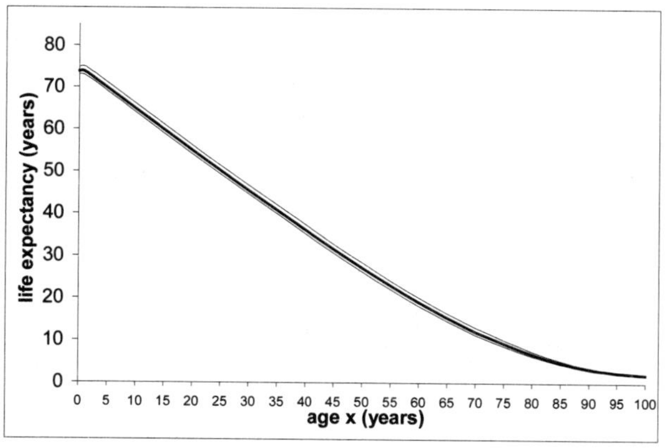

Distribution of deaths by age with uncertainty interval

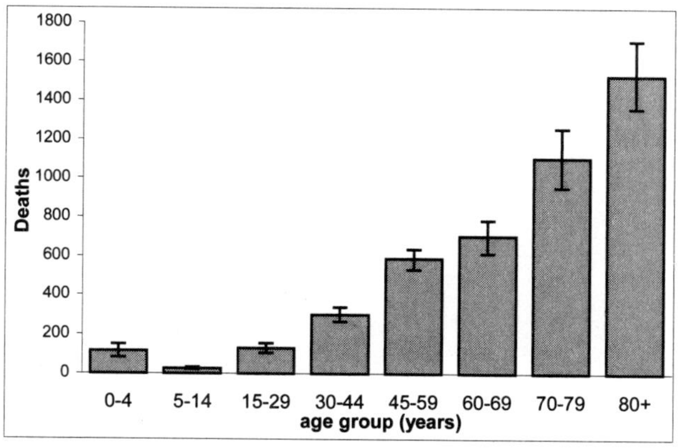

Tunisia | Tunisie | Túnez

Males, 2000 | *Hommes, 2000* | *Hombres, 2000*

Age	$_nM_x$	$_nq_x$ mean	$_nq_x$ 2½	$_nq_x$ 97½	l_x mean	l_x 2½	l_x 97½	$_nd_x$	$_nL_x$	e_x mean	e_x 2½	e_x 97½	Actual deaths	Actual population
<1	0.0231	0.0226	0.0185	0.0254	100,000	100,000	100,000	2,259	97,967	69.2	68.3	70.0	2,002	86,821
1-4	0.0015	0.0061	0.0053	0.0067	97,741	97,456	98,152	592	389,543	69.8	68.8	70.8	522	342,950
5-9	0.0005	0.0023	0.0020	0.0025	97,149	96,806	97,635	223	485,185	66.2	65.2	67.2	224	486,715
10-14	0.0003	0.0014	0.0012	0.0015	96,925	96,561	97,439	136	484,287	61.4	60.4	62.3	148	526,105
15-19	0.0005	0.0027	0.0024	0.0030	96,789	96,412	97,319	264	483,287	56.5	55.4	57.4	291	531,750
20-24	0.0014	0.0070	0.0062	0.0077	96,525	96,123	97,085	679	480,930	51.6	50.6	52.6	688	487,460
25-29	0.0012	0.0059	0.0053	0.0065	95,847	95,380	96,481	569	477,812	47.0	45.9	47.9	502	421,577
30-34	0.0017	0.0085	0.0077	0.0093	95,278	94,759	95,971	812	474,359	42.2	41.2	43.2	631	368,425
35-39	0.0024	0.0119	0.0108	0.0131	94,466	93,873	95,236	1,127	469,511	37.6	36.5	38.5	791	329,694
40-44	0.0036	0.0178	0.0163	0.0195	93,339	92,647	94,208	1,665	462,531	33.0	32.0	33.9	1,038	288,329
45-49	0.0056	0.0276	0.0256	0.0301	91,674	90,841	92,671	2,531	452,040	28.5	27.5	29.5	1,286	229,677
50-54	0.0085	0.0414	0.0382	0.0454	89,142	88,109	90,302	3,692	436,480	24.3	23.3	25.2	1,320	156,029
55-59	0.0120	0.0583	0.0530	0.0644	85,450	84,107	86,855	4,978	414,805	20.2	19.2	21.1	1,501	125,061
60-64	0.0220	0.1043	0.0940	0.1172	80,472	78,690	82,248	8,390	381,385	16.3	15.4	17.2	2,625	119,298
65-69	0.0340	0.1567	0.1401	0.1780	72,082	69,468	74,518	11,294	332,174	12.9	12.1	13.7	3,776	111,059
70-74	0.0565	0.2475	0.2210	0.2804	60,788	57,103	64,078	15,045	266,328	9.9	9.2	10.5	4,316	76,409
75-79	0.0906	0.3693	0.3335	0.4110	45,743	41,092	49,919	16,893	186,484	7.3	6.8	7.8	4,449	49,112
80-84	0.1600	0.5714	0.5315	0.6144	28,850	24,205	33,269	16,486	103,037	5.1	4.8	5.4	3,930	24,562
85-89	0.2600	0.7879	0.7587	0.8183	12,364	9,333	15,588	9,742	37,468	3.6	3.4	3.7	3,361	12,925
90-94	0.4000	0.9091	0.8929	0.9258	2,623	1,696	3,762	2,384	5,961	2.5	2.4	2.6	877	2,192
95-99	0.6000	0.9800	0.9741	0.9853	238	126	403	234	491	2.1	2.1	2.1	157	261
100+	0.7412	1	1	1	5	2	10	5	6	1.3	1.3	1.9	7	10
All ages													*34,439*	*4,776,421*

(Column groups $_nq_x$, l_x and e_x: mean and percentiles 2½, 97½.)

$\ln(_nq_x)$ with uncertainty interval

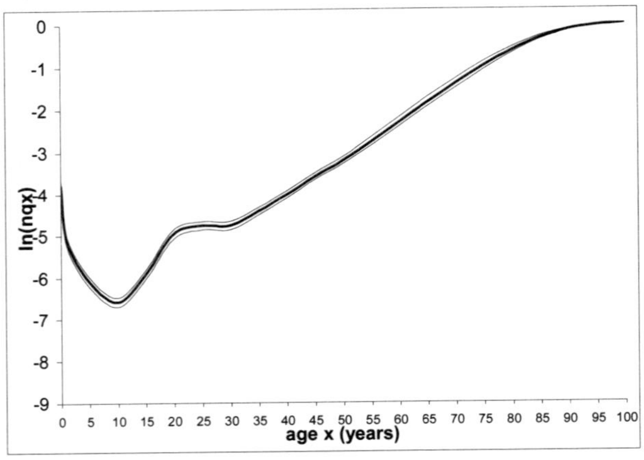

l_x with uncertainty interval

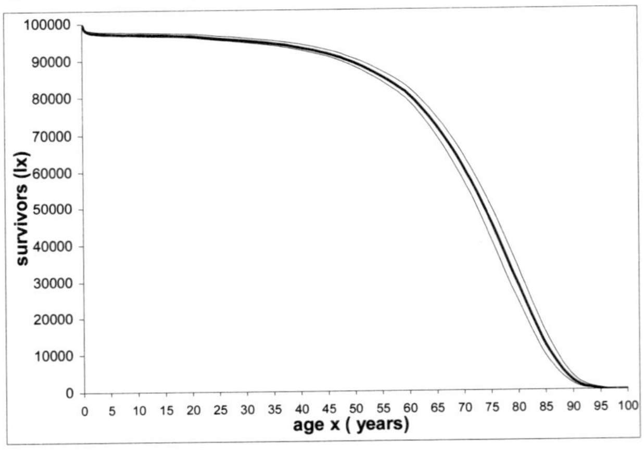

e_x with uncertainty interval

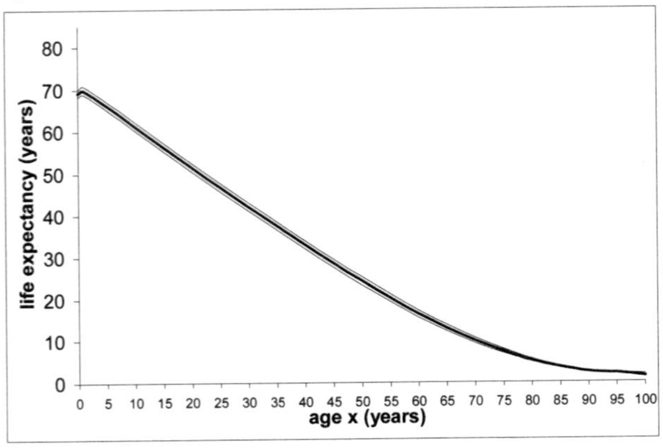

Distribution of deaths by age with uncertainty interval

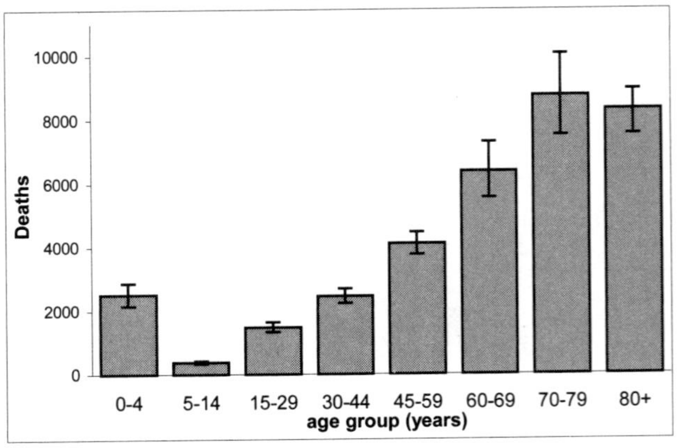

Tunisia — Tunisie — Túnez

Females, 2000 — *Femmes, 2000* — *Mujeres, 2000*

Age	$_nM_x$	$_nq_x$ mean	$_nq_x$ percentiles 2½	$_nq_x$ percentiles 97½	l_x mean	l_x percentiles 2½	l_x percentiles 97½	$_nd_x$	$_nL_x$	e_x mean	e_x percentiles 2½	e_x percentiles 97½	Actual deaths	Actual population
<1	0.0211	0.0207	0.0185	0.0251	100,000	100,000	100,000	2,068	98,138	73.4	72.0	74.3	1,735	82,295
1-4	0.0033	0.0132	0.0122	0.0154	97,932	97,486	98,151	1,298	388,612	73.9	72.7	74.9	1,070	320,385
5-9	0.0005	0.0023	0.0021	0.0027	96,634	95,984	96,955	223	482,611	70.9	69.8	71.9	213	459,336
10-14	0.0004	0.0020	0.0019	0.0023	96,411	95,727	96,750	195	481,566	66.0	64.9	67.1	204	504,315
15-19	0.0005	0.0025	0.0023	0.0029	96,216	95,503	96,571	238	480,484	61.2	60.1	62.2	251	506,906
20-24	0.0006	0.0031	0.0029	0.0036	95,978	95,230	96,351	300	479,138	56.3	55.2	57.3	293	468,029
25-29	0.0007	0.0033	0.0031	0.0038	95,677	94,886	96,074	320	477,588	51.5	50.4	52.5	276	411,996
30-34	0.0011	0.0054	0.0049	0.0062	95,358	94,521	95,781	511	475,513	46.6	45.6	47.6	410	381,880
35-39	0.0013	0.0067	0.0061	0.0077	94,847	93,938	95,310	635	472,649	41.9	40.9	42.9	454	337,789
40-44	0.0021	0.0102	0.0093	0.0117	94,212	93,216	94,725	962	468,658	37.1	36.2	38.1	585	285,317
45-49	0.0031	0.0155	0.0142	0.0177	93,251	92,126	93,840	1,441	462,651	32.5	31.6	33.5	708	227,140
50-54	0.0044	0.0217	0.0199	0.0249	91,810	90,493	92,511	1,989	454,074	28.0	27.1	28.9	714	162,958
55-59	0.0071	0.0348	0.0316	0.0399	89,820	88,243	90,674	3,130	441,276	23.5	22.7	24.5	939	132,343
60-64	0.0118	0.0573	0.0514	0.0654	86,690	84,723	87,806	4,970	421,026	19.3	18.5	20.2	1,462	123,847
65-69	0.0224	0.1062	0.0944	0.1200	81,720	79,184	83,290	8,681	386,899	15.3	14.6	16.1	2,553	113,775
70-74	0.0375	0.1713	0.1515	0.1910	73,039	69,682	75,431	12,510	333,922	11.8	11.2	12.6	2,779	74,178
75-79	0.0649	0.2793	0.2485	0.3080	60,529	56,372	64,005	16,907	260,380	8.8	8.3	9.4	3,270	50,355
80-84	0.1192	0.4593	0.4156	0.4959	43,623	39,008	48,097	20,035	168,025	6.2	5.9	6.7	2,810	23,567
85-89	0.2002	0.6671	0.6249	0.6986	23,588	19,664	28,106	15,736	78,597	4.4	4.1	4.7	2,617	13,071
90-94	0.3074	0.7996	0.7691	0.8208	7,851	5,926	10,544	6,278	20,424	3.1	2.9	3.2	759	2,468
95-99	0.4315	0.9403	0.9260	0.9497	1,574	1,062	2,434	1,480	3,429	2.3	2.3	2.4	119	275
100+	0.5538	1	1	1	94	53	180	94	170	1.8	1.8	1.9	8	15
All ages													*24,226*	*4,682,240*

ln($_nq_x$) with uncertainty interval

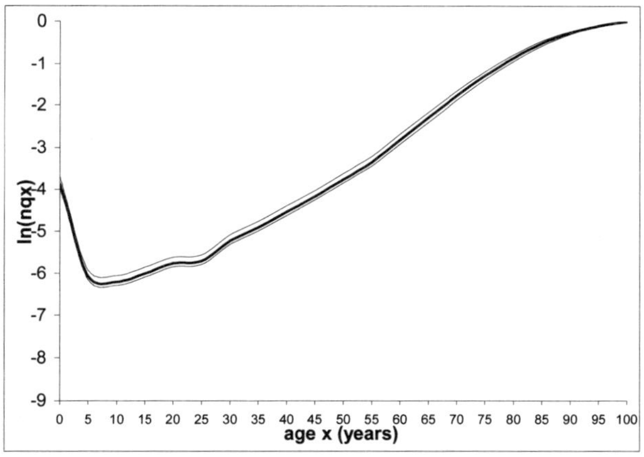

l_x with uncertainty interval

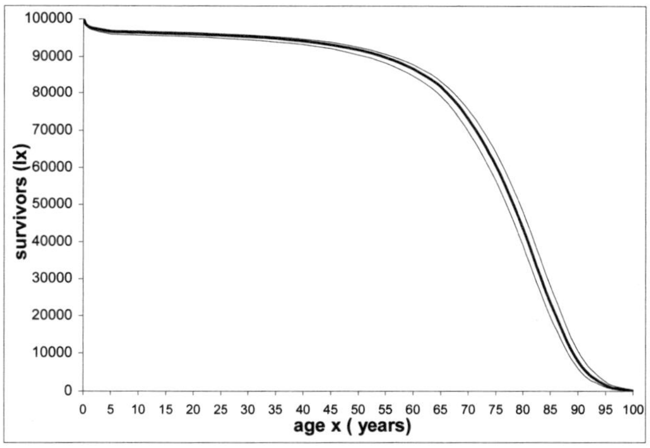

e_x with uncertainty interval

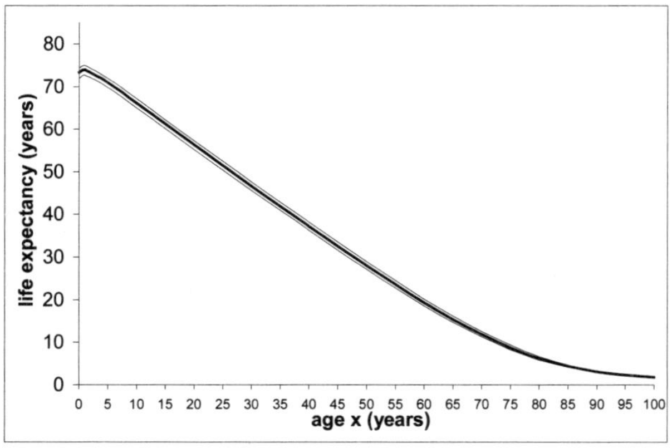

Distribution of deaths by age with uncertainty interval

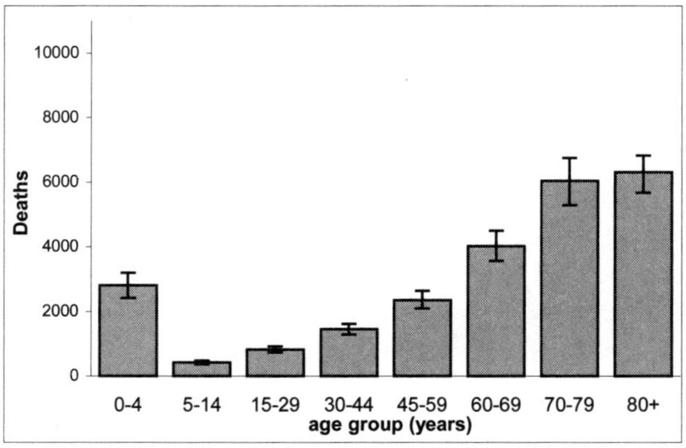

Turkey **Turquie** **Turquía**

Males, 2000 *Hommes, 2000* *Hombres, 2000*

Age	$_nM_x$	$_nq_x$ mean	$_nq_x$ percentiles 2½	$_nq_x$ percentiles 97½	l_x mean	l_x percentiles 2½	l_x percentiles 97½	$_nd_x$	$_nL_x$	e_x mean	e_x percentiles 2½	e_x percentiles 97½	Actual deaths	Actual population
<1	0.0441	0.0424	0.0388	0.0461	100,000	100,000	100,000	4,241	96,183	66.8	65.8	67.8	33,103	750,728
1-4	0.0018	0.0070	0.0065	0.0076	95,759	95,391	96,121	672	381,422	68.7	67.9	69.6	5,046	2,863,174
5-9	0.0008	0.0040	0.0038	0.0044	95,087	94,687	95,486	385	474,471	65.2	64.4	66.1	2,930	3,612,178
10-14	0.0008	0.0042	0.0039	0.0045	94,702	94,283	95,121	397	472,517	60.5	59.6	61.3	2,493	2,968,826
15-19	0.0012	0.0060	0.0056	0.0065	94,305	93,857	94,750	568	470,106	55.7	54.9	56.5	4,190	3,470,351
20-24	0.0014	0.0070	0.0065	0.0075	93,738	93,249	94,219	657	467,045	51.0	50.2	51.8	4,884	3,471,844
25-29	0.0018	0.0089	0.0082	0.0095	93,080	92,546	93,602	825	463,339	46.4	45.6	47.2	5,450	3,060,199
30-34	0.0022	0.0111	0.0103	0.0120	92,255	91,664	92,826	1,028	458,706	41.8	41.0	42.5	5,566	2,483,370
35-39	0.0030	0.0150	0.0138	0.0161	91,227	90,567	91,858	1,365	452,724	37.2	36.5	38.0	7,071	2,345,554
40-44	0.0048	0.0238	0.0220	0.0256	89,862	89,114	90,574	2,141	443,959	32.7	32.0	33.5	9,713	2,013,874
45-49	0.0071	0.0349	0.0322	0.0375	87,721	86,841	88,553	3,059	430,959	28.5	27.8	29.2	12,045	1,696,938
50-54	0.0109	0.0531	0.0491	0.0572	84,662	83,609	85,691	4,497	412,068	24.4	23.8	25.1	14,440	1,323,072
55-59	0.0167	0.0803	0.0745	0.0864	80,165	78,886	81,483	6,440	384,725	20.6	20.1	21.3	16,802	1,003,789
60-64	0.0236	0.1116	0.1035	0.1198	73,725	72,114	75,415	8,230	348,051	17.2	16.7	17.8	19,919	842,370
65-69	0.0330	0.1523	0.1413	0.1625	65,495	63,479	67,578	9,972	302,547	14.1	13.6	14.6	26,108	792,124
70-74	0.0484	0.2160	0.2008	0.2300	55,523	53,166	57,935	11,993	247,634	11.1	10.7	11.6	26,097	538,848
75-79	0.0715	0.3031	0.2832	0.3215	43,530	41,017	46,094	13,196	184,660	8.5	8.2	8.9	19,032	266,323
80-84	0.1234	0.4714	0.4465	0.4941	30,334	28,011	32,931	14,300	115,921	6.1	5.9	6.4	12,344	100,062
85-89	0.1998	0.6663	0.6432	0.6871	16,034	14,311	18,096	10,683	53,464	4.4	4.2	4.5	10,497	52,534
90-94	0.2990	0.7881	0.7707	0.8038	5,351	4,501	6,457	4,217	14,104	3.1	3.0	3.2	5,250	17,557
95-99	0.4174	0.9266	0.9174	0.9348	1,134	883	1,480	1,051	2,518	2.4	2.3	2.4	902	2,161
100+	0.5407	1	1	1	83	58	122	83	154	1.8	1.8	1.9	169	312
All ages													*244,049*	*33,676,188*

ln($_nq_x$) with uncertainty interval

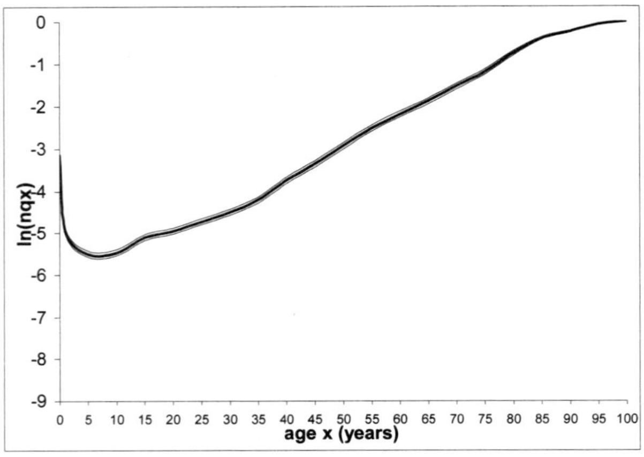

l_x with uncertainty interval

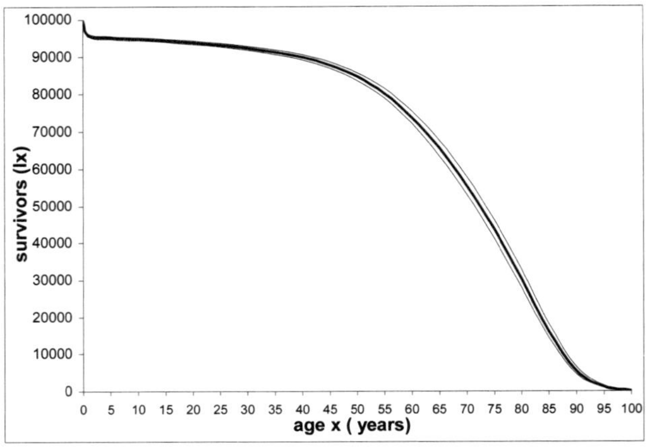

e_x with uncertainty interval

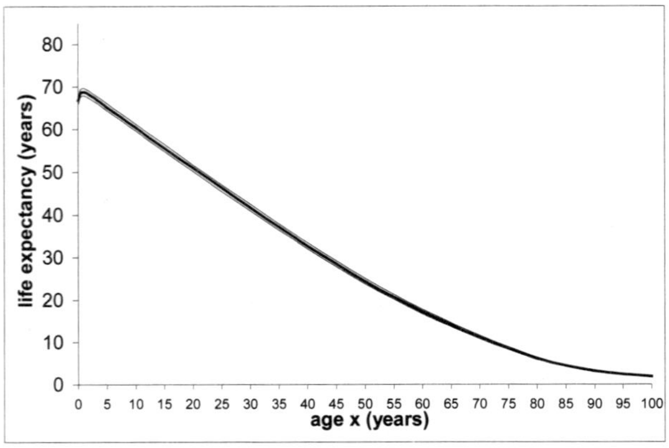

Distribution of deaths by age with uncertainty interval

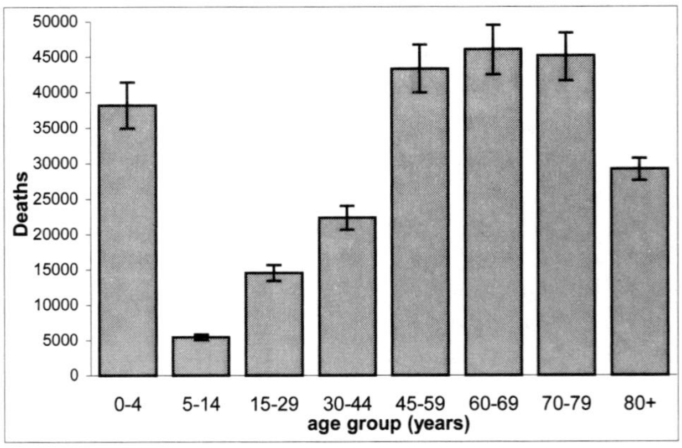

Turkey Turquie Turquía

Females, 2000 *Femmes, 2000* *Mujeres, 2000*

Age	$_nM_x$	$_nq_x$			l_x			$_nd_x$	$_nL_x$	e_x			Actual deaths	Actual population
			percentiles			percentiles					percentiles			
		mean	2½	97½	mean	2½	97½			mean	2½	97½		
<1	0.0407	0.0393	0.0361	0.0424	100,000	100,000	100,000	3,928	96,465	72.5	71.3	74.2	28,932	710,581
1-4	0.0016	0.0063	0.0056	0.0069	96,072	95,755	96,395	603	382,843	74.5	73.4	76.0	4,382	2,783,038
5-9	0.0005	0.0026	0.0023	0.0028	95,470	95,120	95,815	245	476,734	71.0	69.9	72.4	1,800	3,495,277
10-14	0.0004	0.0022	0.0019	0.0024	95,224	94,860	95,584	207	475,603	66.1	65.1	67.6	1,235	2,836,770
15-19	0.0007	0.0034	0.0030	0.0038	95,017	94,640	95,389	325	474,272	61.3	60.2	62.7	2,266	3,304,547
20-24	0.0007	0.0035	0.0031	0.0039	94,692	94,295	95,091	333	472,627	56.5	55.4	57.9	2,296	3,259,136
25-29	0.0009	0.0044	0.0039	0.0049	94,359	93,933	94,787	418	470,750	51.7	50.6	53.1	2,633	2,964,479
30-34	0.0011	0.0056	0.0049	0.0062	93,941	93,474	94,416	525	468,393	46.9	45.9	48.3	2,724	2,432,556
35-39	0.0016	0.0081	0.0071	0.0089	93,416	92,897	93,951	755	465,194	42.1	41.2	43.5	3,736	2,302,641
40-44	0.0024	0.0121	0.0106	0.0134	92,662	92,067	93,284	1,123	460,501	37.5	36.5	38.8	4,691	1,923,740
45-49	0.0036	0.0181	0.0157	0.0201	91,539	90,831	92,294	1,654	453,558	32.9	32.0	34.2	5,986	1,641,084
50-54	0.0055	0.0270	0.0234	0.0300	89,884	89,007	90,841	2,424	443,362	28.4	27.6	29.7	7,119	1,302,097
55-59	0.0090	0.0441	0.0380	0.0492	87,460	86,334	88,719	3,858	427,658	24.2	23.3	25.4	9,172	1,016,732
60-64	0.0134	0.0648	0.0556	0.0723	83,603	82,087	85,347	5,422	404,459	20.2	19.4	21.3	12,623	941,709
65-69	0.0211	0.1003	0.0860	0.1116	78,181	76,149	80,598	7,844	371,296	16.4	15.7	17.4	17,246	816,387
70-74	0.0345	0.1588	0.1371	0.1751	70,337	67,652	73,663	11,169	323,763	12.9	12.3	13.8	22,779	660,304
75-79	0.0534	0.2356	0.2065	0.2582	59,168	55,803	63,566	13,940	260,990	9.9	9.4	10.6	18,182	340,411
80-84	0.0901	0.3678	0.3315	0.3976	45,228	41,506	50,442	16,635	184,552	7.2	6.8	7.6	13,435	149,048
85-89	0.1668	0.5885	0.5532	0.6194	28,593	25,261	33,718	16,828	100,895	4.9	4.7	5.2	12,226	73,304
90-94	0.2703	0.7464	0.7197	0.7689	11,765	9,680	15,064	8,781	32,482	3.4	3.3	3.6	8,958	33,136
95-99	0.3838	0.8920	0.8765	0.9047	2,984	2,237	4,168	2,662	6,935	2.6	2.5	2.6	1,526	3,975
100+	0.4773	1	1	1	322	213	500	322	675	2.1	1.8	2.1	409	857
All ages													*184,354*	*32,991,809*

$\ln(_nq_x)$ with uncertainty interval

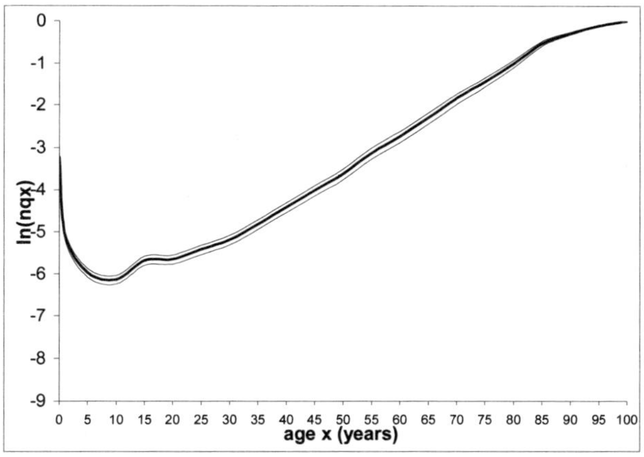

l_x with uncertainty interval

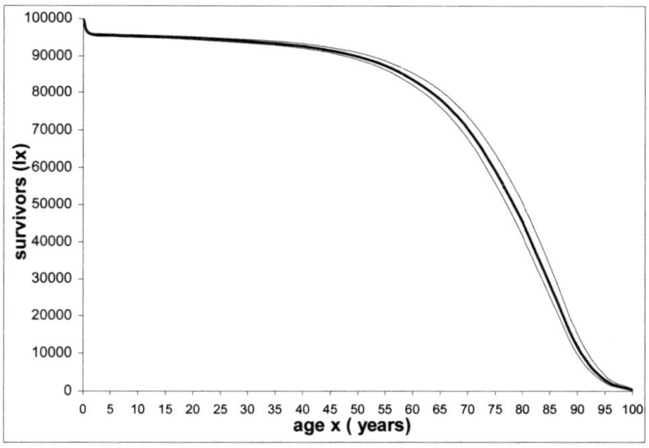

e_x with uncertainty interval

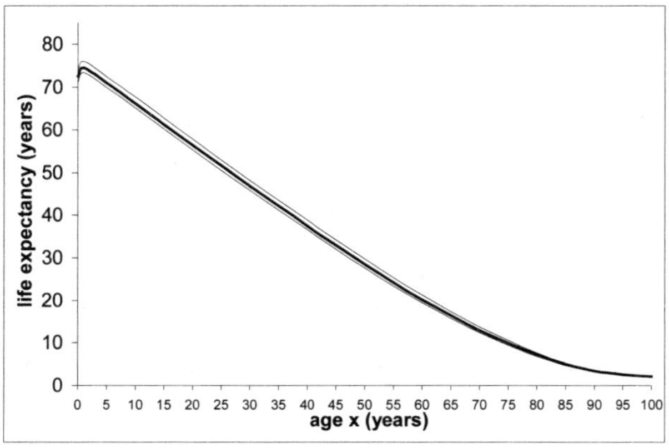

Distribution of deaths by age with uncertainty interval

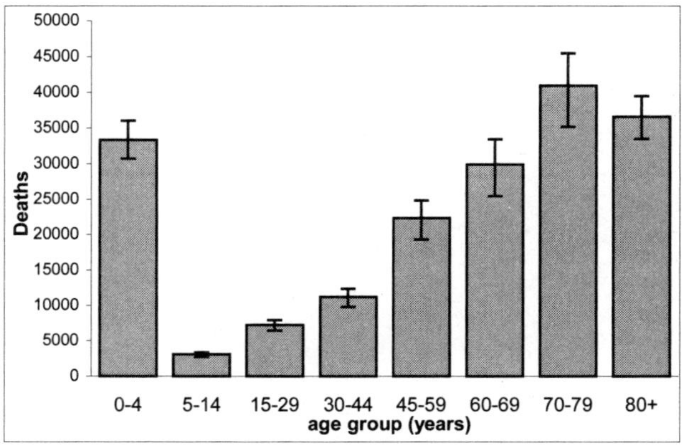

Turkmenistan | Turkménistan | Turkmenistán

Males, 2000 | *Hommes, 2000* | *Hombres, 2000*

Age	$_nM_x$	$_nq_x$			l_x			$_nd_x$	$_nL_x$	e_x			Actual deaths	Actual population
			percentiles			percentiles					percentiles			
		mean	2½	97½	mean	2½	97½			mean	2½	97½		
<1	0.0370	0.0358	0.0313	0.0402	100,000	100,000	100,000	3,581	96,777	60.0	59.3	60.8	2,331	62,997
1-4	0.0060	0.0237	0.0217	0.0257	96,419	95,983	96,871	2,281	380,202	61.3	60.7	61.8	1,455	242,439
5-9	0.0010	0.0050	0.0046	0.0054	94,138	93,540	94,737	470	469,516	58.7	58.1	59.3	304	303,891
10-14	0.0010	0.0050	0.0046	0.0054	93,669	93,037	94,298	467	467,175	54.0	53.4	54.6	295	294,536
15-19	0.0010	0.0050	0.0046	0.0054	93,201	92,538	93,863	465	464,845	49.3	48.7	49.9	248	248,236
20-24	0.0030	0.0149	0.0139	0.0160	92,736	92,043	93,429	1,381	460,231	44.5	43.9	45.1	645	214,847
25-29	0.0030	0.0149	0.0139	0.0159	91,356	90,577	92,134	1,360	453,379	40.1	39.5	40.8	570	190,162
30-34	0.0040	0.0198	0.0186	0.0211	89,996	89,137	90,850	1,782	445,523	35.7	35.1	36.3	690	172,510
35-39	0.0060	0.0296	0.0279	0.0313	88,214	87,257	89,157	2,607	434,550	31.4	30.7	32.0	943	157,093
40-44	0.0090	0.0440	0.0418	0.0463	85,606	84,523	86,659	3,768	418,613	27.2	26.6	27.9	1,196	132,911
45-49	0.0130	0.0630	0.0602	0.0658	81,839	80,607	83,014	5,152	396,314	23.4	22.7	24.0	1,228	94,460
50-54	0.0190	0.0907	0.0864	0.0955	76,687	75,305	77,989	6,955	366,046	19.8	19.1	20.4	1,167	61,396
55-59	0.0260	0.1221	0.1150	0.1296	69,732	68,233	71,173	8,512	327,379	16.5	15.9	17.1	1,059	40,727
60-64	0.0390	0.1777	0.1663	0.1902	61,220	59,689	62,723	10,877	278,906	13.5	12.9	14.0	1,917	49,160
65-69	0.0570	0.2495	0.2329	0.2679	50,343	48,875	51,833	12,558	220,318	10.8	10.3	11.3	1,935	33,950
70-74	0.0830	0.3437	0.3221	0.3675	37,784	35,882	39,575	12,986	156,457	8.6	8.2	9.0	2,169	26,129
75-79	0.1110	0.4344	0.4105	0.4605	24,799	22,747	26,828	10,774	97,059	6.8	6.5	7.1	1,222	11,005
80-84	0.1690	0.5940	0.5694	0.6201	14,025	12,272	15,816	8,331	49,297	5.1	4.9	5.3	799	4,726
85-89	0.2460	0.7616	0.7413	0.7826	5,694	4,662	6,810	4,336	17,628	3.9	3.7	4.0	615	2,502
90-94	0.2950	0.7825	0.7644	0.8014	1,357	1,013	1,762	1,062	3,600	3.2	3.1	3.3	219	744
95-99	0.3750	0.8824	0.8689	0.8961	295	201	415	260	695	2.6	2.5	2.7	63	169
100+	0.4890	1	1	1	35	21	54	35	71	2.0	2.0	2.2	10	21
All ages													*21,079*	*2,344,611*

ln($_nq_x$) with uncertainty interval

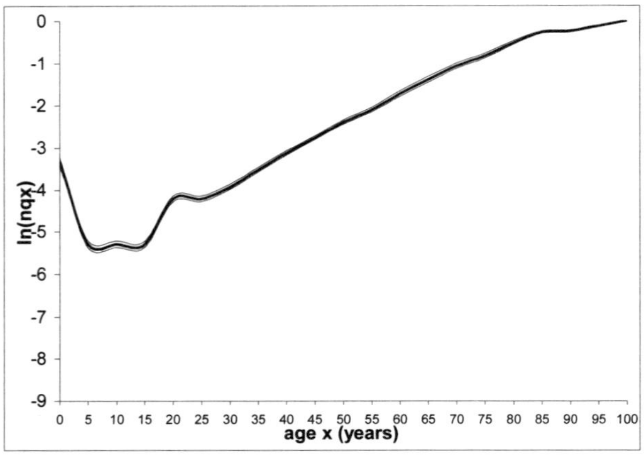

l_x with uncertainty interval

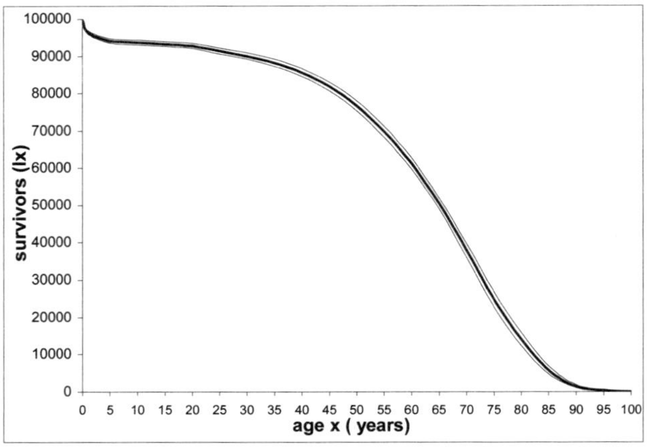

e_x with uncertainty interval

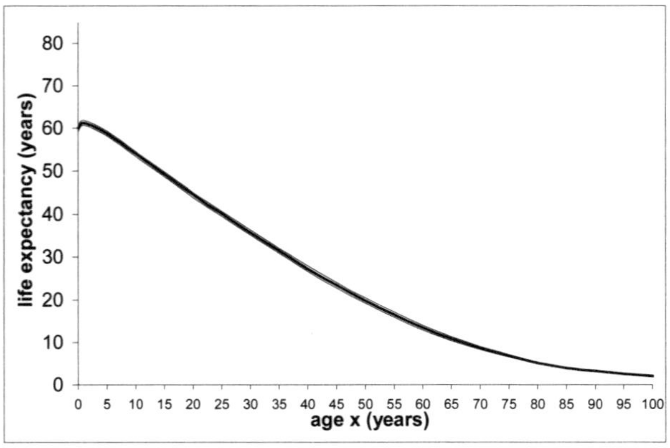

Distribution of deaths by age with uncertainty interval

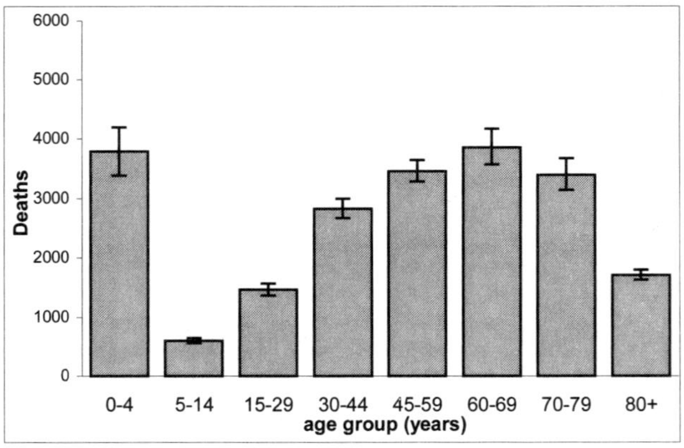

Turkmenistan Turkménistan Turkmenistán

Females, 2000 *Femmes, 2000* *Mujeres, 2000*

Age	$_nM_x$	$_nq_x$ mean	$_nq_x$ percentiles 2½	$_nq_x$ percentiles 97½	l_x mean	l_x percentiles 2½	l_x percentiles 97½	$_nd_x$	$_nL_x$	e_x mean	e_x percentiles 2½	e_x percentiles 97½	Actual deaths	Actual population
<1	0.0300	0.0292	0.0254	0.0329	100,000	100,000	100,000	2,921	97,371	64.9	64.1	65.8	1,803	60,112
1-4	0.0060	0.0237	0.0215	0.0259	97,079	96,711	97,461	2,297	382,803	65.9	65.2	66.7	1,417	236,203
5-9	0.0010	0.0050	0.0046	0.0054	94,782	94,234	95,331	473	472,728	63.5	62.8	64.2	296	296,009
10-14	0.0010	0.0050	0.0046	0.0054	94,309	93,723	94,896	470	470,371	58.8	58.1	59.5	287	287,082
15-19	0.0010	0.0050	0.0046	0.0054	93,839	93,215	94,461	468	468,025	54.0	53.4	54.8	244	244,202
20-24	0.0020	0.0100	0.0092	0.0108	93,371	92,710	94,028	929	464,532	49.3	48.7	50.1	425	212,733
25-29	0.0020	0.0100	0.0092	0.0108	92,442	91,711	93,165	920	459,910	44.8	44.2	45.5	377	188,626
30-34	0.0020	0.0100	0.0092	0.0107	91,522	90,725	92,306	911	455,334	40.2	39.6	40.9	337	168,257
35-39	0.0030	0.0149	0.0139	0.0160	90,611	89,753	91,452	1,349	449,684	35.6	35.0	36.3	503	167,805
40-44	0.0040	0.0198	0.0185	0.0212	89,262	88,317	90,182	1,768	441,893	31.1	30.5	31.8	563	140,774
45-49	0.0060	0.0296	0.0277	0.0316	87,495	86,445	88,508	2,586	431,009	26.7	26.1	27.3	612	102,050
50-54	0.0110	0.0535	0.0500	0.0570	84,909	83,722	86,040	4,545	413,181	22.4	21.8	23.1	730	66,379
55-59	0.0180	0.0861	0.0801	0.0914	80,364	78,965	81,658	6,921	384,515	18.5	18.0	19.1	780	43,351
60-64	0.0270	0.1265	0.1173	0.1347	73,442	71,809	74,932	9,288	343,993	15.0	14.5	15.6	1,516	56,147
65-69	0.0450	0.2022	0.1880	0.2156	64,155	62,352	65,913	12,975	288,335	11.8	11.3	12.4	1,877	41,721
70-74	0.0720	0.3051	0.2846	0.3256	51,180	49,174	53,513	15,614	216,862	9.2	8.8	9.6	2,700	37,495
75-79	0.1060	0.4190	0.3946	0.4445	35,565	33,369	38,213	14,901	140,575	7.2	6.8	7.5	2,434	22,960
80-84	0.1540	0.5560	0.5314	0.5822	20,665	18,703	23,008	11,489	74,601	5.5	5.3	5.8	1,850	12,015
85-89	0.2180	0.7055	0.6840	0.7280	9,176	7,837	10,747	6,474	29,696	4.3	4.1	4.4	1,386	6,358
90-94	0.2670	0.7413	0.7226	0.7609	2,702	2,132	3,387	2,003	7,502	3.5	3.4	3.6	515	1,929
95-99	0.3460	0.8489	0.8343	0.8639	699	510	936	594	1,715	2.8	2.7	2.9	138	400
100+	0.4690	1	1	1	106	69	154	106	225	2.1	2.1	2.2	17	37
All ages													*20,810*	*2,392,645*

ln($_nq_x$) with uncertainty interval

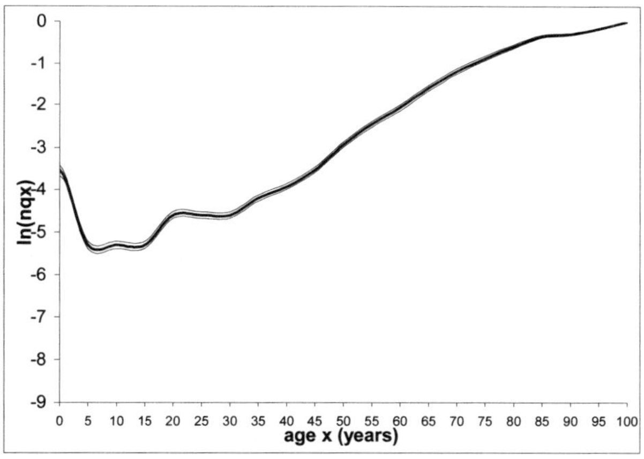

l_x with uncertainty interval

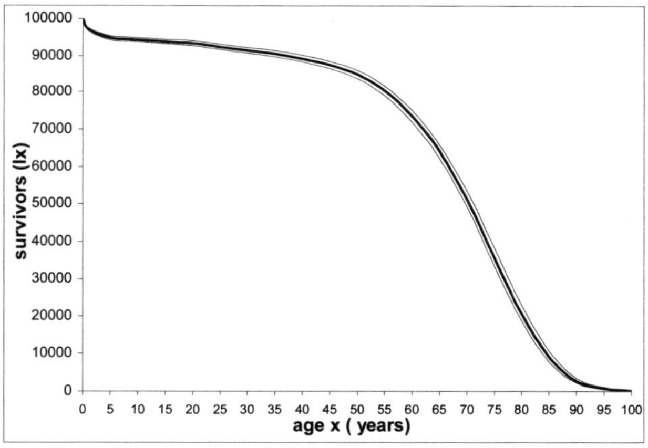

e_x with uncertainty interval

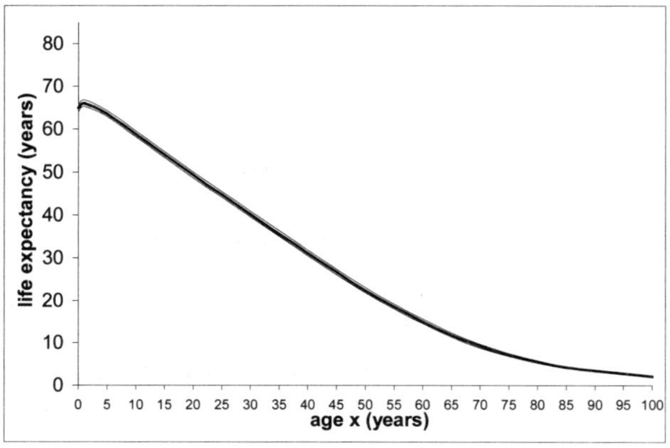

Distribution of deaths by age with uncertainty interval

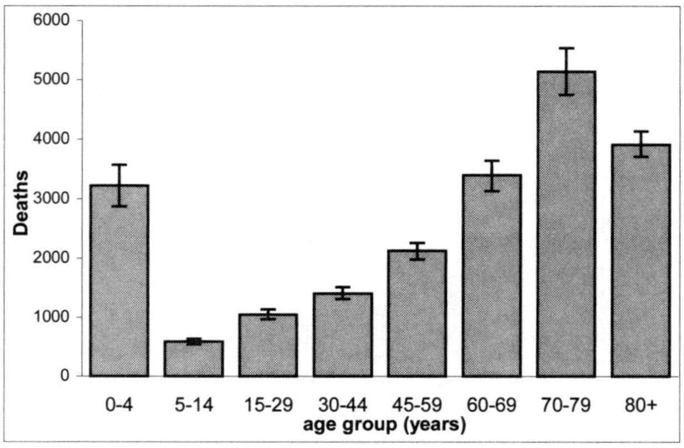

Tuvalu **Tuvalu** **Tuvalu**

Males, 2000 *Hommes, 2000* *Hombres, 2000*

Age	$_nM_x$	$_nq_x$			l_x			$_nd_x$	$_nL_x$	e_x			Actual deaths	Actual population
			percentiles			percentiles					percentiles			
		mean	2½	97½	mean	2½	97½			mean	2½	97½		
<1	0.0376	0.0364	0.0256	0.0474	100,000	100,000	100,000	3,639	96,725	63.6	62.0	64.9	4	117
1-4	0.0018	0.0071	0.0055	0.0086	96,361	95,265	97,438	682	383,807	64.9	64.0	66.0	1	488
5-9	0.0007	0.0034	0.0026	0.0041	95,679	94,442	96,902	321	477,592	61.4	60.4	62.4	0	600
10-14	0.0007	0.0036	0.0029	0.0044	95,358	94,057	96,646	344	475,927	56.6	55.6	57.6	0	631
15-19	0.0019	0.0093	0.0075	0.0111	95,013	93,647	96,369	883	472,858	51.8	50.8	52.8	1	579
20-24	0.0030	0.0150	0.0123	0.0177	94,130	92,607	95,649	1,410	467,125	47.3	46.3	48.3	2	505
25-29	0.0021	0.0105	0.0088	0.0122	92,720	90,972	94,471	972	461,169	42.9	42.0	43.9	1	455
30-34	0.0018	0.0090	0.0077	0.0104	91,748	89,861	93,643	829	456,668	38.4	37.5	39.3	1	399
35-39	0.0025	0.0126	0.0108	0.0144	90,919	88,923	92,926	1,146	451,732	33.7	32.8	34.6	1	340
40-44	0.0051	0.0251	0.0219	0.0283	89,773	87,639	91,921	2,250	443,242	29.1	28.3	30.0	1	282
45-49	0.0092	0.0451	0.0404	0.0500	87,523	85,155	89,904	3,951	427,740	24.8	24.0	25.7	2	237
50-54	0.0135	0.0653	0.0593	0.0715	83,573	80,897	86,241	5,457	404,221	20.8	20.1	21.7	2	181
55-59	0.0216	0.1025	0.0928	0.1122	78,116	75,187	80,971	8,004	370,568	17.1	16.4	17.9	3	154
60-64	0.0352	0.1618	0.1463	0.1765	70,111	67,086	72,980	11,342	322,202	13.8	13.1	14.5	4	122
65-69	0.0485	0.2163	0.1961	0.2358	58,769	55,897	61,731	12,710	262,072	10.9	10.4	11.6	4	92
70-74	0.0810	0.3368	0.3072	0.3665	46,060	42,913	49,375	15,513	191,516	8.3	7.8	8.8	5	59
75-79	0.1260	0.4791	0.4416	0.5143	30,547	27,512	34,059	14,635	116,148	6.2	5.8	6.6	5	36
80-84	0.1890	0.6418	0.6041	0.6792	15,912	13,606	18,714	10,212	54,031	4.6	4.4	4.9	3	14
85-89	0.2700	0.8060	0.7756	0.8353	5,700	4,426	7,295	4,594	17,016	3.5	3.3	3.6	1	5
90-94	0.4050	0.9142	0.8950	0.9320	1,106	752	1,627	1,011	2,497	2.4	2.3	2.5	0	1
95-99	0.5130	0.9800	0.9726	0.9861	95	53	169	93	195	2.1	2.1	2.1	0	0
100+	0.6300	1	1	1	2	1	5	2	3	1.6	1.6	1.7	0	0
All ages													*43*	*5,298*

ln($_nq_x$) with uncertainty interval

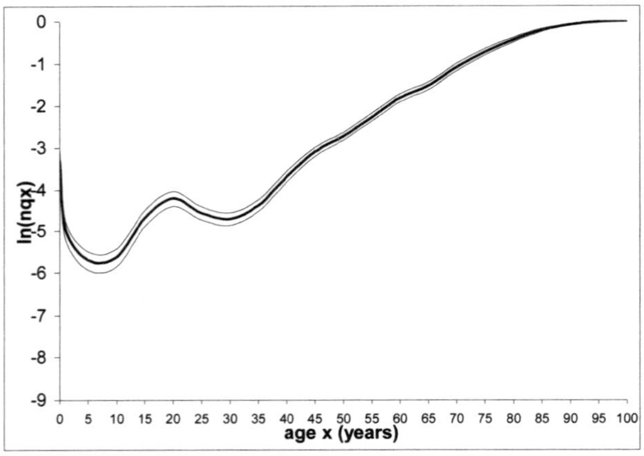

l_x with uncertainty interval

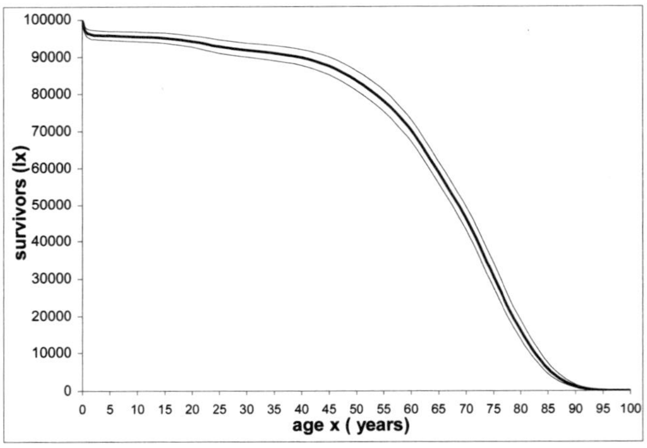

e_x with uncertainty interval

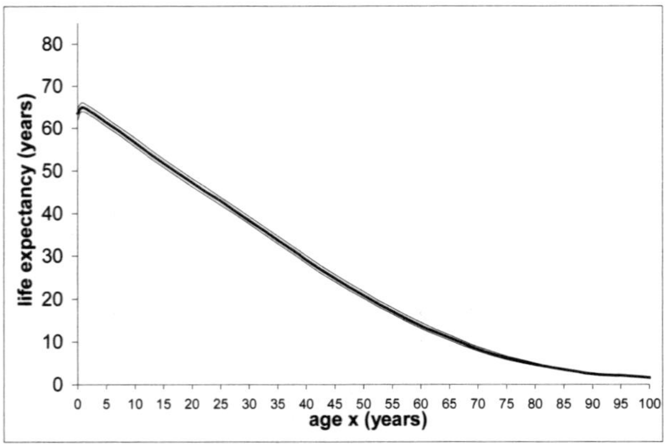

Distribution of deaths by age with uncertainty interval

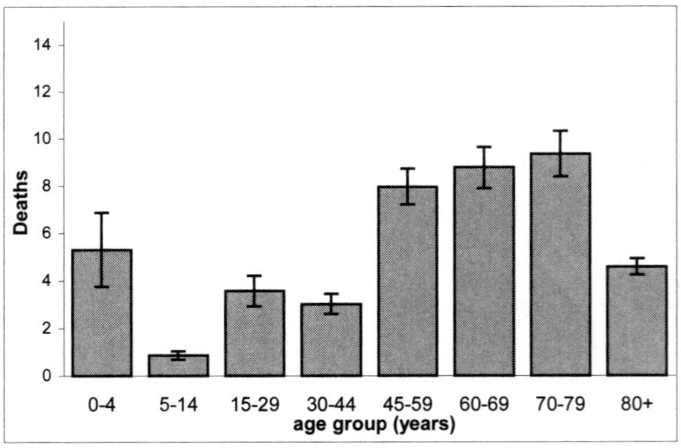

Tuvalu **Tuvalu** **Tuvalu**

Females, 2000 *Femmes, 2000* *Mujeres, 2000*

Age	$_nM_x$	$_nq_x$			l_x			$_nd_x$	$_nL_x$	e_x			Actual deaths	Actual population
			percentiles			percentiles					percentiles			
		mean	2½	97½	mean	2½	97½			mean	2½	97½		
<1	0.0354	0.0343	0.0256	0.0433	100,000	100,000	100,000	3,429	96,914	67.6	66.0	69.2	4	112
1-4	0.0015	0.0059	0.0047	0.0071	96,571	95,670	97,436	568	384,920	69.0	67.7	70.5	1	463
5-9	0.0004	0.0021	0.0017	0.0025	96,003	95,010	96,979	204	479,504	65.4	64.2	66.9	0	566
10-14	0.0012	0.0058	0.0047	0.0069	95,799	94,768	96,814	558	477,601	60.5	59.3	62.0	1	587
15-19	0.0019	0.0096	0.0078	0.0113	95,241	94,111	96,359	913	473,925	55.9	54.7	57.4	1	537
20-24	0.0019	0.0092	0.0076	0.0108	94,329	93,043	95,606	870	469,468	51.4	50.3	52.9	1	457
25-29	0.0017	0.0083	0.0069	0.0096	93,459	92,037	94,878	771	465,366	46.9	45.7	48.3	1	385
30-34	0.0013	0.0063	0.0053	0.0073	92,687	91,151	94,226	584	461,977	42.2	41.1	43.6	0	327
35-39	0.0018	0.0087	0.0074	0.0101	92,103	90,484	93,728	804	458,506	37.5	36.4	38.9	1	300
40-44	0.0035	0.0173	0.0147	0.0198	91,299	89,573	93,038	1,576	452,557	32.8	31.7	34.2	1	253
45-49	0.0062	0.0305	0.0264	0.0347	89,723	87,802	91,667	2,739	441,771	28.3	27.2	29.7	1	219
50-54	0.0101	0.0494	0.0433	0.0556	86,985	84,774	89,236	4,298	424,180	24.1	23.1	25.4	2	167
55-59	0.0159	0.0766	0.0669	0.0859	82,687	80,106	85,333	6,337	397,591	20.3	19.2	21.5	2	144
60-64	0.0234	0.1105	0.0958	0.1239	76,350	73,396	79,390	8,439	360,650	16.7	15.8	17.9	3	121
65-69	0.0332	0.1531	0.1324	0.1727	67,910	64,676	71,178	10,395	313,565	13.5	12.6	14.5	3	94
70-74	0.0546	0.2402	0.2089	0.2712	57,516	53,871	61,456	13,816	253,039	10.5	9.7	11.4	4	66
75-79	0.0878	0.3598	0.3165	0.4040	43,700	39,790	48,538	15,724	179,189	8.0	7.3	8.7	4	46
80-84	0.1268	0.4813	0.4333	0.5302	27,976	23,987	32,969	13,463	106,221	6.1	5.6	6.6	3	22
85-89	0.1950	0.6555	0.6097	0.7004	14,512	11,447	18,464	9,512	48,781	4.5	4.1	4.8	2	10
90-94	0.2925	0.7790	0.7422	0.8139	5,000	3,429	7,192	3,895	13,316	3.2	3.0	3.4	1	3
95-99	0.3900	0.8986	0.8746	0.9204	1,105	638	1,841	993	2,547	2.5	2.4	2.6	0	1
100+	0.4875	1	1	1	112	51	228	112	230	2.1	1.8	2.1	0	0
All ages													*35*	*4,877*

ln($_nq_x$) with uncertainty interval

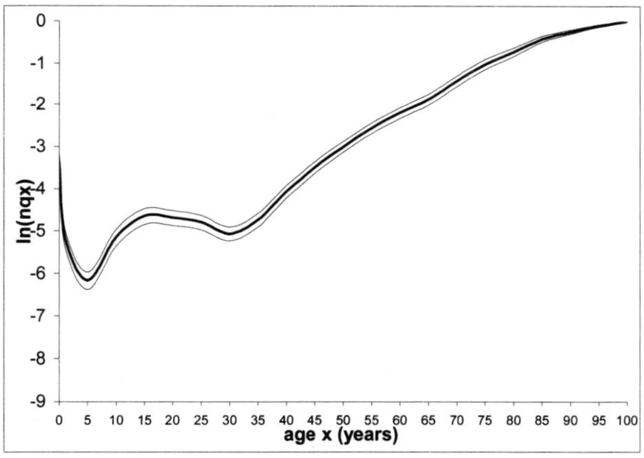

l_x with uncertainty interval

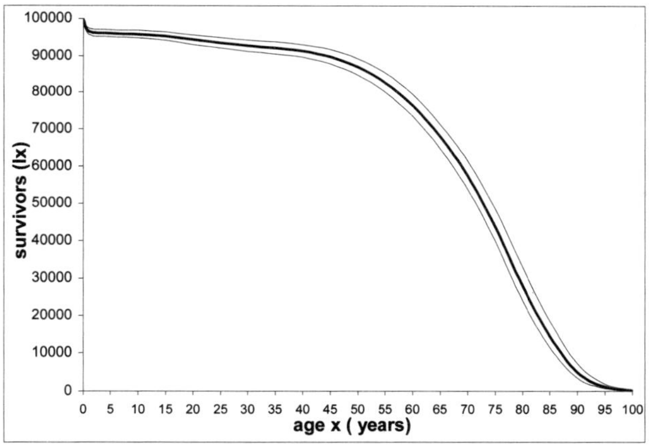

e_x with uncertainty interval

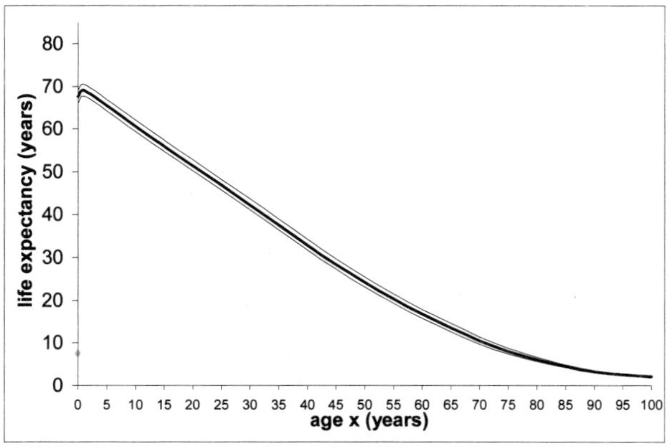

Distribution of deaths by age with uncertainty interval

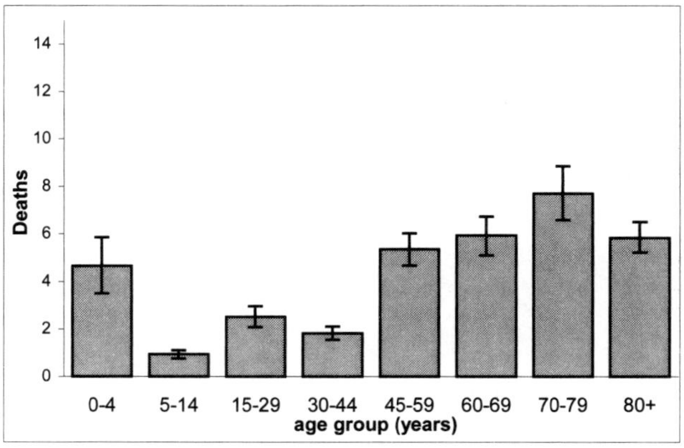

Uganda | Ouganda | Uganda

Males, 2000 | *Hommes, 2000* | *Hombres, 2000*

Age	$_nM_x$	$_nq_x$ mean	$_nq_x$ percentiles 2½	$_nq_x$ percentiles 97½	l_x mean	l_x percentiles 2½	l_x percentiles 97½	$_nd_x$	$_nL_x$	e_x mean	e_x percentiles 2½	e_x percentiles 97½	Actual deaths	Actual population
<1	0.1343	0.1198	0.1040	0.1372	100,000	100,000	100,000	11,979	89,219	43.5	40.1	47.3	72,567	540,464
1-4	0.0131	0.0510	0.0435	0.0587	88,021	86,280	89,599	4,487	341,316	48.4	44.6	53.0	23,893	1,817,668
5-9	0.0031	0.0152	0.0128	0.0179	83,534	82,071	85,005	1,272	414,492	46.9	43.3	51.4	5,702	1,858,226
10-14	0.0021	0.0103	0.0086	0.0123	82,262	80,716	83,899	851	409,184	42.6	39.1	47.0	3,192	1,534,444
15-19	0.0033	0.0165	0.0136	0.0196	81,411	79,790	83,172	1,341	403,703	38.0	34.5	42.4	4,265	1,283,916
20-24	0.0076	0.0372	0.0305	0.0448	80,070	78,330	82,028	2,982	392,897	33.6	30.1	38.0	8,052	1,061,005
25-29	0.0143	0.0691	0.0556	0.0841	77,089	75,002	79,487	5,328	372,124	29.8	26.4	34.1	11,977	836,555
30-34	0.0221	0.1046	0.0824	0.1286	71,761	68,808	74,941	7,508	340,034	26.8	23.6	31.0	14,238	644,789
35-39	0.0253	0.1191	0.0920	0.1474	64,253	59,958	68,478	7,650	302,139	24.7	21.8	28.5	12,246	483,669
40-44	0.0273	0.1276	0.0975	0.1580	56,603	51,123	61,822	7,224	264,956	22.7	20.0	26.2	10,551	387,008
45-49	0.0290	0.1353	0.1029	0.1669	49,379	43,043	55,671	6,680	230,197	20.6	18.3	23.7	9,157	315,566
50-54	0.0304	0.1413	0.1074	0.1742	42,699	35,861	49,822	6,032	198,417	18.5	16.4	21.1	7,809	256,878
55-59	0.0321	0.1487	0.1135	0.1826	36,667	29,655	44,469	5,452	169,707	16.1	14.4	18.4	6,467	201,299
60-64	0.0406	0.1844	0.1424	0.2241	31,215	24,300	39,420	5,757	141,686	13.5	12.0	15.4	5,643	138,884
65-69	0.0545	0.2398	0.1884	0.2867	25,459	18,934	33,809	6,104	112,035	11.0	9.8	12.6	6,244	114,602
70-74	0.0794	0.3313	0.2678	0.3873	19,355	13,599	27,440	6,413	80,742	8.6	7.7	9.9	6,210	78,179
75-79	0.1163	0.4506	0.3773	0.5121	12,942	8,433	20,092	5,832	50,129	6.6	5.9	7.6	5,272	45,314
80-84	0.1697	0.5958	0.5193	0.6560	7,110	4,168	12,497	4,236	24,959	5.1	4.6	5.8	3,370	19,857
85-89	0.2418	0.7535	0.6855	0.8038	2,874	1,440	5,975	2,165	8,955	3.8	3.5	4.3	1,302	5,384
90-94	0.3359	0.8366	0.7804	0.8767	708	285	1,856	593	1,764	2.8	2.6	3.2	285	848
95-99	0.4554	0.9623	0.9365	0.9772	116	35	400	111	245	2.2	2.1	2.3	31	69
100+	0.6020	1	1	1	4	1	25	4	7	1.7	1.6	1.8	2	3
All ages													*218,473*	*11,624,627*

ln($_nq_x$) with uncertainty interval

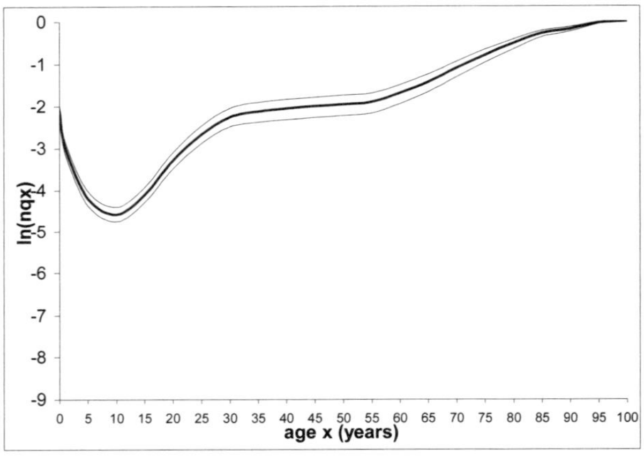

l_x with uncertainty interval

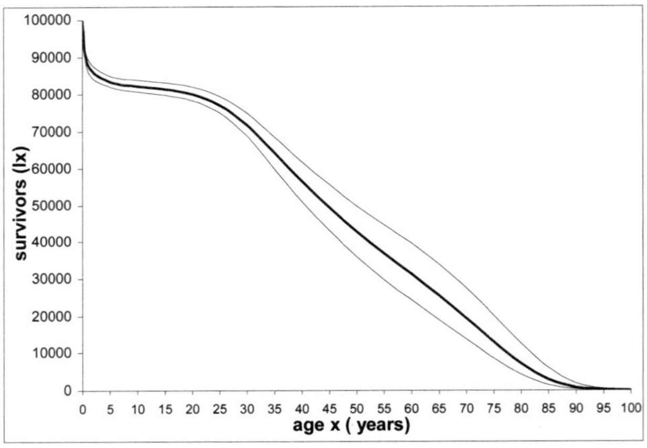

e_x with uncertainty interval

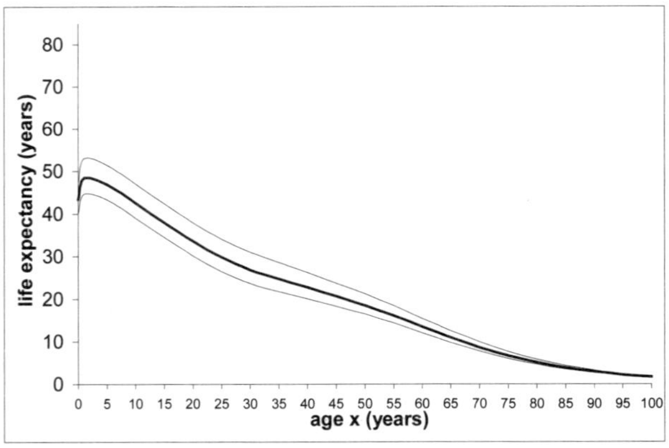

Distribution of deaths by age with uncertainty interval

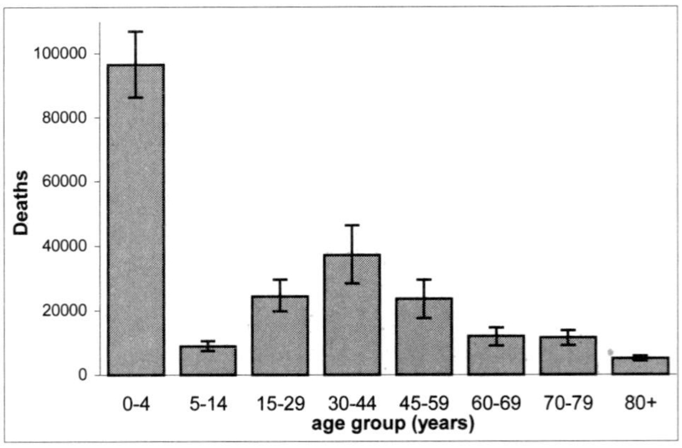

Uganda

Ouganda

Uganda

Females, 2000

Femmes, 2000

Mujeres, 2000

Age	$_nM_x$	$_nq_x$			l_x			$_nd_x$	$_nL_x$	e_x			Actual deaths	Actual population
			percentiles			percentiles					percentiles			
		mean	2½	97½	mean	2½	97½			mean	2½	97½		
<1	0.1096	0.0998	0.0866	0.1152	100,000	100,000	100,000	9,975	91,022	44.6	41.2	48.5	57,506	524,722
1-4	0.0147	0.0567	0.0492	0.0640	90,025	88,484	91,339	5,105	347,847	48.6	44.7	53.2	26,542	1,808,500
5-9	0.0034	0.0171	0.0146	0.0198	84,920	83,576	86,269	1,452	420,968	47.4	43.6	51.9	6,380	1,849,748
10-14	0.0026	0.0127	0.0108	0.0148	83,468	81,970	84,957	1,061	414,686	43.2	39.5	47.7	3,921	1,532,363
15-19	0.0060	0.0294	0.0247	0.0346	82,407	80,779	84,032	2,421	405,982	38.7	35.0	43.2	7,660	1,284,725
20-24	0.0126	0.0610	0.0505	0.0727	79,986	78,064	81,923	4,876	387,739	34.8	31.2	39.2	13,336	1,060,406
25-29	0.0197	0.0937	0.0757	0.1132	75,110	72,611	77,671	7,039	357,952	31.9	28.4	36.2	16,231	825,384
30-34	0.0235	0.1111	0.0879	0.1356	68,071	64,661	71,566	7,566	321,440	29.9	26.7	33.9	14,671	623,339
35-39	0.0221	0.1047	0.0816	0.1285	60,505	55,891	64,959	6,336	286,686	28.4	25.5	32.0	10,379	469,615
40-44	0.0203	0.0964	0.0745	0.1185	54,169	48,708	59,446	5,222	257,792	26.4	23.9	29.6	7,849	387,479
45-49	0.0190	0.0907	0.0698	0.1114	48,947	42,938	54,929	4,442	233,633	23.9	21.7	26.8	6,241	328,303
50-54	0.0206	0.0980	0.0754	0.1203	44,506	38,157	51,006	4,364	211,619	21.1	19.1	23.6	5,671	275,019
55-59	0.0236	0.1115	0.0860	0.1365	40,142	33,586	47,137	4,478	189,516	18.1	16.4	20.3	5,286	223,724
60-64	0.0312	0.1447	0.1124	0.1758	35,664	29,038	43,081	5,161	165,419	15.0	13.6	17.0	5,007	160,493
65-69	0.0439	0.1980	0.1559	0.2375	30,503	23,992	38,237	6,039	137,419	12.2	10.9	13.8	5,950	135,382
70-74	0.0664	0.2846	0.2293	0.3344	24,464	18,376	32,276	6,962	104,915	9.6	8.6	10.9	6,242	94,058
75-79	0.0985	0.3952	0.3289	0.4522	17,502	12,331	24,876	6,917	70,217	7.4	6.6	8.4	5,563	56,464
80-84	0.1484	0.5412	0.4696	0.5995	10,585	6,848	16,695	5,729	38,602	5.5	5.0	6.3	3,862	26,027
85-89	0.2175	0.7044	0.6391	0.7553	4,856	2,756	8,855	3,421	15,728	4.1	3.8	4.6	1,743	8,014
90-94	0.3100	0.8030	0.7494	0.8441	1,435	674	3,196	1,153	3,718	3.0	2.8	3.4	488	1,573
95-99	0.4296	0.9385	0.9088	0.9584	283	105	801	265	618	2.3	2.2	2.4	80	186
100+	0.5788	1	1	1	17	4	73	17	30	1.7	1.7	1.8	6	11
All ages													*210,612*	*11,675,535*

$\ln(_nq_x)$ with uncertainty interval

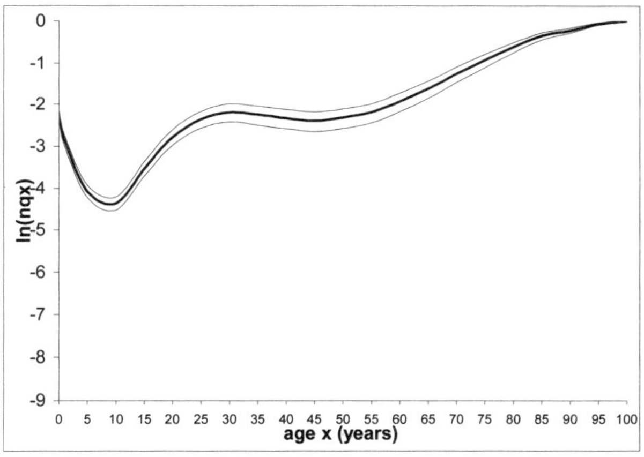

l_x with uncertainty interval

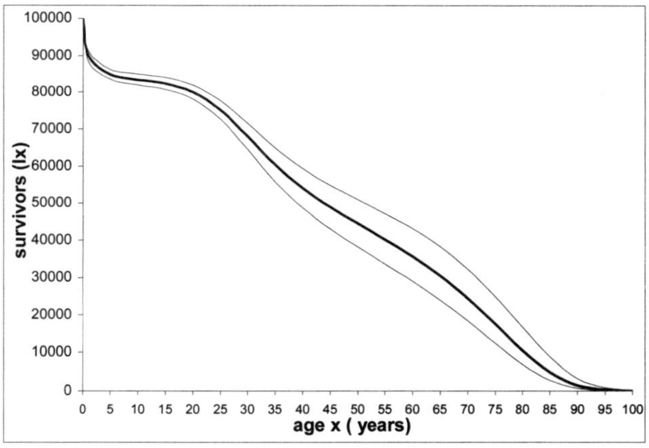

e_x with uncertainty interval

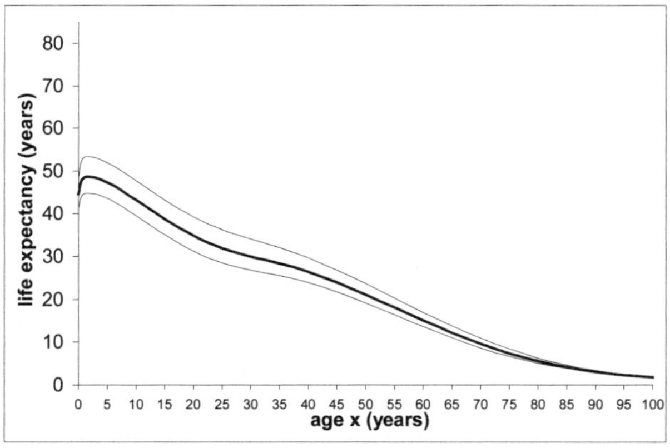

Distribution of deaths by age with uncertainty interval

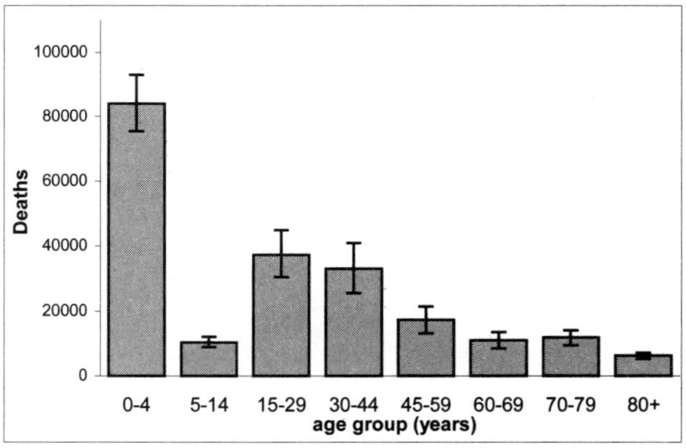

Ukraine **Ukraine** **Ucrania**

Males, 2000 *Hommes, 2000* *Hombres, 2000*

Age	$_nM_x$	$_nq_x$			l_x			$_nd_x$	$_nL_x$	e_x			Actual deaths	Actual population
			percentiles			percentiles					percentiles			
		mean	2½	97½	mean	2½	97½			mean	2½	97½		
<1	0.0139	0.0138	0.0129	0.0146	100,000	100,000	100,000	1,375	98,762	62.6	62.0	63.2	3,081	221,212
1-4	0.0010	0.0040	0.0038	0.0042	98,625	98,540	98,710	392	393,557	62.5	61.9	63.1	897	900,462
5-9	0.0005	0.0022	0.0021	0.0024	98,233	98,133	98,332	221	490,610	58.7	58.1	59.3	672	1,493,055
10-14	0.0005	0.0023	0.0022	0.0024	98,012	97,902	98,120	224	489,498	53.8	53.2	54.4	871	1,903,642
15-19	0.0012	0.0061	0.0058	0.0065	97,788	97,667	97,906	601	487,436	48.9	48.4	49.5	2,356	1,911,085
20-24	0.0026	0.0131	0.0124	0.0138	97,187	97,039	97,334	1,276	482,743	44.2	43.7	44.8	4,799	1,814,933
25-29	0.0034	0.0167	0.0157	0.0176	95,910	95,706	96,126	1,597	475,560	39.8	39.2	40.3	6,121	1,822,662
30-34	0.0050	0.0245	0.0231	0.0258	94,313	94,026	94,614	2,308	465,796	35.4	34.9	35.9	8,004	1,615,216
35-39	0.0069	0.0338	0.0320	0.0356	92,005	91,599	92,426	3,110	452,251	31.2	30.8	31.7	12,690	1,845,463
40-44	0.0098	0.0481	0.0455	0.0506	88,895	88,334	89,471	4,272	433,795	27.3	26.8	27.7	17,933	1,820,771
45-49	0.0141	0.0683	0.0648	0.0718	84,623	83,862	85,398	5,776	408,674	23.5	23.1	23.9	23,440	1,658,453
50-54	0.0193	0.0921	0.0878	0.0965	78,847	77,844	79,862	7,263	376,075	20.0	19.7	20.5	24,312	1,258,838
55-59	0.0284	0.1325	0.1269	0.1383	71,583	70,329	72,851	9,488	334,197	16.8	16.5	17.2	31,058	1,093,936
60-64	0.0386	0.1760	0.1686	0.1828	62,095	60,606	63,601	10,926	283,161	14.0	13.7	14.3	54,330	1,408,039
65-69	0.0538	0.2370	0.2278	0.2457	51,169	49,549	52,813	12,125	225,534	11.5	11.2	11.7	45,183	840,443
70-74	0.0737	0.3112	0.3006	0.3214	39,044	37,460	40,658	12,150	164,847	9.3	9.1	9.5	58,515	793,917
75-79	0.0986	0.3955	0.3839	0.4062	26,894	25,549	28,355	10,636	107,883	7.3	7.2	7.5	35,428	359,373
80-84	0.1495	0.5442	0.5322	0.5553	16,259	15,228	17,445	8,848	59,175	5.4	5.3	5.5	21,359	142,853
85-89	0.2300	0.7302	0.7202	0.7394	7,411	6,779	8,133	5,411	23,528	4.0	3.9	4.0	20,418	88,773
90-94	0.3300	0.8291	0.8216	0.8362	2,000	1,770	2,270	1,658	5,025	2.9	2.8	2.9	7,318	22,177
95-99	0.4600	0.9664	0.9636	0.9689	342	291	405	330	718	2.2	2.1	2.2	1,487	3,232
100+	0.5800	1	1	1	11	9	15	11	20	1.7	1.7	1.7	121	209
All ages													*380,394*	*23,018,744*

$\ln(_nq_x)$ with uncertainty interval

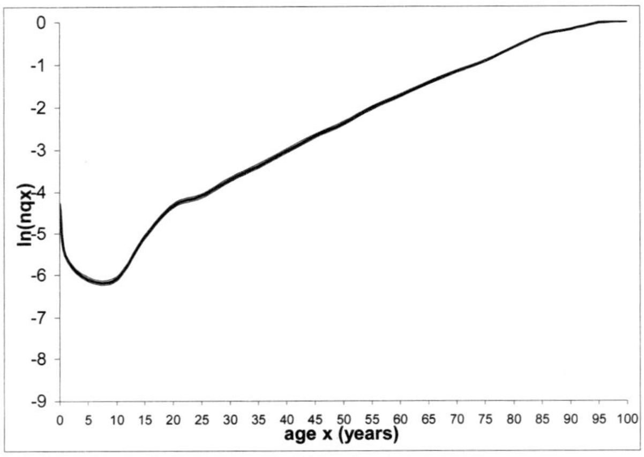

l_x with uncertainty interval

e_x with uncertainty interval

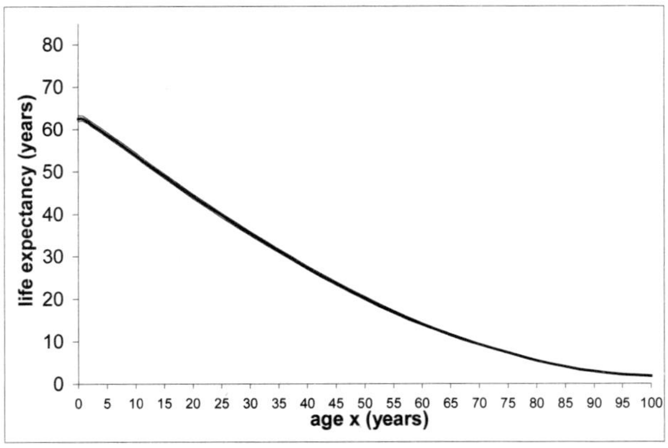

Distribution of deaths by age with uncertainty interval

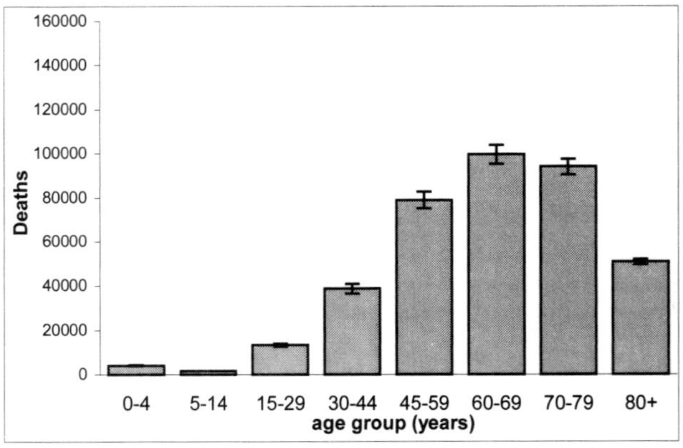

Ukraine Ukraine Ucrania

Females, 2000 *Femmes, 2000* *Mujeres, 2000*

Age	$_nM_x$	$_nq_x$			l_x			$_nd_x$	$_nL_x$	e_x			Actual deaths	Actual population
			percentiles			percentiles					percentiles			
		mean	2 1/2	97 1/2	mean	2 1/2	97 1/2			mean	2 1/2	97 1/2		
<1	0.0097	0.0096	0.0086	0.0106	100,000	100,000	100,000	962	99,134	73.3	72.8	73.8	2,040	210,078
1-4	0.0008	0.0033	0.0030	0.0036	99,038	98,940	99,137	325	395,370	73.0	72.6	73.5	705	857,942
5-9	0.0003	0.0014	0.0013	0.0015	98,712	98,588	98,837	138	493,218	69.2	68.8	69.7	397	1,421,200
10-14	0.0003	0.0013	0.0012	0.0014	98,575	98,439	98,709	129	492,552	64.3	63.9	64.8	479	1,832,809
15-19	0.0005	0.0024	0.0023	0.0026	98,446	98,300	98,590	241	491,627	59.4	59.0	59.9	909	1,852,544
20-24	0.0007	0.0034	0.0032	0.0037	98,205	98,040	98,368	336	490,185	54.5	54.1	55.0	1,222	1,783,948
25-29	0.0010	0.0048	0.0044	0.0051	97,869	97,678	98,057	465	488,183	49.7	49.3	50.2	1,680	1,762,884
30-34	0.0013	0.0064	0.0059	0.0068	97,404	97,179	97,626	619	485,473	44.9	44.6	45.4	2,080	1,631,817
35-39	0.0017	0.0087	0.0081	0.0093	96,785	96,517	97,050	840	481,826	40.2	39.8	40.7	3,364	1,929,003
40-44	0.0028	0.0138	0.0129	0.0147	95,945	95,621	96,265	1,325	476,414	35.5	35.2	36.0	5,468	1,966,397
45-49	0.0043	0.0212	0.0200	0.0225	94,620	94,215	95,021	2,006	468,086	31.0	30.7	31.4	7,979	1,861,687
50-54	0.0065	0.0320	0.0303	0.0337	92,614	92,098	93,125	2,964	455,660	26.6	26.3	27.0	9,748	1,498,426
55-59	0.0103	0.0501	0.0476	0.0526	89,650	88,992	90,301	4,494	437,014	22.4	22.1	22.8	14,548	1,414,701
60-64	0.0156	0.0752	0.0713	0.0787	85,156	84,323	85,980	6,403	409,770	18.5	18.2	18.8	30,126	1,927,818
65-69	0.0255	0.1201	0.1139	0.1252	78,752	77,737	79,754	9,456	370,121	14.8	14.5	15.1	32,576	1,275,027
70-74	0.0414	0.1877	0.1785	0.1944	69,296	68,095	70,548	13,004	313,970	11.4	11.2	11.7	62,339	1,505,102
75-79	0.0697	0.2967	0.2845	0.3070	56,292	54,856	57,955	16,702	239,703	8.5	8.3	8.7	67,044	962,174
80-84	0.1250	0.4761	0.4622	0.4905	39,589	38,142	41,468	18,847	150,831	6.0	5.9	6.2	55,369	443,119
85-89	0.2059	0.6796	0.6669	0.6933	20,743	19,692	22,289	14,096	68,473	4.3	4.2	4.4	64,927	315,385
90-94	0.3116	0.8053	0.7959	0.8154	6,646	6,049	7,369	5,352	17,176	3.0	3.0	3.1	26,706	85,702
95-99	0.4334	0.9421	0.9377	0.9467	1,294	1,117	1,486	1,219	2,813	2.3	2.3	2.3	4,990	11,515
100+	0.5537	1	1	1	75	60	92	75	135	1.8	1.8	1.9	80	145
All ages													*394,776*	*26,549,423*

ln($_nq_x$) with uncertainty interval

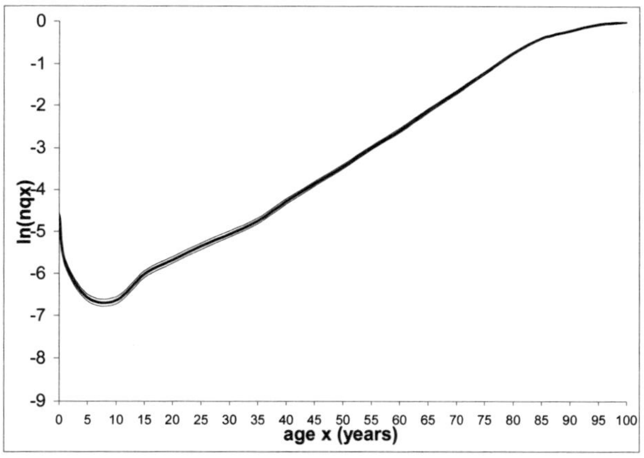

l_x with uncertainty interval

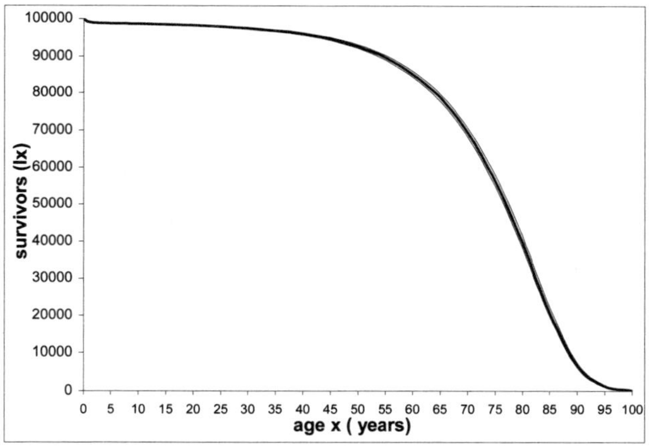

e_x with uncertainty interval

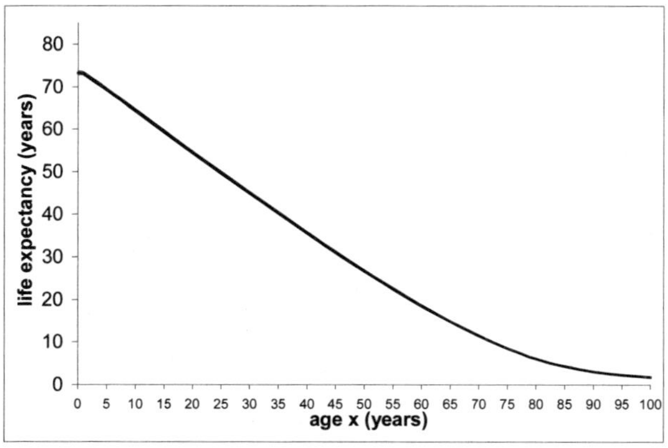

Distribution of deaths by age with uncertainty interval

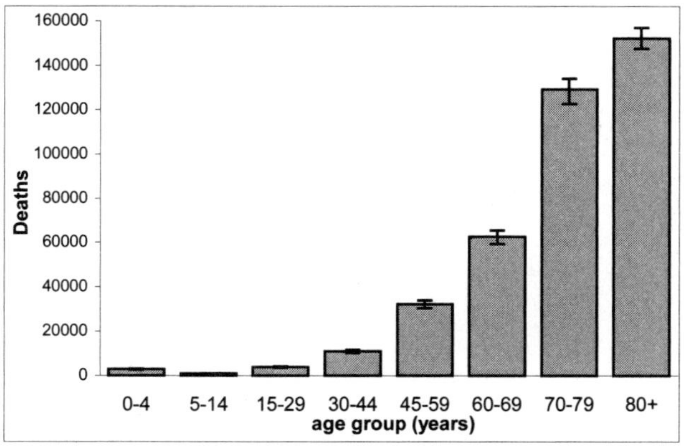

United Arab Emirates **Emirats arabes unis** **Emiratos Arabes Unidos**

Males, 2000 *Hommes, 2000* *Hombres, 2000*

Age	$_nM_x$	$_nq_x$			l_x			$_nd_x$	$_nL_x$	e_x			Actual deaths	Actual population
			percentiles			percentiles					percentiles			
		mean	2 1/2	97 1/2	mean	2 1/2	97 1/2			mean	2 1/2	97 1/2		
<1	0.0136	0.0135	0.0109	0.0162	100,000	100,000	100,000	1,348	98,787	72.3	71.1	73.5	280	20,549
1-4	0.0006	0.0026	0.0022	0.0030	98,652	98,382	98,913	252	394,004	72.2	71.1	73.4	52	81,531
5-9	0.0003	0.0017	0.0014	0.0019	98,400	98,101	98,700	163	491,594	68.4	67.3	69.6	38	114,811
10-14	0.0003	0.0016	0.0014	0.0019	98,237	97,919	98,561	159	490,789	63.5	62.4	64.7	44	135,149
15-19	0.0009	0.0045	0.0038	0.0051	98,078	97,737	98,426	437	489,300	58.6	57.5	59.8	107	119,910
20-24	0.0012	0.0060	0.0051	0.0068	97,642	97,243	98,051	582	486,753	53.9	52.8	55.1	127	106,465
25-29	0.0012	0.0058	0.0050	0.0066	97,059	96,589	97,545	559	483,899	49.2	48.1	50.4	137	118,809
30-34	0.0013	0.0065	0.0056	0.0073	96,500	95,960	97,056	626	480,936	44.5	43.4	45.6	163	125,606
35-39	0.0017	0.0085	0.0074	0.0096	95,874	95,255	96,504	815	477,334	39.7	38.7	40.9	283	165,770
40-44	0.0026	0.0127	0.0112	0.0143	95,059	94,341	95,788	1,210	472,273	35.1	34.0	36.2	568	221,886
45-49	0.0041	0.0204	0.0180	0.0228	93,850	92,994	94,716	1,913	464,467	30.5	29.4	31.6	792	192,404
50-54	0.0068	0.0333	0.0296	0.0371	91,937	90,881	93,008	3,065	452,023	26.1	25.1	27.2	954	140,767
55-59	0.0111	0.0542	0.0480	0.0606	88,872	87,531	90,232	4,820	432,313	21.9	20.9	23.0	906	81,286
60-64	0.0179	0.0857	0.0758	0.0963	84,053	82,333	85,786	7,202	402,259	18.0	17.1	19.0	877	48,997
65-69	0.0287	0.1340	0.1175	0.1506	76,851	74,624	79,133	10,294	358,518	14.4	13.6	15.4	702	24,466
70-74	0.0462	0.2071	0.1818	0.2311	66,557	63,494	69,677	13,786	298,318	11.3	10.6	12.1	544	11,774
75-79	0.0742	0.3130	0.2786	0.3460	52,770	48,832	56,750	16,519	222,554	8.6	8.0	9.2	452	6,092
80-84	0.1199	0.4614	0.4201	0.4987	36,251	32,105	40,940	16,725	139,442	6.3	5.9	6.8	412	3,434
85-89	0.1864	0.6357	0.5961	0.6695	19,526	16,266	23,666	12,412	66,599	4.6	4.4	4.9	300	1,608
90-94	0.2785	0.7586	0.7278	0.7845	7,114	5,395	9,445	5,396	19,379	3.3	3.2	3.5	119	426
95-99	0.4001	0.9092	0.8916	0.9234	1,717	1,172	2,551	1,561	3,903	2.4	2.4	2.5	40	101
100+	0.5527	1	1	1	156	91	277	156	282	1.8	1.8	1.9	14	25
All ages													*7,914*	*1,721,866*

ln($_nq_x$) with uncertainty interval

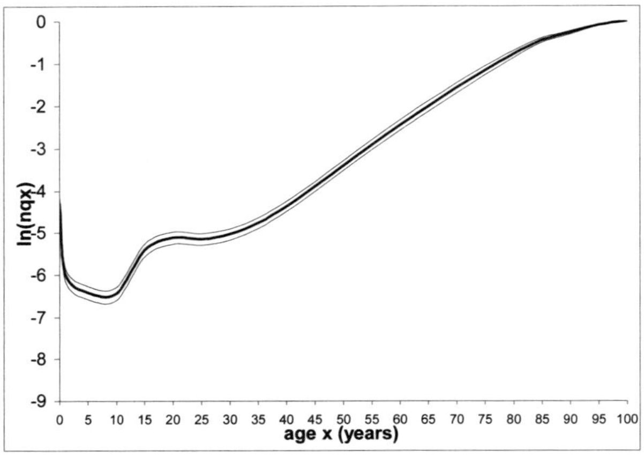

l_x with uncertainty interval

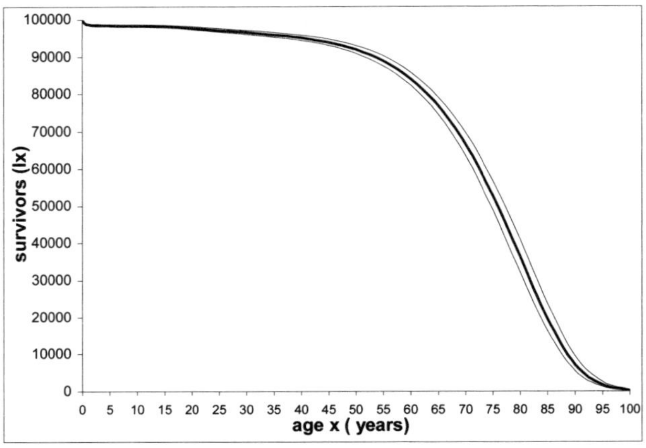

e_x with uncertainty interval

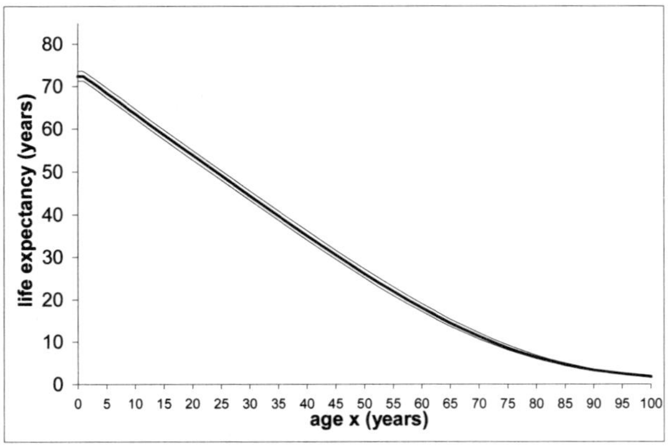

Distribution of deaths by age with uncertainty interval

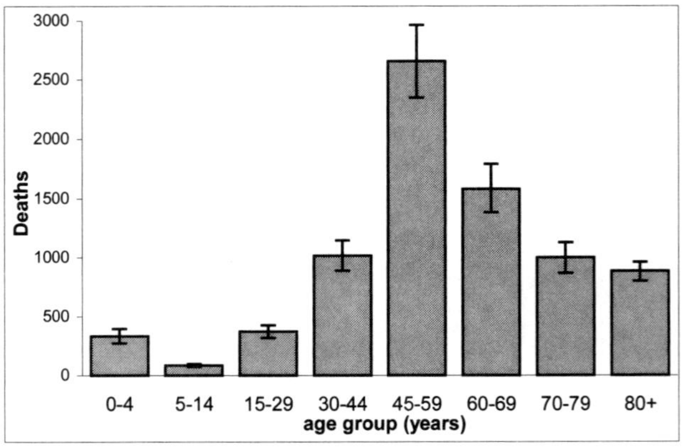

United Arab Emirates — Emirats arabes unis — Emiratos Arabes Unidos

Females, 2000 — *Femmes, 2000* — *Mujeres, 2000*

Age	$_nM_x$	$_nq_x$			l_x			$_nd_x$	$_nL_x$	e_x			Actual deaths	Actual population
			percentiles			percentiles					percentiles			
		mean	2½	97½	mean	2½	97½			mean	2½	97½		
<1	0.0117	0.0115	0.0094	0.0138	100,000	100,000	100,000	1,153	98,962	76.4	75.4	77.7	232	19,892
1-4	0.0006	0.0025	0.0021	0.0028	98,847	98,621	99,062	246	394,795	76.3	75.3	77.7	48	76,556
5-9	0.0003	0.0013	0.0012	0.0015	98,600	98,351	98,850	133	492,670	72.5	71.5	73.9	29	106,747
10-14	0.0002	0.0012	0.0010	0.0014	98,468	98,203	98,735	119	492,042	67.6	66.6	69.0	30	122,499
15-19	0.0004	0.0022	0.0019	0.0025	98,349	98,070	98,631	215	491,209	62.6	61.7	64.1	44	99,922
20-24	0.0005	0.0026	0.0023	0.0029	98,134	97,830	98,443	254	490,038	57.8	56.8	59.2	38	73,464
25-29	0.0006	0.0030	0.0026	0.0034	97,881	97,547	98,219	293	488,672	52.9	51.9	54.3	36	60,526
30-34	0.0008	0.0040	0.0035	0.0045	97,588	97,221	97,959	394	486,956	48.1	47.1	49.5	39	48,002
35-39	0.0012	0.0059	0.0052	0.0066	97,194	96,784	97,608	574	484,536	43.3	42.2	44.7	73	61,331
40-44	0.0018	0.0092	0.0082	0.0102	96,620	96,148	97,094	889	480,879	38.5	37.5	39.9	130	70,418
45-49	0.0029	0.0143	0.0128	0.0158	95,732	95,173	96,292	1,370	475,233	33.8	32.8	35.2	151	52,242
50-54	0.0044	0.0220	0.0198	0.0243	94,362	93,679	95,049	2,075	466,622	29.3	28.2	30.7	161	36,114
55-59	0.0068	0.0334	0.0295	0.0371	92,287	91,421	93,158	3,084	453,726	24.9	23.8	26.3	134	19,756
60-64	0.0109	0.0530	0.0457	0.0591	89,203	88,098	90,312	4,726	434,202	20.7	19.6	22.0	148	13,582
65-69	0.0179	0.0857	0.0726	0.0967	84,477	83,051	85,878	7,236	404,297	16.7	15.7	17.9	161	8,981
70-74	0.0311	0.1443	0.1208	0.1659	77,241	75,106	79,507	11,145	358,345	13.0	12.1	14.1	171	5,506
75-79	0.0547	0.2406	0.2021	0.2774	66,097	63,061	69,902	15,901	290,731	9.8	9.0	10.7	226	4,137
80-84	0.0972	0.3908	0.3374	0.4411	50,196	45,731	55,773	19,619	201,932	7.1	6.5	7.8	250	2,570
85-89	0.1628	0.5785	0.5238	0.6277	30,577	25,559	36,953	17,689	108,663	5.0	4.7	5.5	220	1,351
90-94	0.2572	0.7260	0.6853	0.7625	12,888	9,516	17,597	9,356	36,372	3.5	3.3	3.8	96	374
95-99	0.3834	0.8916	0.8695	0.9114	3,532	2,260	5,538	3,149	8,213	2.5	2.4	2.7	35	92
100+	0.5390	1	1	1	383	200	723	383	711	1.9	1.8	2.0	16	30
All ages													*2,466*	*884,092*

ln($_nq_x$) with uncertainty interval

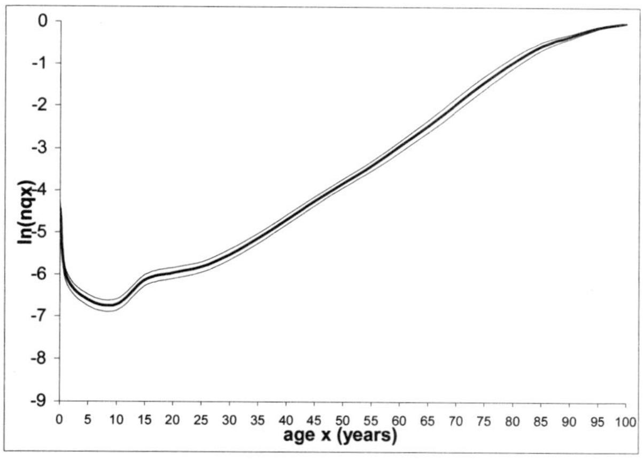

l_x with uncertainty interval

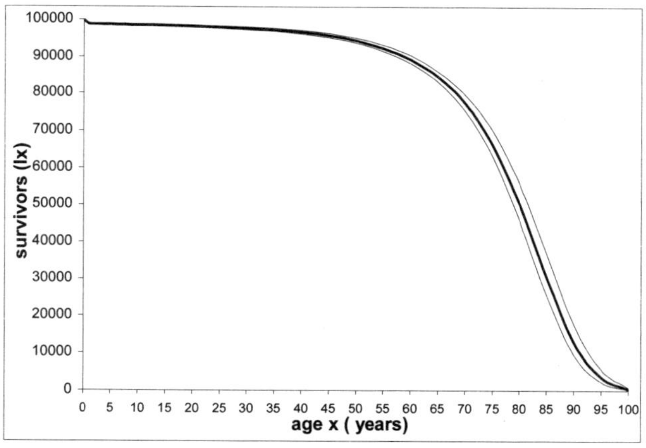

e_x with uncertainty interval

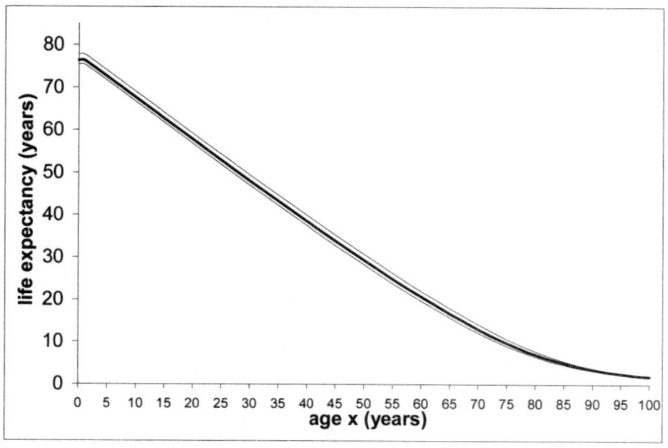

Distribution of deaths by age with uncertainty interval

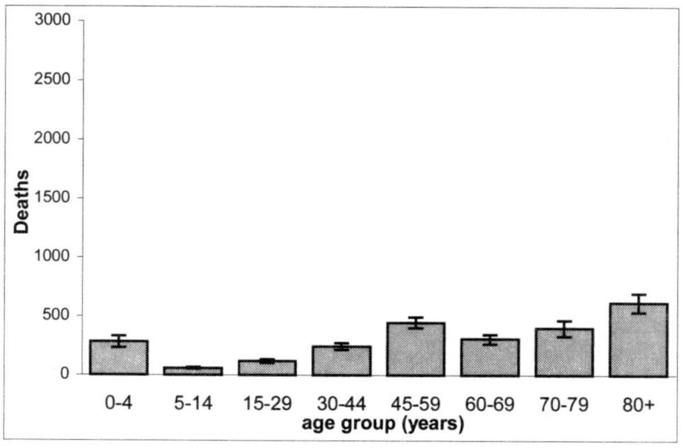

United Kingdom　　Royaume-Uni　　Reino Unido

Males, 2000　　*Hommes, 2000*　　*Hombres, 2000*

Age	$_nM_x$	$_nq_x$			l_x			$_nd_x$	$_nL_x$	e_x			Actual deaths	Actual population
			percentiles			percentiles					percentiles			
		mean	2½	97½	mean	2½	97½			mean	2½	97½		
<1	0.0059	0.0059	0.0054	0.0063	100,000	100,000	100,000	585	99,473	74.8	74.6	75.1	2,077	352,864
1-4	0.0003	0.0012	0.0011	0.0012	99,415	99,372	99,458	114	397,384	74.3	74.1	74.5	418	1,451,009
5-9	0.0001	0.0007	0.0006	0.0007	99,300	99,251	99,350	66	496,336	70.4	70.1	70.6	264	1,978,010
10-14	0.0002	0.0008	0.0008	0.0009	99,234	99,181	99,287	83	495,964	65.4	65.2	65.6	332	1,994,031
15-19	0.0005	0.0027	0.0025	0.0028	99,151	99,094	99,209	264	495,098	60.5	60.2	60.7	1,009	1,893,326
20-24	0.0008	0.0041	0.0039	0.0042	98,888	98,818	98,958	401	493,435	55.6	55.4	55.8	1,456	1,789,059
25-29	0.0009	0.0045	0.0043	0.0047	98,486	98,399	98,575	443	491,324	50.8	50.6	51.1	1,850	2,051,080
30-34	0.0010	0.0050	0.0048	0.0052	98,043	97,938	98,150	488	488,995	46.1	45.8	46.3	2,367	2,369,797
35-39	0.0012	0.0062	0.0060	0.0064	97,555	97,433	97,679	602	486,270	41.3	41.0	41.5	3,014	2,436,351
40-44	0.0019	0.0096	0.0093	0.0099	96,953	96,813	97,097	930	482,442	36.5	36.3	36.7	4,027	2,089,461
45-49	0.0030	0.0151	0.0147	0.0155	96,023	95,859	96,194	1,453	476,485	31.8	31.6	32.1	5,754	1,887,382
50-54	0.0049	0.0243	0.0238	0.0249	94,571	94,375	94,777	2,302	467,097	27.3	27.1	27.5	9,988	2,026,335
55-59	0.0087	0.0425	0.0415	0.0436	92,268	92,039	92,517	3,924	451,531	22.9	22.7	23.1	14,042	1,615,662
60-64	0.0147	0.0711	0.0694	0.0729	88,344	88,065	88,636	6,280	426,020	18.8	18.6	19.0	20,883	1,416,707
65-69	0.0251	0.1183	0.1151	0.1214	82,064	81,674	82,446	9,708	386,052	15.1	14.9	15.3	31,159	1,239,138
70-74	0.0430	0.1940	0.1886	0.1990	72,357	71,791	72,917	14,037	326,690	11.7	11.6	11.9	45,510	1,059,148
75-79	0.0687	0.2932	0.2855	0.3001	58,319	57,521	59,110	17,098	248,852	9.0	8.8	9.1	56,885	827,944
80-84	0.1117	0.4366	0.4268	0.4456	41,222	40,285	42,219	17,996	161,118	6.7	6.6	6.8	50,220	449,615
85-89	0.1737	0.6055	0.5959	0.6145	23,225	22,413	24,152	14,062	80,972	4.9	4.8	5.0	39,822	229,298
90-94	0.2583	0.7276	0.7199	0.7350	9,163	8,667	9,744	6,667	25,815	3.5	3.5	3.6	18,687	72,358
95-99	0.3673	0.8737	0.8689	0.8786	2,496	2,298	2,729	2,181	5,938	2.6	2.6	2.7	4,488	12,220
100+	0.4996	1	1	1	315	279	358	315	631	2.0	2.0	2.0	504	1,008
All ages													*314,754*	*29,241,803*

ln($_nq_x$) with uncertainty interval

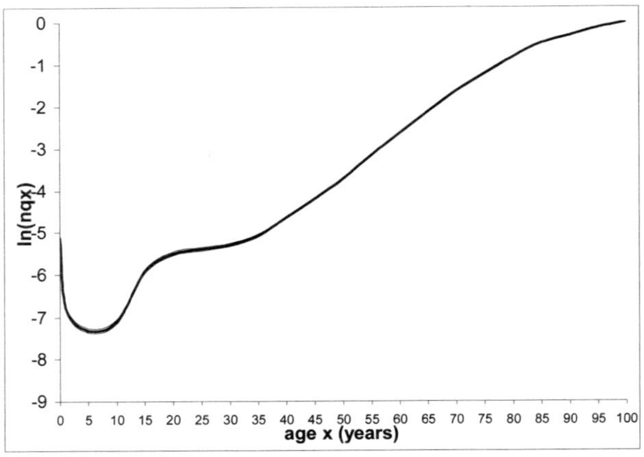

l_x with uncertainty interval

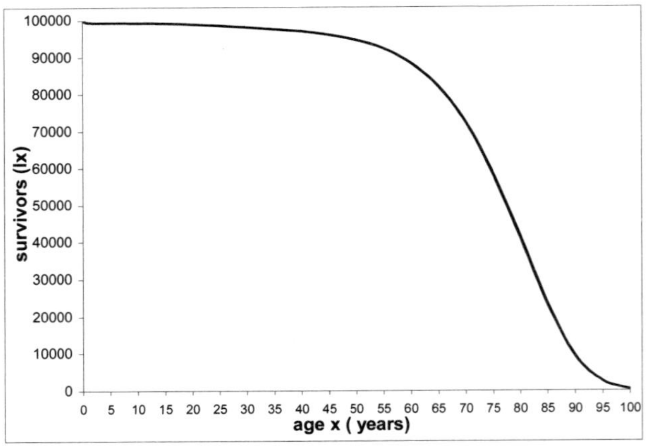

e_x with uncertainty interval

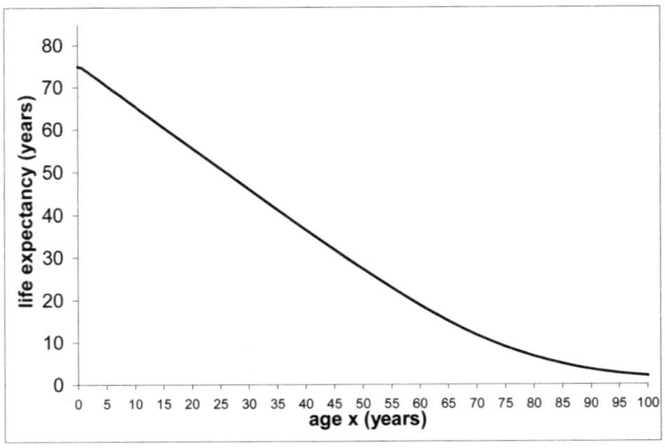

Distribution of deaths by age with uncertainty interval

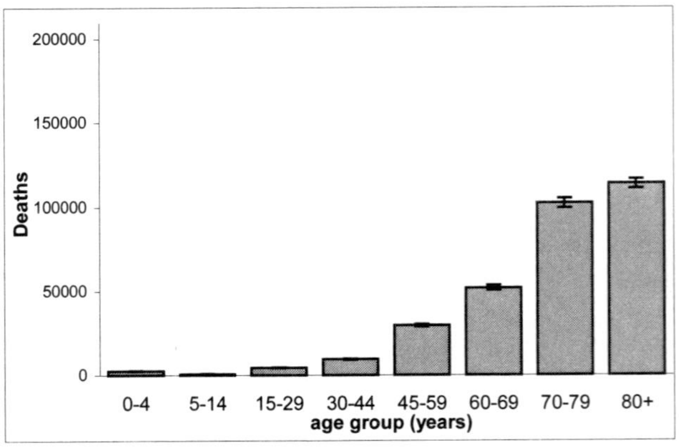

United Kingdom   Royaume-Uni   Reino Unido

Females, 2000   *Femmes, 2000*   *Mujeres, 2000*

Age	$_nM_x$	$_nq_x$ mean	$_nq_x$ percentiles 2½	$_nq_x$ percentiles 97½	l_x mean	l_x percentiles 2½	l_x percentiles 97½	$_nd_x$	$_nL_x$	e_x mean	e_x percentiles 2½	e_x percentiles 97½	Actual deaths	Actual population
<1	0.0048	0.0048	0.0043	0.0053	100,000	100,000	100,000	477	99,570	79.9	79.6	80.2	1,626	339,129
1-4	0.0002	0.0009	0.0009	0.0010	99,523	99,472	99,566	92	397,870	79.3	79.0	79.6	319	1,379,409
5-9	0.0001	0.0005	0.0005	0.0006	99,431	99,373	99,480	53	497,021	75.3	75.1	75.7	202	1,884,210
10-14	0.0001	0.0006	0.0005	0.0006	99,378	99,317	99,430	55	496,749	70.4	70.1	70.7	211	1,893,546
15-19	0.0003	0.0013	0.0012	0.0014	99,322	99,258	99,378	130	496,286	65.4	65.1	65.8	470	1,797,577
20-24	0.0003	0.0016	0.0015	0.0017	99,192	99,120	99,256	157	495,568	60.5	60.2	60.8	538	1,693,534
25-29	0.0004	0.0018	0.0017	0.0019	99,035	98,954	99,106	178	494,729	55.6	55.3	55.9	705	1,955,861
30-34	0.0005	0.0024	0.0023	0.0025	98,857	98,766	98,937	234	493,698	50.7	50.4	51.0	1,081	2,281,114
35-39	0.0007	0.0037	0.0036	0.0039	98,623	98,521	98,713	365	492,200	45.8	45.5	46.2	1,750	2,358,200
40-44	0.0013	0.0063	0.0060	0.0065	98,257	98,142	98,362	615	489,749	41.0	40.7	41.3	2,592	2,064,774
45-49	0.0020	0.0099	0.0097	0.0102	97,642	97,508	97,767	969	485,790	36.2	35.9	36.6	3,766	1,888,186
50-54	0.0032	0.0158	0.0154	0.0162	96,674	96,518	96,822	1,525	479,554	31.5	31.3	31.9	6,494	2,041,460
55-59	0.0052	0.0258	0.0251	0.0265	95,148	94,974	95,319	2,451	469,613	27.0	26.7	27.4	8,603	1,648,461
60-64	0.0085	0.0416	0.0403	0.0429	92,697	92,503	92,880	3,852	453,857	22.7	22.4	23.0	12,557	1,479,565
65-69	0.0146	0.0704	0.0677	0.0727	88,845	88,584	89,092	6,251	428,599	18.5	18.3	18.9	19,775	1,355,817
70-74	0.0254	0.1196	0.1145	0.1237	82,594	82,141	83,009	9,874	388,286	14.8	14.5	15.1	32,736	1,287,276
75-79	0.0413	0.1873	0.1790	0.1946	72,720	71,983	73,496	13,618	329,556	11.4	11.2	11.7	48,674	1,177,945
80-84	0.0726	0.3071	0.2945	0.3186	59,102	58,063	60,336	18,151	250,136	8.5	8.3	8.7	57,823	796,874
85-89	0.1222	0.4681	0.4520	0.4820	40,952	39,634	42,545	19,171	156,831	6.1	6.0	6.3	65,500	535,818
90-94	0.1976	0.6202	0.6050	0.6328	21,780	20,532	23,231	13,508	68,377	4.3	4.2	4.4	47,663	241,260
95-99	0.3063	0.7981	0.7872	0.8071	8,272	7,547	9,154	6,602	21,554	3.0	3.0	3.1	19,622	64,064
100+	0.4556	1	1	1	1,670	1,460	1,948	1,670	3,666	2.2	2.2	2.2	3,991	8,760
All ages													*336,698*	*30,172,840*

ln($_nq_x$) with uncertainty interval

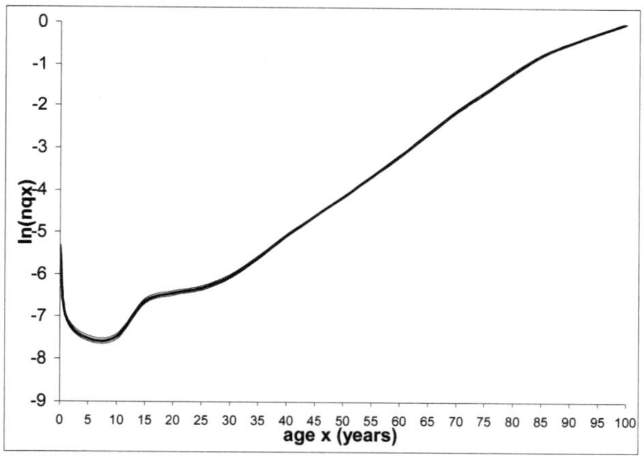

l_x with uncertainty interval

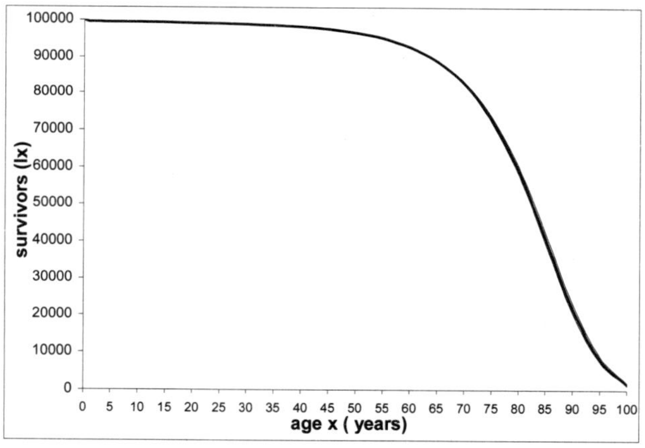

e_x with uncertainty interval

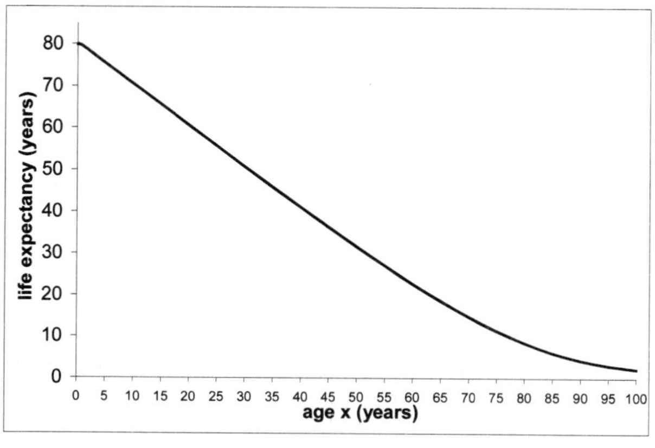

Distribution of deaths by age with uncertainty interval

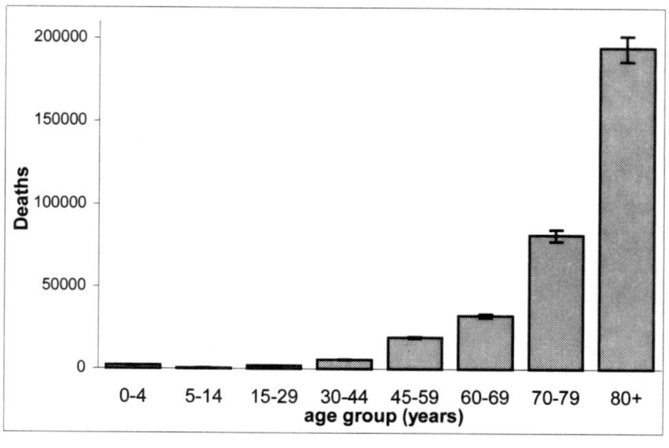

United Republic of Tanzania   République-Unie de Tanzanie   República Unida de Tanzanía

Males, 2000   *Hommes, 2000*   *Hombres, 2000*

Age	$_nM_x$	$_nq_x$			l_x			$_nd_x$	$_nL_x$	e_x			Actual deaths	Actual population
			percentiles			percentiles					percentiles			
		mean	2½	97½	mean	2½	97½			mean	2½	97½		
<1	0.1225	0.1104	0.1054	0.1157	100,000	100,000	100,000	11,035	90,068	45.8	44.6	47.0	81,736	667,124
1-4	0.0116	0.0450	0.0428	0.0472	88,965	88,429	89,464	4,008	346,241	50.4	49.1	51.8	27,171	2,347,464
5-9	0.0028	0.0138	0.0130	0.0146	84,957	84,464	85,456	1,172	421,857	48.7	47.5	50.1	7,288	2,623,726
10-14	0.0020	0.0099	0.0094	0.0105	83,786	83,256	84,329	832	416,848	44.4	43.1	45.7	4,616	2,312,964
15-19	0.0031	0.0154	0.0145	0.0164	82,954	82,394	83,538	1,279	411,571	39.8	38.6	41.1	6,159	1,982,286
20-24	0.0067	0.0330	0.0310	0.0351	81,675	81,059	82,320	2,694	401,640	35.4	34.2	36.7	10,945	1,632,035
25-29	0.0125	0.0605	0.0565	0.0647	78,981	78,248	79,757	4,779	382,958	31.5	30.3	32.8	16,794	1,345,803
30-34	0.0190	0.0908	0.0843	0.0976	74,202	73,267	75,206	6,736	354,170	28.4	27.3	29.6	20,001	1,051,596
35-39	0.0217	0.1031	0.0952	0.1112	67,466	66,200	68,802	6,956	319,940	26.0	24.9	27.1	18,244	839,134
40-44	0.0234	0.1107	0.1020	0.1196	60,510	58,837	62,166	6,699	285,802	23.7	22.7	24.6	15,491	660,906
45-49	0.0252	0.1184	0.1090	0.1280	53,811	51,798	55,780	6,372	253,124	21.3	20.4	22.2	12,908	512,725
50-54	0.0269	0.1259	0.1160	0.1360	47,439	45,170	49,699	5,974	222,258	18.8	18.1	19.6	12,243	455,511
55-59	0.0294	0.1371	0.1265	0.1479	41,465	39,032	43,934	5,683	193,116	16.2	15.5	16.8	10,336	351,236
60-64	0.0406	0.1842	0.1706	0.1981	35,782	33,273	38,378	6,590	162,434	13.3	12.8	13.9	10,457	257,761
65-69	0.0555	0.2438	0.2272	0.2609	29,192	26,709	31,826	7,116	128,169	10.8	10.3	11.2	9,864	177,658
70-74	0.0814	0.3381	0.3178	0.3587	22,076	19,782	24,586	7,463	91,720	8.4	8.1	8.8	9,028	110,949
75-79	0.1200	0.4615	0.4386	0.4841	14,612	12,733	16,754	6,744	56,202	6.5	6.2	6.8	7,146	59,556
80-84	0.1761	0.6113	0.5885	0.6332	7,869	6,601	9,379	4,810	27,317	4.9	4.7	5.1	4,353	24,721
85-89	0.2539	0.7766	0.7577	0.7943	3,058	2,421	3,843	2,375	9,354	3.7	3.6	3.8	1,810	7,126
90-94	0.3537	0.8581	0.8432	0.8718	683	498	926	586	1,657	2.7	2.7	2.8	493	1,395
95-99	0.4762	0.9804	0.9761	0.9840	97	64	145	95	200	2.1	2.1	2.1	87	182
100+	0.6192	1	1	1	2	1	3	2	3	1.6	1.6	1.7	9	14
All ages													*287,180*	*17,421,872*

$\ln({}_nq_x)$ with uncertainty interval

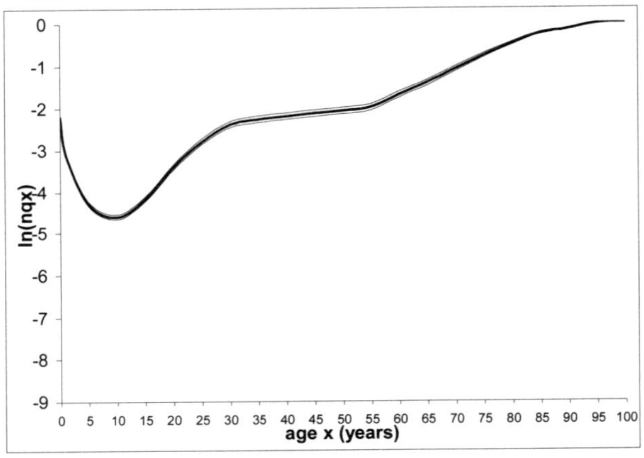

l_x with uncertainty interval

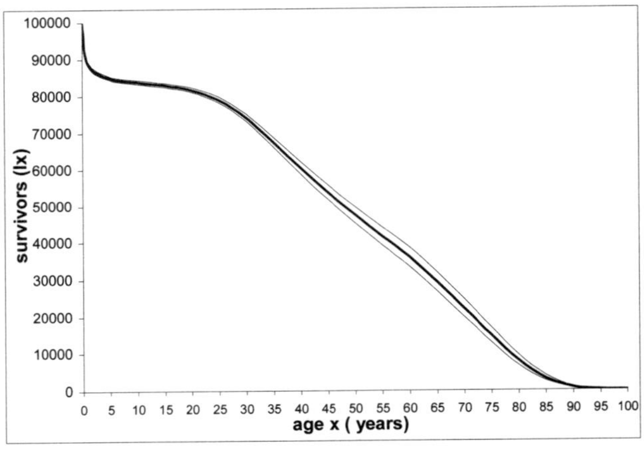

e_x with uncertainty interval

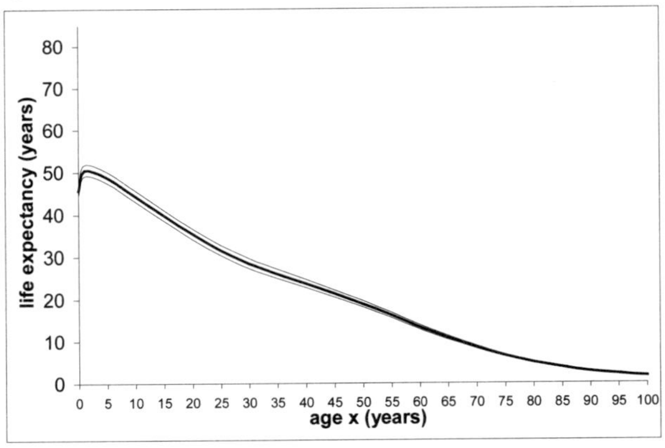

Distribution of deaths by age with uncertainty interval

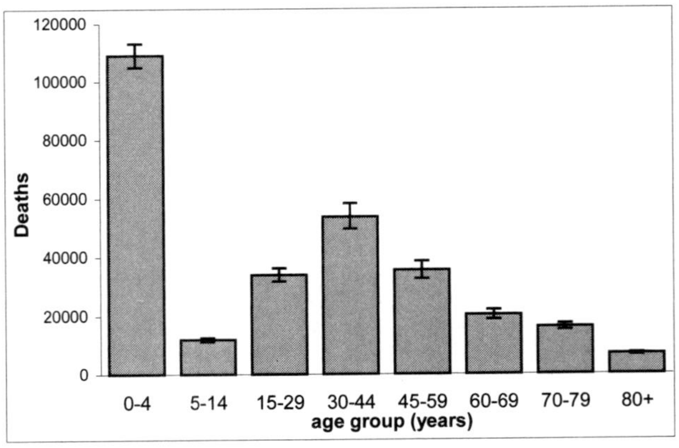

United Republic of Tanzania — République-Unie de Tanzanie — República Unida de Tanzanía

Females, 2000 — *Femmes, 2000* — *Mujeres, 2000*

Age	$_nM_x$	$_nq_x$			l_x			$_nd_x$	$_nL_x$	e_x			Actual deaths	Actual population
			percentiles			percentiles					percentiles			
		mean	2½	97½	mean	2½	97½			mean	2½	97½		
<1	0.1032	0.0945	0.0893	0.0997	100,000	100,000	100,000	9,447	91,498	47.2	46.0	48.4	66,227	641,466
1-4	0.0131	0.0507	0.0482	0.0534	90,553	90,029	91,069	4,595	351,187	51.1	49.8	52.4	30,329	2,318,168
5-9	0.0030	0.0151	0.0142	0.0159	85,959	85,462	86,453	1,294	426,558	49.7	48.5	51.0	7,870	2,593,385
10-14	0.0022	0.0111	0.0104	0.0117	84,664	84,109	85,212	939	420,975	45.4	44.2	46.7	5,119	2,295,473
15-19	0.0051	0.0252	0.0237	0.0267	83,726	83,121	84,320	2,106	413,363	40.9	39.7	42.1	10,076	1,977,742
20-24	0.0108	0.0525	0.0493	0.0557	81,620	80,904	82,322	4,282	397,394	36.9	35.7	38.1	17,725	1,644,967
25-29	0.0169	0.0811	0.0759	0.0862	77,338	76,399	78,257	6,270	371,013	33.8	32.7	35.0	23,616	1,397,339
30-34	0.0202	0.0963	0.0900	0.1028	71,067	69,818	72,292	6,847	338,221	31.6	30.5	32.7	21,909	1,082,316
35-39	0.0190	0.0909	0.0848	0.0973	64,221	62,669	65,756	5,838	306,510	29.7	28.7	30.7	16,459	864,214
40-44	0.0176	0.0841	0.0780	0.0904	58,383	56,615	60,175	4,909	279,644	27.4	26.5	28.4	12,128	690,895
45-49	0.0167	0.0802	0.0742	0.0864	53,474	51,561	55,451	4,289	256,649	24.7	23.9	25.6	9,166	548,483
50-54	0.0185	0.0883	0.0816	0.0951	49,185	47,175	51,295	4,343	235,068	21.6	20.9	22.4	9,077	491,253
55-59	0.0217	0.1029	0.0950	0.1108	44,842	42,756	47,057	4,614	212,675	18.5	17.8	19.2	8,352	385,004
60-64	0.0293	0.1367	0.1264	0.1470	40,228	38,059	42,517	5,500	187,392	15.3	14.7	15.9	8,540	290,968
65-69	0.0422	0.1909	0.1770	0.2046	34,728	32,522	37,047	6,631	157,065	12.3	11.9	12.9	8,804	208,536
70-74	0.0647	0.2785	0.2598	0.2965	28,097	25,945	30,357	7,827	120,921	9.6	9.3	10.1	8,949	138,269
75-79	0.0974	0.3917	0.3690	0.4129	20,271	18,267	22,400	7,941	81,503	7.4	7.1	7.7	7,687	78,899
80-84	0.1472	0.5380	0.5135	0.5602	12,330	10,737	14,135	6,634	45,067	5.6	5.4	5.8	5,258	35,723
85-89	0.2165	0.7023	0.6801	0.7219	5,696	4,732	6,877	4,001	18,480	4.2	4.0	4.3	2,466	11,391
90-94	0.3092	0.8021	0.7839	0.8179	1,696	1,316	2,200	1,360	4,398	3.0	3.0	3.2	770	2,490
95-99	0.4291	0.9380	0.9284	0.9460	336	240	475	315	734	2.3	2.3	2.3	158	369
100+	0.5784	1	1	1	21	13	34	21	36	1.7	1.7	1.8	19	33
All ages													*280,704*	*17,697,383*

ln($_nq_x$) with uncertainty interval

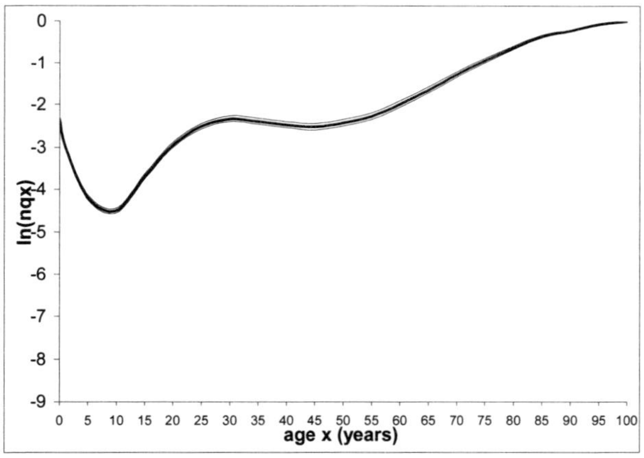

l_x with uncertainty interval

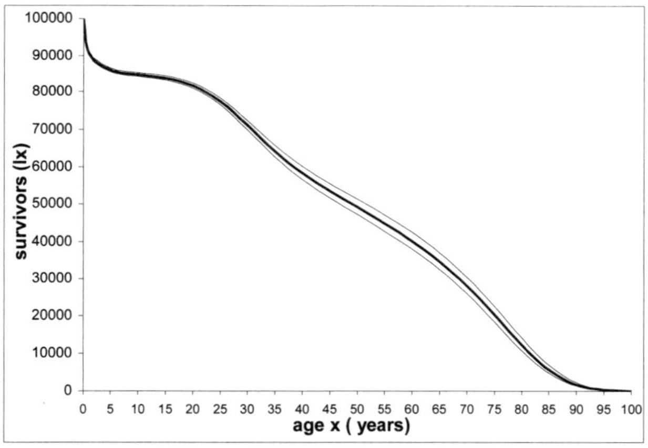

e_x with uncertainty interval

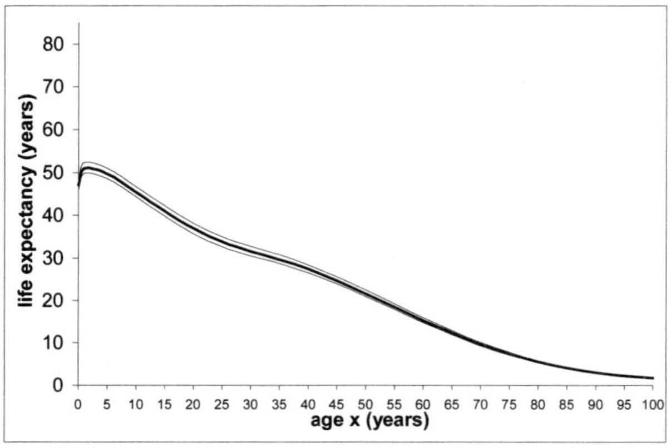

Distribution of deaths by age with uncertainty interval

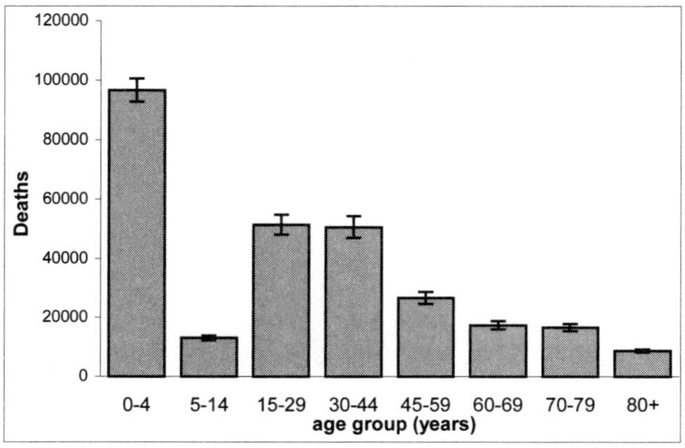

United States of America | Etats-Unis d'Amérique | Estados Unidos de América

Males, 2000 | *Hommes, 2000* | *Hombres, 2000*

Age	$_nM_x$	$_nq_x$			l_x			$_nd_x$	$_nL_x$	e_x			Actual deaths	Actual population
			percentiles			percentiles					percentiles			
		mean	2½	97½	mean	2½	97½			mean	2½	97½		
<1	0.0079	0.0079	0.0075	0.0083	100,000	100,000	100,000	789	99,290	73.9	73.7	74.2	15,999	2,013,918
1-4	0.0004	0.0015	0.0015	0.0016	99,211	99,168	99,255	151	396,482	73.5	73.3	73.8	3,147	8,251,240
5-9	0.0002	0.0010	0.0009	0.0010	99,060	99,010	99,110	94	495,065	69.6	69.4	69.9	2,038	10,716,132
10-14	0.0003	0.0013	0.0012	0.0013	98,966	98,912	99,019	124	494,519	64.7	64.4	64.9	2,636	10,508,545
15-19	0.0010	0.0048	0.0046	0.0050	98,842	98,783	98,900	472	493,030	59.8	59.5	60.0	9,689	10,125,080
20-24	0.0014	0.0068	0.0066	0.0071	98,370	98,293	98,445	671	490,173	55.0	54.8	55.3	12,925	9,445,868
25-29	0.0014	0.0069	0.0066	0.0071	97,699	97,599	97,797	670	486,822	50.4	50.1	50.6	13,103	9,523,566
30-34	0.0016	0.0080	0.0077	0.0082	97,030	96,908	97,148	774	483,212	45.7	45.5	46.0	16,614	10,369,215
35-39	0.0021	0.0104	0.0102	0.0107	96,255	96,111	96,395	1,005	478,764	41.1	40.8	41.3	24,546	11,693,917
40-44	0.0030	0.0149	0.0146	0.0153	95,250	95,080	95,415	1,423	472,694	36.5	36.2	36.7	34,954	11,610,655
45-49	0.0045	0.0224	0.0219	0.0229	93,827	93,624	94,025	2,104	463,876	32.0	31.8	32.2	46,352	10,217,141
50-54	0.0065	0.0321	0.0314	0.0328	91,723	91,479	91,962	2,940	451,264	27.7	27.4	27.9	57,258	8,788,131
55-59	0.0102	0.0498	0.0487	0.0510	88,783	88,495	89,070	4,423	432,857	23.5	23.3	23.7	68,492	6,703,231
60-64	0.0159	0.0765	0.0745	0.0784	84,360	84,025	84,699	6,450	405,675	19.6	19.4	19.8	83,251	5,236,404
65-69	0.0247	0.1163	0.1132	0.1193	77,910	77,476	78,327	9,064	366,891	16.0	15.8	16.2	109,481	4,431,410
70-74	0.0380	0.1736	0.1690	0.1779	68,846	68,252	69,396	11,955	314,343	12.8	12.6	13.0	149,380	3,927,827
75-79	0.0570	0.2495	0.2431	0.2553	56,891	56,113	57,668	14,194	248,970	10.0	9.8	10.1	174,445	3,059,825
80-84	0.0929	0.3771	0.3685	0.3846	42,697	41,809	43,638	16,100	173,234	7.4	7.3	7.6	170,574	1,835,304
85-89	0.1475	0.5389	0.5296	0.5469	26,597	25,777	27,518	14,332	97,153	5.4	5.4	5.5	123,366	836,256
90-94	0.2280	0.6769	0.6690	0.6839	12,264	11,697	12,916	8,302	36,415	3.9	3.8	3.9	64,618	283,429
95-99	0.3431	0.8453	0.8400	0.8501	3,962	3,705	4,275	3,349	9,763	2.8	2.8	2.8	23,344	68,046
100+	0.5026	1	1	1	613	557	684	613	1,219	2.0	2.0	2.1	5,092	10,131
All ages													*1,211,303*	*139,655,271*

ln($_nq_x$) with uncertainty interval

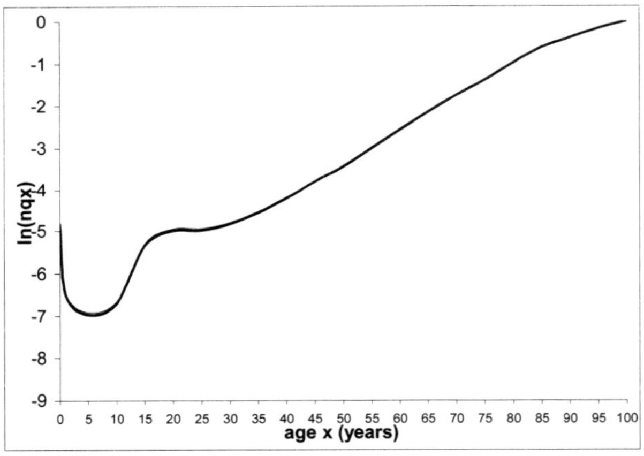

l_x with uncertainty interval

e_x with uncertainty interval

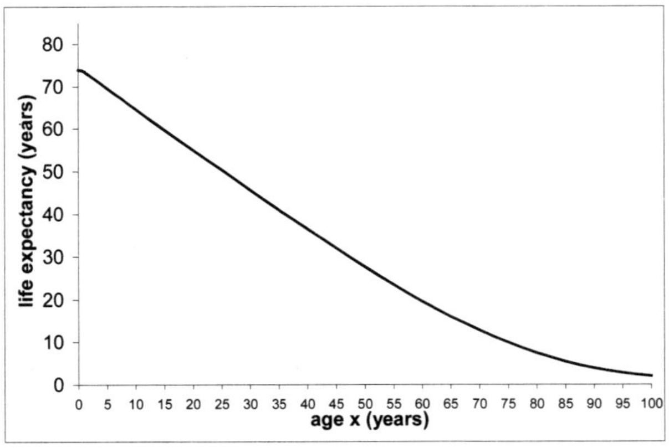

Distribution of deaths by age with uncertainty interval

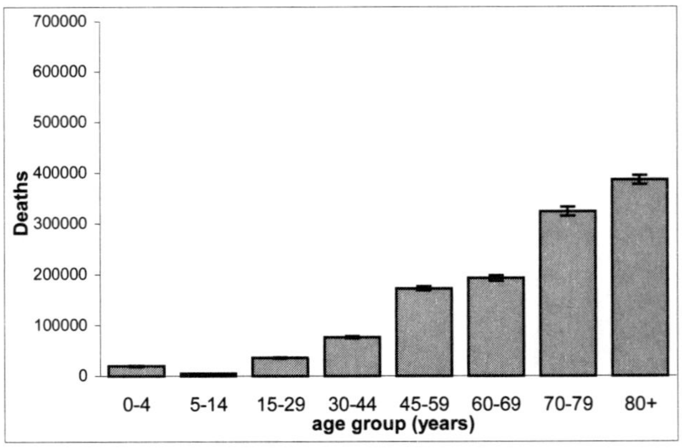

United States of America — Etats-Unis d'Amérique — Estados Unidos de América

Females, 2000 — *Femmes, 2000* — *Mujeres, 2000*

Age	$_nM_x$	$_nq_x$ mean	$_nq_x$ percentiles 2½	$_nq_x$ percentiles 97½	l_x mean	l_x percentiles 2½	l_x percentiles 97½	$_nd_x$	$_nL_x$	e_x mean	e_x percentiles 2½	e_x percentiles 97½	Actual deaths	Actual population
<1	0.0064	0.0063	0.0061	0.0066	100,000	100,000	100,000	633	99,431	79.5	79.3	79.6	12,163	1,911,645
1-4	0.0003	0.0012	0.0011	0.0012	99,367	99,336	99,395	117	397,188	79.0	78.8	79.1	2,324	7,865,462
5-9	0.0002	0.0008	0.0007	0.0008	99,250	99,214	99,282	75	496,062	75.1	74.9	75.2	1,545	10,215,916
10-14	0.0002	0.0008	0.0008	0.0008	99,175	99,137	99,209	80	495,674	70.1	70.0	70.3	1,622	10,024,501
15-19	0.0004	0.0020	0.0020	0.0021	99,095	99,054	99,132	203	494,966	65.2	65.1	65.3	3,964	9,661,241
20-24	0.0005	0.0024	0.0023	0.0024	98,892	98,844	98,935	233	493,877	60.3	60.2	60.5	4,294	9,111,146
25-29	0.0006	0.0028	0.0027	0.0028	98,659	98,603	98,709	272	492,614	55.4	55.3	55.6	5,160	9,335,733
30-34	0.0008	0.0038	0.0037	0.0039	98,387	98,323	98,444	370	491,007	50.6	50.5	50.7	7,698	10,205,165
35-39	0.0011	0.0057	0.0056	0.0058	98,016	97,942	98,084	559	488,685	45.8	45.7	45.9	13,089	11,447,370
40-44	0.0017	0.0084	0.0082	0.0085	97,458	97,370	97,538	814	485,252	41.0	40.9	41.2	19,219	11,454,971
45-49	0.0025	0.0124	0.0122	0.0127	96,643	96,537	96,742	1,199	480,219	36.3	36.2	36.5	25,655	10,275,653
50-54	0.0038	0.0189	0.0186	0.0193	95,444	95,316	95,565	1,806	472,707	31.8	31.7	31.9	34,366	8,995,769
55-59	0.0062	0.0303	0.0298	0.0308	93,639	93,483	93,788	2,839	461,096	27.3	27.2	27.5	43,234	7,022,874
60-64	0.0098	0.0480	0.0472	0.0487	90,800	90,615	90,982	4,354	443,115	23.1	23.0	23.2	55,698	5,668,674
65-69	0.0154	0.0741	0.0729	0.0751	86,446	86,220	86,667	6,403	416,223	19.1	19.1	19.3	77,839	5,059,848
70-74	0.0239	0.1129	0.1111	0.1143	80,043	79,753	80,329	9,036	377,626	15.5	15.4	15.6	116,861	4,883,892
75-79	0.0378	0.1729	0.1703	0.1748	71,007	70,658	71,366	12,275	324,350	12.1	12.1	12.2	162,346	4,289,837
80-84	0.0639	0.2754	0.2719	0.2783	58,733	58,312	59,215	16,177	253,221	9.1	9.1	9.2	195,775	3,064,543
85-89	0.1062	0.4197	0.4156	0.4238	42,556	42,113	43,115	17,859	168,131	6.7	6.6	6.7	195,049	1,836,261
90-94	0.1740	0.5715	0.5670	0.5759	24,697	24,343	25,192	14,115	81,138	4.7	4.6	4.7	152,436	876,258
95-99	0.2806	0.7618	0.7580	0.7653	10,582	10,340	10,879	8,061	28,725	3.2	3.2	3.3	85,005	302,910
100+	0.4459	1	1	1	2,521	2,427	2,618	2,521	5,653	2.2	2.2	2.3	29,118	65,303
All ages													*1,244,461*	*143,574,972*

ln($_nq_x$) with uncertainty interval

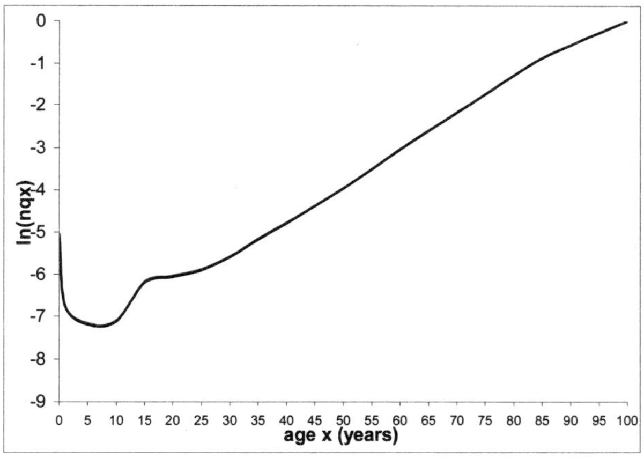

l_x with uncertainty interval

e_x with uncertainty interval

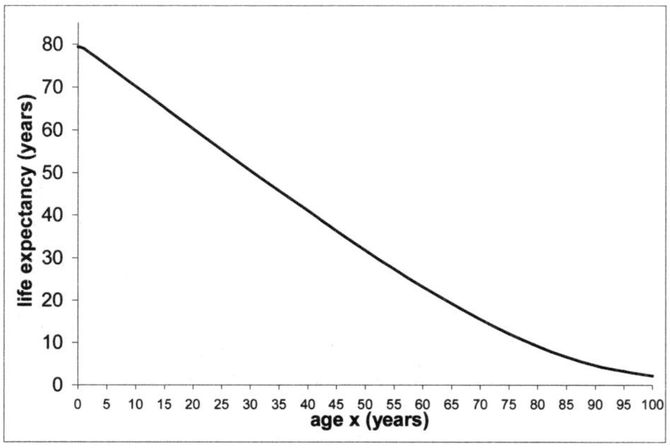

Distribution of deaths by age with uncertainty interval

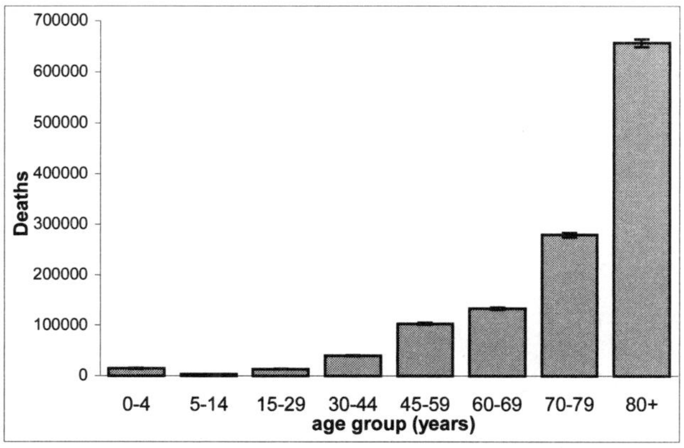

Uruguay **Uruguay** **Uruguay**

Males, 2000 *Hommes, 2000* *Hombres, 2000*

Age	$_nM_x$	$_nq_x$			l_x			$_nd_x$	$_nL_x$	e_x			Actual deaths	Actual population
			percentiles			percentiles					percentiles			
		mean	2½	97½	mean	2½	97½			mean	2½	97½		
<1	0.0172	0.0169	0.0156	0.0183	100,000	100,000	100,000	1,691	98,478	70.0	69.8	70.3	508	29,576
1-4	0.0006	0.0025	0.0024	0.0027	98,309	98,173	98,445	247	392,645	70.2	69.9	70.6	72	114,998
5-9	0.0003	0.0014	0.0013	0.0015	98,063	97,913	98,210	139	489,966	66.4	66.1	66.8	40	142,052
10-14	0.0004	0.0021	0.0020	0.0022	97,924	97,767	98,078	209	489,097	61.5	61.2	61.9	58	135,882
15-19	0.0011	0.0056	0.0054	0.0059	97,715	97,547	97,879	551	487,198	56.6	56.3	57.0	147	130,078
20-24	0.0013	0.0066	0.0063	0.0069	97,164	96,971	97,351	641	484,218	51.9	51.6	52.3	183	137,980
25-29	0.0017	0.0086	0.0083	0.0089	96,523	96,303	96,733	828	480,544	47.2	47.0	47.6	221	128,028
30-34	0.0018	0.0089	0.0086	0.0092	95,695	95,445	95,934	850	476,349	42.6	42.3	43.0	196	110,141
35-39	0.0021	0.0104	0.0100	0.0107	94,845	94,569	95,111	982	471,768	38.0	37.7	38.4	223	107,114
40-44	0.0034	0.0169	0.0165	0.0174	93,862	93,560	94,156	1,590	465,336	33.4	33.0	33.8	343	100,487
45-49	0.0057	0.0281	0.0276	0.0287	92,272	91,935	92,601	2,594	454,875	28.9	28.6	29.3	512	89,848
50-54	0.0087	0.0427	0.0418	0.0436	89,678	89,299	90,048	3,832	438,810	24.7	24.3	25.1	708	81,075
55-59	0.0150	0.0723	0.0701	0.0742	85,846	85,435	86,247	6,209	413,710	20.6	20.3	21.1	1,097	73,077
60-64	0.0219	0.1038	0.0997	0.1070	79,638	79,226	80,038	8,265	377,527	17.1	16.7	17.5	1,414	64,586
65-69	0.0363	0.1665	0.1590	0.1729	71,373	70,910	71,841	11,882	327,162	13.7	13.4	14.1	2,224	61,248
70-74	0.0518	0.2292	0.2184	0.2389	59,492	58,724	60,403	13,637	263,366	11.0	10.7	11.3	2,569	49,609
75-79	0.0802	0.3341	0.3194	0.3476	45,855	44,800	47,212	15,321	190,972	8.5	8.3	8.8	2,687	33,498
80-84	0.1186	0.4574	0.4408	0.4728	30,534	29,229	32,132	13,968	117,752	6.5	6.4	6.7	2,109	17,782
85-89	0.1724	0.6025	0.5866	0.6172	16,567	15,408	17,969	9,981	57,880	4.9	4.8	5.1	1,489	8,632
90-94	0.2465	0.7085	0.6953	0.7208	6,586	5,899	7,428	4,666	18,930	3.7	3.6	3.7	647	2,624
95-99	0.3600	0.8654	0.8565	0.8736	1,920	1,647	2,263	1,661	4,615	2.7	2.6	2.7	188	522
100+	0.5000	1	1	1	258	208	325	258	517	2.0	1.9	2.0	30	60
All ages													*17,665*	*1,618,897*

ln($_nq_x$) with uncertainty interval

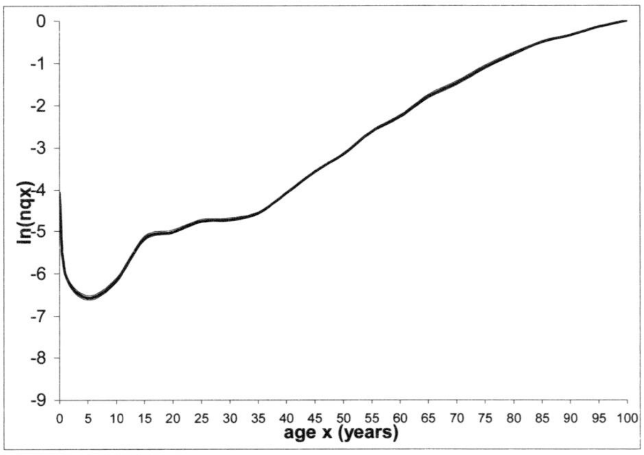

l_x with uncertainty interval

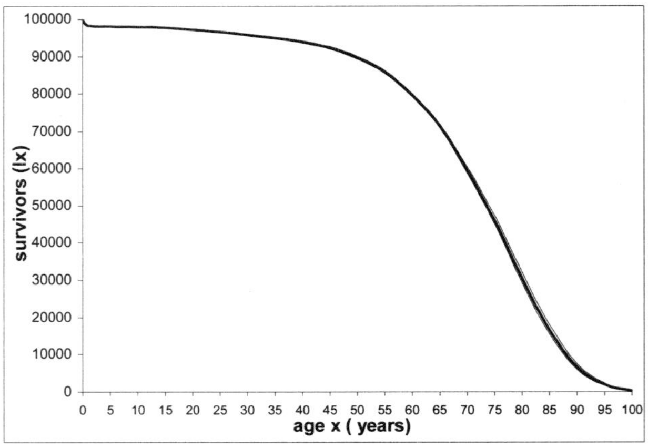

e_x with uncertainty interval

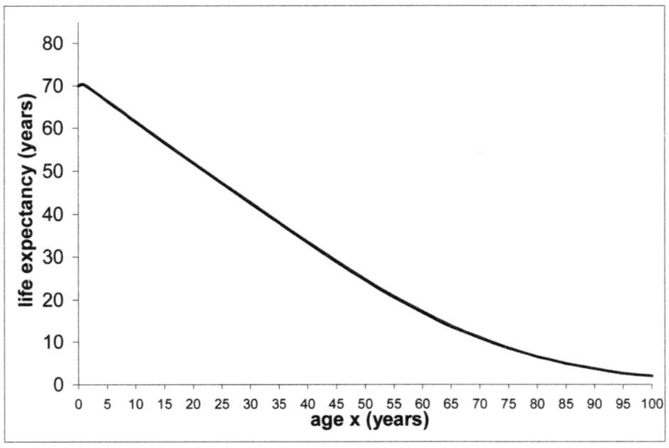

Distribution of deaths by age with uncertainty interval

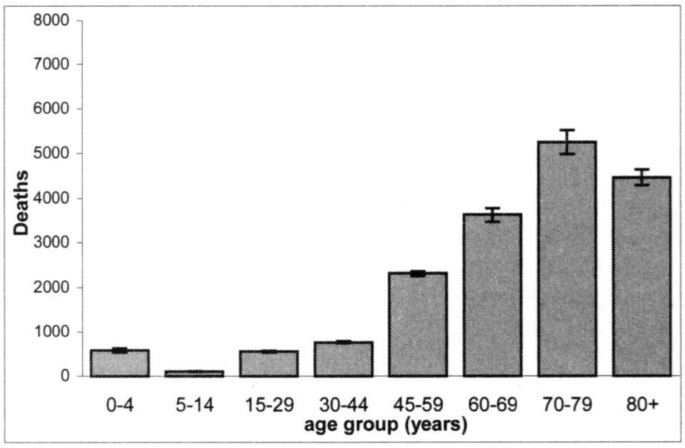

Uruguay

Uruguay

Uruguay

Females, 2000

Femmes, 2000

Mujeres, 2000

Age	$_nM_x$	$_nq_x$			l_x			$_nd_x$	$_nL_x$	e_x			Actual deaths	Actual population
			percentiles			percentiles					percentiles			
		mean	2½	97½	mean	2½	97½			mean	2½	97½		
<1	0.0121	0.0119	0.0108	0.0130	100,000	100,000	100,000	1,193	98,926	77.9	77.6	78.2	337	27,902
1-4	0.0005	0.0021	0.0020	0.0022	98,807	98,697	98,917	208	394,729	77.8	77.5	78.2	58	110,652
5-9	0.0002	0.0010	0.0010	0.0011	98,599	98,476	98,724	102	492,740	74.0	73.7	74.3	28	136,043
10-14	0.0002	0.0011	0.0010	0.0012	98,497	98,367	98,628	108	492,215	69.1	68.7	69.4	29	130,235
15-19	0.0004	0.0022	0.0021	0.0023	98,389	98,253	98,527	217	491,403	64.2	63.8	64.5	55	125,123
20-24	0.0004	0.0021	0.0020	0.0022	98,172	98,023	98,323	204	490,351	59.3	59.0	59.6	56	133,733
25-29	0.0006	0.0029	0.0028	0.0031	97,968	97,808	98,130	286	489,126	54.4	54.1	54.7	73	124,525
30-34	0.0008	0.0038	0.0036	0.0040	97,682	97,507	97,860	371	487,484	49.6	49.2	49.9	85	111,663
35-39	0.0013	0.0067	0.0064	0.0070	97,311	97,118	97,508	654	484,922	44.7	44.4	45.0	151	112,166
40-44	0.0017	0.0085	0.0081	0.0088	96,657	96,435	96,883	817	481,245	40.0	39.7	40.3	182	106,999
45-49	0.0029	0.0145	0.0140	0.0151	95,841	95,587	96,098	1,391	475,724	35.3	35.0	35.7	278	94,897
50-54	0.0043	0.0211	0.0204	0.0219	94,449	94,152	94,752	1,993	467,264	30.8	30.5	31.1	378	88,714
55-59	0.0062	0.0305	0.0295	0.0317	92,456	92,112	92,814	2,824	455,222	26.4	26.2	26.7	507	81,721
60-64	0.0096	0.0469	0.0453	0.0488	89,633	89,229	90,044	4,207	437,645	22.2	21.9	22.5	746	77,593
65-69	0.0152	0.0732	0.0706	0.0761	85,425	84,961	85,888	6,255	411,489	18.2	17.9	18.4	1,157	76,147
70-74	0.0240	0.1133	0.1091	0.1177	79,170	78,493	79,806	8,972	373,423	14.4	14.2	14.6	1,655	68,866
75-79	0.0423	0.1913	0.1841	0.1980	70,199	69,253	71,069	13,430	317,418	10.9	10.7	11.1	2,220	52,458
80-84	0.0794	0.3311	0.3197	0.3415	56,768	55,539	57,885	18,797	236,849	7.9	7.8	8.1	2,545	32,066
85-89	0.1389	0.5154	0.5013	0.5284	37,971	36,646	39,348	19,571	140,927	5.6	5.5	5.7	2,583	18,596
90-94	0.2267	0.6747	0.6622	0.6862	18,400	17,401	19,546	12,414	54,757	3.9	3.8	4.0	1,457	6,428
95-99	0.3452	0.8479	0.8397	0.8556	5,986	5,480	6,593	5,076	14,703	2.8	2.7	2.8	512	1,484
100+	0.4905	1	1	1	910	793	1,054	910	1,856	2.0	2.0	2.1	84	172
All ages													*15,175*	*1,718,183*

$\ln(_nq_x)$ with uncertainty interval

l_x with uncertainty interval

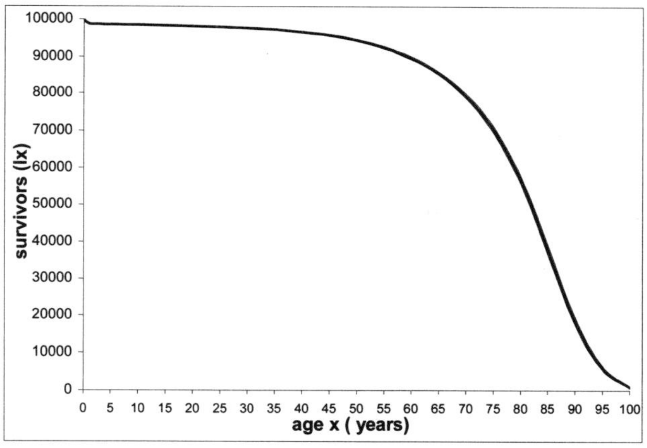

e_x with uncertainty interval

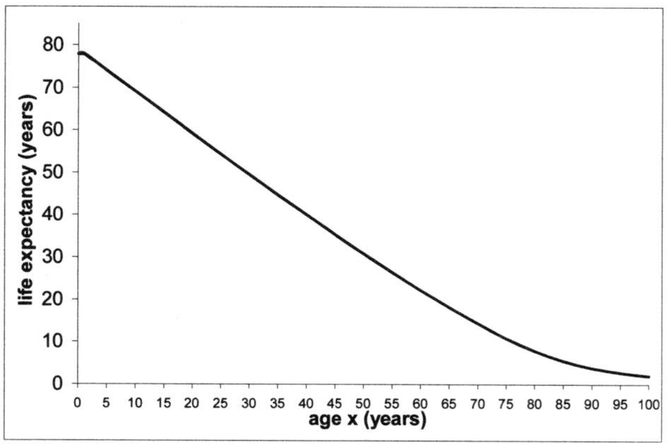

Distribution of deaths by age with uncertainty interval

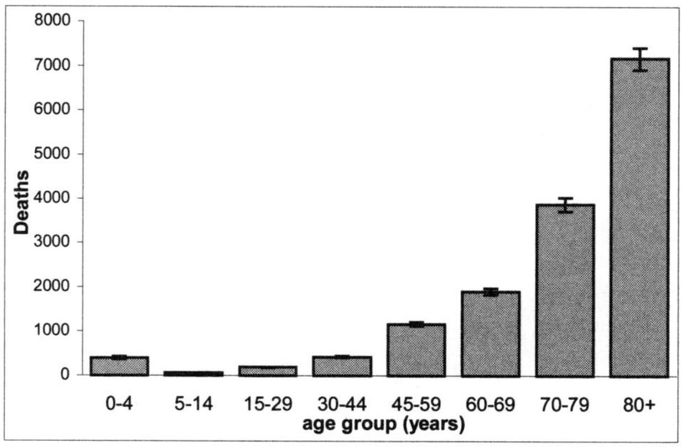

Uzbekistan / Ouzbékistan / Uzbekistán

Males, 2000 / *Hommes, 2000* / *Hombres, 2000*

Age	$_nM_x$	$_nq_x$ mean	$_nq_x$ percentiles 2½	$_nq_x$ percentiles 97½	l_x mean	l_x percentiles 2½	l_x percentiles 97½	$_nd_x$	$_nL_x$	e_x mean	e_x percentiles 2½	e_x percentiles 97½	Actual deaths	Actual population
<1	0.0491	0.0470	0.0436	0.0504	100,000	100,000	100,000	4,700	95,770	62.1	61.4	62.8	14,233	289,994
1-4	0.0058	0.0230	0.0217	0.0242	95,300	94,961	95,637	2,188	375,946	64.2	63.6	64.7	6,504	1,117,319
5-9	0.0010	0.0052	0.0049	0.0054	93,111	92,713	93,511	480	464,357	61.6	61.1	62.2	1,627	1,574,891
10-14	0.0009	0.0045	0.0043	0.0048	92,631	92,217	93,047	420	462,106	56.9	56.4	57.5	1,462	1,606,858
15-19	0.0015	0.0074	0.0070	0.0077	92,211	91,784	92,641	678	459,361	52.2	51.7	52.8	2,016	1,366,315
20-24	0.0024	0.0119	0.0113	0.0125	91,533	91,082	91,994	1,088	454,945	47.6	47.0	48.1	2,749	1,148,884
25-29	0.0029	0.0144	0.0137	0.0151	90,445	89,944	90,955	1,299	448,977	43.1	42.6	43.7	2,958	1,022,360
30-34	0.0036	0.0179	0.0170	0.0187	89,146	88,588	89,711	1,591	441,751	38.7	38.2	39.3	3,137	870,856
35-39	0.0046	0.0226	0.0216	0.0236	87,555	86,929	88,185	1,977	432,831	34.4	33.8	34.9	3,880	849,540
40-44	0.0065	0.0318	0.0304	0.0332	85,578	84,874	86,284	2,719	421,092	30.1	29.6	30.7	4,621	715,727
45-49	0.0100	0.0489	0.0468	0.0510	82,859	82,056	83,661	4,050	404,171	26.0	25.5	26.6	5,045	503,530
50-54	0.0146	0.0704	0.0674	0.0736	78,809	77,874	79,736	5,551	380,170	22.2	21.7	22.7	4,620	316,431
55-59	0.0201	0.0959	0.0912	0.1006	73,259	72,176	74,324	7,023	348,735	18.7	18.3	19.2	4,534	225,106
60-64	0.0308	0.1430	0.1353	0.1503	66,235	65,017	67,424	9,472	307,495	15.4	15.0	15.9	8,698	282,365
65-69	0.0443	0.1992	0.1880	0.2094	56,763	55,449	58,072	11,309	255,542	12.6	12.2	13.0	8,630	195,018
70-74	0.0581	0.2537	0.2396	0.2662	45,454	43,951	46,934	11,533	198,438	10.1	9.8	10.4	9,123	156,972
75-79	0.0830	0.3438	0.3263	0.3590	33,921	32,266	35,676	11,663	140,449	7.6	7.4	7.9	5,259	63,330
80-84	0.1501	0.5457	0.5247	0.5635	22,258	20,713	24,028	12,145	80,927	5.3	5.2	5.5	3,752	25,000
85-89	0.2455	0.7607	0.7435	0.7751	10,113	9,083	11,408	7,693	31,331	3.7	3.6	3.8	3,892	15,852
90-94	0.3636	0.8696	0.8578	0.8793	2,420	2,048	2,922	2,104	5,787	2.7	2.6	2.7	2,651	7,291
95-99	0.4875	0.9899	0.9878	0.9914	316	248	415	312	641	2.0	2.0	2.1	1,268	2,600
100+	0.5916	1	1	1	3	2	5	3	5	1.7	1.7	1.8	354	599
All ages													*101,013*	*12,356,838*

$\ln(_nq_x)$ with uncertainty interval

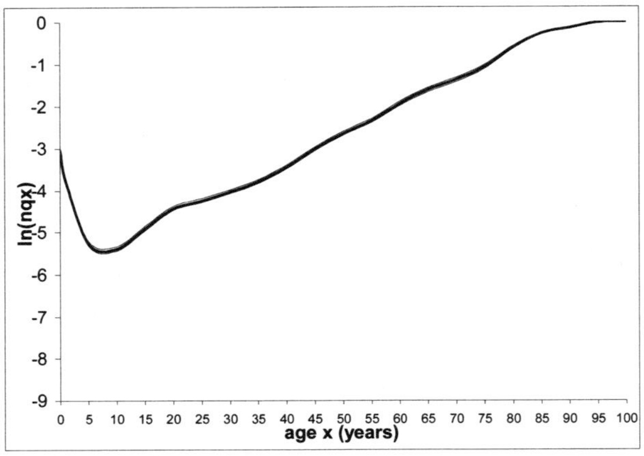

l_x with uncertainty interval

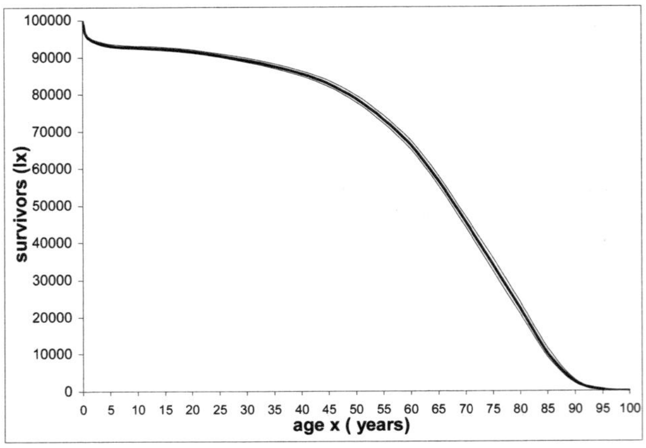

e_x with uncertainty interval

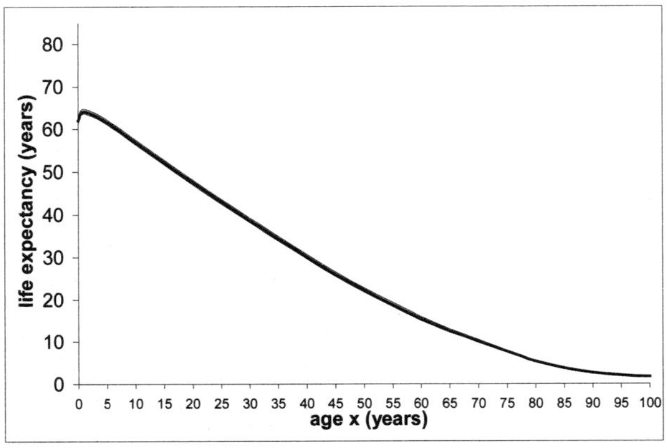

Distribution of deaths by age with uncertainty interval

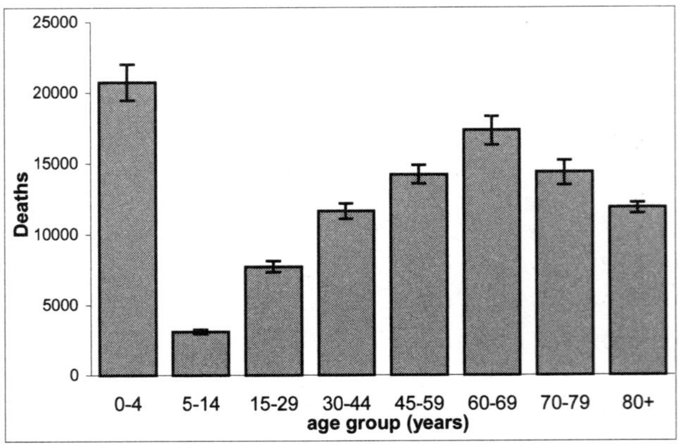

Uzbekistan — Ouzbékistan — Uzbekistán

Females, 2000 — *Femmes, 2000* — *Mujeres, 2000*

Age	$_nM_x$	$_nq_x$			l_x			$_nd_x$	$_nL_x$	e_x			Actual deaths	Actual population
			percentiles			percentiles					percentiles			
		mean	2½	97½	mean	2½	97½			mean	2½	97½		
<1	0.0382	0.0369	0.0339	0.0401	100,000	100,000	100,000	3,692	96,677	68.0	67.2	68.8	10,554	276,317
1-4	0.0052	0.0206	0.0192	0.0219	96,308	95,991	96,607	1,985	380,466	69.6	68.9	70.3	5,621	1,077,462
5-9	0.0008	0.0040	0.0038	0.0043	94,323	93,924	94,720	380	470,663	67.0	66.4	67.7	1,229	1,521,564
10-14	0.0006	0.0031	0.0029	0.0033	93,943	93,524	94,358	293	468,980	62.3	61.6	62.9	974	1,557,413
15-19	0.0009	0.0047	0.0044	0.0050	93,649	93,217	94,079	441	467,145	57.5	56.8	58.1	1,267	1,342,254
20-24	0.0016	0.0080	0.0075	0.0085	93,209	92,755	93,658	749	464,170	52.7	52.1	53.4	1,825	1,130,309
25-29	0.0018	0.0090	0.0084	0.0095	92,459	91,969	92,953	831	460,220	48.1	47.5	48.8	1,830	1,013,988
30-34	0.0021	0.0105	0.0098	0.0111	91,629	91,094	92,171	963	455,737	43.6	43.0	44.2	1,887	893,272
35-39	0.0025	0.0124	0.0116	0.0132	90,666	90,080	91,264	1,126	450,516	39.0	38.4	39.6	2,195	878,703
40-44	0.0035	0.0172	0.0161	0.0183	89,540	88,897	90,201	1,541	443,849	34.5	33.9	35.0	2,547	733,586
45-49	0.0053	0.0261	0.0245	0.0277	87,999	87,280	88,743	2,298	434,251	30.0	29.5	30.6	2,781	525,613
50-54	0.0082	0.0402	0.0377	0.0426	85,701	84,878	86,563	3,441	419,903	25.8	25.3	26.3	2,758	336,501
55-59	0.0127	0.0617	0.0580	0.0655	82,260	81,295	83,285	5,074	398,615	21.7	21.3	22.3	2,893	227,265
60-64	0.0204	0.0972	0.0916	0.1031	77,186	76,046	78,421	7,506	367,164	18.0	17.6	18.5	6,410	313,553
65-69	0.0308	0.1431	0.1348	0.1511	69,680	68,216	71,211	9,975	323,462	14.7	14.3	15.1	7,196	233,355
70-74	0.0447	0.2010	0.1897	0.2117	59,705	57,907	61,556	12,003	268,519	11.7	11.3	12.1	9,282	207,651
75-79	0.0706	0.2999	0.2838	0.3146	47,702	45,694	49,750	14,306	202,746	9.0	8.7	9.3	9,021	127,844
80-84	0.1102	0.4321	0.4126	0.4500	33,396	31,453	35,500	14,430	130,906	6.8	6.6	7.0	7,229	65,580
85-89	0.1670	0.5892	0.5698	0.6070	18,966	17,446	20,753	11,174	66,895	5.0	4.9	5.2	6,559	39,265
90-94	0.2456	0.7070	0.6903	0.7221	7,792	6,893	8,911	5,509	22,434	3.7	3.6	3.8	3,921	15,969
95-99	0.3502	0.8538	0.8421	0.8643	2,283	1,918	2,754	1,950	5,568	2.7	2.7	2.8	1,878	5,363
100+	0.4844	1	1	1	334	261	433	334	689	2.1	2.0	2.1	426	880
All ages													*90,282*	*12,523,707*

ln($_nq_x$) with uncertainty interval

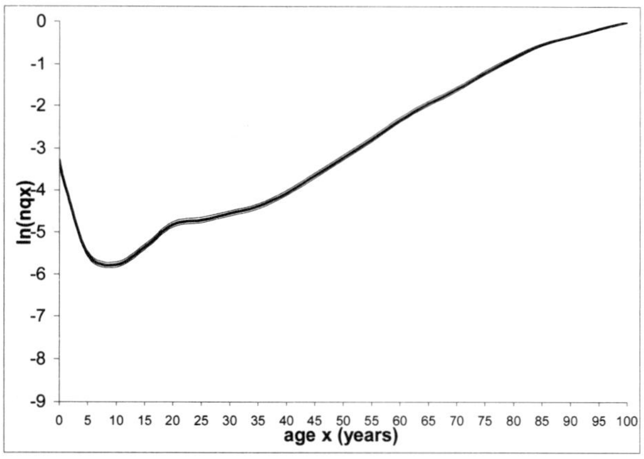

l_x with uncertainty interval

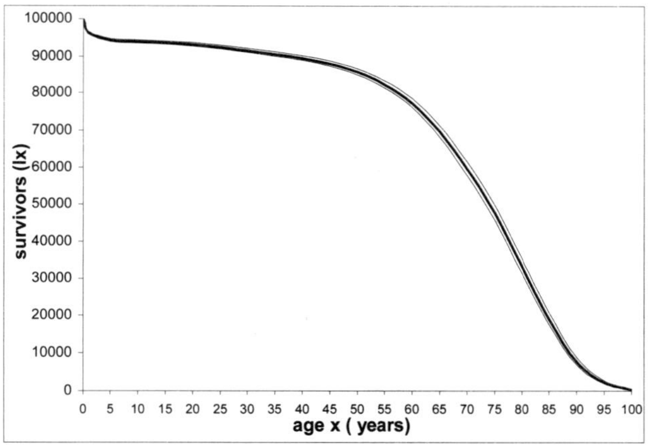

e_x with uncertainty interval

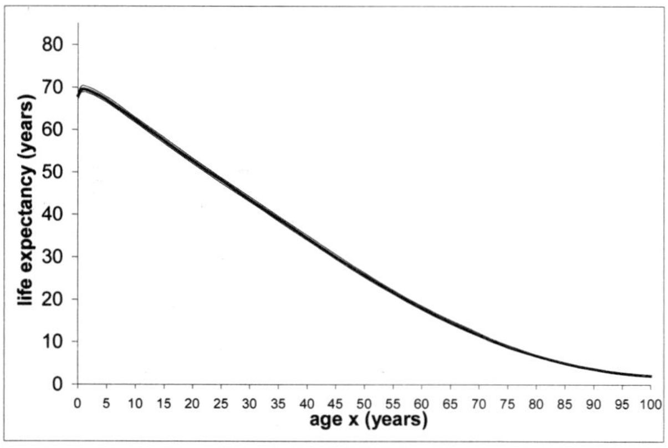

Distribution of deaths by age with uncertainty interval

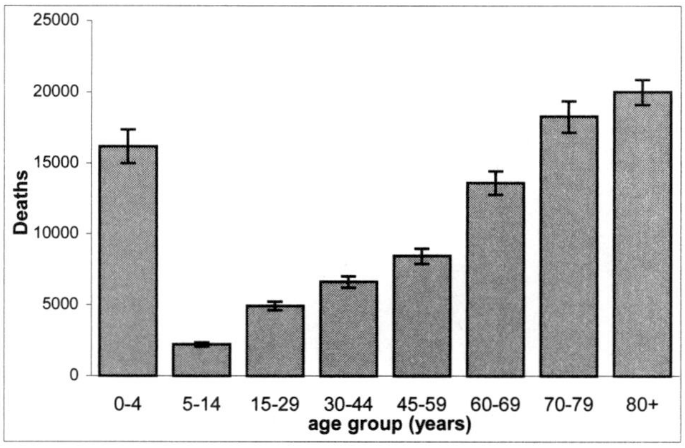

Vanuatu **Vanuatu** **Vanuatu**

Males, 2000 *Hommes, 2000* *Hombres, 2000*

Age	$_nM_x$	$_nq_x$			l_x			$_nd_x$	$_nL_x$	e_x			Actual deaths	Actual population
			percentiles			percentiles					percentiles			
		mean	2½	97½	mean	2½	97½			mean	2½	97½		
<1	0.0473	0.0453	0.0324	0.0585	100,000	100,000	100,000	4,533	95,920	64.2	60.2	69.5	156	3,291
1-4	0.0031	0.0122	0.0088	0.0159	95,467	94,147	96,755	1,165	379,070	66.2	62.6	71.2	38	12,244
5-9	0.0010	0.0052	0.0037	0.0068	94,301	92,803	95,794	486	470,292	63.0	59.6	67.8	15	14,209
10-14	0.0008	0.0042	0.0029	0.0055	93,815	92,208	95,389	390	468,101	58.3	55.0	63.1	11	13,180
15-19	0.0016	0.0082	0.0057	0.0107	93,425	91,705	95,068	763	465,220	53.6	50.3	58.2	19	11,288
20-24	0.0023	0.0114	0.0079	0.0150	92,663	90,722	94,466	1,056	460,674	49.0	45.8	53.6	19	8,324
25-29	0.0024	0.0119	0.0082	0.0158	91,607	89,360	93,667	1,095	455,298	44.5	41.5	49.0	17	7,232
30-34	0.0028	0.0139	0.0094	0.0185	90,512	87,951	92,852	1,260	449,412	40.0	37.1	44.4	18	6,554
35-39	0.0037	0.0181	0.0121	0.0242	89,252	86,334	91,925	1,618	442,217	35.6	32.8	39.8	20	5,396
40-44	0.0052	0.0256	0.0170	0.0343	87,634	84,266	90,743	2,244	432,562	31.2	28.5	35.2	25	4,862
45-49	0.0077	0.0377	0.0247	0.0506	85,390	81,418	89,205	3,220	418,902	26.9	24.4	30.8	28	3,632
50-54	0.0116	0.0566	0.0367	0.0757	82,171	77,379	87,003	4,650	399,227	22.9	20.6	26.5	38	3,263
55-59	0.0176	0.0842	0.0544	0.1117	77,520	71,540	83,809	6,528	371,282	19.1	17.0	22.4	39	2,204
60-64	0.0267	0.1251	0.0813	0.1628	70,992	63,549	79,250	8,880	332,761	15.6	13.7	18.5	48	1,809
65-69	0.0403	0.1830	0.1217	0.2338	62,112	53,202	72,807	11,370	282,137	12.5	10.9	15.0	52	1,279
70-74	0.0624	0.2700	0.1882	0.3384	50,743	40,914	63,944	13,702	219,457	9.7	8.4	11.7	61	974
75-79	0.0964	0.3884	0.2916	0.4717	37,040	27,582	51,907	14,385	149,239	7.4	6.4	9.0	48	502
80-84	0.1482	0.5406	0.4330	0.6243	22,655	15,323	36,773	12,247	82,660	5.5	4.8	6.6	55	372
85-89	0.2200	0.7096	0.6131	0.7775	10,409	5,919	20,395	7,386	33,577	4.1	3.7	4.8	41	185
90-94	0.3155	0.8105	0.7358	0.8617	3,022	1,324	7,389	2,449	7,763	3.0	2.7	3.4	14	45
95-99	0.4372	0.9456	0.9073	0.9678	573	183	1,905	542	1,239	2.3	2.2	2.4	3	6
100+	0.5851	1	1	1	31	6	177	31	53	1.7	1.7	1.8	0	0
All ages													*763*	*100,851*

ln($_nq_x$) with uncertainty interval

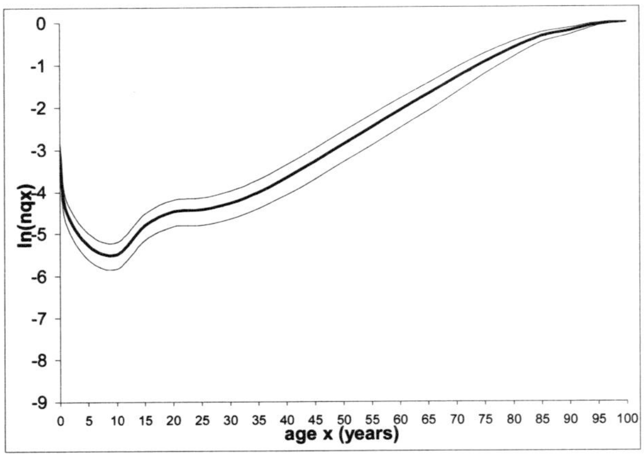

l_x with uncertainty interval

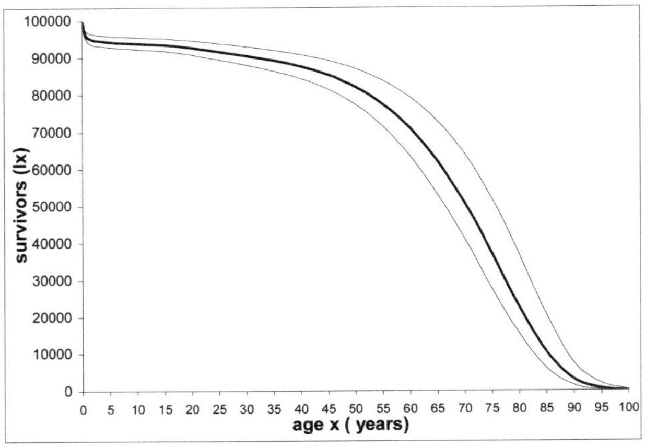

e_x with uncertainty interval

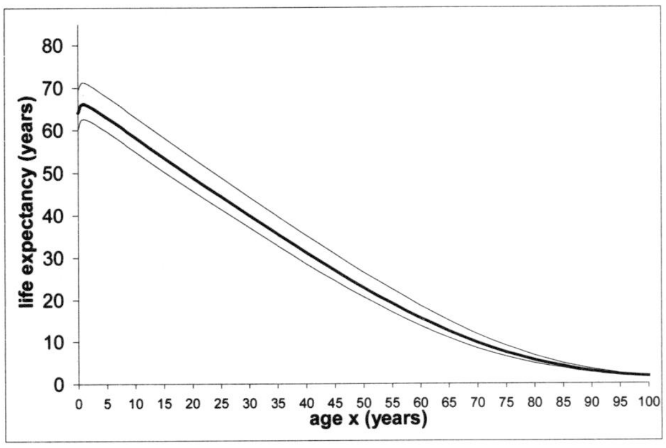

Distribution of deaths by age with uncertainty interval

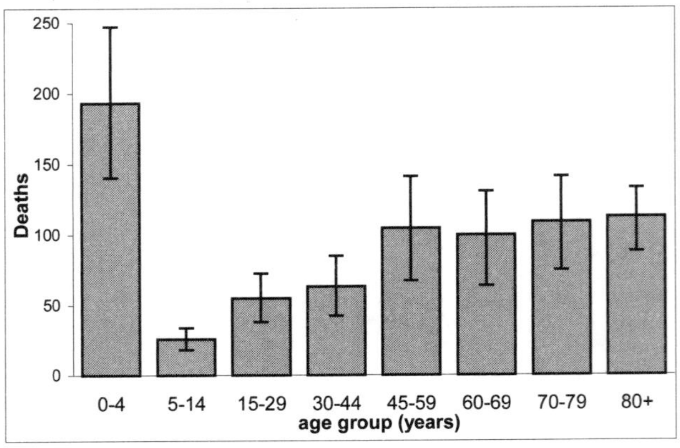

Vanuatu **Vanuatu** **Vanuatu**

Females, 2000 *Femmes, 2000* *Mujeres, 2000*

Age	$_nM_x$	$_nq_x$			l_x			$_nd_x$	$_nL_x$	e_x			Actual deaths	Actual population
			percentiles			percentiles					percentiles			
		mean	2½	97½	mean	2½	97½			mean	2½	97½		
<1	0.0350	0.0339	0.0234	0.0456	100,000	100,000	100,000	3,389	96,950	68.1	64.3	73.1	109	3,131
1-4	0.0029	0.0115	0.0084	0.0149	96,611	95,437	97,661	1,109	383,780	69.5	66.0	74.3	33	11,301
5-9	0.0009	0.0045	0.0033	0.0059	95,501	94,253	96,747	432	476,428	66.3	63.0	70.9	12	13,103
10-14	0.0007	0.0036	0.0026	0.0047	95,070	93,719	96,393	343	474,491	61.6	58.4	66.1	9	12,181
15-19	0.0012	0.0061	0.0043	0.0080	94,727	93,294	96,114	575	472,194	56.8	53.6	61.3	13	10,381
20-24	0.0016	0.0077	0.0055	0.0102	94,151	92,584	95,664	728	468,935	52.1	49.0	56.6	12	7,964
25-29	0.0018	0.0089	0.0062	0.0118	93,423	91,671	95,090	832	465,032	47.5	44.5	51.9	14	7,591
30-34	0.0022	0.0111	0.0076	0.0146	92,590	90,593	94,437	1,024	460,392	42.9	40.0	47.2	15	6,936
35-39	0.0031	0.0152	0.0103	0.0202	91,567	89,268	93,666	1,391	454,356	38.4	35.6	42.5	17	5,700
40-44	0.0042	0.0207	0.0139	0.0277	90,176	87,471	92,639	1,867	446,211	33.9	31.2	37.9	20	4,728
45-49	0.0060	0.0294	0.0194	0.0395	88,309	85,070	91,272	2,596	435,052	29.6	27.1	33.4	21	3,458
50-54	0.0086	0.0422	0.0275	0.0568	85,712	81,755	89,478	3,614	419,524	25.4	23.1	29.0	25	2,863
55-59	0.0124	0.0601	0.0387	0.0807	82,098	77,204	87,022	4,930	398,163	21.4	19.3	24.8	25	2,022
60-64	0.0188	0.0899	0.0577	0.1196	77,168	70,995	83,656	6,940	368,488	17.6	15.7	20.7	30	1,580
65-69	0.0294	0.1368	0.0885	0.1780	70,228	62,504	78,830	9,609	327,115	14.1	12.4	16.8	35	1,179
70-74	0.0482	0.2150	0.1435	0.2739	60,618	51,375	71,851	13,031	270,514	11.0	9.5	13.2	40	829
75-79	0.0783	0.3273	0.2325	0.4066	47,587	37,546	61,539	15,577	198,994	8.3	7.1	10.0	36	465
80-84	0.1265	0.4807	0.3713	0.5700	32,010	22,964	47,229	15,386	121,587	6.1	5.3	7.4	41	327
85-89	0.1961	0.6578	0.5498	0.7353	16,624	10,367	29,503	10,936	55,782	4.4	3.9	5.3	32	161
90-94	0.2911	0.7770	0.6915	0.8352	5,688	2,807	12,687	4,420	15,183	3.2	2.9	3.7	13	44
95-99	0.4142	0.9235	0.8752	0.9525	1,269	463	3,678	1,172	2,829	2.4	2.2	2.6	3	7
100+	0.5649	1	1	1	97	22	459	97	172	1.8	1.7	1.9	1	1
All ages													*554*	*95,952*

$\ln(_nq_x)$ with uncertainty interval

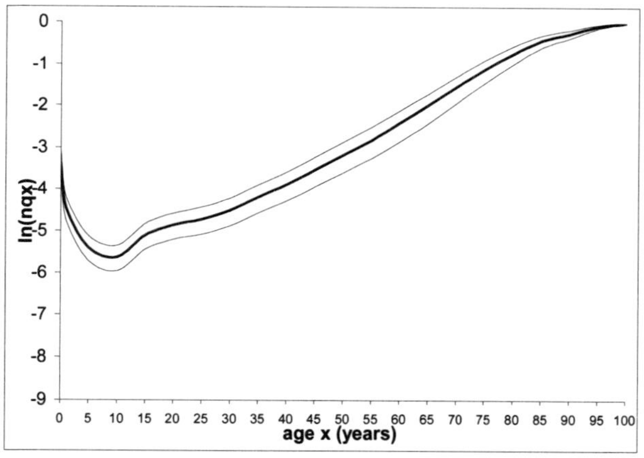

l_x with uncertainty interval

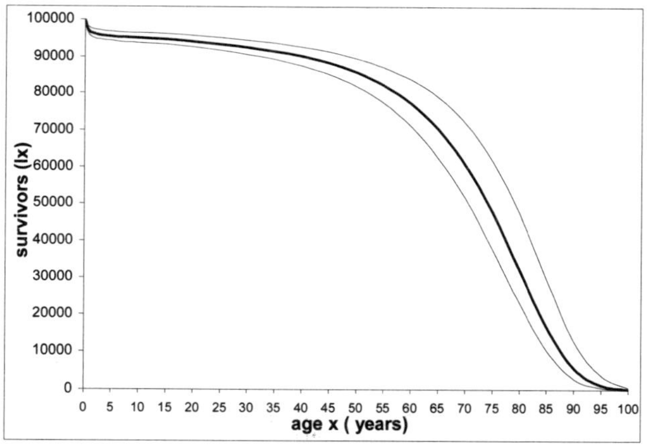

e_x with uncertainty interval

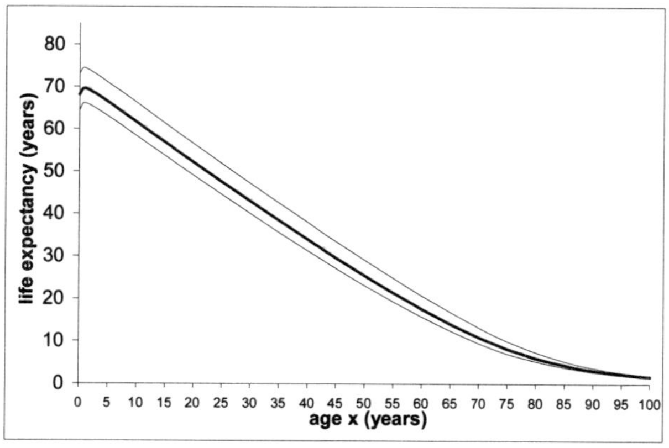

Distribution of deaths by age with uncertainty interval

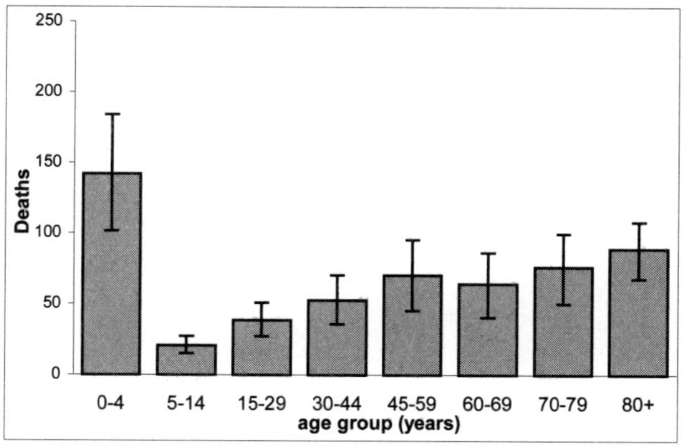

Venezuela (Bolivarian Republic of) Venezuela (République bolivarienne du) Venezuela (República Bolivariana de)

Males, 2000 *Hommes, 2000* *Hombres, 2000*

Age	$_nM_x$	$_nq_x$ mean	$_nq_x$ percentiles 2½	$_nq_x$ percentiles 97½	l_x mean	l_x percentiles 2½	l_x percentiles 97½	$_nd_x$	$_nL_x$	e_x mean	e_x percentiles 2½	e_x percentiles 97½	Actual deaths	Actual population
<1	0.0220	0.0215	0.0181	0.0251	100,000	100,000	100,000	2,154	98,061	70.6	70.0	71.2	6,346	288,833
1-4	0.0012	0.0047	0.0041	0.0052	97,846	97,494	98,192	455	390,289	71.1	70.5	71.8	1,331	1,140,286
5-9	0.0005	0.0023	0.0020	0.0025	97,390	96,992	97,787	223	486,393	67.4	66.8	68.1	644	1,404,446
10-14	0.0005	0.0024	0.0021	0.0026	97,167	96,748	97,587	230	485,260	62.6	61.9	63.3	648	1,366,152
15-19	0.0019	0.0095	0.0085	0.0104	96,937	96,497	97,380	919	482,387	57.7	57.1	58.4	2,385	1,252,235
20-24	0.0027	0.0136	0.0124	0.0147	96,018	95,505	96,543	1,305	476,827	53.3	52.6	54.0	3,125	1,141,582
25-29	0.0026	0.0129	0.0119	0.0138	94,713	94,113	95,340	1,217	470,520	49.0	48.3	49.7	2,528	977,046
30-34	0.0028	0.0139	0.0130	0.0148	93,495	92,816	94,206	1,298	464,233	44.6	43.9	45.3	2,571	919,711
35-39	0.0030	0.0147	0.0139	0.0155	92,198	91,446	92,985	1,357	457,597	40.2	39.5	40.9	2,492	840,640
40-44	0.0041	0.0205	0.0195	0.0215	90,841	90,024	91,696	1,861	449,553	35.7	35.0	36.5	2,852	689,103
45-49	0.0047	0.0230	0.0220	0.0241	88,980	88,090	89,910	2,050	439,777	31.4	30.7	32.2	2,745	588,949
50-54	0.0068	0.0336	0.0319	0.0354	86,930	85,976	87,923	2,921	427,349	27.1	26.4	27.8	3,223	471,556
55-59	0.0106	0.0517	0.0487	0.0549	84,009	82,993	85,058	4,346	409,180	23.0	22.3	23.7	3,651	343,749
60-64	0.0165	0.0793	0.0740	0.0848	79,663	78,602	80,738	6,319	382,517	19.1	18.5	19.8	4,128	249,902
65-69	0.0251	0.1180	0.1092	0.1270	73,344	72,300	74,381	8,657	345,076	15.5	14.9	16.1	4,887	194,794
70-74	0.0400	0.1818	0.1672	0.1958	64,687	63,239	66,080	11,761	294,030	12.2	11.8	12.8	5,803	145,086
75-79	0.0632	0.2730	0.2514	0.2928	52,925	50,854	54,957	14,451	228,499	9.4	9.0	9.9	5,505	87,048
80-84	0.1032	0.4102	0.3819	0.4359	38,474	35,969	41,083	15,781	152,918	7.0	6.7	7.4	4,346	42,115
85-89	0.1617	0.5758	0.5465	0.6021	22,693	20,355	25,392	13,066	80,801	5.1	4.9	5.4	2,388	14,768
90-94	0.2433	0.7033	0.6781	0.7255	9,627	8,167	11,514	6,770	27,824	3.7	3.6	3.9	678	2,785
95-99	0.3516	0.8555	0.8381	0.8705	2,857	2,248	3,697	2,444	6,952	2.7	2.7	2.8	63	178
100+	0.4878	1	1	1	413	292	595	413	846	2.0	2.0	2.1	1	2
All ages													*62,342*	*12,160,966*

$ln(_nq_x)$ with uncertainty interval

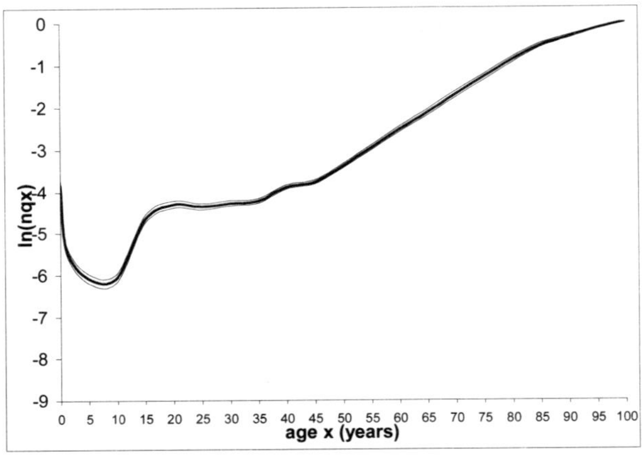

l_x with uncertainty interval

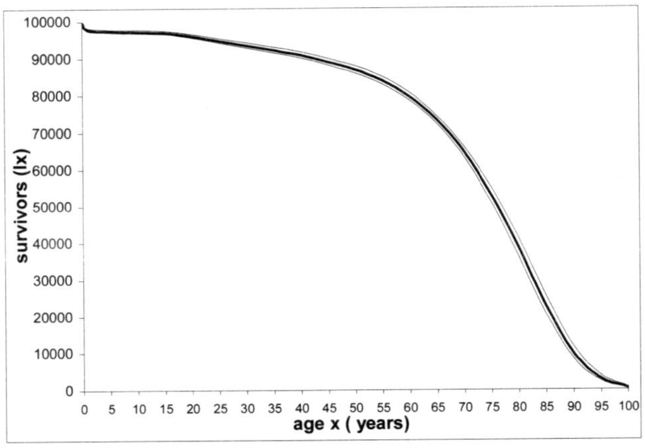

e_x with uncertainty interval

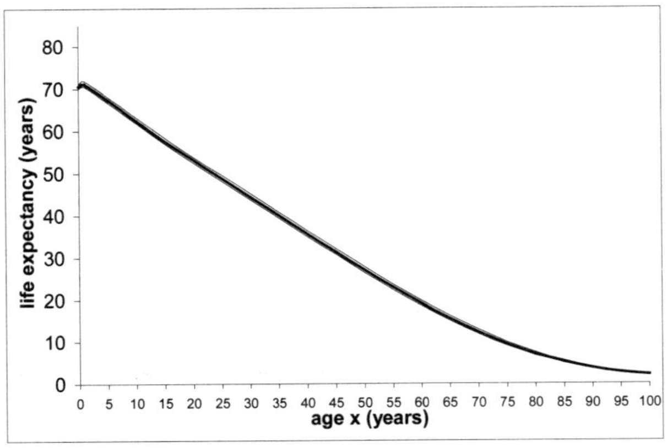

Distribution of deaths by age with uncertainty interval

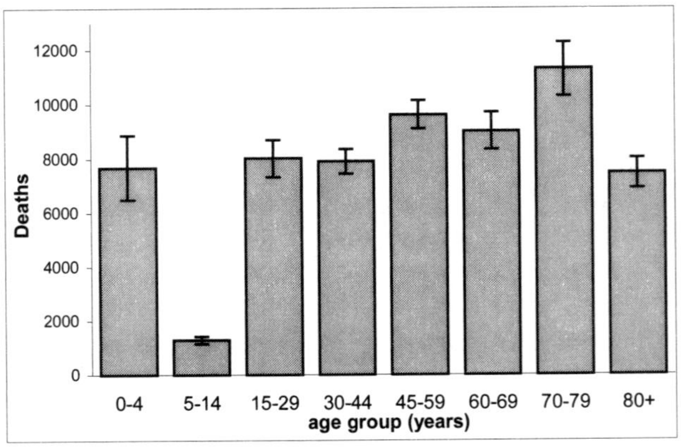

Venezuela (Bolivarian Republic of) / Venezuela (République bolivarienne du) / Venezuela (República Bolivariana de)

Females, 2000 / *Femmes, 2000* / *Mujeres, 2000*

Age	$_nM_x$	$_nq_x$ mean	$_nq_x$ percentiles 2½	$_nq_x$ percentiles 97½	l_x mean	l_x percentiles 2½	l_x percentiles 97½	$_nd_x$	$_nL_x$	e_x mean	e_x percentiles 2½	e_x percentiles 97½	Actual deaths	Actual population
<1	0.0172	0.0169	0.0135	0.0204	100,000	100,000	100,000	1,692	98,478	76.5	75.9	77.1	4,725	275,079
1-4	0.0011	0.0042	0.0037	0.0047	98,308	97,962	98,649	416	392,236	76.8	76.2	77.5	1,159	1,093,369
5-9	0.0003	0.0017	0.0015	0.0019	97,893	97,498	98,289	167	489,045	73.1	72.5	73.8	461	1,346,574
10-14	0.0003	0.0015	0.0013	0.0017	97,725	97,312	98,142	150	488,251	68.3	67.6	68.9	404	1,312,208
15-19	0.0006	0.0030	0.0026	0.0033	97,575	97,147	98,010	291	487,147	63.4	62.7	64.0	722	1,208,139
20-24	0.0007	0.0035	0.0031	0.0038	97,284	96,828	97,751	338	485,573	58.5	57.9	59.2	772	1,108,031
25-29	0.0008	0.0042	0.0038	0.0046	96,945	96,461	97,448	408	483,708	53.7	53.1	54.4	805	955,133
30-34	0.0011	0.0057	0.0051	0.0061	96,538	96,022	97,080	547	481,323	49.0	48.3	49.7	1,030	907,250
35-39	0.0015	0.0075	0.0069	0.0080	95,991	95,439	96,580	717	478,163	44.2	43.6	44.9	1,254	835,926
40-44	0.0021	0.0105	0.0098	0.0111	95,274	94,683	95,916	997	473,877	39.5	38.9	40.3	1,451	689,620
45-49	0.0029	0.0143	0.0136	0.0150	94,277	93,645	94,979	1,349	468,014	34.9	34.3	35.7	1,703	591,114
50-54	0.0042	0.0208	0.0198	0.0219	92,928	92,245	93,690	1,933	459,809	30.4	29.7	31.1	1,995	474,454
55-59	0.0068	0.0337	0.0318	0.0356	90,995	90,256	91,807	3,063	447,320	26.0	25.3	26.7	2,431	354,970
60-64	0.0112	0.0544	0.0512	0.0578	87,932	87,153	88,757	4,788	427,693	21.8	21.2	22.5	3,003	268,298
65-69	0.0171	0.0820	0.0763	0.0880	83,145	82,391	83,873	6,817	398,681	17.9	17.3	18.6	3,705	216,696
70-74	0.0280	0.1308	0.1202	0.1417	76,328	75,232	77,398	9,987	356,672	14.3	13.7	15.0	4,762	170,061
75-79	0.0442	0.1989	0.1812	0.2160	66,341	64,722	67,928	13,197	298,712	11.1	10.5	11.7	4,923	111,429
80-84	0.0800	0.3333	0.3048	0.3618	53,144	50,814	55,426	17,715	221,433	8.2	7.7	8.7	4,778	59,728
85-89	0.1250	0.4762	0.4421	0.5114	35,429	32,573	38,473	16,871	134,969	6.1	5.7	6.4	3,008	24,064
90-94	0.2000	0.6250	0.5927	0.6584	18,558	16,146	21,465	11,599	57,994	4.3	4.0	4.6	1,200	6,002
95-99	0.3000	0.7895	0.7645	0.8144	6,959	5,522	8,737	5,494	18,314	3.1	3.0	3.3	185	615
100+	0.4400	1	1	1	1,465	1,025	2,048	1,465	3,330	2.3	2.2	2.5	7	16
All ages													*44,483*	*12,008,776*

ln($_nq_x$) with uncertainty interval

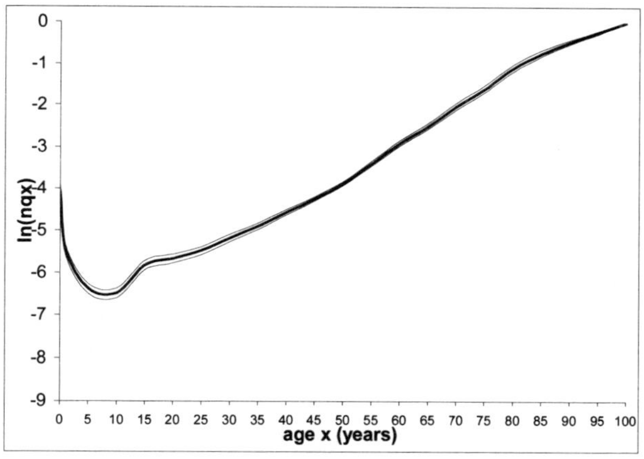

l_x with uncertainty interval

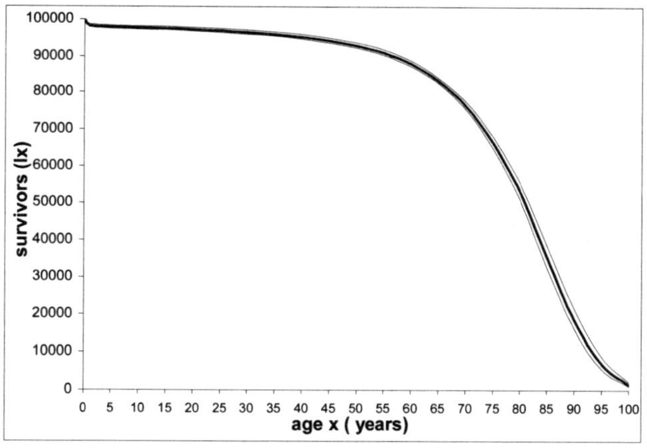

e_x with uncertainty interval

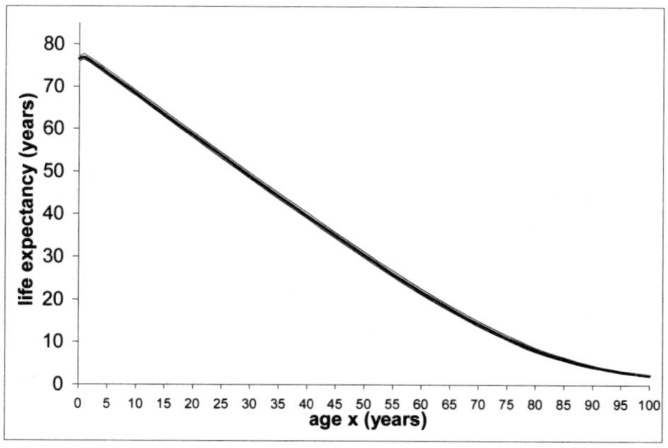

Distribution of deaths by age with uncertainty interval

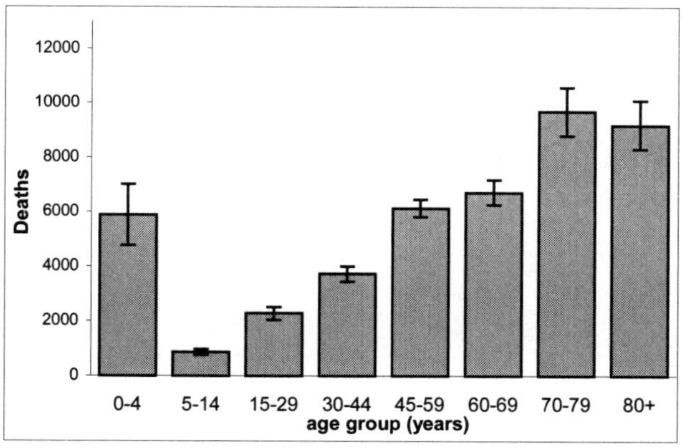

Viet Nam **Viet Nam** **Viet Nam**

Males, 2000 *Hommes, 2000* *Hombres, 2000*

Age	$_nM_x$	$_nq_x$			l_x			$_nd_x$	$_nL_x$	e_x			Actual deaths	Actual population
			percentiles			percentiles					percentiles			
		mean	2½	97½	mean	2½	97½			mean	2½	97½		
<1	0.0263	0.0256	0.0232	0.0280	100,000	100,000	100,000	2,565	97,692	66.7	65.7	67.7	20,662	787,000
1-4	0.0032	0.0128	0.0119	0.0138	97,435	97,201	97,680	1,250	386,740	67.4	66.5	68.4	10,189	3,152,261
5-9	0.0019	0.0093	0.0085	0.0100	96,185	95,859	96,522	891	478,697	64.3	63.4	65.2	8,616	4,626,532
10-14	0.0014	0.0067	0.0062	0.0073	95,294	94,903	95,697	643	474,860	59.9	59.0	60.8	6,321	4,666,677
15-19	0.0016	0.0080	0.0074	0.0086	94,650	94,213	95,102	758	471,357	55.2	54.5	56.1	6,723	4,179,728
20-24	0.0022	0.0109	0.0101	0.0118	93,892	93,401	94,402	1,028	466,893	50.7	49.9	51.5	8,372	3,803,746
25-29	0.0019	0.0097	0.0089	0.0104	92,865	92,300	93,452	900	462,074	46.2	45.5	47.0	6,611	3,395,310
30-34	0.0023	0.0114	0.0104	0.0123	91,965	91,336	92,621	1,045	457,213	41.6	40.9	42.4	6,984	3,055,665
35-39	0.0033	0.0164	0.0150	0.0177	90,920	90,217	91,656	1,489	450,879	37.1	36.4	37.9	8,962	2,714,604
40-44	0.0047	0.0234	0.0215	0.0253	89,432	88,623	90,280	2,095	441,920	32.7	32.0	33.4	10,441	2,202,523
45-49	0.0070	0.0344	0.0315	0.0371	87,337	86,383	88,342	3,004	429,173	28.4	27.7	29.1	11,892	1,698,884
50-54	0.0092	0.0451	0.0414	0.0486	84,332	83,176	85,556	3,803	412,155	24.3	23.7	25.0	9,790	1,061,034
55-59	0.0130	0.0631	0.0580	0.0680	80,529	79,132	82,017	5,083	389,939	20.3	19.8	21.0	11,375	872,586
60-64	0.0228	0.1077	0.0993	0.1160	75,446	73,748	77,258	8,127	356,912	16.5	16.0	17.1	18,116	795,572
65-69	0.0321	0.1485	0.1368	0.1595	67,319	65,190	69,589	9,997	311,602	13.2	12.8	13.7	24,157	752,976
70-74	0.0558	0.2448	0.2268	0.2618	57,322	54,791	60,072	14,032	251,530	10.1	9.7	10.5	29,711	532,601
75-79	0.0883	0.3616	0.3391	0.3832	43,290	40,445	46,445	15,655	177,314	7.6	7.3	7.9	31,994	362,375
80-84	0.1500	0.5455	0.5200	0.5683	27,635	24,947	30,697	15,074	100,492	5.4	5.2	5.7	28,174	187,828
85-89	0.2300	0.7302	0.7094	0.7485	12,561	10,769	14,735	9,172	39,878	3.9	3.8	4.1	16,622	72,268
90-94	0.3400	0.8416	0.8271	0.8544	3,390	2,708	4,282	2,853	8,390	2.8	2.7	2.9	5,362	15,772
95-99	0.4806	0.9841	0.9808	0.9866	537	394	741	528	1,100	2.1	2.1	2.1	991	2,063
100+	0.6976	1	1	1	9	5	14	9	12	1.4	1.4	1.7	70	100
All ages													*282,136*	*38,938,105*

ln($_nq_x$) with uncertainty interval

l_x with uncertainty interval

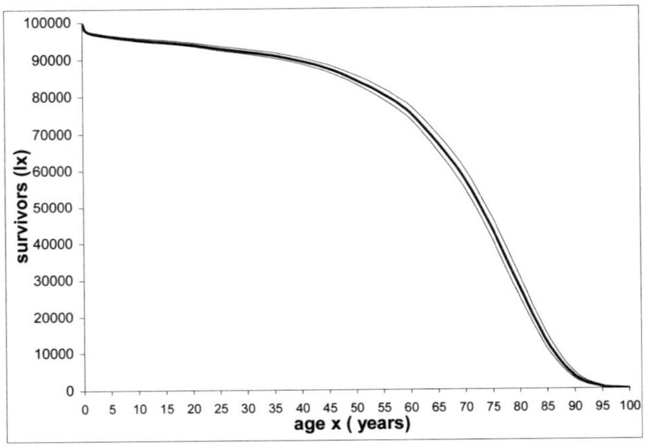

e_x with uncertainty interval

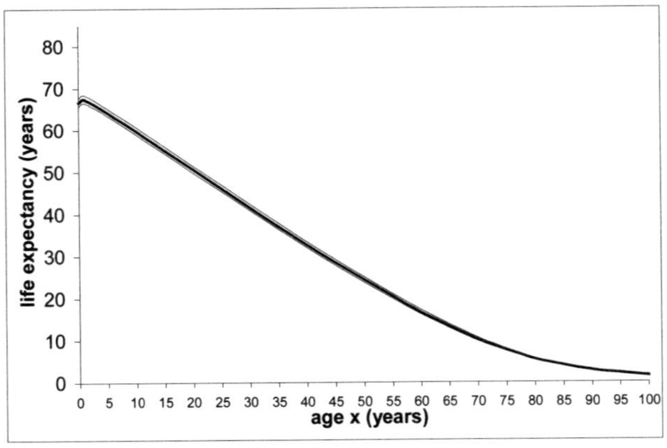

Distribution of deaths by age with uncertainty interval

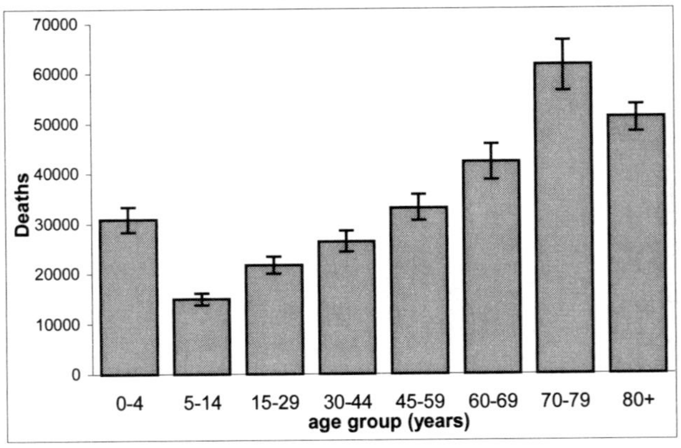

Viet Nam

Females, 2000

Viet Nam

Femmes, 2000

Viet Nam

Mujeres, 2000

Age	$_nM_x$	$_nq_x$ mean	$_nq_x$ percentiles 2½	$_nq_x$ percentiles 97½	l_x mean	l_x percentiles 2½	l_x percentiles 97½	$_nd_x$	$_nL_x$	e_x mean	e_x percentiles 2½	e_x percentiles 97½	Actual deaths	Actual population
<1	0.0185	0.0182	0.0160	0.0203	100,000	100,000	100,000	1,821	98,362	71.0	69.9	72.1	13,873	749,524
1-4	0.0031	0.0121	0.0110	0.0132	98,179	97,973	98,405	1,189	389,863	71.3	70.3	72.4	9,318	3,054,123
5-9	0.0023	0.0114	0.0103	0.0124	96,990	96,680	97,323	1,102	482,195	68.1	67.2	69.2	10,266	4,491,615
10-14	0.0014	0.0072	0.0065	0.0078	95,888	95,481	96,318	687	477,721	63.9	63.1	64.9	6,537	4,542,588
15-19	0.0012	0.0059	0.0054	0.0065	95,200	94,733	95,690	562	474,596	59.3	58.5	60.3	4,854	4,095,556
20-24	0.0016	0.0080	0.0073	0.0088	94,638	94,122	95,177	758	471,295	54.7	53.9	55.6	6,055	3,764,305
25-29	0.0013	0.0063	0.0058	0.0069	93,880	93,297	94,484	594	467,914	50.1	49.3	51.0	4,362	3,434,841
30-34	0.0016	0.0080	0.0073	0.0088	93,286	92,652	93,940	747	464,561	45.4	44.7	46.3	5,023	3,123,005
35-39	0.0019	0.0093	0.0084	0.0101	92,539	91,840	93,256	858	460,549	40.7	40.0	41.6	5,192	2,787,866
40-44	0.0030	0.0151	0.0138	0.0165	91,681	90,909	92,470	1,386	454,938	36.1	35.4	37.0	6,911	2,267,862
45-49	0.0039	0.0193	0.0176	0.0211	90,294	89,405	91,197	1,741	447,120	31.6	31.0	32.5	6,817	1,750,706
50-54	0.0063	0.0312	0.0284	0.0341	88,553	87,521	89,595	2,767	435,849	27.2	26.6	28.0	7,055	1,111,169
55-59	0.0091	0.0447	0.0405	0.0487	85,786	84,536	87,051	3,834	419,346	23.0	22.4	23.7	8,410	919,902
60-64	0.0155	0.0748	0.0677	0.0811	81,952	80,421	83,525	6,130	394,436	18.9	18.5	19.6	13,263	853,342
65-69	0.0233	0.1099	0.0996	0.1184	75,822	73,898	77,870	8,334	358,274	15.3	14.8	15.9	19,393	833,664
70-74	0.0423	0.1911	0.1743	0.2035	67,488	65,148	70,118	12,897	305,196	11.8	11.5	12.4	25,824	611,108
75-79	0.0637	0.2747	0.2538	0.2912	54,591	51,893	57,893	14,997	235,461	9.1	8.7	9.4	27,705	434,983
80-84	0.1109	0.4342	0.4098	0.4564	39,594	36,781	43,200	17,192	154,988	6.5	6.3	6.8	26,620	239,982
85-89	0.1829	0.6275	0.6033	0.6494	22,402	19,994	25,495	14,056	76,866	4.6	4.5	4.8	18,643	101,949
90-94	0.2854	0.7687	0.7504	0.7856	8,345	7,009	10,114	6,415	22,480	3.2	3.1	3.3	7,431	26,041
95-99	0.4215	0.9307	0.9217	0.9386	1,930	1,503	2,525	1,796	4,261	2.3	2.3	2.4	1,837	4,359
100+	0.5894	1	1	1	134	92	198	134	227	1.7	1.7	1.9	187	318
All ages													*235,576*	*39,198,808*

ln($_nq_x$) with uncertainty interval

l_x with uncertainty interval

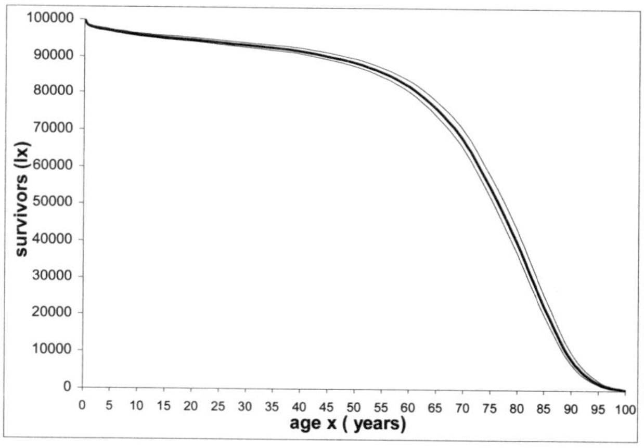

e_x with uncertainty interval

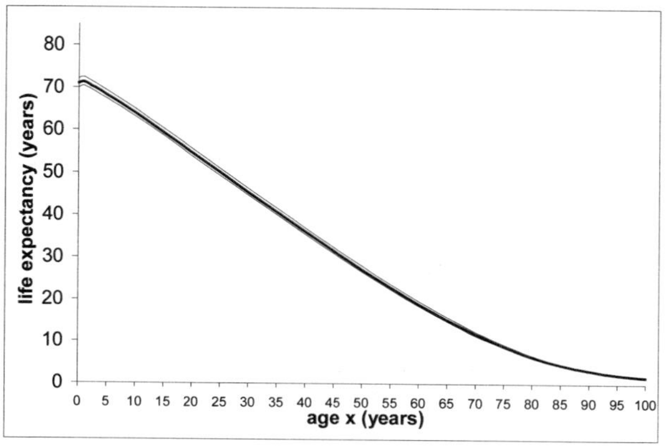

Distribution of deaths by age with uncertainty interval

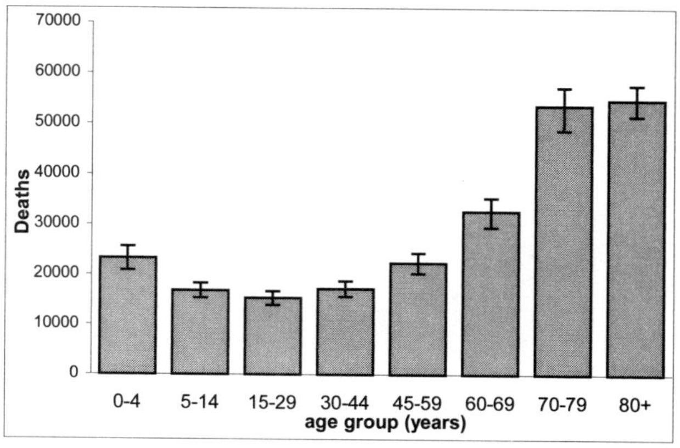

Yemen / Yémen / Yemen

Males, 2000 / *Hommes, 2000* / *Hombres, 2000*

Age	$_nM_x$	$_nq_x$			l_x			$_nd_x$	$_nL_x$	e_x			Actual deaths	Actual population
			percentiles			percentiles					percentiles			
		mean	21/2	971/2	mean	21/2	971/2			mean	21/2	971/2		
<1	0.0877	0.0813	0.0743	0.0887	100,000	100,000	100,000	8,131	92,682	59.3	57.7	61.4	38,078	434,022
1-4	0.0063	0.0246	0.0220	0.0270	91,869	91,131	92,574	2,263	362,043	63.5	61.8	65.6	9,762	1,561,662
5-9	0.0017	0.0085	0.0075	0.0093	89,606	88,856	90,355	762	446,122	61.1	59.4	63.1	2,666	1,560,151
10-14	0.0012	0.0060	0.0052	0.0066	88,843	88,035	89,644	529	442,893	56.6	55.0	58.5	1,366	1,142,604
15-19	0.0016	0.0082	0.0072	0.0090	88,314	87,462	89,157	725	439,758	51.9	50.3	53.8	1,510	916,413
20-24	0.0026	0.0128	0.0112	0.0141	87,589	86,677	88,491	1,119	435,150	47.3	45.7	49.2	2,039	793,015
25-29	0.0030	0.0148	0.0129	0.0164	86,471	85,465	87,483	1,282	429,146	42.9	41.3	44.7	2,069	692,309
30-34	0.0036	0.0178	0.0154	0.0197	85,188	84,072	86,330	1,518	422,146	38.5	37.0	40.3	2,127	591,758
35-39	0.0047	0.0233	0.0200	0.0259	83,671	82,420	84,965	1,946	413,489	34.1	32.7	35.9	2,277	483,894
40-44	0.0065	0.0320	0.0274	0.0359	81,725	80,303	83,216	2,614	402,091	29.9	28.5	31.6	1,513	232,734
45-49	0.0093	0.0455	0.0387	0.0514	79,111	77,462	80,918	3,597	386,565	25.8	24.5	27.4	1,545	166,013
50-54	0.0137	0.0661	0.0561	0.0754	75,514	73,564	77,783	4,995	365,086	21.9	20.6	23.4	2,016	147,342
55-59	0.0201	0.0955	0.0808	0.1095	70,520	68,183	73,419	6,737	335,755	18.3	17.1	19.7	2,524	125,806
60-64	0.0298	0.1386	0.1174	0.1591	63,782	60,938	67,486	8,838	296,818	14.9	13.9	16.2	3,188	107,075
65-69	0.0442	0.1992	0.1692	0.2275	54,945	51,296	59,565	10,943	247,366	11.9	11.1	13.1	3,666	82,864
70-74	0.0681	0.2909	0.2494	0.3275	44,002	39,624	49,432	12,801	188,007	9.3	8.6	10.2	3,730	54,791
75-79	0.1042	0.4132	0.3623	0.4543	31,201	26,646	36,923	12,892	123,776	7.1	6.6	7.8	3,253	31,232
80-84	0.1573	0.5645	0.5116	0.6049	18,309	14,540	23,271	10,336	65,707	5.3	5.0	5.8	2,128	13,530
85-89	0.2301	0.7303	0.6858	0.7630	7,973	5,765	11,325	5,823	25,309	4.0	3.8	4.3	916	3,982
90-94	0.3259	0.8240	0.7898	0.8489	2,150	1,378	3,558	1,772	5,436	2.9	2.8	3.1	247	758
95-99	0.4471	0.9548	0.9391	0.9652	379	211	748	361	808	2.2	2.2	2.3	27	60
100+	0.5940	1	1	1	17	7	46	17	29	1.7	1.7	1.8	1	2
All ages													*86,648*	*9,142,017*

ln($_nq_x$) with uncertainty interval

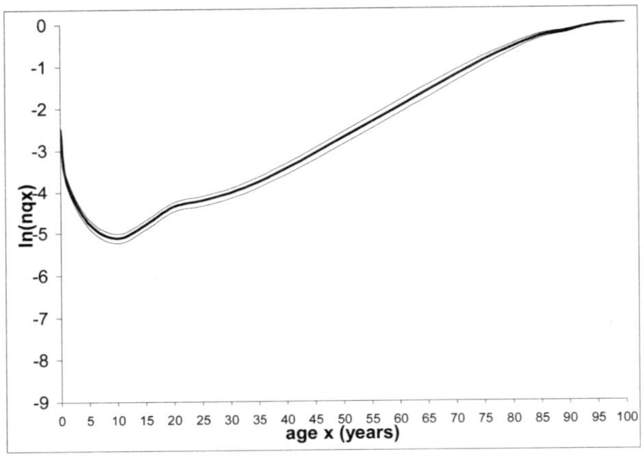

l_x with uncertainty interval

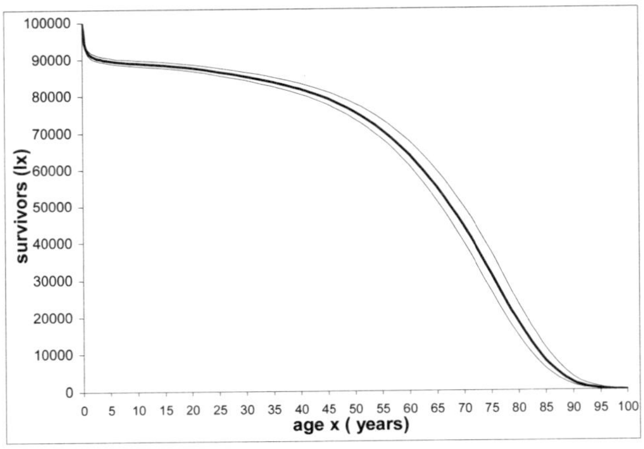

e_x with uncertainty interval

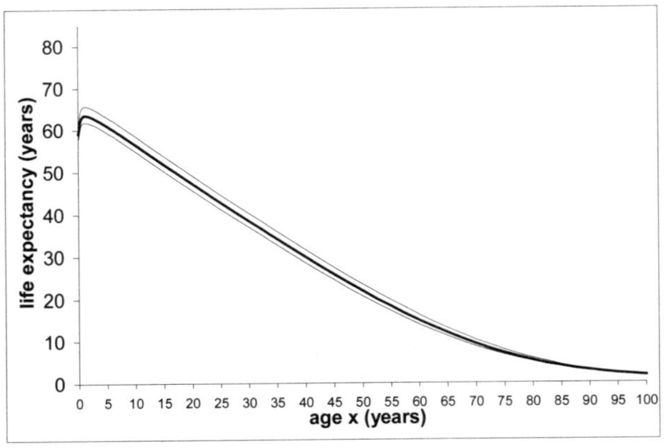

Distribution of deaths by age with uncertainty interval

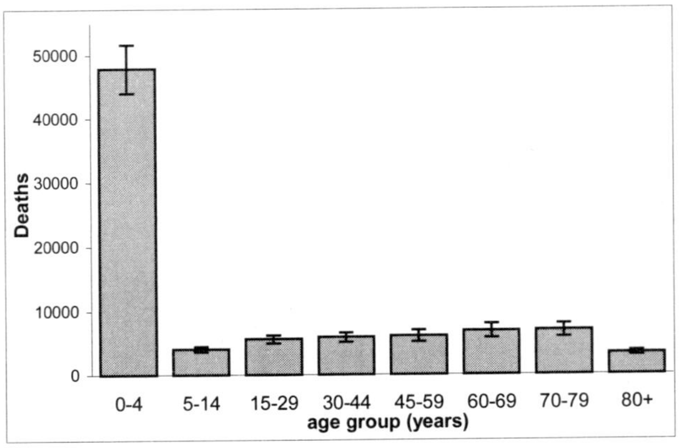

Yemen

Yémen

Yemen

Females, 2000

Femmes, 2000

Mujeres, 2000

Age	$_nM_x$	$_nq_x$			l_x			$_nd_x$	$_nL_x$	e_x			Actual deaths	Actual population
			percentiles			percentiles					percentiles			
		mean	2½	97½	mean	2½	97½			mean	2½	97½		
<1	0.0760	0.0711	0.0655	0.0775	100,000	100,000	100,000	7,112	93,599	62.0	60.4	64.2	31,709	417,329
1-4	0.0068	0.0267	0.0239	0.0291	92,888	92,248	93,449	2,482	365,597	65.7	64.0	68.0	10,158	1,496,449
5-9	0.0017	0.0084	0.0074	0.0093	90,407	89,808	91,005	763	450,127	63.5	61.9	65.6	2,521	1,487,798
10-14	0.0012	0.0059	0.0051	0.0065	89,644	89,005	90,256	525	446,907	59.0	57.4	61.1	1,278	1,087,617
15-19	0.0017	0.0086	0.0075	0.0095	89,119	88,453	89,752	764	443,683	54.4	52.7	56.4	1,496	868,748
20-24	0.0024	0.0118	0.0103	0.0131	88,355	87,645	89,062	1,047	439,156	49.8	48.2	51.8	1,798	754,243
25-29	0.0028	0.0139	0.0120	0.0154	87,308	86,496	88,143	1,209	433,516	45.4	43.8	47.3	1,847	661,923
30-34	0.0032	0.0161	0.0138	0.0179	86,099	85,166	87,085	1,382	427,037	41.0	39.5	42.8	1,827	564,498
35-39	0.0042	0.0209	0.0179	0.0234	84,716	83,645	85,881	1,773	419,148	36.6	35.1	38.4	1,945	459,745
40-44	0.0052	0.0255	0.0217	0.0286	82,943	81,692	84,341	2,112	409,433	32.3	30.9	34.1	1,827	354,067
45-49	0.0070	0.0346	0.0293	0.0392	80,831	79,365	82,512	2,799	397,155	28.1	26.7	29.8	1,948	276,425
50-54	0.0101	0.0492	0.0414	0.0562	78,032	76,284	80,093	3,838	380,562	24.0	22.7	25.6	2,267	224,779
55-59	0.0145	0.0699	0.0586	0.0804	74,193	72,070	76,775	5,185	358,003	20.1	18.9	21.6	2,586	178,570
60-64	0.0222	0.1053	0.0881	0.1218	69,008	66,411	72,278	7,269	326,867	16.5	15.4	17.8	3,088	138,837
65-69	0.0349	0.1604	0.1347	0.1854	61,739	58,552	65,907	9,903	283,936	13.1	12.2	14.3	3,563	102,165
70-74	0.0569	0.2489	0.2101	0.2847	51,836	47,695	57,027	12,902	226,924	10.1	9.4	11.2	3,902	68,630
75-79	0.0906	0.3693	0.3173	0.4131	38,934	34,115	44,933	14,379	158,721	7.7	7.1	8.5	3,610	39,844
80-84	0.1410	0.5211	0.4629	0.5652	24,555	20,021	30,408	12,797	90,782	5.7	5.3	6.3	2,565	18,194
85-89	0.2116	0.6920	0.6401	0.7277	11,758	8,705	16,152	8,136	38,450	4.2	4.0	4.6	1,182	5,587
90-94	0.3064	0.7983	0.7584	0.8249	3,622	2,370	5,813	2,892	9,436	3.1	2.9	3.3	359	1,171
95-99	0.4282	0.9371	0.9164	0.9499	731	415	1,404	685	1,599	2.3	2.2	2.4	45	104
100+	0.5772	1	1	1	46	21	117	46	80	1.7	1.7	1.8	3	6
All ages													*81,523*	*9,206,729*

ln($_nq_x$) with uncertainty interval

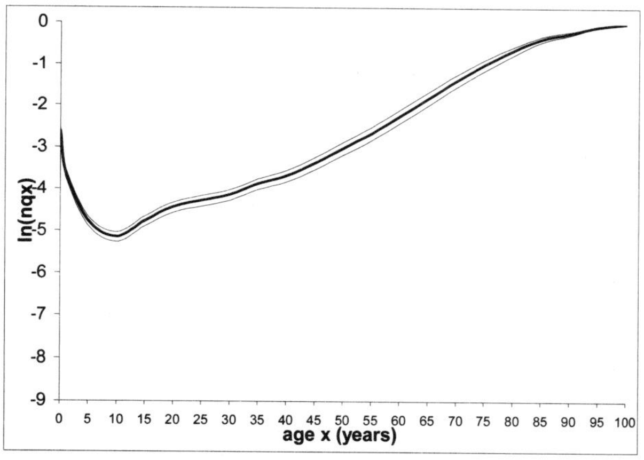

l_x with uncertainty interval

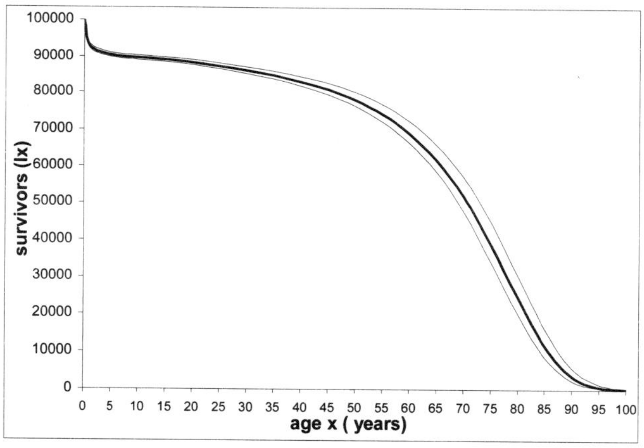

e_x with uncertainty interval

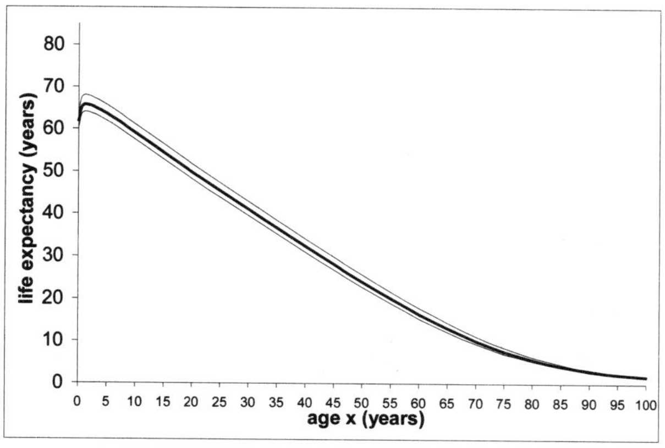

Distribution of deaths by age with uncertainty interval

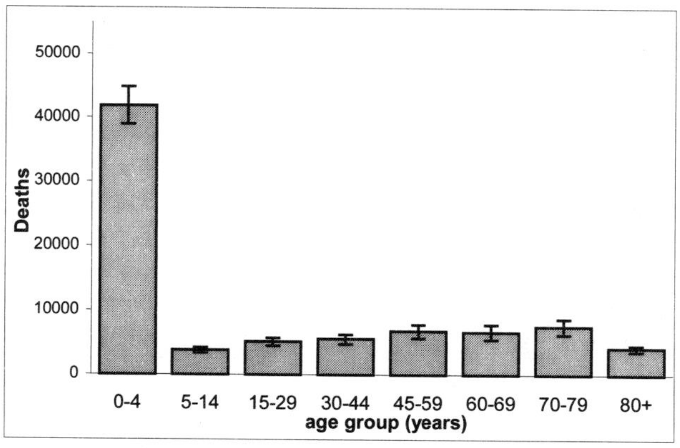

Yugoslavia | Yougoslavie | Yugoslavia

Males, 2000 | *Hommes, 2000* | *Hombres, 2000*

Age	$_nM_x$	$_nq_x$ mean	$_nq_x$ percentiles 21/2	$_nq_x$ percentiles 971/2	l_x mean	l_x percentiles 21/2	l_x percentiles 971/2	$_nd_x$	$_nL_x$	e_x mean	e_x percentiles 21/2	e_x percentiles 971/2	Actual deaths	Actual population
<1	0.0164	0.0162	0.0144	0.0180	100,000	100,000	100,000	1,617	98,545	69.8	69.6	70.0	1,087	66,264
1-4	0.0006	0.0024	0.0022	0.0025	98,383	98,203	98,560	232	392,976	69.9	69.7	70.2	157	265,707
5-9	0.0003	0.0015	0.0014	0.0016	98,151	97,954	98,345	143	490,397	66.1	65.8	66.4	105	359,905
10-14	0.0003	0.0013	0.0012	0.0014	98,008	97,800	98,211	130	489,713	61.2	60.9	61.5	107	402,370
15-19	0.0006	0.0032	0.0030	0.0034	97,877	97,661	98,089	316	488,597	56.2	56.0	56.6	273	422,010
20-24	0.0011	0.0055	0.0051	0.0058	97,561	97,324	97,793	532	486,476	51.4	51.2	51.7	451	412,302
25-29	0.0012	0.0058	0.0055	0.0061	97,029	96,760	97,291	564	483,735	46.7	46.4	47.0	457	391,776
30-34	0.0015	0.0075	0.0071	0.0078	96,465	96,167	96,755	720	480,526	41.9	41.7	42.3	554	369,943
35-39	0.0022	0.0108	0.0104	0.0113	95,745	95,414	96,065	1,036	476,137	37.2	37.0	37.6	794	365,137
40-44	0.0036	0.0180	0.0174	0.0186	94,709	94,338	95,064	1,701	469,294	32.6	32.3	33.0	1,356	374,178
45-49	0.0057	0.0282	0.0275	0.0288	93,008	92,586	93,403	2,619	458,494	28.2	27.9	28.5	2,311	404,523
50-54	0.0094	0.0460	0.0451	0.0469	90,389	89,919	90,818	4,162	441,543	23.9	23.6	24.3	3,052	323,771
55-59	0.0143	0.0689	0.0669	0.0707	86,228	85,732	86,683	5,944	416,278	19.9	19.6	20.3	3,453	241,782
60-64	0.0220	0.1043	0.1003	0.1074	80,283	79,827	80,702	8,371	380,490	16.2	15.9	16.6	5,649	256,783
65-69	0.0340	0.1567	0.1498	0.1626	71,913	71,464	72,373	11,267	331,395	12.8	12.5	13.1	8,228	242,008
70-74	0.0560	0.2456	0.2344	0.2563	60,645	59,919	61,497	14,895	265,988	9.7	9.5	10.0	10,104	180,428
75-79	0.0992	0.3974	0.3812	0.4139	45,750	44,712	47,083	18,179	183,302	7.1	6.9	7.3	10,432	105,193
80-84	0.1600	0.5714	0.5542	0.5898	27,571	26,268	29,137	15,755	98,468	5.1	5.0	5.3	5,494	34,335
85-89	0.2500	0.7692	0.7557	0.7836	11,816	10,775	12,988	9,089	36,357	3.7	3.6	3.8	5,499	21,996
90-94	0.3700	0.8768	0.8678	0.8863	2,727	2,332	3,167	2,391	6,462	2.6	2.6	2.7	2,375	6,418
95-99	0.4900	0.9919	0.9906	0.9932	336	265	417	333	680	2.0	2.0	2.0	482	983
100+	0.6200	1	1	1	3	2	4	3	4	1.6	1.6	1.7	51	83
All ages													*62,471*	*5,247,895*

ln($_nq_x$) with uncertainty interval

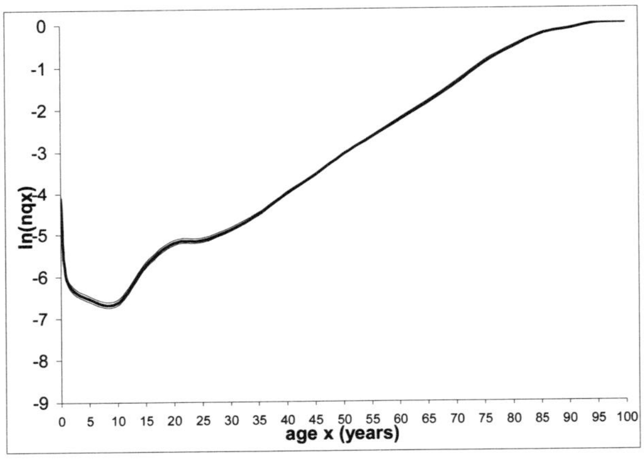

l_x with uncertainty interval

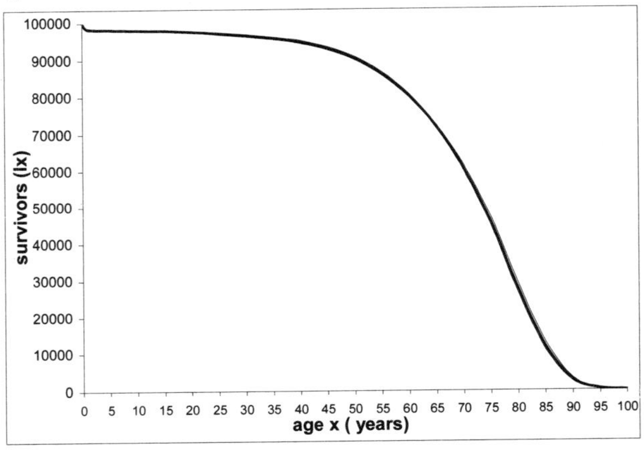

e_x with uncertainty interval

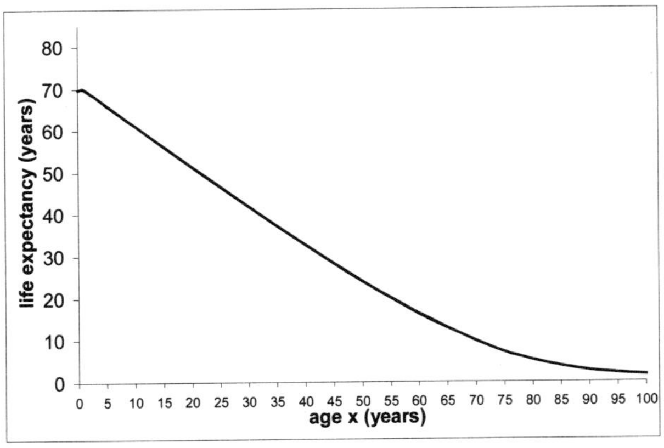

Distribution of deaths by age with uncertainty interval

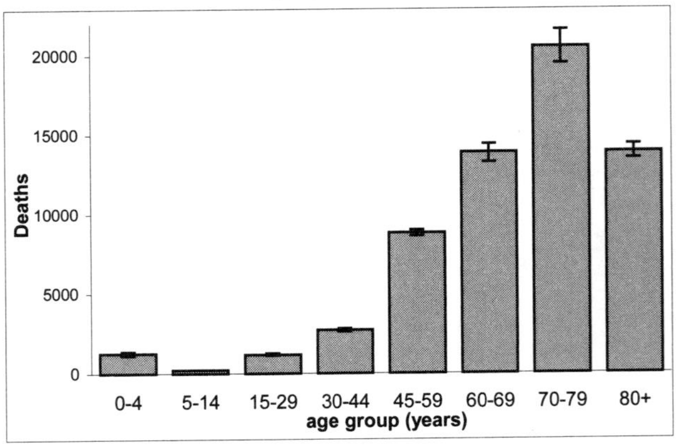

Yugoslavia **Yougoslavie** **Yugoslavia**

Females, 2000 *Femmes, 2000* *Mujeres, 2000*

Age	$_nM_x$	$_nq_x$			l_x			$_nd_x$	$_nL_x$	e_x			Actual deaths	Actual population
			percentiles			percentiles					percentiles			
		mean	2½	97½	mean	2½	97½			mean	2½	97½		
<1	0.0131	0.0130	0.0118	0.0144	100,000	100,000	100,000	1,298	98,832	74.7	74.5	75.0	817	62,220
1-4	0.0006	0.0025	0.0024	0.0027	98,702	98,557	98,816	251	394,206	74.7	74.4	75.0	157	246,238
5-9	0.0002	0.0012	0.0011	0.0013	98,451	98,288	98,579	117	491,963	70.9	70.6	71.1	80	335,618
10-14	0.0002	0.0009	0.0009	0.0010	98,334	98,164	98,469	92	491,441	65.9	65.7	66.2	70	374,794
15-19	0.0003	0.0016	0.0016	0.0017	98,242	98,066	98,381	161	490,808	61.0	60.7	61.3	129	393,371
20-24	0.0004	0.0019	0.0018	0.0021	98,081	97,895	98,228	190	489,929	56.1	55.8	56.4	150	385,523
25-29	0.0005	0.0025	0.0024	0.0027	97,891	97,694	98,047	246	488,839	51.2	50.9	51.5	186	368,793
30-34	0.0007	0.0037	0.0035	0.0039	97,645	97,435	97,812	359	487,326	46.3	46.0	46.6	262	355,706
35-39	0.0012	0.0061	0.0059	0.0064	97,286	97,058	97,468	594	484,942	41.5	41.2	41.8	429	349,987
40-44	0.0020	0.0100	0.0097	0.0104	96,691	96,438	96,896	968	481,035	36.7	36.4	37.0	728	361,887
45-49	0.0031	0.0151	0.0148	0.0155	95,723	95,437	95,957	1,449	474,992	32.1	31.8	32.4	1,220	400,014
50-54	0.0047	0.0231	0.0226	0.0236	94,274	93,954	94,539	2,176	465,929	27.5	27.2	27.8	1,531	327,832
55-59	0.0082	0.0400	0.0389	0.0409	92,098	91,751	92,389	3,680	451,289	23.1	22.8	23.4	2,120	259,935
60-64	0.0132	0.0638	0.0618	0.0657	88,418	88,076	88,713	5,642	427,984	19.0	18.7	19.3	3,859	292,708
65-69	0.0232	0.1095	0.1055	0.1137	82,776	82,442	83,104	9,068	391,209	15.1	14.8	15.4	6,746	291,042
70-74	0.0385	0.1758	0.1681	0.1834	73,708	73,145	74,317	12,958	336,144	11.7	11.4	11.9	9,361	242,842
75-79	0.0641	0.2762	0.2635	0.2879	60,750	59,770	61,727	16,779	261,802	8.6	8.4	8.8	10,310	160,866
80-84	0.1251	0.4763	0.4582	0.4929	43,971	42,562	45,383	20,945	167,491	5.9	5.8	6.1	6,748	53,958
85-89	0.2179	0.7054	0.6890	0.7203	23,026	21,625	24,588	16,242	74,523	4.1	4.0	4.2	7,146	32,789
90-94	0.3393	0.8407	0.8296	0.8504	6,784	6,081	7,645	5,703	16,809	2.8	2.8	2.9	2,545	7,501
95-99	0.4718	0.9766	0.9735	0.9792	1,081	910	1,299	1,055	2,237	2.1	2.1	2.1	401	849
100+	0.5859	1	1	1	25	19	34	25	43	1.7	1.7	1.9	30	52
All ages													*55,025*	*5,304,525*

ln($_nq_x$) with uncertainty interval

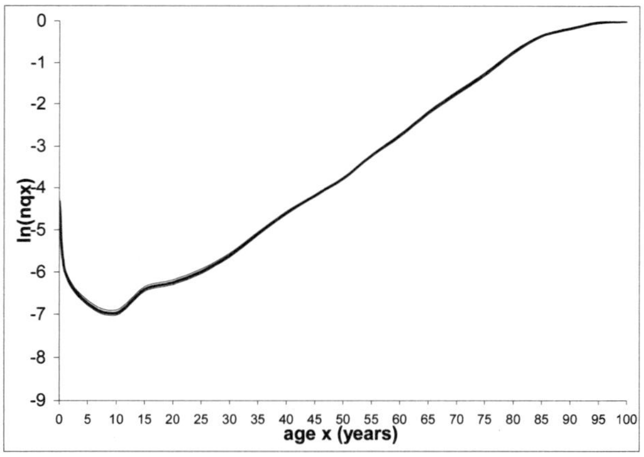

l_x with uncertainty interval

e_x with uncertainty interval

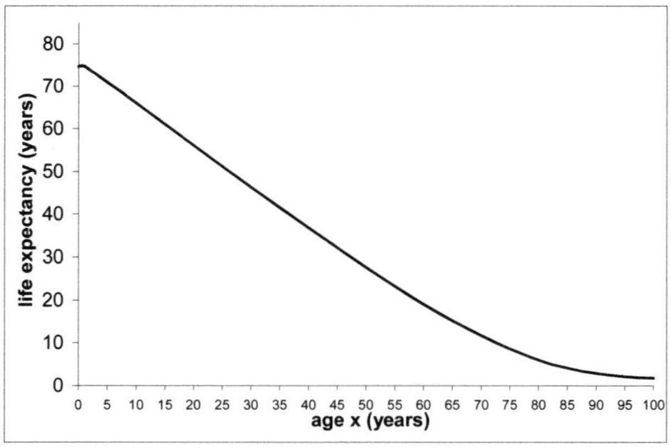

Distribution of deaths by age with uncertainty interval

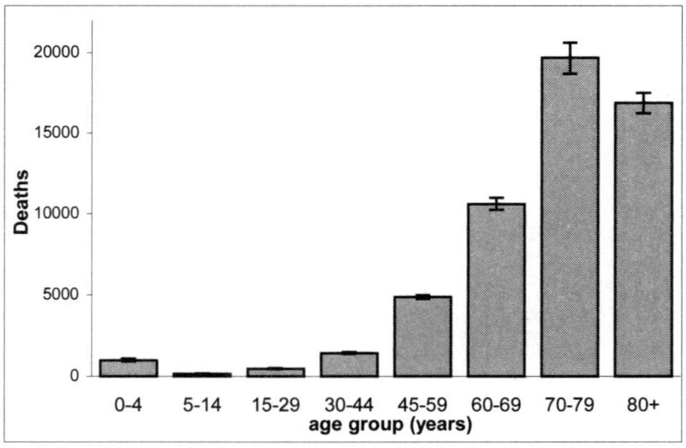

Zambia   Zambie   Zambia

Males, 2000   *Hommes, 2000*   *Hombres, 2000*

Age	$_nM_x$	$_nq_x$			l_x			$_nd_x$	$_nL_x$	e_x			Actual deaths	Actual population
			percentiles			percentiles					percentiles			
		mean	2½	97½	mean	2½	97½			mean	2½	97½		
<1	0.1339	0.1195	0.1012	0.1402	100,000	100,000	100,000	11,954	89,242	39.2	35.4	43.9	28,499	212,757
1-4	0.0147	0.0568	0.0470	0.0670	88,046	85,978	89,881	5,001	340,183	43.6	39.3	49.3	10,896	741,228
5-9	0.0029	0.0142	0.0115	0.0172	83,045	81,301	84,795	1,177	412,284	42.1	38.0	47.7	2,314	810,468
10-14	0.0019	0.0096	0.0077	0.0118	81,868	80,001	83,749	787	407,373	37.6	33.6	43.2	1,328	686,839
15-19	0.0036	0.0178	0.0142	0.0219	81,081	79,132	83,098	1,443	401,798	33.0	29.0	38.5	2,116	589,416
20-24	0.0095	0.0462	0.0364	0.0574	79,638	77,524	81,908	3,679	388,993	28.5	24.5	34.0	4,739	500,981
25-29	0.0218	0.1033	0.0791	0.1301	75,959	73,343	78,878	7,846	360,180	24.8	20.9	30.2	8,952	410,994
30-34	0.0358	0.1643	0.1224	0.2085	68,113	63,842	72,400	11,189	312,593	22.4	18.6	27.6	10,428	291,317
35-39	0.0393	0.1791	0.1309	0.2275	56,924	50,531	62,975	10,196	259,129	21.3	17.8	26.1	8,889	225,895
40-44	0.0394	0.1793	0.1302	0.2273	46,728	39,035	54,521	8,381	212,687	20.4	17.2	24.7	7,230	183,488
45-49	0.0387	0.1766	0.1283	0.2231	38,347	30,204	47,346	6,771	174,808	19.3	16.6	23.0	5,729	147,909
50-54	0.0361	0.1657	0.1210	0.2084	31,576	23,546	41,270	5,232	144,801	17.9	15.6	21.0	4,426	122,476
55-59	0.0330	0.1525	0.1120	0.1908	26,344	18,722	36,277	4,018	121,675	15.9	14.1	18.5	3,245	98,264
60-64	0.0416	0.1885	0.1405	0.2325	22,326	15,222	32,214	4,208	101,111	13.4	11.8	15.6	3,208	77,079
65-69	0.0551	0.2422	0.1845	0.2939	18,118	11,760	27,687	4,388	79,622	10.9	9.6	12.7	3,168	57,485
70-74	0.0797	0.3324	0.2612	0.3937	13,730	8,387	22,578	4,564	57,242	8.6	7.5	10.0	3,206	40,218
75-79	0.1170	0.4527	0.3691	0.5205	9,167	5,165	16,677	4,149	35,459	6.6	5.8	7.7	2,820	24,101
80-84	0.1747	0.6079	0.5192	0.6751	5,017	2,486	10,494	3,050	17,462	4.9	4.4	5.7	1,981	11,340
85-89	0.2524	0.7738	0.6953	0.8288	1,967	812	5,007	1,522	6,031	3.7	3.4	4.2	837	3,318
90-94	0.3521	0.8562	0.7921	0.8989	445	140	1,502	381	1,082	2.7	2.5	3.1	192	545
95-99	0.4748	0.9792	0.9572	0.9897	64	14	305	63	132	2.1	2.0	2.2	21	45
100+	0.6179	1	1	1	1	0	13	1	2	1.6	1.6	1.7	1	2
All ages													*114,224*	*5,236,165*

$\ln({}_nq_x)$ with uncertainty interval

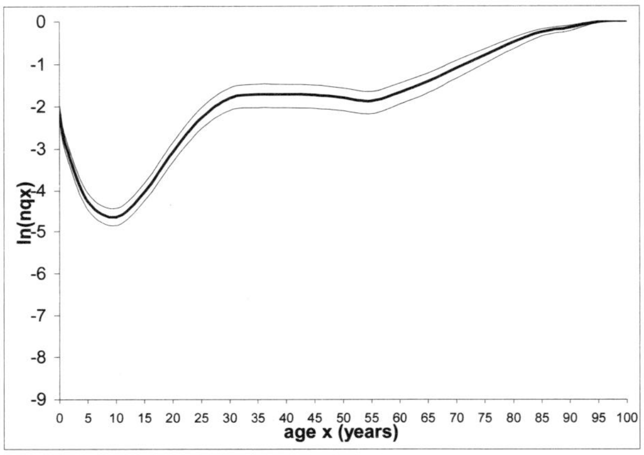

l_x with uncertainty interval

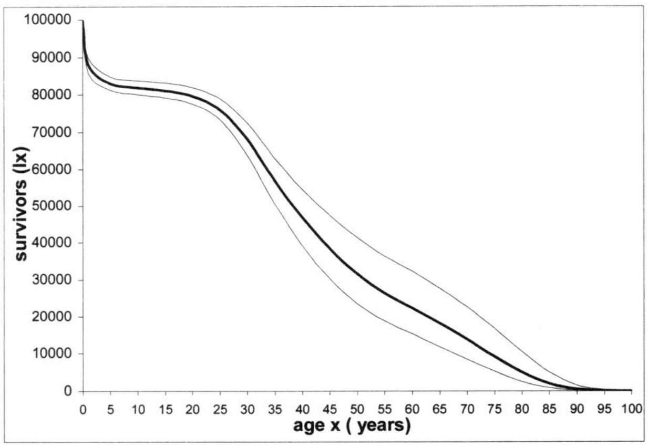

e_x with uncertainty interval

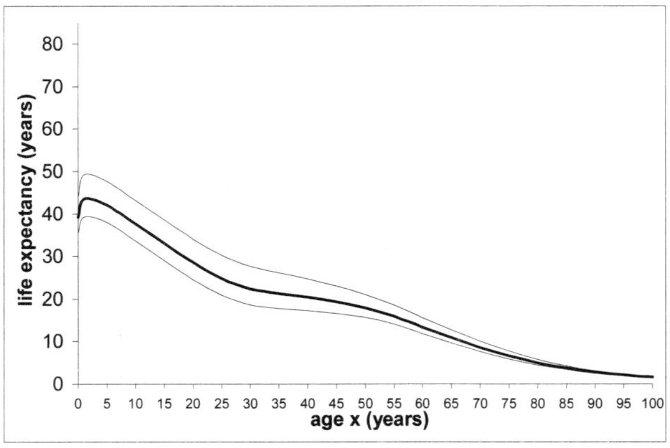

Distribution of deaths by age with uncertainty interval

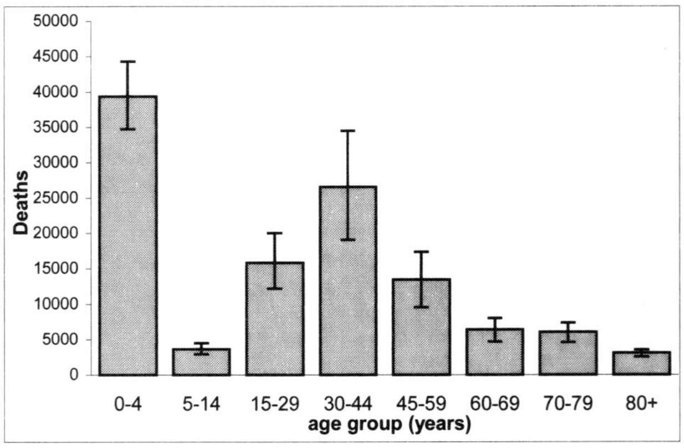

Zambia

Zambie

Zambia

Females, 2000

Femmes, 2000

Mujeres, 2000

Age	$_nM_x$	$_nq_x$			l_x			$_nd_x$	$_nL_x$	e_x			Actual deaths	Actual population
			percentiles			percentiles					percentiles			
		mean	2½	97½	mean	2½	97½			mean	2½	97½		
<1	0.1124	0.1021	0.0870	0.1175	100,000	100,000	100,000	10,207	90,814	39.5	36.0	43.6	22,993	204,574
1-4	0.0155	0.0597	0.0517	0.0674	89,793	88,248	91,301	5,357	346,314	43.0	39.1	48.0	11,267	728,296
5-9	0.0030	0.0147	0.0125	0.0171	84,436	83,073	85,808	1,244	419,067	41.6	37.8	46.6	2,355	793,114
10-14	0.0021	0.0105	0.0089	0.0123	83,191	81,716	84,724	876	413,767	37.2	33.4	42.1	1,423	672,346
15-19	0.0070	0.0346	0.0290	0.0408	82,315	80,758	83,961	2,850	404,452	32.6	28.8	37.5	4,075	578,293
20-24	0.0190	0.0907	0.0746	0.1085	79,465	77,646	81,480	7,209	379,304	28.7	24.9	33.5	9,366	492,762
25-29	0.0331	0.1527	0.1221	0.1849	72,256	69,553	75,203	11,033	333,699	26.3	22.6	31.0	13,065	395,173
30-34	0.0401	0.1824	0.1431	0.2218	61,224	56,693	65,610	11,169	278,196	25.6	22.1	30.0	11,000	273,992
35-39	0.0347	0.1594	0.1236	0.1940	50,055	44,120	55,924	7,981	230,322	25.7	22.6	29.6	7,396	213,432
40-44	0.0291	0.1358	0.1049	0.1659	42,074	35,562	48,885	5,714	196,083	25.1	22.4	28.4	5,278	181,114
45-49	0.0236	0.1114	0.0860	0.1361	36,359	29,703	43,757	4,050	171,672	23.6	21.4	26.5	3,593	152,302
50-54	0.0224	0.1060	0.0820	0.1292	32,309	25,702	39,995	3,423	152,989	21.3	19.4	23.7	2,915	130,270
55-59	0.0228	0.1079	0.0838	0.1310	28,886	22,426	36,717	3,117	136,639	18.5	16.9	20.6	2,513	110,164
60-64	0.0288	0.1343	0.1052	0.1618	25,769	19,533	33,638	3,460	120,196	15.5	14.1	17.3	2,579	89,584
65-69	0.0409	0.1854	0.1474	0.2206	22,309	16,427	30,098	4,137	101,203	12.5	11.3	14.0	2,835	69,354
70-74	0.0631	0.2725	0.2215	0.3184	18,172	12,868	25,661	4,952	78,480	9.7	8.8	11.0	3,109	49,274
75-79	0.0961	0.3874	0.3248	0.4418	13,220	8,851	19,976	5,121	53,297	7.5	6.7	8.4	2,952	30,725
80-84	0.1461	0.5349	0.4662	0.5925	8,099	5,003	13,488	4,332	29,664	5.6	5.1	6.3	2,134	14,612
85-89	0.2156	0.7005	0.6365	0.7523	3,767	2,045	7,199	2,639	12,237	4.2	3.8	4.6	1,022	4,740
90-94	0.3088	0.8015	0.7480	0.8439	1,128	509	2,617	904	2,928	3.1	2.8	3.4	290	938
95-99	0.4291	0.9380	0.9081	0.9586	224	80	659	210	489	2.3	2.2	2.4	47	109
100+	0.5783	1	1	1	14	3	61	14	24	1.7	1.7	1.9	3	6
All ages													*112,210*	*5,185,174*

$\ln({}_nq_x)$ with uncertainty interval

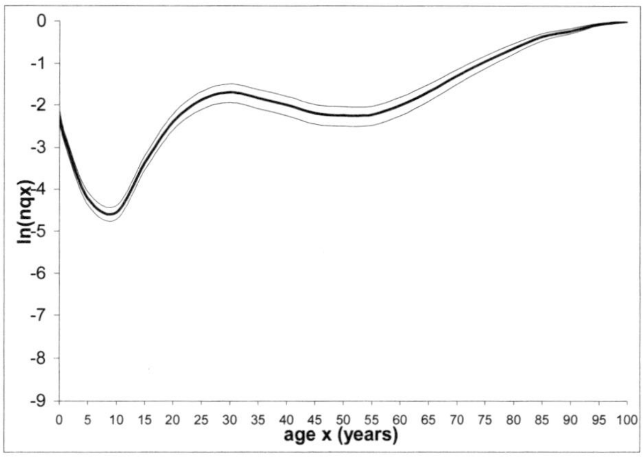

l_x with uncertainty interval

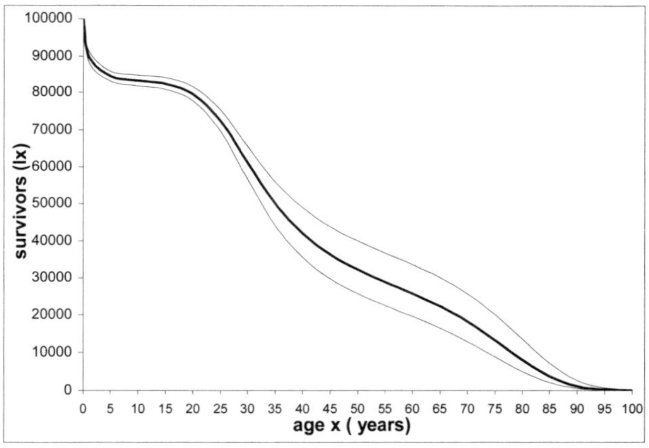

e_x with uncertainty interval

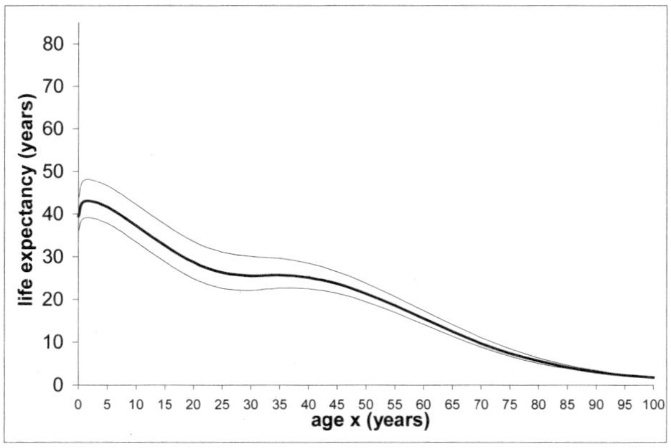

Distribution of deaths by age with uncertainty interval

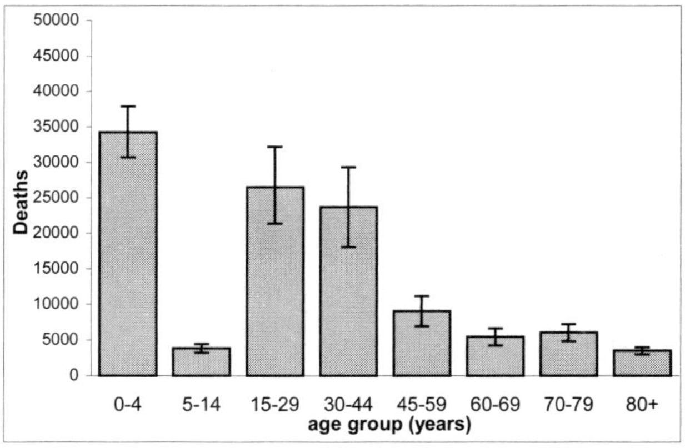

Zimbabwe Zimbabwe Zimbabwe

Males, 2000 *Hommes, 2000* *Hombres, 2000*

Age	$_nM_x$	$_nq_x$			l_x			$_nd_x$	$_nL_x$	e_x			Actual deaths	Actual population
			percentiles			percentiles					percentiles			
		mean	2 1/2	97 1/2	mean	2 1/2	97 1/2			mean	2 1/2	97 1/2		
<1	0.0793	0.0740	0.0695	0.0788	100,000	100,000	100,000	7,403	93,337	45.4	44.0	47.2	17,523	220,929
1-4	0.0092	0.0362	0.0337	0.0384	92,597	92,122	93,052	3,350	362,348	48.0	46.4	49.9	7,483	809,470
5-9	0.0018	0.0091	0.0084	0.0097	89,247	88,747	89,747	811	444,208	45.7	44.2	47.6	1,757	962,414
10-14	0.0014	0.0068	0.0063	0.0073	88,436	87,901	88,977	605	440,667	41.1	39.6	43.0	1,194	869,062
15-19	0.0033	0.0162	0.0148	0.0172	87,831	87,261	88,412	1,420	435,603	36.4	34.9	38.2	2,460	754,398
20-24	0.0078	0.0381	0.0348	0.0407	86,410	85,758	87,085	3,293	423,819	32.0	30.4	33.8	4,847	623,762
25-29	0.0174	0.0836	0.0757	0.0901	83,117	82,275	84,012	6,947	398,217	28.1	26.6	29.9	8,480	486,077
30-34	0.0287	0.1337	0.1205	0.1456	76,170	74,938	77,617	10,187	355,382	25.5	24.0	27.2	10,534	367,499
35-39	0.0314	0.1457	0.1309	0.1595	65,983	64,237	68,263	9,611	305,887	24.0	22.7	25.6	9,175	292,014
40-44	0.0315	0.1461	0.1312	0.1602	56,372	54,070	59,326	8,234	261,275	22.7	21.5	24.1	7,503	238,100
45-49	0.0313	0.1451	0.1302	0.1590	48,138	45,411	51,529	6,987	223,223	21.1	20.1	22.4	5,708	182,341
50-54	0.0298	0.1387	0.1244	0.1516	41,151	38,191	44,779	5,708	191,485	19.3	18.4	20.3	3,666	122,993
55-59	0.0283	0.1323	0.1188	0.1442	35,443	32,401	39,153	4,689	165,493	17.0	16.3	17.9	3,199	112,920
60-64	0.0362	0.1661	0.1500	0.1802	30,754	27,728	34,444	5,110	140,998	14.2	13.6	14.9	3,228	89,071
65-69	0.0493	0.2194	0.1995	0.2362	25,645	22,732	29,206	5,626	114,158	11.5	11.0	12.1	3,445	69,888
70-74	0.0723	0.3060	0.2813	0.3265	20,018	17,363	23,291	6,126	84,776	9.0	8.7	9.5	3,881	53,711
75-79	0.1075	0.4236	0.3951	0.4469	13,892	11,700	16,675	5,885	54,748	6.9	6.6	7.3	3,637	33,838
80-84	0.1624	0.5776	0.5483	0.6012	8,007	6,482	10,087	4,625	28,472	5.2	5.0	5.4	3,109	19,136
85-89	0.2376	0.7454	0.7201	0.7653	3,382	2,595	4,556	2,521	10,607	3.9	3.7	4.0	1,388	5,843
90-94	0.3356	0.8361	0.8160	0.8519	861	613	1,275	720	2,146	2.8	2.8	3.0	391	1,165
95-99	0.4579	0.9646	0.9563	0.9705	141	91	235	136	297	2.2	2.1	2.2	59	129
100+	0.6031	1	1	1	5	3	10	5	8	1.7	1.6	1.7	5	8
All ages													*102,673*	*6,314,768*

ln($_nq_x$) with uncertainty interval

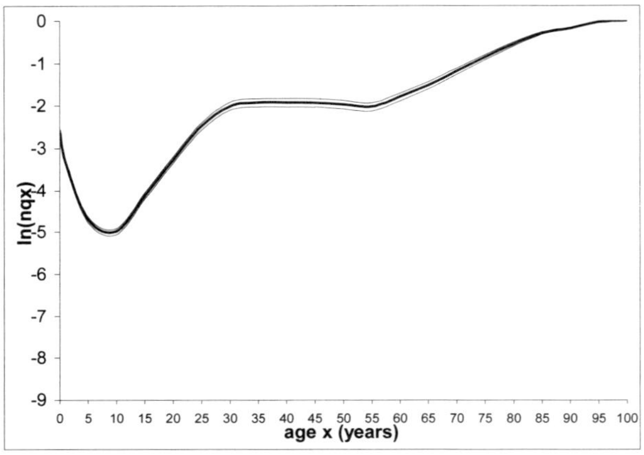

l_x with uncertainty interval

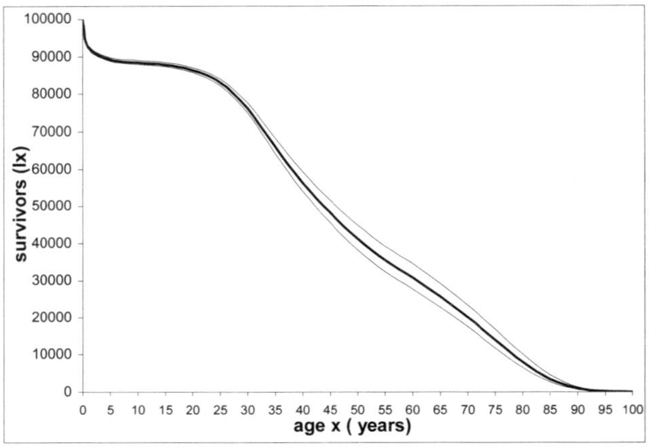

e_x with uncertainty interval

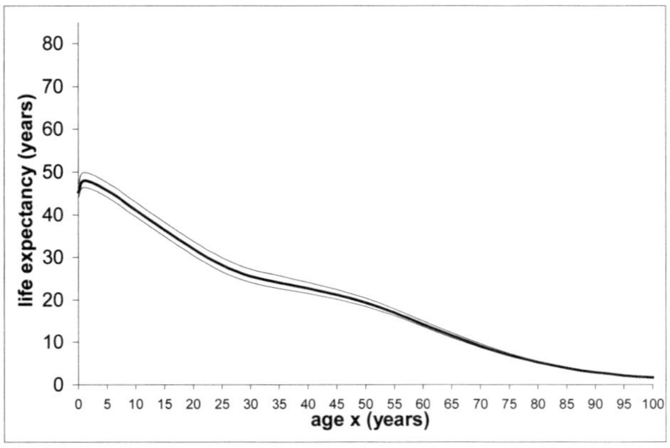

Distribution of deaths by age with uncertainty interval

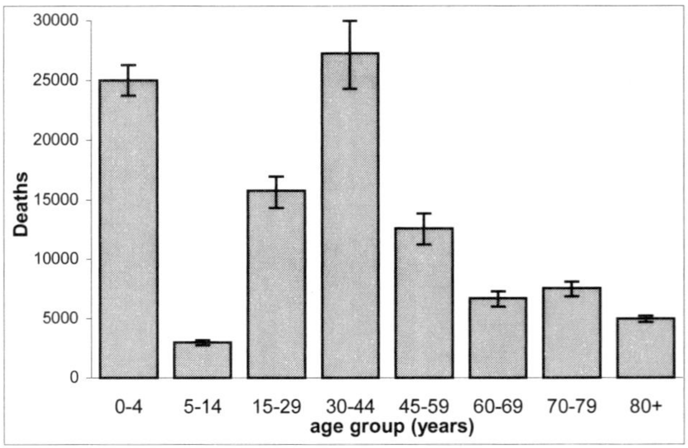

Zimbabwe

Zimbabwe

Zimbabwe

Females, 2000

Femmes, 2000

Mujeres, 2000

Age	$_nM_x$	$_nq_x$			l_x			$_nd_x$	$_nL_x$	e_x			Actual deaths	Actual population
			percentiles			percentiles					percentiles			
		mean	2½	97½	mean	2½	97½			mean	2½	97½		
<1	0.0665	0.0627	0.0585	0.0676	100,000	100,000	100,000	6,273	94,354	46.0	44.3	47.7	14,042	211,213
1-4	0.0096	0.0377	0.0352	0.0400	93,727	93,236	94,154	3,530	366,436	48.0	46.2	50.0	7,801	809,776
5-9	0.0019	0.0093	0.0086	0.0099	90,197	89,697	90,696	834	448,899	45.9	44.0	47.8	1,782	958,447
10-14	0.0014	0.0071	0.0066	0.0076	89,363	88,833	89,901	634	445,228	41.3	39.4	43.2	1,235	867,760
15-19	0.0054	0.0268	0.0248	0.0286	88,729	88,167	89,298	2,378	437,698	36.5	34.7	38.5	4,102	755,051
20-24	0.0149	0.0720	0.0664	0.0773	86,351	85,653	87,069	6,219	416,205	32.5	30.7	34.4	9,255	619,379
25-29	0.0263	0.1235	0.1131	0.1345	80,132	79,058	81,289	9,897	375,915	29.8	28.0	31.7	12,275	466,235
30-34	0.0322	0.1491	0.1350	0.1637	70,234	68,609	72,054	10,470	324,997	28.7	27.0	30.5	11,053	343,104
35-39	0.0278	0.1300	0.1170	0.1434	59,764	57,551	62,168	7,769	279,401	28.2	26.8	29.8	7,744	278,503
40-44	0.0238	0.1124	0.1009	0.1241	51,996	49,298	54,719	5,845	245,368	27.1	25.8	28.5	5,642	236,880
45-49	0.0197	0.0939	0.0843	0.1037	46,151	43,178	49,132	4,334	219,921	25.2	24.1	26.4	3,741	189,802
50-54	0.0191	0.0911	0.0818	0.1004	41,817	38,700	44,993	3,808	199,567	22.6	21.7	23.6	2,525	132,333
55-59	0.0199	0.0946	0.0851	0.1041	38,009	34,815	41,314	3,596	181,056	19.6	18.8	20.5	2,469	124,310
60-64	0.0253	0.1188	0.1071	0.1302	34,413	31,191	37,798	4,088	161,845	16.3	15.7	17.1	2,531	100,199
65-69	0.0360	0.1651	0.1497	0.1798	30,325	27,130	33,745	5,007	139,107	13.2	12.6	13.9	2,920	81,130
70-74	0.0560	0.2457	0.2247	0.2651	25,318	22,251	28,686	6,219	111,040	10.3	9.9	10.9	3,552	63,420
75-79	0.0868	0.3565	0.3303	0.3803	19,098	16,354	22,228	6,808	78,472	7.9	7.5	8.3	3,524	40,618
80-84	0.1355	0.5060	0.4771	0.5319	12,290	10,145	14,867	6,219	45,903	5.9	5.6	6.2	3,265	24,101
85-89	0.2045	0.6766	0.6498	0.7002	6,071	4,761	7,757	4,108	20,085	4.3	4.2	4.5	1,649	8,063
90-94	0.2980	0.7867	0.7648	0.8059	1,963	1,435	2,715	1,544	5,182	3.1	3.0	3.3	564	1,891
95-99	0.4193	0.9285	0.9163	0.9388	419	281	638	389	927	2.3	2.3	2.4	114	273
100+	0.5695	1	1	1	30	17	53	30	53	1.8	1.7	1.8	12	21
All ages													*101,798*	*6,312,509*

$\ln(_nq_x)$ with uncertainty interval

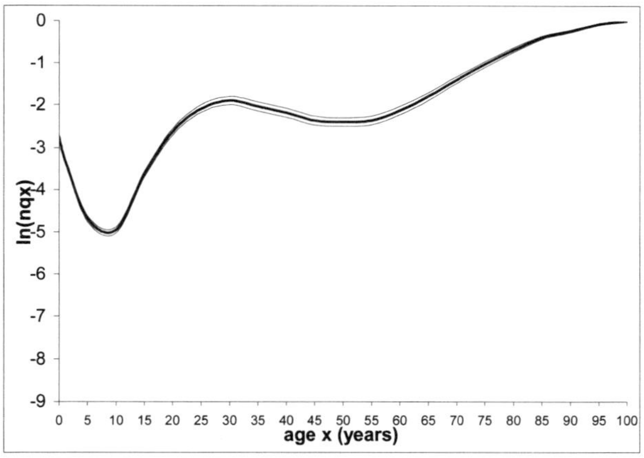

l_x with uncertainty interval

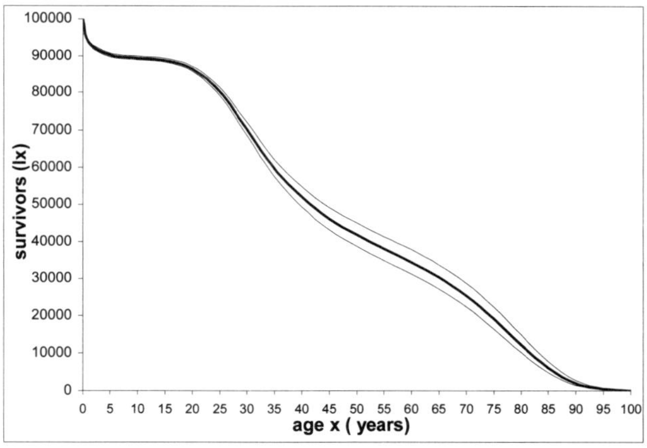

e_x with uncertainty interval

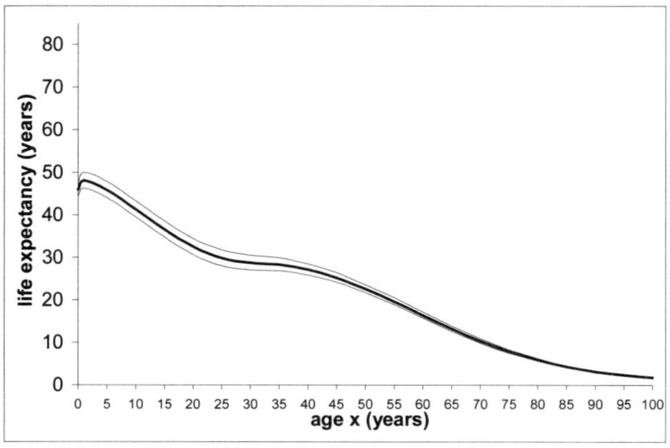

Distribution of deaths by age with uncertainty interval

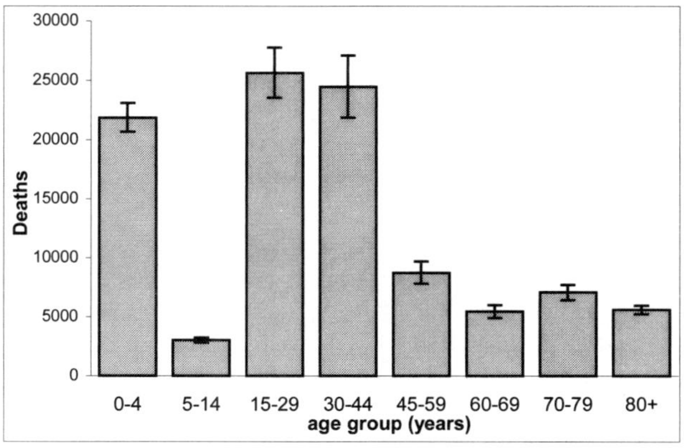

APPENDIX C:
Child Mortality Data, 1960-2000

APPENDICE C:
Données des mortalité des enfants, 1960-2000

Anexo C:
Càlculo de la mortalidad Infantiles, 1960-2000

Reducing levels of child mortality has been the focus of national, regional and global public health initiatives, for several decades. In part to monitor progress with, and better orientate disease control programmes, extensive data collection efforts have been carried out in most countries. In the series of country graphs presented in this appendix, all data available to the World Health Organization as of September 2001 on levels of child mortality have been presented, along with the sources of the data. These include estimates of child mortality calculated from vital registration or demographic surveillance systems, as well as those derived from data collected in demographic surveys or censuses, using direct or indirect techniques to measure child mortality. The data plots present the estimates as calculated from these sources *without* adjustments for underreporting. However, to be truly useful for public health purposes, the best possible estimate of trends and levels of child mortality is required. We have used the data presented for each country to estimate the underlying trend in child mortality, based on knowledge about the reliability of each data source. This trend line has been extrapolated to the year 2000 to yield an estimate of child mortality, with appropriate uncertainty. These estimates were then used for the construction of the life tables reported in Appendices A and B. The data evaluation and estimation methods to estimate national trends in under-five mortality have been reported in detail elsewhere (Ahmad et al. 2000).

As might be expected, the amount, quality and timeliness of data vary substantially among countries and this is reflected in the uncertainty intervals around the child mortality estimates reported in the life tables. We have included countries where only very limited evidence on child mortality was available in order to provide readers with the actual data, however limited, on which the estimates were prepared.

Réduire le niveau de la mortalité des enfants fut au centre des préoccupations nationales, régionales et mondiales des initiatives de santé publique ces dernières décennies. Afin de pouvoir suivre les progrès des programmes de contrôles de maladies d'une part et pour mieux les diriger d'autre part, des efforts considérables dans collecte de données furent menés dans la plupart des pays. Dans la série des graphes par pays présentés dans l'appendice, toutes les données disponibles à l'Organisation mondiale de la Santé en septembre 2001 sur le niveau de la mortalité des enfants ont été exposées, ainsi que leurs sources. Celle-ci inclut les estimations des mortalités des enfants calculées à partir des statistiques de l'état civil ou des systèmes de surveillance démographique , ainsi que celles dérivées des données rassemblées au sein des enquêtes et des recensements, en appliquant des techniques directes ou indirectes de mesure des mortalités des enfants. Les graphiques présentent les estimations calculées à partir de ces données *sans aucun* ajustements pour corriger les sous-enregistrements. Cependant, pour être vraiment utiles à la santé publique, les meilleurs estimations possibles des tendances et des niveaux de la mortalité des enfants sont requises. Nous avons utilisé les renseignements mentionnés pour chaque pays pour estimer la tendance sous-jacente de la mortalité des enfants, en nous basant sur ce que nous connaissons au sujet de la validité de chacune des sources. Cette courbe de tendace a été extrapolée à l'année 2000 pour estimer la mortalité des enfants, avec des incertitudes appropriées. Ces résultats furent alors utilisés pour la construction des tables de mortalité rapportées dans les appendices A et B. Les évaluations et les méthodes des estimations des tendaces des mortalité des moins de cinq ans par pays sont explicitées ailleurs (Ahmad et al. 2000).

Comme on pouvait s'y attendre, la quantité, la qualité et l'ajournement des données varient sensiblement selon les pays et ceci est reflété dans les intervalles d'incertitude autour des estimations des mortalités rapportées dans les tables de mortalité. Nous avons inclus les pays où peu de faits sur la mortalité des enfants sont disponibles pour fournir aux lecteurs les informations réelles, bien que partielles, à partir desquelles les estimations furent préparées.

Reducir los niveles de mortalidad infantil ha sido el propósito, por varias décadas, de iniciativas de salud pública nacionales, regionales y globales. En parte para monitorear el progreso de, y mejores programas orientados de control de enfermedades, extensivos esfuerzos para recolectar datos se han llevado ha cabo en varios países. En la serie de gráficas de países presentados en este apéndice, toda la información disponible a la Organización Mundial de la Salud hasta Septiembre del año 2001 es presentada, así como las fuentes de esta información. Esta incluye estimaciones de mortalidad infantil calculada con datos de registro civil o sistemas de vigilancia demográficos, así como aquellos derivados de recolección de datos en encuestas demográficas o censos, utilizando técnicas directas o indirectas para medir mortalidad infantil. Las gráficas presentan estimaciones calculadas por estas fuentes ante la falta de declaraciones. Sin embargo, para ser verdaderamente útil para propósitos de salud pública, la mejor estimación posible de tendencias y niveles de mortalidad infantil es necesaria. Hemos utilizado la información presentada para cada país para estimar la tendencia subyacente en mortalidad infantil, basada en conocimientos de la fiabilidad de cada fuente de información. Esta tendencia ha sido extrapolada al año 2000 para obtener una estimación de mortalidad infantil, con una apropiada incertidumbre. Estas estimaciones fueron entonces utilizadas para construir las tablas de vida reportadas en los apéndices A y B. La evaluación de información y métodos de estimación para estimar tendencias nacionales en mortalidad (por debajo de 5) han sido reportados en otros lados (Ahmad et al. 2000).

Como puede ser esperado, la cantidad, calidad y disposición en tiempo de la información varia substancialmente entre países y esto se refleja en los intervalos de incertidumbre alrededor de la mortalidad infantil reportados en las tablas de vida. Hemos incluido países donde solamente muy limitada evidencia de mortalidad infantil fue obtenida con el fin de proveer a los lectores de información actual, aunque limitada, con la que las estimaciones fueron calculadas.

Under 5 mortality
Afghanistan

Mortalité des moins de cinq ans
Afghanistan

Mortalidad de menores de cinco años
Afganistán

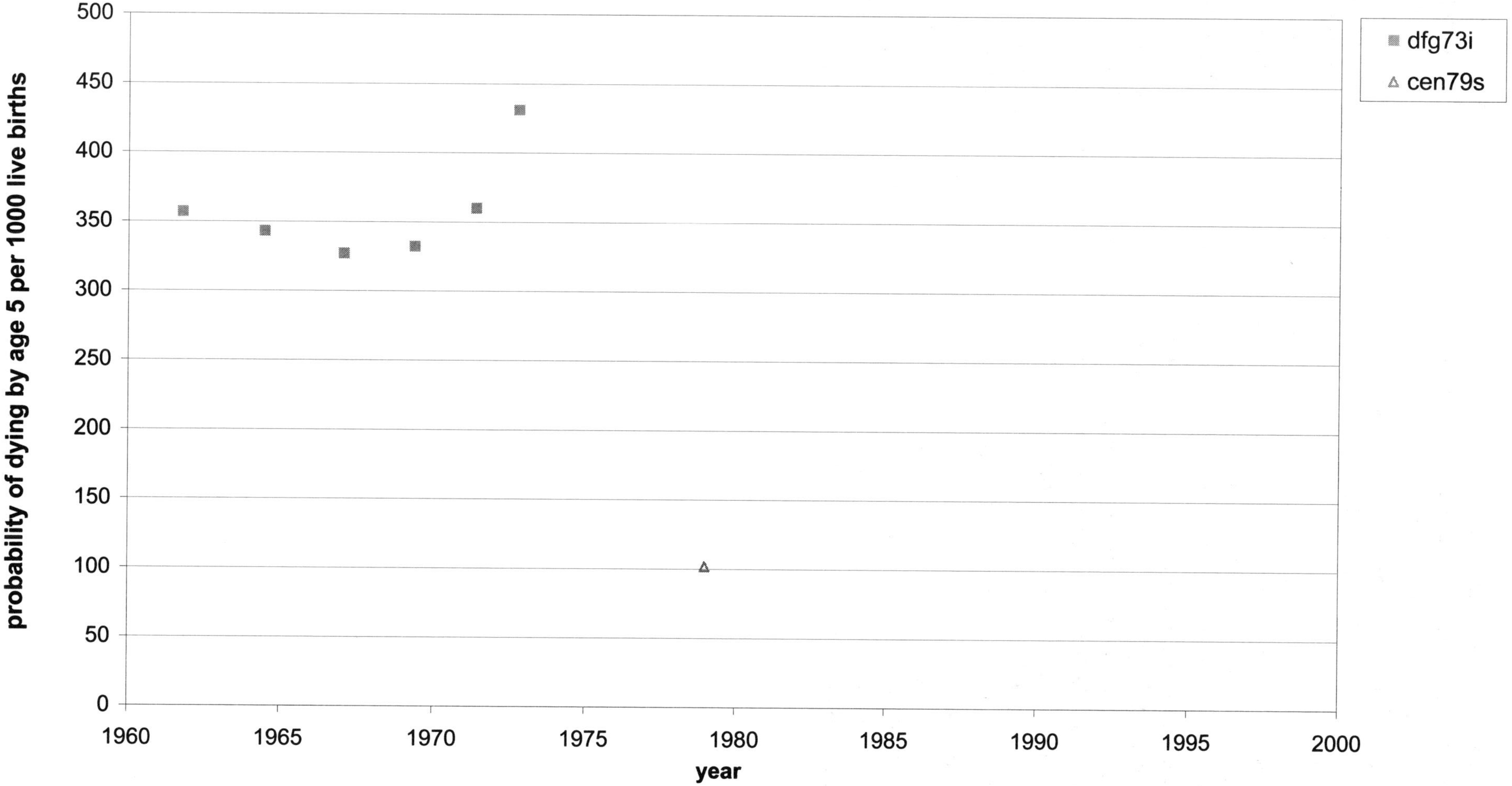

Data sources

Sources des données

Fuentes de los datos

dfg73i : National Demographic and Family Guidance Survey 1972-1974 (indirect method); ***cen79s*** : Census 1979 sample (Mohtasham Shalikar, Estimation of basic demographic parameters and an assessment of the Afghanistan census of 1979, MA Thesis, University of the Philippines, 1983)

Under 5 mortality
Albania

Mortalité des moins de cinq ans
Albanie

Mortalidad de menores de cinco años
Albania

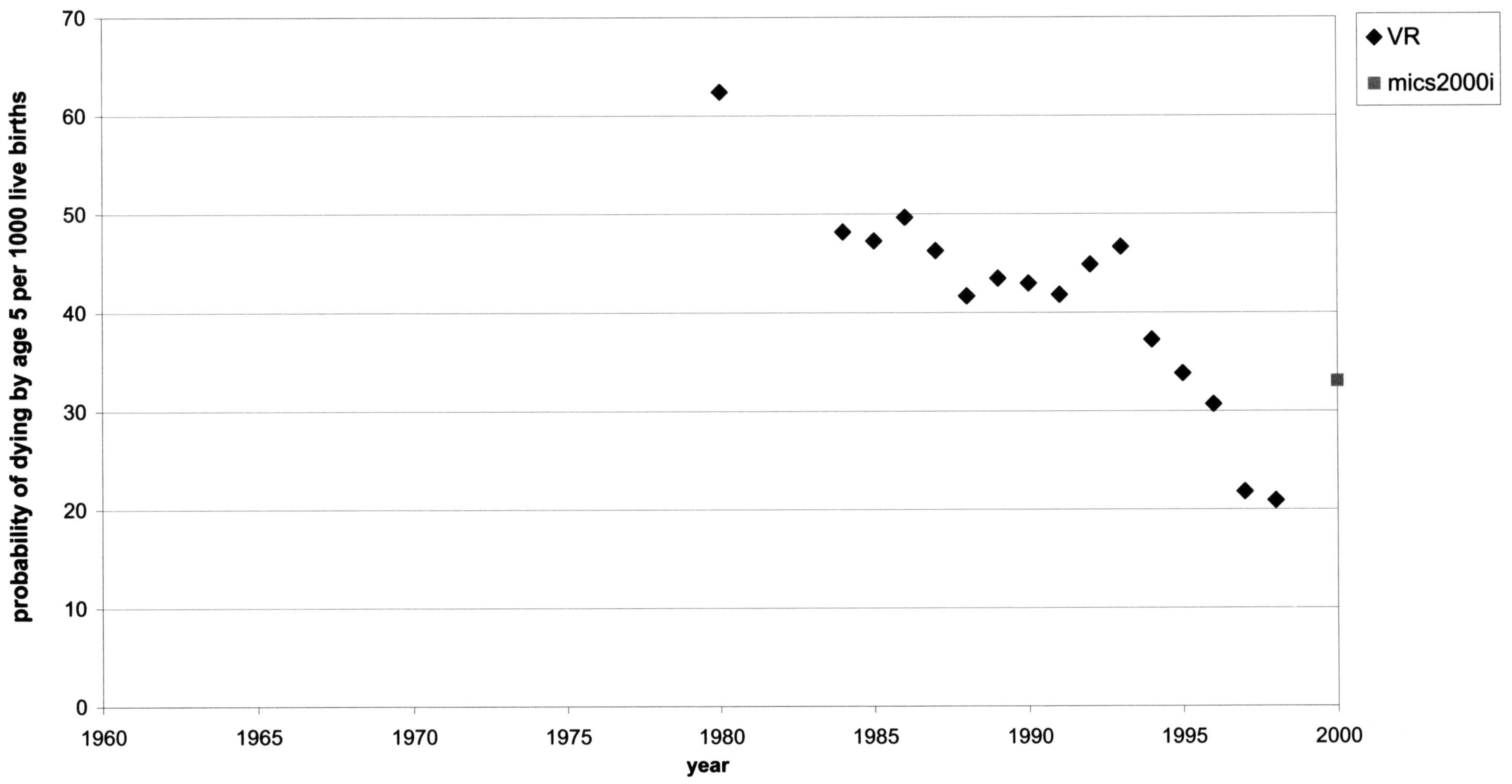

Data sources

Sources des données

Fuentes de los datos

VR : Vital registration; ***mics2000i*** : Mulptiple Indicator Cluster Survey 2000 - draft report (indirect method)

Under 5 mortality

Algeria

Mortalité des moins de cinq ans

Algérie

Mortalidad de menores de cinco años

Argelia

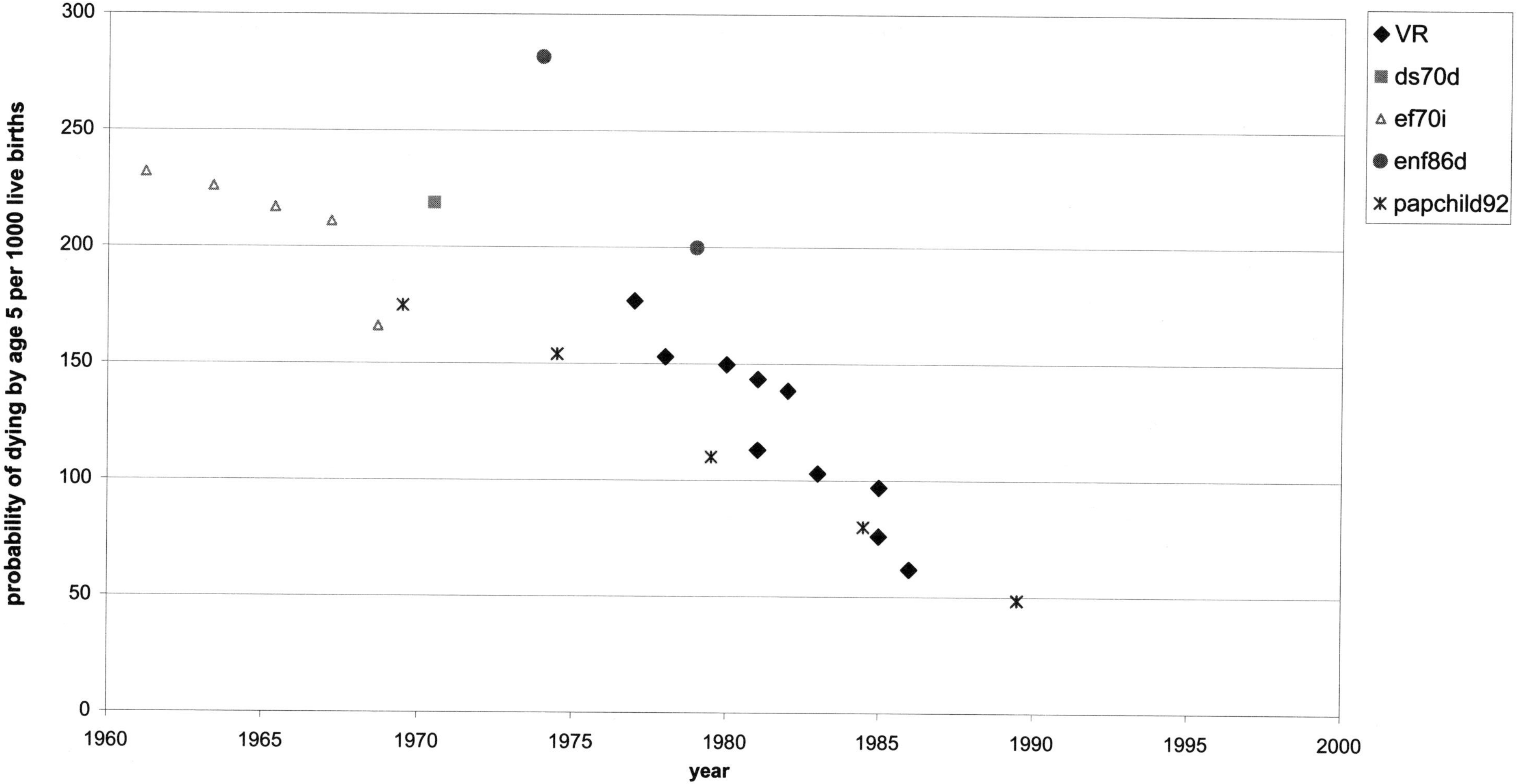

Data sources

Sources des données

Fuentes de los datos

VR: Vital Registration; ***ds70d***: Enquête Démographique, August 1969-March 1971 (direct method); ***ef70i***: Enquête Fécondité, 1970 (indirect method); ***enf86d***: Enquête nationale sur la fécondité 1986 (direct method); ***papchild92***: Maternal and Child Health Survey, 1992

Under 5 mortality
Antigua and Barbuda

Mortalité des moins de cinq ans
Antigua-et-Barbuda

Mortalidad de menores de cinco años
Antigua y Barbuda

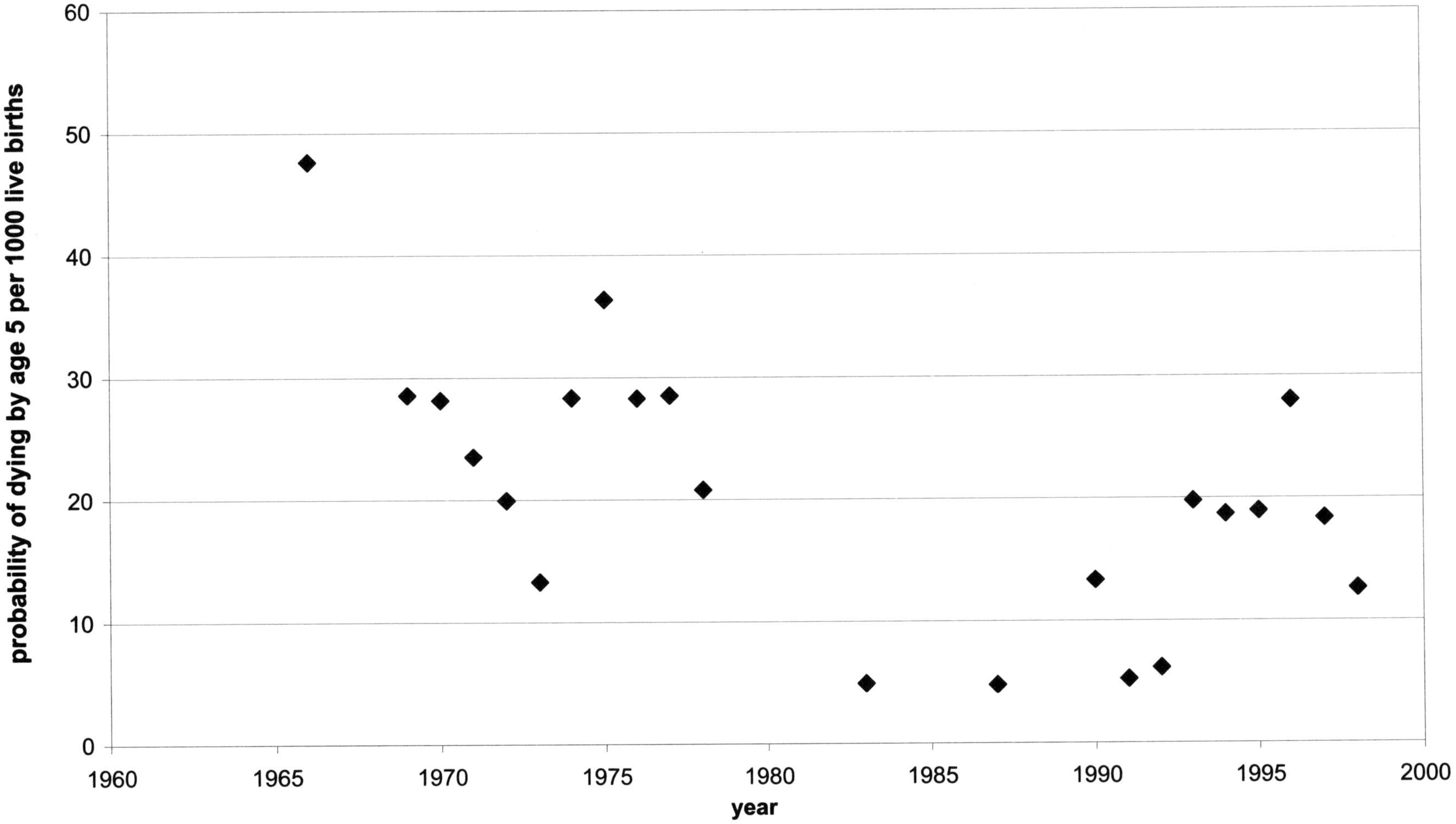

Data sources
Vital Registration

Sources des données

Fuentes de los datos

Under 5 mortality

Argentina

Mortalité des moins de cinq ans

Argentine

Mortalidad de menores de cinco años

Argentina

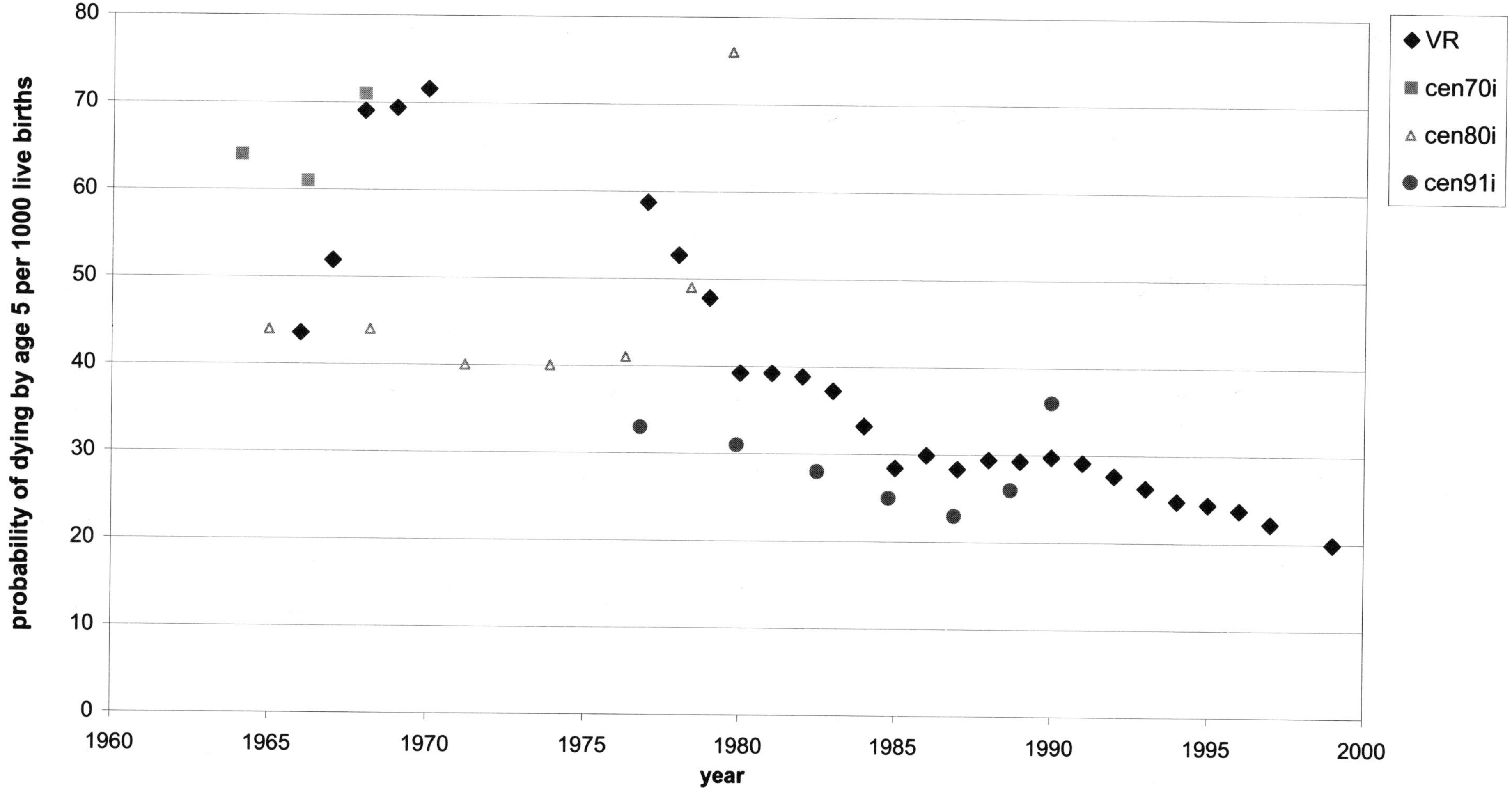

Data sources

Sources des données

Fuentes de los datos

VR: Vital Registration; ***cen70i***: Census, 30 September 1970 (indirect method); ***cen80i***: Census, 22 October 1980 (indirect method); ***cen91i***: Census, 15 May 1991 (indirect method)

Under 5 mortality

Armenia

Mortalité des moins de cinq ans

Arménie

Mortalidad de menores de cinco años

Armenia

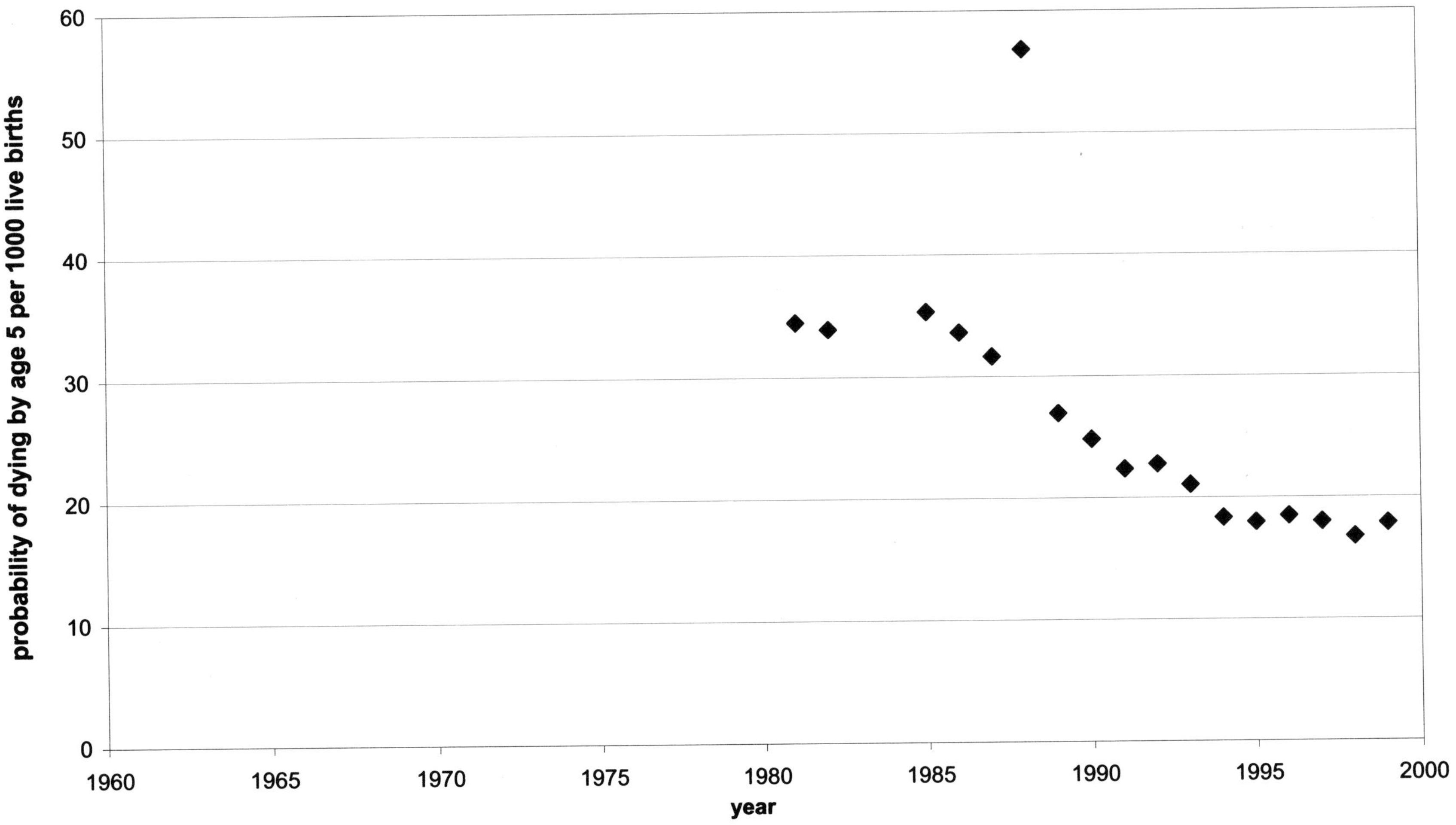

Data sources

Vital Registration

Sources des données

Fuentes de los datos

Under 5 mortality
Australia

Mortalité des moins de cinq ans
Australie

Mortalidad de menores de cinco años
Australia

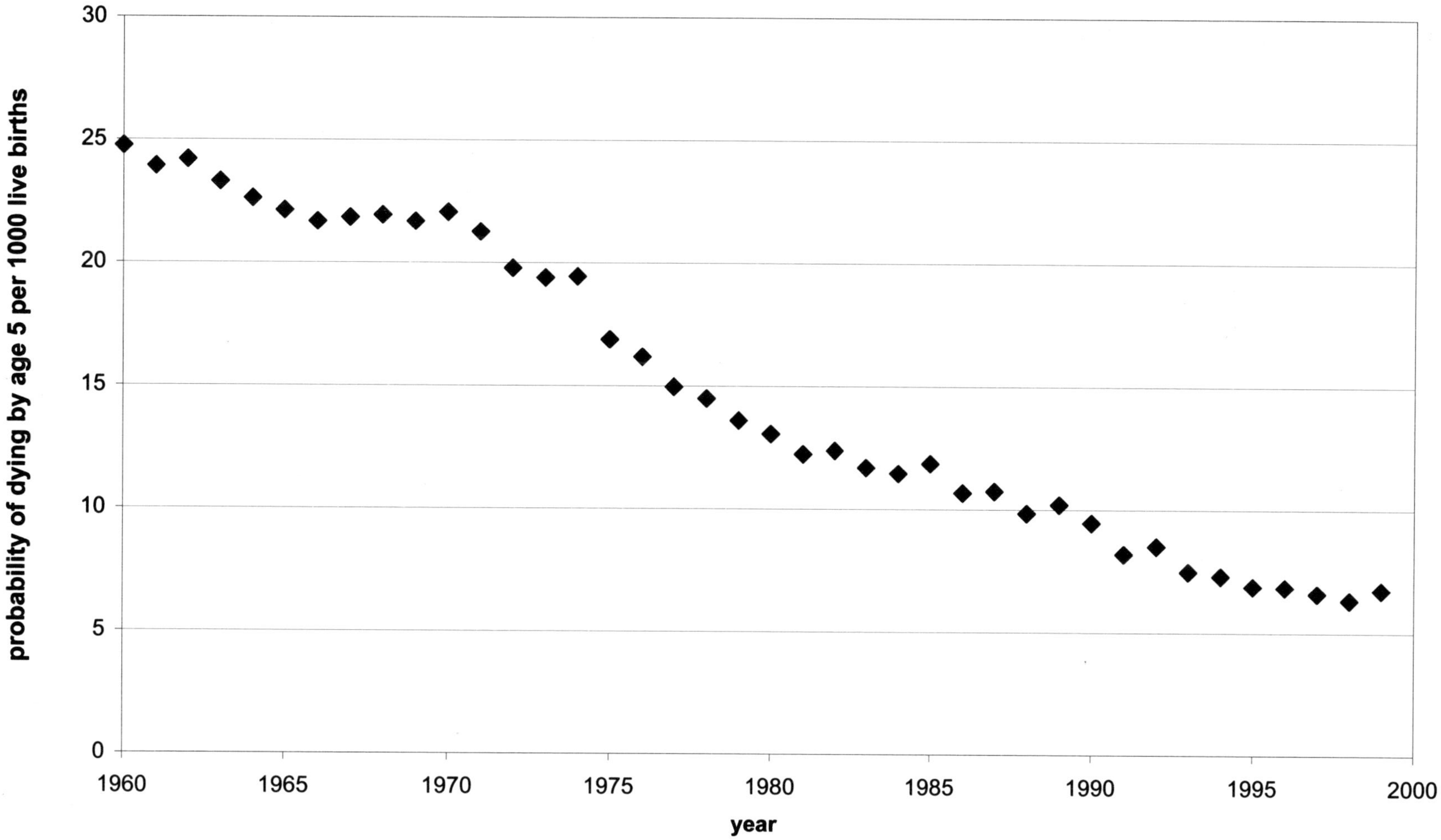

Data sources
Vital Registration

Sources des données

Fuentes de los datos

Under 5 mortality
Austria

Mortalité des moins de cinq ans
Autriche

Mortalidad de menores de cinco años
Austria

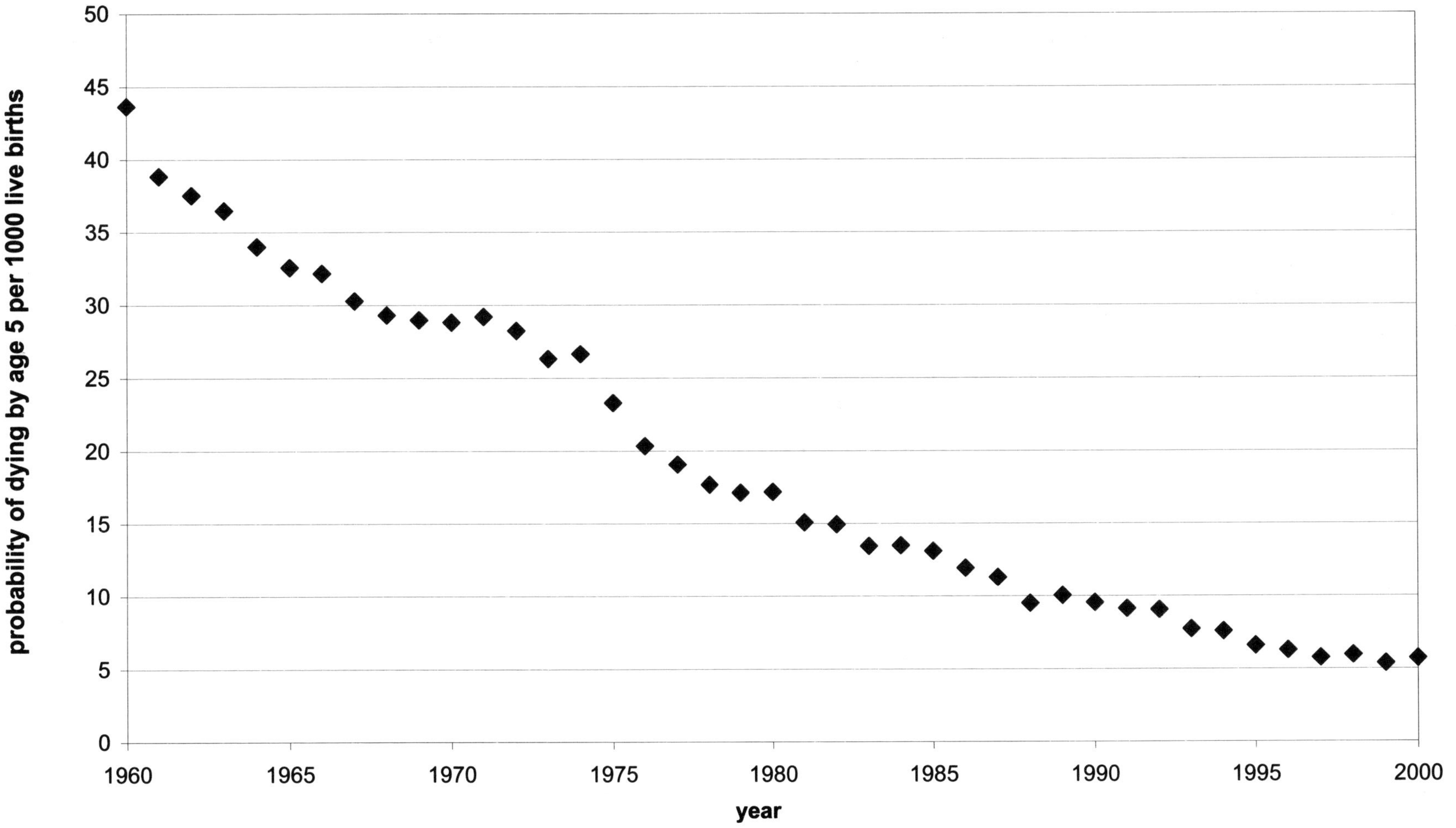

Data sources
Vital Registration

Sources des données

Fuentes de los datos

Under 5 mortality Mortalité des moins de cinq ans Mortalidad de menores de cinco años

Azerbaijan **Azerbaïdjan** **Azerbaiyán**

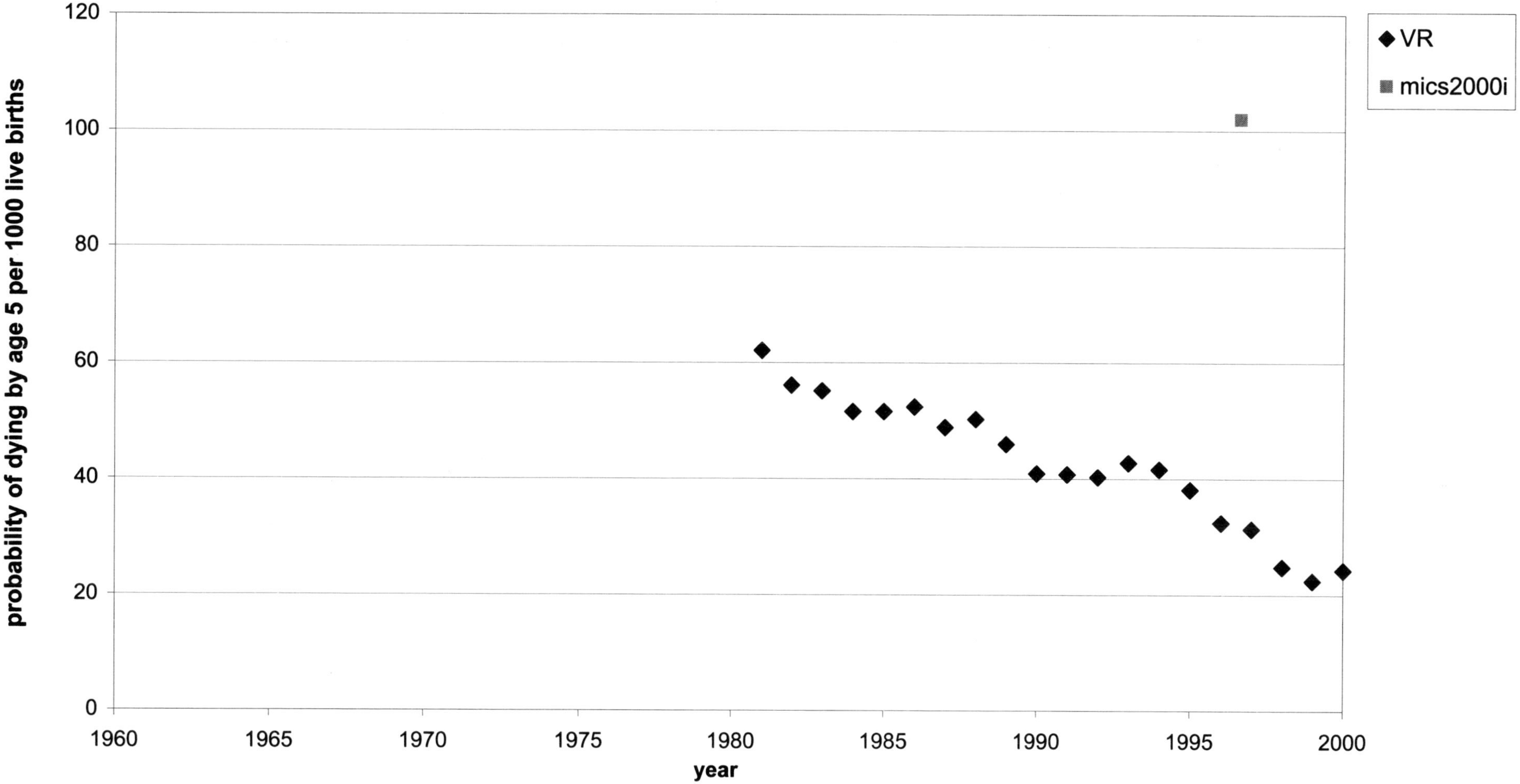

Data sources **Sources des données** **Fuentes de los datos**

VR: Vital Registration; ***mics2000i***: Mulptiple Indicator Cluster Survey, 2000 (indirect method)

Under 5 mortality

Bahamas

Mortalité des moins de cinq ans

Bahamas

Mortalidad de menores de cinco años

Bahamas

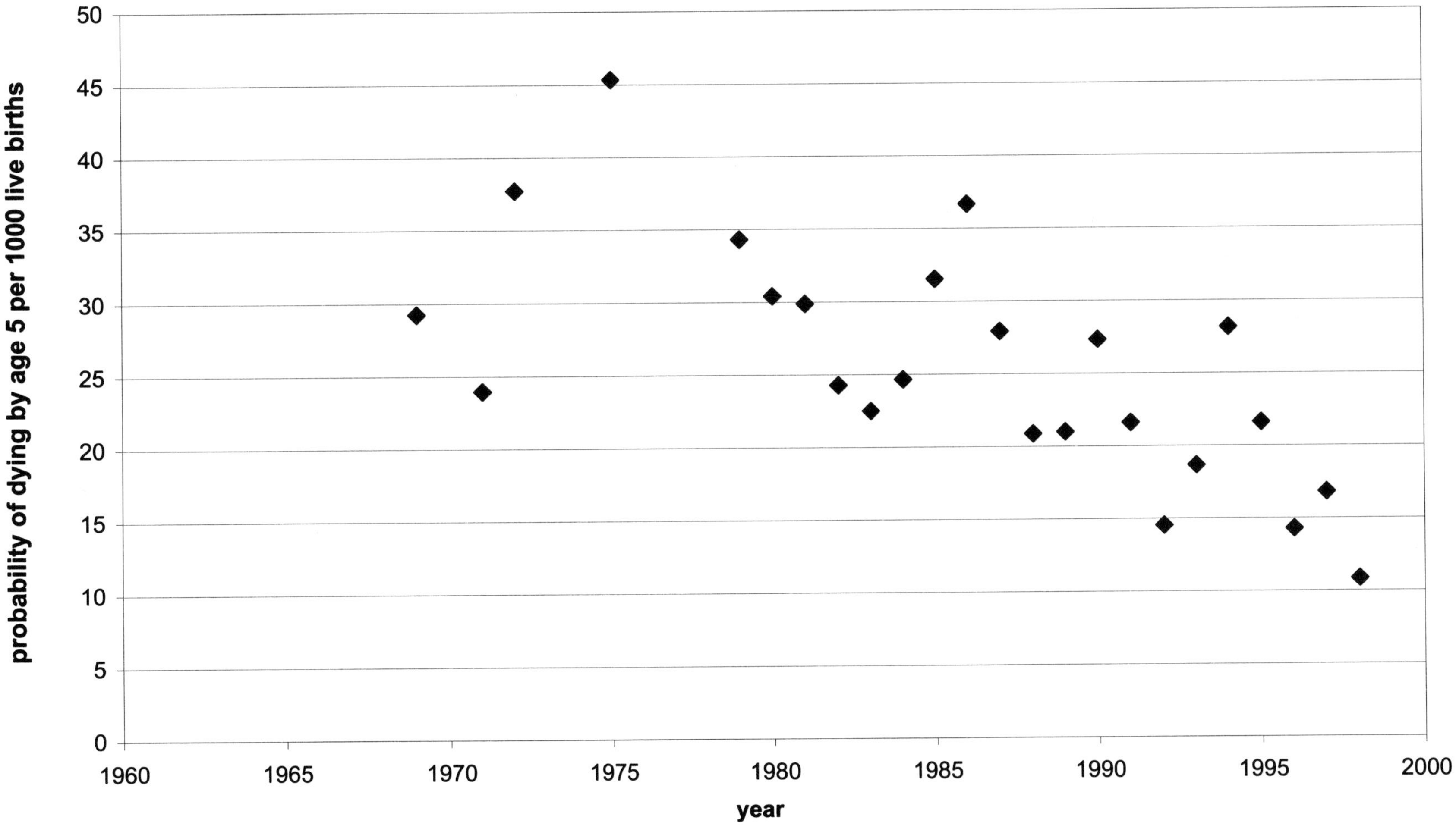

Data sources
Vital Registration

Sources des données

Fuentes de los datos

Under 5 mortality
Bahrain

Mortalité des moins de cinq ans
Bahreïn

Mortalidad de menores de cinco años
Bahrein

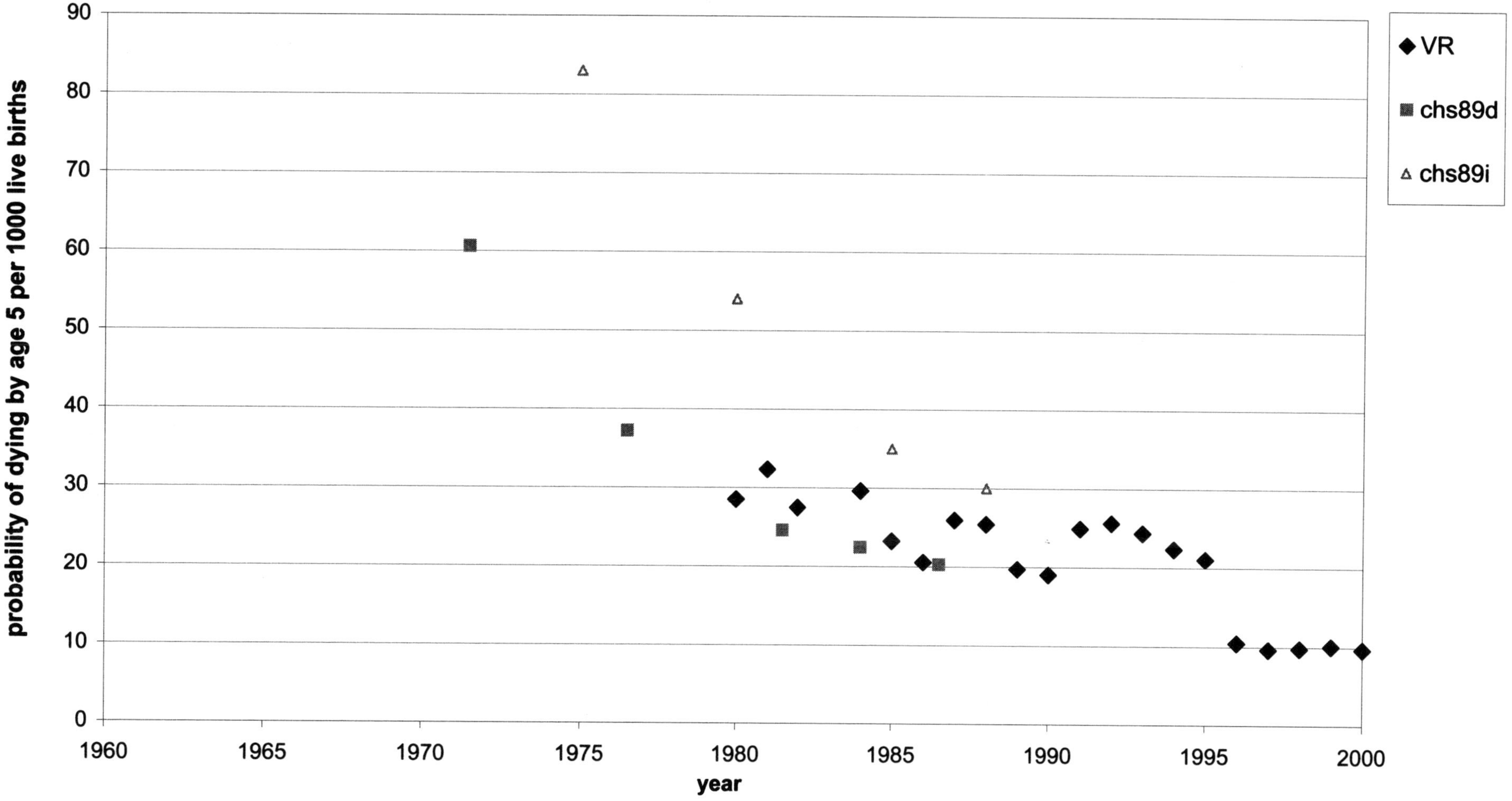

Data sources **Sources des données** **Fuentes de los datos**

VR: Vital Registration; ***chs89d***: Bahrain Child Health Survey 1989 (direct method); ***chs89i***: Child Health Survey 1989 (indirect method)

Under 5 mortality

Bangladesh

Mortalité des moins de cinq ans

Bangladesh

Mortalidad de menores de cinco años

Bangladesh

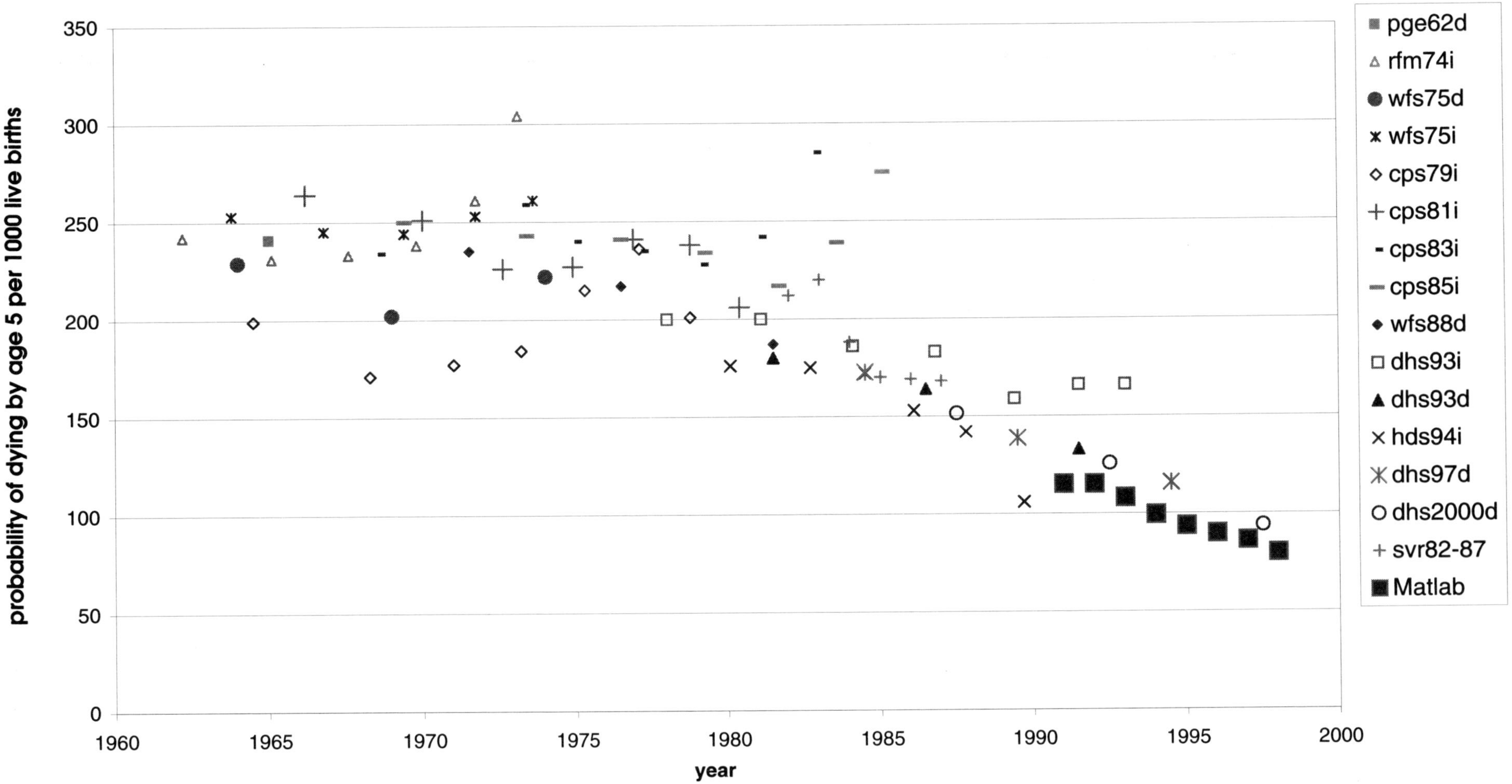

Data sources | **Sources des données** | **Fuentes de los datos**

pge62d: Population Growth Estimation Experiment 1962-1965 (direct method); ***rfm74i***: Retrospective Survey of Fertility and Mortality 1974 (indirect method); ***wfs75d***: Fertility Survey 1975-1976 (direct method); ***wfs75i***: Fertility Survey 1975-1976 (indirect method); ***cps79i***: Contraceptive Prevalence Survey 1979-1980 (indirect method); ***cps81i***: Contraceptive Prevalence Survey 1981 (indirect method); ***cps83i***: Contraceptive Prevalence Survey 1983-1984 (indirect method); ***cps85i***: Contraceptive Prevalence Survey 1985-1986 (indirect method); ***wfs88d:*** Fertility Survey 1988-1989 (direct method); ***dhs93i:*** Demographic and Health Survey 1993-1994 (indirect method); ***dhs93d***: Demographic and Health Survey 1993-1994 (direct method); ***hds94i***: Health and Demographic Survey 1994 (indirect method); ***dhs97d***: Demographic and Health Survey 1997 (direct method); ***dhs2000d***: Demographic and Health Survey 2000 (direct method); ***svr82-87***: Bureau of Statistics, Patterns, levels, trends in mortality and regional life tables for Bangladesh: evidence from sample vital registration system, 1981-88; ***Matlab***: Matlab Surveillance Area Data

Under 5 mortality
Barbados

Mortalité des moins de cinq ans
Barbade

Mortalidad de menores de cinco años
Barbados

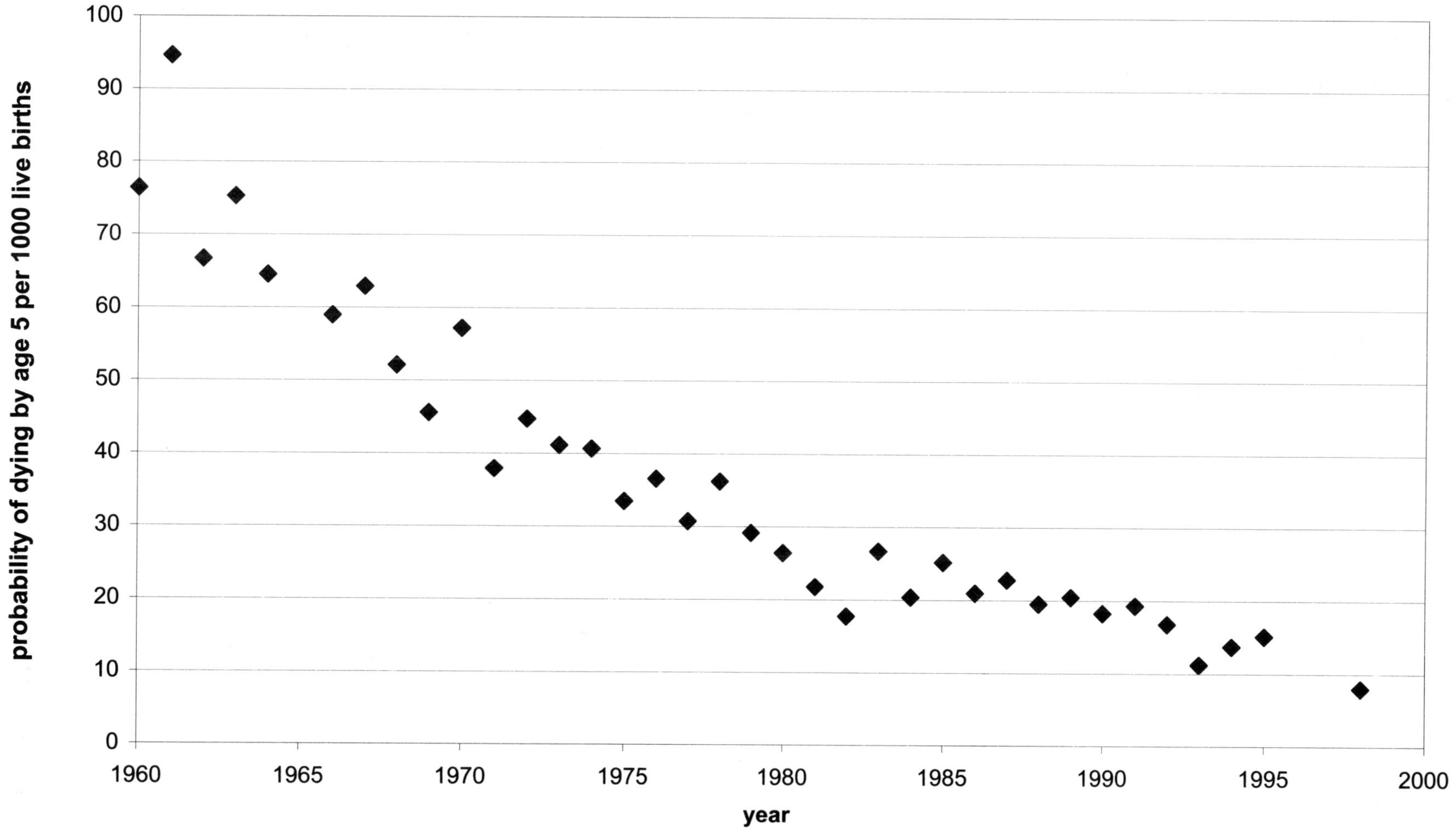

Data sources
Vital Registration

Sources des données

Fuentes de los datos

Under 5 mortality
Belarus

Mortalité des moins de cinq ans
Bélarus

Mortalidad de menores de cinco años
Belarús

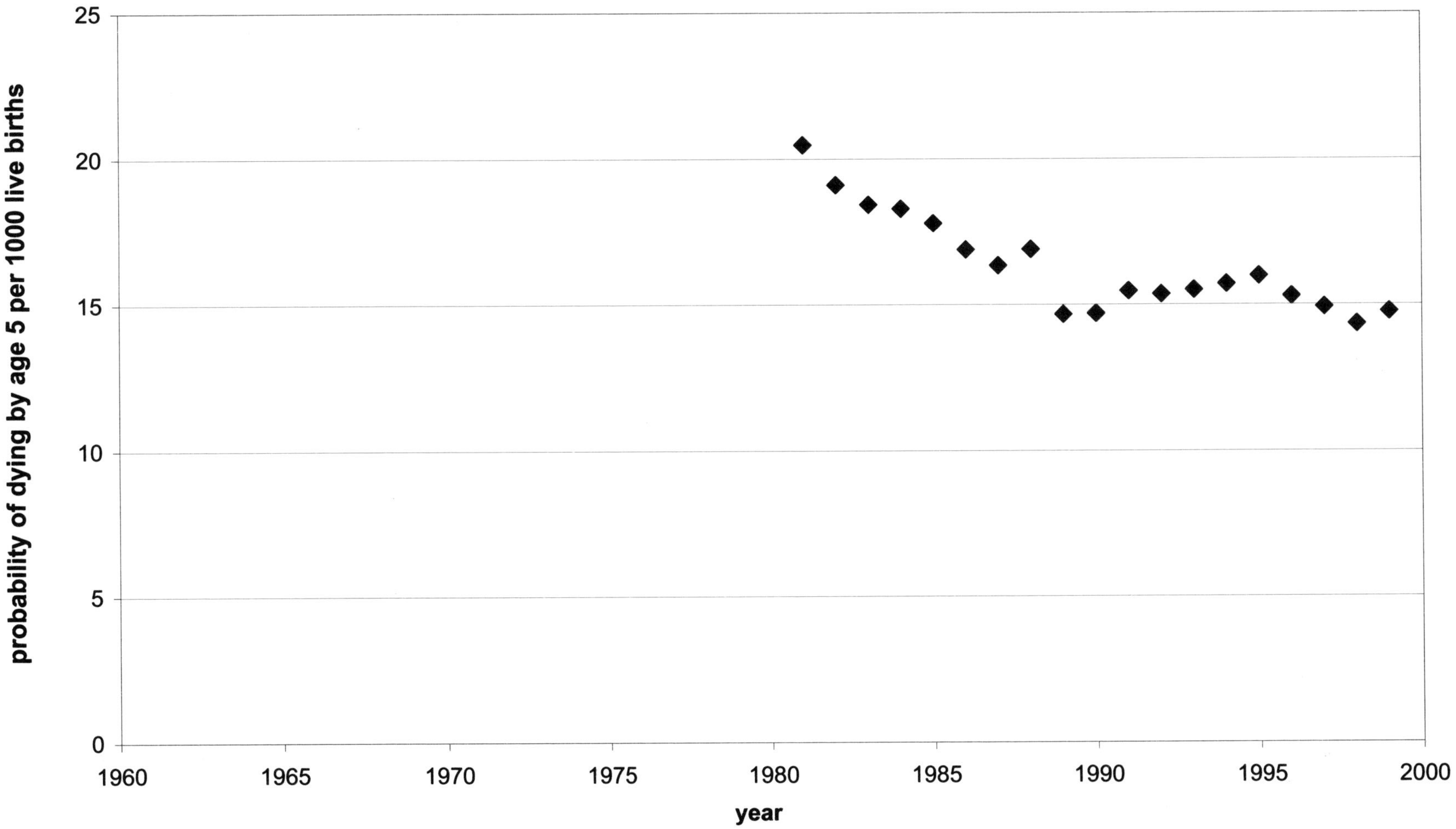

Data sources
Vital Registration

Sources des données

Fuentes de los datos

Under 5 mortality
Belgium

Mortalité des moins de cinq ans
Belgique

Mortalidad de menores de cinco años
Bélgica

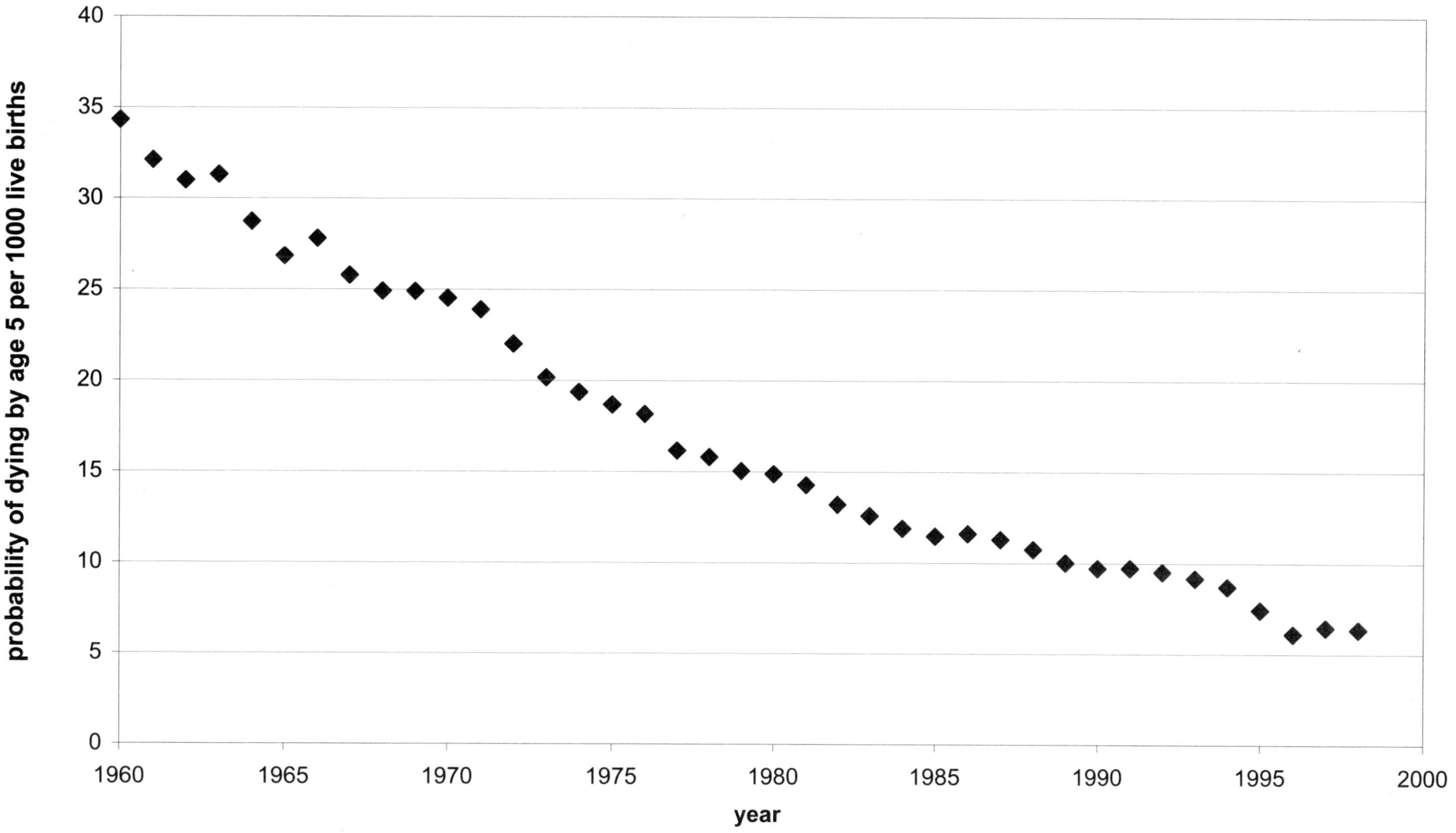

Data sources
Vital Registration

Sources des données

Fuentes de los datos

Under 5 mortality

Belize

Mortalité des moins de cinq ans

Belize

Mortalidad de menores de cinco años

Belice

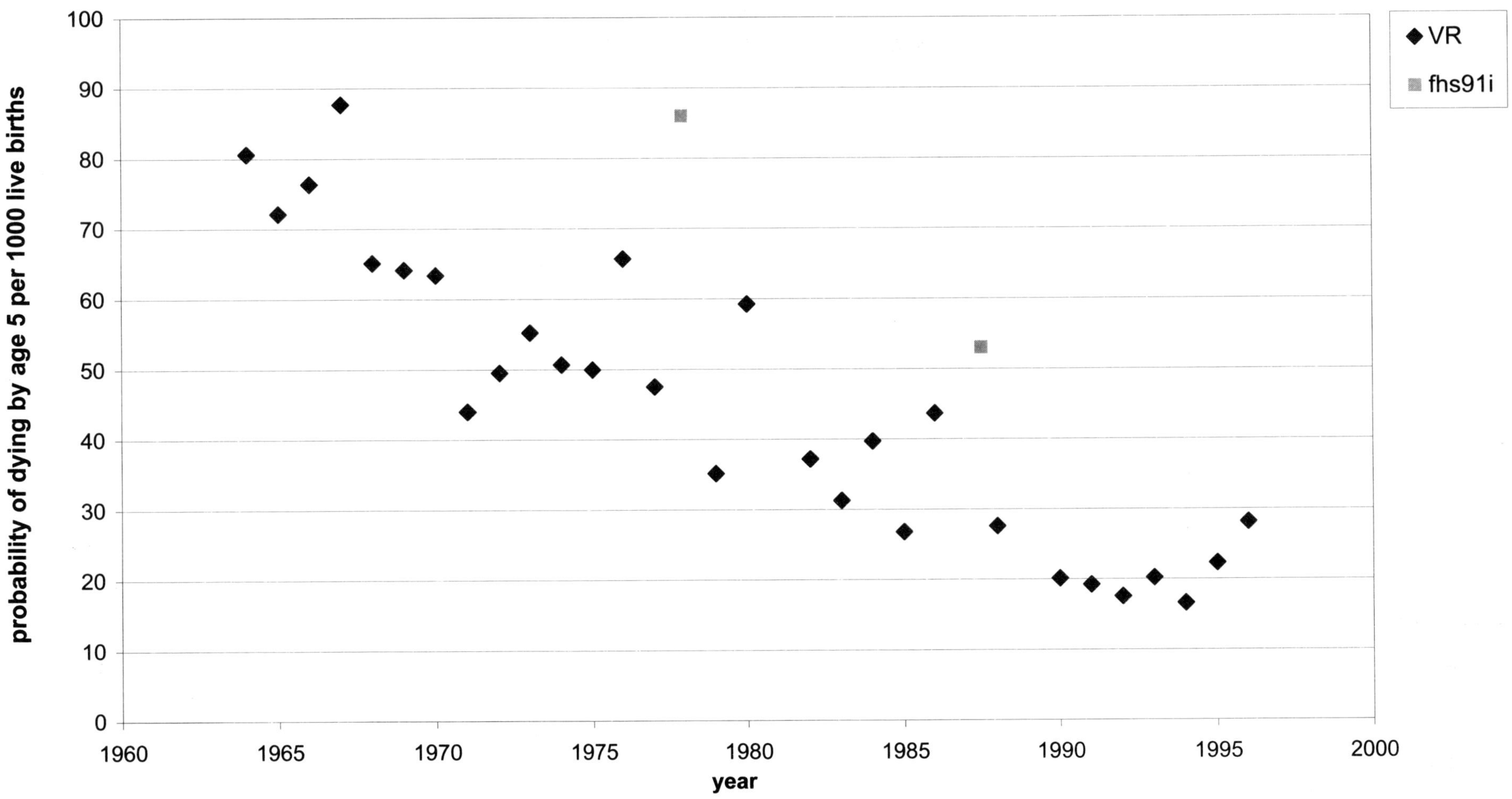

Data sources

Sources des données

Fuentes de los datos

VR: Vital Registration; ***fhs91i***: Belize Family Health Survey 1991 (indirect method)

Under 5 mortality

Benin

Mortalité des moins de cinq ans

Bénin

Mortalidad de menores de cinco años

Benin

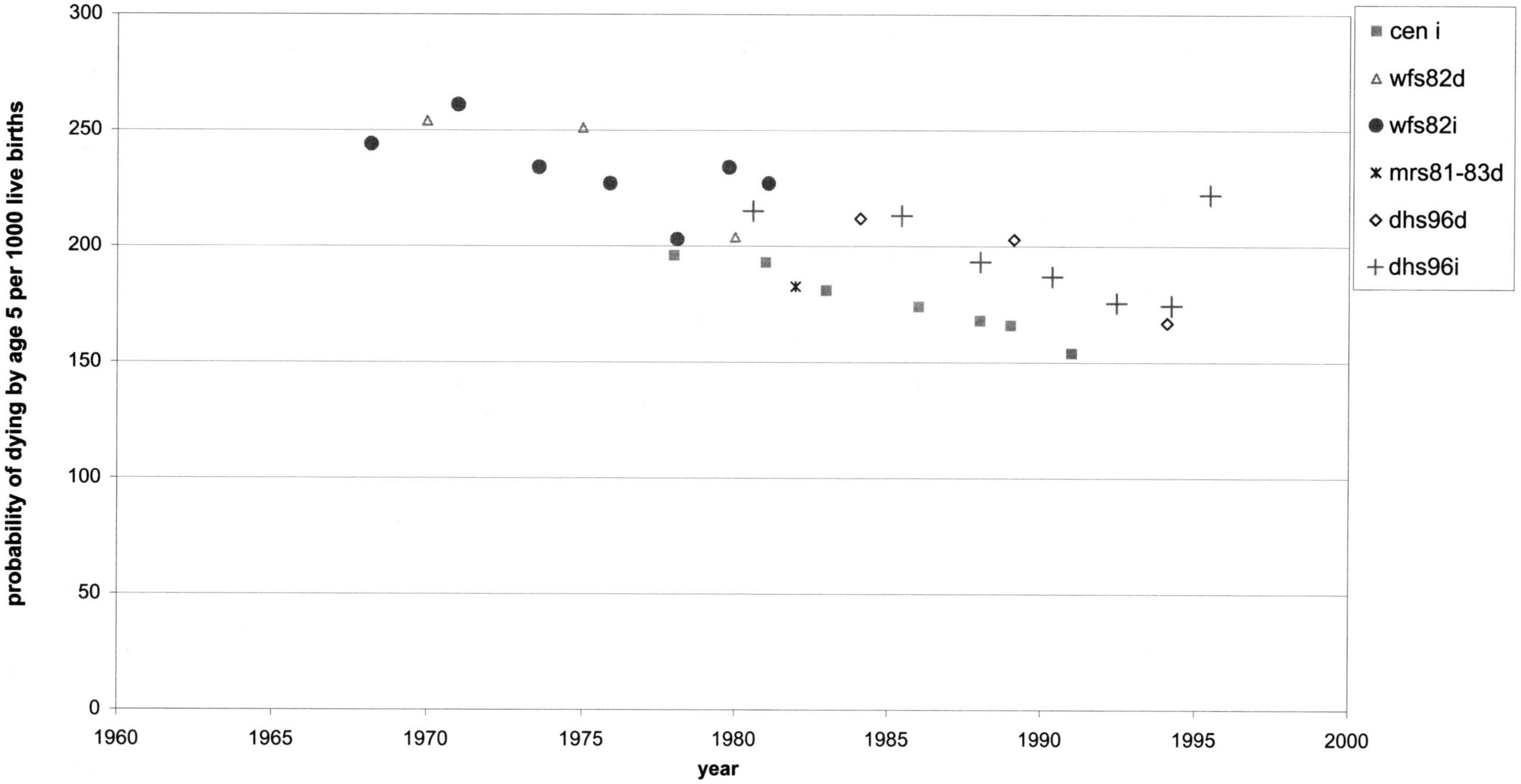

Data sources

Sources des données

Fuentes de los datos

cen i: Census (indirect method); ***wfs82d***: Enquête sur la Fécondité au Benin, Jan-June 1982 (direct method); ***wfs82i***: Enquête sur la Fécondité au Benin, Jan-June 1982 (indirect method); ***mrs81-83d***: Multi-Round Survey 1981-1983 (direct method); ***dhs96d***: Demographic and Health Survey 1996 (direct method); ***dhs96i***: Demographic and Health Survey 1996 (indirect method)

Under 5 mortality
Bhutan

Mortalité des moins de cinq ans
Bhoutan

Mortalidad de menores de cinco años
Bhután

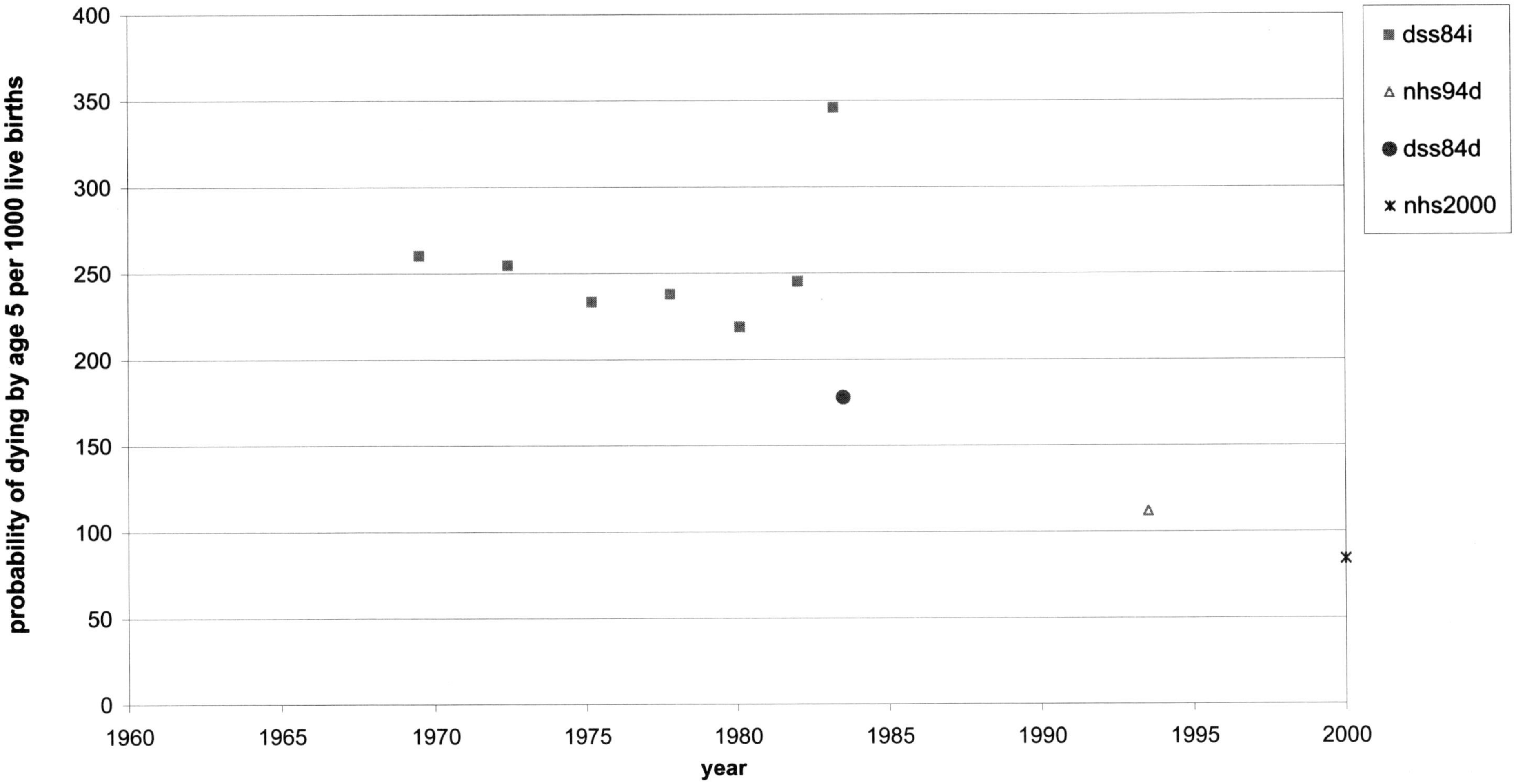

Data sources | **Sources des données** | **Fuentes de los datos**

dss84i: Demographic Sample Survey, 1984 (indirect method); ***nhs94d***: National Health Survey, 1994 (direct method); ***dss84d***: Demographic Sample Survey, 1984 (direct method); ***nhs2000***: Ministry of Health and Education, Report - National Health Survey 2000

Under 5 mortality
Bolivia

Mortalité des moins de cinq ans
Bolivie

Mortalidad de menores de cinco años
Bolivia

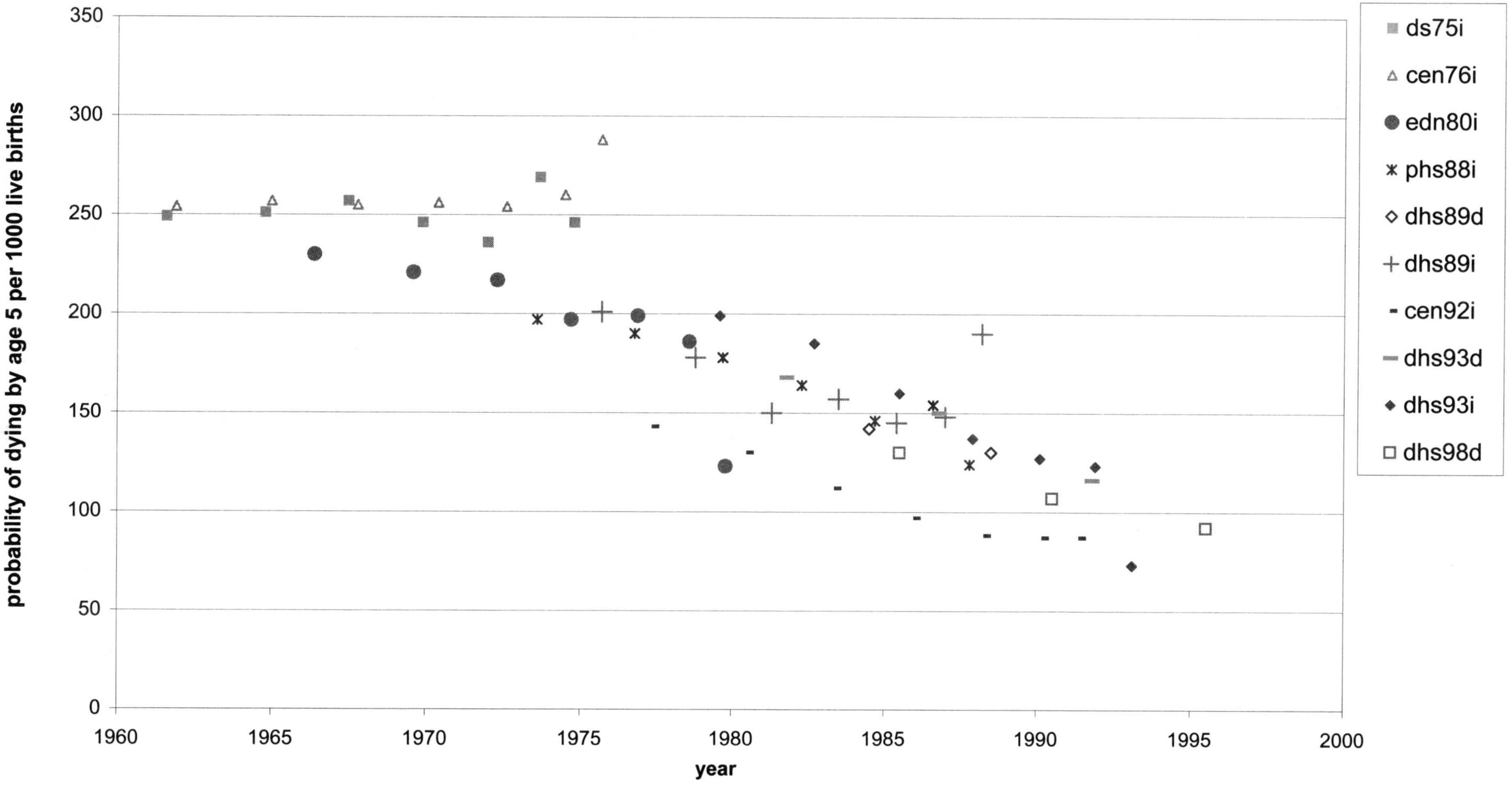

Data sources **Sources des données** **Fuentes de los datos**

ds75i: Encuesta Demografica Nacional, June-October 1975 (indirect method); ***cen76i***: Census, 29 September 1976 (indirect method); ***edn80i***: Encuesta Demografica Nacional, 1980 (indirect method); ***phs88i***: Encuesta Nacional de Poblacion y Vivienda, September 1988 (indirect method); ***dhs89d***: Encuesta Nacional de Demografica y Salud, March-June 1989 (direct method); ***dhs89i***: Encuesta Nacional de Demografica y Salud, March-June 1989 (indirect method); ***cen92i***: Census, 3 June 1992 (indirect method); ***dhs93d***: Demographic and Health Survey 1993-94 (direct method); ***dhs93i***: Demographic and Health Survey 1993-94 (indirect method); ***dhs98d***: Demographic and Health Survey 1998 (direct method)

Under 5 mortality

Bosnia and Herzegovina

Mortalité des moins de cinq ans

Bosnie-Herzégovine

Mortalidad de menores de cinco años

Bosnia y Herzegovina

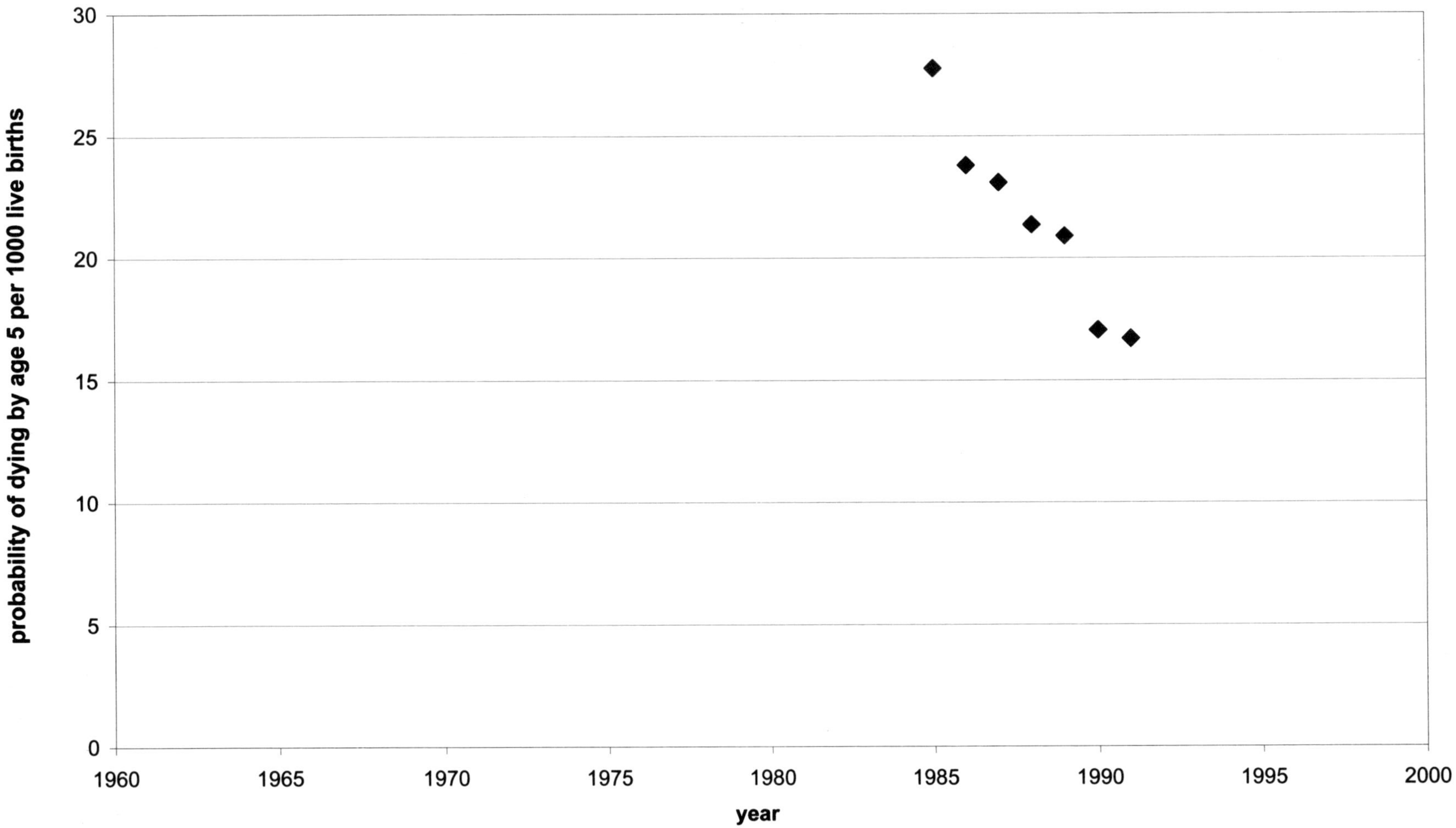

Data sources
Vital Registration

Sources des données

Fuentes de los datos

Under 5 mortality

Botswana

Mortalité des moins de cinq ans

Botswana

Mortalidad de menores de cinco años

Botswana

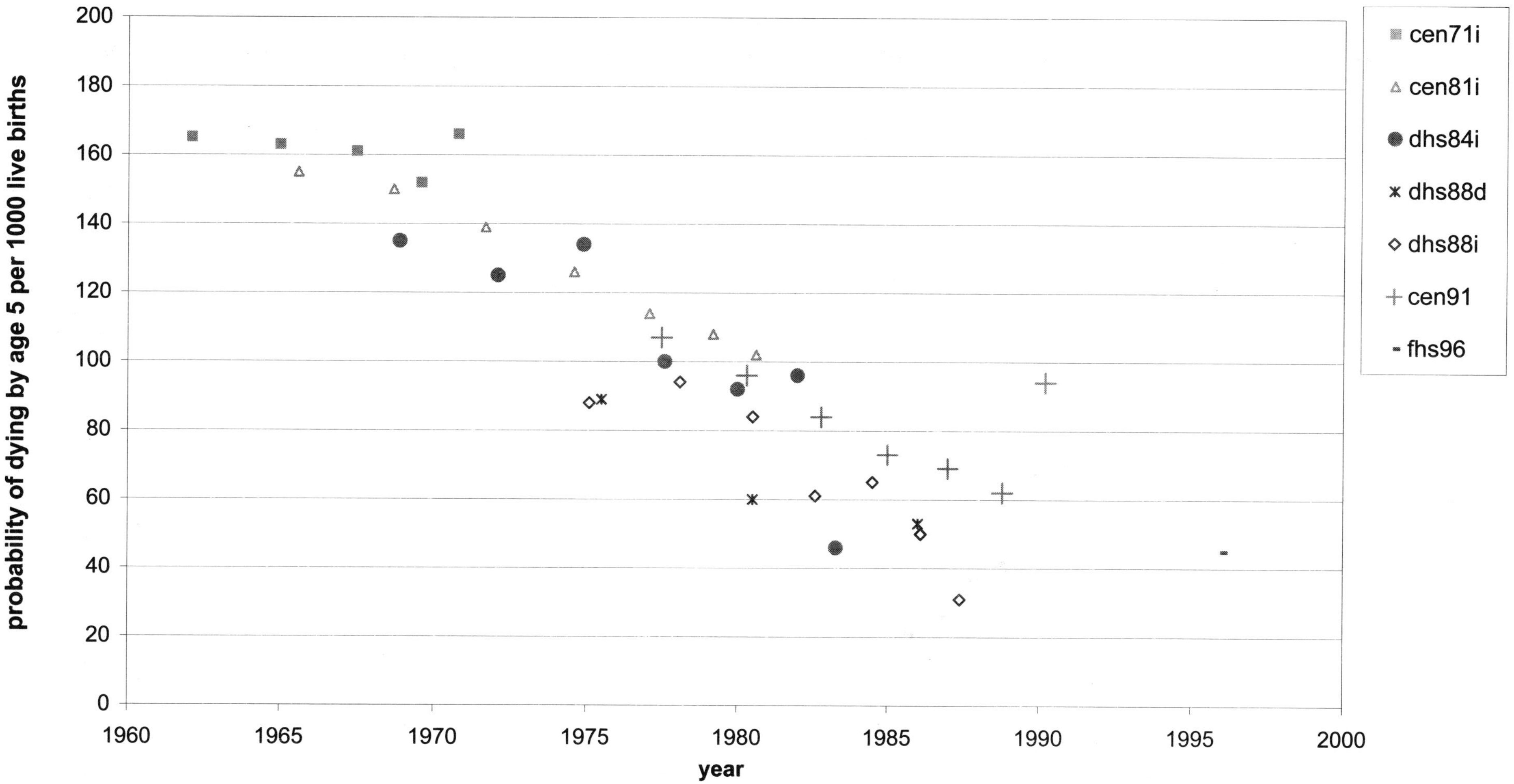

Data sources **Sources des données** **Fuentes de los datos**

cen71i: Census 1971 (indirect method); ***cen81i***: Census 1981 (indirect method); ***dhs84i***: Family Health Survey I (CPS) 1984 (indirect method); ***dhs88d***: Family Health Survey II (DHS) 1988 (direct method); ***dhs88i***: Family Health Survey II (DHS) 1988 (indirect method); ***cen91***: Census 1991; ***fhs96***: Family Health Survey III 1996

Under 5 mortality

Brazil

Mortalité des moins de cinq ans

Brésil

Mortalidad de menores de cinco años

Brasil

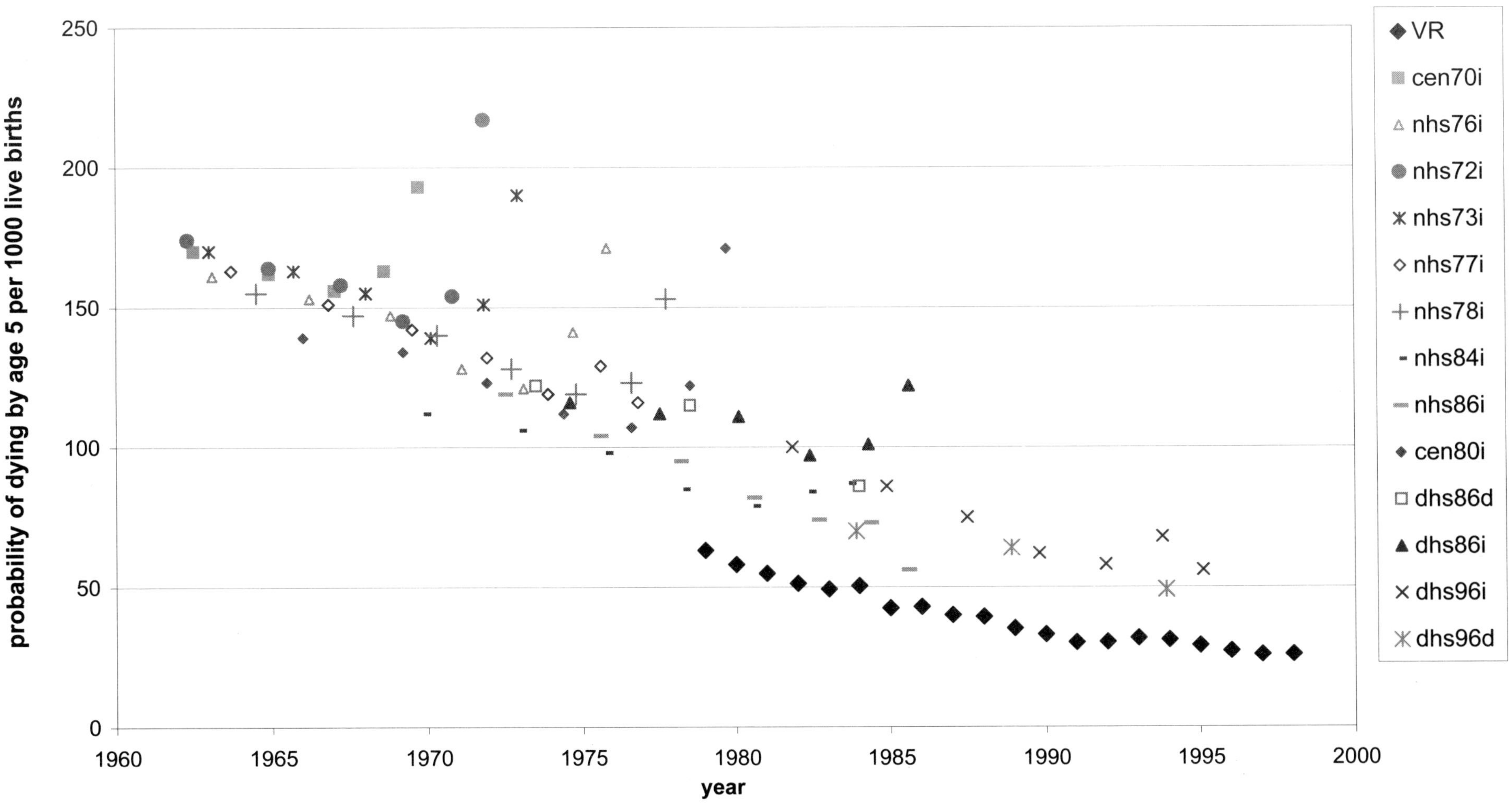

Data sources **Sources des données** **Fuentes de los datos**

VR: Vital Registration; ***cen70i***: Census, 1 September 1970 (indirect method); ***nhs76i***: Pesquisa Nacional por Amostra de Domicilios, 28 November 1976 (indirect method); ***nhs72i***: Pesquisa Nacional por Amostra de Domicilios, October-December 1972 (indirect method); ***nhs73i***: Pesquisa Nacional por Amostra de Domicilios, October-December 1973 (indirect method); ***nhs77i***: Pesquisa Nacional por Amostra de Domicilios, 31 October 1977 (indirect method); ***nhs78i***: Pesquisa Nacional por Amostra de Domicilios, 31 October 1978 (indirect method); ***nhs84i***: Pesquisa Nacional por Amostra de Domicilios, 29 September 1984 (indirect method); ***nhs86i***: Pesquisa Nacional por Amostra de Domicilios, 4 October 1986 (indirect method); ***cen80i***: Census, 1 September 1980 (indirect method); ***dhs86d***: Demographic and Health Survey 1986 (direct method); ***dhs86i***: Demographic and Health Survey 1986 (indirect method); ***dhs96i***: Demographic and Health Survey 1996 (indirect method); ***dhs96d***: Demographic and Health Survey 1996 (direct method)

Under 5 mortality

Brunei Darussalam

Mortalité des moins de cinq ans

Brunéi Darussalam

Mortalidad de menores de cinco años

Brunei Darussalam

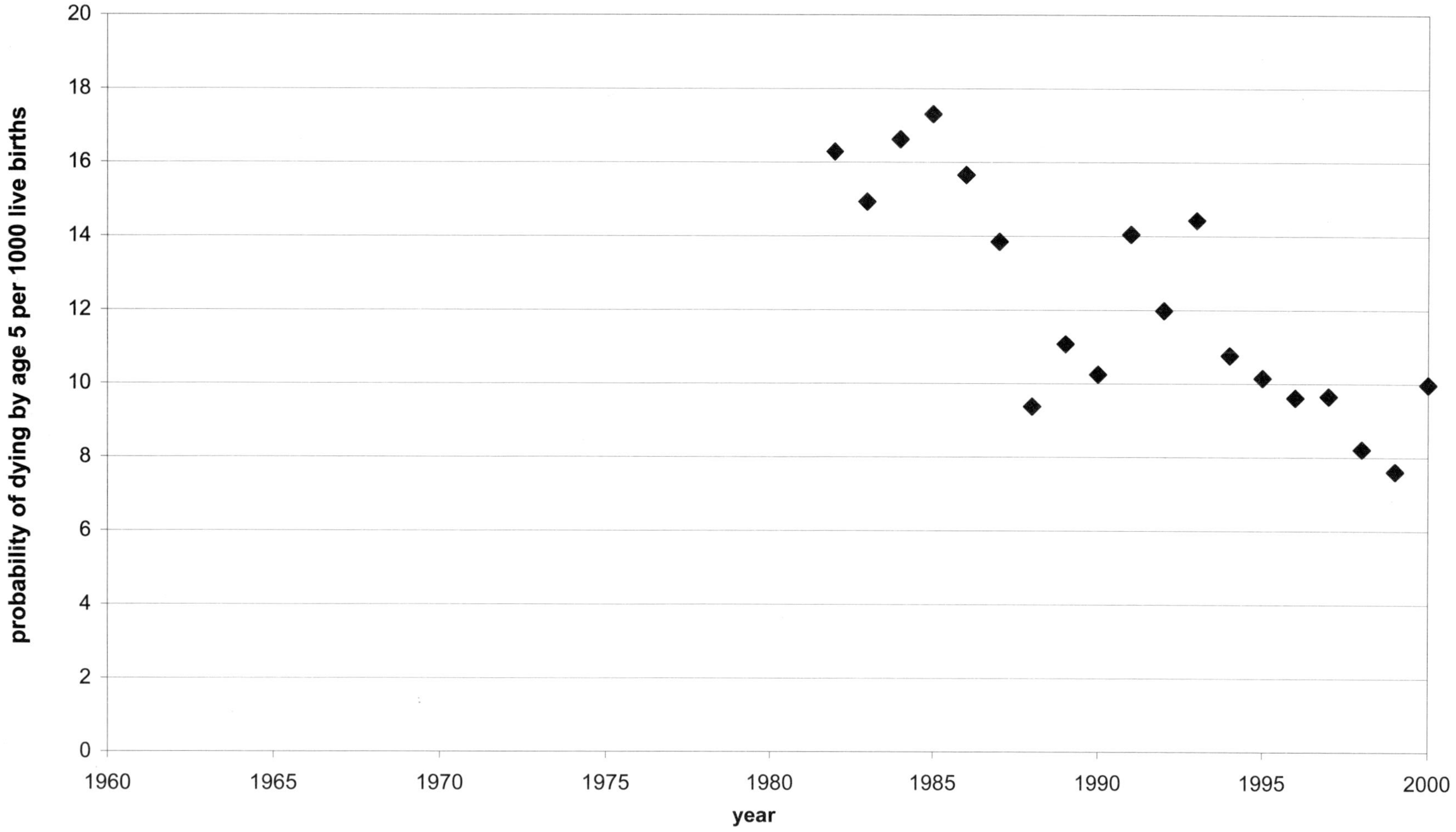

Data sources

Vital Registration

Sources des données

Fuentes de los datos

Under 5 mortality
Bulgaria

Mortalité des moins de cinq ans
Bulgarie

Mortalidad de menores de cinco años
Bulgaria

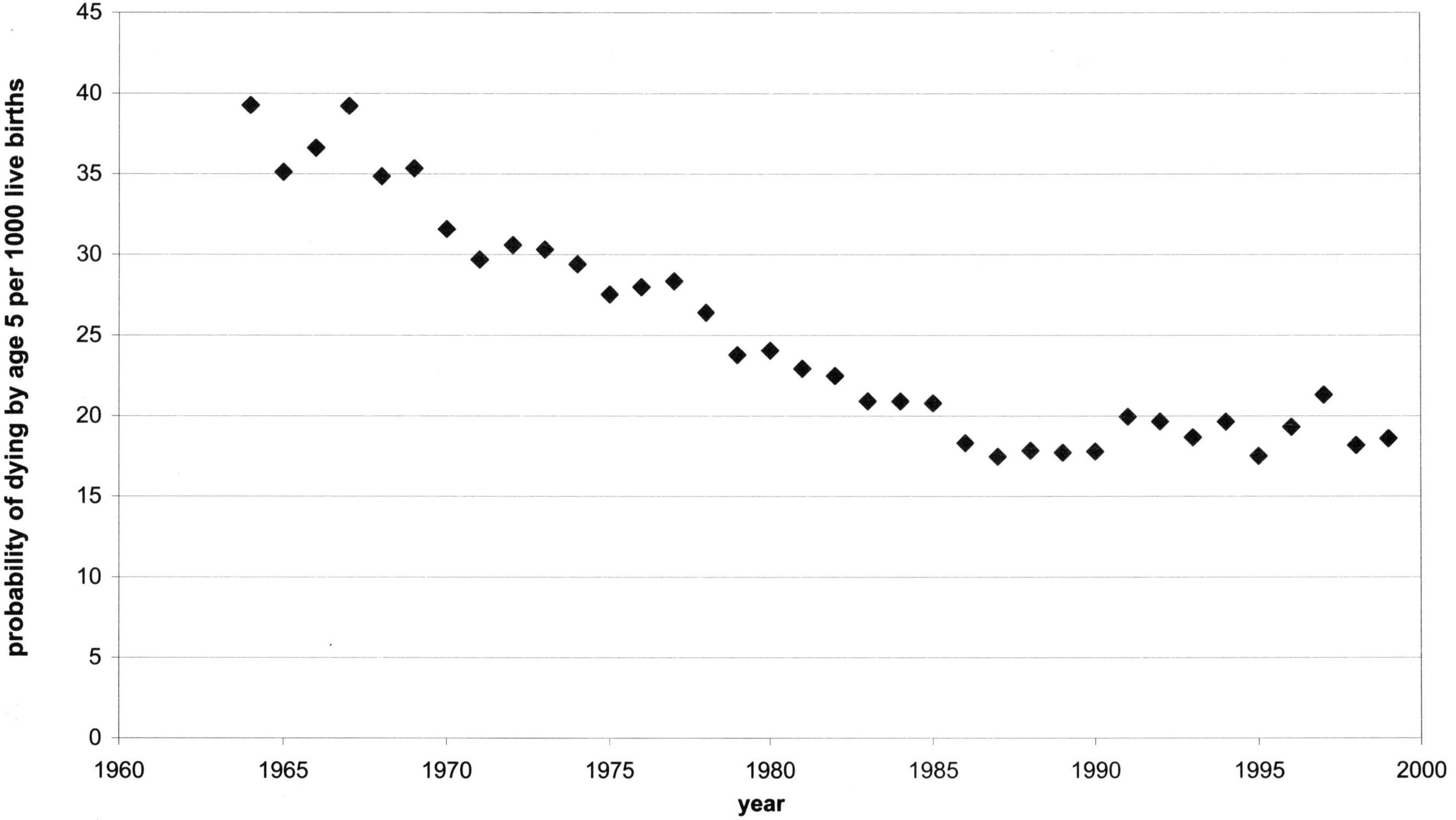

Data sources
Vital Registration

Sources des données

Fuentes de los datos

Under 5 mortality
Burkina Faso

Mortalité des moins de cinq ans
Burkina Faso

Mortalidad de menores de cinco años
Burkina Faso

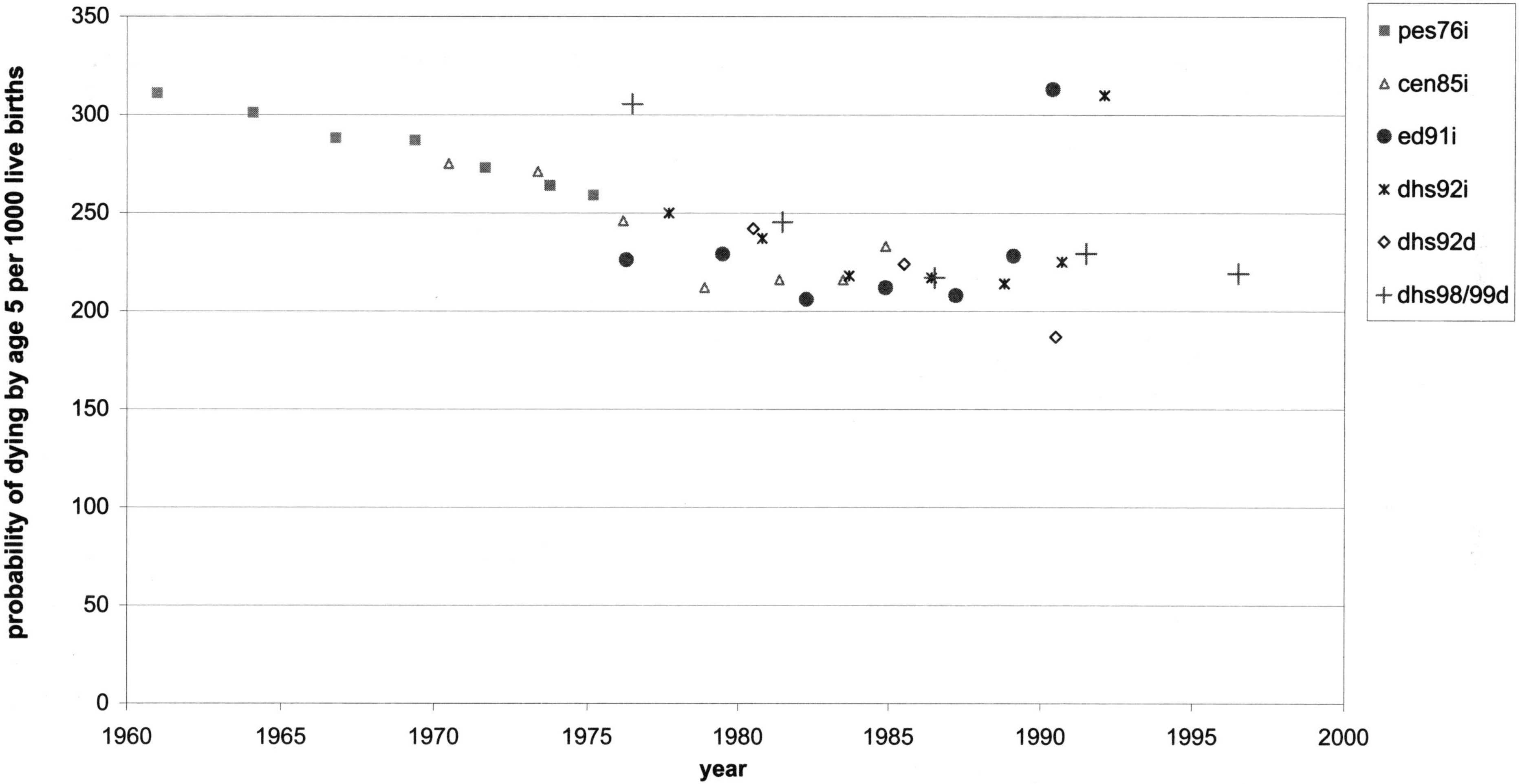

Data sources **Sources des données** **Fuentes de los datos**

pes76i: Post-enumeration Survey 1976 (indirect method); ***cen85i***: Census, 10-20 July 1985 (indirect method); ***ed91i***: Enquête Démographique 1991 (indirect method); ***dhs92i***: Enquête Démographique et de Santé 1993 (indirect method); ***dhs92d***: Enquête Démographique et de Santé 1993 (direct method); ***dhs98/99d***: Demographic and Health Survey 1998-999 (direct method)

Under 5 mortality Mortalité des moins de cinq ans Mortalidad de menores de cinco años

Burundi **Burundi** **Burundi**

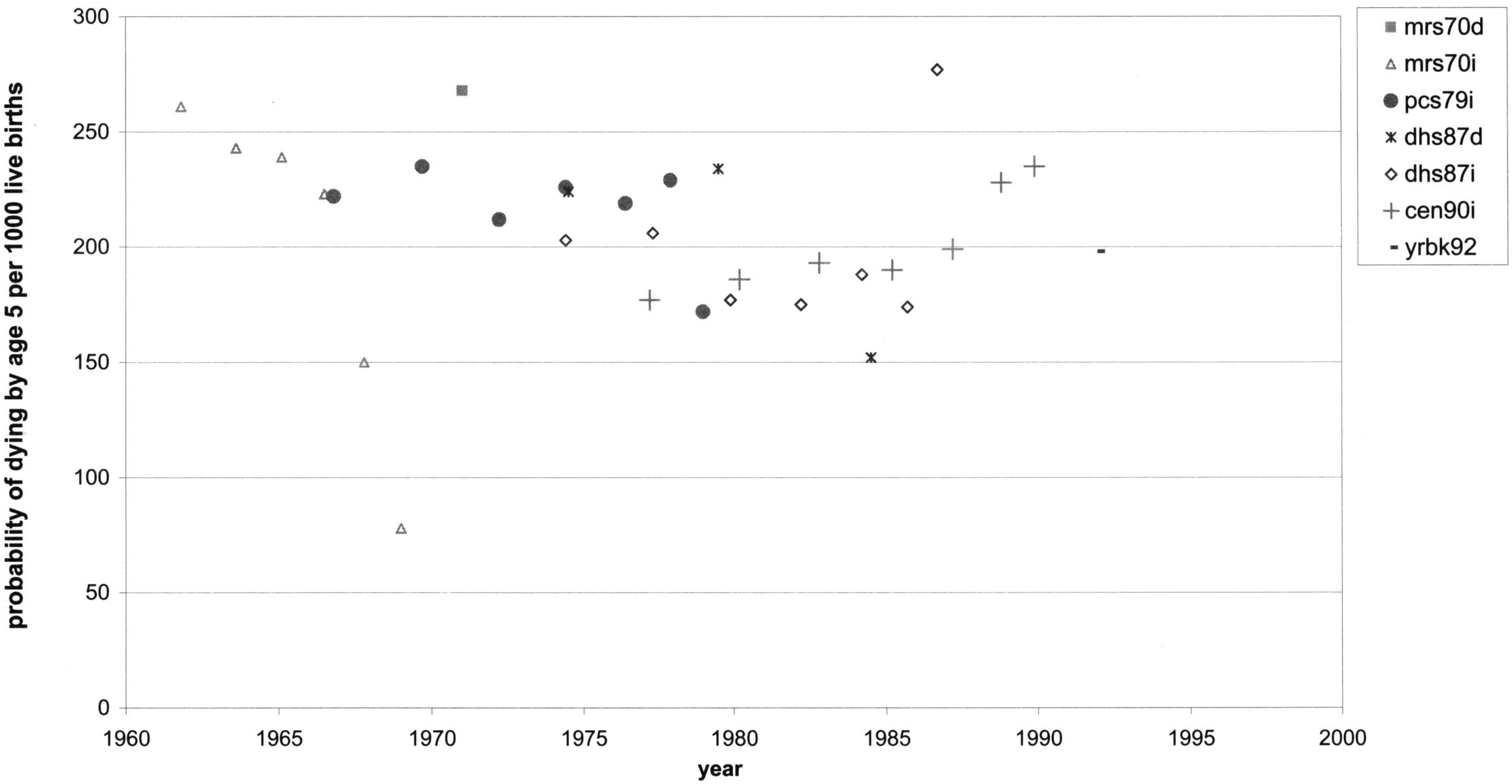

Data sources **Sources des données** **Fuentes de los datos**

mrs70d: Enquête Démographique, May 1970-July 1971 (direct method); ***mrs70i***: Enquête Démographique, May 1970-July 1971 (indirect method); ***pcs79i***: Enquête Post-censitaire, November 1979 (indirect method); ***dhs87d***: Enquête Démographique et de Santé 1987 (direct method); ***dhs87i***: Enquête Démographique et de Santé 1987 (indirect method); ***cen90i***: Census, 16-30 August 1990 (indirect method); ***yrbk92***: Annuaire statistique de Burundi 1992

Under 5 mortality Mortalité des moins de cinq ans Mortalidad de menores de cinco años

Cambodia **Cambodge** **Camboya**

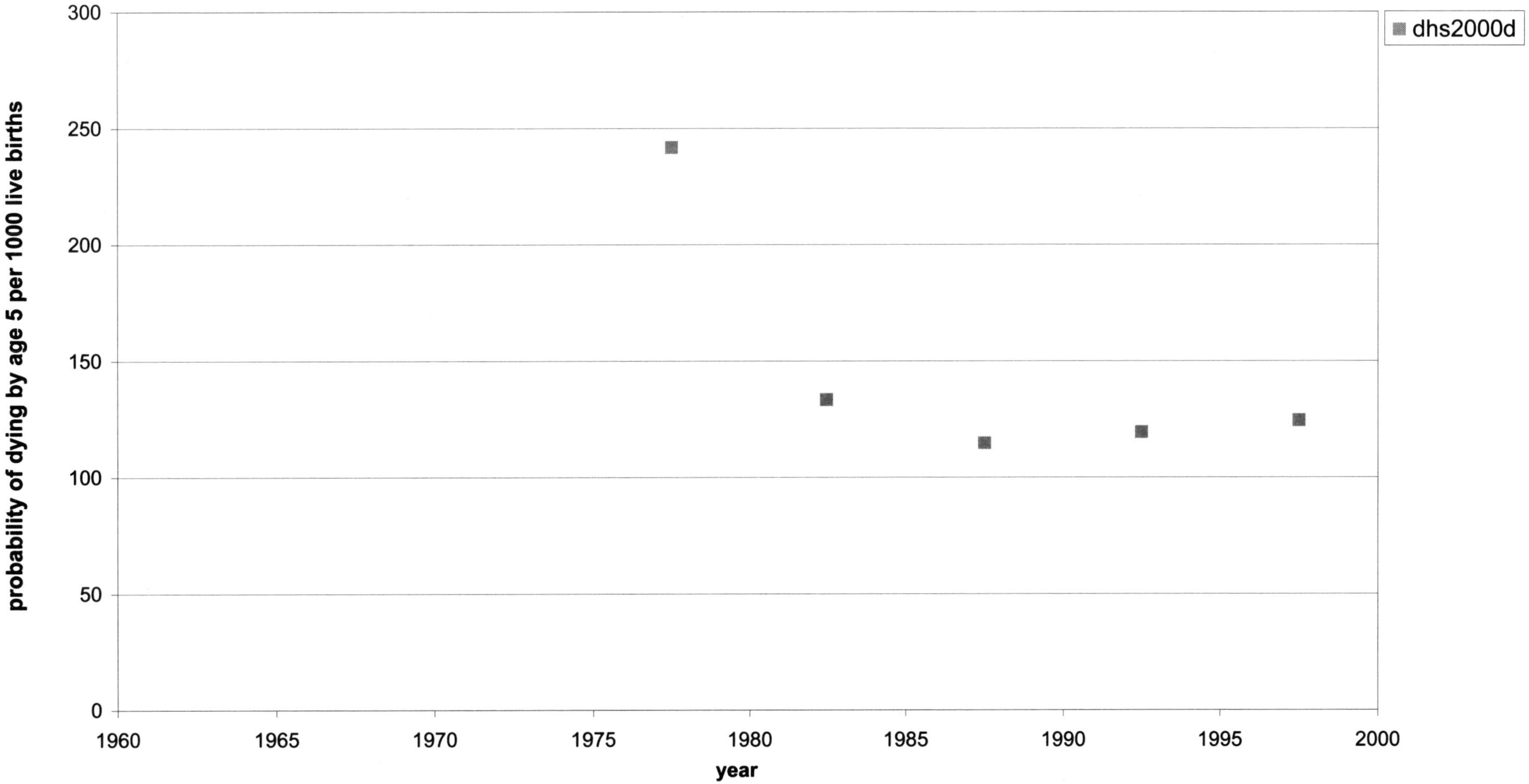

Data sources **Sources des données** **Fuentes de los datos**

dhs2000d Demographic and Health Survey 2000 - preliminary report (direct method)

Under 5 mortality
Cameroon

Mortalité des moins de cinq ans
Cameroun

Mortalidad de menores de cinco años
Camerún

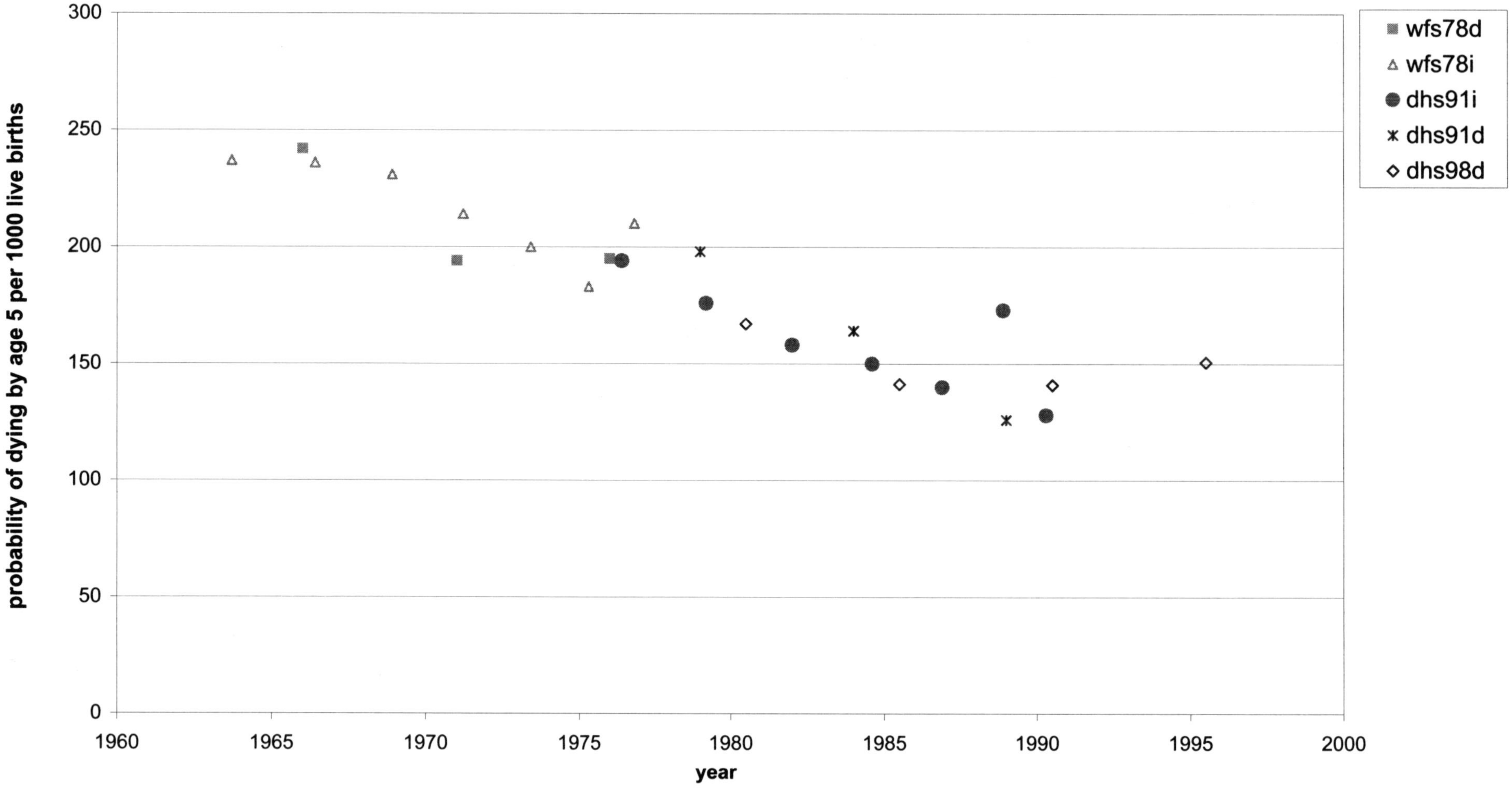

Data sources

Sources des données

Fuentes de los datos

wfs78d : Fertility Survey 1978 (direct method); ***wfs78i*** : Fertility Survey 1978 (indirect method); ***dhs91i*** : Demographic and Health Survey 1991 (indirect method); ***dhs91d*** : Demographic and Health Survey 1991 (direct method); ***dhs98d*** : Demographic and Health Survey 1998 (direct method)

Under 5 mortality

Canada

Mortalité des moins de cinq ans

Canada

Mortalidad de menores de cinco años

Canadá

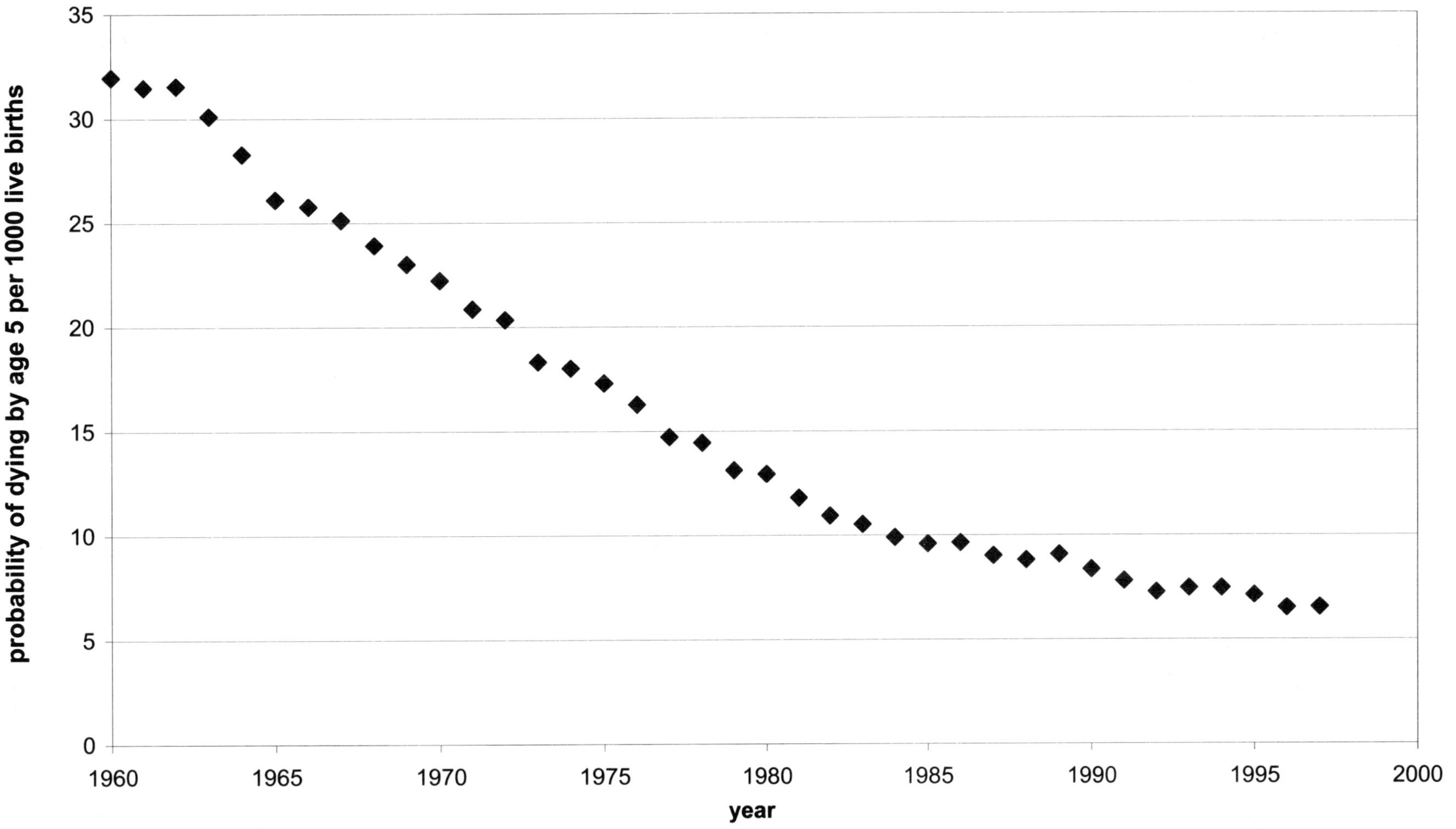

Data sources

VitalRegistration

Sources des données

Fuentes de los datos

Under 5 mortality

Cape Verde

Mortalité des moins de cinq ans

Cap-Vert

Mortalidad de menores de cinco años

Cabo Verde

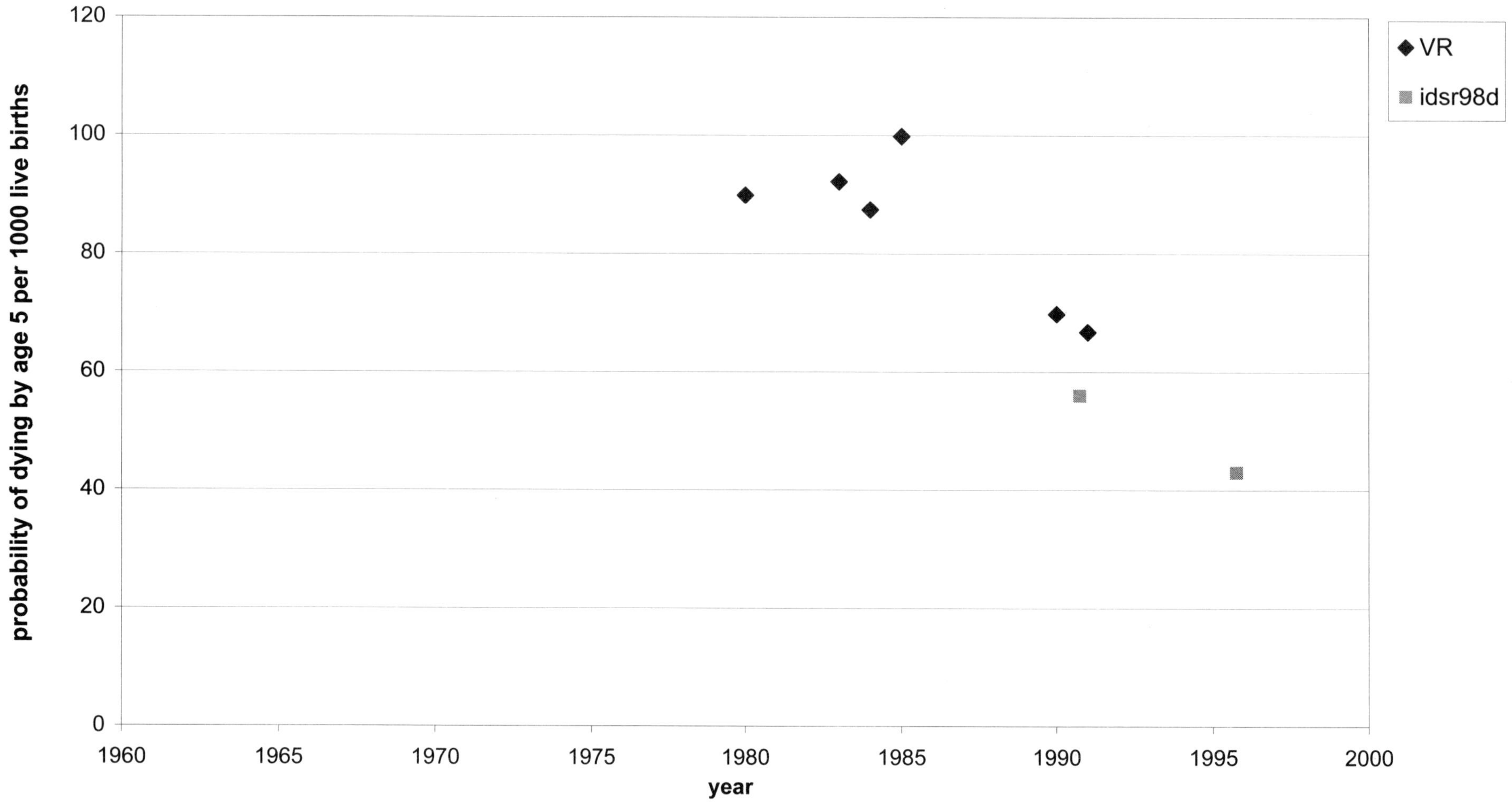

Data sources **Sources des données** **Fuentes de los datos**

VR: Vital Registration; ***idsr98d***: Inquérito demográfico e de saúde reprodutiva 1998 (direct method)

Under 5 mortality
Central African Republic

Mortalité des moins de cinq ans
République centrafricaine

Mortalidad de menores de cinco años
República Centroafricana

Data sources

Sources des données

Fuentes de los datos

cen75i: Census 1975 (indirect method); ***cen88d***: Census 1988 (direct method); ***cen88i***: Census 1988 (indirect method); ***dhs95i***: Demographic and Health Survey 1995 (indirect method); ***dhs95d***: Demographic and Health Survey 1995 (direct method); ***mics2000i***: Enquête à indicateurs multiples en République centrafricaine 2000 - rapport préliminaire (indirect method)

Under 5 mortality

Chad

Mortalité des moins de cinq ans

Tchad

Mortalidad de menores de cinco años

Chad

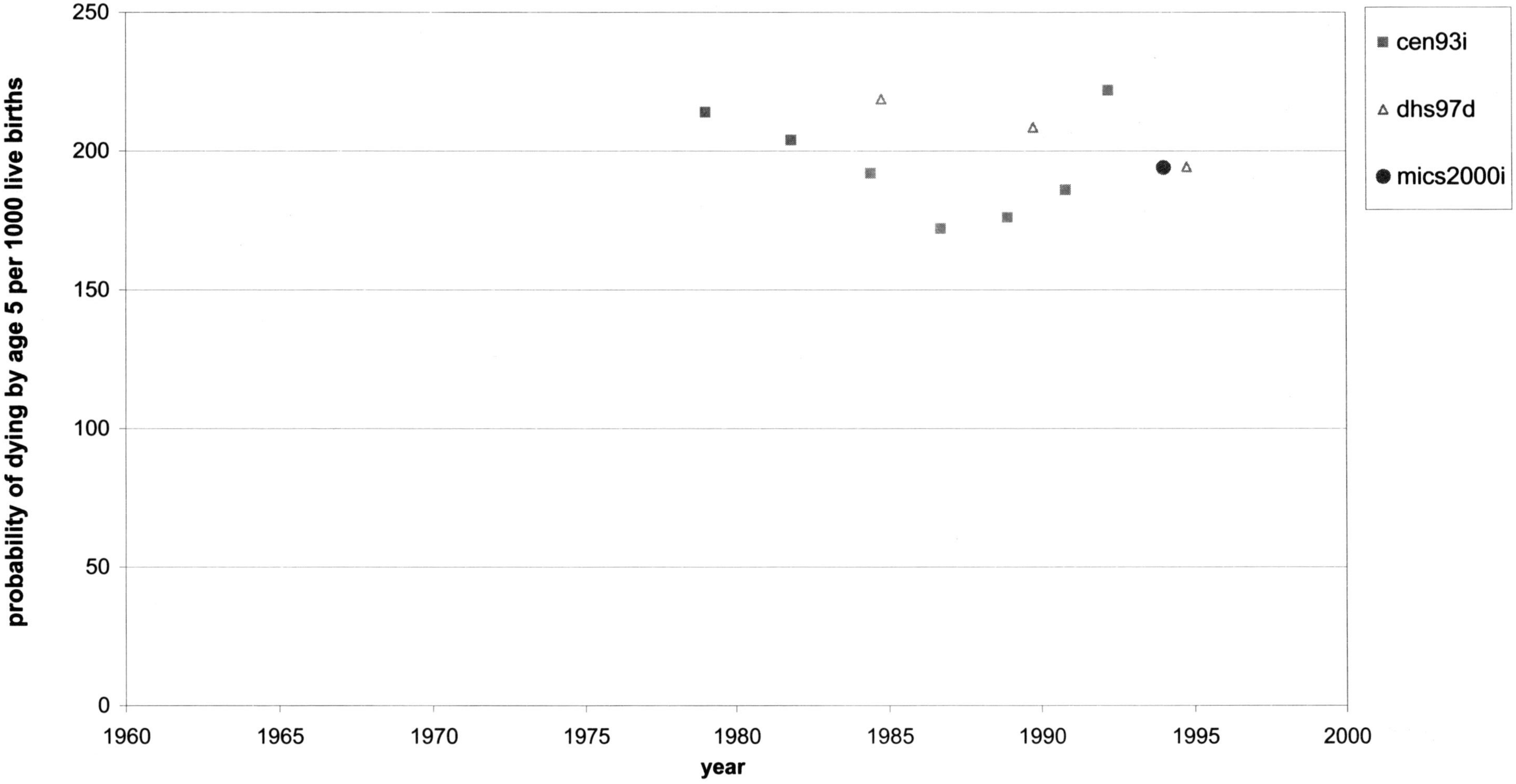

Data sources

Sources des données

Fuentes de los datos

cen93i : Census 1993 (indirect method); ***dhs97d*** : Demographic and Health Survey 1997 (direct method); ***mics2000i*** : Multiple Indicator Cluster Survey 2000 (indirect method)

Under 5 mortality

Chile

Mortalité des moins de cinq ans

Chili

Mortalidad de menores de cinco años

Chile

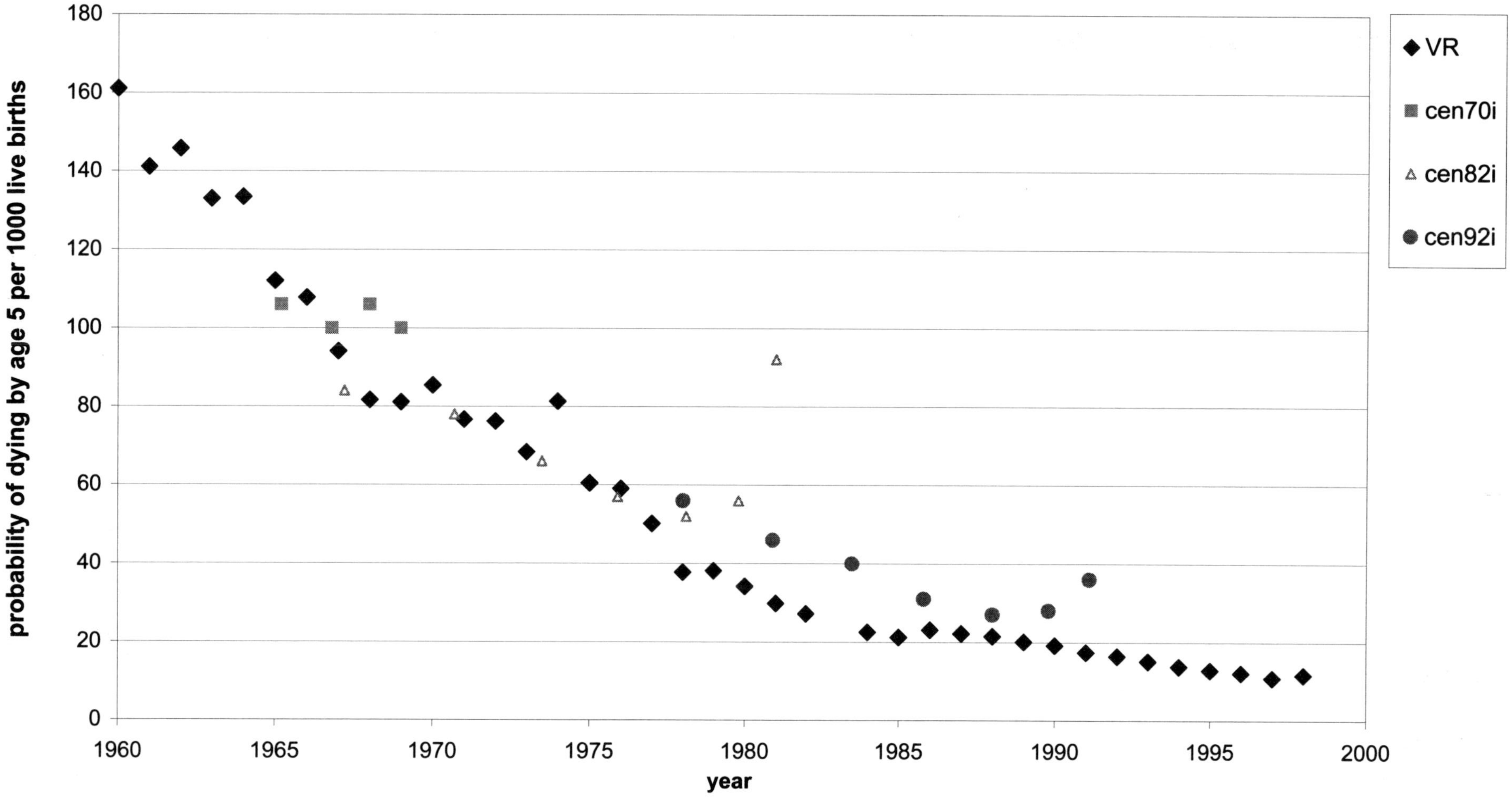

Data sources **Sources des données** **Fuentes de los datos**

VR: Vital Registration; ***cen70i***: Census, 22 April 1970 (indirect method); ***cen82i***: Census, 21 April 1982 (indirect method); ***cen92i***: Census, 23 April 1992 (indirect method)

Under 5 mortality
China

Mortalité des moins de cinq ans
Chine

Mortalidad de menores de cinco años
China

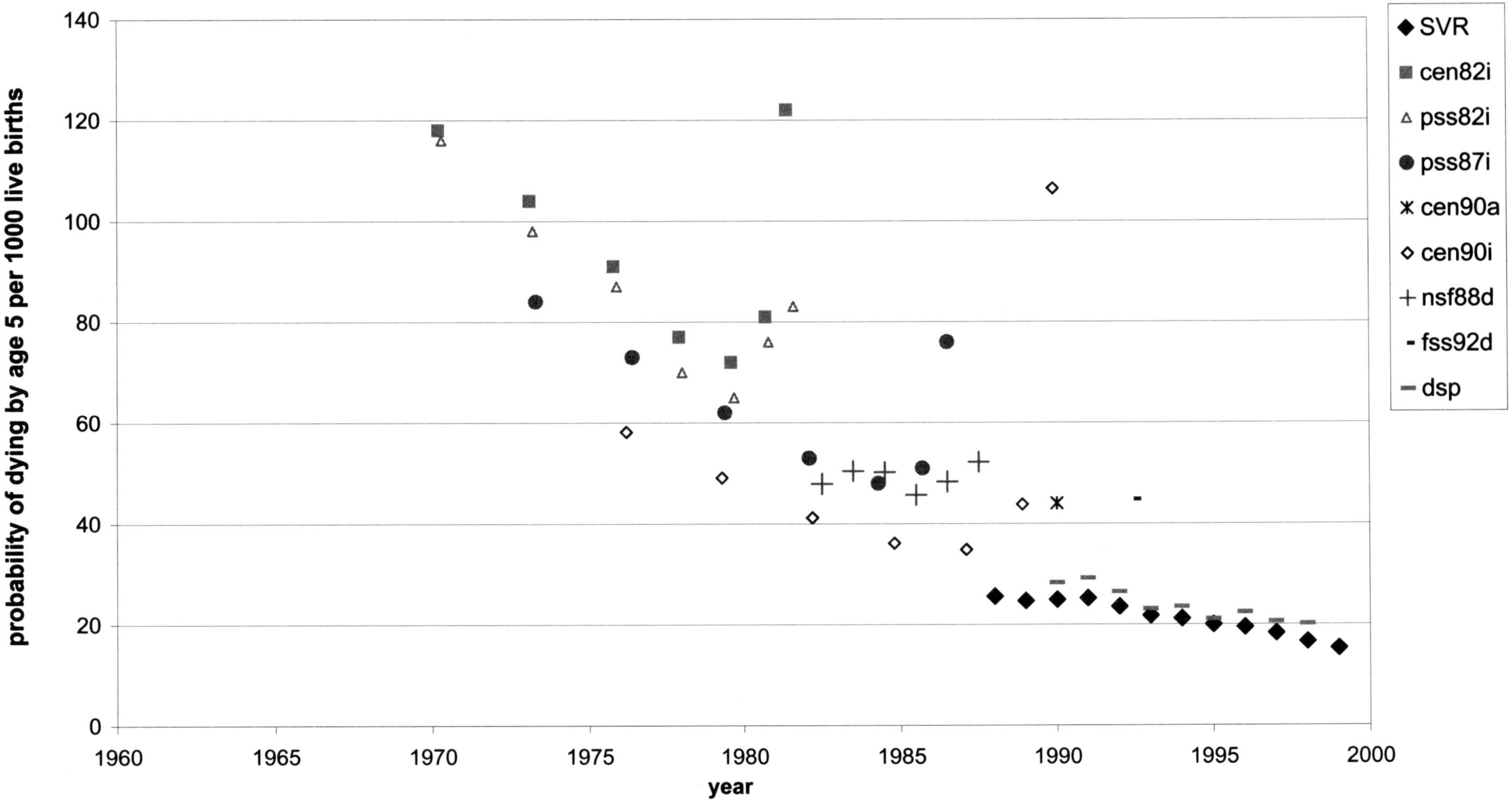

Data sources

Sources des données

Fuentes de los datos

SVR: Sample Vital Registration (MoH); ***cen82i***: Census, 1 July 1982 (indirect method); ***pss82i***: Female Fertility in China: a 1 0/00 Population Survey, 31 June 1982 (indirect method); ***pss87i***: China 1 0/00 Population Sample Survey, 1 July 1987 (indirect method); ***cen90a***: Census, 1 July 1990 (adjusted); ***cen90i***: Census, 1 July 1990 (indirect method); ***nsf88d***: National Survey on Fertility and Birth Control 1988 (direct method); ***fss92d***: Fertility Sampling Survey 1992 (direct method); ***dsp***: Disease surveillance points system

Under 5 mortality
Colombia

Mortalité des moins de cinq ans
Colombie

Mortalidad de menores de cinco años
Colombia

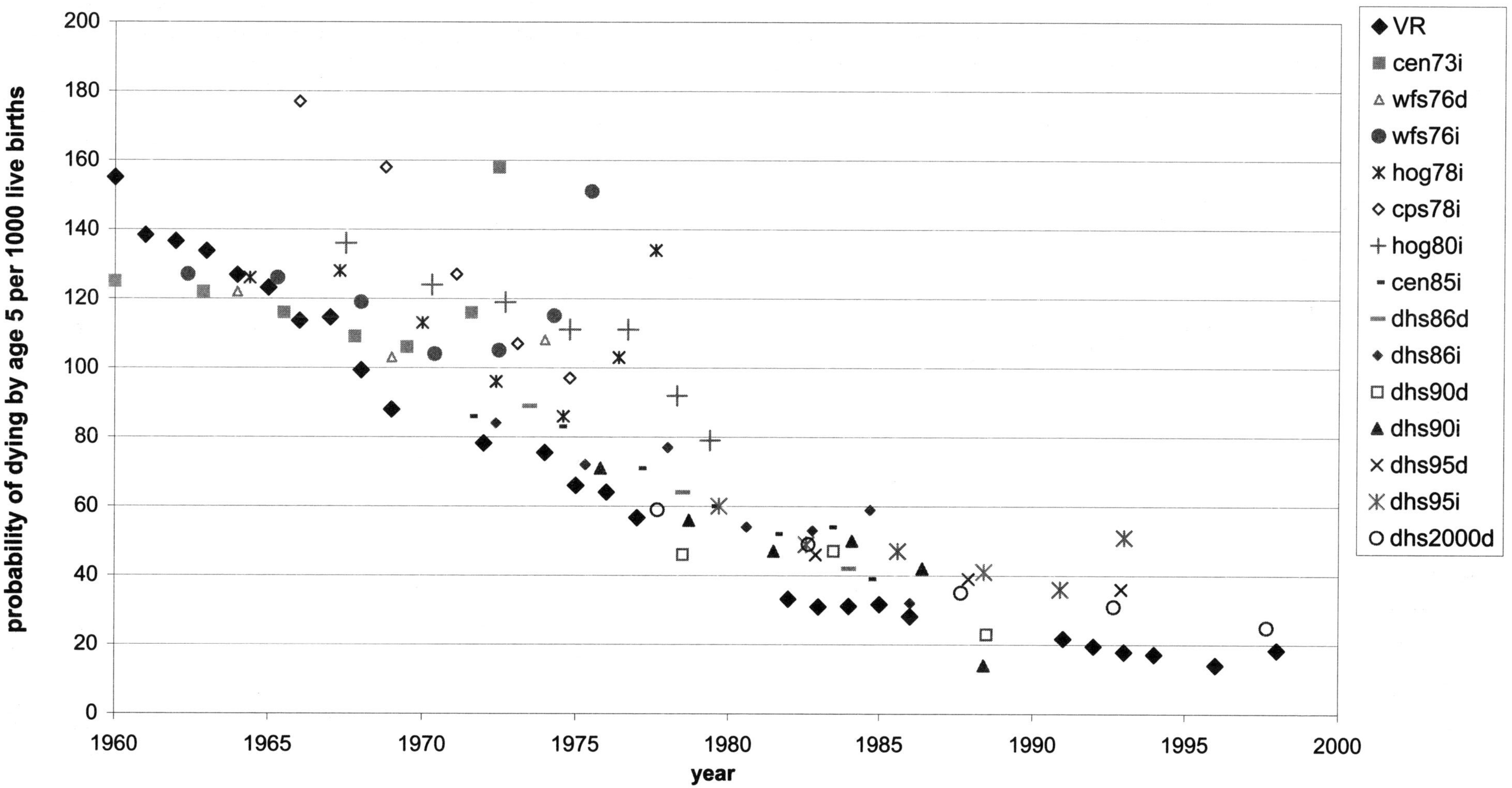

Data sources **Sources des données** **Fuentes de los datos**

VR: Vital Registration; ***cen73i***: Census, 24 October 1973 (indirect method); ***wfs76d***: Encuesta Colombiana de Fecundidad, May-August 1976 (direct method); ***wfs76i***: Encuesta Colombiana de Fecundidad, May-August 1976 (indirect method); ***hog78i***: Encuesta Nacional de Hogares, June-July 1978 (indirect method); ***cps78i***: Encuesta Nacional de Prevalencia de Uso de Anticoncepion, October-December 1978 (indirect method); ***hog80i***: Encuesta Nacional de Hogares, June-July 1980 (indirect method); ***cen85i***: Census, 15 October 1985 (indirect method); ***dhs86d***: Demographic and Health Survey, October-December 1986 (direct method); ***dhs86i***: Demographic and Health Survey, October-December 1986 (indirect method); ***dhs90d***: Demographic and Health Survey, May-August 1990 (direct method); ***dhs90i***: Demographic and Health Survey, May-August 1990 (indirect method); ***dhs95d***: Demographic and Health Survey, March-June 1995 (direct method); ***dhs95i***: Demographic and Health Survey, March-June 1995 (indirect method); ***dhs2000d***: Demographic and Health Survey 2000 (direct method)

Under 5 mortality

Comoros

Mortalité des moins de cinq ans

Comores

Mortalidad de menores de cinco años

Comoras

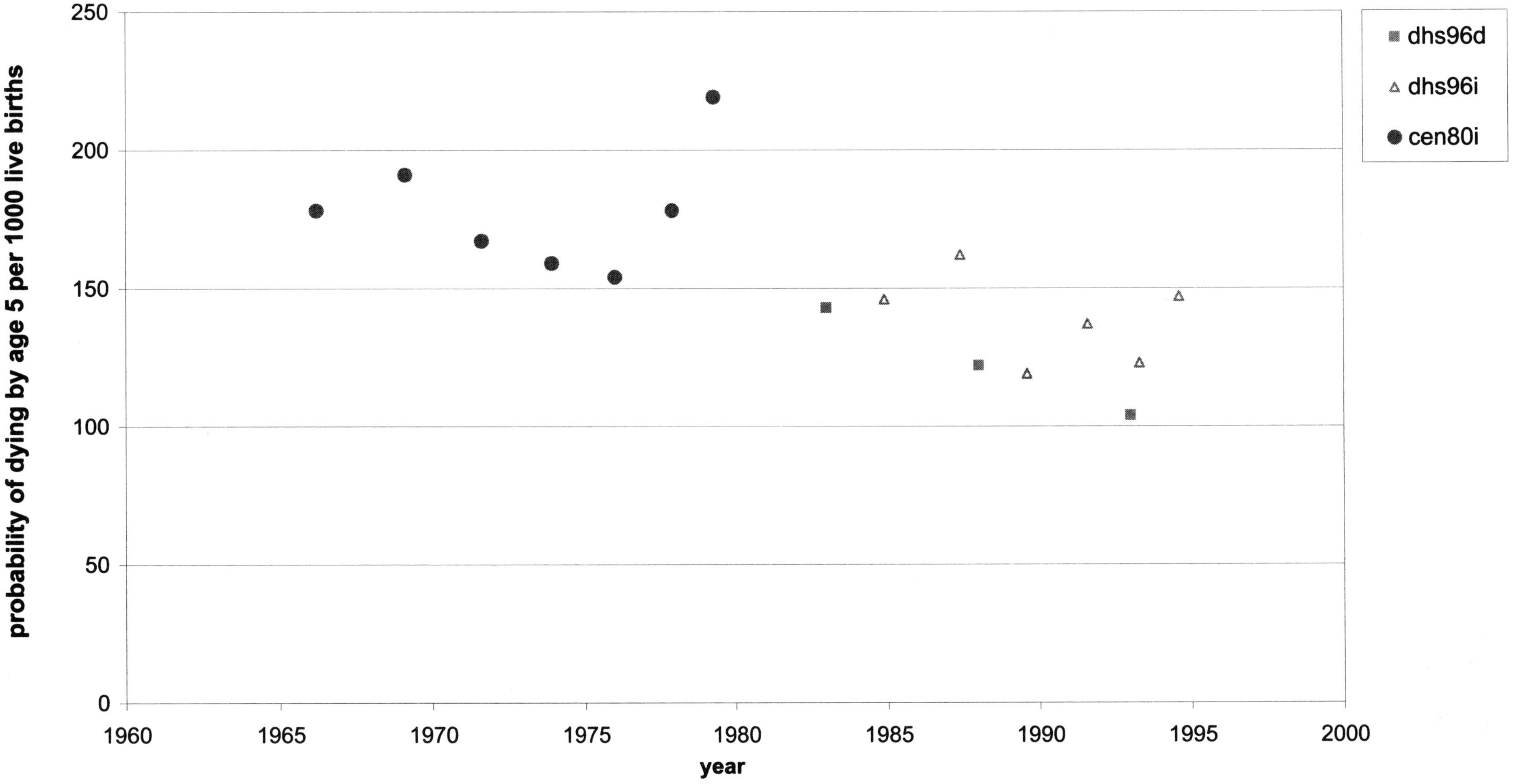

Data sources

Sources des données

Fuentes de los datos

dhs96d : Demographic and Health Survey 1996 (direct method); ***dhs96i*** : Demographic and Health Survey 1996 (indirect method); ***cen80i*** : Census 1980 (indirect method)

Under 5 mortality
Congo

Mortalité des moins de cinq ans
Congo

Mortalidad de menores de cinco años
Congo

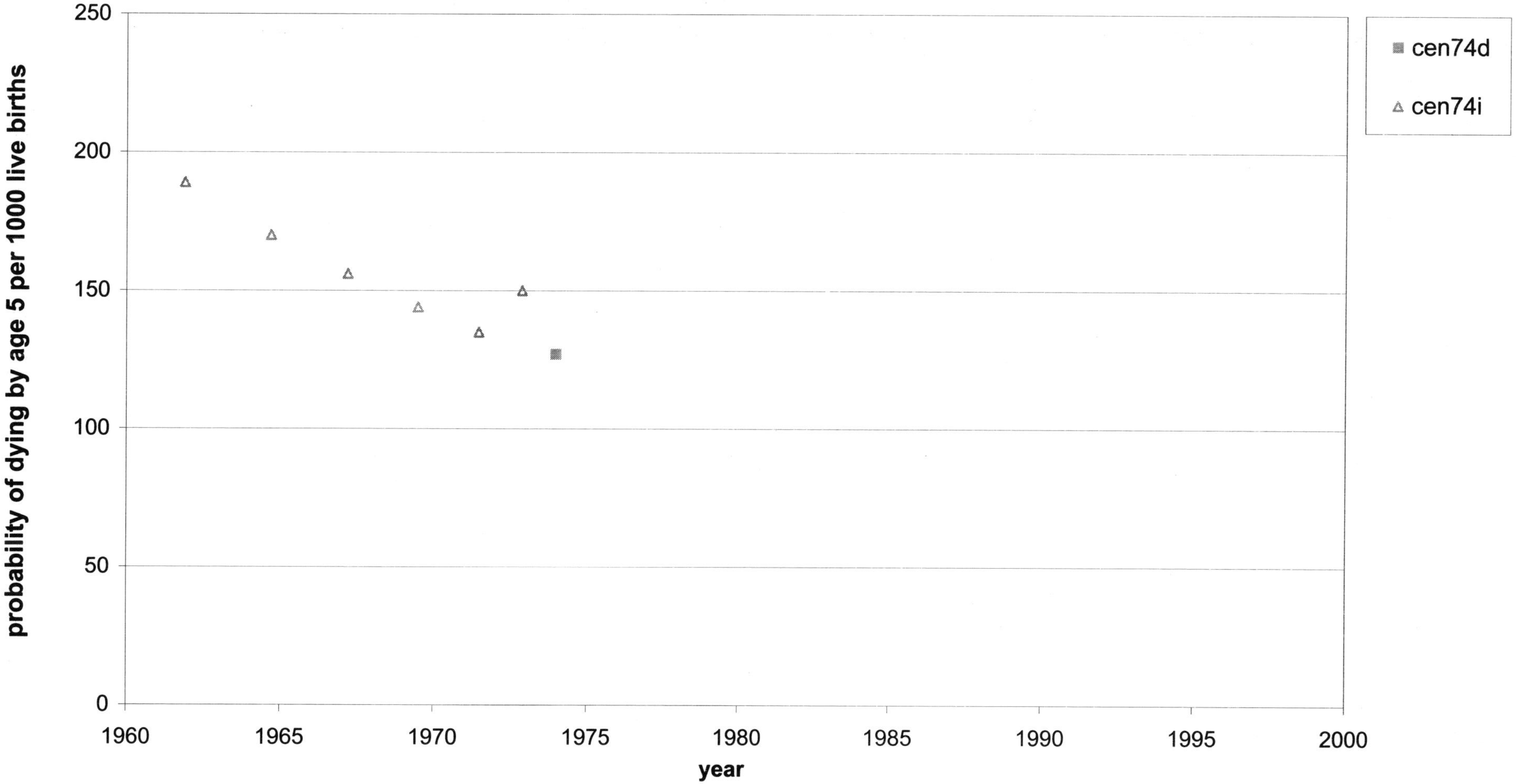

Data sources

Sources des données

Fuentes de los datos

cen74d: Census 1974 (direct method); ***cen74i***: Census 1974 (indirect method)

Under 5 mortality

Cook Islands

Mortalité des moins de cinq ans

Iles Cook

Mortalidad de menores de cinco años

Islas Cook

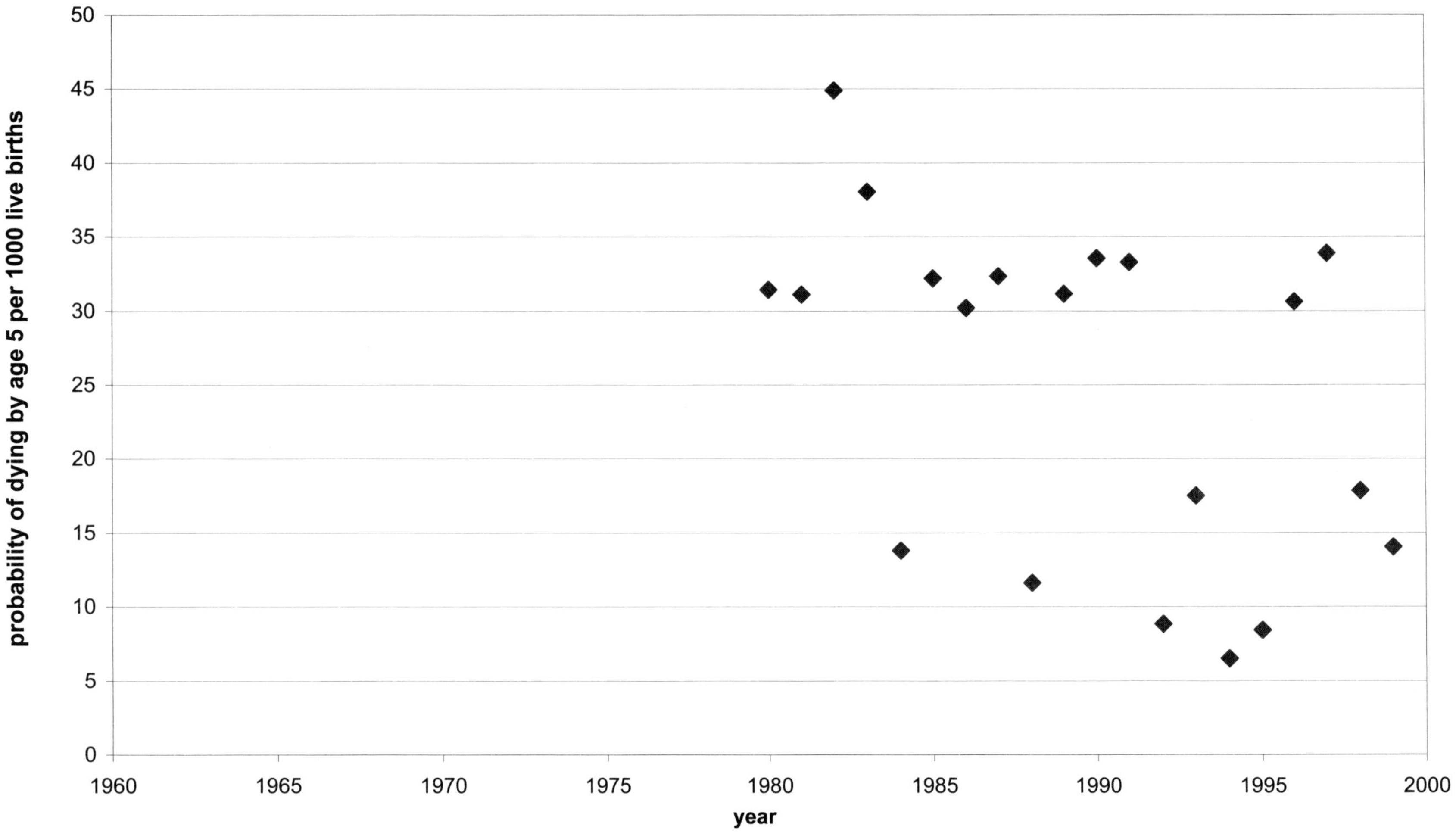

Data sources

Vital Registration

Sources des données

Fuentes de los datos

Under 5 mortality

Costa Rica

Mortalité des moins de cinq ans

Costa Rica

Mortalidad de menores de cinco años

Costa Rica

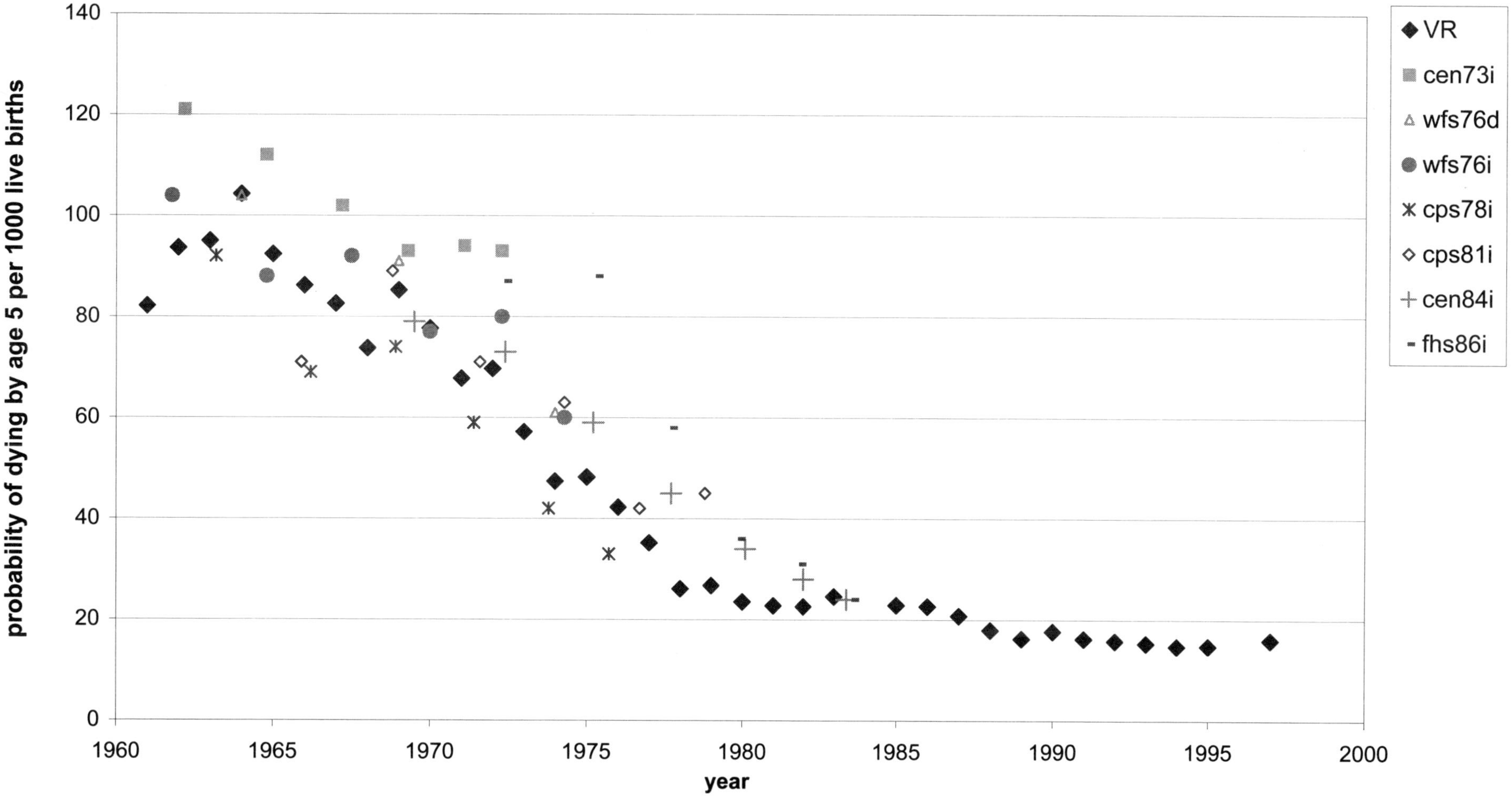

Data sources

Sources des données

Fuentes de los datos

VR: Vital Registration; ***cen73i***: Census, 14-19 May 1973 (indirect method); ***wfs76d***: Encuesta Nacional de Fecundidad, July-December 1976 (direct method); ***wfs76i***: Encuesta Nacional de Fecundidad, July-December 1976 (indirect method); ***cps78i***: Encuesta de Prevalencia Anticonceptiva, March-May 1978 (indirect method); ***cps81i***: Encuesta de Prevalencia Anticonceptiva, January-April 1981 (indirect method); ***cen84i***: Census, 11 June 1984 (indirect method); ***fhs86i***: Encuesta de Fecundidad y Salud, January-May 1986 (indirect method)

Under 5 mortality
Côte d'Ivoire

Mortalité des moins de cinq ans
Côte d'Ivoire

Mortalidad de menores de cinco años
Côte d'Ivoire

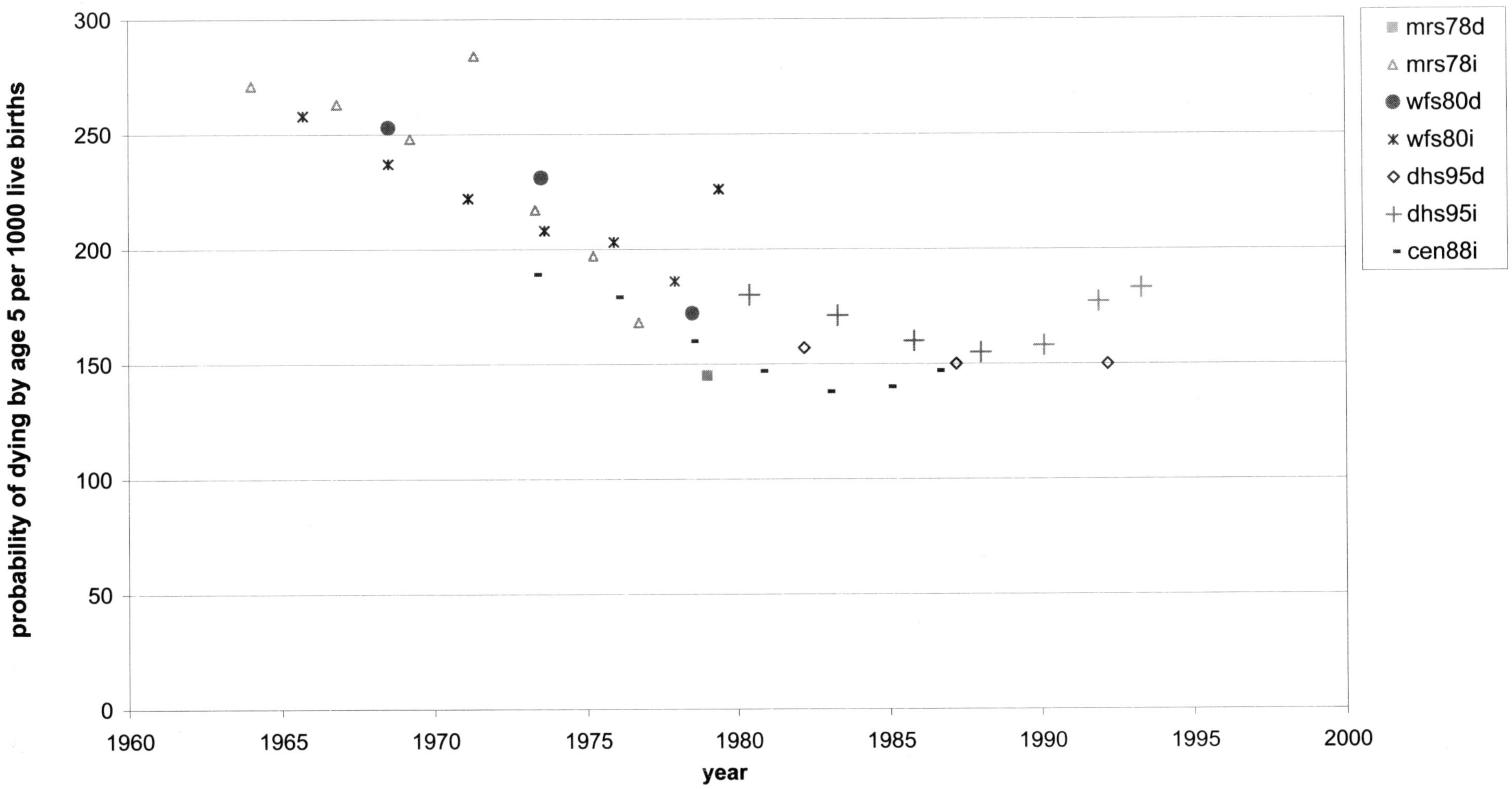

Data sources **Sources des données** **Fuentes de los datos**

mrs78d: Enquête Démographique à Passages Répétés, 1978-1979 (Nationals only) (direct method); ***mrs78i***: Enquête Démographique à Passages Répétés, 1978-1979 (Nationals only) (indirect method); ***wfs80d***: Enquête Ivoirienne sur la Fécondite, August 1980-March 1981 (direct method); ***wfs80i***: Enquête Ivoirienne sur la Fécondite, August 1980-March 1981 (indirect method); ***dhs95d***: Demographic and Health Survey, June-November 1994 (direct method); ***dhs95i***: Demographic and Health Survey, June-November 1994 (indirect method); ***cen88i***: Population and Housing Census, March 1988 (indirect method)

Under 5 mortality

Croatia

Mortalité des moins de cinq ans

Croatie

Mortalidad de menores de cinco años

Croacia

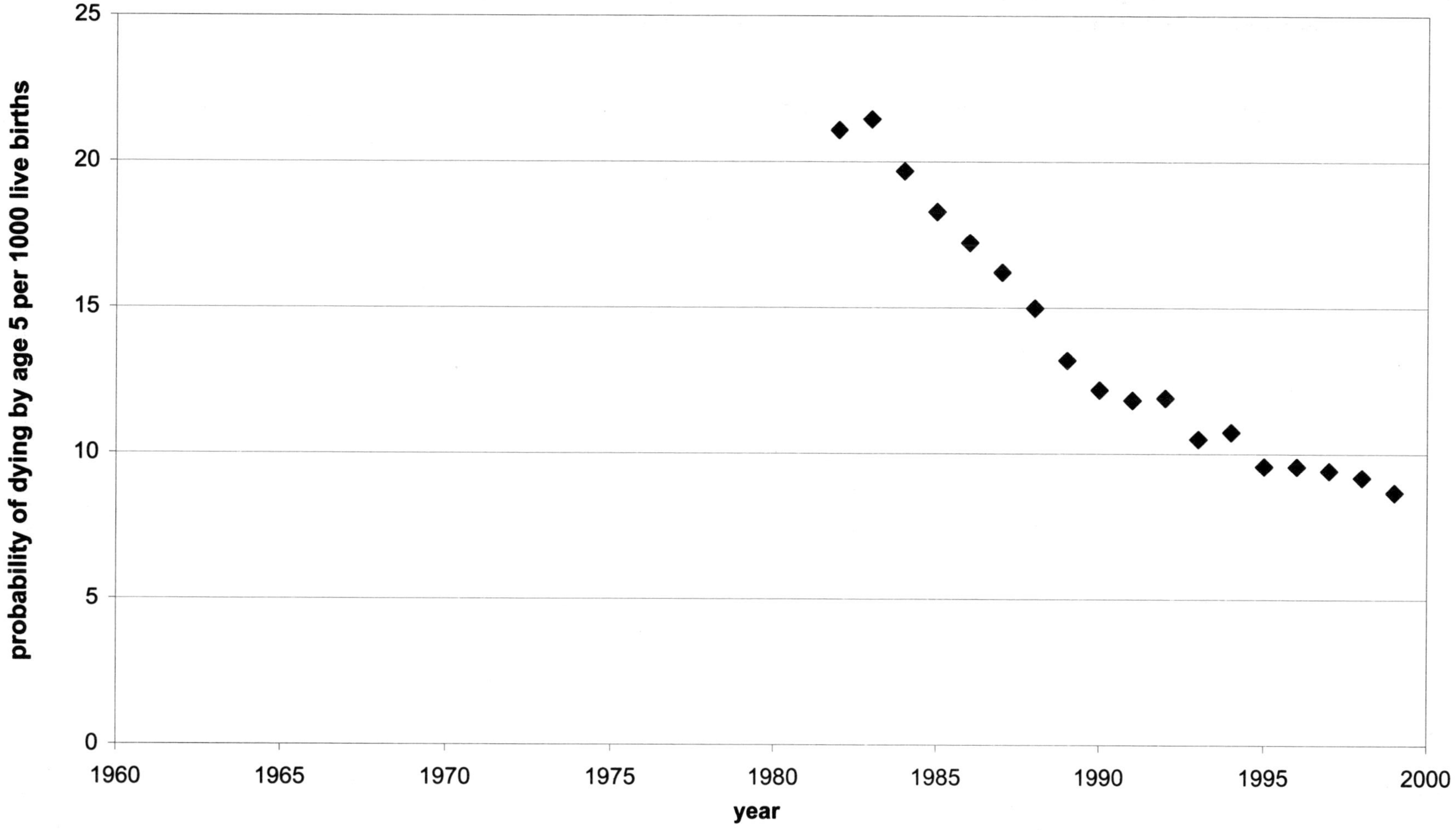

Data sources

Vital Registration

Sources des données

Fuentes de los datos

Under 5 mortality
Cuba

Mortalité des moins de cinq ans
Cuba

Mortalidad de menores de cinco años
Cuba

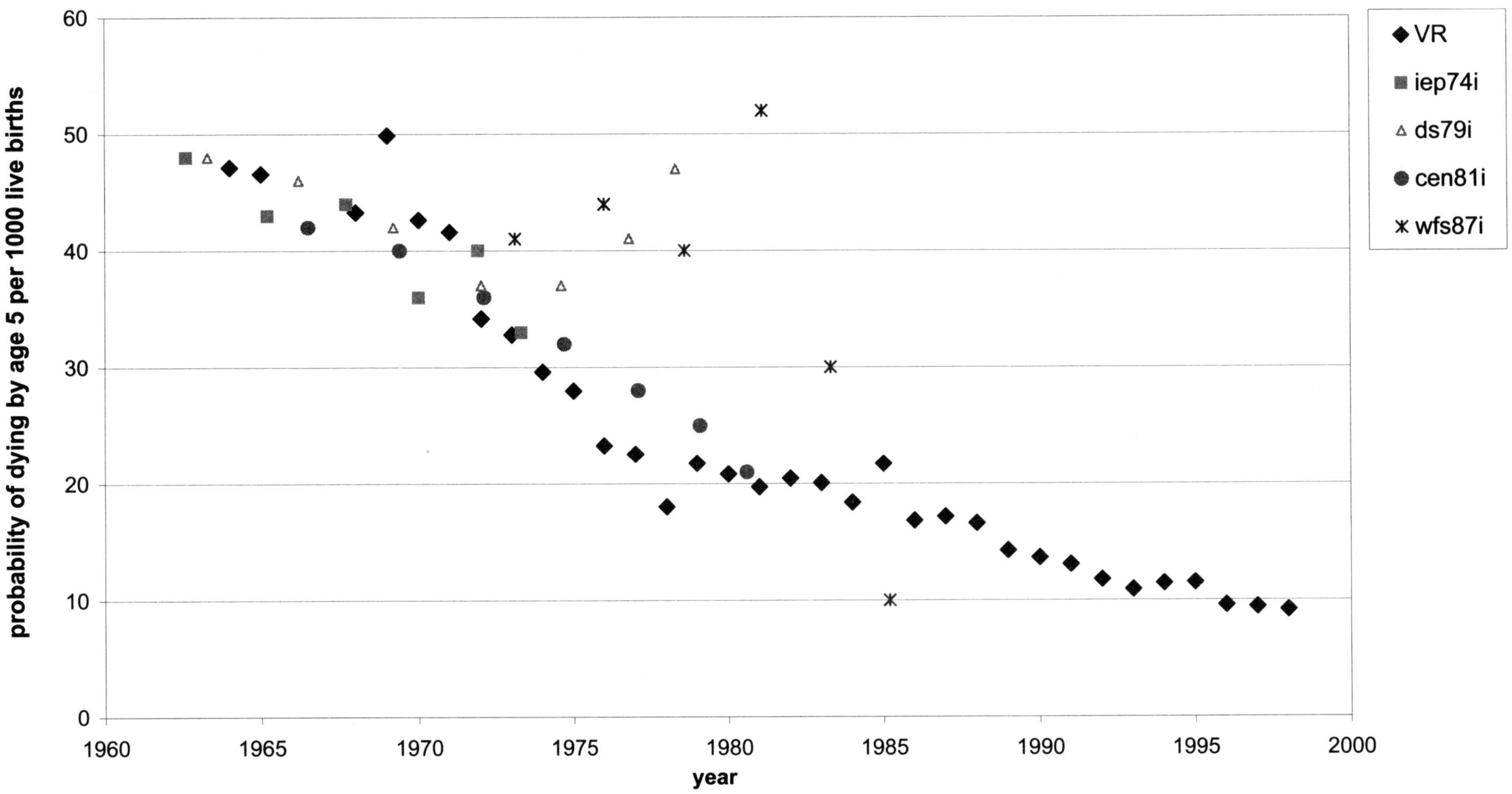

Data sources

Sources des données

Fuentes de los datos

VR: Vital Registration; ***iep74i***: Encuesta Nacional de Ingresos y Egresos de la Poblacion, 1974 (indirect method); ***ds79i***: Encuesta Demografica Nacional, 22-31 March 1979 (indirect method); ***cen81i***: Census, 11 September 1981 (indirect method); ***wfs87i***: Encuesta Nacoinal de Fecundidad, November-December 1987 (indirect method)

Under 5 mortality

Cyprus

Mortalité des moins de cinq ans

Chypre

Mortalidad de menores de cinco años

Chipre

Data sources

Vital Registration

Sources des données

Fuentes de los datos

Under 5 mortality

Czech Republic

Mortalité des moins de cinq ans

République tchèque

Mortalidad de menores de cinco años

República Checa

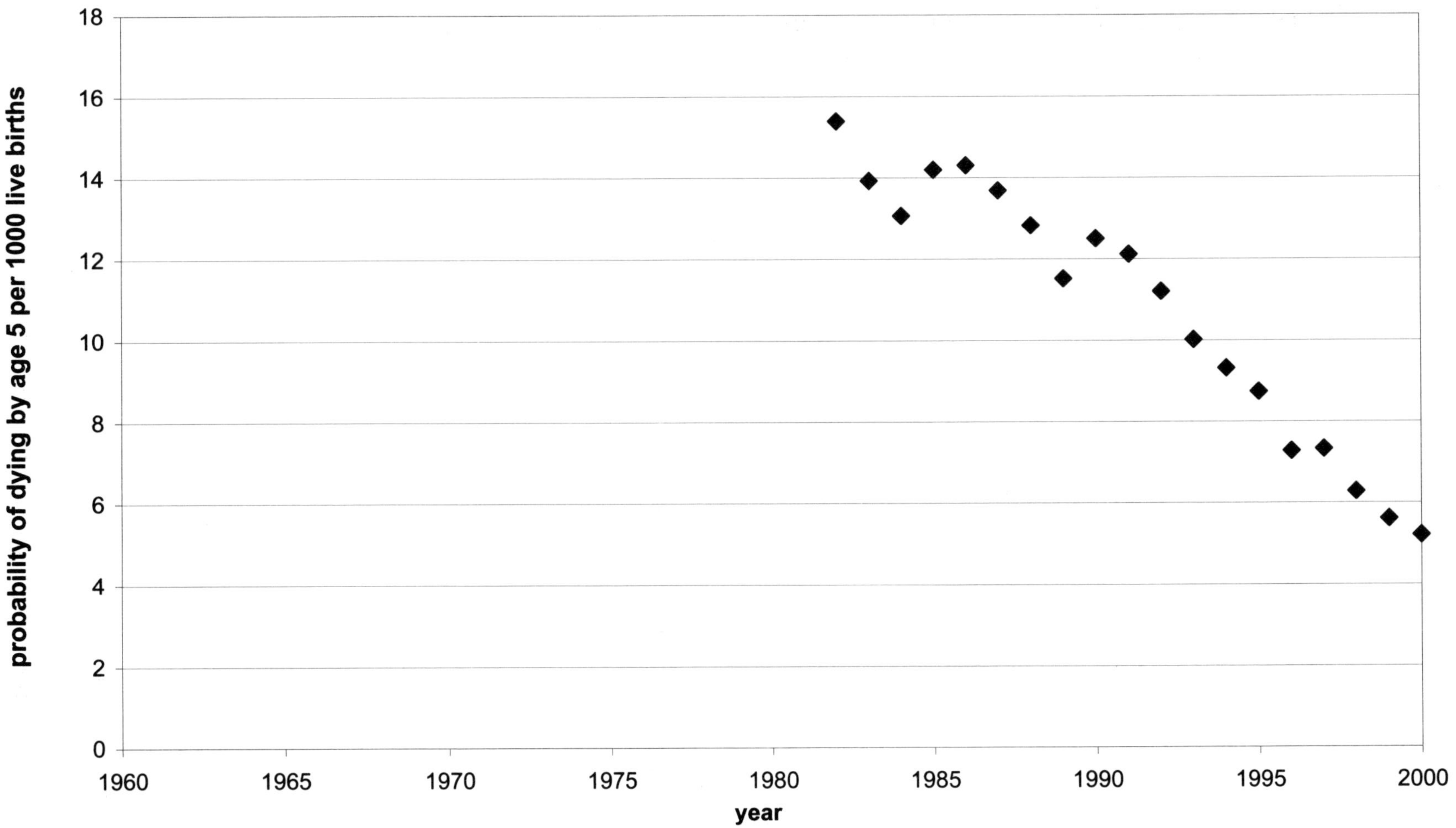

Data sources

Vital Registration

Sources des données

Fuentes de los datos

Under 5 mortality

Democratic People's Republic of Korea

Mortalité des moins de cinq ans

République populaire démocratique de Corée

Mortalidad de menores de cinco años

República Popular Democrática de Corea

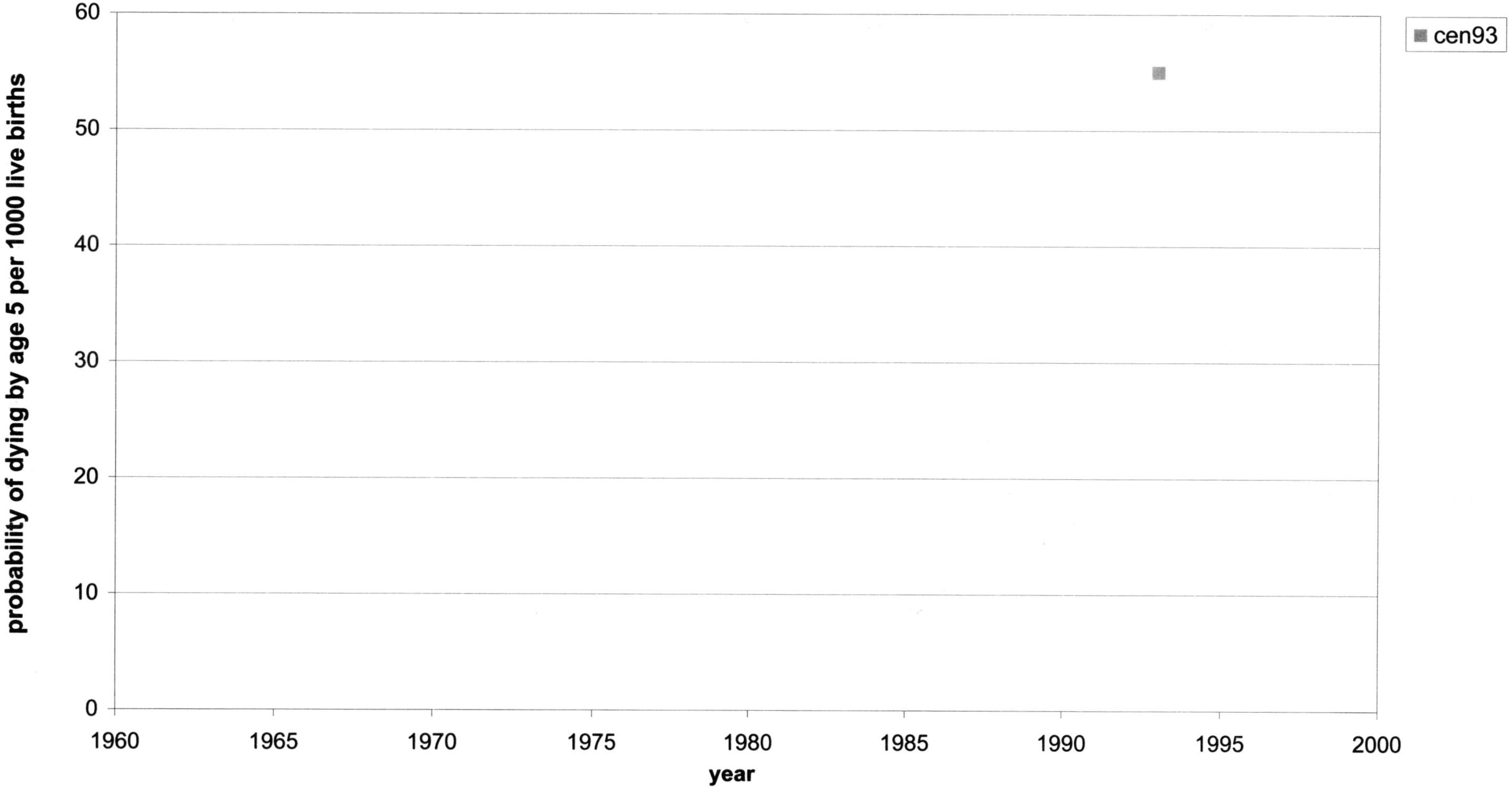

Data sources

cen93: Census 1993

Sources des données

Fuentes de los datos

Under 5 mortality

Democratic Republic of the Congo

Mortalité des moins de cinq ans

République démocratique du Congo

Mortalidad de menores de cinco años

República Democrática del Congo

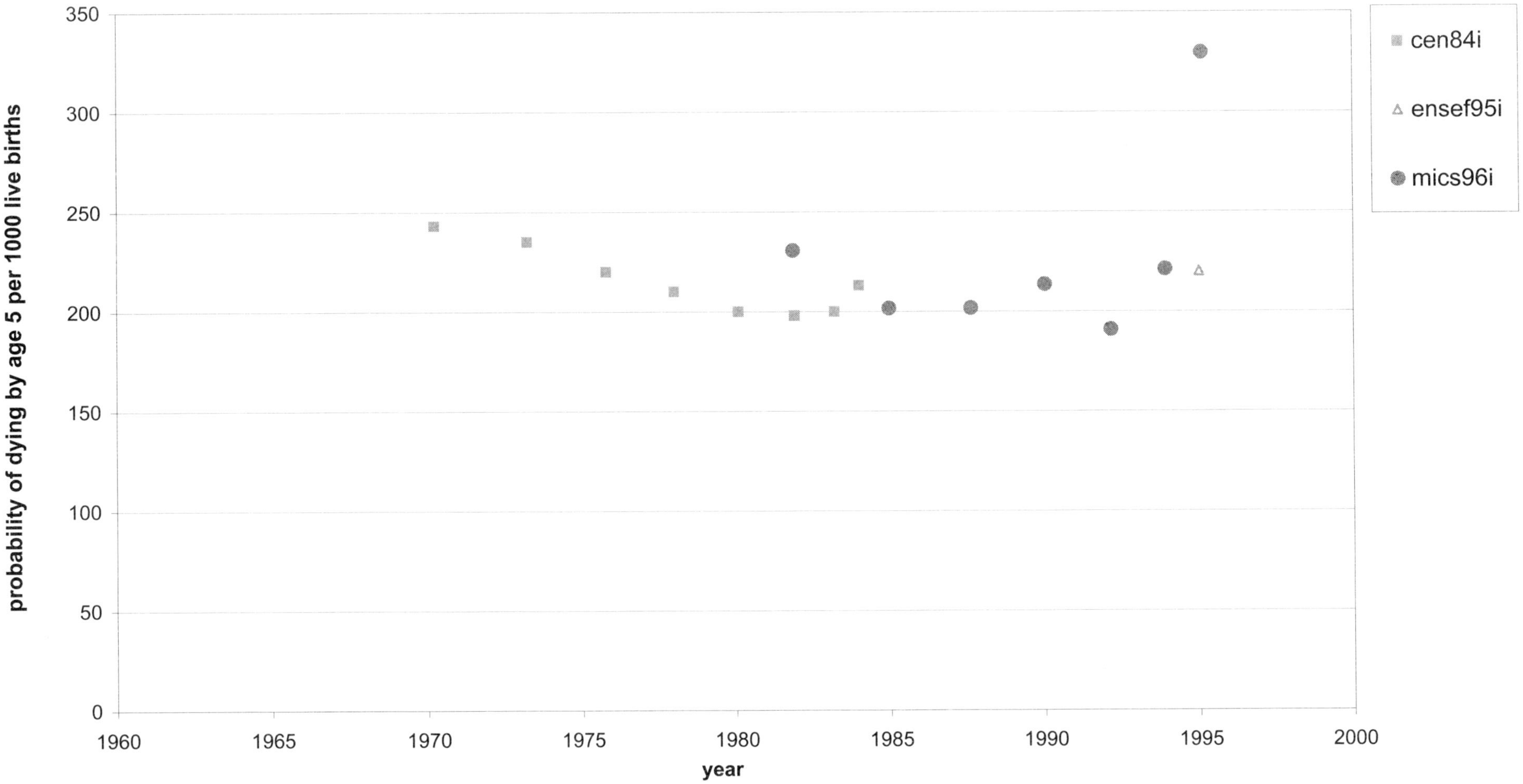

Data sources

Sources des données

Fuentes de los datos

cen84i: Census, 1 July 1984 (indirect method); ***ensef95i***: Enquête nationale sur la situation des enfants et des femmes 1995 (indirect method); ***mics96i***: Multiple Indicator Cluster Survey 1996 (indirect method)

Under 5 mortality

Denmark

Mortalité des moins de cinq ans

Danemark

Mortalidad de menores de cinco años

Dinamarca

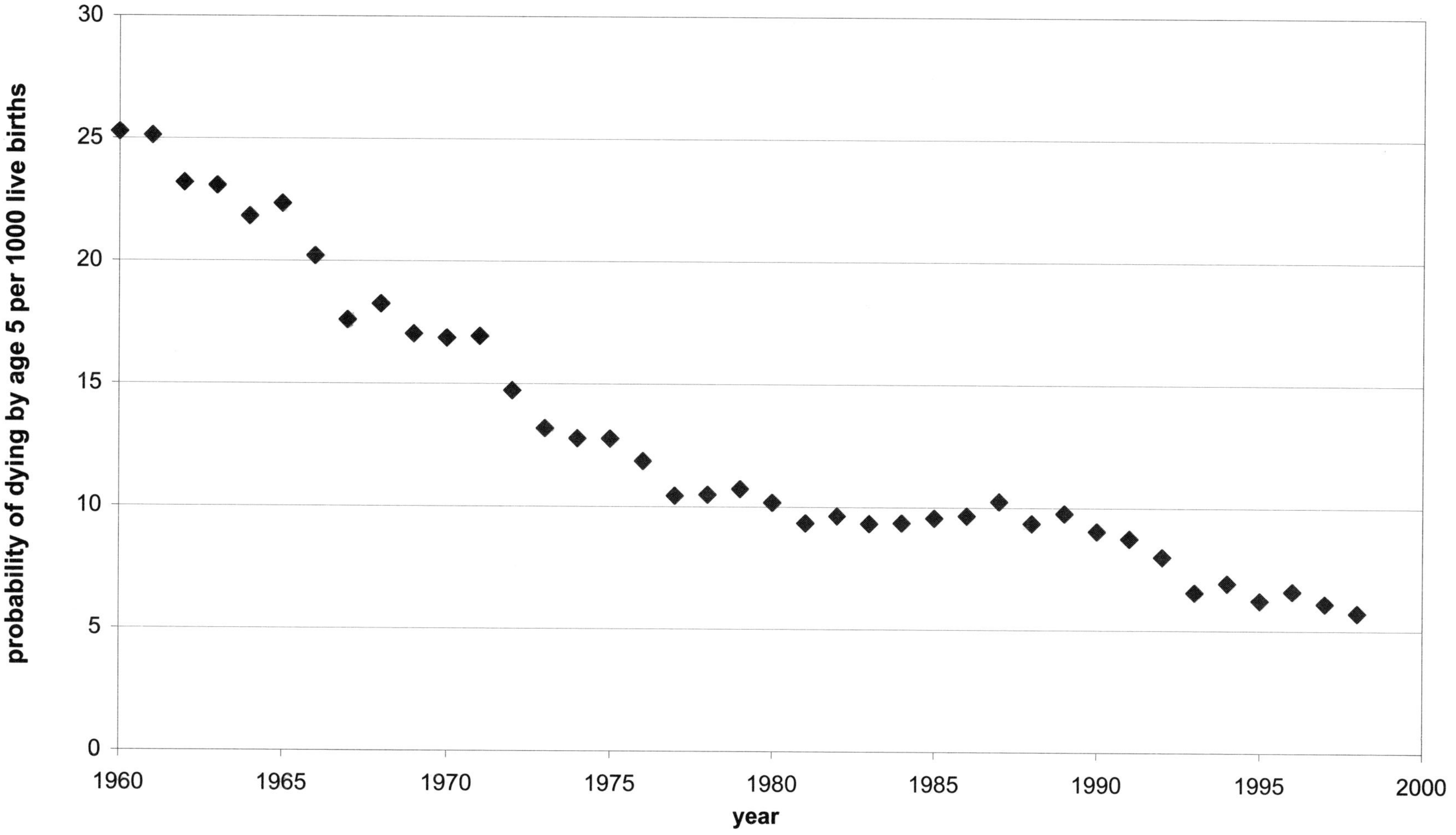

Data sources

Vital Registration

Sources des données

Fuentes de los datos

Under 5 mortality

Dominica

Mortalité des moins de cinq ans

Dominique

Mortalidad de menores de cinco años

Dominica

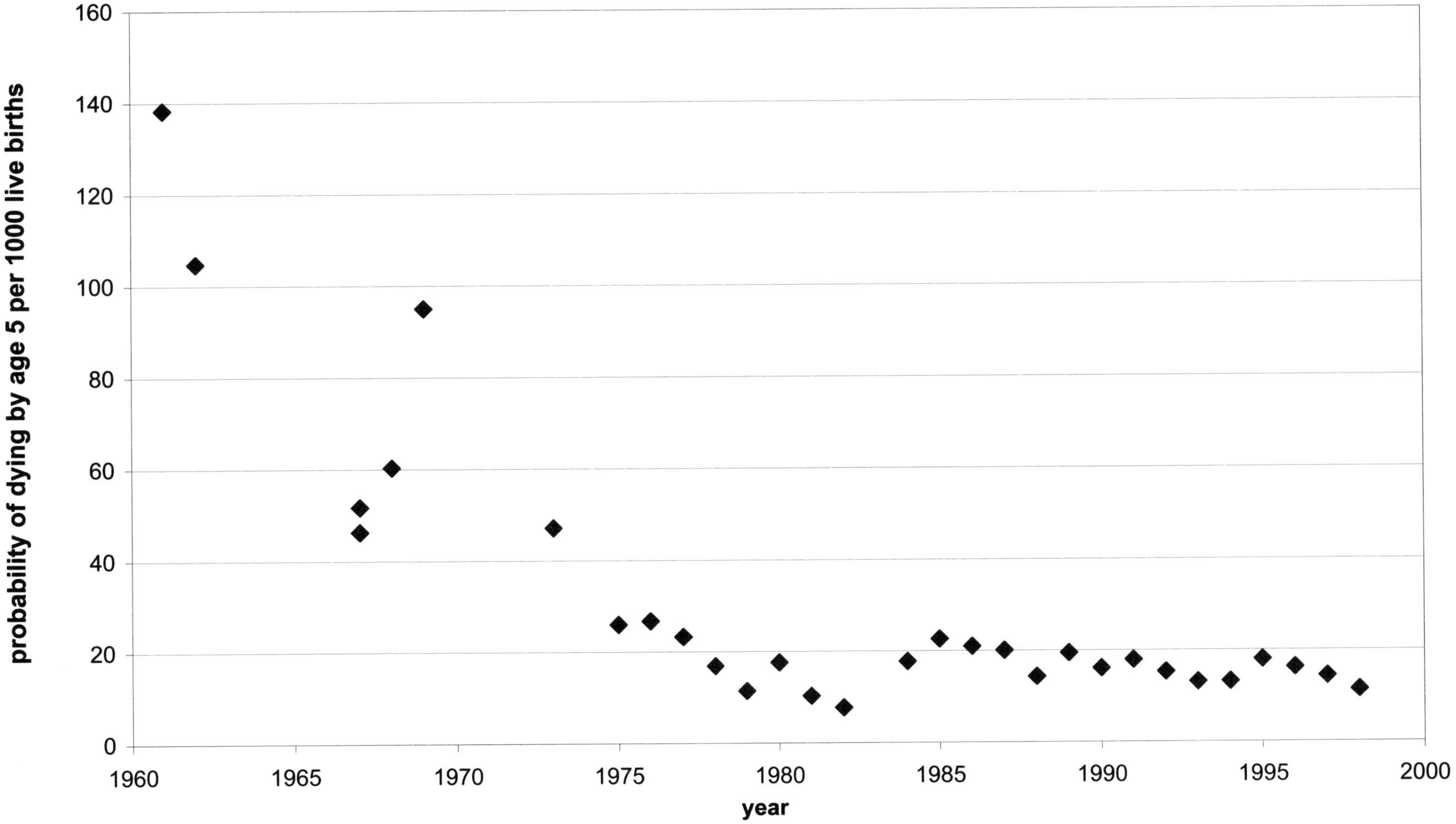

Data sources

Vital Registration

Sources des données

Fuentes de los datos

Under 5 mortality

Dominican Republic

Mortalité des moins de cinq ans

République dominicaine

Mortalidad de menores de cinco años

República Dominicana

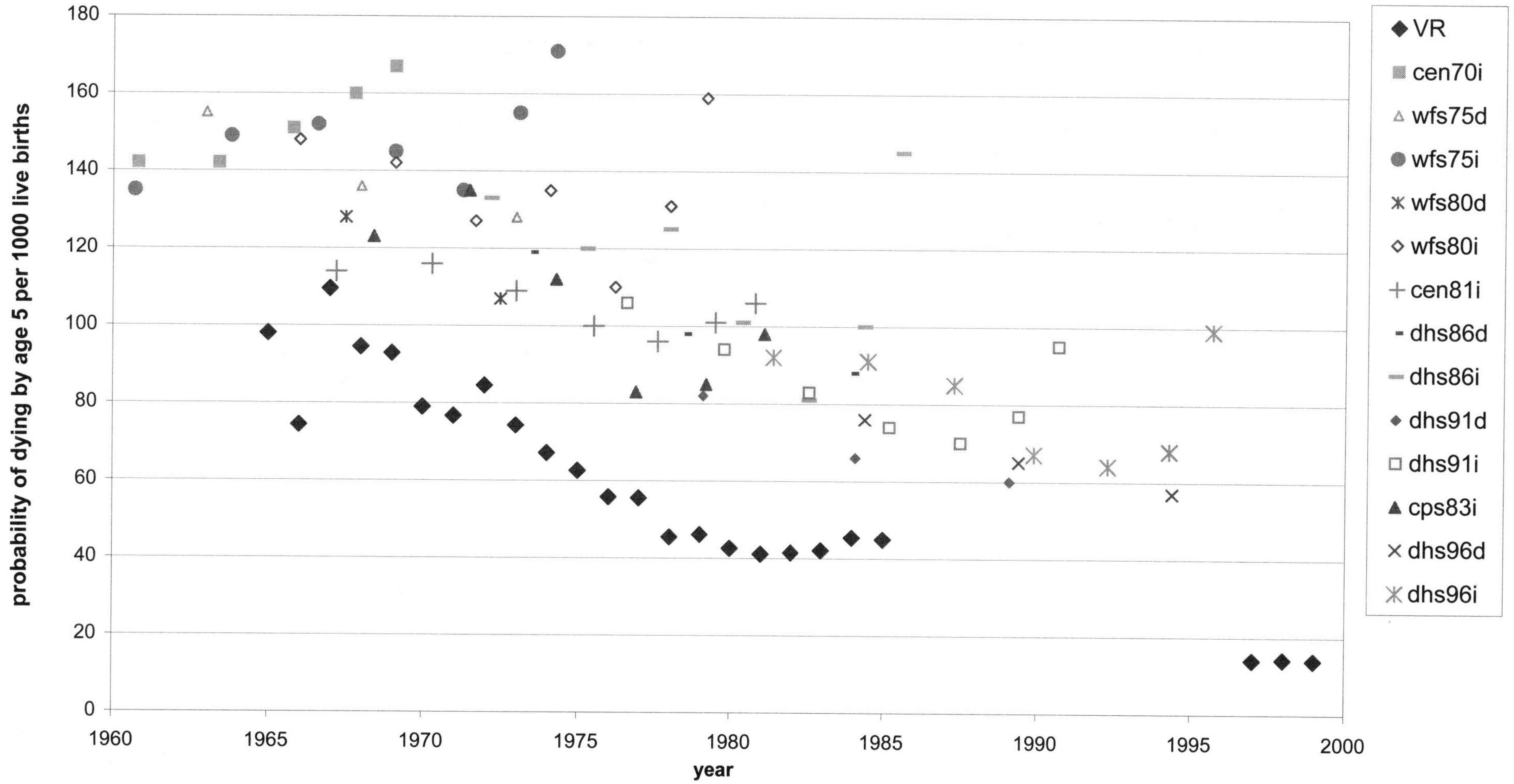

Data sources

Sources des données

Fuentes de los datos

VR: Vital Registration; ***cen70i***: Census, 9-10 January 1970 (indirect method); ***wfs75d***: Encuesta Nacional de Fecundidad, April-June 1975 (direct method); ***wfs75i***: Encuesta Nacional de Fecundidad, April-June 1975 (indirect method); ***wfs80d***: Encuesta Nacional de Fecundidad, February-May 1980 (direct method); ***wfs80i***: Encuesta Nacional de Fecundidad, February-May 1980 (indirect method); ***cen81i***: Census, 12-13 December 1981 (indirect method); ***dhs86d***: Encuesta Demografica y de Salud, September-December 1986 (direct method); ***dhs86i***: Encuesta Demografica y de Salud, September-December 1986 (indirect method); ***dhs91d***: Demographic and Health Survey 1991 (direct method); ***dhs91i***: Demographic and Health Survey 1991 (indirect method); ***cps83i***: Encuesta Nacional de Prevalenceia del Uso de Anticonceptivos, May-July 1983 (indirect method); ***dhs96d***: Demographic and Health Survey 1996 (direct method); ***dhs96i***: Demographic and Health Survey 1996 (indirect method)

Under 5 mortality
Ecuador

Mortalité des moins de cinq ans
Equateur

Mortalidad de menores de cinco años
Ecuador

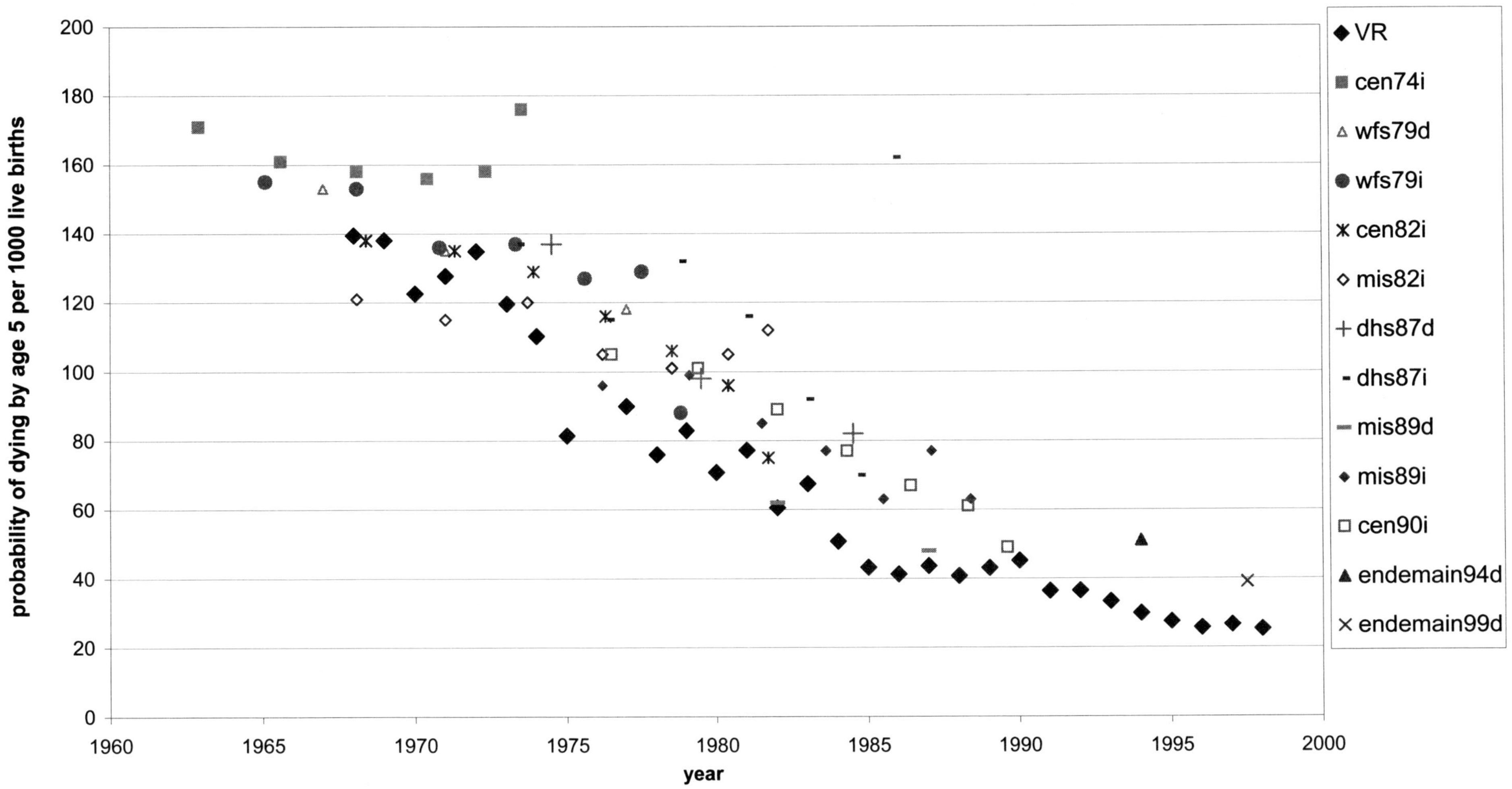

Data sources **Sources des données** **Fuentes de los datos**

VR: Vital Registration; ***cen74i***: Census, 8 June 1974 (indirect method); ***wfs79d***: Encuesta Nacional de Fecundidad, August-December 1979 (direct method); ***wfs79i***: Encuesta Nacional de Fecundidad, August-December 1979 (indirect method); ***cen82i***: Census, 28 November 1982 (indirect method); ***mis82i***: Encuesta National de Salud Materno Infantil y Variables Demograficas, September-November 1982 (indirect method); ***dhs87d***: Encuesta Demografica y de Salud Familiar, January-March 1987 (direct method); ***dhs87i***: Encuesta Demografica y de Salud Familiar, January-March 1987 (indirect method); ***mis89d***: Encuesta Demografica y de Salud Materna e infantil, July-October 1989 (direct method); ***mis89i***: Encuesta Demografica y de Salud Materna e infantil, July-October 1989 (indirect method); ***cen90i***: Census, 25 November 1990 (indirect method); ***endemain94d***: Encuesta Demografica y de Salud Materna e infantil 1994 (direct method); ***endemain99d***: Encuesta Demografica y de Salud Materna e infantil 1999 (direct method)

Under 5 mortality

Egypt

Mortalité des moins de cinq ans

Egypte

Mortalidad de menores de cinco años

Egipto

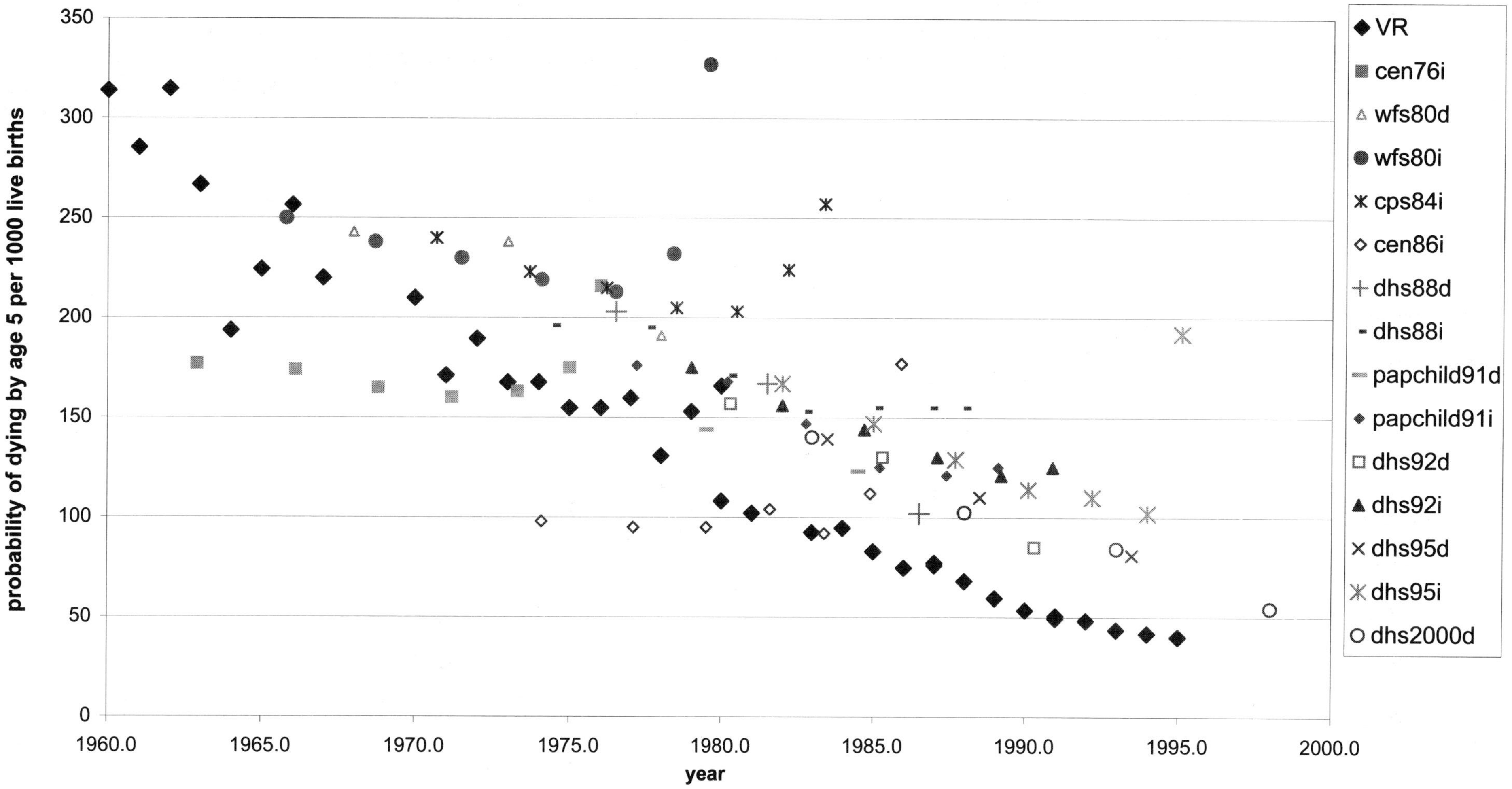

Data sources

Sources des données

Fuentes de los datos

VR: Vital Registration; ***cen76i***: Census, 23 November 1976 (indirect method); ***wfs80d***: Egyptian Fertility Survey, May-June 1980 (direct method); ***wfs80i***: Egyptian Fertility Survey, May-June 1980 (indirect method); ***cps84i***: Contraceptive Prevalence Survey, October-November 1984 (indirect method); ***cen86i***: Census, 17-18 November 1986 (indirect method); ***dhs88d***: Demographic and Health Survey, October 1988 - January 1989 (direct method); ***dhs88i***: Demographic and Health Survey, October 1988 - January 1989 (indirect method); ***papchild91d***: Maternal and Child Health Survey 1991 (direct method); ***papchils91i***: Maternal and Child Health Survey 1991 (indirect method); ***dhs92d***: Demographic and Health Survey, November 1992- February 1993 (direct method); ***dhs92i***: Demographic and Health Survey, November 1992- February 1993 (indirect method); ***dhs95d***: Demographic and Health Survey, November 1995- January 1996 (direct method); ***dhs95i***: Demographic and Health Survey, November 1995- January 1996 (indirect method); ***dhs2000d***: Demographic and Health Survey 2000 (direct method)

Under 5 mortality

El Salvador

Mortalité des moins de cinq ans

El Salvador

Mortalidad de menores de cinco años

El Salvador

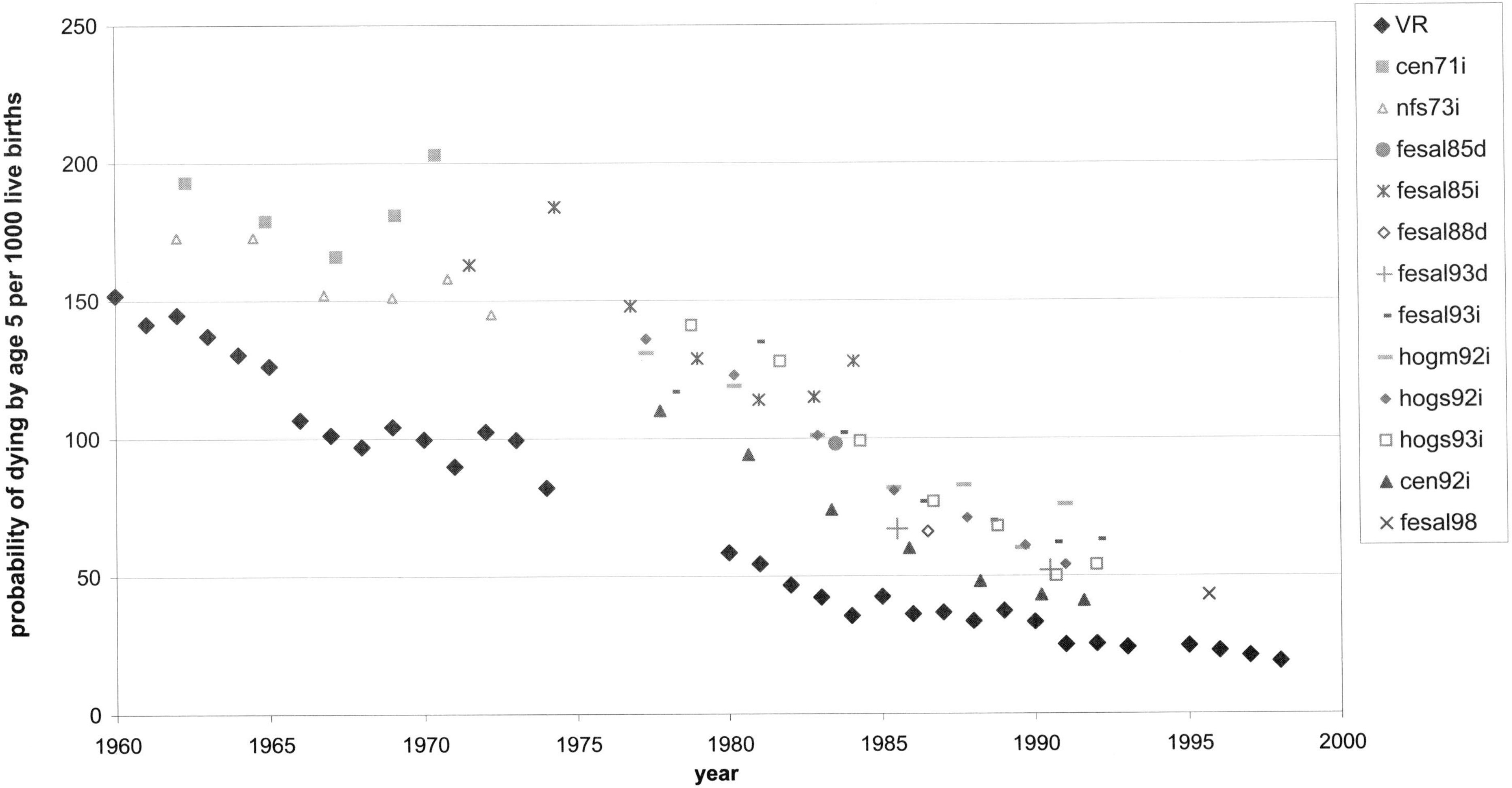

Data sources **Sources des données** **Fuentes de los datos**

VR: Vital Registration; ***cen71i***: Census, 28 June 1971 (indirect method); ***nfs73i***: Encuesta Nacional de Fecundidad, May-June 1973 (indirect method); ***fesal85d***: Encuesta Nacional de Salud Familiar, May-June 1985 (direct method); ***fesal85i***: Encuesta Nacional de Salud Familiar, May-June 1985 (indirect method); ***fesal88d***: Encuesta Nacional de Salud Familiar, May-June 1988 (direct method); ***fesal93d***: Encuesta Nacional de Salud Familiar, 1993 (direct method); ***fesal93i***: Encuesta Nacional de Salud Familiar, 1993 (indirect method); ***hogm92i***: Encuesta de Hongares de Propositos Multiples, March 1992 (indirect method); ***hogs92i***: Encuesta de Hongares de Propositos Multiples, September 1992 (indirect method); ***hogs93i***: Encuesta de Hongares de Propositos Multiples, March 1993 (indirect method); ***cen92i***: Census 1992 (indirect method); ***fesal98***: Encuesta nacional de salud familiar El Salvador 1998

Under 5 mortality
Eritrea

Mortalité des moins de cinq ans
Erythrée

Mortalidad de menores de cinco años
Eritrea

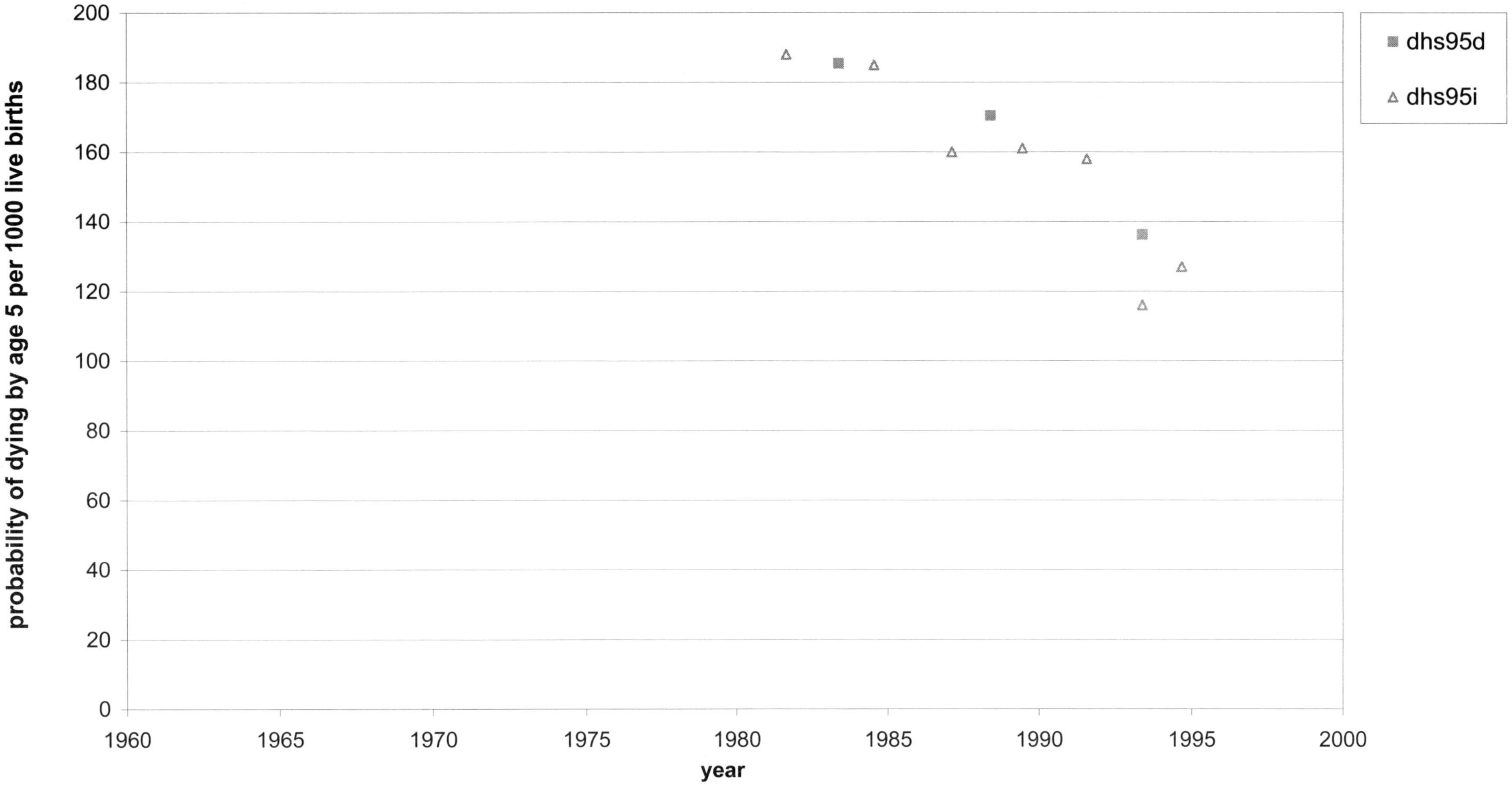

Data sources

Sources des données

Fuentes de los datos

dhs95d: Demographic and Health Survey, September- December 1995 (direct method); ***dhs95i:*** Demographic and Health Survey, September- December 1995 (indirect method)

Under 5 mortality

Estonia

Mortalité des moins de cinq ans

Estonie

Mortalidad de menores de cinco años

Estonia

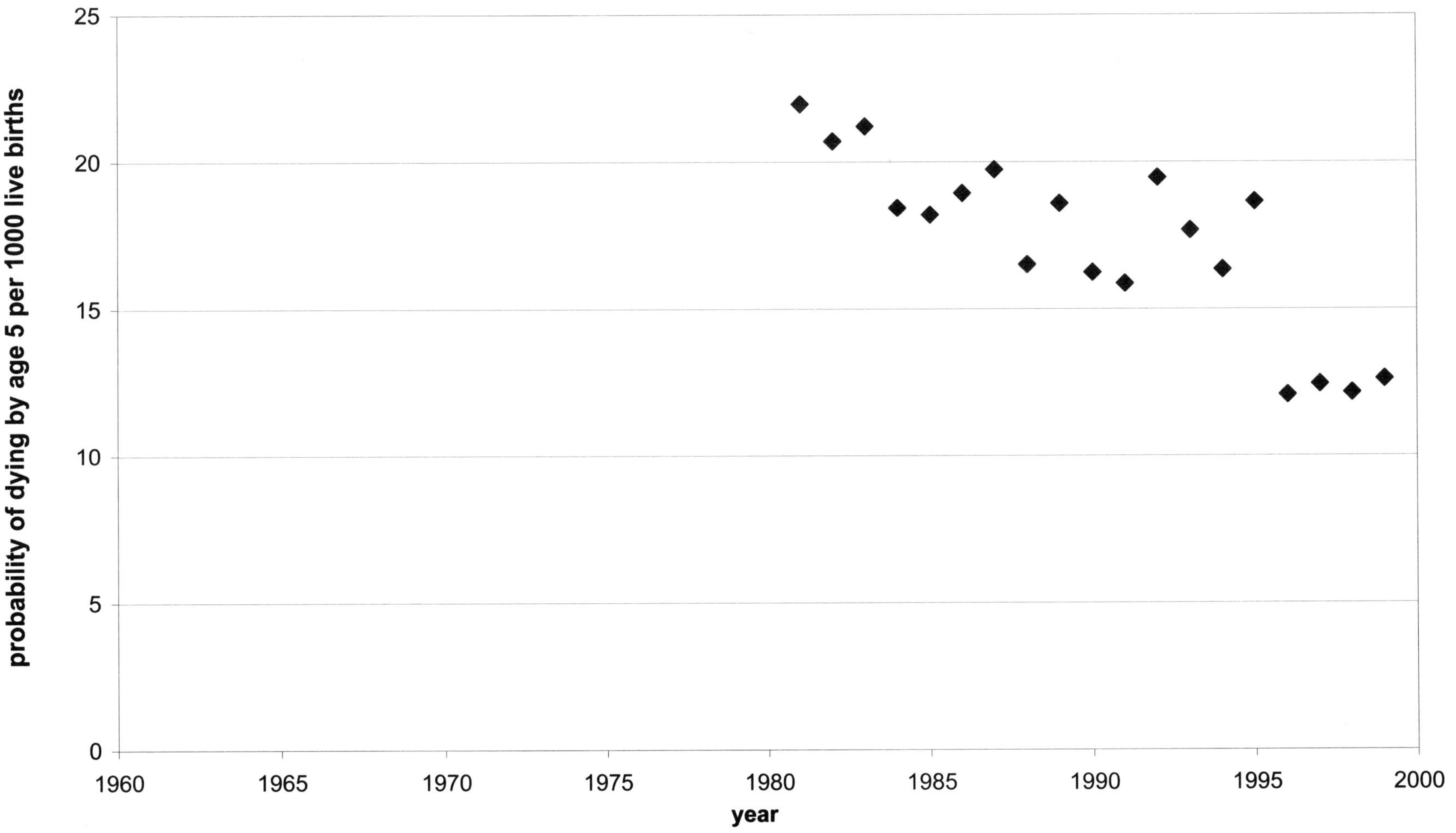

Data sources

Vital Registration

Sources des données

Fuentes de los datos

Under 5 mortality

Ethiopia

Mortalité des moins de cinq ans

Ethiopie

Mortalidad de menores de cinco años

Etiopía

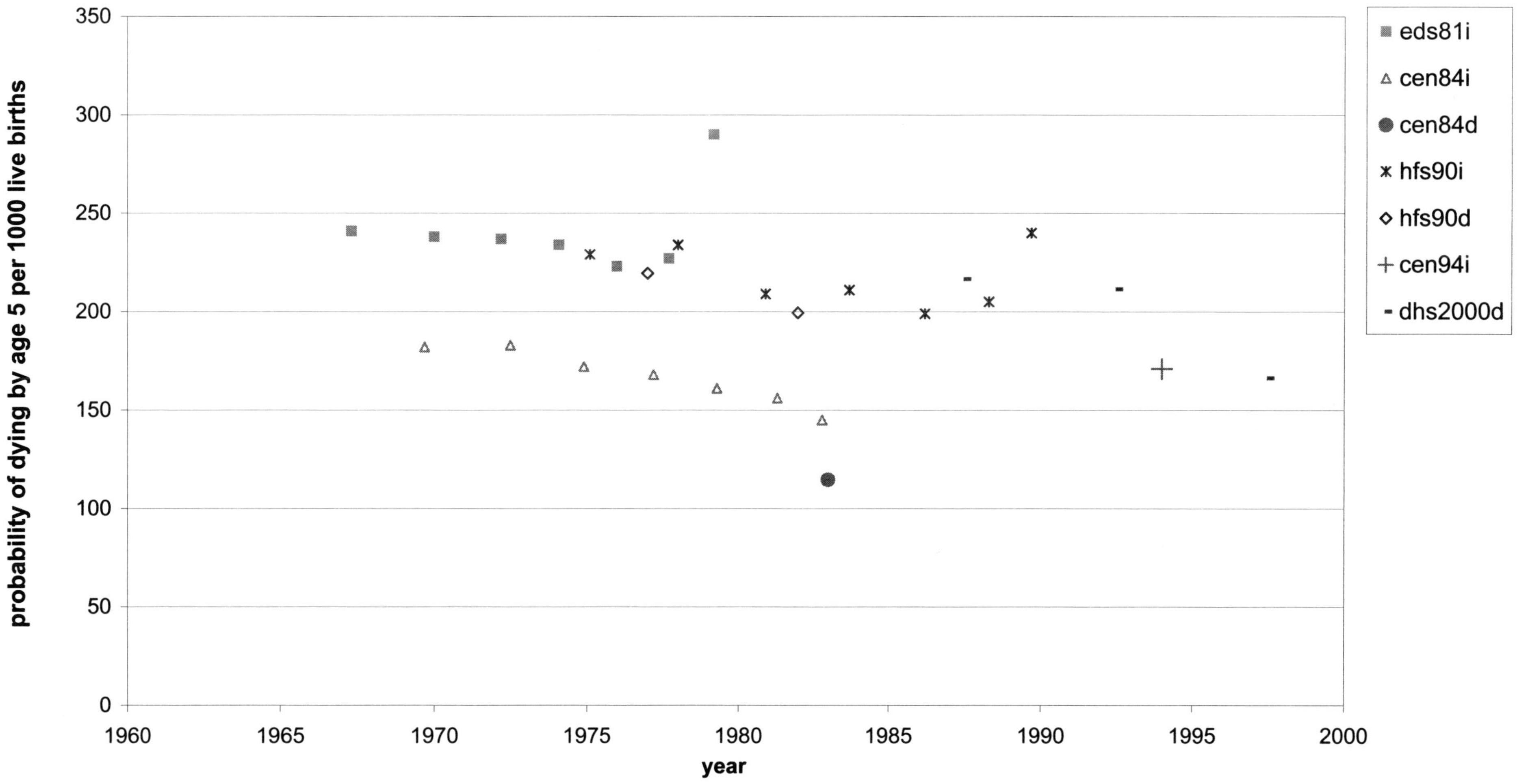

Data sources **Sources des données** **Fuentes de los datos**

eds81i: Demographic Survey 1981 (indirect method); ***cen84i***: Census, 9 May 1984 (indirect method); ***cen84d***: Census, 9 May 1984 (direct method); ***hfs90i***: National Family and Fertility Survey 1990 (indirect method); ***hfs90d***: National Family and Fertility Survey 1990 (direct method); ***cen94i***: Population and Housing Census 1994 (indirect method); ***dhs2000d***: Demographic and Health Survey 2000 (direct method)

Under 5 mortality

Fiji

Mortalité des moins de cinq ans

Fidji

Mortalidad de menores de cinco años

Fiji

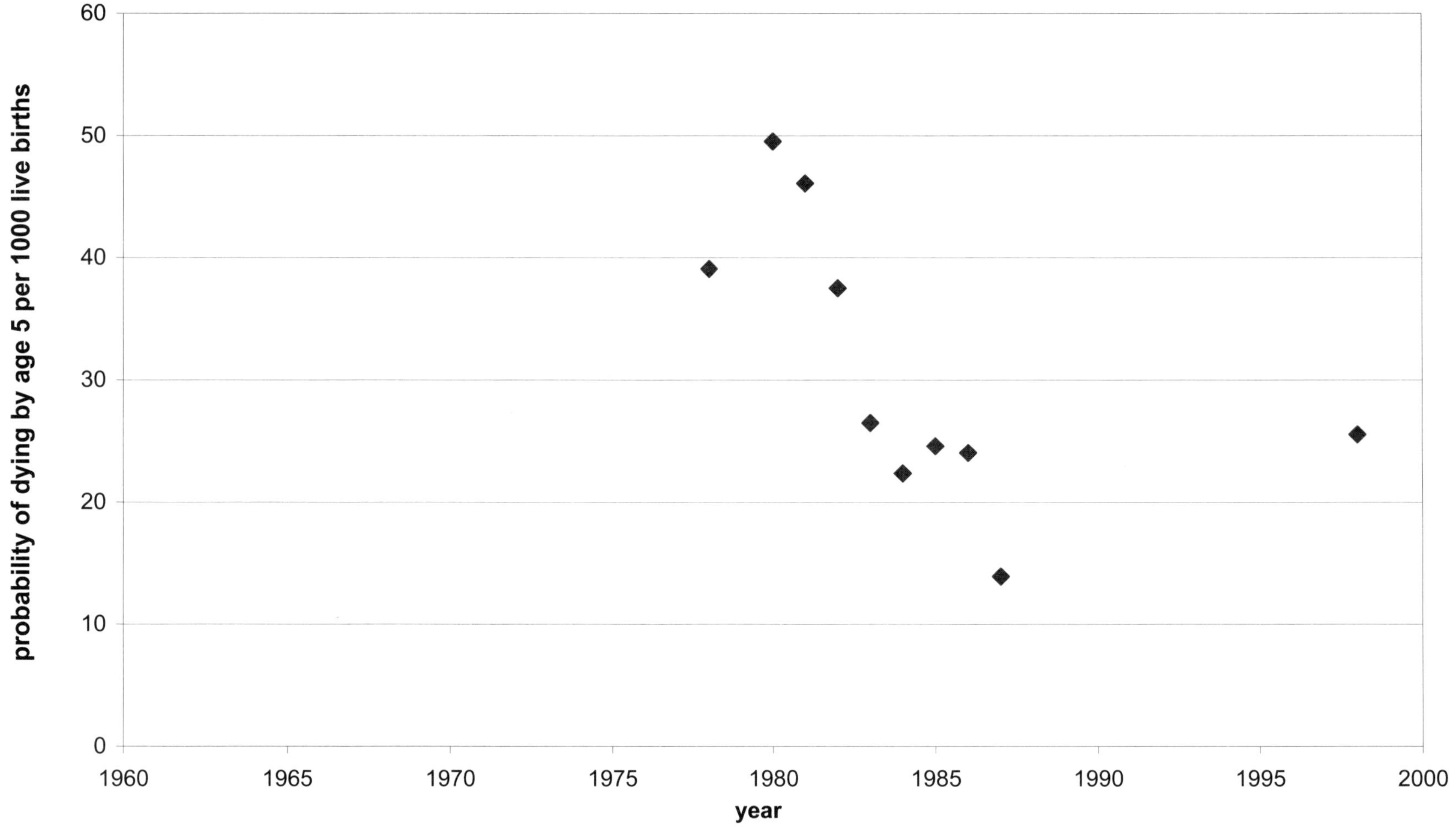

Data sources

Vital Registration

Sources des données

Fuentes de los datos

Under 5 mortality

Finland

Mortalité des moins de cinq ans

Finlande

Mortalidad de menores de cinco años

Finlandia

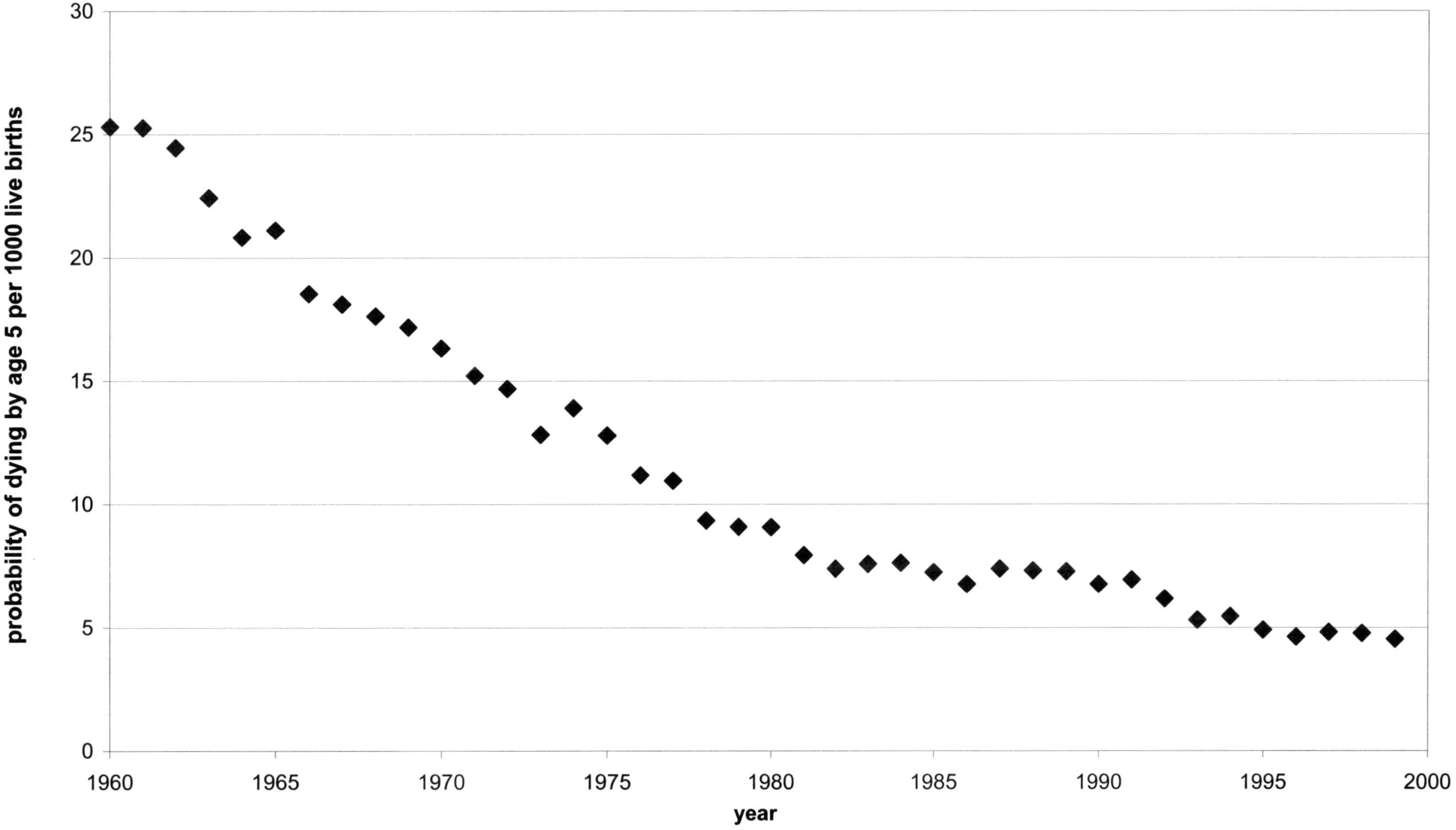

Data sources

Vital Registration

Sources des données

Fuentes de los datos

Under 5 mortality
France

Mortalité des moins de cinq ans
France

Mortalidad de menores de cinco años
Francia

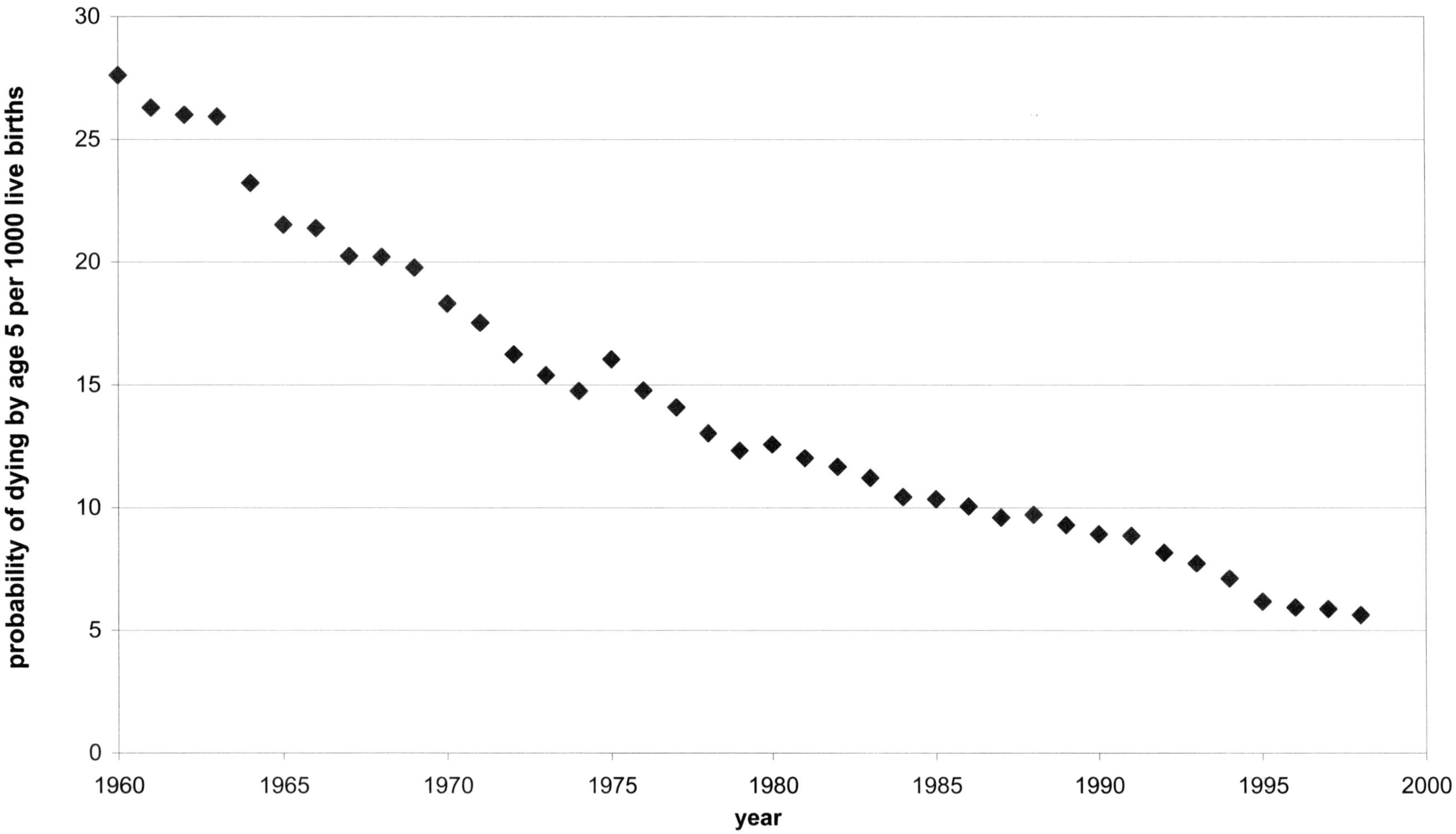

Data sources
Vital Registration

Sources des données

Fuentes de los datos

Under 5 mortality
Gabon

Mortalité des moins de cinq ans
Gabon

Mortalidad de menores de cinco años
Gabón

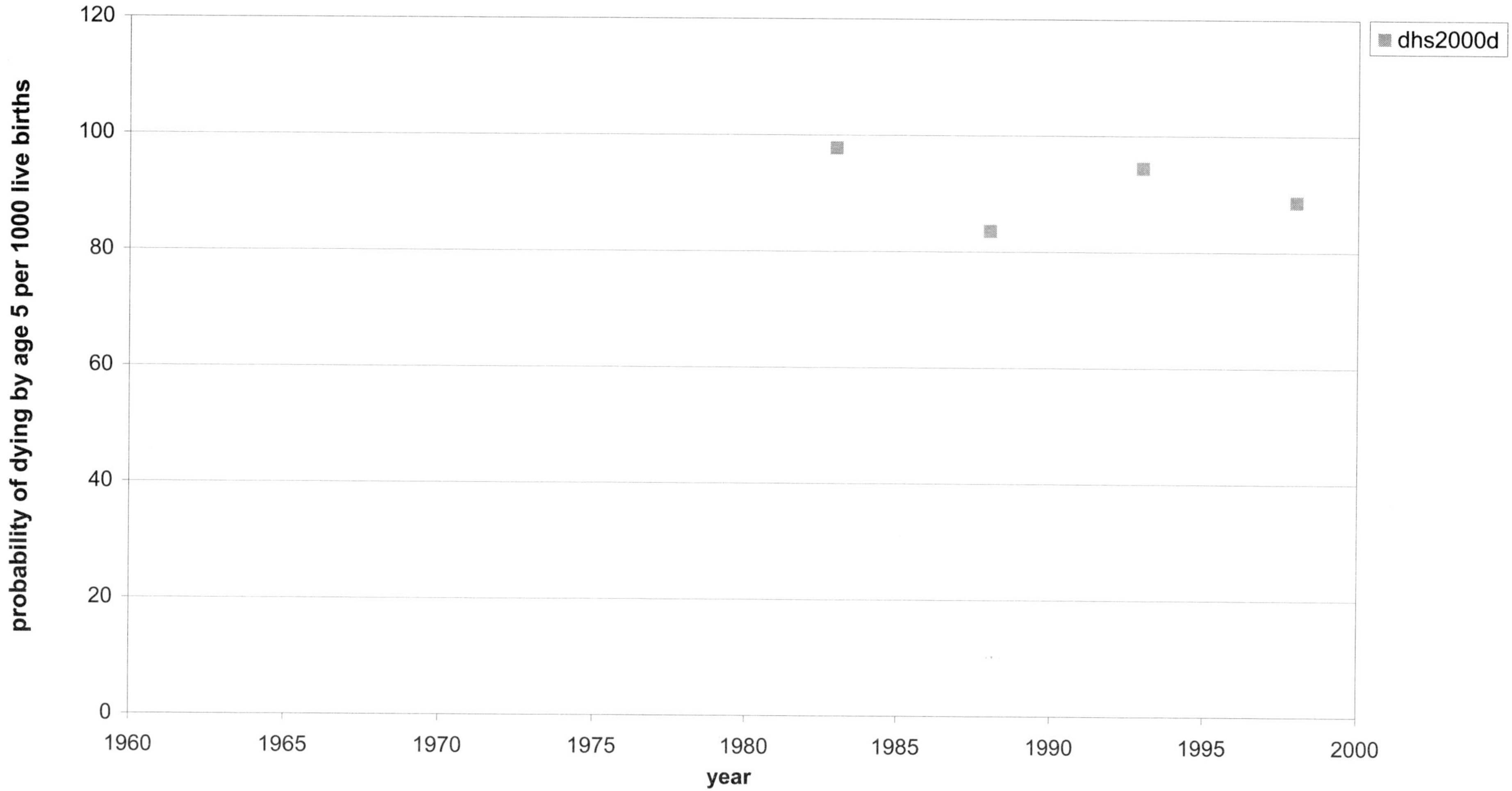

Data sources

Sources des données

Fuentes de los datos

dhs2000d : Demographic and Health Survey 2000 (direct method)

Under 5 mortality

Gambia

Mortalité des moins de cinq ans

Gambie

Mortalidad de menores de cinco años

Gambia

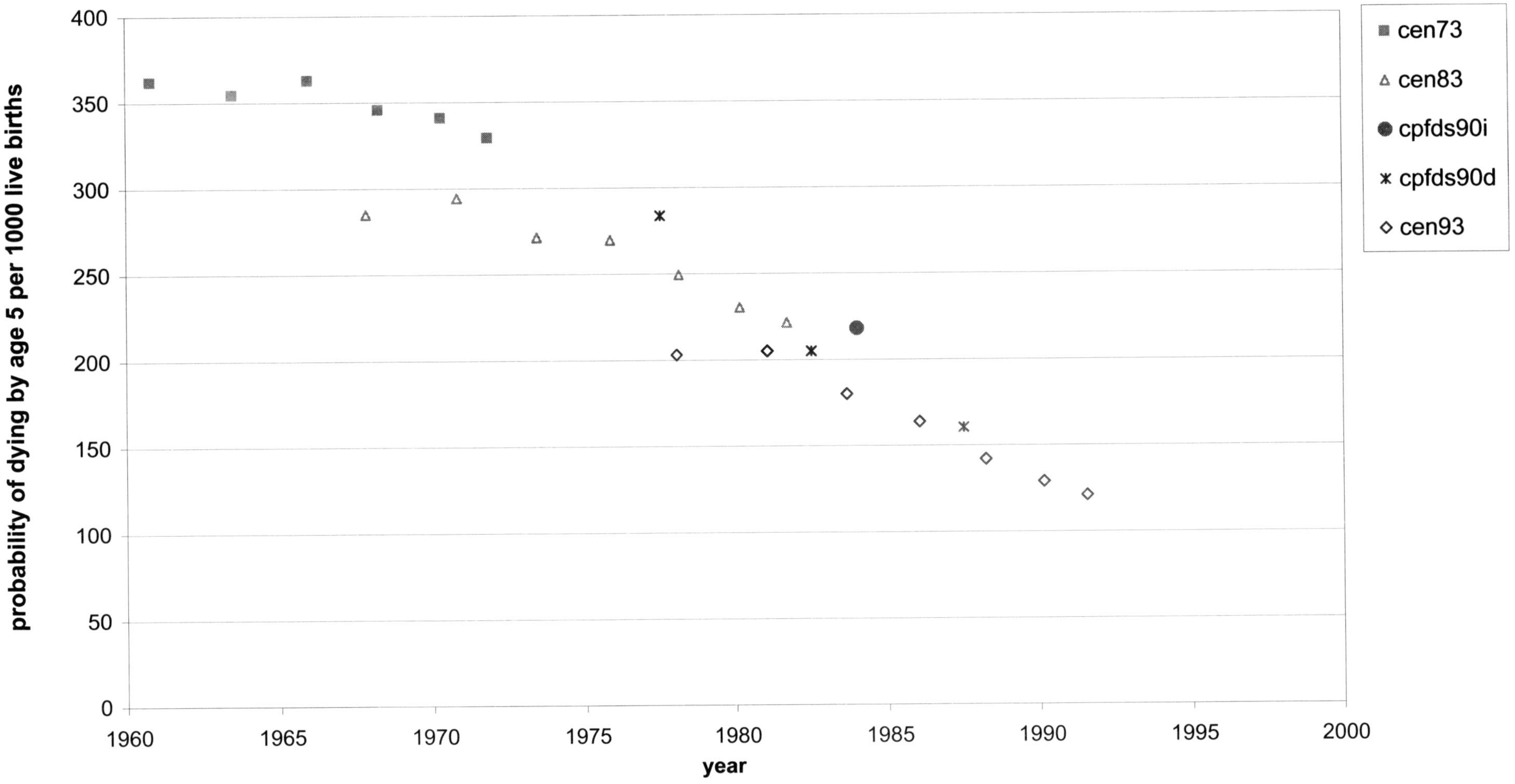

Data sources

Sources des données

Fuentes de los datos

cen73 : Census 1973; ***cen83*** : Census 1983; ***cpfds90*** i: Gambian Contraceptive Prevalence and Fertility Determinants Survey 1990 (indirect method); ***cpfds90d*** : Gambian Contraceptive Prevalence and Fertility Determinants Survey1990 (direct method); ***cen93*** : Census 1993

Under 5 mortality
Georgia

Mortalité des moins de cinq ans
Géorgie

Mortalidad de menores de cinco años
Georgia

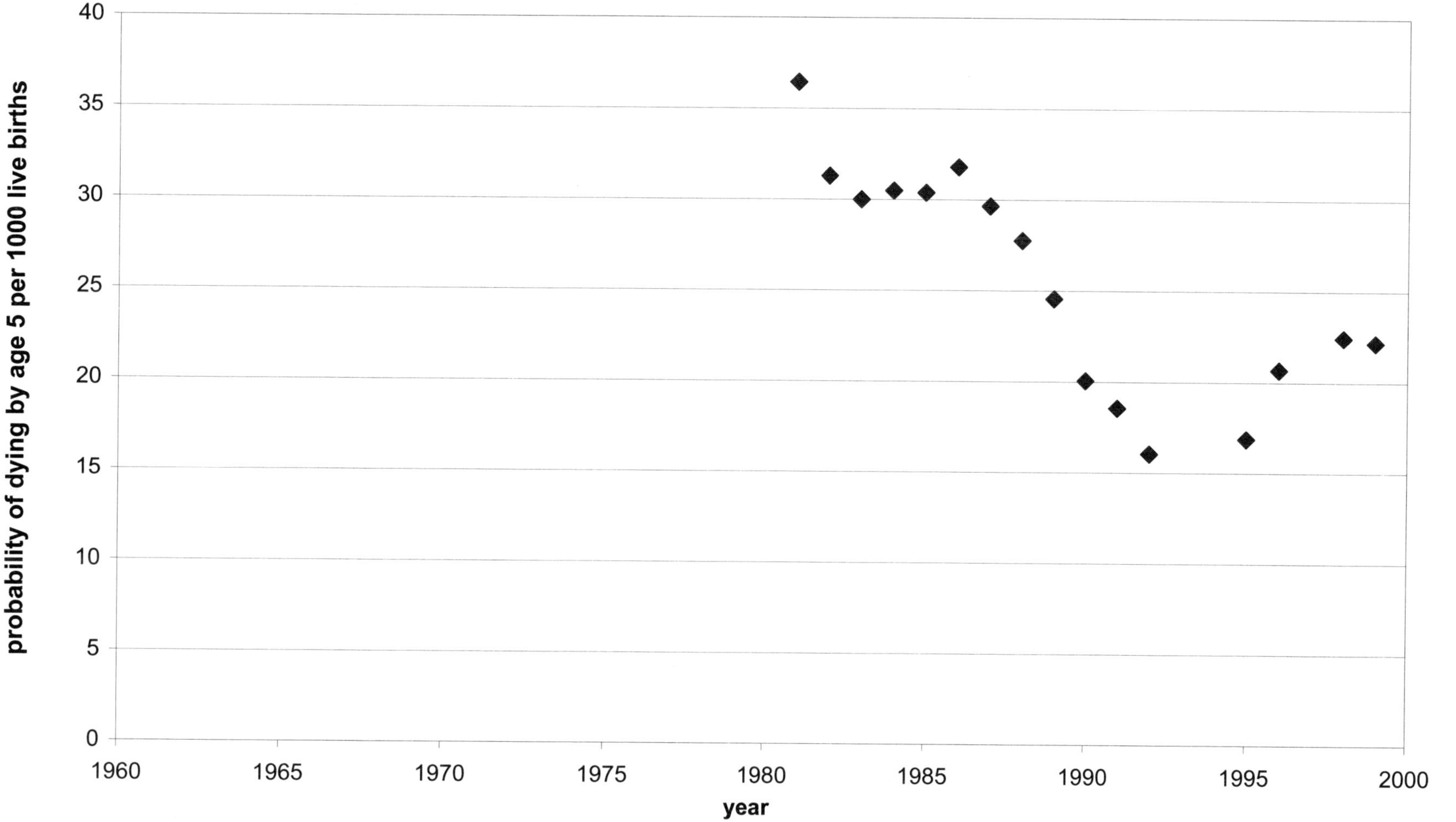

Data sources
Vital Registration

Sources des données

Fuentes de los datos

Under 5 mortality
Germany

Mortalité des moins de cinq ans
Allemagne

Mortalidad de menores de cinco años
Alemania

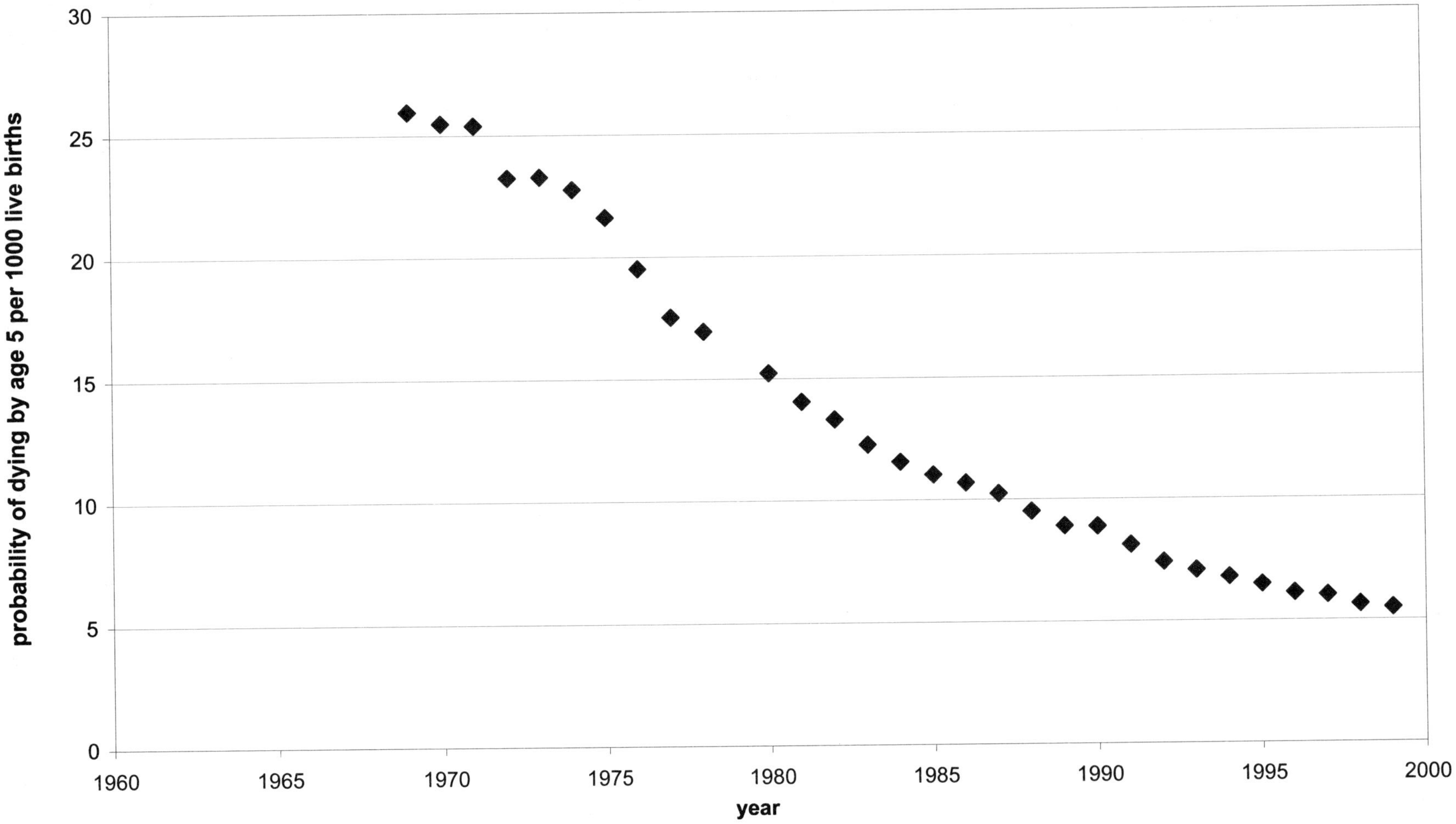

Data sources
Vital Registration

Sources des données

Fuentes de los datos

Under 5 mortality

Ghana

Mortalité des moins de cinq ans

Ghana

Mortalidad de menores de cinco años

Ghana

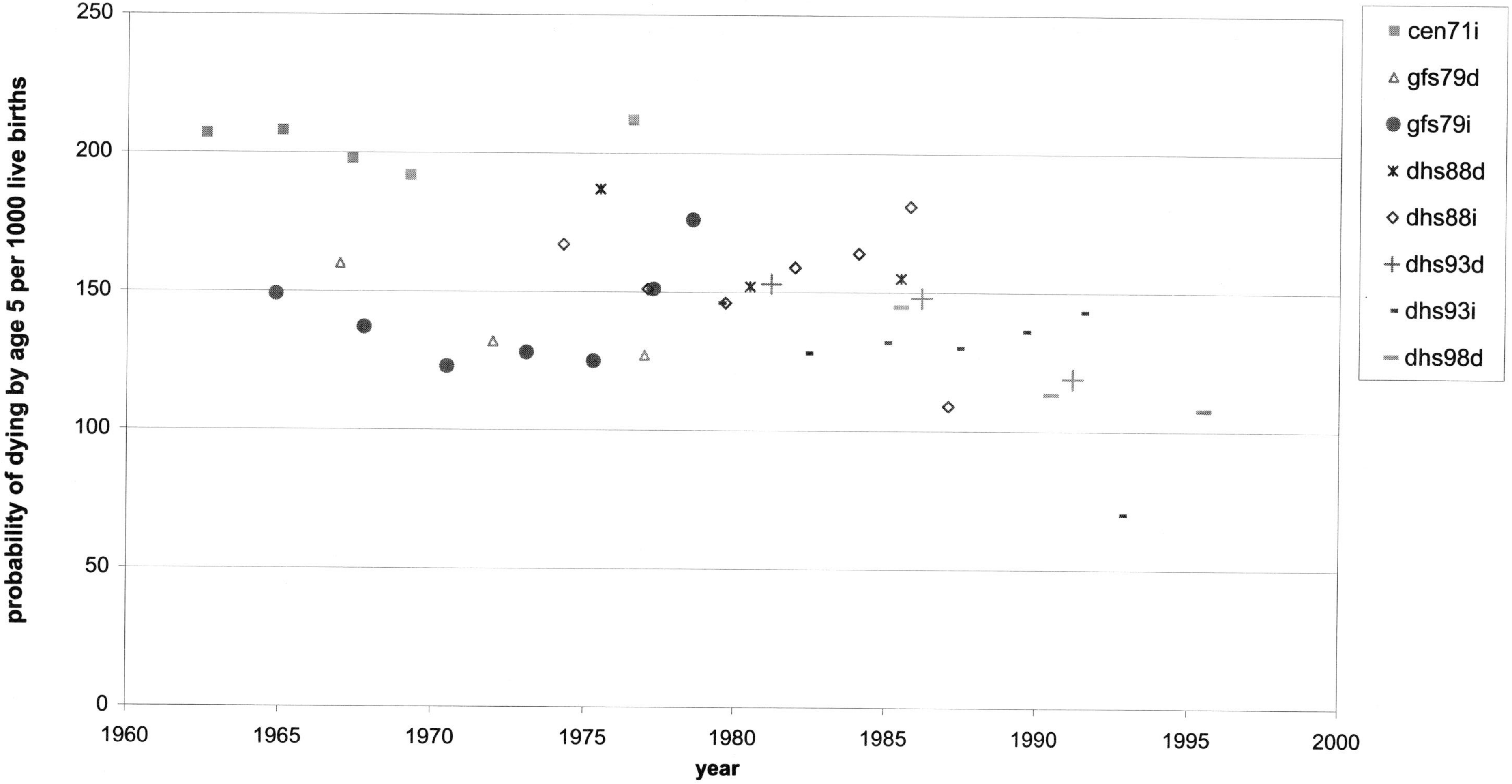

Data sources

Sources des données

Fuentes de los datos

cen71i: Supplementary Inquiry of Ghana, August-September 1971 (indirect method); ***gfs79d***: Ghana Fertility Survey, February 1979-March 1980 (direct method); ***gfs79i***: Ghana Fertility Survey, February 1979-March 1980 (indirect method); ***dhs88d***: Demographic and Health Survey, February-May 1988 (direct method); ***dhs88i***: Demographic and Health Survey, February-May 1988 (indirect method); ***dhs93d***: Demographic and Health Survey 1993 (direct method); ***dhs93i***: Demographic and Health Survey 1993 (indirect method); ***dhs98d***: Demographic and Health Survey 1998 (direct method)

Under 5 mortality
Greece

Mortalité des moins de cinq ans
Grèce

Mortalidad de menores de cinco años
Grecia

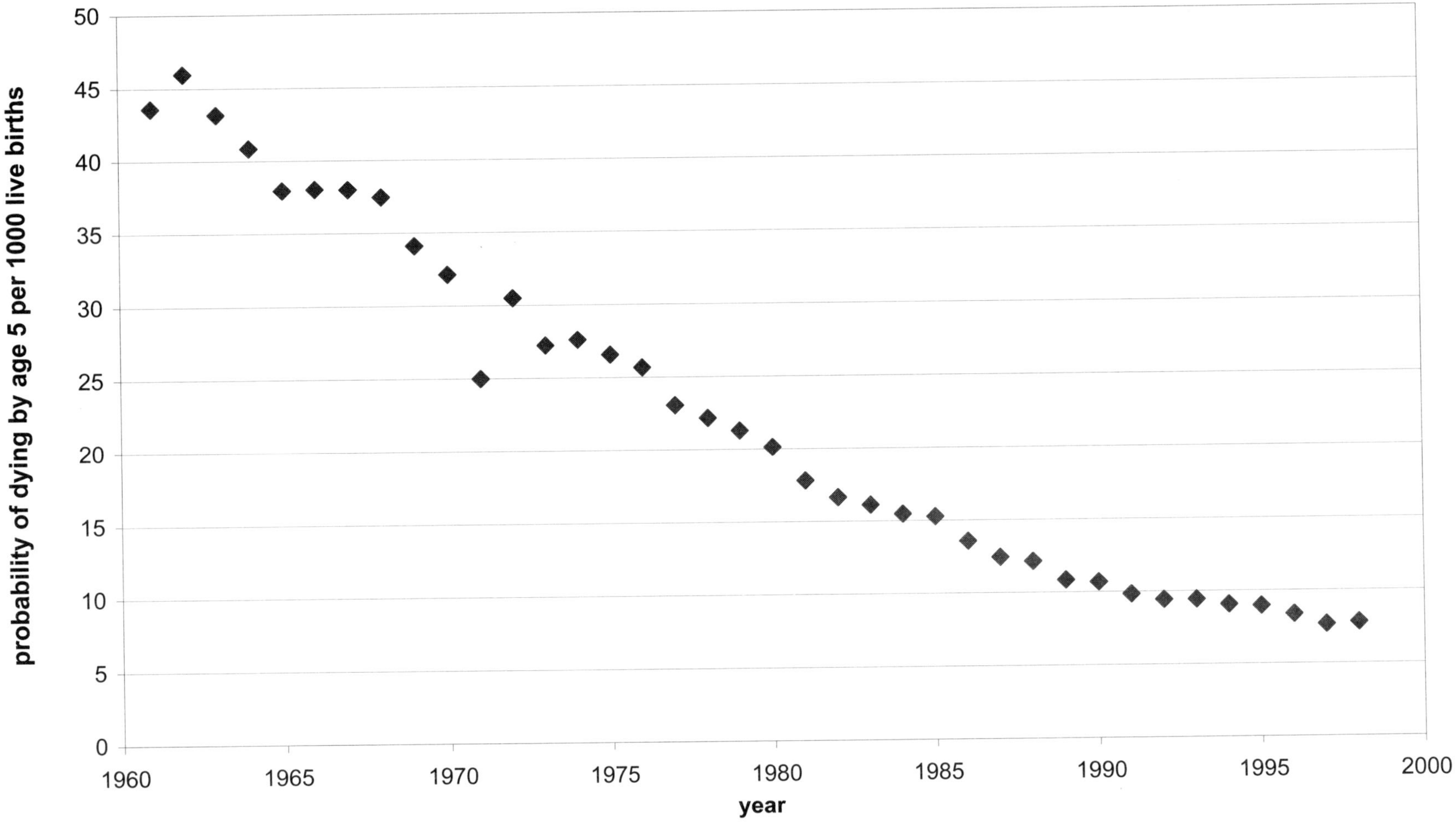

Data sources
Vital Registration

Sources des données

Fuentes de los datos

Under 5 mortality

Guatemala

Mortalité des moins de cinq ans

Guatemala

Mortalidad de menores de cinco años

Guatemala

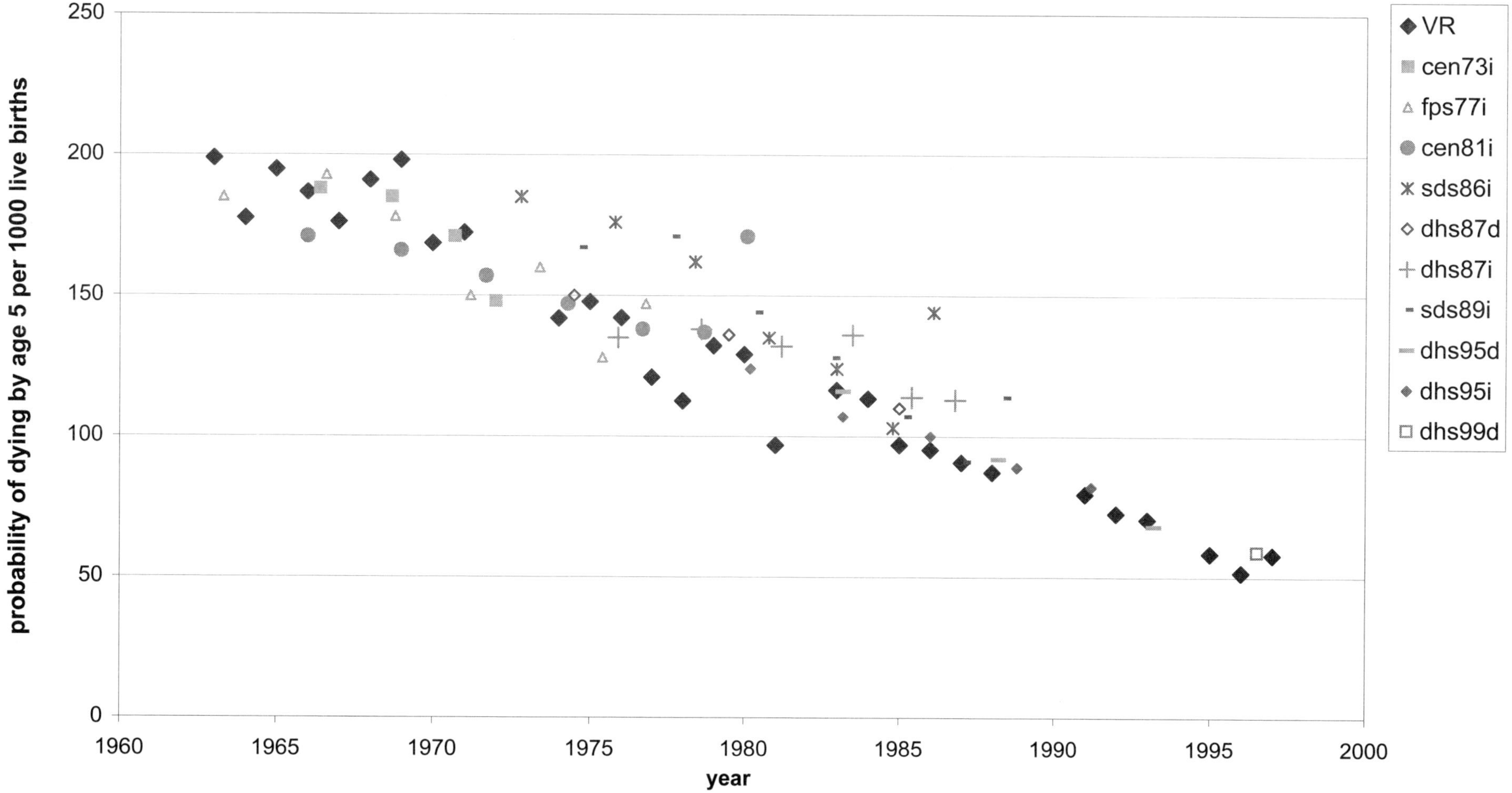

Data sources

Sources des données

Fuentes de los datos

VR: Vital Registration; ***cen73i***: Census, 26 March-7 April 1973 (indirect method); ***fps77i***: Encuesta Nacional de Fecundidad, Planificacion Familiar y Comunicacion, September 1977-August 1978 (indirect method); ***cen81i***: Census, 26 March 1981 (indirect method); ***sds86i***: Encuesta Nacional Socio-Demografica, October 1986-August 1987 (indirect method); ***dhs87d***: Encuesta Nacional de Salud Materno Infantil, September-December 1987 (direct method); ***dhs87i***: Encuesta Nacional de Salud Materno Infantil, September-December 1987 (indirect method); ***sds89i***: Encuesta Nacional Socio-Demografica, April-July 1989 (indirect method); ***dhs95d***: Demographic and Health Survey 1995 (direct method); ***dhs95i***: Demographic and Health Survey 1995 (indirect method); ***dhs99d***: Demographic and Health Survey 1999 - preliminary report (direct method)

Under 5 mortality

Guinea

Mortalité des moins de cinq ans

Guinée

Mortalidad de menores de cinco años

Guinea

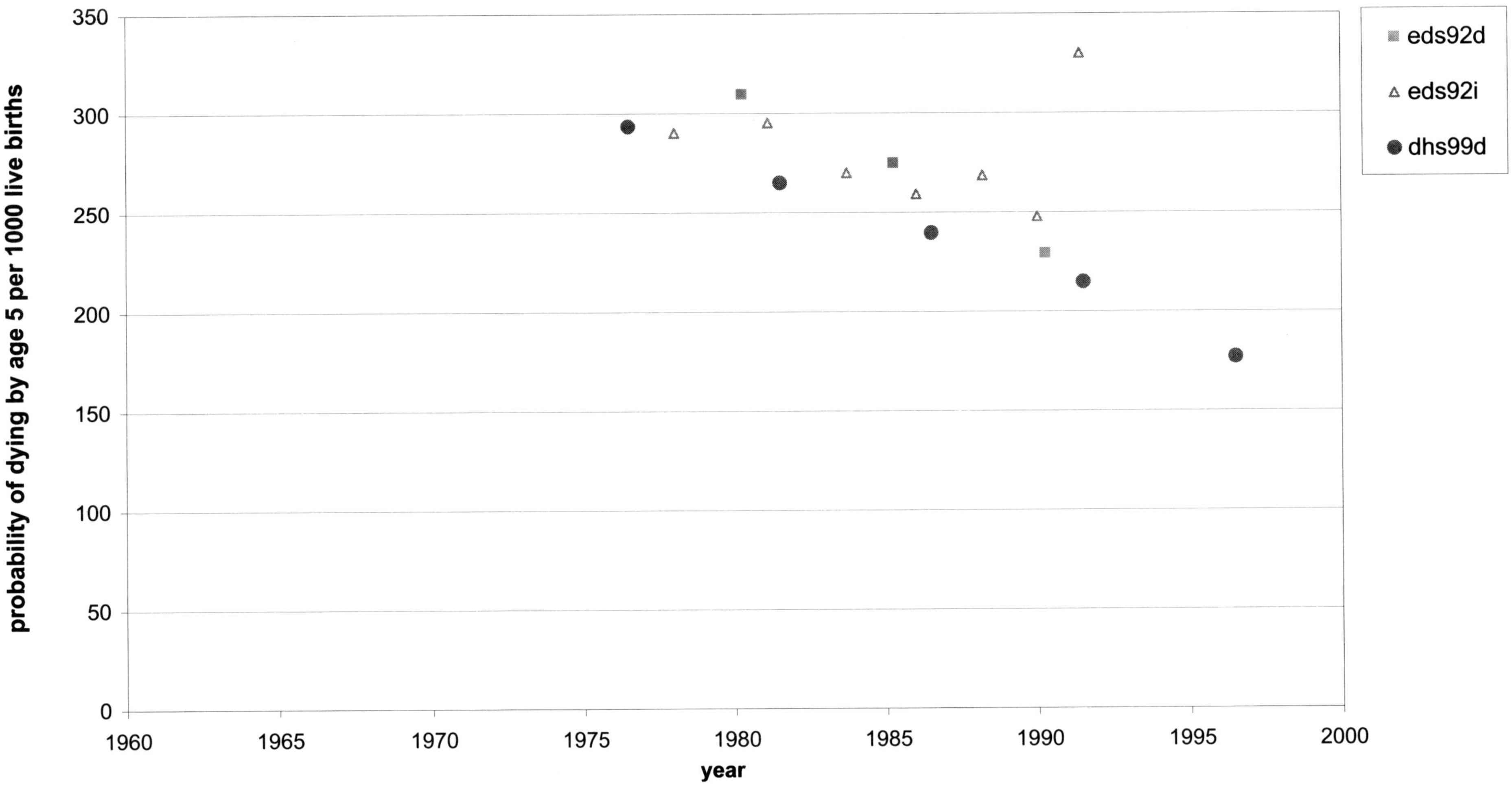

Data sources

Sources des données

Fuentes de los datos

eds92d: Enquête Démographique et de Santé, May 1992-March 1993 (direct method); ***eds92i***: Enquête Démographique et de Santé, May 1992-March 1993 (indirect method); ***dhs99d***: Demographic and Health Survey 1999 (direct method)

Under 5 mortality
Guinea-Bissau

Mortalité des moins de cinq ans
Guinée-Bissau

Mortalidad de menores de cinco años
Guinea-Bissau

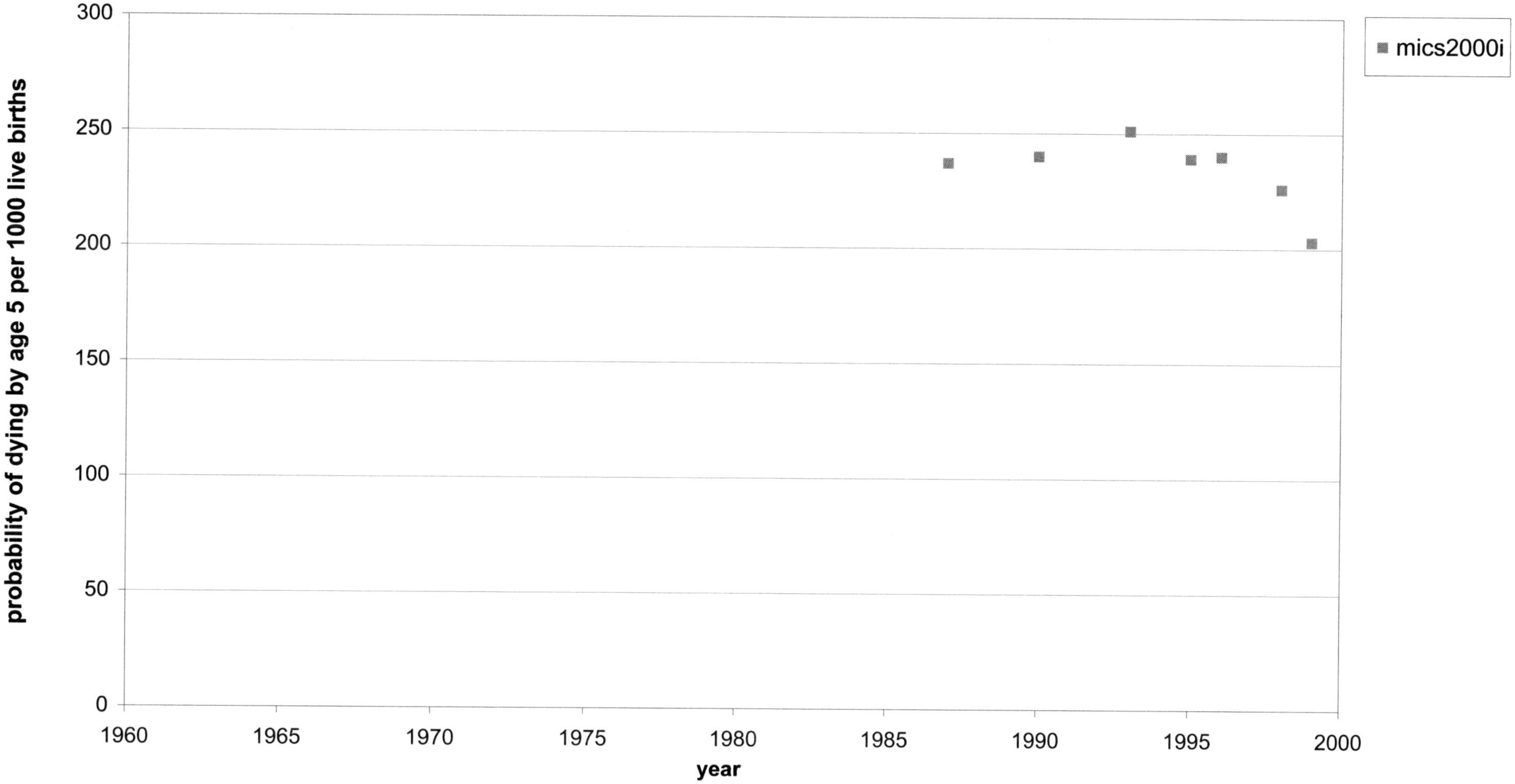

Data sources

Sources des données

Fuentes de los datos

mics2000i : Multiple Indicator Cluster Survey 2000 (indirect method)

Under 5 mortality
Guyana

Mortalité des moins de cinq ans
Guyana

Mortalidad de menores de cinco años
Guyana

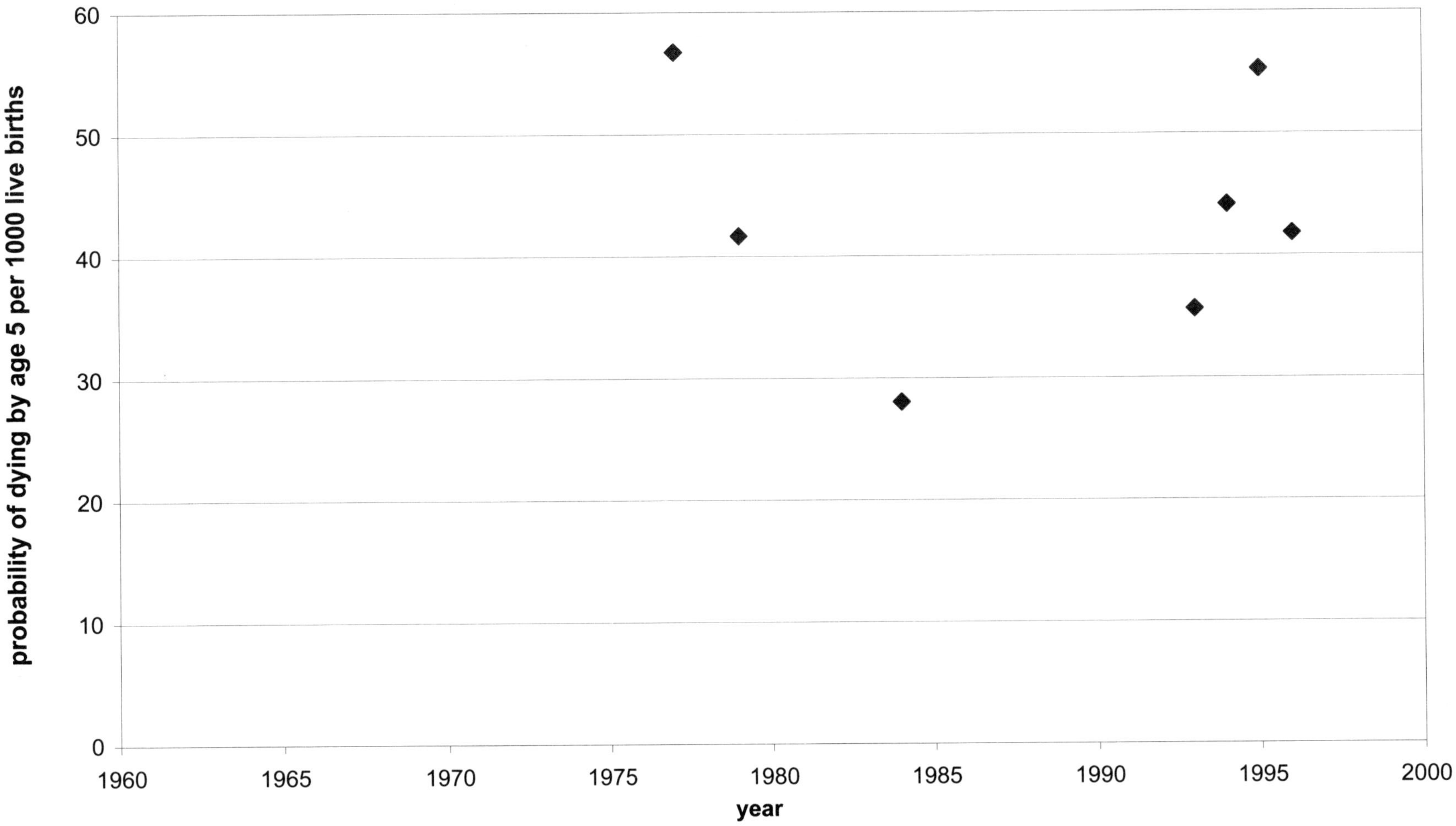

Data sources
Vital Registration

Sources des données

Fuentes de los datos

Under 5 mortality **Haiti** | Mortalité des moins de cinq ans **Haïti** | Mortalidad de menores de cinco años **Haití**

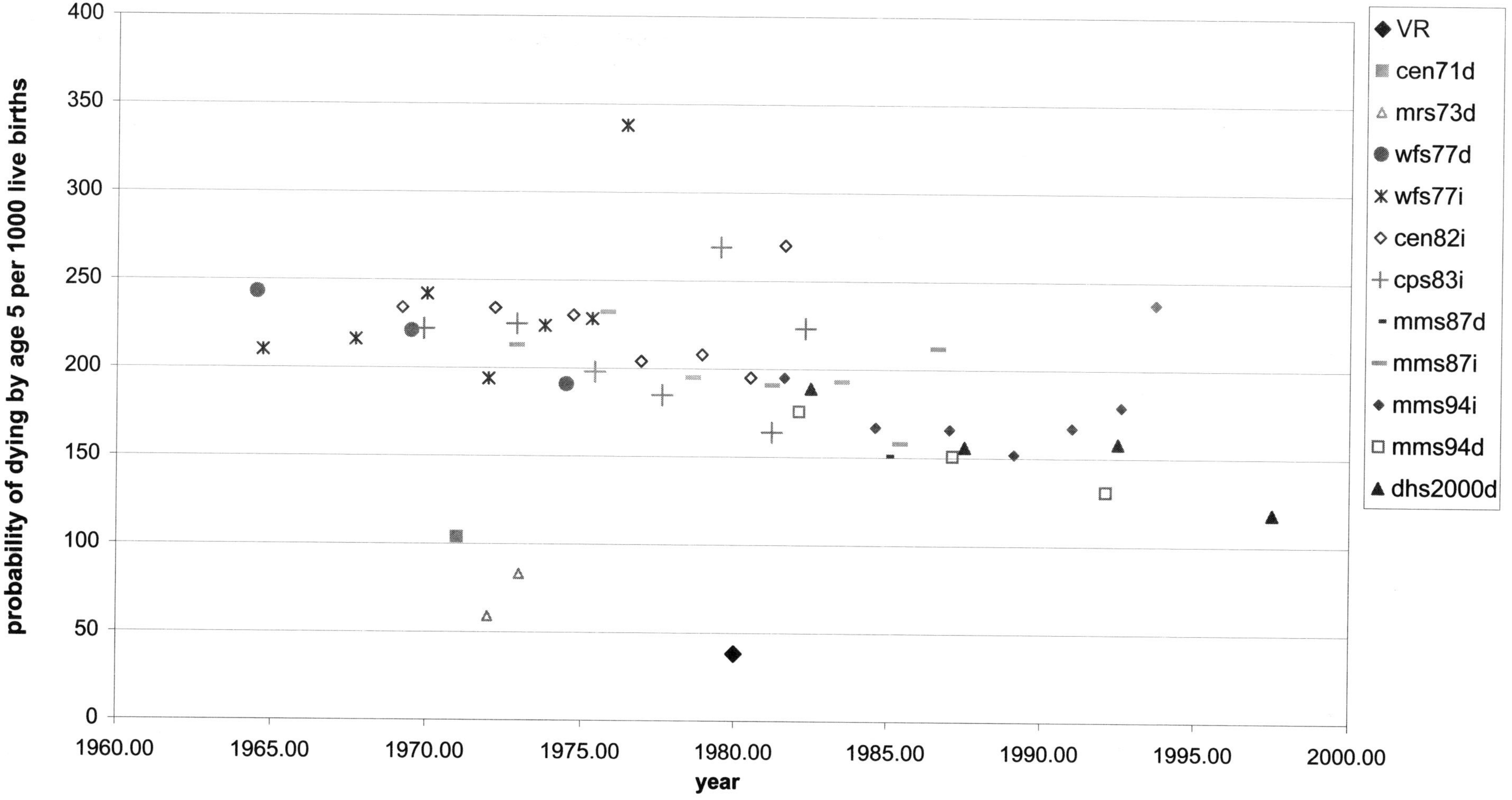

Data sources | **Sources des données** | **Fuentes de los datos**

VR: Vital Registration; ***cen71d***: Census, 31 August 1971 (direct method); ***mrs73d***: Enquête Démographique à Passages Répétés, August 1971-November 1973 (direct method); ***wfs77d***: Enquête Haitienne sur la Fécondité, July-September 1977 (direct method); ***wfs77i***: Enquête Haitienne sur la Fécondité, July-September 1977 (indirect method); ***cen82i***: Census, 30 August 1982 (indirect method); ***cps83i***: Enquête Haitienne sur la Prévalence de la Contraception, 1983 (indirect method); ***mms87d***: Enquête Mortalité, Morbidité et Utilisation des Services, May-September 1987 (direct method); ***mms87i***: Enquête Mortalité, Morbidité et Utilisation des Services, May-September 1987 (indirect method); ***mms94i***: Enquête Mortalité, Morbidité et Utilisation des Services II, July1994-January 1995 (indirect method); ***mms94d***: Enquête Mortalité, Morbidité et Utilisation des Services II, July1994-January 1995 (direct method); ***dhs2000d***: Demographic and Health Survey 2000 (direct method)

Under 5 mortality

Honduras

Mortalité des moins de cinq ans

Honduras

Mortalidad de menores de cinco años

Honduras

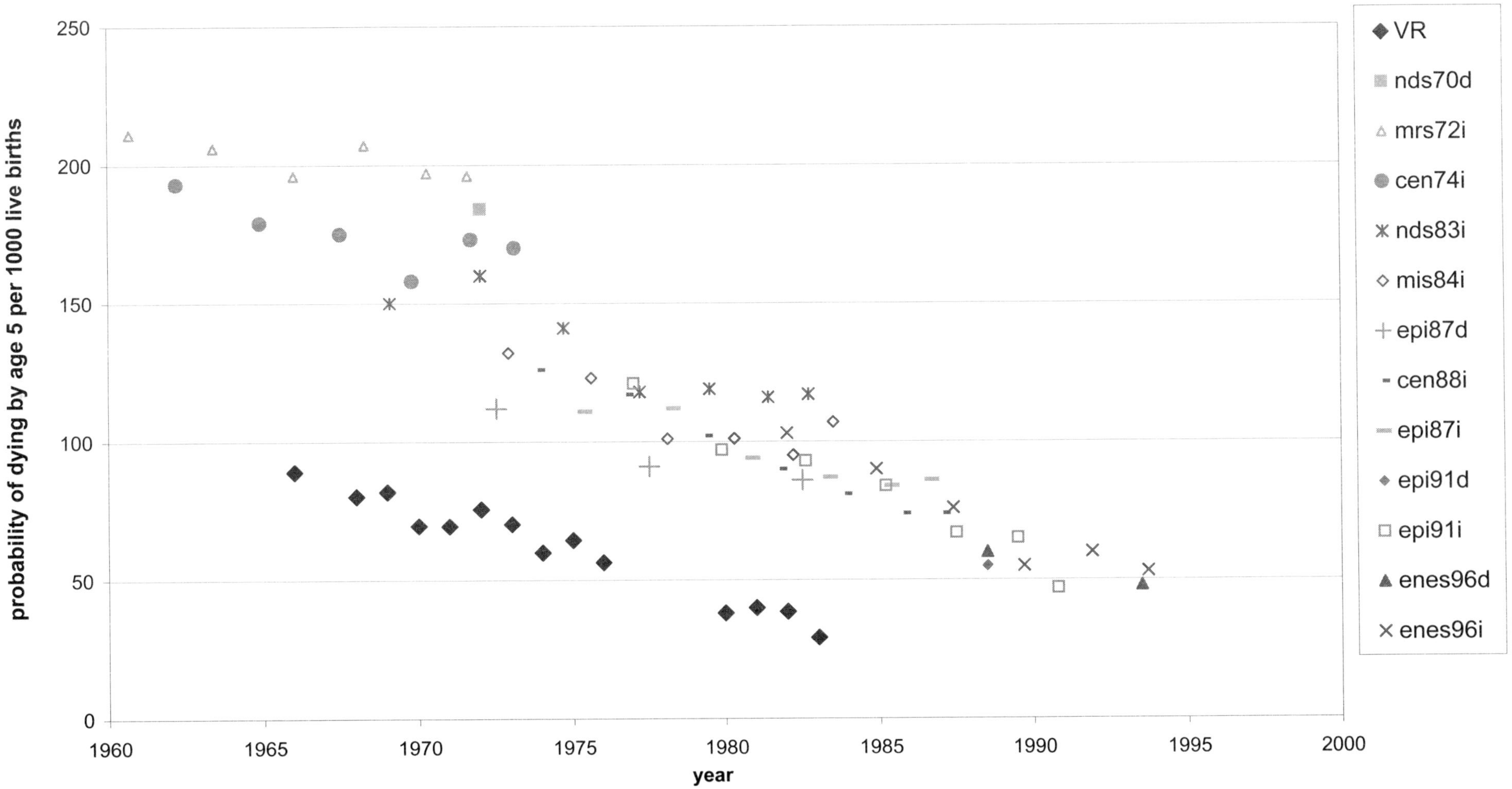

Data sources **Sources des données** **Fuentes de los datos**

VR: Vital Registration; ***nds70d***: Encuesta Demografica Nacional de Honduras, December 1970-October 1972 (direct method); ***mrs72i***: Encuesta Demografica Nacional Retrospectiva, July-October 1972 (indirect method); ***cen74i***: Census, 6 March 1974 (indirect method); ***nds83i***: Encuesta Demografica Nacional de Honduras, July 1983-January 1984 (indirect method); ***mis84i***: Encuesta Nacional de Salud Materno Infantil, February 1984-February 1985 (indirect method); ***epi87d***: Encuesta Nacional de Epidemiologia y Salud Familiar, June-November 1987 (direct method); ***cen88i***: Census, 29 May 1988 (indirect method); ***epi87i***: Encuesta Nacional de Epidemiologia y Salud Familiar, June-November 1987 (indirect method); ***epi91d***: Encuesta Nacional de Epidemiologia y Salud Familiar, September 1991-February 1992 (direct method); ***epi91i***: Encuesta Nacional de Epidemiologia y Salud Familiar, September 1991-February 1992 (indirect method); ***enes96d***: Encuesta nacional de epidemiología y salud familiar 1996 (direct method); ***enes96i***: Encuesta nacional de epidemiología y salud familiar 1996 (indirect method)

Under 5 mortality

Hungary

Mortalité des moins de cinq ans

Hongrie

Mortalidad de menores de cinco años

Hungría

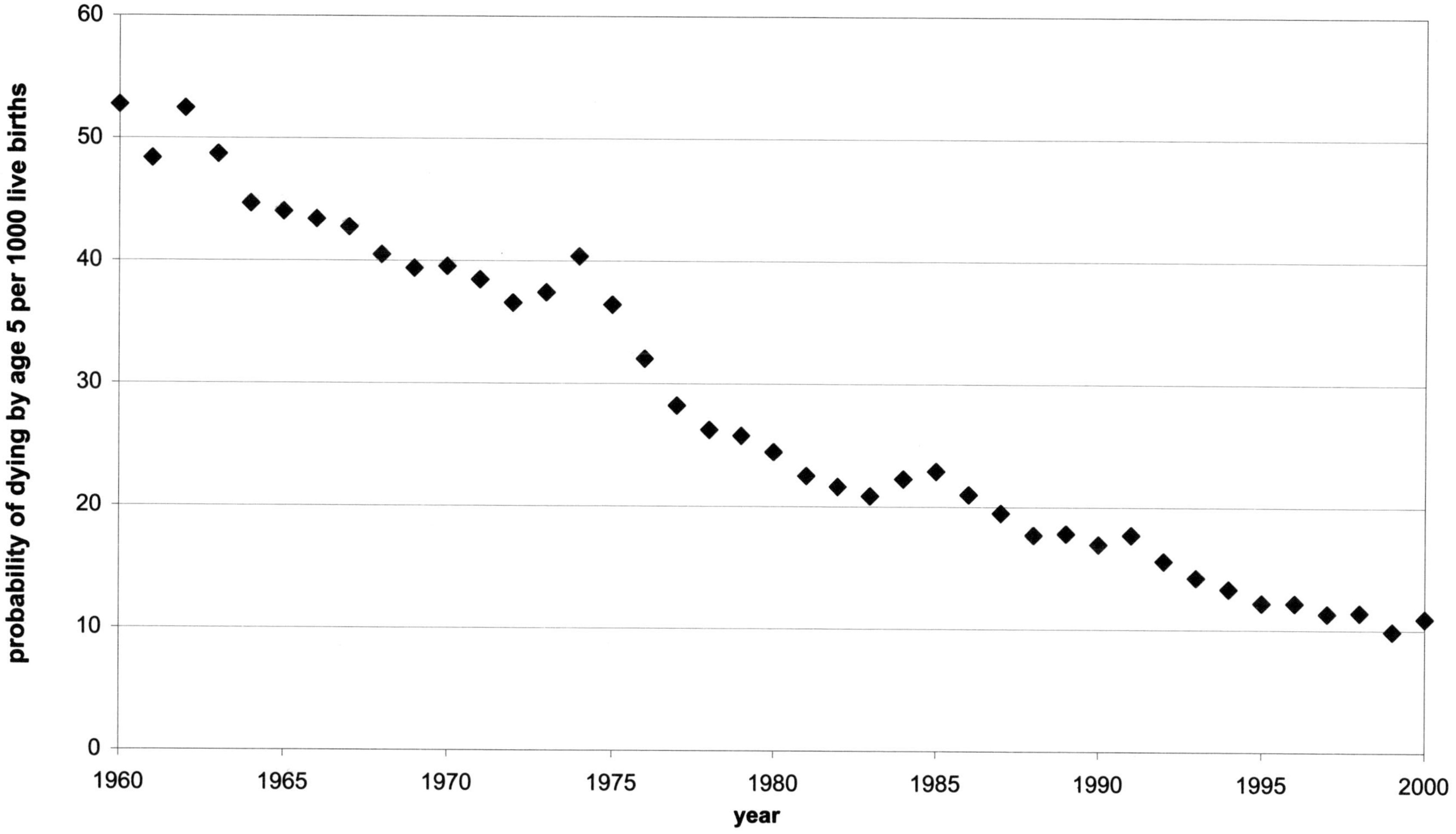

Data sources

Vital Registration

Sources des données

Fuentes de los datos

Under 5 mortality | Mortalité des moins de cinq ans | Mortalidad de menores de cinco años

Iceland | **Islande** | **Islandia**

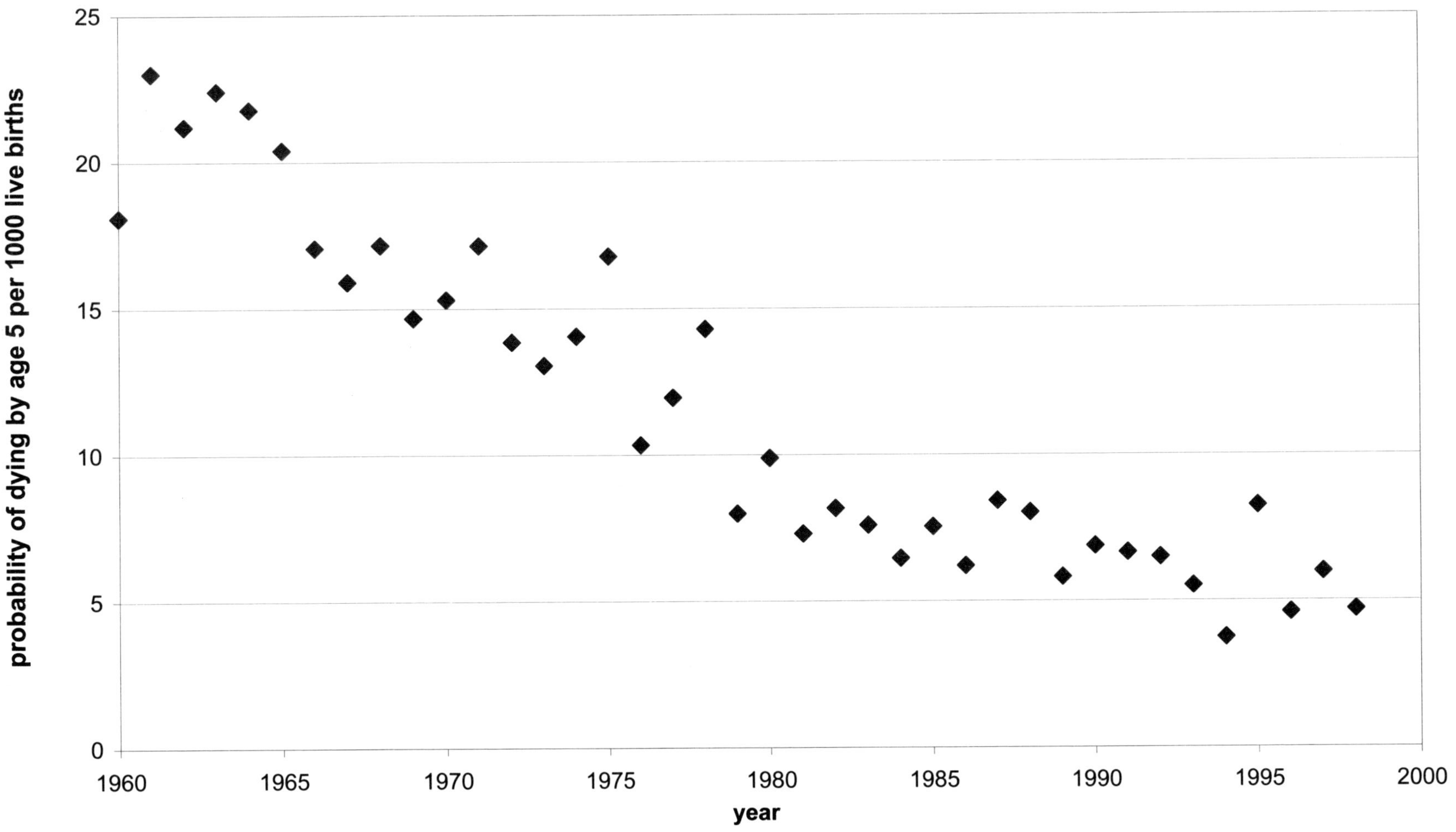

Data sources | **Sources des données** | **Fuentes de los datos**

Vital Registration

Under 5 mortality
India

Mortalité des moins de cinq ans
Inde

Mortalidad de menores de cinco años
India

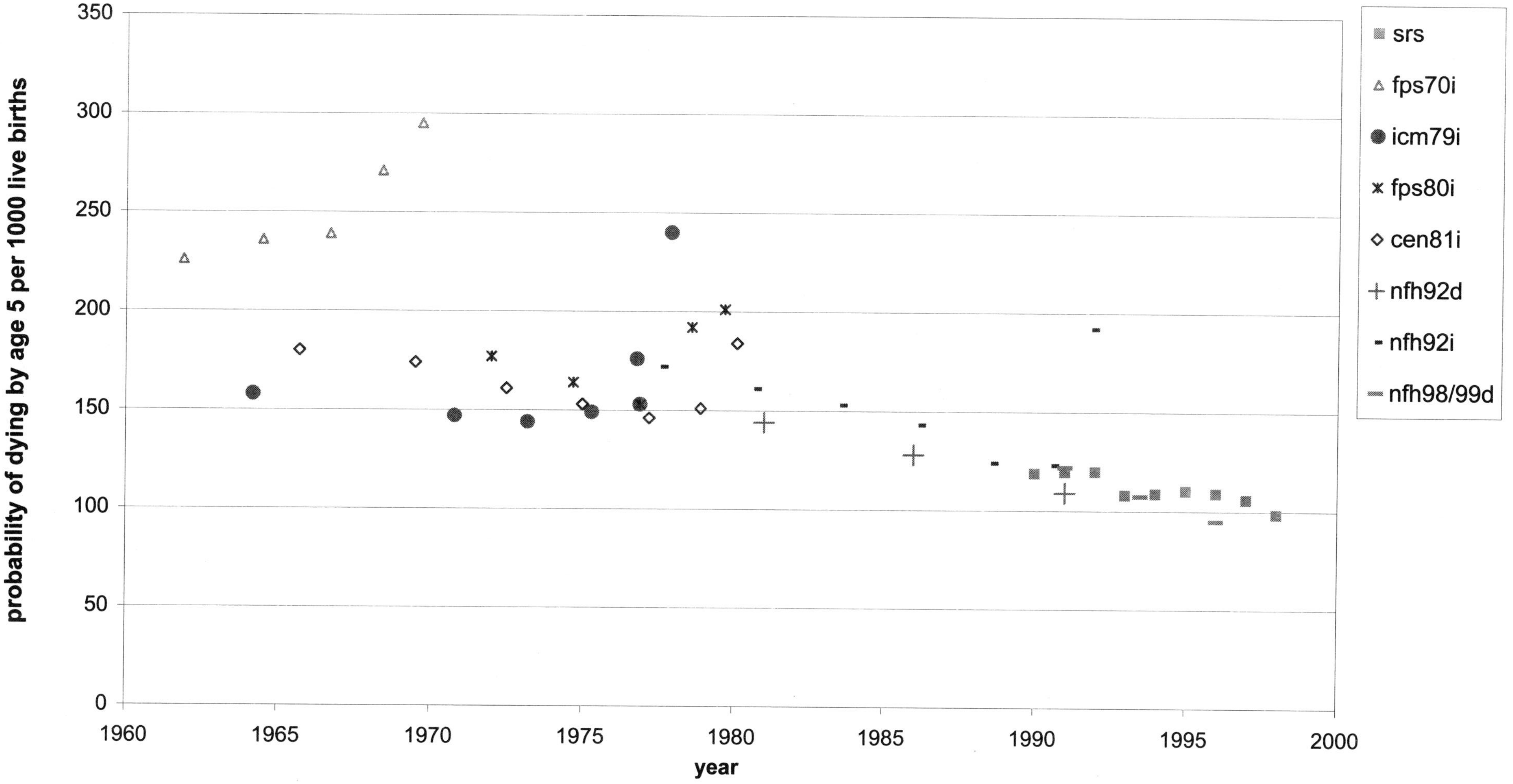

Data sources **Sources des données** **Fuentes de los datos**

srs: Sample Registration System; ***fps70i***: National Family Planning Survey, July 1970-January 1971 (indirect method); ***icm79i***: Survey on Infant and Child Mortality, 1979 (indirect method); ***fps80i***: Second All-India Family Planning Survey, 1980-1981 (indirect method); ***cen81i***: Census, 1 March 1981 (indirect method); ***nfh92d***: National Famliy Health Survey April 1992-September1993 (direct method); ***nfh92i***: National Famliy Health Survey April 1992-September1993 (indirect method); ***nfh98d***: National Famliy Health Survey 1998-1999 (direct method)

Under 5 mortality

Indonesia

Mortalité des moins de cinq ans

Indonésie

Mortalidad de menores de cinco años

Indonesia

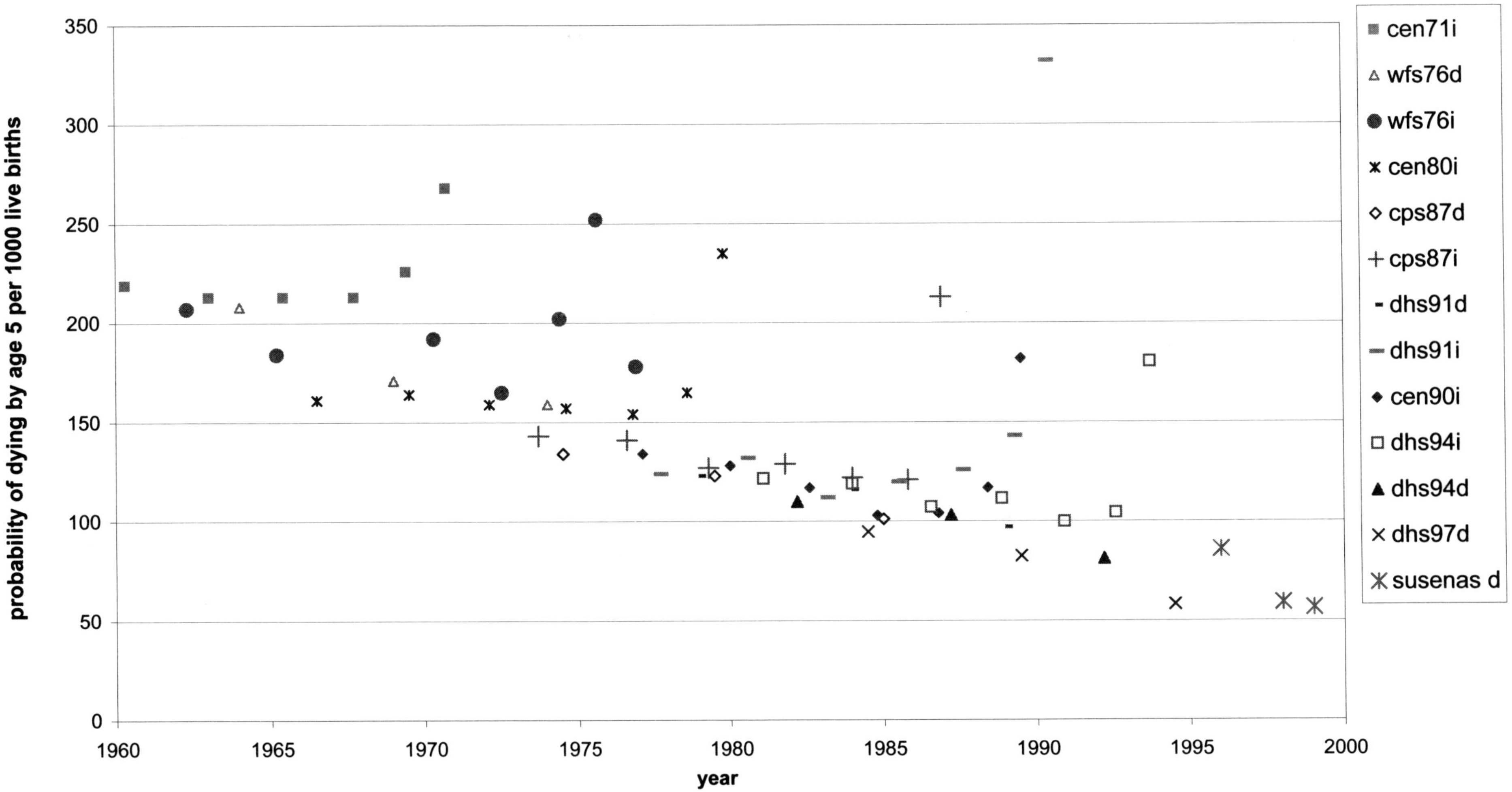

Data sources **Sources des données** **Fuentes de los datos**

cen71i: Census, 24 September 1971 (indirect method); ***wfs76d***: Fertility Survey, April-May 1976 (direct method); ***wfs76i***: Fertility Survey, April-May 1976 (indirect method); ***cen80i***: Census, 31 October 1980 ; ***cps87d***: National Indonesia Contraceptive Prevalence Survey, September-December 1987 (direct method); ***cps87i***: National Indonesia Contraceptive Prevalence Survey, September-December 1987 (indirect method); ***dhs91d***: Demographic and Health Survey, May-June 1991 (direct method); ***dhs91i***: Demographic and Health Survey, May-June 1991 (indirect method); ***cen90i***: Census, 31 December 1990 (indirect method); ***dhs94i***: Demographic and Health Survey, July-November 1994 (indirect method); ***dhs94d***: Demographic and Health Survey, July-November 1994 (direct method); ***dhs97d***: Demographic and Health Survey, 1997 (direct method); ***susenas***: Annual National Socio-Economic Survey (direct method)

Under 5 mortality
Iran (Islamic Republic of)

Mortalité des moins de cinq ans
Iran (République islamique d')

Mortalidad de menores de cinco años
Irán (República Islámica del)

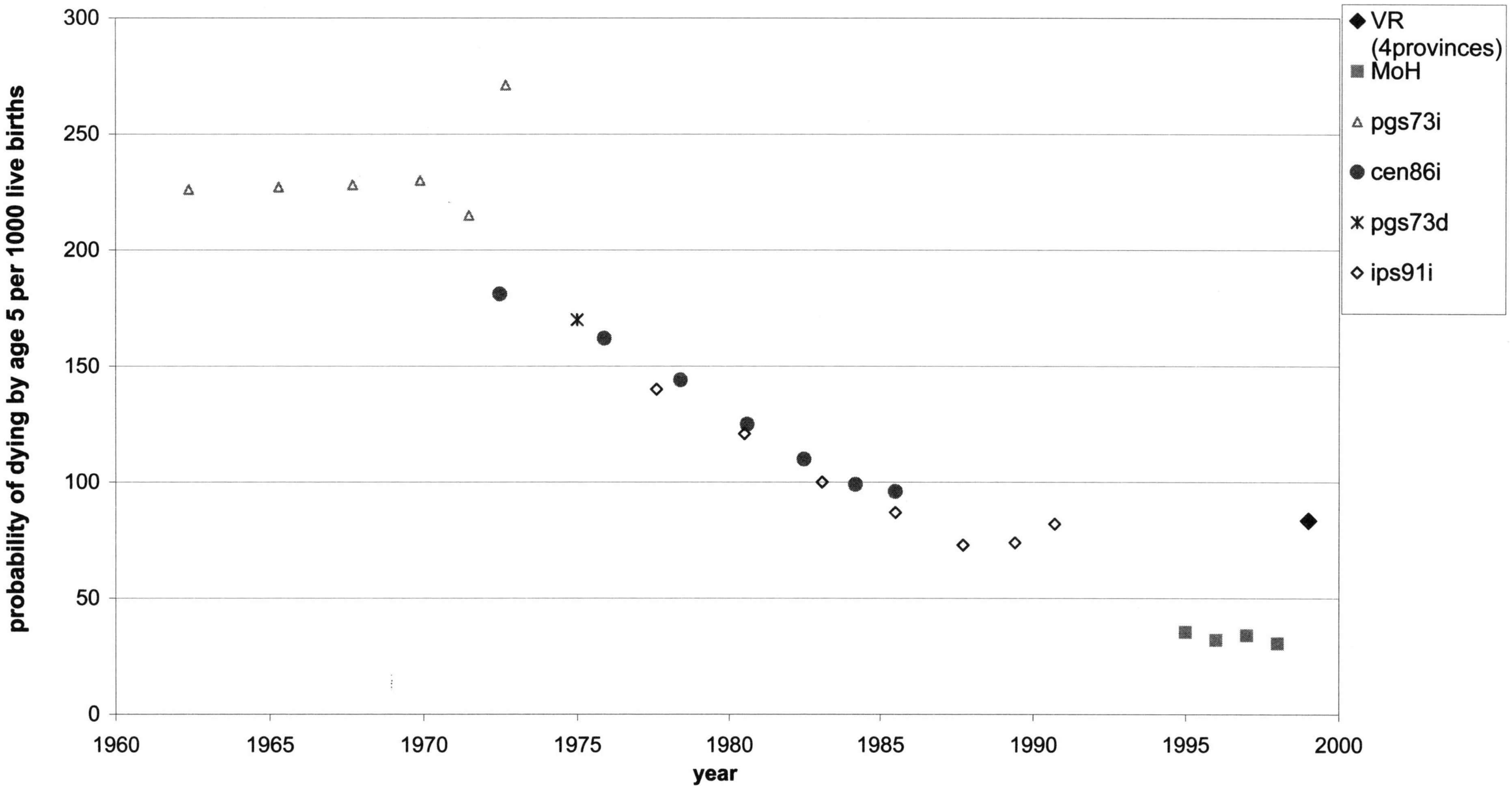

Data sources

Sources des données

Fuentes de los datos

VR (4provinces) : Vital Registration (4 Provinces); ***MoH*** : Ministry of Health rural Vital Registration System (Vital Horoscope); ***pgs73i*** : Baseline Population Growth Survey, October 1973 (indirect method); ***cen86i*** : Census, 22 September 1986 (indirect method); ***pgs73d*** : Population Growth Survey of Iran, 1973-1976 (direct method); ***ips91i*** : Intercensal Population Survey 1991 (indirect method)

Under 5 mortality

Iraq

Mortalité des moins de cinq ans

Iraq

Mortalidad de menores de cinco años

Iraq

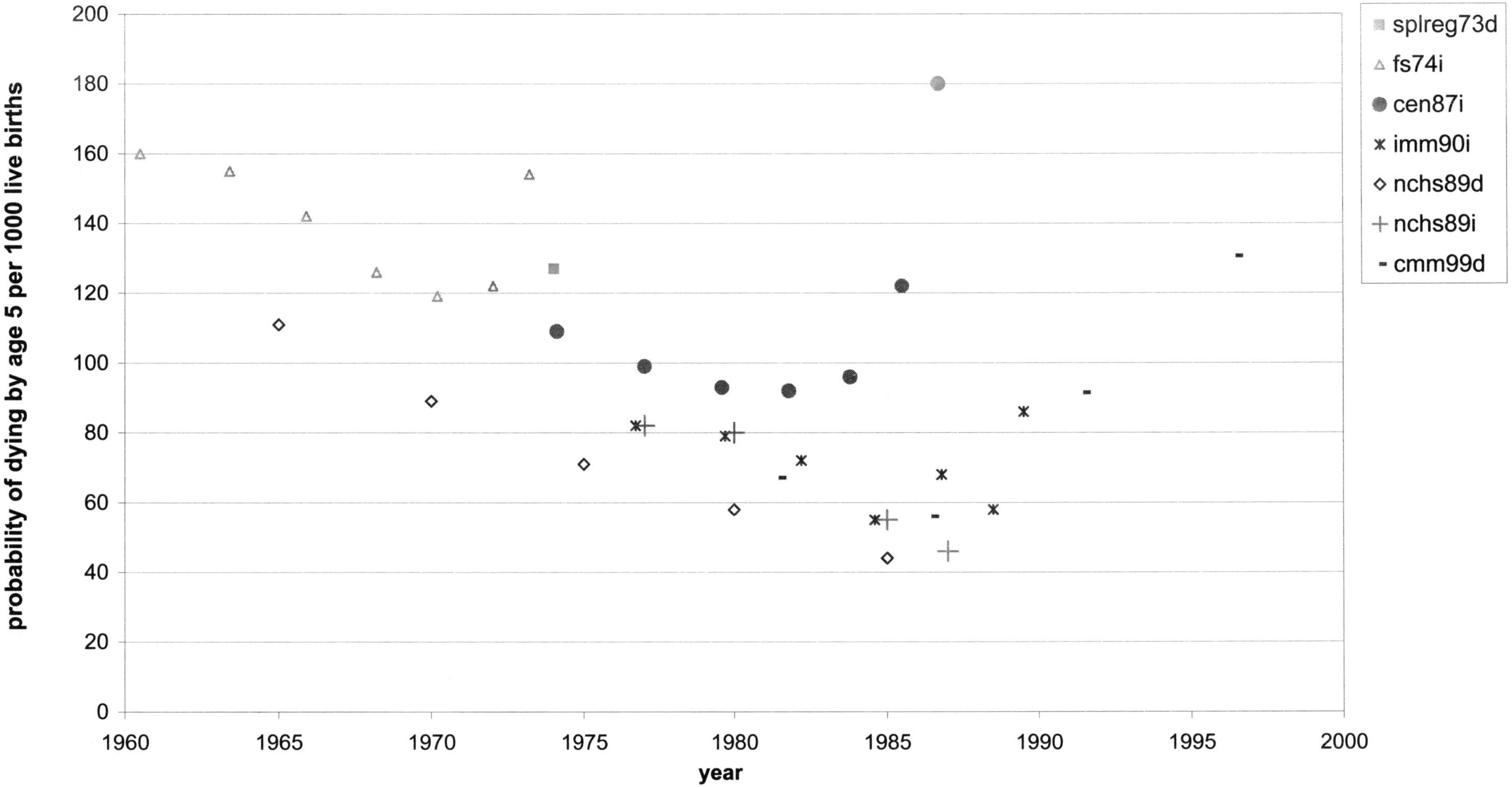

Data sources

Sources des données

Fuentes de los datos

splreg73d: Demographic Sample Survey and Sample Registration System 1973-1974 (direct method); ***fs74i***: Fertility Survey, 1974 (indirect method); ***cen87i***: Census, 17 October 1987 (indirect method); ***imm90i***: Immunization, Diarrhoeal Disease, Maternal and Child Mortality Survey, May-June 1990 (indirect method); ***nchs89d***: National Child Health Survey 1989 (direct method); ***nchs89i***: National Child Health Survey 1989 (indirect method); ***cmm99d***: Child and Maternal Mortality Survey 1999 (direct method)

Under 5 mortality

Ireland

Mortalité des moins de cinq ans

Irlande

Mortalidad de menores de cinco años

Irlanda

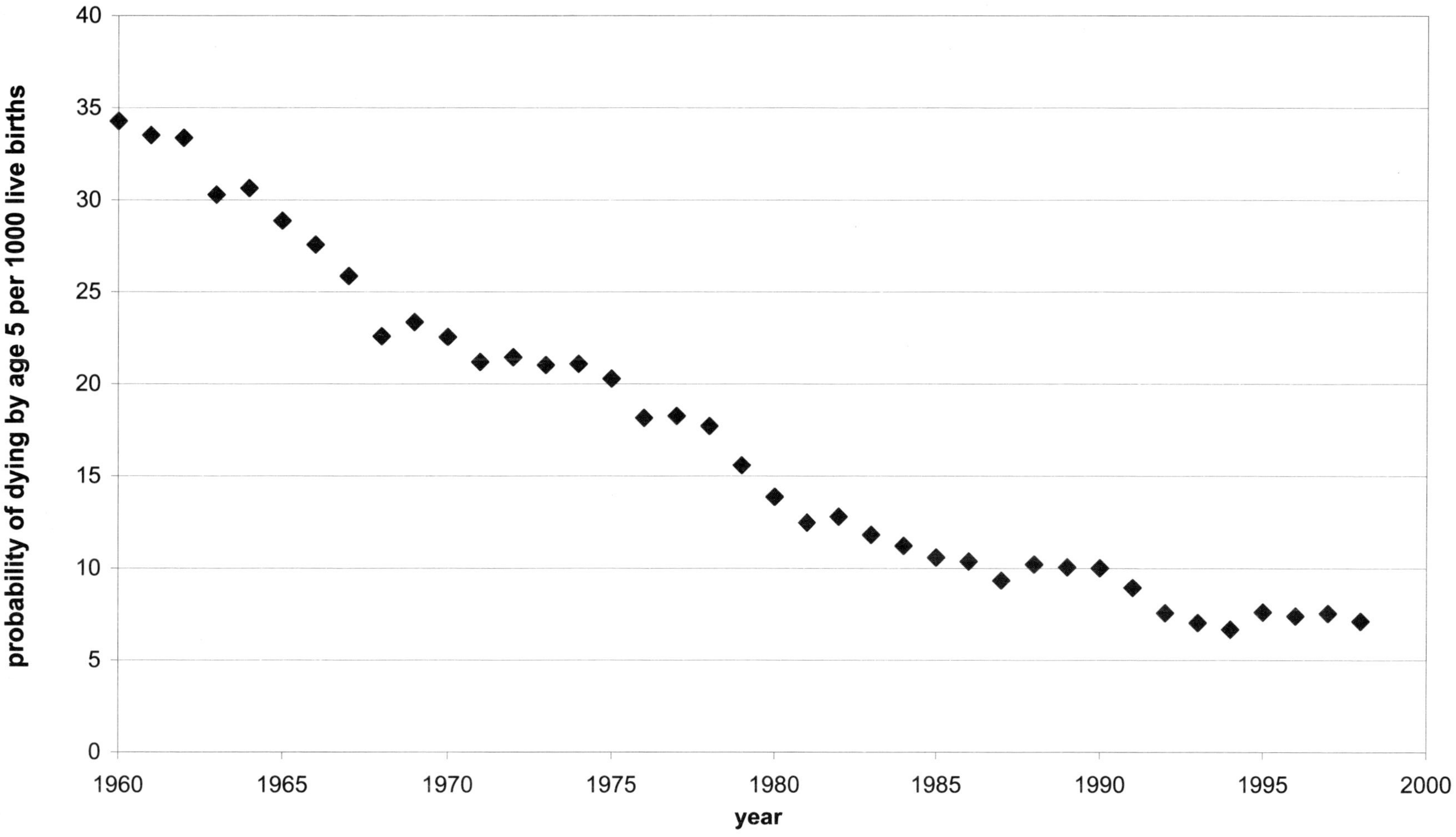

Data sources

Vital Registration

Sources des données

Fuentes de los datos

Under 5 mortality

Israel

Mortalité des moins de cinq ans

Israël

Mortalidad de menores de cinco años

Israel

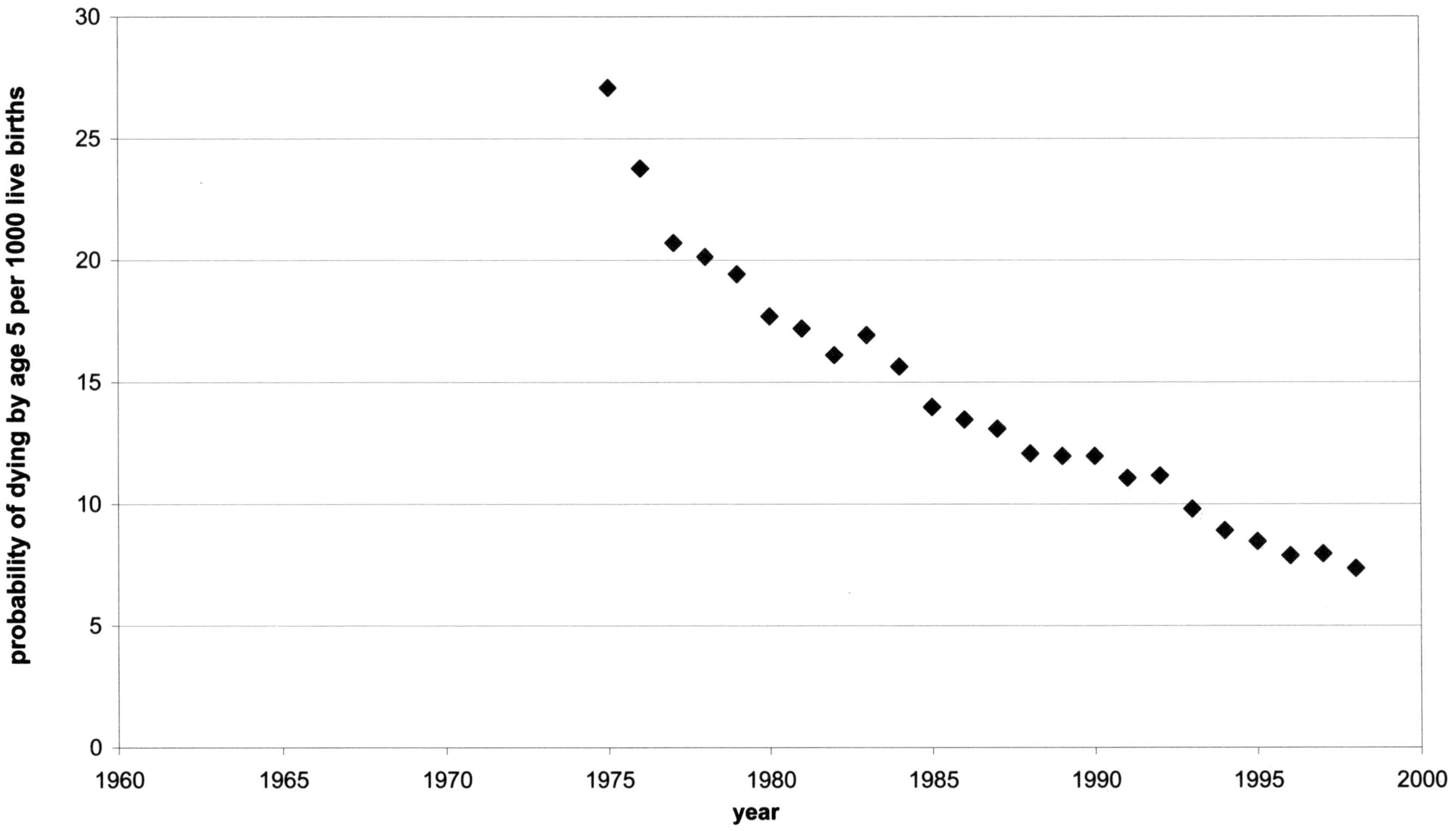

Data sources

Vital Registration

Sources des données

Fuentes de los datos

Under 5 mortality

Italy

Mortalité des moins de cinq ans

Italie

Mortalidad de menores de cinco años

Italia

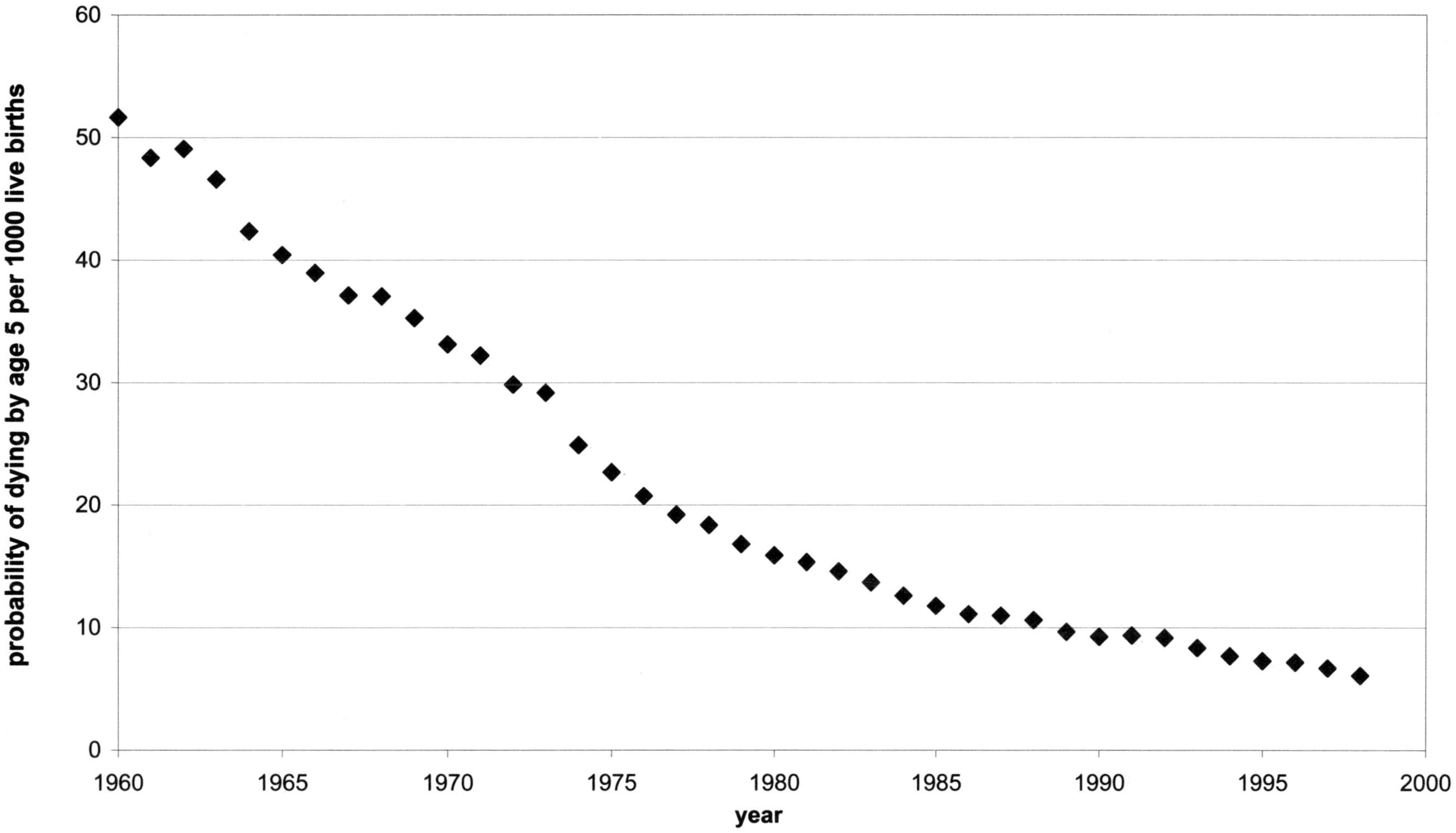

Data sources

Vital Registration

Sources des données

Fuentes de los datos

Under 5 mortality

Jamaica

Mortalité des moins de cinq ans

Jamaïque

Mortalidad de menores de cinco años

Jamaica

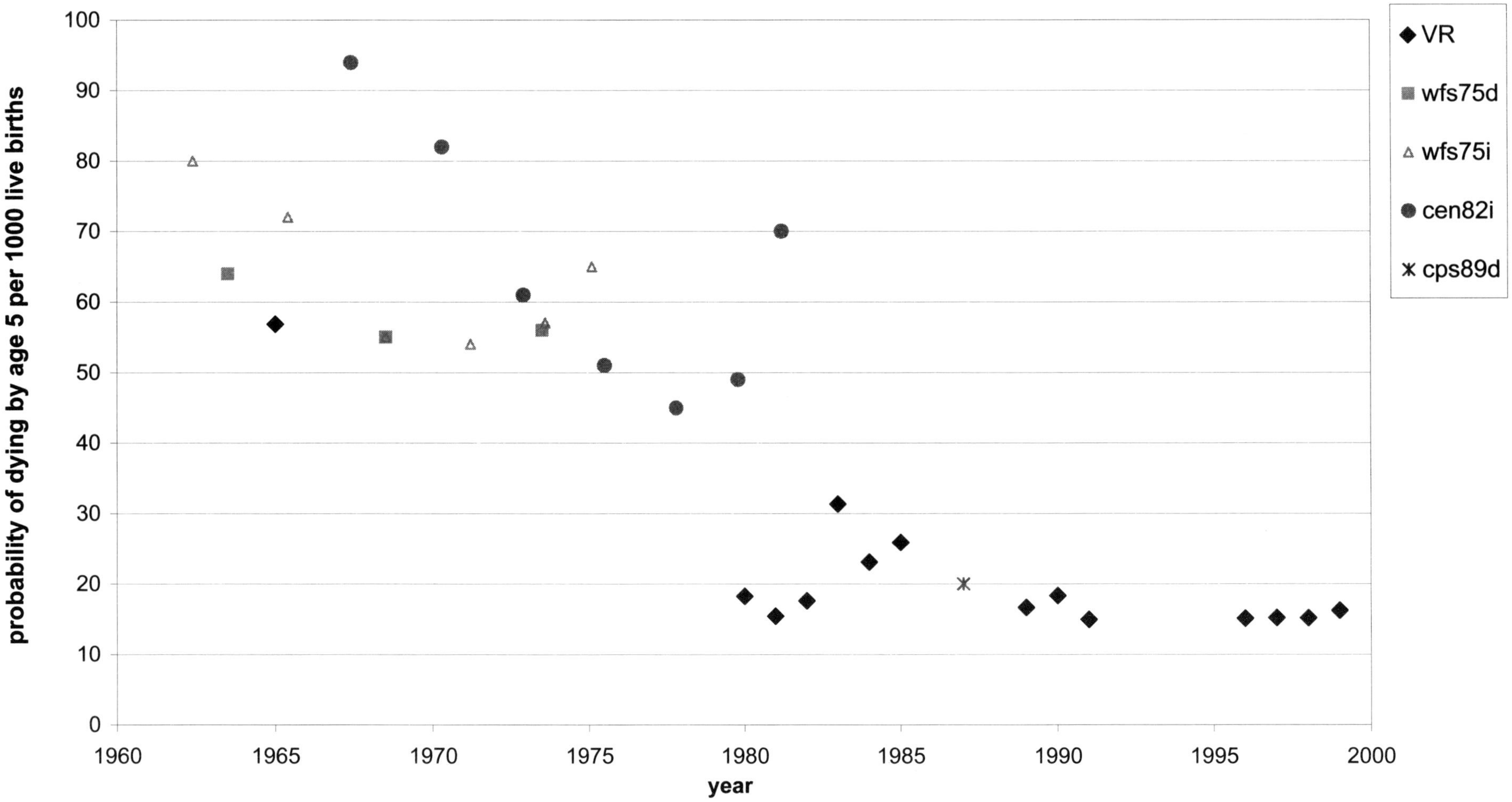

Data sources

VR: Vital Registration; ***wfs75d***: Fertility Survey, November 1975-January 1976 (direct method); ***wfs75i***: Fertility Survey, November 1975-January 1976 (indirect method); ***cen82i***: Census, 8 June 1982 (indirect method); ***cps89d***: Contraceptive Prevalence Survey 1989 (direct method)

Under 5 mortality

Japan

Mortalité des moins de cinq ans

Japon

Mortalidad de menores de cinco años

Japón

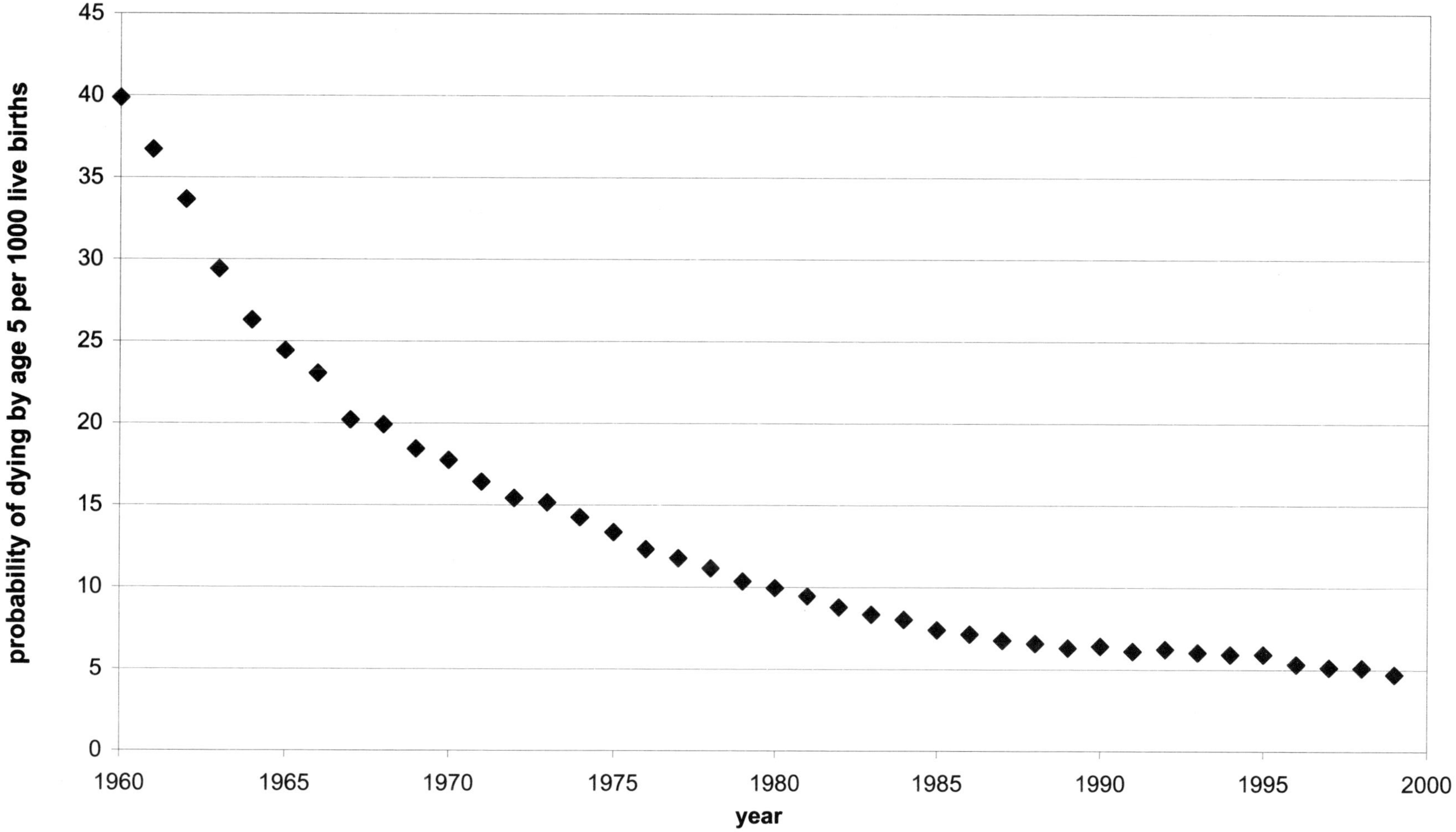

Data sources

Vital Registration

Sources des données

Fuentes de los datos

Under 5 mortality
Jordan

Mortalité des moins de cinq ans
Jordanie

Mortalidad de menores de cinco años
Jordania

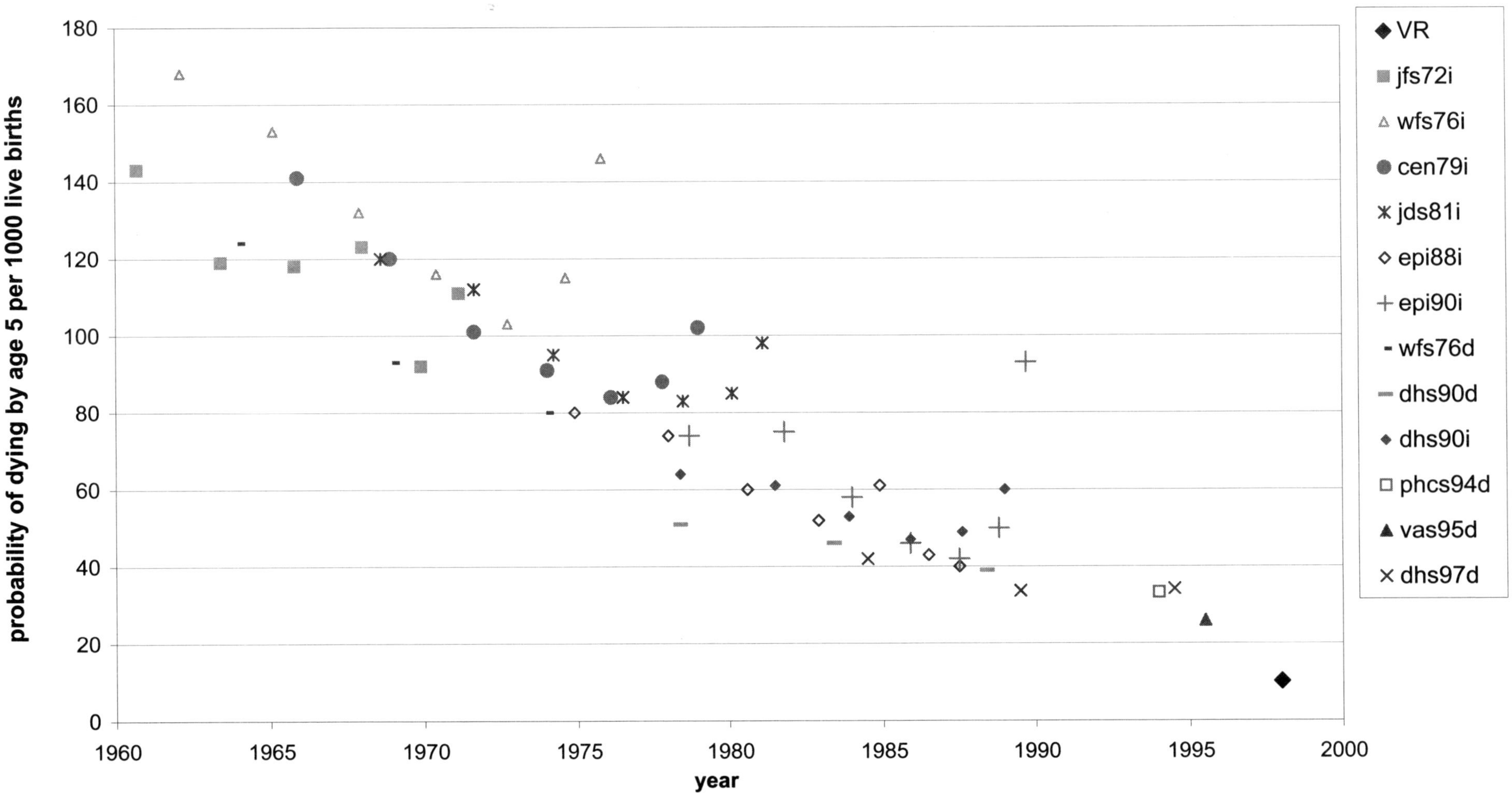

Data sources **Sources des données** **Fuentes de los datos**

VR: Vital Registration; ***jfs72i***: National Fertility Survey in Jordan, January 1972 (indirect method); ***wfs76i***: Jordan Fertitlity Survey, 1976 (indirect method); ***cen79i***: Census, 10 November 1979 ; ***jds81i***: Jordan Demographic Survey, 1981 (indirect method); ***epi88i***: Epi/Cdd and Child Mortality Survey 1988 (indirect method); ***epi90i***: Epi/Cdd and Child Mortality Survey 1990 (indirect method); ***wfs76d***: Jordan Fertitlity Survey, 1976 (direct method); ***dhs90d***: Demographic and Health Survey 1990 (direct method); ***dhs90i***: Demographic and Health Survey 1990 (indirect method); ***phcs94d***: Population and Housing Census Survey 1994 (direct method); ***vas95d***: MoH, Verbal Autopsy Study 1995-1996 (direct method); ***dhs97d***: Demographic and Health Survey 1997 (direct method)

Under 5 mortality

Kazakhstan

Mortalité des moins de cinq ans

Kazakhstan

Mortalidad de menores de cinco años

Kazajstán

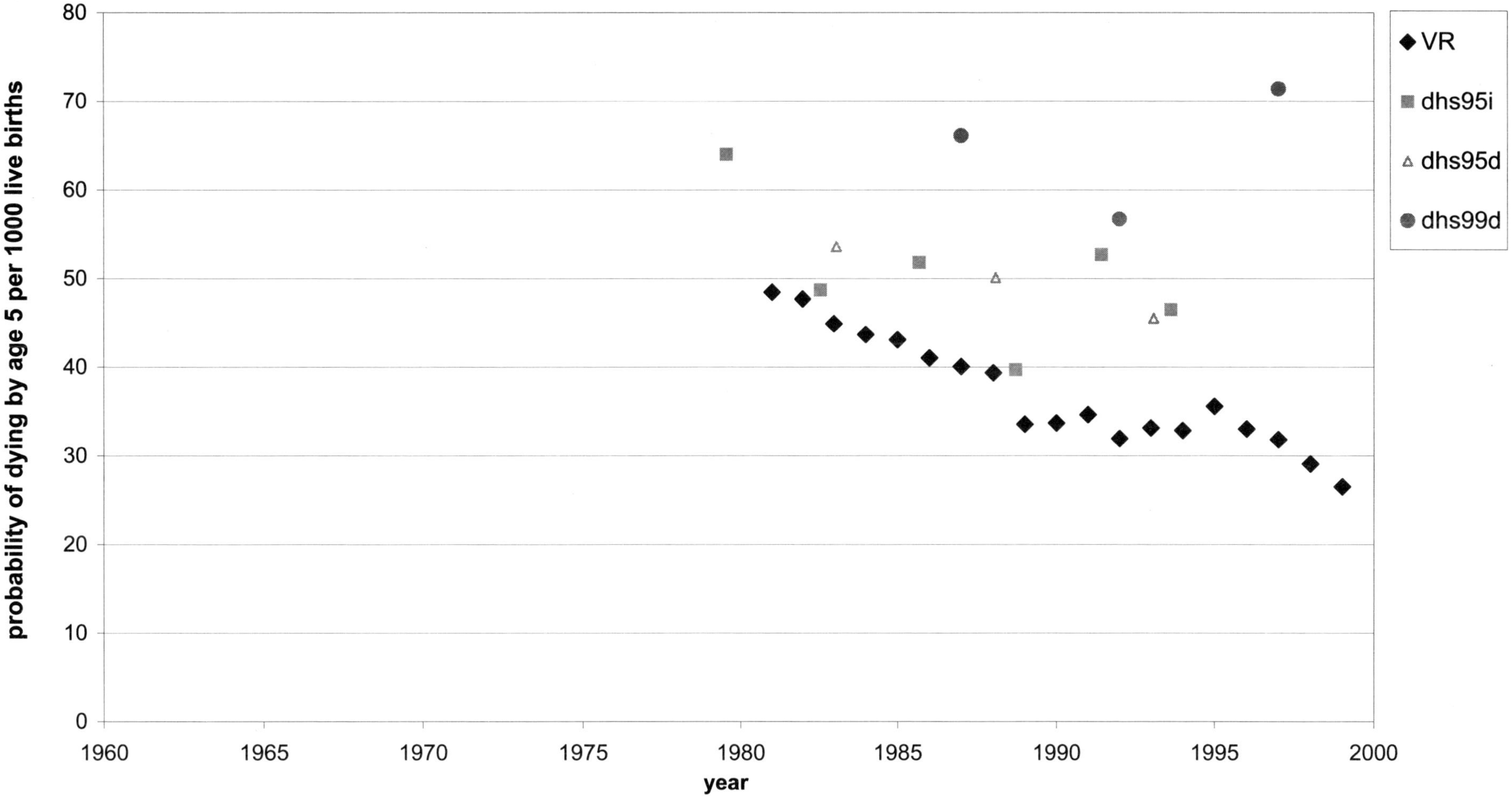

Data sources | **Sources des données** | **Fuentes de los datos**

VR : Vital Registration; ***dhs95i*** : Demographic and Health Survey 1995 (indirect method); ***dhs95d*** : Demographic and Health Survey 1995 (direct method); ***dhs99d*** : Demographic and Health Survey 1999 (direct method)

Under 5 mortality | Mortalité des moins de cinq ans | Mortalidad de menores de cinco años

Kenya | **Kenya** | **Kenya**

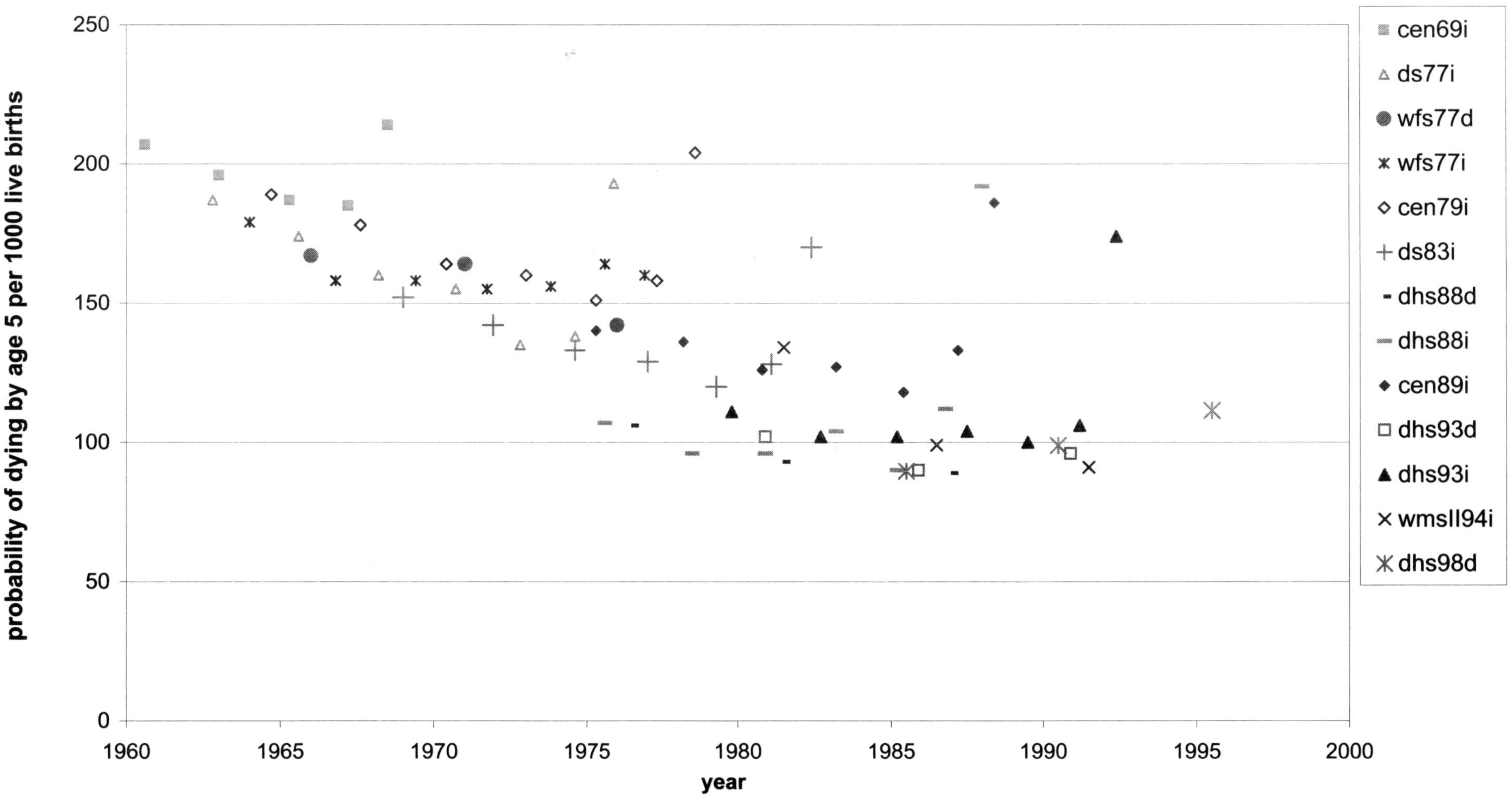

Data sources | **Sources des données** | **Fuentes de los datos**

cen69i: Census, 24-25 August 1969 (indirect method); ***ds77i***: National Demographic Survey, January 1977 (indirect method); ***wfs77d***: Fertility Survey, August 1977-May 1978 (direct method); ***wfs77i***: Fertility Survey, August 1977-May 1978 (indirect method); ***cen79i***: Census, 24-25 August 1979 (indirect method); ***ds83i***: National Demographic Survey, 1983 (indirect method); ***dhs88d***: Demographic and Health Survey, October 1988 -May 1989 (direct method); ***dhs88i***: Demographic and Health Survey, October 1988 -May 1989 (indirect method); ***cen89i***: Census 1989 (indirect method); ***dhs93d***: Demographic and Health Survey, February-August 1993 (direct method); ***dhs93i***: Demographic and Health Survey, February-August 1993 (indirect method); ***wmsII94i***: Welfare monitoring survey II 1994 (indirect method); ***dhs98d***: Demographic and Health Survey 1998 (direct method)

Under 5 mortality

Kuwait

Mortalité des moins de cinq ans

Koweït

Mortalidad de menores de cinco años

Kuwait

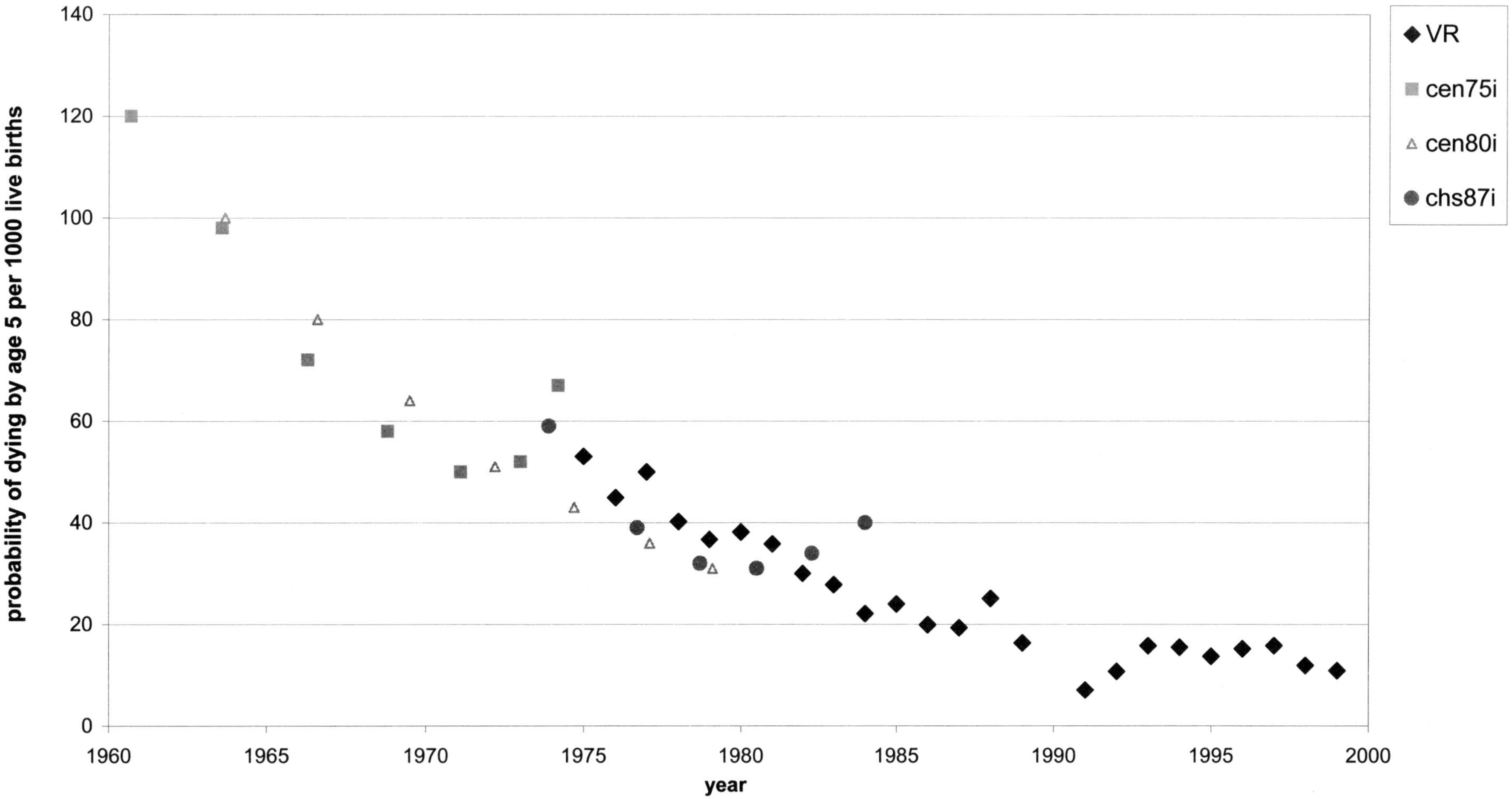

Data sources

Sources des données

Fuentes de los datos

VR: Vital Registration; ***cen75i***: Census, 20-21 April 1975 (indirect method); ***cen80i***: Census, 20-21 April 1980 (indirect method); ***chs87i***: Child Health Survey, March-April 1987 (indirect method)

Under 5 mortality Mortalité des moins de cinq ans Mortalidad de menores de cinco años

Kyrgyzstan **Kirghizistan** **Kirguistán**

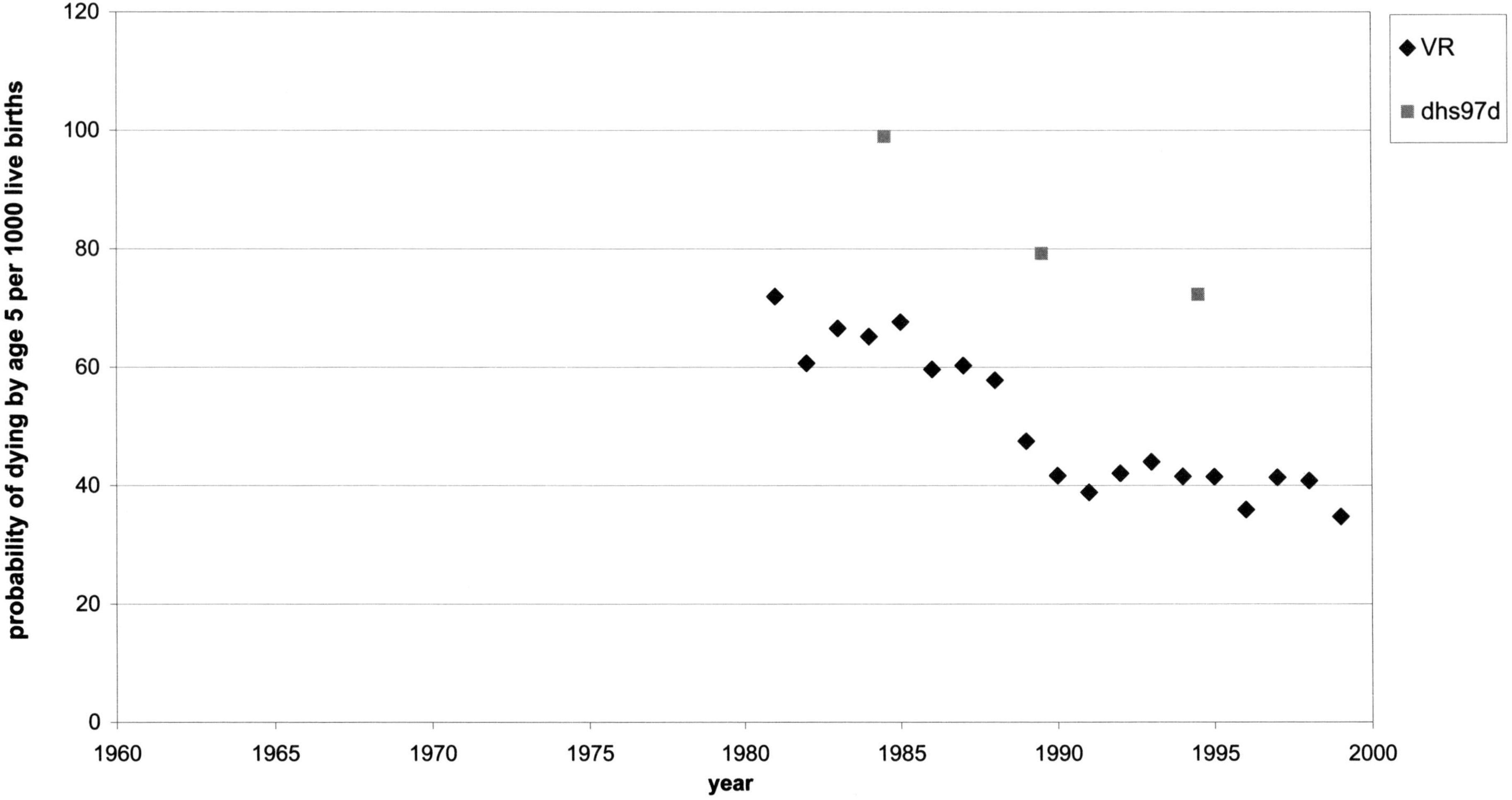

Data sources **Sources des données** **Fuentes de los datos**

VR: Vital Registration; ***dhs97d***: Demographic and Health Survey 1997 (direct method)

Under 5 mortality

Lao People's Democratic Republic

Mortalité des moins de cinq ans

République démocratique populaire lao

Mortalidad de menores de cinco años

República Democrática Popular Lao

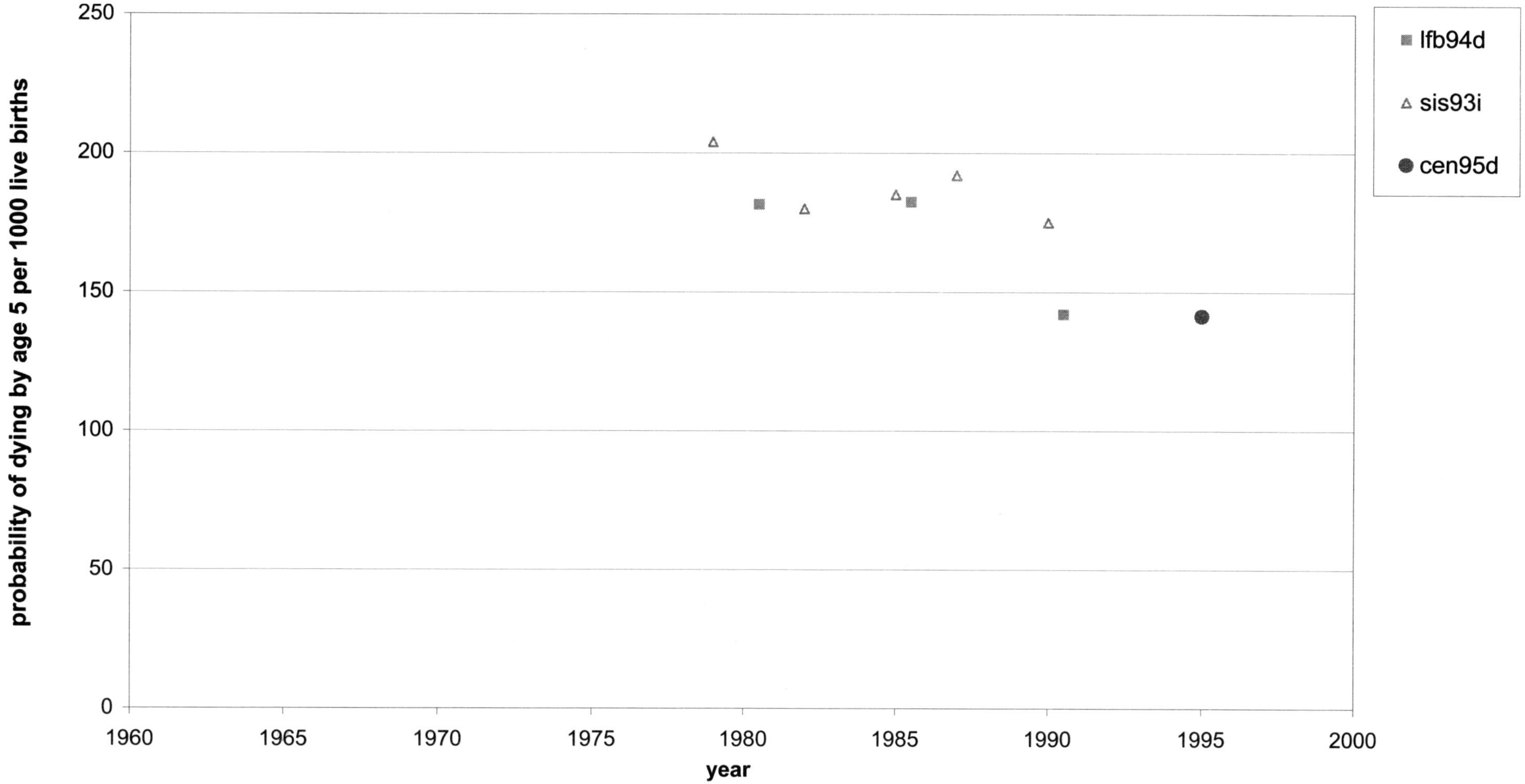

Data sources

Sources des données

Fuentes de los datos

lfb94d : Fertility and Birth Spacing Survey September 1994-March 1995 (direct method); ***sis93i*** : Living Conditions Survey 1992-1993 (indirect method); ***cen95d*** : Census 1995 (direct method)

Under 5 mortality
Latvia

Mortalité des moins de cinq ans
Lettonie

Mortalidad de menores de cinco años
Letonia

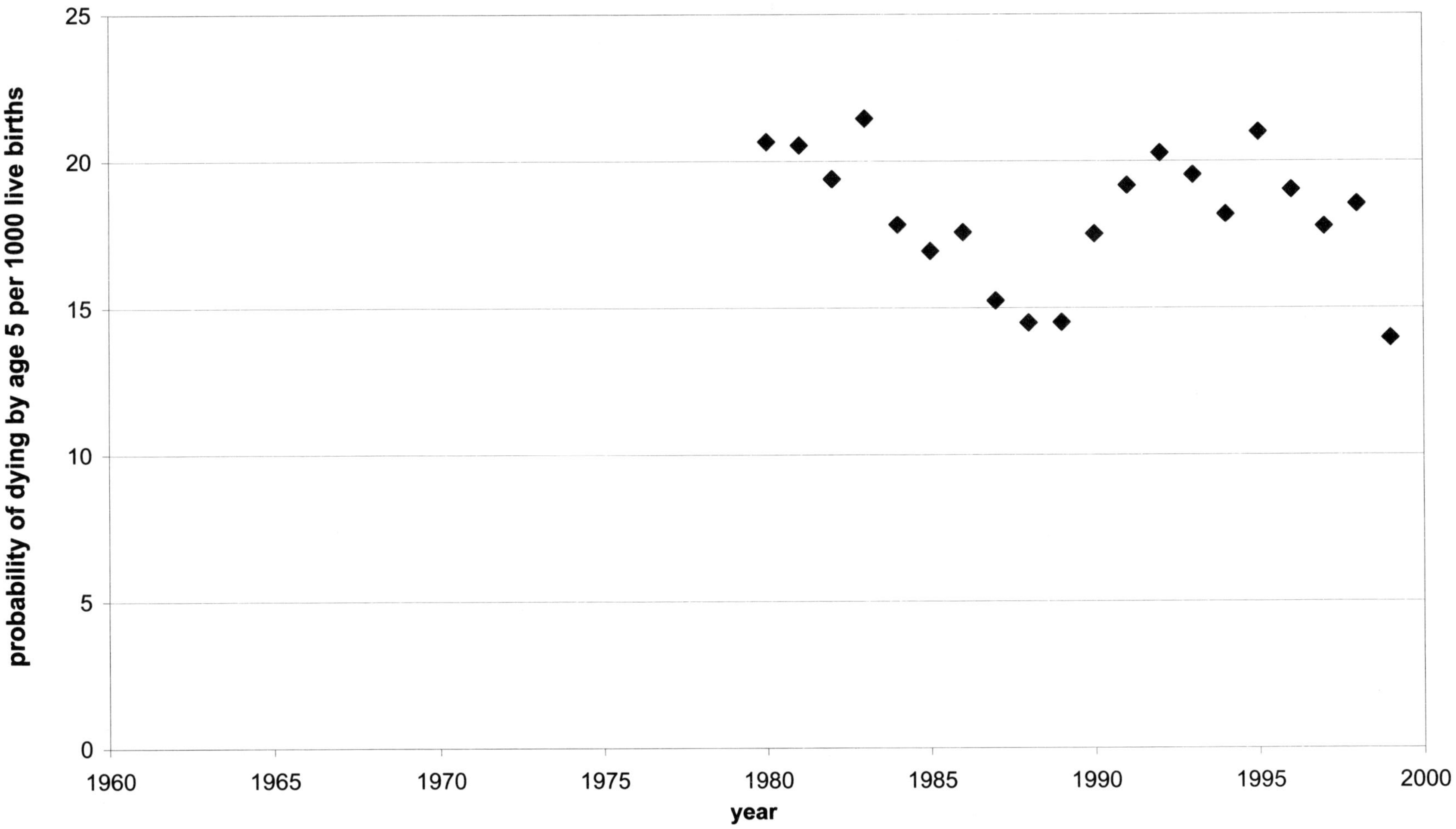

Data sources
Vital Registration

Sources des données

Fuentes de los datos

Under 5 mortality
Lebanon

Mortalité des moins de cinq ans
Liban

Mortalidad de menores de cinco años
Líbano

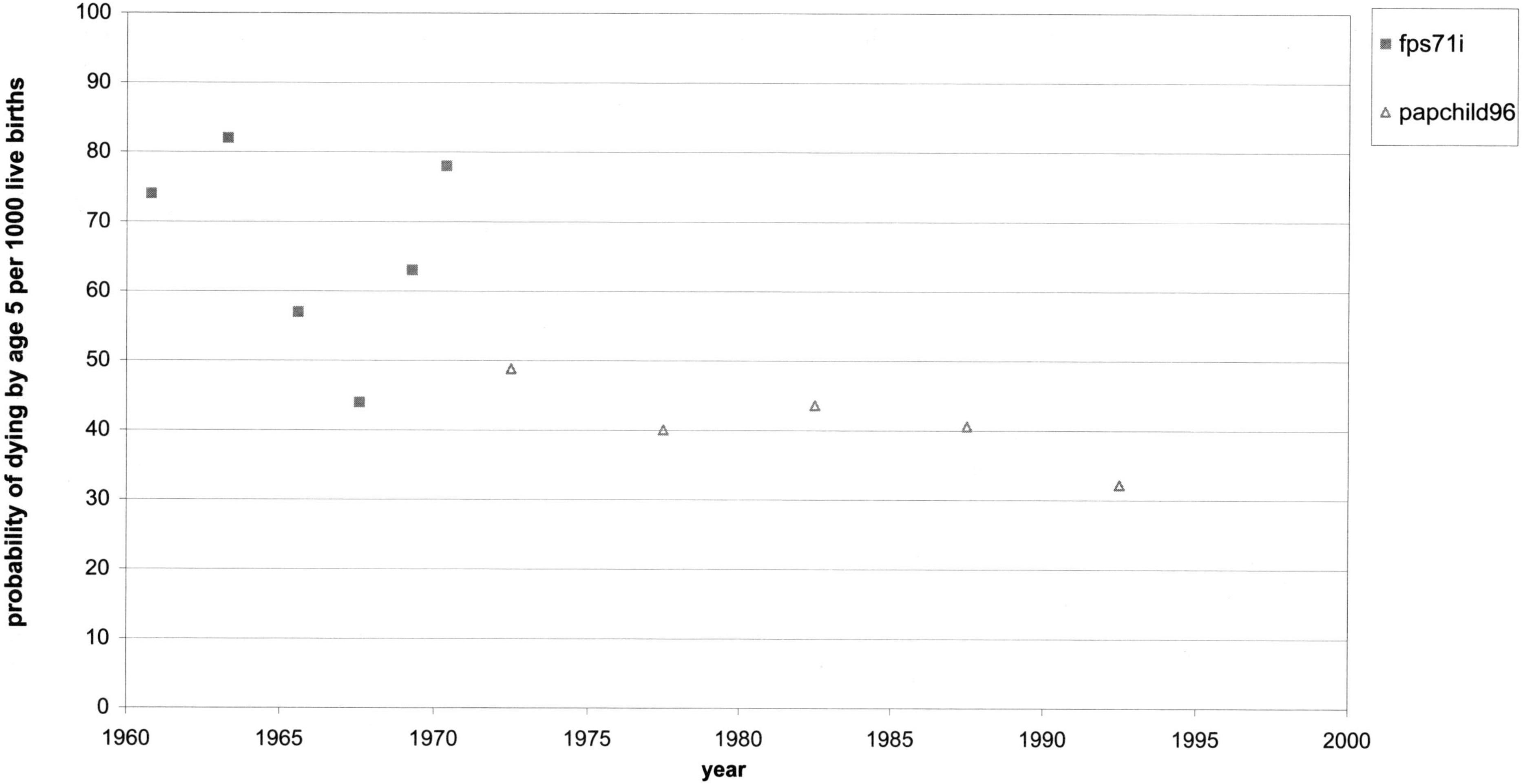

Data sources

Sources des données

Fuentes de los datos

fps71i : National Fertility and Family Planning Survey, 1971 (indirect method); ***papchild96*** : Maternal and Child Health Survey 1996

Under 5 mortality
Lesotho

Mortalité des moins de cinq ans
Lesotho

Mortalidad de menores de cinco años
Lesotho

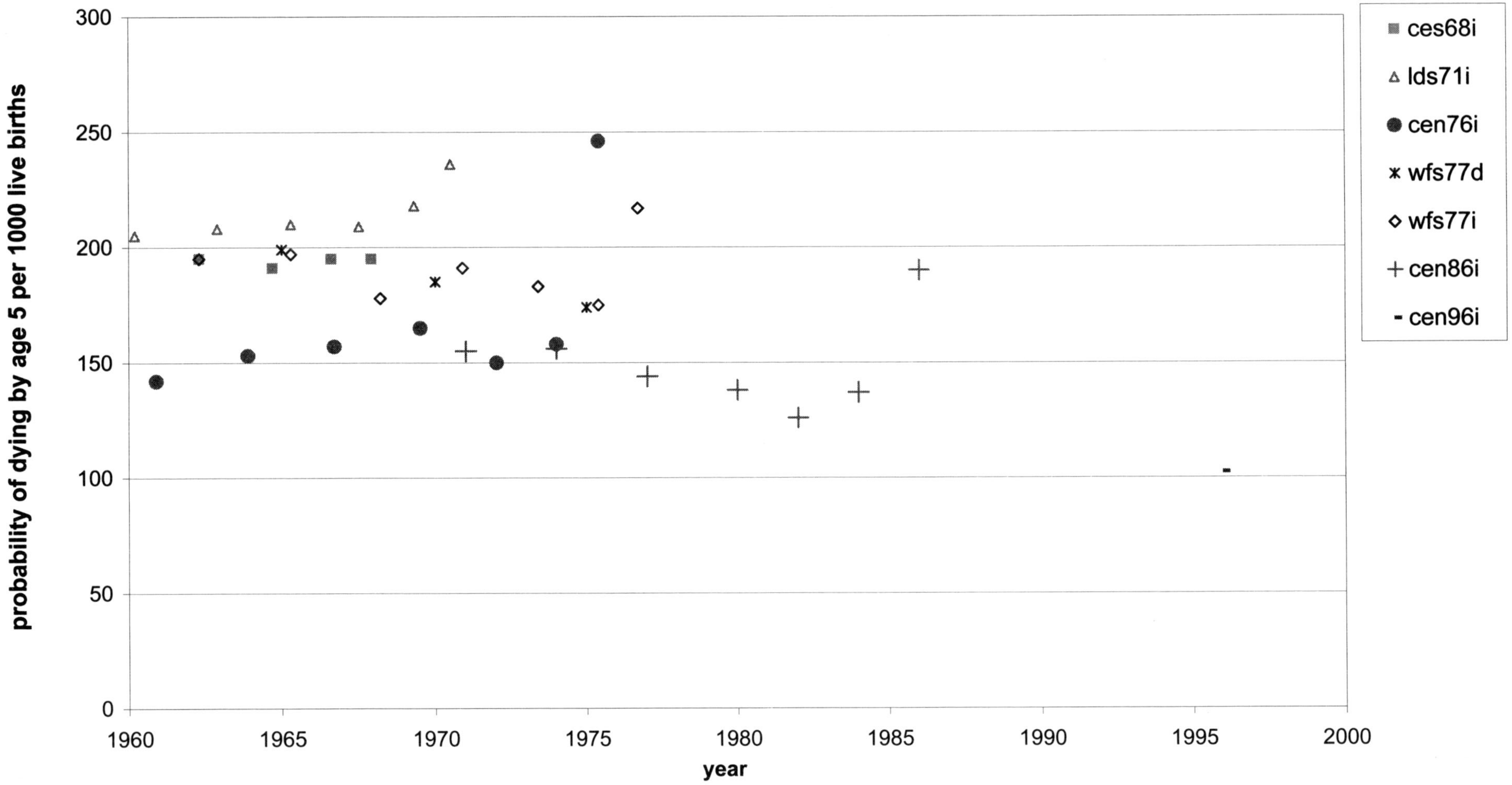

Data sources | **Sources des données** | **Fuentes de los datos**

ces68i: Rural Household Consumption and Expenditure Survey, March 1968-May 1969 (indirect method); ***lds71i***: Lesotho Demographic Survey, May 1971 (indirect method); ***cen76i***: Census, 12 April 1976 (indirect method); ***wfs77d***: Lesotho Fertility Survey, April-December 1977 (direct method); ***wfs77i***: Lesotho Fertility Survey, April-December 1977 (indirect method); ***cen86i***: Census, 12 April 1986 (indirect method); ***cen96i***: Census 1996 (indirect method)

Under 5 mortality Mortalité des moins de cinq ans Mortalidad de menores de cinco años

Liberia **Libéria** **Liberia**

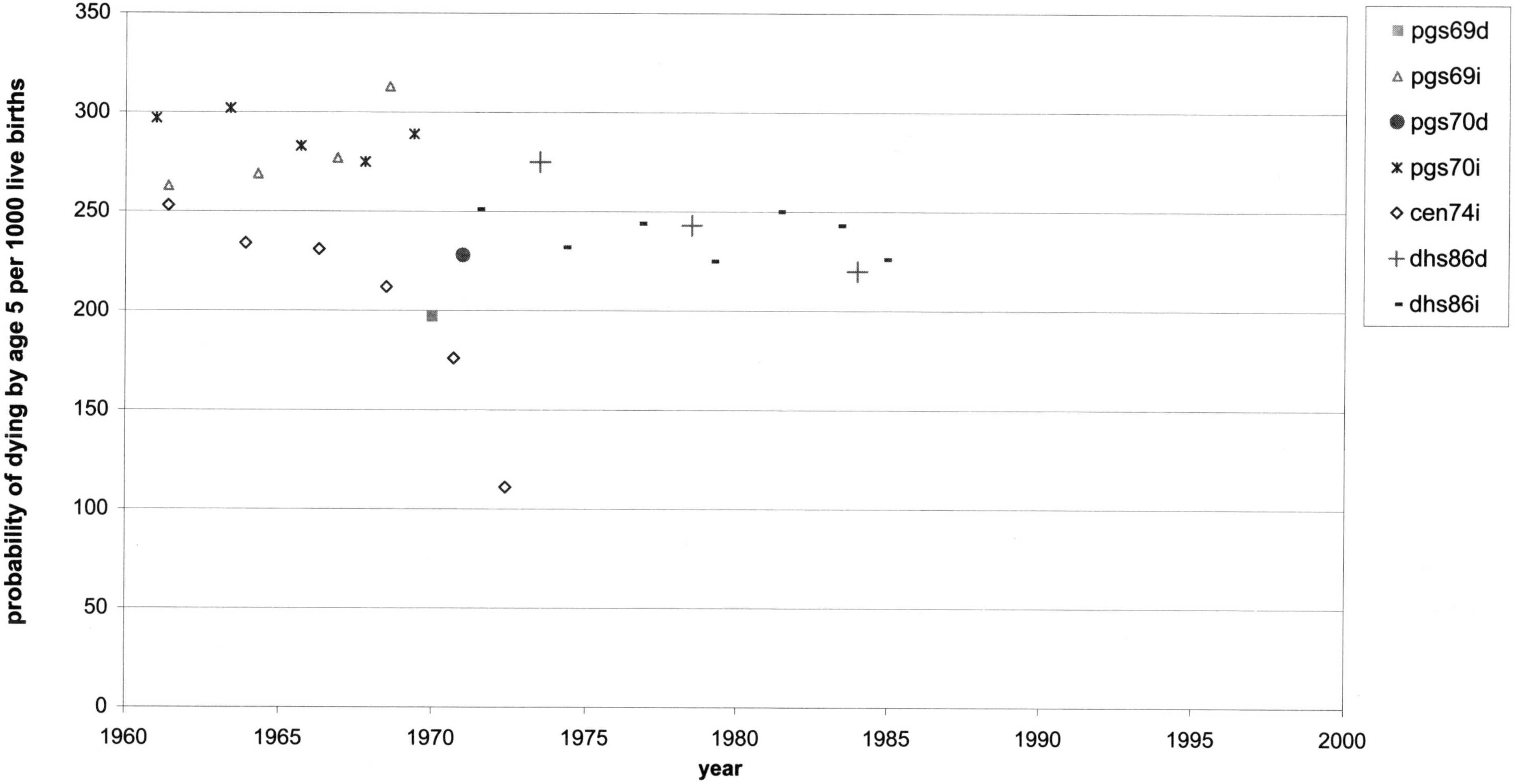

Data sources **Sources des données** **Fuentes de los datos**

pgs69d: Population Growth Survey, May 1969-April 1970 (direct method); ***pgs69i***: Population Growth Survey, May 1969-April 1970 (indirect method); ***pgs70d***: Population Growth Survey, May 1970-April 1971 (direct method); ***pgs70i***: Population Growth Survey, May 1970-April 1971 (indirect method); ***cen74i***: Census, 1 February 1974 (indirect method); ***dhs86d***: Demographic and Health Survey, February-July 1986 (direct method); ***dhs86i***: Demographic and Health Survey, February-July 1986 (indirect method)

Under 5 mortality

Libyan Arab Jamahiriya

Mortalité des moins de cinq ans

Jamahiriya arabe libyenne

Mortalidad de menores de cinco años

Jamahiriya Arabe Libia

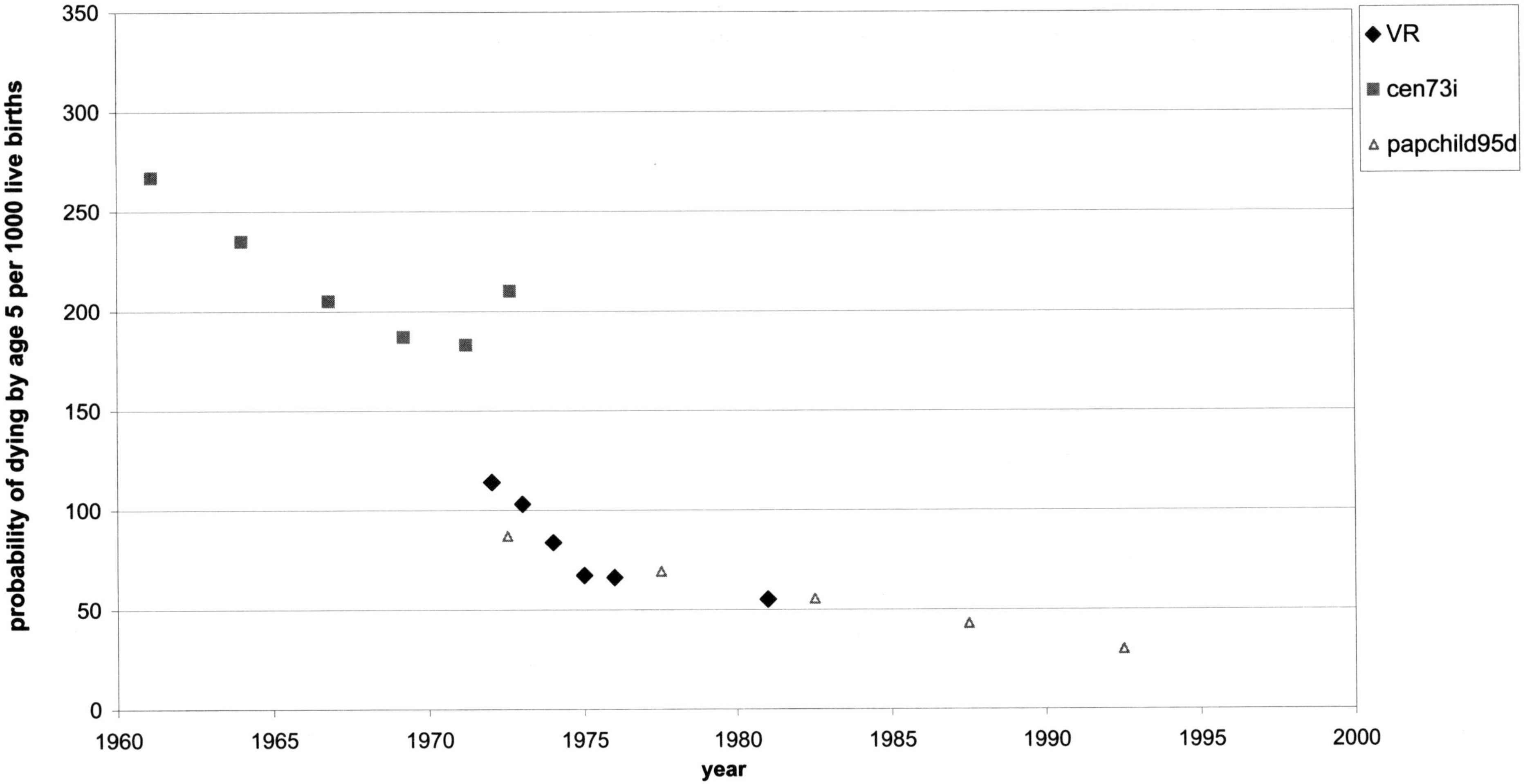

Data sources **Sources des données** **Fuentes de los datos**

VR: Vital Registration; ***cen73i***: Census, 31 July 1973 (indirect method); ***papchild95d***: Maternal and Child Health Survey 1995 (direct method)

Under 5 mortality

Lithuania

Mortalité des moins de cinq ans

Lituanie

Mortalidad de menores de cinco años

Lituania

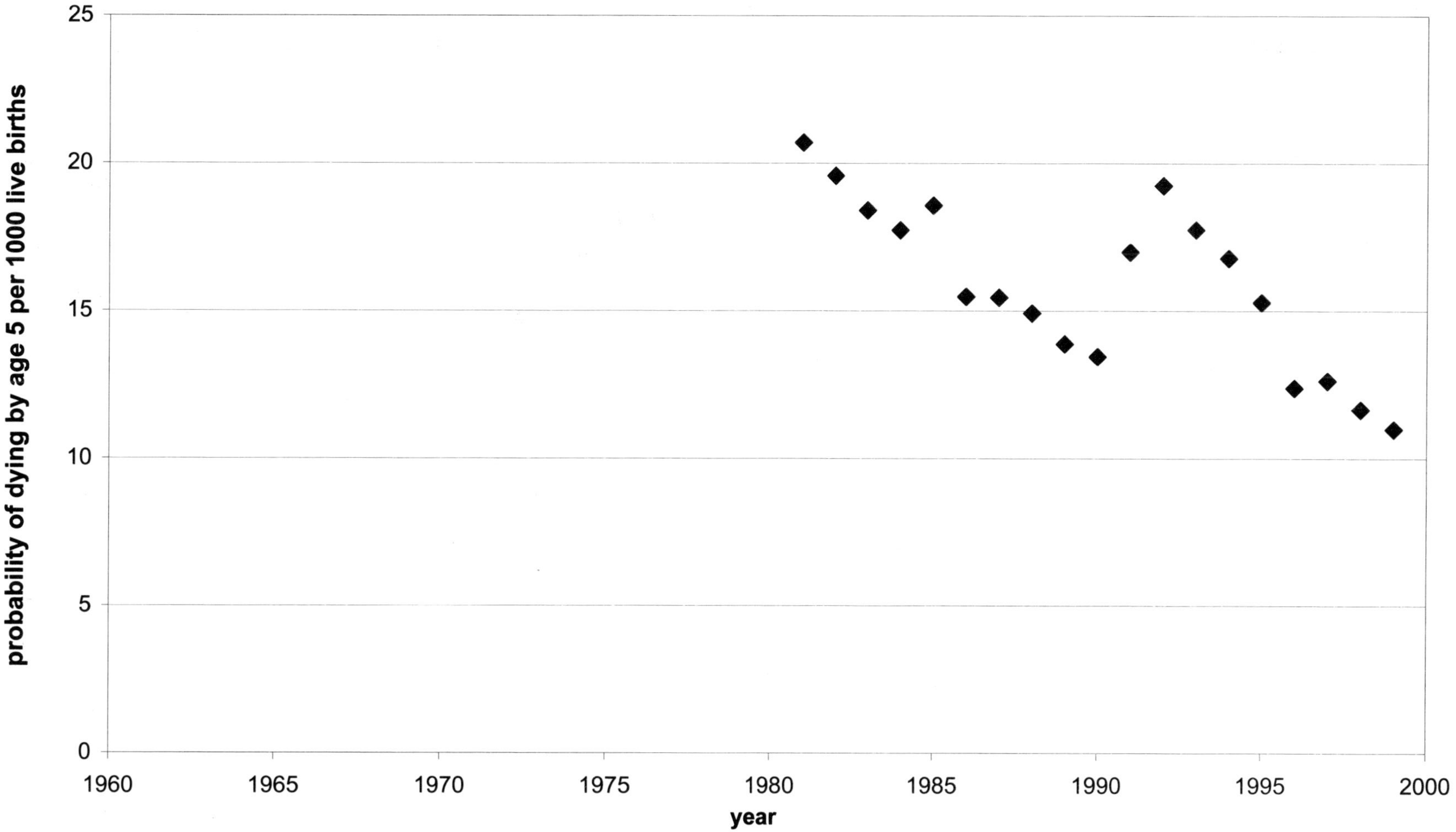

Data sources

Vital Registration

Sources des données

Fuentes de los datos

Under 5 mortality
Luxembourg

Mortalité des moins de cinq ans
Luxembourg

Mortalidad de menores de cinco años
Luxemburgo

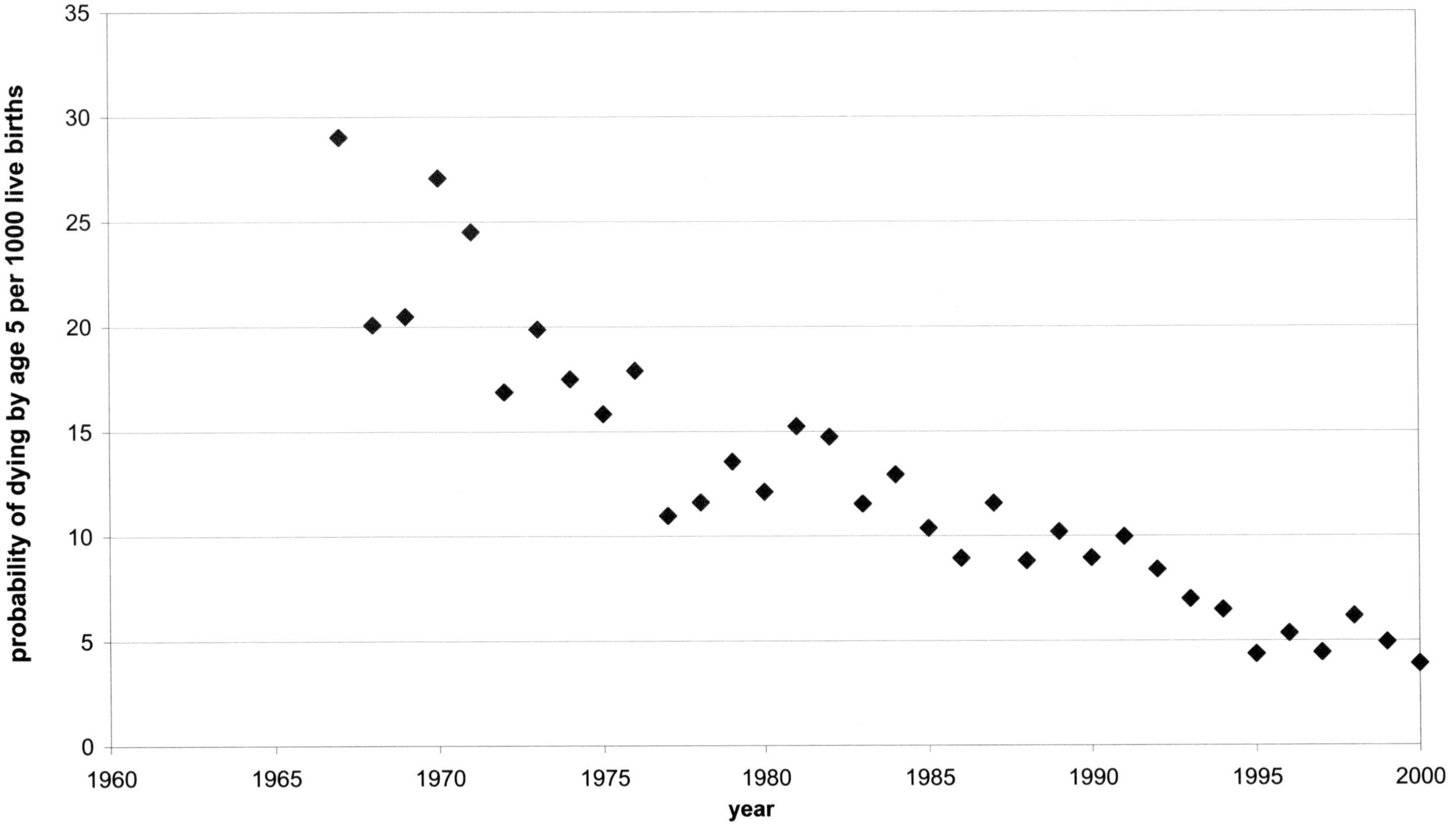

Data sources
Vital Registration

Sources des données

Fuentes de los datos

Under 5 mortality

Madagascar

Mortalité des moins de cinq ans

Madagascar

Mortalidad de menores de cinco años

Madagascar

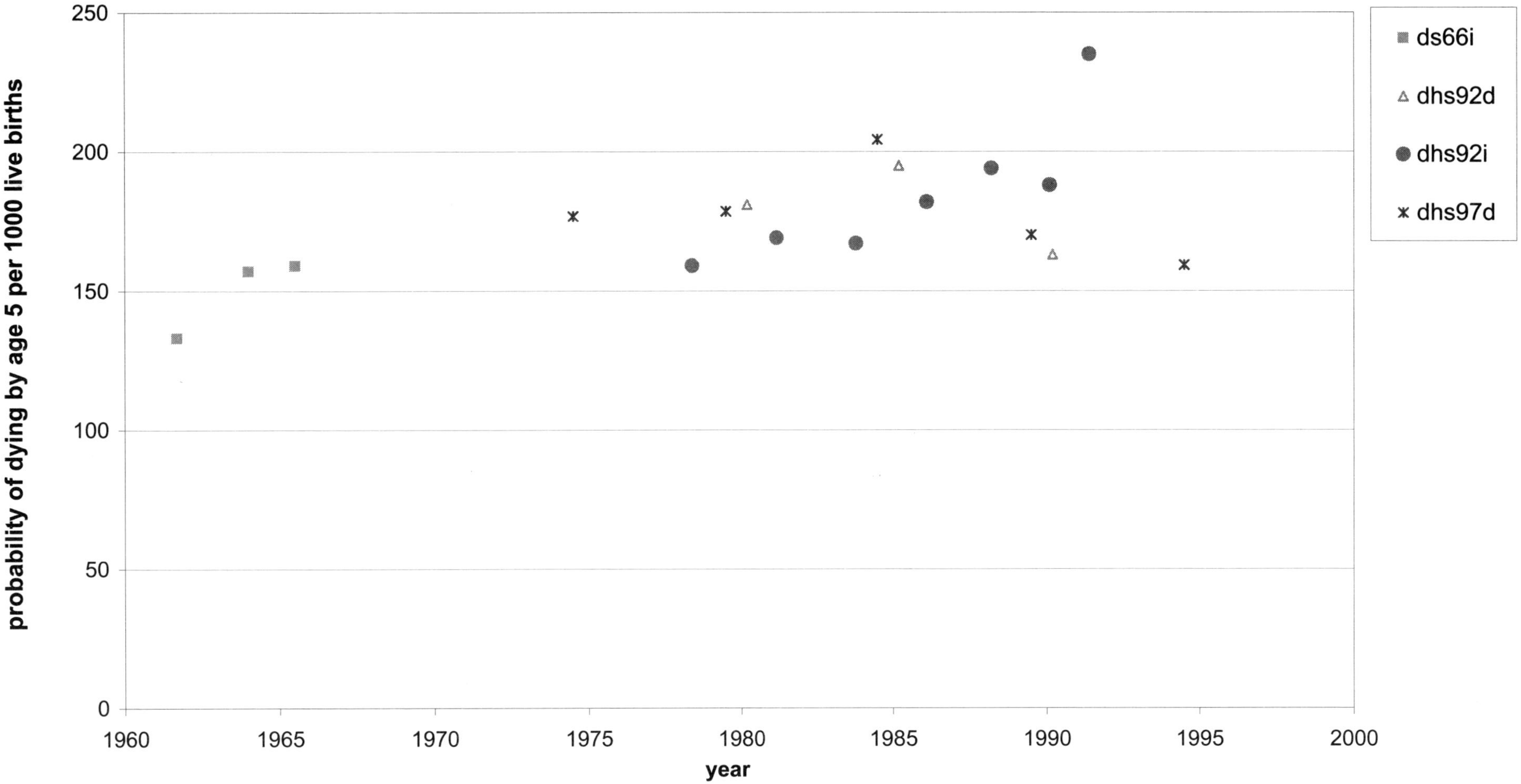

Data sources

Sources des données

Fuentes de los datos

ds66i: Enquête Démographique, 9 May-11 November 1966 (indirect method); ***dhs92d***: Demographic and Health Survey 1992 (direct method); ***dhs92i***: Demographic and Health Survey 1992 (indirect method); ***dhs97d***: Demographic and Health Survey 1997 (direct method)

Under 5 mortality | Mortalité des moins de cinq ans | Mortalidad de menores de cinco años

Malawi | **Malawi** | **Malawi**

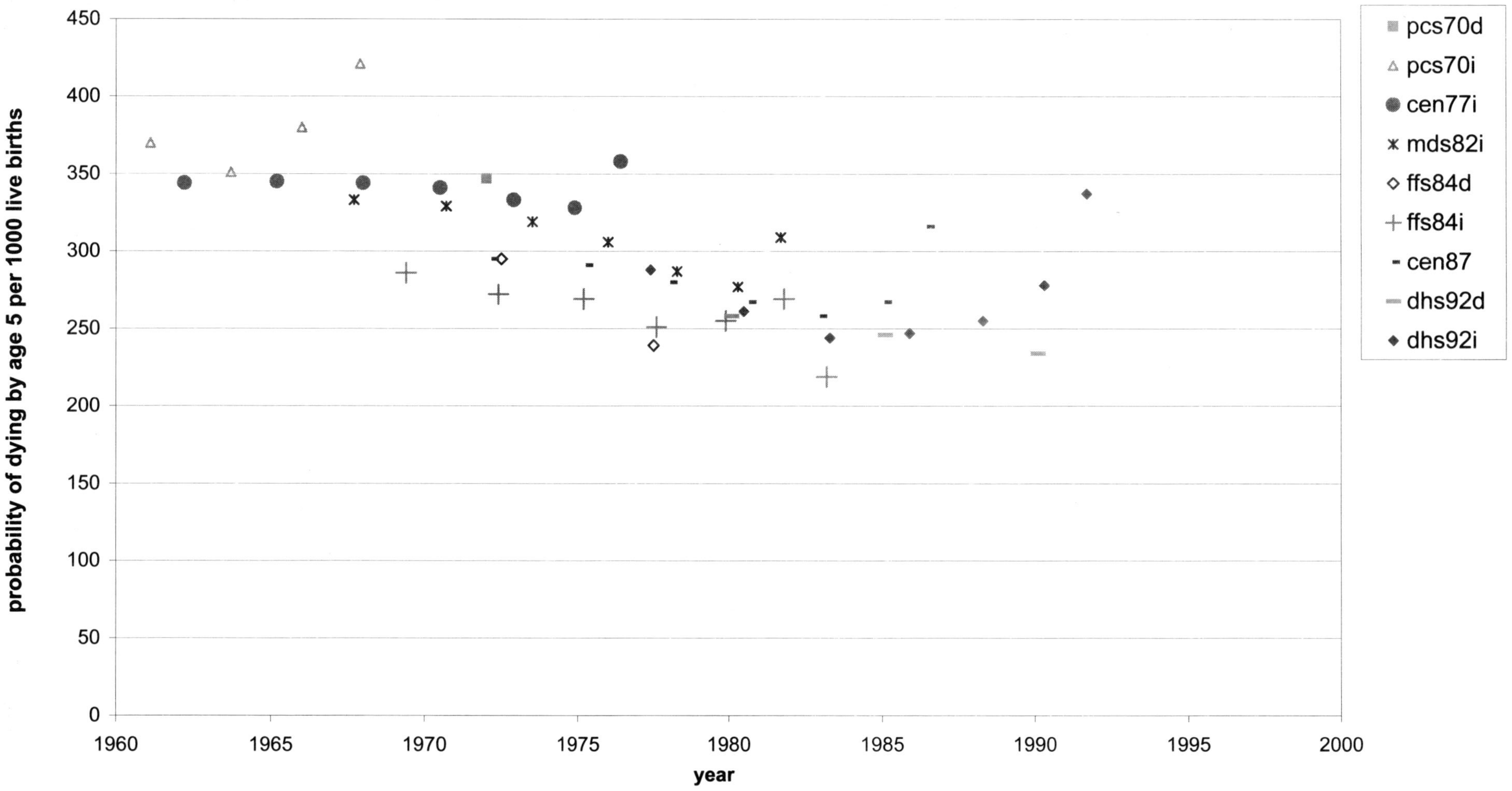

Data sources | **Sources des données** | **Fuentes de los datos**

pcs70d: Population Change Survey, February 1970-January 1972 (direct method); ***pcs70i***: Population Change Survey, February 1970-January 1972 (indirect method); ***cen77i***: Census, 20 September-10 October 1977 (indirect method); ***mds82i***: Demographic Survey, November-December 1982 (indirect method); ***ffs84d***: Family Formation Survey, May-June 1984 (direct method); ***ffs84i***: Family Formation Survey, May-June 1984 (indirect method); ***cen87***: Census 1987; ***dhs92d***: Demographic and Health Survey, September- November 1992 (direct method); ***dhs92i***: Demographic and Health Survey, September- November 1992 (indirect method)

Under 5 mortality
Malaysia

Mortalité des moins de cinq ans
Malaisie

Mortalidad de menores de cinco años
Malasia

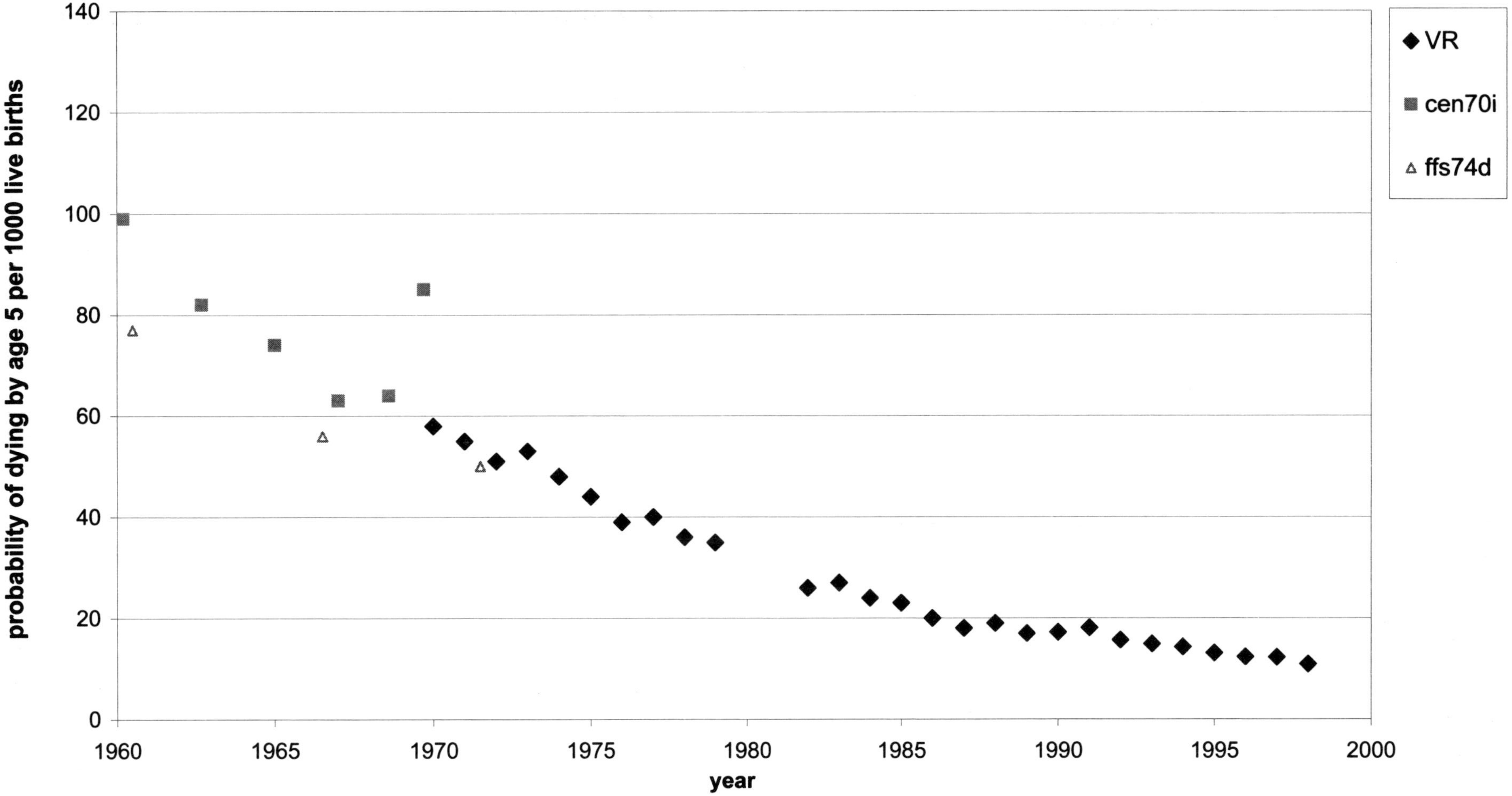

Data sources

Sources des données

Fuentes de los datos

VR: Vital Registration; ***cen70i***: Census, 24 August 1970 (indirect method); ***ffs74d***: Malaysian Fertility and Family Survey, August-December 1974 (direct method)

Under 5 mortality
Maldives

Mortalité des moins de cinq ans
Maldives

Mortalidad de menores de cinco años
Maldivas

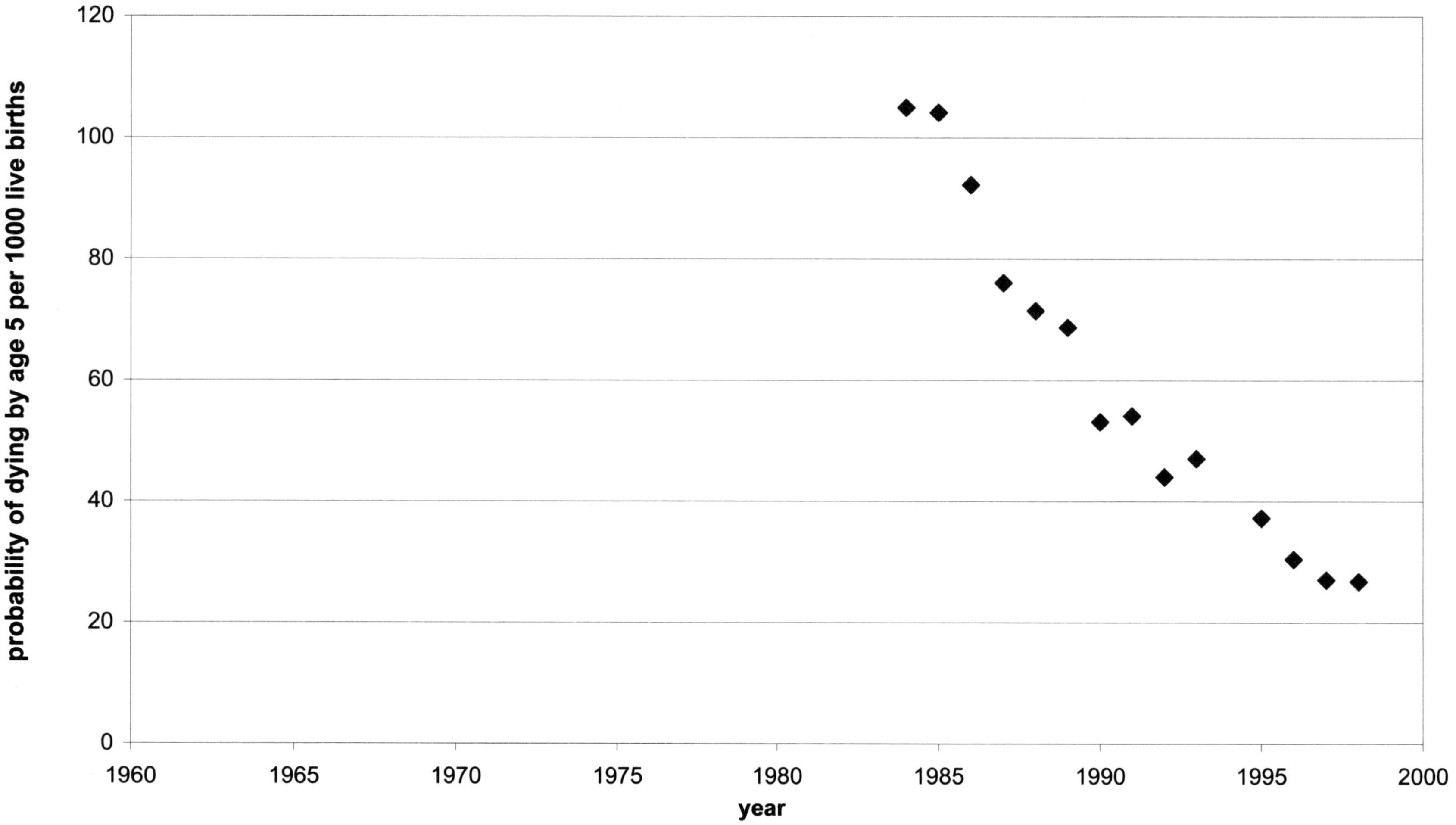

Data sources
Vital Registration

Sources des données

Fuentes de los datos

Under 5 mortality **Mali** | Mortalité des moins de cinq ans **Mali** | Mortalidad de menores de cinco años **Malí**

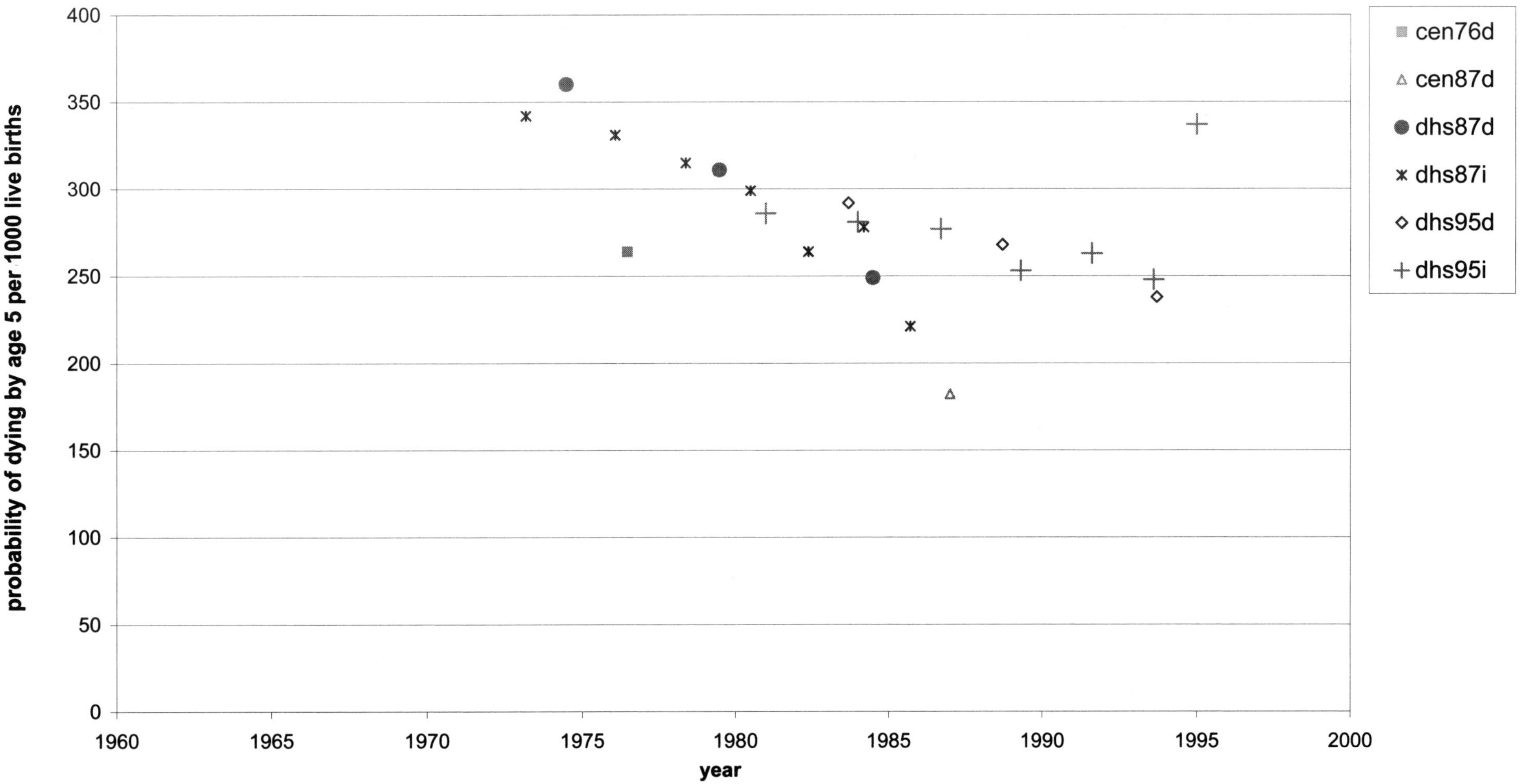

Data sources | **Sources des données** | **Fuentes de los datos**

cen76d: Census, 1-6 December 1976 (direct method); cen87d: Census 1987 (direct method); ***dhs87d***: Enquête Démographique et de Santé au Mali, March-August 1987 (direct method); ***dhs87i***: Enquête Démographique et de Santé au Mali, March-August 1987 (indirect method); ***dhs95d***: Enquête Démographique et de Santé au Mali, November 1995 - May 1996 (direct method); ***dhs95i***: Enquête Démographique et de Santé au Mali, November 1995 - May 1996 (indirect method)

Under 5 mortality

Malta

Mortalité des moins de cinq ans

Malte

Mortalidad de menores de cinco años

Malta

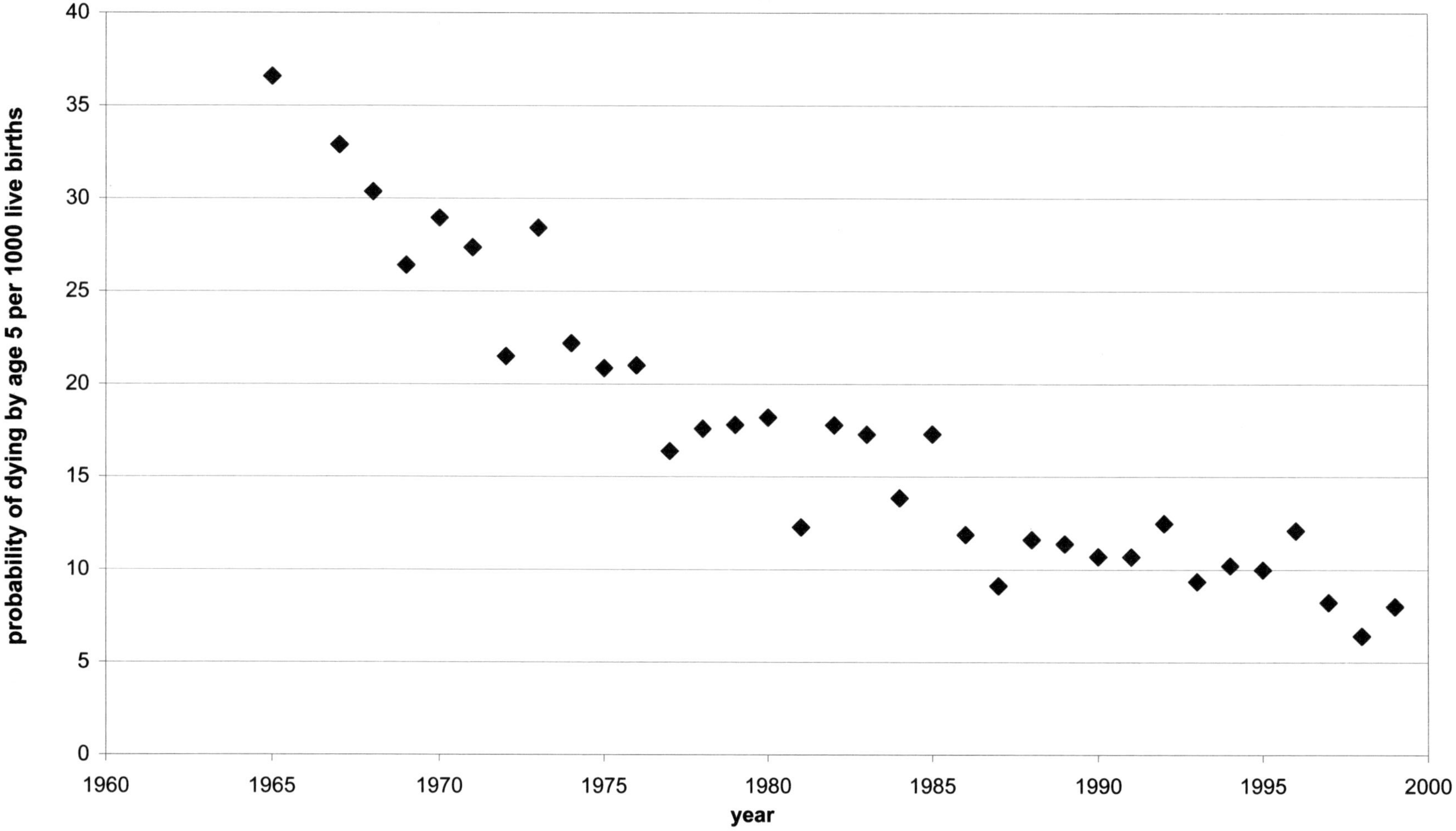

Data sources

Vital Registration

Sources des données

Fuentes de los datos

Under 5 mortality

Marshall Islands

Mortalité des moins de cinq ans

Iles Marshall

Mortalidad de menores de cinco años

Islas Marshall

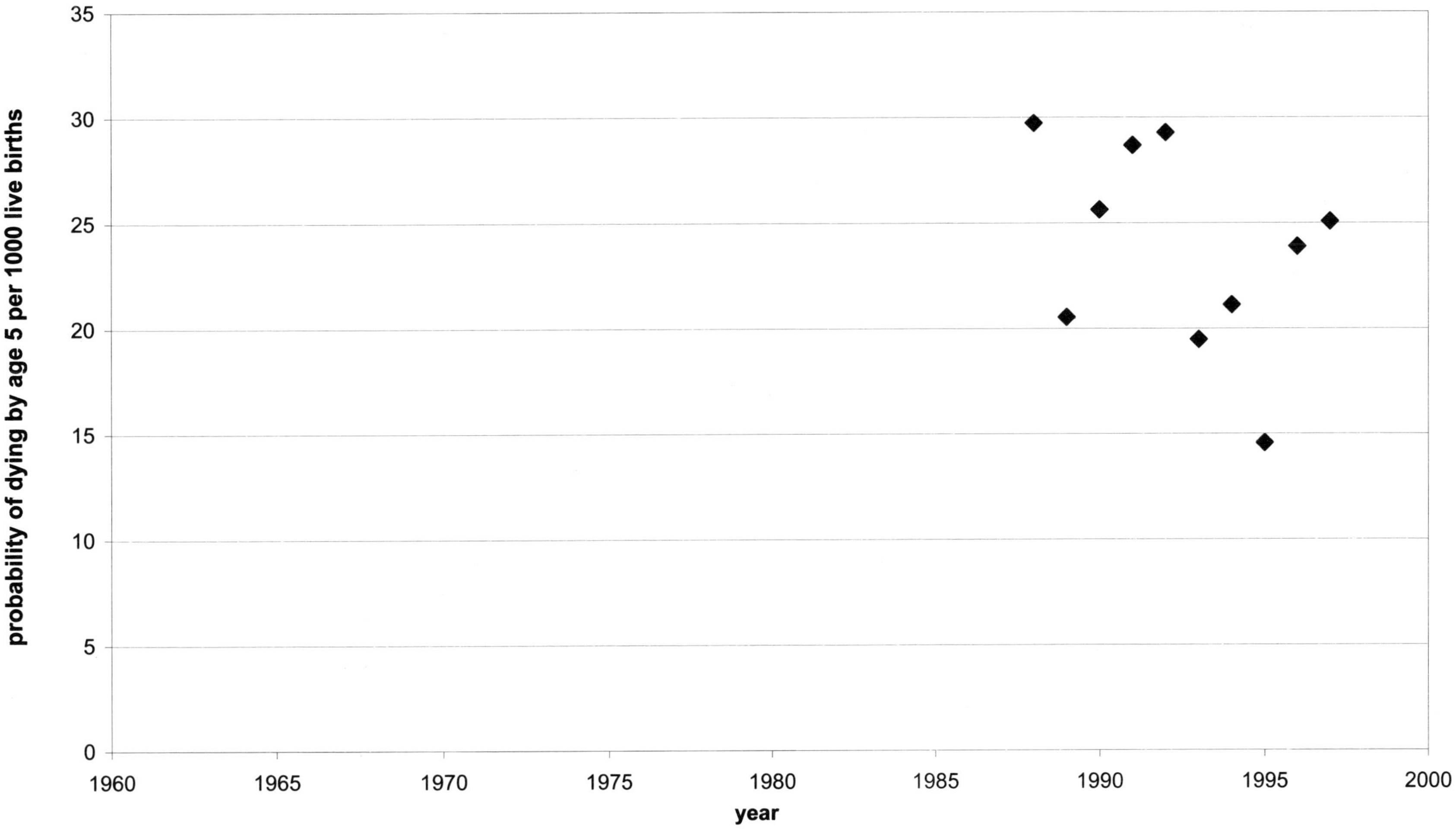

Data sources

Vital Registration

Sources des données

Fuentes de los datos

Under 5 mortality
Mauritius

Mortalité des moins de cinq ans
Maurice

Mortalidad de menores de cinco años
Mauricio

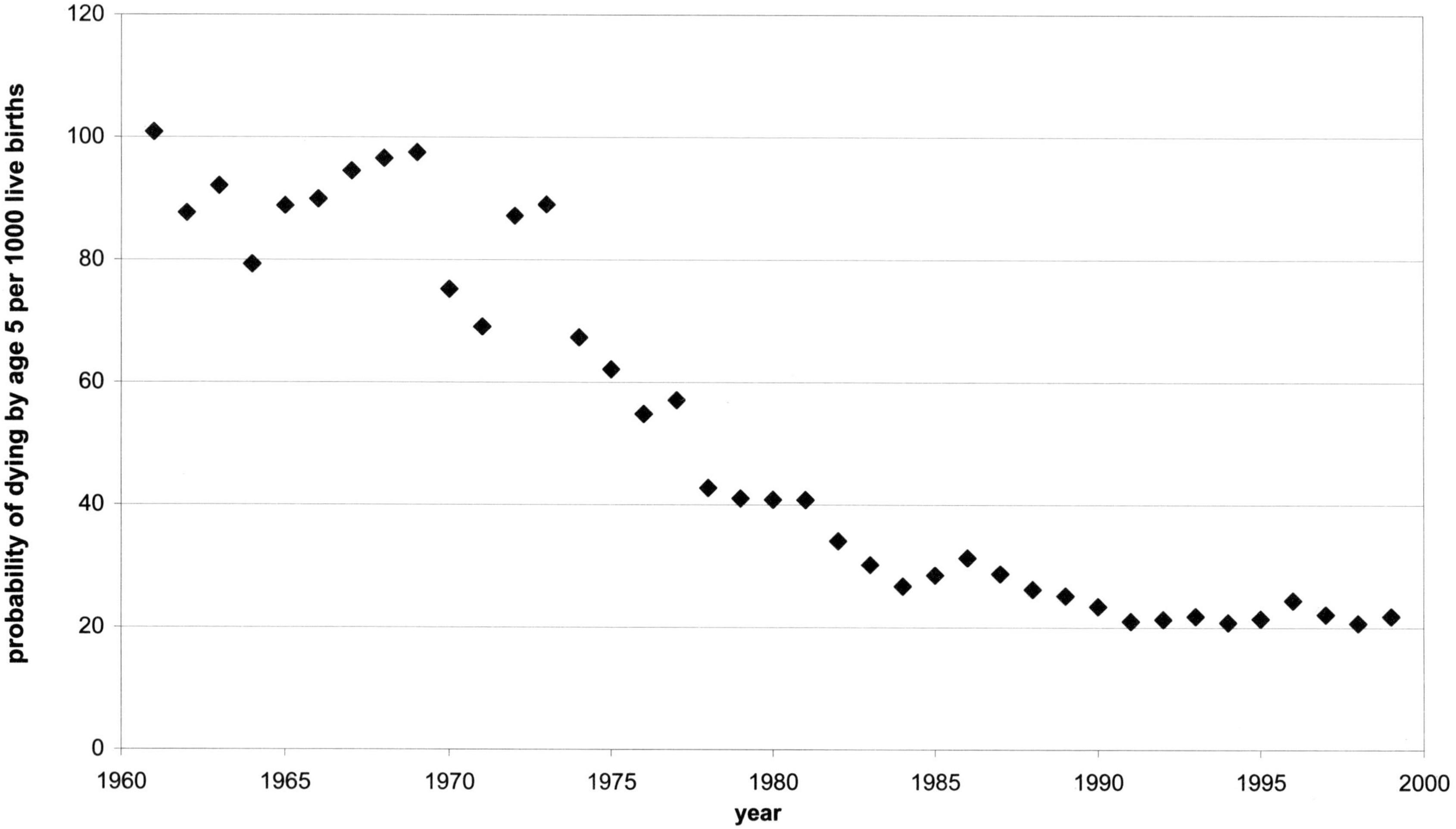

Data sources
Vital Registration

Sources des données

Fuentes de los datos

Under 5 mortality
Mauritania

Mortalité des moins de cinq ans
Mauritanie

Mortalidad de menores de cinco años
Mauritania

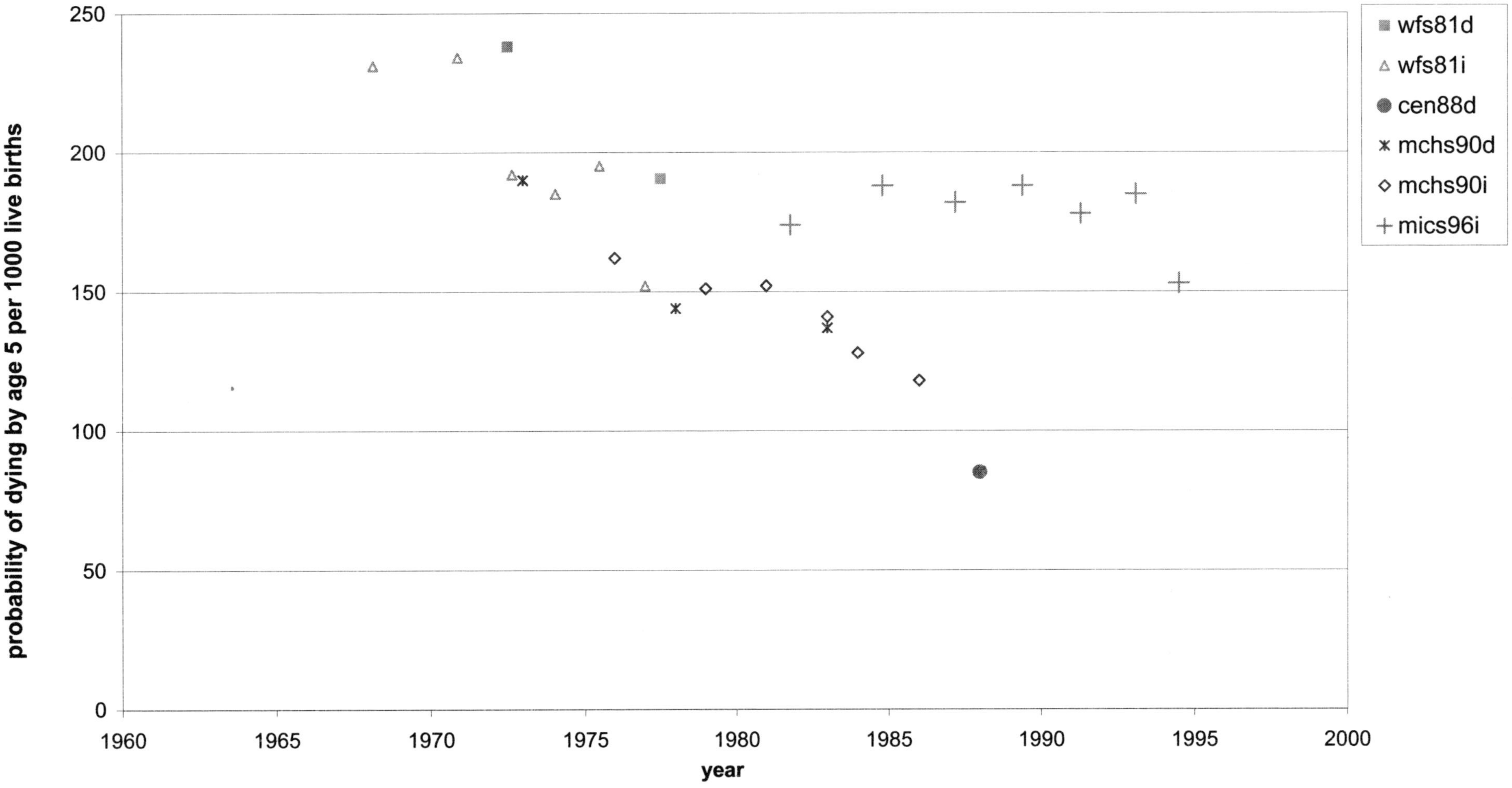

Data sources

Sources des données

Fuentes de los datos

wfs81d: Fertility Survey 1981-1982 (direct method); ***wfs81i***: Fertility Survey 1981-1982 (indirect method); ***cen88d***: Census 1988 (direct method); ***mchs90d***: Maternal and Child Health Survey 1990-1991 (direct method); ***mchs90i***: Maternal and Child Health Survey 1990-1991 (indirect method); ***mics96i***: Enquête Nationale sur les Indicateurs des objectifs à Mi-terme 1996 (indirect method)

Under 5 mortality

Mexico

Mortalité des moins de cinq ans

Mexique

Mortalidad de menores de cinco años

México

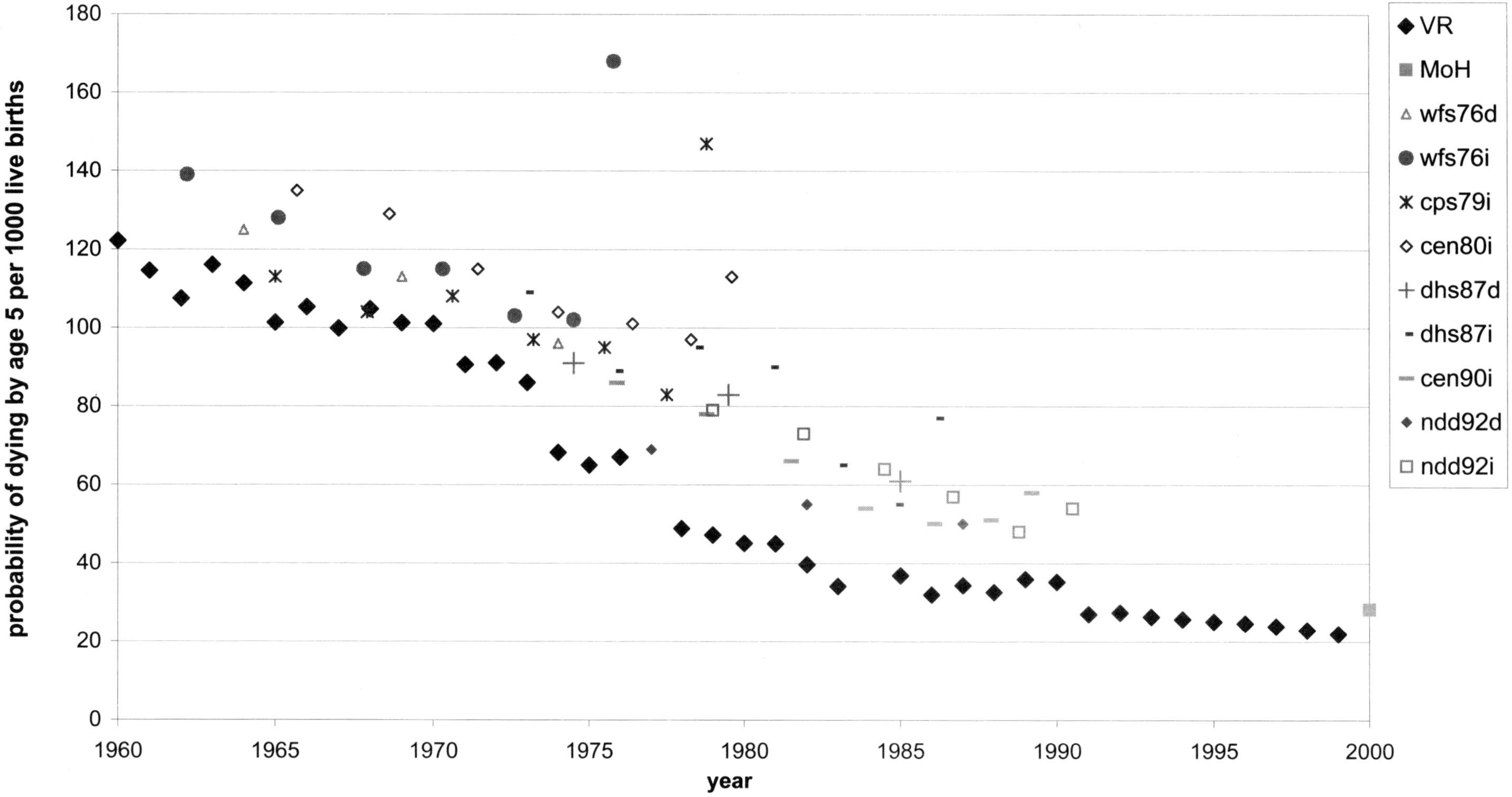

Data sources **Sources des données** **Fuentes de los datos**

VR: Vital Registration; ***MoH***: Ministry of Health estimate; ***wfs76d***: Encuesta Nactional de Fecundidad, July 1976-March 1977 (direct method); ***wfs76i***: Encuesta Nactional de Fecundidad, July 1976-March 1977 (indirect method); ***cps79i***: Encuesta Nacional de Prevalencia en el Usa de Metodos Anticonceptivos, September-December 1979 (indirect method); ***cen80i***: Census, 4-11 June 1980 (indirect method); ***dhs87d***: Encuesta Nactional Sobre Fecundidad y Salud, February-May 1987 (direct method); ***dhs87i***: Encuesta Nactional Sobre Fecundidad y Salud, February-May 1987 (indirect method); cen90i: Census, 12 March 1990 (indirect method); ***ndd92d***: Encuesta Nacional de la Dinamica Demografica 1992 (direct method); ***ndd92i***: Encuesta Nacional de la Dinamica Demografica 1992 (indirect method)

Under 5 mortality

Mongolia

Mortalité des moins de cinq ans

Mongolie

Mortalidad de menores de cinco años

Mongolia

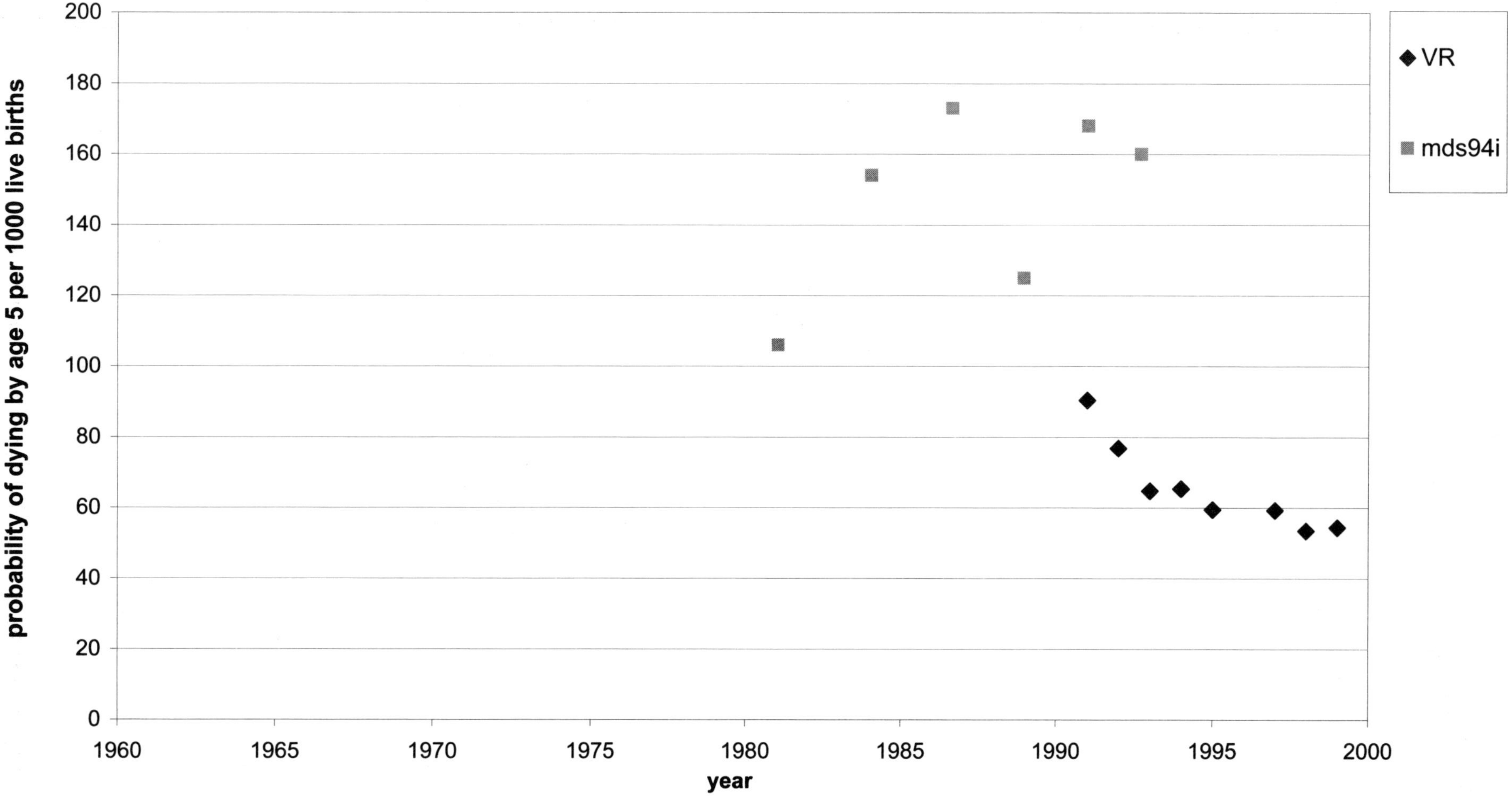

Data sources

Sources des données

Fuentes de los datos

VR: Vital Registration; ***mds94i***: Mongolian Demographic Survey 1994 (indirect method)

Under 5 mortality | Mortalité des moins de cinq ans | Mortalidad de menores de cinco años

Morocco | **Maroc** | **Marruecos**

Data sources | **Sources des données** | **Fuentes de los datos**

VR: Vital Registration; ***wfs80d***: Enquête Nationale sur la Fécondite et la Planification Familiale 1979-80 (direct method); ***wfs80i***: Enquête Nationale sur la Fécondite et la Planification Familiale 1979-80 (indirect method); ***cen82i***: Census, 3-21 September 1982 (indirect method); ***cps83i***: Planification Familiale, Fécondite et Santé Familiale 1983-84 (indirect method); ***dhs87d***: Enquête Nationale sur la Planification Familiale, la Fécondite et la Santé de la Population, April-July 1987 (direct method); ***dhs87i***: Enquête Nationale sur la Planification Familiale, la Fécondite et la Santé de la Population, April-July 1987 (indirect method); ***dhs92d***: Enquête Nationale sur la Population et la Santé, January-March 1992 (direct method); ***dhs92i***: Enquête Nationale sur la Population et la Santé, January-March 1992 (indirect method); ***dhs95d***: Enquête de Panel sur la Population et la Santé, April-May 1995 (direct method); ***dhs95i***: Enquête de Panel sur la Population et la Santé, April-May 1995 (indirect method); ***papchild97***: Maternal and Child Health Survey 1997

Under 5 mortality

Mozambique

Mortalité des moins de cinq ans

Mozambique

Mortalidad de menores de cinco años

Mozambique

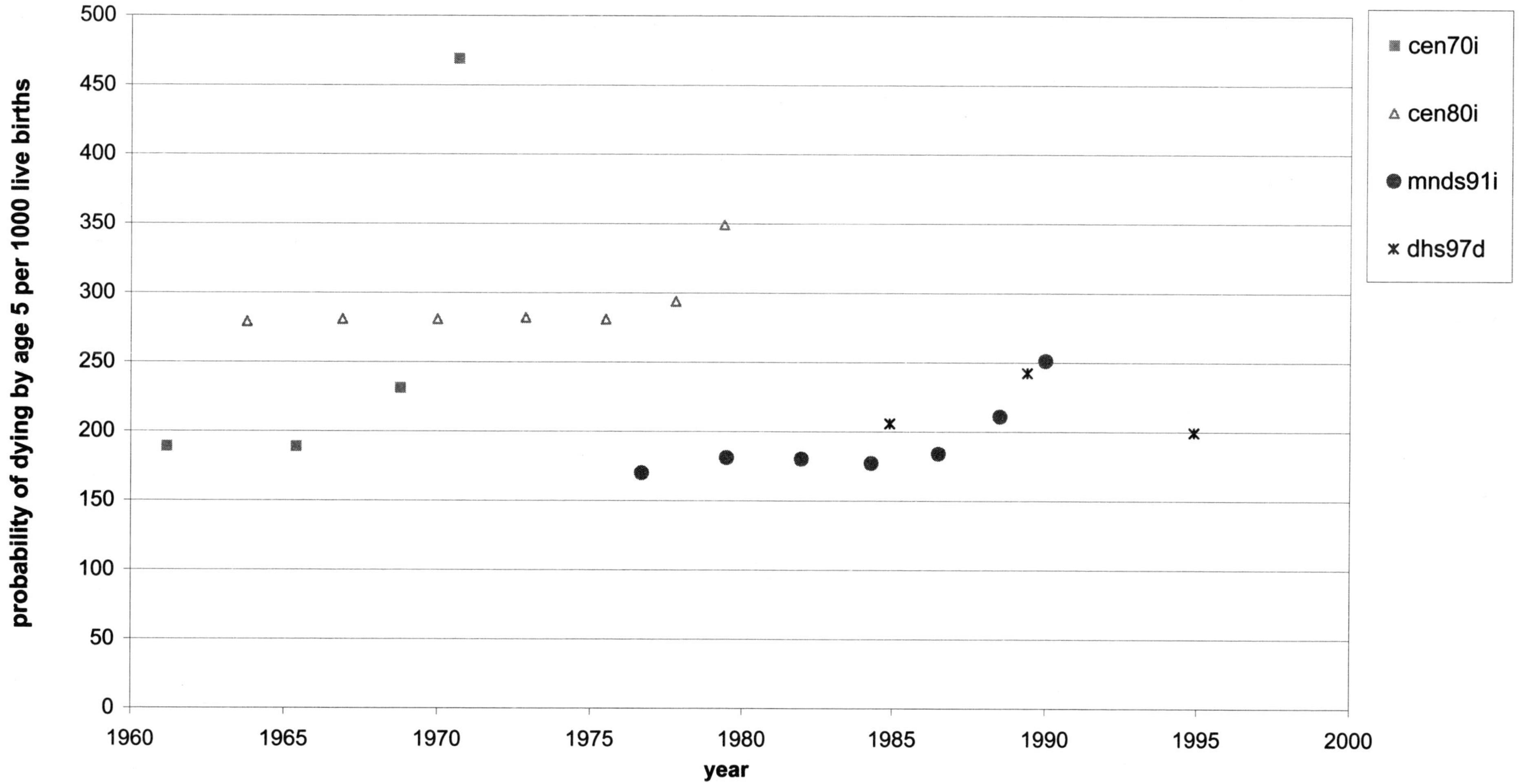

Data sources **Sources des données** **Fuentes de los datos**

cen70i : Census, 15 September 1970 (indirect method); ***cen80i*** : Census, 1-15 August 1980 (indirect method); ***mnds91i*** : Mozambique National Demographic Survey 1991 (indirect method); ***dhs97d*** : Demographic and Health Survey 1996 (direct method)

Under 5 mortality

Myanmar

Mortalité des moins de cinq ans

Myanmar

Mortalidad de menores de cinco años

Myanmar

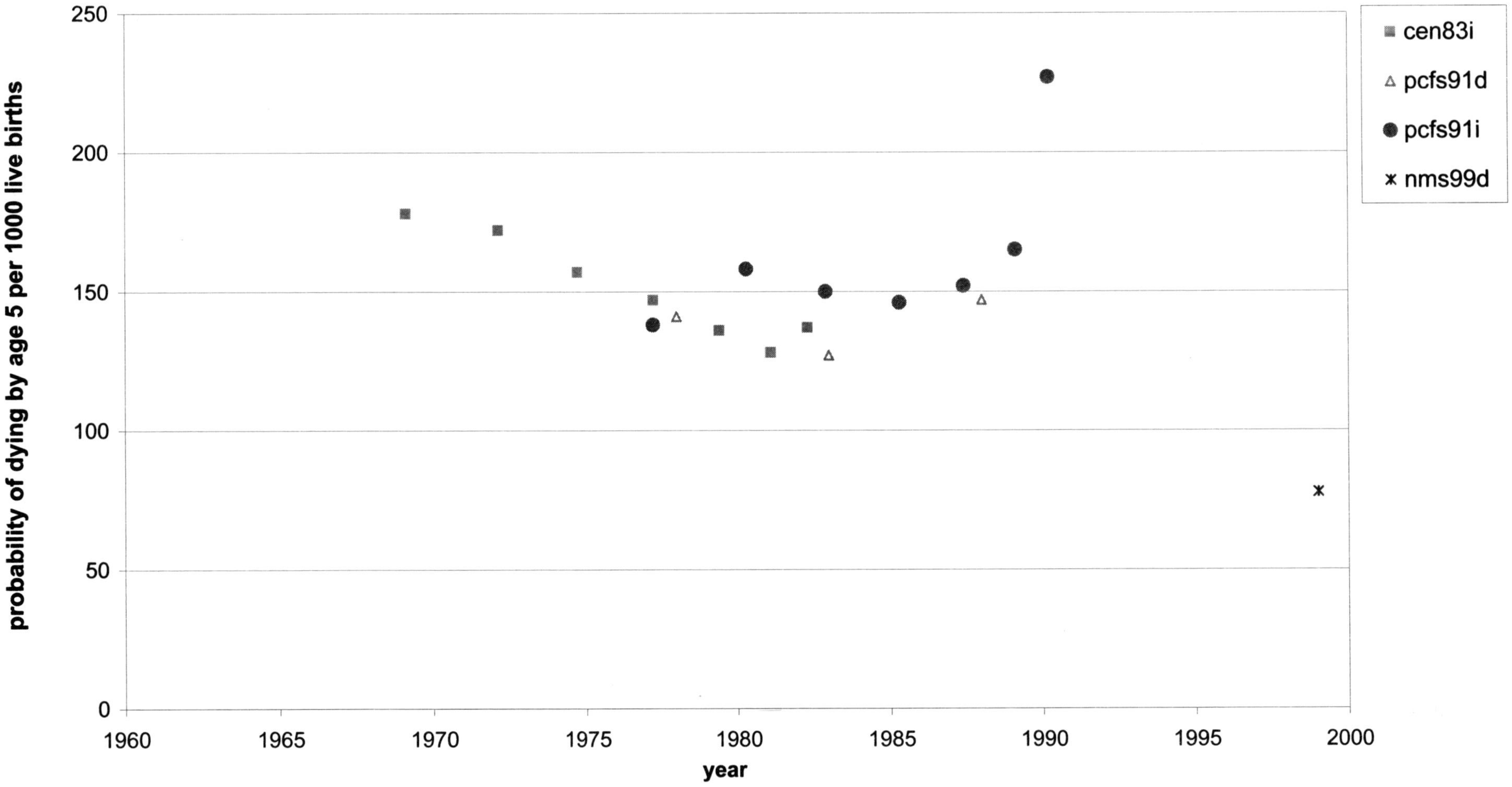

Data sources **Sources des données** **Fuentes de los datos**

cen83i: Census, 31 March 1983 (indirect method); ***pcfs91d***: Population Changes and Fertility Survey, January 1991 (direct method); ***pcfs91i***: Population Changes and Fertility Survey, January 1991 (indirect method); ***nms99d***: National Mortality Survey 1999 (direct method)

Under 5 mortality
Namibia

Mortalité des moins de cinq ans
Namibie

Mortalidad de menores de cinco años
Namibia

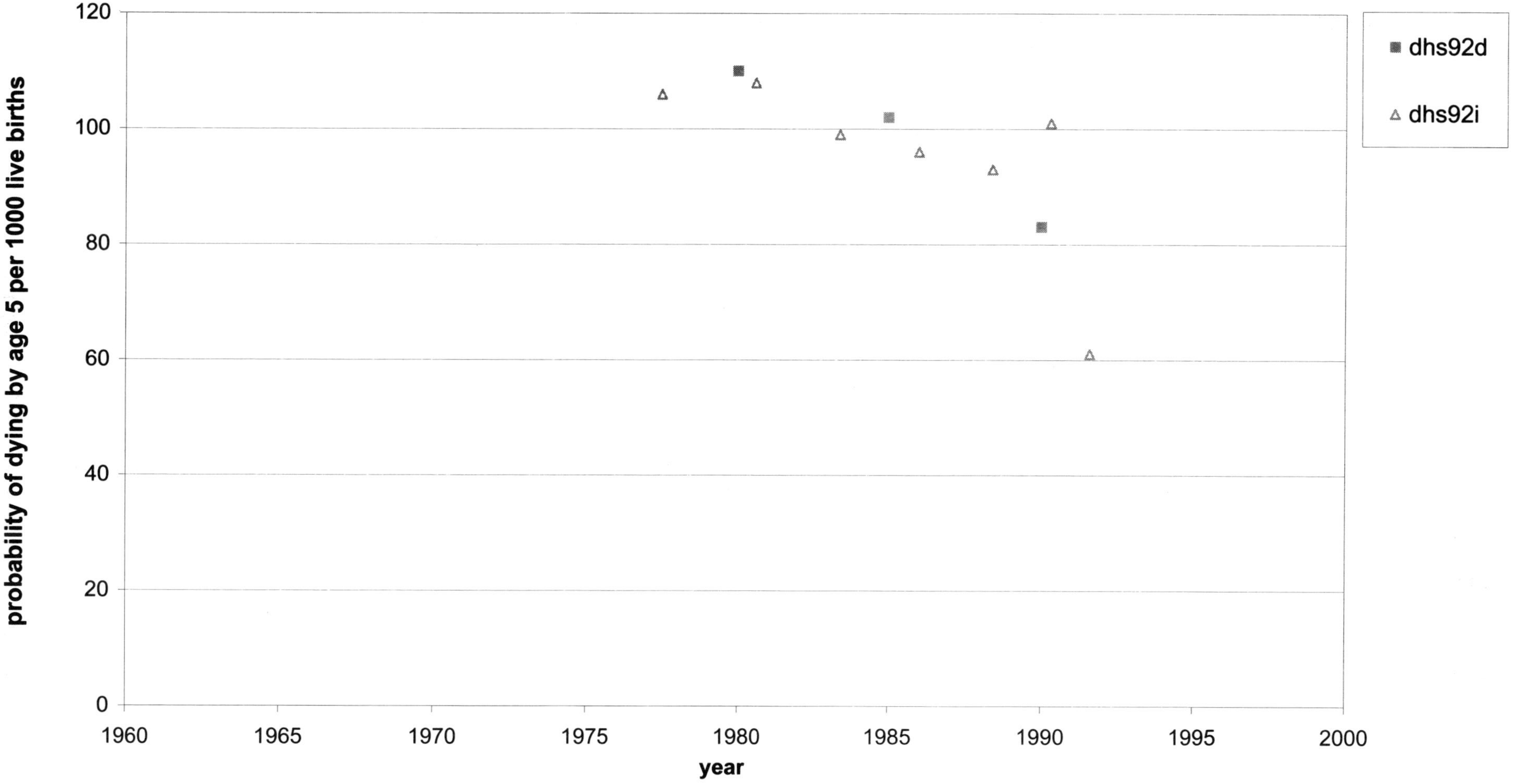

Data sources

Sources des données

Fuentes de los datos

dhs92d : Demographic and Health Survey, July-November 1992 (direct method); ***dhs92i*** : Demographic and Health Survey, July-November 1992 (indirect method)

Under 5 mortality

Nauru

Mortalité des moins de cinq ans

Nauru

Mortalidad de menores de cinco años

Nauru

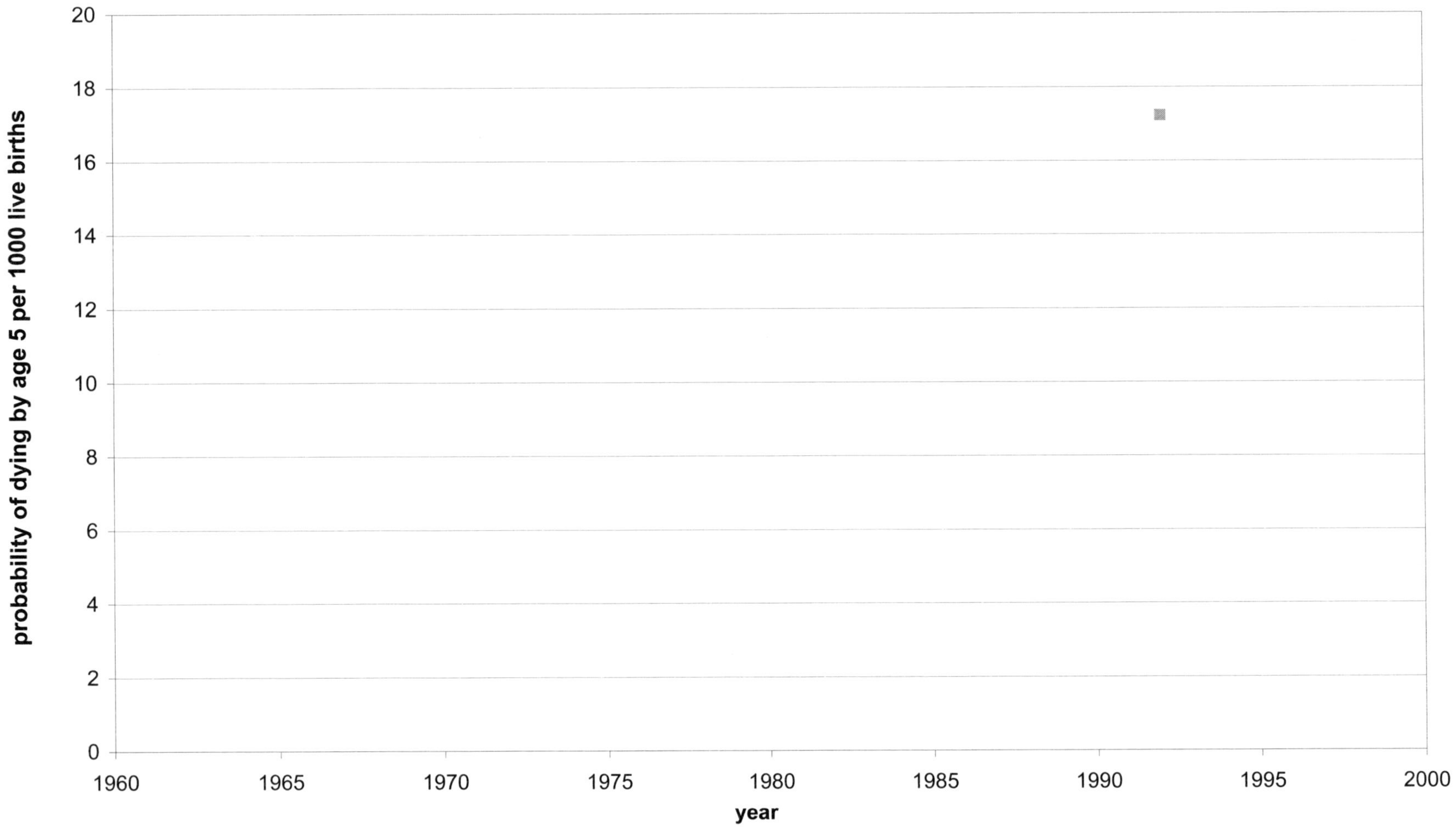

Data sources

Census 1992 (indirect method)

Sources des données

Fuentes de los datos

Under 5 mortality

Nepal

Mortalité des moins de cinq ans

Népal

Mortalidad de menores de cinco años

Nepal

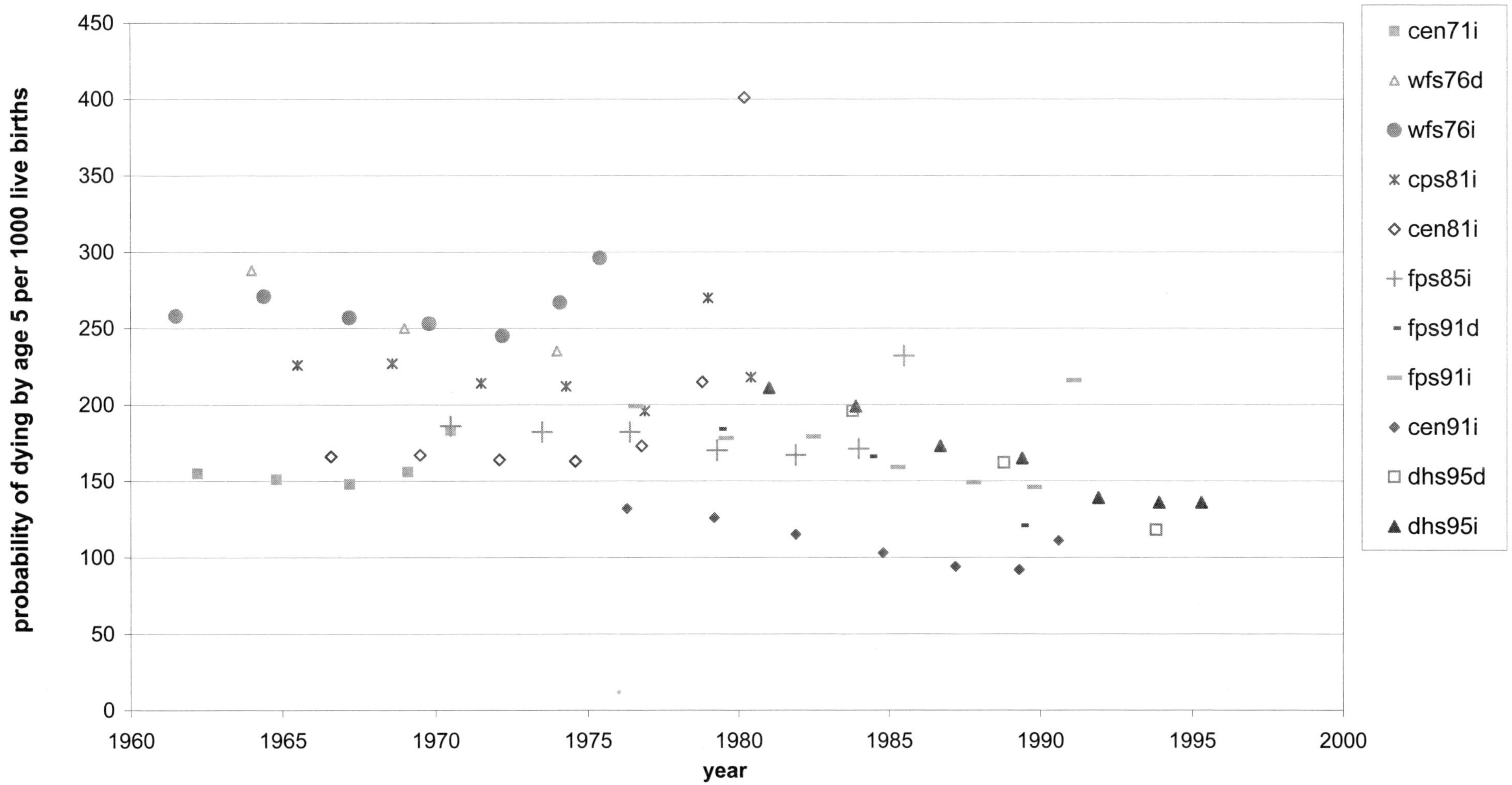

Data sources **Sources des données** **Fuentes de los datos**

cen71i: Census, 22 June 1971 (indirect method); ***wfs76d***: Fertility Survey, 1 April-20 June 1976 (direct method); ***wfs76i***: Fertility Survey, 1 April-20 June 1976 (indirect method); ***cps81i***: Contraceptive Prevalence Survey, 27 January-22 June 1981 (indirect method); ***cen81i***: Census, 22 June 1981 (indirect method); ***fps85i***: Fertility and Family Planning Survey, 28 December 1985-30 April 1986 (indirect method); ***fps91d***: Fertility, Family Planning and Health Survey, November 1991 (direct method); ***fps91i***: Fertility, Family Planning and Health Survey, November 1991 (indirect method); ***cen91i***: Census, 22 June 1991 (indirect method); ***dhs95d***: Demographic and Health Survey 1995 (direct method); ***dhs95i***: Demographic and Health Survey 1995 (indirect method)

Under 5 mortality
Netherlands

Mortalité des moins de cinq ans
Pays-Bas

Mortalidad de menores de cinco años
Países Bajos

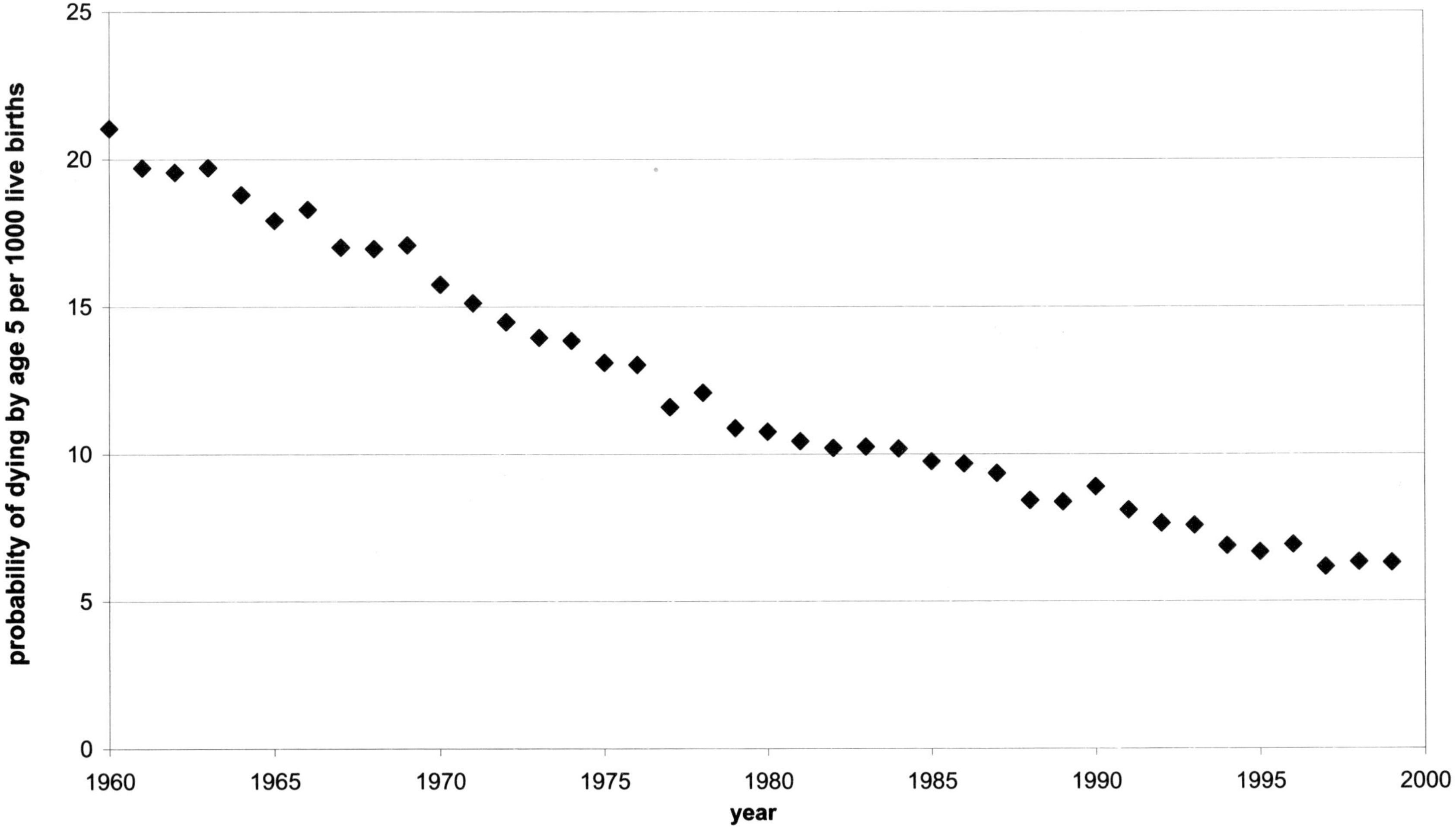

Data sources
Vital Registration

Sources des données

Fuentes de los datos

Under 5 mortality

New Zealand

Mortalité des moins de cinq ans

Nouvelle-Zélande

Mortalidad de menores de cinco años

Nueva Zelandia

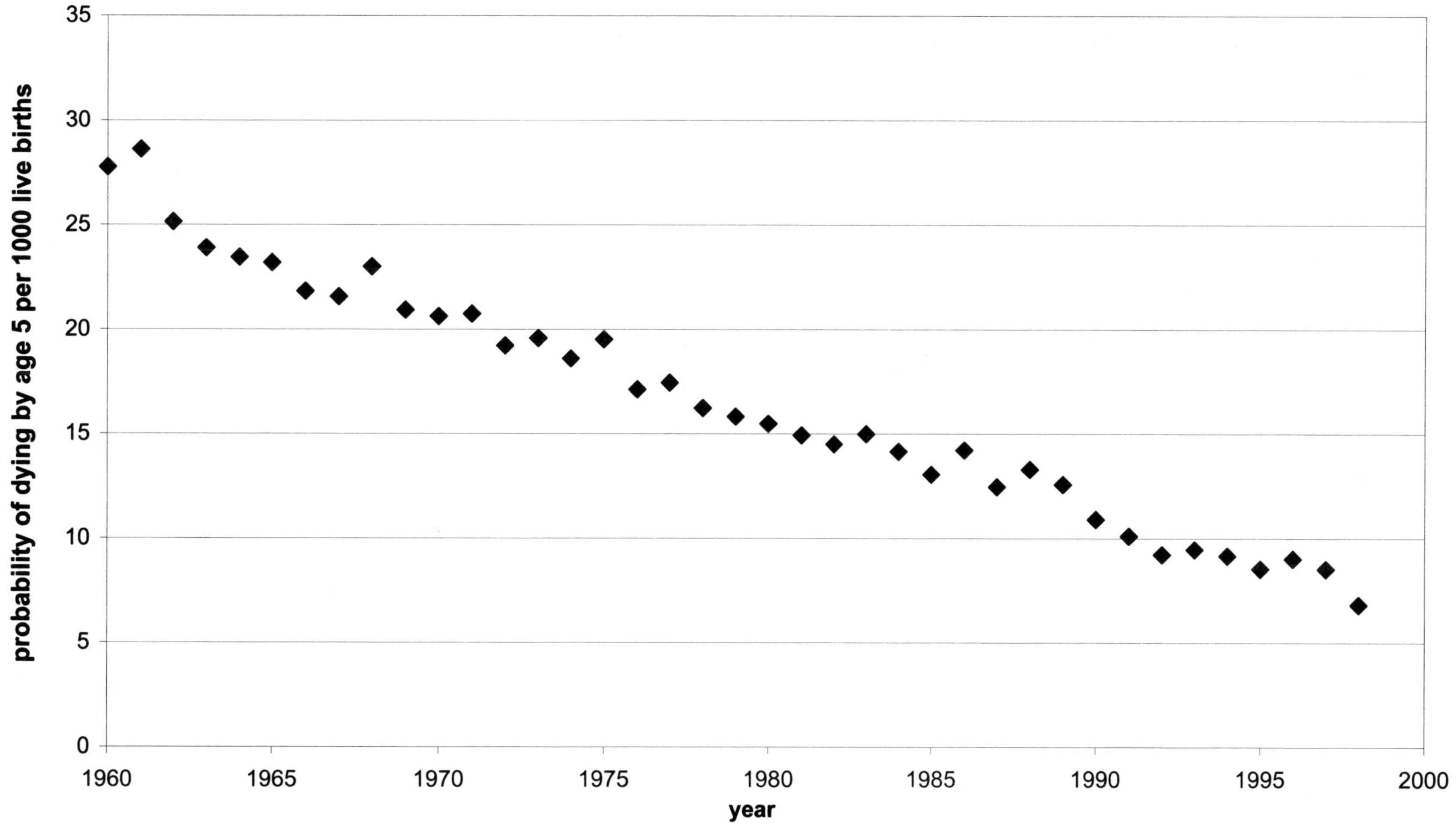

Data sources

Vital Registration

Sources des données

Fuentes de los datos

Under 5 mortality
Nicaragua

Mortalité des moins de cinq ans
Nicaragua

Mortalidad de menores de cinco años
Nicaragua

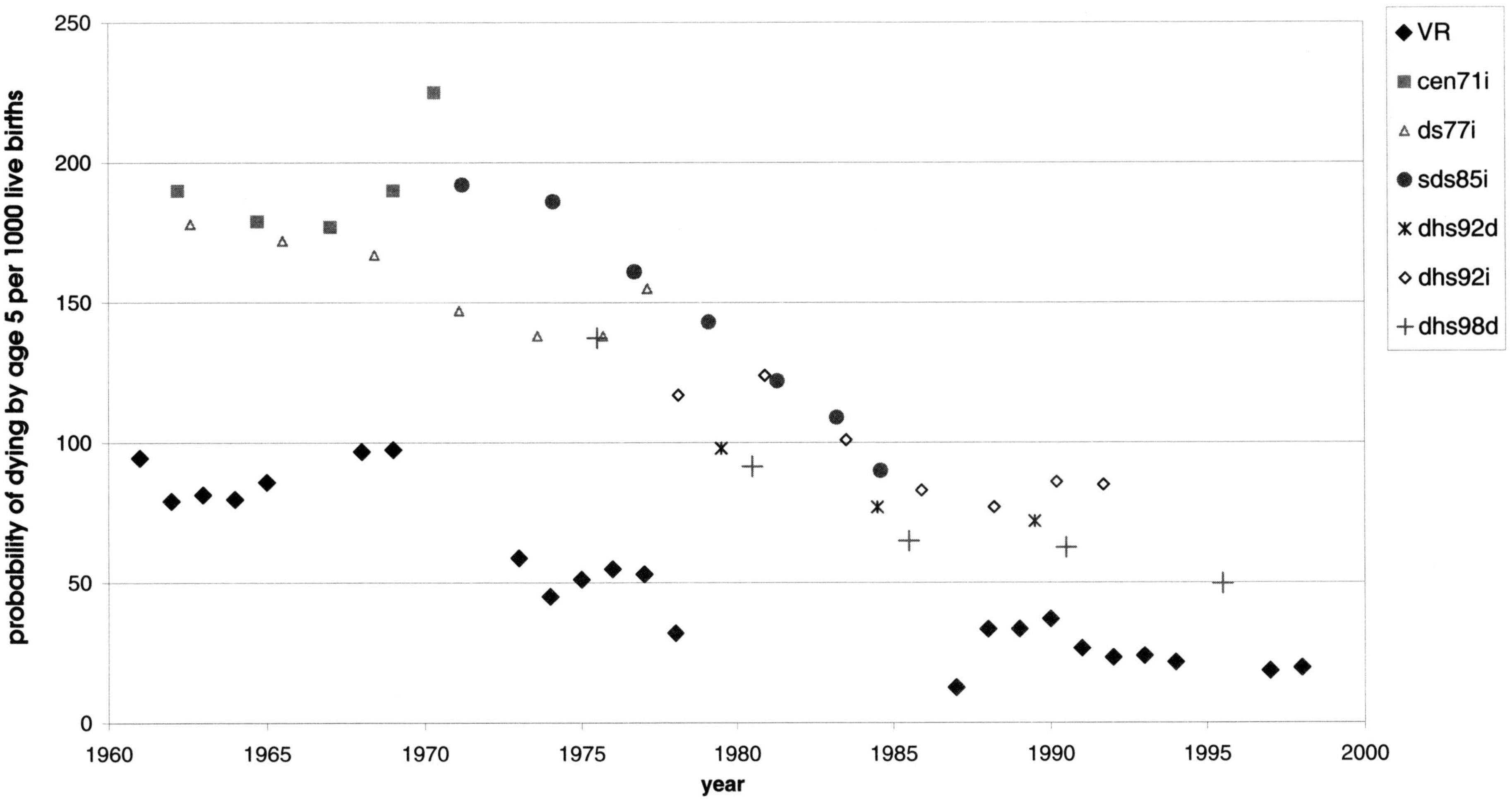

Data sources

Sources des données

Fuentes de los datos

VR: Vital Registration; ***cen71i***: Census, 20 April 1971 (indirect method); ***ds77i***: Encuesta Retrospectiva Demografica Nacional, December 1977-February 1978 (indirect method); ***sds85i***: Encuesta Socio-Demografica Nicarguense, July 1985-March 1986 (indirect method); ***dhs92d***: Encuesta Socio-Demografica Nicarguense, November 1992-February 1993 (direct method); ***dhs92i***: Encuesta Socio-Demografica Nicarguense, November 1992-February 1993 (indirect method); ***dhs98d***: Demographic and Health Survey 1998 (direct method)

Under 5 mortality

Niger

Mortalité des moins de cinq ans

Niger

Mortalidad de menores de cinco años

Níger

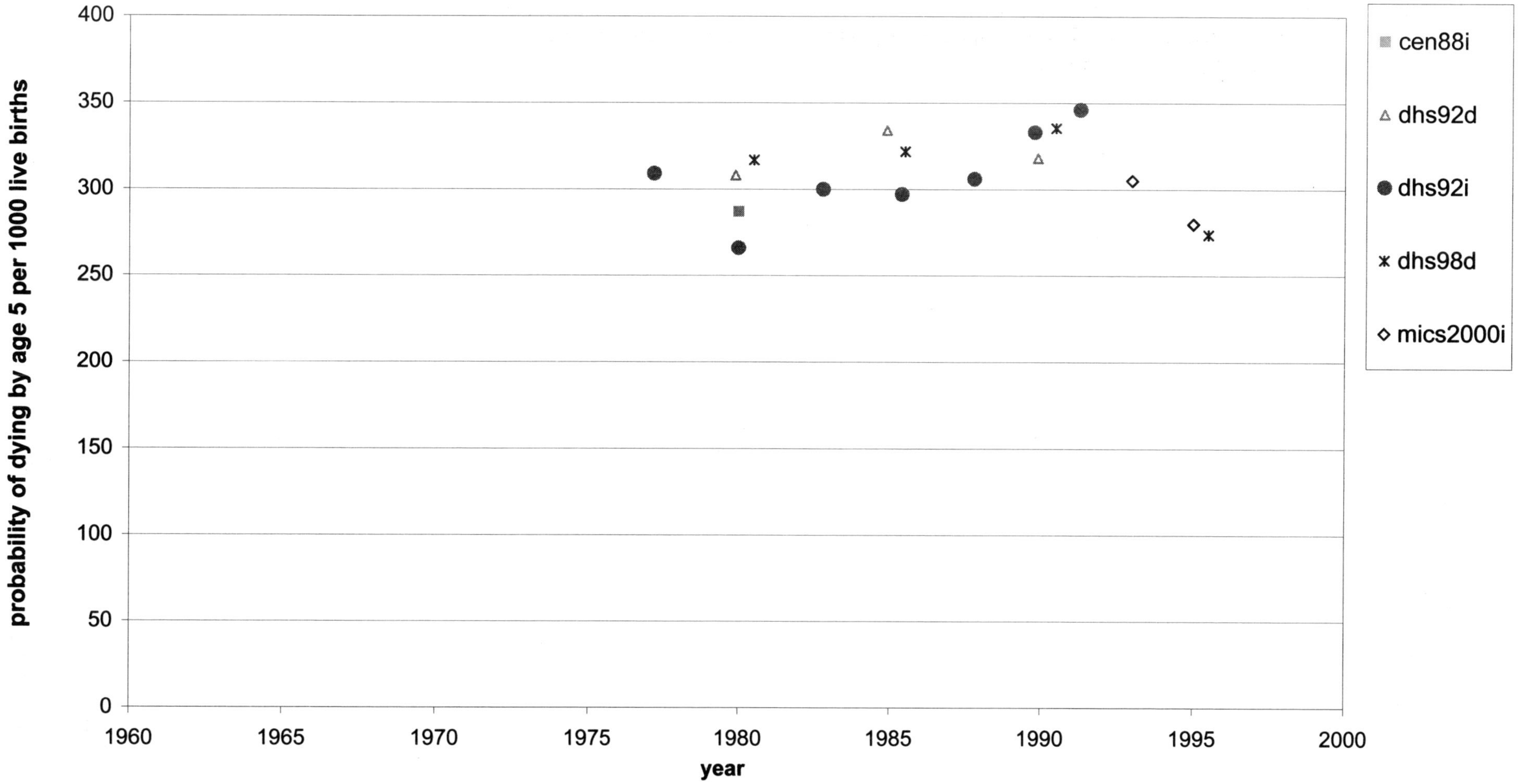

Data sources

Sources des données

Fuentes de los datos

cen88i : Census 1988 (indirect method); ***dhs92d*** : Demographic and Health Survey, March-June 1992 (direct method); ***dhs92i*** : Demographic and Health Survey, March-June 1992 (indirect method); ***dhs98d*** : Demographic and Health Survey 1998 (direct method); ***mics2000i*** : Enquête à indicateurs multiples de la fin de la décennie - draft report (indirect method)

Under 5 mortality

Nigeria

Mortalité des moins de cinq ans

Nigéria

Mortalidad de menores de cinco años

Nigeria

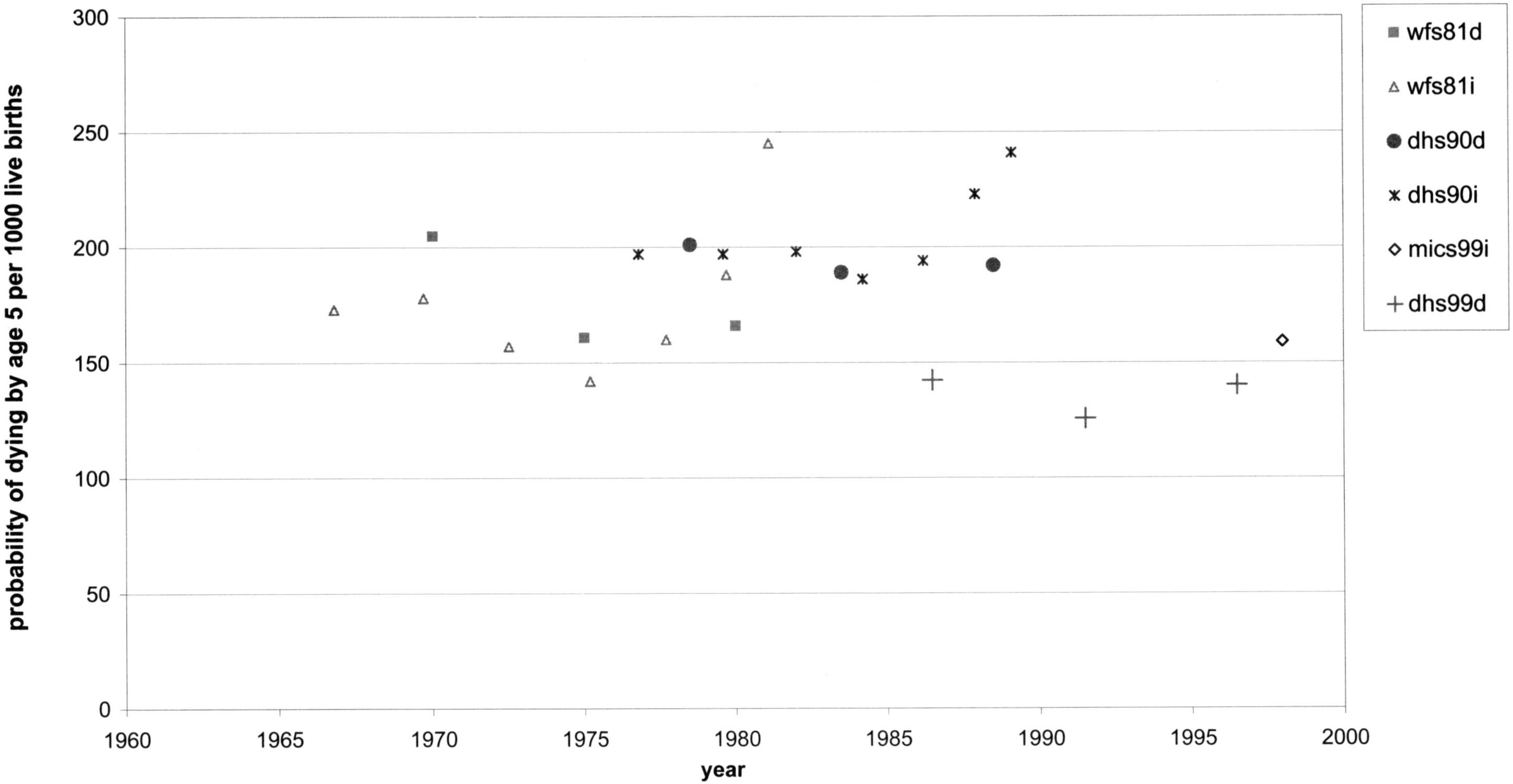

Data sources **Sources des données** **Fuentes de los datos**

wfs81d: Fertility Survey, October 1981-August 1982 (direct method); ***wfs81i***: Fertility Survey, October 1981-August 1982 (indirect method); ***dhs90d***: Demographic and Health Survey, April-October 1990 (direct method); ***dhs90i***: Demographic and Health Survey, April-October 1990 (indirect method); ***mics99i***: Multiple Indicator Cluster Survey (indirect method); ***dhs99d***: Demographic and Health Survey 1999 (direct method)

Under 5 mortality
Niue

Mortalité des moins de cinq ans
Nioué

Mortalidad de menores de cinco años
Niue

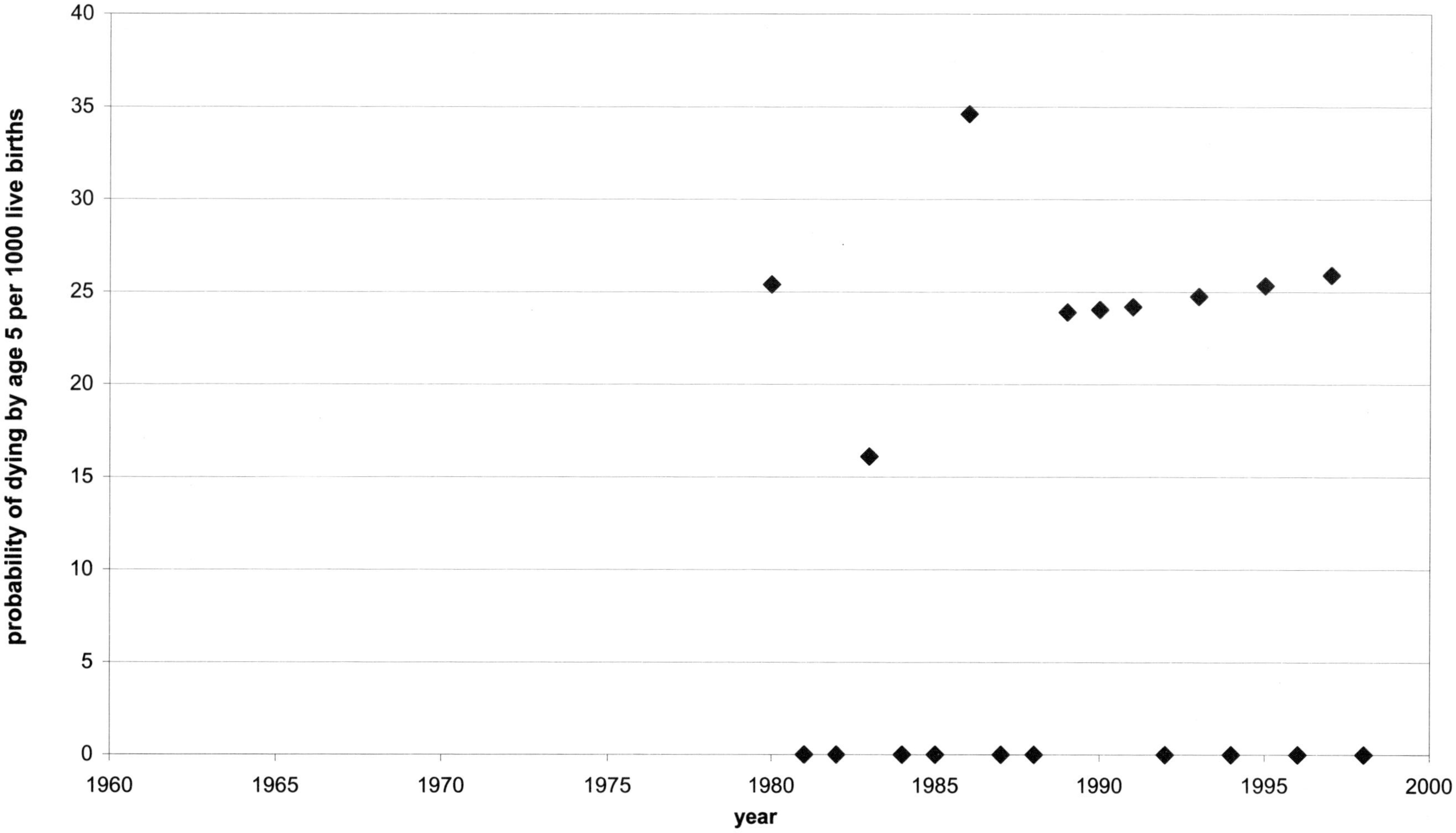

Data sources
Vital Registration

Sources des données

Fuentes de los datos

Under 5 mortality

Norway

Mortalité des moins de cinq ans

Norvège

Mortalidad de menores de cinco años

Noruega

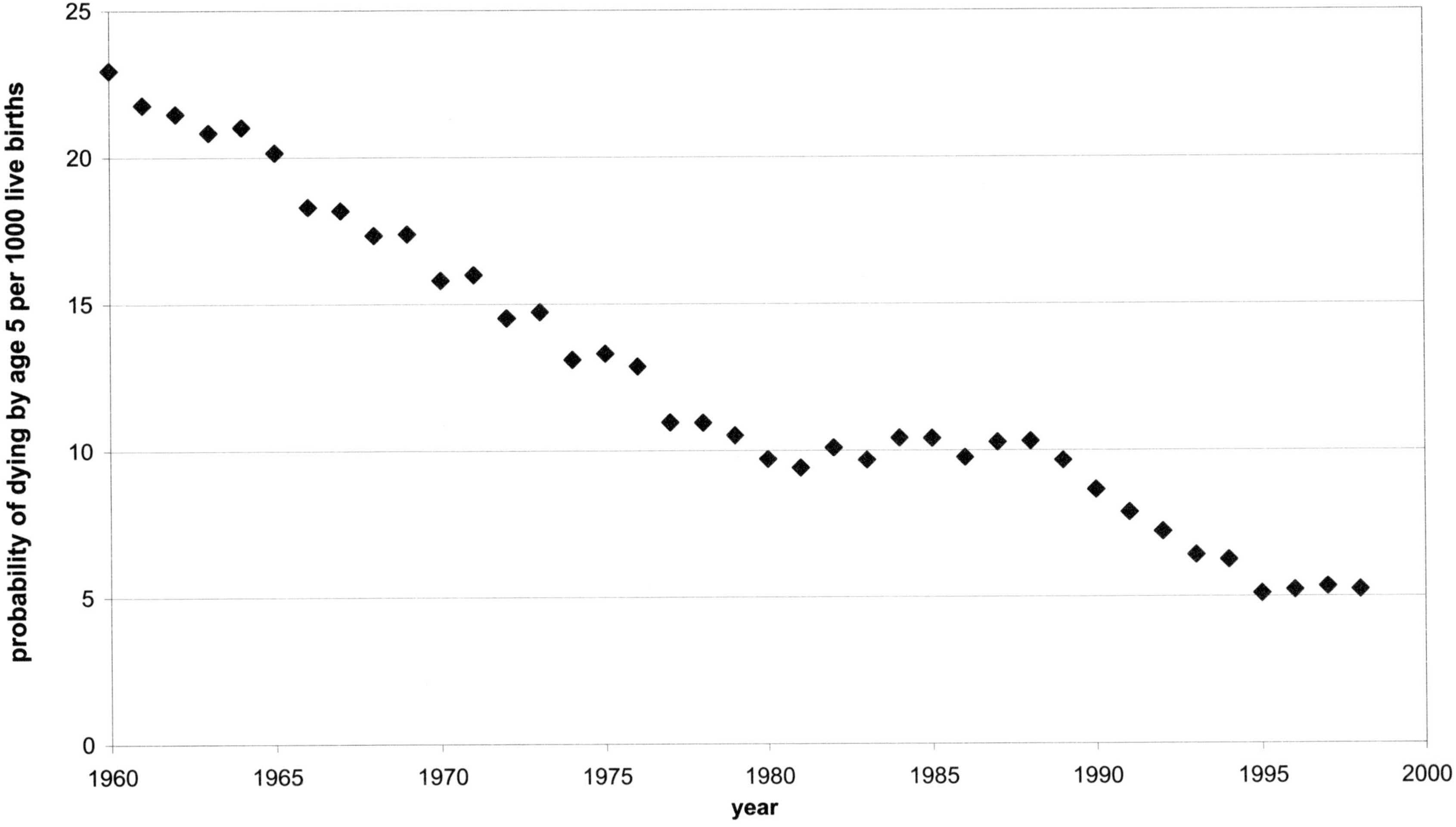

Data sources

Vital Registration

Sources des données

Fuentes de los datos

Under 5 mortality Mortalité des moins de cinq ans Mortalidad de menores de cinco años

Oman **Oman** **Omán**

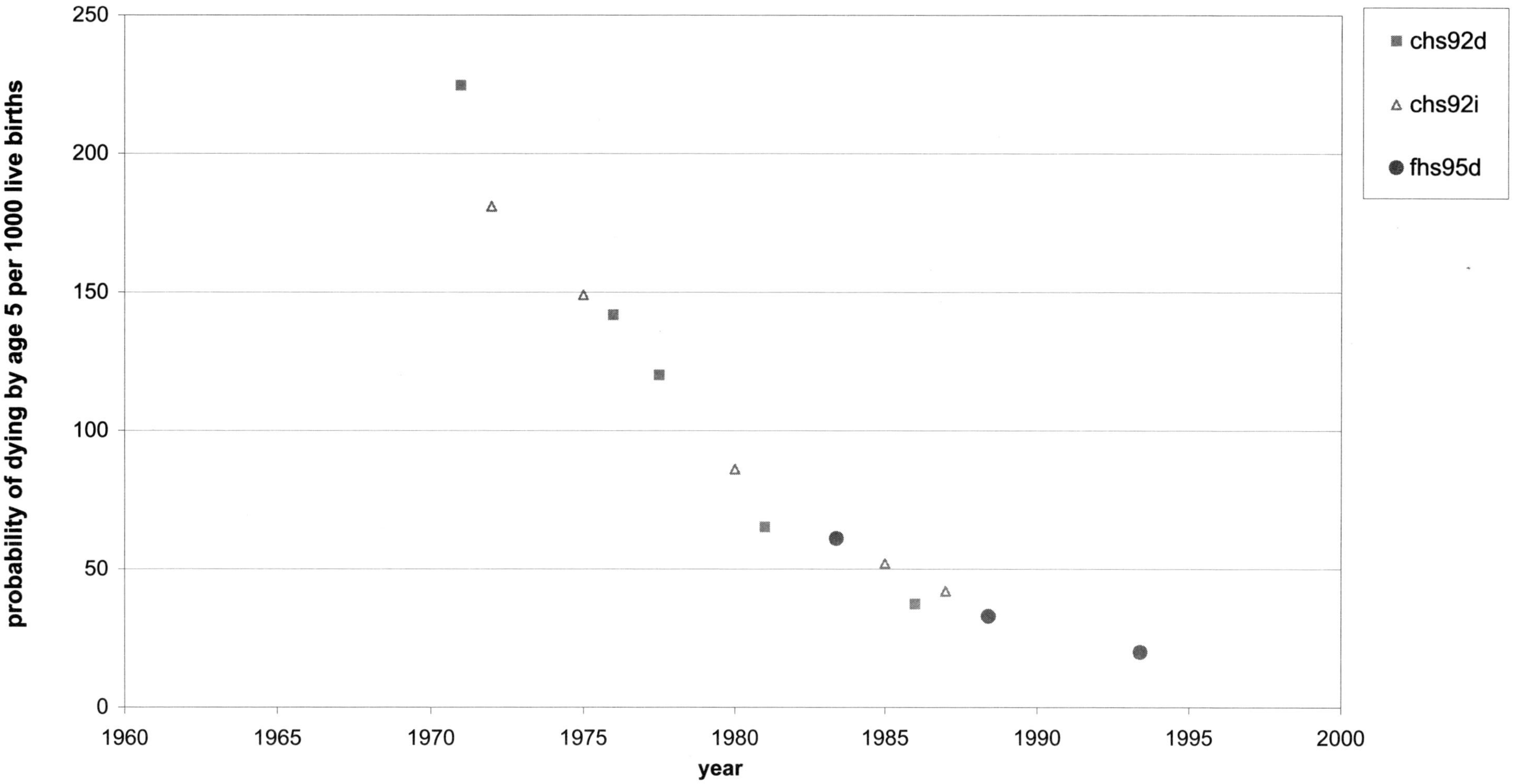

Data sources **Sources des données** **Fuentes de los datos**

chs92d : Child Health Survey 1992 (direct method); ***chs92i*** : Child Health Survey 1992 (indirect method); ***fhs95d*** : Family Health Survey 1995 (direct method)

Under 5 mortality

Pakistan

Mortalité des moins de cinq ans

Pakistan

Mortalidad de menores de cinco años

Pakistán

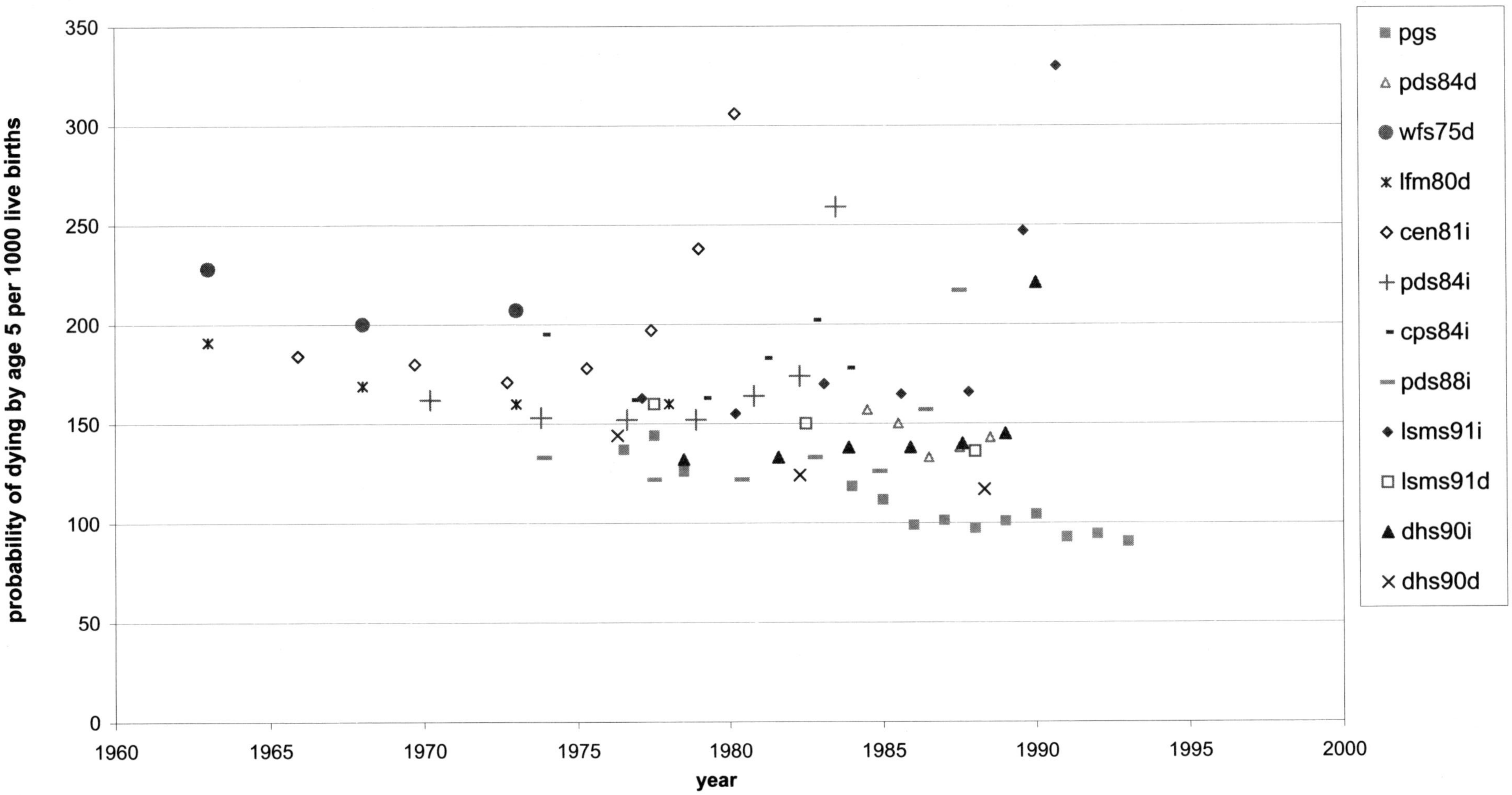

Data sources

Sources des données

Fuentes de los datos

pgs: Population Growth Survey, various years; ***pds84d***: Demographic Survey 1984-1988 (direct method); ***wfs75d***: Fertility Survey, May-December 1975 (direct method); ***lfm80d***: Labour Force and Migration Survey 1980 (direct method); ***cen81i***: Census, 1 March 1981 (indirect method); ***pds84i***: Demographic Survey, August 1984 (indirect method); ***cps84i***: Contraceptive Prevalence Survey, October 1984-March 1985 (indirect method); ***pds88i***: Demographic Survey, July 1988 (indirect method); ***lsms91i***: Living Standards Survey 1991 (indirect method); ***lsms91d***: Living Standards Survey 1991 (direct method); ***dhs90i***: Demographic and Health Survey, December 1990-May1991 (indirect method); ***dhs90d***: Demographic and Health Survey, December 1990-May1991 (direct method)

Under 5 mortality
Panama

Mortalité des moins de cinq ans
Panama

Mortalidad de menores de cinco años
Panamá

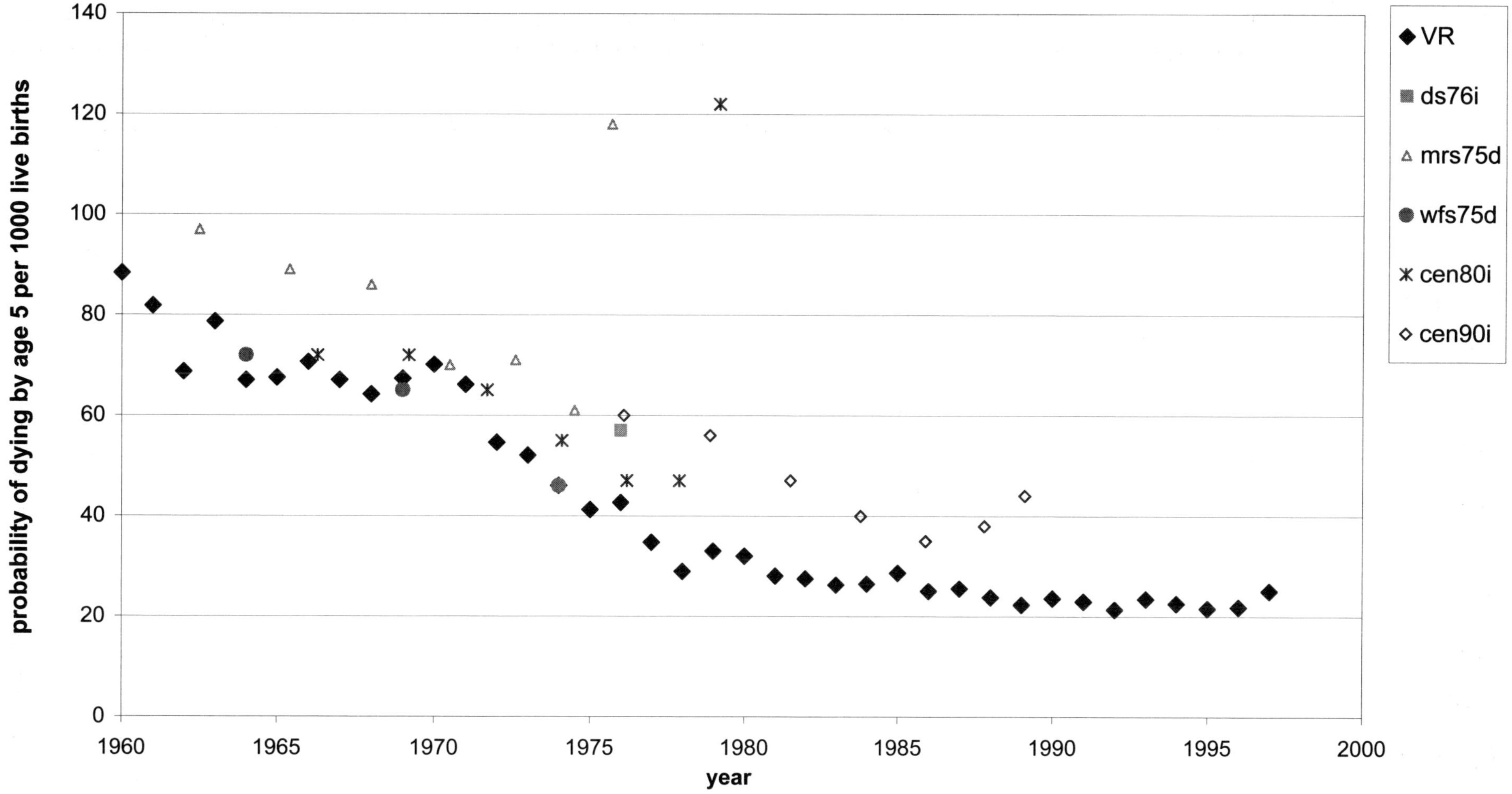

Data sources

Sources des données

Fuentes de los datos

VR: Vital Registration; ***ds76i***: Encuesta Demografica Nacional, June 1975-February 1977 (indirect method); ***mrs75d***: Encuesta Deomgrafica Nacional Retrospectiva, September 1976- February 1977 (direct method); ***wfs75d***: Encuesta Nacional de Fecundidad, December 1975-April 1976 (direct method); ***cen80i***: Census, 11 May 1980 (indirect method); ***cen90i***: Census, 13 May 1990 (indirect method)

Under 5 mortality

Papua New Guinea

Mortalité des moins de cinq ans

Papouasie-Nouvelle-Guinée

Mortalidad de menores de cinco años

Papua Nueva Guinea

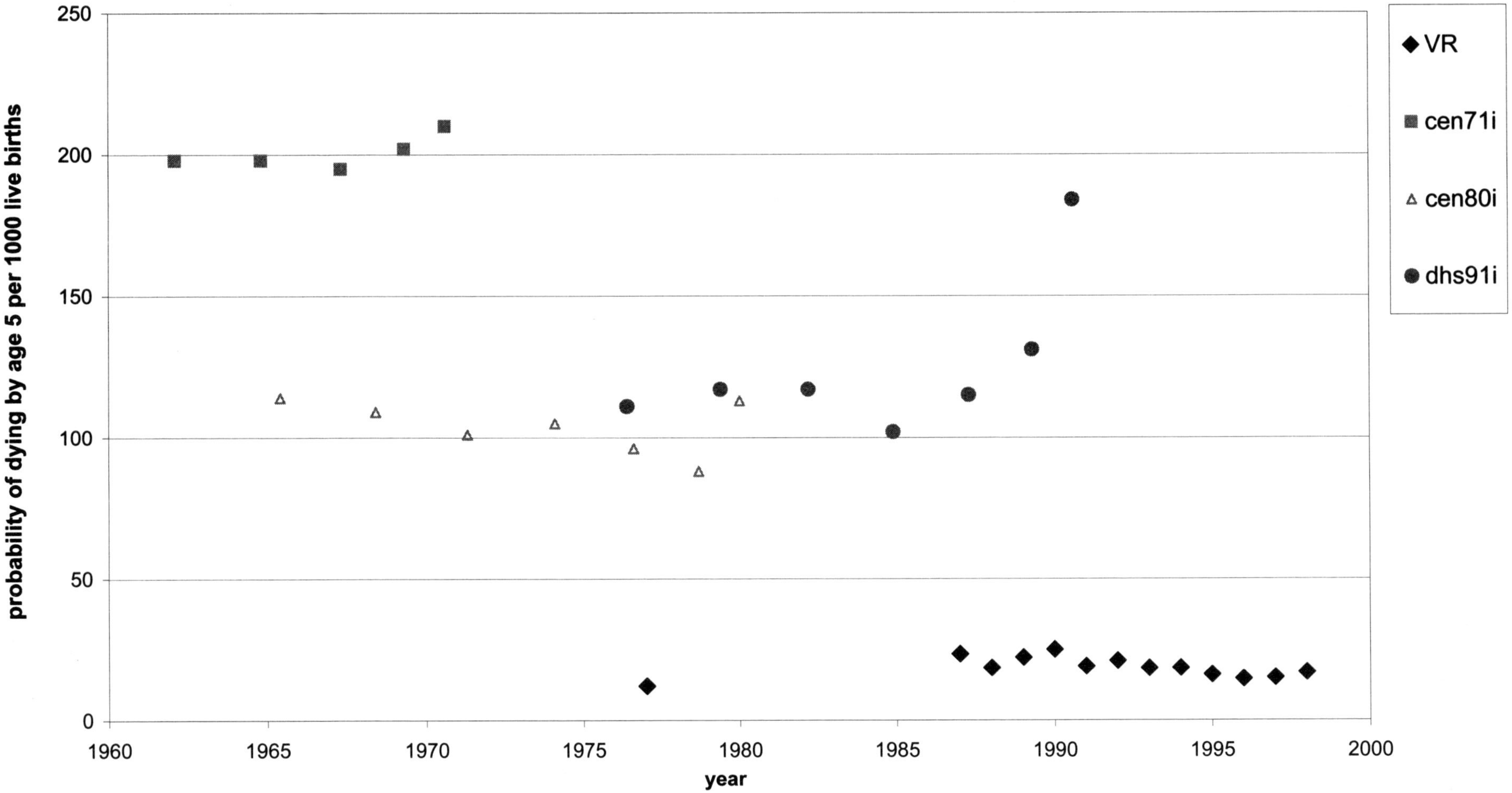

Data sources

Sources des données

Fuentes de los datos

VR: Vital Registration; ***cen71i***: Census, 7 July 1971 (indirect method); ***cen80i***: Census, 1 September 1980 (indirect method); ***dhs91i***: Demographic and Health Survey 1991 (indirect method)

Under 5 mortality

Paraguay

Mortalité des moins de cinq ans

Paraguay

Mortalidad de menores de cinco años

Paraguay

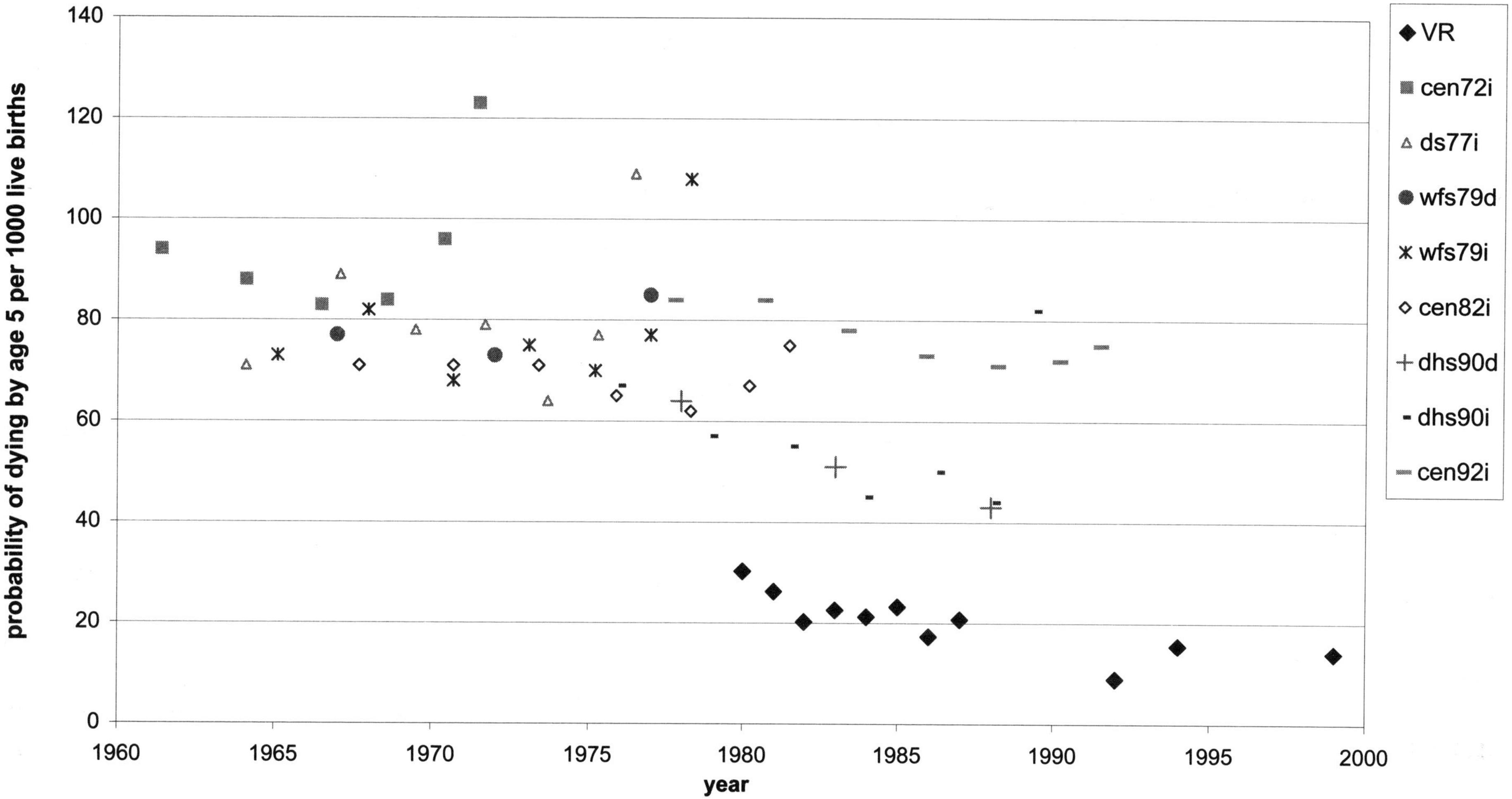

Data sources

Sources des données

Fuentes de los datos

VR: Vital Registration; ***cen72i***: Census, 9 July 1972 (indirect method); ***ds77i***: Encuesta Demografica Nacional, July-October 1977 (indirect method); ***wfs79d***: Encuesta Nacional de Fecundidad, February-May 1979 (direct method); ***wfs79i***: Encuesta Nacional de Fecundidad, February-May 1979 (indirect method); ***cen82i***: Census, 11 July 1982 ; ***dhs90d***: Encuesta Nacional de Demografia y Salud, May-August 1990 (direct method); ***dhs90i***: Encuesta Nacional de Demografia y Salud, May-August 1990 (indirect method); ***cen92i***: Census, 26 August 1992 (indirect method)

Under 5 mortality
Peru

Mortalité des moins de cinq ans
Pérou

Mortalidad de menores de cinco años
Perú

Data sources **Sources des données** **Fuentes de los datos**

VR: Vital Registration; ***cen72i***: Census, 4 June 1972 (indirect method); ***ds74d***: Encuesta Demografica Nacional, November 1974-December 1976 (direct method); ***ednr76i***: Encuesta Demografica Nacional Retrospectiva, April-December 1976 (indirect method); ***wfs77d***: Encuesta Nacional de Fecundidad, July 1977-June 1978 (direct method); ***wfs77i***: Encuesta Nacional de Fecundidad, July 1977-June 1978 (indirect method); ***cen81i***: Census, 12 July 1981 (indirect method); ***cps81i***: Encuesta Nacional de Prevalencia de Anticonceptivos, August-December 1981 (indirect method); ***dhs86d***: Encuesta Demografica y de Salud Familiar, September-December 1986 (direct method); ***dhs86i***: Encuesta Demografica y de Salud Familiar, September-December 1986 (indirect method); ***dhs91d***: Encuesta Demografica y de Salud Familiar, October 1991-March 1992 (direct method); ***dhs91i***: Encuesta Demografica y de Salud Familiar, October 1991-March 1992 (indirect method); ***cen93i***: Census, July 1993 (indirect method); ***dhs96d***: Demographic and Health Survey 1996 (direct method)

Under 5 mortality Mortalité des moins de cinq ans Mortalidad de menores de cinco años

Philippines **Philippines** **Filipinas**

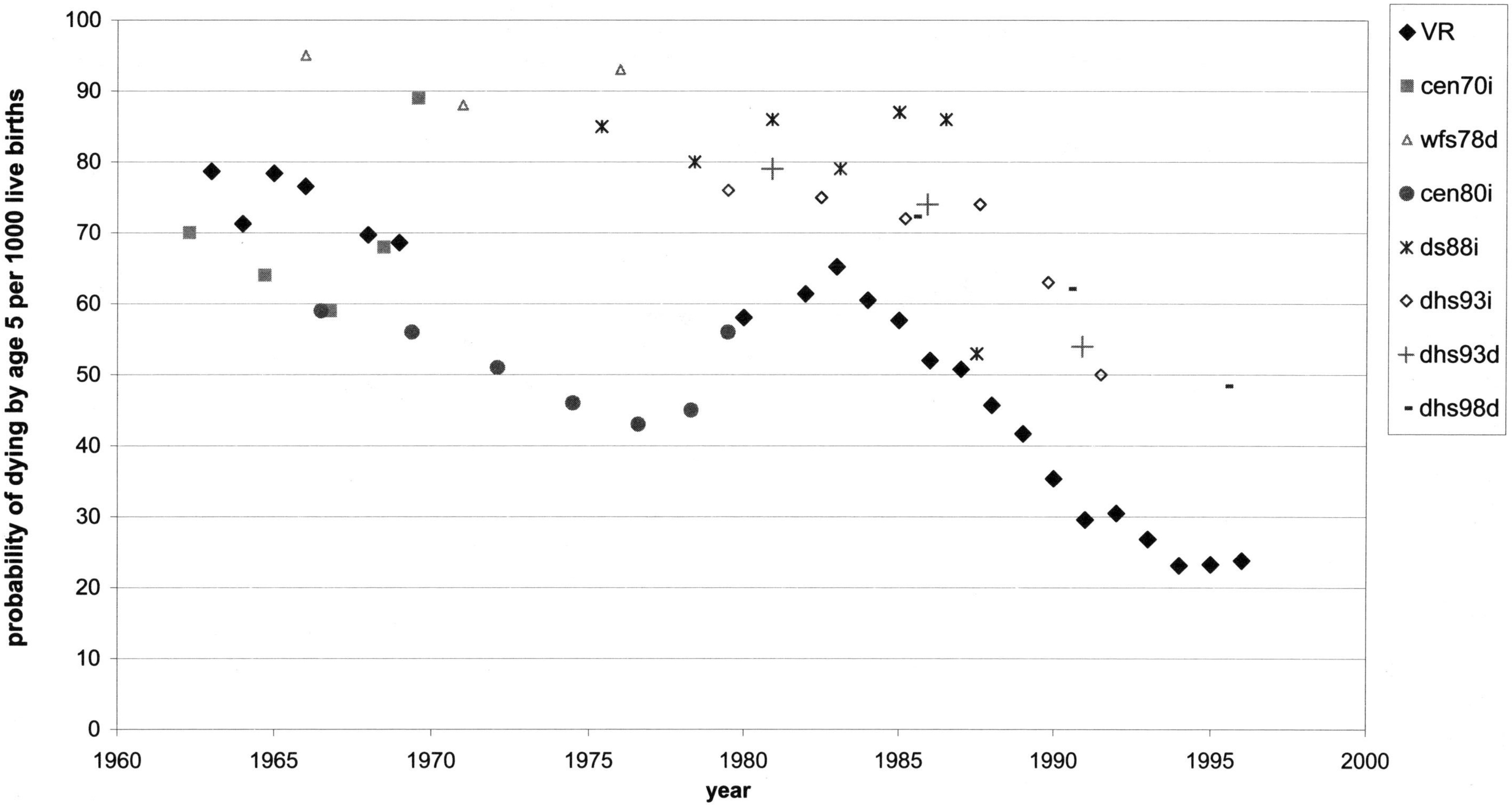

Data sources **Sources des données** **Fuentes de los datos**

VR: Vital Registration; ***cen70i***: Census, 6 May 1970 (indirect method); ***wfs78d***: Fertility Survey, 27 February-18 June 1978 (direct method); ***cen80i***: Census, 1 May 1980 (indirect method); ***ds88i***: Demographic Survey, 1988 (indirect method); ***dhs93i***: National Demographic Survey, April-June 1993 (indirect method); ***dhs93d***: National Demographic Survey, April-June 1993 (direct method); ***dhs98d***: Demographic and Health Survey 1998 (direct method)

Under 5 mortality
Poland

Mortalité des moins de cinq ans
Pologne

Mortalidad de menores de cinco años
Polonia

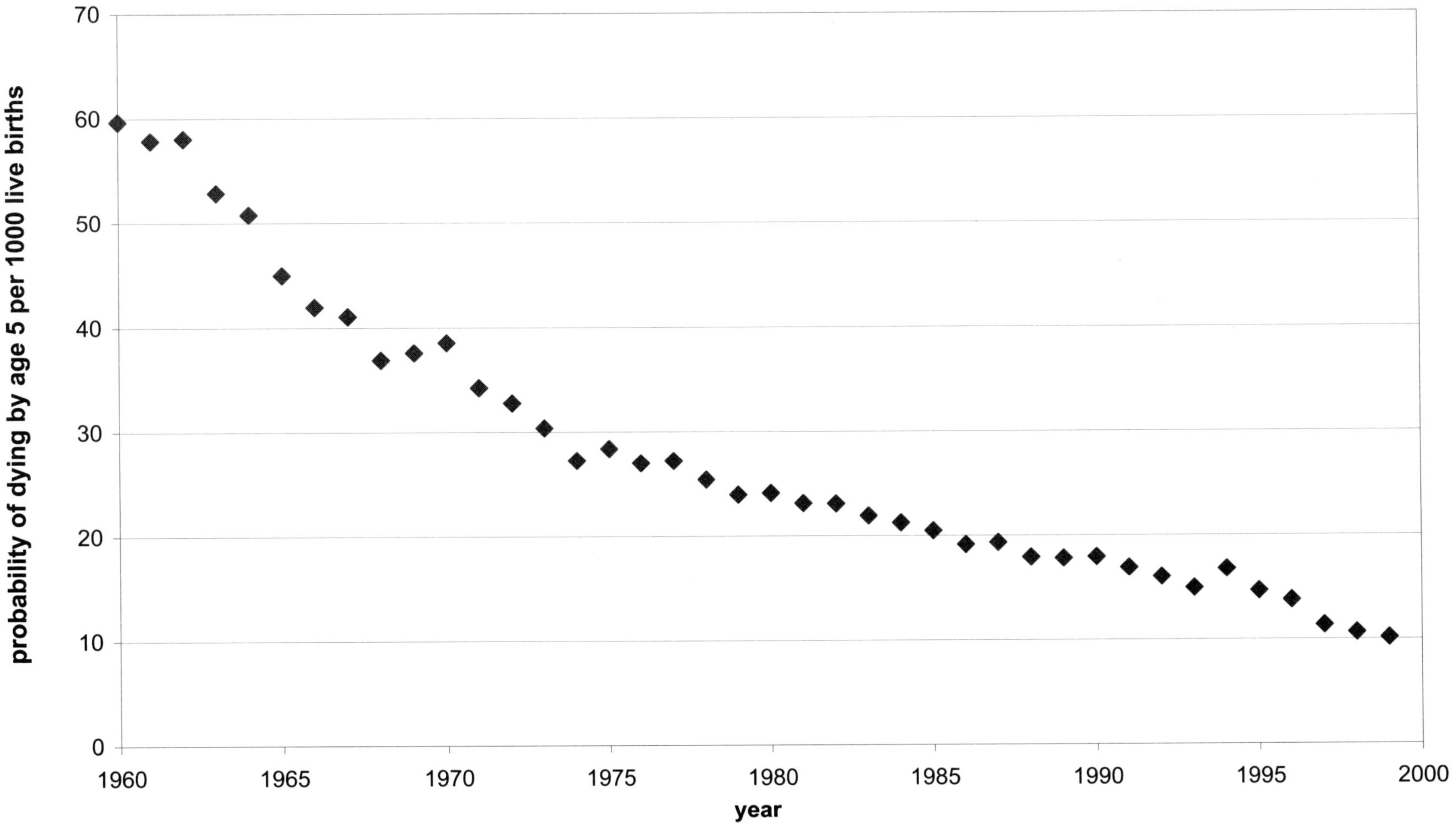

Data sources
Vital Registration

Sources des données

Fuentes de los datos

Under 5 mortality

Portugal

Mortalité des moins de cinq ans

Portugal

Mortalidad de menores de cinco años

Portugal

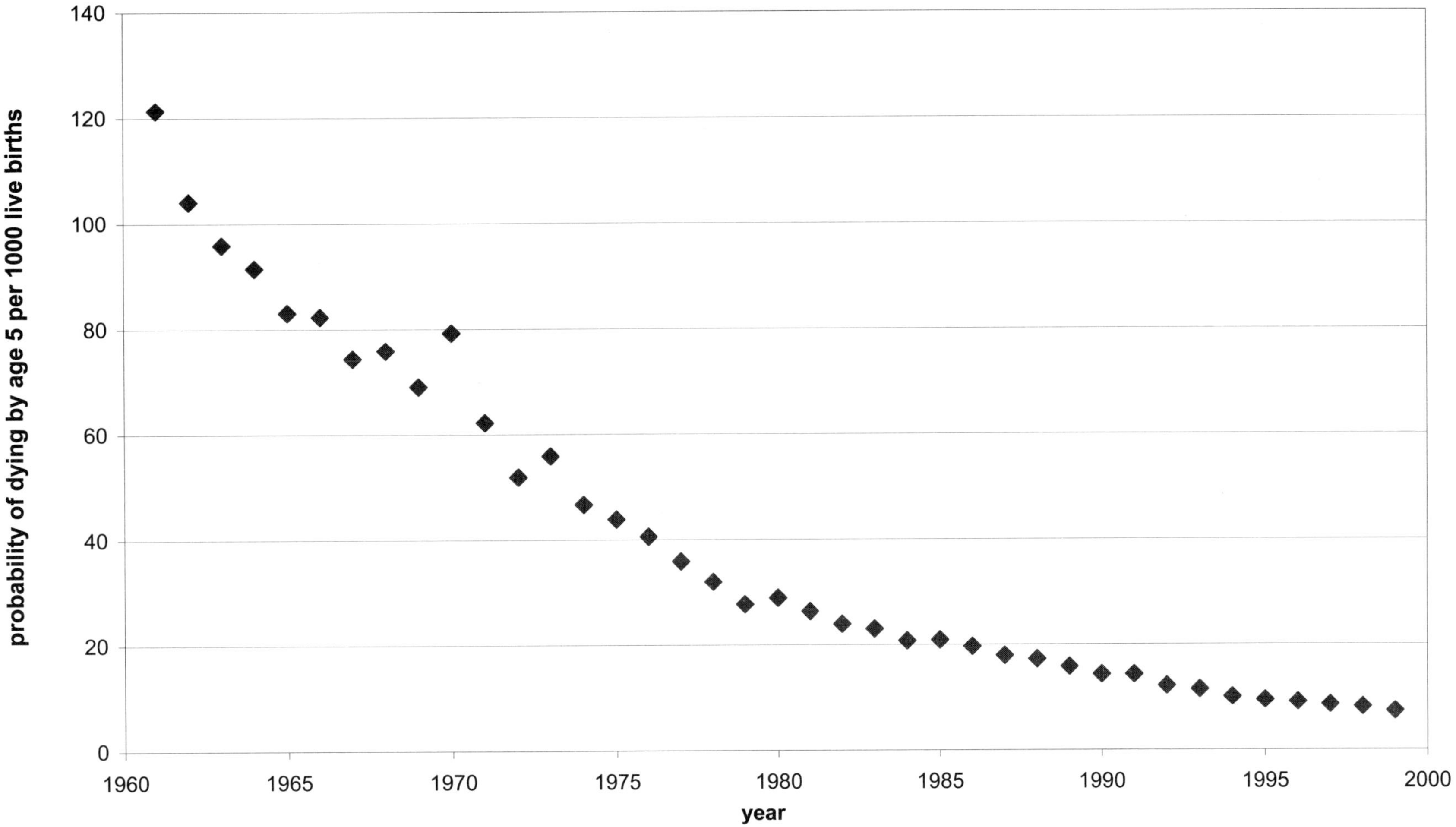

Data sources

Vital Registration

Sources des données

Fuentes de los datos

Under 5 mortality

Qatar

Mortalité des moins de cinq ans

Qatar

Mortalidad de menores de cinco años

Qatar

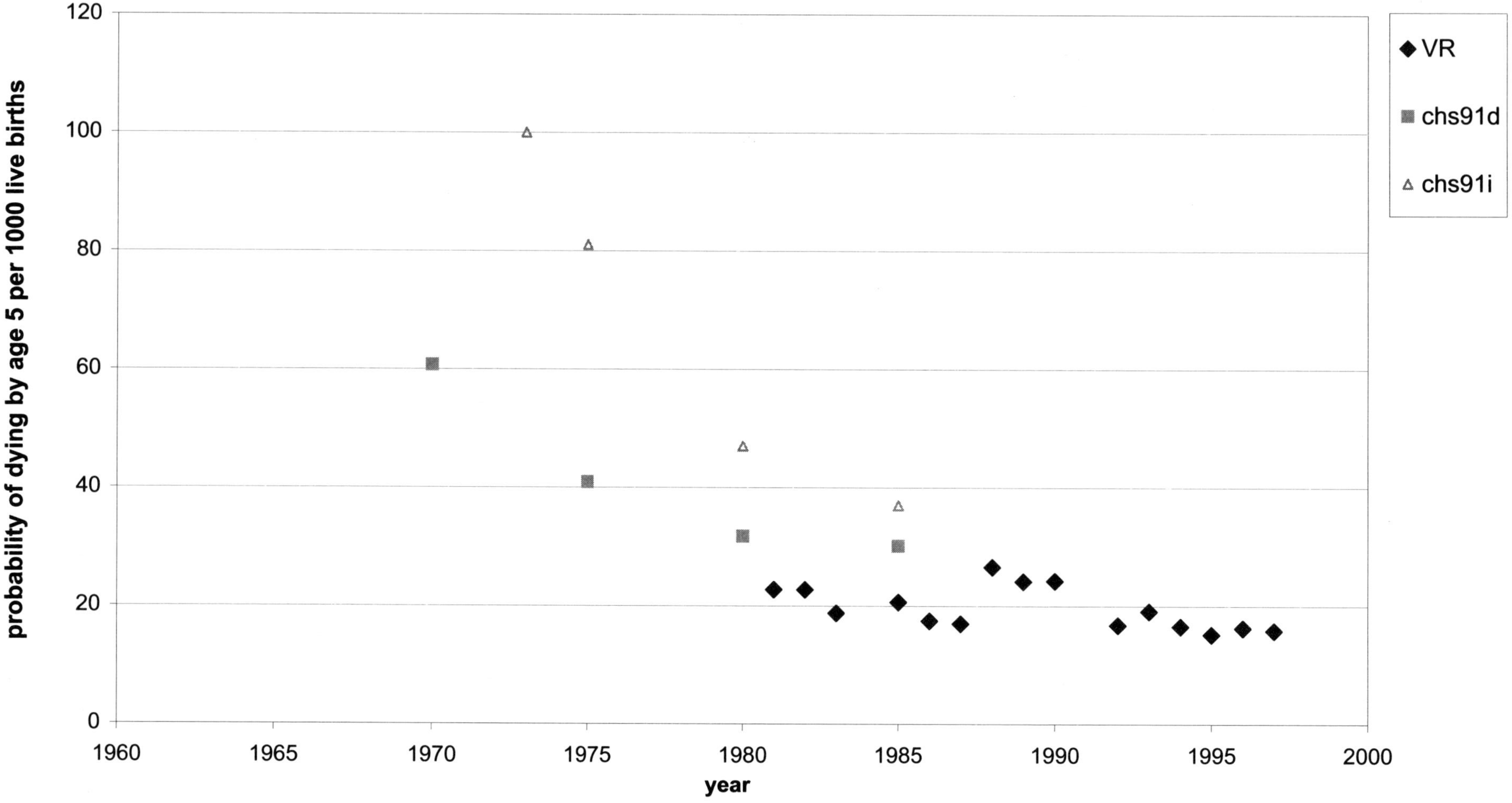

Data sources

Sources des données

Fuentes de los datos

VR: Vital Registration; ***chs91d***: Child Health Survey 1991 (direct method); ***chs91i***: Child Health Survey 1991 (indirect method)

Under 5 mortality
Republic of Korea

Mortalité des moins de cinq ans
République de Corée

Mortalidad de menores de cinco años
República de Corea

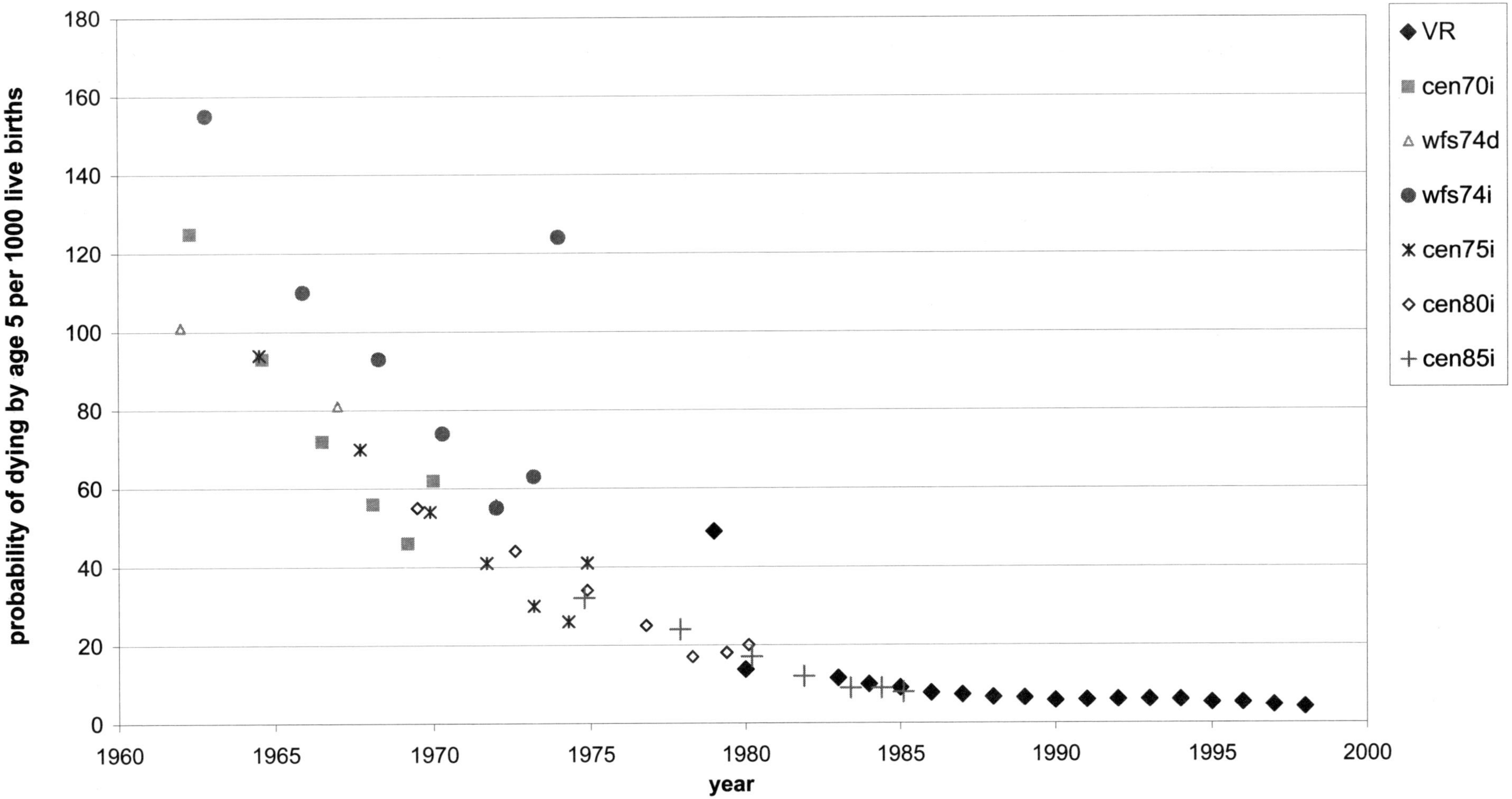

Data sources | **Sources des données** | **Fuentes de los datos**

VR: Vital Registration; ***cen70i***: Census, 1 October 1970 (indirect method); ***wfs74d***: National Fertility Survey, 16 September-5 December 1974 (direct method); ***wfs74i***: National Fertility Survey, 16 September-5 December 1974 (indirect method); ***cen75i***: Census, 1 October 1975 (indirect method); ***cen80i***: Census, 1 November 1980 (indirect method); ***cen85i***: Census, 1 November 1985 (indirect method)

Under 5 mortality
Republic of Moldova

Mortalité des moins de cinq ans
République de Moldova

Mortalidad de menores de cinco años
República de Moldova

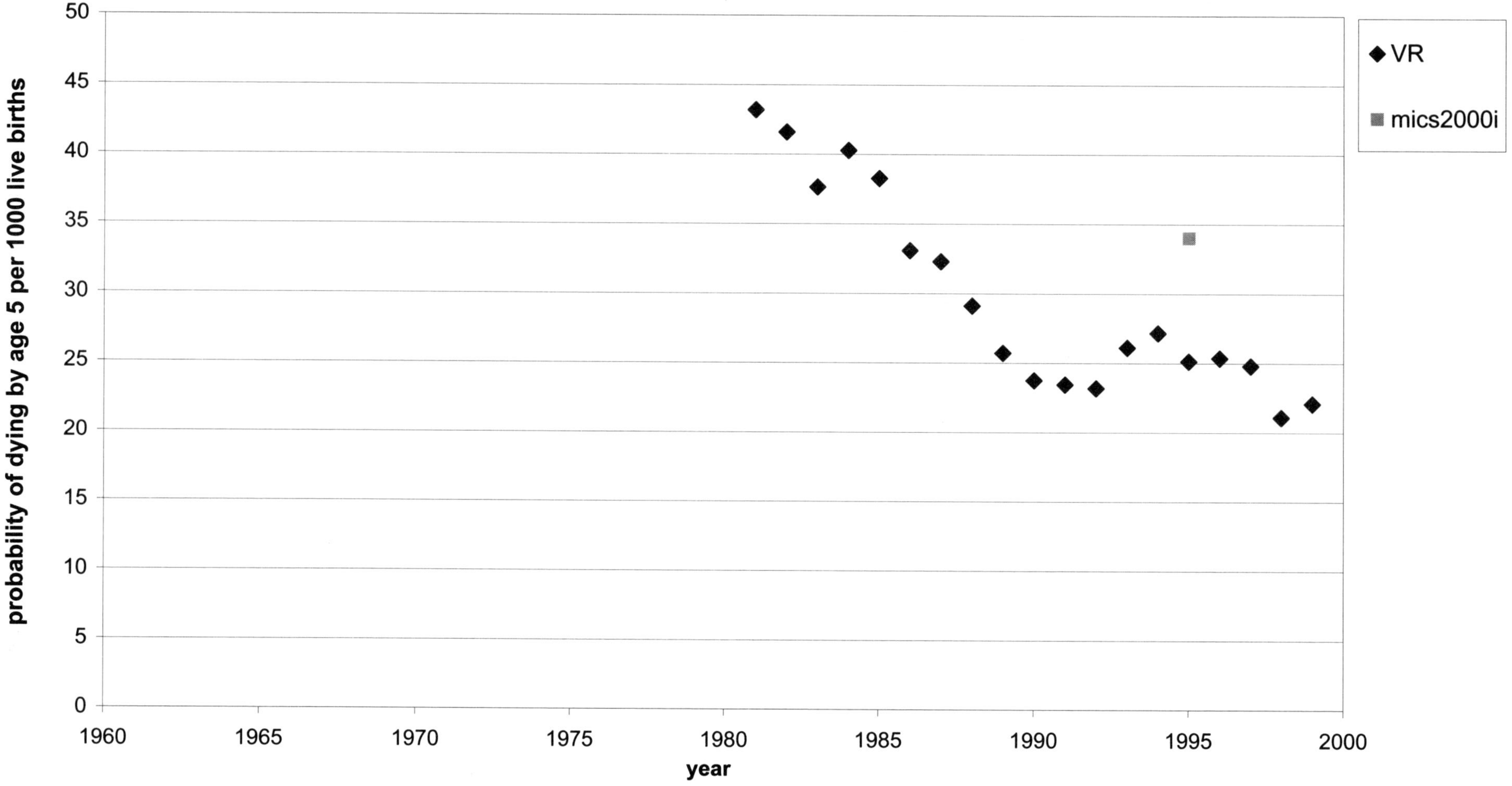

Data sources

Sources des données

Fuentes de los datos

VR : Vital Registration; ***mics2000i*** : Multiple Indicator Cluster Survey 2000 (indirect method)

Under 5 mortality

Romania

Mortalité des moins de cinq ans

Roumanie

Mortalidad de menores de cinco años

Rumania

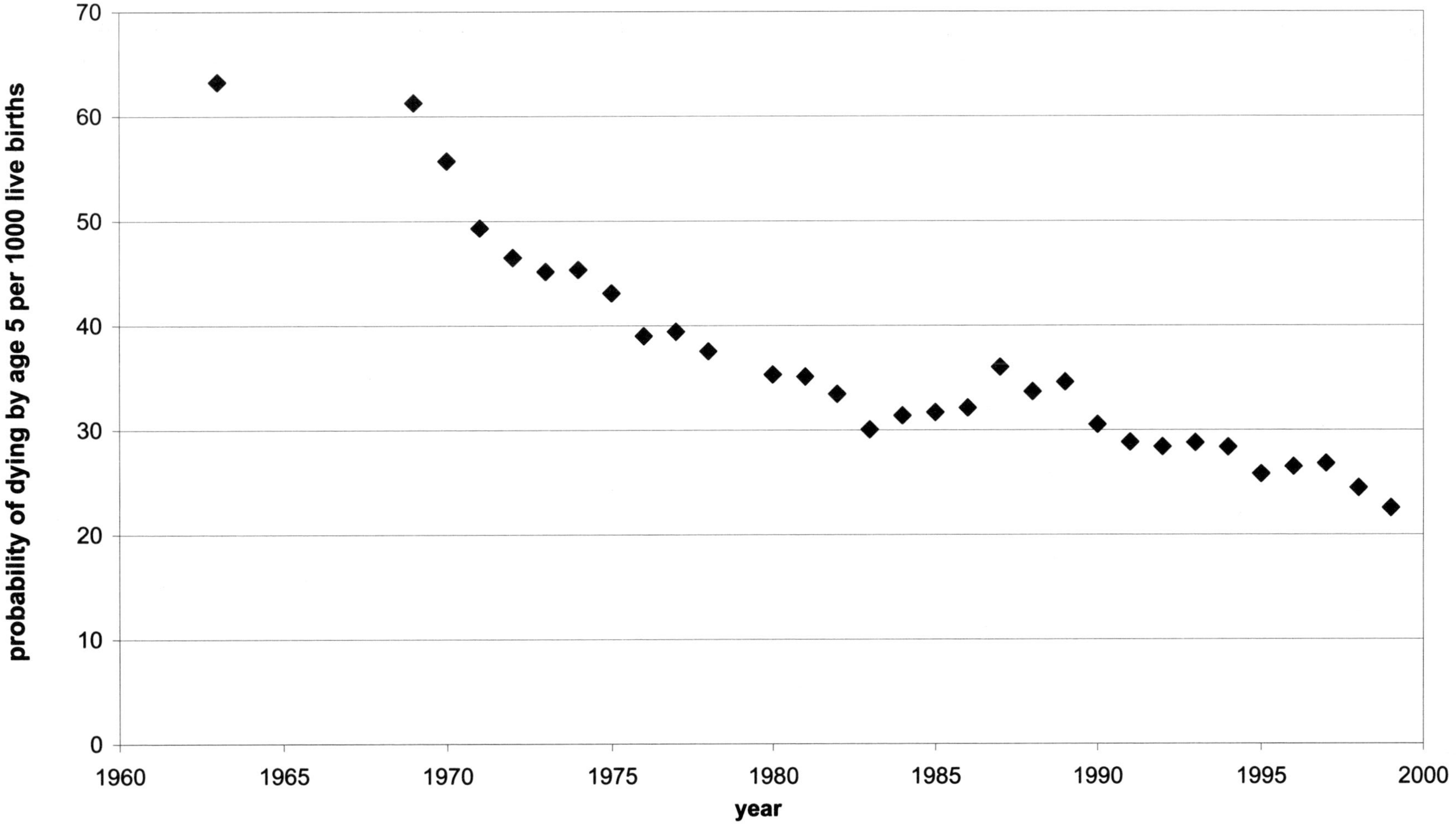

Data sources

Vital Registration

Sources des données

Fuentes de los datos

Under 5 mortality
Russian Federation

Mortalité des moins de cinq ans
Fédération de Russie

Mortalidad de menores de cinco años
Federación de Rusia

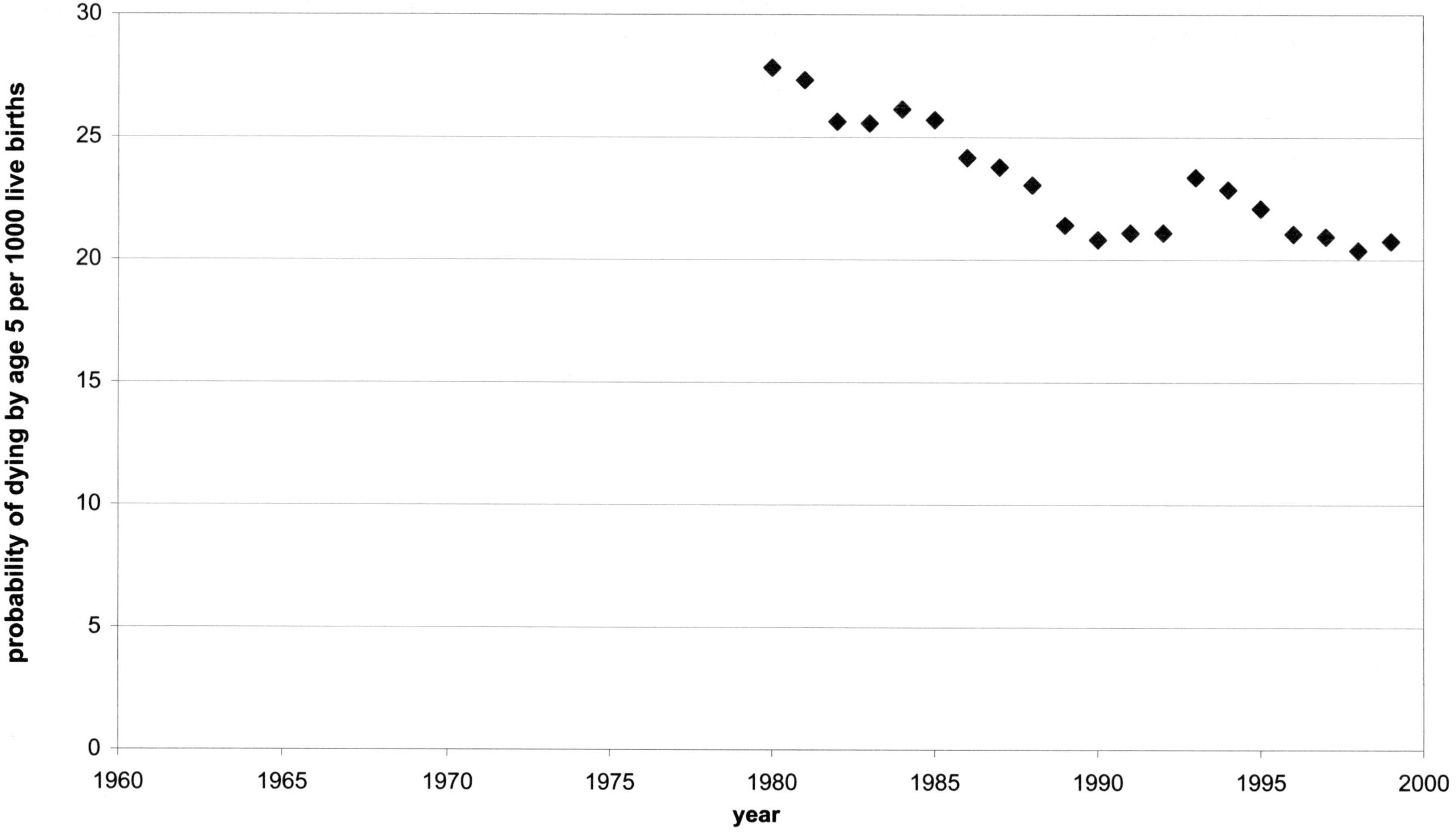

Data sources
Vital Registration

Sources des données

Fuentes de los datos

Under 5 mortality | Mortalité des moins de cinq ans | Mortalidad de menores de cinco años

Rwanda | **Rwanda** | **Rwanda**

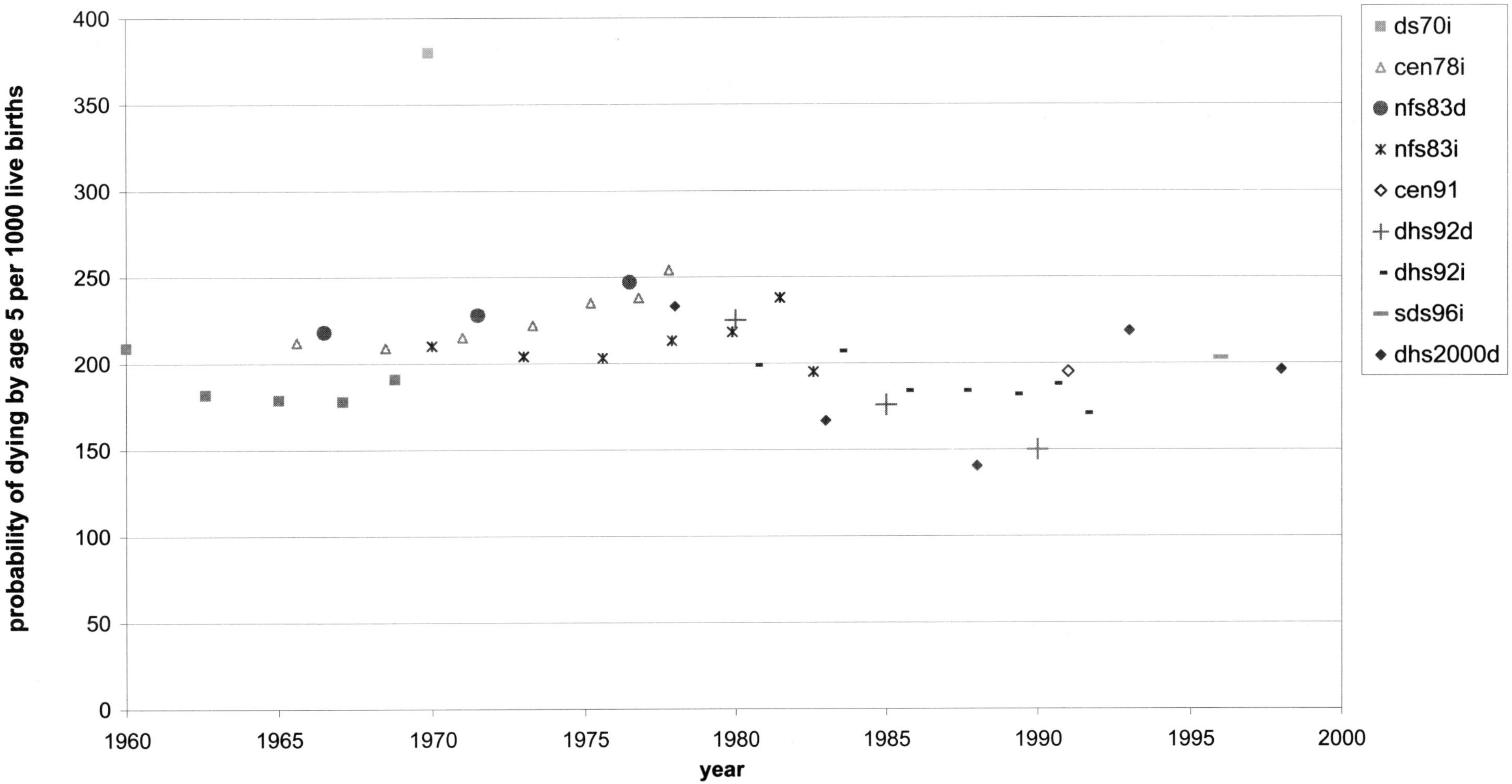

Data sources | **Sources des données** | **Fuentes de los datos**

ds70i: Enquête Démographique, 30 September 1970 (indirect method); ***cen78i***: Census, 15-16 August 1978 (indirect method); ***nfs83d***: Enquête National sur la Fécondite, August-December 1983 (direct method); ***nfs83i***: Enquête National sur la Fécondite, August-December 1983 (indirect method); ***cen91***: Census 1991; ***dhs92d***: Enquête Démographique et de Santé, June-October 1992 (direct method); ***dhs92i***: Enquête Démographique et de Santé, June-October 1992 (indirect method); ***sds96***: Socio-Demographic Survey 1996; ***dhs2000d***: Demographic and Health Survey 2000 (direct method)

Under 5 mortality

Saint Kitts and Nevis

Mortalité des moins de cinq ans

Saint-Kitts-et-Nevis

Mortalidad de menores de cinco años

Saint Kitts y Nevis

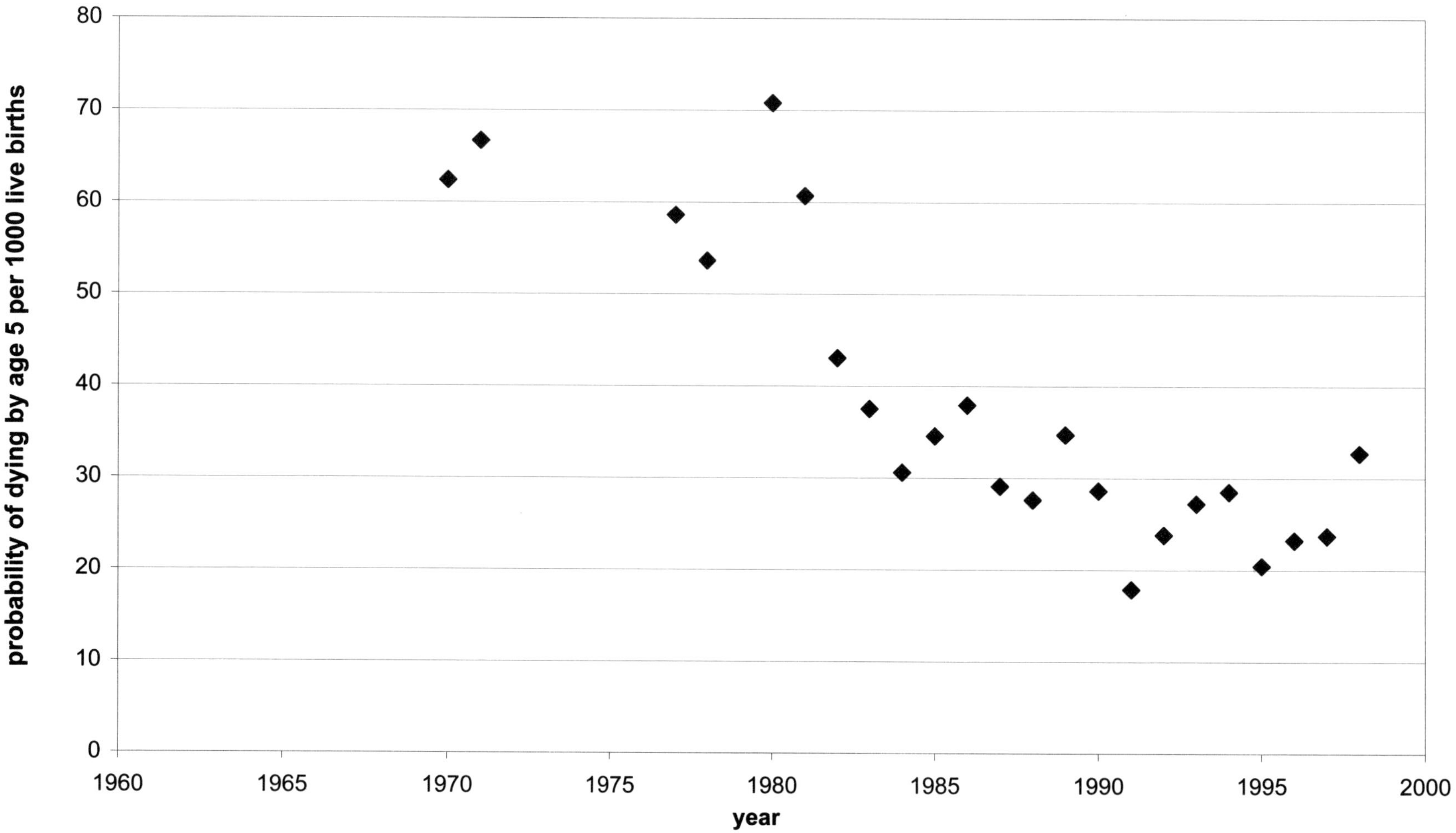

Data sources

Vital Registration

Sources des données

Fuentes de los datos

Under 5 mortality

Saint Lucia

Mortalité des moins de cinq ans

Sainte-Lucie

Mortalidad de menores de cinco años

Santa Lucía

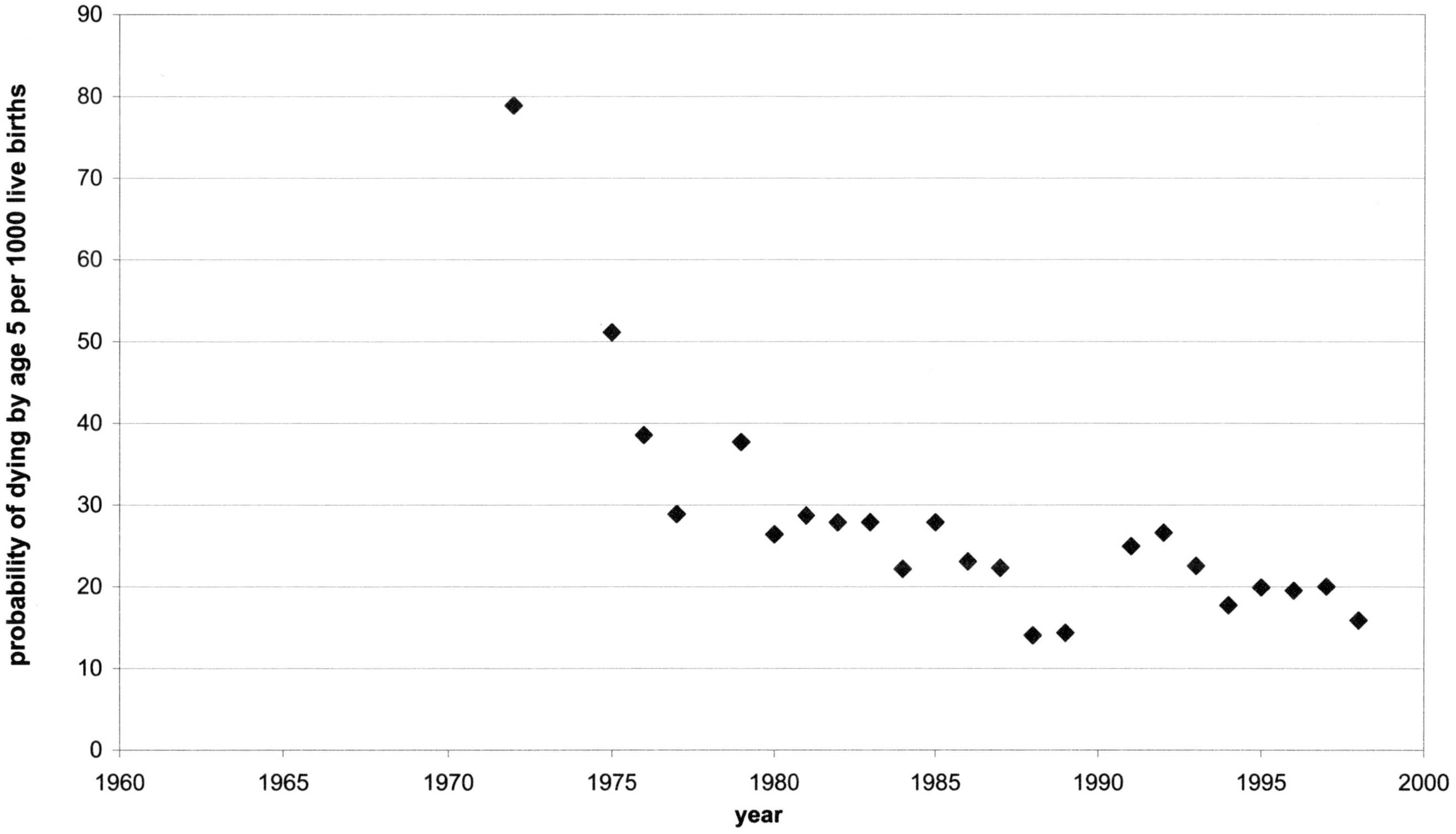

Data sources

Vital Registration

Sources des données

Fuentes de los datos

Under 5 mortality

Saint Vincent and the Grenadines

Mortalité des moins de cinq ans

Saint-Vincent-et-Grenadines

Mortalidad de menores de cinco años

San Vicente y las Granadinas

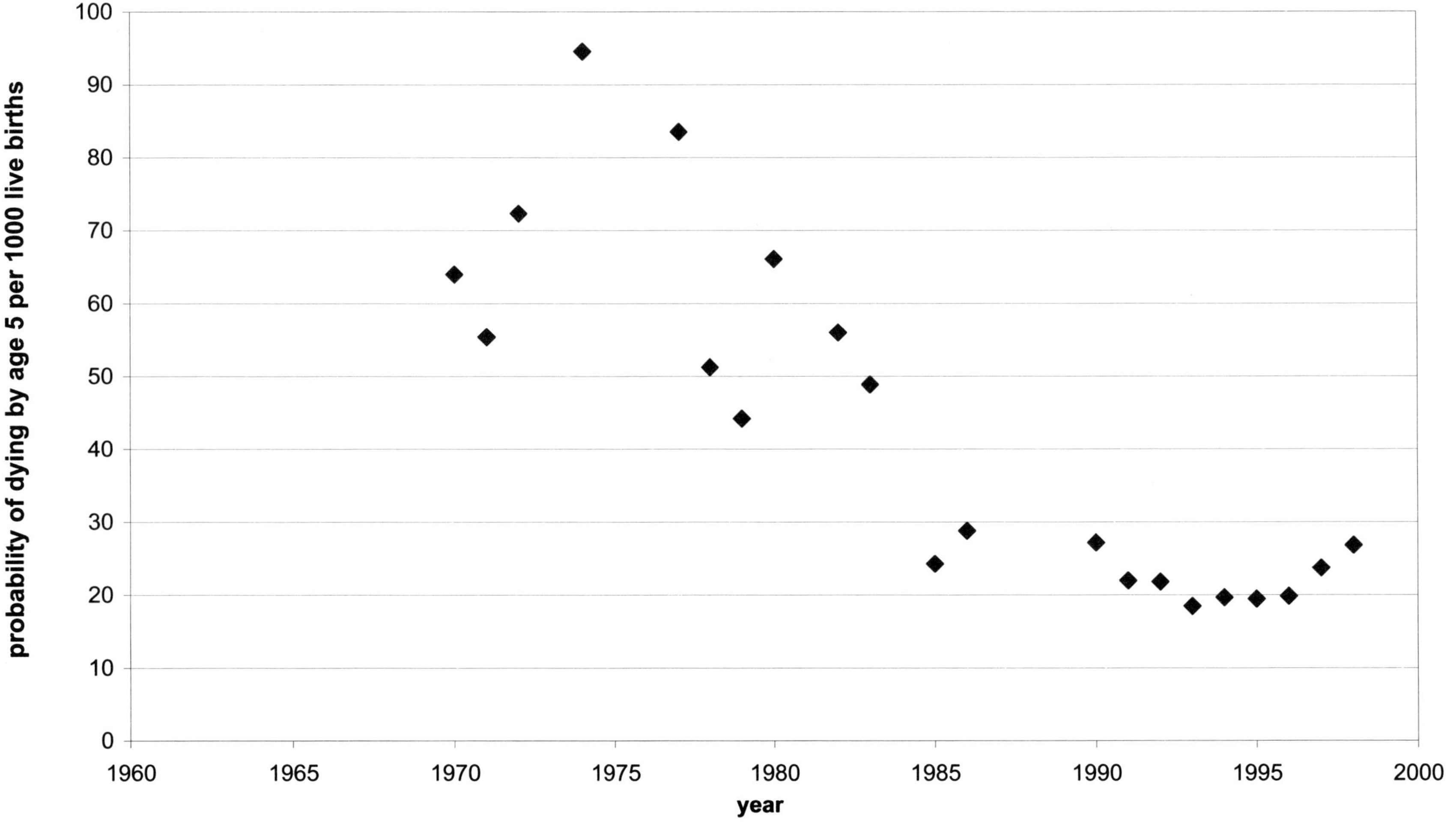

Data sources

Vital Registration

Sources des données

Fuentes de los datos

Under 5 mortality
Samoa

Mortalité des moins de cinq ans
Samoa

Mortalidad de menores de cinco años
Samoa

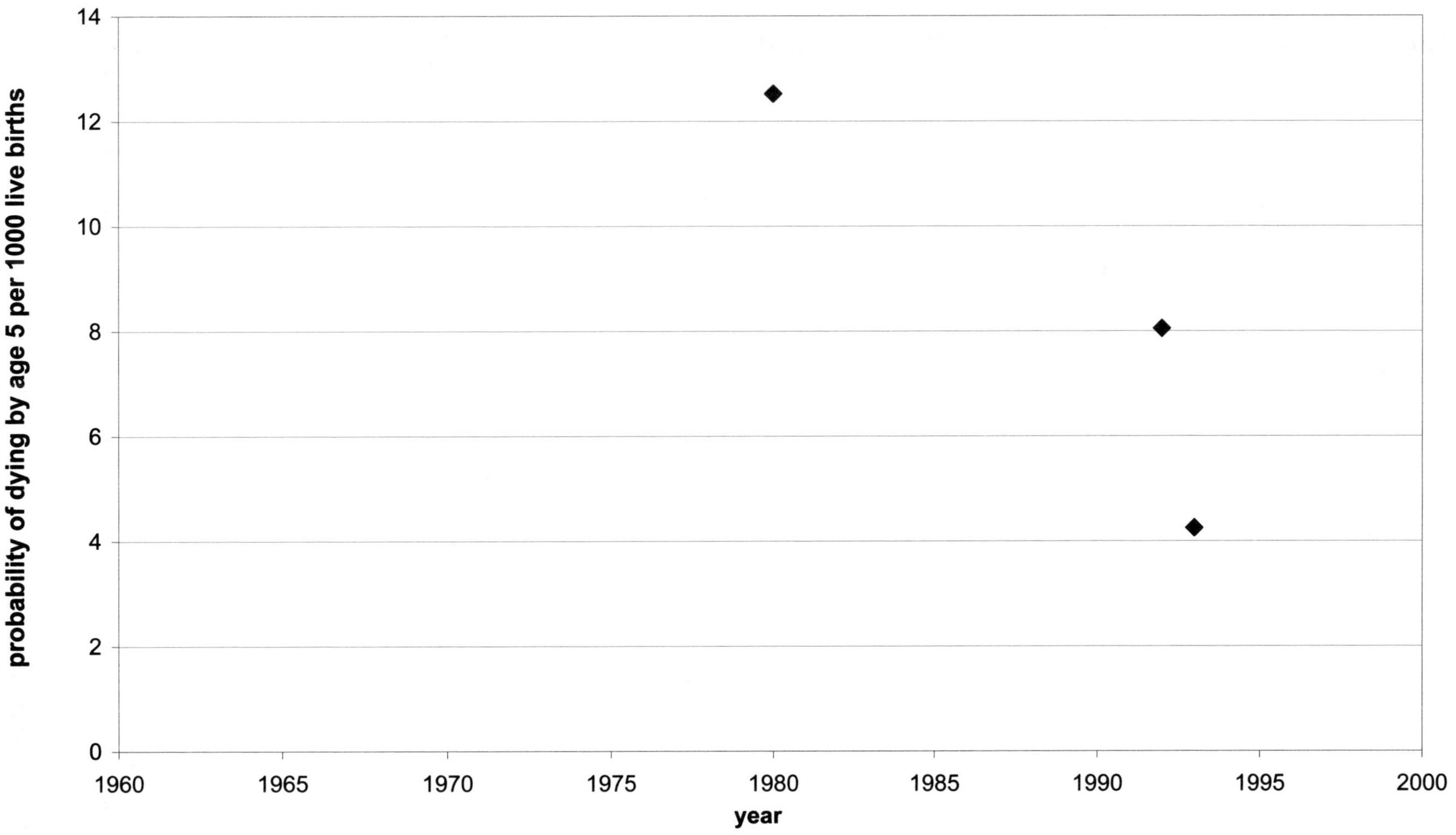

Data sources
Vital Registration

Sources des données

Fuentes de los datos

Under 5 mortality
San Marino

Mortalité des moins de cinq ans
Saint-Marin

Mortalidad de menores de cinco años
San Marino

Data sources
Vital Registration

Sources des données

Fuentes de los datos

Under 5 mortality
Sao Tome and Principe

Mortalité des moins de cinq ans
Sao Tomé-et-Principe

Mortalidad de menores de cinco años
Santo Tomé y Príncipe

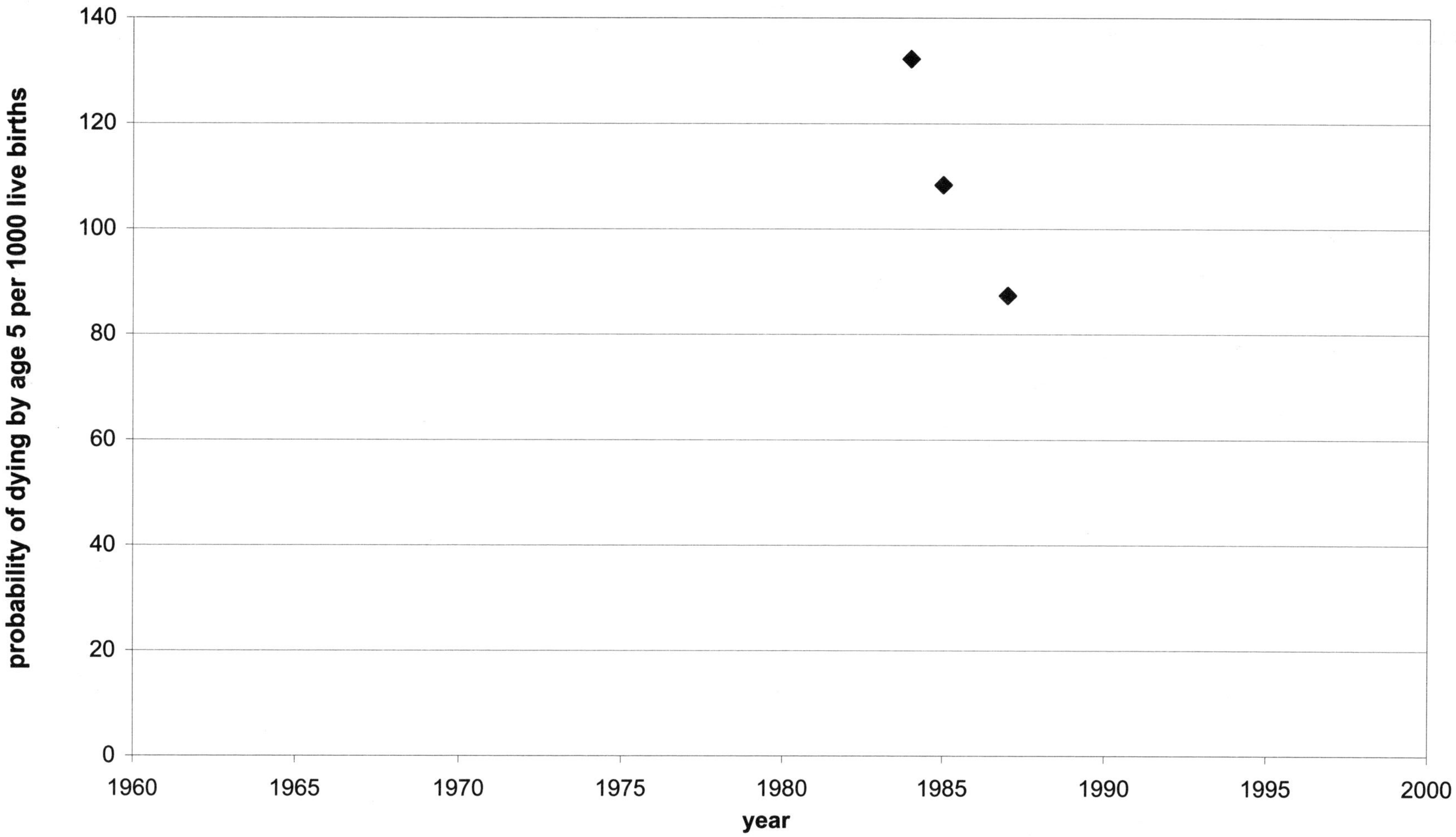

Data sources
Vital Registration

Sources des données

Fuentes de los datos

Under 5 mortality
Saudi Arabia

Mortalité des moins de cinq ans
Arabie saoudite

Mortalidad de menores de cinco años
Arabia Saudita

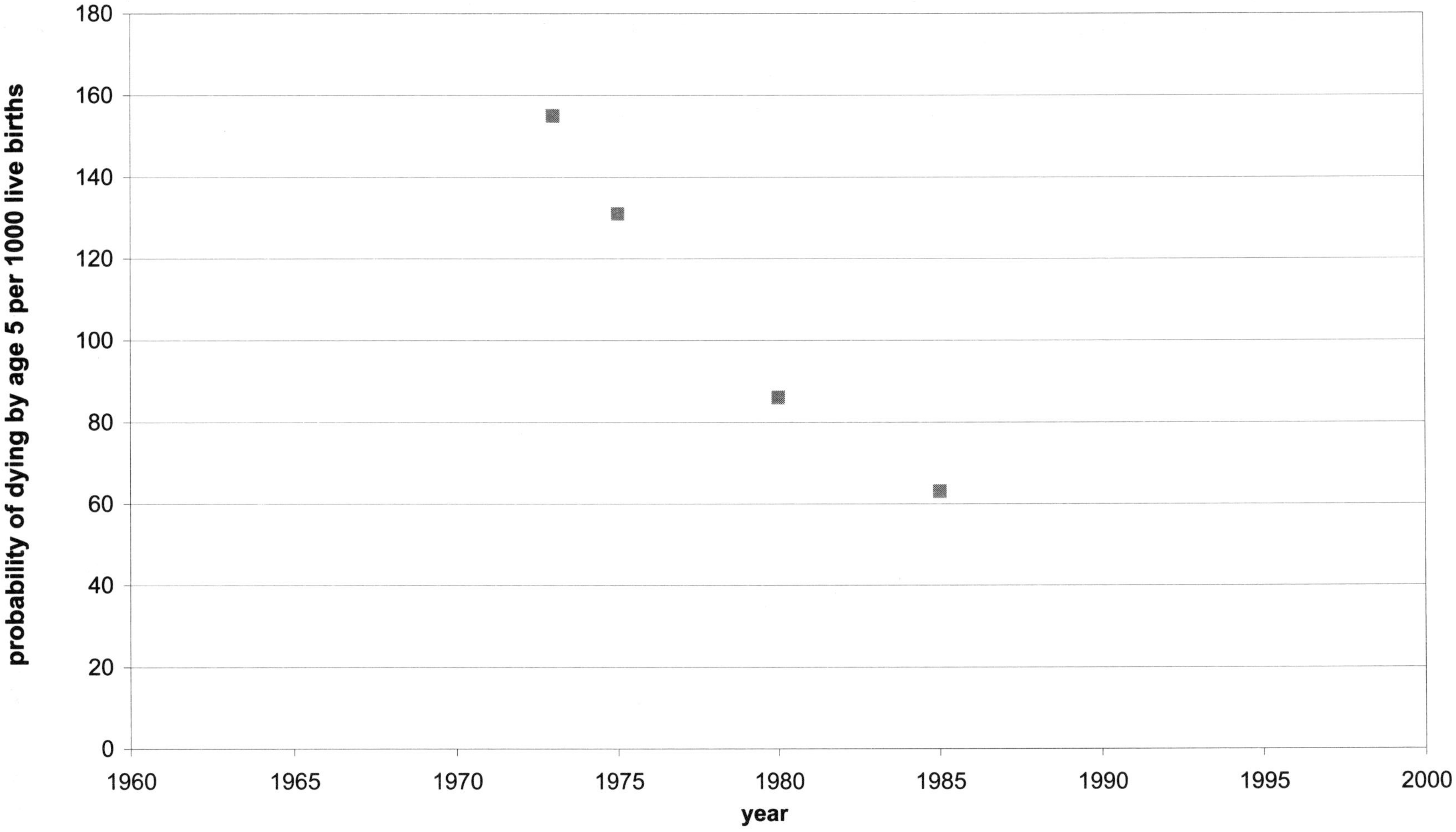

Data sources
Child Health Survey 1991 (indirect method)

Sources des données

Fuentes de los datos

Under 5 mortality | Mortalité des moins de cinq ans | Mortalidad de menores de cinco años

Senegal | **Sénégal** | **Senegal**

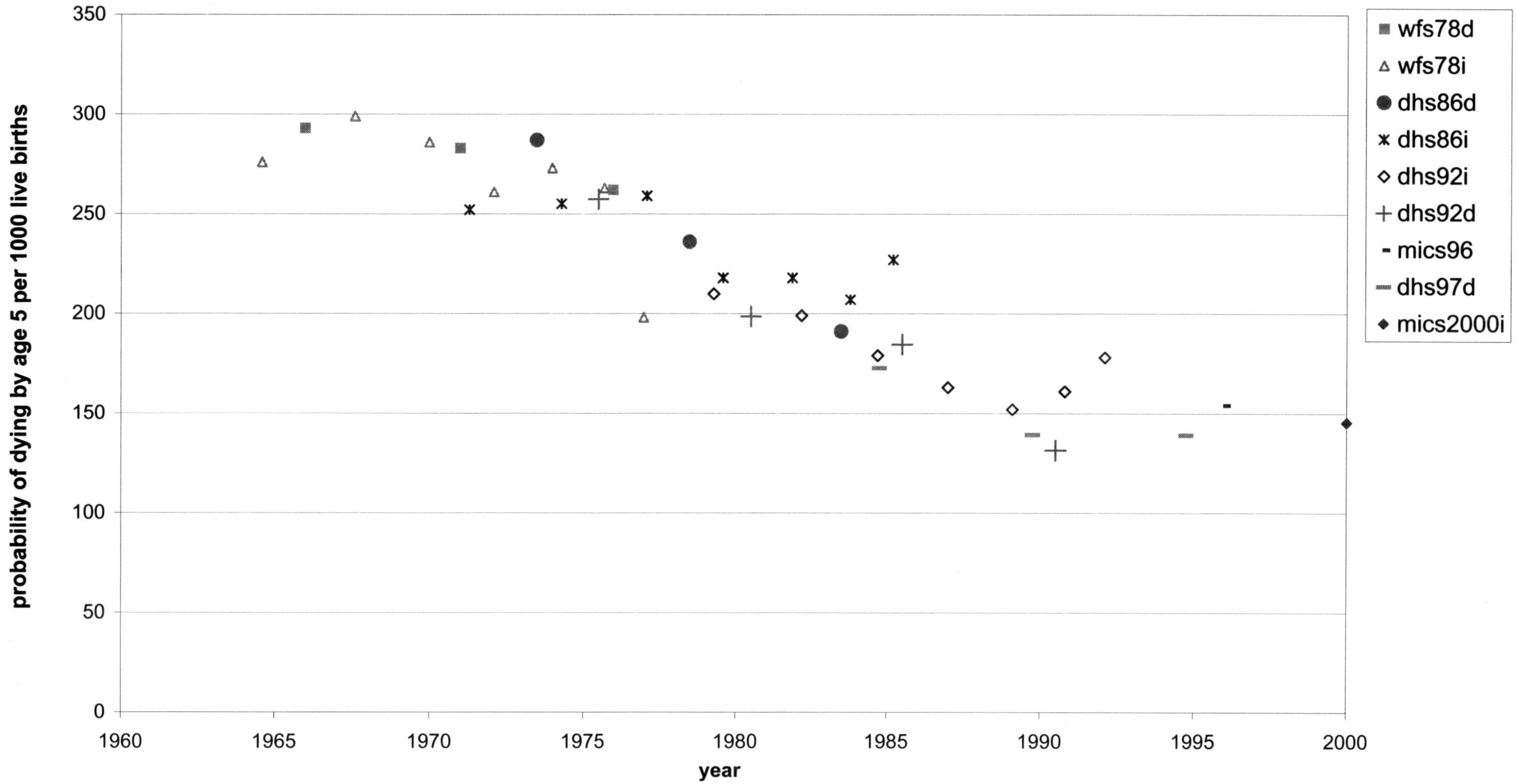

Data sources | **Sources des données** | **Fuentes de los datos**

wfs78d : Enquête Sénégalaise sur la Fécondité, April-October 1978 (direct method); ***wfs78i*** : Enquête Sénégalaise sur la Fécondité, April-October 1978 (indirect method); ***dhs86d*** : Enquête Démographique et de Santé, April-July 1986 (direct method); ***dhs86i*** : Enquête Démographique et de Santé, April-July 1986 (indirect method); ***dhs92i*** : Enquête Démographique et de Santé, November 1992-March 1993 (indirect method); ***dhs92d*** : Enquête Démographique et de Santé, November 1992-March 1993 (direct method); ***mics96*** : Evaluation des Objectives intermédiaires 1996; ***dhs97d*** : Demographic and Health Survey 1997 (direct method); ***mics2000i*** : Multiple Indicator Cluster Survey 2000 (indirect method)

Under 5 mortality
Seychelles

Mortalité des moins de cinq ans
Seychelles

Mortalidad de menores de cinco años
Seychelles

Data sources
Vital Registration

Sources des données

Fuentes de los datos

Under 5 mortality

Sierra Leone

Mortalité des moins de cinq ans

Sierra Leone

Mortalidad de menores de cinco años

Sierra Leona

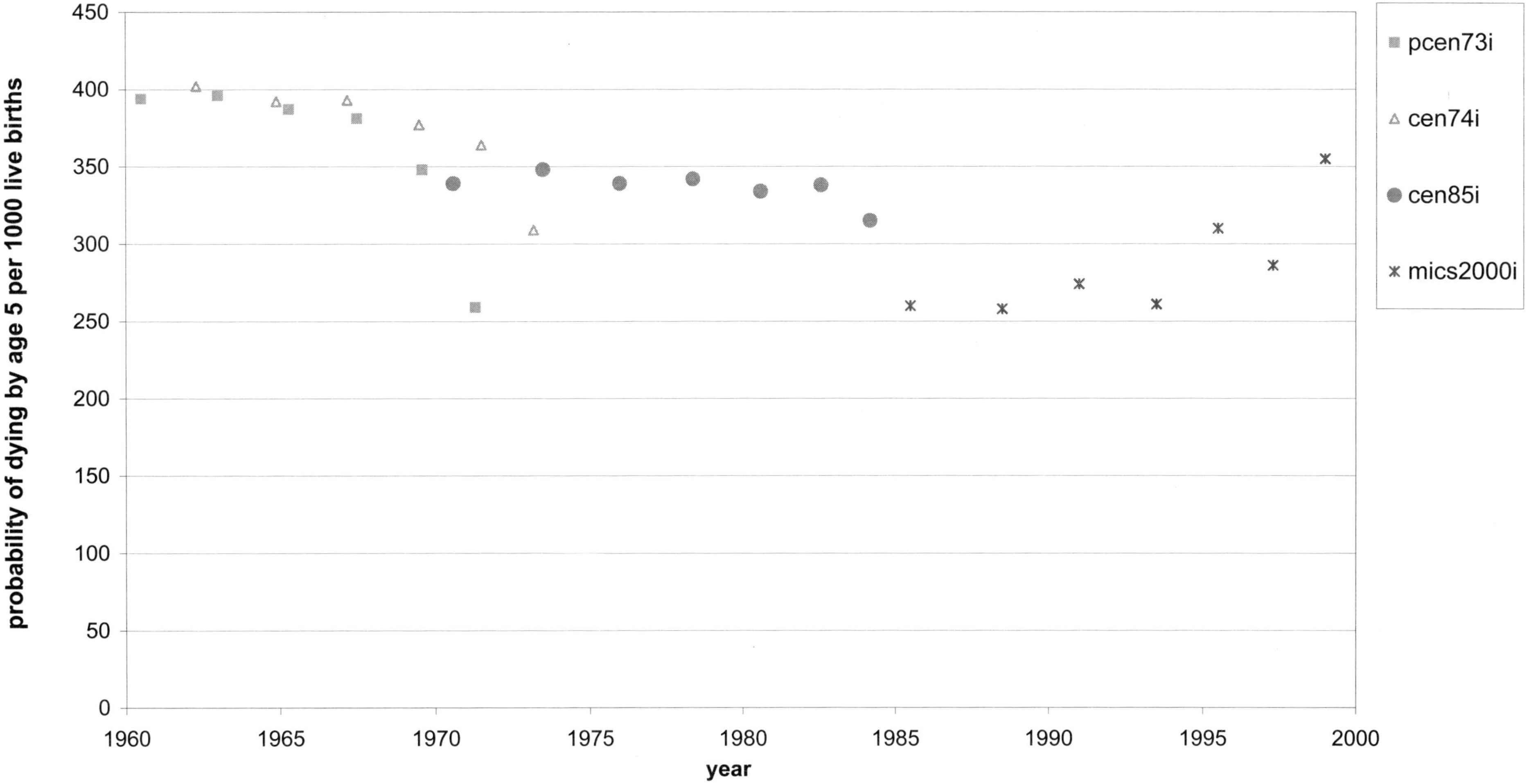

Data sources

Sources des données

Fuentes de los datos

pcen73i: Pilot Census, 8-24 April 1973 (indirect method); ***cen74i***: Census, 8 December 1974 (indirect method); ***cen85i***: Census, 15 July 1985 (indirect method); ***mics2000i***: Multiple Indicator Cluster Survey 2000 (indirect method)

Under 5 mortality
Singapore

Mortalité des moins de cinq ans
Singapour

Mortalidad de menores de cinco años
Singapur

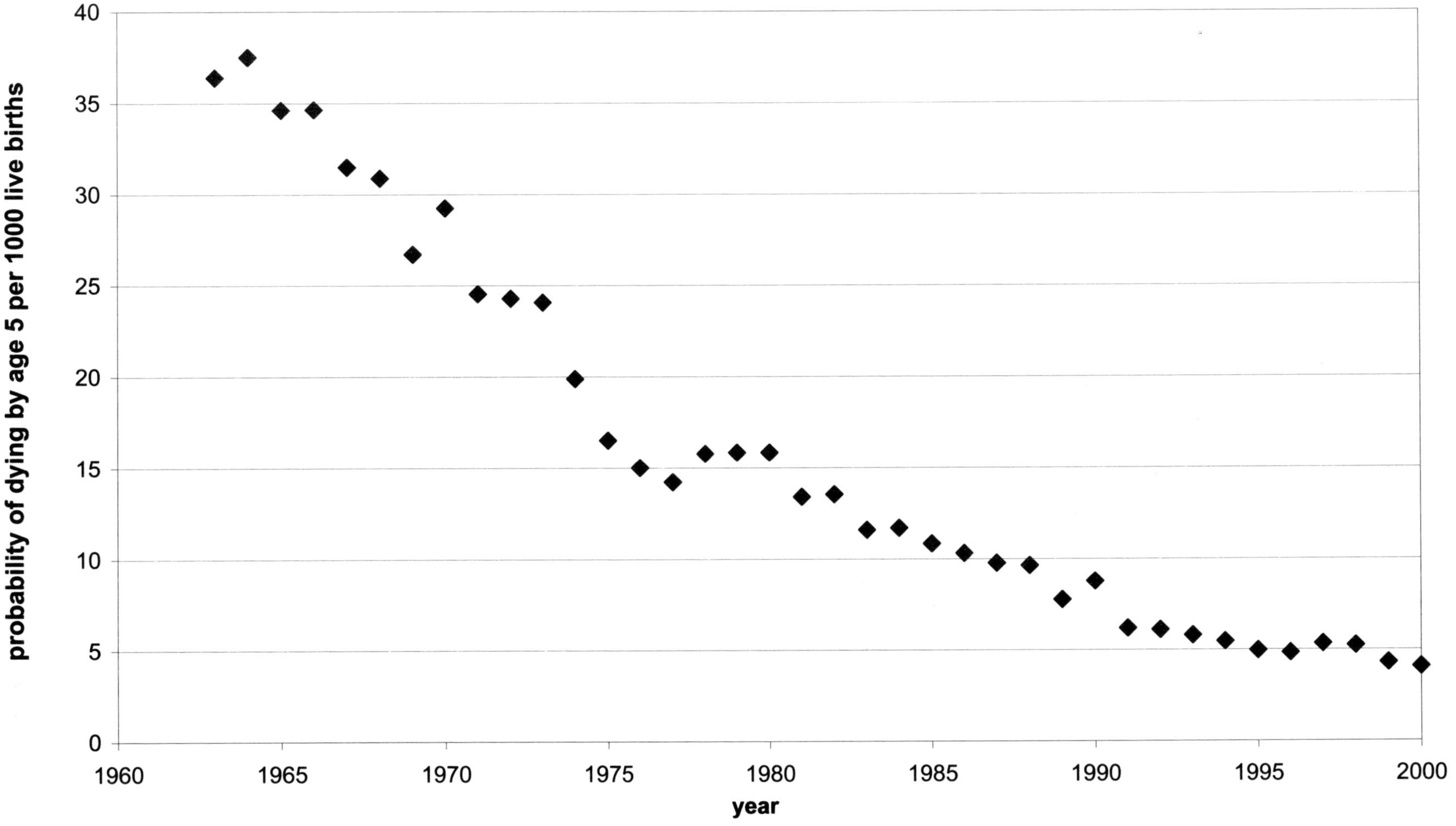

Data sources
Vital Registration

Sources des données

Fuentes de los datos

Under 5 mortality | Mortalité des moins de cinq ans | Mortalidad de menores de cinco años

Slovakia | **Slovaquie** | **Eslovaquia**

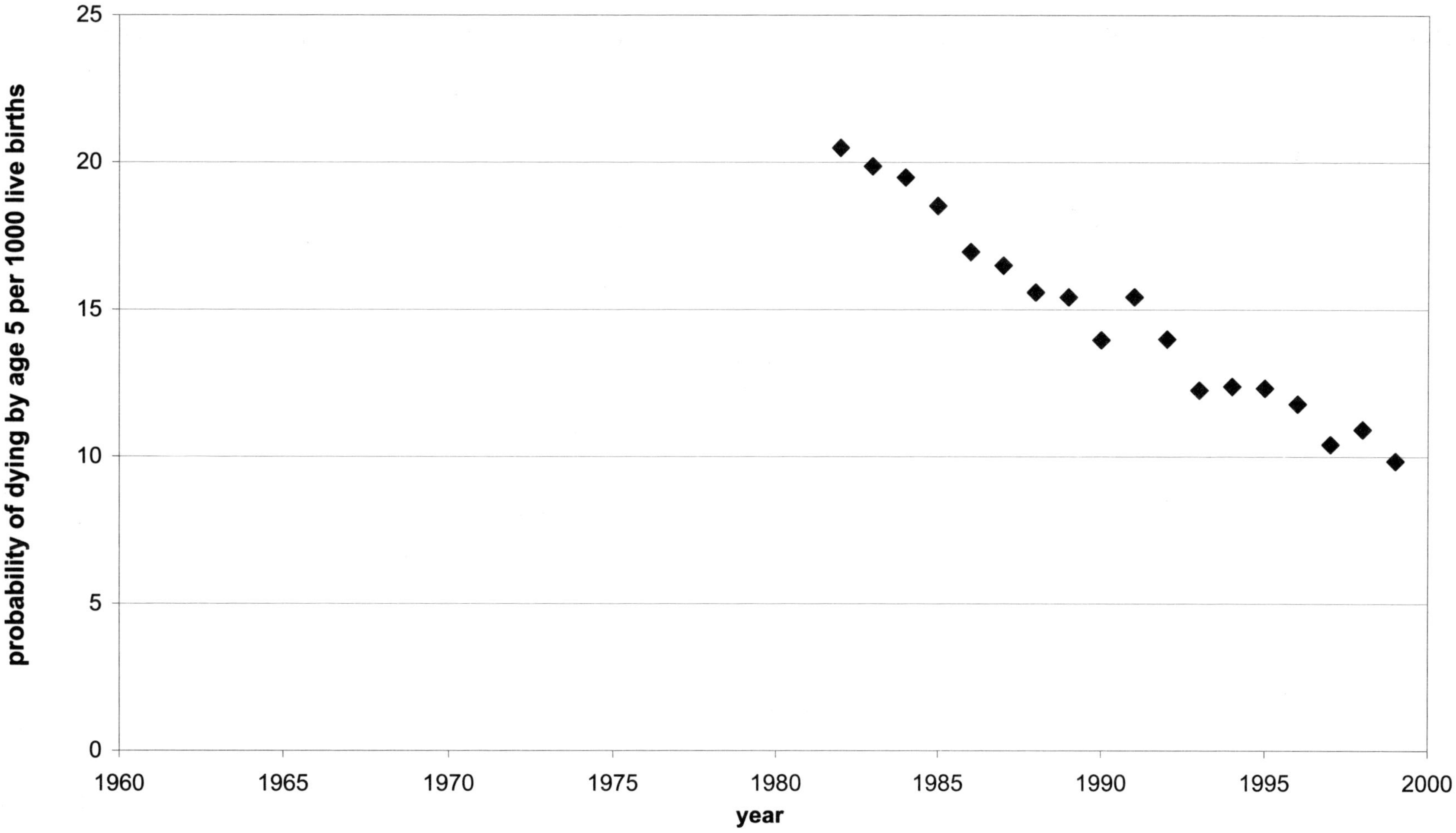

Data sources | **Sources des données** | **Fuentes de los datos**

Vital Registration

Under 5 mortality
Slovenia

Mortalité des moins de cinq ans
Slovénie

Mortalidad de menores de cinco años
Eslovenia

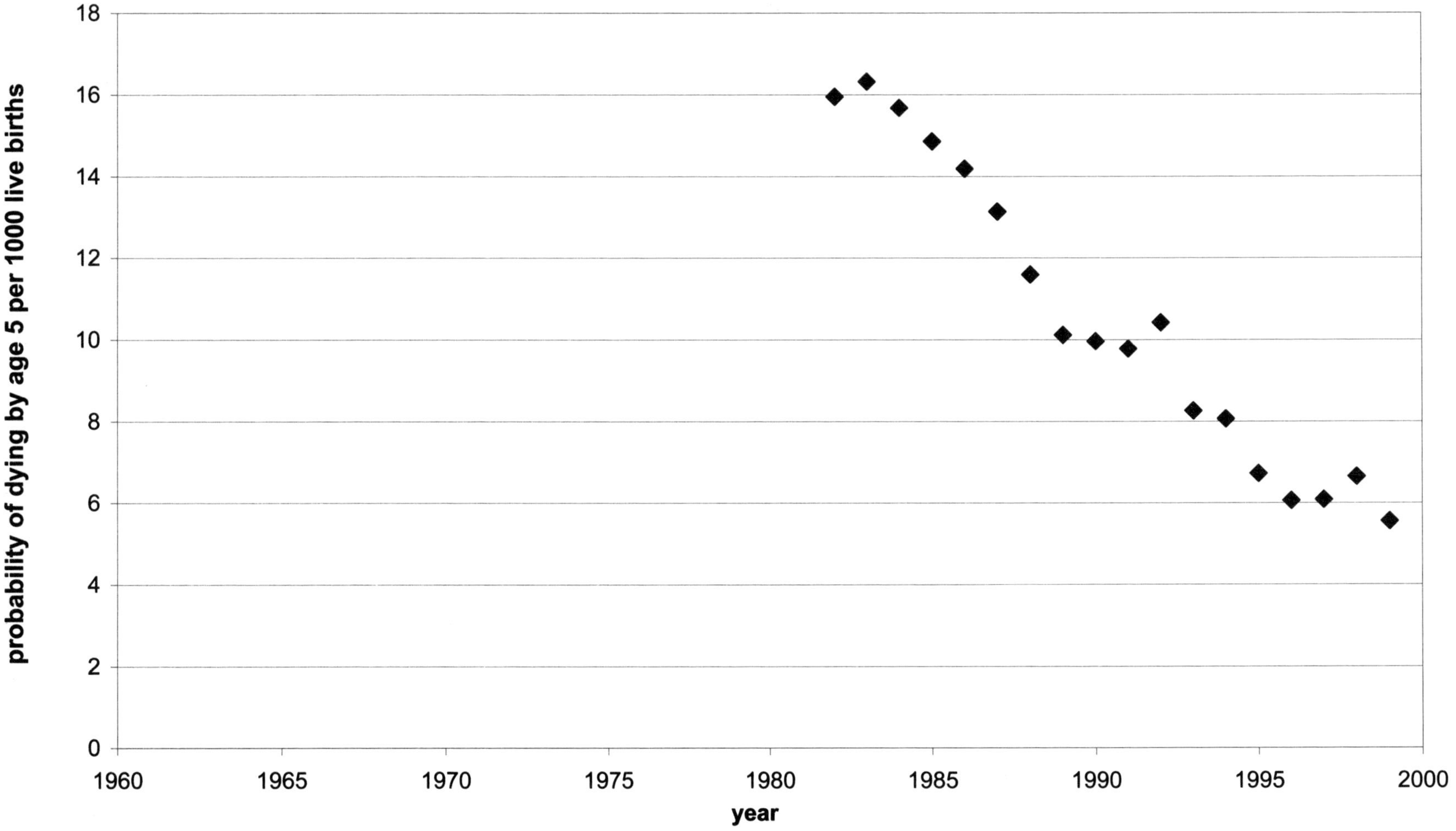

Data sources
Vital Registration

Sources des données

Fuentes de los datos

Under 5 mortality
Somalia

Mortalité des moins de cinq ans
Somalie

Mortalidad de menores de cinco años
Somalia

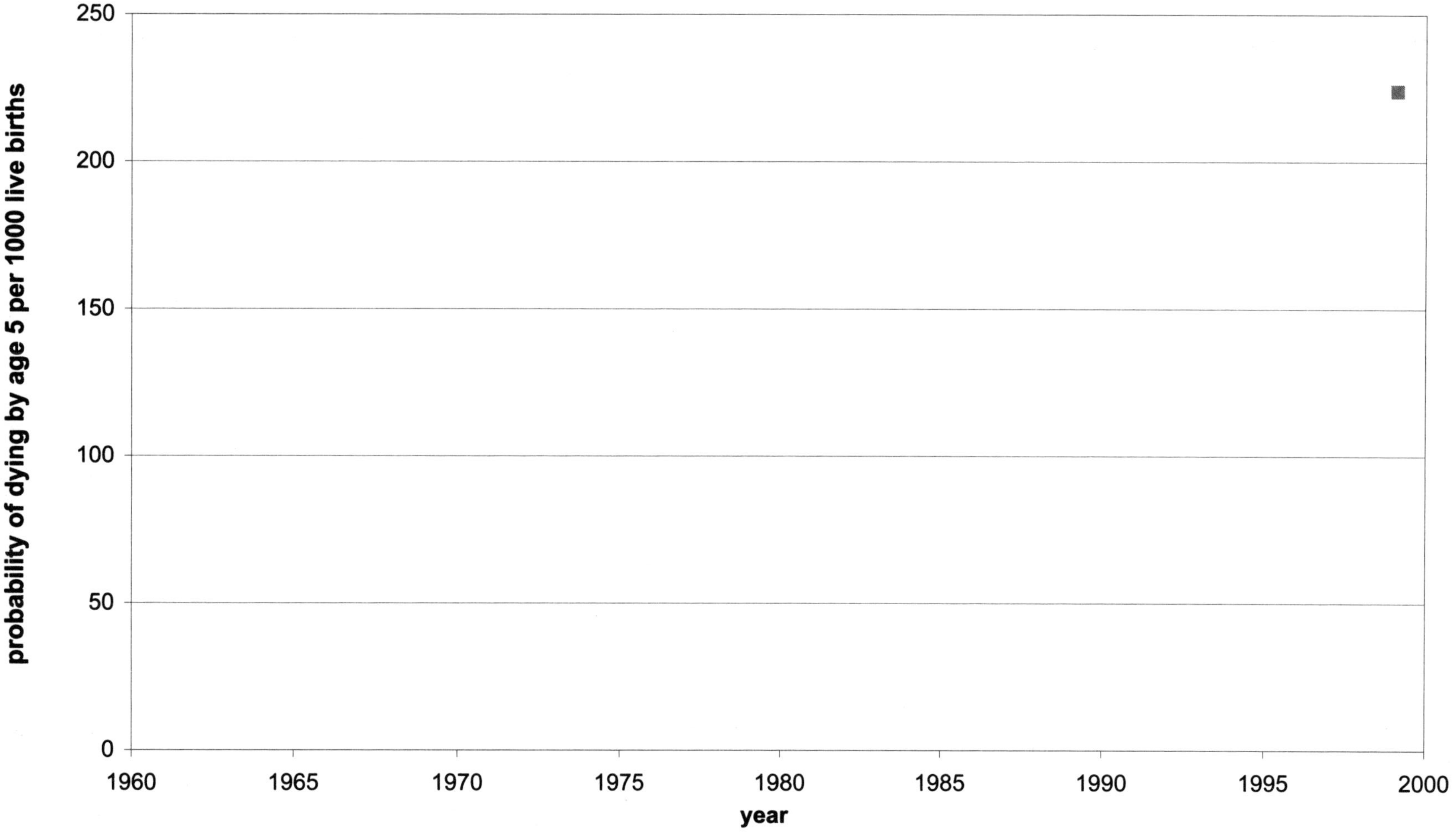

Data sources
Multiple Indicator Cluster Survey

Sources des données

Fuentes de los datos

Under 5 mortality | Mortalité des moins de cinq ans | Mortalidad de menores de cinco años

South Africa | **Afrique du Sud** | **Sudáfrica**

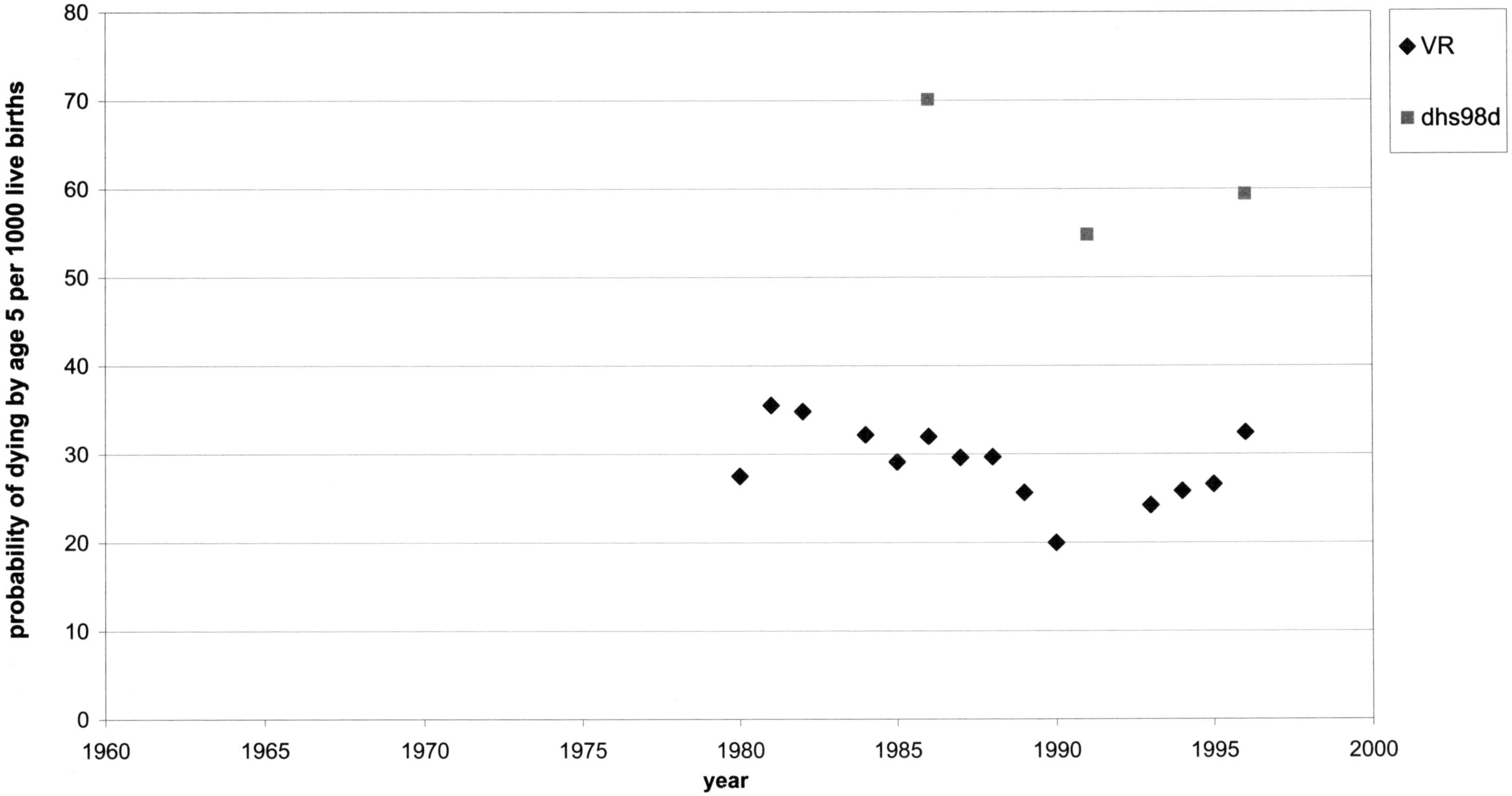

Data sources | **Sources des données** | **Fuentes de los datos**

VR: Vital Registration; ***dhs98d***: Demographic and Health Survey 1998 - preliminary report (direct method)

Under 5 mortality

Spain

Mortalité des moins de cinq ans

Espagne

Mortalidad de menores de cinco años

España

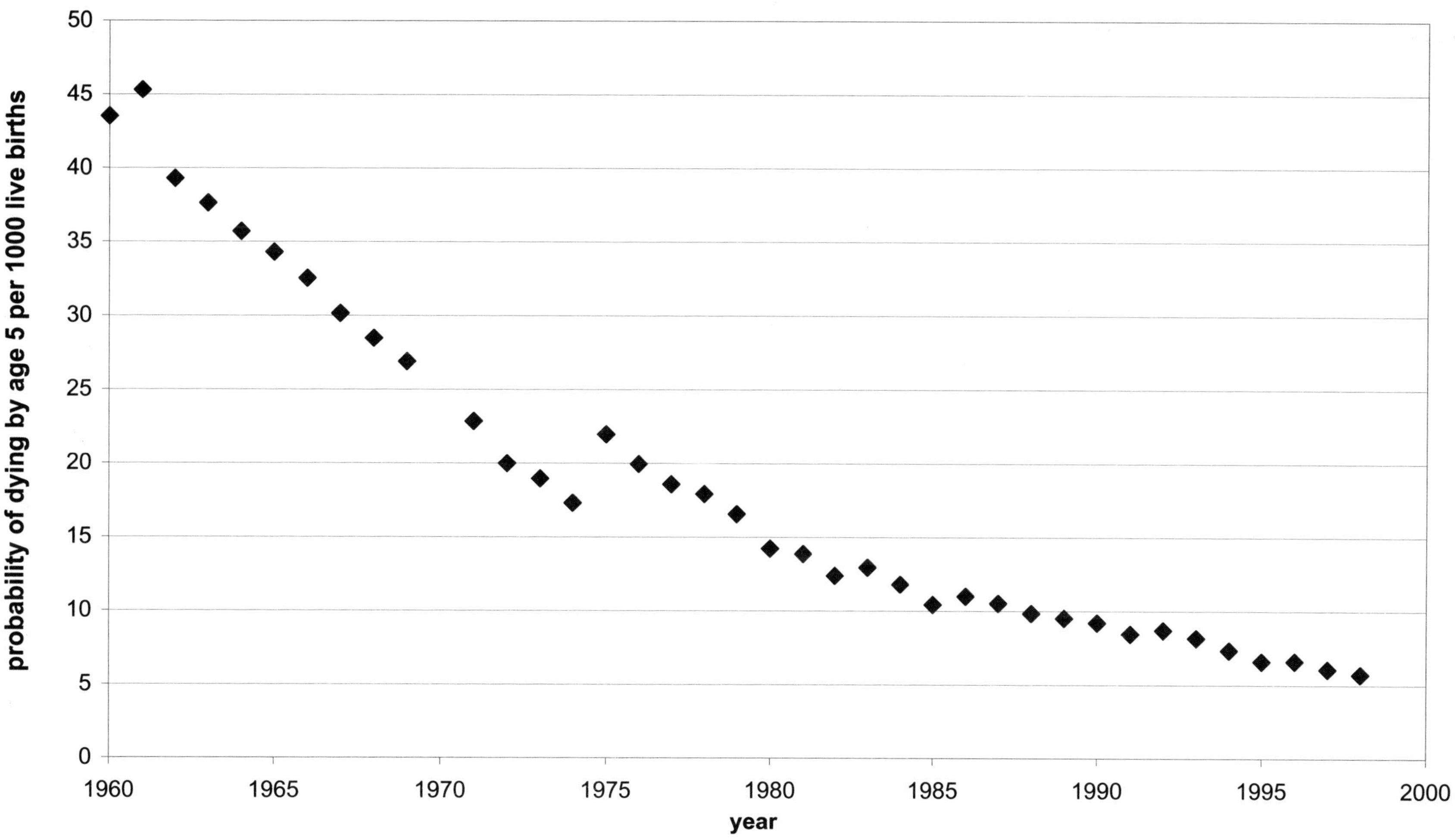

Data sources

Vital Registration

Sources des données

Fuentes de los datos

Under 5 mortality

Sri Lanka

Mortalité des moins de cinq ans

Sri Lanka

Mortalidad de menores de cinco años

Sri Lanka

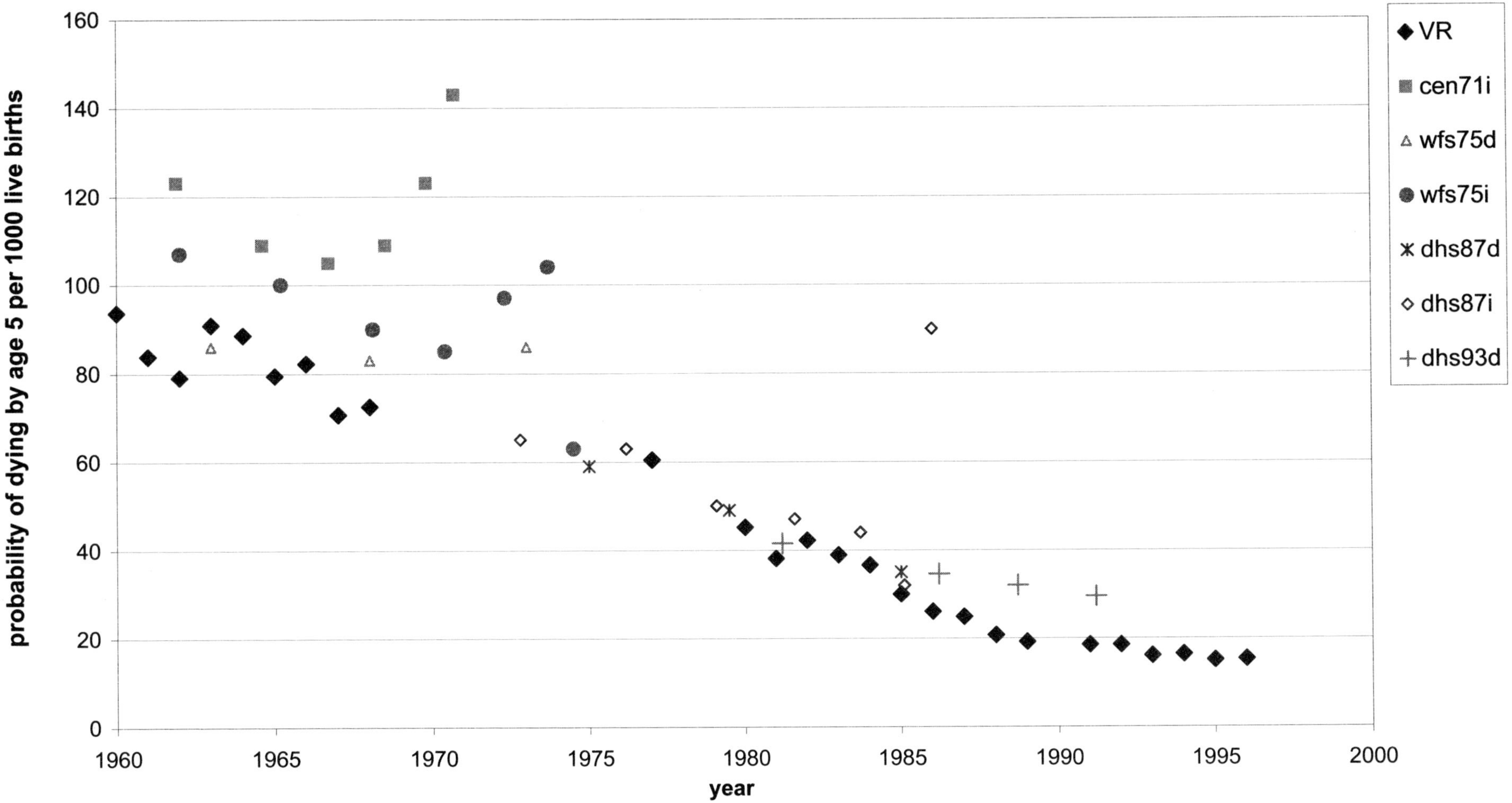

Data sources **Sources des données** **Fuentes de los datos**

VR: Vital Registration; ***cen71i***: Census, 9 October 1971 (indirect method); ***wfs75d***: World Fertility Survey, August-November 1975 (direct method); ***wfs75i***: World Fertility Survey, August-November 1975 (indirect method); ***dhs87d***: Demographic and Health Survey, January-March 1987 (direct method); ***dhs87i***: Demographic and Health Survey, January-March 1987 (indirect method); ***dhs93d***: Demographic and Health Survey, July-September 1993 (direct method)

Under 5 mortality
Sudan

Mortalité des moins de cinq ans
Soudan

Mortalidad de menores de cinco años
Sudán

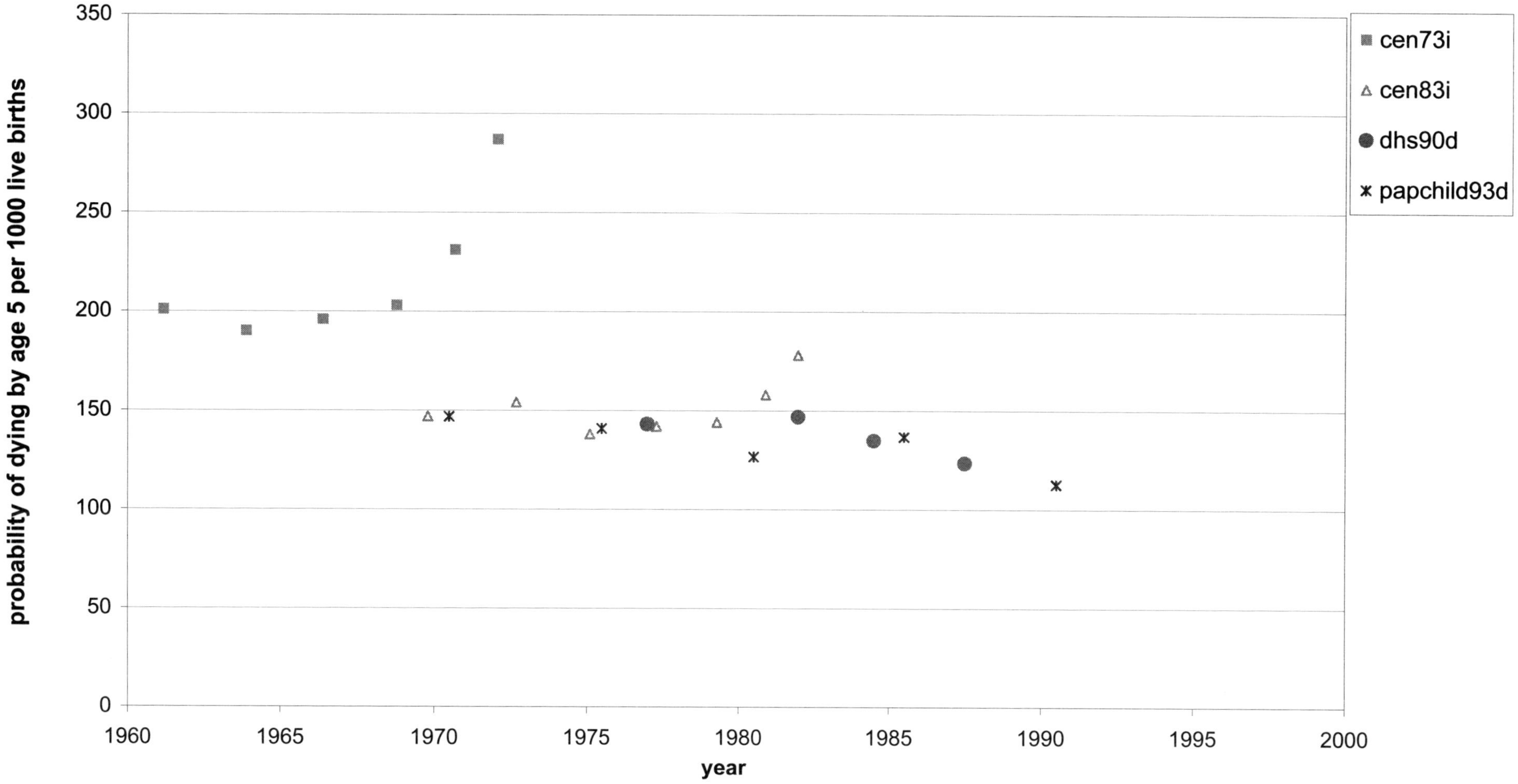

Data sources

Sources des données

Fuentes de los datos

cen73i : Census, 3 April 1973 (indirect method); ***cen83i*** : Census, 1 February 1983 (indirect method); ***dhs90d*** : Demographic and Health Survey 1989-1990 (direct method); ***papchild93d*** : Maternal and Child Health Survey 1993 (direct method)

Under 5 mortality

Suriname

Mortalité des moins de cinq ans

Suriname

Mortalidad de menores de cinco años

Suriname

Data sources

Vital Registration

Sources des données

Fuentes de los datos

Under 5 mortality
Swaziland

Mortalité des moins de cinq ans
Swaziland

Mortalidad de menores de cinco años
Swazilandia

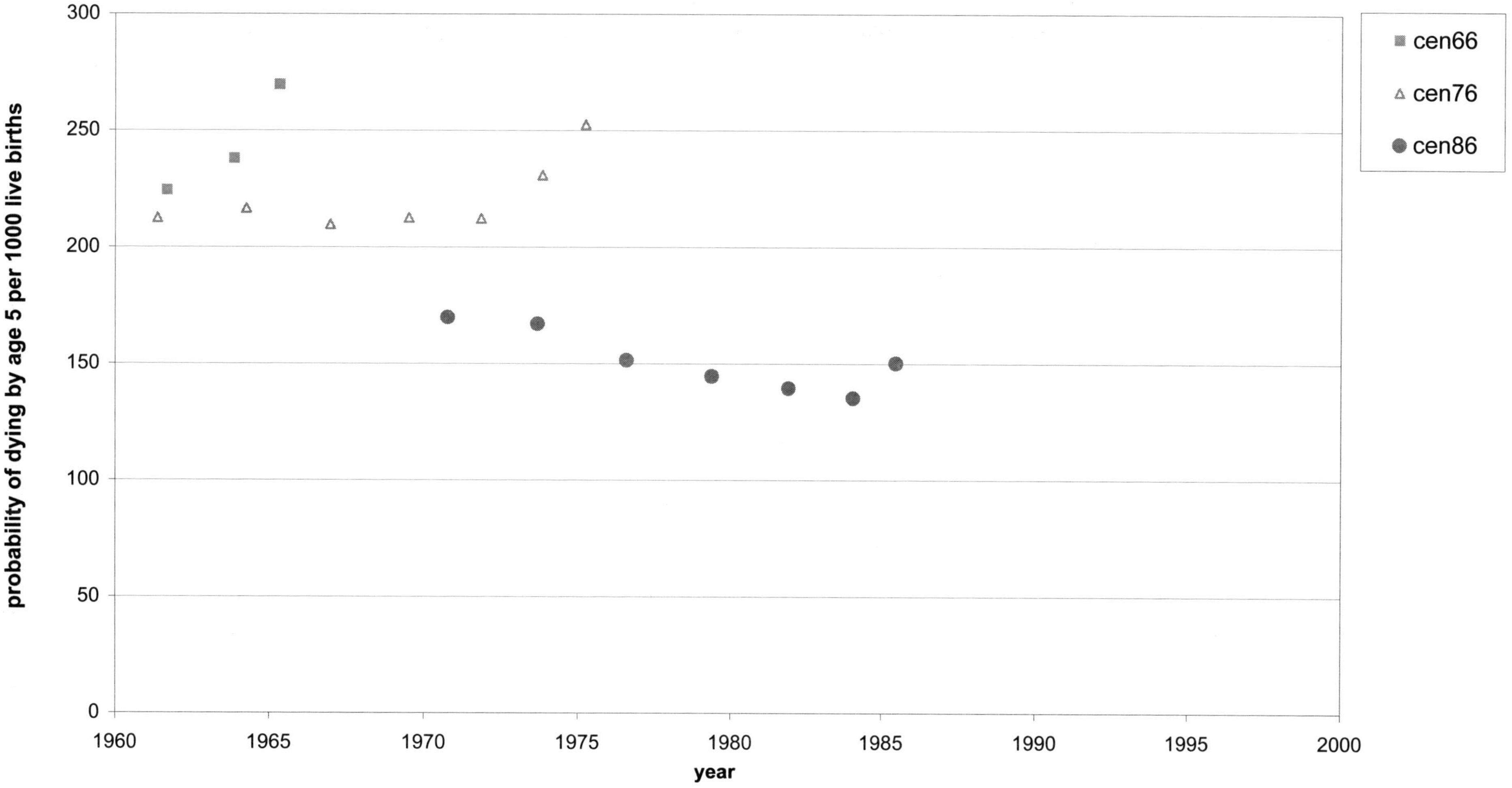

Data sources

Sources des données

Fuentes de los datos

cen66d: Census 1966 (direct method); ***cen76d***: Census 1976 (direct method); ***cen86d***: Census 1986 (direct method)

Under 5 mortality

Sweden

Mortalité des moins de cinq ans

Suède

Mortalidad de menores de cinco años

Suecia

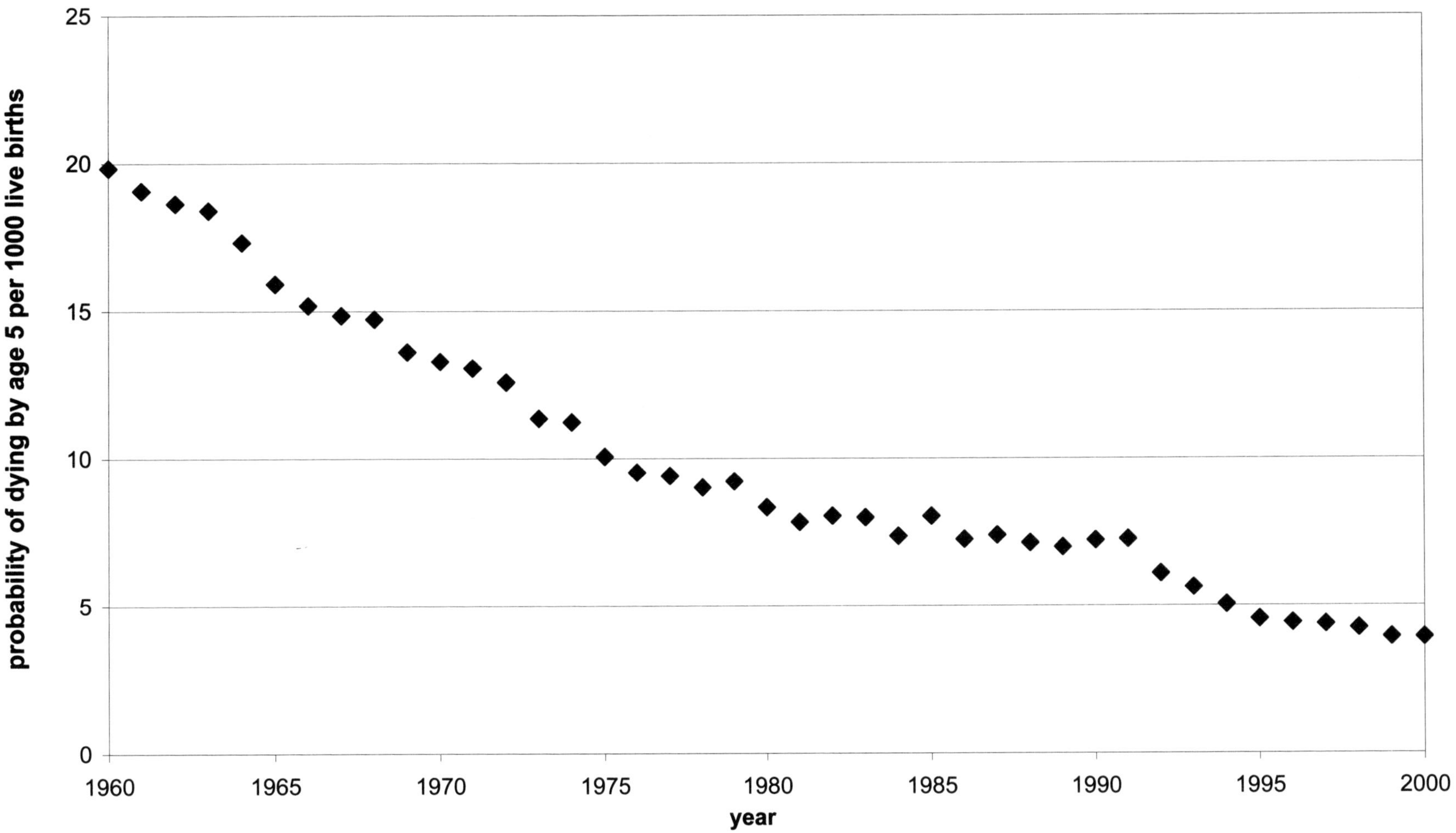

Data sources

Vital Registration

Sources des données

Fuentes de los datos

Under 5 mortality | Mortalité des moins de cinq ans | Mortalidad de menores de cinco años

Switzerland | **Suisse** | **Suiza**

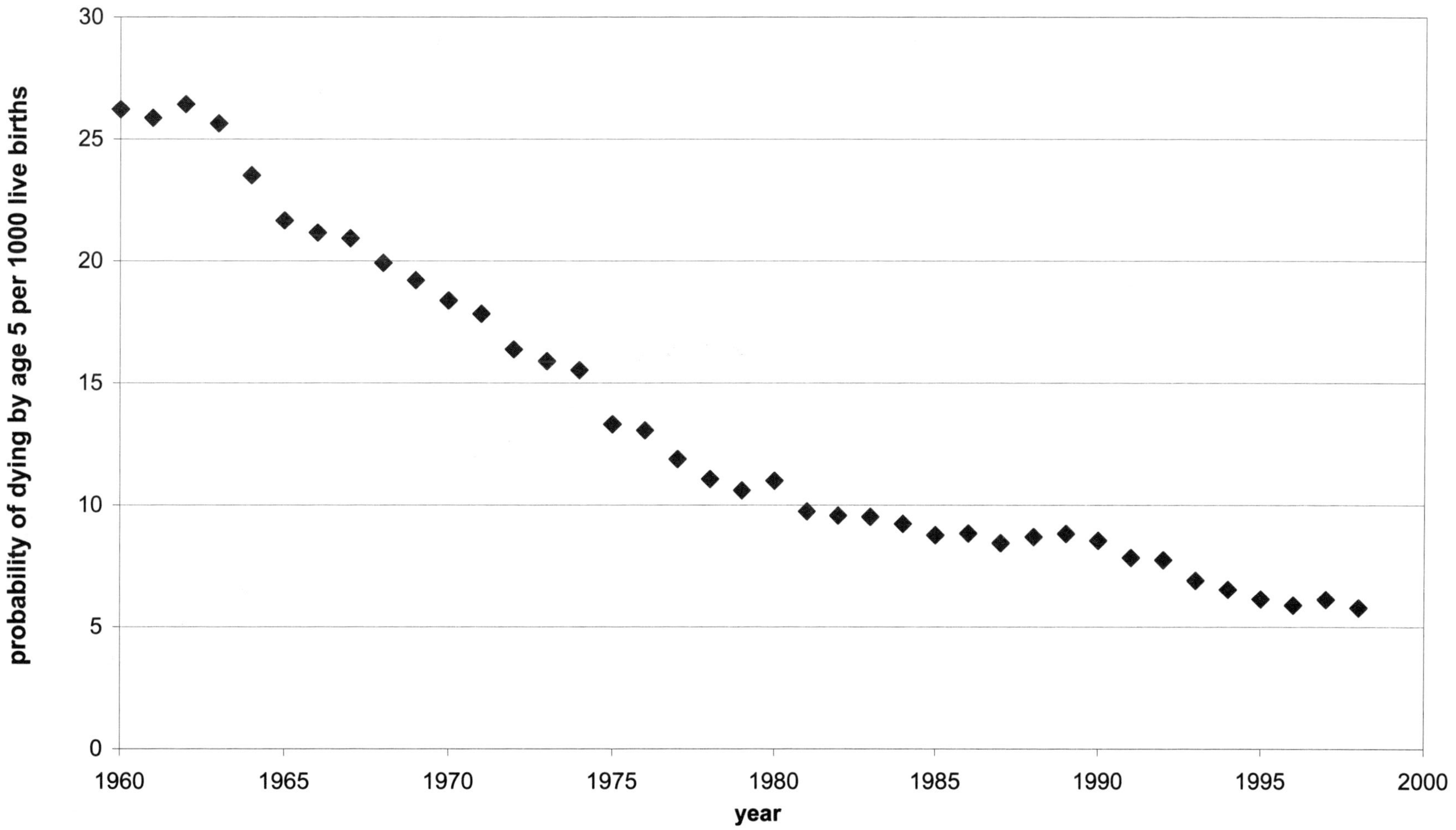

Data sources
Vital Registration

Sources des données

Fuentes de los datos

Under 5 mortality

Syrian Arab Republic

Mortalité des moins de cinq ans

République arabe syrienne

Mortalidad de menores de cinco años

República Arabe Siria

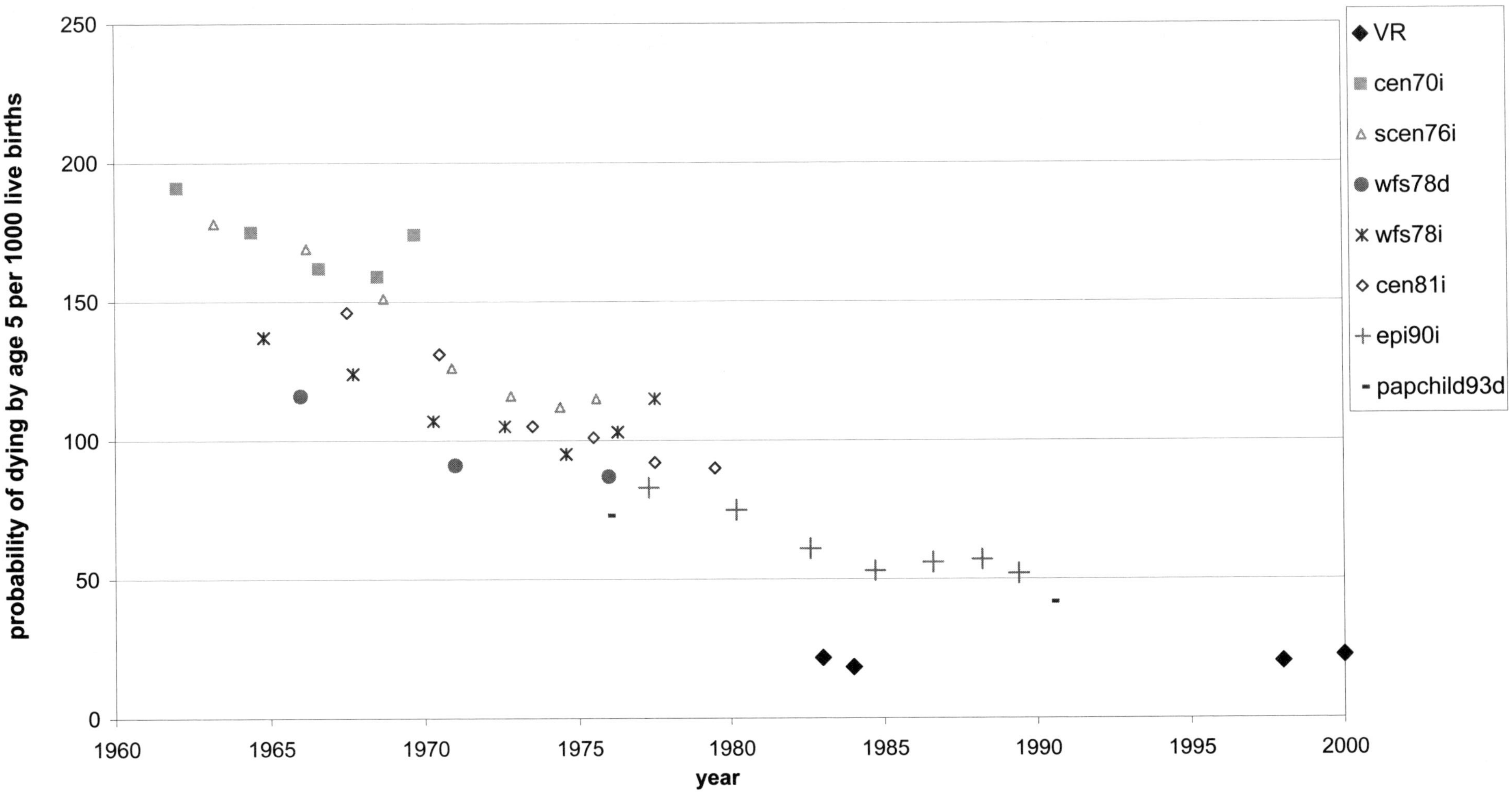

Data sources

Sources des données

Fuentes de los datos

VR: Vital Registration; ***cen70i***: Census, 23 September 1970 (indirect method); ***scen76i***: Sample Census, September 1976 (indirect method); ***wfs78d***: Syria Fertility Survey, June-August 1978 (direct method); ***wfs78i***: Syria Fertility Survey, June-August 1978 (indirect method); ***cen81i***: Census, 8 September 1981 (indirect method); ***epi90i***: Syria Epi/Cdd and Child Mortality Survey, September 1990 (indirect method); ***papchild93d***: Maternal and Child Health Survey 1993 (direct)

Under 5 mortality
Tajikistan

Mortalité des moins de cinq ans
Tadjikistan

Mortalidad de menores de cinco años
Tayikistán

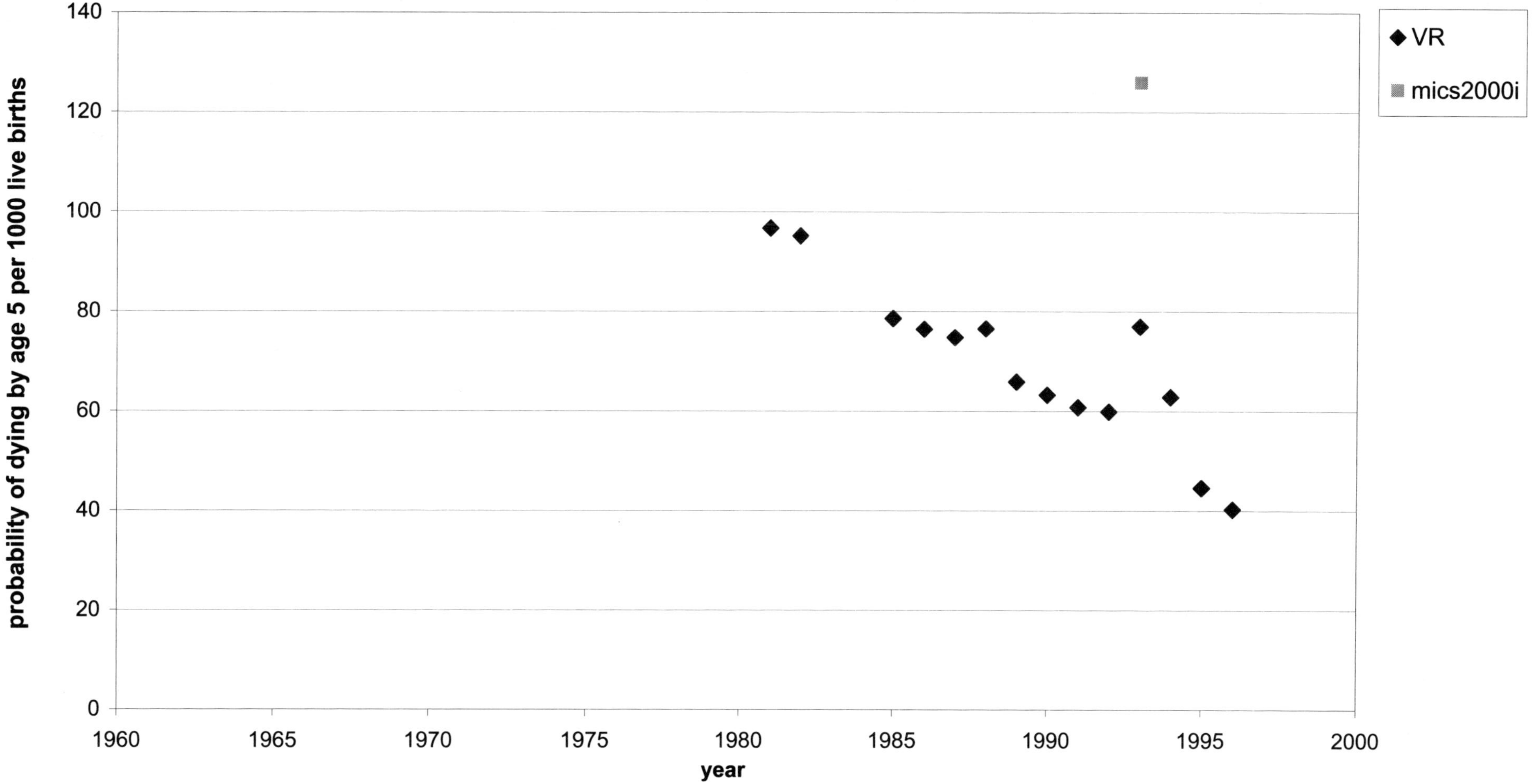

Data sources

Sources des données

Fuentes de los datos

VR : Vital Registration; ***mics2000i*** : Multiple Indicator Cluster Survey 2000 (indirect method)

Under 5 mortality

Thailand

Mortalité des moins de cinq ans

Thaïlande

Mortalidad de menores de cinco años

Tailandia

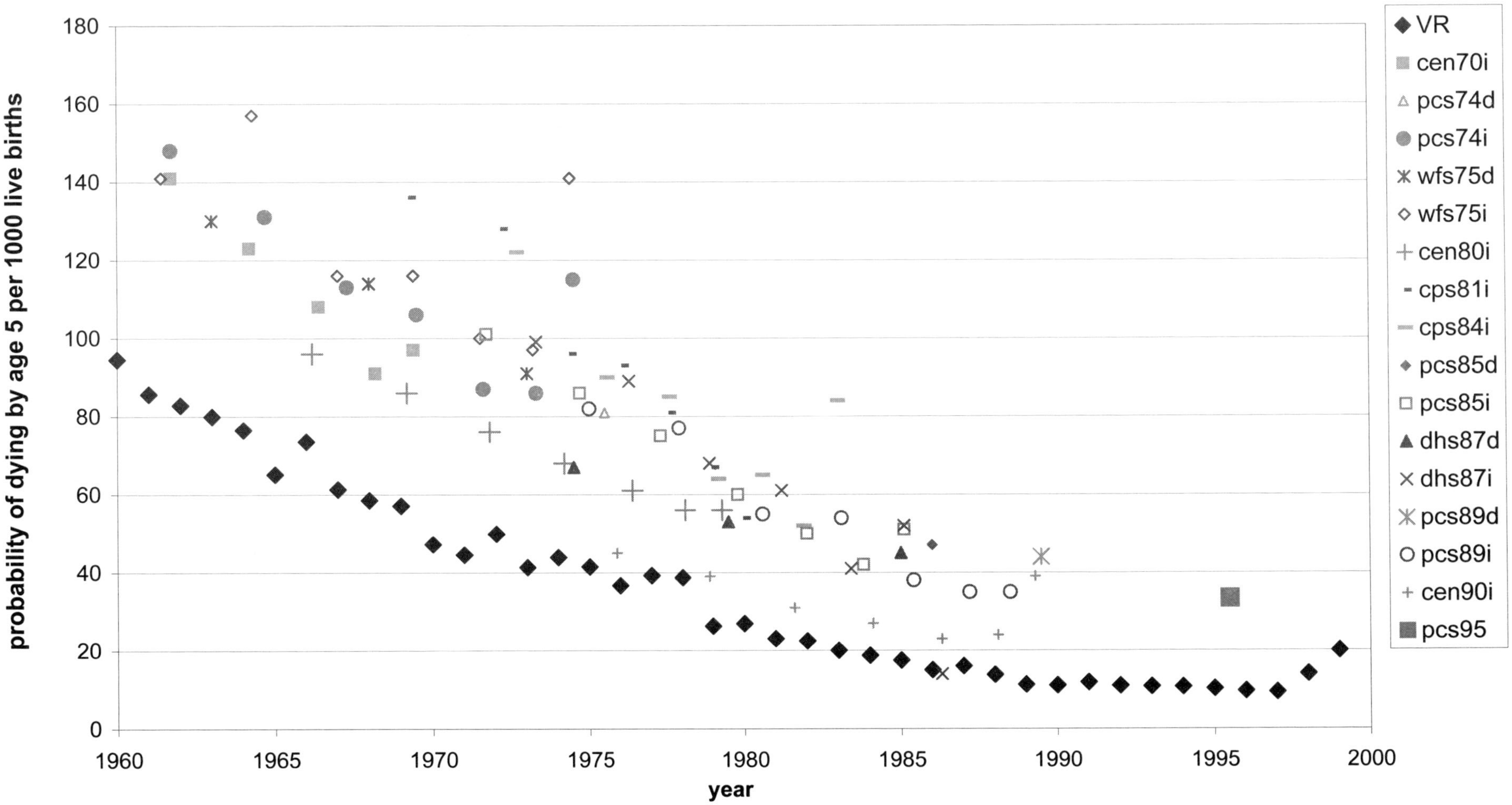

Data sources **Sources des données** **Fuentes de los datos**

VR: Vital Registration; ***cen70i***: Census, 1 April 1970 (indirect method); ***pcs74d***: Survey of Population Change, July 1974-June 1976 (direct method); ***pcs74i***: Survey of Population Change, July 1974-June 1976 (indirect method); ***wfs75d***: Survey of Fertility, April-June 1975 (direct method); ***wfs75i***: Survey of Fertility, April-June 1975 (indirect method); ***cen80i***: Census, 1 April 1980 (indirect method); ***cps81i***: Contraceptive Prevalence Survey, 25 March-10 June 1981 (indirect method); ***cps84i***: Contraceptive Prevalence Survey 1984 (indirect method); ***pcs85d***: Survey of Population Change, July 1985-September 1986 (direct method); ***pcs85i***: Survey of Population Change, July 1985-September 1986 (indirect method); ***dhs87d***: Demographic and Health Survey, March-June 1987 (direct method); ***dhs87i***: Demographic and Health Survey, March-June 1987 (indirect method); ***pcs89d***: Survey of Population Change 1989 (direct method); ***pcs89i***: Survey of Population Change 1989 (indirect method); ***cen90i***: Census 1 April 1990 (indirect method); ***pcs95***: Survey of Population Change 1995

Under 5 mortality

The former Yugoslav Republic of Macedonia

Mortalité des moins de cinq ans

Ex-République yougoslave de Macédoine

Mortalidad de menores de cinco años

Ex República Yugoslava de Macedonia

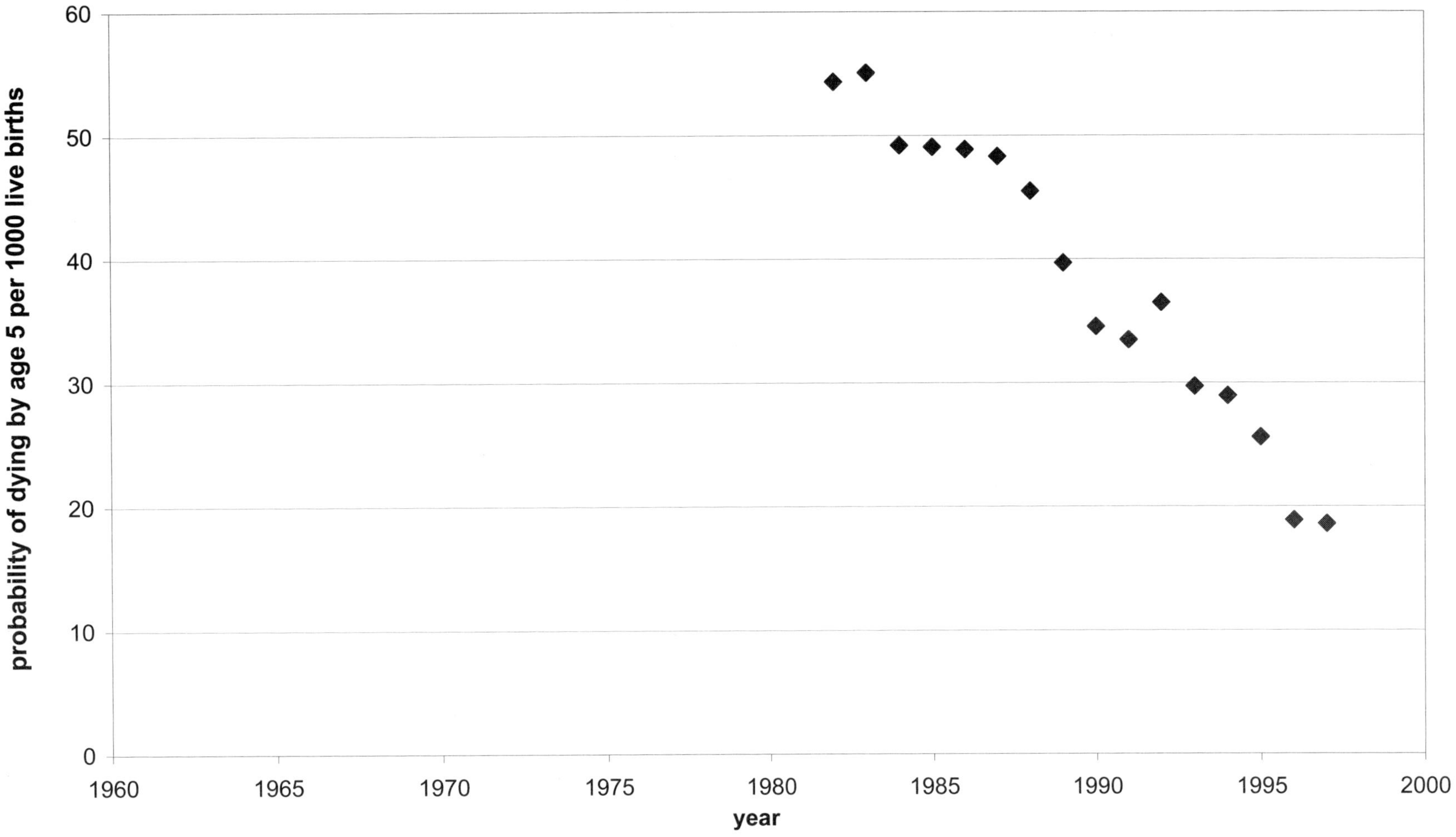

Data sources

Vital Registration

Sources des données

Fuentes de los datos

Under 5 mortality
Togo

Mortalité des moins de cinq ans
Togo

Mortalidad de menores de cinco años
Togo

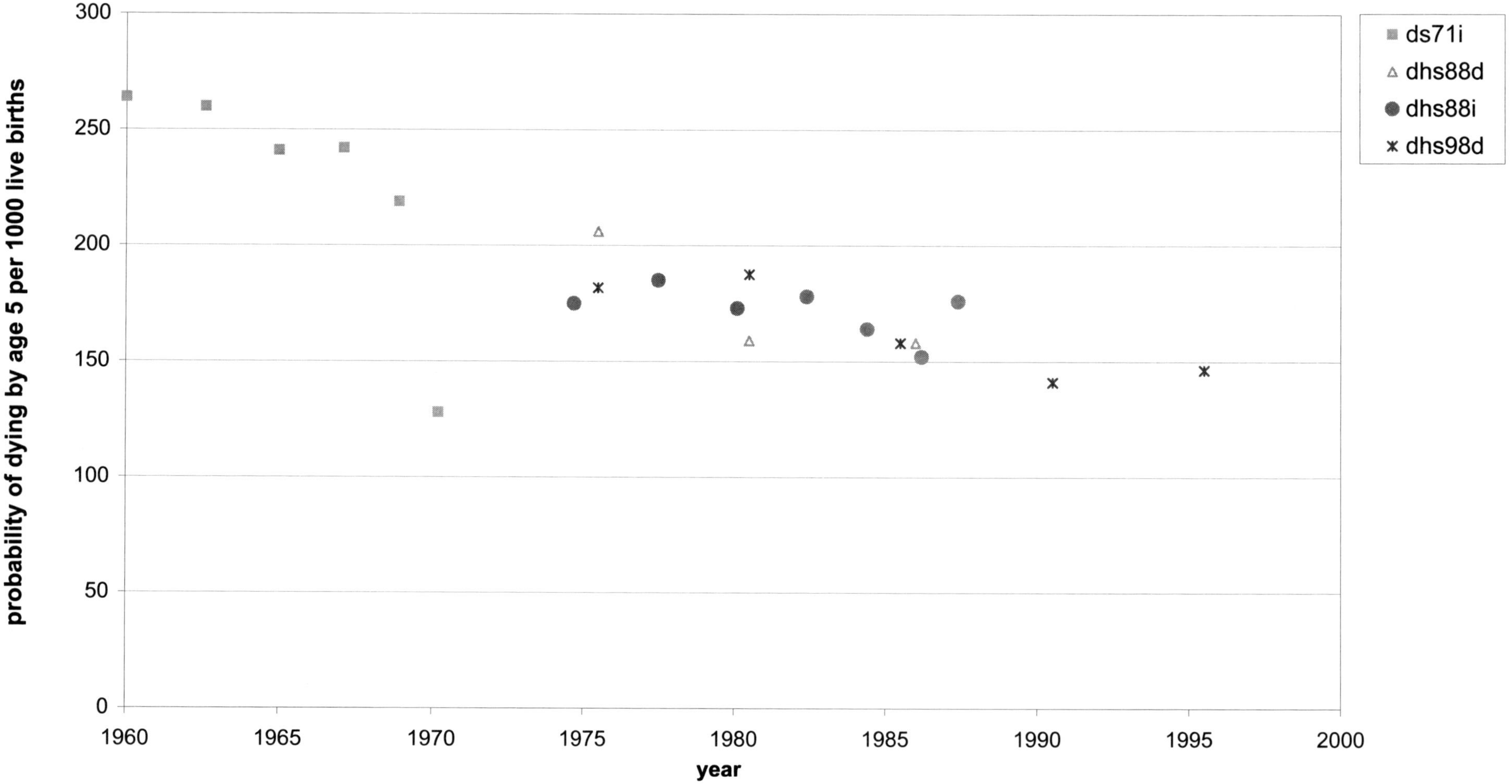

Data sources **Sources des données** **Fuentes de los datos**

ds71i : Enquête Démographique, March-April 1971 (indirect method); ***dhs88d*** : Demographic and Health Survey 1988 (direct method); ***dhs88i*** : Demographic and Health Survey 1988 (indirect method); ***dhs98d*** : Demographic and Health Survey 1998 (direct method)

Under 5 mortality

Tonga

Mortalité des moins de cinq ans

Tonga

Mortalidad de menores de cinco años

Tonga

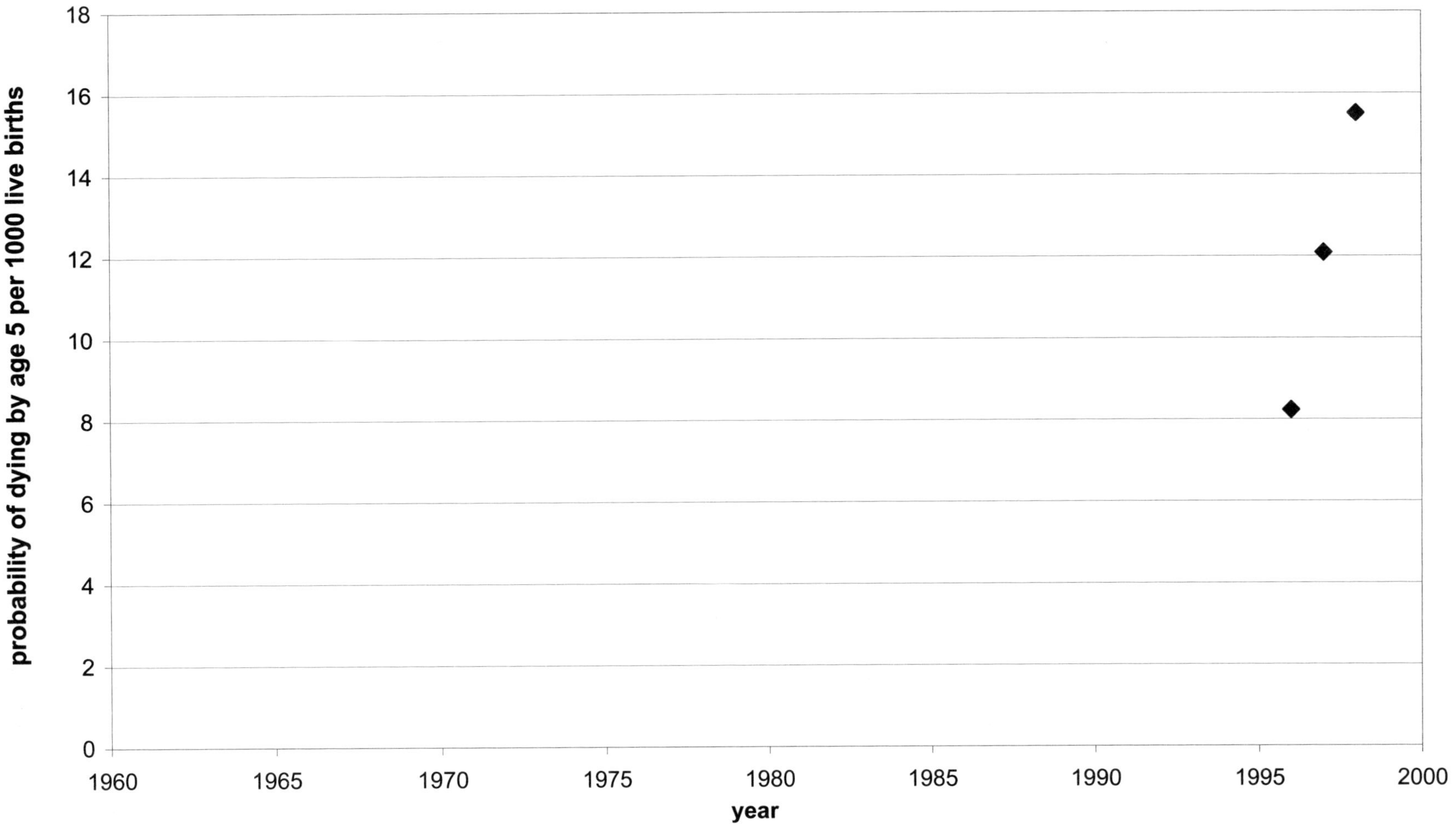

Data sources

Vital Registration

Sources des données

Fuentes de los datos

Under 5 mortality

Trinidad and Tobago

Mortalité des moins de cinq ans

Trinité-et-Tobago

Mortalidad de menores de cinco años

Trinidad y Tabago

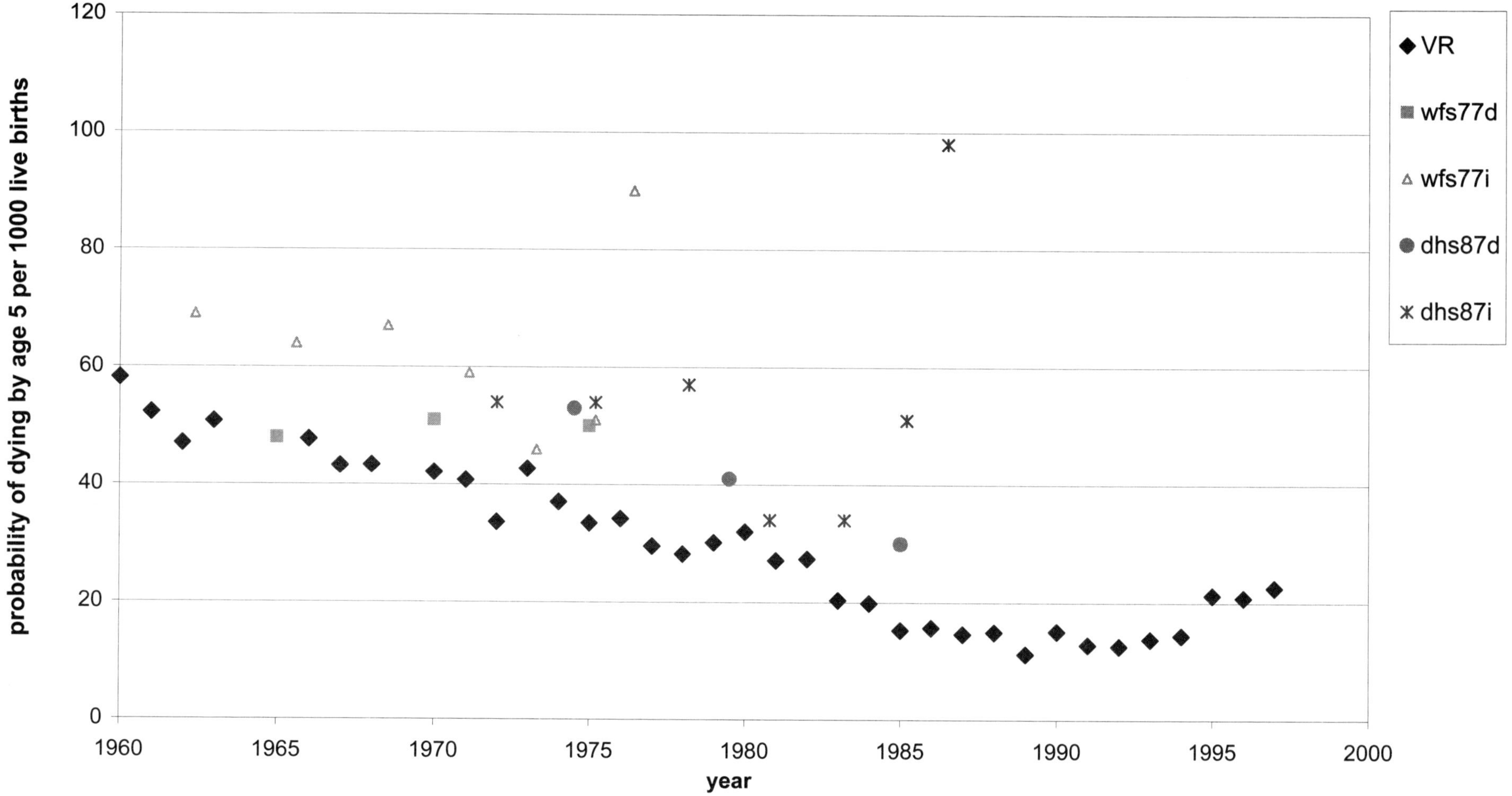

Data sources

Sources des données

Fuentes de los datos

VR: Vital Registration; ***wfs77d***: Fertility Survey, 26 March-30 June 1977 (direct method); ***wfs77i***: Fertility Survey, 26 March-30 June 1977 (indirect method); ***dhs87d***: Demographic and Health Survey, May-September 1987 (direct method); ***dhs87i***: Demographic and Health Survey, May-September 1987 (indirect method)

Under 5 mortality

Tunisia

Mortalité des moins de cinq ans

Tunisie

Mortalidad de menores de cinco años

Túnez

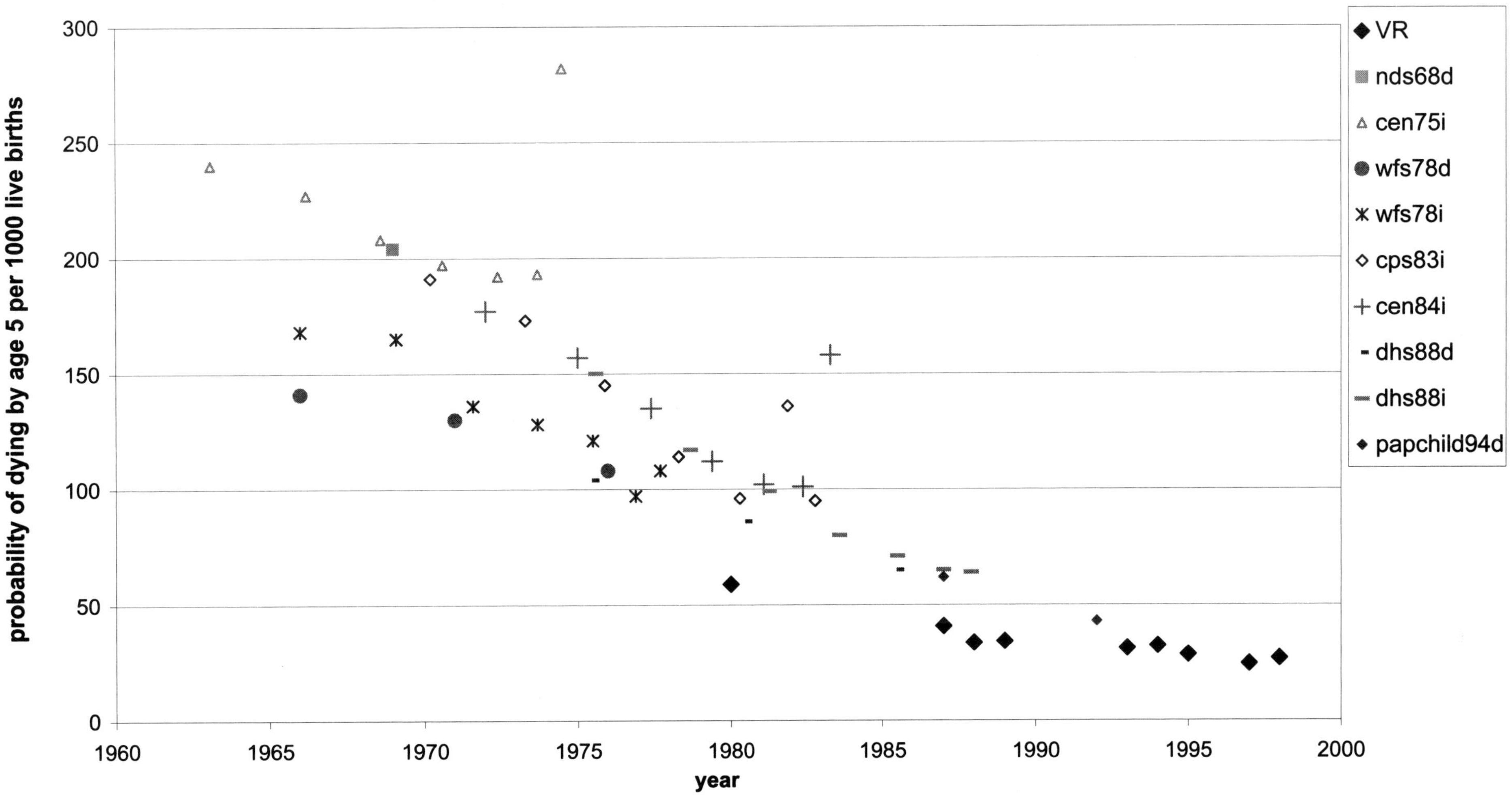

Data sources **Sources des données** **Fuentes de los datos**

VR: Vital Registration; ***nds68d***: Enquête Nationale Démographique, 1968-69 (direct method); ***cen75i***: Census, 8 May 1975 (indirect method); ***wfs78d***: Enquête Tunisienne sur la Fécondite, May-August 1978 (direct method); ***wfs78i***: Enquête Tunisienne sur la Fécondite, May-August 1978 (indirect method); ***cps83i***: Enquête Tunisienne sur la Prévalence de la Contraception 1983 (indirect method); ***cen84i***: Census, 30 March 1984 (indirect method); ***dhs88d***: Enquête Démographique et de Santé, June-October 1988 (direct method); ***dhs88i***: Enquête Démographique et de Santé, June-October 1988 (indirect method); ***papchild94d***: Maternal and Child Health Survey 1994 (direct method)

Under 5 mortality

Turkey

Mortalité des moins de cinq ans

Turquie

Mortalidad de menores de cinco años

Turquía

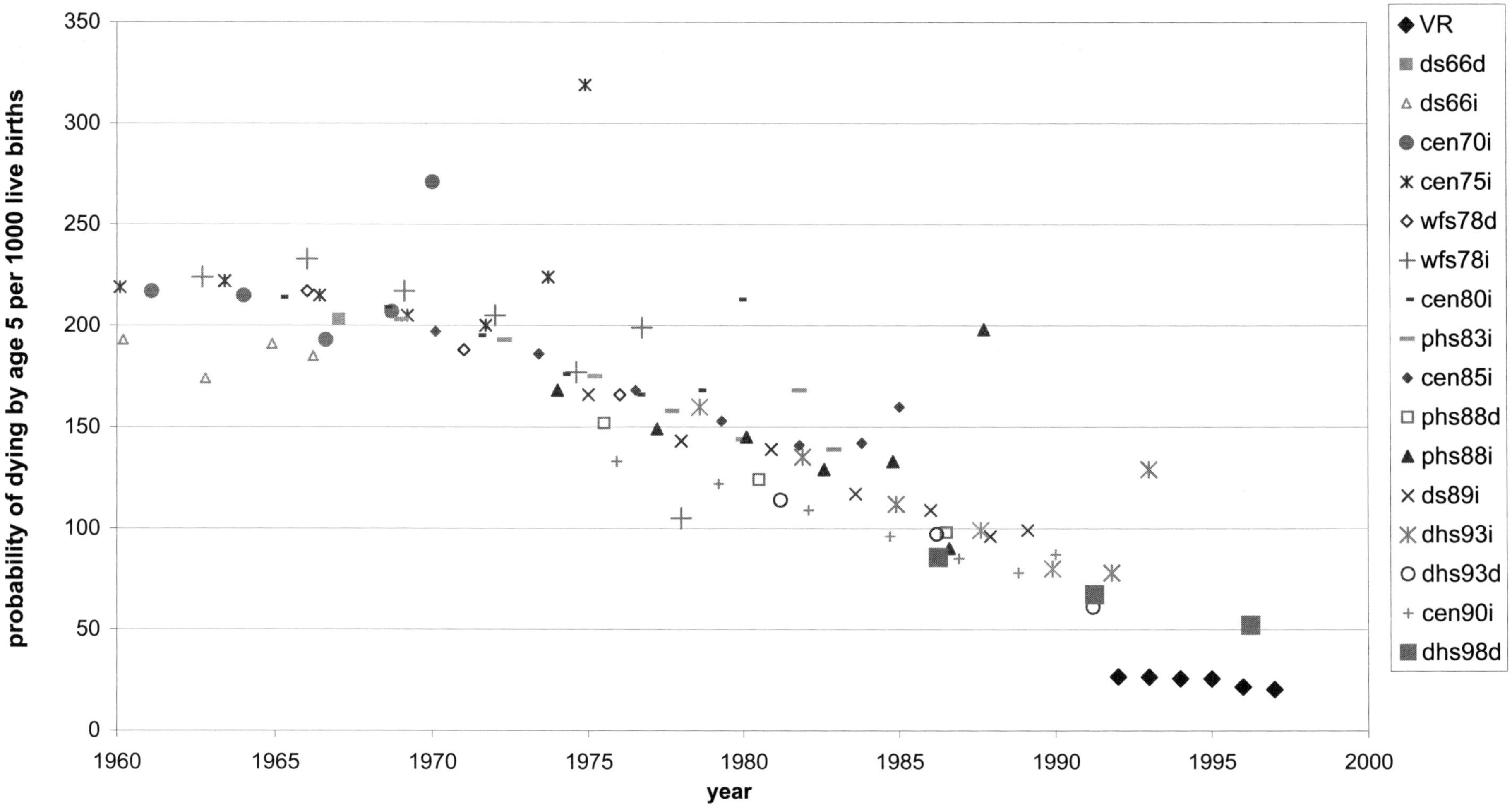

Data sources **Sources des données** **Fuentes de los datos**

VR: Vital Registration (urban area)***; ds66d***: Demographic Survey, 1966-1967 (direct method); ***ds66i***: Demographic Survey, 1966-1967 (indirect method); ***cen70i***: Census, 25 October 1970 (indirect method); ***cen75i***: Census, 26 October 1975 (indirect method); ***wfs78d***: Fertility Survey, September-October 1978 (direct method); ***wfs78i***: Fertility Survey, September-October 1978 (indirect method); ***cen80i***: Census, 12 October 1980 (indirect method); ***phs83i***: Population and Health Survey, August-November 1983 (indirect method); ***cen85i***: Census, 20 October 1985 (indirect method); ***phs88d***: Population and Health Survey, August-September 1988 (direct method); ***phs88i***: Population and Health Survey, August-September 1988 (indirect method); ***ds89i***: Demographic Survey 1989 (indirect method); ***dhs93i***: Demographic and Health Survey, August-October 1993 (indirect method); ***dhs93d***: Demographic and Health Survey, August-October 1993 (direct method); ***cen90i***: Census, 21 October 1990 (indirect method); ***dhs98d***: Demographic and Health Survey 1998 (direct method)

Under 5 mortality
Turkmenistan

Mortalité des moins de cinq ans
Turkménistan

Mortalidad de menores de cinco años
Turkmenistán

Data sources
Vital Registration

Sources des données

Fuentes de los datos

Under 5 mortality

Tuvalu

Mortalité des moins de cinq ans

Tuvalu

Mortalidad de menores de cinco años

Tuvalu

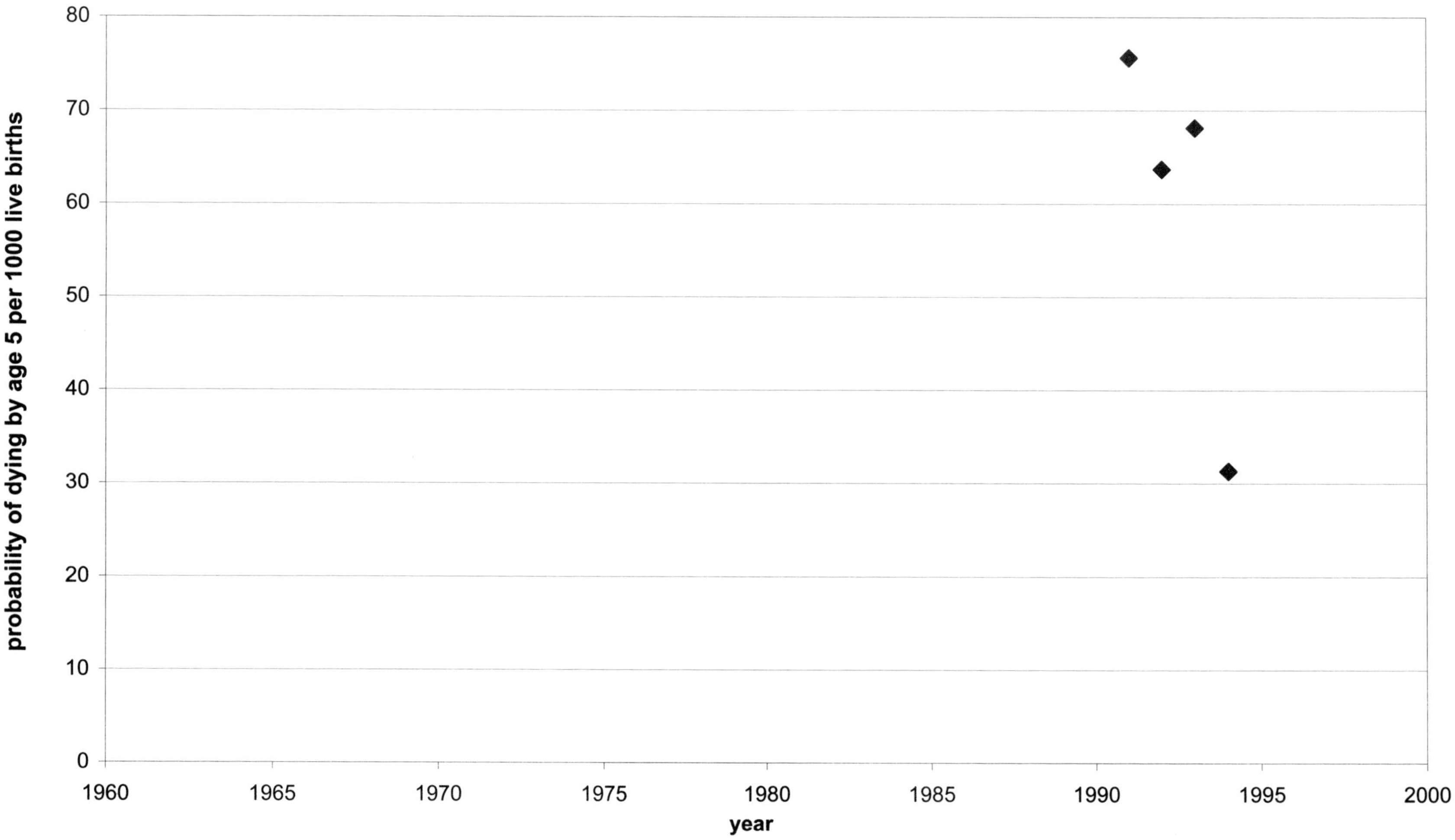

Data sources

Vital Registration

Sources des données

Fuentes de los datos

Under 5 mortality

Uganda

Mortalité des moins de cinq ans

Ouganda

Mortalidad de menores de cinco años

Uganda

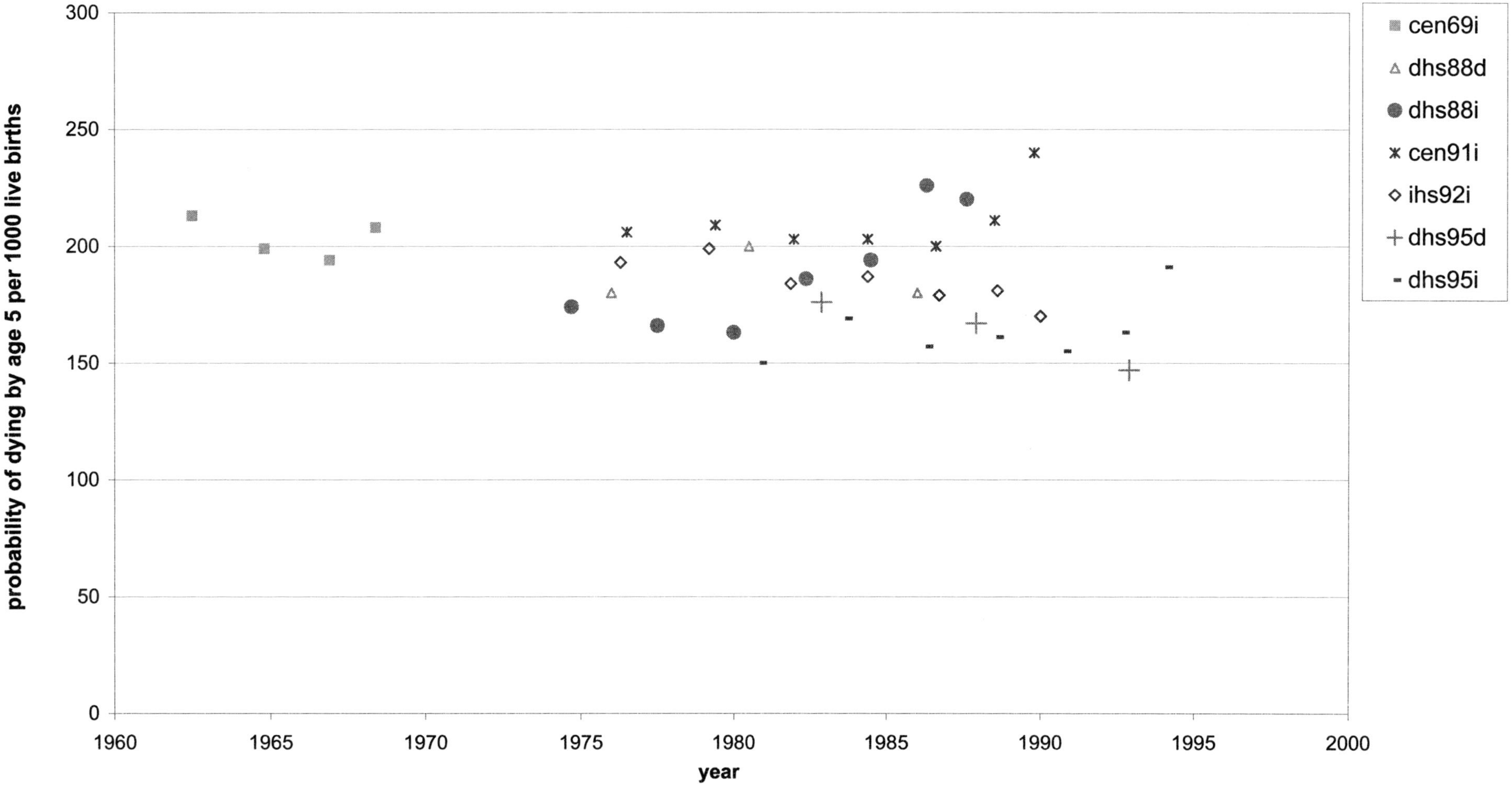

Data sources **Sources des données** **Fuentes de los datos**

cen69i: Census, 18 August 1969 (indirect method); ***dhs88d***: Demographic and Health Survey, September 1988-February 1989 (direct method); ***dhs88i***: Demographic and Health Survey, September 1988-February 1989 (indirect method); ***cen91i***: Census, 1-19 January 1991 (indirect method); ***ihs92i***: National Integrated Household Survey 1992 (indirect method); ***dhs95d***: Demographic and Health Survey, March-August 1995 (direct method); ***dhs95i***: Demographic and Health Survey, March-August 1995 (indirect method)

Under 5 mortality
Ukraine

Mortalité des moins de cinq ans
Ukraine

Mortalidad de menores de cinco años
Ucrania

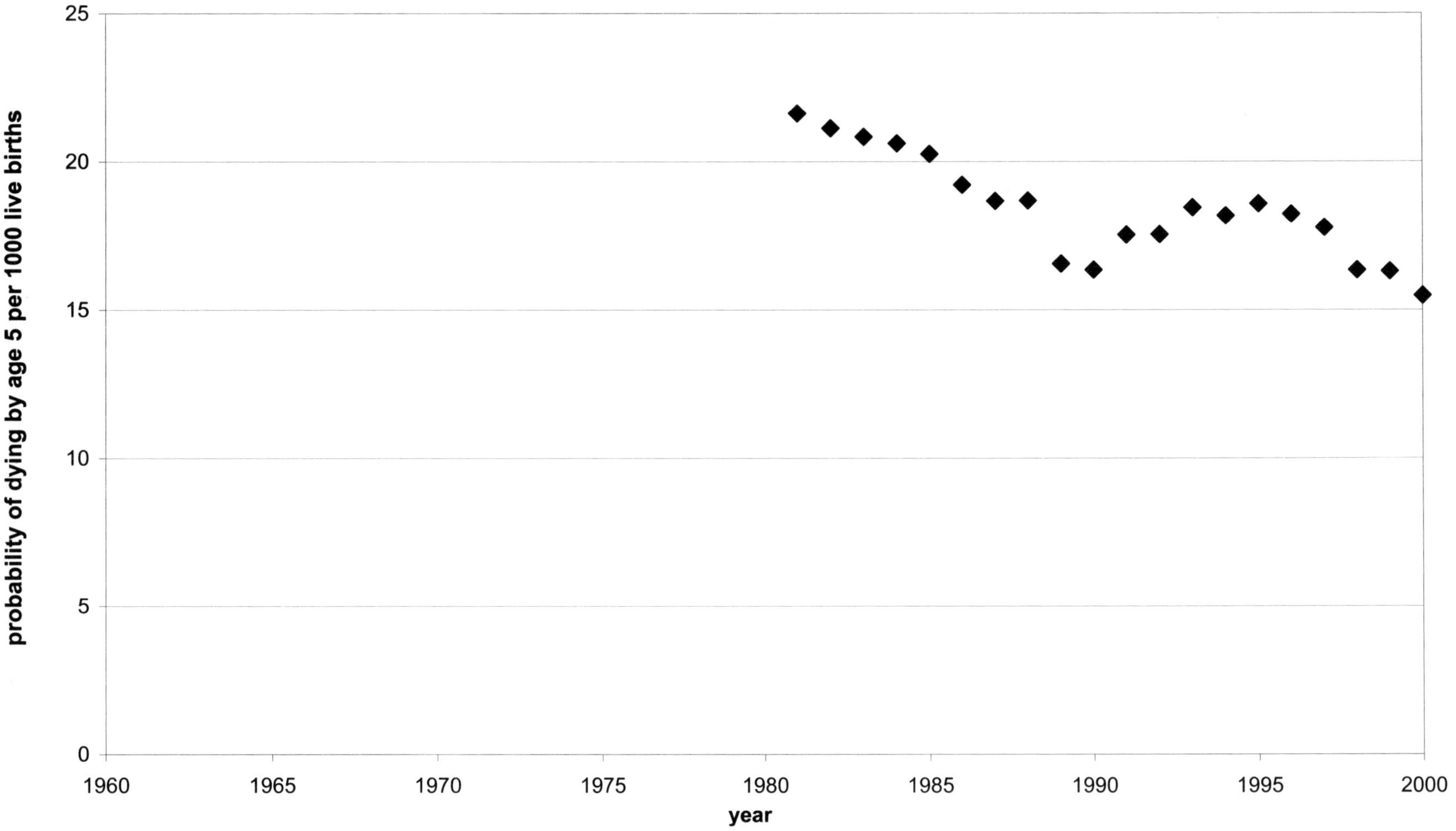

Data sources
Vital Registration

Sources des données

Fuentes de los datos

Under 5 mortality

United Arab Emirates

Mortalité des moins de cinq ans

Emirats arabes unis

Mortalidad de menores de cinco años

Emiratos Arabes Unidos

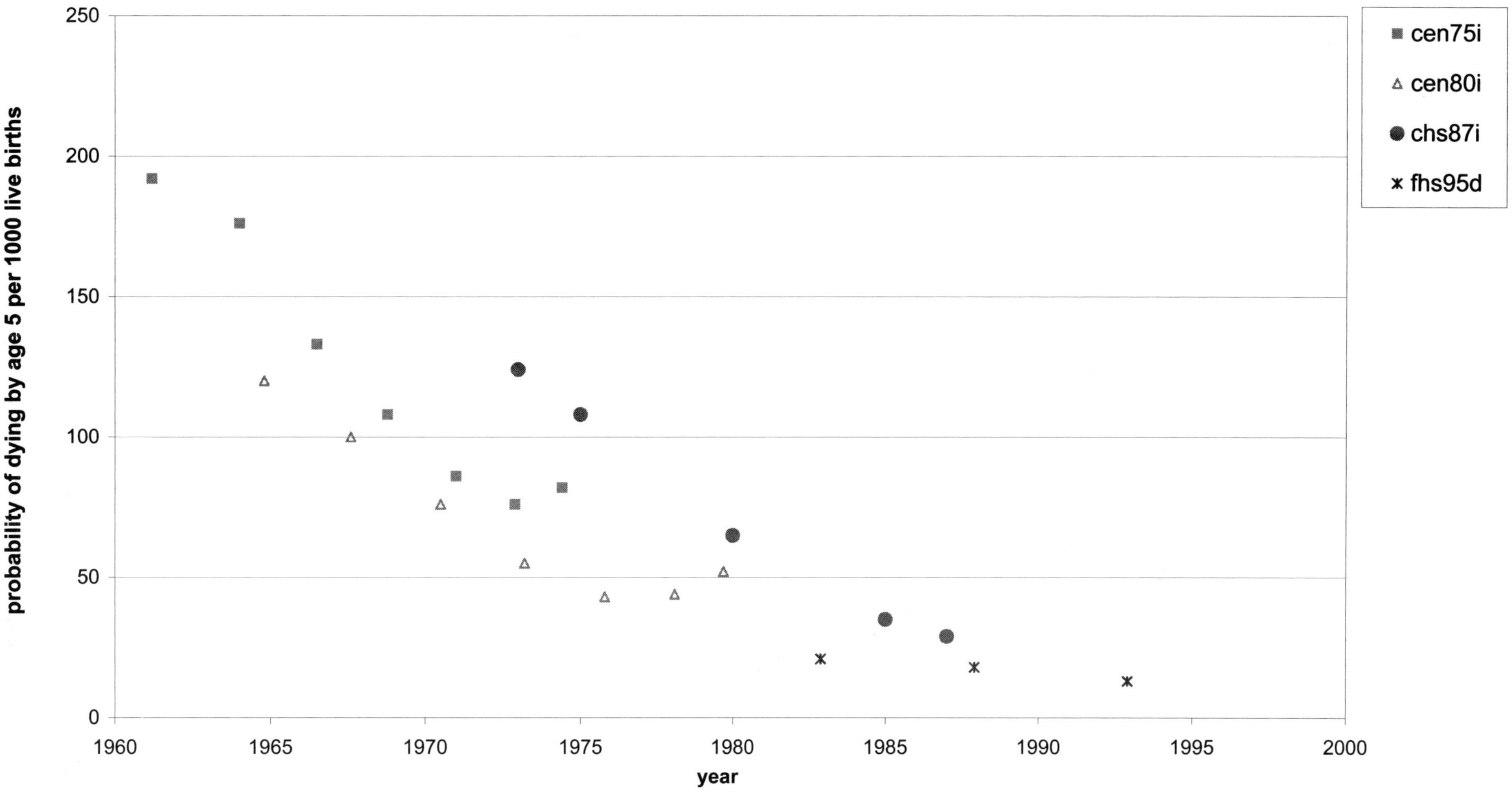

Data sources

Sources des données

Fuentes de los datos

cen75i: Census, 31 December 1975 (indirect method); ***cen80i***: Census, 15 December 1980 (indirect method); ***chs87i***: Child Health Survey, March 1987-May 1988 (indirect method); ***fhs95d***: Family Health Survey 1995 (direct method)

Under 5 mortality

United Kingdom

Mortalité des moins de cinq ans

Royaume-Uni

Mortalidad de menores de cinco años

Reino Unido

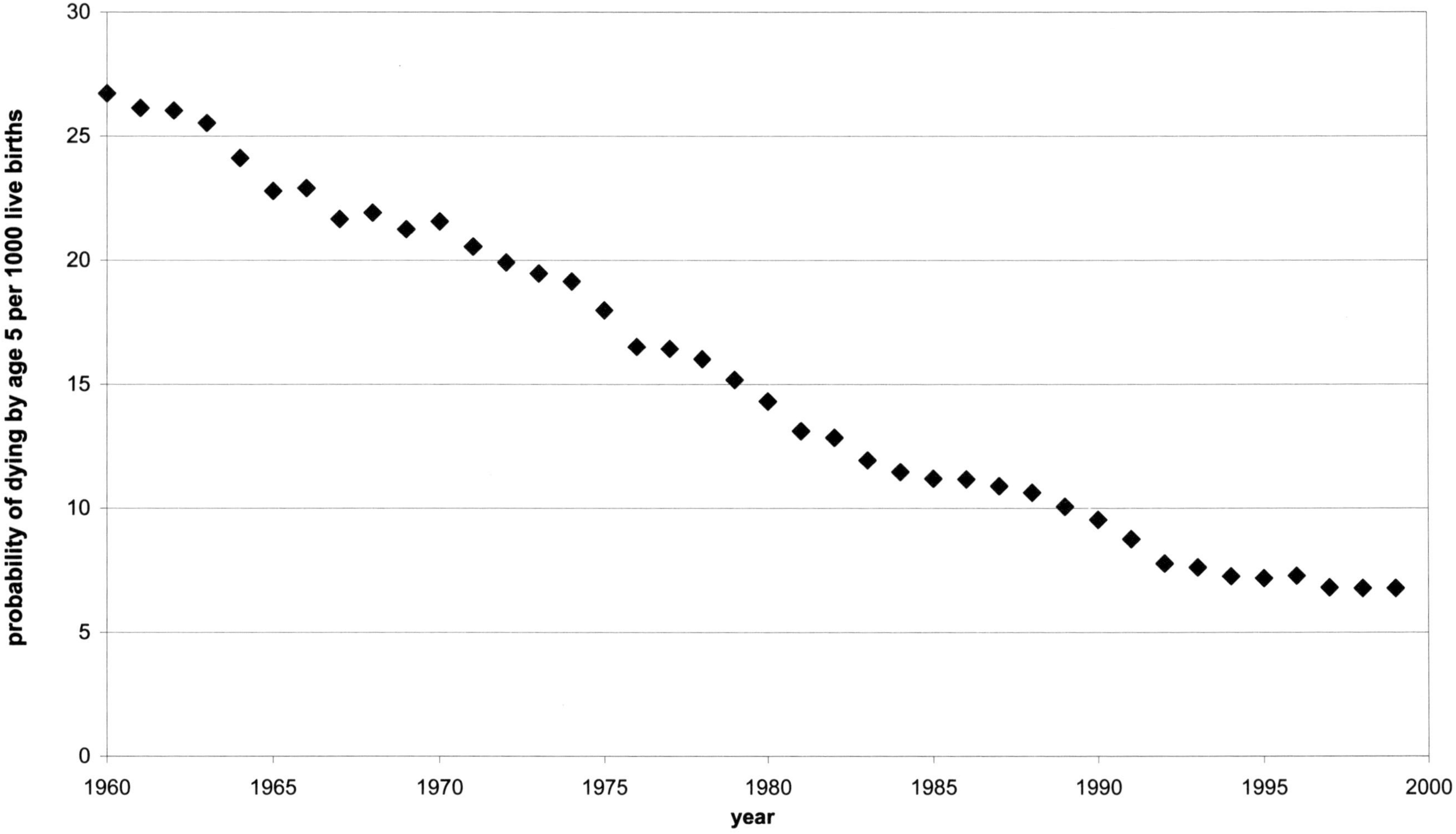

Data sources

Vital Registration

Sources des données

Fuentes de los datos

Under 5 mortality
United Republic of Tanzania

Mortalité des moins de cinq ans
République-Unie de Tanzanie

Mortalidad de menores de cinco años
República Unida de Tanzanía

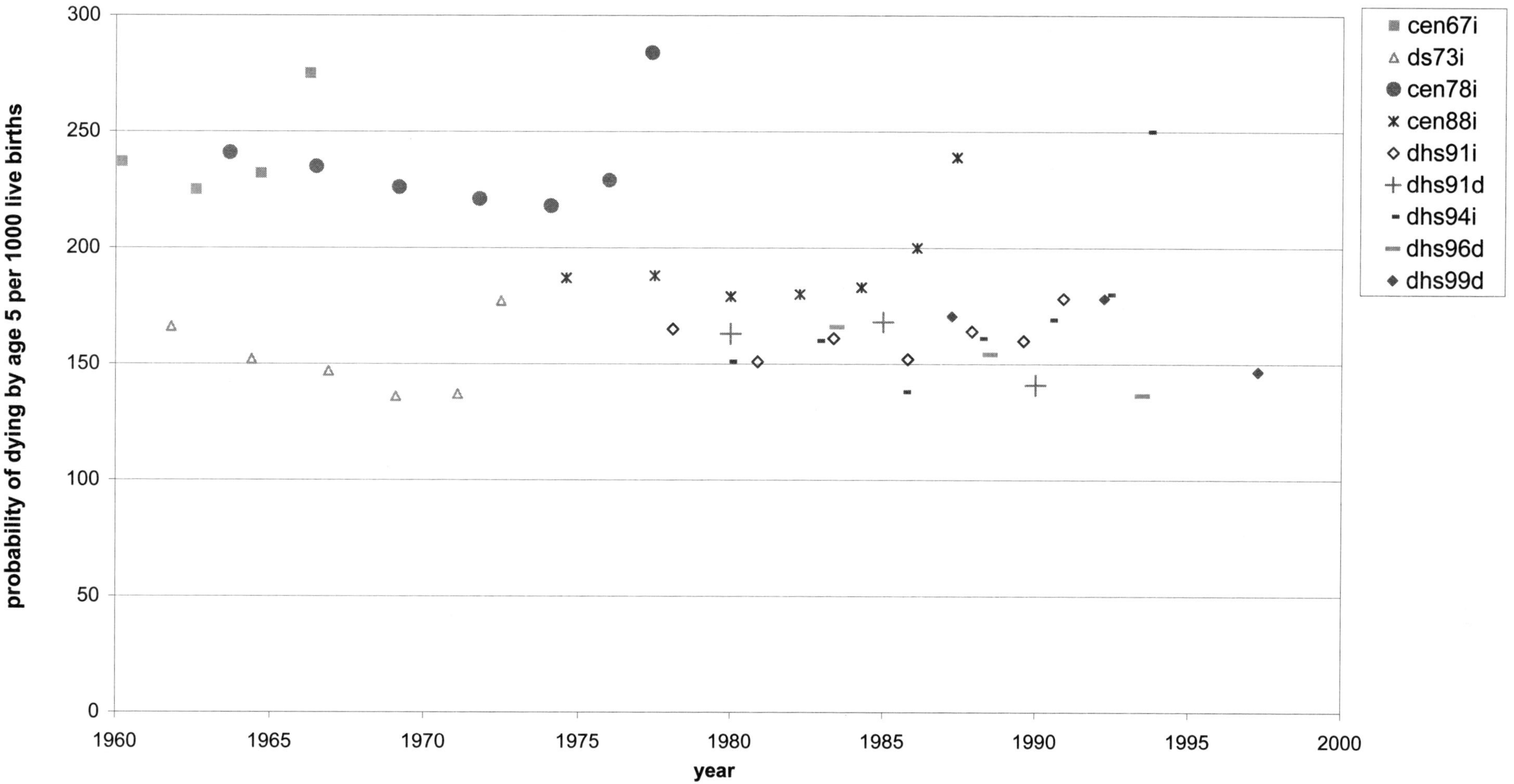

Data sources **Sources des données** **Fuentes de los datos**

cen67i : Census, 26 August 1967 (indirect method); ***ds73i*** : National Demographic Survey of Tanzania, July-December 1973 (indirect method); ***cen78i*** : Census, 26-27 August 1978 (indirect method); ***cen88i*** : Census, 28 August 1988 (indirect method); ***dhs91i*** : Demographic and Health Survey, October 1991-March 1992 (indirect method); ***dhs91d*** : Demographic and Health Survey, October 1991-March 1992 (direct method); ***dhs94i*** : Demographic and Health Survey 1994 (indirect method); ***dhs96d*** : Demographic and Health Survey 1996 (direct method); ***dhs99d*** : Demographic and Health Survey 1999 (direct method)

Under 5 mortality

United States of America

Mortalité des moins de cinq ans

Etats-Unis d'Amérique

Mortalidad de menores de cinco años

Estados Unidos de América

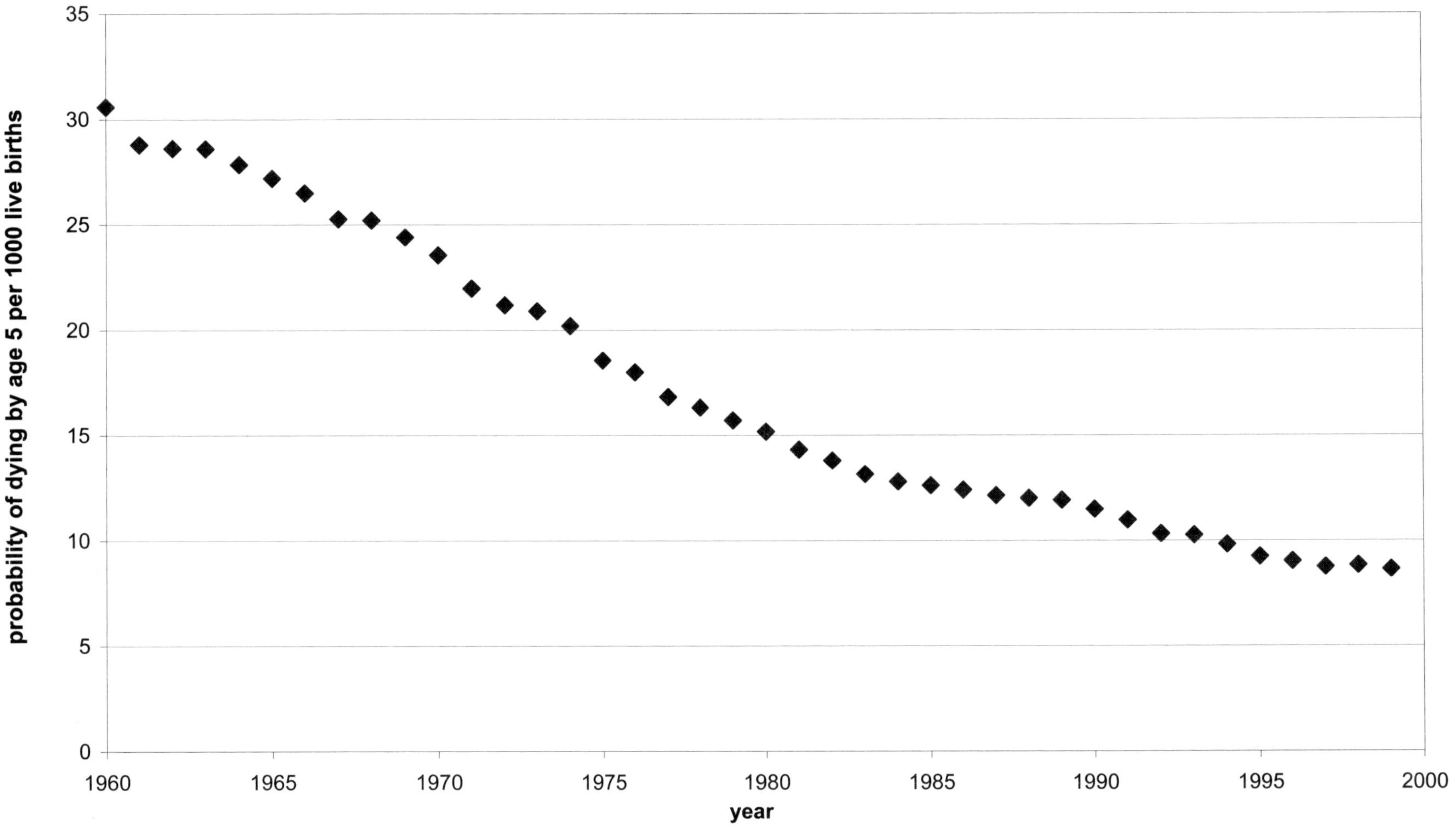

Data sources

Vital Registration

Sources des données

Fuentes de los datos

Under 5 mortality

Uruguay

Mortalité des moins de cinq ans

Uruguay

Mortalidad de menores de cinco años

Uruguay

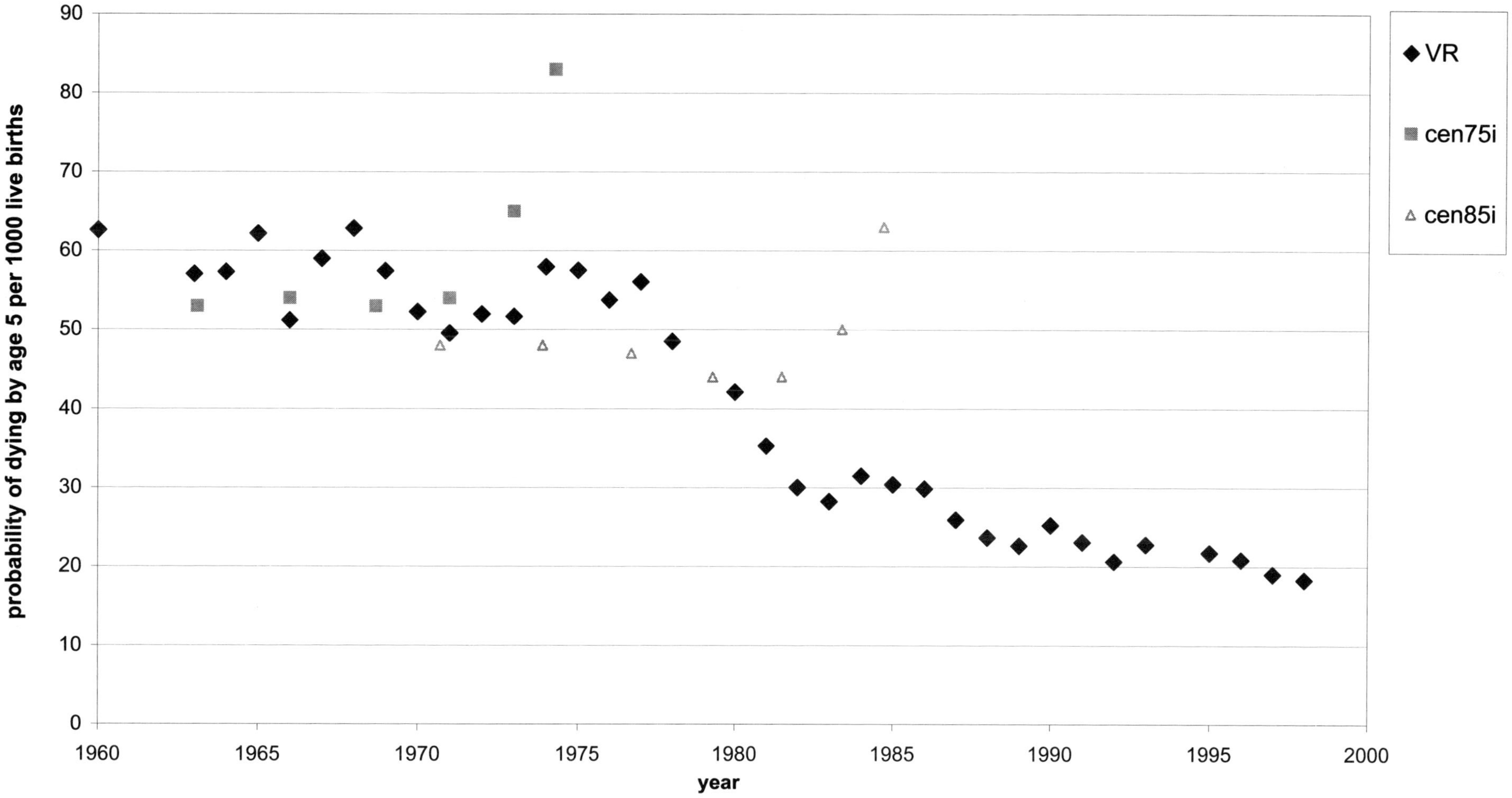

Data sources

Sources des données

Fuentes de los datos

VR: Vital Registration; ***cen75i***: Census 1975 (indirect method); ***cen85i***: Census 1985 (indirect method)

Under 5 mortality
Uzbekistan

Mortalité des moins de cinq ans
Ouzbékistan

Mortalidad de menores de cinco años
Uzbekistán

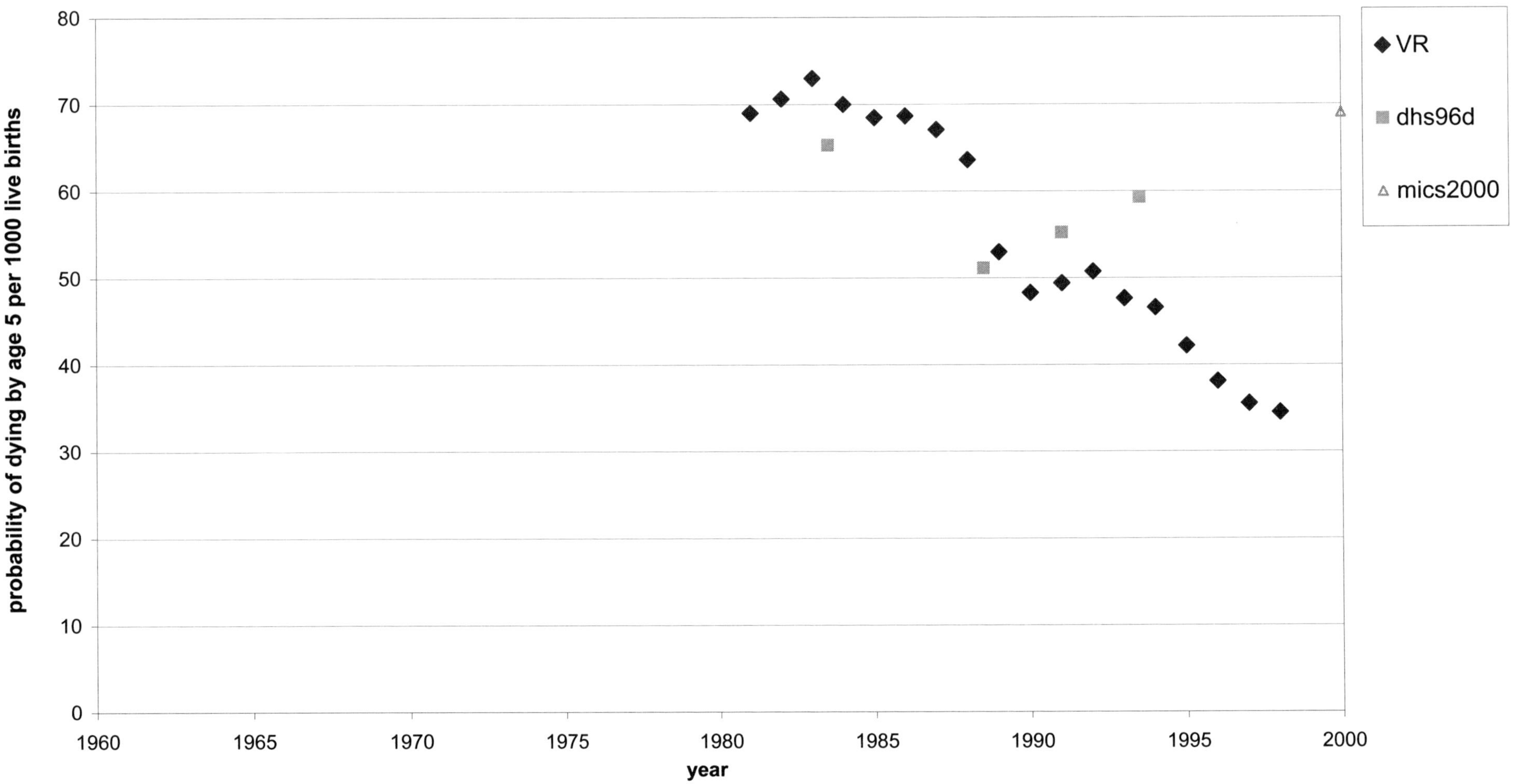

Data sources **Sources des données** **Fuentes de los datos**

VR: Vital Registration; ***dhs96d***: Demographic and Health Survey 1996 (direct method); ***mics2000***: Multiple Indicator Cluster Survey 2000

Under 5 mortality
Vanuatu

Mortalité des moins de cinq ans
Vanuatu

Mortalidad de menores de cinco años
Vanuatu

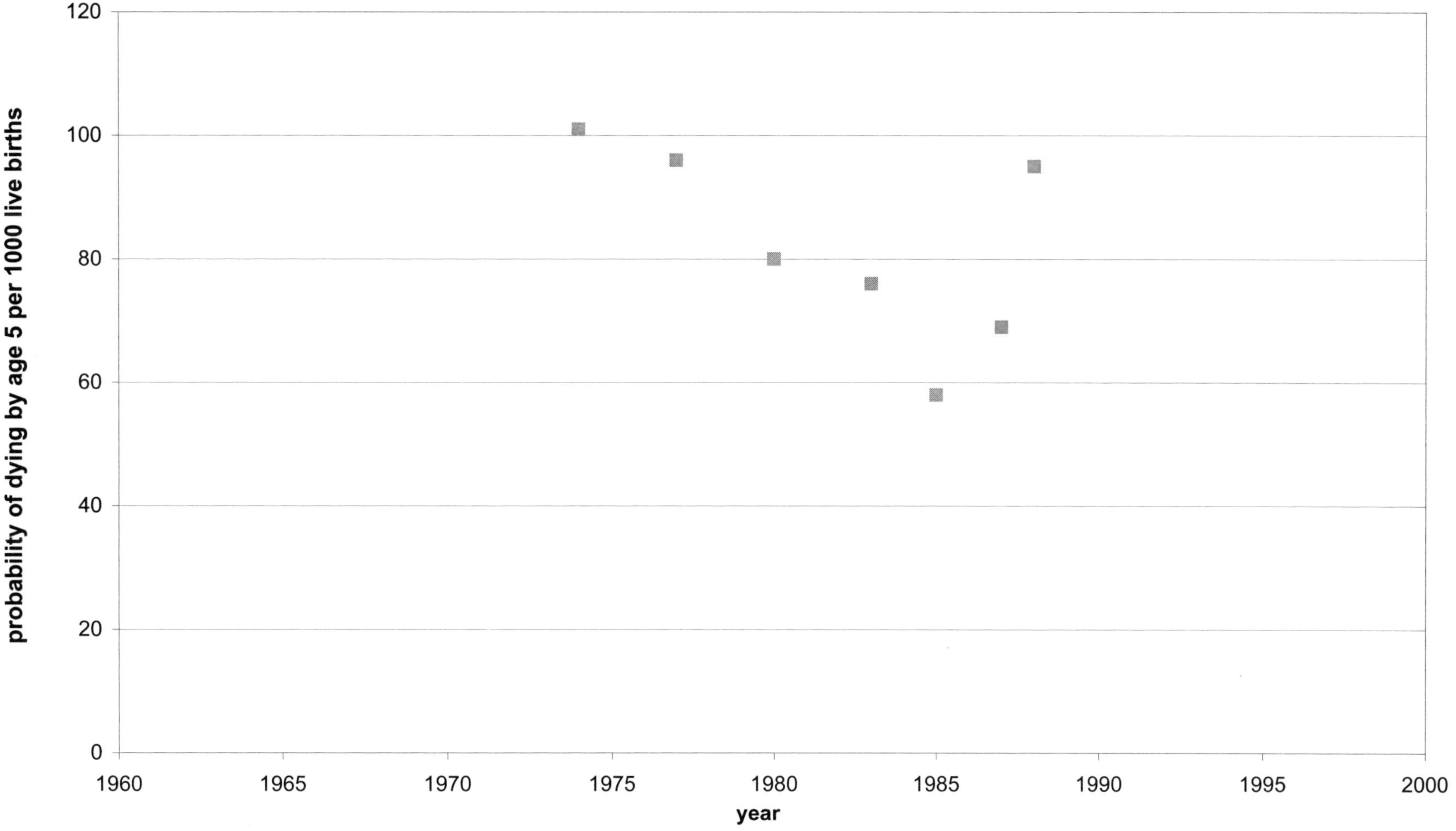

Data sources
Census 1989 (indirect method)

Sources des données

Fuentes de los datos

Under 5 mortality
Venezuela (Bolivarian Republic of)

Mortalité des moins de cinq ans
Venezuela (République bolivarienne du)

Mortalidad de menores de cinco años
Venezuela (Reùplica Bolivariana de)

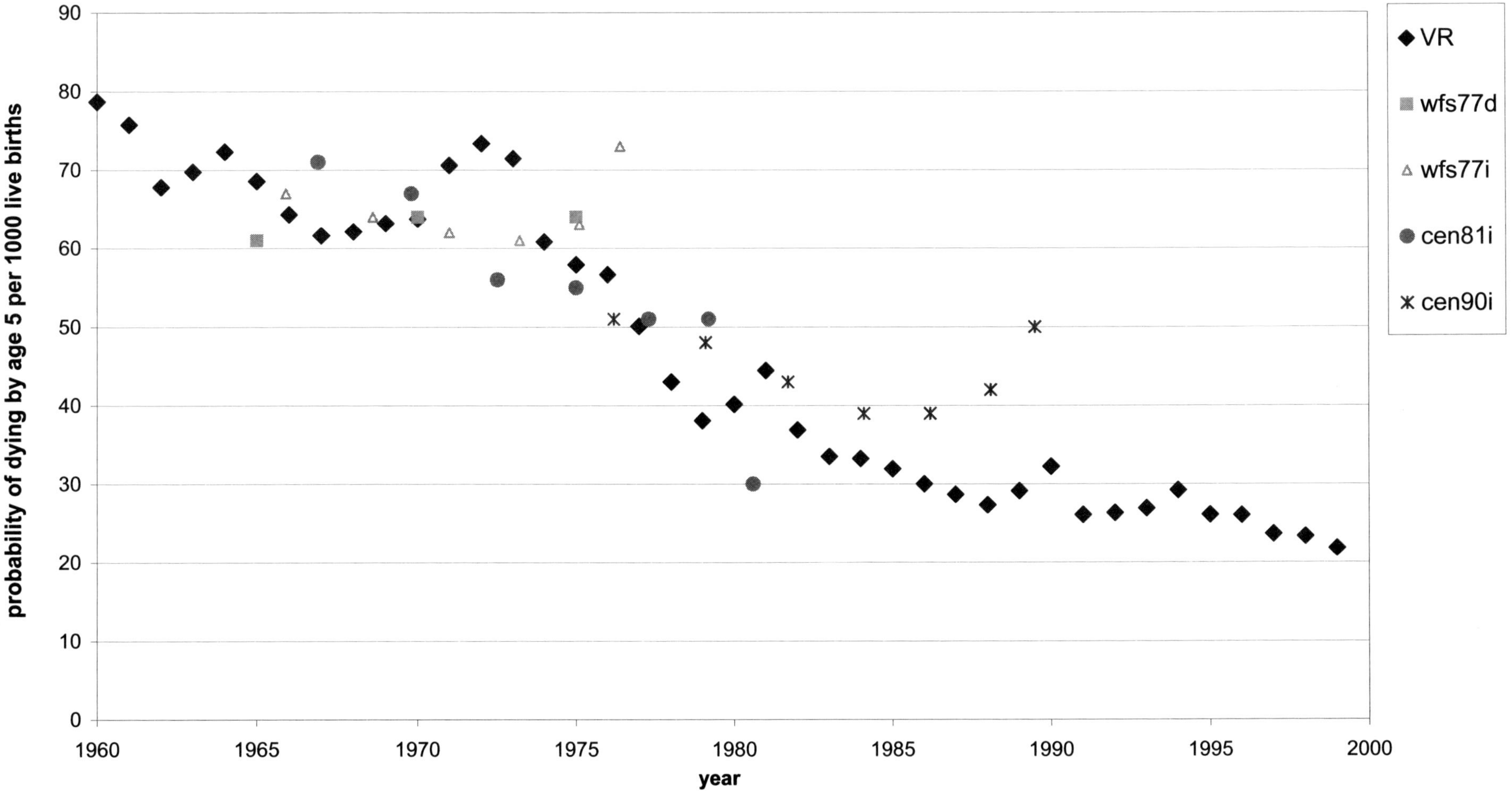

Data sources

Sources des données

Fuentes de los datos

VR: Vital Registration; ***wfs77d***: Encuesta Nacional de Fecundidad, March-August 1977 (direct method); ***wfs77i***: Encuesta Nacional de Fecundidad, March-August 1977 (indirect method); ***cen81i***: Census, 20 October 1981 (indirect method); ***cen90i***: Census, 21 October 1990 (indirect method)

Under 5 mortality
Viet Nam

Mortalité des moins de cinq ans
Viet Nam

Mortalidad de menores de cinco años
Viet Nam

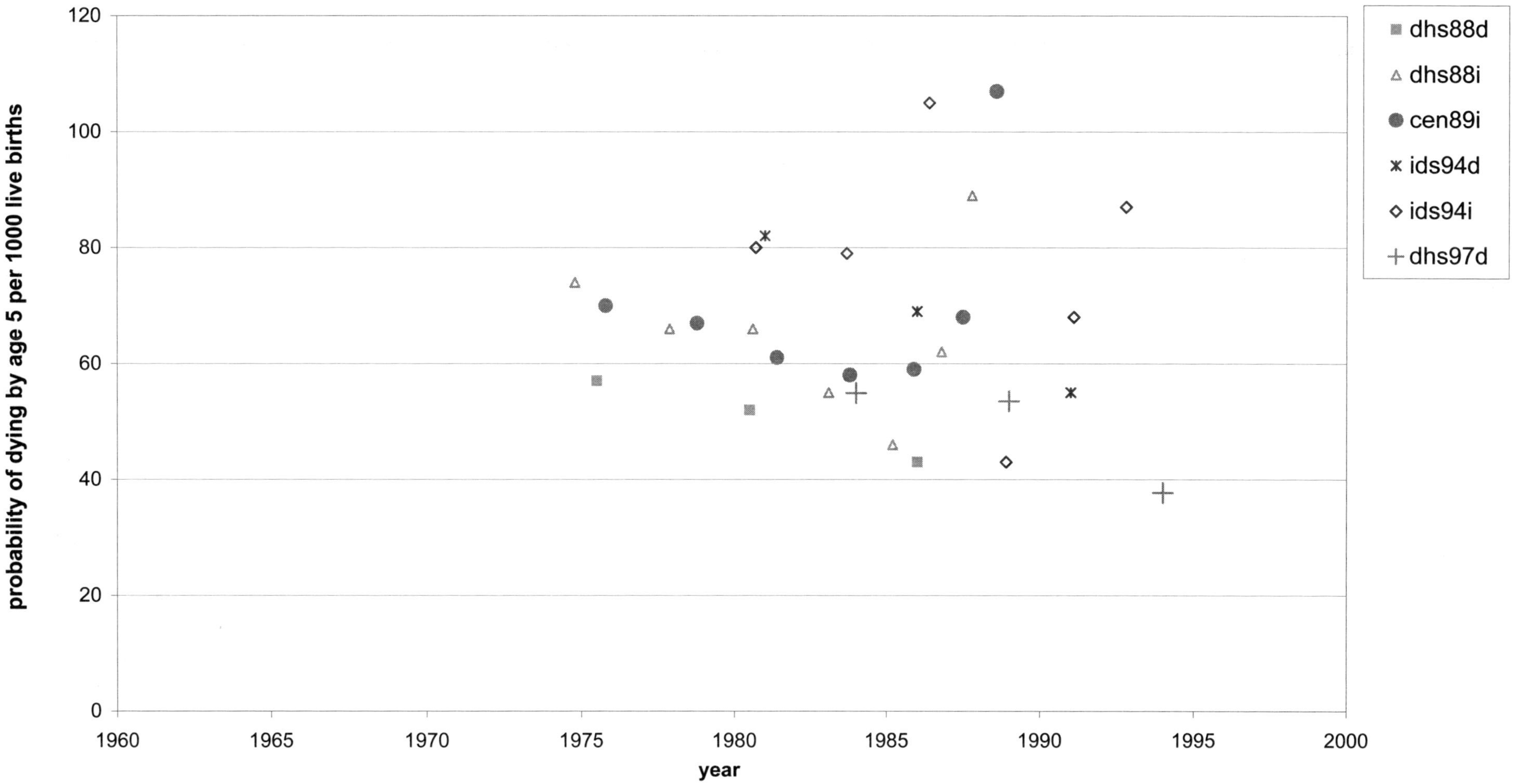

Data sources **Sources des données** **Fuentes de los datos**

dhs88d : Demographic and Health Survey, May-June 1988 (direct method); ***dhs88i*** : Demographic and Health Survey, May-June 1988 (indirect method); ***cen89i*** : Census, 31 March 1989 (indirect method); ***ids94d*** : Intercensal Demographic Survey, 1994 (direct method); ***ids94i*** : Intercensal Demographic Survey, 1994 (indirect method); ***dhs97d*** : Demographic and Health Survey 1997 (direct method)

Under 5 mortality
Yemen

Mortalité des moins de cinq ans
Yémen

Mortalidad de menores de cinco años
Yemen

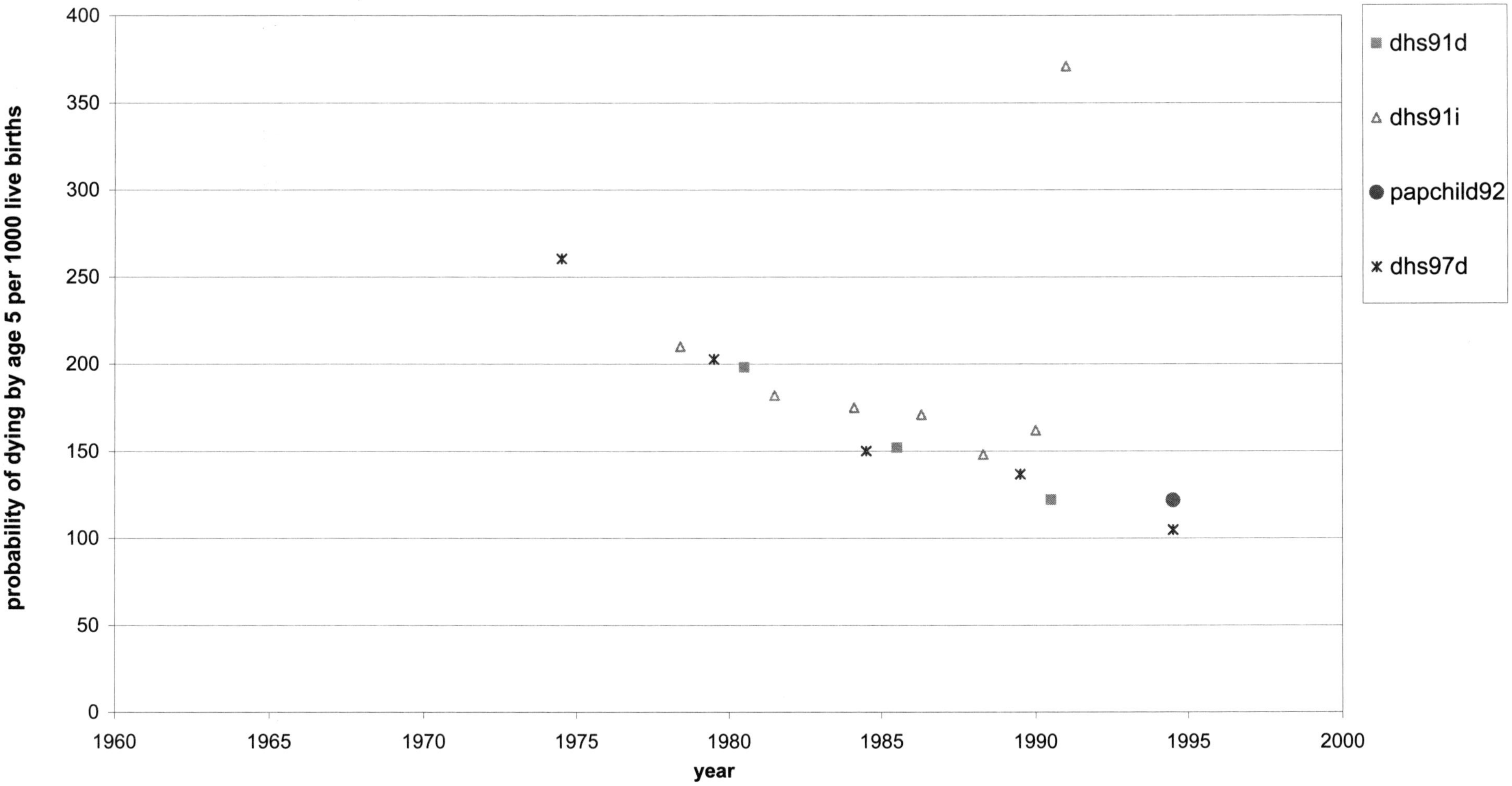

Data sources

Sources des données

Fuentes de los datos

dhs91d : Demographic and Health Survey 1991 (direct method); ***dhs91i*** : Demographic and Health Survey 1991 (indirect method); ***papchild92*** : Maternal and Child Health Survey 1992; ***dhs97d*** : Demographic and Health Survey 1997 (direct method)

Under 5 mortality

Yugoslavia

Mortalité des moins de cinq ans

Yougoslavie

Mortalidad de menores de cinco años

Yugoslavia

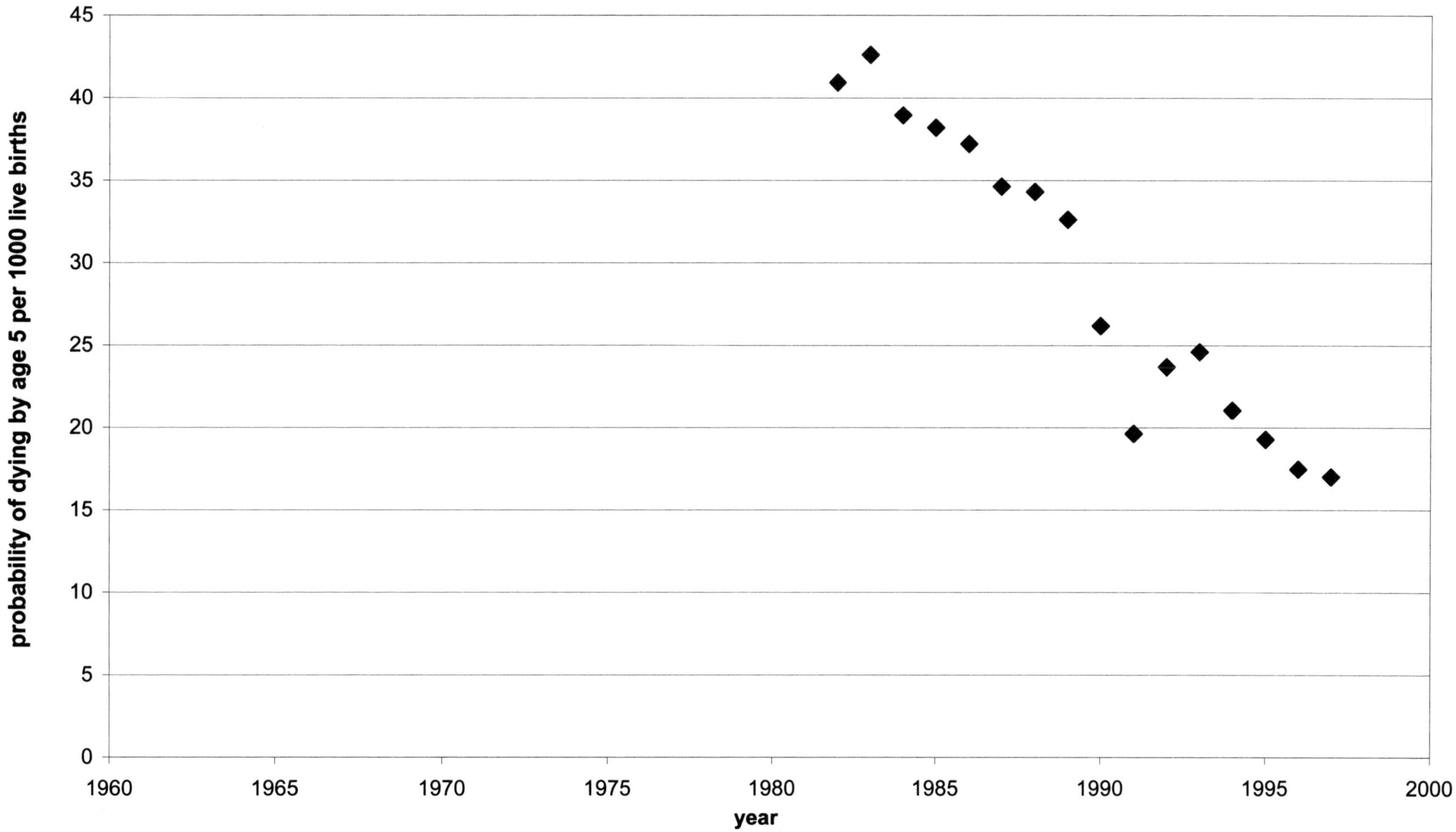

Data sources

Vital Registration

Sources des données

Fuentes de los datos

Under 5 mortality

Zambia

Mortalité des moins de cinq ans

Zambie

Mortalidad de menores de cinco años

Zambia

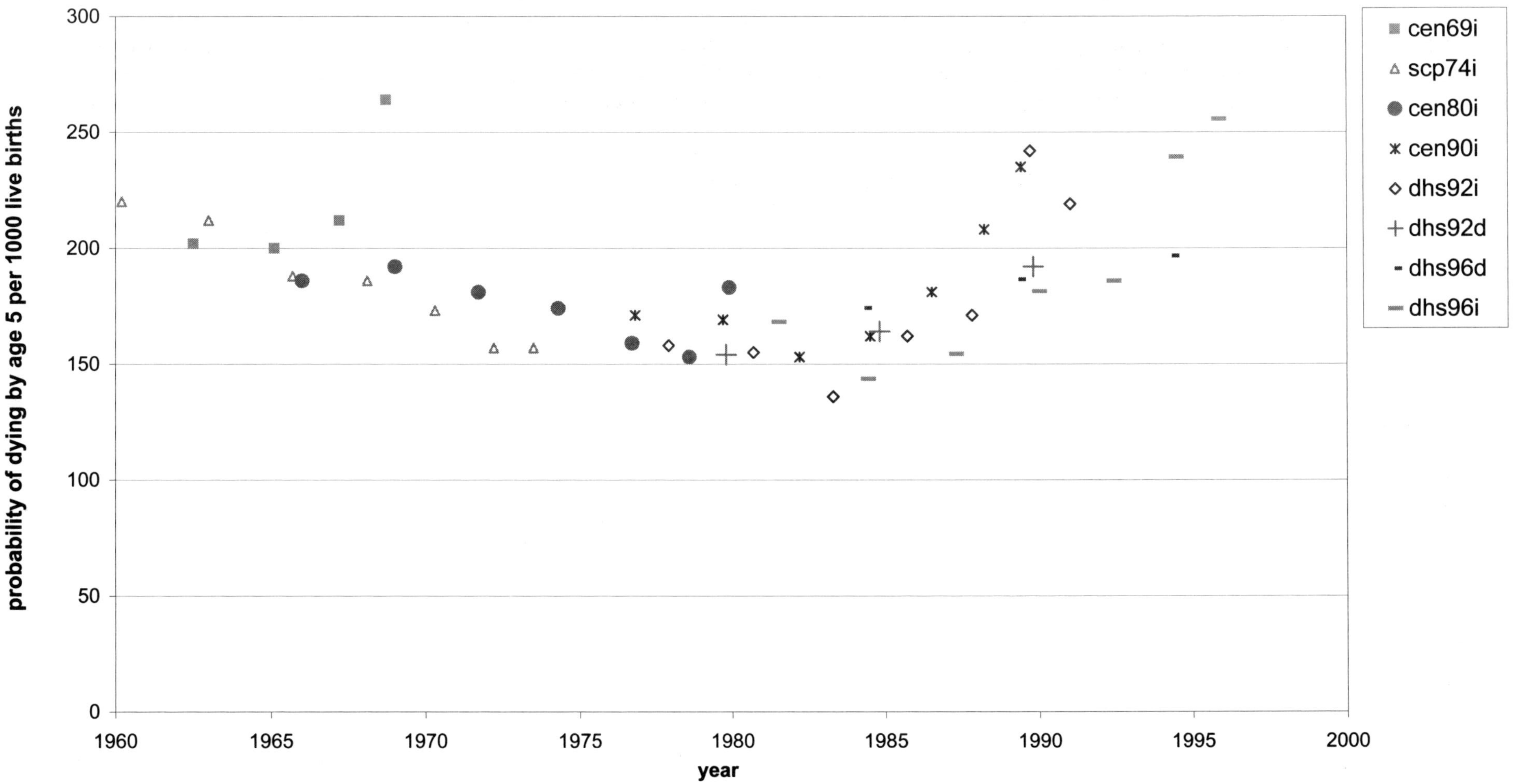

Data sources **Sources des données** **Fuentes de los datos**

cen69i : Census, 22-30 August 1969 (indirect method); ***scp74i*** : Sample Census of Population, 26 August-7 September 1974 (indirect method); ***cen80i*** : Census, 25 August 1980 (indirect method); ***cen90i*** : Census, 1990 (indirect method); ***dhs92i*** : Demographic and Health Survey, January-May 1992 (indirect method); ***dhs92d*** : Demographic and Health Survey, January-May 1992 (direct method); ***dhs96d*** : Demographic and Health Survey, July 1996-January 1997 (direct method); ***dhs96i*** : Demographic and Health Survey, July 1996-January 1997 (indirect method)

Under 5 mortality

Zimbabwe

Mortalité des moins de cinq ans

Zimbabwe

Mortalidad de menores de cinco años

Zimbabwe

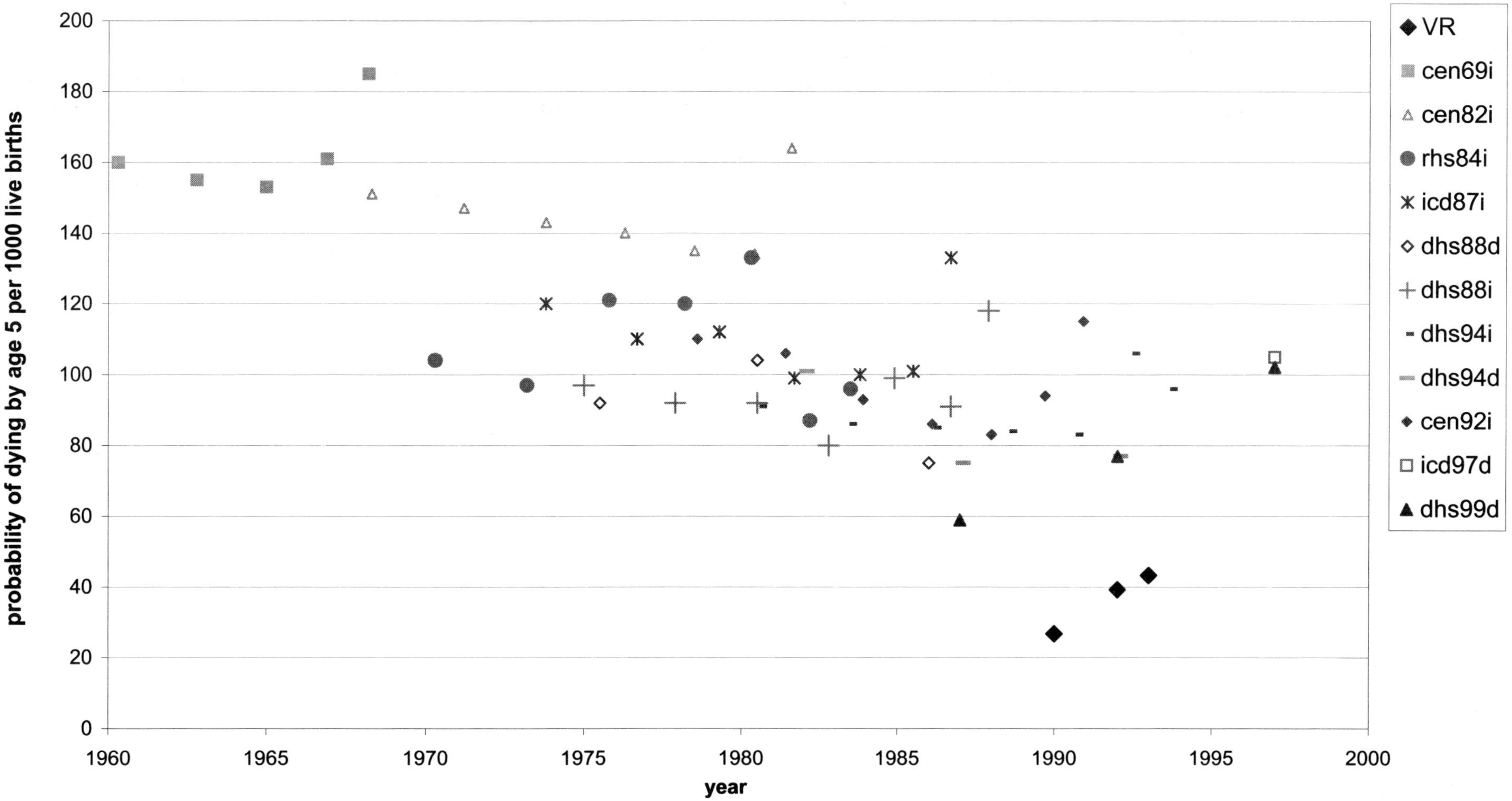

Data sources **Sources des données** **Fuentes de los datos**

VR: Vital Registration; ***cen69i***: Census, 20 March 1969 (indirect method); ***cen82i***: Census, 16 August 1982 (indirect method); ***rhs84i***: Reproductive Health Survey, July-October 1984 (indirect method); ***icd87i***: Intercensal Demographic Survey - Round 1, August 1987 (indirect method); ***dhs88d***: Demographic and Health Survey, September 1988-January 1989 (direct method); ***dhs88i***: Demographic and Health Survey, September 1988-January 1989 (indirect method); ***dhs94i***: Demographic and Health Survey, July-November 1994 (indirect method); ***dhs94d***: Demographic and Health Survey, July-November 1994 (direct method); ***cen92i***: Census, August 1992 (indirect method); ***icd97d***: Intercensal Demographic Survey 1997 (direct method); ***dhs99d***: Demographic and Health Survey 1999 (direct method)